COMMON FUNCTIONAL GROUPS

Type of Compound	General Structure	Example	Functional Group	Type of Compound	General Structure	Example	Functional Group
Acid chloride	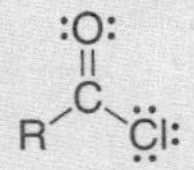	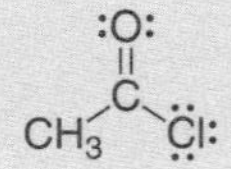	$-COCl$	Aromatic compound	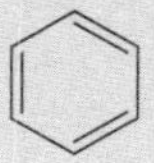	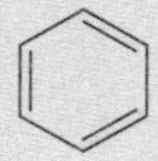	phenyl group
Acyl phosphate	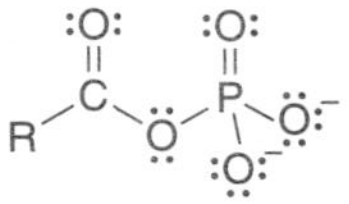	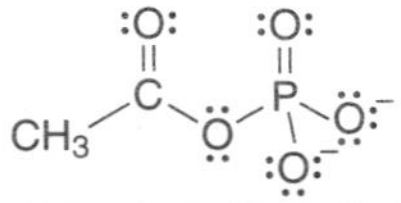	$-CO_2PO_3^{2-}$	Carboxylic acid	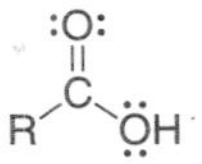	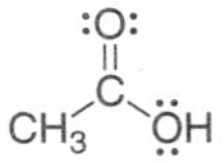	$-CO_2H$ carboxy group
Alcohol	$R-OH$	CH_3-OH	$-OH$ hydroxy group	Disulfide	$R-S-S-R$	$CH_3-S-S-CH_3$	$-SS-$
Aldehyde	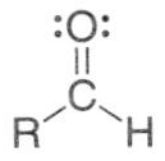	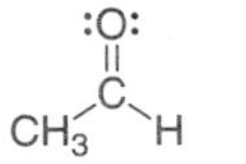	$C=O$ carbonyl group	Ester	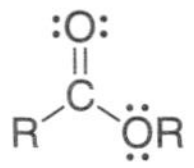	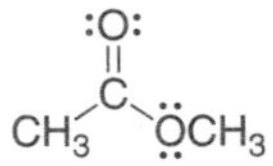	$-CO_2R$
Alkane	$R-H$	CH_3CH_3	– –	Ether	$R-O-R$	CH_3-O-CH_3	$-OR$ alkoxy group
Alkene	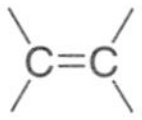	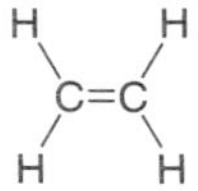	double bond	Ketone	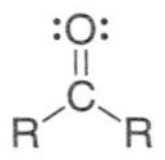	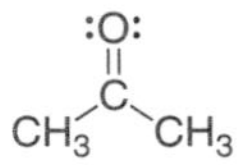	$C=O$ carbonyl group
Alkyl halide	$R-X$: (X = F, Cl, Br, I)	CH_3-Br:	$-X$ halo group	Monophosphate	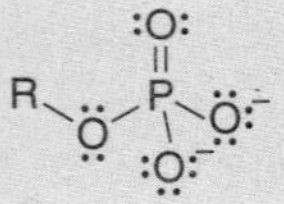	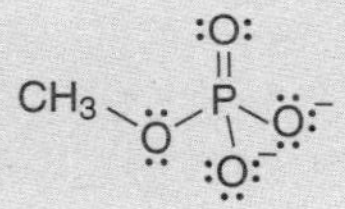	$-OPO_3^{2-}$
Alkyne	$-C\equiv C-$	$H-C\equiv C-H$	triple bond	Nitrile	$R-C\equiv N$:	$CH_3-C\equiv N$:	$-C\equiv N$ cyano gr
Amide	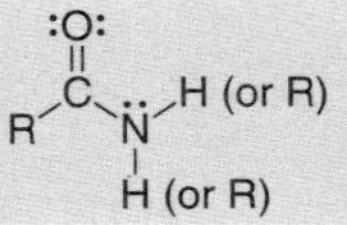	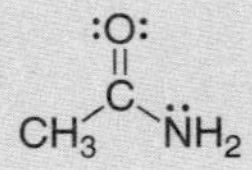	$-CONH_2$, $-CONHR$, $-CONR_2$	Sulfide	$R-S-R$	CH_3-S-CH_3	$-SR$ alkylthio
Amine	$R-NH_2$ or R_2NH or R_3N	CH_3-NH_2	$-NH_2$ amino group	Thiol	$R-SH$	CH_3-SH	$-SH$ mercapto
Anhydride	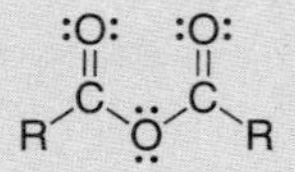	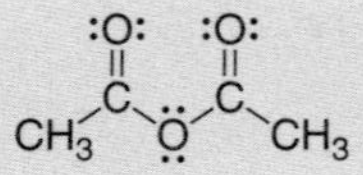	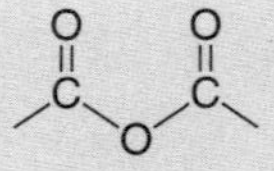	Thioester	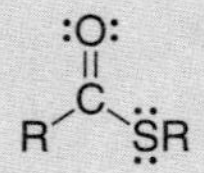	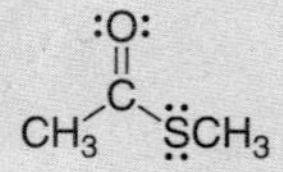	$-C$

Organic Chemistry with Biological Topics

Fifth Edition

Janice Gorzynski Smith
University of Hawaiʻi at Mānoa

Heidi R. Vollmer–Snarr
Stanford University

ORGANIC CHEMISTRY WITH BIOLOGICAL TOPICS, FIFTH EDITION

This book is printed on acid-free paper.

1 2 3 4 5 6 7 8 9 LWI 21 20 19 18 17

ISBN 978-1-259-92001-1
MHID 1-259-92001-1

Chief Product Officer, SVP Products & Markets: *G. Scott Virkler*
Vice President, General Manager, Products & Markets: *Marty Lange*
Vice President, Content Design & Delivery: *Betsy Whalen*
Managing Director: *Thomas Timp*
Director: *David Spurgeon, Ph.D.*
Brand Manager: *Andrea Pellerito, Ph.D.*
Director, Product Development: *Rose M. Koos*
Product Developer: *Mary Hurley*
Director of Marketing: *Tamara Hodge*
Marketing Manager: *Matthew Garcia*
Market Development Manager: *Shannon O'Donnell*
Digital Product Developer: *Joan Weber*
Director of Digital Content: *Philip Janowicz, Ph.D.*
Director, Content Design & Delivery: *Linda Avenarius*
Program Manager: *Lora Neyens*
Content Project Managers: *Sherry Kane/Tammy Juran*
Buyer: *Laura M. Fuller*
Design: *Matt Backhaus*
Content Licensing Specialists: *Carrie Burger/Shannon Manderscheid*
Cover Image: © *Adam Gault/Getty Images*
Compositor: *Lachina Publishing*
Printer: *LSC Communications*

Library of Congress Cataloging-in-Publication Data

Names: Smith, Janice G. | Vollmer-Snarr, Heidi R. | Smith, Janice G. Organic chemistry.
Title: Organic chemistry with biological topics / Janice Gorzynski Smith, Heidi R. Vollmer-Snarr.
Description: 5e [5th edition, updated]. | New York, NY : McGraw-Hill Education, 2018. | Previous edition: Organic chemistry / Janice Gorzynski Smith (New York, NY : McGraw-Hill, 2014). | Includes index.
Identifiers: LCCN 2016042232 | ISBN 9781259920011 (hardcover)
Subjects: LCSH: Chemistry, Organic—Textbooks.
Classification: LCC QD253.2 .S6325 2018 | DDC 547—dc23

mheducation.com/highered

About the Authors

Janice Gorzynski Smith was born in Schenectady, New York. She received an A.B. degree *summa cum laude* in chemistry from Cornell University and a Ph.D. in organic chemistry from Harvard University under the direction of Nobel Laureate E. J. Corey. After a postdoctoral fellowship, Jan joined the faculty of Mount Holyoke College, where she was employed for 21 years, teaching organic chemistry and conducting a research program in organic synthesis. After spending two sabbaticals in Hawai'i in the 1990s, Jan and her family moved there permanently in 2000, and she became a faculty member at the University of Hawai'i at Mānoa. She has four children and four grandchildren. When not teaching, writing, or enjoying her family, Jan bikes, hikes, snorkels, and scuba dives, and time permitting, enjoys travel and quilting.

Heidi R. Vollmer-Snarr was born in Pittsburgh, Pennsylvania. She received a B.S. degree in chemistry and a B.A. degree in German from the University of Utah and a Ph.D. in organic chemistry from Oxford University under the direction of Sir Jack Baldwin. As an NIH Postdoctoral Fellow, she worked for Koji Nakanishi at Columbia University and was an Assistant Professor at Brigham Young University, where her research involved the synthesis and photochemistry of ocular retinoid age pigments. Heidi now focuses on curriculum development at Stanford University and serves on the NIH Small Business Sensory Technologies study section and ACS Committee on Chemistry and Public Affairs. She also loves to spend time skiing, biking, and hiking with her husband, Trent, and three children, Zach, Grady, and Elli.

For Megan Sarah Smith and Charles J. Vollmer

Contents in Brief

Prologue 1
1 Structure and Bonding 7
2 Acids and Bases 61
3 Introduction to Organic Molecules and Functional Groups 91
4 Alkanes 134
5 Stereochemistry 180
6 Understanding Organic Reactions 219
7 Alkyl Halides and Nucleophilic Substitution 255
8 Alkyl Halides and Elimination Reactions 305
9 Alcohols, Ethers, and Related Compounds 339
10 Alkenes 391
11 Alkynes 434
12 Oxidation and Reduction 463
13 Mass Spectrometry and Infrared Spectroscopy 503
14 Nuclear Magnetic Resonance Spectroscopy 535
15 Radical Reactions 578
16 Conjugation, Resonance, and Dienes 612
17 Benzene and Aromatic Compounds 649
18 Reactions of Aromatic Compounds 686
19 Carboxylic Acids and the Acidity of the O–H Bond 738
20 Introduction to Carbonyl Chemistry; Organometallic Reagents; Oxidation and Reduction 774
21 Aldehydes and Ketones—Nucleophilic Addition 827
22 Carboxylic Acids and Their Derivatives—Nucleophilic Acyl Substitution 878
23 Substitution Reactions of Carbonyl Compounds at the α Carbon 934
24 Carbonyl Condensation Reactions 972
25 Amines 1010
26 Amino Acids and Proteins 1063
27 Carbohydrates 1109
28 Lipids 1155
29 Carbon–Carbon Bond-Forming Reactions in Organic Synthesis 1185
30 Pericyclic Reactions 1212
31 Synthetic Polymers 1242 (Available online)
Appendices A-1
Glossary G-1
Credits C-1
Index I-1

Contents

Preface xiii

Acknowledgments xxi

List of *How To's* xxiii

List of Mechanisms xxiv

List of Selected Applications xxvii

Prologue 1

What Is Organic Chemistry? 1

Some Representative Organic Molecules 2

Organic Chemistry and Malaria 4

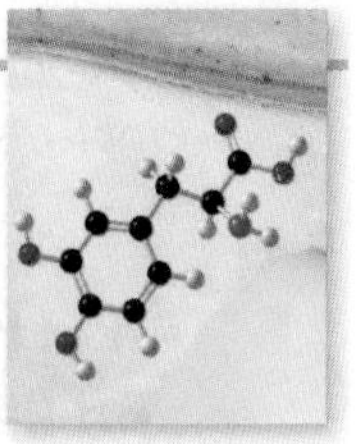

1 Structure and Bonding 7

1.1 The Periodic Table 8

1.2 Bonding 11

1.3 Lewis Structures 13

1.4 Isomers 18

1.5 Exceptions to the Octet Rule 19

1.6 Resonance 19

1.7 Determining Molecular Shape 25

1.8 Drawing Organic Structures 30

1.9 Hybridization 36

1.10 Ethane, Ethylene, and Acetylene 40

1.11 Bond Length and Bond Strength 45

1.12 Electronegativity and Bond Polarity 47

1.13 Polarity of Molecules 49

1.14 L-Dopa—A Representative Organic Molecule 50

Key Concepts 52

Problems 53

2 Acids and Bases 61

2.1 Brønsted–Lowry Acids and Bases 62

2.2 Reactions of Brønsted–Lowry Acids and Bases 63

2.3 Acid Strength and pK_a 66

2.4 Predicting the Outcome of Acid–Base Reactions 68

2.5 Factors That Determine Acid Strength 70

2.6 Common Acids and Bases 78

2.7 Aspirin 80

2.8 Lewis Acids and Bases 81

Key Concepts 84

Problems 85

3 Introduction to Organic Molecules and Functional Groups 91

3.1 Functional Groups 92

3.2 An Overview of Functional Groups 93

3.3 Intermolecular Forces 101

3.4 Physical Properties 105

3.5 Application: Vitamins 111

3.6 Application of Solubility: Soap 112

3.7 Application: The Cell Membrane 114

3.8 Functional Groups and Reactivity 117

3.9 Biomolecules 119

Key Concepts 125

Problems 126

4 Alkanes 134

4.1 Alkanes—An Introduction 135

4.2 Cycloalkanes 138

4.3 An Introduction to Nomenclature 138

4.4 Naming Alkanes 139

4.5 Naming Cycloalkanes 144

4.6 Common Names 147

4.7 Fossil Fuels 147

4.8 Physical Properties of Alkanes 149

4.9 Conformations of Acyclic Alkanes—Ethane 150

4.10 Conformations of Butane 154

4.11 An Introduction to Cycloalkanes 157

4.12 Cyclohexane 158

4.13 Substituted Cycloalkanes 162

4.14 Oxidation of Alkanes 167

4.15 Lipids—Part 1 170

Key Concepts 172

Problems 173

13.5 Electromagnetic Radiation 514
13.6 Infrared Spectroscopy 516
13.7 IR Absorptions 518
13.8 IR and Structure Determination 525
Key Concepts 527
Problems 528

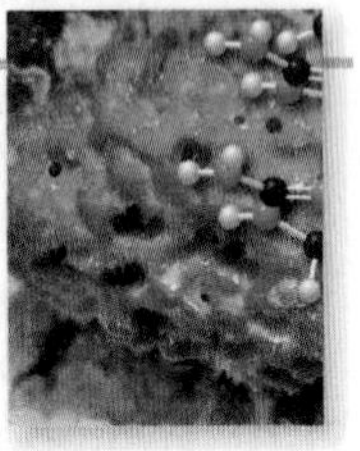

14 Nuclear Magnetic Resonance Spectroscopy 535

14.1 An Introduction to NMR Spectroscopy 536
14.2 ^{1}H NMR: Number of Signals 539
14.3 ^{1}H NMR: Position of Signals 543
14.4 The Chemical Shift of Protons on sp^2 and *sp* Hybridized Carbons 547
14.5 ^{1}H NMR: Intensity of Signals 549
14.6 ^{1}H NMR: Spin–Spin Splitting 550
14.7 More Complex Examples of Splitting 554
14.8 Spin–Spin Splitting in Alkenes 557
14.9 Other Facts About ^{1}H NMR Spectroscopy 559
14.10 Using ^{1}H NMR to Identify an Unknown 561
14.11 ^{13}C NMR Spectroscopy 564
14.12 Magnetic Resonance Imaging (MRI) 568
Key Concepts 569
Problems 569

15 Radical Reactions 578

15.1 Introduction 579
15.2 General Features of Radical Reactions 580
15.3 Halogenation of Alkanes 582
15.4 The Mechanism of Halogenation 583
15.5 Chlorination of Other Alkanes 586
15.6 Chlorination Versus Bromination 586
15.7 Halogenation as a Tool in Organic Synthesis 589
15.8 The Stereochemistry of Halogenation Reactions 590
15.9 Application: The Ozone Layer and CFCs 592
15.10 Radical Halogenation at an Allylic Carbon 593
15.11 Application: Oxidation of Unsaturated Lipids 596
15.12 Application: Antioxidants 597
15.13 Radical Addition Reactions to Double Bonds 598
15.14 Polymers and Polymerization 601
Key Concepts 603
Problems 604

16 Conjugation, Resonance, and Dienes 612

16.1 Conjugation 613
16.2 Resonance and Allylic Carbocations 615
16.3 Common Examples of Resonance 616
16.4 The Resonance Hybrid 618
16.5 Electron Delocalization, Hybridization, and Geometry 620
16.6 Conjugated Dienes 621
16.7 Interesting Dienes and Polyenes 622
16.8 The Carbon–Carbon σ Bond Length in Buta-1,3-diene 622
16.9 Stability of Conjugated Dienes 623
16.10 Electrophilic Addition: 1,2- Versus 1,4-Addition 624
16.11 Kinetic Versus Thermodynamic Products 626
16.12 The Diels–Alder Reaction 629
16.13 Specific Rules Governing the Diels–Alder Reaction 631
16.14 Other Facts About the Diels–Alder Reaction 635
16.15 Conjugated Dienes and Ultraviolet Light 638
Key Concepts 640
Problems 642

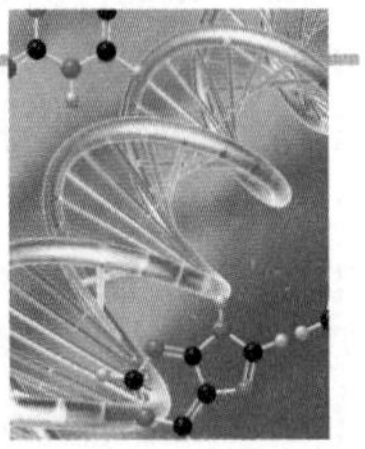

17 Benzene and Aromatic Compounds 649

17.1 Background 650
17.2 The Structure of Benzene 651
17.3 Nomenclature of Benzene Derivatives 653
17.4 Spectroscopic Properties 655
17.5 Benzene's Unusual Stability 656
17.6 The Criteria for Aromaticity—Hückel's Rule 657
17.7 Examples of Aromatic Compounds 660
17.8 Aromatic Heterocycles 664
17.9 What Is the Basis of Hückel's Rule? 669
17.10 The Inscribed Polygon Method for Predicting Aromaticity 672
17.11 Application: Aromatase Inhibitors for Estrogen-Dependent Cancer Treatment 674
Key Concepts 676
Problems 677

18 Reactions of Aromatic Compounds 686

18.1 Electrophilic Aromatic Substitution 687
18.2 The General Mechanism 688
18.3 Halogenation 690
18.4 Nitration and Sulfonation 691
18.5 Friedel–Crafts Alkylation and Friedel–Crafts Acylation 693
18.6 Substituted Benzenes 700
18.7 Electrophilic Aromatic Substitution of Substituted Benzenes 703
18.8 Why Substituents Activate or Deactivate a Benzene Ring 705
18.9 Orientation Effects in Substituted Benzenes 707
18.10 Limitations on Electrophilic Substitution Reactions with Substituted Benzenes 710
18.11 Disubstituted Benzenes 712
18.12 Synthesis of Benzene Derivatives 714
18.13 Nucleophilic Aromatic Substitution 715
18.14 Halogenation of Alkyl Benzenes 718
18.15 Oxidation and Reduction of Substituted Benzenes 720
18.16 Multistep Synthesis 724
Key Concepts 727
Problems 730

19 Carboxylic Acids and the Acidity of the O–H Bond 738

19.1 Structure and Bonding 739
19.2 Nomenclature 739
19.3 Physical Properties 742
19.4 Spectroscopic Properties 743
19.5 Interesting Carboxylic Acids 745
19.6 Aspirin, Arachidonic Acid, and Prostaglandins 745
19.7 Preparation of Carboxylic Acids 747
19.8 Reactions of Carboxylic Acids—General Features 748
19.9 Carboxylic Acids—Strong Organic Brønsted–Lowry Acids 749
19.10 The Henderson–Hasselbalch Equation 752
19.11 Inductive Effects in Aliphatic Carboxylic Acids 754
19.12 Substituted Benzoic Acids 756
19.13 Extraction 758
19.14 Organic Acids Containing Sulfur and Phosphorus 760
19.15 Amino Acids 761
Key Concepts 765
Problems 766

20 Introduction to Carbonyl Chemistry; Organometallic Reagents; Oxidation and Reduction 774

20.1 Introduction 775
20.2 General Reactions of Carbonyl Compounds 776
20.3 A Preview of Oxidation and Reduction 779
20.4 Reduction of Aldehydes and Ketones 781
20.5 The Stereochemistry of Carbonyl Reduction 783
20.6 Enantioselective Carbonyl Reductions 784
20.7 Reduction of Carboxylic Acids and Their Derivatives 787
20.8 Oxidation of Aldehydes 792
20.9 Organometallic Reagents 792
20.10 Reaction of Organometallic Reagents with Aldehydes and Ketones 796
20.11 Retrosynthetic Analysis of Grignard Products 800
20.12 Protecting Groups 802
20.13 Reaction of Organometallic Reagents with Carboxylic Acid Derivatives 804
20.14 Reaction of Organometallic Reagents with Other Compounds 807
20.15 α,β-Unsaturated Carbonyl Compounds 809
20.16 Summary—The Reactions of Organometallic Reagents 812
20.17 Synthesis 812
Key Concepts 815
Problems 818

21 Aldehydes and Ketones—Nucleophilic Addition 827

21.1 Introduction 828
21.2 Nomenclature 829
21.3 Physical Properties 832
21.4 Spectroscopic Properties 833
21.5 Interesting Aldehydes and Ketones 835

21.6 Preparation of Aldehydes and Ketones 836
21.7 Reactions of Aldehydes and Ketones—General Considerations 838
21.8 Nucleophilic Addition of H^- and R^-—A Review 841
21.9 Nucleophilic Addition of ^-CN 843
21.10 The Wittig Reaction 845
21.11 Addition of 1° Amines 850
21.12 Addition of 2° Amines 852
21.13 Addition of H_2O—Hydration 854
21.14 Addition of Alcohols—Acetal Formation 857
21.15 Acetals as Protecting Groups 861
21.16 Cyclic Hemiacetals 862
21.17 An Introduction to Carbohydrates 865
Key Concepts 866
Problems 868

22 Carboxylic Acids and Their Derivatives—Nucleophilic Acyl Substitution 878

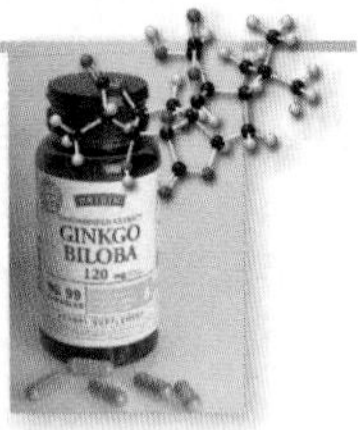
GINKGO
BILOBA

22.1 Introduction 879
22.2 Structure and Bonding 881
22.3 Nomenclature 883
22.4 Physical Properties 888
22.5 Spectroscopic Properties 889
22.6 Interesting Esters and Amides 891
22.7 Introduction to Nucleophilic Acyl Substitution 892
22.8 Reactions of Acid Chlorides 896
22.9 Reactions of Anhydrides 897
22.10 Reactions of Carboxylic Acids 898
22.11 Reactions of Esters 903
22.12 Application: Lipid Hydrolysis 905
22.13 Reactions of Amides 908
22.14 Application: The Mechanism of Action of β-Lactam Antibiotics 909
22.15 Summary of Nucleophilic Acyl Substitution Reactions 910
22.16 Acyl Phosphates—Biological Anhydrides 911
22.17 Reactions of Thioesters—Biological Acylation Reactions 914
22.18 Nitriles 916
Key Concepts 921
Problems 924

23 Substitution Reactions of Carbonyl Compounds at the α Carbon 934

AWARENESS
FIGHT
LIKE A
GIRL

23.1 Introduction 935
23.2 Enols 936
23.3 Enolates 938
23.4 Enolates of Unsymmetrical Carbonyl Compounds 944
23.5 Racemization at the α Carbon 946
23.6 A Preview of Reactions at the α Carbon 947
23.7 Halogenation at the α Carbon 947
23.8 Direct Enolate Alkylation 952
23.9 Malonic Ester Synthesis 955
23.10 Acetoacetic Ester Synthesis 959
Key Concepts 962
Problems 963

24 Carbonyl Condensation Reactions 972

24.1 The Aldol Reaction 973
24.2 Crossed Aldol Reactions 978
24.3 Directed Aldol Reactions 981
24.4 Intramolecular Aldol Reactions 984
24.5 The Claisen Reaction 986
24.6 The Crossed Claisen and Related Reactions 987
24.7 The Dieckmann Reaction 990
24.8 Biological Carbonyl Condensation Reactions 991
24.9 The Michael Reaction 994
24.10 The Robinson Annulation 996
Key Concepts 1000
Problems 1001

25 Amines 1010

25.1 Introduction 1011
25.2 Structure and Bonding 1011
25.3 Nomenclature 1013
25.4 Physical Properties 1015
25.5 Spectroscopic Properties 1016
25.6 Interesting and Useful Amines 1018
25.7 Preparation of Amines 1021
25.8 Reactions of Amines—General Features 1028

31 Synthetic Polymers (Available online) 1242

31.1 Introduction 1243
31.2 Chain-Growth Polymers—Addition Polymers 1244
31.3 Anionic Polymerization of Epoxides 1251
31.4 Ziegler–Natta Catalysts and Polymer Stereochemistry 1252
31.5 Natural and Synthetic Rubbers 1254
31.6 Step-Growth Polymers—Condensation Polymers 1255
31.7 Polymer Structure and Properties 1260
31.8 Green Polymer Synthesis 1261
31.9 Polymer Recycling and Disposal 1264
Key Concepts 1267
Problems 1268

Appendix A pK_a Values for Selected Compounds A-1
Appendix B Nomenclature A-3
Appendix C Bond Dissociation Energies for Some Common Bonds [A–B → A• + •B] A-7
Appendix D Reactions That Form Carbon–Carbon Bonds A-8
Appendix E Characteristic IR Absorption Frequencies A-9
Appendix F Characteristic NMR Absorptions A-10
Appendix G General Types of Organic Reactions A-12
Appendix H How to Synthesize Particular Functional Groups A-14

Glossary G-1
Credits C-1
Index I-1

Preface

Since the publication of *Organic Chemistry* in 2005, chemistry has witnessed a rapid growth in its understanding of the biological world. The molecular basis of many complex biological processes is now known with certainty, and can be explained by applying the basic principles of organic chemistry. Because of the close relationship between chemistry and many biological phenomena, *Organic Chemistry with Biological Topics* presents an approach to traditional organic chemistry that incorporates the discussion of biological applications that are understood using the fundamentals of organic chemistry.

The Basic Features

Organic Chemistry with Biological Topics continues the successful student-oriented approach used in *Organic Chemistry* by Janice Gorzynski Smith. This text uses less prose and more diagrams and bulleted summaries for today's students, who rely more heavily on visual imagery to learn than ever before. Each topic is broken down into small chunks of information that are more manageable and easily learned. Sample Problems illustrate stepwise problem solving, and relevant examples from everyday life are used to illustrate topics. New concepts are introduced one at a time so that the basic themes are kept in focus.

The organization of *Organic Chemistry with Biological Topics* provides the student with a logical and accessible approach to an intense and fascinating subject. The text begins with a healthy dose of review material in Chapters 1 and 2 to ensure that students have a firm grasp of the fundamentals. Stereochemistry, the three-dimensional structure of molecules, is introduced early (Chapter 5) and reinforced often. Certain reaction types with unique characteristics and terminology are grouped together. These include acid–base reactions (Chapter 2), oxidation and reduction (Chapters 12 and 20), radical reactions (Chapter 15), and reactions of organometallic reagents (Chapter 20). Each chapter ends with Key Concepts, end-of-chapter summaries that succinctly organize the main concepts and reactions.

New to *Organic Chemistry with Biological Topics*

While there is no shortage of biological applications that can be added to an organic chemistry text, we have chosen to concentrate on the following areas.

- **Chapter 3** on functional groups now includes an expanded section on four types of biomolecules—amino acids and proteins, monosaccharides and carbohydrates, nucleotides and nucleic acids, and lipids. This material augments the discussions of vitamins and the cell membrane, topics already part of *Organic Chemistry* in past editions. Phosphorus-containing compounds such as ATP (adenosine triphosphate), the key intermediate used in energy transfer in cells, are also introduced in this chapter.
- **Chapter 6** now uses biological examples to illustrate the basic types of organic reactions, and the energetics of coupled reactions in metabolism is presented. The discussion of enzymes as biological catalysts is expanded, and a specific example of an enzyme's active site is shown.
- **Chapter 17** now applies the discussion of aromatic heterocycles to the bases in DNA, the high molecular weight molecule that holds the encrypted genetic instructions for our development and cellular processes. In addition, new material has been added on the synthesis of female sex hormones with the aromatase enzyme, which has resulted in the development of drugs used to treat estrogen-dependent breast cancers.

- **Chapter 19** contains a section on the Henderson–Hasselbalch equation, a mathematical expression that allows us to tell whether a compound exists as an uncharged compound or ion at the cellular pH of 7.4. A section on phosphoric acid esters has been added, and the ionization of amino acids is now explained using the Henderson–Hasselbalch equation.
- **Chapter 22** contains additional material on two common carboxylic acid derivatives—acyl phosphates and thioesters. The role of these functional groups in the biosynthesis of amino acids and the metabolism of fatty acids is discussed.
- **Chapter 24** contains a new section on biological carbonyl condensation reactions. Topics include the biological aldol reaction in the citric acid cycle, the retro-aldol reaction in the metabolism of glucose, and the biological Claisen reaction in the biosynthesis of fatty acids.

In addition, the later chapters of the text are now reorganized to emphasize the connection of biomolecules to prior sections. The chapter on Amino Acids and Proteins (Chapter 26) now directly follows the chapter on Amines (Chapter 25), followed by the remaining chapters on biomolecules, Carbohydrates (Chapter 27) and Lipids (Chapter 28).

Tools to Make Learning Organic Chemistry Easier

Illustrations

Organic Chemistry with Biological Topics is supported by a well-developed illustration program. Besides traditional skeletal (line) structures and condensed formulas, there are numerous ball-and-stick molecular models and electrostatic potential maps to help students grasp the three-dimensional structure of molecules (including stereochemistry) and to better understand the distribution of electronic charge.

Figure 21.9 The key reaction in the chemistry of vision

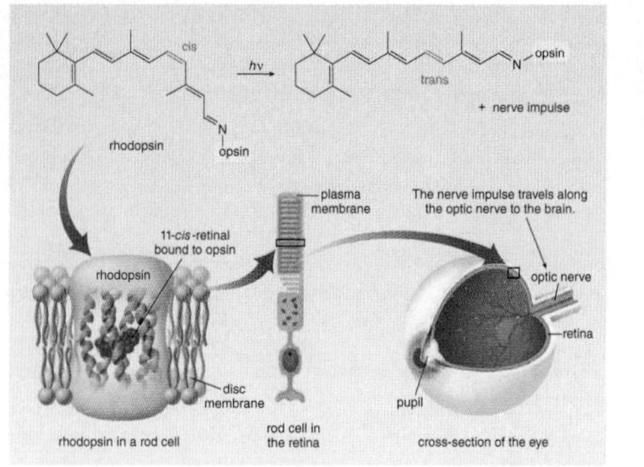

- Rhodopsin is a light-sensitive compound located in the membrane of the rod cells in the retina of the eye. Rhodopsin contains the protein opsin bonded to 11-*cis*-retinal via an imine linkage. When light strikes this molecule, the crowded 11-cis double bond isomerizes to the 11-trans isomer, and a nerve impulse is transmitted to the brain by the optic nerve.

Micro-to-Macro Illustrations

Unique to *Organic Chemistry with Biological Topics* are micro-to-macro illustrations, where line art and photos combine with chemical structures to reveal the underlying molecular structures giving rise to macroscopic properties of common phenomena. Examples include starch and cellulose (Chapter 5), adrenaline (Chapter 7), partial hydrogenation of vegetable oil (Chapter 12), and dopamine (Chapter 25).

Figure 12.4 Partial hydrogenation of the double bonds in a vegetable oil

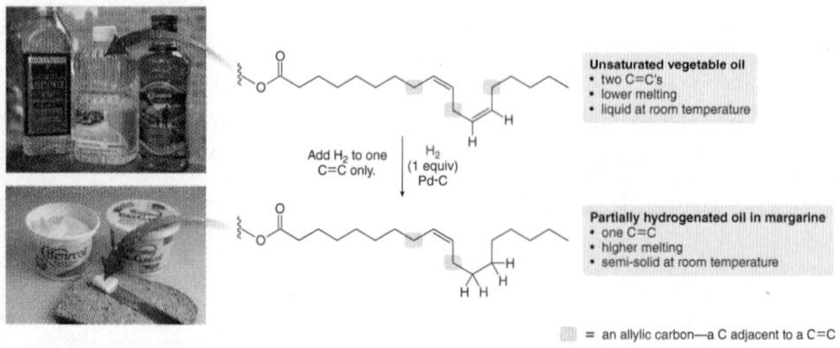

- **Decreasing** the number of degrees of unsaturation **increases** the melting point. Only one long chain of the triacylglycerol is drawn.
- When an oil is *partially* hydrogenated, some double bonds react with H_2, whereas some double bonds remain in the product.
- Partial hydrogenation **decreases** the number of allylic sites (shown in blue), making a triacylglycerol **less** susceptible to oxidation, thereby increasing its shelf life.

Spectra

Over 100 spectra created specifically for *Organic Chemistry with Biological Topics* are presented throughout the text. The spectra are color-coded by type and generously labeled. Mass spectra are green; infrared spectra are red; and proton and carbon nuclear magnetic resonance spectra are blue.

Figure 13.5 Identifying fragments in the mass spectrum of hexane

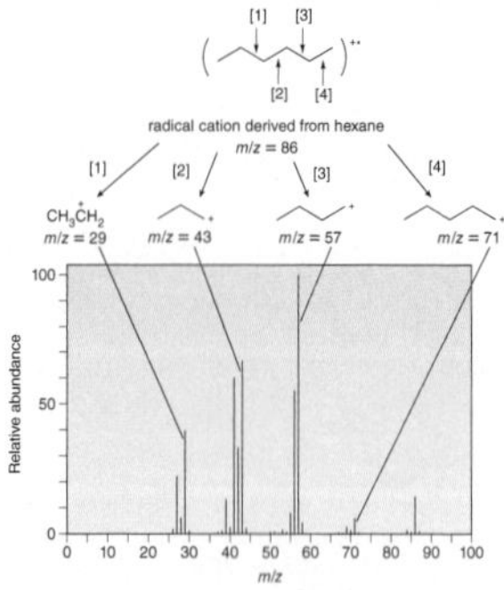

- Cleavage of C–C bonds (labeled [1]–[4]) in hexane forms lower molecular weight fragments that correspond to lines in the mass spectrum. Although the mass spectrum is complex, possible structures can be assigned to some of the fragments, as shown.

Mechanisms

Curved arrow notation is used extensively to help students follow the movement of electrons in reactions.

Mechanism 9.2 Dehydration of a 1° ROH—An E2 Mechanism

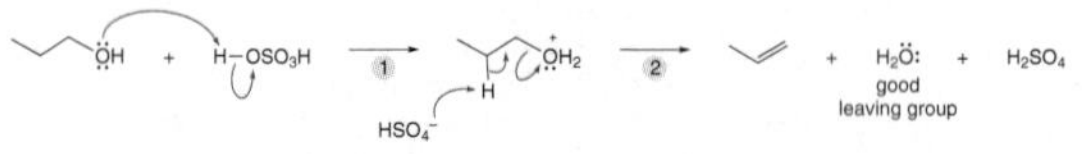

1 Protonation of the oxygen atom converts the poor leaving group (^-OH) into a good leaving group (H_2O).
2 Two bonds are broken and two bonds are formed. The base (HSO_4^- or H_2O) removes a proton from the β carbon; the electron pair in the β C–H bond forms the new π bond and the leaving group (H_2O) departs.

Problem Solving

Sample Problems

Sample Problems show students how to solve organic chemistry problems in a logical, stepwise manner. More than 800 follow-up problems are located throughout the chapters to test whether students understand concepts covered in the Sample Problems.

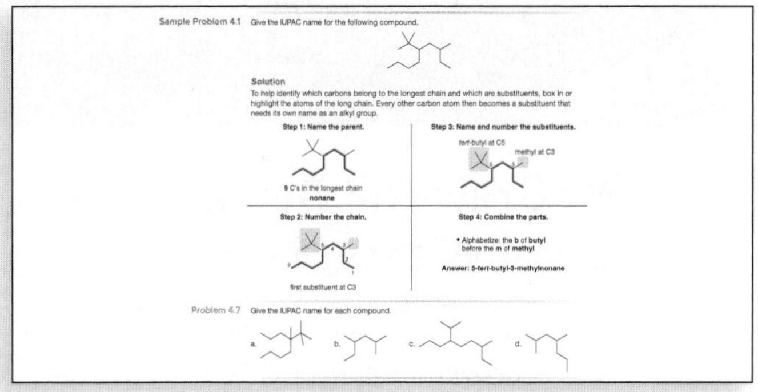

How To's

How To's provide students with detailed instructions on how to work through key processes.

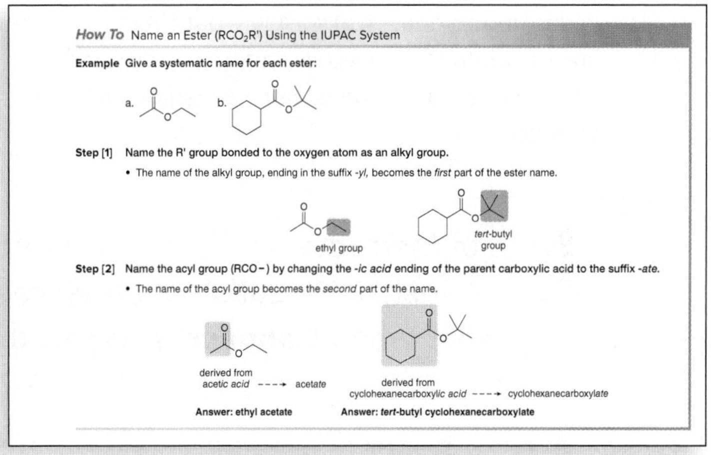

Applications and Summaries

Key Concept Summaries

Succinct summary tables reinforcing important principles and concepts are provided at the end of each chapter.

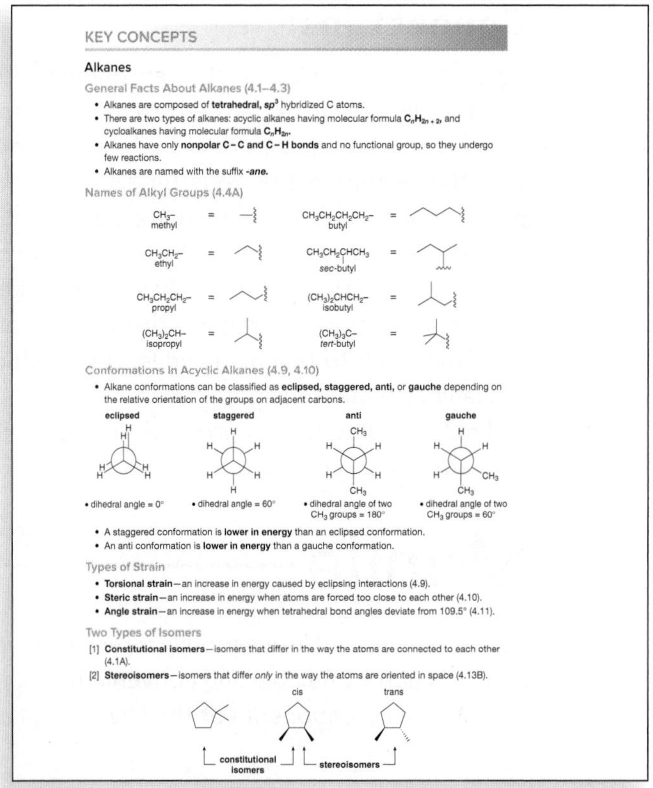

Margin Notes

Margin notes are placed carefully throughout the chapters, providing interesting information relating to topics covered in the text. Some margin notes are illustrated with photos to make the chemistry more relevant.

All soaps are salts of fatty acids. The main difference between soaps is the addition of other ingredients that do not alter their cleaning properties: dyes for color, scents for a pleasing odor, and oils for lubrication. Soaps that float are aerated, so that they are less dense than water.

Mc Graw Hill Education **connect**®

Required=Results

McGraw-Hill Connect®
Learn Without Limits

Connect is a teaching and learning platform that is proven to deliver better results for students and instructors.

Connect empowers students by continually adapting to deliver precisely what they need, when they need it, and how they need it, so your class time is more engaging and effective.

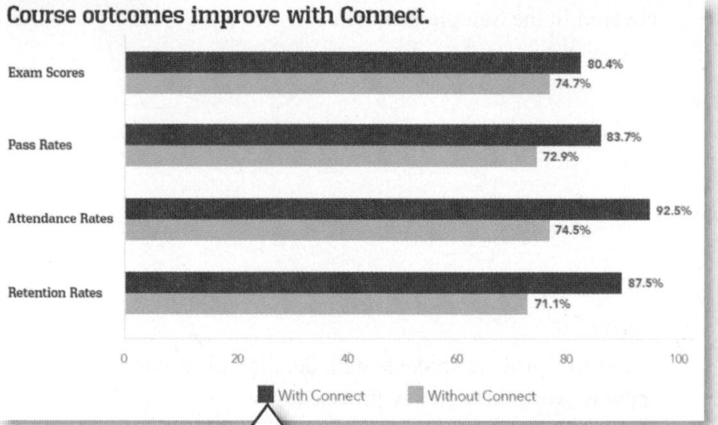

Using **Connect** improves passing rates by **10.8%** and retention by **16.4%.**

88% of instructors who use **Connect** require it; instructor satisfaction **increases** by 38% when **Connect** is required.

Analytics

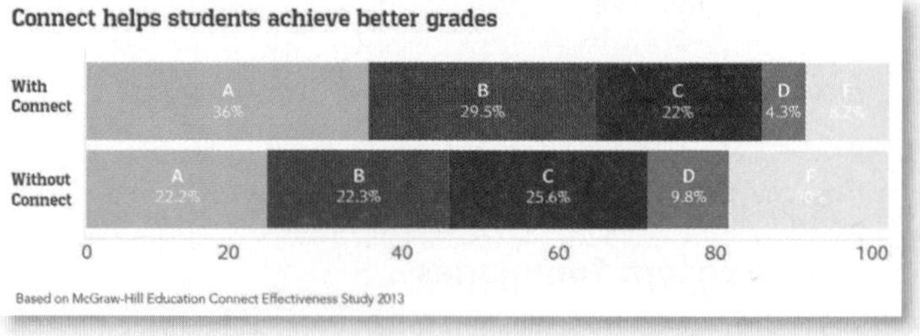

Connect Insight®

Connect Insight is Connect's new one-of-a-kind visual analytics dashboard—now available for both instructors and students—that provides at-a-glance information regarding student performance, which is immediately actionable. By presenting assignment, assessment, and topical performance results together with a time metric that is easily visible for aggregate or individual results, Connect Insight gives the user the ability to take a just-in-time approach to teaching and learning, which was never before available. Connect Insight presents data that empowers students and helps instructors improve class performance in a way that is efficient and effective.

Students can view their results for any **Connect** course.

Mobile

Connect's new, intuitive mobile interface gives students and instructors flexible and convenient, anytime–anywhere access to all components of the Connect platform.

Adaptive

THE FIRST AND ONLY **ADAPTIVE READING EXPERIENCE** DESIGNED TO TRANSFORM THE WAY STUDENTS READ

More students earn **A's** and **B's** when they use McGraw-Hill Education **Adaptive** products.

SmartBook®

Proven to help students improve grades and study more efficiently, SmartBook contains the same content within the print book, but actively tailors that content to the needs of the individual. SmartBook's adaptive technology provides precise, personalized instruction on what the student should do next, guiding the student to master and remember key concepts, targeting gaps in knowledge and offering customized feedback, and driving the student toward comprehension and retention of the subject matter. Available on smartphones and tablets, SmartBook puts learning at the student's fingertips—anywhere, anytime.

Over **4 billion questions** have been answered, making McGraw-Hill Education products more intelligent, reliable, and precise.

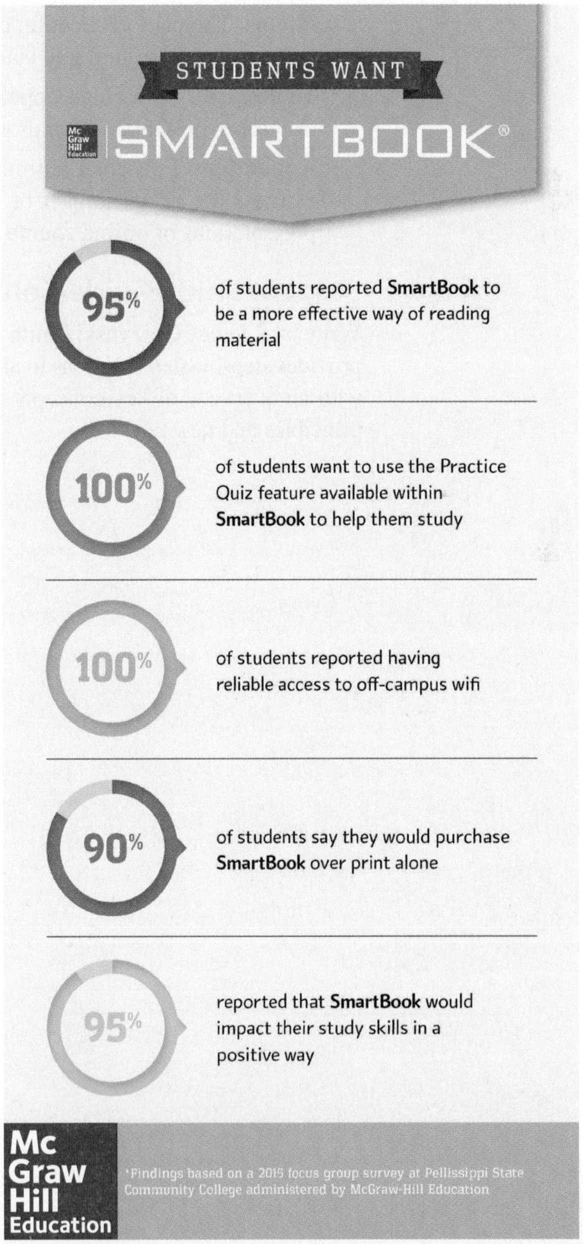

STUDENTS WANT
SMARTBOOK®
95%
of students reported SmartBook to be a more effective way of reading material
100%
of students want to use the Practice Quiz feature available within SmartBook to help them study
100%
of students reported having reliable access to off-campus wifi
90%
of students say they would purchase SmartBook over print alone
95%
reported that SmartBook would impact their study skills in a positive way
Mc Graw Hill Education
*Findings based on a 2015 focus group survey at Pellissippi State Community College administered by McGraw-Hill Education

Learning Resources for Instructors and Students

The following items may accompany this text. Please consult your McGraw-Hill representative for policies, prices, and availability as some restrictions may apply.

Presentation Tools

Within the Instructor's Presentation Tools, instructors have access to editable PowerPoint lecture outlines, which appear as ready-made presentations that combine art and lecture notes for each chapter of the text. For instructors who prefer to create their lecture notes from scratch, all illustrations, photos, tables, *How To*'s, and Sample Problems are pre-inserted by chapter into a separate set of PowerPoint slides. They are also available as individual .jpg files.

Photos, artwork, animations, and other media types can be used to create customized lectures, visually enhanced tests and quizzes, compelling course websites, or attractive printed support materials. All assets are copyrighted by McGraw-Hill Higher Education, but can be used by instructors for classroom purposes. The visual resources in this collection include:

- **Art** Full-color digital files of all illustrations in the book can be readily incorporated into lecture presentations, exams, or custom-made classroom materials.
- **Photos** The photo collection contains digital files of photographs from the text, which can be reproduced for multiple classroom uses.
- **Tables** Every table that appears in the text has been saved in electronic form for use in classroom presentations and/or quizzes.
- **Animations** Full-color animations illustrating important processes are also provided. Harness the visual impact of concepts in motion by importing these files into classroom presentations or online course materials.

Student Study Guide/Solutions Manual

Written by Janice Gorzynski Smith and Erin R. Smith, the Student Study Guide/Solutions Manual provides step-by-step solutions to all in-chapter and end-of-chapter problems. Each chapter begins with an overview of key concepts and includes a short-answer practice test on the fundamental principles and new reactions.

Acknowledgments

Organic Chemistry with Biological Topics is an outgrowth of many fruitful discussions with McGraw-Hill personnel about how best to meld biological applications with basic organic chemistry. Special thanks go to Brand Manager Andrea Pellerito, an organic chemist with extensive teaching experience, who understood the need to maintain the integrity and rigor of organic chemistry in this approach, and devised a method to bring this plan to reality.

Special thanks are also due to Senior Product Developer Mary Hurley, who skillfully navigated the logistics involved with integrating a new project within the framework of an existing text. Much appreciation also goes to Production Manager Sherry Kane, who managed an aggressive but workable production schedule. In truth, this new text is the result of an entire team of publishing professionals, beginning with manuscript preparation and culminating with publication of the completed text that is brought to the chemistry community through the dedicated work of the marketing and sales team. Our sincere appreciation goes out to all of them.

JGS: I especially thank my husband Dan and the other members of my immediate family, who have experienced the day-do-day demands of living with a busy author. The joys and responsibilities of the family have always kept me grounded during the rewarding but sometimes all-consuming process of writing a textbook. This book, like prior editions of *Organic Chemistry*, is dedicated to my wonderful daughter Megan, who passed away after a nine-year battle with cystic fibrosis.

HVS: I am honored to be working with Jan Smith and have already learned so much from her. Thanks to my colleagues Steve Wood, Megan Brennan, Charlie Cox, Jen Schwartz Poehlmann, Chris Chidsey, Dan Stack, and Justin Du Bois for many great conversations about using biological examples to teach the fundamental concepts of organic chemistry. Work on this book would not have been possible without the support of my husband Trent and our three energetic children, Zach, Grady, and Elli. I am also grateful for the encouragement of my mother and brother, Jeanette and Devin Vollmer. This book is dedicated to my father, Chuck Vollmer, who could not have been prouder of my work on this book, but passed away before it was published.

Among the many others that go unnamed but who have profoundly affected this work are the thousands of students we have been lucky to teach over many years. We have learned so much from our daily interactions with them, and we hope that the wider chemistry community can benefit from this experience.

This edition has evolved based on the helpful feedback of many people who reviewed the fourth edition text and digital products, class-tested the book, and attended focus groups or symposiums. These many individuals have collectively provided constructive improvements to the project.

Listed below are the reviewers of the *Organic Chemistry,* fourth edition text:

Steven Castle, *Brigham Young University*
Ihsan Erden, *San Francisco State University*
Andrew Frazer, *University of Central Florida, Orlando*
Tiffany Gierasch, *University of Maryland, Baltimore County*
Anne Gorden, *Auburn University*
Michael Lewis, *Saint Louis University*
Eugene A. Mash, Jr., *University of Arizona*
Mark McMills, *Ohio University*
Joan Mutanyatta–Comar, *Georgia State University*
Felix Ngassa, *Grand Valley State University*
Michael Rathke, *Michigan State University*
Jacob Schroeder, *Clemson University*
Keith Schwartz, *Portland State University*
John Selegue, *University of Kentucky*
Paul J. Toscano, *University at Albany, SUNY*
Jane E. Wissinger, *University of Minnesota*

The following contributed to the editorial direction of *Organic Chemistry,* fifth edition, by responding to our survey on the MCAT and the organic chemistry course student population:

Chris Abelt, *College of William and Mary*
Orlando Acevedo, *Auburn University*
Kim Albizati, *University of California, San Diego*
Merritt Andrus, *Brigham Young University*
Ardeshir Azadnia, *Michigan State University*
Susan Bane, *Binghamton University*
Russell Barrows, *Metropolitan State University of Denver*
Peter Beak, *University of Illinois, Urbana—Champaign*
Phil Beauchamp, *Cal Poly, Pomona*
Michael Berg, *Virginia Tech*
K. Darrell Berlin, *Oklahoma State University*
Thomas Bertolini, *University of South Carolina*
Ned Bowden, *University of Iowa*
David W. Brown, *Florida Gulf Coast University*
Rebecca Broyer, *University of Southern California*
Arthur Bull, *Oakland University*
K. Nolan Carter, *University of Central Arkansas*
Steven Castle, *Brigham Young University*
Victor Cesare, *St. John's University*
Manashi Chatterjee, *University of Nebraska, Lincoln*
Melissa Cichowicz, *West Chester University*

Jeff Corkill, *Eastern Washington University, Cheney*
Sulekha Coticone, *Florida Gulf Coast University*
Michael Crimmins, *University of North Carolina at Chapel Hill*
Eric Crumpler, *Valencia College*
David Dalton, *Temple University*
Rick Danheiser, *Massachusetts Institute of Technology*
Tammy Davidson, *University of Florida*
Brenton DeBoef, *University of Rhode Island*
Amy Deveau, *University of New England*
Kenneth M. Doxsee, *University of Oregon*
Larissa D'Souza, *Johns Hopkins University*
Philip Egan, *Texas A&M University, Corpus Christi*
Seth Elsheimer, *University of Central Florida*
John Esteb, *Butler University*
Steve Fleming, *Temple University*
Marion Franks, *North Carolina A&T State University*
Andy Frazer, *University of Central Florida*
Brian Ganley, *University of Missouri, Columbia*
Robert Giuliano, *Villanova University*
Anne Gorden, *Auburn University*
Carlos G. Gutierrez, *California State University, Los Angeles*
Scott Handy, *Middle Tennessee State University*
Rick Heldrich, *College of Charleston*
James Herndon, *New Mexico State University*
Kathleen Hess, *Brown University*
Sean Hickey, *University of New Orleans*
Carl Hoeger, *University of California, San Diego*
Javier Horta, *University of Massachusetts, Lowell*
Bob A. Howell, *Central Michigan University*
Jennifer Irvin, *Texas State University*
Phil Janowicz, *Cal State, Fullerton*
Mohammad Karim, *Tennessee State University*
Mark L. Kearley, *Florida State University*
Amy Keirstead, *University of New England*
Margaret Kerr, *Worcester State University*
James Kiddle, *Western Michigan University*
Jisook Kim, *University of Tennessee at Chattanooga*
Angela King, *Wake Forest University*
Margaret Kline, *Santa Monica College*
Dalila G. Kovacs, *Grand Valley State University*
Deborah Lieberman, *University of Cincinnati*
Carl Lovely, *University of Texas, Arlington*
Kristina Mack, *Grand Valley State University*
Daniel Macks, *Towson University*
Vivian Mativo, *Georgia Perimeter College, Clarkston*
Mark McMills, *Ohio University*
Stephen Mills, *Xavier University*
Robert Minto, *Indiana University–Purdue University, Indianapolis*
Debbie Mohler, *James Madison University*
Kathleen Morgan, *Xavier University of Louisiana*
Paul Morgan, *Butler University*
James C. Morris, *Georgia Institute of Technology*
Linda Munchausen, *Southeastern Louisiana University*
Toby Nelson, *Oklahoma State University*
Felix Ngassa, *Grand Valley State University*
George A. O'Doherty, *Northeastern University*
Anne Padias, *University of Arizona*
Dan Paschal, *Georgia Perimeter College*
Richard Pennington, *Georgia Gwinnett College*
John Pollard, *University of Arizona*
Gloria Proni, *John Jay College*
Khalilah Reddie, *University of Massachusetts, Lowell*
Joel M. Ressner, *West Chester University of Pennsylvania*
Christine Rich, *University of Louisville*
Carmelo Rizzo, *Vanderbilt University*
Harold R. Rogers, *California State University, Fullerton*
Paul B. Savage, *Brigham Young University*
Deborah Schwyter, *Santa Monica College*
Holly Sebahar, *University of Utah*
Laura Serbulea, *University of Virginia*
Abid Shaikh, *Georgia Southern University*
Kevin Shaughnessy, *The University of Alabama*
Joel Shulman, *University of Cincinnati*
Joseph M. Simard, *University of New England*
Rhett Smith, *Clemson University*
Priyantha Sugathapala, *University at Albany, SUNY*
Claudia Taenzler, *University of Texas at Dallas*
Robin Tanke, *University of Wisconsin, Stevens Point*
Richard T. Taylor, *Miami University, Oxford*
Edward Turos, *University of South Florida*
Ted Wood, *Pierce College*
Kana Yamamoto, *University of Toledo*

The following individuals helped write and review learning goal-oriented content for **LearnSmart for Organic Chemistry:** David G. Jones, Vistamar School; and Adam I. Keller, Columbus State Community College. Andrea Leonard of the University of Louisiana, Lafayette, revised the PowerPoint Lectures, and Elizabeth Clizbe, University at Buffalo, SUNY, revised the Test Bank for Organic Chemistry with Biological Topics, fifth edition.

Although every effort has been made to make this text and its accompanying Student Study Guide/Solutions Manual as error-free as possible, some errors undoubtedly remain. Please feel free to email one of the authors about any inaccuracies, so that subsequent editions may be further improved.

With much aloha,

Janice Gorzynski Smith
jgsmith@hawaii.edu

Heidi R. Vollmer–Snarr
hrvsnarr@stanford.edu

List of *How To's*

How To boxes provide detailed instructions for key procedures that students need to master. Below is a list of each *How To* and where it is presented in the text.

Chapter 1 **Structure and Bonding**
How To Draw a Lewis Structure 14
How To Interpret a Skeletal Structure 33

Chapter 2 **Acids and Bases**
How To Determine the Relative Acidity of Protons 77

Chapter 4 **Alkanes**
How To Name an Alkane Using the IUPAC System 141
How To Name a Cycloalkane Using the IUPAC System 145
How To Draw a Newman Projection 151
How To Draw the Chair Form of Cyclohexane 160
How To Draw the Two Conformations for a Substituted Cyclohexane 162
How To Draw Two Conformations for a Disubstituted Cyclohexane 165

Chapter 5 **Stereochemistry**
How To Assign *R* or *S* to a Stereogenic Center 193
How To Find and Draw All Possible Stereoisomers for a Compound with Two Stereogenic Centers 197

Chapter 7 **Alkyl Halides and Nucleophilic Substitution**
How To Name an Alkyl Halide Using the IUPAC System 257

Chapter 9 **Alcohols, Ethers, and Related Compounds**
How To Name an Alcohol Using the IUPAC System 342

Chapter 10 **Alkenes**
How To Name an Alkene 395
How To Assign the Prefixes *E* and *Z* to an Alkene 397

Chapter 11 **Alkynes**
How To Develop a Retrosynthetic Analysis 453

Chapter 13 **Mass Spectrometry and Infrared Spectroscopy**
How To Use MS and IR for Structure Determination 526

Chapter 14 **Nuclear Magnetic Resonance Spectroscopy**
How To Use ^{1}H NMR Data to Determine a Structure 562

Chapter 16 **Conjugation, Resonance, and Dienes**
How To Draw the Product of a Diels–Alder Reaction 630

Chapter 17 **Benzene and Aromatic Compounds**
How To Use the Inscribed Polygon Method to Determine the Relative Energies of MOs for Cyclic, Completely Conjugated Compounds 672

Chapter 18 **Reactions of Aromatic Compounds**
How To Determine the Directing Effects of a Particular Substituent 707

Chapter 21 **Aldehydes and Ketones—Nucleophilic Addition**
How To Determine the Starting Materials for a Wittig Reaction Using Retrosynthetic Analysis 848

Chapter 22 **Carboxylic Acids and Their Derivatives—Nucleophilic Acyl Substitution**
How To Name an Ester (RCO_2R') Using the IUPAC System 884
How To Name a Thioester (RCOSR') Using the IUPAC System 884
How To Name a 2° or 3° Amide 885

Chapter 24 **Carbonyl Condensation Reactions**
How To Synthesize a Compound Using the Aldol Reaction 978
How To Synthesize a Compound Using the Robinson Annulation 999

Chapter 25 **Amines**
How To Name 2° and 3° Amines with Different Alkyl Groups 1013

Chapter 26 **Amino Acids and Proteins**
How To Use (*R*)-α-Methylbenzylamine to Resolve a Racemic Mixture of Amino Acids 1072
How To Synthesize a Dipeptide from Two Amino Acids 1084
How To Synthesize a Peptide Using the Merrifield Solid Phase Technique 1089

Chapter 27 **Carbohydrates**
How To Draw a Haworth Projection from an Acyclic Aldohexose 1122

List of Mechanisms

Mechanisms are the key to understanding the reactions of organic chemistry. For this reason, great care has been given to present mechanisms in a detailed, step-by-step fashion. The list below indicates when each mechanism in the text is presented for the first time.

Chapter 7 Alkyl Halides and Nucleophilic Substitution

7.1 The S_N2 Mechanism 272
7.2 The S_N1 Mechanism 277

Chapter 8 Alkyl Halides and Elimination Reactions

8.1 The E2 Mechanism 312
8.2 The E1 Mechanism 318

Chapter 9 Alcohols, Ethers, and Related Compounds

9.1 Dehydration of 2° and 3° ROH—An E1 Mechanism 354
9.2 Dehydration of a 1° ROH—An E2 Mechanism 355
9.3 A 1,2-Methyl Shift—Carbocation Rearrangement During Dehydration 357
9.4 Dehydration Using $POCl_3$ + Pyridine—An E2 Mechanism 359
9.5 Reaction of a 1° ROH with HX—An S_N2 Mechanism 361
9.6 Reaction of 2° and 3° ROH with HX—An S_N1 Mechanism 362
9.7 Reaction of ROH with $SOCl_2$ + Pyridine—An S_N2 Mechanism 364
9.8 Reaction of ROH with PBr_3—An S_N2 Mechanism 365
9.9 Mechanism of Ether Cleavage in Strong Acid—
$(CH_3)_3COCH_3 + HI \rightarrow (CH_3)_3CI + CH_3I + H_2O$ 371

Chapter 10 Alkenes

10.1 Electrophilic Addition of HX to an Alkene 407
10.2 Electrophilic Addition of H_2O to an Alkene—Hydration 412
10.3 Addition of X_2 to an Alkene—Halogenation 414
10.4 Addition of X and OH—Halohydrin Formation 416
10.5 Addition of H and BH_2—Hydroboration 419

Chapter 11 Alkynes

11.1 Electrophilic Addition of HX to an Alkyne 443
11.2 Addition of X_2 to an Alkyne—Halogenation 444
11.3 Tautomerization in Acid 446
11.4 Hydration of an Alkyne 446

Chapter 12 Oxidation and Reduction

12.1 Addition of H_2 to an Alkene—Hydrogenation 467
12.2 Dissolving Metal Reduction of an Alkyne to a Trans Alkene 473
12.3 Reduction of RX with $LiAlH_4$ 475
12.4 Epoxidation of an Alkene with a Peroxyacid 477
12.5 Oxidation of an Alcohol with CrO_3 486
12.6 Oxidation of a 1° Alcohol to a Carboxylic Acid 486

Chapter 15 Radical Reactions

15.1 Radical Halogenation of Alkanes 584
15.2 Allylic Bromination with NBS 594
15.3 Radical Addition of HBr to an Alkene 599
15.4 Radical Polymerization of $CH_2{=}CHZ$ 603

Chapter 16 Conjugation, Resonance, and Dienes

16.1 Biological Formation of Geranyl Diphosphate 616
16.2 Electrophilic Addition of HBr to a 1,3-Diene—1,2- and 1,4-Addition 625

Chapter 18 Reactions of Aromatic Compounds

18.1 General Mechanism—Electrophilic Aromatic Substitution 688
18.2 Bromination of Benzene 690
18.3 Formation of the Nitronium Ion ($^+NO_2$) for Nitration 691
18.4 Formation of the Electrophile $^+SO_3H$ for Sulfonation 692

18.5 Formation of the Electrophile in Friedel–Crafts Alkylation—Two Possibilities 694
18.6 Friedel–Crafts Alkylation Using a 3° Carbocation 694
18.7 Formation of the Electrophile in Friedel–Crafts Acylation 695
18.8 Friedel–Crafts Alkylation Involving Carbocation Rearrangement 696
18.9 A Rearrangement Reaction Beginning with a 1° Alkyl Chloride 697
18.10 Nucleophilic Aromatic Substitution by Addition–Elimination 716
18.11 Nucleophilic Aromatic Substitution by Elimination–Addition: Benzyne 717
18.12 Benzylic Bromination 719

Chapter 20 Introduction to Carbonyl Chemistry; Organometallic Reagents; Oxidation and Reduction
20.1 Nucleophilic Addition—A Two-Step Process 777
20.2 Nucleophilic Substitution—A Two-Step Process 778
20.3 $LiAlH_4$ Reduction of RCHO and $R_2C{=}O$ 782
20.4 Reduction of RCOCl and RCOOR' with a Metal Hydride Reagent 788
20.5 Reduction of an Amide to an Amine with $LiAlH_4$ 790
20.6 Nucleophilic Addition of R"MgX to RCHO and $R_2C{=}O$ 796
20.7 Reaction of R"MgX or R"Li with RCOCl and RCOOR' 805
20.8 Carboxylation—Reaction of RMgX with CO_2 808
20.9 1,2-Addition to an α,β-Unsaturated Carbonyl Compound 810
20.10 1,4-Addition to an α,β-Unsaturated Carbonyl Compound 810

Chapter 21 Aldehydes and Ketones—Nucleophilic Addition
21.1 General Mechanism—Nucleophilic Addition 839
21.2 General Mechanism—Acid-Catalyzed Nucleophilic Addition 839
21.3 Nucleophilic Addition of $^-$CN—Cyanohydrin Formation 843
21.4 The Wittig Reaction 847
21.5 Imine Formation from an Aldehyde or Ketone 851
21.6 Enamine Formation from an Aldehyde or Ketone 853
21.7 Base-Catalyzed Addition of H_2O to a Carbonyl Group 856
21.8 Acid-Catalyzed Addition of H_2O to a Carbonyl Group 856
21.9 Acetal Formation 859
21.10 Acid-Catalyzed Cyclic Hemiacetal Formation 863
21.11 A Cyclic Acetal from a Cyclic Hemiacetal 864

Chapter 22 Carboxylic Acids and Their Derivatives—Nucleophilic Acyl Substitution
22.1 General Mechanism—Nucleophilic Acyl Substitution 892
22.2 Conversion of Acid Chlorides to Anhydrides 896
22.3 Conversion of an Anhydride to an Amide 898
22.4 Conversion of Carboxylic Acids to Acid Chlorides 899
22.5 Fischer Esterification—Acid-Catalyzed Conversion of Carboxylic Acids to Esters 900
22.6 Conversion of Carboxylic Acids to Amides with DCC 902
22.7 Acid-Catalyzed Hydrolysis of an Ester to a Carboxylic Acid 904
22.8 Base-Promoted Hydrolysis of an Ester to a Carboxylic Acid 904
22.9 Amide Hydrolysis in Base 908
22.10 Biological Conversion of a Carboxylate to an Acyl Phosphate 911
22.11 Biological Conversion of an Acyl Phosphate to a Thioester 913
22.12 Hydrolysis of a Nitrile in Base 918
22.13 Reduction of a Nitrile with $LiAlH_4$ 919
22.14 Reduction of a Nitrile with DIBAL-H 919
22.15 Addition of Grignard and Organolithium Reagents (R–M) to Nitriles 920

Chapter 23 Substitution Reactions of Carbonyl Compounds at the α Carbon
23.1 Tautomerization in Acid 937
23.2 Tautomerization in Base 937
23.3 Acid-Catalyzed Halogenation at the α Carbon 948
23.4 Halogenation at the α Carbon in Base 949
23.5 The Haloform Reaction 950

Chapter 24 Carbonyl Condensation Reactions
24.1 The Aldol Reaction 974
24.2 The Retro-Aldol Reaction 976
24.3 Dehydration of β-Hydroxy Carbonyl Compounds with Base 977
24.4 The Intramolecular Aldol Reaction 984
24.5 The Claisen Reaction 986

24.6 The Dieckmann Reaction 990
24.7 The Michael Reaction 995
24.8 The Robinson Annulation 997

Chapter 25 Amines
25.1 The E2 Mechanism for the Hofmann Elimination 1039
25.2 Formation of a Diazonium Salt from a 1° Amine 1042
25.3 Formation of an *N*-Nitrosamine from a 2° Amine 1043
25.4 Azo Coupling 1048

Chapter 26 Amino Acids and Proteins
26.1 Formation of an α-Amino Nitrile 1070
26.2 Edman Degradation 1081

Chapter 27 Carbohydrates
27.1 Glycoside Formation 1128
27.2 Glycoside Hydrolysis 1129

Chapter 28 Lipids
28.1 Biological Formation of Farnesyl Diphosphate 1172
28.2 Isomerization of Geranyl Diphosphate to Neryl Diphosphate 1173

Chapter 29 Carbon–Carbon Bond-Forming Reactions in Organic Synthesis
29.1 Suzuki Reaction 1191
29.2 Heck Reaction 1194
29.3 Formation of Dichlorocarbene 1195
29.4 Addition of Dichlorocarbene to an Alkene 1196
29.5 Simmons–Smith Reaction 1198
29.6 Olefin Metathesis: $2\ RCH{=}CH_2 \rightarrow RCH{=}CHR + CH_2{=}CH_2$ 1200

Chapter 31 Synthetic Polymers (Available online)
31.1 Radical Polymerization of $CH_2{=}CHPh$ 1245
31.2 Forming Branched Polyethylene During Radical Polymerization 1247
31.3 Cationic Polymerization of $CH_2{=}CHZ$ 1248
31.4 Anionic Polymerization of $CH_2{=}CHZ$ 1250
31.5 Ziegler–Natta Polymerization of $CH_2{=}CH_2$ 1253

List of Selected Applications

Applications make any subject seem more relevant and interesting—for nonmajors and majors alike. The following is a list of the biological, medicinal, and environmental applications that have been integrated throughout *Organic Chemistry with Biological Topics.* Each chapter opener showcases an interesting and current application relating to the chapter's topic. (Code: G = general; M = medicinal; B = biological; E = environmental)

Prologue

G Methane, the main component of natural gas
G Ethanol, the alcohol in beverages
E Trichlorofluoromethane, a CFC responsible for destroying the stratospheric ozone layer
M Amoxicillin, a widely used antibiotic
M Fluoxetine, the antidepressant Prozac
M AZT, a drug used to treat HIV
M Capsaicin, a compound found in topical pain relief creams
E DDT, a nonspecific pesticide that persists in the environment
M The antimalarial drugs quinine, chloroquine, and artemisinin

Chapter 1 Structure and Bonding

M L-Dopa, a drug used to treat Parkinson's disease (Chapter opener and Section 1.14)
M Alendronic acid (Fosamax), a drug used to prevent osteoporosis (Section 1.5)
B Enanthotoxin, a poisonous compound isolated from hemlock water dropwort (Section 1.7)
G Vanillin, the principal component in the extract of the vanilla bean (Section 1.8B)
M Structures of active ingredients in common sunscreens (Section 1.8B)
G Ethane, a component of natural gas (Section 1.10A)
G Ethylene, a hydrocarbon used to make the plastic polyethylene (Section 1.10B)
G Acetylene, a gas used in welding torches (Section 1.10C)
G Cucumber aldehyde, the compound responsible for the odor of freshly cut cucumbers (Section 1.10C)
M Sinemet, a drug used to treat Parkinson's disease that combines L-dopa and carbidopa (Section 1.14)
B Vitamin B_6 (Section 1.14)

Chapter 2 Acids and Bases

M Aspirin, a common analgesic and antipyretic (Chapter opener and Section 2.7)
M The acid–base chemistry of morphine (Section 2.1)
M The nasal decongestant pseudoephedrine (Section 2.5, Problem 2.17)
M Glycolic acid, an α-hydroxy acid used in skin care products (Section 2.5, Problem 2.20)
E Sulfuric acid, a major contributor to acid rain (Section 2.6)
M Salicin, an analgesic found in willow bark (Section 2.7)

Chapter 3 Introduction to Organic Molecules and Functional Groups

B Vitamin C, a water-soluble vitamin that is important in the formation of collagen (Chapter opener and Section 3.5B)
E Hemibrevetoxin B, a neurotoxin produced by algal blooms ("red tides") (Section 3.2B)
M Diethyl ether, the first common general anesthetic (Section 3.2B)
B Sucrose and the antibiotic amoxicillin (Section 3.2B, Problem 3.3)
M Dexamethasone, a synthetic steroid (Section 3.2B, Problem 3.5)
B Spermine, isolated from semen, and meperidine, the narcotic Demerol (Section 3.2B, Problem 3.6)
M The anticancer agent doxorubicin (Adriamycin) (Section 3.2B, Problem 3.7)
M Thyrotropin-releasing hormone (Section 3.2C, Problem 3.8)
M Tamiflu, an antiviral drug used to treat influenza (Section 3.2C, Problem 3.9)
B Pyruvic acid, lipoic acid, and folic acid as examples of biological molecules with multiple functional groups (Section 3.2C, Problem 3.10)
B Biological phosphorus compounds (Section 3.2D)
G How geckos use van der Waals forces to stick to walls (Section 3.3B)
B Ionic, water-soluble biological compounds: isopentenyl diphosphate and acetylcholine (Section 3.4C)
G MTBE, a high-octane additive in unleaded gasoline, and 4,4'-dichlorobiphenyl, a PCB (Section 3.4C)
B Phenylalanine and 11-*cis*-retinal (Section 3.4C, Sample Problem 3.4)

B Adrenaline and estrone (Section 3.4C, Problem 3.17)
B Progesterone and testosterone (Section 3.4C, Sample Problem 3.5)
B Norethindrone, an oral contraceptive, and arachidonic acid, a fatty acid (Section 3.4C, Problem 3.18)
B Vitamin A (retinol), a fat-soluble vitamin found in the vision receptors of the eyes (Section 3.5A)
B β-Carotene, a precursor to vitamin A (Section 3.5A)
B Vitamin B_3 and vitamin K_1 (Section 3.5B, Problem 3.19)
B Avocados as a source of pantothenic acid, vitamin B_5 (Section 3.5B, Problem 3.20)
M Morphine and heroin (Section 3.7A, Problem 3.23)
M The antibiotics nonactin and valinomycin (Section 3.7B)
B The reactive features of isopentenyl diphosphate and pyruvic acid (Section 3.8)
B The nucleophilic thiol of coenzyme A (Section 3.8)
B Methionine, ATP, and *S*-adenosylmethionine (Section 3.8, Problem 3.28)
B Amino acids and proteins (Section 3.9A)
B Monosaccharides and carbohydrates (Section 3.9B)
B Nucleotides and nucleic acids (Section 3.9C)
B Lipids (Section 3.9D)
M, B End-of-chapter problems: 3.33–3.35, 3.37, 3.38, 3.40, 3.48–3.57, 3.60, 3.61, 3.63–3.65, and 3.67

Chapter 4 Alkanes

E Oil slicks that result from crude petroleum being spilled into the ocean from oil tankers or oil wells (Chapter opener)
B The cockroach pheromone undecane (Section 4.1)
B Cyclohexane, one component of mangoes (Section 4.1)
B Allicin, a compound responsible for the odor of garlic (Section 4.3)
M Systematic names, generic names, and trade names in over-the-counter drugs like Motrin (Section 4.3)
G Fossil fuels such as natural gas and petroleum (Section 4.7)
E The combustion of alkanes and how it contributes to climate change (Section 4.14B)
B Lipids such as fat-soluble vitamins, phospholipids, waxes, prostaglandins, and steroids (Section 4.15)
B Pristane, a high molecular weight alkane found in shark liver oil (Section 4.15, Problem 4.33)
B End-of-chapter problems: 4.66 and 4.69

Chapter 5 Stereochemistry

M, B Paclitaxel (Taxol), a drug used to treat ovarian, breast, and other cancers (Chapter opener)
B How differences in the three-dimensional structure of starch and cellulose affect their shape and function (Section 5.1)
M, B Identifying stereogenic centers in Darvon (an analgesic), ephedrine (a decongestant), and fructose (a simple sugar) (Section 5.4A)
M The three-dimensional structure of thalidomide, an anti-nausea drug that caused catastrophic birth defects (Section 5.5)
M, B Identifying stereogenic centers in paclitaxel (anticancer agent) and sucrose (Section 5.5)
M Identifying stereogenic centers in gabapentin (a drug used to treat seizures and chronic pain), gabapentin enacarbil, cholesterol, and Zocor (cholesterol-lowering drug) (Section 5.5, Problems 5.9 and 5.10)
M Assigning *R* and *S* configurations in the drugs Plavix and Zestril (Section 5.6, Problems 5.14 and 5.15)
B The sweetener sorbitol (Section 5.9, Problem 5.24)
B The specific rotation of MSG, a common flavor enhancer (Section 5.12D, Problem 5.32)
M Chiral drugs and how mirror image isomers can have drastically different properties—the analgesic ibuprofen, the antidepressant fluoxetine, and the anti-inflammatory agent naproxen (Section 5.13A)
B The sense of smell and how mirror image isomers (e.g., carvone and celery ketone) can smell different (Section 5.13B and Problem 5.35)
M, B End-of-chapter problems: 5.36, 5.43, 5.49, 5.50, 5.53, 5.55, 5.60, and 5.65–5.71

Chapter 6 Understanding Organic Reactions

B Entropy changes in the metabolism of glucose (Chapter opener and Section 6.4)
B A biological substitution reaction: the hydrolysis of a triacylglycerol to glycerol + fatty acids (Section 6.2A)
B A biological elimination reaction in the citric acid cycle (Section 6.2B)
B A biological addition reaction with a thioester, a key step in fatty acid metabolism (Section 6.2C)
B Four enzyme-catalyzed steps in the citric acid cycle (Section 6.2C, Problem 6.2)
B The air oxidation of vegetable oils (Section 6.3C, Sample Problem 6.1)
B Examples of exothermic reactions: the hydrolysis of ATP and the oxidation of glucose (Section 6.4)
B Coupled reactions in metabolism (Section 6.5C)
G The reaction of gasoline with O_2 (Section 6.9A)
G Refrigeration and spoilage (Section 6.9A)
B Enzymes, biological catalysts (Section 6.11)
B End-of-chapter problems: 6.27, 6.28, 6.32, 6.39, 6.41, 6.52, and 6.56

Chapter 7 Alkyl Halides and Nucleophilic Substitution

M Flonase, a synthetic steroid used to treat seasonal allergies (Chapter opener)

B, M Telfairine (insecticide) and halomon (antitumor agent), halogenated compounds isolated from red algae (Section 7.1, Problem 7.1)

B, M Simple alkyl halides—chloromethane (found in emissions from volcanoes), dichloromethane (once used to decaffeinate coffee), and halothane (a general anesthetic) (Section 7.4)

E CFCs and DDT, two polyhalogenated compounds once widely used, now discontinued because of adverse environmental effects (Section 7.4)

B, M Ma'ilione and plocoralide B, halogenated compounds isolated from red algae (Section 7.4)

B Chondrocole A, a marine natural product isolated from red seaweed (Section 7.4, Problem 7.5)

M The antiseptic CPC (Section 7.6)

M Nucleophilic substitutions in the syntheses of Myambutol (used to treat tuberculosis) and Prozac (an antidepressant) (Section 7.11)

M The synthesis of imatinib, an anticancer drug, by a nucleophilic substitution reaction (Section 7.11, Problem 7.22)

B, M Biological nucleophilic substitution reactions: phosphate leaving groups and *S*-adenosylmethionine (SAM) (Section 7.16)

B The biological synthesis of adrenaline using SAM (Section 7.16)

B The synthesis of nicotine using SAM (Section 7.16, Problem 7.36)

M The importance of organic synthesis in preparing useful drugs such as aspirin (Section 7.18)

B, M End-of-chapter problems: 7.64–7.66, 7.70, 7.76

Chapter 8 Alkyl Halides and Elimination Reactions

E DDE, a degradation product of the pesticide DDT (Chapter opener and Section 8.1)

B Ethylene, a hormone that regulates plant growth and fruit ripening (Section 8.2)

B Classifying alkenes using vitamins A and D (Section 8.2, Problem 8.2)

B Identifying stereoisomerism in alkenes using (*E*)-ocimene, found in lilacs (Section 8.2, Problem 8.4)

B, M Elimination reactions in the syntheses of a prostaglandin, quinine, and estradiol (Section 8.4)

B, M End-of-chapter problems: 8.29 and 8.66

Chapter 9 Alcohols, Ethers, and Related Compounds

B Linalool, an alcohol used in scented soaps and lotions and as an insecticide for controlling fleas and cockroaches (Chapter opener)

B Classifying alcohols using cortisol (Section 9.1)

B Classifying ethers and alcohols using brevenal, a marine natural product formed in red tides (Section 9.1, Problem 9.1)

G, E Ethanol, a gasoline additive and renewable fuel source that can be produced from the fermentation of carbohydrates in grains (Section 9.5A)

G Useful simple alcohols: methanol (wood alcohol), isopropanol (rubbing alcohol), and ethylene glycol (antifreeze) (Section 9.5A)

M Diethyl ether, a general anesthetic (Section 9.5B)

M Sevoflurane, a halogenated ether currently used as a general anesthetic (Section 9.5B)

M Medicinal epoxides: eplerenone (a drug that reduces cardiovascular risk in patients who have already had a heart attack) and tiotropium bromide (a bronchodilator) (Section 9.5C)

M A Williamson ether synthesis in the preparation of paroxetine (antidepressant) (Section 9.6, Problem 9.9)

G The syntheses of vitamin A and patchouli alcohol (used in perfumery) using a dehydration reaction (Section 9.10)

G, B The unpleasant odors related to skunks, onions, and human sweat (Section 9.15A)

B The oxidation of a thiol to a disulfide using grapefruit mercaptan (Section 9.15A, Problem 9.31)

B The synthesis of SAM from methionine and ATP by an S_N2 reaction (Section 9.15B)

M The syntheses of salmeterol and albuterol (two bronchodilators) by the opening of an epoxide ring (Section 9.16)

M The design of asthma drugs that block the synthesis of leukotrienes, highly potent molecules that contribute to the asthmatic response (Section 9.17)

B The metabolism of polycyclic aromatic hydrocarbons (PAHs) to carcinogens that disrupt normal cell function, resulting in cancer or cell death (Section 9.18)

M End-of-chapter problems: 9.49, 9.73, and 9.81

Chapter 10 Alkenes

B The unsaturated fatty acids found in kukui nuts (Chapter opener)

M Degrees of unsaturation in the drugs Ambien and mefloquine (Section 10.2, Problem 10.3)

B 11-*cis*-Retinal, the light-sensitive aldehyde involved in the vision of all vertebrates, arthropods, and mollusks (Section 10.3B, Problem 10.7)

B The sex pheromone of the codling moth (Section 10.3B, Problem 10.9)
G Ethylene, the starting material for preparing polyethylene and a variety of other polymers (Section 10.5)
B The naturally occurring alkenes β-carotene, zingiberene, (*R*)-limonene, and α-farnesene (Section 10.5)
B Triacylglycerols, fatty acids, fats, and oils (Section 10.6)
B Omega-3 fatty acids (Section 10.6, Problem 10.11)
B The synthesis of the female sex hormone estrone (Section 10.15B)
M The synthesis of artemisinin, an antimalarial drug, by a hydroboration–oxidation step (Section 10.16B)
B, M End-of-chapter problems: 10.37, 10.43, 10.44, 10.45, 10.69, and 10.71

Chapter 11 Alkynes
M Oral contraceptives (Chapter opener and Section 11.4)
B Nepheliosyne B, a novel acetylenic fatty acid (Section 11.1, Problem 11.1)
M Synthetic hormones Mifepristone and Plan B, drugs that prevent pregnancy (Section 11.4)
B Histrionicotoxin, a diyne isolated from the skin of a frog, used as a poison on arrow tips by the Choco tribe of South America (Section 11.4)
B Acetylide anion reactions in the synthesis of two marine natural products (Section 11.11)
M, B End-of-chapter problems: 11.25 and 11.43

Chapter 12 Oxidation and Reduction
B The metabolism of ethanol, the alcohol in alcoholic beverages (Chapter opener and Section 12.14)
B The partial hydrogenation of vegetable oils and the formation of "trans fats" (Section 12.4)
B The reduction of an alkyne to form *cis*-jasmone, a component of perfume (Section 12.5B, Problem 12.10)
B The use of disparlure, a sex hormone, in controlling the spread of gypsy moths (Section 12.8B)
G The production of ozone from O_2 during electrical storms (Section 12.10)
G Blood alcohol screening (Section 12.12)
E Green chemistry—environmentally benign oxidation reactions (Section 12.13)
B Biological oxidations (Section 12.14)
B The synthesis of insect pheromones using asymmetric epoxidation (Section 12.15)
B, M End-of-chapter problems: 12.37, 12.41, 12.51, 12.53, 12.55, 12.56, 12.60, and 12.61

Chapter 13 Mass Spectrometry and Infrared Spectroscopy
M Infrared spectroscopy and the structure determination of penicillin (Chapter opener and Section 13.8)
M Applying the nitrogen rule to 3-methylfentanyl and MPPP, two drugs that mimic the effects of heroin (Section 13.1)
B Determining the molecular formula of nootkatone (found in grapefruit) (Section 13.1, Problem 13.3)
M Using instrumental analysis to detect THC, the active compound in marijuana, and other drugs (Section 13.4B)
B Mass spectrometry and high molecular weight biomolecules (Section 13.4C)
B End-of-chapter problems: 13.29, 13.30, 13.44, and 13.62

Chapter 14 Nuclear Magnetic Resonance Spectroscopy
B Modern spectroscopic methods and the structure of palau'amine, a complex natural product isolated from a sea sponge (Chapter opener and Problem 14.23)
E The high-octane gasoline additive MTBE, which has contaminated water supplies (Section 14.1B)
B Esters of chrysanthemic acid (from chrysanthemum flowers) as insecticides (Section 14.11, Problem 14.29)
M Magnetic resonance imaging (Section 14.12)
B End-of-chapter problem: 14.37

Chapter 15 Radical Reactions
G Polystyrene, a common synthetic polymer used in packaging materials and beverage cups (Chapter opener)
E Ozone destruction and CFCs (Section 15.9)
B The oxidation of unsaturated lipids by radical reactions (Section 15.11)
M, B Two antioxidants—naturally occurring vitamin E and synthetic BHT (Section 15.12)
B The antioxidant rosmarinic acid (Section 15.12)
G The formation of useful polymers from monomers by radical reactions (Section 15.14)
B, G, M End-of-chapter problems: 15.63, 15.66–15.70, and 15.79

Chapter 16 Conjugation, Resonance, and Dienes
M The laboratory synthesis of morphine by a Diels–Alder reaction (Chapter opener)
B Allylic carbocations in biological reactions, such as the formation of geranyl diphosphate (Section 16.2B)
B Isoprene, a conjugated compound that helps plants tolerate heat stress (Section 16.7)
M The antioxidant lycopene (Sections 16.7 and 16.15A)
M Simvastatin (Zocor) and calcitriol (Rocaltrol), two drugs with conjugated double bonds (Section 16.7)
B The synthesis of tetrodotoxin (found in Japanese puffer fish) by a Diels–Alder reaction (Section 16.12)

B	The trienes zingiberene and β-sesquiphellandrene found in ginger root (Section 16.13A, Problem 16.21)
B	The Diels–Alder reaction in the synthesis of steroids (Section 16.14C)
G	Why lycopene and other highly conjugated compounds are colored (Section 16.15A)
G	How sunscreens work (Section 16.15B)
B, M	End-of-chapter problems: 16.54, 16.61, 16.69, 16.73, and 16.75
Chapter 17	**Benzene and Aromatic Compounds**
B	Hydrogen bonding in complementary DNA base pairs (Chapter opener)
B	Novocain, a disubstituted aniline derivative (Section 17.3B)
B	Histamine and scombroid fish poisoning (Section 17.8A)
M	Quinine, an antimalarial drug (Section 17.8A, Problem 17.18)
M	Januvia, a drug used to treat type 2 diabetes (Section 17.8A, Problem 17.19)
B	DNA (Section 17.8B)
B	Aromatase inhibitors for estrogen-dependent cancer treatment (Section 17.11)
M, B	End-of-chapter problems: 17.35–17.37 and 17.59–17.64
Chapter 18	**Reactions of Aromatic Compounds**
B	Vitamin K_1, a fat-soluble vitamin that regulates the synthesis of proteins needed for blood to clot (Chapter opener and Section 18.5E)
M, E	Biologically active aryl chlorides: the drugs bupropion and chlorpheniramine, and 2,4-D and 2,4,5-T, herbicide components of the defoliant Agent Orange (Section 18.3)
M	Intramolecular Friedel–Crafts acylation in the synthesis of LSD (Section 18.5D)
M	The synthesis of sertraline (Zoloft), an SSRI antidepressant (Section 18.5D, Problem 18.10)
B	A biological Friedel–Crafts reaction (Section 18.5E)
M	Nucleophilic aromatic substitution by addition–elimination in the synthesis of Prozac (Section 18.13A, Problem 18.25)
M	Benzocaine, the active ingredient in the over-the-counter topical anesthetic Orajel (Section 18.15C)
M, G, B	End-of-chapter problems: 18.42–18.44, 18.61, 18.63, 18.67, 18.68, 18.70, 18.73, and 18.77
Chapter 19	**Carboxylic Acids and the Acidity of the O–H Bond**
B	The essential amino acid lysine (Chapter opener)
B	Hexanoic acid, the foul-smelling carboxylic acid in ginkgo seeds (Section 19.2B)
B	Biologically significant diacids: succinic acid, malic acid, and α-ketoglutaric acid (Section 19.2C)
M	Depakote (used to treat seizures) (Section 19.2C, Problem 19.5)
B	Biologically significant carboxylic acids: formic acid (ant stings), acetic acid (vinegar), butanoic acid (body odor), oxalic acid (spinach), and lactic acid (sour milk) (Section 19.5)
B	GHB (4-hydroxybutanoic acid), an illegal recreational intoxicant used as a "date rape" drug (Section 19.5)
M, B	How NSAIDs block the synthesis of prostaglandins to prevent inflammation (Section 19.6)
B	Mandelic acid, a naturally occurring carboxylic acid in plums and peaches (Section 19.9, Problem 19.13)
B	Applications of the Henderson–Hasselbalch equation to biological molecules (Section 19.10)
M	The irritant urushiol in poison ivy (Section 19.12, Problem 19.19)
B	Organic acids containing S and P: sulfonic acids and phosphoric acid esters (Section 19.14)
B	An introduction to amino acids, the building blocks of proteins; why vegetarians must have a balanced diet (Section 19.15)
B, M	End-of-chapter problems: 19.39, 19.51, 19.59–19.65, 19.68, and 19.69
Chapter 20	**Introduction to Carbonyl Chemistry; Organometallic Reagents; Oxidation and Reduction**
B	The use of a reduction reaction to synthesize the marine neurotoxin ciguatoxin CTX3C (Chapter opener and Section 20.7A)
B	The aldehyde α-sinensal, a component of mandarin oil (Section 20.1, Problem 20.1)
M	The anticancer drug Taxol and nucleophilic substitution (Section 20.2, Problem 20.2)
B, M	Reduction reactions in the synthesis of the analgesic ibuprofen and the perfume component muscone (Section 20.4)
M	The synthesis of the long-acting bronchodilator salmeterol (Section 20.6A)
M	The use of CBS reagents in the synthesis of cholesterol-lowering drugs (Section 20.6A, Problem 20.9)
B	Biological oxidation–reduction reactions with the coenzymes NADH and NAD^+ (Section 20.6B)
B	The synthesis of NAD^+ from the vitamin niacin (Section 20.6B)
M	The use of organometallic reagents to synthesize the oral contraceptive ethynylestradiol (Section 20.10C)
B	The use of Grignard reagents in the synthesis of C_{18} juvenile hormones and the use of juvenile hormone mimics to regulate the life cycle of insects (Section 21.10C)
B	The use of organolithium reagents in the synthesis of two components of lavender oil (Section 20.11, Problem 20.24)

M The use of protecting groups in the conversion of estrone to ethynylestradiol (Section 20.12, Problem 20.26)
M, B End-of-chapter problems: 20.50, 20.56, 20.61, 20.68, 20.75, and 20.78

Chapter 21 Aldehydes and Ketones—Nucleophilic Addition

M Digoxin, a naturally occurring drug isolated from the woolly foxglove plant and used to treat congestive heart failure (Chapter opener and Problem 21.37)
B Determining the IUPAC names of neral (from lemongrass) and cucumber aldehyde (Section 21.2E, Problem 21.7)
G Formaldehyde and acetone, an industrially useful aldehyde and ketone (Section 21.5)
B Examples of naturally occurring compounds that contain aldehydes or ketones—vanillin, citronellal, cinnamaldehyde, and geranial (Section 21.5)
M Cortisone and prednisone, steroids that contain ketones (Section 21.5)
B Naturally occurring cyanohydrin derivatives: linamarin, from cassava root; and amygdalin, from apricot, peach, and wild cherry pits (Section 21.9B)
B The use of the Wittig reaction in the synthesis of β-carotene, the orange pigment in carrots (Section 21.10B)
B The role of rhodopsin in the chemistry of vision (Section 21.11B)
B The acid-catalyzed hydrolysis of safrole, a carcinogen once used in root beer (Section 21.14B, Problem 21.33)
B, M The acid-catalyzed hydrolysis of the acetal in oleandrin (Section 21.14B, Problem 21.34)
B The carbohydrates glucose and lactose (Section 21.17)
M The role of carbohydrates in diabetes (Section 21.17)
B The carbohydrate galactose (Section 21.17, Problem 21.39)
M, B End-of-chapter problems: 21.52, 21.65, 21.69–21.71, 21.79, 21.80, 21.82, and 21.84–21.86

Chapter 22 Carboxylic Acids and Their Derivatives—Nucleophilic Acyl Substitution

B, M Ginkgolide B, a major constituent of the extracts of the ginkgo tree, *Ginkgo biloba* (Chapter opener and Problem 22.21)
B Acyl phosphates, common intermediates in biological pathways (Section 22.1)
B, M Oxytocin, a naturally occuring hormone used to stimulate uterine contractions and induce labor (Section 22.1, Problem 22.1)
B The esters responsible for the odors of banana, mango, and pineapple (Section 22.6A)
M, B Compounds that contain an ester: vitamin C and cocaine (Section 22.6A)
M, B Useful amides: proteins and met-enkephalin (Section 22.6B)
B Mechanism for the synthesis of blattellaquinone, the sex pheromone of the female German cockroach (Section 22.8, Problem 22.13)
M The cholesterol-lowering drug fenofibrate (Section 22.11B, Problem 22.20)
B The hydrolysis of triacylglycerols in the metabolism of lipids (Section 22.12A)
G Olestra, a synthetic fat (Section 22.12A)
G The synthesis of soap (Section 22.12B)
M The mechanism of action of β-lactam antibiotics like penicillin (Section 22.14)
B Acyl phosphates as biological anhydrides (Section 22.16)
B Biological acylation reactions (Section 22.17)
M Cholesteryl esters in plaque, the deposits that form on the inside walls of arteries (Section 22.17)
B The acylation of glucosamine to form NAG, the monomer in chitin (Section 22.17, Problem 22.33)
B, M End-of-chapter problems: 22.49, 22.50, 22.53, 22.54, 22.56, 22.57, 22.59–22.64, 22.71, 22.72, 22.74, 22.79, and 22.85–22.87

Chapter 23 Substitution Reactions of Carbonyl Compounds at the α Carbon

M The synthesis of the anticancer drug tamoxifen (Chapter opener and Section 23.8C)
B Keto–enol tautomerizations in glycolysis (Section 23.2A, Problem 23.2)
M The synthesis of the antimalarial drug quinine by an intramolecular substitution reaction (Section 23.7C)
M The heterocyclic ring system in some antitumor agents (Section 23.8C, Problem 23.19)
M The use of the acetoacetic ester synthesis in the synthesis of illudin-S, an antitumor agent (Section 23.10, Problem 23.27)
M Retrosynthesis of the pain reliever nabumetone (Section 23.10, Problem 23.28)
B, M End-of-chapter problems: 23.38, 23.40, 23.45, 23.53, 23.54, 23.61, 23.64, 23.68, 23.72, and 23.74

Chapter 24 Carbonyl Condensation Reactions

B Stearic acid, a key component of cocoa butter (Chapter opener)
B The E1cB mechanism in biological pathways (Section 24.1C)
B The perfume component flosal, an α,β-unsaturated aldehyde (Section 24.2B, Problem 24.6)
B The synthesis of periplanone B, sex pheromone of the female American cockroach (Section 24.3)
B The synthesis of *ar*-turmerone, a component of turmeric, a principal ingredient in curry powder (Section 24.3)
B The conversion of zingerone to gingerol, components of ginger, using a directed aldol reaction (Section 24.3, Problem 24.10)
M A directed aldol reaction in the synthesis of the drug donepezil (for treating dementia) (Section 24.3, Problem 24.11)

B The synthesis of the steroid progesterone by an intramolecular aldol reaction (Section 24.4)
M Avobenzone, a common ingredient in commercial sunscreens (Section 24.6A, Problem 24.16)
M The synthesis of ibuprofen (Section 24.6B, Problem 24.18)
B Biological carbonyl condensation reactions (Section 24.8)
M, B End-of-chapter problems: 24.34, 24.42–24.46, 24.49, 24.53, 24.55–24.58, 24.60, 24.72, and 24.73

Chapter 25 Amines

M Scopolamine, an alkaloid used to treat the nausea and vomiting associated with motion sickness (Chapter opener)
M The stereogenic centers in dobutamine, an amine used in stress tests (Section 25.2, Problem 25.1)
B Poisonous diamines with putrid odors: putrescine and cadaverine (Section 25.6A)
B Naturally occurring alkaloids: atropine, nicotine, and coniine (Section 25.6A)
M Histamine, antihistamines, and antiulcer drugs like Tagamet (cimetidine) (Section 25.6B)
B, M Biologically active derivatives of 2-phenylethanamine: adrenaline, noradrenaline, methamphetamine, mescaline, and dopamine (Section 25.6C)
B, M The neurotransmitter serotonin and SSRI antidepressants (Section 25.6C)
B Bufotenin and psilocin (hallucinogens) (Section 25.6C)
M The synthesis of methamphetamine (Section 25.7C)
M The synthesis of enalapril, an antihypertensive, by reductive amination (Section 25.7C, Problem 25.14)
M The synthesis of the drugs rimantadine and pseudoephedrine by reductive amination (Section 25.7C, Problem 25.15)
M The systematic name of a component of the diet drug fen–phen (Section 25.7C, Problem 25.16)
M Drugs, such as the antihistamine diphenhydramine, sold as water-soluble ammonium salts (Section 25.9)
M Hybridization effects on the basicity of nicotine (Section 25.10E, Problem 25.22)
M Acid–base properties of the drugs chloroquine, matrine, tacrine, and quinine (Section 25.10F, Problem 25.23)
G Azo dyes (Section 25.15)
G Perkin's mauveine and synthetic dyes (Section 25.16A)
M Sulfa drugs (Section 25.16B)
M End-of-chapter problems: 25.37, 25.42, 25.44, 25.54, 25.57, 25.58, 25.68, 25.70, 25.77, and 25.78

Chapter 26 Amino Acids and Proteins

B Myoglobin, the protein that stores oxygen in tissues (Chapter opener and Section 26.10C)
B The naturally occurring amino acids (Section 26.1)
M L-Thyroxine, used to treat thyroid hormone deficiency (Section 26.1B, Problem 26.4)
B The structures of the hormones bradykinin, oxytocin, and vasopressin (Section 26.5C)
B The artificial sweetener aspartame (Section 26.5C)
B The amino acid sequence of leu-enkephalin, an analgesic and opiate (Section 26.5C, Problem 26.17)
B The structure of glutathione, a powerful antioxidant in cells (Section 26.5C, Problem 26.18)
B The Merrifield method of automated protein synthesis (Section 26.8)
B The structures of lysozyme and spider silk (Section 26.9B)
M The structure of insulin (Section 26.9C)
B α-Keratin, the protein in hair, hooves, nails, skin, and wool (Section 26.10A)
B Collagen, the protein in connective tissue (Section 26.10B)
B, M Hemoglobin and the structure of sickle cell hemoglobin (Section 26.10C)
M, B End-of-chapter problems: 26.32, 26.46, 26.48, 26.50, 26.54, 26.56, 26.67, and 26.70

Chapter 27 Carbohydrates

B Solanine, the defensive toxin found in the leaves, stems, and green spots on the skin of potatoes (Chapter opener and Section 27.7C)
B The use of fructose in "lite" foods (Section 27.2)
B Dihydroxyacetone, the active ingredient in many artificial tanning agents (Section 27.2)
B Glucose, the most common simple sugar (Section 27.6)
G Honey, a mixture of D-fructose and D-glucose (Section 27.6D)
B, M The naturally occurring glycosides salicin and solanine (Section 27.7C)
G Rebaudioside A, a sweet glycoside from the stevia plant (Section 27.7C, Problem 27.19)
B Glucitol (sorbitol), a sucrose substitute (Section 27.9A)
B The common disaccharides maltose, lactose, and sucrose (Section 27.11)
M Lactose intolerance (Section 27.11B)
G Artificial sweeteners (Section 27.11C)
B The common polysaccharides cellulose, starch, and glycogen (Section 27.12)
B, M Glucosamine, an over-the-counter remedy for osteoarthritis, and chitin, the carbohydrate that gives rigidity to crab shells (Section 27.13A)
B *N*-Glycosides and the structure of DNA (Section 27.13B)
B, M End-of-chapter problems: 27.66 and 27.69

Chapter 28	**Lipids**
B	Cholesterol, the most prominent steroid (Chapter opener and Section 28.8B)
B	Structure of spermaceti wax (Section 28.2)
B	Waxes obtained from jojoba seeds that are used in cosmetics and personal care products (Section 28.2, Problem 28.1)
B	Triacylglycerols, the components of fats and oils (Section 28.3)
B	Essential fatty acids (Section 28.3)
B	The saturated versus unsaturated fatty acid content of fats and oils (Section 28.3)
B	Energy storage and the metabolism of fats (Section 28.3)
B	The phospholipids in cell membranes (Section 28.4)
B	Fat-soluble vitamins: A, D, E, and K (Section 28.5)
B	The eicosanoids, a group of biologically active lipids that includes the prostaglandins and leukotrienes (Section 28.6)
M	Misoprostol, an analogue of PGE_1 used to prevent gastric ulcers, and unoprostone isopropyl, a prostaglandin analogue used to treat glaucoma (Section 28.6)
M	NSAIDs like aspirin and ibuprofen and the COX-1 and COX-2 enzymes (Section 28.6)
M	The anti-inflammatory drugs Vioxx, Bextra, and Celebrex (Section 28.6)
B	Essential oils that are terpenes and terpenoids (Section 28.7)
B	Locating isoprene units in geraniol, vitamin A, grandisol (pheromone), and camphor (Section 28.7, Problem 28.10)
B	Biformene, a terpenoid from amber (Section 28.7, Problem 28.11)
B, M	The structures of steroids: cholesterol, sex hormones (female and male), adrenal cortical steroids, anabolic steroids, and oral contraceptives (Section 28.8)
M	Cholesterol and the cholesterol-lowering drugs Lipitor and Zocor (Section 28.8B)
B, M	Anabolic steroids (Section 28.8C)
B, M	End-of-chapter problems: 28.20, 28.26–25.28, 28.30, 28.31, 28.35, 28.36, and 28.39
Chapter 29	**Carbon–Carbon Bond-Forming Reactions in Organic Synthesis**
M	Ingenol mebutate, used to treat the skin condition actinic keratosis (Chapter opener and Section 29.6, Problem 29.16)
B	The synthesis of C_{18} juvenile hormone (Section 29.1A, Problem 29.2)
B, E	Use of the Suzuki reaction to prepare bombykol, the sex pheromone of the female silkworm moth, and humulene, a lipid isolated from hops (Section 29.2B)
E	Pyrethrin I, a biodegradable insecticide isolated from chrysanthemums, and decamethrin, a synthetic analogue (Section 29.4)
M	Ring-closing metathesis and the synthesis of epothilone A, an anticancer drug, and Sch38516, an antiviral agent (Section 29.6)
M, B, G	End-of-chapter problems: 29.25, 29.26, 29.33, 29.37, 29.38, 29.50
Chapter 30	**Pericyclic Reactions**
B	One synthesis of periplanone B (sex pheromone of the female American cockroach) using pericyclic reactions (Chapter opener and Section 30.5B, Problem 30.22)
B	The role of photochemical electrocyclic ring opening and sigmatropic rearrangements in the formation of vitamin D_3 from 7-dehydrocholesterol (Section 30.3C, Problem 30.9)
M	The synthesis of the alkaloid reserpine by a [4 + 2] cycloaddition reaction (Section 30.4B, Problem 30.15)
M	Garsubellin A and the synthesis of the neurotransmitter acetylcholine (Section 30.5B, Problem 30.25)
B	End-of-chapter problems: 30.43, 30.48, and 30.62
Chapter 31	**Synthetic Polymers (Available online)**
G	Polyethylene terephthalate, an easily recycled synthetic polymer used in transparent soft drink containers (Chapter opener and Sections 31.6B and 31.9A)
G	Consumer products made from Lexan, nylon 6,6, rubber, and polyethylene (Section 31.1)
G	Polyethylene, the plastic in milk jugs and plastic bags, and other chain-growth polymers (Section 31.2)
G	ABS, a copolymer used in crash helmets, small appliances, and toys (Section 31.2, Problem 31.11)
G	Using Ziegler–Natta catalysts to make high-density polyethylene (Section 31.4)
G	Dyneema, a strong fiber made of ultra high-density polyethylene (Section 31.4)
B	Natural and synthetic rubber (Section 31.5)
G	The synthesis of the step-growth polymers nylon, Kevlar, Dacron, spandex, and Lexan (Section 31.6)
M	Dissolving sutures (Section 31.6B)
E	Polyethylene furanoate, a polymer synthesized from renewable resources (Section 31.6B, Problem 31.16)

G	Spandex for active wear (Section 31.6C)
G	Lexan for bike helmets, goggles, catcher's masks, and bulletproof glass (Section 31.6D)
G	Epoxy resins (Section 31.6E)
G	Bakelite for bowling balls (Section 31.7)
E	Green polymer synthesis: environmentally benign methods for preparing polymers (Section 31.8)
E	Polymer recycling (Section 31.9A)
E	Biodegradable polymers (Section 31.9B)
G, E, M	End-of-chapter problems: 31.34, 31.35, 31.50, 31.52, and 31.56–31.58

Prologue

What is organic chemistry?
Some representative organic molecules
Organic chemistry and malaria

Organic chemistry. You might wonder how a discipline that conjures up images of eccentric old scientists working in basement laboratories is relevant to you, a student in the twenty-first century.

Consider for a moment the activities that occupied your past 24 hours. You likely showered with soap, drank a caffeinated beverage, ate at least one form of starch, took some medication, listened to a CD, and traveled in a vehicle that had rubber tires and was powered by fossil fuels. If you did any *one* of these, your life was touched by organic chemistry.

What Is Organic Chemistry?

- **Organic chemistry is the chemistry of compounds that contain the element carbon.**

It is one branch in the entire field of chemistry, which encompasses many classical subdisciplines including inorganic, physical, and analytical chemistry, and newer fields such as bioinorganic chemistry, physical biochemistry, polymer chemistry, and materials science.

Some compounds that contain the element carbon are *not* organic compounds. Examples include carbon dioxide (CO_2), sodium carbonate (Na_2CO_3), and sodium bicarbonate ($NaHCO_3$).

Organic chemistry was singled out as a separate discipline for historical reasons. Originally, it was thought that compounds in living things, termed ***organic compounds,*** were fundamentally different from those in nonliving things, called ***inorganic compounds.*** Although we have known for more than 150 years that this distinction is artificial, the name *organic* persists. Today the term refers to the study of the compounds that contain carbon, many of which, incidentally, are found in living organisms.

It may seem odd that a whole discipline is devoted to the study of a single element in the periodic table, when more than 100 elements exist. It turns out, though, that there are far more organic compounds than any other type. **Organic chemicals affect virtually every facet of our lives, and for this reason, it is important and useful to know something about them.**

Clothes, foods, medicines, gasoline, refrigerants, and soaps are composed almost solely of organic compounds. Some, like cotton, wool, or silk, are naturally occurring; that is, they can be isolated directly from natural sources. Others, such as nylon and polyester, are synthetic, meaning they are produced by chemists in the laboratory. By studying the principles and concepts of organic chemistry, you can learn more about compounds such as these and how they affect the world around you.

Realize, too, what organic chemistry has done for us. Organic chemistry has made available both comforts and necessities that were previously nonexistent, or reserved for only the wealthy. We have seen an enormous increase in life span, from 47 years in 1900 to over 70 years currently. To a large extent this is due to the isolation and synthesis of new drugs to fight infections and the availability of vaccines for childhood diseases. Chemistry has also given us the tools to control insect populations that spread disease, and there is more food for all because of fertilizers, pesticides, and herbicides. Our lives would be vastly different today without the many products that result from organic chemistry (Figure 1).

Figure 1
Products of organic chemistry used in medicine

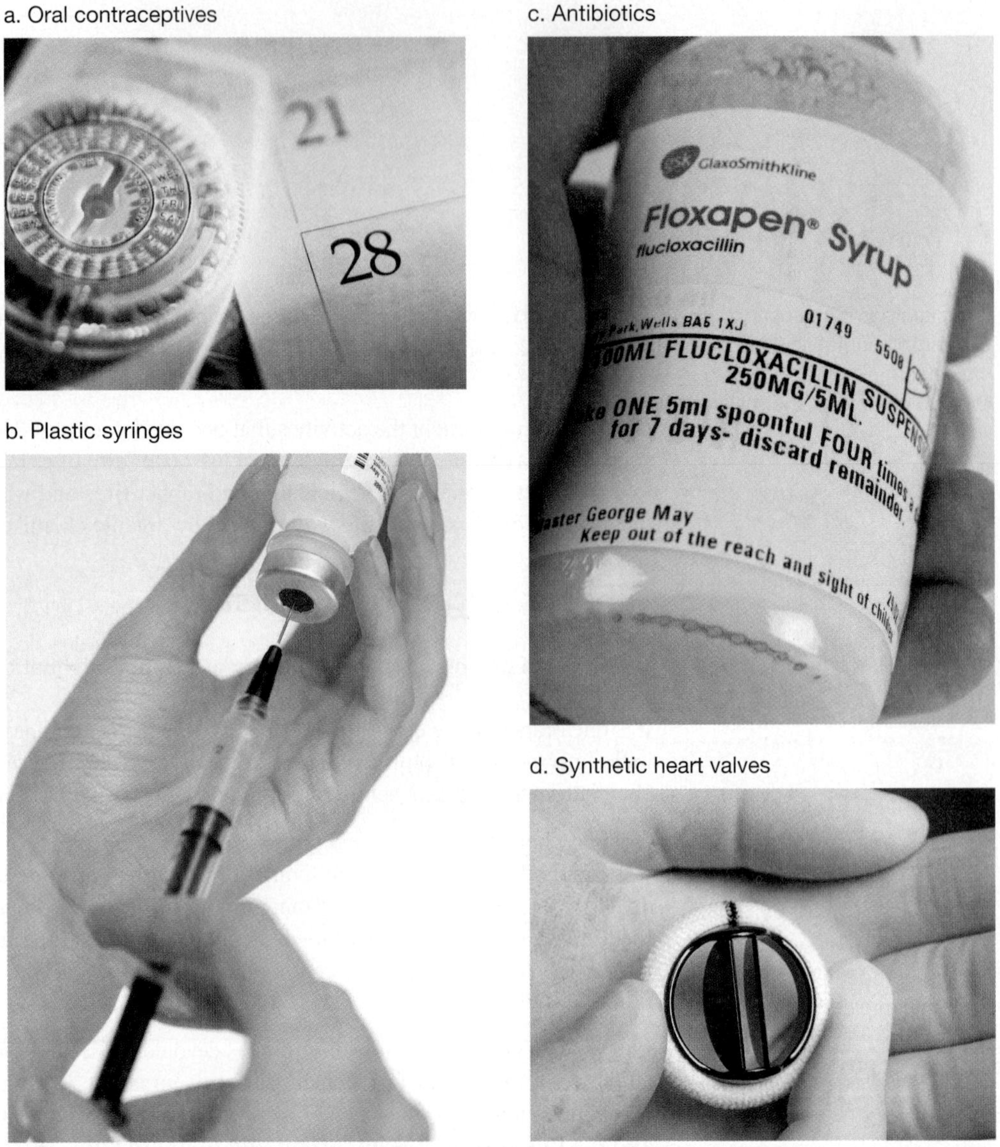

• Organic chemistry has given us contraceptives, plastics, antibiotics, and the knitted material used in synthetic heart valves.

Some Representative Organic Molecules

Perhaps the best way to appreciate the variety of organic molecules is to look at a few. Three simple organic compounds are **methane, ethanol,** and **trichlorofluoromethane.**

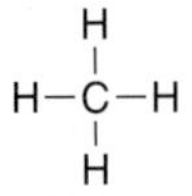

methane

- **Methane,** the simplest of all organic compounds, contains one carbon atom. Methane—the main component of natural gas—occurs widely in nature. Like other **hydrocarbons**—organic compounds that contain only carbon and hydrogen—methane is combustible; that is, it burns in the presence of oxygen. Methane is the product of the anaerobic (without air) decomposition of organic matter by bacteria. The natural gas we use today was formed by the decomposition of organic material millions of years ago. Hydrocarbons such as methane are discussed in Chapter 4.

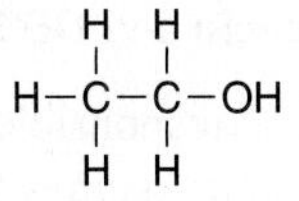

ethanol

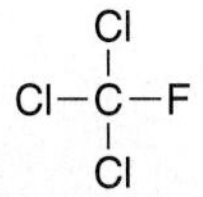

trichlorofluoromethane

- **Ethanol,** the alcohol present in beer, wine, and other alcoholic beverages, is formed by the fermentation of sugar, quite possibly the oldest example of organic synthesis. Ethanol can also be made in the lab by a totally different process, but **the ethanol produced in the lab is *identical* to the ethanol produced by fermentation.** Alcohols including ethanol are discussed in Chapter 9.
- **Trichlorofluoromethane** is a member of a class of molecules called **chlorofluorocarbons,** or **CFCs,** which contain one or two carbon atoms and several halogens. Trichlorofluoromethane is an unusual organic molecule in that **it contains no hydrogen atoms.** Because it has a low molecular weight and is easily vaporized, trichlorofluoromethane has been used as an aerosol propellant and refrigerant. It and other CFCs have been implicated in the destruction of the stratospheric ozone layer, a topic discussed in Chapter 15.

Three complex organic molecules that are important medications are **amoxicillin, fluoxetine,** and **AZT.**

- **Amoxicillin** is one of the most widely used antibiotics in the penicillin family. The discovery and synthesis of such antibiotics in the twentieth century have made routine the treatment of infections that were formerly fatal. You were likely given some amoxicillin to treat an ear infection when you were a child. The penicillin antibiotics are discussed in Chapter 22.

Complex organic structures are drawn with shorthand conventions described in Chapter 1.

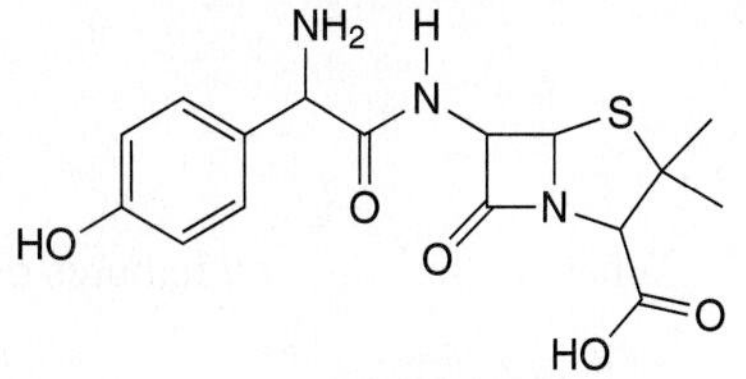

amoxicillin

- **Fluoxetine** is the generic name for the antidepressant **Prozac.** Prozac was designed and synthesized by chemists in the laboratory, and is now produced on a large scale in chemical factories. Because it is safe and highly effective in treating depression, Prozac is widely prescribed. Over 40 million individuals worldwide have used Prozac since 1986.

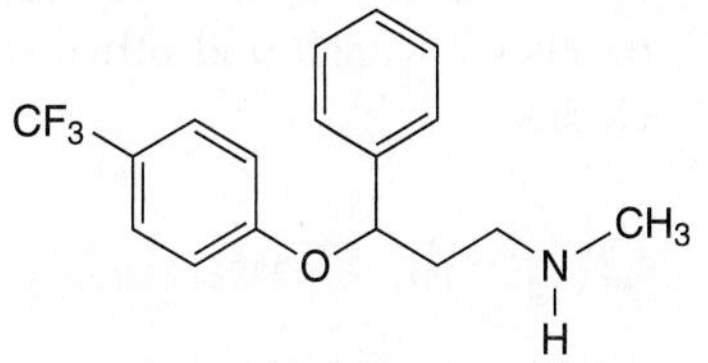

fluoxetine

- **AZT, az**ido**d**eoxy**t**hymidine, is a drug that treats human immunodeficiency virus (HIV), the virus that causes acquired immune deficiency syndrome (AIDS). Also known by its generic name **zidovudine,** AZT represents a chemical success to a different challenge: synthesizing agents that combat viral infections.

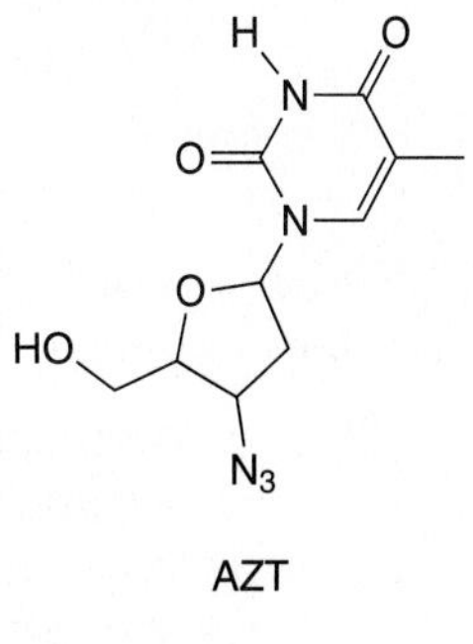

AZT

Other complex organic compounds having interesting properties are **capsaicin** and **DDT.**

- **Capsaicin,** one member of a group of compounds called *vanilloids,* is responsible for the characteristic spiciness of hot peppers. It is the active ingredient in pepper sprays used for personal defense and topical creams used for pain relief.

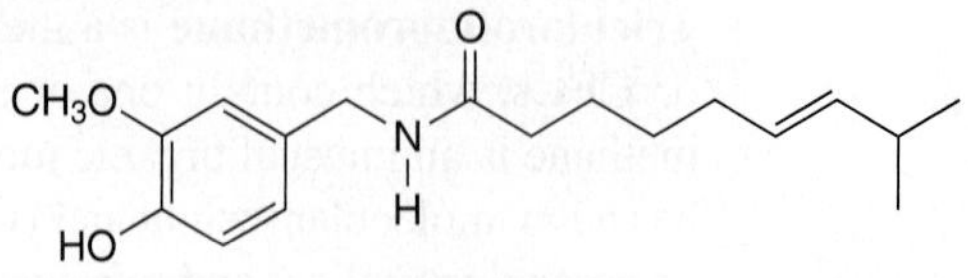

capsaicin

- **DDT, d**ichloro**d**iphenyl**t**richloroethane, is a pesticide once called "miraculous" by Winston Churchill because of the many lives it saved by killing disease-carrying mosquitoes. DDT use is now banned in the United States and many developed countries because it is a nonspecific insecticide that persists in the environment.

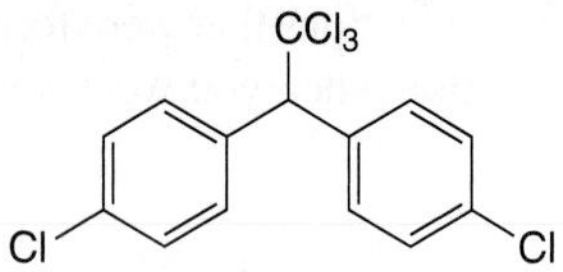

DDT

What are the common features of these organic compounds?

- All organic compounds contain carbon atoms and most contain hydrogen atoms.
- All the carbon atoms have four bonds. A stable carbon atom is said to be *tetravalent*.
- Other elements may also be present. Any atom that is not carbon or hydrogen is called a *heteroatom*. Common heteroatoms include N, O, S, P, and the halogens.
- Some compounds have chains of atoms and some compounds have rings.

These features explain why there are so many organic compounds: **Carbon forms four strong bonds with itself and other elements. Carbon atoms combine together to form rings and chains.**

Organic Chemistry and Malaria

A vast array of organic compounds is now available to fight malaria, a mosquito-borne infectious disease that affects an estimated 200 million people worldwide. Antimalarial medications include organic compounds isolated from natural sources or those synthesized by chemists in the laboratory. Two common antimalarial drugs shown in Figure 2 are **quinine,** a centuries-old remedy obtained from the bark of the cinchona tree native to the Andes Mountains, and **chloroquine,** a synthetic drug introduced in the late 1940s.

Figure 2
Antimalarial drugs

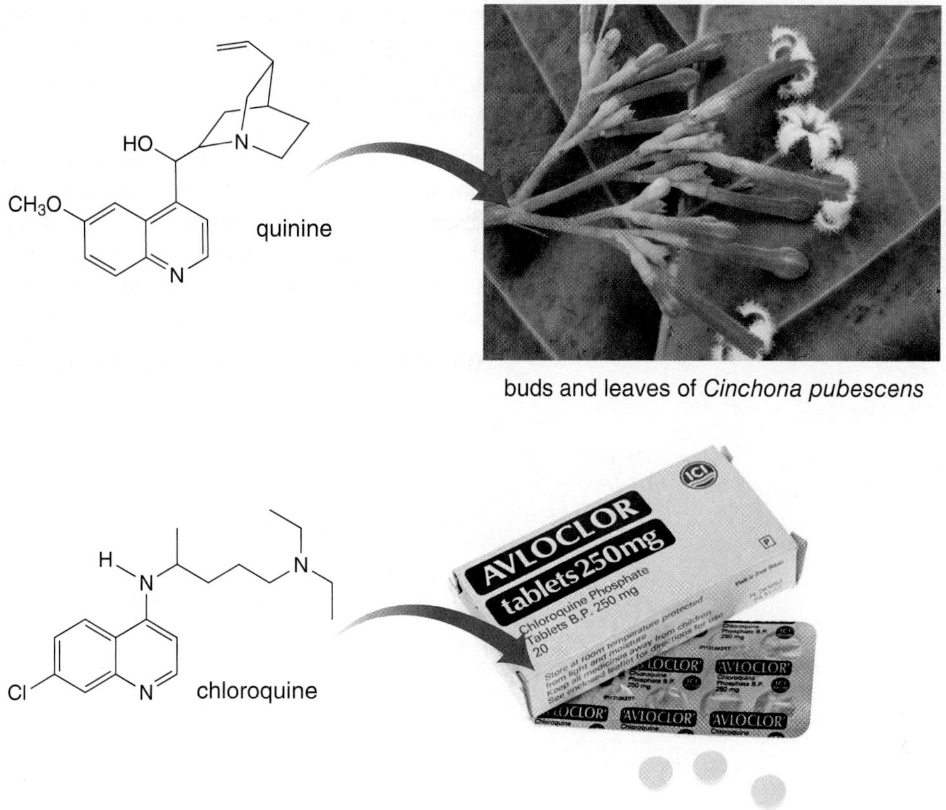

Because malaria is caused by a variety of closely related parasitic microorganisms and drug-resistant strains have developed, currently recommended therapy consists of a combination of drugs that includes **artemisinin** or a related compound. Artemisinin is a complex compound isolated from sweet wormwood, *Artemisia annua,* a plant used for hundreds of years in traditional Chinese medicine. Although artemisinin can be obtained by extracting the active drug from the dried leaves of *Artemisia annua,* this process does not meet the worldwide demand. As a result, artemisinin can now be obtained using genetic engineering and fermentation processes.

The 2015 Nobel Prize in Physiology or Medicine was awarded to Youyou Tu for her discovery of artemisinin as an antimalarial drug.

Artemisia annua, sweet wormwood

Malaria continues to present a major public health challenge for chemists, health professionals, and biologists. Despite extensive efforts to prevent and control the disease in the equatorial regions of Asia, Africa, and Latin America, it is estimated that malaria was responsible for over 450,000 deaths in 2012.

In this introduction, we have seen a variety of molecules that have diverse structures. They represent a miniscule fraction of the organic compounds currently known and the many thousands that are newly discovered or synthesized each year. The principles you learn in organic chemistry will apply to all of these molecules, from simple ones like methane and ethanol, to complex ones like capsaicin and artemisinin. It is these beautiful molecules, their properties, and their reactions that we will study in organic chemistry.

WELCOME TO THE WORLD OF ORGANIC CHEMISTRY!

Structure and Bonding

1

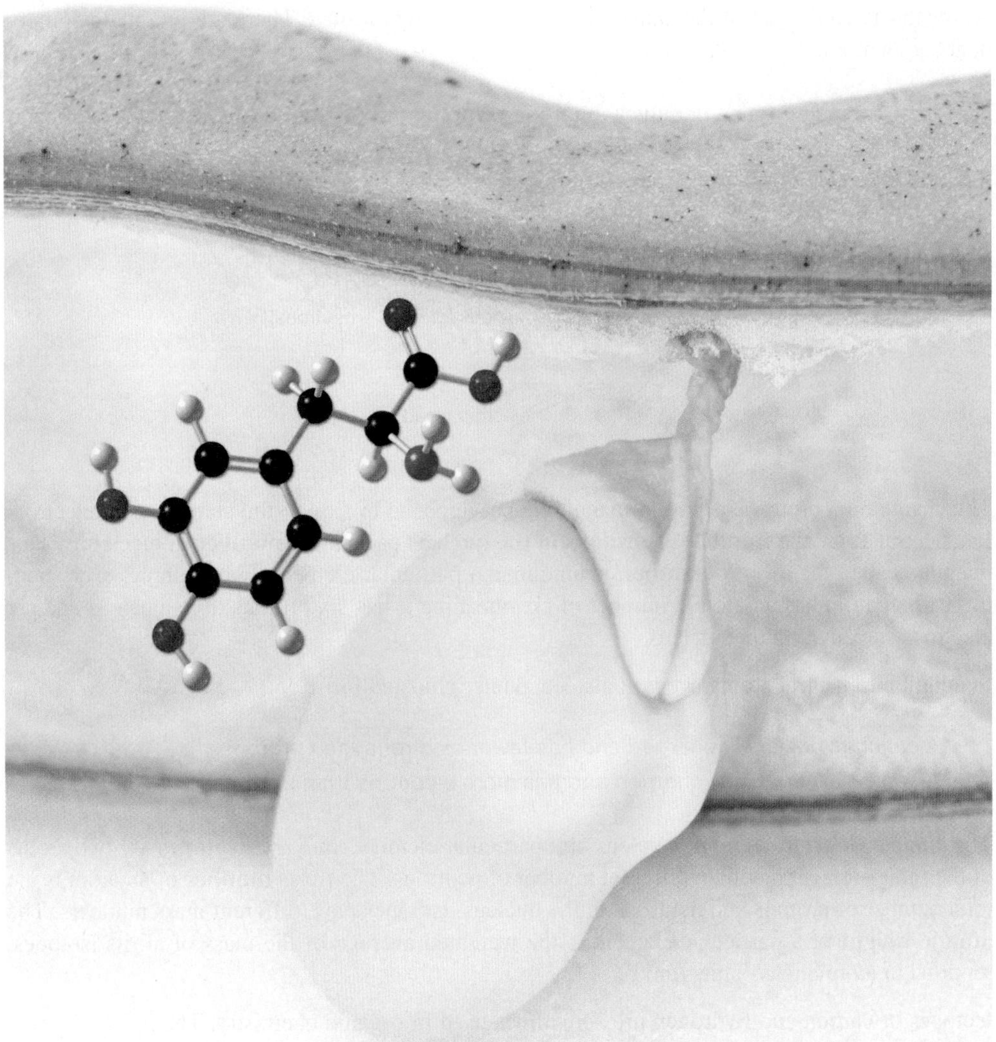

1.1 The periodic table
1.2 Bonding
1.3 Lewis structures
1.4 Isomers
1.5 Exceptions to the octet rule
1.6 Resonance
1.7 Determining molecular shape
1.8 Drawing organic structures
1.9 Hybridization
1.10 Ethane, ethylene, and acetylene
1.11 Bond length and bond strength
1.12 Electronegativity and bond polarity
1.13 Polarity of molecules
1.14 L-Dopa—A representative organic molecule

L-Dopa, also called levodopa, was first isolated from seeds of the broad bean plant *Vicia faba* in 1913. Since 1967 it has been the drug of choice for the treatment of Parkinson's disease, a debilitating illness that results from the degeneration of neurons that produce the neurotransmitter dopamine in the brain. L-Dopa is an oral medication that is transported to the brain by the bloodstream, where it is converted to dopamine. Since L-dopa must be taken in large doses with some serious side effects, today it is often given with other drugs that lessen its negative impact. In Chapter 1, we learn about the structure, bonding, and properties of organic molecules like L-dopa.

Before examining organic molecules in detail, we must review topics about structure and bonding learned in previous chemistry courses. We will discuss these concepts primarily from an organic chemist's perspective, and spend time on only the particulars needed to understand organic compounds.

Important topics in Chapter 1 include drawing Lewis structures, predicting the shape of molecules, determining what orbitals are used to form bonds, and how electronegativity affects bond polarity. Equally important is Section 1.8 on drawing organic molecules, both shorthand methods routinely used for simple and complex compounds, as well as three-dimensional representations that allow us to more clearly visualize them.

1.1 The Periodic Table

All matter is composed of the same building blocks called **atoms.** There are two main components of an atom.

- The **nucleus** contains positively charged **protons** and uncharged **neutrons.** Most of the mass of the atom is contained in the nucleus.
- The **electron cloud** is composed of negatively charged **electrons.** The electron cloud comprises most of the volume of the atom.

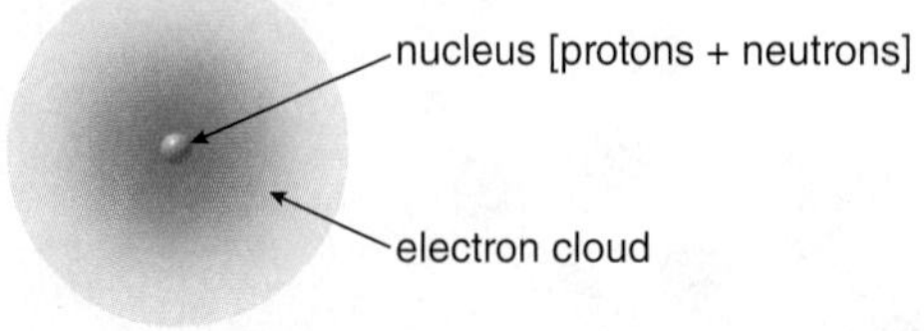

The charge on a proton is equal in magnitude but opposite in sign to the charge on an electron. In a neutral atom, the **number of protons in the nucleus equals the number of electrons.** This quantity, called the **atomic number,** is unique to a particular element. For example, every neutral carbon atom has an atomic number of six, meaning it has six protons in its nucleus and six electrons surrounding the nucleus.

In addition to neutral atoms, we will also encounter **charged ions.**

- A *cation* is positively charged and has fewer electrons than protons.
- An *anion* is negatively charged and has more electrons than protons.

The number of neutrons in the nucleus of a particular element can vary. **Isotopes** are two atoms of the same element having a different number of neutrons. The **mass number** of an atom is the total number of protons and neutrons in the nucleus. Isotopes have different mass numbers. The **atomic weight** of a particular element is the weighted average of the mass of all its isotopes, reported in atomic mass units (amu).

Isotopes of carbon and hydrogen are sometimes used in organic chemistry. The most common isotope of hydrogen has one proton and no neutrons in the nucleus, but 0.02% of hydrogen atoms have one proton and one neutron. This isotope of hydrogen is called **deuterium,** and is sometimes symbolized by the letter **D.**

Each atom is identified by a one- or two-letter abbreviation that is the characteristic symbol for that element. Carbon is identified by the single letter **C.** Sometimes the atomic number is indicated as a subscript to the left of the element symbol, and the mass number is indicated as a superscript. Using this convention, the most common isotope of carbon, which contains six protons and six neutrons, is designated as ${}^{12}_{6}C$.

A **row** in the periodic table is also called a ***period,*** and a **column** is also called a ***group.*** A periodic table is located on the inside front cover for your reference.

Long ago it was realized that groups of elements have similar properties, and that these atoms could be arranged in a schematic way called the **periodic table.** There are more than 100 known elements, arranged in the periodic table in order of increasing atomic number. The periodic table is composed of rows and columns.

- **Elements in the same row are similar in *size.***
- **Elements in the same column have similar *electronic and chemical properties.***

Each column in the periodic table is identified by a **group number,** an Arabic (1 to 8) or Roman (I to VIII) numeral followed by the letter A or B. Carbon is located in group **4A** in the periodic table in this text.

Although more than 100 elements exist, most are not common in organic compounds. Figure 1.1 contains a truncated periodic table, indicating the handful of elements that are routinely seen in this text. **Most of these elements are located in the first and second rows of the periodic table.**

Figure 1.1
A periodic table of the common elements seen in organic chemistry

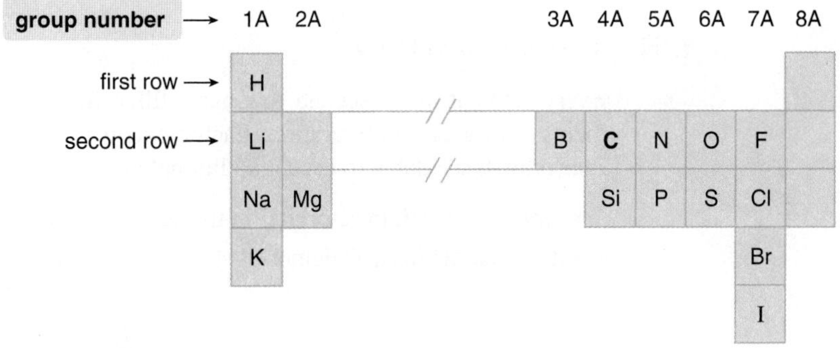

- Carbon is located in the second row, group 4A.

Carbon's entry in the periodic table:

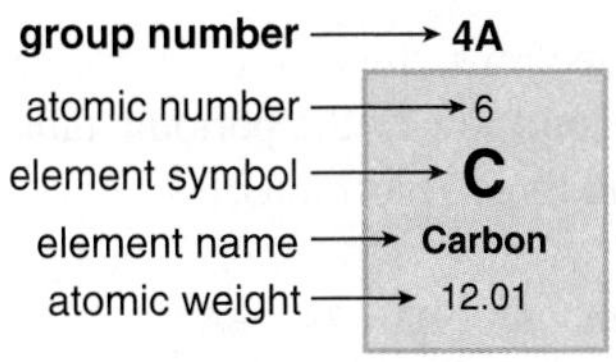

Across each row of the periodic table, electrons are added to a particular shell of orbitals around the nucleus. The shells are numbered 1, 2, 3, and so on. Adding electrons to the first shell forms the first row. Adding electrons to the second shell forms the second row. **Electrons are first added to the shells closest to the nucleus.**

Each shell contains a certain number of **orbitals.** An orbital is a region of space that is high in electron density. There are four different kinds of orbitals, called ***s, p, d,*** and ***f.*** The first shell has only one orbital, called an *s* orbital. The second shell has two kinds of orbitals, ***s*** and ***p,*** and so on. Each type of orbital has a particular shape.

For the first- and second-row elements, we must consider only ***s*** **orbitals** and ***p*** **orbitals.**

- **An *s* orbital has a sphere of electron density. It is *lower in energy* than other orbitals of the same shell, because electrons are kept closer to the positively charged nucleus.**
- **A *p* orbital has a dumbbell shape. It contains a node of electron density at the nucleus. A node means there is *no* electron density in this region. A *p* orbital is *higher in energy* than an *s* orbital (in the same shell) because its electron density is farther away from the nucleus.**

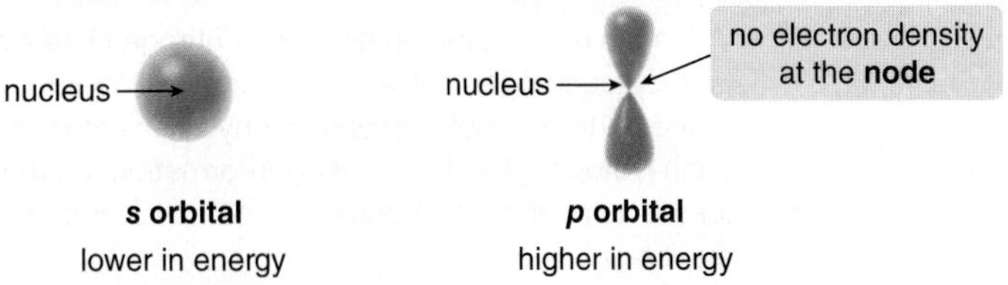

An *s* orbital is filled with electrons before a *p* orbital in the same shell.

1.1A The First Row

The first row of the periodic table is formed by adding electrons to the first shell of orbitals around the nucleus. There is only one orbital in the first shell, called the **1*s* orbital.**

- **Each orbital can have a maximum of two electrons.**

As a result, there are **two elements in the first row,** one having one electron added to the 1*s* orbital, and one having two. The element hydrogen (H) has what is called a $1s^1$ configuration with one electron in the 1*s* orbital, and helium (He) has a $1s^2$ configuration with two electrons in the 1*s* orbital.

	H	He
first row →	H	He
	$1s^1$	$1s^2$
	one electron in the 1*s* orbital	two electrons in the 1*s* orbital

1.1B The Second Row

Every element in the second row has a filled first shell of electrons. Thus, all second-row elements have a $1s^2$ configuration. Each element in the second row of the periodic table also has four orbitals available to accept additional electrons:

- **one 2*s* orbital,** the *s* orbital in the second shell
- **three 2*p* orbitals,** all dumbbell-shaped and perpendicular to each other along the *x*, *y*, and *z* axes

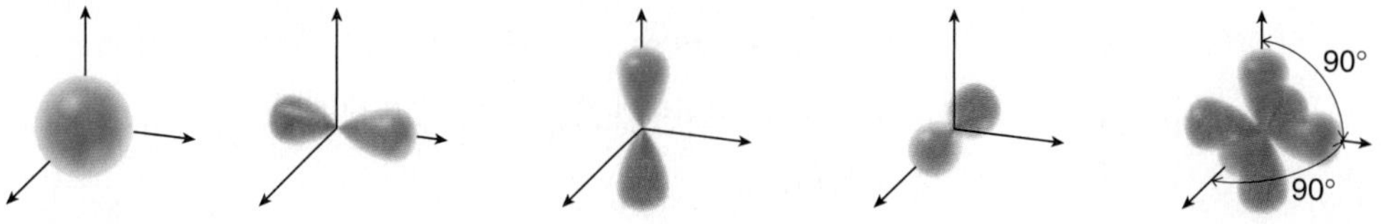

2*s* orbital — $2p_x$ orbital — $2p_y$ orbital — $2p_z$ orbital — all three 2*p* orbitals drawn on the same set of axes

Because each of the four orbitals in the second shell can hold two electrons, there is a **maximum capacity of *eight* electrons** for elements in the second row. The second row of the periodic table consists of eight elements, obtained by adding electrons to the 2*s* and three 2*p* orbitals.

group number →	1A	2A	3A	4A	5A	6A	7A	8A
second row →	Li	Be	B	**C**	N	O	F	Ne
number of valence electrons →	1	2	3	4	5	6	7	8

The outermost electrons are called **valence electrons.** The valence electrons are more loosely held than the electrons closer to the nucleus, and as such, they participate in chemical reactions. **The group number of a second-row element reveals its number of valence electrons.** For example, carbon in group 4A has four valence electrons, and oxygen in group 6A has six.

Problem 1.1 While the most common isotope of nitrogen has a mass number of 14 (nitrogen-14), a radioactive isotope of nitrogen has a mass number of 13 (nitrogen-13). Nitrogen-13 is used in PET (positron emission tomography) scans by physicians to monitor brain activity and diagnose dementia. For each isotope, give the following information: (a) the number of protons; (b) the number of neutrons; (c) the number of electrons in the neutral atom; (d) the group number; and (e) the number of valence electrons.

1.2 Bonding

Until now our discussion has centered on individual atoms, but it is more common in nature to find two or more atoms joined together.

- ***Bonding*** **is the joining of two atoms in a stable arrangement.**

Bonding is a favorable process because it always leads to ***lowered energy and increased stability.*** Joining two or more elements forms **compounds.** Although only about 100 elements exist, more than 50 million compounds are known. Examples of compounds include hydrogen gas (H_2), formed by joining two hydrogen atoms, and methane (CH_4), the simplest organic compound, formed by joining a carbon atom with four hydrogen atoms.

One general rule governs the bonding process.

- **Through bonding, atoms attain a complete outer shell of valence electrons.**

Alternatively, because the noble gases in group 8A of the periodic table are especially stable as atoms having a filled shell of valence electrons, the general rule can be restated.

- **Through bonding, atoms gain, lose, or share electrons to attain the electronic configuration of the noble gas closest to them in the periodic table.**

What does this mean for first- and second-row elements? **A first-row element like hydrogen can accommodate *two electrons* around it.** This would make it like the noble gas helium at the end of the same row. **A second-row element is generally most stable with *eight valence electrons* around it** like neon. Elements that behave in this manner are said to follow the **octet rule.**

There are two different kinds of bonding: **ionic bonding** and **covalent bonding.**

- ***Ionic bonds*** **result from the** ***transfer*** **of electrons from one element to another.**
- ***Covalent bonds*** **result from the** ***sharing*** **of electrons between two nuclei.**

Atoms readily form ionic bonds when they can attain a noble gas configuration by gaining or losing just one or two electrons. NaCl and KI are ionic compounds.

The type of bonding is determined by the location of an element in the periodic table. An ionic bond generally occurs when elements on the **far left** side of the periodic table combine with elements on the **far right** side, ignoring the noble gases, which form bonds only rarely. **The resulting ions are held together by extremely strong electrostatic interactions.** A positively charged cation formed from the element on the left side attracts a negatively charged anion formed from the element on the right side. Examples of ionic inorganic compounds include sodium chloride (NaCl), common table salt, and potassium iodide (KI), an essential nutrient added to make iodized salt.

Ionic compounds form extended crystal lattices that maximize the positive and negative electrostatic interactions. In NaCl, each positively charged Na^+ ion is surrounded by six negatively charged Cl^- ions, and each Cl^- ion is surrounded by six Na^+ ions.

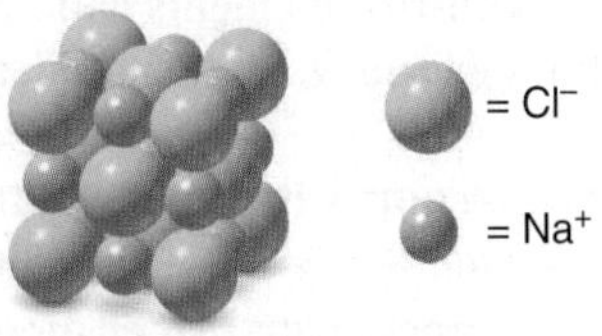

NaCl

Lithium fluoride, LiF, is an example of an ionic compound.

- The element **lithium,** located in group 1A of the periodic table, has one valence electron in its second shell. Loss of this electron forms the cation Li^+ having no electrons in the second shell and two electrons in the first shell like helium.
- The element **fluorine,** located in group 7A of the periodic table, has seven valence electrons. By gaining one it forms the anion F^-, which has a filled valence shell (an octet of electrons), like neon.
- Thus, lithium fluoride is a stable ionic compound composed of Li^+ cations and F^- anions.

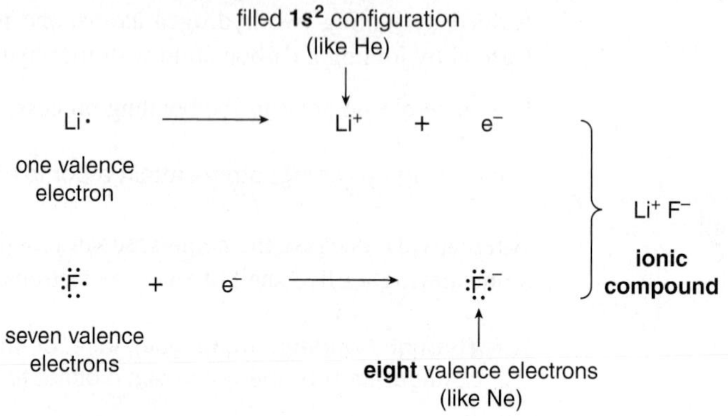

- **The transfer of electrons forms stable salts composed of cations and anions.**

A **compound** may have either ionic or covalent bonds. A **molecule** has only covalent bonds.

The second type of bonding, **covalent bonding,** occurs with elements like carbon in the middle of the periodic table, which would otherwise have to gain or lose several electrons to form an ion with a complete valence shell. **A covalent bond is a two-electron bond,** and a compound with covalent bonds is called a **molecule.** Covalent bonds also form between two elements from the same side of the table, such as two hydrogen atoms or two chlorine atoms. H_2, Cl_2, and CH_4 are all examples of covalent molecules.

Problem 1.2 Label each bond in the following compounds as ionic or covalent.

a. F_2 b. LiBr c. CH_3CH_3 d. $NaNH_2$ e. $NaOCH_3$

How many covalent bonds will a particular atom typically form? As you might expect, it depends on the location of the atom in the periodic table. In the first row, **hydrogen forms one covalent bond** using its one valence electron. When two hydrogen atoms are joined in a bond, each has a filled valence shell of two electrons.

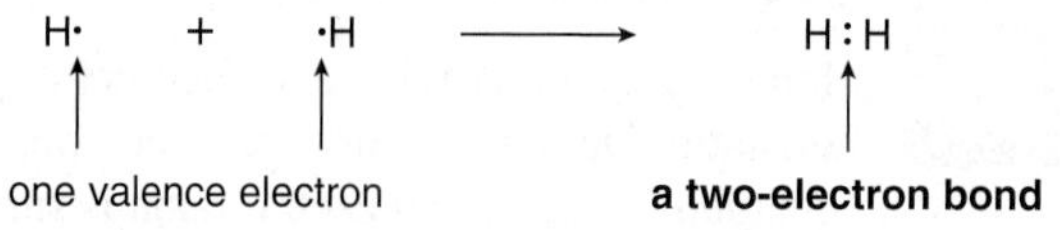

Second-row elements can have no more than eight valence electrons around them. For neutral molecules, two consequences result.

- **Atoms with one, two, three, or four valence electrons form one, two, three, or four bonds,** respectively, in neutral molecules.
- **Atoms with five or more valence electrons** form enough bonds to give an octet. In this case, **the predicted number of bonds = 8 – the number of valence electrons.**

For example, B has three valence electrons, so it forms three bonds, as in BF_3. N has five valence electrons, so it also forms three bonds ($8 - 5 = 3$ bonds), as in NH_3.

Nonbonded pair of electrons = unshared pair of electrons = lone pair

These guidelines are used in Figure 1.2 to summarize the usual number of bonds formed by the common atoms in organic compounds. When second-row elements form fewer than four bonds their octets consist of both **bonding (shared) electrons** and **nonbonding (unshared) electrons.** Unshared electrons are also called **lone pairs.**

Problem 1.3 How many covalent bonds are predicted for each atom?

a. O b. Al c. Br d. Si

Figure 1.2 The usual number of bonds of common neutral atoms

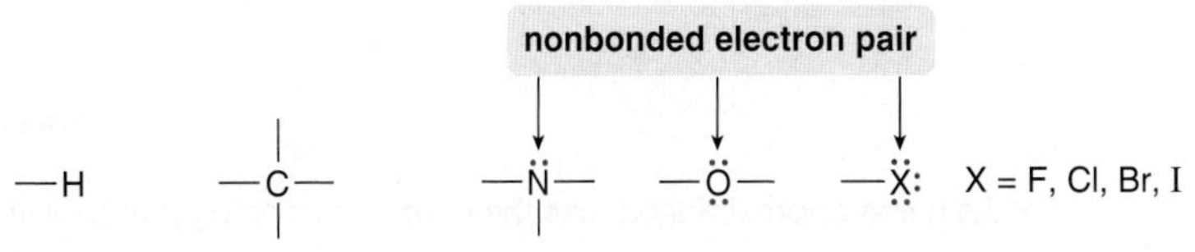

	—H	C	N	O	X
number of bonds →	1	4	3	2	1
number of nonbonded electron pairs →	0	0	1	2	3

1.3 Lewis Structures

***Lewis structures* are electron dot representations for molecules.** Three rules are used for drawing Lewis structures.

1. Draw only the valence electrons.
2. Give every second-row element no more than eight electrons.
3. Give each hydrogen two electrons.

The letter **X** is often used to represent one of the halogens in group 7A: F, Cl, Br, or I.

To draw a Lewis structure for a diatomic molecule like **HF,** recall that hydrogen has one valence electron and fluorine has seven. H and F each donate one electron to form a two-electron bond. The resulting molecule gives both H and F a filled valence shell. In a Lewis structure, **a *solid line* indicates a two-electron covalent bond.**

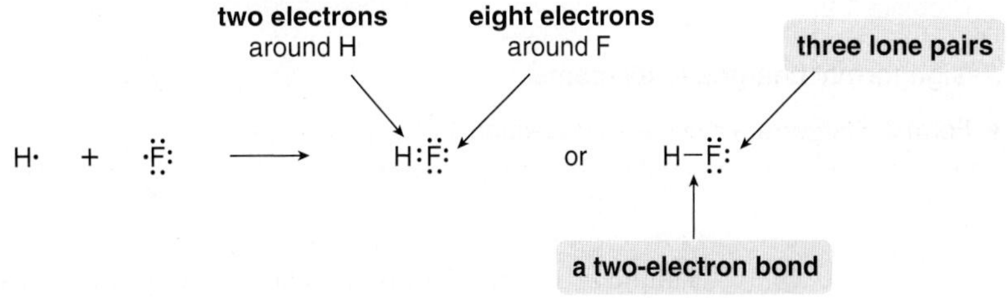

1.3A A Procedure for Drawing Lewis Structures

Drawing a Lewis structure for larger molecules is easier if you follow a stepwise procedure.

How To Draw a Lewis Structure

Step [1] Arrange atoms next to each other that you think are bonded together.

- Always place hydrogen atoms and halogen atoms on the periphery because H and X (X = F, Cl, Br, and I) form only one bond each.

For $\mathbf{CH_4}$:

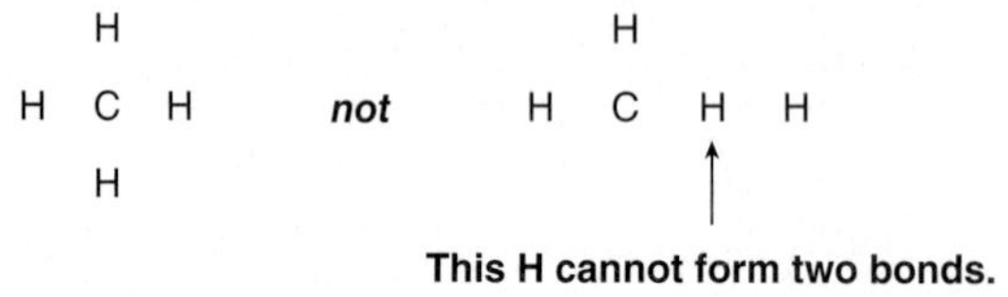

- As a first approximation, use the common bonding patterns in Figure 1.2 to arrange the atoms.

For $\mathbf{CH_5N}$:

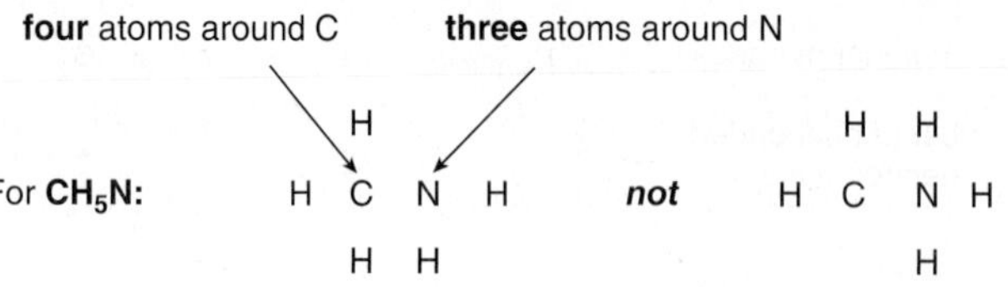

- In truth, the proper arrangement of atoms may not be obvious, or more than one arrangement may be possible (Section 1.4). Even in many simple molecules, the connectivity between atoms must be determined experimentally.

Step [2] Count the electrons.

- Count the number of valence electrons from all atoms.
- Add one electron for each negative charge.
- Subtract one electron for each positive charge.
- This sum gives the total number of electrons that must be used in drawing the Lewis structure.

Step [3] Arrange the electrons around the atoms.

- Place a bond between every two atoms, giving two electrons to each H and no more than eight to any second-row atom.
- Use all remaining electrons to fill octets with lone pairs.
- If all valence electrons are used and an atom does not have an octet, form multiple bonds, as shown in Sample Problem 1.2.

Step [4] Assign formal charges to all atoms.

- Formal charges are discussed in Section 1.3C.

Sample Problem 1.1 illustrates how to draw a Lewis structure of a simple organic molecule.

Sample Problem 1.1 Draw a Lewis structure for methanol, a compound with molecular formula CH_4O.

Solution

Step [1] Arrange the atoms.

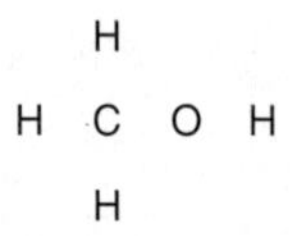

- Place the second-row elements, C and O, in the middle.
- Place three H's around C to surround C by four atoms.
- Place one H next to O to surround O by two atoms.

Step [2] **Count the electrons.**

$$\begin{aligned} 1\ \text{C} \times 4\ e^- &= 4\ e^- \\ 1\ \text{O} \times 6\ e^- &= 6\ e^- \\ 4\ \text{H} \times 1\ e^- &= \underline{4\ e^-} \\ &\ \mathbf{14\ e^-\ total} \end{aligned}$$

Step [3] **Add the bonds and lone pairs.**

- Add five two-electron bonds to form the C–H, C–O, and O–H bonds, using 10 of the 14 electrons.
- Place two lone pairs on the O atom to use the remaining four electrons and give the O atom an octet.

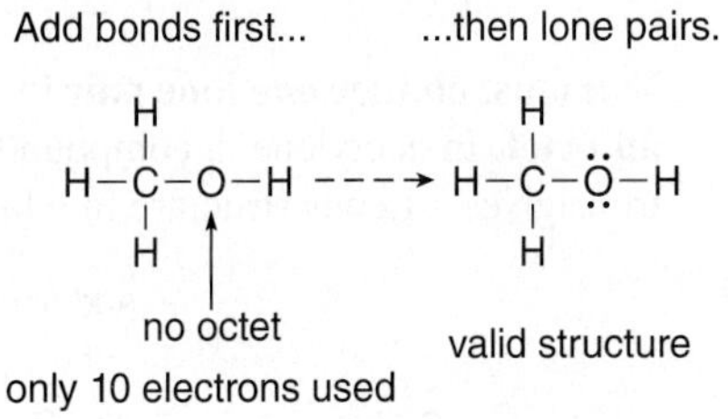
Add bonds first...
...then lone pairs.
no octet
only 10 electrons used
valid structure

This Lewis structure is valid because it uses all 14 electrons, each H is surrounded by two electrons, and each second-row element is surrounded by no more than eight electrons.

Problem 1.4 Draw a valid Lewis structure for each species.

a. CH_3CH_3 b. CH_5N c. CH_3Cl

1.3B Multiple Bonds

Sample Problem 1.2 illustrates an example of a Lewis structure with a double bond.

Sample Problem 1.2 Draw a Lewis structure for ethylene, a compound of molecular formula C_2H_4, in which each carbon is bonded to two hydrogens.

Solution

Follow Steps [1] to [3] to draw a Lewis structure.

Step [1] **Arrange the atoms.**

H C C H
H H

- Each C gets 2 H's.

Step [2] **Count the electrons.**

$$\begin{aligned} 2\ \text{C} \times 4\ e^- &= 8\ e^- \\ 4\ \text{H} \times 1\ e^- &= \underline{4\ e^-} \\ &\ \mathbf{12\ e^-\ total} \end{aligned}$$

Step [3] **Add the bonds and lone pairs.**

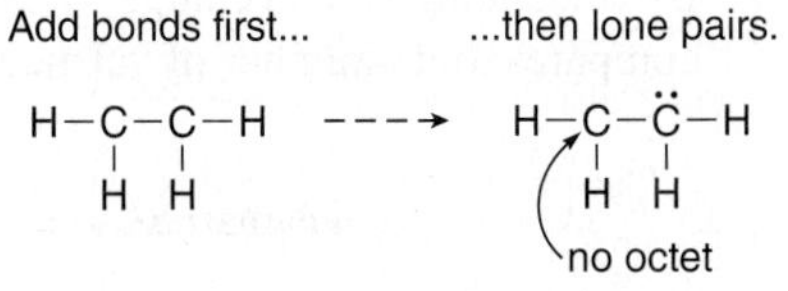
Add bonds first...
...then lone pairs.
no octet

After placing five bonds between the atoms and adding the two remaining electrons as a lone pair, one C still has no octet.

To give both C's an octet, **change *one* lone pair into *one* bonding pair of electrons between the two C's, forming a double bond.**

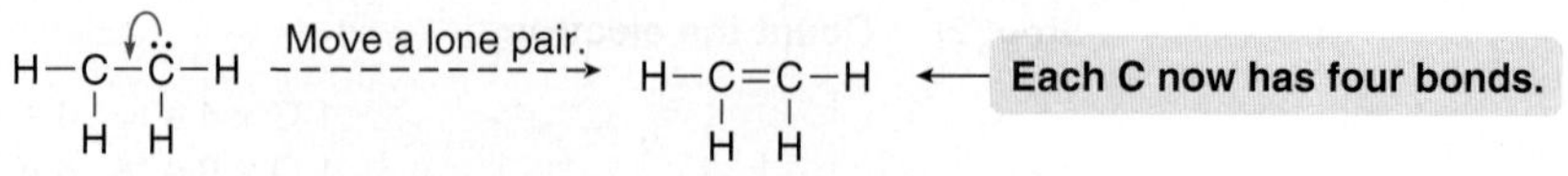
Move a lone pair.
Each C now has four bonds.

ethylene
a valid Lewis structure

This uses all 12 electrons, each C has an octet, and each H has two electrons. The Lewis structure is valid. **Ethylene contains a carbon–carbon double bond.**

- **After placing all electrons in bonds and lone pairs, use a lone pair to form a multiple bond if an atom does not have an octet.**

Carbon always forms four bonds in stable organic molecules. Carbon forms single, double, and triple bonds to itself and other elements.

You must change *one* lone pair into *one* new bond for each *two* electrons needed to complete an octet. In acetylene, a compound with molecular formula C_2H_2, placing the 10 valence electrons gives a Lewis structure in which one or both of the C's lack an octet.

C_2H_2

acetylene

10 valence electrons from 2 C's and 2 H's

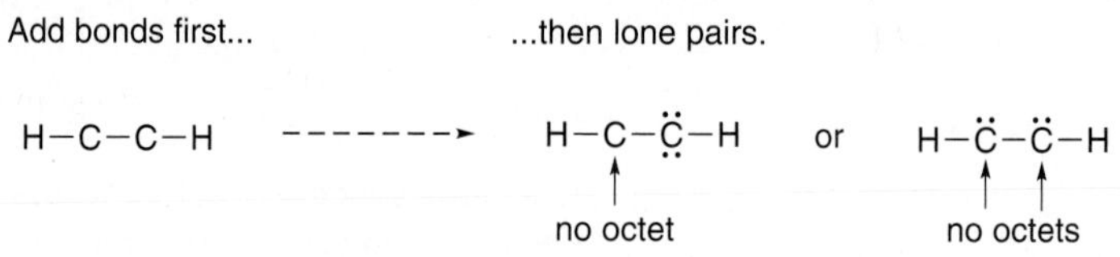
Add bonds first...
...then lone pairs.
or
no octet
no octets

In this case, **change *two* lone pairs into *two* bonding pairs of electrons, forming a triple bond.**

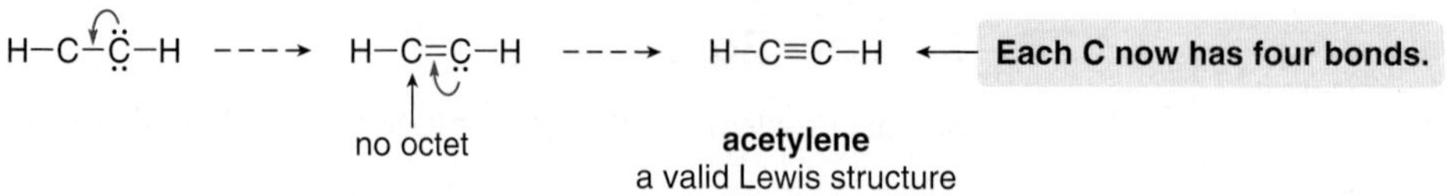
no octet
Each C now has four bonds.

acetylene
a valid Lewis structure

Problem 1.5 Draw an acceptable Lewis structure for each compound, assuming the atoms are connected as arranged. Hydrogen cyanide (HCN) is a poison, formaldehyde (H_2CO) is a preservative, and glycolic acid ($HOCH_2CO_2H$) is used to make dissolving sutures.

a. HCN   H C N

b. H_2CO H C O (with H below C)

c. $HOCH_2CO_2H$ H O C C O H (with H above and H below the first C, O above the second C)

1.3C Formal Charge

To manage electron bookkeeping in a Lewis structure, chemists use **formal charge.**

- ***Formal charge* is the charge assigned to individual atoms in a Lewis structure.**

By calculating formal charge, we determine how the number of electrons around a particular atom compares to its number of valence electrons. Formal charge is calculated as follows:

formal charge = **number of valence electrons** – **number of electrons an atom "owns"**

The number of electrons "owned" by an atom is determined by its number of bonds and lone pairs.

- **An atom "owns" *all* of its unshared electrons and *half* of its shared electrons.**

$$\textbf{number of electrons owned} = \textbf{number of unshared electrons} + \frac{1}{2}\left[\textbf{number of shared electrons}\right]$$

The number of electrons "owned" by different carbon atoms is indicated in the following examples:

- C shares eight electrons.
- C "owns" **four** electrons.

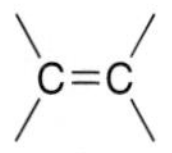

- Each C shares eight electrons.
- Each C "owns" **four** electrons.

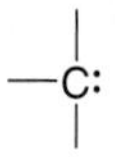

- C shares six electrons.
- C has two unshared electrons.
- C "owns" **five** electrons.

Sample Problem 1.3 illustrates how formal charge is calculated on the atoms of a polyatomic ion. **The sum of the formal charges on the individual atoms equals the net charge on the molecule or ion.**

Sample Problem 1.3

Determine the formal charge on each atom in the ion H_3O^+.

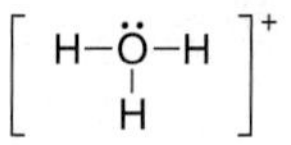

Solution

To calculate the formal charge on each atom:

- Determine the number of valence electrons from the group number.
- Determine the number of electrons an atom "owns" from the number of bonding and nonbonding electrons it has.
- Subtract the second quantity from the first to give the formal charge.

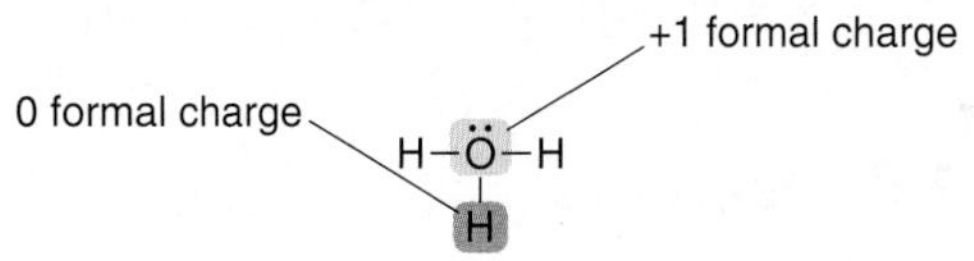

For the O atom (group 6A):

- number of valence electrons = 6
- number of bonding electrons = 6
- number of nonbonding electrons = 2

$$\text{formal charge} = 6 - \left[2 + \frac{1}{2}(6)\right]$$
$$= +1$$

For each H atom (group 1A):

- number of valence electrons = 1
- number of bonding electrons = 2
- number of nonbonding electrons = 0

$$\text{formal charge} = 1 - \left[0 + \frac{1}{2}(2)\right]$$
$$= 0$$

The formal charge on the O atom is +1 and the formal charge on each H is 0. The overall charge on the ion H_3O^+ is the sum of all of the formal charges on the atoms: 1 + 0 + 0 + 0 = +1.

Problem 1.6

Calculate the formal charge on each second-row atom.

a. 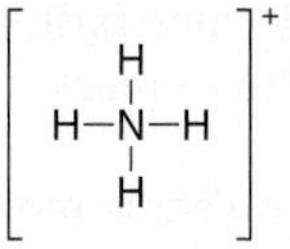

b. CH_3—N≡C:

c. :Ö=Ö—Ö:

Problem 1.7 Draw a Lewis structure for each ion.

a. CH_3O^- b. HC_2^- c. $(CH_3NH_3)^+$ d. $(CH_3NH)^-$

When you first add formal charges to Lewis structures, use the procedure in Sample Problem 1.3. With practice, you will notice that certain bonding patterns always result in the same formal charge. For example, any N atom with four bonds (and, thus no lone pairs) has a +1 formal charge. Table 1.1 lists the bonding patterns and resulting formal charges for carbon, nitrogen, and oxygen.

Table 1.1 Formal Charge Observed with Common Bonding Patterns for C, N, and O

Atom	Number of valence electrons	Formal charge		
		+1	0	−1
C	4	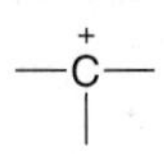		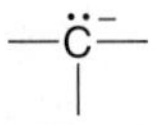
N	5		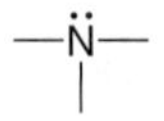	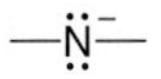
O	6	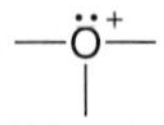	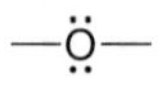	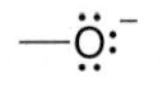

Problem 1.8 What is the formal charge on the O atom in each of the following species that contains a multiple bond to O?

a. ≡O: b. =Ö— c. =Ö:

1.4 Isomers

In drawing a Lewis structure for a molecule with several atoms, sometimes more than one arrangement of atoms is possible for a given molecular formula. For example, there are two acceptable arrangements of atoms for the molecular formula C_2H_6O.

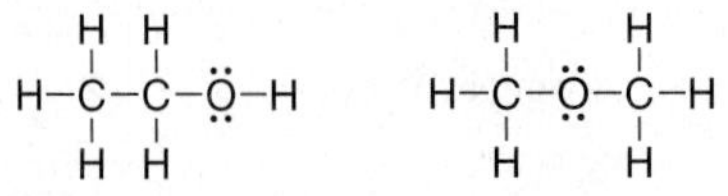

ethanol dimethyl ether

same molecular formula
C_2H_6O
isomers

Both are valid Lewis structures, and both molecules exist. One is called ethanol, and the other, dimethyl ether. These two compounds are called **isomers.**

- *Isomers* are different molecules having the same molecular formula.

Ethanol and dimethyl ether are **constitutional isomers** because they have the same molecular formula, but the ***connectivity of their atoms is different.*** Ethanol has one C–C bond and one O–H bond, whereas dimethyl ether has two C–O bonds. A second class of isomers, called **stereoisomers,** is introduced in Section 4.13B.

Problem 1.9 Draw Lewis structures for each molecular formula.

a. $C_2H_4Cl_2$ (two isomers) b. C_3H_8O (three isomers) c. C_3H_6 (two isomers)

1.5 Exceptions to the Octet Rule

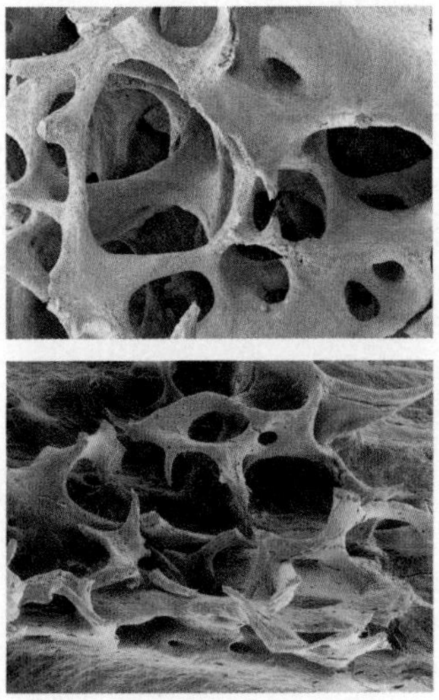

Alendronic acid, sold as a sodium salt under the trade name of **Fosamax,** is used to prevent osteoporosis in women. Osteoporosis decreases bone density, as shown by comparing normal bone (top) with brittle bone (bottom).

Most of the common elements in organic compounds—**C, N, O, and the halogens**—follow the octet rule. Hydrogen is a notable exception, because it accommodates only two electrons in bonding. Additional exceptions include boron and beryllium (second-row elements in groups 3A and 2A, respectively), and elements in the third row (particularly phosphorus and sulfur).

Elements in groups 2A and 3A of the periodic table, such as beryllium and boron, do not have enough valence electrons to form an octet in a neutral molecule. Lewis structures for BeH_2 and BF_3 show that these atoms have only four and six electrons, respectively, around the central atom. There simply aren't enough electrons to form an octet. Because the Be and B atoms each have less than an octet of electrons, these molecules are highly reactive.

four electrons around Be

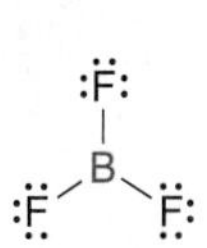

six electrons around B

A second exception to the octet rule occurs with some elements located in the third row and later in the periodic table. These elements have empty *d* orbitals available to accept electrons, and thus they may have ***more than eight*** **electrons** around them. For organic chemists, the two most common elements in this category are **phosphorus** and **sulfur,** which can have 10 or even 12 electrons around them, as shown in dimethyl sulfoxide, sulfuric acid, and alendronic acid.

10 electrons around S

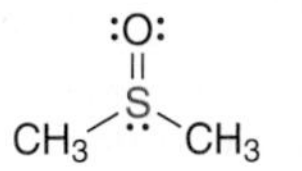

dimethyl sulfoxide
DMSO

12 electrons around S

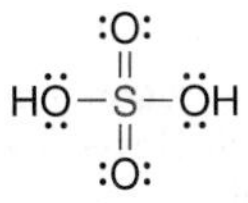

sulfuric acid

10 electrons around each P

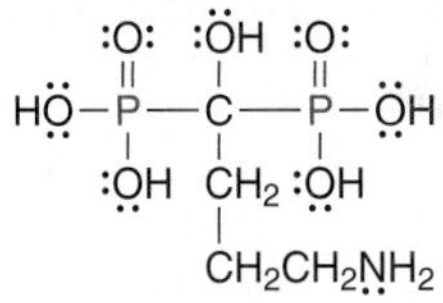

alendronic acid

1.6 Resonance

Some molecules can't be adequately represented by a single Lewis structure. For example, two valid Lewis structures can be drawn for the anion $(HCONH)^-$. One structure has a negatively charged N atom and a C–O double bond; the other has a negatively charged O atom and a C–N double bond. These structures are called **resonance structures** or **resonance forms.** A **double-headed arrow** is used to separate two resonance structures.

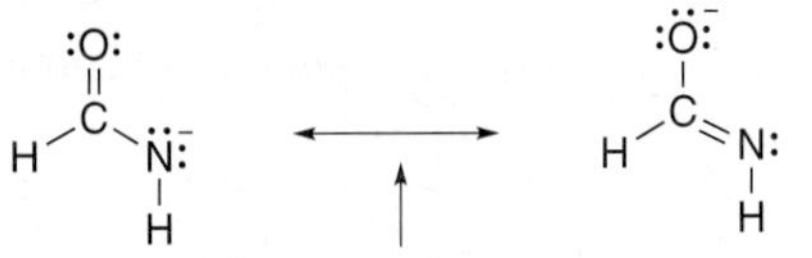

double-headed arrow

- *Resonance structures* are two Lewis structures having the *same* placement of atoms but a *different* arrangement of electrons.

Which resonance structure is an accurate representation for $(HCONH)^-$? **The answer is *neither* of them.** The true structure is a composite of both resonance forms, and is called a **resonance hybrid.** The hybrid shows characteristics of ***both*** resonance structures.

Each resonance structure implies that electron pairs are localized in bonds or on atoms. In actuality, resonance allows certain electron pairs to be ***delocalized*** over two or more atoms, and this delocalization of electron density adds stability. **A molecule with two or more resonance structures is said to be *resonance stabilized.***

1.6A An Introduction to Resonance Theory

Keep in mind the following basic principles of resonance theory.

- Resonance structures are not real. An individual resonance structure does not accurately represent the structure of a molecule or ion.
- Resonance structures are *not* in equilibrium with each other. There is no movement of electrons from one form to another.
- Resonance structures are *not* isomers. Two isomers differ in the arrangement of *both* atoms and electrons, whereas resonance structures differ *only* in the *arrangement of electrons.*

For example, ions **A** and **B** are resonance structures because the atom position is the same in both compounds, but the location of an electron pair is different. In contrast, compounds **C** and **D** are isomers because the atom placement is different; **C** has an O–H bond, and **D** has an additional C–H bond.

A **B** **C** **D**

- **A** and **B** are **resonance** structures.
- The position of one electron pair (in red) is different.

- **C** and **D** are **isomers.**
- The position of a H atom (in red) is different.

Problem 1.10 Classify each pair of compounds as isomers or resonance structures.

a. 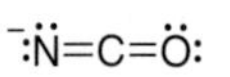 and 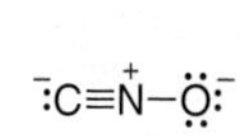 b. 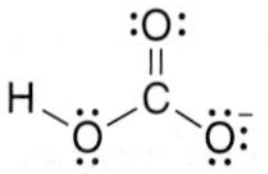 and 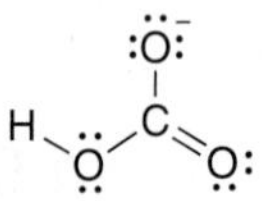

Problem 1.11 Considering structures **A–D,** classify each pair of compounds as isomers, resonance structures, or neither: (a) **A** and **B;** (b) **A** and **C;** (c) **A** and **D;** (d) **B** and **D.**

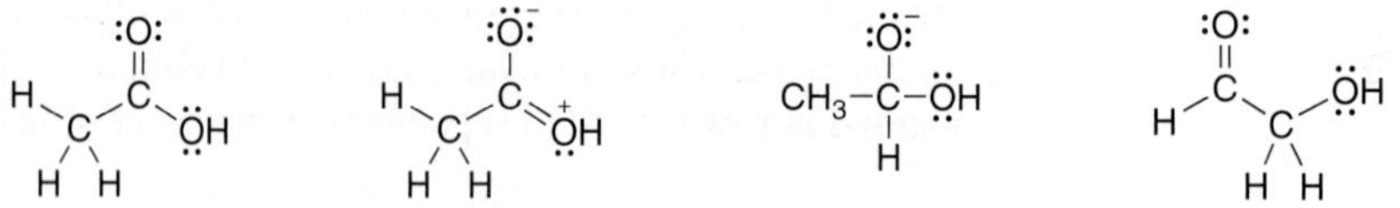

A **B** **C** **D**

1.6B Drawing Resonance Structures

To draw resonance structures, use three rules.

Rule [1] **Two resonance structures differ in the position of multiple bonds and nonbonded electrons. The placement of atoms and single bonds always stays the same.**

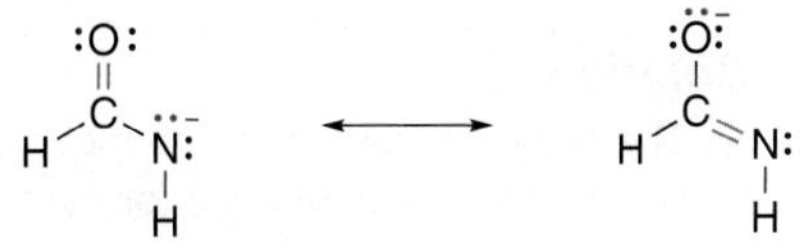

A **B**

- The position of a double bond (in blue) is different.
- The position of a lone pair (in red) is different.

Rule [2] **Two resonance structures must have the same number of unpaired electrons.**

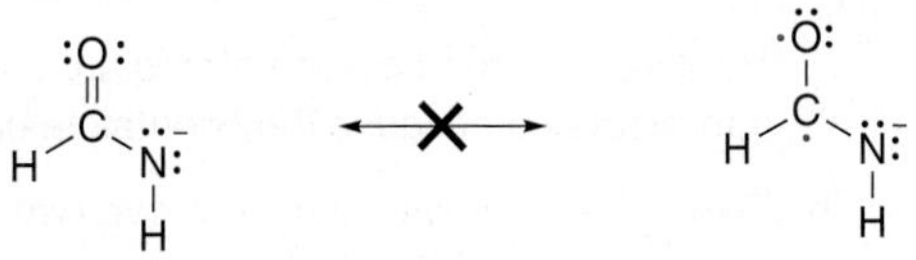

A — no unpaired electrons

C — two unpaired electrons
not a resonance structure of **A** (or **B**)

Rule [3] **Resonance structures must be valid Lewis structures. Hydrogen must have two electrons and no second-row element can have more than eight electrons.**

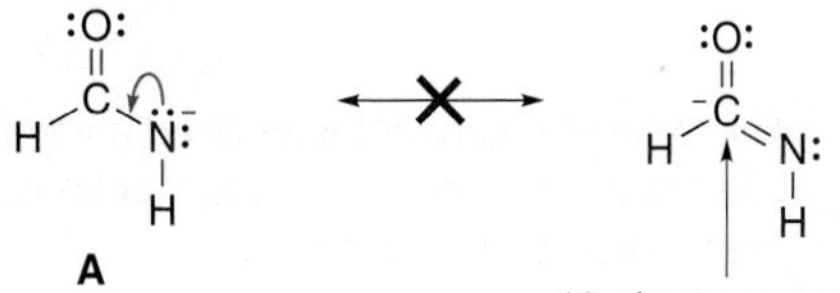

A

10 electrons around C
not a valid resonance structure of **A**

Curved arrow notation is a convention that shows how electron position differs between the two resonance forms.

- *Curved arrow notation shows the movement of an electron pair.* The tail of the arrow always begins at an electron pair, either in a bond or lone pair. The head points to where the electron pair "moves."

A curved arrow always begins at an electron pair. It ends at an atom or a bond.

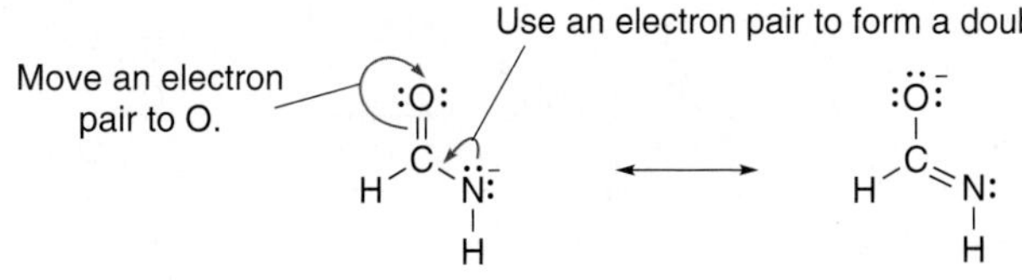
Use an electron pair to form a double bond.
Move an electron pair to O.

A **B**

Resonance structures **A** and **B** differ in the location of two electron pairs, so two curved arrows are needed. To convert **A** to **B,** take the lone pair on N and form a double bond between C and N. Then, move an electron pair in the C–O double bond to form a lone pair on O. Curved arrows thus show how to reposition the electrons in converting one resonance form to another. **The electrons themselves do not actually move.** Sample Problem 1.4 illustrates the use of curved arrows to convert one resonance structure to another.

Sample Problem 1.4 Follow the curved arrows to draw a second resonance structure for each ion.

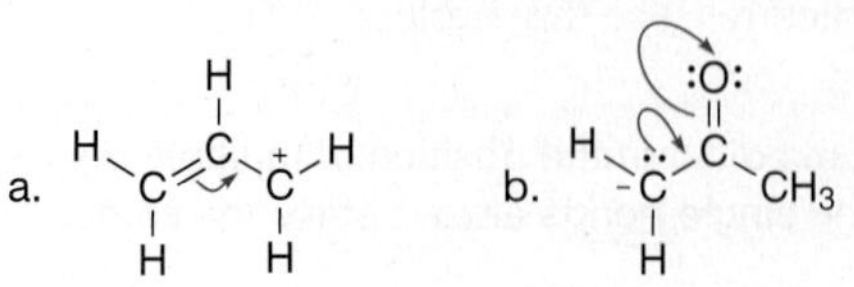

Solution

a. The curved arrow tells us to move **one** electron pair in the double bond to the adjacent C – C bond. Then determine the formal charge on any atom whose bonding is different.

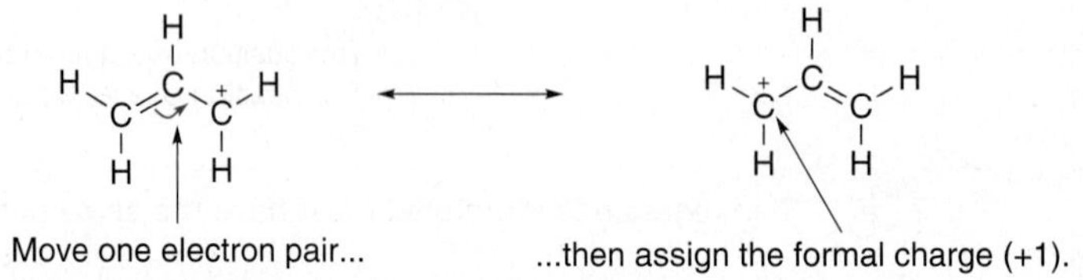
Move one electron pair...
...then assign the formal charge (+1).

Positively charged carbon atoms are called **carbocations.** Carbocations are unstable intermediates because they contain a carbon atom that is lacking an octet of electrons.

b. **Two** curved arrows tell us to move **two** electron pairs. The second resonance structure has a formal charge of (–1) on O.

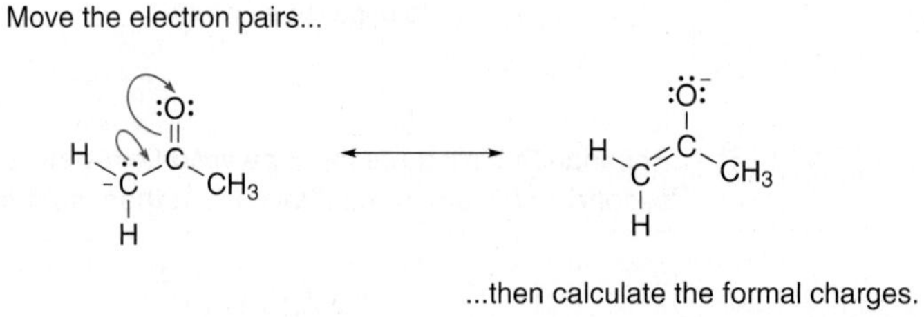
Move the electron pairs...
...then calculate the formal charges.

This type of resonance-stabilized anion is called an **enolate anion.** Enolates are important intermediates in many organic reactions, and all of Chapters 23 and 24 is devoted to their preparation and reactions.

Problem 1.12 Follow the curved arrows to draw a second resonance structure for each species.

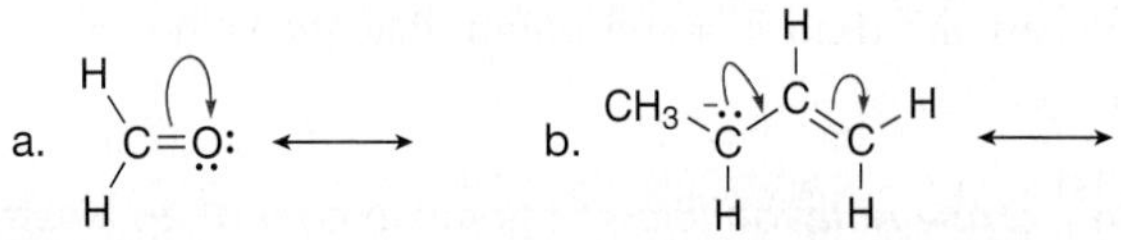

Problem 1.13 Use curved arrow notation to show how the first resonance structure can be converted to the second.

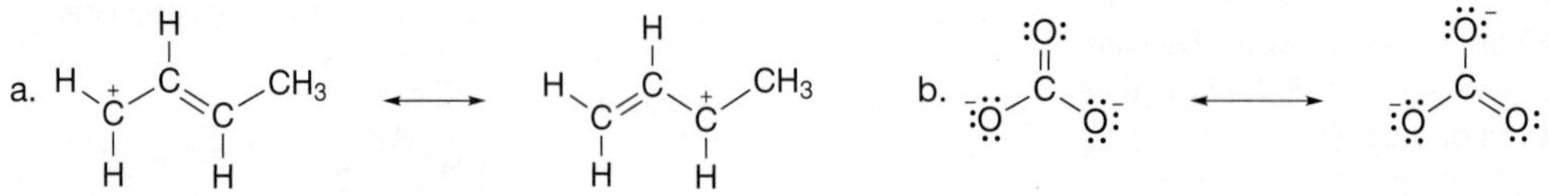

Two resonance structures can have exactly the same kinds of bonds, as they do in the carbocation in Sample Problem 1.4a, or they may have different types of bonds, as they do in the enolate in Sample Problem 1.4b. Either possibility is fine as long as the individual resonance structures are valid Lewis structures.

A resonance structure can have an atom with *fewer* than eight electrons around it. **B** is a resonance structure of **A** even though the carbon atom is surrounded by only six electrons.

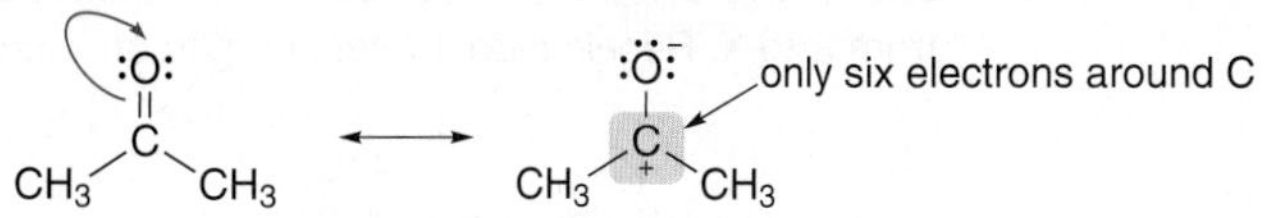

A **B**

valid resonance structure

In contrast, a resonance structure can *never* have a second-row element with more than eight electrons. **C** is *not* a resonance structure of **A** because the carbon atom is now surrounded by 10 electrons.

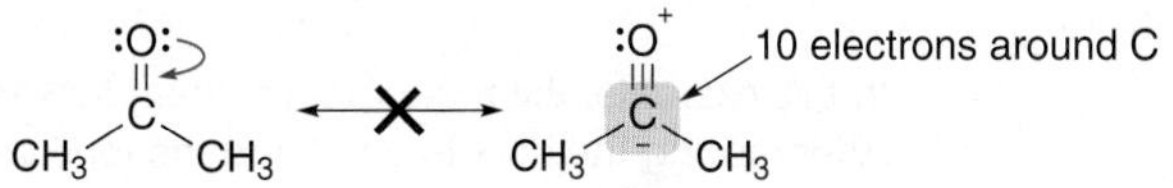

A **C**

not a valid resonance structure

We will learn much more about resonance in Chapter 16.

The ability to draw and manipulate resonance structures is a necessary skill that will be used throughout your study of organic chemistry. With practice, you will begin to recognize certain common bonding patterns for which more than one Lewis structure can be drawn. For instance, both the carbocation in Sample Problem 1.4a and the enolate anion in Sample Problem 1.4b are specific examples of one general type of resonance observed in certain three-atom systems.

- **In a group of three atoms having a multiple bond X=Y joined to an atom Z having a *p* orbital with zero, one, or two electrons, two resonance structures can be drawn.**

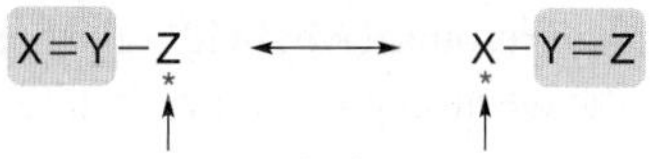

The * corresponds to a charge, a lone pair, or a single electron.

* = +, –, ·, or :

Recall from the Prologue that a ***heteroatom* is an atom other than carbon or hydrogen.**

X, Y, and Z may all be carbon atoms or they may be **heteroatoms** such as nitrogen or oxygen. The atom Z can be charged (positive or negative) or neutral (with a lone pair or a single electron), corresponding to the [*] in the general structure X=Y–Z*. The two resonance structures differ in the location of the multiple bond and the [*].

In the enolate anion in Sample Problem 1.4b, X corresponds to oxygen and [*] is a lone pair, which gives carbon a net negative charge. Moving the double bond and the lone pair and readjusting charges gives the second resonance structure.

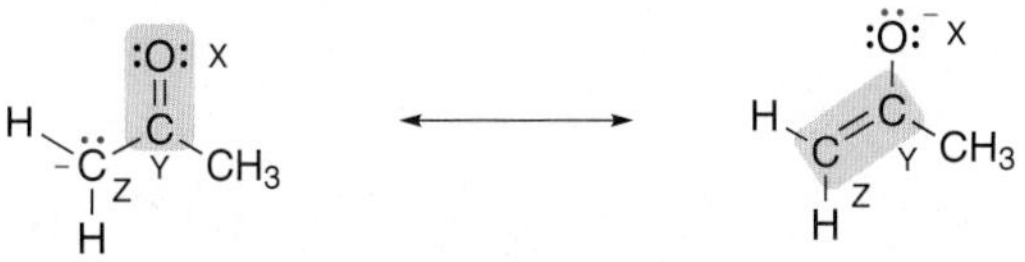

- The position of the double bond changes.
- The location of a lone pair changes.

In Chapter 16, we learn more about the orbitals involved in this type of resonance.

Sample Problem 1.5 Draw a second resonance structure for acetamide.

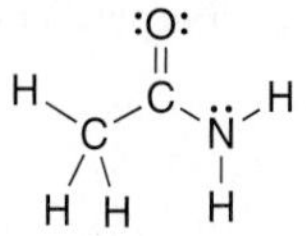

acetamide

Solution

Always look for a three-atom system that contains a multiple bond joined to an atom Z with zero, one, or two electrons in a *p* orbital. Move the double bond (from X = Y to Y = Z) and move the [*] from Z to X. Recalculate formal charges on X and Z.

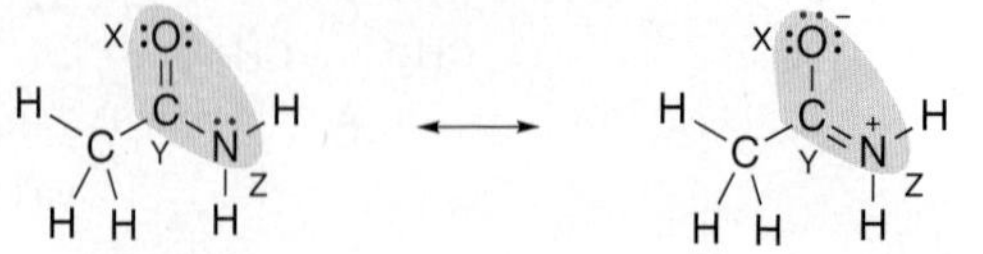

acetamide
- C=O
- lone pair on N

second resonance structure
- C=N
- additional lone pair on O

In this example, the three-atom system for resonance (X = Y – Z*) is O = C – N with a lone pair on N. After moving the double bond and the lone pair, the formal charges on O and N are –1 and +1, respectively, calculated using the procedure for determining formal charges.

Problem 1.14 Draw a second resonance structure for each species in parts (a) and (b). Draw two additional resonance structures for the ion in part (c).

a. 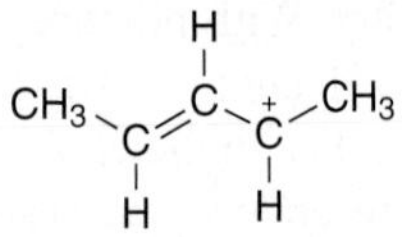 b. 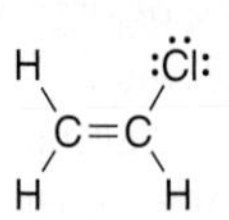 c. 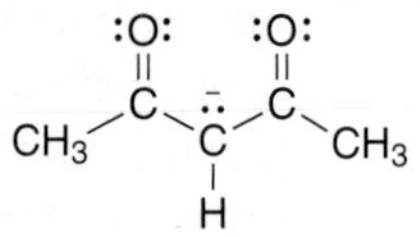

1.6C The Resonance Hybrid

The **resonance hybrid** is the composite of all possible resonance structures. In the resonance hybrid, the electron pairs drawn in different locations in individual resonance structures are ***delocalized.***

- **The resonance hybrid is more stable than any resonance structure because it delocalizes electron density over a larger volume.**

What does the hybrid look like? When all resonance forms are identical, as they were in the carbocation in Sample Problem 1.4a, each resonance form contributes **equally** to the hybrid.

When two resonance structures are different, the hybrid looks more like the "better" resonance structure. The "better" resonance structure is called the **major contributor** to the hybrid, and all others are **minor contributors.** The hybrid is the weighted average of the contributing resonance structures. What makes one resonance structure "better" than another? There are many factors, but for now, we will learn one fact.

- **A "better" resonance structure is one that has *more bonds* and *fewer charges.***

X
more bonds
fewer charges

major contributor

Y
minor contributor

Comparing resonance structures **X** and **Y, X** is the major contributor because it has more bonds and fewer charges. Thus, the hybrid looks more like **X** than **Y.**

How can we draw a hybrid, which has delocalized electron density? First, we must determine what is different in the resonance structures. Two differences commonly seen are the **position of a multiple bond** and the **site of a charge.** The anion $(HCONH)^-$ illustrates two conventions for drawing resonance hybrids.

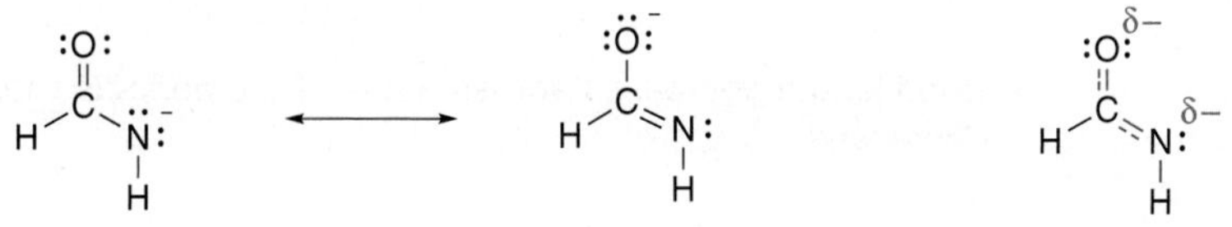

A B

individual resonance structures — resonance hybrid

- The (–) charge is delocalized on N and O.
- The double bond is delocalized between O, C, and N.

Common symbols and conventions used in organic chemistry are listed on the inside back cover.

- **Double bond position. Use a dashed line for a bond that is single in one resonance structure and double in another.**
- **Location of charge. Use a δ– (partial negative charge) or δ+ (partial positive charge) for an atom that is neutral in one resonance structure and charged in another.**

The hybrid for $(HCONH)^-$ shows two dashed bonds, indicating that both the C–O and C–N bonds have partial double bond character. Both the O and N atoms bear a partial negative charge (δ–) because these atoms are neutral in one resonance structure and negatively charged in the other.

This discussion of resonance is meant to serve as an introduction only. You will learn many more facets of resonance theory in later chapters. In Chapter 2, for example, the enormous effect of resonance on acidity is discussed.

Problem 1.15 Label the resonance structures in each pair as major, minor, or equal contributors to the hybrid. Then draw the hybrid.

a. 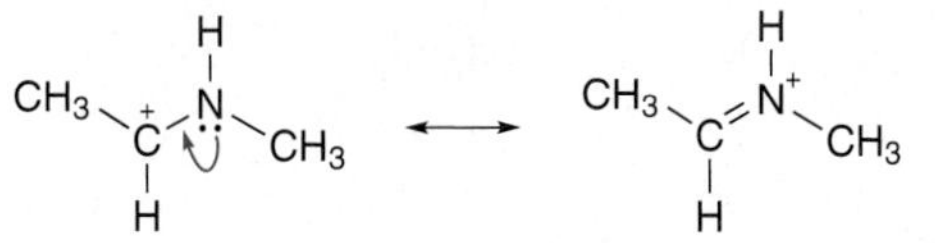

b. 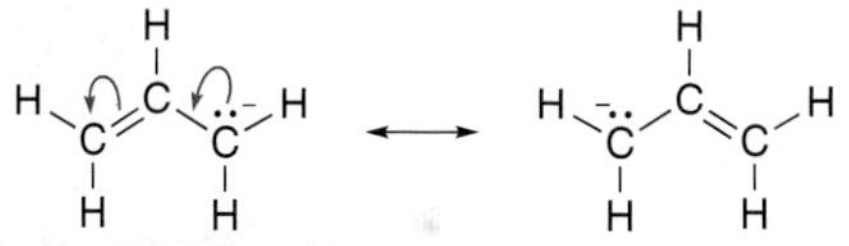

Problem 1.16 (a) Draw a second resonance structure for **A.** (b) Why can't a second resonance structure be drawn for **B?**

A **B**

1.7 Determining Molecular Shape

Consider the H_2O molecule. The Lewis structure tells us which atoms are connected to each other, but it implies nothing about the geometry. What does the overall molecule look like? Is H_2O a bent or linear molecule? Two variables define a molecule's structure: **bond length** and **bond angle.**

1.7A Bond Length

Although the SI unit for bond length is the picometer (pm), the angstrom (Å) is still widely used in the chemical literature; 1 Å = 10^{-10} m. As a result, 1 pm = 10^{-2} Å, and 95.8 pm = 0.958 Å.

***Bond length* is the average distance between the centers of two bonded nuclei.** Bond lengths are typically reported in picometers (pm), where 1 pm = 10^{-12} m. For example, the O–H bond length in H_2O is 95.8 pm. Average bond lengths for common bonds are listed in Table 1.2.

- **Bond length *decreases* across a row of the periodic table as the size of the atom *decreases*.**

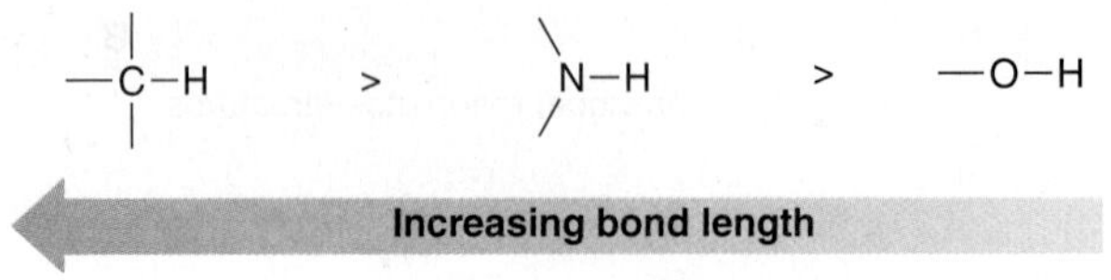

- **Bond length *increases* down a column of the periodic table as the size of an atom *increases*.**

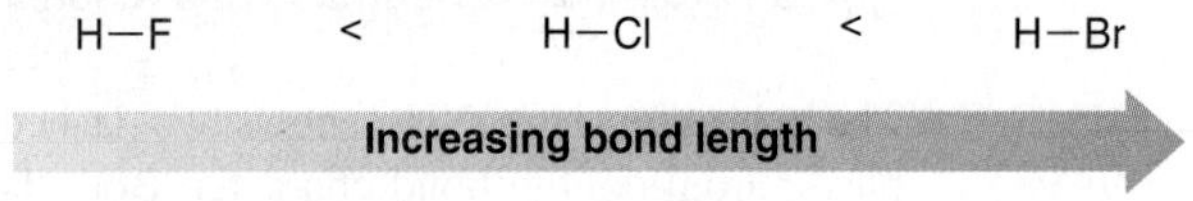

Table 1.2 Average Bond Lengths

Bond	Length (pm)	Bond	Length (pm)	Bond	Length (pm)
H–H	74	H–F	92	C–F	133
C–H	109	H–Cl	127	C–Cl	177
N–H	101	H–Br	141	C–Br	194
O–H	96	H–I	161	C–I	213

1.7B Bond Angle

Bond angle determines the shape around any atom bonded to two other atoms. To determine the bond angle and shape around a given atom, first count how many groups surround the atom. **A group is either an atom or a lone pair of electrons.** Then use the **valence shell electron pair repulsion (VSEPR) theory** to determine the shape. VSEPR is based on the fact that electron pairs repel each other; thus:

- **The most stable arrangement keeps these groups as far away from each other as possible.**

A second-row element has only three possible arrangements, defined by the number of groups surrounding it.

To determine geometry: [1] Draw a valid Lewis structure; [2] count groups around a given atom.

Number of groups	Geometry	Bond angle
• two groups	linear	180°
• three groups	trigonal planar	120°
• four groups	tetrahedral	109.5°

Let's examine several molecules to illustrate this phenomenon. We first need a valid Lewis structure, and then we count groups around a given atom to predict its geometry.

Two Groups Around an Atom

Any atom surrounded by only two groups is linear and has a bond angle of 180°. Two examples illustrating this geometry are **BeH_2** (beryllium hydride) and **HC≡CH** (acetylene). We consider each carbon atom in acetylene *separately.* Because each C is surrounded by two atoms and no lone pairs, each H–C–C bond angle in acetylene is 180°, and therefore all four atoms are linear.

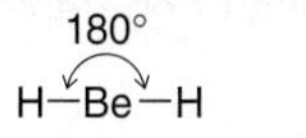

two atoms around Be

two groups
linear molecule

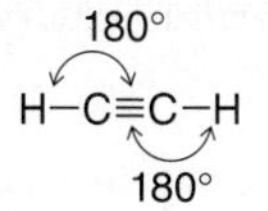

two atoms around each C

two groups
linear carbons

=

ball-and-stick model

Acetylene illustrates another important feature: ***ignore multiple bonds in predicting geometry.* Count only atoms and lone pairs.**

Most students in organic chemistry find that building models helps them visualize the shape of molecules. Invest in a set of models *now*.

Common element colors are also shown on the inside back cover.

We will represent molecules with models having balls for atoms and sticks for bonds, as in the ball-and-stick model of acetylene just shown. These representations are analogous to a set of molecular models. Balls are color-coded using accepted conventions: carbon (black), hydrogen (white or gray), oxygen (red), and so forth, as shown.

Three Groups Around an Atom

Any atom surrounded by three groups is trigonal planar and has bond angles of 120°. Two examples illustrating this geometry are **BF_3** (boron trifluoride) and **$CH_2{=}CH_2$** (ethylene). All three B–F bonds lie in one plane. *Each* carbon atom of ethylene is surrounded by three atoms and no lone pairs, making *each* H–C–C bond angle 120°. All six atoms of $CH_2{=}CH_2$ lie in one plane.

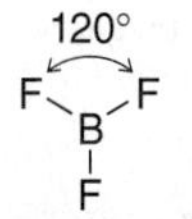

three atoms around B

three groups
trigonal planar molecule

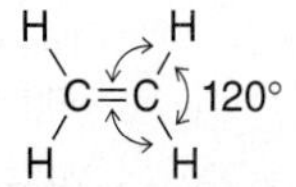

three atoms around each C

three groups
trigonal planar carbons

=

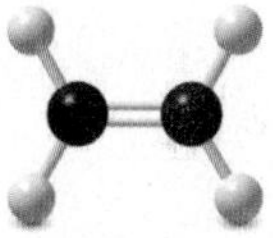

ethylene

Four Groups Around an Atom

Any atom surrounded by four groups is tetrahedral and has bond angles of approximately 109.5°. The simple organic compound methane, **CH_4,** has a central carbon atom with bonds to four hydrogen atoms, each pointing to a corner of a tetrahedron. This arrangement keeps four groups farther apart than a square planar arrangement in which all bond angles would be only 90°.

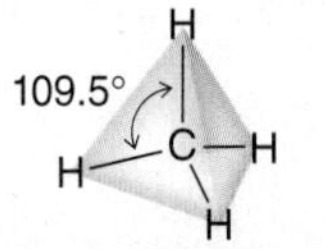

preferred geometry
larger H–C–H bond angle

four groups
tetrahedral molecule

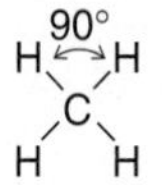

square planar arrangement
This geometry does ***not*** occur.

How can we represent the three-dimensional geometry of a tetrahedron on a two-dimensional piece of paper? **Place two of the bonds in the plane of the paper, one bond in front and one bond behind,** using the following conventions:

- A *solid line* is used for a bond *in* the plane.
- A *wedge* is used for a bond in *front* of the plane.
- A *dashed wedge* is used for a bond *behind* the plane.

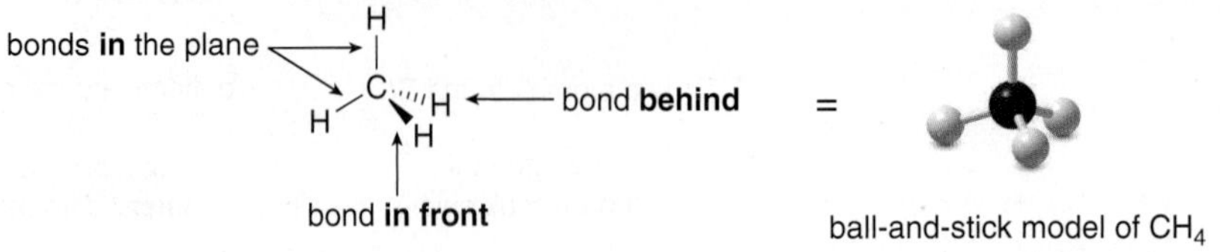

This is just one way to draw a tetrahedron for CH_4. We can turn the molecule in many different ways, generating many equivalent representations. All of the following are acceptable drawings for CH_4, because each drawing has two solid lines, one wedge, and one dashed wedge.

Finally, **wedges and dashed wedges are used for groups that are really *aligned one behind another.*** It does not matter in the following two drawings whether the wedge or dashed wedge is skewed to the left or right, because the two H atoms are really aligned as shown in the three-dimensional model.

All carbons in stable molecules are ***tetravalent,*** but the geometry varies with the number of groups around the particular carbon.

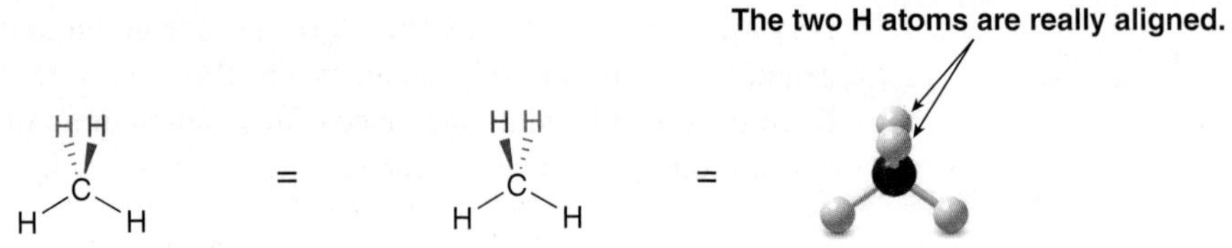

- These representations are equivalent.
- The wedge can be skewed to the left or the right of the dashed wedge.

Ammonia (NH_3) and water (H_2O) both have atoms surrounded by four groups, some of which are lone pairs. In $\mathbf{NH_3}$**,** the three H atoms and one lone pair around N point to the corners of a tetrahedron. The H–N–H bond angle of 107° is close to the theoretical tetrahedral bond angle of 109.5°. This molecular shape is referred to as **trigonal pyramidal,** because one of the groups around the N is a nonbonded electron pair, not another atom.

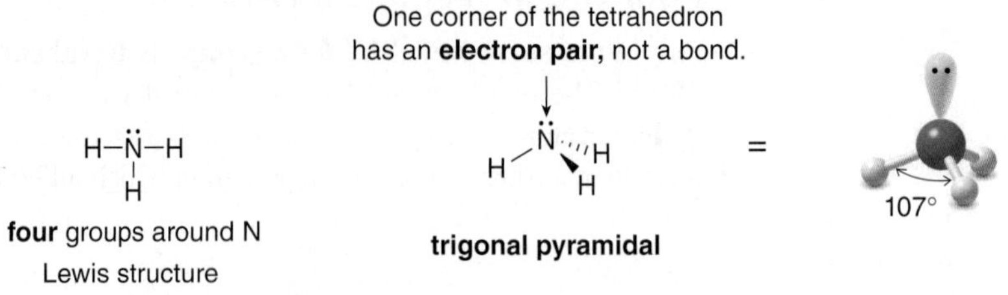

In H_2O**,** the two H atoms and two lone pairs around O point to the corners of a tetrahedron. The H–O–H bond angle of 105° is close to the theoretical tetrahedral bond angle of 109.5°. Water has a **bent** molecular shape, because two of the groups around oxygen are lone pairs of electrons.

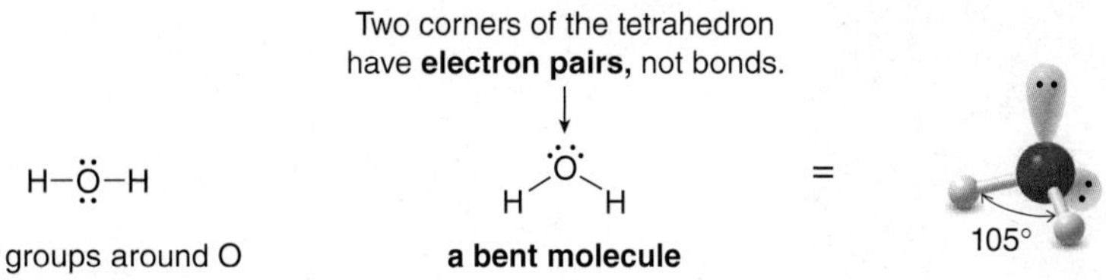

four groups around O
Lewis structure

a bent molecule

In both NH_3 and H_2O the bond angle is somewhat smaller than the theoretical tetrahedral bond angle because of repulsion of the lone pairs of electrons. The bonded atoms are compressed into a smaller space with a smaller bond angle.

Predicting geometry based on counting groups is summarized in Table 1.3.

Table 1.3 Summary: Determining Geometry Based on the Number of Groups

Number of groups around an atom	Geometry	Bond angle	Examples
2	linear	180°	BeH_2, HC≡CH
3	trigonal planar	120°	BF_3, $CH_2{=}CH_2$
4	tetrahedral	109.5°	CH_4, NH_3, H_2O

Sample Problem 1.6

Determine the geometry around the highlighted atom in each species.

a. 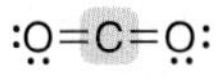 b. 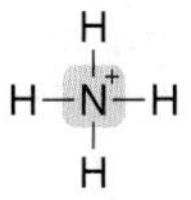

Solution

a.

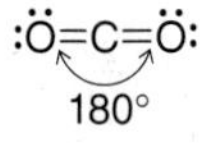

two atoms around C
no lone pairs

two groups

linear

b.

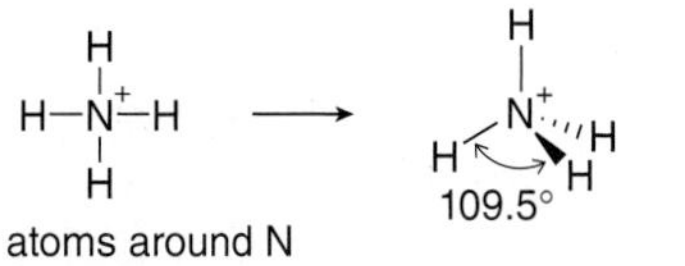

four atoms around N
no lone pairs

four groups ⟶ **tetrahedral**

Problem 1.17 Determine the geometry around all second-row elements in each compound drawn as a Lewis structure with no implied geometry.

a. 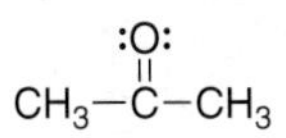 b. 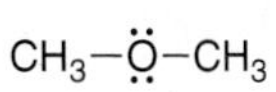 c. 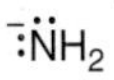 d. 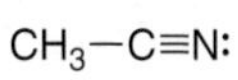

Problem 1.18 Predict the indicated bond angles in each compound drawn as a Lewis structure with no implied geometry.

a. 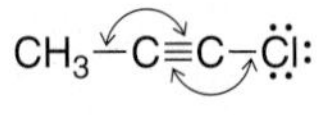 b. 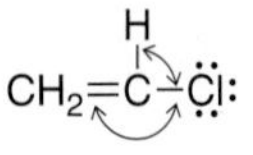 c. 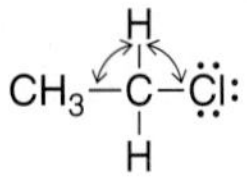

Problem 1.19 Using the principles of VSEPR theory, you can predict the geometry around any atom in any molecule, no matter how complex. Enanthotoxin is a poisonous compound isolated from a common variety of hemlock grown in England. Predict the geometry around the highlighted atoms in enanthotoxin.

The hemlock water dropwort is the source of the poison enanthotoxin.

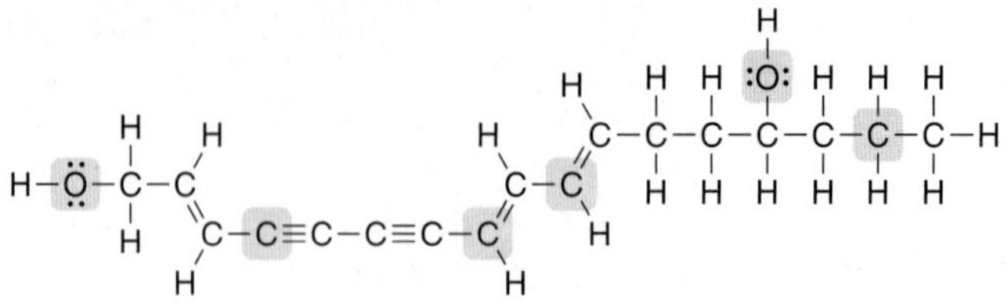

enanthotoxin

1.8 Drawing Organic Structures

Drawing organic molecules presents a special challenge. Because they often contain many atoms, we need shorthand methods to simplify their structures. The two main types of shorthand representations used for organic compounds are **condensed structures** and **skeletal structures.**

1.8A Condensed Structures

Condensed structures can be used for compounds having a chain of atoms bonded together. The following conventions are used:

- **All of the atoms are drawn in, but the two-electron bond lines are generally omitted.**
- **Atoms are usually drawn next to the atoms to which they are bonded.**
- **Parentheses are used around similar groups bonded to the same atom.**
- **Lone pairs are omitted.**

To interpret a condensed formula, it is usually best to start at the *left side* of the molecule and remember that the ***carbon atoms must be tetravalent.*** A carbon bonded to three H atoms becomes $\mathbf{CH_3}$; a carbon bonded to two H atoms becomes $\mathbf{CH_2}$; and a carbon bonded to one H atom becomes **CH.**

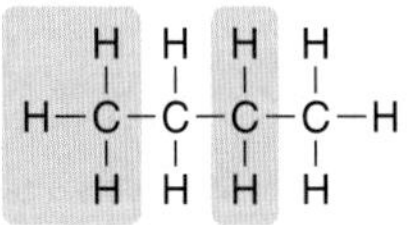

= $CH_3CH_2CH_2CH_3$ or $CH_3(CH_2)_2CH_3$

2 CH_2 groups bonded together

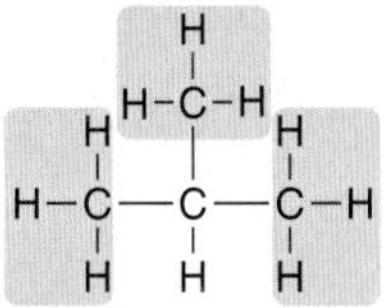

= $(CH_3)_3CH$

Other examples of condensed structures with heteroatoms and carbon–carbon multiple bonds are given in Figure 1.3.

Figure 1.3 Examples of condensed structures

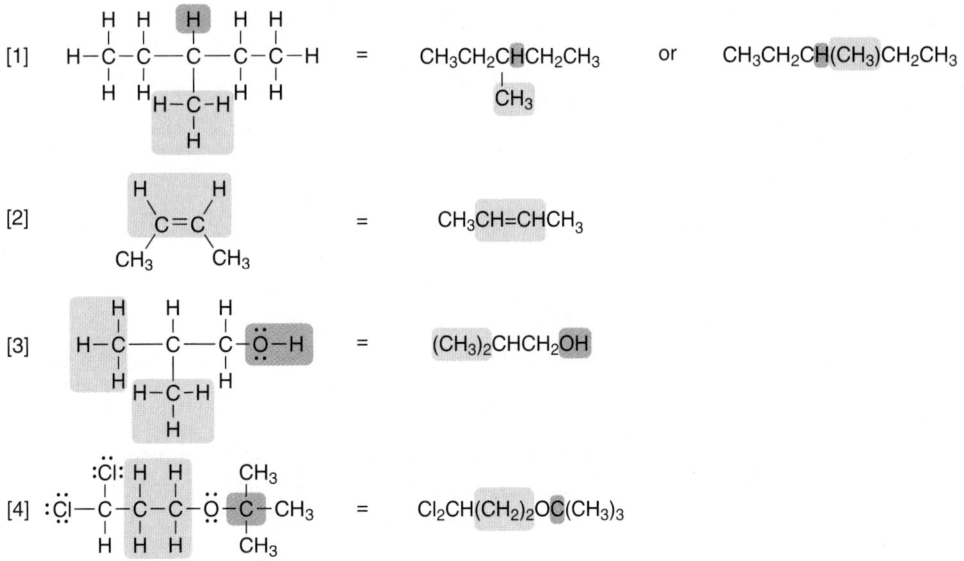
[1]
=
$CH_3CH_2CHCH_2CH_3$ with CH_3 below CH
or
$CH_3CH_2CH(CH_3)CH_2CH_3$
[2]
=
$CH_3CH{=}CHCH_3$
[3]
=
$(CH_3)_2CHCH_2OH$
[4]
=
$Cl_2CH(CH_2)_2OC(CH_3)_3$

- Entry [1]: Draw the H atom next to the C to which it is bonded, and use parentheses around CH_3 to show it is bonded to the carbon chain.
- Entry [2]: Keep the carbon–carbon double bond and draw the H atoms after each C to which they are bonded.
- Entry [3]: Omit the lone pairs on the O atom in the condensed structure.
- Entry [4]: Omit the lone pairs on Cl and O and draw the two CH_2 groups as $(CH_2)_2$.

Translating some condensed formulas is not obvious, and it will come only with practice. This is especially true for compounds containing a carbon–oxygen double bond. Some noteworthy examples in this category are given in Figure 1.4. While carbon–carbon double bonds are generally drawn in condensed structures, carbon–oxygen double bonds are usually omitted.

Figure 1.4 Condensed structures containing a C–O double bond

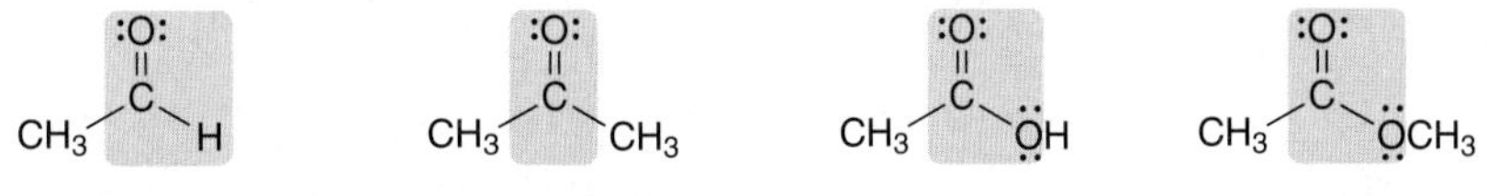

CH_3CHO	CH_3COCH_3	CH_3CO_2H	$CH_3CO_2CH_3$
A	**B**	**C**	**D**

- In **A,** the H atom is bonded to C, *not* O.
- In **B,** each CH_3 group is bonded to C, *not* O.
- In **C** and **D,** the C atom is doubly bonded to one O and singly bonded to the other O.

Sample Problem 1.7 Convert each condensed formula to a Lewis structure.

a. $(CH_3)_2CHOCH_2CH_2CH_2OH$ b. $CH_3(CH_2)_2CO_2C(CH_3)_3$

Solution

Start at the left and proceed to the right, making sure that each carbon has four bonds. Give each O atom two lone pairs to have an octet.

a. $(CH_3)_2CHOCH_2CH_2CH_2OH$

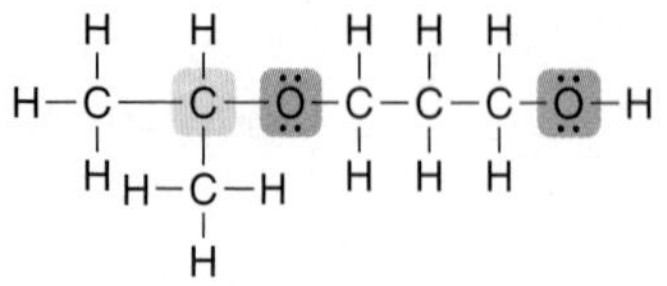

One C atom (labeled in blue) is bonded to 2 CH_3's, 1 H, and 1 O.

b. $CH_3(CH_2)_2CO_2C(CH_3)_3$

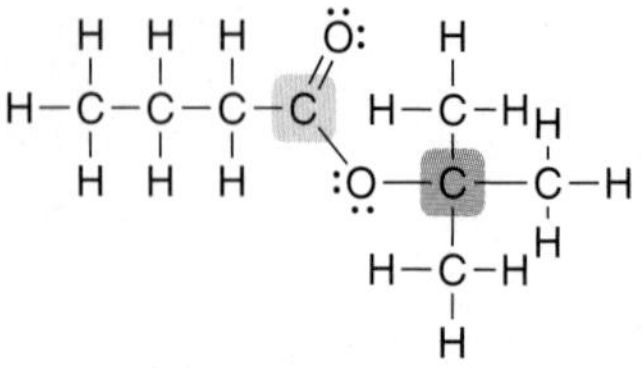

One C atom (labeled in blue) is bonded to both O's.

Problem 1.20 Convert each condensed formula to a Lewis structure.

a. $CH_3(CH_2)_4CH(CH_3)_2$
b. $(CH_3)_3CCH(OH)CH_2CH_3$
c. $(CH_3)_2CHCHO$
d. $(HOCH_2)_2CH(CH_2)_3C(CH_3)_2CH_2CH_3$

Problem 1.21 During periods of strenuous exercise, the buildup of lactic acid $[CH_3CH(OH)CO_2H]$ causes the aching feeling in sore muscles. Convert this condensed structure to a Lewis structure of lactic acid.

1.8B Skeletal Structures

Skeletal structures are used for organic compounds containing both rings and chains of atoms. Three rules are used to draw them.

- **Assume a carbon atom is located at the junction of any two lines or at the end of any line.**
- **Assume each carbon has enough hydrogens to make it tetravalent.**
- **Draw in all heteroatoms and the hydrogens directly bonded to them.**

Carbon chains are drawn in a zigzag fashion, and rings are drawn as polygons, as shown for hexane and cyclohexane.

Lewis structure

skeletal structure

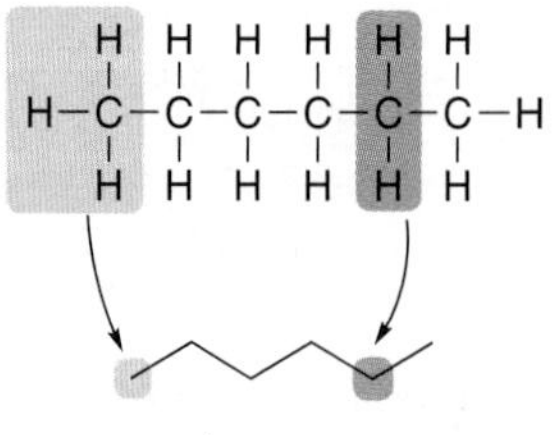

hexane

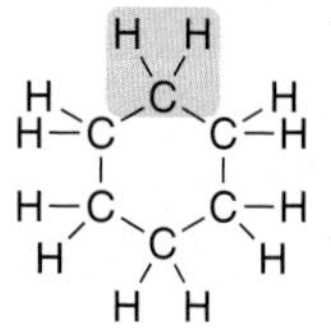

Lewis structure

C_6H_{12}

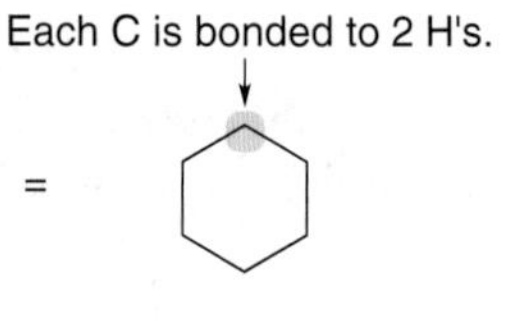
Each C is bonded to 2 H's.
=

skeletal structure

cyclohexane

How To Interpret a Skeletal Structure

Example Draw in all C atoms, H atoms, and lone pairs in the following molecule:

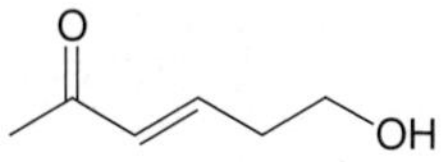
O
OH

Step [1] Place a C atom at the intersection of any two lines and at the end of any line.

- This molecule has six carbons, including the C labeled in red at the left end of the chain.
- There are two C's (labeled in green) between the C = C and the OH group.

Step [2] Add enough H's to make each C tetravalent.

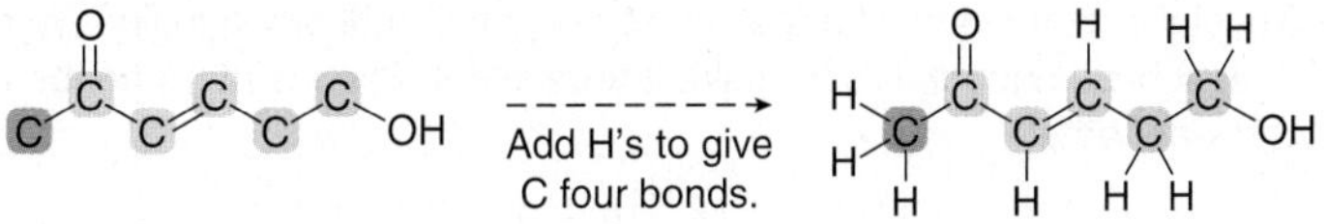
Add H's to give
C four bonds.

- The end C labeled in red needs three H's to be tetravalent.
- Each C on the C = C has three bonds already, so only one H must be drawn.
- There are two CH_2 groups between the C = C and the OH group.

Step [3] Add lone pairs to give each heteroatom an octet.

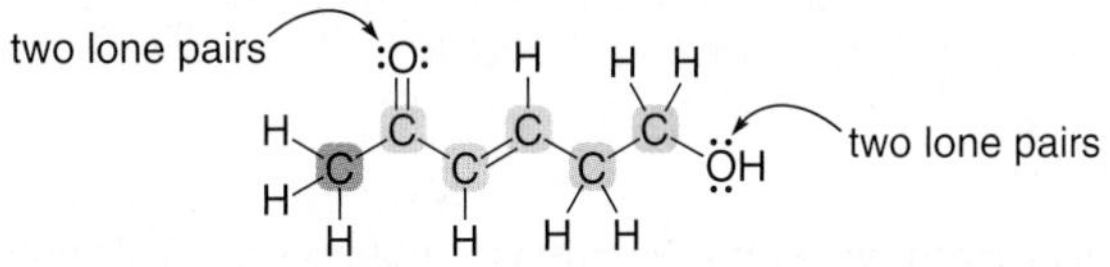
two lone pairs
two lone pairs

Each O needs **two** lone pairs for an octet.

Figure 1.5 shows other examples of skeletal structures, and Sample Problem 1.8 illustrates how to interpret the skeletal structure for a more complex cyclic compound.

Figure 1.5
Interpreting skeletal structures

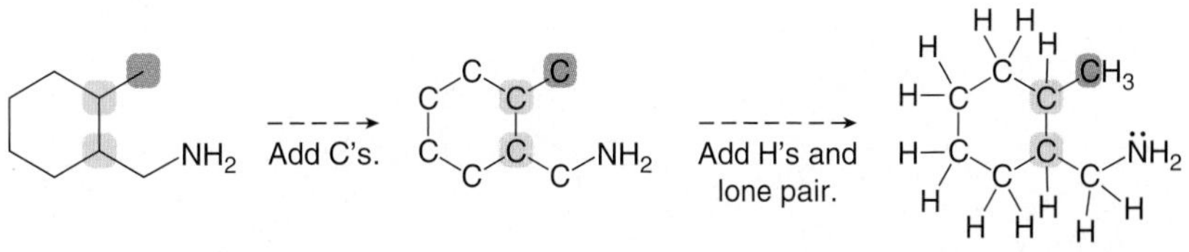
Add C's.
Add H's and
lone pair.

- C labeled in red needs 3 H's.
- C's labeled in blue need 1 H.
- All other C's need 2 H's.
- N needs one lone pair.

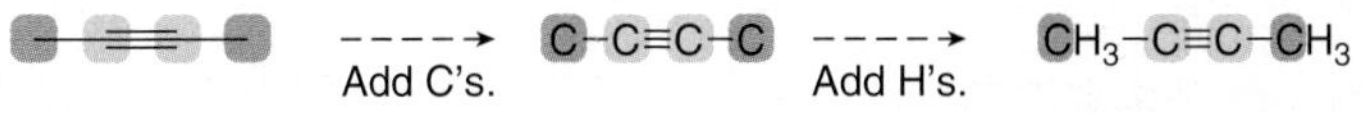
Add C's.
Add H's.

- C's labeled in red need 3 H's.
- C's labeled in green have four bonds to C.

Sample Problem 1.8 Draw a complete structure for vanillin showing all C atoms, H atoms, and lone pairs, and give the molecular formula. Vanillin is the principal component of the extract of the vanilla bean.

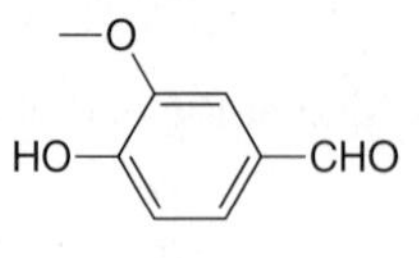

vanillin

vanilla bean

Solution

Skeletal structures have a C atom at the junction of any two lines and at the end of any line. Each C must have enough H's to make it tetravalent. Each O atom needs two lone pairs to have a complete octet.

- C in blue has three H's.
- C in green has four bonds to other C's.
- C in red is doubly bonded to O.

$C_8H_8O_3$

Problem 1.22 How many hydrogen atoms are present around each highlighted carbon atom in the following molecules? What is the molecular formula for each molecule? Both compounds are active ingredients in some common sunscreens.

a. 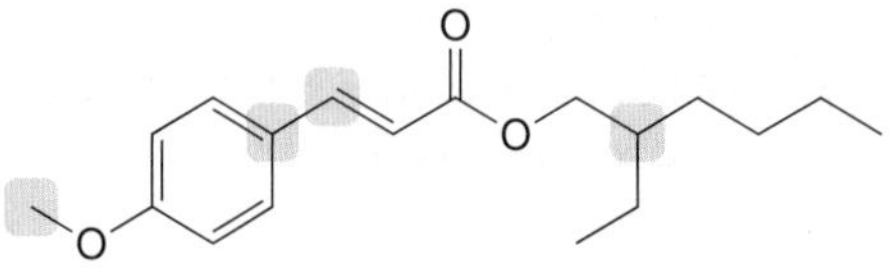

octinoxate
(2-ethylhexyl 4-methoxycinnamate)

b. 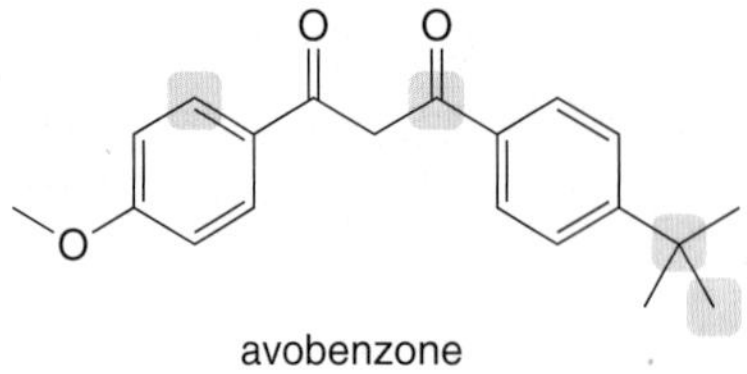

avobenzone

Problem 1.23 Convert each skeletal structure to a complete structure with all C's, H's, and lone pairs drawn in.

a. 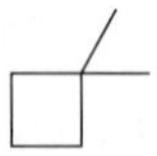 b.  c. 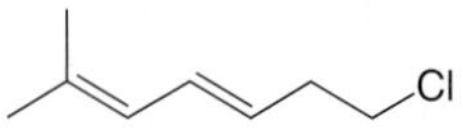 d. 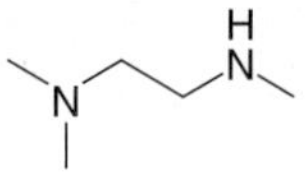

Problem 1.24 What is the molecular formula of quinine, the antimalarial medication whose skeletal structure appears in Figure 2 of the Prologue?

When heteroatoms are bonded to a carbon skeleton, the heteroatom is joined *directly* to the carbon to which it is bonded, with no H atoms in between. Thus, an OH group is drawn as OH or HO depending on where the OH is located. In contrast, when carbon appendages are bonded to a carbon skeleton, the H atoms will be drawn to the *right* of the carbon to which they are bonded regardless of the location.

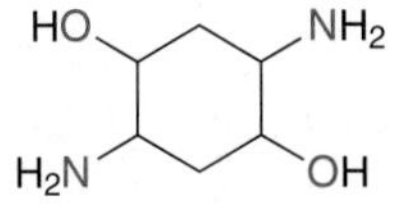
HO
NH2
H2N
OH

Place the O and N atoms directly joined to the ring.

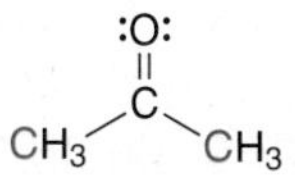
:O:
C
CH3
CH3

Two C atoms in red are bonded to the middle C.

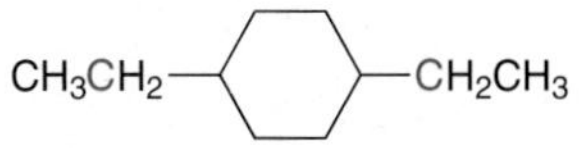
CH3CH2
CH2CH3

Two C atoms in red are bonded to the ring.

1.8C Skeletal Structures with Charged Carbon Atoms

Take care in interpreting skeletal structures for positively and negatively charged carbon atoms, because *both* the hydrogen atoms *and* the lone pairs are omitted. Keep in mind the following:

- **A charge on a carbon atom takes the place of one hydrogen atom.**
- **The charge determines the number of lone pairs. Negatively charged carbon atoms have one lone pair and positively charged carbon atoms have none.**

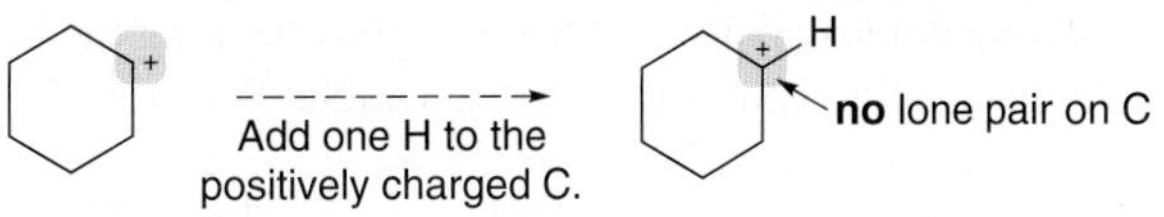
Add one H to the positively charged C.
H
no lone pair on C

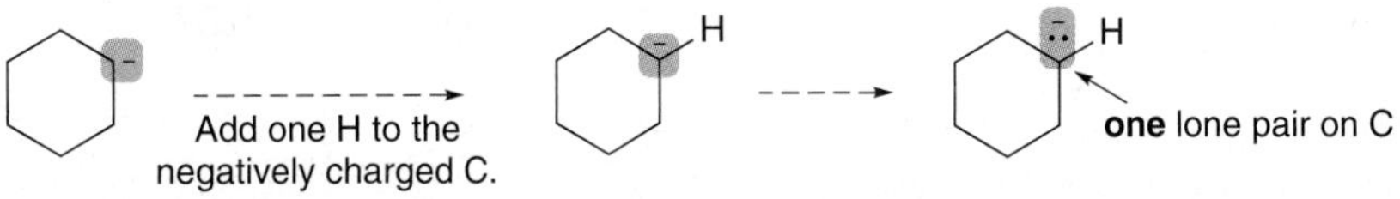
Add one H to the negatively charged C.
H
H
one lone pair on C

Skeletal structures often leave out lone pairs on heteroatoms, but *don't forget about them.* Use the formal charge on an atom to determine the number of lone pairs. For example, a neutral O atom with two bonds needs two additional lone pairs, and a positively charged O atom with three bonds needs only one lone pair.

OH
=
Ö
H
ÖH2
=
H
O
H

neutral O atom
two lone pairs

positively charged O atom
one lone pair

Problem 1.25 Draw in all hydrogens and lone pairs on the charged carbons in each ion.

a. 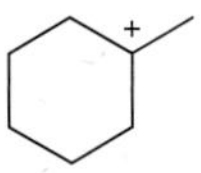

b. 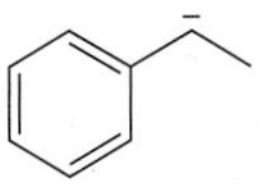

c. 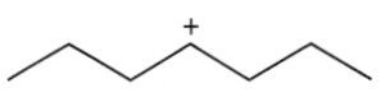

d. 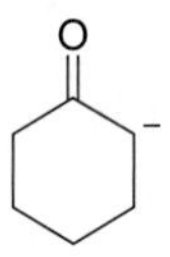
O

Problem 1.26 Draw a skeletal structure for the molecules in parts (a) and (b), and a condensed structure for the molecules in parts (c) and (d).

a. $(CH_3)_2C{=}CH(CH_2)_4CH_3$

b. 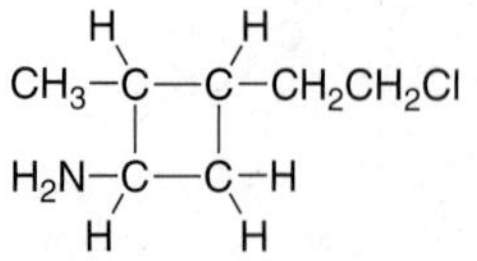

c. 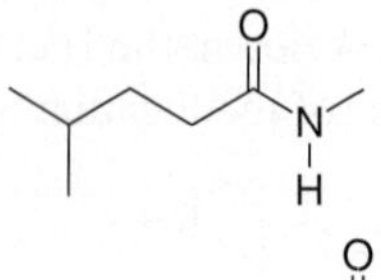

d. 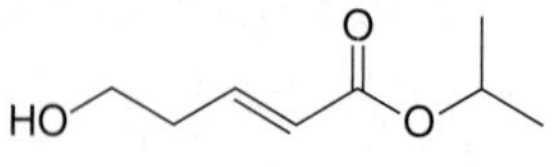

1.9 Hybridization

What orbitals do the first- and second-row atoms use to form bonds?

1.9A Hydrogen

Recall from Section 1.2 that two hydrogen atoms share each of their electrons to form H_2. Thus, the 1*s* orbital on one H overlaps with the 1*s* orbital on the other H to form a bond that concentrates electron density between the two nuclei. This type of bond, called a **σ** (sigma) **bond,** is cylindrically symmetrical because the electrons forming the bond are distributed symmetrically about an imaginary line connecting the two nuclei.

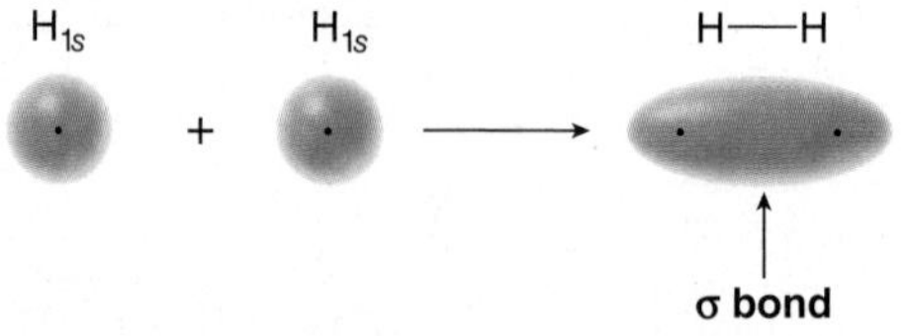

- **A σ bond concentrates electron density on the axis that joins two nuclei. All single bonds are σ bonds.**

1.9B Bonding in Methane

To account for the bonding patterns observed in more complex molecules, we must take a closer look at the 2*s* and 2*p* orbitals of atoms of the second row. Let's illustrate this with methane, CH_4.

Carbon has **four valence electrons.** To fill atomic orbitals in the most stable arrangement, electrons are placed in the orbitals of lowest energy. For carbon, this places two electrons in the 2*s* orbital and one each in two 2*p* orbitals.

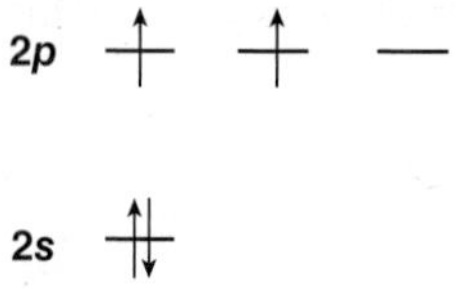

ground state for carbon's four valence electrons

- **This lowest energy arrangement of electrons for an atom is called its ground state.**

In this description, **carbon should form *only two bonds*** because it has only two unpaired valence electrons, and CH_2 should be a stable molecule. In reality, however, CH_2 is a highly reactive species because carbon does not have an octet of electrons.

Alternatively, promotion of an electron from a $2s$ to a vacant $2p$ orbital would form **four** unpaired electrons for bonding. This process requires energy because it moves an electron to a higher energy orbital. This higher energy electron configuration is called an electronically **excited state.**

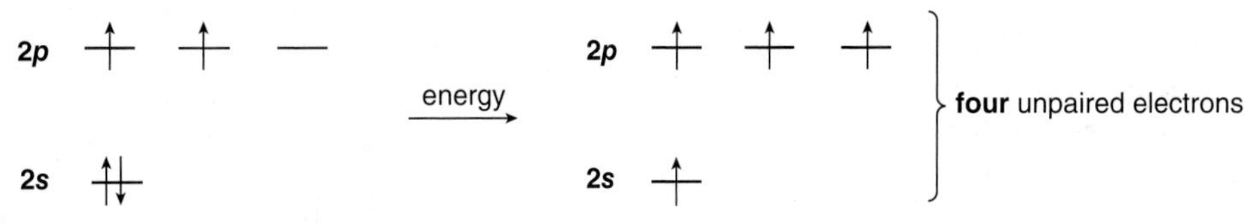

ground state for carbon **excited state for carbon**

In this description carbon would form two different types of bonds: three with $2p$ orbitals and one with a $2s$ orbital. **But experimental evidence points to carbon forming *four identical bonds* in methane.**

To solve this dilemma, chemists have proposed that atoms like carbon do not use pure *s* and pure *p* orbitals in forming bonds. Instead, atoms use a set of new orbitals called **hybrid orbitals.** The mathematical process by which these orbitals are formed is called **hybridization.**

- ***Hybridization* is the combination of two or more atomic orbitals to form the same number of hybrid orbitals, each having the same shape and energy.**

Hybridization of one $2s$ orbital and three $2p$ orbitals for carbon forms four hybrid orbitals, each with one electron. These new hybrid orbitals are intermediate in energy between the $2s$ and $2p$ orbitals.

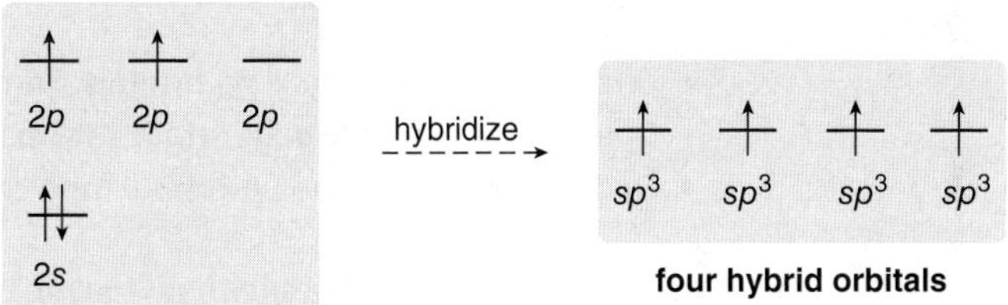

four hybrid orbitals

four atomic orbitals

- **These hybrid orbitals are called sp^3 *hybrids* because they are formed from *one s* orbital and *three p* orbitals.**

***p* orbital** sp^3 hybrid orbital

What do these new hybrid orbitals look like? Mixing a spherical $2s$ orbital and three dumbbell-shaped $2p$ orbitals together produces four orbitals having one large lobe and one small lobe, oriented toward the corners of a tetrahedron. Each large lobe concentrates electron density in the bonding direction between two nuclei. This makes bonds formed from hybrid orbitals ***stronger*** than bonds formed from pure *p* orbitals.

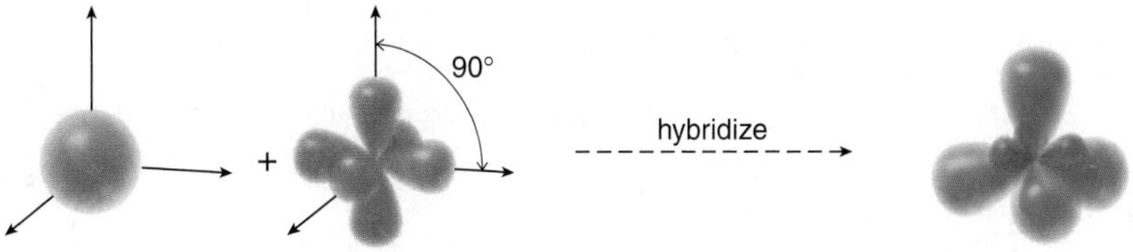

2s orbital three 2p orbitals together

tetrahedron
four sp^3 hybrid orbitals

The four hybrid orbitals form four equivalent bonds. We can now explain the observed bonding in CH_4.

- **Each bond in CH_4 is formed by overlap of an sp^3 hybrid orbital of carbon with a 1*s* orbital of hydrogen. These four bonds point to the corners of a tetrahedron.**

All four C–H bonds in methane are **σ bonds,** because the electron density is concentrated on the axis joining C and H. An orbital picture of the bonding in CH_4 is given in Figure 1.6.

Figure 1.6
Bonding in CH_4 using sp^3 hybrid orbitals

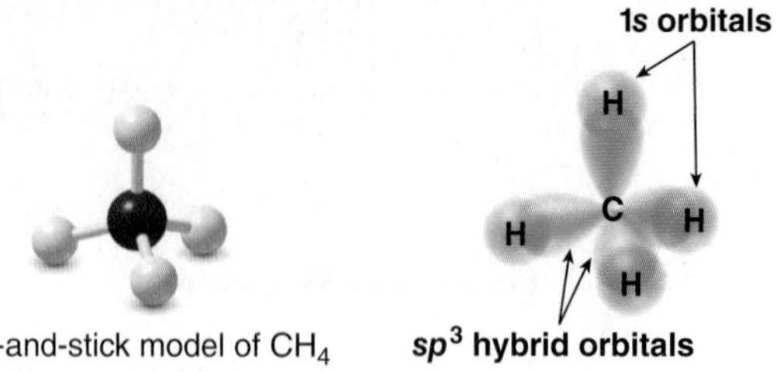

- All four C–H bonds are σ bonds.

Problem 1.27 What orbitals are used to form each of the C–C and C–H bonds in $CH_3CH_2CH_3$ (propane)? How many σ bonds are present in this molecule?

1.9C Other Hybridization Patterns—*sp* and *sp*^{2} Hybrid Orbitals

Forming sp^3 hybrid orbitals is just one way that 2*s* and 2*p* orbitals can hybridize. Three common modes of hybridization are seen in organic molecules. The number of orbitals is always conserved in hybridization; that is, a **given number of atomic orbitals hybridizes to form an equivalent number of hybrid orbitals.**

- ***One*** **2s orbital and *three* 2p orbitals form *four* sp^3 hybrid orbitals.**
- ***One*** **2s orbital and *two* 2p orbitals form *three* sp^2 hybrid orbitals.**
- ***One*** **2s orbital and *one* 2p orbital form *two* sp hybrid orbitals.**

We have already seen pictorially how four sp^3 hybrid orbitals are formed from one 2*s* and three 2*p* orbitals. Figures 1.7 and 1.8 illustrate the same process for *sp* and sp^2 hybrids. Each *sp* and sp^2 hybrid orbital has one large and one small lobe, much like an sp^3 hybrid orbital. Note, however, that both sp^2 and *sp* hybridization **leave one and two 2*p* orbitals *unhybridized,*** respectively, on each atom.

Figure 1.7
Forming two *sp* hybrid orbitals

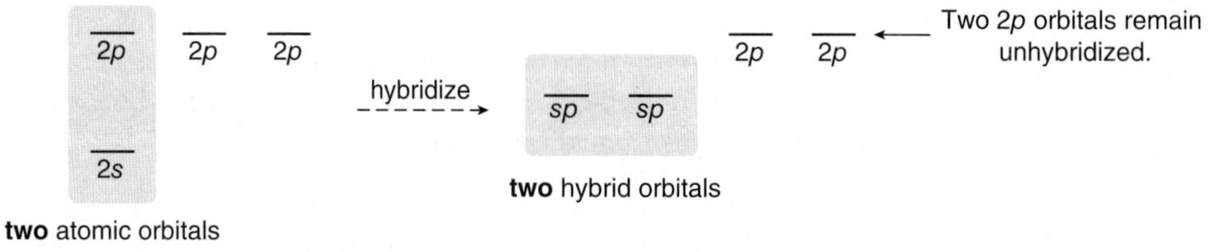

- Forming **two *sp* hybrid orbitals** uses **one 2*s*** and **one 2*p* orbital,** leaving **two 2*p* orbitals *unhybridized.***

Figure 1.8
Forming three sp^2 hybrid orbitals

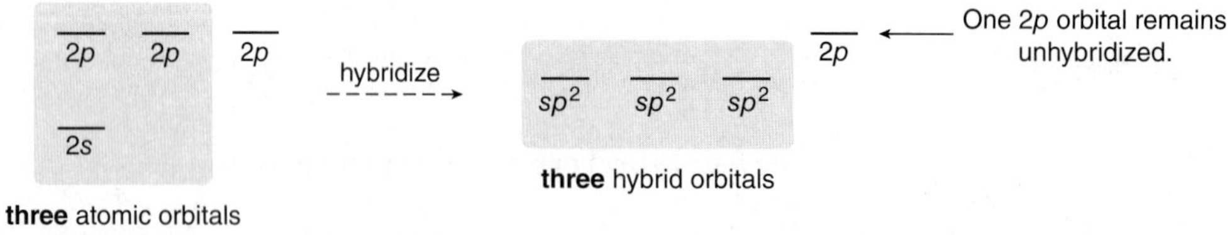

- Forming **three sp^2 hybrid orbitals** uses **one 2*s*** and **two 2*p* orbitals,** leaving **one 2*p* orbital *unhybridized.***

The **superscripts** for hybrid orbitals correspond to the **number of atomic orbitals** used to form them. The number "1" is understood.

For example: $sp^3 = s^1p^3$

one 2*s* + **three** 2*p* orbitals used to make each hybrid orbital

To determine the hybridization of an atom in a molecule, we count groups (atoms and lone pairs) around the atom, just as we did in determining geometry.

- **The number of groups around an atom equals the number of atomic orbitals that are hybridized to form hybrid orbitals (Table 1.4).**

Table 1.4 Three Types of Hybrid Orbitals

Number of groups	Number of orbitals used	Type of hybrid orbital
2	2	two *sp* hybrid orbitals
3	3	three sp^2 hybrid orbitals
4	4	four sp^3 hybrid orbitals

Let's illustrate this phenomenon with BeH_2, BF_3, NH_3, and H_2O, whose geometry was determined by counting groups in Section 1.7.

In **BeH_2, the Be atom is *sp* hybridized** because it is surrounded by two groups (two H atoms). Each Be–H bond is formed by overlap of an *sp* hybrid orbital from Be and a 1*s* orbital from H. The *sp* hybrid orbitals are oriented 180° away from each other.

In **BF_3, the B atom is sp^2 hybridized** because it is surrounded by three groups (three F atoms). Each B–F bond is formed by overlap of an sp^2 hybrid orbital from B and a 2*p* orbital from F. The sp^2 hybrid orbitals all lie in a plane, and are oriented 120° apart. The B atom also has a vacant unhybridized 2*p* orbital. This orbital is located ***above and below the plane*** of the BF_3 molecule.

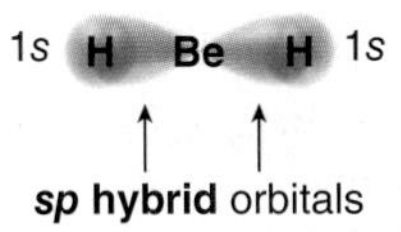

two Be–H bonds

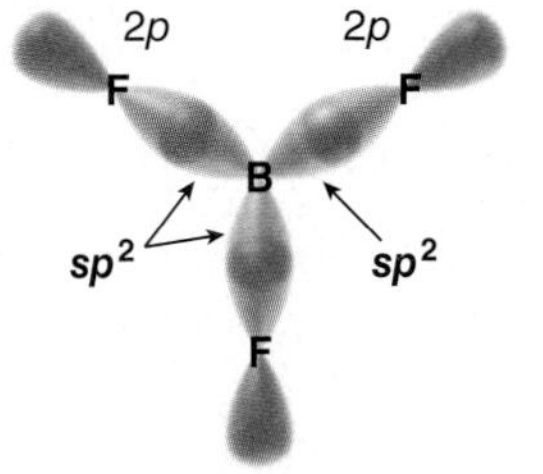

BF_3 from above

The three B–F bonds all lie in a plane, 120° apart.

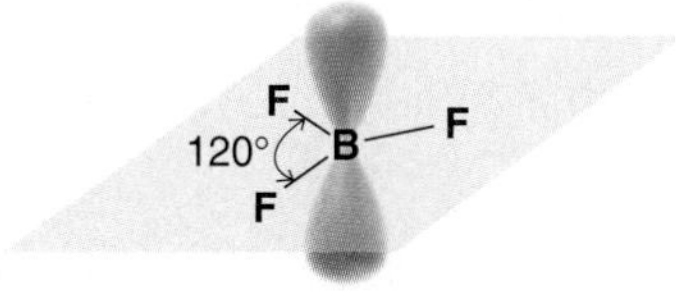

BF_3 from the side

The **unhybridized *p* orbital** extends above and below the plane.

The N atom in **NH_3** and the O atom in **H_2O** are both surrounded by four groups, making them sp^3 hybridized. Each N–H and O–H bond in these molecules is formed by overlap of an sp^3 hybrid orbital with a 1*s* orbital from H. The lone pairs of electrons on N and O also occupy sp^3 hybrid orbitals, as shown in Figure 1.9.

Figure 1.9
Hybrid orbitals of NH_3 and H_2O

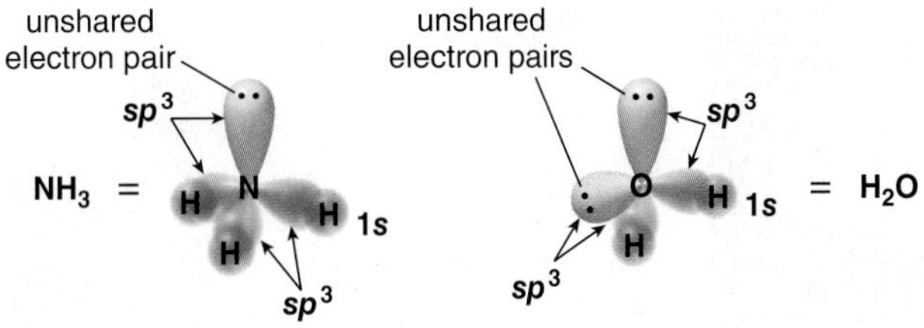

Sample Problem 1.9 What orbitals are used to form each bond in methanol, CH_3OH?

Solution

To solve this problem, draw a valid Lewis structure and count groups around each atom. Then, use the rule to determine hybridization: two groups = *sp*, three groups = sp^2, and four groups = sp^3.

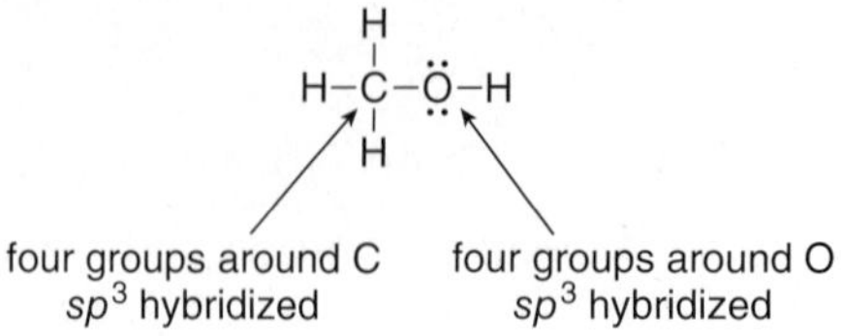

- All C–H bonds are formed from C_{sp^3}–H_{1s}.
- The C–O bond is formed from C_{sp^3}–O_{sp^3}.
- The O–H bond is formed from O_{sp^3}–H_{1s}.

Problem 1.28 What orbitals are used to form each bond in the following molecules?

a. CH_3BeH b. $(CH_3)_3B$ c. CH_3OCH_3

1.10 Ethane, Ethylene, and Acetylene

The principles of hybridization determine the type of bonds in **ethane, ethylene,** and **acetylene.**

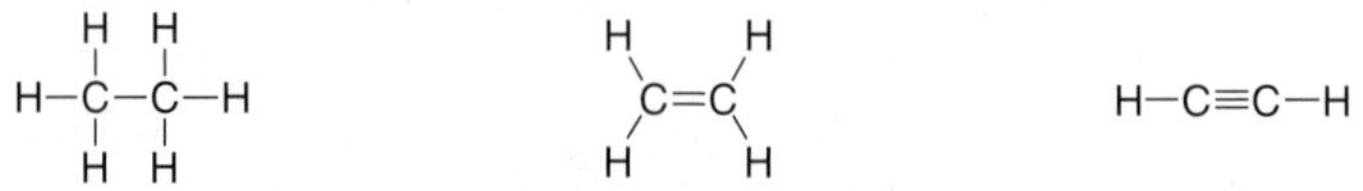

ethane ethylene acetylene

1.10A Ethane—CH_3CH_3

According to the Lewis structure for **ethane, CH_3CH_3,** each carbon atom is singly bonded to four other atoms. As a result:

- **Each carbon is tetrahedral.**
- **Each carbon is sp^3 hybridized.**

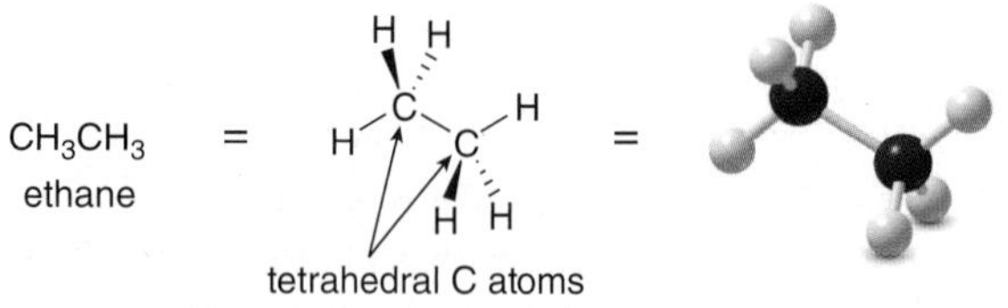

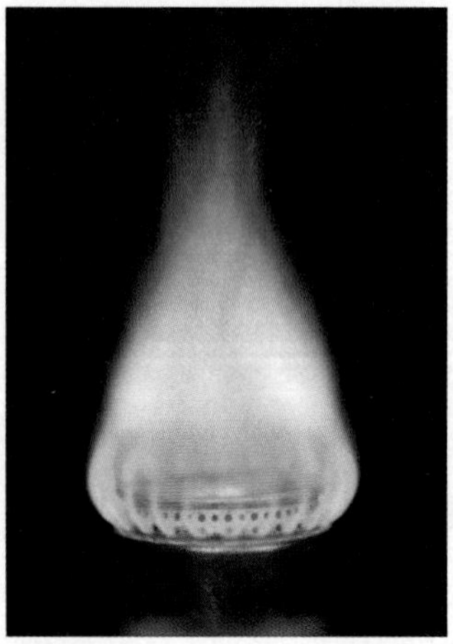

Ethane is a constituent of natural gas.

All of the bonds in ethane are σ bonds. The C–H bonds are formed from the overlap of one of the three sp^3 hybrid orbitals on each carbon atom with the 1*s* orbital on hydrogen. The C–C bond is formed from the overlap of an sp^3 hybrid orbital on each carbon atom.

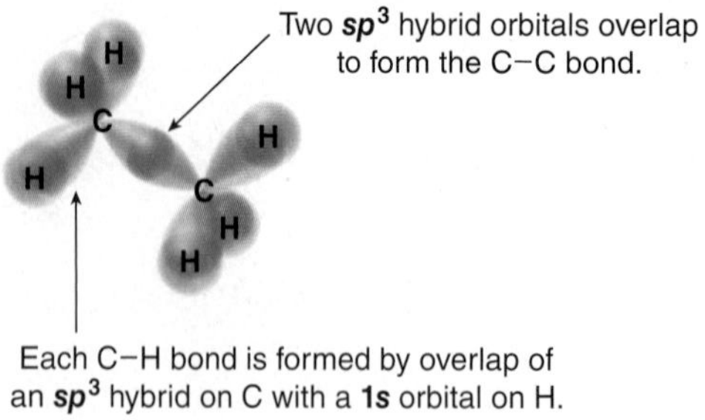

A model of ethane shows that **rotation can occur around the central C–C σ bond.** The relative position of the H atoms on the adjacent CH_3 groups changes with bond rotation, as seen in the location of the labeled red H atom before and after rotation. This process is discussed in greater detail in Chapter 4.

1.10B Ethylene—C_2H_4

Ethylene is an important starting material in the preparation of the plastic polyethylene.

Based on the Lewis structure of **ethylene, $CH_2{=}CH_2$,** each carbon atom is singly bonded to two H atoms and doubly bonded to the other C atom, so each C is surrounded by three groups. As a result:

- **Each carbon is trigonal planar (Section 1.7B).**
- **Each carbon is sp^2 hybridized.**

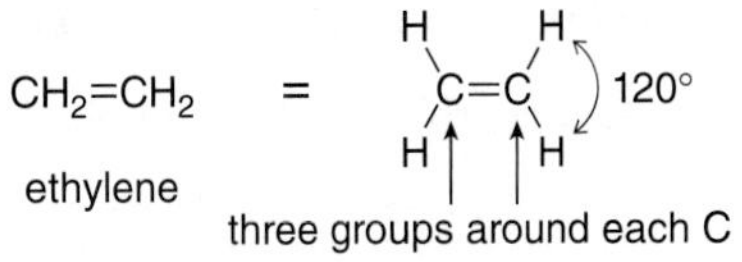

What orbitals are used to form the two bonds of the C–C double bond? Recall from Section 1.9 that ***sp*² hybrid orbitals** are formed from **one 2*s* and two 2*p* orbitals,** leaving one **2*p* orbital unhybridized.** Because carbon has four valence electrons, **each of these orbitals has one electron** that can be used to form a bond.

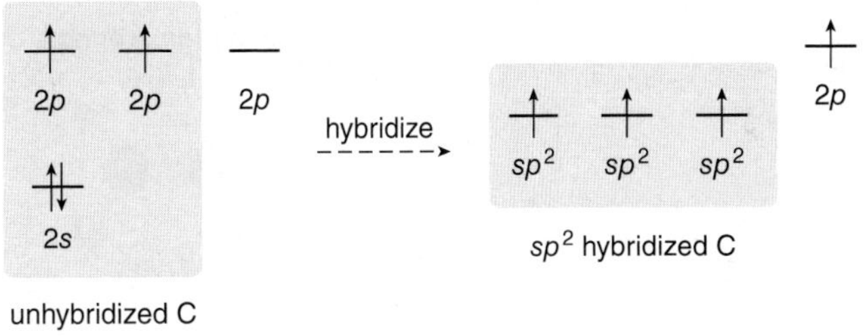

Each C–H bond results from the end-on overlap of an sp^2 hybrid orbital on carbon and the 1*s* orbital on hydrogen. Similarly, one of the C–C bonds results from the end-on overlap of an sp^2 hybrid orbital on each carbon atom. Each of these bonds is a **σ bond.** All five σ bonds lie in the same plane, viewed from above in the following representation, and from the side in Figure 1.10a.

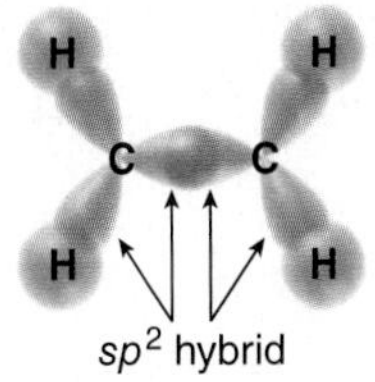

- Each C has three sp^2 hybrid orbitals.
- The C–H bonds and the C–C bond are σ bonds.

The second C–C bond results from the side-by-side overlap of the 2*p* orbitals on each carbon. Side-by-side overlap creates an area of electron density above and below the plane containing the

sp^2 hybrid orbitals (that is, the plane containing the six atoms in the σ bonding system), as shown in Figure 1.10b.

Figure 1.10
The σ and π bonds in ethylene

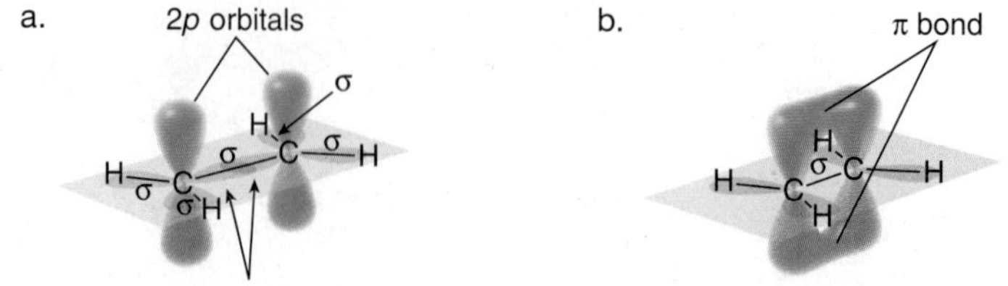

a.
- Overlap of the two sp^2 hybrid orbitals forms the C–C σ bond.

b.
- Overlap of the two 2*p* orbitals forms the C–C π bond.
- The π bond extends **above and below the plane** of the molecule.

In this second bond, the electron density is ***not*** concentrated on the axis joining the two nuclei. This new type of bond is called a **π bond.** Because the electron density in a π bond is farther from the two nuclei, **π bonds are usually weaker and therefore more easily broken than σ bonds.**

Thus, a carbon–carbon double bond has two components:

- **a σ bond, formed by end-on overlap of two sp^2 hybrid orbitals;**
- **a π bond, formed by side-by-side overlap of two 2*p* orbitals.**

Unlike the C–C single bond in ethane, rotation about the C–C double bond in ethylene is **restricted.** It can occur only if the π bond first breaks and then re-forms, a process that requires considerable energy.

All double bonds are composed of one σ and one π bond.

Rotation around a C=C bond does *not* occur.

1.10C Acetylene—C_2H_2

Because acetylene produces a very hot flame on burning, it is often used in welding torches. The fire is very bright, too, so it was once used in the lamps worn by spelunkers—people who study and explore caves.

Based on the Lewis structure of **acetylene, HC≡CH,** each carbon atom is singly bonded to one hydrogen atom and triply bonded to the other carbon atom, so each carbon atom is surrounded by two groups. As a result:

- **Each carbon is linear (Section 1.7B).**
- **Each carbon is *sp* hybridized.**

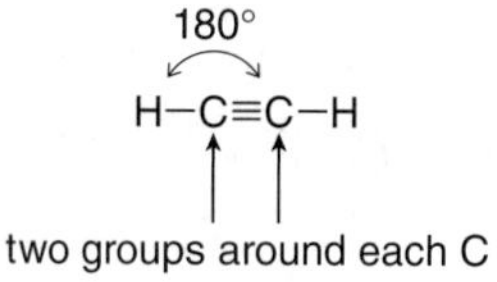

acetylene

What orbitals are used to form the bonds of the C–C triple bond? Recall from Section 1.9 that ***sp* hybrid orbitals** are formed from **one 2*s* and one 2*p* orbital,** leaving **two 2*p* orbitals**

unhybridized. Because carbon has four valence electrons, **each of these orbitals has one electron** that can be used to form a bond.

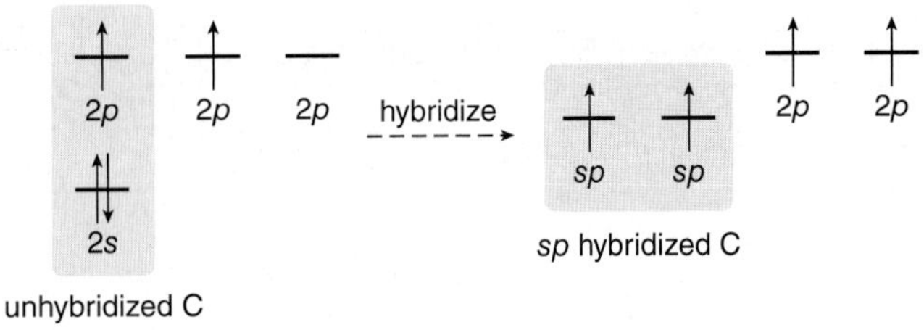

Each C−H bond results from the end-on overlap of an *sp* hybrid orbital on carbon and the 1*s* orbital on hydrogen. Similarly, one of the C−C bonds results from the end-on overlap of an *sp* hybrid orbital on each carbon atom. Each of these bonds is a **σ bond.**

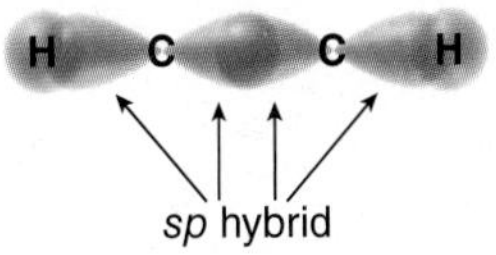

- Each C has two *sp* hybrid orbitals.
- The C−H bonds and C−C bond are σ bonds.

Each carbon atom also has two **unhybridized 2*p* orbitals** that are perpendicular to each other and to the *sp* hybrid orbitals (Figure 1.11a). Side-by-side overlap between the two 2*p* orbitals on one carbon with the two 2*p* orbitals on the other carbon creates the second and third bonds of the C−C triple bond (Figure 1.11b). The electron density from one of these two bonds is above and below the axis joining the two nuclei, and the electron density from the second of these two bonds is in front of and behind the axis, so both of these bonds are **π bonds.**

Figure 1.11
The σ and π bonds in acetylene

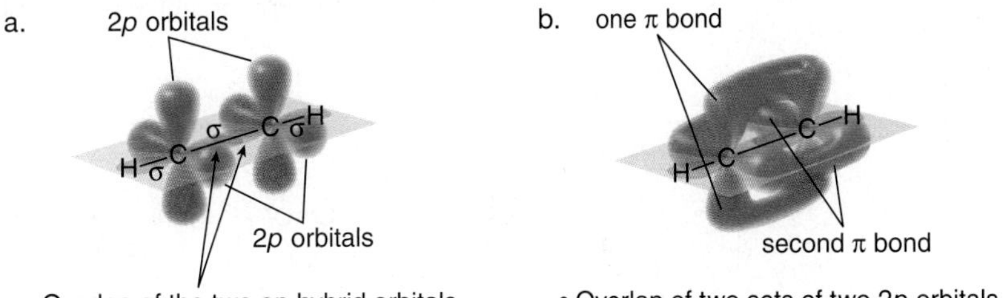

- Overlap of the two *sp* hybrid orbitals forms the C−C σ bond.

- Overlap of two sets of two 2*p* orbitals forms two C−C π bonds.
- Two π bonds extend out from the axis of the linear molecule.

The side-by-side overlap of two *p* orbitals always forms a π bond.

All triple bonds are composed of one σ and two π bonds.

Thus, a carbon–carbon triple bond has three components:

- **a σ bond, formed by end-on overlap of two *sp* hybrid orbitals;**
- **two π bonds, formed by side-by-side overlap of two sets of 2*p* orbitals.**

Table 1.5 summarizes the three possible types of bonding in carbon compounds.

Table 1.5 A Summary of Covalent Bonding Seen in Carbon Compounds

Number of groups bonded to C	Hybridization	Bond angle	Example	Observed bonding
4	sp^3	109.5°	**CH_3CH_3** ethane	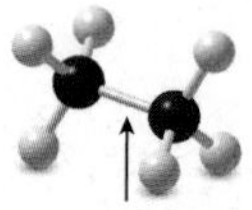 **one σ bond** $C_{sp^3}–C_{sp^3}$
3	sp^2	120°	**$CH_2{=}CH_2$** ethylene	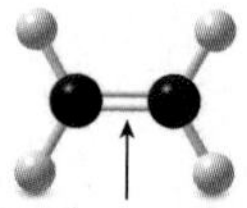 **one σ bond** + **one π bond** $C_{sp^2}–C_{sp^2}$ $C_{2p}–C_{2p}$
2	sp	180°	**HC≡CH** acetylene	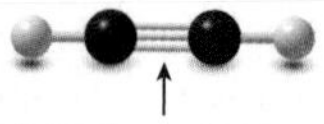 **one σ bond** + **two π bonds** $C_{sp}–C_{sp}$ $C_{2p}–C_{2p}$ $C_{2p}–C_{2p}$

Sample Problem 1.10 Answer each question for the molecule acetone.

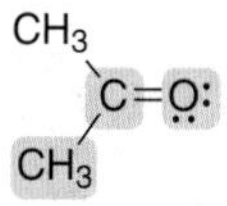
acetone

a. Determine the hybridization of the highlighted atoms.
b. What orbitals are used to form the C–O double bond?
c. In what type of orbital does each lone pair reside?

Solution

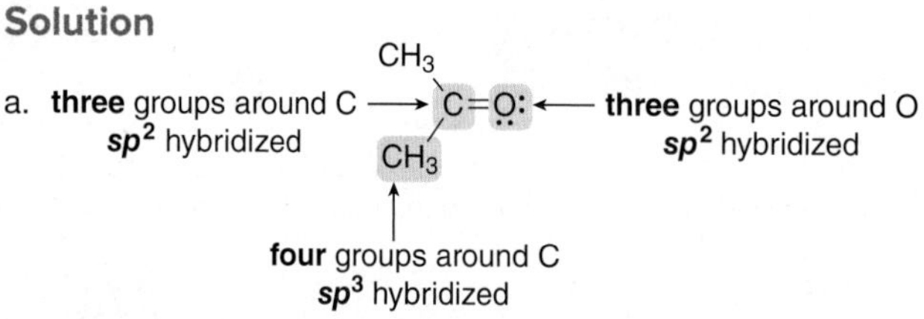
a. **three** groups around C
sp^2 hybridized
three groups around O
sp^2 hybridized
four groups around C
sp^3 hybridized

b. • The σ bond is formed from the end-on overlap of $C_{sp^2}–O_{sp^2}$.
 • The π bond is formed from the side-by-side overlap of $C_{2p}–O_{2p}$.

c. The O atom has three sp^2 hybrid orbitals.
 • One is used for the σ bond of the double bond.
 • The remaining two sp^2 hybrids are occupied by the lone pairs.

Problem 1.29 Determine the hybridization around the highlighted atoms in each molecule.

a. $CH_3{-}C{\equiv}CH$ b. 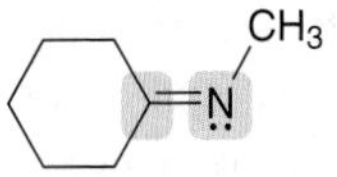

c. $CH_2{=}C{=}CH_2$

Problem 1.30 The unmistakable odor of a freshly cut cucumber is largely due to cucumber aldehyde. (a) How many sp^2 hybridized carbon atoms does cucumber aldehyde contain? (b) What is the hybridization of the O atom? (c) What orbitals are used to form the carbon–oxygen double bond? (d) How many σ bonds does cucumber aldehyde contain? (e) How many π bonds does it contain?

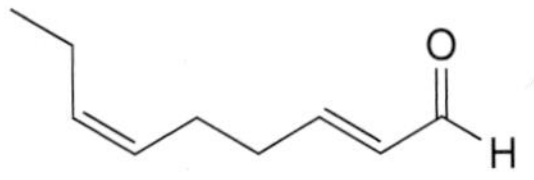

cucumber aldehyde

1.11 Bond Length and Bond Strength

Let's now examine the relative bond length and bond strength of the C–C and C–H bonds in ethane, ethylene, and acetylene.

1.11A A Comparison of Carbon–Carbon Bonds

While the SI unit of energy is the **joule** (J), organic chemists often report energy values in **calories** (cal). For this reason, energy values in the tables in this text are reported in joules, followed by the number of calories in parentheses.
1 cal = 4.18 J

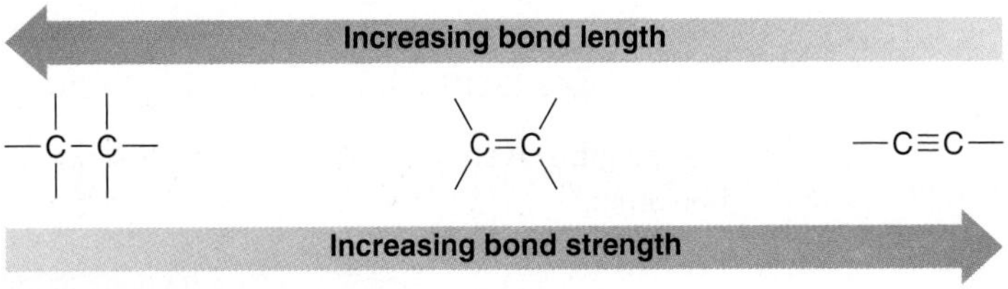

- **As the number of electrons between two nuclei *increases*, bonds become shorter and stronger.**
- **Triple bonds are shorter and stronger than double bonds, which are shorter and stronger than single bonds.**

An inverse relationship exists between bond length and bond strength. The shorter the bond, the closer the electron density is kept to the nucleus, and the harder the bond is to break. ***Shorter*** **bonds are** ***stronger*** **bonds.**

Values for bond lengths and bond strengths for CH_3CH_3, $CH_2{=}CH_2$, and HC≡CH are listed in Table 1.6. Be careful not to confuse two related but different principles regarding multiple bonds such as C–C double bonds. **Double bonds, consisting of both a σ and a π bond, are *strong*.** The **π component** of the double bond, however, is usually much ***weaker*** **than the σ component.** This is a particularly important consideration when studying alkenes in Chapter 10.

Table 1.6 Bond Lengths and Bond Strengths for Ethane, Ethylene, and Acetylene

Compound	C–C bond length (pm)		Bond strength kJ/mol (kcal/mol)	
$CH_3{-}CH_3$	153	↑	368 (88)	↓
$CH_2{=}CH_2$	134	Increasing bond length	635 (152)	Increasing bond strength
HC≡CH	121		837 (200)	
Compound	**C–H bond length (pm)**		**Bond strength kJ/mol (kcal/mol)**	
$CH_3CH_2{-}H$	111	↑	410 (98)	↓
$CH_2{=}CH_2$ (C–H)	110	Increasing bond length	435 (104)	Increasing bond strength
HC≡C–H	109		523 (125)	

1.11B A Comparison of Carbon–Hydrogen Bonds

The length and strength of a C–H bond vary slightly depending on the hybridization of the carbon atom.

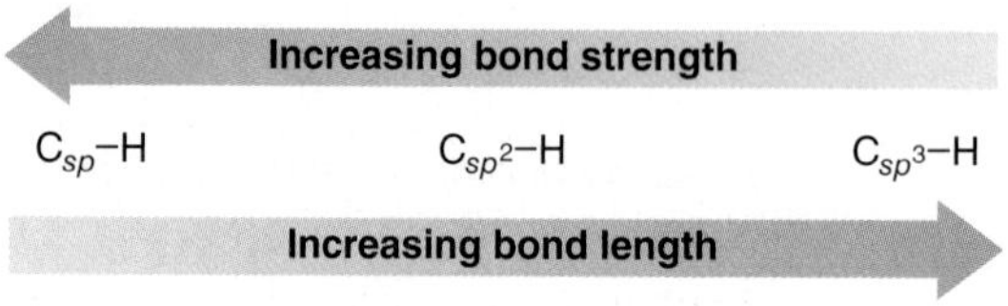

To understand why this is so, we must look at the atomic orbitals used to form each type of hybrid orbital. A single 2*s* orbital is always used, but the number of 2*p* orbitals varies with the type of hybridization. The **percent *s*-character** indicates the fraction of a hybrid orbital due to the 2*s* orbital used to form it.

sp **hybrid** $\dfrac{\text{one 2}s\text{ orbital}}{\text{two hybrid orbitals}}$ = **50% *s*-character**

***sp*²** **hybrid** $\dfrac{\text{one 2}s\text{ orbital}}{\text{three hybrid orbitals}}$ = **33% *s*-character**

***sp*³** **hybrid** $\dfrac{\text{one 2}s\text{ orbital}}{\text{four hybrid orbitals}}$ = **25% *s*-character**

Why should the percent *s*-character of a hybrid orbital affect the length of a C–H bond? A 2*s* orbital keeps electron density closer to a nucleus compared to a 2*p* orbital. As the **percent *s*-character increases,** a hybrid orbital holds its electrons closer to the nucleus, and the **bond becomes shorter and stronger.**

- **Increased percent *s*-character ---→ Increased bond strength ---→ Decreased bond length**

Problem 1.31 Which of the bonds shown in red in each compound or pair of compounds is shorter?

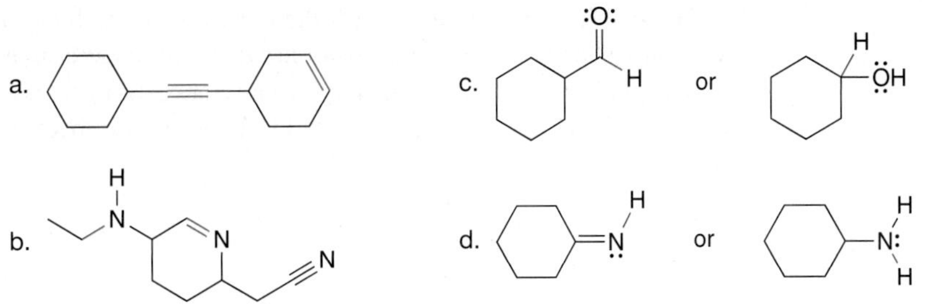

1.12 Electronegativity and Bond Polarity

***Electronegativity* is a measure of an atom's attraction for electrons in a bond.** Electronegativity indicates how much a particular atom "*wants*" electrons.

- **Electronegativity *increases* across a row of the periodic table as the nuclear charge increases (excluding the noble gases).**
- **Electronegativity *decreases* down a column of the periodic table as the atomic radius increases, pushing the valence electrons farther from the nucleus.**

As a result, the *most* electronegative elements are located at the **upper right-hand corner** of the periodic table, and the *least* electronegative elements in the **lower left-hand corner.** A scale has been established to represent electronegativity values arbitrarily, from 0 to 4, as shown in Figure 1.12.

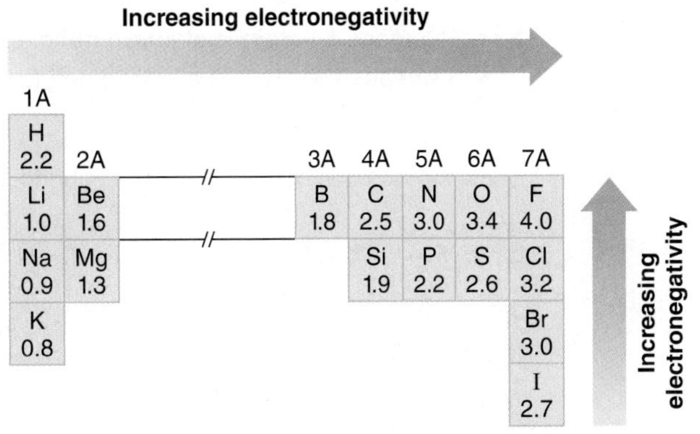

Figure 1.12 Electronegativity values for some common elements

Electronegativity values are relative, so they can be used for comparison purposes only. When comparing two different elements, one is **more electronegative** than the other if it attracts electron density toward itself. One is less electronegative—**more electro*positive***—if it gives up electron density to the other element.

Problem 1.32 Rank the following atoms in order of increasing electronegativity. Label the most electronegative and most electropositive atom in each group.

a. Se, O, S b. P, Na, Cl c. Cl, S, F d. O, P, N

Electronegativity values are used as a guideline to indicate whether the electrons in a bond are **equally shared** or **unequally shared** between two atoms. Whenever two identical atoms are bonded together, each atom attracts the electrons in the bond to the same extent. The electrons are equally shared, and the **bond is *nonpolar*.** Thus, a **carbon–carbon bond is nonpolar.** Whenever two different atoms having similar electronegativities are bonded together, the bond is also **nonpolar. C–H bonds are considered to be nonpolar,** because the electronegativity difference between C (2.5) and H (2.2) is small.

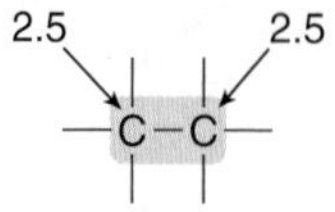

nonpolar bond

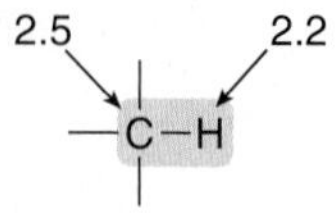

nonpolar bond

The small electronegativity difference between C and H is ignored.

Bonding between atoms of different electronegativity values results in the **unequal sharing** of electrons. In a C–O bond, the electrons are pulled away from C (2.5) toward O (3.4), the element of higher electronegativity. **The bond is *polar,*** or ***polar covalent.*** The bond is said to have a **dipole;** that is, **a partial separation of charge.**

A C–O bond is a **polar** bond.

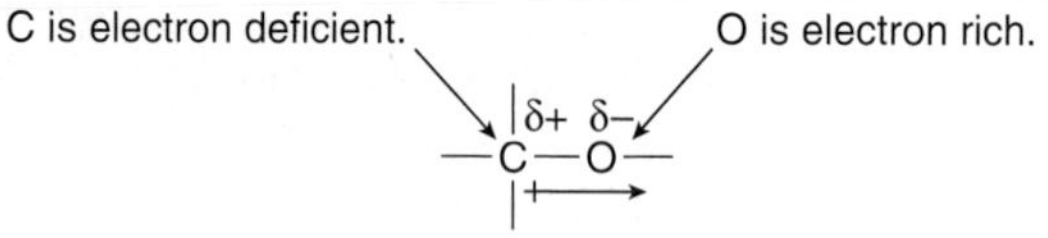

a bond dipole

The direction of polarity in a bond is often indicated by an arrow, with the head of the arrow pointing toward the more electronegative element. The tail of the arrow, with a perpendicular line drawn through it, is positioned at the less electronegative element. Alternatively, the symbols δ+ and δ– indicate this unequal sharing of electron density.

- **δ+ means an atom is electron deficient (has a partial positive charge).**
- **δ– means the atom is electron rich (has a partial negative charge).**

Problem 1.33 Show the direction of the dipole in each bond. Label the atoms with δ+ and δ–.

a. H–F b. 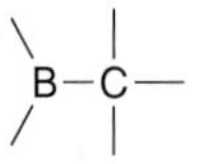

c. 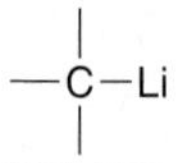

d. 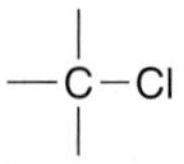

Students often wonder how large an electronegativity difference must be to consider a bond polar. That's hard to say. We will set an arbitrary value for this difference and use it as an *approximation.* **Usually, a polar bond will be one in which the electronegativity difference between two atoms is ≥ 0.5 units.**

The distribution of electron density in a molecule can be shown using an **electrostatic potential map.** These maps are color coded to illustrate areas of high and low electron density. Electron-rich regions are indicated in red, and electron-deficient sites are indicated in blue. Regions of intermediate electron density are shown in orange, yellow, and green.

An electrostatic potential map of CH_3Cl indicates the polar nature of the C–Cl bond (Figure 1.13). The more electronegative Cl atom pulls electron density toward it, making it electron rich. This is indicated by the red around the Cl in the plot. The carbon is electron deficient, and this is shown with blue. When comparing two maps, the comparison is useful only if they are plotted *using the same scale* of color gradation. For this reason, whenever we compare two plots in this text, they will be drawn side by side using the same scale.

Figure 1.13
Electrostatic potential plot of CH_3Cl

a. Color scheme used for electron density

b. Electrostatic potential plot

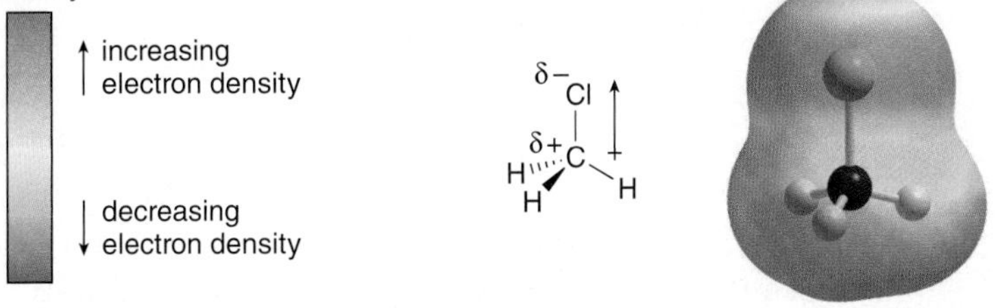

1.13 Polarity of Molecules

A **polar molecule** has either one polar bond, or two or more bond dipoles that reinforce. A **nonpolar molecule** has either no polar bonds, or two or more bond dipoles that cancel.

Whenever C or H is bonded to N, O, and all halogens, the bond is ***polar.*** Thus, the C–I bond is considered polar even though the electronegativity difference between C and I is small. Remember, electronegativity is just an approximation.

Thus far, we have been concerned with the polarity of one bond. To determine whether a molecule has a net dipole, use the following two-step procedure:

[1] Use electronegativity differences to **identify all of the polar bonds and the directions of the bond dipoles.**

[2] **Determine the geometry** around individual atoms by counting groups, and decide if individual dipoles **cancel** or **reinforce each other in space.**

The two molecules $\mathbf{H_2O}$ and $\mathbf{CO_2}$ illustrate different outcomes of this process. In H_2O, each O–H bond is polar because the electronegativity difference between O (3.4) and H (2.2) is large. Since H_2O is a **bent** molecule, the two dipoles reinforce (both point *up*). Thus, **$\mathbf{H_2O}$ has a net dipole, making it a polar molecule.** CO_2 also has polar C–O bonds because the electronegativity difference between O (3.4) and C (2.5) is large. However, CO_2 is a **linear** molecule, so the two dipoles, which are equal and opposite in direction, **cancel.** Thus, CO_2 is a **nonpolar molecule** with **no net dipole.**

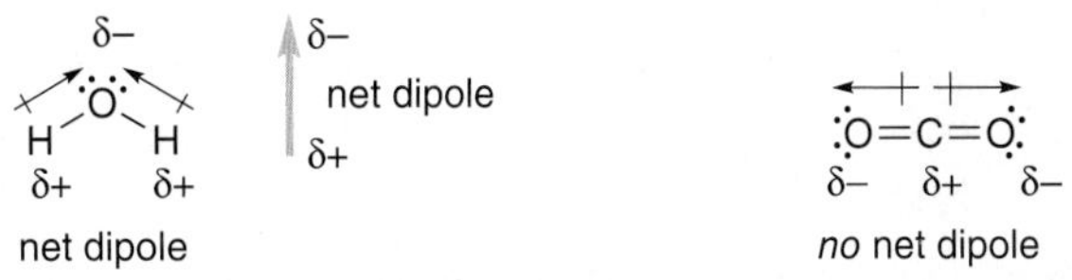

Electrostatic potential plots for H_2O and CO_2 appear in Figure 1.14. Additional examples of polar and nonpolar molecules are given in Figure 1.15.

Problem 1.34 Indicate which of the following molecules is polar because it possesses a net dipole. Show the direction of the net dipole if one exists.

a. CH_3Br b. CH_2Br_2 c. CF_4 d. 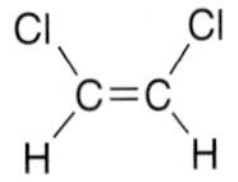

e. 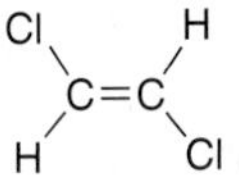

Figure 1.14
Electrostatic potential plots for H_2O and CO_2

- The electron-rich (red) region is concentrated on the more electronegative O atom. Both H atoms are electron deficient (blue-green).
- Both electronegative O atoms are electron rich (red) and the central C atom is electron deficient (blue).

Figure 1.15
Examples of polar and nonpolar molecules

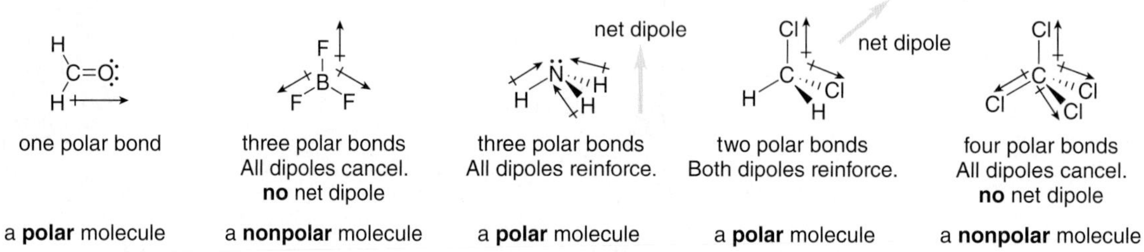

1.14 L-Dopa—A Representative Organic Molecule

The principles learned in this chapter apply to all organic molecules regardless of size or complexity. We now know a great deal about the structure of **L-dopa,** a drug used to treat Parkinson's disease described in the chapter opener.

Sample Problem 1.11 Answer each question about L-dopa.

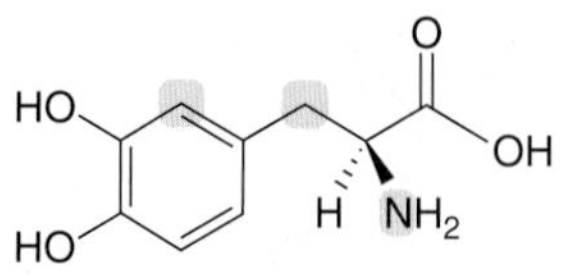

L-dopa

a. Convert the skeletal structure to a Lewis structure.
b. What is the hybridization and geometry around each labeled atom?
c. Label three polar bonds.

Solution

a, b. Convert the skeletal structure to a Lewis structure as in Sample Problem 1.8. Each O atom needs two lone pairs and each N atom needs one for an octet. Count groups around an atom to determine hybridization and shape.

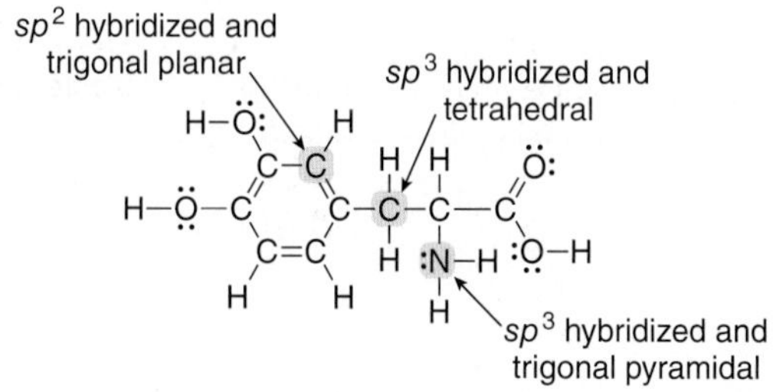

Lewis structure of L-dopa

c. All C–O, O–H, N–H, and C–N bonds are polar because the electronegativity difference between the atoms is large.

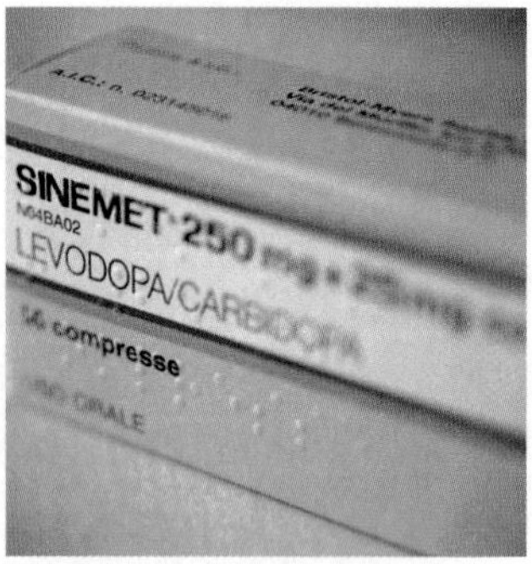

Sinemet, the trade name of a drug used to treat Parkinson's disease, contains a combination of L-dopa (Sample Problem 1.11) and carbidopa (Problem 1.36). Carbidopa increases the effectiveness of L-dopa by inhibiting its metabolism prior to crossing the blood–brain barrier and entering the brain.

Problem 1.35 Provide the following information about L-dopa.

a. Label all sp^3 hybridized C atoms.
b. Label all H atoms that bear a partial positive charge ($\delta+$).
c. Draw another resonance structure.

Sample Problem 1.12 illustrates how to derive structural information from a ball-and-stick model.

Sample Problem 1.12 Use the ball-and-stick model of vitamin B_6 to answer each question.

vitamin B_6

a. Draw a skeletal structure of vitamin B_6.
b. How many sp^2 hybridized carbons are present?
c. What is the hybridization of the N atom in the ring?

Solution

Use the element colors shown in Section 1.7B to convert the 3-D model to a skeletal structure [black (C), gray (H), red (O), blue (N)]. H atoms on carbon are omitted, but H atoms on heteroatoms are drawn. Count groups to determine hybridization. Each O atom needs two lone pairs and each N needs one to give an octet of electrons.

a. 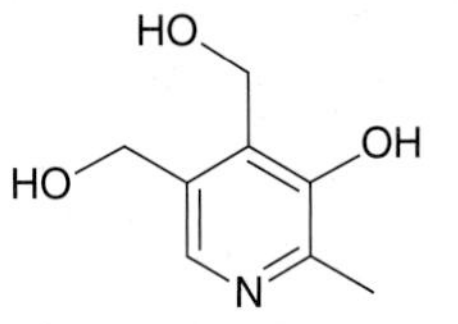

skeletal structure of vitamin B_6

b, c. 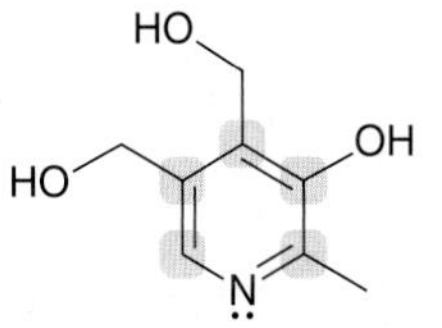

- Each C labeled in blue is sp^2 hybridized.
- The N atom is surrounded by three groups (two atoms and one lone pair), making it sp^2 hybridized.

Problem 1.36 Use the ball-and-stick model to answer each question about carbidopa, a drug used in combination with L-dopa to treat Parkinson's disease

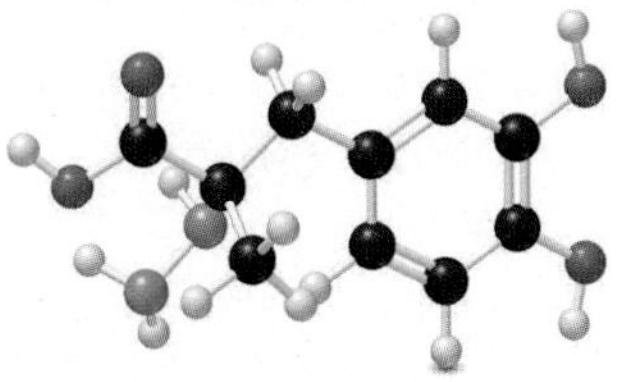

carbidopa

a. Draw a skeletal structure of carbidopa.
b. Determine the hybridization around each carbon atom.
c. What is the hybridization and geometry around each N atom?
d. How many polar bonds are present?

KEY CONCEPTS

Structure and Bonding

Important Facts

- **The general rule of bonding:** Atoms "strive" to attain a complete outer shell of valence electrons (1.2). H "wants" two electrons. Second-row elements "want" eight electrons.
- **Formal charge** is the difference between the number of valence electrons of an atom and the number of electrons it "owns" (1.3C). See Sample Problem 1.3 for a stepwise example.
- **Curved arrow notation** shows the movement of an electron pair. The tail of the arrow always begins at an electron pair, either in a bond or a lone pair. The head points to where the electron pair "moves" (1.6).
- **Electrostatic potential plots** are color-coded maps of electron density, indicating electron-rich and electron-deficient regions (1.12).

The Importance of Lewis Structures (1.3–1.5)

A properly drawn Lewis structure shows the number of bonds and lone pairs present around each atom in a molecule. In a valid Lewis structure, each H has two electrons, and each second-row element has no more than eight. This is the first step needed to determine many properties of a molecule.

Lewis structure ---→
- **Geometry:** linear, trigonal planar, or tetrahedral (1.7)
- **Hybridization:** *sp*, *sp*^{2}, or *sp*^{3} (1.9)
- **Types of bonds:** single, double, or triple (1.3, 1.10)

Resonance (1.6)

The basic principles:

- Resonance exists when a compound cannot be represented by a single Lewis structure.
- Resonance structures differ in the position of only nonbonded electrons and π bonds, not atoms.
- The resonance hybrid is the only accurate representation for a resonance-stabilized compound. A hybrid represents the compound better than any single resonance structure because electron density is delocalized.

The difference between resonance structures and isomers:

- Two **isomers** differ in the arrangement of *both* atoms and electrons.
- **Resonance structures** differ *only* in the arrangement of electrons.

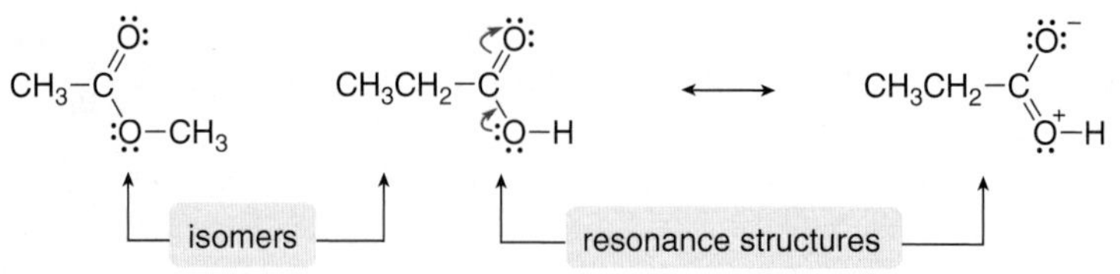

Geometry and Hybridization

The number of groups around an atom determines both its geometry (1.7) and hybridization (1.9).

Number of groups	Geometry	Bond angle (°)	Hybridization
2	linear	180	*sp*
3	trigonal planar	120	*sp*^{2}
4	tetrahedral	109.5	*sp*^{3}

Drawing Organic Molecules (1.8)

- Shorthand methods are used to abbreviate the structure of organic molecules.

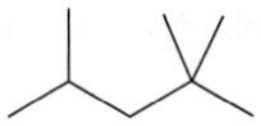

skeletal structure

=

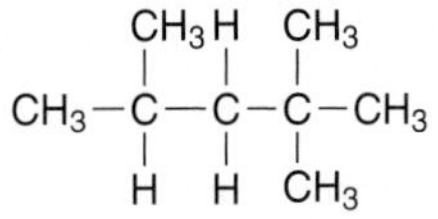

isooctane

= $(CH_3)_2CHCH_2C(CH_3)_3$

condensed structure

- A carbon bonded to four atoms is tetrahedral. The best way to represent a tetrahedron is to draw two bonds in the plane, one bond in front, and one bond behind.

Bond Length

- Bond length decreases across a row and increases down a column of the periodic table (1.7A).
- Bond length decreases as the number of electrons between two nuclei increases (1.11A).
- Bond length decreases as the percent *s*-character increases (1.11B).
- Bond length and bond strength are inversely related. In general, shorter bonds are stronger bonds (1.11).
- Sigma (σ) bonds are generally stronger than π bonds (1.10).

Electronegativity and Polarity (1.12, 1.13)

- Electronegativity increases across a row and decreases down a column of the periodic table.
- A polar bond results when two atoms of different electronegativity values are bonded together. Whenever C or H is bonded to N, O, or any halogen, the bond is polar.
- A polar molecule has either one polar bond, or two or more bond dipoles that reinforce.

PROBLEMS

Problems Using Three-Dimensional Models

1.37 Citric acid is responsible for the tartness of citrus fruits, especially lemons and limes.

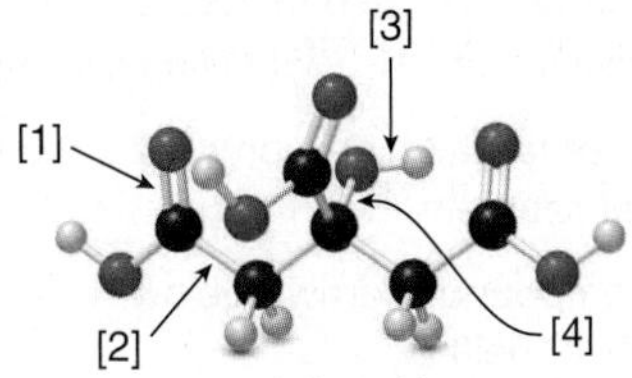

citric acid

a. What is the molecular formula for citric acid?
b. How many lone pairs are present?
c. Draw a skeletal structure.
d. How many sp^2 hybridized carbons are present?
e. What orbitals are used to form each indicated bond ([1]–[4])?

1.38 Zingerone gives ginger its pungent taste.

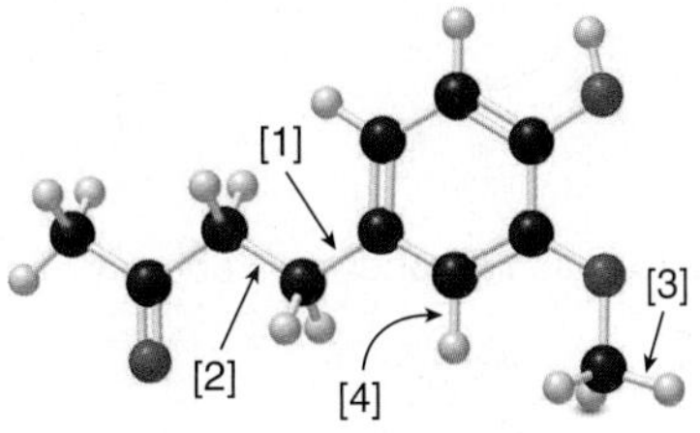

zingerone

a. What is the molecular formula for zingerone?
b. How many lone pairs are present?
c. Draw a skeletal structure.
d. How many sp^2 hybridized carbons are present?
e. What orbitals are used to form each indicated bond ([1]–[4])?

Lewis Structures and Formal Charge

1.39 Assign formal charges to each carbon atom in the given species. All lone pairs have been drawn in.

a. $CH_2{=}\ddot{C}H$ b. $H{-}\ddot{C}{-}H$ c. 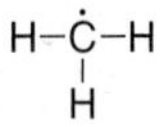 d. 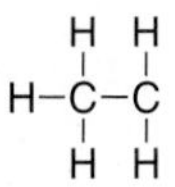

1.40 Assign formal charges to each N and O atom in the given molecules. All lone pairs have been drawn in.

a. 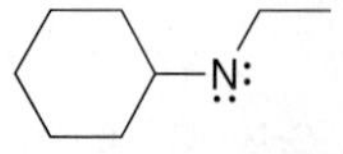 b. $:\ddot{N}=N=\ddot{N}:$ c. 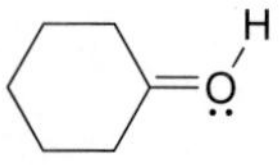 d. 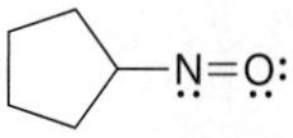

1.41 Draw one valid Lewis structure for each compound. Assume the atoms are arranged as drawn.

a. CH_2N_2

```
H C N N
  H
```

c. CH_3CNO

```
  H
H C C N O
  H
```

b. CH_3NO_2

```
  H
H C N O
  H O
```

d. $(CH_2CN)^-$

```
  H
H C C N
  H
```

1.42 Draw an acceptable Lewis structure from each condensed structure, such that all atoms have zero formal charge.

a. diethyl ether, $(CH_3CH_2)_2O$, the first general anesthetic used in medical procedures
b. acrylonitrile, CH_2CHCN, starting material used to manufacture synthetic Orlon fibers
c. dihydroxyacetone, $(HOCH_2)_2CO$, an ingredient in sunless tanning products
d. acetic anhydride, $(CH_3CO)_2O$, a reagent used to synthesize aspirin

Isomers and Resonance Structures

1.43 Creatine is a dietary supplement used by some athletes to boost their athletic performance. (a) Draw in all lone pairs in creatine. (b) Draw two additional resonance structures showing all lone pairs and formal charges.

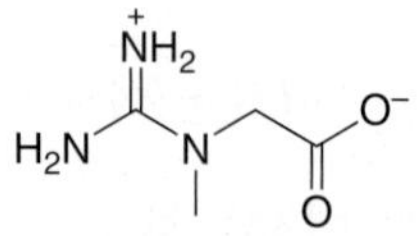

creatine

1.44 Draw all possible isomers for each molecular formula.

a. C_3H_7Cl (two isomers) b. C_2H_4O (three isomers) c. C_3H_9N (four isomers)

1.45 Draw Lewis structures for the nine isomers having molecular formula C_3H_6O, with all atoms having a zero formal charge.

1.46 With reference to compound **A** drawn below, label each compound as an isomer, a resonance structure, or neither.

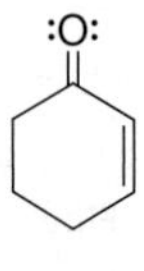

A

a. 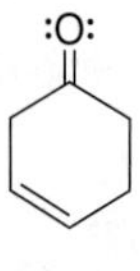 b. 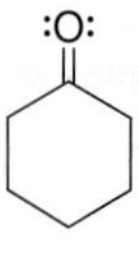 c. 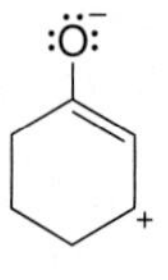 d. 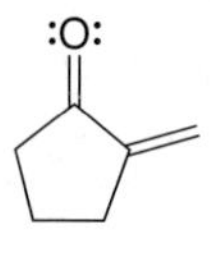

1.47 With reference to cation **B,** label each species as an isomer, a resonance structure, or neither.

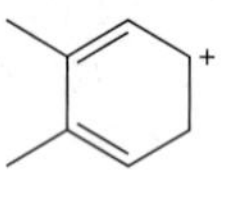

B

a. 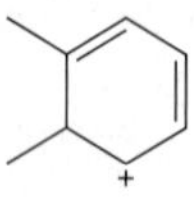 b. 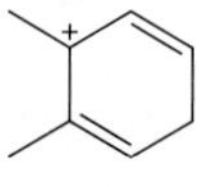 c. 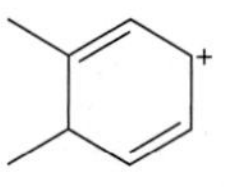 d. 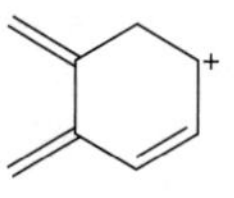

1.48 How are the molecules or ions in each pair related? Classify them as resonance structures, isomers, or neither.

a. 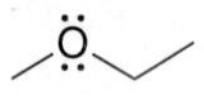 and 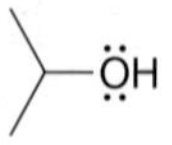

c. 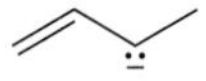 and 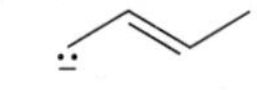

b.  and 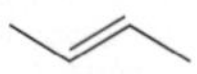

d. $CH_3CH_2CH_3$ and $CH_3CH_2\overset{-}{\ddot{C}}H_2$

1.49 Add curved arrows to show how the first resonance structure can be converted into the second.

a. 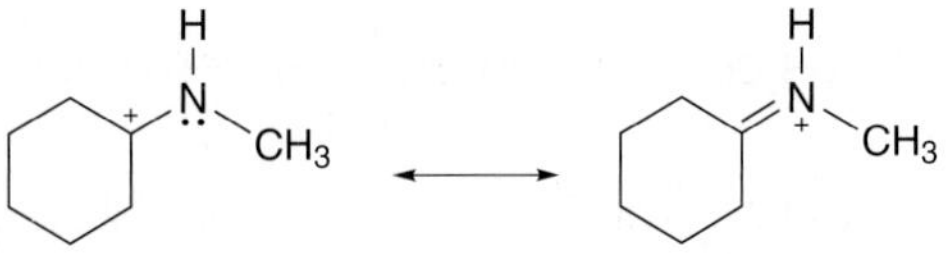

b. 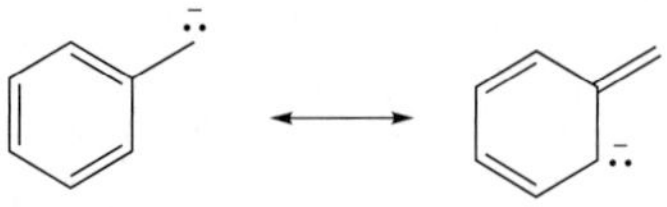

1.50 Follow the curved arrows to draw a second resonance structure for each species.

a. 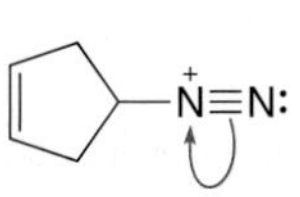
b. 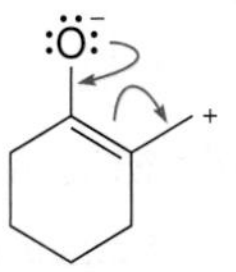
c. 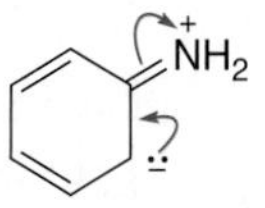

1.51 Draw a second resonance structure for each ion. Then, draw the resonance hybrid.

a. 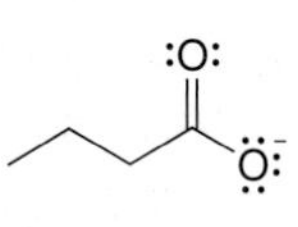
b. 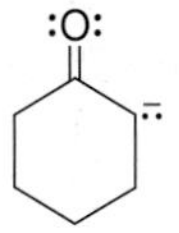
c. 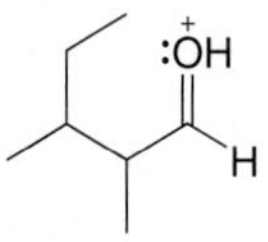

1.52 Draw all reasonable resonance structures for each species.

a. O_3 b. NO_3^- (a central N atom) c. N_3^- d.  e. 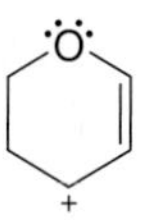

1.53 Consider compounds **A–D,** which contain both a heteroatom and a double bond. (a) For which compounds are no additional Lewis structures possible? (b) When two or more Lewis structures can be drawn, draw all additional resonance structures.

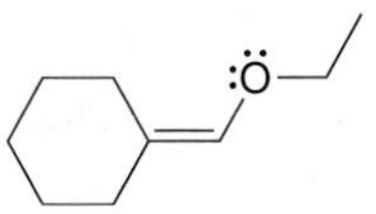
A

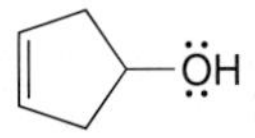
B

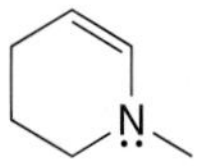
C

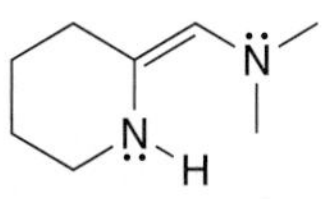
D

1.54 Draw four additional resonance structures for the following cation. Then draw the resonance hybrid.

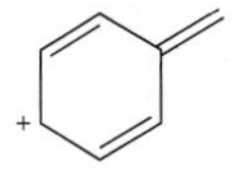

1.55 Which of the given resonance structures (**A, B,** or **C**) contributes most to the resonance hybrid? Which contributes least?

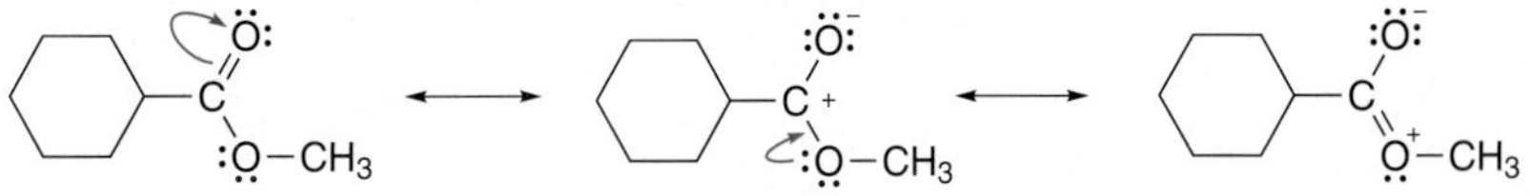
A **B** **C**

1.56 Consider the compounds and ions with curved arrows drawn below. When the curved arrows give a second valid resonance structure, draw the resonance structure. When the curved arrows generate an invalid Lewis structure, explain why the structure is unacceptable.

a. 
b. 
c. $CH_3CH_2{-}C{\equiv}N{:}$
d. 

Geometry

1.57 Predict all bond angles in each compound.

a. CH_3Cl b. NH_2OH c. $CH_2{=}NCH_3$ d. $HC{\equiv}CCH_2OH$ e. 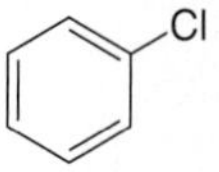

1.58 Predict the geometry around each highlighted atom.

a.  b. $(CH_3)_2N^-$ c.  d. 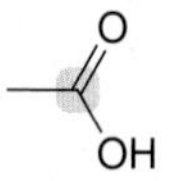

e. $(CH_3)_3N$

Drawing Organic Molecules

1.59 How many hydrogens are present around each carbon atom in the following molecules?

a. 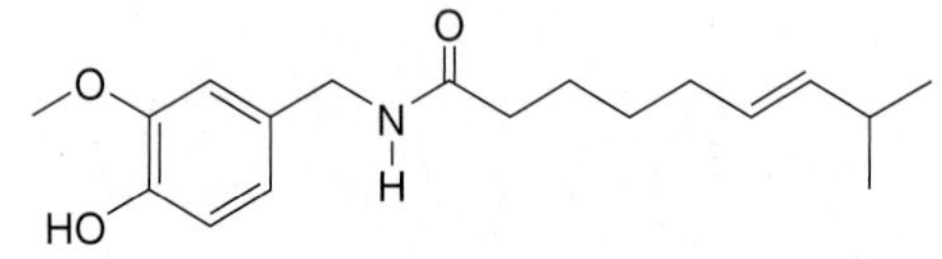

capsaicin
(spicy component of hot peppers)

b. 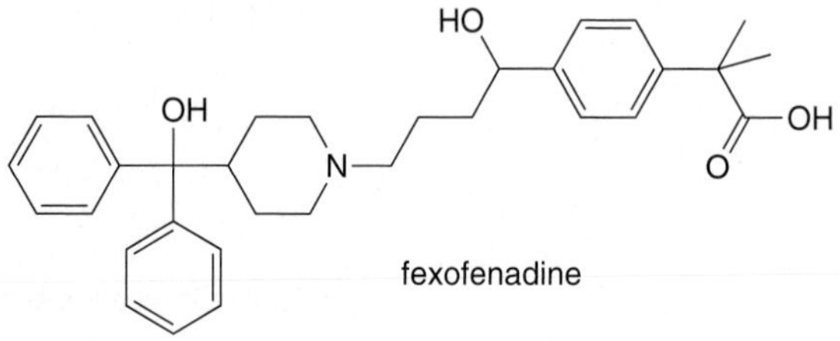

fexofenadine

1.60 Draw in all the carbon and hydrogen atoms in each molecule.

a. 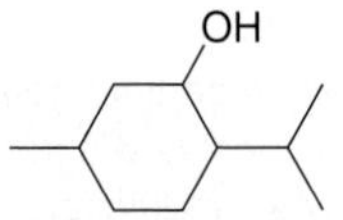

menthol
(isolated from peppermint oil)

b. 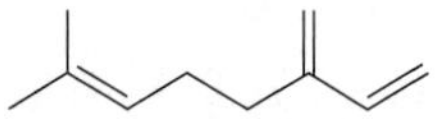
myrcene
(isolated from bayberry)

c. 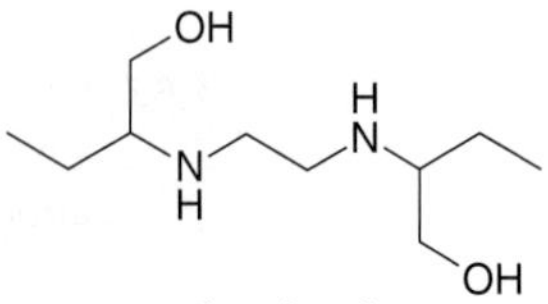

ethambutol
(drug used to treat tuberculosis)

d. 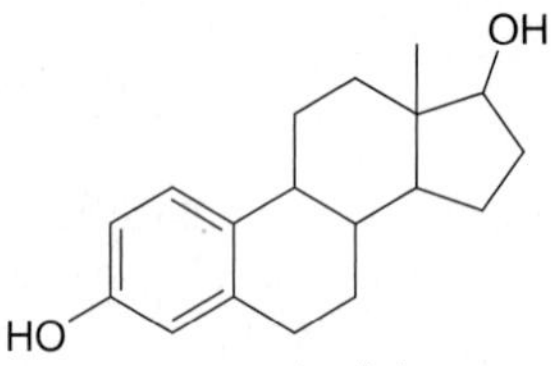

estradiol
(a female sex hormone)

1.61 Convert each molecule into a skeletal structure.

a. $(CH_3)_2CHCH_2CH_2CH(CH_3)_2$

b. $CH_3CH(Cl)CH(OH)CH_3$

c. $CH_3(CH_2)_2C(CH_3)_2CH(CH_3)CH(CH_3)CH(Br)CH_3$

d. 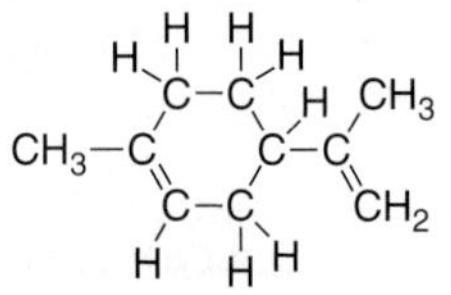

limonene
(oil of lemon)

1.62 Convert the following condensed formulas into skeletal structures.

a. $CH_3CONHCH_3$
b. CH_3COCH_2Br
c. $(CH_3)_3COH$
d. CH_3COCl
e. $CH_3COCH_2CO_2H$
f. $HO_2CCH(OH)CO_2H$

1.63 Draw in all the hydrogen atoms and nonbonded electron pairs in each ion.

a. 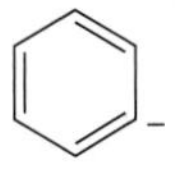 b. 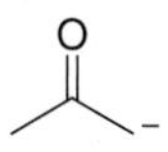 c. 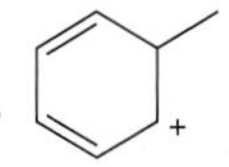 d. 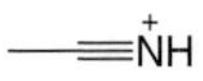 e. 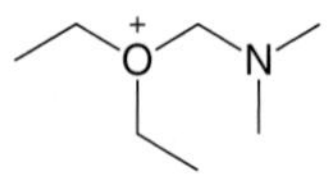

Hybridization

1.64 Predict the hybridization and geometry around each highlighted atom.

a. 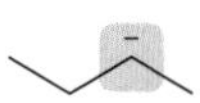 b. 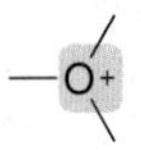 c. 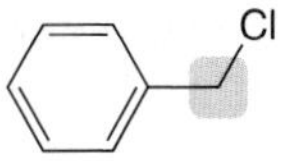 d.  e. 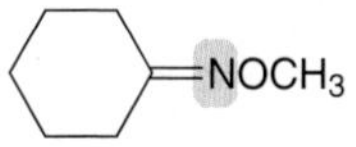

1.65 What orbitals are used to form each highlighted bond? For multiple bonds, indicate the orbitals used in individual bonds.

a. 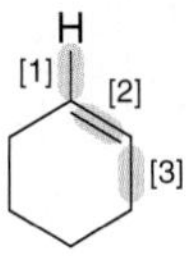 b. 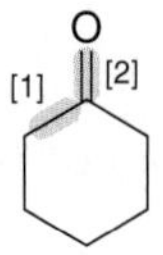 c. 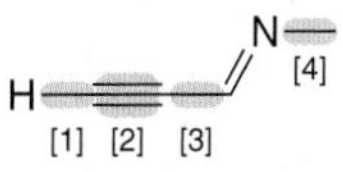

1.66 Ketene, $CH_2=C=O$, is an unusual organic molecule that has a single carbon atom doubly bonded to two different atoms. Determine the hybridization of both C atoms and the O in ketene. Then, draw a diagram showing what orbitals are used to form each bond (similar to Figures 1.10 and 1.11).

Bond Length and Strength

1.67 Rank the following bonds in order of *increasing* bond length.

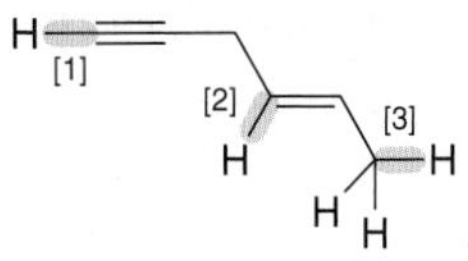

1.68 Answer the following questions about compound **A**.

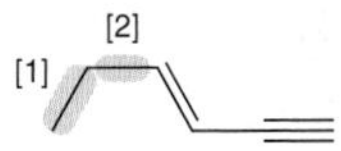

A

a. Label the shortest C–C single bond.
b. Label the longest C–C single bond.
c. Considering all the bonds, label the shortest C–C bond.
d. Label the weakest C–C bond.
e. Label the strongest C–H bond.
f. Explain why bond [1] and bond [2] are different in length, even though they are both C–C single bonds.

1.69 Two useful organic compounds that contain Cl atoms are vinyl chloride ($CH_2=CHCl$) and chloroethane (CH_3CH_2Cl). Vinyl chloride is the starting material used to prepare poly(vinyl chloride), a plastic used in insulation, pipes, and bottles. Chloroethane (ethyl chloride) is a local anesthetic. Why is the C–Cl bond of vinyl chloride stronger than the C–Cl bond in chloroethane?

Bond Polarity

1.70 Use the symbols δ+ and δ– to indicate the polarity of the highlighted bonds.

a. NH_2–OH b. 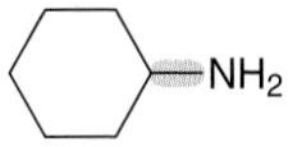 c. 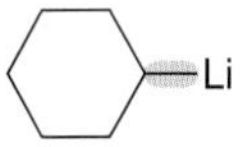

1.71 Label the polar bonds in each molecule. Indicate the direction of the net dipole (if there is one).

a. $CHBr_3$
b. $CH_3CH_2OCH_2CH_3$
c. 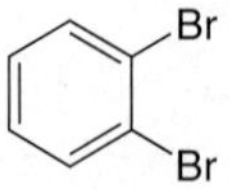
d. 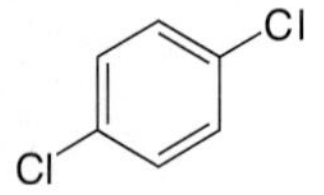

General Questions

1.72 Anacin is an over-the-counter pain reliever that contains aspirin and caffeine. Answer the following questions about each compound.

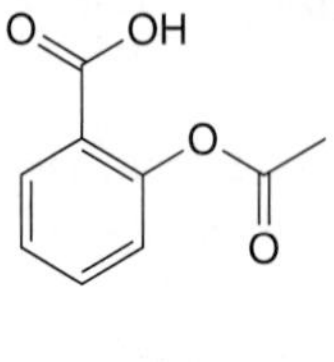

aspirin
(acetylsalicylic acid)

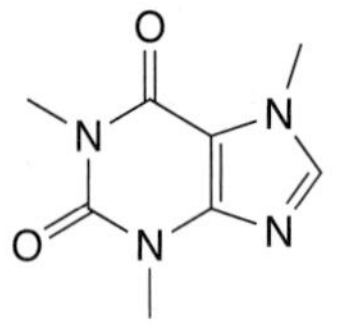

caffeine

a. What is the molecular formula?
b. How many lone pairs are present on heteroatoms?
c. Label the hybridization state of each carbon.
d. Draw three additional resonance structures.

1.73 Answer the following questions about acetonitrile ($CH_3C{\equiv}N$:).

a. Determine the hybridization of both C atoms and the N atom.
b. Label all bonds as σ or π.
c. In what type of orbital does the lone pair on N reside?
d. Label all bonds as polar or nonpolar.

1.74 Benzene is the simplest member of a whole class of compounds called aromatic hydrocarbons.

benzene

a. How is each carbon atom hybridized?
b. What is the geometry around each carbon atom? What is the overall geometry of the benzene ring?
c. Follow the indicated curved arrow notation to draw a second resonance structure.

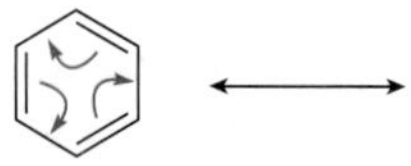

d. Benzene and other aromatic hydrocarbons are shown in Chapter 17 to be very stable. Offer an explanation.

1.75 The principles of this chapter can be applied to organic molecules of any size. Answer the following questions about amoxicillin, an antibiotic from the penicillin family.

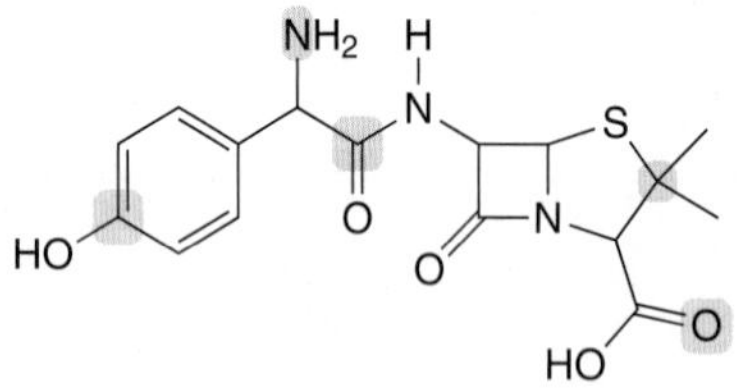

amoxicillin

a. Predict the hybridization and geometry around each highlighted atom.
b. Label five polar bonds using the symbols $\delta+$ and $\delta-$.
c. How many π bonds does amoxicillin have? Label them.
d. Find a C–H bond containing a carbon atom having a hybrid orbital with 33% *s*-character.

1.76

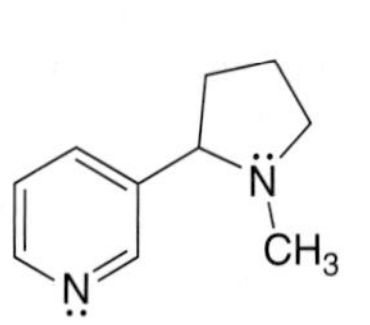

nicotine

a. What is the hybridization of each N atom in nicotine?
b. What is the geometry around each N atom?
c. In what type of orbital does the lone pair on each N atom reside?
d. Draw a constitutional isomer of nicotine.
e. Draw a resonance structure of nicotine.

1.77 Stalevo is the trade name for a medication used for Parkinson's disease, which contains L-dopa, carbidopa, and entacapone.

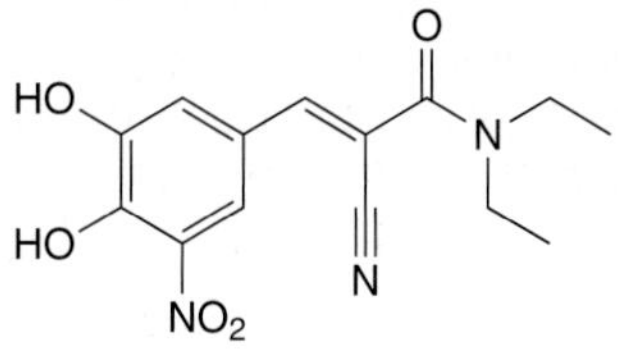

entacapone

a. Draw a Lewis structure for entacapone.
b. Which C–C bond in entacapone is the longest?
c. Which C–C single bond is the shortest?
d. Which C–N bond is the longest?
e. Which C–N bond is the shortest?
f. Use curved arrows to draw a resonance structure that is an equal contributor to the resonance hybrid.
g. Use curved arrows to draw a resonance structure that is a minor contributor to the resonance hybrid.

1.78 CH_3^+ and CH_3^- are two highly reactive carbon species.

a. What is the predicted hybridization and geometry around each carbon atom?
b. Two electrostatic potential plots are drawn for these species. Which ion corresponds to which diagram and why?

A B

Challenge Problems

1.79 The N atom in CH_3CONH_2 (acetamide) is sp^2 hybridized, even though it is surrounded by four groups. Using this information, draw a diagram that shows the orbitals used by the atoms in the $-CONH_2$ portion of acetamide, and offer an explanation as to the observed hybridization.

1.80 Use the observed bond lengths to answer each question. (a) Why is bond [1] longer than bond [2] (143 pm versus 136 pm)? (b) Why are bonds [3] and [4] equal in length (127 pm), and shorter than bond [2]?

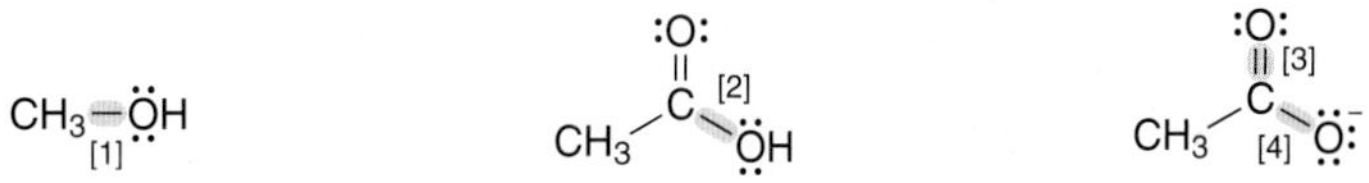

1.81 Draw at least 10 more resonance structures for acetaminophen, the active pain reliever in Tylenol.

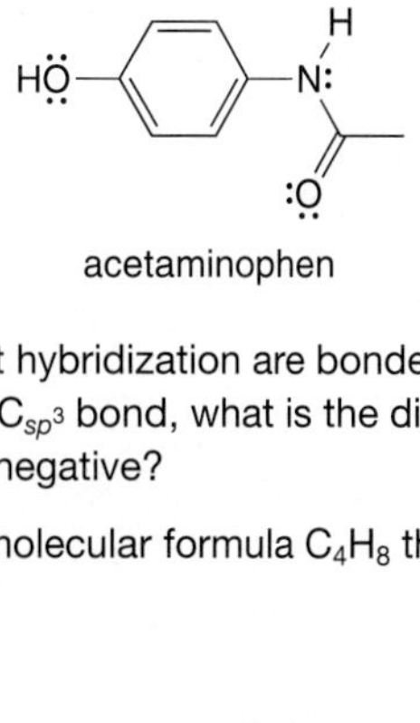

acetaminophen

1.82 When two carbons having different hybridization are bonded together, the C–C bond contains a slight dipole. In a $C_{sp^2}-C_{sp^3}$ bond, what is the direction of the dipole? Which carbon is considered more electronegative?

1.83 Draw all possible isomers having molecular formula C_4H_8 that contain one π bond.

1.84 Use the principles of resonance theory to explain why carbocation **A** is more stable than carbocation **B.**

A **B**

1.85 The curved arrow notation introduced in Section 1.6 is a powerful method used by organic chemists to show the movement of electrons not only in resonance structures, but also in chemical reactions. Since each curved arrow shows the movement of two electrons, following the curved arrows illustrates what bonds are broken and formed in a reaction. Consider the following three-step process. (a) Add curved arrows in Step [1] to show the movement of electrons. (b) Use the curved arrows drawn in Step [2] to identify the structure of **X. X** is converted in Step [3] to phenol and HCl.

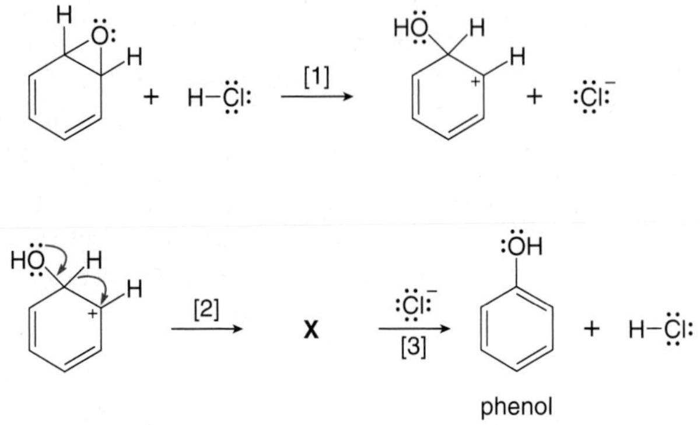

Acids and Bases

2

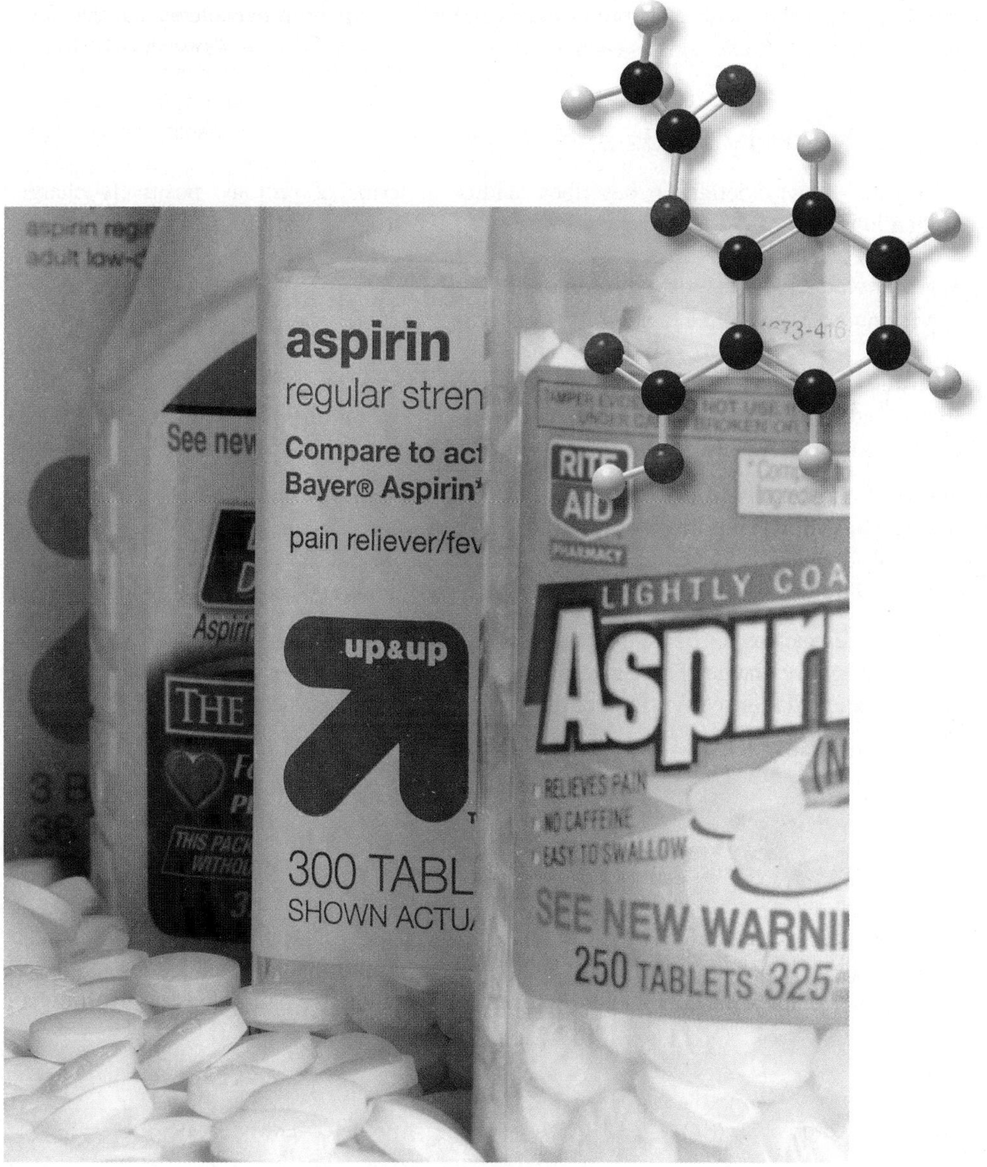

2.1 Brønsted–Lowry acids and bases

2.2 Reactions of Brønsted–Lowry acids and bases

2.3 Acid strength and pK_a

2.4 Predicting the outcome of acid–base reactions

2.5 Factors that determine acid strength

2.6 Common acids and bases

2.7 Aspirin

2.8 Lewis acids and bases

Aspirin is one of the most widely used over-the-counter drugs. Whether you purchase Anacin, Bufferin, Bayer, or a generic, the active ingredient is the same—**acetylsalicylic acid.** Aspirin, a synthetic compound that does not occur in nature, was first marketed to the general public in 1899, and is now used regularly by over 100 million people throughout the world. Like many drugs, aspirin undergoes a proton transfer reaction after ingestion. In Chapter 2, we learn about acidity and the role of acid–base reactions in aspirin's chemistry.

Chemical terms such as *anion* and *cation* may be unfamiliar to most nonscientists, but *acid* has found a place in everyday language. Commercials advertise the latest remedy for the heartburn caused by excess stomach *acid.* The nightly news may report the latest environmental impact of *acid* rain. Wine lovers know that wine sours because its alcohol has turned to *acid. Acid* comes from the Latin word *acidus,* meaning sour, because when tasting compounds was a routine method of identification, these compounds were sour.

In Chapter 2, we concentrate on two definitions of acids and bases: the **Brønsted–Lowry** definition, which describes acids as **proton donors** and bases as **proton acceptors,** and the **Lewis** definition, which describes acids as **electron pair acceptors** and bases as **electron pair donors.**

2.1 Brønsted–Lowry Acids and Bases

The general words "acid" and "base" usually mean a *Brønsted–Lowry* acid and *Brønsted–Lowry* base.

H^+ = proton.

HA = Brønsted–Lowry acid.
B: = Brønsted–Lowry base.

The Brønsted–Lowry definition describes acidity in terms of protons: positively charged **hydrogen ions, H^+.**

- A Brønsted–Lowry acid is a *proton donor.*
- A Brønsted–Lowry base is a *proton acceptor.*

A Brønsted–Lowry acid must contain a *hydrogen* atom. This definition of an acid is often familiar to students, because many inorganic acids in general chemistry are Brønsted–Lowry acids. The symbol **HA** is used for a general Brønsted–Lowry acid.

A Brønsted–Lowry base must be able to form a bond to a proton. Because a proton has no electrons, **a base must contain an "available" electron pair** that can be easily donated to form a new bond. These include **lone pairs** or electron pairs in **π bonds.** The symbol **B:** is used for a general Brønsted–Lowry base. Examples of Brønsted–Lowry acids and bases are given in Figure 2.1.

Charged species such as ^-OH and $^-NH_2$ are used as **salts,** with cations such as Li^+, Na^+, or K^+ to balance the negative charge. These cations are called **counterions** or **spectator ions,** and their **identity is usually inconsequential.** For this reason, the counterion is often omitted.

salt		counterion	base
NaOH	=	Na^+	^-OH
KOH	=	K^+	^-OH

Compounds like H_2O and CH_3OH that contain both hydrogen atoms and lone pairs may be either an acid or a base, depending on the particular reaction. These fundamental principles are true no matter how complex the compound. For example, the addictive pain reliever **morphine** is a

Figure 2.1 Examples of Brønsted–Lowry acids and bases

a. **Brønsted–Lowry acids (HA)**

HCl
H_2SO_4
HSO_4^-
H_2O
H_3O^+

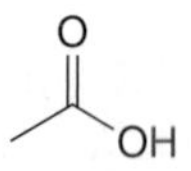

acetic acid

- All Brønsted–Lowry acids contain a proton.
- The net charge may be zero, (+), or (–).

b. **Brønsted–Lowry bases (B:)**

$H_2\ddot{O}$: $\quad$ $^-$:$\ddot{O}$H $\quad$ $CH_3\ddot{O}$:$^-$

:NH_3 $\quad$ $^-$:$\ddot{N}H_2$ $\quad$ $CH_3\ddot{N}H_2$

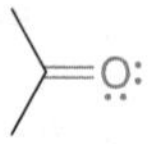

- All Brønsted–Lowry bases contain a lone pair of electrons or a π bond.
- The net charge may be zero or (–).

Morphine is obtained from the opium poppy.

Brønsted–Lowry acid because it contains many hydrogen atoms. It is also a Brønsted–Lowry base because it has lone pairs on O and N, and four π bonds.

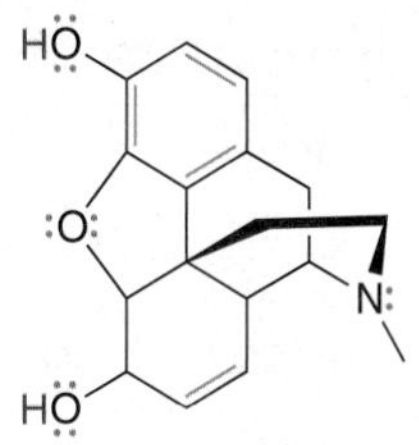
HÖ
:O:
N:
HÖ
morphine

- H atoms on O make morphine an acid.
- Lone pairs and π bonds (in blue) make morphine a base.

Problem 2.1

a. Which compounds are Brønsted–Lowry acids: HBr, NH_3, CCl_4?
b. Which compounds are Brønsted–Lowry bases: CH_3CH_3, $(CH_3)_3CO^-$, HC≡CH?
c. Classify each compound as an acid, a base, or both: CH_3CH_2OH, $CH_3CH_2CH_2CH_3$, $CH_3CO_2CH_3$.

2.2 Reactions of Brønsted–Lowry Acids and Bases

A Brønsted–Lowry acid–base reaction results in transfer of a proton from an acid to a base. These acid–base reactions, also called ***proton transfer reactions,*** are fundamental to the study of organic chemistry.

Consider, for example, the reaction of the acid HA with the base :B. **In an acid–base reaction, one bond is broken and one is formed.**

- **The electron pair of the base B: forms a new bond to the proton of the acid.**
- **The acid HA loses a proton, leaving the electron pair in the HA bond on A.**

Recall from Section 1.6 that a curved arrow shows the movement of an **electron pair.** The tail of the arrow always begins at an electron pair and the head points to where that electron pair "moves."

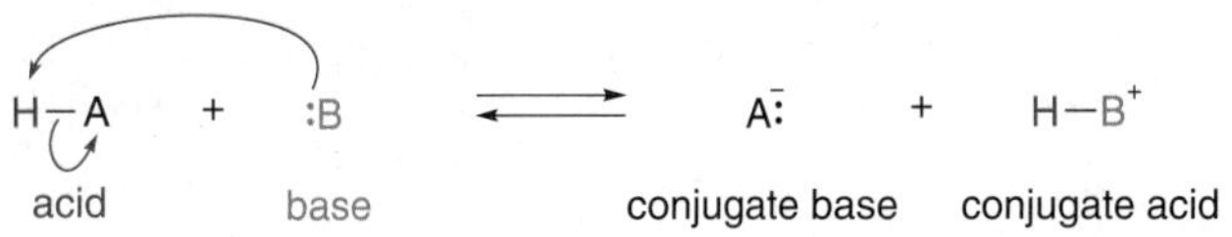
H–A
+
:B
A:⁻
+
H–B⁺
acid
base
conjugate base
conjugate acid

This "movement" of electrons in reactions can be illustrated using curved arrow notation. Because **two electron pairs** are involved in this reaction, **two curved arrows** are needed. Two products are formed.

- Loss of a proton from an acid forms its *conjugate base.*
- Gain of a proton by a base forms its *conjugate acid.*

The **net charge must be the same** on both sides of any equation. In this example, the net charge on each side is zero. Individual charges can be calculated using formal charges. **A double reaction arrow** is used between starting materials and products to indicate that the reaction can proceed in the forward and reverse directions. These are **equilibrium arrows.**

Two examples of proton transfer reactions are drawn here with curved arrow notation.

A double reaction arrow indicates equilibrium.

equilibrium arrows

Remove H^+ from an acid to form its conjugate base. Add H^+ to a base to form its conjugate acid.

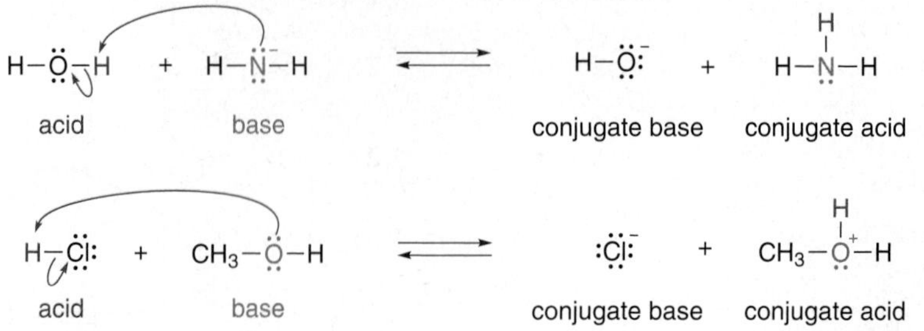

The acid loses H^+. **The base gains H^+.**

- **Brønsted–Lowry acid–base reactions always result in the transfer of a proton from an acid to a base.**

The ability to identify and draw a conjugate acid or base from a given starting material is illustrated in Sample Problems 2.1 and 2.2.

Sample Problem 2.1

a. What is the conjugate acid of CH_3O^-?
b. What is the conjugate base of NH_3?

Solution

a. **Add H^+ to CH_3O^- to form its conjugate acid.**

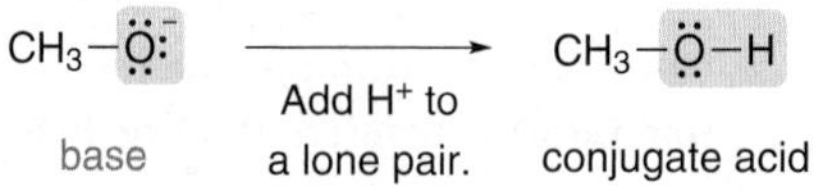

b. **Remove H^+ from NH_3 to form its conjugate base.**

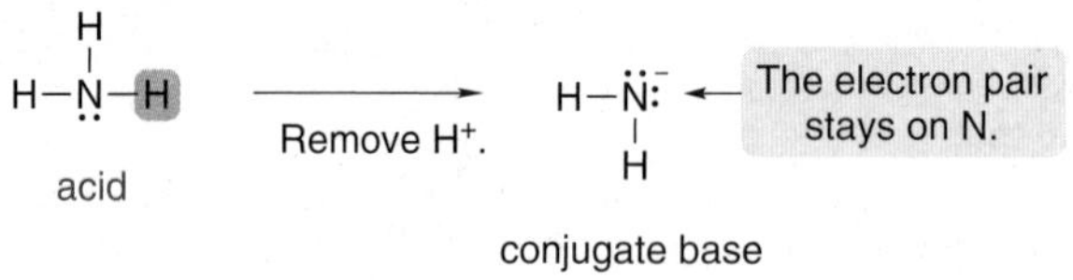

Problem 2.2

a. Draw the conjugate acid of each base: NH_3, Cl^-, $(CH_3)_2C{=}O$.
b. Draw the conjugate base of each acid: HBr, HSO_4^-, CH_3OH.

Problem 2.3

Label each statement as True or False.

a. $CH_3CH_2^+$ is the conjugate acid of $CH_2{=}CH_2$.
b. $CH_3CH_2^-$ is the conjugate base of $CH_3CH_2^+$.
c. $CH_2{=}CH_2$ is the conjugate base of $CH_3CH_2^-$.
d. $CH_2{=}CH^-$ is the conjugate base of $CH_2{=}CH_2$.
e. CH_3CH_3 is the conjugate acid of $CH_3CH_2^-$.

Sample Problem 2.2

Label the acid and base, and the conjugate acid and base in the following reaction. Use curved arrow notation to show the movement of electron pairs.

A **B** **C** **D**

Solution

A is the base because it accepts a proton, forming its conjugate acid, **C. B** is the acid because it donates a proton, forming its conjugate base, **D.** Two curved arrows are needed. One shows that the lone pair on **A** bonds to a proton of **B,** and the second shows that the electron pair in the O–H bond remains on O.

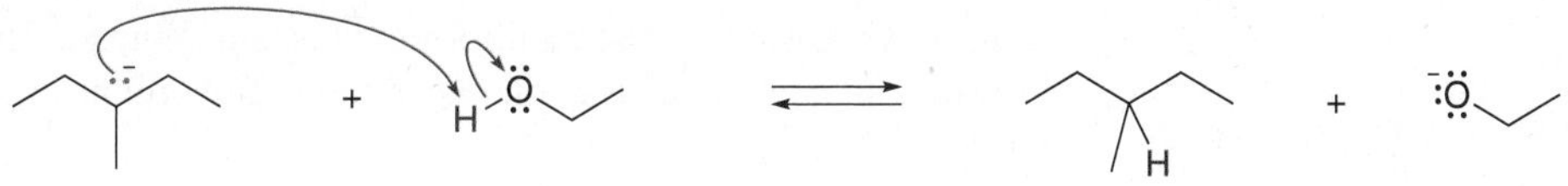

A base **The base gains H⁺.**

B acid **The acid loses H⁺.**

C conjugate acid

D conjugate base

Problem 2.4 Label the acid and base, and the conjugate acid and base in the following reactions. Use curved arrows to show the movement of electron pairs.

a. 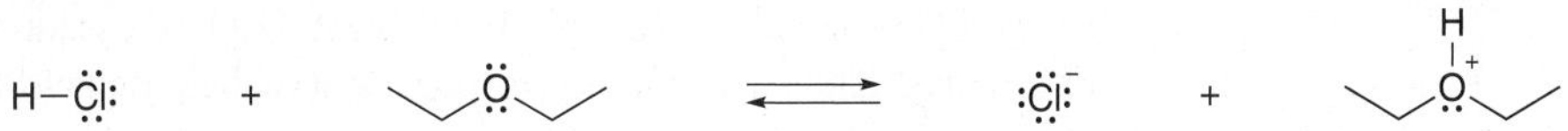

b. 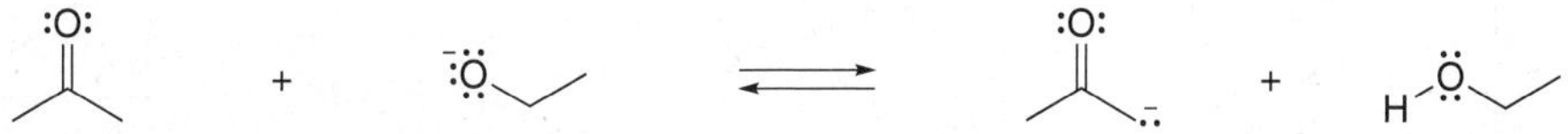

In all proton transfer reactions, the **electron-rich base** donates an electron pair to the acid, which usually has a polar HA bond. The H of the acid bears a partial positive charge, making it **electron deficient.** This is the first example of a general pattern of reactivity.

- **Electron-rich species react with electron-deficient ones.**

Given two starting materials, how do you know which is the acid and which is the base in a proton transfer reaction? Use the following generalizations:

[1] Common acids and bases introduced in general chemistry are used in the same way in organic reactions. HCl and H_2SO_4 are strong acids, and ^-OH is a strong base.

[2] When only one starting material contains a hydrogen, it must be the acid. If only one starting material has a lone pair or a π bond, it must be the base.

[3] A starting material with a net positive charge is usually the acid. A starting material with a negative charge is usually the base.

Problem 2.5 Decide which compound is the acid and which is the base, and draw the products of each proton transfer reaction.

a. 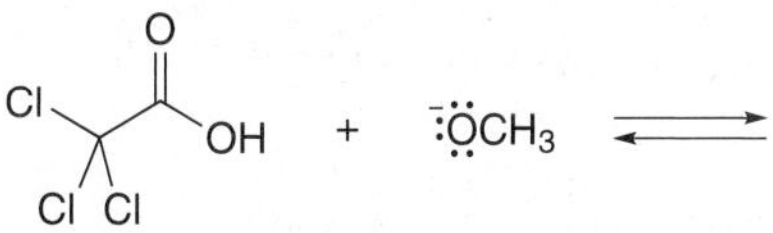

c. 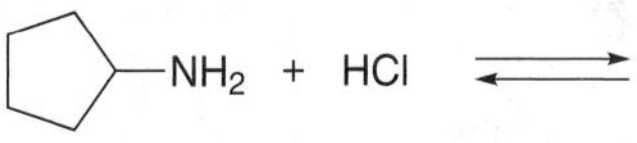

b. 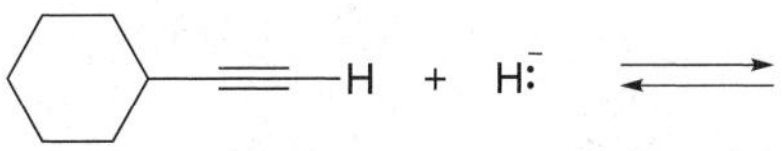

d. 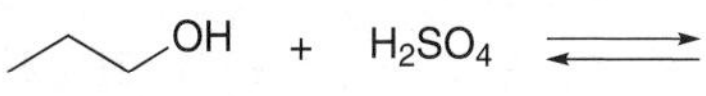

Problem 2.6 Draw the products formed from the acid–base reaction of HCl with each compound.

a. 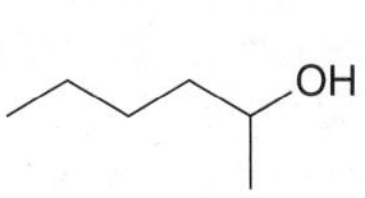

b. 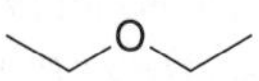

c. 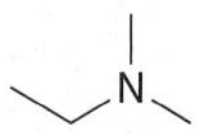

d. 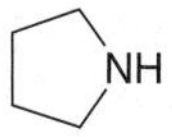

2.3 Acid Strength and pK_a

Acid strength is the tendency of an acid to donate a proton.

- **The more readily a compound donates a proton, the stronger the acid.**

Acidity is measured by an equilibrium constant. When a Brønsted–Lowry acid HA is dissolved in water, an acid–base reaction occurs, and an equilibrium constant K_{eq} can be written for the reaction.

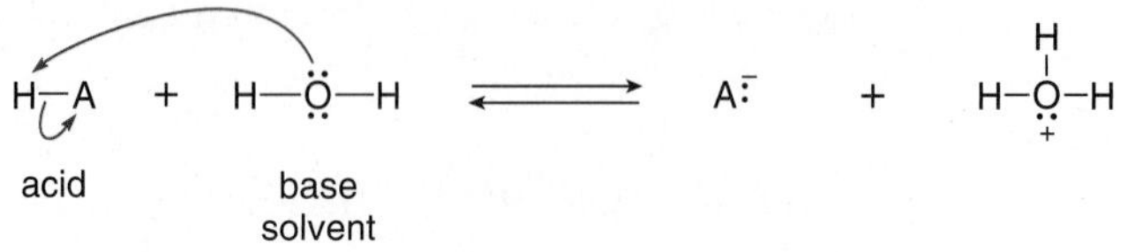

$$K_{eq} = \frac{[\text{products}]}{[\text{starting materials}]} = \frac{[H_3O^+][A\!:^-]}{[HA][H_2O]}$$

Because the concentration of the solvent H_2O is essentially constant, the equation can be re-arranged and a new equilibrium constant, called the **acidity constant, K_a,** can be defined.

$$K_a = [H_2O]K_{eq} = \frac{[H_3O^+][A\!:^-]}{[HA]}$$

How is the magnitude of K_a related to acid strength?

- **The *stronger the acid,* the further the equilibrium lies to the right, and the *larger the* K_a.**

For most organic compounds, K_a is small, typically 10^{-5} to 10^{-50}. This contrasts with the K_a values for many inorganic acids, which range from 10^0 to 10^{10}. Because using exponents can be cumbersome, it is often more convenient to use pK_a values instead of K_a values.

$$pK_a = -\log K_a$$

How does pK_a relate to acid strength?

Recall that a **log** is an **exponent;** for example, log $10^{-5} = -5$.

K_a values of typical organic acids		**pK_a values** of typical organic acids	
10^{-5} →	10^{-50}	+5 →	+50
larger number stronger acid	smaller number weaker acid	smaller number stronger acid	larger number weaker acid

- **The *smaller* the pK_a, the *stronger* the acid.**

Problem 2.7 Which compound in each pair is the stronger acid?

a. $CH_3CH_2CH_3$ (pK_a = 50) or CH_3CH_2OH (pK_a = 16)

b. 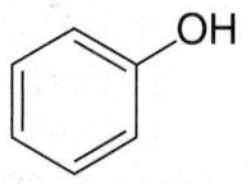

$K_a = 10^{-10}$

or

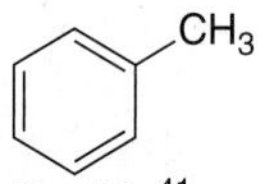

$K_a = 10^{-41}$

Problem 2.8 Use a calculator when necessary to answer the following questions.

a. What is the pK_a for each K_a: 10^{-10}, 10^{-21}, and 5.2×10^{-5}?

b. What is the K_a for each pK_a: 7, 11, and 3.2?

An inverse relationship exists between acidity and basicity.

- A *strong acid* readily donates a proton, forming a *weak conjugate base.*
- A *strong base* readily accepts a proton, forming a *weak conjugate acid.*

Table 2.1 is a brief list of pK_a values for some common compounds, ranked in order of ***increasing* pK_a** and therefore ***decreasing* acidity.** Because strong acids form weak conjugate bases, this list also ranks their conjugate bases, in order of ***increasing* basicity.** CH_4 is the weakest acid in the list, because it has the highest pK_a (50). Its conjugate base, CH_3^-, is therefore the strongest conjugate base. An extensive pK_a table is located in Appendix A.

Table 2.1 Selected pK_a Values

Acid	pK_a	Conjugate base
H–Cl	–7	Cl^-
CH_3CO_2–H	4.8	$CH_3CO_2^-$
HO–H	15.7	HO^-
CH_3CH_2O–H	16	$CH_3CH_2O^-$
HC≡CH	25	$HC{\equiv}C^-$
H–H	35	H^-
H_2N–H	38	H_2N^-
$CH_2{=}CH_2$	44	$CH_2{=}\bar{C}H$
CH_3–H	50	CH_3^-

Increasing acidity (upward, acid column) — Increasing basicity (downward, conjugate base column)

Comparing pK_a values tells us the **relative acidity of two acids,** and the **relative basicity of their conjugate bases,** as shown in Sample Problem 2.3.

Sample Problem 2.3 Rank the following compounds in order of increasing acidity, and then rank their conjugate bases in order of increasing basicity.

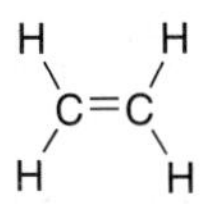

H–Cl

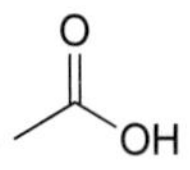

Solution

Use the pK_a values in Table 2.1 and the rule: **the lower the pK_a, the stronger the acid.**

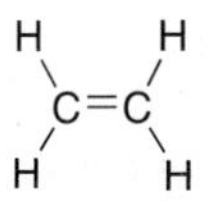

pK_a = 44

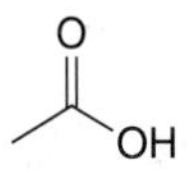

pK_a = 4.8

H–Cl

pK_a = –7

Increasing acidity →

Because strong acids form weak conjugate bases, the **basicity of conjugate bases increases with increasing pK_a** of their acids.

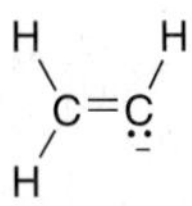

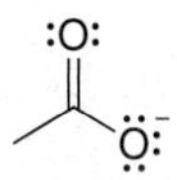

← **Increasing basicity**

Problem 2.9 Rank the conjugate bases of each group of acids in order of increasing basicity.

a. NH_3, H_2O, CH_4 b. $CH_2{=}CH_2$, $HC{\equiv}CH$, CH_4

Problem 2.10 Consider two acids: HCO_2H (formic acid, $pK_a = 3.8$) and pivalic acid [$(CH_3)_3CCO_2H$, $pK_a = 5.0$]. (a) Which acid has the larger K_a? (b) Which acid is the stronger acid? (c) Which acid forms the stronger conjugate base? (d) When each acid is dissolved in water, for which acid does the equilibrium lie further to the right?

The pK_a values in Table 2.1 span a large range (–7 to 50). The pK_a scale is logarithmic, so a small difference in pK_a translates into a large numerical difference. The difference between the pK_a values of NH_3 (38) and $CH_2{=}CH_2$ (44) is six pK_a units, so NH_3 is 10^6 or *one million times more acidic* than $CH_2{=}CH_2$.

Although Table 2.1 is abbreviated, it is a useful tool for *estimating* the pK_a of a compound similar though not identical to one in the table. Suppose you are asked to estimate the pK_a of the N–H bond of CH_3NH_2. Although CH_3NH_2 is not listed in the table, we have enough information to *approximate* its pK_a. Because the pK_a of the N–H bond of NH_3 is 38, we can estimate the pK_a of the N–H bond of CH_3NH_2 to be 38. Its actual pK_a is 40, so this is a good first approximation.

Problem 2.11 Estimate the pK_a of each of the indicated bonds.

a. 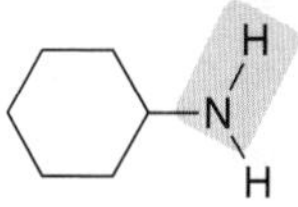
H
N
H

b. 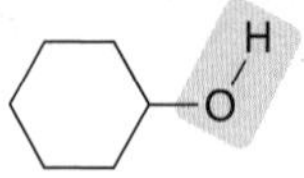
H
O

c. 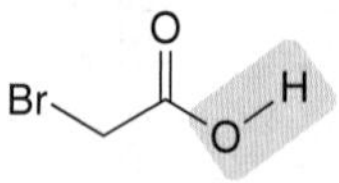
Br
O
O
H

2.4 Predicting the Outcome of Acid–Base Reactions

In a proton transfer reaction, the **stronger acid reacts with the stronger base** to form the weaker acid and the weaker base.

A proton transfer reaction represents an equilibrium. Because an acid donates a proton to a base, forming a conjugate acid and conjugate base, there are always two acids and two bases in the reaction mixture. Which pair of acids and bases is favored at equilibrium? **The position of the equilibrium depends on the relative strengths of the acids and bases.**

- Equilibrium always favors formation of the weaker acid and base.

Because a strong acid readily donates a proton and a strong base readily accepts one, these two species react to form a weaker conjugate acid and base that do not donate or accept a proton as readily. Comparing pK_a values allows us to determine the position of equilibrium, as illustrated in Sample Problem 2.4.

Sample Problem 2.4 Determine the direction of equilibrium when acetylene ($HC{\equiv}CH$) reacts with $^-NH_2$ in a proton transfer reaction.

Solution

Follow three steps to determine the position of equilibrium:

Step [1] Identify the acid and base in the starting materials.

- $^-NH_2$ is the base because it bears a net negative charge, so $HC{\equiv}CH$ is the acid.

Step [2] Draw the products of proton transfer and identify the conjugate acid and base in the products.

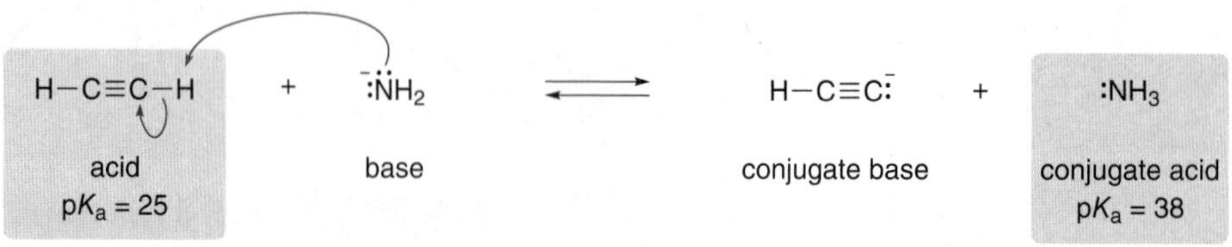
H–C≡C–H
+
:NH2
H–C≡C:
+
:NH3
acid
pKa = 25
base
conjugate base
conjugate acid
pKa = 38

Step [3] Compare the pK_a values of the acid and the conjugate acid. Equilibrium favors formation of the weaker acid with the higher pK_a.

Use unequal equilibrium arrows (⇌ or ⇌) for a reversible reaction in which products or reactants are favored at equilibrium.

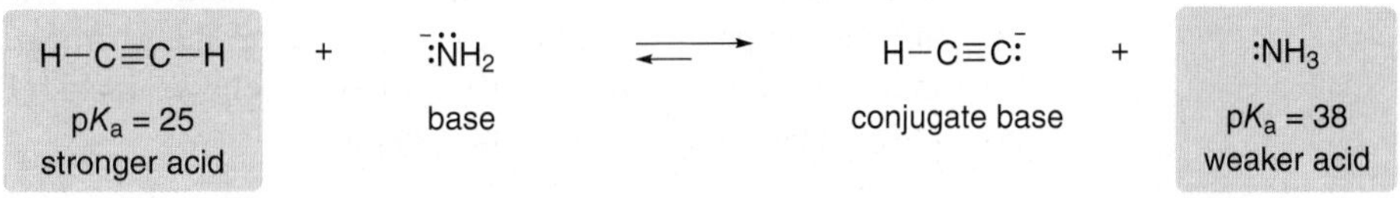
H−C≡C−H
pK_a = 25
stronger acid
+
:N̈H₂
base
H−C≡C:
conjugate base
+
:NH₃
pK_a = 38
weaker acid

Equilibirum favors the products, forming the weaker acid.

- Because the pK_a of the starting acid (25) is ***lower*** than the pK_a of the conjugate acid (38), HC≡CH is a ***stronger*** acid and equilibrium favors the products.

Problem 2.12 Draw the products of each reaction and determine the direction of equilibrium.

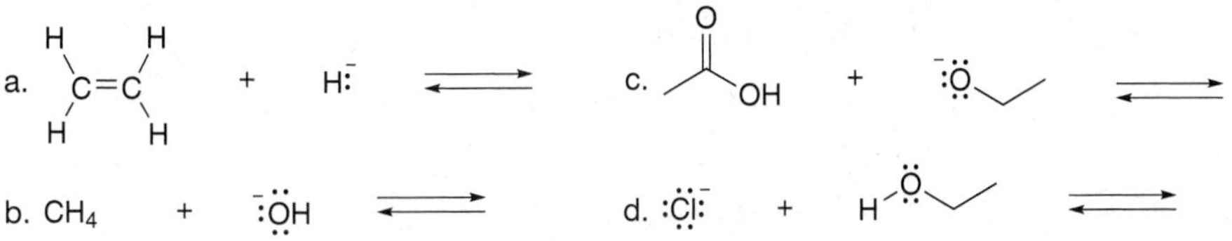
a.
H
H
C=C
H
H
+
H:
c.
O
OH
+
:Ö
b. CH_4
+
:ÖH
d. :C̈l:
+
H
Ö

How can we know if a particular base is strong enough to deprotonate a given acid, so that the equilibrium lies to the right? The pK_a table readily gives us this information, as shown in Sample Problem 2.5.

Sample Problem 2.5 Which of the following bases is strong enough to deprotonate *N,N*-dimethylacetamide [$CH_3CON(CH_3)_2$, pK_a = 30], so that equilibrium favors the products: (a) $NaNH_2$; (b) NaOH?

Solution

- Draw the structure of the conjugate acid of each base, and determine its pK_a from Table 2.1 or Appendix A.
- Compare the pK_a values of the starting acid and the conjugate acid. If the conjugate acid has a *higher* pK_a than the starting acid, the conjugate acid is the *weaker* acid and equilibrium favors the *products.* **The base is strong enough** to deprotonate the acid.
- If the conjugate acid has a *lower* pK_a than the starting acid, the conjugate acid is the *stronger* acid and equilibrium favors the *starting materials.* **The base is not strong enough** to deprotonate the acid.

a. Na^+ is a counterion and $^-NH_2$ is the base in $NaNH_2$.

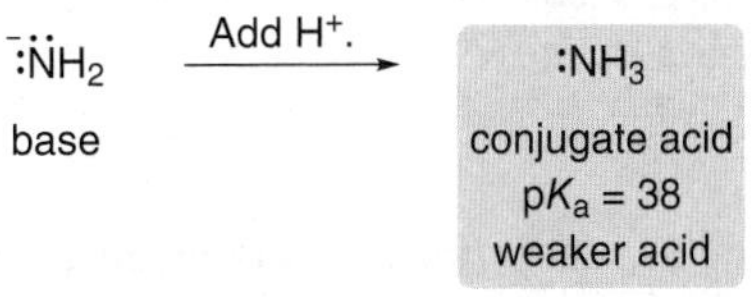
:N̈H₂
base
Add H⁺.
:NH₃
conjugate acid
pK_a = 38
weaker acid

The conjugate acid (NH_3) of the base is a weaker acid than $CH_3CON(CH_3)_2$ (pK_a = 30), so the base is strong enough to deprotonate the acid, and equilibrium favors the products.

b. Na^+ is a counterion and ^-OH is the base in NaOH.

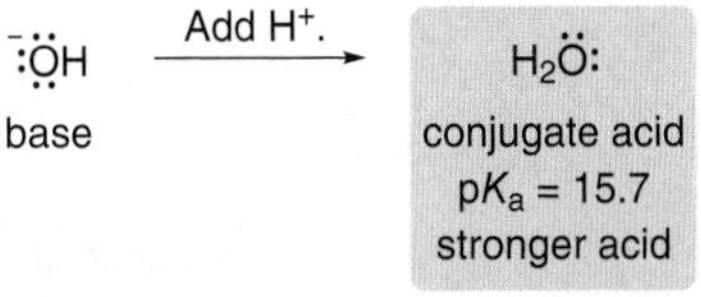
:ÖH
base
Add H⁺.
H₂Ö:
conjugate acid
pK_a = 15.7
stronger acid

The conjugate acid (H_2O) of the base is a stronger acid than $CH_3CON(CH_3)_2$ (pK_a = 30), so the base is *not* strong enough to deprotonate the acid, and equilibrium favors the starting materials.

Because Table 2.1 is arranged from low to high pK_a, **an acid can be deprotonated by the conjugate base of any acid below it in the table.**

Sample Problem 2.5 illustrates a fundamental principle in acid–base reactions.

- **An acid can be deprotonated by the conjugate base of any acid having a *higher* pK_a.**

Problem 2.13 Using the data in Appendix A, determine which of the following bases is strong enough to deprotonate acetonitrile (CH_3CN), so that equilibrium favors the products: (a) NaH; (b) Na_2CO_3; (c) NaOH; (d) $NaNH_2$; (e) $NaHCO_3$.

2.5 Factors That Determine Acid Strength

The wide range of pK_a values in Table 2.1 illustrates that a tremendous difference in acidity exists among compounds. HCl (pK_a < 0) is an extremely strong acid, water (pK_a = 15.7) is moderate in acidity, and CH_4 (pK_a = 50) is an extremely weak acid. How are these differences explained? One general rule governs acid strength.

- **Anything that stabilizes a conjugate base A:$^-$ makes the starting acid HA more acidic.**

Four factors affect the acidity of HA:

[1] Element effects
[2] Inductive effects
[3] Resonance effects
[4] Hybridization effects

No matter which factor is discussed, follow the same procedure. To compare the acidity of any two acids:

- **Draw the conjugate bases.**
- **Determine which conjugate base is more stable.**
- **The *more stable* the conjugate base, the *more acidic* the acid.**

2.5A Element Effects—Trends in the Periodic Table

The most important factor determining the acidity of HA is the location of A in the periodic table.

Comparing Elements in the Same Row of the Periodic Table

To examine acidity trends **across a row** of the periodic table, we compare CH_4 and H_2O, two compounds having H atoms bonded to a second-row element. We know from Table 2.1 that **H_2O has a much *lower* pK_a and therefore is much *more acidic* than CH_4,** but why is this the case?

To answer this question, first draw both conjugate bases and then determine which is more stable. Each conjugate base has a net negative charge, but the negative charge in $^-$OH is on oxygen and in CH_3^- it is on carbon.

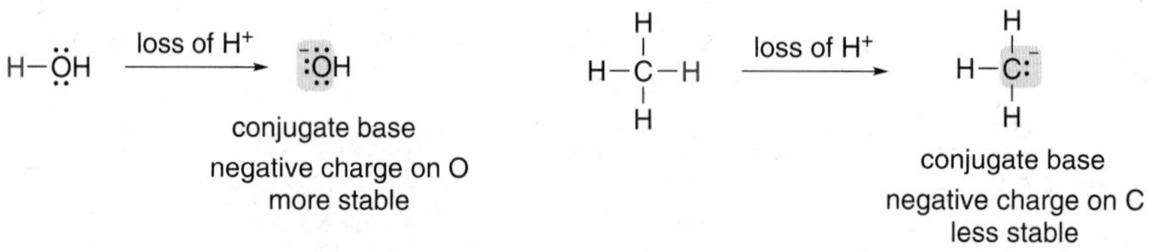
H–ÖH
loss of H+
:ÖH
conjugate base
negative charge on O
more stable
H–C–H
loss of H+
H–C:
conjugate base
negative charge on C
less stable

Because the oxygen atom is much **more electronegative** than carbon, oxygen more readily accepts a negative charge, making $^-$OH much more stable than CH_3^-. **H_2O is a stronger acid than CH_4 because $^-$OH is a more stable conjugate base than CH_3^-.** This is a specific example of a general trend.

- **Across a row of the periodic table, the acidity of HA increases as the electronegativity of A increases.**

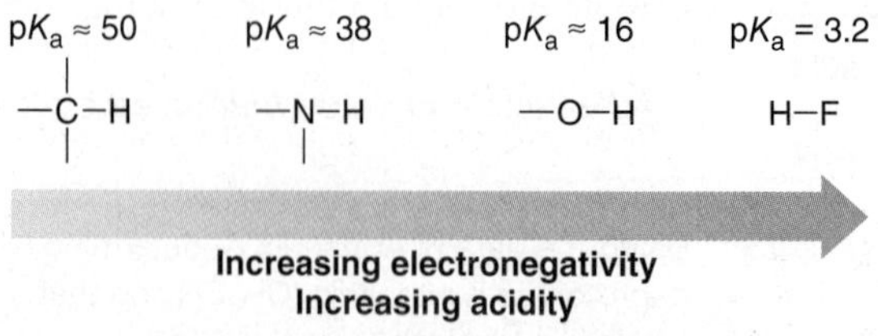
pKa ≈ 50
pKa ≈ 38
pKa ≈ 16
pKa = 3.2
–C–H
–N–H
–O–H
H–F
Increasing electronegativity
Increasing acidity

The enormity of this effect is evident by comparing the pK_a values for these bonds. **A C–H bond is approximately 10^{47} times *less acidic* than H–F.**

Comparing Elements Down a Column of the Periodic Table

To examine acidity trends down a column of the periodic table, we compare H–F and H–Br. Draw both conjugate bases and then determine which is more stable. In this case, removal of a proton forms F^- and Br^-.

loss of H+
conjugate base
loss of H+
conjugate base

There are two important differences between F^- and Br^-—electronegativity and size. In this case, **size is more important than electronegativity.** The size of an atom or ion increases down a column of the periodic table, so Br^- is much larger than F^-, and this stabilizes the negative charge.

- **Positive or negative charge is stabilized when it is spread over a larger volume.**

Because Br^- is larger than F^-, Br^- is more stable than F^-, and H–Br is a stronger acid than H–F.

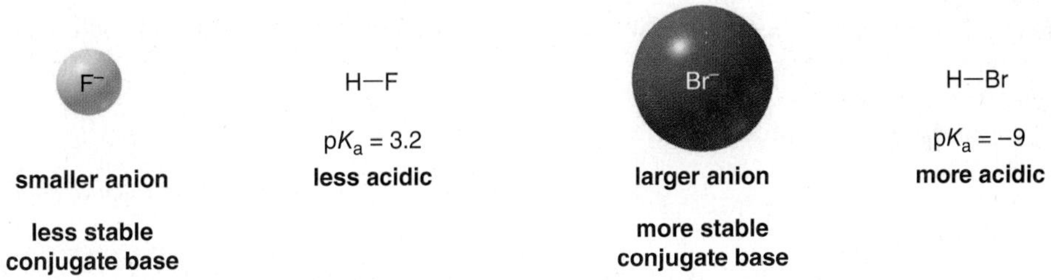
F⁻
smaller anion
less stable
conjugate base
H–F
pKa = 3.2
less acidic
Br⁻
larger anion
more stable
conjugate base
H–Br
pKa = –9
more acidic

This again is a specific example of a general trend.

- **Down a column of the periodic table, the acidity of HA increases as the size of A increases.**

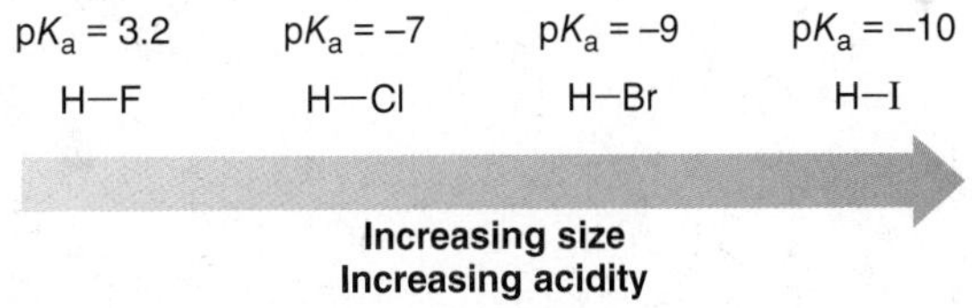
pKa = 3.2
H–F
pKa = –7
H–Cl
pKa = –9
H–Br
pKa = –10
H–I
Increasing size
Increasing acidity

Because of carbon's position in the periodic table (in the second row and to the left of O, N, and the halogens), **C–H bonds are usually the *least acidic* bonds in a molecule.**

This is ***opposite*** to what would be expected on the basis of electronegativity differences between F and Br, because F is more electronegative than Br. **Size and *not* electronegativity determines acidity down a column.** Combining both trends together:

- **The acidity of HA increases both left-to-right across a row and down a column of the periodic table.**

Sample Problem 2.6

Without reference to a pK_a table, decide which compound in each pair is the stronger acid:

a. H_2O or HF   b. H_2S or H_2O

Solution

a. H_2O and HF both have H atoms bonded to a second-row element. Because the acidity of HA increases across a row of the periodic table, the H–F bond is more acidic than the H–O bond. **HF is a stronger acid than H_2O.**

b. H_2O and H_2S both have H atoms bonded to elements in the same column. Because the acidity of HA increases down a column of the periodic table, the H–S bond is more acidic than the H–O bond. **H_2S is a stronger acid than H_2O.**

Problem 2.14 Without reference to a pK_a table, decide which compound in each pair is the stronger acid.

a. 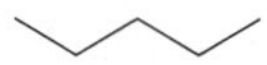 or H_2O b. 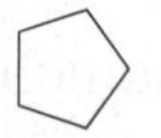 or H_2S

Problem 2.15 Rank the labeled H atoms in the following compound in order of increasing acidity.

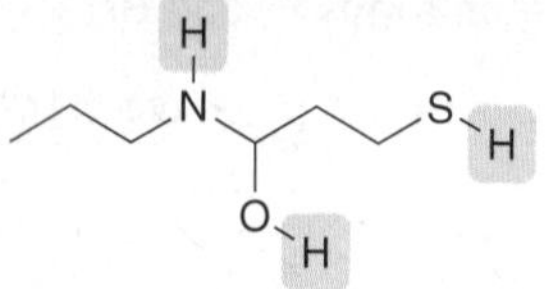

Because the pseudoephedrine (Problem 2.17) in Sudafed can be readily converted to the illegal, addictive drug methamphetamine, products that contain pseudoephedrine are now stocked behind the pharmacy counter, so their sale can be more closely monitored. Sudafed PE is a related product that contains a decongestant less easily converted to methamphetamine.

When discussing acidity, the most acidic proton in a compound is the one removed first by a base. Although four factors determine the overall acidity of a particular hydrogen atom, **the element effect—the identity of A—is the single most important factor in determining the acidity of the HA bond.**

To decide which hydrogen is most acidic, **first determine what element each hydrogen is bonded to and then decide its acidity based on periodic trends.** For example, $CH_3NHCH_2CH_2CH_2CH_3$ contains only C−H and N−H bonds. Because the acidity of HA increases across a row of the periodic table, the single H on N is the most acidic H in this compound.

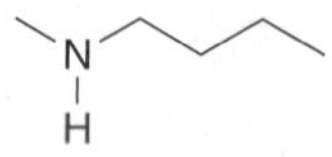

most acidic H shown in red

Problem 2.16 Which hydrogen in each molecule is most acidic?

a. 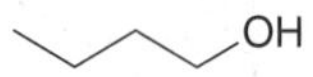 b. 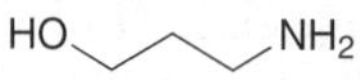 c. 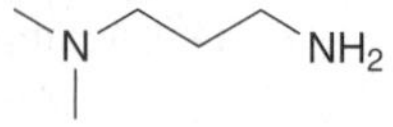

Problem 2.17 Which hydrogen in pseudoephedrine, the nasal decongestant in the commercial medication Sudafed, is most acidic?

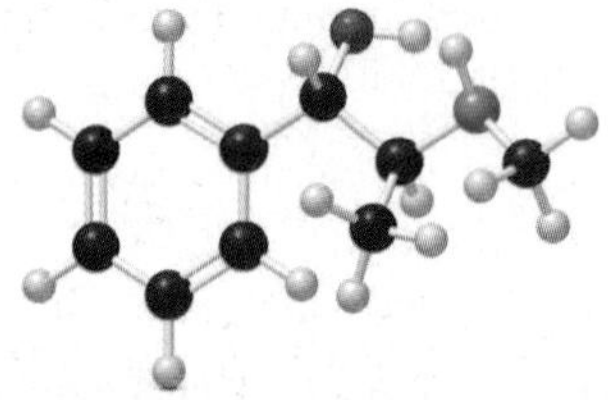

pseudoephedrine

Problem 2.18 Which compound in each pair of isomers is the stronger acid?

a. 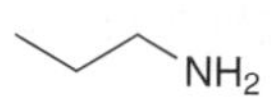 or 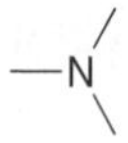 b. 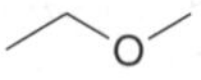 or 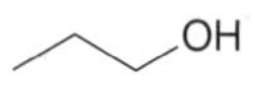

2.5B Inductive Effects

A second factor affecting the acidity of HA is the presence of atoms more electronegative than carbon. To illustrate this phenomenon, compare ethanol (CH_3CH_2OH) and 2,2,2-trifluoroethanol (CF_3CH_2OH), two compounds containing O−H bonds. The pK_a table in Appendix A indicates that CF_3CH_2OH is a stronger acid than CH_3CH_2OH. We are comparing the acidity of the O−H bond in both compounds, so what causes the difference?

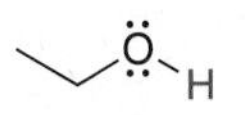

ethanol
pK_a = 16

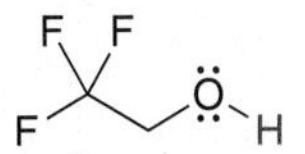

2,2,2-trifluoroethanol
pK_a = 12.4
stronger acid

Draw both conjugate bases and then determine which is more stable. Both bases have a negative charge on an electronegative oxygen, but the second anion has three very electronegative fluorine atoms. These fluorine atoms withdraw electron density from the carbon to which they are bonded, making it electron deficient. Furthermore, this electron-deficient carbon pulls electron density through σ bonds from the negatively charged oxygen atom, stabilizing the negative charge. This is called an **inductive effect.**

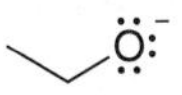

No additional electronegative atoms stabilize the conjugate base.

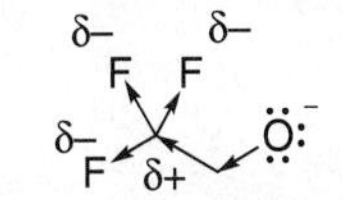

CF_3 withdraws electron density, stabilizing the conjugate base.

- **An *inductive effect* is the pull of electron density through σ bonds caused by electronegativity differences of atoms.**

In this case, the electron density is pulled away from the negative charge through σ bonds by the very electronegative fluorine atoms, so it is called an **electron-*withdrawing* inductive effect. Thus, the three very electronegative fluorine atoms stabilize the negatively charged conjugate base $CF_3CH_2O^-$, making CF_3CH_2OH a stronger acid than CH_3CH_2OH.** We have learned two important principles from this discussion:

- **More electronegative atoms stabilize regions of high electron density by an electron-withdrawing inductive effect.**
- **The acidity of HA increases with the presence of electron-withdrawing groups in A.**

Inductive effects result because an electronegative atom stabilizes the negative charge of the conjugate base. **The *more electronegative* the atom and the *closer* it is to the site of the negative charge, the greater the effect.** This effect is discussed in greater detail in Chapter 19.

Electrostatic potential plots in Figure 2.2 compare the electron density around the oxygen atoms in $CH_3CH_2O^-$ and $CF_3CH_2O^-$. The darker red region around the O atom of $CH_3CH_2O^-$ indicates a higher concentration of electron density compared to the O atom of $CF_3CH_2O^-$.

Figure 2.2
Electrostatic potential plots of $CH_3CH_2O^-$ and $CF_3CH_2O^-$

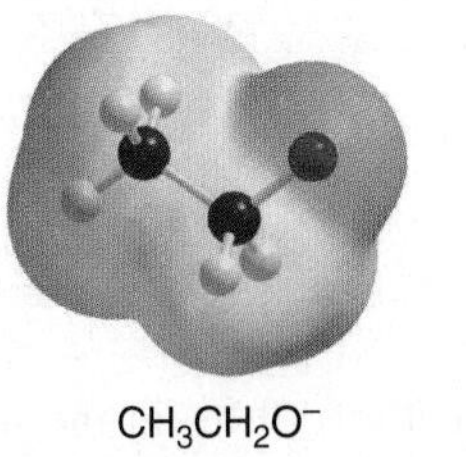

$CH_3CH_2O^-$
The dark red of the O atom indicates a region of high electron density.

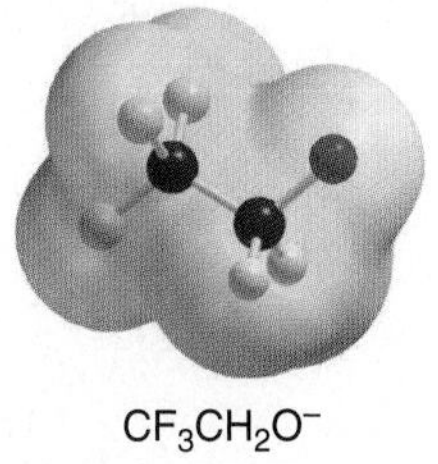

$CF_3CH_2O^-$
The O atom is yellow, indicating it is less electron rich.

Problem 2.19 Which compound in each pair is the stronger acid?

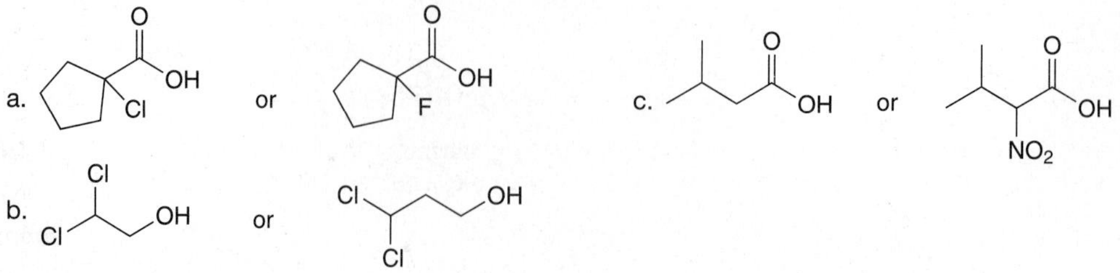

Problem 2.20 Glycolic acid, $HOCH_2CO_2H$, is the simplest member of a group of compounds called α-hydroxy acids, ingredients in skin care products that have an OH group on the carbon adjacent to a CO_2H group. Would you expect $HOCH_2CO_2H$ to be a stronger or weaker acid than acetic acid, CH_3CO_2H?

Problem 2.21 Explain the apparent paradox. HBr is a stronger acid than HCl, but HOCl is a stronger acid than HOBr.

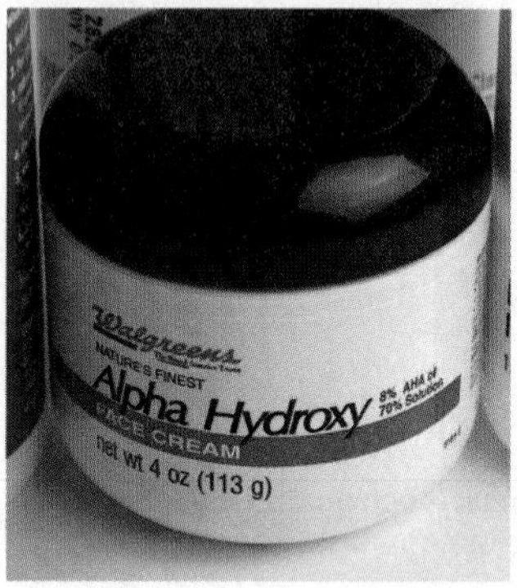

α-Hydroxy acids (Problem 2.20) are used in skin care products that purportedly smooth fine lines and improve skin texture by reacting with the outer layer of skin cells, causing them to loosen and flake off.

Resonance structures are two Lewis structures having the same placement of atoms but a different arrangement of electrons.

2.5C Resonance Effects

A third factor that determines acidity is resonance. Recall from Section 1.6 that resonance occurs whenever two or more different Lewis structures can be drawn for the same arrangement of atoms. To illustrate this phenomenon, compare ethanol (CH_3CH_2OH) and acetic acid (CH_3CO_2H), two compounds containing O–H bonds. Based on Table 2.1, CH_3CO_2H is a stronger acid than CH_3CH_2OH.

ethanol
pK_a = 16

acetic acid
pK_a = 4.8
stronger acid

Draw the conjugate bases of these acids to illustrate the importance of resonance. For ethoxide ($CH_3CH_2O^-$), the conjugate base of ethanol, only one Lewis structure can be drawn. The negative charge of this conjugate base is *localized* on the O atom.

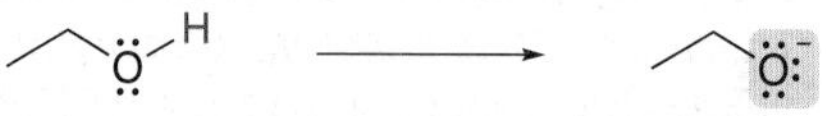

ethanol
acid

ethoxide
conjugate base
only **one** Lewis structure

With acetate ($CH_3CO_2^-$), however, two resonance structures can be drawn.

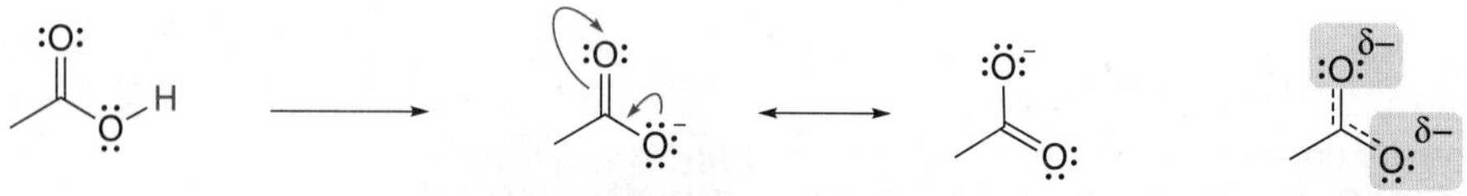

acetic acid

acetate
conjugate base
two resonance structures

hybrid
resonance-stabilized conjugate base

These two resonance structures differ in the **position of a π bond** and a **lone pair.** Although each resonance structure of acetate implies that the negative charge is localized on an O atom, in actuality, charge is ***delocalized*** over both O atoms. **Delocalization of electron density stabilizes acetate, making it a weaker base.**

Resonance delocalization often produces a larger effect on pK_a than the inductive effects discussed in Section 2.5B. Resonance makes CH_3CO_2H (pK_a = 4.8) a much stronger acid than CH_3CH_2OH (pK_a = 16), while the inductive effects due to three electronegative F atoms make CF_3CH_2OH (pK_a = 12.4) a somewhat stronger acid than CH_3CH_2OH.

Remember that neither resonance form adequately represents acetate. The true structure is a **hybrid** of both structures. In the hybrid, the electron pairs drawn in different locations in individual resonance structures are ***delocalized.*** With acetate, a dashed line is used to show that each C–O bond has partial double bond character. The symbol δ– (partial negative) indicates that the charge is delocalized on both O atoms in the hybrid.

Thus, **resonance delocalization makes $CH_3CO_2^-$ more stable than $CH_3CH_2O^-$, so CH_3CO_2H is a stronger acid than CH_3CH_2OH.** This is another example of a general rule.

- **The acidity of HA increases when the conjugate base A:⁻ is resonance stabilized.**

Electrostatic potential plots of $CH_3CH_2O^-$ and $CH_3CO_2^-$ in Figure 2.3 indicate that the negative charge is concentrated on a single O in $CH_3CH_2O^-$, but delocalized over the O atoms in $CH_3CO_2^-$.

Figure 2.3
Electrostatic potential plots of $CH_3CH_2O^-$ and $CH_3CO_2^-$

$CH_3CH_2O^-$
The negative charge is concentrated on the single oxygen atom, making this anion *less stable*.

$CH_3CO_2^-$
The negative charge is delocalized over both oxygen atoms, making this anion *more stable*.

Problem 2.22 The C–H bond in acetone, $(CH_3)_2C{=}O$, has a pK_a of 19.2. Draw two resonance structures for its conjugate base. Then, explain why acetone is much more acidic than propane, $CH_3CH_2CH_3$ (pK_a = 50).

Problem 2.23 Acetonitrile (CH_3CN) has a pK_a of 25, making it more acidic than many other compounds having only C–H bonds. Draw Lewis structures for acetonitrile and its conjugate base. Use resonance structures to account for the acidity of acetonitrile.

2.5D Hybridization Effects

The final factor affecting the acidity of HA is the hybridization of A. To illustrate this phenomenon, compare ethane (CH_3CH_3), ethylene ($CH_2{=}CH_2$), and acetylene ($HC{\equiv}CH$). Appendix A indicates that there is a considerable difference in the pK_a values of these compounds.

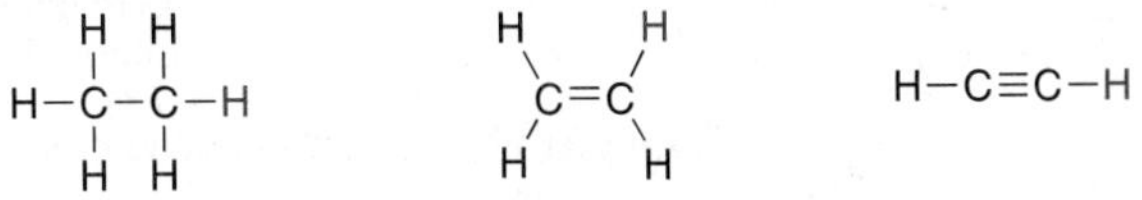

ethane pK_a = 50 | ethylene pK_a =44 | acetylene pK_a = 25

Increasing acidity →

The conjugate bases formed by removing a proton from ethane, ethylene, and acetylene are **carbanions—species with a negative charge on carbon.**

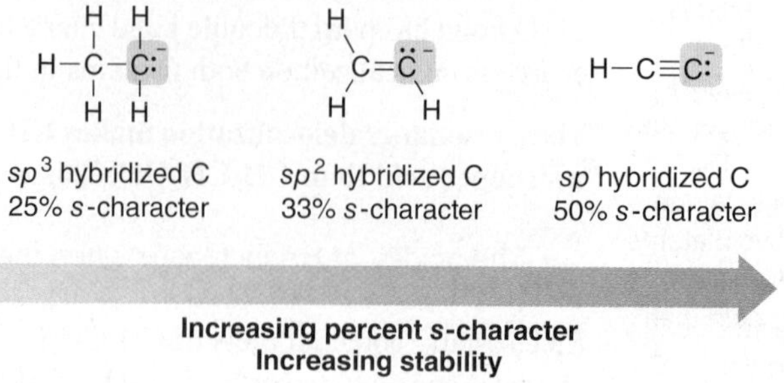

sp^3 hybridized C
25% s-character
sp^2 hybridized C
33% s-character
sp hybridized C
50% s-character
Increasing percent s-character
Increasing stability

The hybridization of the carbon bearing the negative charge is different in each anion, so the lone pair of electrons occupies an orbital with a different percent *s*-character in each case. A higher percent *s*-character means a hybrid orbital has a larger fraction of the lower energy *s* orbital.

- The *higher* the percent s-character of the hybrid orbital, the **more stable** the conjugate base.

Thus, **acidity increases from CH_3CH_3 to $CH_2=CH_2$ to $HC\equiv CH$ as the negative charge of the conjugate base is stabilized by increasing percent *s*-character.** Once again this is a specific example of a general trend.

- **The acidity of HA increases as the percent s-character of A:⁻ increases.**

Electrostatic potential plots of these carbanions appear in Figure 2.4.

Figure 2.4 Electrostatic potential plots of three carbanions

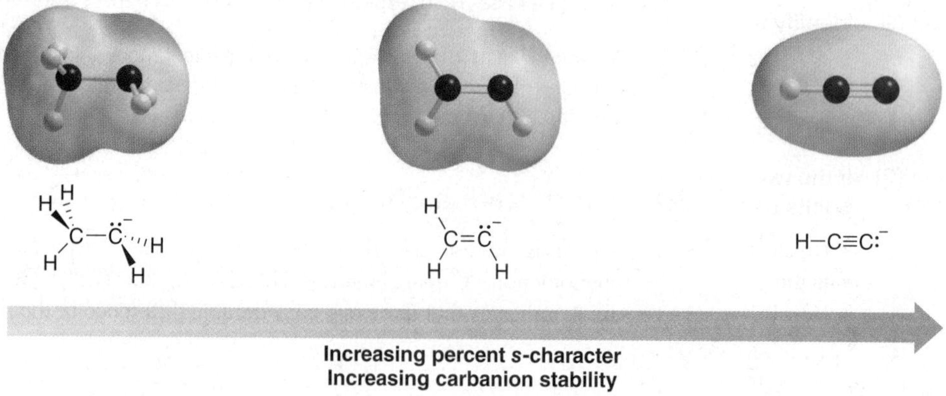

Increasing percent s-character
Increasing carbanion stability

- As the lone pair of electrons is pulled closer to the nucleus, the negatively charged carbon appears less intensely red.

Problem 2.24 For each pair of compounds: [1] Which indicated H is more acidic? [2] Draw the conjugate base of each acid. [3] Which conjugate base is stronger?

a. 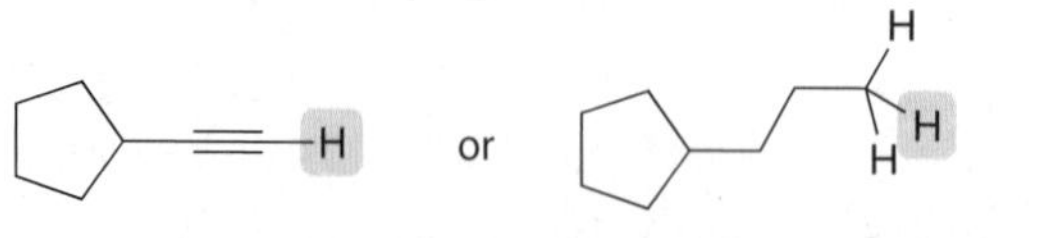

b. 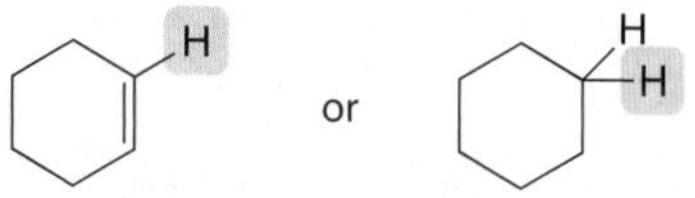

2.5E Summary of Factors Determining Acid Strength

The ability to recognize the most acidic site in a molecule will be important throughout the study of organic chemistry. All the factors that determine acidity are therefore summarized in Figure 2.5. The following two-step procedure shows how these four factors can be used to determine the relative acidity of protons.

Figure 2.5 Summary of the factors that determine acidity

Factor	Example		
1. **Element effects:** The acidity of HA increases both left-to-right across a row and down a column of the periodic table.	CH_4	and	H_2O **more acidic**
2. **Inductive effects:** The acidity of HA increases with the presence of electron-withdrawing groups in A.	$CH_3CH_2O{-}H$	and	$CF_3CH_2O{-}H$ **more acidic**
3. **Resonance effects:** The acidity of HA increases when the conjugate base A:⁻ is resonance stabilized.	$CH_3CH_2O{-}H$	and	$CH_3CO_2{-}H$ **more acidic**
4. **Hybridization effects:** The acidity of HA increases as the percent *s*-character of A:⁻ increases.	$CH_2{=}CH_2$	and	$H{-}C{\equiv}C{-}H$ **more acidic**

How To Determine the Relative Acidity of Protons

Step [1] Identify the atoms bonded to hydrogen, and use periodic trends to assign relative acidity.

- The most common HA bonds in organic compounds are C – H, N – H, and O – H. Because acidity increases left-to-right across a row, the relative acidity of these bonds is C – H < N – H < O – H. Therefore, H atoms bonded to C atoms are usually *less acidic* than H atoms bonded to any heteroatom.

Step [2] If the two H atoms in question are bonded to the same element, draw the conjugate bases and look for other points of difference. Ask three questions:

- Do electron-withdrawing groups stabilize the conjugate base?
- Is the conjugate base resonance stabilized?
- How is the conjugate base hybridized?

Sample Problem 2.7 shows how to apply this procedure to actual compounds.

Sample Problem 2.7 Rank the following compounds in order of increasing acidity of their most acidic hydrogen atom.

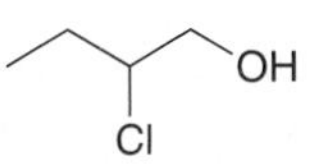

A

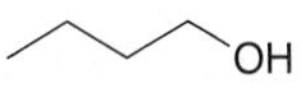

B

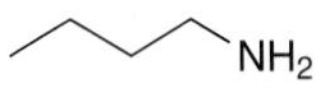

C

Solution

[1] Compounds **A, B,** and **C** contain C–H, N–H, and O–H bonds. Because acidity increases left-to-right across a row of the periodic table, the O–H bonds are most acidic. Compound **C** is thus the least acidic because it has *no* O–H bonds.

[2] The only difference between compounds **A** and **B** is the presence of an electronegative Cl in **A.** The Cl atom stabilizes the conjugate base of **A,** making it more acidic than **B.** Thus,

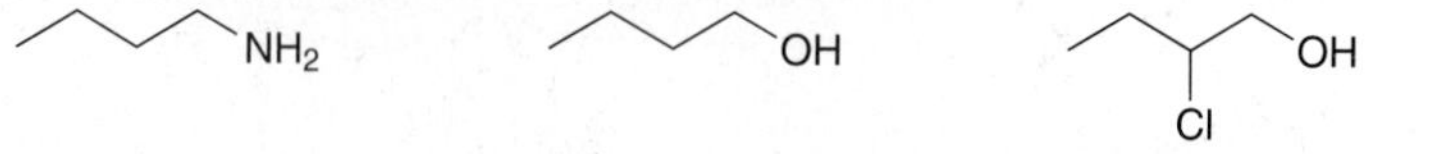

C **B** **A**

Increasing acidity →

Problem 2.25 Rank the compounds in each group in order of increasing acidity.

a. 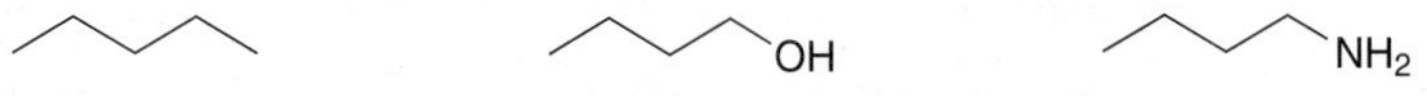

b. 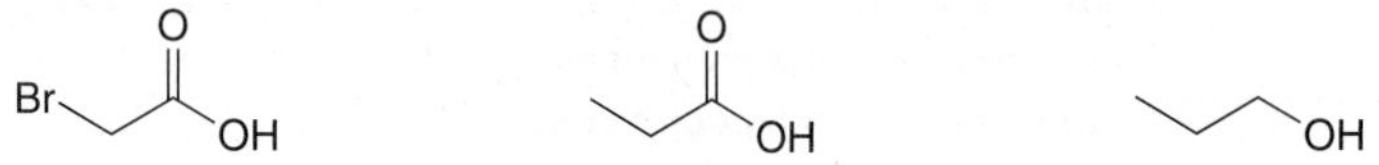

c. 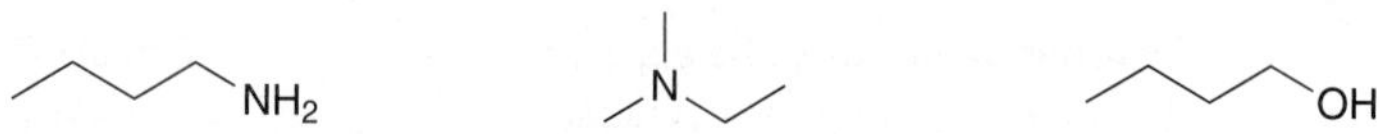

Problem 2.26 Which proton in each of the following drugs is most acidic? THC is the active component in marijuana, and ketoprofen is an anti-inflammatory agent.

a. 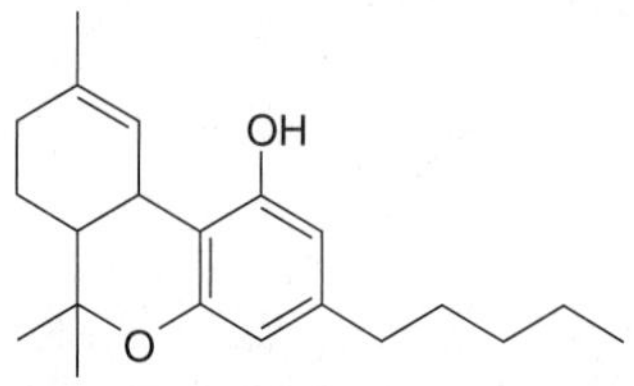

THC
tetrahydrocannabinol

b. 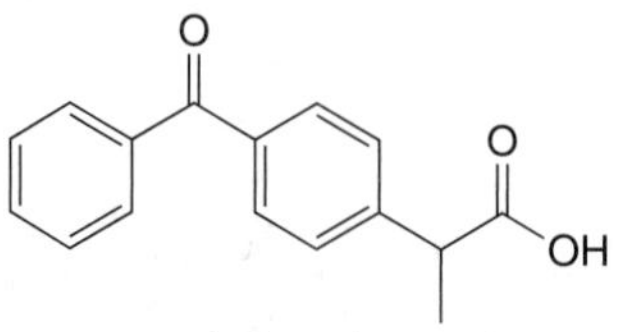

ketoprofen

Problem 2.27 Which anion (**A** or **B**) is the stronger base?

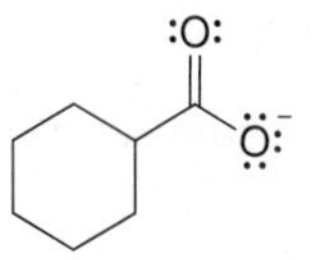

A

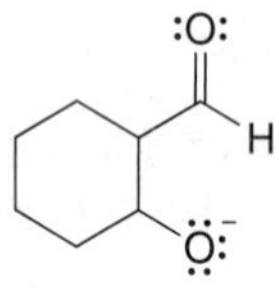

B

2.6 Common Acids and Bases

Many strong or moderately strong acids and bases are used as reagents in organic reactions.

2.6A Common Acids

Several organic reactions are carried out in the presence of strong inorganic acids, most commonly **HCl** and **H_2SO_4.** These strong acids, with **pK_a values ≤ 0,** should be familiar from previous chemistry courses.

Sulfuric acid is the most widely produced industrial chemical. It is also formed when sulfur oxides, emitted into the atmosphere by burning fossil fuels high in sulfur content, dissolve in water. This makes rainwater acidic, forming acid rain, which has destroyed acres of forests worldwide.

Two organic acids are also commonly used, namely **acetic acid** and ***p*-toluenesulfonic acid** (usually abbreviated as **TsOH**). Although acetic acid has a higher pK_a than the inorganic acids, making it a weaker acid, it is more acidic than most organic compounds. *p*-Toluenesulfonic acid is similar in acidity to the strong inorganic acids. Because it is a solid, small quantities can be easily weighed on a balance and then added to a reaction mixture.

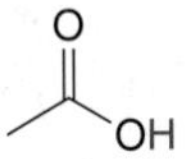

acetic acid
$pK_a = 4.8$

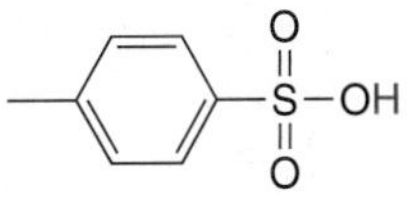

= TsOH

p-toluenesulfonic acid
$pK_a = -7$

2.6B Common Bases

Three common kinds of strong bases include:

[1] Negatively charged oxygen bases: **^-OH** (hydroxide) and its organic derivatives
[2] Negatively charged nitrogen bases: **$^-NH_2$** (amide) and its organic derivatives
[3] Hydride **(H^-)**

Figure 2.6 gives examples of these strong bases. Each negatively charged base is used as a salt with a spectator ion (usually Li^+, Na^+, or K^+) that serves to balance charge.

- **Strong bases have weak conjugate acids with high pK_a values, usually > 12.**

Figure 2.6
Some common negatively charged bases

Oxygen bases		Nitrogen bases	
$Na^+\ {}^-\!:\ddot{O}H$	sodium hydroxide	$Na^+\ {}^-\!:\ddot{N}H_2$	sodium amide
$Na^+\ {}^-\!:\ddot{O}CH_3$	sodium methoxide	$Li^+\ {}^-\!:\ddot{N}[CH(CH_3)_2]_2$	lithium diisopropylamide
$Na^+\ {}^-\!:\ddot{O}CH_2CH_3$	sodium ethoxide		
$K^+\ {}^-\!:\ddot{O}C(CH_3)_3$	potassium *tert*-butoxide	Hydride	
		$Na^+\ H\!:^-$	sodium hydride

Strong bases have a net negative charge, but not all negatively charged species are strong bases. For example, none of the halides, F^-, Cl^-, Br^-, or I^-, is a strong base. These anions have very strong conjugate acids and have little affinity for donating their electron pairs to a proton.

Carbanions, negatively charged carbon atoms discussed in Section 2.5D, are especially strong bases. Perhaps the most common example is **butyllithium.** Butyllithium and related compounds are discussed in greater detail in Chapter 20.

$$CH_3CH_2CH_2CH_2^-\ Li^+$$

butyllithium

Two other weaker organic bases are **triethylamine** and **pyridine.** These compounds have a lone pair on nitrogen, making them basic, but they are considerably weaker than the amide bases because they are neutral, not negatively charged.

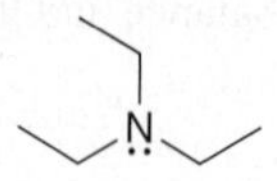

triethylamine

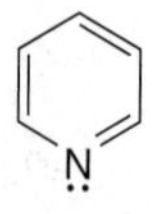

pyridine

Problem 2.28 Draw the products formed when propan-2-ol [$(CH_3)_2CHOH$], the main ingredient in rubbing alcohol, is treated with each acid or base: (a) NaH; (b) H_2SO_4; (c) $Li^{+-}N[CH(CH_3)_2]_2$; (d) CH_3CO_2H.

2.7 Aspirin

Aspirin, or acetylsalicylic acid, is the most well known member of a group of compounds called **salicylates.** Although aspirin was first used in medicine for its analgesic (pain-relieving), anti-pyretic (fever-reducing), and anti-inflammatory properties, today it is commonly used as an anti-platelet agent in the treatment and prevention of heart attacks and strokes. **Aspirin is a synthetic compound;** it does not occur in nature, although some related salicylates are found in willow bark and meadowsweet blossoms (Figure 2.7).

Figure 2.7
Salicin, an analgesic in willow bark

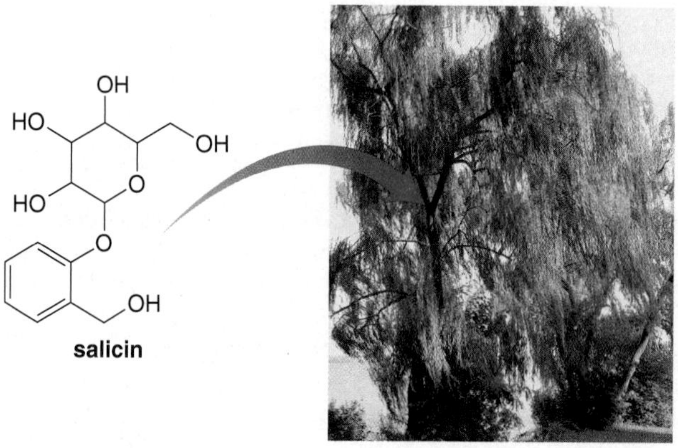

willow tree

- The modern history of aspirin dates back to 1763 when Reverend Edmund Stone reported on the analgesic effect of chewing on the bark of the willow tree. Willow bark is now known to contain *salicin,* which is structurally related to aspirin.

Like many drugs, aspirin undergoes a proton transfer reaction. Its most acidic proton is the H bonded to O, and in the presence of base, this H is readily removed.

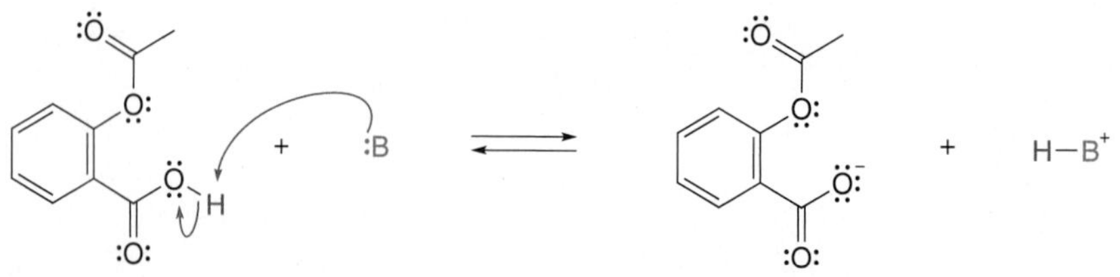

aspirin
acetylsalicylic acid
neutral form
This form exists in the **stomach.**

conjugate base
ionic form
This form exists in the **intestines.**

Why is this acid–base reaction important? After ingestion, aspirin first travels into the stomach and then the intestines. In the acidic environment of the stomach, aspirin remains in its neutral form, but in the basic environment of the small intestine, aspirin is deprotonated to form its conjugate base, an ion. Likewise, in the slightly basic environment of the blood, aspirin exists primarily as its ionic conjugate base.

We will learn more about solubility and the cell membrane in Section 3.7.

Whether aspirin is a neutral acid or an ionic conjugate base affects its transport throughout the body and its ability to pass through a cell membrane. In its ionic form, aspirin is readily soluble in the aqueous environment of the blood, so it is transported in the bloodstream to tissues. Once aspirin has reached its target location, however, its conjugate base must be re-protonated to form the neutral acid that can pass through the nonpolar interior of a cell membrane where it inhibits prostaglandin synthesis, as we will learn in Chapter 19. Thus, in the body, aspirin undergoes acid–base reactions and these reactions are crucial in determining its properties and action.

Problem 2.29

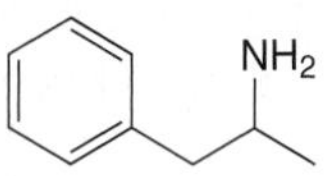

amphetamine

Compounds like amphetamine that contain nitrogen atoms are protonated by the HCl in the gastric juices of the stomach, and the resulting salt is then deprotonated in the basic environment of the intestines to regenerate the neutral form. Write proton transfer reactions for both of these processes. In which form will amphetamine pass through a cell membrane?

2.8 Lewis Acids and Bases

The Lewis definition of acids and bases is more general than the Brønsted–Lowry definition.

- A Lewis acid is an *electron pair acceptor.*
- A Lewis base is an *electron pair donor.*

All Brønsted–Lowry bases are Lewis bases.

Lewis bases are structurally the same as Brønsted–Lowry bases. Both have an **available electron pair**—a lone pair or an electron pair in a π bond. A Brønsted–Lowry base always donates this electron pair to a proton, but a Lewis base donates this electron pair to anything that is electron deficient. Simple Lewis bases are shown in Figure 2.8.

A Lewis acid must be able to accept an electron pair, but there are many ways for this to occur. **All Brønsted–Lowry acids are also Lewis acids, but the reverse is not necessarily true.** Any species that is electron deficient and capable of accepting an electron pair is also a Lewis acid, as shown in Figure 2.8.

Common examples of Lewis acids (which are not Brønsted–Lowry acids) include BF_3 and $AlCl_3$. These compounds contain elements in group 3A of the periodic table that can accept an electron pair because they do not have filled valence shells of electrons.

Figure 2.8
Simple Lewis acids and bases

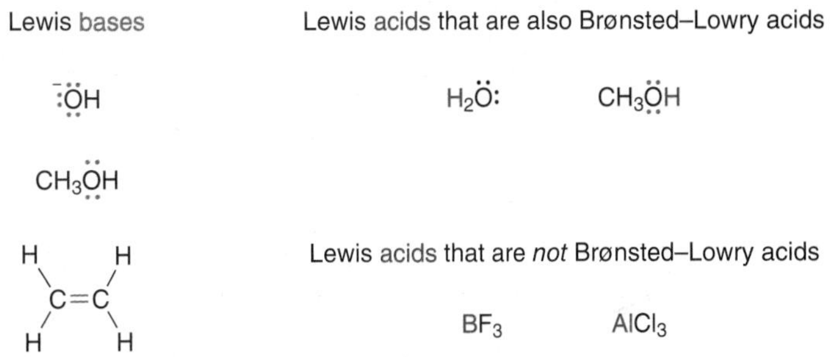

Problem 2.30 Which species are Lewis bases?

a. NH_3 b. $CH_3CH_2CH_3$ c. H^- d. $H-C\equiv C-H$

Problem 2.31 Which species are Lewis acids?

a. BBr_3 b. CH_3CH_2OH c. $(CH_3)_3C^+$ d. Br^-

Any reaction in which one species donates an electron pair to another species is a Lewis acid–base reaction.

In a Lewis acid–base reaction, a Lewis base donates an electron pair to a Lewis acid. Most reactions in organic chemistry involving movement of electron pairs can be classified as Lewis acid–base reactions. Lewis acid–base reactions illustrate a general pattern of reactivity.

- **Electron-rich species react with electron-poor species.**

In the simplest Lewis acid–base reaction one bond is formed and no bonds are broken. This is illustrated with the reaction of BF_3 with H_2O. BF_3 has only six electrons around B, so it is the electron-deficient Lewis acid. H_2O has two lone pairs on O, so it is the electron-rich Lewis base.

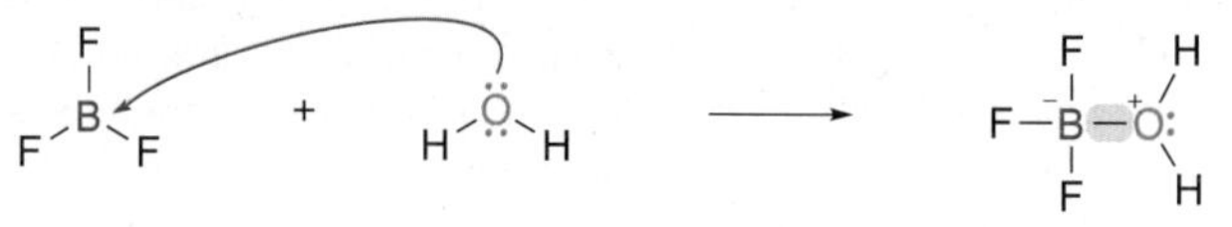

H_2O donates an electron pair to BF_3 to form one new bond. The electron pair in the new B–O bond comes from the oxygen atom, and a single product, a Lewis acid–base complex, is formed. Both B and O bear formal charges in the product, but the overall product is neutral.

Nucleophile = nucleus loving.
Electrophile = electron loving.

- **A Lewis acid is called an *electrophile.***
- **When a Lewis base reacts with an electrophile other than a proton, the Lewis base is called a *nucleophile.***

In this Lewis acid–base reaction, **BF_3 is the electrophile** and **H_2O is the nucleophile.**

Two other examples are drawn. In each reaction the **electron pair is not removed from the Lewis base;** instead, the electron pair is donated to an atom of the Lewis acid, and one new covalent bond is formed.

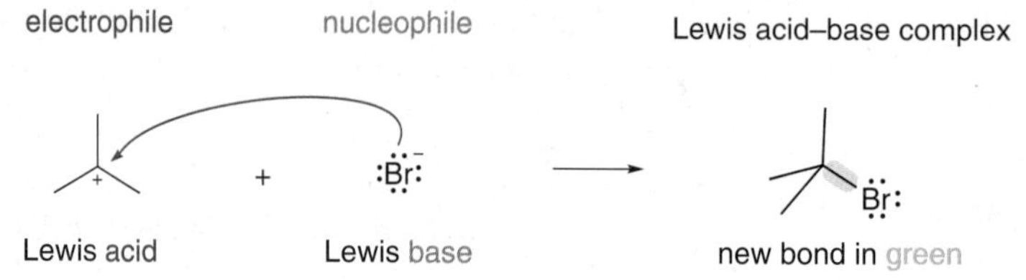

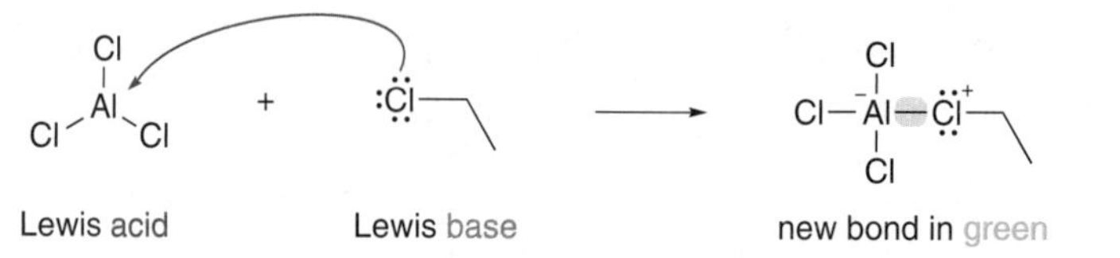

Problem 2.32 For each reaction, label the Lewis acid and base. Use curved arrow notation to show the movement of electron pairs.

a. 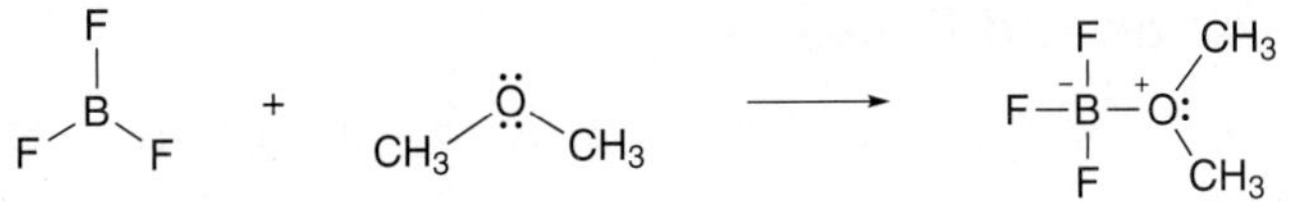

b. 

Problem 2.33 Draw the products of each reaction, and label the nucleophile and electrophile.

a. 

b. 

Problem 2.34 Draw the product formed when $(CH_3CH_2)_3N$:, a Lewis base, reacts with each Lewis acid: (a) $B(CH_3)_3$; (b) $(CH_3)_3C^+$; (c) $AlCl_3$.

In some Lewis acid–base reactions, one bond is formed and one bond is broken. To draw the products of these reactions, keep the following steps in mind.

> **[1] Always identify the Lewis acid and base first.**
> **[2] Draw a curved arrow from the electron pair of the base to the electron-deficient atom of the acid.**
> **[3] Count electron pairs and break a bond when needed to keep the correct number of valence electrons.**

Recall from Chapter 1 that a positively charged carbon atom is called a **carbocation.**

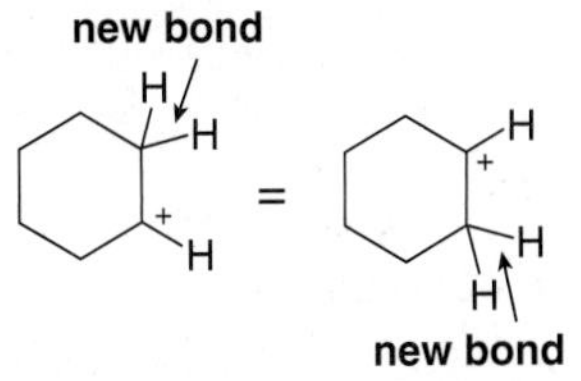

In the reaction of cyclohexene with HCl, the new bond to H could form at **either carbon of the double bond,** because the same carbocation results.

For example, draw the Lewis acid–base reaction between cyclohexene and H−Cl. The Brønsted–Lowry acid HCl is also a Lewis acid, and cyclohexene, having a π bond, is the Lewis base.

cyclohexene
Lewis base

H−Cl
δ+ δ−

Lewis acid

To draw the product of this reaction, the electron pair in the π bond of the Lewis base forms a new bond to the proton of the Lewis acid, forming a carbocation. The H−Cl bond must break, giving its two electrons to Cl, forming Cl^-. Because two electron pairs are involved, two curved arrows are needed.

Lewis base | Lewis acid | carbocation new bond in green

The Lewis acid–base reaction of cyclohexene with HCl is a specific example of a fundamental reaction of compounds containing C−C double bonds, as discussed in Chapter 10.

Problem 2.35 Label the Lewis acid and base. Use curved arrow notation to show the movement of electron pairs.

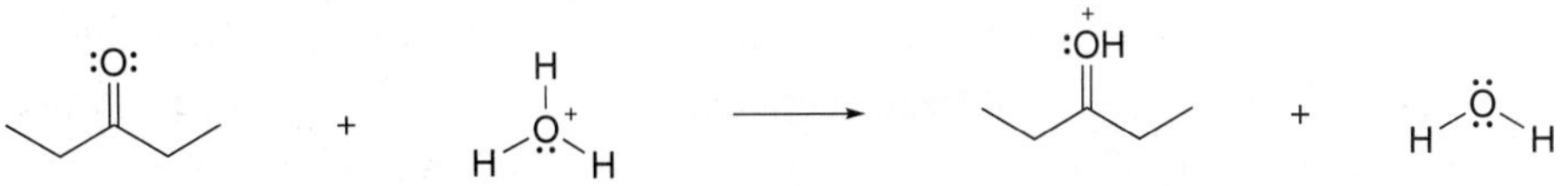

KEY CONCEPTS

Acids and Bases

A Comparison of Brønsted–Lowry and Lewis Acids and Bases

Type	Definition	Structural feature	Examples
Brønsted–Lowry acid (2.1)	proton donor	a proton	HCl, H_2SO_4, H_2O, CH_3CO_2H, TsOH
Brønsted–Lowry base (2.1)	proton acceptor	a lone pair *or* a π bond	^-OH, $^-OCH_3$, H^-, $^-NH_2$, NH_3, $CH_2{=}CH_2$
Lewis acid (2.8)	electron pair acceptor	a proton, *or* an unfilled valence shell, *or* a partial (+) charge	BF_3, $AlCl_3$, HCl, CH_3CO_2H, H_2O
Lewis base (2.8)	electron pair donor	a lone pair *or* a π bond	^-OH, $^-OCH_3$, H^-, $^-NH_2$, NH_3, $CH_2{=}CH_2$

Acid–Base Reactions

[1] A Brønsted–Lowry acid donates a proton to a Brønsted–Lowry base (2.2).

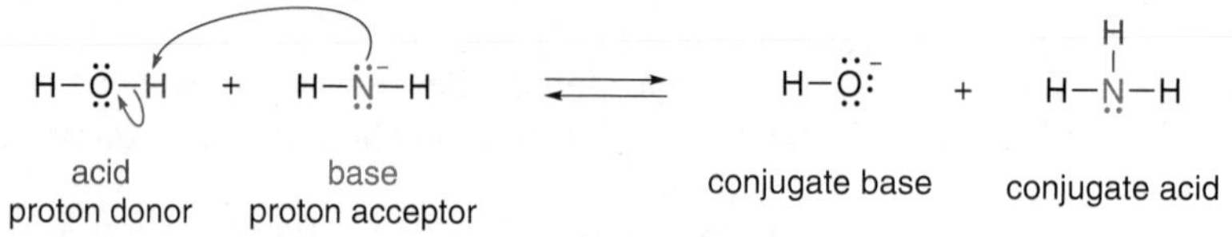

[2] A Lewis base donates an electron pair to a Lewis acid (2.8).

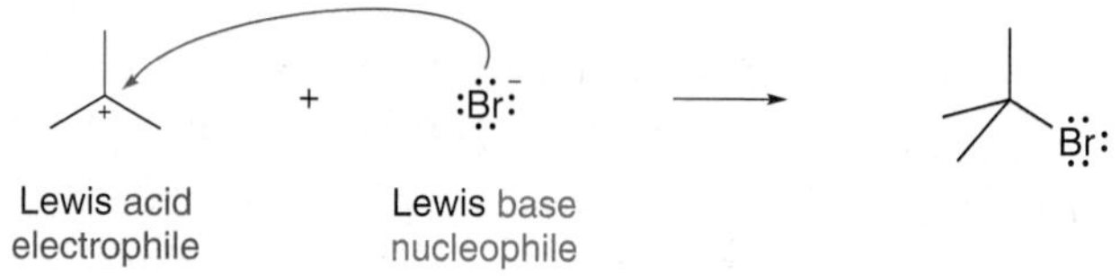

- Electron-rich species react with electron-poor ones.
- Nucleophiles react with electrophiles.

Important Facts

- Definition: $\mathbf{pK_a = -log\ K_a}$. **The lower the pK_a, the stronger the acid** (2.3).
- The stronger the acid, the weaker the conjugate base (2.3).
- In proton transfer reactions, equilibrium favors the weaker acid and weaker base (2.4).
- An acid can be deprotonated by the conjugate base of any acid having a higher pK_a (2.4).

Periodic Trends in Acidity and Basicity (2.5A)

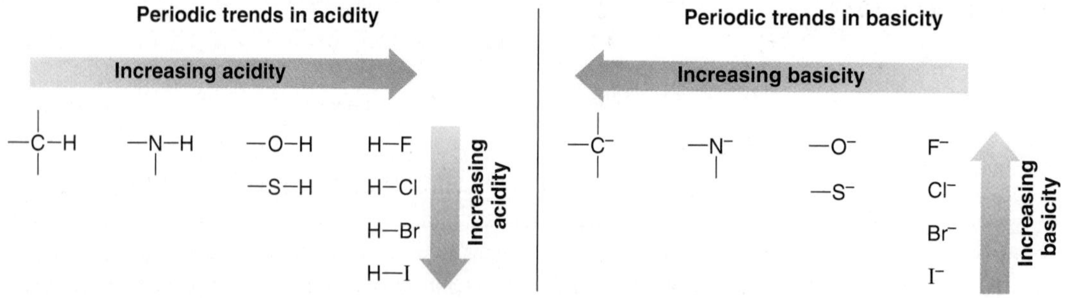

Factors That Determine Acidity (2.5)

[1] **Element effects** (2.5A)	The acidity of HA increases both left-to-right across a row and down a column of the periodic table.
[2] **Inductive effects** (2.5B)	The acidity of HA increases with the presence of electron-withdrawing groups in A.
[3] **Resonance effects** (2.5C)	The acidity of HA increases when the conjugate base A:$^-$ is resonance stabilized.
[4] **Hybridization effects** (2.5D)	The acidity of HA increases as the percent *s*-character of A:$^-$ increases.

PROBLEMS

Problems Using Three-Dimensional Models

2.36 Propranolol is an antihypertensive agent—that is, it lowers blood pressure. (a) Which proton in propranolol is most acidic? (b) What products are formed when propranolol is treated with NaH? (c) Which atom is most basic? (d) What products are formed when propranolol is treated with HCl?

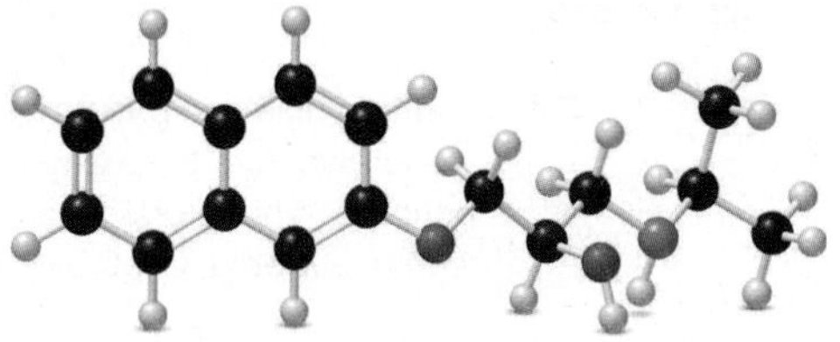

propranolol

2.37 Amphetamine is a powerful stimulant of the central nervous system. (a) Which proton in amphetamine is most acidic? (b) What products are formed when amphetamine is treated with NaH? (c) What products are formed when amphetamine is treated with HCl?

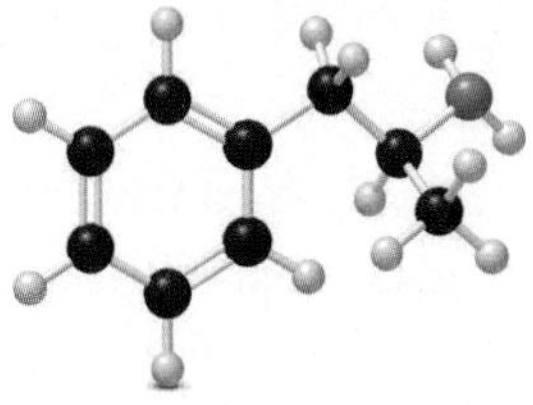

amphetamine

Brønsted–Lowry Acids and Bases

2.38 What is the conjugate acid of each base?

a. HCO_3^- b. 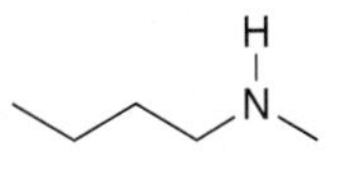 c. 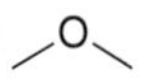 d. 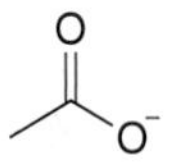

2.39 What is the conjugate base of each acid?

a. HCO_3^- b. 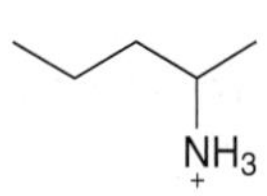 c. 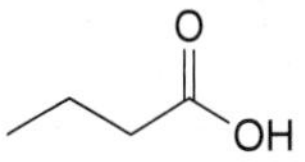 d. 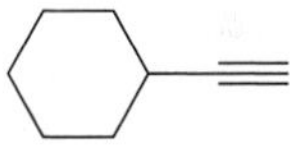

Reactions of Brønsted–Lowry Acids and Bases

2.40 Draw the products formed from the acid–base reaction of H_2SO_4 with each compound.

a. 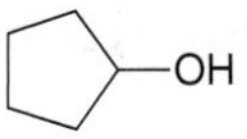 b. 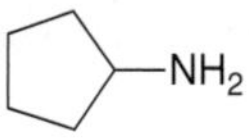 c. 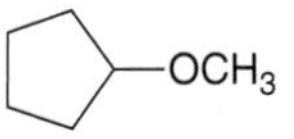 d. 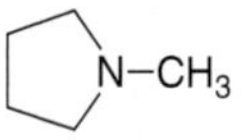

2.41 Draw the products formed from the acid–base reaction of KOH with each compound.

a. 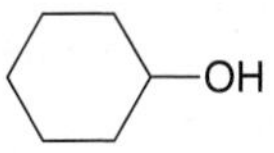 b. 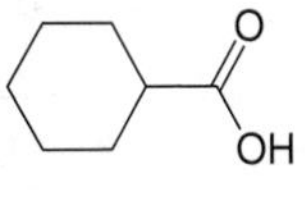 c. 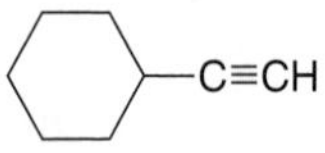 d. 

2.42 Draw the products of each proton transfer reaction. Label the acid and base in the starting materials, and the conjugate acid and base in the products.

a. 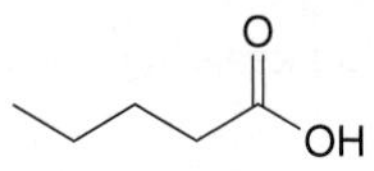 + $CH_3\ddot{O}:^-$ ⇌

b. 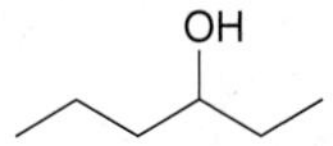 + HBr ⇌

c. 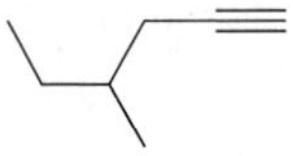 + $NaNH_2$ ⇌

d. 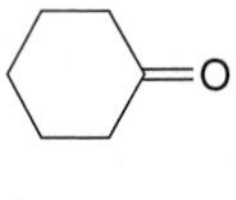 + H_2SO_4 ⇌

2.43 Draw the products of each acid–base reaction.

a. 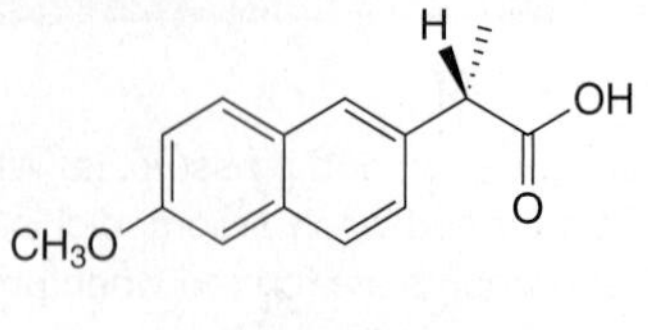 + NaOH $\rightleftharpoons$

naproxen
anti-inflammatory agent

b. 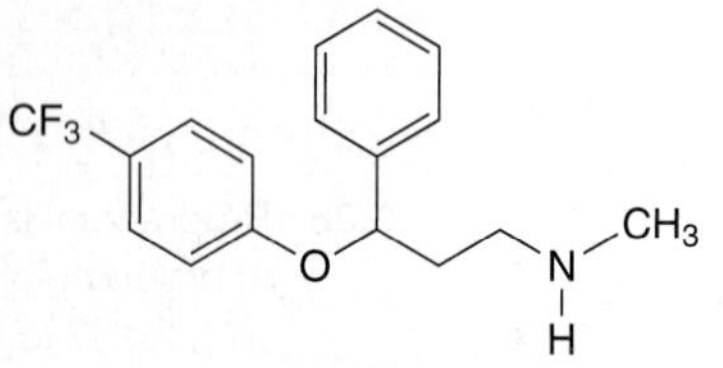 + HCl $\rightleftharpoons$

fluoxetine
antidepressant

2.44 Fenfluramine and phentermine are two components of fen–phen, an appetite suppressant withdrawn from the market in 1997 after it was shown to damage the heart valves in some patients. What products are formed when fenfluramine and phentermine are each treated with acetic acid (CH_3CO_2H)?

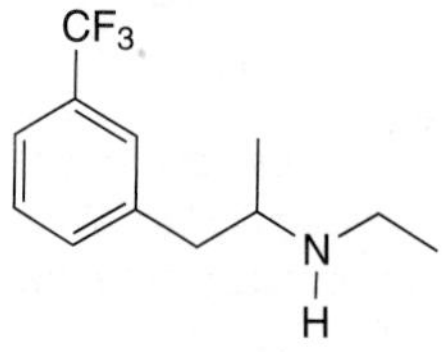

fenfluramine

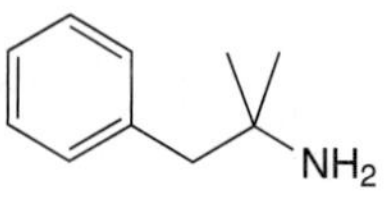

phentermine

pK_a, K_a, and the Direction of Equilibrium

2.45 What is K_a for each compound? Use a calculator when necessary.

a. H_2S pK_a = 7.0

b. $ClCH_2COOH$ pK_a = 2.8

c. HCN pK_a = 9.1

2.46 What is the pK_a for each compound?

a. 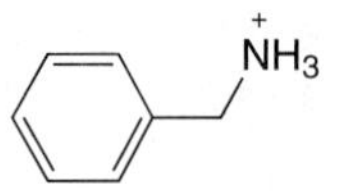 $K_a = 4.7 \times 10^{-10}$

b. 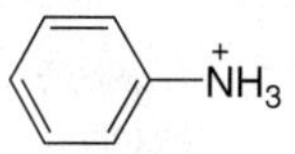 $K_a = 2.3 \times 10^{-5}$

c. 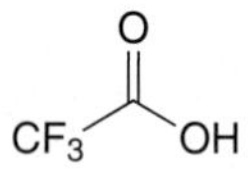 $K_a = 5.9 \times 10^{-1}$

2.47 Which of the following bases are strong enough to deprotonate $CH_3CH_2CH_2C{\equiv}CH$ (pK_a = 25), so that equilibrium favors the products: (a) H_2O; (b) NaOH; (c) $NaNH_2$; (d) NH_3; (e) NaH; (f) CH_3Li?

2.48 Which compounds can be deprotonated by ^{-}OH, so that equilibrium favors the products? Refer to the pK_a table in Appendix A.

a. HCOOH b. H_2S c. 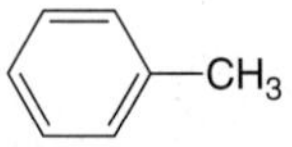 d. CH_3NH_2

2.49 Draw the products of each reaction. Use the pK_a table in Appendix A to decide if the equilibrium favors the starting materials or products.

a. CH_3NH_2 + H_2SO_4 $\rightleftharpoons$

b. 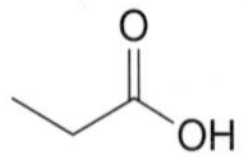 + NaCl $\rightleftharpoons$

c. 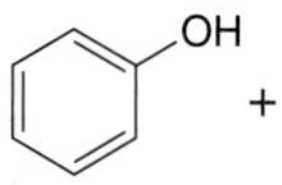 + $NaHCO_3$ $\rightleftharpoons$

d. H–C≡C–H + $CH_3CH_2^{-}\ Li^{+}$ $\rightleftharpoons$

Relative Acid Strength

2.50 Rank the following compounds in order of increasing acidity.

a. 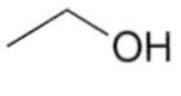 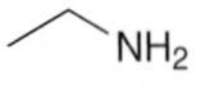 

b. 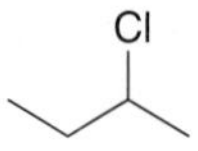 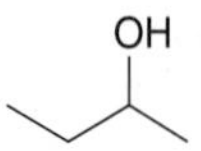 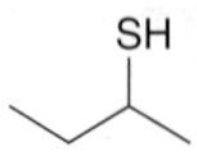

c.  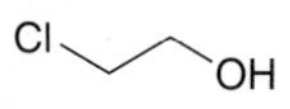 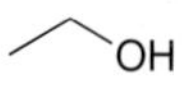

d. 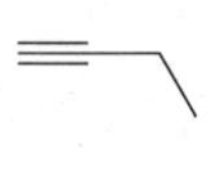  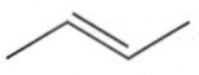

2.51 Rank the following ions in order of increasing basicity.

a. $CH_3\bar{C}H_2$, $CH_3\bar{O}$, $CH_3\bar{N}H$

b. CH_3^-, HO^-, Br^-

c. 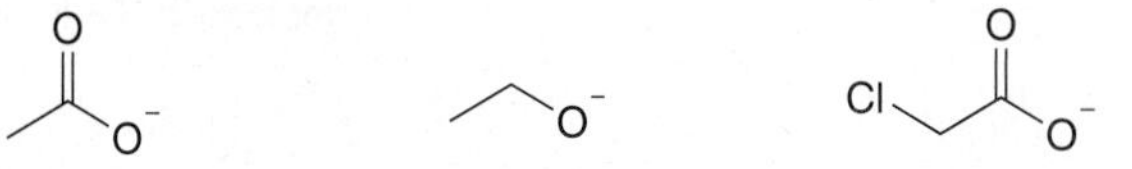

d. 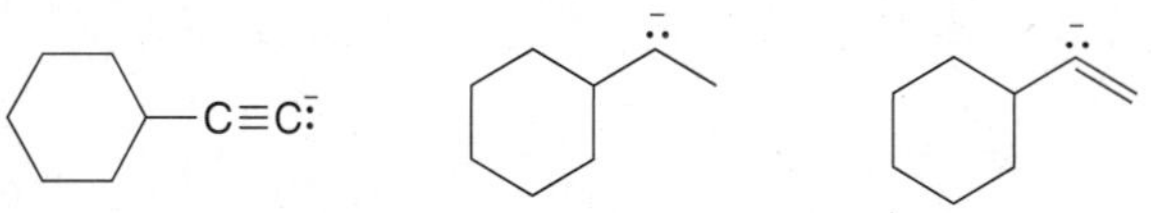

2.52 The pK_a values of the two ammonium cations drawn below are 8.33 and 11.1. Which pK_a corresponds to which cation? Explain your choice.

2.53 Explain why the C – H_a bond is much more acidic than the C – H_b bond in pentan-2-one.

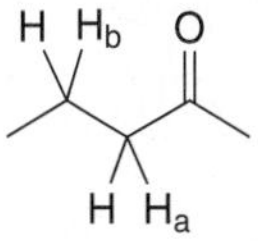

pentan-2-one

2.54 The pK_a of three C – H bonds is given below.

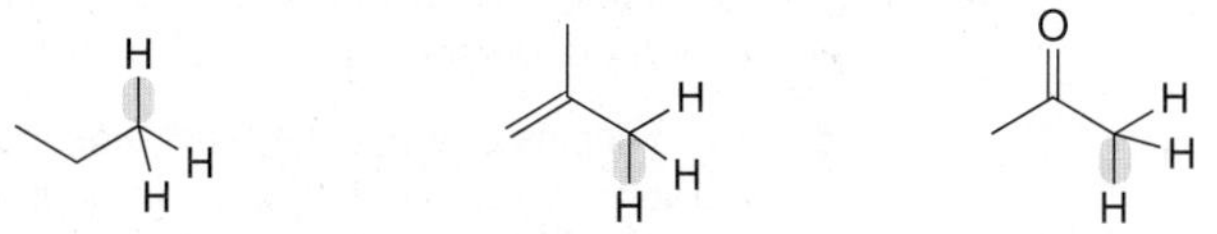

pK_a = 50 pK_a = 43 pK_a = 19.2

a. For each compound, draw the conjugate base, including all possible resonance structures.

b. Explain the observed trend in pK_a.

2.55 a. What is the conjugate acid of **A?**

b. What is the conjugate base of **A?**

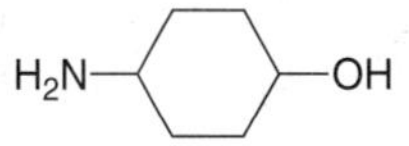

A

2.56 Draw the structure of a constitutional isomer of compound **B** that fits each description.

a. an isomer that is at least 10^5 times more acidic than **B**

b. an isomer that is at least 10^5 times less acidic than **B**

c. an isomer that is comparable in acidity to **B**

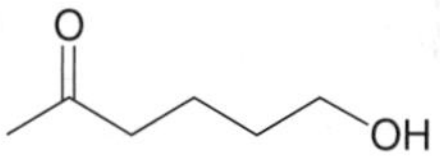

B

2.57 Many drugs are Brønsted–Lowry acids or bases.

a. What is the most acidic proton in the analgesic ibuprofen? Draw the conjugate base.

b. What is the most basic electron pair in cocaine? Draw the conjugate acid.

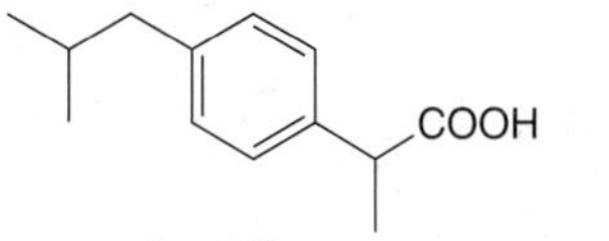

ibuprofen

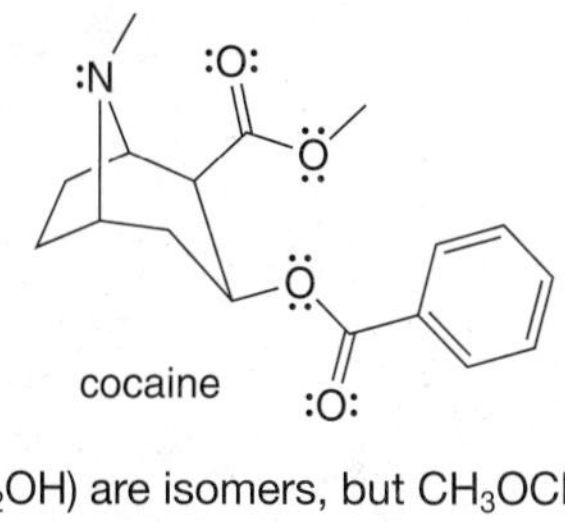

cocaine

2.58 Dimethyl ether (CH_3OCH_3) and ethanol (CH_3CH_2OH) are isomers, but CH_3OCH_3 has a pK_a of 40 and CH_3CH_2OH has a pK_a of 16. Why are these pK_a values so different?

2.59 Atenolol is a β (beta) blocker, a drug used to treat high blood pressure. Which of the indicated N–H bonds is more acidic? Explain your reasoning.

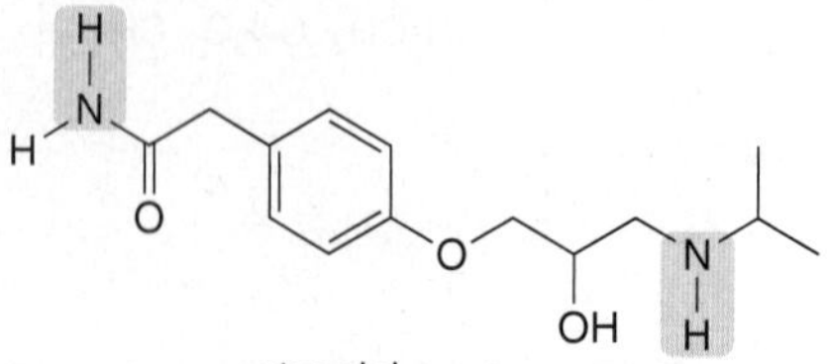

atenolol

2.60 Use the principles in Section 2.5 to label the most acidic hydrogen in each drug. Explain your choice.

a. 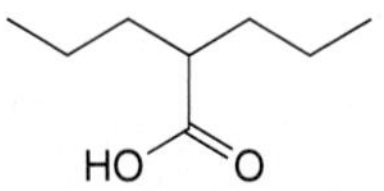

valproic acid
(used to treat epilepsy)

b. 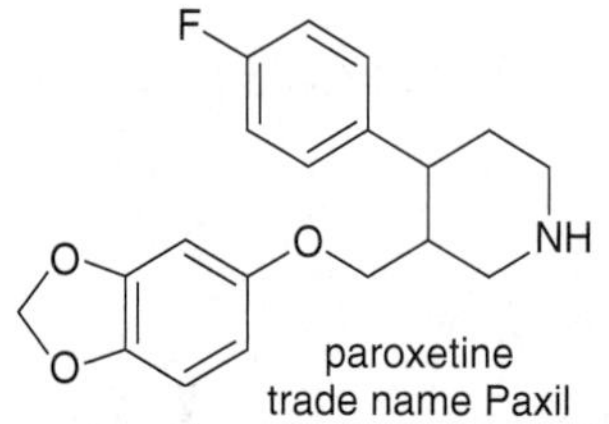

paroxetine
trade name Paxil
(used to treat depression)

c. 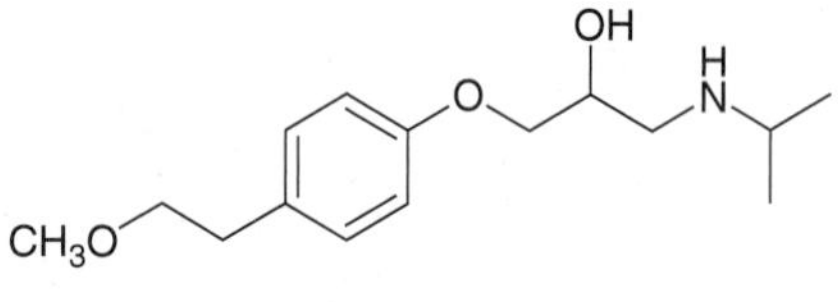

metoprolol
(used to treat high blood pressure)

2.61 Label the three most acidic hydrogen atoms in lactic acid, $CH_3CH(OH)CO_2H$, and rank them in order of decreasing acidity. Explain your reasoning.

2.62 Bupivacaine (trade name Marcaine) is a quick-acting anesthetic often used during labor and delivery. Which nitrogen atom in bupivacaine is more basic? Explain your reasoning.

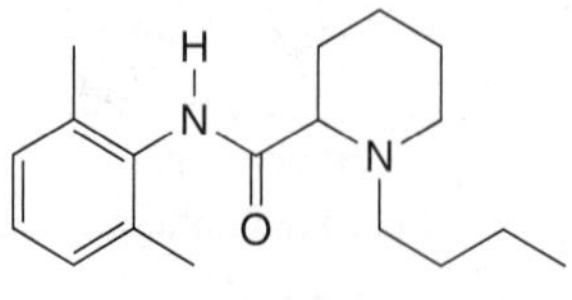

bupivacaine

Lewis Acids and Bases

2.63 Classify each compound as a Lewis base, a Brønsted–Lowry base, both, or neither.

a. 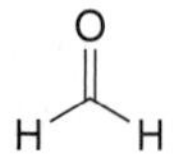 b.  c.  d. 

2.64 Classify each species as a Lewis acid, a Brønsted–Lowry acid, both, or neither.

a. H_3O^+ b. Cl_3C^+ c. BCl_3 d. BF_4^-

Lewis Acid–Base Reactions

2.65 Label the Lewis acid and Lewis base in each reaction. Use curved arrows to show the movement of electron pairs.

a. 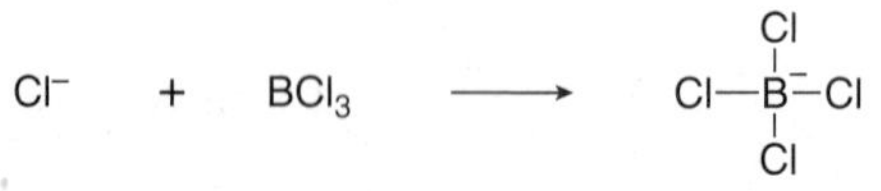

b. 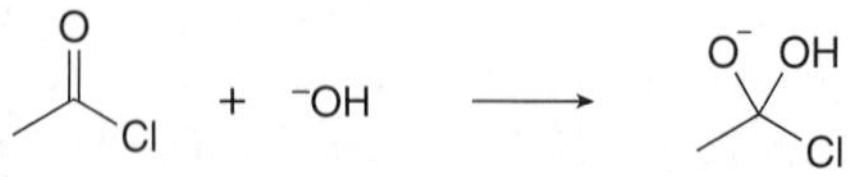

2.66 Draw the products of each Lewis acid–base reaction. Label the electrophile and nucleophile.

a. 

c. 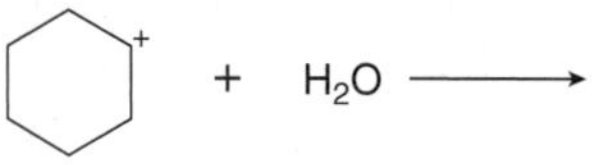

b. 

2.67 Draw the product formed when the Lewis acid $(CH_3CH_2)_3C^+$ reacts with each Lewis base: (a) CH_3OH; (b) $(CH_3)_2O$; (c) $(CH_3)_2NH$.

General Problems

2.68 Answer the following questions about the four species **A–D**.

A **B** **C** **D**

a. Which two species represent a conjugate acid–base pair?

b. Which two species represent resonance structures?

c. Which two species represent constitutional isomers?

2.69 Classify each reaction as either a proton transfer reaction, or a reaction of a nucleophile with an electrophile. Use curved arrows to show how the electron pairs move.

a. 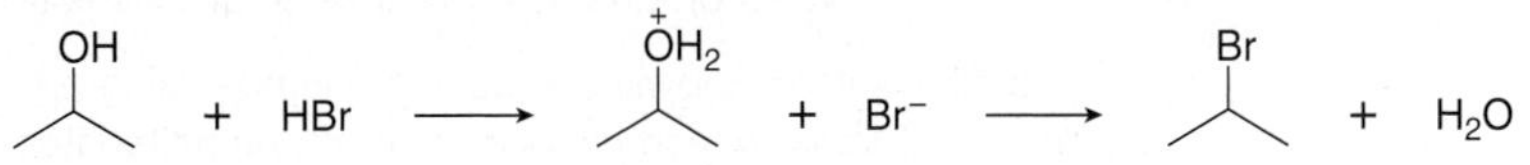

b. 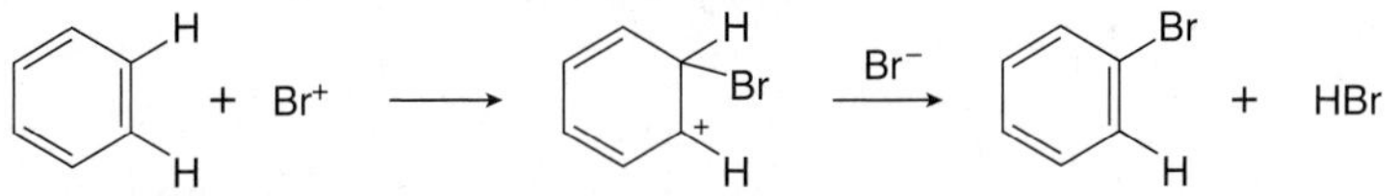

2.70 Hydroxide (⁻OH) can react as a Brønsted–Lowry base (and remove a proton), or a Lewis base (and attack a carbon atom). (a) What organic product is formed when ⁻OH reacts with the carbocation $(CH_3)_3C^+$ as a Brønsted–Lowry base? (b) What organic product is formed when ⁻OH reacts with $(CH_3)_3C^+$ as a Lewis base?

2.71 Answer the following questions about esmolol, a drug used to treat high blood pressure sold under the trade name Brevibloc.

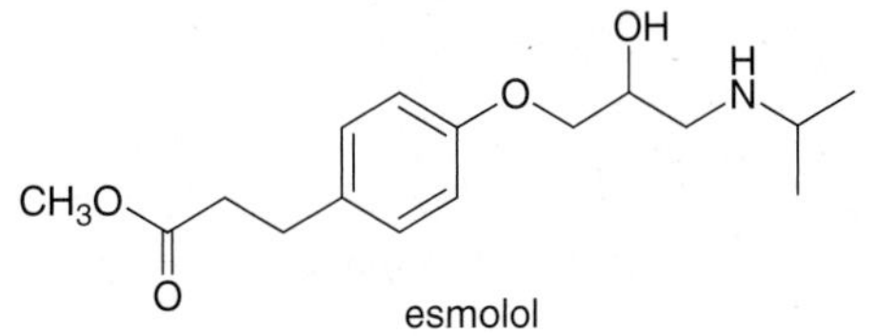

esmolol

a. Label the most acidic hydrogen atom in esmolol.

b. What products are formed when esmolol is treated with NaH?

c. What products are formed when esmolol is treated with HCl?

d. Label all sp^2 hybridized C atoms.

e. Label the only trigonal pyramidal atom.

f. Label all C's that bear a δ+ charge.

Challenge Problems

2.72 DBU, 1,8-diazabicyclo[5.4.0]undec-7-ene, is a base we will encounter in elimination reactions in Chapter 8. Which N atom is more basic in DBU? Explain your choice.

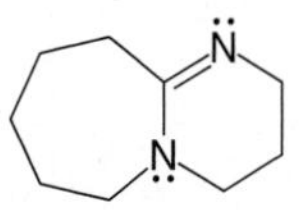

DBU

2.73 Molecules like acetamide (CH_3CONH_2) can be protonated on either their O or N atoms when treated with a strong acid like HCl. Which site is more readily protonated and why?

2.74 Two pK_a values are reported for malonic acid, a compound with two COOH groups. Explain why one pK_a is lower and one pK_a is higher than the pK_a of acetic acid (CH_3COOH, pK_a = 4.8).

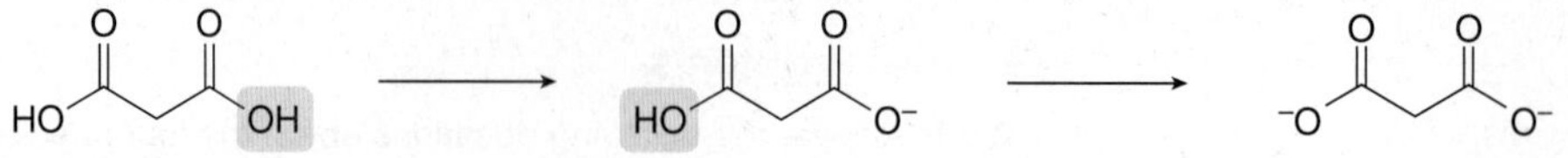

malonic acid
pK_a = 2.86 pK_a = 5.70

2.75 Amino acids such as glycine are the building blocks of large molecules called proteins that give structure to muscle, tendon, hair, and nails.

glycine zwitterion form

a. Explain why glycine does not actually exist in the form with all atoms uncharged, but actually exists as a salt called a zwitterion.
b. What product is formed when glycine is treated with concentrated HCl?
c. What product is formed when glycine is treated with NaOH?

2.76 Write a stepwise reaction sequence using proton transfer reactions to show how the following reaction occurs. (Hint: As a first step, use ^{-}OH to remove a proton from the CH_2 group between the C=O and C=C.)

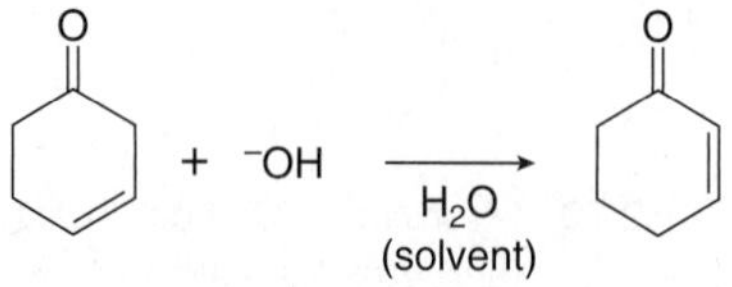

2.77 Which H atom in vitamin C (ascorbic acid) is most acidic?

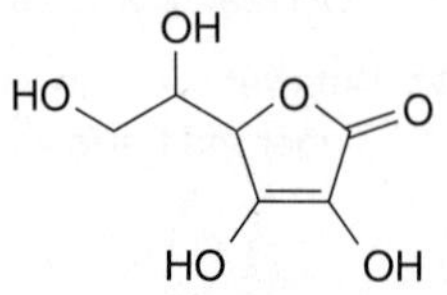

vitamin C
ascorbic acid

2.78 Which compound, **M** or **N,** is the stronger acid? Explain your choice.

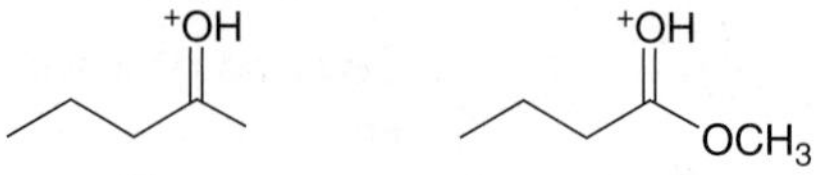

M **N**

Introduction to Organic Molecules and Functional Groups

3

3.1 Functional groups
3.2 An overview of functional groups
3.3 Intermolecular forces
3.4 Physical properties
3.5 Application: Vitamins
3.6 Application of solubility: Soap
3.7 Application: The cell membrane
3.8 Functional groups and reactivity
3.9 Biomolecules

Vitamin C, or **ascorbic acid,** is important in the formation of collagen, a protein that holds together the connective tissues of skin, muscle, and blood vessels. Citrus fruits—oranges, grapefruit, and lemons—are well known sources of vitamin C. A deficiency of vitamin C causes scurvy, a common disease of sailors in the 1600s when they had no access to fresh fruits on long voyages. Early Arctic explorers, however, were able to obtain vitamin C by eating scurvy grass, a low-growing plant found along the coasts of the Arctic region of the northern hemisphere. In Chapter 3, we learn why some vitamins like vitamin A can be stored in the fat cells in the body, whereas others like vitamin C are excreted in urine.

Having learned some basic concepts about structure, bonding, and acid–base chemistry in Chapters 1 and 2, we will now concentrate on organic molecules.

- What are the characteristic features of an organic compound?
- What determines the properties of an organic compound?

After these questions are answered, we can understand some common phenomena. Why do we store some vitamins in the body and readily excrete others? How does soap clean away dirt? Moreover, learning about the structure and properties of organic molecules will give us an understanding of the organic compounds found in biological systems.

3.1 Functional Groups

What are the characteristic features of an organic compound? Most organic molecules have C–C and C–H σ bonds. These bonds are strong, nonpolar, and not readily broken. Organic molecules may have the following structural features as well:

- **Heteroatoms—atoms other than carbon or hydrogen.** Common heteroatoms are nitrogen, oxygen, sulfur, phosphorus, and the halogens.
- **π Bonds.** The most common π bonds occur in C–C and C–O double bonds.

These structural features distinguish one organic molecule from another. They determine a molecule's geometry, physical properties, and reactivity, and comprise what is called a **functional group.**

- A *functional group* is an atom or a group of atoms with characteristic chemical and physical properties. It is the *reactive part* of the molecule.

Why do heteroatoms and π bonds confer reactivity on a particular molecule?

- Heteroatoms have lone pairs and create electron-deficient sites on carbon.
- π Bonds are easily broken in chemical reactions. A π bond makes a molecule a base and a nucleophile.

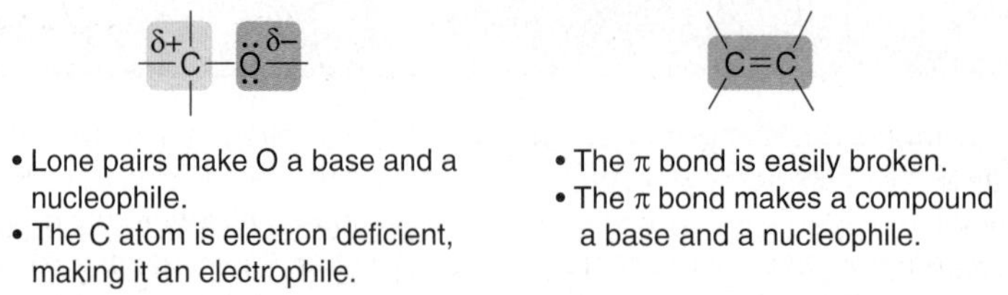
δ+
δ–
C=C
• Lone pairs make O a base and a nucleophile.
• The C atom is electron deficient, making it an electrophile.
• The π bond is easily broken.
• The π bond makes a compound a base and a nucleophile.

Don't think, though, that the C–C and C–H σ bonds are unimportant. They form the **carbon backbone** or **skeleton** to which the functional groups are bonded. A functional group usually behaves the same whether it is bonded to a carbon skeleton having as few as two or as many as 20 carbons. For this reason, we often abbreviate the carbon and hydrogen portion of the molecule by a capital letter **R,** and draw the **R** bonded to a particular functional group.

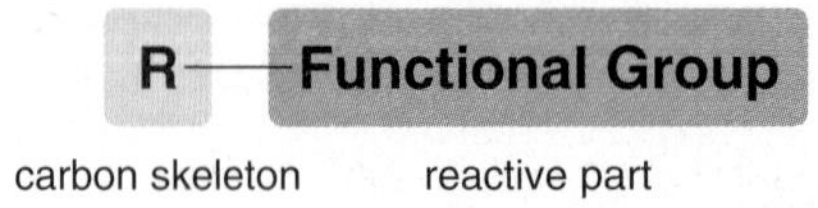
R
Functional Group
carbon skeleton
reactive part

Ethane, for example, has only C–C and C–H σ bonds, so it has *no* functional group. Ethane has no polar bonds, no lone pairs, and no π bonds, so it has **no reactive sites.** Because of this, ethane and molecules like it are very unreactive.

Ethanol, on the other hand, has two carbons and five hydrogens in its carbon backbone, as well as an OH group, a functional group called a **hydroxy** group. Ethanol has lone pairs and polar bonds that make it reactive with a variety of reagents, including the acids and bases discussed in Chapter 2. The hydroxy group makes the properties of ethanol very different from the properties of ethane. Moreover, any organic molecule containing a hydroxy group has properties similar to ethanol.

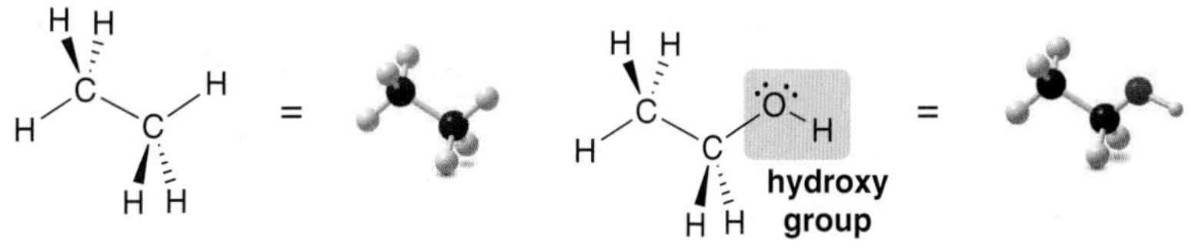

ethane

- all C–C and C–H σ bonds
- no functional group

ethanol

- polar C–O and O–H bonds
- two lone pairs

Most organic compounds can be grouped into a relatively small number of categories, based on the structure of their functional group. Ethane, for example, is an **alkane,** whereas ethanol is a simple **alcohol.**

Problem 3.1 What reaction occurs when CH_3CH_2OH is treated with (a) H_2SO_4? (b) NaH? What happens when CH_3CH_3 is treated with these same reagents?

3.2 An Overview of Functional Groups

The most common functional groups in organic compounds can be subdivided into three types: hydrocarbons, compounds containing a C–Z σ bond (where Z = an electronegative element), and compounds containing a C=O group (Tables 3.1–3.3). In addition, Table 3.4 lists functional groups with phosphorus–oxygen bonds that are found in several biological molecules.

3.2A Hydrocarbons

To review the structure and bonding of the simple aliphatic hydrocarbons, return to Section 1.10.

The word *aliphatic* is derived from the Greek word *aleiphas* meaning *fat.* Aliphatic compounds have physical properties similar to fats.

***Hydrocarbons* are compounds made up of only the elements carbon and hydrogen.** They may be **aliphatic** or **aromatic.**

[1] Aliphatic hydrocarbons. Aliphatic hydrocarbons can be divided into three subgroups.

- **Alkanes** have only C–C σ bonds and no functional group. Ethane, $\mathbf{CH_3CH_3}$, is a simple alkane.
- **Alkenes** have a C–C double bond as a functional group. Ethylene, $\mathbf{CH_2{=}CH_2}$, is a simple alkene.
- **Alkynes** have a C–C triple bond as a functional group. Acetylene, **HC≡CH,** is a simple alkyne.

[2] Aromatic hydrocarbons. This class of hydrocarbons was so named because many of the earliest known aromatic compounds had strong, characteristic odors.

The simplest aromatic hydrocarbon is **benzene.** The six-membered ring and three π bonds of benzene comprise a *single* functional group. Benzene is a component of the **BTX** mixture (**B** for **b**enzene) added to gasoline to boost octane ratings.

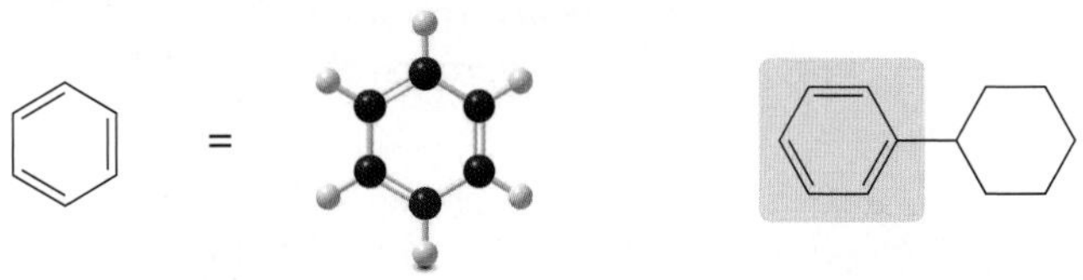

benzene
molecular formula C_6H_6

phenyl group
C_6H_5–
phenylcyclohexane

When a benzene ring is bonded to another group, it is called a **phenyl group.** In phenylcyclohexane, for example, a phenyl group is bonded to the six-membered cyclohexane ring. Table 3.1 summarizes the four different types of hydrocarbons.

Table 3.1 Hydrocarbons

Type of compound	General structure	Example	Functional group
Alkane	R—H	CH_3CH_3	—
Alkene	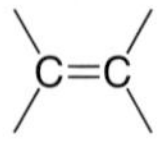	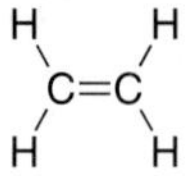	double bond
Alkyne	—C≡C—	H—C≡C—H	triple bond
Aromatic compound			phenyl group

Polyethylene is a synthetic plastic first produced in the 1930s, and initially used as insulating material for radar during World War II. It is now a plastic used in milk containers, sandwich bags, and plastic wrapping. Over 100 billion pounds of polyethylene are manufactured each year.

Alkanes, which have no functional groups, are notoriously unreactive except under very drastic conditions. For example, **polyethylene** is a synthetic plastic and high molecular weight alkane, consisting of chains of $-CH_2-$ groups bonded together, hundreds or even thousands of atoms long. Because it is an alkane with no reactive sites, it is a very stable compound that does not readily degrade and thus persists for years in landfills.

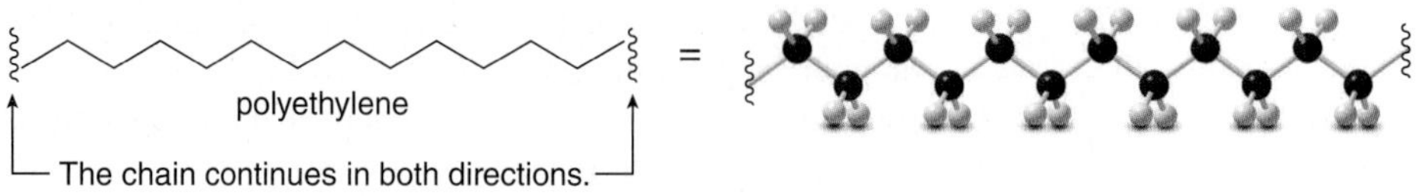

Carbon atoms in alkanes and other organic compounds are classified by the number of other carbons directly bonded to them.

- A *primary carbon* (1° carbon) is bonded to *one* other C atom.
- A *secondary carbon* (2° carbon) is bonded to *two* other C atoms.
- A *tertiary carbon* (3° carbon) is bonded to *three* other C atoms.
- A *quaternary carbon* (4° carbon) is bonded to *four* other C atoms.

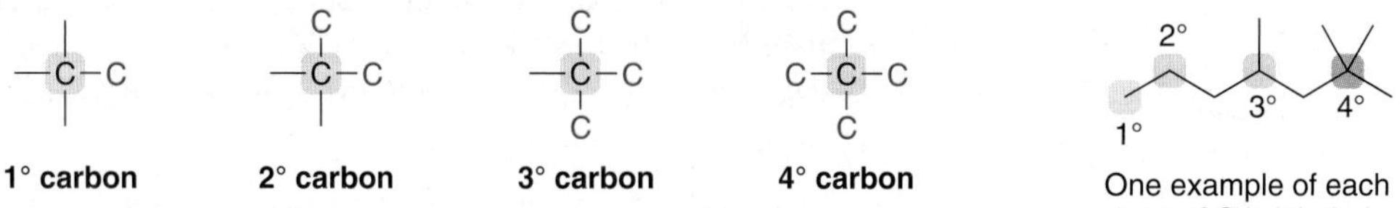

One example of each type of C is labeled.

Hydrogen atoms are classified as **primary (1°), secondary (2°),** or **tertiary (3°)** depending on the **type of carbon atom** to which they are bonded.

- A *primary hydrogen* (1° H) is on a C bonded to one other C atom.
- A *secondary hydrogen* (2° H) is on a C bonded to two other C atoms.
- A *tertiary hydrogen* (3° H) is on a C bonded to three other C atoms.

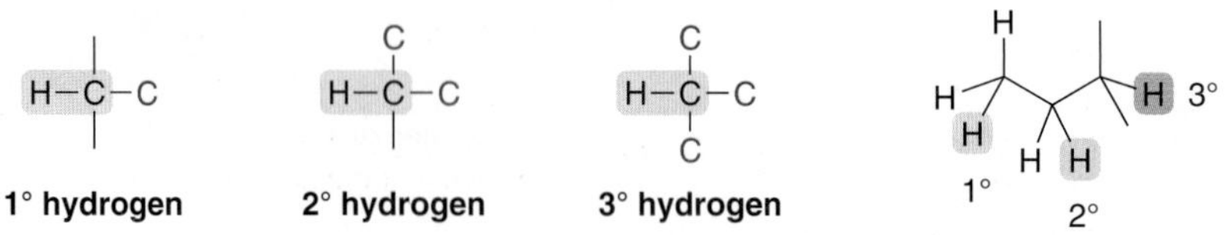

One example of each type of H is labeled.

Sample Problem 3.1 Classify the designated carbon atoms in **A** as 1°, 2°, 3°, or 4°. Classify the designated hydrogen atoms in **B** as 1°, 2°, or 3°.

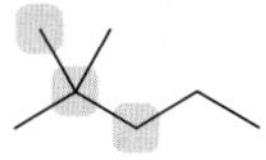

A

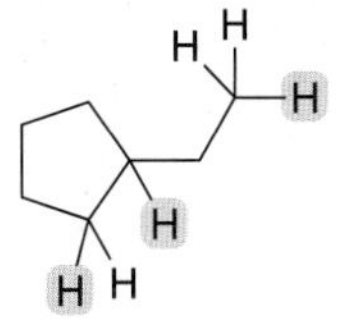

B

Solution

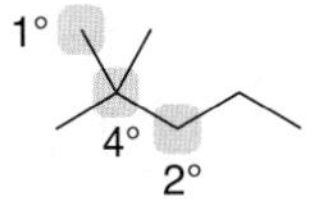

A

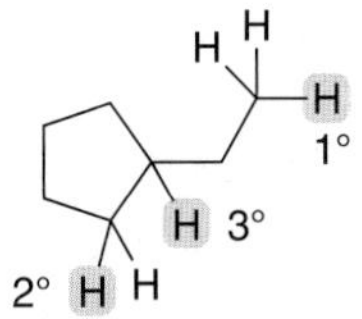

B

Problem 3.2 (a) Classify the carbon atoms in each compound as 1°, 2°, 3°, or 4°. (b) Classify the hydrogen atoms in each compound as 1°, 2°, or 3°.

[1]  [2]  [3] 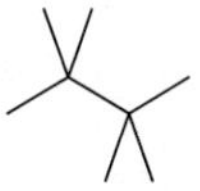 [4] 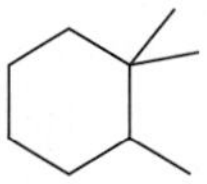

Problem 3.3 Classifying a carbon atom by the number of carbons to which it is bonded can also be done in more complex molecules that contain heteroatoms. Classify each sp^3 hybridized carbon atom in the carbohydrate sucrose (table sugar) and the antibiotic amoxicillin as 1°, 2°, 3°, or 4°.

a. 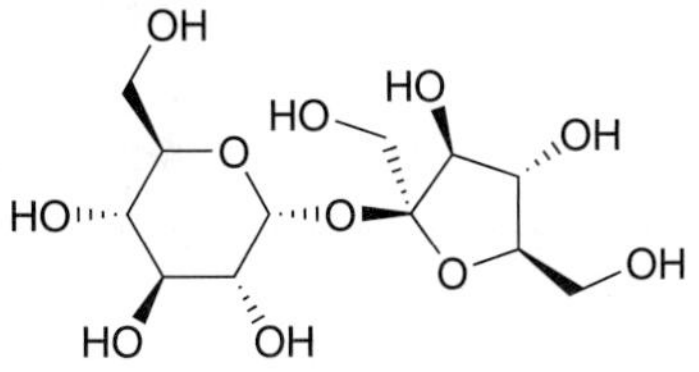

sucrose

b. 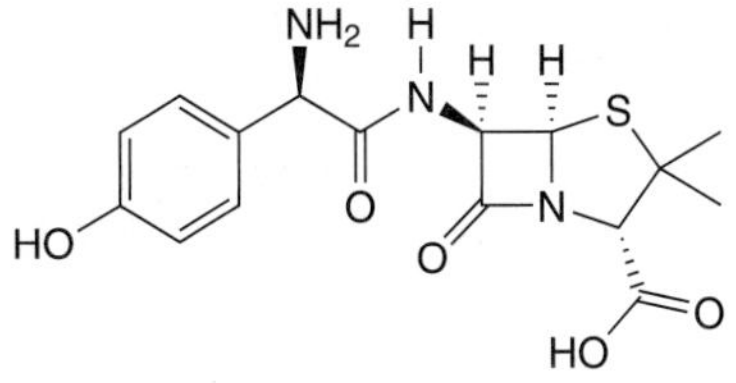

amoxicillin

3.2B Compounds Containing C–Z σ Bonds

Functional groups that contain C–Z σ bonds include alkyl halides, alcohols, ethers, amines, thiols, sulfides, and disulfides (Table 3.2). The electronegative heteroatom Z creates a polar bond, making carbon electron deficient. The lone pairs on Z are available for reaction with protons and other electrophiles, especially when Z = N or O.

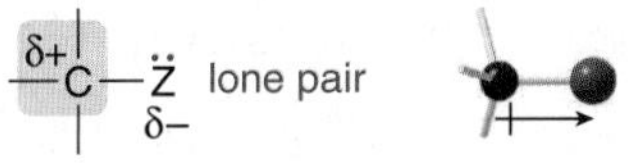

Molecules containing these functional groups may be simple or very complex. Diethyl ether, the first common general anesthetic, is a simple ether because it contains a single O atom, depicted in red, bonded to two C atoms. Hemibrevetoxin B, on the other hand, contains four ether groups, in addition to other functional groups.

Hemibrevetoxin B is a neurotoxin produced by algal blooms referred to as "red tides," because of the color often seen in shallow ocean waters when these algae proliferate.

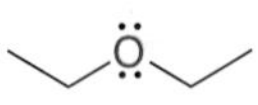

diethyl ether

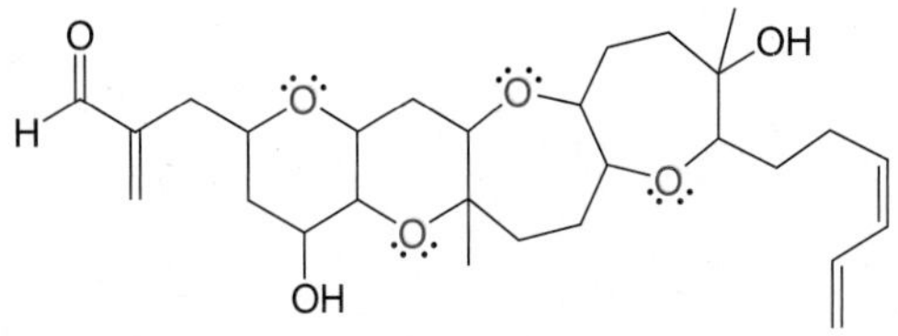

hemibrevetoxin B

Table 3.2 Compounds Containing C–Z σ Bonds

Type of compound	General structure	Example	3-D structure	Functional group
Alkyl halide	$R-\ddot{\underset{..}{X}}:$ (X = F, Cl, Br, I)	$CH_3-\ddot{\underset{..}{Br}}:$	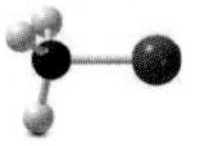	**–X** halo group
Alcohol	$R-\ddot{\underset{..}{O}}H$	$CH_3-\ddot{\underset{..}{O}}-H$	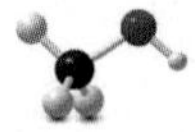	**–OH** hydroxy group
Ether	$R-\ddot{\underset{..}{O}}-R$	$CH_3-\ddot{\underset{..}{O}}-CH_3$		**–OR** alkoxy group
Amine	$R-\ddot{N}H_2$ or $R_2\ddot{N}H$ or $R_3\ddot{N}$	$CH_3-\ddot{N}H-H$		**$-NH_2$** amino group
Thiol	$R-\ddot{\underset{..}{S}}H$	$CH_3-\ddot{\underset{..}{S}}-H$	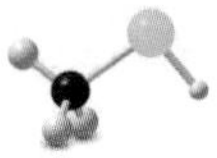	**–SH** mercapto group
Sulfide	$R-\ddot{\underset{..}{S}}-R$	$CH_3-\ddot{\underset{..}{S}}-CH_3$		**–SR** alkylthio group
Disulfide	$R-\ddot{\underset{..}{S}}-\ddot{\underset{..}{S}}-R$	$CH_3-\ddot{\underset{..}{S}}-\ddot{\underset{..}{S}}-CH_3$	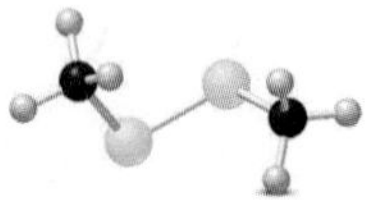	**—SS—**

Alkyl halides and alcohols are classified as **primary (1°)**, **secondary (2°)**, or **tertiary (3°)** based on the number of carbon atoms bonded to the carbon bearing the halogen or OH group. The classification of four functional groups in sucralose, the synthetic sweetener sold as Splenda, is shown.

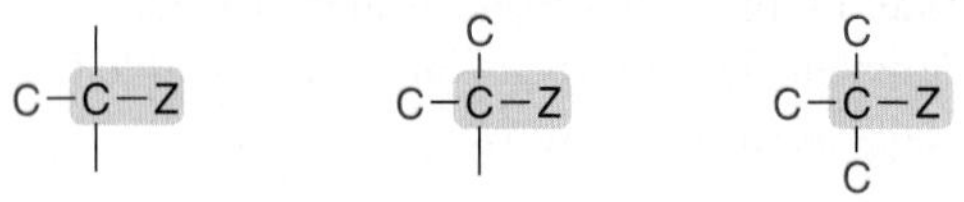

Z = X	1° alkyl halide	2° alkyl halide	3° alkyl halide
Z = OH	1° alcohol	2° alcohol	3° alcohol

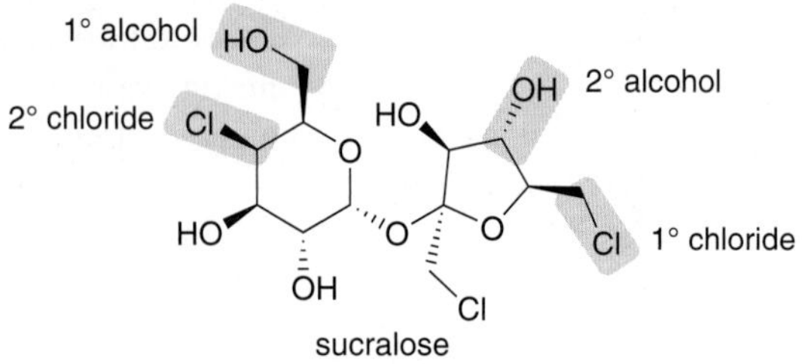

sucralose

Problem 3.4 Classify each alkyl halide and alcohol as 1°, 2°, or 3°.

a. 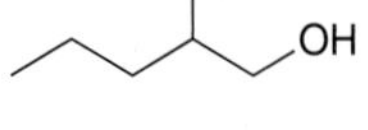

b. 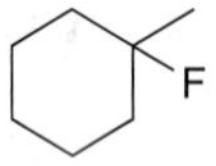

c. 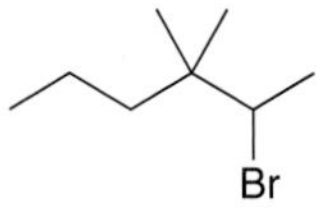

d. 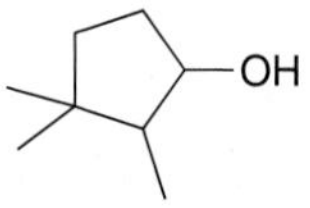

Problem 3.5 Classify each OH group and halogen in dexamethasone, a synthetic steroid, as 1°, 2°, or 3°.

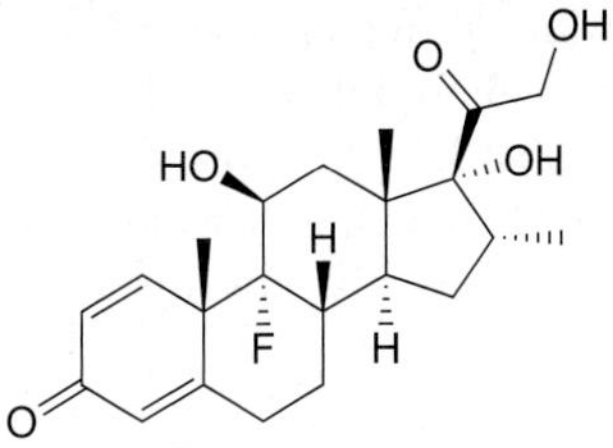

dexamethasone

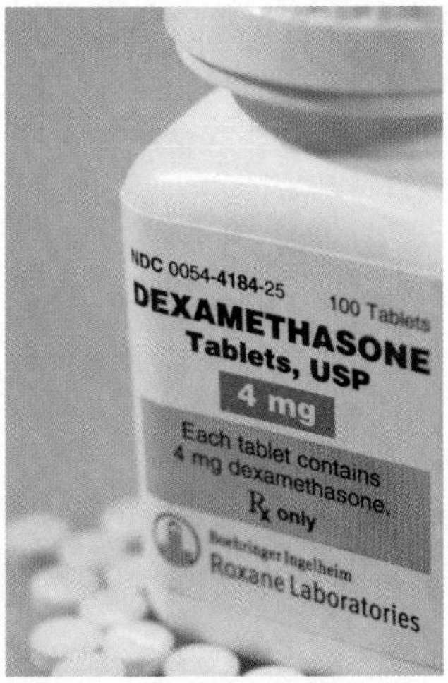

Dexamethasone relieves inflammation and is used to treat some forms of arthritis, skin conditions, and asthma.

Amines are classified as **primary (1°), secondary (2°),** or **tertiary (3°)** based on the number of carbon atoms bonded to the nitrogen atom.

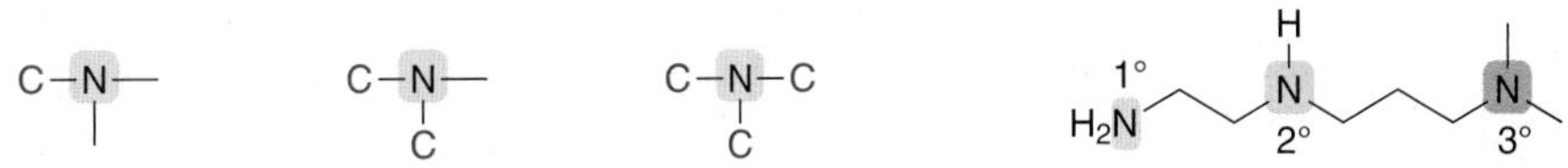

1° amine 2° amine 3° amine

Classifying amines is different from classifying alcohols and alkyl halides as primary (1°), secondary (2°), or tertiary (3°). Amines are classified by the number of carbon–*nitrogen* bonds, whereas alkyl halides and alcohols are classified by the type of *carbon* bonded to the halogen or hydroxy group.

Problem 3.6 Classify each amine in the following compounds as 1°, 2°, or 3°.

a. 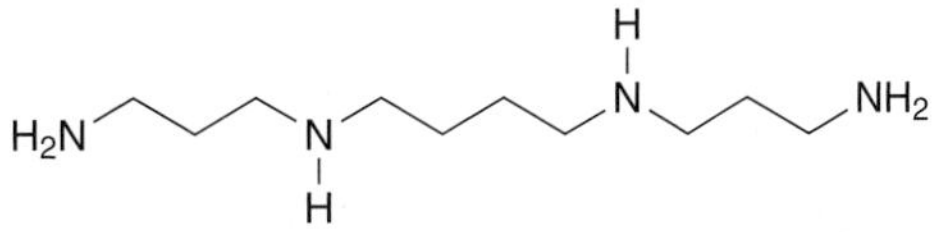

spermine
(isolated from semen)

b. 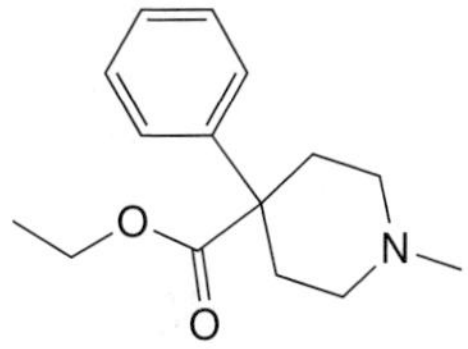

meperidine
(a narcotic)
trade name Demerol

Problem 3.7 Classify the designated hydroxy groups and amine in the anticancer agent doxorubicin (trade name Adriamycin) as 1°, 2°, or 3°.

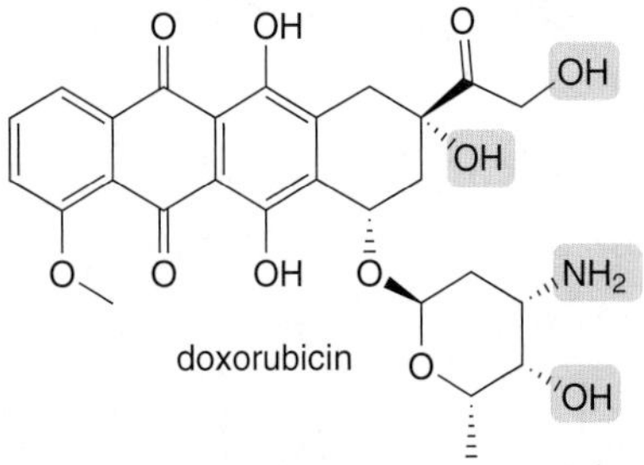

doxorubicin

Sulfur-containing functional groups are especially prevalent in the chemistry of proteins, because cysteine and methionine, two common amino acids, contain a thiol and sulfide, respectively. We will learn more about amino acids and proteins in Section 3.9.

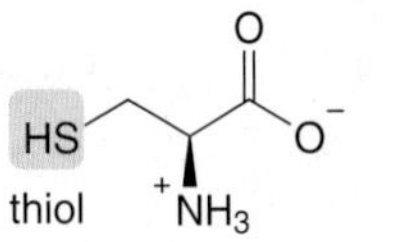

thiol

cysteine

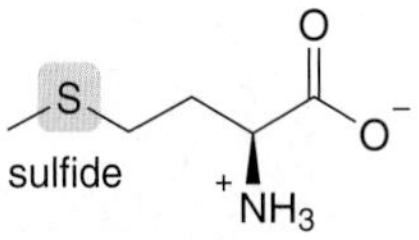

sulfide

methionine

3.2C Compounds Containing a C=O Group

Many different types of functional groups possess a C−O double bond (a **carbonyl group**), including aldehydes, ketones, carboxylic acids, esters, thioesters, amides, and acid chlorides (Table 3.3). The polar C−O bond makes the carbonyl carbon an electrophile, while the lone pairs on O allow it to react as a nucleophile and base. The carbonyl group also contains a π bond that is more easily broken than a C−O σ bond.

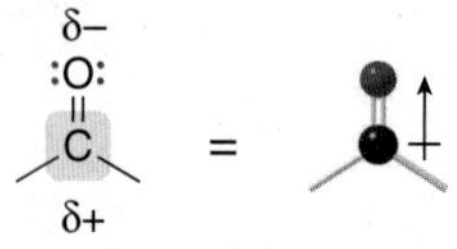

carbonyl group

Table 3.3 Compounds Containing a C=O Group

Type of compound	General structure	Example	Condensed structure	3-D structure	Functional group
Aldehyde	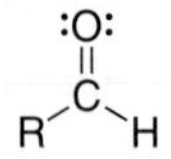	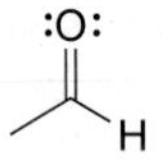	CH_3CHO	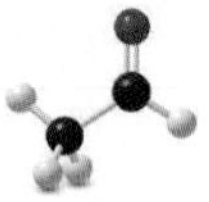	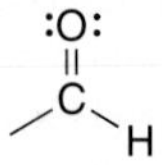
Ketone	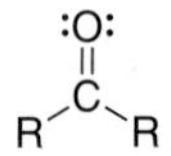	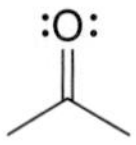	$(CH_3)_2CO$	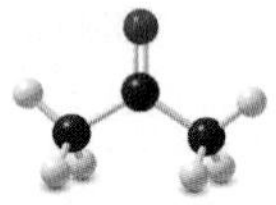	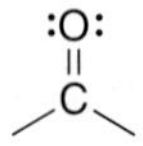 carbonyl group
Carboxylic acid	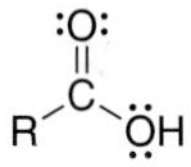	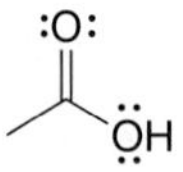	CH_3CO_2H	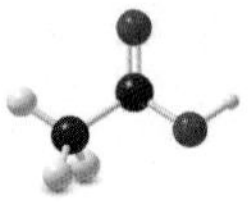	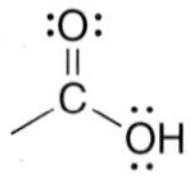 carboxy group
Ester	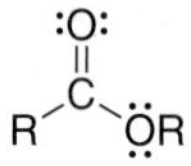	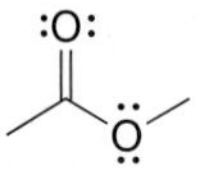	$CH_3CO_2CH_3$	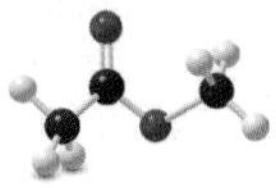	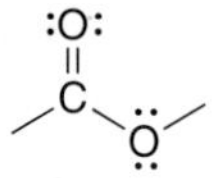
Thioester	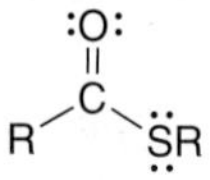	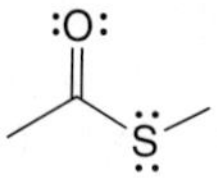	CH_3COSCH_3	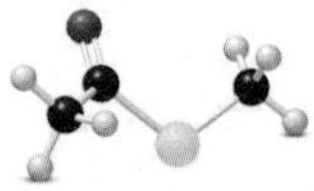	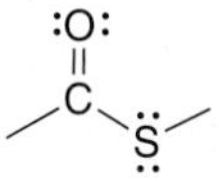
Amide	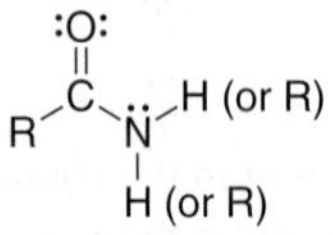	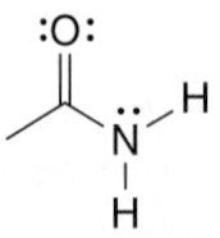	CH_3CONH_2	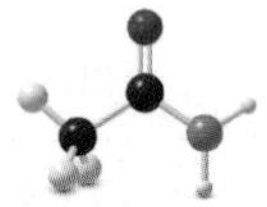	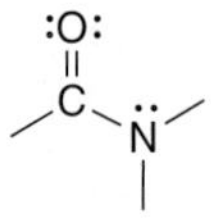
Acid chloride	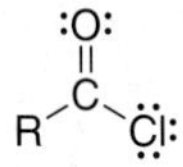	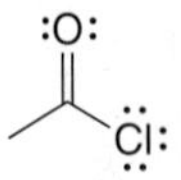	CH_3COCl	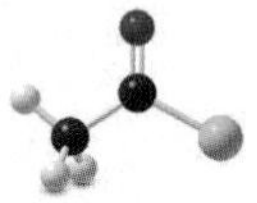	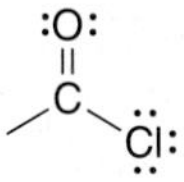

Amides, compounds that contain a nitrogen atom bonded directly to the carbonyl carbon, are classified as **primary (1°), secondary (2°),** or **tertiary (3°)** based on the number of carbon atoms bonded to the nitrogen atom.

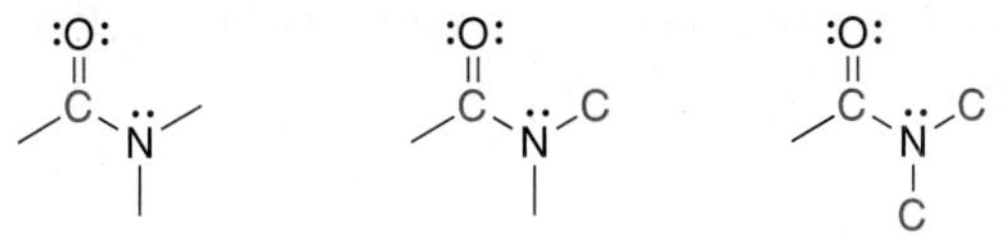

1° amide   2° amide   3° amide

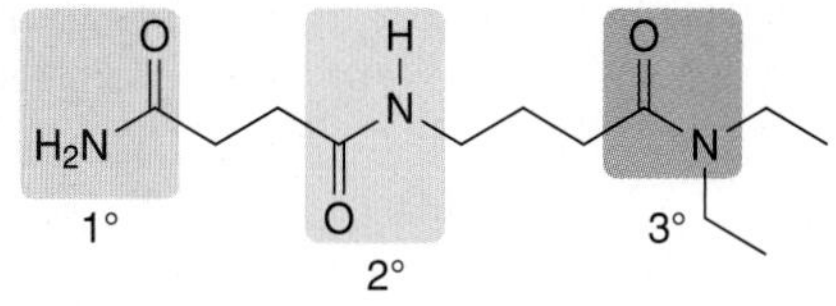

Problem 3.8 Classify the amides in thyrotropin-releasing hormone (TRH), a hormone produced by the hypothalamus, as 1°, 2°, or 3°.

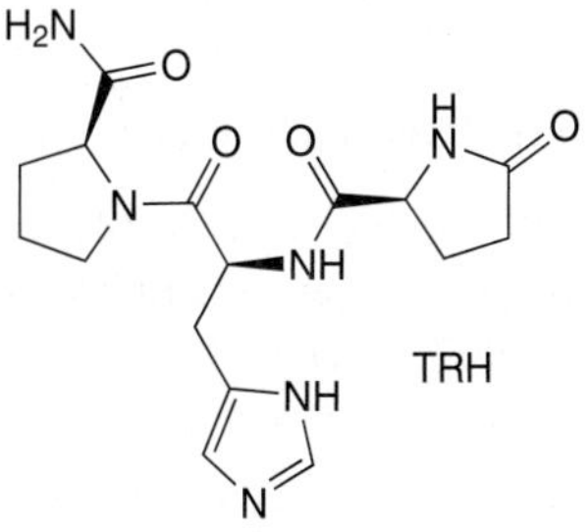

TRH

The importance of a functional group cannot be overstated. A functional group determines a molecule's bonding and shape, type and strength of intermolecular forces, physical properties, nomenclature, and chemical reactivity.

Problem 3.9 Oseltamivir can be prepared in 10 steps from shikimic acid. Identify the functional groups in oseltamivir and shikimic acid.

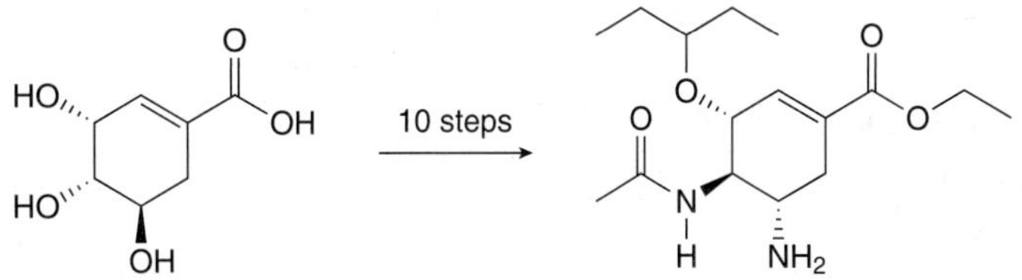

shikimic acid   oseltamivir

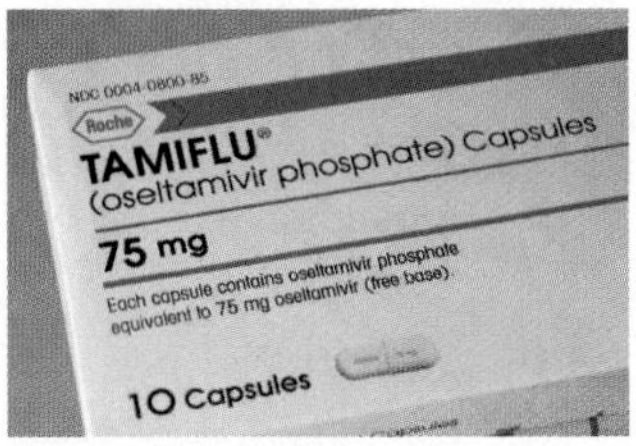

Tamiflu is the trade name for oseltamivir, an antiviral drug used to treat influenza.

Leukotriene C_4 and glutathione are two biological molecules that contain a variety of functional groups. Leukotriene C_4 is a major contributor to the inflammation associated with asthma, whereas glutathione is an antioxidant that destroys harmful oxidizing agents in cells.

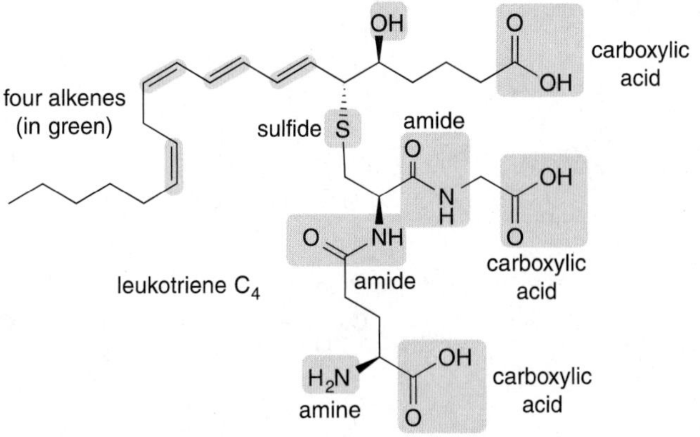

leukotriene C_4

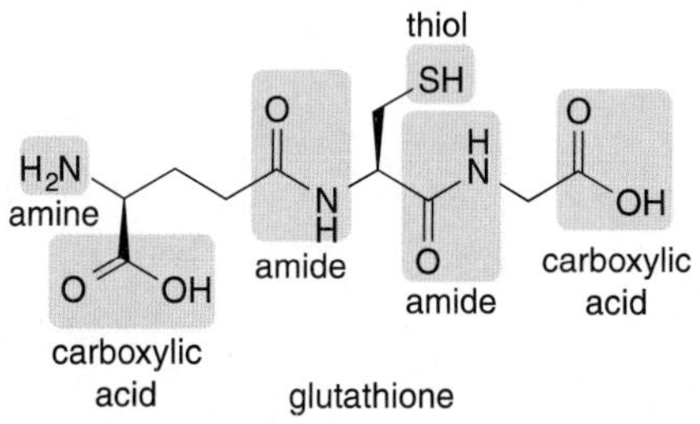

glutathione

Problem 3.10 Pyruvic acid, lipoic acid, and folic acid are three biological molecules with more than one functional group. Pyruvic acid is formed during carbohydrate metabolism; lipoic acid is utilized in several metabolic reactions; and folic acid is a vitamin. (a) Identify all the functional groups in pyruvic acid and lipoic acid and the designated functional groups in folic acid. (b) Draw an isomer of pyruvic acid that contains at least one different functional group.

[1]
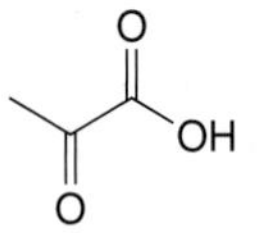
pyruvic acid

[2]
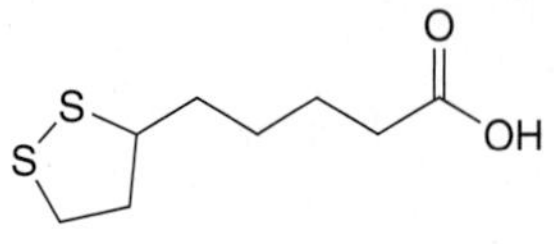
lipoic acid

[3]
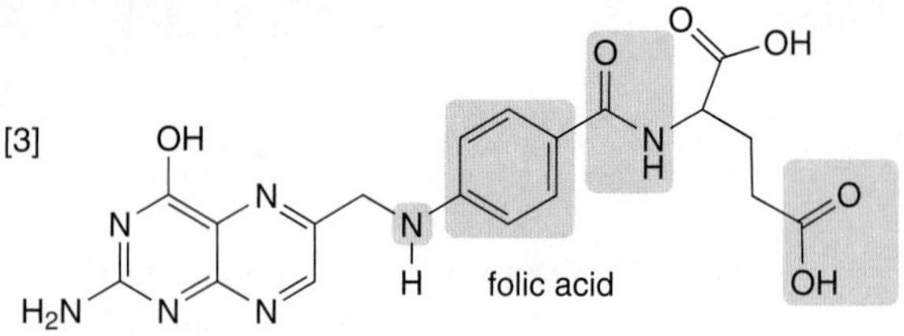
folic acid

3.2D Compounds Containing P–O Bonds

Many biological molecules contain functional groups with one or more phosphorus atoms bonded to four oxygens, including monophosphates, diphosphates, triphosphates, and acyl phosphates (Table 3.4). These functional groups bear net negative charges at the pH of cells, a topic discussed in more detail in Chapter 19.

The role of phosphates in biological substitution reactions is discussed in Section 7.16.

The bond between one oxygen of phosphate to an alkyl group results in a phosphorus analogue of an ester—a **phosphate ester.** When two oxygens of phosphate are bonded to alkyl groups, the compound becomes a **phosphate diester** or **phosphodiester.** Phosphodiesters are key elements of nucleic acids, as we will see in Section 3.9.

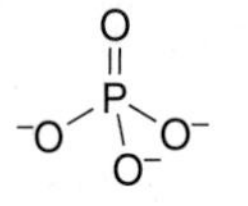
phosphate

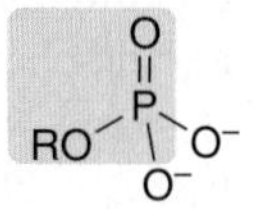
phosphate ester

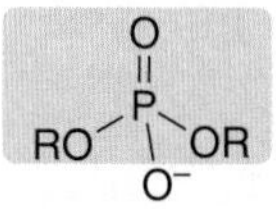
phosphate diester
phosphodiester

Table 3.4 Compounds Containing P–O Bonds

Type of compound	General structure	Example	Condensed structure
Monophosphate	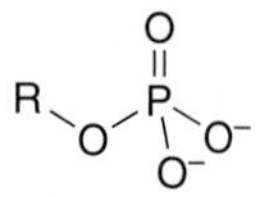	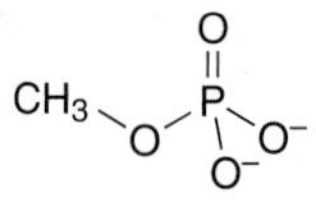	$CH_3OPO_3^{2-}$
Diphosphate	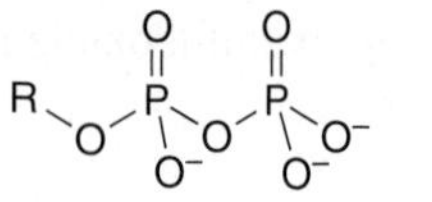	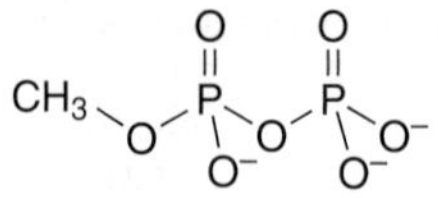	$CH_3OP_2O_6^{3-}$
Triphosphate	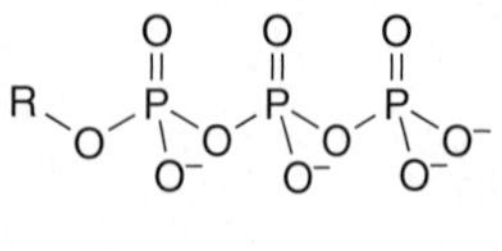	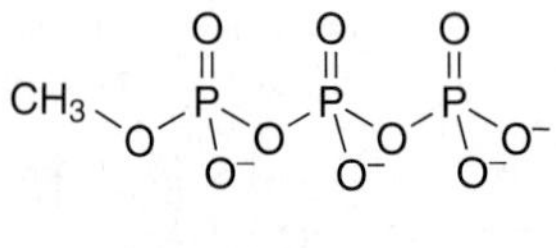	$CH_3OP_3O_9^{4-}$
Acyl phosphate	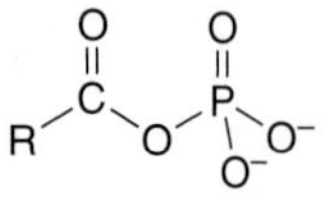	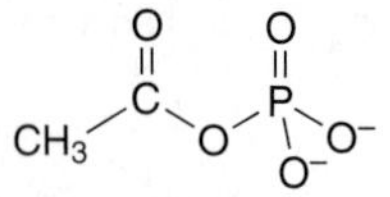	$CH_3CO_2PO_3^{2-}$

Adenosine triphosphate (ATP), the key compound used in energy transfer in metabolism in cells, contains a triphosphate.

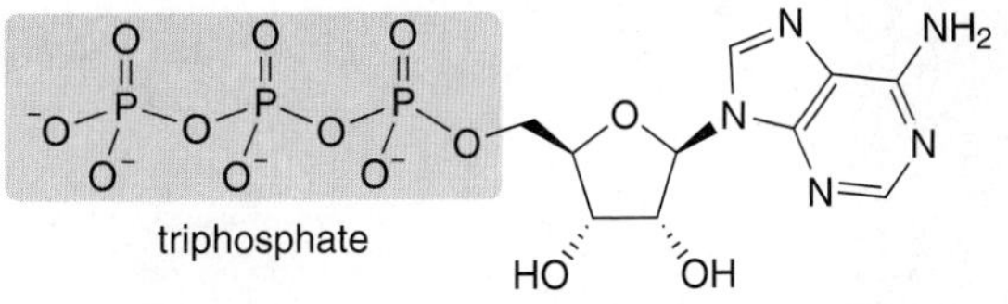

adenosine triphosphate
ATP

Problem 3.11 Label all the functional groups in **A** and **B**, intermediates formed during the metabolism of the simple sugar glucose.

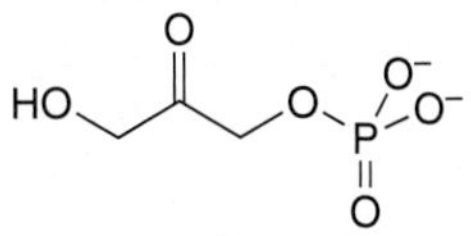

A

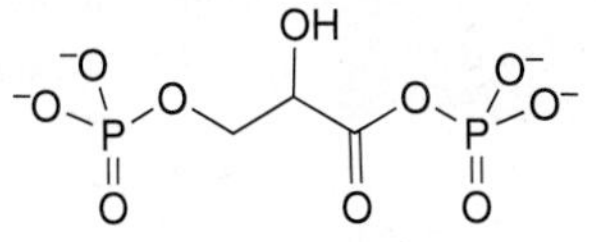

B

3.3 Intermolecular Forces

Intermolecular forces are also referred to as **noncovalent interactions** or **nonbonded interactions.**

strong electrostatic interaction between Na^+ and Cl^-

Intermolecular forces are the interactions that exist *between* molecules. A functional group determines the type and strength of these interactions.

3.3A Ionic Compounds

Ionic compounds, such as NaCl, contain oppositely charged particles held together by **extremely strong electrostatic interactions.** These ionic interactions are much *stronger* than the intermolecular forces present between covalent molecules, so it takes a great deal of energy to separate oppositely charged ions from each other.

3.3B Covalent Compounds

Covalent compounds are composed of discrete molecules. The nature of the forces between the molecules depends on the functional group present. There are three different types of interactions, presented here in order of *increasing strength:*

- **van der Waals forces**
- **dipole–dipole interactions**
- **hydrogen bonding**

Van der Waals Forces

Van der Waals forces, also called **London forces,** are very weak interactions caused by the **momentary changes in electron density in a molecule.** Van der Waals forces are the only attractive forces present in nonpolar compounds.

For example, although a nonpolar CH_4 molecule has no net dipole, at any one instant its electron density may not be completely symmetrical, creating a *temporary* dipole. This can induce a temporary dipole in another CH_4 molecule, with the partial positive and negative charges arranged close to each other. **The weak interaction of these temporary dipoles constitutes van der Waals forces.** All compounds exhibit van der Waals forces.

Van der Waals interactions occur between temporary dipoles.

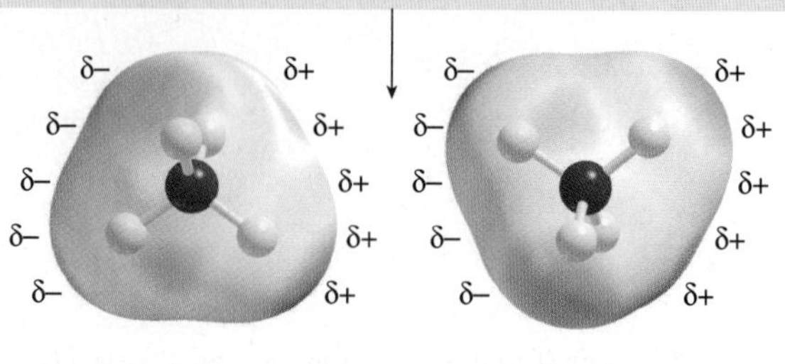

The surface area of a molecule determines the strength of the van der Waals interactions. **The larger the surface area, the larger the attractive force between two molecules, and the stronger the intermolecular forces.** Long, sausage-shaped molecules such as $CH_3CH_2CH_2CH_2CH_3$ (pentane) have stronger van der Waals interactions than compact, spherical ones like $C(CH_3)_4$ (2,2-dimethylpropane), as shown in Figure 3.1.

Another factor affecting the strength of van der Waals forces is **polarizability.**

- *Polarizability* is a measure of how the electron cloud around an atom responds to changes in its electronic environment.

Larger atoms like iodine, which have more loosely held valence electrons, are more polarizable than smaller atoms like fluorine, which have more tightly held electrons. Because larger atoms have more easily induced dipoles, compounds containing them possess stronger intermolecular interactions.

Although any single van der Waals interaction is weak, a large number of van der Waals interactions creates a strong force. For example, geckos stick to walls and ceilings by van der Waals interactions of the surfaces with the 500,000 tiny hairs on each foot.

Dipole–Dipole Interactions

***Dipole–dipole interactions* are the attractive forces between the permanent dipoles of two polar molecules.** In acetone, $(CH_3)_2C{=}O$, for example, the dipoles in adjacent molecules align so that the partial positive and partial negative charges are in close proximity. These attractive forces caused by permanent dipoles are much stronger than weak van der Waals forces.

Figure 3.1 Surface area and van der Waals forces

a. $CH_3CH_2CH_2CH_2CH_3$ (pentane)

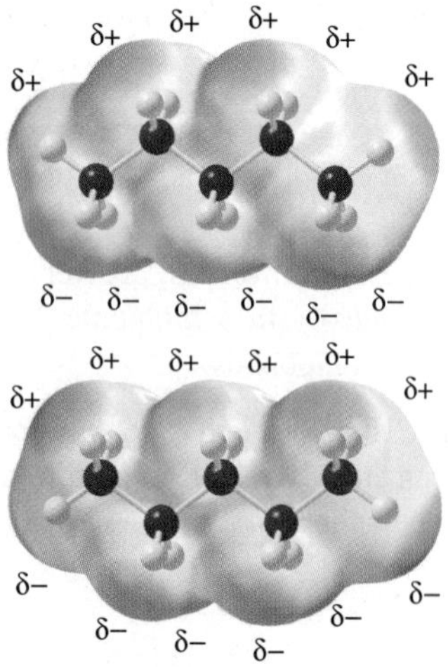

- A long, cylindrical molecule like pentane has a larger surface area, causing stronger van der Waals interactions.

b. $C(CH_3)_4$ (2,2-dimethylpropane)

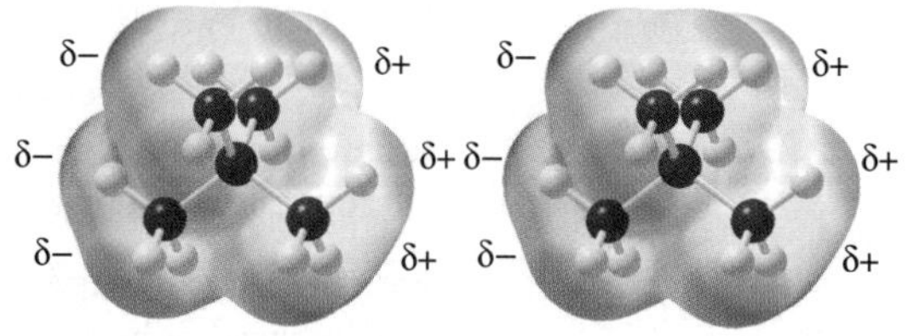

- A compact, spherical molecule like 2,2-dimethylpropane has a smaller surface area, causing weaker van der Waals interactions.

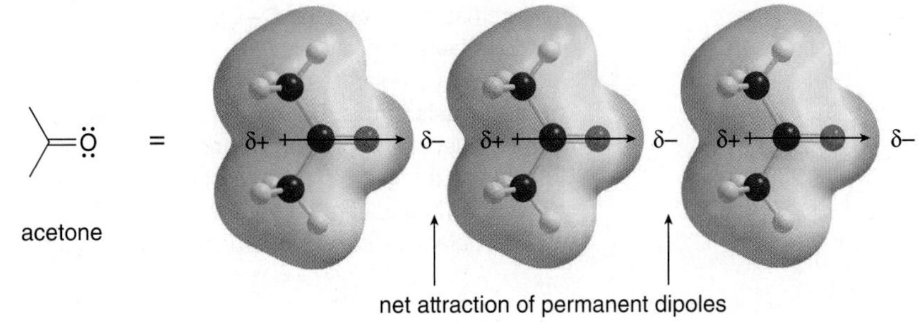

Hydrogen Bonding

Hydrogen bonding helps determine the three-dimensional shape of large biomolecules such as proteins and carbohydrates. See Chapters 26 and 27 for details.

***Hydrogen bonding* typically occurs when a hydrogen atom bonded to O, N, or F is electrostatically attracted to a lone pair of electrons on an O, N, or F atom in another molecule.** Thus, H_2O molecules can hydrogen bond to each other. When they do, a H atom covalently bonded to O in one water molecule is attracted to a lone pair of electrons on the O in another water molecule. Hydrogen bonds are the strongest of the three types of intermolecular forces, though they are still much weaker than any covalent bond.

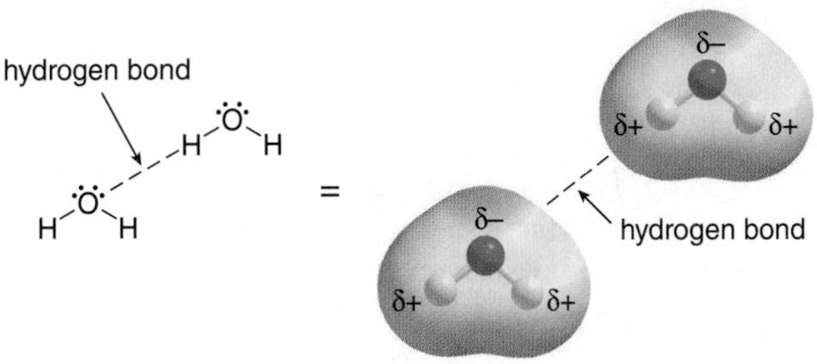

Sample Problem 3.2 illustrates how to determine the relative strength of intermolecular forces for a group of compounds. Table 3.5 summarizes the four types of interactions that affect the properties of all compounds.

Table 3.5 Summary of Types of Intermolecular Forces

Type of force	Relative strength	Exhibited by	Example
van der Waals	weak	all molecules	$CH_3CH_2CH_2CH_2CH_3$ $CH_3CH_2CH_2CHO$ $CH_3CH_2CH_2CH_2OH$
dipole–dipole	moderate	molecules with a net dipole	$CH_3CH_2CH_2CHO$ $CH_3CH_2CH_2CH_2OH$
hydrogen bonding	strong	molecules with an O–H, N–H, or H–F bond	$CH_3CH_2CH_2CH_2OH$
ion–ion	very strong	ionic compounds	NaCl, LiF

Sample Problem 3.2 Rank the following compounds in order of increasing strength of intermolecular forces: $CH_3CH_2CH_2CH_2CH_3$ (pentane), $CH_3CH_2CH_2CH_2OH$ (butan-1-ol), and $CH_3CH_2CH_2CHO$ (butanal).

Solution

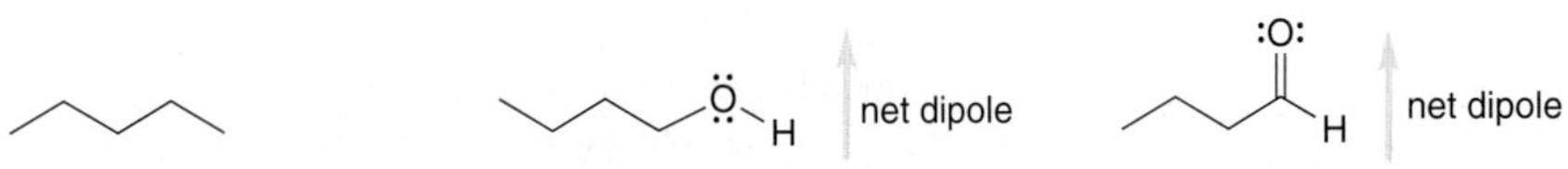

pentane
- nonpolar molecule

butan-1-ol
- bent molecule around O
- polar C–O and O–H bonds
- O–H bond for hydrogen bonding

butanal
- trigonal planar C
- polar C=O bond

- Pentane has only nonpolar C–C and C–H bonds, so its molecules are held together by only **van der Waals** forces.
- Butan-1-ol is a polar bent molecule, so it can have **dipole–dipole** interactions in addition to **van der Waals** forces. Because it has an O–H bond, butan-1-ol molecules are held together by intermolecular **hydrogen bonds** as well.
- Butanal has a trigonal planar carbon with a polar C=O bond, so it exhibits **dipole–dipole** interactions in addition to **van der Waals** forces. There is *no* H atom bonded to O, so two butanal molecules *cannot* hydrogen bond to each other.

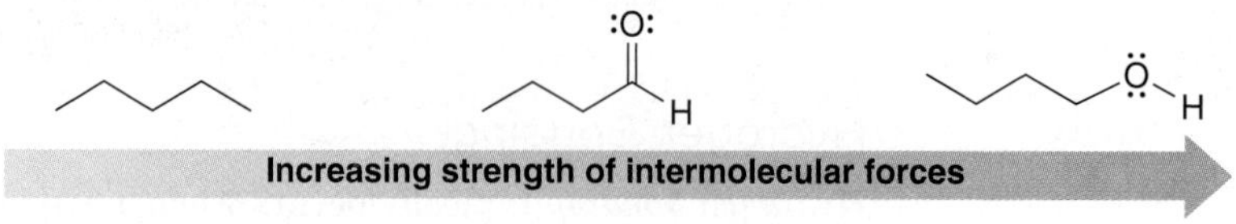

Problem 3.12 What types of intermolecular forces are present in each compound?

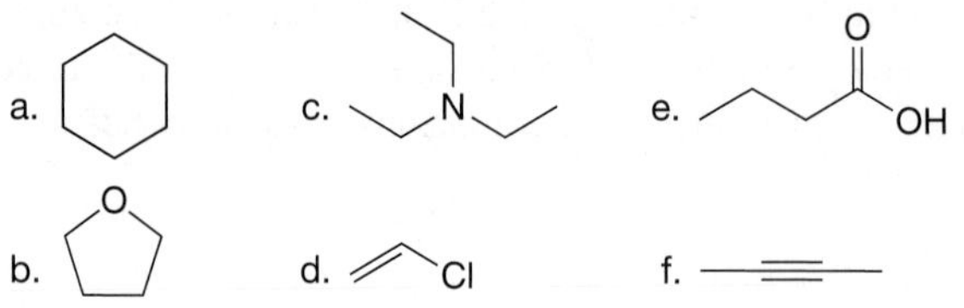

Hydrogen bonding is important in DNA, the high molecular weight compound that stores the genetic information of an organism. As we will learn in Section 3.9, DNA is composed of two long strands of atoms that are held together by an extensive network of hydrogen bonds, as shown in Figure 3.2.

Figure 3.2 Hydrogen bonding and DNA

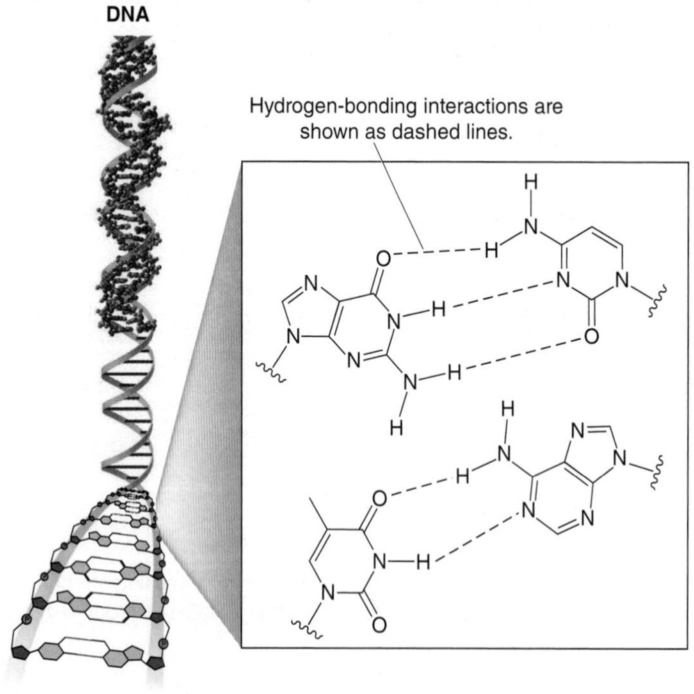

- DNA, which is contained in the chromosomes of the nucleus of the cell, is composed of two long strands of atoms held together by hydrogen bonding.

3.4 Physical Properties

The strength of a compound's intermolecular forces determines many of its physical properties, including its boiling point, melting point, and solubility.

3.4A Boiling Point (bp)

The *boiling point* of a compound is the temperature at which a liquid is converted to a gas. In boiling, energy is needed to overcome the attractive forces in the more ordered liquid state.

- **The stronger the intermolecular forces, the higher the boiling point.**

Because **ionic compounds** are held together by extremely strong interactions, they have **very high boiling points.** The boiling point of NaCl, for example, is 1413 °C. **With covalent molecules, the boiling point depends on the identity of the functional group.** For compounds of approximately the same molecular weight:

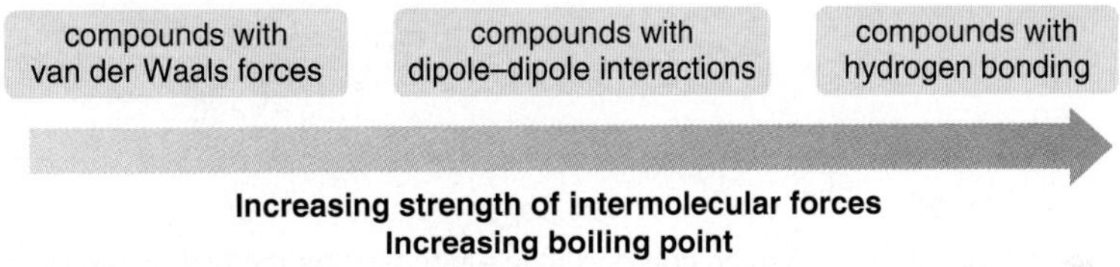
compounds with van der Waals forces
compounds with dipole–dipole interactions
compounds with hydrogen bonding
Increasing strength of intermolecular forces
Increasing boiling point

Recall from Sample Problem 3.2, for example, that the relative strength of the intermolecular forces increases from pentane to butanal to butan-1-ol. The boiling points of these compounds increase in the same order.

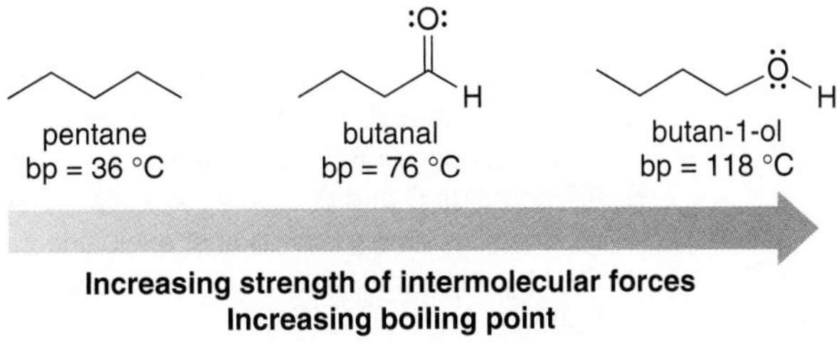
:O:
H
Ö
H
pentane
bp = 36 °C
butanal
bp = 76 °C
butan-1-ol
bp = 118 °C
Increasing strength of intermolecular forces
Increasing boiling point

Because surface area and polarizability affect the strength of intermolecular forces, they also affect the boiling point. For two compounds with similar functional groups:

- **The larger the surface area, the higher the boiling point.**
- **The more polarizable the atoms, the higher the boiling point.**

Examples of each phenomenon are illustrated in Figure 3.3. In comparing two ketones that differ in size, pentan-3-one has a higher boiling point than acetone because it has a greater molecular weight and larger surface area. In comparing two alkyl halides having the same number of carbon atoms, CH_3I has a higher boiling point than CH_3F because I is more polarizable than F.

Figure 3.3 Effect of surface area and polarizability on boiling point

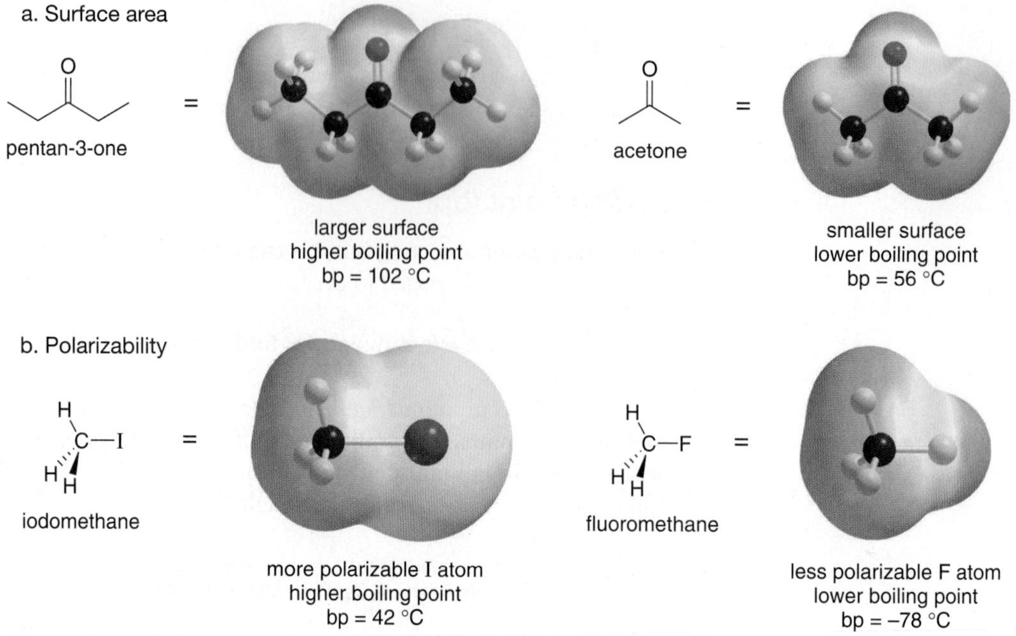
a. Surface area
pentan-3-one
larger surface
higher boiling point
bp = 102 °C
acetone
smaller surface
lower boiling point
bp = 56 °C
b. Polarizability
iodomethane
more polarizable I atom
higher boiling point
bp = 42 °C
fluoromethane
less polarizable F atom
lower boiling point
bp = –78 °C

Sample Problem 3.3 Which compound in each pair has the higher boiling point?

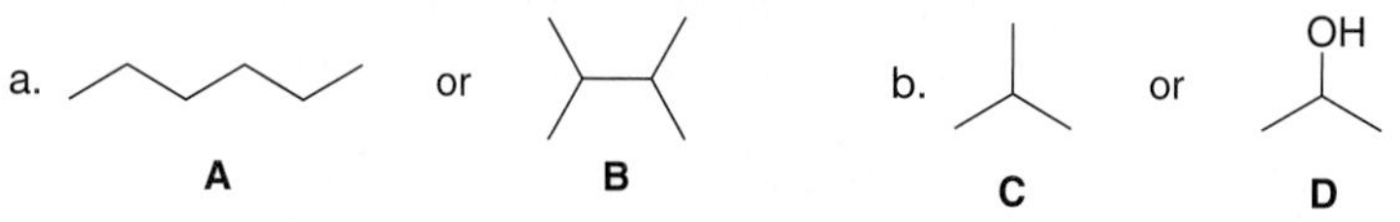
a.
or
A
B
b.
or
C
D

Solution

a. Isomers **A** and **B** have only nonpolar C–C and C–H bonds, so they exhibit only van der Waals forces. Because **B** is more compact, it has less surface area and a lower boiling point.

b. Compounds **C** and **D** have approximately the same molecular weight but different functional groups. **C** is a nonpolar alkane, exhibiting only van der Waals forces. **D** is an alcohol with an O–H group available for hydrogen bonding, so it has stronger intermolecular forces and a higher boiling point.

Problem 3.13 Which compound in each pair has the higher boiling point?

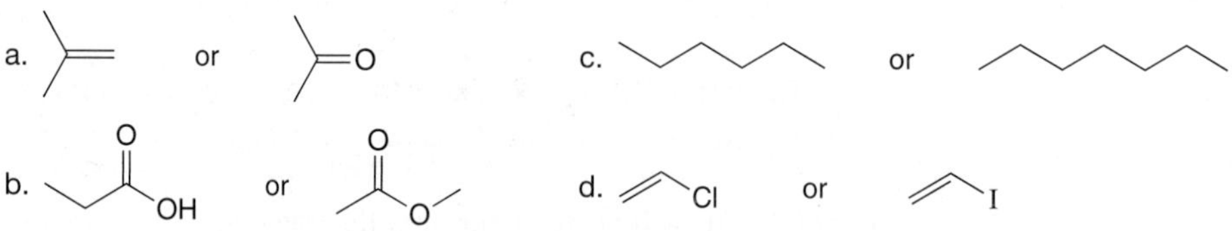
a.
or
c.
or
b.
or
d.
or

Problem 3.14 Explain why the boiling point of propanamide, $CH_3CH_2CONH_2$, is considerably higher than the boiling point of *N,N*-dimethylformamide, $HCON(CH_3)_2$ (213 °C vs. 153 °C), even though both compounds are isomeric amides.

3.4B Melting Point (mp)

The *melting point* is the temperature at which a solid is converted to its liquid phase. In melting, energy is needed to overcome the attractive forces in the more ordered crystalline solid. Two factors determine the melting point of a compound.

- **The stronger the intermolecular forces, the higher the melting point.**
- **Given the same functional group, the more symmetrical the compound, the higher the melting point.**

Because **ionic compounds** are held together by extremely strong interactions, they have **very high melting points.** For example, the melting point of NaCl is 801 °C. With covalent molecules, the melting point once again depends on the identity of the functional group. For compounds of approximately the same molecular weight:

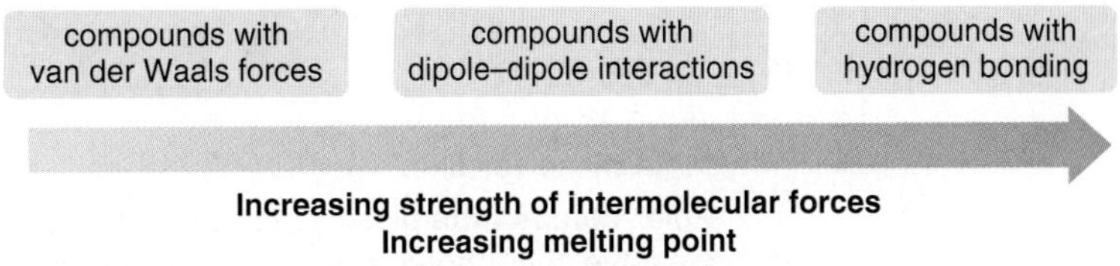
compounds with van der Waals forces
compounds with dipole–dipole interactions
compounds with hydrogen bonding
Increasing strength of intermolecular forces
Increasing melting point

The trend in the melting points of pentane, butanal, and butan-1-ol parallels the trend observed in their boiling points.

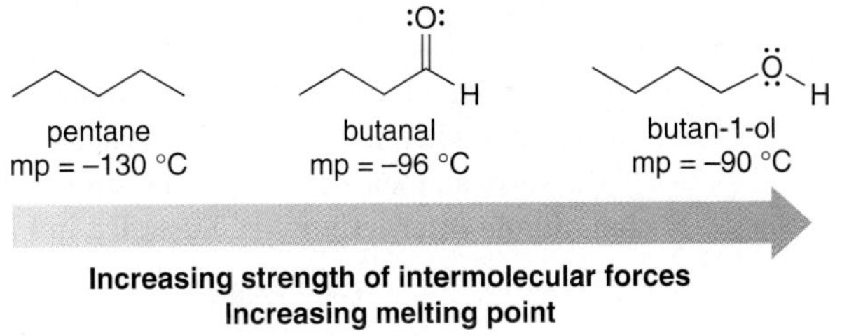
:O:
H
Ö
H
pentane
mp = –130 °C
butanal
mp = –96 °C
butan-1-ol
mp = –90 °C
Increasing strength of intermolecular forces
Increasing melting point

Symmetry also plays a role in determining the melting points of compounds having the same functional group and similar molecular weights, but very different shapes. A compact symmetrical molecule like 2,2-dimethylpropane packs well into a crystalline lattice whereas 2-methylbutane, which has a CH_3 group dangling from a four-carbon chain, does not. Thus, 2,2-dimethylpropane has a much higher melting point.

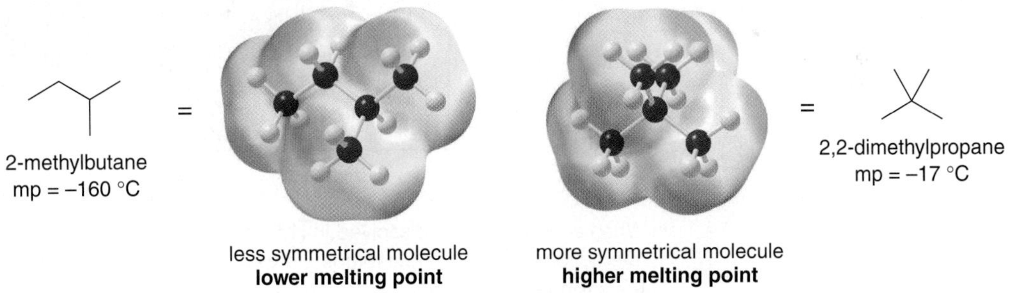
=
2-methylbutane
mp = –160 °C
=
2,2-dimethylpropane
mp = –17 °C
less symmetrical molecule
lower melting point
more symmetrical molecule
higher melting point

Problem 3.15 Predict which compound in each pair has the higher melting point.

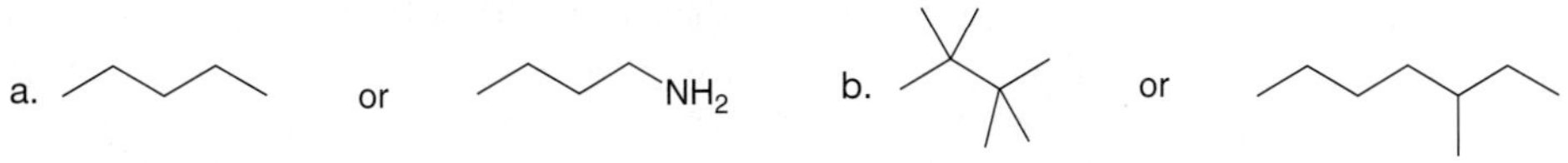
a.
or
NH2
b.
or

Problem 3.16 Consider acetic acid (CH_3CO_2H) and its conjugate base, sodium acetate (CH_3CO_2Na). (a) What intermolecular forces are present in each compound? (b) Explain why the melting point of sodium acetate (324 °C) is considerably higher than the melting point of acetic acid (17 °C).

3.4C Solubility

Quantitatively, a compound may be considered soluble when 3 g of solute dissolves in 100 mL of solvent.

Solubility* is the extent to which a compound, called the *solute,* dissolves in a liquid, called the *solvent. In dissolving a compound, the energy needed to break up the interactions between the molecules or ions of the solute comes from new interactions between the solute and the solvent.

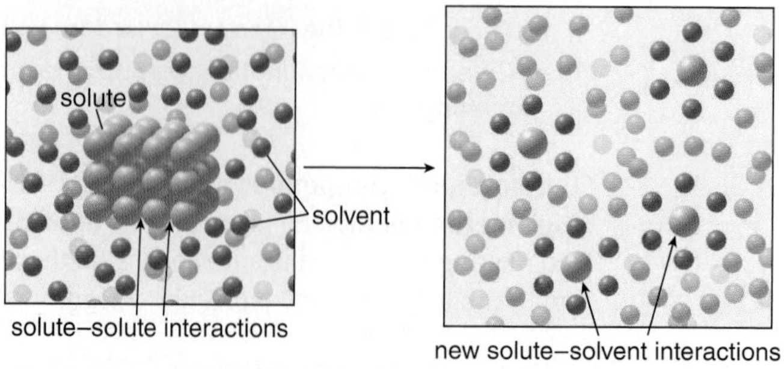

Compounds dissolve in solvents having similar kinds of intermolecular forces.

- "Like dissolves like."
- Polar compounds dissolve in polar solvents. Nonpolar or weakly polar compounds dissolve in nonpolar or weakly polar solvents.

Water and **organic liquids** are two different kinds of solvents. Water is very polar because it is capable of hydrogen bonding with a solute. Many organic solvents are either nonpolar, like carbon tetrachloride (CCl_4) and hexane [$CH_3(CH_2)_4CH_3$], or weakly polar like diethyl ether ($CH_3CH_2OCH_2CH_3$).

Ionic compounds are held together by strong electrostatic forces, so they need very polar solvents to dissolve. **Most ionic compounds are soluble in water, but are insoluble in organic solvents.** To dissolve an ionic compound, the strong ion–ion interactions must be replaced by many weaker **ion–dipole interactions,** as illustrated in Figure 3.4.

Figure 3.4
Dissolving an ionic compound in H_2O

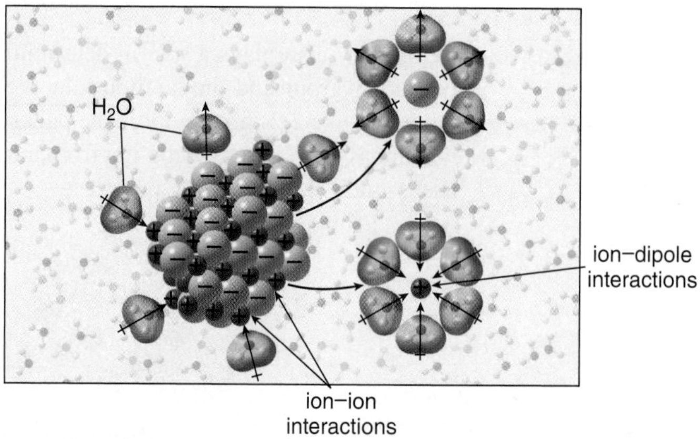

- When an ionic solid is dissolved in H_2O, the ion–ion interactions are replaced by ion–dipole interactions. Though these forces are weaker, there are so many of them that they compensate for the stronger ionic bonds.

Because water is the solvent in cells and many body fluids, many biological organic compounds contain ionic functional groups, so that they are water soluble. Isopentenyl diphosphate, a precursor of cholesterol, and acetylcholine, a neurotransmitter, are examples of ionic, water-soluble biological compounds.

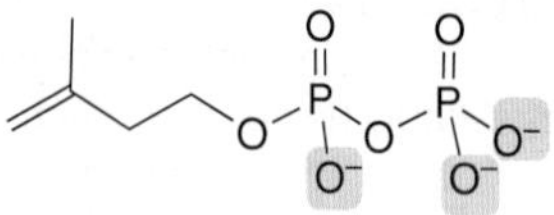

isopentenyl diphosphate

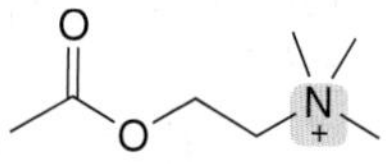

acetylcholine

Most uncharged organic compounds are soluble in organic solvents (remember, *like dissolves like*). **An organic compound is water soluble only if it contains one polar functional group capable of hydrogen bonding with the solvent for every five C atoms it contains.** In other words, a water-soluble organic compound has an O- or N-containing functional group that solubilizes its nonpolar carbon backbone.

Compare, for example, the solubility of butane and acetone in H_2O and CCl_4.

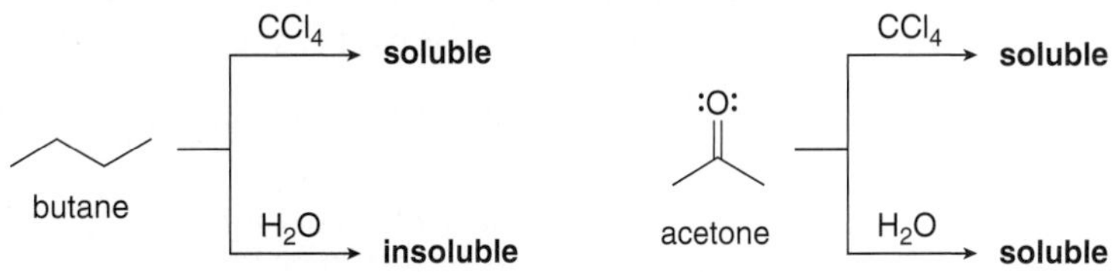

Because butane and acetone are both organic compounds having a C−C and C−H backbone, they are soluble in the organic solvent CCl_4. Butane, a nonpolar molecule, is insoluble in the polar solvent H_2O. Acetone, however, is H_2O soluble because it contains only three C atoms and its O atom can hydrogen bond with one H atom of H_2O. In fact, acetone is so soluble in water that acetone and water are **miscible**—they form solutions in all proportions with each other.

$(CH_3)_2C=O$ molecules cannot hydrogen bond to each other because they have no OH group. However, $(CH_3)_2C=O$ can hydrogen bond to H_2O because its O atom can hydrogen bond to one of the H atoms of H_2O.

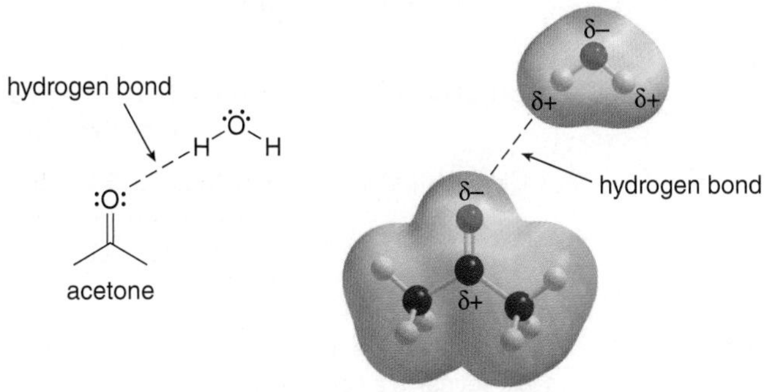

For an organic compound with one functional group, **a compound is water soluble only if it has ≤ five C atoms and contains an O or N atom.**

The size of an organic molecule with a polar functional group determines its water solubility. A low molecular weight alcohol like **ethanol is water soluble** because it has a small carbon skeleton (≤ five C atoms) compared to the size of its polar OH group. Cholesterol, on the other hand, has 27 carbon atoms and only one OH group. Its carbon skeleton is too large for the OH group to solubilize by hydrogen bonding, so **cholesterol is insoluble in water.**

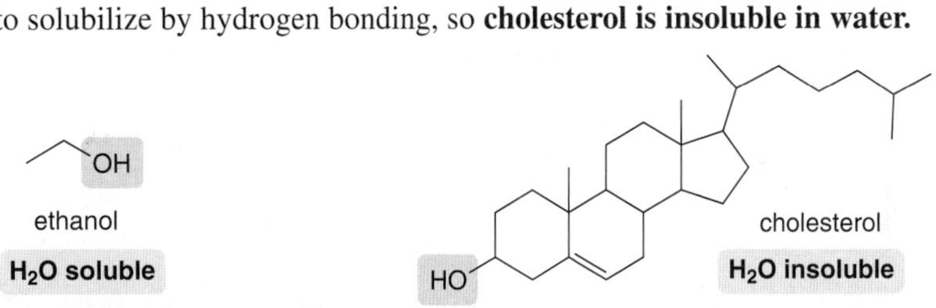

Hydrophobic = afraid of H_2O.
Hydrophilic = H_2O loving.

- The nonpolar part of a molecule that is not attracted to H_2O is said to be *hydrophobic.*
- The polar part of a molecule that can hydrogen bond to H_2O is said to be *hydrophilic.*

In cholesterol, for example, the **hydroxy group is hydrophilic,** whereas the **carbon skeleton is hydrophobic.**

MTBE (*tert*-butyl methyl ether) and 4,4'-dichlorobiphenyl (a polychlorinated biphenyl, abbreviated as PCB) demonstrate that solubility properties can help determine the fate of organic compounds in the environment.

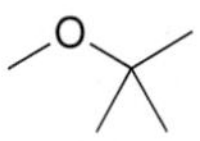

MTBE
tert-butyl methyl ether

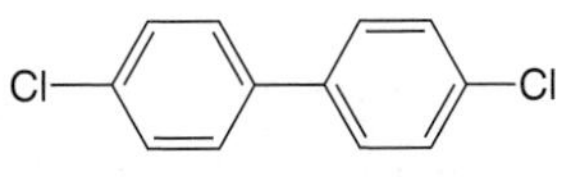

4,4'-dichlorobiphenyl
(a polychlorinated biphenyl, PCB)

Using **MTBE** as a high-octane additive in unleaded gasoline has had a negative environmental impact. Although MTBE is not toxic or carcinogenic, it has a distinctive, nauseating odor, and **it is water soluble.** Small amounts of MTBE have contaminated the drinking water in several communities, making it unfit for consumption. For this reason, the use of MTBE as a gasoline additive has steadily declined in the United States since 1999.

4,4'-Dichlorobiphenyl is a polychlorinated biphenyl **(PCB),** a compound that contains two benzene rings joined by a C−C bond, and substituted by one or more chlorine atoms on each ring. PCBs have been used as plasticizers in polystyrene coffee cups and coolants in transformers. They have been released into the environment during production, use, storage, and disposal, making them one of the most widespread organic pollutants. **PCBs are insoluble in H_2O, but very soluble in**

organic media, so they are soluble in fatty tissue, including that found in all types of fish and birds around the world. Although PCBs are not acutely toxic, frequently ingesting large quantities of fish contaminated with PCBs has been shown to retard growth and memory retention in children.

Solubility properties of some representative compounds are summarized in Table 3.6.

Table 3.6 Summary of Solubility

Type of compound	Solubility in H_2O	Solubility in organic solvents (such as CCl_4)
Ionic		
NaCl	**soluble**	**insoluble**
Covalent		
$CH_3CH_2CH_2CH_3$	**insoluble** (no N or O atom to hydrogen bond to H_2O)	**soluble**
$CH_3CH_2CH_2OH$	**soluble** ($\leq$ 5 C's and an O atom for hydrogen bonding to H_2O)	**soluble**
$CH_3(CH_2)_{10}OH$	**insoluble** (> 5 C's; too large to be soluble even though it has an O atom for hydrogen bonding to H_2O)	**soluble**

Sample Problem 3.4 Which compounds are water soluble? Phenylalanine is an amino acid that occurs in proteins, and 11-*cis*-retinal is a light-sensitive aldehyde that plays a key role in the chemistry of vision.

a. 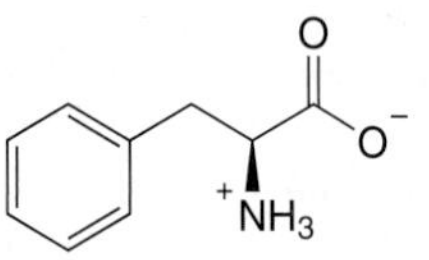

phenylalanine

b. 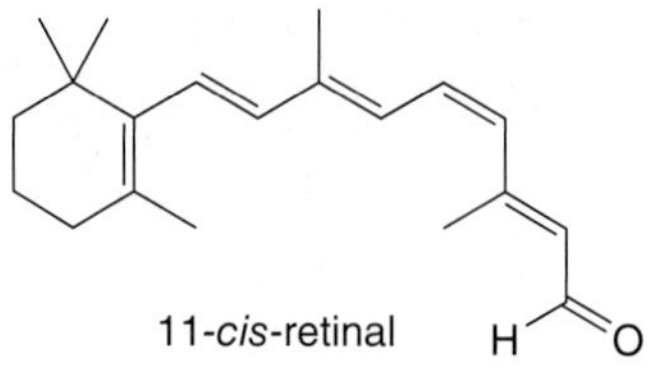

11-*cis*-retinal

Solution

a. Phenylalanine is ionic, so it is water soluble.
b. 11-*cis*-Retinal has 20 carbons and only one polar functional group. Its carbon skeleton is too large for the C=O to solubilize by hydrogen bonding with water, so it is water insoluble.

Problem 3.17 Which compounds are water soluble? Adrenaline is a hormone synthesized in the adrenal glands, and estrone is a female sex hormone.

a. 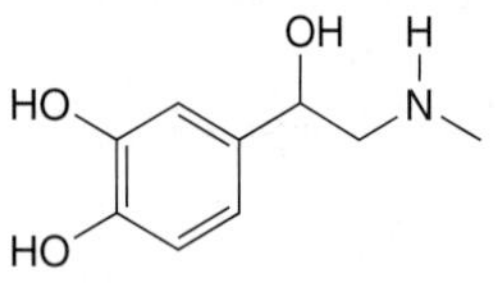

adrenaline

b. 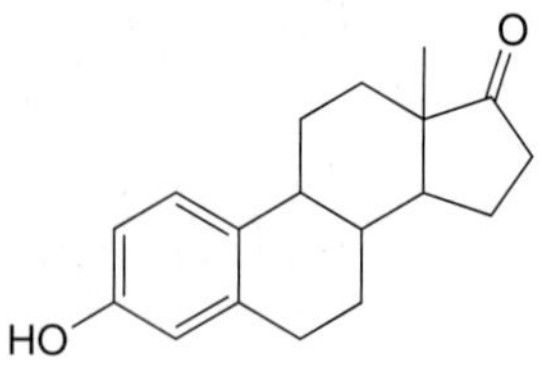

estrone

Sample Problem 3.5 Consider the female sex hormone progesterone (**A**) and the male sex hormone testosterone (**B**). (a) Which of these molecules can hydrogen bond to another molecule like itself? (b) Which of these molecules can hydrogen bond with water?

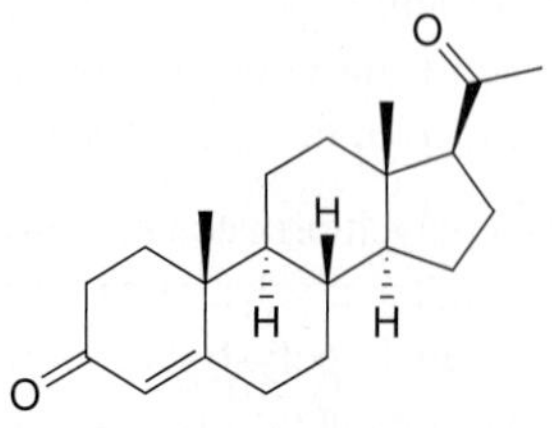

progesterone

A

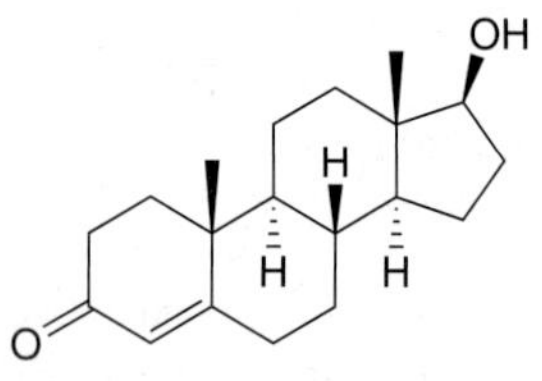

testosterone

B

Solution

a. Compounds **A** and **B** have polar C–O bonds but **A** does not have an O–H bond, so it cannot hydrogen bond to another **A** molecule. **B,** on the other hand, has an O–H bond, so two **B** molecules can hydrogen bond with each other.
b. Because **A** and **B** have electronegative O atoms, both **A** and **B** can hydrogen bond with a hydrogen atom in water.

Problem 3.18 (a) Label the hydrophobic and hydrophilic portions of each molecule. (b) At which sites can **C** hydrogen bond to another molecule like itself? (c) At which sites can **D** hydrogen bond to water?

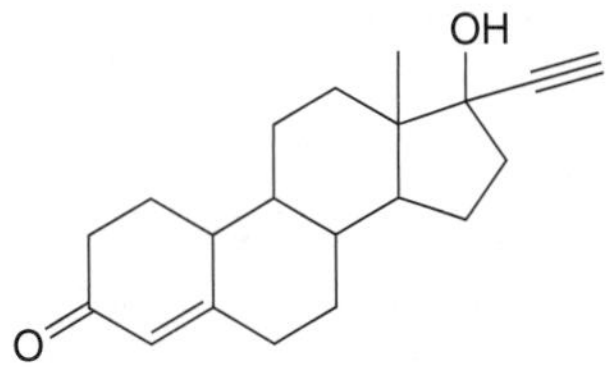

C
norethindrone
(oral contraceptive component)

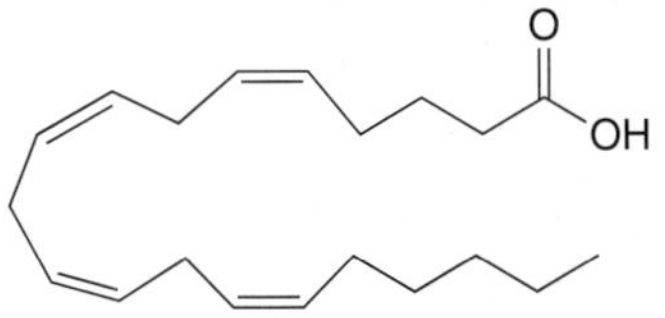

D
arachidonic acid
(fatty acid)

3.5 Application: Vitamins

Vitamins **are organic compounds needed in small amounts for normal cell function.** Our bodies cannot synthesize these compounds, so they must be obtained in the diet. Most vitamins are identified by a letter, such as A, C, D, E, and K. There are several different B vitamins, though, so a subscript is added to distinguish them: for example, B_1, B_2, and B_{12}.

Whether a vitamin is **fat soluble** (it dissolves in organic media) or **water soluble** can be determined by applying the solubility principles discussed in Section 3.4C. Vitamins A and C illustrate the differences between fat-soluble and water-soluble vitamins.

3.5A Vitamin A

The name **vitamin** was first used in 1912 by the Polish chemist Casimir Funk, who called them *vitamines,* because he thought that they all contained an *amine* functional group. Later the word was shortened to vitamin, because some are amines but others, like vitamins A and C, are not.

Vitamin A, or **retinol,** is an essential component of the vision receptors in the eyes. It also helps to maintain the health of mucous membranes and the skin, so many anti-aging creams contain vitamin A. A deficiency of this vitamin leads to a loss of night vision.

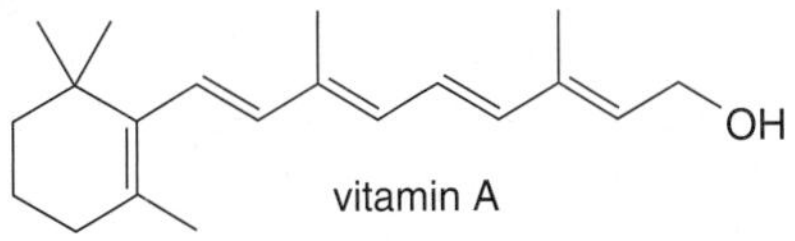

vitamin A

Vitamin A contains 20 carbons and a single OH group, making it **water insoluble.** As a result, vitamin A is insoluble in bodily fluids such as blood, gastric juices in the stomach, and urine, which are largely water with dissolved ions such as Na^+ and K^+. There are also fat cells composed of organic compounds having C–C and C–H bonds. Vitamin A is soluble in this organic environment because it is an uncharged organic compound, and thus it is readily stored in these fat cells, particularly in the liver.

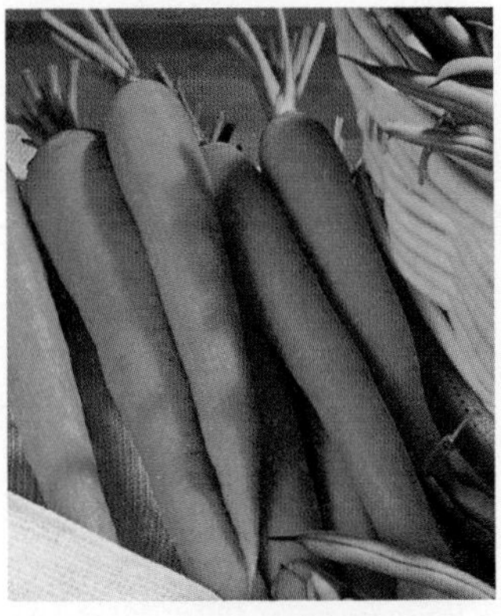

Vitamin A is synthesized from β-carotene, the orange pigment in carrots.

Vitamin A may be obtained directly from the diet. In addition, β-carotene, the orange pigment found in many plants including carrots, is readily converted to vitamin A in our bodies.

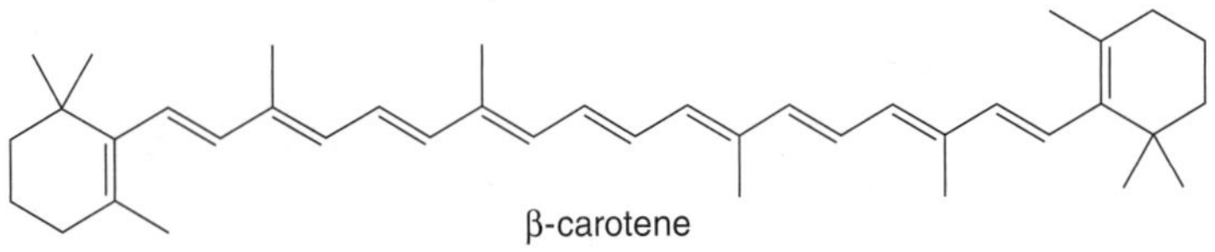

β-carotene

Eating too many carrots does not result in an excess of stored vitamin A. If you consume more β-carotene than you need, your body stores this precursor until it needs more vitamin A. Some β-carotene reaches the surface tissues of the skin and eyes, giving them an orange color.

This phenomenon may look odd, but it is harmless and reversible. When stored β-carotene is converted to vitamin A and is no longer in excess, these tissues will return to their normal hue.

Vitamin C is obtained by eating citrus fruits and a wide variety of other fruits and vegetables. Individuals can also obtain the recommended daily dose of vitamin C by taking tablets that contain vitamin C prepared in the laboratory. Both the "natural" vitamin C in oranges and the "synthetic" vitamin C in vitamin supplements are identical.

3.5B Vitamin C

Although most animal species can synthesize vitamin C, humans, guinea pigs, the Indian fruit bat, and the bulbul bird must obtain this vitamin from dietary sources. Citrus fruits, strawberries, tomatoes, and sweet potatoes are all excellent sources of vitamin C.

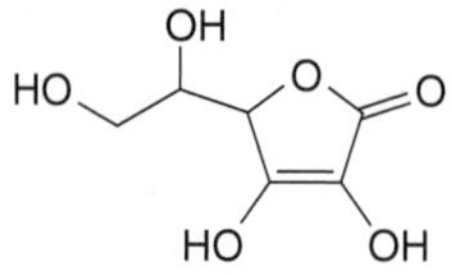

vitamin C
(ascorbic acid)

Vitamin C has six carbon atoms, each bonded to an oxygen atom that is capable of hydrogen bonding, making it **water soluble.** Vitamin C thus dissolves in urine. Although it has been acclaimed as a deterrent for all kinds of diseases, from the common cold to cancer, the consequences of taking large amounts of vitamin C are not really known, because any excess of the minimum daily requirement is excreted in the urine.

Problem 3.19 Predict the water solubility of each vitamin.

a. 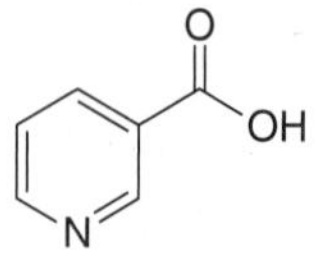

vitamin B_3
(niacin)

b. 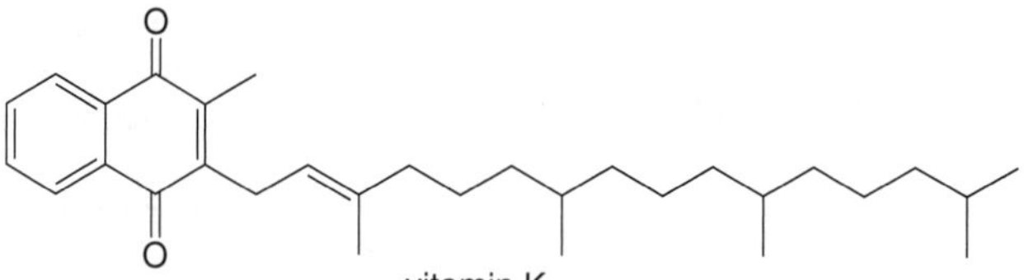

vitamin K_1
(phylloquinone)

Problem 3.20 (a) Identify the functional groups in the ball-and-stick model of pantothenic acid, vitamin B_5. (b) At which sites can pantothenic acid hydrogen bond to water? (c) Predict the water solubility of pantothenic acid.

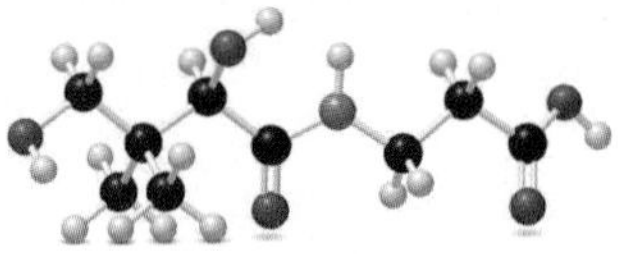

pantothenic acid
vitamin B_5

Avocados are an excellent dietary source of pantothenic acid, vitamin B_5.

3.6 Application of Solubility: Soap

Soap has been used by humankind for some 2000 years. Historical records describe its manufacture in the first century and document the presence of a soap factory in Pompeii. Before this time clothes were cleaned by rubbing them on rocks in water, or by forming soapy lathers from the roots, bark, and leaves of certain plants. These plants produced natural materials called *saponins,* which act in much the same way as modern soaps.

On a molecular level, soap has two distinct parts:

- **a hydrophilic portion composed of ions called the *polar head***
- **a hydrophobic carbon chain of nonpolar C–C and C–H bonds, called the *nonpolar tail***

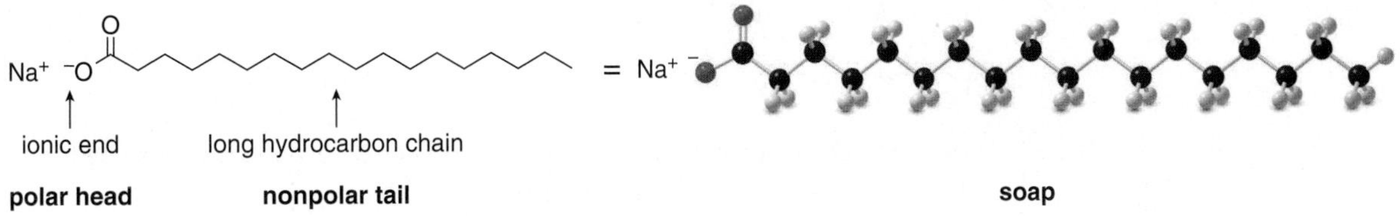

Dissolving soap in water forms ***micelles,* spherical droplets having the ionic heads on the surface and the nonpolar tails packed together in the interior,** as shown in Figure 3.5. In this arrangement, the ionic heads are solvated by the polar solvent water, thus solubilizing the nonpolar, "greasy" hydrocarbon portion of the soap.

Figure 3.5
Dissolving soap in water

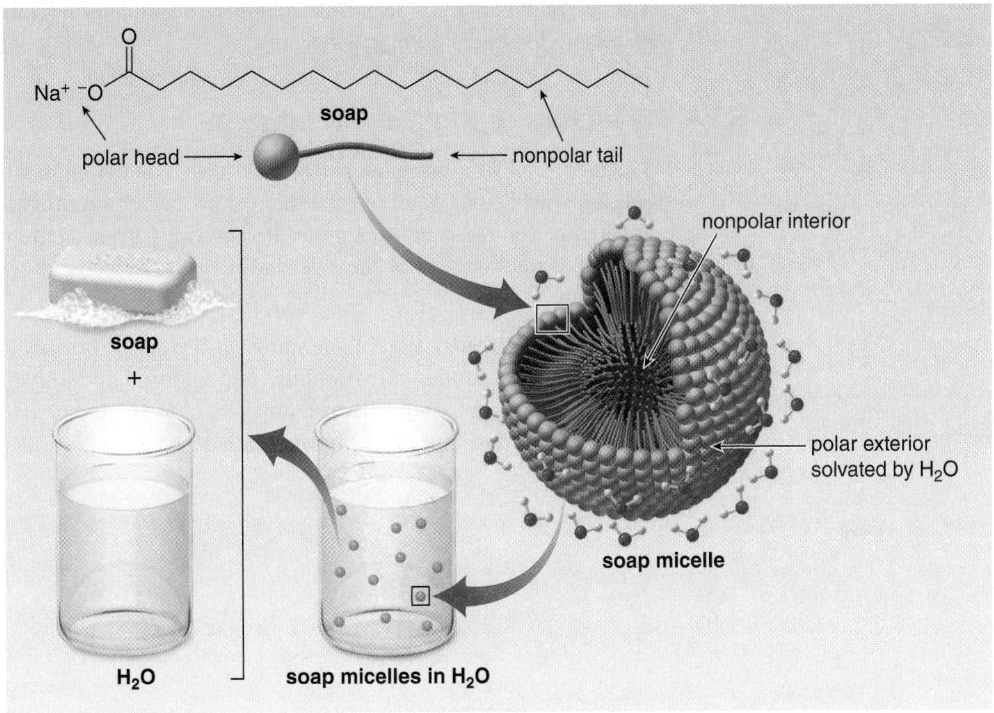

- When soap is dissolved in H_2O, it forms micelles with the nonpolar tails in the interior and the polar heads on the surface. The polar heads are solvated by ion–dipole interactions with H_2O molecules.

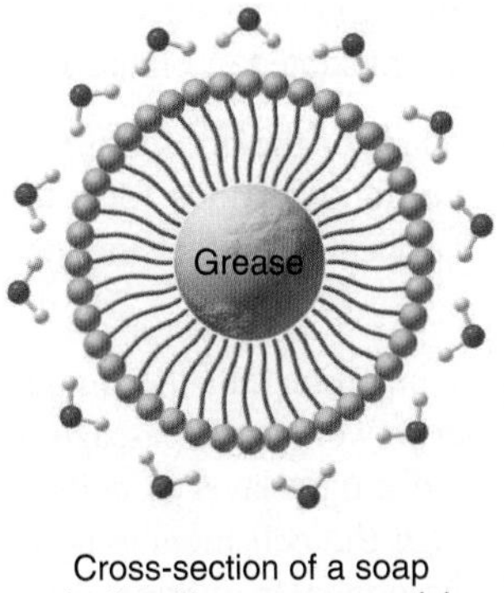

Cross-section of a soap micelle with a grease particle dissolved in the interior

How does soap dissolve grease and oil? Water alone cannot dissolve dirt, which is composed largely of nonpolar hydrocarbons. When soap is mixed with water, however, the nonpolar hydrocarbon tails dissolve the dirt in the interior of the micelle. The polar head of the soap remains on the surface of the micelle to interact with water. The nonpolar tails of the soap are so well sealed off from the water by the polar head groups that the micelles are water soluble, allowing them to separate from the fibers of our clothes and be washed down the drain with water. In this way, soaps do a seemingly impossible task: they remove nonpolar hydrocarbon material from skin and clothes, by solubilizing it in the polar solvent water.

Problem 3.21 Which of the following structures represent soaps? Explain your answers.

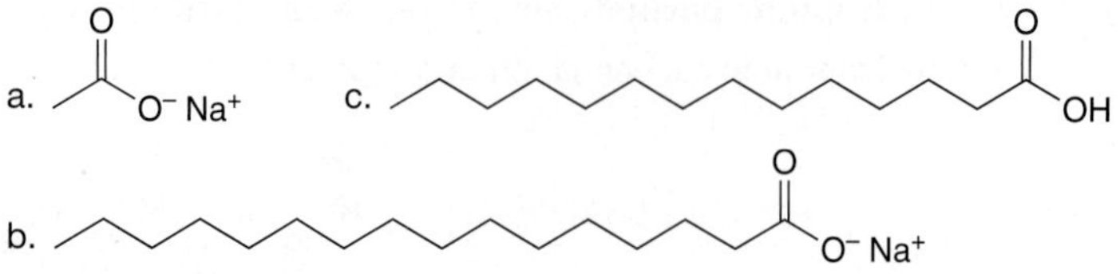
a.
O
O⁻ Na⁺
c.
O
OH
b.
O
O⁻ Na⁺

Problem 3.22 Today, synthetic detergents like the compound drawn here, not soaps, are used to clean clothes. Explain how this detergent cleans away dirt.

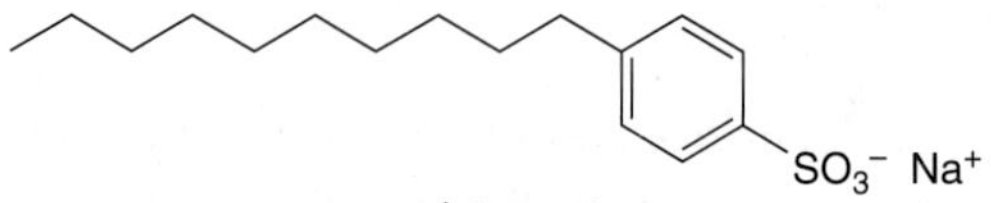
SO3⁻ Na⁺

a detergent

3.7 Application: The Cell Membrane

The cell membrane is a beautifully complex example of how the principles of organic chemistry come into play in a biological system.

3.7A Structure of the Cell Membrane

The basic unit of living organisms is the **cell.** The cytoplasm is the aqueous medium inside the cell, separated from water outside the cell by the **cell membrane.** The cell membrane serves two apparently contradictory functions. It acts as a barrier to the passage of ions, water, and other molecules into and out of the cell, and it is also selectively permeable, letting nutrients in and waste out.

A major component of the cell membrane is a group of organic compounds called **phospholipids.** Like soap, they contain a hydrophilic ionic portion and a hydrophobic hydrocarbon portion, in this case two long carbon chains composed of C–C and C–H bonds. **Phospholipids thus contain a polar head and *two* nonpolar tails.**

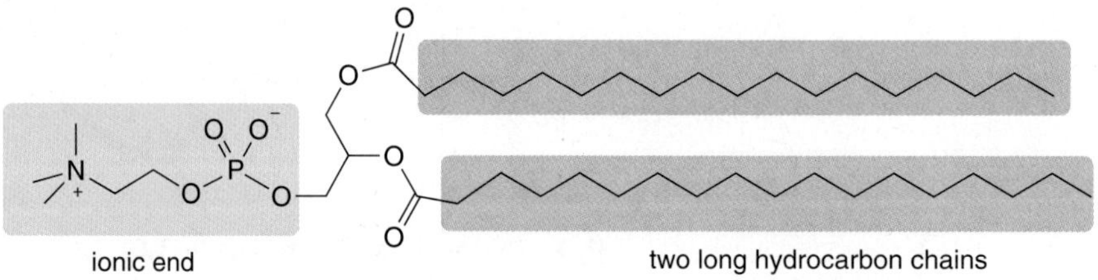
ionic end
polar head
two long hydrocarbon chains
nonpolar tails

phospholipid

When phospholipids are mixed with water, they assemble in an arrangement called a **lipid bilayer,** with the ionic heads oriented on the outside and the nonpolar tails on the inside. The polar heads electrostatically interact with the polar solvent H_2O, while the nonpolar tails are held in close proximity by numerous van der Waals interactions. This is schematically illustrated in Figure 3.6.

Cell membranes are composed of these lipid bilayers. The charged heads of the phospholipids are oriented toward the aqueous interior and exterior of the cell. The nonpolar tails form the hydrophobic interior of the membrane, thus serving as an insoluble barrier that protects the cell from the outside.

The nonpolar interior of the cell membrane is especially important in protecting the human brain from fluctuation in the concentration of compounds in the blood, as well as the passage of unwanted substances into the brain. The blood–brain barrier consists of a tight layer of cells in the blood capillaries of the brain, and all substances must pass through the cell membrane of these capillaries to enter the brain. Because ions are not soluble in the nonpolar interior of the cell membrane, the blood–brain barrier is only slightly permeable to ions. On the other hand,

Figure 3.6 The cell membrane

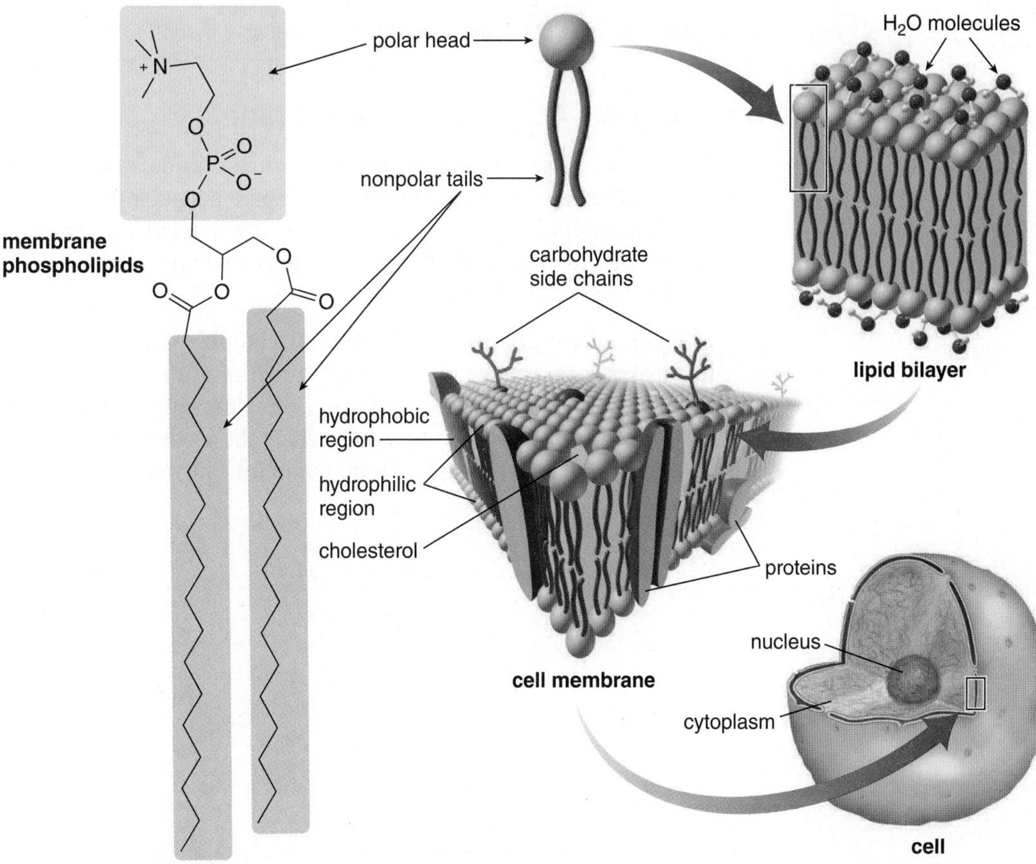

- Phospholipids contain an ionic or polar head, and two long nonpolar hydrocarbon tails. In an aqueous environment, phospholipids form a lipid bilayer, with the polar heads oriented toward the aqueous exterior and the nonpolar tails forming a hydrophobic interior. Cell membranes are composed largely of this lipid bilayer.

uncharged organic molecules like nicotine, caffeine, and heroin are very soluble in the interior of the cell membrane, so they readily pass into the brain.

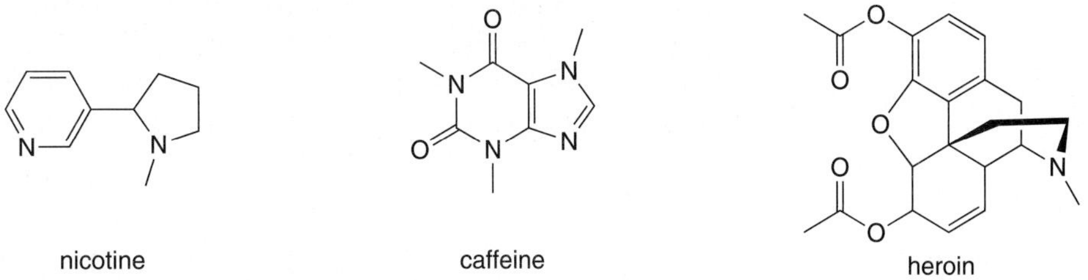

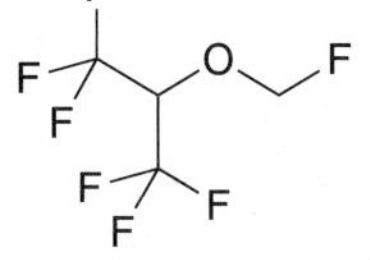

sevoflurane

General anesthetics such as sevoflurane are also weakly polar compounds that can penetrate the blood–brain barrier because they are soluble in the lipid bilayer of the blood capillaries.

Problem 3.23 (a) What types of intermolecular forces do morphine and heroin each possess? (b) Which compound can cross the blood–brain barrier more readily, and therefore serve as the more potent pain reliever?

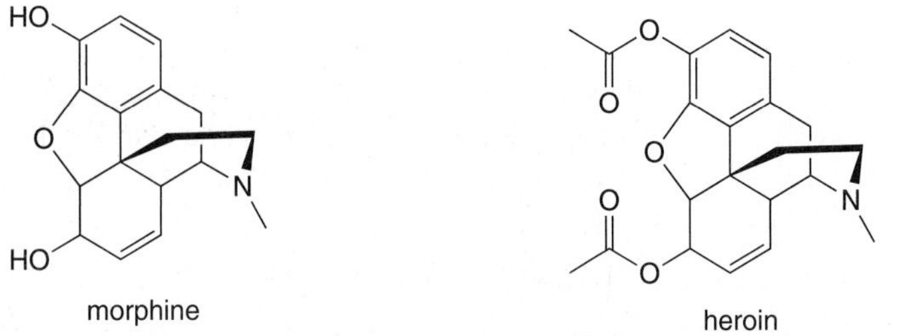

Problem 3.24 Explain why the noble gas xenon is a general anesthetic.

3.7B Transport Across a Cell Membrane

How does a polar molecule or ion in the water outside a cell pass through the nonpolar interior of the cell membrane and enter the cell? Some nonpolar molecules like O_2 are small enough to enter and exit the cell by diffusion. Polar molecules and ions, on the other hand, may be too large or too polar to diffuse efficiently. Some ions are transported across the membrane with the help of molecules called **ionophores.**

***Ionophores* are organic molecules that complex cations.** They have a hydrophobic exterior that makes them soluble in the nonpolar interior of the cell membrane, and a central cavity with several oxygen atoms whose lone pairs complex with a given ion. The size of the cavity determines the identity of the cation with which the ionophore complexes. Two naturally occurring antibiotics that act as ionophores are **nonactin** and **valinomycin.**

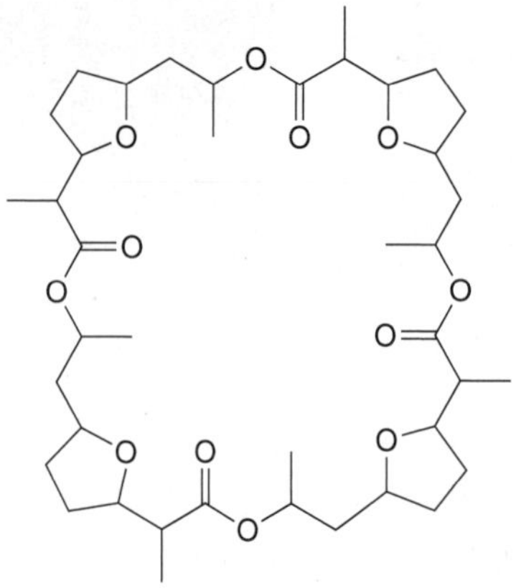

nonactin

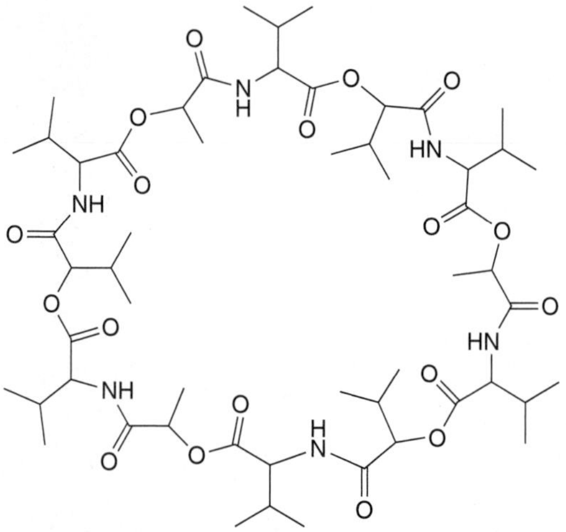

valinomycin

Several synthetic ionophores have also been prepared, including one group called **crown ethers.** ***Crown ethers* are cyclic ethers containing several oxygen atoms that bind specific cations depending on the size of their cavity.** Crown ethers are named according to the general format ***x*-crown-*y*,** where ***x*** is the total number of atoms in the ring and ***y*** is the number of oxygen atoms. For example, 18-crown-6 contains 18 atoms in the ring, including 6 O atoms. This crown ether binds potassium ions. Sodium ions are too small to form a tight complex with the O atoms, and larger cations do not fit in the cavity.

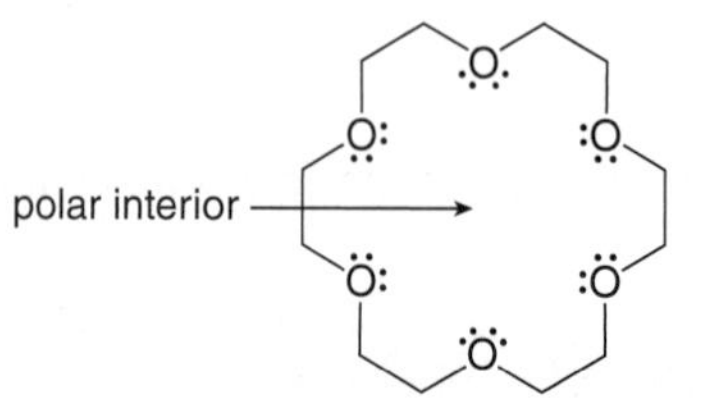

18-crown-6

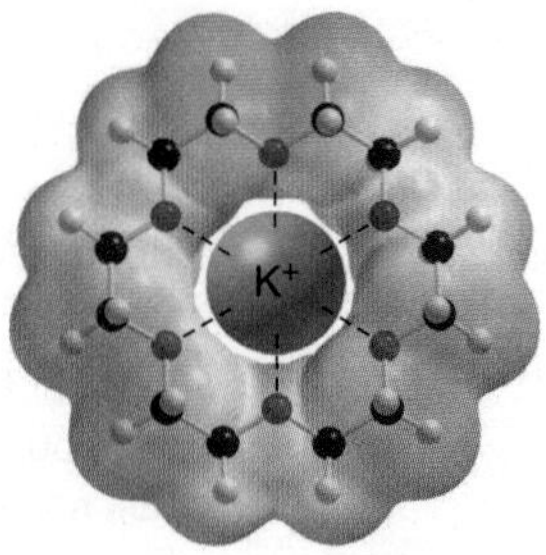

complex with K^+

How does an ionophore transfer an ion across a membrane? The ionophore binds the ion on one side of the membrane in its polar interior. It can then move across the membrane because its

Figure 3.7
Transport of ions across a cell membrane

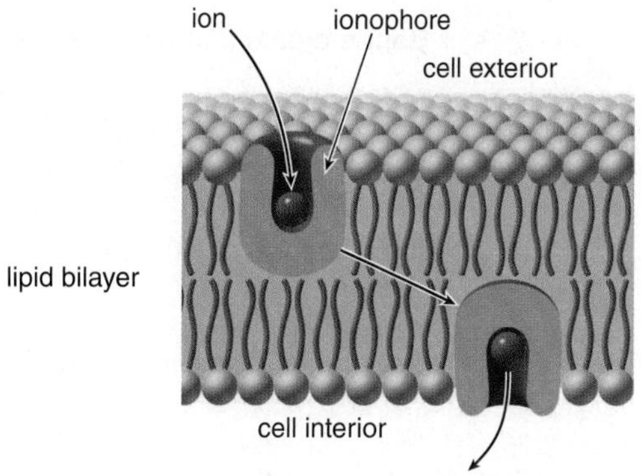

- By binding an ion on one side of a lipid bilayer (where the concentration of the ion is high) and releasing it on the other side of the bilayer (where the concentration of the ion is low), an ionophore transports an ion across a cell membrane.

hydrophobic exterior interacts with the hydrophobic tails of the phospholipid. The ionophore then releases the ion on the other side of the membrane. This ion-transfer role is essential for normal cell function. This process is illustrated in Figure 3.7.

In this manner, antibiotic ionophores like nonactin transport ions across a cell membrane of bacteria. This disrupts the normal ionic balance in the cell, thus interfering with cell function and causing the bacteria to die.

Problem 3.25 Now that you have learned about solubility, explain why aspirin (Section 2.7) crosses a cell membrane as a neutral carboxylic acid rather than an ionic conjugate base.

3.8 Functional Groups and Reactivity

Much of Chapter 3 has been devoted to how a functional group determines the strength of intermolecular forces and, consequently, the physical properties of molecules. A functional group also determines reactivity. What type of reaction does a particular kind of organic compound undergo? Begin by recalling two fundamental concepts.

- **Functional groups create reactive sites in molecules.**
- **Electron-rich sites react with electron-poor sites.**

All functional groups contain a heteroatom, a π bond, or both, and these features make electron-deficient (or electrophilic) sites and electron-rich (or nucleophilic) sites in a molecule. To predict reactivity, first locate the functional group and then determine the resulting electron-rich or electron-deficient sites it creates. Keep three guidelines in mind.

- **An electronegative heteroatom like N, O, or X makes a carbon atom *electrophilic*.**

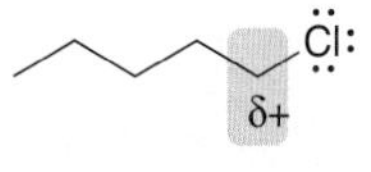

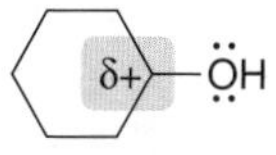

- **A lone pair on a heteroatom makes it basic and nucleophilic.**

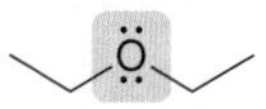

base
nucleophile

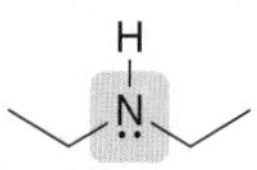

base
nucleophile

- **π Bonds create *nucleophilic* sites and are more easily broken than σ bonds.**

Problem 3.26 Label the electrophilic and nucleophilic sites in each molecule.

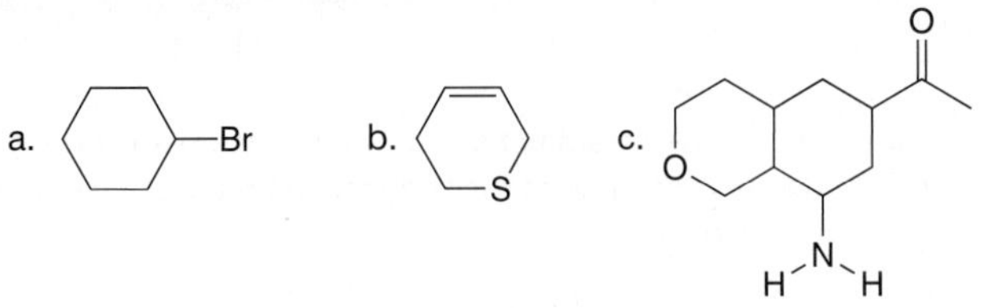

By identifying the nucleophilic and electrophilic sites in a compound, you can begin to understand how it will react. In general, electron-rich sites react with electron-deficient sites:

:Nu⁻ = a nucleophile;
E⁺ = an electrophile.

- **An electron-deficient carbon atom reacts with a nucleophile, symbolized as :Nu⁻.**
- **An electron-rich carbon reacts with an electrophile, symbolized as E⁺.**

At this point we don't know enough organic chemistry to draw the products of many reactions with confidence. We do know enough, however, to begin to predict if two compounds might react together based solely on electron density arguments, and at what atoms that reaction is most likely to occur.

For example, the carbon–carbon double bond of isopentenyl diphosphate is electron rich, so it reacts with electrophiles, E^+. This reaction is a key step in the early stages of the biosynthesis of cholesterol. On the other hand, the carbonyl carbon of pyruvic acid is electrophilic, so it reacts with electron-rich nucleophiles. This type of reaction is a key step in glucose metabolism.

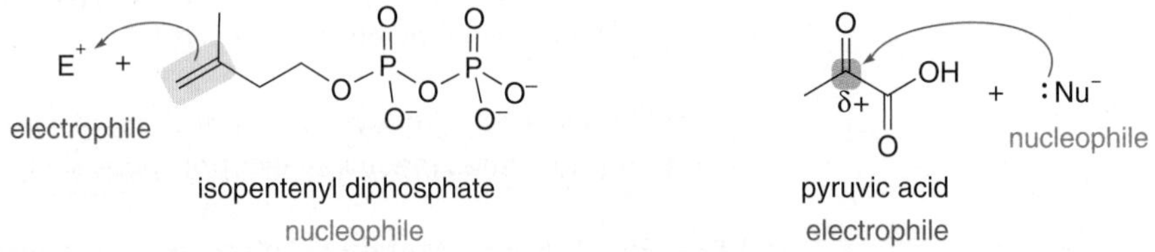

Complex molecules in biological systems may contain many functional groups, but often one of those functional groups determines many of the reactions of the molecule. For example, most reactions of **coenzyme A,** a key molecule in many biological pathways, occur when its electron-rich thiol reacts with electrophiles. As a result, we often abbreviate the structures of these compounds to emphasize the reacting functional group. Coenzyme A is written as **HS–CoA** to emphasize its nucleophilic thiol.

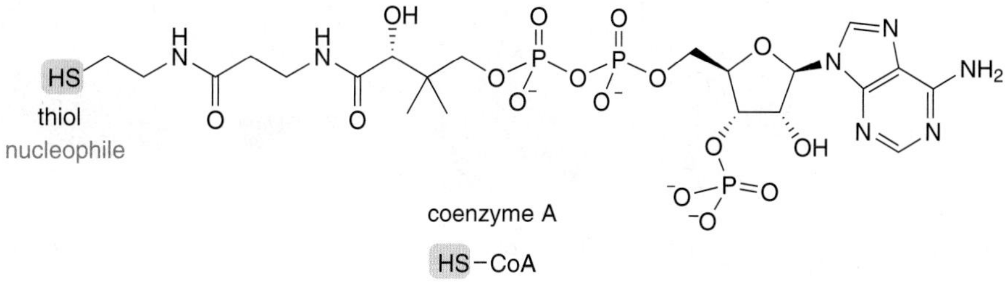
HS
thiol
nucleophile
coenzyme A
HS–CoA

Problem 3.27 Considering only electron density, state whether the following reactions will occur.

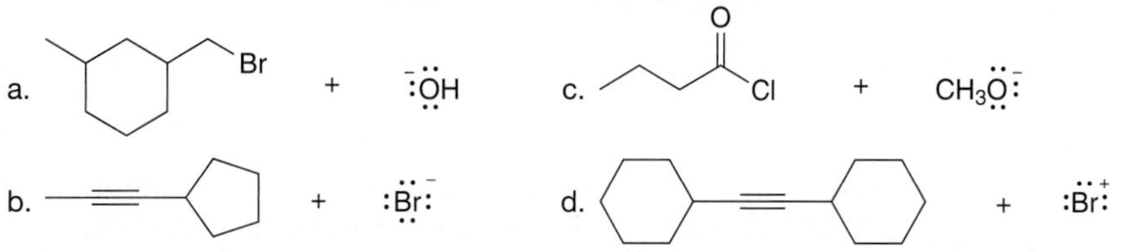
a.
b.
c.
d.

Problem 3.28 Which atom acts as the nucleophile and which atom acts as the electrophile in the starting materials of the following reaction, a key step in the synthesis of *S*-adenosylmethionine from ATP? We will learn more about the preparation and reactions of this biological molecule in Sections 7.16 and 9.15.

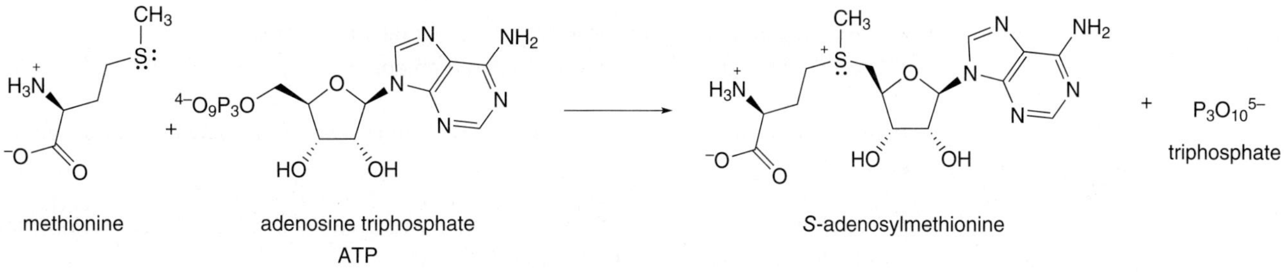
methionine
adenosine triphosphate
ATP
S-adenosylmethionine
triphosphate

3.9 Biomolecules

Biomolecules **are organic compounds found in biological systems.** Many are relatively small, with molecular weights of less than 1000 g/mol. There are four main families of these small molecules—amino acids, simple sugars or monosaccharides, nucleotides, and lipids. Many simple biomolecules are used to synthesize larger compounds that have important cellular functions.

3.9A Amino Acids and Proteins

Of the four major types of biomolecules, proteins have the widest range of functions. Proteins like collagen form long insoluble fibers in connective tissue, cartilage, and blood vessels (Figure 3.8), whereas enzymes are proteins that regulate all aspects of cellular function. Hemoglobin, which transports oxygen from the lungs to the tissues, and insulin, a hormone that regulates blood glucose levels, are both proteins.

- **Proteins are polyamides, formed by joining amino acids together.**

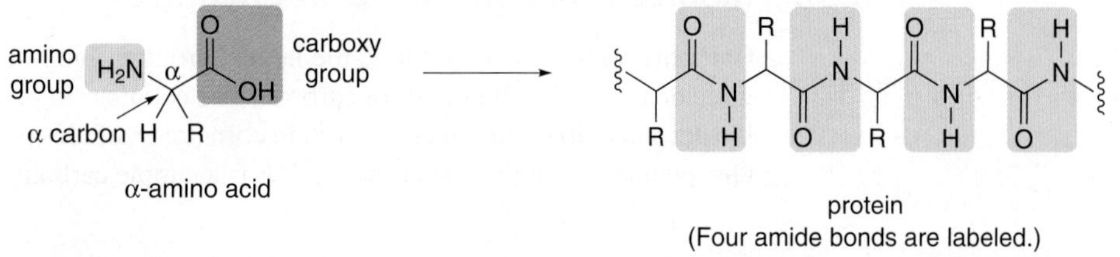
amino group
carboxy group
α carbon
α-amino acid
protein
(Four amide bonds are labeled.)

Figure 3.8 Proteins, large molecules formed from amino acids

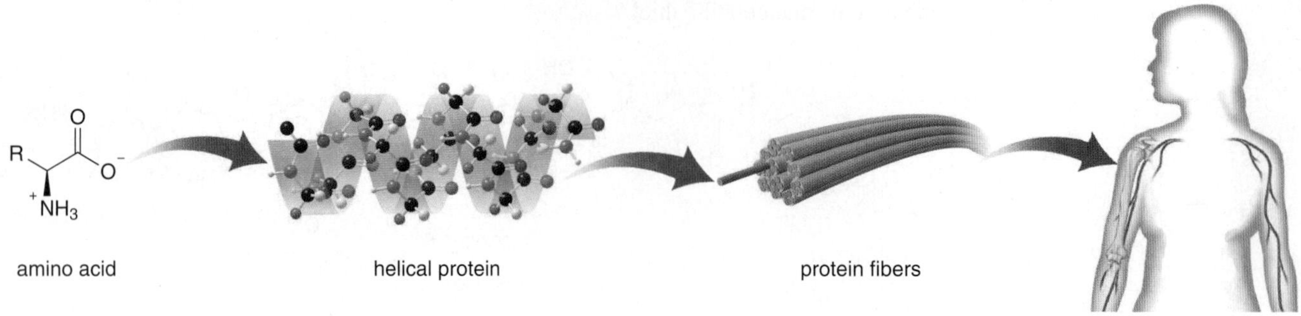

- The fibrous tissue in muscle, cartilage, and tendons is composed of proteins, which are formed from amino acids joined together by amide bonds.

A listing of the 20 naturally occurring amino acids is given in Figure 26.2.

Amino acids contain two functional groups—an **amino group** (NH_2) and a **carboxy group** (CO_2H)—bonded to the same carbon, called the **α carbon.** The 20 naturally occurring amino acids that form proteins differ in the identity of the R group bonded to the α carbon. The R group is called the **side chain** of the amino acid. An R group can be hydrogen, carbon groups with only C–C and C–H bonds, phenyl, or have additional functional groups such as SH, NH_2, OH, or CO_2H.

An amino acid is both an acid and a base.

- **The NH_2 group has a lone pair of electrons, making it a base.**
- **The CO_2H has an acidic proton, making it an acid.**

Even though we sometimes draw amino acids with all neutral atoms, in reality, amino acids exist as salts.

- **Proton transfer from the acid to the base forms an ionic salt called a zwitterion, which contains both a positive and negative charge.**

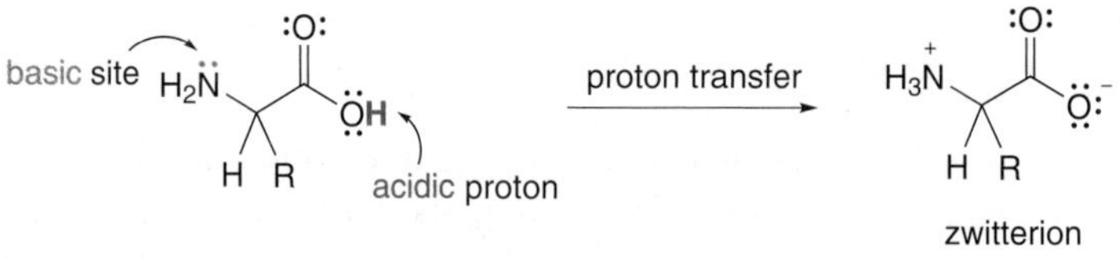

We will learn more about the acid–base chemistry of amino acids in Chapter 19.

Problem 3.29 Draw the zwitterionic form of each amino acid.

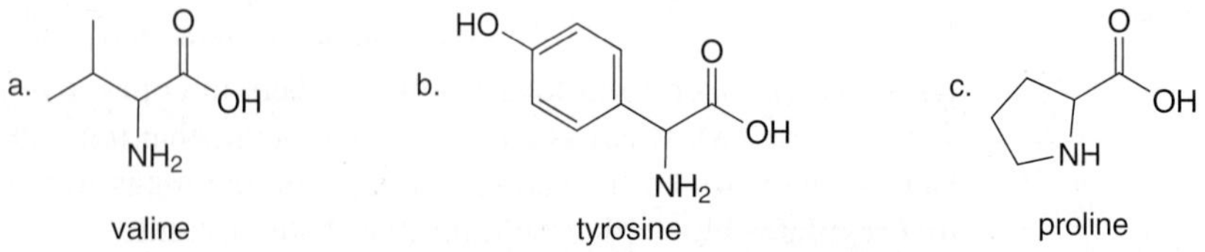

3.9B Monosaccharides and Carbohydrates

Carbohydrates, which constitute the largest group of biomolecules in nature, may have as few as three or as many as thousands of carbons. The cellulose in plant stems and tree trunks, the chitin in lobster and crab shells, and the starch in corn are examples of complex carbohydrates (Figure 3.9). One component of the nucleic acid DNA is a simple carbohydrate, as we will see in Section 3.9C.

Figure 3.9 Glucose and starch, a complex carbohydrate

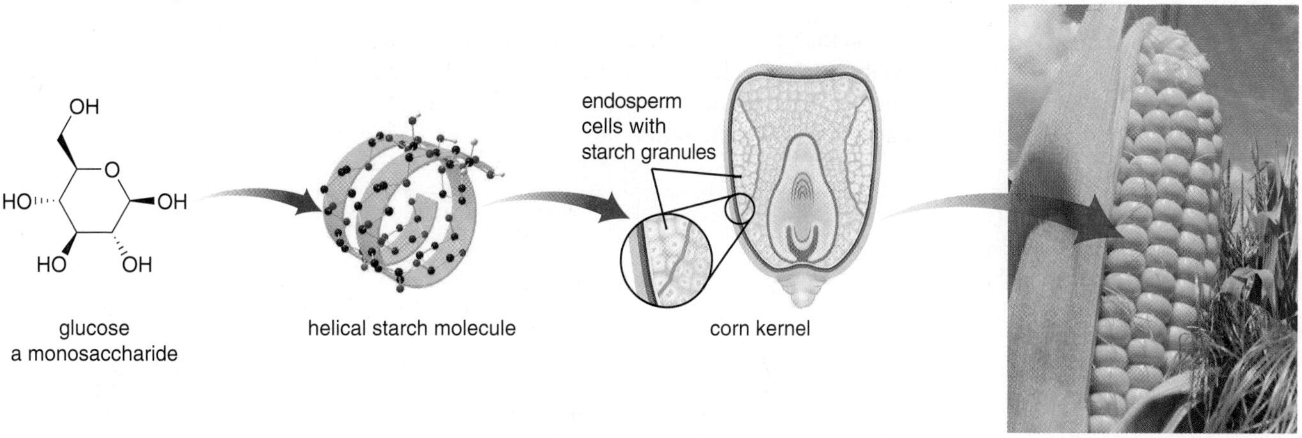

- Kernels of corn contain starch, a complex carbohydrate formed from the simple sugar glucose.

- **Carbohydrates, commonly called sugars and starches, are polyhydroxy aldehydes and ketones, or compounds that can be converted to them by reaction with water.**

The simplest carbohydrates are called **monosaccharides** or **simple sugars.** Monosaccharides have three to six carbon atoms in a chain, with a carbonyl group at either the terminal carbon or the carbon adjacent to it. In most monosaccharides, each remaining carbon has a hydroxy group. Glyceraldehyde and glucose are monosaccharides. As we will learn in Chapter 21, glucose, the most prevalent monosaccharide, can be drawn as an acyclic compound with an aldehyde or as a cyclic compound with a six-membered ring that contains an oxygen.

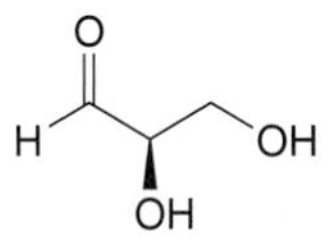

glyceraldehyde

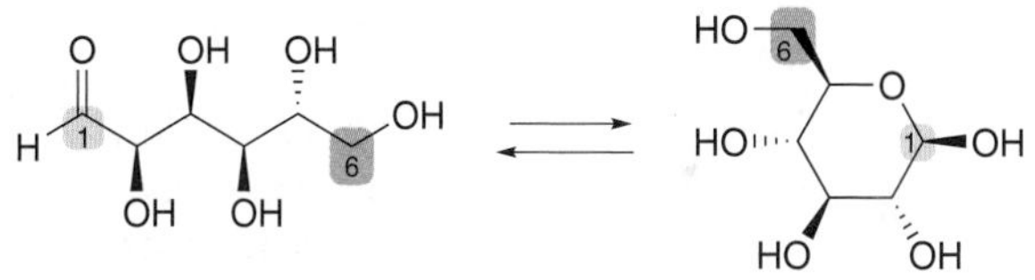

glucose
acyclic form

glucose
cyclic form

Joining two monosaccharides together forms **disaccharides.** Lactose is a disaccharide found in milk. Joining three or more monosaccharides together forms **polysaccharides.** Starch is a common polysaccharide found in corn, rice, and wheat. Disaccharides and polysaccharides are discussed in Chapter 27.

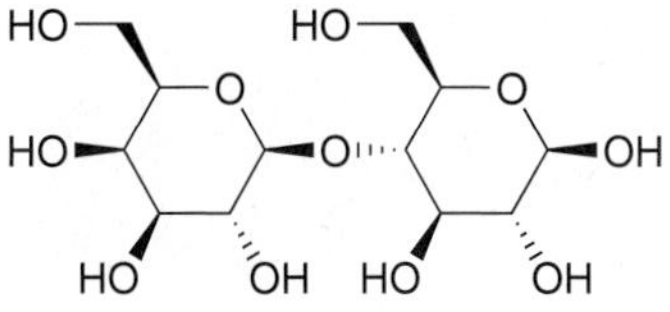

lactose
a disaccharide

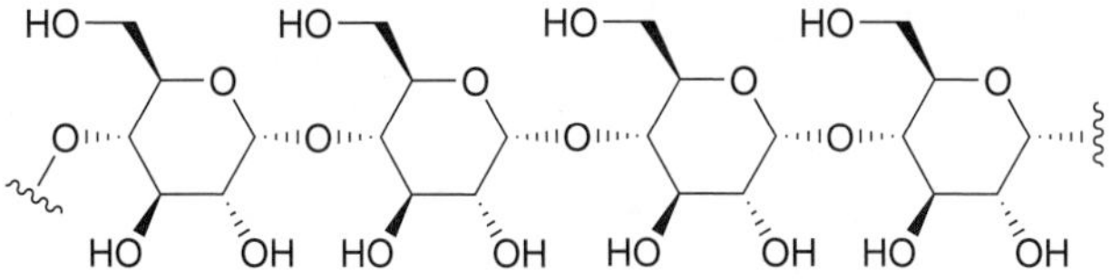

amylose
(one form of starch)
a polysaccharide

Problem 3.30 Why are glucose and lactose water soluble, even though they each contain more than five carbon atoms?

3.9C Nucleotides and Nucleic Acids

Nucleic acids are large molecules composed of repeating units called **nucleotides. DNA, deoxyribonucleic acid,** stores the genetic information of an organism and transmits that information from one generation to another (Figure 3.10), whereas **RNA, ribonucleic acid,** translates that information into proteins needed for all cellular functions.

Figure 3.10
Nucleotides, the building blocks of DNA, a nucleic acid

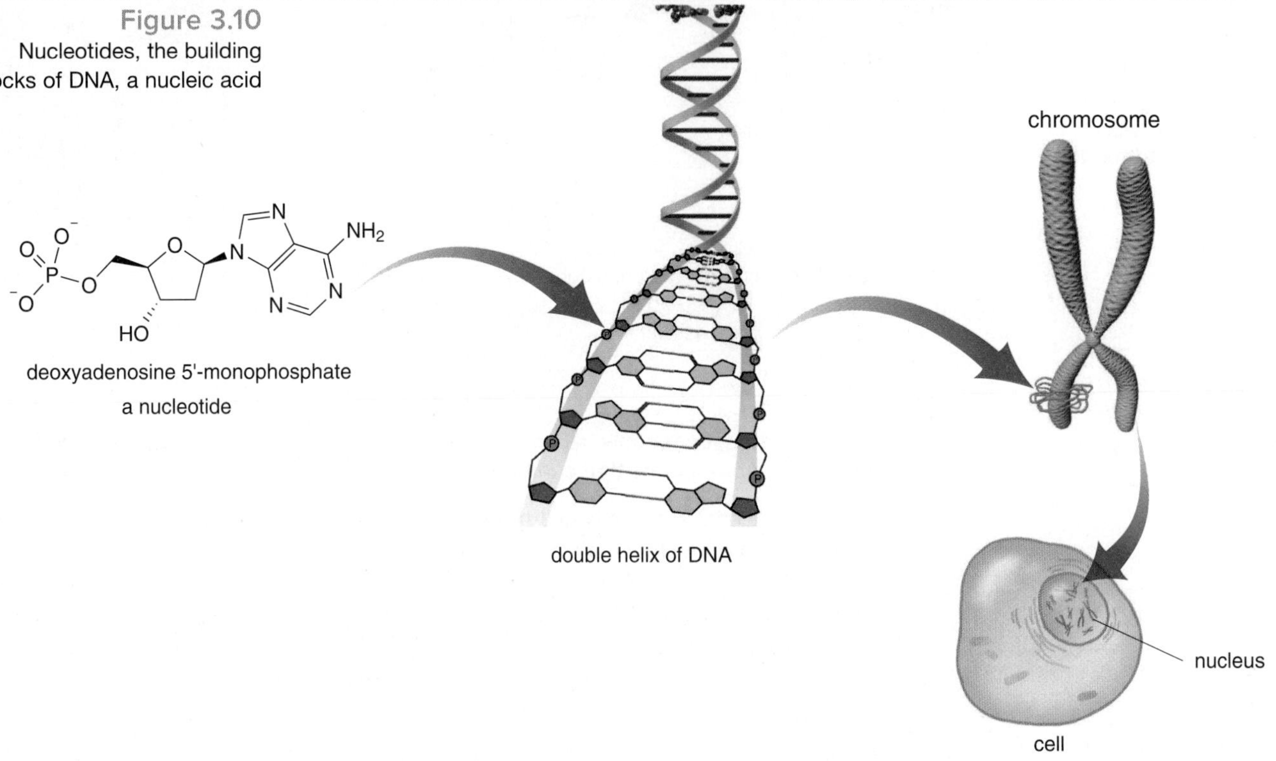

- DNA, which is formed from smaller units called nucleotides, is contained in the chromosomes of the nucleus. Humans have 46 chromosomes (23 pairs). An individual chromosome is composed of many genes. A gene is a portion of the DNA molecule responsible for the synthesis of a single protein.

- **A nucleotide consists of three components—a monosaccharide, a nitrogen-containing base, and a phosphate group.**

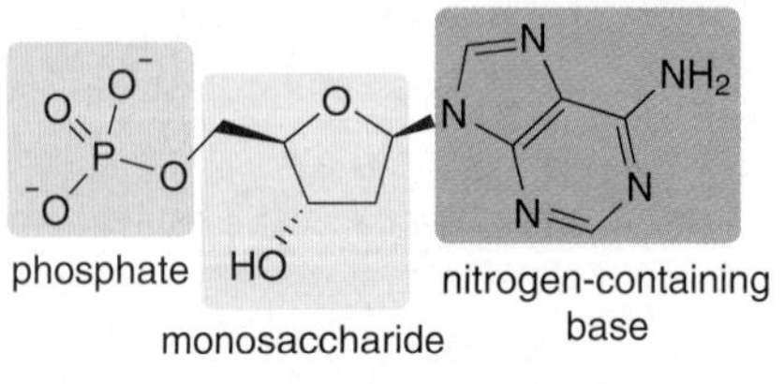

nucleotide

In RNA, the monosaccharide is ribose, whereas in DNA, the monosaccharide is 2-deoxyribose, a compound that lacks a hydroxy group at C2.

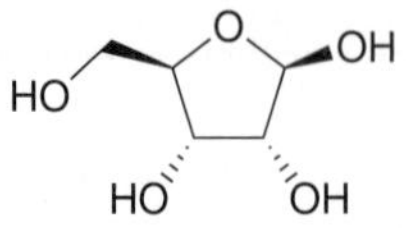

ribose

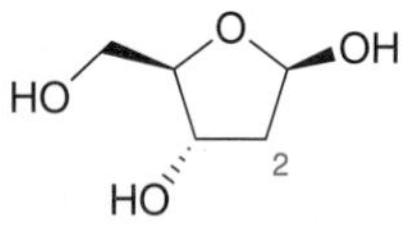

2-deoxyribose

Nucleic acids are polynucleotides, formed by joining an OH group of one nucleotide with the phosphate of another in a **phosphodiester linkage.** A nucleic acid contains a backbone consisting of alternating sugar and phosphate groups. The identity and order of the bases distinguish one nucleic acid from another. Figure 3.11 shows a portion of a polynucleotide.

Figure 3.11
A polynucleotide

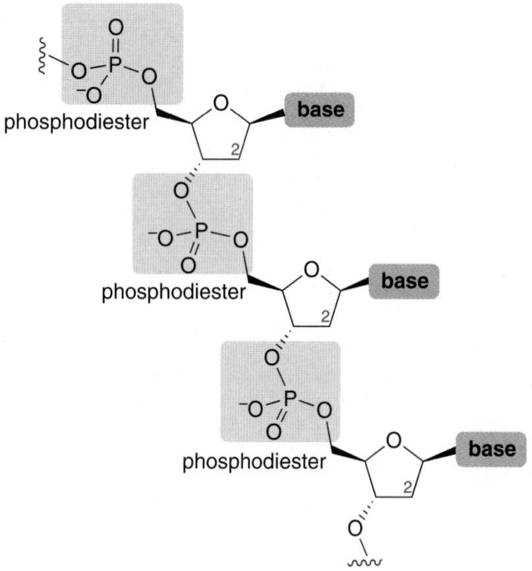

- Phosphodiester bonds that hold the polynucleotide together are highlighted in blue. All atoms except the nitrogen-containing bases form the sugar–phosphate backbone. Because no OH group is present on C2 in the monosaccharide rings, the polynucleotide is formed from 2-deoxyribose and represents a segment of DNA.

DNA consists of two polynucleotide chains that wind into a right-handed double helix held together by hydrogen bonding interactions of the nitrogen-containing bases (Figure 3.2). RNA consists of a single chain of nucleotides that can fold back on itself and form loops.

Problem 3.31 (a) Label each of the following in dinucleotide **A,** formed by joining two nucleotides together: the monosaccharide units, the phosphodiester, and the nitrogen-containing bases. (b) Can **A** be part of DNA or RNA?

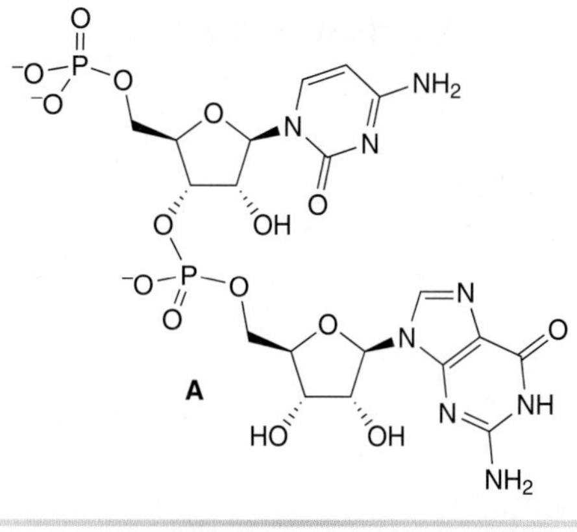

A

3.9D Lipids

- **Lipids are biomolecules that are soluble in organic solvents and insoluble in water.**

Because lipids are composed mainly of carbon–carbon and carbon–hydrogen bonds, their properties resemble those of the alkanes and other hydrocarbons. Lipids have a wide variety of shapes and sizes. The fat-soluble vitamins like vitamin A (Section 3.5) and the phospholipids in cell membranes (Section 3.7) are two groups of lipids. Triacylglycerols, the most common lipids, compose animal fat and vegetable oil (Figure 3.12).

Figure 3.12 Fatty acids and triacylglycerols, the most common lipids

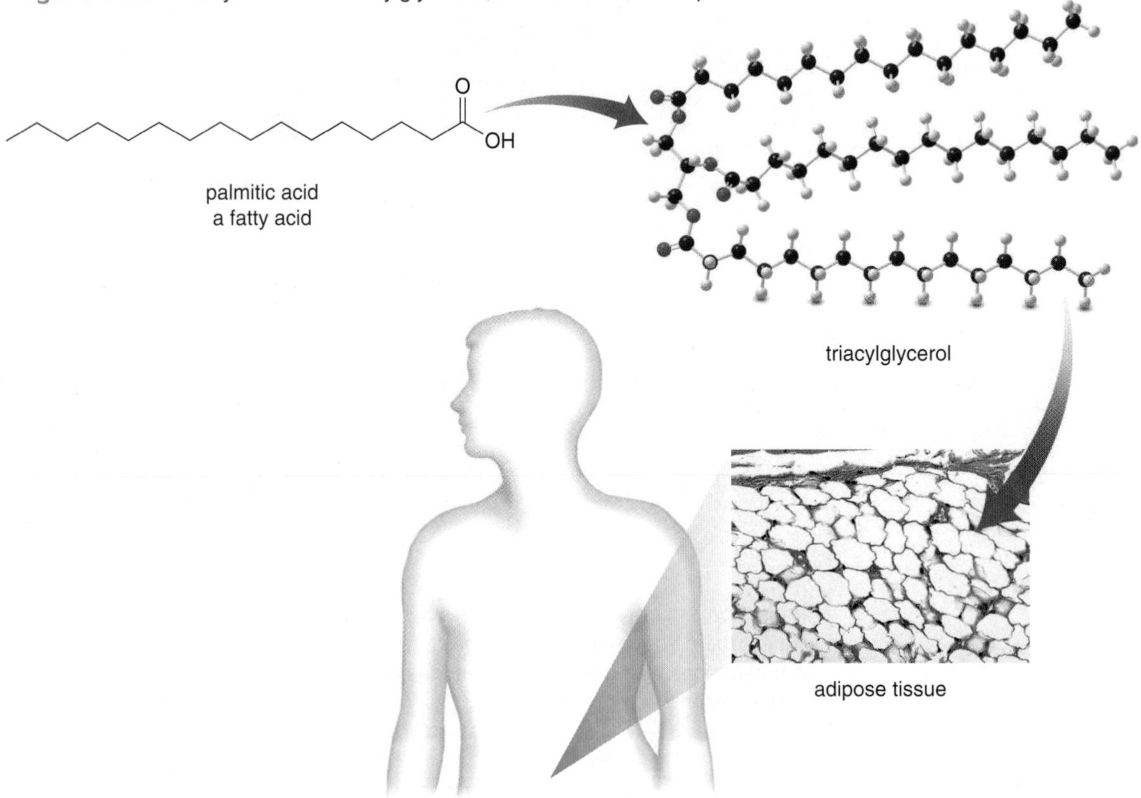

- Triacylglycerols are stored in adipose cells below the skin and are concentrated in some regions of the body. Triacylglycerols are formed from fatty acids like palmitic acid.

- **Triacylglycerols are triesters, formed from glycerol and three fatty acids (water-insoluble carboxylic acids).**

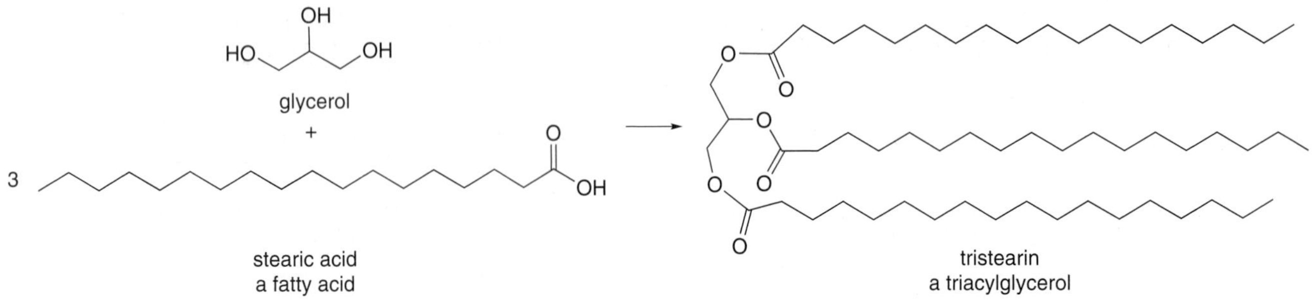

We will learn more about lipids in Sections 4.15 and 10.6 and Chapter 28.

Problem 3.32 (a) Label the hydrophobic and hydrophilic regions of tristearin. (b) Can two molecules of tristearin hydrogen bond to each other? (c) Can tristearin hydrogen bond to water?

KEY CONCEPTS

Introduction to Organic Molecules and Functional Groups

Classifying Carbon Atoms, Hydrogen Atoms, Alcohols, Alkyl Halides, Amines, and Amides (3.2)

- Carbon atoms are classified by the number of carbons bonded to them; a 1° carbon is bonded to one other carbon, and so forth.
- Hydrogen atoms are classified by the type of carbon atom to which they are bonded; a 1° hydrogen is bonded to a 1° carbon, and so forth.
- Alkyl halides and alcohols are classified by the type of carbon to which the OH or X group is bonded; a 1° alcohol has an OH group bonded to a 1° carbon, and so forth.
- Amines and amides are classified by the number of carbons bonded to the nitrogen atom; a 1° amine has one carbon–nitrogen bond, and so forth.

Types of Intermolecular Forces (3.3)

Increasing strength ↓

Type of force	Cause
van der Waals	Caused by the interaction of temporary dipoles • Larger surface area, stronger forces • Larger, more polarizable atoms, stronger forces
dipole–dipole	Caused by the interaction of permanent dipoles
hydrogen bonding	Caused by the electrostatic interaction of a H atom in an O–H, N–H, or H–F bond with the lone pair of another N, O, or F atom
ion–ion	Caused by the charge attraction of two ions

Physical Properties

Property	Observation
Boiling point (3.4A)	• For compounds of comparable molecular weight, the stronger the intermolecular forces the higher the bp.

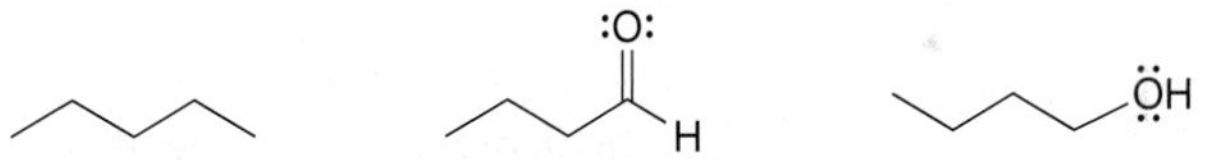

VDW bp = 36 °C | VDW, DD bp = 76 °C | VDW, DD, HB bp = 118 °C

Increasing strength of intermolecular forces
Increasing boiling point

• For compounds with similar functional groups, the larger the surface area, the higher the bp.

bp = 0 °C | bp = 36 °C

Increasing surface area
Increasing boiling point

• For compounds with similar functional groups, the more polarizable the atoms, the higher the bp.

CH_3F bp = –78 °C | CH_3I bp = 42 °C

Increasing polarizability
Increasing boiling point

Melting point (3.4B)	• For compounds of comparable molecular weight, the stronger the intermolecular forces the higher the mp. 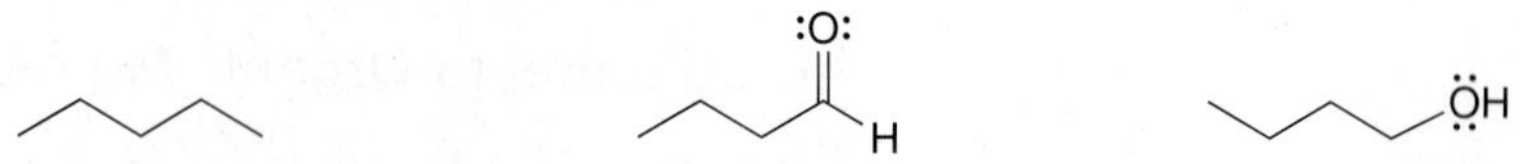 VDW, mp = −130 °C; VDW, DD, mp = −96 °C; VDW, DD, HB, mp = −90 °C **Increasing strength of intermolecular forces** **Increasing melting point** • For compounds with similar functional groups, the more symmetrical the compound, the higher the mp.  mp = −160 °C; mp = −17 °C **Increasing symmetry** **Increasing melting point**
Solubility (3.4C)	Types of water-soluble compounds: • Ionic compounds • Organic compounds having ≤ 5 C's, and an O or N atom for hydrogen bonding (for a compound with one functional group) Types of compounds soluble in organic solvents: • Organic compounds regardless of size or functional group

Key: VDW = van der Waals, DD = dipole–dipole, HB = hydrogen bonding

Reactivity (3.8)

- **Nucleophiles react with electrophiles.**
- Electronegative heteroatoms create electrophilic carbon atoms that react with nucleophiles.
- Lone pairs and π bonds are nucleophilic sites that react with electrophiles.

PROBLEMS

Problems with Three-Dimensional Models

3.33

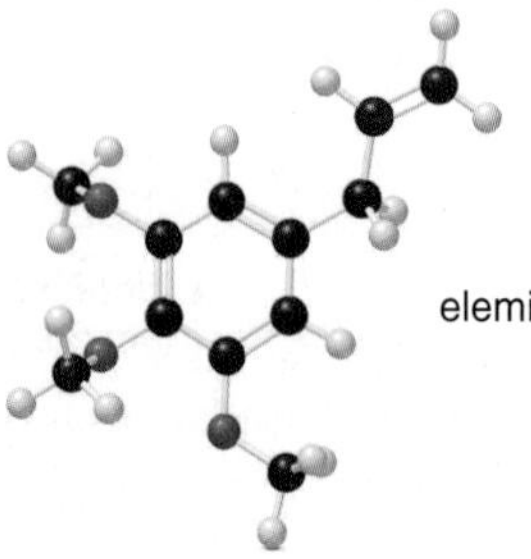

elemicin

a. Identify the functional groups in the ball-and-stick model of elemicin, a compound partly responsible for the flavor and fragrance of nutmeg.
b. Draw a skeletal structure of a constitutional isomer of elemicin that should have a higher boiling point and melting point.
c. Label all electrophilic carbon atoms.

3.34

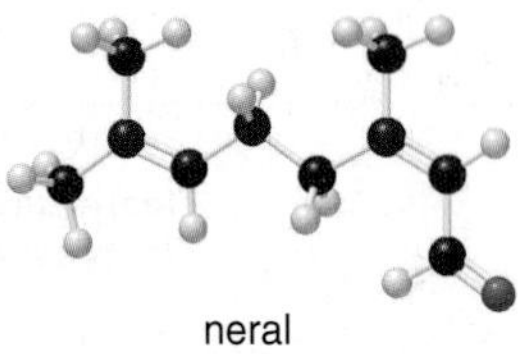

neral

a. Identify the functional groups in the ball-and-stick model of neral, a compound with a lemony odor isolated from lemongrass.
b. Draw a skeletal structure of a constitutional isomer of neral that should be more water soluble.
c. Label the most electrophilic carbon atom.

3.35

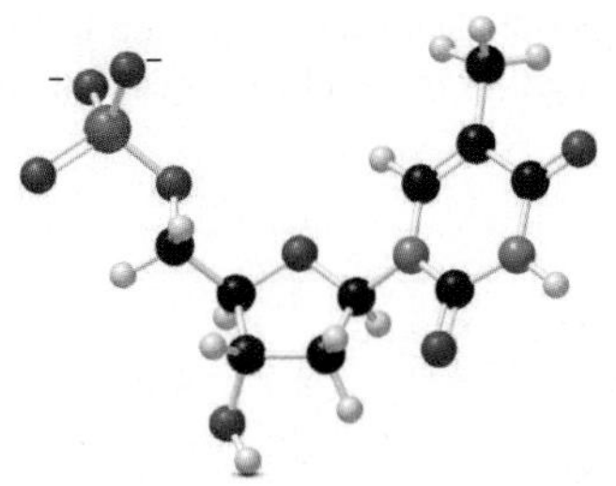

X

a. Convert the ball-and-stick model of nucleotide **X** into a skeletal structure.
b. Identify the phosphorus-containing functional group and classify any hydroxy group as 1°, 2°, or 3°.
c. Can this nucleotide be a component of DNA or RNA?

Functional Groups

3.36 For each alkane: (a) classify each carbon atom as 1°, 2°, 3°, or 4°; (b) classify each hydrogen atom as 1°, 2°, or 3°.

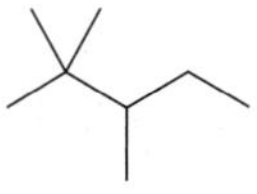

A

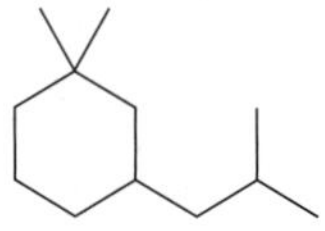

B

3.37 Identify the functional groups in each molecule. Classify each alcohol, alkyl halide, amide, and amine as 1°, 2°, or 3°.

a.

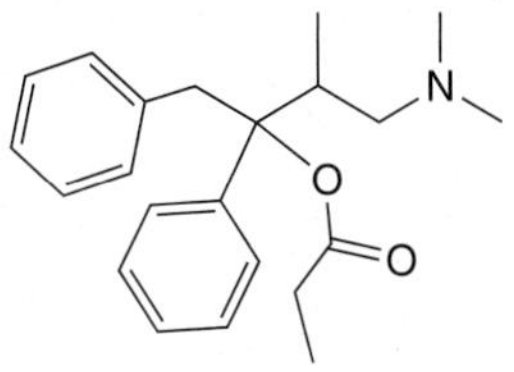

Darvon
(analgesic)

c.

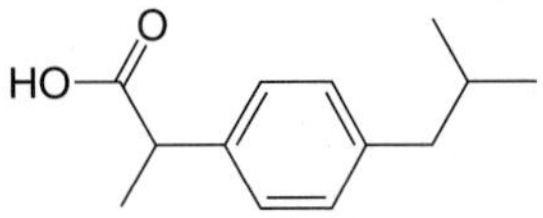

ibuprofen
(analgesic)

e.

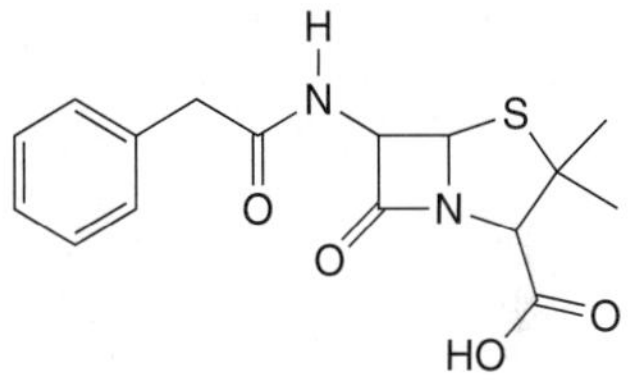

penicillin G
(an antibiotic)

b.

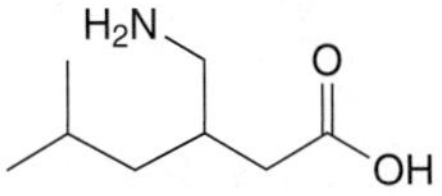

pregabalin
trade name Lyrica
(used in treating chronic pain)

d.

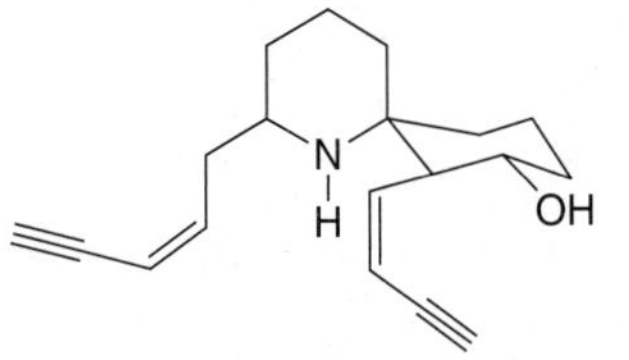

histrionicotoxin
(poison secreted by a South American frog)

f.

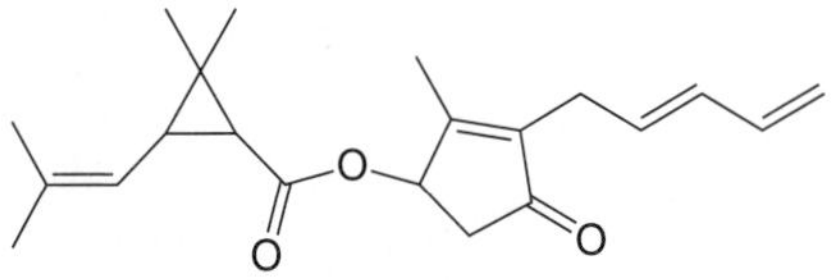

pyrethrin I
(potent insecticide from chrysanthemums)

3.38 (a) Identify the functional groups in salinosporamide A, an anticancer agent isolated from marine sediment. (b) Classify each alcohol, alkyl halide, amide, and amine as 1°, 2°, or 3°.

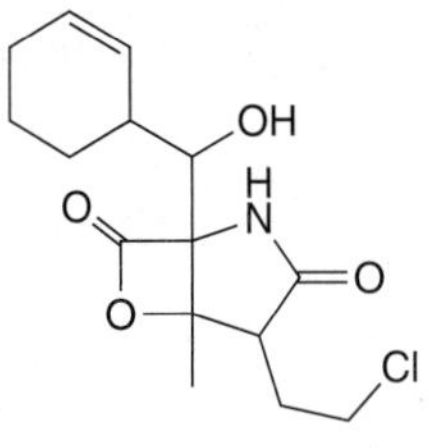

salinosporamide A

Intermolecular Forces

3.39 Rank the compounds in each group in order of increasing strength of intermolecular forces.

a.

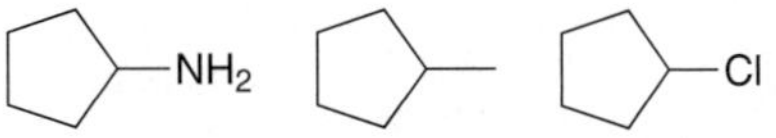

b.

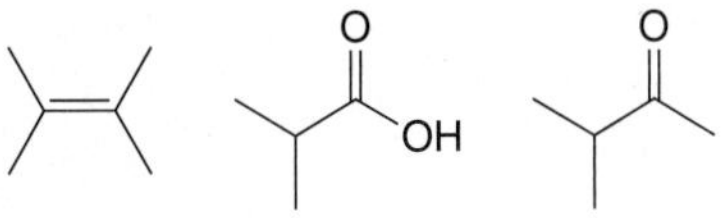

3.40 Indinavir (trade name Crixivan) is a drug used to treat HIV. (a) At which sites can indinavir hydrogen bond to another molecule like itself? (b) At which sites can indinavir hydrogen bond to water?

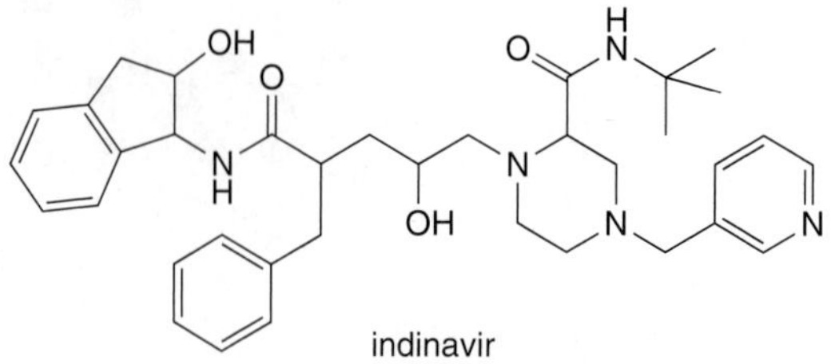

indinavir

3.41 Intramolecular forces of attraction are often important in holding large molecules together. For example, some proteins fold into compact shapes, held together by attractive forces between nearby functional groups. A schematic of a folded protein is drawn here, with the protein backbone indicated by a blue-green ribbon, and various appendages drawn dangling from the chain. What types of intramolecular forces occur at each labeled site (**A–F**)?

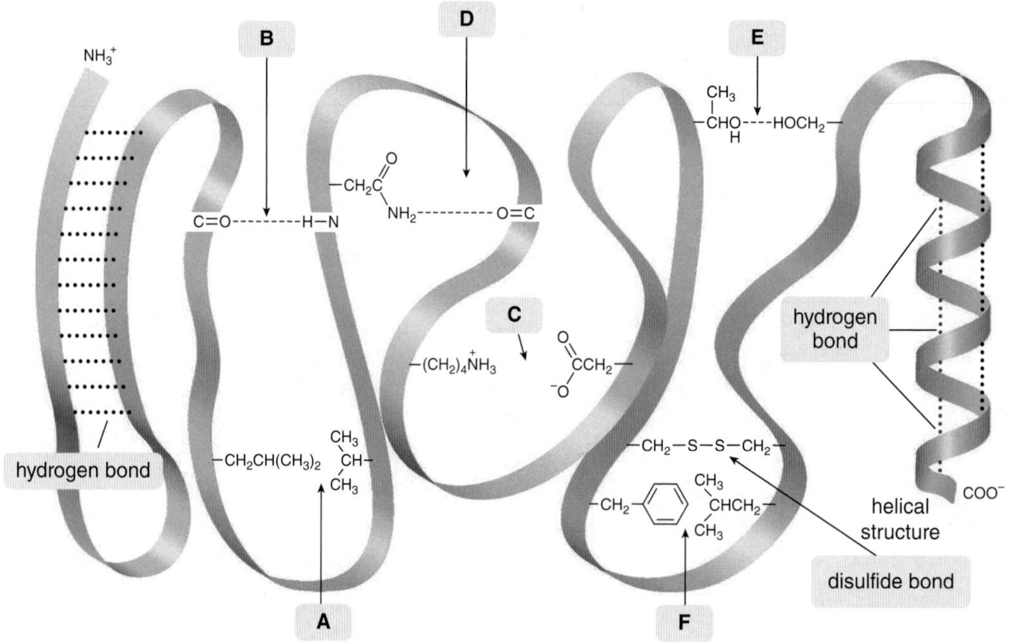

Physical Properties

3.42 Rank the compounds in each group in order of increasing boiling point.

a.

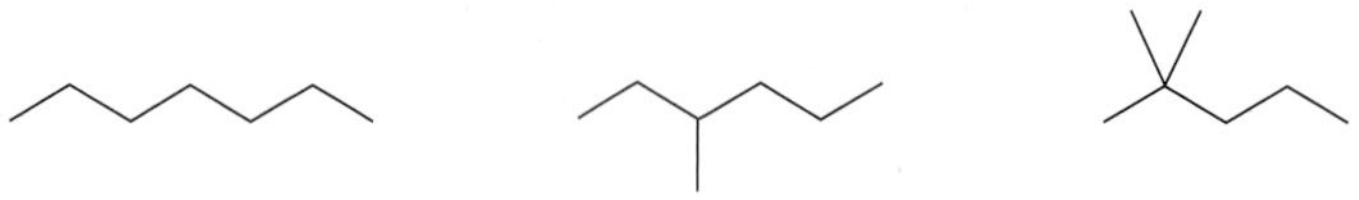

b.

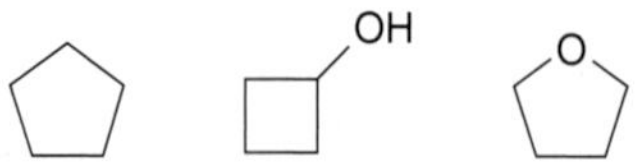

3.43 Explain why $CH_3CH_2NHCH_3$ has a higher boiling point than $(CH_3)_3N$, even though they have the same molecular weight.

3.44 Explain why benzene has a lower boiling point but much higher melting point than toluene.

benzene
bp = 80 °C
mp = 5 °C

and

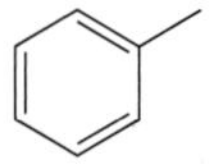

toluene
bp = 111 °C
mp = −93 °C

3.45 Rank the following compounds in order of increasing water solubility.

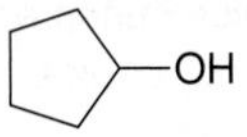

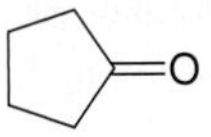

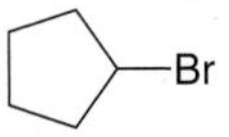

3.46 Which of the following molecules can hydrogen bond to another molecule of itself? Which can hydrogen bond with water?

a. 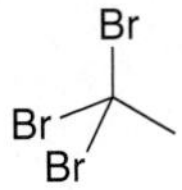

b. 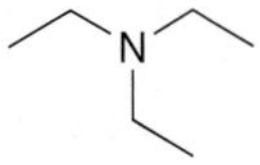

c. 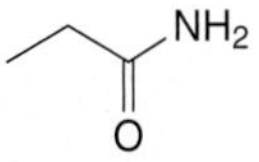

d. 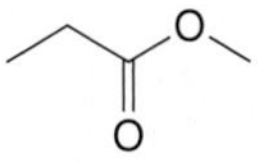

3.47 Explain why diethyl ether ($CH_3CH_2OCH_2CH_3$) and butan-1-ol ($CH_3CH_2CH_2CH_2OH$) have similar solubility properties in water, but butan-1-ol has a much higher boiling point.

3.48 Predict the water solubility of each of the following organic molecules.

a. 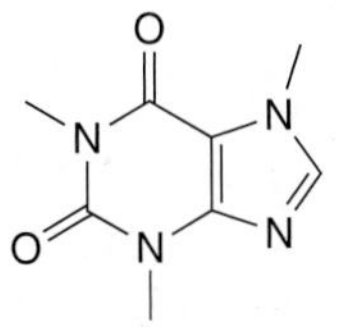

caffeine
(stimulant in coffee, tea, and many soft drinks)

c. 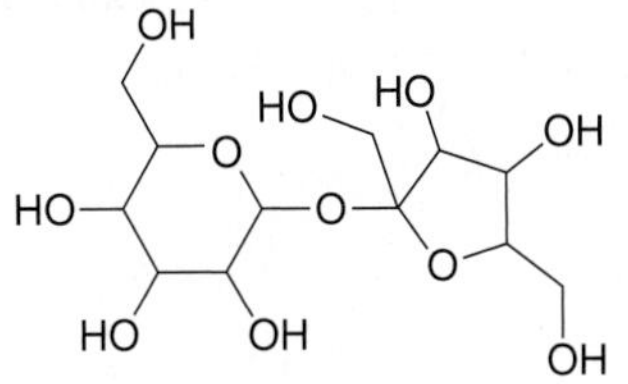

sucrose
(table sugar)

b. 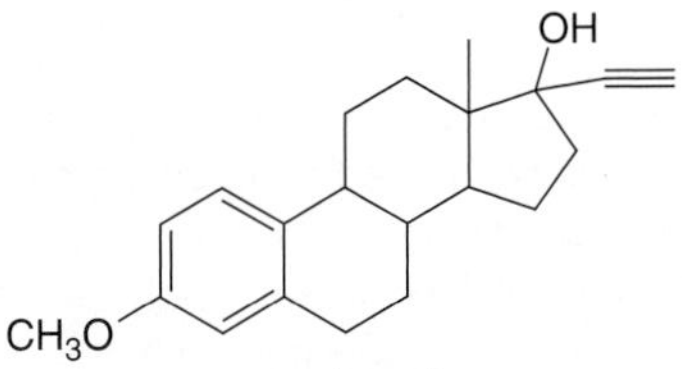

mestranol
(component in oral contraceptives)

d. 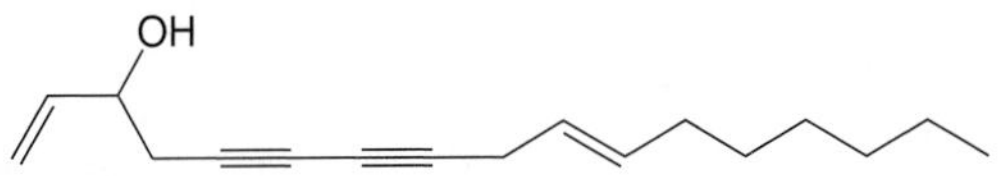

carotatoxin
(neurotoxin isolated from carrots)

Applications and Biomolecules

3.49 Predict the solubility of each of the following vitamins in water and in organic solvents.

a. 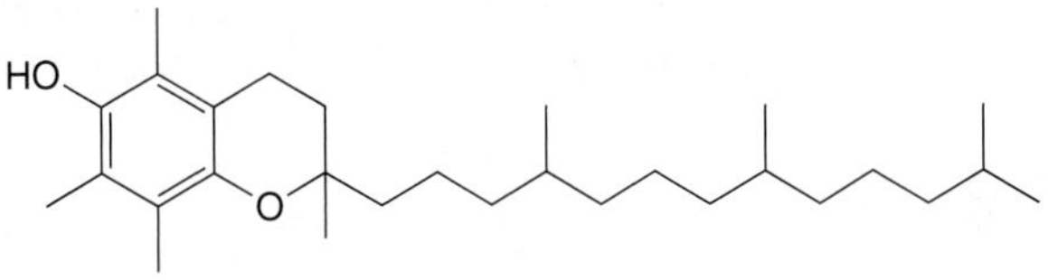

vitamin E

b. 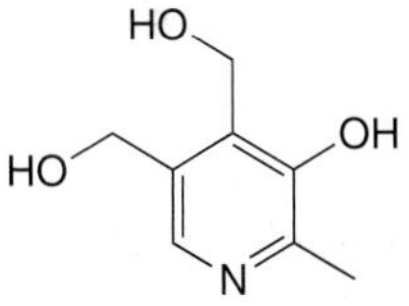

pyridoxine
vitamin B_6

3.50 Use the structure of keto acid **A** and amino acid **B** to answer each question. (a) Which compound has the higher melting point? (b) Which compound is more soluble in water? (c) Which compound is more soluble in diethyl ether [$(CH_3CH_2)_2O$]?

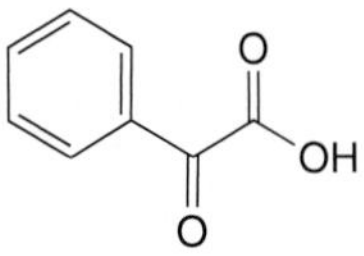

A

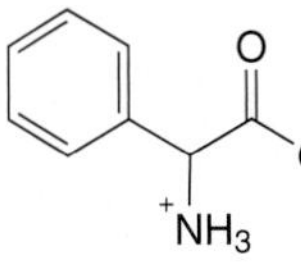

B

3.51 Avobenzone and dioxybenzone are two commercial sunscreens. Using the principles of solubility, predict which sunscreen is more readily washed off when an individual goes swimming. Explain your choice.

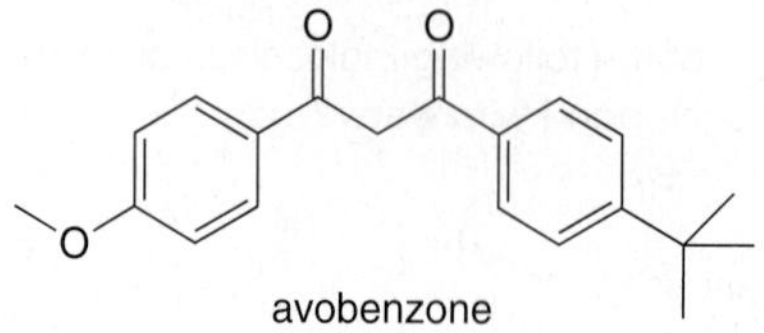

avobenzone

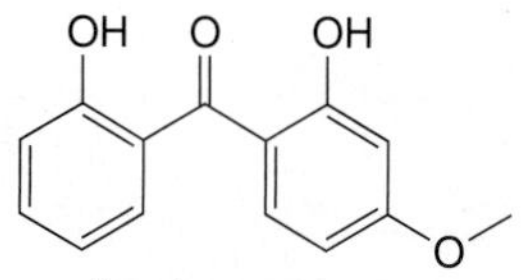

dioxybenzone

3.52 Prostaglandin E_2 (PGE_2) is a lipid that relaxes the smooth muscles of the uterus, so it is sometimes administered when labor is induced. (a) Identify the functional groups in PGE_2. (b) At which sites can PGE_2 hydrogen bond to another molecule like itself? (c) At which sites can PGE_2 hydrogen bond to water?

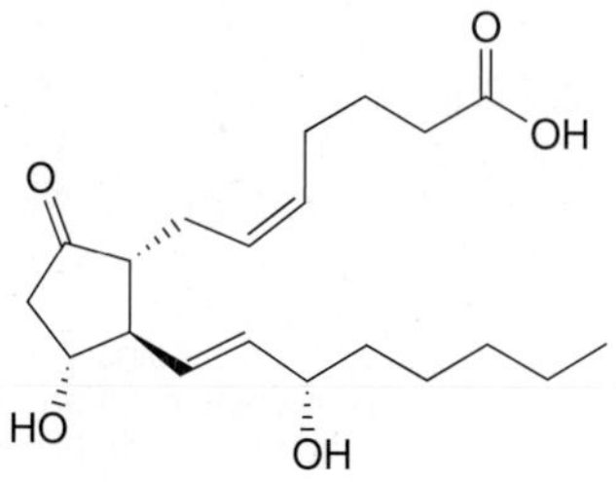

PGE_2

3.53 Poly(ethylene glycol) (PEG) and poly(vinyl chloride) (PVC) are examples of polymers, large organic molecules composed of repeating smaller units covalently bonded together. Polymers have very different properties depending (in part) on their functional groups. Discuss the water solubility of each polymer and suggest why PEG is used in shampoos, whereas PVC is used to make garden hoses and pipes. Synthetic polymers are discussed in detail in Chapters 15 and 31.

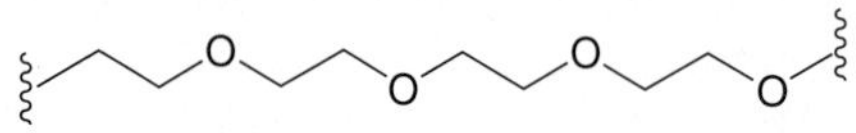

poly(ethylene glycol)
PEG

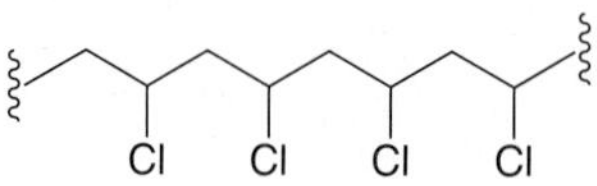

poly(vinyl chloride)
PVC

3.54 THC is the active component in marijuana, and ethanol is the alcohol in alcoholic beverages. Explain why drug screenings are able to detect the presence of THC but not ethanol weeks after these substances have been introduced into the body.

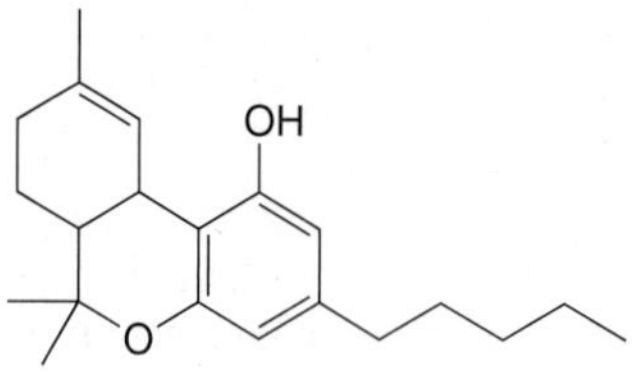

tetrahydrocannabinol
THC

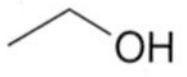

ethanol

3.55 Cocaine is a widely abused, addicting drug. Cocaine is usually obtained as its hydrochloride salt (cocaine hydrochloride) but can be converted to crack (the neutral molecule) by treatment with base. Which of the two compounds here has a higher boiling point? Which is more soluble in water? How does the relative solubility explain why crack is usually smoked but cocaine hydrochloride is injected directly into the bloodstream?

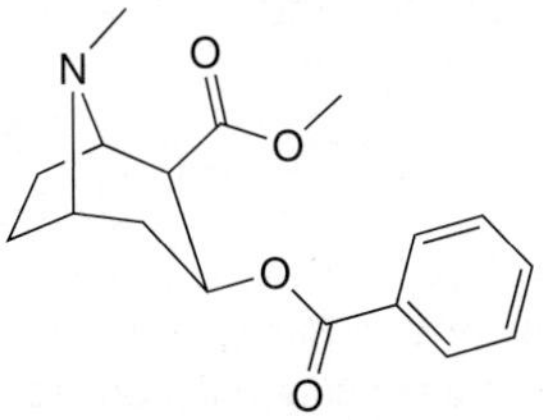

cocaine (crack)
neutral organic molecule

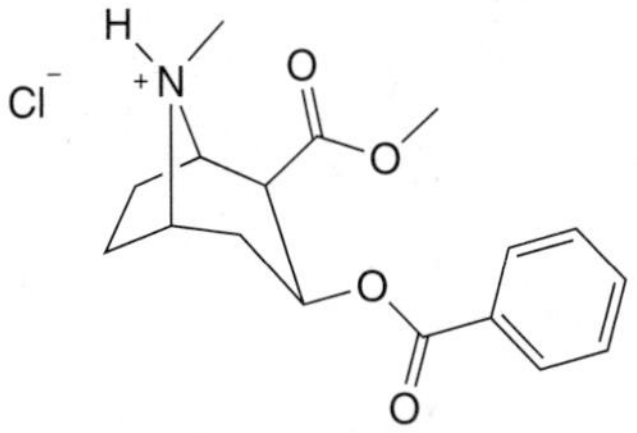

cocaine hydrochloride
a salt

3.56 Many drugs are sold as their hydrochloride salts ($R_2NH_2^+\ Cl^-$), formed by reaction of an amine (R_2NH) with HCl.

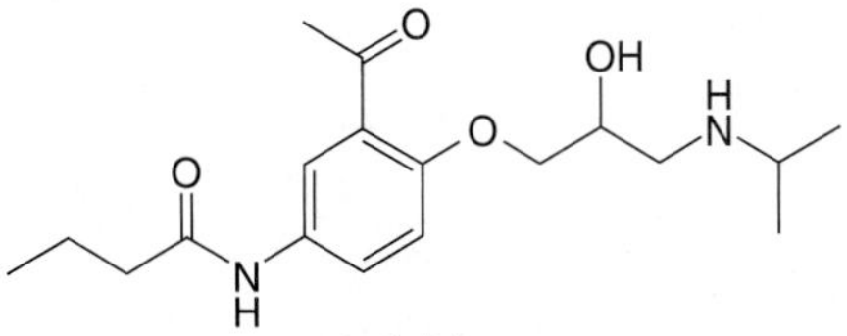

acebutolol

a. Draw the product (a hydrochloride salt) formed by reaction of acebutolol with HCl. Acebutolol is a β blocker used to treat high blood pressure.

b. Discuss the solubility of acebutolol and its hydrochloride salt in water.

c. Offer a reason as to why the drug is marketed as a hydrochloride salt rather than a neutral amine.

3.57 As we will learn in Chapter 22, acetyl coenzyme A (acetyl CoA) is a biological compound involved in reactions in which an acyl group ($CH_3CO—$) is transferred from one species to another. (a) Identify the functional groups in the highlighted portions of acetyl CoA. (b) Which functional groups serve as the nucleophile and electrophile in the given reaction, which synthesizes the neurotransmitter acetylcholine?

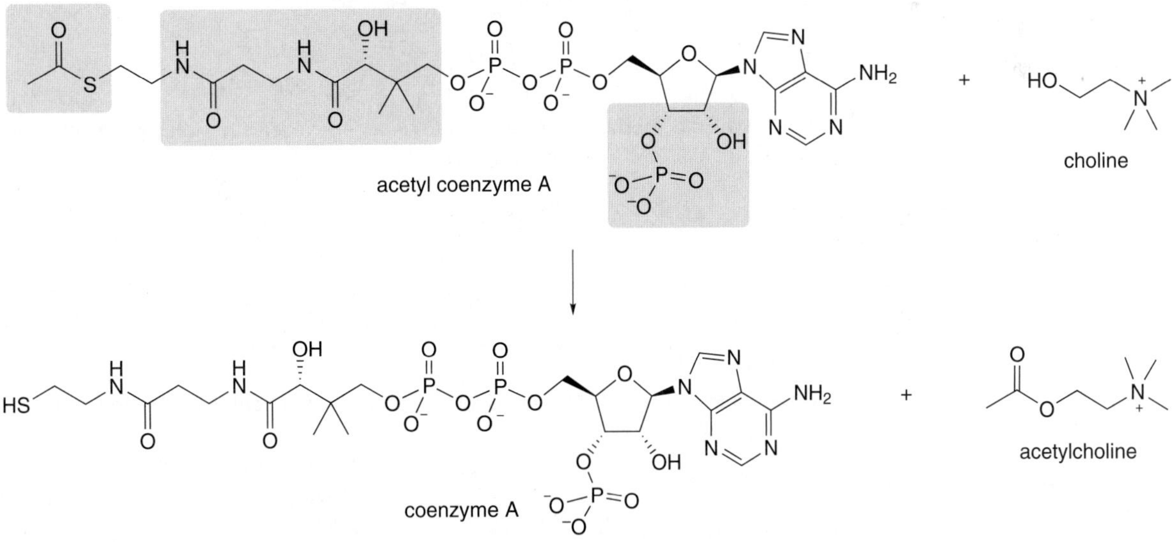

acetyl coenzyme A + choline → coenzyme A + acetylcholine

Reactivity of Organic Molecules

3.58 Label the electrophilic and nucleophilic sites in each molecule.

a. 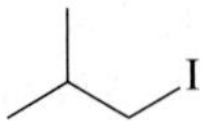 b. 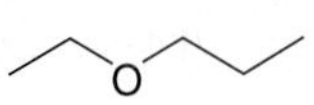 c. 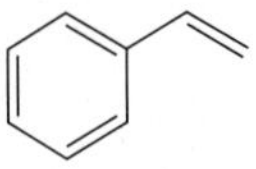 d. 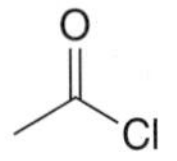

3.59 By using only electron density arguments, determine whether the following reactions will occur.

a.  + Br^- ⟶

b. 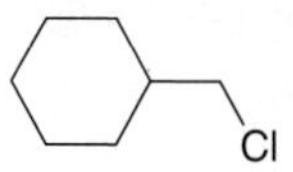 + ^-CN ⟶

c.  + ^-OH ⟶

d.  + H_3O^+ ⟶

Cell Membrane

3.60 The composition of a cell membrane is not uniform for all types of cells. Some cell membranes are more rigid than others. Rigidity is determined by a variety of factors, one of which is the structure of the carbon chains in the phospholipids that comprise the membrane. One example of a phospholipid was drawn in Section 3.7A, and another, having C–C double bonds in its carbon chains, is drawn here. Which phospholipid would be present in the more rigid cell membrane and why?

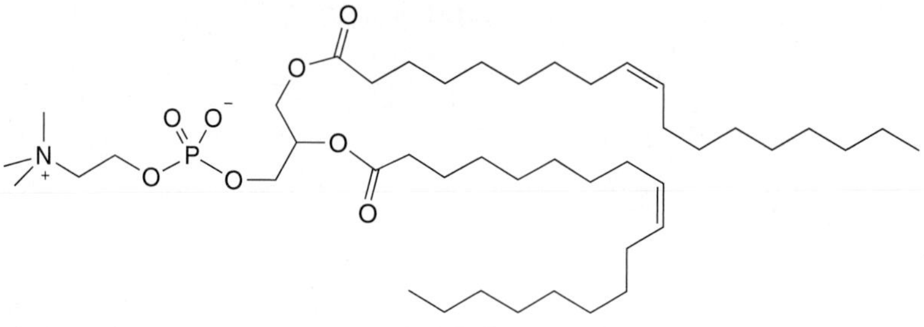

phospholipid

3.61 Sphingomyelins, a group of lipids that resemble the membrane phospholipids discussed in Section 3.7, are a major component of the myelin sheath, the insulating layer that surrounds a nerve fiber. (a) What functional groups are present in sphingomyelin **X?** (b) Classify any alcohol, amine, and amide as 1°, 2°, or 3°. (c) Label the polar head and nonpolar tails of **X.**

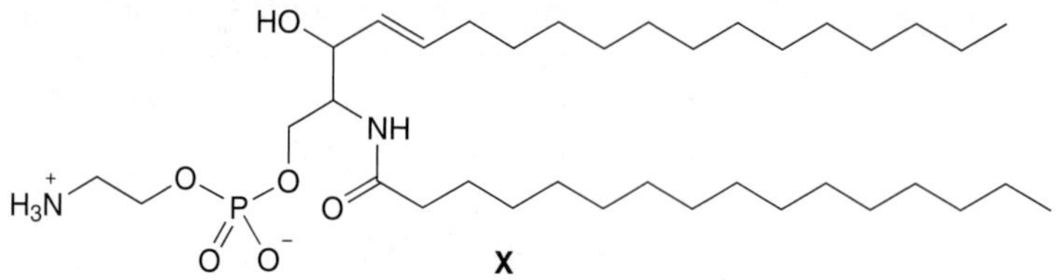

X

a sphingomyelin

3.62 Which compound is more likely to be a general anesthetic? Explain your choice.

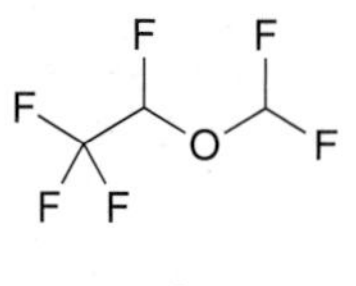

A

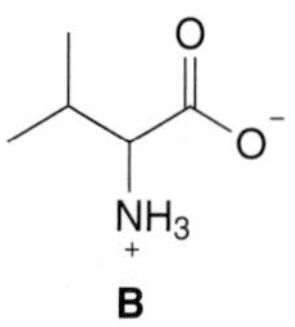

B

General Questions

3.63 Quinapril (trade name Accupril) is a drug used to treat hypertension and congestive heart failure.

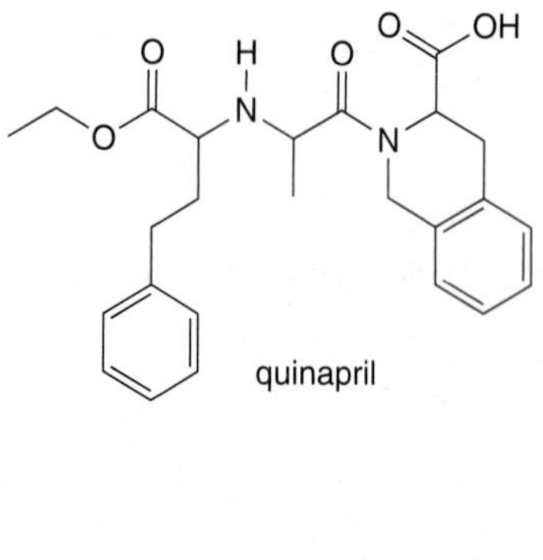

quinapril

a. Identify the functional groups in quinapril.
b. Classify any alcohol, amide, or amine as 1°, 2°, or 3°.
c. At which sites can quinapril hydrogen bond to water?
d. At which sites can quinapril hydrogen bond to acetone $[(CH_3)_2CO]$?
e. Label the most acidic hydrogen atom.
f. Which site is most basic?

3.64 Answer each question about oxycodone, a narcotic analgesic used for severe pain.

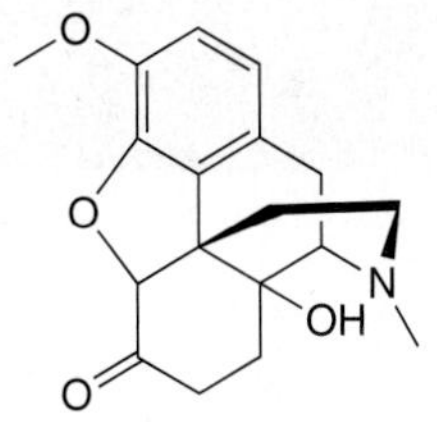

oxycodone

a. Identify the functional groups in oxycodone.
b. Classify any alcohol, amide, or amine as 1°, 2°, or 3°.
c. Which proton is most acidic?
d. Which site is most basic?
e. What is the hybridization of the N atom?
f. How many sp^2 hybridized C atoms does oxycodone contain?

Challenge Problems

3.65 Answer the following questions by referring to the ball-and-stick model of fentanyl, a potent narcotic analgesic used in surgical procedures.

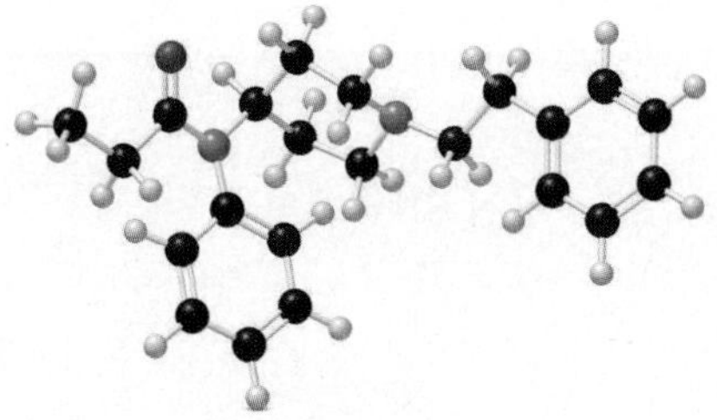

fentanyl

a. Identify the functional groups.
b. Label the most acidic proton.
c. Label the most basic atom.
d. What types of intermolecular forces are present between two molecules of fentanyl?
e. Draw an isomer predicted to have a higher boiling point.
f. Which sites in the molecule can hydrogen bond to water?
g. Label all electrophilic carbons.

3.66 Explain why **A** is less water soluble than **B,** even though both compounds have the same functional groups.

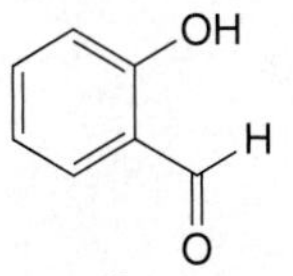

A

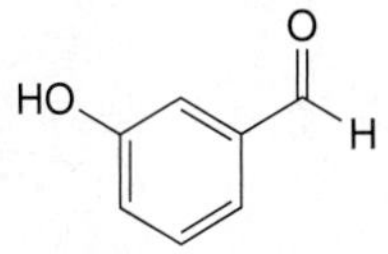

B

3.67 Recall from Section 1.10B that there is restricted rotation around carbon–carbon double bonds. Maleic acid and fumaric acid are two isomers with vastly different physical properties and pK_a values for loss of both protons. Explain why each of these differences occurs.

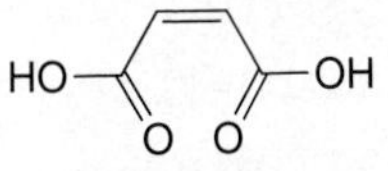

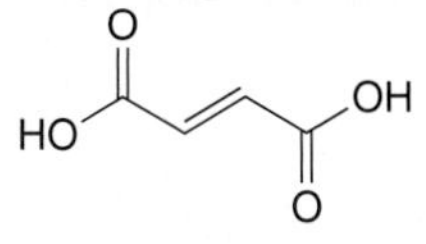

	maleic acid	fumaric acid
mp (°C)	130	286
solubility (g/L) in H_2O at 25 °C	788	7
pK_{a1}	1.9	3.0
pK_{a2}	6.5	4.5

4 Alkanes

4.1 Alkanes—An introduction
4.2 Cycloalkanes
4.3 An introduction to nomenclature
4.4 Naming alkanes
4.5 Naming cycloalkanes
4.6 Common names
4.7 Fossil fuels
4.8 Physical properties of alkanes
4.9 Conformations of acyclic alkanes—Ethane
4.10 Conformations of butane
4.11 An introduction to cycloalkanes
4.12 Cyclohexane
4.13 Substituted cycloalkanes
4.14 Oxidation of alkanes
4.15 Lipids—Part 1

Alkanes, the simplest hydrocarbons, are found in all shapes and sizes and occur widely in nature. They are the major constituents of petroleum, a complex mixture of compounds that includes hydrocarbons such as **hexane** and **decane.** Crude petroleum spilled into the sea from a ruptured oil tanker or offshore oil well creates an insoluble oil slick on the surface. Petroleum is refined to produce gasoline, diesel fuel, home heating oil, and a myriad of other useful compounds. In Chapter 4, we learn about the properties of alkanes, how to name them (nomenclature), and oxidation—one of their important reactions.

In Chapter 4, we apply the principles of bonding, shape, and reactivity discussed in Chapters 1–3 to our first family of organic compounds, the **alkanes.** Because alkanes have no functional group, they are much less reactive than other organic compounds, and for this reason, much of Chapter 4 is devoted to learning how to name and draw them, as well as to understanding what happens when rotation occurs about their carbon–carbon single bonds.

Studying alkanes also provides an opportunity to learn about **lipids,** a group of biomolecules similar to alkanes, in that they are composed mainly of nonpolar carbon–carbon and carbon–hydrogen σ bonds. Section 4.15 serves as a brief introduction only, so we will return to lipids in Chapters 10 and 28.

4.1 Alkanes—An Introduction

Secretion of **undecane** by a cockroach causes other members of the species to aggregate. Undecane is a ***pheromone,* a chemical substance used for communication** in an animal species, most commonly an insect population.

Cyclohexane is one component of the mango, the most widely consumed fruit in the world.

Recall from Section 3.2 that **alkanes are aliphatic hydrocarbons having only C–C and C–H σ bonds.** Because their carbon atoms can be joined together in chains or rings, they can be categorized as acyclic or cyclic.

- **Acyclic alkanes** have the molecular formula C_nH_{2n+2} (where n = an integer) and contain only linear and branched chains of carbon atoms. Acyclic alkanes are also called **saturated hydrocarbons** because they have the maximum number of hydrogen atoms per carbon.
- **Cycloalkanes** contain carbons joined in one or more rings. Because their general formula is C_nH_{2n}, they have two fewer H atoms than an acyclic alkane with the same number of carbons.

Undecane, an acyclic alkane, and cyclohexane, a cycloalkane, are two naturally occurring alkanes.

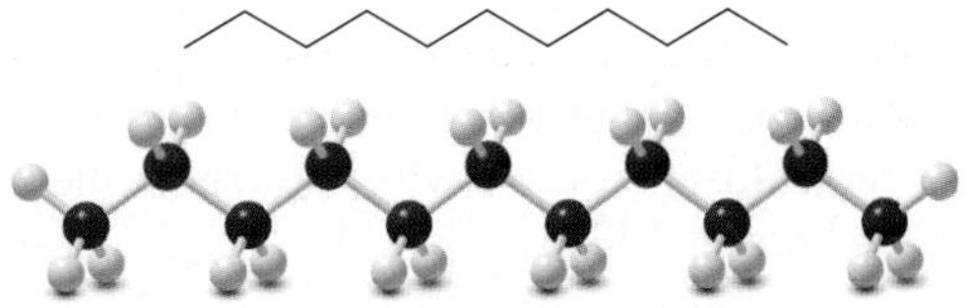

undecane
$C_{11}H_{24}$

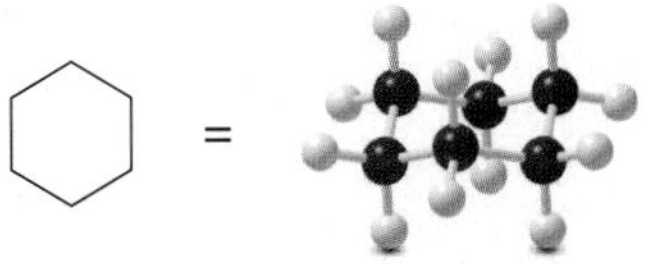

cyclohexane
C_6H_{12}

4.1A Acyclic Alkanes Having One to Five C Atoms

Structures for the two simplest acyclic alkanes were given in Chapter 1. **Methane, CH_4,** has a single carbon atom, and **ethane, CH_3CH_3,** has two. All C atoms in an alkane are surrounded by four groups, making them ***sp*³ hybridized** and **tetrahedral,** and all bond angles are 109.5°.

CH_4
methane
=

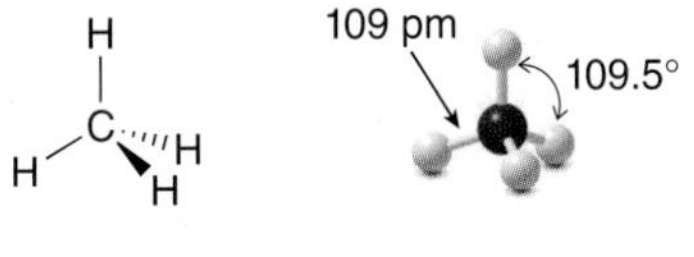
H
C
H
H
H
109 pm
109.5°

CH_3CH_3
ethane
=

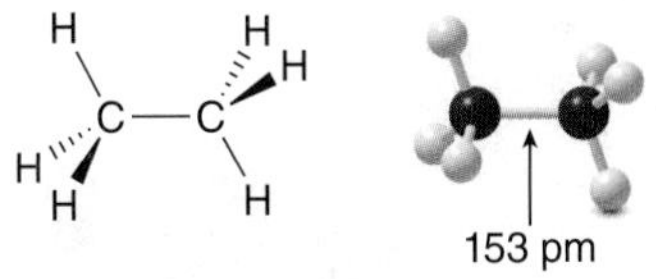
H
H
H
C
C
H
H
H
153 pm

The three-carbon alkane **$CH_3CH_2CH_3$, propane,** has molecular formula C_3H_8. Each carbon in the three-dimensional drawing has two bonds in the plane (solid lines), one bond in front (on a wedge), and one bond behind the plane (on a dashed wedge).

To draw the structure of an alkane, join the carbon atoms together with single bonds, and add enough H atoms to make each C tetravalent.

$CH_3CH_2CH_3$ =
propane

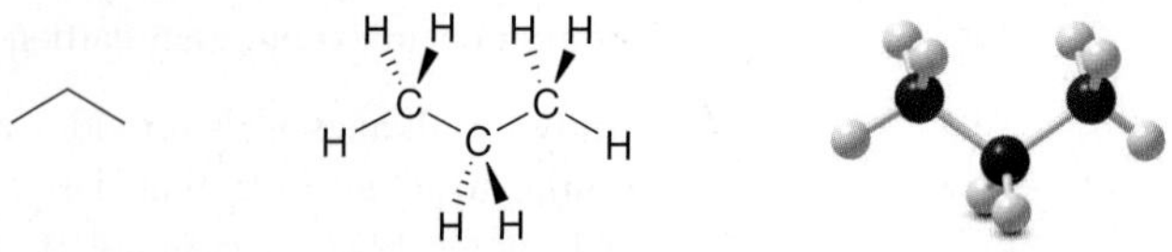

Problem 4.1 Three components of the sex pheromone of the female sand bee (*Ophrys sphegodes*) are saturated hydrocarbons containing 23, 25, and 27 carbon atoms. How many H atoms does each of these alkanes contain? Interestingly, the early spider orchid emits a similar hydrocarbon mixture to attract male sand bees to pollinate its flowers.

The molecular formulas for methane, ethane, and propane fit into the general molecular formula for an alkane, $\mathbf{C_nH_{2n+2}}$.

- Methane = $CH_4 = C_1H_{2(1)+2}$
- Ethane = $C_2H_6 = C_2H_{2(2)+2}$
- Propane = $C_3H_8 = C_3H_{2(3)+2}$

There are two different ways to arrange four carbons, giving two compounds with molecular formula C_4H_{10}, named **butane** and **2-methylpropane** (or isobutane).

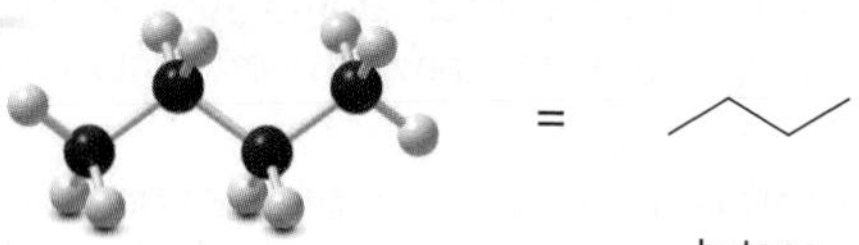

butane

4 C's in a row

straight-chain alkane

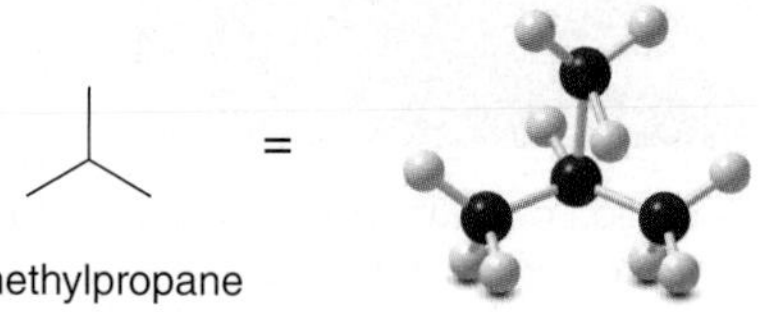

2-methylpropane
(or isobutane)

3 C's in a row with a one-carbon branch

branched-chain alkane

Butane and 2-methylpropane are ***isomers,*** **two different compounds with the same molecular formula** (Section 1.4). They belong to one of the two major classes of isomers called **constitutional** or **structural isomers.** We will learn about the second major class of isomers, called **stereoisomers,** in Section 4.13B.

- *Constitutional isomers* differ in the way the atoms are connected to each other.

Butane, which has four carbons in a row, is a **straight-chain** or **normal alkane** (an ***n*-alkane**). 2-Methylpropane, on the other hand, is a **branched-chain alkane.**

With alkanes having more than four carbons, the names of the straight-chain isomers are systematic and derive from Greek roots: ***pent***ane for five C atoms, ***hex***ane for six, and so on. There are three constitutional isomers for the five-carbon alkane, each having molecular formula C_5H_{12}: **pentane, 2-methylbutane** (or isopentane), and **2,2-dimethylpropane** (or neopentane).

pentane

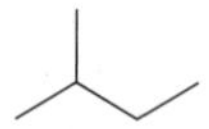

2-methylbutane
(isopentane)

2,2-dimethylpropane
(neopentane)

Take care in interpreting skeletal structures. Although pentane is typically drawn using a zigzag structure, the carbon skeleton can be drawn in a variety of ways, and still represent the same compound. Each of the following representations has five carbon atoms in a row, so each represents pentane, not an isomer of pentane.

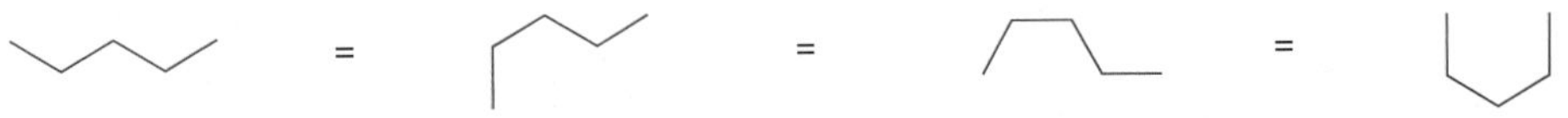

pentane
C_5H_{12} — 5 C's in a row — 5 C's in a row — 5 C's in a row

Problem 4.2 Which of the following is *not* another representation for 2-methylbutane?

a. 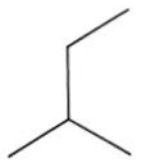 b. 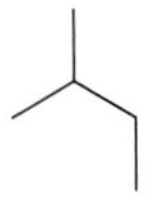 c. 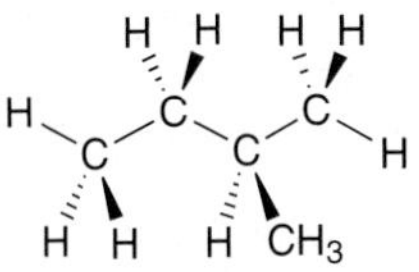

d. 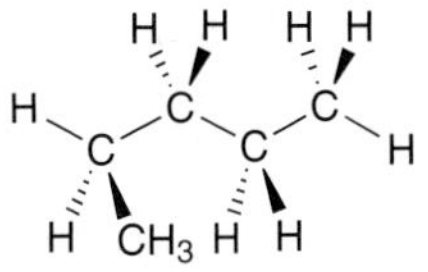

4.1B Acyclic Alkanes Having More Than Five C Atoms

The maximum number of possible constitutional isomers increases dramatically as the number of carbon atoms in the alkane increases, as shown in Table 4.1. For example, there are 75 possible isomers for an alkane having 10 carbon atoms, and 366,319 possible isomers for one having 20 carbons.

Each entry in Table 4.1 is formed from the preceding entry by adding a CH_2 group. **A CH_2 group is called a *methylene group.* A group of compounds that differ by only a CH_2 group is called a *homologous series.*** The names of all alkanes end in the suffix ***-ane,*** and the syllables preceding the suffix identify the number of carbon atoms in the chain.

Table 4.1 Summary: Straight-Chain Alkanes

Number of C atoms	Molecular formula	Name (*n*-alkane)	Number of constitutional isomers	Number of C atoms	Molecular formula	Name (*n*-alkane)	Number of constitutional isomers
1	CH_4	methane	—	9	C_9H_{20}	nonane	35
2	C_2H_6	ethane	—	10	$C_{10}H_{22}$	decane	75
3	C_3H_8	propane	—	11	$C_{11}H_{24}$	undecane	159
4	C_4H_{10}	butane	2	12	$C_{12}H_{26}$	dodecane	355
5	C_5H_{12}	pentane	3	13	$C_{13}H_{28}$	tridecane	802
6	C_6H_{14}	hexane	5	14	$C_{14}H_{30}$	tetradecane	1858
7	C_7H_{16}	heptane	9	15	$C_{15}H_{32}$	pentadecane	4347
8	C_8H_{18}	octane	18	20	$C_{20}H_{42}$	icosane	366,319

Problem 4.3 Draw the five constitutional isomers having molecular formula C_6H_{14}.

Problem 4.4 Review classifying carbons and hydrogens in Section 3.2, and draw the structure of an alkane with molecular formula C_7H_{16} that contains (a) one 4° carbon; (b) only 1° and 2° carbons; (c) 1°, 2°, and 3° hydrogens.

Problem 4.5 (a) Which compounds (**B–F**) are identical to **A?** (b) Which compounds (**B–F**) represent an isomer of **A?**

A **B** **C** **D** **E** **F**

4.2 Cycloalkanes

Cycloalkanes have molecular formula C_nH_{2n} and contain carbon atoms arranged in a ring. Think of a cycloalkane as being formed by removing two H atoms from the end carbons of a chain, and then bonding the two carbons together. Simple cycloalkanes are named by adding the prefix **cyclo-** to the name of the acyclic alkane having the same number of carbons.

Cycloalkanes with three to six carbon atoms are shown.

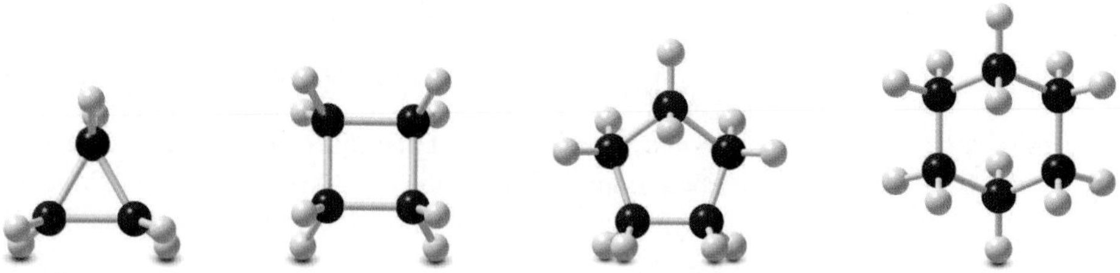

cyclopropane C_3H_6 | cyclobutane C_4H_8 | cyclopentane C_5H_{10} | cyclohexane C_6H_{12}

Problem 4.6 Draw the five constitutional isomers that have molecular formula C_5H_{10} and contain one ring.

Garlic has been a valued commodity throughout history. It has been used in Chinese herbal medicine for more than 4000 years, as a form of currency in Siberia, and as a repellent for witches by the Saxons. Today it is used as a dietary supplement because of its reported health benefits. **Allicin,** the molecule largely responsible for garlic's odor, is not stored in the garlic bulb, but instead is produced by the action of enzymes when the bulb is crushed or bruised.

4.3 An Introduction to Nomenclature

How are organic compounds named? Long ago, the name of a compound was often based on the plant or animal source from which it was obtained. For example, the name for **formic acid,** a caustic compound isolated from certain ants, comes from the Latin word *formica,* meaning *ant;* and **allicin,** the pungent principle of garlic, is derived from the botanical name for garlic, *Allium sativum.* Other compounds were named by their discoverer for more personal reasons. Adolf von Baeyer supposedly named barbituric acid after a woman named Barbara, although speculation continues on Barbara's identity—a lover, a Munich waitress, or even St. Barbara.

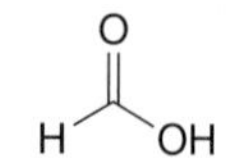

formic acid
(obtained from certain ants)

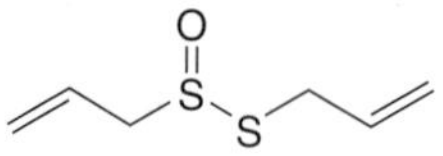

allicin
(odor of garlic)

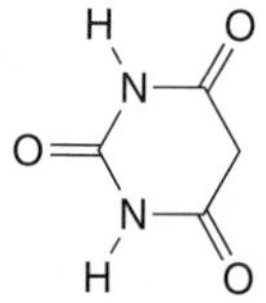

barbituric acid
(named for Barbara ?)

With the isolation and preparation of thousands of new organic compounds it became clear that each organic compound must have an unambiguous name, derived from a set of easily remembered rules. A systematic method of naming compounds was developed by the ***I****nternational* ***U****nion of* ***P****ure and* ***A****pplied* ***C****hemistry*. It is referred to as the **IUPAC system of nomenclature;** how it can be used to name alkanes is explained in Sections 4.4 and 4.5.

The IUPAC system of nomenclature has been regularly revised since it was first adopted in 1892. Revisions in 1979 and 1993 and extensive recommendations in 2004 have given chemists a variety of acceptable names for compounds. Many changes are minor. For example, the 1979 nomenclature rules assign the name 1-butene to $CH_2{=}CHCH_2CH_3$, while the 1993 rules assign the name but-1-ene; that is, only the position of the number differs. In this text, more recent IUPAC conventions will be used, and often a margin note will be added to mention the differences between past and recent recommendations.

Naming organic compounds has become big business for drug companies. The IUPAC name of an organic compound can be long and complex, and may be comprehensible only to a chemist. As a result, most drugs have three names:

- **Systematic:** The systematic name follows the accepted rules of nomenclature and indicates the compound's chemical structure; this is the IUPAC name.
- **Generic:** The generic name is the official, internationally approved name for the drug.
- **Trade:** The trade name for a drug is assigned by the company that manufactures it. Trade names are often "catchy" and easy to remember. Companies hope that the public will continue to purchase a drug with an easily recalled trade name long after a cheaper generic version becomes available.

In the world of over-the-counter anti-inflammatory agents, the compound a chemist calls 2-[4-(2-methylpropyl)phenyl]propanoic acid has the generic name ibuprofen. It is marketed under a variety of trade names including Motrin and Advil.

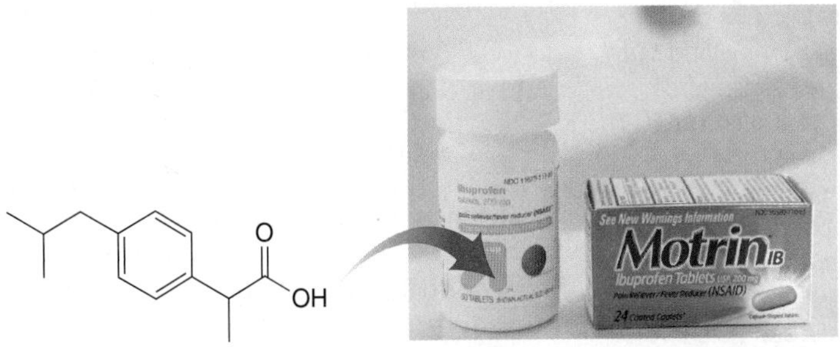

Systematic name: 2-[4-(2-methylpropyl)phenyl]propanoic acid
Generic name: ibuprofen
Trade name: Motrin or Advil

4.4 Naming Alkanes

The name of every organic molecule has three parts.

- The **parent name** indicates the number of carbons in the longest continuous carbon chain in the molecule.
- The **suffix** indicates what functional group is present.
- The **prefix** reveals the identity, location, and number of substituents attached to the carbon chain.

prefix + **parent** + **suffix**

What and where are the substituents? | What is the longest carbon chain? | What is the functional group?

The names listed in Table 4.1 of Section 4.1B for the simple *n*-alkanes consist of the parent name, which indicates the number of carbon atoms in the longest carbon chain, and the suffix ***-ane,*** which indicates that the compounds are alkanes. The parent name for **one carbon is *meth-,*** for **two carbons is *eth-,*** and so on. Thus, we are already familiar with two parts of the name of an organic compound.

To determine the third part of a name, the prefix, we must learn how to name the carbon groups or *substituents* that are bonded to the longest carbon chain.

4.4A Naming Substituents

Carbon substituents bonded to a long carbon chain are called **alkyl groups.**

- **An *alkyl group* is formed by removing one hydrogen from an alkane.**

An alkyl group is a part of a molecule that is now able to bond to another atom or a functional group. **To name an alkyl group, change the *-ane* ending of the parent alkane to *-yl.*** Thus, methane (CH_4) becomes methyl (CH_3-) and ethane (CH_3CH_3) becomes ethyl (CH_3CH_2-). As we learned in Section 3.1, **R** denotes a general carbon group bonded to a functional group. **R** thus denotes any alkyl group.

Naming three- and four-carbon alkyl groups is more complicated because the parent hydrocarbons have more than one type of hydrogen atom. Propane has both 1° and 2° H atoms, and removal of each of these H atoms forms a different alkyl group, **propyl** or **isopropyl.**

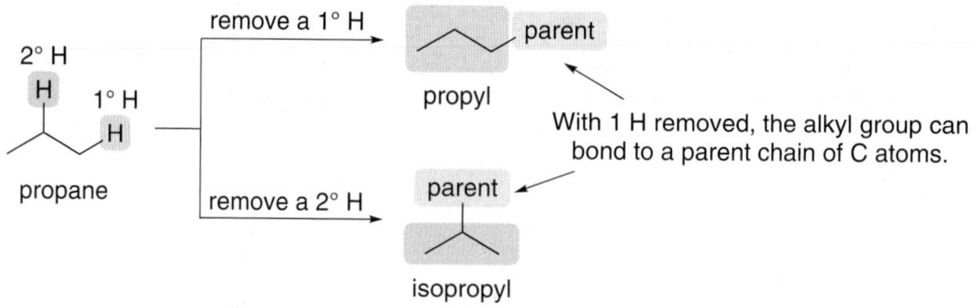

The prefix **iso-** is part of the words *propyl* and *butyl*, forming a single word: **isopropyl** and **isobutyl.** The prefixes ***sec-*** and ***tert-*** are separated from the word *butyl* by a hyphen: ***sec*-butyl** and ***tert*-butyl.**

The prefix *sec-* is short for *secondary.* A *sec*-butyl group is formed by removal of a **2° H.** The prefix *tert-* is short for *tertiary.* A *tert*-butyl group is formed by removal of a **3° H.**

Because there are two different butane isomers to begin with, each with two different kinds of H atoms, there are four possible alkyl groups containing four carbon atoms: **butyl, *sec*-butyl, isobutyl,** and ***tert*-butyl.**

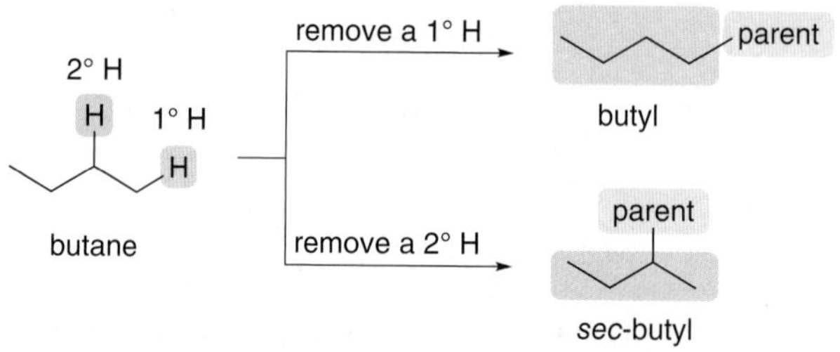

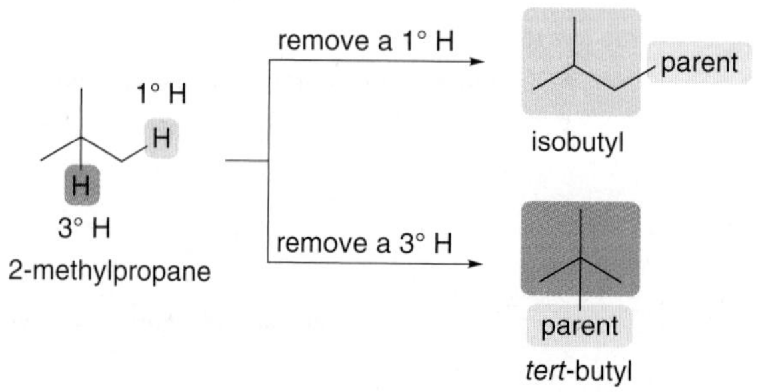

Abbreviations are sometimes used for certain common alkyl groups.

- methyl (Me)
- ethyl (Et)
- butyl (Bu)
- *tert*-butyl (*t*-Bu)

The names isopropyl, *sec*-butyl, isobutyl, and *tert*-butyl are recognized as acceptable substituent names in both the 1979 and 1993 revisions of IUPAC nomenclature. A general method to name these substituents, as well as alkyl groups that contain five or more carbon atoms, is described in Appendix B.

4.4B Naming an Acyclic Alkane

Four steps are needed to name an alkane.

How To Name an Alkane Using the IUPAC System

Step [1] Find the parent carbon chain and add the suffix.

- Find the longest continuous carbon chain, and name the molecule by using the parent name for that number of carbons, given in Table 4.1. To the name of the parent, add the suffix ***-ane*** for an alkane. Each functional group has its own characteristic suffix.

Correct

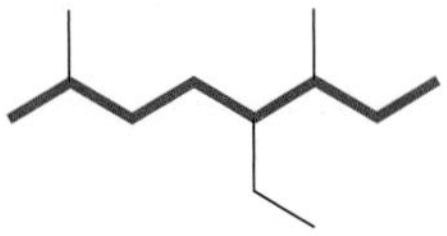

8 C's in the longest chain
8 C's ---→ **octane**

Incorrect

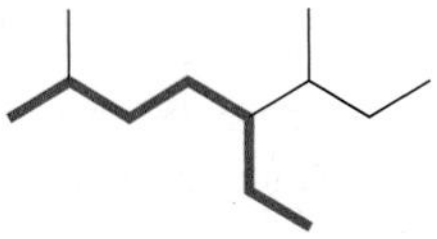

7 C's in the longest chain

- Finding the longest chain is a matter of trial and error. Place your pencil on one end of the chain, go to the other end without picking it up, and count carbons. Repeat this procedure until you have found the chain with the largest number of carbons.
- It does not matter if the chain is *straight* or has *bends.* All of the following representations are equivalent, and each longest chain has eight carbons.

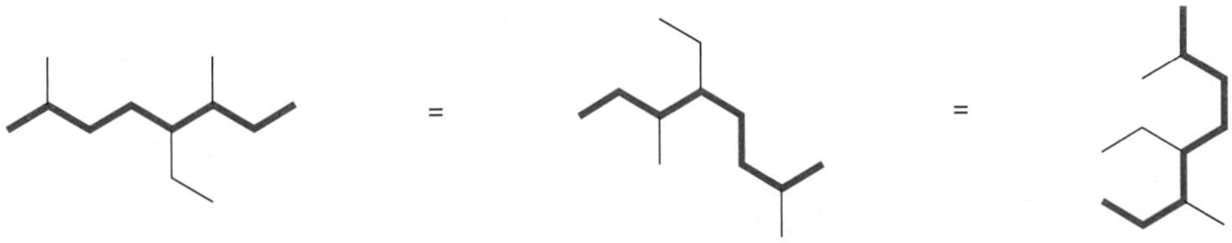

- If there are two chains of equal length, pick the chain with more substituents. In the following example, two different chains in the same alkane contain 7 C's, but the compound on the left has two alkyl groups attached to its long chain, whereas the compound to the right has only one.

Correct

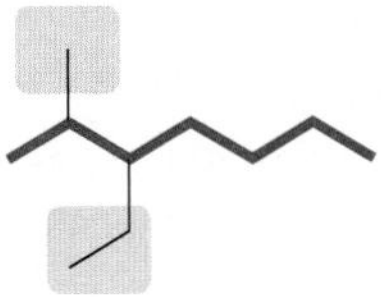

7 atoms in the longest chain
2 substituents

more substituents

Incorrect

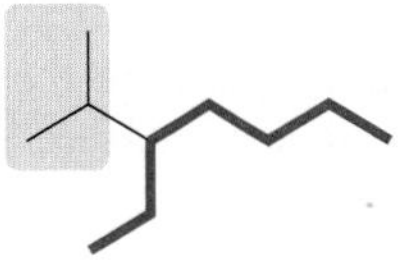

7 atoms in the longest chain
only 1 substituent

fewer substituents

Step [2] Number the atoms in the carbon chain.

- Number the longest chain to give the *first* substituent the lower number.

Correct

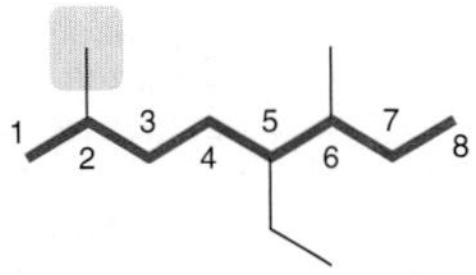

first substituent at C2

Incorrect

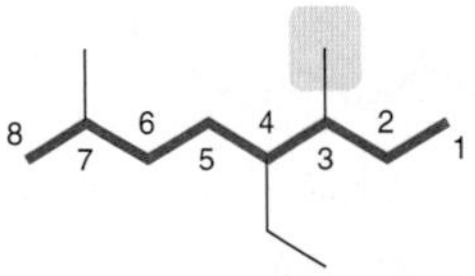

first substituent at C3

—Continued

How To, continued . . .

- If the first substituent is the same distance from both ends, number the chain to give the *second* substituent the lower number. Always look for the first point of difference in numbering from each end of the longest chain.

Correct

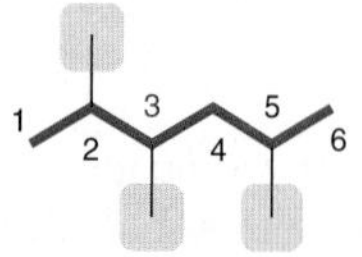

CH_3 groups at C2, C3, and C5
The second CH_3 group has the lower number (C3).

Incorrect

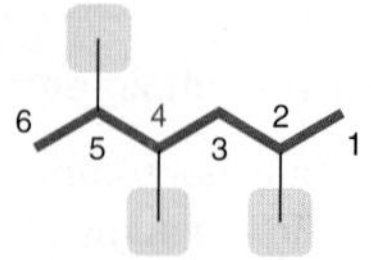

CH_3 groups at C2, C4, and C5
The second CH_3 group has the higher number (C4).

- When numbering a carbon chain results in the same numbers from either end of the chain, assign the lower number *alphabetically* to the first substituent.

Correct

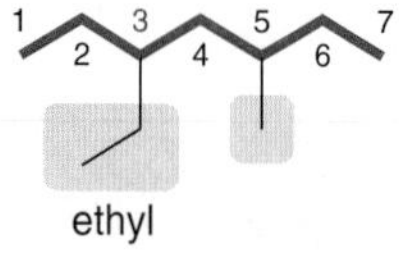

- **e**thyl at **C3**
- **m**ethyl at **C5**

Incorrect

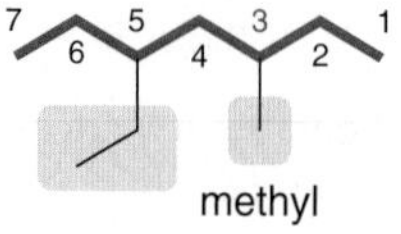

- **m**ethyl at **C3**
- **e**thyl at **C5**

Earlier letter ⟶ lower number

Step [3] Name and number the substituents.

methyl at C2 methyl at C6

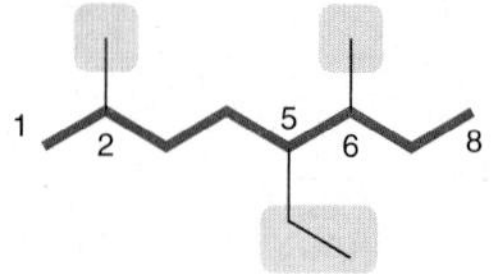

ethyl at C5
8 C's in the longest chain

- Name the substituents as alkyl groups, and use the numbers from Step 2 to designate their location.
- Every carbon belongs to *either* the longest chain or a substituent, but *not both.*
- Each substituent needs its *own* number.
- If two or more identical substituents are bonded to the longest chain, use prefixes to indicate how many: **di-** for two groups, **tri-** for three groups, **tetra-** for four groups, and so forth. This molecule has two methyl substituents, so its name contains the prefix di-before the word methyl → *di*methyl.

Step [4] Combine substituent names and numbers + parent + suffix.

- Precede the name of the parent by the names of the substituents.
- Alphabetize the names of the substituents, ignoring all prefixes except *iso,* as in isopropyl and isobutyl.
- Precede the name of each substituent by the number that indicates its location. There must be **one number for each substituent.**
- Separate numbers by commas and separate numbers from letters by hyphens. The name of an alkane is a single word, with no spaces after hyphens or commas.

[1] Identify all the pieces of a compound, using Steps 1–3.

2,6-dimethyl

methyl at C2 methyl at C6

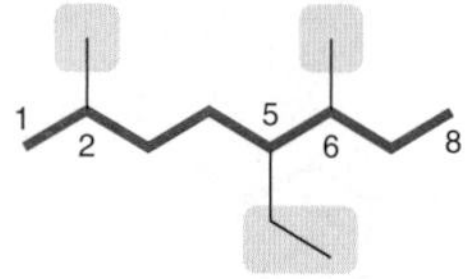

8 C's in the longest chain

octane

ethyl at C5

5-ethyl

[2] Then, put the pieces of the name together.

substituent names and numbers	+	parent	+	suffix
5-ethyl-2,6-dimethyl	+	**oct**	+	**ane**
Alphabetize: **e** for **ethyl,** then **m** for **methyl**		**8** C's		an **alkane**

Answer: 5-ethyl-2,6-dimethyloctane

Several additional examples of alkane nomenclature are given in Figure 4.1.

Figure 4.1
Examples of alkane nomenclature

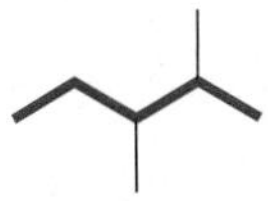

2,3-dimethylpentane

Number to give the 1st methyl group the lower number.

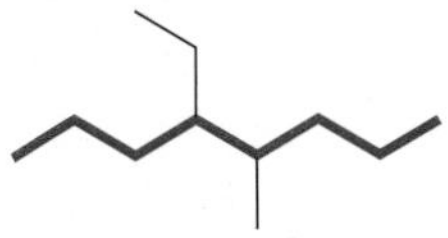

4-ethyl-5-methyloctane

Assign the lower number to the 1st substituent alphabetically: the **e** of **e**thyl before the **m** of **m**ethyl.

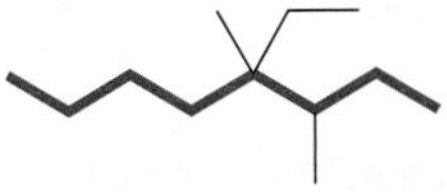

4-ethyl-3,4-dimethyloctane

Alphabetize the **e** of **e**thyl before the **m** of **m**ethyl.

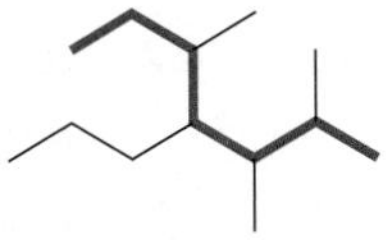

2,3,5-trimethyl-4-propylheptane

Pick the long chain with more substituents.

- The carbon atoms of each long chain are drawn in **red.**

Sample Problem 4.1 Give the IUPAC name for the following compound.

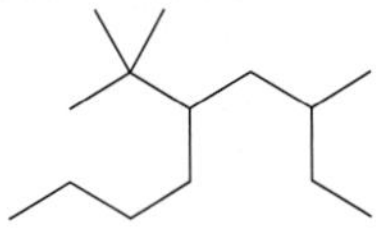

Solution

To help identify which carbons belong to the longest chain and which are substituents, box in or highlight the atoms of the long chain. Every other carbon atom then becomes a substituent that needs its own name as an alkyl group.

Step 1: Name the parent.

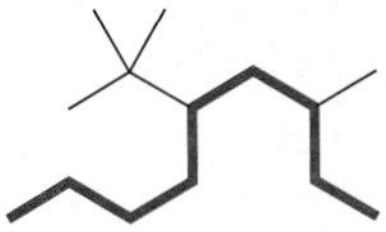

9 C's in the longest chain
nonane

Step 2: Number the chain.

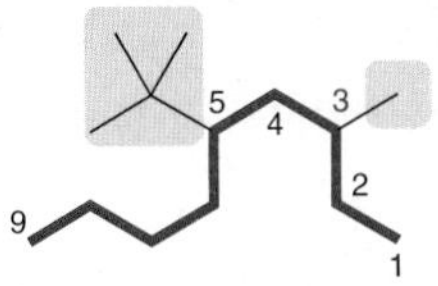

first substituent at C3

Step 3: Name and number the substituents.

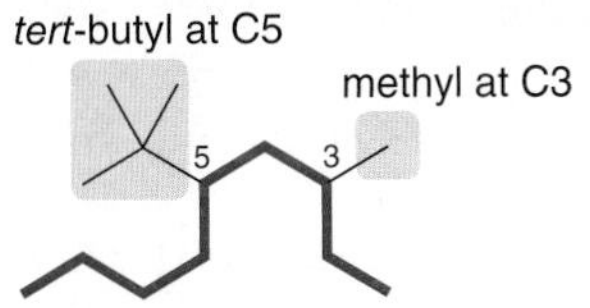

Step 4: Combine the parts.

- Alphabetize: the **b** of **butyl** before the **m** of **methyl**

Answer: 5-*tert*-butyl-3-methylnonane

Problem 4.7 Give the IUPAC name for each compound.

a. 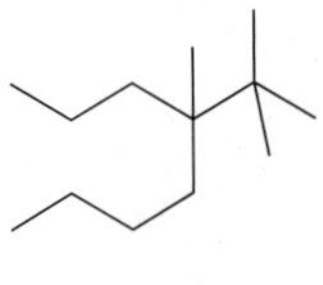
b. 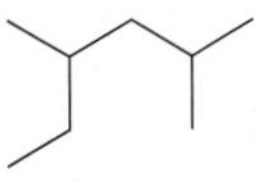
c. 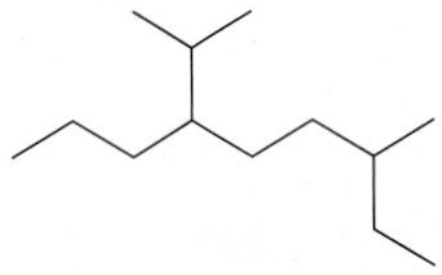
d. 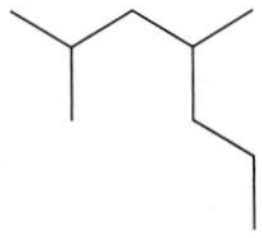

Problem 4.8 Give the IUPAC name for each compound.

a. $(CH_3)_3CCH_2CH(CH_2CH_3)_2$

c. $CH_3(CH_2)_3CH(CH_2CH_2CH_3)CH(CH_3)_2$

b. 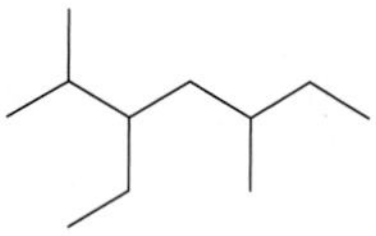

d. 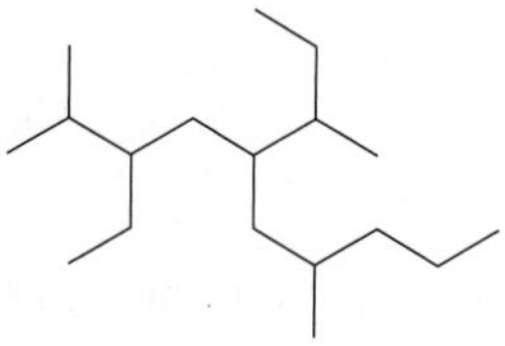

You must also know how to derive a structure from a given name. Sample Problem 4.2 illustrates a stepwise method.

Sample Problem 4.2 Give the structure corresponding to the following IUPAC name: 6-isopropyl-3,3,7-trimethyldecane.

Solution

Follow three steps to derive a structure from a name.

Step [1] Identify the parent name and functional group found at the *end* of the name.

decane ---→ **10** C's ---→ 

Step [2] Number the carbon skeleton in *either* direction.

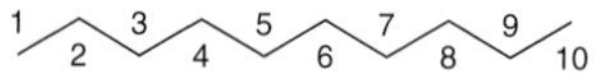

Step [3] Add the substituents at the appropriate carbons.

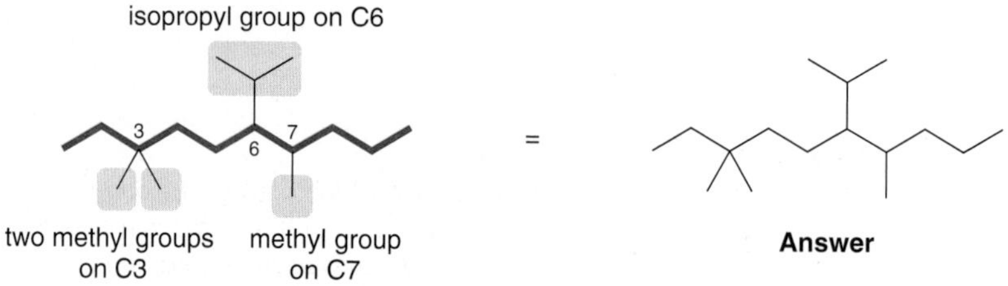

Problem 4.9 Give the structure corresponding to each IUPAC name.

a. 3-methylhexane
b. 3,3-dimethylpentane
c. 3,5,5-trimethyloctane
d. 3-ethyl-4-methylhexane
e. 3-ethyl-5-isobutylnonane

Problem 4.10 Give the IUPAC name for each of the five constitutional isomers of molecular formula C_6H_{14} in Problem 4.3.

4.5 Naming Cycloalkanes

Cycloalkanes are named by using similar rules, but the prefix **cycloꟷ** immediately precedes the name of the parent.

prefix	+	**cyclo-**	+	**parent**	+	**suffix**
What and where are the substituents?				How many C's are in the ring?		What is the functional group?

How To Name a Cycloalkane Using the IUPAC System

Step [1] Find the parent cycloalkane.

- Count the number of carbon atoms in the ring and use the parent name for that number of carbons. Add the prefix **cyclo-** and the suffix ***-ane*** to the parent name.

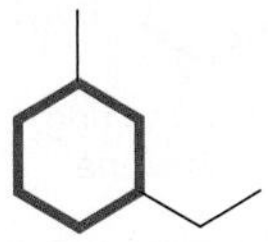

6 C's in the ring
cyclohexane

Step [2] Name and number the substituents.

- No number is needed to indicate the location of a single substituent.

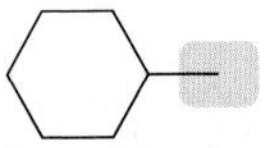

methylcyclohexane

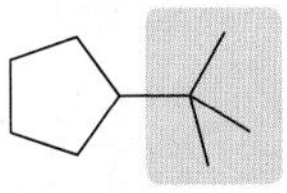

tert-butylcyclopentane

- For rings with more than one substituent, begin numbering at one substituent and proceed around the ring clockwise or counterclockwise to give the second substituent the lower number.

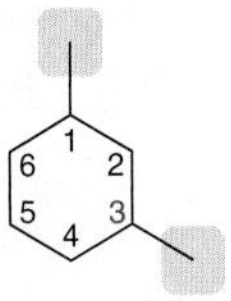

CH_3 groups at C1 and C3
The 2nd substituent has a lower number.

Correct: 1,3-dimethylcyclohexane

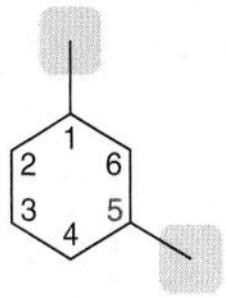

CH_3 groups at C1 and C5

Incorrect: 1,5-dimethylcyclohexane

- With two different substituents, number the ring to assign the lower number to the substituents *alphabetically.*

ethyl at C1

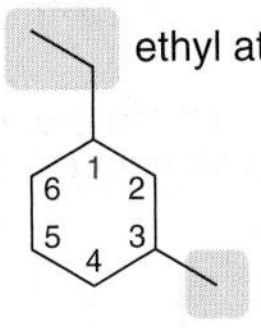

methyl at C3

earlier letter ⟶ lower number

Correct: 1-ethyl-3-methylcyclohexane

ethyl at C3

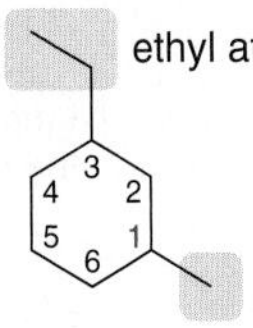

methyl at C1

Incorrect: 3-ethyl-1-methylcyclohexane

When an alkane is composed of both a ring and a long chain, what determines whether a compound is named as an acyclic alkane or a cycloalkane? If the number of carbons in the ring is greater than or equal to the number of carbons in the longest chain, the compound is named as a **cycloalkane,** as shown in Figure 4.2. Several examples of cycloalkane nomenclature are given in Figure 4.3.

Figure 4.2
Naming compounds containing both a ring and a chain of carbon atoms

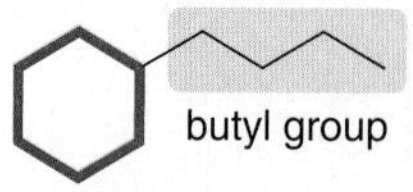
butyl group
6 C's in the ring
cyclohexane

Name as a **cyclohexane** with a substituent.

butylcyclohexane

- Name the molecule as a substituted cycloalkane when it has more C's in the ring than any single alkyl substituent.

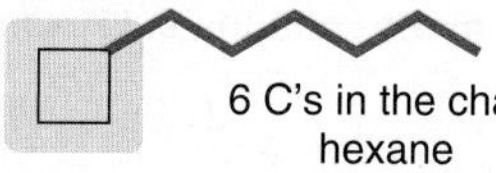
6 C's in the chain
hexane
4 C's in the ring
cyclobutyl group

Name as a **hexane** with a substituent.

1-cyclobutylhexane

- Name the molecule as a substituted alkane when it has a carbon chain with more C's than the ring.

Figure 4.3
Examples of cycloalkane nomenclature

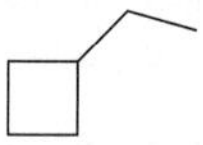

ethylcyclobutane

No number is needed with only one substituent.

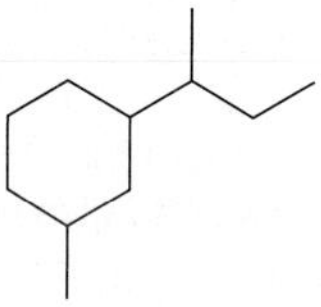

1-sec-butyl-3-methylcyclohexane

Assign the lower number to the 1st substituent alphabetically: the **b** of **b**utyl before the **m** of **m**ethyl.

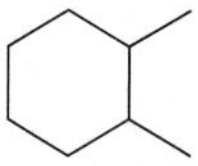

1,2-dimethylcyclohexane

Number to give the 2nd CH_3 group the lower number: 1,2- not 1,6-.

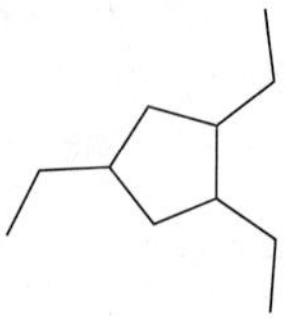

1,2,4-triethylcyclopentane

Number to give the 2nd CH_3CH_2 group the lower number: 1,2,4- not 1,3,4- or 1,3,5-.

Problem 4.11 Give the IUPAC name for each compound.

a. 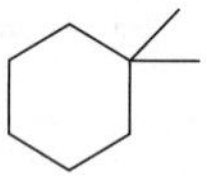

c. 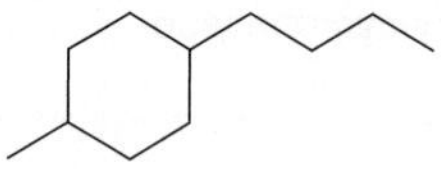

e. 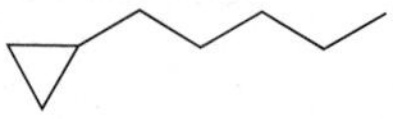

b. 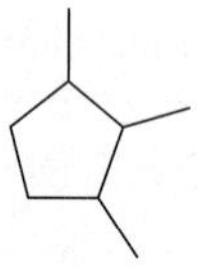

d. 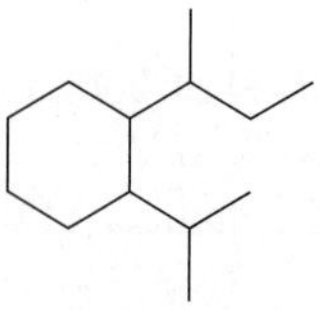

f. 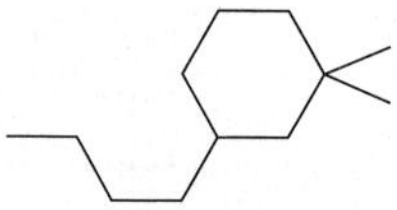

Problem 4.12 Give the structure corresponding to each IUPAC name.

a. 1,2-dimethylcyclobutane
b. 1,1,2-trimethylcyclopropane
c. 4-ethyl-1,2-dimethylcyclohexane
d. 1-*sec*-butyl-3-isopropylcyclopentane
e. 1,1,2,3,4-pentamethylcycloheptane

4.6 Common Names

Some organic compounds are identified using **common names** that do not follow the IUPAC system of nomenclature. Many of these names were given to molecules long ago, before the IUPAC system was adopted. These names are still widely used. For example, isopentane, an older name for 2-methylbutane, is still allowed by IUPAC rules. We will follow the IUPAC system except in cases in which a common name is widely accepted.

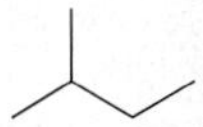

isopentane or 2-methylbutane

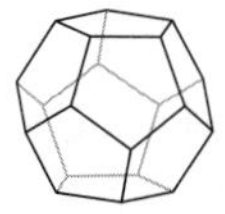

dodecahedrane

In the past several years, organic chemists have attempted to synthesize some unusual cycloalkanes not found in nature. **Dodecahedrane,** a beautifully symmetrical compound composed of 12 five-membered rings, is one such molecule. It was first prepared at The Ohio State University in 1982. The IUPAC name for dodecahedrane is undecacyclo$[9.9.0.0^{2,9}.0^{3,7}.0^{4,20}.0^{5,18}.0^{6,16}.0^{8,15}.0^{10,14}.0^{12,19}.0^{13,17}]$-icosane, a name so complex that few trained organic chemists would be able to identify its structure.

Because these systematic names are so unwieldy, organic chemists often assign a name to a polycyclic compound that is more descriptive of its shape and structure. Dodecahedrane is named because its 12 five-membered rings resemble a dodecahedron. Figure 4.4 shows the names and structures of several other cycloalkanes whose names were inspired by the shape of their carbon skeletons. All the names end in the suffix ***-ane,*** indicating that they refer to alkanes.

Figure 4.4
Common names for some polycyclic alkanes

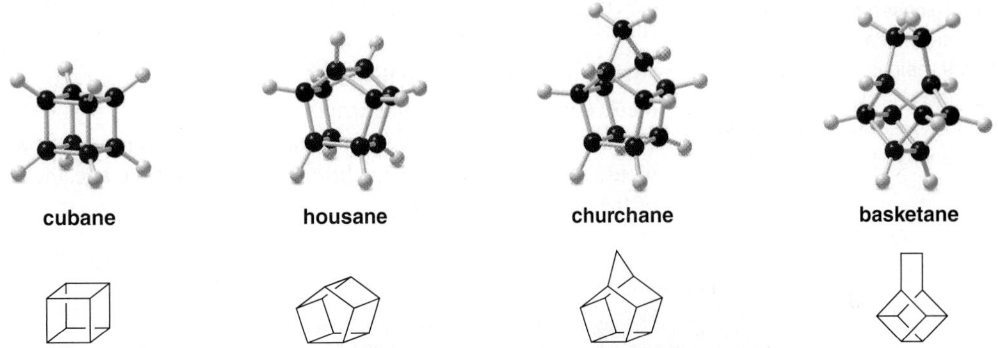
cubane
housane
churchane
basketane

For a more comprehensive list of unusual polycyclic alkanes (including windowpane, davidane, catenane, propellane, and many others), see *Organic Chemistry: The Name Game* by Alex Nickon and Ernest Silversmith, Pergamon Press, 1987.

4.7 Fossil Fuels

Natural gas is odorless. The smell observed in a gas leak is due to minute amounts of a sulfur additive such as methanethiol, CH_3SH, which provides an odor for easy detection.

Many alkanes occur in nature, primarily in natural gas and petroleum. Both of these fossil fuels serve as energy sources, formed from the degradation of organic material long ago.

Natural gas is composed largely of **methane** (60% to 80% depending on its source), with lesser amounts of ethane, propane, and butane. These organic compounds burn in the presence of oxygen, releasing energy for cooking and heating.

Petroleum is a complex mixture of compounds, most of which are hydrocarbons containing 1–40 carbon atoms. Distilling crude petroleum, a process called **refining,** separates it into usable

Figure 4.5
Refining crude petroleum into usable fuel and other petroleum products

a. **An oil refinery.** At an oil refinery, crude petroleum is separated into fractions of similar boiling point by the process of **distillation.**

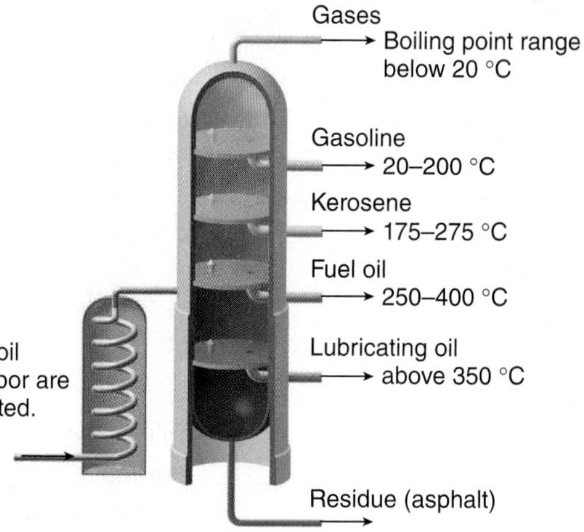

b. **Schematic of a refinery tower.** As crude petroleum is heated, the lower boiling, more volatile components distill first, followed by fractions of progressively higher boiling point.

Methane is formed and used in a variety of ways. The CH_4 released from decaying vegetable matter in New York City's main landfill is used for heating homes. CH_4 generators in China convert cow manure into energy in rural farming towns.

fractions that differ in boiling point (Figure 4.5). Most products of petroleum refining provide fuel for home heating, automobiles, diesel engines, and airplanes. Each fuel type has a different composition of hydrocarbons: gasoline ($C_5H_{12}-C_{12}H_{26}$), kerosene ($C_{12}H_{26}-C_{16}H_{34}$), and diesel fuel ($C_{15}H_{32}-C_{18}H_{38}$).

Petroleum provides more than fuel. About 3% of crude oil is used to make plastics and other synthetic compounds including drugs, fabrics, dyes, and pesticides. These products are responsible for many of the comforts we now take for granted in industrialized countries. Imagine what life would be like without air conditioning, refrigeration, anesthetics, and pain relievers, all products of the petroleum industry.

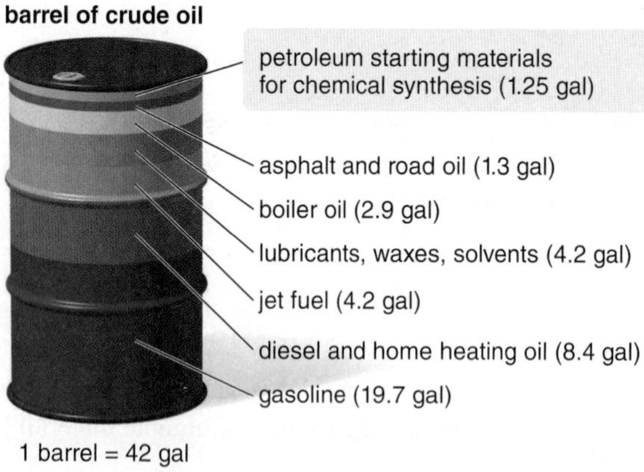

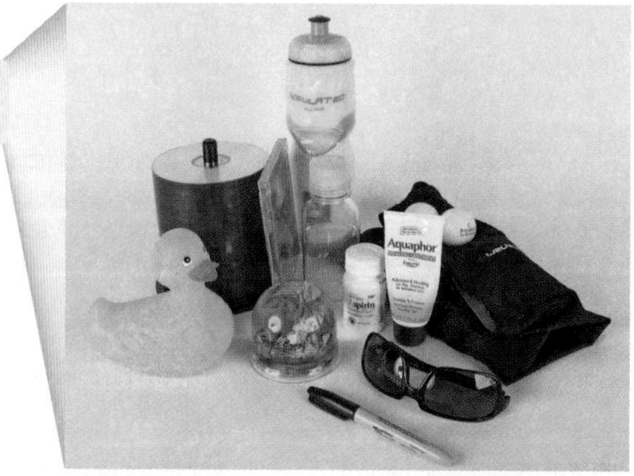

products made from petroleum

Energy from petroleum is ***nonrenewable,*** and the remaining known oil reserves are limited. Given our dependence on petroleum, not only for fuel, but also for the many necessities of modern society, it becomes clear that we must both conserve what we have and find alternate energy sources.

4.8 Physical Properties of Alkanes

Alkanes contain only nonpolar C–C and C–H bonds, and as a result they exhibit only **weak van der Waals forces.** Table 4.2 summarizes how these intermolecular forces affect the physical properties of alkanes.

Table 4.2 Physical Properties of Alkanes

Property	Observation
Boiling point and melting point	• Alkanes have low bp's and mp's compared to more polar compounds of comparable size.  VDW bp = –42 °C mp = –190 °C 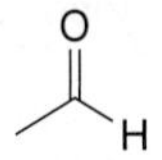 VDW, DD bp = 21 °C mp = –121 °C **Increasing strength of intermolecular forces** **Increasing boiling point and melting point**
	• Bp and mp increase as the number of carbons increases because of increased surface area.  bp = 0 °C mp = –138 °C 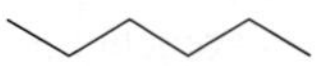 bp = 69 °C mp = –95 °C **Increasing surface area** **Increasing boiling point and melting point**
	• The bp of isomers decreases with branching because of decreased surface area.  bp = 10 °C 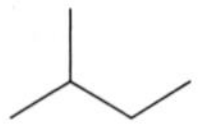 bp = 30 °C  bp = 36 °C **Increasing surface area** **Increasing boiling point**
	• Mp increases with increased symmetry. 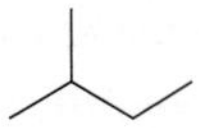 mp = –160 °C  mp = –17 °C **Increasing symmetry** **Increasing melting point**
Solubility	• Alkanes are soluble in organic solvents. • Alkanes are insoluble in water.

Key: bp = boiling point; mp = melting point; VDW = van der Waals; DD = dipole–dipole; HB = hydrogen bonding

The gasoline industry exploits the dependence of boiling point and melting point on alkane size by seasonally changing the composition of gasoline in locations where it gets very hot in the summer and very cold in the winter. Gasoline is refined to contain a larger fraction of higher boiling hydrocarbons in warmer weather, so it evaporates less readily. In colder weather, it is refined to contain more lower boiling hydrocarbons, so it freezes less readily.

The mutual insolubility of nonpolar oil and very polar water leads to the common expression, "Oil and water don't mix."

Because nonpolar alkanes are not water soluble, crude petroleum that leaks into the sea from an oil tanker or offshore oil well creates an insoluble oil slick on the surface. The insoluble hydrocarbon oil poses a special threat to birds whose feathers are coated with natural nonpolar oils for insulation. Because these hydrophobic oils dissolve in the crude petroleum, birds lose their layer of natural protection and many die.

Problem 4.13 Arrange the following compounds in order of increasing boiling point.

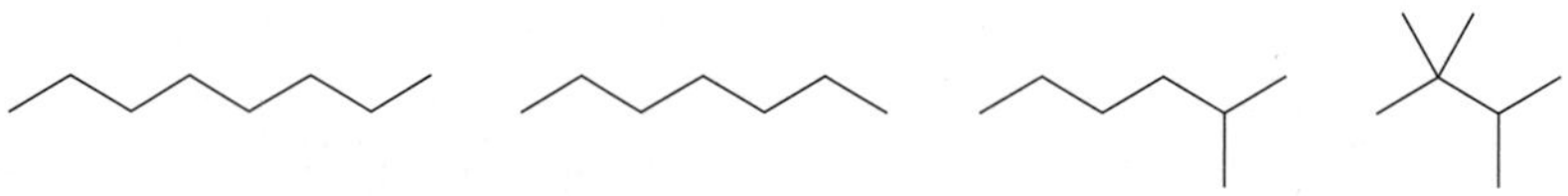

4.9 Conformations of Acyclic Alkanes—Ethane

Let's now take a closer look at the three-dimensional structure of alkanes. The three-dimensional structure of molecules is called **stereochemistry.** In Chapter 4, we examine the effect of rotation around single bonds. In Chapter 5, we will learn about other aspects of stereochemistry.

Recall from Section 1.10A that **rotation occurs around carbon–carbon σ bonds.** Thus, the two CH_3 groups of ethane rotate, allowing the hydrogens on one carbon to adopt different orientations relative to the hydrogens on the other carbon. These arrangements are called **conformations.**

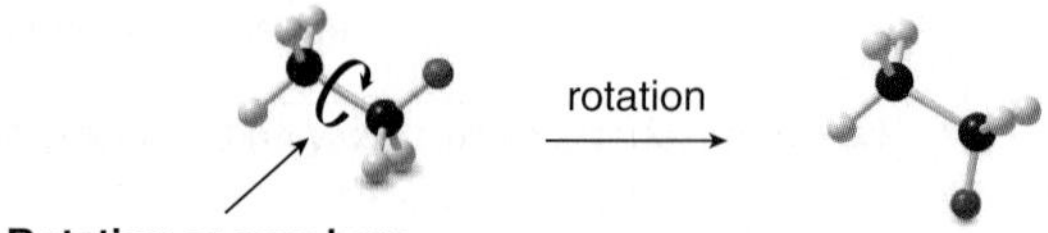

two different conformations

- *Conformations* are different arrangements of atoms that are interconverted by rotation about single bonds.

Two different arrangements are the **eclipsed conformation** and the **staggered conformation.**

- In the *eclipsed conformation,* the C – H bonds on one carbon are directly aligned with the C – H bonds on the adjacent carbon.
- In the *staggered conformation,* the C – H bonds on one carbon bisect the H – C – H bond angle on the adjacent carbon.

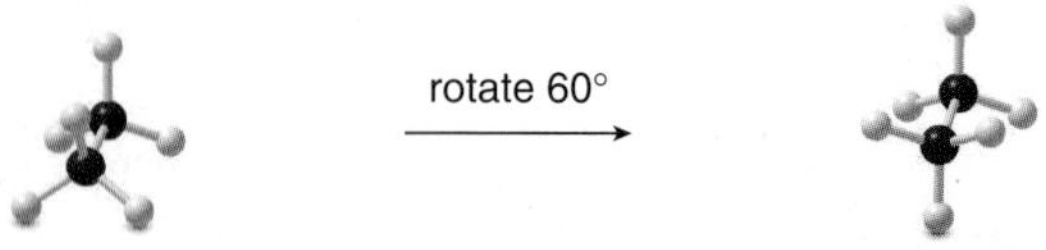

eclipsed conformation

The C–H bonds are all **aligned.**

staggered conformation

The C–H bonds in front **bisect** the H–C–H bond angles in back.

Rotating the atoms on one carbon by 60° converts an eclipsed conformation into a staggered conformation, and vice versa. These conformations are often viewed end-on—that is, looking directly down the carbon–carbon bond. The angle that separates a bond on one atom from a bond on an adjacent atom is called a **dihedral angle.** For ethane in the staggered conformation, the dihedral angle for the C–H bonds is **60°.** For eclipsed ethane, it is **0°.**

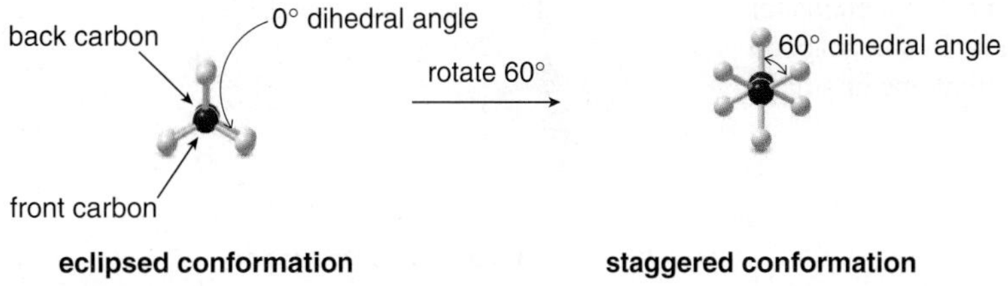

End-on representations for conformations are commonly drawn using a convention called a **Newman projection.** A Newman projection is a graphic that shows the three groups bonded to each carbon atom in a particular C–C bond, as well as the dihedral angle that separates them.

How To Draw a Newman Projection

Step [1] **Look directly down the C–C bond (end-on), and draw a circle with a dot in the center to represent the carbons of the C–C bond.**

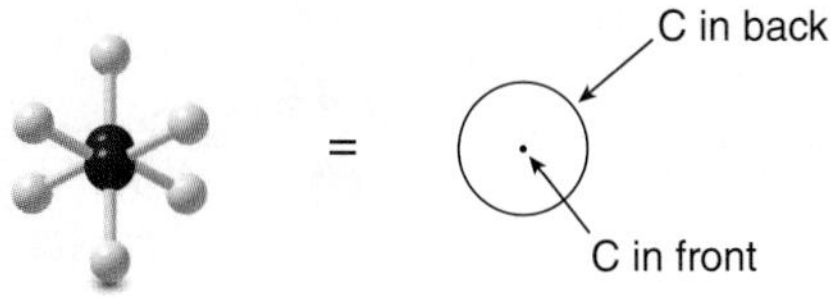

- The circle represents the back carbon and the dot represents the front carbon.

Step [2] **Draw in the bonds.**

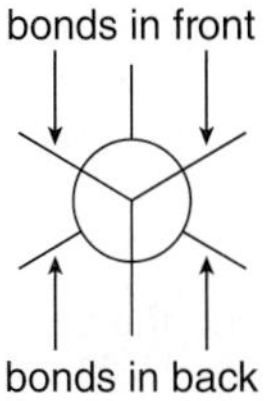

- Draw the bonds on the **front** C as three lines **meeting at the center** of the circle.
- Draw the bonds on the **back** C as three lines coming **out of the edge** of the circle.

Step [3] **Add the atoms on each bond.**

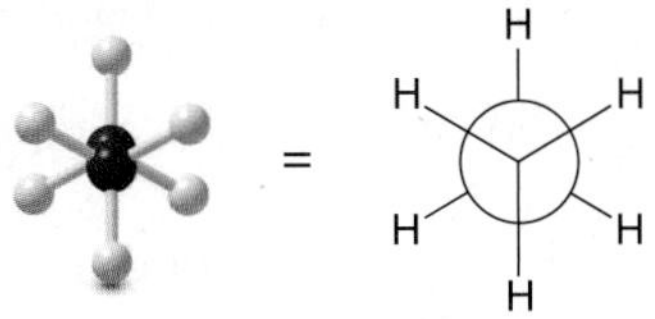

- Each C has 3 H's in ethane.

Figure 4.6 illustrates the Newman projections for both the staggered and eclipsed **conformations** for ethane.

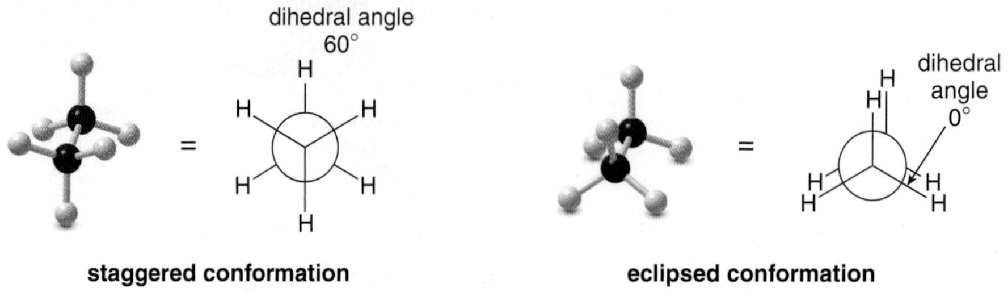

Figure 4.6 Newman projections for the staggered and eclipsed conformations of ethane

Follow this procedure for any C–C bond. With a Newman projection, **always consider *one* C–C bond only and draw the atoms bonded to the carbon atoms, *not* the carbon atoms in the bond itself.** Newman projections for the staggered and eclipsed conformations of propane are drawn in Figure 4.7.

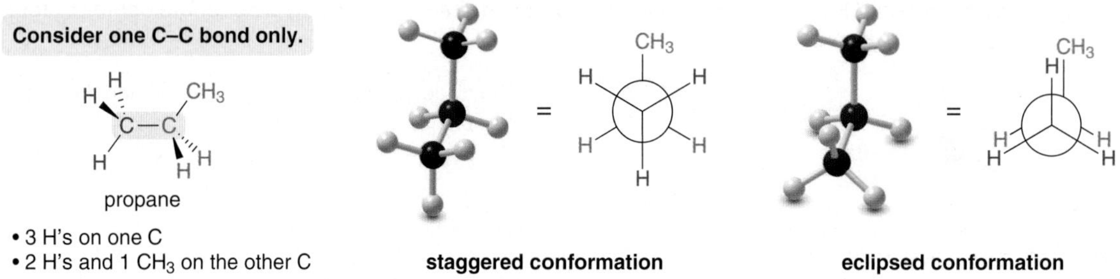

Figure 4.7 Newman projections for the staggered and eclipsed conformations of propane

Problem 4.14 Draw the staggered and eclipsed conformations that result from rotation around the C–C bond in $CH_3–CH_2Br$.

Problem 4.15 Which of the following is (are) possible Newman projections for 2-methylpentane?

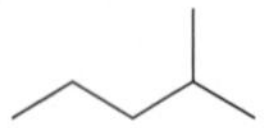

2-methylpentane

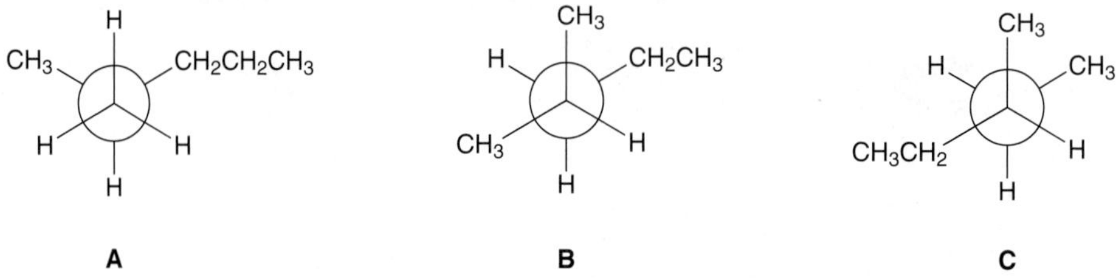

The staggered and eclipsed conformations of ethane interconvert at room temperature, but **each conformation is *not* equally stable.**

- **The staggered conformations are more stable (lower in energy) than the eclipsed conformations.**

The cause of this stability difference is the subject of some debate in the chemical literature. A contributing factor may be increased electron–electron repulsion between the bonds in the eclipsed conformation compared to the staggered conformation, where the bonding electrons are farther apart.

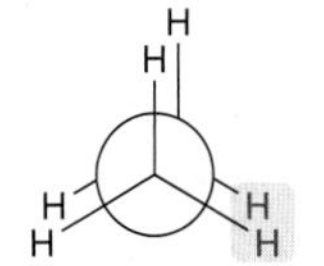

Each H,H eclipsing increases energy by 4.0 kJ/mol.

The difference in energy between the staggered and eclipsed conformations is 12 kJ/mol (2.9 kcal/mol), a small enough difference that the rotation is still very rapid at room temperature, and the conformations cannot be separated. Because three eclipsed C–H bonds increase the energy of a conformation by 12 kJ/mol, **each eclipsed C–H bond results in an increase in energy of 4.0 kJ/mol (1.0 kcal/mol).** The energy difference between the staggered and eclipsed conformations is called **torsional energy.** Thus, eclipsing introduces **torsional strain** into a molecule.

- *Torsional strain* is an increase in energy caused by eclipsing interactions.

The graph in Figure 4.8 shows how the potential energy of ethane changes with dihedral angle as one CH_3 group rotates relative to the other. **The staggered conformation is the most stable arrangement, so it is at an *energy minimum.*** As the C–H bonds on one carbon are rotated relative to the C–H bonds on the other carbon, the energy increases as the C–H bonds get closer until a **maximum is reached after 60° rotation to the eclipsed conformation.** As rotation continues, the energy decreases until after 60° rotation, when the staggered conformation is reached once again. At any given moment, all ethane molecules do not exist in the more stable staggered conformation; rather, a higher percentage of molecules is present in the more stable staggered conformation than any other possible arrangement.

Strain results in an **increase in energy.** Torsional strain is the first of three types of strain discussed in this text. The other two are steric strain (Section 4.10) and angle strain (Section 4.11).

- An energy minimum and maximum occur every 60° as the conformation changes from staggered to eclipsed. Conformations that are neither staggered nor eclipsed are intermediate in energy.

Problem 4.16 The torsional energy in propane is 14 kJ/mol (3.4 kcal/mol). Because each H,H eclipsing interaction is worth 4.0 kJ/mol (1.0 kcal/mol) of destabilization, how much is one H,CH_3 eclipsing interaction worth in destabilization? (See Section 4.10 for an alternate way to arrive at this value.)

Figure 4.8
Graph: Energy versus dihedral angle for ethane

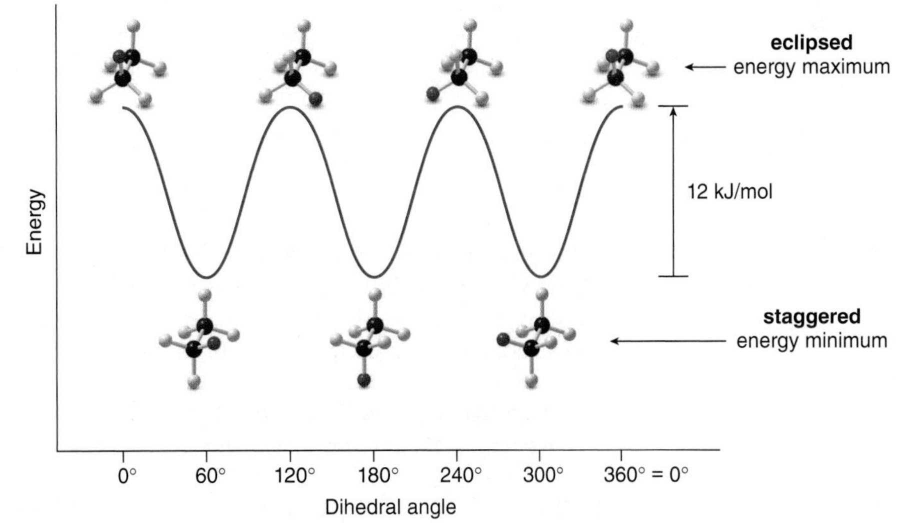

- Note the position of the labeled H atom after each 60° rotation. All three staggered conformations are identical (except for the position of the label), and the same is true for all three eclipsed conformations.

4.10 Conformations of Butane

Butane and higher molecular weight alkanes have several carbon–carbon bonds, all capable of rotation.

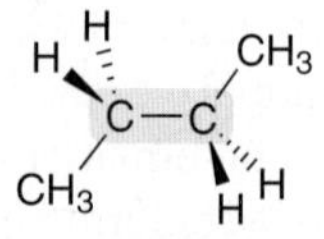

butane
Consider rotation at C2–C3.
Each C is bonded to 2 H's and 1 CH_3 group.

To analyze the different conformations that result from rotation about the C2–C3 bond, begin arbitrarily with one—for example, the staggered conformation that places two CH_3 groups 180° from each other—then,

It takes six 60° rotations to return to the original conformation.

- **Rotate one carbon atom in 60° increments either clockwise or counterclockwise, while keeping the other carbon fixed. Continue until you return to the original conformation.**

Figure 4.9 illustrates the six possible conformations that result from this process.

Figure 4.9 Six different conformations of butane

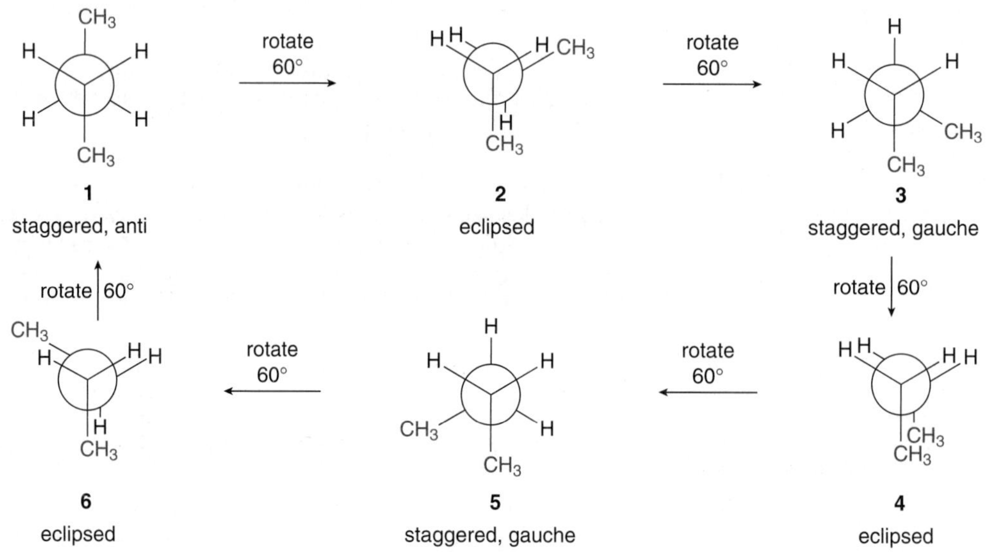

Although each 60° bond rotation converts a staggered conformation into an eclipsed conformation (or vice versa), neither all the staggered conformations nor all the eclipsed conformations are the same. For example, the dihedral angle between the methyl groups in staggered conformations **3** and **5** are both 60°, whereas it is 180° in staggered conformation **1.**

- **A staggered conformation with two larger groups 180° from each other is called *anti.***
- **A staggered conformation with two larger groups 60° from each other is called *gauche.***

Similarly, the methyl groups in conformations **2** and **6** both eclipse hydrogen atoms, whereas they eclipse each other in conformation **4.**

The staggered conformations (**1, 3,** and **5**) are lower in energy than the eclipsed conformations (**2, 4,** and **6**), but how do the energies of the individual staggered and eclipsed conformations compare to each other? The relative energies of the individual staggered conformations (or the individual eclipsed conformations) depend on their **steric strain.**

- *Steric* strain is an increase in energy resulting when atoms are forced too close to one another.

The methyl groups are farther apart in the anti conformation (**1**) than in the gauche conformations (**3** and **5**), so among the staggered conformations, **1** is lower in energy (more stable) than **3** and **5.** In fact, the anti conformation is 3.8 kJ/mol (0.9 kcal/mol) lower in energy than either gauche conformation because of the steric strain that results from the proximity of the methyl groups in **3** and **5.**

Anti conformation

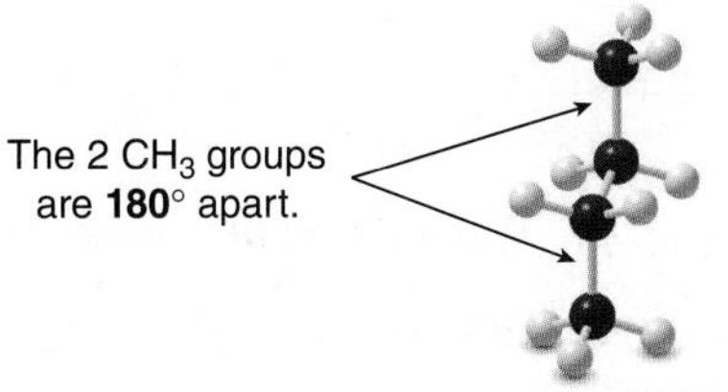

lower in energy
1

Gauche conformation

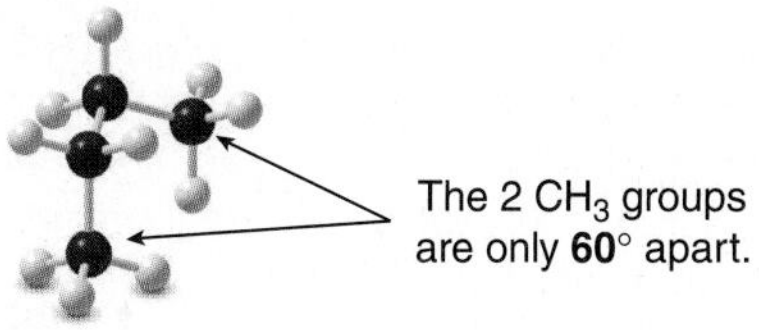

higher in energy
3

- Gauche conformations are generally higher in energy than anti conformations because of steric strain.

Steric strain also affects the relative energies of eclipsed conformations. Conformation **4** is higher in energy than **2** or **6,** because the two larger CH_3 groups are forced close to each other, introducing considerable steric strain.

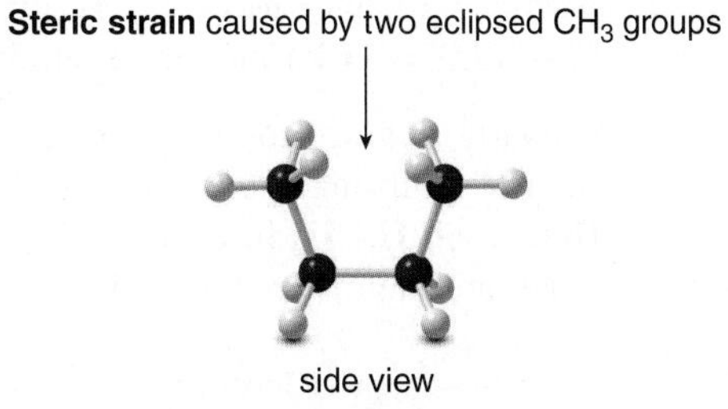

4

To graph energy versus dihedral angle, keep in mind two considerations:

- Staggered conformations are at energy minima and eclipsed conformations are at energy maxima.
- Unfavorable steric interactions increase energy.

For butane, this means that anti conformation **1** is lowest in energy, and conformation **4** with two eclipsed CH_3 groups is the highest in energy. The relative energy of other conformations is depicted in the energy versus rotation diagram for butane in Figure 4.10.

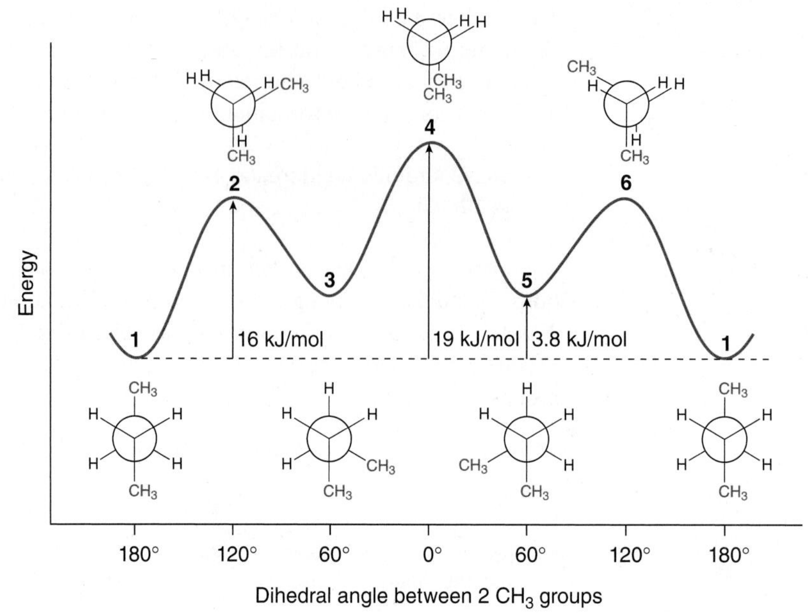

Figure 4.10
Graph: Energy versus dihedral angle for butane

- Staggered conformations **1, 3,** and **5** are at energy minima.
- Anti conformation **1** is lower in energy than gauche conformations **3** and **5,** which possess steric strain.
- Eclipsed conformations **2, 4,** and **6** are at energy maxima.
- Eclipsed conformation **4,** which has additional steric strain due to two eclipsed CH_3 groups, is highest in energy.

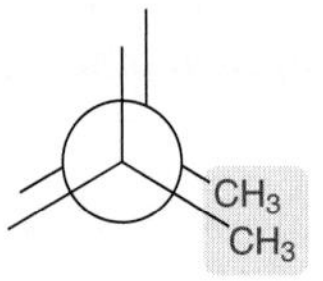

CH_3,CH_3 eclipsing
11 kJ/mol destabilization

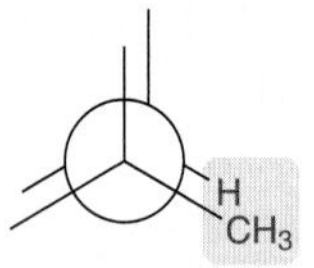

H,CH_3 eclipsing
6.0 kJ/mol destabilization

We can now use the values in Figure 4.10 to estimate the destabilization caused by other eclipsed groups. For example, conformation **4** is 19 kJ/mol less stable than the anti conformation **1.** Conformation **4** possesses two H,H eclipsing interactions, worth 4.0 kJ/mol each in destabilization (Section 4.9), and one CH_3,CH_3 eclipsing interaction. Thus, the **CH_3,CH_3 interaction** is worth 19 – 2(4.0) = **11 kJ/mol** of destabilization.

Similarly, conformation **2** is 16 kJ/mol less stable than the anti conformation **1,** and possesses one H,H eclipsing interaction (worth 4.0 kJ/mol of destabilization) and two H,CH_3 interactions. Thus, **each H,CH_3 interaction** is worth 1/2(16 – 4.0) = **6.0 kJ/mol** of destabilization. These values are summarized in Table 4.3.

- **The energy difference between the lowest and highest energy conformations is called the *barrier to rotation.***

Table 4.3 Summary: Torsional and Steric Strain Energies in Acyclic Alkanes

	Energy increase	
Type of interaction	kJ/mol	kcal/mol
H,H eclipsing	4.0	1.0
H,CH_3 eclipsing	6.0	1.4
CH_3,CH_3 eclipsing	11	2.6
gauche CH_3 groups	3.8	0.9

We can use these same principles to determine conformations and relative energies for any acyclic alkane. Because the lowest energy conformation has all bonds staggered and all large groups anti, alkanes are often drawn in zigzag skeletal structures to indicate this.

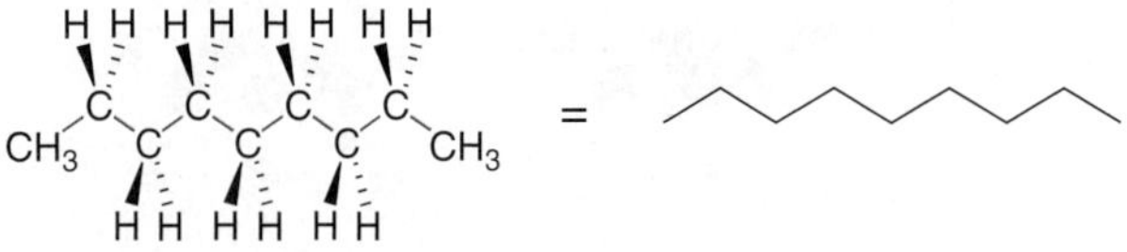

Problem 4.17

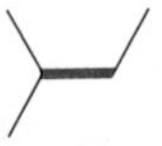

a. Draw the three staggered and three eclipsed conformations that result from rotation around the bond labeled in red using Newman projections.
b. Label the most stable and least stable conformation.

Problem 4.18 Rank the following conformations in order of increasing energy.

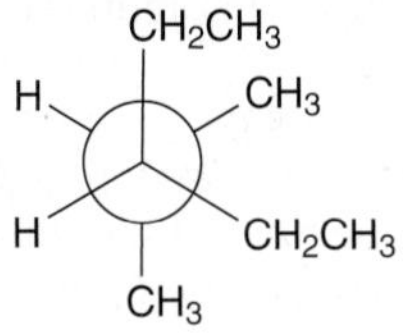

A

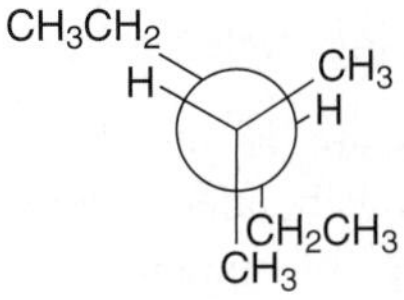

B

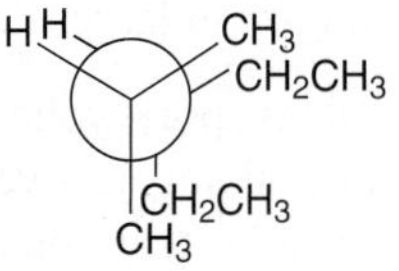

C

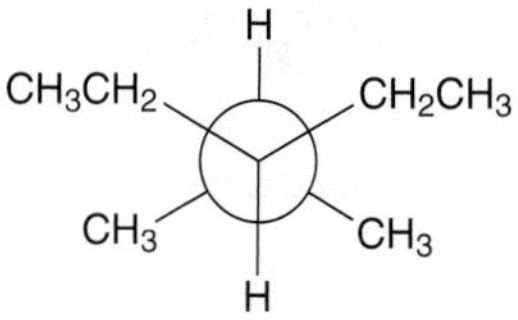

D

Problem 4.19 Consider rotation around the carbon–carbon bond in 1,2-dichloroethane ($ClCH_2CH_2Cl$).

a. Using Newman projections, draw all of the staggered and eclipsed conformations that result from rotation around this bond.
b. Graph energy versus dihedral angle for rotation around this bond.

Problem 4.20 Calculate the destabilization present in each eclipsed conformation.

a. 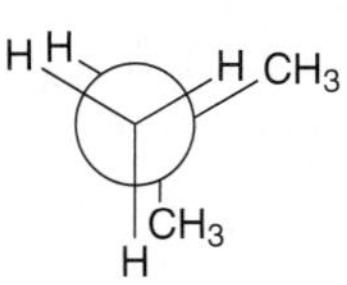

b. 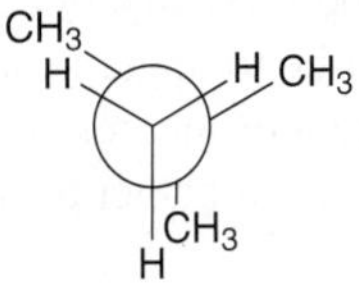

4.11 An Introduction to Cycloalkanes

Besides torsional strain and steric strain, the conformations of cycloalkanes are also affected by **angle strain.**

- ***Angle*** **strain is an increase in energy when tetrahedral bond angles deviate from the optimum angle of 109.5°.**

Originally cycloalkanes were thought to be flat rings, with the bond angles between carbon atoms determined by the size of the ring. For example, a flat cyclopropane ring would have 60° internal bond angles, a flat cyclobutane ring would have 90° angles, and large flat rings would have very large angles. It was assumed that rings with bond angles so different from the tetrahedral bond angle would be very strained and highly reactive. This is called the Baeyer strain theory.

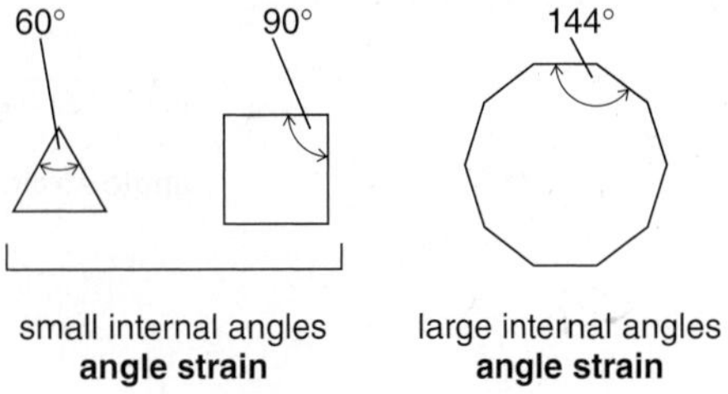
60°
90°
144°
small internal angles
angle strain
large internal angles
angle strain

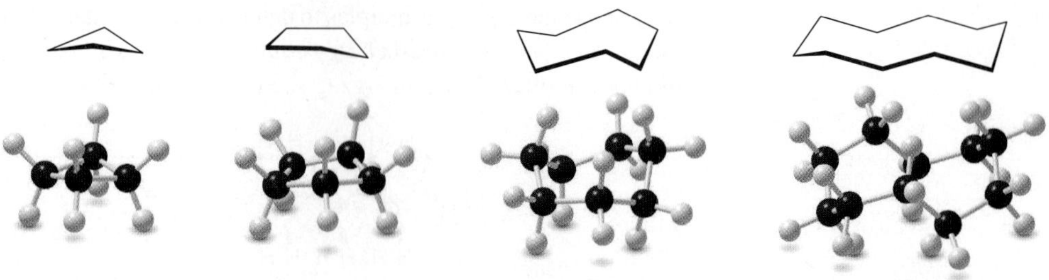

cyclobutane cyclopentane cycloheptane cyclodecane

Figure 4.11 Three-dimensional structure of some cycloalkanes

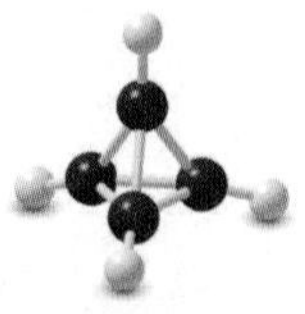

tetrahedrane

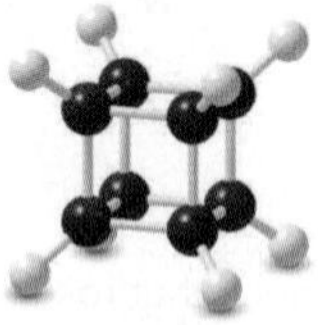

cubane

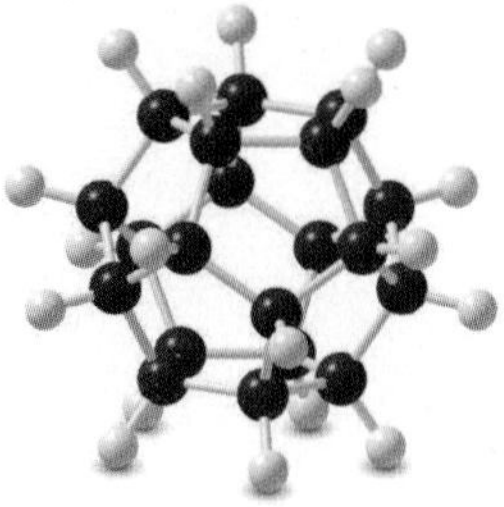

dodecahedrane

It turns out, though, that **cycloalkanes with more than three C atoms in the ring are not flat molecules.** They are puckered to **reduce strain,** both angle strain and torsional strain. The three-dimensional structures of some simple cycloalkanes are shown in Figure 4.11. Three- and four-membered rings still possess considerable angle strain, but puckering reduces the internal bond angles in larger rings, thus reducing angle strain.

Many polycyclic hydrocarbons are of interest to chemists. For example, **dodecahedrane,** containing 12 five-membered rings bonded together, is one member of a family of three hydrocarbons that contain several rings of one size joined together. The two other members of this family are **tetrahedrane,** consisting of four three-membered rings, and **cubane,** consisting of six four-membered rings. These compounds are the simplest regular polyhedra whose structures resemble three of the highly symmetrical Platonic solids: the tetrahedron, the cube, and the dodecahedron.

How stable are these compounds? Tetrahedrane (with internal 60° bond angles) is so strained that all attempts to prepare it have been thus far unsuccessful. Although cubane is also highly strained because of its 90° bond angles, it was first synthesized in 1964 and is a stable molecule at room temperature. Finally, dodecahedrane is very stable because it has bond angles very close to the tetrahedral bond angle (108° versus 109.5°). Its synthesis eluded chemists for years not because of its strain or inherent instability, but because of the enormous challenge of joining 12 five-membered rings together to form a sphere.

4.12 Cyclohexane

Let's now examine the conformation of **cyclohexane,** the most common ring size in naturally occurring compounds.

4.12A The Chair Conformation

A planar cyclohexane ring would experience angle strain, because the internal bond angle between the carbon atoms would be 120°, and torsional strain, because all of the hydrogens on adjacent carbon atoms would be eclipsed.

If a cyclohexane ring were flat...

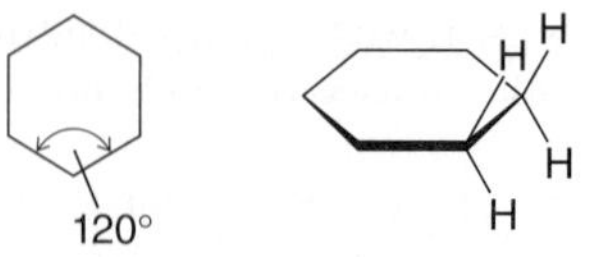

The internal bond angle is >109.5°. **angle strain**

All H's are aligned. **torsional strain**

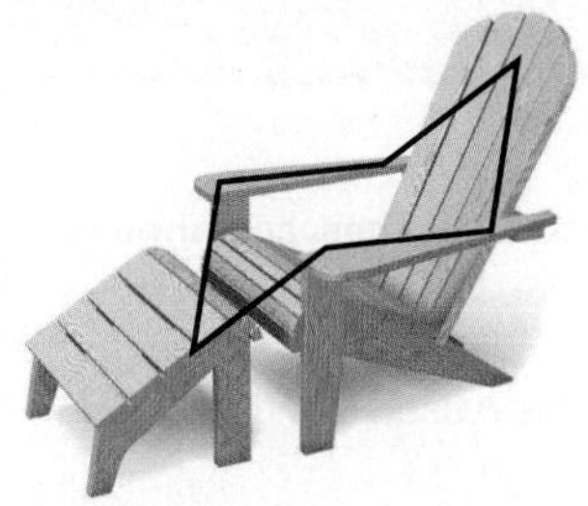

Visualizing the chair. If the cyclohexane chair conformation is tipped downward, we can more easily view it as a chair with a back, seat, and foot support.

In reality, cyclohexane adopts a puckered conformation, called the **chair** form, which is more stable than any other possible conformation.

The chair conformation is so stable because it eliminates angle strain **(all C−C−C bond angles are 109.5°)** and torsional strain (all hydrogens on adjacent carbon atoms are **staggered,** not eclipsed).

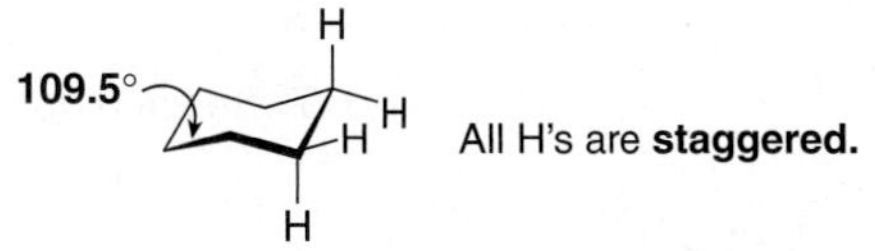

- In cyclohexane, three C atoms pucker up and three C atoms pucker down, alternating around the ring. These C atoms are called *up* C's and *down* C's.

Each cyclohexane carbon atom has one axial and one equatorial hydrogen.

Each carbon in cyclohexane has two different kinds of hydrogens.

- *Axial* hydrogens are located above and below the ring (along a perpendicular axis).
- *Equatorial* hydrogens are located in the plane of the ring (around the equator).

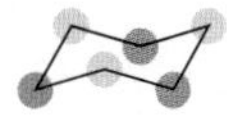

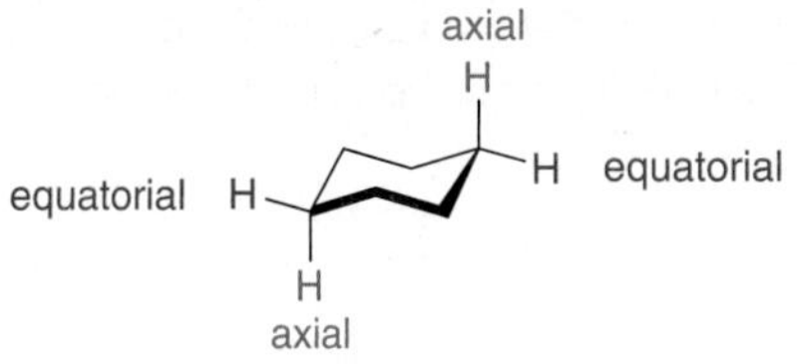

A three-dimensional representation of the chair form is shown in Figure 4.12.

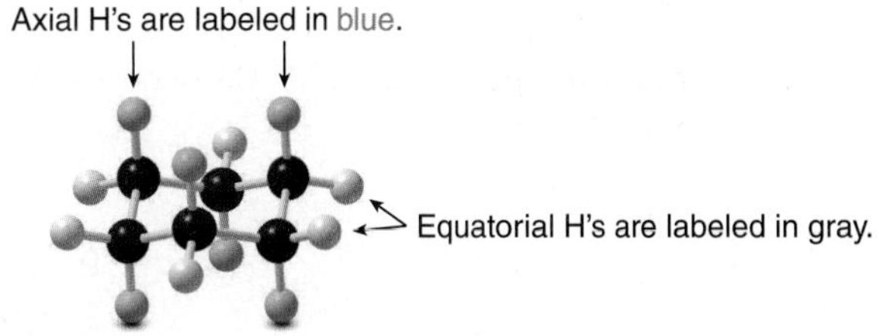

Figure 4.12
A three-dimensional model of the chair form of cyclohexane with all H atoms drawn

- Cyclohexane has **six axial H's** and **six equatorial H's.**

How To Draw the Chair Form of Cyclohexane

Step [1] Draw the carbon skeleton.

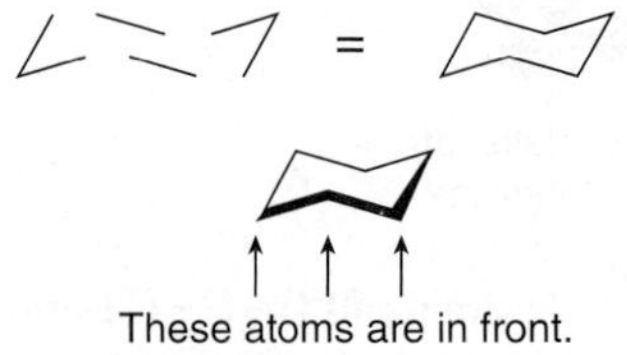
=
These atoms are in front.

- Draw three parts of the chair: **a wedge, a set of parallel lines,** and **another wedge.**
- Then, join them together.
- The bottom 3 C's come out of the page, and for this reason, bonds to them are sometimes highlighted in bold.

Step [2] Label the *up* C's and *down* C's on the ring.

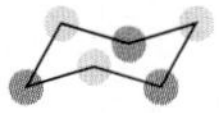

● = *up* C ● = *down* C

- There are 3 *up* and 3 *down* C's, and they alternate around the ring.

Step [3] Draw in the axial H atoms.

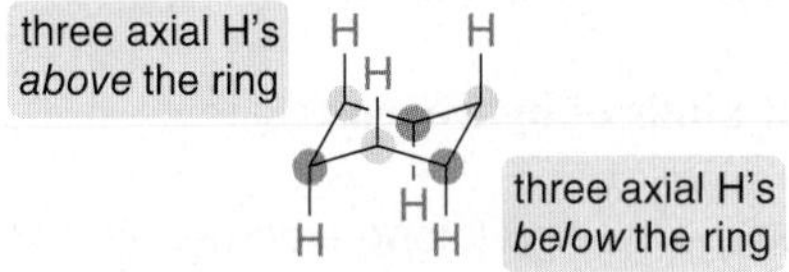
three axial H's *above* the ring
three axial H's *below* the ring
H
H
H
H
H
H

- On an *up* C the axial H is *up.*
- On a *down* C the axial H is *down.*

Step [4] Draw in the equatorial H atoms.

- The axial H is **down** on a down C, so the equatorial H must be up.
- The axial H is **up** on an up C, so the equatorial H must be down.

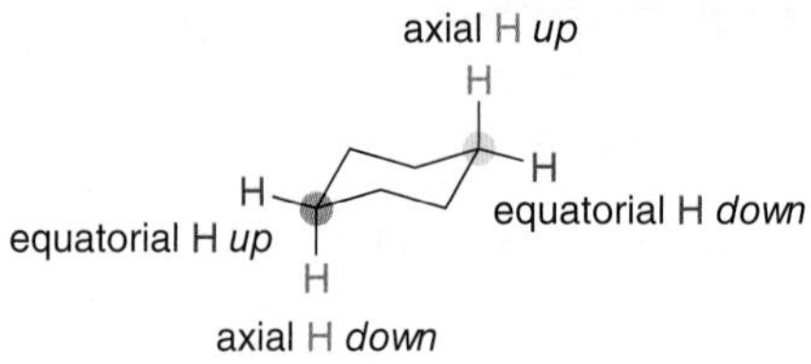
axial H *up*
equatorial H *up*
equatorial H *down*
axial H *down*
H
H
H
H

All equatorial H's drawn in.

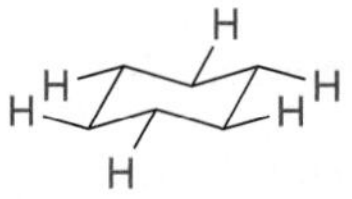

All H's drawn in.

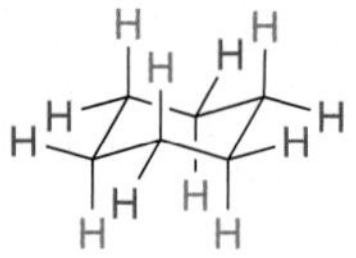

Problem 4.21 Classify the ring carbons as *up* C's or *down* C's. Identify the bonds highlighted in bold as axial or equatorial.

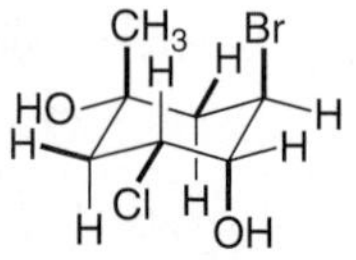
CH3
Br
HO
OH
Cl
H

Problem 4.22 Using the cyclohexane with the C's numbered as shown, draw a chair form that fits each description.

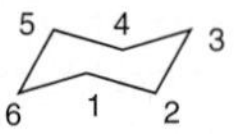
5
4
3
6
1
2

a. The ring has an axial CH_3 group at C1 and an equatorial OH on C2.
b. The ring has an equatorial CH_3 group on C6 and an axial OH group on C4.
c. The ring has equatorial OH groups on C1, C2, and C5.

4.12B Ring-Flipping

Like acyclic alkanes, **cyclohexane does not remain in a single conformation.** The bonds twist and bend, resulting in new arrangements, but the movement is more restricted. One conformational change involves **ring-flipping,** which can be viewed as a two-step process.

- **A *down* carbon flips up.** This forms a new conformation of cyclohexane called a **boat.** The boat form has two carbons oriented above a plane containing the other four carbons.
- The boat form can flip in two possible ways. The carbon labeled with a red circle can flip down, re-forming the initial conformation; or the **second *up* carbon,** labeled with a blue circle, **can flip down. This forms a second chair conformation.**

Because of ring-flipping, the ***up* carbons become *down* carbons and the *down* carbons become *up* carbons.** Thus, cyclohexane exists as two different chair conformations of equal stability, which rapidly interconvert at room temperature.

The process of ring-flipping also affects the orientation of cyclohexane's hydrogen atoms.

- **Axial and equatorial H atoms are interconverted during a ring flip. Axial H atoms become equatorial H atoms, and equatorial H atoms become axial H atoms (Figure 4.13).**

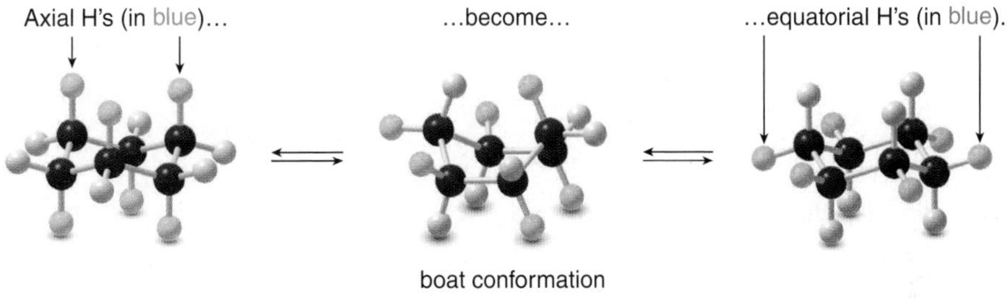

Figure 4.13 Ring-flipping interconverts axial and equatorial hydrogens in cyclohexane.

The chair forms of cyclohexane are 30 kJ/mol more stable than the boat forms. The boat conformation is destabilized by torsional strain because the hydrogens on the four carbon atoms in the plane are eclipsed. Additionally, there is steric strain because two hydrogens at either end of the boat—the **flagpole hydrogens**—are forced close to each other, as shown in Figure 4.14.

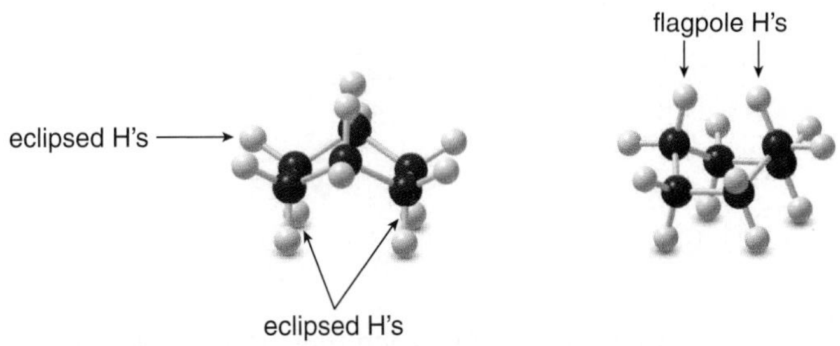

Figure 4.14 Two views of the boat conformation of cyclohexane

The boat form of cyclohexane is less stable than the chair forms for two reasons.

- Eclipsing interactions between H's cause torsional strain.
- The proximity of the flagpole H's causes steric strain.

4.13 Substituted Cycloalkanes

What happens when one hydrogen on cyclohexane is replaced by a larger substituent? Is there a difference in the stability of the two cyclohexane conformations? To answer these questions, remember one rule.

- **The equatorial position has more room than the axial position, so larger substituents are more stable in the equatorial position.**

4.13A Cyclohexane with One Substituent

There are two possible chair conformations of a monosubstituted cyclohexane, such as methylcyclohexane, as shown in the following *How To.*

How To Draw the Two Conformations for a Substituted Cyclohexane

Step [1] Draw one chair form and add the substituents.

- Arbitrarily pick a ring carbon, classify it as an *up* or *down* carbon, and draw the bonds. Each C has one axial and one equatorial bond.
- Add the substituents, in this case H and CH_3, arbitrarily placing one axial and one equatorial. In this example, the CH_3 group is drawn equatorial.
- This forms one of the two possible chair conformations, labeled **A.**

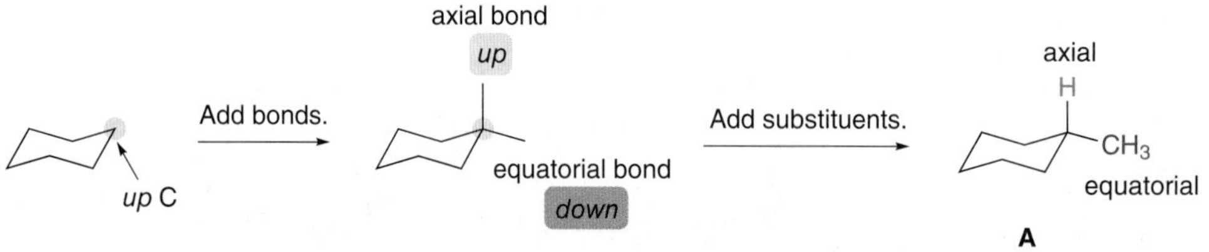

A

Step [2] Ring-flip the cyclohexane ring.

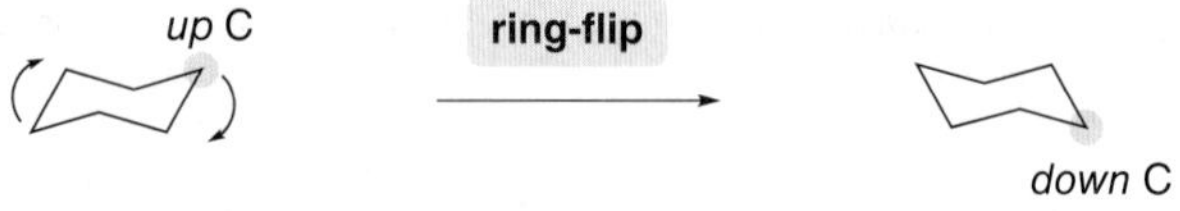

- Convert *up* C's to *down* C's and vice versa. The chosen *up* C now puckers down.

Step [3] Add the substituents to the second conformation.

- Draw axial and equatorial bonds. On a *down* C the axial bond is *down.*
- Ring-flipping converts axial bonds to equatorial bonds, and vice versa. The equatorial methyl becomes axial.
- This forms the other possible chair conformation, labeled **B.**

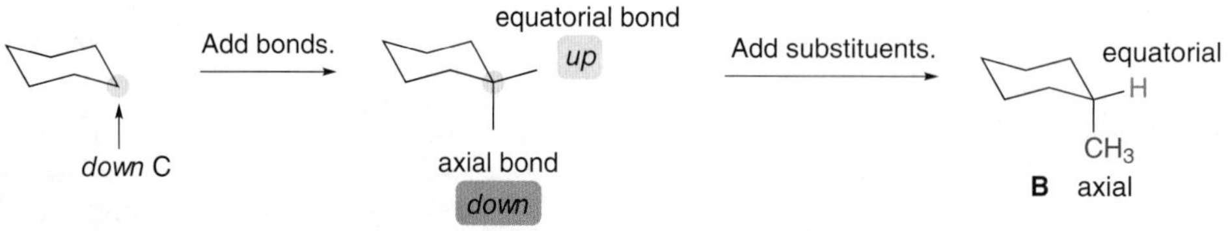

B

Although the CH_3 group flips from equatorial to axial, it starts on a down bond, and stays on a down bond. It *never* flips from below the ring to above the ring.

- **A substituent always stays on the same side of the ring—either below or above—during the process of ring-flipping.**

Each carbon atom has one *up* and one *down* bond. An *up* bond can be either axial or equatorial, depending on the carbon to which it is attached. On an *up* C, the axial bond is *up*, but on a *down* C, the equatorial bond is *up*.

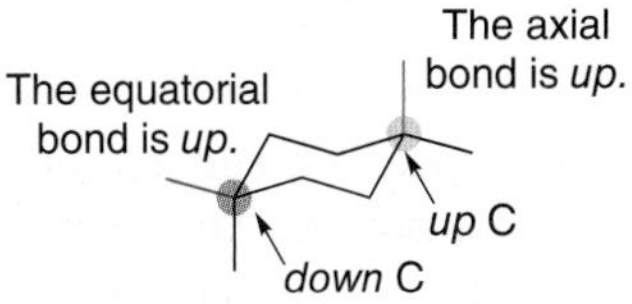

The two conformations of methylcyclohexane are different, so they are not equally stable. In fact, **A,** which places the larger methyl group in the roomier equatorial position, is considerably more stable than **B,** which places it axial.

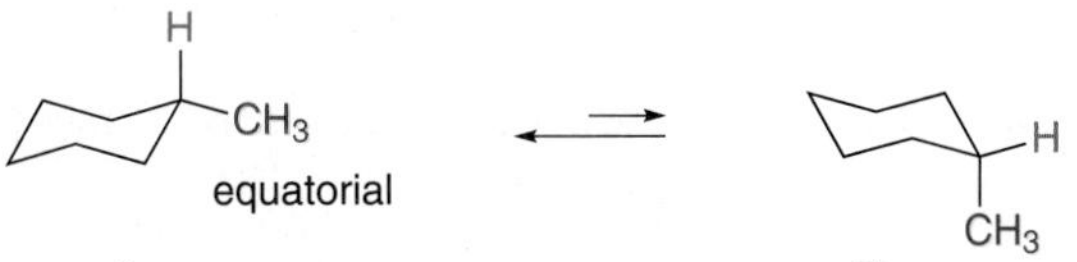

A The larger CH_3 group is equatorial. **more stable** 95%

B less stable 5%

Why is a substituted cyclohexane ring more stable with a larger group in the equatorial position? Figure 4.15 shows that with an equatorial CH_3 group, steric interactions with nearby groups are minimized. An axial CH_3 group, however, is close to two other axial H atoms, creating two destabilizing steric interactions called **1,3-diaxial interactions.** Each unfavorable H,CH_3 interaction destabilizes the conformation by 3.8 kJ/mol, so **B** is 7.6 kJ/mol less stable than **A.**

Figure 4.15 Three-dimensional representations for the two conformations of methylcyclohexane

Equatorial CH_3 group **Axial CH_3 group**

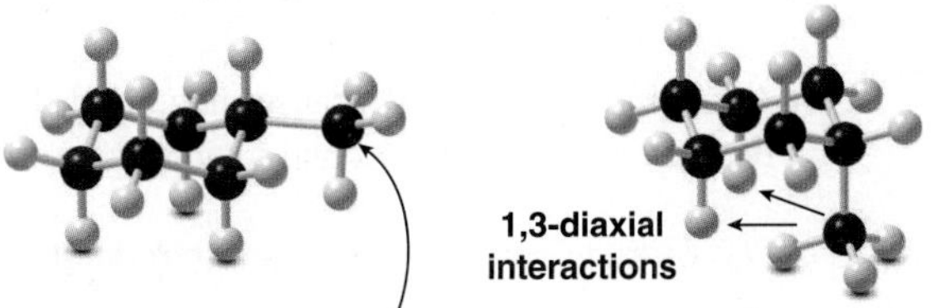

A The CH_3 has more room. **preferred**

B An axial CH_3 group has unfavorable steric interactions.

- **Larger axial substituents create unfavorable 1,3-diaxial interactions, destabilizing a cyclohexane conformation.**

The larger the substituent on the six-membered ring, the higher the percentage of the conformation containing the equatorial substituent at equilibrium. With a very large substituent like *tert*-butyl [$(CH_3)_3C-$], essentially none of the conformation containing an axial *tert*-butyl group is present at room temperature, so **the ring is essentially anchored in a single conformation having an equatorial *tert*-butyl group.** This is illustrated in Figure 4.16.

Figure 4.16 The two conformations of *tert*-butylcyclohexane

C highly destabilized

D 100%

- The large *tert*-butyl group anchors the cyclohexane ring in conformation **D.**

Problem 4.23 Draw a second chair conformation for each cyclohexane. Then decide which conformation is present in higher concentration at equilibrium.

a. 

b. 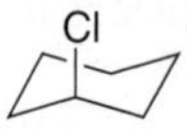

c. 

Problem 4.24 Draw both conformations for 1-ethyl-1-methylcyclohexane and decide which conformation (if any) is more stable.

4.13B A Disubstituted Cycloalkane

Rotation around the C–C bonds in the ring of a cycloalkane is restricted, so **a group on one side of the ring can *never* rotate to the other side of the ring.** As a result, there are two different 1,2-dimethylcyclopentanes—one having two CH_3 groups on the **same side** of the ring and one having them on **opposite sides** of the ring.

Wedges indicate bonds in front of the plane of the ring, and dashed wedges indicate bonds behind. For a review of this convention, see Section 1.7B. If a ring carbon is bonded to a CH_3 group in **front** of the ring (on a wedge), it is *assumed* that the other atom bonded to this carbon is hydrogen, located **behind** the ring (on a dashed wedge).

2 CH_3's above the ring
A
cis isomer

1 CH_3 above and 1 CH_3 below
B
trans isomer

A and **B** are **isomers,** because they are different compounds with the same molecular formula, but they represent the second major class of isomers called **stereoisomers.**

- ***Stereoisomers* are isomers that differ *only* in the way the atoms are oriented in space.**

The prefixes **cis** and **trans** are used to distinguish these stereoisomers.

- **The cis isomer has two groups on the *same side* of the ring.**
- **The trans isomer has two groups on *opposite sides* of the ring.**

Cis and **trans** isomers are named by adding the prefixes *cis* and *trans* to the name of the cycloalkane. Thus, **A** is *cis*-1,2-dimethylcyclopentane, and **B** is *trans*-1,2-dimethylcyclopentane.

Problem 4.25 Draw the structure for each compound using wedges and dashed wedges.
a. *cis*-1,2-dimethylcyclopropane b. *trans*-1-ethyl-2-methylcyclopentane

Problem 4.26 For *cis*-1,3-diethylcyclobutane, draw (a) a stereoisomer; (b) a constitutional isomer.

4.13C A Disubstituted Cyclohexane

A disubstituted cyclohexane like 1,4-dimethylcyclohexane also has cis and trans stereoisomers. In addition, each of these stereoisomers has two possible chair conformations.

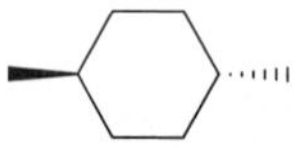

***trans*-1,4-dimethylcyclohexane**

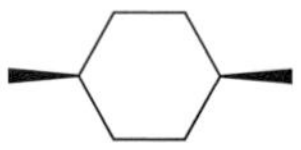

***cis*-1,4-dimethylcyclohexane**

All disubstituted cycloalkanes with two groups bonded to *different* atoms have cis and trans isomers.

To draw both conformations for each stereoisomer, follow the procedure in Section 4.13A for a monosubstituted cyclohexane, keeping in mind that two substituents must now be added to the ring.

How To Draw Two Conformations for a Disubstituted Cyclohexane

Step [1] Draw one chair form and add the substituents.

- For *trans*-1,4-dimethylcyclohexane, arbitrarily pick two C's located 1,4- to each other, classify them as *up* or *down* C's, and draw in the substituents.
- The trans isomer must have one group *above* the ring (on an *up* bond) and one group *below* the ring (on a *down* bond). The substituents can be either axial or equatorial, as long as one is up and one is down. The easiest trans isomer to visualize has two axial CH_3 groups. This arrangement is said to be **diaxial.**
- This forms one of the two possible chair conformations, labeled **A.**

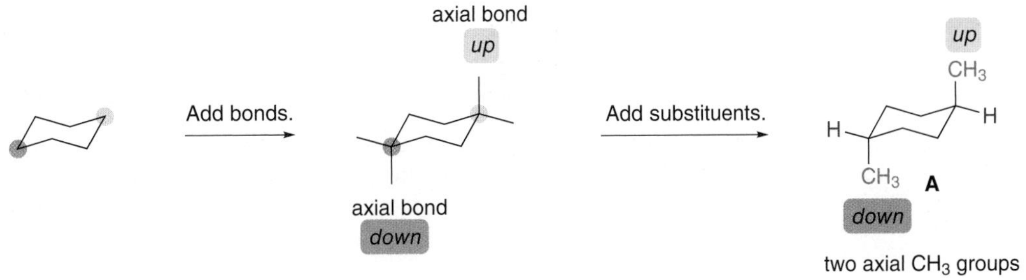

Step [2] Ring-flip the cyclohexane ring.

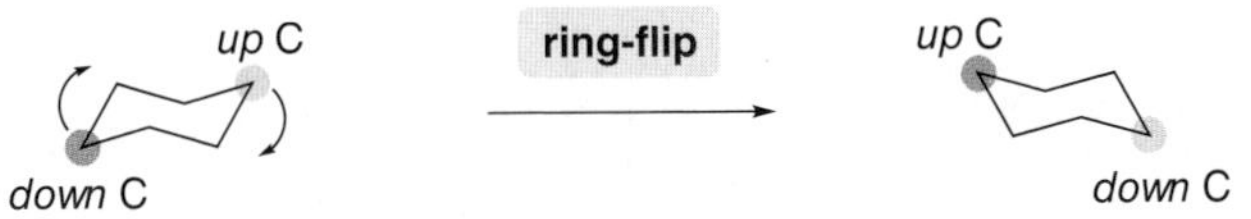

- The *up* C flips down, and the *down* C flips up.

Step [3] Add the substituents to the second conformation.

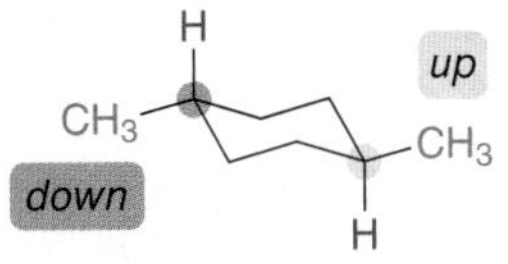

two equatorial CH_3 groups

B

- Ring-flipping converts axial bonds to equatorial bonds, and vice versa. The diaxial CH_3 groups become diequatorial. This trans conformation is less obvious to visualize. It is still trans, because one CH_3 group is above the ring (on an *up* bond), and one is below (on a *down* bond).

Conformations **A** and **B** are not equally stable. **Because B has both larger CH_3 groups in the roomier equatorial position, B is lower in energy.**

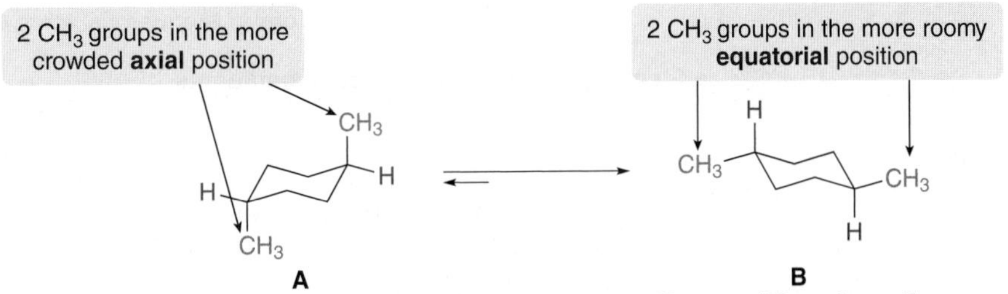

A
diaxial conformation

B
diequatorial conformation
more stable

The cis isomer of 1,4-dimethylcyclohexane also has two conformations, as shown in Figure 4.17. Because each conformation has one CH_3 group axial and one equatorial, they are **identical in energy.** At room temperature, therefore, the two conformations exist in a 50:50 mixture at equilibrium.

Figure 4.17
The two conformations of *cis*-1,4-dimethylcyclohexane

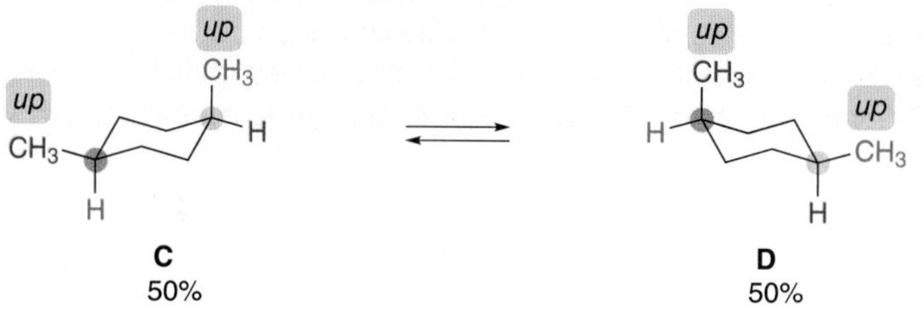

- A cis isomer has two groups on the same side of the ring, either both *up* or both *down*. In this example, Conformations **C** and **D** have two CH_3 groups drawn up.
- Both conformations have one CH_3 group axial and one equatorial, making them equally stable.

The relative stability of the two conformations of any disubstituted cyclohexane can be analyzed using this procedure.

- **A cis isomer has two substituents on the same side, either both on *up* bonds or both on *down* bonds.**
- **A trans isomer has two substituents on opposite sides, one *up* and one *down.***
- **Whether substituents are axial or equatorial depends on the relative location of the two substituents (on carbons 1,2-, 1,3-, or 1,4-).**

Sample Problem 4.3 Draw both chair conformations for *trans*-1,3-dimethylcyclohexane.

Solution

Step [1] Draw one chair form and add substituents.

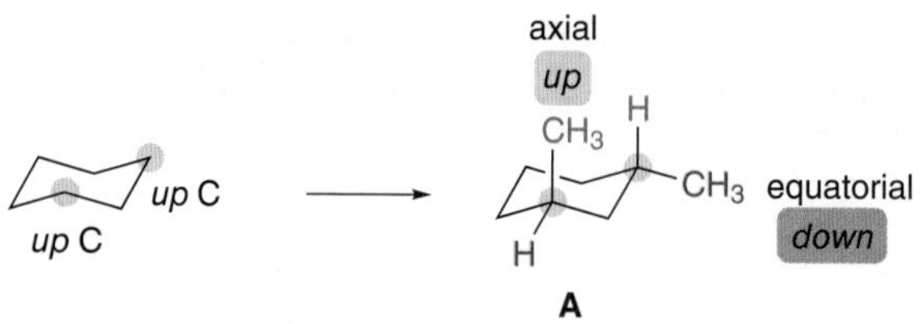

- Pick two C's 1,3- to each other.
- The trans isomer has two groups on **opposite sides.** In Conformation **A,** one CH_3 is axial (on an *up* bond), and one group is equatorial (on a *down* bond).

Steps [2–3] Ring-flip and add substituents.

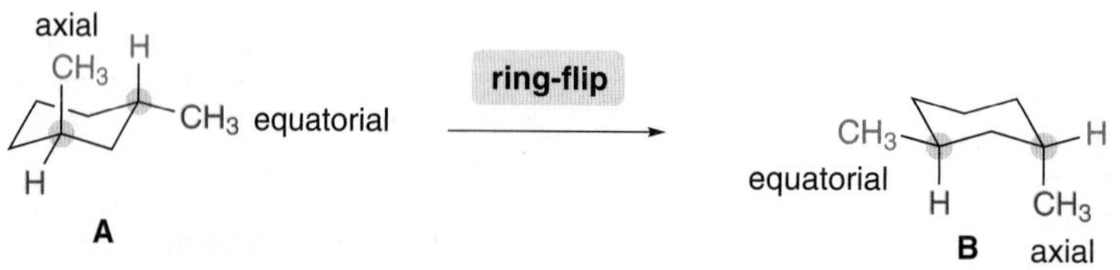

- The two *up* C's flip down.
- The axial CH_3 flips equatorial (still an *up* bond) and the equatorial CH_3 flips axial (still a *down* bond). Conformation **B** is trans because the two CH_3's are still on opposite sides.
- **Conformations A and B are equally stable** because each has one CH_3 equatorial and one axial.

Problem 4.27 Label each compound as cis or trans. Then draw the second chair conformation.

a. 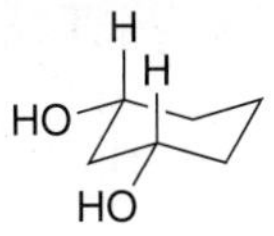 b. 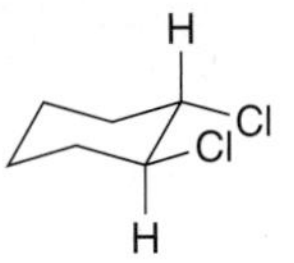 c. 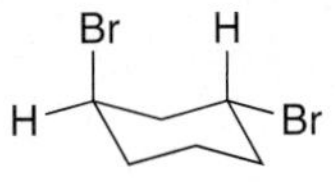

Problem 4.28 Consider 1,2-dimethylcyclohexane.

a. Draw structures for the cis and trans isomers using a hexagon for the six-membered ring.
b. Draw the two possible chair conformations for the cis isomer. Which conformation, if either, is more stable?
c. Draw the two possible chair conformations for the trans isomer. Which conformation, if either, is more stable?
d. Which isomer, cis or trans, is more stable and why?

Problem 4.29 Draw a chair conformation of cyclohexane with one CH_3CH_2 group and one CH_3 group that fits each description.

a. a 1,1-disubstituted cyclohexane with an axial CH_3CH_2 group
b. a cis-1,2-disubstituted cyclohexane with an axial CH_3 group
c. a trans-1,3-disubstituted cyclohexane with an equatorial CH_3 group
d. a trans-1,4-disubstituted cyclohexane with an equatorial CH_3CH_2 group

4.14 Oxidation of Alkanes

In Chapter 3, we learned that a functional group contains a heteroatom or π bond and constitutes **the reactive part of a molecule.** Alkanes are the only family of organic molecules that has no functional group, and therefore, **alkanes undergo few reactions.** In fact, alkanes are inert to reaction unless forcing conditions are used.

In Chapter 4, we consider only one reaction of alkanes—**combustion.** Combustion is an **oxidation–reduction** reaction.

4.14A Oxidation and Reduction Reactions

- *Oxidation* is the *loss* of electrons.
- *Reduction* is the *gain* of electrons.

Compounds that contain many C–H bonds and few C–Z bonds are said to be in a ***reduced state,*** whereas those that contain few C–H bonds and more C–Z bonds are in a ***more oxidized state.*** CH_4 is highly reduced, while CO_2 is highly oxidized.

Oxidation and reduction are opposite processes. As in acid–base reactions, there are always two components in these reactions. **One component is oxidized and one is reduced.**

To determine if an organic compound undergoes oxidation or reduction, we concentrate on the carbon atoms of the starting material and product, and **compare the relative number of C–H and C–Z bonds,** where Z = an element *more electronegative* than carbon (usually O, N, or X). Oxidation and reduction are then defined in two complementary ways.

- *Oxidation* results in an *increase* in the number of C–Z bonds; *or*
- *Oxidation* results in a *decrease* in the number of C–H bonds.

- *Reduction* results in a *decrease* in the number of C–Z bonds; *or*
- *Reduction* results in an *increase* in the number of C–H bonds.

Because Z is more electronegative than C, replacing C–H bonds with C–Z bonds decreases the electron density around C. Loss of electron density = oxidation.

Figure 4.18 illustrates the oxidation of CH_4 by replacing C–H bonds with C–O bonds (from left to right). The symbol **[O]** indicates oxidation. Because reduction is the reverse of oxidation, the molecules in Figure 4.18 are progressively reduced moving from right to left, from CO_2 to CH_4. The symbol **[H]** indicates reduction.

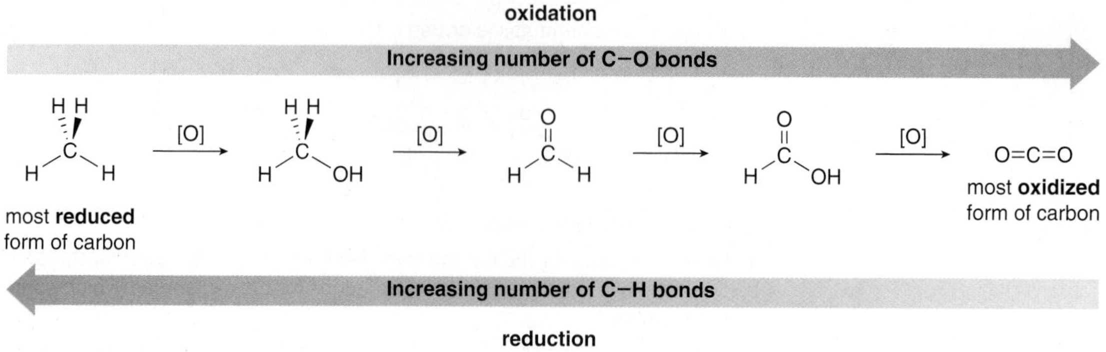

Figure 4.18 The oxidation and reduction of a carbon compound

Sample Problem 4.4 Determine whether the organic compound is oxidized or reduced in each transformation.

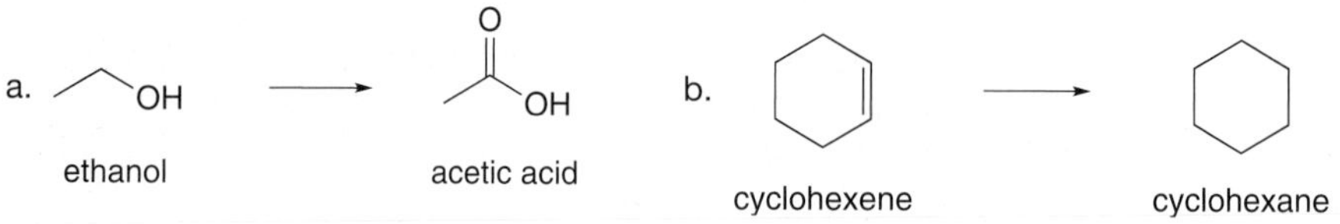

Solution

a. The conversion of ethanol to acetic acid is an **oxidation** because the number of C–O bonds increases: CH_3CH_2OH has one C–O bond and CH_3COOH has three C–O bonds.

b. The conversion of cyclohexene (C_6H_{10}) to cyclohexane (C_6H_{12}) is a **reduction** because the number of C–H bonds increases: cyclohexane has two more C–H bonds than cyclohexene.

Problem 4.30 Classify each transformation as an oxidation, reduction, or neither.

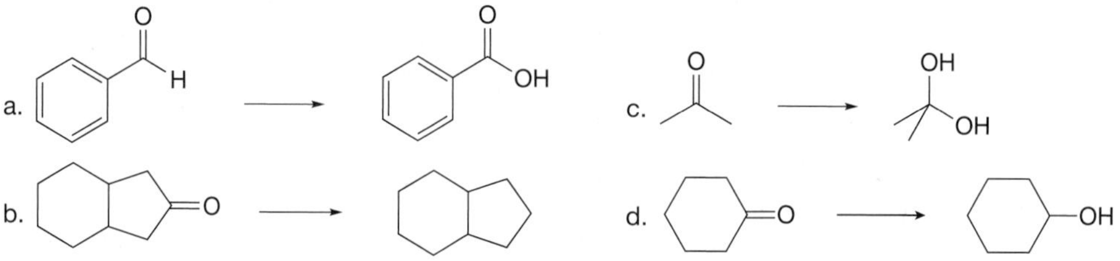

4.14B Combustion of Alkanes

When an organic compound is *oxidized* by a reagent, the reagent itself is *reduced.* Similarly, when an organic compound is *reduced* by a reagent, the reagent is *oxidized.* **Organic chemists identify a reaction as an oxidation or reduction by what happens to the *organic* component of the reaction.**

Alkanes undergo **combustion**—that is, **they burn in the presence of oxygen to form carbon dioxide and water.** This is a practical example of oxidation. Every C–H and C–C bond in the starting material is converted to a C–O bond in the product. The reactions drawn show the combustion of two different alkanes. The products, $\mathbf{CO_2 + H_2O}$, are the same, regardless of the identity of the starting material. Combustion of alkanes in the form of natural gas, gasoline, or heating oil releases energy for heating homes, powering vehicles, and cooking food.

$$\underset{\text{methane}}{CH_4} + 2\,O_2 \xrightarrow[\text{flame}]{} CO_2 + 2\,H_2O + \text{(heat) energy}$$

2 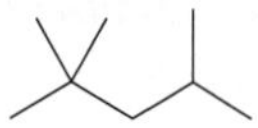 + 25 O_2 $\xrightarrow[\text{flame}]{}$ 16 CO_2 + 18 H_2O + (heat) energy

2,2,4-trimethylpentane (isooctane)

reduced starting material

oxidized product

Combustion requires a spark or a flame to initiate the reaction. Gasoline, therefore, which is composed largely of alkanes, can be safely handled and stored in the air, but the presence of a spark or match causes immediate and violent combustion.

Driving an automobile 10,000 miles at 25 miles per gallon releases ~10,000 lb of CO_2 into the atmosphere.

The combustion of alkanes and other hydrocarbons obtained from fossil fuels adds a tremendous amount of CO_2 to the atmosphere each year. Quantitatively, data show over a 20% increase in the atmospheric concentration of CO_2 in the last 56 years (from 315 parts per million in 1958 to 395 parts per million in 2014; Figure 4.19). Although the composition of the atmosphere has changed over the lifetime of the earth, this may be the first time that the actions of humankind have altered that composition significantly and so quickly.

Figure 4.19 The changing concentration of CO_2 in the atmosphere since 1958

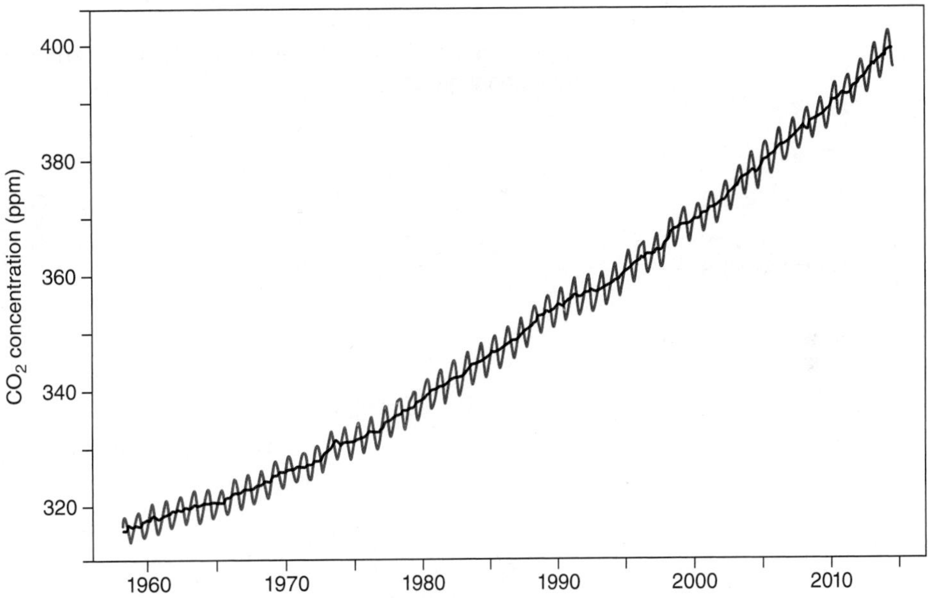

- The increasing level of atmospheric CO_2 is clearly evident on the graph. Two data points are recorded each year. The sawtooth nature of the graph is due to seasonal variation of CO_2 level with the seasonal variation in photosynthesis. (Data recorded at Mauna Loa, Hawaii)

An increased CO_2 concentration in the atmosphere may have long-range and far-reaching effects. CO_2 absorbs thermal energy that normally radiates from the earth's surface, and redirects it back to the surface. Higher levels of CO_2 may therefore contribute to an increase in the average temperature of the earth's atmosphere. The global climate change resulting from these effects may lead to melting of the polar ice caps, a rise in sea level, and many more unforeseen consequences. How great a role CO_2 plays in this process is hotly debated.

Problem 4.31 Draw the products of each combustion reaction.

a. 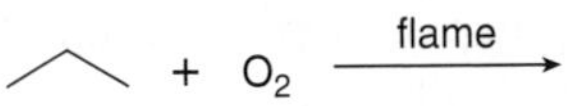

b. 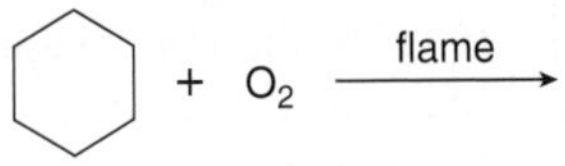

4.15 Lipids—Part 1

Lipids that contain carbon–carbon double bonds are discussed in Section 10.6.

Lipids are biomolecules whose properties resemble those of alkanes and other hydrocarbons. They are unlike any other class of biomolecules, because they are defined by a **physical property,** not by the presence of a particular functional group.

- **Lipids are biomolecules that are soluble in organic solvents and insoluble in water.**

Lipids have varied sizes and shapes, and a diverse number of functional groups. Fat-soluble vitamins like vitamin A and the phospholipids that comprise cell membranes are two examples of lipids that were presented in Sections 3.5 and 3.7. Other examples are shown in Figure 4.20. One unifying feature accounts for their solubility.

- **Lipids are composed of many nonpolar C–H and C–C bonds, and have few polar functional groups.**

Figure 4.20 Three representative lipid molecules

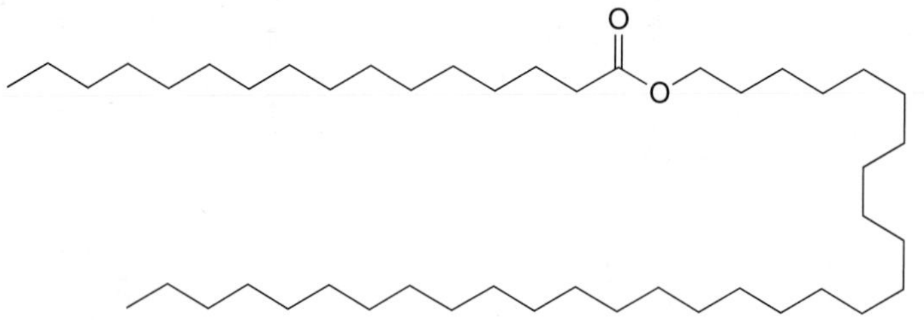

a component of beeswax

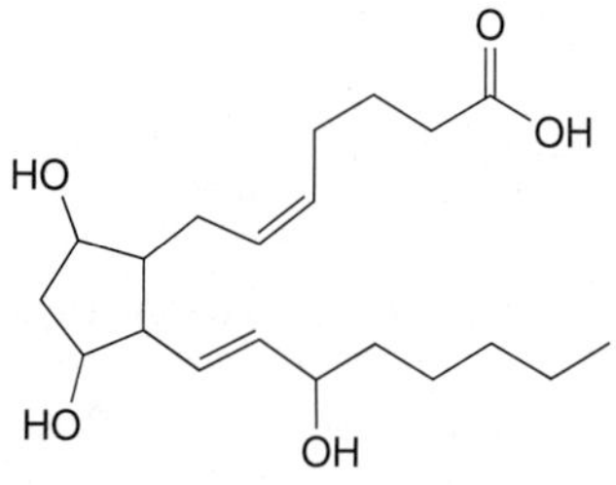

$PGF_{2\alpha}$

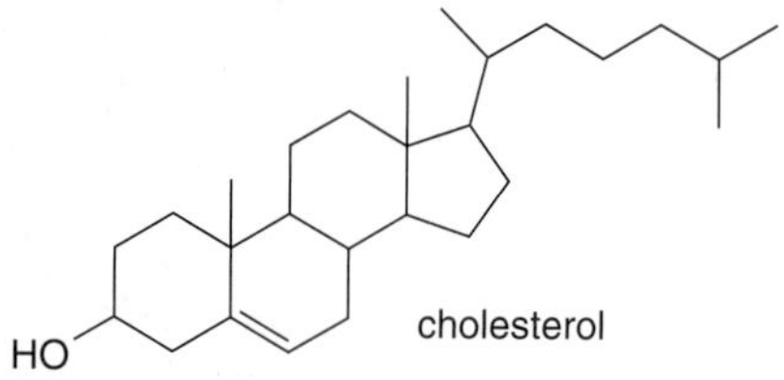

cholesterol

Waxes are lipids having two long alkyl chains joined by a single oxygen-containing functional group. Because of their many C–C and C–H bonds, **waxes are hydrophobic.** They form a protective coating on the feathers of birds to make them water repellent, and on leaves to prevent water evaporation. Bees secrete $\mathbf{CH_3(CH_2)_{14}COO(CH_2)_{29}CH_3}$, a wax that forms the honeycomb in which they lay eggs.

$\mathbf{PGF_{2\alpha}}$ belongs to a class of lipids called **prostaglandins.** Prostaglandins contain many C–C and C–H bonds and a single COOH group (a **carboxy group**). Prostaglandins possess a wide range of biological activities. They control inflammation, affect blood-platelet aggregation, and stimulate uterine contractions. Nonsteroidal anti-inflammatory drugs such as ibuprofen operate by blocking the synthesis of prostaglandins, as discussed in Section 19.6.

Cholesterol is a member of the steroid family, a group of lipids having four rings joined together. Because it has just one polar OH group, cholesterol is insoluble in the aqueous medium of the blood. It is synthesized in the liver and transported to other cells bound to water-soluble organic molecules. Elevated cholesterol levels can lead to coronary artery disease.

Figure 4.21
Cholesterol embedded in a lipid bilayer of a cell membrane

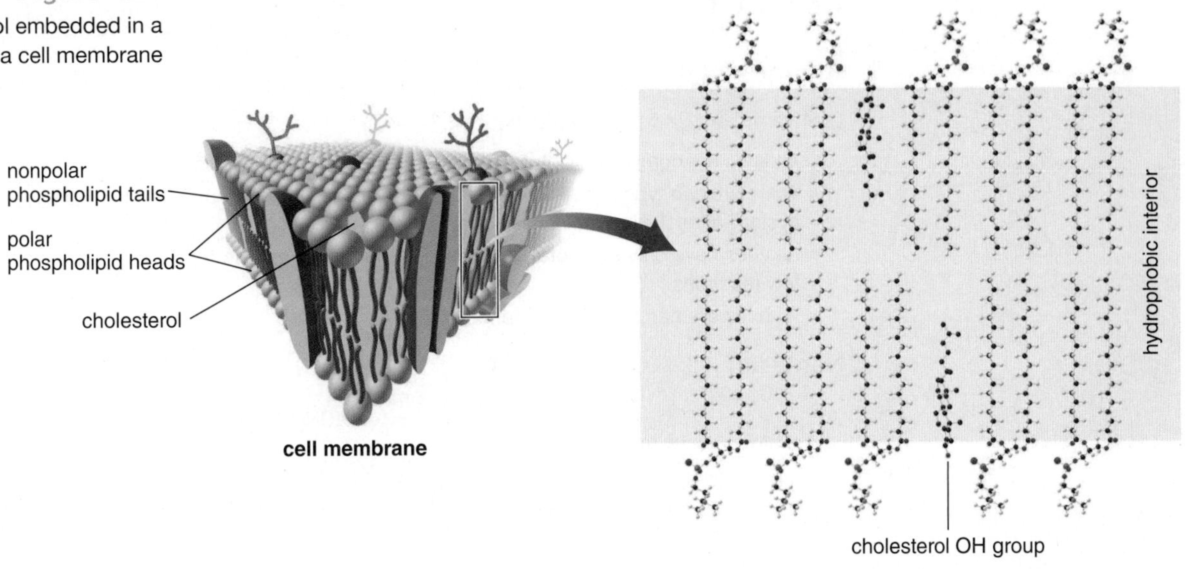

- The nonpolar hydrocarbon skeleton of cholesterol is embedded in the nonpolar interior of the cell membrane. Its rigid carbon skeleton stiffens the fluid lipid bilayer, giving it strength.
- Cholesterol's polar OH group is oriented toward the aqueous media inside and outside the cell.

Cholesterol is a vital component of the cell membrane. Its hydrophobic carbon chain is embedded in the interior of the lipid bilayer, and its hydrophilic hydroxy group is oriented toward the aqueous exterior (Figure 4.21). Because its tetracyclic carbon skeleton is quite rigid compared to the long floppy side chains of a phospholipid, cholesterol stiffens the cell membrane somewhat, giving it more strength.

Lipids have a high energy content, meaning that much energy is released on their metabolism. Because lipids are composed mainly of C–C and C–H bonds, they are oxidized with the release of energy, just like alkanes are. In fact, lipids are the most efficient biomolecules for the storage of energy. **The combustion of alkanes provides heat for our homes, and the metabolism of lipids provides energy for our bodies.**

Problem 4.32 Explain why beeswax is insoluble in H_2O, slightly soluble in ethanol (CH_3CH_2OH), and soluble in chloroform ($CHCl_3$).

Problem 4.33 Pristane is a high molecular weight alkane present in shark liver oil, a lipid used as a folk remedy and lubricant by coastal societies in northern Europe for centuries. What is the IUPAC name for pristane?

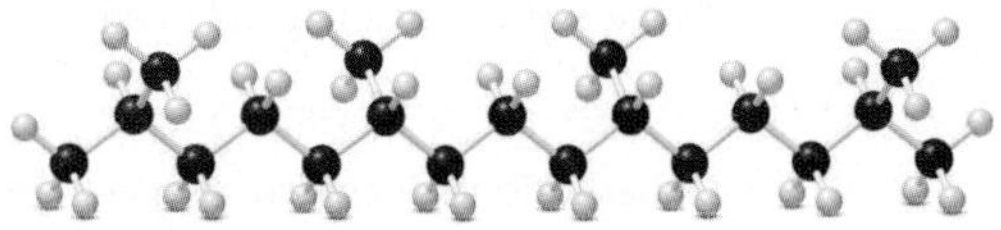

pristane

Shark liver oil, the source of pristane and other lipids, is obtained from sharks that inhabit the cold waters of the northern Atlantic and southern Pacific Oceans.

KEY CONCEPTS

Alkanes

General Facts About Alkanes (4.1–4.3)

- Alkanes are composed of **tetrahedral, sp^3 hybridized** C atoms.
- There are two types of alkanes: acyclic alkanes having molecular formula $\mathbf{C_nH_{2n+2}}$, and cycloalkanes having molecular formula $\mathbf{C_nH_{2n}}$.
- Alkanes have only **nonpolar C–C and C–H bonds** and no functional group, so they undergo few reactions.
- Alkanes are named with the suffix ***-ane.***

Names of Alkyl Groups (4.4A)

$CH_3–$ methyl = 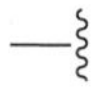

$CH_3CH_2–$ ethyl = 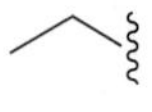

$CH_3CH_2CH_2–$ propyl = 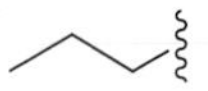

$(CH_3)_2CH–$ isopropyl = 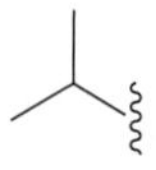

$CH_3CH_2CH_2CH_2–$ butyl = 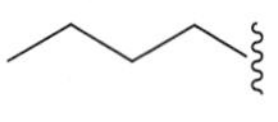

$CH_3CH_2CHCH_3$ (attached at CH) *sec*-butyl = 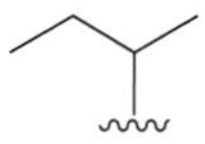

$(CH_3)_2CHCH_2–$ isobutyl = 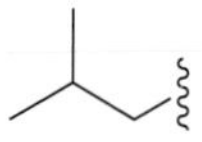

$(CH_3)_3C–$ *tert*-butyl = 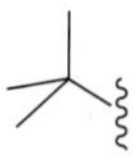

Conformations in Acyclic Alkanes (4.9, 4.10)

- Alkane conformations can be classified as **eclipsed, staggered, anti,** or **gauche** depending on the relative orientation of the groups on adjacent carbons.

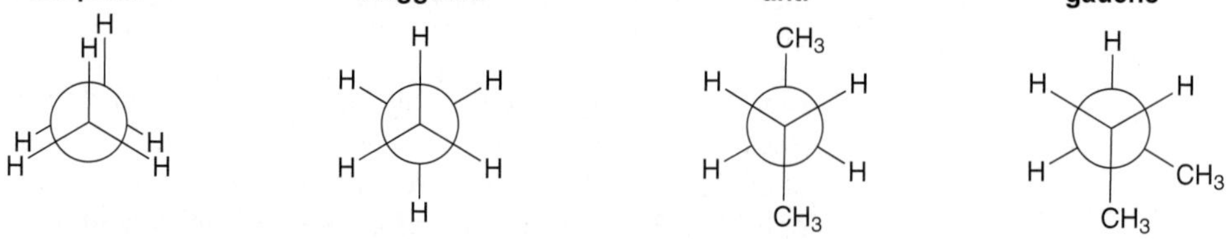

- dihedral angle = 0° (eclipsed)
- dihedral angle = 60° (staggered)
- dihedral angle of two CH_3 groups = 180° (anti)
- dihedral angle of two CH_3 groups = 60° (gauche)

- A staggered conformation is **lower in energy** than an eclipsed conformation.
- An anti conformation is **lower in energy** than a gauche conformation.

Types of Strain

- **Torsional strain**—an increase in energy caused by eclipsing interactions (4.9).
- **Steric strain**—an increase in energy when atoms are forced too close to each other (4.10).
- **Angle strain**—an increase in energy when tetrahedral bond angles deviate from 109.5° (4.11).

Two Types of Isomers

[1] **Constitutional isomers**—isomers that differ in the way the atoms are connected to each other (4.1A).

[2] **Stereoisomers**—isomers that differ *only* in the way the atoms are oriented in space (4.13B).

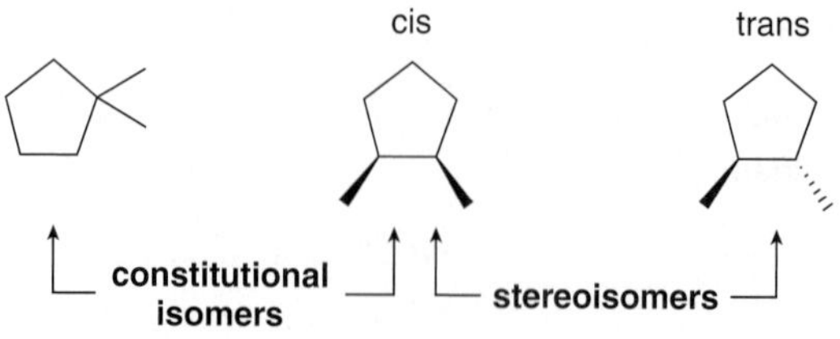

Conformations in Cyclohexane (4.12, 4.13)

- Cyclohexane exists as **two chair conformations** in rapid equilibrium at room temperature.
- Each carbon atom on a cyclohexane ring has **one axial** and **one equatorial hydrogen.** Ring-flipping converts axial H's to equatorial H's, and vice versa.

- In substituted cyclohexanes, groups larger than hydrogen are more stable in the **roomier equatorial position.**
- Disubstituted cyclohexanes with substituents on different atoms exist as two possible stereoisomers.
  - The **cis** isomer has two groups on the **same side** of the ring, either both up or both down.
  - The **trans** isomer has two groups on **opposite sides** of the ring, one up and one down.

Oxidation–Reduction Reactions (4.14)

- **Oxidation** results in an **increase in the number of C – Z bonds** or a **decrease in the number of C – H bonds.**
- **Reduction** results in a **decrease in the number of C – Z bonds** or an **increase in the number of C – H bonds.**

PROBLEMS

Problems Using Three-Dimensional Models

4.34 Name each alkane using the ball-and-stick model, and classify each carbon as 1°, 2°, 3°, or 4°.

a. 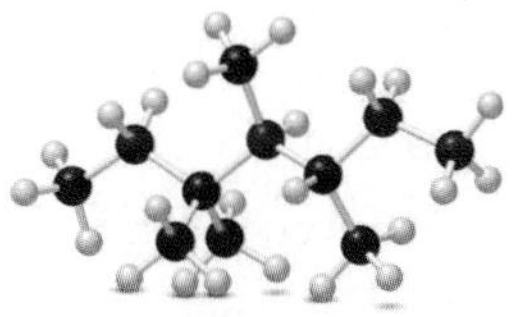 b. 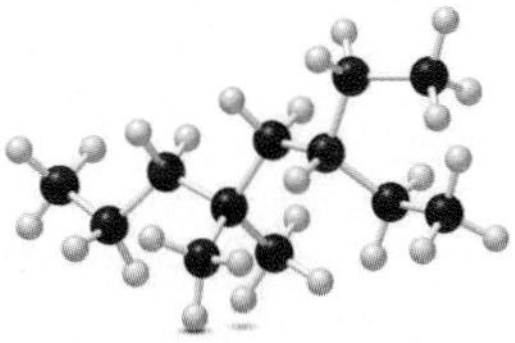

4.35 Consider the substituted cyclohexane shown in the ball-and-stick model.

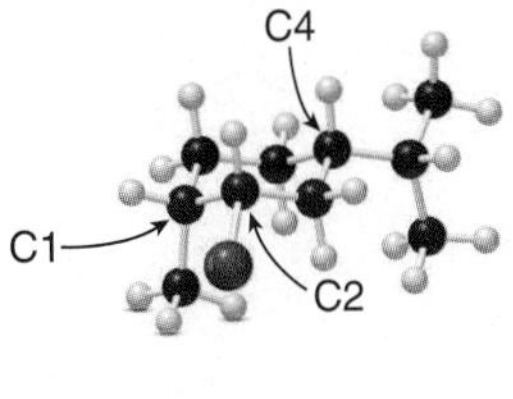

a. Label the substituents on C1, C2, and C4 as axial or equatorial.
b. Are the substituents on C1 and C2 cis or trans to each other?
c. Are the substituents on C2 and C4 cis or trans to each other?
d. Draw the second possible conformation in the chair form, and classify it as more stable or less stable than the conformation shown in the three-dimensional model.

4.36 Convert each three-dimensional model to a Newman projection around the indicated bond.

Constitutional Isomers

4.37 Draw the structure of all compounds that fit the following descriptions.
a. five constitutional isomers having the molecular formula C_4H_8
b. nine constitutional isomers having the molecular formula C_7H_{16}
c. twelve constitutional isomers having the molecular formula C_6H_{12} and containing one ring

IUPAC Nomenclature

4.38 Give the IUPAC name for each compound.

a. 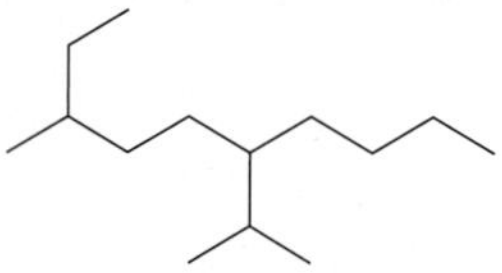

b. 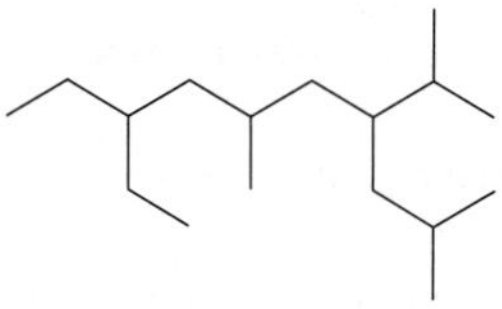

c. 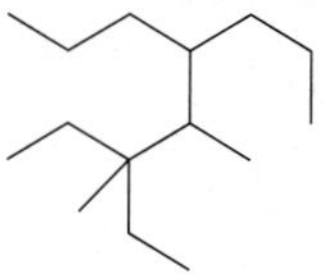

d. 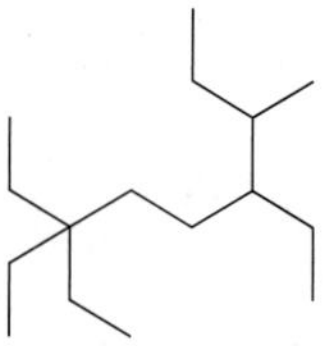

e. 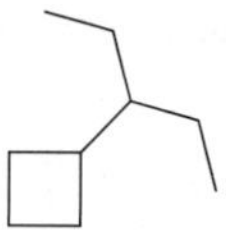

f. 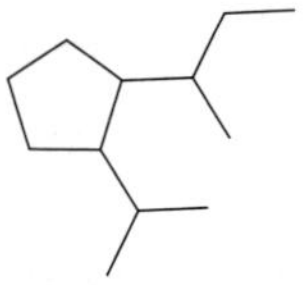

g. 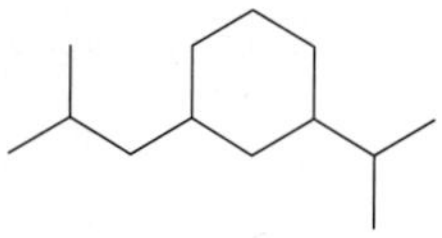

h. 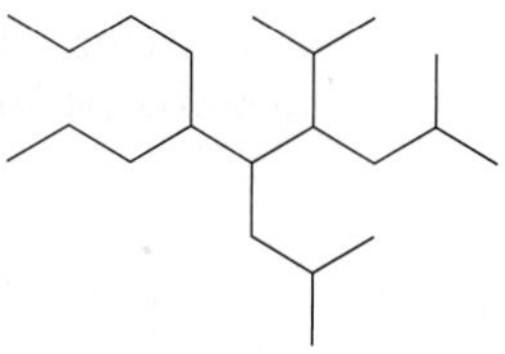

i. 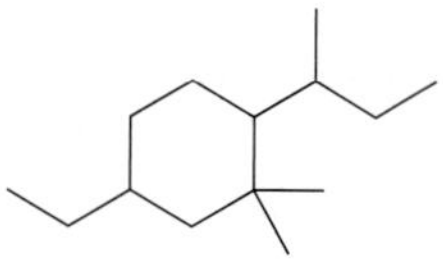

j. 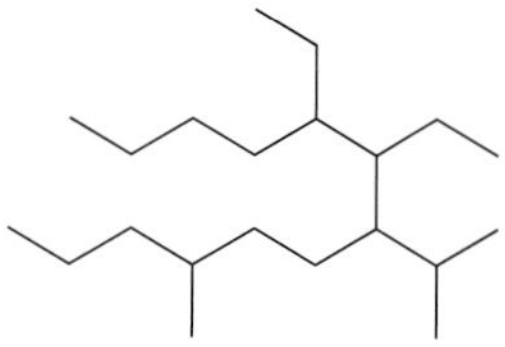

4.39 Give the structure and IUPAC name for each of the nine isomers having molecular formula C_9H_{20} that contains seven carbons in the longest chain and two methyl groups as substituents.

4.40 Draw the structure corresponding to each IUPAC name.

a. 3-ethyl-2-methylhexane
b. *sec*-butylcyclopentane
c. 4-isopropyl-2,4,5-trimethylundecane
d. cyclobutylcycloheptane
e. 3-ethyl-1,1-dimethylcyclohexane
f. 4-butyl-1,1-diethylcyclooctane
g. 6-isopropyl-2,3-dimethyldodecane
h. 2,2,6,6,7-pentamethyloctane
i. *cis*-1-ethyl-3-methylcyclopentane
j. *trans*-1-*tert*-butyl-4-ethylcyclohexane

4.41 Each of the following IUPAC names is incorrect. Explain why it is incorrect and give the correct IUPAC name.

a. 2,2-dimethyl-4-ethylheptane
b. 5-ethyl-2-methylhexane
c. 2-methyl-2-isopropylheptane
d. 1,5-dimethylcyclohexane
e. 1-ethyl-2,6-dimethylcycloheptane
f. 5,5,6-trimethyloctane
g. 3-butyl-2,2-dimethylhexane
h. 1,3-dimethylbutane

4.42 Give the IUPAC name for each compound.

a. 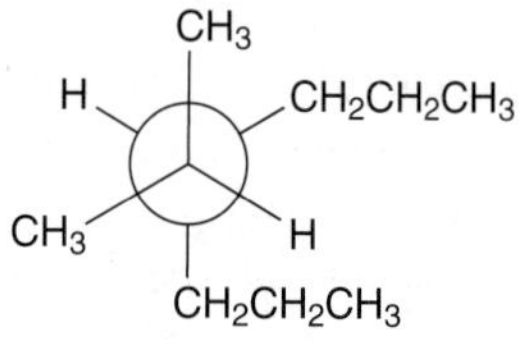

b. 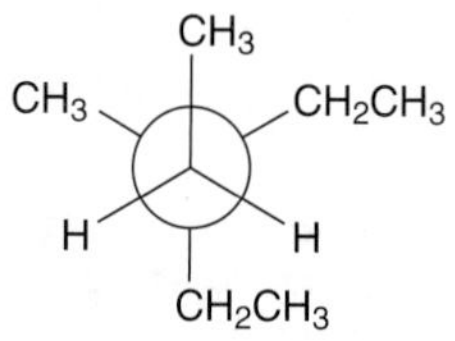

c. 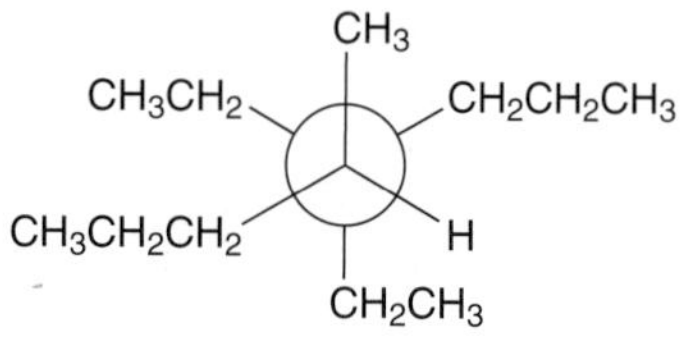

Physical Properties

4.43 Rank the following alkanes in order of increasing boiling point.

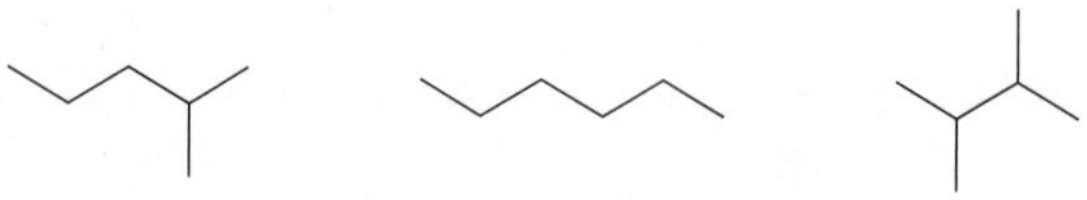

4.44 The melting points and boiling points of two isomeric alkanes are as follows: $CH_3(CH_2)_6CH_3$, mp = –57 °C and bp = 126 °C; $(CH_3)_3CC(CH_3)_3$, mp = 102 °C and bp = 106 °C. (a) Explain why one isomer has a lower melting point but higher boiling point. (b) Explain why there is a small difference in the boiling points of the two compounds, but a huge difference in their melting points.

Conformation of Acyclic Alkanes

4.45 Which conformation in each pair is *higher* in energy? Calculate the energy difference between the two conformations using the values given in Table 4.3.

a. 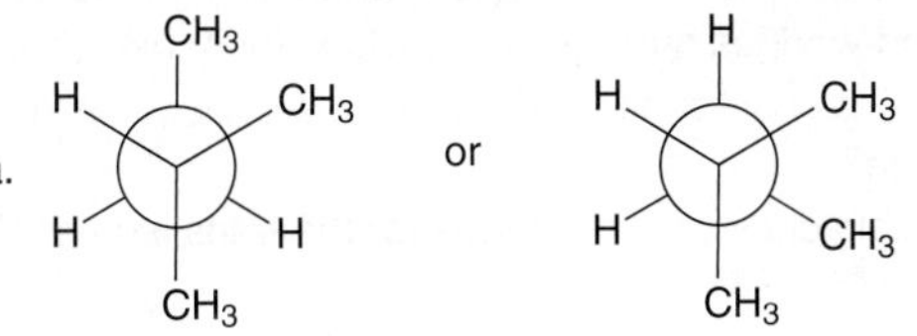

b. 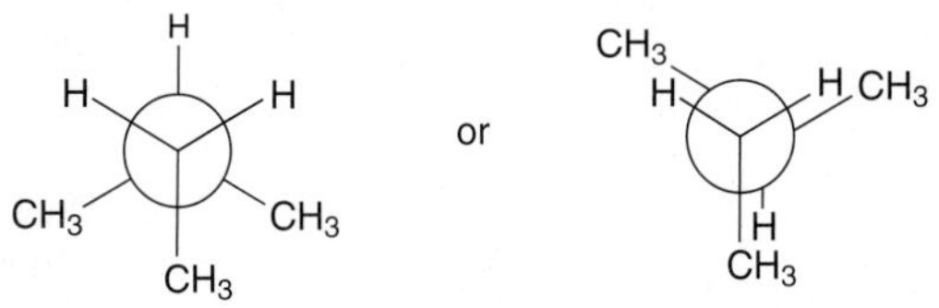

4.46 Considering rotation around the bond highlighted in red in each compound, draw Newman projections for the most stable and least stable conformations.

a.  b. 

4.47 Convert each structure to a Newman projection around the bond highlighted in red.

a. 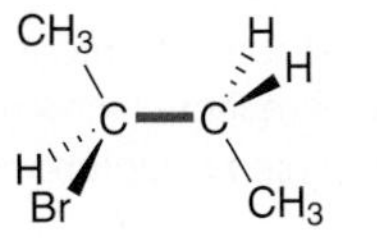 b. 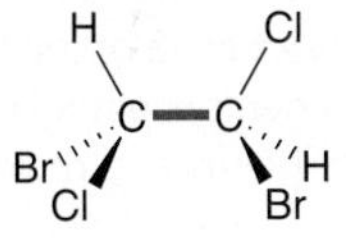 c. 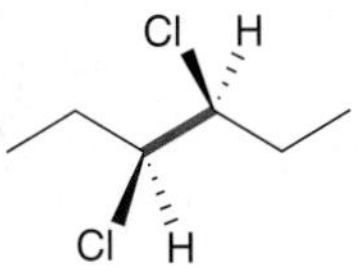

4.48 (a) Using Newman projections, draw all staggered and eclipsed conformations that result from rotation around the bond highlighted in red in each molecule; (b) draw a graph of energy versus dihedral angle for rotation around this bond.

[1]  [2] 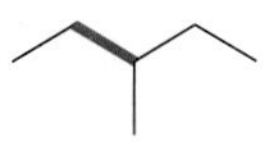

4.49 Label the sites of torsional and steric strain in each conformation.

a. 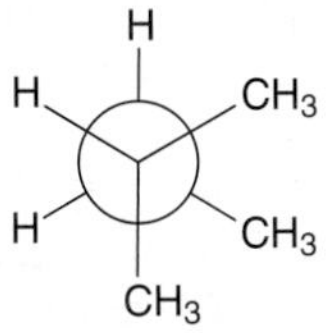 b. 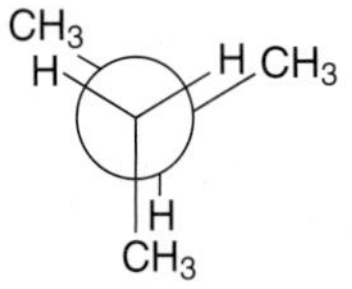 c. 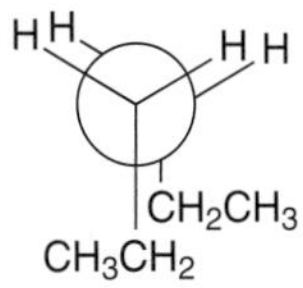

4.50 Calculate the barrier to rotation for each bond highlighted in red.

a. 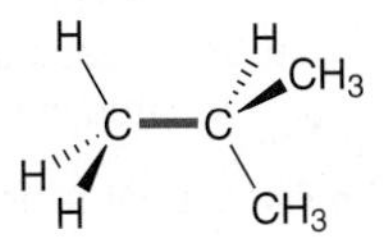 b. 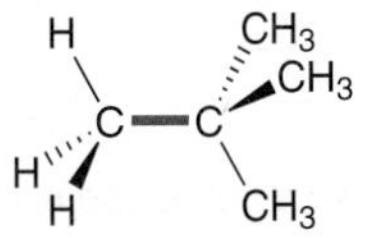

4.51 The eclipsed conformation of CH_3CH_2Cl is 15 kJ/mol less stable than the staggered conformation. How much is the H,Cl eclipsing interaction worth in destabilization?

4.52 (a) Draw the anti and gauche conformations for ethylene glycol ($HOCH_2CH_2OH$). (b) Ethylene glycol is unusual in that the gauche conformation is more stable than the anti conformation. Offer an explanation.

Conformations and Stereoisomers in Cycloalkanes

4.53 For each compound drawn below:

a. Label each OH, Br, and CH_3 group as axial or equatorial.
b. Classify each conformation as cis or trans.
c. Translate each structure into a representation with a hexagon for the six-membered ring, and wedges and dashed wedges for groups above and below the ring.
d. Draw the second possible chair conformation for each compound.

[1] 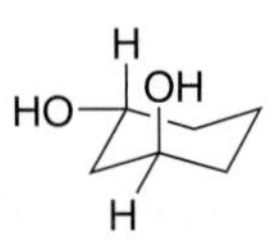 [2] 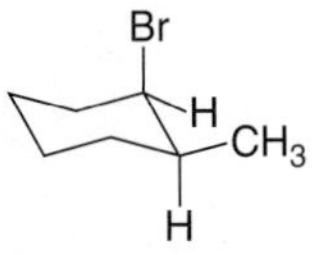 [3] 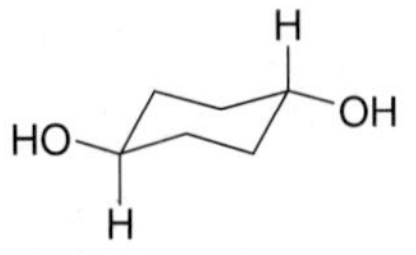

4.54 Draw the two possible chair conformations for *cis*-1,3-dimethylcyclohexane. Which conformation, if either, is more stable?

4.55 For each compound drawn below:

a. Draw representations for the cis and trans isomers using a hexagon for the six-membered ring, and wedges and dashed wedges for substituents.

b. Draw the two possible chair conformations for the cis isomer. Which conformation, if either, is more stable?

c. Draw the two possible chair conformations for the trans isomer. Which conformation, if either, is more stable?

d. Which isomer, cis or trans, is more stable and why?

[1] 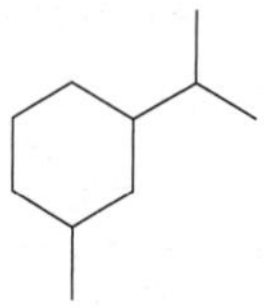

[2] 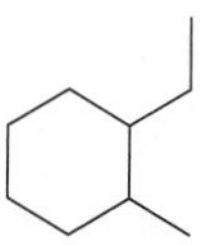

[3] 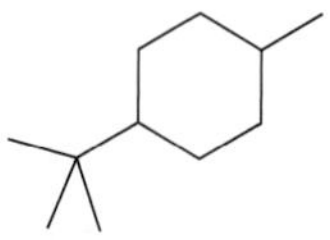

4.56 Convert each of the following structures into its more stable chair form. One structure represents menthol and one represents isomenthol. Menthol, the more stable isomer, is used in lip balms and mouthwash. Which structure corresponds to menthol?

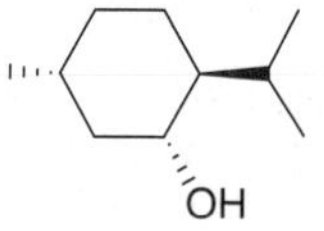

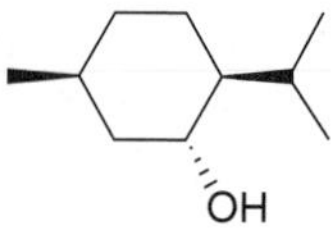

4.57 Answer the following questions about compound **A,** which contains a CH_3 group and OH group bonded to the carbon skeleton that consists of three six-membered rings in the conformation shown.

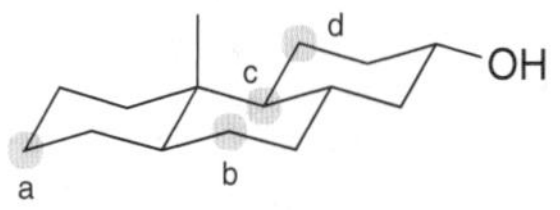

A

a. Are the CH_3 and OH groups oriented cis or trans to each other?

b. Is a substituent on C_a that is cis to the CH_3 group located in the axial or equatorial position?

c. Is an equatorial Br at C_b oriented cis or trans to the OH group?

d. Is the H atom on C_c located cis or trans to the OH group?

e. Is a substituent on C_d that is trans to the OH group located in the axial or equatorial position?

4.58 Glucose is a simple sugar with five substituents bonded to a six-membered ring.

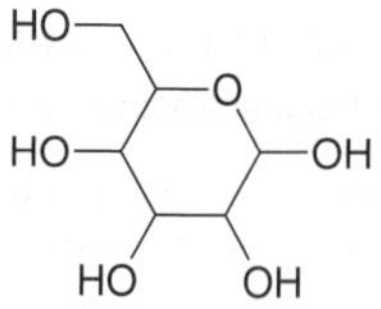

glucose

a. Using a chair representation, draw the most stable arrangement of these substituents on the six-membered ring.

b. Convert this representation into one that uses a hexagon with wedges and dashed wedges.

c. Draw a constitutional isomer of glucose.

d. Draw a stereoisomer that has an axial OH group on one carbon.

Constitutional Isomers and Stereoisomers

4.59 Classify each pair of compounds as constitutional isomers, stereoisomers, identical molecules, or not isomers of each other.

a. 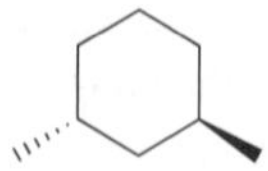 and 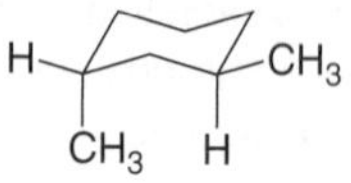

c. 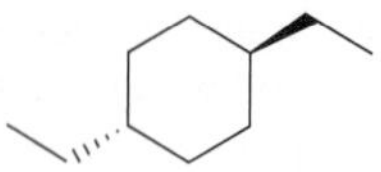 and 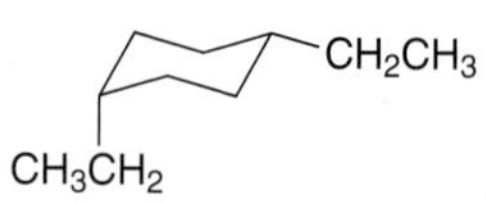

b. 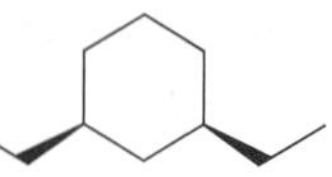 and 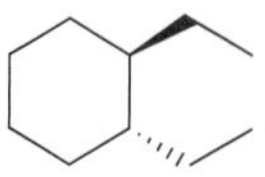

d. 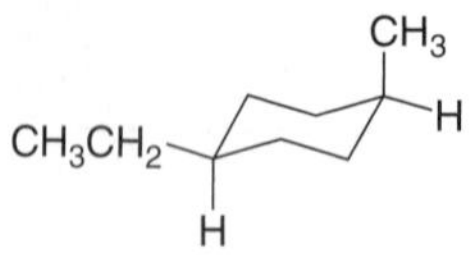

and 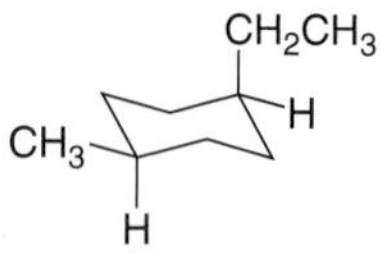

4.60 Classify each pair of compounds as constitutional isomers or identical molecules.

a. 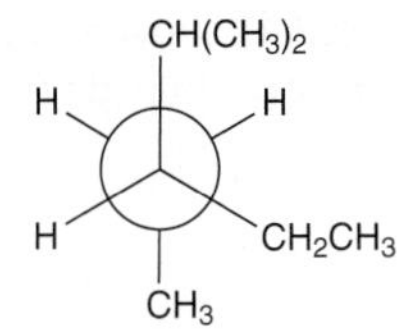 and 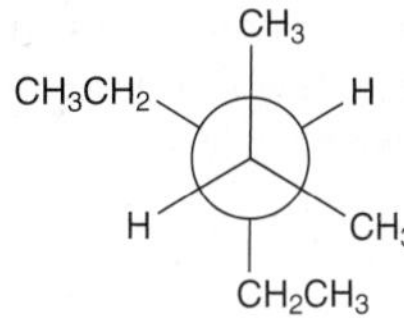

b. 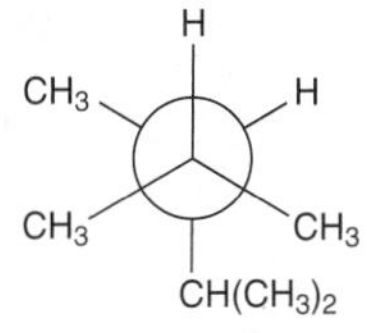 and 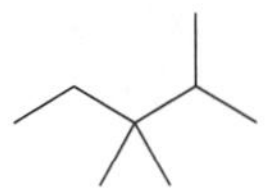

4.61 Answer the following questions about compounds **A–D.**

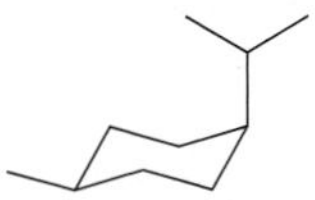

A

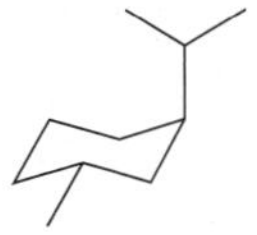

B

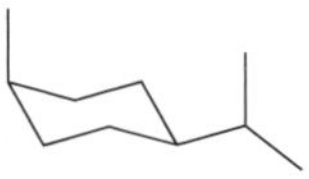

C

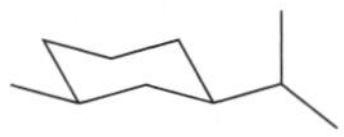

D

a. How are the compounds in each pair related? Choose from constitutional isomers, stereoisomers, or identical molecules: **A** and **B; A** and **C; B** and **D.**

b. Label each compound as a cis or trans isomer.

c. Draw **B** as a hexagon with wedges and dashed wedges to show the stereochemistry of substituents.

d. Draw a stereoisomer of **A** as a hexagon using wedges and dashed wedges to show the orientation of substituents.

4.62 Draw the three constitutional isomers having molecular formula C_7H_{14} that contain a five-membered ring and two methyl groups as substituents. For each constitutional isomer that can have cis and trans isomers, draw the two stereoisomers.

Oxidation and Reduction

4.63 Classify each reaction as oxidation, reduction, or neither.

a. 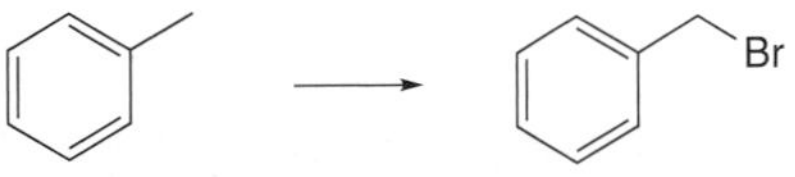

b. 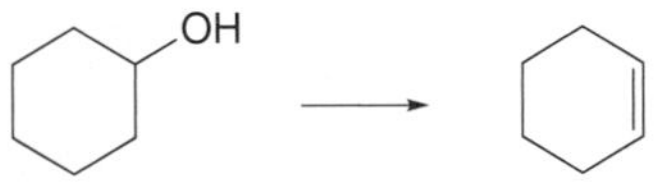

4.64 Draw the products of combustion of each alkane.

a. 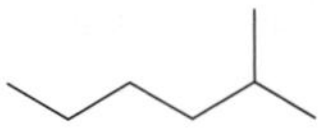 b. 

4.65 Hydrocarbons like benzene are metabolized in the body to arene oxides, which rearrange to form phenols. This is an example of a general process in the body, in which an unwanted compound (benzene) is converted to a more water-soluble derivative called a *metabolite,* so that it can be excreted more readily from the body.

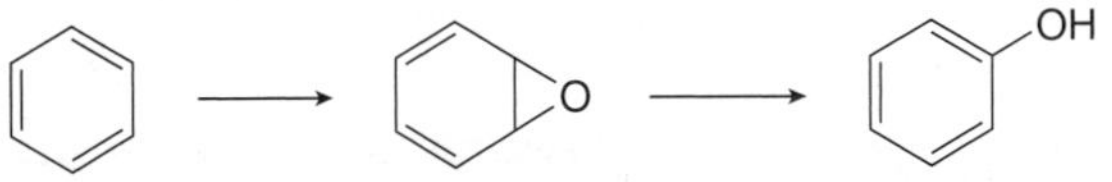

benzene arene oxide phenol

a. Classify each of these reactions as oxidation, reduction, or neither.

b. Explain why phenol is more water soluble than benzene. This means that phenol dissolves in urine, which is largely water, to a greater extent than benzene.

Lipids

4.66 Cholic acid, a compound called a **bile acid,** is converted to a **bile salt** in the body. Bile salts have properties similar to soaps, and they help transport lipids through aqueous solutions. Explain why this is so.

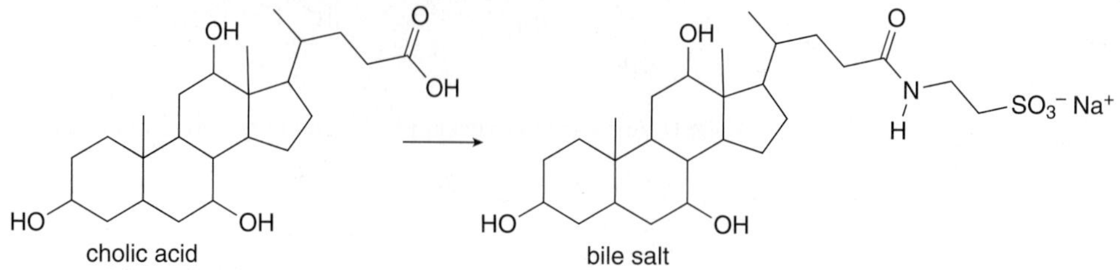

cholic acid
a bile acid

bile salt

4.67 Mineral oil, a mixture of high molecular weight alkanes, is sometimes used as a laxative. Why are individuals who use mineral oil for this purpose advised to avoid taking it at the same time they consume foods rich in fat-soluble vitamins such as vitamin A?

Challenge Problems

4.68 Cyclopropane and cyclobutane have similar strain energy despite the fact that the C–C–C bond angles of cyclopropane are much smaller than those of cyclobutane. Suggest an explanation for this observation, considering all sources of strain discussed in Chapter 4.

4.69 Although penicillin G has two amide functional groups, one is much more reactive than the other. Which amide is more reactive and why?

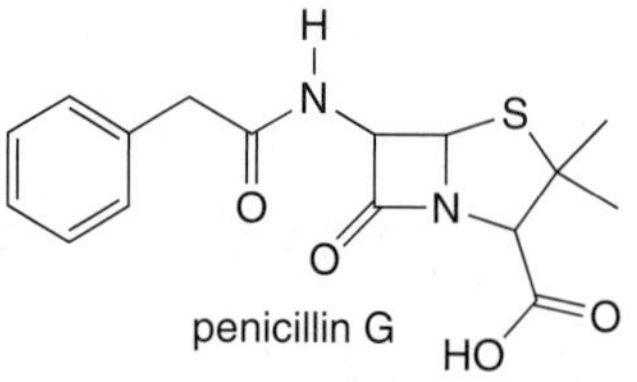

penicillin G

4.70 Haloethanes (CH_3CH_2X, X = Cl, Br, I) have similar barriers to rotation (13.4–15.5 kJ/mol) despite the fact that the size of the halogen increases, Cl → Br → I. Offer an explanation.

4.71 When two six-membered rings share a C–C bond, this bicyclic system is called a **decalin.** There are two possible arrangements: *trans*-decalin having two hydrogen atoms at the ring fusion on opposite sides of the rings, and *cis*-decalin having the two hydrogens at the ring fusion on the same side.

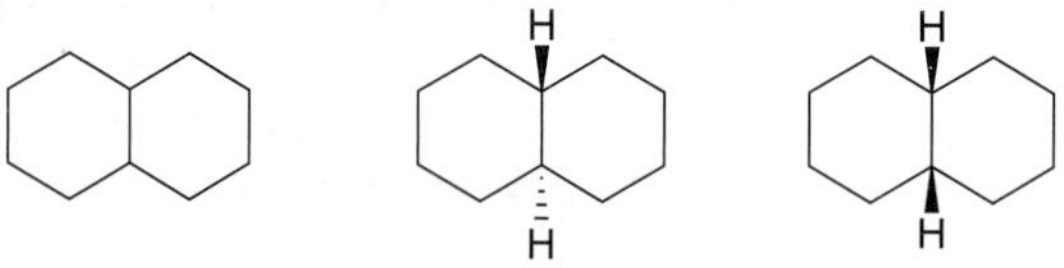

decalin *trans*-decalin *cis*-decalin

a. Draw *trans*- and *cis*-decalin using the chair form for the cyclohexane rings.
b. The trans isomer is more stable. Explain why.

4.72 Consider the tricyclic structure **A.** (a) Label each substituent on the rings as axial or equatorial. (b) Draw a skeletal structure for **A,** using wedges and dashed wedges to show whether the substituents are located above or below the rings.

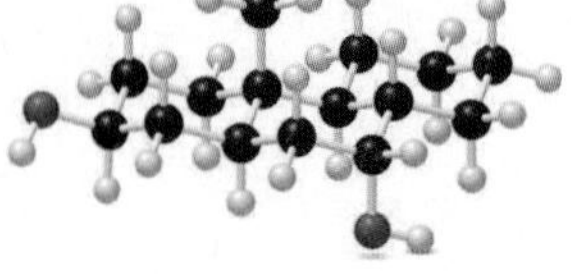

A

4.73 Consider the tricyclic structure **B.** (a) Label each substituent on the rings as axial or equatorial. (b) Draw **B** using chair conformations for each six-membered ring. (c) Label the atoms on the ring fusions (the carbons that join each set of two rings together) as cis or trans to each other.

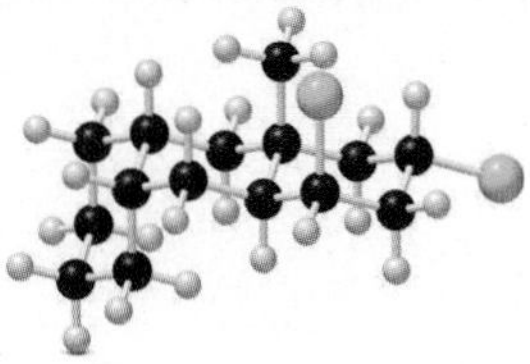

B

4.74 Read Appendix B on naming branched alkyl substituents, and draw all possible alkyl groups having the formula $C_5H_{11}-$. Give the IUPAC names for the eight compounds of molecular formula $C_{10}H_{20}$ that contain a cyclopentane ring with each of these alkyl groups as a substituent.

4.75 Read Appendix B on naming bicyclic compounds. Then give the IUPAC name for each of the following compounds.

a. 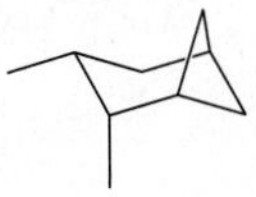

b. 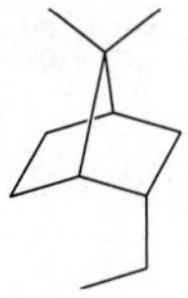

c. 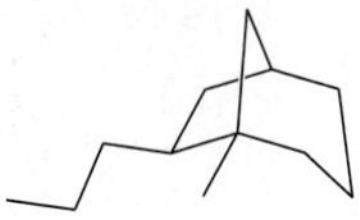

d. 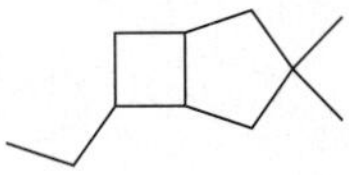

5 Stereochemistry

5.1 Starch and cellulose
5.2 The two major classes of isomers
5.3 Looking glass chemistry—Chiral and achiral molecules
5.4 Stereogenic centers
5.5 Stereogenic centers in cyclic compounds
5.6 Labeling stereogenic centers with *R* or *S*
5.7 Diastereomers
5.8 Meso compounds
5.9 *R* and *S* assignments in compounds with two or more stereogenic centers
5.10 Disubstituted cycloalkanes
5.11 Isomers—A summary
5.12 Physical properties of stereoisomers
5.13 Chemical properties of enantiomers

Paclitaxel (trade name Taxol), a potent anticancer agent active against ovarian, breast, and several other cancers, was discovered in 1962 and approved for use by the Food and Drug Administration in 1992. Initial studies with paclitaxel were carried out with material isolated from the bark of the Pacific yew tree, but stripping the bark killed these magnificent trees. Paclitaxel was synthesized in the laboratory in 1994, and is now produced by a plant cell fermentation process. Like other widely used drugs, paclitaxel is biologically active because of its complex structure and the particular three-dimensional arrangement of its functional groups. In Chapter 5 we learn about the stereochemistry of molecules like paclitaxel.

Are you left-handed or right-handed? If you're right-handed, you've probably spent little time thinking about your hand preference. If you're left-handed, though, you probably learned at an early age that many objects—like scissors and baseball gloves—"fit" for righties, but are "backwards" for lefties. **Hands, like many objects in the world around us, are mirror images that are *not* identical.**

In Chapter 5 we examine the "handedness" of molecules, and ask, "How important is the three-dimensional shape of a molecule?"

5.1 Starch and Cellulose

Recall from Chapter 4 that ***stereochemistry* is the three-dimensional structure of a molecule.** How important is stereochemistry? Two biomolecules—starch and cellulose—illustrate how apparently minute differences in structure can result in vastly different properties.

Starch and **cellulose** are two polymers that belong to the family of biomolecules called **carbohydrates** (Figure 5.1). **A *polymer* is a large molecule composed of repeating smaller units—called monomers—that are covalently bonded together.**

Figure 5.1
Starch and cellulose—Two common carbohydrates

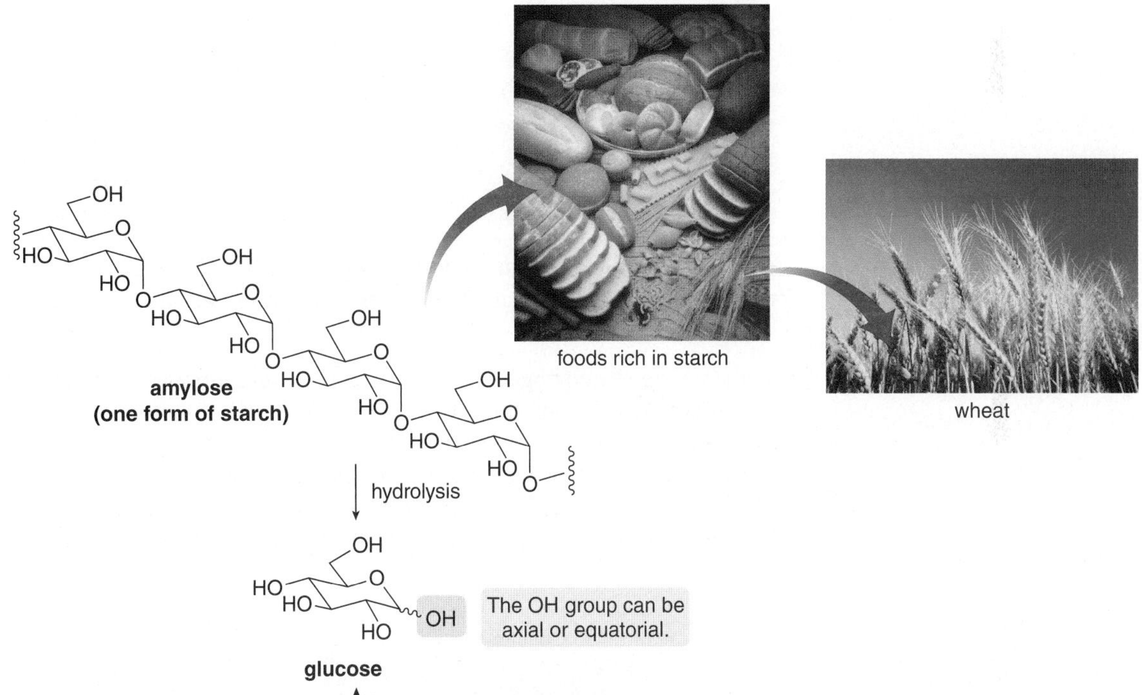

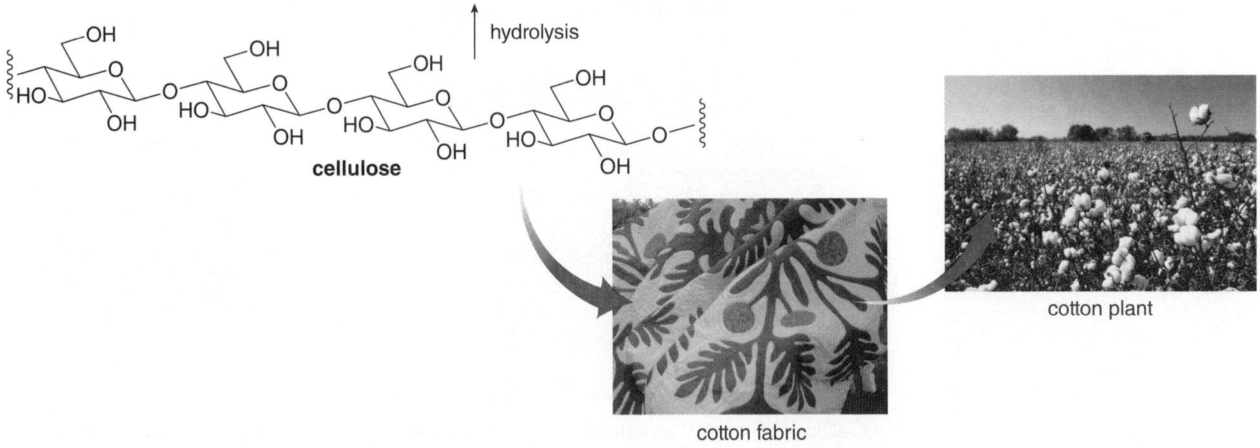

Starch is the main carbohydrate in the seeds and roots of plants. When we humans ingest wheat, rice, or potatoes, we consume starch, which is then hydrolyzed to the simple sugar **glucose,** one of the compounds our bodies use for energy. **Cellulose,** nature's most abundant organic material, gives rigidity to tree trunks and plant stems. Wood, cotton, and flax are composed largely of cellulose. Complete hydrolysis of cellulose also forms glucose, but unlike starch, humans cannot metabolize cellulose to glucose. In other words, we can digest starch but not cellulose.

Cellulose and starch are both composed of the same repeating unit—a six-membered ring containing an oxygen atom and three OH groups—joined by an oxygen atom. They differ in the position of the O atom joining the rings together.

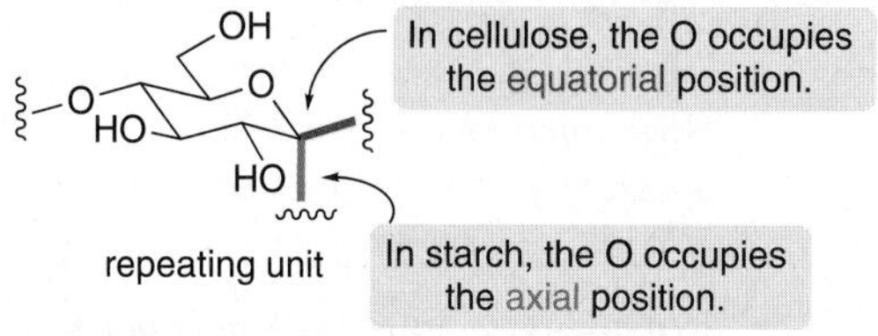

- **In cellulose, the O atom joins two rings using two equatorial bonds.**
- **In starch, the O atom joins two rings using one equatorial and one axial bond.**

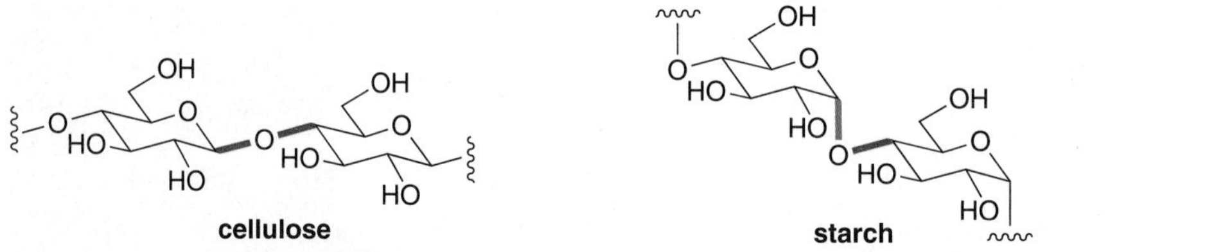

cellulose
two equatorial bonds (in red)

starch
one axial bond (in blue) and one equatorial bond (in red)

Starch and cellulose are **isomers** because they are different compounds with the same molecular formula $(C_6H_{10}O_5)_n$. They are **stereoisomers** because only the three-dimensional arrangement of atoms is different.

How the six-membered rings are joined together has an enormous effect on the shape and properties of these carbohydrate molecules. Cellulose is composed of long chains held together by intermolecular hydrogen bonds, forming sheets that stack in an extensive three-dimensional network. The axial–equatorial ring junction in starch creates chains that fold into a helix (Figure 5.2). Moreover, the human digestive system contains the enzyme necessary to hydrolyze starch by cleaving its axial C–O bond, but not an enzyme to hydrolyze the equatorial C–O bond in cellulose.

Figure 5.2
Three-dimensional structure of cellulose and starch

Cellulose **Starch**

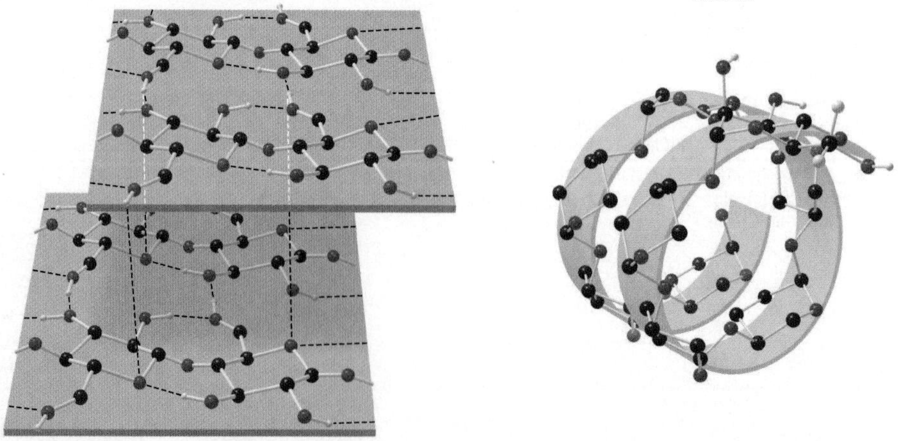

- Cellulose consists of an extensive three-dimensional network held together by hydrogen bonds.
- The starch polymer is composed of chains that wind into a helix.

Thus, an **apparently minor difference in the three-dimensional arrangement of atoms confers very different properties on starch and cellulose.**

Problem 5.1 Cellulose is water insoluble, despite its many OH groups. Considering its three-dimensional structure, why do you think this is so?

5.2 The Two Major Classes of Isomers

Because an understanding of isomers is integral to the discussion of stereochemistry, let's begin with an overview of isomers.

- **Isomers are different compounds with the same molecular formula.**

There are two major classes of isomers: **constitutional isomers** and **stereoisomers.** ***Constitutional (or structural) isomers*** **differ in the way the atoms are connected to each other.** Constitutional isomers have:

- different IUPAC names;
- the same or different functional groups;
- different physical properties, so they are separable by physical techniques such as distillation; and
- different chemical properties. They behave differently or give different products in chemical reactions.

Stereoisomers **differ *only* in the way atoms are oriented in space.** Stereoisomers have identical IUPAC names (except for a prefix like cis or trans). Because they differ only in the three-dimensional arrangement of atoms, stereoisomers always have the same functional group(s).

A particular three-dimensional arrangement is called a *configuration.* Thus, stereoisomers differ in configuration. The cis and trans isomers in Section 4.13B and the biomolecules starch and cellulose in Section 5.1 are two examples of stereoisomers.

Figure 5.3 illustrates examples of both types of isomers. Chapter 5 concentrates on the types and properties of stereoisomers.

Problem 5.2 Classify each pair of compounds as constitutional isomers or stereoisomers.

a. 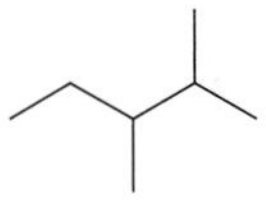 and 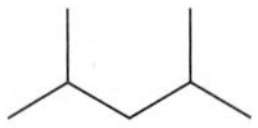

b. 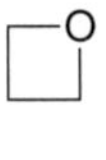 and 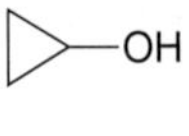

c.  and 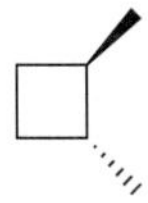

d.  and 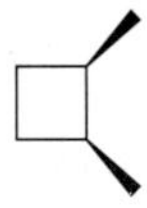

Figure 5.3 A comparison of constitutional isomers and stereoisomers

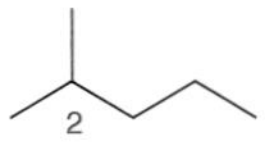

and
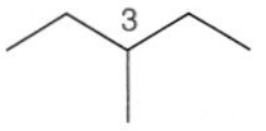

2-methylpentane C_6H_{14}

3-methylpentane C_6H_{14}

same molecular formula
different names
constitutional isomers

and

cis-1,2-dimethylcyclopentane

trans-1,2-dimethylcyclopentane

same molecular formula
same name except for the prefix
stereoisomers

5.3 Looking Glass Chemistry—Chiral and Achiral Molecules

Despite the dominance of right-handedness over left-handedness, even identical twins can exhibit differences in hand preference. Pictured are Matthew (right-handed) and Zachary (left-handed), identical twin sons of one of the authors.

Everything has a mirror image. What's significant is **whether a molecule is *identical* to or *different* from its mirror image.**

Some molecules are like hands. **Left and right hands are mirror images of each other, but they are *not* identical.** If you try to mentally place one hand inside the other hand, you can never superimpose either all the fingers, or the tops and palms. To *superimpose* an object on its mirror image means to align *all* parts of the object with its mirror image. With molecules, this means aligning all atoms and all bonds.

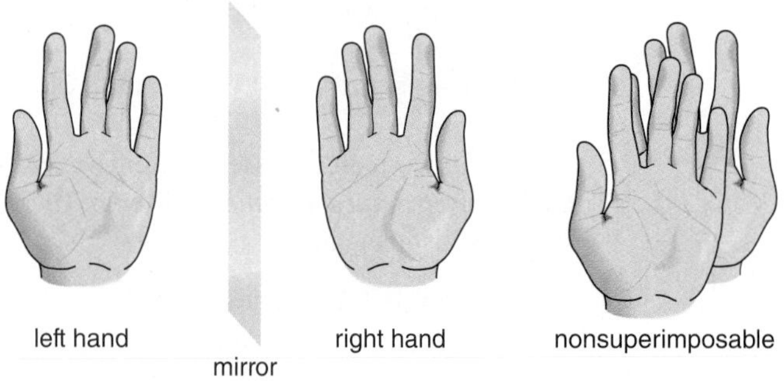

- A molecule (or object) that is *not* superimposable on its mirror image is said to be *chiral.*

Other molecules are like socks. **Two socks from a pair are mirror images that *are* superimposable.** One sock can fit inside another, aligning toes and heels, and tops and bottoms. A sock and its mirror image are *identical.*

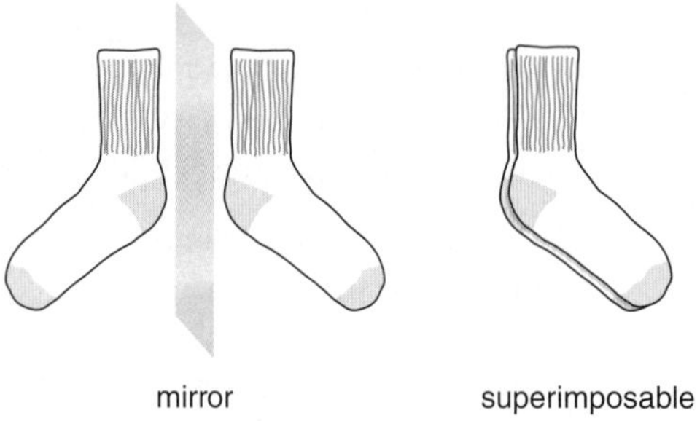

- A molecule (or object) that *is* superimposable on its mirror image is said to be *achiral.*

Let's determine whether three molecules—H_2O, CH_2BrCl, and CHBrClF—are superimposable on their mirror images; that is, **are H_2O, CH_2BrCl, and CHBrClF chiral or achiral?**

The adjective ***chiral*** comes from the Greek *cheir,* meaning "hand." Left and right hands are **chiral:** they are mirror images that do not superimpose on each other.

To test chirality:

- Draw the molecule in three dimensions.
- Draw its mirror image.
- Try to align all bonds and atoms. To superimpose a molecule and its mirror image, you can perform any rotation but **you cannot break bonds.**

Few beginning students of organic chemistry can readily visualize whether a compound and its mirror image are superimposable by looking at drawings on a two-dimensional page. Molecular models can help a great deal in this process.

Following this procedure, H_2O and CH_2BrCl are both **achiral** molecules because each molecule is superimposable on its mirror image.

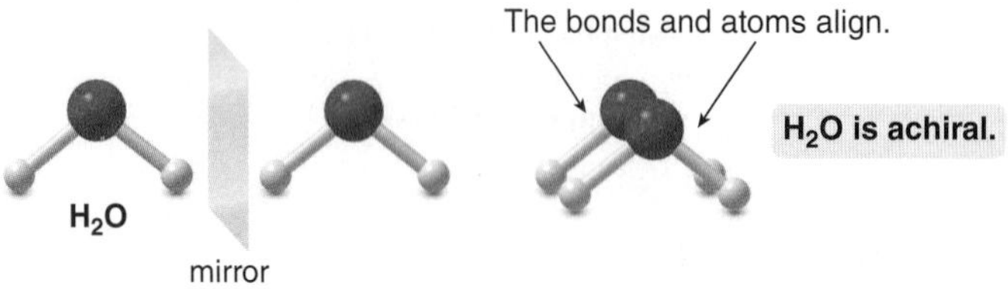

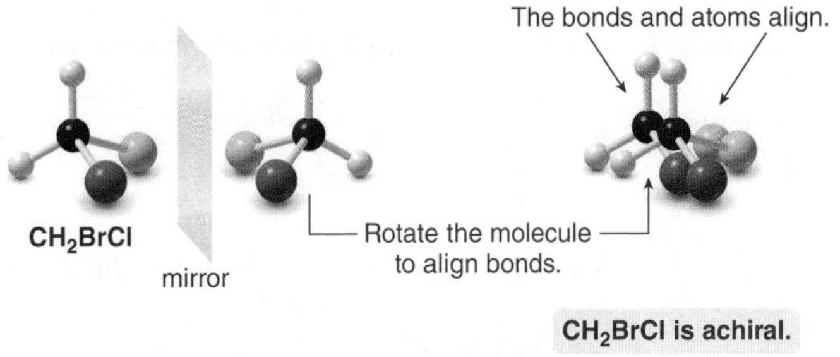
The bonds and atoms align.
CH2BrCl
mirror
Rotate the molecule
to align bonds.
CH2BrCl is achiral.

With CHBrClF, the result is different. The molecule (labeled **A**) and its mirror image (labeled **B**) are not superimposable. No matter how you rotate **A** and **B,** all the atoms never align. **CHBrClF is thus a chiral molecule,** and **A** and **B** are different compounds.

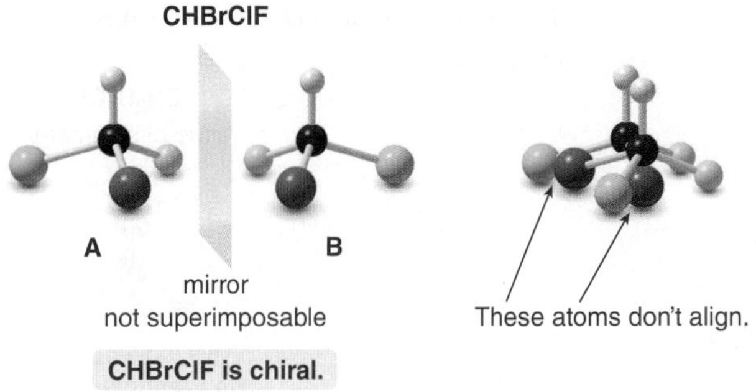
CHBrClF
A
B
mirror
not superimposable
These atoms don't align.
CHBrClF is chiral.

A and **B** are **stereoisomers** because they are isomers differing only in the three-dimensional arrangement of substituents. These stereoisomers are called **enantiomers.**

- *Enantiomers* are mirror images that are not superimposable.

CHBrClF contains a carbon atom bonded to four different groups. **A carbon atom bonded to four different groups is called a tetrahedral** ***stereogenic center.*** Most chiral molecules contain one or more stereogenic centers.

Naming a carbon atom with four different groups is a topic that currently has no firm agreement among organic chemists. The IUPAC recommends the term *chirality center,* but the term has not gained wide acceptance among organic chemists since it was first suggested in 1996. Other terms in common use are chiral center, chiral carbon, asymmetric carbon, stereocenter, and stereogenic center, the term used in this text.

The general term *stereogenic center* refers to any site in a molecule at which the interchange of two groups forms a stereoisomer. A **carbon atom with four different groups is a** ***tetrahedral*** **stereogenic center,** because the interchange of two groups converts one enantiomer into another. We will learn about another type of stereogenic center in Section 8.2B.

We have now learned two related but different concepts, and it is necessary to distinguish between them.

- A molecule that is not superimposable on its mirror image is a *chiral molecule.*
- A carbon atom bonded to four different groups is a *stereogenic center.*

Molecules can contain zero, one, or more stereogenic centers.

- **With no stereogenic centers, a molecule generally is not chiral.** H_2O and CH_2BrCl have *no* stereogenic centers and are *achiral* molecules. (There are a few exceptions to this generalization, as we will learn in Section 26.4.)
- **With one tetrahedral stereogenic center, a molecule is** ***always*** **chiral.** CHBrClF is a *chiral* molecule containing *one* stereogenic center.
- **With two or more stereogenic centers, a molecule** ***may*** **or** ***may not*** **be chiral,** as we will learn in Section 5.8.

Problem 5.3 Draw the mirror image of each compound. Label each molecule as chiral or achiral.

a. 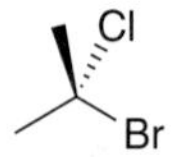 b. 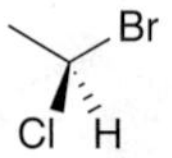 c.  d. 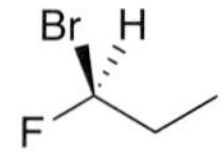

When trying to distinguish between chiral and achiral compounds, keep in mind:

- **A *plane of symmetry* is a mirror plane that cuts a molecule in half, so that one half of the molecule is a reflection of the other half.**

- **Achiral molecules usually contain a plane of symmetry but chiral molecules do not.**

The achiral molecule CH_2BrCl has a plane of symmetry, but the chiral molecule CHBrClF does not.

CH_2BrCl
plane of symmetry

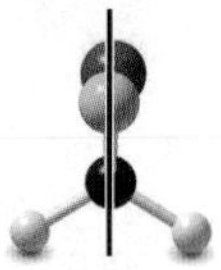

This molecule has
two identical halves.

CH_2BrCl is achiral.

CHBrClF
NO plane of symmetry

CHBrClF is chiral.

Figure 5.4 summarizes the main facts about chirality we have learned thus far.

Figure 5.4 The basic principles of chirality

- Everything has a mirror image. The fundamental question is whether a molecule and its mirror image are superimposable.
- If a molecule and its mirror image are *not* superimposable, the molecule and its mirror image are *chiral.*
- The terms *stereogenic center* and *chiral molecule* are related but distinct. In general, a chiral molecule must have one or more stereogenic centers.
- The presence of a *plane of symmetry* makes a molecule achiral.

Problem 5.4 Draw in a plane of symmetry for each molecule.

a.  b. 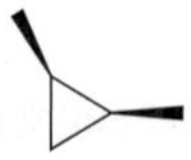 c. 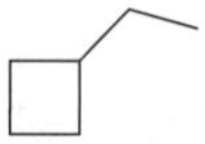 d. 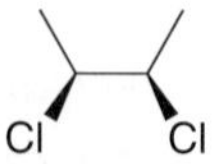

Problem 5.5 A molecule is achiral if it has a plane of symmetry in *any* conformation. Each of the following conformations does not have a plane of symmetry, but rotation around a carbon–carbon bond forms a conformation that does have a plane of symmetry. Draw this conformation for each molecule.

a. 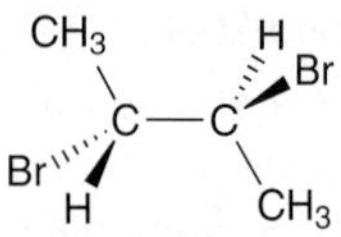 b. 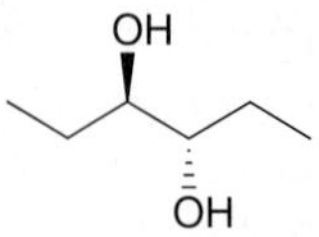

When a right-handed shell is held in the right hand with the thumb pointing toward the wider end, the opening is on the right side.

Stereochemistry may seem esoteric, but chirality pervades our very existence. On a molecular level, many biomolecules fundamental to life are chiral. On a macroscopic level, many naturally occurring objects possess handedness. Examples include chiral helical seashells shaped like right-handed screws, and plants such as honeysuckle that wind in a chiral left-handed helix. The human body is chiral, and hands, feet, and ears are not superimposable.

5.4 Stereogenic Centers

A necessary skill in the study of stereochemistry is the ability to locate and draw tetrahedral stereogenic centers.

5.4A Stereogenic Centers on Carbon Atoms That Are Not Part of a Ring

Recall from Section 5.3 that any carbon atom bonded to four different groups is a tetrahedral stereogenic center. To locate a stereogenic center, examine each *tetrahedral* carbon atom in a molecule, and look at the four ***groups***—not the four *atoms*—bonded to it. CBrClFI has one stereogenic center because its central carbon atom is bonded to four different elements. 3-Bromohexane also has one stereogenic center because one carbon is bonded to H, Br, CH_2CH_3, and $CH_2CH_2CH_3$. We consider all atoms in a group as a *whole unit,* not just the atom directly bonded to the carbon in question. Although C3 of 3-bromohexane is bonded to two carbon atoms, one is part of an ethyl group and one is part of a propyl group.

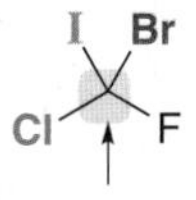

stereogenic center

stereogenic center

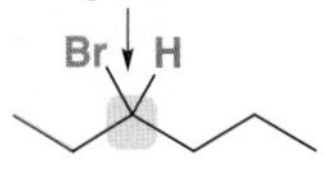

3-bromohexane

The stereogenic center is bonded to: H
Br
CH_2CH_3
$CH_2CH_2CH_3$

Ephedrine is isolated from ma huang, an herb used to treat respiratory ailments in traditional Chinese medicine. Once a popular drug to promote weight loss and enhance athletic performance, ephedrine has now been linked to episodes of sudden death, heart attack, and stroke.

Always omit from consideration all C atoms that can't be tetrahedral stereogenic centers. These include:

- CH_2 and CH_3 groups (more than one H bonded to C)
- any *sp* or *sp*^{2} hybridized C (less than four groups around C)

Larger organic molecules can have two, three, or even hundreds of stereogenic centers. **Propoxyphene** and **ephedrine** each contain two stereogenic centers, and **fructose,** a simple carbohydrate, has three.

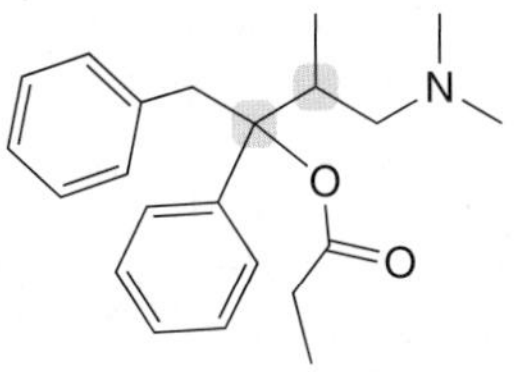

propoxyphene
Trade name Darvon
(analgesic)

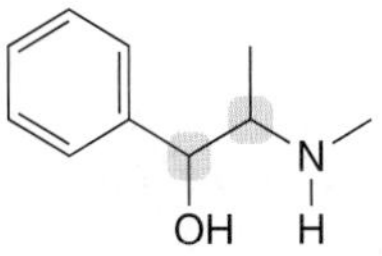

ephedrine
(bronchodilator, decongestant)

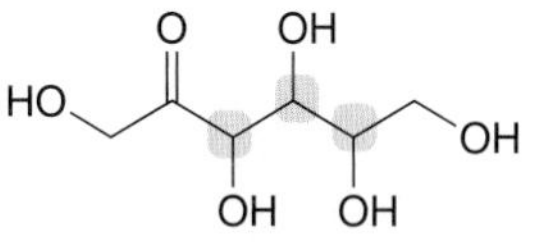

fructose
(a simple sugar)

 = stereogenic center

Sample Problem 5.1 Locate the stereogenic centers in each drug. Albuterol is a bronchodilator—that is, it widens airways—so it is used to treat asthma. Chloramphenicol is an antibiotic used extensively in developing countries because of its low cost.

a. 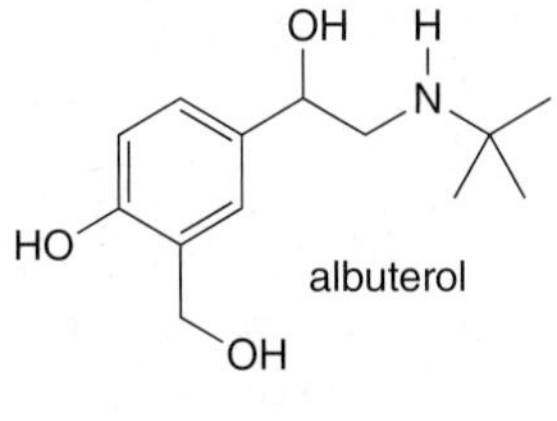

albuterol

b. 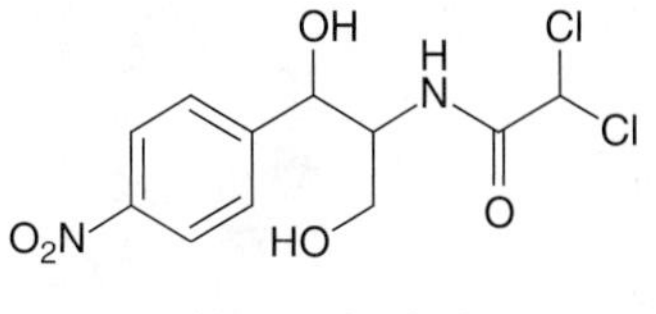

chloramphenicol

Heteroatoms surrounded by four different groups are also stereogenic centers. Stereogenic N atoms are discussed in Chapter 25.

Solution

Omit all CH_2 and CH_3 groups and all doubly bonded (*sp*^{2} hybridized) C's. In albuterol, one C has three CH_3 groups bonded to it, so it can be eliminated as well. Draw in H atoms on tetrahedral C's in skeletal structures to more clearly see the groups. This leaves one C in albuterol and two C's in chloramphenicol surrounded by four different groups, making them stereogenic centers.

a. 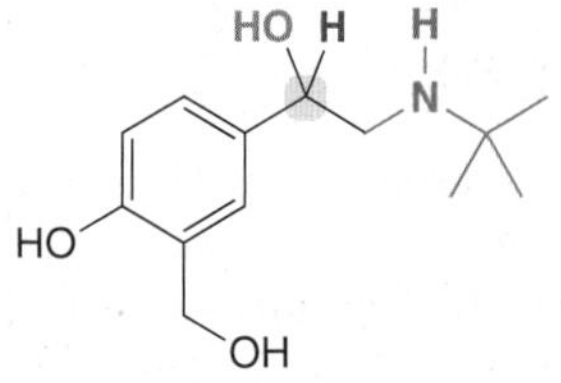

one stereogenic center

b. 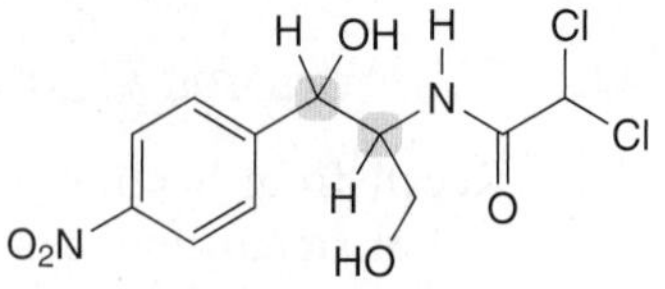

two stereogenic centers

Problem 5.6 Locate the stereogenic centers in each molecule. Compounds may have one or more stereogenic centers.

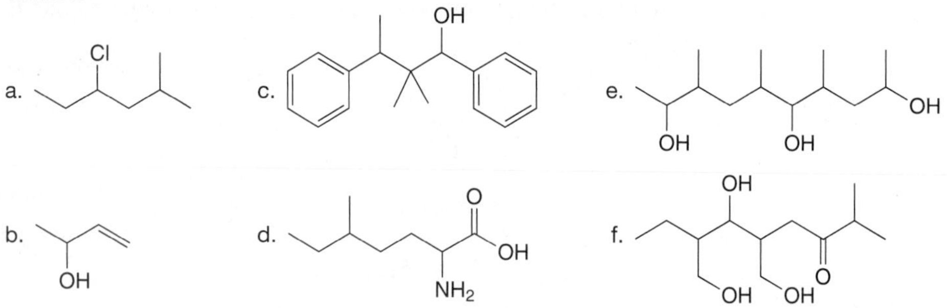

a.
b.
c.
d.
e.
f.

Problem 5.7 The facts in Section 5.4A can be used to locate stereogenic centers in any molecule, no matter how complicated. Always look for carbons surrounded by four different groups. With this in mind, locate the four stereogenic centers in aliskiren, a drug introduced in 2007 for the treatment of hypertension.

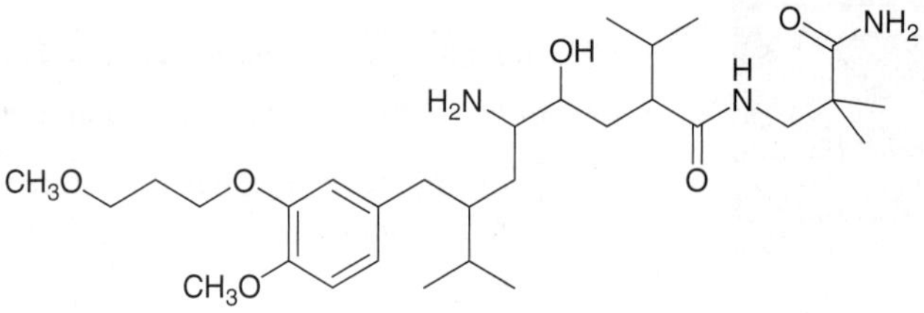

aliskiren

5.4B Drawing a Pair of Enantiomers

- **Any molecule with one tetrahedral stereogenic center is a chiral compound and exists as a pair of enantiomers.**

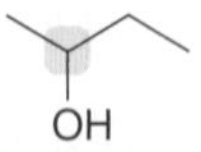

butan-2-ol
one stereogenic center

Butan-2-ol, for example, has one stereogenic center. To draw both enantiomers, use the typical convention for depicting a tetrahedron: **place two bonds in the plane, one in front of the plane on a wedge, and one behind the plane on a dashed wedge.** Then, to form the first enantiomer **A,** arbitrarily place the four groups—H, OH, CH_3, and CH_2CH_3—on any bond to the stereogenic center.

In Section 27.2, we will learn about Fischer projection formulas, an older convention used for drawing stereogenic centers utilized mainly in carbohydrate chemistry.

Draw the molecule...then the mirror image.

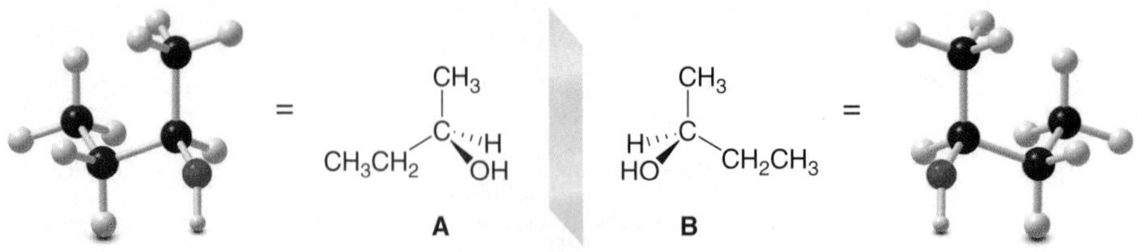

A
B
mirror
not superimposable
enantiomers

Then, draw a mirror plane and arrange the substituents in the mirror image so that they are a reflection of the groups in the first molecule, forming **B.** No matter how **A** and **B** are rotated, it is impossible to align all of their atoms. Because **A** and **B** are mirror images and not superimposable, **A** and **B** are a pair of **enantiomers.** Two other pairs of enantiomers are drawn in Figure 5.5.

Figure 5.5 Three-dimensional representations for pairs of enantiomers

a. Alanine, an amino acid

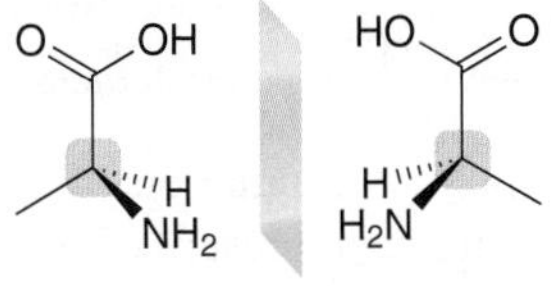

enantiomers

b. 3-Bromohexane

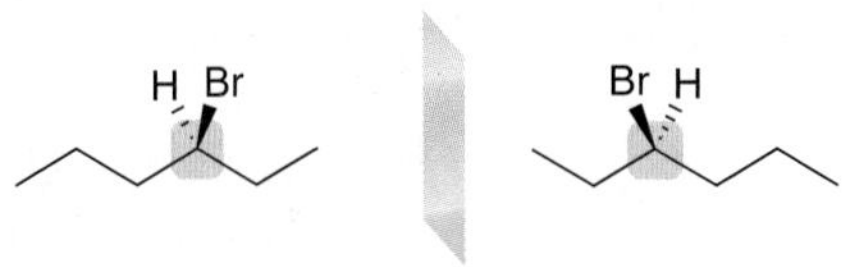

enantiomers

- Remember: H and Br are directly aligned, one behind the other.

Problem 5.8 Locate the stereogenic center in each compound and draw both enantiomers.

a. 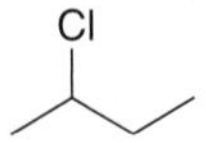

b. 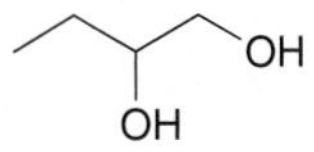

c. 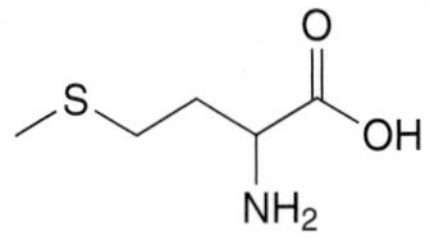

5.5 Stereogenic Centers in Cyclic Compounds

In drawing a tetrahedron using solid lines, wedges, and dashed wedges, always draw the two solid lines first; then draw the wedge and the dashed wedge on the *opposite side* of the solid lines.

If you draw the two solid lines *down*...

...then add the wedge and dashed wedge *above*.

If you draw the two solid lines to the *left*...

...then add the wedge and dashed wedge to the *right*.

Stereogenic centers may also occur at carbon atoms that are part of a ring. To find stereogenic centers on ring carbons, always draw the rings as flat polygons, and look for tetrahedral carbons that are bonded to four different groups, as usual. Each ring carbon is bonded to two other atoms in the ring, as well as two substituents attached to the ring. When the two substituents on the ring are *different,* we must compare the ring atoms equidistant from the atom in question.

Does methylcyclopentane have a stereogenic center? All of the carbon atoms are bonded to two or three hydrogen atoms except for C1, the ring carbon bonded to the methyl group. Next, compare the ring atoms and bonds on both sides equidistant from C1, and continue until a point of difference is reached, or until both sides meet, either at an atom or in the middle of a bond. In this case, there is no point of difference on either side, so C1 is bonded to identical alkyl groups that happen to be part of a ring. **C1, therefore, is *not* a stereogenic center.**

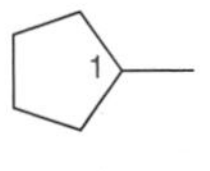

methylcyclopentane

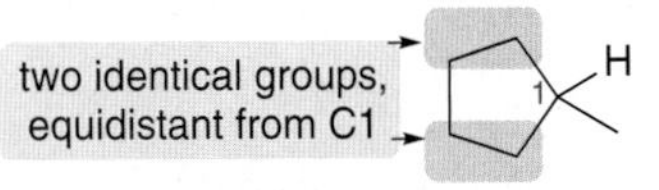

C1 is *not* a stereogenic center.

With 3-methylcyclohexene, the result is different. All carbon atoms are bonded to two or three hydrogen atoms or are sp^2 hybridized except for C3, the ring carbon bonded to the methyl group. In this case, the atoms equidistant from C3 are different, so C3 is bonded to *different* alkyl groups in the ring. **C3 is therefore bonded to four different groups, making it a stereogenic center.**

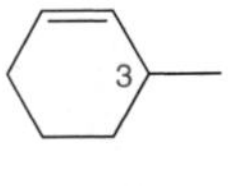

3-methylcyclohexene

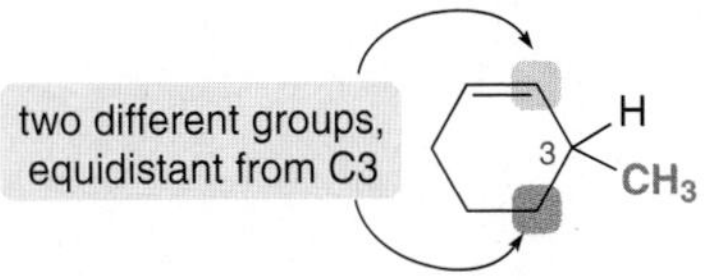

C3 is a stereogenic center.

Two enantiomers are *different* compounds. To convert one enantiomer to another, you must **switch the position of two atoms.** This amounts to breaking bonds.

Because 3-methylcyclohexene has one tetrahedral stereogenic center, it is a chiral compound and exists as a pair of enantiomers.

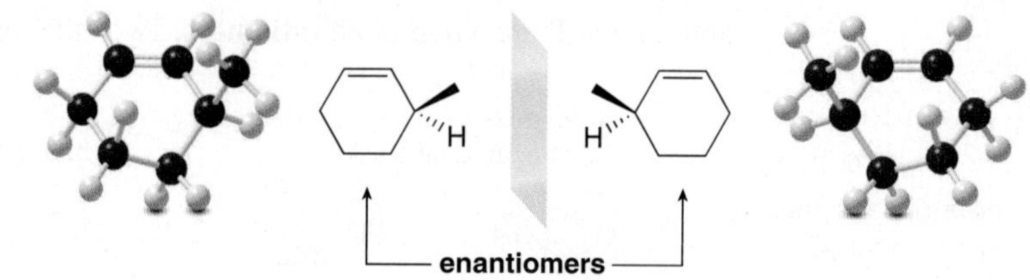

Many biologically active compounds contain one or more stereogenic centers on ring carbons. For example, **thalidomide,** a drug once prescribed as a sedative and anti-nausea agent for pregnant women in Great Britain and Europe, contains one stereogenic center, so it exists as a pair of enantiomers.

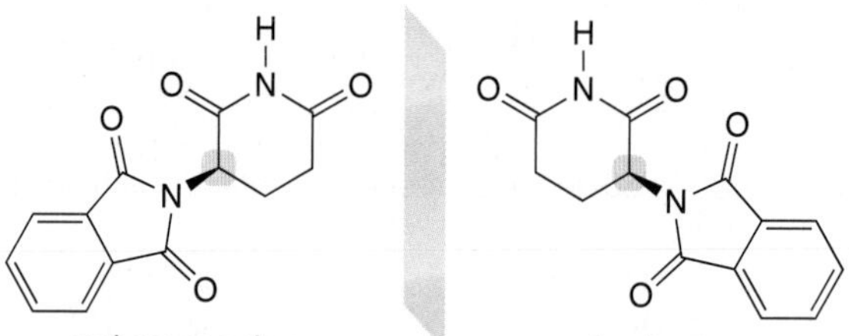

thalidomide
enantiomers

Today, thalidomide is prescribed under strict controls for the treatment of Hansen's disease (leprosy). Because it was once thought to be highly contagious, individuals in Hawai'i with Hansen's disease were sent to Kalaupapa, a remote and inaccessible peninsula on the north shore of the Hawaiian island of Moloka'i. Hansen's disease is now known to be a curable bacterial infection, which is treated by the sulfa drugs discussed in Section 25.16.

Unfortunately thalidomide was sold as a mixture of its two enantiomers, and each of these stereoisomers has a different biological activity. This is a property not uncommon in chiral drugs, as we will see in Section 5.13. Although one enantiomer was an effective sedative and anti-nausea drug, the other enantiomer was responsible for thousands of catastrophic birth defects in children born to women who took the drug during pregnancy. Thalidomide was never approved for use in the United States due to the diligence of Frances Oldham Kelsey, a medical reviewing officer for the Food and Drug Administration, who insisted that the safety data on thalidomide were inadequate.

Sucrose and **paclitaxel** (the chapter-opening molecule) are two useful compounds with several stereogenic centers at ring carbons. Identify the stereogenic centers in these more complicated compounds in exactly the same way, looking at one carbon at a time. **Sucrose,** with nine stereogenic centers on two rings, is the carbohydrate used as table sugar. **Paclitaxel,** with 11 stereogenic centers, is an anticancer agent active against ovarian, breast, and some lung tumors.

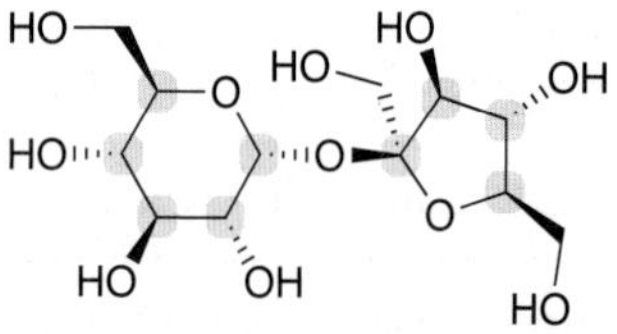

sucrose
(table sugar)

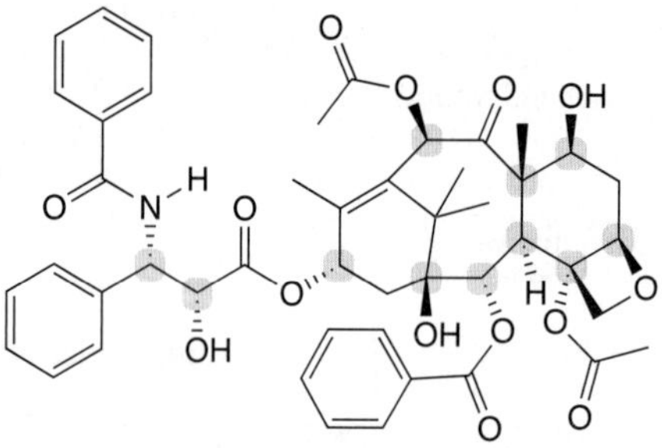

paclitaxel
Trade name Taxol
(anticancer agent)

Problem 5.9 Locate the stereogenic centers in each compound. A molecule may have zero, one, or more stereogenic centers. Gabapentin [part (d)] is used clinically to treat seizures and certain types of chronic pain. Gabapentin enacarbil [part (e)] is a related compound that is three times more potent.

a. 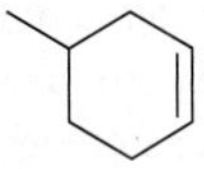

c. 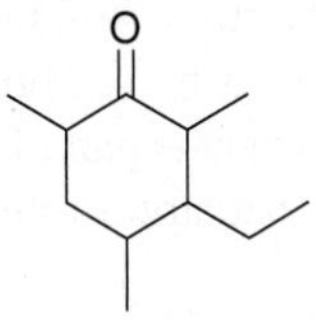

e. 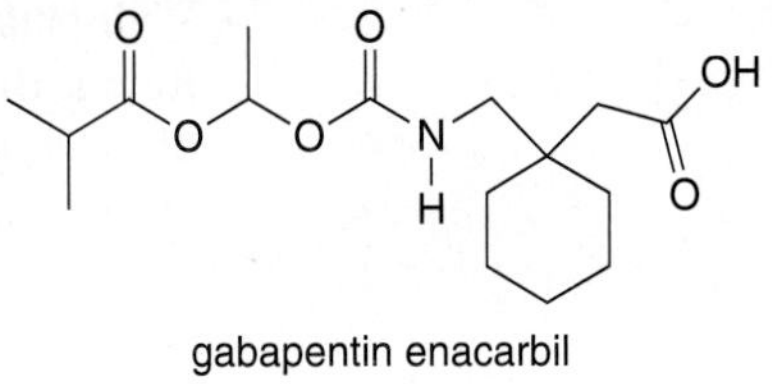

gabapentin enacarbil

b. 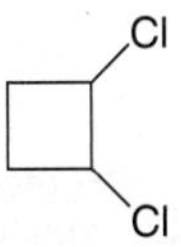

d. 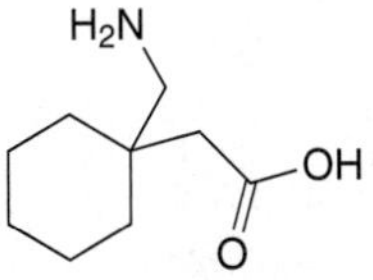

gabapentin

f. 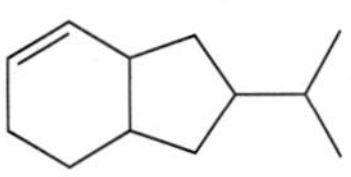

Problem 5.10 Locate the stereogenic centers in each compound.

a. 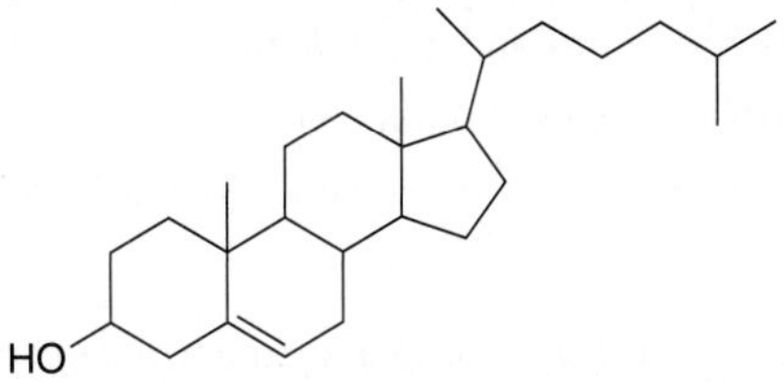

cholesterol

b. 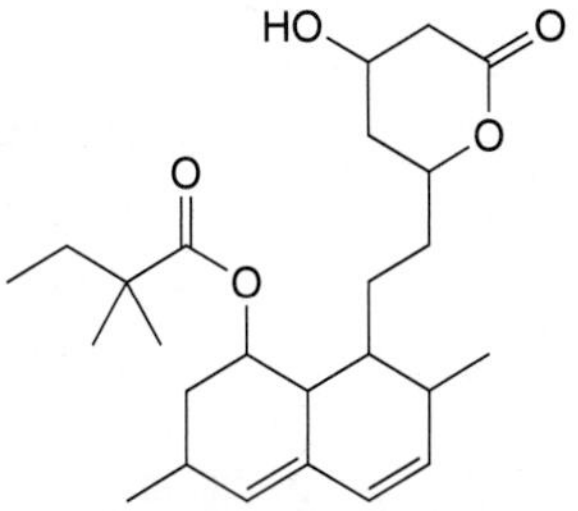

simvastatin
Trade name Zocor
(cholesterol-lowering drug)

5.6 Labeling Stereogenic Centers with *R* or *S*

Naming enantiomers with the prefixes *R* or *S* is called the Cahn–Ingold–Prelog system after the three chemists who devised it.

Because enantiomers are two different compounds, we need a method to distinguish them by name. This is done by adding the prefix *R* or *S* to the IUPAC name of the enantiomer. To designate an enantiomer as *R* or *S*, first **assign a priority** (1, 2, 3, or 4) to each group bonded to the stereogenic center, and then use these priorities to label one enantiomer *R* and one *S*.

Rules Needed to Assign Priority

Rule 1 **Assign priorities (1, 2, 3, or 4) to the atoms directly bonded to the stereogenic center in order of decreasing atomic number. The atom of *highest* atomic number gets the *highest* priority (1).**

- In CHBrClF, priorities are assigned as follows: Br (1, highest) → Cl (2) → F (3) → H (4, lowest). In many molecules the lowest priority group will be H.

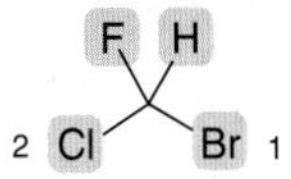

Rule 2 **If two atoms on a stereogenic center are the *same,* assign priority based on the atomic number of the atoms bonded to these atoms. *One* atom of higher atomic number determines a higher priority.**

- With butan-2-ol, the O atom gets highest priority (1) and H gets lowest priority (4) using Rule 1. Butan-2-ol also has two carbon atoms bonded to the stereogenic center, one that is part of a CH_3 group and one that is part of a CH_2CH_3 group. To assign priority (either 2 or 3) to the two C atoms, look at what atoms (other than the stereogenic center) are bonded to each C.

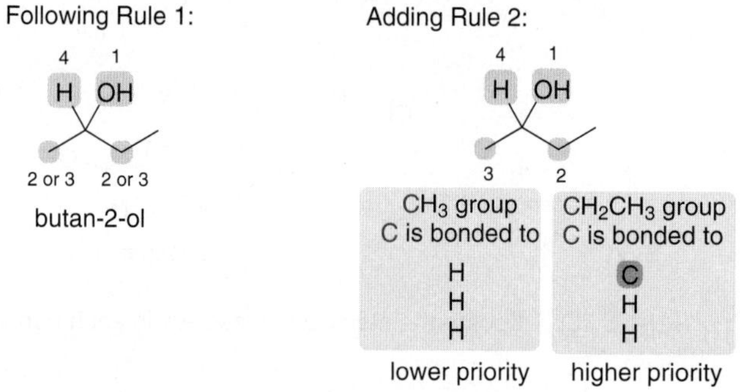

- The CH_2CH_3 gets higher priority (2) than the CH_3 group (priority 3) because the carbon of the ethyl group is bonded to another carbon.
- The order of priority of groups in butan-2-ol is $-OH$ **(1)**, $-CH_2CH_3$ **(2)**, $-CH_3$ **(3)**, and $-H$ **(4)**.
- If priority still cannot be assigned, continue along a chain until a point of difference is reached.

Rule 3 **If two isotopes are bonded to the stereogenic center, assign priorities in order of decreasing *mass* number.**

- In comparing two isotopes of the element hydrogen, deuterium, which has a mass number of two (one proton and one neutron), has a higher priority than hydrogen, which has a mass number of one (one proton only).

Rule 4 **To assign a priority to an atom that is part of a multiple bond, treat a multiply bonded atom as an equivalent number of singly bonded atoms.**

- The C of a C=O is considered to be bonded to two O atoms.

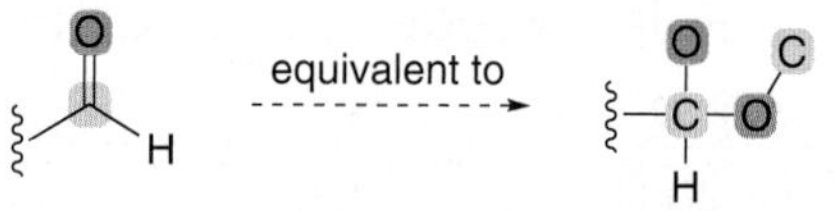

- C is bonded to 2 O's.
- O is bonded to 2 C's.

- Other common multiple bonds are drawn below.

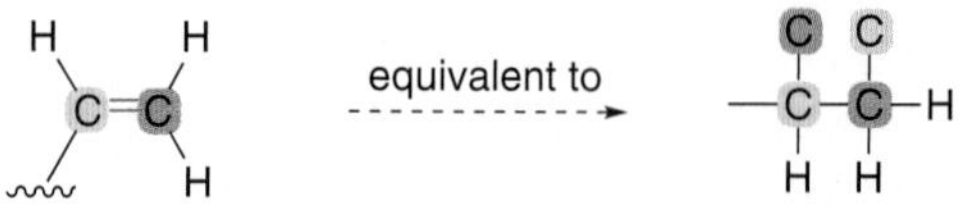

- Each C in the double bond is drawn twice.

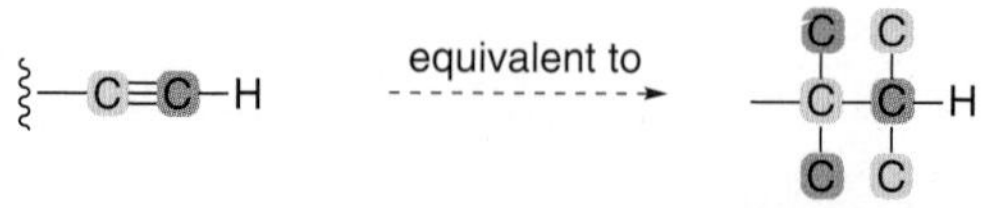

- Each C in the triple bond is drawn three times.

Figure 5.6 gives examples of priorities assigned to stereogenic centers.

Figure 5.6 Examples of assigning priorities to stereogenic centers

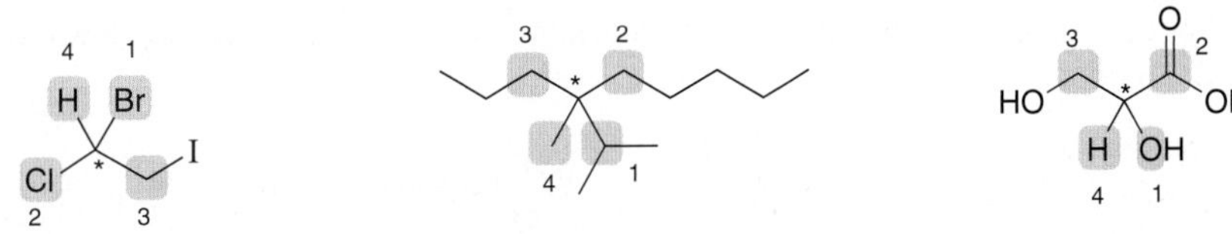

- The stereogenic center is bonded to Br, Cl, C, and H.
- The stereogenic center is *not* bonded directly to I.

- $CH(CH_3)_2$ gets the highest priority because the C is bonded to 2 other C's.

- OH gets the highest priority because O has the highest atomic number.
- CO_2H (three bonds to O) gets higher priority than CH_2OH (one bond to O).

[* = stereogenic center]

Problem 5.11 Which group in each pair is assigned the *higher* priority?

a. $-CH_3$, $-CH_2CH_3$
b. $-I$, $-Br$
c. $-H$, $-D$
d. $-CH_2Br$, $-CH_2CH_2Br$
e. $-CH_2CH_2Cl$, $-CH_2CH(CH_3)_2$
f. $-CH_2OH$, $-CHO$

Problem 5.12 Rank the following groups in order of *decreasing* priority.

a. $-COOH$, $-H$, $-NH_2$, $-OH$
b. $-H$, $-CH_3$, $-Cl$, $-CH_2Cl$
c. $-CH_2CH_3$, $-CH_3$, $-H$, $-CH(CH_3)_2$
d. $-CH{=}CH_2$, $-CH_3$, $-C{\equiv}CH$, $-H$

R is derived from the Latin word *rectus* meaning "right" and *S* is from the Latin word *sinister* meaning "left."

Once priorities are assigned to the four groups around a stereogenic center, we can use three steps to designate the center as either *R* or *S*.

How To Assign *R* or *S* to a Stereogenic Center

Example Label each enantiomer as *R* or *S*.

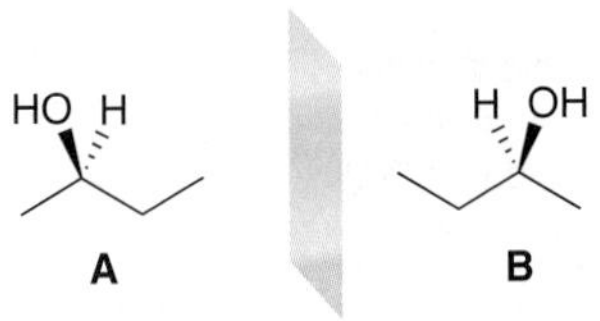

two enantiomers of butan-2-ol

Step [1] **Assign priorities from 1 to 4 to each group bonded to the stereogenic center.**

- The priorities for the four groups around the stereogenic center in butan-2-ol were given in Rule 2, on page 192.

$-OH$	$-CH_2CH_3$	$-CH_3$	$-H$
1	2	3	4
highest			lowest

Decreasing priority →

—Continued

How To, continued . . .

Step [2] Orient the molecule with the lowest priority group (4) *back* (on a *dashed wedge*), and visualize the relative positions of the remaining three groups (priorities 1, 2, and 3).

- For each enantiomer of butan-2-ol, look toward the lowest priority group, drawn behind the plane, down the C – H bond.

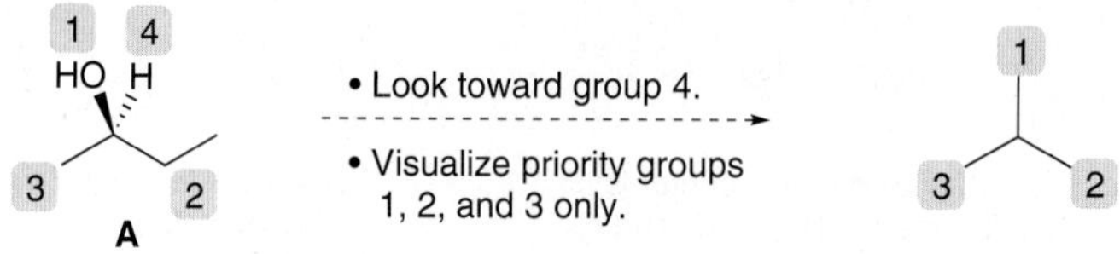

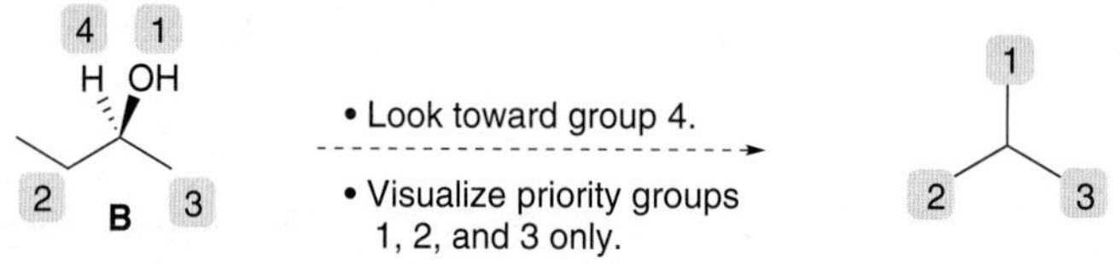

Step [3] Trace a circle from priority group 1 → 2 → 3.

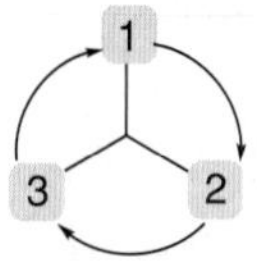

clockwise
***R* isomer**

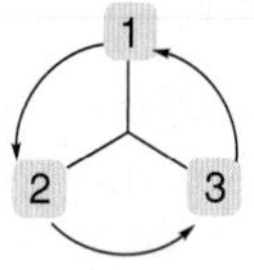

counterclockwise
***S* isomer**

- If tracing the circle goes in the **clockwise** direction—to the right from the noon position—the isomer is named ***R***.
- If tracing the circle goes in the **counterclockwise** direction—to the left from the noon position—the isomer is named ***S***.

- The letter *R* or *S* precedes the IUPAC name of the molecule. For the enantiomers of butan-2-ol:

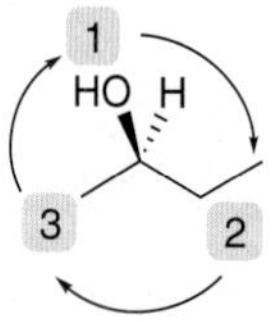

A

clockwise
***R* isomer**
(*R*)-butan-2-ol

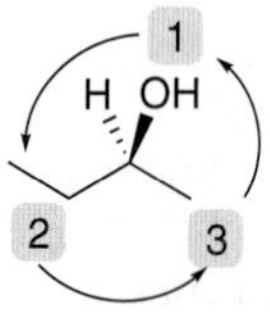

B

counterclockwise
***S* isomer**
(*S*)-butan-2-ol

Sample Problem 5.2

Label the following compound as *R* or *S*.

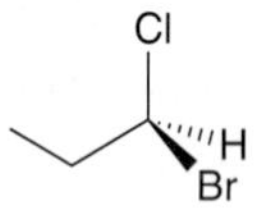

Solution

[1] Assign priorities.

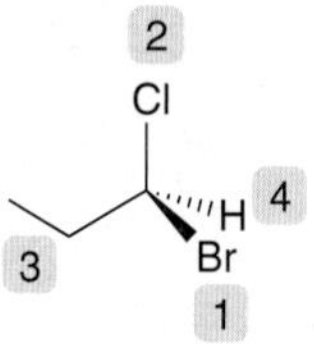

[2] Looking toward the lowest priority group (H), trace a circle, 1→2→3.

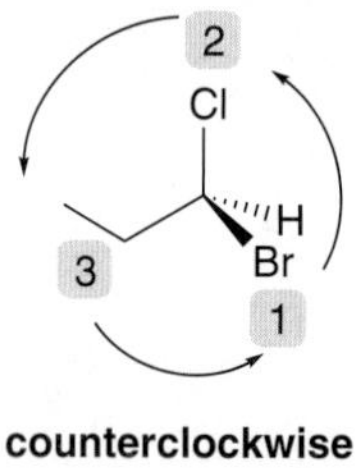

counterclockwise

Answer: ***S*** isomer

How do you assign *R* or *S* to a molecule when the lowest priority group is not oriented toward the back, on a dashed wedge? You could rotate and flip the molecule until the lowest priority group is in the back, as shown in Figure 5.7; then follow the stepwise procedure for assigning the configuration. Or, if manipulating and visualizing molecules in three dimensions is difficult for you, try the procedure suggested in Sample Problem 5.3.

Figure 5.7 Orienting the lowest priority group in back

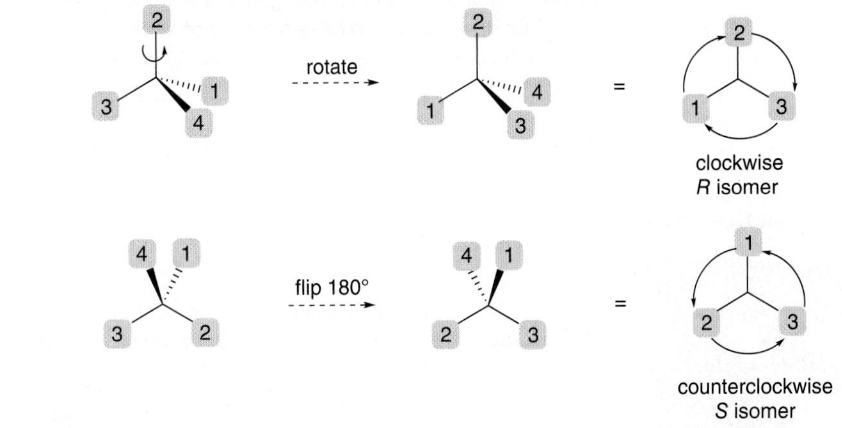

- In rotating a molecule about a single bond, the position of three groups changes.
- In flipping a molecule 180°, the position of all four groups changes.

Sample Problem 5.3 Label the following compound as *R* or *S*.

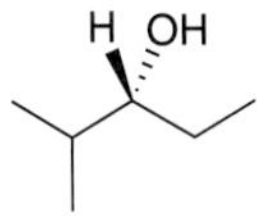

Solution

In this problem, the lowest priority group (H) is oriented in **front** of, not behind, the page. To assign *R* or *S* in this case:

- **Switch** the position of the lowest priority group (H) with the group located *behind* the page (OH).
- Determine *R* or *S* in the usual manner.
- **Reverse the answer.** Because we switched the position of two groups on the stereogenic center to begin with, and there are only two possibilities, the answer is **opposite** to the correct answer.

[1] Assign priorities.

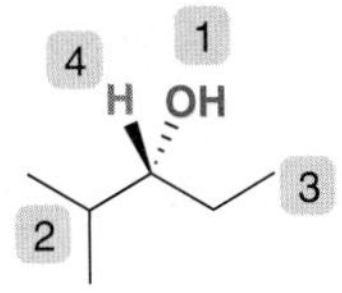

[2] Switch groups 4 and 1.

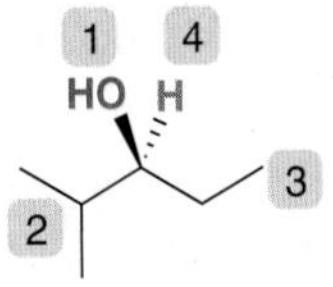

[3] Trace a circle, 1 → 2 → 3, and reverse the answer.

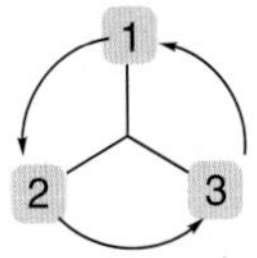

counterclockwise
It looks like an *S* isomer, but we must reverse the answer because we switched groups 1 and 4, *S* → *R*.

Answer: ***R*** isomer

Problem 5.13 Label each compound as *R* or *S*.

a. 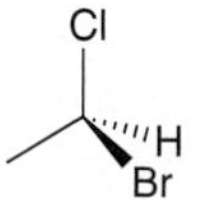

b. 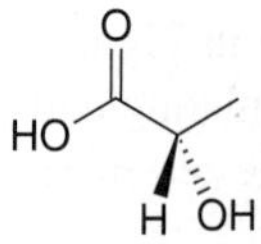

c. 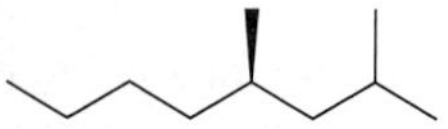

d. 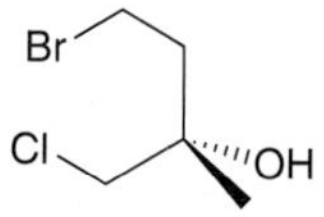

Problem 5.14 Draw both enantiomers of clopidogrel (trade name Plavix), a drug given to prevent the formation of blood clots in persons who have a history of stroke or coronary artery disease. Plavix is sold as a single enantiomer with the *S* configuration. Which enantiomer is Plavix?

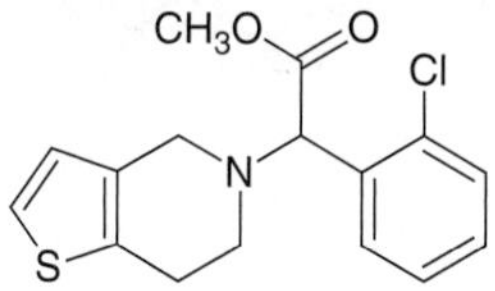

clopidogrel

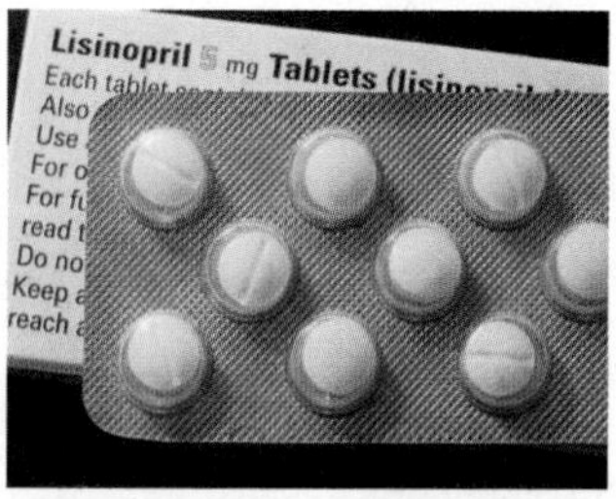

Lisinopril (trade name Zestril) is an ACE inhibitor, a drug that lowers blood pressure by decreasing the amount of angiotensin in the blood. Angiotensin is a polyamide that narrows blood vessels, thus increasing blood pressure.

Problem 5.15 (a) Locate the stereogenic centers in the ball-and-stick model of lisinopril, a drug used to treat high blood pressure. (b) Label each stereogenic center as *R* or *S*.

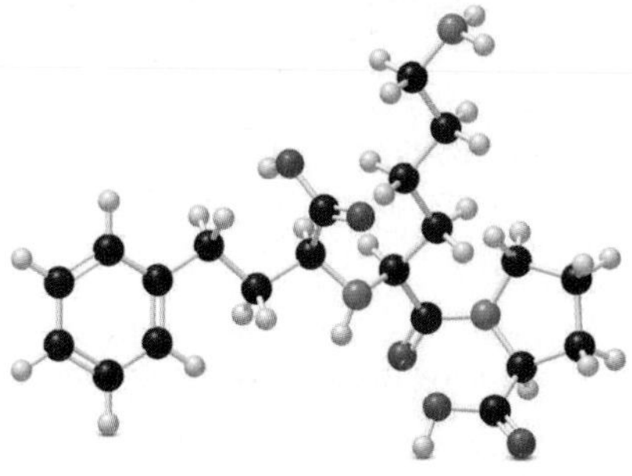

lisinopril

5.7 Diastereomers

We have now seen many examples of compounds containing one tetrahedral stereogenic center. The situation is more complex for compounds with two stereogenic centers, because more stereoisomers are possible. Moreover, a molecule with two or more stereogenic centers ***may* or *may not* be chiral.**

- **For *n* stereogenic centers, the maximum number of stereoisomers is 2^n.**

- When $n = 1$, $2^1 = 2$. With one stereogenic center, there are always two stereoisomers and they are **enantiomers.**
- When $n = 2$, $2^2 = 4$. With two stereogenic centers, the maximum number of stereoisomers is four, although sometimes there are *fewer* than four.

Problem 5.16 What is the maximum number of stereoisomers possible for a compound with: (a) three stereogenic centers; (b) eight stereogenic centers?

Let's illustrate a stepwise procedure for finding all possible stereoisomers using 2,3-dibromopentane. Because 2,3-dibromopentane has two stereogenic centers, the maximum number of stereoisomers is four.

In testing to see if one compound is superimposable on another, rotate atoms and flip the entire molecule, but **do not break any bonds.**

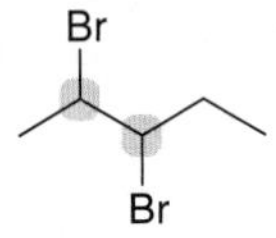

2,3-dibromopentane

● = stereogenic center

To find stereoisomers, use the *eclipsed* conformation.

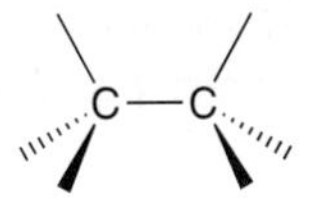

eclipsed conformation

easier conformation to visualize

rapid interconversion

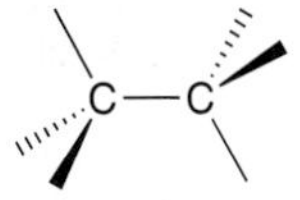

staggered conformation

more stable

How To Find and Draw All Possible Stereoisomers for a Compound with Two Stereogenic Centers

Step [1] Draw one stereoisomer by arbitrarily arranging substituents around the stereogenic centers. Then draw its mirror image.

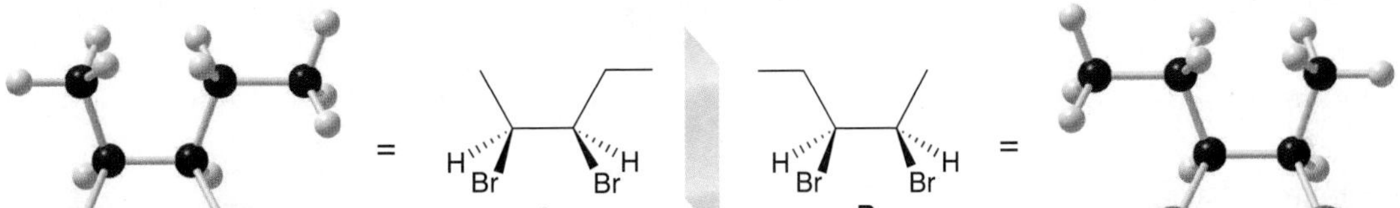

A **B**

- Arbitrarily add the H, Br, CH_3, and CH_2CH_3 groups to the stereogenic centers, forming **A.** Then draw the mirror image **B** so that substituents in **B** are a reflection of the substituents in **A.**
- Determine whether **A** and **B** are superimposable by flipping or rotating one molecule to see if all the atoms align.
- If you have drawn the compound and the mirror image in the described manner, you only have to do two operations to see if the atoms align. Place **B** directly on top of **A** (either in your mind or use models); and rotate **B** 180° and place it on top of **A** to see if the atoms align.

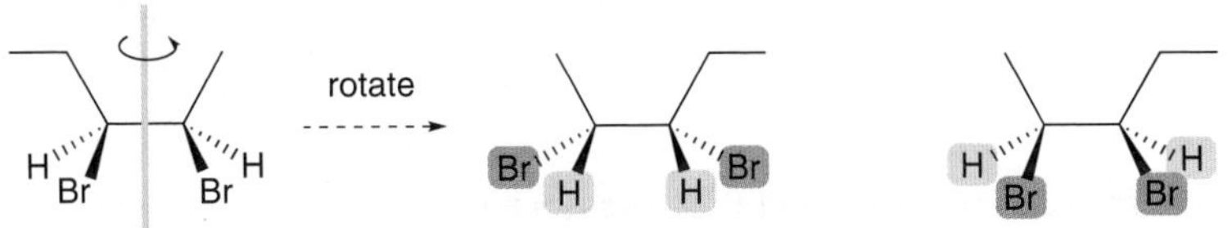

180°

B **B** **A**

- H and Br do *not* align.
- **A** and **B** are *different* compounds.

- In this case, the atoms of **A** and **B** do not align, making **A** and **B** nonsuperimposable mirror images—**enantiomers. A** and **B** are two of the four possible stereoisomers for 2,3-dibromopentane.

—Continued

How To, continued...

Step [2] Draw a third possible stereoisomer by switching the positions of any two groups on only *one* stereogenic center. Then draw its mirror image.

- Switching the positions of H and Br (or any two groups) on one stereogenic center of either **A** or **B** forms a new stereoisomer (labeled **C** in this example), which is different from both **A** and **B.** Then draw the mirror image of **C,** labeled **D. C** and **D** are nonsuperimposable mirror images—**enantiomers.** We have now drawn four stereoisomers for 2,3-dibromopentane, the maximum number possible.

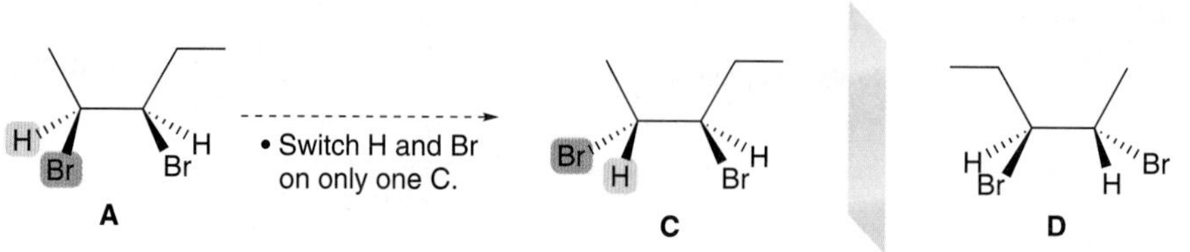

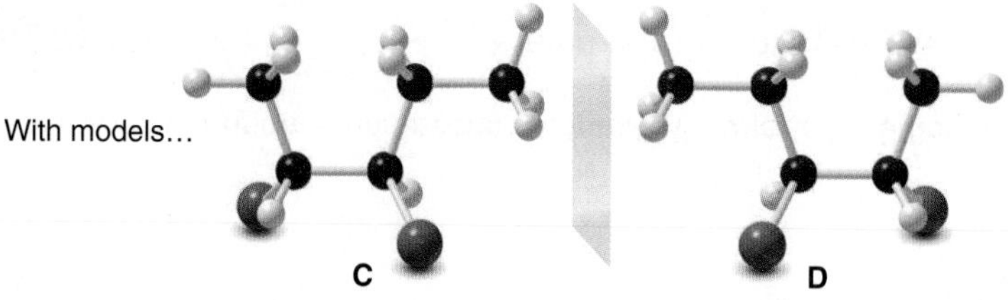

There are only two types of stereoisomers: ***Enantiomers* are stereoisomers that *are* mirror images. *Diastereomers* are stereoisomers that are *not* mirror images.**

There are four stereoisomers for 2,3-dibromopentane: enantiomers **A** and **B,** and enantiomers **C** and **D.** What is the relationship between two stereoisomers like **A** and **C?** **A** and **C** represent the second class of stereoisomers, called **diastereomers. *Diastereomers* are stereoisomers that are *not* mirror images of each other. A** and **B** are diastereomers of **C** and **D,** and vice versa. Figure 5.8 summarizes the relationships between the stereoisomers of 2,3-dibromopentane.

Problem 5.17 Label the two stereogenic centers in each compound and draw all possible stereoisomers.

a. 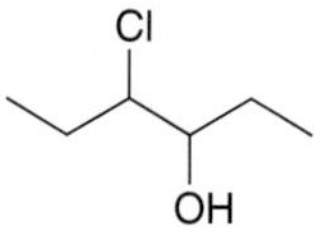

b. 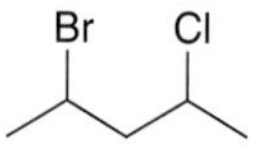

Problem 5.18 Compounds **E** and **F** are two isomers of 2,3-dibromopentane drawn in staggered conformations. Which compounds (**A–D**) in Figure 5.8 are identical to **E** and **F?**

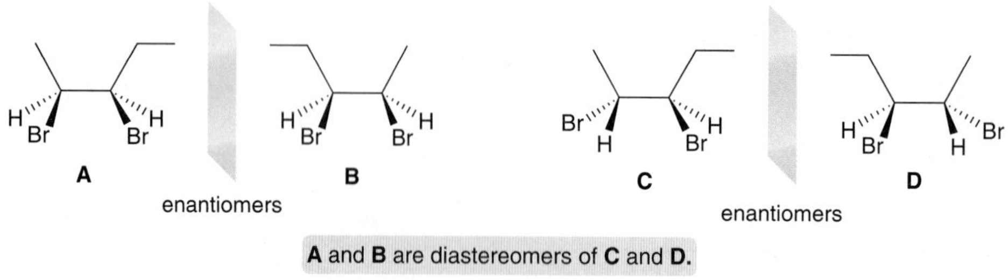

Figure 5.8 The four stereoisomers of 2,3-dibromopentane

- Pairs of enantiomers: **A** and **B; C** and **D.**
- Pairs of diastereomers: **A** and **C; A** and **D; B** and **C; B** and **D.**

5.8 Meso Compounds

Whereas 2,3-dibromopentane has two stereogenic centers and the maximum of four stereoisomers, **2,3-dibromobutane** has two stereogenic centers but fewer than the maximum number of stereoisomers.

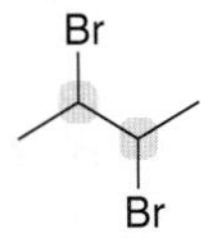

2,3-dibromobutane

= stereogenic center

To find and draw all the stereoisomers of 2,3-dibromobutane, follow the same stepwise procedure outlined in Section 5.7. Arbitrarily add the H, Br, and CH_3 groups to the stereogenic centers, forming one stereoisomer **A,** and then draw its mirror image **B. A** and **B** are nonsuperimposable mirror images—**enantiomers.**

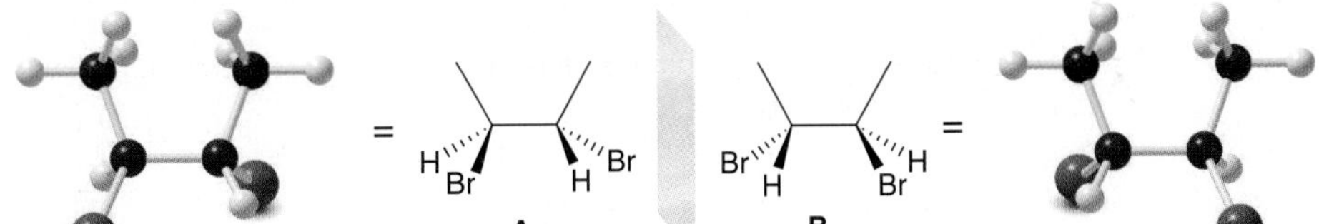

A **B**

enantiomers

To find the other two stereoisomers (if they exist), switch the position of two groups on *one* stereogenic center of only *one* enantiomer. In this case, switching the positions of H and Br on one stereogenic center of **A** forms **C,** which is different from both **A** and **B** and is thus a new stereoisomer.

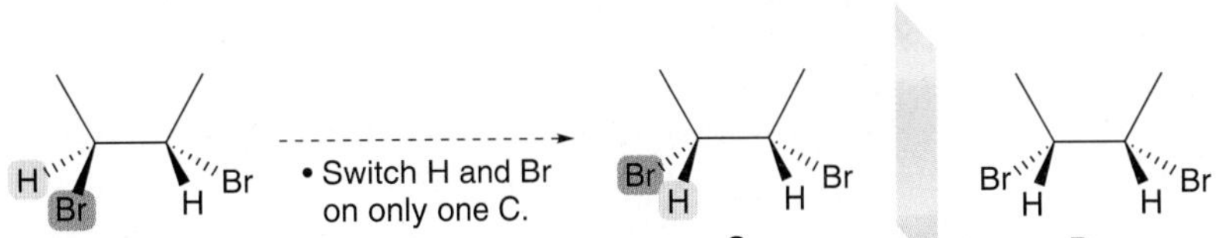

A **C** **D**

identical
C = D

With models...

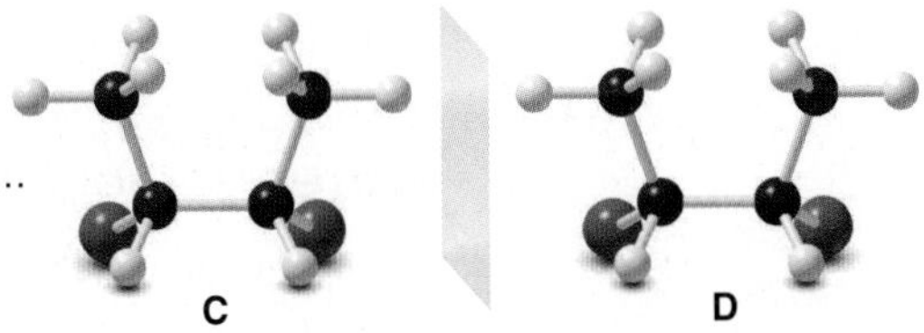

C **D**

However, the mirror image of **C,** labeled **D,** is superimposable on **C,** so **C** and **D** are *identical.* Thus, **C** is achiral, even though it has two stereogenic centers. **C** is a **meso compound.**

- A *meso compound* is an achiral compound that contains tetrahedral stereogenic centers.

C contains a **plane of symmetry. Meso compounds generally have a plane of symmetry,** so they possess two identical halves.

plane of symmetry

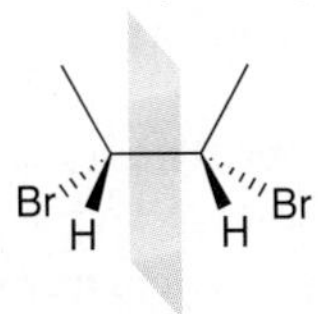

C

two identical halves

Because one stereoisomer of 2,3-dibromobutane is superimposable on its mirror image, there are only three stereoisomers and not four, as summarized in Figure 5.9.

Figure 5.9 The three stereoisomers of 2,3-dibromobutane

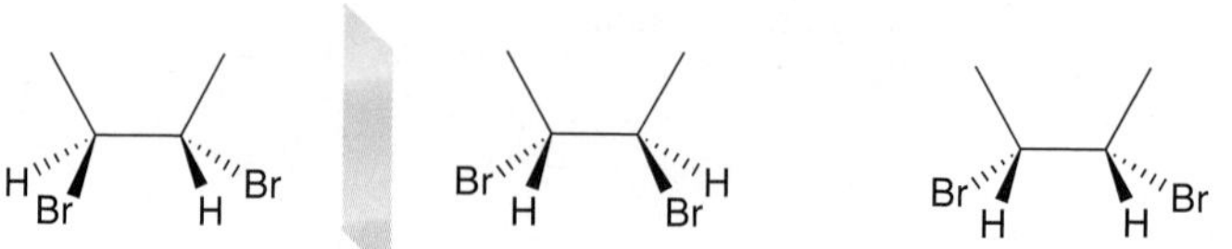

A **B** enantiomers

C meso compound

A and **B** are diastereomers of **C**.

- Pair of enantiomers: **A** and **B.**
- Pairs of diastereomers: **A** and **C; B** and **C.**

Problem 5.19 Draw all the possible stereoisomers for each compound, and label pairs of enantiomers and diastereomers.

a. 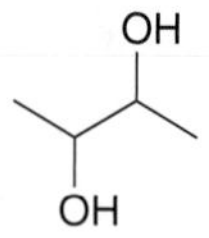 b. 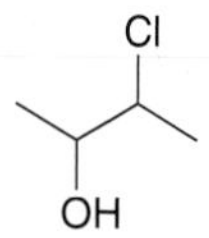

Problem 5.20 Which compounds are meso compounds?

a. 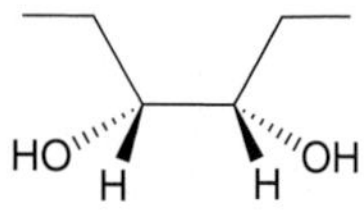 b. 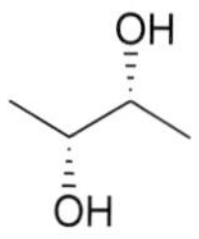 c. 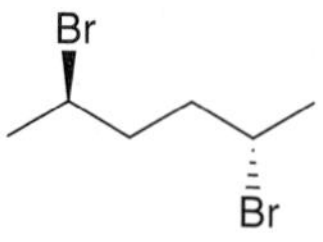

Problem 5.21 Draw a meso compound for each of the following molecules.

a. 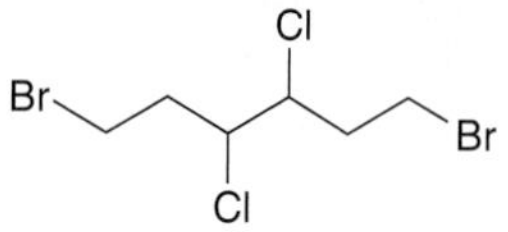 b. 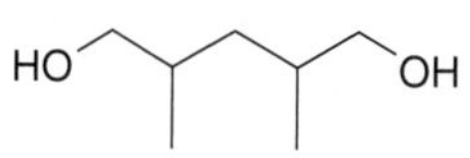 c. 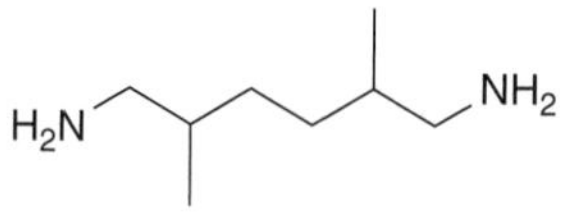

5.9 *R* and *S* Assignments in Compounds with Two or More Stereogenic Centers

When a compound has more than one stereogenic center, the *R* or *S* configuration must be assigned to each of them. In the stereoisomer of 2,3-dibromopentane drawn here, C2 has the *S* configuration and C3 has the *R*, so the complete name of the compound is (2*S*,3*R*)-2,3-dibromopentane.

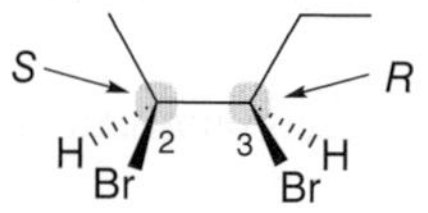

(2*S*,3*R*)-2,3-dibromopentane

R,*S* configurations can be used to determine whether two compounds are identical, enantiomers, or diastereomers.

- Identical compounds have the *same R,S* designations at every tetrahedral stereogenic center.
- Enantiomers have exactly *opposite R,S* designations.
- Diastereomers have the *same R,S* designation for at least one stereogenic center and the *opposite* for at least one of the other stereogenic centers.

For example, if a compound has two stereogenic centers, both with the *R* configuration, then its enantiomer is *S,S* and the diastereomers are either *R,S* or *S,R*.

Problem 5.22 If the two stereogenic centers of a compound are *R,S* in configuration, what are the *R,S* assignments for its enantiomer and two diastereomers?

Problem 5.23 Without drawing out the structures, label each pair of compounds as enantiomers or diastereomers.

a. (2*R*,3*S*)-hexane-2,3-diol and (2*R*,3*R*)-hexane-2,3-diol
b. (2*R*,3*R*)-hexane-2,3-diol and (2*S*,3*S*)-hexane-2,3-diol
c. (2*R*,3*S*,4*R*)-hexane-2,3,4-triol and (2*S*,3*R*,4*R*)-hexane-2,3,4-triol

Problem 5.24 (a) Label the four stereogenic centers in sorbitol as *R* or *S*. (b) How are sorbitol and **A** related? (c) How are sorbitol and **B** related?

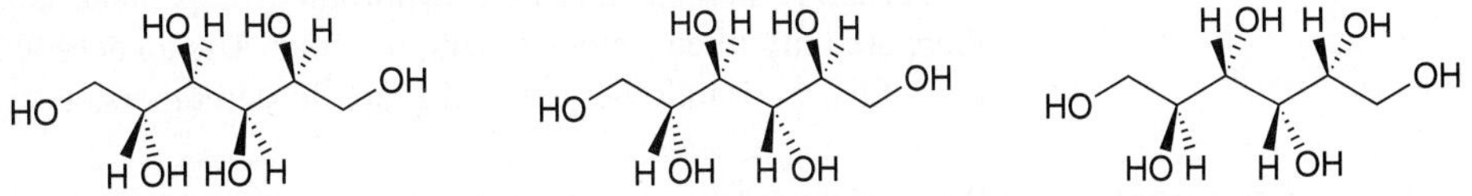
HO H HO H
HO
OH
H OH HO H
HO H HO H
HO
OH
H OH H OH
H OH H OH
HO
OH
HO H H OH

sorbitol A B

NEW!
Trident
Splash
Citrus Blackberry

Sorbitol (Problem 5.24) occurs naturally in some berries and fruits. It is used as a substitute sweetener in sugar-free—that is, sucrose-free—candy and gum.

5.10 Disubstituted Cycloalkanes

Let us now turn our attention to disubstituted cycloalkanes, and draw all possible stereoisomers for **1,3-dibromocyclopentane.** Because 1,3-dibromocyclopentane has two stereogenic centers (labeled in blue), it has a maximum of four stereoisomers.

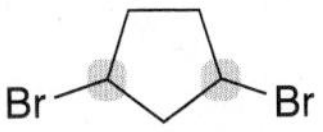
Br
Br

1,3-dibromocyclopentane

In determining chirality in substituted cycloalkanes, always draw the rings as **flat polygons.** This is especially true for cyclohexane derivatives, where having two chair forms that interconvert can make analysis especially difficult.

To draw all possible stereoisomers, remember that a disubstituted cycloalkane can have two substituents on the same side of the ring (**cis isomer,** labeled **A**) or on opposite sides of the ring (**trans isomer,** labeled **B**). These compounds are **stereoisomers but not mirror images of each other,** making them **diastereomers.** **A** and **B** are two of the four possible stereoisomers.

Br
Br

A
cis isomer

Br
Br

B
trans isomer

A and **B** are diastereomers.

cis-1,3-Dibromocyclopentane contains a plane of symmetry.

plane of symmetry

two identical halves

To find the other two stereoisomers (if they exist), draw the mirror image of each compound and determine whether the compound and its mirror image are superimposable.

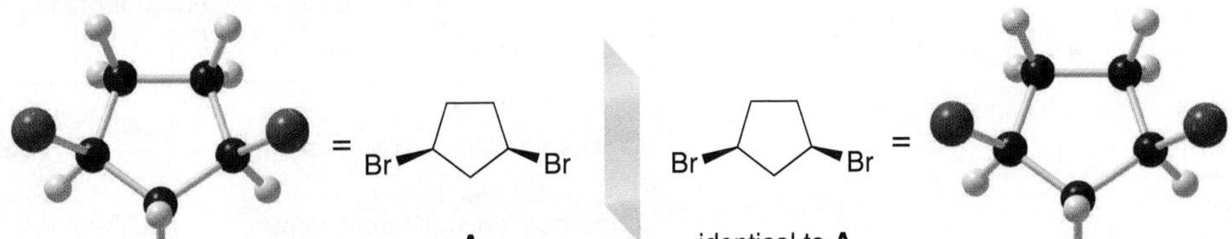
=
Br
Br
Br
Br
=

A
cis isomer

identical to **A**

- The cis isomer is superimposable on its mirror image, making them *identical.* Thus, **A** is an **achiral meso compound.**

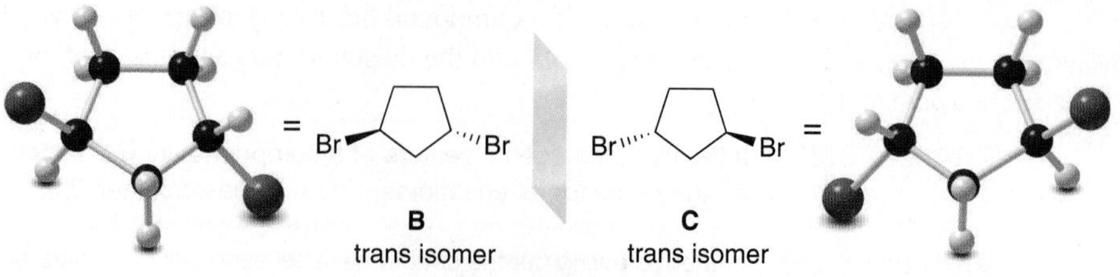

B and **C** are enantiomers.

- The trans isomer **B** is *not* superimposable on its mirror image, labeled **C,** making **B** and **C** different compounds. Thus, **B** and **C** are **enantiomers.**

Because one stereoisomer of 1,3-dibromocyclopentane is superimposable on its mirror image, there are only three stereoisomers, not four. **A** is an achiral meso compound, and **B** and **C** are a pair of chiral enantiomers. **A** and **B** are diastereomers, as are **A** and **C.**

Problem 5.25 Which of the following cyclic molecules are *meso* compounds?

a. 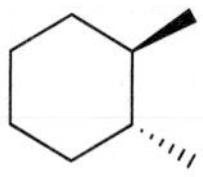 b. 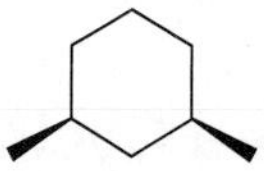 c. 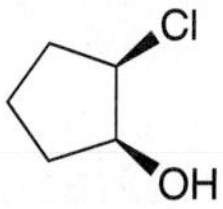

Problem 5.26 Draw all possible stereoisomers for each compound. Label pairs of enantiomers and diastereomers.

a. 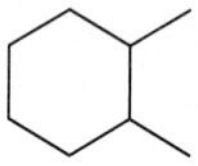 b. 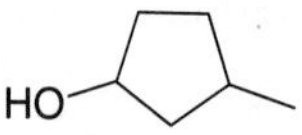

c. 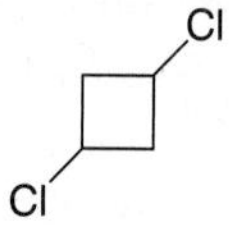

5.11 Isomers—A Summary

Before moving on to other aspects of stereochemistry, take the time to review Figures 5.10 and 5.11. Keep in mind the following facts, and use Figure 5.10 to summarize the types of isomers.

- **There are two major classes of isomers: constitutional isomers and stereoisomers.**
- **There are only two kinds of stereoisomers: enantiomers and diastereomers.**

Then, to determine the relationship between two nonidentical molecules, refer to the flowchart in Figure 5.11.

Figure 5.10
Summary—Types of isomers

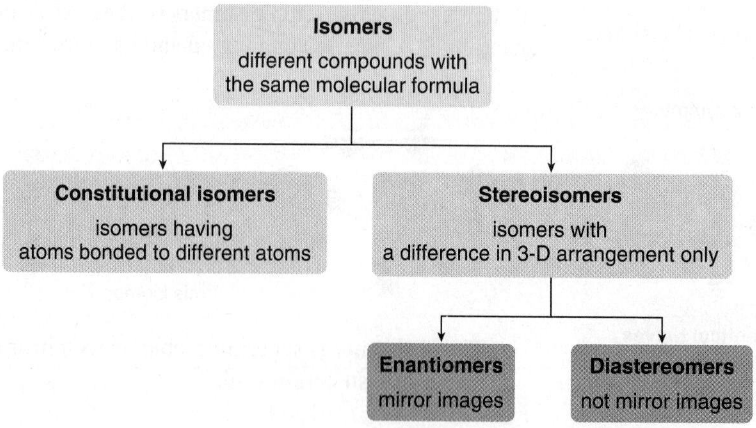

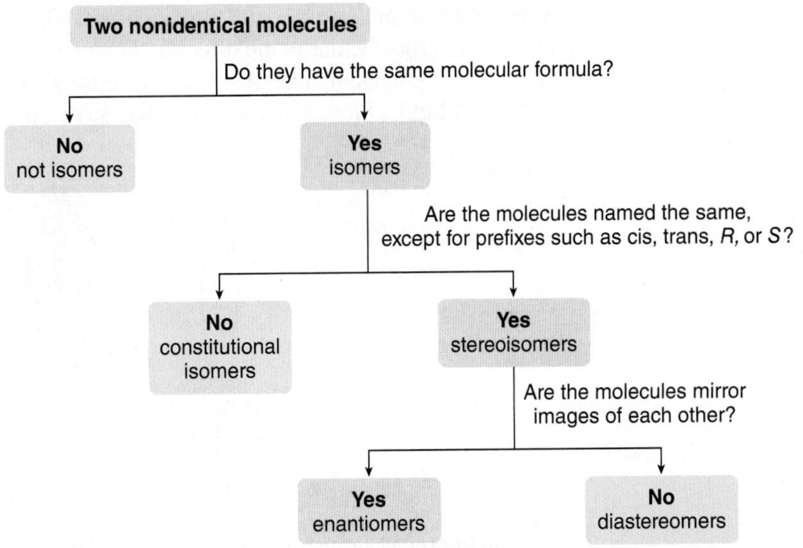

Figure 5.11 Determining the relationship between two nonidentical molecules

Problem 5.27 State how each pair of compounds is related. Are they enantiomers, diastereomers, constitutional isomers, or identical?

a. 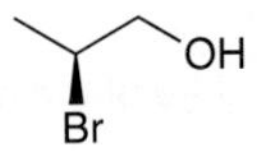 and 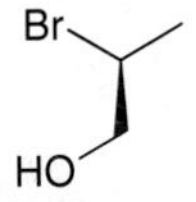

b. 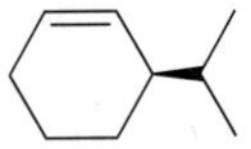 and 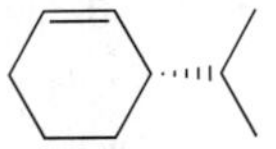

c. 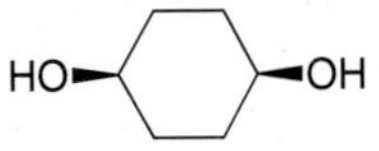 and 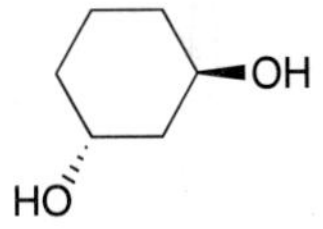

d. 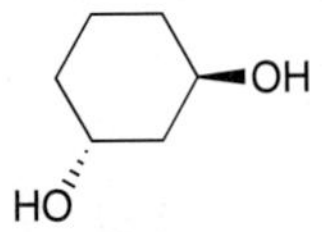 and 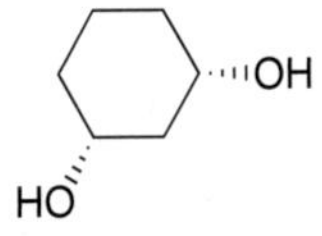

5.12 Physical Properties of Stereoisomers

Recall from Section 5.2 that constitutional isomers have different physical and chemical properties. How, then, do the physical and chemical properties of enantiomers compare?

- **The chemical and physical properties of two enantiomers are *identical* except in their interaction with *chiral* substances.**

5.12A Optical Activity

Two enantiomers have identical physical properties—melting point, boiling point, solubility—except for how they interact with plane-polarized light.

What is plane-polarized light? Ordinary light consists of electromagnetic waves that oscillate in all planes perpendicular to the direction in which the light travels. Passing light through a polarizer allows light in only one plane to come through, resulting in **plane-polarized light** (or simply **polarized light**). Plane-polarized light has an electric vector that oscillates in a single plane.

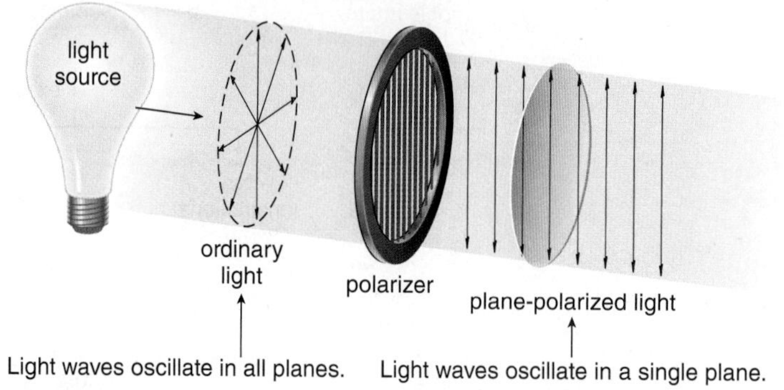

A **polarimeter** is an instrument that allows plane-polarized light to travel through a sample tube containing an organic compound. After the light exits the sample tube, an analyzer slit is rotated to determine the direction of the plane of the exiting polarized light.

With **achiral compounds,** the light exits the sample tube *unchanged,* and the plane of the polarized light is in the same position it was before entering the sample tube. **A compound that does not change the plane of polarized light is said to be *optically inactive.***

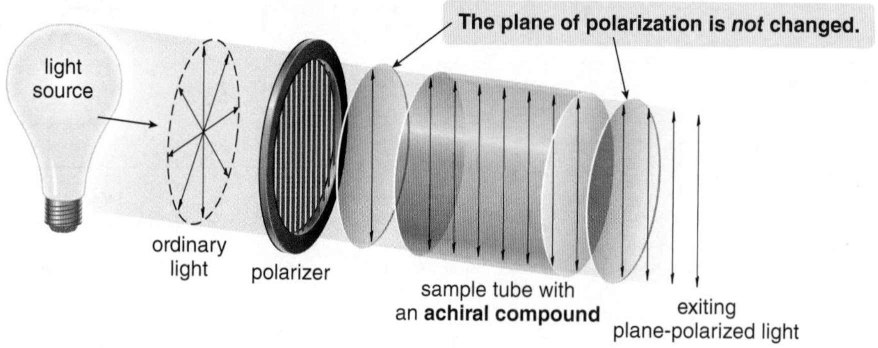

With **chiral compounds,** the plane of the polarized light is rotated through an angle α. The angle α, measured in degrees (°), is called the **observed rotation. A compound that rotates the plane of polarized light is said to be *optically active.***

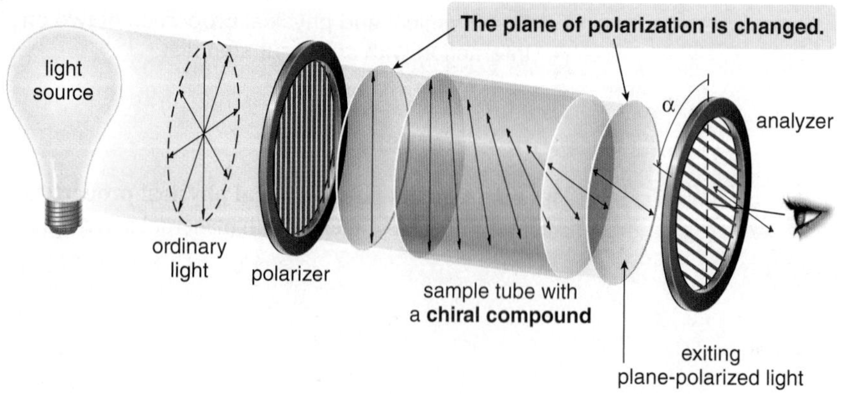

The achiral compound CH_2BrCl is optically inactive, whereas a single enantiomer of CHBrClF, a chiral compound, is optically active.

The rotation of polarized light can be in the **clockwise** or **counterclockwise** direction.

- If the rotation is *clockwise* (to the right from the noon position), the compound is called *dextrorotatory.* The rotation is labeled *d* or (+).
- If the rotation is *counterclockwise* (to the left from noon), the compound is called *levorotatory.* The rotation is labeled *l* or (–).

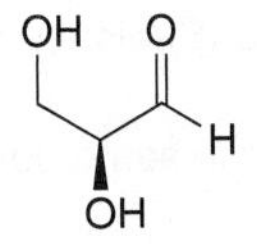

(*S*)-(–)-glyceraldehyde

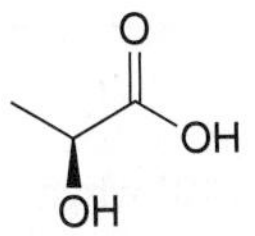

(*S*)-(+)-lactic acid

No relationship exists between the *R* and *S* prefixes that designate configuration and the (+) and (–) designations indicating optical rotation. For example, the *S* enantiomer of lactic acid is dextrorotatory (+), whereas the *S* enantiomer of glyceraldehyde is levorotatory (–).

How does the rotation of two enantiomers compare?

- Two enantiomers rotate plane-polarized light *to an equal extent* but in the *opposite* direction.

Thus, if enantiomer **A** rotates polarized light +5°, then the same concentration of enantiomer **B** rotates it –5°.

5.12B Racemic Mixtures

What is the observed rotation of an equal amount of two enantiomers? Because **two enantiomers rotate plane-polarized light to an equal extent but in opposite directions, the rotations cancel,** and no rotation is observed.

- An equal amount of two enantiomers is called a *racemic mixture* or a *racemate.* A racemic mixture is optically inactive.

Besides optical rotation, other physical properties of a racemate are not readily predicted. The melting point and boiling point of a racemic mixture are not necessarily the same as either pure enantiomer, and this fact is not easily explained. The physical properties of two enantiomers and their racemic mixture are summarized in Table 5.1.

Table 5.1 The Physical Properties of Enantiomers **A** and **B** Compared

Property	A alone	B alone	Racemic A + B
Melting point	identical to **B**	identical to **A**	may be different from **A** and **B**
Boiling point	identical to **B**	identical to **A**	may be different from **A** and **B**
Optical rotation	equal in magnitude but opposite in sign to **B**	equal in magnitude but opposite in sign to **A**	0°

5.12C Specific Rotation

The observed rotation depends on the number of chiral molecules that interact with polarized light. This in turn depends on the concentration of the sample and the length of the sample tube. To standardize optical rotation data, the quantity **specific rotation** ([α]) is defined using a specific sample tube length (usually 1 dm), concentration, temperature (25 °C), and wavelength (589 nm, the D line emitted by a sodium lamp).

$$\text{specific rotation} = [\alpha] = \frac{\alpha}{l \times c}$$

α = observed rotation (°)
l = length of sample tube (dm)
c = concentration (g/mL)

[dm = decimeter
1 dm = 10 cm]

Specific rotations are physical constants just like melting points or boiling points, and are reported in chemical reference books for a wide variety of compounds.

Problem 5.28 The amino acid (*S*)-alanine has the physical characteristics listed under the structure.

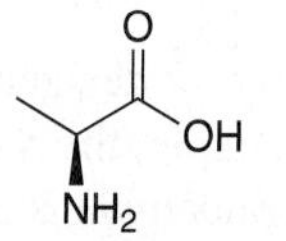

(*S*)-alanine
[α] = +8.5
mp = 297 °C

a. What is the melting point of (*R*)-alanine?
b. How does the melting point of a racemic mixture of (*R*)- and (*S*)-alanine compare to the melting point of (*S*)-alanine?
c. What is the specific rotation of (*R*)-alanine, recorded under the same conditions as the reported rotation of (*S*)-alanine?
d. What is the optical rotation of a racemic mixture of (*R*)- and (*S*)-alanine?
e. Label each of the following as optically active or inactive: a solution of pure (*S*)-alanine; an equal mixture of (*R*)- and (*S*)-alanine; a solution that contains 75% (*S*)- and 25% (*R*)-alanine.

Problem 5.29 A natural product was isolated in the laboratory, and its observed rotation was +10° when measured in a 1 dm sample tube containing 1.0 g of compound in 10 mL of water. What is the specific rotation of this compound?

5.12D Enantiomeric Excess

Sometimes in the laboratory we have neither a pure enantiomer nor a racemic mixture, but rather a mixture of two enantiomers in which one enantiomer is present in excess of the other. The **enantiomeric excess (*ee*),** also called the **optical purity,** tells how much more there is of one enantiomer.

- **Enantiomeric excess = *ee* = % of one enantiomer – % of the other enantiomer.**

Enantiomeric excess tells how much one enantiomer is present in excess of the racemic mixture. For example, if a mixture contains 75% of one enantiomer and 25% of the other, the enantiomeric excess is 75% – 25% = 50%. There is a 50% excess of one enantiomer over the racemic mixture.

Problem 5.30 What is the *ee* for each of the following mixtures of enantiomers **A** and **B?**

a. 95% **A** and 5% **B** b. 85% **A** and 15% **B**

Knowing the *ee* of a mixture makes it possible to calculate the amount of each enantiomer present, as shown in Sample Problem 5.4.

Sample Problem 5.4 If the enantiomeric excess is 95%, how much of each enantiomer is present?

Solution

Label the two enantiomers **A** and **B** and assume that **A** is in excess. A 95% *ee* means that the solution contains an excess of 95% of **A,** and 5% of the racemic mixture of **A** and **B.** Because a racemic mixture is an equal amount of both enantiomers, it has 2.5% of **A** and 2.5% of **B.**

- Total amount of **A** = 95% + 2.5% = 97.5%
- Total amount of **B** = 2.5% (or 100% – 97.5%)

Problem 5.31 For the given *ee* values, calculate the percentage of each enantiomer present.

a. 90% *ee* b. 99% *ee* c. 60% *ee*

The enantiomeric excess can also be calculated if two quantities are known—the specific rotation [α] of a mixture and the specific rotation [α] of a pure enantiomer.

$$ee = \frac{[\alpha]\text{ mixture}}{[\alpha]\text{ pure enantiomer}} \times 100\%$$

Sample Problem 5.5 Pure cholesterol has a specific rotation of –32. A sample of cholesterol prepared in the lab had a specific rotation of –16. What is the enantiomeric excess of this sample of cholesterol?

Solution

Calculate the *ee* of the mixture using the given formula.

$$ee = \frac{[\alpha]\text{ mixture}}{[\alpha]\text{ pure enantiomer}} \times 100\% = \frac{-16}{-32} \times 100\% = 50\%\ ee$$

Answer

Problem 5.32 Pure MSG, a common flavor enhancer, exhibits a specific rotation of +24. (a) Calculate the *ee* of a solution whose [α] is +10. (b) If the *ee* of a solution of MSG is 80%, what is [α] for this solution?

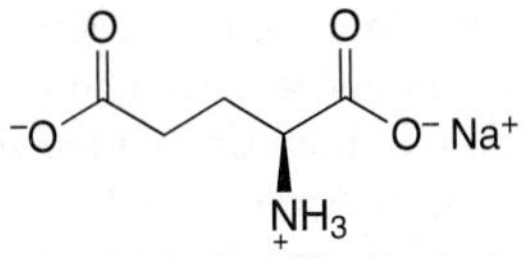

MSG
monosodium glutamate

Problem 5.33 (*S*)-Lactic acid has a specific rotation of +3.8. (a) If the *ee* of a solution of lactic acid is 60%, what is [α] for this solution? (b) How much of the dextrorotatory and levorotatory isomers does the solution contain?

5.12E The Physical Properties of Diastereomers

Diastereomers are not mirror images of each other, and as such, **their physical properties are different, including optical rotation.** Figure 5.12 compares the physical properties of the three stereoisomers of tartaric acid, consisting of a meso compound that is a diastereomer of a pair of enantiomers.

Figure 5.12 The physical properties of the three stereoisomers of tartaric acid

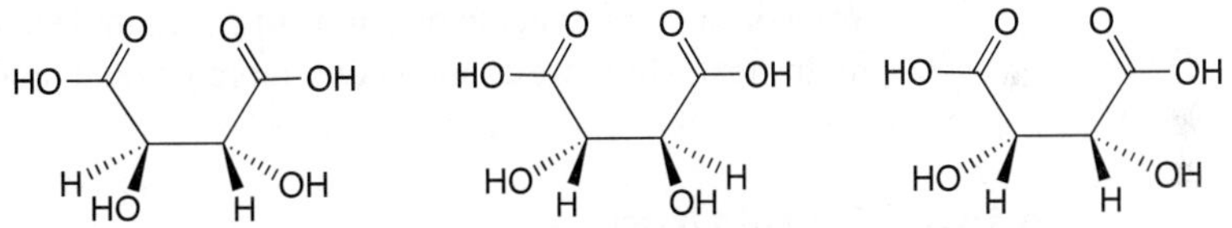

A B C

- **A** and **B** are enantiomers.
- **A** and **B** are diastereomers of **C**.

Property	A	B	C	A + B (1:1)
melting point (°C)	171	171	146	206
solubility (g/100 mL H_2O)	139	139	125	139
[α]	+13	–13	0	0
R,S designation	*R,R*	*S,S*	*R,S*	—
d,l designation	*d*	*l*	none	*d,l*

- The physical properties of **A** and **B** differ from their diastereomer **C**.
- The physical properties of a racemic mixture of **A** and **B** (last column) can also differ from either enantiomer and diastereomer **C**.
- **C** is an achiral meso compound, so it is optically inactive; [α] = 0.

Whether the physical properties of a set of compounds are the same or different has practical applications in the lab. Physical properties characterize a compound's physical state, and two compounds can usually be separated only if their physical properties are different.

Two enantiomers can be separated by the process of **resolution,** as described in Section 26.3.

- **Because two enantiomers have identical physical properties, they cannot be separated by common physical techniques like distillation.**
- **Diastereomers and constitutional isomers have different physical properties, and therefore they can be separated by common physical techniques.**

Problem 5.34 Compare the physical properties of the three stereoisomers of 1,3-dimethylcyclopentane.

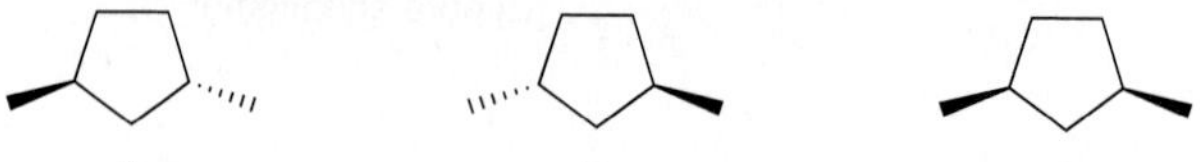

three stereoisomers of 1,3-dimethylcyclopentane

a. How do the boiling points of **A** and **B** compare? What about those of **A** and **C?**
b. Characterize a solution of each of the following as optically active or optically inactive: pure **A;** pure **B;** pure **C;** an equal mixture of **A** and **B;** an equal mixture of **A** and **C.**
c. A reaction forms a 1:1:1 mixture of **A, B,** and **C.** If this mixture is distilled, how many fractions would be obtained? Which fractions would be optically active and which would be optically inactive?

5.13 Chemical Properties of Enantiomers

When two enantiomers react with an achiral reagent, they react at the same rate, but when they react with a chiral, non-racemic reagent, they react at different rates.

- **Two enantiomers have exactly the same chemical properties except for their reaction with chiral, non-racemic reagents.**

For an everyday analogy, consider what happens when you are handed an achiral object like a pen and a chiral object like a right-handed glove. Your left and right hands are enantiomers, but they can both hold the achiral pen in the same way. With the glove, however, only your right hand can fit inside it, not your left.

We will examine specific reactions of chiral molecules with both chiral and achiral reagents later in this text. Here, we examine two more general applications.

5.13A Chiral Drugs

A living organism is a sea of chiral molecules. Many drugs are chiral, and often they must interact with a chiral receptor or a chiral enzyme to be effective. One enantiomer of a drug may treat a disease whereas its mirror image may be ineffective. Alternatively, one enantiomer may trigger one biochemical response and its mirror image may elicit a totally different response.

Although (*R*)-ibuprofen shows no anti-inflammatory activity itself, it is slowly converted to the *S* enantiomer in vivo.

The drugs ibuprofen and fluoxetine each contain one stereogenic center, and thus exist as a pair of enantiomers, only one of which exhibits biological activity. **(*S*)-Ibuprofen** is the active component of the anti-inflammatory agents Motrin and Advil, and **(*R*)-fluoxetine** is the active component in the antidepressant Prozac.

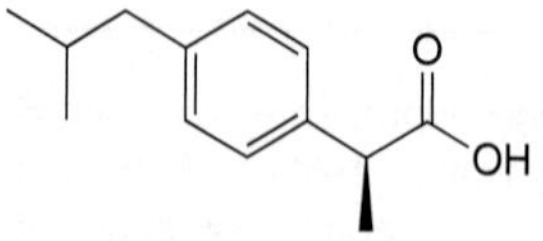

(*S*)-ibuprofen
anti-inflammatory agent

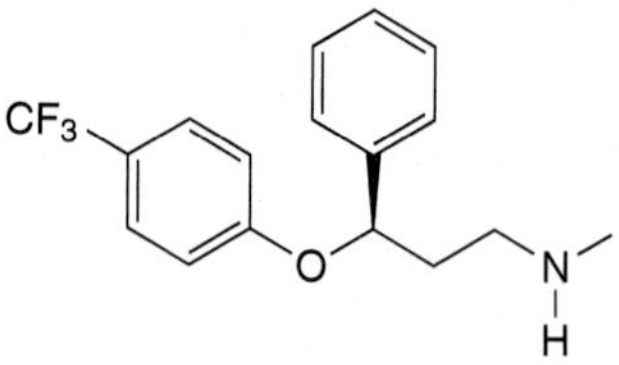

(*R*)-fluoxetine
antidepressant

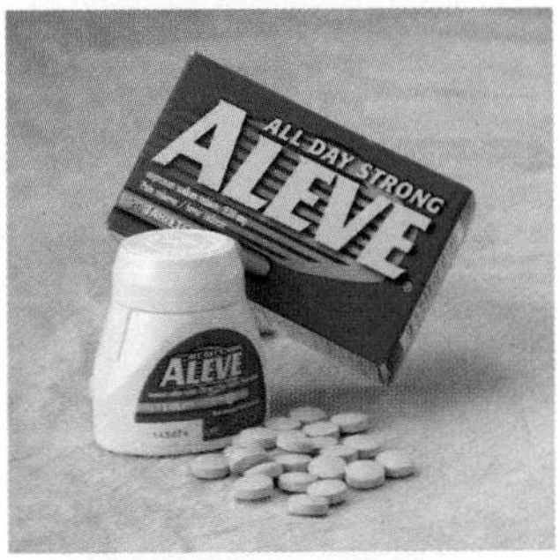

(*S*)-Naproxen is the active ingredient in the widely used pain relievers Naprosyn and Aleve.

For more examples of two enantiomers that exhibit very different biochemical properties, see *Journal of Chemical Education*, **1996,** *73,* 481–484.

Changing the orientation of two substituents to form a mirror image can also alter biological activity to produce an undesirable side effect in the other enantiomer. The *S* enantiomer of **naproxen** is an active anti-inflammatory agent, but the *R* enantiomer is a harmful liver toxin.

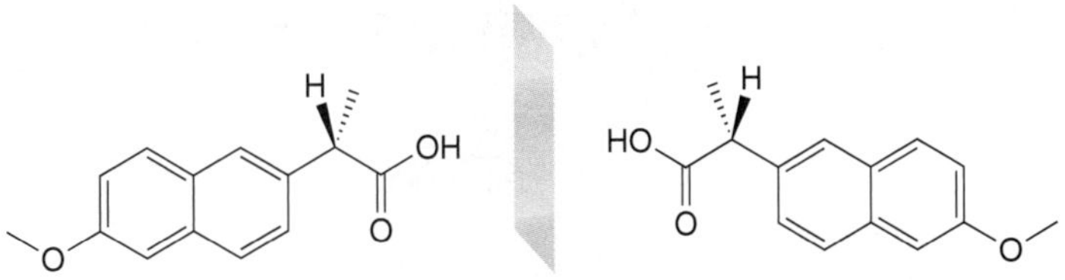

(*S*)-naproxen anti-inflammatory agent

(*R*)-naproxen liver toxin

If a chiral drug could be sold as a single active enantiomer, it should be possible to use smaller doses with fewer side effects. Many chiral drugs continue to be sold as racemic mixtures, however, because it is more difficult and therefore more costly to obtain a single enantiomer. An enantiomer is not easily separated from a racemic mixture because the two enantiomers have the same physical properties. In Chapter 12 we will study a reaction that can form a single active enantiomer, an important development in making chiral drugs more readily available.

5.13B Enantiomers and the Sense of Smell

Research suggests that the odor of a particular molecule is determined more by its shape than by the presence of a particular functional group. For example, hexachloroethane (Cl_3CCCl_3) and cyclooctane have no obvious structural similarities, but they both have a camphor-like odor, a fact attributed to their similar spherical shape. Each molecule binds to spherically shaped olfactory receptors present on the nerve endings in the nasal passage, resulting in similar odors (Figure 5.13).

Figure 5.13
The shape of molecules and the sense of smell

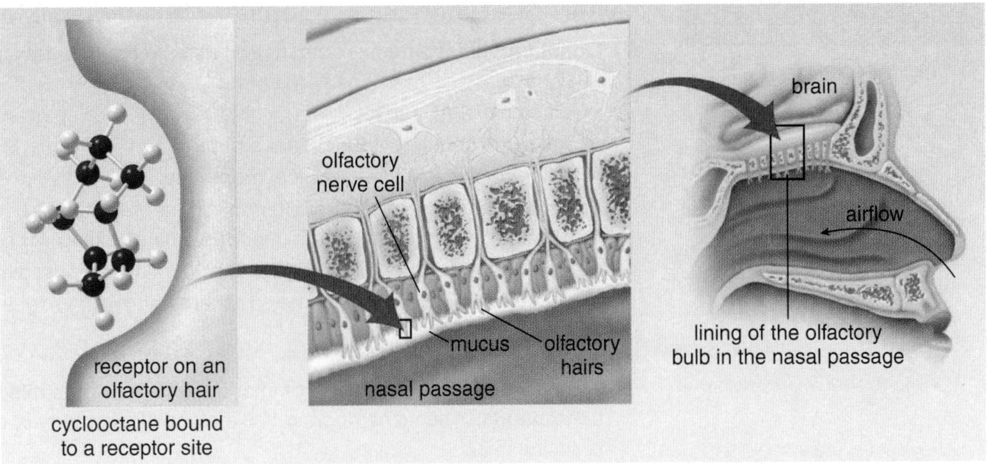

- Cyclooctane and other molecules similar in shape bind to a particular olfactory receptor on the nerve cells that lie at the top of the nasal passage. Binding results in a nerve impulse that travels to the brain, which interprets impulses from particular receptors as specific odors.

Because enantiomers interact with chiral smell receptors, some enantiomers have different odors. There are a few well-characterized examples of this phenomenon in nature. For example,

(*S*)-carvone is responsible for the odor of caraway, whereas (*R*)-carvone is responsible for the odor of spearmint.

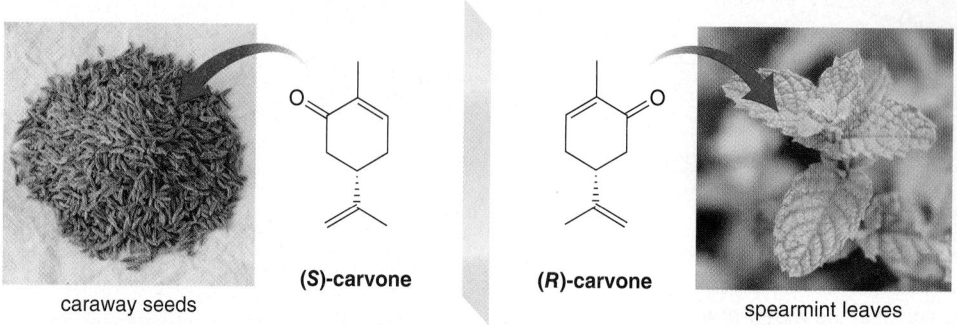
(*S*)-carvone
(*R*)-carvone
caraway seeds
spearmint leaves

These examples demonstrate that understanding the three-dimensional structure of a molecule is very important in organic chemistry.

Problem 5.35 Like carvone, the two enantiomers of celery ketone smell different. The *R* enantiomer smells like celery leaves, whereas the *S* enantiomer smells like licorice. Draw each enantiomer and assign its odor.

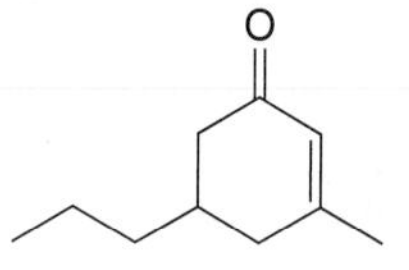

celery ketone

(*R*)-Celery ketone (Problem 5.35) has an odor reminiscent of celery leaves.

KEY CONCEPTS

Stereochemistry

Isomers Are Different Compounds with the Same Molecular Formula (5.2, 5.11).

[1] **Constitutional isomers**—isomers that differ in the way the atoms are connected to each other. They have:
- different IUPAC names
- the same or different functional groups
- different physical and chemical properties

[2] **Stereoisomers**—isomers that differ only in the way atoms are oriented in space**.** They have the same functional group and the same IUPAC name except for prefixes such as cis, trans, *R,* and *S*.
- **Enantiomers**—stereoisomers that are nonsuperimposable mirror images of each other (5.4).
- **Diastereomers**—stereoisomers that are not mirror images of each other (5.7).

Some Basic Principles

- When a compound and its mirror image are **superimposable,** they are **identical achiral compounds.** When a compound has a plane of symmetry in one conformation, the compound is achiral (5.3).
- When a compound and its mirror image are **not superimposable,** they are **different chiral compounds** called **enantiomers.** A chiral compound has no plane of symmetry in any conformation (5.3).
- A **tetrahedral stereogenic center** is a carbon atom bonded to four different groups (5.4, 5.5).
- For ***n*** **stereogenic centers,** the maximum number of stereoisomers is $\mathbf{2^n}$ (5.7).

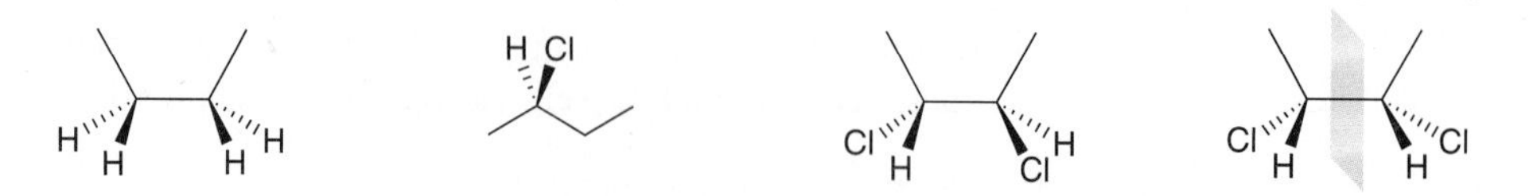
plane of symmetry

no stereogenic centers achiral	one stereogenic center chiral	two stereogenic centers chiral	two stereogenic centers achiral

Optical Activity Is the Ability of a Compound to Rotate Plane-Polarized Light (5.12).

- An optically active solution contains a chiral compound.
- An optically inactive solution contains one of the following:
  - an achiral compound with no stereogenic centers
  - a meso compound—an achiral compound with two or more stereogenic centers
  - a racemic mixture—an equal amount of two enantiomers

The Prefixes *R* and *S* Compared with *d* and *l*

The prefixes *R* and *S* are labels used in nomenclature. Rules on assigning *R,S* are found in Section 5.6.

- An enantiomer has every stereogenic center opposite in configuration. If a compound with two stereogenic centers has the *R,R* configuration, its enantiomer has the *S,S* configuration.
- A diastereomer of this same compound has either the *R,S* or *S,R* configuration; one stereogenic center has the same configuration and one is opposite.

The prefixes *d* (or +) and *l* (or –) tell the direction a compound rotates plane-polarized light (5.12).

- Dextrorotatory (*d* or +) compounds rotate polarized light clockwise.
- Levorotatory (*l* or –) compounds rotate polarized light counterclockwise.
- There is no relation between whether a compound is *R* or *S* and whether it is *d* or *l*.

The Physical Properties of Isomers Compared (5.12)

Type of isomer	Physical properties
Constitutional isomers	Different
Enantiomers	Identical except for the direction polarized light is rotated
Diastereomers	Different
Racemic mixture	Possibly different from either enantiomer

Equations

- Specific rotation (5.12C):

$$\text{specific rotation} = [\alpha] = \frac{\alpha}{l \times c}$$

α = observed rotation (°)
l = length of sample tube (dm)
c = concentration (g/mL)

[dm = decimeter
1 dm = 10 cm]

- Enantiomeric excess (5.12D):

$$ee = \%\text{ of one enantiomer} - \%\text{ of the other enantiomer} = \frac{[\alpha]\text{ mixture}}{[\alpha]\text{ pure enantiomer}} \times 100\%$$

PROBLEMS

Problems Using Three-Dimensional Models

5.36 (a) Locate the stereogenic centers in the ball-and-stick model of ezetimibe (trade name Zetia), a cholesterol-lowering drug. (b) Label each stereogenic center as *R* or *S*.

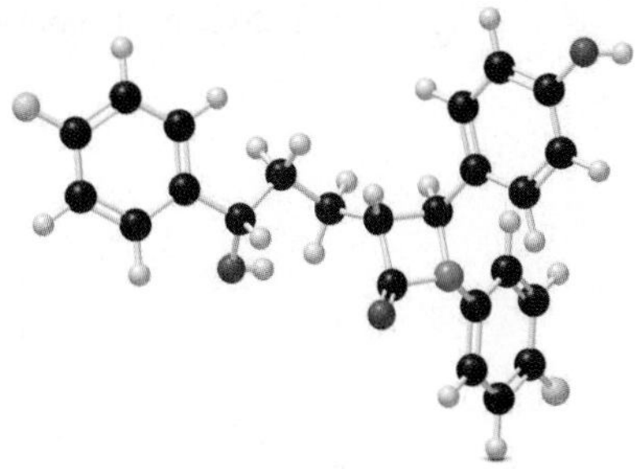

ezetimibe

5.37 Consider the ball-and-stick models **A–D.** How is each pair of compounds related: (a) **A** and **B;** (b) **A** and **C;** (c) **A** and **D;** (d) **C** and **D?** Choose from identical molecules, enantiomers, or diastereomers.

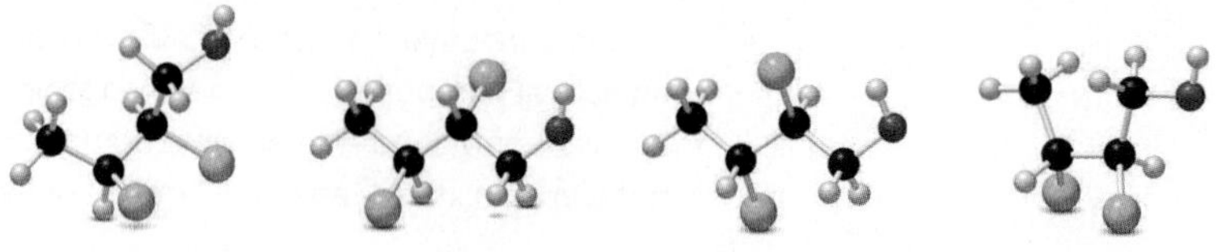

A **B** **C** **D**

Constitutional Isomers Versus Stereoisomers

5.38 Label each pair of compounds as constitutional isomers, stereoisomers, or not isomers of each other.

a. 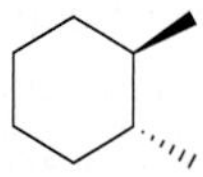 and 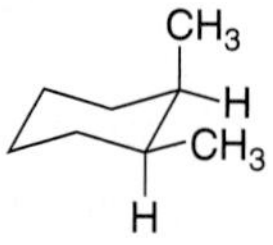

c. 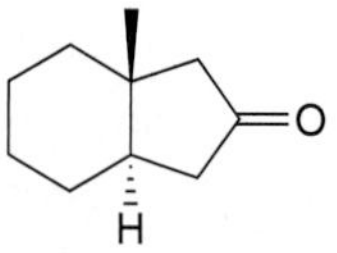 and 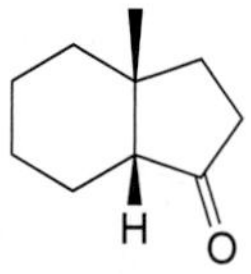

b. 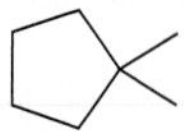 and 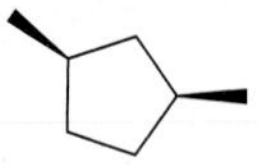

d. 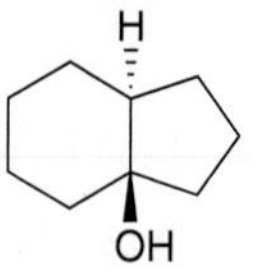 and 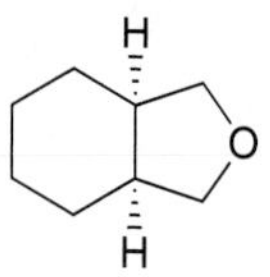

Mirror Images and Chirality

5.39 Label each compound as chiral or achiral.

a. 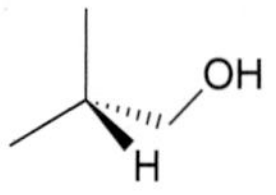 b. 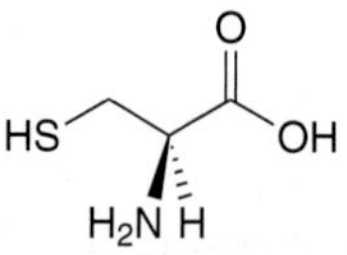 c. 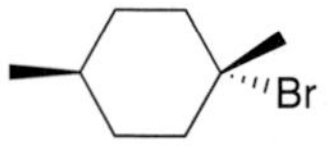 d. 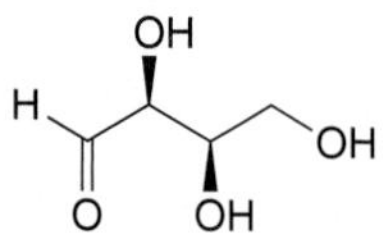

5.40 Determine if each compound is identical to or an enantiomer of **A.**

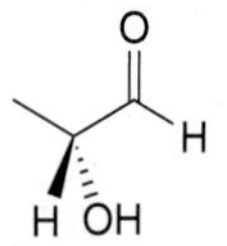

A

a. 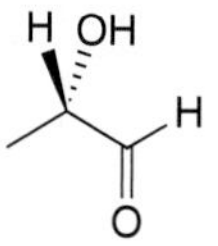 b. 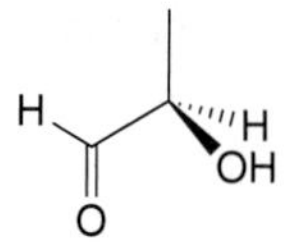 c. 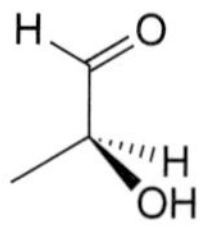

5.41 Indicate a plane of symmetry for each molecule that contains one. A molecule may require rotation around a carbon–carbon bond to see the plane of symmetry.

a. 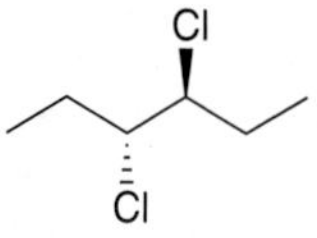 b. 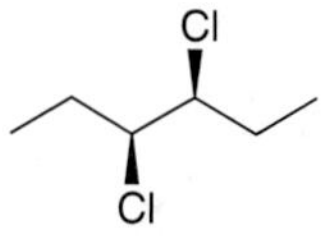 c. 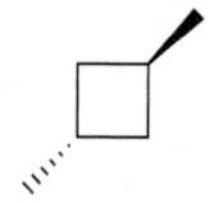 d. 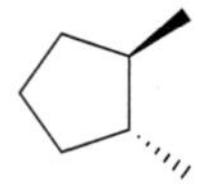

Finding and Drawing Stereogenic Centers

5.42 Locate the tetrahedral stereogenic center(s) in each compound. A molecule may have one or more stereogenic centers.

a. 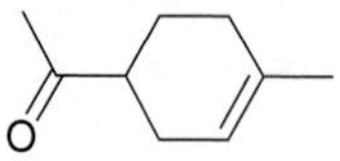 c. 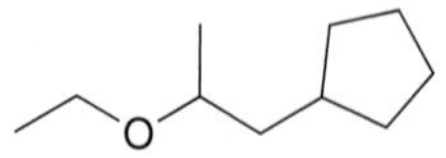 e. 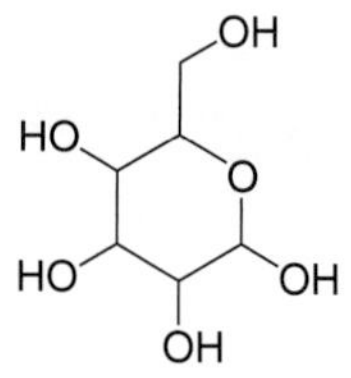

b. 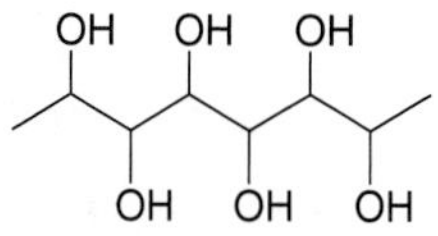 d. 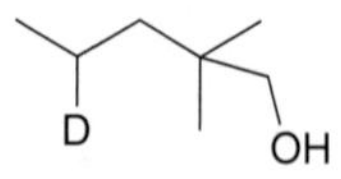 f. 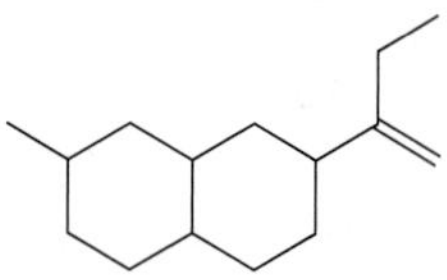

5.43 Locate the stereogenic centers in each drug.

a. 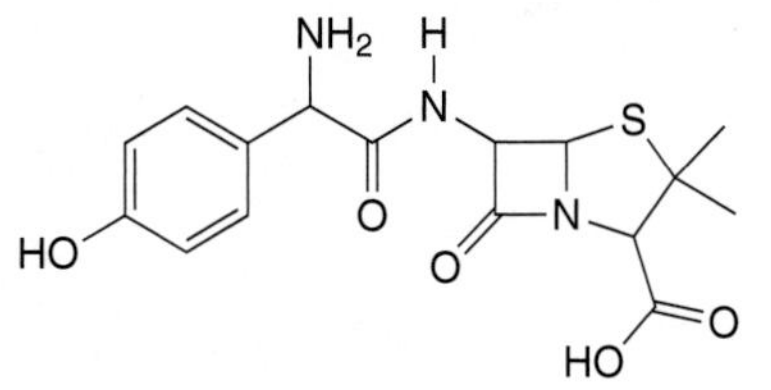

amoxicillin
(an antibiotic)

b. 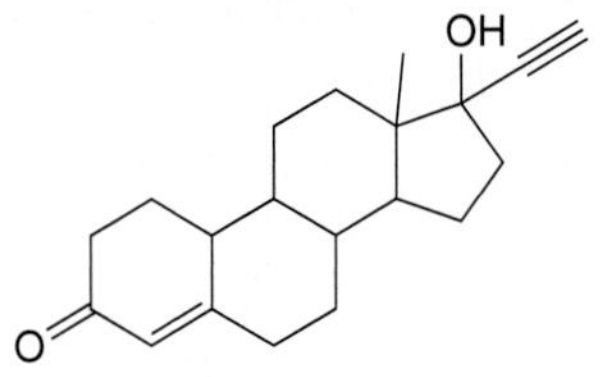

norethindrone
(oral contraceptive component)

c. 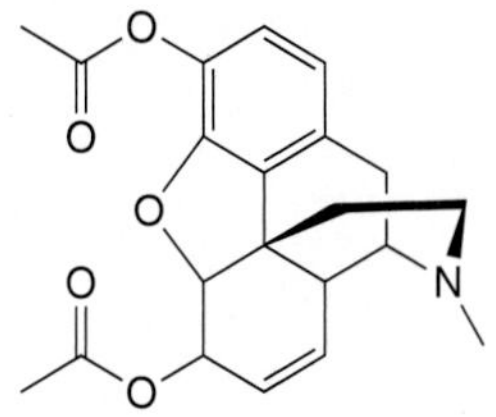

heroin
(an opiate)

5.44 Draw both enantiomers for each biologically active compound.

a. 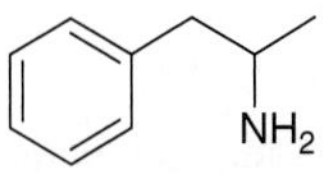

amphetamine
(a powerful central nervous stimulant)

b. 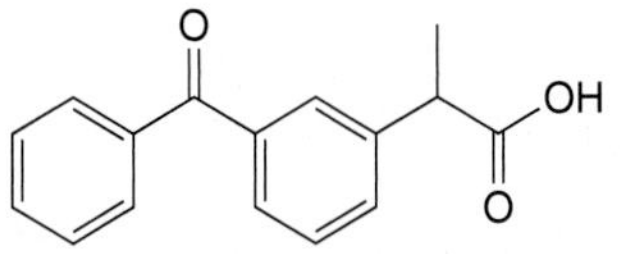

ketoprofen
(analgesic and anti-inflammatory agent)

Nomenclature

5.45 Which group in each pair is assigned the higher priority in *R,S* nomenclature?

a. $-CD_3$, $-CH_3$

b. $-CH(CH_3)_2$, $-CH_2OH$

c. $-CH_2Cl$, $-CH_2CH_2CH_2Br$

d. $-CH_2NH_2$, $-NHCH_3$

5.46 Rank the following groups in order of decreasing priority.

a. $-F$, $-NH_2$, $-CH_3$, $-OH$

b. $-CH_3$, $-CH_2CH_3$, $-CH_2CH_2CH_3$, $-(CH_2)_3CH_3$

c. $-NH_2$, $-CH_2NH_2$, $-CH_3$, $-CH_2NHCH_3$

d. $-COOH$, $-CH_2OH$, $-H$, $-CHO$

e. $-Cl$, $-CH_3$, $-SH$, $-OH$

f. $-C{\equiv}CH$, $-CH(CH_3)_2$, $-CH_2CH_3$, $-CH{=}CH_2$

5.47 Label each stereogenic center as *R* or *S*.

a. 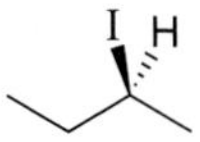

b. 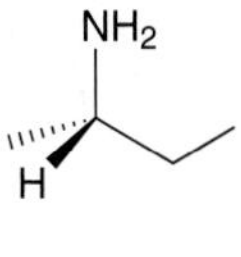

c. 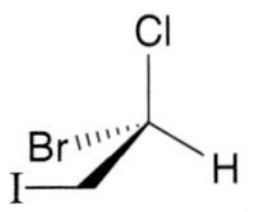

d. 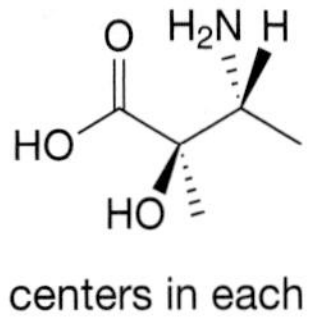

e. 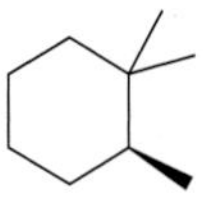

f. 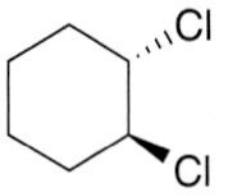

g. 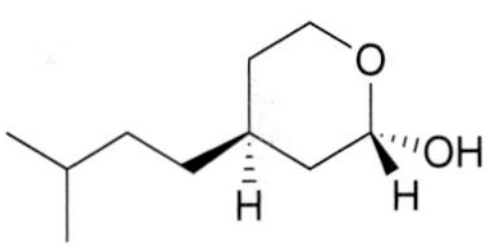

h. 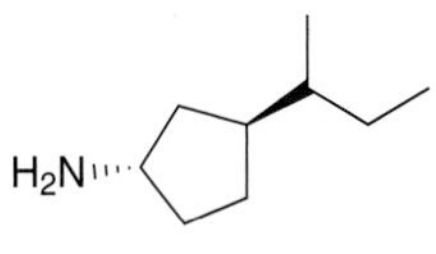

5.48 Locate the stereogenic centers in each Newman projection and label each center as *R* or *S*.

a. 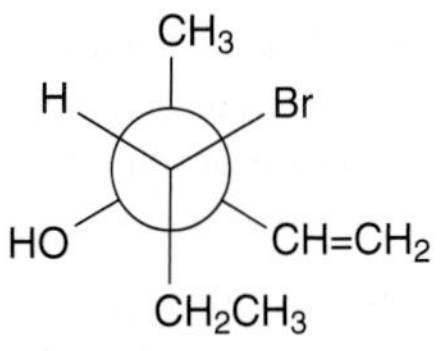

b. 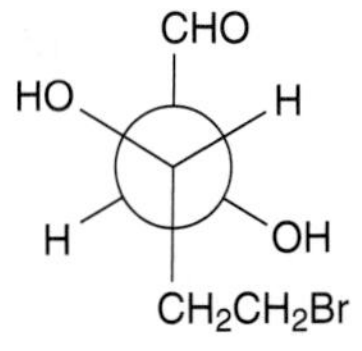

5.49 Draw the structure of (*S,S*)-ethambutol, a drug used to treat tuberculosis that is 10 times more potent than any of its other stereoisomers.

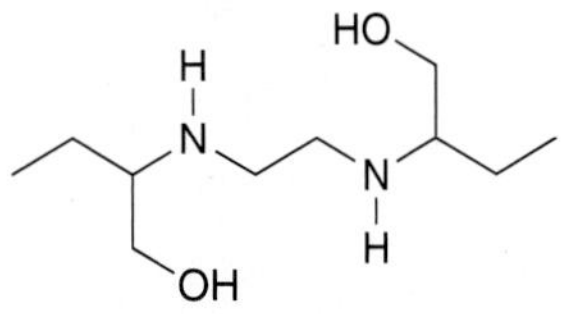

ethambutol

5.50 Draw the structure of (S)-citalopram, a drug used to treat depression and anxiety that is much more potent than its *R* enantiomer.

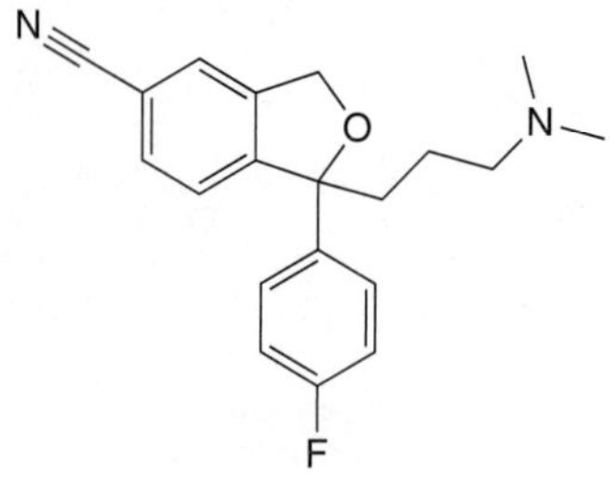

citalopram

5.51 Draw the structure for each compound.

a. (*R*)-3-methylhexane
b. (4*R*,5*S*)-4,5-diethyloctane
c. (3*R*,5*S*,6*R*)-5-ethyl-3,6-dimethylnonane
d. (3*S*,6*S*)-6-isopropyl-3-methyldecane

5.52 Give the IUPAC name for each compound, including the *R*,*S* designation for each stereogenic center.

a. 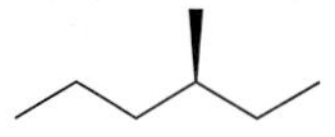

b. 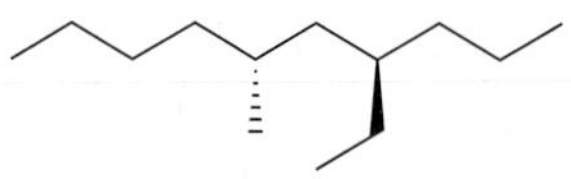

c. 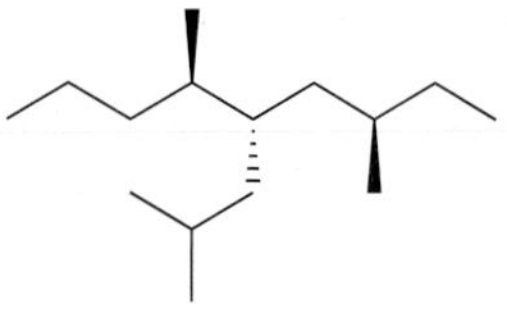

5.53 Label the stereogenic centers in paclitaxel, the chapter-opening anticancer drug, as *R* or *S*.

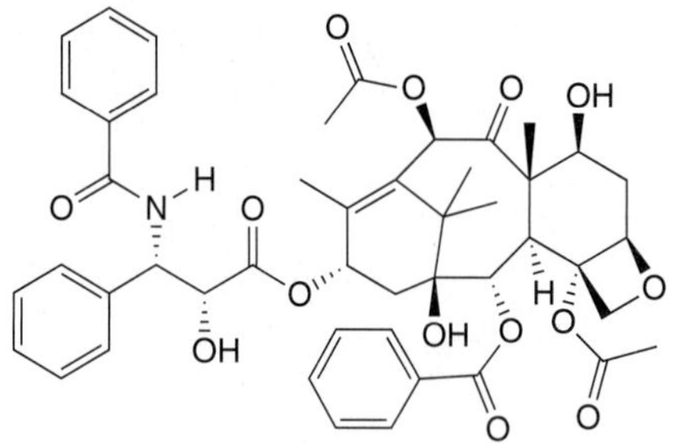

paclitaxel

Compounds with More Than One Stereogenic Center

5.54 What is the maximum number of stereoisomers possible for each compound?

a. 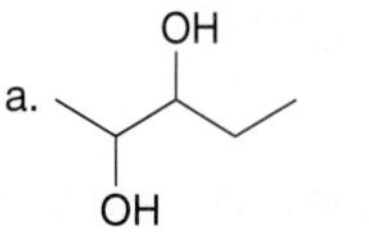

b. 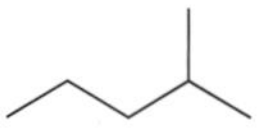

c. 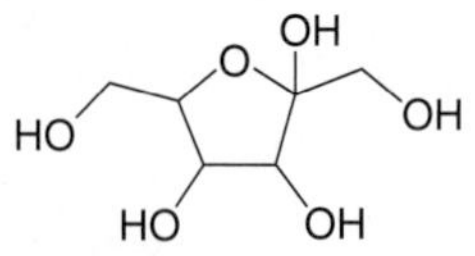

5.55 The shrub ma huang (Section 5.4A) contains two biologically active stereoisomers—ephedrine and pseudoephedrine—with two stereogenic centers as shown in the given structure. Ephedrine is one component of a once popular combination drug used by body builders to increase energy and alertness, while pseudoephedrine is a nasal decongestant.

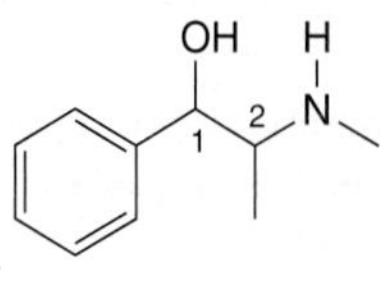

isolated from ma huang

a. Draw the structure of naturally occurring (–)-ephedrine, which has the 1*R*,2*S* configuration.
b. Draw the structure of naturally occurring (+)-pseudoephedrine, which has the 1*S*,2*S* configuration.
c. How are ephedrine and pseudoephedrine related?
d. Draw all other stereoisomers of (–)-ephedrine and (+)-pseudoephedrine, and give the *R*,*S* designation for all stereogenic centers.
e. How is each compound drawn in part (d) related to (–)-ephedrine?

5.56 Draw all possible stereoisomers for each compound. Label pairs of enantiomers and diastereomers. Label any meso compound.

a. 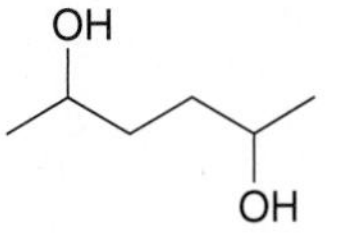 b. 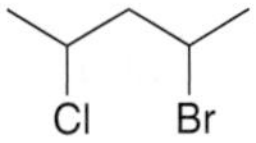 c. 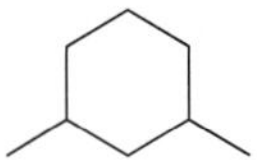 d. 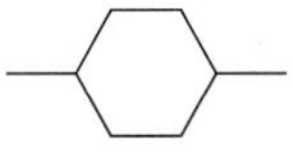

5.57 Draw the enantiomer and a diastereomer for each compound.

a. 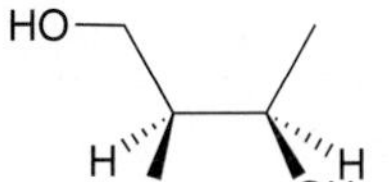 b. 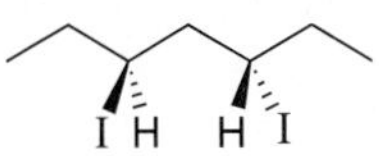 c. 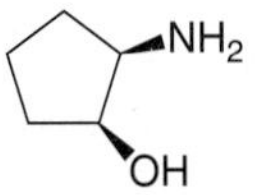

5.58 Draw all possible constitutional isomers and stereoisomers for a compound of molecular formula C_6H_{12} having a cyclobutane ring and two methyl groups as substituents. Label each compound as chiral or achiral.

5.59 Explain each statement by referring to compounds **A–E**.

A

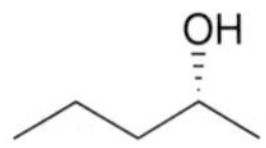
B

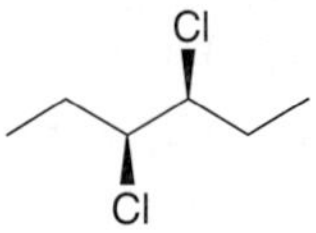
C

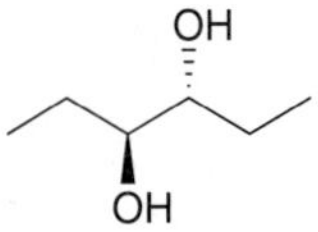
D

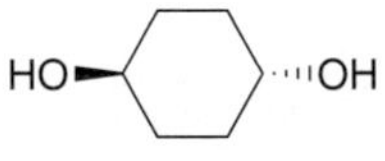
E

a. **A** has a mirror image but no enantiomer.
b. **B** has an enantiomer and no diastereomer.
c. **C** has both an enantiomer and a diastereomer.
d. **D** has a diastereomer but no enantiomer.
e. **E** has a diastereomer but no enantiomer.

Comparing Compounds: Enantiomers, Diastereomers, and Constitutional Isomers

5.60 How is each compound related to the simple sugar D-erythrose? Is it an enantiomer, diastereomer, or identical?

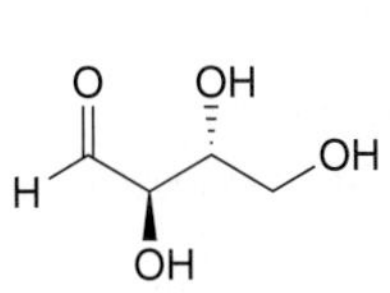
D-erythrose

a. 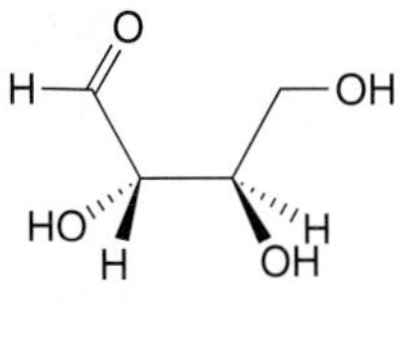 b. 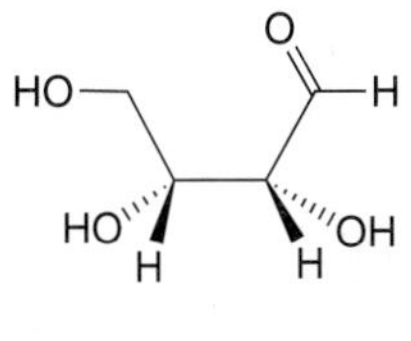 c. 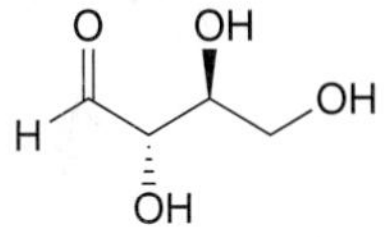

5.61 Consider Newman projections (**A–D**) for four-carbon carbohydrates. How is each pair of compounds related: (a) **A** and **B;** (b) **A** and **C;** (c) **A** and **D;** (d) **C** and **D?** Choose from identical molecules, enantiomers, or diastereomers.

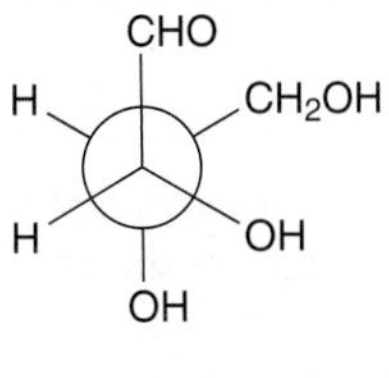
A

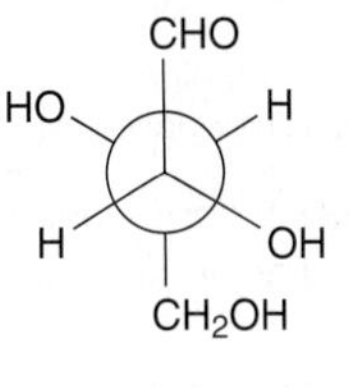
B

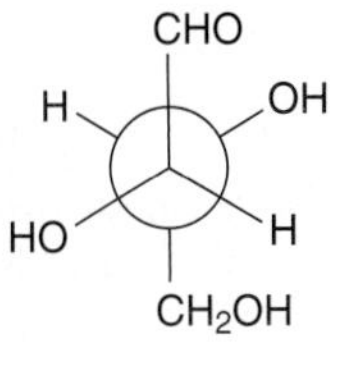
C

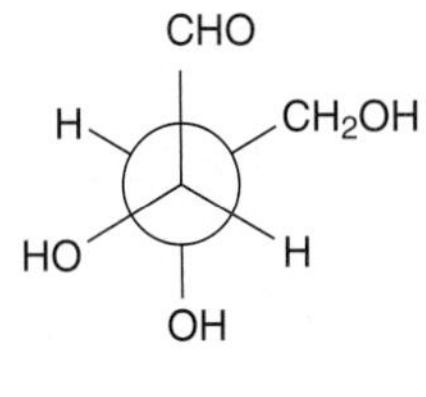
D

5.62 How is compound **A** related to compounds **B–E?** Choose from enantiomers, diastereomers, constitutional isomers, or identical molecules.

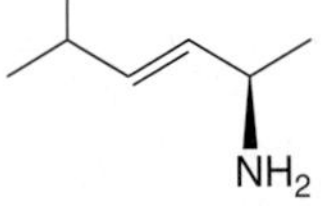
A

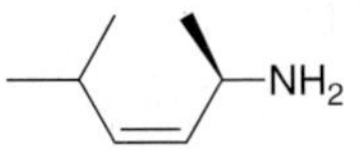
B

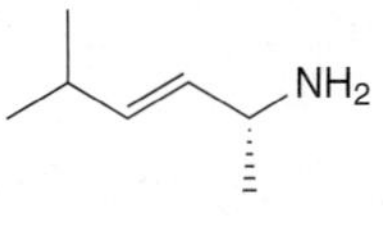
C

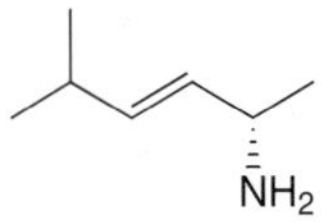
D

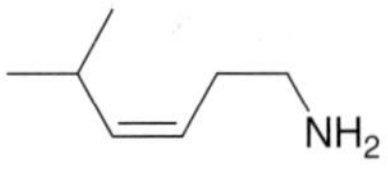
E

5.63 How are the compounds in each pair related to each other? Are they identical, enantiomers, diastereomers, constitutional isomers, or not isomers of each other?

a. 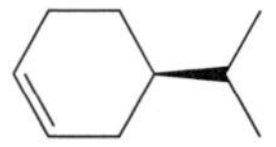 and 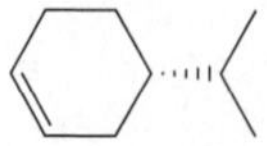

d. 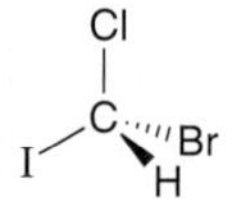 and 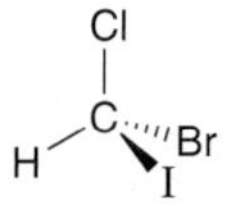

b. 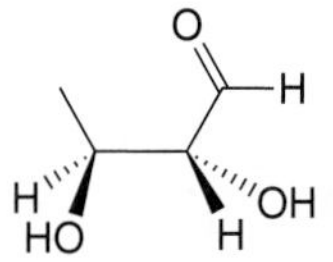 and 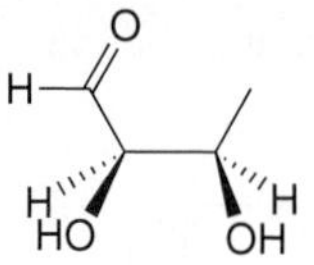

e. 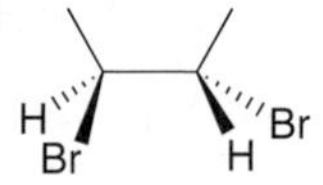 and 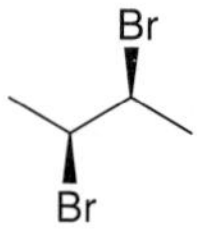

c. 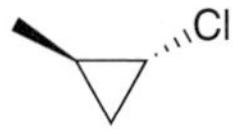 and 

f. 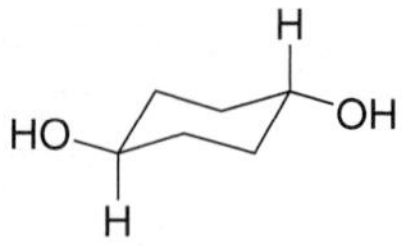 and 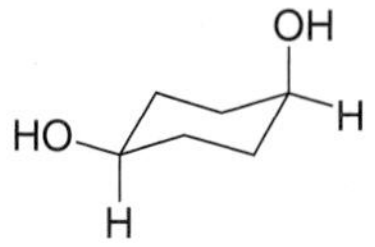

Physical Properties of Isomers

5.64 Drawn are four isomeric dimethylcyclopropanes.

A **B** **C** **D**

a. How are the compounds in each pair related (enantiomers, diastereomers, constitutional isomers): **A** and **B; A** and **C; B** and **C; C** and **D?**
b. Label each compound as chiral or achiral.
c. Which compounds, alone, would be optically active?
d. Which compounds have a plane of symmetry?
e. How do the boiling points of the compounds in each pair compare: **A** and **B; B** and **C; C** and **D?**
f. Which of the compounds are meso compounds?
g. Would an equal mixture of compounds **C** and **D** be optically active? What about an equal mixture of **B** and **C?**

5.65 The [α] of pure quinine, an antimalarial drug, is –165.

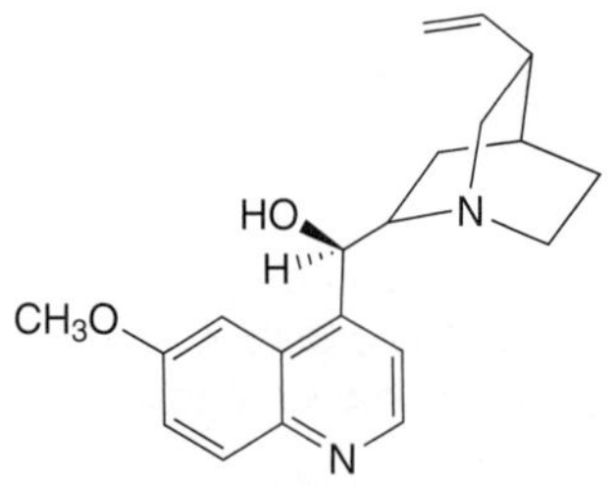

quinine
(antimalarial drug)

a. Calculate the *ee* of a solution with the following [α] values: –50, –83, and –120.
b. For each *ee,* calculate the percent of each enantiomer present.
c. What is [α] for the enantiomer of quinine?
d. If a solution contains 80% quinine and 20% of its enantiomer, what is the *ee* of the solution?
e. What is [α] for the solution described in part (d)?

5.66 Amygdalin, a compound isolated from the pits of apricots, peaches, and wild cherries, has been used as an unsanctioned anticancer drug both within and outside of the United States. One hydrolysis product formed from amygdalin is mandelic acid, used in treating common skin problems caused by photo-aging and acne.

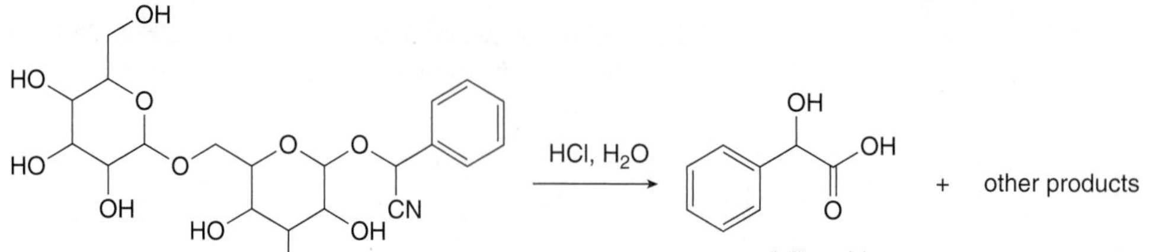
HCl, H2O
+ other products

amygdalin

mandelic acid

a. How many stereogenic centers are present in amygdalin? What is the maximum number of stereoisomers possible?
b. Draw both enantiomers of mandelic acid, and label each stereogenic center as *R* or *S*.
c. Pure (*R*)-mandelic acid has a specific rotation of –154. If a sample contains 60% of the *R* isomer and 40% of its enantiomer, what is [α] of this solution?
d. Calculate the *ee* of a solution of mandelic acid having [α] = +50. What is the percentage of each enantiomer present?

General Problems

5.67 Artemisinin and mefloquine are widely used antimalarial drugs. A ball-and-stick model of artemisinin appears on the cover of this text.

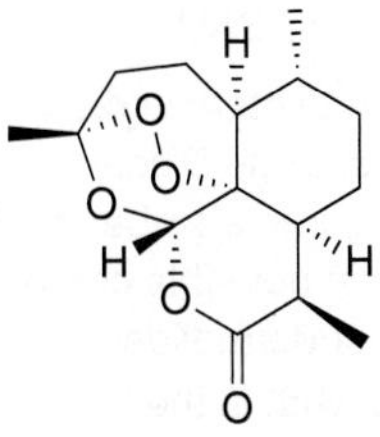

artemisinin

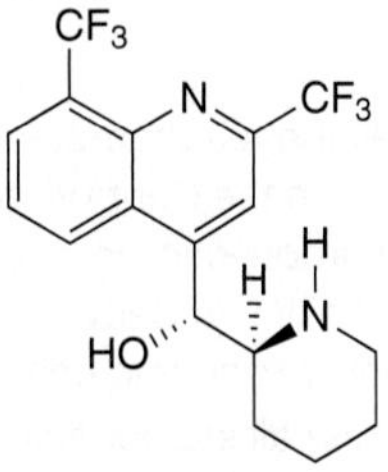

mefloquine

a. Locate the stereogenic centers in both drugs.
b. Label each stereogenic center in mefloquine as *R* or *S*.
c. What is the maximum number of stereoisomers possible for artemisinin?
d. How are the N atoms in mefloquine hybridized?
e. Can two molecules of artemisinin intermolecularly hydrogen bond to each other?
f. What product is formed when mefloquine is treated with HCl?

5.68 Saquinavir (trade name Invirase) is a protease inhibitor, used to treat HIV (human immunodeficiency virus).

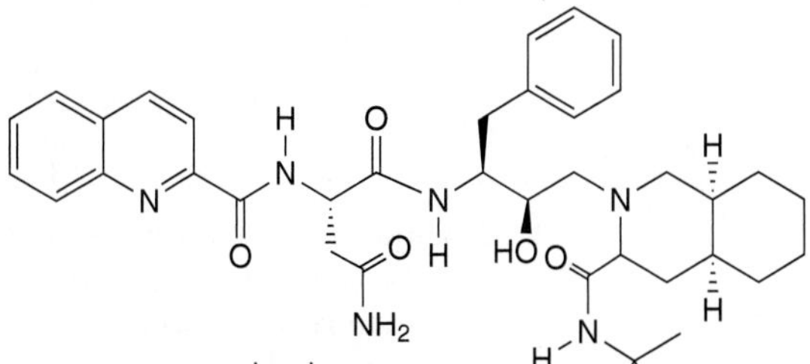

saquinavir
Trade name Invirase

a. Locate all stereogenic centers in saquinavir, and label each stereogenic center as *R* or *S*.
b. Draw the enantiomer of saquinavir.
c. Draw a diastereomer of saquinavir.
d. Draw a constitutional isomer that contains at least one different functional group.

Challenge Problems

5.69 A limited number of chiral compounds having no stereogenic centers exist. For example, although **A** is achiral, constitutional isomer **B** is chiral. Make models and explain this observation. Compounds containing two double bonds that share a single carbon atom are called *allenes.* Locate the allene in the antibiotic mycomycin and decide whether mycomycin is chiral or achiral.

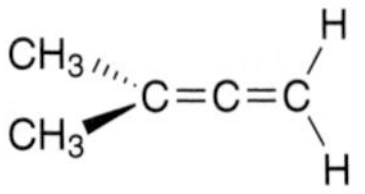

achiral
A

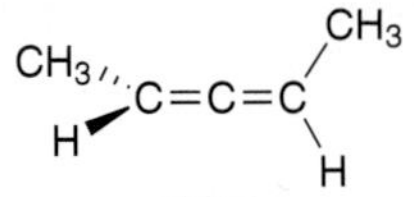

chiral
B

$HC{\equiv}C{-}C{\equiv}C{-}CH{=}C{=}CH{-}CH{=}CH{-}CH{=}CH{-}CH_2CO_2H$

mycomycin

5.70 A sulfonium ion (R_3S^+) is a stereogenic center if three different alkyl groups are bonded to sulfur because sulfur is surrounded by four different groups, including its lone pair. In assigning an *R* or *S* designation to sulfur, the lone pair is always assigned the lowest priority (4). SAM, *S*-adenosylmethionine, is a biologically active sulfonium ion that we will learn about in Section 7.16. Locate all the stereogenic centers in SAM, and assign an *R,S* designation to each center.

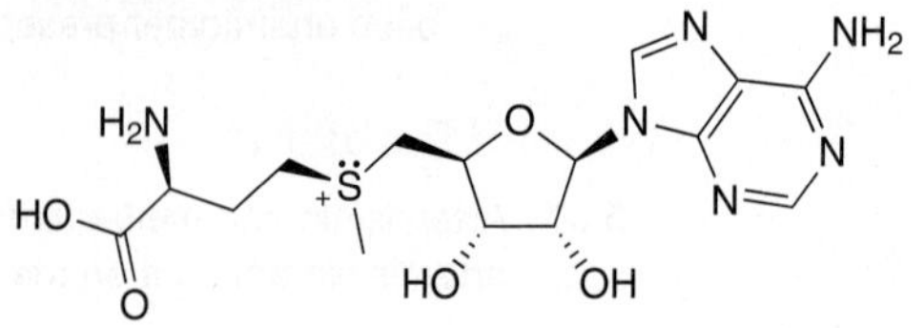

SAM

5.71 a. Locate all the tetrahedral stereogenic centers in discodermolide, a tumor inhibitor isolated from the Caribbean marine sponge *Discodermia dissoluta.*

b. Certain carbon–carbon double bonds can also be stereogenic centers. With reference to the definition in Section 5.3, explain how this can occur, and then locate the three additional stereogenic centers in discodermolide.

c. Considering all stereogenic centers, what is the maximum number of stereoisomers possible for discodermolide?

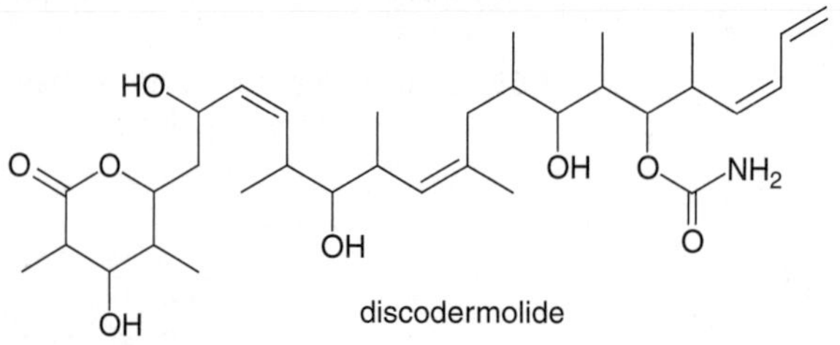

discodermolide

5.72 Label each compound as chiral or achiral. Compounds that contain a single carbon common to two rings are called spiro compounds. Because carbon is tetrahedral, the two rings are perpendicular to each other.

a. 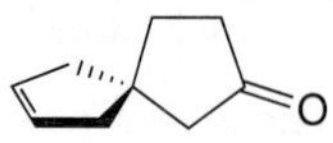 b.  c. 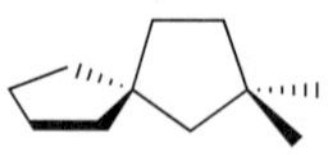 d. 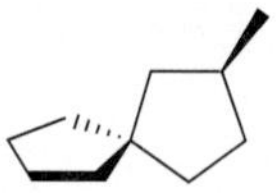

5.73 An acid–base reaction of (*R*)-*sec*-butylamine with a racemic mixture of 2-phenylpropanoic acid forms two products having different melting points and somewhat different solubilities. Draw the structure of these two products. Assign *R* and *S* to any stereogenic centers in the products. How are the two products related? Choose from enantiomers, diastereomers, constitutional isomers, or not isomers.

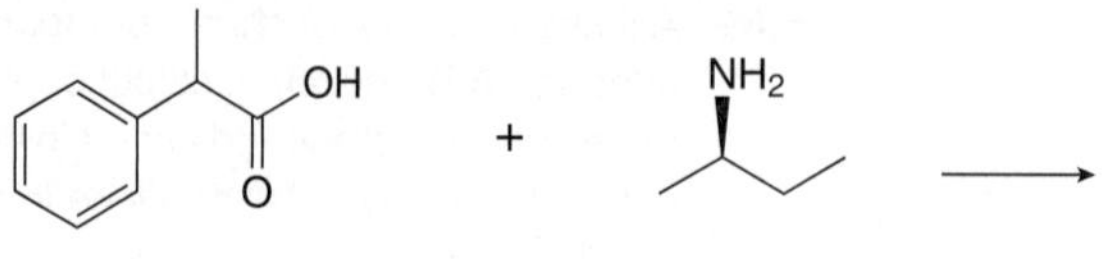

2-phenylpropanoic acid (racemic mixture) (*R*)-*sec*-butylamine

Understanding Organic Reactions 6

6.1 Writing equations for organic reactions
6.2 Kinds of organic reactions
6.3 Bond breaking and bond making
6.4 Bond dissociation energy
6.5 Thermodynamics
6.6 Enthalpy and entropy
6.7 Energy diagrams
6.8 Energy diagram for a two-step reaction mechanism
6.9 Kinetics
6.10 Catalysts
6.11 Enzymes

Glucose, the most abundant simple carbohydrate, is the building block for starch and cellulose and a major sweet-tasting component of honey. Glucose is used as an energy source by most organisms. In humans, when glucose levels are high after a meal is digested, the body stores glucose as glycogen, which is then hydrolyzed when glucose levels fall and energy demands increase. Glucose is transported in the bloodstream and metabolized aerobically to carbon dioxide and water and a great deal of energy. In Chapter 6, we learn about energy changes that accompany chemical reactions.

Why do certain reactions occur when two compounds are mixed together whereas others do not? To answer this question we must learn how and why organic compounds react.

Reactions are at the heart of organic chemistry. The mastery of chemical transformations is essential to our understanding of living organisms as well as laboratory reactions. The most fundamental biological processes, including vision and metabolism, occur because of enzyme-catalyzed organic reactions. Furthermore, our knowledge of these reactions has made possible the conversion of natural substances into new compounds with different, and sometimes superior, properties. Aspirin, ibuprofen, nylon, and polyethylene are all products of chemical reactions between substances derived from petroleum.

Reactions are difficult to learn when each reaction is considered a unique and isolated event. *Avoid this tendency.* Virtually all chemical reactions are woven together by a few basic themes. After we learn the general principles, specific reactions then fit neatly into a general pattern.

In our study of organic reactions we will begin with the functional groups, looking for electron-rich and electron-deficient sites, and bonds that might be broken easily. These reactive sites give us a clue as to the general type of reaction a particular class of compound undergoes. Finally, we will learn about how a reaction occurs. Does it occur in one step or in a series of steps? Understanding the details of an organic reaction allows us to determine when it might be used in preparing interesting and useful organic compounds.

6.1 Writing Equations for Organic Reactions

Like other reactions, equations for organic reactions are usually drawn with a single reaction arrow (→) between the starting material and product, but other conventions make these equations look different from those encountered in general chemistry.

Although chemical reactions are equilibria, which are designated by double reaction arrows (⇄), single reaction arrows (→) are often used instead.

The **reagent,** the chemical substance with which an organic compound reacts, is sometimes drawn on the left side of the equation with the other reactants. At other times, the reagent is drawn above or below the reaction arrow itself, to focus attention on the organic starting material by itself on the left side. The solvent and temperature of a reaction may be added above or below the arrow. **The symbols "*hν*" and "Δ" are used for reactions that require *light* or *heat,* respectively.** Figure 6.1 presents an organic reaction in different ways.

Often the solvent and temperature of a reaction are omitted from chemical equations, to further focus attention on the main substances involved in the reaction.

When two sequential reactions are carried out without drawing any intermediate compound, the steps are usually numbered above or below the reaction arrow. This convention signifies that the first step occurs *before* the second, and the reagents are added *in sequence,* not at the same time.

Most organic reactions take place in a **liquid solvent.** Solvents solubilize key reaction components and serve as heat reservoirs to maintain a given temperature. Chapter 7 presents the two major types of reaction solvents and how they affect substitution reactions.

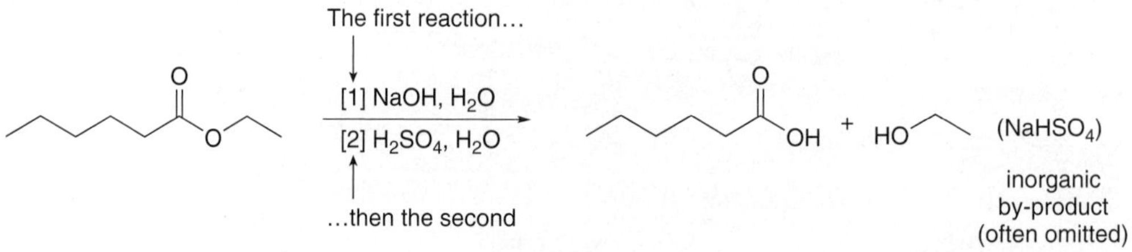
The first reaction...
[1] NaOH, H2O
[2] H2SO4, H2O
...then the second
O
O
O
OH
+
HO
(NaHSO4)
inorganic
by-product
(often omitted)

In this equation only the organic products are drawn on the right side of the arrow. Although the reagents NaOH and H_2SO_4 contain both Na and S, these elements do not appear in the organic

Figure 6.1 Different ways of writing organic reactions

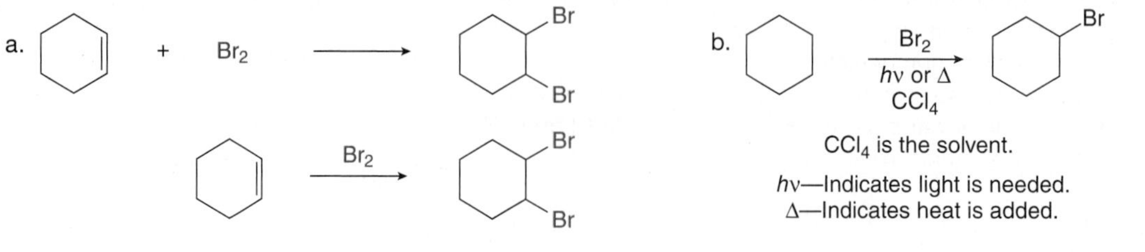
a.
+
Br2
Br
Br
Br2
Br
Br
b.
Br2
hν or Δ
CCl4
Br
CCl4 is the solvent.
hν—Indicates light is needed.
Δ—Indicates heat is added.

• The reagent (Br_2) can be on the left side or above the arrow.

• Other reaction parameters can be indicated.

products, and they are often omitted on the product side of the equation. These elements have not disappeared. They are part of an inorganic by-product ($NaHSO_4$ in this case), and are often of little interest to an organic chemist.

6.2 Kinds of Organic Reactions

Like other compounds, organic molecules undergo acid–base and oxidation–reduction reactions, as discussed in Chapters 2 and 4. Organic molecules also undergo **substitution, elimination,** and **addition** reactions.

6.2A Substitution Reactions

- ***Substitution* is a reaction in which an atom or a group of atoms is *replaced* by another atom or group of atoms.**

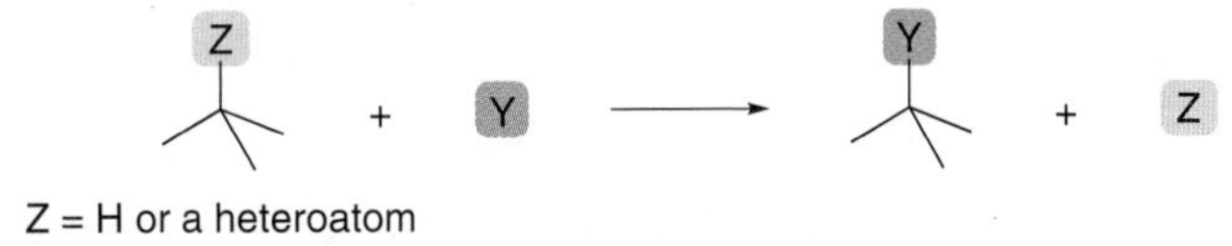

In a general substitution reaction, Y *replaces* Z on a carbon atom. **Substitution reactions involve σ bonds: one σ bond breaks and another forms at the same carbon atom.** The most common examples of substitution occur when Z is hydrogen or a heteroatom that is more electronegative than carbon.

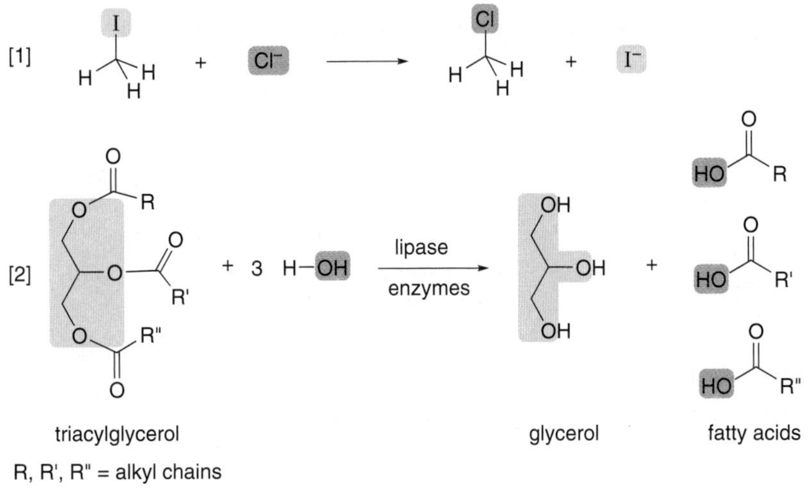

With a complex starting material, concentrate on the functional groups that *change.* The conversion of the esters in the triacylglycerol to glycerol and three fatty acids is a substitution reaction, because the OH in H_2O replaces the glycerol portion of the ester.

6.2B Elimination Reactions

- ***Elimination* is a reaction in which elements of the starting material are "lost" and a π bond is formed.**

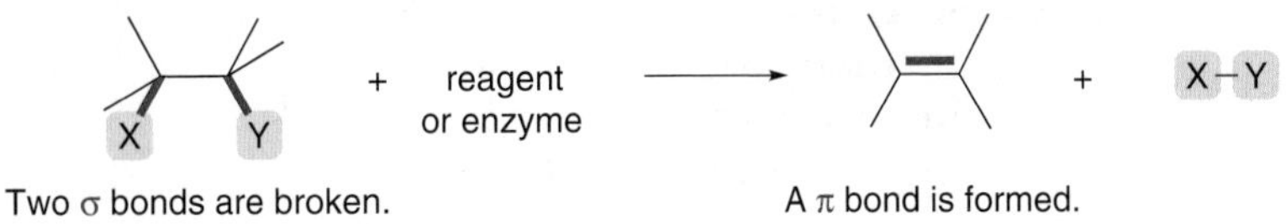

In an elimination reaction, two groups X and Y are removed from a starting material. **Two σ bonds are broken, and a π bond is formed between adjacent atoms.** The most common examples of elimination occur when X = H and Y is a heteroatom more electronegative than carbon.

[1]

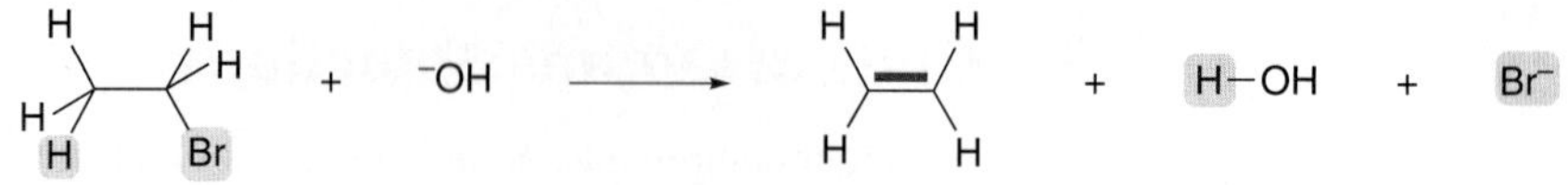

−OH
H−OH
Br−

loss of HBr

new π bond

Citrate loses H_2O during the citric acid cycle, an enzyme-catalyzed pathway that occurs during metabolism.

[2]

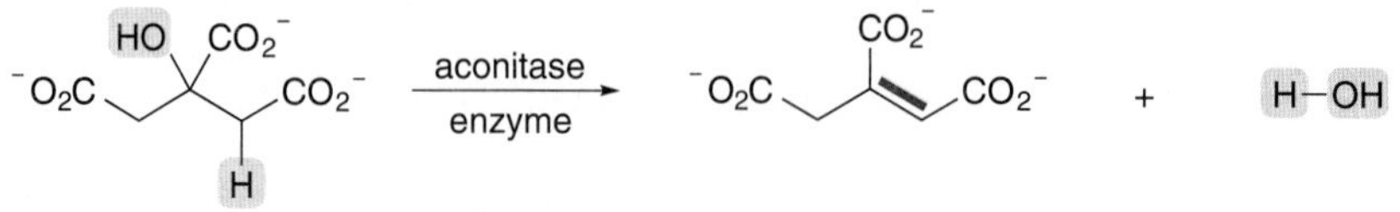

aconitase
enzyme
H−OH

citrate

loss of H_2O

new π bond

6.2C Addition Reactions

- *Addition* is a reaction in which elements are added to a starting material.

X−Y
X
Y

A π bond is broken.

Two σ bonds are formed.

In an addition reaction, new groups X and Y are added to a starting material. **A π bond is broken and two σ bonds are formed.**

[1]

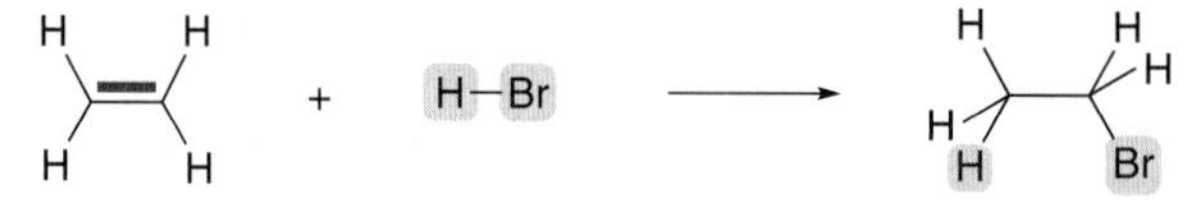

H−Br
Br

A π bond is broken.

HBr is added.

The addition of H_2O to thioesters derived from coenzyme A (HS–CoA) is a key step in fatty acid metabolism.

[2]

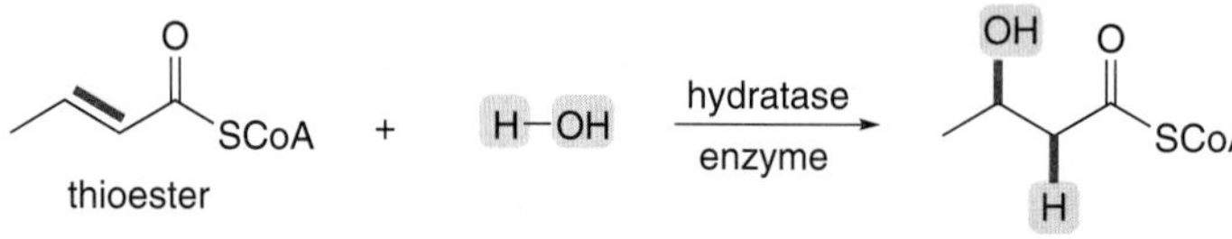

SCoA
H−OH
hydratase
enzyme
OH
SCoA

thioester

A π bond is broken.

H_2O is added.

Addition and elimination reactions are exactly opposite. A π bond is *formed* in elimination reactions, whereas a π bond is *broken* in addition reactions.

A summary of the general types of organic reactions is given in Appendix G.

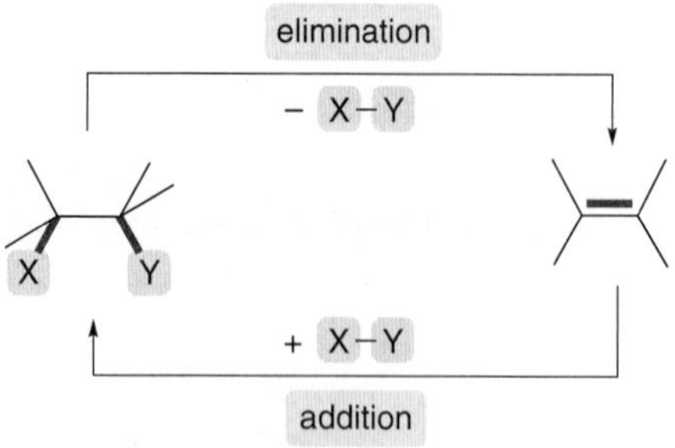

elimination
− X−Y
X
Y
+ X−Y
addition

Problem 6.1 Classify each transformation as substitution, elimination, or addition.

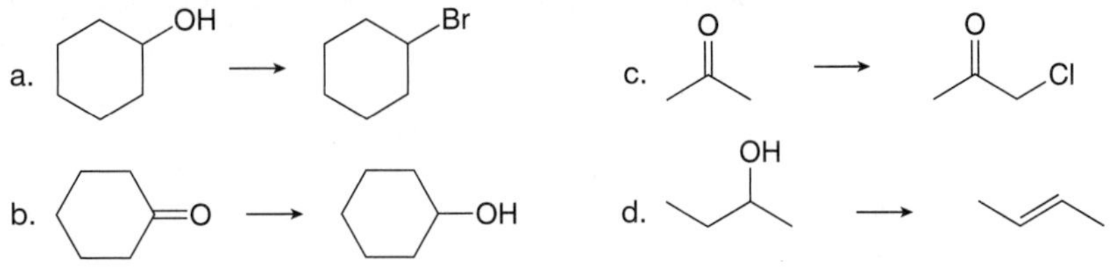

Problem 6.2 The following enzyme-catalyzed reactions illustrate the last four steps in the citric acid cycle, a critical part of metabolism. Classify each reaction as a substitution, elimination, or addition.

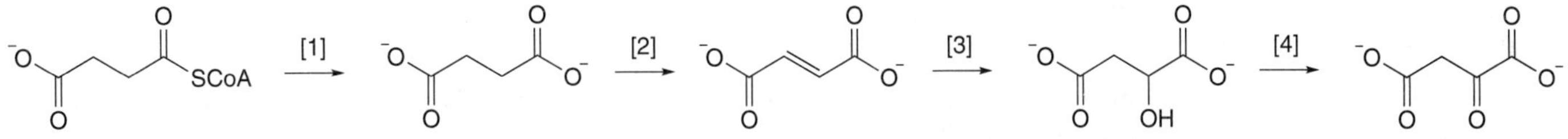

succinyl CoA   succinate   fumarate   malate   oxaloacetate

6.3 Bond Breaking and Bond Making

Having now learned how to write and identify some common kinds of organic reactions, we can turn to a discussion of **reaction mechanism.**

- **A *reaction mechanism* is a detailed description of how bonds are broken and formed as a starting material is converted to a product.**

A reaction mechanism describes the relative order and rate of bond cleavage and formation. It explains all the known facts about a reaction and accounts for all products formed, and it is subject to modification or refinement as new details are discovered.

A reaction can occur either in one step or in a series of steps.

- **A one-step reaction is called a *concerted reaction.*** No matter how many bonds are broken or formed, a starting material is converted *directly* to a product.

A ⟶ **B**

- **A stepwise reaction** involves more than one step. A starting material is first converted to an unstable intermediate, called a **reactive intermediate,** which then goes on to form the product.

A ⟶ **reactive intermediate** ⟶ **B**

6.3A Bond Cleavage

Bonds are broken and formed in all chemical reactions. When a bond is broken, the electrons in the bond can be divided **equally** or **unequally** between the two atoms of the bond.

- Breaking a bond by **equally dividing the electrons** between the two atoms in the bond is called **homolysis** or **homolytic cleavage.**

$$\text{A–B} \xrightarrow{\text{homolysis}} \text{A}\cdot \;+\; \cdot\text{B}$$

Each atom gets one electron.

- **Breaking a bond by unequally dividing the electrons** between the two atoms in the bond is called **heterolysis or heterolytic cleavage.**

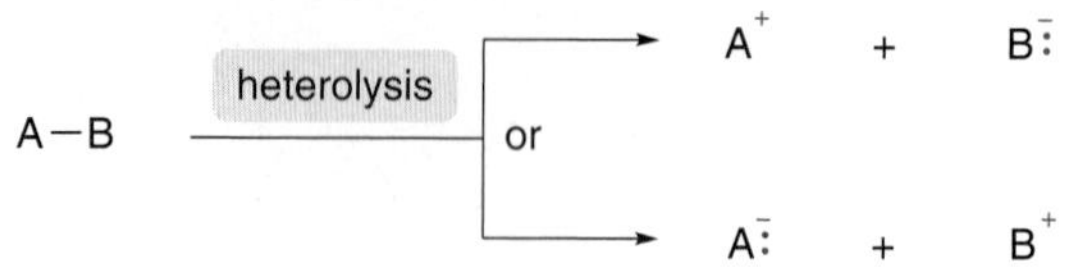

A or **B** gets two electrons.

Heterolysis of a bond between **A** and **B** can give either **A** or **B** the two electrons in the bond. When **A** and **B** have different electronegativities, the ***electrons normally end up on the more electronegative atom.***

Homolysis and heterolysis require energy. Both processes generate reactive intermediates, but the products are different in each case.

- **Homolysis generates uncharged reactive intermediates with *unpaired* electrons.**
- **Heterolysis generates *charged* intermediates.**

Each of these reactive intermediates has a very short lifetime and reacts quickly to form a stable organic product.

6.3B Radicals, Carbocations, and Carbanions

The curved arrow notation first discussed in Section 1.6 works fine for heterolytic bond cleavage because it illustrates the movement of an **electron pair.** For homolytic cleavage, however, one electron moves to one atom in the bond and one electron moves to the other, so a different kind of curved arrow is needed.

- **To illustrate the movement of a single electron, use a half-headed curved arrow, sometimes called a *fishhook.***

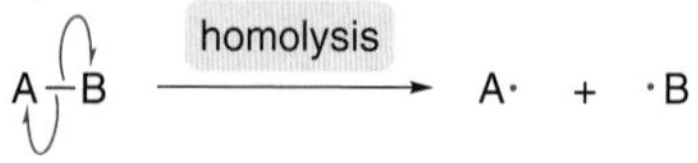

- Two **half-headed** curved arrows are needed for two **single** electrons.

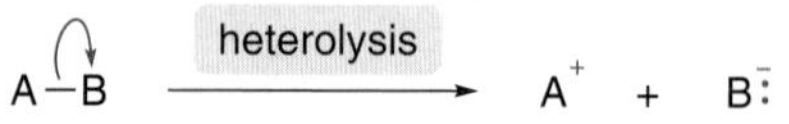

- One **full-headed** curved arrow is needed for one electron **pair.**

A full-headed curved arrow (⌒) shows the movement of an electron *pair.* **A half-headed curved arrow** (⌒) shows the movement of a *single* electron.

Figure 6.2 illustrates homolysis and two different heterolysis reactions for a carbon compound using curved arrows. Three different reactive intermediates are formed.

Homolysis of the C–Z bond generates two uncharged products with unpaired electrons.

- **A reactive intermediate with a single unpaired electron is called a *radical.***

Most radicals are highly unstable because they contain an atom that does not have an octet of electrons. Radicals typically have **no charge. They are intermediates in a group of reactions called *radical reactions,*** which are discussed in detail in Chapter 15.

Heterolysis of the C–Z bond can generate a **carbocation** or a **carbanion.**

- **Giving two electrons to Z and none to carbon generates a positively charged carbon intermediate called a *carbocation.***
- **Giving two electrons to C and none to Z generates a negatively charged carbon species called a *carbanion.***

Figure 6.2 Three reactive intermediates resulting from homolysis and heterolysis of a C–Z bond

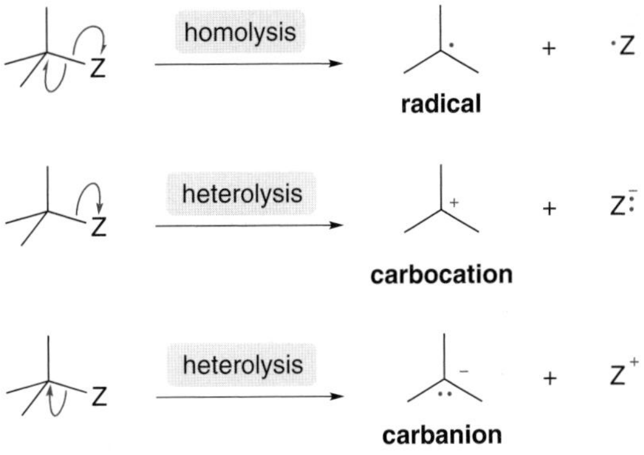

- Radicals are intermediates in **radical** reactions.
- **Ionic intermediates** are seen in **polar** reactions.

The chemistry of **carbenes,** another type of organic reactive intermediate, is discussed in Section 29.4.

Both carbocations and carbanions are unstable reactive intermediates: A carbocation contains a carbon atom surrounded by only six electrons. A carbanion has a negative charge on carbon, which is not a very electronegative atom. **Carbocations (electrophiles)** and **carbanions (nucleophiles)** can be intermediates in ***polar reactions*****—reactions in which a nucleophile reacts with an electrophile.**

Problem 6.3 By taking into account electronegativity differences, draw the products formed by heterolysis of the carbon–heteroatom bond in each molecule. Classify the organic reactive intermediate as a carbocation or a carbanion.

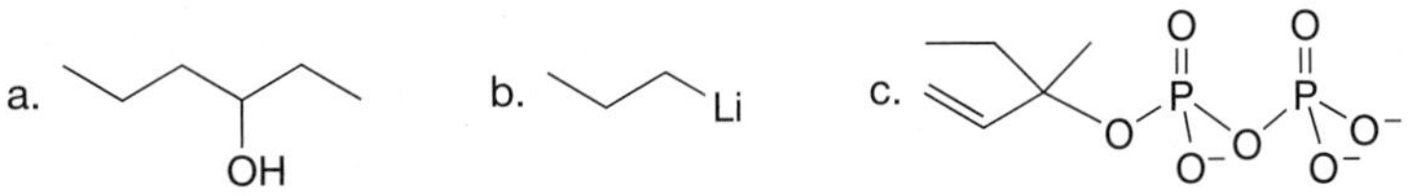

6.3C Bond Formation

Like bond cleavage, bond formation occurs in two different ways. Two radicals can each donate **one electron** to form a two-electron bond. Alternatively, two ions with unlike charges can come together, with the negatively charged ion donating **both electrons** to form the resulting two-electron bond. **Bond formation always releases energy.**

With two radicals...

...one electron comes from each atom.

With two ions...

...both electrons come from one atom.

6.3D All Kinds of Arrows

Table 6.1 summarizes the many kinds of arrows used in describing organic reactions. Curved arrows are especially important because they explicitly show what electrons are involved in a reaction, how these electrons move in forming and breaking bonds, and if a reaction proceeds via a radical or polar pathway.

A more complete summary of the arrows used in organic chemistry is given in the table Common Abbreviations, Arrows, and Symbols, located on the inside back cover.

Table 6.1 A Summary of Arrow Types in Chemical Reactions

Arrow	Name	Use
$\longrightarrow$	Reaction arrow	Drawn between the starting materials and products in an equation (6.1)
$\rightleftharpoons$	Double reaction arrows (equilibrium arrows)	Drawn between the starting materials and products in an equilibrium equation (2.2)
	Coupled reaction arrows	Shows reactions that occur together because energy is needed from one to drive the other (6.5)
$\longleftrightarrow$	Double-headed arrow	Drawn between resonance structures (1.6)
	Full-headed curved arrow	Shows movement of an electron pair (1.6, 2.2)
	Half-headed curved arrow (fishhook)	Shows movement of a single electron (6.3)

Sample Problem 6.1 Use full-headed or half-headed curved arrows to show the movement of electrons in each equation.

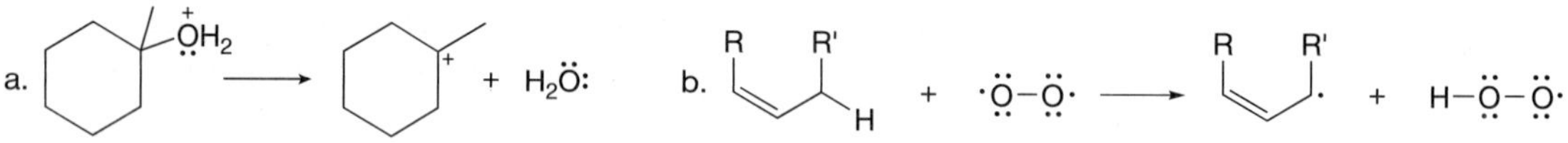

Solution

Look for all bonds that are broken or formed. Use full-headed curved arrows for electron pairs and half-headed curved arrows for single electrons.

Sample Problem 6.1b depicts the first step in the air oxidation of vegetable oils, a process that turns the oil rancid (Section 15.11).

a. In this reaction, the C–O bond is broken heterolytically. Because only one electron pair is involved, **one full-headed curved arrow** is needed. The electron pair in the C–O bond ends up on O.

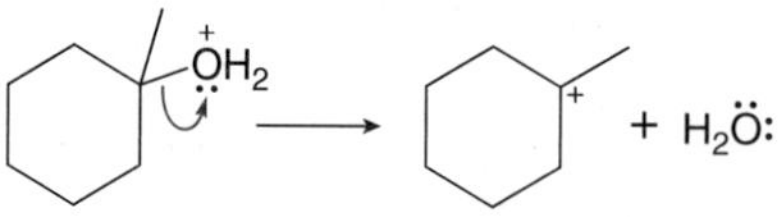

b. This reaction involves radicals, so half-headed curved arrows are needed to show the movement of single electrons. One new two-electron bond is formed between H and O, and an unpaired electron is left on C. Because a total of three electrons are involved, three half-headed curved arrows are needed.

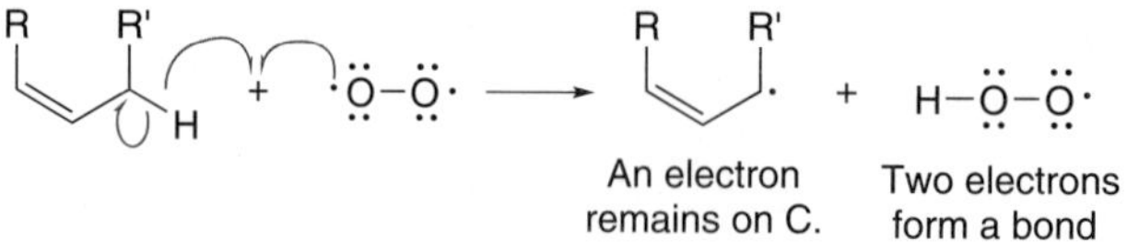
An electron remains on C.
Two electrons form a bond

Problem 6.4 Use curved arrows to show the movement of electrons in each equation.

a. 

b. 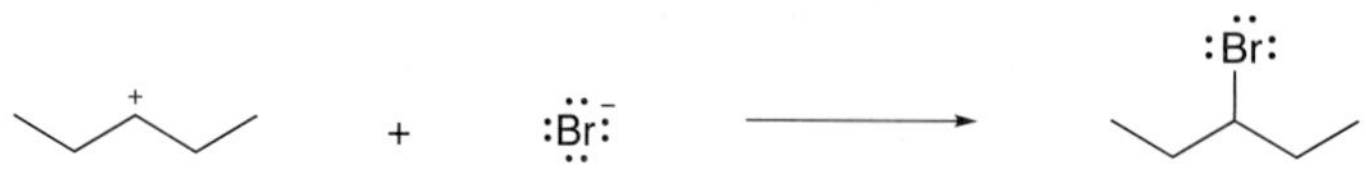

Sample Problem 6.2 Follow the curved arrows and draw the products of the following reaction.

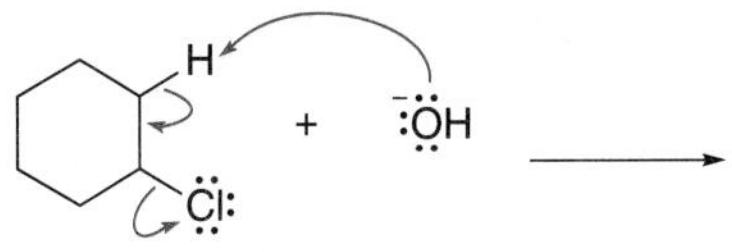

Solution

Three full-headed curved arrows are drawn, so three electron *pairs* take part in the reaction. Arrow **1** shows that a lone pair on $^{-}$OH forms a new bond to H, forming H_2O. Arrow **2** indicates that the electron pair in the C – H bond forms a carbon–carbon double bond. Arrow **3** shows that the electron pair in the C – Cl bond ends up on Cl, forming Cl^{-}. After breaking and making bonds, formal charges on the atoms involved in the reaction are adjusted when necessary.

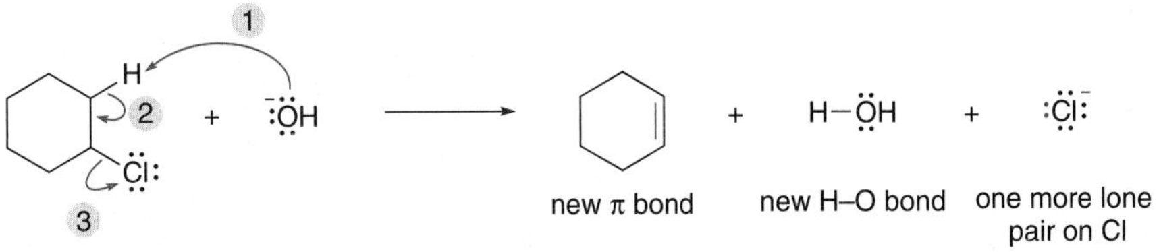
1
2
3
new π bond
new H–O bond
one more lone pair on Cl

Problem 6.5 Follow the curved arrows and draw the products of the following reaction.

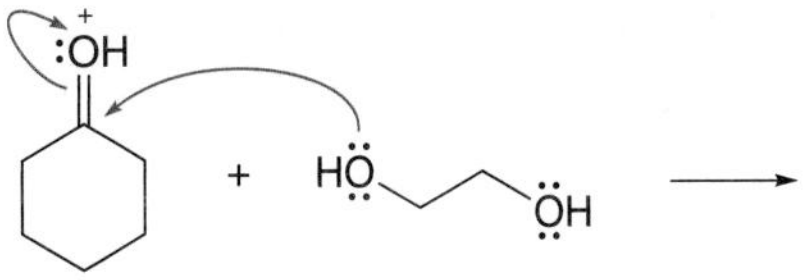

6.4 Bond Dissociation Energy

Bond breaking can be quantified using the bond dissociation energy.

- **The *bond dissociation energy* is the energy needed to homolytically cleave a covalent bond.**

Bond dissociation ***energy*** is also called *bond dissociation* ***enthalpy*** because it refers to the heat absorbed when bonds are cleaved.

A–B ⟶ A· + ·B $\Delta H°$ = bond dissociation energy

Homolysis requires energy.

The superscript (°) means that values are determined under standard state conditions (pure compounds in their most stable state at a specified temperature and pressure).

The heat energy absorbed or released in any reaction, symbolized by $\Delta H°$, is called the **enthalpy change** or **heat of reaction.**

- **When $\Delta H°$ is positive (+), energy is absorbed and the reaction is *endothermic.***
- **When $\Delta H°$ is negative (–), energy is released and the reaction is *exothermic.***

Additional bond dissociation energies for C – C multiple bonds are given in Table 1.6.

A bond dissociation energy is the $\Delta H°$ for a specific kind of reaction—the homolysis of a covalent bond to form two radicals. Because bond breaking requires energy, **bond dissociation energies are always *positive* numbers,** and homolysis is always **endothermic.** Conversely, **bond formation always *releases* energy,** so this reaction is always **exothermic.** The H – H bond requires +435 kJ/mol to cleave and releases –435 kJ/mol when formed. Table 6.2 contains a representative list of bond dissociation energies for many common bonds.

H–H ⟶ H· + ·H
ΔH = +435 kJ/mol
endothermic reaction

H· + ·H ⟶ H–H
ΔH = –435 kJ/mol
exothermic reaction

A table of bond dissociation energies also appears in Appendix C.

Table 6.2 Bond Dissociation Energies for Some Common Bonds [$A–B \rightarrow A\cdot + \cdot B$]

Bond	$\Delta H°$ kJ/mol	(kcal/mol)
H–Z bonds		
$H-F$	569	(136)
$H-Cl$	431	(103)
$H-Br$	368	(88)
$H-I$	297	(71)
$H-OH$	498	(119)
Z–Z bonds		
$H-H$	435	(104)
$F-F$	159	(38)
$Cl-Cl$	242	(58)
$Br-Br$	192	(46)
$I-I$	151	(36)
$HO-OH$	213	(51)
R–H bonds		
CH_3-H	435	(104)
CH_3CH_2-H	410	(98)
$CH_3CH_2CH_2-H$	410	(98)
$(CH_3)_2CH-H$	397	(95)
$(CH_3)_3C-H$	381	(91)
$CH_2=CH-H$	435	(104)
$HC\equiv C-H$	523	(125)
$CH_2=CHCH_2-H$	364	(87)
C_6H_5-H	460	(110)
$C_6H_5CH_2-H$	356	(85)
R–R bonds		
CH_3-CH_3	368	(88)
$CH_3-CH_2CH_3$	356	(85)
$CH_3-CH=CH_2$	385	(92)
$CH_3-C\equiv CH$	489	(117)
R–X bonds		
CH_3-F	456	(109)
CH_3-Cl	351	(84)
CH_3-Br	293	(70)
CH_3-I	234	(56)
CH_3CH_2-F	448	(107)
CH_3CH_2-Cl	339	(81)
CH_3CH_2-Br	285	(68)
CH_3CH_2-I	222	(53)
$(CH_3)_2CH-F$	444	(106)
$(CH_3)_2CH-Cl$	335	(80)
$(CH_3)_2CH-Br$	285	(68)
$(CH_3)_2CH-I$	222	(53)
$(CH_3)_3C-F$	444	(106)
$(CH_3)_3C-Cl$	331	(79)
$(CH_3)_3C-Br$	272	(65)
$(CH_3)_3C-I$	209	(50)
R–Z bonds		
CH_3-OH	389	(93)
CH_3CH_2-OH	393	(94)
$CH_3CH_2CH_2-OH$	385	(92)
$(CH_3)_2CH-OH$	401	(96)
$(CH_3)_3C-OH$	401	(96)
CH_3-NH_2	331	(79)
CH_3-SH	305	(73)

Comparing bond dissociation energies is equivalent to comparing **bond strength.**

- The *stronger* the bond, the *higher* its bond dissociation energy.

For example, the H–H bond is stronger than the Cl–Cl bond because its bond dissociation energy is higher [Table 6.2: 435 kJ/mol (H_2) versus 242 kJ/mol (Cl_2)]. The data in Table 6.2 demonstrate that **bond dissociation energies *decrease* down a column of the periodic table as the valence electrons used in bonding are farther from the nucleus.** Bond dissociation energies for a group of methyl–halogen bonds exemplify this trend.

Increasing size of the halogen →

CH_3-F	CH_3-Cl	CH_3-Br	CH_3-I
$\Delta H°$ = 456 kJ/mol	351 kJ/mol	293 kJ/mol	234 kJ/mol

← **Increasing bond strength**

Because bond length increases down a column of the periodic table, bond dissociation energies are a quantitative measure of the general phenomenon noted in Chapter 1—***shorter* bonds are *stronger* bonds.**

Problem 6.6 Which bond in each pair has the higher bond dissociation energy?

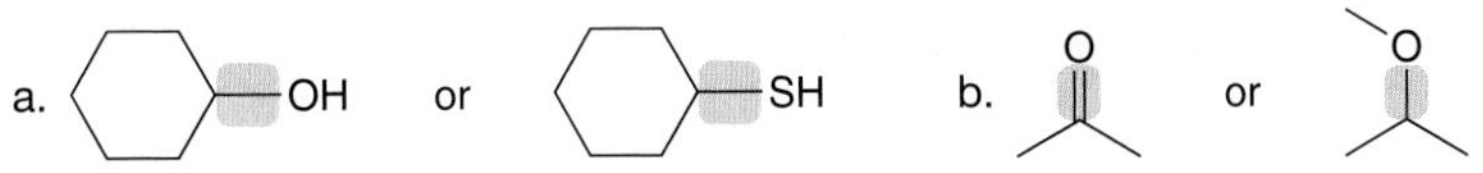
a.
OH
or
SH
b.
O
or
O

Bond dissociation energies are also used to calculate the enthalpy change ($\Delta H°$) in a reaction in which several bonds are broken and formed. **$\Delta H°$ indicates the relative strength of bonds broken and formed in a reaction.**

- **When $\Delta H°$ is positive, more energy is needed to break bonds than is released in forming bonds. The bonds broken in the starting material are *stronger* than the bonds formed in the product.**

- **When $\Delta H°$ is negative, more energy is released in forming bonds than is needed to break bonds. The bonds formed in the product are *stronger* than the bonds broken in the starting material.**

To determine the overall $\Delta H°$ for a reaction:

[1] Beginning with a *balanced* equation, add the bond dissociation energies for all bonds broken in the starting materials. This (+) value represents the **energy needed** to break bonds.

[2] Add the bond dissociation energies for all bonds formed in the products. This (–) value represents the **energy released** in forming bonds.

[3] **The overall $\Delta H°$ is the sum in Step [1] *plus* the sum in Step [2].**

$\Delta H°$ overall enthalpy change = sum of $\Delta H°$ of bonds broken + (–) sum of $\Delta H°$ of bonds formed

Sample Problem 6.3 Use the values in Table 6.2 to determine $\Delta H°$ for the following reaction.

Cl
+
H
O
H
OH
+
H–Cl

Solution

[1] Bonds broken

	$\Delta H°$ (kJ/mol)
$(CH_3)_3C–Cl$	+331
H–OH	+498
Total	+829 kJ/mol

Energy needed to break bonds.

[2] Bonds formed

	$\Delta H°$ (kJ/mol)
$(CH_3)_3C–OH$	–401
H–Cl	–431
Total	–832 kJ/mol

Energy released in forming bonds.

[3] Overall $\Delta H°$ =

sum in Step [1] + sum in Step [2]

+829 kJ/mol
–832 kJ/mol

Answer: –3 kJ/mol

Because $\Delta H°$ is a negative value, this reaction is **exothermic** and energy is released. **The bonds broken in the starting material are weaker than the bonds formed in the product.**

Problem 6.7 Use the values in Table 6.2 to calculate $\Delta H°$, and classify the reaction as endothermic or exothermic.

Br
+
H2O
OH
+
HBr

Certain metabolic compounds, such as adenosine triphosphate (**ATP,** Section 3.2D), are called "high-energy" molecules, because they undergo highly exothermic reactions. Processes such as walking, running, swallowing, and breathing are fueled by the energy released from ATP hydrolysis.

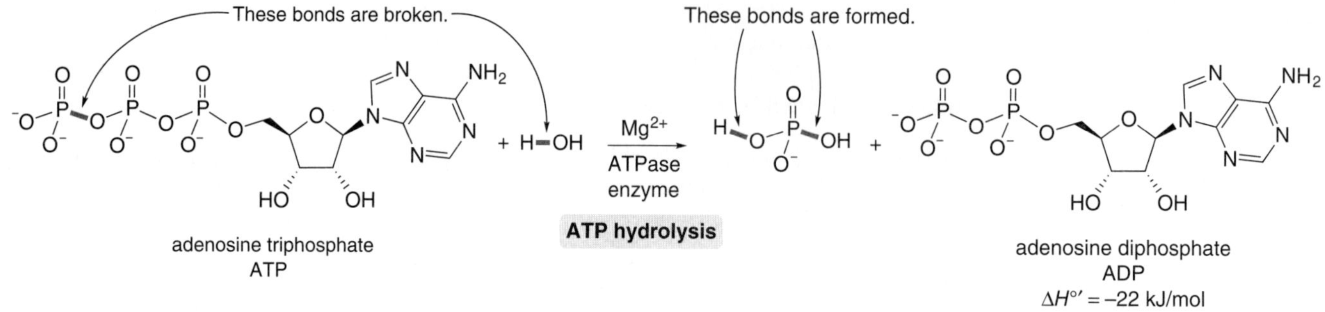

Multiple enzymes exist in humans and other organisms to carry out phosphate hydrolysis reactions in the presence of Mg^{2+}.

In the hydrolysis of ATP with H_2O to form ADP and $H_2PO_4^-$, a P–O bond and H–O bond are broken, and a P–O bond and H–O bond are formed. Because more energy is released in bond formation than is absorbed in bond cleavage, the reaction is **exothermic** (–22 kJ/mol).

- **Because a substantial amount of heat is given off during ATP hydrolysis, ATP is an excellent energy source in biological systems.**

The superscript (′) means the reaction was run at a specified concentration of a species. In biochemical reactions H^+ is often either consumed or produced, so (′) is added to designate that the H^+ concentration (pH) remains constant.

In a series of chemical reactions in metabolic biochemical pathways, glucose, the chapter-opening molecule, is oxidized to form CO_2 and H_2O. Several bonds break and form in this process.

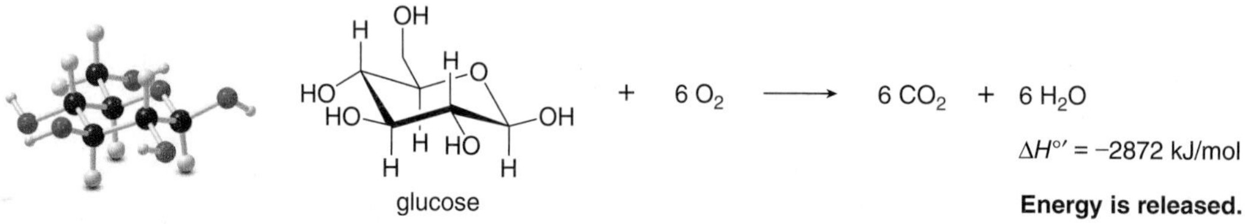

$\Delta H°'$ is negative for this oxidation, so the overall reaction is exothermic. **Glucose releases heat on oxidation because the reaction series results in a net gain in bond strength in the products versus the reactants.**

Bond dissociation energies have two important limitations. They present *overall* energy changes only. They reveal nothing about the reaction mechanism or how fast a reaction proceeds. Moreover, bond dissociation energies are determined for reactions in the gas phase, whereas most organic reactions are carried out in a liquid solvent where solvation energy contributes to the overall enthalpy of a reaction. As such, bond dissociation energies are imperfect indicators of energy changes in a reaction. Despite these limitations, using bond dissociation energies to calculate $\Delta H°$ gives a useful approximation of the energy changes that occur when bonds are broken and formed in a reaction.

Problem 6.8 Calculate $\Delta H°$ for each oxidation reaction. Each equation is balanced as written; remember to take into account the coefficients in determining the number of bonds broken or formed.
[$\Delta H°$ for O_2 = 497 kJ/mol; $\Delta H°$ for one C=O in CO_2 = 535 kJ/mol]

a. $CH_4 + 2\ O_2 \longrightarrow CO_2 + 2\ H_2O$ b. $2\ CH_3CH_3 + 7\ O_2 \longrightarrow 4\ CO_2 + 6\ H_2O$

6.5 Thermodynamics

For a reaction to be practical, the equilibrium must favor the products, *and* the reaction rate must be fast enough to form them in a reasonable time. These two conditions depend on the **thermodynamics** and the **kinetics** of a reaction, respectively.

- *Thermodynamics* describes energy and equilibrium. How do the *energies* of the reactants and the products compare? What are the relative *amounts* of reactants and products at equilibrium?

Reaction kinetics are discussed in Section 6.9.

- *Kinetics* describes reaction rates. How *fast* are reactants converted to products?

6.5A Equilibrium Constant and Free Energy Changes

K_{eq} was first defined in Section 2.3 for acid–base reactions.

The **equilibrium constant, K_{eq},** is a mathematical expression that relates the amount of starting material and product at equilibrium. For example, when starting materials **A** and **B** react to form products **C** and **D,** the equilibrium constant is given by the following expression.

$$\mathbf{A} + \mathbf{B} \rightleftharpoons \mathbf{C} + \mathbf{D}$$

$$K_{eq} = \frac{[\text{products}]}{[\text{starting materials}]} = \frac{[\text{C}][\text{D}]}{[\text{A}][\text{B}]}$$

The size of K_{eq} tells about the position of equilibrium; that is, it expresses whether the starting materials or products predominate once equilibrium has been reached.

- When $K_{eq} > 1$, equilibrium favors the products (**C** and **D**) and the equilibrium lies to the right as the equation is written.
- When $K_{eq} < 1$, equilibrium favors the starting materials (**A** and **B**) and the equilibrium lies to the left as the equation is written.

- **For a reaction to be useful, the equilibrium must favor the products, and $K_{eq} > 1$.**

What determines whether equilibrium favors the products in a given reaction? **The position of equilibrium is determined by the relative energies of the reactants and products.** The free energy of a molecule, also called its **Gibbs free energy,** is symbolized by $G°$. The **change in free energy** between reactants and products, symbolized by $\Delta G°$, determines whether the starting materials or products are favored at equilibrium.

- **$\Delta G°$ is the overall energy difference between reactants and products.**

$$\Delta G° = G°_{products} - G°_{reactants}$$

$G°_{products}$: free energy of the products
$G°_{reactants}$: free energy of the reactants

$\Delta G°$ is related to the equilibrium constant K_{eq} by the following equation:

$$\Delta G° = -2.303RT \log K_{eq}$$

[R = 8.314 J/(K•mol), the gas constant
T = Kelvin temperature (K)]

Using this expression we can determine the relationship between the equilibrium constant and the free energy change between reactants and products.

At 25 °C, $2.303RT = 5.7$ kJ/mol; thus, $\Delta G° = -5.7 \log K_{eq}$.

$K_{eq} > 1$ when $\Delta G° < 0$, and equilibrium favors the products. $K_{eq} < 1$ when $\Delta G° > 0$, and equilibrium favors the starting materials.

- **When $K_{eq} > 1$, log K_{eq} is positive, making $\Delta G°$ negative, and energy is released. Thus, equilibrium favors the products when the energy of the products is *lower* than the energy of the reactants.**
- **When $K_{eq} < 1$, log K_{eq} is negative, making $\Delta G°$ positive, and energy is absorbed. Thus, equilibrium favors the reactants when the energy of the products is *higher* than the energy of the reactants.**

Compounds that are lower in energy have increased stability. Thus, **equilibrium favors the products when they are more stable (lower in energy) than the starting materials of a reaction.** This is summarized in Figure 6.3.

Figure 6.3
Summary of the relationship between $\Delta G°$ and K_{eq}

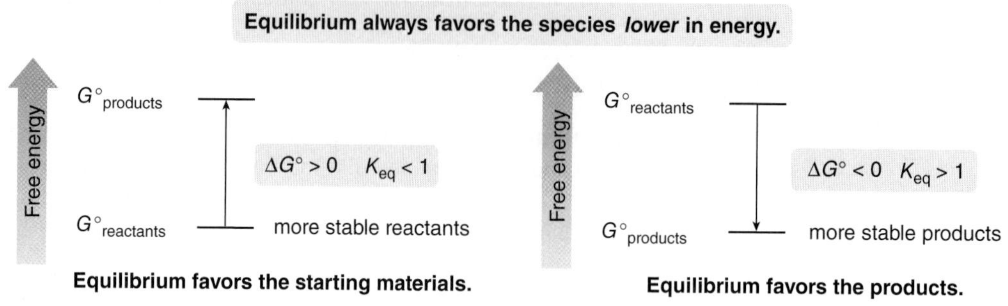

Because $\Delta G°$ depends on the logarithm of K_{eq}, a **small change in energy corresponds to a large difference in the relative amount of starting material and product at equilibrium.** Several values of $\Delta G°$ and K_{eq} are given in Table 6.3. For example, a difference in energy of only ~6 kJ/mol means that there is 10 times as much of the more stable species at equilibrium. A difference in energy of ~18 kJ/mol means that there is essentially only one compound, either starting material or product, at equilibrium.

The symbol ~ means *approximately.*

Table 6.3 Representative Values for $\Delta G°$ and K_{eq} at 25 °C, for a Reaction A → B

$\Delta G°$ (kJ/mol)	K_{eq}	Relative amount of A and B at equilibrium
+18	**10^{-3}**	**Essentially all A (99.9%)**
+12	10^{-2}	100 times as much **A** as **B**
+6	10^{-1}	10 times as much **A** as **B**
0	1	Equal amounts of **A** and **B**
−6	10^{1}	10 times as much **B** as **A**
−12	10^{2}	100 times as much **B** as **A**
−18	**10^{3}**	**Essentially all B (99.9%)**

Increasing [product] ↓

Problem 6.9 a. Which K_{eq} corresponds to a negative value of $\Delta G°$, $K_{eq} = 1000$ or $K_{eq} = .001$?
b. Which K_{eq} corresponds to a lower value of $\Delta G°$, $K_{eq} = 10^{-2}$ or $K_{eq} = 10^{-5}$?

Problem 6.10 Given each of the following values, is the starting material or product favored at equilibrium?
a. $K_{eq} = 5.5$ b. $\Delta G° = 40$ kJ/mol

Problem 6.11 Given each of the following values, is the starting material or product lower in energy?
a. $\Delta G° = 8.0$ kJ/mol b. $K_{eq} = 10$ c. $\Delta G° = -12$ kJ/mol d. $K_{eq} = 10^{-3}$

6.5B Energy Changes and Conformational Analysis

These equations can be used for any process with two states in equilibrium. As an example, the cyclic form of glucose (Section 3.9B) exists as two different chair conformations that rapidly interconvert at room temperature, with the conformation having the substituents in the roomier equatorial positions favored (Section 4.13). Knowing the energy difference between the two conformations allows us to calculate the amount of each at equilibrium.

For example, the energy difference between the two chair conformations of glucose (**A** and **B**) is –36 kJ/mol, as shown in the accompanying equation. Extrapolating from the values in Table 6.3, this corresponds to an equilibrium constant of ~10^6, meaning that there is approximately 1,000,000 times more **B** (all equatorial substituents) than **A** (all axial substituents) at equilibrium.

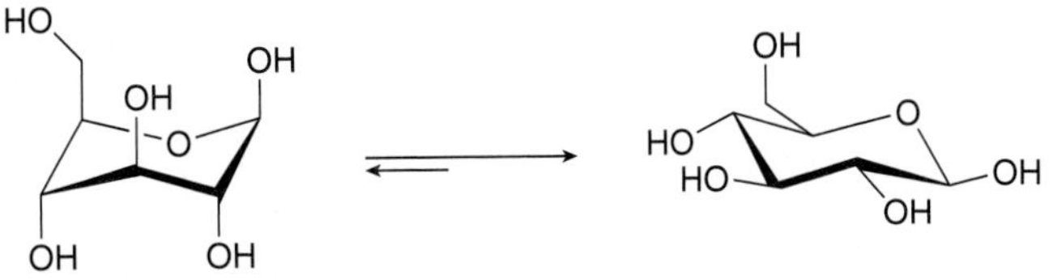

$\Delta G°$ = –36 kJ/mol

glucose — all axial substituents — **A**

glucose — all equatorial substituents — **B**

$$\Delta G° = -2.303RT \log K_{eq}$$

$$-36 \text{ kJ/mol} \longrightarrow K_{eq} \approx 10^6$$

There is ~1,000,000 times more **B** than **A** at equilibrium.

Problem 6.12 The equilibrium constant for the conversion of the axial to the equatorial conformation of methoxycyclohexane is 2.7.

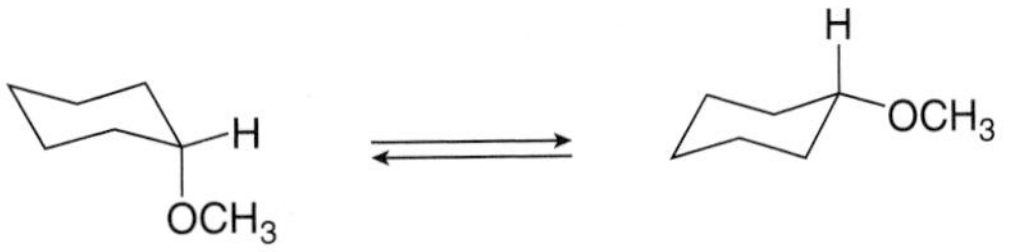

K_{eq} = 2.7

a. Given these data, which conformation is present in the larger amount at equilibrium?
b. Is $\Delta G°$ for this process positive or negative?
c. From the values in Table 6.3, approximate the size of $\Delta G°$.

6.5C Coupled Reactions in Metabolism

As we learned in Section 6.4, the hydrolysis of ATP to ADP is highly exothermic. Because the overall energy difference between the reactants and products ($\Delta G°'$) is also negative (–), the energy released in ATP hydrolysis can be used to drive a reaction that has an unfavorable energy change by **coupling** the two reactions together.

- **Coupled reactions are reactions that are paired together to drive an unfavorable process. The energy released by one reaction provides the energy to drive the other reaction.**

Consider the oxidation of glucose to CO_2 and H_2O, resulting from several biochemical reactions. Although the overall conversion releases a great deal of energy, the energy change associated with each individual step is much smaller. In some reactions energy is released, and in others energy is absorbed. The hydrolysis of ATP provides energy to drive reactions that require energy.

Coupled reactions that involve ATP or other reagents are often drawn using a combination of horizontal and curved arrows. The principal reactants and products for a given pathway are drawn from left to right with a reaction arrow as usual, and the other reagents are drawn using a curved arrow.

In the first step of glucose metabolism, glucose is phosphorylated to form glucose 6-phosphate.

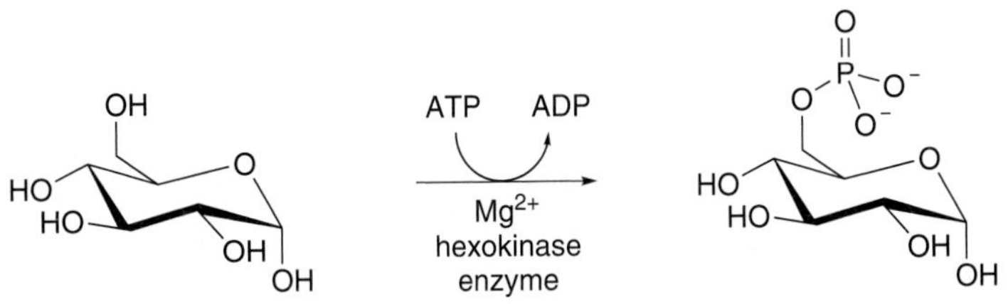
ATP ADP
Mg^{2+}
hexokinase
enzyme

glucose

glucose 6-phosphate
$\Delta G°'$ = –16.7 kJ/mol

When glucose reacts directly with hydrogen phosphate (HPO_4^{2-}) to form glucose 6-phosphate, 13.8 kJ/mol of energy is required, so this process is energetically unfavorable. By coupling glucose phosphorylation with energetically favorable ATP hydrolysis (30.5 kJ/mol of energy released), the coupled reaction becomes an energetically favorable process (–16.7 kJ/mol of energy released).

					$\Delta G^{\circ\prime}$	
Phosphorylation	glucose + HPO_4^{2-}	$\longrightarrow$	glucose 6-phosphate + H_2O		+13.8 kJ/mol	← energy required
Hydrolysis	ATP + H_2O	$\longrightarrow$	ADP + HPO_4^{2-}		–30.5 kJ/mol	← energy released
Coupled reaction	glucose + ATP	$\longrightarrow$	glucose 6-phosphate + ADP		–16.7 kJ/mol	

Energy is released.

- The coupled reaction is the **net reaction**, written by summing the substances in both equations and eliminating those compounds that appear on both sides of the reaction arrows.
- The overall energy change is found by summing the energies for the individual steps.

Thus, in this example:

- The hydrolysis of ATP provides the energy for the phosphorylation of glucose.

Although the coupled reactions are written as two separate equations for emphasis in the preceding example, in reality a single reaction takes place. ATP transfers a phosphate to glucose, forming glucose 6-phosphate and ADP, while both molecules are held in close proximity at the active site of hexokinase, as shown in Figure 6.4.

Figure 6.4
A coupled reaction—Phosphorylation of glucose by ATP

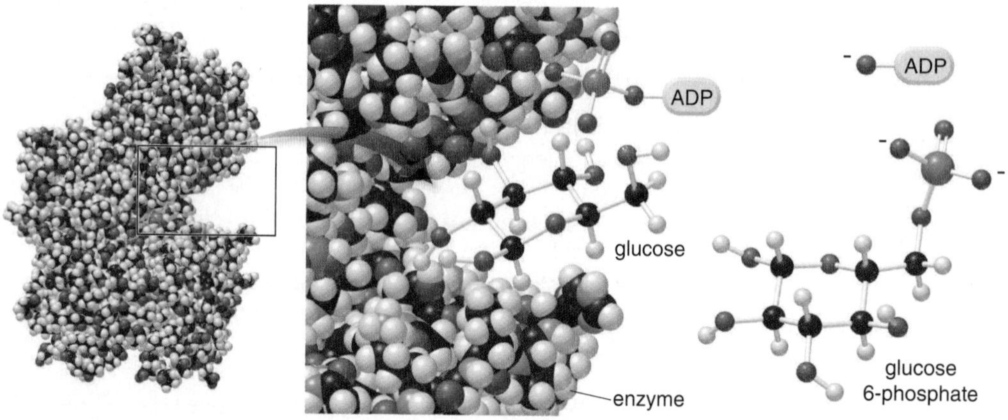

- The enzyme hexokinase binds glucose at its active site and ATP is also bound in close proximity. ATP transfers a phosphate directly to the glucose molecule to form ADP and glucose 6-phosphate. This coupled reaction is energetically favorable.

Problem 6.13 The phosphorylation of fructose to fructose 6-phosphate requires 15.9 kJ/mol of energy. In fructose metabolism, this unfavorable reaction is driven by the hydrolysis of ATP to ADP in the presence of Mg^{2+} and the fructokinase enzyme.

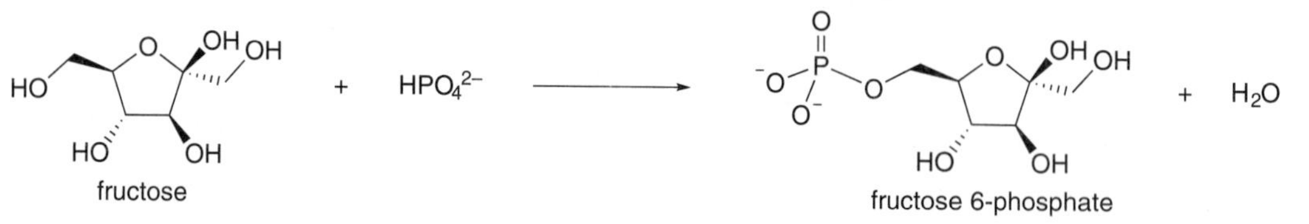

$\Delta G^{\circ\prime} = +15.9$ kJ/mol

a. Write the net (coupled) reaction by summing the substances in both the phosphorylation and hydrolysis equations.
b. How much energy is released in the coupled reaction?
c. Write the equation for the coupled reaction using coupled reaction arrows.

6.6 Enthalpy and Entropy

The **free energy change ($\Delta G°$)** depends on the **enthalpy change ($\Delta H°$)** and the **entropy change ($\Delta S°$).** $\Delta H°$ indicates relative bond strength, but what does $\Delta S°$ measure?

Entropy is a rather intangible concept that comes up again and again in chemistry courses. One way to remember the relation between entropy and disorder is to consider a handful of chopsticks. Dropped on the floor, they are arranged randomly (a state of high entropy). Placed end-to-end in a straight line, they are arranged intentionally (a state of low entropy). The more disordered, random arrangement is favored and easier to achieve energetically.

***Entropy* ($S°$) is a measure of the randomness in a system.** The more freedom of motion or the more disorder present, the higher the entropy. Gas molecules move more freely than liquid molecules and are higher in entropy. Cyclic molecules have more restricted bond rotation than similar acyclic molecules and are lower in entropy.

The *entropy change* ($\Delta S°$) is the change in the amount of disorder between reactants and products. $\Delta S°$ is positive (+) when the products are more disordered than the reactants. $\Delta S°$ is negative (–) when the products are less disordered (more ordered) than the reactants.

- **Reactions resulting in an increase in entropy are favored.**

$\Delta G°$ is related to $\Delta H°$ and $\Delta S°$ by the following equation:

$$\Delta G° = \Delta H° - T\Delta S° \qquad [T = \text{Kelvin temperature}]$$

$\Delta G°$	$\Delta H°$	$T\Delta S°$
total energy change	change in bonding enthalpy	change in disorder

This equation tells us that the total energy change in a reaction is due to two factors: the change in the **bonding enthalpy** and the change in **disorder.** The change in bonding enthalpy can be calculated from bond dissociation energies (Section 6.4). Entropy changes, on the other hand, are more difficult to assess, but they are important when the number of molecules of starting material *differs* from the number of molecules of product in the balanced chemical equation. The entropy of a system also changes when an acyclic molecule is *cyclized* to a cyclic one, or a cyclic molecule is converted to an acyclic one.

For example, **when a single starting material forms two products,** as in the homolytic cleavage of a bond to form two radicals, **entropy increases** and favors formation of the products. In contrast, **entropy decreases when an acyclic compound forms a ring,** because a ring has fewer degrees of freedom. In this case, therefore, entropy does *not* favor formation of the product.

single reactant

two products
Entropy *increases.*

Entropy favors the products.

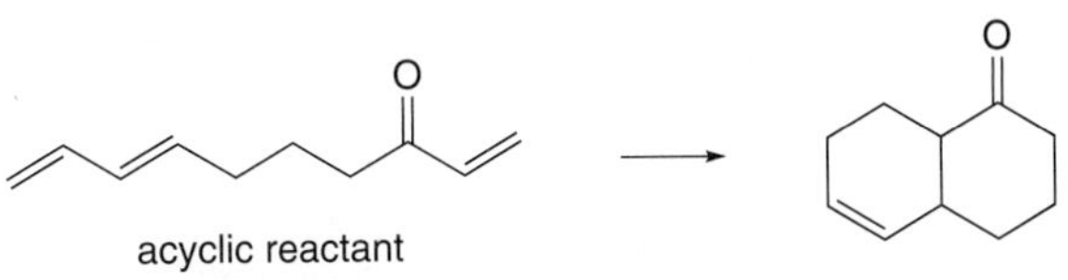

acyclic reactant

Entropy favors the reactants.

cyclic product
Entropy *decreases.*

The metabolism of glucose (Section 6.4) is favored by entropy because the number of molecules of products formed (6 CO_2 and 6 H_2O) is greater than the number of molecules of reactants ($C_6H_{12}O_6$ and 6 O_2). Moreover, a cyclic reactant is cleaved to form 12 acyclic product molecules.

In many reactions that are not carried out at high temperature, the entropy term ($T\Delta S°$) is small compared to the enthalpy term ($\Delta H°$) and it can be neglected. Thus, **we will often approximate the overall free energy change of a reaction by the change in the bonding enthalpy only.** Keep in mind that this is an approximation, but it gives us a starting point from which to decide if the reaction is energetically favorable.

Recall from Section 6.4 that a reaction is endothermic when $\Delta H°$ is positive and exothermic when $\Delta H°$ is negative. A reaction is **endergonic when $\Delta G°$ is positive** and **exergonic when $\Delta G°$ is negative.** $\Delta G°$ is usually approximated by $\Delta H°$ in this text, so the terms endergonic and exergonic are rarely used.

$$\Delta G° \approx \Delta H°$$

According to this approximation:

- **The product is favored when $\Delta H°$ is a *negative* value; that is, the reaction results in a net gain in bond strength in the products versus the reactants.**

- **The starting material is favored when $\Delta H°$ is a *positive* value; that is, the reaction results in a net loss in bond strength in the products versus the reactants.**

Problem 6.14 Considering each of the following values and neglecting entropy, tell whether the starting material or product is favored at equilibrium: (a) $\Delta H°$ = 80 kJ/mol; (b) $\Delta H°$ = –40 kJ/mol.

6.7 Energy Diagrams

An **energy diagram** is a schematic representation of the energy changes that take place as reactants are converted to products. An energy diagram indicates how readily a reaction proceeds, how many steps are involved, and how the energies of the reactants, products, and intermediates compare.

Consider a concerted reaction between molecule **A–B** with anion **C:⁻** to form products **A:⁻** and **B–C.** If the reaction occurs in a single step, the bond between **A** and **B** is broken *as* the bond between **B** and **C** is formed. Let's assume that the products are lower in energy than the reactants in this hypothetical reaction.

An energy diagram plots **energy on the *y* axis** versus the progress of reaction, often labeled the **reaction coordinate,** on the ***x* axis.** As the starting materials **A–B** and **C:⁻** approach one another, their electron clouds feel some repulsion, causing an increase in energy, until a maximum value is reached. This unstable energy maximum is called the **transition state.** In the transition state the bond between **A** and **B** is partially broken, and the bond between **B** and **C** is partially formed. Because it is at the top of an energy "hill," **a transition state can never be isolated.**

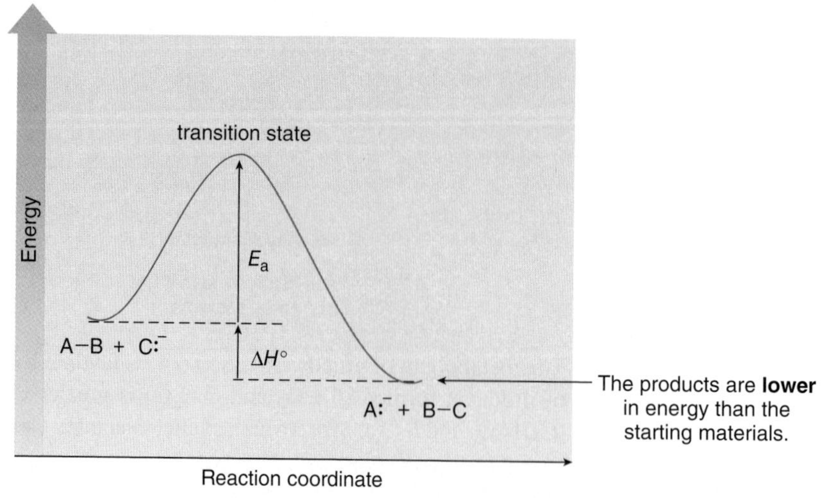

At the transition state, the bond between **A** and **B** can re-form to regenerate starting material, *or* the bond between **B** and **C** can form to generate product. As the bond forms between **B** and **C,** the energy decreases until some stable energy minimum of the products is reached.

- **The energy difference between the reactants and products is $\Delta H°$. Because the products are at lower energy than the reactants, this reaction is exothermic and energy is released.**

- **The energy difference between the transition state and the starting material is called the *energy of activation,* symbolized by E_a.**

The *energy of activation* is the minimum amount of energy needed to break bonds in the reactants. It represents an **energy barrier** that must be overcome for a reaction to occur. The size of E_a tells us about the reaction rate.

A slow reaction has a large E_a. A fast reaction has a low E_a.

- **The larger the E_a, the greater the amount of energy that is needed to break bonds, and the slower the reaction rate.**

How can we draw the structure of the unstable transition state? The structure of the transition state is somewhere in between the structures of the starting material and product. Any bond that is partially broken or formed is drawn with a dashed line. Any atom that gains or loses a charge contains a partial charge in the transition state. Transition states are drawn in brackets, with a superscript double dagger (‡).

In the hypothetical reaction between **A – B** and **C:**$^-$ to form **A:**$^-$ and **B – C,** the bond between **A** and **B** is partially broken, and the bond between **B** and **C** is partially formed. Because **A** gains a negative charge and **C** loses a charge in the course of the reaction, each atom bears a partial negative charge in the transition state.

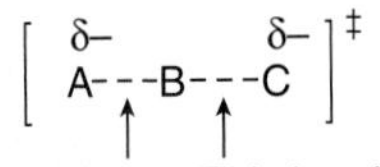

This bond is partially broken. This bond is partially formed.

Several energy diagrams are drawn in Figure 6.5. For any energy diagram:

- **E_a determines the height of the energy barrier.**
- **$\Delta H°$ determines the relative position of the reactants and products.**

The two variables, **E_a and $\Delta H°$, are independent** of each other. Two reactions can have identical values for $\Delta H°$ but very different E_a values. For two exothermic reactions with the same negative value of $\Delta H°$ but different E_a values, the reaction with the lower E_a is faster.

Problem 6.15 Draw an energy diagram for a reaction in which the products are higher in energy than the starting materials and E_a is large. Clearly label all of the following on the diagram: the axes, the starting materials, the products, the transition state, $\Delta H°$, and E_a.

Problem 6.16 Draw the structure for the transition state in the following reaction.

Figure 6.5 Some representative energy diagrams

Example [1]

- **Large E_a ⟶ slow reaction**
- **(+) $\Delta H°$ ⟶ endothermic reaction**

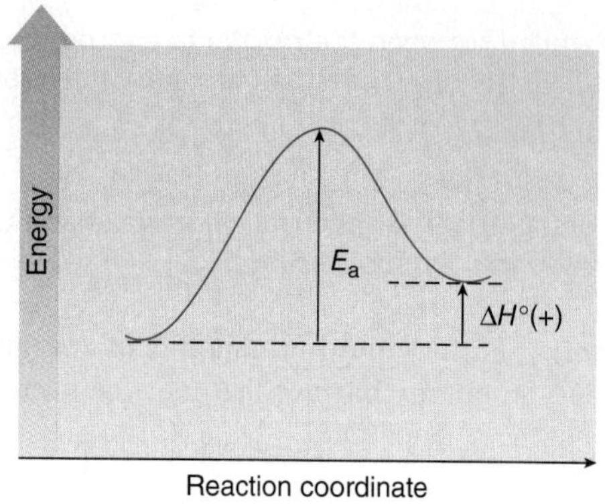

Example [3]

- **Low E_a ⟶ fast reaction**
- **(+) $\Delta H°$ ⟶ endothermic reaction**

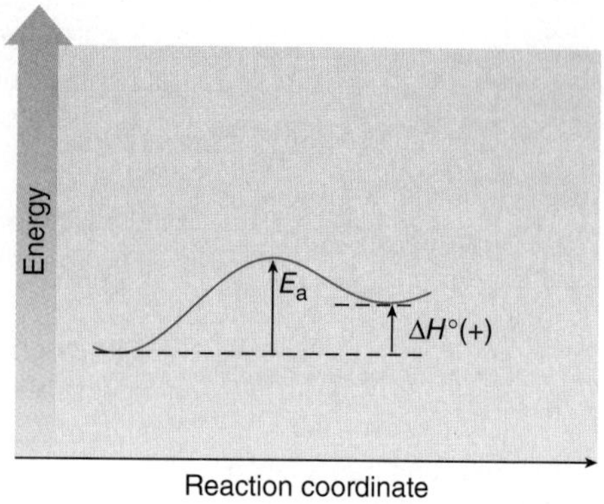

Example [2]

- **Large E_a ⟶ slow reaction**
- **(–) $\Delta H°$ ⟶ exothermic reaction**

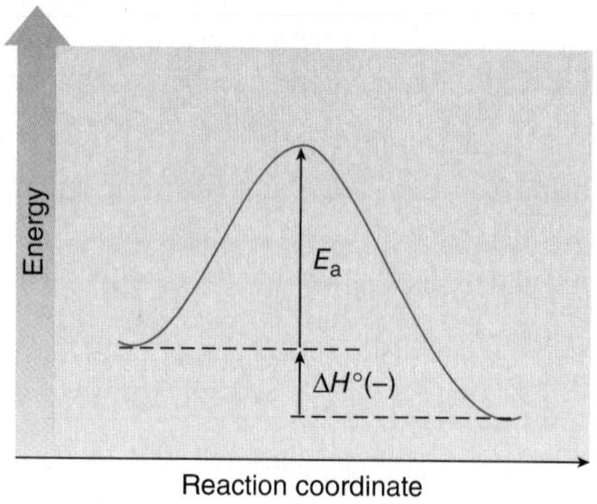

Example [4]

- **Low E_a ⟶ fast reaction**
- **(–) $\Delta H°$ ⟶ exothermic reaction**

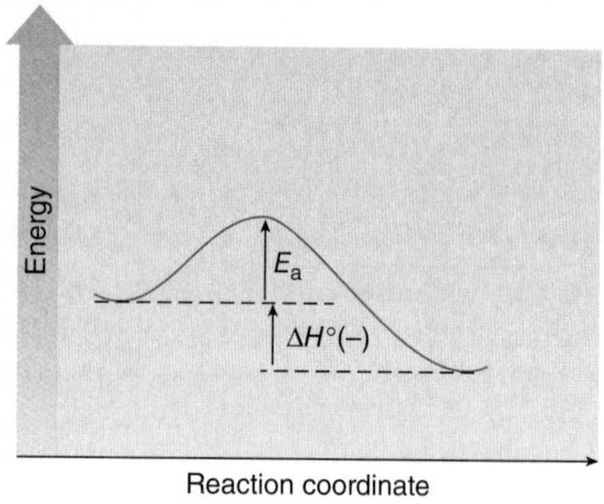

Problem 6.17 Compound **A** can be converted to either **B** or **C.** The energy diagrams for both processes are drawn on the graph below.

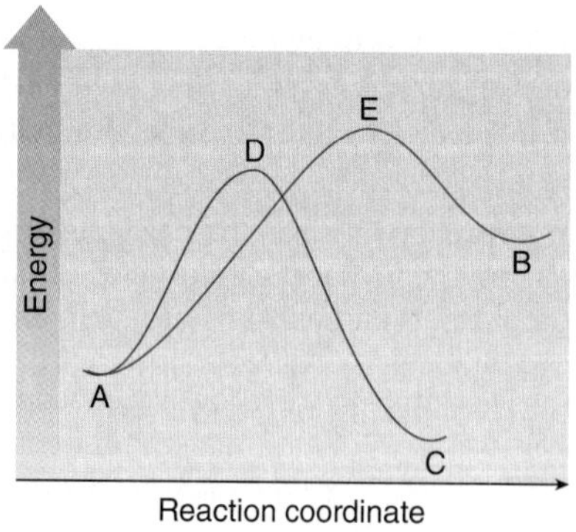

a. Label each reaction as endothermic or exothermic.
b. Which reaction is faster?
c. Which reaction generates the product lower in energy?
d. Which points on the graphs correspond to transition states?
e. Label the energy of activation for each reaction.
f. Label the $\Delta H°$ for each reaction.

6.8 Energy Diagram for a Two-Step Reaction Mechanism

Although the hypothetical reaction in Section 6.7 is concerted, many reactions involve more than one step with formation of a reactive intermediate. Consider the same overall reaction, **A–B + C:$^-$** to form products **A:$^-$ + B–C,** but in this case begin with the assumption that the reaction occurs by a *stepwise* pathway—that is, bond breaking occurs *before* bond making. Once again, assume that the overall process is exothermic.

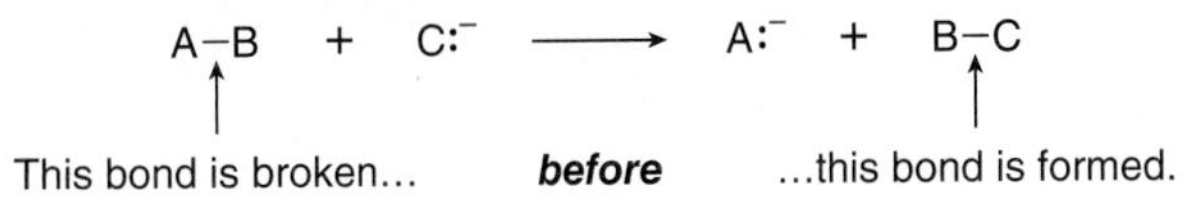

One possible stepwise mechanism involves heterolysis of the **A–B** bond to form two ions **A:$^-$** and **B$^+$**, followed by reaction of **B$^+$** with anion **C:$^-$** to form product **B–C,** as outlined in the accompanying equations. Species **B$^+$** is a **reactive intermediate. B$^+$** is a product in Step [1] that reacts with **C:$^-$** in Step [2].

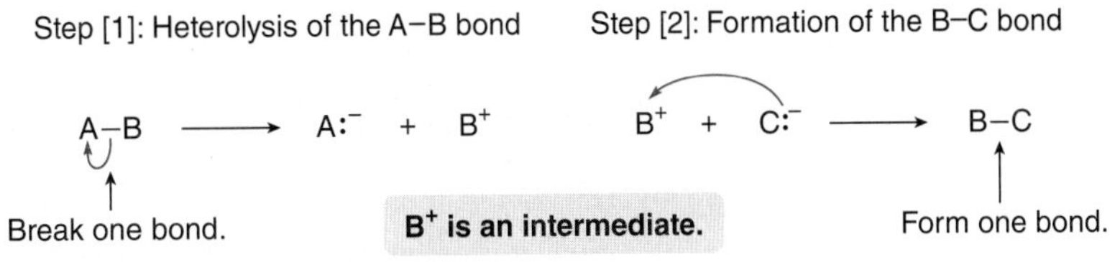

To draw an energy diagram for a two-step mechanism, we must draw an energy diagram for each step, and then combine them. Each step has its own energy barrier, with a transition state at the energy maximum.

Step [1] is endothermic because energy is needed to cleave the **A–B** bond, making $\Delta H°$ a positive value and placing the products of Step [1] at higher energy than the starting materials. In the transition state, the **A–B** bond is partially broken.

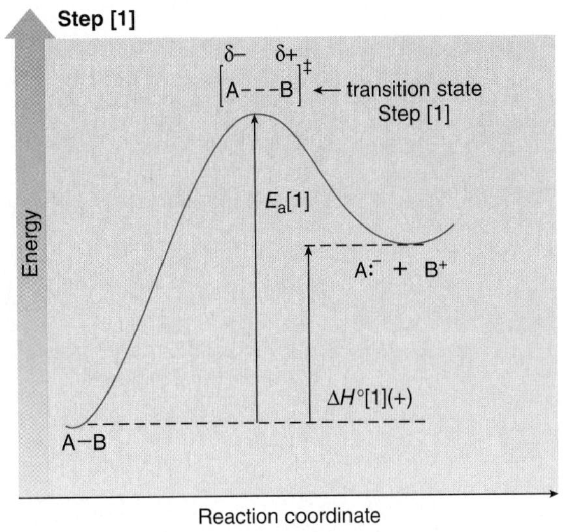

Step [2] is exothermic because energy is released in forming the **B–C** bond, making $\Delta H°$ a negative value and placing the products of Step [2] at lower energy than the starting materials of Step [2]. In the transition state, the **B–C** bond is partially formed.

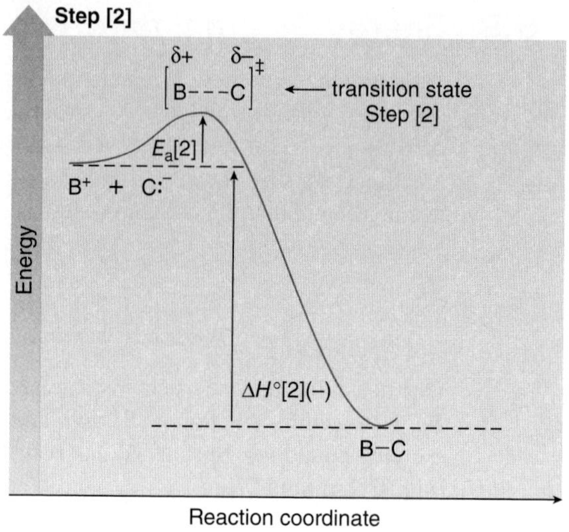

The overall process is shown in Figure 6.6 as a single energy diagram that combines both steps. Because the reaction has two steps, there are two transition states, each corresponding to an energy barrier. The transition states are separated by an energy minimum, at which the reactive intermediate $\mathbf{B^+}$ is located. Because we made the assumption that the overall two-step process is exothermic, the overall energy difference between the reactants and products, labeled $\Delta H^\circ_{overall}$, has a negative value, and the final products are at a lower energy than the starting materials.

The energy barrier for Step [1], labeled $E_a[1]$, is higher than the energy barrier for Step [2], labeled $E_a[2]$. This is because bond cleavage (Step [1]) is more difficult (requires more energy) than bond formation (Step [2]). A higher energy transition state for Step [1] makes it the slower step of the mechanism.

- **In a multistep mechanism, the step with the highest energy transition state is called the rate-determining step.**

In this reaction, the rate-determining step is Step [1].

Figure 6.6 Complete energy diagram for the two-step conversion of A–B + C:$^-$ → A:$^-$ + B–C

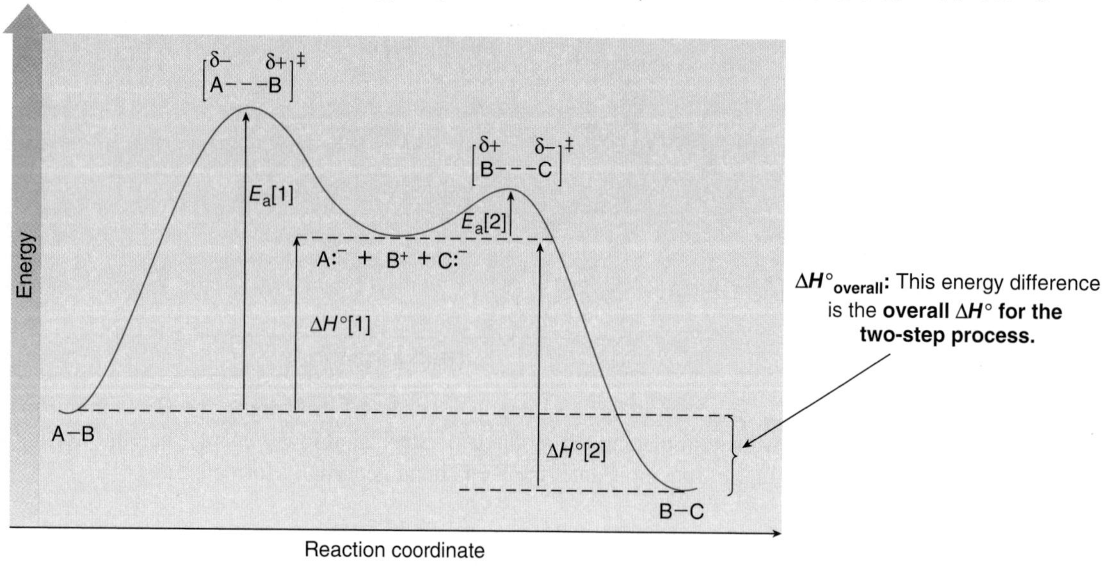

- **The transition states are located at energy maxima, while the reactive intermediate B^+ is located at an energy minimum.**
- Each step has its own value of ΔH° and E_a.
- The overall energy difference between starting material and products is called $\Delta H^\circ_{overall}$. In this example, the products of the two-step sequence are at lower energy than the starting materials.
- Since Step [1] has the higher energy transition state, it is the **rate-determining step.**

Problem 6.18 Consider the following energy diagram.

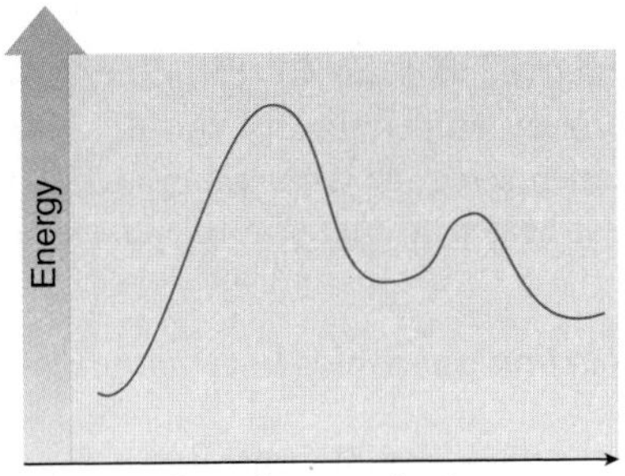

a. How many steps are involved in this reaction?
b. Label $\Delta H°$ and E_a for each step, and label $\Delta H°_{overall}$.
c. Label each transition state.
d. Which point on the graph corresponds to a reactive intermediate?
e. Which step is rate-determining?
f. Is the overall reaction endothermic or exothermic?

Problem 6.19 Draw an energy diagram for a two-step reaction, A → B → C, where the relative energy of these compounds is C < A < B, and the conversion of B → C is rate-determining.

6.9 Kinetics

We now turn to a more detailed discussion of **reaction rate**—that is, how fast a particular reaction proceeds. **The study of reaction rates is called *kinetics*.**

The rate of chemical processes affects many facets of our lives. Aspirin is an effective anti-inflammatory agent because it rapidly inhibits the synthesis of prostaglandins (Section 19.6). Butter turns rancid with time because its lipids are only slowly oxidized by oxygen in the air to undesirable by-products (Section 15.11). DDT (Section 7.4) is a persistent environmental pollutant because it does not react appreciably with water, oxygen, or any other chemical with which it comes into contact. All of these processes occur at different rates, resulting in beneficial or harmful effects.

Some reactions have a very favorable equilibrium constant (K_{eq} >> 1), but the rate is very slow. Gasoline can be safely handled in the air because its reaction with O_2 is slow unless there is a spark to provide energy to initiate the reaction.

6.9A Energy of Activation

As we learned in Section 6.7, the energy of activation, E_a, is the energy difference between the reactants and the transition state. It is the **energy barrier** that must be exceeded for reactants to be converted to products.

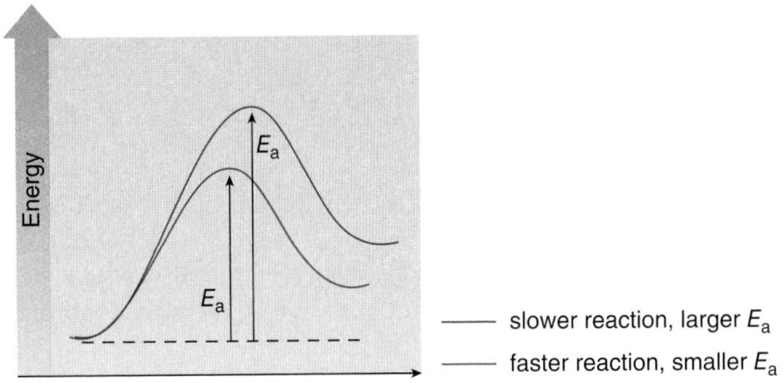

- The *larger* the E_a, the *slower* the reaction.

Concentration and temperature also affect reaction rate.

- The *higher* the concentration, the *faster* the rate. Increasing concentration increases the number of collisions between reacting molecules, which in turn increases the rate.

- The *higher* the temperature, the *faster* the rate in laboratory reactions. Increasing temperature increases the average kinetic energy of the reacting molecules. Because the kinetic energy of colliding molecules is used for bond cleavage, increasing the average kinetic energy increases the rate.

Practically, the effect of temperature on reaction rate is used to an advantage in the kitchen. Food is stored in a cold refrigerator to slow the reactions that cause spoilage.

In contrast to laboratory reactions, enzyme-catalyzed reactions must be run at the specific temperature of the organism, and increasing temperatures could deactivate the enzyme.

The E_a values of most organic reactions are 40–150 kJ/mol. When E_a < 80 kJ/mol, the reaction occurs readily at or below room temperature. When E_a > 80 kJ/mol, higher temperatures are needed. As a rule of thumb, increasing the temperature by 10 °C doubles the reaction rate. Thus, reactions in the lab are often heated to increase their rates so that they occur in a reasonable amount of time.

Keep in mind that certain **reaction quantities have *no effect* on reaction rate.**

- **$\Delta G°$, $\Delta H°$, and K_{eq} do *not* determine the rate of a reaction. These quantities indicate the direction of equilibrium and the relative energy of reactants and products.**

Problem 6.20 Which value (if any) corresponds to a faster reaction: (a) E_a = 40 kJ/mol or E_a = 4 kJ/mol; (b) a reaction temperature of 0 °C or a reaction temperature of 25 °C; (c) K_{eq} = 10 or K_{eq} = 100; (d) $\Delta H°$ = –10 kJ/mol or $\Delta H°$ = 10 kJ/mol?

Problem 6.21 For a reaction with K_{eq} = 0.8 and E_a = 80 kJ/mol, decide which of the following statements is (are) true. Correct any false statement to make it true. Ignore entropy considerations. (a) The reaction is faster than a reaction with K_{eq} = 8 and E_a = 80 kJ/mol. (b) The reaction is faster than a reaction with K_{eq} = 0.8 and E_a = 40 kJ/mol. (c) $\Delta G°$ for the reaction is a positive value. (d) The starting materials are lower in energy than the products of the reaction. (e) The reaction is exothermic.

6.9B Rate Equations

The rate of a chemical reaction is determined by measuring the decrease in the concentration of the reactants over time, or the increase in the concentration of the products over time. A **rate law** (or **rate equation**) is an equation that shows the relationship between the rate of a reaction and the concentration of the reactants. A rate law is determined *experimentally,* and it depends on the mechanism of the reaction.

A rate law has two important terms: the **rate constant symbolized by *k*,** and the **concentration of the reactants.** Not all reactant concentrations may appear in the rate equation, as we shall soon see.

$$\text{rate} = k[\text{reactants}]$$

k = the rate constant

A rate constant k and the energy of activation E_a are inversely related. **A high E_a corresponds to a small k.**

A rate constant k is a fundamental characteristic of a reaction. It is a complex mathematical term that takes into account the dependence of a reaction rate on temperature and the energy of activation.

- **Fast reactions have large rate constants.**
- **Slow reactions have small rate constants.**

What concentration terms appear in the rate equation? That depends on the mechanism. For the organic reactions we will encounter:

- **A rate equation contains concentration terms for *all* reactants involved in a *one-step* mechanism.**
- **A rate equation contains concentration terms for *only* the reactants involved in the *rate-determining step* in a multistep reaction.**

In the one-step reaction of **A–B + C:⁻** to form **A:⁻ + B–C,** *both* reactants appear in the transition state of the only step of the mechanism. The **concentration of *both* reactants affects the**

reaction rate and *both* terms appear in the rate equation. This type of reaction involving two reactants is said to be **bimolecular.**

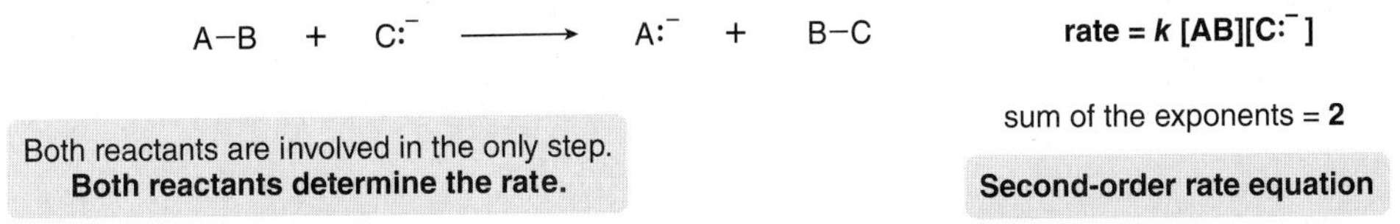

The ***order*** **of a rate equation equals the sum of the exponents of the concentration terms** in the rate equation. In the rate equation for the concerted reaction of **A–B** + **C:⁻**, there are two concentration terms, each with an exponent of one. Thus, the sum of the exponents is two and the **rate equation is *second order*** (the reaction follows second-order kinetics).

Because the rate of the reaction depends on the concentration of both reactants, doubling the concentration of *either* **A–B** or **C:⁻** doubles the rate of the reaction. Doubling the concentration of *both* **A–B** and **C:⁻** increases the reaction rate by a factor of *four.*

The situation is different in the stepwise conversion of **A–B** + **C:⁻** to form **A:⁻** + **B–C.** The mechanism shown in Section 6.8 has two steps: a slow step (the **rate-determining** step) in which the **A–B** bond is broken, and a fast step in which the **B–C** bond is formed.

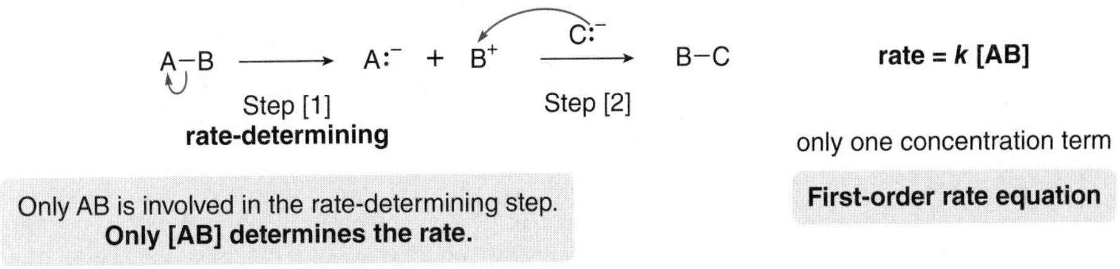

In a multistep mechanism, a reaction can occur no faster than its rate-determining step. **Only the concentrations of the reactants in the rate-determining step appear in the rate equation.** In this example, the rate depends on the concentration of **A–B** *only,* because only **A–B** appears in the rate-determining step. A reaction involving only one reactant is said to be **unimolecular.** Because there is only one concentration term (raised to the first power), the **rate equation is *first order*** (the reaction follows first-order kinetics).

Because the rate of the reaction depends on the concentration of only *one* reactant, doubling the concentration of **A–B** doubles the rate of the reaction, but **doubling the concentration of C:⁻ has *no effect* on the reaction rate.**

This might seem like a puzzling result. If **C:⁻** is involved in the reaction, why doesn't it affect the overall rate of the reaction? Not only can you change the concentration of **C:⁻** and not affect the rate, but you also can replace it by a different anion without affecting the rate. How can this be? **C:⁻ is not involved in the slow step of the reaction, so neither its concentration nor its identity affects the reaction rate.**

The following analogy is useful. Let's say three students must make 20 peanut butter and jelly sandwiches for a class field trip. Student (**1**) spreads the peanut butter on the bread. Student (**2**) spreads on the jelly, and student (**3**) cuts the sandwiches in half. Suppose student (**2**) is very slow in spreading the jelly. It doesn't matter how fast students (**1**) and (**3**) are; they can't finish making sandwiches any faster than student (**2**) can add the jelly. Five more students can spread on the peanut butter, or an entirely different individual can replace student (**3**), and this doesn't speed up the process. How fast the sandwiches are made is determined entirely by the rate-determining step—that is, spreading the jelly.

Rate equations provide very important information about the mechanism of a reaction. Rate laws for new reactions with unknown mechanisms are determined by a set of experiments that measure how a reaction's rate changes with concentration. Then, a mechanism is suggested based on which reactants affect the rate.

Problem 6.22 For each rate equation, what effect does the indicated concentration change have on the overall rate of the reaction?

[1] rate = $k[CH_3CH_2Br][^{-}OH]$
- a. tripling the concentration of CH_3CH_2Br only
- b. tripling the concentration of ^{-}OH only
- c. tripling the concentration of both CH_3CH_2Br and ^{-}OH

[2] rate = $k[(CH_3)_3COH]$
- a. doubling the concentration of $(CH_3)_3COH$
- b. increasing the concentration of $(CH_3)_3COH$ by a factor of 10

Problem 6.23 Write a rate equation for each reaction, given the indicated mechanism.

a.
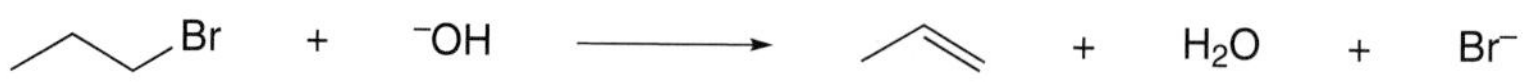

b.
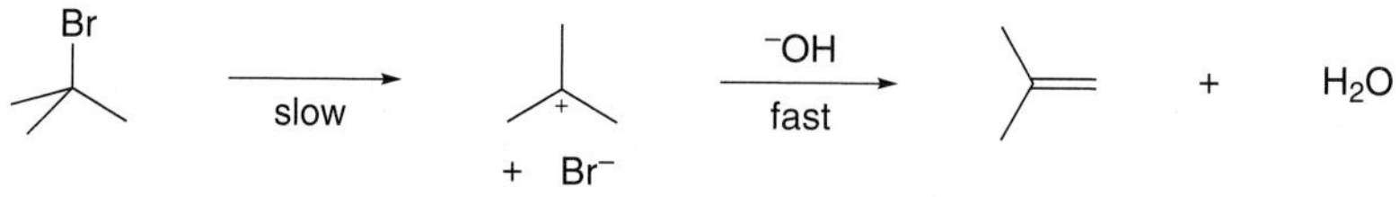

6.10 Catalysts

Some reactions do not occur in a reasonable time unless a **catalyst** is added.

- **A *catalyst* is a substance that speeds up the rate of a reaction. A catalyst is recovered unchanged in a reaction, and it does not appear in the product.**

Common catalysts in organic reactions are **acids** and **metals.** Two examples are shown with the catalyst drawn in red.

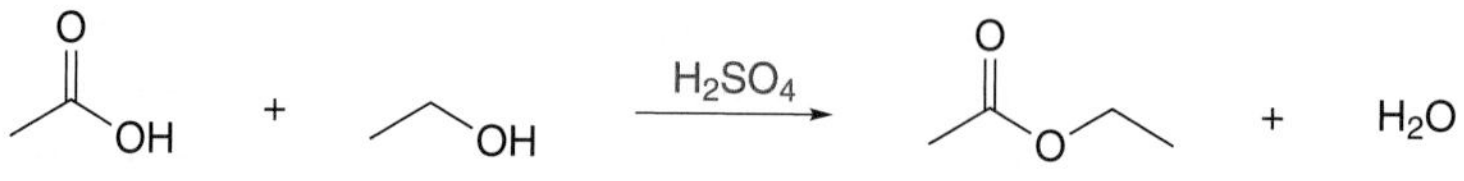

acetic acid ethanol ethyl acetate

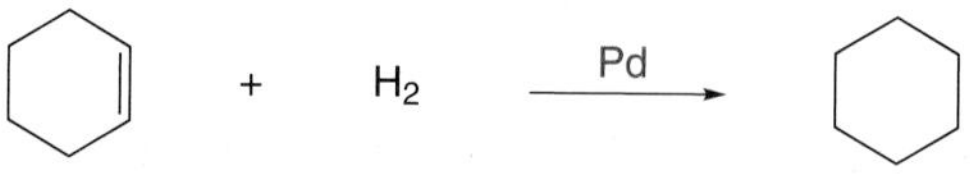

cyclohexene cyclohexane

The reaction of acetic acid with ethanol to yield ethyl acetate and water occurs in the presence of an acid catalyst. The acid catalyst is written over or under the arrow to emphasize that it is not part of the starting materials or the products. The details of this reaction are discussed in Chapter 22.

The reaction of cyclohexene with hydrogen to form cyclohexane occurs only in the presence of a metal catalyst such as palladium, platinum, or nickel. The metal provides a surface that binds both the cyclohexene and the hydrogen, and in doing so, facilitates the reaction. We return to this mechanism in Chapter 12.

Catalysts accelerate a reaction by lowering the energy of activation (Figure 6.7). They have no effect on the equilibrium constant, so they do not change the amount of reactant and product at equilibrium. Thus, catalysts affect how *quickly* equilibrium is achieved, but not the relative amounts of reactants and products at equilibrium. If a catalyst is somehow used up in one step of a reaction sequence, it must be regenerated in another step. Because only a small amount of a catalyst is needed relative to starting material, it is said to be present in a **catalytic amount.**

Problem 6.24 Identify the catalyst in each equation.

a. $CH_2{=}CH_2 \xrightarrow[H_2SO_4]{H_2O} CH_3CH_2OH$ b. $CH_3Cl \xrightarrow[^{-}OH]{I^{-}} CH_3OH + Cl^{-}$

Figure 6.7

The effect of a catalyst on a reaction

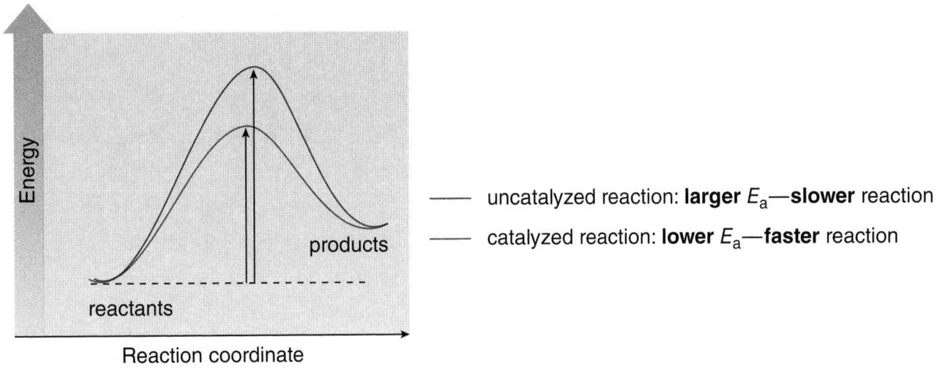

- The catalyst lowers the energy of activation, thus **increasing the rate of the catalyzed reaction.**
- The energy of the reactants and products is the same in both the uncatalyzed and catalyzed reactions, so the **position of equilibrium is unaffected.**

6.11 Enzymes

The catalysts that synthesize and break down biomolecules in living organisms are governed by the same principles as the acids and metals in organic reactions. The catalysts in living organisms, however, are usually protein molecules called **enzymes.**

- *Enzymes* are biochemical catalysts composed of amino acids held together in a very specific three-dimensional shape.

An enzyme contains a region called its **active site,** which binds an organic reactant, called a **substrate.** When bound, this unit is called the **enzyme–substrate complex,** as shown schematically in Figure 6.8 for the enzyme lactase, the enzyme that binds lactose, the principal carbohydrate in milk. Once bound, the organic substrate undergoes a very specific reaction at an enhanced rate. In this example, lactose is converted into two simpler sugars, glucose and galactose. When individuals lack adequate amounts of lactase, a condition called lactose intolerance, they are unable to digest lactose, causing abdominal cramping and diarrhea.

β-Galactosidase is shown as a ribbon diagram, a convention used to illustrate protein structure explained in Figure 26.11.

An enzyme speeds up a biological reaction in a variety of ways. It may hold reactants in the proper conformation to facilitate reaction, or it may provide acidic and basic sites needed for a particular transformation. Once the reaction is completed, the enzyme releases the substrate and it is then able to catalyze another reaction.

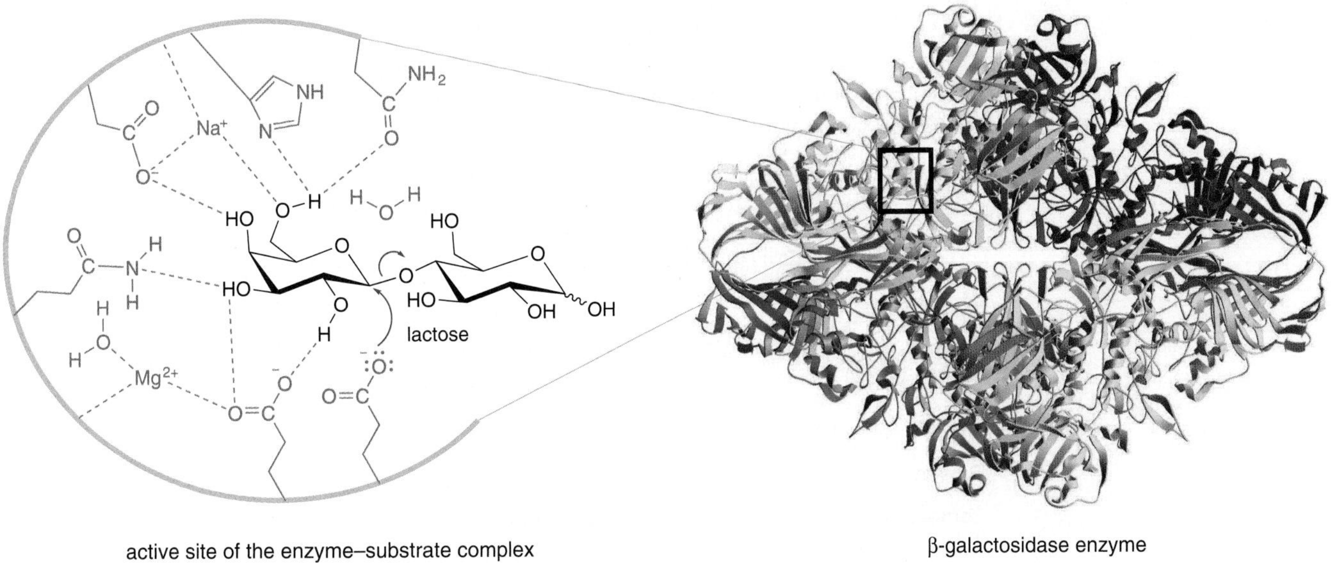

Figure 6.8 Lactase, an example of a biological catalyst

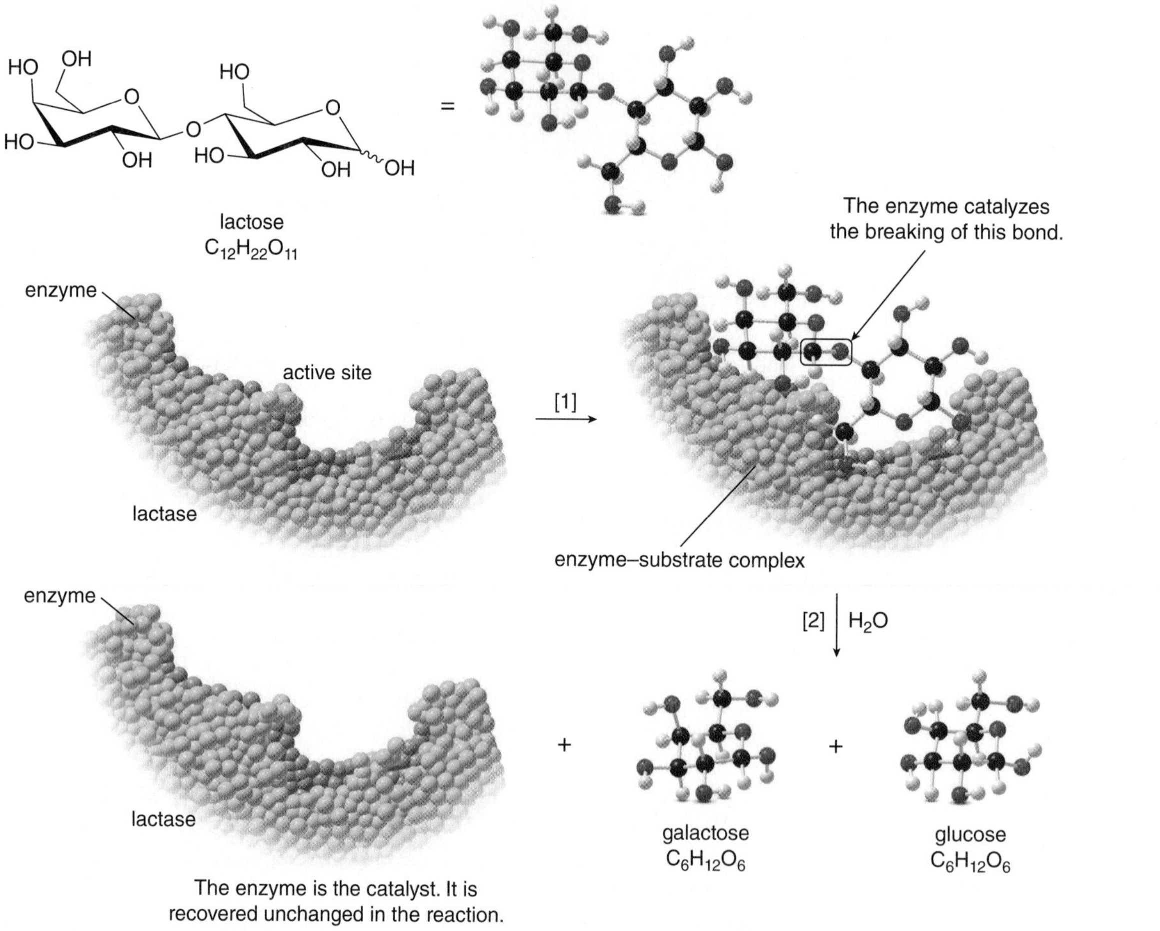

- The enzyme lactase binds the carbohydrate lactose ($C_{12}H_{22}O_{11}$) in its active site in Step [1]. Lactose then reacts with water to break a bond and form two simpler sugars, galactose and glucose, in Step [2]. This process is the first step in digesting lactose, the principal carbohydrate in milk.

The enzyme β-galactosidase is used to produce lactose-free dairy products by catalyzing the hydrolysis of lactose to glucose and galactose prior to human consumption.

A closer look at the active site of the β-galactosidase complex with lactose reveals a selection of amino acids and metal ions that bind lactose. These amino acids hydrogen bond to lactose so that another amino acid in the enzyme can initiate hydrolysis. The curved arrows designate the making and breaking of covalent bonds.

Key distinctions between enzymatic and laboratory reactions are summarized in Table 6.4.

Table 6.4 Comparison of Enzymatic and Laboratory Reactions

	Enzymatic reaction	Laboratory reaction
Size	Small part(s) of a large biomolecule	Small molecule
Catalyst	Amino acids and metal ions	Acids, bases, and metal ions
Specificity	Highly specific	Less specific
Solvent	Water	Organic solvent
Temperature and pH	Specific to an organism	Wide range

KEY CONCEPTS

Understanding Organic Reactions

Writing Equations for Organic Reactions (6.1)

- Use curved arrows to show the movement of electrons. Full-headed arrows are used for electron pairs and half-headed arrows are used for single electrons.
- Reagents can be drawn either on the left side of an equation or over the reaction arrow. Catalysts are drawn over or under the reaction arrow.

Types of Reactions (6.2)

[1] Substitution

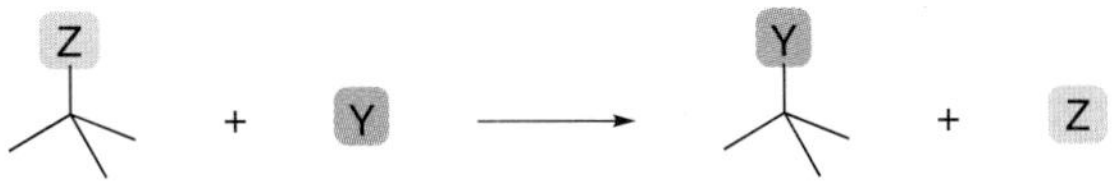

Z = H or a heteroatom

[2] Elimination

Two σ bonds are broken. A π bond is formed.

[3] Addition

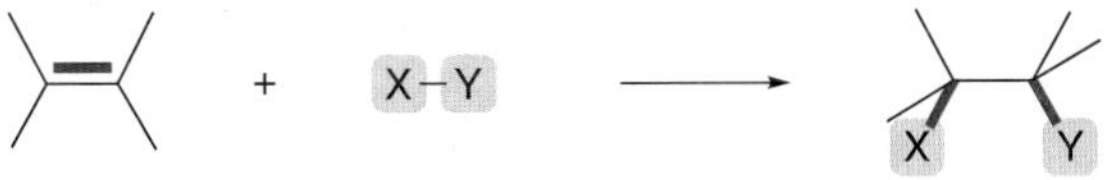

A π bond is broken. Two σ bonds are formed.

Important Trends

Values compared	Trend
Bond dissociation energy and **bond strength**	The ***higher*** the bond dissociation energy, the ***stronger*** the bond (6.4).
Energy and **stability**	The ***higher*** the energy, the ***less stable*** the species (6.5A).
E_a and **reaction rate**	The ***larger*** the energy of activation, the ***slower*** the reaction (6.9A).
E_a and **rate constant**	The ***larger*** the energy of activation, the ***smaller*** the rate constant (6.9B).

Reactive Intermediates (6.3)

- Breaking bonds generates reactive intermediates.
- Homolysis generates radicals with unpaired electrons.
- Heterolysis generates ions.

Reactive intermediate	General structure	Reactive feature	Reactivity
Radical		Unpaired electron	Electrophilic
Carbocation	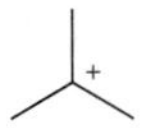	Positive charge; only six electrons around C	Electrophilic
Carbanion	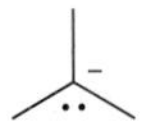	Net negative charge; lone electron pair on C	Nucleophilic

Energy Diagrams (6.7, 6.8)

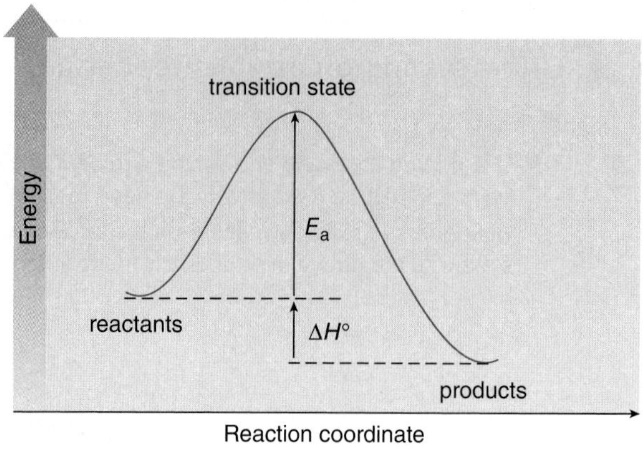

- E_a determines the rate.
- $\Delta H°$ is the difference in bonding energy between the reactants and products.

Conditions Favoring Product Formation (6.5, 6.6)

Variable	Value	Meaning
K_{eq}	$K_{eq} > 1$	More products than reactants are present at equilibrium.
$\Delta G°$	$\Delta G° < 0$	The free energy of the products is **lower** than the energy of the reactants.
$\Delta H°$	$\Delta H° < 0$	Bonds in the products are **stronger** than bonds in the reactants.
$\Delta S°$	$\Delta S° > 0$	The products are **more disordered** than the reactants.

Equations (6.5, 6.6)

$$\Delta G° = -2.303RT \log K_{eq}$$

K_{eq} depends on the energy difference between reactants and products.

[R = 8.314 J/(K•mol), the gas constant
T = Kelvin temperature (K)]

$$\Delta G° = \Delta H° - T\Delta S°$$

$\Delta G°$: free energy change; $\Delta H°$: change in bonding enthalpy; $T\Delta S°$: change in disorder

[T = Kelvin temperature (K)]

Factors Affecting Reaction Rate (6.9)

Factor	Effect
Energy of activation	Larger E_a ---→ slower reaction
Concentration	Higher concentration ---→ faster reaction
Temperature	Higher temperature ---→ faster reaction

PROBLEMS

Problems Using Three-Dimensional Models

6.25 Draw the products of homolysis or heterolysis of each indicated bond. Use electronegativity differences to decide on the location of charges in the heterolysis reaction. Classify each carbon reactive intermediate as a radical, carbocation, or carbanion.

a. 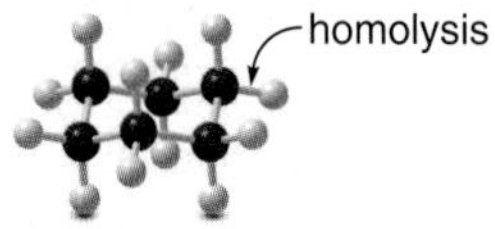

b. 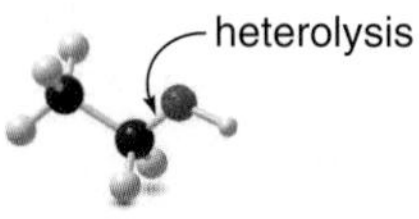

6.26 Explain why the bond dissociation energy for bond (a) is lower than the bond dissociation energy for bond (b).

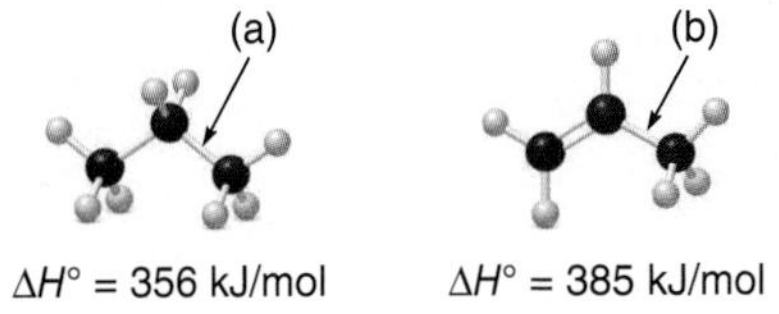
(a)
(b)
ΔH° = 356 kJ/mol
ΔH° = 385 kJ/mol

Types of Reactions

6.27 Classify each transformation as substitution, elimination, or addition.

a. 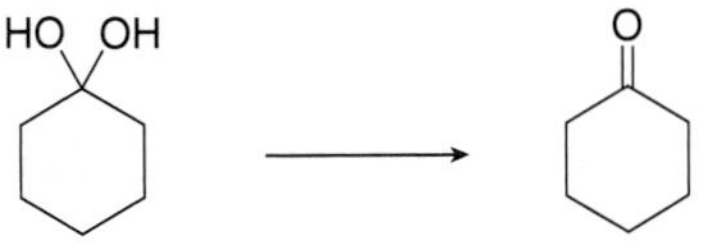

b. 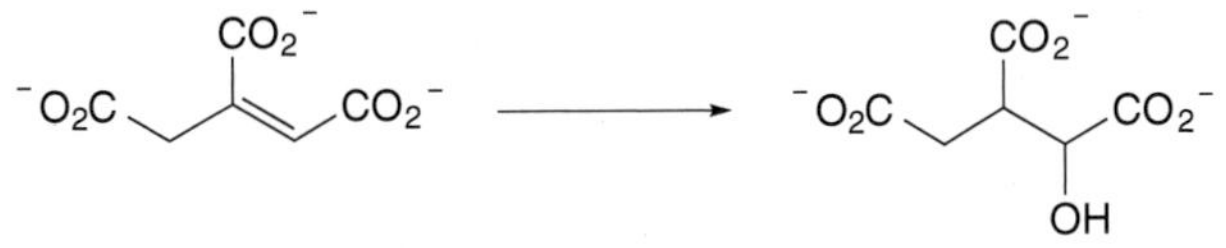

Curved Arrows

6.28 Use full-headed or half-headed curved arrows to show the movement of electrons in each reaction.

a. 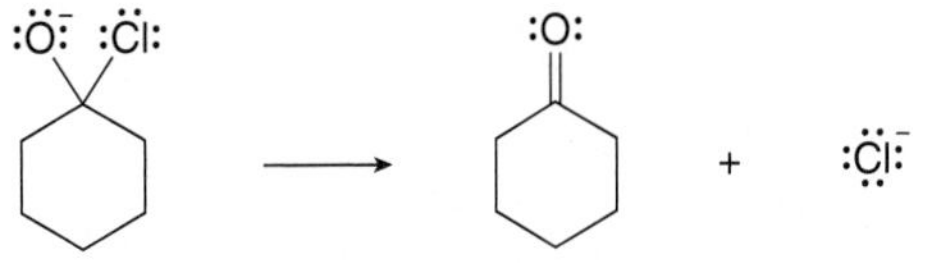

b. 

c. 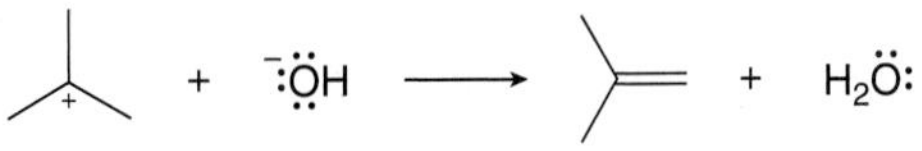

d. 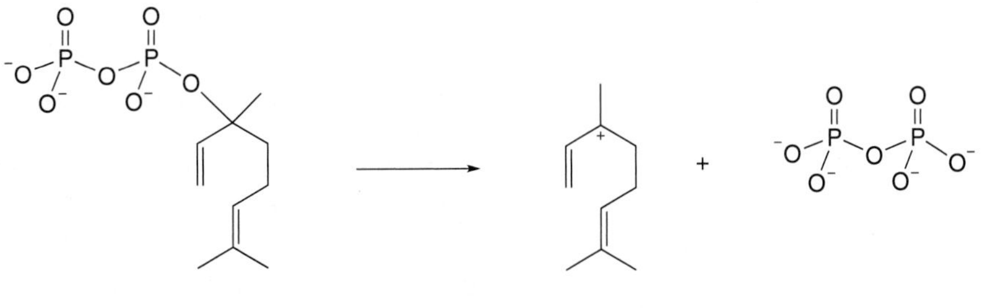

6.29 Draw the products of each reaction by following the curved arrows.

a. 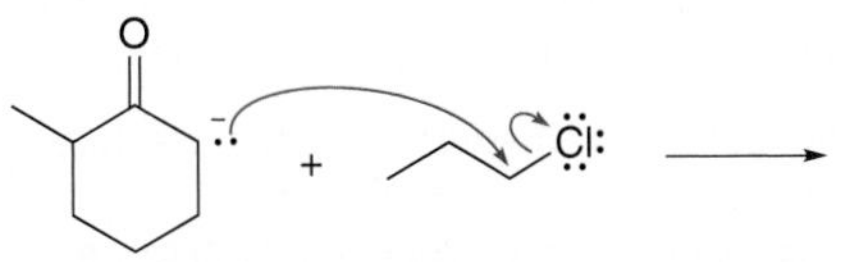

b. 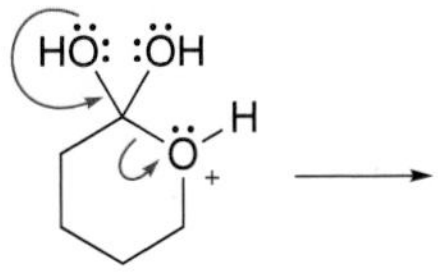

6.30 (a) Add curved arrows for each step to show how **A** is converted to the epoxy ketone **C.** (b) Classify the conversion of **A** to **C** as a substitution, elimination, or addition. (c) Draw one additional resonance structure for **B.**

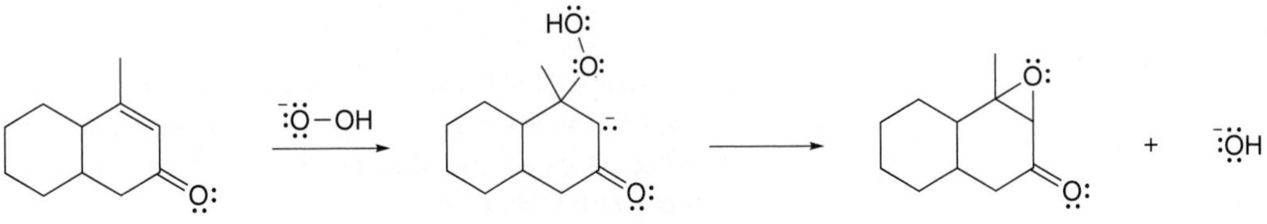

A **B** **C**

6.31 (a) Draw in the curved arrows to show how **A** is converted to **B** in Step [1]. (b) Identify **X,** using the curved arrows drawn for Step [2].

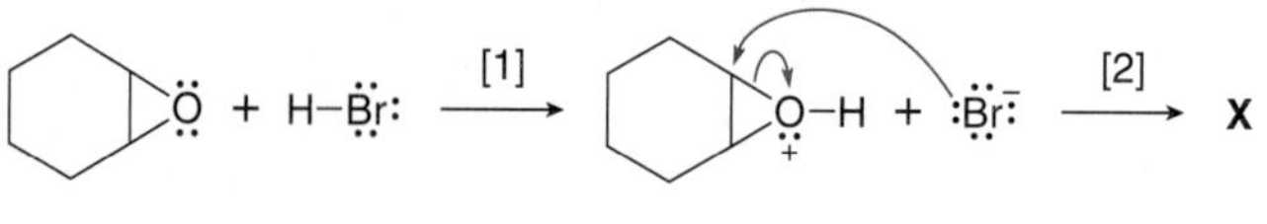

A **B**

6.32 $PGF_{2\alpha}$ (Section 4.15) is synthesized in cells using a cyclooxygenase enzyme that catalyzes a multistep radical pathway. Two steps in the pathway are depicted in the accompanying equations. (a) Draw in curved arrows to illustrate how **C** is converted to **D** in Step [1]. (b) Identify **Y,** the product of Step [2], using the curved arrows that are drawn on compound **D.**

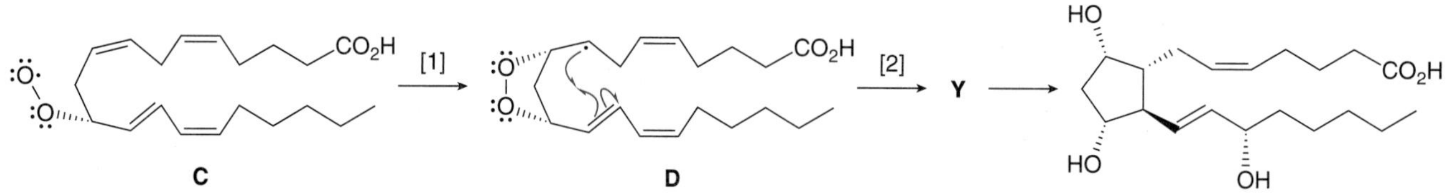

C **D** $PGF_{2\alpha}$

Bond Dissociation Energy and Calculating $\Delta H°$

6.33 Rank the indicated bonds in order of increasing bond dissociation energy.

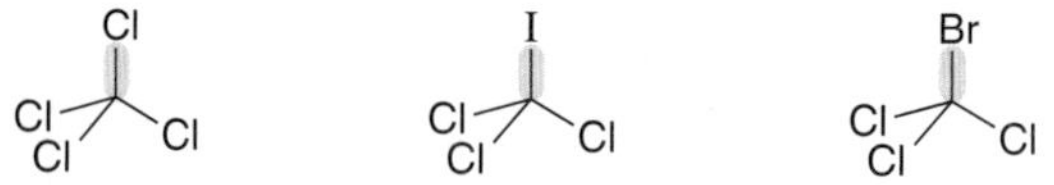

6.34 Calculate $\Delta H°$ for each reaction.

a. $HO\cdot + CH_4 \longrightarrow \cdot CH_3 + H_2O$

b. $CH_3OH + HBr \longrightarrow CH_3Br + H_2O$

6.35 As we learned in Chapter 4, propane ($CH_3CH_2CH_3$) has both 1° and 2° hydrogens.

a. Draw the carbon radical formed by homolysis of each type of C–H bond.

b. Use the values in Table 6.2 to determine which C–H bond is stronger.

c. Explain how this information can be used to determine the relative stability of the two radicals formed. Which radical formed from propane is more stable?

Thermodynamics, $\Delta G°$, $\Delta H°$, $\Delta S°$, and K_{eq}

6.36 Given each value, determine whether the starting material or product is favored at equilibrium.

a. $K_{eq} = 0.5$

b. $\Delta G° = -100$ kJ/mol

c. $\Delta H° = 8.0$ kJ/mol

d. $K_{eq} = 16$

e. $\Delta G° = 2.0$ kJ/mol

f. $\Delta H° = 200$ kJ/mol

g. $\Delta S° = 8$ J/(K•mol)

h. $\Delta S° = -8$ J/(K•mol)

6.37 a. Which value corresponds to a negative value of $\Delta G°$: $K_{eq} = 10^{-2}$ or $K_{eq} = 10^{2}$?

b. In a unimolecular reaction with five times as much starting material as product at equilibrium, what is the value of K_{eq}? Is $\Delta G°$ positive or negative?

c. Which value corresponds to a larger K_{eq}: $\Delta G° = -8$ kJ/mol or $\Delta G° = 20$ kJ/mol?

6.38 As we learned in Chapter 4, monosubstituted cyclohexanes exist as an equilibrium mixture of two conformations having either an axial or equatorial substituent. When R = CH_2CH_3, K_{eq} for this process is 23. When R = $C(CH_3)_3$, K_{eq} for this process is 4000.

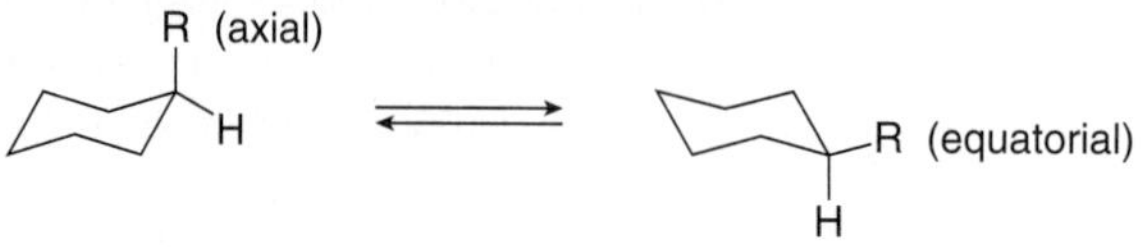

a. When R = CH_2CH_3, which conformation is present in higher concentration?

b. Which R shows the higher percentage of equatorial conformation at equilibrium?

c. Which R shows the higher percentage of axial conformation at equilibrium?

d. For which R is $\Delta G°$ more negative?

e. How is the size of R related to the amount of axial and equatorial conformations at equilibrium?

6.39 In solution, glucose equilibrates between two stereoisomers, α-glucose and β-glucose. The energy difference between the two isomers is 1.46 kJ/mol.

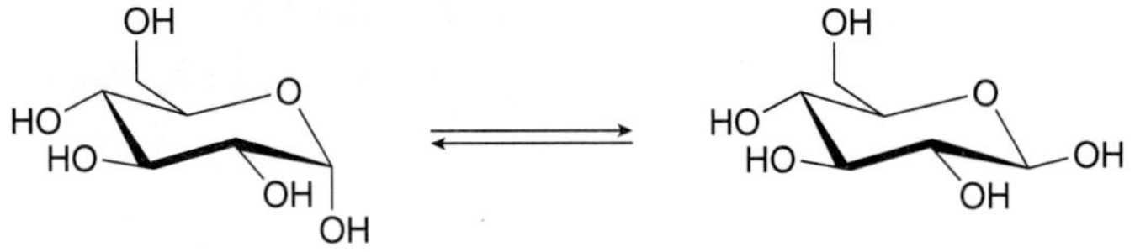

α-glucose β-glucose

a. Which isomer is lower in energy and why?
b. Is the $\Delta G^{\circ\prime}$ positive or negative as drawn?
c. Is the K_{eq} greater or less than 1?

6.40 For which of the following reactions is ΔS° a positive value?

a. 

b. 

6.41 The hydrolysis of phosphoenolpyruvate releases 61.9 kJ/mol of energy. In glucose metabolism, this highly spontaneous reaction drives the energetically unfavorable phosphorylation of ADP to ATP in the presence of Mg^{2+} and the pyruvate kinase enzyme. (Hint: See Section 6.5C.)

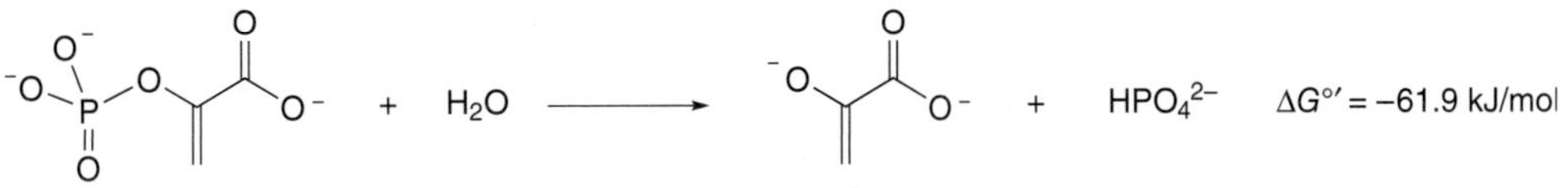

phosphoenolpyruvate pyruvate

a. Write the net (coupled) reaction by summing the substances in both the phosphorylation and hydrolysis equations.
b. How much energy is released in the coupled reaction?
c. Write the equation for the coupled reaction using coupled reaction arrows.

Energy Diagrams and Transition States

6.42 Draw the transition state for the following reaction.

6.43 Draw an energy diagram for each reaction. Label the axes, the starting material, product, transition state, ΔH°, and E_a.

a. a concerted reaction with $\Delta H^\circ = -80$ kJ/mol and $E_a = 16$ kJ/mol
b. a two-step reaction, **A** → **B** → **C,** in which the relative energy of the compounds is **A** < **C** < **B,** and the step **A** → **B** is rate-determining

6.44 Consider the following energy diagram for the conversion of **A** → **G.**

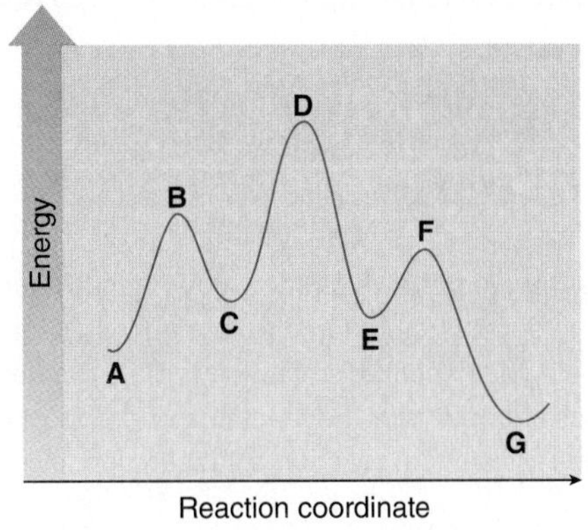

a. Which points on the graph correspond to transition states?
b. Which points on the graph correspond to reactive intermediates?
c. How many steps are present in the reaction mechanism?
d. Label each step of the mechanism as endothermic or exothermic.
e. Label the overall reaction as endothermic or exothermic.

6.45 Draw an energy diagram for the Brønsted–Lowry acid–base reaction of CH_3CO_2H with $^-OC(CH_3)_3$ to form $CH_3CO_2^-$ and $(CH_3)_3COH$. Label the axes, starting materials, products, $\Delta H°$, and E_a. Draw the structure of the transition state.

6.46 Consider the following two-step reaction:

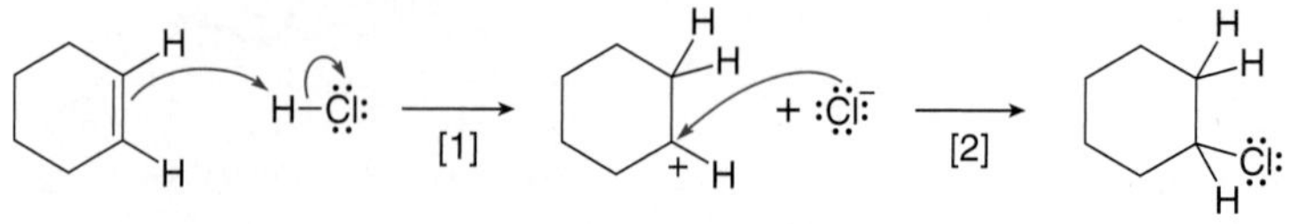

a. How many bonds are broken and formed in Step [1]? Would you predict $\Delta H°$ of Step [1] to be positive or negative?
b. How many bonds are broken and formed in Step [2]? Would you predict the $\Delta H°$ of Step [2] to be positive or negative?
c. Which step is rate-determining?
d. Draw the structure for the transition state in both steps of the mechanism.
e. If $\Delta H°_{overall}$ is negative for this two-step reaction, draw an energy diagram illustrating all of the information in parts (a)–(d).

Kinetics and Rate Laws

6.47 The following is a concerted, bimolecular reaction: $CH_3Br + NaCN \rightarrow CH_3CN + NaBr$.
a. What is the rate equation for this reaction?
b. What happens to the rate of the reaction if $[CH_3Br]$ is doubled?
c. What happens to the rate of the reaction if [NaCN] is halved?
d. What happens to the rate of the reaction if $[CH_3Br]$ and [NaCN] are both increased by a factor of five?

6.48 The conversion of acetyl chloride to methyl acetate occurs via the following two-step mechanism:

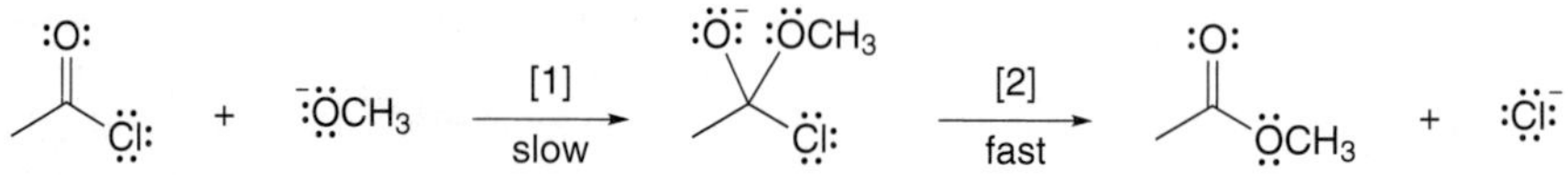

acetyl chloride | methyl acetate

a. Add curved arrows to show the movement of the electrons in each step.
b. Write the rate equation for this reaction, assuming the first step is rate-determining.
c. If the concentration of $^-OCH_3$ were increased 10 times, what would happen to the rate of the reaction?
d. If the concentrations of both CH_3COCl and $^-OCH_3$ were increased 10 times, what would happen to the rate of the reaction?
e. Classify the conversion of acetyl chloride to methyl acetate as an addition, elimination, or substitution.

General Problems

6.49 The Diels–Alder reaction, a powerful reaction discussed in Chapter 16, occurs when a 1,3-diene such as **A** reacts with an alkene such as **B** to form the six-membered ring in **C.**

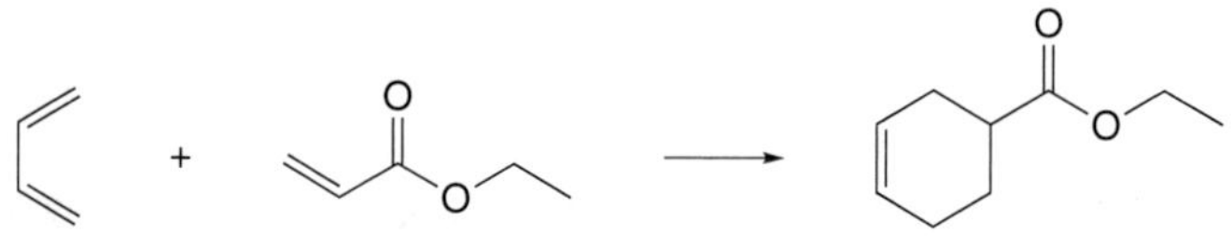

A **B** **C**

a. Draw curved arrows to show how **A** and **B** react to form **C.**
b. What bonds are broken and formed in this reaction?
c. Would you expect this reaction to be endothermic or exothermic?
d. Does entropy favor the reactants or products?
e. Is the Diels–Alder reaction a substitution, elimination, or addition?

6.50 The conversion of $(CH_3)_3CI$ to $(CH_3)_2C=CH_2$ can occur by either a one-step or a two-step mechanism, as shown in Equations [1] and [2].

[1] 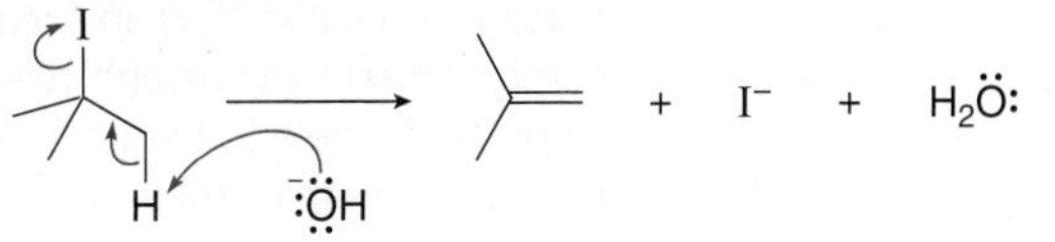

[2] 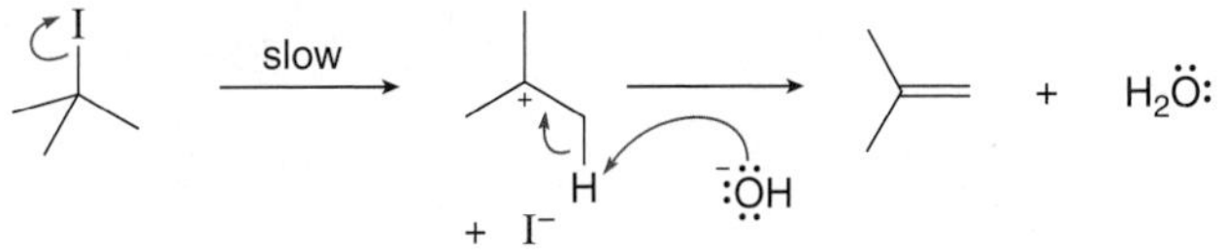

a. What rate equation would be observed for the mechanism in Equation [1]?
b. What rate equation would be observed for the mechanism in Equation [2]?
c. What is the order of each rate equation (i.e., first, second, and so forth)?
d. How can these rate equations be used to show which mechanism is the right one for this reaction?
e. Assume Equation [1] represents an endothermic reaction and draw an energy diagram for the reaction. Label the axes, reactants, products, E_a, and $\Delta H°$. Draw the structure for the transition state.
f. Assume Equation [2] represents an endothermic reaction and that the product of the rate-determining step is higher in energy than the reactants or products. Draw an energy diagram for this two-step reaction. Label the axes, reactants and products for each step, and the E_a and $\Delta H°$ for each step. Label $\Delta H°_{overall}$. Draw the structure for both transition states.

Challenge Problems

6.51 Explain why $HC{\equiv}CH$ is more acidic than CH_3CH_3, even though the C–H bond in $HC{\equiv}CH$ has a higher bond dissociation energy than the C–H bond in CH_3CH_3.

6.52 The use of curved arrows is a powerful tool that illustrates even complex reactions.

a. Add curved arrows to show how carbocation **A** is converted to carbocation **B.** Label each new σ bond formed. Similar reactions have been used in elegant syntheses of steroids.

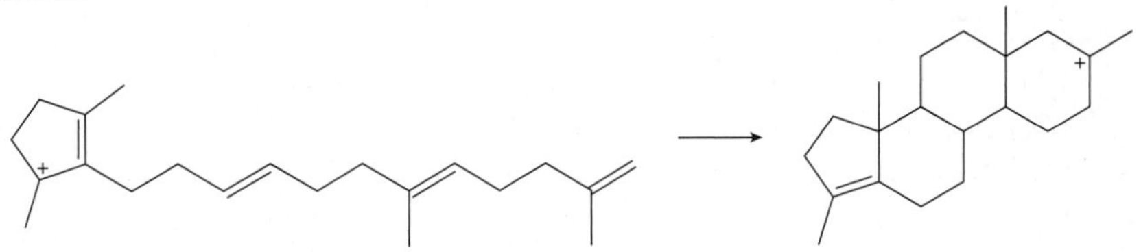

A **B**

b. Draw the product by following the curved arrows. This reaction is an example of a [3,3] sigmatropic rearrangement, as we will learn in Chapter 30.

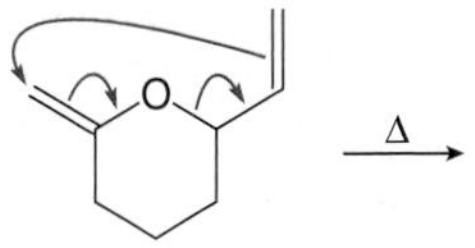

6.53

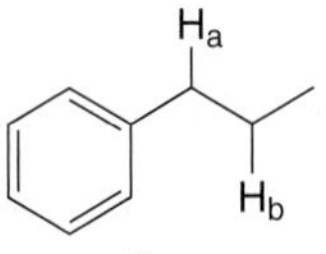

propylbenzene

a. What carbon radical is formed by homolysis of the $C-H_a$ bond in propylbenzene? Draw all reasonable resonance structures for this radical.
b. What carbon radical is formed by homolysis of the $C-H_b$ bond in propylbenzene? Draw all reasonable resonance structures for this radical.
c. The bond dissociation energy of one of the C–H bonds is considerably less than the bond dissociation energy of the other. Which C–H bond is weaker? Offer an explanation.

6.54 Esterification is the reaction of a carboxylic acid (RCOOH) with an alcohol (R′OH) to form an ester (RCOOR′) with loss of water. Equation [1] is an example of an *intermolecular* esterification reaction. Equation [2] is an example of an *intramolecular* esterification reaction; that is, the carboxylic acid and alcohol are contained in the same starting material, forming a cyclic ester as product. The equilibrium constants for both reactions are given. Explain why K_{eq} is different for these two apparently similar reactions.

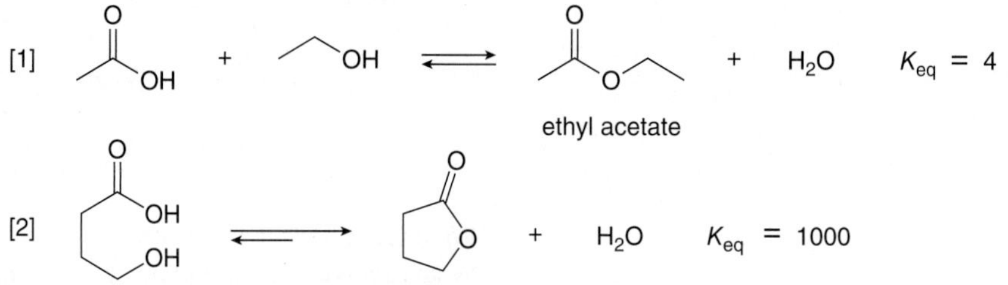

6.55 Although K_{eq} of Equation [1] in Problem 6.54 does not greatly favor formation of the product, it is sometimes possible to use Le Châtelier's principle to increase the yield of ethyl acetate. Le Châtelier's principle states that if an equilibrium is disturbed, a system will react to counteract this disturbance. How can Le Châtelier's principle be used to drive the equilibrium to increase the yield of ethyl acetate? Another example of Le Châtelier's principle is given in Section 9.8.

6.56 As we will learn in Section 15.12, many antioxidants—compounds that prevent unwanted radical oxidation reactions from occurring—are phenols, compounds that contain an OH group bonded directly to a benzene ring.

a. Explain why homolysis of the O–H bond in phenol requires considerably less energy than homolysis of the O–H bond in ethanol (362 kJ/mol vs. 438 kJ/mol).

b. Why is the C–O bond in phenol shorter than the C–O bond in ethanol?

Alkyl Halides and Nucleophilic Substitution

7

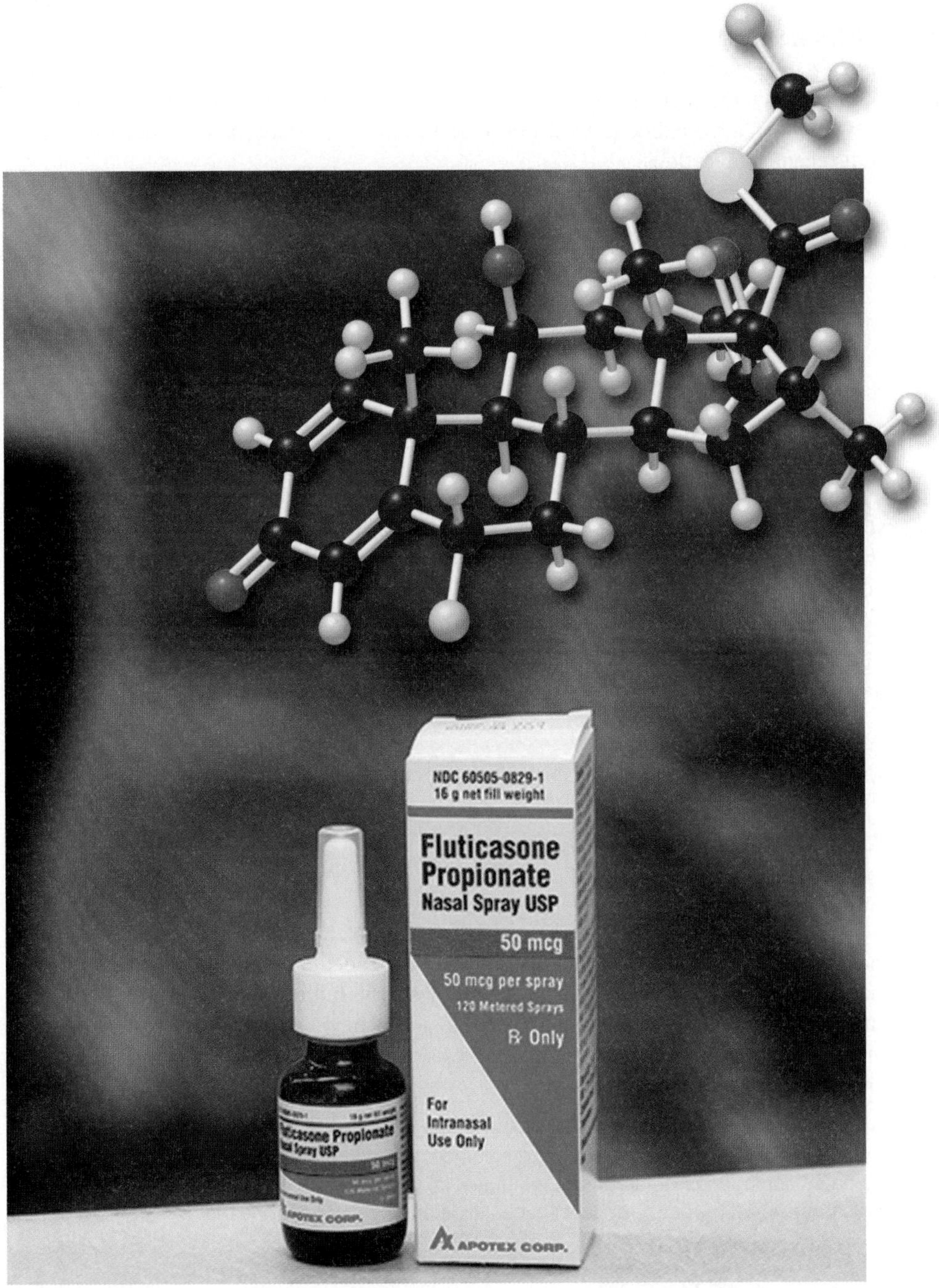

7.1 Introduction to alkyl halides
7.2 Nomenclature
7.3 Physical properties
7.4 Interesting alkyl halides
7.5 The polar carbon–halogen bond
7.6 General features of nucleophilic substitution
7.7 The leaving group
7.8 The nucleophile
7.9 Possible mechanisms for nucleophilic substitution
7.10 Two mechanisms for nucleophilic substitution
7.11 The S_N2 mechanism
7.12 The S_N1 mechanism
7.13 Carbocation stability
7.14 The Hammond postulate
7.15 When is the mechanism S_N1 or S_N2?
7.16 Biological nucleophilic substitution
7.17 Vinyl halides and aryl halides
7.18 Organic synthesis

Fluticasone is a prominent member of a growing number of useful drugs that contain halogens. Sold under the trade name Flonase, fluticasone is a synthetic steroid that is used as an aerosol inhalation for the treatment of seasonal and recurrent nasal allergies. Because it acts by decreasing inflammation, fluticasone is also used to combat the chronic lung inflammation associated with asthma. In Chapter 7 we learn about alkyl halides like fluticasone and one of their characteristic reactions, nucleophilic substitution.

This is the first of three chapters dealing with an in-depth study of the organic reactions of compounds containing C–Z σ bonds, where Z is an element more electronegative than carbon. In Chapter 7, we learn about **alkyl halides** and one of their characteristic reactions, **nucleophilic substitution**. In Chapter 8, we look at **elimination,** a second general reaction of alkyl halides. We conclude this discussion in Chapter 9 by examining other molecules that also undergo nucleophilic substitution and elimination reactions.

7.1 Introduction to Alkyl Halides

Alkyl halides are organic molecules containing a halogen atom X bonded to an sp^3 hybridized carbon atom. As we learned in Section 3.2, alkyl halides are classified as **primary (1°), secondary (2°),** or **tertiary (3°)** depending on the number of carbons bonded to the carbon with the halogen. Whether an alkyl halide is 1°, 2°, or 3° is the *most important factor* in determining the course of its chemical reactions.

Alkyl halides have the general molecular formula $\mathbf{C_nH_{2n+1}X}$, and are formally derived from an alkane by replacing a hydrogen atom with a halogen.

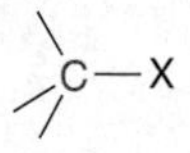

alkyl halide
X = F, Cl, Br, I
C is sp^3 hybridized.

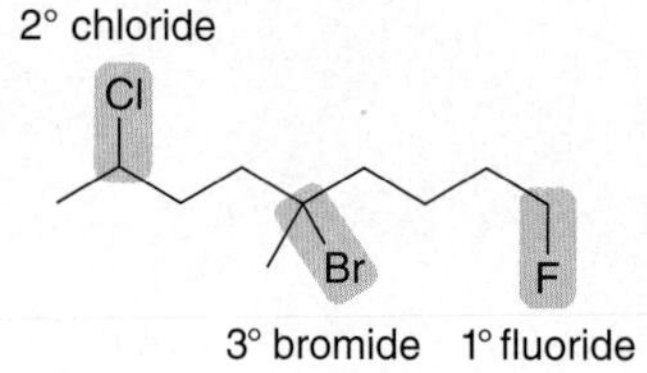

Four types of organic halides having the halogen atom in close proximity to a π bond are illustrated in Figure 7.1. **Vinyl halides** have a halogen atom bonded to a carbon–carbon double bond, and **aryl halides** have a halogen atom bonded to a benzene ring. These two types of organic halides with X bonded directly to an sp^2 hybridized carbon atom do *not* undergo the reactions presented in Chapter 7, as discussed in Section 7.17.

Figure 7.1 Four types of organic halides (RX) having X near a π bond

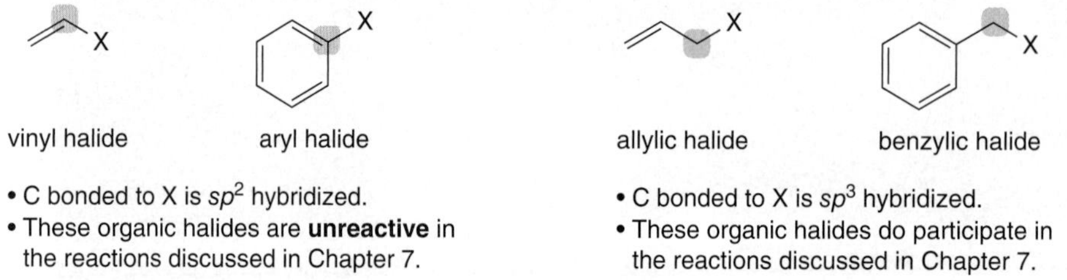

Allylic halides and benzylic halides have halogen atoms bonded to sp^3 hybridized carbon atoms and *do* undergo the reactions described in Chapter 7. **Allylic halides** have X bonded to the carbon atom *adjacent* to a carbon–carbon double bond, and **benzylic halides** have X bonded to the carbon atom *adjacent* to a benzene ring. The synthesis of allylic and benzylic halides is discussed in Sections 15.10 and 18.14, respectively.

Problem 7.1 Telfairine, a naturally occurring insecticide, and halomon, an antitumor agent, are two polyhalogenated compounds isolated from red algae. (a) Classify each halide bonded to an sp^3 hybridized carbon as 1°, 2°, or 3°. (b) Label each halide as vinyl, allylic, or neither.

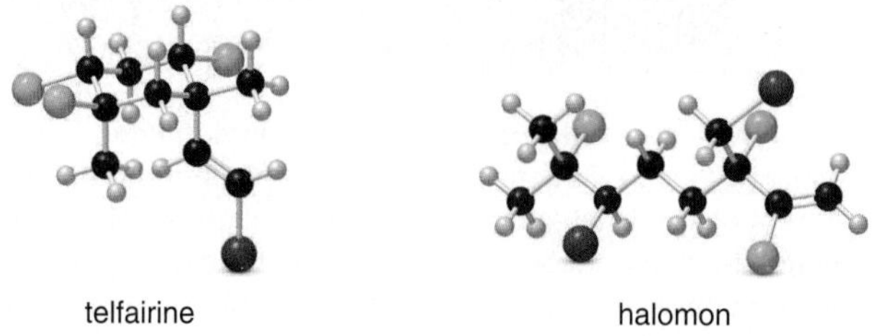

7.2 Nomenclature

The systematic (IUPAC) method for naming alkyl halides follows from the basic rules described in Chapter 4. Common names are also discussed in Section 7.2B, because many low molecular weight alkyl halides are often referred to by their common names.

7.2A IUPAC System

An alkyl halide is named as an alkane with a halogen substituent—that is, as a ***halo alkane.*** To name a halogen substituent, change the ***-ine*** ending of the name of the halogen to the suffix ***-o*** (chlor*ine* → chlor*o*).

How To Name an Alkyl Halide Using the IUPAC System

Example Give the IUPAC name of the following alkyl halide:

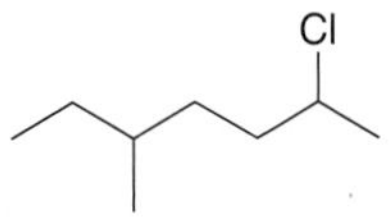

Step [1] Find the parent carbon chain and name it as an alkane.

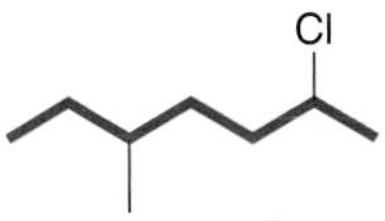

7 C's in the longest chain

7 C's ---→ heptane

- Name the parent chain as an ***alkane,*** with the halogen as a substituent bonded to the longest chain.

Step [2] Apply all other rules of nomenclature.

a. **Number** the chain.

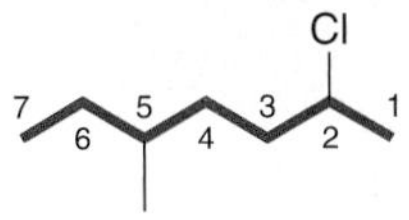

- Begin at the end nearest the first substituent, either alkyl or halogen.

b. **Name** and **number** the substituents.

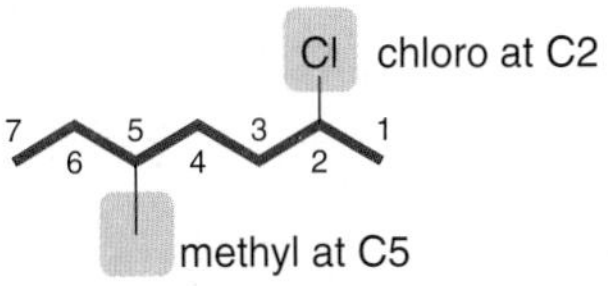

c. **Alphabetize: c** for **c**hloro, then **m** for **m**ethyl.

Answer: 2-chloro-5-methylheptane

7.2B Common Names

Common names for alkyl halides are used only for simple alkyl halides. To assign a common name:

- **Name all the carbon atoms of the molecule as a single alkyl group.**
- **Name the halogen bonded to the alkyl group. To name the halogen, change the *-ine* ending of the halogen name to the suffix *-ide;* for example, brom*ine* → brom*ide.***
- **Combine the names of the alkyl group and halide, separating the words with a space.**

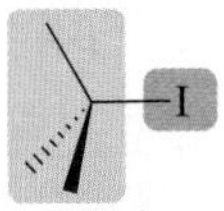

iodine → iodide

tert-butyl group

***tert*-butyl iodide**

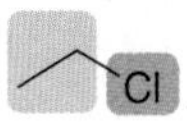

chlorine → chloride

ethyl group

ethyl chloride

Other examples of alkyl halide nomenclature are given in Figure 7.2.

Figure 7.2
Examples: Nomenclature of alkyl halides

IUPAC: **1-chloro-2-methylpropane**
Common: isobutyl chloride

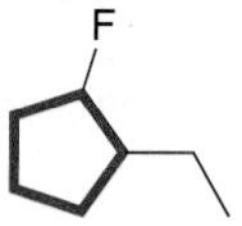

- **e**thyl group at **C1**
- **f**luoro group at **C2**

IUPAC: **1-ethyl-2-fluorocyclopentane**
earlier letter --> lower number
[too complex to use a common name]

Problem 7.2 Give the IUPAC name for each compound.

a. 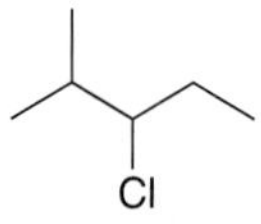
b. 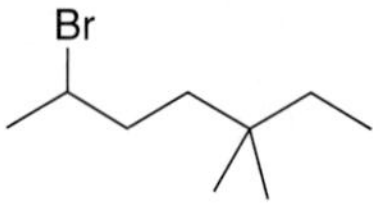
c. 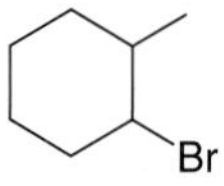
d. 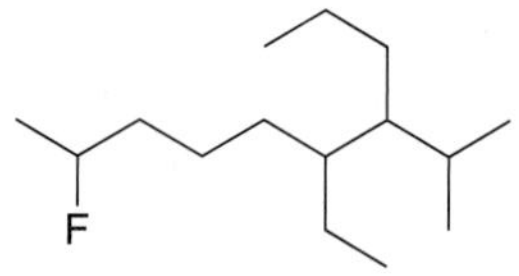

Problem 7.3 Give the structure corresponding to each name.

a. 3-chloro-2-methylhexane
b. 4-ethyl-5-iodo-2,2-dimethyloctane
c. *cis*-1,3-dichlorocyclopentane
d. 1,1,3-tribromocyclohexane
e. *sec*-butyl bromide
f. 6-ethyl-3-iodo-3,5-dimethylnonane

7.3 Physical Properties

Alkyl halides are weakly polar molecules. They exhibit **dipole–dipole** interactions because of their polar C–X bond, but because the rest of the molecule contains only C–C and C–H bonds they are incapable of intermolecular hydrogen bonding. How this affects their physical properties is summarized in Table 7.1.

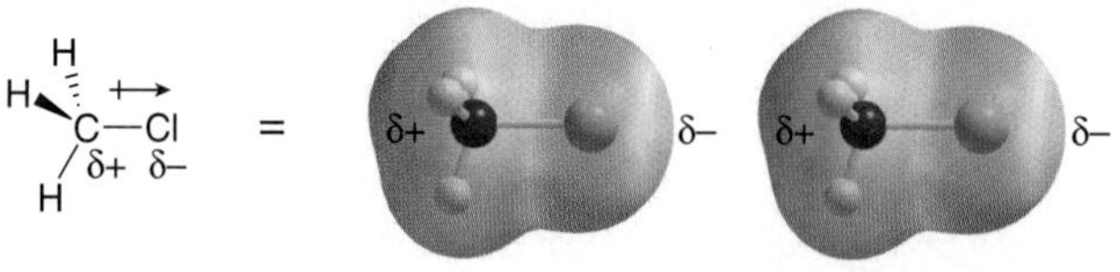

Opposite ends of the dipoles interact.

Table 7.1 Physical Properties of Alkyl Halides

Property	Observation		
Boiling point and melting point	• Alkyl halides have higher bp's and mp's than alkanes having the same number of carbons.	 bp = –42 °C	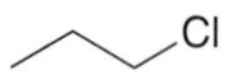 bp = 47 °C
	• Bp's and mp's increase as the size of R increases.	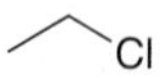 mp = –136 °C bp = 12 °C	and 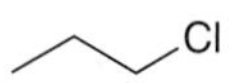 mp = –123 °C bp = 47 °C larger surface area higher mp and bp
	• Bp's and mp's increase as the size of X increases.	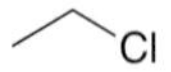 mp = –136 °C bp = 12 °C	and 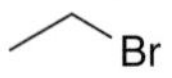 mp = –119 °C bp = 39 °C more polarizable halogen higher mp and bp
Solubility	• RX is soluble in organic solvents. • RX is insoluble in water.		

Problem 7.4 An sp^3 hybridized C – Cl bond is more polar than an sp^2 hybridized C – Cl bond. (a) Explain why this phenomenon arises. (b) Rank the following compounds in order of increasing boiling point.

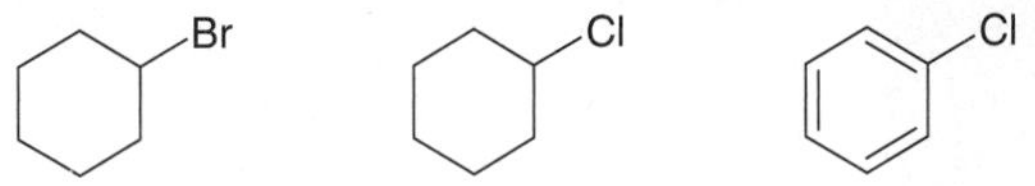

7.4 Interesting Alkyl Halides

Many simple alkyl halides make excellent solvents because they are not flammable and dissolve a wide variety of organic compounds. Compounds in this category include **$CHCl_3$** (chloroform or trichloromethane) and **CCl_4** (carbon tetrachloride or tetrachloromethane). Large quantities of these solvents are produced industrially each year, but like many chlorinated organic compounds, both chloroform and carbon tetrachloride are toxic if inhaled or ingested. Other simple alkyl halides are shown in Figure 7.3.

Figure 7.3 Some simple alkyl halides

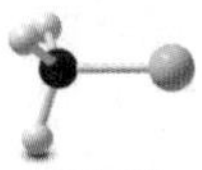
CH_3Cl

- **Chloromethane (CH_3Cl)** is produced by giant kelp and algae and also found in emissions from volcanoes such as Hawai'i's Kilauea. Almost all of the atmospheric chloromethane results from these natural sources.

CH_2Cl_2

- **Dichloromethane (or methylene chloride, CH_2Cl_2)** is an important solvent, once used to decaffeinate coffee. Coffee is now decaffeinated by using supercritical CO_2 due to concerns over the possible ill effects of trace amounts of residual CH_2Cl_2 in the coffee. Subsequent studies on rats have shown, however, that no cancers occurred when animals ingested the equivalent of over 100,000 cups of decaffeinated coffee per day.

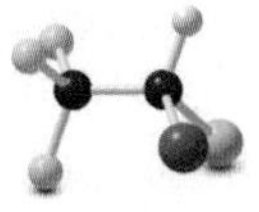
$CF_3CHClBr$

- **Halothane ($CF_3CHClBr$)** is a safe general anesthetic compared to other organic anesthetics such as $CHCl_3$, which causes liver and kidney damage, and $CH_3CH_2OCH_2CH_3$ (diethyl ether), which is very flammable.

Synthetic organic halides are also used in insulating materials, plastic wrap, and coatings. Two such compounds are **Teflon** and **poly(vinyl chloride) (PVC).**

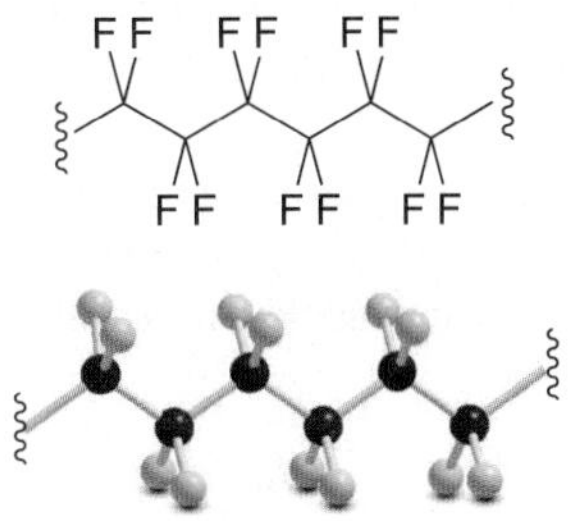

Teflon
(nonstick coating)

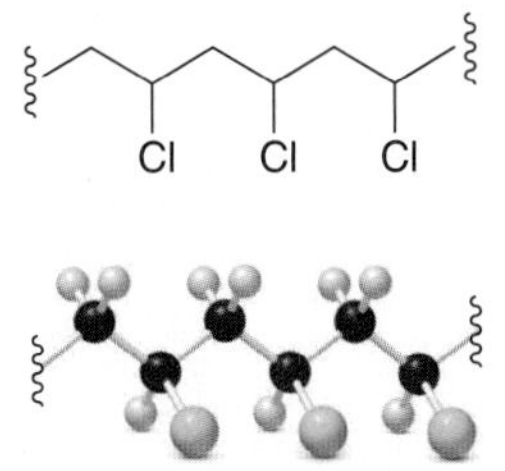

poly(vinyl chloride) (PVC)
(plastic used in films, pipes, and insulation)

Thousands of organic halides have now been isolated from a variety of marine organisms. The characteristic smell of the ocean is due in part to simple alkyl halides found in seaweed. While many sponges and corals use these compounds as a chemical defense against predators and parasites, recent research has shown that some may have novel medicinal properties that offer

Hundreds of organic halides with diverse structures and biological activities have been isolated from red algae of the genus *Laurencia,* seaweed that grows in shallow water at the edges of reefs.

treatment for cancer and other diseases. Two halogenated compounds are ma'ilione, isolated from the Hawaiian red algae *Laurencia cartilaginea,* and plocoralide B, which kills esophageal cancer cells.

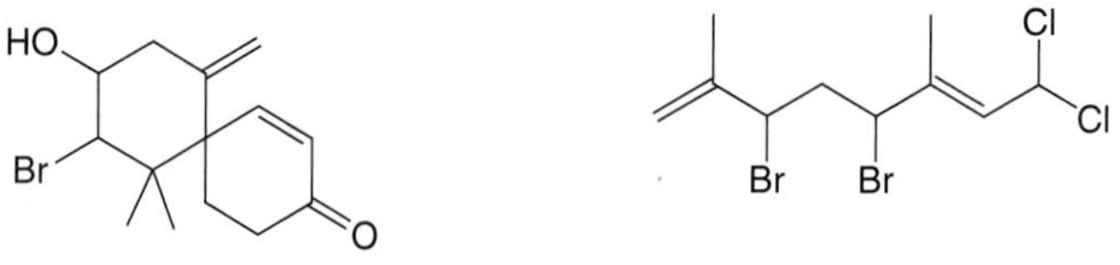

ma'ilione plocoralide B

Although the beneficial effects of many organic halides are undisputed, certain synthetic chlorinated organics such as the **chlorofluorocarbons** and the pesticide **DDT** have caused lasting harm to the environment.

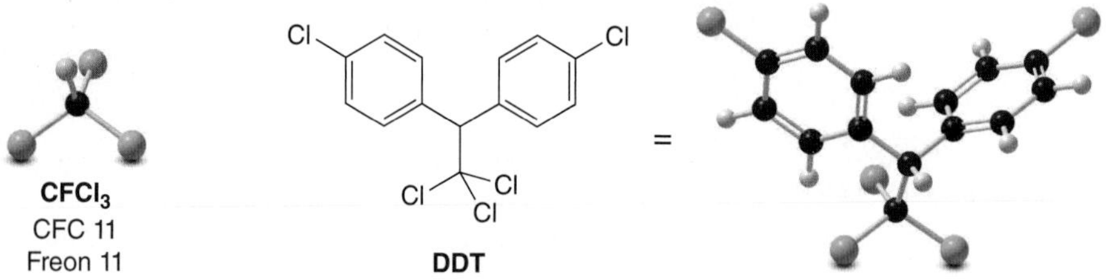

$CFCl_3$
CFC 11
Freon 11

DDT

Chlorofluorocarbons (CFCs) have the general molecular structure **CF_xCl_{4-x}**. Trichlorofluoromethane [$CFCl_3$, CFC 11, or Freon 11 (trade name)] is an example of these easily vaporized compounds, having been extensively used as a refrigerant and an aerosol propellant. CFCs slowly rise to the stratosphere, where sunlight catalyzes their decomposition, a process that contributes to the destruction of the ozone layer, the thin layer of atmosphere that shields the earth's surface from harmful ultraviolet radiation (Section 15.9). Although it is now easy to second-guess the extensive use of CFCs, it is also easy to see why they were used so widely. **CFCs made refrigeration available to the general public.** Would you call your refrigerator a comfort or a necessity?

DDT, a nonbiodegradable pesticide, has been labeled both a "miraculous" discovery by Winston Churchill in 1945 and the "elixir of death" by Rachel Carson in her 1962 book *Silent Spring.* DDT use was banned in the United States in 1973, but because of its effectiveness and low cost, it is still widely used to control insect populations in developing countries.

The story of the insecticide **DDT** (**d**ichloro**d**iphenyl**t**richloroethane) follows the same theme: DDT is an organic molecule with valuable short-term effects that has caused long-term problems. DDT kills insects that spread diseases such as malaria and typhus, and in controlling insect populations, DDT has saved millions of lives worldwide. DDT is a weakly polar organic compound that persists in the environment for years. Because DDT is soluble in organic media, it accumulates in fatty tissues. Most adults in the United States have low concentrations of DDT (or a degradation product of DDT) in their bodies. DDT is acutely toxic to many types of marine life (crayfish, sea shrimp, and some fish), but the long-term effect on humans is not known.

Problem 7.5 Chondrocole A is a marine natural product isolated from red seaweed that grows in regions of heavy surf in the Pacific Ocean. (a) Predict the solubility of chondrocole A in water and CH_2Cl_2. (b) Locate the stereogenic centers and label each as *R* or *S*. (c) Draw a stereoisomer and a constitutional isomer of chondrocole A.

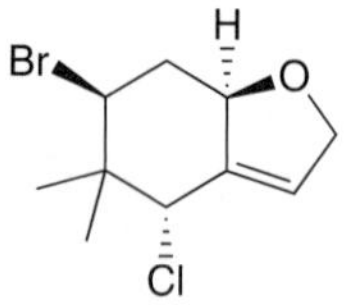

chondrocole A

7.5 The Polar Carbon–Halogen Bond

The properties of alkyl halides dictate their reactivity. The electrostatic potential maps of four simple alkyl halides in Figure 7.4 illustrate that the electronegative halogen X creates a polar C−X bond, making the carbon atom electron deficient. **The chemistry of alkyl halides is determined by this polar C−X bond.**

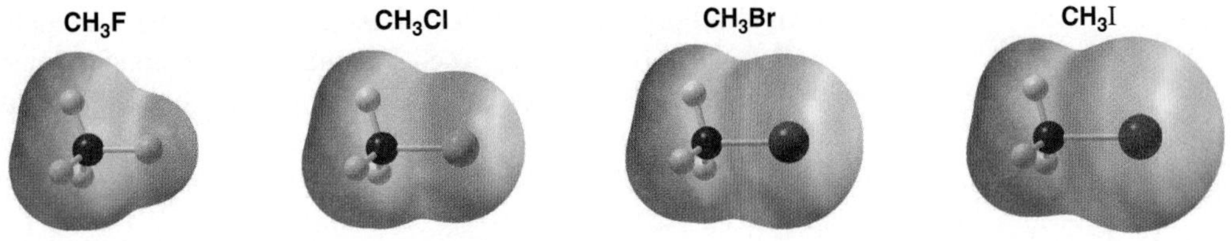

Figure 7.4
Electrostatic potential maps of four halomethanes (CH_3X)

- The polar C–X bond makes the carbon atom *electron deficient* in each CH_3X molecule.

What kind of reactions do alkyl halides undergo? **The characteristic reactions of alkyl halides are substitution and elimination.** Because alkyl halides contain an electrophilic carbon, they react with electron-rich reagents—Lewis bases (nucleophiles) and Brønsted–Lowry bases.

- **Alkyl halides undergo substitution reactions with nucleophiles.**

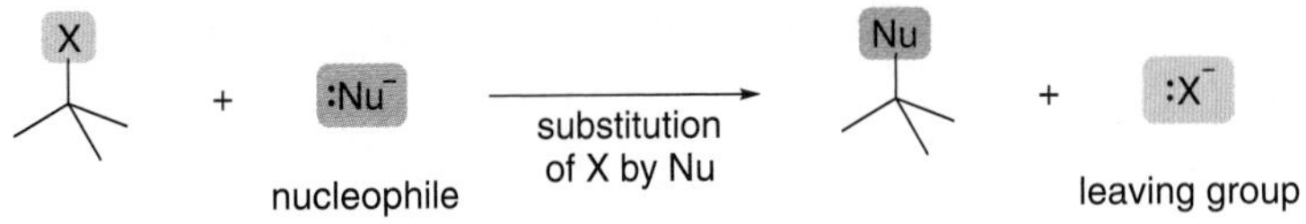

In a substitution reaction of an alkyl halide, **the halogen X is replaced by an electron-rich nucleophile :Nu⁻.** The C–X σ bond is broken and the C–Nu σ bond is formed.

- **Alkyl halides undergo elimination reactions with Brønsted–Lowry bases.**

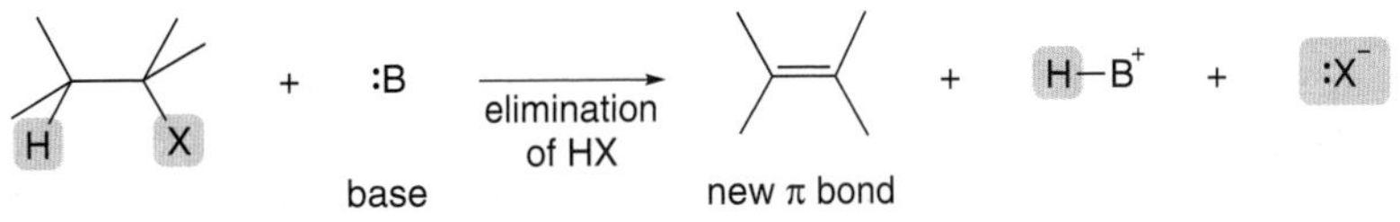

In an elimination reaction of an alkyl halide, the **elements of HX are removed by a Brønsted–Lowry base :B.**

The remainder of Chapter 7 is devoted to a discussion of the substitution reactions of alkyl halides. Elimination reactions are discussed in Chapter 8.

7.6 General Features of Nucleophilic Substitution

Three components are necessary in any substitution reaction.

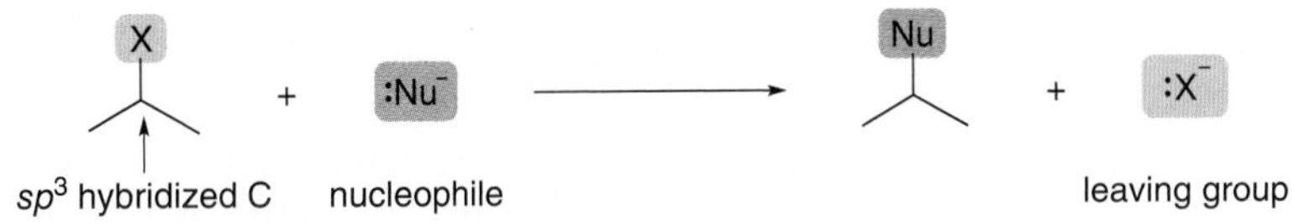

[1] An alkyl group containing an ***sp*³** hybridized carbon bonded to X.

[2] **X**—An atom X (or a group of atoms) called **a leaving group,** which is able to accept the electron density in the C–X bond. The most common leaving groups are halide anions (X^-), but H_2O (from ROH_2^+) and N_2 (from RN_2^+) are also encountered.

[3] **:Nu⁻**—A **nucleophile.** Nucleophiles contain a lone pair or a π bond but not necessarily a negative charge.

Because these substitution reactions involve electron-rich nucleophiles, they are called ***nucleophilic* substitution reactions.** Examples are shown in Equations [1]–[3]. **Nucleophilic**

substitutions are Lewis acid–base reactions. The nucleophile donates its electron pair, the alkyl halide (Lewis acid) accepts it, and the C−X bond is heterolytically cleaved. Curved arrow notation can be used to show the movement of electron pairs, as shown in Equation [3].

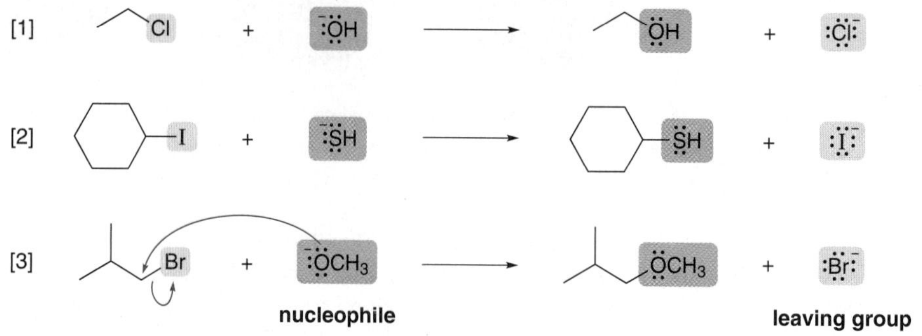

Negatively charged nucleophiles like ^{-}OH and ^{-}SH are used as **salts** with Li^+, Na^+, or K^+ counterions to balance charge. The identity of the cation is usually inconsequential, and therefore it is often omitted from the chemical equation.

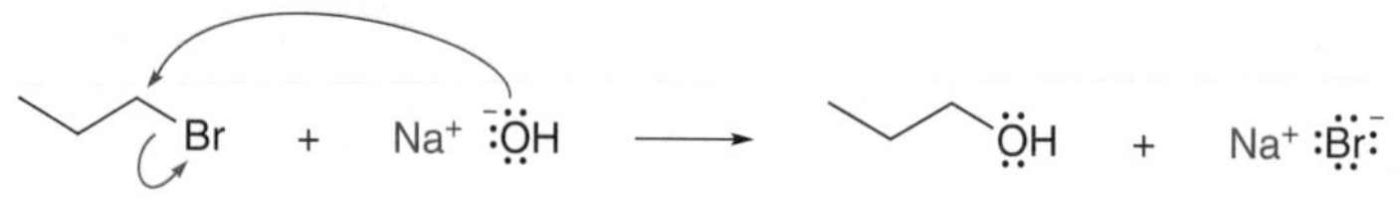

When a neutral nucleophile is used, the substitution product bears a positive charge. **All atoms originally bonded to the nucleophile stay bonded to it after substitution occurs.** All three CH_3 groups stay bonded to the N atom in the given example.

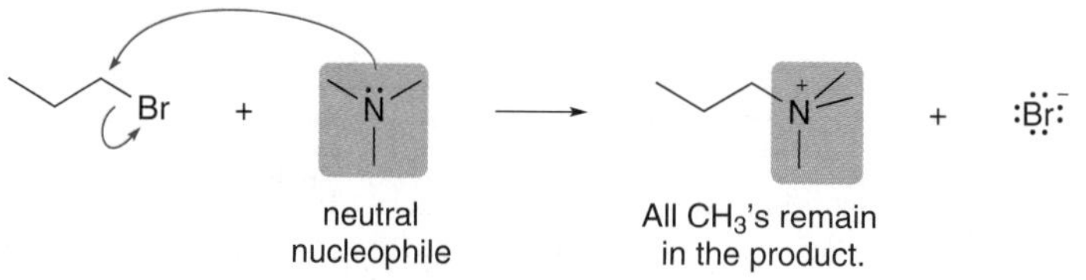

The reaction of alkyl halides with NH_3 to form amines (RNH_2) is discussed in Chapter 25.

Furthermore, when the substitution product bears a positive charge and also contains a *proton* bonded to O or N, the initial substitution product readily loses a proton in a Brønsted–Lowry acid–base reaction, forming a neutral product.

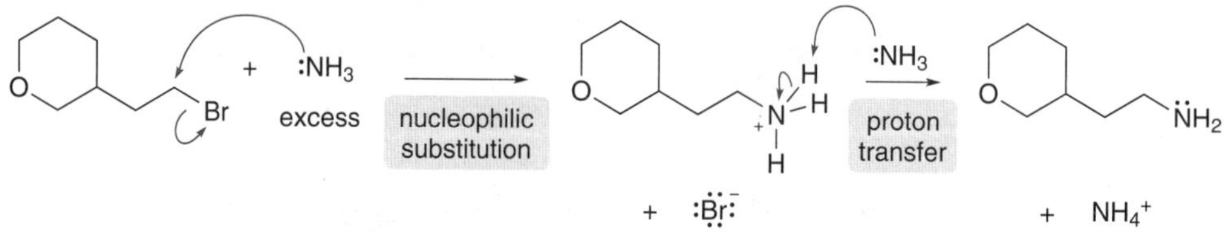

All of these reactions are nucleophilic substitutions and have the same overall result—**replacement of the leaving group by the nucleophile,** regardless of the identity or charge of the nucleophile. To draw any nucleophilic substitution product:

- **Find the *sp*^{3} hybridized carbon with the leaving group.**
- **Identify the nucleophile, the species with a lone pair or π bond.**
- **Substitute the nucleophile for the leaving group and assign charges (if necessary) to any atom that is involved in bond breaking or bond formation.**

Problem 7.6 Identify the nucleophile and leaving group and draw the products of each substitution reaction.

a. 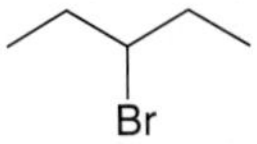 + $^{-}OCH_2CH_3$ ⟶

b. 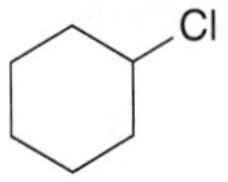 + NaOH ⟶

c.  + N_3^- ⟶

d. 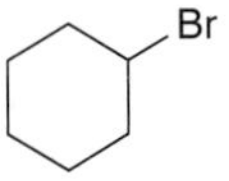 + NaCN ⟶

Problem 7.7 Draw the product of nucleophilic substitution with each neutral nucleophile. When the initial substitution product can lose a proton to form a neutral product, draw the product after proton transfer.

a. 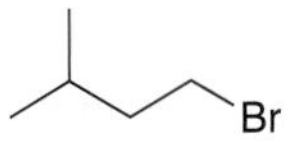 + $:N(CH_2CH_3)_3$ ⟶

b. 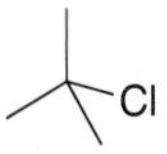 + $H_2\ddot{O}:$ ⟶

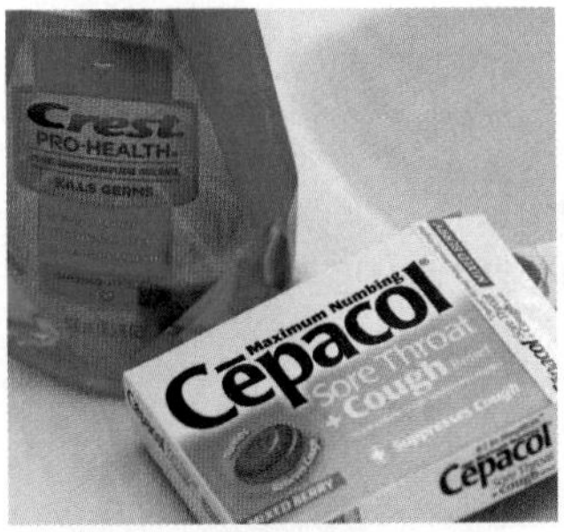

Cepacol throat lozenges and Crest Pro-Health mouthwash contain the antiseptic CPC, which is prepared by nucleophilic substitution (Problem 7.8).

Problem 7.8 CPC (cetylpyridinium chloride), an antiseptic found in throat lozenges and mouthwash, is synthesized by the following reaction. Draw the structure of CPC.

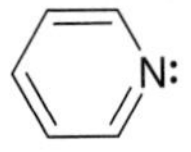 +  ⟶ CPC

Problem 7.9 What neutral nucleophile is needed to convert dihalide **A** to ticlopidine, an antiplatelet drug used to reduce the risk of strokes?

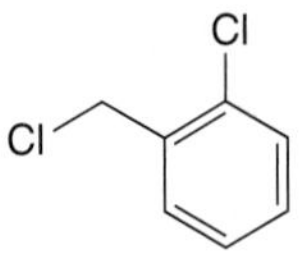 ⟶ 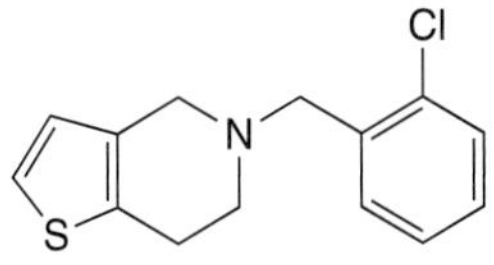

A ticlopidine

7.7 The Leaving Group

Nucleophilic substitution is a general reaction of organic compounds. Why, then, are alkyl halides the most common substrates, and halide anions the most common leaving groups? To answer this question, we must understand leaving group ability. **What makes a good leaving group?**

In a nucleophilic substitution reaction of R–X, the C–X bond is heterolytically cleaved, and the leaving group departs with the electron pair in that bond, forming $X:^-$. **The more stable the leaving group $X:^-$, the better able it is to accept an electron pair,** giving rise to the following generalization:

- **In comparing two leaving groups, the better leaving group is the weaker base.**

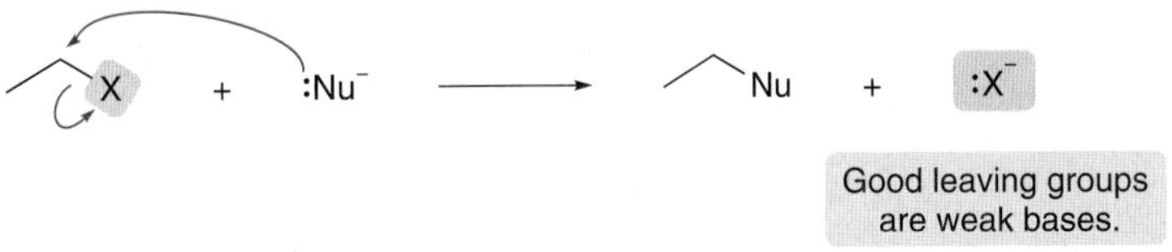

For example, H_2O is a better leaving group than ^{-}OH because H_2O is a weaker base. Moreover, the periodic trends in basicity can now be used to identify **periodic trends in leaving group ability:**

- **Left-to-right across a row of the periodic table, basicity *decreases* so leaving group ability *increases*.**

Increasing basicity (←)

With second-row elements: $:NH_3$ $H_2\ddot{O}:$ (better leaving group)

Increasing leaving group ability (→)

- Down a column of the periodic table, basicity *decreases* so leaving group ability *increases.*

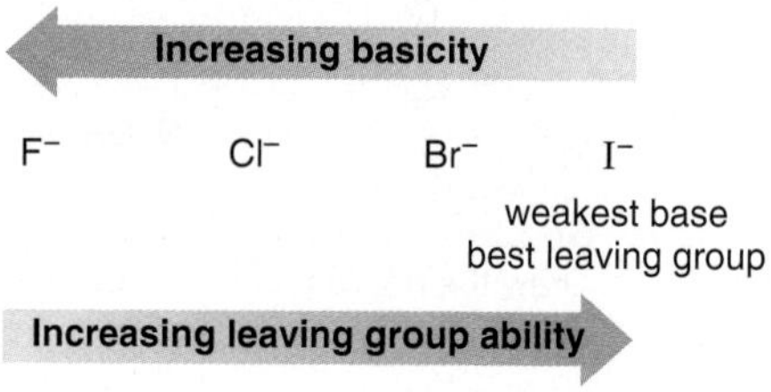

All good leaving groups are weak bases with strong conjugate acids having low pK_a values. Thus, all halide anions except F^- are good leaving groups because their conjugate acids (HCl, HBr, and HI) have low pK_a values. Tables 7.2 and 7.3 list good and poor leaving groups for nucleophilic substitution reactions, respectively. Nucleophilic substitution does not occur with any of the leaving groups in Table 7.3 because these leaving groups are strong bases.

Table 7.2 Good Leaving Groups for Nucleophilic Substitution

Starting material	Leaving group	Conjugate acid	pK_a
R–Cl	$:\ddot{\underset{..}{Cl}}:^-$	HCl	–7
R–Br	$:\ddot{\underset{..}{Br}}:^-$	HBr	–9
R–I	$:\ddot{\underset{..}{I}}:^-$	HI	–10
R–OH_2^+	$H_2\ddot{O}:$	H_3O^+	–1.7

Table 7.3 Poor Leaving Groups for Nucleophilic Substitution

Starting material	Leaving group	Conjugate acid	pK_a
R–F	$:\ddot{\underset{..}{F}}:^-$	HF	3.2
R–OH	$^-:\ddot{\underset{..}{O}}H$	H_2O	15.7
R–NH_2	$^-:\ddot{N}H_2$	NH_3	38
R–H	$H:^-$	H_2	35
R–R	$R:^-$	RH	50

Problem 7.10 Which is the better leaving group in each pair?

a. Cl^-, I^- b. NH_3, $^-NH_2$ c. H_2O, H_2S

Problem 7.11 Which molecules contain good leaving groups?

a. 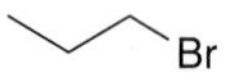
b. 
c. 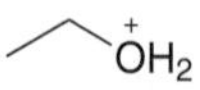
d. 

Given a particular nucleophile and leaving group, how can we determine whether the equilibrium will favor products in a nucleophilic substitution? We can often correctly predict the direction of equilibrium by comparing the basicity of the nucleophile and the leaving group.

- **Equilibrium favors the products of nucleophilic substitution when the leaving group is a weaker base than the nucleophile.**

Sample Problem 7.1 illustrates how to apply this general rule.

Sample Problem 7.1 Will the following substitution reaction favor formation of the products?

Solution

Determine the basicity of the nucleophile (^-OH) and the leaving group (Cl^-) by comparing the pK_a values of their conjugate acids. **The stronger the conjugate acid, the weaker the base, and the better the leaving group.**

			conjugate acids	
nucleophile	^-OH	⟶	H_2O	pK_a = 15.7
leaving group	Cl^-	⟶	HCl	pK_a = −7
	weaker base		**stronger acid**	

Because Cl^-, the leaving group, is a weaker base than ^-OH, the nucleophile, the reaction favors the products.

Problem 7.12 Does the equilibrium favor the reactants or products in each substitution reaction?

a. 

b. 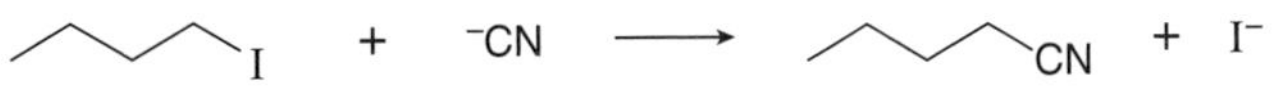

7.8 The Nucleophile

We use the word base *to mean* Brønsted–Lowry *base and the word* nucleophile *to mean a* Lewis base *that reacts with electrophiles* other than *protons.*

Nucleophiles and bases are structurally similar: both have a lone pair or a π bond. They differ in what they attack.

- **Bases attack protons. Nucleophiles attack other electron-deficient atoms (usually carbons).**

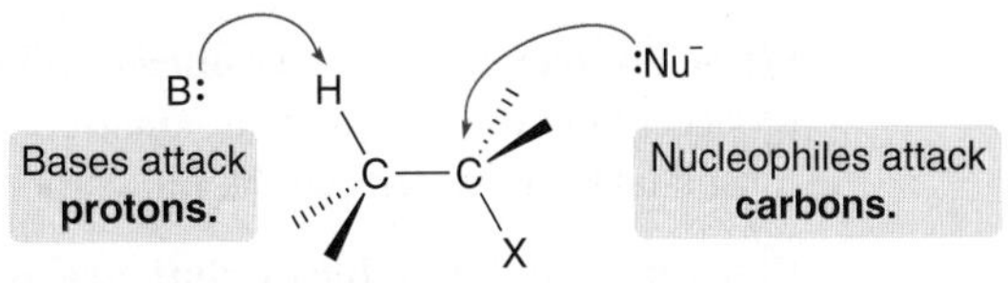

7.8A Nucleophilicity Versus Basicity

How is **nucleophilicity** (nucleophile strength) related to basicity? Although it is generally true that **a strong base is a strong nucleophile,** nucleophile size and steric factors can sometimes change this relationship.

Nucleophilicity parallels basicity in three instances.

[1] For two nucleophiles with the same nucleophilic atom, the stronger base is the stronger nucleophile.

- The relative nucleophilicity of ^-OH and $CH_3CO_2^-$, two oxygen nucleophiles, is determined by comparing the pK_a values of their conjugate acids (H_2O and CH_3CO_2H). CH_3CO_2H (pK_a = 4.8) is a stronger acid than H_2O (pK_a = 15.7), so ^-OH is a stronger base and stronger nucleophile than $CH_3CO_2^-$.

[2] A negatively charged nucleophile is always stronger than its conjugate acid.

- ^{-}OH is a stronger base and stronger nucleophile than H_2O, its conjugate acid.

[3] Right-to-left across a row of the periodic table, nucleophilicity increases as basicity increases.

CH_3^{-} $\quad$ $^{-}NH_2$ $\quad$ ^{-}OH $\quad$ F^{-}

← **Increasing basicity**
Increasing nucleophilicity

Problem 7.13 Identify the stronger nucleophile in each pair.

a. NH_3, $^{-}NH_2$ b. CH_3NH_2, CH_3OH c. $CH_3CO_2^{-}$, $CH_3CH_2O^{-}$

7.8B Steric Effects and Nucleophilicity

All steric effects arise because two atoms cannot occupy the same space. In Chapter 4, for example, we learned that **steric strain** is an increase in energy when big groups (occupying a large volume) are forced close to each other.

Nucleophilicity does not parallel basicity when **steric hindrance** becomes important. ***Steric hindrance* is a decrease in reactivity resulting from the presence of bulky groups at the site of a reaction.**

For example, although pK_a tables indicate that *tert*-butoxide $[(CH_3)_3CO^{-}]$ is a stronger base than ethoxide $(CH_3CH_2O^{-})$, **ethoxide is the *stronger* nucleophile.** The three CH_3 groups around the O atom of *tert*-butoxide create steric hindrance, making it more difficult for this big, bulky base to attack a tetravalent carbon atom.

ethoxide
stronger nucleophile

tert-butoxide
stronger base
Three CH_3 groups crowd the O.
weaker nucleophile

Steric hindrance decreases nucleophilicity but *not* basicity. Because bases pull off small, easily accessible protons, they are unaffected by steric hindrance. Nucleophiles, on the other hand, must attack a crowded tetrahedral carbon, so bulky groups decrease reactivity.

Sterically hindered bases that are poor nucleophiles are called *nonnucleophilic bases.* Potassium *tert*-butoxide $[K^{+}\,^{-}OC(CH_3)_3]$ is a strong, nonnucleophilic base.

7.8C Comparing Nucleophiles of Different Size—Solvent Effects

Atoms vary greatly in size down a column of the periodic table, and in this case, **nucleophilicity depends on the solvent used in a substitution reaction.** Although solvent has thus far been ignored, most organic reactions take place in a liquid solvent that dissolves all reactants to some extent. Because substitution reactions involve polar starting materials, polar solvents are used to dissolve them. There are two main kinds of polar solvents—**polar *protic* solvents and polar *aprotic* solvents.**

Polar Protic Solvents

In addition to dipole–dipole interactions, **polar *protic* solvents are capable of intermolecular hydrogen bonding,** because they contain an O–H or N–H bond. The most common polar protic solvents are water and alcohols (ROH) (Figure 7.5). **Polar protic solvents solvate *both* cations and anions well.**

- **Cations are solvated by ion–dipole interactions.**
- **Anions are solvated by hydrogen bonding.**

Figure 7.5
Polar protic solvents

H_2O	CH_3OH methanol	CH_3CH_2OH ethanol	$(CH_3)_3COH$ *tert*-butanol	CH_3CO_2H acetic acid

For example, if the salt NaBr is used as a source of the nucleophile Br^- in H_2O, the Na^+ cations are solvated by ion–dipole interactions with H_2O molecules, and the Br^- anions are solvated by strong hydrogen bonding interactions.

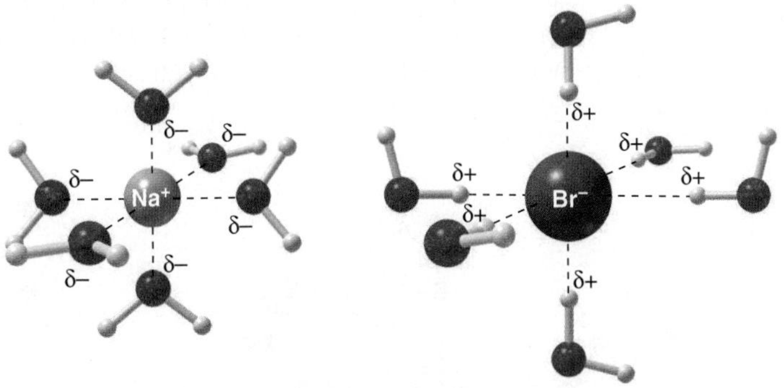

How do polar protic solvents affect nucleophilicity? **In polar protic solvents, nucleophilicity *increases* down a column of the periodic table as the size of the anion increases. This is *opposite* to basicity.** A small electronegative anion like F^- is very well solvated by hydrogen bonding, effectively shielding it from reaction. On the other hand, a large, less electronegative anion like I^- does not hold onto solvent molecules as tightly. The solvent does not "hide" a large nucleophile as well, and the nucleophile is much more able to donate its electron pairs in a reaction. Thus, **nucleophilicity increases down a column** even though basicity decreases, giving rise to the following trend in polar protic solvents:

F^- Cl^- Br^- I^- →

Increasing nucleophilicity in polar protic solvents

I^- is a weak base but a strong nucleophile in polar protic solvents.

Polar Aprotic Solvents

Polar *aprotic* solvents also exhibit dipole–dipole interactions, but they have no O−H or N−H bond so they are **incapable of hydrogen bonding.** Examples of polar aprotic solvents are shown in Figure 7.6. **Polar aprotic solvents solvate only cations well.**

- Cations are solvated by ion–dipole interactions.
- Anions are not well solvated because the solvent cannot hydrogen bond to them.

Figure 7.6
Polar aprotic solvents

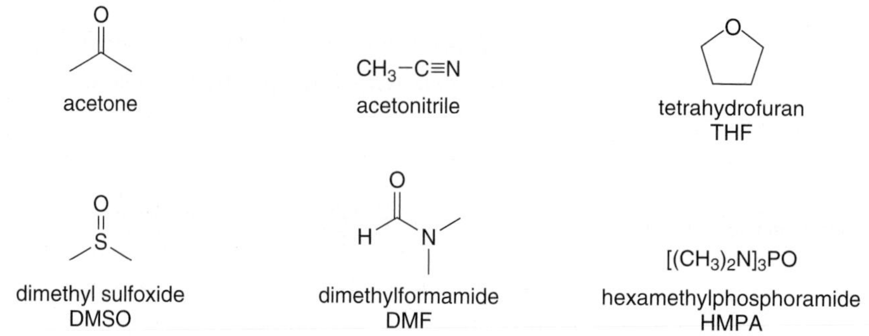

Abbreviations are often used in organic chemistry, instead of a compound's complete name. A list of common abbreviations is given on the inside back cover.

When the salt NaBr is dissolved in acetone, $(CH_3)_2C{=}O$, the Na^+ cations are solvated by ion–dipole interactions with the acetone molecules, but, with no possibility for hydrogen bonding, the Br^- anions are not well solvated. Often these anions are called **naked anions** because they are not bound by tight interactions with solvent.

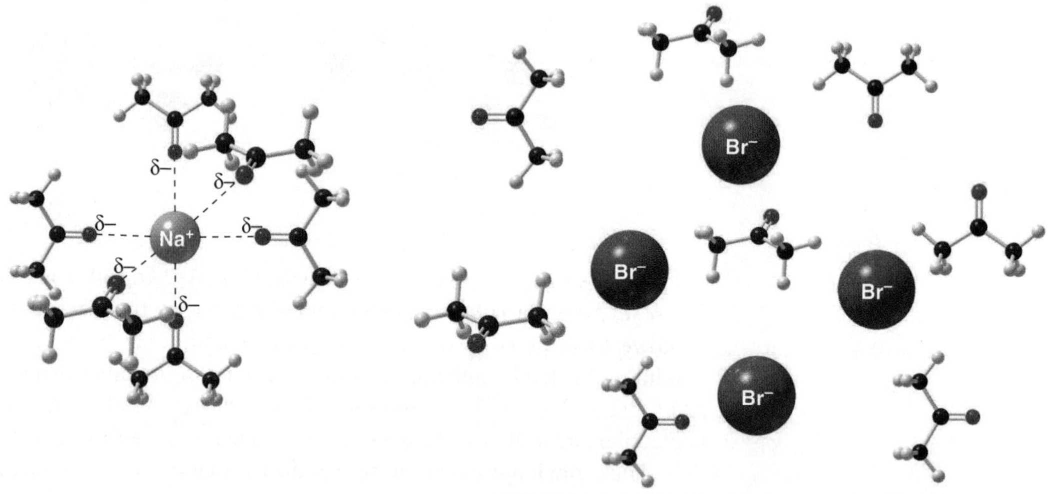

Br^- anions are surrounded by solvent but not well solvated by the $(CH_3)_2C{=}O$ molecules.

How do polar aprotic solvents affect nucleophilicity? Because anions are not well solvated in polar aprotic solvents, there is no need to consider whether solvent molecules more effectively hide one anion than another. Nucleophilicity parallels basicity and **the stronger base is the stronger nucleophile.** Because basicity decreases with size down a column, nucleophilicity decreases as well:

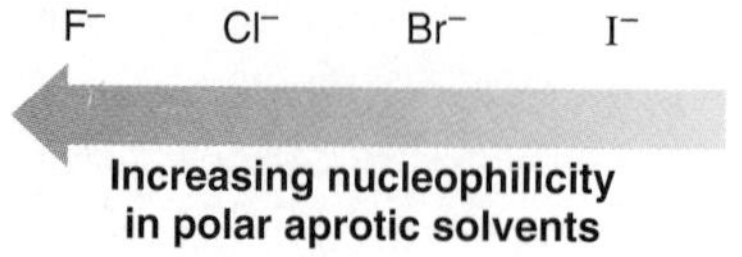

Problem 7.14 Classify each solvent as protic or aprotic.

a. 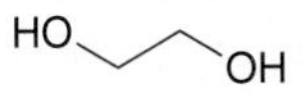 b. 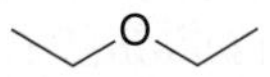 c. 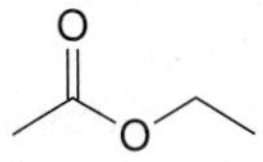

Problem 7.15 Identify the stronger nucleophile in each pair of anions.

a. Br^- or Cl^- in a polar protic solvent
b. HO^- or Cl^- in a polar aprotic solvent
c. HS^- or F^- in a polar protic solvent

7.8D Summary

This discussion of nucleophilicity has brought together many new concepts, such as steric hindrance and solvent effects, both of which we will meet again in our study of organic chemistry. Keep in mind the central relationship between nucleophilicity and basicity in comparing two nucleophiles.

- **It is generally true that the *stronger* base is the *stronger* nucleophile.**
- **In polar protic solvents, however, nucleophilicity increases with increasing size of an anion (opposite to basicity).**
- **Steric hindrance decreases nucleophilicity without decreasing basicity, making $(CH_3)_3CO^-$ a stronger base but a weaker nucleophile than $CH_3CH_2O^-$.**

Table 7.4 lists some common nucleophiles used in nucleophilic substitution reactions.

Problem 7.16 Rank the nucleophiles in each group in order of increasing nucleophilicity.

a. ^-OH, $^-NH_2$, H_2O
b. ^-OH, Br^-, F^- (polar aprotic solvent)
c. H_2O, ^-OH, $CH_3CO_2^-$

Problem 7.17 What nucleophile is needed to convert $(CH_3)_2CHCH_2CH_2Br$ to each product?

a. 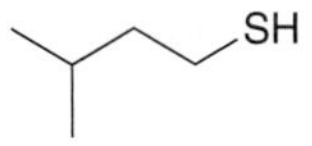 b. 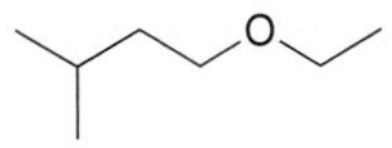 c. 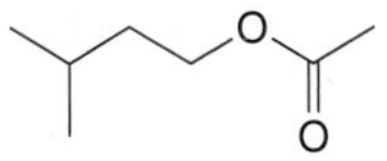 d. 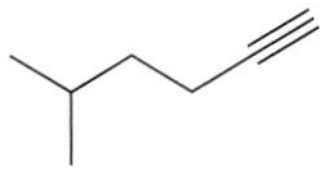

Table 7.4 Common Nucleophiles in Organic Chemistry

	Negatively charged nucleophiles			Neutral nucleophiles	
Oxygen	^-OH	^-OR	$CH_3CO_2^-$	H_2O	ROH
Nitrogen	N_3^-			NH_3	RNH_2
Carbon	^-CN	$HC{\equiv}C^-$			
Halogen	Cl^-	Br^-	I^-		
Sulfur	HS^-	RS^-		H_2S	RSH

7.9 Possible Mechanisms for Nucleophilic Substitution

Now that you know something about the general features of nucleophilic substitution, you can begin to understand the mechanism.

$$R{-}X \quad + \quad :Nu^- \longrightarrow R{-}Nu \quad + \quad :X^-$$

The σ bond is broken. The σ bond is formed.

Nucleophilic substitution at an sp^3 hybridized carbon involves two σ bonds: the bond to the leaving group is broken and the bond to the nucleophile is formed. To understand the mechanism of this reaction, though, we must know the timing of these two events; that is, **what is the order of bond breaking and bond making?** Do they happen at the same time, or does one event precede the other? There are three possibilities.

[1] **The mechanism has one step, and bond breaking and bond making occur at the *same* time.**

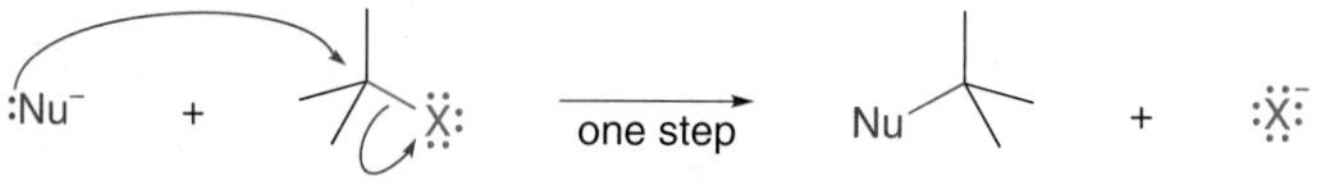

The C–X σ bond is broken... *as* ...the C–Nu σ bond is formed.

- If the C–X bond is broken *as* the C–Nu bond is formed, the mechanism has **one step.** As we learned in Section 6.9, the rate of such a bimolecular reaction depends on the concentration of both reactants; that is, the rate equation is **second order,** and **rate =** $\mathbf{k[RX][:Nu^-]}$**.**

[2] **The mechanism has two steps, and bond breaking occurs *before* bond making.**

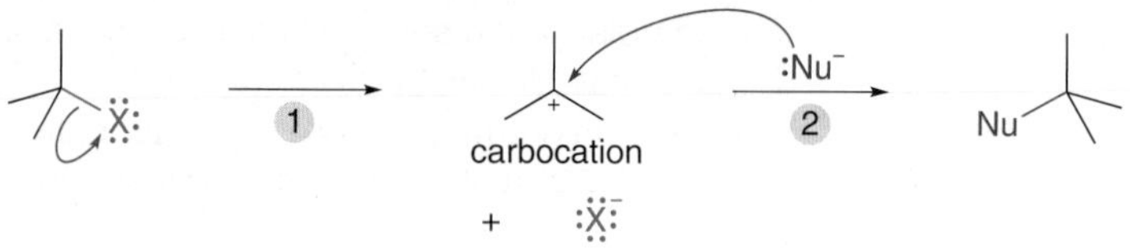

The C–X σ bond is broken... *before* ...the C–Nu σ bond is formed.

- If the C–X bond is broken *first* and then the C–Nu bond is formed, the mechanism has **two steps** and a **carbocation** is formed as an intermediate. Because the first step is rate-determining, the rate depends on the concentration of RX only; that is, the rate equation is **first order,** and **rate =** $\mathbf{k[RX]}$**.**

[3] **The mechanism has two steps and bond making occurs *before* bond breaking.**

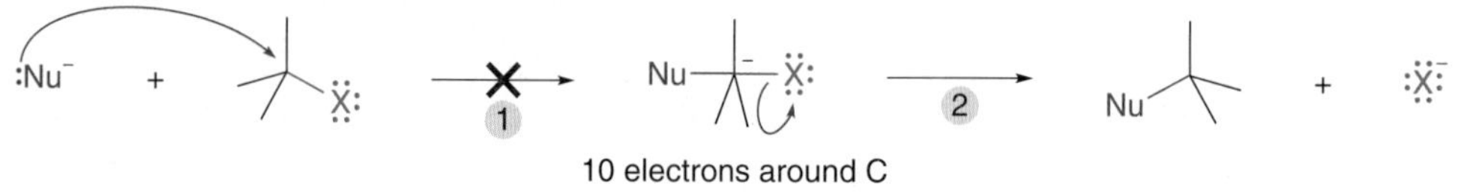

The C–X σ bond is broken... *after* ...the C–Nu σ bond is formed.

- If the C–Nu bond is formed *first* and then the C–X bond is broken, the mechanism has **two steps,** but this mechanism has an inherent problem. The intermediate generated in the first step has 10 electrons around carbon, violating the octet rule, so this last possibility can be disregarded.

Thus, there are two possible mechanisms for nucleophilic substitution: **a one-step mechanism in which bond breaking and bond making are simultaneous, and a two-step mechanism in which bond breaking comes before bond making.** In Section 7.10 we look at data for two specific nucleophilic substitution reactions and see if those data fit either of these proposed mechanisms.

7.10 Two Mechanisms for Nucleophilic Substitution

Rate equations for two different reactions give us insight into the possible mechanism for nucleophilic substitution.

Reaction of bromomethane (CH_3Br) with acetate ($CH_3CO_2^-$) affords the substitution product methyl acetate with loss of Br^- as the leaving group (Equation [1]). Kinetic data show that the reaction rate depends on the concentration of *both* reactants; that is, the rate equation is **second order.** This suggests a **bimolecular reaction with a one-step mechanism** in which the C−X bond is broken *as* the C−Nu bond is formed.

[1] 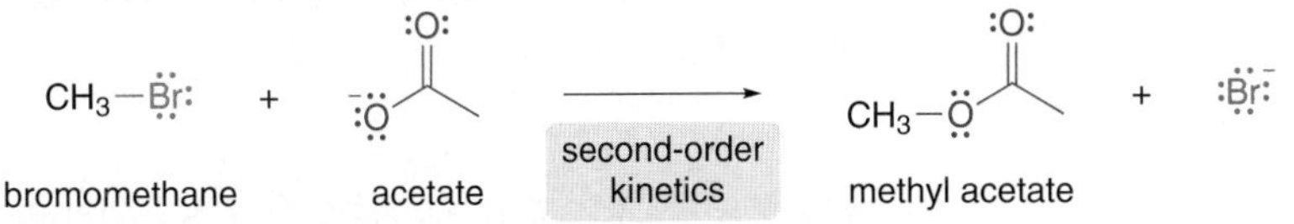
CH3−Br:
+
acetate
second-order kinetics
CH3−O
methyl acetate
bromomethane
+
:Br:−

Equation [2] illustrates a similar nucleophilic substitution reaction with a different alkyl halide, $(CH_3)_3CBr$, which also leads to substitution of Br^- by $CH_3CO_2^-$. Kinetic data show that this reaction rate depends on the concentration of only *one* reactant, the alkyl halide; that is, the rate equation is **first order.** This suggests a **two-step mechanism in which the rate-determining step involves the alkyl halide only.**

[2] 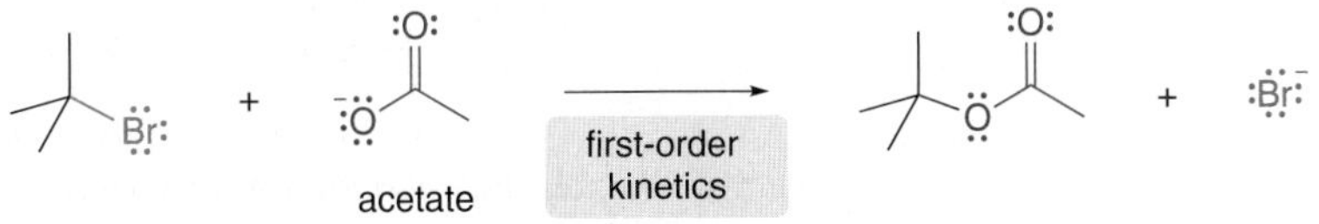
Br:
+
acetate
first-order kinetics
+
:Br:−

The numbers **1** and **2** in the names S_N1 and S_N2 refer to the kinetic order of the reactions. For example, S_N**2** means that the kinetics are **second** order. The number *2* does *not* refer to the number of steps in the mechanism.

How can these two different results be explained? Although these two reactions have the same nucleophile and leaving group, **there must be two different mechanisms** because there are two different rate equations. These equations are specific examples of two well-known mechanisms for nucleophilic substitution at an sp^3 hybridized carbon:

- **S_N2 mechanism (substitution nucleophilic bimolecular).**
- **S_N1 mechanism (substitution nucleophilic unimolecular).**

The reaction in Equation [1] illustrates an S_N2 mechanism, whereas the reaction in Equation [2] illustrates an S_N1 mechanism.

7.11 The S_N2 Mechanism

The reaction of CH_3Br with $CH_3CO_2^-$ is an example of an **S_N2 reaction.** What are the general features of this mechanism?

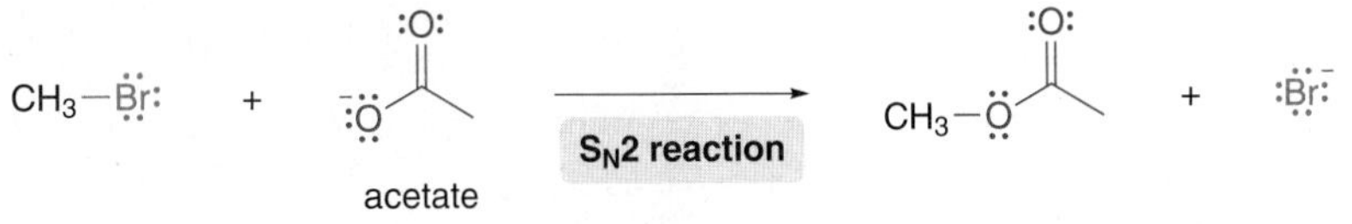
CH3−Br:
+
acetate
SN2 reaction
CH3−O
+
:Br:−

7.11A Kinetics

An S_N2 reaction exhibits **second-order kinetics;** that is, the reaction is **bimolecular** and both the alkyl halide and the nucleophile appear in the rate equation.

- rate = $k[CH_3Br][CH_3CO_2^-]$

Changing the concentration of *either* reactant affects the rate. For example, doubling the concentration of *either* the nucleophile or the alkyl halide doubles the rate. Doubling the concentration of *both* reactants increases the rate by a factor of four.

Problem 7.18 What happens to the rate of an S_N2 reaction under each of the following conditions?

a. [RX] is tripled, and [:Nu⁻] stays the same.
b. Both [RX] and [:Nu⁻] are tripled.
c. [RX] is halved, and [:Nu⁻] stays the same.
d. [RX] is halved, and [:Nu⁻] is doubled.

7.11B A One-Step Mechanism

The most straightforward explanation for the observed second-order kinetics is a **concerted reaction—bond breaking and bond making occur at the *same* time,** as shown in Mechanism 7.1.

Mechanism 7.1 The S_N2 Mechanism

One step The C – Br bond breaks as the C – O bond forms.

An energy diagram for the reaction of $CH_3Br + CH_3CO_2^-$ is shown in Figure 7.7. The reaction has one step, so there is one energy barrier between reactants and products. Because the equilibrium for this S_N2 reaction favors the products, the products are drawn at lower energy than the starting materials.

Problem 7.19 Draw an energy diagram for the following S_N2 reaction. Label the axes, the starting materials, and the product. Draw the structure of the transition state.

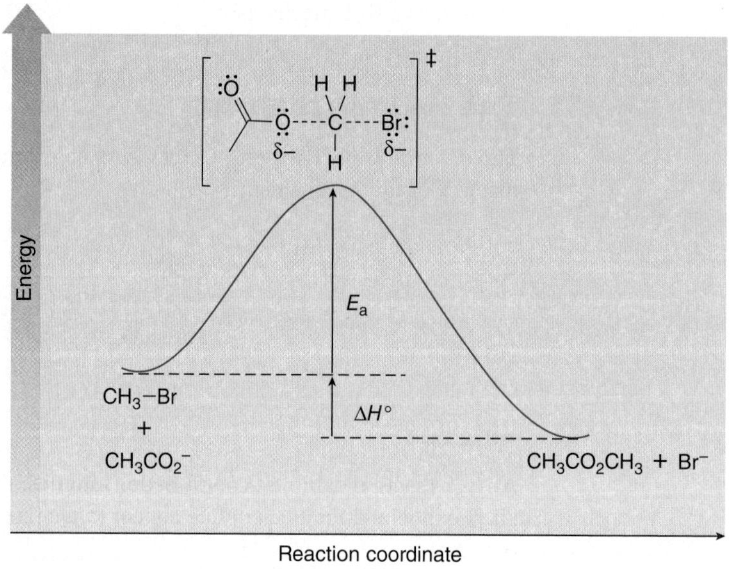

Figure 7.7
An energy diagram for the S_N2 reaction:
$CH_3Br + CH_3CO_2^- \rightarrow CH_3CO_2CH_3 + Br^-$

- In the transition state, the C–Br bond is partially broken, the C–O bond is partially formed, and both the attacking nucleophile and the departing leaving group bear a partial negative charge.

7.11C Stereochemistry of the S_N2 Reaction

From what direction does the nucleophile approach the substrate in an S_N2 reaction? There are two possibilities.

- **Frontside attack:** The nucleophile approaches from the ***same*** side as the leaving group.
- **Backside attack:** The nucleophile approaches from the side ***opposite*** the leaving group.

The results of frontside and backside attack of a nucleophile are illustrated with $CH_3CH(D)Br$ as substrate and the general nucleophile $:Nu^-$. This substrate has the leaving group bonded to a stereogenic center, thus allowing us to see the structural difference that results when the nucleophile attacks from two different directions.

In **frontside attack,** the nucleophile approaches from the **same** side as the leaving group, forming **A.** In this example, the leaving group was drawn on the right, so the nucleophile attacks from the right, and all other groups remain in their original positions. Because the nucleophile and leaving group are in the same position relative to the other three groups on carbon, frontside attack results in **retention of configuration** around the stereogenic center.

Recall from Section 1.1 that D stands for the isotope deuterium (^2H).

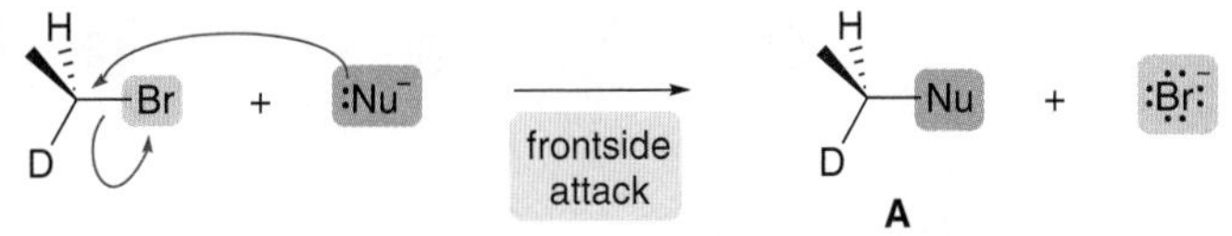

A

Nu replaces Br on the same side.

In **backside attack,** the nucleophile approaches from the **opposite** side to the leaving group, forming **B.** In this example, the leaving group was drawn on the right, so the nucleophile attacks from the left. Because the nucleophile and leaving group are in the opposite position relative to the other three groups on carbon, backside attack results in **inversion of configuration** around the stereogenic center.

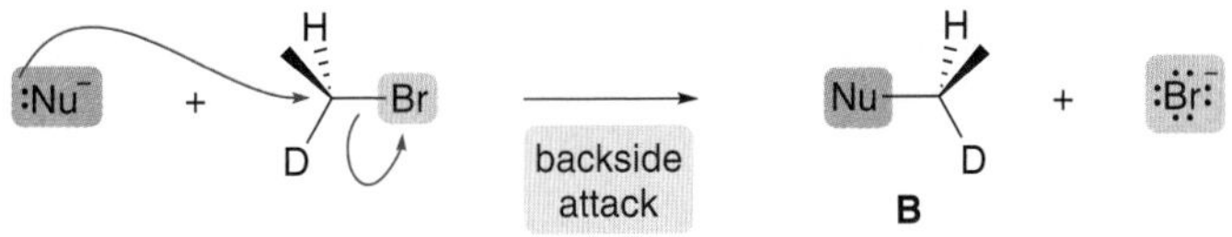

B

Nu replaces Br on the opposite side.

The products of frontside and backside attack are *different* compounds. **A** and **B** are stereoisomers that are nonsuperimposable—they are **enantiomers.**

Inversion of configuration in an S_N2 reaction is often called **Walden inversion,** after Latvian chemist Dr. Paul Walden, who first observed this process in 1896.

A **B**

only S_N2 product

enantiomers

product with retention of configuration

product with inversion of configuration

Backside attack occurs in all S_N2 reactions, but we can observe this change only when the leaving group is bonded to a stereogenic center.

Which product is formed in an S_N2 reaction? When the stereochemistry of the product is determined, **only B, the product of backside attack, is formed.**

- **All S_N2 reactions proceed with backside attack of the nucleophile, resulting in *inversion* of configuration at a stereogenic center.**

One explanation for backside attack is based on an electronic argument. Both the nucleophile and leaving group are electron rich and these like charges repel each other. Backside attack keeps these two groups as far away from each other as possible. In the transition state, the nucleophile and leaving group are 180° away from each other, and the other three groups around carbon occupy a plane, as illustrated in Figure 7.8.

Figure 7.8
Stereochemistry of the S_N2 reaction

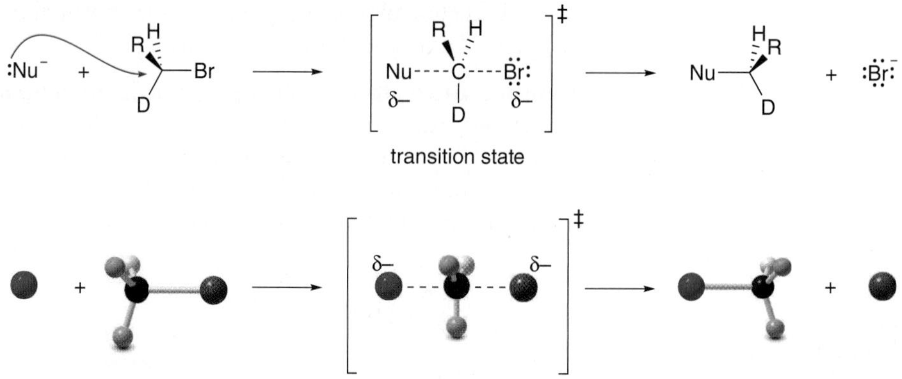

- :Nu⁻ and Br⁻ are 180° away from each other, on either side of a plane containing R, H, and D.

Two additional examples of inversion of configuration in S_N2 reactions are given in Figure 7.9.

Figure 7.9
Two examples of inversion of configuration in the S_N2 reaction

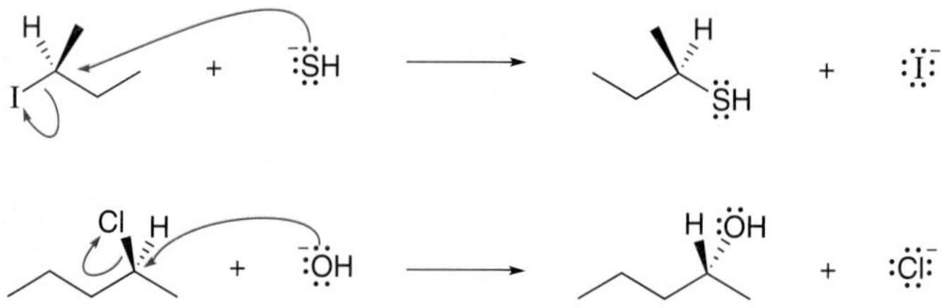

- The bond to the nucleophile in the product is always on the **opposite side** compared to the bond to the leaving group in the starting material. If the leaving group is drawn to the left, the nucleophile approaches from the right. If the leaving group is drawn in front of the plane (on a wedge), the nucleophile approaches from the back and ends up on a dashed wedge.

Sample Problem 7.2 Label the nucleophile and leaving group, and draw the product (including stereochemistry) of the following S_N2 reaction.

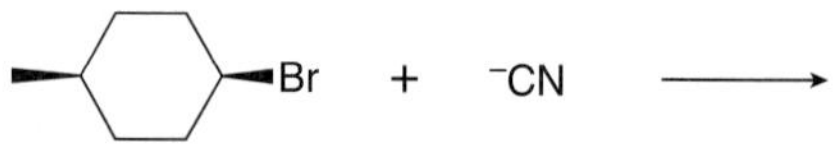

Solution

Br⁻ is the leaving group and ⁻CN is the nucleophile. Because S_N2 reactions proceed with **inversion** of configuration and the leaving group is drawn above the ring (on a wedge), the nucleophile must come in from below (ending up on a dashed wedge).

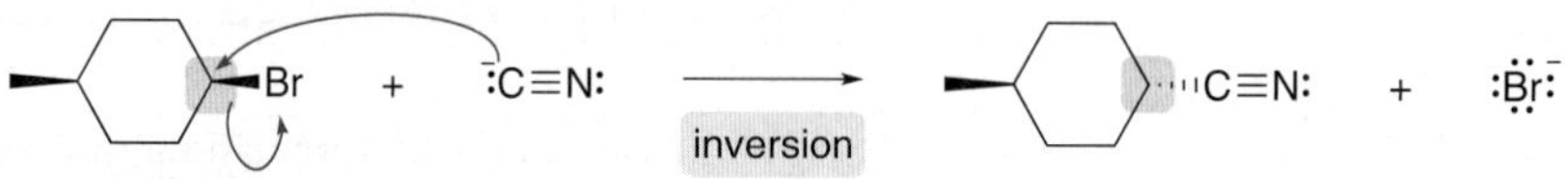

- **Inversion** of configuration occurs at the C – Br bond.
- Backside attack converts the **cis** starting material to a **trans** product because the nucleophile (⁻CN) attacks from below the plane of the ring.

Problem 7.20 Draw the product of each S_N2 reaction and indicate stereochemistry.

a. 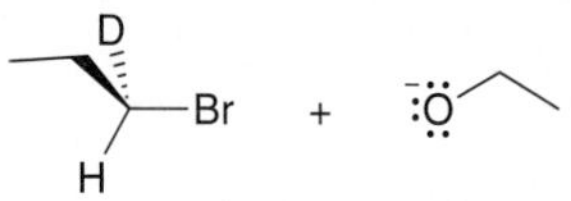

b. 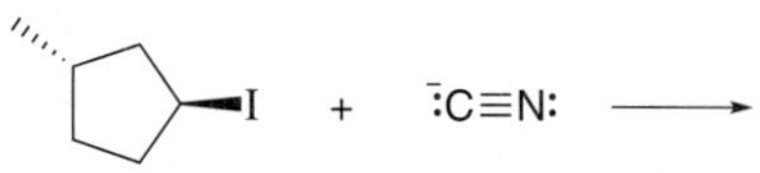

7.11D The Identity of the R Group

How does the rate of an S_N2 reaction change as the alkyl group in the substrate alkyl halide changes from CH_3 --→ 1° --→ 2° --→ 3°?

- As the number of R groups on the carbon with the leaving group *increases,* the rate of an S_N2 reaction *decreases.*

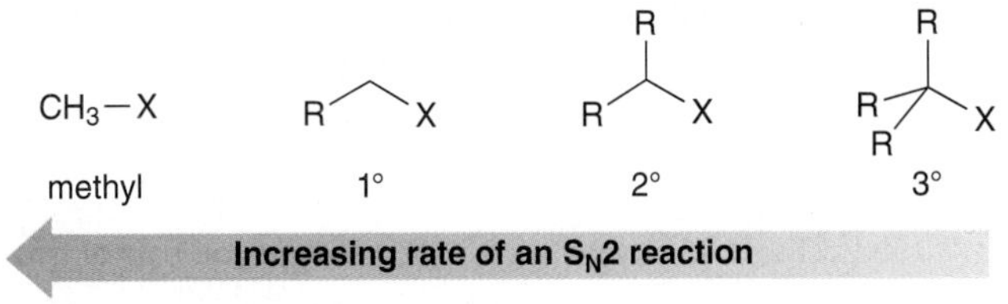

- Methyl and 1° alkyl halides undergo S_N2 reactions with ease.
- 2° Alkyl halides react more slowly.
- 3° Alkyl halides *do not* undergo S_N2 reactions.

This order of reactivity can be explained by steric effects. As small H atoms are replaced by larger alkyl groups, **steric hindrance caused by bulky R groups makes nucleophilic attack from the back side more difficult,** slowing the reaction rate. Figure 7.10 illustrates the effect of increasing steric hindrance in a series of alkyl halides.

Figure 7.10
Steric effects in the S_N2 reaction

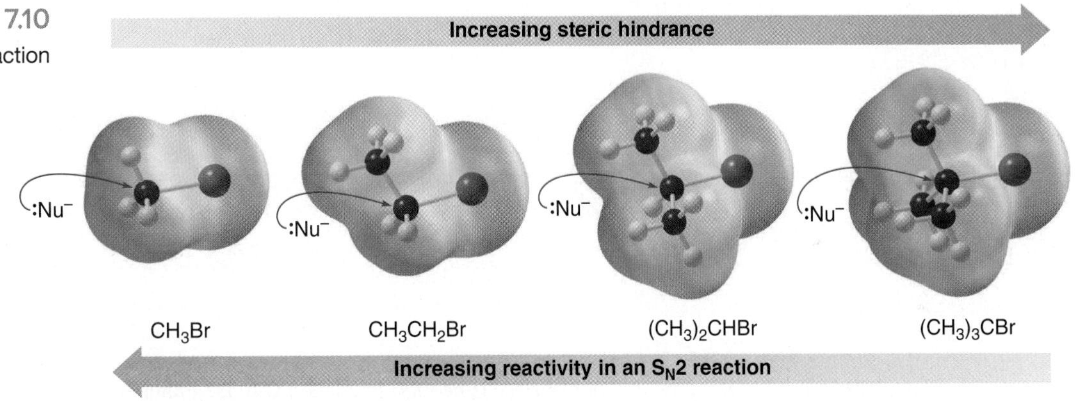

- The S_N2 reaction is fastest with unhindered halides.

Table 7.5 summarizes what we have learned thus far about an S_N2 mechanism.

Table 7.5 Characteristics of the S_N2 Mechanism

Characteristic	Result
Kinetics	• Second-order kinetics; rate = $k[RX][:Nu^-]$
Mechanism	• One step
Stereochemistry	• Backside attack of the nucleophile • Inversion of configuration at a stereogenic center
Identity of R	• Unhindered halides react fastest. • Rate: $CH_3X > RCH_2X > R_2CHX > R_3CX$

Problem 7.21 Which compound in each pair undergoes a faster S_N2 reaction?

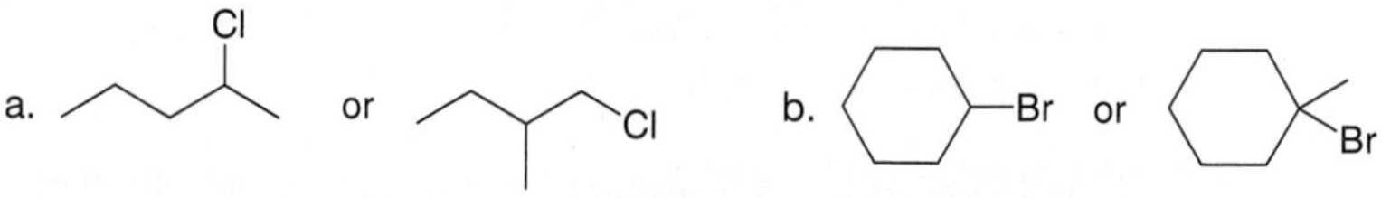

The S_N2 reaction is a key step in the laboratory synthesis of many drugs including **ethambutol** (trade name Myambutol), used in the treatment of tuberculosis, and **fluoxetine** (trade name Prozac), an antidepressant, as illustrated in Figure 7.11.

Figure 7.11 Nucleophilic substitution in the synthesis of two useful drugs

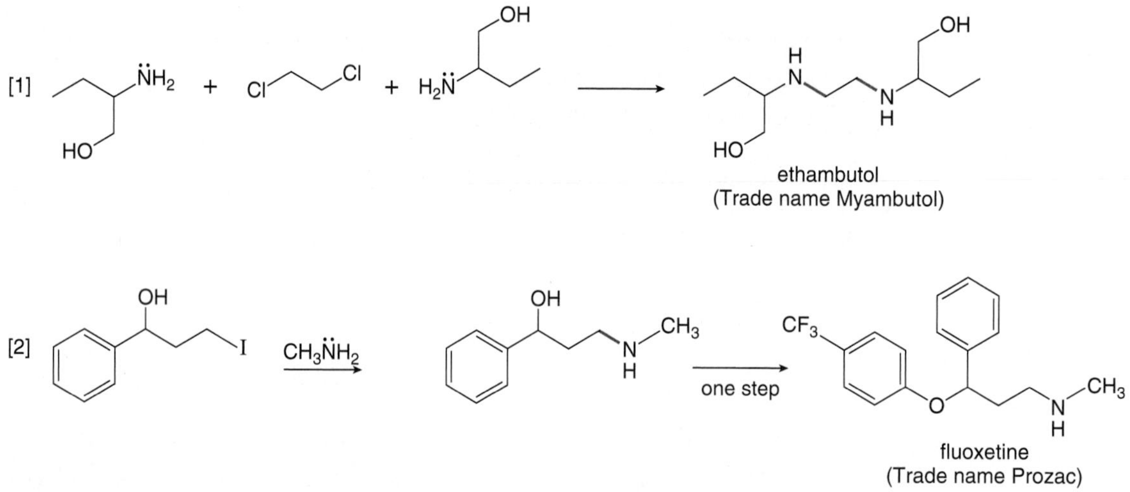

- In both examples, the initial substitution product bears a positive charge and goes on to lose a proton to form the product drawn.
- The NH_2 group serves as a neutral nucleophile to displace halogen in each synthesis. The new bonds formed by nucleophilic substitution are drawn in red in the products.

Problem 7.22 What neutral nucleophile is needed to convert **C** to **D?** Imatinib, an effective treatment for certain cancers, is prepared in one step from **D.**

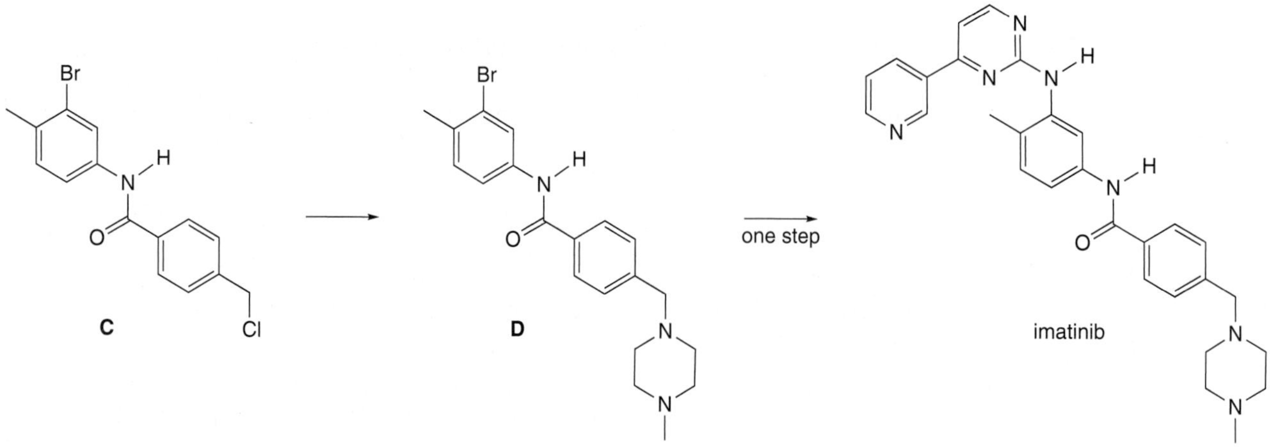

7.12 The S_N1 Mechanism

The reaction of $(CH_3)_3CBr$ with $CH_3CO_2^-$ is an example of the second mechanism for nucleophilic substitution, the **S_N1 mechanism.** What are the general features of this mechanism?

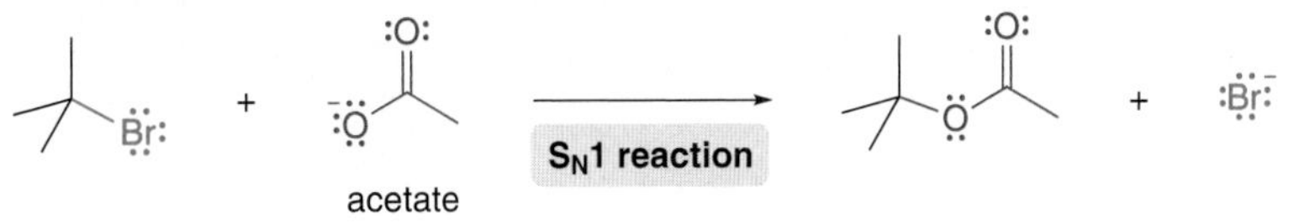

7.12A Kinetics

The S_N1 reaction exhibits **first-order kinetics.**

- rate = $k[(CH_3)_3CBr]$

As we learned in Section 7.10, the kinetics suggest that the S_N1 mechanism involves **more than one step,** and that the slow step is **unimolecular,** involving *only* the alkyl halide. **The identity and concentration of the nucleophile have no effect on the reaction rate.** Doubling the concentration of $(CH_3)_3CBr$ doubles the rate, but doubling the concentration of the nucleophile has no effect.

Problem 7.23 What happens to the rate of an S_N1 reaction under each of the following conditions?

a. [RX] is tripled, and [:Nu⁻] stays the same.
b. Both [RX] and [:Nu⁻] are tripled.
c. [RX] is halved, and [:Nu⁻] stays the same.
d. [RX] is halved, and [:Nu⁻] is doubled.

7.12B A Two-Step Mechanism

The most straightforward explanation for the observed first-order kinetics is a **two-step mechanism** in which **bond breaking occurs *before* bond making,** as shown in Mechanism 7.2.

Mechanism 7.2 The S_N1 Mechanism

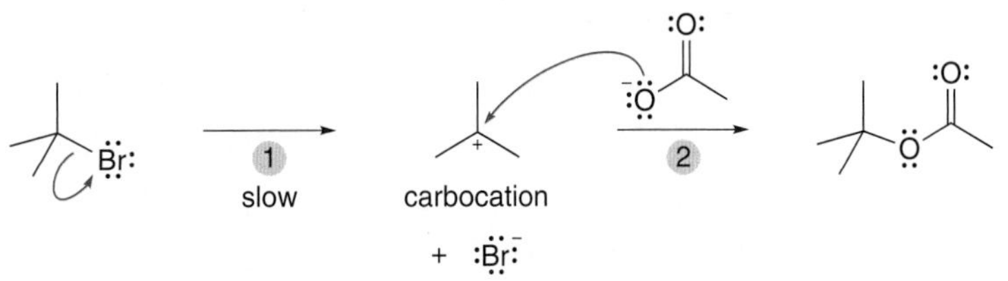

1 Heterolysis of the C–Br bond forms a carbocation in the rate-determining step.

2 Nucleophilic attack of acetate (a Lewis base) on the carbocation (a Lewis acid) forms the new C–O bond.

The key features of the S_N1 mechanism are:

- The mechanism has two steps.
- Carbocations are formed as reactive intermediates.

An energy diagram for the reaction of $(CH_3)_3CBr + CH_3CO_2^-$ is shown in Figure 7.12. Each step has its own energy barrier, with a transition state at each energy maximum. Because the transition state for Step [1] is at higher energy, Step [1] is rate-determining. $\Delta H°$ for Step [1] has a positive value because only bond breaking occurs, whereas $\Delta H°$ of Step [2] has a negative value because only bond making occurs. The overall reaction is assumed to be exothermic, so the final product is drawn at lower energy than the initial starting material.

Figure 7.12
An energy diagram for the S_N1 reaction: $(CH_3)_3CBr + CH_3CO_2^- \rightarrow (CH_3)_3COCOCH_3 + Br^-$

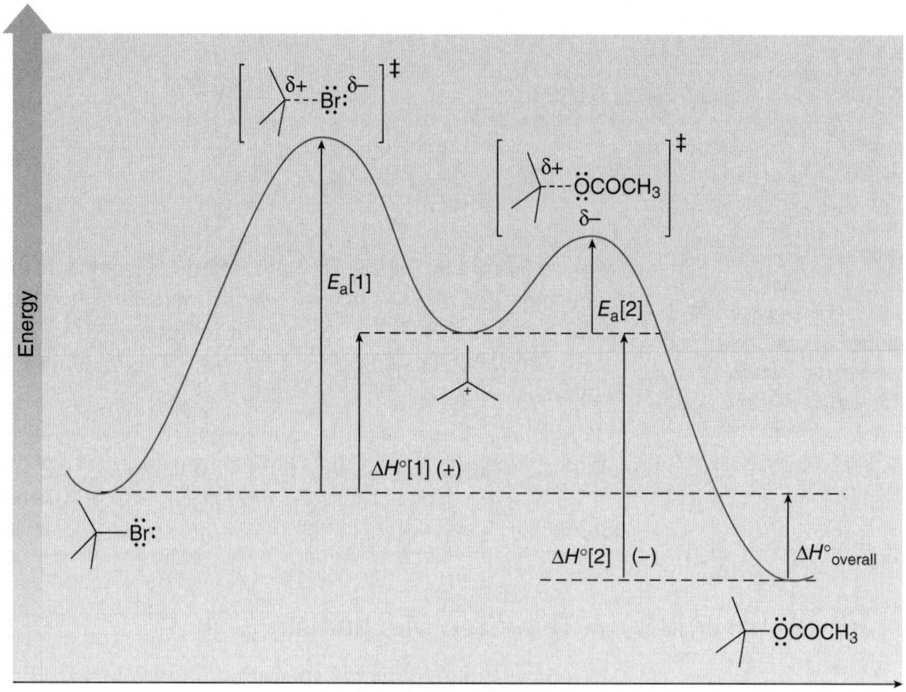

- The S_N1 mechanism has two steps, so there are two energy barriers.
- $E_a[1] > E_a[2]$ because Step [1] involves bond breaking and Step [2] involves bond formation.
- In each step only one bond is broken or formed, so the transition state for each step has one partial bond.

7.12C Stereochemistry of the S_N1 Reaction

To understand the stereochemistry of the S_N1 reaction, we must examine the geometry of the carbocation intermediate.

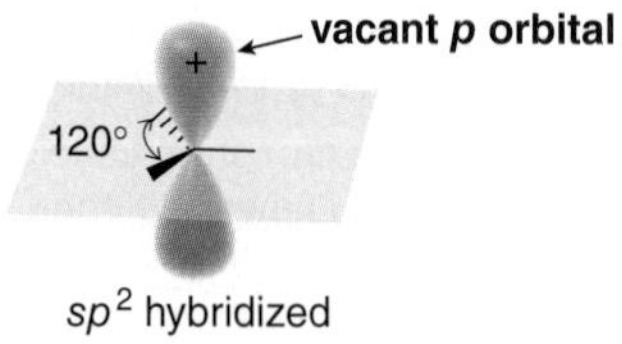

- **A carbocation (with three groups around C) is sp^2 hybridized and trigonal planar, and contains a vacant *p* orbital extending above and below the plane.**

To illustrate the consequences of having a trigonal planar carbocation formed as a reactive intermediate, we examine the S_N1 reaction of a 3° alkyl halide **A** having the leaving group bonded to a stereogenic center.

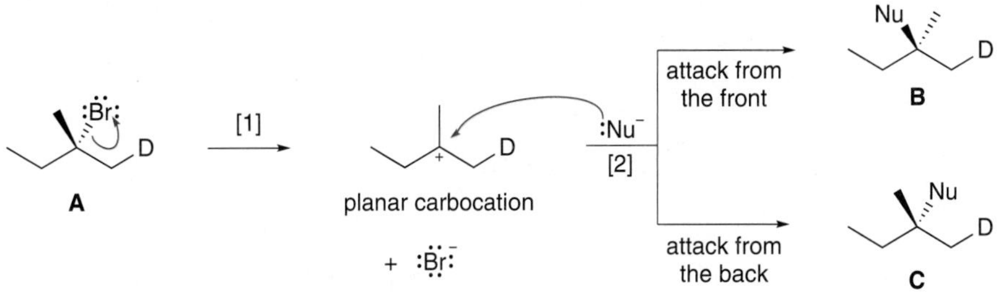

Loss of the leaving group in Step [1] generates a planar carbocation that is now achiral. Attack of the nucleophile in Step [2] can occur from either the front or the back to afford two products, **B** and **C.** These two products are *different* compounds containing one stereogenic center. **B** and **C** are stereoisomers that are not superimposable—they are **enantiomers.** Because there is no preference for nucleophilic attack from either direction, an equal amount of the two enantiomers is formed—a **racemic mixture.** We say that ***racemization*** has occurred.

Nucleophilic attack from both sides of a planar carbocation occurs in S_N1 reactions, but we see the result of this phenomenon only when the leaving group is bonded to a stereogenic center.

- ***Racemization* is the formation of equal amounts of two enantiomeric products from a single starting material.**
- **S_N1 reactions proceed with *racemization* at a single stereogenic center.**

Two additional examples of racemization in S_N1 reactions are given in Figure 7.13.

Figure 7.13
Two examples of racemization in the S_N1 reaction

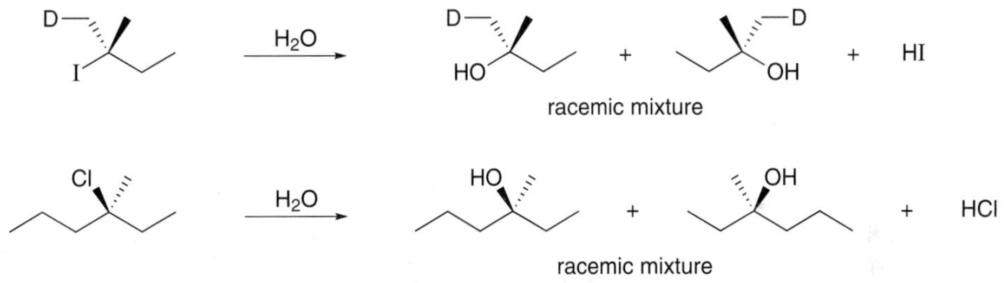

- Nucleophilic substitution of each starting material by an S_N1 mechanism forms a **racemic mixture** of two products.
- With H_2O, a neutral nucleophile, the initial product of nucleophilic substitution (ROH_2^+) loses a proton to form the final neutral product, ROH (Section 7.6).

Sample Problem 7.3 Label the nucleophile and leaving group, and draw the products (including stereochemistry) of the following S_N1 reaction.

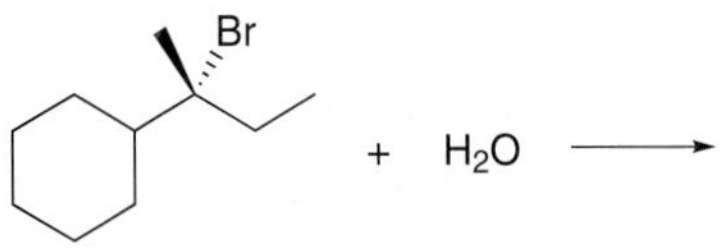

Solution

Br^- is the leaving group and H_2O is the nucleophile. Loss of the leaving group generates a trigonal planar carbocation, which can react with the nucleophile from either direction to form two products.

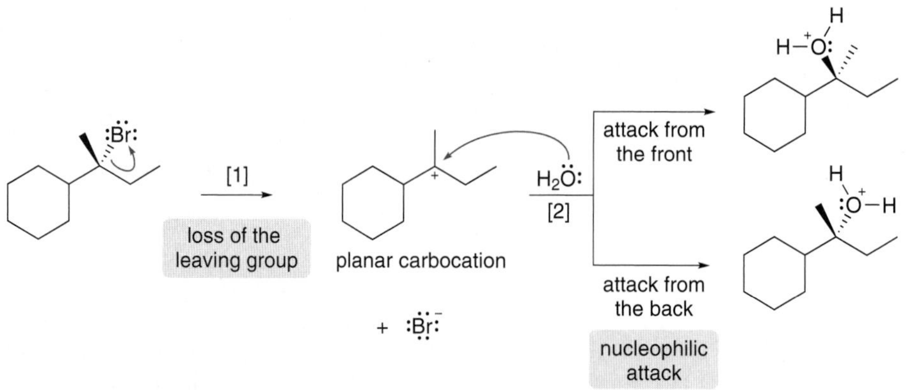

In this example, the initial products of nucleophilic substitution bear a positive charge. They readily lose a proton to form neutral products. The overall process with a neutral nucleophile thus has **three steps:** the first two constitute the **two-step S_N1 mechanism** (loss of the leaving group and attack of the nucleophile), and the third is a **Brønsted–Lowry acid–base reaction** leading to a neutral organic product.

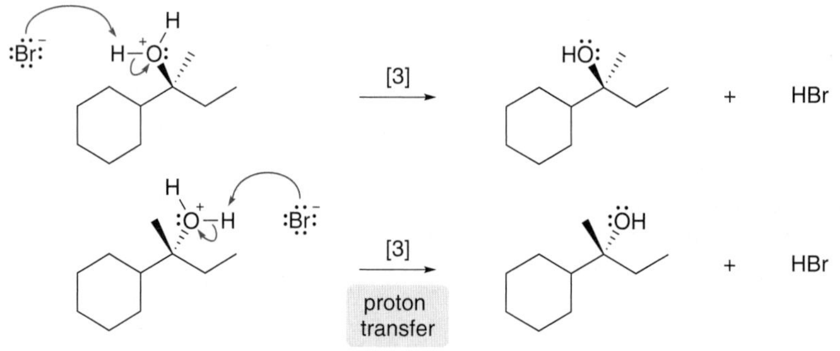

The two products in this reaction are nonsuperimposable mirror images—**enantiomers.** Because nucleophilic attack on the trigonal planar carbocation occurs with equal frequency from both directions, a **racemic mixture is formed.**

Problem 7.24 Draw the products of each S_N1 reaction and indicate the stereochemistry of any stereogenic centers.

a. 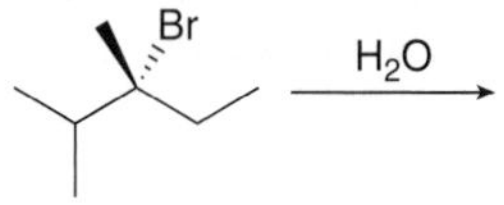

b. 

7.12D The Identity of the R Group

How does the rate of an S_N1 reaction change as the alkyl group in the substrate alkyl halide changes from CH_3 → 1° → 2° → 3°?

- **As the number of R groups on the carbon with the leaving group *increases*, the rate of an S_N1 reaction *increases*.**

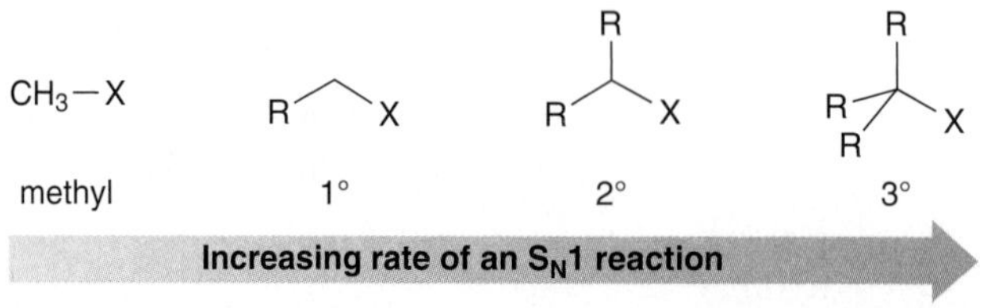

- 3° Alkyl halides undergo S_N1 reactions rapidly.
- 2° Alkyl halides react more slowly.
- Methyl and 1° alkyl halides do *not* undergo S_N1 reactions.

This trend is exactly opposite to that observed for the S_N2 mechanism. To explain this result, we must examine the rate-determining step, the formation of the carbocation, and learn about the effect of alkyl groups on **carbocation stability.** Table 7.6 summarizes the characteristics of the S_N1 mechanism.

Table 7.6 Characteristics of the S_N1 Mechanism

Characteristic	Result
Kinetics	• First-order kinetics; rate = k[RX]
Mechanism	• Two steps
Stereochemistry	• Trigonal planar carbocation intermediate • Racemization at a single stereogenic center
Identity of R	• More substituted halides react fastest. • Rate: $R_3CX > R_2CHX > RCH_2X > CH_3X$

7.13 Carbocation Stability

Carbocations are classified as **primary (1°), secondary (2°),** or **tertiary (3°)** by the number of R groups bonded to the charged carbon atom. As the number of R groups on the positively charged carbon atom increases, the stability of the carbocation **increases.**

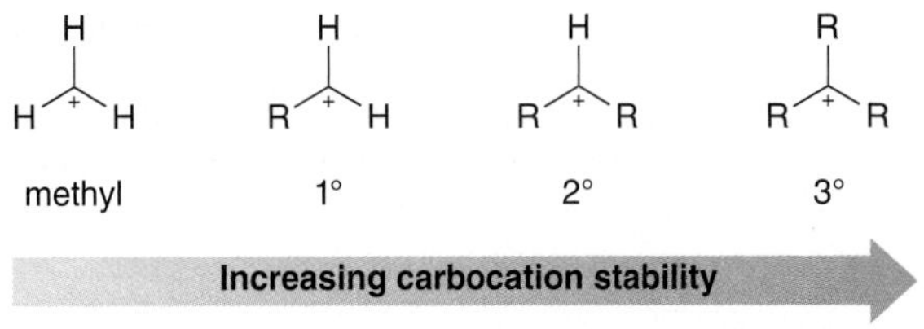

When we speak of carbocation stability, we really mean *relative* stability. Tertiary carbocations are too unstable to isolate, but they are more stable than secondary carbocations. We will examine the reason for this order of stability by invoking two different principles: **inductive effects** and **hyperconjugation.**

Problem 7.25 Classify each carbocation as 1°, 2°, or 3°.

a. 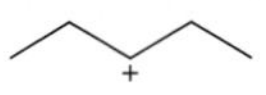 b. 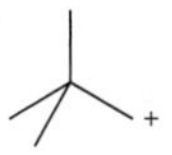 c. 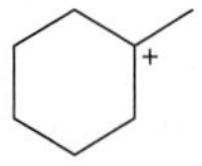 d. 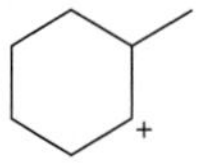

Problem 7.26 Draw the structure of a 1°, 2°, and 3° carbocation, each having molecular formula $C_4H_9^+$. Rank the three carbocations in order of increasing stability.

7.13A Inductive Effects

Inductive effects are electronic effects that occur through σ bonds. In Section 2.5B, for example, we learned that more electronegative atoms stabilize a negative charge by an **electron-withdrawing inductive effect.**

Electron-donor groups (Z) stabilize a (+) charge; Z→Y^+. Electron-withdrawing groups (W) stabilize a (–) charge; W←Y^-.

To stabilize a positive charge, **electron-donating groups** are needed. **Alkyl groups are electron-donor groups that stabilize a positive charge.** An alkyl group with several σ bonds is more polarizable than a hydrogen atom, and more able to donate electron density. Thus, as R groups successively replace the H atoms in CH_3^+, **the positive charge is more dispersed on the electron-donor R groups, and the carbocation is more stabilized.**

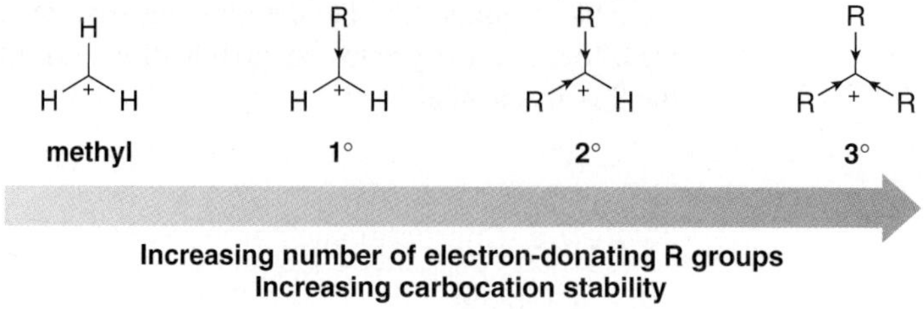

Electrostatic potential maps for four carbocations in Figure 7.14 illustrate the effect of increasing alkyl substitution on the positive charge of the carbocation.

Figure 7.14 Electrostatic potential maps for different carbocations

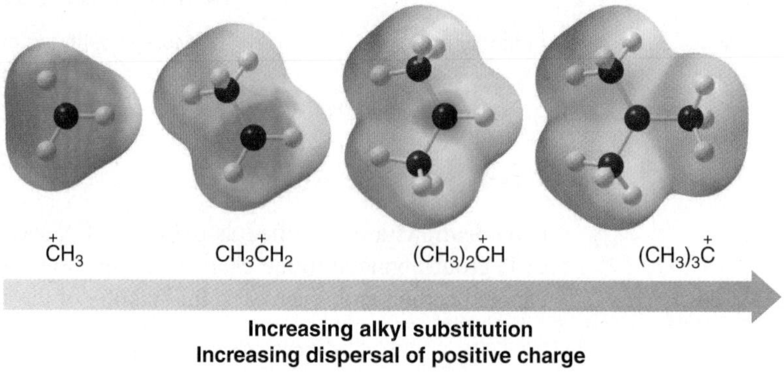

- Dark blue areas in electrostatic potential plots indicate regions low in electron density. As alkyl substitution increases, the region of positive charge is less concentrated on carbon.

Problem 7.27 Rank the following carbocations in order of increasing stability.

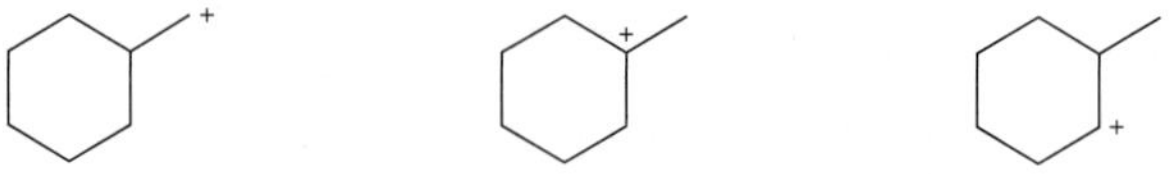

7.13B Hyperconjugation

A second explanation for the observed trend in carbocation stability is based on orbital overlap. A 3° carbocation is more stable than a 2°, 1°, or methyl carbocation because the positive charge is delocalized over more than one atom.

- **Spreading out charge by the overlap of an empty *p* orbital with an adjacent σ bond is called *hyperconjugation.***

For example, CH_3^+ cannot be stabilized by hyperconjugation, but $(CH_3)_2CH^+$ can:

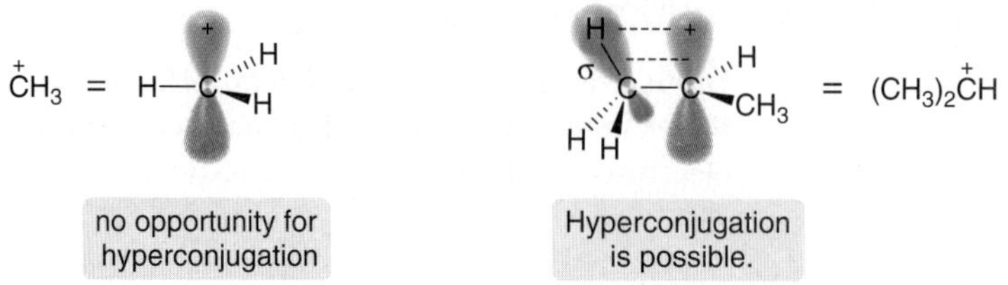

Both carbocations contain an sp^2 hybridized carbon, so both are trigonal planar with a vacant *p* orbital extending above and below the plane. There are no adjacent C–H σ bonds with which the *p* orbital can overlap in CH_3^+, but there *are* adjacent C–H σ bonds in $(CH_3)_2CH^+$. This overlap (the **hyperconjugation**) delocalizes the positive charge on the carbocation, spreading it over a larger volume, and this stabilizes the carbocation.

The larger the number of alkyl groups on the adjacent carbons, the greater the possibility for hyperconjugation, and the larger the stabilization. Hyperconjugation thus provides an alternate way of explaining why **carbocations with a larger number of R groups are more stabilized**.

7.14 The Hammond Postulate

The rate of an S_N1 reaction depends on the rate of formation of the carbocation (the product of the rate-determining step) via heterolysis of the C–X bond.

- **The rate of an S_N1 reaction *increases* as the number of R groups on the carbon with the leaving group *increases*.**
- **The stability of a carbocation *increases* as the number of R groups on the positively charged carbon *increases*.**

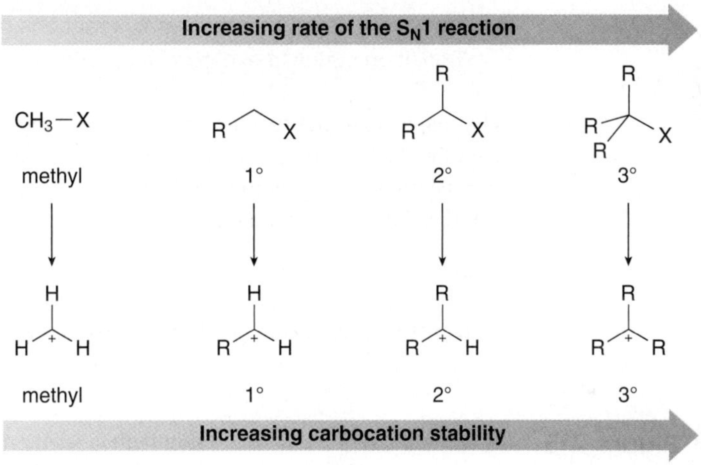

- **Thus, the rate of an S_N1 reaction *increases* as the stability of the carbocation *increases*.**

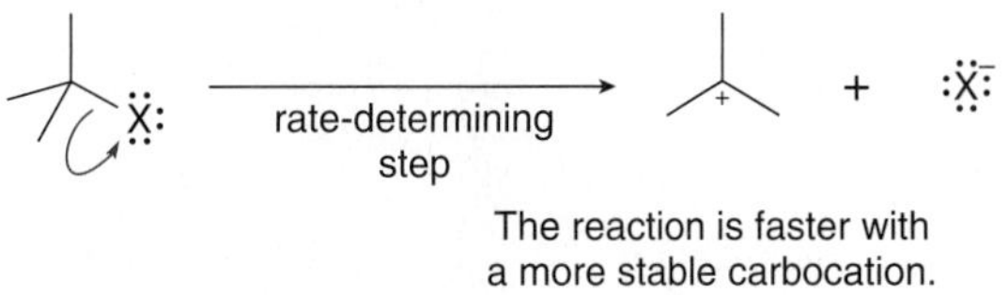

The rate of a reaction depends on the magnitude of E_a, and the stability of a product depends on $\Delta G°$. The **Hammond postulate,** first proposed in 1955, **relates rate to stability.**

7.14A The General Features of the Hammond Postulate

The Hammond postulate provides a qualitative estimate of the energy of a transition state. Because the energy of the transition state determines the energy of activation and therefore the reaction rate, predicting the relative energy of two transition states allows us to determine the relative rates of two reactions.

According to the Hammond postulate, the transition state of a reaction resembles the structure of the species (reactant or product) to which it is closer in energy. In **endothermic reactions,** the transition state is closer in energy to the **products.** In **exothermic reactions,** the transition state is closer in energy to the **reactants.**

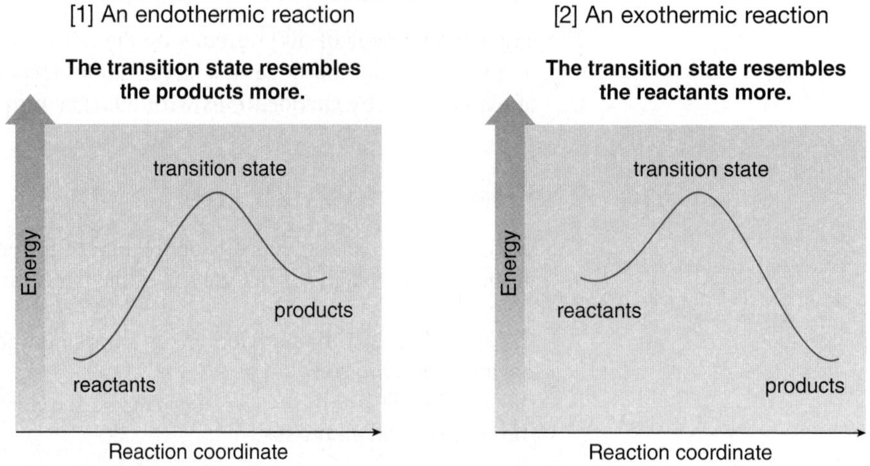

- **Transition states in endothermic reactions resemble the products.**
- **Transition states in exothermic reactions resemble the reactants.**

What happens to the reaction rate if the energy of the product is lowered? In an **endothermic reaction,** the transition state resembles the products, so anything that stabilizes the product stabilizes the transition state, too. **Lowering the energy of the transition state *decreases* the energy of activation (E_a), which *increases* the reaction rate.**

Suppose there are two possible products of an endothermic reaction, but one is more stable (lower in energy) than the other (Figure 7.15). According to the Hammond postulate, **the transition state to form the more stable product is lower in energy, so this reaction should occur faster.**

Figure 7.15
An endothermic reaction—how the energies of the transition state and products are related

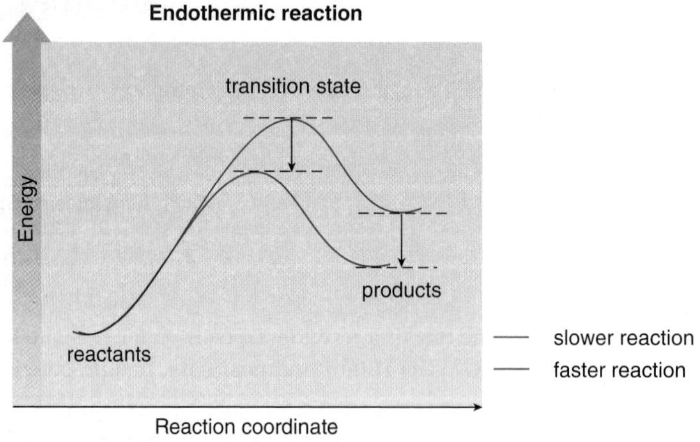

• The lower energy transition state leads to the lower energy product.

- **In an endothermic reaction, the more stable product forms faster.**

What happens to the reaction rate of an **exothermic reaction** if the energy of the product is lowered? The transition state resembles the reactants, so **lowering the energy of the products has little or no effect on the energy of the transition state.** If E_a is unaffected, then the reaction rate is unaffected, too, as shown in Figure 7.16.

Figure 7.16
An exothermic reaction—how the energies of the transition state and products are related

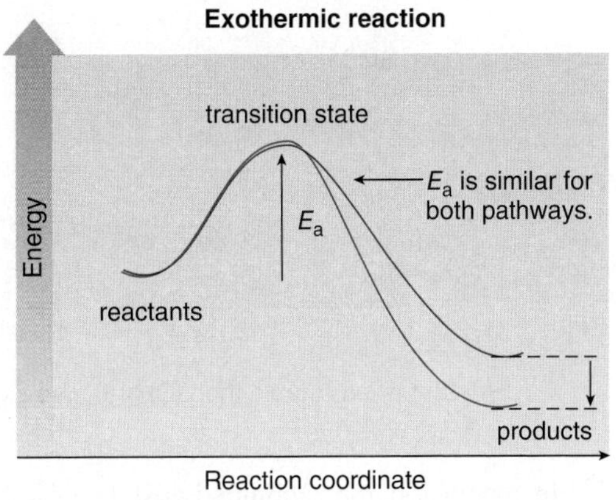

• Decreasing the energy of the product often has little effect on the energy of the transition state.

- **In an exothermic reaction, the more stable product may or may not form faster because E_a is similar for both products.**

7.14B The Hammond Postulate and the S_N1 Reaction

In the S_N1 reaction, the rate-determining step is the formation of the carbocation, an *endothermic* reaction. According to the Hammond postulate, the **stability of the carbocation determines the rate of its formation.**

For example, heterolysis of the C–Cl bond in $(CH_3)_2CHCl$ affords a less stable 2° carbocation, $(CH_3)_2CH^+$ (Equation [1]), whereas heterolysis of the C–Cl bond in $(CH_3)_3CCl$ affords a more stable 3° carbocation, $(CH_3)_3C^+$ (Equation [2]). The Hammond postulate states that Reaction [2] is faster than Reaction [1], because the transition state to form the more stable 3° carbocation is lower in energy. Figure 7.17 depicts an energy diagram comparing these two endothermic reactions.

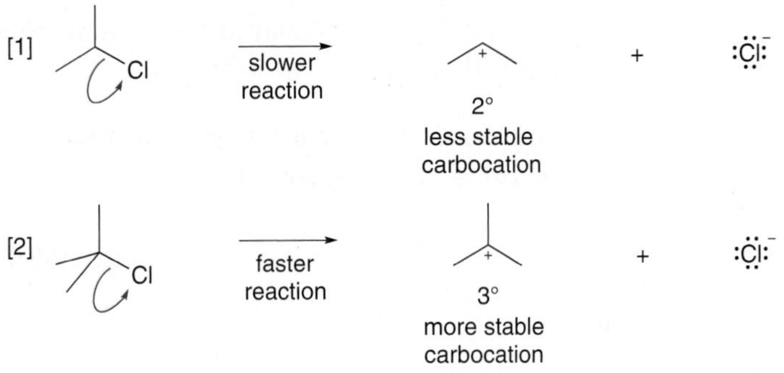

Figure 7.17
Energy diagram for carbocation formation in two different S_N1 reactions

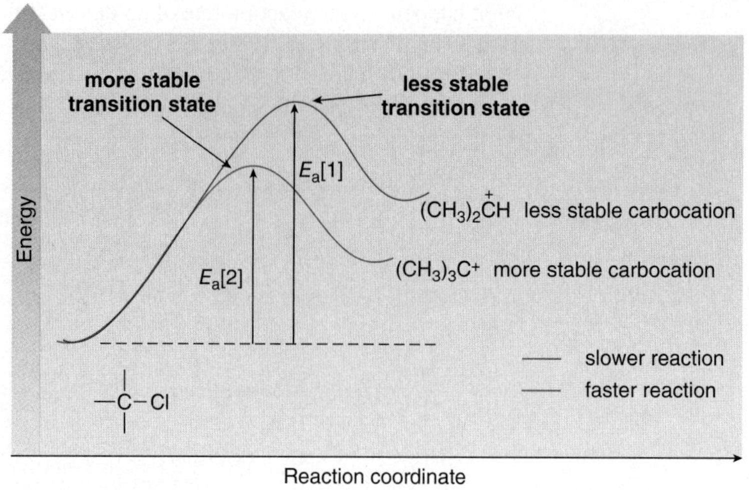

- $(CH_3)_2CH^+$ is less stable than $(CH_3)_3C^+$, so $E_a[1] > E_a[2]$, and Reaction [1] is slower.

In conclusion, the Hammond postulate can be used to predict the relative rates of two reactions. In the S_N1 reaction the rate-determining step is endothermic, so the more stable carbocation is formed faster.

Problem 7.28 Which alkyl halide in each pair reacts faster in an S_N1 reaction?

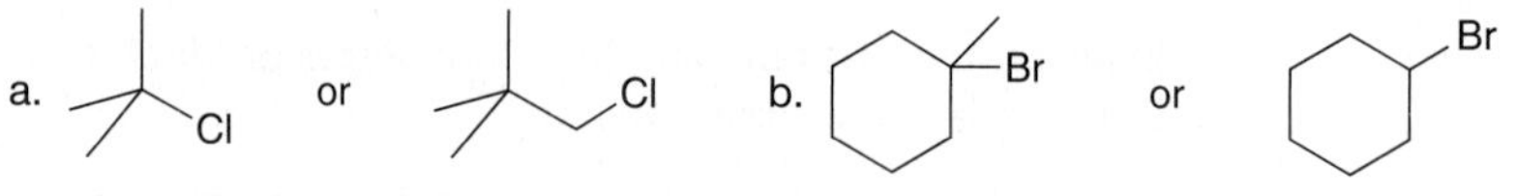

7.15 When Is the Mechanism S_N1 or S_N2?

Given a particular starting material and nucleophile, how do we know whether a reaction occurs by the S_N1 or S_N2 mechanism? Four factors are examined:

- **The alkyl halide—CH_3X, RCH_2X, R_2CHX, or R_3CX**
- **The nucleophile—strong or weak**
- **The leaving group—good or poor**
- **The solvent—protic or aprotic**

7.15A The Alkyl Halide—The Most Important Factor

The most important factor in determining whether a reaction follows the S_N1 or S_N2 mechanism is the *identity of the alkyl halide.*

- Increasing alkyl substitution favors S_N1.
- Decreasing alkyl substitution favors S_N2.

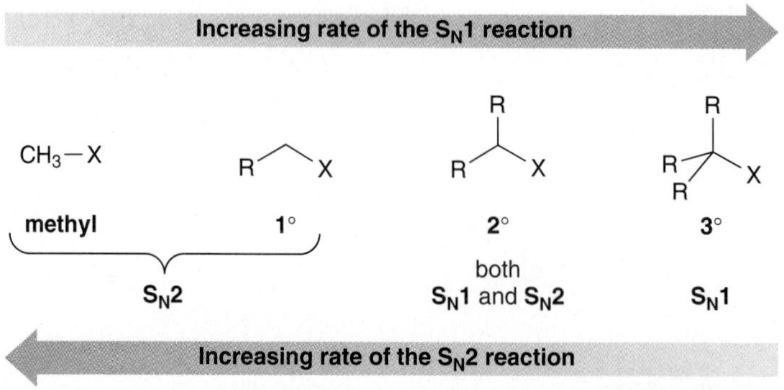

- **Methyl and 1° halides (CH_3X and RCH_2X) undergo S_N2 reactions only.**
- **3° Alkyl halides (R_3CX) undergo S_N1 reactions only.**
- **2° Alkyl halides (R_2CHX) undergo both S_N1 and S_N2 reactions. Other factors determine the mechanism.**

Examples are given in Figure 7.18.

Figure 7.18
The identity of RX and the mechanism of nucleophilic substitution

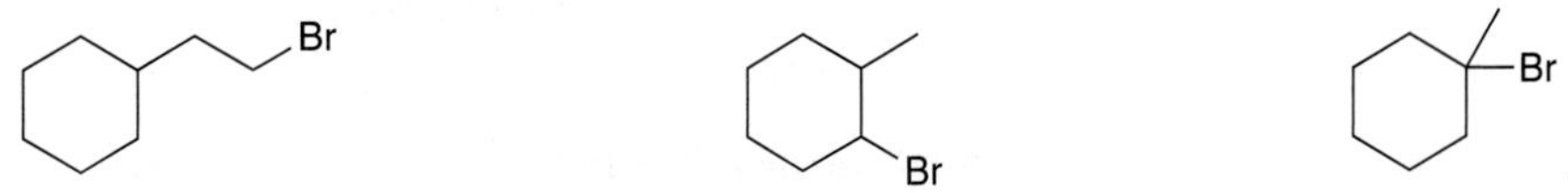
Br
1° halide
S_N2
Br
2° halide
Both S_N2 and S_N1 are possible.
Br
3° halide
S_N1

Problem 7.29 What is the likely mechanism of nucleophilic substitution for each alkyl halide?

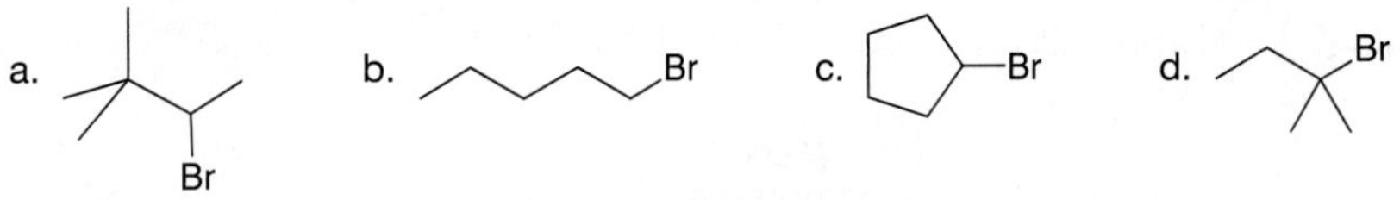
a.
Br
b.
Br
c.
Br
d.
Br

7.15B The Nucleophile

How does the strength of the nucleophile affect an S_N1 or S_N2 mechanism? The rate of the S_N1 reaction is unaffected by the identity of the nucleophile because the nucleophile does not appear in the rate equation (rate = k[RX]). The identity of the nucleophile *is* important for the S_N2 reaction, however, because the nucleophile does appear in the rate equation for this mechanism (rate = $k[RX][:Nu^-]$).

- **Strong nucleophiles present in high concentration favor S_N2 reactions.**
- **Weak nucleophiles favor S_N1 reactions by decreasing the rate of any competing S_N2 reaction.**

The most common nucleophiles in S_N2 reactions bear a net negative charge. The most common nucleophiles in S_N1 reactions are weak nucleophiles such as H_2O and ROH. The identity of the nucleophile is especially important in determining the mechanism and therefore the stereochemistry of nucleophilic substitution when 2° alkyl halides are starting materials.

Let's compare the substitution products formed when the 2° alkyl halide **A** (*cis*-1-bromo-4-methylcyclohexane) is treated with either the strong nucleophile ^-OH or the weak nucleophile H_2O. Because a 2° alkyl halide can react by either mechanism, the strength of the nucleophile determines which mechanism takes place.

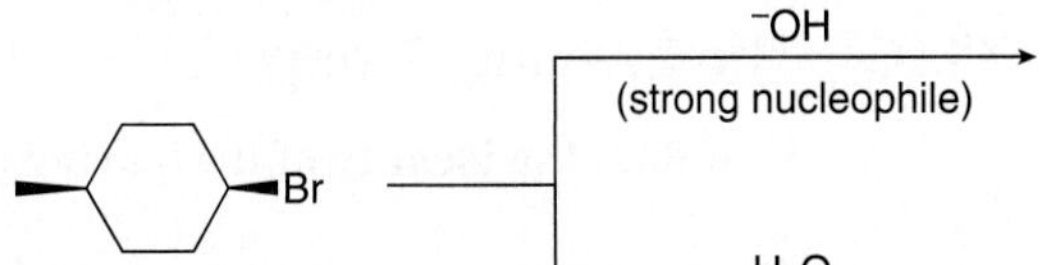
Br
^-OH
(strong nucleophile)
H_2O
(weak nucleophile)

cis-1-bromo-4-methyl-cyclohexane
A

The **strong nucleophile $^{-}$OH favors an S_N2 reaction,** which occurs with **backside** attack of the nucleophile, resulting in **inversion of configuration.** Because the leaving group Br^- is above the plane of the ring, the nucleophile attacks from below, and a single product **B** is formed.

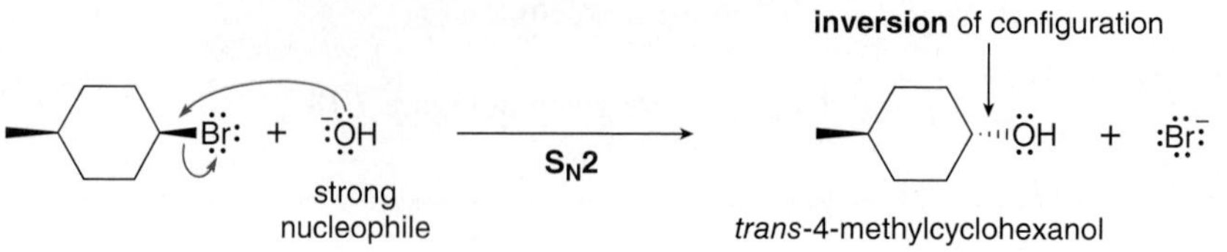

A B

The **weak nucleophile H_2O favors an S_N1 reaction,** which occurs by way of an intermediate carbocation. Loss of the leaving group in **A** forms the carbocation, which undergoes nucleophilic attack from both above and below the plane of the ring to afford two products, **C** and **D.** Loss of a proton by proton transfer forms the final products, **B** and **E. B** and **E** are diastereomers of each other (**B** is a trans isomer and **E** is a cis isomer).

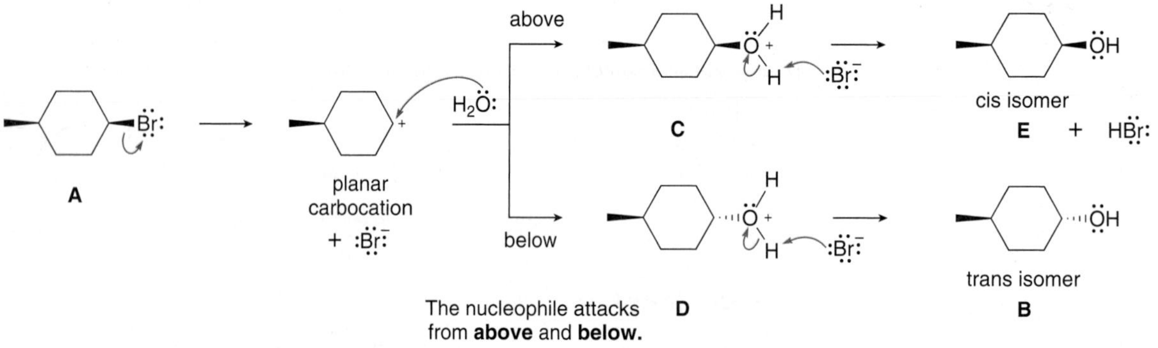

Thus, the mechanism of nucleophilic substitution determines the stereochemistry of the products formed.

Problem 7.30 For each alkyl halide and nucleophile: [1] Draw the product of nucleophilic substitution; [2] determine the likely mechanism (S_N1 or S_N2) for each reaction.

a. 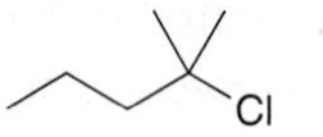 + CH_3OH

c. 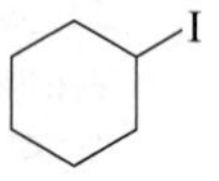 + $CH_3CH_2O^-$

b. 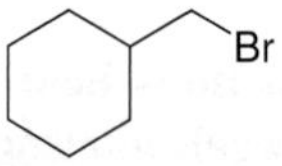 + ^{-}SH

d. 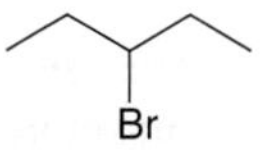 + CH_3OH

Problem 7.31 Draw the products (including stereochemistry) for each reaction.

a. 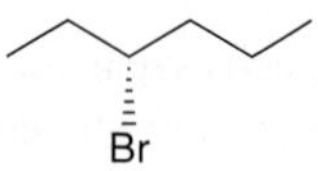 + H_2O ⟶

b. 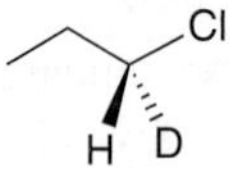 + $^{-}$:C≡C–H ⟶

7.15C The Leaving Group

How does the identity of the leaving group affect an S_N1 or S_N2 reaction?

- A better leaving group increases the rate of *both* S_N1 and S_N2 reactions.

Because the bond to the leaving group is partially broken in the transition state of the only step of the S_N2 mechanism and the slow step of the S_N1 mechanism, **a better leaving group increases the rate of both reactions.** The better the leaving group, the more willing it is to accept the electron pair in the C–X bond, and the faster the reaction.

For alkyl halides, the following order of reactivity is observed for the S_N1 and the S_N2 mechanisms:

R–F R–Cl R–Br R–I

Increasing leaving group ability
Increasing rate of S_N1 and S_N2 reactions

Problem 7.32 Which compound in each pair reacts faster in nucleophilic substitution?

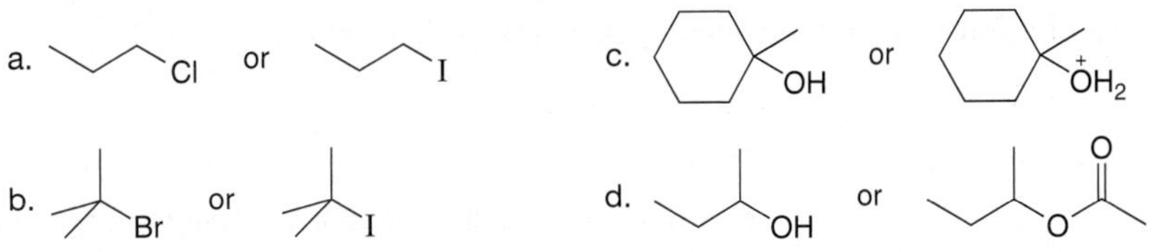
a.
Cl
or
I
c.
OH
or
$\overset{+}{O}H_2$
b.
Br
or
I
d.
OH
or
O
O

7.15D The Solvent

Polar protic solvents and polar aprotic solvents affect the rates of S_N1 and S_N2 reactions differently.

- **Polar *protic* solvents are especially good for S_N1 reactions.**
- **Polar *aprotic* solvents are especially good for S_N2 reactions.**

Polar protic solvents like H_2O and ROH solvate both cations and anions well, and this characteristic is important for the S_N1 mechanism, in which two ions (a carbocation and a leaving group) are formed by heterolysis of the C–X bond. The carbocation is solvated by ion–dipole interactions with the polar solvent, and the leaving group is solvated by hydrogen bonding, in much the same way that Na^+ and Br^- are solvated in Section 7.8C. These interactions stabilize the reactive intermediate. In fact, a polar protic solvent is generally needed for an S_N1 reaction.

Summary of solvent effects:
- Polar protic solvents favor S_N1 reactions because the ionic intermediates are stabilized by solvation.
- Polar aprotic solvents favor S_N2 reactions because nucleophiles are not well solvated, and therefore are more nucleophilic.

Polar aprotic solvents exhibit dipole–dipole interactions but not hydrogen bonding, and as a result, they do not solvate anions well. This has a pronounced effect on the nucleophilicity of anionic nucleophiles. Because these nucleophiles are not "hidden" by strong interactions with the solvent, they are **more nucleophilic.** Because stronger nucleophiles favor S_N2 reactions, **polar aprotic solvents are especially good for S_N2 reactions.**

Problem 7.33 Which solvents favor S_N1 reactions and which favor S_N2 reactions?

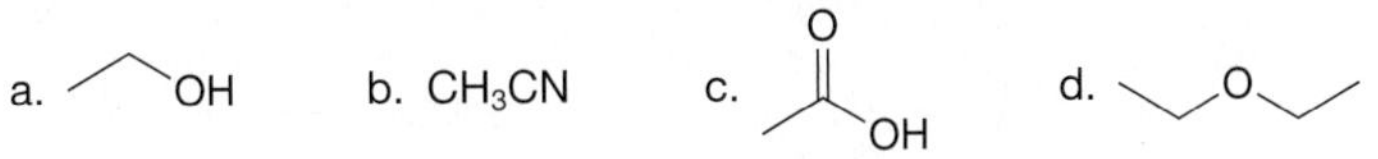
a.
OH
b. CH_3CN
c.
O
OH
d.
O

Problem 7.34 For each reaction, use the identity of the alkyl halide and nucleophile to determine which substitution mechanism occurs. Then determine which solvent affords the faster reaction.

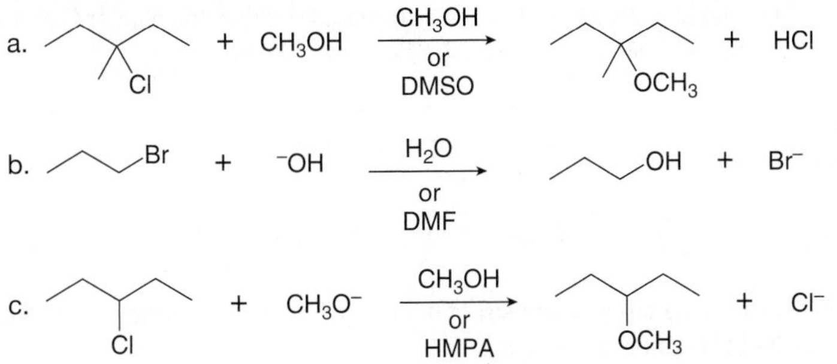
a.
Cl
+ CH_3OH
CH_3OH
or
DMSO
OCH_3
+ HCl
b.
Br
+ ^-OH
H_2O
or
DMF
OH
+ Br^-
c.
Cl
+ CH_3O^-
CH_3OH
or
HMPA
OCH_3
+ Cl^-

7.15E Summary of Factors That Determine Whether the S_N1 or S_N2 Mechanism Occurs

Table 7.7 summarizes the factors that determine whether a reaction occurs by the S_N1 or S_N2 mechanism. Sample Problems 7.4 and 7.5 illustrate how these factors are used to determine the mechanism of a given reaction.

Table 7.7 Summary of Factors That Determine the S_N1 or S_N2 Mechanism

Alkyl halide	Mechanism	Other factors
CH_3X RCH_2X (1°)	**S_N2**	Favored by • **strong nucleophiles** (usually a net negative charge) • polar **aprotic** solvents
R_3CX (3°)	**S_N1**	Favored by • **weak nucleophiles** (usually neutral) • polar **protic** solvents
R_2CHX (2°)	**S_N1 or S_N2**	The mechanism depends on the conditions. • **Strong nucleophiles favor the S_N2 mechanism over the S_N1 mechanism.** RO^- is a stronger nucleophile than ROH, so RO^- favors the S_N2 reaction and ROH favors the S_N1 reaction. • **Protic solvents favor the S_N1 mechanism and aprotic solvents favor the S_N2 mechanism.** H_2O and CH_3OH are polar protic solvents that favor the S_N1 mechanism, whereas acetone $[(CH_3)_2C{=}O]$ and DMSO $[(CH_3)_2S{=}O]$ are polar aprotic solvents that favor the S_N2 mechanism.

Sample Problem 7.4 Determine the mechanism of nucleophilic substitution for each reaction and draw the products.

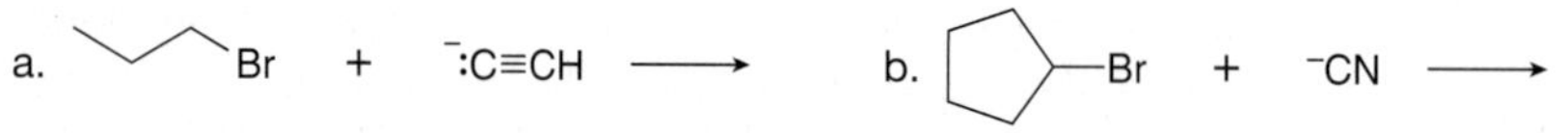

Solution

a. The alkyl halide is 1°, so it must react by an S_N2 mechanism with the nucleophile $^-$:C≡CH.

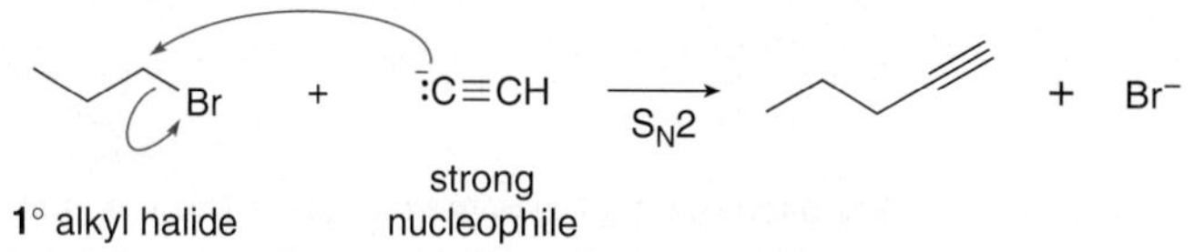

b. The alkyl halide is 2°, so it can react by either the S_N1 or S_N2 mechanism. The strong nucleophile ($^-$CN) favors the S_N2 mechanism.

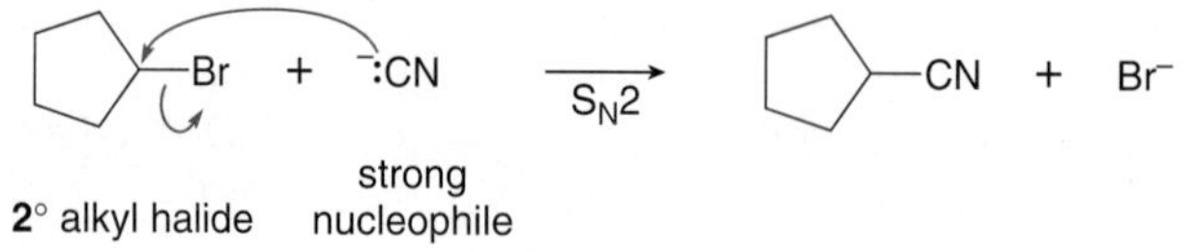

Sample Problem 7.5 Determine the mechanism of nucleophilic substitution for each reaction and draw the products, including stereochemistry.

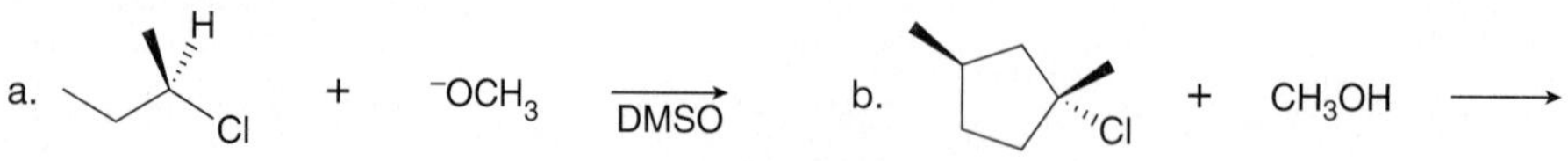

Solution

a. The 2° alkyl halide can react by either the S_N1 or S_N2 mechanism. The strong nucleophile ($^-OCH_3$) favors the S_N2 mechanism, as does the polar aprotic solvent (DMSO). S_N2 reactions proceed with inversion of configuration.

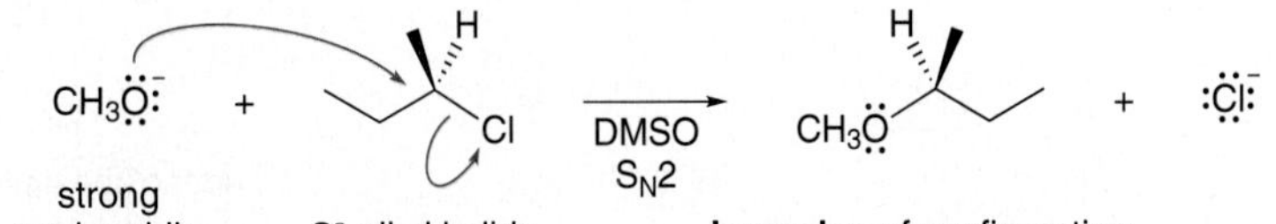

strong nucleophile — 2° alkyl halide — **inversion** of configuration

b. The alkyl halide is 3°, so it reacts by an S_N1 mechanism with the weak nucleophile CH_3OH. S_N1 reactions proceed with racemization at a single stereogenic center, so two products are formed.

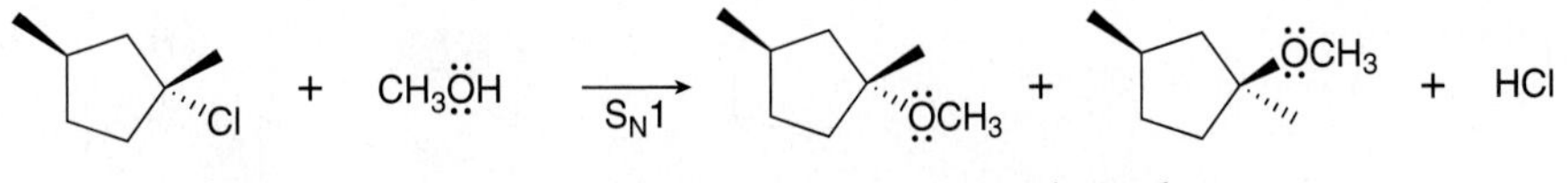

3° alkyl halide — weak nucleophile — two products of nucleophilic substitution

Problem 7.35 Determine the mechanism and draw the products of each reaction. Include the stereochemistry at all stereogenic centers.

a. 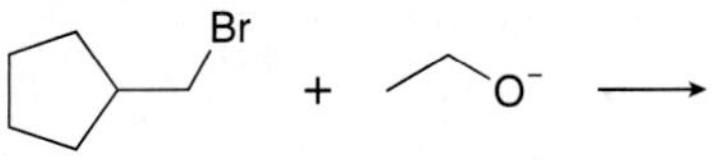

c. 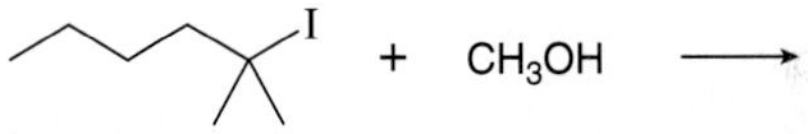

b. 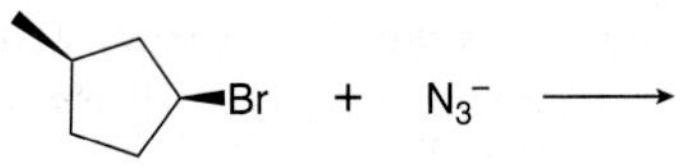

d. 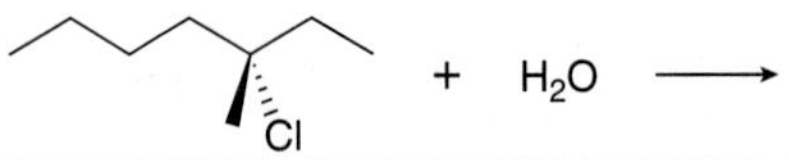

7.16 Biological Nucleophilic Substitution

Nucleophilic substitution occurs in a wide variety of biological reactions.

7.16A Leaving Groups Derived from Phosphorus

In contrast to nucleophilic substitutions run in the laboratory that use alkyl halides as substrates and halide anions as leaving groups, biological substitutions often occur with phosphorus leaving groups, such as phosphate (PO_4^{3-}, abbreviated as P_i for inorganic phosphate), diphosphate ($P_2O_7^{4-}$, abbreviated as PP_i), and triphosphate ($P_3O_{10}^{5-}$, abbreviated as PPP_i). These anions are excellent leaving groups because they are weak, resonance-stabilized bases.

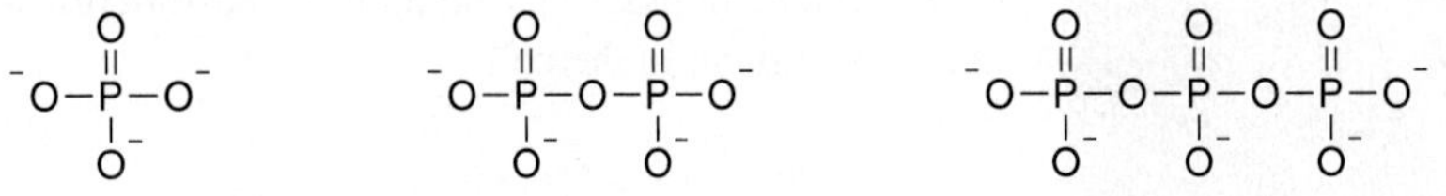

phosphate P_i — diphosphate PP_i — triphosphate PPP_i

When an organic compound contains a carbon bonded to one of these leaving groups, the compound is called an organic monophosphate, diphosphate, or triphosphate.

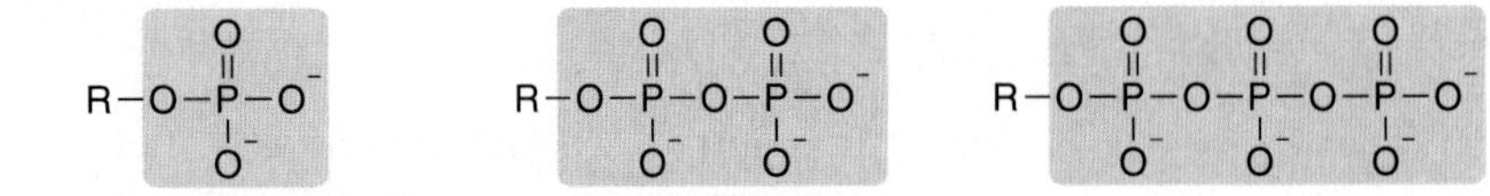

organic monophosphate — organic diphosphate — organic triphosphate

R–OP — R–OPP — R–OPPP

Adenosine triphosphate (ATP) is an organic triphosphate.

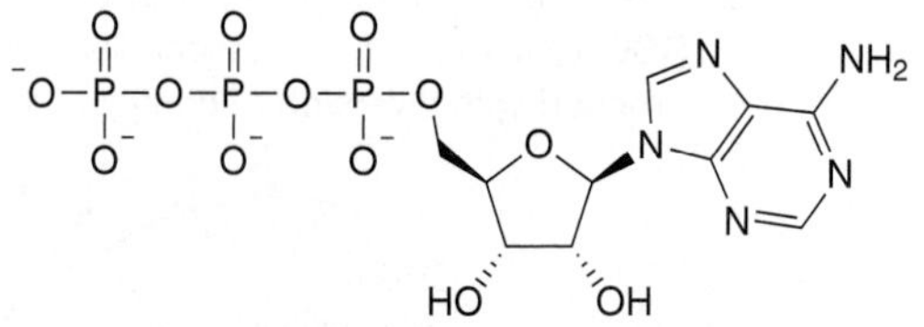

adenosine triphosphate
ATP

Nucleophilic substitutions with these substrates may proceed by either an S_N2 or S_N1 pathway, as shown with the general diphosphate R_2CHOPP.

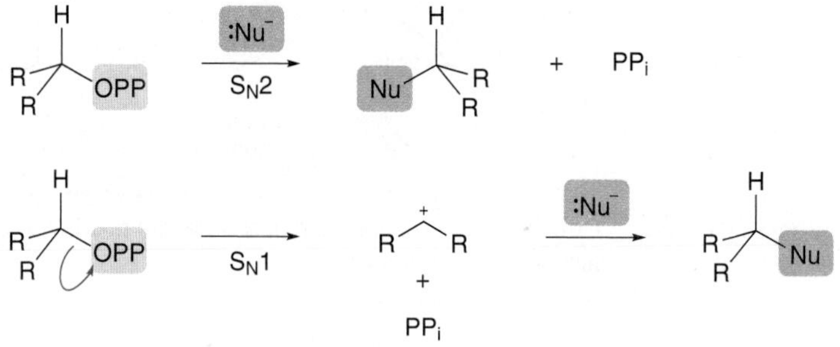

The final step in the biosynthesis of geraniol, a component of rose oil used in perfumery, is an S_N1 reaction of geranyl diphosphate with water. This reaction occurs by way of a resonance-stabilized carbocation. We will learn more about reactions of diphosphates in Chapter 16.

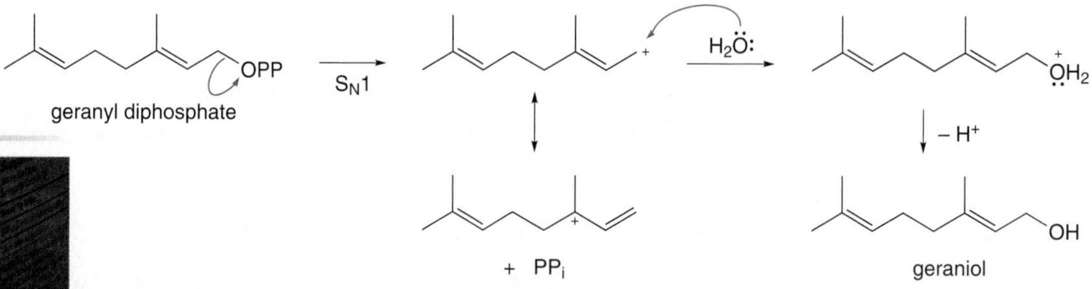

SAM, a nutritional supplement sold under the name SAM-e (pronounced sammy), has been used in Europe to treat depression and arthritis for over 20 years. In cells, SAM is used in nucleophilic substitutions that synthesize key amino acids, hormones, and neurotransmitters.

7.16B *S*-Adenosylmethionine

A common nucleophilic substitution occurs with *S*-adenosylmethionine, or **SAM.** SAM is the cell's equivalent of CH_3I. The many polar functional groups in SAM make it soluble in the aqueous environment in the cell.

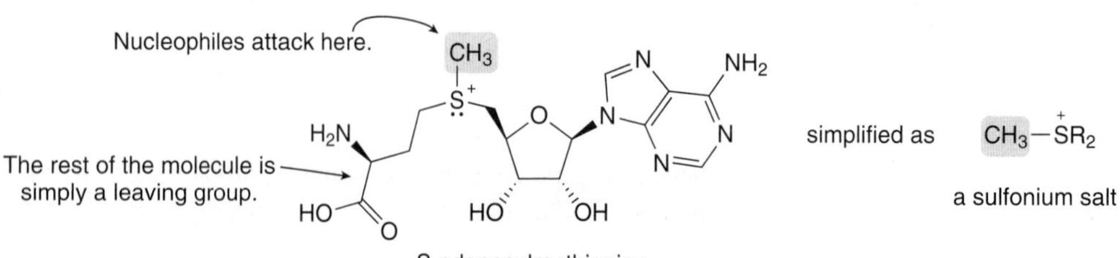

S-adenosylmethionine
SAM

The CH_3 group in SAM [abbreviated as $(CH_3SR_2)^+$] is part of a sulfonium salt, a positively charged sulfur species that contains a good leaving group. Nucleophilic attack at the CH_3 group of SAM displaces R_2S, a good neutral leaving group. This reaction is called **methylation,** because a CH_3 group is transferred from one compound (SAM) to another ($:Nu^-$).

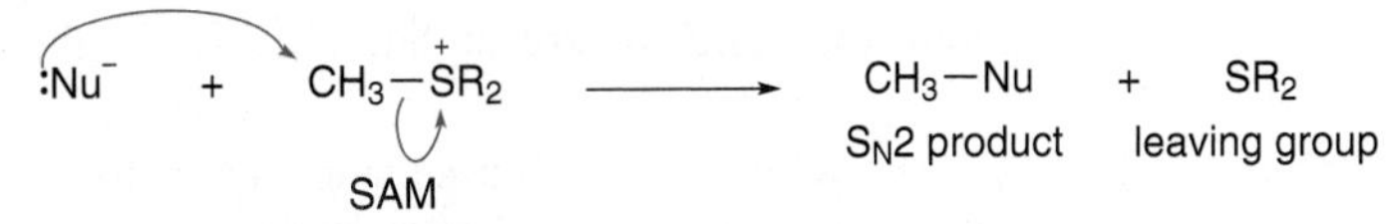

For example, **adrenaline** (epinephrine) is a hormone synthesized in the adrenal glands from noradrenaline (norepinephrine) by nucleophilic substitution using SAM (Figure 7.19). When an individual senses danger or is confronted by stress, the hypothalamus region of the brain signals the adrenal glands to synthesize and release adrenaline, which enters the bloodstream and then

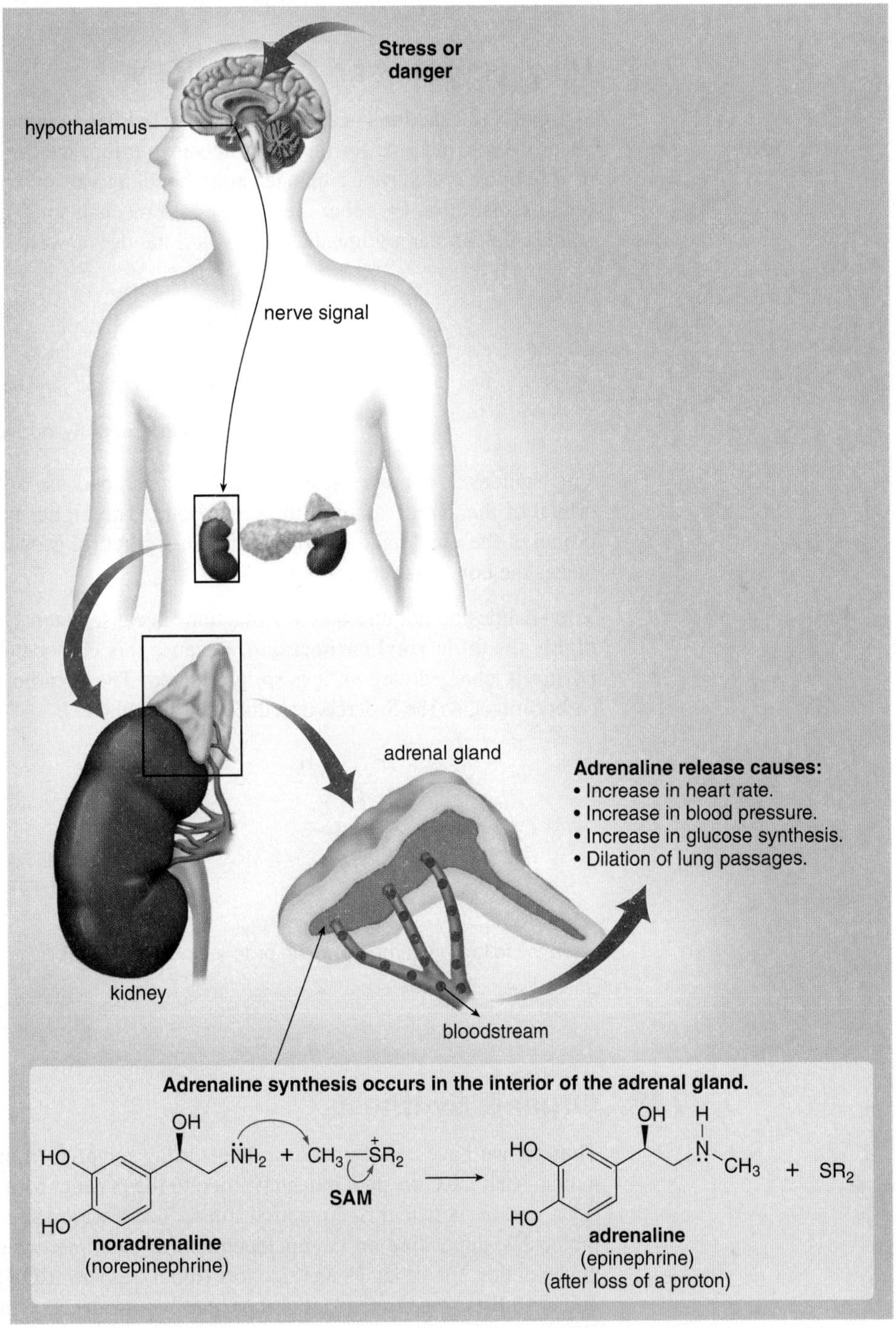

Figure 7.19 Adrenaline synthesis from noradrenaline in response to stress

stimulates a response in many organs. Stored carbohydrates are metabolized in the liver to form glucose, which is further metabolized to provide an energy boost. Heart rate and blood pressure increase, and lung passages are dilated. These physiological changes result from the "rush of adrenaline," and prepare an individual for "fight or flight."

Problem 7.36 Nicotine, a toxic and addictive component of tobacco, is synthesized from **A** using SAM. Write out the reaction that converts **A** into nicotine.

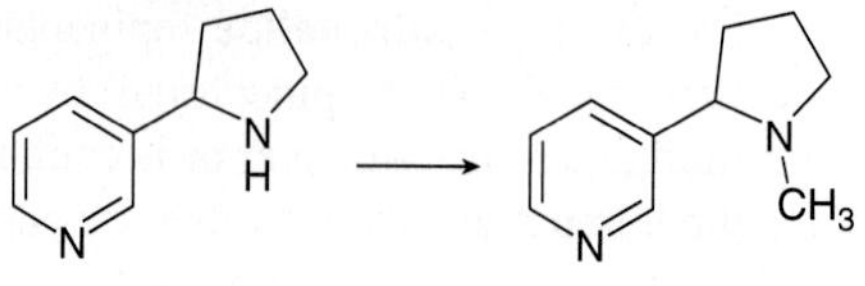

7.17 Vinyl Halides and Aryl Halides

S_N1 and S_N2 reactions occur only at *sp^3* hybridized carbon atoms. Now that we have learned about the mechanisms for nucleophilic substitution, we can understand why **vinyl halides** and **aryl halides,** which have a halogen atom bonded to an sp^2 hybridized C, do not undergo nucleophilic substitution by either the S_N1 or S_N2 mechanism. The discussion here centers on vinyl halides, but similar arguments hold for aryl halides as well.

C bonded to X is sp^2 hybridized.

Vinyl halides do not undergo S_N2 reactions in part because of the percent *s*-character in the hybrid orbital of the carbon atom in the C−X bond. The higher percent *s*-character in the sp^2 hybrid orbital of the vinyl halide compared to the sp^3 hybrid orbital of the alkyl halide (33% vs. 25%) makes the bond shorter and stronger.

Vinyl halides do not undergo S_N1 reactions because heterolysis of the C−X bond would form a **highly unstable vinyl carbocation.** Because this carbocation has only two groups around the positively charged carbon, it is ***sp*** hybridized. These carbocations are even less stable than 1° carbocations, so the S_N1 reaction does not take place.

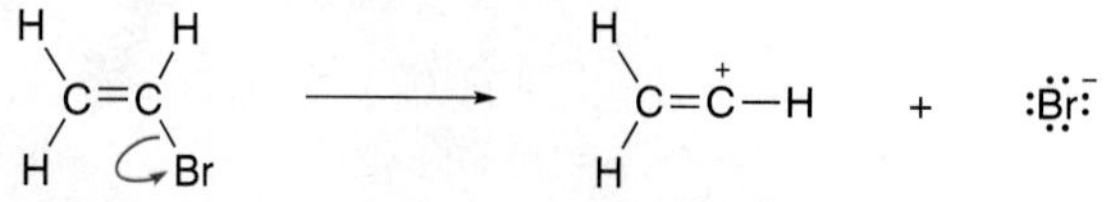

vinyl carbocation
highly unstable

Problem 7.37 Rank the following carbocations in order of increasing stability.

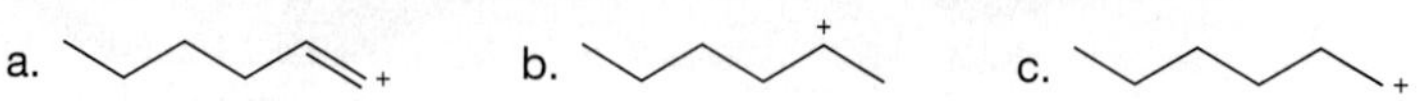

7.18 Organic Synthesis

Thus far we have concentrated on the starting material in nucleophilic substitution—the alkyl halide—and have not paid much attention to the product formed. Nucleophilic substitution reactions, and in particular S_N2 reactions, introduce a wide variety of different functional groups in molecules, depending on the nucleophile. For example, when ^-OH, ^-OR, and ^-CN are used as nucleophiles, the products are alcohols (ROH), ethers (ROR), and nitriles (RCN), respectively. Table 7.8 lists some functional groups readily introduced using nucleophilic substitution.

Table 7.8 Molecules Synthesized from R–X by the S_N2 Reaction

	Nucleophile (:Nu^-)	Product	Name
Oxygen compounds	^-OH	R–OH	alcohol
	$^-OR'$	R–OR'	ether
	$^-O-C(=O)-R'$	$R-O-C(=O)-R'$	ester
Carbon compounds	^-CN	R–CN	nitrile
	$^-$:C≡C–H	R–C≡C–H	alkyne
Nitrogen compounds	N_3^-	$R-N_3$	azide
	$:NH_3$	$R-NH_2$	amine
Sulfur compounds	^-SH	R–SH	thiol
	$^-SR'$	R–SR'	sulfide

By thinking of **nucleophilic substitution as a reaction that *makes* a particular kind of organic compound,** we begin to think about ***synthesis.***

- **Organic synthesis is the systematic preparation of a compound from a readily available starting material by one or many steps.**

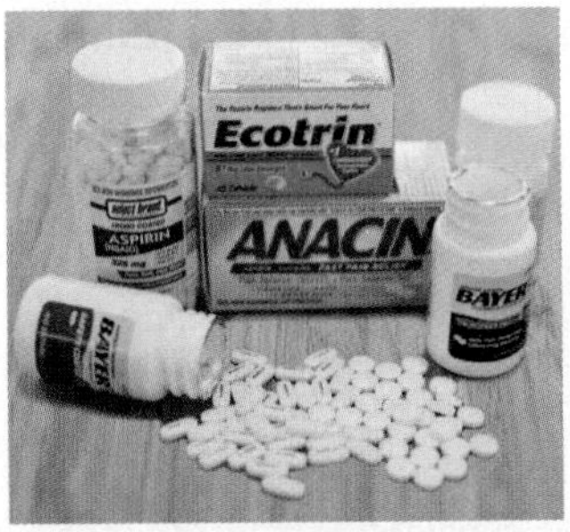

Aspirin is synthesized by a two-step procedure from simple, cheap starting materials.

Phenol, the starting material for the aspirin synthesis, is a petroleum product, like most of the starting materials used in large quantities in industrial syntheses. A shortage of petroleum reserves thus affects the availability not only of fuels for transportation, but also of raw materials needed for most chemical synthesis.

7.18A Background on Organic Synthesis

Chemists synthesize molecules for many reasons. Sometimes a **natural product,** a compound isolated from natural sources, has useful medicinal properties, but is produced by an organism in only minute quantities. Synthetic chemists then prepare this molecule from simpler starting materials, so that it can be made available to a large number of people.

Sometimes, chemists prepare molecules that do not occur in nature (although they may be similar to those in nature), because these molecules have superior properties to their naturally occurring relatives. **Aspirin, or acetylsalicylic acid** (Section 2.7), is a well-known example. Acetylsalicylic acid is prepared from phenol, a product of the petroleum industry, by a two-step procedure (Figure 7.20). Aspirin has become one of the most popular and widely used drugs in the world because it has excellent analgesic and anti-inflammatory properties, *and* it is cheap and readily available.

Figure 7.20 Synthesis of aspirin

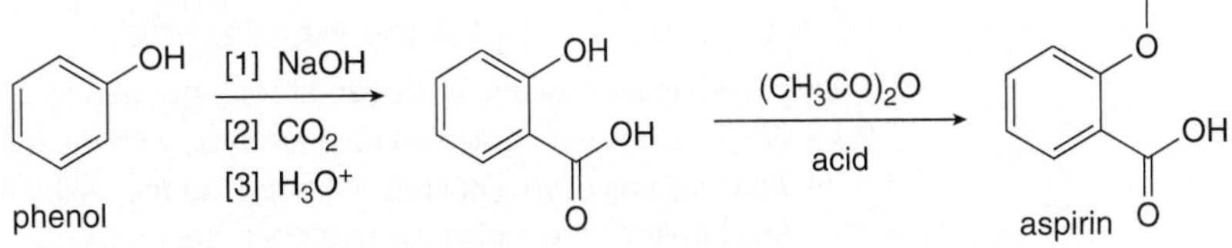

7.18B Nucleophilic Substitution and Organic Synthesis

To carry out synthesis we must think backwards. We examine a compound and ask: **What starting material and reagent are needed to make it?** If we are using nucleophilic substitution, we must determine what alkyl halide and what nucleophile can be used to form a specific product. This is the simplest type of synthesis because it involves only one step. In Chapter 11 we will learn about multistep syntheses.

Suppose, for example, that we are asked to prepare $(CH_3)_2CHCH_2OH$ (2-methylpropan-1-ol) from an alkyl halide and any required reagents. To accomplish this synthesis, we must "fill in the boxes" for the starting material and reagent in the accompanying equation.

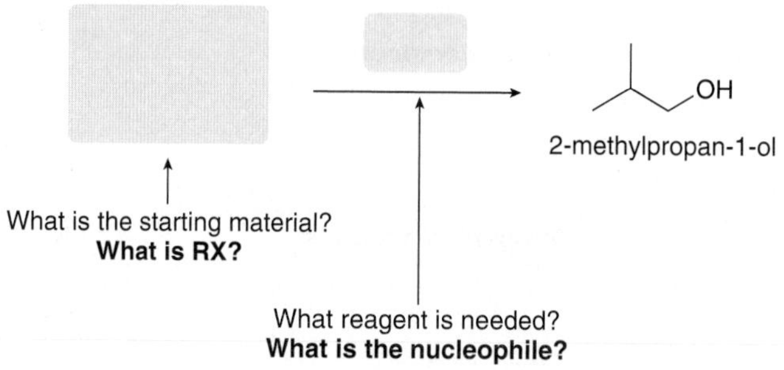

To determine the two components needed for the synthesis, remember that the carbon atoms come from the organic starting material, in this case a 1° alkyl halide $[(CH_3)_2CHCH_2Br]$. The functional group comes from the nucleophile, ^-OH in this case. With these two components, we can "fill in the boxes" to complete the synthesis.

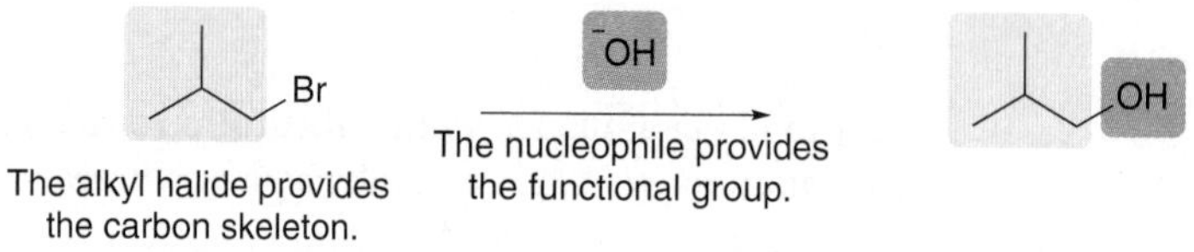

After any synthesis is proposed, check to see if it is reasonable, given what we know about reactions. Will the reaction written give a high yield of product? The synthesis of $(CH_3)_2CHCH_2OH$ is reasonable, because the starting material is a 1° alkyl halide and the nucleophile (^-OH) is strong, and both facts contribute to a successful S_N2 reaction.

Problem 7.38 What alkyl halide and nucleophile are needed to prepare each compound?

a. 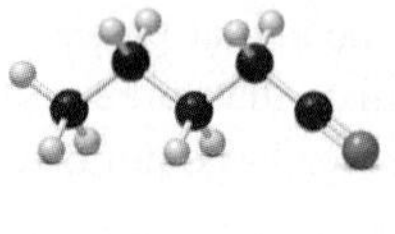 b. 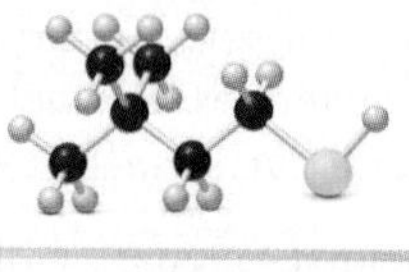 c.  d. 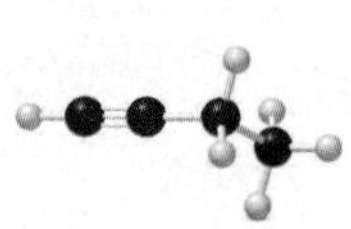

Problem 7.39 The ether, $CH_3OCH_2CH_3$, can be prepared by two different nucleophilic substitution reactions, one using CH_3O^- as nucleophile and the other using $CH_3CH_2O^-$ as nucleophile. Draw both routes.

KEY CONCEPTS

Alkyl Halides and Nucleophilic Substitution

General Facts about Alkyl Halides

- Alkyl halides contain a halogen atom X bonded to an sp^3 hybridized carbon (7.1).
- Alkyl halides are named as halo alkanes, with the halogen as a substituent (7.2).
- Alkyl halides have a polar C–X bond, so they exhibit dipole–dipole interactions but are incapable of intermolecular hydrogen bonding (7.3).

- The polar C – X bond containing an electrophilic carbon makes alkyl halides reactive toward nucleophiles and bases (7.5).

The Central Theme (7.6)

- Nucleophilic substitution is one of the two main reactions of alkyl halides. A nucleophile replaces a leaving group on an sp^3 hybridized carbon.

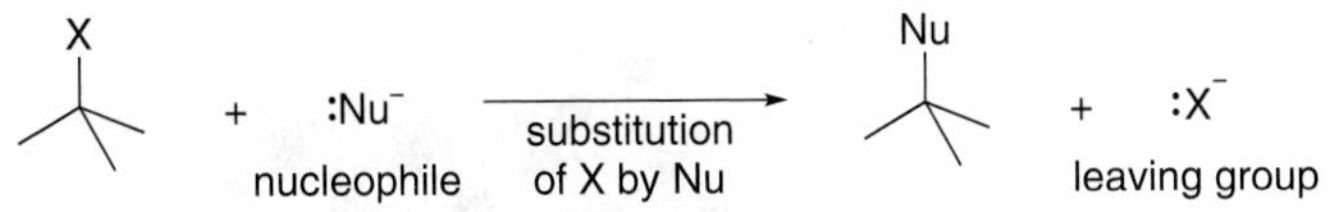

- One σ bond is broken and one σ bond is formed.
- There are two possible mechanisms: S_N1 and S_N2.

S_N1 and S_N2 Mechanisms Compared

	S_N2 mechanism	**S_N1 mechanism**
[1] Mechanism	• One step (7.11B)	• Two steps (7.12B)
[2] Alkyl halide	• Order of reactivity: $CH_3X > RCH_2X > R_2CHX > R_3CX$ (7.11D)	• Order of reactivity: $R_3CX > R_2CHX > RCH_2X > CH_3X$ (7.12D)
[3] Rate equation	• Rate = k[RX][:Nu^-] • Second-order kinetics (7.11A)	• Rate = k[RX] • First-order kinetics (7.12A)
[4] Stereochemistry	• Backside attack of the nucleophile (7.11C) • Inversion of configuration at a stereogenic center	• Trigonal planar carbocation intermediate (7.12C) • Racemization at a single stereogenic center
[5] Nucleophile	• Favored by stronger nucleophiles (7.15B)	• Favored by weaker nucleophiles (7.15B)
[6] Leaving group	• Better leaving group ⇢ faster reaction (7.15C)	• Better leaving group ⇢ faster reaction (7.15C)
[7] Solvent	• Favored by polar aprotic solvents (7.15D)	• Favored by polar protic solvents (7.15D)

Important Trends

- The best leaving group is the weakest base. Leaving group ability increases left-to-right across a row and down a column of the periodic table (7.7).
- Nucleophilicity decreases left-to-right across a row of the periodic table (7.8A).
- Nucleophilicity decreases down a column of the periodic table in polar aprotic solvents (7.8C).
- Nucleophilicity increases down a column of the periodic table in polar protic solvents (7.8C).
- The stability of a carbocation increases as the number of R groups bonded to the positively charged carbon increases (7.13).

Important Principles

Principle	**Example**
• Electron-donating groups (such as R groups) stabilize a positive charge (7.13A).	• 3° Carbocations (R_3C^+) are more stable than 2° carbocations (R_2CH^+), which are more stable than 1° carbocations (RCH_2^+).
• Steric hindrance decreases nucleophilicity but not basicity (7.8B).	• $(CH_3)_3CO^-$ is a stronger base but a weaker nucleophile than $CH_3CH_2O^-$.
• Hammond postulate: In an endothermic reaction, the more stable product is formed faster. In an exothermic reaction, this is not necessarily true (7.14).	• S_N1 reactions are faster when more stable (more substituted) carbocations are formed, because the rate-determining step is endothermic.
• Planar, sp^2 hybridized atoms react with reagents from both sides of the plane (7.12C).	• A trigonal planar carbocation reacts with nucleophiles from both sides of the plane.

PROBLEMS

Problems Using Three-Dimensional Models

7.40 Give the IUPAC name for each compound, including any *R,S* designation.

a. 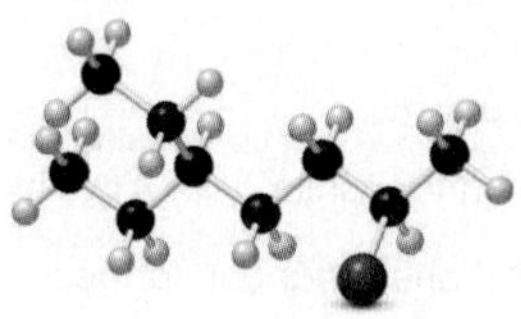

b. 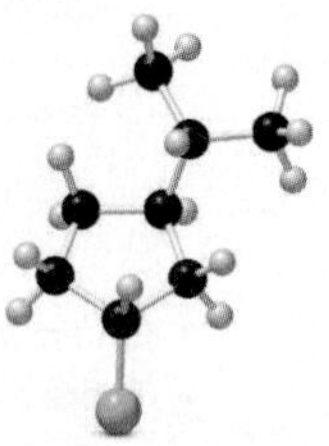

7.41 Draw the products formed when each alkyl halide is treated with NaCN.

a. 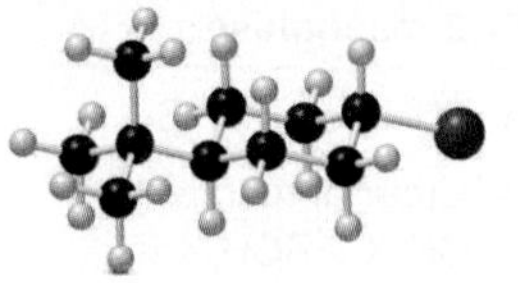

b. 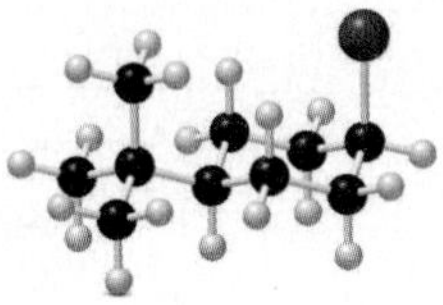

Nomenclature

7.42 Give the IUPAC name for each compound.

a. 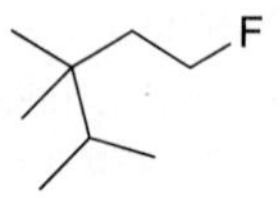

b. 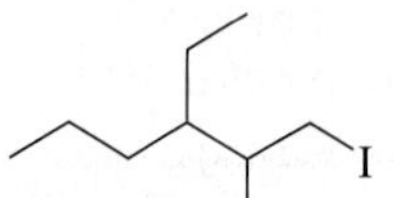

c. 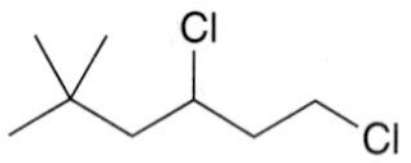

d. 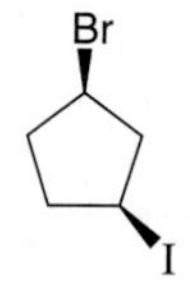

e. 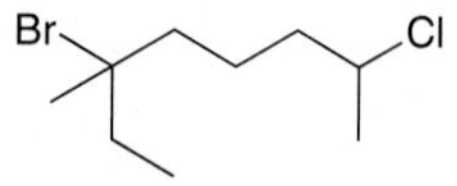

f. 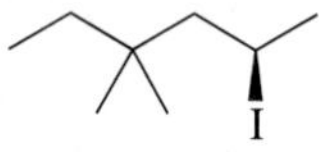

Include the *R,S* designation in the name.

7.43 Give the structure corresponding to each name.

a. 3-bromo-4-ethylheptane
b. 1,1-dichloro-2-methylcyclohexane
c. 1-bromo-4-ethyl-3-fluorooctane
d. (*S*)-3-iodo-2-methylnonane
e. (1*R*,2*R*)-*trans*-1-bromo-2-chlorocyclohexane
f. (*R*)-4,4,5-trichloro-3,3-dimethyldecane

7.44 Draw the eight constitutional isomers having the molecular formula $C_5H_{11}Cl$.

a. Give the IUPAC name for each compound (ignoring *R* and *S* designations).
b. Classify each alkyl halide as 1°, 2°, or 3°.
c. Label any stereogenic centers.
d. For each constitutional isomer that contains a stereogenic center, draw all possible stereoisomers, and label each stereogenic center as *R* or *S*.

Physical Properties

7.45 Which compound in each pair has the higher boiling point?

a. 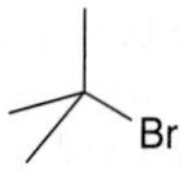

or
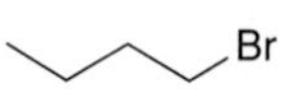

b. 

or

General Nucleophilic Substitution, Leaving Groups, and Nucleophiles

7.46 Draw the products of each nucleophilic substitution reaction.

a. 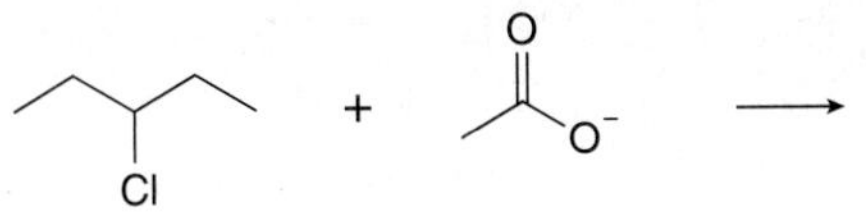

b. 

c. 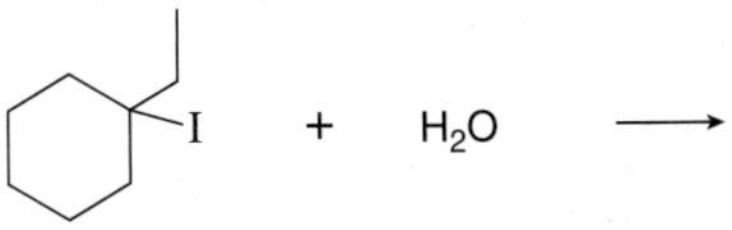

d. 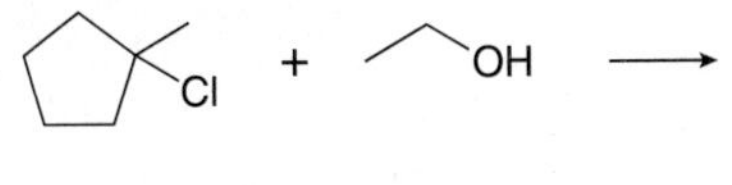

e. 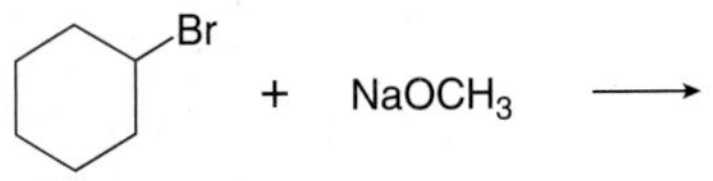

f. 

7.47 Which of the following molecules contain a good leaving group?

a. 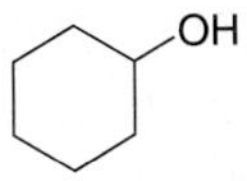 b. 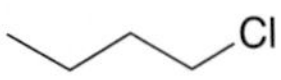 c.  d. 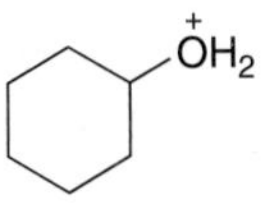

7.48 Rank the species in each group in order of increasing leaving group ability.

a. Br^-, Cl^-, I^- b. NH_3, H_2S, H_2O

7.49 Which of the following nucleophilic substitution reactions will take place?

a. 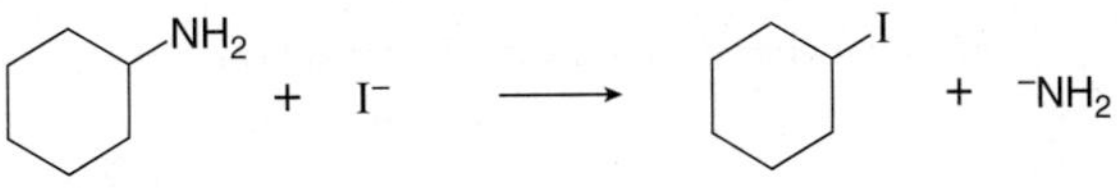

b. 

7.50 Rank the species in each group in order of increasing nucleophilicity.

a. $CH_3CH_2S^-$, $CH_3CH_2O^-$, $CH_3CO_2^-$ in CH_3OH

b. CH_3NH_2, CH_3SH, CH_3OH in acetone

c. ^-OH, F^-, Cl^- in acetone

d. HS^-, F^-, Cl^- in CH_3OH

7.51 Classify each solvent as protic or aprotic.

a. $(CH_3)_2CHOH$ b. CH_3NO_2 c. CH_2Cl_2 d. NH_3 e. $N(CH_3)_3$ f. $HCONH_2$

7.52 Why is the amine N atom more nucleophilic than the amide N atom in $CH_3CONHCH_2CH_2CH_2NHCH_3$?

The S_N2 Reaction

7.53 Consider the following S_N2 reaction.

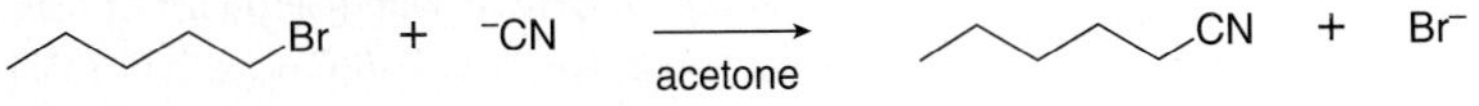

a. Draw a mechanism using curved arrows.

b. Draw an energy diagram. Label the axes, the reactants, products, E_a, and $\Delta H°$. Assume that the reaction is exothermic.

c. Draw the structure of the transition state.

d. What is the rate equation?

e. What happens to the reaction rate in each of the following instances? [1] The leaving group is changed from Br^- to I^-; [2] The solvent is changed from acetone to CH_3CH_2OH; [3] The alkyl halide is changed from $CH_3(CH_2)_4Br$ to $CH_3CH_2CH_2CH(Br)CH_3$; [4] The concentration of ^-CN is increased by a factor of five; and [5] The concentrations of both the alkyl halide and ^-CN are increased by a factor of five.

7.54 Pick the reactant or solvent in each part that gives the faster S_N2 reaction.

a. reaction of ^-OH with CH_3CH_2Br or CH_3CH_2Cl

b. reaction of $CH_3CH_2CH_2Cl$ with NaOH or $NaOCOCH_3$

c. reaction of $CH_3CH_2CH_2I$ with $^-OCH_3$ in CH_3OH or DMSO

7.55 Draw the products of each S_N2 reaction and indicate the stereochemistry where appropriate.

a. 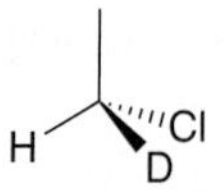 + $^{-}OCH_3$ ⟶

b.  + $^{-}OCH_2CH_3$ ⟶

c. 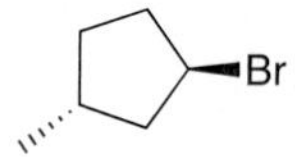 + ^{-}CN ⟶

Carbocations

7.56 Classify the carbocations as 1°, 2°, or 3°, and rank the carbocations in each group in order of increasing stability.

a. 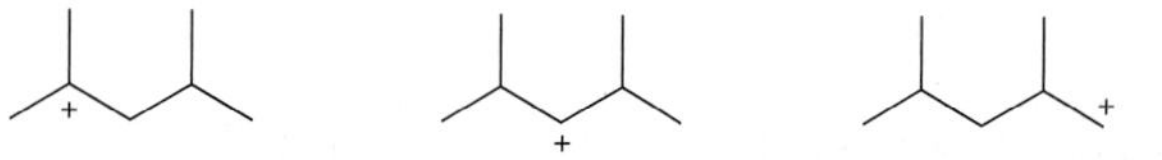

b. 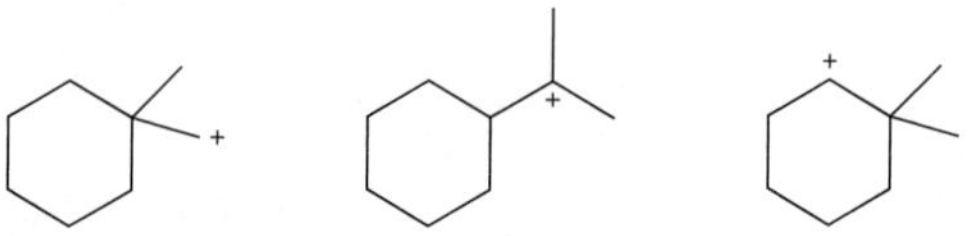

7.57 Which of the following carbocations (**A** or **B**) is more stable? Explain your choice.

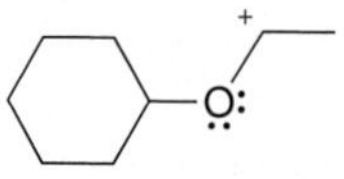

A

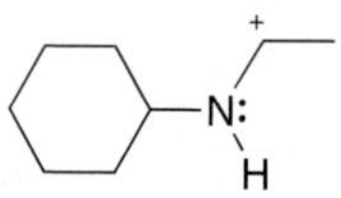

B

The S_N1 Reaction

7.58 Consider the following S_N1 reaction.

a. Draw a mechanism for this reaction using curved arrows.

b. Draw an energy diagram. Label the axes, starting material, product, E_a, and $\Delta H°$. Assume that the starting material and product are equal in energy.

c. Draw the structure of any transition states.

d. What is the rate equation for this reaction?

e. What happens to the reaction rate in each of the following instances? [1] The leaving group is changed from I^- to Cl^-; [2] The solvent is changed from H_2O to DMF; [3] The alkyl halide is changed from $(CH_3)_2C(I)CH_2CH_3$ to $(CH_3)_2CHCH(I)CH_3$; [4] The concentration of H_2O is increased by a factor of five; and [5] The concentrations of both the alkyl halide and H_2O are increased by a factor of five.

7.59 Pick the reactant or solvent in each part that gives the faster S_N1 reaction.

a. reaction of H_2O with $(CH_3)_3CCl$ or $(CH_3)_3CI$

b. reaction of CH_3OH with $(CH_3)_3CBr$ or $(CH_3)_2CHCH_2Br$

c. reaction of $CH_3CH_2CH(I)CH_3$ with CH_3CH_2OH in water or DMSO

7.60 Draw the products of each S_N1 reaction and indicate the stereochemistry when necessary.

a. 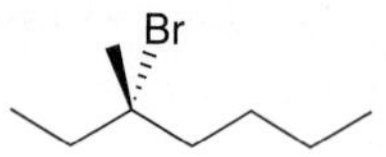 + CH_3CH_2OH ⟶

b. 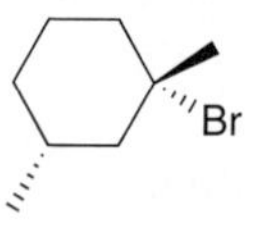 + H_2O ⟶

7.61 Draw a stepwise mechanism for the following reaction that illustrates how two substitution products are formed. Explain why 1-bromohex-2-ene reacts rapidly with a weak nucleophile (CH_3OH) under S_N1 reaction conditions, even though it is a 1° alkyl halide.

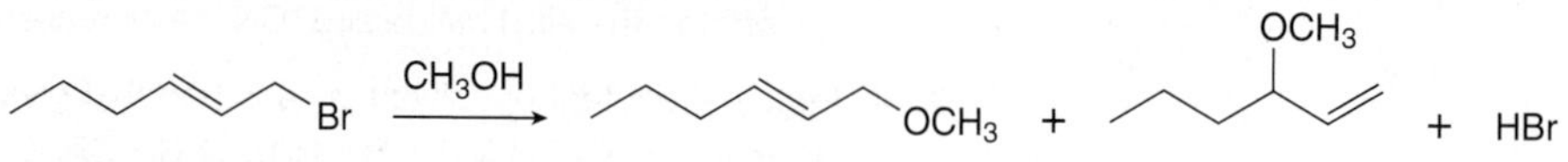

1-bromohex-2-ene

S_N1 and S_N2 Reactions

7.62 (a) Rank **A, B,** and **C** in order of increasing S_N2 reactivity. (b) Rank **A, B,** and **C** in order of increasing S_N1 reactivity.

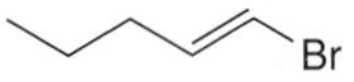
Br

A

Br

B

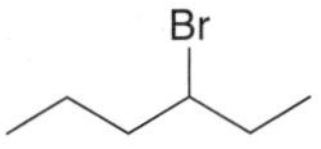
Br

C

7.63 Determine the mechanism of nucleophilic substitution of each reaction and draw the products, including stereochemistry.

a. 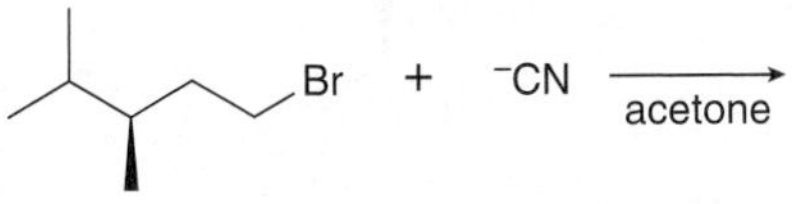
Br
+
⁻CN
acetone

d. 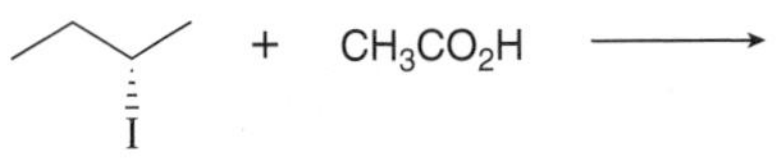
I
+
CH_3CO_2H

b. 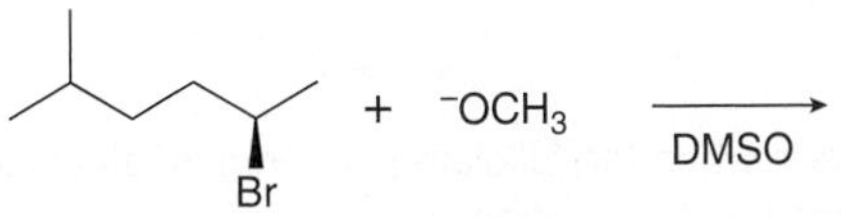
Br
+
⁻OCH$_3$
DMSO

e. 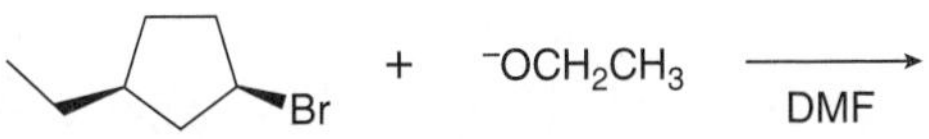
Br
+
⁻OCH$_2$CH$_3$
DMF

c. 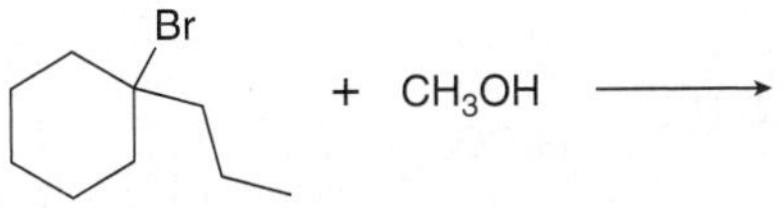
Br
+ CH_3OH

f. 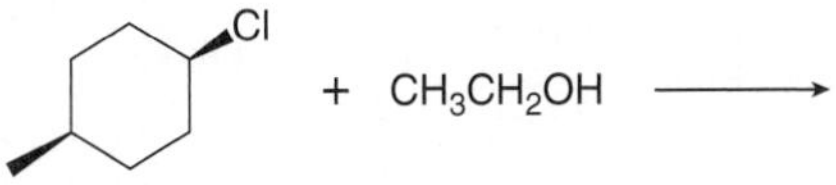
Cl
+ CH_3CH_2OH

7.64 Fluticasone, the chapter-opening molecule, can be prepared by the following reaction. Draw a stepwise mechanism for this reaction.

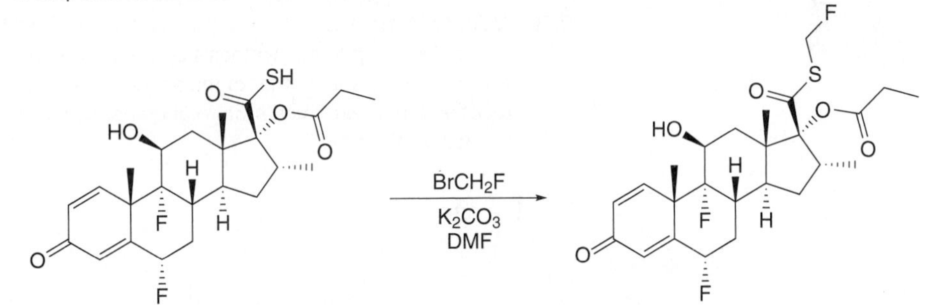
SH
O
HO
H
F
H
F
$BrCH_2F$
K_2CO_3
DMF
S
F

A fluticasone

7.65 Uridine monophosphate (UMP) is one of the four nucleotides that compose RNA, the nucleic acid that translates the genetic information of DNA into proteins needed by cells for proper function and development. A key step in the synthesis of UMP is the S_N1 reaction of **A** with **B** to form **C,** which is then converted to UMP in one step. Draw a stepwise mechanism for this S_N1 reaction.

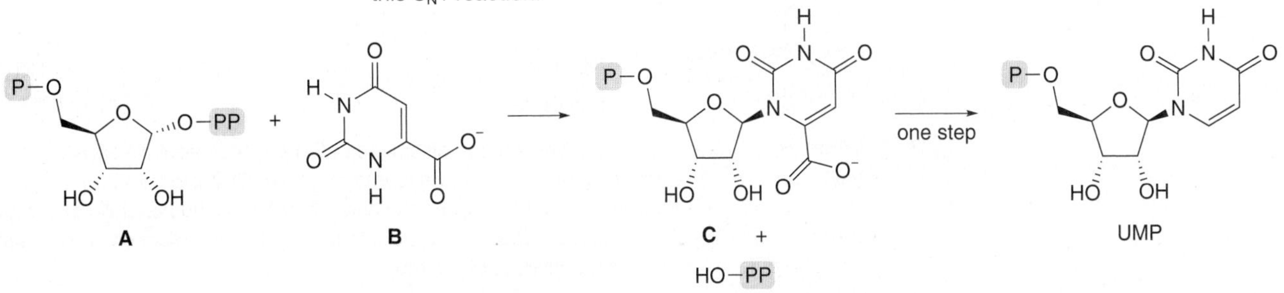
P–O
O–PP
HO
OH
A
+
B
C
+
HO–PP
one step
UMP

7.66 Diphenhydramine, the antihistamine in Benadryl, can be prepared by the following two-step sequence. What is the structure of diphenhydramine?

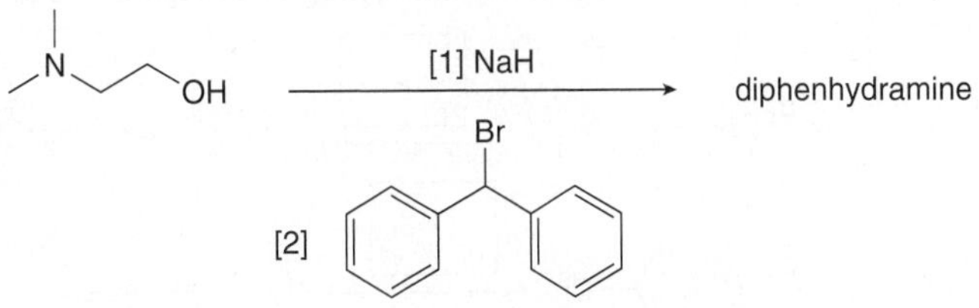
N
OH
[1] NaH
[2]
Br
diphenhydramine

7.67 Draw a stepwise, detailed mechanism for the following reaction. Use curved arrows to show the movement of electrons.

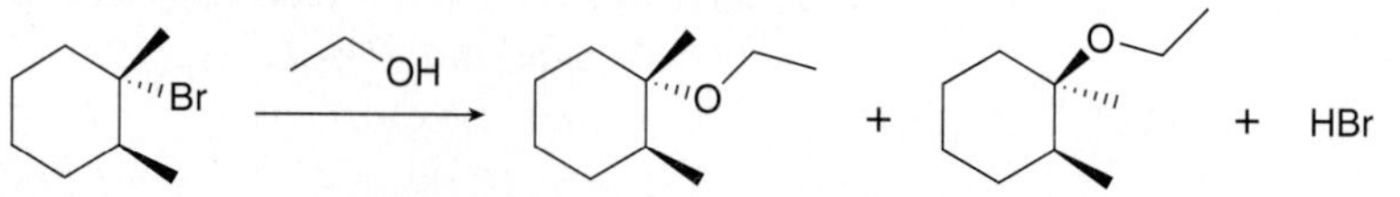
Br
OH
O
+
O
+ HBr

7.68 When a single compound contains both a nucleophile and a leaving group, an **intramolecular** reaction may occur. With this in mind, draw the product of the following reaction.

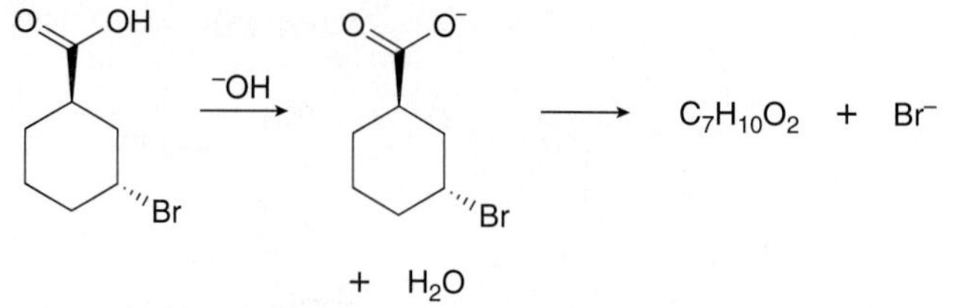
O OH
Br
⁻OH
O O⁻
Br
C7H10O2 + Br⁻
+ H2O

7.69 Nicotine can be made when the following ammonium salt is treated with Na_2CO_3. Draw a stepwise mechanism for this reaction.

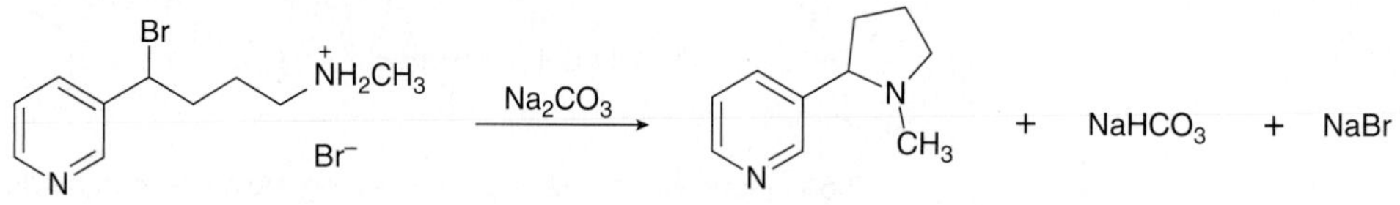
Br
NH2CH3
Br⁻
N
Na2CO3
N
CH3
N
+ NaHCO3 + NaBr

nicotine

7.70 Quinapril (trade name Accupril) is used to treat high blood pressure and congestive heart failure. One step in the synthesis of quinapril involves reaction of the racemic alkyl bromide **A** with a single enantiomer of the amino ester **B.** (a) What two products are formed in this reaction? (b) Given the structure of quinapril, which one of these two products is needed to synthesize the drug?

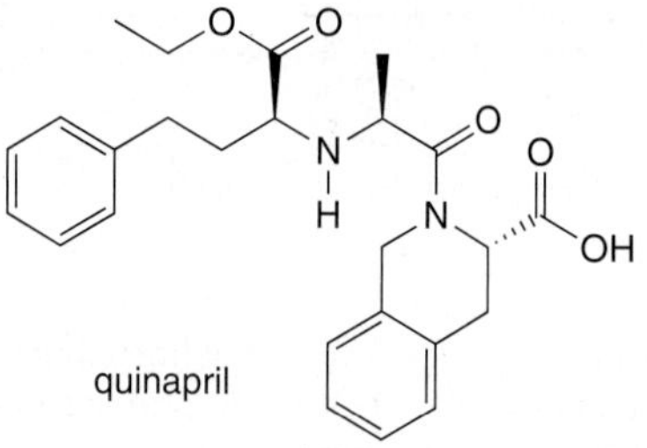
O O
N
H
O
N
O
OH

quinapril

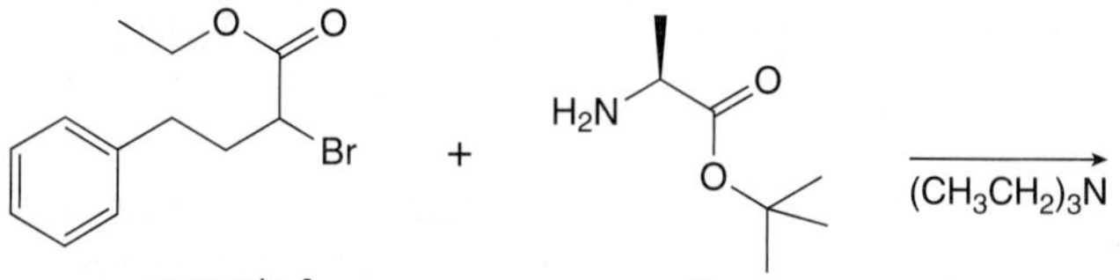
O O
Br
+
H2N
O
O
(CH3CH2)3N

racemic **A** **B**

7.71 Draw a stepwise, detailed mechanism for the following reaction.

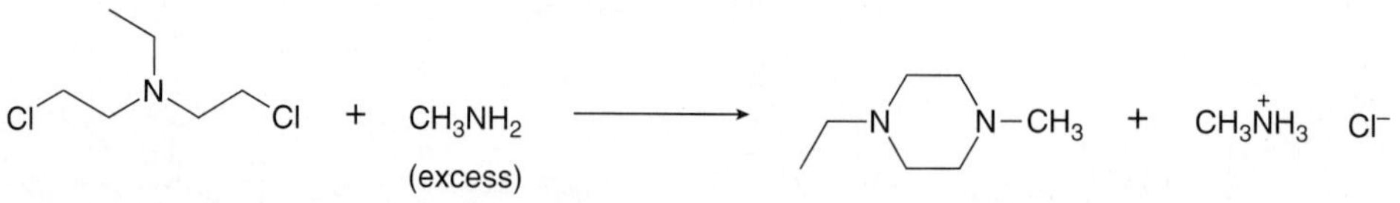
N
Cl Cl
+ CH3NH2
(excess)
N N−CH3
+ CH3NH3 Cl⁻

7.72 When (*R*)-6-bromo-2,6-dimethylnonane is dissolved in CH_3OH, nucleophilic substitution yields an optically inactive solution. When the isomeric halide (*R*)-2-bromo-2,5-dimethylnonane is dissolved in CH_3OH under the same conditions, nucleophilic substitution forms an optically active solution. Draw the products formed in each reaction, and explain why the difference in optical activity is observed.

Synthesis

7.73 Fill in the appropriate reagent or starting material in each of the following reactions.

a. 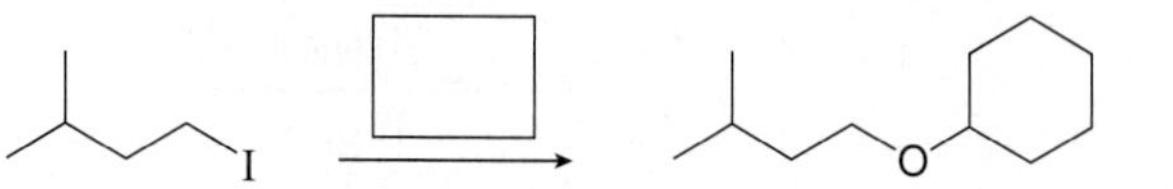
I
O

c. 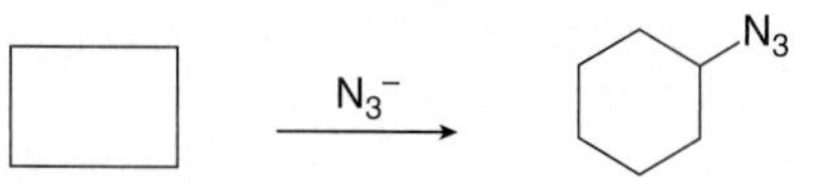
N3⁻
N3

b. 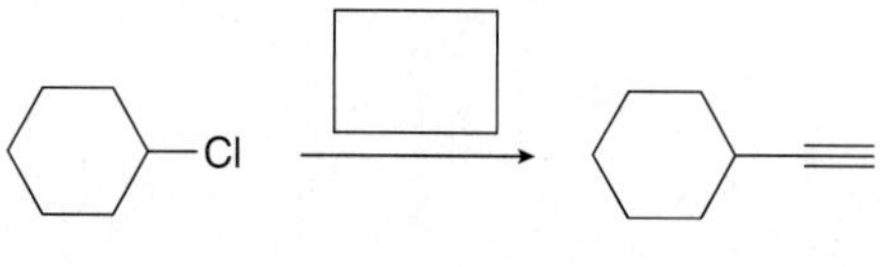
Cl

d. 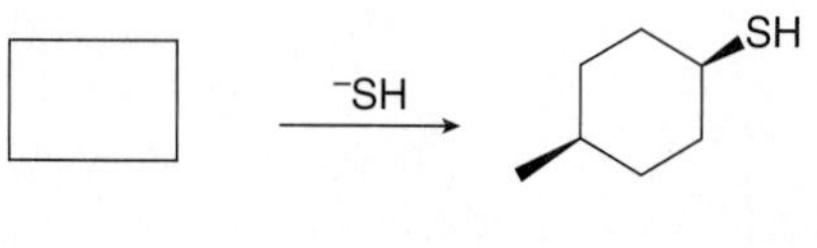
⁻SH
SH

7.74 Devise a synthesis of each compound from an alkyl halide using any other organic or inorganic reagents.

a. 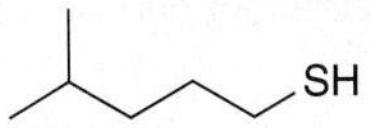 b. 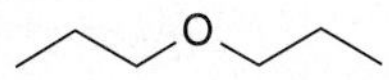 c. 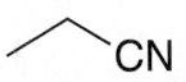 d. 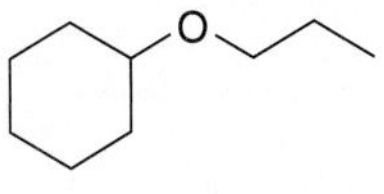 e. 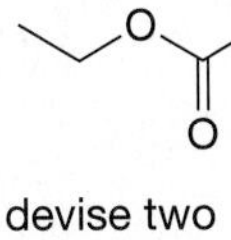

7.75 Suppose you have compounds **A–D** at your disposal. Using these compounds, devise two different ways to make **E.** Which one of these methods is preferred, and why?

CH_3I 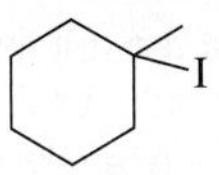 $NaOCH_3$ 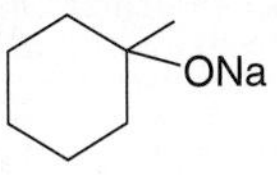 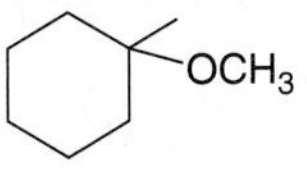

A **B** **C** **D** **E**

7.76 Muscalure, the sex pheromone of the common housefly, can be prepared by a reaction sequence that uses two nucleophilic substitutions. Identify compounds **A–D** in the following synthesis of muscalure.

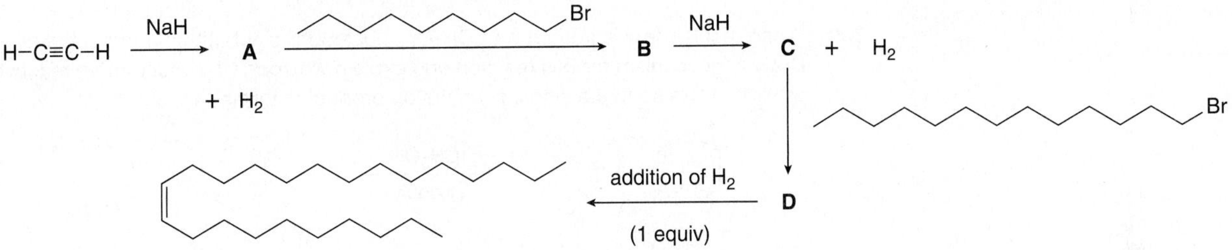
NaH
H–C≡C–H
A
+ H2
Br
B
NaH
C + H2
Br
addition of H2
(1 equiv)
D

muscalure

Challenge Problems

7.77 We will return often to nucleophilic substitution, in particular the S_N2 reaction, in subsequent chapters and concentrate on the *nucleophile,* rather than the alkyl halide. By using different nucleophiles, nucleophilic substitution allows the synthesis of a wide variety of organic compounds with many different functional groups. With this in mind, draw the products of each two-step sequence. (Hint: Step [1] in each part involves an acid–base reaction that removes the most acidic hydrogen from the starting material.)

a. 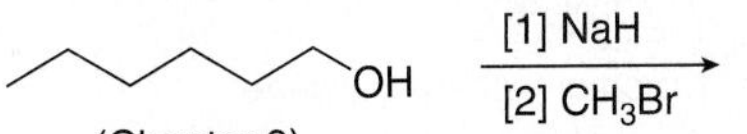
OH
[1] NaH
[2] CH3Br

(Chapter 9)

b. 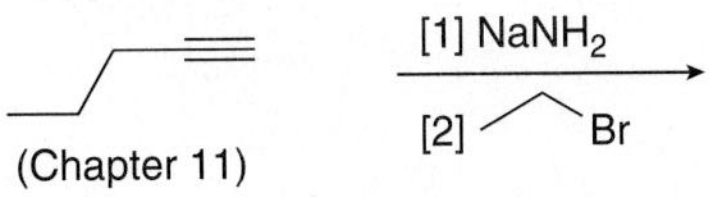
[1] NaNH2
[2]
Br

(Chapter 11)

c. 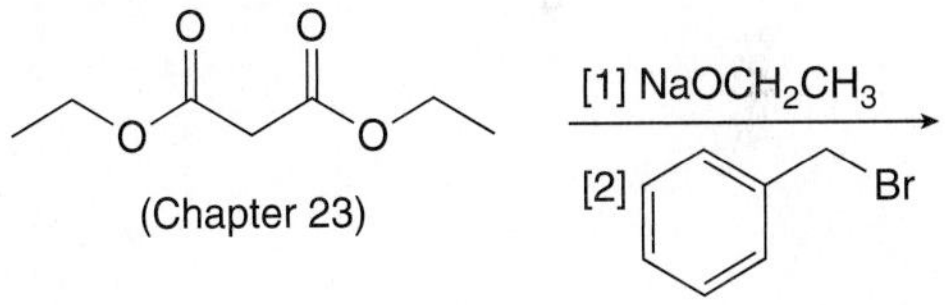
O
O
O
O
[1] NaOCH2CH3
[2]
Br

(Chapter 23)

7.78 Explain why quinuclidine is a much more reactive nucleophile than triethylamine, even though both compounds have N atoms surrounded by three R groups.

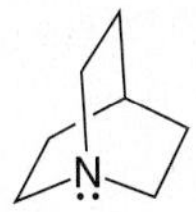

quinuclidine

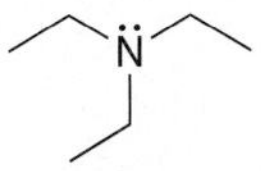

triethylamine

7.79 Draw a stepwise mechanism for the following reaction sequence.

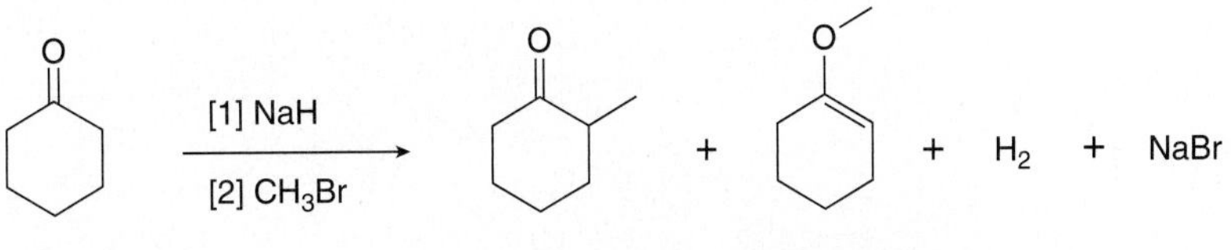
O
[1] NaH
[2] CH3Br
O
+
O
+ H2 + NaBr

major product minor product

7.80 As we will learn in Chapter 9, an epoxide is an ether with an oxygen atom in a three-membered ring. Epoxides can be made by intramolecular S_N2 reactions of intermediates that contain a nucleophile and a leaving group on adjacent carbons, as shown.

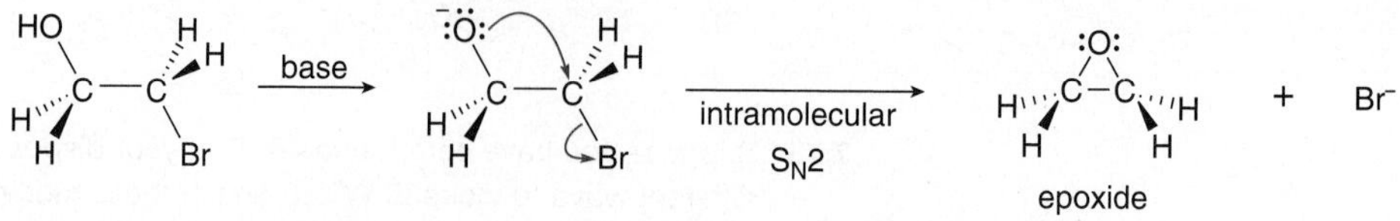

Assume that each of the following starting materials can be converted to an epoxide by this reaction. Draw the product formed (including stereochemistry) from each starting material. Why might some of these reactions be more difficult than others in yielding nucleophilic substitution products?

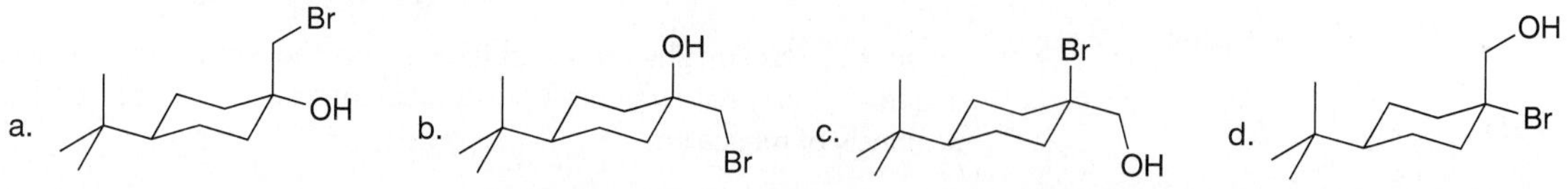

7.81 When trichloride **J** is treated with CH_3OH, nucleophilic substitution forms the dihalide **K.** Draw a mechanism for this reaction and explain why one Cl is much more reactive than the other two Cl's so that a single substitution product is formed.

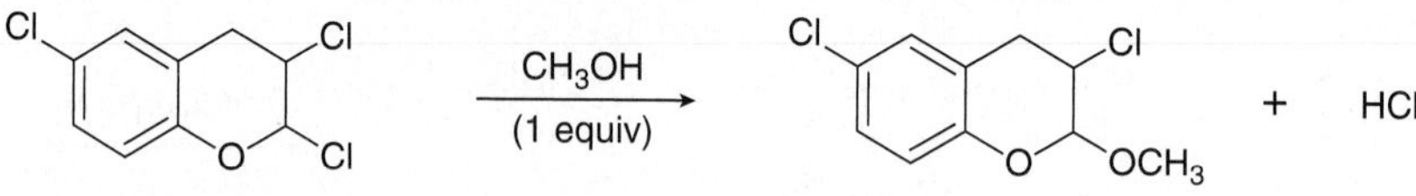

J **K**

7.82 In some nucleophilic substitutions under S_N1 conditions, complete racemization does not occur and a small excess of one enantiomer is present. For example, treatment of optically pure 1-bromo-1-phenylpropane with water forms 1-phenylpropan-1-ol. (a) Calculate how much of each enantiomer is present using the given optical rotation data. (b) Which product predominates—the product of inversion or the product of retention of configuration? (c) Suggest an explanation for this phenomenon.

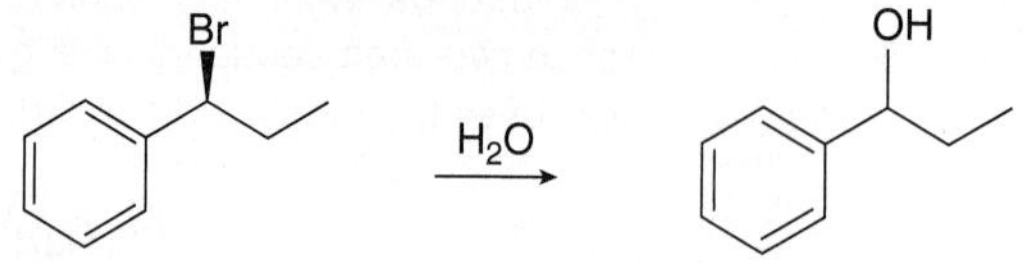

1-bromo-1-phenylpropane

1-phenylpropan-1-ol
observed [α] = +5.0
optically pure *S* isomer, [α] = −48

Alkyl Halides and Elimination Reactions

8

8.1 General features of elimination
8.2 Alkenes—The products of elimination reactions
8.3 The mechanisms of elimination
8.4 The E2 mechanism
8.5 The Zaitsev rule
8.6 The E1 mechanism
8.7 S_N1 and E1 reactions
8.8 Stereochemistry of the E2 reaction
8.9 When is the mechanism E1 or E2?
8.10 E2 reactions and alkyne synthesis
8.11 When is the reaction S_N1, S_N2, E1, or E2?

DDE, dichlorodiphenyl**d**ichloro**e**thylene, is formed by the elimination of HCl from the pesticide DDT. DDE and DDT accumulate in the fatty tissues of predator birds such as osprey that feed on fish contaminated with DDT. When DDE and DDT concentration is high, mother osprey produce eggs with very thin shells that are easily crushed, so fewer osprey chicks hatch. In Chapter 8, we learn about **elimination reactions,** the second general reaction of alkyl halides, which form alkenes like DDE.

Elimination reactions introduce π bonds into organic compounds, so they can be used to synthesize **alkenes** and **alkynes**—hydrocarbons that contain one and two π bonds, respectively. Like nucleophilic substitution, elimination reactions can occur by two different pathways, depending on the conditions. By the end of Chapter 8, therefore, you will have learned four different reaction mechanisms, two for nucleophilic substitution (S_N1 and S_N2) and two for elimination (E1 and E2).

The biggest challenge with this material is learning how to sort out two different reactions that follow four different mechanisms. **Will a particular alkyl halide undergo substitution or elimination with a given reagent, and by which of the four possible mechanisms?** To answer this question, we conclude Chapter 8 with a summary that allows you to predict which reaction and mechanism are likely for a given substrate.

8.1 General Features of Elimination

All **elimination reactions** involve loss of elements from the starting material to form a new π bond in the product.

- **Alkyl halides undergo elimination reactions with Brønsted–Lowry bases. The elements of HX are lost and an alkene is formed.**

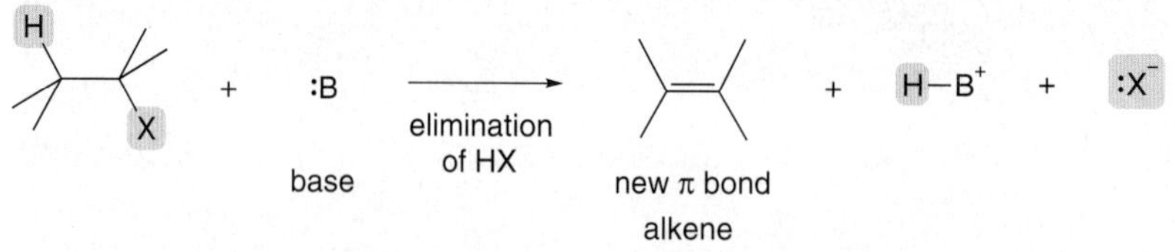

Equations [1] and [2] illustrate examples of elimination reactions. In both reactions a base removes the elements of an acid, HBr or HCl, from the organic starting material.

[1]
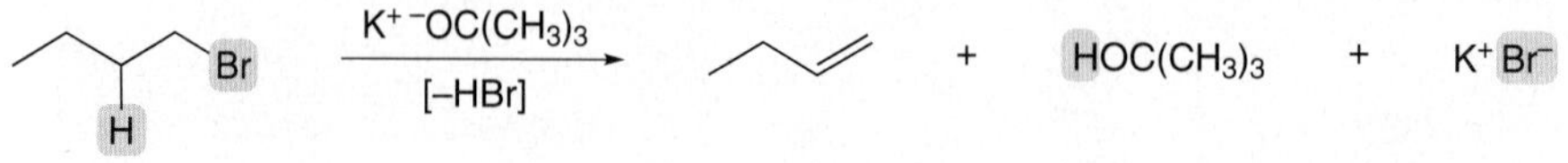

[2]
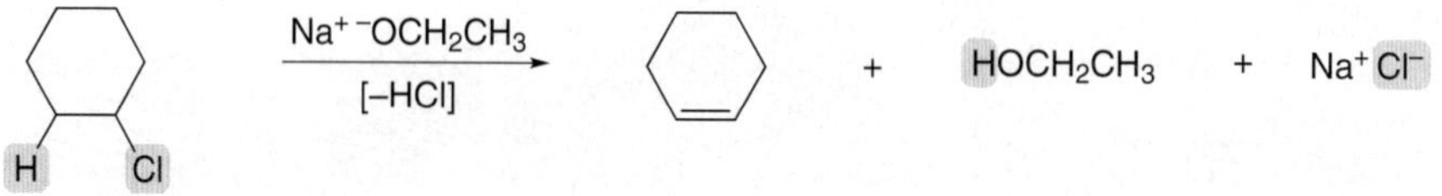

Removal of the elements of HX, called **dehydrohalogenation,** is one of the most common methods to introduce a π bond and prepare an alkene. Dehydrohalogenation is an example of **β elimination,** because it involves loss of elements from two adjacent atoms: the **α carbon** bonded to the leaving group X, and the **β carbon** adjacent to it. Three curved arrows illustrate how four bonds are broken or formed in the process.

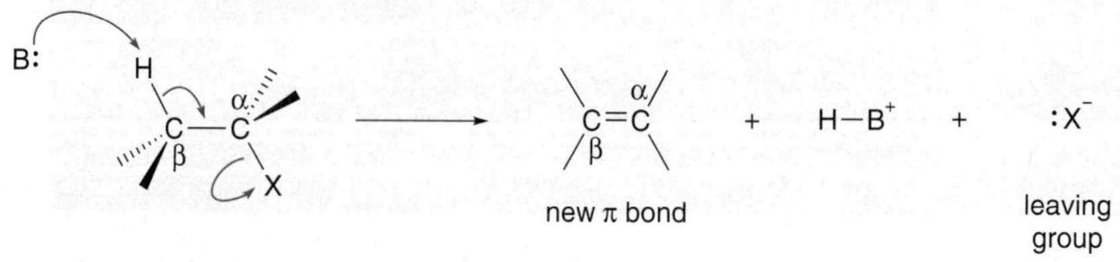

- The base (B:) removes a proton on the β carbon, thus forming $H{-}B^+$.
- The electron pair in the β C–H bond forms the new π bond between the α and β carbons.
- The electron pair in the C–X bond ends up on halogen, forming the leaving group $:X^-$.

The most common bases used in elimination reactions are negatively charged oxygen compounds such as ^-OH and its alkyl derivatives, ^-OR, called **alkoxides,** listed in Table 8.1. Potassium *tert*-butoxide, $K^+\ ^-OC(CH_3)_3$, a bulky nonnucleophilic base, is especially useful (Section 7.8B).

Table 8.1
Common Bases Used in Dehydrohalogenation

Structure	Name
$Na^+\ ^-OH$	Sodium hydroxide
$K^+\ ^-OH$	Potassium hydroxide
$Na^+\ ^-OCH_3$	Sodium methoxide
$Na^+\ ^-OCH_2CH_3$	Sodium ethoxide
$K^+\ ^-OC(CH_3)_3$	Potassium *tert*-butoxide

To draw any product of dehydrohalogenation:

- **Find the α carbon—the sp^3 hybridized carbon bonded to the leaving group.**
- **Identify all β carbons with H atoms.**
- **Remove the elements of H and X from the α and β carbons and form a π bond.**

For example, 2-bromo-2-methylpropane has three β carbons (three CH_3 groups), but because all three are *identical,* only *one* alkene is formed upon elimination of HBr. In contrast, 2-bromobutane has two *different* β carbons (labeled $β_1$ and $β_2$), so elimination affords *two* constitutional isomers by loss of HBr across either the α and $β_1$ carbons, or the α and $β_2$ carbons. We learn about which product predominates and why in Section 8.5.

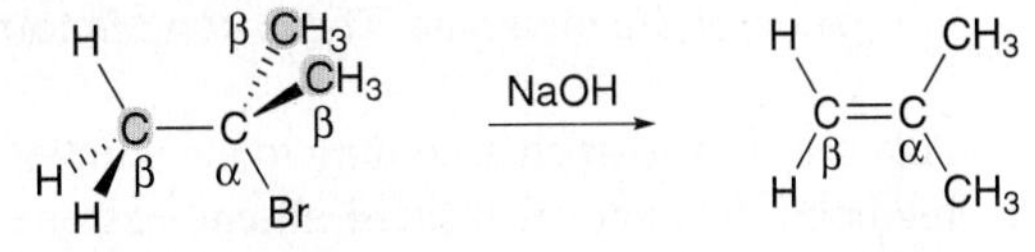

2-bromo-2-methylpropane — one alkene

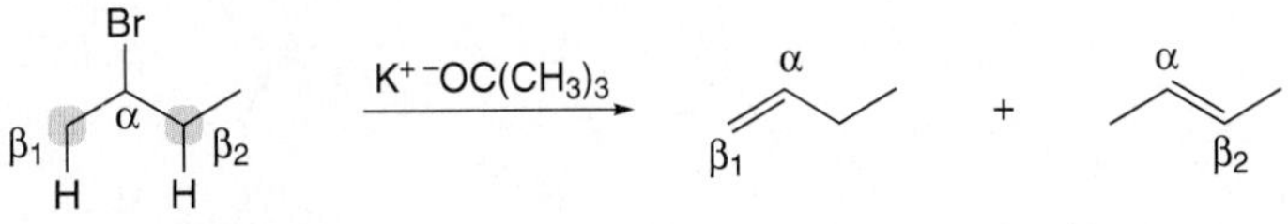

2-bromobutane — two constitutional isomers

An elimination reaction is the first step in the slow degradation of the **pesticide DDT** (Chapter 8 opening paragraph and Section 7.4). Elimination of HCl from DDT forms the degradation product **DDE** (dichlorodiphenyldichloroethylene). This stable alkene is found in minute concentration in the fatty tissues of most adults in the United States.

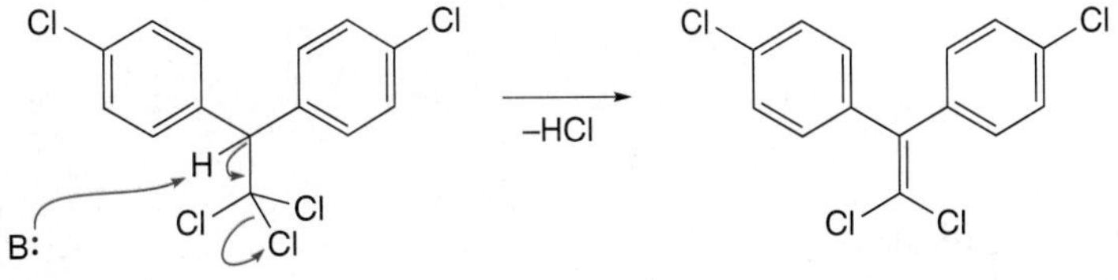

DDT — DDE

Problem 8.1 Label the α and β carbons in each alkyl halide. Draw all possible elimination products formed when each alkyl halide is treated with $K^{+}\,^{-}OC(CH_3)_3$.

a. 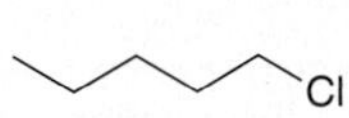 b. 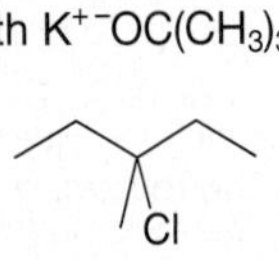 c. 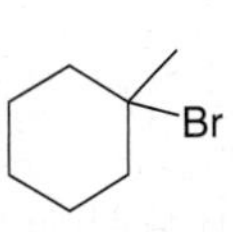

8.2 Alkenes—The Products of Elimination Reactions

Because elimination reactions of alkyl halides form alkenes, let's review earlier material on alkene structure and learn some additional facts as well.

8.2A Bonding in a Carbon–Carbon Double Bond

Recall from Section 1.10B that alkenes are hydrocarbons containing a carbon–carbon double bond. Each carbon of the double bond is sp^2 hybridized and trigonal planar, and all bond angles are 120°.

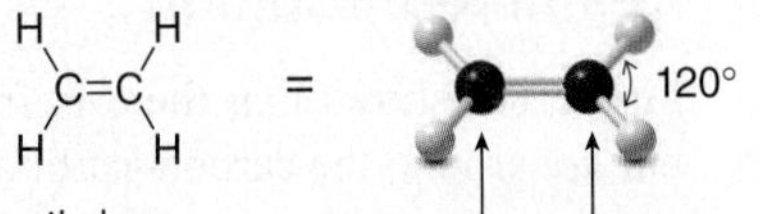

ethylene

sp^2 hybridized

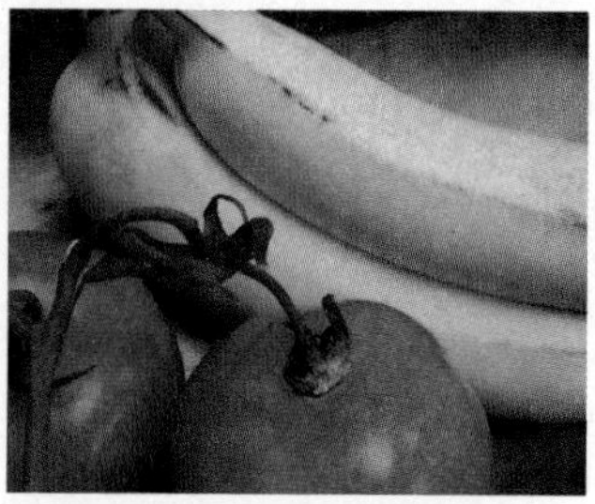

Ethylene, the simplest alkene, is a hormone that regulates plant growth and fruit ripening. A ripe banana speeds up the ripening of green tomatoes because the banana gives off ethylene.

The double bond of an alkene consists of a σ bond and a π bond.

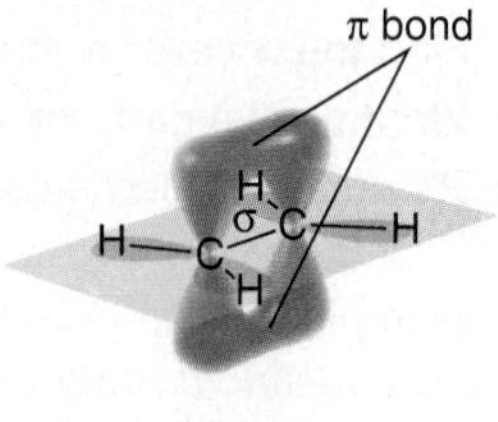

- **The σ bond, formed by end-on overlap of the two sp^2 hybrid orbitals, lies in the plane of the molecule.**
- **The π bond, formed by side-by-side overlap of two 2*p* orbitals, lies perpendicular to the plane of the molecule. The π bond is formed during elimination.**

Alkenes are classified according to the number of carbon atoms bonded to the carbons of the double bond. A **monosubstituted alkene** has one carbon atom bonded to the carbons of the double bond. A **disubstituted alkene** has two carbon atoms bonded to the carbons of the double bond, and so forth.

monosubstituted (*one* R group) | disubstituted (*two* R groups) | trisubstituted (*three* R groups) | tetrasubstituted (*four* R groups)

Figure 8.1 shows several alkenes and how they are classified. You must be able to classify alkenes in this way to determine the major and minor products of elimination reactions, when a mixture of alkenes is formed.

Figure 8.1 Classifying alkenes by the number of R groups bonded to the double bond

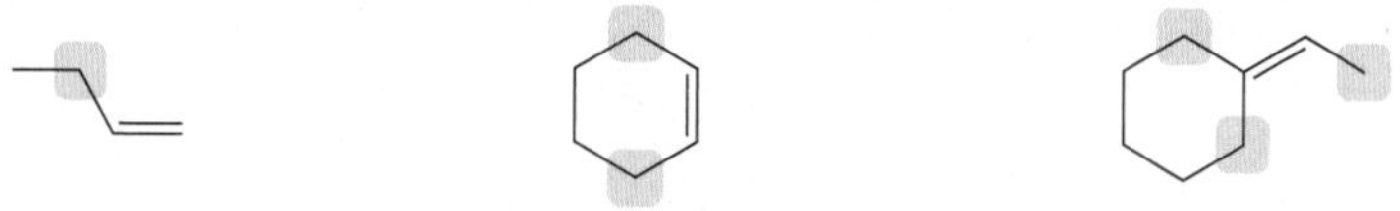

monosubstituted | disubstituted | trisubstituted

- Carbon atoms bonded to the double bond are screened in blue.

Problem 8.2 Classify each alkene in the following vitamins by the number of carbon substituents bonded to the double bond.

a.

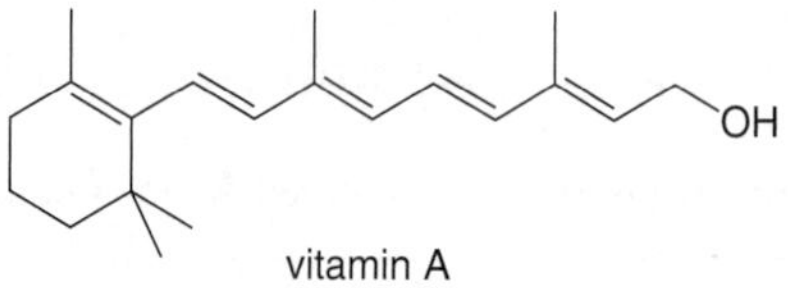

vitamin A

b.

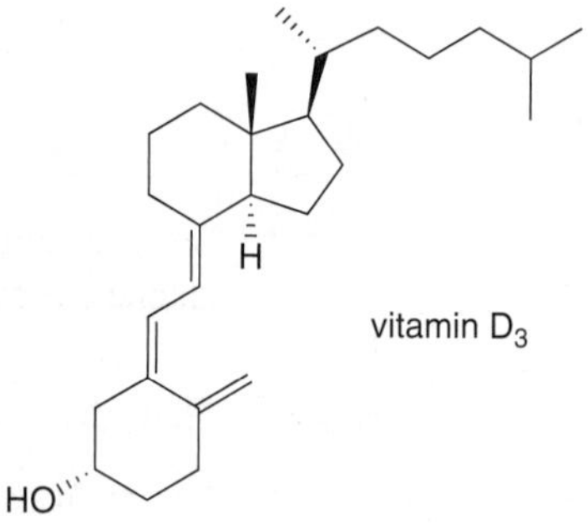

vitamin D_3

8.2B Restricted Rotation

Figure 8.2 shows that there is free rotation about the carbon–carbon single bonds of butane, but *not* around the carbon–carbon double bond of but-2-ene. Because of restricted rotation, two stereoisomers of but-2-ene are possible.

Figure 8.2
Rotation around C–C and C=C compared

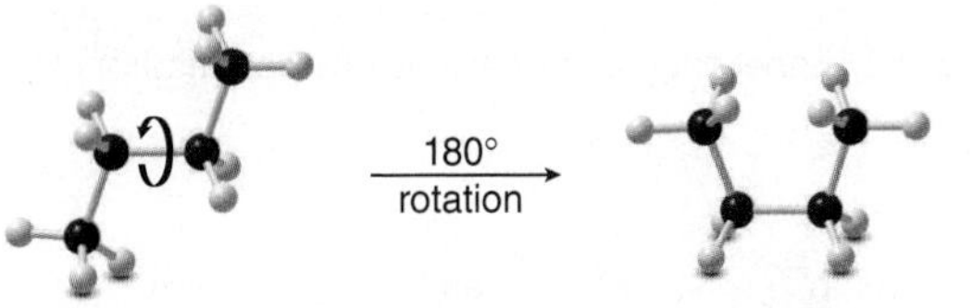

butane
anti conformation

butane
eclipsed conformation

These conformations **interconvert** by rotation.
They represent the **same** molecule.

cis-but-2-ene

trans-but-2-ene

These molecules **do not interconvert** by rotation.
They are **different** molecules.

The concept of cis and trans isomers was first introduced for disubstituted cycloalkanes in Chapter 4. In both cases, a ring or a double bond restricts motion, preventing the rotation of a group from one side of the ring or double bond to the other.

- **The cis isomer has two groups on the *same side* of the double bond.**
- **The trans isomer has two groups on *opposite sides* of the double bond.**

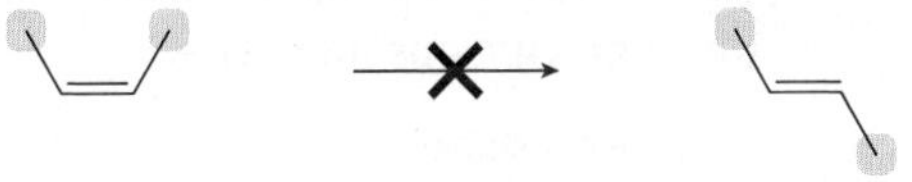

cis-but-2-ene

two R groups
on the **same** side

trans-but-2-ene

two R groups
on **opposite** sides

cis-But-2-ene and *trans*-but-2-ene are stereoisomers, but not mirror images of each other, so they are **diastereomers.**

The cis and trans isomers of but-2-ene are a specific example of a general type of stereoisomer occurring at carbon–carbon double bonds. **Whenever the two groups on *each* end of a carbon–carbon double bond are *different from each other,* two diastereomers are possible.**

X and X' must be different from *each other...*

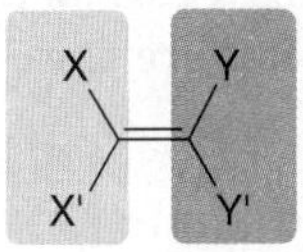

...and Y and Y' must be different from *each other.*

Problem 8.3 For which double bonds are stereoisomers possible?

a. 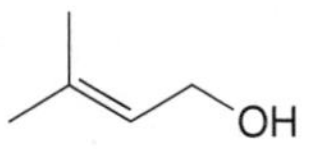

b. 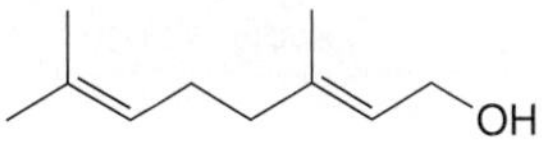

c. 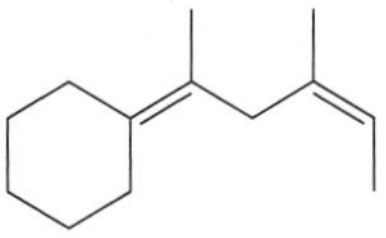

Problem 8.4 (a) Which double bonds in (*E*)-ocimene, a major component of the odor of lilac flowers, can exhibit stereoisomerism? (b) Draw a diastereomer of (*E*)-ocimene.

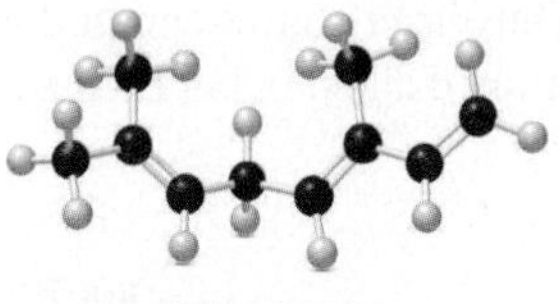

(*E*)-ocimene

The characteristic fragrance of lilac is a mixture of (*E*)-ocimene and other volatile ethers, aldehydes, and alcohols.

Problem 8.5 Label each pair of alkenes as constitutional isomers, stereoisomers, or identical.

a. 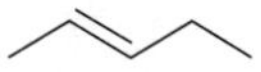 and 

b. 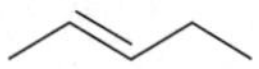 and 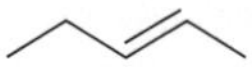

c.  and 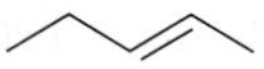

d. 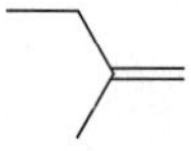 and 

8.2C Stability of Alkenes

Some alkenes are more stable than others. For example, **trans alkenes are generally more stable than cis alkenes** because the larger groups bonded to the double bond carbons are farther apart, reducing steric interactions.

The trans isomer has the CH_3 groups farther away from each other.

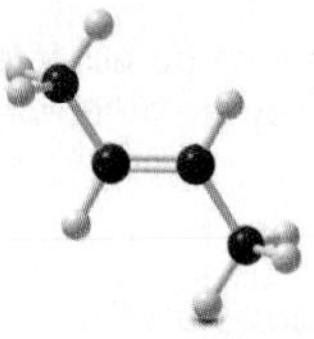

more stable

Steric interactions of the CH_3 groups destabilize the cis isomer.

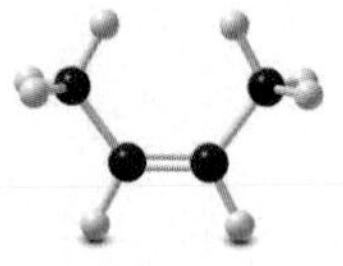

less stable

The stability of an alkene increases, moreover, as the **number of R groups bonded to the double bond carbons increases.**

least stable → most stable

$CH_2{=}CH_2$ < $RCH{=}CH_2$ < $R_2C{=}CH_2$ ~ $RCH{=}CHR$ < $R_2C{=}CHR$ < $R_2C{=}CR_2$

Increasing number of R groups
Increasing stability

R groups increase the stability of an alkene because R groups are sp^3 hybridized, whereas the carbon atoms of the double bond are sp^2 hybridized. Recall from Sections 1.11B and 2.5D that the percent *s*-character of a hybrid orbital increases from 25% to 33% in going from sp^3 to sp^2. The higher the percent *s*-character, the more readily an atom accepts electron density. Thus, ***sp*^{2} hybridized carbon atoms are more able to *accept* electron density,** and ***sp*^{3} hybridized carbon atoms are more able to *donate* electron density.**

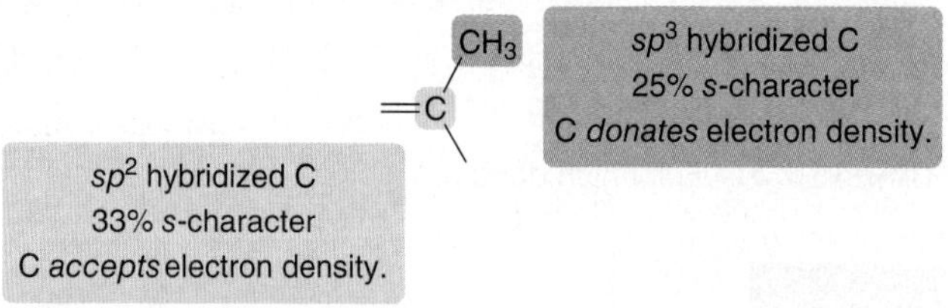

- **As a result, increasing the number of electron-donating R groups on a carbon atom able to accept electron density makes the alkene more stable.**

Thus, *trans*-but-2-ene (a disubstituted alkene) is more stable than *cis*-but-2-ene (another disubstituted alkene), but both are more stable than but-1-ene (a monosubstituted alkene).

but-1-ene — *cis*-but-2-ene — *trans*-but-2-ene (fewer steric interactions)

Increasing stability

In summary:

- **Trans alkenes are more stable than cis alkenes because they have fewer steric interactions.**
- **Increasing alkyl substitution stabilizes an alkene by an electron-donating inductive effect.**

Problem 8.6 Which alkene in each pair is more stable?

a.  or 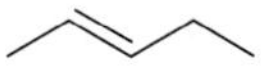

c. 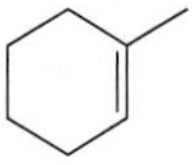 or 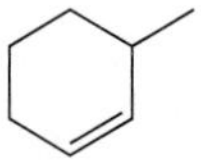

b.  or 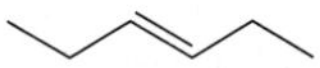

Problem 8.7 Several factors can affect alkene stability. Explain why alkene **A** is more stable than alkene **B** even though both contain disubstituted carbon–carbon double bonds.

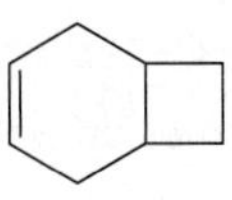

A

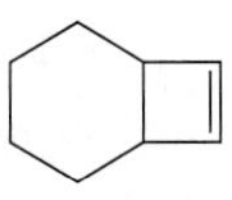

B

8.3 The Mechanisms of Elimination

What is the mechanism for elimination? What is the order of bond breaking and bond making? Is the reaction a one-step process or does it occur in many steps?

There are two mechanisms for elimination—E2 and E1—just as there are two mechanisms for nucleophilic substitution—S_N2 and S_N1.

- The **E2 mechanism (bimolecular elimination)**
- The **E1 mechanism (unimolecular elimination)**

The E2 and E1 mechanisms differ in the timing of bond cleavage and bond formation, analogous to the S_N2 and S_N1 mechanisms. In fact, E2 and S_N2 reactions have some features in common, as do E1 and S_N1 reactions.

8.4 The E2 Mechanism

The most common mechanism for dehydrohalogenation is the E2 mechanism. For example, $(CH_3)_3CBr$ reacts with ^-OH to form $(CH_3)_2C{=}CH_2$ via an E2 mechanism.

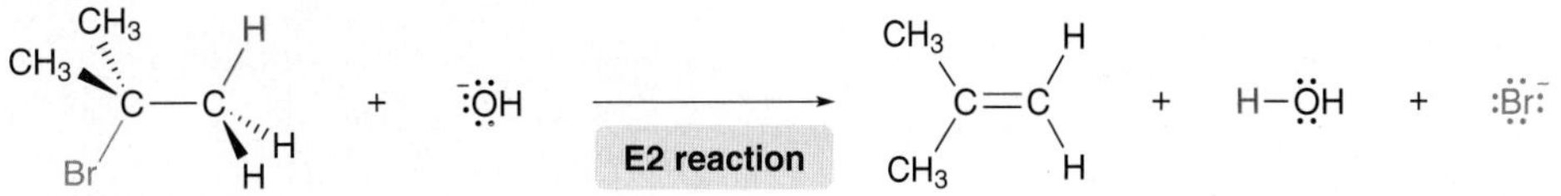

8.4A Kinetics

An E2 reaction exhibits **second-order kinetics;** that is, the reaction is **bimolecular** and both the alkyl halide and the base appear in the rate equation.

- rate = $k[(CH_3)_3CBr][^-OH]$

8.4B A One-Step Mechanism

The most straightforward explanation for the second-order kinetics is a **concerted reaction: all bonds are broken and formed in a single step,** as shown in Mechanism 8.1.

Mechanism 8.1 The E2 Mechanism

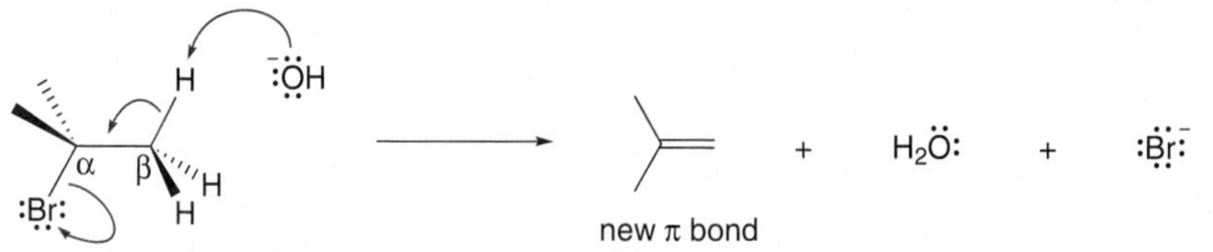

- The base ^-OH removes a proton from the β carbon, forming H_2O (a by-product).
- The electron pair in the β C–H bond forms the new π bond.
- The leaving group Br^- comes off with the electron pair in the C–Br bond.

An energy diagram for the reaction of $(CH_3)_3CBr$ with ^-OH is shown in Figure 8.3. The reaction has one step, so there is one energy barrier between reactants and products. Two bonds are broken (C–H and C–Br) and two bonds are formed (H–OH and the π bond) in a single step, so the transition state contains four partial bonds, with the negative charge distributed over the base and the leaving group. Entropy favors the products of an E2 reaction because two molecules of starting material form three molecules of product.

Figure 8.3
An energy diagram for an E2 reaction:
$(CH_3)_3CBr + {}^-OH \rightarrow (CH_3)_2C{=}CH_2 + H_2O + Br^-$

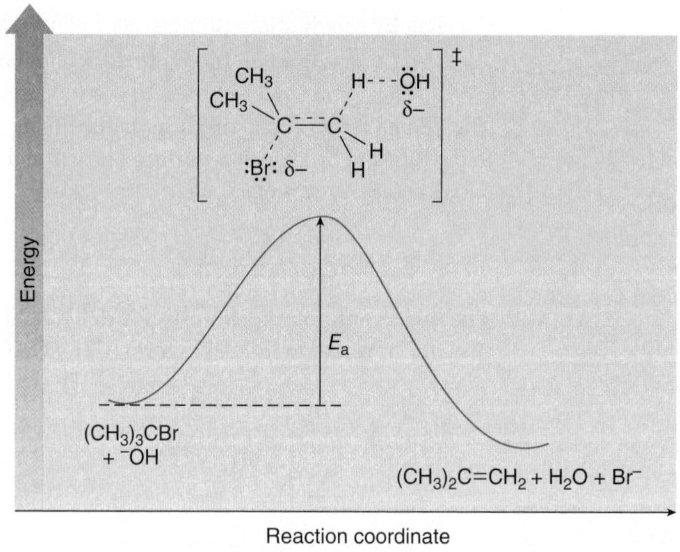

- In the transition state, the C–H and C–Br bonds are partially broken, the O–H and π bonds are partially formed, and both the base and the departing leaving group bear a partial negative charge.

Problem 8.8 Use curved arrows to show the movement of electrons in the following E2 mechanism. Draw the structure of the transition state.

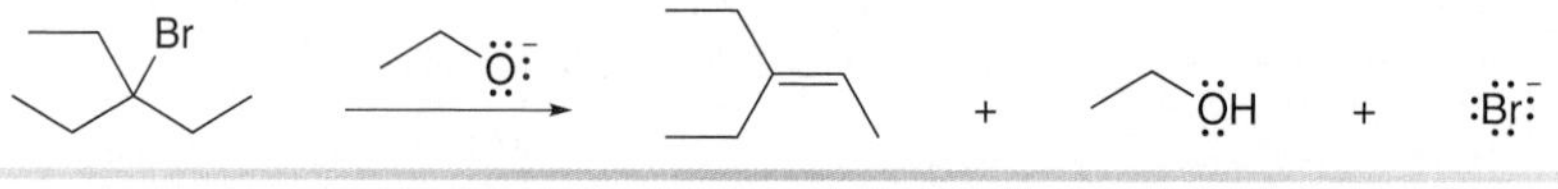

There are close parallels between the E2 and S_N2 mechanisms in how the identity of the base, the leaving group, and the solvent affect the rate.

The Base

- **The base appears in the rate equation, so the rate of the E2 reaction increases as the strength of the base increases.**

The IUPAC names for **DBN** and **DBU** are rarely used because the names are complex. **DBN** stands for 1,5-**d**iaza**b**icyclo[4.3.0]-**n**on-5-ene, and **DBU** stands for 1,8-**d**iaza**b**icyclo[5.4.0]-**u**ndec-7-ene.

E2 reactions are generally run with strong, negatively charged bases like ^{-}OH and ^{-}OR. Two strong, sterically hindered nitrogen bases, called **DBN** and **DBU,** are also sometimes used. An example of an E2 reaction with DBN is shown in Figure 8.4.

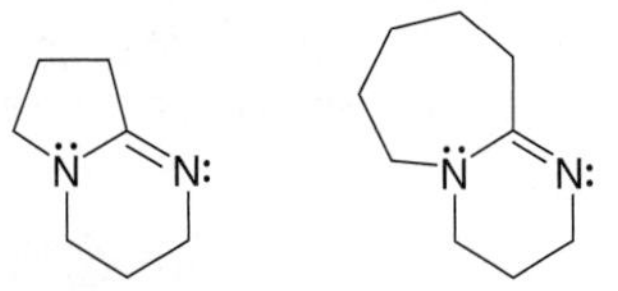

DBN DBU

Figure 8.4 An E2 elimination with DBN used as the base

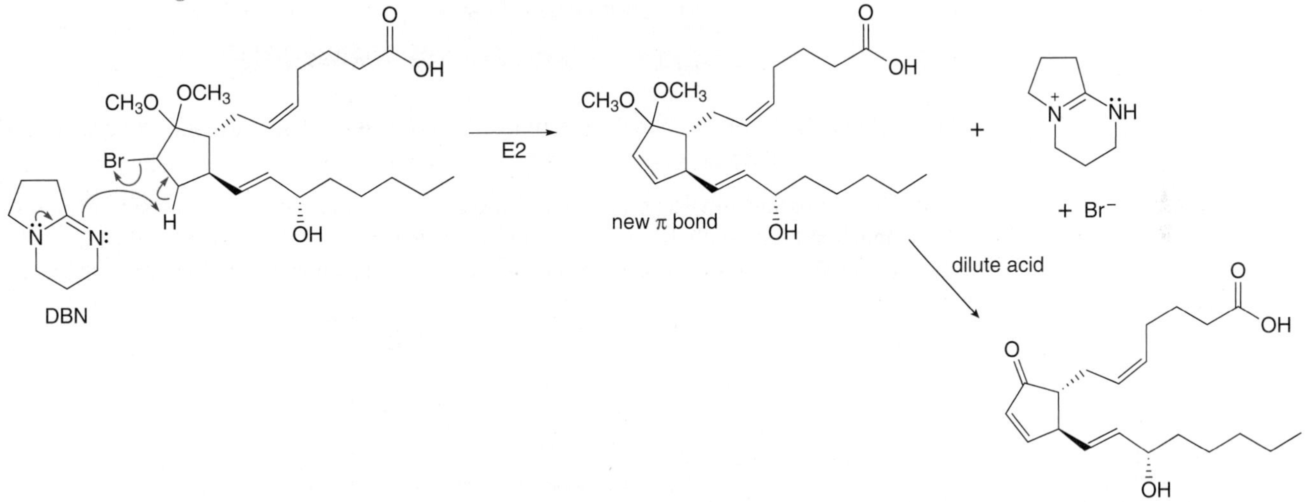

PGA$_2$
a prostaglandin
(Section 19.6)

The Leaving Group

- **Because the bond to the leaving group is partially broken in the transition state, the better the leaving group the faster the E2 reaction.**

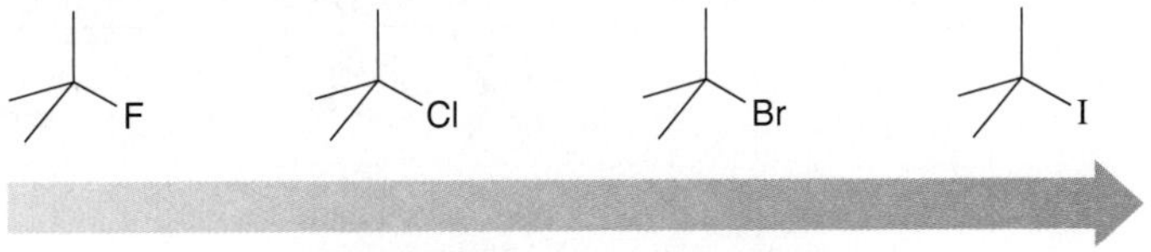

Increasing leaving group ability
Increasing rate of the E2 reaction

The Solvent

- **Polar aprotic solvents increase the rate of E2 reactions.**

Because **polar aprotic solvents** like $(CH_3)_2C{=}O$ do not solvate anions well, a negatively charged base is not "hidden" by strong interactions with the solvent (Section 7.15D), and the base is stronger. **A stronger base increases the reaction rate.**

Problem 8.9 Consider an E2 reaction between CH_3CH_2Br and $KOC(CH_3)_3$. What effect does each of the following changes have on the rate of elimination? (a) The base is changed to KOH. (b) The alkyl halide is changed to CH_3CH_2Cl.

8.4C The Identity of the Alkyl Halide

The S_N2 and E2 mechanisms differ in how the R group affects the reaction rate.

- **As the number of R groups on the carbon with the leaving group increases, the rate of the E2 reaction increases.**

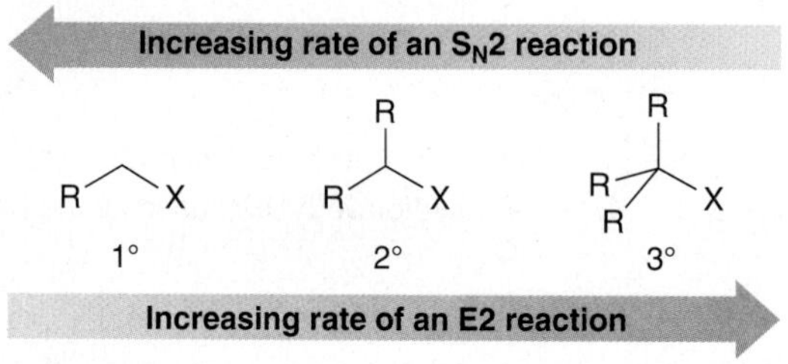

This trend is exactly *opposite* to the reactivity of alkyl halides in S_N2 reactions, where increasing alkyl substitution decreases the rate of reaction (Section 7.11D).

Why does increasing alkyl substitution increase the rate of an E2 reaction? In the transition state, the double bond is partially formed, so increasing the stability of the double bond with alkyl substituents stabilizes the transition state (i.e., it lowers E_a), which increases the rate of the reaction.

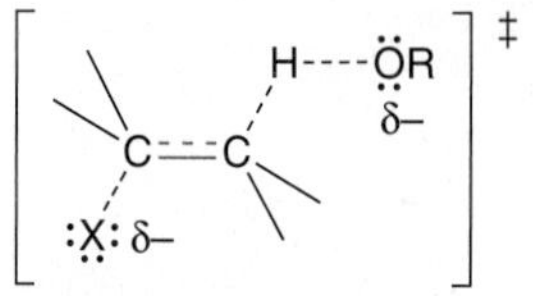

The double bond is partially formed.

- **Increasing the number of R groups on the carbon with the leaving group forms more highly substituted, more stable alkenes in E2 reactions.**

For example, the E2 reaction of a 1° alkyl halide (1-bromobutane) forms a monosubstituted alkene, whereas the E2 reaction of a 3° alkyl halide (2-bromo-2-methylpropane) forms a disubstituted alkene. The disubstituted alkene is more stable, so the 3° alkyl halide reacts faster than the 1° alkyl halide.

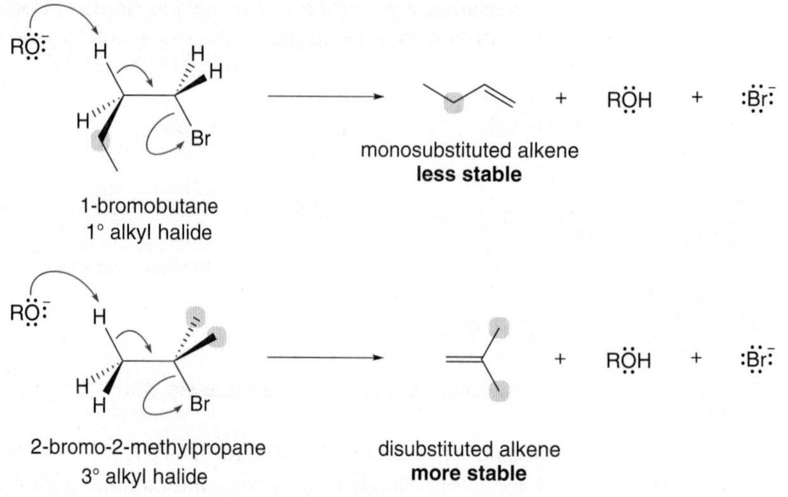

Problem 8.10 Rank the alkyl halides in each group in order of increasing reactivity in an E2 reaction.

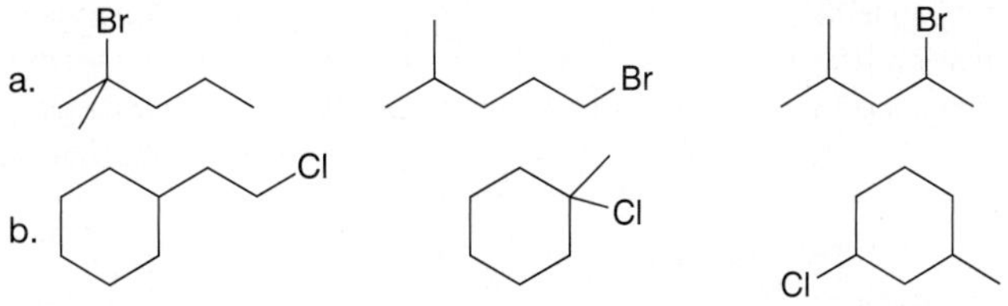

Table 8.2 summarizes the characteristics of the E2 mechanism, and Figure 8.5 illustrates two examples of E2 reactions used in the synthesis of **quinine** and **estradiol,** two naturally occurring compounds.

Table 8.2 Characteristics of the E2 Mechanism

Characteristic	Result
Kinetics	• Second order
Mechanism	• One step
Identity of R	• More substituted halides react faster. • Rate: $R_3CX > R_2CHX > RCH_2X$
Base	• Favored by strong bases
Leaving group	• Better leaving group ---→ faster reaction
Solvent	• Favored by polar aprotic solvents

Figure 8.5 Two examples of the E2 reaction used in organic synthesis

Quinine, a natural product isolated from the bark of the cinchona tree native to the Andes Mountains, is a powerful antipyretic—that is, it reduces fever—and for centuries, it was the only effective treatment for malaria.

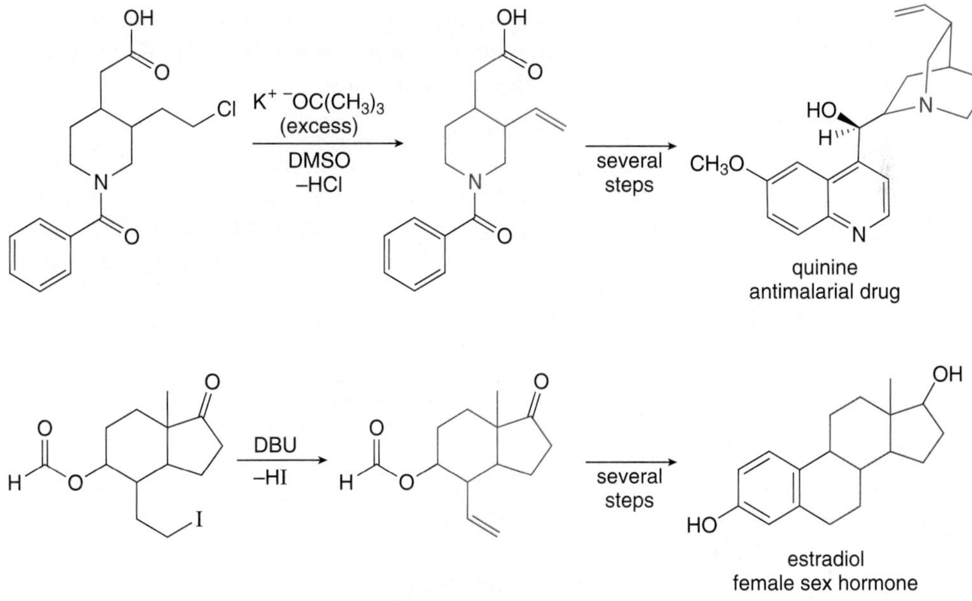

• Bonds and atoms in quinine and estradiol that originate in the alkene intermediate are shown in red.

Problem 8.11 How does each of the following changes affect the rate of an E2 reaction?

a. tripling [RX]
b. halving [B:]
c. changing the solvent from CH_3OH to DMSO
d. changing the leaving group from I^- to Br^-
e. changing the base from ^-OH to H_2O
f. changing the alkyl halide from CH_3CH_2Br to $(CH_3)_2CHBr$

8.5 The Zaitsev Rule

Recall from Section 8.1 that a mixture of alkenes can form from the dehydrohalogenation of alkyl halides having two or more different β carbon atoms. When this occurs, one of the products usually predominates. The **major product is the more stable product—the one with the more substituted double bond.** For example, elimination of the elements of H and I from 1-iodo-1-methylcyclohexane yields two constitutional isomers: the trisubstituted alkene **A** (the major product) and the disubstituted alkene **B** (the minor product).

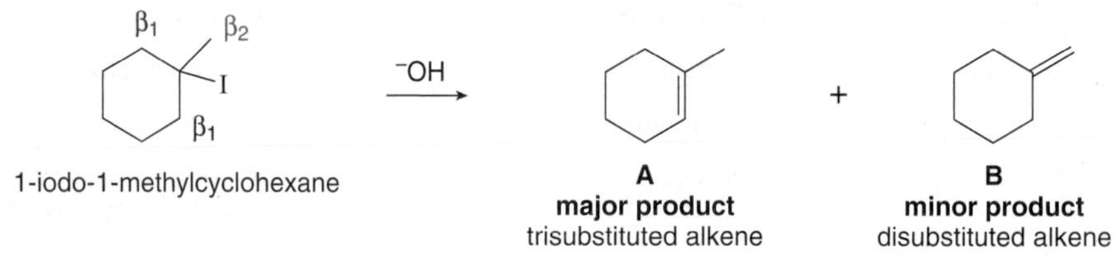

This phenomenon is called the **Zaitsev rule** (also called the **Saytzeff rule,** depending on the translation) for the Russian chemist who first noted this trend.

- **The Zaitsev rule: The major product in β elimination has the more substituted double bond.**

A reaction is *regioselective* when it yields predominantly or exclusively one constitutional isomer when more than one is possible. The E2 reaction is **regioselective** because the more substituted alkene predominates.

The Zaitsev rule results because the double bond is partially formed in the transition state for the E2 reaction. Thus, increasing the stability of the double bond by adding R groups lowers the energy of the transition state, which increases the reaction rate. E2 elimination of HBr from 2-bromo-2-methylbutane yields alkenes **C** and **D. D,** having the more substituted double bond, is the major product, because the transition state leading to its formation is lower in energy.

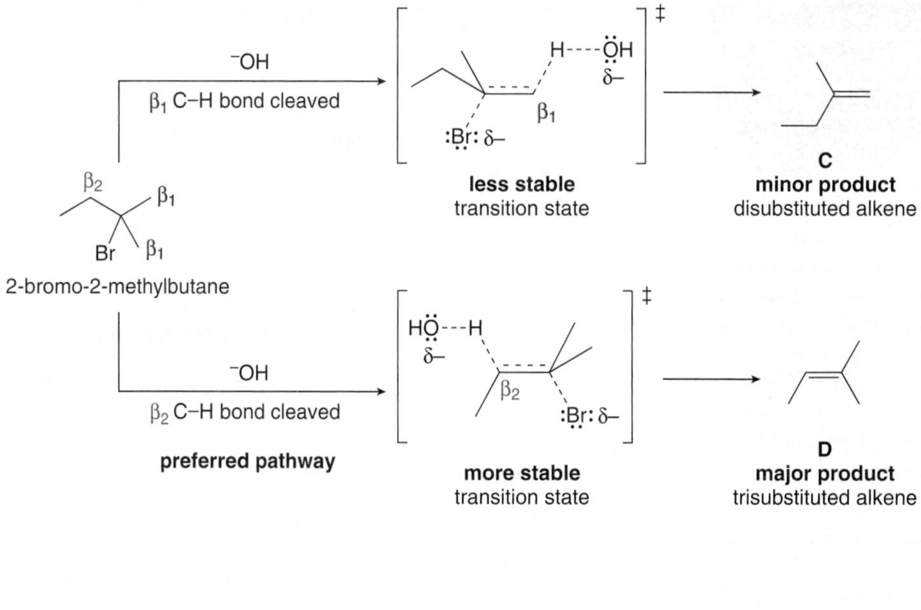

When a mixture of stereoisomers is possible from dehydrohalogenation, the **major product is the more stable stereoisomer.** Dehydrohalogenation of alkyl halide **X** forms a mixture of trans and cis alkenes, **Y** and **Z.** The trans alkene **Y** is the major product because it is more stable.

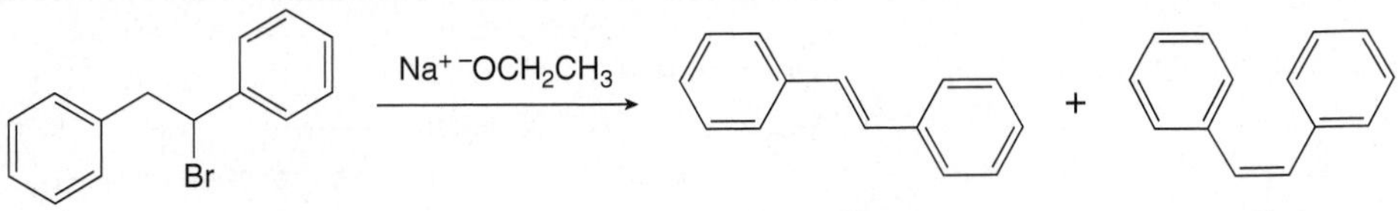

X — **Y** trans alkene **major product** — **Z** cis alkene **minor product**

A reaction is *stereoselective* when it forms predominantly or exclusively one stereoisomer when two or more are possible. The **E2 reaction is stereoselective** because one stereoisomer is formed preferentially.

Sample Problem 8.1 Predict the major product in the following E2 reaction.

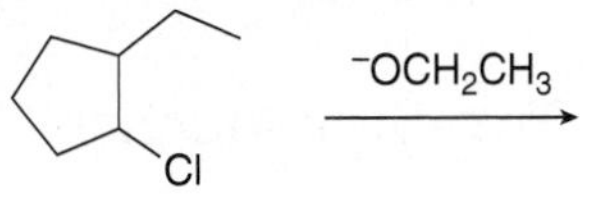

Solution

The alkyl halide has two different β C atoms (labeled β_1 and β_2), so two different alkenes are possible: one formed by removal of HCl across the α and β_1 carbons, and one formed by removal of HCl across the α and β_2 carbons. Using the Zaitsev rule, the major product should be **A,** because it has the more substituted double bond.

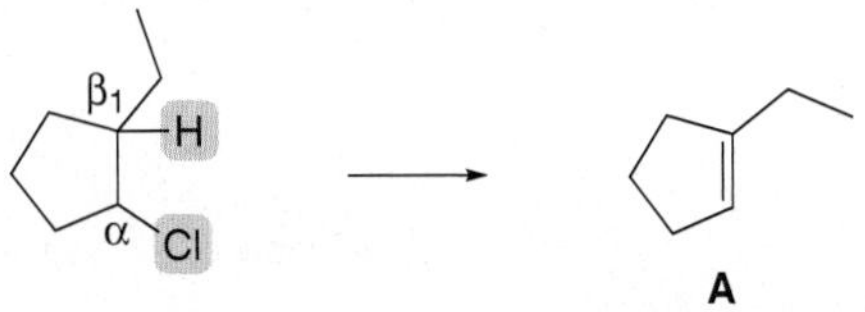

A
trisubstituted alkene
major product

B
disubstituted alkene
minor product

Problem 8.12 What alkenes are formed from each alkyl halide by an E2 reaction? Use the Zaitsev rule to predict the major product.

a. 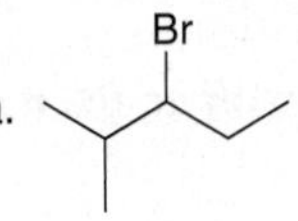

b. 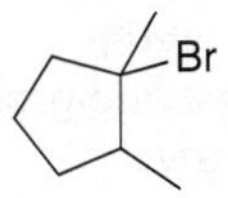

c. 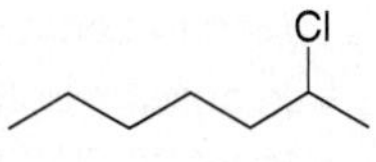

d. 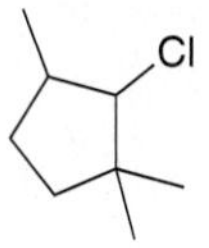

8.6 The E1 Mechanism

The dehydrohalogenation of $(CH_3)_3CI$ with H_2O to form $(CH_3)_2C{=}CH_2$ can be used to illustrate the second general mechanism of elimination, the **E1 mechanism.**

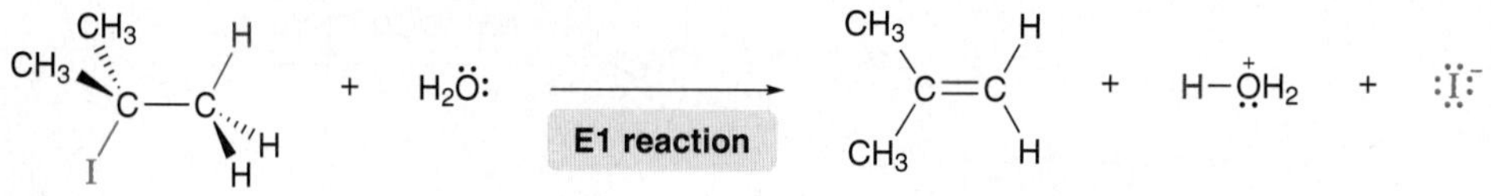

8.6A Kinetics

An E1 reaction exhibits first-order kinetics.

- rate = $k[(CH_3)_3CI]$

Like the S_N1 mechanism, the kinetics suggest that the reaction mechanism involves more than one step, and that the slow step is **unimolecular,** involving *only* the alkyl halide.

8.6B A Two-Step Mechanism

The most straightforward explanation for the observed first-order kinetics is a **two-step reaction: the bond to the leaving group breaks first *before* the π bond is formed,** as shown in Mechanism 8.2.

Mechanism 8.2 The E1 Mechanism

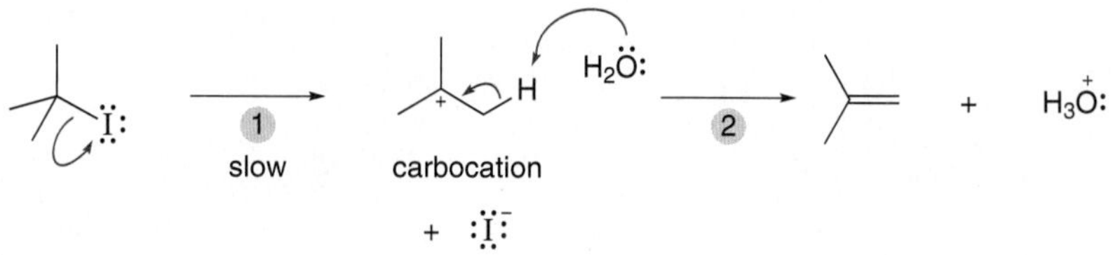

1 Heterolysis of the C–I bond forms a carbocation in the rate-determining step.

2 A base (either H_2O or I^-) removes a proton from a carbon adjacent to the carbocation, and the electron pair in the C–H bond forms the π bond.

The E1 and E2 mechanisms both involve the same number of bonds broken and formed. **The only difference is the timing.**

- **In an E1 reaction, the leaving group comes off *before* the β proton is removed, and the reaction occurs in two steps.**
- **In an E2 reaction, the leaving group comes off *as* the β proton is removed, and the reaction occurs in one step.**

An energy diagram for the reaction of $(CH_3)_3CI + H_2O$ is shown in Figure 8.6. Each step has its own energy barrier, with a transition state at each energy maximum. Because its transition state is higher in energy, Step [1] is rate-determining. $\Delta H°$ for Step [1] is positive because only bond breaking occurs, whereas $\Delta H°$ of Step [2] is negative because two bonds are formed and only one is broken.

Figure 8.6 Energy diagram for an E1 reaction: $(CH_3)_3CI + H_2O \rightarrow (CH_3)_2C{=}CH_2 + H_3O^+ + I^-$

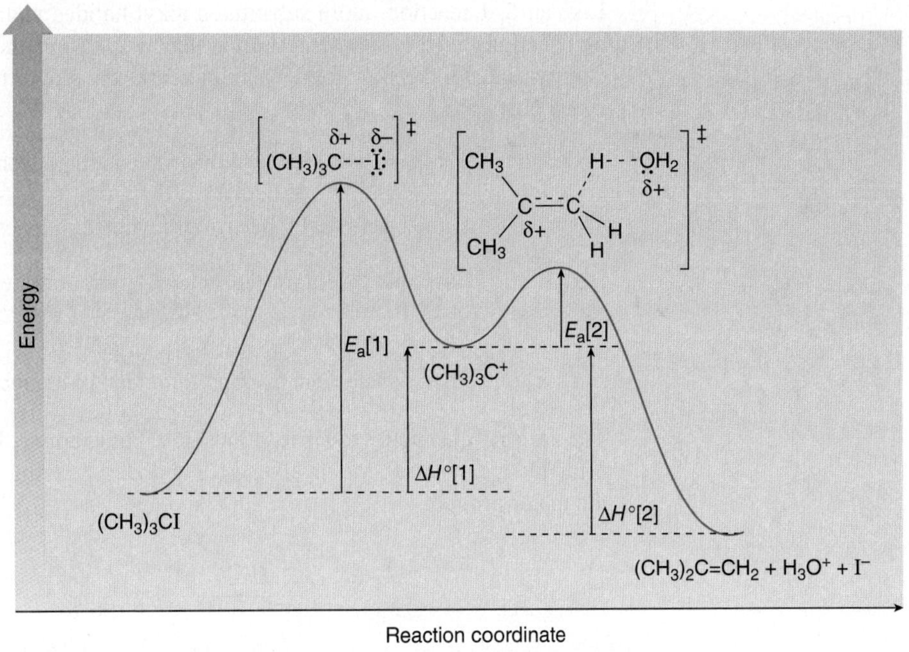

- The E1 mechanism has two steps, so there are two energy barriers.
- Step [1] is rate-determining.

Problem 8.13 Draw an E1 mechanism for the following reaction. Draw the structure of the transition state for each step.

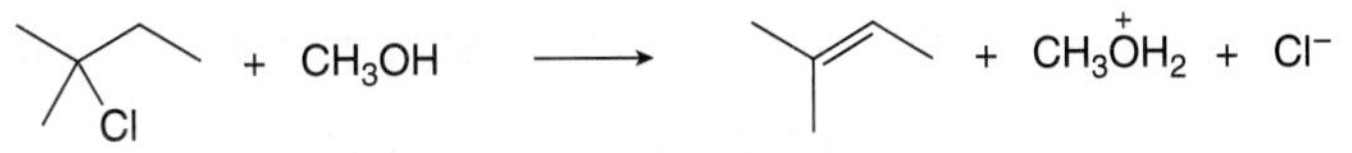

8.6C Other Characteristics of E1 Reactions

Three other features of E1 reactions are worthy of note.

[1] The rate of an E1 reaction increases as the number of R groups on the carbon with the leaving group increases.

Increasing alkyl substitution has the same effect on the rate of *both* an E1 and E2 reaction; increasing rate of the E1 and E2 reactions: RCH_2X (1°) < R_2CHX (2°) < R_3CX (3°).

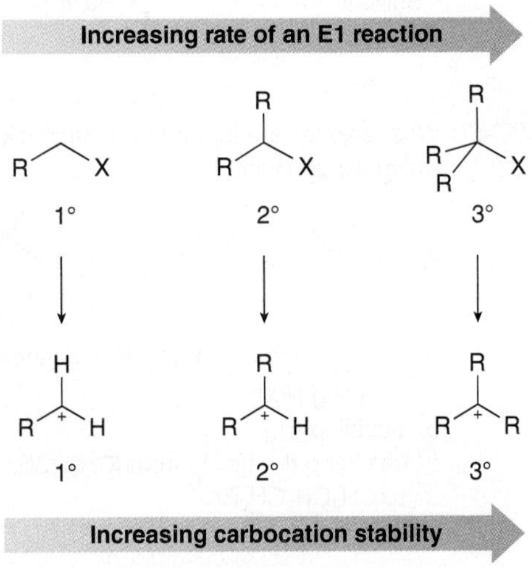

Like an S_N1 reaction, more substituted alkyl halides yield more substituted (and more stable) carbocations in the rate-determining step. Increasing the stability of a carbocation, in turn, decreases E_a for the slow step, which increases the rate of the E1 reaction according to the Hammond postulate.

[2] Because the base does not appear in the rate equation, weak bases favor E1 reactions.

The strength of the base usually determines whether a reaction follows the E1 or E2 mechanism.

- **Strong bases like ^-OH and ^-OR favor E2 reactions, whereas weaker bases like H_2O and ROH favor E1 reactions.**

[3] E1 reactions are regioselective, favoring formation of the more substituted, more stable alkene.

The Zaitsev rule applies to E1 reactions, too. For example, E1 elimination of HBr from 1-bromo-1-methylcyclopentane yields alkenes **A** and **B. A,** having the more substituted double bond, is the major product.

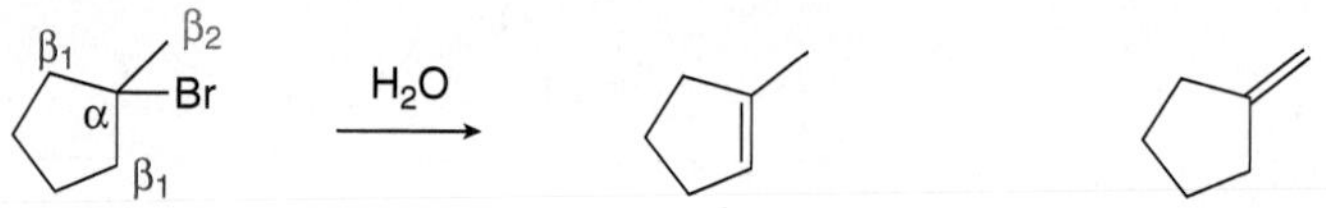

1-bromo-1-methyl-cyclopentane | **A** trisubstituted alkene **major product** | **B** disubstituted alkene **minor product**

Table 8.3 summarizes the characteristics of E1 reactions.

Table 8.3 Characteristics of the E1 Mechanism

Characteristic	Result
Kinetics	• First order
Mechanism	• Two steps
Identity of R	• More substituted halides react faster. • Rate: $R_3CX > R_2CHX > RCH_2X$
Base	• Favored by weaker bases such as H_2O and ROH
Leaving group	• A better leaving group makes the reaction faster because the bond to the leaving group is partially broken in the rate-determining step.
Solvent	• Polar protic solvents that solvate the ionic intermediates are needed.

Problem 8.14 What alkenes are formed from each alkyl halide by an E1 reaction? Use the Zaitsev rule to predict the major product.

a. 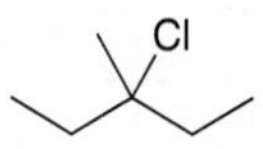

b. 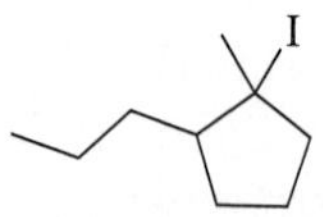

Problem 8.15 How does each of the following changes affect the rate of an E1 reaction?

a. doubling [RX]
b. doubling [B:]
c. changing the halide from $(CH_3)_3CBr$ to $CH_3CH_2CH_2Br$
d. changing the leaving group from Cl^- to Br^-
e. changing the solvent from DMSO to CH_3OH

8.7 S_N1 and E1 Reactions

S_N1 and E1 reactions have exactly the same first step—formation of a carbocation. They differ in what happens to the carbocation.

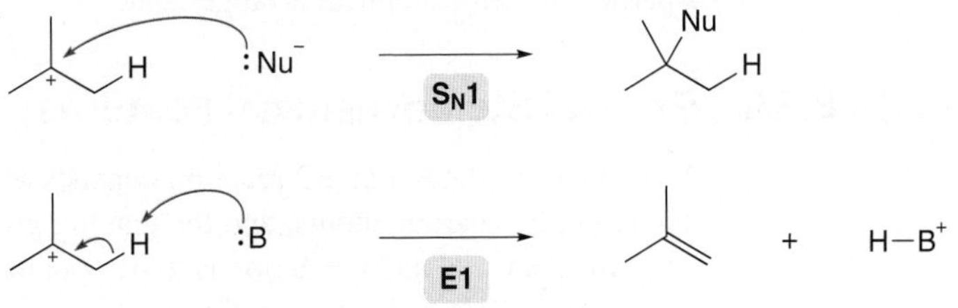

- **In an S_N1 reaction, a nucleophile attacks the carbocation, forming a substitution product.**
- **In an E1 reaction, a base removes a proton, forming a new π bond.**

The same conditions that favor substitution by an S_N1 mechanism also favor elimination by an E1 mechanism: a 3° alkyl halide as substrate, a weak nucleophile or base as reagent, and a polar protic solvent. As a result, both reactions usually occur in the same reaction mixture to afford a mixture of products, as illustrated in Sample Problem 8.2.

Sample Problem 8.2 Draw the S_N1 and E1 products formed in the reaction of $(CH_3)_3CBr$ with H_2O.

Solution

The first step in both reactions is heterolysis of the C–Br bond to form a **carbocation.**

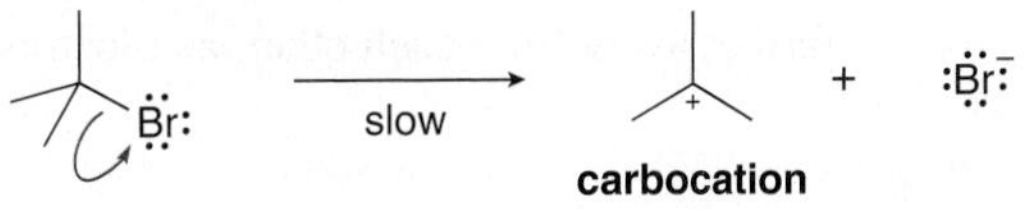

Reaction of the carbocation with H_2O as a nucleophile affords the substitution product (Reaction [1]). Alternatively, H_2O acts as a base to remove a proton, affording the elimination product (Reaction [2]). **Two products are formed.**

[1] 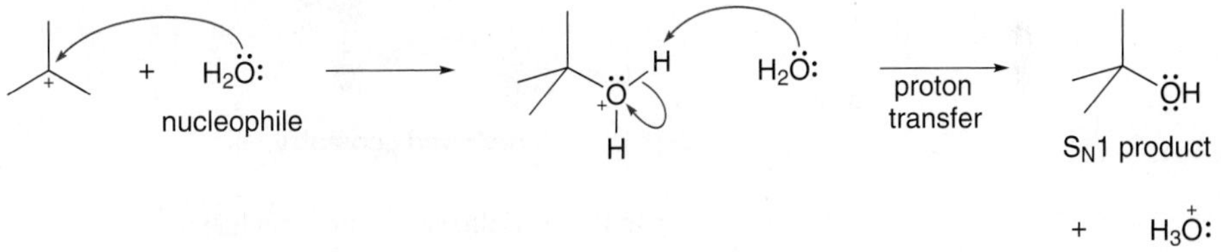

[2] 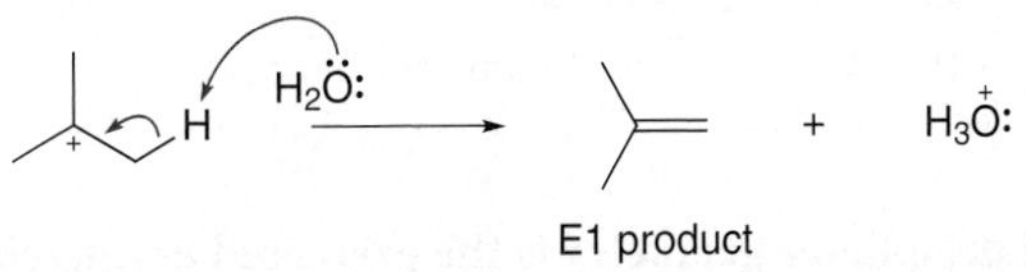

Because E1 reactions often occur with a competing S_N1 reaction, **E1 reactions of alkyl halides are *much less useful* than E2 reactions.**

Problem 8.16 Draw both the S_N1 and E1 products of each reaction.

a. 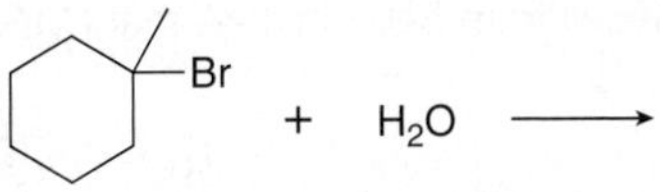

b. 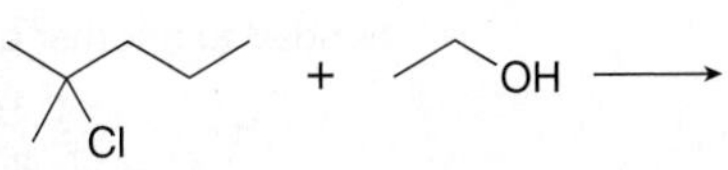

8.8 Stereochemistry of the E2 Reaction

Although the E2 reaction does not produce products with tetrahedral stereogenic centers, its transition state consists of four atoms that react at the same time, and they react only if they possess a particular stereochemical arrangement.

8.8A General Stereochemical Features

The transition state of an E2 reaction consists of **four atoms** from the alkyl halide—one hydrogen atom, two carbon atoms, and the leaving group (X)—**all aligned in a plane.** There are two ways for the C–H and C–X bonds to be coplanar.

The dihedral angle for the C–H and C–X bonds equals 0° for the syn periplanar arrangement and 180° for the anti periplanar arrangement.

- **The H and X atoms can be oriented on the same side of the molecule. This geometry is called *syn periplanar.***
- **The H and X atoms can be oriented on opposite sides of the molecule. This geometry is called *anti periplanar.***

All evidence suggests that **E2 elimination occurs most often in the anti periplanar geometry.** This arrangement allows the molecule to react in the lower energy *staggered* conformation. It also allows two electron-rich species, the incoming base and the departing leaving group, to be farther away from each other, as illustrated in Figure 8.7.

Figure 8.7
Two possible geometries for the E2 reaction

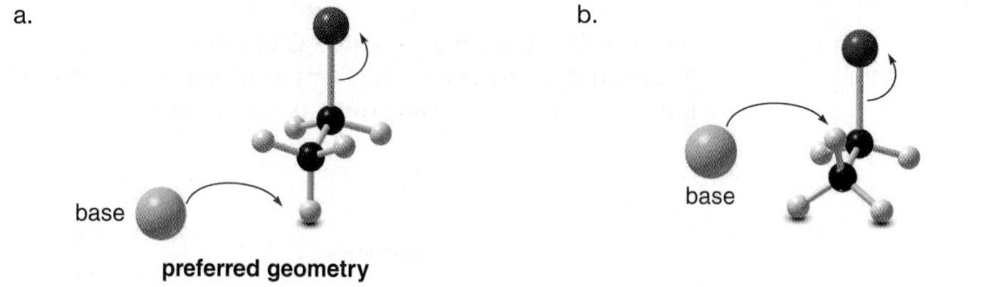

a.
- An **anti periplanar** arrangement has a **staggered** conformation.
- The two electron-rich groups are far apart.

b.
- A **syn periplanar** arrangement has an **eclipsed** conformation.
- The two electron-rich groups are close.

Anti periplanar geometry is the preferred arrangement for any alkyl halide undergoing E2 elimination, regardless of whether it is cyclic or acyclic. This stereochemical requirement has important consequences for compounds containing six-membered rings.

Problem 8.17 Given that an E2 reaction proceeds with anti periplanar stereochemistry, draw the products of each elimination. The alkyl halides in (a) and (b) are diastereomers of each other. How are the products of these two reactions related? Recall from Section 3.2A that C_6H_5- is a phenyl group, a benzene ring bonded to another group.

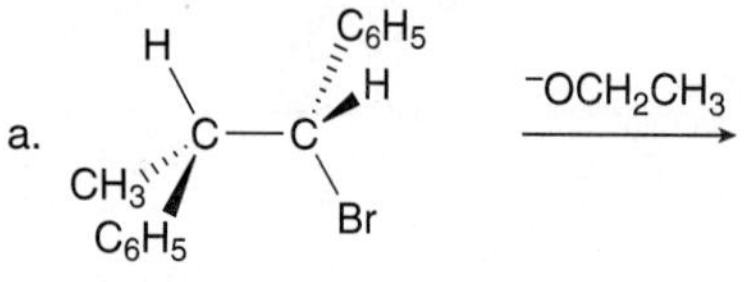

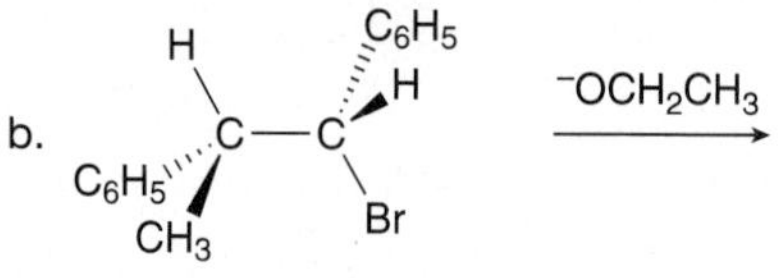

8.8B Anti Periplanar Geometry and Halocyclohexanes

Recall from Section 4.13 that cyclohexane exists as two chair conformations that rapidly interconvert, and that substituted cyclohexanes are more stable with substituents in the roomier equatorial position. Chlorocyclohexane exists as two chair conformations, but **X** is preferred because the Cl group is equatorial.

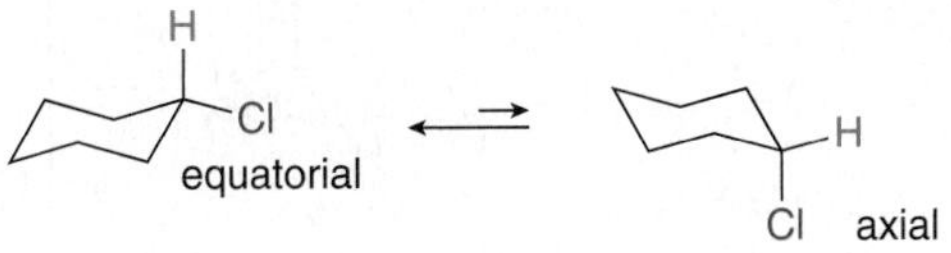

more stable
X

less stable
Y

chlorocyclohexane

For E2 elimination, **the C–Cl bond must be anti periplanar to a C–H bond on a β carbon,** and this occurs only when the H and Cl atoms are both in the **axial** position. This requirement for **trans diaxial geometry** means that E2 elimination must occur from the less stable conformation **Y,** as shown in Figure 8.8.

Figure 8.8

The trans diaxial geometry for the E2 elimination in chlorocyclohexane

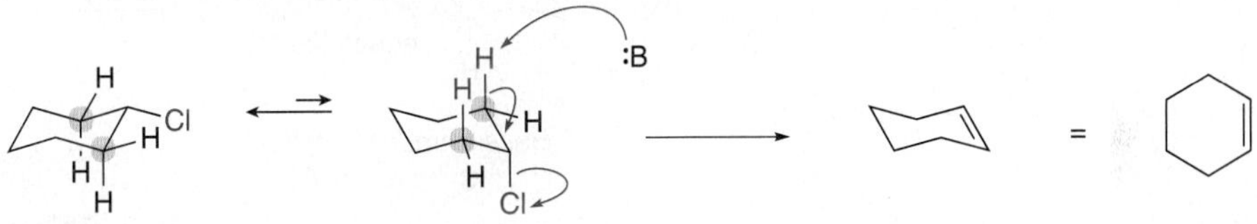

X

This conformation does *not* react.

Y

This conformation reacts.

- In conformation **X** (**equatorial** Cl group), a β C–H bond and a C–Cl bond are never anti periplanar; therefore, no E2 elimination can occur. β Carbons are highlighted in blue.
- In conformation **Y** (**axial** Cl group), two β C–H bonds and the C–Cl bond are **trans diaxial;** therefore, E2 elimination occurs. Axial H's on β carbons that can react are shown in red.

Sometimes this rigid stereochemical requirement affects the regioselectivity of the E2 reaction of substituted cyclohexanes. Dehydrohalogenation of *cis*- and *trans*-1-chloro-2-methylcyclohexane via an E2 mechanism illustrates this phenomenon.

cis-1-chloro-2-methylcyclohexane

trans-1-chloro-2-methylcyclohexane

The **cis isomer** exists as two conformations (**A** and **B**), each of which has one group axial and one group equatorial. E2 reaction must occur from conformation **B,** which contains an axial Cl atom.

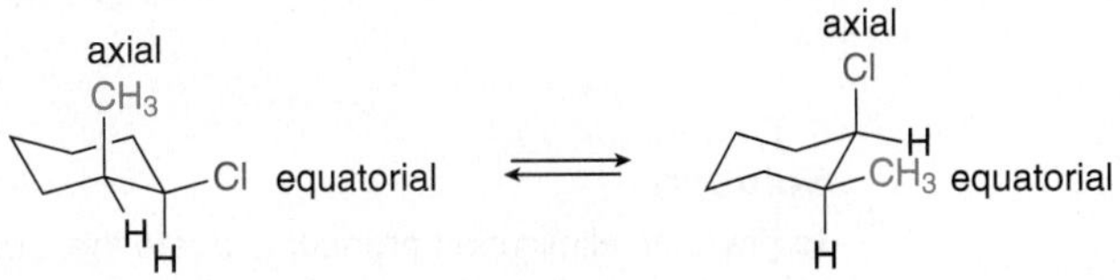

cis isomer
A

cis isomer
B
This conformation reacts.

Because conformation **B** has two different axial β H atoms, labeled H_a and H_b, E2 reaction occurs in two different directions to afford two alkenes. **The major product contains the more stable trisubstituted double bond, as predicted by the Zaitsev rule.**

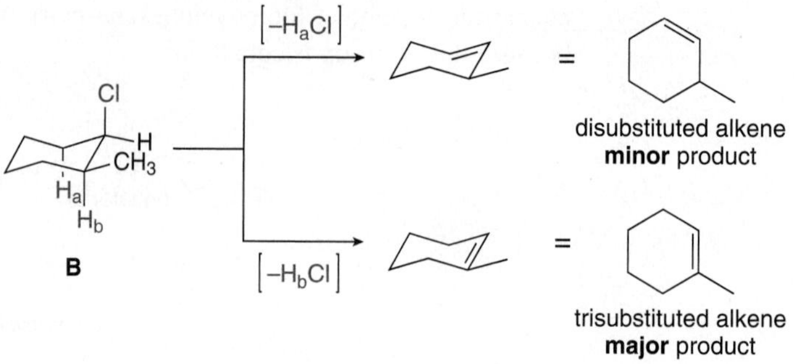

The **trans isomer** exists as two conformations, **C,** having two equatorial substituents, and **D,** having two axial substituents. E2 reaction must occur from conformation **D,** which contains an axial Cl atom.

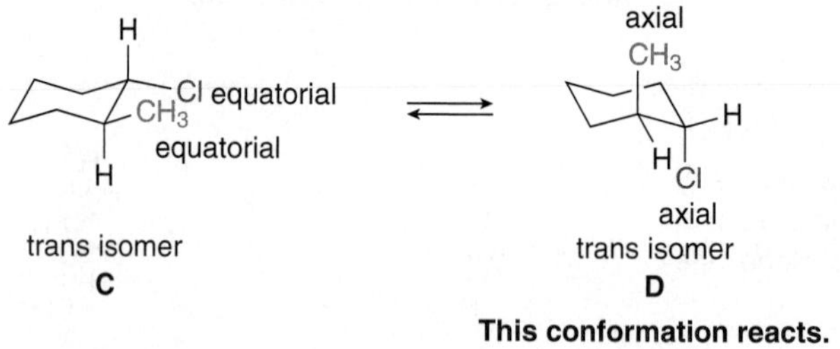

Because conformation **D** has **only one axial β H,** E2 reaction occurs in only one direction to afford a **single product,** having the disubstituted double bond. This is *not* predicted by the Zaitsev rule. E2 reaction requires H and Cl to be trans and diaxial, and with the trans isomer, this is possible only when the less stable alkene is formed as product.

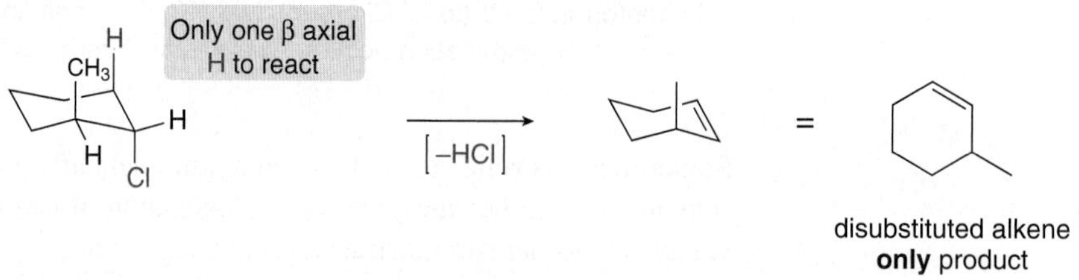

- **In conclusion, with substituted cyclohexanes, E2 elimination must occur with a trans diaxial arrangement of H and X, and as a result of this requirement, the more substituted alkene is not necessarily the major product.**

Sample Problem 8.3 Draw the major E2 elimination product formed from the following alkyl halide.

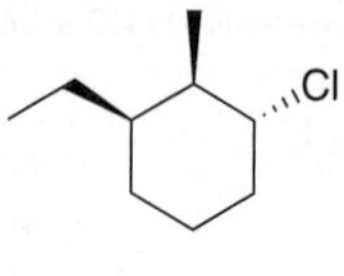

Solution

To draw the elimination products, locate the β carbons and look for H atoms that are trans to the leaving group. The given alkyl chloride has two different β carbons, labeled β_1 and β_2. Elimination can occur only when the leaving group (Cl) and a H atom on the β carbon are **trans.**

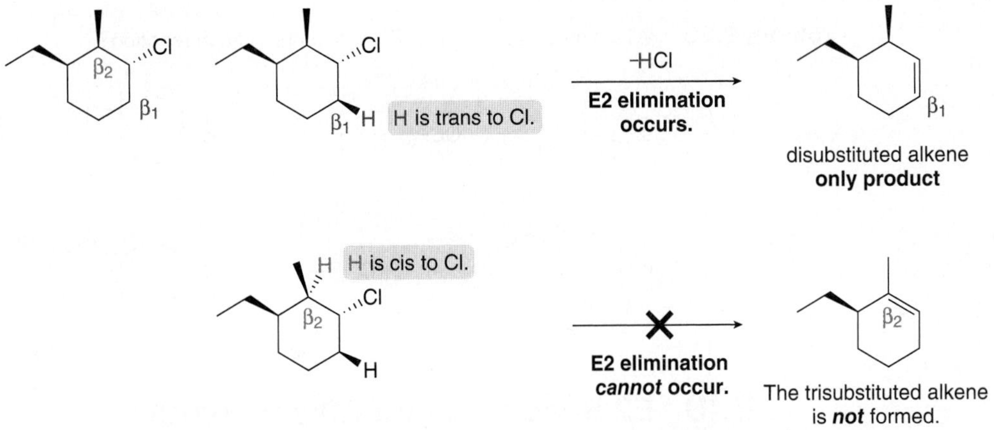

The β_1 C has a H atom trans to Cl, so E2 elimination occurs to form a disubstituted alkene. Because there is no trans H on the β_2 C, E2 elimination cannot occur in this direction, and the more stable trisubstituted alkene is *not* formed. Although this result is not predicted by the Zaitsev rule, it is consistent with the requirement that the H and X atoms in an E2 elimination must be located trans to each other.

Problem 8.18 Draw the major E2 elimination products from each of the following alkyl halides.

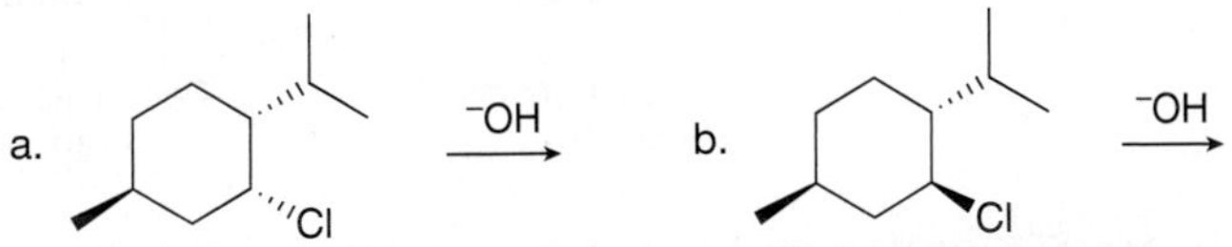

Problem 8.19 Explain why *cis*-1-chloro-2-methylcyclohexane undergoes E2 elimination much faster than its trans isomer.

8.9 When Is the Mechanism E1 or E2?

Given a particular starting material and base, how do we know whether a reaction occurs by the E1 or E2 mechanism?

Because the rate of *both* the E1 and E2 reactions increases as the number of R groups on the carbon with the leaving group increases, you cannot use the identity of the alkyl halide to decide which elimination mechanism occurs.

- **The strength of the base is the most important factor in determining the mechanism for elimination. Strong bases favor the E2 mechanism. Weak bases favor the E1 mechanism.**

Table 8.4 compares the E1 and E2 mechanisms.

Table 8.4 A Comparison of the E1 and E2 Mechanisms

Mechanism	Comment
E2 mechanism	• Much more common and useful • Favored by strong, negatively charged bases, especially ^-OH and ^-OR • The reaction occurs with 1°, 2°, and 3° alkyl halides. Order of reactivity: $R_3CX > R_2CHX > RCH_2X$.
E1 mechanism	• Much less useful because a mixture of S_N1 and E1 products usually results • Favored by weaker, neutral bases, such as H_2O and ROH • This mechanism does not occur with 1° RX because they form highly unstable 1° carbocations.

Problem 8.20 Which mechanism, E1 or E2, will occur in each reaction?

a. 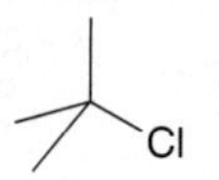 + $^{-}OCH_3$ ⟶

b. 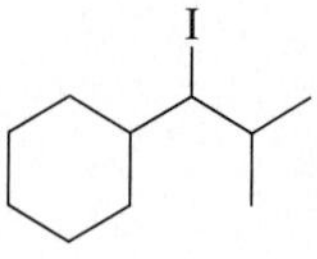 + H_2O ⟶

c. 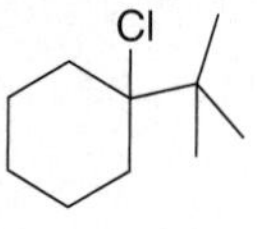 + CH_3OH ⟶

d. 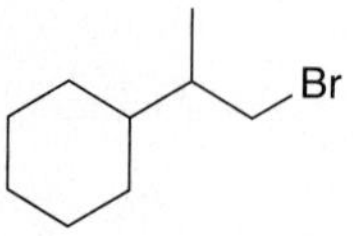 + $^{-}OC(CH_3)_3$ ⟶

8.10 E2 Reactions and Alkyne Synthesis

Recall from Section 1.10C that the carbon–carbon triple bond of alkynes consists of one σ and two π bonds.

A single elimination reaction produces the π bond of an alkene. Two consecutive elimination reactions produce the two π bonds of an alkyne.

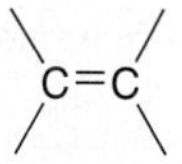

alkene
one π bond

One elimination reaction is needed.

—C≡C—

alkyne
two π bonds

Two elimination reactions are needed.

- **Alkynes are prepared by two successive dehydrohalogenation reactions.**

Two elimination reactions are needed to remove two moles of HX from a **dihalide** as substrate. Two different starting materials can be used.

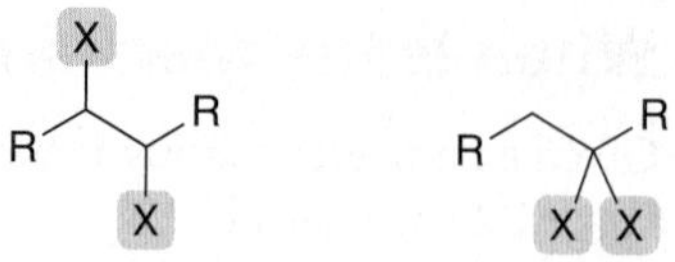

vicinal dihalide   geminal dihalide

- **A vicinal dihalide has two X atoms on *adjacent* carbon atoms.**
- **A geminal dihalide has two X atoms on the *same* carbon atom.**

The word *geminal* comes from the Latin *geminus,* meaning "twin."

Equations [1] and [2] illustrate how two moles of HX can be removed from these dihalides with base. Two equivalents of strong base are used and each step follows an E2 mechanism.

[1] 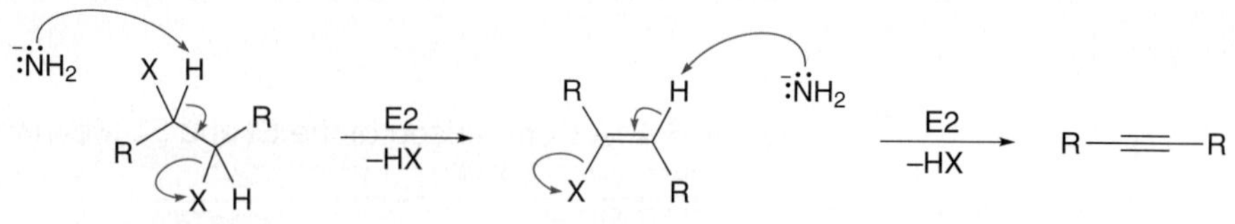

vicinal dihalide   vinyl halide

[2] 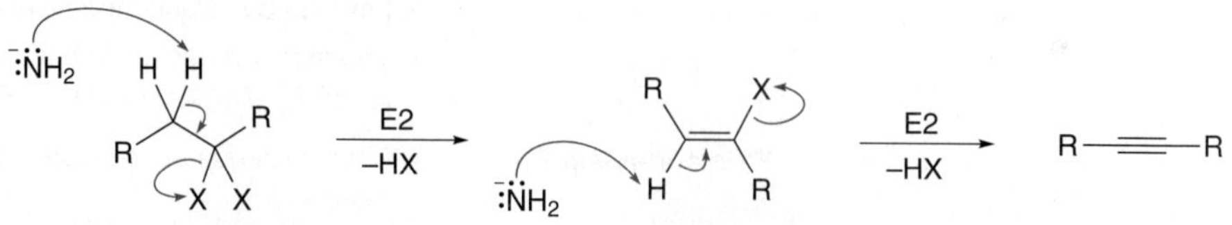

geminal dihalide   vinyl halide

The relative strength of C–H bonds depends on the hybridization of the carbon atom: *sp* > *sp*^{2} > *sp*^{3}. For more information, review Section 1.11B.

Stronger bases are needed to synthesize alkynes by dehydrohalogenation than are needed to synthesize alkenes. The typical base is **amide ($^-NH_2$)**, used as the sodium salt $NaNH_2$ (sodium amide). $KOC(CH_3)_3$ can also be used with DMSO as solvent. Because DMSO is a polar aprotic solvent, the anionic base is not well solvated, thus **increasing its basicity** and making it strong enough to remove two equivalents of HX. Examples are given in Figure 8.9.

The strongly basic conditions needed for alkyne synthesis result from the difficulty of removing the second equivalent of HX from the intermediate vinyl halide, RCH=C(R)X. Because H and X are both bonded to *sp*^{2} hybridized carbons, these bonds are shorter and stronger than the *sp*^{3} hybridized C–H and C–X bonds of an alkyl halide, necessitating the use of a stronger base.

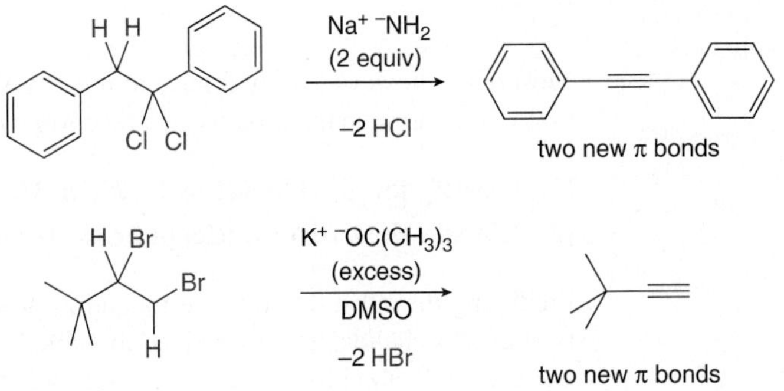

Figure 8.9 Examples of dehydrohalogenation of dihalides to afford alkynes

Problem 8.21 Draw the alkynes formed when each dihalide is treated with excess base.

a. 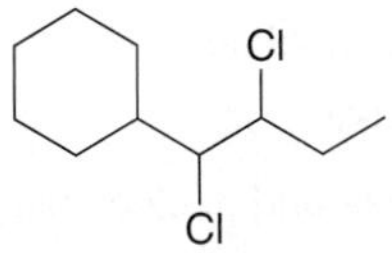

b. 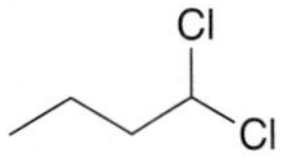

c. 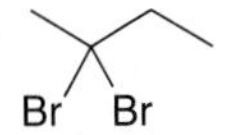

d. 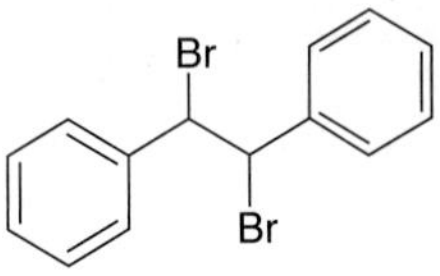

8.11 When Is the Reaction S_N1, S_N2, E1, or E2?

We have now considered two different kinds of reactions (substitution and elimination) and four different mechanisms (S_N1, S_N2, E1, and E2) that begin with one class of compounds (alkyl halides). How do we know if a given alkyl halide will undergo substitution or elimination with a given base or nucleophile, and by what mechanism?

Unfortunately, there is no easy answer, and often mixtures of products result. Two generalizations help to determine whether substitution or elimination occurs.

[1] Good nucleophiles that are weak bases favor substitution over elimination.

Certain anions generally give products of substitution because they are good nucleophiles but weak bases. These include: **I^-, Br^-, HS^-, ^-CN,** and **$CH_3CO_2^-$.**

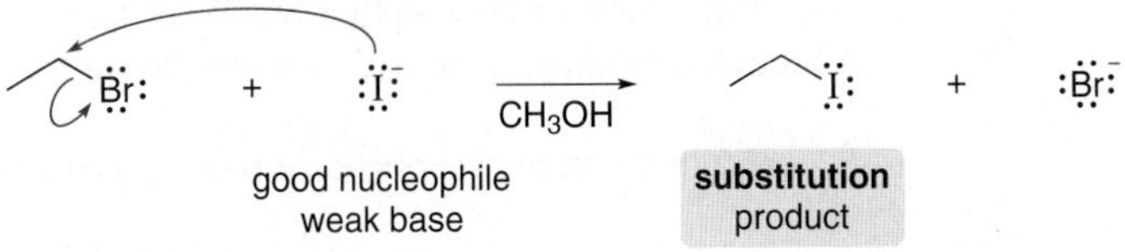

[2] Bulky, nonnucleophilic bases favor elimination over substitution.

$KOC(CH_3)_3$, DBU, and **DBN** are too sterically hindered to attack a tetravalent carbon, but are able to remove a small proton, favoring elimination over substitution.

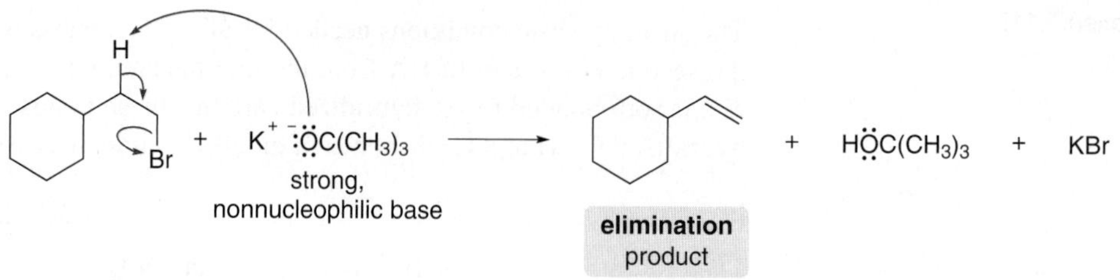

Most often, however, we will have to rely on other criteria to predict the outcome of these reactions. To determine the product of a reaction with an alkyl halide:

[1] Classify the alkyl halide as 1°, 2°, or 3°.

[2] Classify the base or nucleophile as strong, weak, or bulky.

Predicting the substitution and elimination products of a reaction can then be organized by the type of alkyl halide, as summarized in Table 8.5. The explanation that follows the table is organized with 2° alkyl halides last, because their reactions can follow any of the four mechanisms and product mixtures often result.

Table 8.5 Summary of Alkyl Halides and S_N1, S_N2, E1, and E2 Mechanisms

Alkyl halide type	Reaction with	Mechanism
1° RCH_2X	• Strong nucleophile	S_N2
	• Strong bulky base	E2
2° R_2CHX	• Strong base and nucleophile	S_N2 and E2
	• Strong bulky base	E2
	• Weak base and nucleophile	S_N1 and E1
3° R_3CX	• Weak base and nucleophile	S_N1 and E1
	• Strong base	E2

8.11A Tertiary Alkyl Halides

Tertiary alkyl halides react by all mechanisms *except* S_N2.

• **With strong bases, elimination occurs by an E2 mechanism.**

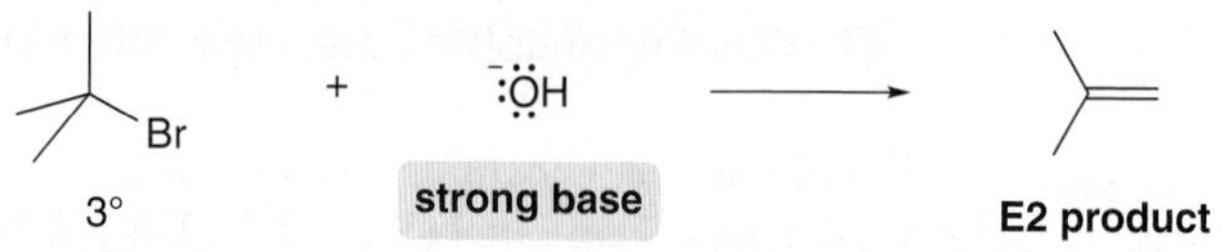

A strong base or nucleophile favors an S_N2 or E2 mechanism, but 3° halides are too sterically hindered to undergo an S_N2 reaction, so only E2 elimination occurs.

• **With weak nucleophiles or bases, a mixture of S_N1 and E1 products results.**

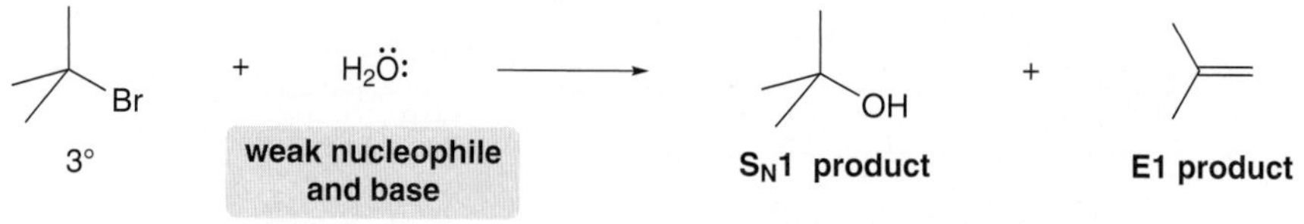

A weak base or nucleophile favors S_N1 and E1 mechanisms and both occur.

8.11B Primary Alkyl Halides

Primary alkyl halides react by S_N2 and E2 mechanisms.

- **With strong nucleophiles, substitution occurs by an S_N2 mechanism.**

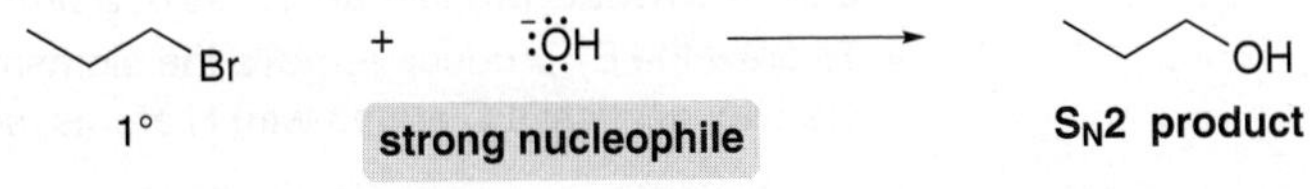

A strong base or nucleophile favors S_N2 or E2, but 1° halides are the least reactive halide type in elimination, so only S_N2 reaction occurs.

- **With strong, bulky bases, elimination occurs by an E2 mechanism.**

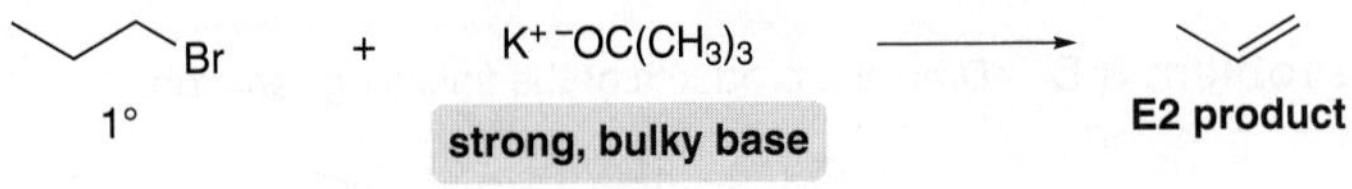

A strong, bulky base cannot act as a nucleophile, so elimination occurs and the mechanism is E2.

8.11C Secondary Alkyl Halides

Secondary alkyl halides react by *all* mechanisms.

- **With strong bases and nucleophiles, a mixture of S_N2 and E2 products results.**

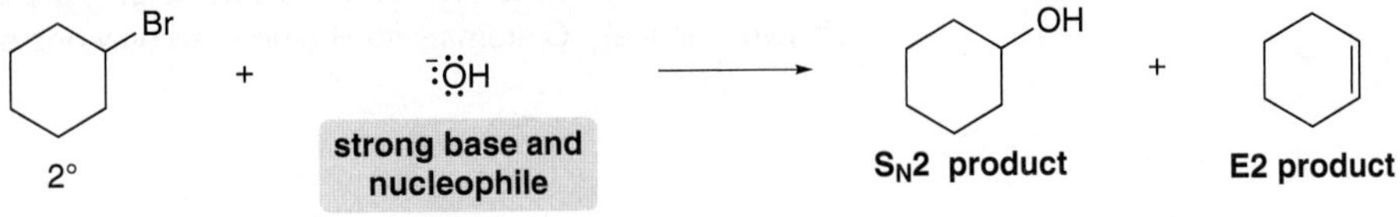

A strong base that is also a strong nucleophile gives a mixture of S_N2 and E2 products.

- **With strong, bulky bases, elimination occurs by an E2 mechanism.**

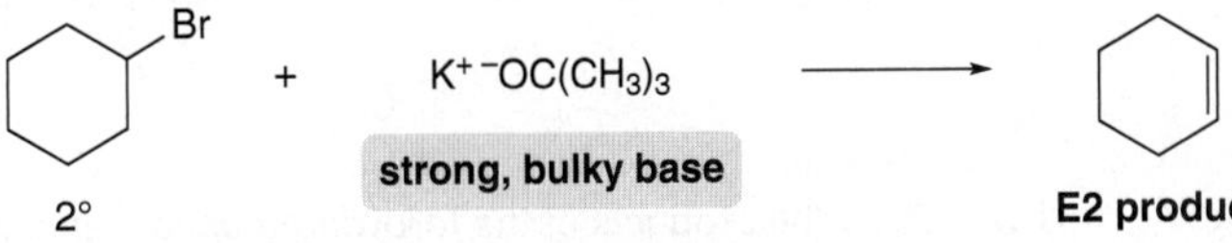

A strong, bulky base cannot act as a nucleophile, so elimination occurs and the mechanism is E2.

- **With weak nucleophiles or bases, a mixture of S_N1 and E1 products results.**

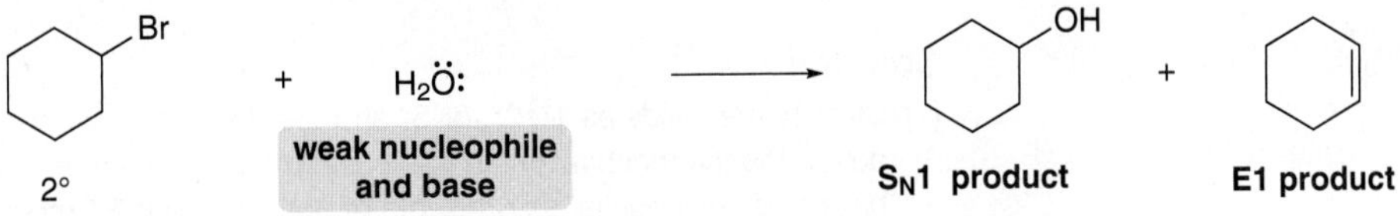

A weak base or nucleophile favors S_N1 and E1 mechanisms and both occur.

Sample Problems 8.4–8.6 illustrate how to apply the information in Table 8.5 to specific alkyl halides.

Sample Problem 8.4 Draw the products of the following reaction.

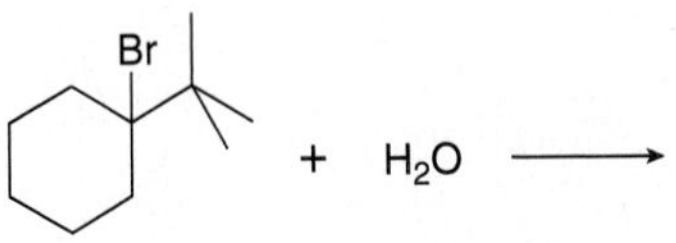

Solution

- Classify the halide as 1°, 2°, or 3° and the reagent as a strong or weak base (and nucleophile) to determine the mechanism. In this case, the alkyl halide is 3° and the reagent (H_2O) is a weak base and nucleophile, so products of both **S_N1** and **E1** mechanisms are formed.
- To draw the S_N1 product, substitute the nucleophile (H_2O) for the leaving group (Br^-), and draw the neutral product after loss of a proton.
- To draw the E1 product, remove the elements of H and Br from the α and β carbons. There are two identical β C atoms with H atoms, so only one elimination product is possible.

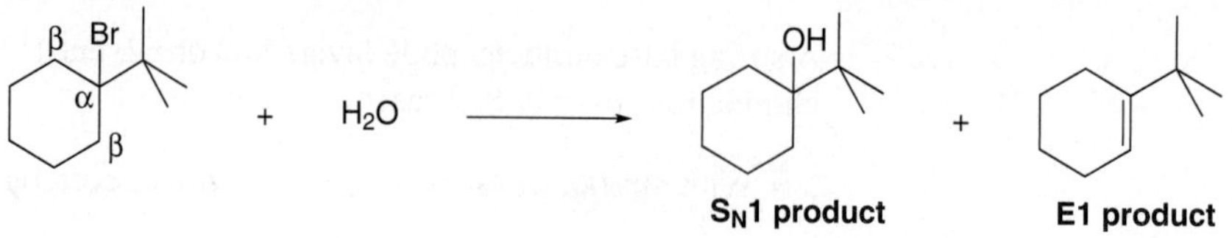

Sample Problem 8.5 Draw the products of the following reaction.

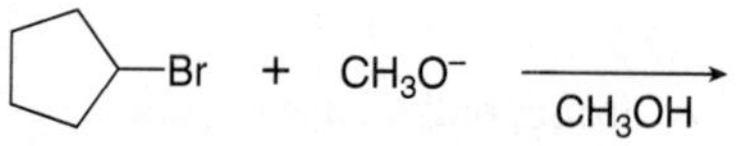

Solution

- Classify the halide as 1°, 2°, or 3° and the reagent as a strong or weak base (and nucleophile) to determine the mechanism. In this case, the alkyl halide is 2° and the reagent (CH_3O^-) is a strong base and nucleophile, so products of both **S_N2** and **E2** mechanisms are formed.
- To draw the S_N2 product, substitute the nucleophile (CH_3O^-) for the leaving group (Br^-).
- To draw the E2 product, remove the elements of H and Br from the α and β carbons. There are two identical β C atoms with H atoms, so only one elimination product is possible.

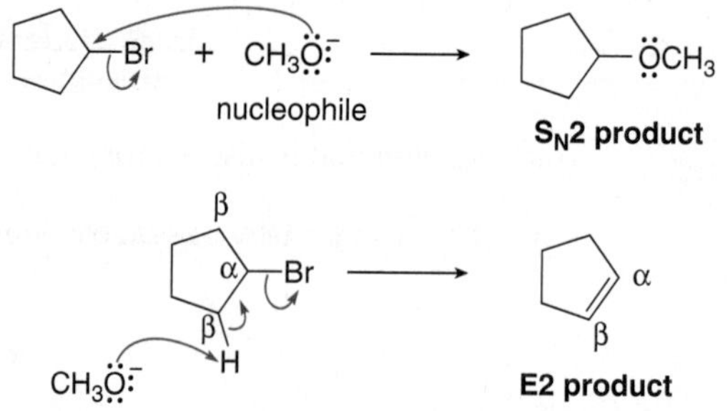

Sample Problem 8.6 Draw the products of the following reaction, and include the mechanism showing how each product is formed.

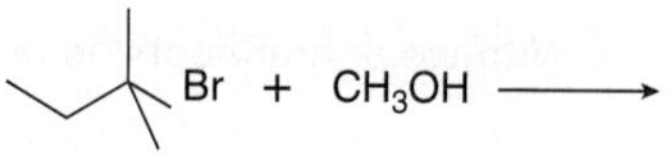

Solution

[1] Classify the halide as 1°, 2°, or 3° and the reagent as a strong or weak base (and nucleophile) to determine the mechanism. In this case, the alkyl halide is 3° and the reagent (CH_3OH) is a weak base and nucleophile, so products of both **S_N1** and **E1** mechanisms are formed.

[2] Draw the steps of the mechanisms to give the products. Both mechanisms begin with the same first step: loss of the leaving group to form a carbocation.

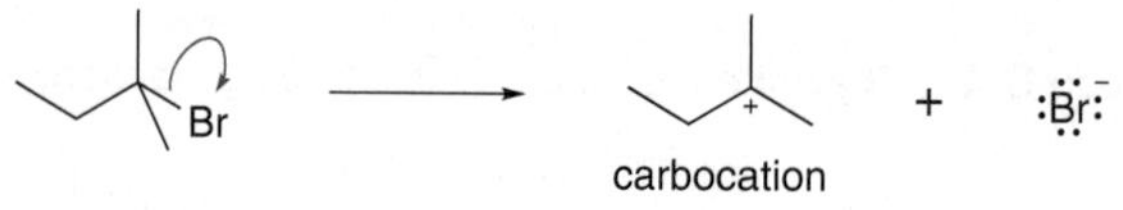

- **For S_N1:** The carbocation reacts with a nucleophile. Nucleophilic attack of CH_3OH on the carbocation generates a positively charged intermediate that loses a proton to afford the neutral S_N1 product.

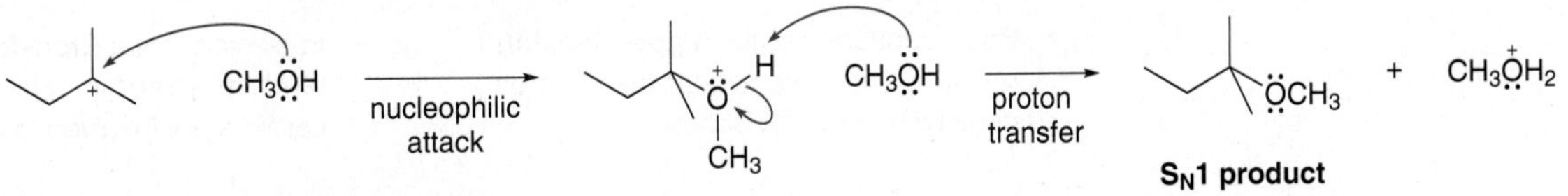

- **For E1:** The carbocation reacts with a base (CH_3OH or Br^-). Two different products of elimination can form because the carbocation has two different β carbons.

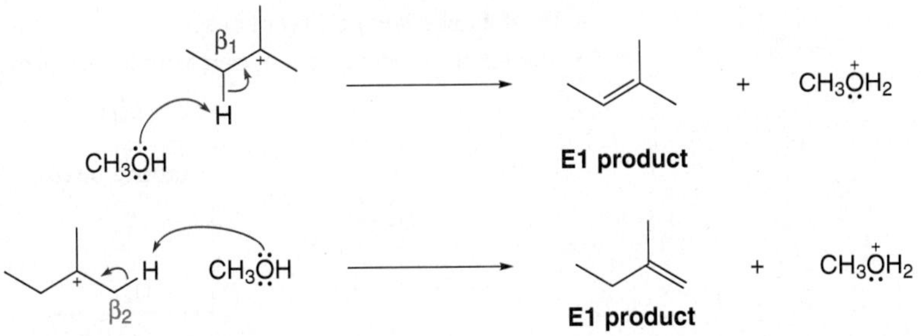

In this problem, three products are formed: one from an S_N1 reaction and two from E1 reactions.

Problem 8.22 Draw the products in each reaction.

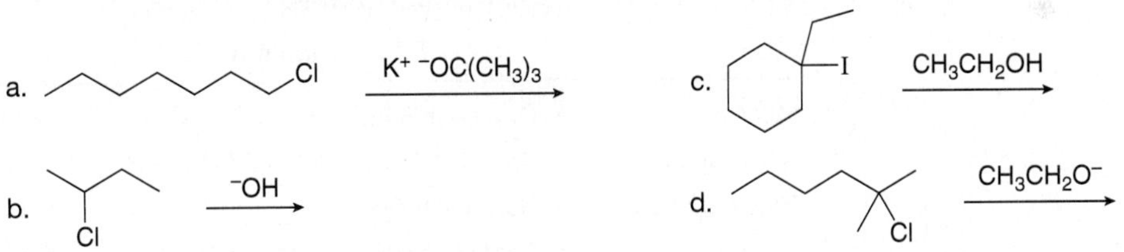

Problem 8.23 Draw a stepwise mechanism for the following reaction.

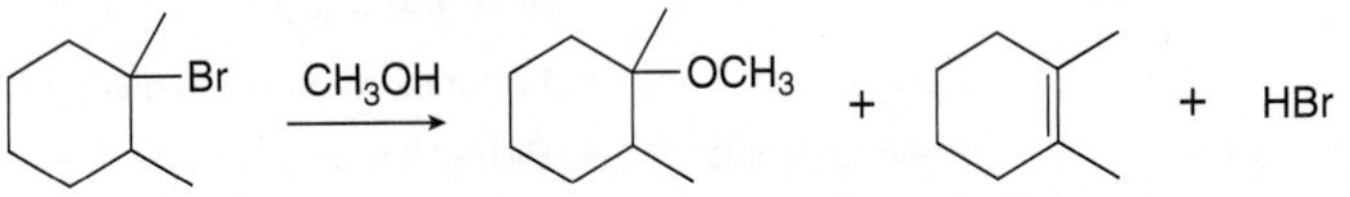

KEY CONCEPTS

Alkyl Halides and Elimination Reactions

A Comparison Between Nucleophilic Substitution and β Elimination

Nucleophilic substitution—A nucleophile attacks a carbon atom (7.6).

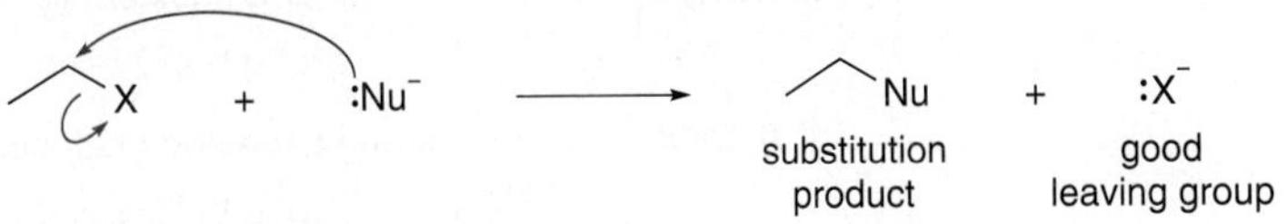

β Elimination—A base attacks a proton (8.1).

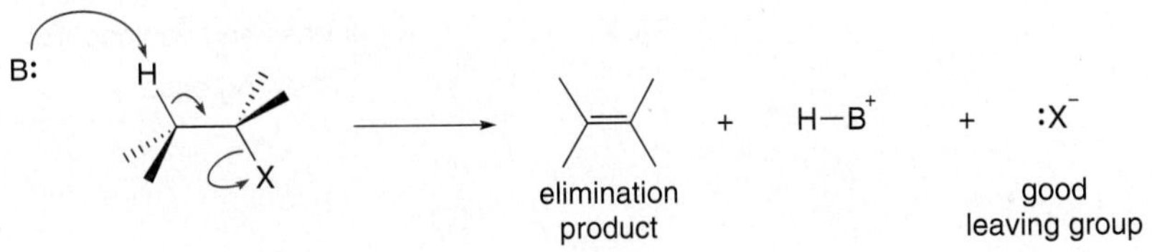

Similarities	Differences
• In both reactions RX acts as an electrophile, reacting with an electron-rich reagent.	• In substitution, a nucleophile attacks a single carbon atom.
• Both reactions require a **good leaving group X:⁻** that can accept the electron density in the C–X bond.	• In elimination, a Brønsted–Lowry base removes a proton to form a π bond, and two carbons are involved in the reaction.

The Importance of the Base in E2 and E1 Reactions (8.9)

The strength of the base determines the mechanism of elimination.

- Strong bases favor E2 reactions.
- Weak bases favor E1 reactions.
- The same organic product can form in two eliminations by different mechanisms.

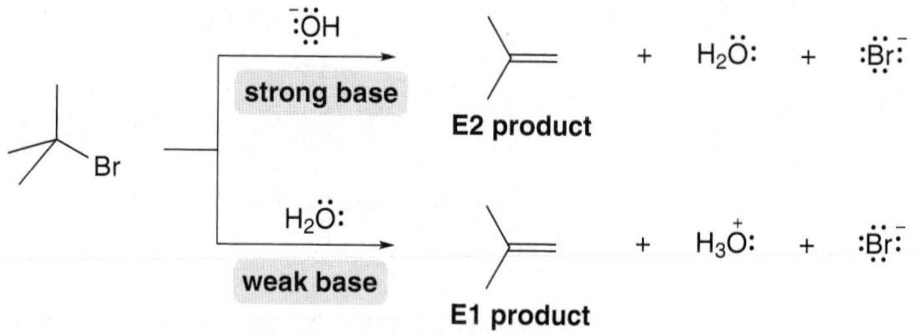

E1 and E2 Mechanisms Compared

	E2 mechanism	E1 mechanism
Mechanism	• One step (8.4B)	• Two steps (8.6B)
Alkyl halide	• Rate: $R_3CX > R_2CHX > RCH_2X$ (8.4C)	• Rate: $R_3CX > R_2CHX > RCH_2X$ (8.6C)
Rate equation	• Rate = k[RX][B:]	• Rate = k[RX]
	• Second-order kinetics (8.4A)	• First-order kinetics (8.6A)
Stereochemistry	• Anti periplanar arrangement of H and X (8.8)	• Trigonal planar carbocation intermediate (8.6B)
Base	• Favored by strong bases (8.4B)	• Favored by weak bases (8.6C)
Leaving group	• Better leaving group ⇢ faster reaction (8.4B)	• Better leaving group ⇢ faster reaction (Table 8.3)
Solvent	• Favored by polar aprotic solvents (8.4B)	• Favored by polar protic solvents (Table 8.3)
Product	• More substituted alkene favored (Zaitsev rule, 8.5)	• More substituted alkene favored (Zaitsev rule, 8.6C)

Summary Chart on the Four Mechanisms: S_N1, S_N2, E1, or E2 (8.11)

Alkyl halide type	Conditions		Mechanism
1° RCH_2X	strong nucleophile	⇢	**S_N2**
	strong bulky base	⇢	**E2**
2° R_2CHX	strong base and nucleophile	⇢	**S_N2 + E2**
	strong bulky base	⇢	**E2**
	weak base and nucleophile	⇢	**S_N1 + E1**
3° R_3CX	weak base and nucleophile	⇢	**S_N1 + E1**
	strong base	⇢	**E2**

PROBLEMS

Problems Using Three-Dimensional Models

8.24 Rank the alkenes shown in the ball-and-stick models (**A–C**) in order of increasing stability.

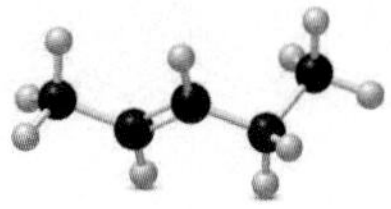

A

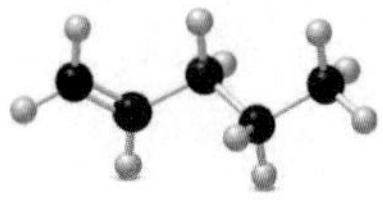

B

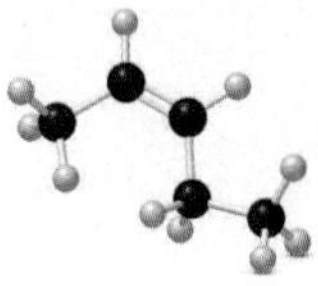

C

8.25 Name each compound and decide which stereoisomer will react faster in an E2 elimination reaction. Explain your choice.

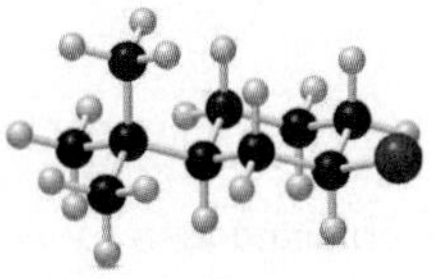

D

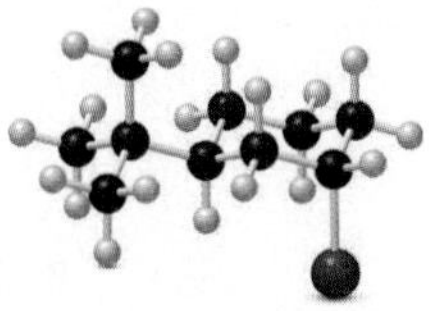

E

8.26 What is the major E2 elimination product formed from each alkyl halide?

a. 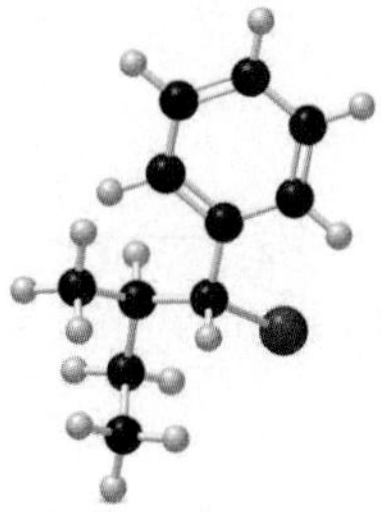

b. 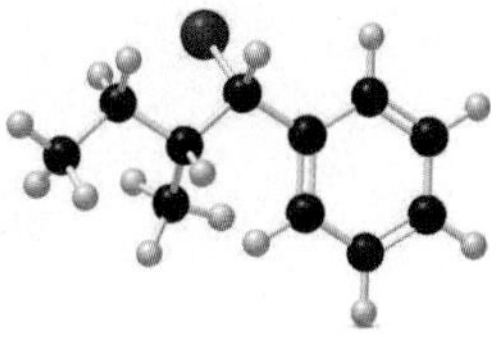

General Elimination

8.27 Draw all possible constitutional isomers formed by dehydrohalogenation of each alkyl halide.

a. 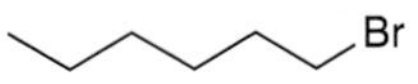

b. 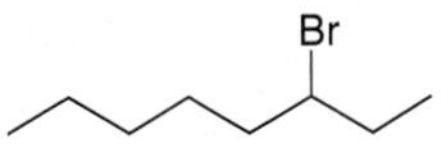

c. 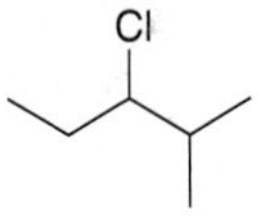

d. 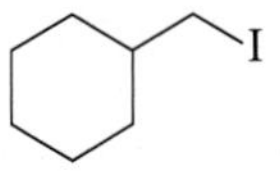

8.28 What alkyl halide forms each of the following alkenes as the *only* product in an elimination reaction?

a. 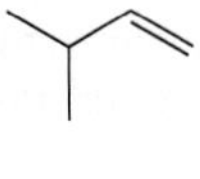

b. 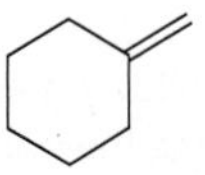

c. 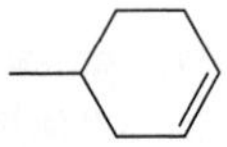

d. 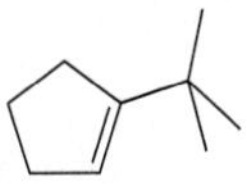

Alkenes

8.29 Which double bonds in the following natural products can exhibit stereoisomerism? Nerolidol is isolated from the angel's trumpet plant, caryophyllene is present in hemp, and humulene comes from hops.

a. 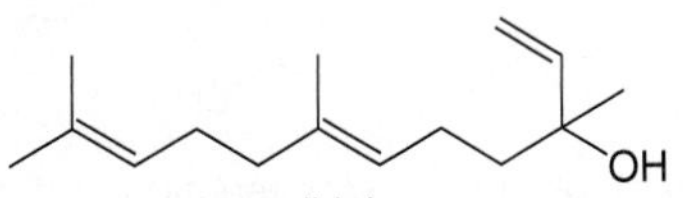

nerolidol

b. 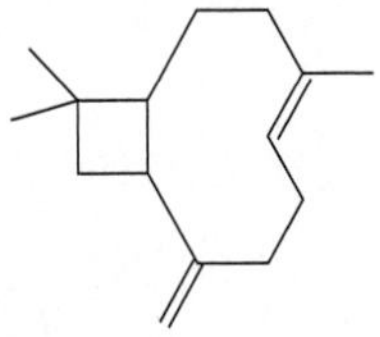

caryophyllene

c. 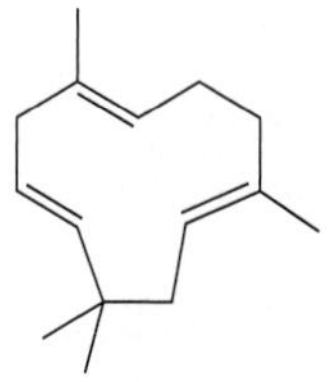

humulene

8.30 Label each pair of alkenes as constitutional isomers, stereoisomers, or identical.

a. 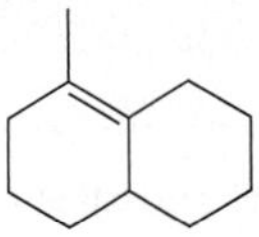 and 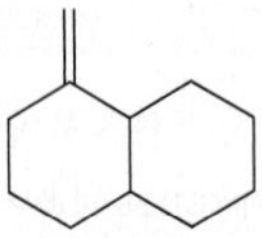

b. 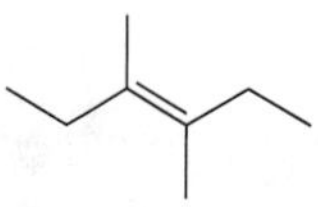 and 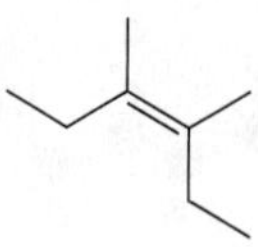

c. 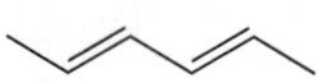 and 

d. 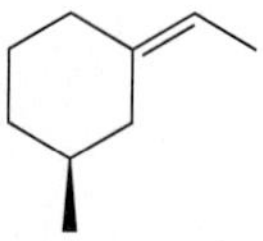 and 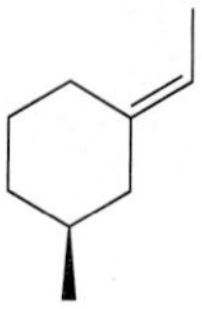

8.31 Rank the following alkenes in order of increasing stability:

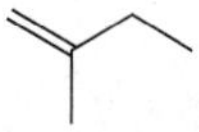

A

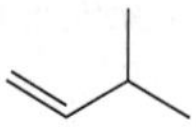

B

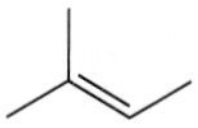

C

8.32 $\Delta H°$ values obtained for a series of similar reactions are one set of experimental data used to determine the relative stability of alkenes. Explain how the following data suggest that *cis*-but-2-ene is more stable than but-1-ene (Section 12.3A).

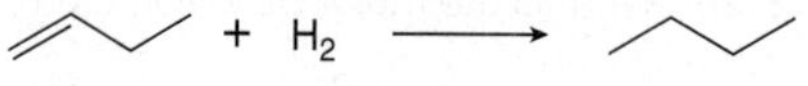

but-1-ene

$\Delta H° = -127$ kJ/mol

cis-but-2-ene

$\Delta H° = -120$ kJ/mol

E2 Reaction

8.33 Draw all constitutional isomers formed in each E2 reaction and predict the major product using the Zaitsev rule.

a. 

b. 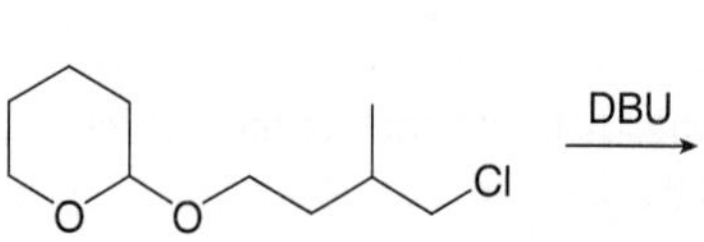

c. 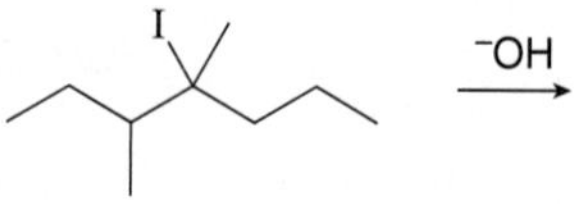

d. 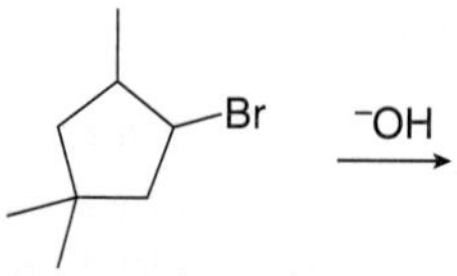

8.34 For each of the following alkenes, draw the structure of two different alkyl halides that yield the given alkene as the only product of dehydrohalogenation.

a. 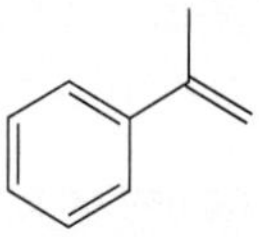

b. 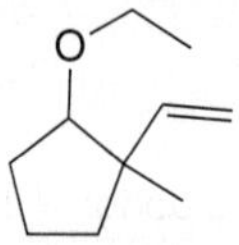

c. 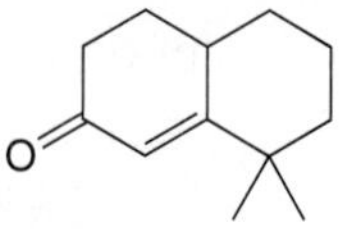

8.35 Consider the following E2 reaction.

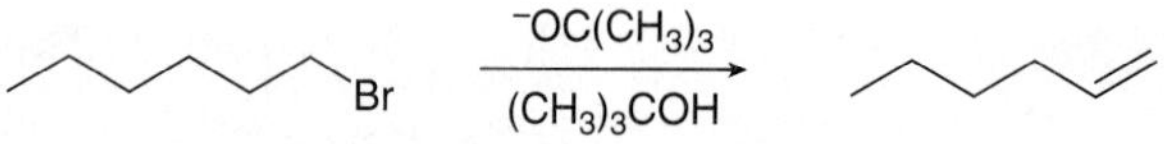

a. Draw the by-products of the reaction and use curved arrows to show the movement of electrons.

b. What happens to the reaction rate with each of the following changes? [1] The solvent is changed to DMF. [2] The concentration of $^{-}OC(CH_3)_3$ is decreased. [3] The base is changed to ^{-}OH. [4] The halide is changed to $CH_3CH_2CH_2CH_2CH(Br)CH_3$. [5] The leaving group is changed to I^{-}.

8.36 Dehydrohalogenation of 1-chloro-1-methylcyclopropane affords two alkenes (**A** and **B**) as products. Explain why **A** is the major product despite the fact that it contains the less substituted double bond.

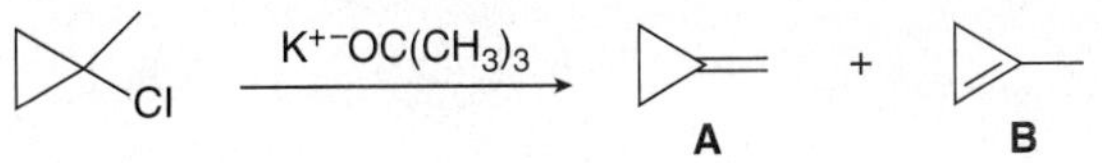

1-chloro-1-methylcyclopropane

8.37 What is the major stereoisomer formed when each alkyl halide is treated with $KOC(CH_3)_3$?

a. 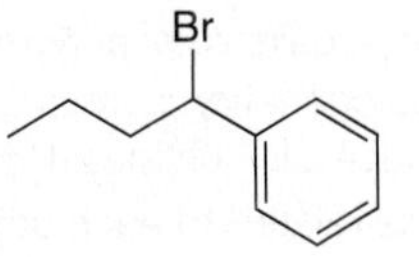

b. 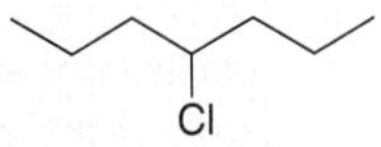

E1 Reaction

8.38 What alkene is the major product formed from each alkyl halide in an E1 reaction?

a. 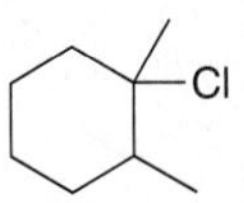

b. 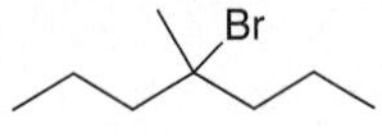

c. 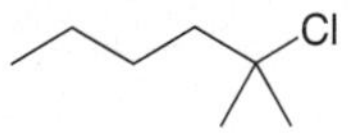

E1 and E2

8.39 Draw all constitutional isomers formed in each elimination reaction. Label the mechanism as E2 or E1.

a. 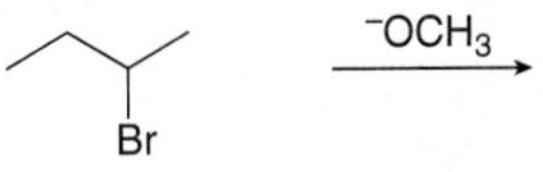

b. 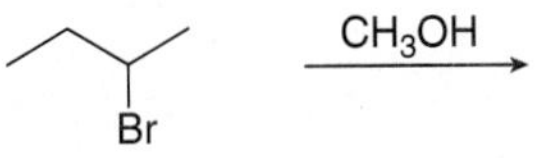

c. 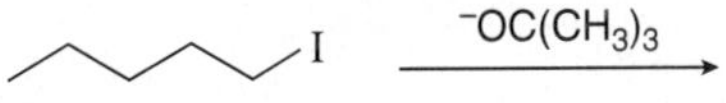

d. 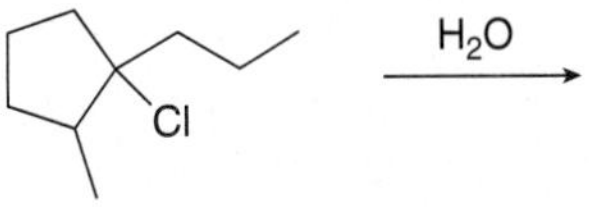

e. 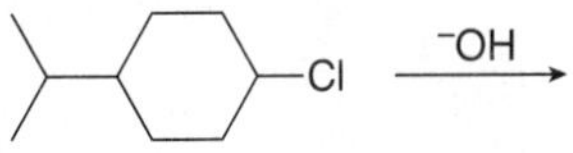

f. 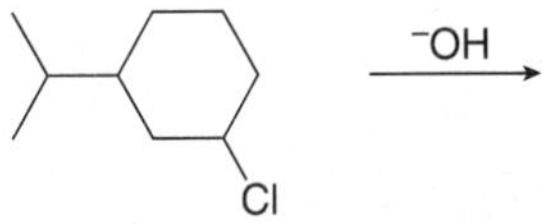

8.40 Rank the alkyl halides in order of increasing E2 reactivity. Then do the same for E1 reactivity.

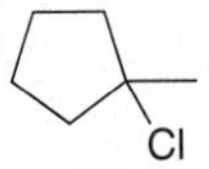

A

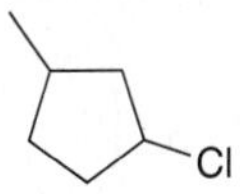

B

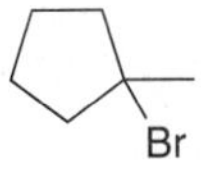

C

8.41 Pick the reactant or solvent in each part that gives the faster elimination reaction.

a. reaction of ^-OH with 1-chloro-1-methylcyclohexane or 1-chloro-3-methylcyclohexane
b. reaction of H_2O with $CH_3CH(Cl)CH_2CH_3$ or $(CH_3)_2C(Cl)CH_2CH_3$
c. reaction of $(CH_3)_3CCl$ with ^-OH in H_2O or DMSO

8.42 In the dehydrohalogenation of bromocyclodecane, the major product is *cis*-cyclodecene rather than *trans*-cyclodecene. Offer an explanation.

8.43 Elimination of HBr from 2-bromobutane affords a mixture of but-1-ene and but-2-ene. With sodium ethoxide as base, but-2-ene constitutes 81% of the alkene products, but with potassium *tert*-butoxide, but-2-ene constitutes only 67% of the alkene products. Offer an explanation for this difference.

Stereochemistry and the E2 Reaction

8.44 What is the major E2 elimination product formed from each halide?

a. 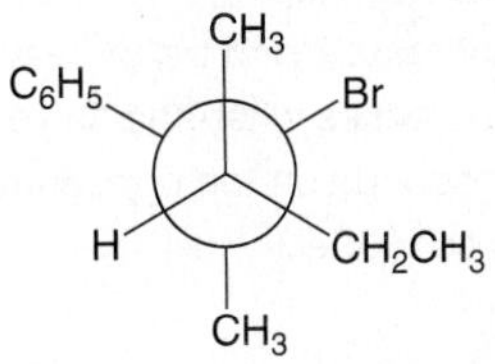

b. 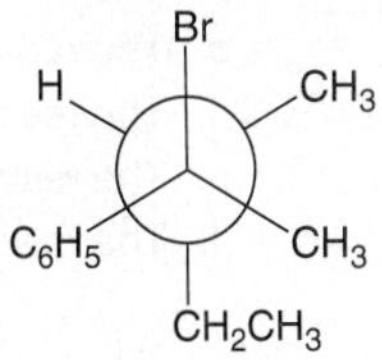

c. 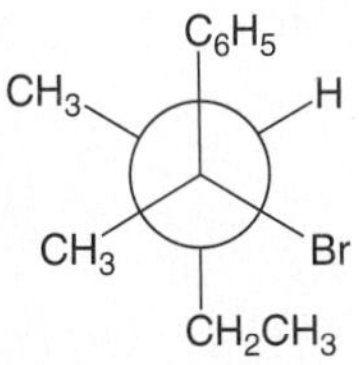

8.45 Taking into account anti periplanar geometry, predict the major E2 product formed from each starting material.

a. 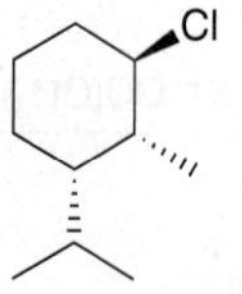 b. 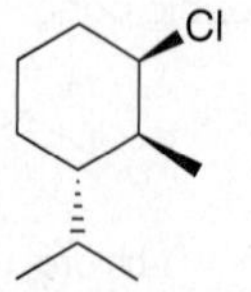

8.46 Does *cis*- or *trans*-1-bromo-4-*tert*-butylcylohexane react faster in an E2 reaction?

8.47 a. Draw three-dimensional representations for all stereoisomers of 2-chloro-3-methylpentane, and label pairs of enantiomers.

b. Considering dehydrohalogenation across C2 and C3 only, draw the E2 product that results from each of these alkyl halides. How many different products have you drawn?

c. How are these products related to each other?

8.48 Explain why compound **A** does not undergo an E2 elimination with strong base.

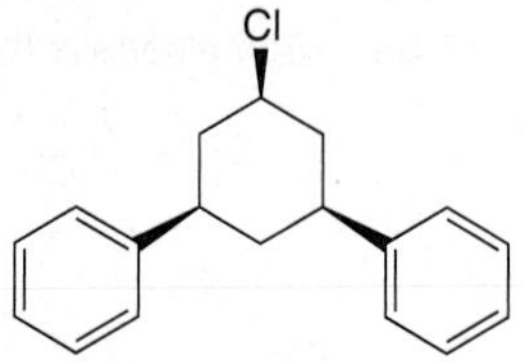

A

8.49 What alkyl chloride affords the following alkene exclusively under E2 reaction conditions?

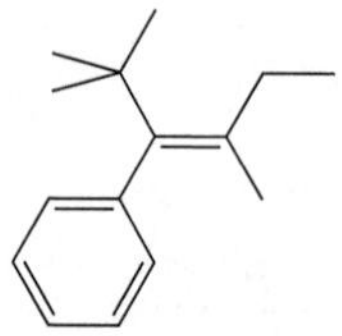

Alkynes

8.50 Draw the products formed when each dihalide is treated with excess $NaNH_2$.

a. 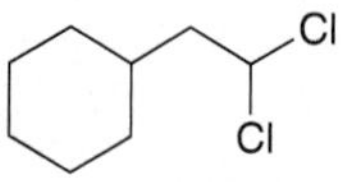 b. 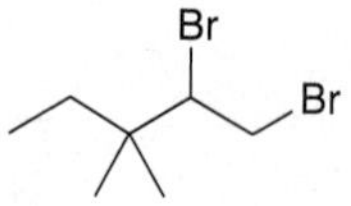 c. 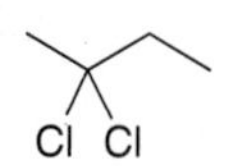 d. 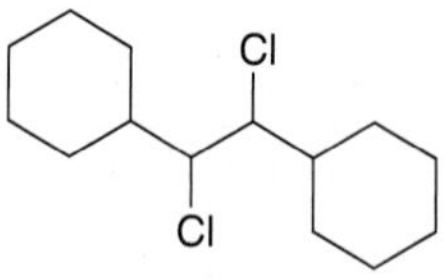

8.51 Draw the structure of a dihalide that could be used to prepare each alkyne. There may be more than one possible dihalide.

a. b. 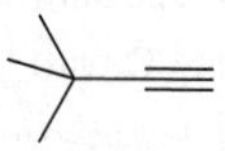 c. 

8.52 Under certain reaction conditions, 2,3-dibromobutane reacts with two equivalents of base to give three products, each of which contains two new π bonds. Product **A** has two *sp* hybridized carbon atoms, product **B** has one *sp* hybridized carbon atom, and product **C** has none. What are the structures of **A, B,** and **C?**

S_N1, S_N2, E1, and E2 Mechanisms

8.53 For which reaction mechanisms—S_N1, S_N2, E1, or E2—are each of the following statements true? A statement may be true for one or more mechanisms.

a. The mechanism involves carbocation intermediates.

b. The mechanism has two steps.

c. The reaction rate increases with better leaving groups.

d. The reaction rate increases when the solvent is changed from CH_3OH to $(CH_3)_2SO$.

e. The reaction rate depends on the concentration of the alkyl halide only.

f. The mechanism is concerted.

g. The reaction of CH_3CH_2Br with NaOH occurs by this mechanism.
h. Racemization at a stereogenic center occurs.
i. Tertiary (3°) alkyl halides react faster than 2° or 1° alkyl halides.
j. The reaction follows a second-order rate equation.

8.54 Draw the organic products formed in each reaction.

a. 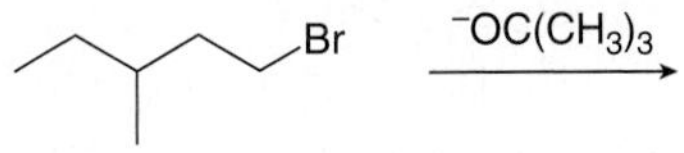

d. 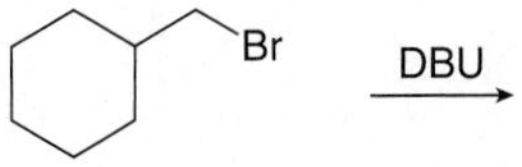

g. 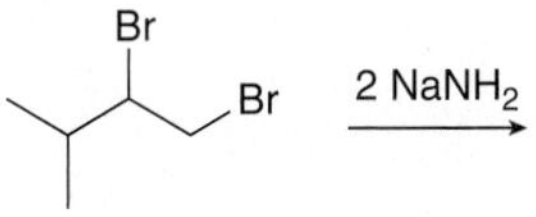

b. 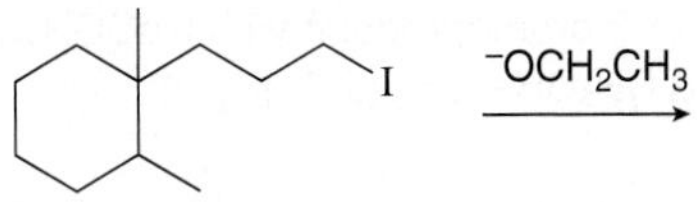

e. 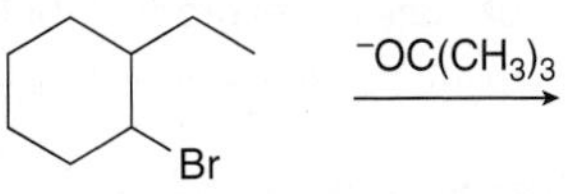

h. 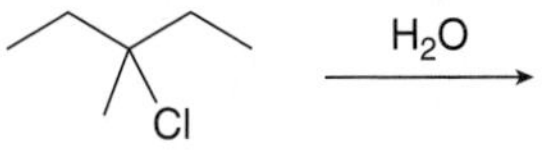

c. 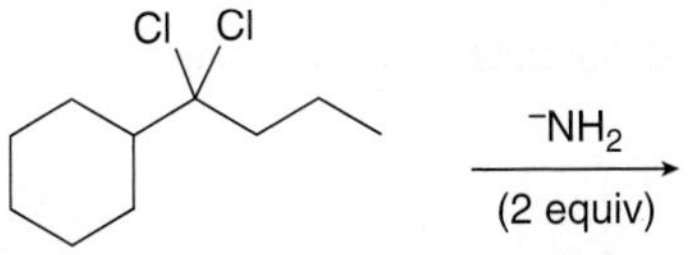

f. 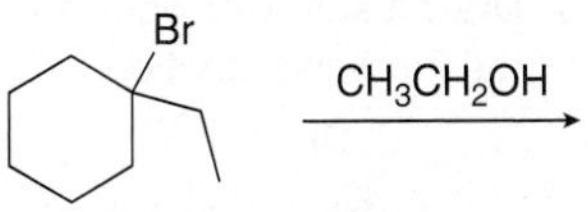

8.55 What is the major product formed when each alkyl halide is treated with each of the following reagents: [1] $NaOCOCH_3$; [2] $NaOCH_3$; [3] $KOC(CH_3)_3$? If it is not possible to predict the major product, identify the products in the mixture and the mechanism by which each is formed.

a. 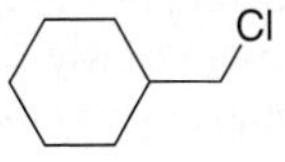

b. 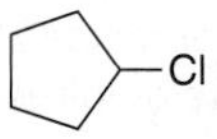

c. 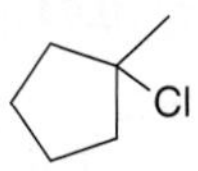

8.56 Draw all products, including stereoisomers, in each reaction.

a. 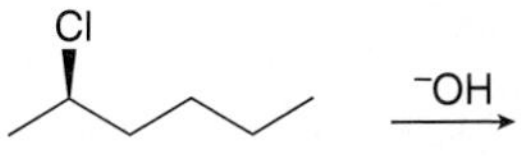

c. 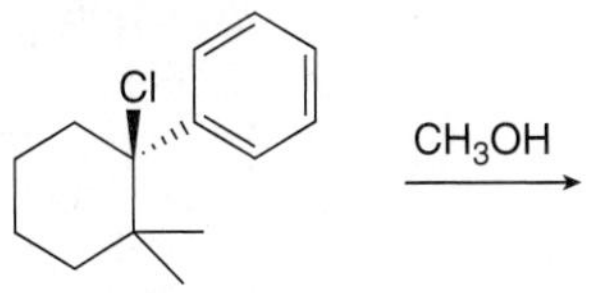

b. 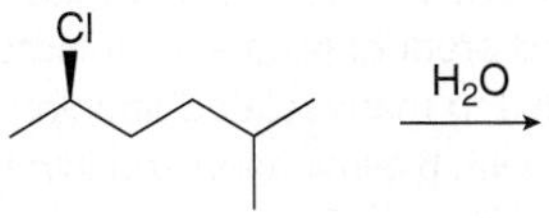

d. 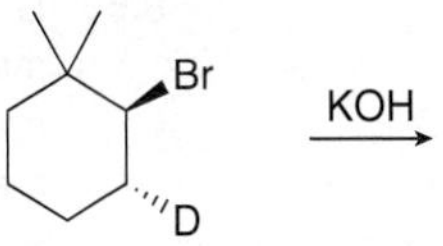

8.57 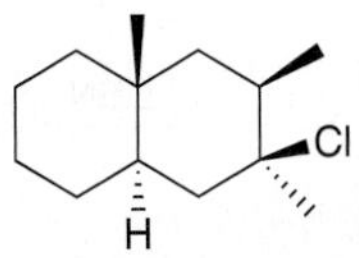

Draw all of the substitution and elimination products formed from the given alkyl halide with each reagent: (a) CH_3OH; (b) KOH. Indicate the stereochemistry around the stereogenic centers present in the products, as well as the mechanism by which each product is formed.

8.58 The following reactions do not afford the major product that is given. Explain why this is so, and draw the structure of the major product actually formed.

a. 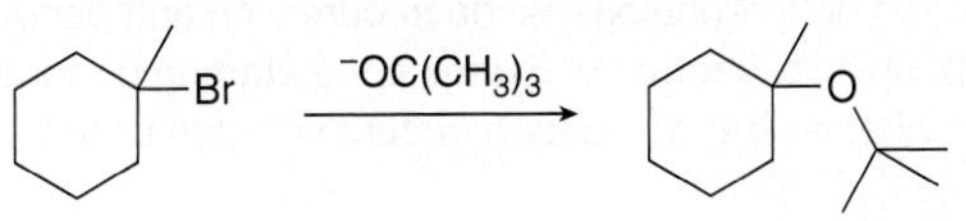

c. 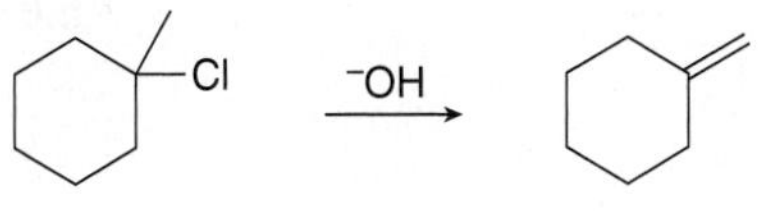

b. 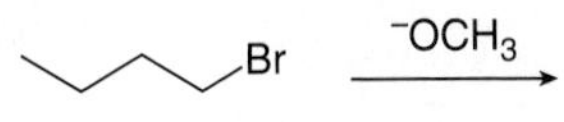

d. 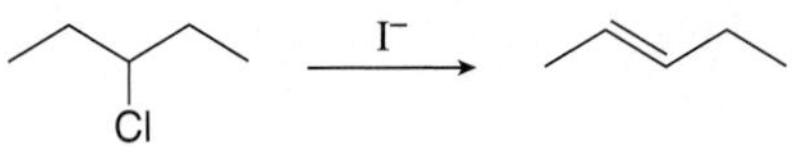

8.59 Draw a stepwise, detailed mechanism for each reaction.

a. 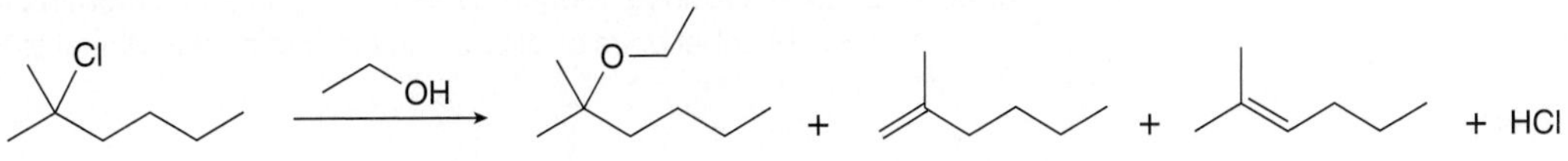

b. 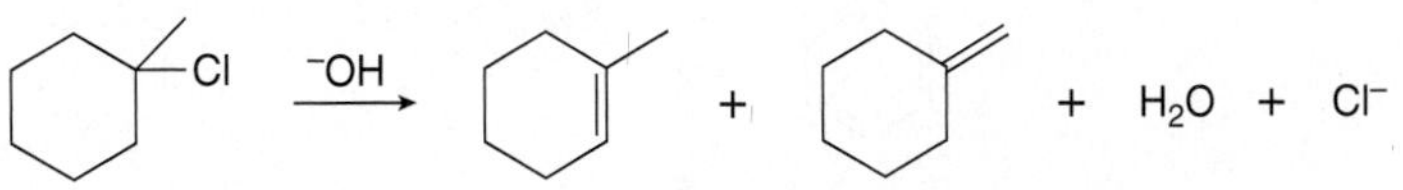

8.60 Draw the major product formed when (*R*)-1-chloro-3-methylpentane is treated with each reagent: (a) $NaOCH_2CH_3$; (b) KCN; (c) DBU.

8.61 Draw a stepwise, detailed mechanism for the following reaction.

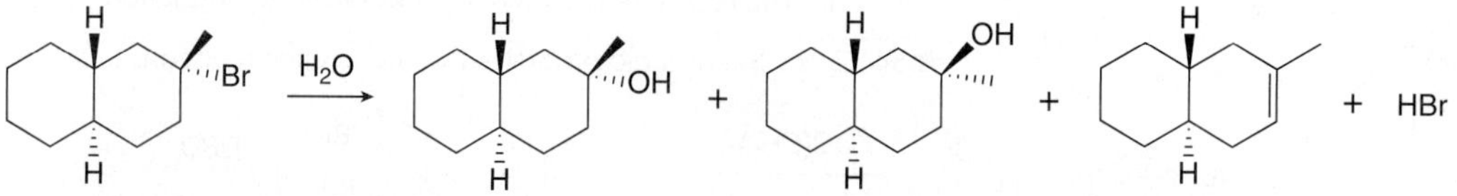

8.62 Explain why the reaction of 2-bromopropane with $NaOCOCH_3$ gives $(CH_3)_2CHOCOCH_3$ exclusively as product, but the reaction of 2-bromopropane with $NaOCH_2CH_3$ gives a mixture of $(CH_3)_2CHOCH_2CH_3$ (20%) and $CH_3CH{=}CH_2$ (80%).

Challenge Problems

8.63 Draw a stepwise detailed mechanism that illustrates how four organic products are formed in the following reaction.

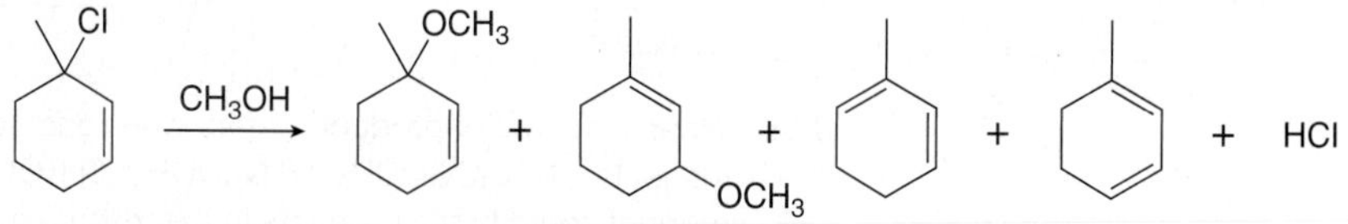

8.64 Although there are nine stereoisomers of 1,2,3,4,5,6-hexachlorocyclohexane, one stereoisomer reacts 7000 times more slowly than any of the others in an E2 elimination. Draw the structure of this isomer and explain why this is so.

8.65 Explain the selectivity observed in the following reactions.

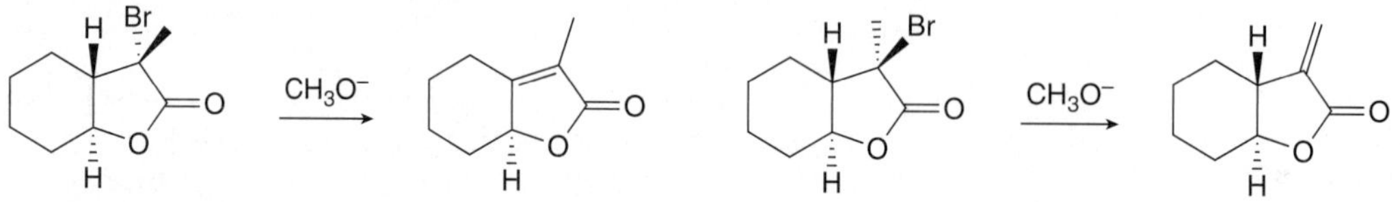

8.66 Draw a stepwise mechanism for the following reaction. The four-membered ring in the starting material and product is called a β-lactam. This functional group confers biological activity on penicillin and many related antibiotics, as is discussed in Chapter 22. (Hint: The mechanism begins with β elimination and involves only two steps.)

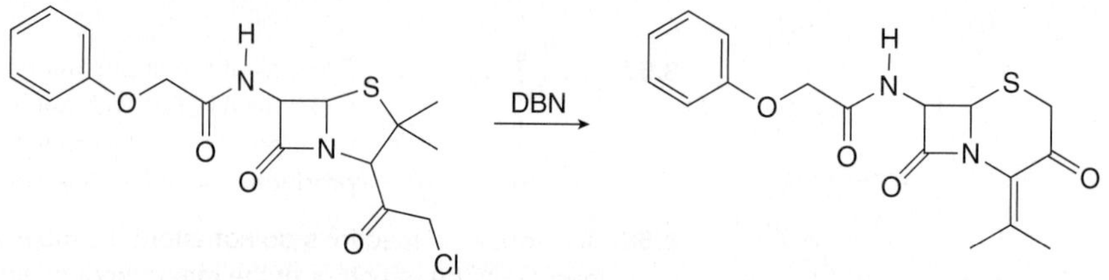

8.67 Although dehydrohalogenation occurs with anti periplanar geometry, some eliminations have syn periplanar geometry. Examine the starting material and product of each elimination, and state whether the elimination occurs with syn or anti periplanar geometry.

a. 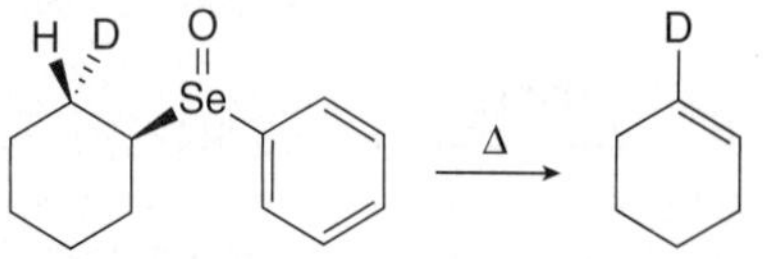

b. 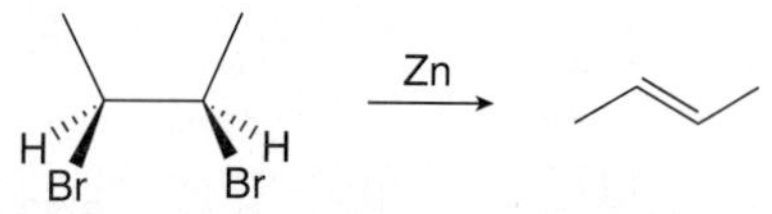

8.68 (a) Draw all products formed by treatment of each dibromide (**A** and **B**) with one equivalent of $NaNH_2$. (b) Label pairs of diastereomers and constitutional isomers.

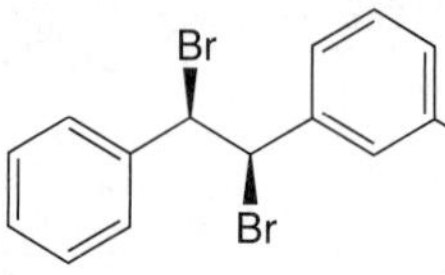

A

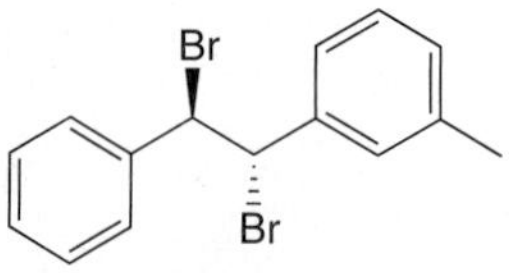

B

Alcohols, Ethers, and Related Compounds

9

9.1 Introduction
9.2 Structure and bonding
9.3 Nomenclature
9.4 Physical properties
9.5 Interesting alcohols, ethers, and epoxides
9.6 Preparation of alcohols, ethers, and epoxides
9.7 General features—Reactions of alcohols, ethers, and epoxides
9.8 Dehydration of alcohols to alkenes
9.9 Carbocation rearrangements
9.10 Dehydration using $POCl_3$ and pyridine
9.11 Conversion of alcohols to alkyl halides with HX
9.12 Conversion of alcohols to alkyl halides with $SOCl_2$ and PBr_3
9.13 Tosylate—Another good leaving group
9.14 Reaction of ethers with strong acid
9.15 Thiols and sulfides
9.16 Reactions of epoxides
9.17 Application: Epoxides, leukotrienes, and asthma
9.18 Application: Benzo[*a*]-pyrene, epoxides, and cancer

Linalool is a 10-carbon alcohol found in a wide variety of scented herbs, laurels, and citrus fruits. The *R* enantiomer is one of the two main components of lavender oil, while the *S* enantiomer is found in coriander and sweet orange flowers. Because of its pleasant odor, linalool is used commercially in scented soaps and lotions, and as an insecticide for controlling fleas and cockroaches. In Chapter 9, we learn about alcohols like linalool, as well as related oxygen- and sulfur-containing functional groups.

In Chapter 9, we take the principles learned in Chapters 7 and 8 about leaving groups, nucleophiles, and bases, and apply them to **alcohols, ethers,** and **epoxides,** three new functional groups that contain polar C–O bonds. In the process, you will discover that all of the reactions in Chapter 9 follow one of the four mechanisms introduced in Chapters 7 and 8—S_N1, S_N2, E1, or E2—so there are **no new general mechanisms to learn.**

Later in the chapter, we will also examine **thiols (RSH)** and **sulfides (R_2S),** sulfur analogues of alcohols and ethers, respectively. These functional groups play a key role in the chemistry of biomolecules, especially the proteins discussed in Chapter 26.

9.1 Introduction

Alcohols, ethers, and **epoxides** are three functional groups that contain carbon–oxygen σ bonds.

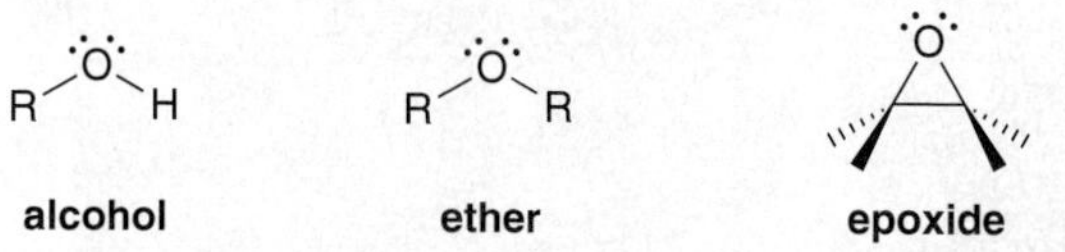
alcohol
ether
epoxide

***Alcohols* contain a hydroxy group (OH** group) bonded to an sp^3 hybridized carbon atom. As we learned in Section 3.2, alcohols are classified as **primary (1°), secondary (2°),** or **tertiary (3°)** based on the number of carbon atoms bonded to the carbon with the OH group.

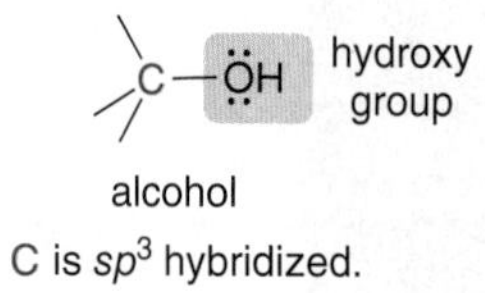
hydroxy group
alcohol
C is sp^3 hybridized.

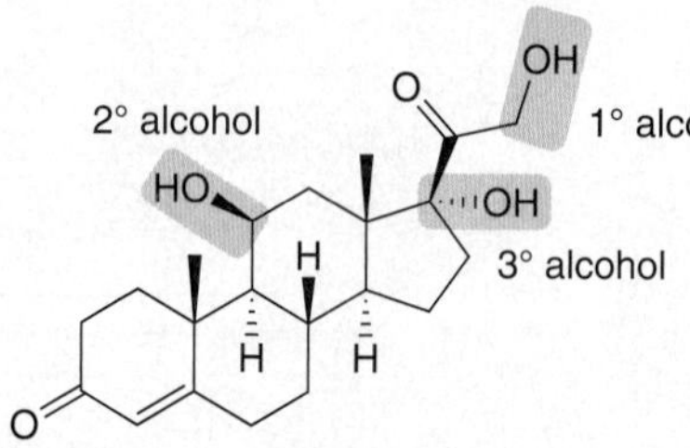
2° alcohol
1° alcohol
3° alcohol

cortisol
anti-inflammatory steroid

Compounds having a hydroxy group on an sp^2 hybridized carbon atom—**enols** and **phenols**—undergo different reactions than alcohols and are discussed in Chapters 11 and 19, respectively. **Enols** have an OH group on a carbon of a C–C double bond. **Phenols** have an OH group on a benzene ring.

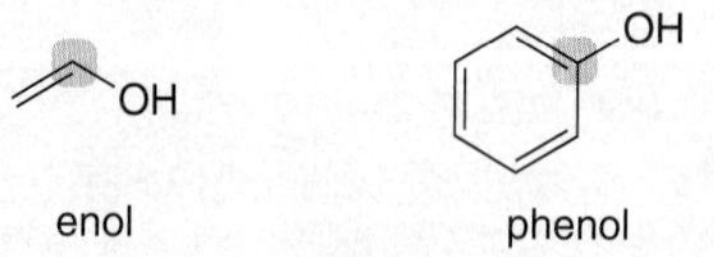
enol
phenol

• C bonded to OH is sp^2 hybridized.

***Ethers* have two alkyl groups bonded to an oxygen atom.** An ether is **symmetrical** if the two alkyl groups are the same, and **unsymmetrical** if they are different. Both alcohols and ethers are organic derivatives of H_2O, formed by replacing one or both of the hydrogens on the oxygen atom by R groups, respectively.

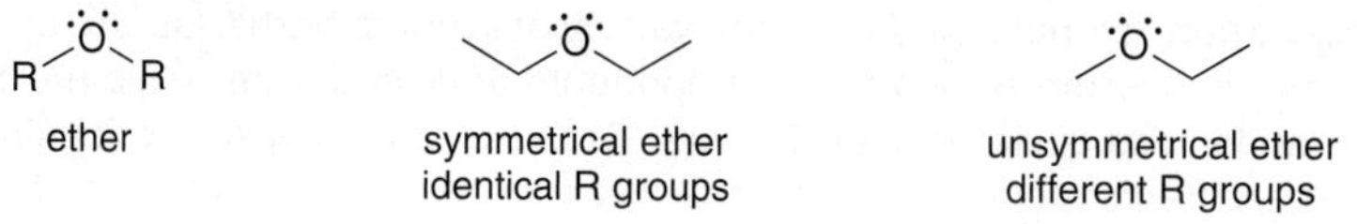
ether
symmetrical ether
identical R groups
unsymmetrical ether
different R groups

Epoxides **are ethers having the oxygen atom in a three-membered ring.** Epoxides are also called **oxiranes.**

epoxide or oxirane

Problem 9.1 Label each ether and alcohol in brevenal, a marine natural product. Classify each alcohol as 1°, 2°, or 3°.

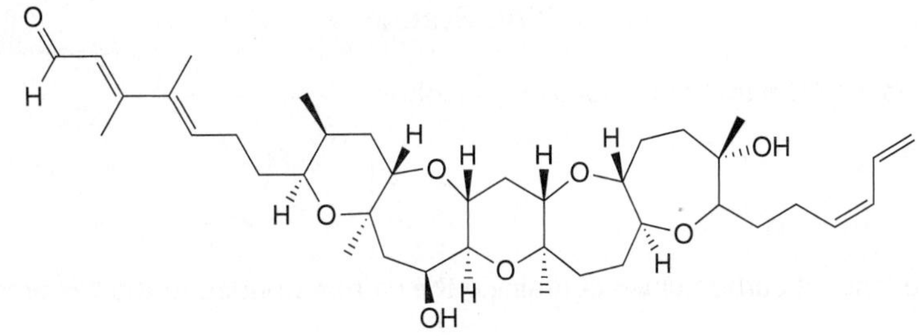

brevenal

Brevenal is a nontoxic marine polyether produced by *Karenia brevis,* a single-celled organism that proliferates during red tides, vast algal blooms that turn the ocean water red, brown, or green.

9.2 Structure and Bonding

Alcohols, ethers, and epoxides each contain an oxygen atom surrounded by two atoms and two nonbonded electron pairs, making the O atom **tetrahedral** and sp^3 hybridized. Because only two of the four groups around O are atoms, alcohols and ethers have a **bent** shape like H_2O.

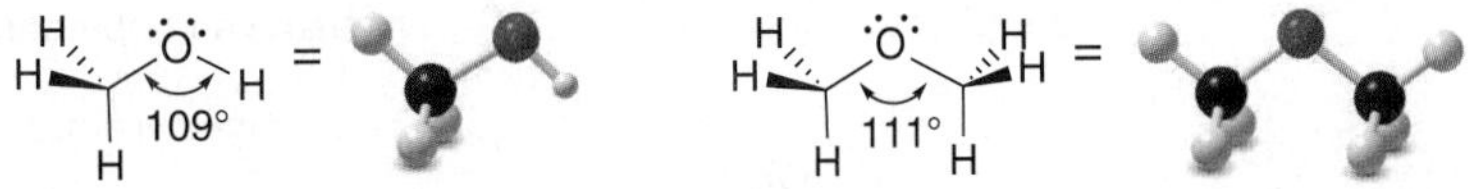

The bond angle around the O atom in an alcohol or ether is similar to the tetrahedral bond angle of **109.5°.** In contrast, the C–O–C bond angle of an epoxide must be **60°,** a considerable deviation from the tetrahedral bond angle. For this reason, **epoxides have angle strain,** making them much more reactive than other ethers.

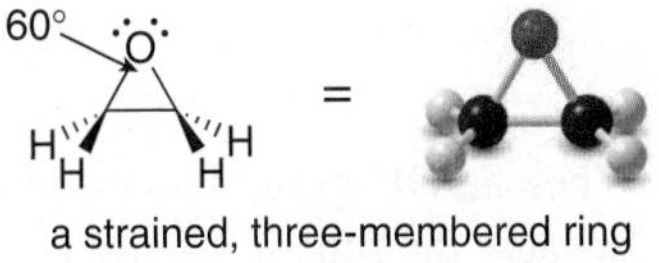

a strained, three-membered ring

Because oxygen is much more electronegative than carbon or hydrogen, the C–O and O–H bonds are all polar, with the O atom electron rich and the C and H atoms electron poor. The electrostatic potential maps in Figure 9.1 show these polar bonds for all three functional groups.

Figure 9.1 Electrostatic potential maps for a simple alcohol, ether, and epoxide

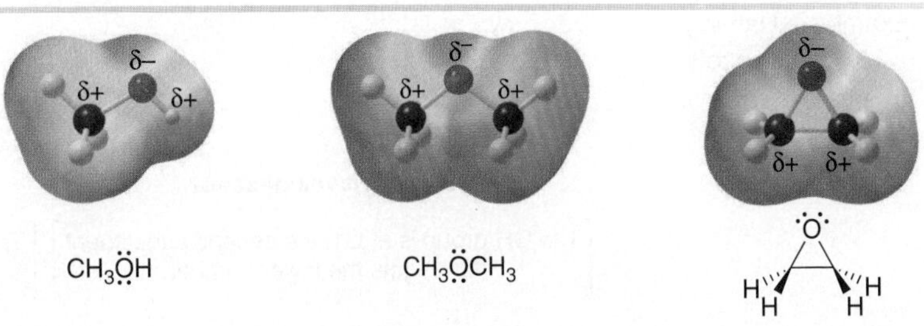

- Electron-rich regions are shown by the red around the O atoms.

9.3 Nomenclature

To name an alcohol, ether, or epoxide using the IUPAC system, we must learn how to name the functional group either as a substituent or by using a suffix added to the parent name.

9.3A Naming Alcohols

- **In the IUPAC system, alcohols are identified by the suffix *-ol.***

How To Name an Alcohol Using the IUPAC System

Example Give the IUPAC name of the following alcohol:

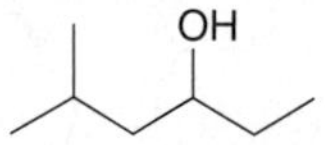

Step [1] Find the longest carbon chain containing the carbon bonded to the OH group.

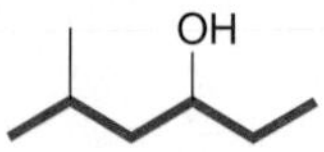

6 C's in the longest chain

6 C's → hexane → hexano*l*

- Change the **-*e*** ending of the parent alkane to the suffix ***-ol.***

Step [2] Number the carbon chain to give the OH group the lower number, and apply all other rules of nomenclature.

a. **Number** the chain.

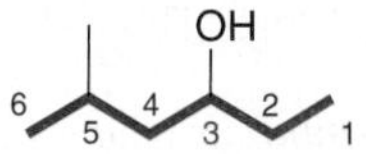

- Number the chain to put the OH group at C3, not C4.

hexan-3-ol

b. **Name** and **number** the substituents.

methyl at C5

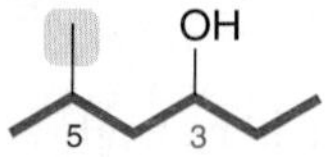

Answer: 5-methylhexan-3-ol

$CH_3CH_2CH_2CH_2OH$ is named as 1-butanol using the 1979 IUPAC recommendations and butan-1-ol using the 1993 IUPAC recommendations.

When an OH group is bonded to a ring, the **ring is numbered beginning with the OH group.** Because the functional group is always at C1, the "1" is usually omitted from the name. The ring is then numbered in a clockwise or counterclockwise fashion to give the next substituent the lower number. Representative examples are given in Figure 9.2.

Figure 9.2
Examples: Naming cyclic alcohols

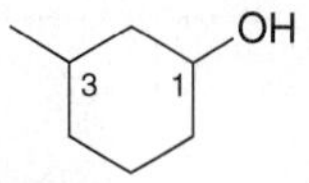

3-methylcyclohexanol

[The OH group is at C1; the second substituent (CH_3) gets the lower number.]

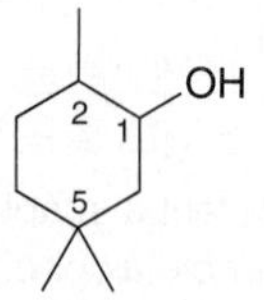

2,5,5-trimethylcyclohexanol

[The OH group is at C1; the second substituent (CH_3) gets the lower number.]

Common names are often used for simple alcohols. To assign a common name:

- Name all the carbon atoms of the molecule as a single **alkyl group.**
- Add the word **alcohol,** separating the words with a space.

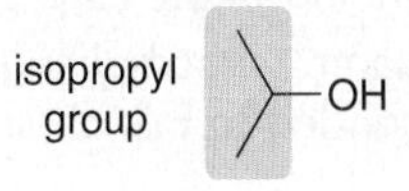

isopropyl alcohol

Compounds with two hydroxy groups are called **diols** (using the IUPAC system) or **glycols.** Compounds with three hydroxy groups are called **triols,** and so forth. To name a diol, for example, the suffix ***-diol*** is added to the name of the parent alkane, and numbers are used to indicate the location of the two OH groups.

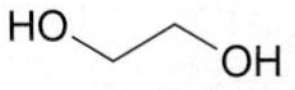

ethylene glycol
(ethane-1,2-diol)

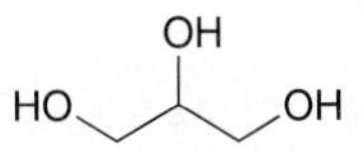

glycerol
(propane-1,2,3-triol)

Common names are usually used for these simple compounds.

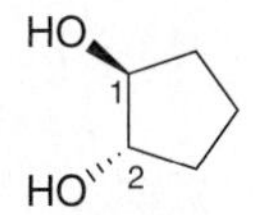

***trans*-cyclopentane-1,2-diol**

Numbers are needed to show the location of **two** OH groups.

Problem 9.2 Give the IUPAC name for each compound.

a. 

b. 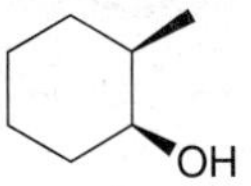

c. 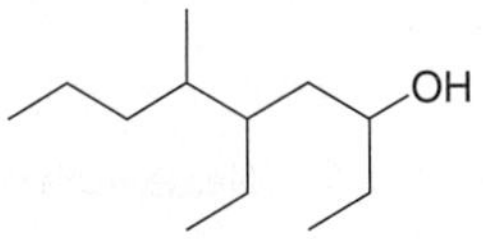

Problem 9.3 Give the structure corresponding to each name.

a. 7,7-dimethyloctan-4-ol
b. 5-methyl-4-propylheptan-3-ol
c. 2-*tert*-butyl-3-methylcyclohexanol
d. *trans*-cyclohexane-1,2-diol

9.3B Naming Ethers

Simple ethers are usually assigned common names. To do so, **name both alkyl groups** bonded to the oxygen, arrange these names alphabetically, and add the word **ether.** For symmetrical ethers, name the alkyl group and add the prefix **di-.**

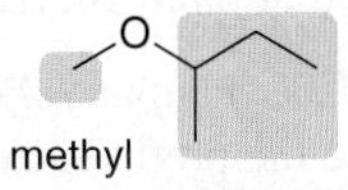

sec-butyl methyl ether

Alphabetize the **b** of **b**utyl before the **m** of **m**ethyl.

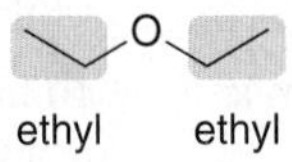

diethyl ether

More complex ethers are named using the IUPAC system. One alkyl group is named as a hydrocarbon chain, and the other is named as part of a substituent bonded to that chain.

- **Name the simpler alkyl group + O atom as an alkoxy substituent by changing the *-yl* ending of the alkyl group to *-oxy.***
- **Name the remaining alkyl group as an alkane, with the alkoxy group as a substituent bonded to this chain.**

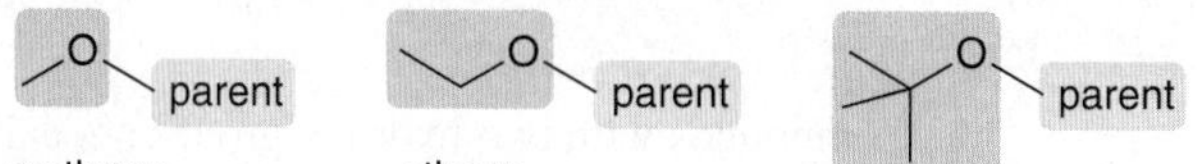

methoxy ethoxy *tert*-butoxy

Sample Problem 9.1 Give the IUPAC name for the following ether.

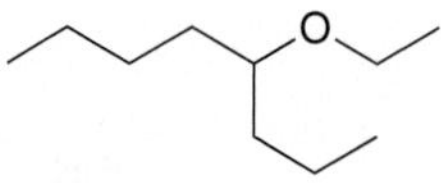

Solution

[1] Name the longer chain as an alkane and the shorter chain as an alkoxy group.

ethoxy group

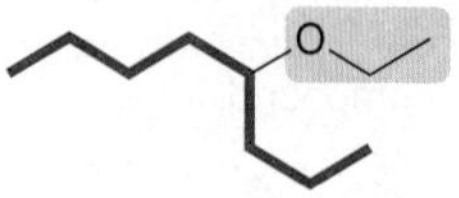

8 C's ----→ octane

[2] Apply the other nomenclature rules to complete the name.

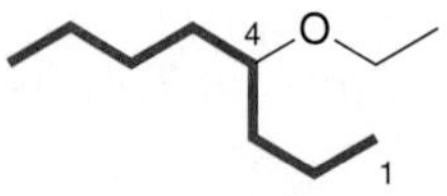

Answer: 4-ethoxyoctane

Problem 9.4 Name each of the following ethers.

a. 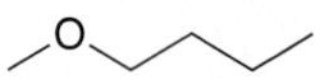 b. 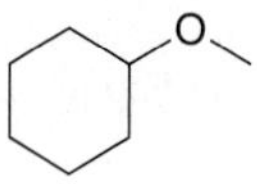 c. 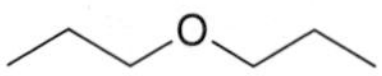

tetrahydrofuran
THF

Cyclic ethers have an O atom in a ring. A common cyclic ether is **tetrahydrofuran (THF),** a polar aprotic solvent used in nucleophilic substitution (Section 7.8C) and many other organic reactions.

9.3C Naming Epoxides

Any cyclic compound containing a heteroatom is called a *heterocycle.*

Epoxides are named in three different ways—**epoxyalkanes, oxiranes,** or **alkene oxides.**

To name an epoxide as an **epoxyalkane,** first name the alkane chain or ring to which the oxygen is attached, and use the prefix ***epoxy*** to name the epoxide as a substituent. Use two numbers to designate the location of the atoms to which the O's are bonded.

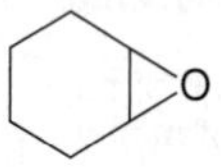

1,2-epoxycyclohexane

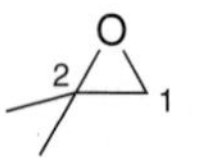

1,2-epoxy-2-methylpropane

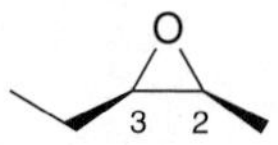

***cis*-2,3-epoxypentane**

Epoxides bonded to a chain of carbon atoms can also be named as derivatives of **oxirane,** the simplest epoxide having two carbons and one oxygen atom in a ring. The oxirane ring is numbered to put the O atom at position "1," and the first substituent at position "2." No number is used for a substituent in a monosubstituted oxirane.

oxirane **2,2-dimethyloxirane**

Epoxides are also named as **alkene oxides,** because they are often prepared by adding an O atom to an alkene (Chapter 12). To name an epoxide this way, mentally replace the epoxide oxygen by a double bond, name the alkene (Section 10.3), and then add the word ***oxide.*** For example, the common name for oxirane is ethylene oxide, because it is an epoxide derived from the alkene ethylene. We will use this method of naming epoxides after the details of alkene nomenclature are presented in Chapter 10.

ethylene **ethylene oxide** oxirane

Problem 9.5 Name each epoxide.

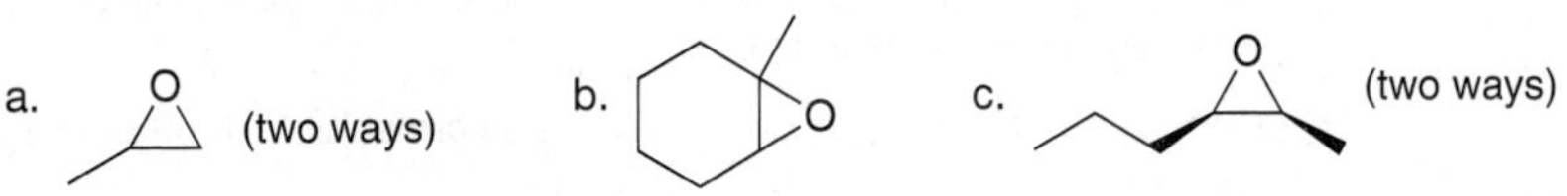

9.4 Physical Properties

Alcohols, ethers, and epoxides exhibit dipole–dipole interactions because they have a bent structure with two polar bonds. **Alcohols are also capable of intermolecular hydrogen bonding,** because they possess a hydrogen atom on an oxygen, making alcohols much more polar than ethers and epoxides.

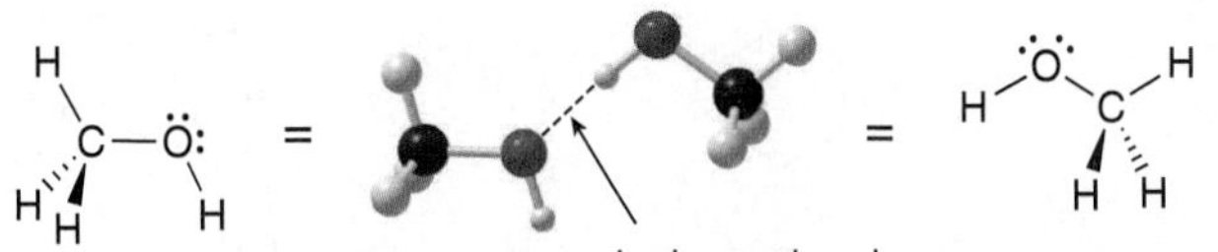

Steric factors affect the extent of hydrogen bonding. Although all alcohols can hydrogen bond, increasing the number of R groups around the carbon atom bearing the OH group decreases the extent of hydrogen bonding. Thus, 3° alcohols are least able to hydrogen bond, whereas 1° alcohols are most able to.

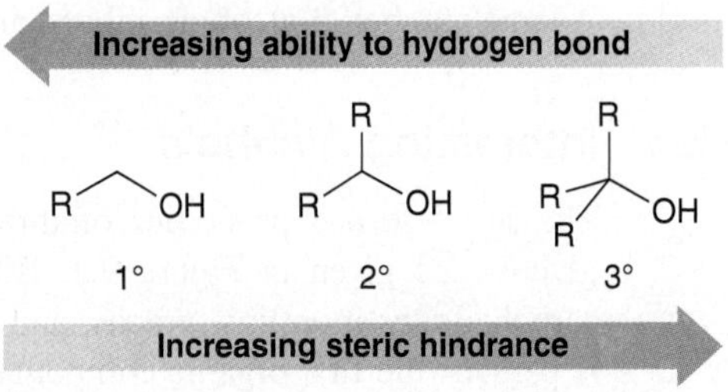

Table 9.1 Physical Properties of Alcohols, Ethers, and Epoxides

Property	Observation
Boiling point and melting point	• For compounds of comparable molecular weight, the **stronger** the intermolecular forces, the **higher** the bp or mp. 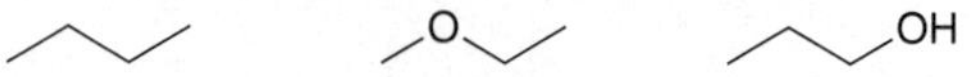 VDW bp 0 °C — VDW, DD bp 11 °C — VDW, DD, HB bp 97 °C **Increasing boiling point**
	• Bp's **increase** as the extent of hydrogen bonding **increases.** 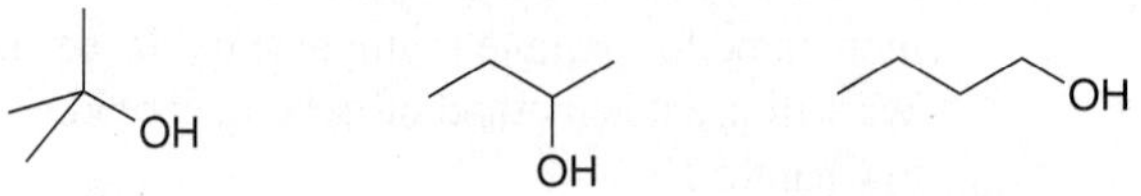 3° bp 83 °C — 2° bp 98 °C — 1° bp 118 °C **Increasing ability to hydrogen bond** **Increasing boiling point**
Solubility	• Alcohols, ethers, and epoxides having ≤ 5 C's are H_2O soluble because they each have an oxygen atom capable of hydrogen bonding to H_2O (Section 3.4C). • Alcohols, ethers, and epoxides having > 5 C's are H_2O insoluble because the nonpolar alkyl portion is too large to dissolve in H_2O. • Alcohols, ethers, and epoxides of any size are soluble in organic solvents.

Key: VDW = van der Waals forces; DD = dipole–dipole; HB = hydrogen bonding

How these factors affect the physical properties of alcohols, ethers, and epoxides is summarized in Table 9.1.

Problem 9.6 Rank the following compounds in order of increasing boiling point.

a. 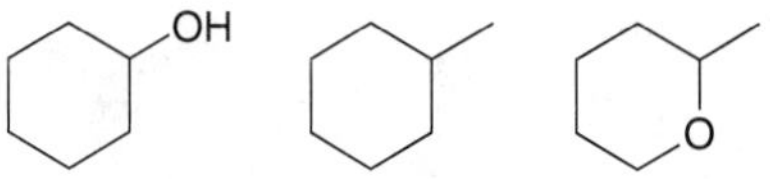

b. 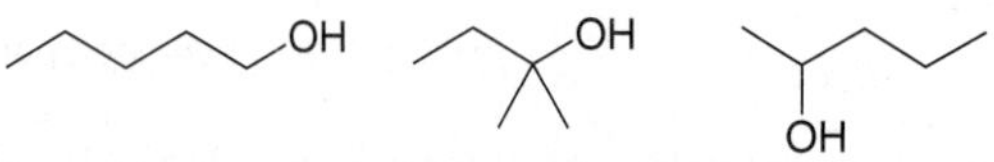

9.5 Interesting Alcohols, Ethers, and Epoxides

A large number of alcohols, ethers, and epoxides have interesting and useful properties.

9.5A Interesting Alcohols

The structure and properties of three simple alcohols—methanol, propan-2-ol, and ethylene glycol—are given in Figure 9.3. **Ethanol** (CH_3CH_2OH), formed by the fermentation of the carbohydrates in grains, grapes, and potatoes, is the alcohol present in alcoholic beverages. It is perhaps the first organic compound synthesized by humans, because alcohol production has been known for at least 4000 years. Ethanol depresses the central nervous system, increases the production of stomach acid, and dilates blood vessels, producing a flushed appearance. Ethanol is also a common laboratory solvent, which is sometimes made unfit to ingest by adding small amounts of benzene or methanol (both of which are toxic).

Figure 9.3
Some simple alcohols

CH_3OH

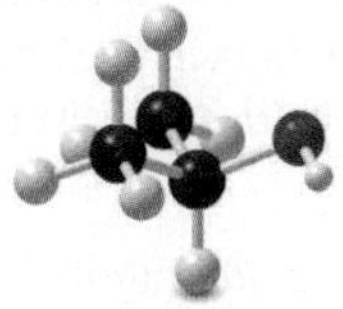

$(CH_3)_2CHOH$

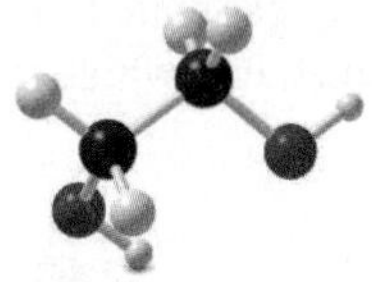

$HOCH_2CH_2OH$

- **Methanol (CH_3OH)** is also called wood alcohol, because it can be obtained by heating wood at high temperatures in the absence of air. Methanol is extremely toxic because of the oxidation products formed when it is metabolized in the liver (Section 12.14). Ingestion of as little as 15 mL causes blindness, and 100 mL causes death.
- **Propan-2-ol [$(CH_3)_2CHOH$]** is the major component of rubbing alcohol. When rubbed on the skin it evaporates readily, producing a pleasant cooling sensation. Because it has weak antibacterial properties, propan-2-ol is used to clean skin before minor surgery and to sterilize medical instruments.
- **Ethylene glycol ($HOCH_2CH_2OH$)** is the major component of antifreeze. It is readily prepared from ethylene oxide by reactions discussed in Section 9.16. It is sweet tasting but toxic.

Ethanol is a common gasoline additive, widely touted as an environmentally friendly fuel source. Two common gasoline–ethanol fuels are gasohol, which contains 10% ethanol, and E-85, which contains 85% ethanol. Ethanol is now routinely prepared from the carbohydrates in corn (Figure 9.4). Starch, a complex carbohydrate polymer, can be hydrolyzed to the simple sugar glucose, which forms ethanol by the process of fermentation. Combining ethanol with gasoline forms a usable fuel, which combusts to form CO_2, H_2O, and a great deal of energy.

Since green plants use sunlight to convert CO_2 and H_2O to carbohydrates during photosynthesis, next year's corn crop removes CO_2 from the atmosphere to make new molecules of starch as the corn grows. While in this way ethanol is a renewable fuel source, the need for large-scale farm equipment and the heavy reliance on fertilizers and herbicides make ethanol expensive to produce. Moreover, many criticize the use of valuable farmland for an energy-producing crop rather than for food production. As a result, discussion continues on ethanol as an alternative to fossil fuels.

Figure 9.4 Ethanol from corn, a renewable fuel source

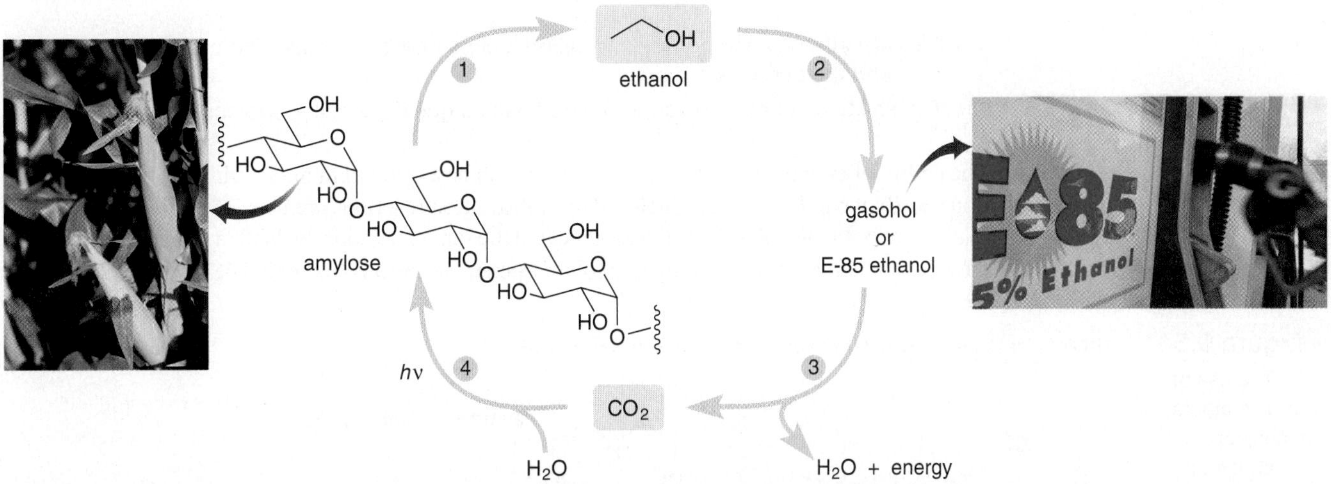

- Hydrolysis of amylose (one form of starch) and fermentation of the resulting simple sugars (Step [1]) yield ethanol, which is mixed with hydrocarbons from petroleum refining (Step [2]) to form usable fuels.
- Combustion of this ethanol–hydrocarbon fuel forms CO_2 and releases a great deal of energy (Step [3]).
- Photosynthesis converts atmospheric CO_2 back to plant carbohydrates in Step [4], and the cycle continues.

9.5B Interesting Ethers

The discovery that **diethyl ether** ($CH_3CH_2OCH_2CH_3$) is a general anesthetic revolutionized surgery in the nineteenth century. For years, a heated controversy existed over who first discovered diethyl ether's anesthetic properties and recognized the enormous benefit in its use. Early experiments were performed by a dentist, Dr. William Morton, resulting in a public demonstration of diethyl ether as an anesthetic in Boston in 1846. In fact, Dr. Crawford Long, a Georgia physician, had been using diethyl ether in surgery and obstetrics for several years, but had not presented his findings to a broader audience.

Diethyl ether is an imperfect anesthetic, but considering the alternatives in the nineteenth century, it was a miracle drug. It is safe, easy to administer, and causes little patient mortality, but it is highly flammable and causes nausea in many patients. For these reasons, it has largely been replaced by sevoflurane and other halogenated ethers, which are non-flammable and cause little patient discomfort.

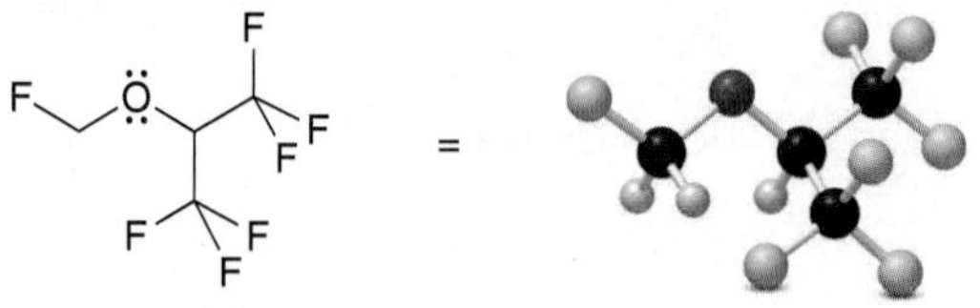

sevoflurane

Recall from Section 3.7B that crown ethers are named as ***x*-crown-*y*,** where ***x*** is the total number of atoms in the ring and ***y*** is the number of O atoms.

Recall from Section 3.7B that some cyclic **polyethers**—compounds with two or more ether linkages—contain cavities that can complex specific-sized cations. For example, 18-crown-6 binds K^+, whereas 12-crown-4 binds Li^+.

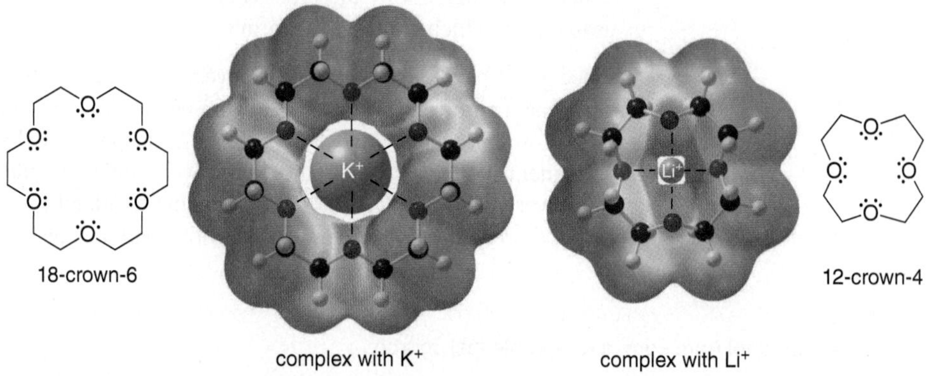
18-crown-6
K+
Li+
12-crown-4
complex with K+
complex with Li+

- **A crown ether–cation complex is called a *host–guest* complex. The crown ether is the *host* and the cation is the *guest.***
- **The ability of a host molecule to bind specific guests is called *molecular recognition.***

The ability of crown ethers to complex cations can be exploited in nucleophilic substitution reactions, as shown in Figure 9.5. Nucleophilic substitution reactions are usually run in polar solvents to dissolve both the polar organic substrate and the ionic nucleophile. With a crown ether, though, the reaction can be run in a nonpolar solvent under conditions that enhance nucleophilicity.

Figure 9.5 The use of crown ethers in nucleophilic substitution reactions

KCN is insoluble in nonpolar solvents alone, but with 18-crown-6:

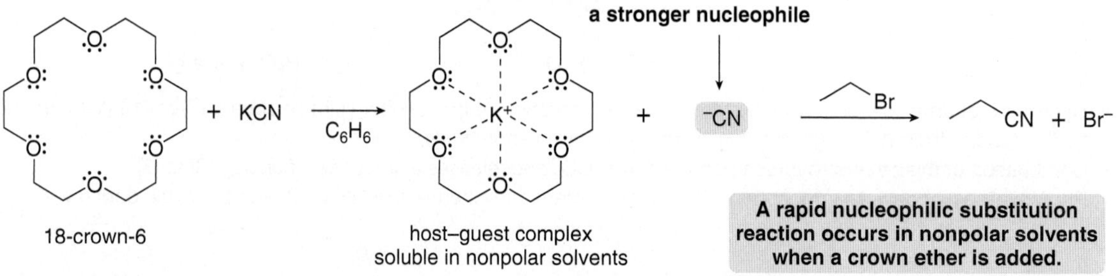
a stronger nucleophile
+ KCN
C6H6
K+
+
−CN
Br
CN + Br−
18-crown-6
host–guest complex
soluble in nonpolar solvents
A rapid nucleophilic substitution reaction occurs in nonpolar solvents when a crown ether is added.

When 18-crown-6 is added to the reaction of CH_3CH_2Br with KCN, for example, the crown ether forms a tight complex with K^+ that has nonpolar C–H bonds on the outside, making the complex soluble in nonpolar solvents like benzene (C_6H_6) or hexane. When the crown ether/K^+ complex dissolves in the nonpolar solvent, it carries the ^-CN along with it to maintain electrical neutrality. The result is a solution of tightly complexed cation and relatively unsolvated anion (nucleophile). The anion, therefore, is extremely nucleophilic because it is not hidden from the substrate by solvent molecules.

Problem 9.7 Which mechanism is favored by the use of crown ethers in nonpolar solvents, S_N1 or S_N2?

9.5C Interesting Epoxides

Although epoxides occur less widely in natural products than alcohols or ethers, interesting and useful epoxides are also known. Examples include two useful drugs, eplerenone and tiotropium bromide. Eplerenone (trade name Inspra) is prescribed to reduce cardiovascular risk in patients who have already had a heart attack. Tiotropium bromide (trade name Spiriva) is a long-acting bronchodilator used to treat the chronic obstructive pulmonary disease of smokers and those routinely exposed to secondhand smoke.

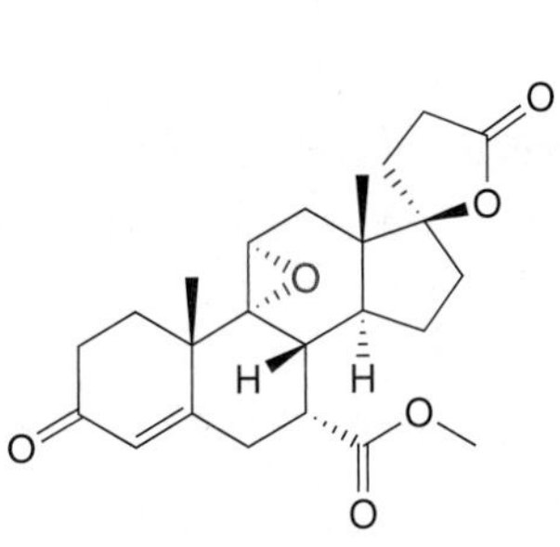

eplerenone

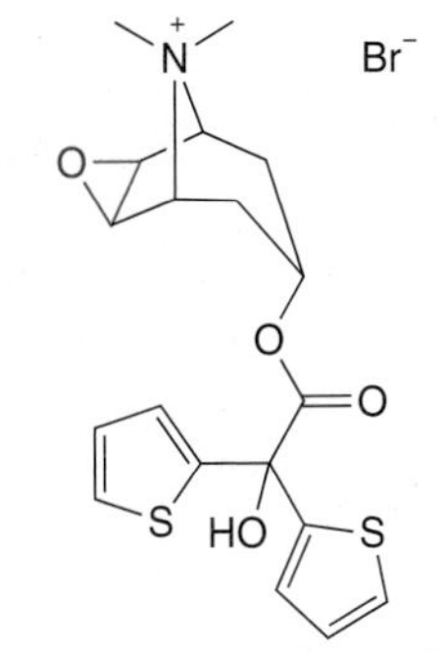

tiotropium bromide

9.6 Preparation of Alcohols, Ethers, and Epoxides

Alcohols and ethers are both common products of nucleophilic substitution. They are synthesized from alkyl halides by S_N2 reactions using strong nucleophiles. As in all S_N2 reactions, highest yields of products are obtained with unhindered methyl and 1° alkyl halides.

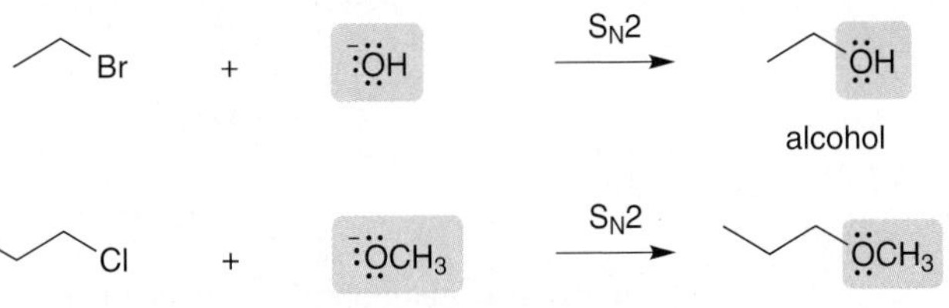

The preparation of ethers by this method is called the **Williamson ether synthesis,** and, although it was first reported in the 1800s, it is still the most general method to prepare an ether. Unsymmetrical ethers can be synthesized in two different ways, but often one path is preferred.

For example, ethyl isopropyl ether can be prepared from $CH_3CH_2O^-$ and 2-bromopropane (Path [a]), or from $(CH_3)_2CHO^-$ and bromoethane (Path [b]). Because the mechanism is S_N2, **the preferred path uses the less sterically hindered halide,** CH_3CH_2Br—Path [b].

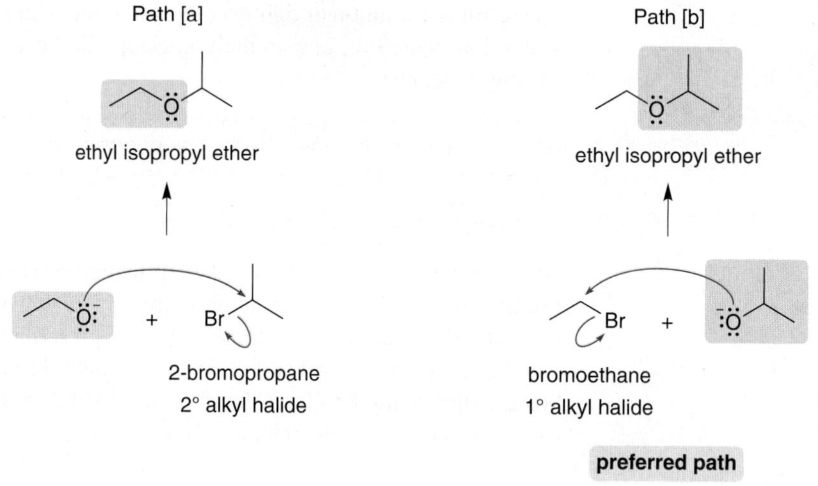
Path [a]
ethyl isopropyl ether
Br
2-bromopropane
2° alkyl halide
Path [b]
ethyl isopropyl ether
Br
bromoethane
1° alkyl halide
preferred path

Problem 9.8 Draw the organic product of each reaction.

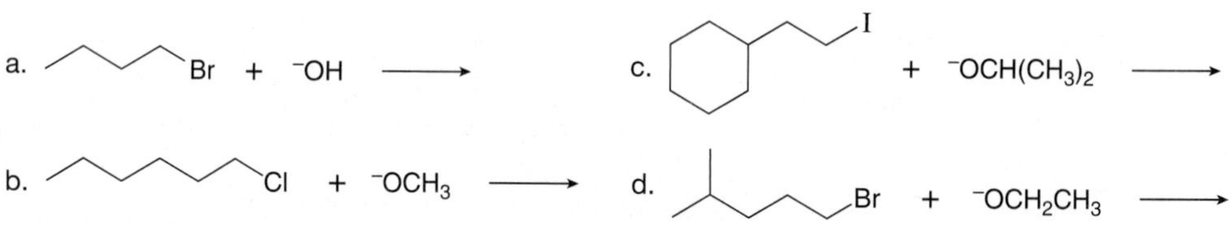
a. Br + ⁻OH
b. Cl + ⁻OCH3
c. I + ⁻OCH(CH3)2
d. Br + ⁻OCH2CH3

Problem 9.9 A key step in the synthesis of the antidepressant paroxetine (trade name Paxil) involves a Williamson ether synthesis of the acyclic ether in **X.** Draw two different routes to this ether and state which is preferred.

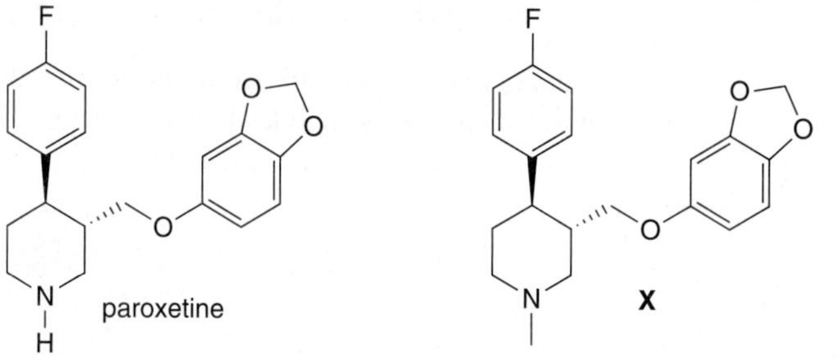
F
O
O
O
N
H
F
N

paroxetine **X**

A **hydroxide** nucleophile is needed to synthesize an alcohol, and salts such as **NaOH** and **KOH** are inexpensive and commercially available. An **alkoxide** salt is needed to make an ether. Simple alkoxides such as sodium methoxide ($NaOCH_3$) can be purchased, but others are prepared from alcohols by a Brønsted–Lowry acid–base reaction. For example, **sodium ethoxide** (**$NaOCH_2CH_3$**) is prepared by treating ethanol with NaH.

NaH is an especially good base for forming an alkoxide, because the by-product of the reaction, H_2, is a gas that just bubbles out of the reaction mixture.

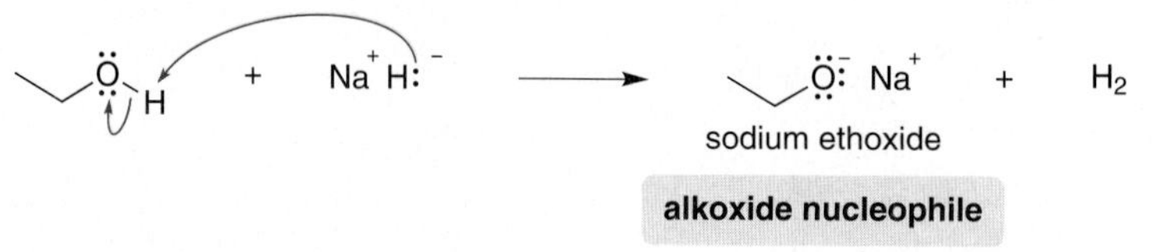
H
Na⁺ H:⁻
Na⁺
H2
sodium ethoxide
alkoxide nucleophile

Sample Problem 9.2 Draw the product of the following two-step reaction sequence.

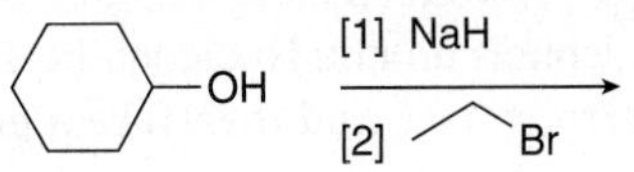
OH
[1] NaH
[2]
Br

Solution

[1] The base removes a proton from the OH group, forming an alkoxide.

[2] The alkoxide acts as a nucleophile in an S_N2 reaction, forming an ether.

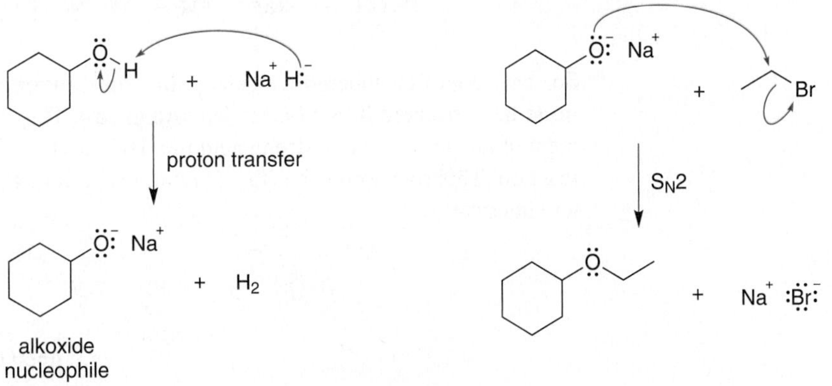
H
+
Na⁺ H:⁻
proton transfer
Na⁺
+ H₂
alkoxide
nucleophile
Na⁺
+
Br
SN2
+
Na⁺ :Br:⁻

- This two-step sequence converts an alcohol to an ether.

When an organic compound contains both a hydroxy group and a halogen atom on adjacent carbon atoms, an *intramolecular* version of this reaction forms an epoxide. The starting material for this two-step sequence, a **halohydrin,** is prepared from an alkene, as we will learn in Chapter 10.

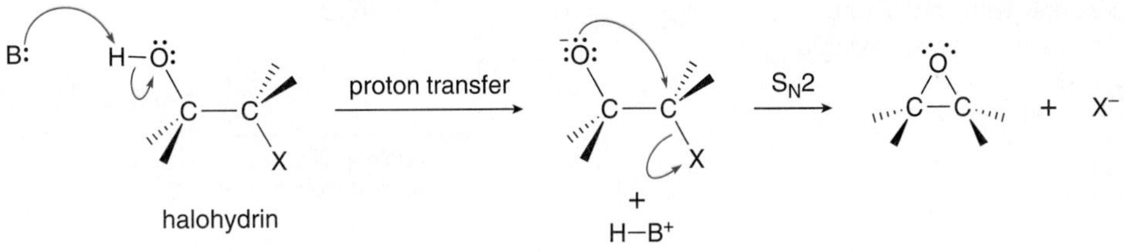
B:
H—O:
C—C
X
halohydrin
proton transfer
C—C
X
+
H—B⁺
SN2
C—C
+ X⁻

Problem 9.10 Draw the products of each reaction.

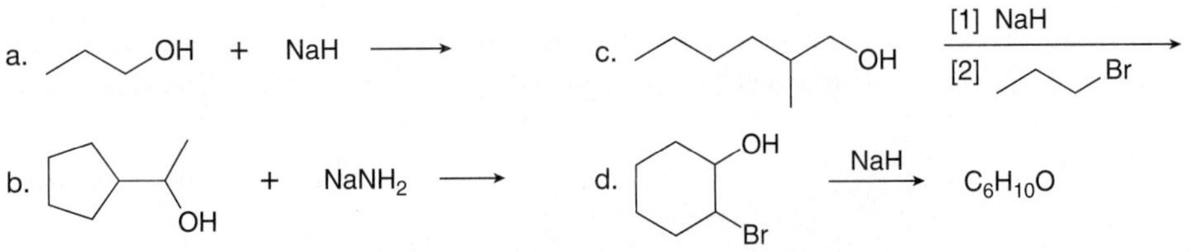
a.
OH + NaH
b.
OH
+ NaNH₂
c.
OH
[1] NaH
[2]
Br
d.
OH
Br
NaH
C₆H₁₀O

9.7 General Features—Reactions of Alcohols, Ethers, and Epoxides

We begin our discussion of the chemical reactions of alcohols, ethers, and epoxides with a look at the general reactive features of each functional group.

9.7A Alcohols

Unlike many families of molecules, the reactions of alcohols do *not* fit neatly into a single reaction class. In Chapter 9, we discuss only the substitution and β elimination reactions of alcohols. Alcohols are also key starting materials in oxidation reactions (Chapter 12), and their polar O–H bond makes them more acidic than many other organic compounds, a feature we will explore in Chapter 19.

Alcohols are similar to alkyl halides in that both contain an electronegative element bonded to an sp^3 hybridized carbon atom. Alkyl halides contain a good leaving group (X^-), however, whereas alcohols do not. Nucleophilic substitution with ROH as starting material would displace **$^-$OH, a strong base and therefore a poor leaving group.**

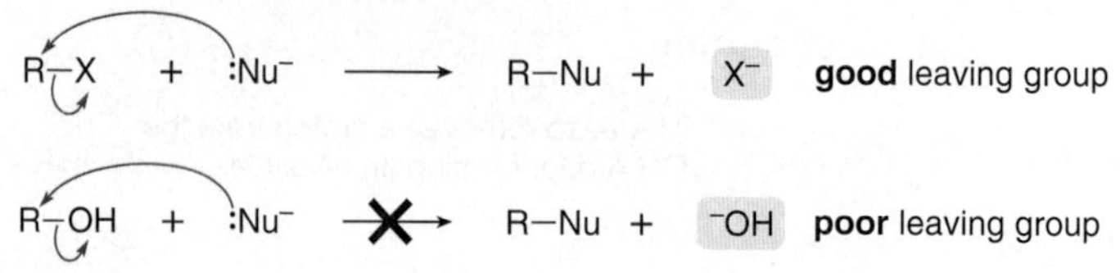
R–X + :Nu⁻ → R–Nu + X⁻ good leaving group
R–OH + :Nu⁻ ✕→ R–Nu + ⁻OH poor leaving group

For an alcohol to undergo a nucleophilic substitution or elimination reaction, the **OH group must be converted into a better leaving group.** This can be done by reaction with acid. Treatment of an alcohol with a strong acid like HCl or H_2SO_4 protonates the O atom via an acid–base reaction. This transforms the $^-$OH leaving group into **H_2O, a weak base and therefore a good leaving group.**

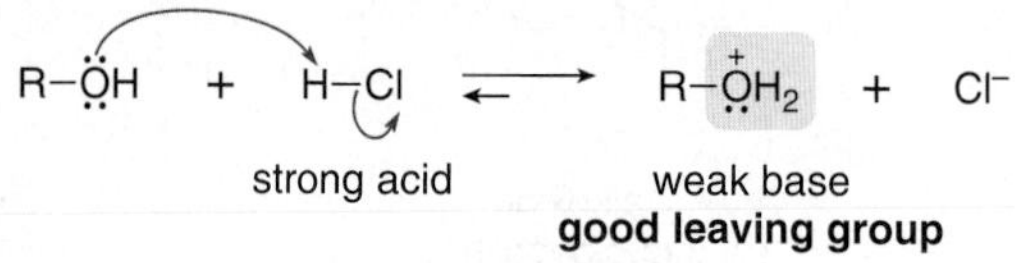
R–ÖH + H–Cl ⇌ R–ÖH₂⁺ + Cl⁻
strong acid
weak base
good leaving group

If the OH group of an alcohol is made into a good leaving group, alcohols *can* undergo β elimination and nucleophilic substitution, as described in Sections 9.8–9.12.

Because the pK_a of $(ROH_2)^+$ is ~–2, protonation of an alcohol occurs only with very strong acids—namely, those having a $pK_a \leq -2$.

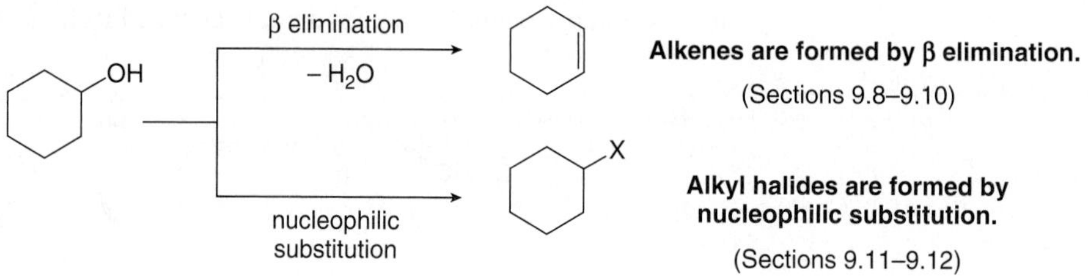
OH
β elimination
– H₂O
Alkenes are formed by β elimination.
(Sections 9.8–9.10)
X
nucleophilic
substitution
Alkyl halides are formed by
nucleophilic substitution.
(Sections 9.11–9.12)

9.7B Ethers and Epoxides

Like alcohols, **ethers do not contain a good leaving group,** which means that nucleophilic substitution and β elimination do not occur directly. Ethers undergo fewer useful reactions than alcohols.

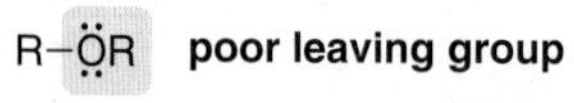
R–ÖR poor leaving group

Epoxides don't have a good leaving group either, but they have one characteristic that neither alcohols nor ethers have: **the "leaving group" is contained in a strained three-membered ring.** Nucleophilic attack opens the three-membered ring and relieves angle strain, making nucleophilic attack a favorable process that occurs even with the poor leaving group. Specific examples are presented in Section 9.16.

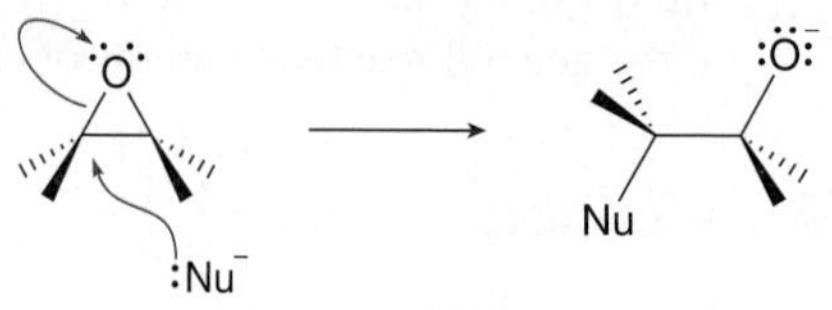
:Nu⁻
Nu

9.8 Dehydration of Alcohols to Alkenes

The dehydrohalogenation of alkyl halides, discussed in Chapter 8, is one way to introduce a π bond into a molecule. Another way is to eliminate water from an alcohol in a **dehydration** reaction.

- **Dehydration is a β elimination reaction in which the elements of OH and H are removed from the α and β carbon atoms, respectively.**

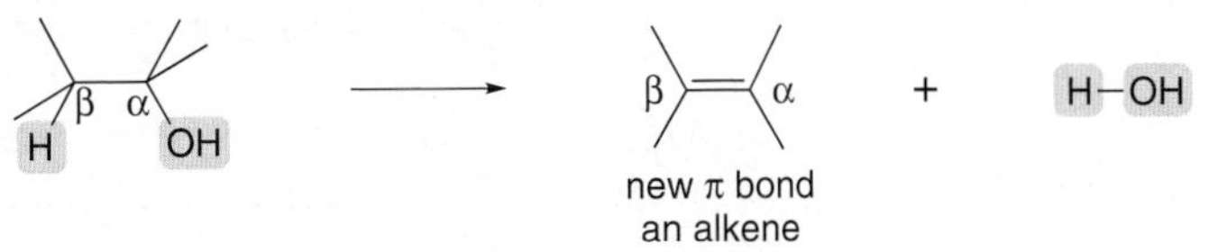

Dehydration is typically carried out using H_2SO_4 and other strong acids, or phosphorus oxychloride ($POCl_3$) in the presence of an amine base. We consider dehydration in acid first, followed by dehydration with $POCl_3$ in Section 9.10.

9.8A General Features of Dehydration in Acid

Alcohols undergo dehydration in the presence of strong acid to afford alkenes, as illustrated in Equations [1] and [2]. Typical acids used for this conversion are H_2SO_4 or *p*-toluenesulfonic acid (abbreviated as TsOH).

Recall from Section 2.6 that *p*-toluenesulfonic acid is a strong organic acid ($pK_a = -7$).

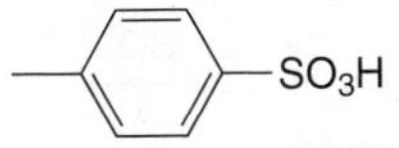

p-toluenesulfonic acid
TsOH

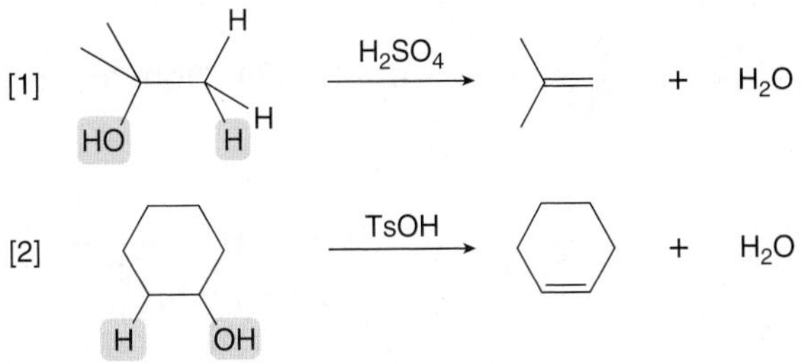

More substituted alcohols dehydrate more readily, giving rise to the following order of reactivity:

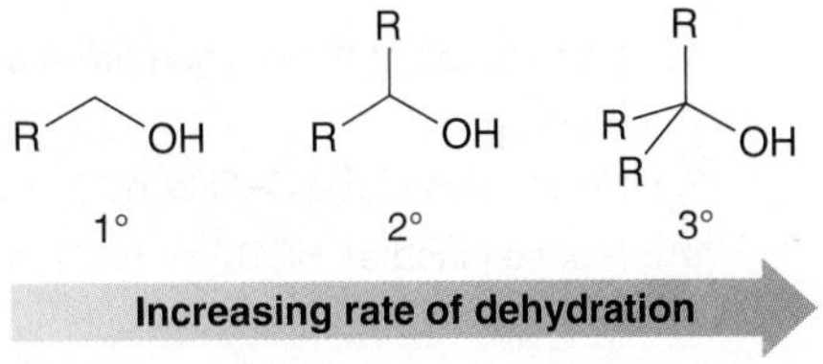

When an alcohol has two or three different β carbons, dehydration is regioselective and follows the Zaitsev rule. **The more substituted alkene is the major product when a mixture of constitutional isomers is possible.** For example, elimination of H and OH from 2-methylbutan-2-ol yields two constitutional isomers: the trisubstituted alkene **A** as major product and the disubstituted alkene **B** as minor product.

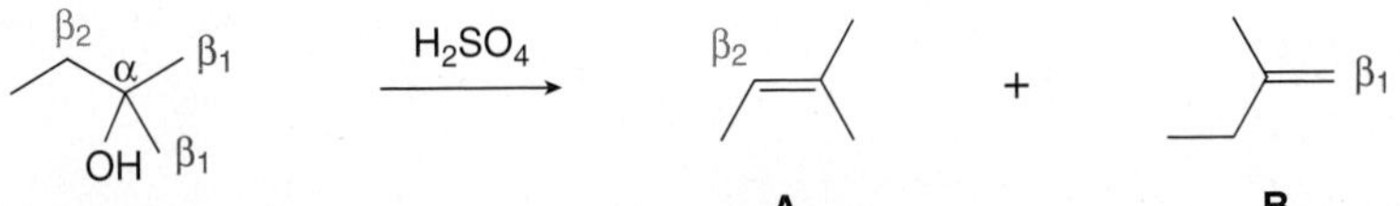

2-methylbutan-2-ol

A
major product
trisubstituted alkene

B
minor product
disubstituted alkene

Problem 9.11 Draw the products formed when each alcohol undergoes dehydration with TsOH, and label the major product when a mixture results.

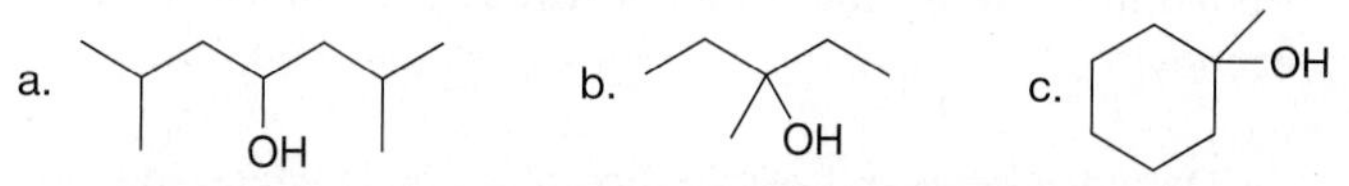

Problem 9.12 Rank the alcohols in order of increasing reactivity when dehydrated with H_2SO_4.

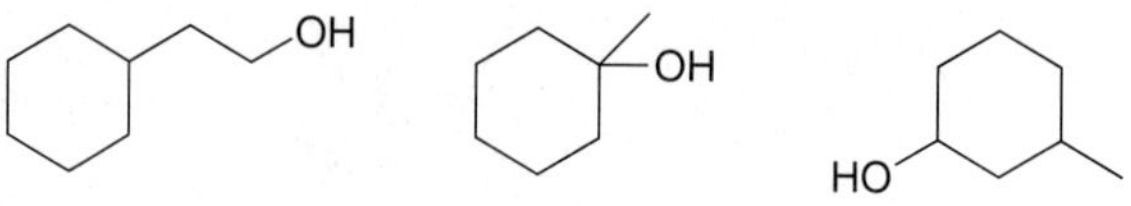

9.8B The E1 Mechanism for the Dehydration of 2° and 3° Alcohols

The mechanism of dehydration depends on the structure of the alcohol: **2° and 3° alcohols react by an E1 mechanism, whereas 1° alcohols react by an E2 mechanism.** Regardless of the type of alcohol, however, strong acid is *always* needed to protonate the O atom to form a good leaving group.

The E1 dehydration of 2° and 3° alcohols is illustrated with $(CH_3)_3COH$ (a 3° alcohol) as starting material to form $(CH_3)_2C{=}CH_2$ as product (Mechanism 9.1). The mechanism consists of **three steps.**

Mechanism 9.1 Dehydration of 2° and 3° ROH—An E1 Mechanism

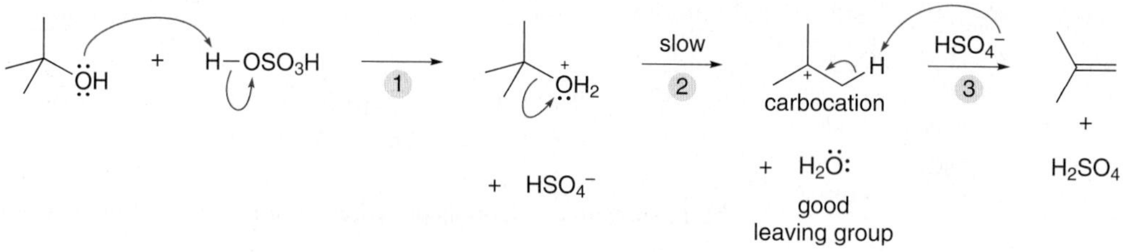

1 Protonation of the oxygen atom converts the poor leaving group (^-OH) into a good leaving group (H_2O).

2 Heterolysis of the C–O bond forms a carbocation in the rate-determining step.

3 A base (such as HSO_4^- or H_2O) removes a proton from a carbon adjacent to the carbocation to form the new π bond.

Thus, **dehydration of 2° and 3° alcohols occurs via an E1 mechanism with an added first step.** Step [1] protonates the OH group to make a good leaving group. Steps [2] and [3] are the two steps of an E1 mechanism: loss of a leaving group (H_2O in this case) to form a carbocation, followed by removal of a β proton to form a π bond. The acid used to protonate the alcohol in Step [1] is regenerated upon removal of the proton in Step [3], so dehydration is **acid-catalyzed.**

The E1 dehydration of 2° and 3° alcohols with acid gives clean elimination products without by-products formed from an S_N1 reaction. This makes the E1 dehydration of alcohols much more synthetically useful than the E1 dehydrohalogenation of alkyl halides (Section 8.7). Clean elimination takes place because the reaction mixture contains no good nucleophile to react with the intermediate carbocation, so no competing S_N1 reaction occurs.

9.8C The E2 Mechanism for the Dehydration of 1° Alcohols

Because 1° carbocations are highly unstable, the dehydration of 1° alcohols cannot occur by an E1 mechanism involving a carbocation intermediate. With 1° alcohols, therefore, **dehydration follows an E2 mechanism.** The two-step process for the conversion of $CH_3CH_2CH_2OH$ (a 1° alcohol) to $CH_3CH{=}CH_2$ with H_2SO_4 as acid catalyst is shown in Mechanism 9.2.

Mechanism 9.2 Dehydration of a 1° ROH—An E2 Mechanism

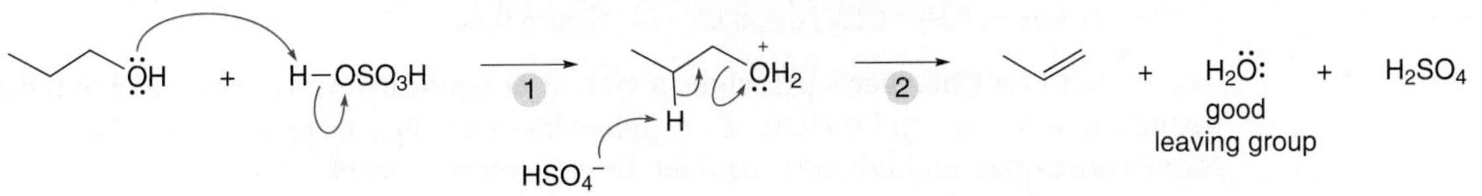

1 Protonation of the oxygen atom converts the poor leaving group (^-OH) into a good leaving group (H_2O).

2 Two bonds are broken and two bonds are formed. The base (HSO_4^- or H_2O) removes a proton from the β carbon; the electron pair in the β C–H bond forms the new π bond and the leaving group (H_2O) departs.

The dehydration of a 1° alcohol begins with the protonation of the OH group to form a good leaving group, just as in the dehydration of a 2° or 3° alcohol. With 1° alcohols, however, loss of the leaving group and removal of a β proton occur at the *same* time, so that no highly unstable 1° carbocation is generated.

Problem 9.13 Draw the structure of each transition state in the two-step mechanism for the reaction, $CH_3CH_2CH_2OH + H_2SO_4 \rightarrow CH_3CH{=}CH_2 + H_2O$.

9.8D Le Châtelier's Principle

Although entropy favors product formation in dehydration (one molecule of reactant forms two molecules of products), enthalpy does not, because the two σ bonds broken in the reactant are

Figure 9.6
The dehydration of CH_3CH_2OH to $CH_2{=}CH_2$—An endothermic reaction

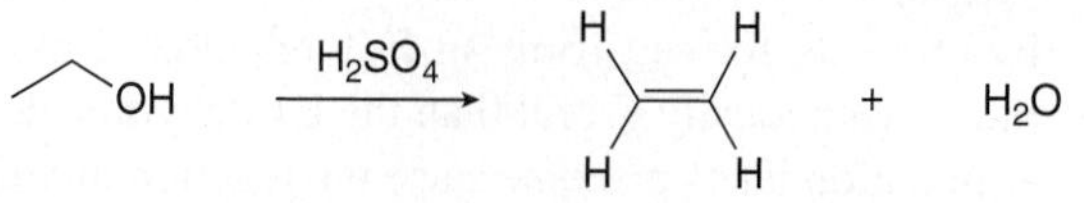

ΔH° calculation:

[1] Bonds broken

	ΔH° (kJ/mol)
CH_3CH_2–OH	+393
$HOCH_2CH_2$–H	+410
Total	+803 kJ/mol

Energy needed to break bonds.

[2] Bonds formed

	ΔH° (kJ/mol)
$CH_2{=}CH_2$ π bond	–267
H–OH	–498
Total	–765 kJ/mol

Energy released in forming bonds.

[3] Overall ΔH° =

sum in Step [1]
+
sum in Step [2]

+803 kJ/mol
–765 kJ/mol

ΔH° = +38 kJ/mol

The reaction is endothermic.

[Values taken from Appendix C.]

stronger than the σ and π bonds formed in the products. For example, $\Delta H°$ for the dehydration of CH_3CH_2OH to $CH_2{=}CH_2$ is +38 kJ/mol (Figure 9.6).

According to **Le Châtelier's principle, a system at equilibrium will react to counteract any disturbance to the equilibrium.** Thus, removing a product from a reaction mixture as it is formed drives the equilibrium to the right, forming more product.

Le Châtelier's principle can be used to favor products in dehydration reactions because the alkene product has a lower boiling point than the alcohol reactant. Thus, the alkene can be distilled from the reaction mixture as it is formed, leaving the alcohol and acid to react further, forming more product.

9.9 Carbocation Rearrangements

Sometimes "unexpected" products are formed in dehydration; that is, the carbon skeletons of the starting material and product might be different, or the double bond might be in an unexpected location. For example, the dehydration of 3,3-dimethylbutan-2-ol yields two alkenes, whose carbon skeletons do not match the carbon framework of the starting material.

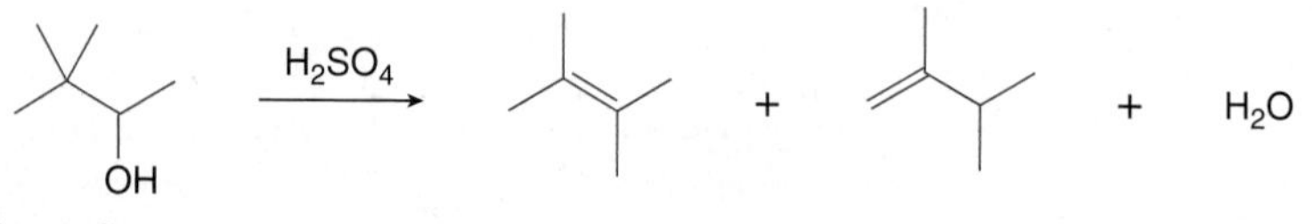

3,3-dimethylbutan-2-ol

This phenomenon sometimes occurs when carbocations are reactive intermediates. **A less stable carbocation can rearrange to a more stable carbocation by shift of a hydrogen atom or an alkyl group.** These **1,2-shifts** involve migration of an alkyl group or hydrogen atom from one carbon to an adjacent carbon atom. The migrating group moves with the two electrons that bonded it to the carbon skeleton.

Because the migrating group in a 1,2-shift moves with two bonding electrons, the carbon it leaves behind now has only three bonds (six electrons), giving it a net positive (+) charge.

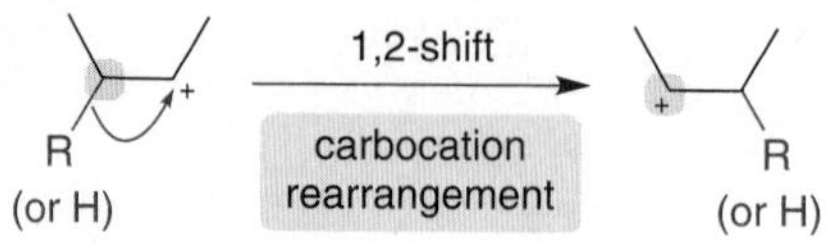

- Movement of a hydrogen atom is called a 1,2-hydride shift.
- Movement of an alkyl group is called a 1,2-alkyl shift.

The dehydration of 3,3-dimethylbutan-2-ol illustrates the rearrangement of a 2° to a 3° carbocation by a **1,2-methyl shift,** as shown in Mechanism 9.3. The carbocation rearrangement occurs in Step [3] of the four-step mechanism.

Mechanism 9.3 A 1,2-Methyl Shift—Carbocation Rearrangement During Dehydration

Part [1] Formation of a 2° carbocation and rearrangement

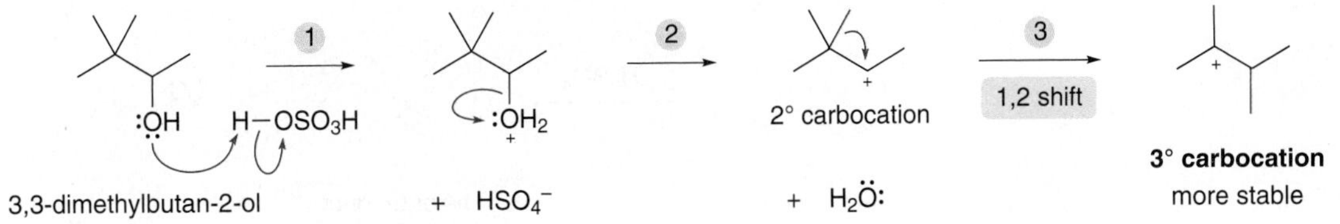

1
:OH
H—OSO3H
3,3-dimethylbutan-2-ol
2
:OH2
+ HSO4−
2° carbocation
+ H2Ö:
3
1,2 shift
3° carbocation
more stable

1. Protonation of the oxygen atom converts the poor leaving group (^{-}OH) into a good leaving group (H_2O).
2. Heterolysis of the C–O bond forms a 2° carbocation.
3. 1,2-Shift of a CH_3 group converts a 2° carbocation to a more stable 3° carbocation.

Part [2] Loss of a proton to form the π bond

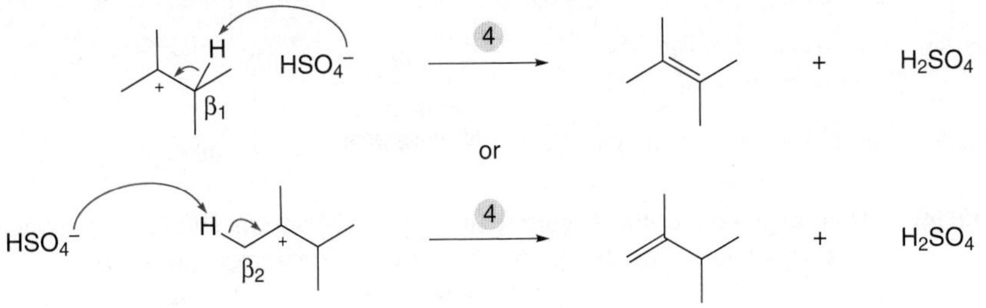

H
HSO4−
β1
4
+
H2SO4
or
HSO4−
H
β2
4
+
H2SO4

4. Loss of a proton from a β carbon ($β_1$ or $β_2$) forms two different alkenes.

Steps [1], [2], and [4] in the mechanism for the dehydration of 3,3-dimethylbutan-2-ol are exactly the same steps previously seen in dehydration: protonation, loss of H_2O, and loss of a proton. Only Step [3], rearrangement of the less stable 2° carbocation to the more stable 3° carbocation, is new.

- **1,2-Shifts convert a less stable carbocation to a more stable carbocation.**

For example, 2° carbocation **A** rearranges to the more stable 3° carbocation by a 1,2-hydride shift, whereas carbocation **B** does not rearrange because it is 3° to begin with.

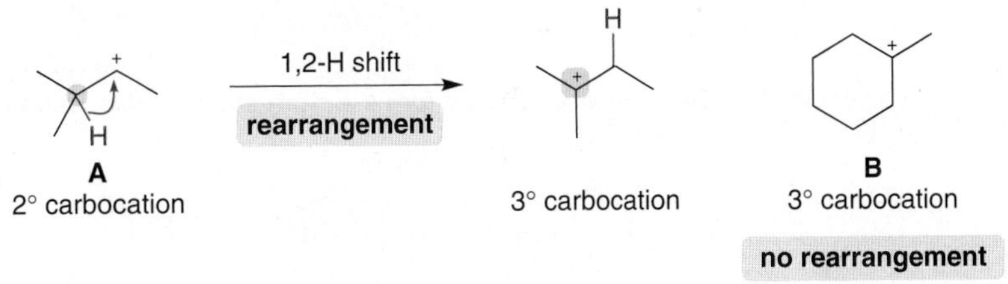

H
A
2° carbocation
1,2-H shift
rearrangement
H
3° carbocation
B
3° carbocation
no rearrangement

Sample Problem 9.3 illustrates a dehydration reaction that occurs with a **1,2-hydride** shift.

Sample Problem 9.3 Show how the dehydration of alcohol **X** forms alkene **Y** using a 1,2-hydride shift.

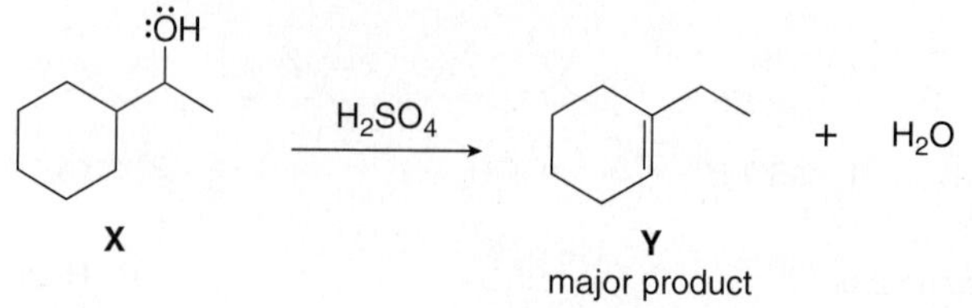
:ÖH
H_2SO_4
+ H_2O
X
Y
major product

Solution

Steps [1] and [2] Protonation of **X** and loss of H_2O form a 2° carbocation.

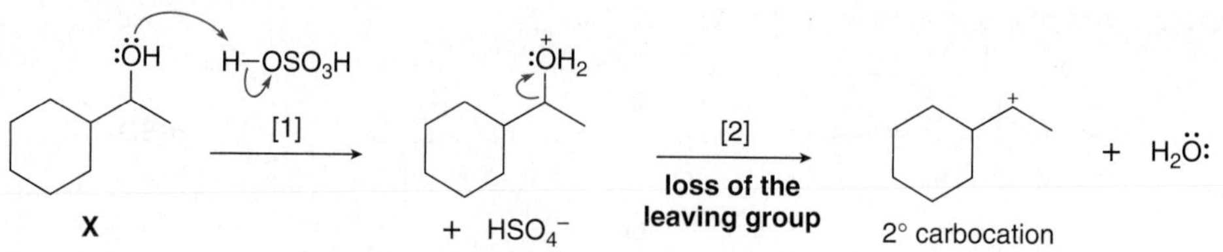
:ÖH
H–OSO₃H
[1]
X
:ÖH₂ +
+ HSO₄⁻
[2]
loss of the
leaving group
2° carbocation
+ H₂Ö:

Steps [3] and [4] Rearrangement of the 2° carbocation by a 1,2-hydride shift forms a more stable 3° carbocation. Loss of a proton from a β carbon forms alkene **Y**.

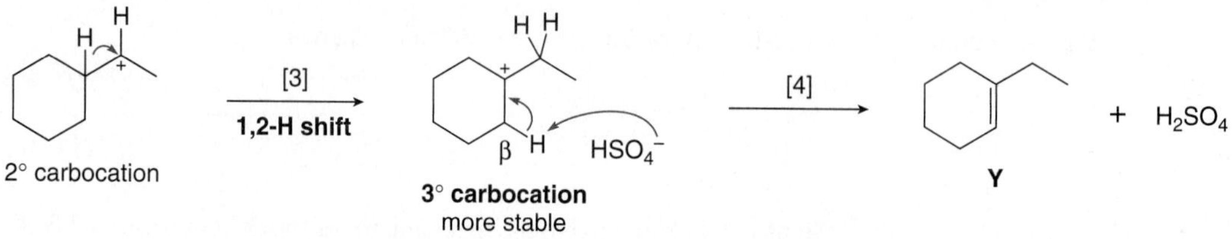
H
H
2° carbocation
[3]
1,2-H shift
H H
β H
HSO₄⁻
3° carbocation
more stable
[4]
Y
+ H₂SO₄

Problem 9.14 What other alkene is also formed along with **Y** in Sample Problem 9.3? What alkenes would form from **X** if no carbocation rearrangement occurred?

Rearrangements are not unique to dehydration reactions. **Rearrangements can occur whenever a carbocation is formed as reactive intermediate,** meaning any S_N1 or E1 reaction. In fact, the formation of rearranged products often indicates the presence of a carbocation intermediate.

Problem 9.15 Show how a 1,2-shift forms a more stable carbocation from each intermediate.

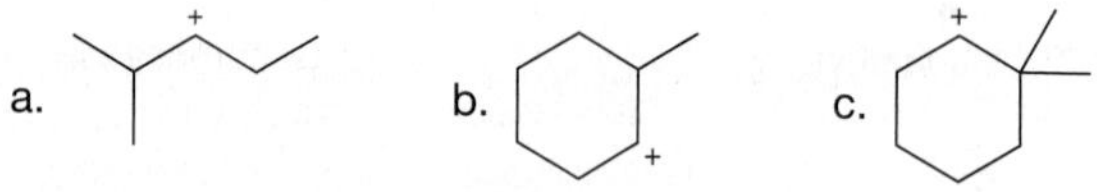
a.
b.
c.

Problem 9.16 Explain why two substitution products are formed in the following reaction.

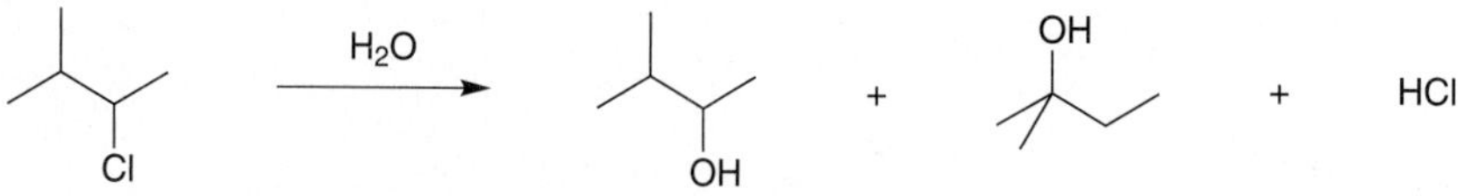
Cl
H₂O
OH
+
OH
+
HCl

9.10 Dehydration Using $POCl_3$ and Pyridine

Because some organic compounds decompose in the presence of strong acid, other methods that avoid strong acid have been developed to convert alcohols to alkenes. A common method uses phosphorus oxychloride ($POCl_3$) and pyridine (an amine base) in place of H_2SO_4 or TsOH. For example, the treatment of cyclohexanol with $POCl_3$ and pyridine forms cyclohexene in good yield.

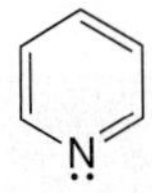

pyridine

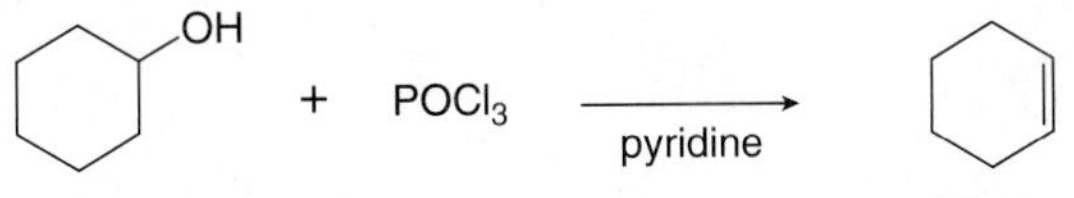

cyclohexanol cyclohexene

$POCl_3$ serves much the same role as strong acid does in acid-catalyzed dehydration. **It converts a poor leaving group (^-OH) into a good leaving group.** Dehydration then proceeds by an **E2 mechanism,** as shown in Mechanism 9.4. Pyridine is the base that removes a β proton during elimination.

Mechanism 9.4 Dehydration Using $POCl_3$ + Pyridine—An E2 Mechanism

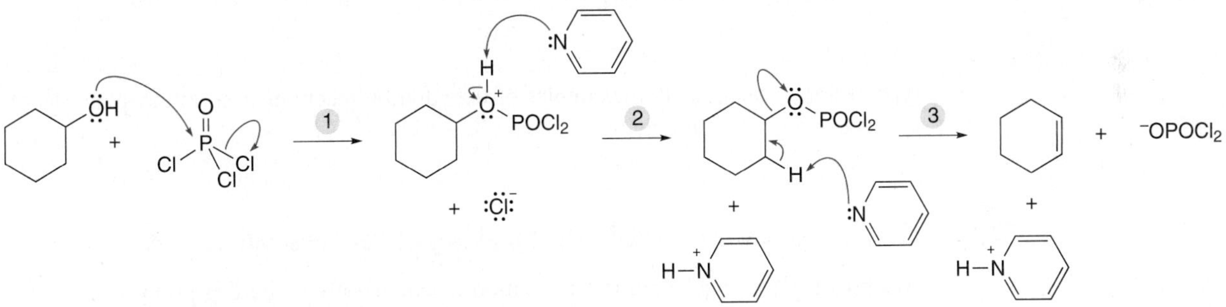

1–2 Reaction of the OH with $POCl_3$ followed by loss of a proton converts a poor leaving group (^-OH) into a good leaving group ($^-OPOCl_2$).

3 Two bonds are broken and two bonds are formed. The base (pyridine) removes a proton; the electron pair in the β C–H bond forms the π bond, and the leaving group ($^-OPOCl_2$) departs.

No rearrangements occur during dehydration with $POCl_3$, suggesting that carbocations are *not* formed as intermediates in this reaction. Steps [1] and [2] of the mechanism convert the OH group to a good leaving group. In Step [3], the C–H and C–O bonds are broken and the π bond is formed.

We have now learned about two different reagents for alcohol dehydration—strong acid (H_2SO_4 or TsOH) and $POCl_3$ + pyridine. The best dehydration method for a given alcohol is often hard to know ahead of time, and this is why organic chemists develop more than one method for a given

Figure 9.7
Dehydration reactions in the synthesis of two natural products

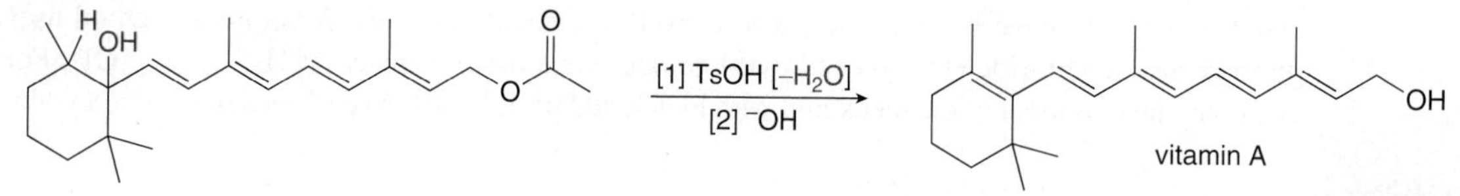

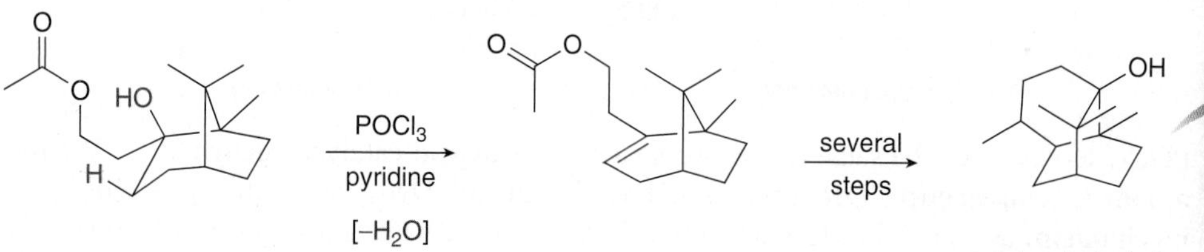

patchouli alcohol

patchouli plant

- New double bonds formed by dehydration are shown in red.
- **Patchouli alcohol,** obtained from the patchouli plant native to Malaysia, has been used in perfumery because of its exotic fragrance. In the 1800s, shawls imported from India were often packed with patchouli leaves to ward off insects, thus permeating the clothing with the distinctive odor.

type of transformation. Two examples of dehydration reactions used in the synthesis of natural products are given in Figure 9.7.

9.11 Conversion of Alcohols to Alkyl Halides with HX

Alcohols undergo nucleophilic substitution reactions only if the OH group is converted into a better leaving group before nucleophilic attack. Thus, substitution does *not* occur when an alcohol is treated with X^- because $^{-}$OH is a poor leaving group (Reaction [1]), but substitution *does* occur on treatment of an alcohol with HX because H_2O is now the leaving group (Reaction [2]).

[1] $R{-}\ddot{\underset{..}{O}}H + X^- \not\longrightarrow R{-}X + {}^{-}OH$ — poor leaving group **Reaction does *not* occur.**

[2] $R{-}\ddot{\underset{..}{O}}H + H{-}X \longrightarrow R{-}X + H_2O$ — good leaving group **Reaction occurs.**

alkyl halide

- **The reaction of alcohols with HX (X = Cl, Br, I) is a general method to prepare 1°, 2°, and 3° alkyl halides.**

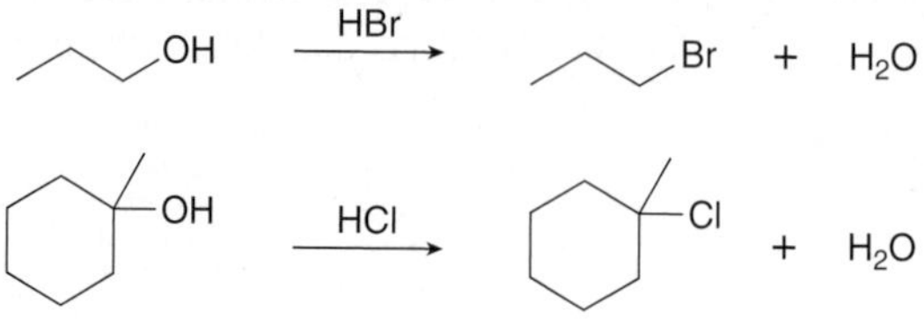

More substituted alcohols usually react more rapidly with HX:

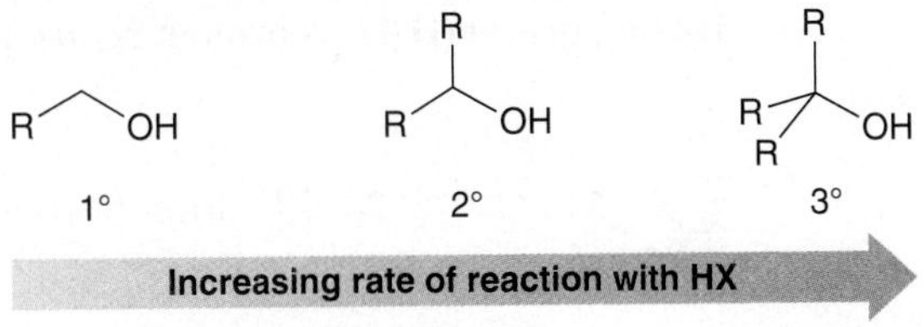
R OH
1°
R OH R
2°
R R R OH
3°
Increasing rate of reaction with HX

Problem 9.17 Draw the products of each reaction.

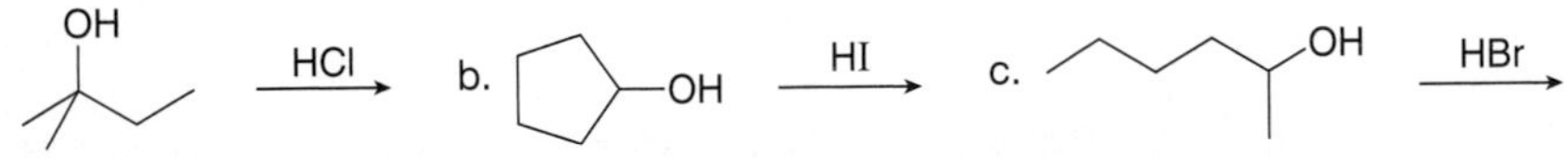
a.
OH
HCl
b.
OH
HI
c.
OH
HBr

9.11A Two Mechanisms for the Reaction of ROH with HX

How does the reaction of ROH with HX occur? Acid–base reactions are very fast, so the strong acid HX protonates the OH group of the alcohol, forming a **good leaving group** (H_2O) and a **good nucleophile** (the conjugate base, X^-). Both components are needed for nucleophilic substitution. The mechanism of substitution of X^- for H_2O then depends on the structure of the R group.

When there is an oxygen-containing reactant and a strong acid, generally the first step in the mechanism is protonation of the oxygen atom.

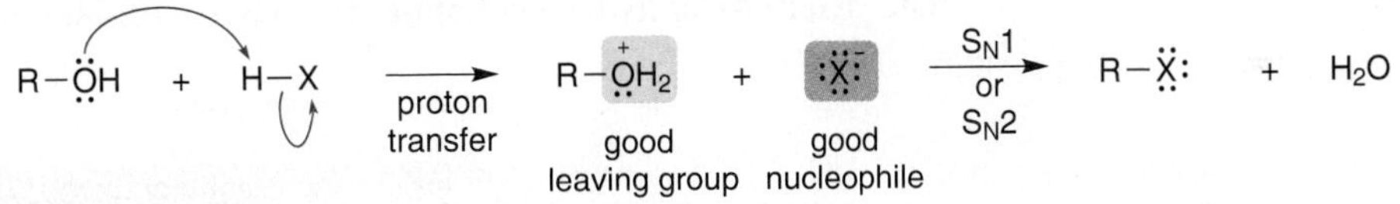
R–ÖH
+
H–X
proton
transfer
R–OH2
+
:X:
good
leaving group
good
nucleophile
SN1
or
SN2
R–X:
+
H2O

- **Methyl and 1° ROH form RX by an S_N2 mechanism.**
- **Secondary (2°) and 3° ROH form RX by an S_N1 mechanism.**

The reaction of CH_3CH_2OH with HBr illustrates the S_N2 mechanism of a 1° alcohol (Mechanism 9.5). Nucleophilic attack on the protonated alcohol occurs in one step: **the bond to the nucleophile X^- is formed *as* the bond to the leaving group (H_2O) is broken.**

Mechanism 9.5 Reaction of a 1° ROH with HX—An S_N2 Mechanism

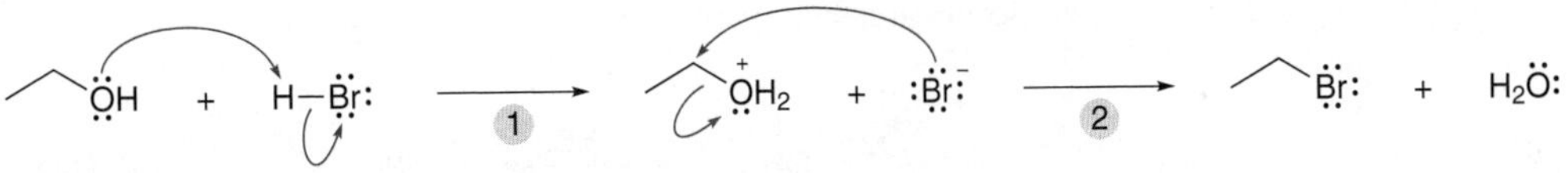
ÖH
+
H–Br:
1
OH2
+
:Br:
2
Br:
+
H2Ö:

1 Protonation of the OH group forms a good leaving group (H_2O).

2 The bond to the nucleophile forms as the leaving group departs.

The reaction of $(CH_3)_3COH$ with HBr illustrates the S_N1 mechanism of a 3° alcohol (Mechanism 9.6). Nucleophilic attack on the protonated alcohol occurs in two steps: **the bond to the leaving group (H_2O) is broken *before* the bond to the nucleophile X^- is formed.**

Mechanism 9.6 Reaction of 2° and 3° ROH with HX—An S_N1 Mechanism

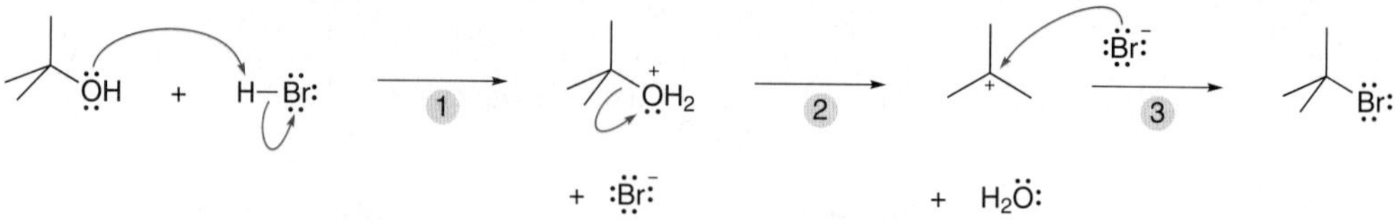

1 Protonation of the OH group forms a good leaving group (H_2O).

2 Loss of the leaving group forms a carbocation.

3 Nucleophilic attack of Br^- forms the substitution product.

Both mechanisms begin with the same first step—protonation of the O atom to form a good leaving group—and both mechanisms give an alkyl halide (RX) as product. The mechanisms differ only in the timing of bond breaking and bond making.

The reactivity of hydrogen halides increases with increasing acidity:

H–Cl H–Br H–I

Increasing reactivity toward ROH

Because Cl^- is a poorer nucleophile than Br^- or I^-, the reaction of 1° alcohols with HCl occurs only when an additional Lewis acid catalyst, usually $\mathbf{ZnCl_2}$, is added. $ZnCl_2$ complexes with the O atom of the alcohol in a Lewis acid–base reaction, making a leaving group and facilitating the S_N2 reaction.

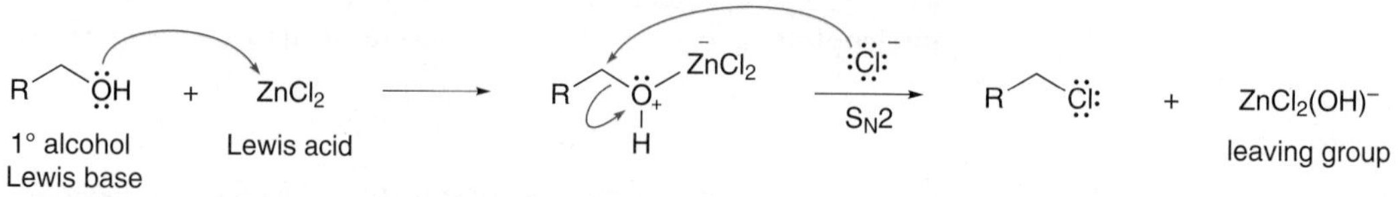

Knowing the mechanism allows us to predict the stereochemistry of the products when reaction occurs at a stereogenic center.

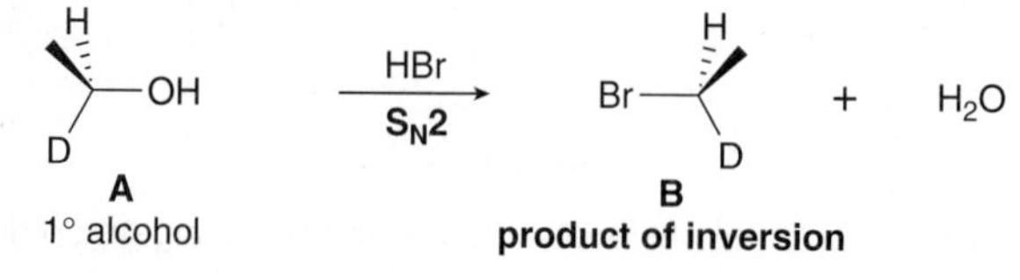

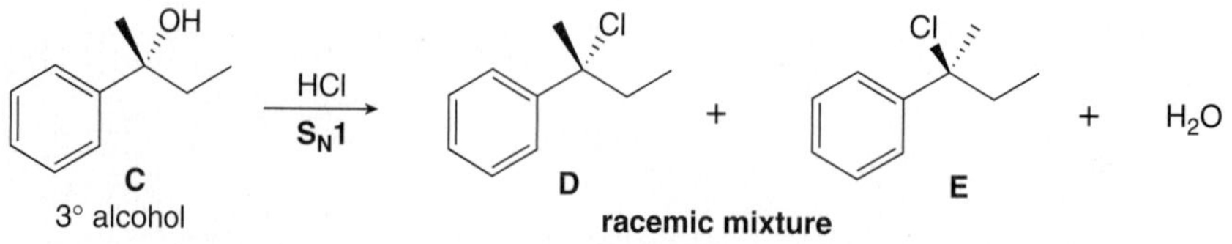

- The 1° alcohol **A** reacts with HBr via an S_N2 mechanism to yield the alkyl bromide **B** with **inversion** of stereochemistry at the stereogenic center.
- The 3° alcohol **C** reacts with HCl via an S_N1 mechanism to yield a **racemic mixture** of alkyl chlorides **D** and **E,** because a trigonal planar carbocation intermediate is formed.

Problem 9.18 Draw the products of each reaction, indicating the stereochemistry around any stereogenic centers.

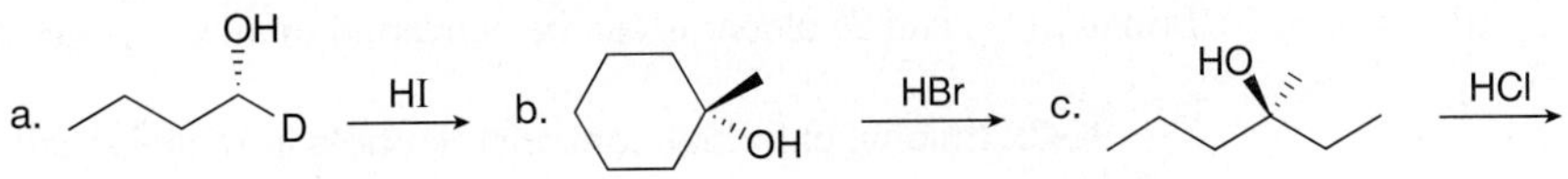

9.11B Carbocation Rearrangement in the S_N1 Reaction

Because carbocations are formed in the S_N1 reaction of 2° and 3° alcohols with HX, **carbocation rearrangements are possible,** as illustrated in Sample Problem 9.4.

Sample Problem 9.4 Draw a stepwise mechanism for the following reaction.

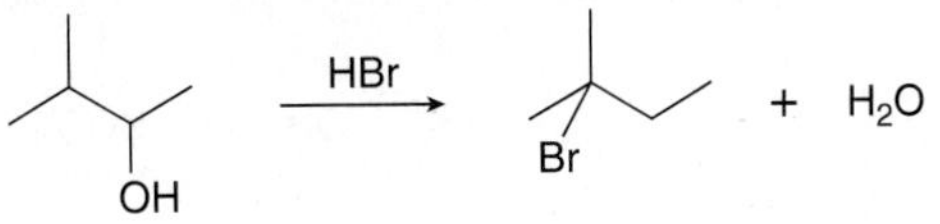

Solution

A 2° alcohol reacts with HBr by an S_N1 mechanism. Because substitution converts a 2° alcohol to a 3° alkyl halide in this example, a carbocation rearrangement must occur.

Steps [1] and [2] Protonation of the O atom and then loss of H_2O form a 2° carbocation.

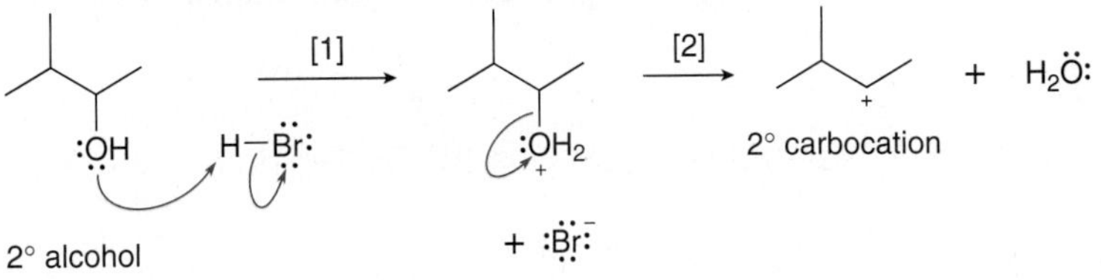

Steps [3] and [4] **Rearrangement of the 2° carbocation by a 1,2-hydride shift forms a more stable 3° carbocation.** Nucleophilic attack forms the substitution product.

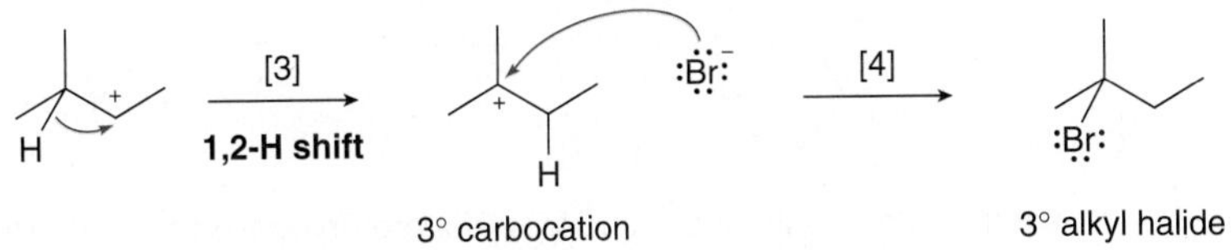

Problem 9.19 What is the major product formed when each alcohol is treated with HCl?

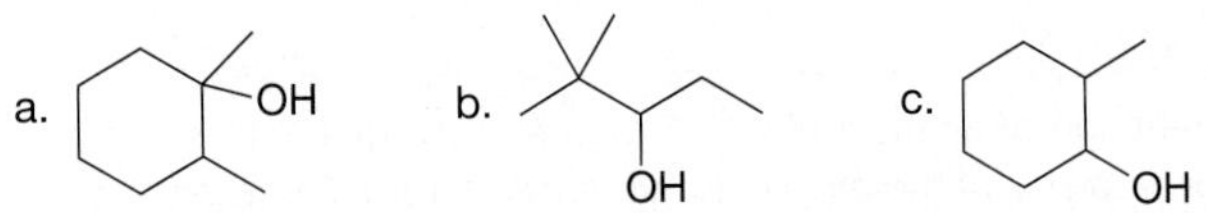

9.12 Conversion of Alcohols to Alkyl Halides with $SOCl_2$ and PBr_3

Primary (1°) and 2° alcohols can be converted to alkyl halides using $SOCl_2$ and PBr_3.

- **$SOCl_2$ (thionyl chloride) converts alcohols into alkyl chlorides.**
- **PBr_3 (phosphorus tribromide) converts alcohols into alkyl bromides.**

Both reagents convert $^-$OH into a good leaving group *in situ*—that is, directly in the reaction mixture—as well as provide the **nucleophile,** either Cl^- or Br^-, to displace the leaving group.

9.12A Reaction of ROH with $SOCl_2$

The treatment of a 1° or 2° alcohol with thionyl chloride, $SOCl_2$, and pyridine forms an alkyl chloride, with SO_2 and HCl as by-products.

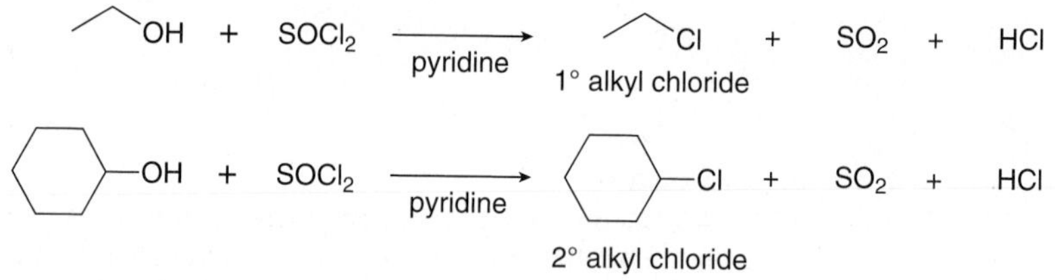

The mechanism for this reaction consists of two parts: **conversion of the OH group into a better leaving group, and nucleophilic attack by Cl^- via an S_N2 reaction,** as shown in Mechanism 9.7.

Mechanism 9.7 Reaction of ROH with $SOCl_2$ + Pyridine—An S_N2 Mechanism

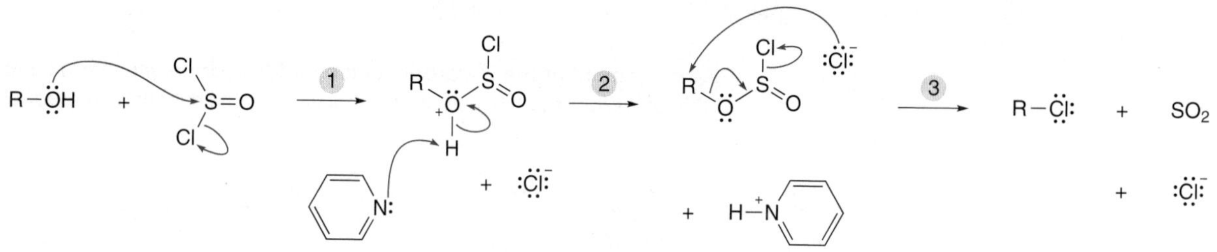

1–2 Reaction of the alcohol with $SOCl_2$ and loss of a proton convert the OH group to OSOCl, a good leaving group.

3 Nucleophilic attack of chloride and loss of the leaving group (SO_2 and Cl^-) form RCl in a single step.

Problem 9.20 If the reaction of an alcohol with $SOCl_2$ and pyridine follows an S_N2 mechanism, what is the stereochemistry of the alkyl chloride formed from (*R*)-butan-2-ol?

9.12B Reaction of ROH with PBr_3

In a similar fashion, the treatment of a 1° or 2° alcohol with phosphorus tribromide, PBr_3, forms an alkyl bromide.

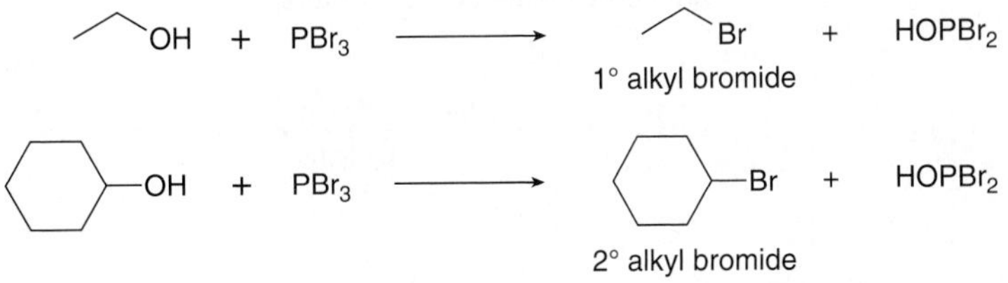
OH
+
PBr3
Br
+
HOPBr2
1° alkyl bromide
OH
+
PBr3
Br
+
HOPBr2
2° alkyl bromide

The mechanism for this reaction also consists of two parts: **conversion of the OH group into a better leaving group, and nucleophilic attack by Br^- via an S_N2 reaction,** as shown in Mechanism 9.8.

Mechanism 9.8 Reaction of ROH with PBr_3—An S_N2 Mechanism

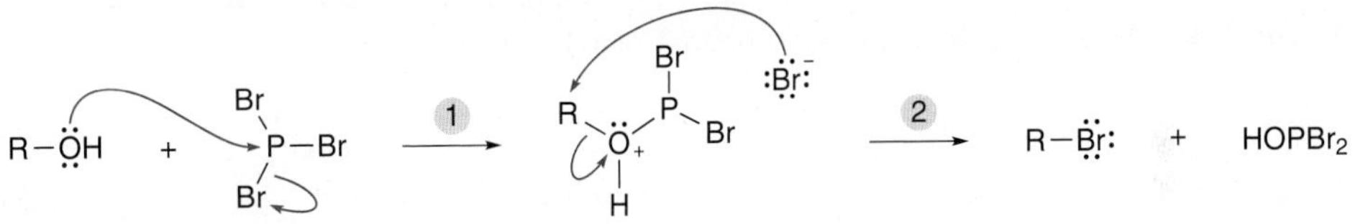
R–ÖH
+
Br
P–Br
Br
1
R
Br
P
Br
O+
H
:Br:
2
R–Br:
+
HOPBr2

1 Reaction of the alcohol with PBr_3 converts the OH group to $OPBr_2$, a good leaving group, and generates the nucleophile, Br^-.

2 Nucleophilic attack of bromide and loss of the leaving group form RBr in a single step.

Table 9.2 summarizes the methods for converting an alcohol to an alkyl halide presented in Sections 9.11 and 9.12.

Table 9.2 Summary of Methods for ROH → RX

Overall reaction	Reagent	Comment
ROH → RCl	HCl	• Useful for all ROH • An S_N1 mechanism for 2° and 3° ROH; an S_N2 mechanism for CH_3OH and 1° ROH
	$SOCl_2$	• Best for CH_3OH, and 1° and 2° ROH • An S_N2 mechanism
ROH → RBr	HBr	• Useful for all ROH • An S_N1 mechanism for 2° and 3° ROH; an S_N2 mechanism for CH_3OH and 1° ROH
	PBr_3	• Best for CH_3OH, and 1° and 2° ROH • An S_N2 mechanism
ROH → RI	HI	• Useful for all ROH • An S_N1 mechanism for 2° and 3° ROH; an S_N2 mechanism for CH_3OH and 1° ROH

Problem 9.21 If the reaction of an alcohol with PBr_3 follows an S_N2 mechanism, what is the stereochemistry of the alkyl bromide formed from (*R*)-butan-2-ol?

Problem 9.22 Draw the organic products formed in each reaction, and indicate the stereochemistry of products that contain stereogenic centers.

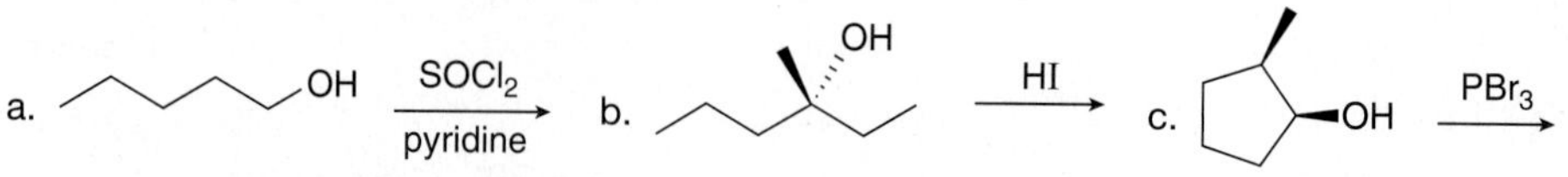

9.12C The Importance of Making RX from ROH

We have now learned two methods to prepare an alkyl chloride and two methods to prepare an alkyl bromide from an alcohol. If there is one good way to carry out a reaction, why search for more? A particular reagent might work well for one starting material, but not so well for another, so organic chemists try to devise several different ways to perform the same overall reaction.

Why are there so many ways to convert an alcohol to an alkyl halide? Alkyl halides are versatile starting materials in organic synthesis, as shown in Sample Problem 9.5.

Sample Problem 9.5 Convert propan-1-ol to butanenitrile (**A**).

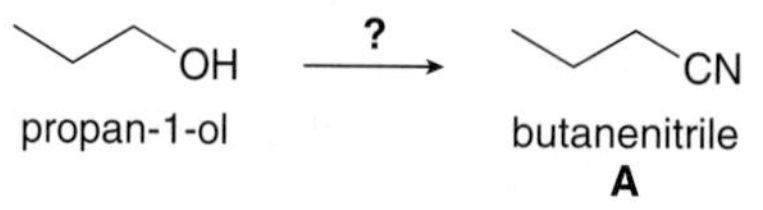

Solution

Direct conversion of propan-1-ol to **A** using ^-CN as a nucleophile is not possible because ^-OH is a poor leaving group. However, conversion of the OH group to a Br atom forms a good leaving group, which can then readily undergo an S_N2 reaction with ^-CN to yield **A.** The overall result of this two-step sequence is the substitution of ^-OH by ^-CN.

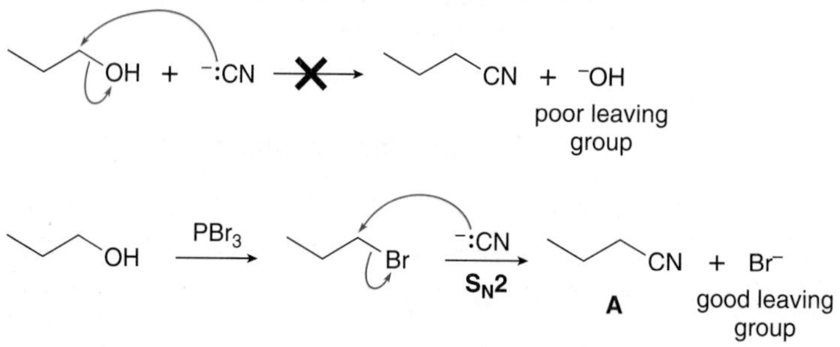

Problem 9.23 Draw two steps to convert $(CH_3)_2CHOH$ into each of the following compounds: $(CH_3)_2CHN_3$ and $(CH_3)_2CHOCH_2CH_3$.

9.13 Tosylate—Another Good Leaving Group

We have now learned two methods to convert the OH group of an alcohol to a better leaving group: treatment with strong acids (Section 9.8), and conversion to an alkyl halide (Sections 9.11–9.12). Alcohols can also be converted to **alkyl tosylates.**

Recall from Section 1.5 that a third-row element like sulfur can have 10 or 12 electrons around it in a valid Lewis structure.

An alkyl tosylate

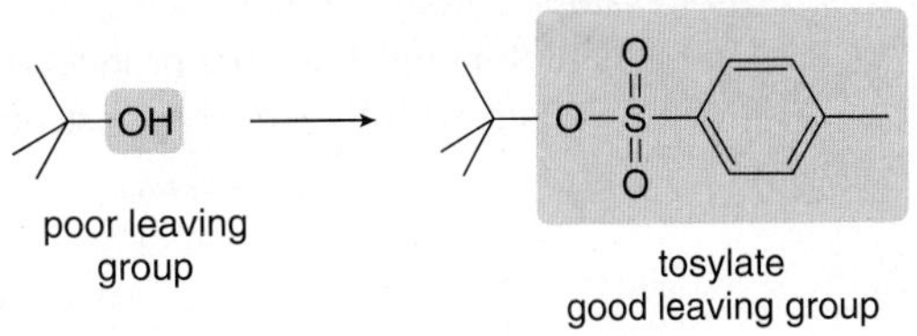

An alkyl tosylate is often called simply a **tosylate.**

An **alkyl tosylate** is composed of two parts: the **alkyl group R,** derived from an alcohol; and the **tosylate** (short for ***p*-toluenesulfonate**), which is a good leaving group. A tosyl group, $CH_3C_6H_4SO_2-$, is abbreviated as **Ts,** so an alkyl tosylate becomes **ROTs.**

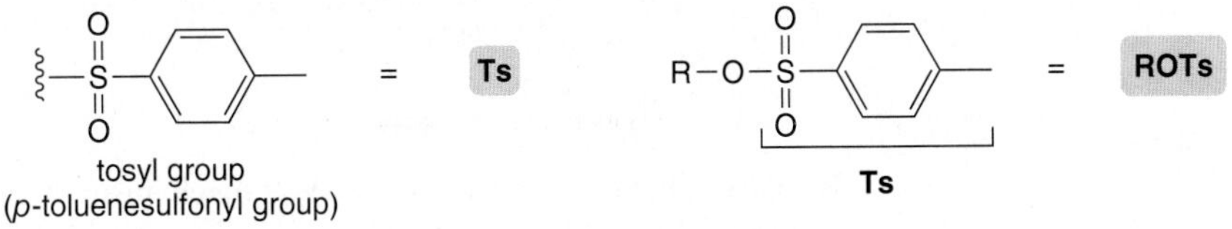

9.13A Conversion of Alcohols to Alkyl Tosylates

A tosylate (TsO⁻) is similar to I⁻ in leaving group ability.

Alcohols are converted to alkyl tosylates by treatment with *p*-toluenesulfonyl chloride (TsCl) in the presence of pyridine. This overall process converts a poor leaving group (⁻OH) into a good one (⁻OTs). A tosylate is a good leaving group because its conjugate acid, *p*-toluenesulfonic acid ($CH_3C_6H_4SO_3H$, TsOH), is a strong acid ($pK_a = -7$, Section 2.6).

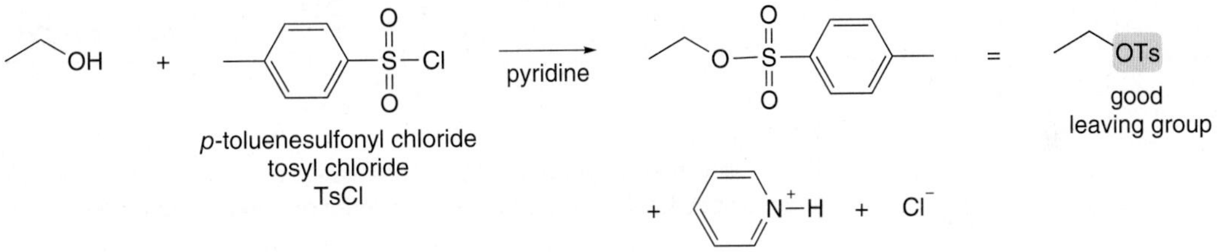

(*S*)-Butan-2-ol is converted to its tosylate with **retention of configuration** at the stereogenic center. Thus, the C–O bond of the alcohol must *not* be broken when the tosylate is formed.

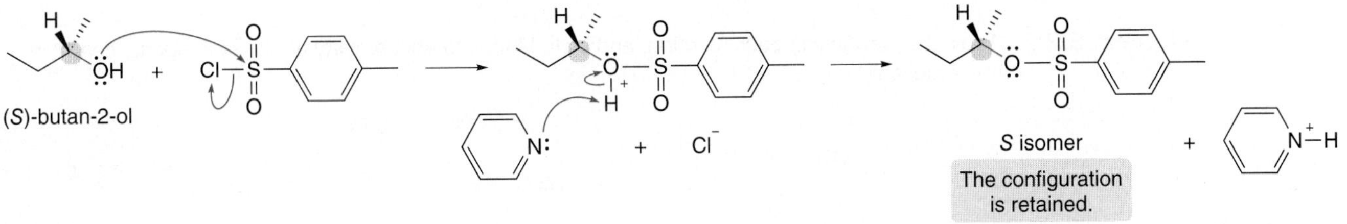

Problem 9.24 Draw the products of each reaction, and indicate the stereochemistry at any stereogenic center.

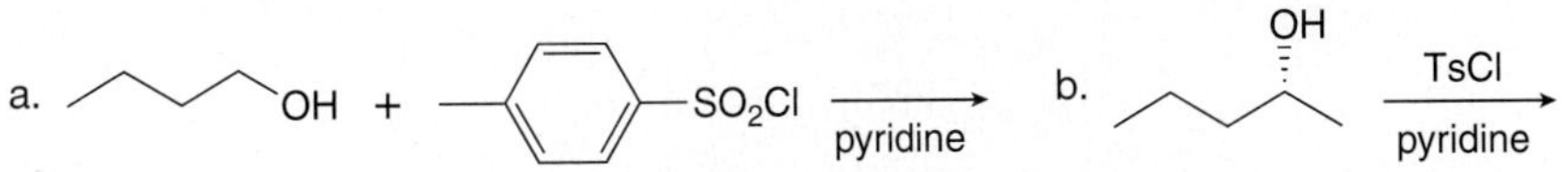

9.13B Reactions of Alkyl Tosylates

Because alkyl tosylates have good leaving groups, **they undergo both nucleophilic substitution and β elimination,** exactly as alkyl halides do. Generally, alkyl tosylates are treated with strong nucleophiles and bases, so that the mechanism of substitution is **S_N2** and the mechanism of elimination is **E2.**

For example, propyl tosylate, which has the leaving group on a 1° carbon, reacts with $NaOCH_3$ to yield methyl propyl ether, the product of nucleophilic substitution by an S_N2 mechanism. Propyl tosylate reacts with $KOC(CH_3)_3$, a strong bulky base, to yield propene by an E2 mechanism.

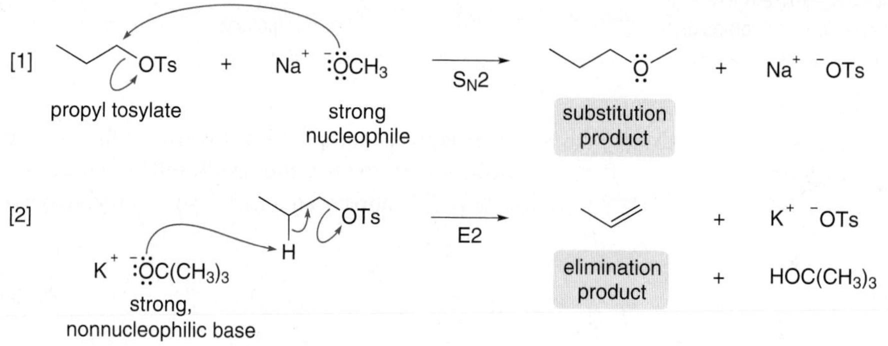

Because substitution occurs via an **S_N2** mechanism, **inversion of configuration** results when the leaving group is bonded to a stereogenic center.

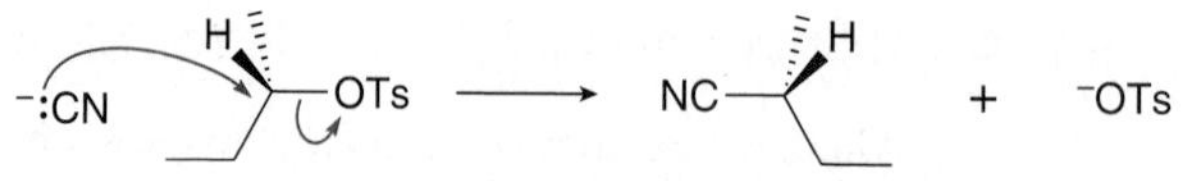

Inversion of configuration

Sample Problem 9.6 Draw the product of the following reaction, including stereochemistry.

Solution

The 1° alkyl tosylate and the strong nucleophile both favor substitution by an **S_N2** mechanism, which proceeds by backside attack, resulting in **inversion** of configuration at the stereogenic center.

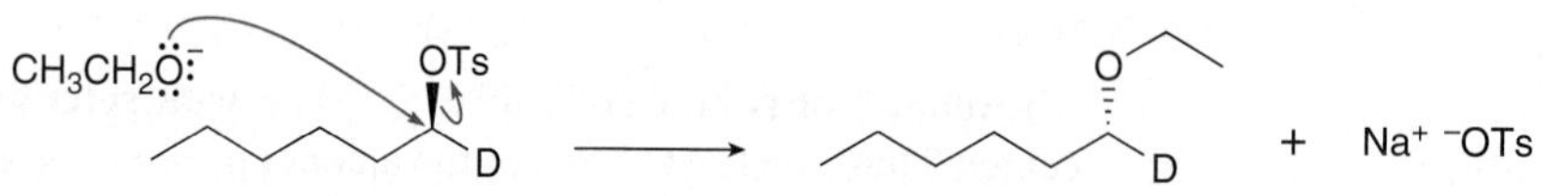

1° tosylate

Problem 9.25 Draw the products of each reaction, and include the stereochemistry at any stereogenic center in the products.

a. 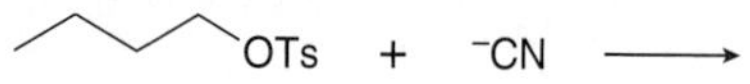

c. 

b. 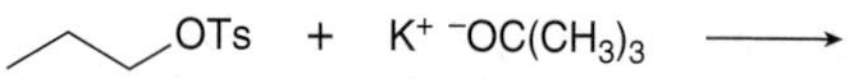

9.13C The Two-Step Conversion of an Alcohol to a Substitution Product

We now have another **two-step method to convert an alcohol to a substitution product:** reaction of an alcohol with TsCl and pyridine to form an alkyl tosylate (Step [1]), followed by nucleophilic attack on the tosylate (Step [2]).

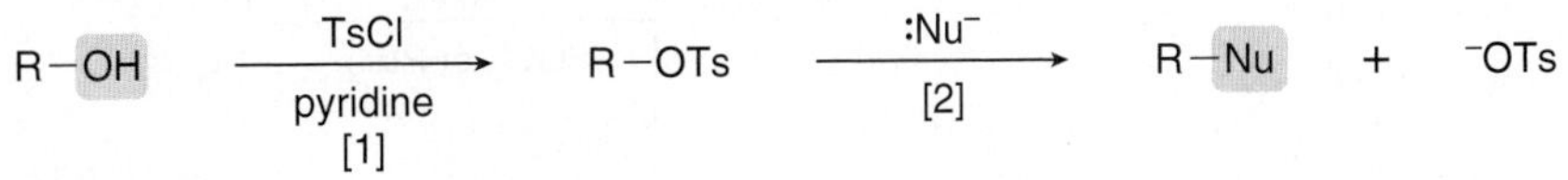

Let's look at the stereochemistry of this two-step process.

- Step [1], formation of the tosylate, proceeds with **retention** of configuration at a stereogenic center because the C–O bond remains intact.
- Step [2] is an S_N2 reaction, so it proceeds with **inversion of configuration** because the nucleophile attacks from the back side.
- Overall there is a **net inversion of configuration** at a stereogenic center.

For example, the treatment of *cis*-3-methylcyclohexanol with *p*-toluenesulfonyl chloride and pyridine forms a cis tosylate **A,** which undergoes backside attack by the nucleophile $^{-}OCH_3$ to yield the trans ether **B.**

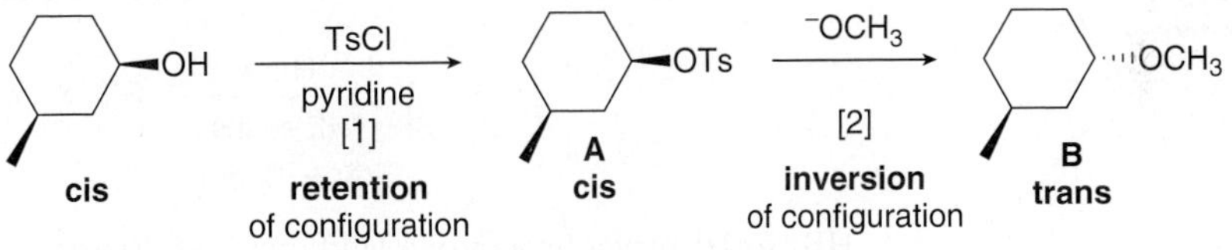

Problem 9.26 Draw the products formed when (*S*)-butan-2-ol is treated with TsCl and pyridine, followed by NaOH. Label the stereogenic center in each compound as *R* or *S*. What is the stereochemical relationship between the starting alcohol and the final product?

9.13D A Summary of Substitution and Elimination Reactions of Alcohols

The reactions of alcohols in Sections 9.8–9.13C share two similarities:

- **The OH group is converted into a better leaving group by treatment with acid or another reagent.**
- **The resulting product undergoes either elimination or substitution, depending on the reaction conditions.**

Figure 9.8 summarizes these reactions with cyclohexanol as starting material.

Problem 9.27 Draw the product formed when $(CH_3)_2CHOH$ is treated with each reagent.

a. $SOCl_2$, pyridine
b. TsCl, pyridine
c. H_2SO_4
d. HBr
e. PBr_3, then NaCN
f. $POCl_3$, pyridine

Figure 9.8 Summary: Nucleophilic substitution and β elimination reactions of alcohols

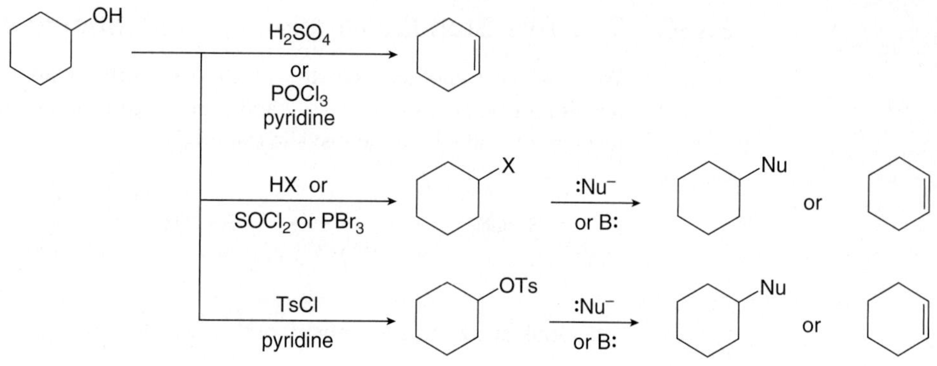

9.14 Reaction of Ethers with Strong Acid

Because ethers are so unreactive, diethyl ether and tetrahydrofuran (THF) are often used as solvents for organic reactions.

Recall from Section 9.7B that ethers have a poor leaving group, so they cannot undergo nucleophilic substitution or β elimination reactions directly. Instead, they must first be converted to a good leaving group by reaction with strong acids. Only **HBr** and **HI** can be used, though, because they are strong acids that are also sources of good nucleophiles (Br^- and I^-, respectively). **When ethers react with HBr or HI, both C–O bonds are cleaved and two alkyl halides are formed as products.**

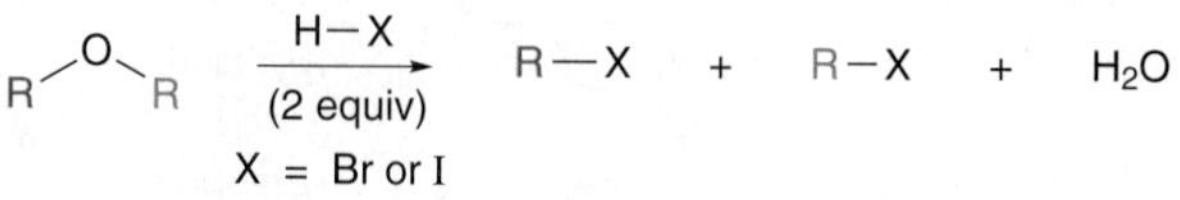

HBr or HI serves as a strong acid that both protonates the O atom of the ether and is the source of a good nucleophile (Br^- or I^-). Because both C–O bonds in the ether are broken, **two successive nucleophilic substitution reactions occur.**

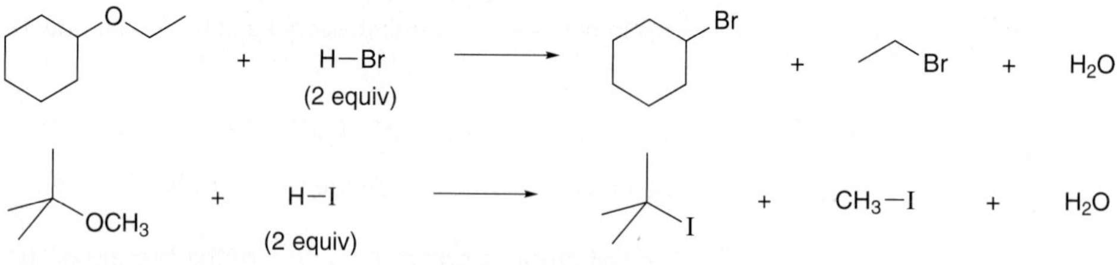

- **The mechanism of ether cleavage is S_N1 or S_N2, depending on the identity of R.**
- **With 2° or 3° alkyl groups bonded to the ether oxygen, the C–O bond is cleaved by an S_N1 mechanism involving a carbocation; with methyl or 1° R groups, the C–O bond is cleaved by an S_N2 mechanism.**

For example, cleavage of $(CH_3)_3COCH_3$ with HI occurs at two bonds, as shown in Mechanism 9.9. The 3° alkyl group undergoes nucleophilic substitution by an $\mathbf{S_N1}$ mechanism, resulting in the cleavage of one C–O bond. The methyl group undergoes nucleophilic substitution by an $\mathbf{S_N2}$ mechanism, resulting in the cleavage of the second C–O bond.

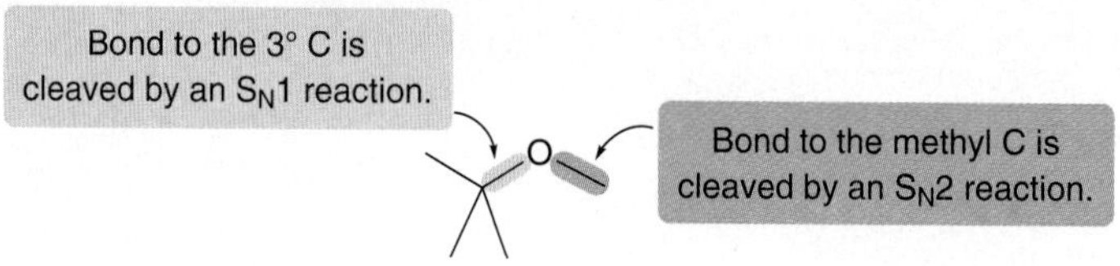

Mechanism 9.9 Mechanism of Ether Cleavage in Strong Acid—$(CH_3)_3COCH_3 + HI \rightarrow (CH_3)_3CI + CH_3I + H_2O$

Part [1] Cleavage of the 3° C–O bond by an S_N1 mechanism

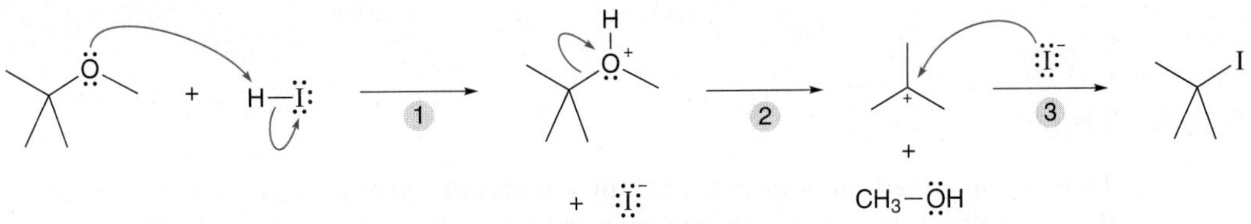

1. Protonation of the ether O atom forms a good leaving group.
2. Cleavage of the C–O bond to the 3° carbon forms a 3° carbocation and CH_3OH.
3. Nucleophilic attack of I^- forms the substitution product.

Part [2] Cleavage of the CH_3–O bond by an S_N2 mechanism

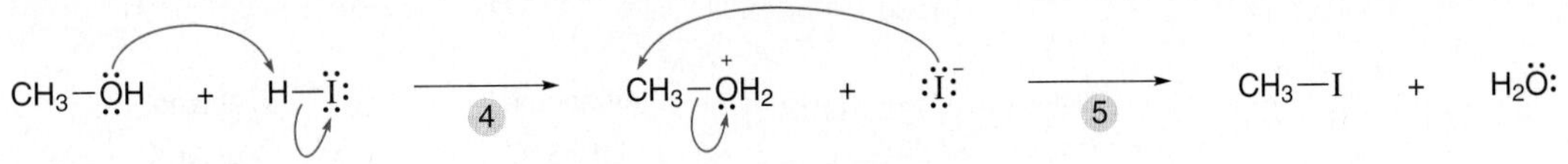

4. Protonation of the OH group forms a good leaving group (H_2O).
5. Nucleophilic attack of iodide forms the second alkyl halide, CH_3I. Because the mechanism is S_N2, the C–O bond is broken as the C–I bond is formed.

The mechanism illustrates the central role of HX in the reaction:

- **HX protonates the ether oxygen, thus making a good leaving group.**
- **HX provides a source of X^- for nucleophilic attack.**

Problem 9.28 What alkyl halides are formed when each ether is treated with HBr?

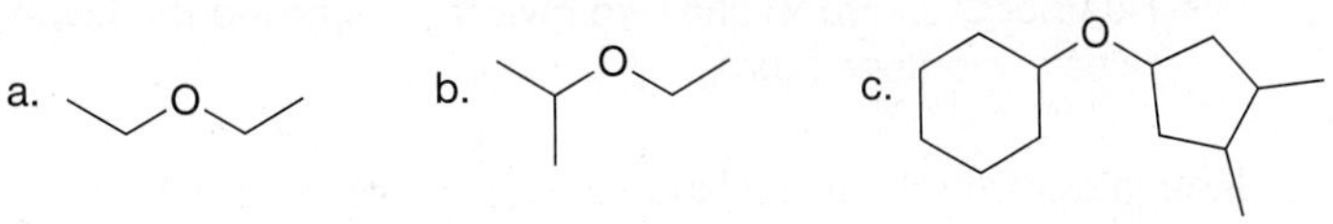

Problem 9.29 Explain why the treatment of anisole with HBr yields phenol and CH_3Br, but not bromobenzene.

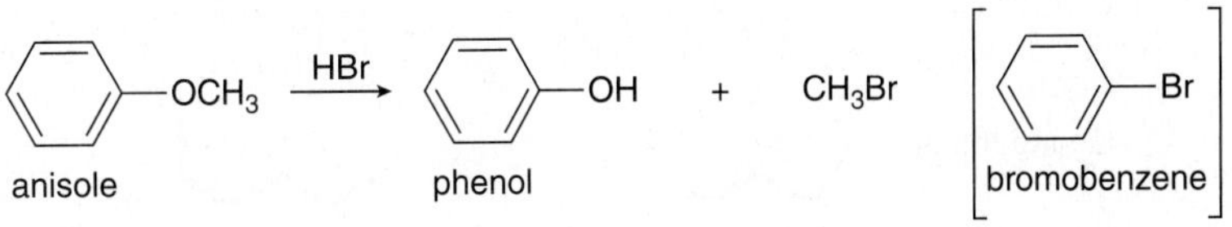

9.15 Thiols and Sulfides

Thiols and sulfides are sulfur analogues of alcohols and ethers, respectively.

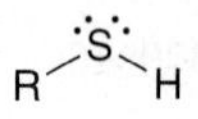

thiol

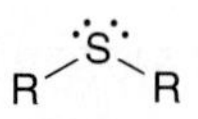

sulfide

9.15A Thiols

Thiols, also called mercaptans, contain a mercapto group (SH) bonded to a carbon atom. Because sulfur is below oxygen in the periodic table, the sulfur atom is surrounded by two atoms and two lone pairs, giving thiols a bent shape. Unlike alcohols, however, thiols are incapable of intermolecular hydrogen bonds, so thiols have lower boiling points and melting points than alcohols with a similar number of carbons.

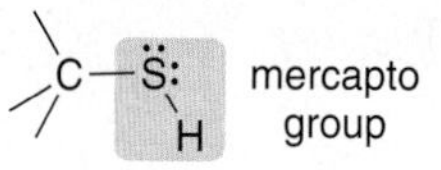

thiol

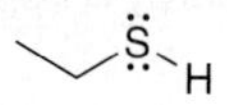

ethanethiol
bp 35 °C

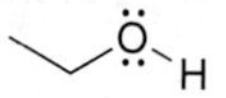

ethanol
bp 78 °C

Many simple thiols have pungent and disagreeable odors. Skunks, onions, and human sweat all contain thiols.

propane-1-thiol
onion odor

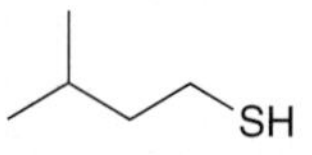

3-methylbutane-1-thiol
skunk odor

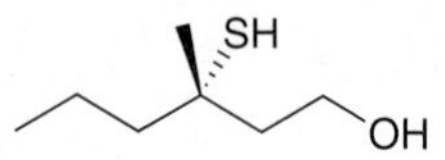

(*S*)-3-methyl-3-sulfanylhexan-1-ol
onion-like odor in human sweat

Thiols are named in a similar method to alcohols, using the suffix ***-thiol*** instead of the suffix ***-ol.*** To name a thiol in the IUPAC system:

- **Name the parent carbon chain and add the suffix *-thiol*.**
- **Number the carbon chain to give the SH group the lower number and apply the other rules of nomenclature.**

Examples of thiol nomenclature are given in Figure 9.9.

Figure 9.9
Naming thiols

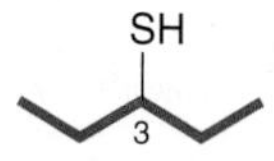

pentane-3-thiol

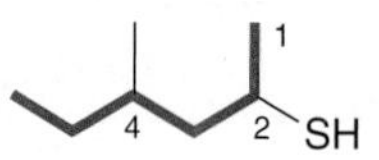

4-methylhexane-2-thiol

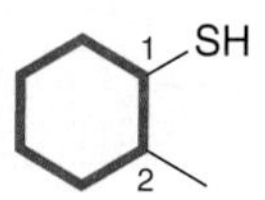

2-methylcyclohexanethiol

Problem 9.30 Name each thiol.

a. 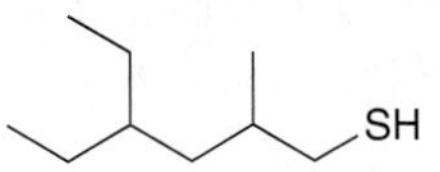

b. 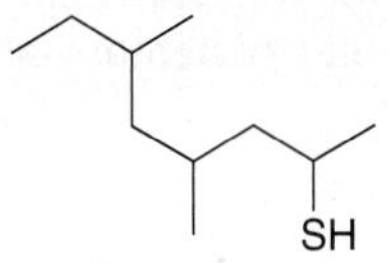

Thiols are prepared by S_N2 reactions of alkyl halides with ^-SH, a good nucleophile.

Thiols are easily oxidized with Br_2 or I_2 to disulfides (RSSR), compounds that contain a sulfur–sulfur bond. This reaction is an oxidation (Section 4.14) because H atoms are removed from the thiol in forming the disulfide. Disulfides are reduced to thiols with Zn and acid.

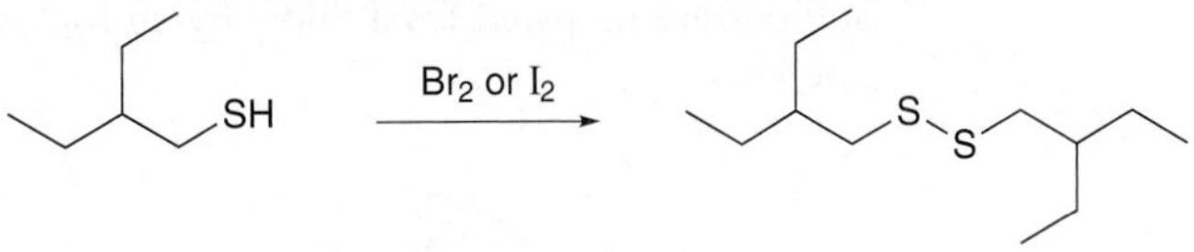

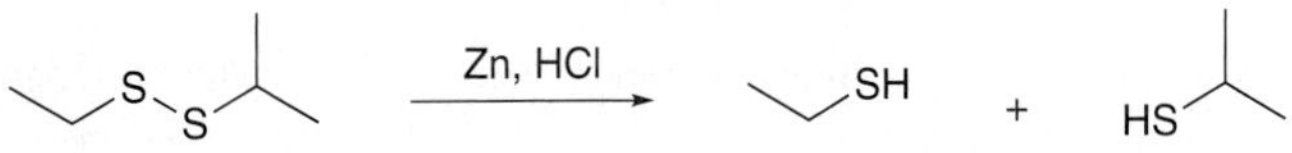

Disulfide formation is especially important in determining the shape and properties of some proteins that contain the amino acid cysteine, as we will learn in Chapter 26.

Problem 9.31 Draw the product of each reaction.

a. 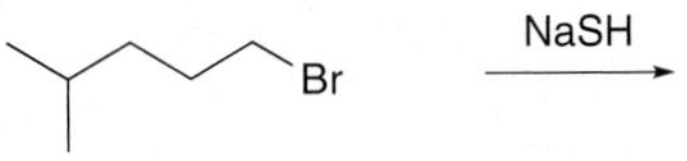

c. 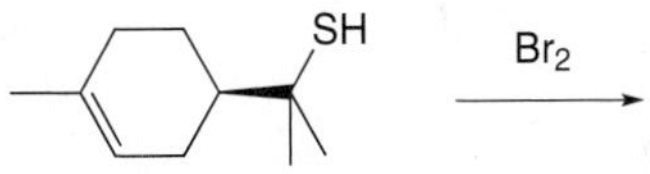

grapefruit mercaptan

b. 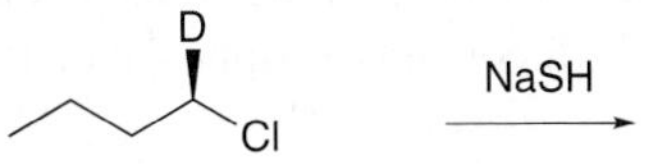

d. 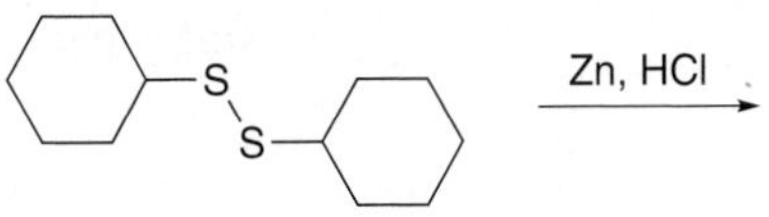

The potent odor of grapefruit mercaptan (Problem 9.31c) contributes to the characteristic aroma of grapefruit.

9.15B Sulfides

Sulfides contain two alkyl groups bonded to a sulfur atom. Sulfides are named with the same rules used to name ethers. The suffix ***sulfide*** is used instead of *ether* for simple compounds.

sulfide

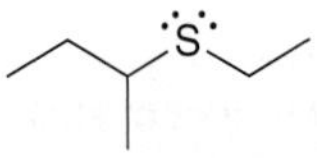

sec-butyl ethyl sulfide

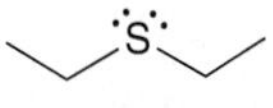

diethyl sulfide

To name more complex sulfides using the IUPAC system, one alkyl group is named as a parent chain and the other is named as part of a substituent bonded to that chain.

- Name the simpler alkyl group + S atom as an *alkylthio* substituent.
- Name the remaining alkyl group as an alkane with an alkylthio substituent using the usual rules of nomenclature.

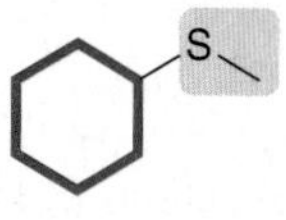

methylthiocyclohexane

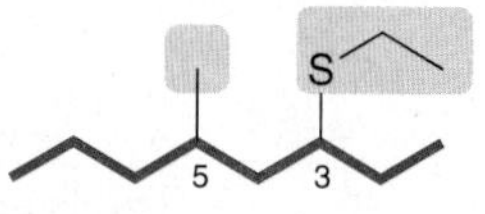

3-ethylthio-5-methyloctane

Problem 9.32 Give the IUPAC name for each sulfide.

a. 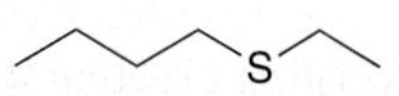

b. 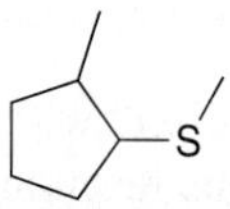

Sulfides are prepared from thiols by an S_N2 reaction that is analogous to the Williamson ether synthesis.

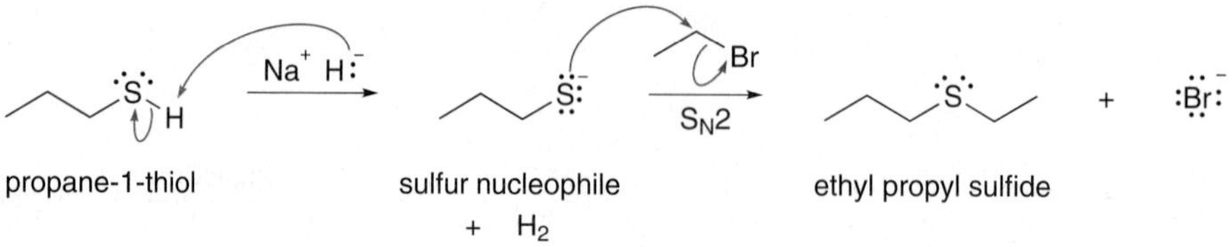

Sulfides contain a nucleophilic sulfur atom that reacts readily with unhindered alkyl halides to form sulfonium ions.

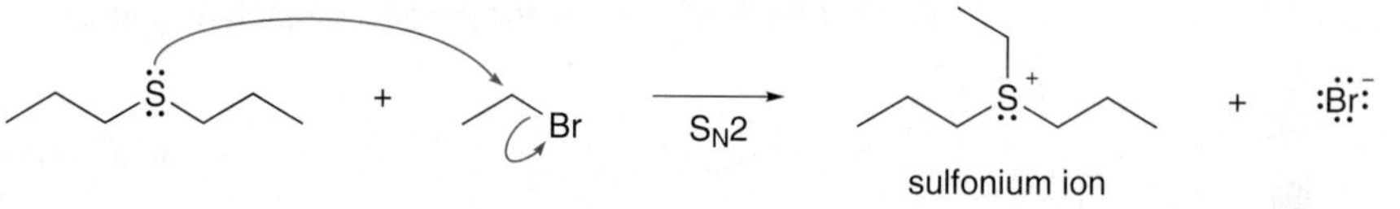

S-Adenosylmethionine (SAM), a biological sulfonium ion that was introduced in Section 7.16, is synthesized from the amino acid methionine, which contains a nucleophilic sulfide, and adenosine triphosphate (ATP), which contains a triphosphate leaving group (Section 7.16).

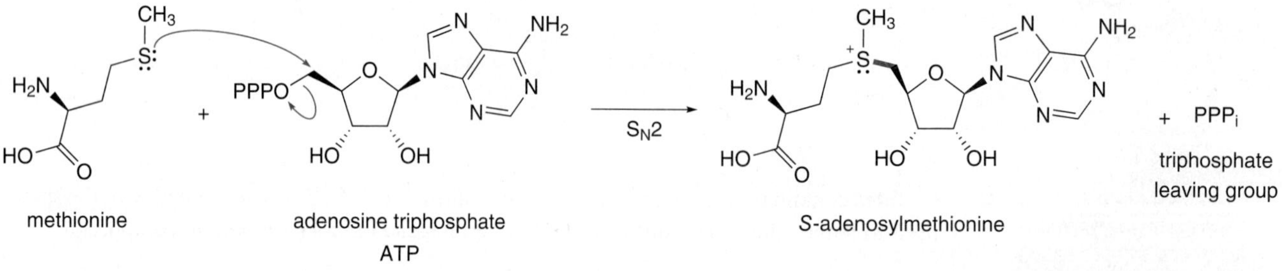

Problem 9.33 Draw the product of each reaction.

a. 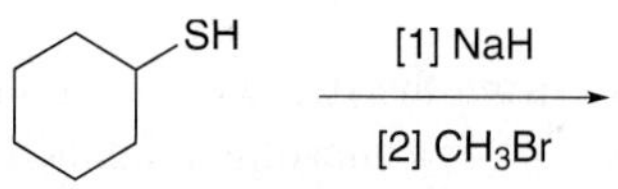

b. 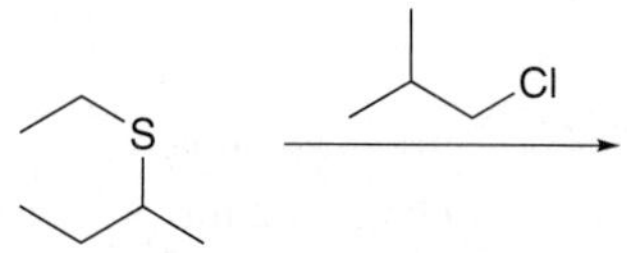

9.16 Reactions of Epoxides

Although epoxides do not contain a good leaving group, they contain a strained three-membered ring with two polar bonds. **Nucleophilic attack opens the strained three-membered ring,** making it a favorable process even with the poor leaving group.

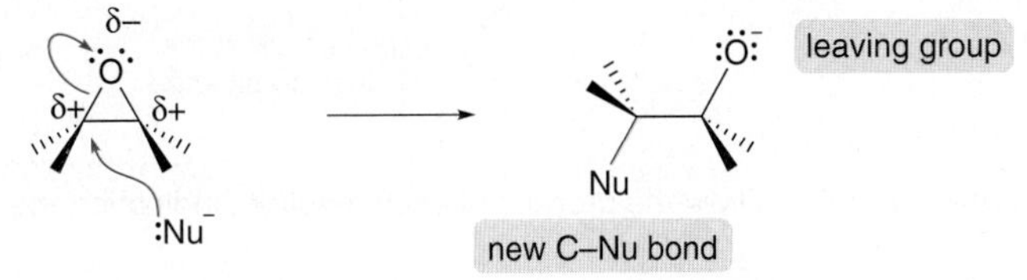

This reaction occurs readily with strong nucleophiles like ⁻CN, and with acids like HZ, where Z is a nucleophilic atom.

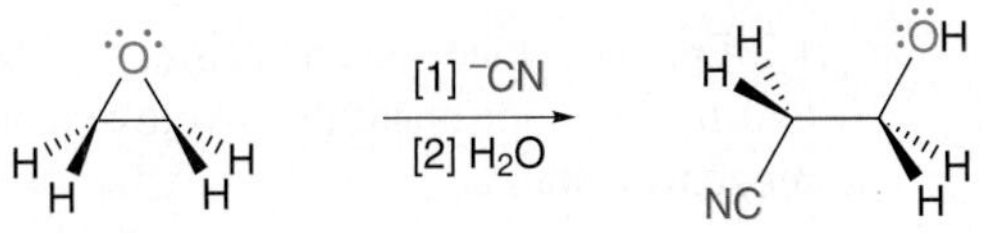

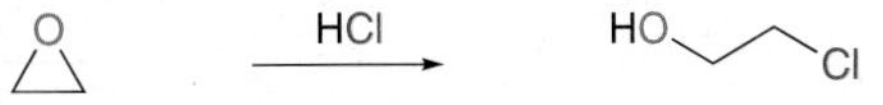

9.16A Opening of Epoxide Rings with Strong Nucleophiles

Virtually all strong nucleophiles open an epoxide ring by a two-step reaction sequence.

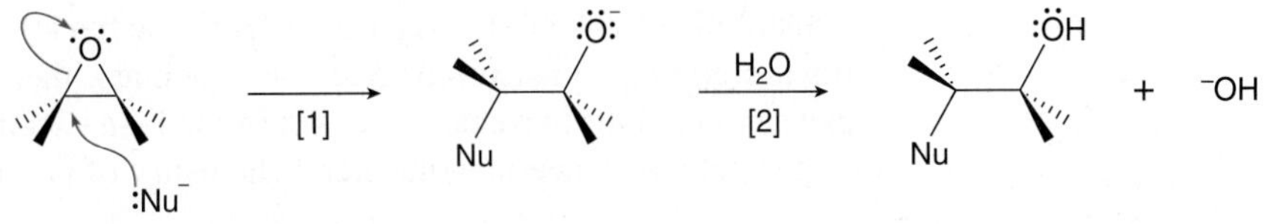

- **Step [1]:** The nucleophile attacks an electron-deficient carbon of the epoxide, cleaving a C−O bond and relieving the strain of the three-membered ring.
- **Step [2]:** Protonation of the alkoxide with water generates a neutral product with two functional groups on adjacent atoms.

Common nucleophiles that open epoxide rings include ⁻OH, ⁻OR, ⁻CN, ⁻SR, and NH_3. With these strong nucleophiles, the reaction occurs via an **S_N2 mechanism,** resulting in two consequences:

- **The nucleophile opens the epoxide ring from the back side.**

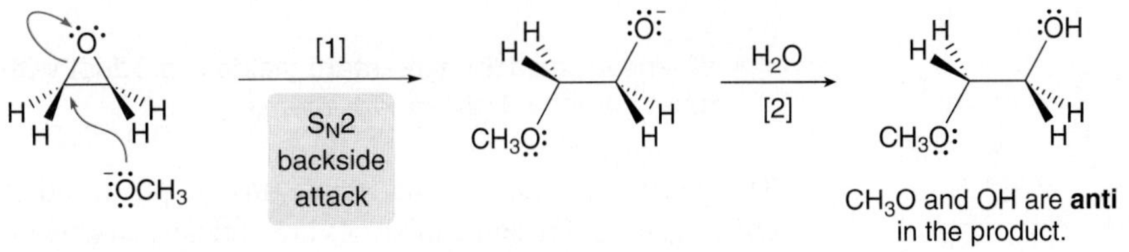

Other examples of the nucleophilic opening of epoxide rings are presented in Sections 12.6 and 20.14.

- **In an unsymmetrical epoxide, the nucleophile attacks at the less substituted carbon atom.**

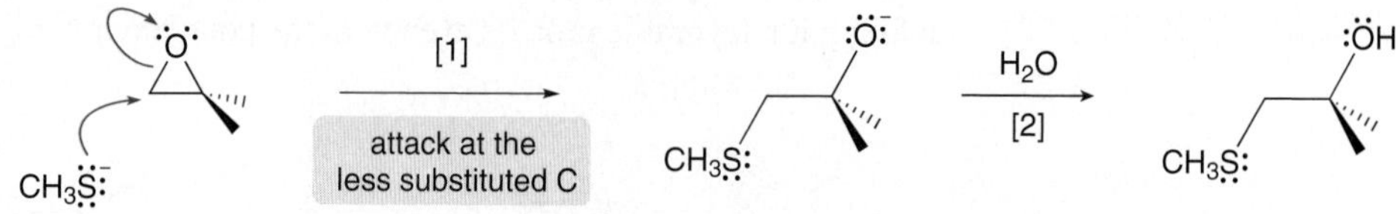
[1]
attack at the less substituted C
H_2O
[2]
$CH_3\ddot{S}$:
:ÖH

Problem 9.34 Draw the product of each reaction, indicating the stereochemistry at any stereogenic center.

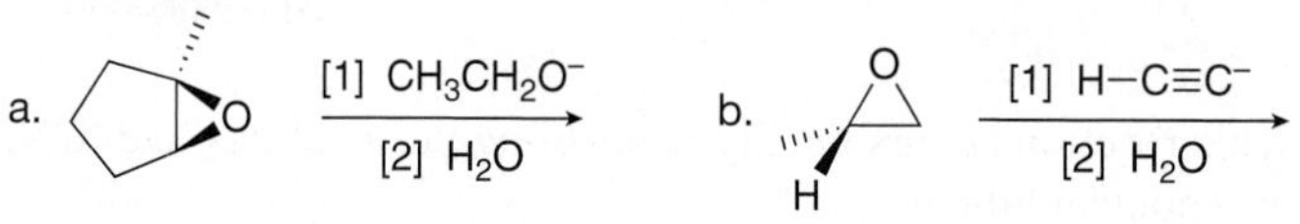
a.
[1] $CH_3CH_2O^-$
[2] H_2O
b.
[1] H–C≡C⁻
[2] H_2O

1,2-Epoxycyclohexane, an achiral epoxide with a plane of symmetry, reacts with $^-OCH_3$ to yield two *trans*-1,2-disubstituted cyclohexanes, **A** and **B,** which are **enantiomers;** each has two stereogenic centers.

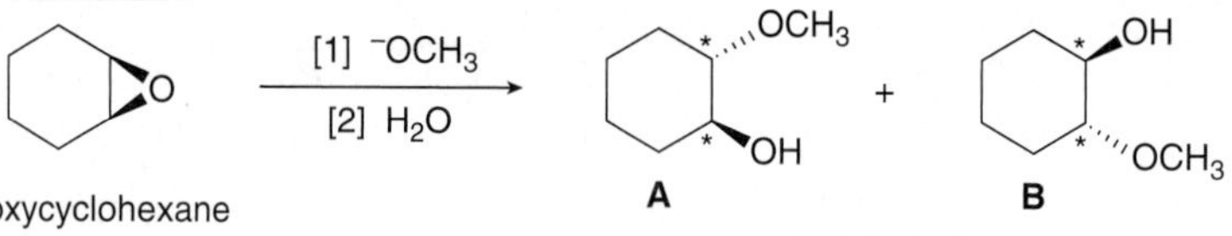
[1] $^-OCH_3$
[2] H_2O
OCH_3
OH
+
OH
OCH_3

1,2-epoxycyclohexane
achiral starting material

A **B**
enantiomers
[* denotes a stereogenic center]

Nucleophilic attack of $^-OCH_3$ occurs from the back side at *either* C–O bond, because both ends are equally substituted. Because attack at either side occurs with equal probability, an equal amount of the two enantiomers is formed—**a racemic mixture.** This is a specific example of a general rule concerning the stereochemistry of products obtained from an achiral reactant.

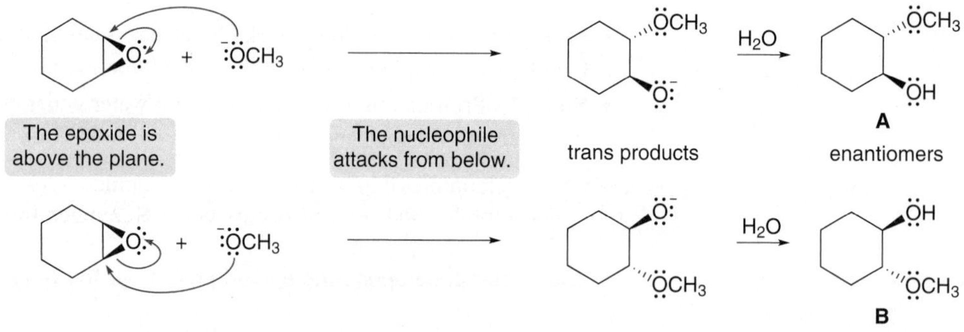
The epoxide is above the plane.
The nucleophile attacks from below.
trans products
H_2O
A
enantiomers
H_2O
B

- **Whenever an achiral reactant yields a product with stereogenic centers, the product must be achiral (meso) or racemic.**

This general rule can be restated in terms of optical activity. Recall from Section 5.12 that achiral compounds and racemic mixtures are optically inactive.

- **Optically inactive starting materials give optically inactive products.**

Problem 9.35 The cis and trans isomers of 2,3-dimethyloxirane both react with ^{-}OH to give butane-2,3-diol. One stereoisomer gives a single achiral product, and one gives two chiral enantiomers. Which epoxide gives one product and which gives two?

cis-2,3-dimethyloxirane

one enantiomer of *trans*-2,3-dimethyloxirane

9.16B Reaction with Acids HZ

Acids **HZ** that contain a nucleophile **Z** also open epoxide rings by a two-step reaction sequence.

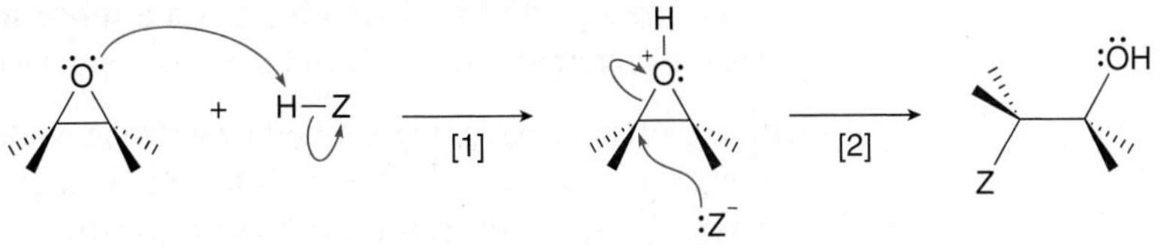

- **Step [1]:** Protonation of the epoxide oxygen with HZ makes the epoxide oxygen into a good leaving group (OH). It also provides a source of a good nucleophile (Z^-) to open the epoxide ring.
- **Step [2]:** The nucleophile Z^- then opens the protonated epoxide ring by **backside** attack.

These two steps—**protonation followed by nucleophilic attack**—are the exact reverse of the opening of epoxide rings with strong nucleophiles, where nucleophilic attack precedes protonation.

HCl, HBr, and **HI** all open an epoxide ring in this manner. $\mathbf{H_2O}$ and **ROH** can, too, but acid must also be added. Regardless of the reaction, the product has an OH group from the epoxide on one carbon and a new functional group Z from the nucleophile on the adjacent carbon. With epoxides fused to rings, ***trans*-1,2-disubstituted cycloalkanes** are formed.

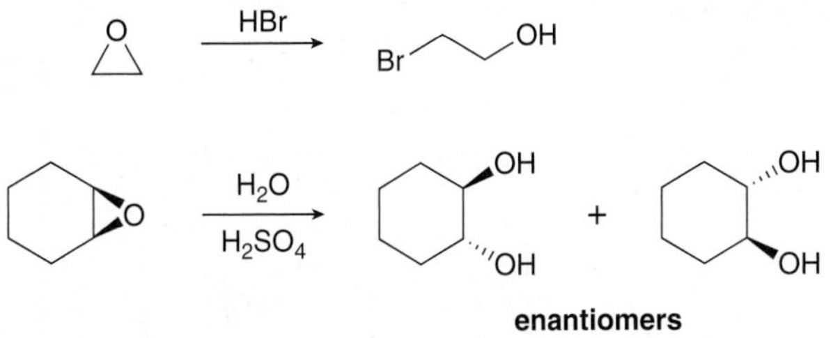

Although backside attack of the nucleophile suggests that this reaction follows an S_N2 mechanism, the regioselectivity of the reaction with unsymmetrical epoxides does not.

- **With unsymmetrical epoxides, nucleophilic attack occurs at the *more* substituted carbon atom.**

For example, the treatment of 2,2-dimethyloxirane with HCl results in nucleophilic attack at the carbon with two methyl groups.

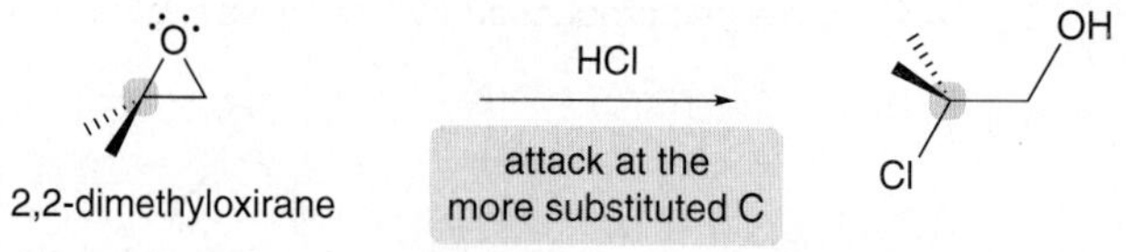

Backside attack of the nucleophile suggests an S_N2 mechanism, but attack at the more substituted carbon suggests an S_N1 mechanism. To explain these results, the **mechanism of nucleophilic attack is thought to be somewhere in between S_N1 and S_N2.**

Figure 9.10 illustrates two possible pathways for the reaction of 2,2-dimethyloxirane with HCl. Backside attack of Cl^- at the more substituted carbon proceeds via transition state **A,** whereas backside attack of Cl^- at the less substituted carbon proceeds via transition state **B. Transition state A has a partial positive charge on a more substituted carbon, making it more stable.** Thus, the preferred reaction path takes place by way of the lower energy transition state **A.**

Opening of an epoxide ring with either a strong nucleophile $:Nu^-$ or an acid HZ is **regioselective,** because one constitutional isomer is the major or exclusive product. The **site selectivity of these two reactions, however, is *exactly the opposite*.**

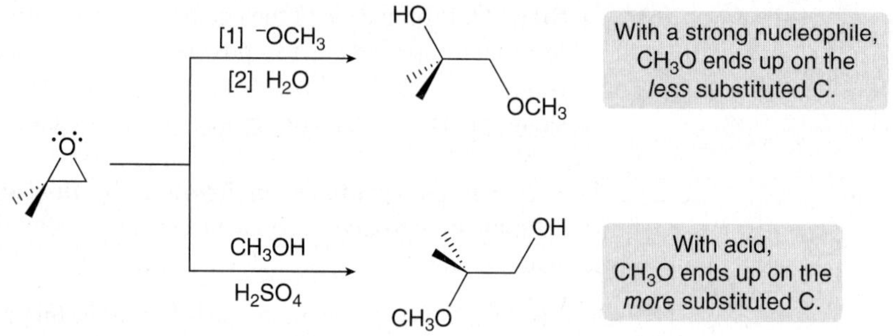

Figure 9.10
Opening of an unsymmetrical epoxide ring with HCl

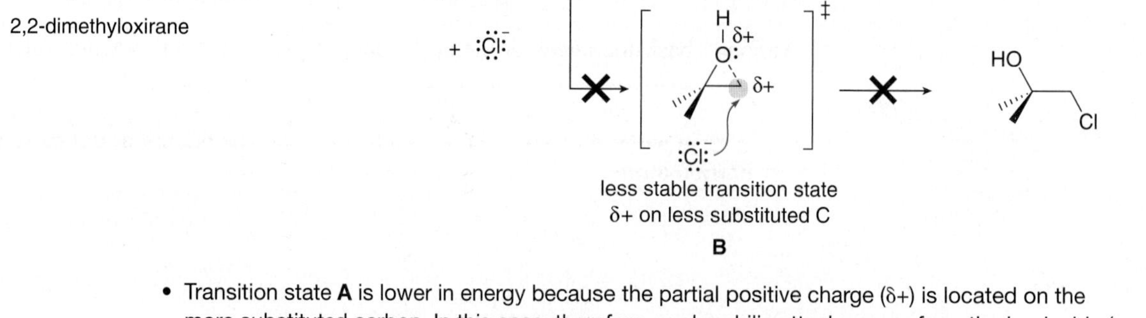

- Transition state **A** is lower in energy because the partial positive charge (δ+) is located on the more substituted carbon. In this case, therefore, nucleophilic attack occurs from the back side (an S_N2 characteristic) at the more substituted carbon (an S_N1 characteristic).

Figure 9.11
The synthesis of two bronchodilators uses the opening of an epoxide ring.

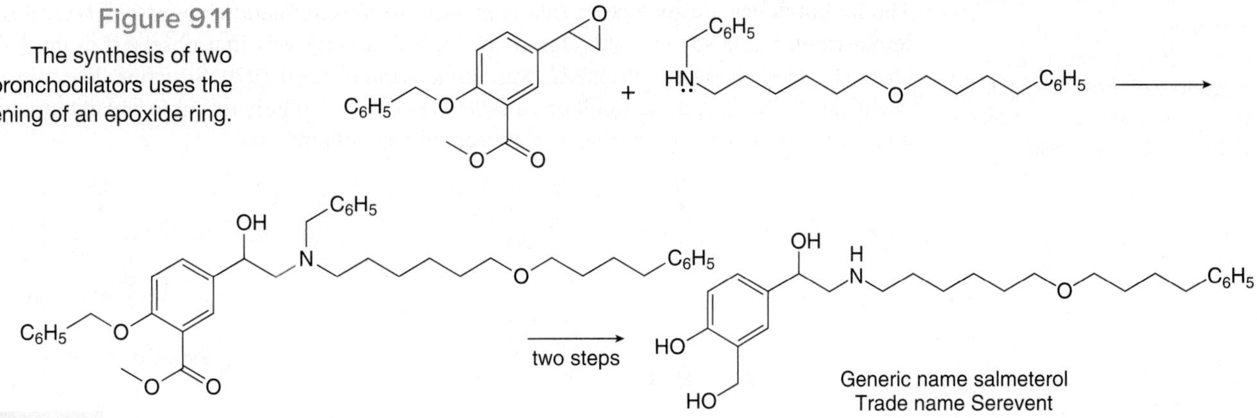

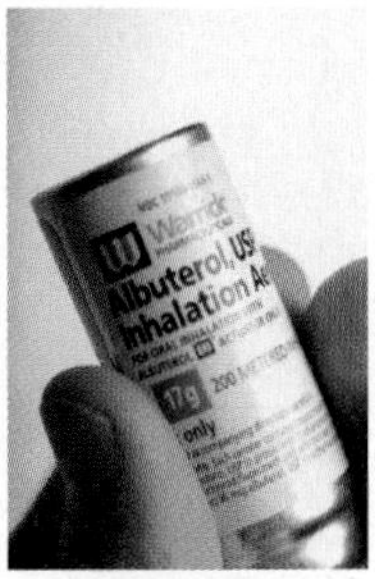

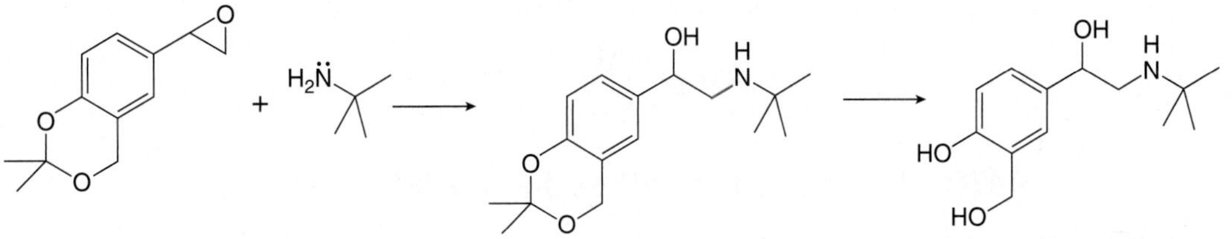

- A key step in each synthesis is the opening of an epoxide ring with a nitrogen nucleophile to form a new C – N bond, shown in red.

- **With a strong nucleophile, :Nu⁻ attacks at the less substituted carbon.**
- **With an acid HZ, the nucleophile attacks at the more substituted carbon.**

The reaction of epoxide rings with nucleophiles is important for the synthesis of many biologically active compounds, including **salmeterol** and **albuterol,** two bronchodilators used in the treatment of asthma (Figure 9.11).

Problem 9.36 Draw the product of each reaction.

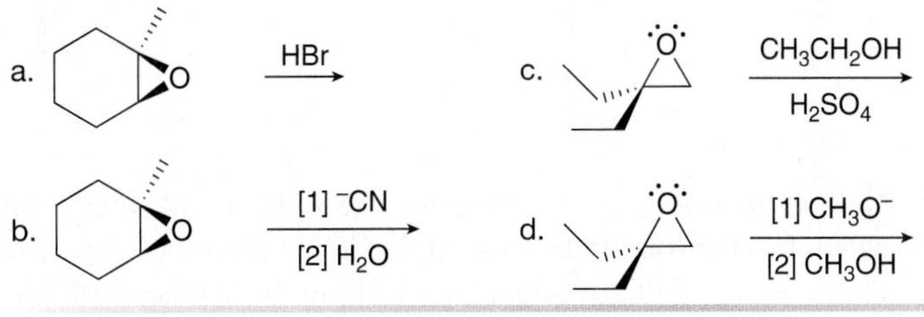

9.17 Application: Epoxides, Leukotrienes, and Asthma

The opening of epoxide rings with nucleophiles is a key step in some important biological processes.

9.17A Asthma and Leukotrienes

Asthma is an obstructive lung disease that affects millions of Americans. Because it involves episodic constriction of small airways, bronchodilators such as albuterol (Figure 9.11) are used to treat symptoms by widening airways. Because asthma is also characterized by chronic inflammation, inhaled steroids that reduce inflammation are also commonly used.

Leukotrienes were first synthesized in 1980 in the laboratory of Professor E. J. Corey, the 1990 recipient of the Nobel Prize in Chemistry.

The **leukotrienes** are molecules that contribute to the asthmatic response. A typical example, **leukotriene C_4,** is shown. Although its biological activity was first observed in the 1930s, the chemical structure of leukotriene C_4 was not determined until 1979. Structure determination and chemical synthesis were difficult because leukotrienes are highly unstable and extremely potent, and are therefore present in tissues in exceedingly small amounts.

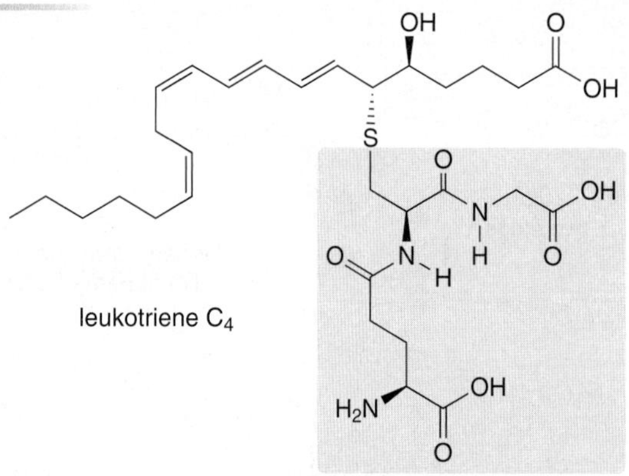

leukotriene C_4

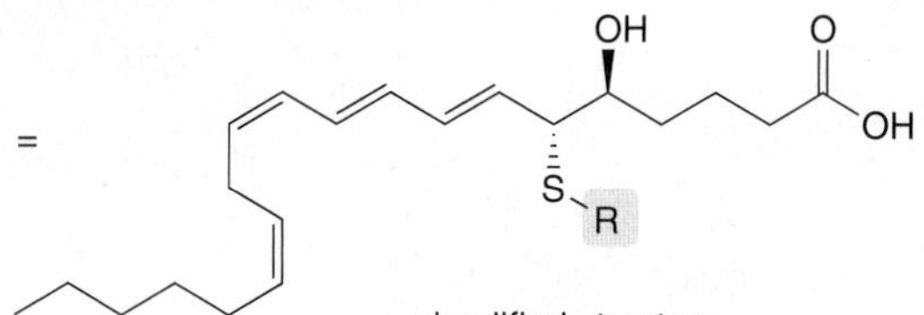

simplified structure

9.17B Leukotriene Synthesis and Asthma Drugs

Leukotrienes are synthesized in cells by the oxidation of **arachidonic acid** to 5-HPETE, which is then converted to an epoxide, **leukotriene A_4.** Opening of the epoxide ring with a sulfur nucleophile **RSH** yields leukotriene C_4.

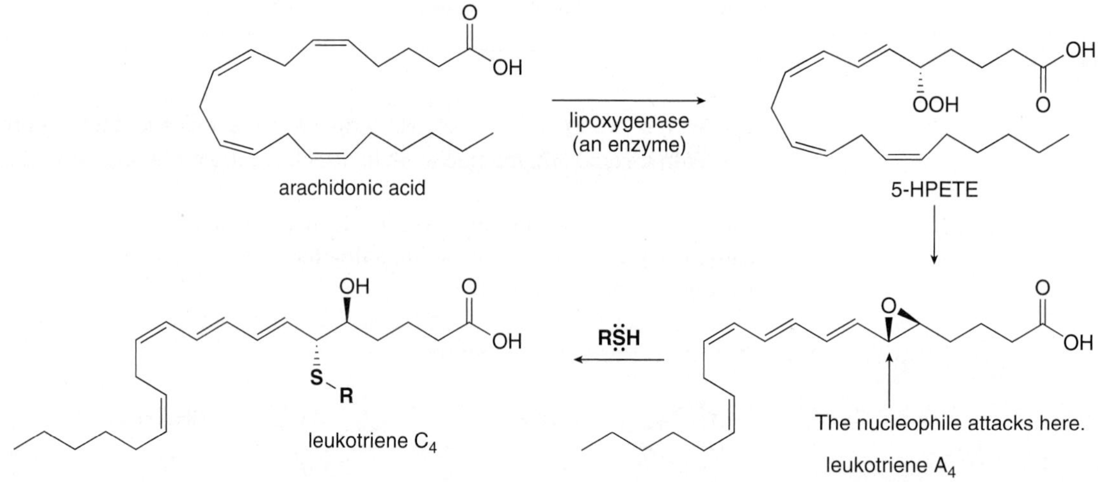

New asthma drugs act by blocking the synthesis of leukotriene C_4 from arachidonic acid. For example, **zileuton** (trade name Zyflo CR) inhibits the enzyme (called a lipoxygenase) needed for the first step of this process. By blocking the synthesis of leukotriene C_4, a compound responsible for the disease, zileuton treats the **cause of asthma,** not just its symptoms.

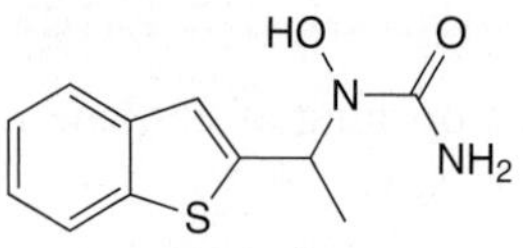

Generic name zileuton
Trade name Zyflo CR
anti-asthma drug

9.18 Application: Benzo[*a*]pyrene, Epoxides, and Cancer

The sooty exhaust from trucks and buses contains PAHs such as benzo[*a*]pyrene.

Benzo[*a*]pyrene is a widespread environmental pollutant, produced during the combustion of all types of organic material—gasoline, fuel oil, wood, garbage, and cigarettes. It is a **polycyclic aromatic hydrocarbon (PAH),** a class of compounds that is discussed further in Chapter 17.

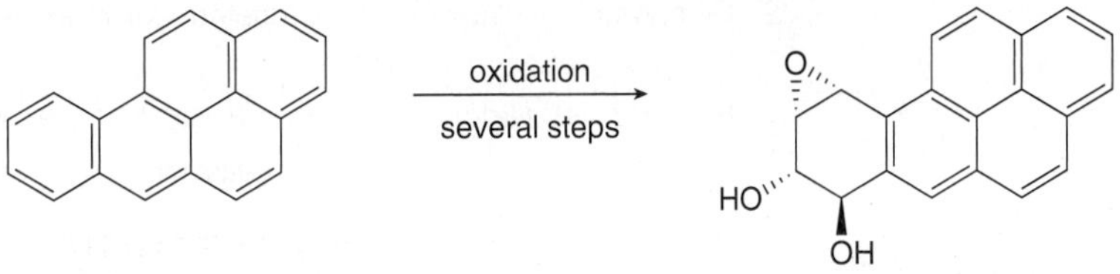

benzo[*a*]pyrene — **water insoluble**

a diol epoxide — **more water soluble**

After this nonpolar and water-insoluble hydrocarbon is inhaled or ingested, it is oxidized in the liver to a diol epoxide. Oxidation is a common fate of foreign substances that are not useful nutrients for the body. The oxidation product has three oxygen-containing functional groups, making it much more water soluble, and more readily excreted in urine. It is also a potent carcinogen. The strained three-membered ring of the epoxide reacts readily with biological nucleophiles $:Nu^-$ (such as DNA or an enzyme), leading to ring-opened products that often disrupt normal cell function, causing cancer or cell death.

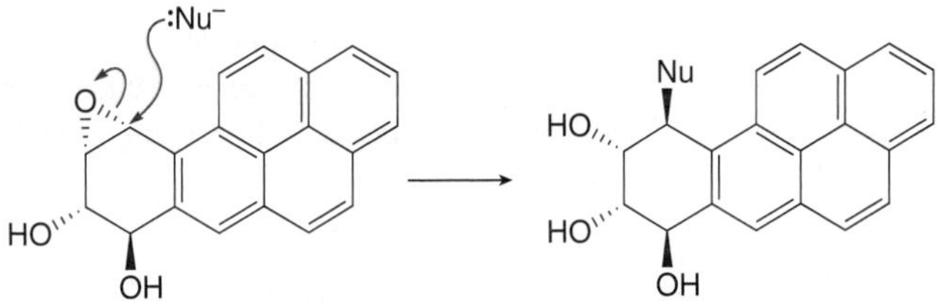

carcinogen

These examples illustrate the central role of the nucleophilic opening of epoxide rings in two well-defined cellular processes.

KEY CONCEPTS

Alcohols, Ethers, and Related Compounds

General Facts about ROH, ROR, and Epoxides

- All three compounds contain an O atom that is sp^3 hybridized and tetrahedral (9.2).
- All three compounds have polar C – O bonds, but only alcohols have an O – H bond for intermolecular hydrogen bonding (9.4).
- Alcohols and ethers do not contain a good leaving group. Nucleophilic substitution can occur only after the OH (or OR) group is converted to a better leaving group (9.7).
- Epoxides have a leaving group located in a strained three-membered ring, making them reactive to strong nucleophiles and acids HZ that contain a nucleophilic atom Z (9.16).

A New Reaction of Carbocations (9.9)

- Less stable carbocations rearrange to more stable carbocations by the shift of a hydrogen atom or an alkyl group.

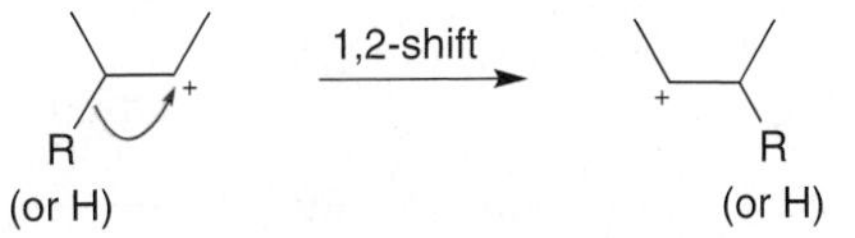

- Besides rearranging, a carbocation can also react with a nucleophile (7.12) and a base (8.6).

Preparation of Alcohols, Ethers, and Epoxides (9.6)

[1] Preparation of alcohols

$$R\text{–}X \;+\; {}^{-}OH \longrightarrow R\text{–}OH \;+\; X^{-}$$

- The mechanism is S_N2.
- The reaction works best for CH_3X and 1° RX.

[2] Preparation of alkoxides—A Brønsted–Lowry acid–base reaction

$$R\text{–}O\text{–}H \;+\; Na^{+}H^{-} \longrightarrow \underset{\text{alkoxide}}{R\text{–}O^{-}\,Na^{+}} \;+\; H_2$$

[3] Preparation of ethers (Williamson ether synthesis)

$$R\text{–}X \;+\; {}^{-}OR' \longrightarrow R\text{–}OR' \;+\; X^{-}$$

- The mechanism is S_N2.
- The reaction works best for CH_3X and 1° RX.

[4] Preparation of epoxides—Intramolecular S_N2 reaction

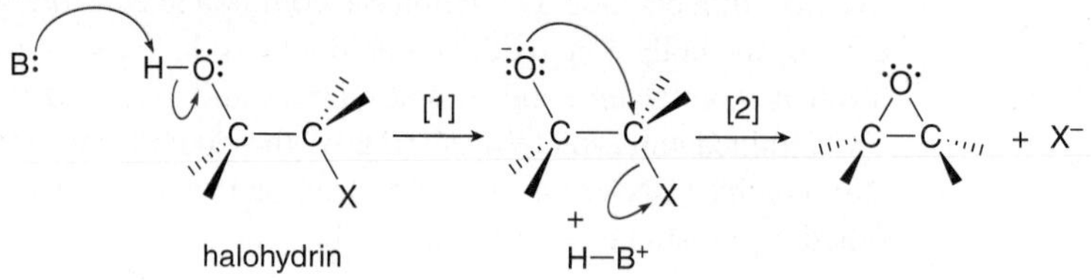
B:
[1]
[2]
+ X⁻
halohydrin
H–B⁺

- A two-step reaction sequence:
  [1] The removal of a proton with base forms an alkoxide.
  [2] An intramolecular S_N2 reaction forms the epoxide.

Reactions of Alcohols

[1] Dehydration to form alkenes

a. Using strong acid (9.8, 9.9)

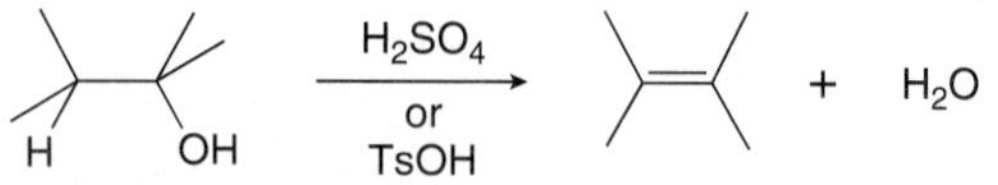
H₂SO₄
or
TsOH
+ H₂O

- Order of reactivity: $R_3COH > R_2CHOH > RCH_2OH$.
- The mechanism for 2° and 3° ROH is E1—carbocations are intermediates and rearrangements occur.
- The mechanism for 1° ROH is E2.
- The Zaitsev rule is followed.

b. Using $POCl_3$ and pyridine (9.10)

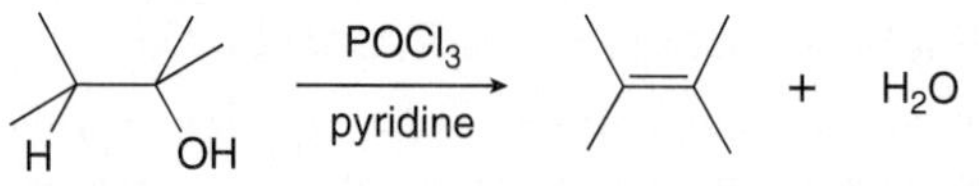
POCl₃
pyridine
+ H₂O

- The mechanism is E2.
- No carbocation rearrangements occur.

[2] Reaction with HX to form RX (9.11)

$$R\text{–}OH \;+\; H\text{–}X \longrightarrow R\text{–}X \;+\; H_2O$$

- Order of reactivity: $R_3COH > R_2CHOH > RCH_2OH$.
- The mechanism for 2° and 3° ROH is S_N1—carbocations are intermediates and rearrangements occur.
- The mechanism for CH_3OH and 1° ROH is S_N2.

[3] Reaction with other reagents to form RX (9.12)

$$R\text{–}OH \;+\; SOCl_2 \xrightarrow[\text{pyridine}]{} R\text{–}Cl$$

$$R\text{–}OH \;+\; PBr_3 \longrightarrow R\text{–}Br$$

- Reactions occur with CH_3OH and 1° and 2° ROH.
- The reactions follow an S_N2 mechanism.

[4] Reaction with tosyl chloride to form alkyl tosylates (9.13A)

$$R{-}OH + Ts{-}Cl \xrightarrow{\text{pyridine}} R{-}OTs$$

- The C–O bond is not broken, so the configuration at a stereogenic center is retained.

Reactions of Alkyl Tosylates

Alkyl tosylates undergo either substitution or elimination, depending on the reagent (9.13B).

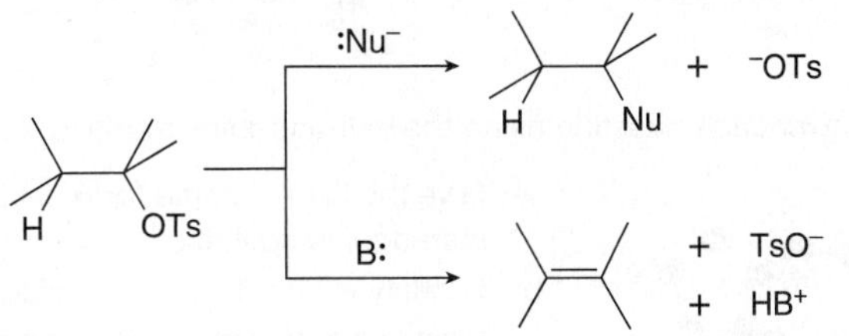

- Substitution is carried out with a strong :Nu⁻, so the mechanism is S_N2.
- Elimination is carried out with a strong base, so the mechanism is E2.

Reactions of Ethers

Only one reaction is useful: cleavage with strong acids (9.14).

$$R{-}O{-}R' + H{-}X \longrightarrow R{-}X + R'{-}X + H_2O$$

(2 equiv)
[X = Br or I]

- With 2° and 3° R groups, the mechanism is S_N1.
- With CH_3 and 1° R groups, the mechanism is S_N2.

Reactions Involving Thiols and Sulfides (9.15)

[1] Preparation of thiols

$$R{-}X + {}^{-}SH \longrightarrow R{-}SH + X^-$$

- The mechanism is S_N2.
- The reaction works best for CH_3X and 1° RX.

[2] Oxidation and reduction involving thiols

a. Oxidation of thiols to disulfides

$$R{-}SH \xrightarrow{Br_2 \text{ or } I_2} RS{-}SR$$

b. Reduction of disulfides to thiols

$$RS{-}SR \xrightarrow[\text{HCl}]{\text{Zn}} R{-}SH$$

[3] Preparation of sulfides

$$R{-}X + {}^{-}SR' \longrightarrow R{-}SR' + X^-$$

- The mechanism is S_N2.
- The reaction works best for CH_3X and 1° RX.

[4] Reaction of sulfides to form sulfonium ions

$$R'_2S + R{-}X \longrightarrow R'_2\overset{+}{S}{-}R + X^-$$

- The mechanism is S_N2.
- The reaction works best for CH_3X and 1° RX.

Reactions of Epoxides

Epoxide rings are opened with nucleophiles :Nu⁻ and acids HZ (9.16).

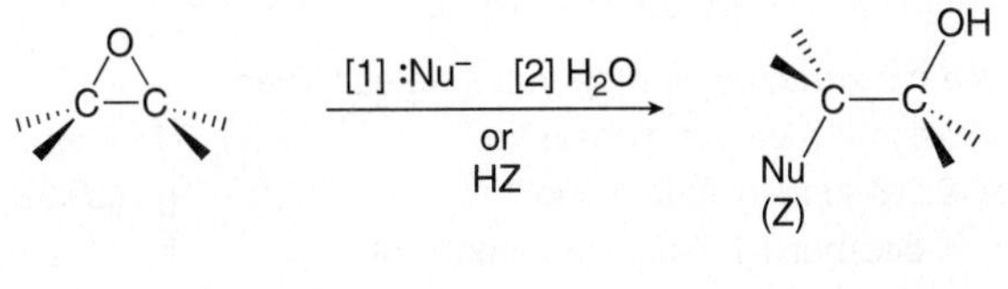

- The reaction occurs with backside attack, resulting in trans or anti products.
- With :Nu⁻, the mechanism is S_N2, and nucleophilic attack occurs at the less substituted C.
- With HZ, the mechanism is between S_N1 and S_N2, and attack of Z^- occurs at the more substituted C.

PROBLEMS

Problems Using Three-Dimensional Models

9.37 Name each compound depicted in the ball-and-stick models.

a. 

b. 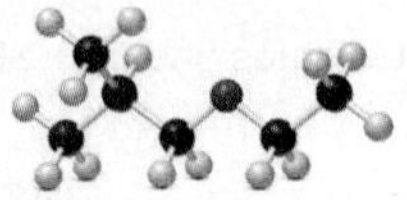

c. 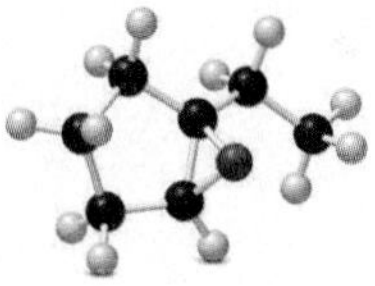

9.38 Answer each question using the ball-and-stick model of compound **A.**

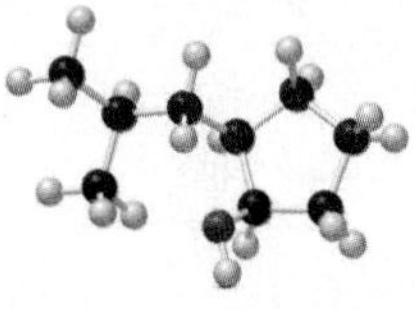

A

a. Give the IUPAC name for **A,** including *R,S* designations for stereogenic centers.
b. Classify **A** as a 1°, 2°, or 3° alcohol.
c. Draw a stereoisomer for **A** and give its IUPAC name.
d. Draw a constitutional isomer that contains an OH group and give its IUPAC name.
e. Draw a constitutional isomer that contains an ether and give its IUPAC name.
f. Draw the products formed (including stereochemistry) when **A** is treated with each reagent: [1] NaH; [2] H_2SO_4; [3] $POCl_3$, pyridine; [4] HCl; [5] $SOCl_2$, pyridine; [6] TsCl, pyridine.

9.39 Draw the product and indicate the stereochemistry when the given alcohol is treated with each reagent: (a) HBr; (b) PBr_3; (c) HCl; (d) $SOCl_2$ and pyridine.

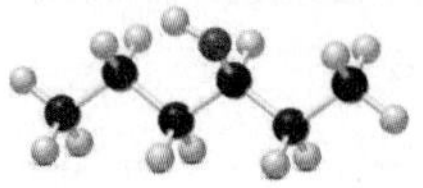

Nomenclature

9.40 Give the IUPAC name for each alcohol.

a. 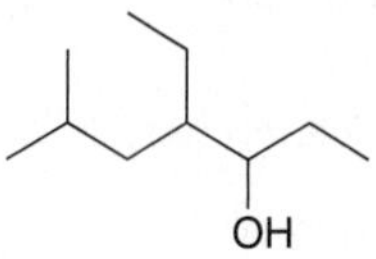

c. 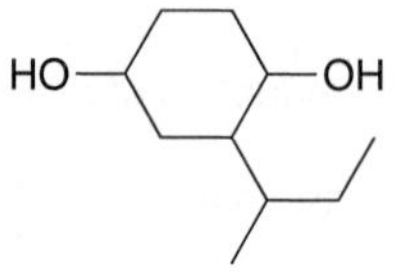

e. 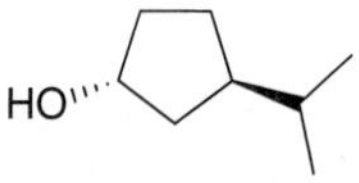

b. 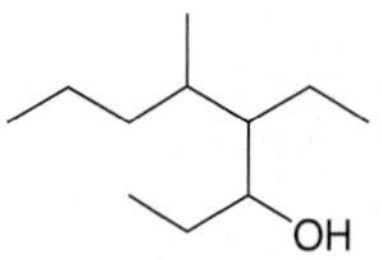

d. 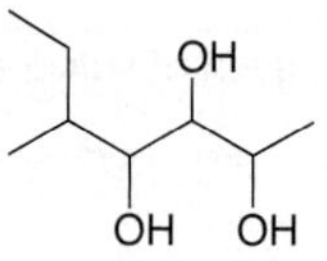

9.41 Name each ether, epoxide, thiol, and sulfide.

a. 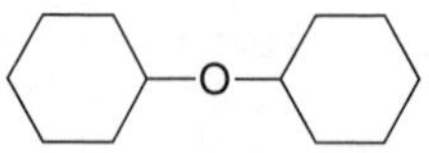

c. 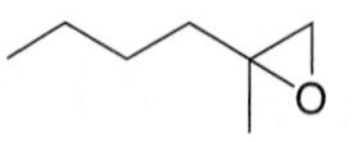

e. 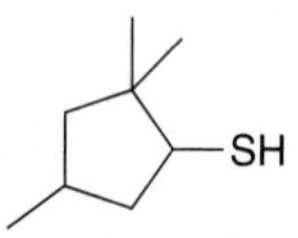

b. 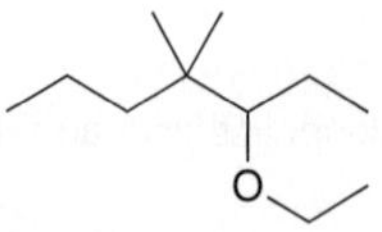

d. 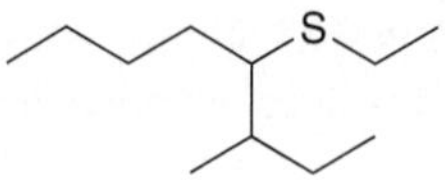

9.42 Give the structure corresponding to each name.

a. *trans*-2-methylcyclohexanol
b. 2,3,3-trimethylbutan-2-ol
c. 6-*sec*-butyl-7,7-diethyldecan-4-ol
d. 3-chloropropane-1,2-diol
e. 1,2-epoxy-1,3,3-trimethylcyclohexane
f. 1-ethoxy-3-ethylheptane
g. (2*R*,3*S*)-3-isopropylhexan-2-ol
h. (*S*)-2-ethoxy-1,1-dimethylcyclopentane
i. 4-ethylheptane-3-thiol
j. 1-isopropylthio-2-methylcyclohexane

Physical Properties

9.43 Explain the observed trend in the melting points of the following three isomeric alcohols: $(CH_3)_2CHCH_2OH$ (–108 °C), $CH_3CH_2CH_2CH_2OH$ (–90 °C), $(CH_3)_3COH$ (26 °C).

9.44 Why is the boiling point of propane-1,3-diol ($HOCH_2CH_2CH_2OH$) higher than the boiling point of propane-1,2-diol [$HOCH_2CH(OH)CH_3$] (215 °C vs. 187 °C)? Why do both diols have a higher boiling point than butan-1-ol ($CH_3CH_2CH_2CH_2OH$, 118 °C)?

Alcohols

9.45 Draw the organic product(s) formed when $CH_3CH_2CH_2OH$ is treated with each reagent.

a. H_2SO_4
b. NaH
c. HCl + $ZnCl_2$
d. HBr
e. $SOCl_2$, pyridine
f. PBr_3
g. TsCl, pyridine
h. [1] NaH; [2] CH_3CH_2Br
i. [1] TsCl, pyridine; [2] NaSH
j. $POCl_3$, pyridine

9.46 What alkenes are formed when each alcohol is dehydrated with TsOH? Label the major product when a mixture results.

a. 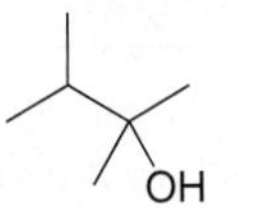

b. 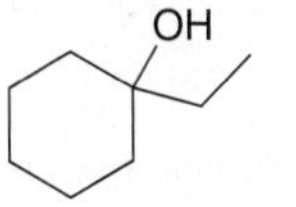

c. 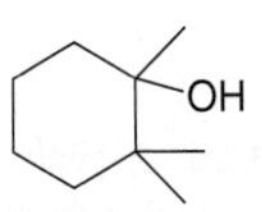

d. 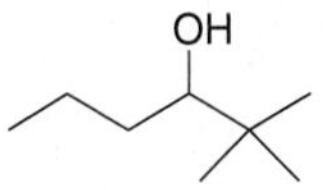

9.47 Draw the products formed when $CH_3CH_2CH_2CH_2OTs$ is treated with each reagent.

a. CH_3SH b. $NaOCH_2CH_3$ c. NaOH d. $KOC(CH_3)_3$

9.48 Draw the products of each reaction and indicate stereochemistry around stereogenic centers.

a. 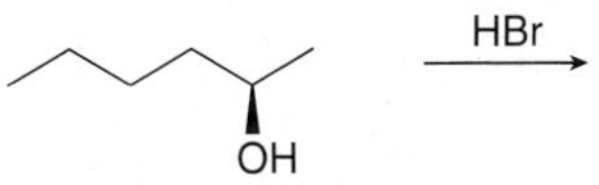

b. 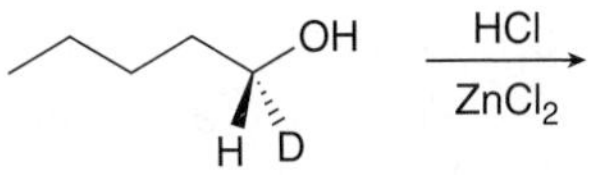

c. 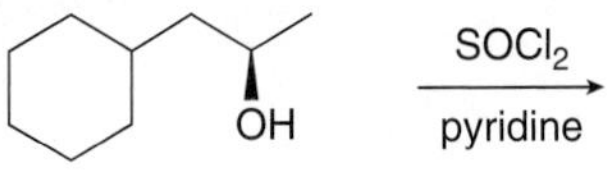

d. 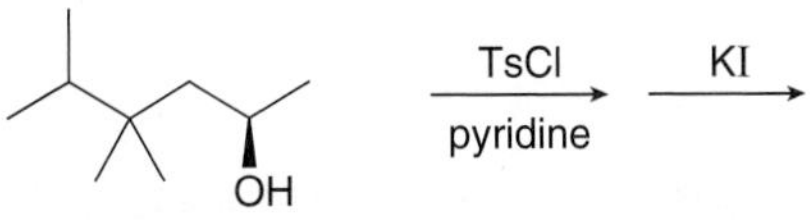

9.49 Draw the product of the following reaction, one step in the synthesis of the antiplatelet agent clopidogrel used to reduce the risk of strokes.

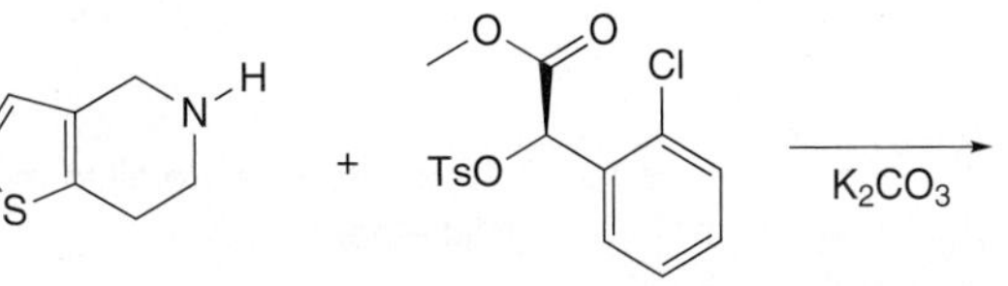

9.50 Draw the substitution product formed (including stereochemistry) when (*R*)-hexan-2-ol is treated with each series of reagents: (a) NaH, followed by CH_3I; (b) TsCl and pyridine, followed by $NaOCH_3$; (c) PBr_3, followed by $NaOCH_3$. Which two routes produce identical products?

9.51 (a) What is the major alkene formed when **A** is dehydrated with H_2SO_4? (b) What is the major alkene formed when **A** is treated with $POCl_3$ and pyridine? Explain why the major product is different in these reactions.

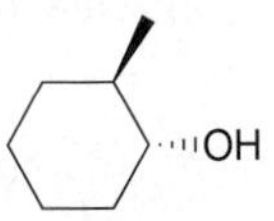

A

9.52 Draw a stepwise mechanism for the following reaction.

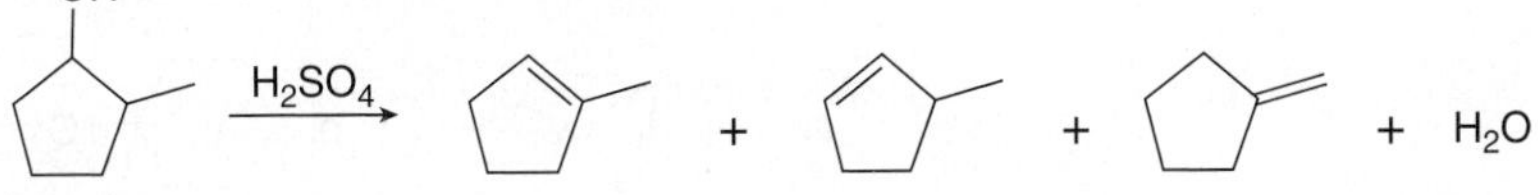

9.53 Although alcohol **V** gives a single alkene **W** when treated with $POCl_3$ and pyridine, three isomeric alkenes (**X–Z**) are formed on dehydration with H_2SO_4. Draw a stepwise mechanism for each reaction and explain why the difference occurs.

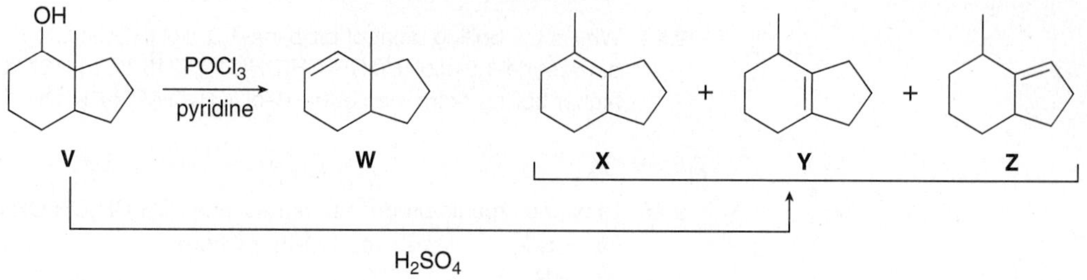

9.54 Sometimes carbocation rearrangements can change the size of a ring. Draw a stepwise, detailed mechanism for the following reaction.

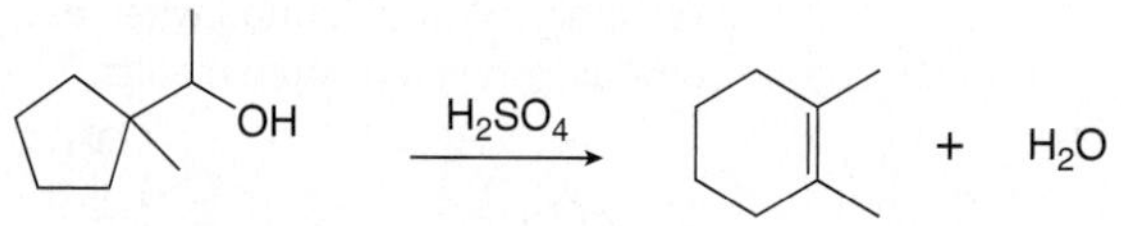

9.55 Explain the following observation. When 3-methylbutan-2-ol is treated with HBr, a single alkyl bromide is isolated, resulting from a 1,2-shift. When 2-methylpropan-1-ol is treated with HBr, no rearrangement occurs to form an alkyl bromide.

9.56 An allylic alcohol contains an OH group on a carbon atom adjacent to a C–C double bond. Treatment of allylic alcohol **A** with HCl forms a mixture of two allylic chlorides, **B** and **C.** Draw a stepwise mechanism that illustrates how both products are formed.

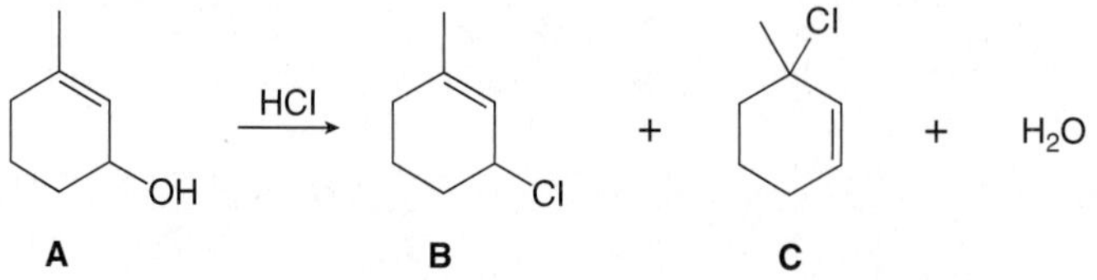

9.57 Draw a stepwise, detailed mechanism for the following reaction.

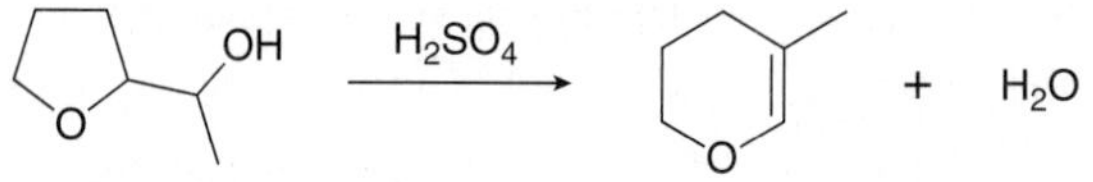

9.58 Draw a stepwise, detailed mechanism for the following intramolecular reaction that forms a cyclic ether.

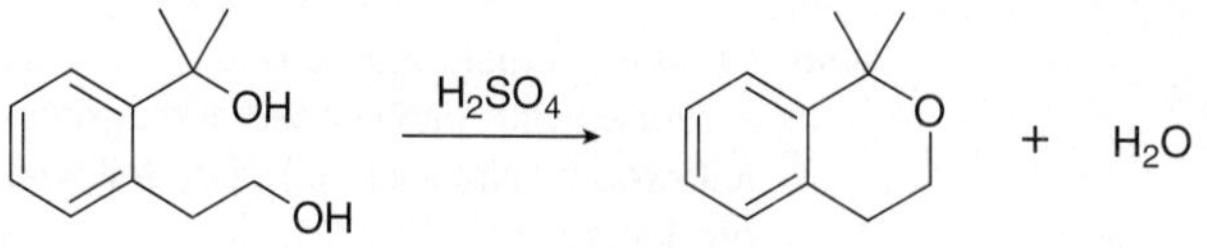

Ethers

9.59 Draw two different routes to each of the following ethers using a Williamson ether synthesis. Indicate the preferred route (if there is one).

a. 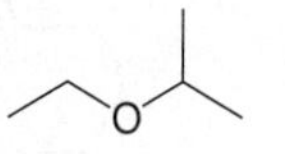

b. 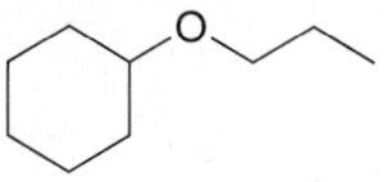

c. 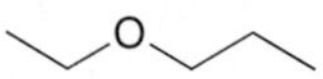

9.60 Explain why it is not possible to prepare *tert*-butyl phenyl ether using a Williamson ether synthesis.

9.61 Draw the products formed when each ether is treated with two equivalents of HBr.

a. 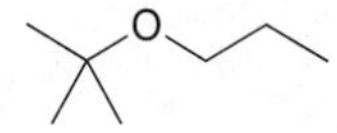

b. 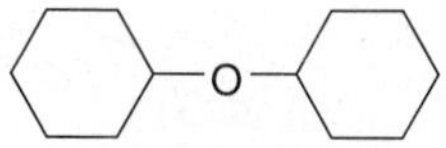

c. 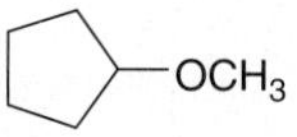

9.62 Draw a stepwise mechanism for each reaction.

a. 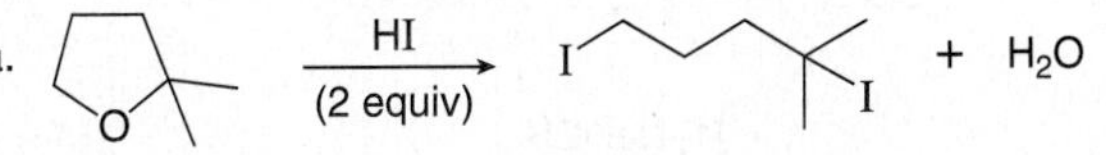
HI
(2 equiv)
I
I
+ H_2O

b. 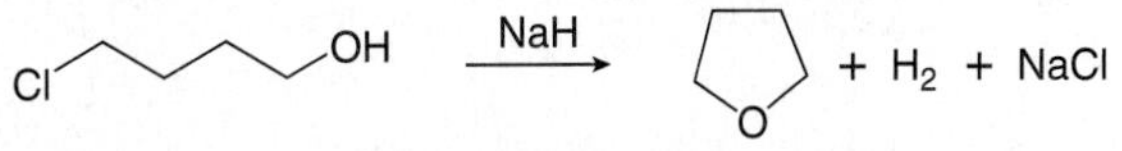
Cl
OH
NaH
O
+ H_2 + NaCl

9.63 Draw a stepwise, detailed mechanism for the following reaction.

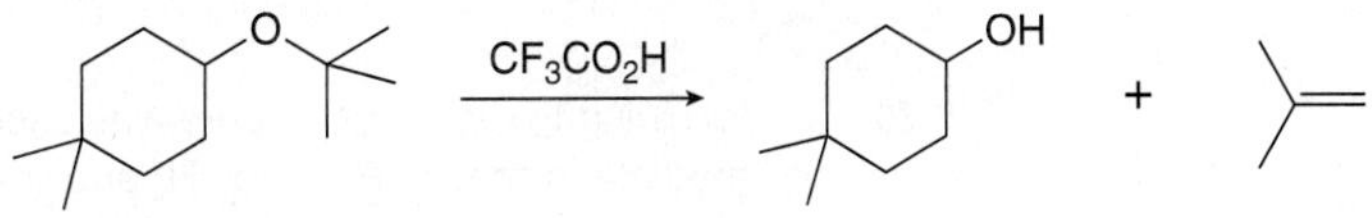
O
CF_3CO_2H
OH
+

Epoxides

9.64 Draw the products formed when ethylene oxide is treated with each reagent.

a. HBr
b. H_2O (H_2SO_4)
c. [1] $CH_3CH_2O^-$; [2] H_2O
d. [1] $HC{\equiv}C^-$; [2] H_2O
e. [1] ^-OH; [2] H_2O
f. [1] CH_3S^-; [2] H_2O

9.65 Draw the products of each reaction.

a. 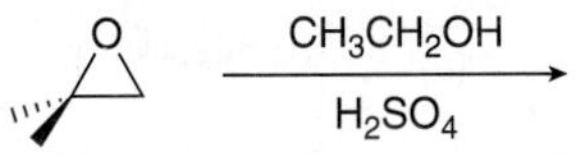
O
CH_3CH_2OH
H_2SO_4

c. 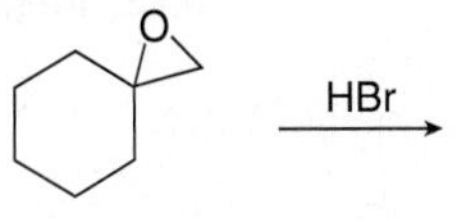
O
HBr

b. 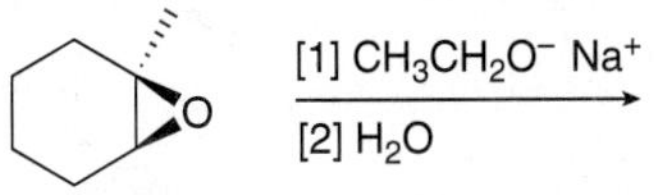
O
[1] $CH_3CH_2O^-$ Na^+
[2] H_2O

d. 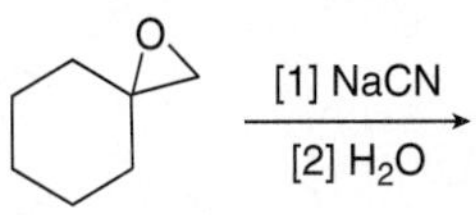
O
[1] NaCN
[2] H_2O

9.66 When each halohydrin is treated with NaH, a product of molecular formula C_4H_8O is formed. Draw the structure of the product and indicate its stereochemistry.

a. 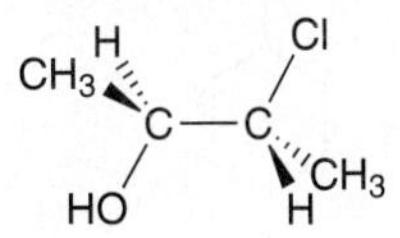
H
CH_3
C
C
Cl
HO
H
CH_3

b. 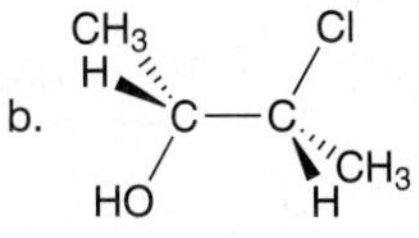
CH_3
H
C
C
Cl
HO
H
CH_3

c. 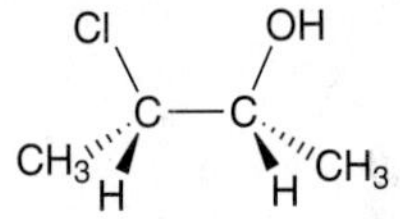
Cl
OH
C
C
CH_3
H
H
CH_3

9.67 (a) What reaction conditions are needed to convert (*R*)-2-ethyl-2-methyloxirane to (*R*)-2-methylbutane-1,2-diol? (b) What reaction conditions are needed to convert (*R*)-2-ethyl-2-methyloxirane to (*S*)-2-methylbutane-1,2-diol?

9.68 Draw a stepwise mechanism for the following reaction, and explain the stereochemistry of the product.

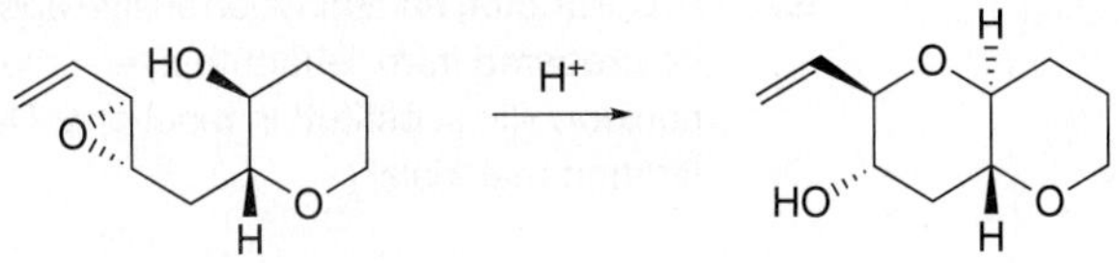
HO
O
H
O
H^+
H
O
HO
H
O

General Problems

9.69 Draw the products of each reaction, and indicate the stereochemistry where appropriate.

a. 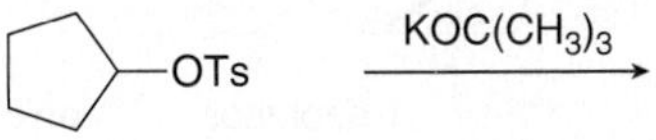
OTs
$KOC(CH_3)_3$

d. 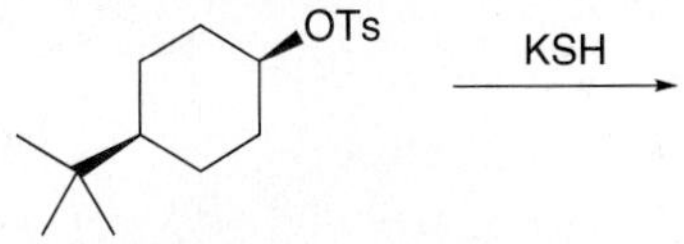
OTs
KSH

b. 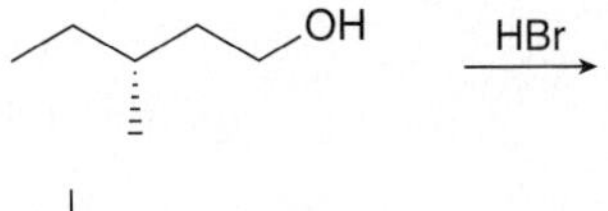
OH
HBr

e. 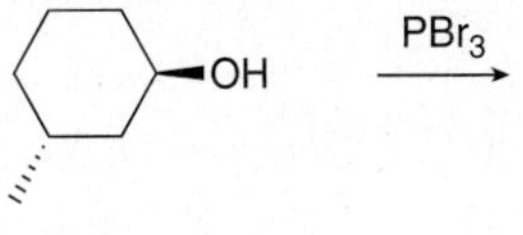
OH
PBr_3

c. 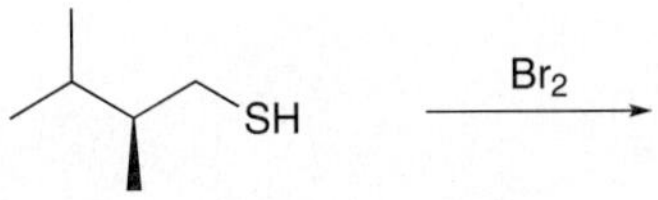
SH
Br_2

f. 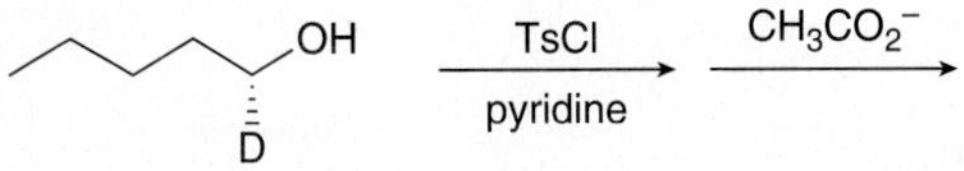
OH
D
TsCl
pyridine
$CH_3CO_2^-$

g. 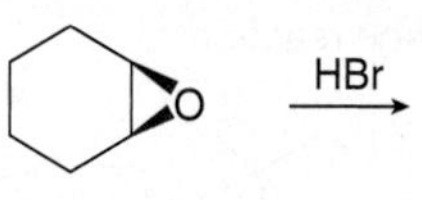

j. 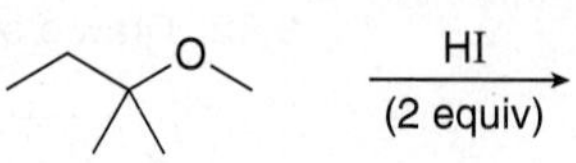

h. 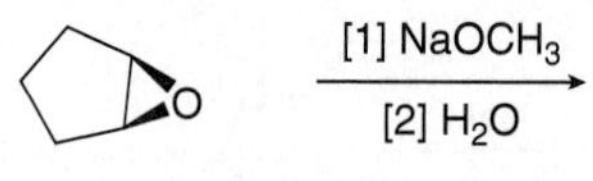

k. 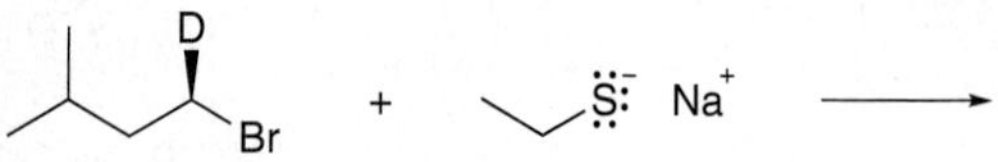

i. 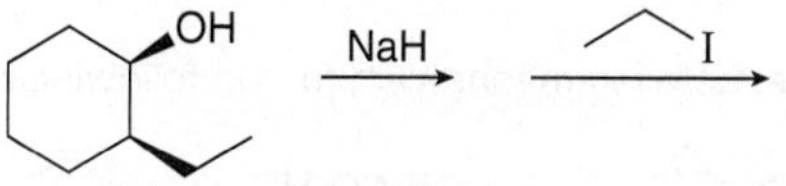

l. 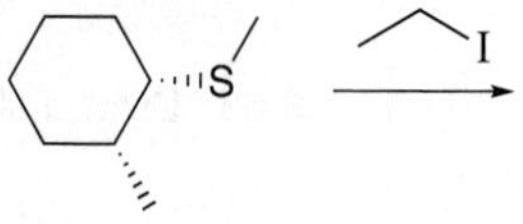

9.70 The following two-step procedure was used to prepare a sulfide from a diol. Draw the intermediate formed in Reaction [1] and draw a mechanism for Reaction [2].

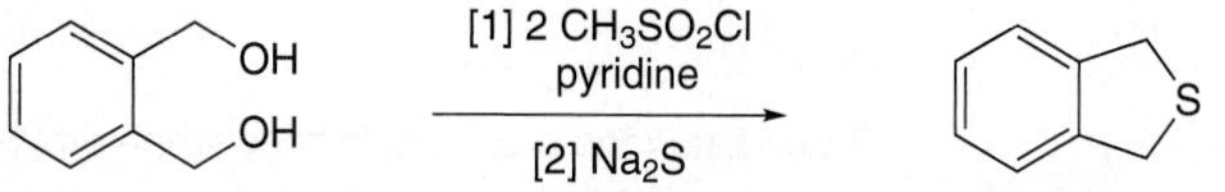

9.71 Prepare each compound from cyclopentanol. More than one step may be needed.

a. 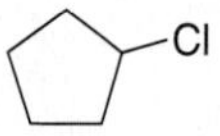

b. 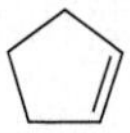

c. 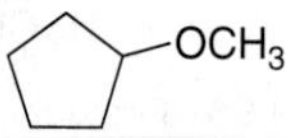

d. 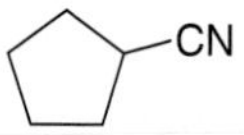

9.72 Identify the reagents (a–h) needed to carry out each reaction.

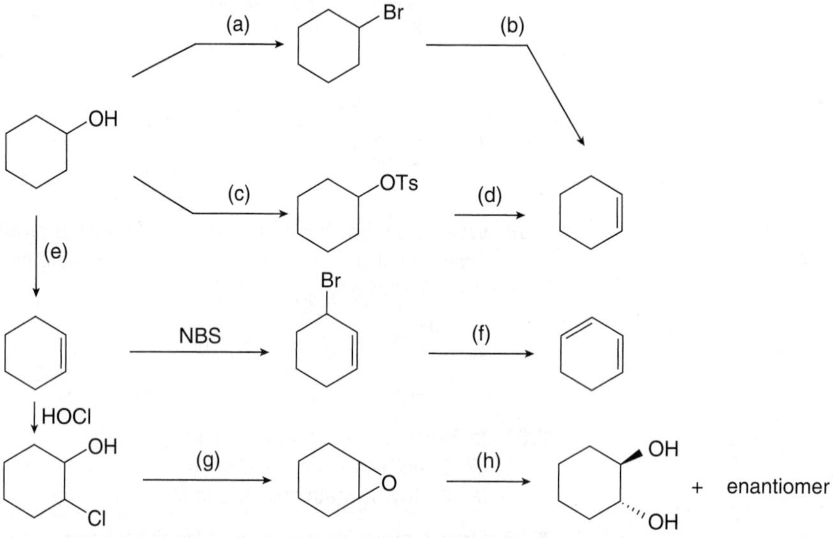

9.73 **Propranolol,** an antihypertensive agent used in the treatment of high blood pressure, can be prepared from 1-naphthol, epichlorohydrin, and isopropylamine using two successive nucleophilic substitution reactions. Devise a stepwise synthesis of propranolol from these starting materials.

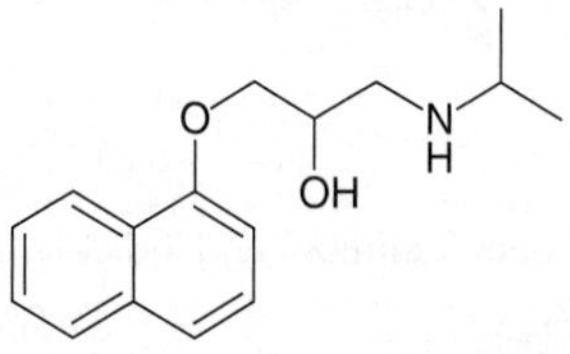

propranolol

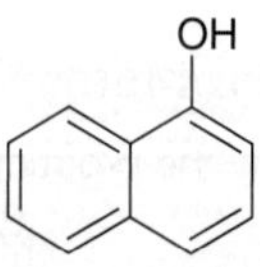

1-naphthol

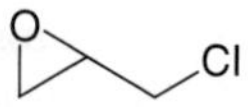

epichlorohydrin

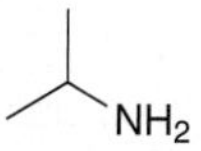

isopropylamine

Challenge Problems

9.74 Treatment of *cis*-4-bromocyclohexanol with HO^- affords compound **A** and cyclohex-3-en-1-ol. Treatment of *trans*-4-bromocyclohexanol under the same conditions forms compound **B** and cyclohex-3-en-1-ol. **A** and **B** contain different functional groups and are not isomers of each other. Propose structures for **A** and **B** and offer an explanation for their formation.

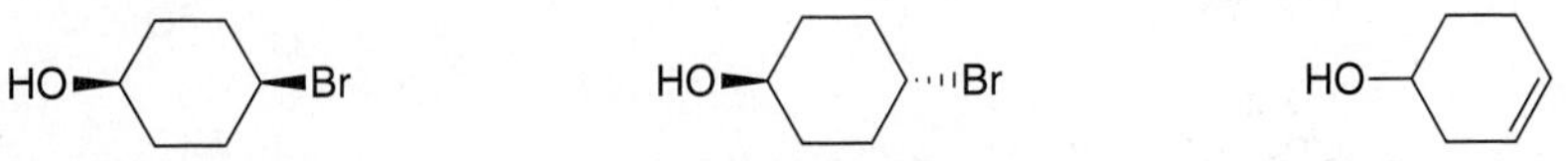

cis-4-bromocyclohexanol *trans*-4-bromocyclohexanol cyclohex-3-en-1-ol

9.75 Epoxides are converted to allylic alcohols with nonnucleophilic bases such as lithium diethylamide [$LiN(CH_2CH_3)_2$]. Draw a stepwise mechanism for the conversion of 1,2-epoxycyclohexane to cyclohex-2-en-1-ol with this base. Explain why a strong bulky base must be used in this reaction.

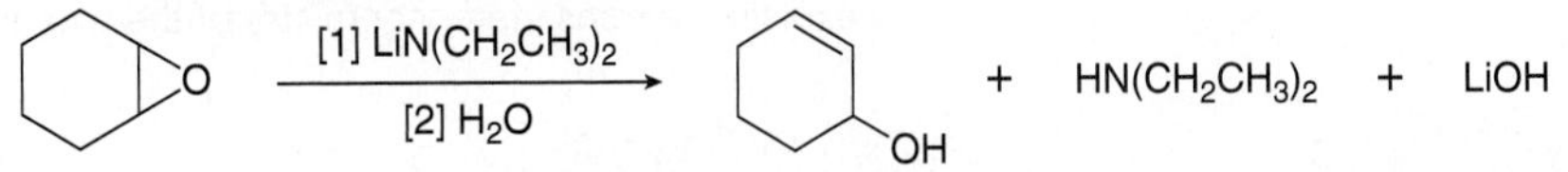

cyclohex-2-en-1-ol

9.76 Rearrangements can occur during the dehydration of 1° alcohols even though no 1° carbocation is formed—that is, a 1,2-shift occurs as the $C-OH_2^+$ bond is broken, forming a more stable 2° or 3° carbocation, as shown. Using this information, draw a stepwise mechanism that shows how $CH_3CH_2CH_2CH_2OH$ is dehydrated with H_2SO_4 to form a mixture of $CH_3CH_2CH=CH_2$ and the cis and trans isomers of $CH_3CH=CHCH_3$. We will see another example of this type of rearrangement in Section 18.5C.

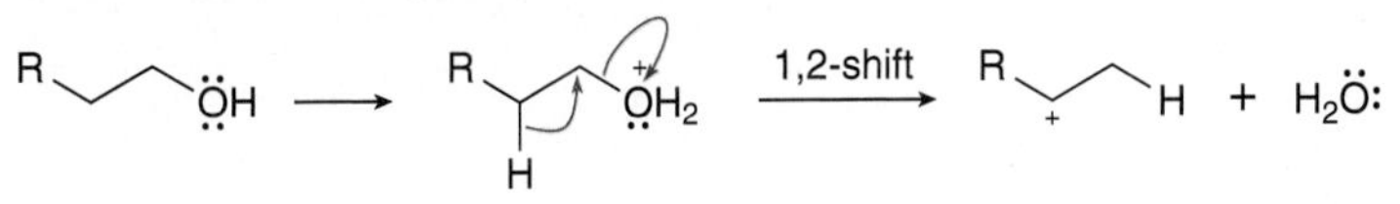

1° alcohol no 1° carbocation at this step 2° carbocation

9.77 Draw a stepwise, detailed mechanism for the following reaction.

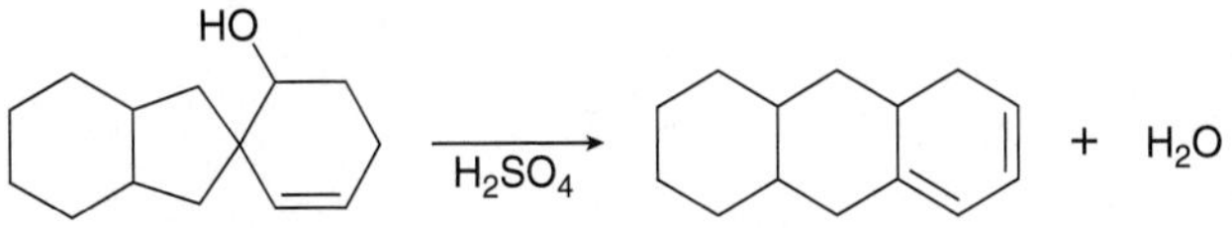

9.78 Dehydration of 1,2,2-trimethylcyclohexanol with H_2SO_4 affords 1-*tert*-butylcyclopentene as a minor product. (a) Draw a stepwise mechanism that shows how this alkene is formed. (b) Draw other alkenes formed in this dehydration. At least one must contain a five-membered ring.

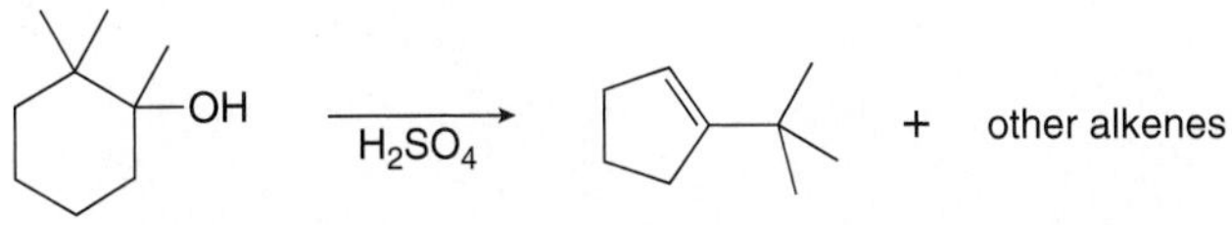

1,2,2-trimethylcyclohexanol 1-*tert*-butylcyclopentene

9.79 1,2-Diols are converted to carbonyl compounds when treated with strong acids, in a reaction called the *pinacol rearrangement.* (a) Draw a stepwise mechanism for this reaction. (Hint: The reaction proceeds by way of carbocation intermediates.) (b) Assuming that the pinacol rearrangement occurs via the more stable carbocation, draw the rearrangement product formed from diol **D.**

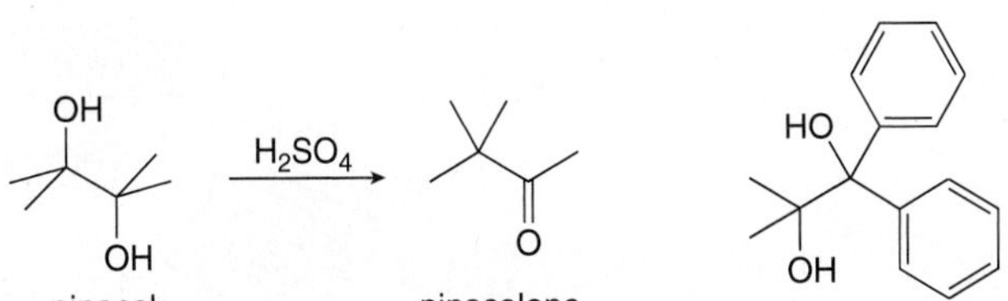

pinacol pinacolone **D**

9.80 Draw a stepwise mechanism for the following reaction.

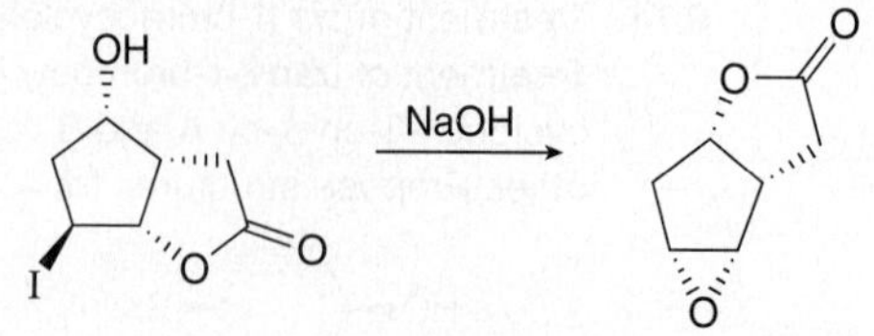

9.81 Aziridines are heterocycles that contain an N atom in a three-membered ring. Like epoxides, aziridines are strained and reactive because the 60° bond angles of the three-membered ring deviate greatly from the theoretical tetrahedral bond angle. One step in the synthesis of the drug oseltamivir (trade name Tamiflu, Section 3.2) involves the conversion of amine **X** to diamine **Y,** a reaction that occurs by way of an intermediate aziridine. Draw a stepwise mechanism for the conversion of **X** to **Y.** Indicate the structure of the aziridine intermediate, and explain the trans stereochemistry of the two amines in **Y.**

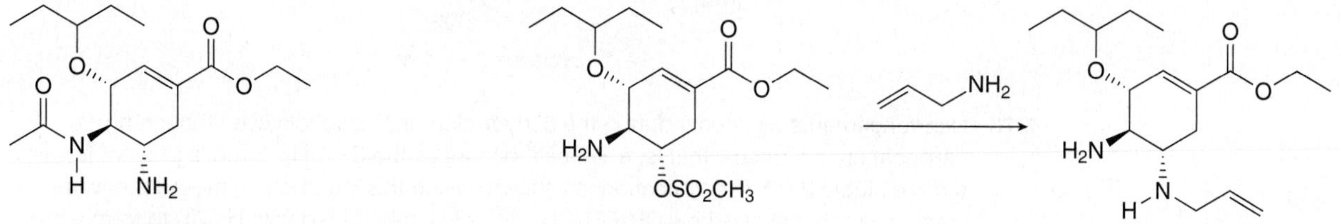

oseltamivir

X

Y

Alkenes

10

10.1 Introduction
10.2 Calculating degrees of unsaturation
10.3 Nomenclature
10.4 Physical properties
10.5 Interesting alkenes
10.6 Lipids—Part 2
10.7 Preparation of alkenes
10.8 Introduction to addition reactions
10.9 Hydrohalogenation—Electrophilic addition of HX
10.10 Markovnikov's rule
10.11 Stereochemistry of electrophilic addition of HX
10.12 Hydration—Electrophilic addition of water
10.13 Halogenation—Addition of halogen
10.14 Stereochemistry of halogenation
10.15 Halohydrin formation
10.16 Hydroboration–oxidation
10.17 Keeping track of reactions
10.18 Alkenes in organic synthesis

Candlenuts, known as kukui nuts in Hawai'i, are rich in **linoleic acid** and **linolenic acid,** two unsaturated fatty acids. **Unsaturated fatty acids** have a carboxy group (CO_2H) bonded to a carbon chain that contains one or more carbon–carbon double bonds. The presence of the double bonds greatly affects the chemical and physical properties of these compounds. In Chapter 10, we learn about alkenes, organic compounds that contain carbon–carbon double bonds.

In Chapters 10 and 11 we turn our attention to **alkenes** and **alkynes,** compounds that contain one and two π bonds, respectively. Because π bonds are easily broken, alkenes and alkynes undergo **addition,** the third general type of organic reaction. These multiple bonds also make carbon atoms electron rich, so alkenes and alkynes react with a wide variety of electrophilic reagents in addition reactions.

In Chapter 10 we review the properties and synthesis of alkenes first, and then concentrate on reactions. **Every new reaction in Chapter 10 is an *addition reaction.*** The most challenging part is learning the reagents, mechanism, and stereochemistry that characterize each individual reaction.

10.1 Introduction

Alkenes are also called **olefins.**

Alkenes are compounds that contain a carbon–carbon double bond. **Terminal alkenes** have the double bond at the end of the carbon chain, whereas **internal alkenes** have at least one carbon atom bonded to each end of the double bond. **Cycloalkenes** contain a double bond in a ring.

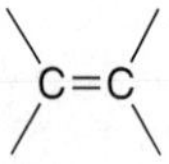

alkene

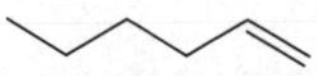

terminal alkene

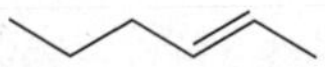

internal alkene

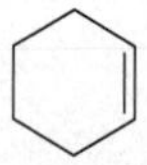

cycloalkene

The double bond of an alkene consists of one σ bond and one π bond. Each carbon is sp^2 hybridized and trigonal planar, and all bond angles are approximately 120° (Section 8.2A).

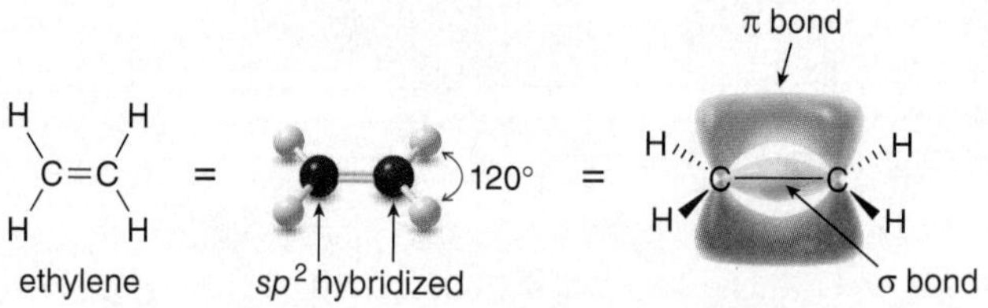

Bond dissociation energies of the C–C bonds in ethane (a σ bond only) and ethylene (one σ and one π bond) can be used to estimate the strength of the π component of the double bond. If we assume that the σ bond in ethylene is similar in strength to the σ bond in ethane (368 kJ/mol), then the π bond is worth 267 kJ/mol.

$CH_2{=}CH_2$		$CH_3{-}CH_3$		
635 kJ/mol (σ + π bond)	–	368 kJ/mol (σ bond)	=	267 kJ/mol **π bond only**

- **The π bond is much weaker than the σ bond of a C–C double bond, making it much more easily broken. As a result, alkenes undergo many reactions that alkanes do not.**

Other features of the carbon–carbon double bond, which were presented in Chapter 8, are summarized in Table 10.1.

Table 10.1 Properties of the Carbon–Carbon Double Bond

Property	Result
Restricted rotation	• **The rotation around the C–C double bond is restricted.** Rotation can occur only if the π bond breaks and then re-forms, a process that is unfavorable (Section 8.2B).
Stereoisomerism	• Whenever the two groups on each end of a C=C are different from each other, two diastereomers are possible. *Cis*- and *trans*-but-2-ene (drawn at the bottom of Table 10.1) are diastereomers (Section 8.2B).
Stability	• **Trans** alkenes are generally more stable than **cis** alkenes. • **The stability of an alkene increases as the number of R groups on the C=C increases** (Section 8.2C).

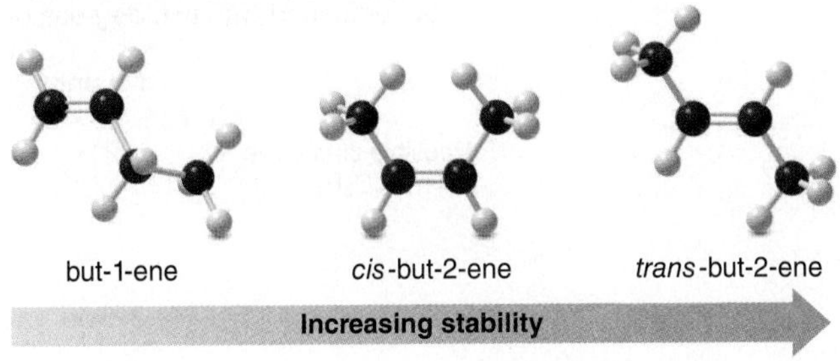
but-1-ene
cis-but-2-ene
trans-but-2-ene
Increasing stability

Cycloalkenes having fewer than eight carbon atoms have a cis geometry. A trans cycloalkene must have a carbon chain long enough to connect the ends of the double bond without introducing too much strain. *trans*-Cyclooctene is the smallest, isolable trans cycloalkene, but it is considerably less stable than *cis*-cyclooctene, making it one of the few alkenes having a higher energy trans isomer.

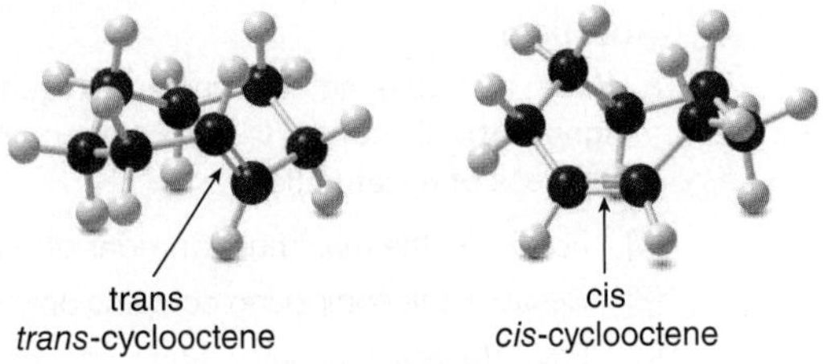
trans
trans-cyclooctene
cis
cis-cyclooctene

Problem 10.1 Draw the six alkenes of molecular formula C_5H_{10}. Label one pair of diastereomers.

10.2 Calculating Degrees of Unsaturation

An acyclic alkene has the general molecular formula C_nH_{2n}, giving it two fewer hydrogens than an acyclic alkane with the same number of carbons.

• Alkenes are *unsaturated hydrocarbons* because they have fewer than the maximum number of hydrogen atoms per carbon.

In Chapter 12 we will learn how to use the hydrogenation of π bonds to determine how many degrees of unsaturation result from π bonds and how many result from rings.

Cycloalkanes also have the general molecular formula C_nH_{2n}. Thus, **each π bond or ring removes two hydrogen atoms from a molecule, and this introduces one *degree of unsaturation.*** The number of degrees of unsaturation for a given molecular formula can be calculated by comparing the actual number of H atoms in a compound and the maximum number of H atoms possible. Remember that for n carbons, the maximum number of H atoms is $2n + 2$ (Section 4.1). This procedure gives the total number of rings and π bonds in a molecule.

Sample Problem 10.1 Calculate the number of degrees of unsaturation in a compound of molecular formula C_4H_6, and propose possible structures.

Solution

[1] Calculate the maximum number of H's possible.

- For n carbons, the maximum number of H's is $2n + 2$; in this example, $2n + 2 = 2(4) + 2 = 10$.

[2] Subtract the actual number of H's from the maximum number and divide by two.

- 10 H's (maximum) – 6 H's (actual) = 4 H's fewer than the maximum number.

$$\frac{\text{4 H's fewer than the maximum}}{\text{2 H's removed for each degree of unsaturation}} =$$

Answer: two degrees of unsaturation

A compound with two degrees of unsaturation has:

	two rings	*or*	**two π bonds**	*or*	**one ring and one π bond**
Possible structures for C_4H_6:	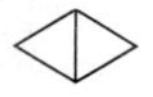		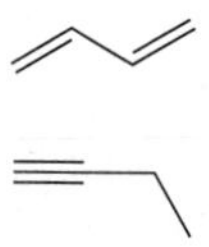		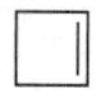

This procedure can be extended to compounds that contain heteroatoms such as oxygen, nitrogen, and halogen, as illustrated in Sample Problem 10.2.

Sample Problem 10.2 Calculate the number of degrees of unsaturation for each molecular formula: (a) C_5H_8O; (b) $C_6H_{11}Cl$; (c) C_8H_9N. Propose one possible structure for each compound.

Solution

a. When a compound contains an oxygen atom, **use the given number of C's and H's and ignore the O atom** in the calculation; that is, C_5H_8O is equivalent to C_5H_8 when calculating degrees of unsaturation.

[1] For 5 C's, the maximum number of H's = $2n + 2 = 2(5) + 2 = 12$.

[2] Because the compound contains only 8 H's, it has 12 – 8 = 4 H's fewer than the maximum number.

[3] Each degree of unsaturation removes 2 H's, so the answer in Step [2] must be divided by 2.
Answer: two degrees of unsaturation

b. A compound with a halogen atom is equivalent to a hydrocarbon having one more H; that is, $C_6H_{11}Cl$ is equivalent to C_6H_{12} when calculating degrees of unsaturation.

[1] For 6 C's, the maximum number of H's = $2n + 2 = 2(6) + 2 = 14$.

[2] Because the compound contains only 12 H's, it has 14 – 12 = 2 H's fewer than the maximum number.

[3] Each degree of unsaturation removes 2 H's, so the answer in Step [2] must be divided by 2.
Answer: one degree of unsaturation

c. A compound with a nitrogen atom is equivalent to a hydrocarbon having one fewer H; that is, C_8H_9N is equivalent to C_8H_8 when calculating degrees of unsaturation.

[1] For 8 C's, the maximum number of H's = $2n + 2 = 2(8) + 2 = 18$.

[2] Because the compound contains only 8 H's, it has 18 – 8 = 10 H's fewer than the maximum number.

[3] Each degree of unsaturation removes 2 H's, so the answer in Step [2] must be divided by 2.
Answer: five degrees of unsaturation

Possible structures:

a. 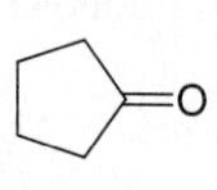

b. 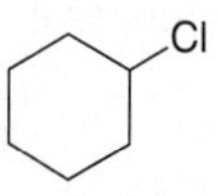

c. 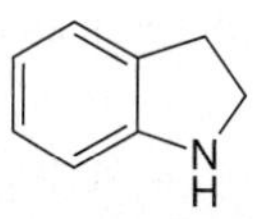

Problem 10.2 How many degrees of unsaturation are present in each compound?

a. C_6H_6 b. C_8H_{18} c. C_7H_8O d. $C_7H_{11}Br$ e. C_5H_9N

Problem 10.3 How many degrees of unsaturation does each of the following drugs contain?

a. zolpidem (sleep aid sold as Ambien), $C_{19}H_{21}N_3O$
b. mefloquine (antimalarial drug), $C_{17}H_{16}F_6N_2O$

10.3 Nomenclature

- **In the IUPAC system, an alkene is identified by the suffix *-ene*.**

10.3A General IUPAC Rules

How To Name an Alkene

Example Give the IUPAC name of the following alkene:

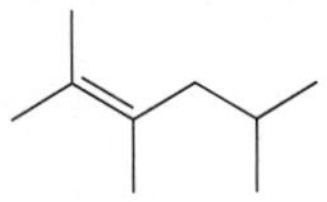

Step [1] Find the longest chain that contains *both* carbon atoms of the double bond.

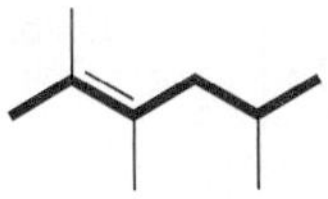

6 C's in the longest chain

hex*ane* ---→ hex*ene*

- Change the ***-ane*** ending of the parent alkane to ***-ene***.

Step [2] Number the carbon chain to give the double bond the lower number, and apply all other rules of nomenclature.

a. **Number** the chain, and name using the ***first* number** assigned to the C=C.

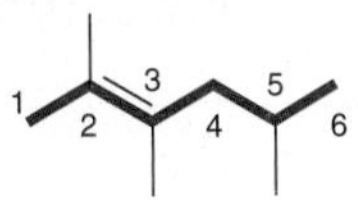

- Number the chain to put the C=C at C2, not C4.

hex-2-ene

b. **Name** and **number** the substituents.

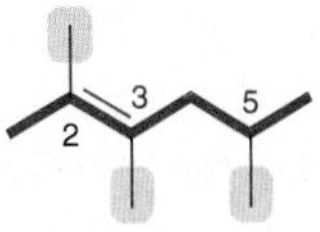

three methyl groups at C2, C3, and C5

Answer: 2,3,5-trimethylhex-2-ene

Compounds with two double bonds are named as **dienes** by changing the ***-ane*** ending of the parent alkane to the suffix ***-adiene.*** Compounds with three double bonds are named as **trienes,** and so forth. Always choose the longest chain that contains *both* atoms of the double bond. In

Figure 10.1, the alkene is named as a derivative of heptene because the seven-carbon chain contains both atoms of the double bond, but the eight-carbon chain does not.

Figure 10.1 Naming an alkene in which the longest carbon chain does not contain both atoms of the double bond

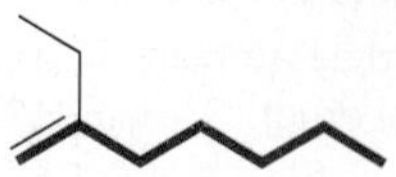

7 C's ---→ heptene

Both C's of the C=C are contained in this long chain.

Correct: 2-ethylhept-1-ene

8 C's

Both C's of the C=C are NOT contained in this long chain.

Incorrect

In naming cycloalkenes, the double bond is located between C1 and C2, and the "1" is usually omitted in the name. The ring is numbered clockwise or counterclockwise to give the first substituent the lower number. Representative examples are given in Figure 10.2.

Figure 10.2 Examples of cycloalkene nomenclature

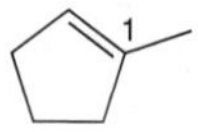

1-methylcyclopentene

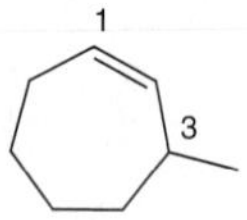

3-methylcycloheptene

[Number clockwise beginning at the C=C and place the CH_3 at C3.]

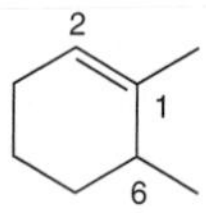

1,6-dimethylcyclohexene

[Number counterclockwise beginning at the C=C and place the first CH_3 at C1.]

$CH_3CH_2CH=CH_2$ is named as 1-butene using the 1979 IUPAC recommendations and but-1-ene using the 1993 IUPAC recommendations.

Compounds that contain both a double bond and a hydroxy group are named as **alkenols,** and the chain (or ring) is numbered to give the OH group the lower number.

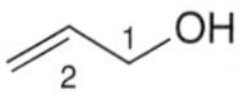

prop-2-en-1-ol

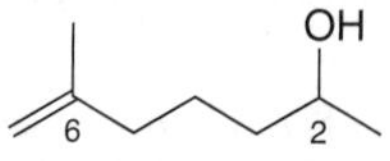

6-methylhept-6-en-2-ol

Problem 10.4 Give the IUPAC name for each alkene.

a. 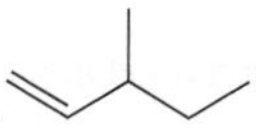 b. 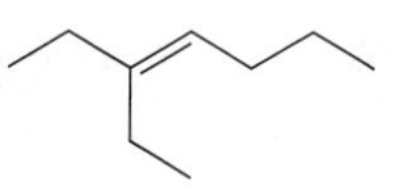 c. 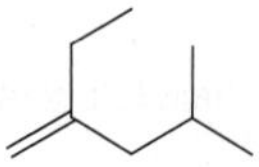 d. 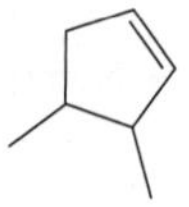 e. 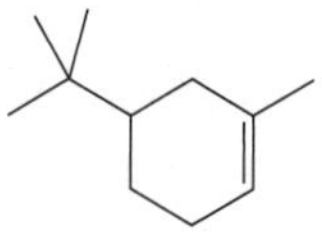

Problem 10.5 Give the IUPAC name for each polyfunctional compound.

a. 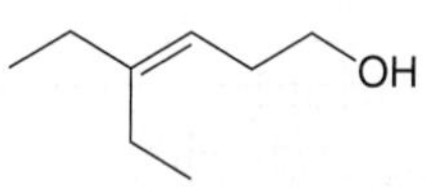

b. 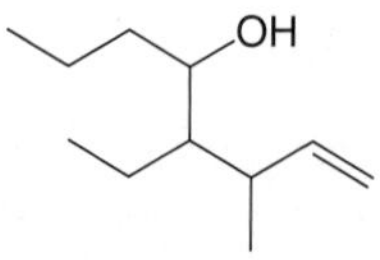

c. 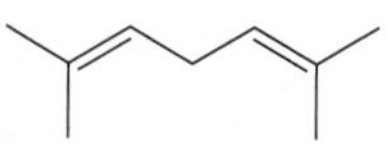

10.3B Naming Stereoisomers

A prefix is needed to distinguish two alkenes when diastereomers are possible.

Using Cis and Trans as Prefixes

An alkene having one alkyl group bonded to each carbon atom can be named using the prefixes **cis** and **trans** to designate the relative location of the two alkyl groups. For example, *cis*-hex-3-ene has two ethyl groups on the **same side** of the double bond, whereas *trans*-hex-3-ene has two ethyl groups on **opposite sides** of the double bond.

2 R's on the same side
cis-hex-3-ene

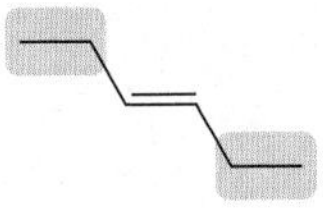

2 R's on opposite sides
trans-hex-3-ene

Using the Prefixes *E* and *Z*

E stands for the German word ***entgegen*** meaning "opposite." *Z* stands for the German word ***zusammen,*** meaning "together." Using *E,Z* nomenclature, a cis isomer has the *Z* configuration and a trans isomer has the *E* configuration.

Although the prefixes cis and trans can be used to distinguish diastereomers when two alkyl groups are bonded to the C=C, they cannot be used when there are three or four alkyl groups bonded to the C=C.

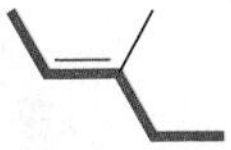

A
3-methylpent-2-ene

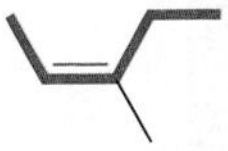

B
3-methylpent-2-ene

For example, alkenes **A** and **B** are two *different* compounds that are both called 3-methylpent-2-ene. In **A** the two CH_3 groups are cis, whereas in **B** the CH_3 and CH_2CH_3 groups are cis. The ***E,Z* system of nomenclature** has been devised to unambiguously name these kinds of alkenes.

How To Assign the Prefixes *E* and *Z* to an Alkene

Step [1] Assign priorities to the two substituents on each end of the C=C by using the priority rules for *R,S* nomenclature (Section 5.6).

- Divide the double bond in half, and assign the numbers **1** and **2** to indicate the relative priority of the two groups on each end—the higher priority group is labeled **1,** and the lower priority group is labeled **2.**

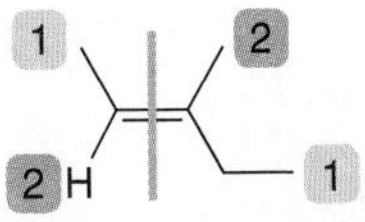

Assign priorities to each side separately.

Step [2] Assign *E* or *Z* based on the location of the two higher priority groups (1).

Two higher priority groups on **opposite sides**

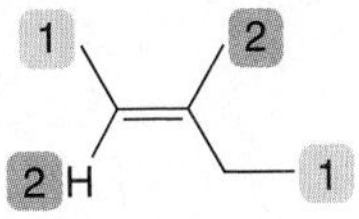

***E* isomer**
(*E*)-3-methylpent-2-ene

Two higher priority groups on the **same side**

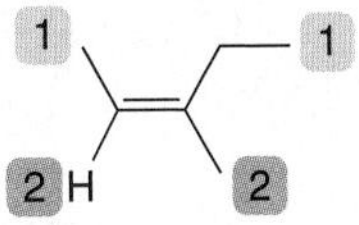

***Z* isomer**
(*Z*)-3-methylpent-2-ene

- The ***E*** isomer has the two higher priority groups on the **opposite sides.**
- The ***Z*** isomer has the two higher priority groups on the **same side.**

Problem 10.6 Label each C–C double bond as *E* or *Z*. Kavain is a naturally occurring relaxant isolated from kava root.

a. 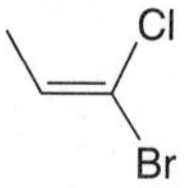
Cl
Br

b. 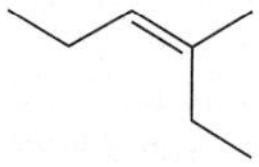

c. 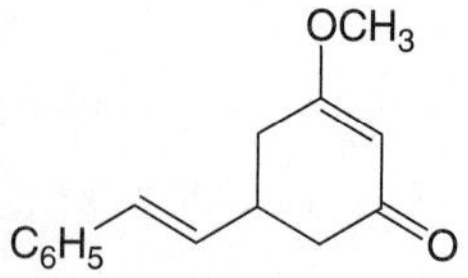
OCH3
C6H5
O

kavain

Problem 10.7 Label each carbon–carbon double bond in 11-*cis*-retinal as *E* or *Z*. As we will learn in Section 21.11, the isomerization of one double bond in this compound to a less crowded stereoisomer takes place when light strikes the retina of the eye.

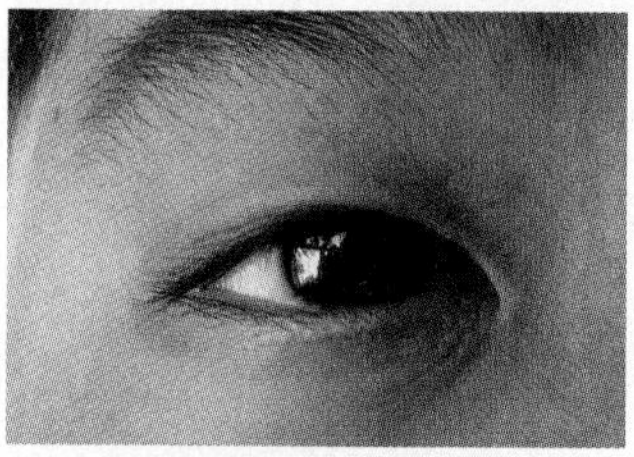

11-*cis*-Retinal is the light-sensitive aldehyde that plays a key role in the chemistry of vision for all vertebrates, arthropods, and mollusks.

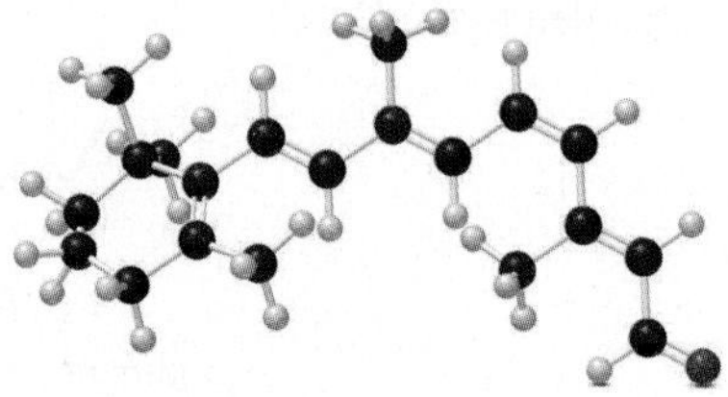

11-*cis*-retinal

Problem 10.8 Draw the structure corresponding to each IUPAC name.

a. (*Z*)-4-ethylhept-3-ene b. (*E*)-3,5,6-trimethyloct-2-ene c. (*Z*)-2-bromo-1-iodohex-1-ene

Problem 10.9 Draw the structure of (2*Z*,6*E*)-3-ethyl-7-methyldeca-2,6-dien-1-ol, the sex pheromone of the codling moth, a common agricultural pest that destroys apple crops. Spraying an apple orchard with this pheromone disrupts the mating of male and female moths, and moth populations can be controlled.

10.3C Common Names

The simplest alkene, $\mathbf{CH_2{=}CH_2}$, named in the IUPAC system as **ethene,** is often called **ethylene,** its common name. The common names for three **alkyl groups** derived from alkenes are also used. Three examples of naming organic molecules using these common names are shown in Figure 10.3.

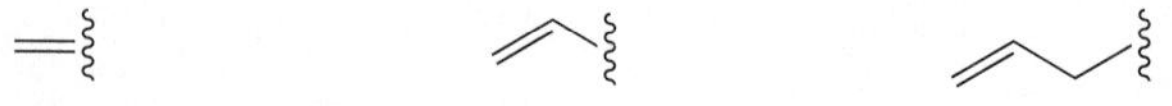
methylene group
vinyl group
allyl group

Figure 10.3 Naming alkenes with common substituent names

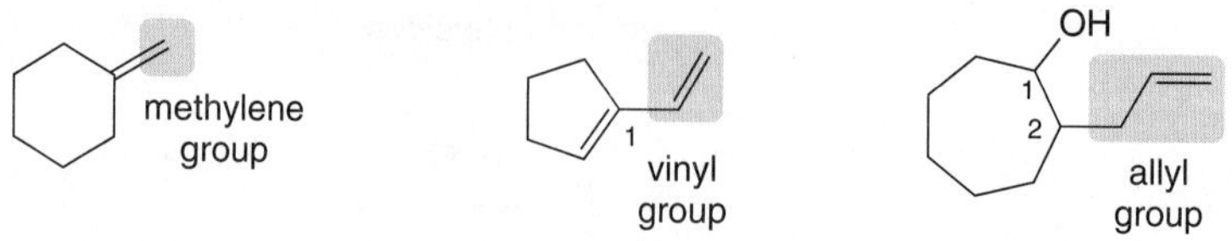
methylene
group
1
vinyl
group
OH
1
2
allyl
group

methylenecyclohexane 1-vinylcyclopentene 2-allylcycloheptanol

10.4 Physical Properties

Most alkenes exhibit only weak van der Waals interactions, so their physical properties are similar to those of alkanes of comparable molecular weight.

- **Alkenes have low melting points and boiling points.**
- **Melting points and boiling points increase as the number of carbons increases because of increased surface area.**
- **Alkenes are soluble in organic solvents and insoluble in water.**

Cis and trans alkenes often have somewhat different physical properties. For example, *cis*-but-2-ene has a higher boiling point (4 °C) than *trans*-but-2-ene (1 °C). This difference arises because the C–C single bond between an alkyl group and one of the double-bond carbons of an alkene is slightly polar. **The sp^3 hybridized alkyl carbon donates electron density to the sp^2 hybridized alkenyl carbon.**

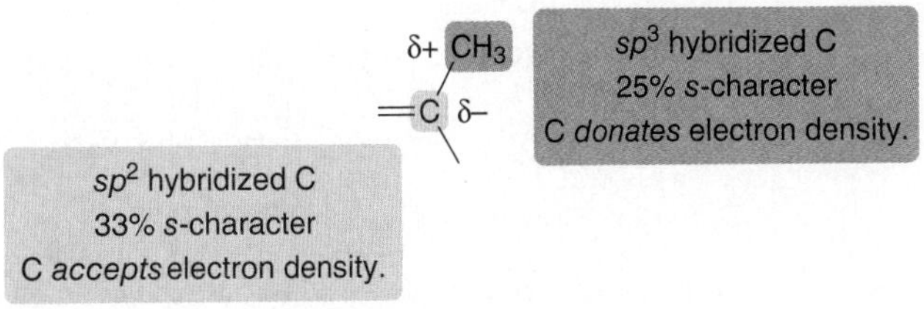

Related arguments involving $C_{sp^3}-C_{sp^2}$ bonds were used in Section 8.2C to explain why the stability of an alkene increases with increasing alkyl substitution.

The bond dipole places a partial negative charge on the alkenyl carbon (sp^2) relative to the alkyl carbon (sp^3) because an sp^2 hybridized orbital has greater percent *s*-character (33%) than an sp^3 hybridized orbital (25%). **In a cis isomer, the two $C_{sp^3}-C_{sp^2}$ bond dipoles reinforce each other, yielding a small net molecular dipole. In a trans isomer, the two bond dipoles cancel.**

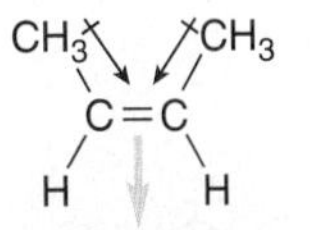

a small net dipole
more polar isomer

cis-but-2-ene

higher bp

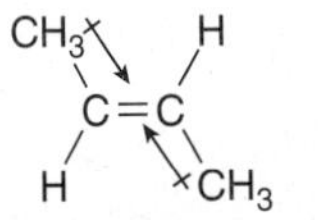

no net dipole
less polar isomer

trans-but-2-ene

lower bp

- **A cis alkene is more polar than a trans alkene, giving it a slightly higher boiling point and making it more soluble in polar solvents.**

Problem 10.10 Rank the following isomers in order of increasing boiling point.

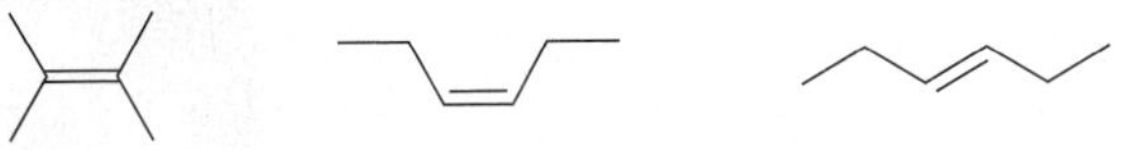

10.5 Interesting Alkenes

Ethylene is prepared from petroleum by a process called **cracking.** Ethylene is the most widely produced organic chemical, serving as the starting material not only for the polymer **polyethylene,** a widely used plastic, but also for many other useful organic compounds, as shown in Figure 10.4.

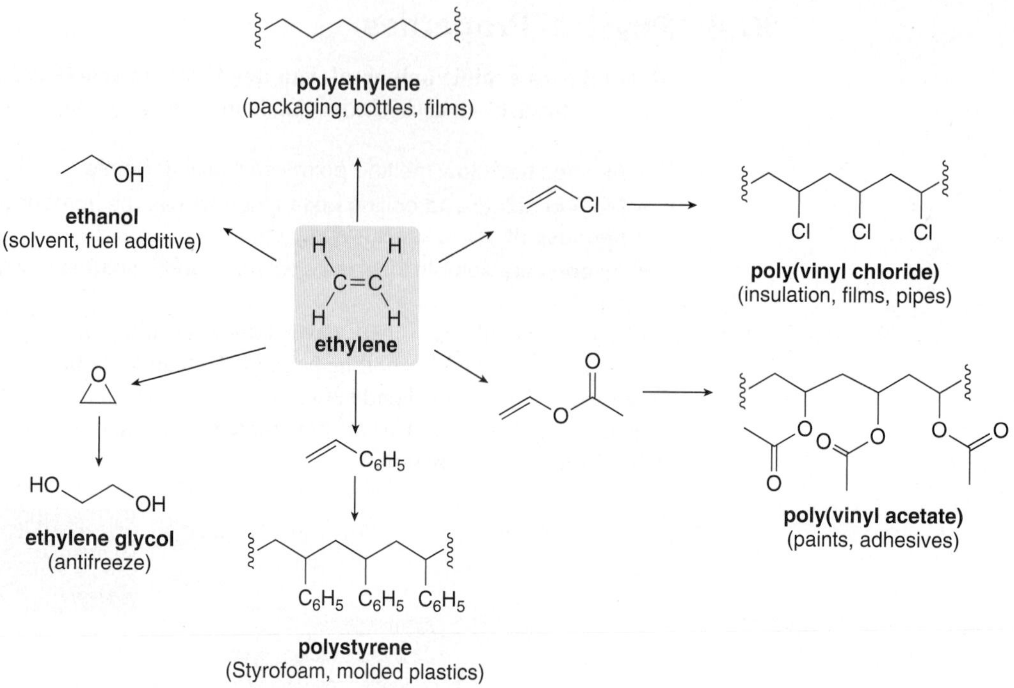

Figure 10.4
Ethylene, an industrial starting material for many useful products

Numerous organic compounds containing carbon–carbon double bonds have been isolated from natural sources (Figure 10.5).

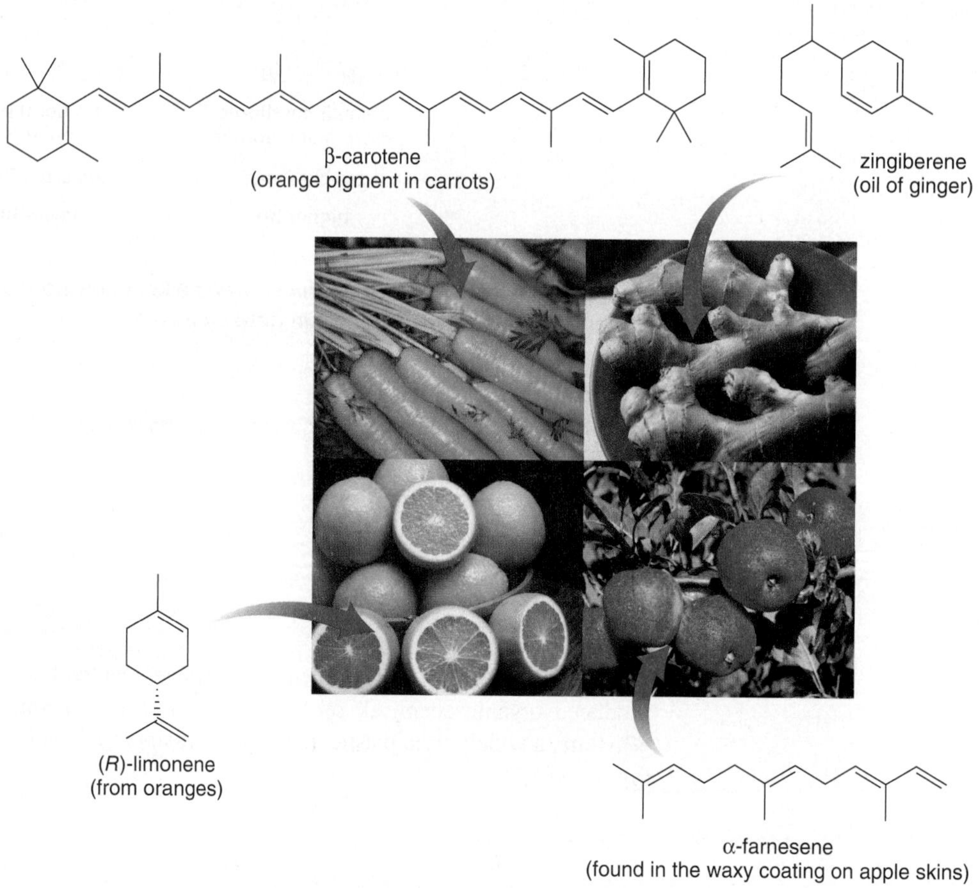

Figure 10.5
Four naturally occurring alkenes

10.6 Lipids—Part 2

Lipids are water-insoluble biomolecules composed largely of nonpolar C – C and C – H bonds (Section 4.15).

General structure of an ester:

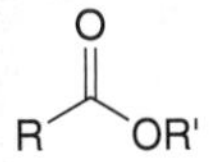

Understanding the geometry of C–C double bonds provides an insight into the properties of **triacylglycerols,** the most abundant lipids. Triacylglycerols contain three ester groups, each having a long carbon chain (abbreviated as R, R', and R") bonded to a carbonyl group (C=O).

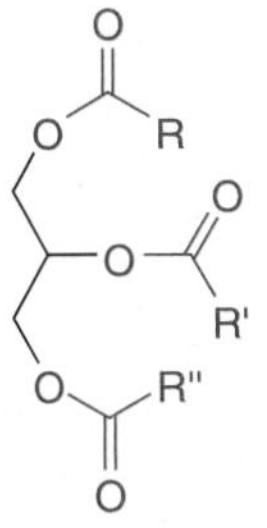

triacylglycerol

R groups have 11–19 C's.

[Three ester groups are labeled in red.]

10.6A Fatty Acids

Linoleic and linolenic acids are essential fatty acids, meaning they cannot be synthesized in the body and must therefore be obtained in the diet. A common source of these essential fatty acids is whole milk. Babies fed a diet of nonfat milk in their early months do not thrive because they do not obtain enough of these essential fatty acids.

Triacylglycerols are hydrolyzed to glycerol (a triol), and three **fatty acids** of general structure RCO_2H. Naturally occurring fatty acids contain 12–20 carbon atoms, with a carboxy group (CO_2H) at one end.

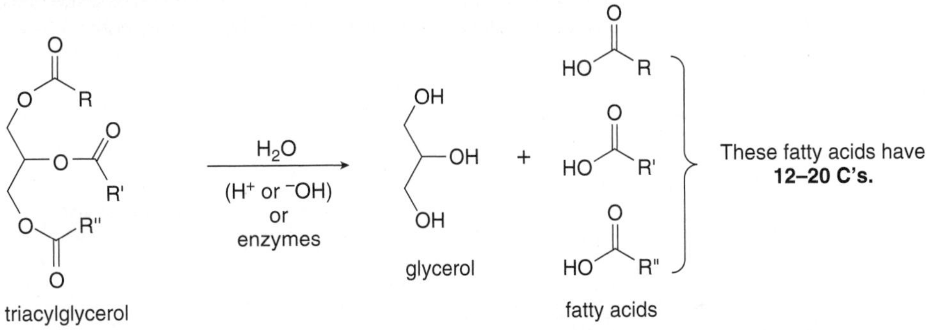

- **Saturated fatty acids have no double bonds in their long hydrocarbon chains, and unsaturated fatty acids have one or more double bonds in their hydrocarbon chains.**
- **Double bonds in naturally occurring fatty acids have the *Z* configuration.**

Table 10.2 lists the structure and melting point of four fatty acids containing 18 carbon atoms. Stearic acid is one of the two most common saturated fatty acids, and oleic and linoleic acids are the most common unsaturated ones. The data show the effect of *Z* double bonds on the melting point of fatty acids.

- **As the number of double bonds in the fatty acid *increases,* the melting point *decreases.***

The three-dimensional structures of the fatty acids in Figure 10.6 illustrate how *Z* double bonds introduce kinks in the long hydrocarbon chain, decreasing the ability of the fatty acid to pack well in a crystalline lattice. **The larger the number of *Z* double bonds, the more kinks in the hydrocarbon chain, and the lower the melting point.**

Table 10.2 The Effect of Double Bonds on the Melting Point of Fatty Acids

Name	Structure	Mp (°C)
Stearic acid (0 C=C)	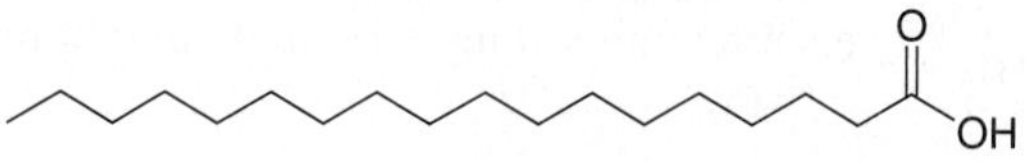	69
Oleic acid (1 C=C)	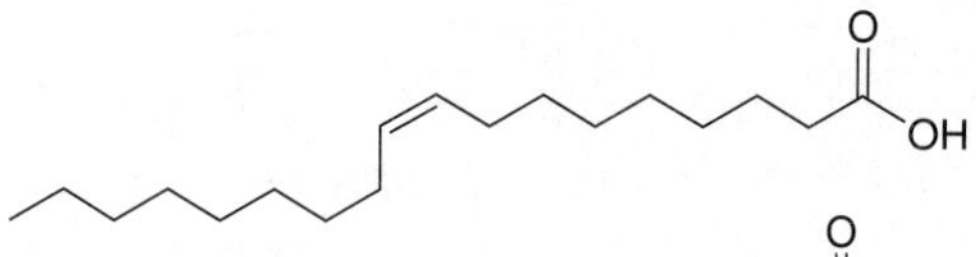	4
Linoleic acid (2 C=C)	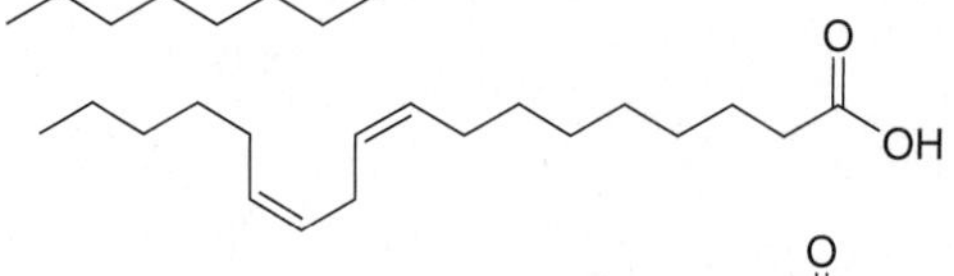	−5
Linolenic acid (3 C=C)	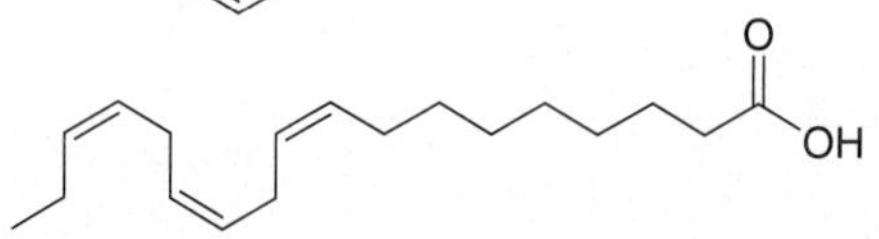	−11

Increasing number of double bonds

Problem 10.11 Linolenic acid (Table 10.2) and stearidonic acid are omega-3 fatty acids, unsaturated fatty acids that contain the first double bond located at C3, when numbering begins at the methyl end of the chain. Predict how the melting point of stearidonic acid compares with the melting points of linolenic and stearic acids. A current avenue of research is examining the use of soybean oil enriched in stearidonic acid as a healthier alternative to vegetable oils that contain fewer degrees of unsaturation.

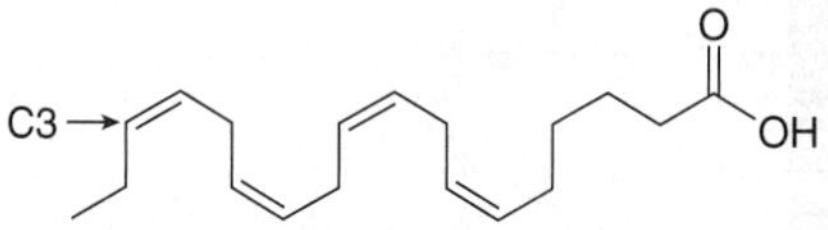
C3
O
OH

stearidonic acid

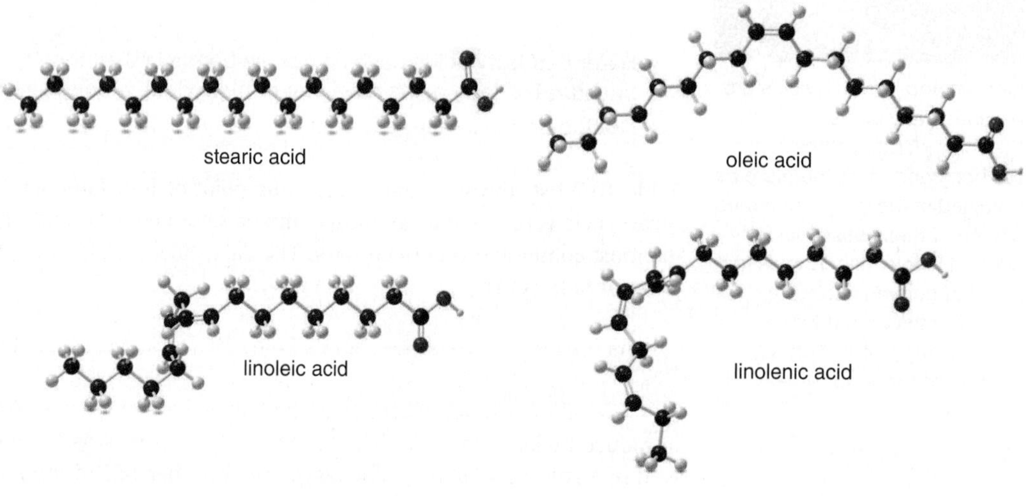
stearic acid
oleic acid
linoleic acid
linolenic acid

Figure 10.6 Three-dimensional structure of four C_{18} fatty acids

10.6B Fats and Oils

Canola, soybeans, and flaxseed are excellent dietary sources of linolenic acid, an essential fatty acid. Oils derived from omega-3 fatty acids (Problem 10.11) are currently thought to be especially beneficial for individuals at risk of developing coronary artery disease.

Fats and **oils** are triacylglycerols with different physical properties.

- Fats have higher melting points—they are *solids* at room temperature.
- Oils have lower melting points—they are *liquids* at room temperature.

The identity of the three fatty acids in the triacylglycerol determines whether it is a fat or an oil. **Increasing the number of double bonds in the fatty acid side chains decreases the melting point of the triacylglycerol.**

- Fats are derived from fatty acids having few double bonds.
- Oils are derived from fatty acids having a larger number of double bonds.

Saturated fats are typically obtained from animal sources, whereas unsaturated oils are common in vegetable sources. Thus, butter and lard are high in saturated triacylglycerols, and olive oil and safflower oil are high in unsaturated triacylglycerols. An exception to this generalization is coconut oil, which is composed largely of saturated alkyl side chains.

Considerable evidence suggests that an elevated cholesterol level is linked to an increased risk of heart disease. Saturated fats stimulate cholesterol synthesis in the liver, thus increasing the cholesterol concentration in the blood.

10.7 Preparation of Alkenes

Recall from Chapters 8 and 9 that alkenes can be prepared from alkyl halides and alcohols via elimination reactions. For example, **dehydrohalogenation of alkyl halides with strong base yields alkenes via an E2 mechanism** (Sections 8.4 and 8.5).

- Typical bases include ^-OH and ^-OR [especially $^-OC(CH_3)_3$], and nonnucleophilic bases such as DBU and DBN.
- Alkyl tosylates can also be used as starting materials under similar reaction conditions (Section 9.13).

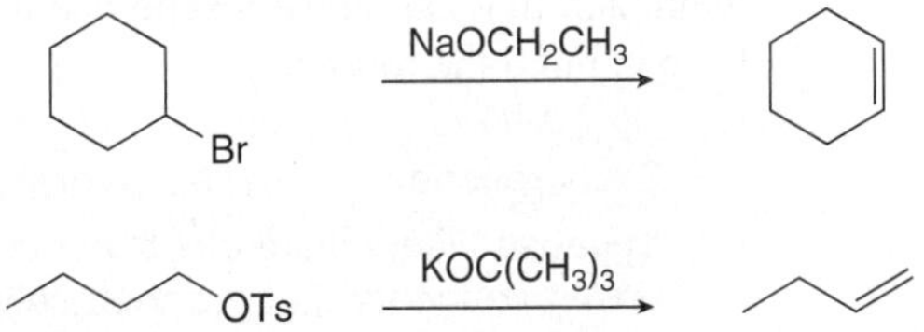

The acid-catalyzed dehydration of alcohols with H_2SO_4 or TsOH yields alkenes, too (Sections 9.8 and 9.9). The reaction occurs via an E1 mechanism for 2° and 3° alcohols, and an E2 mechanism for 1° alcohols. E1 reactions involve carbocation intermediates, so rearrangements are possible. **Dehydration can also be carried out with $POCl_3$ and pyridine** by an E2 mechanism (Section 9.10).

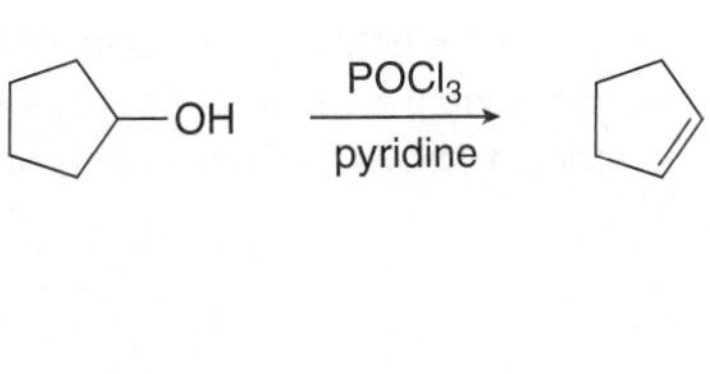

These elimination reactions are stereoselective and regioselective, so the most stable alkene is usually formed as the major product.

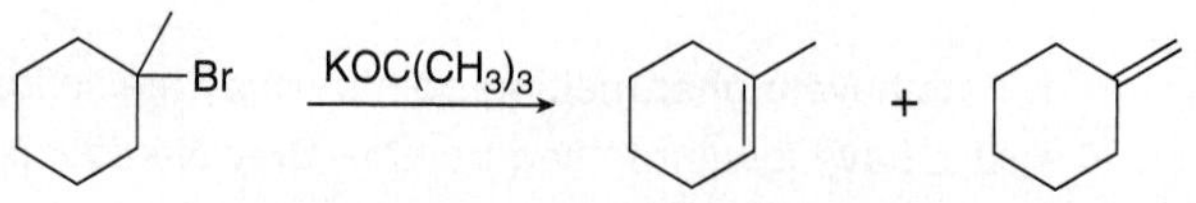

major product

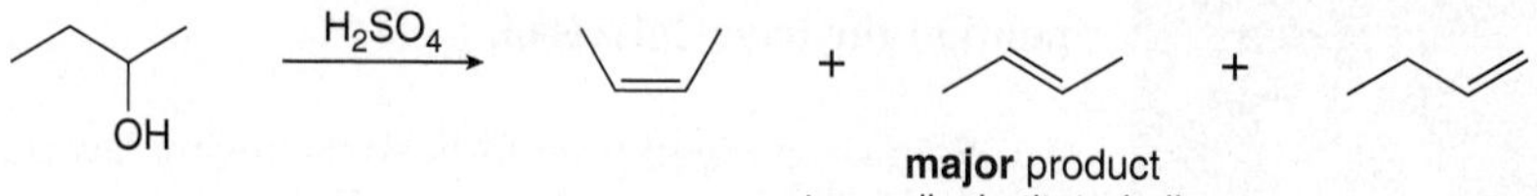

major product
trans disubstituted alkene

Problem 10.12 Draw the products of each elimination reaction.

a. 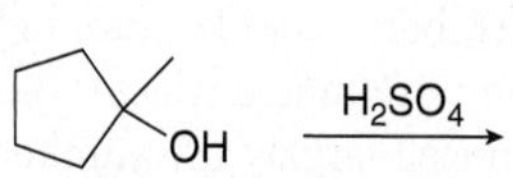

b. 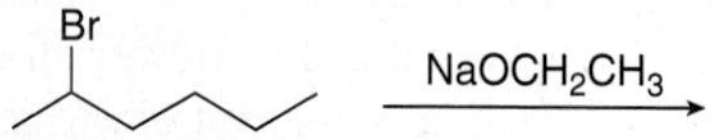

10.8 Introduction to Addition Reactions

Because the C–C π bond of an alkene is much weaker than a C–C σ bond, the characteristic reaction of alkenes is **addition: the π bond is broken and two new σ bonds are formed.**

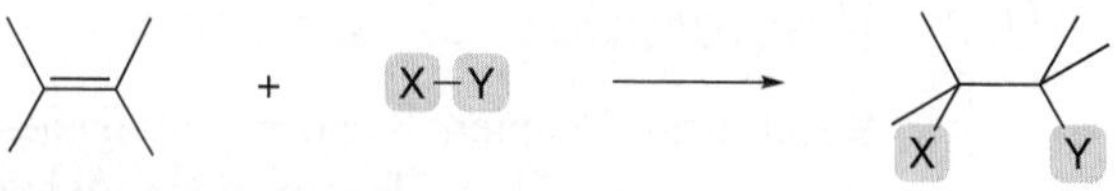

A π bond is broken. Two σ bonds are formed.

Alkenes are electron rich, as seen in the electrostatic potential plot in Figure 10.7. The electron density of the π bond is concentrated above and below the plane of the molecule, making the π bond more exposed than the σ bond.

What kinds of reagents add to the weak, electron-rich π bond of alkenes? There are many of them, and that can make alkene chemistry challenging. To help you organize this information, keep in mind the following:

The addition reactions of alkenes are discussed in Sections 10.9–10.16 and in Chapter 12 (Oxidation and Reduction).

- **Every reaction of alkenes involves addition: the π bond is always broken.**
- **Because alkenes are electron rich, simple alkenes do *not* react with nucleophiles or bases, reagents that are themselves electron rich. Alkenes react with *electrophiles.***

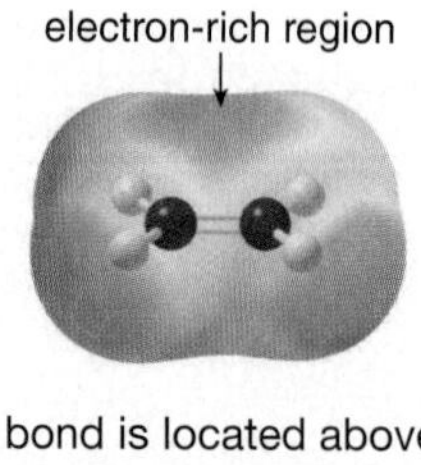

Figure 10.7 Electrostatic potential plot of ethylene

- The red electron-rich region of the π bond is located above and below the plane of the molecule. Because the plane of the alkene depicted in this electrostatic potential plot is tipped, only the red region above the molecule is visible.

The stereochemistry of addition is often important in delineating a reaction's mechanism. Because the carbon atoms of a double bond are both trigonal planar, the elements of X and Y can be added to them from the **same side** or from **opposite sides.**

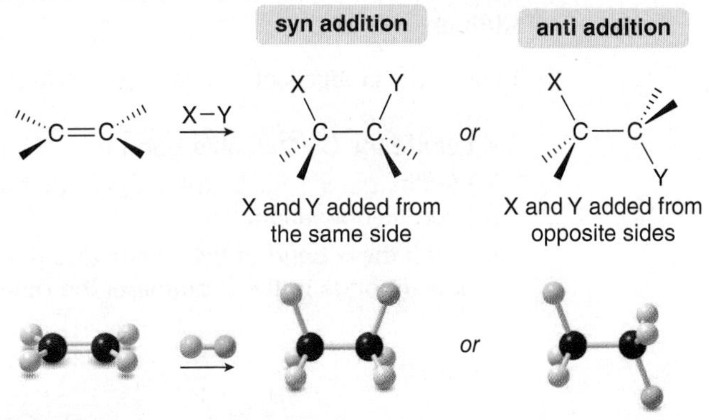
syn addition
anti addition
X–Y
or
X and Y added from the same side
X and Y added from opposite sides
or

- *Syn addition* takes place when both X and Y are added from the *same* side.
- *Anti addition* takes place when X and Y are added from *opposite* sides.

Five reactions of alkenes are discussed in Chapter 10 and each is illustrated in Figure 10.8, using cyclohexene as the starting material.

Figure 10.8
Five addition reactions of cyclohexene

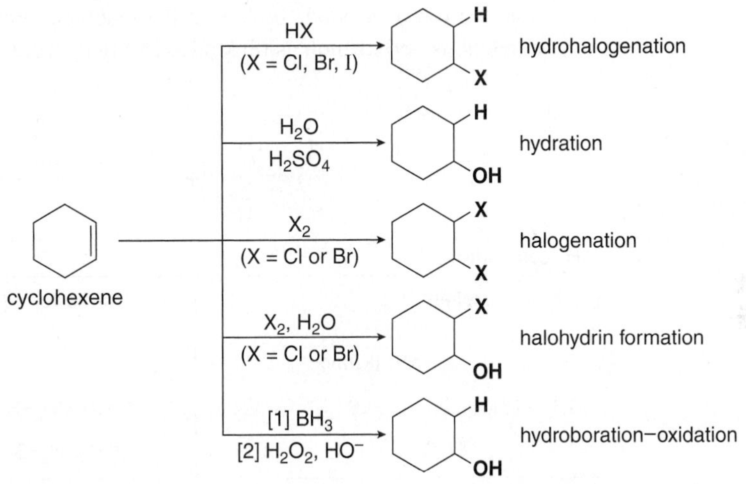
HX
(X = Cl, Br, I)
hydrohalogenation
H2O
H2SO4
hydration
X2
(X = Cl or Br)
halogenation
cyclohexene
X2, H2O
(X = Cl or Br)
halohydrin formation
[1] BH3
[2] H2O2, HO–
hydroboration–oxidation

- In each reaction, the π bond is broken and two new σ bonds are formed.

10.9 Hydrohalogenation—Electrophilic Addition of HX

Hydrohalogenation of an alkene to form an alkyl halide is the reverse of the dehydrohalogenation of an alkyl halide to form an alkene, a reaction discussed in detail in Sections 8.4 and 8.5.

Hydrohalogenation results in the addition of hydrogen halides HX (X = Cl, Br, and I) to alkenes to form alkyl halides.

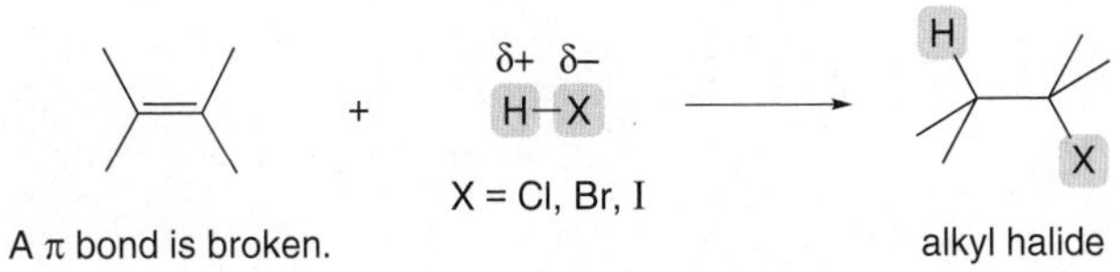
δ+ δ–
H–X
X = Cl, Br, I
A π bond is broken.
alkyl halide

Two bonds are broken in this reaction—the weak π bond of the alkene and the HX bond—and two new σ bonds are formed—one to H and one to X. Because X is more electronegative than H, the H–X bond is polarized, with a partial positive charge on H. Because the electrophilic (H) end of HX is attracted to the electron-rich double bond, these reactions are called **electrophilic additions.**

To draw the products of an addition reaction:

- **Locate the C–C double bond.**
- **Identify the σ bond of the reagent that breaks—namely, the H–X bond in hydrohalogenation.**
- **Break the π bond of the alkene and the σ bond of the reagent, and form two new σ bonds to the C atoms of the double bond.**

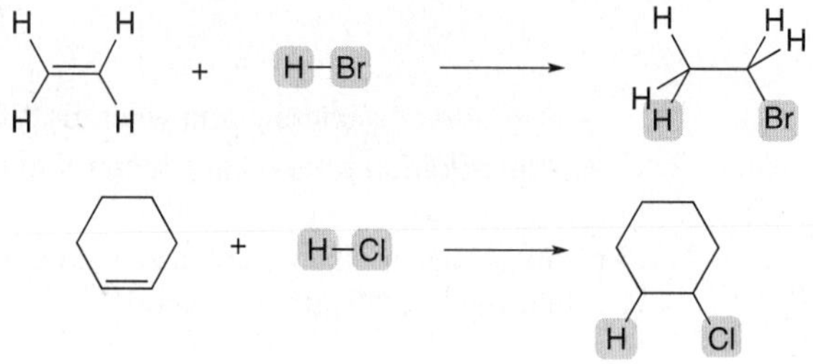

Addition reactions are exothermic because the two σ bonds formed in the product are stronger than the σ and π bonds broken in the reactants. For example, $\Delta H°$ for the addition of HBr to ethylene is –60 kJ/mol, as illustrated in Figure 10.9.

Figure 10.9
The addition of HBr to $CH_2{=}CH_2$, an exothermic reaction

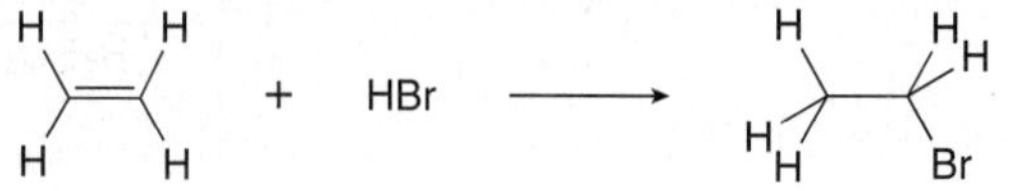

$\Delta H°$ calculation:

[1] Bonds broken	$\Delta H°$ (kJ/mol)	[2] Bonds formed	$\Delta H°$ (kJ/mol)	[3] Overall $\Delta H°$ =
$CH_2{=}CH_2$ π bond	+267	$BrCH_2CH_2$–H	–410	sum in Step [1] + sum in Step [2]
H–Br	+368	CH_3CH_2–Br	–285	
Total	+635 kJ/mol	Total	–695 kJ/mol	+635 kJ/mol –695 kJ/mol
Energy needed to break bonds.		**Energy released in forming bonds.**		$\Delta H°$ = –60 kJ/mol
				The reaction is exothermic.

[Values taken from Appendix C.]

The mechanism of electrophilic addition of HX consists of **two steps:** addition of H^+ to form a carbocation, followed by nucleophilic attack of X^-. The mechanism is illustrated for the reaction of *cis*-but-2-ene with HBr in Mechanism 10.1.

Mechanism 10.1 Electrophilic Addition of HX to an Alkene

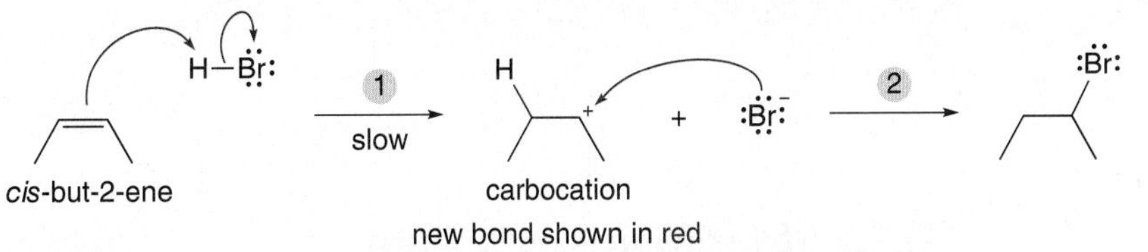

1 The π bond of the alkene attacks the H of HBr to form a new C–H bond and a carbocation in the rate-determining in step.

2 Nucleophilic attack of Br^- on the carbocation forms the new C–Br bond.

The mechanism of electrophilic addition consists of two successive Lewis acid–base reactions. In Step [1], the **alkene is the Lewis base** that donates an electron pair to **H–Br, the Lewis acid,** while in Step [2], **Br^- is the Lewis base** that donates an electron pair to the **carbocation, the Lewis acid.**

An energy diagram for the reaction of $CH_3CH{=}CHCH_3$ with HBr is given in Figure 10.10. Each step has its own energy barrier with a transition state at each energy maximum. Because Step [1] has a higher energy transition state, it is rate-determining. $\Delta H°$ for Step [1] is positive because more bonds are broken than formed, whereas $\Delta H°$ for Step [2] is negative because only bond making occurs.

Figure 10.10
Energy diagram for electrophilic addition: $CH_3CH{=}CHCH_3 + HBr \rightarrow CH_3CH_2CH(Br)CH_3$

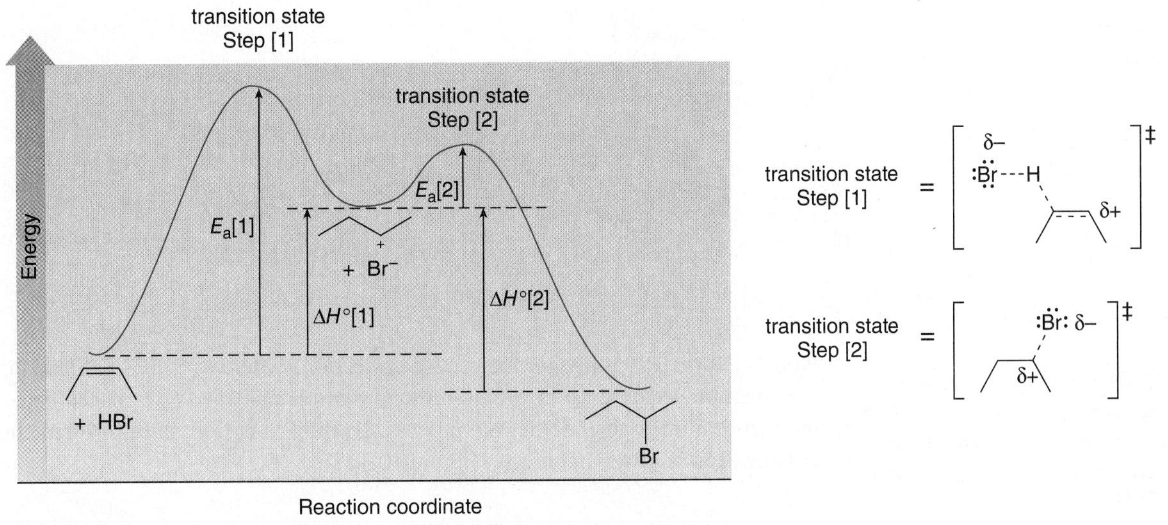

- The mechanism has two steps, so there are two energy barriers.
- Step [1] is rate-determining.

Problem 10.13 What product is formed when each alkene is treated with HCl?

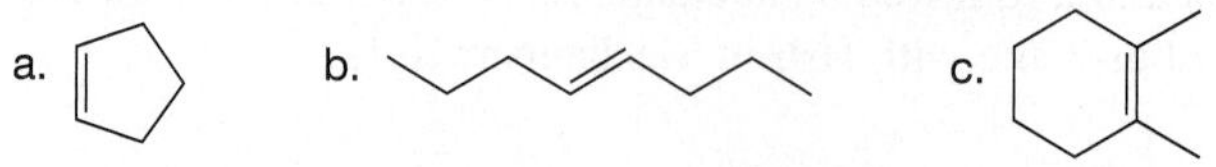

Problem 10.14 Draw a stepwise mechanism for the following reaction. Draw the transition state for each step.

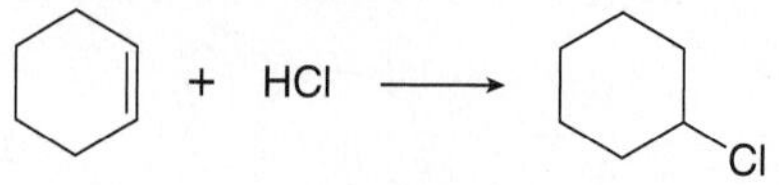

10.10 Markovnikov's Rule

With an unsymmetrical alkene, HX can add to the double bond to give two constitutional isomers.

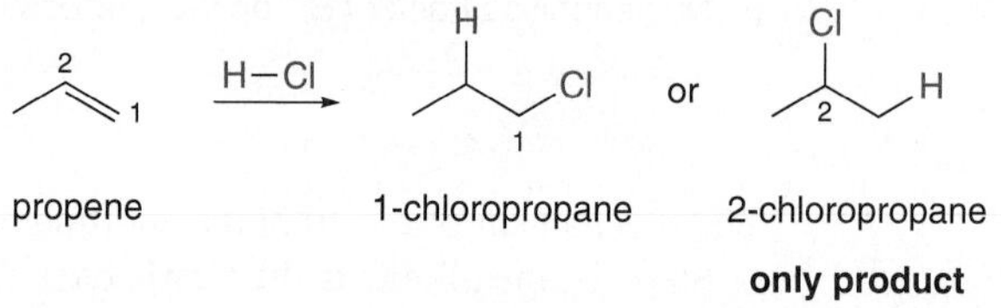

For example, HCl addition to propene could in theory form 1-chloropropane by addition of H and Cl to C2 and C1, respectively, and 2-chloropropane by addition of H and Cl to C1 and C2, respectively. In fact, **electrophilic addition forms *only* 2-chloropropane.** This is a specific example of a general trend called **Markovnikov's rule,** named for the Russian chemist who first determined the regioselectivity of electrophilic addition of HX.

- **Markovnikov's rule: In the addition of HX to an unsymmetrical alkene, the H atom bonds to the less substituted carbon atom—that is, the carbon that has more H atoms to begin with.**

The basis of Markovnikov's rule is the formation of a carbocation in the rate-determining step of the mechanism. With propene, there are two possible paths for this first step, depending on which carbon atom of the double bond forms the new bond to hydrogen.

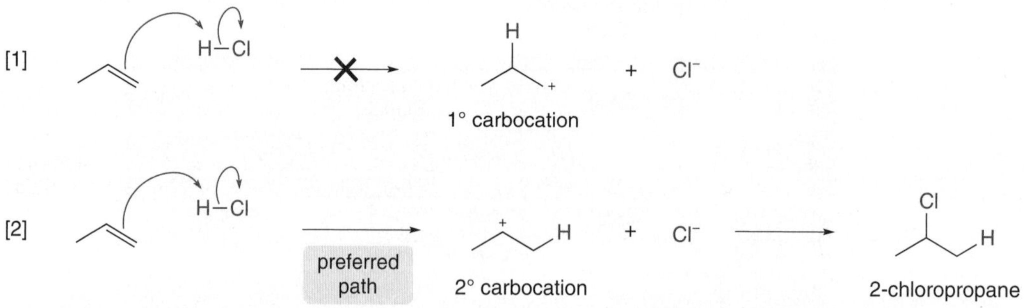

The Hammond postulate was first introduced in Section 7.14 to explain the relative rate of S_N1 reactions with 1°, 2°, and 3° RX.

Path [1] forms a highly unstable 1° carbocation, whereas Path [2] forms a **more stable 2° carbocation.** According to the Hammond postulate, Path [2] is faster because formation of the carbocation is an endothermic process, so **the transition state to form the more stable 2° carbocation is lower in energy** (Figure 10.11).

Figure 10.11
Electrophilic addition and the Hammond postulate

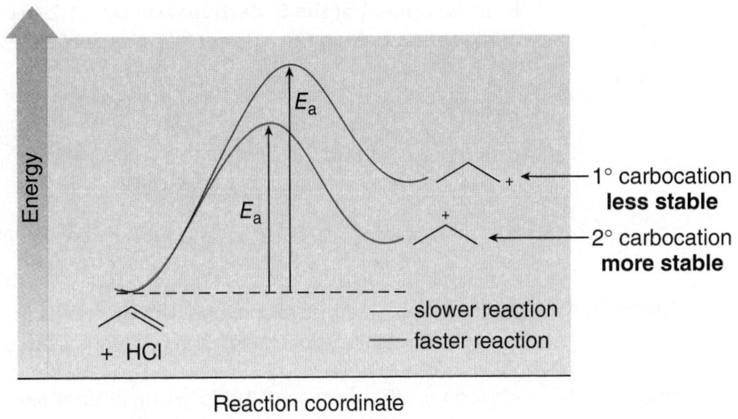

- The E_a for formation of the more stable 2° carbocation is lower than the E_a for formation of the 1° carbocation. The 2° carbocation is formed faster.

- **In the addition of HX to an unsymmetrical alkene, the H atom is added to the less substituted carbon to form the more stable, more substituted carbocation.**

Similar results are seen in any electrophilic addition involving an intermediate carbocation: **the more stable, more substituted carbocation is formed by addition of the electrophile to the less substituted carbon.**

Problem 10.15 Draw the products formed when each alkene is treated with HCl.

a. b. c.

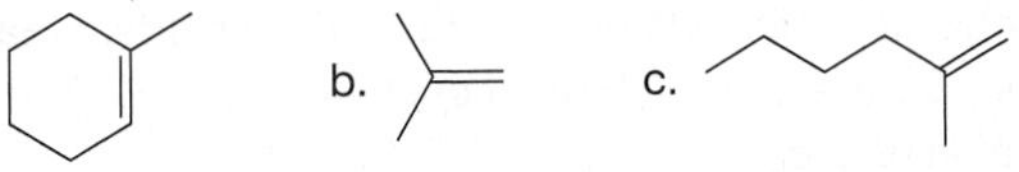

Problem 10.16 Use the Hammond postulate to explain why $(CH_3)_2C{=}CH_2$ reacts faster than $CH_3CH{=}CH_2$ in electrophilic addition of HX.

Because carbocations are formed as intermediates in hydrohalogenation, carbocation rearrangements can occur, as illustrated in Sample Problem 10.3.

Sample Problem 10.3 Draw a stepwise mechanism for the following reaction.

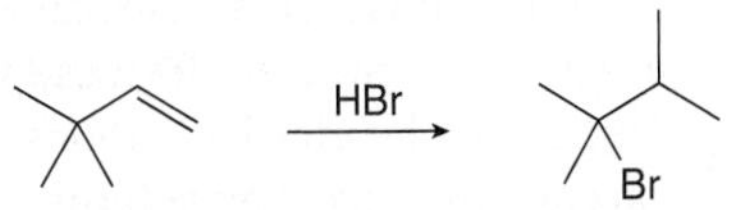

Solution

Because the carbon skeletons of the starting material and product are different—the alkene reactant has a 4° carbon and the product alkyl halide does not—a carbocation rearrangement must have occurred.

Markovnikov addition of HBr adds H^+ to the less substituted end of the double bond, forming a 2° carbocation in Step [1].

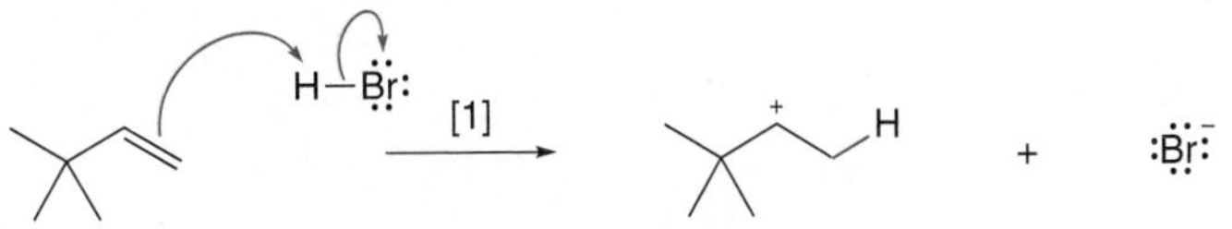

—Continued

Rearrangement of the 2° carbocation by a 1,2-methyl shift forms a more stable 3° carbocation in Step [2]. Nucleophilic attack of Br^- forms the product, a 3° alkyl halide, in Step [3].

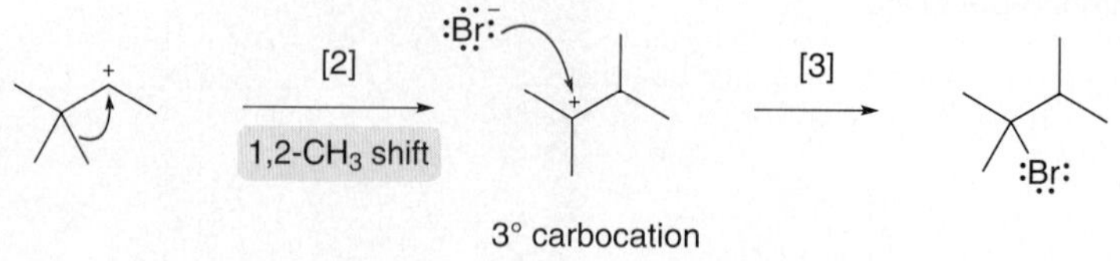

Problem 10.17 Treatment of 3-methylcyclohexene with HCl yields two products, 1-chloro-3-methylcyclohexane and 1-chloro-1-methylcyclohexane. Draw a mechanism to explain this result.

Problem 10.18 Addition of HBr to which of the following alkenes will lead to a rearrangement?

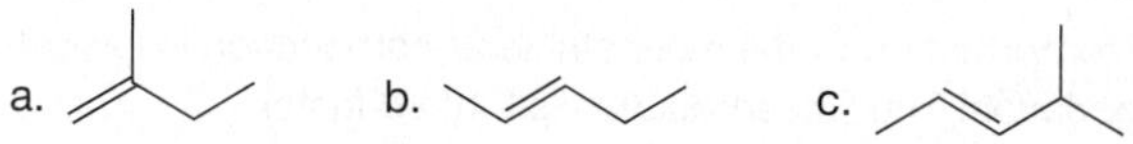

10.11 Stereochemistry of Electrophilic Addition of HX

To understand the stereochemistry of electrophilic addition, recall two stereochemical principles learned in Chapters 7 and 9.

- **Trigonal planar atoms react with reagents from two directions with equal probability (Section 7.12C).**
- **Achiral starting materials yield achiral or racemic products (Section 9.16).**

Many hydrohalogenation reactions begin with an **achiral reactant** and form an **achiral product.** For example, the addition of HBr to cyclohexene, an achiral alkene, forms bromocyclohexane, an achiral alkyl halide.

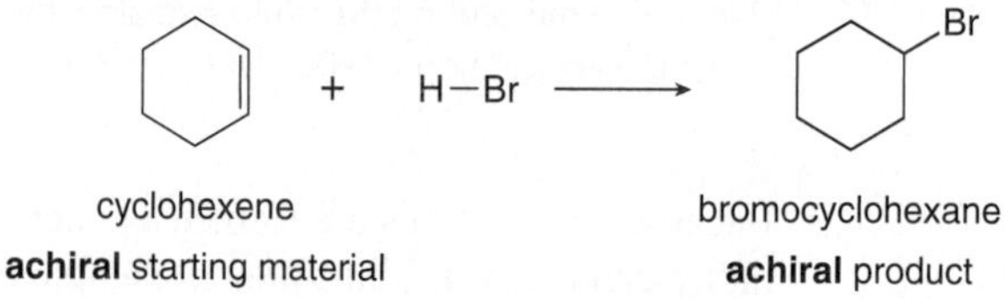

Because addition converts sp^2 hybridized carbons to sp^3 hybridized carbons, sometimes new stereogenic centers are formed from hydrohalogenation. Markovnikov addition of HCl to 1,3,3-trimethylcyclohexene, an achiral alkene, forms one constitutional isomer, 1-chloro-1,3,3-trimethylcyclohexane. Because this product now has a stereogenic center at one of the newly formed sp^3 hybridized carbons (labeled in blue), **an equal amount of two enantiomers—a racemic mixture**—must form.

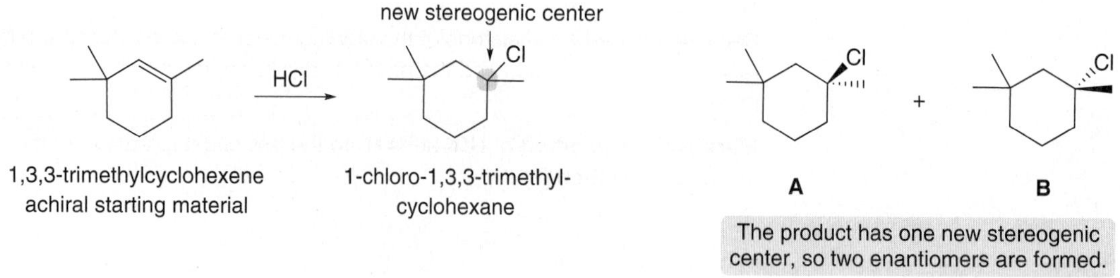

The mechanism of hydrohalogenation illustrates why two enantiomers are formed. Initial addition of the electrophile H^+ (from HCl) occurs from **either side of the planar double bond** to form a carbocation. Both modes of addition (from above and below) generate the same **achiral carbocation.** Either representation of this carbocation can then be used to draw the second step of the mechanism.

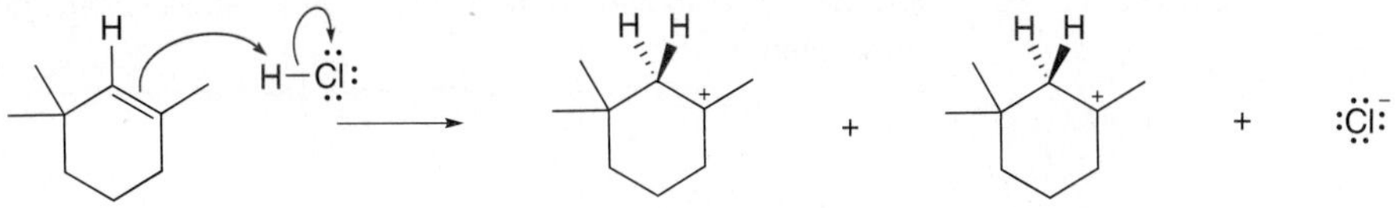

H^+ from above H^+ from below

identical carbocations

Nucleophilic attack of Cl^- on the trigonal planar carbocation also occurs from two different directions, forming two products, **A** and **B,** having a new stereogenic center. **A** and **B** are not superimposable, so they are **enantiomers.** Because attack from either direction occurs with equal probability, a **racemic mixture** of **A** and **B** is formed.

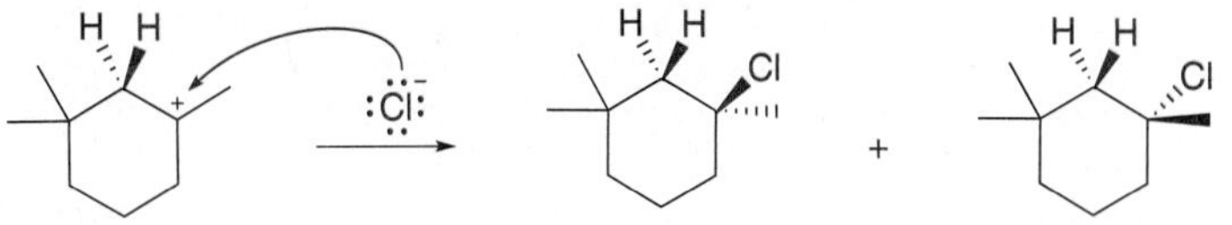

Cl^- from above	Cl^- from below
A	**B**
syn addition	**anti addition**
H and Cl are added from the same side.	H and Cl are added from opposite sides.

The terms **cis** and **trans** refer to the arrangement of groups in a particular compound, usually an alkene or a disubstituted cycloalkane. The terms **syn** and **anti** describe the stereochemistry of a process—for example, how two groups are added to a double bond.

Because hydrohalogenation begins with a **planar** double bond and forms a **planar** carbocation, addition of H and Cl occurs in two different ways. The elements of H and Cl can both be added from the same side of the double bond—that is, **syn addition**—or they can be added from opposite sides—that is, **anti addition.** *Both* modes of addition occur in this two-step reaction mechanism.

- **Hydrohalogenation occurs with syn and anti addition of HX.**

Table 10.3 summarizes the characteristics of electrophilic addition of HX to alkenes.

Table 10.3 Summary: Electrophilic Addition of HX to Alkenes

	Observation
Mechanism	• The mechanism involves two steps. • The rate-determining step forms a carbocation. • Rearrangements can occur.
Regioselectivity	• Markovnikov's rule is followed. In unsymmetrical alkenes, H bonds to the less substituted C to form the more stable carbocation.
Stereochemistry	• Syn and anti addition occur.

Problem 10.19 Draw the products, including stereochemistry, of each reaction.

a. 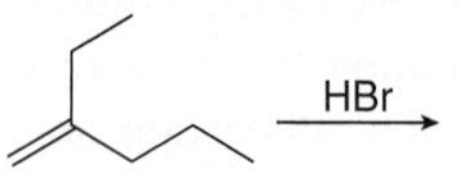

b. 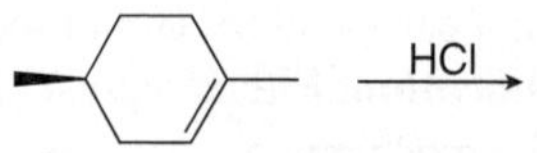

Problem 10.20 Draw all stereoisomers formed when 1,2-dimethylcyclohexene is treated with HCl. Label pairs of enantiomers.

10.12 Hydration—Electrophilic Addition of Water

Hydration results in the addition of water to an alkene to form an alcohol. H_2O itself is too weak an acid to protonate an alkene, but with added H_2SO_4, H_3O^+ is formed and addition readily occurs.

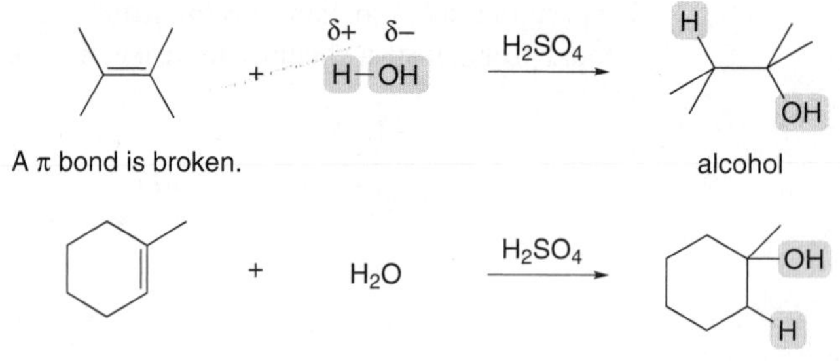

Hydration is simply another example of **electrophilic addition.** The first two steps of the mechanism are similar to those of electrophilic addition of HX—that is, addition of H^+ (from H_3O^+) to generate a carbocation, followed by nucleophilic attack of H_2O. Mechanism 10.2 illustrates the addition of H_2O to cyclohexene to form cyclohexanol.

Mechanism 10.2 Electrophilic Addition of H_2O to an Alkene—Hydration

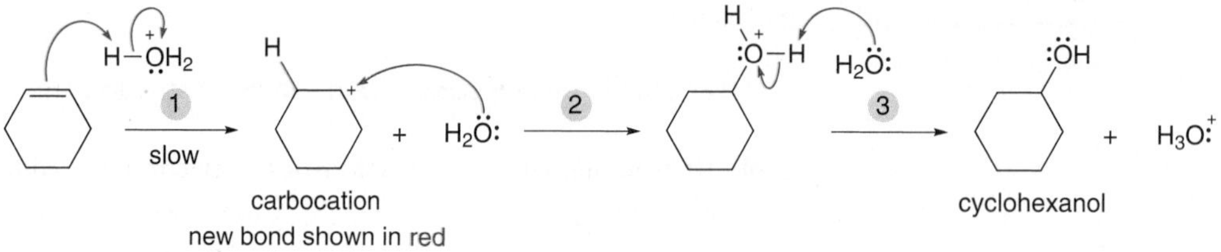

1 The π bond of the alkene attacks the H of H_3O^+ to form a new C–H bond and a carbocation in the rate-determining step.

2 Nucleophilic attack of H_2O on the carbocation forms the new C–O bond.

3 Removal of a proton with H_2O forms a neutral alcohol. Because the acid used in Step [1] is regenerated in Step [3], the reaction is acid-catalyzed.

Hydration of an alkene to form an alcohol is the reverse of the dehydration of an alcohol to form an alkene, a reaction discussed in detail in Section 9.8.

There are three consequences to the formation of carbocation intermediates.

- **In unsymmetrical alkenes, H adds to the less substituted carbon to form the more stable carbocation; that is, Markovnikov's rule holds.**
- **Addition of H and OH occurs in both a syn and anti fashion.**
- **Carbocation rearrangements can occur.**

Alcohols add to alkenes, forming ethers, using the same mechanism. Addition of CH_3OH to 2-methylpropene, for example, forms *tert*-butyl methyl ether (**MTBE**), a high octane fuel additive described in Section 3.4C.

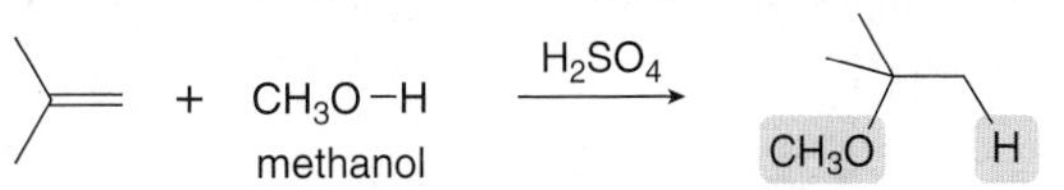

tert-butyl methyl ether
MTBE

Problem 10.21 What two alkenes give rise to each alcohol as the major product of acid-catalyzed hydration?

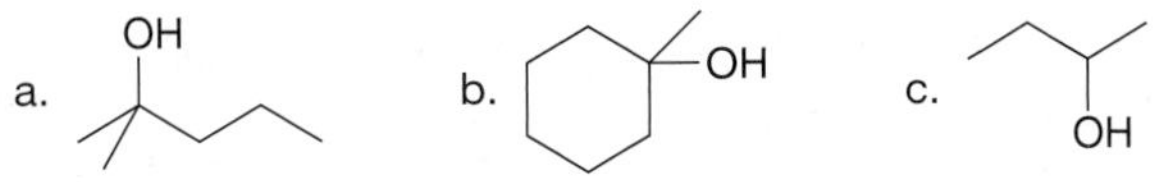

Problem 10.22 What stereoisomers are formed when pent-1-ene is treated with H_2O and H_2SO_4?

10.13 Halogenation—Addition of Halogen

Halogenation results in the addition of halogen X_2 (X = Cl or Br) **to an alkene, forming a vicinal dihalide.**

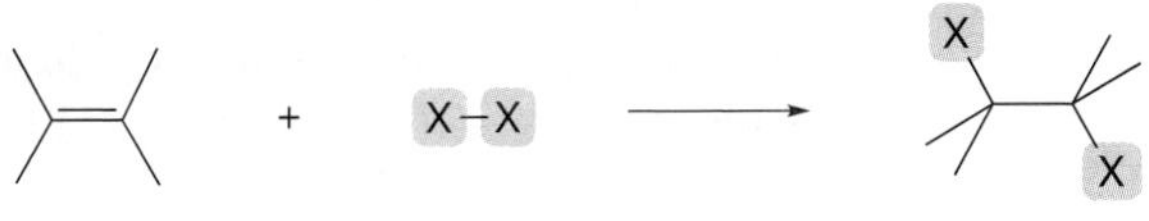

A π bond is broken. vicinal dihalide

Bromination is a simple chemical test for the presence of π bonds. When an alkene such as hex-1-ene is treated with red liquid bromine, the Br_2 adds to the double bond to give a colorless product, as seen in the right test tube. With the alkane hexane (test tube on the left), no reaction occurs, and the red color of Br_2 remains.

Halogenation is synthetically useful only with Cl_2 and Br_2. The dichlorides and dibromides formed in this reaction serve as starting materials for the synthesis of alkynes, as we learned in Section 8.10.

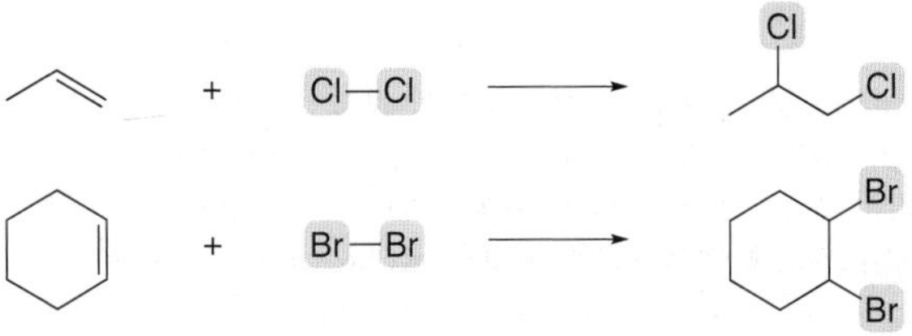

Halogens add to π bonds because halogens are **polarizable.** The electron-rich double bond induces a dipole in an approaching halogen molecule, making one halogen atom electron deficient and the other electron rich ($X^{\delta+}-X^{\delta-}$). **The electrophilic halogen atom is then attracted to the nucleophilic double bond,** making addition possible.

Two facts demonstrate that halogenation follows a different mechanism from that of hydrohalogenation or hydration. First, no rearrangements occur, and second, only anti addition of X_2 is observed. For example, treatment of cyclohexene with Br_2 yields two **trans** enantiomers formed by **anti addition.**

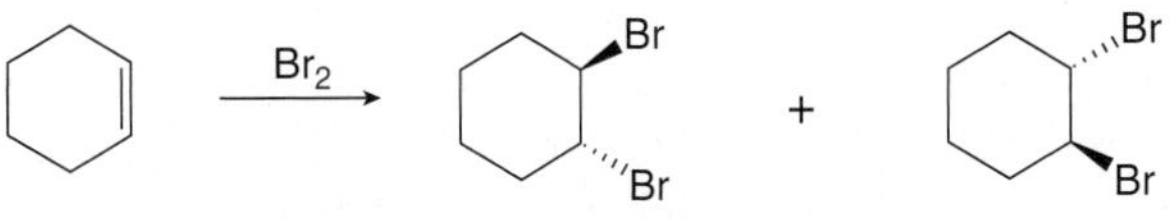

enantiomers

These facts suggest that **carbocations are *not* intermediates in halogenation.** Unstable carbocations rearrange, and both syn and anti addition is possible with carbocation intermediates. The accepted mechanism for halogenation comprises **two steps,** but it does *not* proceed with formation of a carbocation, as shown in Mechanism 10.3.

Mechanism 10.3 Addition of X_2 to an Alkene—Halogenation

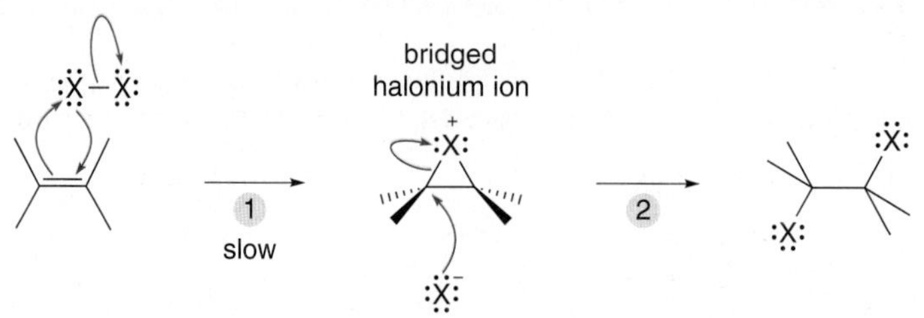

1 Four bonds are broken or formed to generate an unstable bridged halonium ion that contains a three-membered ring. The electron pair in the π bond and a lone pair on a halogen are used to form two new C–X bonds, and the X–X bond is cleaved.

2 Nucleophilic attack of X^- ring opens the bridged halonium ion and forms a new C–X bond.

Bridged halonium ions resemble carbocations in that they are short-lived intermediates that react readily with nucleophiles. Carbocations are inherently unstable because only six electrons surround carbon, whereas **halonium ions are unstable because they contain a strained three-membered ring** with a positively charged halogen atom.

carbocation — C has no octet.

bridged halonium ion — The ring has angle strain.

Problem 10.23 Draw the products of each reaction, including stereochemistry.

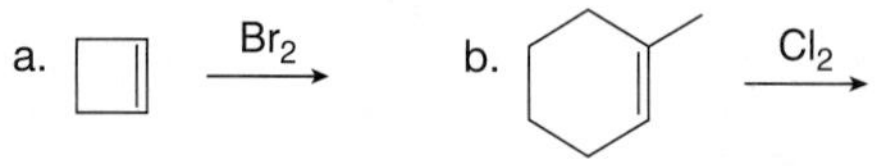

10.14 Stereochemistry of Halogenation

How does the proposed mechanism invoking a bridged halonium ion intermediate explain the observed **trans products of halogenation?** For example, chlorination of cyclopentene affords both enantiomers of *trans*-1,2-dichlorocyclopentane, with *no* cis products.

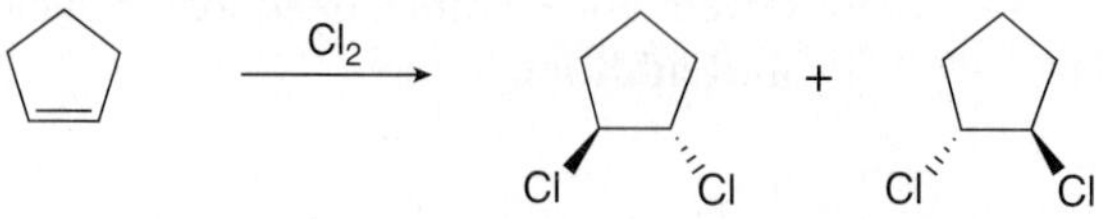

trans enantiomers

Initial addition of the electrophile Cl^+ (from Cl_2) occurs from either side of the planar double bond to form the bridged chloronium ion. In this example, both modes of addition (from above and below) generate the same **achiral** intermediate, so either representation can be used to draw the second step.

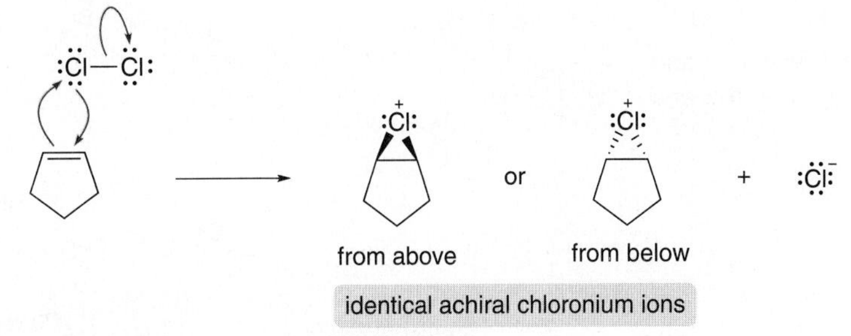
from above
from below
identical achiral chloronium ions

The opening of bridged halonium ion intermediates resembles the opening of epoxide rings with nucleophiles discussed in Section 9.16A.

In the second step, **nucleophilic attack of Cl⁻ must occur from the back side**—that is, from the side of the five-membered ring opposite to the side having the bridged chloronium ion. Because the nucleophile attacks from below in this example and the leaving group departs from above, the two Cl atoms in the product are oriented **trans** to each other. Backside attack occurs with equal probability at either carbon of the three-membered ring to yield an equal amount of two enantiomers—**a racemic mixture.**

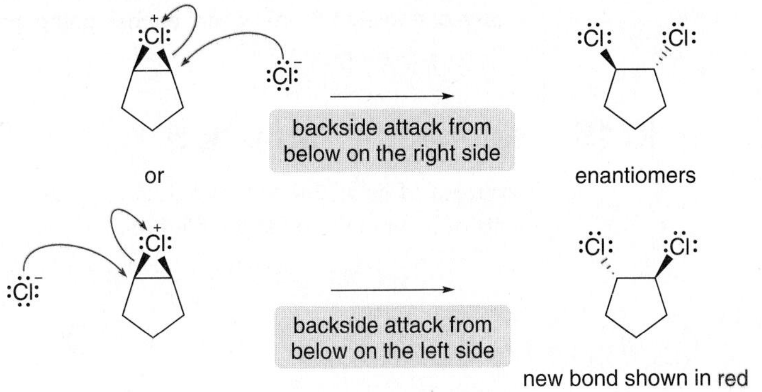
backside attack from below on the right side
or
enantiomers
backside attack from below on the left side
new bond shown in red

In summary, the mechanism for halogenation of alkenes occurs in two steps:

- **Addition of X^+ forms an unstable bridged halonium ion in the rate-determining step.**
- **Nucleophilic attack of X^- occurs from the back side to form trans products. The overall result is anti addition of X_2 across the double bond.**

Because halogenation occurs exclusively in an anti fashion, cis and trans alkenes yield different stereoisomers. Halogenation of alkenes is a **stereospecific reaction.**

- **A reaction is stereospecific when each of two specific stereoisomers of a starting material yields a particular stereoisomer of a product.**

cis-But-2-ene yields two enantiomers, whereas *trans*-but-2-ene yields a single achiral meso compound, as shown in Figure 10.12.

Problem 10.24 Draw all stereoisomers formed in each reaction.

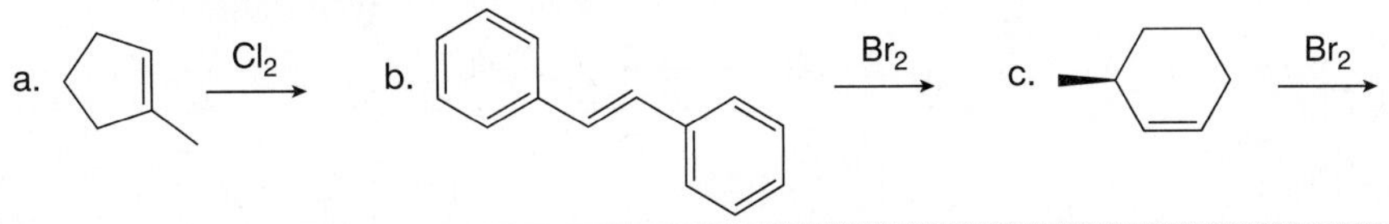
a.
Cl_2
b.
Br_2
c.
Br_2

Figure 10.12
Halogenation of *cis*- and *trans*-but-2-ene

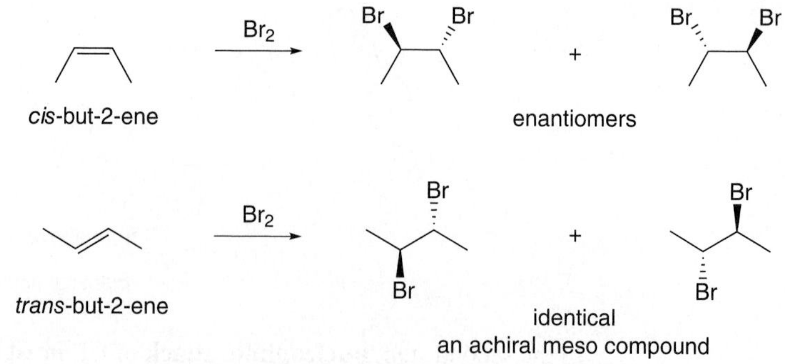

To draw the products of halogenation:

- Add Br_2 in an **anti** fashion across the double bond, leaving all other groups in their original orientations. With the alkene drawn in the plane of the page, one Br adds from the front (ending up on a wedge), and one Br adds from the back (ending up on a dashed wedge).
- Sometimes this reaction produces two stereoisomers, as in the case of *cis*-but-2-ene, which forms an equal amount of two enantiomers. Sometimes it produces a single compound, as in the case of *trans*-but-2-ene, where a meso compound is formed.

10.15 Halohydrin Formation

Treatment of an alkene with a halogen $\mathbf{X_2}$ and $\mathbf{H_2O}$ forms a **halohydrin** by addition of the elements of **X** and **OH** to the double bond.

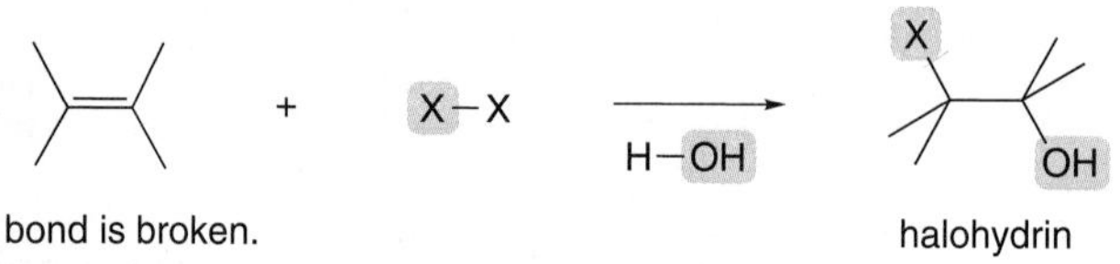

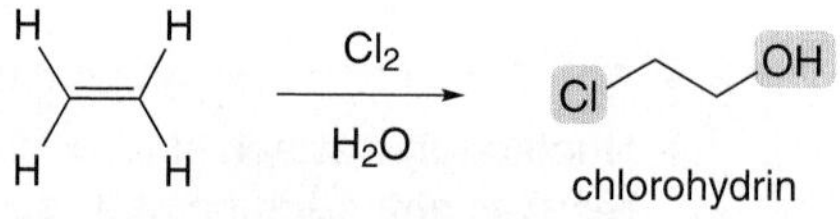

The mechanism for halohydrin formation is similar to the mechanism for halogenation: addition of the electrophile X^+ (from X_2) to form a **bridged halonium ion,** followed by nucleophilic attack by H_2O from the back side on the three-membered ring (Mechanism 10.4). Even though X^- is formed in Step [1] of the mechanism, its concentration is small compared to H_2O (often the solvent), so H_2O and *not* X^- is the nucleophile.

Mechanism 10.4 Addition of X and OH—Halohydrin Formation

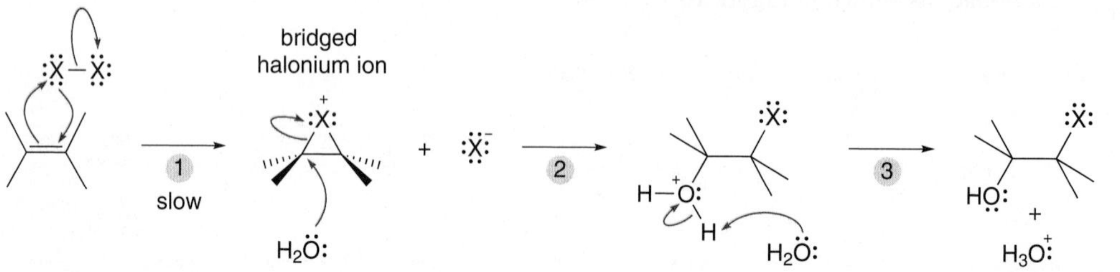

1. Four bonds are broken or formed to generate an unstable bridged halonium ion that contains a three-membered ring. The electron pair in the π bond and a lone pair on a halogen are used to form two new C–X bonds, and the X–X bond is cleaved.
2. Nucleophilic attack of H_2O ring opens the bridged halonium ion and forms a new C–O bond.
3. Loss of a proton forms the halohydrin.

Recall from Section 7.8C that DMSO (dimethyl sulfoxide) is a polar aprotic solvent.

Although the combination of Br_2 and H_2O effectively forms **bromohydrins** from alkenes, other reagents can also be used. Bromohydrins are also formed with ***N*-bromosuccinimide** (abbreviated as **NBS**) in **aqueous DMSO** [$(CH_3)_2S{=}O$]. NBS serves as a source of Br_2, which then goes on to form a bromohydrin by the same reaction mechanism.

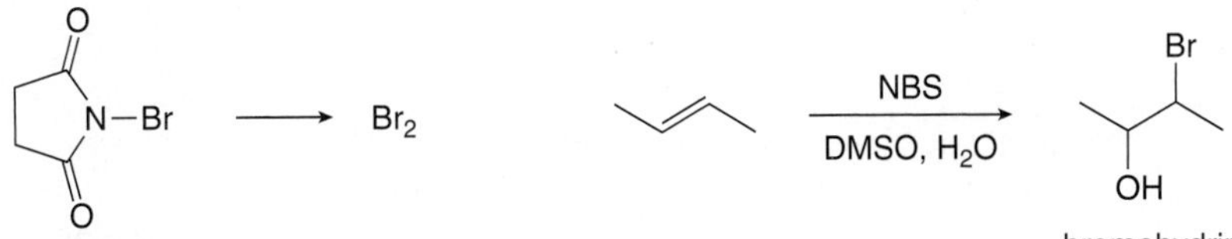

N-bromosuccinimide
NBS

bromohydrin

10.15A Stereochemistry and Regioselectivity of Halohydrin Formation

Because the bridged halonium ion ring is opened by backside attack of H_2O, addition of X and OH occurs in an **anti** fashion and **trans** products are formed.

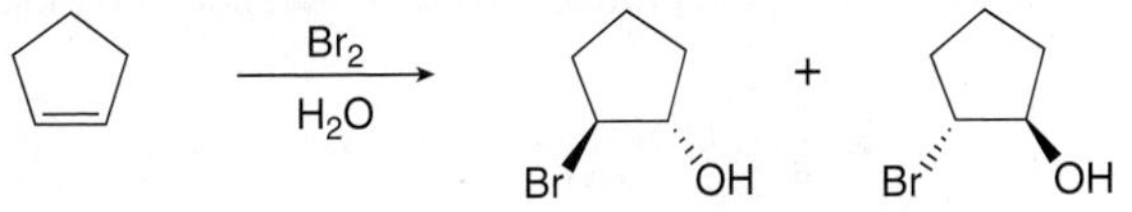

trans enantiomers

Sample Problem 10.4 Draw the products of the following reaction, including stereochemistry.

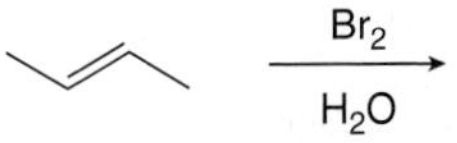

trans-but-2-ene

Solution

The reagent (Br_2 + H_2O) adds the elements of Br and OH to a double bond in an **anti** fashion—that is, from **opposite** sides. To draw two products of anti addition: add Br from above and OH from below in one product; then add Br from below and OH from above in the other product. In this example, the two products are nonsuperimposable mirror images—**enantiomers.**

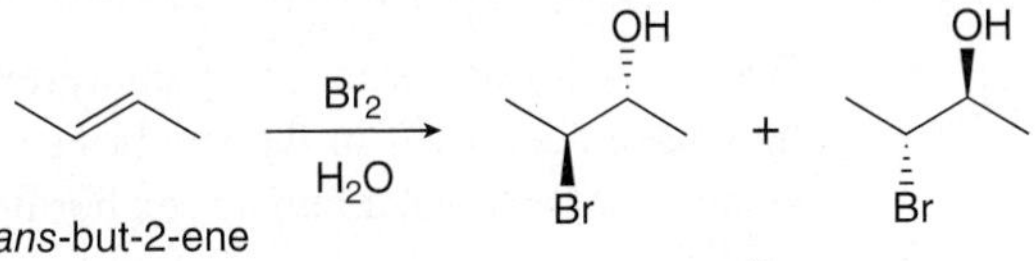

trans-but-2-ene

enantiomers

With unsymmetrical alkenes, two constitutional isomers are possible from addition of X and OH, but only one is formed. **The preferred product has the electrophile X^+ bonded to the less substituted carbon atom**—that is, the carbon that has more H atoms to begin with in the reacting alkene. Thus, the **nucleophile (H_2O) bonds to the more substituted carbon.**

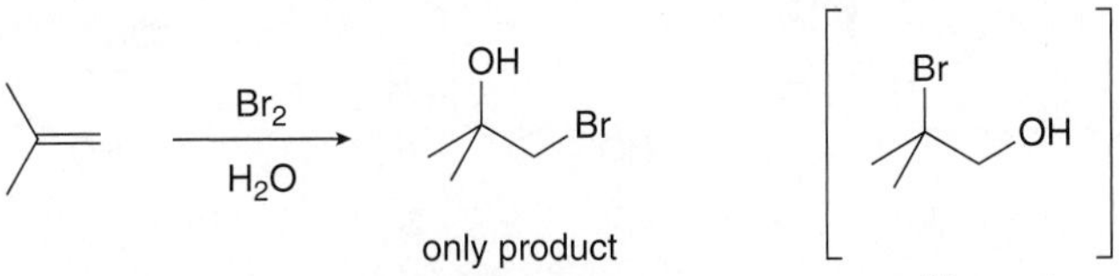

only product

Br ends up on the less substituted C.

not formed

This result is reminiscent of the opening of epoxide rings with acids HZ (Z = a nucleophile), which we encountered in Section 9.16B. As in the opening of an epoxide ring, **nucleophilic attack occurs at the more substituted carbon end of the bridged halonium ion** because that carbon is better able to accommodate a partial positive charge in the transition state.

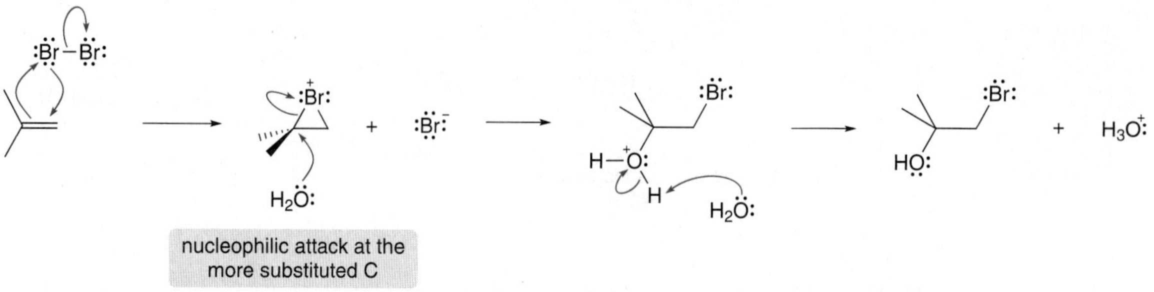

Table 10.4 summarizes the characteristics of halohydrin formation.

Problem 10.25 Draw the products of each reaction and indicate their stereochemistry.

Table 10.4 Summary: Conversion of Alkenes to Halohydrins

	Observation
Mechanism	• The mechanism involves three steps. • The rate-determining step forms a bridged halonium ion. • No rearrangements can occur.
Regioselectivity	• The electrophile X^+ bonds to the less substituted carbon.
Stereochemistry	• Anti addition occurs.

10.15B Halohydrins: Useful Compounds in Organic Synthesis

Because halohydrins are easily converted to epoxides by intramolecular S_N2 reaction (Section 9.6), they have been used in the synthesis of many naturally occurring compounds. Key steps in the synthesis of estrone, a female sex hormone, are illustrated in Figure 10.13.

Figure 10.13 The synthesis of estrone from a chlorohydrin

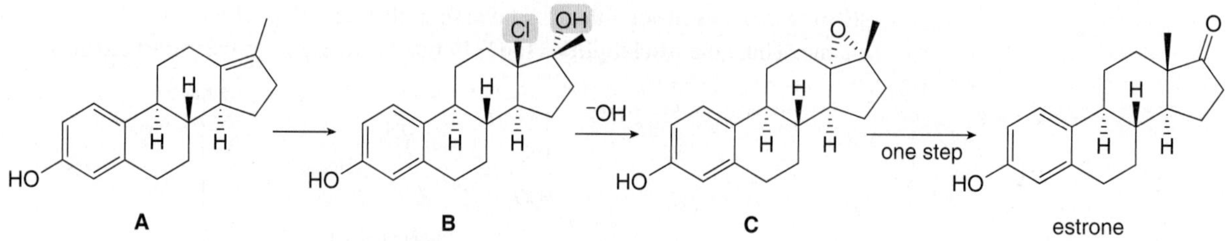

- Chlorohydrin **B,** prepared from alkene **A** by addition of Cl and OH, is converted to epoxide **C** with base. **C** is converted to estrone in one step.

10.16 Hydroboration–Oxidation

Hydroboration–oxidation is a two-step reaction sequence that converts an alkene to an alcohol.

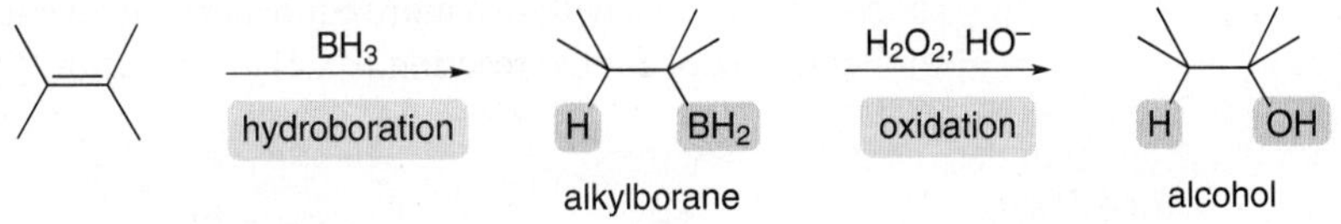

- *Hydroboration* is the addition of borane (BH_3) to an alkene, forming an alkylborane.
- *Oxidation* converts the C – B bond of the alkylborane to a C – O bond.

Hydroboration–oxidation results in **addition of H_2O** to an alkene.

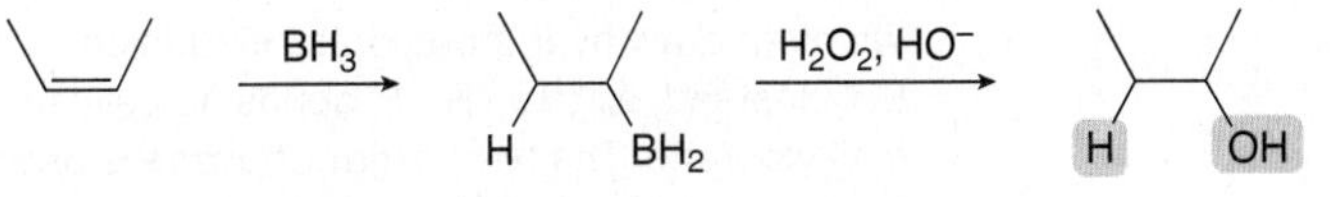

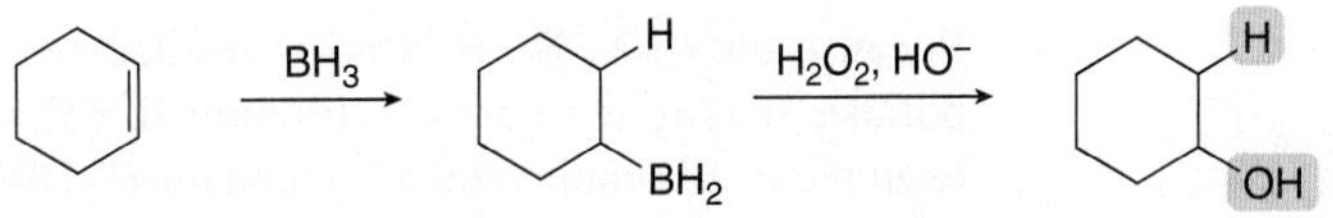

Borane (BH_3) is a reactive gas that exists mostly as the dimer, diborane (B_2H_6). Borane is a strong **Lewis acid** that reacts readily with Lewis bases. For ease in handling in the laboratory, it is commonly used as a complex with tetrahydrofuran (THF).

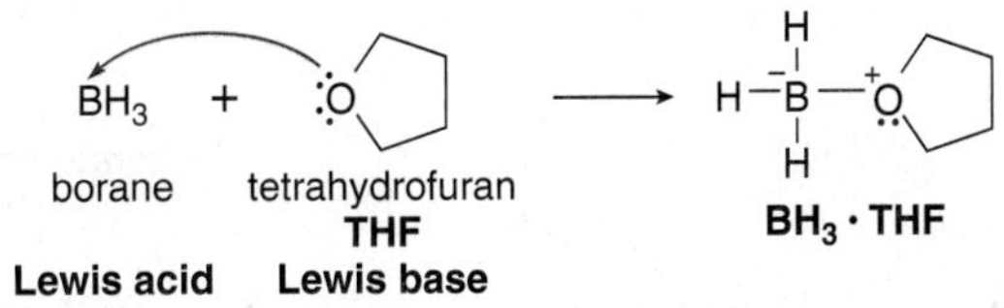

10.16A Hydroboration

The first step in hydroboration–oxidation is **addition of the elements of H and BH_2** to the π bond of the alkene, forming an intermediate alkylborane.

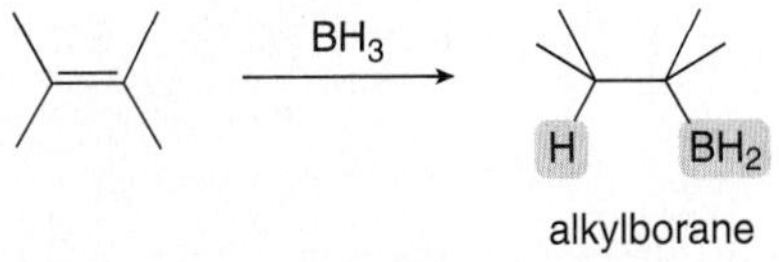

Because syn addition to the double bond occurs and no carbocation rearrangements are observed, carbocations are *not* formed during hydroboration, as shown in Mechanism 10.5. The proposed

Mechanism 10.5 Addition of H and BH_2—Hydroboration

One step The π bond and H – BH_2 bonds break as the C – H and C – B bonds form.

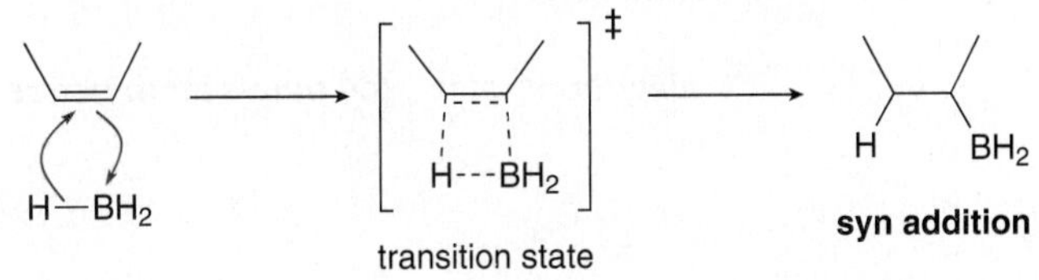

mechanism involves a **concerted addition of H and BH_2 from the same side of the planar double bond:** the π bond and $H-BH_2$ bond are broken as two new σ bonds are formed. Because four atoms are involved, the transition state is said to be **four-centered.**

Because the alkylborane formed by reaction with one equivalent of alkene still has two B–H bonds, it can react with two more equivalents of alkene to form a trialkylborane. This is illustrated in Figure 10.14 for the reaction of $CH_2=CH_2$ with BH_3.

Figure 10.14 Conversion of BH_3 to a trialkylborane with three equivalents of $CH_2=CH_2$

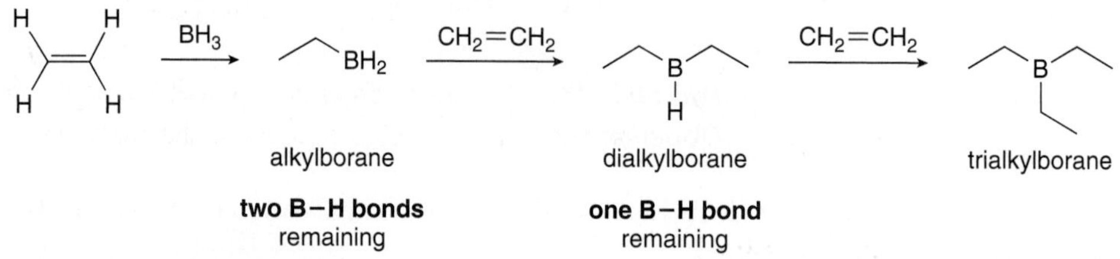

- We often draw hydroboration as if addition stopped after one equivalent of alkene reacts with BH_3. Instead, all three B–H bonds actually react with three equivalents of an alkene to form a trialkylborane. The term **organoborane** is used for any compound with a carbon–boron bond.

Because only one B–H bond is needed for hydroboration, commercially available dialkylboranes having the general structure $\mathbf{R_2BH}$ are sometimes used instead of BH_3. A common example is 9-borabicyclo[3.3.1]nonane **(9-BBN).** 9-BBN undergoes hydroboration in the same manner as BH_3.

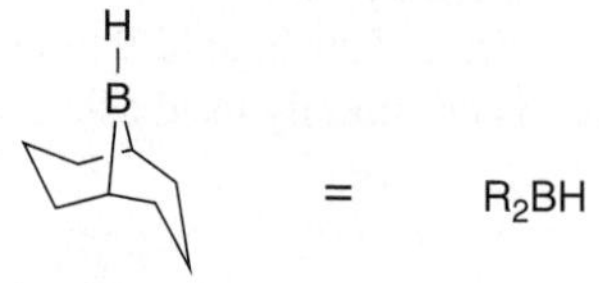

9-borabicyclo[3.3.1]nonane
9-BBN

Hydroboration is regioselective. **With unsymmetrical alkenes, the boron atom bonds to the less substituted carbon atom.** For example, addition of BH_3 to propene forms an alkylborane with the B bonded to the terminal carbon atom.

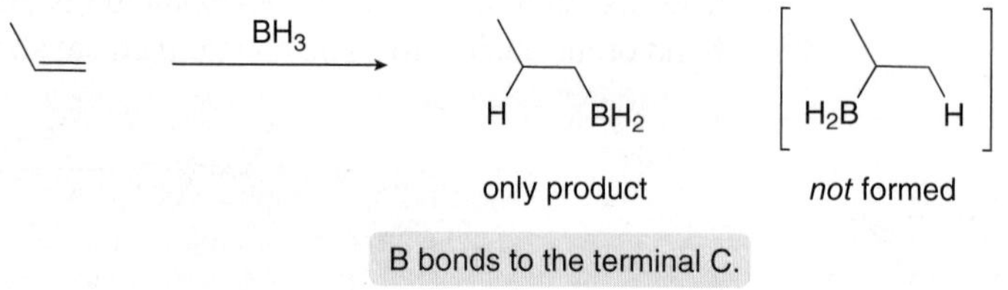

Because H is more electronegative than B, the B–H bond is polarized to give boron a partial positive charge ($H^{\delta-}-B^{\delta+}$), making BH_2 the electrophile in hydroboration.

Steric factors explain this regioselectivity. The larger boron atom bonds to the less sterically hindered, more accessible carbon atom.

Electronic factors are also used to explain this regioselectivity. If bond breaking and bond making are not completely symmetrical, boron bears a partial negative charge in the transition state and carbon bears a partial positive charge. Because alkyl groups stabilize a positive charge, the more stable transition state has the partial positive charge on the more substituted carbon, as illustrated in Figure 10.15.

- **In hydroboration, the boron atom bonds to the less substituted carbon.**

Figure 10.15
Hydroboration of an unsymmetrical alkene

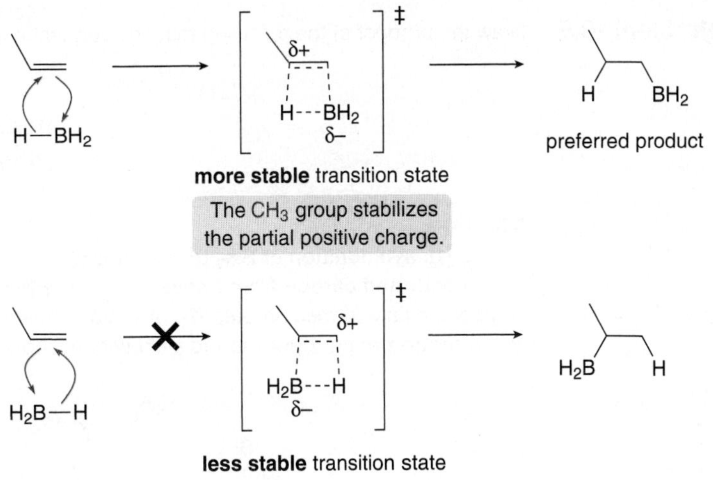

Problem 10.26 What alkylborane is formed from hydroboration of each alkene?

a.  b.  c. 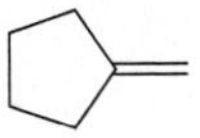

10.16B Oxidation of the Alkylborane

Because alkylboranes react rapidly with water and spontaneously burn when exposed to the air, they are oxidized, without isolation, with basic hydrogen peroxide (H_2O_2, HO^-). **Oxidation replaces the C–B bond with a C–O bond, forming a new OH group with retention of configuration;** that is, the **OH group replaces the BH_2 group in the same position** relative to the other three groups on carbon.

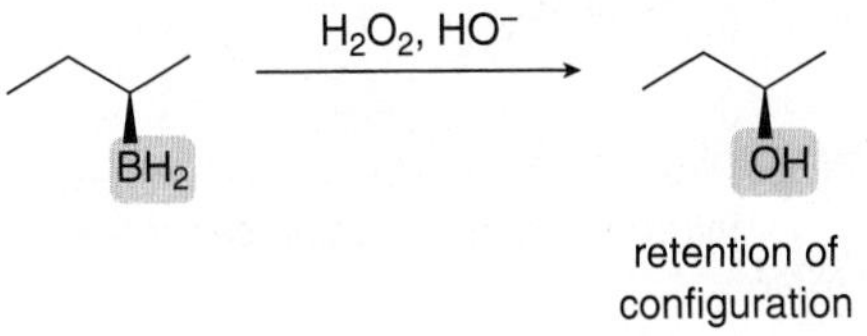

Thus, to draw the product of a hydroboration–oxidation reaction, keep in mind two stereochemical facts:

- **Hydroboration occurs with syn addition.**
- **Oxidation occurs with retention of configuration.**

The overall result of this two-step sequence is **syn addition of the elements of H and OH** to a double bond, as illustrated in Sample Problem 10.5. The OH group bonds to the *less* substituted carbon.

Sample Problem 10.5 Draw the product of the following reaction sequence, including stereochemistry.

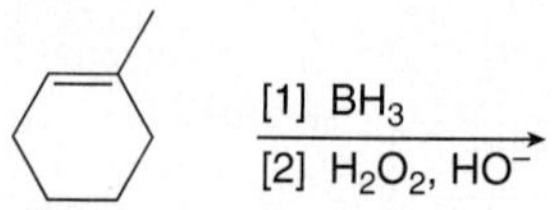

Solution

In Step [1], **syn addition of BH_3 to the unsymmetrical alkene adds the BH_2 group to the less substituted carbon from above and below the planar double bond.** Two enantiomeric alkylboranes are formed. In Step [2], oxidation replaces the BH_2 group with OH in each enantiomer with **retention of configuration** to yield two alcohols that are also enantiomers.

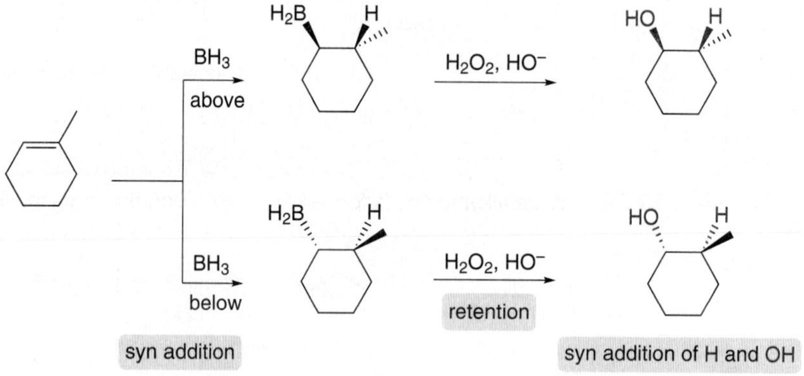

Hydroboration–oxidation results in the **addition of H and OH in a syn fashion** across the double bond. The achiral alkene is converted to an equal mixture of two enantiomers—that is, a **racemic mixture of alcohols.**

Problem 10.27 Draw the products formed when each alkene is treated with BH_3 followed by H_2O_2, HO^-. Include the stereochemistry at all stereogenic centers.

a.  b. 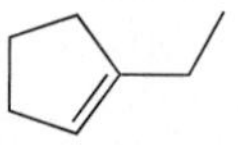 c. 

Problem 10.28 What alkene can be used to prepare each alcohol as the exclusive product of a two-step hydroboration–oxidation sequence?

a. 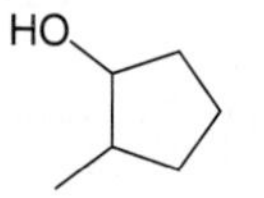

b. 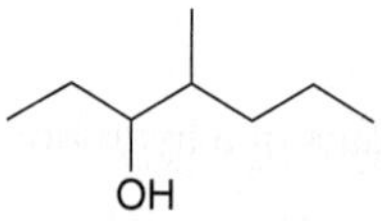

c. 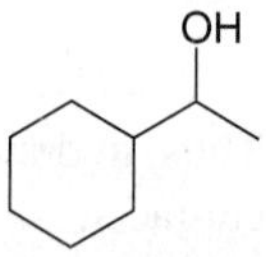

Table 10.5 summarizes the features of hydroboration–oxidation.

Table 10.5 Summary: Hydroboration–Oxidation of Alkenes

	Observation
Mechanism	• The addition of H and BH_2 occurs in one step. • No rearrangements can occur.
Regioselectivity	• The OH group bonds to the less substituted carbon atom.
Stereochemistry	• Syn addition occurs. • OH replaces BH_2 with retention of configuration.

Hydroboration–oxidation is a very common method for adding H_2O across a double bond. One example is shown in the synthesis of **artemisinin** (or **qinghaosu**), the active component of **qinghao,** a Chinese herbal remedy used for the treatment of malaria (Figure 10.16). As mentioned in the Prologue, currently recommended antimalarial therapy consists of a combination of drugs

Figure 10.16 An example of hydroboration–oxidation in synthesis

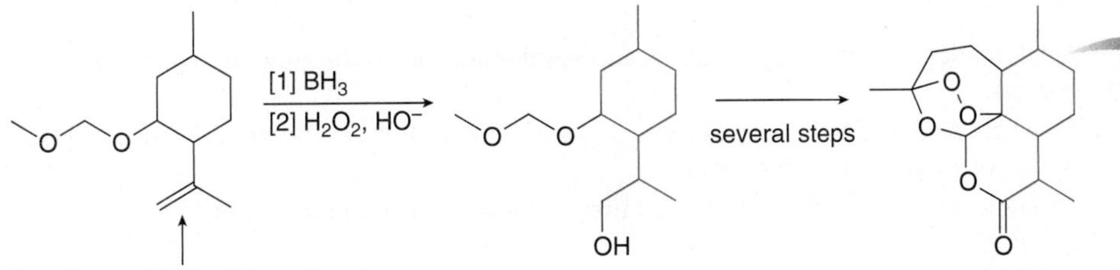

Hydroboration−oxidation takes place here. **A** artemisinin (antimalarial drug)

Artemisia annua, source of the antimalarial agent artemisinin

• The carbon atoms of artemisinin that come from alcohol **A** are indicated in red.

that includes artemisinin or a related compound. A ball-and-stick model of artemisinin appears on the cover of this text.

10.16C A Comparison of Hydration Methods

Hydration (H_2O, H^+) and hydroboration–oxidation (BH_3 followed by H_2O_2, HO^-) both add the elements of H_2O across a double bond. Despite their similarities, these reactions often form different constitutional isomers, as shown in Sample Problem 10.6.

Sample Problem 10.6 Draw the product formed when $CH_3CH_2CH_2CH_2CH{=}CH_2$ is treated with (a) H_2O, H_2SO_4; and (b) BH_3 followed by H_2O_2, HO^-.

Solution

With $H_2O + H_2SO_4$, electrophilic addition of H and OH places the **H atom on the less substituted carbon** of the alkene to yield a **2° alcohol.** In contrast, addition of BH_3 gives an alkylborane with the **BH_2 group on the less substituted terminal carbon** of the alkene. Oxidation replaces BH_2 by OH to yield a **1° alcohol.**

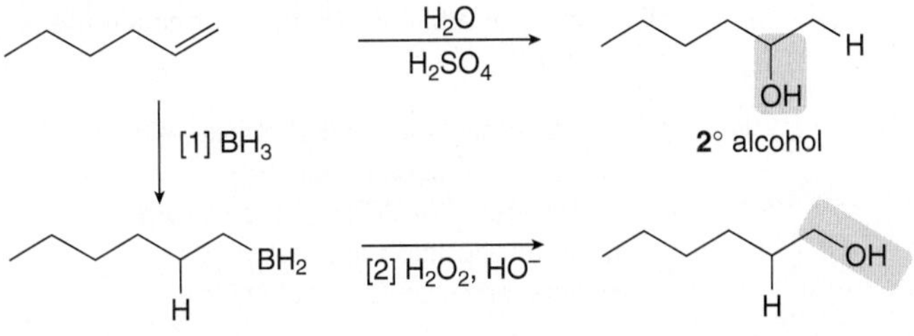

2° alcohol

1° alcohol

Problem 10.29 Draw the constitutional isomer formed when the following alkenes are treated with each set of reagents: [1] H_2O, H_2SO_4; or [2] BH_3 followed by H_2O_2, ^-OH.

a. 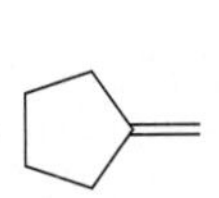 b. 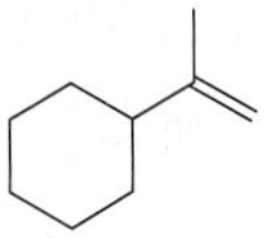 c. 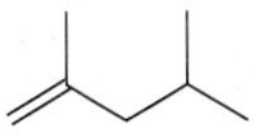

10.17 Keeping Track of Reactions

Chapters 7–10 have introduced three basic kinds of organic reactions: **nucleophilic substitution, β elimination,** and **addition.** In the process, many specific reagents have been discussed and the stereochemistry that results from many different mechanisms has been examined. **How can we keep track of all the reactions?**

To make the process easier, **remember that most organic molecules undergo only one or two different kinds of reactions.** For example:

- **Alkyl halides undergo substitution and elimination because they have good leaving groups.**
- **Alcohols also undergo substitution and elimination, but can do so only when OH is made into a good leaving group.**
- **Alkenes undergo addition because they have easily broken π bonds.**

You must still learn many reaction details, and in truth, there is no one method to learn them. ***You must practice these reactions over and over again, not by merely looking at them, but by writing them.*** Some students do this by making a list of specific reactions for each functional group, and then rewriting them with different starting materials. Others make flash cards: index cards that have the starting material and reagent on one side and the product on the other. Whatever method you choose, **the details must become second nature,** much like the answers to simple addition problems, such as, what is the sum of 2 + 2?

Learning reactions is really a two-step process.

- **First, learn the basic type of reaction for a functional group. This provides an overall organization to the reactions.**
- **Then, learn the specific reagents for each reaction. It helps to classify the reagent according to its properties. Is it an acid or a base? Is it a nucleophile or an electrophile? Is it an oxidizing agent or a reducing agent?**

Sample Problem 10.7 illustrates this process.

Sample Problem 10.7 Draw the product of each reaction.

a. 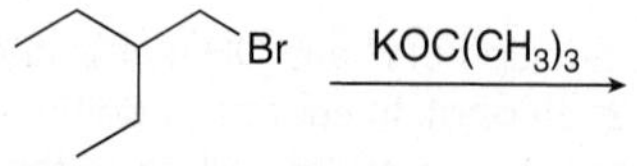

b. 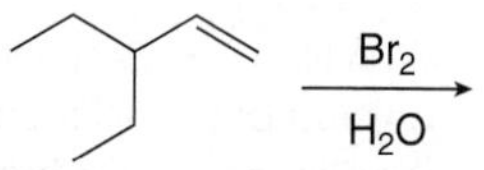

Solution

In each problem, identify the functional group to determine the general reaction type—substitution, elimination, or addition. Then, determine if the reagent is an electrophile, nucleophile, acid, base, and so forth.

a. The reactant is a **1° alkyl halide,** which can undergo substitution and elimination. The reagent [$KOC(CH_3)_3$] is a **strong nonnucleophilic base,** favoring elimination by an E2 mechanism.

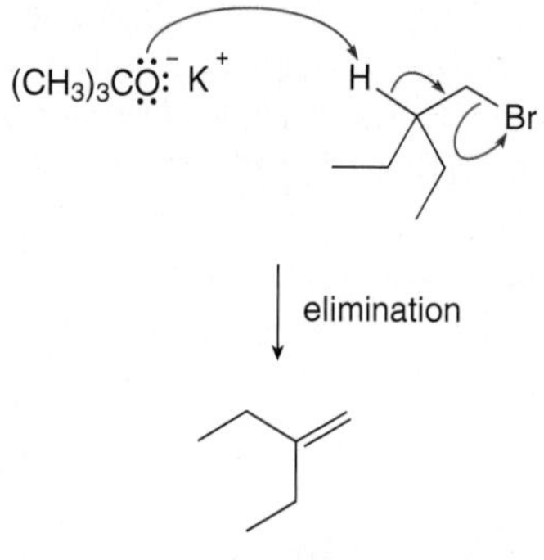

E2 product

b. The reactant is an **alkene,** which undergoes addition reactions to its π bond. The reagent (Br_2 + H_2O) serves as the source of the **electrophile Br^+,** resulting in **addition** of Br and OH to the double bond (Section 10.15).

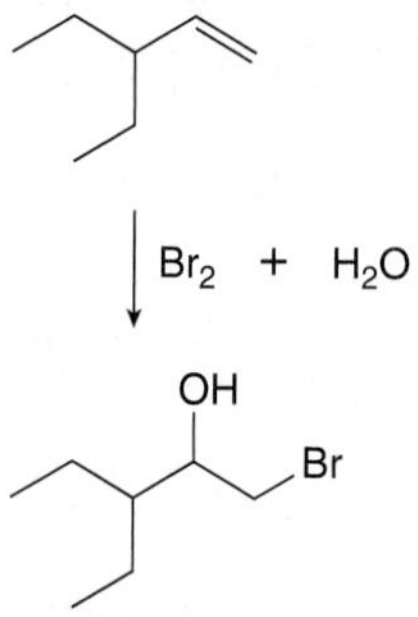

addition product

Problem 10.30 Draw the products of each reaction using the two-part strategy from Sample Problem 10.7.

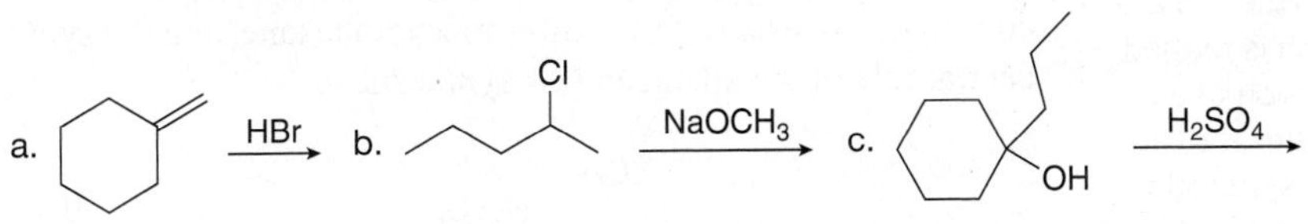
a.
HBr
b.
Cl
NaOCH3
c.
OH
H2SO4

10.18 Alkenes in Organic Synthesis

Alkenes are a central functional group in organic chemistry. **Alkenes are easily prepared by elimination reactions** such as dehydrohalogenation and dehydration. **Because their π bond is easily broken, they undergo many addition reactions** to prepare a variety of useful compounds.

Suppose, for example, that we must synthesize 1,2-dibromocyclohexane from cyclohexanol, a cheap and readily available starting material. Because there is no way to accomplish this transformation in one step, this synthesis must have at least two steps.

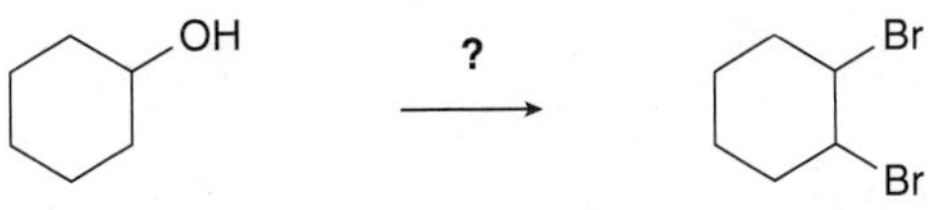
OH
?
Br
Br

cyclohexanol **starting material** 1,2-dibromocyclohexane **product**

To solve this problem we must:

- **Work backwards from the product by asking: What type of reactions introduce the functional groups in the product?**
- **Work forwards from the starting material by asking: What type of reactions does the starting material undergo?**

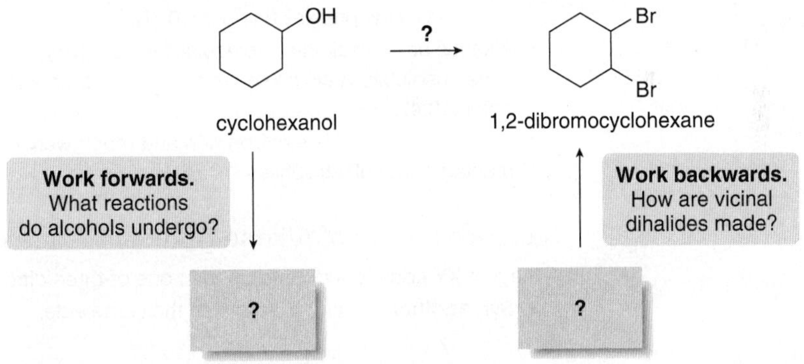
OH
?
Br
Br
cyclohexanol
1,2-dibromocyclohexane
Work forwards.
What reactions do alcohols undergo?
Work backwards.
How are vicinal dihalides made?
?
?

In Chapter 11 we will learn about retrosynthetic analysis in more detail.

Working backwards from the product to determine the starting material from which it is made is called *retrosynthetic analysis*.

We know reactions that answer each of these questions.

Working backwards:

[1] 1,2-Dibromocyclohexane, a vicinal dibromide, can be prepared by the addition of Br_2 to **cyclohexene.**

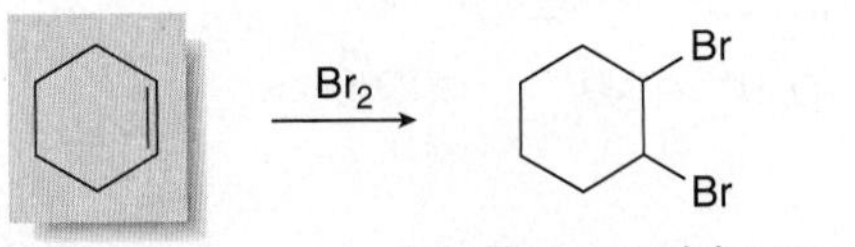
Br2
Br
Br

cyclohexene 1,2-dibromocyclohexane

Working forwards:

[2] Cyclohexanol can undergo acid-catalyzed dehydration to form **cyclohexene.**

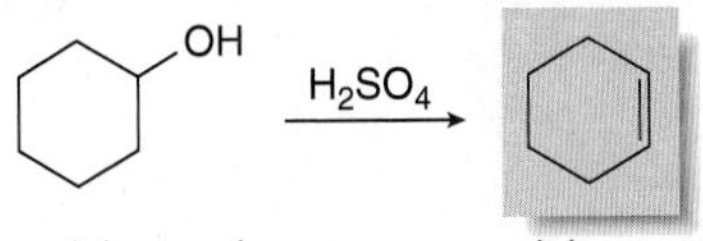
OH
H2SO4

cyclohexanol cyclohexene

A **reactive intermediate** is an unstable intermediate like a carbocation, which is formed during the conversion of a stable starting material to a stable product. A **synthetic intermediate** is a stable compound that is the product of one step and the starting material of another in a multistep synthesis.

Cyclohexene is called a **synthetic intermediate,** or simply an **intermediate,** because it is the **product of one step and the starting material of another.** We now have a two-step sequence to convert cyclohexanol to 1,2-dibromocyclohexane, and the synthesis is complete. Take note of the central role of the alkene in this synthesis.

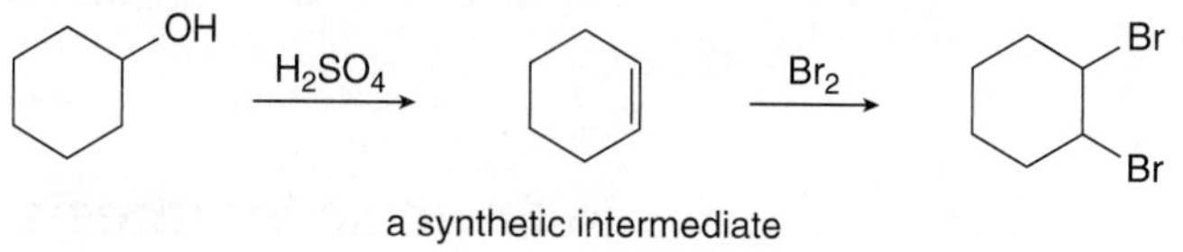

Problem 10.31 Devise a synthesis of each compound from the indicated starting material.

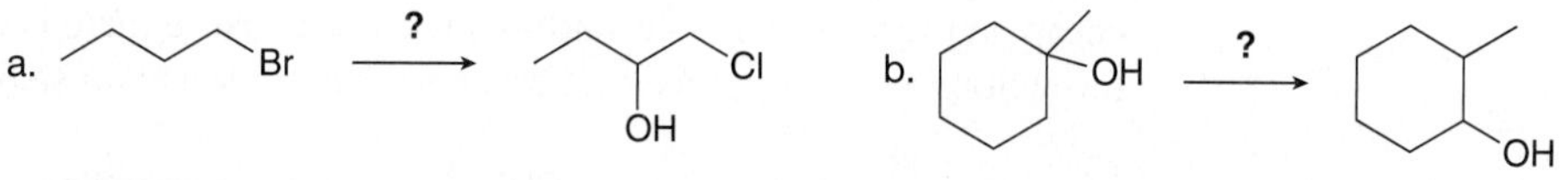

KEY CONCEPTS

Alkenes

General Facts About Alkenes

- Alkenes contain a carbon–carbon double bond consisting of a stronger σ bond and a weaker π bond. Each carbon is sp^2 hybridized and trigonal planar (10.1).
- Alkenes are named using the suffix ***-ene*** (10.3).
- Alkenes with different groups on each end of the double bond exist as a pair of diastereomers, identified by the prefixes *E* and *Z* (10.3B).
- Alkenes have weak intermolecular forces, giving them low mp's and bp's, and making them water insoluble. A cis alkene is more polar than a trans alkene, giving it a slightly higher boiling point (10.4).
- Because a π bond is electron rich and much weaker than a σ bond, alkenes undergo addition reactions with electrophiles (10.8).

Stereochemistry of Alkene Addition Reactions (10.8)

A reagent XY adds to a double bond in one of three different ways:

- **Syn addition**—X and Y add from the same side.

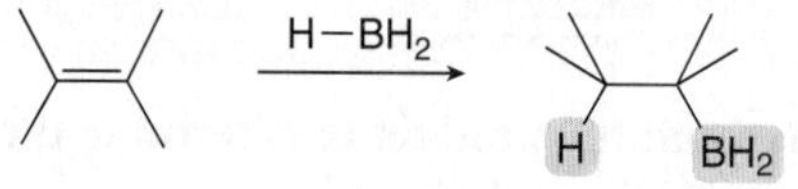

- Syn addition occurs in **hydroboration.**

- **Anti addition**—X and Y add from opposite sides.

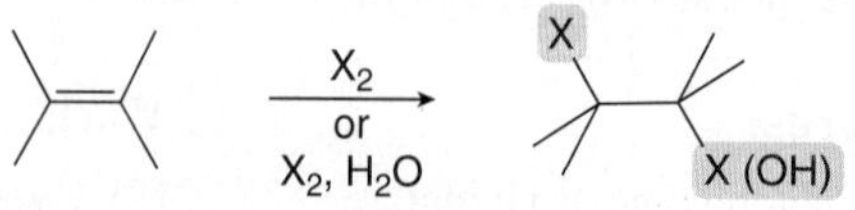

- Anti addition occurs in **halogenation** and **halohydrin formation.**

- **Both syn and anti addition** occur when carbocations are intermediates.

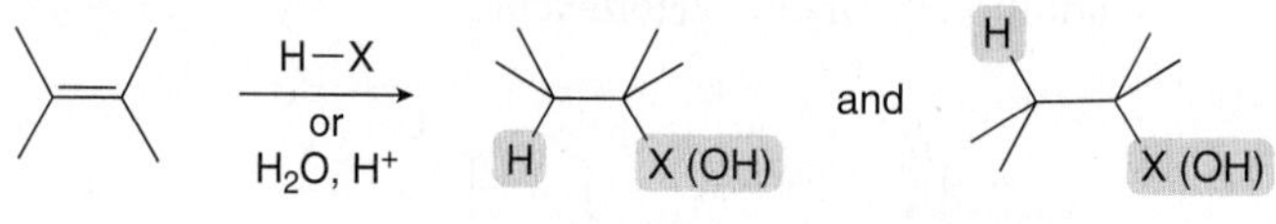

- Syn and anti addition occur in **hydrohalogenation** and **hydration.**

Addition Reactions of Alkenes

[1] Hydrohalogenation (Addition of HX; X = Cl, Br, I) (10.9–10.11)

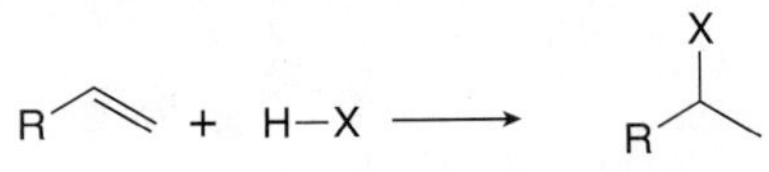

alkyl halide

- The mechanism has two steps.
- Carbocations are formed as intermediates.
- Carbocation rearrangements are possible.
- Markovnikov's rule is followed. H bonds to the less substituted C to form the more stable carbocation.
- Syn and anti addition occur.

[2] Hydration and related reactions (Addition of H_2O or ROH) (10.12)

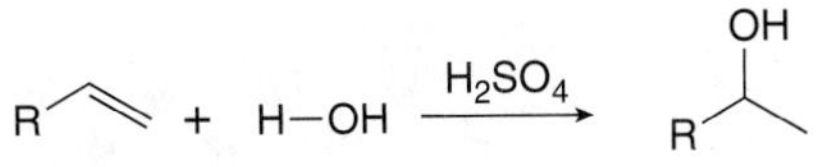

alcohol

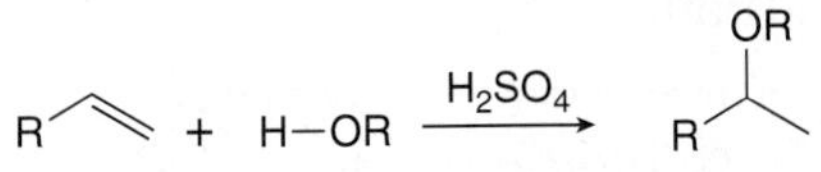

ether

For both reactions:

- The mechanism has three steps.
- Carbocations are formed as intermediates.
- Carbocation rearrangements are possible.
- Markovnikov's rule is followed. H bonds to the less substituted C to form the more stable carbocation.
- Syn and anti addition occur.

[3] Halogenation (Addition of X_2; X = Cl or Br) (10.13–10.14)

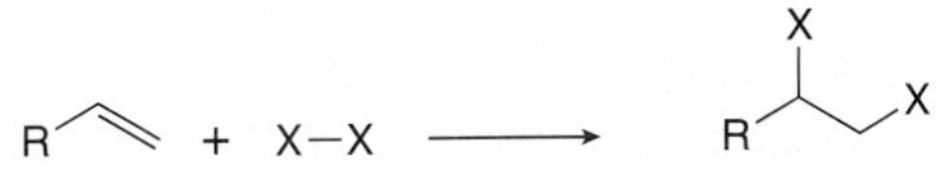

vicinal dihalide

- The mechanism has two steps.
- Bridged halonium ions are formed as intermediates.
- No rearrangements can occur.
- Anti addition occurs.

[4] Halohydrin formation (Addition of OH and X; X = Cl, Br) (10.15)

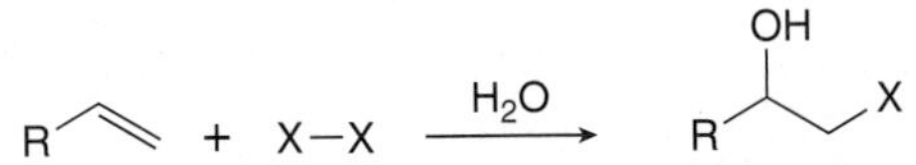

halohydrin

- The mechanism has three steps.
- Bridged halonium ions are formed as intermediates.
- No rearrangements can occur.
- X bonds to the less substituted C.
- Anti addition occurs.
- NBS in DMSO and H_2O adds Br and OH in the same fashion.

[5] Hydroboration–oxidation (Addition of H_2O) (10.16)

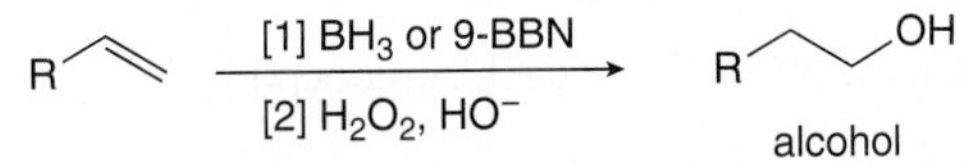

alcohol

- Hydroboration has a one-step mechanism.
- No rearrangements can occur.
- OH bonds to the less substituted C.
- Syn addition of H_2O results.

PROBLEMS

Problems Using Three-Dimensional Models

10.32 Give the IUPAC name for each compound.

a. 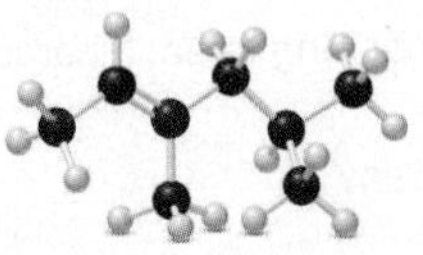

b. 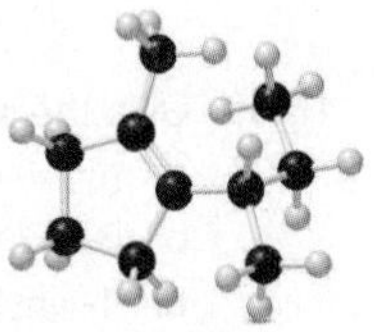

10.33 (a) Label the carbon–carbon double bond in **A** as *E* or *Z*. (b) Draw the products (including stereoisomers) formed when **A** is treated with H_2O in the presence of H_2SO_4.

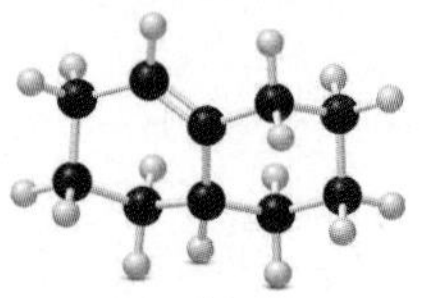

A

10.34 Name the alkene depicted in the ball-and-stick model and draw the constitutional isomers formed when the alkene is treated with each reagent: (a) Br_2; (b) Br_2 in H_2O; (c) Br_2 in CH_3OH.

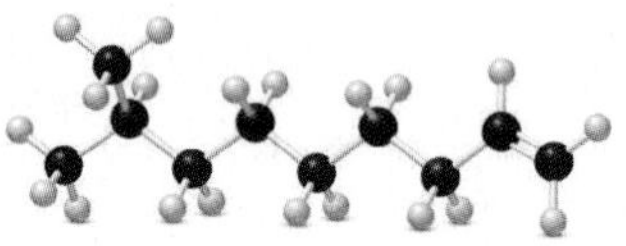

Degrees of Unsaturation

10.35 Calculate the number of degrees of unsaturation for each molecular formula.

a. C_6H_8
b. $C_{40}H_{56}$
c. $C_{10}H_{16}O_2$
d. C_8H_9Br
e. C_8H_9ClO
f. $C_7H_{11}N$
g. C_4H_8BrN
h. $C_{10}H_{18}ClNO$

10.36 How many rings and π bonds does a compound with molecular formula $C_{10}H_{14}$ possess? List all possibilities.

Nomenclature and Stereochemistry

10.37 Label the alkene in each drug as *E* or *Z*. Enclomiphene is one component of the fertility drug Clomid. Tamoxifen is an anticancer drug. Clavulanic acid is sold in combination with the antibiotic amoxicillin under the trade name Augmentin.

a.
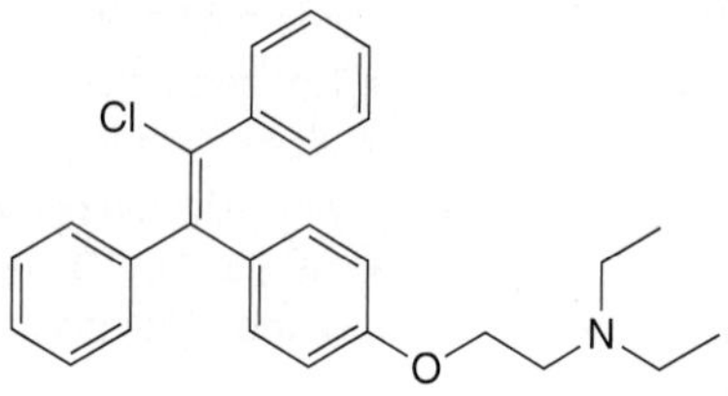

enclomiphene

b.
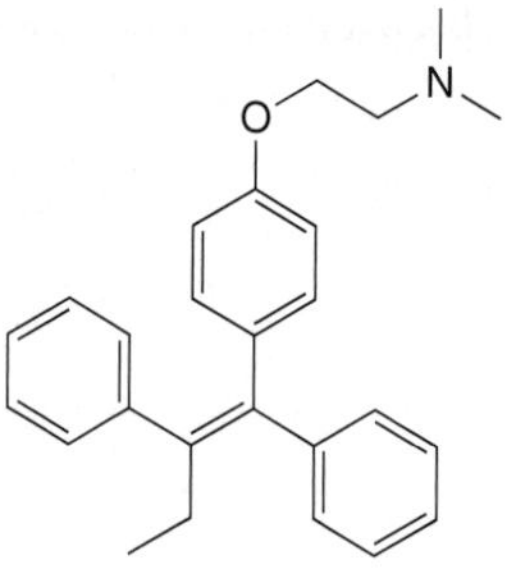

tamoxifen

c.
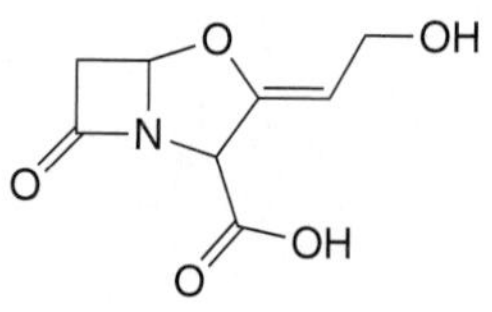

clavulanic acid

10.38 Give the IUPAC name for each compound.

a.
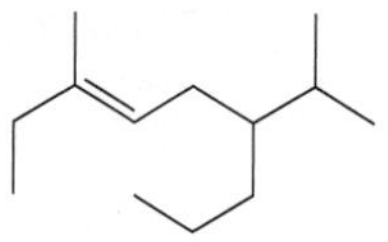

b.
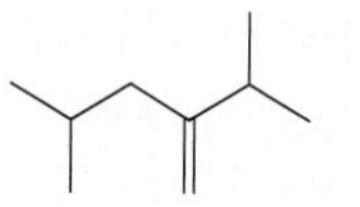

c.
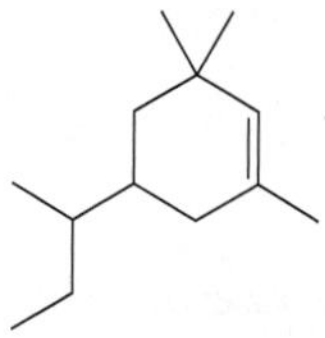

d.
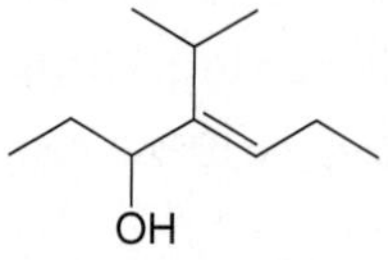

e.
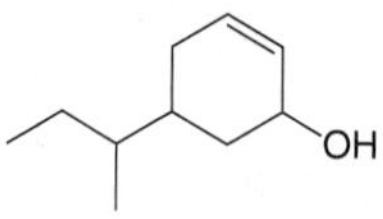

f.
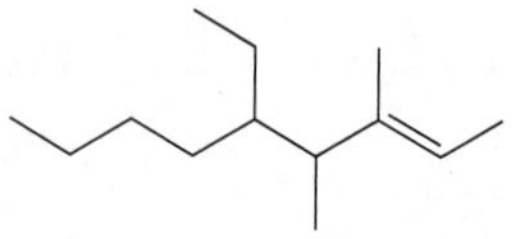

10.39 Give the structure corresponding to each name.

a. (*E*)-4-ethylhept-3-ene
b. 3,3-dimethylcyclopentene
c. 4-vinylcyclopentene
d. (*Z*)-3-isopropylhept-2-ene
e. *cis*-3,4-dimethylcyclopentene
f. 1-isopropyl-4-propylcyclohexene
g. 3,4-dimethylcyclohex-2-enol
h. 3,5-diethylhex-5-en-3-ol

10.40 (a) Draw all possible stereoisomers of 4-methylnon-2-ene, and name each isomer, including its *E,Z* and *R,S* prefixes. (b) Label two pairs of enantiomers. (c) Label four pairs of diastereomers.

10.41 (a) Draw the structure of (1*E*,4*R*)-1,4-dimethylcyclodecene. (b) Draw the enantiomer and name it, including its *E,Z* and *R,S* prefixes. (c) Draw two diastereomers and name them, including the *E,Z* and *R,S* prefixes.

10.42 Now that you have learned how to name alkenes in Section 10.3, name each of the following epoxides as an alkene oxide, as described in Section 9.3.

a. 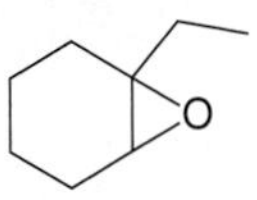
b. 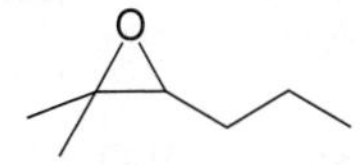
c. 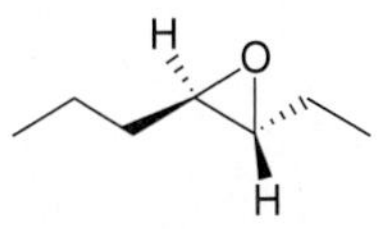
d. 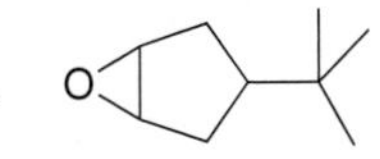

10.43 Iejimalide B, an anticancer agent with a 24-membered ring, is isolated from a tunicate found off Ie Island in Okinawa. (a) Label each double bond in iejimalide B as *E* or *Z*. (b) Label each tetrahedral stereogenic center as *R* or *S*. (c) How many stereoisomers are possible for iejimalide B?

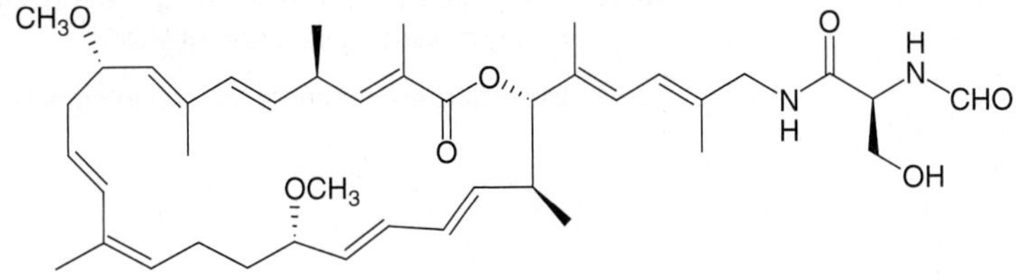

iejimalide B

Lipids

10.44 Although naturally occurring unsaturated fatty acids generally have the *Z* configuration, elaidic acid, a C_{18} fatty acid having an *E* double bond, is present in processed foods such as margarine and cooking oils. Predict how the melting point of elaidic acid compares with the melting points of stearic and oleic acids (Table 10.2).

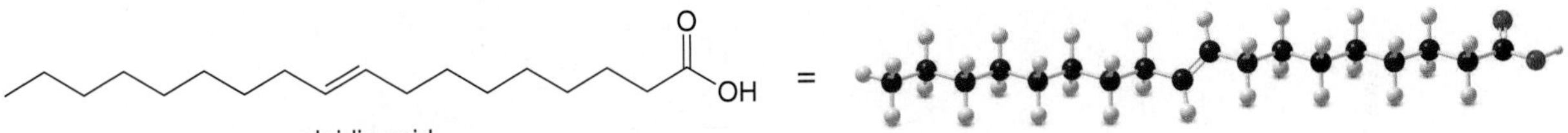

elaidic acid

10.45 Eleostearic acid is an unsaturated fatty acid obtained from the seeds of the tung oil tree (*Aleurites fordii*), a deciduous tree native to China. (a) Draw the structure of a stereoisomer that has a higher melting point than eleostearic acid. (b) Draw the structure of a stereoisomer that has a lower melting point.

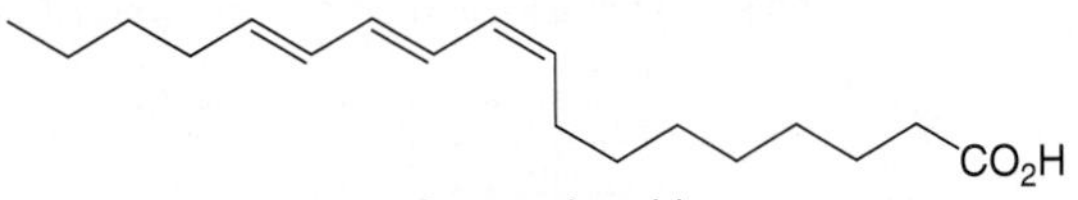

eleostearic acid

Reactions of Alkenes

10.46 Draw the products formed when $(CH_3)_2C{=}CH_2$ is treated with each reagent.

a. HBr
b. H_2O, H_2SO_4
c. CH_3CH_2OH, H_2SO_4
d. Cl_2
e. Br_2, H_2O
f. NBS (aqueous DMSO)
g. [1] BH_3; [2] H_2O_2, HO^-

10.47 What alkene can be used to prepare each alkyl halide or dihalide as the exclusive or major product of an addition reaction?

a. 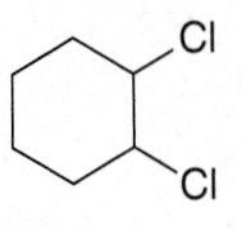
b. 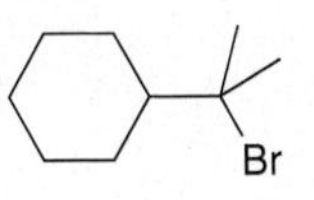
c. 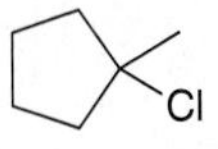
d. 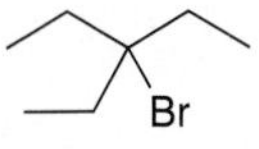

10.48 Which alcohols can be prepared as a single product by hydroboration–oxidation of an alkene? Which alcohols can be prepared as a single product by the acid-catalyzed addition of H_2O to an alkene?

a. 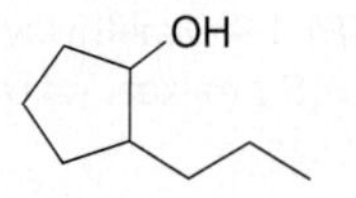

b. 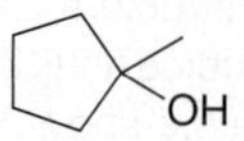

c. 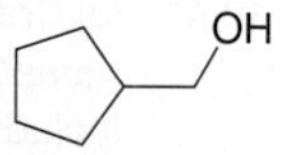

d. 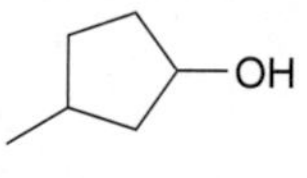

10.49 Draw the constitutional isomer formed in each reaction.

a. 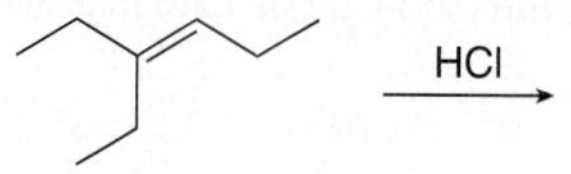

d. 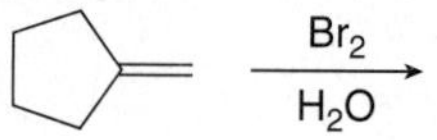

b. 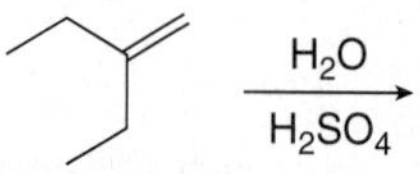

e. 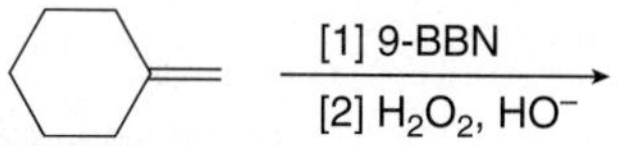

c. 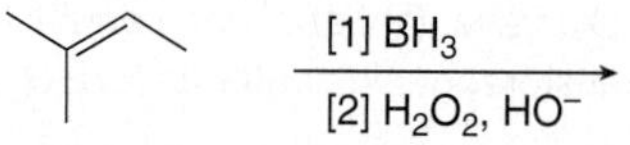

f. 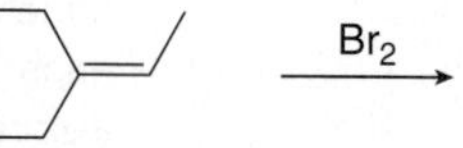

10.50 What three alkenes (excluding stereoisomers) can be used to prepare 3-chloro-3-methylhexane by addition of HCl?

10.51 Draw all stereoisomers formed in each reaction.

a. 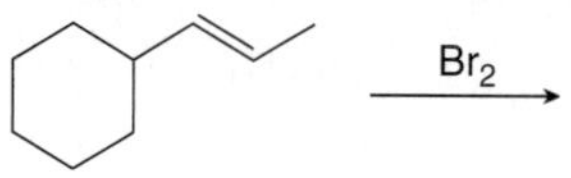

c. 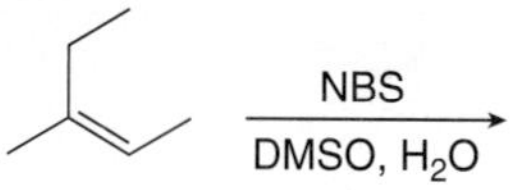

b. 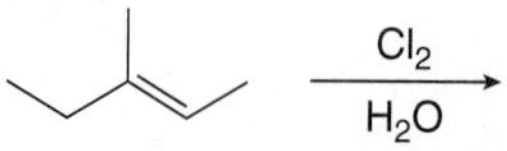

10.52 Draw the products of each reaction, including stereoisomers.

a. 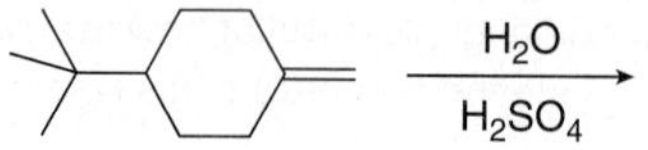

d. 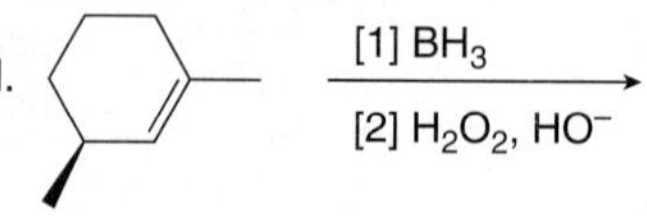

b. 

e. 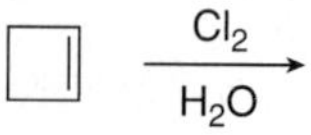

c. 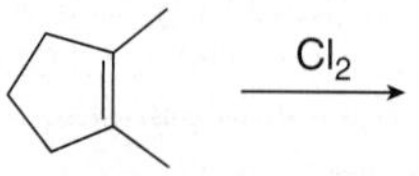

f. 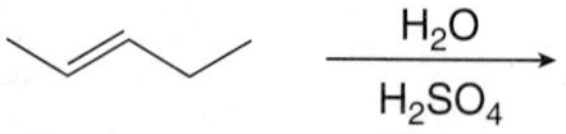

10.53 Which alkene reacts faster with HBr? Explain your choice.

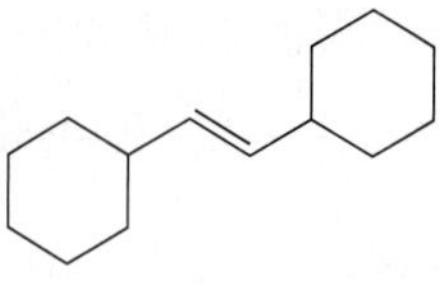
A

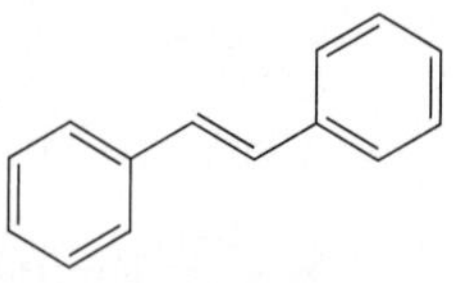
B

10.54 (a) Which diastereomer of oct-4-ene yields a mixture of two enantiomers, (4*R*,5*R*)- and (4*S*,5*S*)-4,5-dibromooctane on reaction with Br_2? (b) Which diastereomer of oct-4-ene yields a single meso compound, (4*R*,5*S*)-4,5-dibromooctane?

10.55 Using *cis*- and *trans*-hex-3-ene, demonstrate that the addition of HCl is not a stereospecific reaction. Draw the structure of the stereoisomers formed from each alkene.

Mechanisms

10.56 Draw a stepwise mechanism for the following reaction.

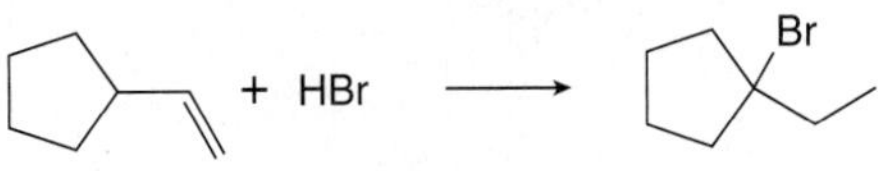

10.57 Draw a stepwise mechanism for the following reaction, which results in ring expansion of a six-membered ring to a seven-membered ring.

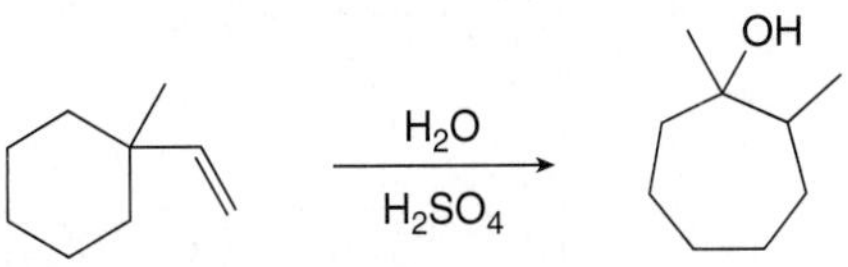

10.58 Draw a stepwise mechanism for the conversion of hex-5-en-1-ol to the cyclic ether **A**.

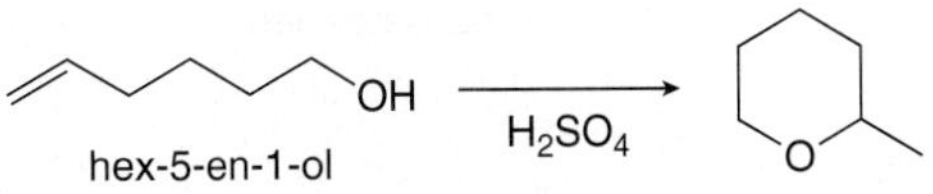

A

10.59 Draw a stepwise mechanism that shows how all three alcohols are formed from the bicyclic alkene.

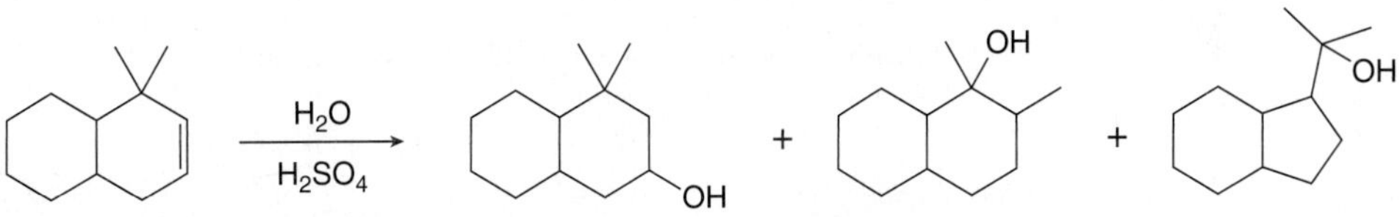

10.60 Less stable alkenes can be isomerized to more stable alkenes by treatment with strong acid. For example, 2,3-dimethylbut-1-ene is converted to 2,3-dimethylbut-2-ene when treated with H_2SO_4. Draw a stepwise mechanism for this isomerization process.

10.61 When buta-1,3-diene ($CH_2=CH-CH=CH_2$) is treated with HBr, two constitutional isomers are formed, $CH_3CHBrCH=CH_2$ and $BrCH_2CH=CHCH_3$. Draw a stepwise mechanism that accounts for the formation of both products.

10.62 Explain why the addition of HBr to alkenes **A** and **C** is regioselective, forming addition products **B** and **D,** respectively.

A **B** **C** **D**

10.63 Bromoetherification, the addition of the elements of Br and OR to a double bond, is a common method for constructing rings containing oxygen atoms. This reaction has been used in the synthesis of the polyether antibiotic monensin (Problem 21.37). Draw a stepwise mechanism for the following intramolecular bromoetherification reaction.

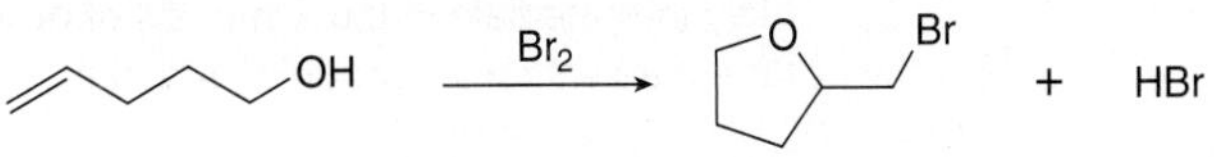

Synthesis

10.64 Devise a synthesis of each product from the given starting material. More than one step is required.

a. 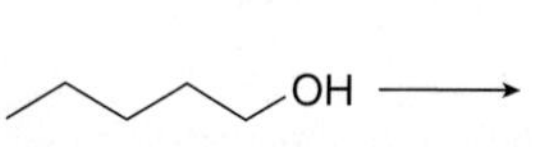

d. 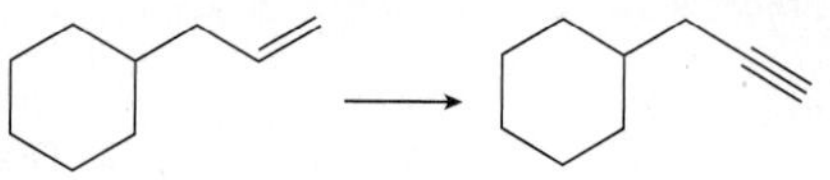

b. 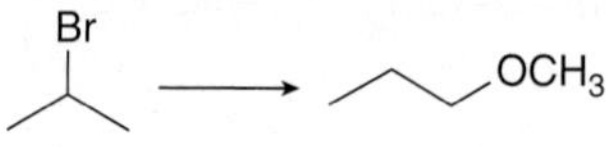

e. 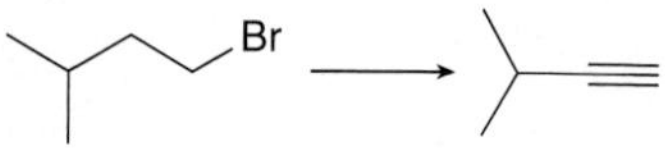

c. 

f. 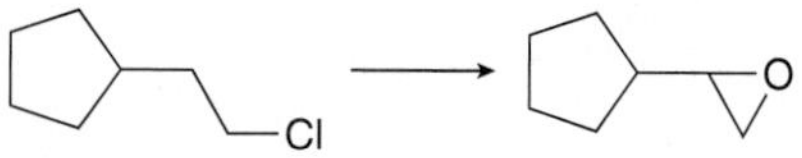

10.65 Devise a synthesis of each compound from cyclohexene as the starting material. More than one step is needed.

a. 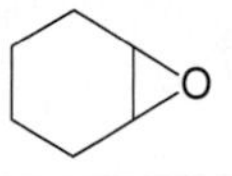

b. 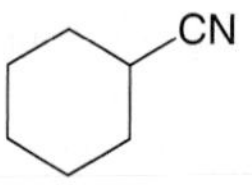

c. 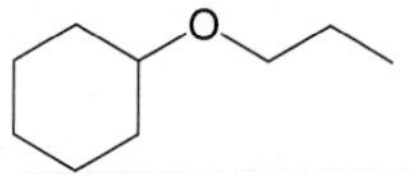

d. 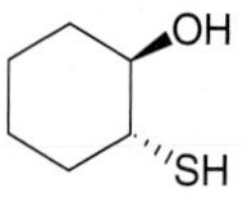

+ enantiomer

Challenge Problems

10.66 Explain why **A** is a stable compound but **B** is not.

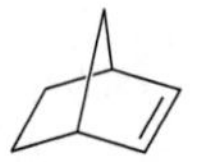

A

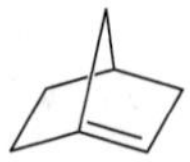

B

10.67 (a) What product(s) are formed when the *E* isomer of $C_6H_5CH{=}CHC_6H_5$ is treated with Br_2, followed by one equivalent of KOH? Label the resulting alkene(s) as *E* or *Z*. (b) What product(s) are formed when the *Z* isomer of $C_6H_5CH{=}CHC_6H_5$ is subjected to the same reaction sequence? (c) How are the compounds in parts (a) and (b) related to each other?

10.68 Alkene **A** can be isomerized to isocomene, a natural product isolated from goldenrod, by treatment with TsOH. Draw a stepwise mechanism for this conversion. (Hint: Look for a carbocation rearrangement.)

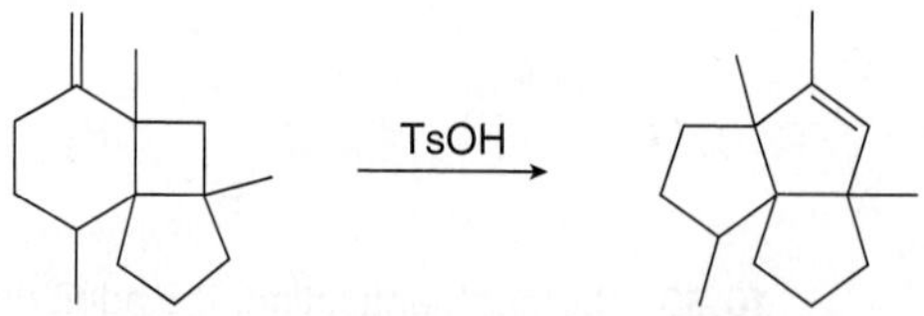

A isocomene

10.69 Lactones, cyclic esters such as compound **A,** are prepared by **halolactonization,** an addition reaction to an alkene. For example, iodolactonization of **B** forms lactone **C,** a key intermediate in the synthesis of prostaglandin $PGF_{2\alpha}$ (Section 4.15). Draw a stepwise mechanism for this addition reaction.

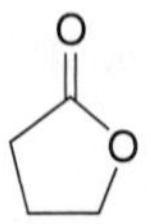

A

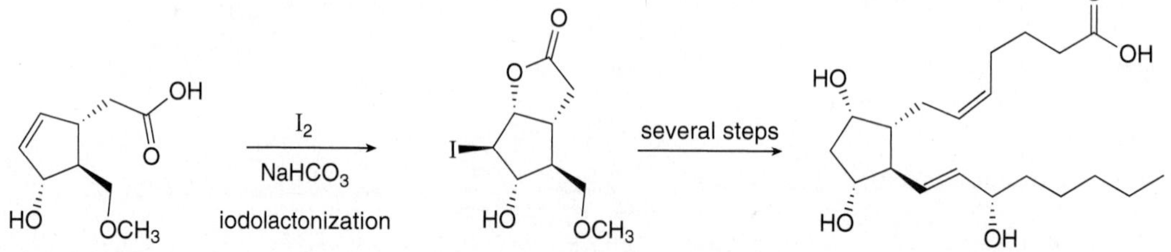

B **C** $PGF_{2\alpha}$

10.70 Draw a stepwise mechanism for the following reaction.

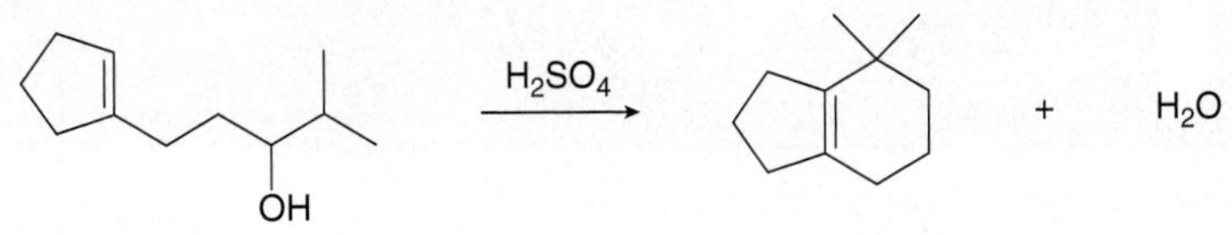

10.71 Like other electrophiles, carbocations add to alkenes to form new carbocations, which can then undergo substitution or elimination reactions depending on the reaction conditions. With this in mind, consider the following reactions of nerol, a natural product isolated from lemongrass and other plant sources. Treatment of nerol with TsOH forms α-terpineol as the major product, whereas treatment of nerol with chlorosulfonic acid, HSO_3Cl, forms a constitutional isomer, α-cyclogeraniol. Write stepwise mechanisms for both processes. Each mechanism involves the addition of an electrophile—a carbocation—to a double bond.

General reaction

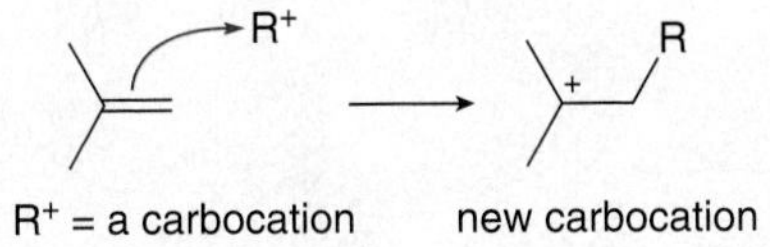

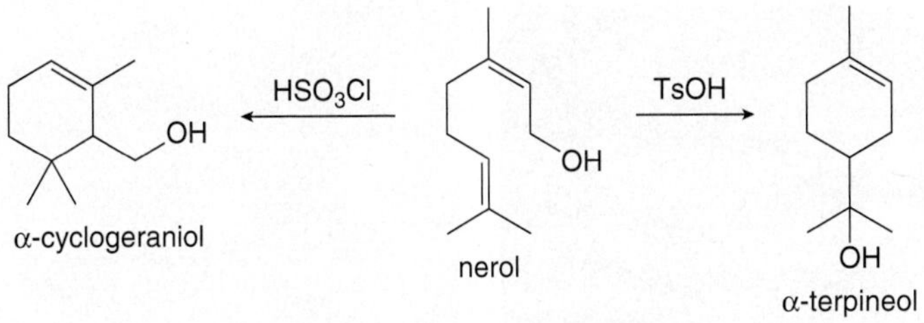

10.72 Draw a stepwise mechanism for the following reaction. This reaction combines two processes: the opening of an epoxide ring with a nucleophile and the addition of an electrophile to a carbon–carbon double bond. (Hint: Begin the mechanism by protonating the epoxide ring.)

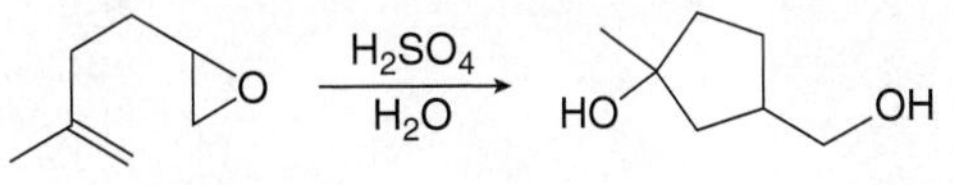

11 Alkynes

11.1 Introduction
11.2 Nomenclature
11.3 Physical properties
11.4 Interesting alkynes
11.5 Preparation of alkynes
11.6 Introduction to alkyne reactions
11.7 Addition of hydrogen halides
11.8 Addition of halogen
11.9 Addition of water
11.10 Hydroboration–oxidation
11.11 Reaction of acetylide anions
11.12 Synthesis

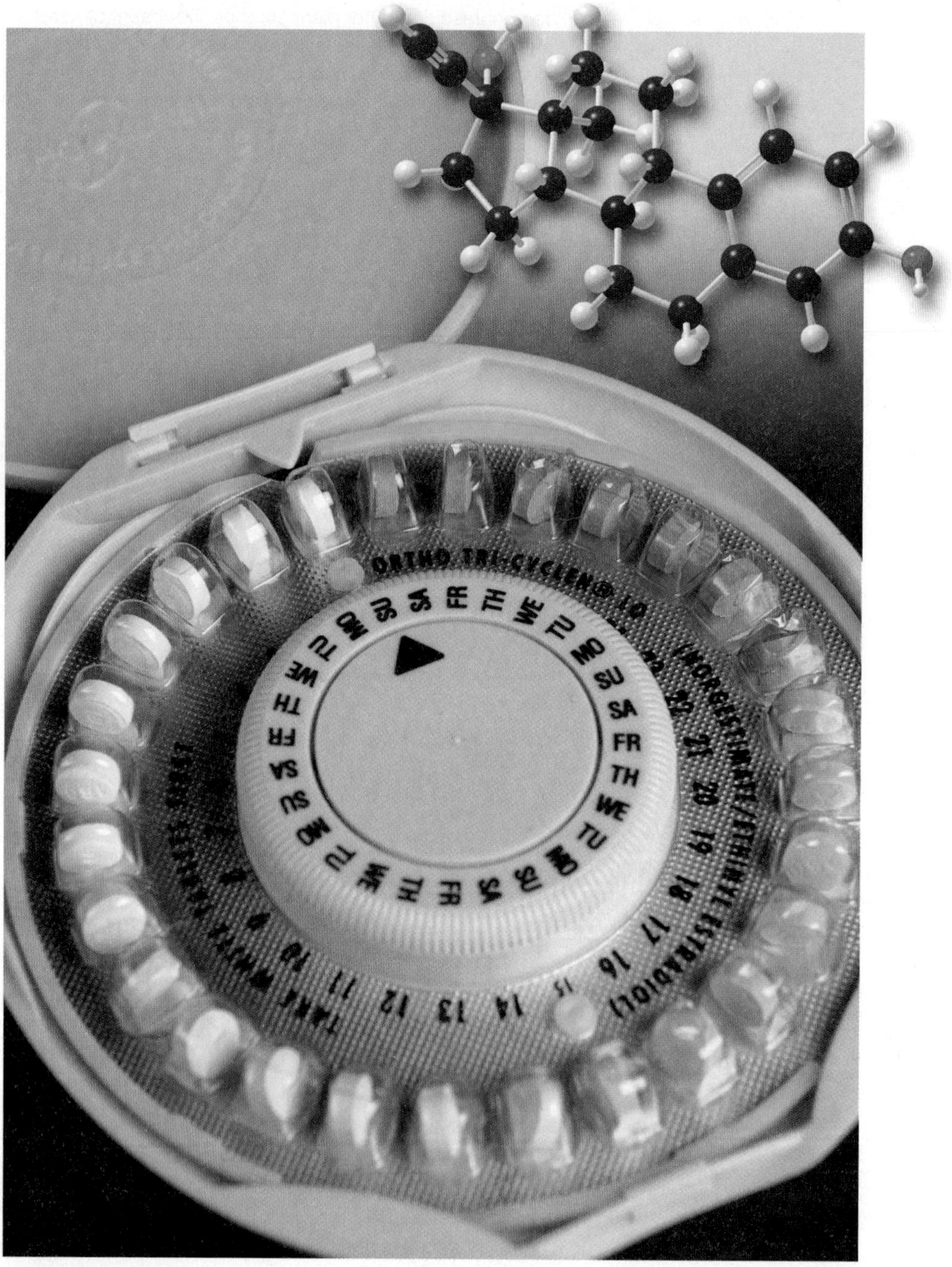

Ethynylestradiol is a synthetic compound whose structure closely resembles the carbon skeleton of female estrogen hormones. Because it is more potent than its naturally occurring analogues, it is a component of several widely used oral contraceptives. Ethynylestradiol and related compounds with similar biological activity contain a carbon–carbon triple bond. In Chapter 11 we learn about alkynes, hydrocarbons that contain triple bonds.

In Chapter 11 we continue our focus on organic molecules with electron-rich functional groups by examining ***alkynes,* compounds that contain a carbon–carbon triple bond.** Like alkenes, **alkynes are nucleophiles with easily broken π bonds,** and as such, they undergo **addition** reactions with electrophilic reagents.

Alkynes also undergo a reaction that has no analogy in alkene chemistry. Because a C−H bond of an alkyne is more acidic than a C−H bond in an alkene or an alkane, alkynes are readily deprotonated with strong base. The resulting nucleophiles react with electrophiles to form new carbon–carbon σ bonds, so that complex molecules can be prepared from simple starting materials. The study of alkynes thus affords an opportunity to learn more about organic synthesis.

11.1 Introduction

Alkynes contain a carbon–carbon triple bond. A **terminal alkyne** has the triple bond at the end of the carbon chain, so that a hydrogen atom is directly bonded to a carbon atom of the triple bond. An **internal alkyne** has a carbon atom bonded to each carbon atom of the triple bond.

—C≡C—

alkyne

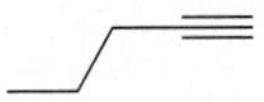

terminal alkyne

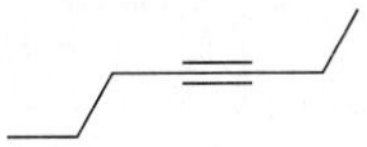

internal alkyne

An alkyne has the general molecular formula $\mathbf{C_nH_{2n-2}}$, giving it four fewer hydrogens than the maximum number possible. Because every degree of unsaturation removes two hydrogens, a **triple bond introduces two degrees of unsaturation.**

Each carbon of a triple bond is ***sp*** hybridized and **linear,** and all bond angles are **180°** (Section 1.10C).

H−C≡C−H
acetylene
=

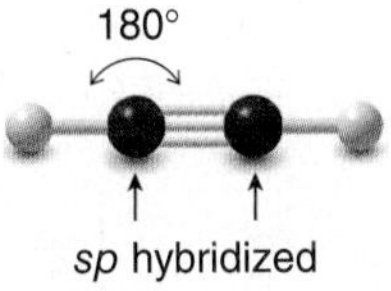

The triple bond of an alkyne consists of one σ bond and two π bonds.

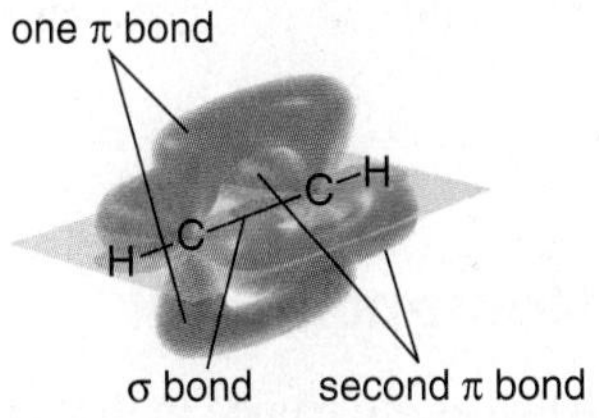

- The σ bond is formed by end-on overlap of the two *sp* hybrid orbitals.
- Each π bond is formed by side-by-side overlap of two 2*p* orbitals.

Bond dissociation energies of the C−C bonds in ethylene (one σ and one π bond) and acetylene (one σ and two π bonds) can be used to estimate the strength of the second π bond of the triple bond. If we assume that the σ bond and first π bond in acetylene are similar in strength to the σ and π bonds in ethylene (368 and 267 kJ/mol, respectively), then the second π bond is worth 202 kJ/mol.

HC≡CH		CH_2=CH_2		
837 kJ/mol	−	635 kJ/mol	=	202 kJ/mol
(σ + two π bonds)		(σ + π bond)		↑ **second π bond**

- **Both π bonds of a C–C triple bond are weaker than a C–C σ bond, making them much more easily broken. As a result, alkynes undergo many addition reactions.**
- **Alkynes are more polarizable than alkenes because the electrons in their π bonds are more loosely held.**

Skeletal structures for alkynes may look somewhat unusual, but they follow the customary convention: a carbon atom is located at the intersection of any two lines and at the end of any line; thus,

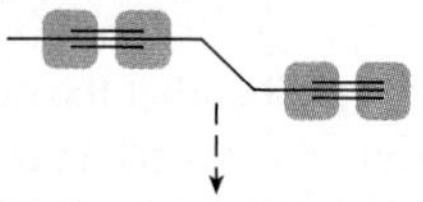

$CH_3C{\equiv}CCH_2CH_2C{\equiv}CH$

Like trans cycloalkenes, cycloalkynes with small rings are unstable. The carbon chain must be long enough to connect the two ends of the triple bond without introducing too much strain. Cyclooctyne is the smallest isolated cycloalkyne, though it decomposes upon standing at room temperature after a short time. To accommodate the triple bond in the ring, bending occurs around the *sp* hybridized C's, destabilizing cyclooctyne.

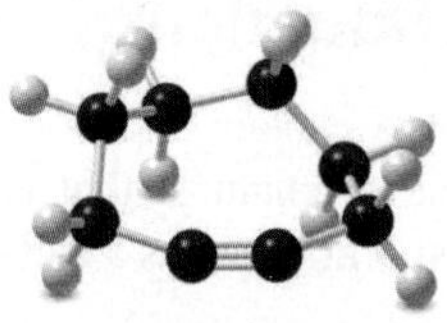

cyclooctyne

Problem 11.1 Nepheliosyne B is a novel acetylenic fatty acid isolated from a New Caledonian marine sponge. (a) Label the most acidic H atom. (b) Which carbon–carbon σ bond is shortest? (c) How many degrees of unsaturation does nepheliosyne B contain? (d) How many bonds are formed from $C_{sp}-C_{sp^3}$? (e) Label each triple bond as internal or terminal.

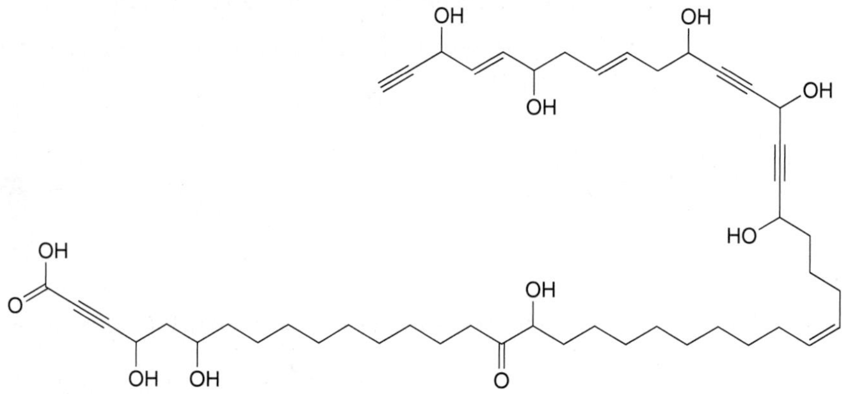

nepheliosyne B

11.2 Nomenclature

Alkynes are named in the same way that alkenes were named in Section 10.3.

- **In the IUPAC system, change the *-ane* ending of the parent alkane to the suffix *-yne.***
- **Choose the longest carbon chain that contains both atoms of the triple bond and number the chain to give the triple bond the lower number.**
- **Compounds with two triple bonds are named as *diynes,* those with three are named as *triynes,* and so forth.**
- **Compounds with both a double and a triple bond are named as *enynes.* The chain is numbered to give the first site of unsaturation (either C=C or C≡C) the lower number.**

Sample Problem 11.1 Give the IUPAC name for the following alkyne.

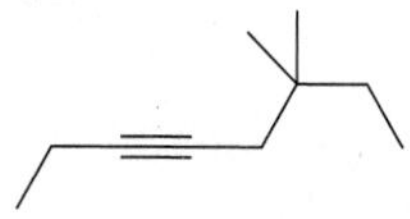

Solution

[1] Find the longest chain that contains both carbons of the triple bond.

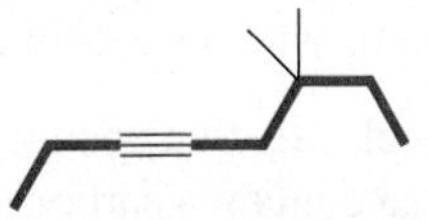

8 C's in the longest chain

oct*ane* ---→ oct*yne*

[2] Number the long chain; then name and number the substituents.

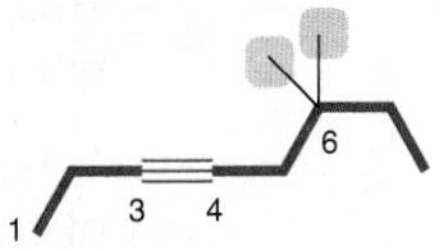

two methyl groups at C6

Answer: 6,6-dimethyloct-3-yne

The simplest alkyne, HC≡CH, named in the IUPAC system as **ethyne,** is more often called **acetylene,** its common name. The two-carbon alkyl group derived from acetylene is called an **ethynyl group** (HC≡C−). Examples of alkyne nomenclature are shown in Figure 11.1.

Problem 11.2 Give the IUPAC name for each compound.

a.

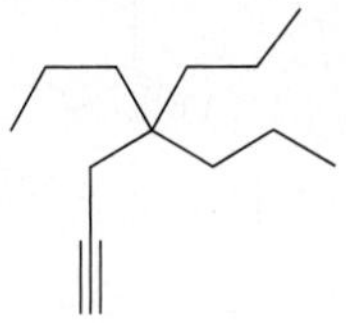

c.

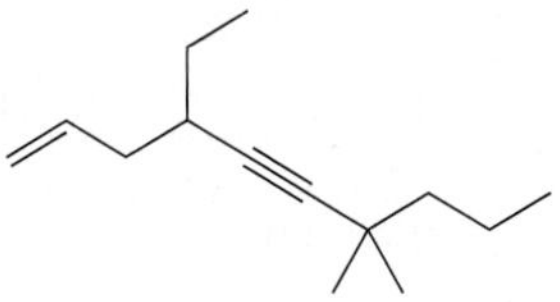

b.

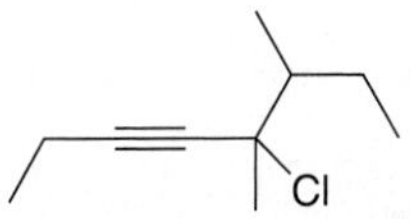

d.

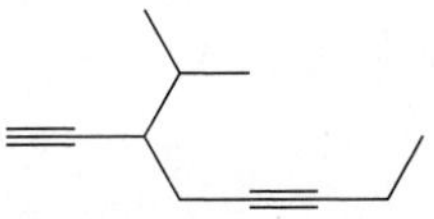

Problem 11.3 Give the structure corresponding to each of the following names.

a. *trans*-2-ethynylcyclopentanol b. 4-*tert*-butyldec-5-yne c. 3,3,5-trimethylcyclononyne

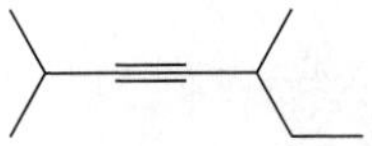

2,5-dimethylhept-3-yne

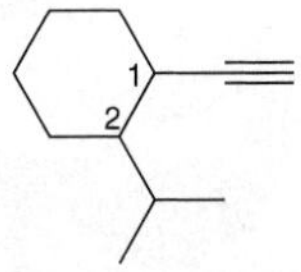

1-ethynyl-2-isopropylcyclohexane

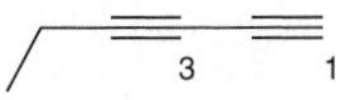

hexa-1,3-diyne

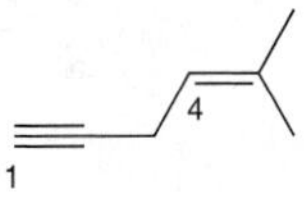

5-methylhex-4-en-1-yne

Figure 11.1 Examples of alkyne nomenclature

11.3 Physical Properties

The physical properties of alkynes resemble those of hydrocarbons having a similar shape and molecular weight.

- **Alkynes have low melting points and boiling points.**
- **Melting points and boiling points increase as the number of carbons increases.**
- **Alkynes are soluble in organic solvents and insoluble in water.**

Problem 11.4 Explain why an alkyne often has a slightly higher boiling point than an alkene of similar molecular weight. For example, the bp of pent-1-yne is 39°C, and the bp of pent-1-ene is 30°C.

11.4 Interesting Alkynes

Acetylene, HC≡CH, is a colorless gas with an ethereal odor that burns in oxygen to form CO_2 and H_2O. Because the combustion of acetylene releases more energy per mole of product formed than other hydrocarbons, it burns with a very hot flame, making it an excellent fuel for welding torches.

Ethynylestradiol, the molecule that opened Chapter 11, and **norethindrone** are two components of oral contraceptives that contain a carbon–carbon triple bond (Figure 11.2). Both molecules are synthetic analogues of the naturally occurring female hormones estradiol and progesterone, but are more potent so they can be administered in lower doses. Most oral contraceptives contain two of these synthetic hormones. They act by artificially elevating hormone levels in a woman, thereby preventing pregnancy.

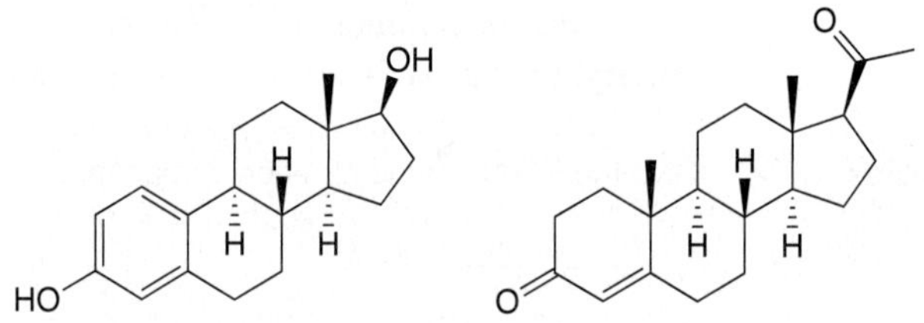

estradiol progesterone

Figure 11.2 How oral contraceptives work

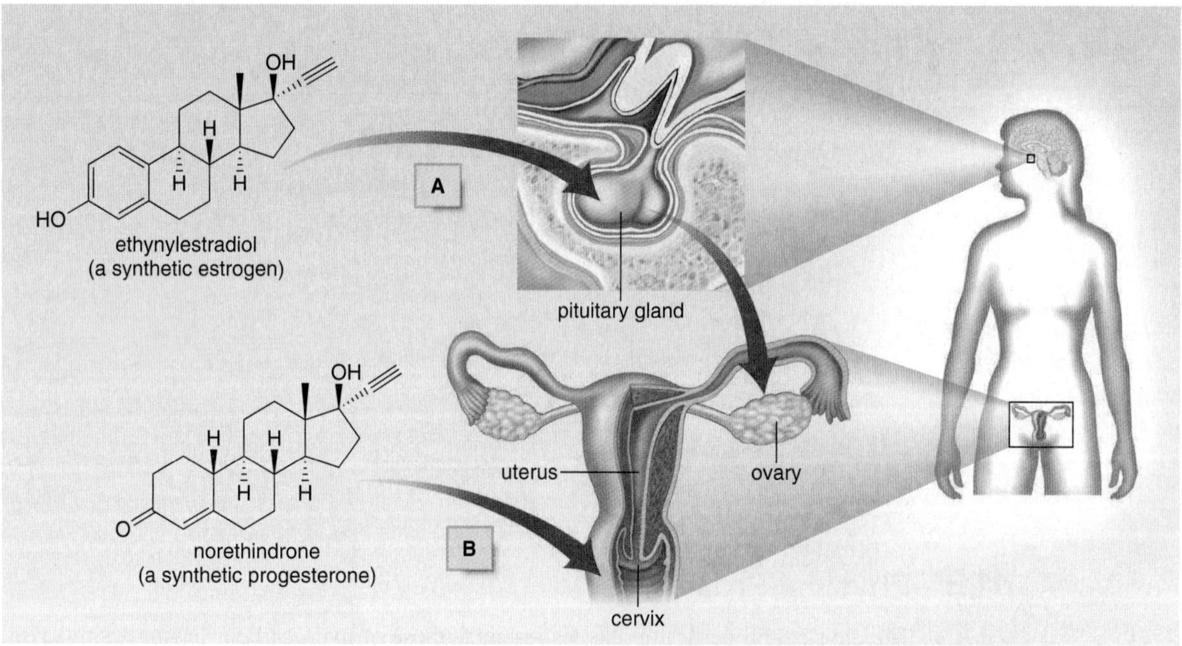

Monthly cycles of hormones from the pituitary gland cause ovulation, the release of an egg from an ovary. To prevent pregnancy, the two synthetic hormones in many oral contraceptives have different effects on the female reproductive system.

A: The elevated level of **ethynylestradiol,** a synthetic estrogen, "fools" the pituitary gland into thinking a woman is pregnant, so ovulation does not occur.

B: The elevated level of **norethindrone,** a synthetic progesterone, stimulates the formation of a thick layer of mucus in the cervix, making it difficult for sperm to reach the uterus.

Two other synthetic hormones with alkynyl appendages are **RU 486** and **levonorgestrel.** RU 486 blocks the effects of progesterone, and because of this, prevents implantation of a fertilized egg. RU 486 is used to induce abortions within the first few weeks of pregnancy. Levonorgestrel interferes with ovulation, and so it prevents pregnancy if taken within a few days of unprotected sex.

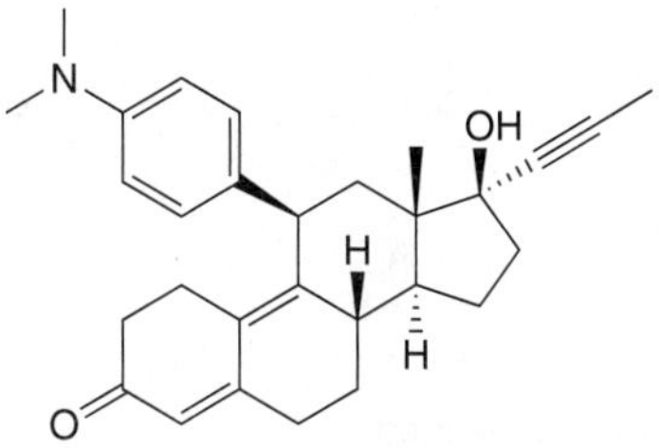

RU 486
(Trade name Mifepristone)

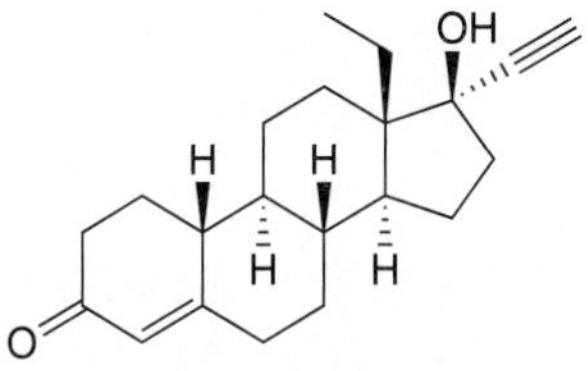

levonorgestrel
(Trade name Plan B)

Histrionicotoxin is a diyne isolated in small quantities from the skin of *Dendrobates histrionicus,* a colorful South American frog (Figure 11.3). This toxin, secreted by the frog as a natural defense mechanism, was used as a poison on arrow tips by the Choco tribe of South America.

Figure 11.3
Histrionicotoxin

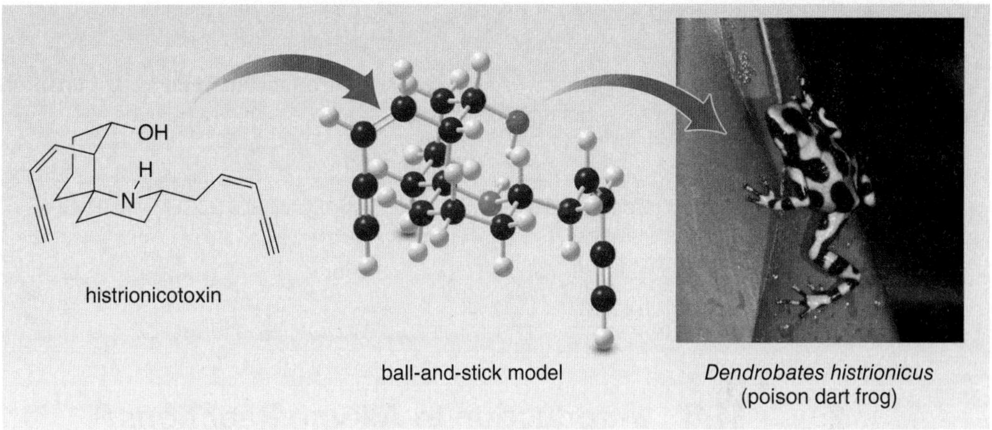

- Histrionicotoxin is a defensive toxin that protects *Dendrobates histrionicus* from potential predators. These small "poison dart" frogs inhabit the moist humid floor of tropical rainforests, and are commonly found in western Ecuador and Colombia. Histrionicotoxin acts by interfering with nerve transmission in mammals, resulting in prolonged muscle contraction.

11.5 Preparation of Alkynes

Alkynes are prepared by elimination reactions, as discussed in Section 8.10. **A strong base removes two equivalents of HX from a vicinal or geminal dihalide to yield an alkyne** by two successive E2 eliminations.

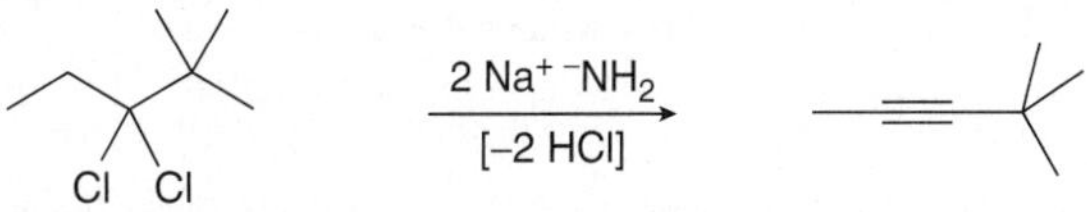

geminal dichloride

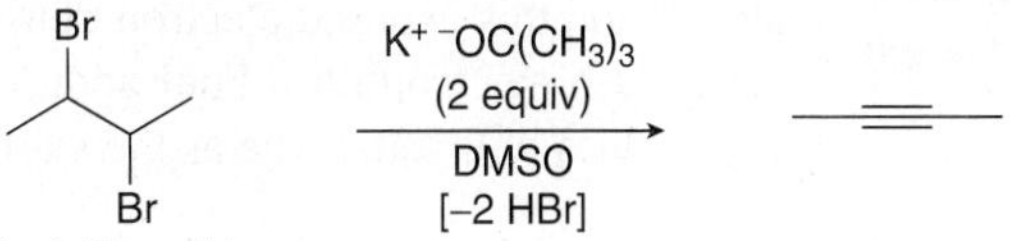

vicinal dibromide

Because vicinal dihalides are synthesized by adding halogens to alkenes, an alkene can be converted to an alkyne by the two-step process illustrated in Sample Problem 11.2.

Sample Problem 11.2 Convert alkene **A** into alkyne **B** by a stepwise method.

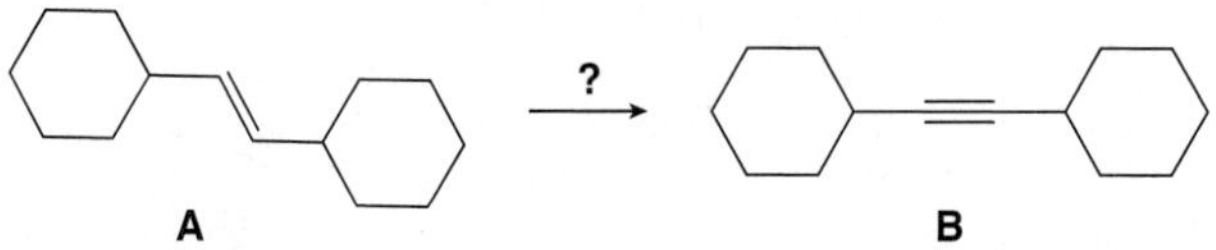

A **B**

Solution

A two-step method is needed:

- **Addition of X_2** forms a vicinal dihalide.
- **Elimination** of two equivalents of HX forms two π bonds.

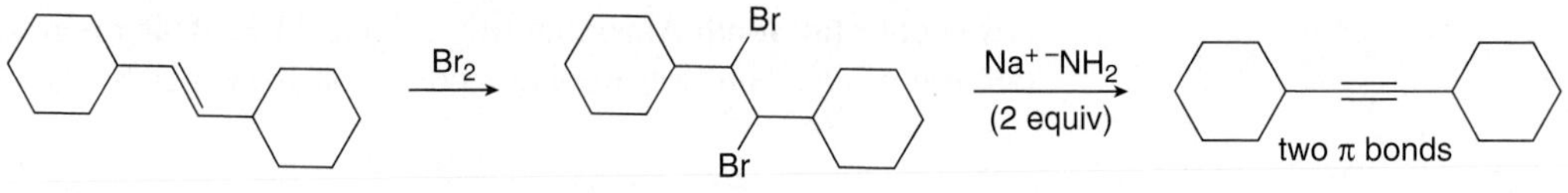

A vicinal dibromide **B**

- **This two-step process introduces one degree of unsaturation:** an alkene with one π bond is converted to an alkyne with two π bonds.

Problem 11.5 Convert each compound to hex-1-yne, $HC{\equiv}CCH_2CH_2CH_2CH_3$.

a. 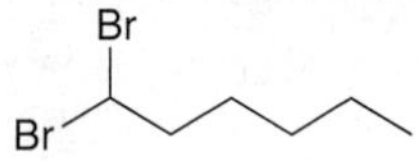

b. 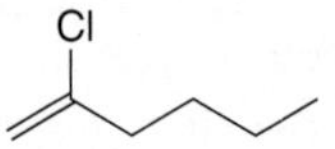

c. 

11.6 Introduction to Alkyne Reactions

All reactions of alkynes occur because they contain **easily broken π bonds** or, in the case of terminal alkynes, an **acidic, *sp* hybridized C–H bond.**

11.6A Addition Reactions

Like alkenes, **alkynes undergo addition reactions because they contain weak π bonds.** Two sequential reactions take place: addition of one equivalent of reagent forms an alkene, which then adds a second equivalent of reagent to yield a product having **four new bonds.**

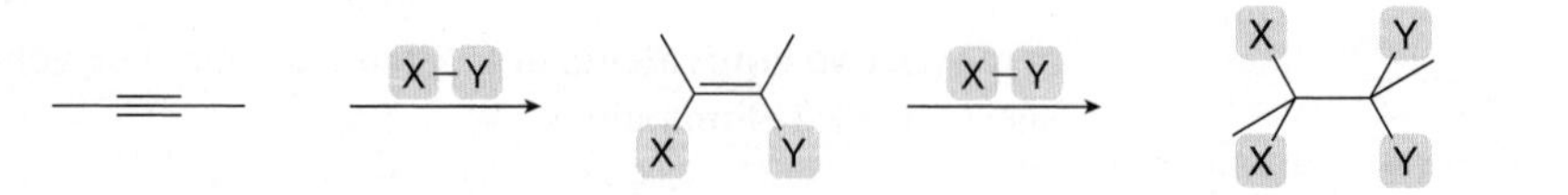

two weak π bonds — weak π bond (*E* or *Z* product) — Four σ bonds are formed.

The oxidation and reduction of alkynes, reactions that also involve addition, are discussed in Chapter 12.

Alkynes are electron rich, as shown in the electrostatic potential map of acetylene in Figure 11.4. The two π bonds form a cylinder of electron density between the two *sp* hybridized carbon atoms, and this exposed electron density makes a triple bond nucleophilic. As a result, **alkynes react with electrophiles.** Four addition reactions are discussed in Chapter 11 and illustrated in Figure 11.5 with but-1-yne as the starting material.

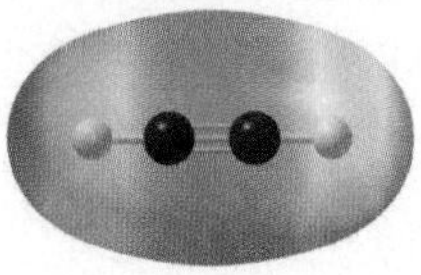

Figure 11.4
Electrostatic potential map of acetylene

- The red electron-rich region is located between the two carbon atoms, forming a cylinder of electron density.

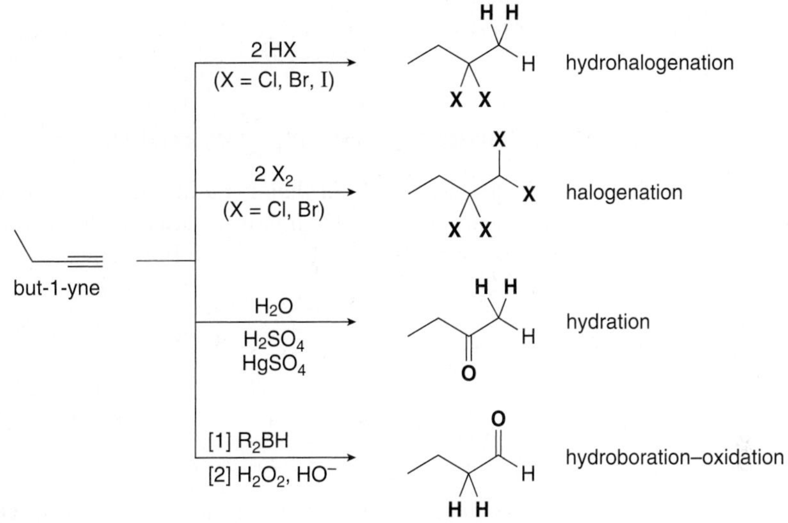

Figure 11.5
Four addition reactions of but-1-yne

- In each addition, both π bonds of the triple bond are broken, and four new bonds are formed.

11.6B Terminal Alkynes—Reaction as an Acid

Because ***sp*** **hybridized C–H bonds are more acidic than** $\boldsymbol{sp^2}$ **and** $\boldsymbol{sp^3}$ **hybridized C–H bonds,** terminal alkynes are readily deprotonated with strong base in a Brønsted–Lowry acid–base reaction. The resulting anion is called an **acetylide anion.**

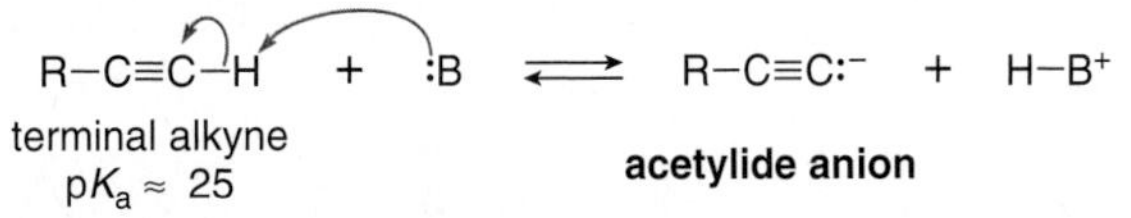

What bases can be used for this reaction? Because an acid–base equilibrium favors the weaker acid and base, only **bases having conjugate acids with pK_a values *higher* than the terminal alkyne—that is, pK_a values > 25—are strong enough** to form a significant concentration of acetylide anion. As shown in Table 11.1, $^-NH_2$ and H^- are strong enough to deprotonate a terminal alkyne, but ^-OH and ^-OR are not.

Recall from Section 2.5D that the acidity of a C–H bond increases as the percent *s*-character of C increases. Thus, the following order of relative acidity results: C_{sp^3}–H < C_{sp^2}–H < C_{sp}–H.

Table 11.1 A Comparison of Bases for Alkyne Deprotonation

	Base	pK_a of the conjugate acid
These bases are **strong** enough to deprotonate an alkyne.	$^-NH_2$	38
	H^-	35
These bases are ***not*** strong enough to deprotonate an alkyne.	^-OH	15.7
	^-OR	15.5–18

Why is this reaction useful? The acetylide anions formed by deprotonating terminal alkynes are **strong nucleophiles** that can react with a variety of electrophiles, as shown in Section 11.11.

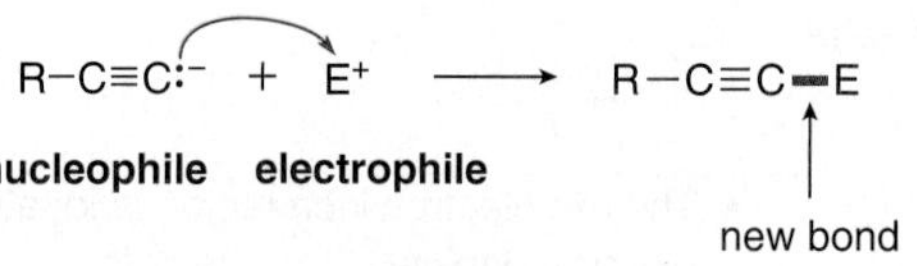

nucleophile
electrophile
new bond

Problem 11.6 Which bases can deprotonate acetylene? The pK_a values of the conjugate acids are given in parentheses.

a. CH_3NH^- (pK_a = 40) b. CO_3^{2-} (pK_a = 10.2) c. $CH_2=CH^-$ (pK_a = 44) d. $(CH_3)_3CO^-$ (pK_a = 18)

11.7 Addition of Hydrogen Halides

Alkynes undergo **hydrohalogenation, the addition of hydrogen halides, HX** (X = Cl, Br, I). Two equivalents of HX are usually used: addition of one mole forms a **vinyl halide,** which then reacts with a second mole of HX to form a **geminal dihalide.**

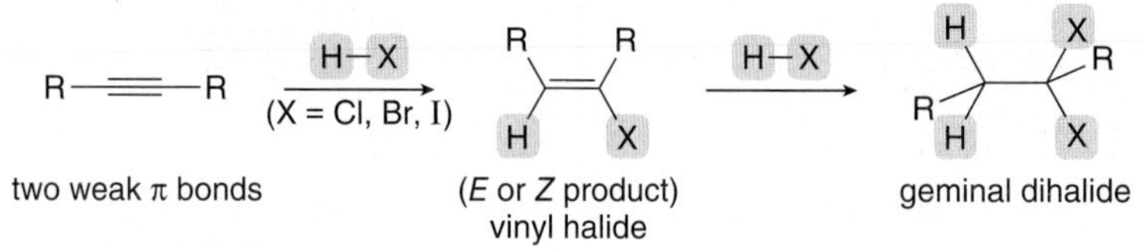

H–X
(X = Cl, Br, I)
H–X
two weak π bonds
(E or Z product)
vinyl halide
geminal dihalide

Addition of HX to an alkyne is another example of **electrophilic addition,** because the electrophilic (H) end of the reagent is attracted to the electron-rich triple bond.

- **With two equivalents of HX, both H atoms bond to the *same* carbon.**
- **With a terminal alkyne, both H atoms bond to the *terminal* carbon; that is, the hydrohalogenation of alkynes follows Markovnikov's rule.**

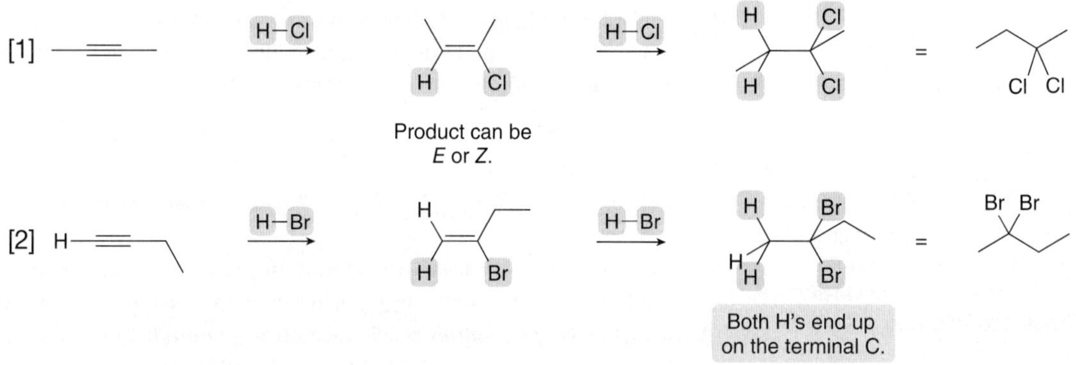

[1]
H–Cl
Product can be
E or Z.
H–Cl
[2]
H–Br
H–Br
Both H's end up
on the terminal C.

- **With only one equivalent of HX, the reaction stops with formation of the vinyl halide.**

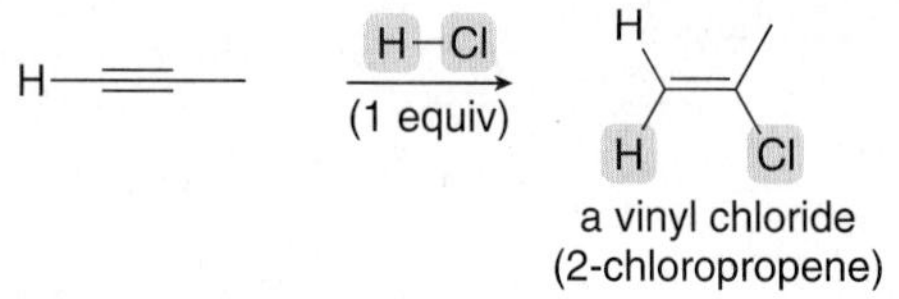

H–Cl
(1 equiv)
a vinyl chloride
(2-chloropropene)

Because of the instability of a vinyl carbocation, other mechanisms for HX addition that avoid formation of a discrete carbocation have been proposed. It is likely that more than one mechanism occurs, depending in part on the identity of the alkyne substrate.

One proposed mechanism for the addition of two equivalents of HX to an alkyne involves **two steps for each addition of HX:** addition of H^+ (from HX) to form a carbocation, followed by nucleophilic attack of X^-. Mechanism 11.1 illustrates the addition of HBr to but-1-yne to yield 2,2-dibromobutane. Each two-step mechanism is similar to the two-step addition of HBr to *cis*-but-2-ene discussed in Section 10.9.

Mechanism 11.1 Electrophilic Addition of HX to an Alkyne

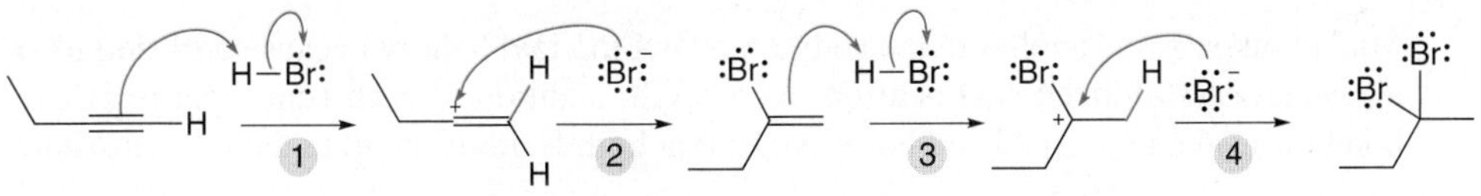

but-1-yne — vinyl carbocation, new bond shown in red — vinyl bromide — carbocation, new bond shown in red — 2,2-dibromobutane

1. Addition of H^+ forms a vinyl carbocation and follows Markovinkov's rule. The H atom bonds to the terminal C to form the more substituted carbocation.
2. Nucleophilic attack of Br^- forms a vinyl bromide. One equivalent of HBr adds in two steps.
3. The addition of a second equivalent of HBr follows in the same two-step manner. Addition of H^+ to the vinyl bromide forms a carbocation.
4. Nucleophilic attack of Br^- forms a geminal dibromide, 2,2-dibromobutane.

The formation of both carbocations (in Steps [1] and [3]) deserves additional scrutiny. **The vinyl carbocation formed in Step [1] is *sp* hybridized and therefore less stable than a 2° *sp*^{2} hybridized carbocation** (Section 7.18). This makes electrophilic addition of HX to an alkyne ***slower*** than electrophilic addition of HX to an alkene, even though alkynes are more polarizable and have more loosely held π electrons than alkenes.

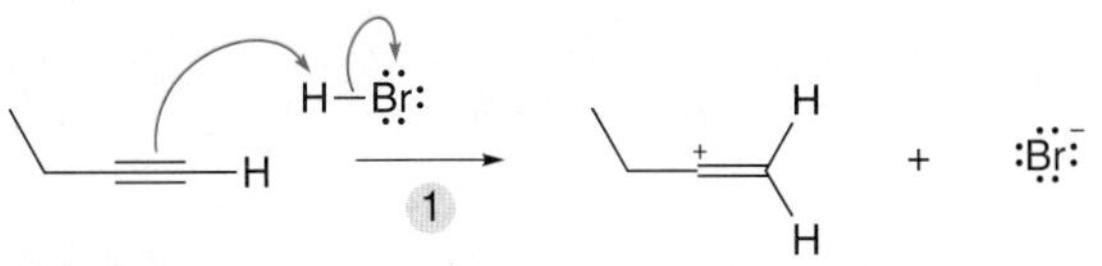

sp hybridized
vinyl carbocation

In Step [3] two carbocations are possible but only one is formed. Markovnikov addition in Step [3] places the H on the terminal carbon (C1) to form the more substituted carbocation **A,** rather than the less substituted carbocation **B.** Because the more stable carbocation is formed faster—another example of the Hammond postulate—carbocation **A** must be more stable than carbocation **B.**

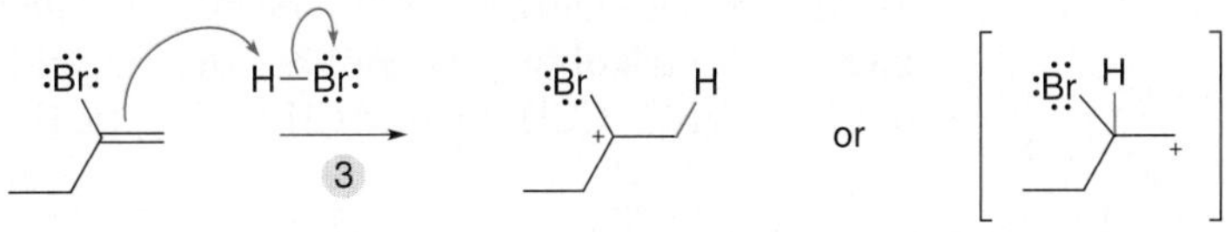

A — more stable carbocation, new bond shown in red

B — *not* formed, new bond shown in red

Why is carbocation **A,** having a positive charge on a carbon that also has a Br atom, more stable? Shouldn't the electronegative Br atom withdraw electron density from the positive charge, and thus destabilize it? It turns out that **A is stabilized by resonance** but **B** is not. Two resonance structures can be drawn for carbocation **A,** but only one Lewis structure can be drawn for carbocation **B.**

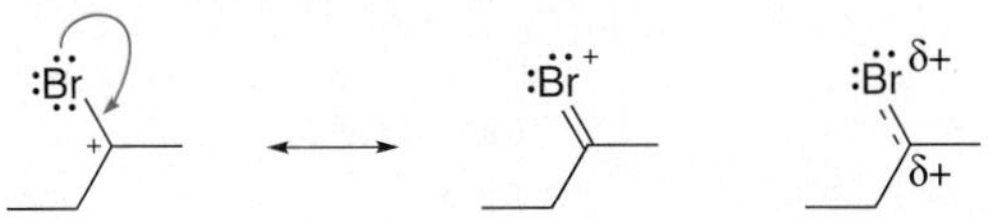

two resonance structures for **A**

hybrid
The positive charge is delocalized.

- **Resonance stabilizes a molecule by delocalizing charge and electron density.**
- **Thus, halogens stabilize an adjacent positive charge by resonance.**

Markovnikov's rule applies to the addition of HX to vinyl halides because **addition of H^+ forms a resonance-stabilized carbocation.** As a result, addition of each equivalent of HX to a triple bond forms the more stable carbocation, so that both H atoms bond to the less substituted C.

Problem 11.7 Draw the organic products formed when each alkyne is treated with two equivalents of HBr.

a. 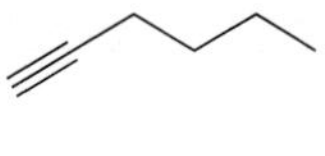 b. 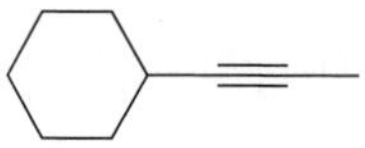 c. 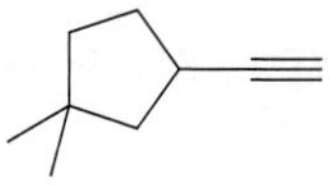

Problem 11.8 Draw additional resonance structures for each cation.

a. 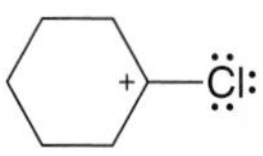 b. 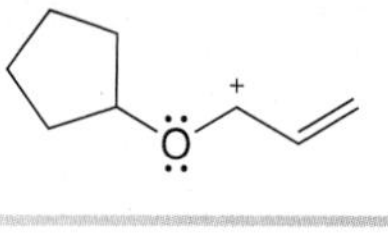 c. 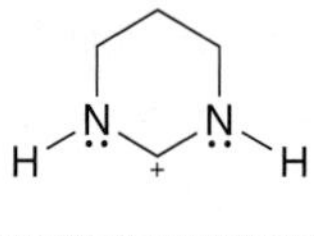

11.8 Addition of Halogen

Halogens, X_2 (X = Cl or Br), add to alkynes in much the same way they add to alkenes (Section 10.13). Addition of one mole of X_2 forms a **trans dihalide,** which can then react with a second mole of X_2 to yield a **tetrahalide.**

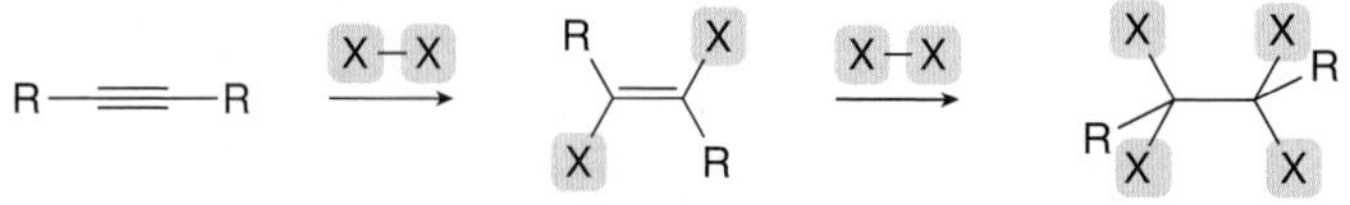

trans dihalide tetrahalide

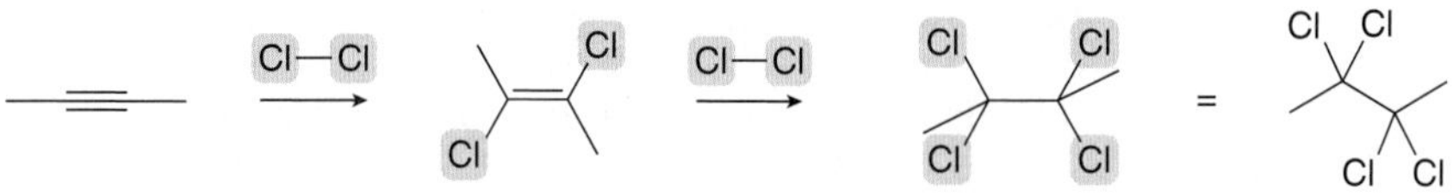

Each addition of X_2 involves a two-step process with a **bridged halonium ion** intermediate, reminiscent of the addition of X_2 to alkenes (Section 10.13). A trans dihalide is formed after addition of one equivalent of X_2 because the intermediate **halonium ion ring is opened upon backside attack of the nucleophile.** Mechanism 11.2 illustrates the addition of two equivalents of Cl_2 to $CH_3C{\equiv}CCH_3$ to form $CH_3CCl_2CCl_2CH_3$.

Mechanism 11.2 Addition of X_2 to an Alkyne—Halogenation

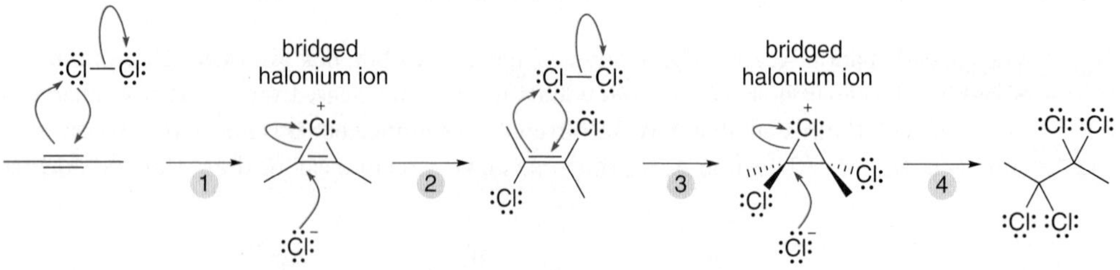

trans dihalide tetrahalide

1 Four bonds are broken or formed to generate a bridged halonium ion that contains a three-membered ring.

2 Nucleophilic attack of Cl^- ring opens the bridged halonium ion from the back side and forms the trans dihalide.

3 Electrophilic addition of Cl^+ forms the bridged halonium ion.

4 Nucleophilic attack of Cl^- ring opens the bridged halonium ion and forms the tetrahalide.

Problem 11.9 Draw the products formed when $CH_3CH_2C{\equiv}CCH_2CH_3$ is treated with each reagent: (a) Br_2 (2 equiv); (b) Cl_2 (1 equiv).

Problem 11.10 Explain the following result. Although alkenes are generally more reactive than alkynes toward electrophiles, the reaction of Cl_2 with but-2-yne can be stopped after one equivalent of Cl_2 has been added.

11.9 Addition of Water

Although the addition of H_2O to an alkyne resembles the acid-catalyzed addition of H_2O to an alkene in some ways, an important difference exists. In the presence of strong acid or Hg^{2+} catalyst, the **elements of H_2O add to the triple bond,** but the initial addition product, an **enol,** is unstable and rearranges to a product containing a **carbonyl group**—that is, a **C=O.** A carbonyl compound having two alkyl groups bonded to the C=O carbon is called a **ketone.**

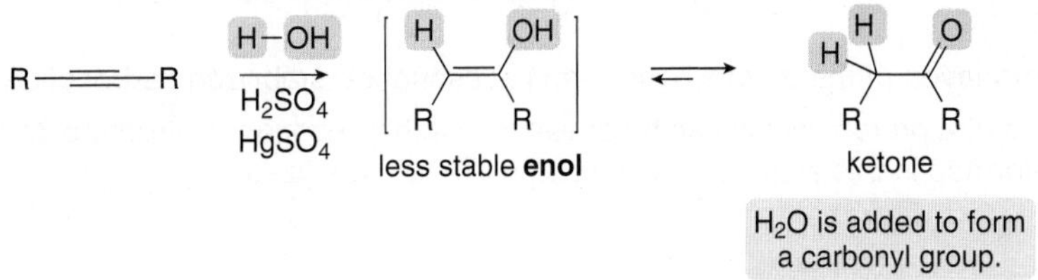

Internal alkynes undergo hydration with concentrated acid, whereas terminal alkynes require the presence of an additional Hg^{2+} catalyst—usually $HgSO_4$—to yield methyl ketones by **Markovnikov addition of H_2O.**

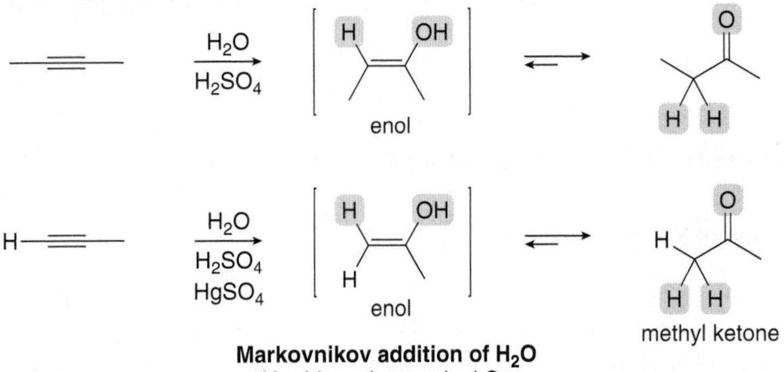

Because an enol contains both a C=C and a hydroxy group, the name **enol** comes from alk**ene** + alcoh**ol.**

$HgSO_4$ is often used in the hydration of internal alkynes as well, because hydration can be carried out under milder reaction conditions.

Let's first examine the conversion of a general enol **A** to the carbonyl compound **B. A** and **B** are called **tautomers: A is the *enol form* and B is the *keto form* of the tautomer.**

- ***Tautomers* are constitutional isomers that differ in the location of a double bond and a hydrogen atom. Two tautomers are in equilibrium with each other.**

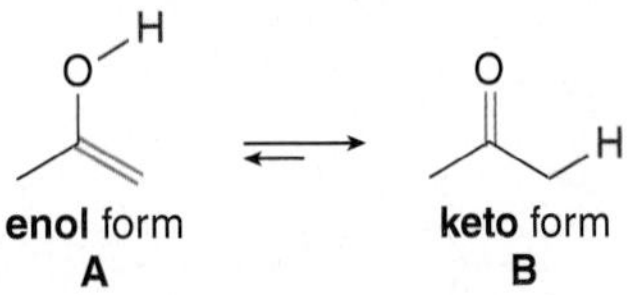

Tautomers differ in the position of a double bond and a hydrogen atom. In Chapter 23 an in-depth discussion of keto–enol tautomers is presented.

- An enol tautomer has an O–H group bonded to a C=C.
- A keto tautomer has a C=O and an additional C–H bond.

Equilibrium favors the keto form largely because a C=O is much stronger than a C=C. Tautomerization, the process of converting one tautomer into another, is catalyzed by both acid and base. Under the strongly acidic conditions of hydration, tautomerization of the enol to the keto form occurs rapidly by a two-step process: **protonation,** followed by **deprotonation** as shown in Mechanism 11.3.

Mechanism 11.3 Tautomerization in Acid

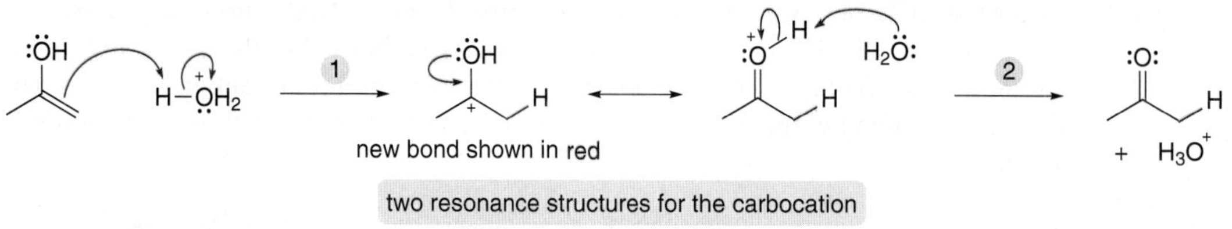

:ÖH
H–ÖH2
1
new bond shown in red
2
+ H3O+
two resonance structures for the carbocation

1 Protonation of the double bond forms a resonance-stabilized carbocation.

2 Loss of a proton, which can be drawn with either resonance structure, forms the carbonyl group. Because acid is re-formed in this step, tautomerizataion is acid-catalyzed.

Hydration of an internal alkyne with strong acid forms an enol by a mechanism similar to that of the acid-catalyzed hydration of an alkene (Section 10.12). Mechanism 11.4 illustrates the hydration of but-2-yne with H_2O and H_2SO_4. Once formed, the enol then tautomerizes to the more stable keto form by protonation followed by deprotonation.

Mechanism 11.4 Hydration of an Alkyne

Part [1] Addition of H_2O to form an enol

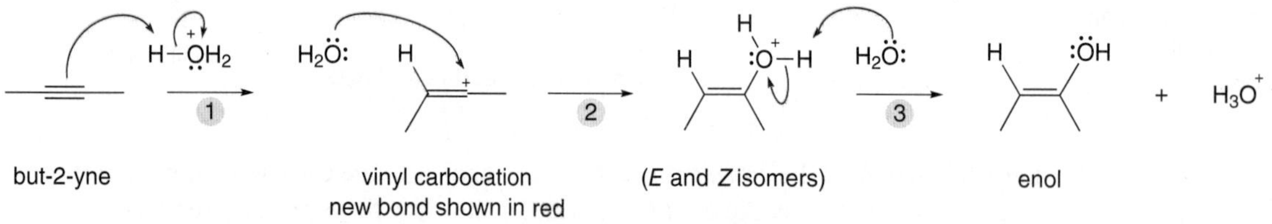

H–ÖH2
1
H2Ö:
2
3
+ H3O+
but-2-yne
vinyl carbocation
new bond shown in red
(E and Z isomers)
enol

1 Addition of H^+ forms a vinyl carbocation.

2 - 3 Nucleophilic attack followed by loss of a proton forms the enol.

Part [2] Tautomerization

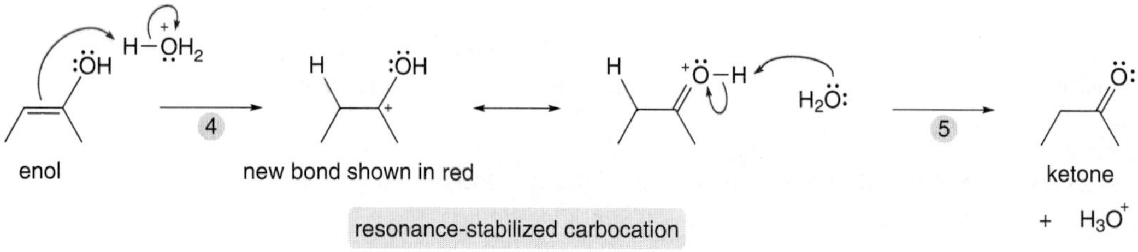

H–ÖH2
:ÖH
4
H2Ö:
5
enol
new bond shown in red
ketone
+ H3O+
resonance-stabilized carbocation

4 Tautomerization of the enol to the keto form begins with protonation of the double bond to form a carbocation.

5 Loss of a proton, which can be drawn with either resonance structure, forms the ketone.

Sample Problem 11.3 Draw the enol intermediate and the ketone product formed in the following reaction.

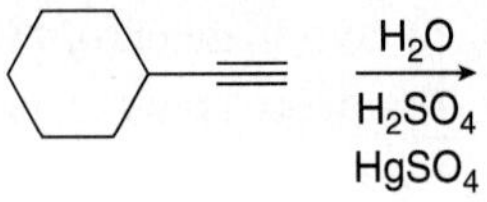

Solution

First, form the enol by adding H_2O to the triple bond with the **H bonded to the less substituted terminal carbon,** according to Markovnikov's rule.

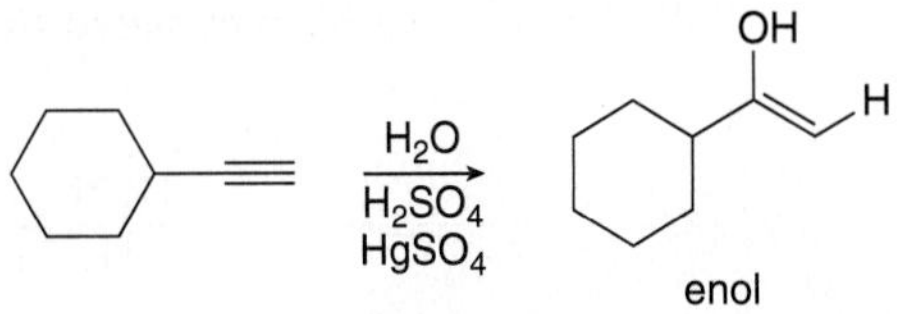

To convert the enol to the keto tautomer, add a proton to the C=C and remove a proton from the OH group. In tautomerization, the C–OH bond is converted to a C=O, and a new C–H bond is formed on the other enol carbon.

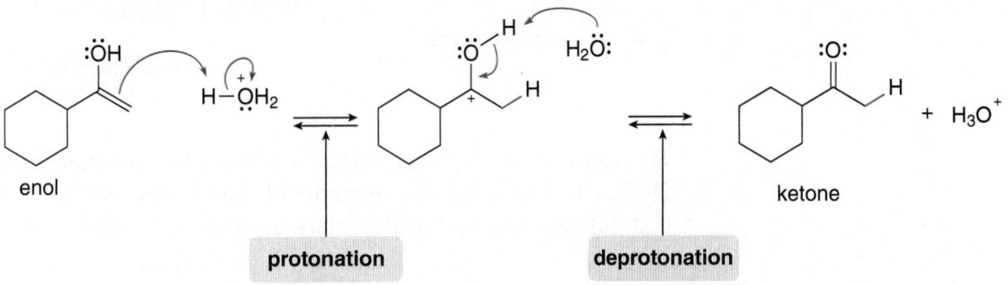

- **The overall result is the addition of H_2O to a triple bond to form a ketone.**

Problem 11.11 Draw the keto tautomer of each enol.

a. 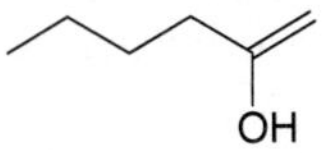

b. 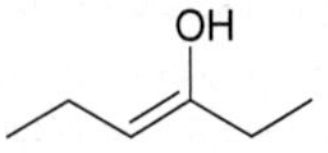

c. 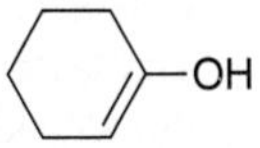

Problem 11.12 What two enols are formed when pent-2-yne is treated with H_2O, H_2SO_4, and $HgSO_4$? Draw the ketones formed from these enols after tautomerization.

Problem 11.13 (a) Draw two different enol tautomers of 2-methylcyclohexanone. (b) Draw two constitutional isomers that are not tautomers, but contain a C=C and an OH group.

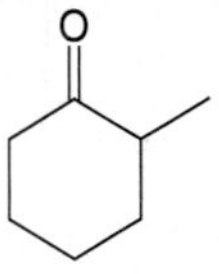

2-methylcyclohexanone

11.10 Hydroboration–Oxidation

Hydroboration–oxidation is a two-step reaction sequence that converts an alkyne to a carbonyl compound.

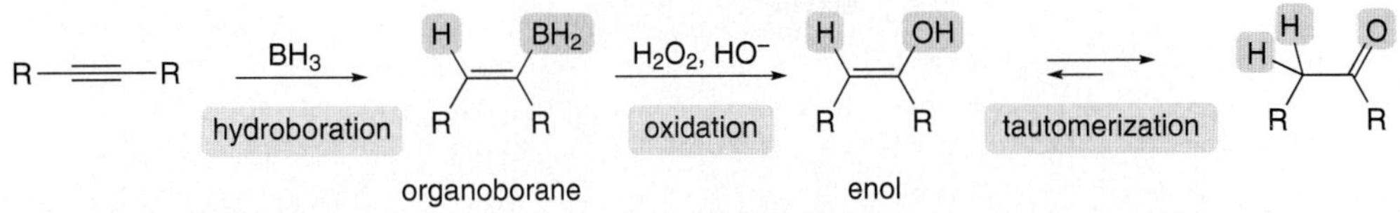

- **Addition of borane forms an organoborane.**
- **Oxidation with basic H_2O_2 forms an enol.**
- **Tautomerization of the enol forms a carbonyl compound.**
- **The overall result is addition of H_2O to a triple bond.**

Hydroboration–oxidation of an *internal* alkyne forms a **ketone. Hydroboration of a *terminal* alkyne adds boron to the less substituted, terminal carbon.** After oxidation to the enol, tautomerization yields an **aldehyde,** a carbonyl compound having a hydrogen atom bonded to the carbonyl carbon. Hydroboration of a terminal alkyne is generally carried out with a dialkylborane (R_2BH), which has been prepared from BH_3 (Section 10.16).

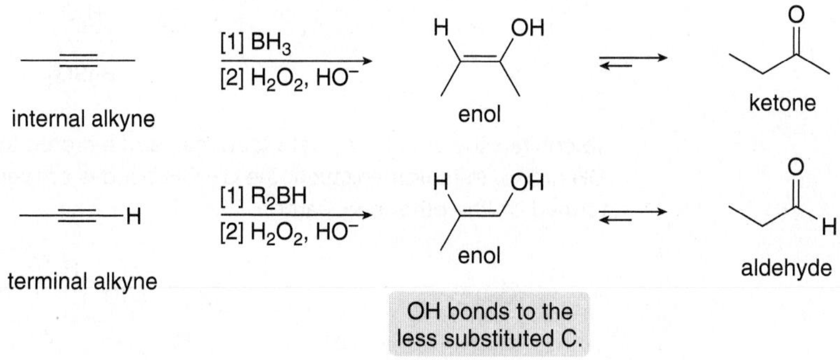

Hydration (H_2O, H_2SO_4, and $HgSO_4$) and **hydroboration–oxidation** (BH_3 or R_2BH followed by H_2O_2, HO^-) both **add the elements of H_2O across a triple bond.** Sample Problem 11.4 shows that different constitutional isomers are formed from terminal alkynes in these two reactions.

Sample Problem 11.4 Draw the product formed when $CH_3CH_2C{\equiv}CH$ is treated with each of the following sets of reagents: (a) H_2O, H_2SO_4, $HgSO_4$; and (b) R_2BH, followed by H_2O_2, HO^-.

Solution

(a) With $H_2O + H_2SO_4 + HgSO_4$, electrophilic addition of H and OH places the **H atom on the less substituted carbon** of the alkyne to form a **ketone** after tautomerization. (b) In contrast, addition of R_2BH places the **R_2B group on the less substituted terminal carbon** of the alkyne. Oxidation and tautomerization yield an **aldehyde.** The ketone and aldehyde formed in these reactions are constitutional isomers.

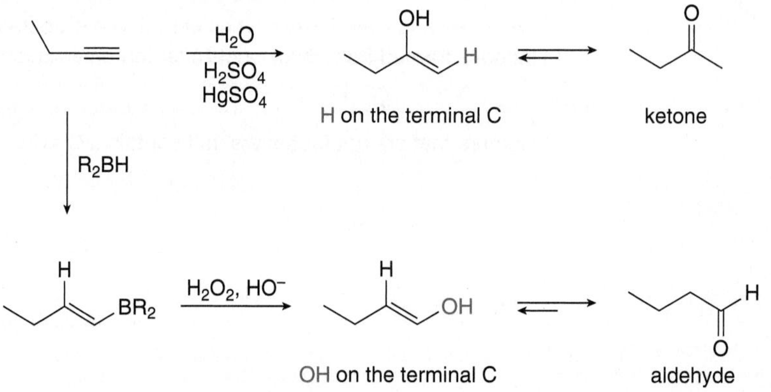

- **Addition of H_2O using H_2O, H_2SO_4, and $HgSO_4$ forms methyl ketones from terminal alkynes.**
- **Addition of H_2O using an organoborane, then H_2O_2, HO^- forms aldehydes from terminal alkynes.**

Problem 11.14 Draw the products formed when the following alkynes are treated with each set of reagents: [1] H_2O, H_2SO_4, $HgSO_4$; or [2] R_2BH followed by H_2O_2, ^-OH.

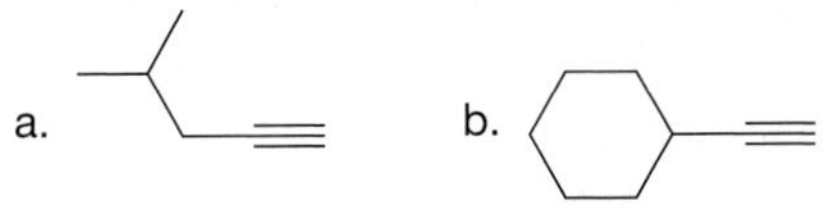
a.
b.

11.11 Reaction of Acetylide Anions

Terminal alkynes are readily converted to acetylide anions with strong bases such as $NaNH_2$ and NaH. These anions are strong nucleophiles, capable of reacting with electrophiles such as alkyl halides and epoxides.

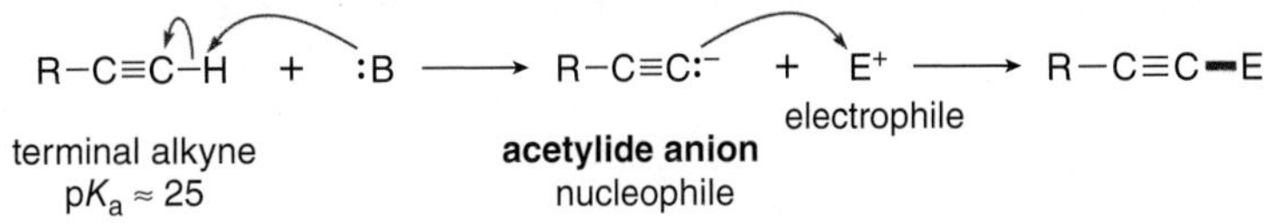
R–C≡C–H + :B → R–C≡C:⁻ + E⁺ → R–C≡C–E
terminal alkyne
pKa ≈ 25
acetylide anion
nucleophile
electrophile

11.11A Reaction of Acetylide Anions with Alkyl Halides

Acetylide anions react with unhindered alkyl halides to yield products of nucleophilic substitution.

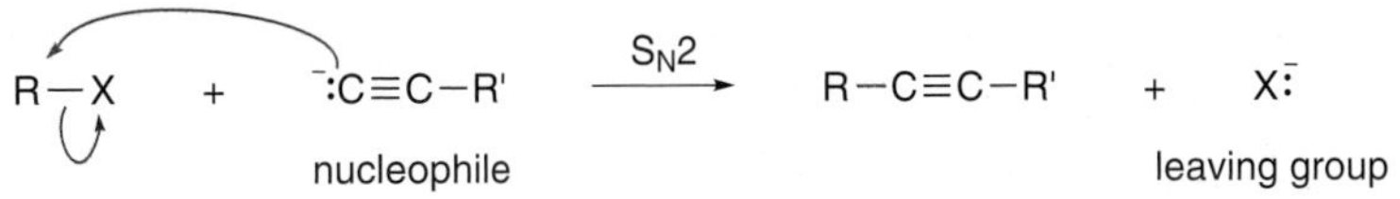
R–X + ⁻:C≡C–R'
SN2
R–C≡C–R' + X:⁻
nucleophile
leaving group

Because acetylide anions are strong nucleophiles, the mechanism of nucleophilic substitution is **S_N2,** and thus the **reaction is fastest with CH_3X and 1° alkyl halides.** Terminal alkynes (Reaction [1]) or internal alkynes (Reaction [2]) can be prepared depending on the identity of the acetylide anion.

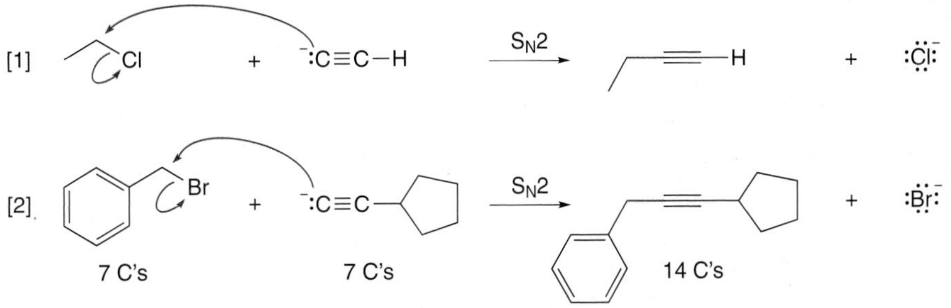
[1]
Cl
⁻:C≡C–H
SN2
H
:Cl:⁻
[2]
Br
⁻:C≡C
SN2
:Br:⁻
7 C's
7 C's
14 C's
new bond drawn in red

- **Nucleophilic substitution with acetylide anions forms new carbon–carbon bonds.**

Because organic compounds consist of a carbon framework, reactions that form carbon–carbon bonds are especially useful. In Reaction [2], for example, nucleophilic attack of a seven-carbon acetylide anion on a seven-carbon alkyl halide yields a 14-carbon alkyne as product.

Although nucleophilic substitution with acetylide anions is a very valuable carbon–carbon bond-forming reaction, it has the same limitations as any S_N2 reaction. **Steric hindrance around the leaving group causes 2° and 3° alkyl halides to undergo elimination by an E2 mechanism,** as shown with 2-bromo-2-methylpropane. Thus, nucleophilic substitution with acetylide anions forms new carbon–carbon bonds in high yield only with unhindered CH_3X and 1° alkyl halides.

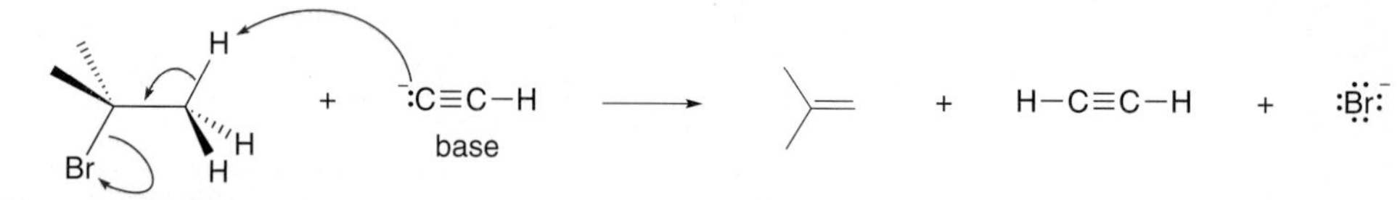
H
H
H
Br
+
⁻:C≡C−H
base
+
H−C≡C−H
+
:Br:⁻

2-bromo-2-methylpropane
3° alkyl halide

E2 product

Steric hindrance prevents an S_N2 reaction.

Sample Problem 11.5 Draw the organic products formed in each reaction.

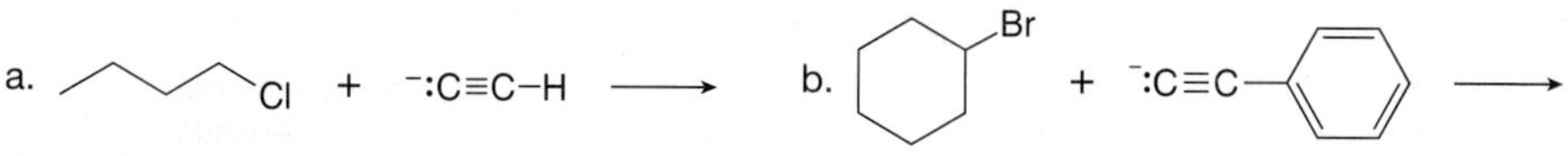
a.
Cl
+ ⁻:C≡C−H
b.
Br
+ ⁻:C≡C−

Solution

a. Because the alkyl halide is **1°** and the acetylide anion is a strong nucleophile, substitution occurs by an $\mathbf{S_N2}$ mechanism, resulting in a new C–C bond.

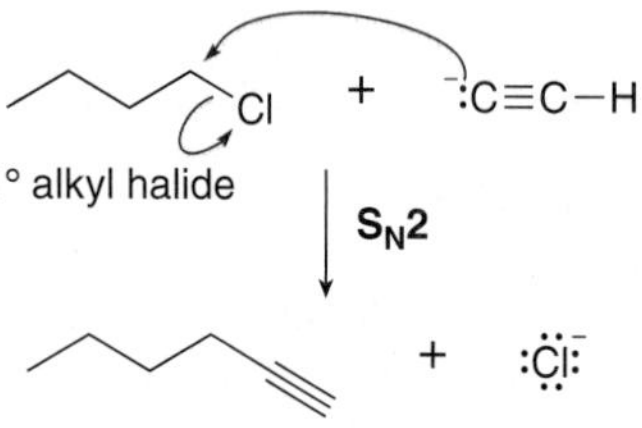
Cl
+
⁻:C≡C−H
1° alkyl halide
SN2
+
:Cl:⁻

b. Because the alkyl halide is **2°,** the major product is formed from elimination by an **E2** mechanism.

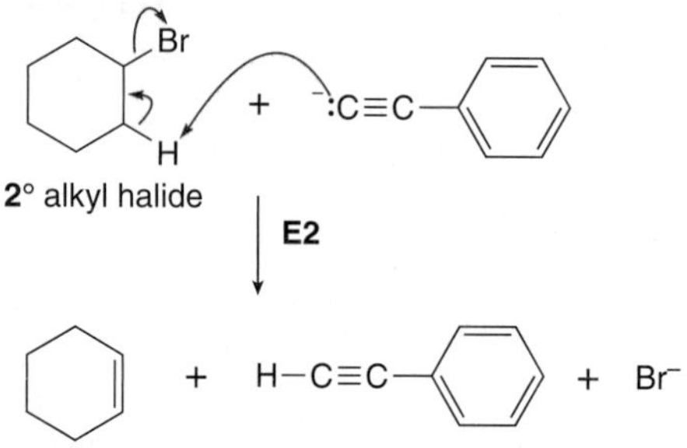
Br
H
+
⁻:C≡C−
2° alkyl halide
E2
+
H−C≡C−
+ Br⁻
major product

Problem 11.15 Draw the organic products formed in each reaction.

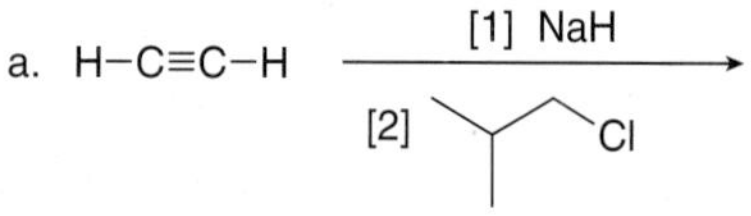
a. H−C≡C−H
[1] NaH
[2]
Cl

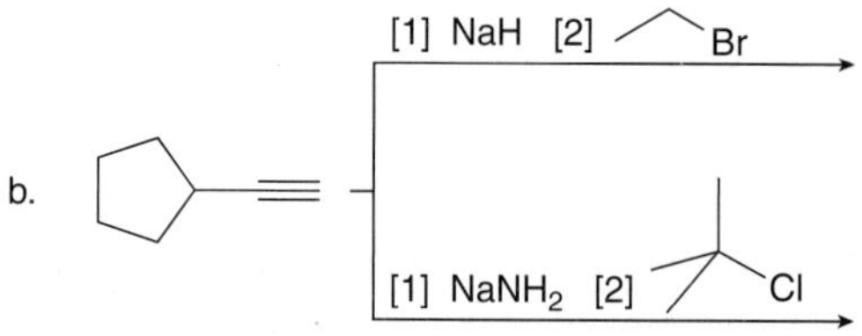
b.
[1] NaH [2]
Br
[1] NaNH2 [2]
Cl

Problem 11.16 What acetylide anion and alkyl halide can be used to prepare each alkyne? Indicate all possibilities when more than one route will work.

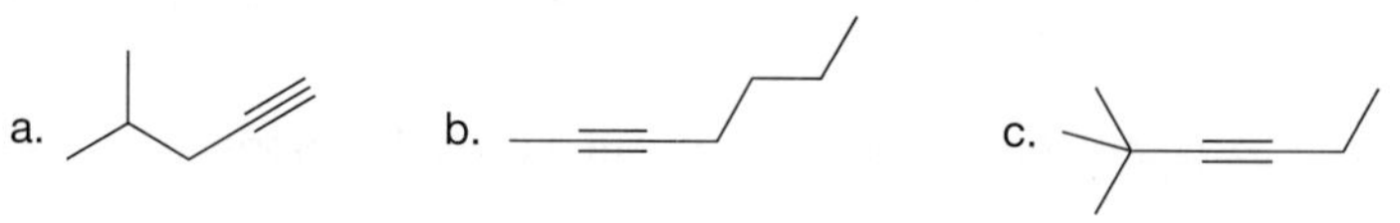
a.
b.
c.

Because acetylene has two *sp* hybridized C–H bonds, two sequential reactions can occur to form **two new carbon–carbon bonds,** as shown in Sample Problem 11.6.

Sample Problem 11.6 Identify the terminal alkyne **A** and the internal alkyne **B** in the following reaction sequence.

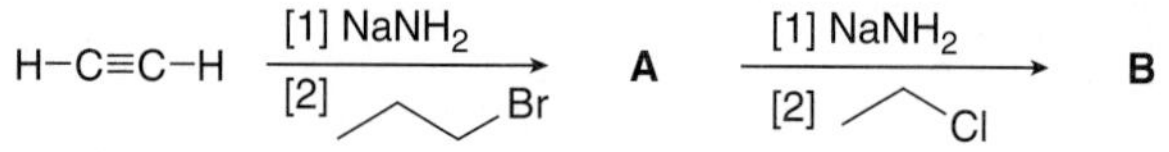
H−C≡C−H
[1] NaNH2
[2]
Br
A
[1] NaNH2
[2]
Cl
B

Solution

In each step, the base $^{-}NH_2$ removes a proton on an *sp* hybridized carbon, and the resulting acetylide anion reacts as a nucleophile with an alkyl halide to yield an S_N2 product. The first two-step reaction sequence forms the **terminal alkyne A** by nucleophilic attack of the acetylide anion on $CH_3CH_2CH_2Br$.

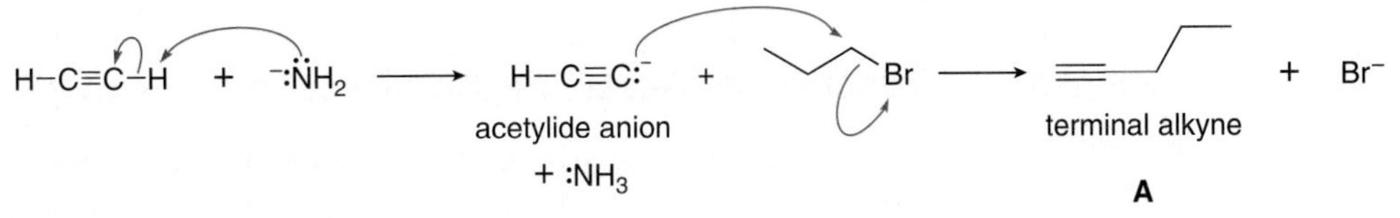

The second two-step reaction sequence forms the **internal alkyne B** by nucleophilic attack of the acetylide anion on CH_3CH_2Cl.

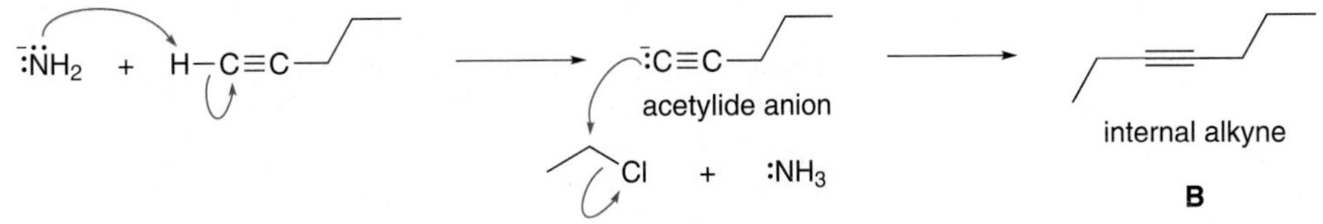

The soft coral *Capnella imbricata* is the source of the natural product capnellene.

Sample Problem 11.6 illustrates how a seven-carbon product can be prepared from three smaller molecules by forming two new carbon–carbon bonds.

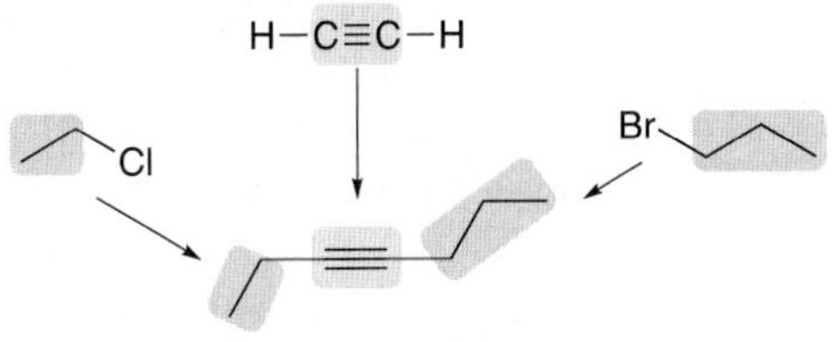

new bonds shown in red

Carbon–carbon bond formation with acetylide anions is a valuable reaction used in the synthesis of numerous natural products. Two examples include **capnellene,** isolated from the soft coral *Capnella imbricata,* and **niphatoxin B,** isolated from a red sea sponge, as shown in Figure 11.6.

Figure 11.6 Use of acetylide anion reactions in the synthesis of two marine natural products

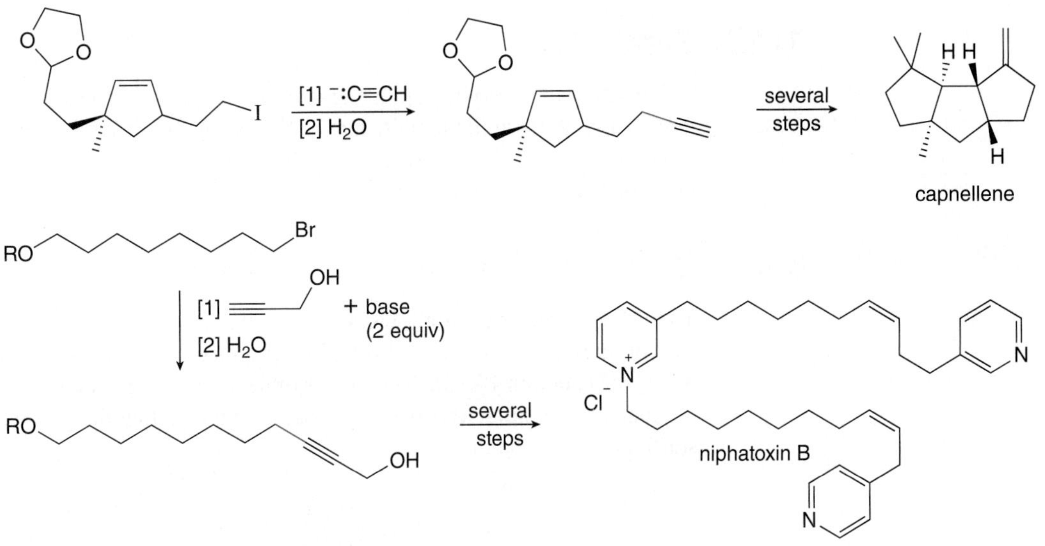

- New carbon–carbon bonds formed from acetylide anions are shown in red.

Problem 11.17 Show how $HC{\equiv}CH$, CH_3CH_2Br, and $(CH_3)_2CHCH_2CH_2Br$ can be used to prepare $CH_3CH_2C{\equiv}CCH_2CH_2CH(CH_3)_2$. Show all reagents, and use curved arrows to show movement of electron pairs.

Problem 11.18 Explain why 2,2,5,5-tetramethylhex-3-yne can't be made using acetylide anions.

11.11B Reaction of Acetylide Anions with Epoxides

Acetylide anions are strong nucleophiles that open epoxide rings by an S_N2 mechanism. This reaction also results in the formation of a **new carbon–carbon bond.** Backside attack occurs at the less substituted end of the epoxide.

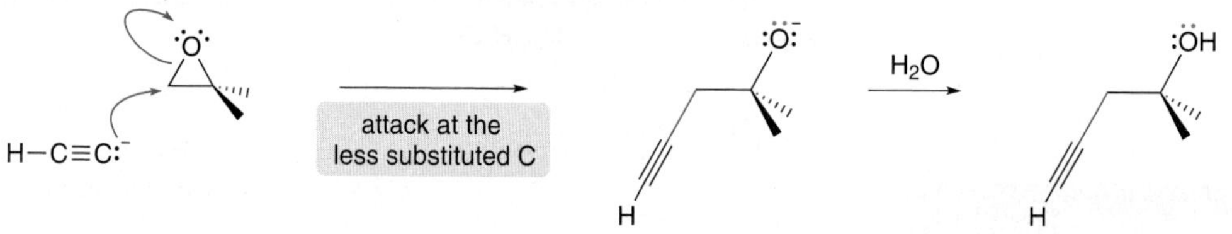

Opening of epoxide rings with strong nucleophiles was first discussed in Section 9.16A.

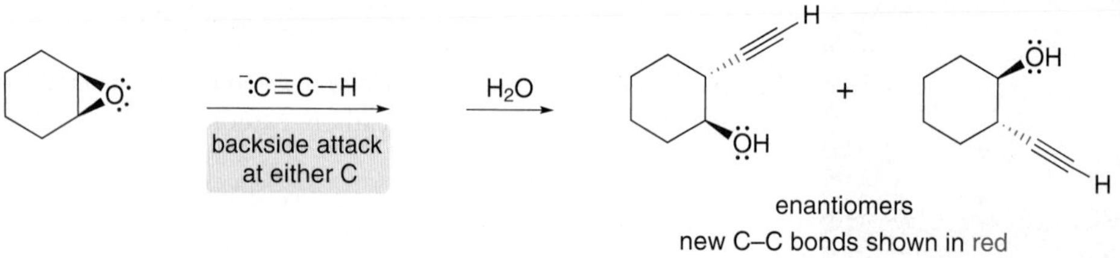

Problem 11.19 Draw the products of each reaction.

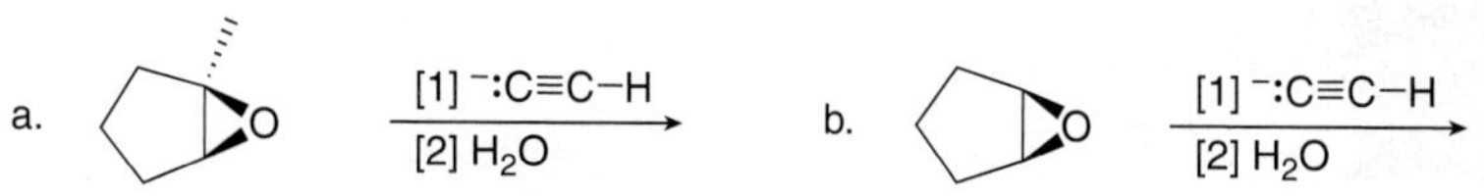

Problem 11.20 Draw the products formed when $CH_3CH_2C{\equiv}C^-Na^+$ reacts with each compound.

a. $CH_3CH_2CH_2Br$
b. $(CH_3)_2CHCH_2CH_2Cl$
c. $(CH_3CH_2)_3CCl$
d. $BrCH_2CH_2CH_2CH_2OH$
e. ethylene oxide followed by H_2O
f. propene oxide followed by H_2O

11.12 Synthesis

The reactions of acetylide anions give us an opportunity to examine organic synthesis more systematically. Performing a multistep synthesis can be difficult. Not only must you know the reactions for a particular functional group, but you must also put these reactions in a logical order, a process that takes much practice to master.

11.12A General Terminology and Conventions

To plan a synthesis of more than one step, we use the process of **retrosynthetic analysis**—that is, working backwards from the desired product to determine the starting materials from which it is made (Section 10.18). To write a synthesis working backwards from the product to the starting material, an **open arrow** (⇒) is used to indicate that the product is drawn on the left and the starting material on the right.

Carefully read the directions for each synthesis problem. Sometimes a starting material is specified, whereas at other times you must begin with a compound that meets a particular criterion; for example, you may be asked to synthesize a compound from alcohols having five or fewer carbon atoms. These limitations are meant to give you some direction in planning a multistep synthesis.

The product of a synthesis is often called the **target compound.** Using retrosynthetic analysis, we must determine what compound can be converted to the target compound by a single reaction. That is, **what is the immediate precursor of the target compound?** After an appropriate precursor is identified, this process is continued until we reach a specified starting material. Sometimes multiple retrosynthetic pathways are examined before a particular route is decided upon.

Target compound $\Longrightarrow$ **Precursor** $\Longrightarrow$ **Precursor** $\Longrightarrow$ **Starting material**

(Target compound: **final product**; $\Longrightarrow$: open arrow)

In designing a synthesis, reactions are often divided into two categories:

- **Those that form new carbon–carbon bonds.**
- **Those that convert one functional group into another—that is, functional group interconversions.**

Appendix D lists the carbon–carbon bond-forming reactions encountered in this text.

Carbon–carbon bond-forming reactions are central to organic synthesis because simpler and less valuable starting materials can be converted to more complex products. Keep in mind that whenever the product of a synthesis has more carbon–carbon bonds than the starting material, the synthesis must contain at least one of these reactions.

How To Develop a Retrosynthetic Analysis

Step [1] Compare the carbon skeletons of the starting material and product.

- If the product has more carbon–carbon σ bonds than the starting material, the synthesis must form one or more C–C bonds. If not, only functional group interconversion occurs.
- Match the carbons in the starting material with those in the product, to see where new C–C bonds must be added or where functional groups must be changed.

Step [2] Concentrate on the functional groups in the starting material and product and ask:

- What methods introduce the functional groups in the product?
- What kind of reactions does the starting material undergo?

Step [3] Work backwards from the product and forwards from the starting material.

- Ask: What is the immediate precursor of the product?
- Compare each precursor to the starting material to determine if there is a one-step reaction that converts one to the other. Continue this process until the starting material is reached.
- Always generate simpler precursors when working backwards.
- Use fewer steps when multiple routes are possible.
- Keep in mind that you may need to evaluate several different precursors for a given compound.

Step [4] Check the synthesis by writing it in the synthetic direction.

- To check a retrosynthetic analysis, write out the steps beginning with the starting material, indicating all necessary reagents.

11.12B Examples of Multistep Synthesis

Retrosynthetic analysis with acetylide anions is illustrated in Sample Problems 11.7 and 11.8.

Sample Problem 11.7 Devise a synthesis of $HC\equiv CCH_2CH_2CH_3$ from $HC\equiv CH$ and any other organic or inorganic reagents.

Retrosynthetic Analysis

The two C's in the starting material match up with the two *sp* hybridized C's in the product, so a three-carbon unit must be added.

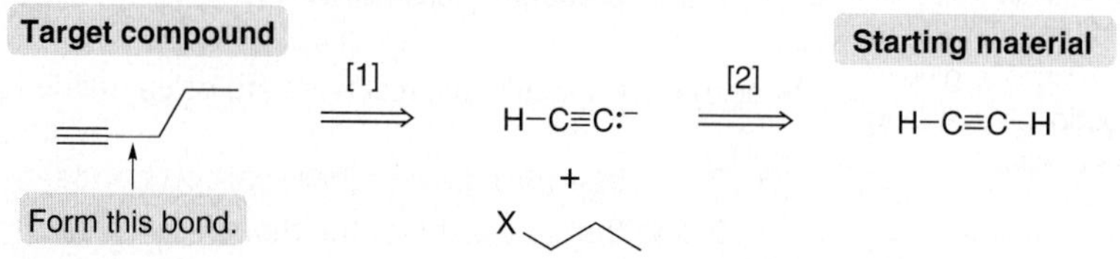

Target compound
Starting material
[1]
[2]
H–C≡C:⁻
+
X
H–C≡C–H
Form this bond.

Thinking backwards . . .

[1] Form a new C–C bond using an acetylide anion and a 1° alkyl halide.
[2] Prepare the acetylide anion from acetylene by treatment with base.

Synthesis

Deprotonation of $HC\equiv CH$ with NaH forms the acetylide anion, which undergoes S_N2 reaction with an alkyl halide to form the target compound, a five-carbon alkyne.

A two-step process:

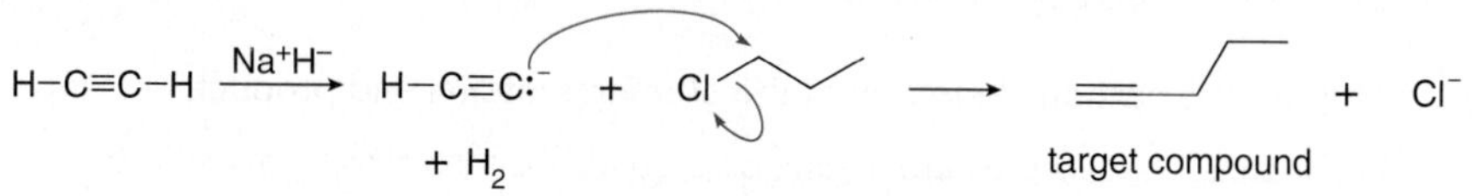

H–C≡C–H
Na⁺H⁻
H–C≡C:⁻
+
Cl
+ H₂
+
Cl⁻
target compound

Sample Problem 11.8 Devise a synthesis of the following compound from starting materials having two or fewer carbons.

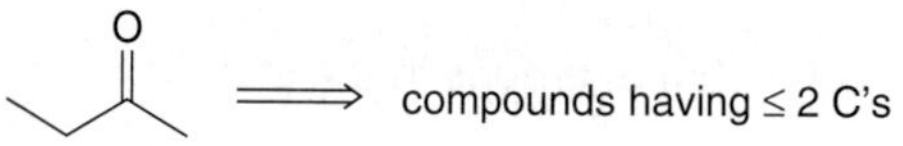

O
compounds having ≤ 2 C's

Retrosynthetic Analysis

A carbon–carbon bond-forming reaction must be used to convert the two-carbon starting materials to the four-carbon product.

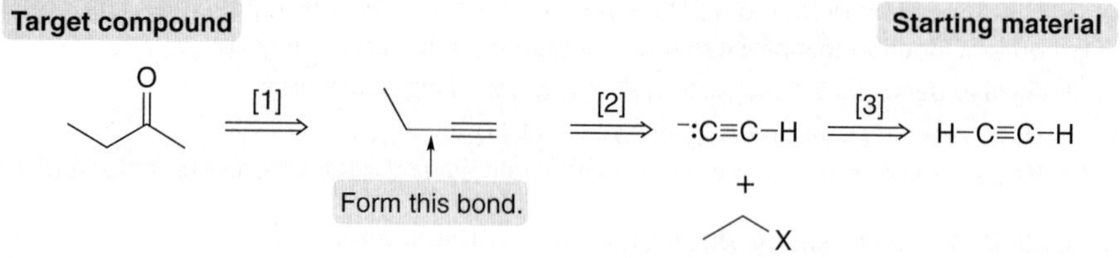

Target compound
Starting material
O
[1]
[2]
[3]
⁻:C≡C–H
H–C≡C–H
+
X
Form this bond.

Thinking backwards . . .

[1] Form the carbonyl group by hydration of a triple bond.
[2] Form a new C–C bond using an acetylide anion and a 1° alkyl halide.
[3] Prepare the acetylide anion from acetylene by treatment with base.

Synthesis

Three steps are needed to complete the synthesis. Treatment of HC≡CH with NaH forms the acetylide anion, which undergoes an S_N2 reaction with an alkyl halide to form a four-carbon terminal alkyne. Hydration of the alkyne with H_2O, H_2SO_4, and $HgSO_4$ yields the target compound.

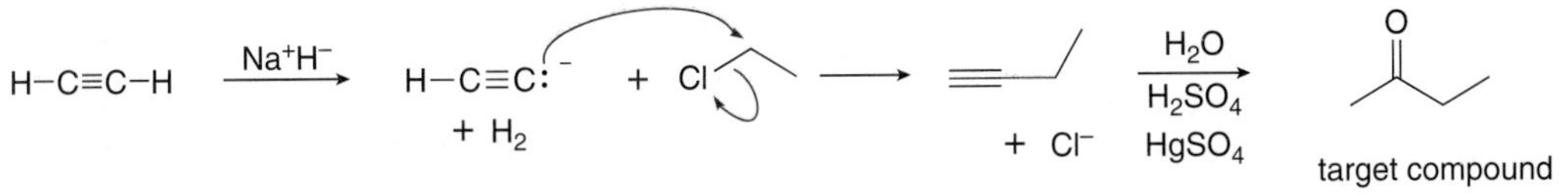

These examples illustrate the synthesis of organic compounds by multistep routes. In Chapter 12 we will learn other useful reactions that expand our capability to do synthesis.

Problem 11.21 Use retrosynthetic analysis to show how hex-3-yne can be prepared from acetylene and any other organic and inorganic compounds. Then draw the synthesis in the synthetic direction, showing all needed reagents.

Problem 11.22 Devise a synthesis of $CH_3CH_2CH_2CHO$ from two-carbon starting materials.

KEY CONCEPTS

Alkynes

General Facts About Alkynes

- Alkynes contain a carbon–carbon triple bond consisting of a strong σ bond and two weak π bonds. Each carbon is *sp* hybridized and linear (11.1).
- Alkynes are named using the suffix ***-yne*** (11.2).
- Alkynes have weak intermolecular forces, giving them low mp's and low bp's, and making them water insoluble (11.3).
- Because its weaker π bonds make an alkyne electron rich, alkynes undergo addition reactions with electrophiles (11.6).

Addition Reactions of Alkynes

[1] Hydrohalogenation—Addition of HX (X = Cl, Br, I) (11.7)

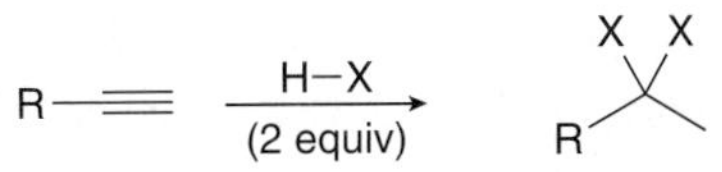

geminal dihalide

- Markovnikov's rule is followed. H bonds to the less substituted C to form the more stable carbocation.

[2] Halogenation—Addition of X_2 (X = Cl or Br) (11.8)

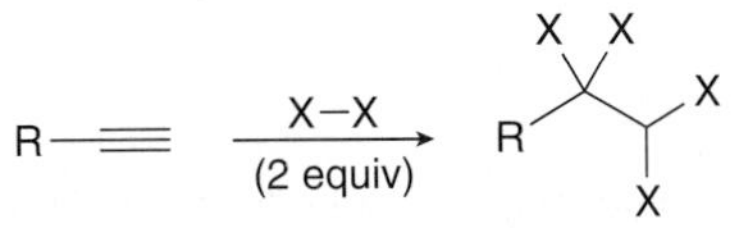

tetrahalide

- Bridged halonium ions are formed as intermediates.
- Anti addition of X_2 occurs.

[3] Hydration—Addition of H_2O (11.9)

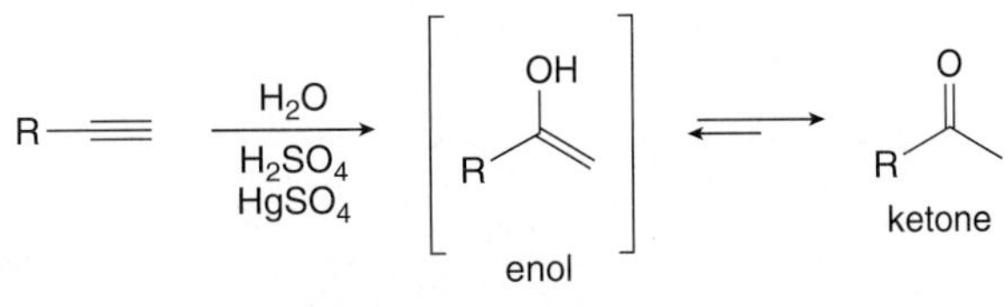

- Markovnikov's rule is followed. H bonds to the less substituted C to form the more stable carbocation.
- An unstable enol is first formed, which rearranges to a carbonyl group.

[4] Hydroboration–oxidation—Addition of H_2O (11.10)

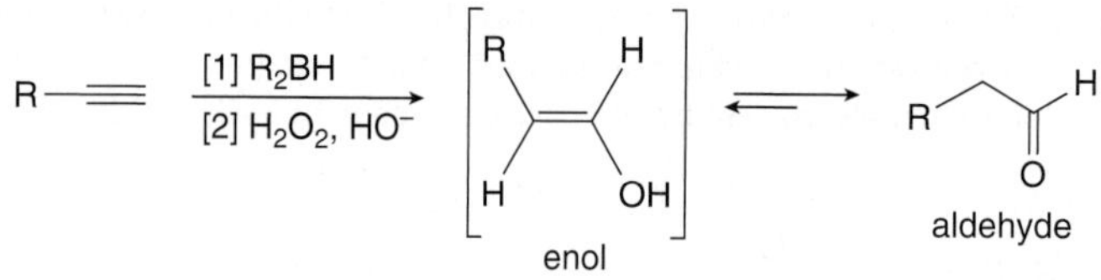
[1] R_2BH
[2] H_2O_2, HO^-
enol
aldehyde

- The unstable enol, first formed after oxidation, rearranges to a carbonyl group.

Reactions Involving Acetylide Anions

[1] Formation of acetylide anions from terminal alkynes (11.6B)

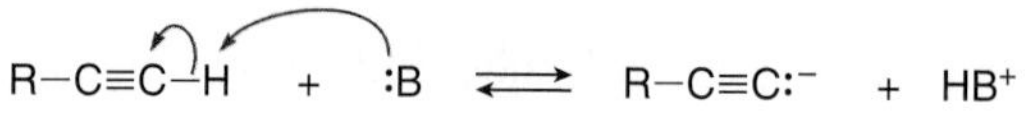

- Typical bases used for the reaction are $NaNH_2$ and NaH.

[2] Reaction of acetylide anions with alkyl halides (11.11A)

- The reaction follows an S_N2 mechanism.
- The reaction works best with CH_3X and RCH_2X.

[3] Reaction of acetylide anions with epoxides (11.11B)

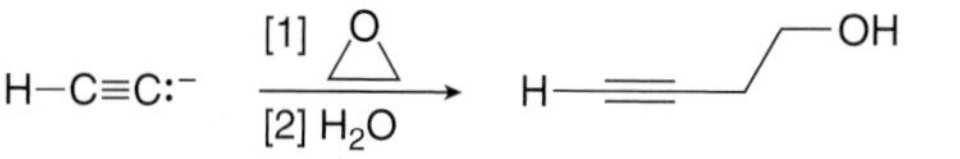
[1]
[2] H_2O

- The reaction follows an S_N2 mechanism.
- Opening of the ring occurs from the back side at the less substituted end of the epoxide.

PROBLEMS

Problems Using Three-Dimensional Models

11.23 Give the IUPAC name for each compound.

a. 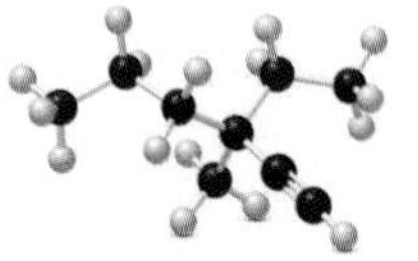

b. 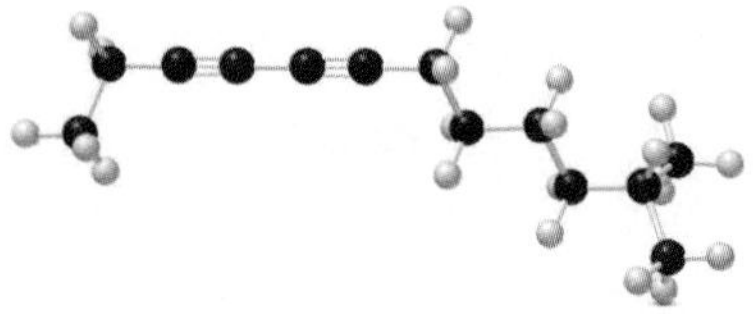

11.24 Draw the enol tautomer of (a) and the keto tautomer of (b).

a. 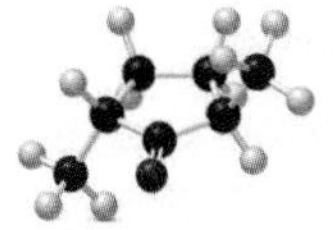

b. 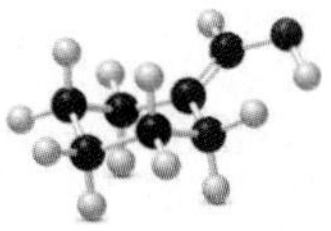

Structure and Nomenclature

11.25 Answer the following questions about erlotinib and phomallenic acid C. Erlotinib, sold under the trade name Tarceva, was introduced in 2004 for the treatment of lung cancer. Phomallenic acid C is an inhibitor of bacterial fatty acid synthesis.

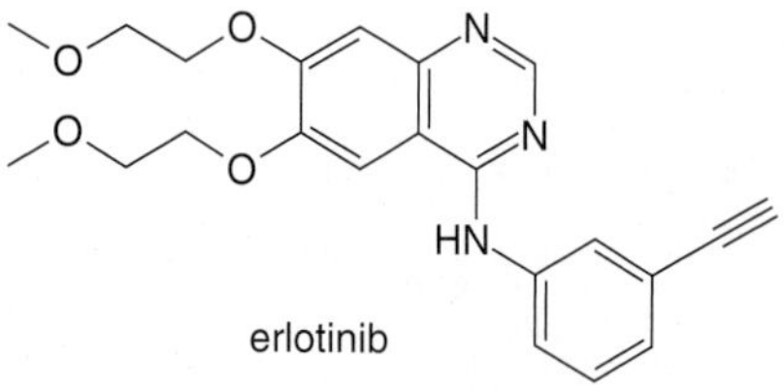

erlotinib

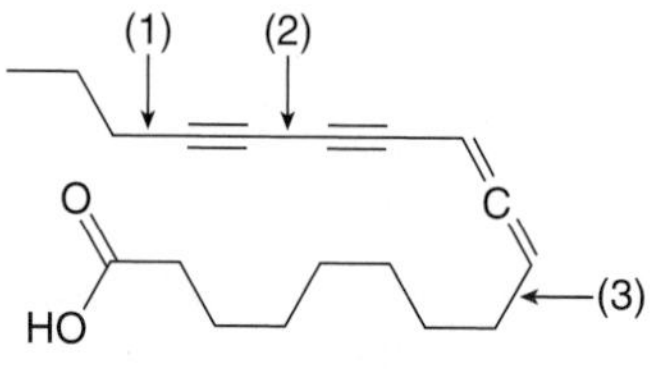
(1)
(2)
(3)

phomallenic acid C

a. Which C–H bond in erlotinib is most acidic?
b. What orbitals are used to form the shortest C–C single bond in erlotinib?
c. Which H atom in phomallenic acid C is most acidic?
d. How many *sp* hybridized carbons are contained in phomallenic acid C?
e. Rank the labeled bonds in phomallenic acid C in order of increasing bond strength.

11.26 Give the IUPAC name for each alkyne.

a. 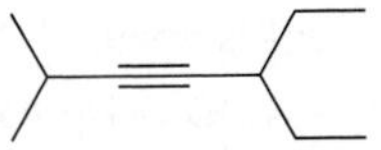

b. 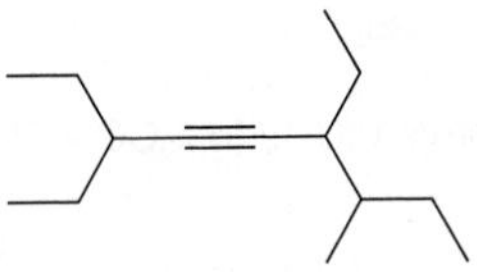

c. 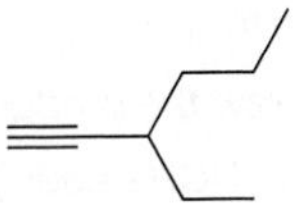

d. 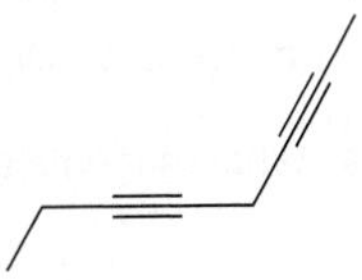

e. 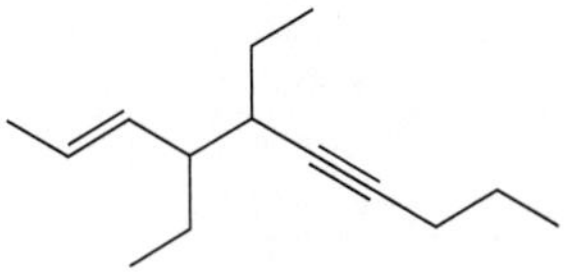

f. 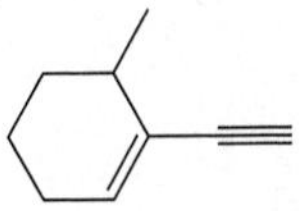

11.27 Give the structure corresponding to each name.

a. 5,6-dimethylhept-2-yne
b. 5-*tert*-butyl-6,6-dimethylnon-3-yne
c. (*S*)-4-chloropent-2-yne
d. *cis*-1-ethynyl-2-methylcyclopentane
e. 3,4-dimethylocta-1,5-diyne
f. (*Z*)-6-methyloct-6-en-1-yne

Tautomers

11.28 Which of the following pairs of compounds represent keto–enol tautomers?

a. 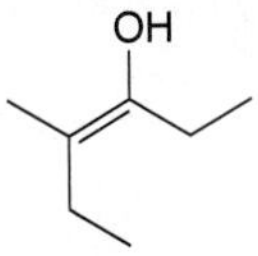 and 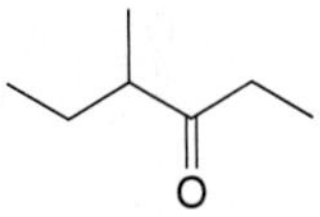

b. 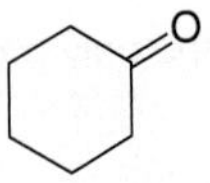 and 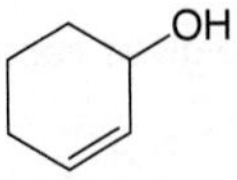

c. 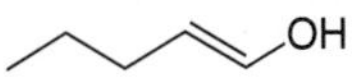 and 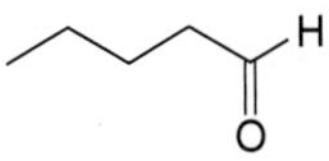

d. 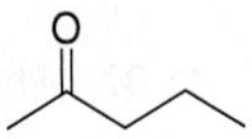 and 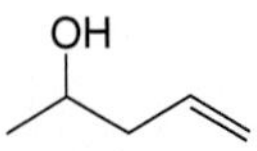

11.29 Draw the enol form of each keto tautomer in parts (a) and (b), and the keto form of each enol tautomer in parts (c) and (d).

a. 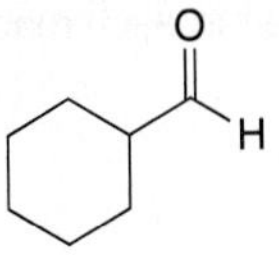

b. 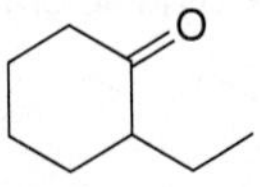

(two different enols)

c. 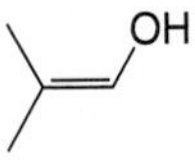

d. 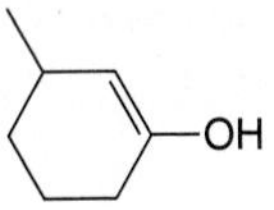

11.30 How is each compound related to **A?** Choose from tautomers, constitutional isomers but not tautomers, or neither.

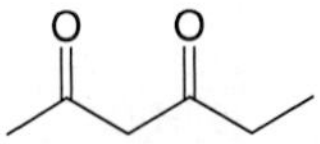

A

a. 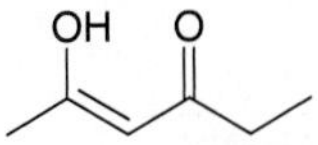

b. 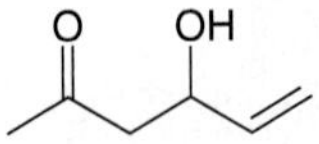

c. 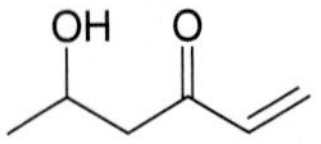

d. 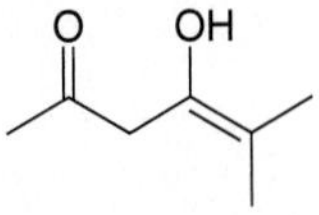

11.31 Ignoring stereoisomers, draw the two possible enols for butan-2-one ($CH_3COCH_2CH_3$), and predict which one is more stable.

11.32 Enamines and imines are tautomers that contain N atoms. Draw a stepwise mechanism for the acid-catalyzed conversion of enamine **X** to imine **Y.**

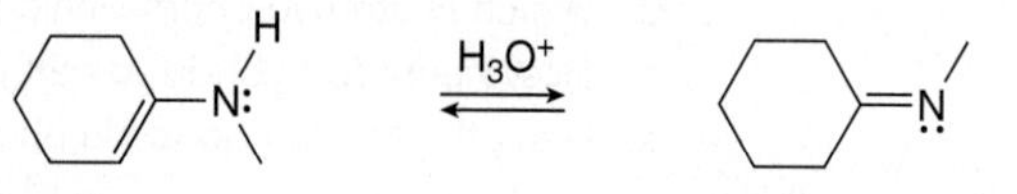

X enamine **Y** imine

Reactions

11.33 Draw the products formed when hex-1-yne is treated with each reagent.

a. HCl (2 equiv)
b. HBr (2 equiv)
c. Cl_2 (2 equiv)
d. $H_2O + H_2SO_4 + HgSO_4$
e. [1] R_2BH; [2] H_2O_2, HO^-
f. NaH
g. [1] $^-NH_2$; [2] CH_3CH_2Br
h. [1] $^-NH_2$; [2]  ; [3] H_2O

11.34 What reagents are needed to convert $(CH_3CH_2)_3CC\equiv CH$ to each compound?

a. 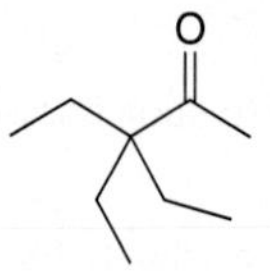 b. 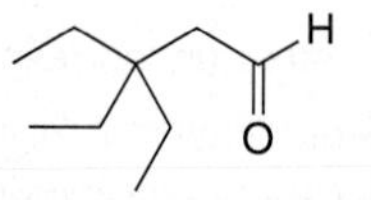 c. 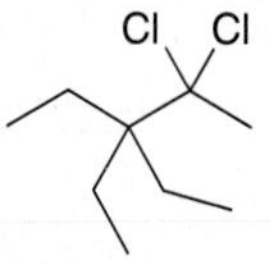 d. 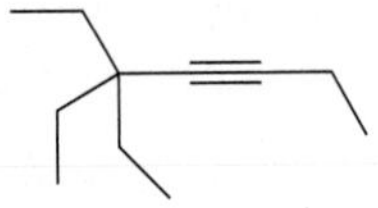

11.35 Explain the apparent paradox. Although the addition of one equivalent of HX to an alkyne is more exothermic than the addition of HX to an alkene, an alkene reacts faster with HX.

11.36 What alkynes give each of the following ketones as the only product after hydration with H_2O, H_2SO_4, and $HgSO_4$?

a. 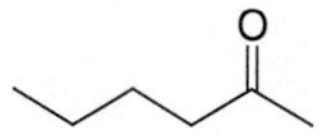 b. 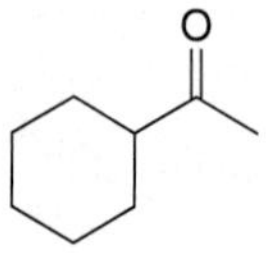 c. 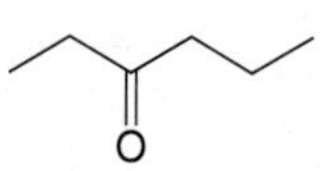 d. 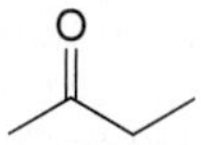

11.37 What alkyne gives each compound as the only product after hydroboration–oxidation?

a. 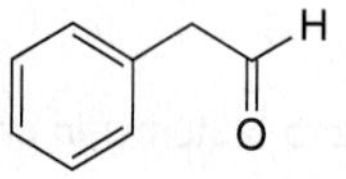 b. 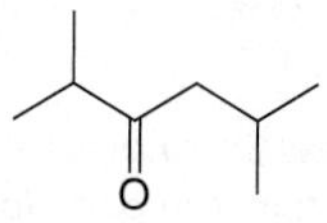

11.38 Draw the organic products formed in each reaction.

a. 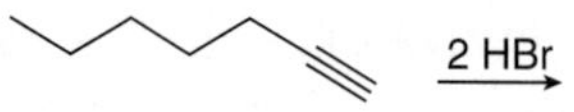

b. 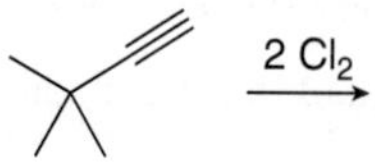

c. 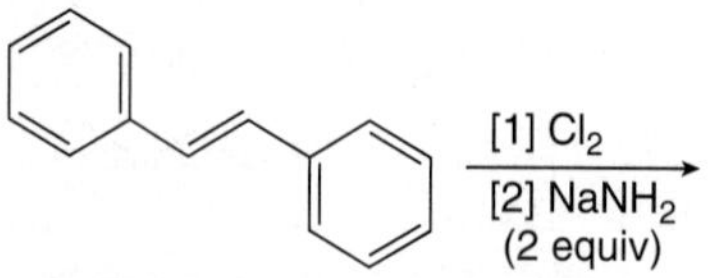

d. 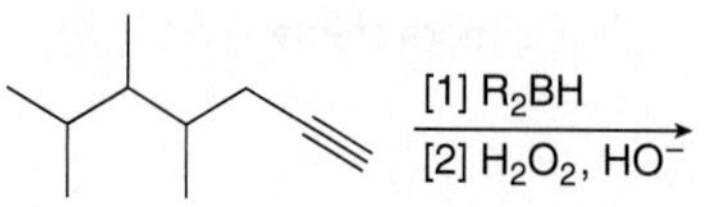

e. $HC\equiv C^- + D_2O \longrightarrow$

f. 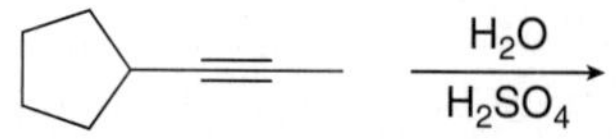

g. 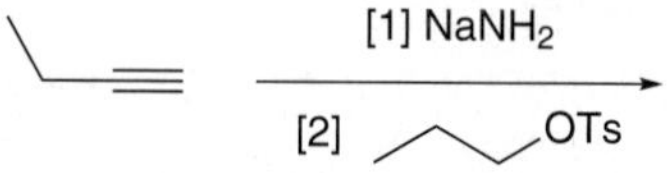

h. 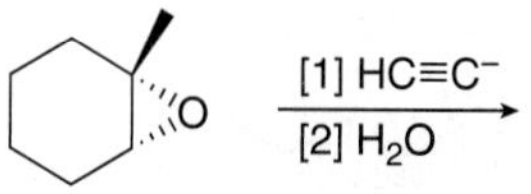

i. 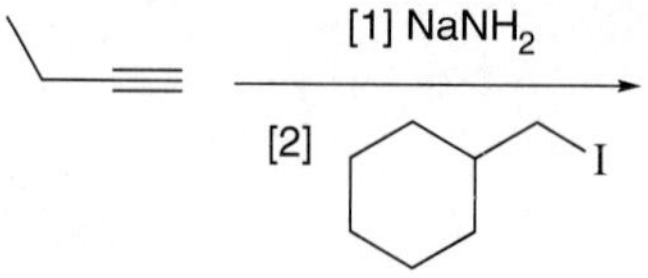

j. 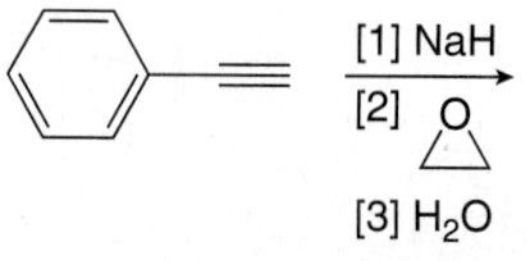

11.39 Draw the structure of compounds **A–E** in the following reaction scheme.

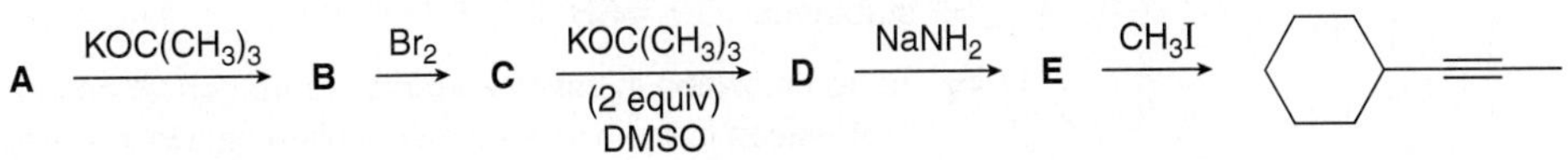
A
KOC(CH3)3
B
Br2
C
KOC(CH3)3
(2 equiv)
DMSO
D
NaNH2
E
CH3I

11.40 When alkyne **A** is treated with $NaNH_2$ followed by CH_3I, a product having molecular formula $C_6H_{10}O$ is formed, but it is *not* compound **B.** What is the structure of the product and why is it formed?

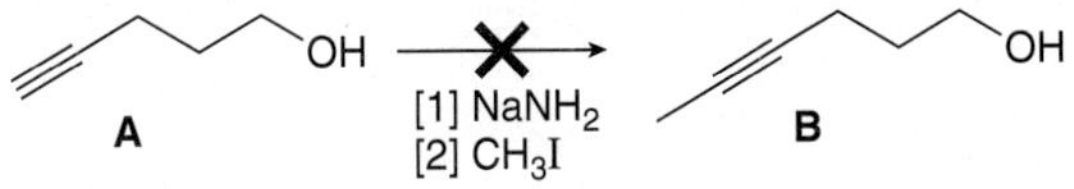
OH
A
[1] NaNH2
[2] CH3I
OH
B

11.41 Draw the products formed in each reaction and indicate stereochemistry.

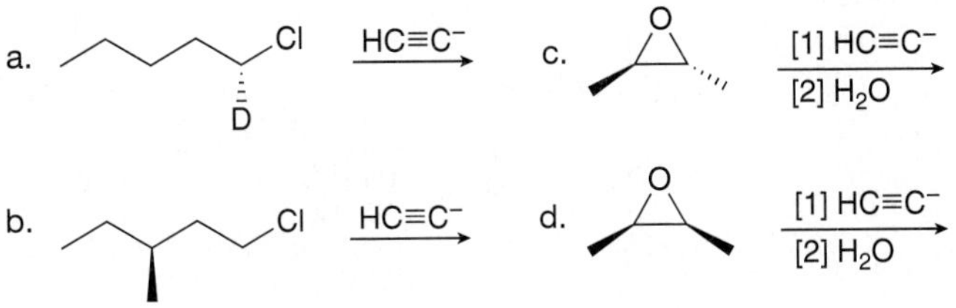
a.
Cl
D
HC≡C⁻
c.
O
[1] HC≡C⁻
[2] H2O
b.
Cl
HC≡C⁻
d.
O
[1] HC≡C⁻
[2] H2O

11.42 What reactions are needed to convert alcohol **A** into either alkyne **B** or alkyne **C?**

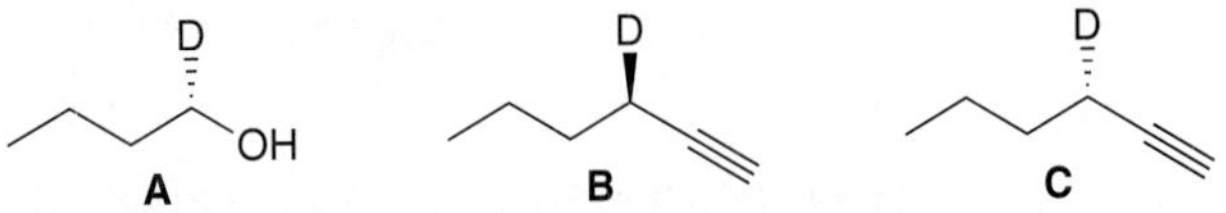
D
OH
A
D
B
D
C

11.43 Identify the lettered compounds in the following reaction schemes. Each reaction sequence was used in the synthesis of a natural product.

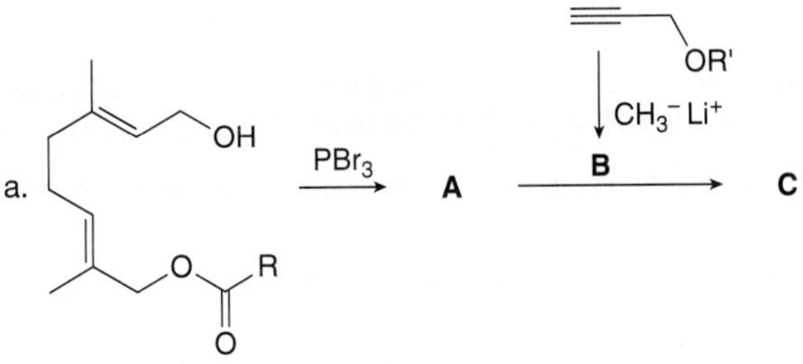
a.
OH
O
R
O
PBr3
A
OR'
CH3⁻ Li⁺
B
C

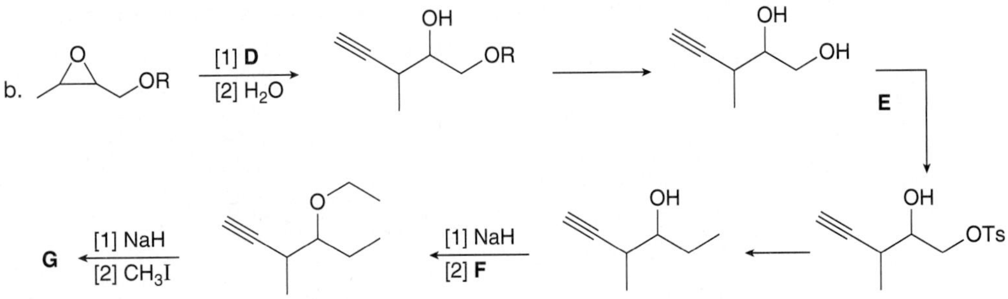
b.
O
OR
[1] D
[2] H2O
OH
OR
OH
OH
E
OH
OTs
OH
[1] NaH
[2] F
O
[1] NaH
[2] CH3I
G

Mechanisms

11.44 Treatment of 2,2-dibromobutane with two equivalents of strong base affords but-1-yne and but-2-yne, as well as a small amount of buta-1,2-diene. Draw a mechanism showing how each compound is formed. Which alkyne should be the major product?

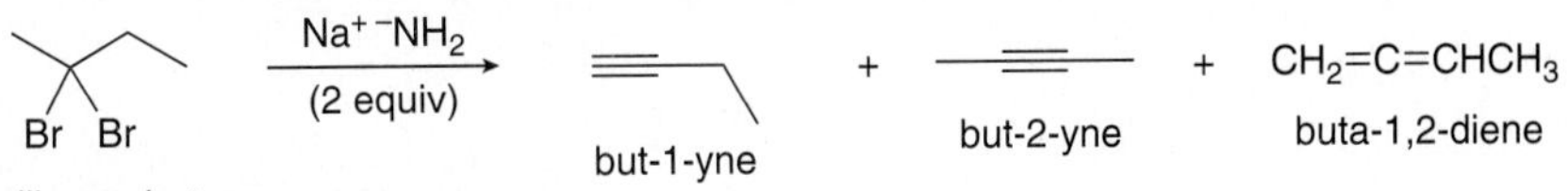
Br Br
Na⁺ ⁻NH2
(2 equiv)
but-1-yne
+
but-2-yne
+
CH2=C=CHCH3
buta-1,2-diene

2,2-dibromobutane

11.45 Explain this statement: Although $HC{\equiv}C^-$ is more stable than $CH_2{=}CH^-$, $HC{\equiv}C^+$ is less stable than $CH_2{=}CH^+$.

11.46 Tautomerization in base resembles tautomerization in acid, but deprotonation precedes protonation in the two-step mechanism. (a) Draw a stepwise mechanism for the following tautomerization. (b) Then draw a stepwise mechanism for the reverse reaction, the conversion of the keto form to the enol.

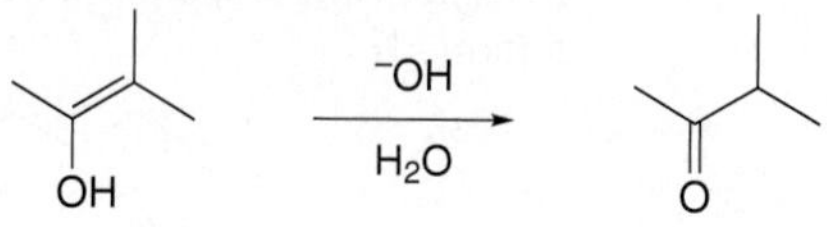

OH
$^-$OH
H_2O
O

enol form keto form

11.47 Draw a stepwise mechanism for each reaction.

a. 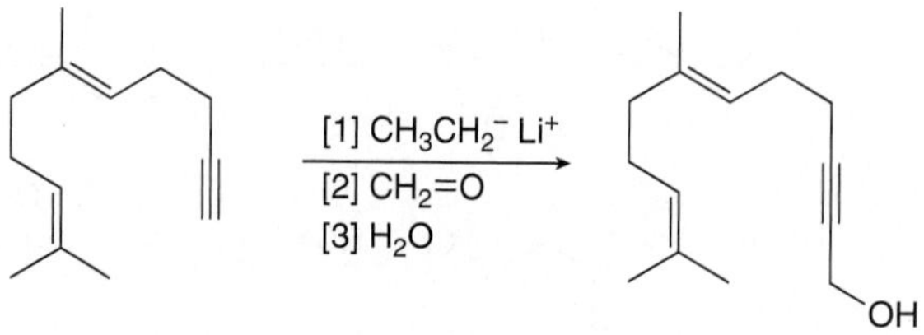

[1] $CH_3CH_2^-$ Li^+
[2] $CH_2{=}O$
[3] H_2O
OH

b. 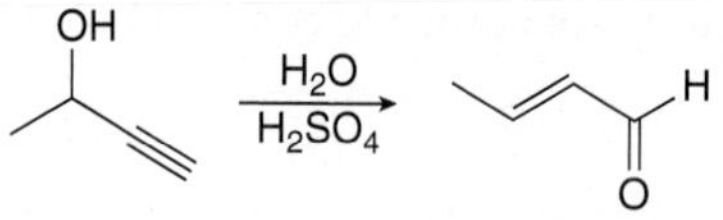

OH
H_2O
H_2SO_4
H
O

11.48 From what you have learned about enols and the hydration of alkynes, predict what product is formed by the acid-catalyzed hydration of $CH_3CH_2CH_2C{\equiv}COCH_3$. Draw a stepwise mechanism that illustrates how it is formed.

Synthesis

11.49 What steps are needed to prepare phenylacetylene, $C_6H_5C{\equiv}CH$, from each compound: (a) $C_6H_5CH_2CHBr_2$; (b) $C_6H_5CHBrCH_3$; (c) $C_6H_5CH_2CH_2OH$?

11.50 What acetylide anion and alkyl halide are needed to synthesize each alkyne?

a. 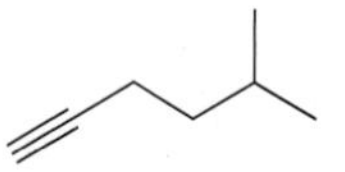 b. 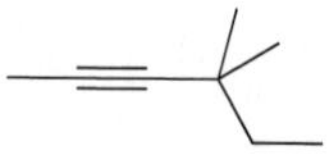 c. 

11.51 Synthesize each compound from acetylene. You may use any other organic or inorganic reagents.

a. 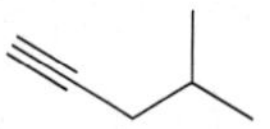 c. 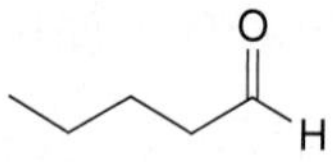

O
H

e. 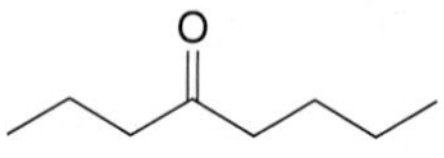
O

b. 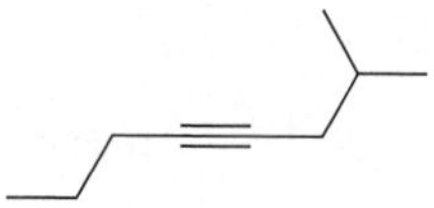 d. 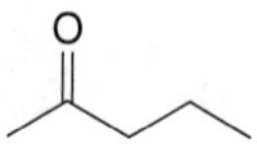
O

11.52 Devise a synthesis of each compound using $CH_3CH_2CH{=}CH_2$ as the starting material. You may use any other organic compounds or inorganic reagents.

a. 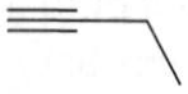 c. 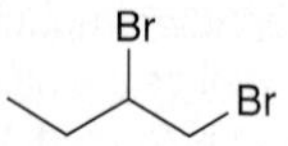

Br
Br

e. 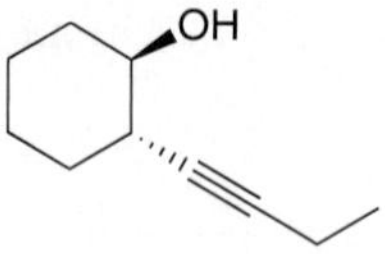
OH

(+ enantiomer)

b. 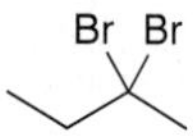

Br Br

d. 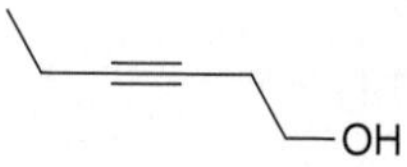
OH

11.53 Devise a synthesis of the following compound from cyclohexene and acetylene. You may use any other inorganic reagents.

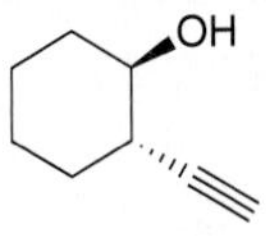

+ enantiomer

11.54 Devise a synthesis of each compound. You may use HC≡CH, ethylene oxide, and alkyl halides as organic starting materials and any inorganic reagents.

a. 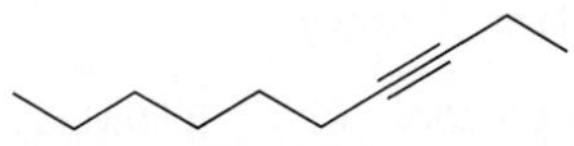

b. 

11.55 Devise a synthesis of the ketone hexan-3-one, $CH_3CH_2COCH_2CH_2CH_3$, from CH_3CH_2Br as the only organic starting material; that is, all the carbon atoms in hexan-3-one must come from CH_3CH_2Br. You may use any other needed reagents.

11.56 Devise a synthesis of each compound using $CH_3CH_2CH_2OH$ as the only organic starting material: (a) $CH_3C{\equiv}CCH_2CH_2CH_3$; (b) $CH_3C{\equiv}CCH_2CH(OH)CH_3$. You may use any other needed inorganic reagents.

11.57 Devise a synthesis of $CH_3CH_2C{\equiv}CCH_2CH_2OH$ from CH_3CH_2OH as the only organic starting material. You may use any other needed reagents.

Challenge Problems

11.58 Explain why the C=C of an enol is more nucleophilic than the C=C of an alkene, despite the fact that the electronegative oxygen atom of the enol inductively withdraws electron density from the carbon–carbon double bond.

11.59 *N*-Chlorosuccinimide (NCS) serves as a source of Cl^+ in electrophilic addition reactions to alkenes and alkynes. Keeping this in mind, draw a stepwise mechanism for the following addition to but-2-yne.

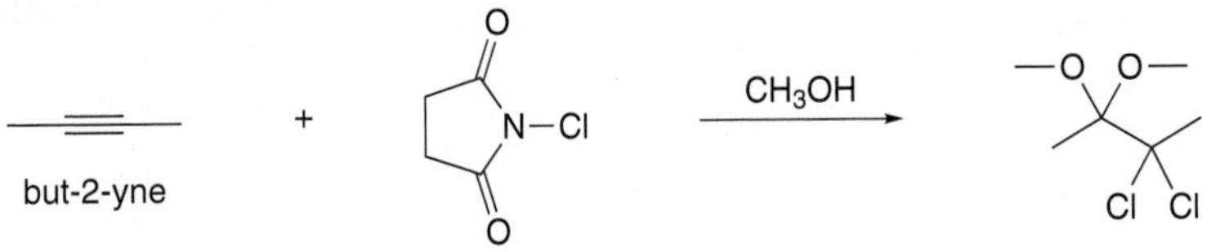

but-2-yne

N-chlorosuccinimide

11.60 Draw a stepwise mechanism for the following reaction.

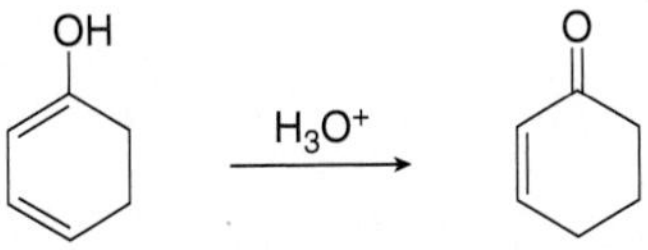

11.61 Draw a stepwise mechanism for the following reaction.

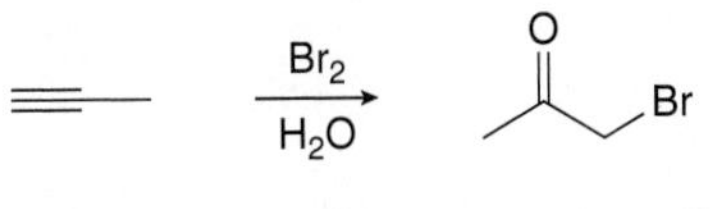

11.62 Why is compound **X** formed in the following reaction, instead of its constitutional isomer **Y**?

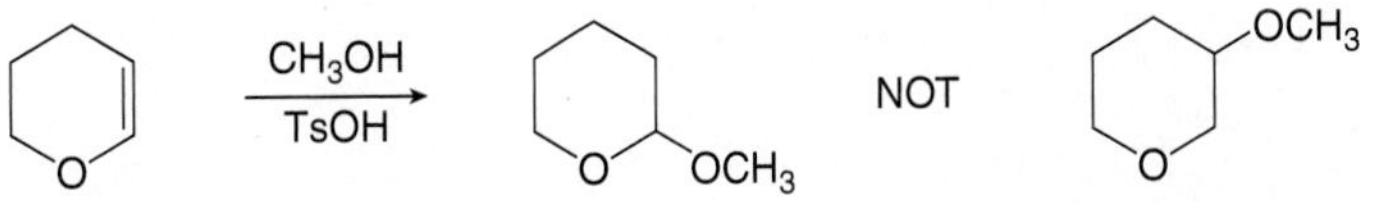

X NOT **Y**

11.63 Write a stepwise mechanism for each of the following reactions. Explain why a more stable alkyne (but-2-yne) is isomerized to a less stable alkyne (but-1-yne), but under similar conditions, 2,5-dimethylhex-3-yne forms 2,5-dimethylhexa-2,3-diene.

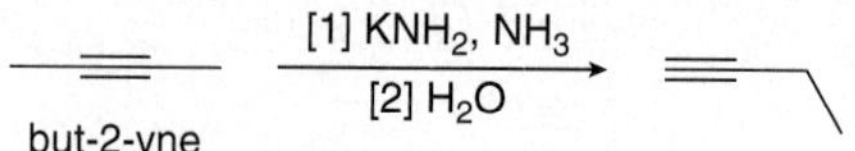
[1] KNH$_2$, NH$_3$
[2] H$_2$O

but-2-yne → but-1-yne

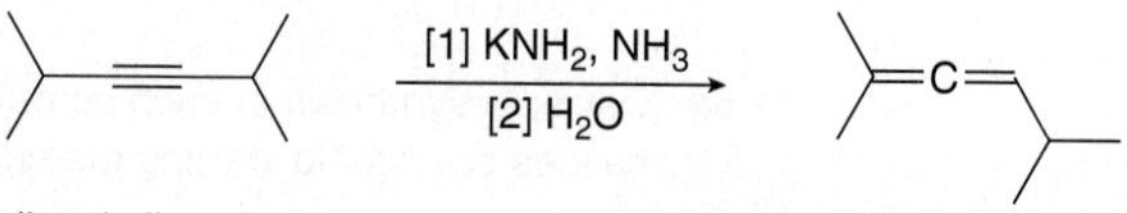
[1] KNH$_2$, NH$_3$
[2] H$_2$O
C

2,5-dimethylhex-3-yne → 2,5-dimethylhexa-2,3-diene

11.64 Draw a stepwise mechanism for the following intramolecular reaction.

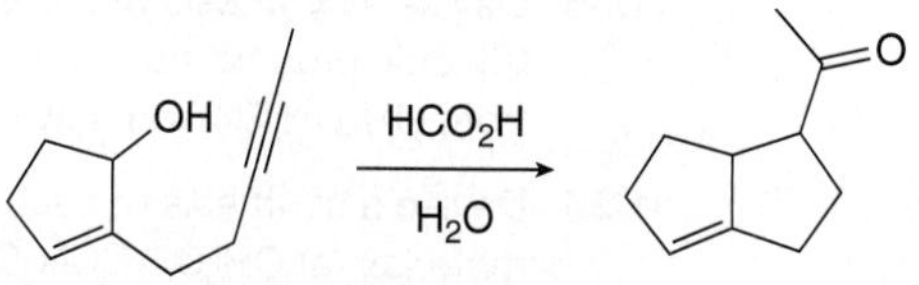
OH
HCO$_2$H
H$_2$O
O

11.65 Explain why an optically active solution of (*R*)-α-methylbutyrophenone loses its optical activity when dilute acid is added to the solution.

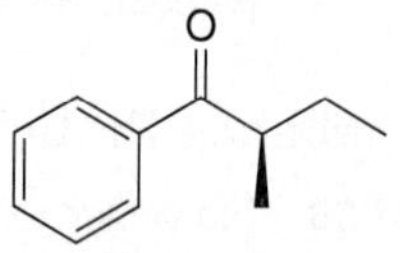
O

(*R*)-α-methylbutyrophenone

Oxidation and Reduction

12

12.1 Introduction
12.2 Reducing agents
12.3 Reduction of alkenes
12.4 Application: Hydrogenation of oils
12.5 Reduction of alkynes
12.6 The reduction of polar C–X σ bonds
12.7 Oxidizing agents
12.8 Epoxidation
12.9 Dihydroxylation
12.10 Oxidative cleavage of alkenes
12.11 Oxidative cleavage of alkynes
12.12 Oxidation of alcohols
12.13 Green chemistry
12.14 Biological oxidation
12.15 Sharpless epoxidation

Throughout history, humans have ingested alcoholic beverages for their pleasant taste and the feeling of euphoria they impart. Wine, beer, and similar products contain **ethanol** (CH_3CH_2OH), a 1° alcohol that is quickly absorbed in the stomach and small intestines and rapidly transported in the bloodstream to other organs. Like other 1° alcohols, ethanol is easily oxidized, and as a result, ethanol is metabolized in the body by a series of enzyme-catalyzed oxidation reactions that take place in the liver. In Chapter 12, we learn about oxidation and reduction reactions of organic molecules like ethanol.

In Chapter 12, we discuss the oxidation and reduction of **alkenes** and **alkynes,** as well as compounds with **polar C–X σ bonds**—alcohols, alkyl halides, and epoxides. Although there will be many different reagents and mechanisms, discussing these reactions as a group allows us to more easily compare and contrast them.

The word *mechanism* will often be used loosely here. In contrast to the S_N1 reaction of alkyl halides or the electrophilic addition reactions of alkenes, the details of some of the mechanisms presented in Chapter 12 are known with less certainty. For example, although the identity of a particular intermediate might be confirmed by experiment, other details of the mechanism are suggested by the structure or stereochemistry of the final product.

Oxidation and reduction reactions are very versatile, and knowing them allows us to design many more complex organic syntheses.

12.1 Introduction

Recall from Section 4.14 that the way to determine whether an organic compound has been oxidized or reduced is to compare the **relative number of C–H and C–Z bonds** (Z = an element *more electronegative* than carbon) in the starting material and product.

Two components are always present in an oxidation or reduction reaction—**one component is oxidized and one is reduced.** When an organic compound is *oxidized* by a reagent, the reagent itself must be *reduced.* Similarly, when an organic compound is *reduced* by a reagent, the reagent becomes *oxidized.*

- *Oxidation* results in an *increase* in the number of C–Z bonds (usually C–O bonds) *or* a *decrease* in the number of C–H bonds.
- *Reduction* results in a *decrease* in the number of C–Z bonds (usually C–O bonds) *or* an *increase* in the number of C–H bonds.

Thus, an organic compound such as CH_4 can be oxidized by replacing C–H bonds with C–O bonds, as shown in Figure 12.1. Reduction is the opposite of oxidation, so Figure 12.1 also shows how a compound can be reduced by replacing C–O bonds with C–H bonds. The symbols **[O]** and **[H]** indicate oxidation and reduction, respectively.

Figure 12.1
A general scheme for the oxidation and reduction of a carbon compound

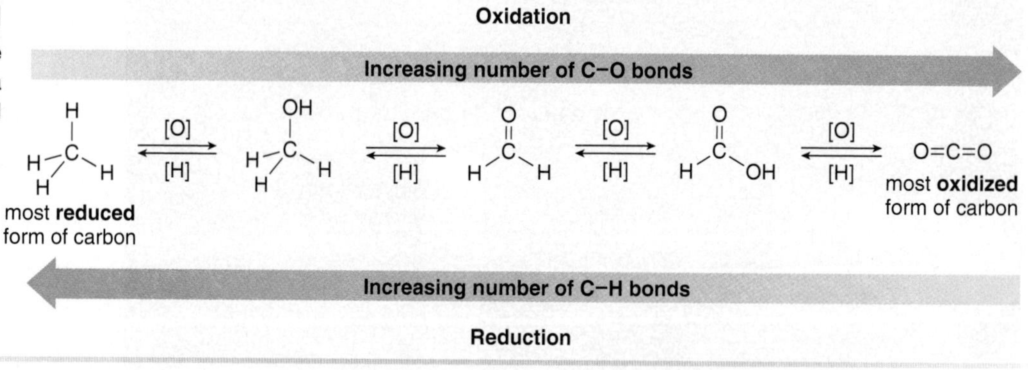

Sometimes two carbon atoms are involved in a single oxidation or reduction reaction, and the net change in the number of C–H or C–Z bonds at *both* atoms must be taken into account. The conversion of an **alkyne to an alkene** and an **alkene to an alkane** are examples of reduction, because each process adds two new C–H bonds to the starting material, as shown in Figure 12.2.

Figure 12.2
Oxidation and reduction of hydrocarbons

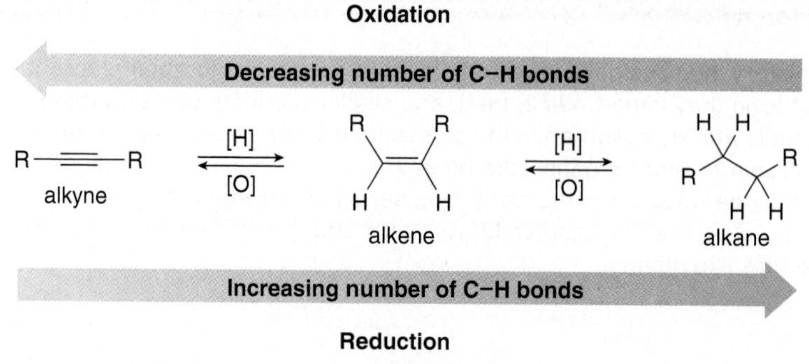

Problem 12.1 Classify each reaction as oxidation, reduction, or neither.

a. 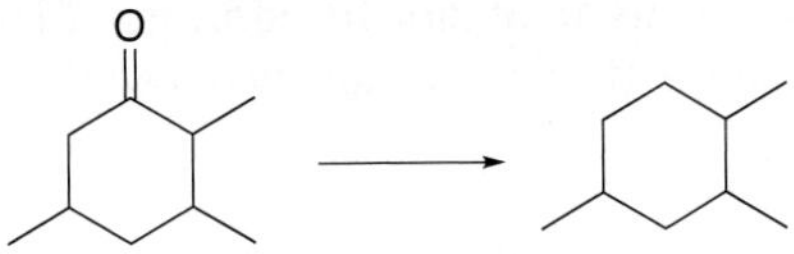

c. 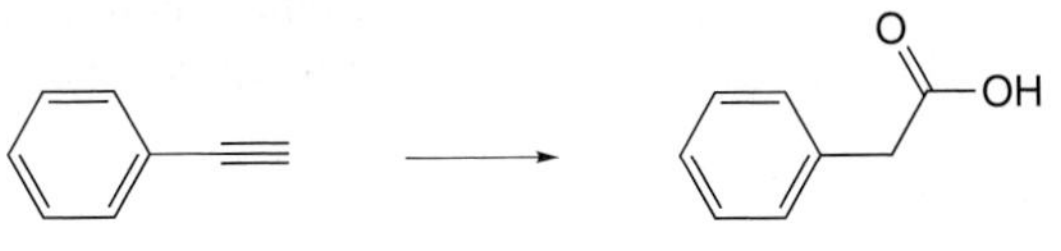

b. 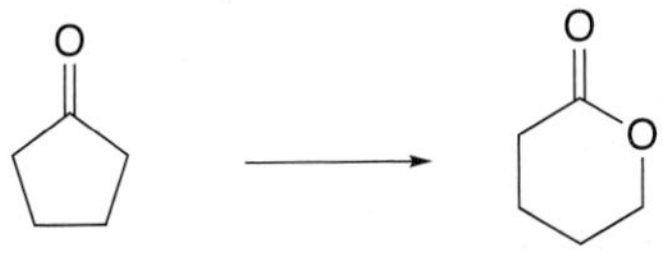

d. 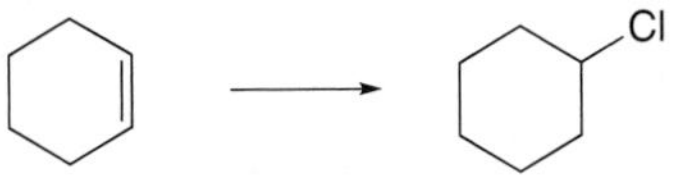

12.2 Reducing Agents

Reducing agents provide the equivalent of two hydrogen atoms, but **there are three types of reductions,** differing in how H_2 is added. The simplest reducing agent is molecular $\mathbf{H_2}$**.** Reductions of this sort are carried out in the presence of a metal catalyst that acts as a surface on which the reaction occurs.

The second way to deliver H_2 in a reduction is to add two protons and two electrons to a substrate—that is, $\mathbf{H_2 = 2\ H^+ + 2\ e^-}$. Reducing agents of this sort use alkali metals as a source of electrons and liquid ammonia (NH_3) as a source of protons. Reductions with **Na in** $\mathbf{NH_3}$ are called **dissolving metal reductions.**

$$2\ Na \longrightarrow 2\ Na^+ + 2\ e^-$$

$$2\ NH_3 \longrightarrow 2\ {}^-NH_2 + 2\ H^+$$

$2\ e^-$ and $2\ H^+$: an equivalent of H_2 for reduction

The third way to deliver the equivalent of two hydrogen atoms is to add **hydride ($\mathbf{H^-}$)** and a **proton ($\mathbf{H^+}$).** The most common hydride reducing agents contain a hydrogen atom bonded to boron or aluminum. Simple examples include **sodium borohydride ($\mathbf{NaBH_4}$)** and **lithium aluminum hydride ($\mathbf{LiAlH_4}$).** These reagents deliver H^- to a substrate, and then a proton is added from H_2O or an alcohol.

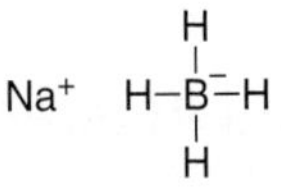

sodium borohydride
$\mathbf{NaBH_4}$

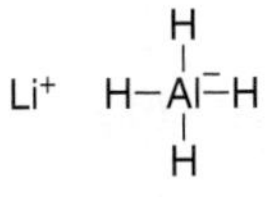

lithium aluminum hydride
$\mathbf{LiAlH_4}$

- **Metal hydride reagents act as a source of $\mathbf{H^-}$ because they contain polar metal–hydrogen bonds that place a partial negative charge on hydrogen.**

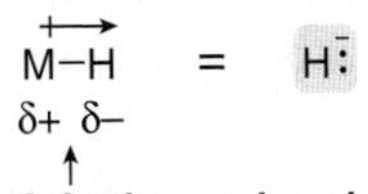

a polar metal–hydrogen bond

M = B or Al

12.3 Reduction of Alkenes

Reduction of an alkene forms an alkane by addition of H_2. Two bonds are broken—the **weak π bond** of the alkene and the H_2 σ bond—and two new C−H σ bonds are formed.

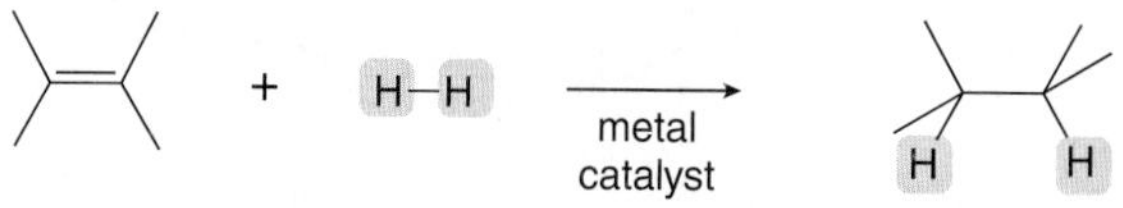

A π bond is broken. Two C–H σ bonds are formed.

> Hydrogenation catalysts are insoluble in common solvents, thus creating a **heterogeneous** reaction mixture. This insolubility has a practical advantage. These catalysts contain expensive metals, but they can be filtered away from the other reactants after the reaction is complete, and then reused.

The addition of H_2 occurs only in the presence of a **metal catalyst,** and thus, the reaction is called **catalytic hydrogenation.** The catalyst consists of a metal—usually Pd, Pt, or Ni—adsorbed onto a finely divided inert solid, such as charcoal. For example, the catalyst 10% Pd on carbon is composed of 10% Pd and 90% carbon, by weight. H_2 adds in a **syn** fashion, as shown in Equation [2].

[1]
[2]

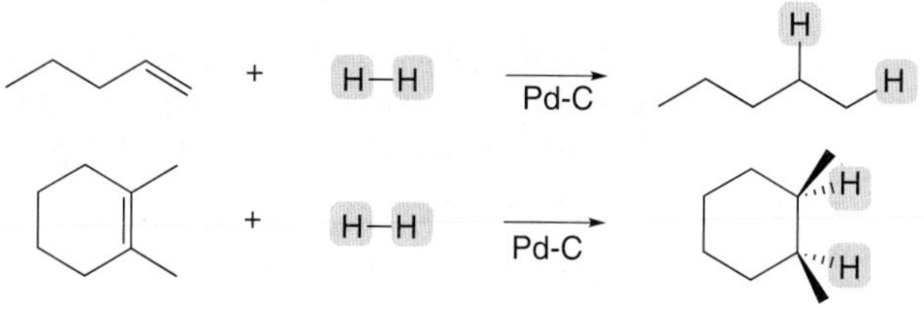

syn addition

Problem 12.2 What alkane is formed when each alkene is treated with H_2 and a Pd catalyst?

a. 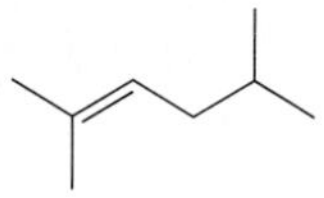 b. 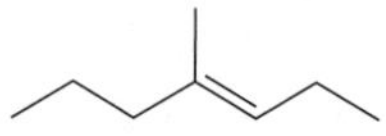 c. 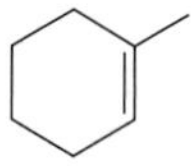

Problem 12.3 Draw all alkenes that react with one equivalent of H_2 in the presence of a palladium catalyst to form each alkane. Consider constitutional isomers only.

a. 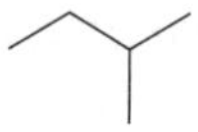 b. 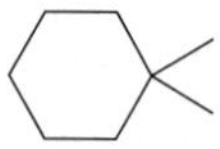 c. 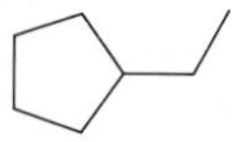

12.3A Hydrogenation and Alkene Stability

Hydrogenation reactions are **exothermic** because the bonds in the product are stronger than the bonds in the starting materials, making them similar to other alkene addition reactions. The $\Delta H°$ for hydrogenation, called the **heat of hydrogenation,** can be used as a measure of the relative stability of two different alkenes that are hydrogenated to the same alkane.

> Recall from Chapter 8 that **trans alkenes are generally more stable than cis alkenes.**

For example, both *cis*- and *trans*-but-2-ene are hydrogenated to butane, and the heat of hydrogenation for the trans isomer is less than that for the cis isomer. **Because less energy is released in converting the trans alkene to butane, it must be lower in energy (more stable) to begin with.** The relative energies of the butene isomers are illustrated in Figure 12.3.

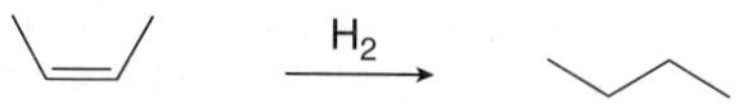

cis-but-2-ene → butane $\Delta H° = -120$ kJ/mol

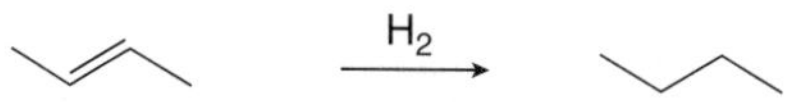

trans-but-2-ene → butane $\Delta H° = -115$ kJ/mol

Less energy is released.

more stable starting material

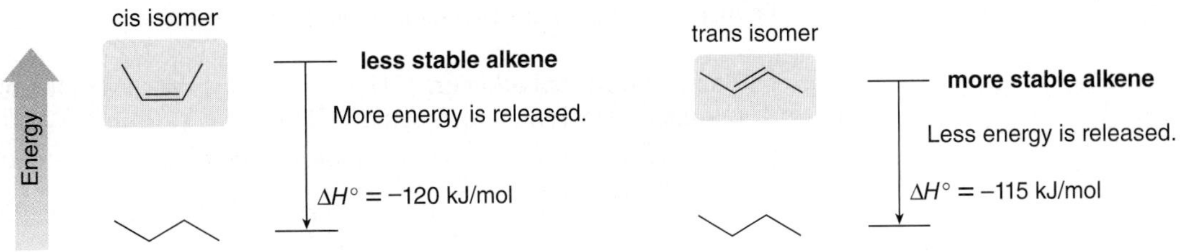

Figure 12.3
Relative energies of *cis*- and *trans*-but-2-ene

- **When hydrogenation of two alkenes gives the same alkane, the more stable alkene has the *smaller* heat of hydrogenation.**

Problem 12.4 Which alkene in each pair has the larger heat of hydrogenation?

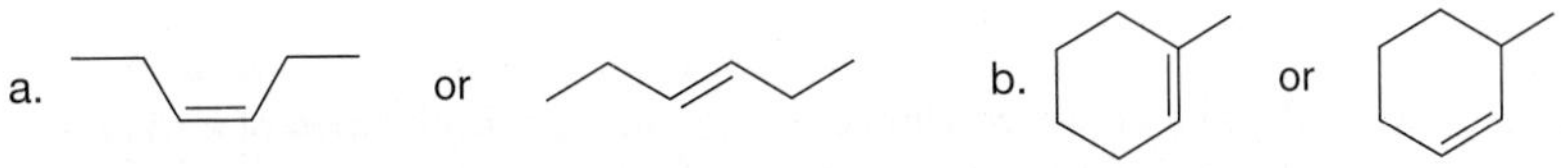

Problem 12.5 Explain why heats of hydrogenation cannot be used to determine the relative stability of 2-methylpent-2-ene and 3-methylpent-1-ene.

12.3B The Mechanism of Catalytic Hydrogenation

In the generally accepted mechanism for catalytic hydrogenation, the surface of the metal catalyst binds both H_2 and the alkene, and H_2 is transferred to the π bond in a rapid but stepwise process (Mechanism 12.1).

Mechanism 12.1 Addition of H_2 to an Alkene—Hydrogenation

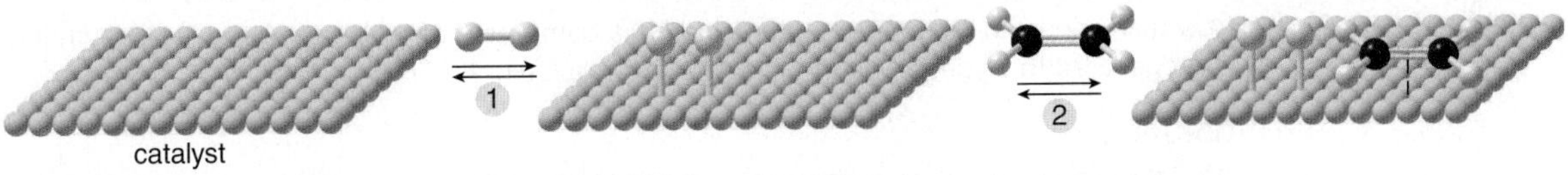

1 **H_2 adsorbs to the catalyst surface** with partial or complete cleavage of the H − H bond.

2 The π bond of the alkene complexes with the metal.

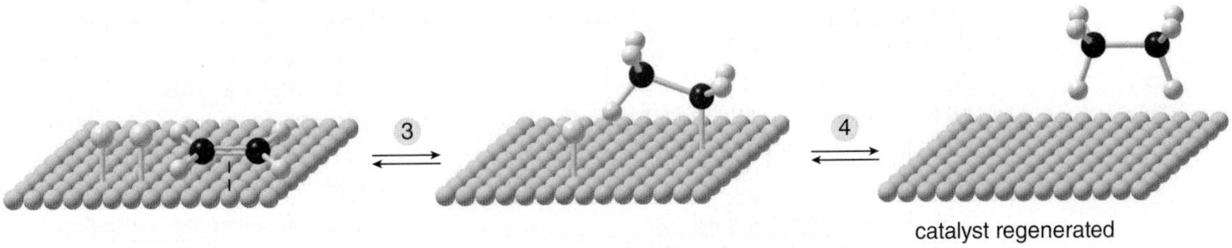

3 − 4 **Two H atoms are transferred sequentially** to the π bond in Steps [3] and [4], forming the alkane. Because the product alkane no longer has a π bond with which to complex to the metal, it is released from the catalyst surface.

The mechanism explains two facts about hydrogenation:

- **Rapid, sequential addition of H_2 occurs from the side of the alkene complexed to the metal surface, resulting in syn addition.**
- **Less crowded double bonds complex more readily to the catalyst surface, resulting in faster reaction.**

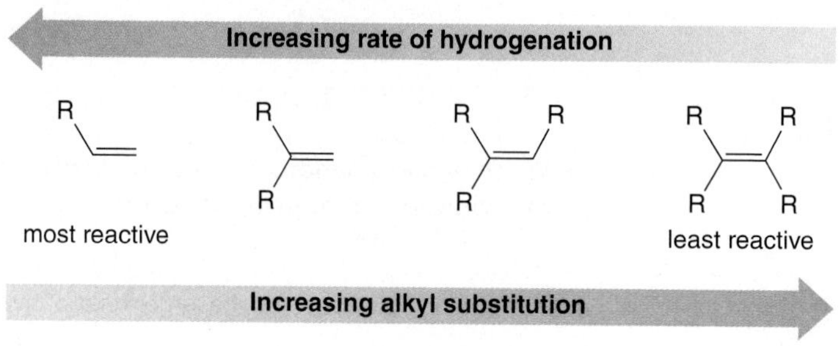

Problem 12.6 Given that syn addition of H_2 occurs from both sides of a trigonal planar double bond, draw all stereoisomers formed when each alkene is treated with H_2.

a. 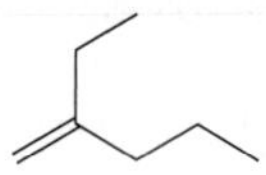 b. 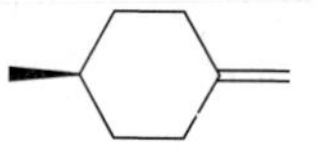 c. 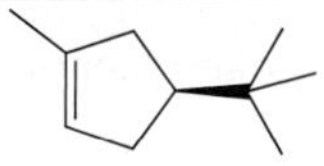

12.3C Hydrogenation Data and Degrees of Unsaturation

Recall from Section 10.2 that the **number of degrees of unsaturation gives the *total* number of rings and π bonds in a molecule.** Because H_2 adds to π bonds but does *not* add to the C–C σ bonds of rings, hydrogenation allows us to determine how many degrees of unsaturation are due to π bonds and how many are due to rings. This is done by comparing the number of degrees of unsaturation before and after a molecule is treated with H_2, as illustrated in Sample Problem 12.1.

Sample Problem 12.1 How many rings and π bonds are contained in a compound of molecular formula C_8H_{12} that is hydrogenated to a compound of molecular formula C_8H_{14}?

Solution

[1] Determine the number of degrees of unsaturation in the compounds before and after hydrogenation.

Before H_2 addition—C_8H_{12}

- The maximum number of H's possible for *n* C's is $2n + 2$; in this example, $2n + 2 = 2(8) + 2 = 18$.
- 18 H's (maximum) – 12 H's (actual) = 6 H's fewer than the maximum number.

$$\frac{\text{6 H's fewer than the maximum}}{\text{2 H's removed for each degree of unsaturation}} =$$

three degrees of unsaturation

After H_2 addition—C_8H_{14}

- The maximum number of H's possible for *n* C's is $2n + 2$; in this example, $2n + 2 = 2(8) + 2 = 18$.
- 18 H's (maximum) – 14 H's (actual) = 4 H's fewer than the maximum number.

$$\frac{\text{4 H's fewer than the maximum}}{\text{2 H's removed for each degree of unsaturation}} =$$

two degrees of unsaturation

[2] Assign the number of degrees of unsaturation to rings or π bonds as follows:

- The number of degrees of unsaturation that remain in the product after H_2 addition = the **number of rings** in the starting material.
- The number of degrees of unsaturation that react with H_2 = the **number of π bonds.**

In this example, **two** degrees of unsaturation remain after hydrogenation, so the starting material has **two** rings. Thus:

Before H_2 addition:		After H_2 addition:			
three degrees of unsaturation	–	**two** degrees of unsaturation	=	**one** degree of unsaturation that reacted with H_2	
three rings or π bonds in C_8H_{12}	=	**two rings**	+	**one π bond**	**ANSWER**

Problem 12.7 Complete the missing information for compounds **A, B,** and **C,** each subjected to hydrogenation. The number of rings and π bonds refers to the reactant (**A, B,** or **C**) prior to hydrogenation.

Compound	Molecular formula before hydrogenation	Molecular formula after hydrogenation	Number of rings	Number of π bonds
A	$C_{10}H_{12}$	$C_{10}H_{16}$	?	?
B	?	C_4H_{10}	0	1
C	C_6H_8	?	1	?

12.3D Hydrogenation of Other Double Bonds

Compounds that contain a carbonyl group also react with H_2 and a metal catalyst. For example, **aldehydes and ketones are reduced to 1° and 2° alcohols,** respectively. We return to this reaction in Chapter 20.

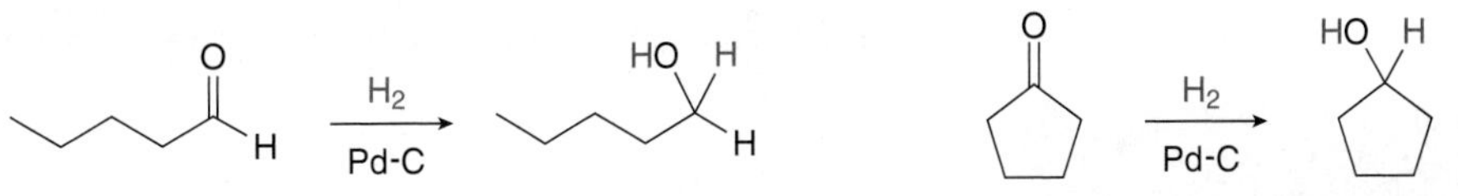

aldehyde 1° alcohol ketone 2° alcohol

12.4 Application: Hydrogenation of Oils

Many processed foods, such as peanut butter, margarine, and some brands of crackers, contain *partially hydrogenated* vegetable oils. These oils are produced by hydrogenating the long hydrocarbon chains of triacylglycerols.

In Section 10.6 we learned that **fats and oils are triacylglycerols that differ in the number of degrees of unsaturation** in their long alkyl side chains.

- **Fats—usually animal in origin—are solids with triacylglycerols having few degrees of unsaturation.**
- **Oils—usually vegetable in origin—are liquids with triacylglycerols having a larger number of degrees of unsaturation.**

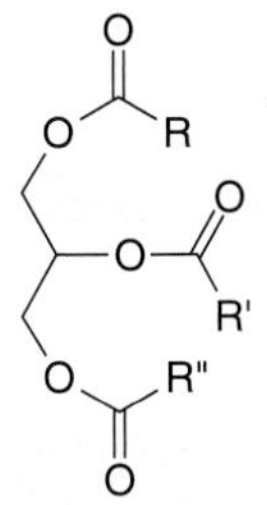

triacylglycerol

The number of double bonds in the R groups of the triacylglycerol determines whether it is a fat or an oil.

Peanut butter is a common consumer product that contains partially hydrogenated vegetable oil.

When an unsaturated vegetable oil is treated with hydrogen, some (or all) of the π bonds add H_2, decreasing the number of degrees of unsaturation (Figure 12.4). This increases the melting point of the oil. For example, margarine is prepared by partially hydrogenating vegetable oil to give a product having a semi-solid consistency that more closely resembles butter. This process is sometimes called ***hardening.***

If unsaturated oils are healthier than saturated fats, why does the food industry hydrogenate oils? There are two reasons—aesthetics and shelf life. Consumers prefer the semi-solid consistency of margarine to a liquid oil. Imagine pouring vegetable oil on a piece of toast or pancakes.

Furthermore, unsaturated oils are more susceptible than saturated fats to oxidation at the **allylic carbon atoms**—the carbons adjacent to the double bond carbons—a process discussed in Chapter 15. Oxidation makes the oil rancid and inedible. Hydrogenating the double bonds reduces the number of allylic carbons (also illustrated in Figure 12.4), thus reducing the likelihood of oxidation and increasing the shelf life of the food product. This process reflects a delicate balance between providing consumers with healthier food products, while maximizing shelf life to prevent spoilage.

One other fact is worthy of note. Because the steps in hydrogenation are reversible and H atoms are added in a sequential rather than concerted fashion, a cis double bond can be isomerized to a trans double bond. After addition of one H atom (Step [3] in Mechanism 12.1), an intermediate can lose a hydrogen atom to re-form a double bond with either the cis or trans configuration.

As a result, some of the cis double bonds in vegetable oils are converted to trans double bonds during hydrogenation, forming so-called "trans fats." The shape of the resulting fatty acid chain is very different, closely resembling the shape of a *saturated* fatty acid chain. Consequently, trans

Figure 12.4 Partial hydrogenation of the double bonds in a vegetable oil

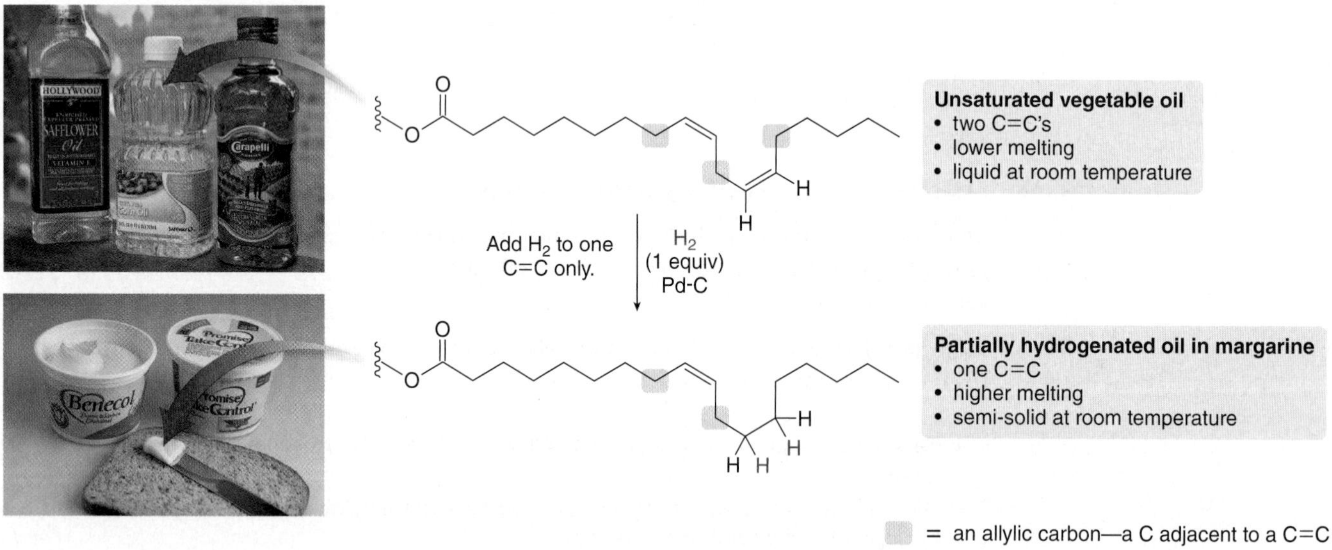

- **Decreasing** the number of degrees of unsaturation **increases** the melting point. Only one long chain of the triacylglycerol is drawn.
- When an oil is *partially* hydrogenated, some double bonds react with H_2, whereas some double bonds remain in the product.
- Partial hydrogenation **decreases** the number of allylic sites (shown in blue), making a triacylglycerol **less** susceptible to oxidation, thereby increasing its shelf life.

fats are thought to have the same negative effects on blood cholesterol levels as saturated fats; that is, trans fats stimulate cholesterol synthesis in the liver, thus increasing blood cholesterol levels, a factor linked to increased risk of heart disease.

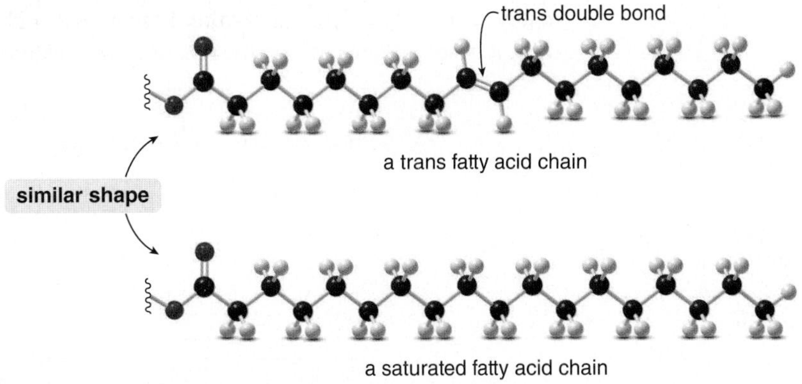
trans double bond
a trans fatty acid chain
similar shape
a saturated fatty acid chain

Problem 12.8 Draw the products formed when triacylglycerol **A** is treated with each reagent, forming compounds **B** and **C.** Rank **A, B,** and **C** in order of increasing melting point.

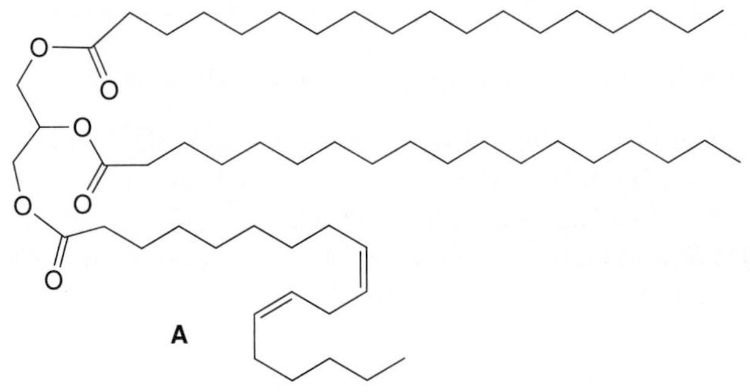

A

a. H_2 (excess), Pd-C (Compound **B**)
b. H_2 (1 equiv), Pd-C (Compound **C**)

12.5 Reduction of Alkynes

Reduction of an alkyne adds H_2 to one or both of the π bonds. There are three different ways by which the elements of H_2 can be added to a triple bond.

- Adding two equivalents of H_2 forms an alkane.

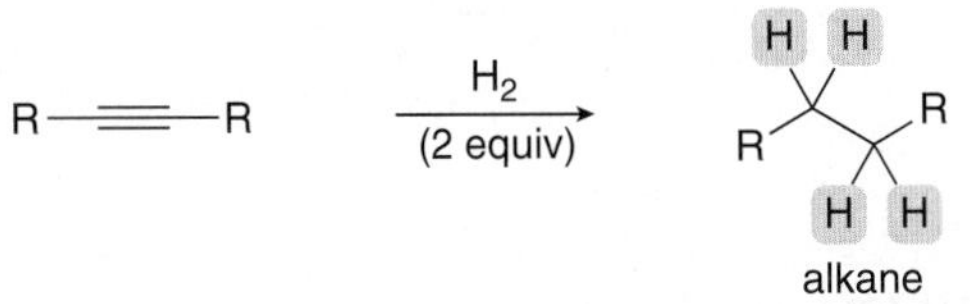
R—≡—R
H_2
(2 equiv)
H H
R
R
H H
alkane

- Adding one equivalent of H_2 in a syn fashion forms a cis alkene.

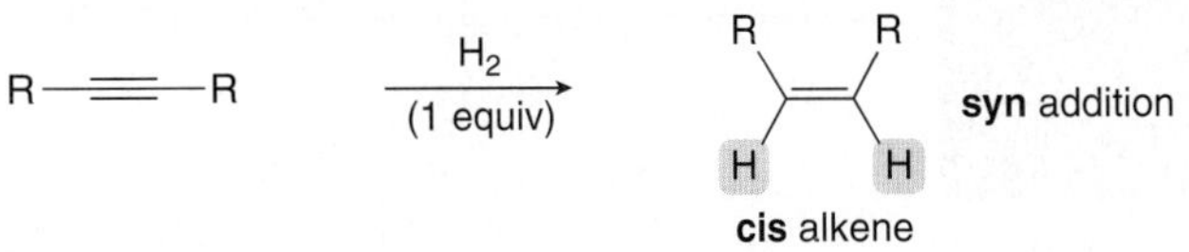
R—≡—R
H_2
(1 equiv)
R R
H H
cis alkene
syn addition

- Adding one equivalent of H_2 in an anti fashion forms a trans alkene.

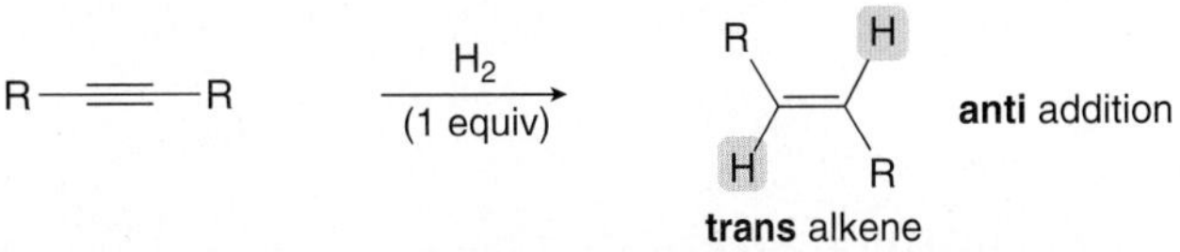
R—≡—R
H_2
(1 equiv)
R H
H R
trans alkene
anti addition

12.5A Reduction of an Alkyne to an Alkane

When an alkyne is treated with two or more equivalents of H_2 and a Pd catalyst, reduction of *both* π bonds occurs. **Syn addition** of one equivalent of H_2 forms a cis alkene, which adds a second equivalent of H_2 to form an **alkane. Four new C–H bonds are formed.** By using a Pd-C catalyst, it is not possible to stop the reaction after addition of only one equivalent of H_2.

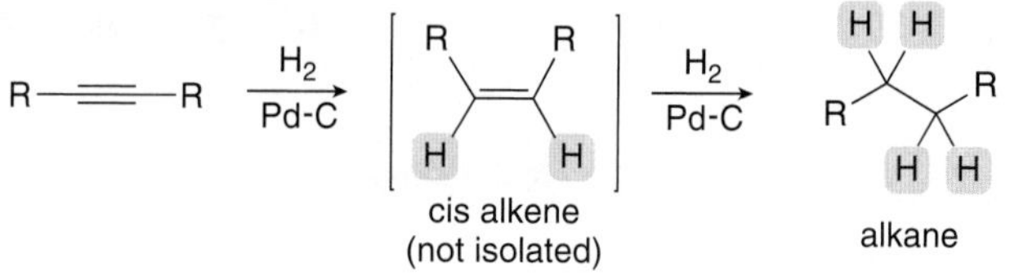

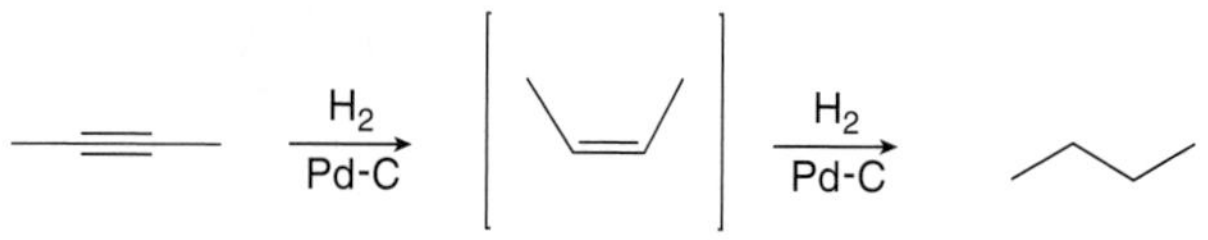

Problem 12.9 Which alkyne has the smaller heat of hydrogenation, $HC{\equiv}CCH_2CH_2CH_3$ or $CH_3C{\equiv}CCH_2CH_3$? Explain your choice.

12.5B Reduction of an Alkyne to a Cis Alkene

Palladium metal is too active a catalyst to allow the hydrogenation of an alkyne to stop after one equivalent of H_2. To prepare a cis alkene from an alkyne and H_2, a less active Pd catalyst is used—Pd adsorbed onto $CaCO_3$ with added lead(II) acetate and quinoline. This catalyst is called the **Lindlar catalyst** after the chemist who first prepared it. Compared to Pd metal, the **Lindlar catalyst is deactivated or "poisoned."**

Pd on $CaCO_3$
+ $Pb(OCOCH_3)_2$ + quinoline

Lindlar catalyst

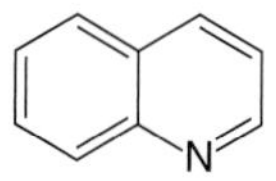

quinoline

Reduction of an alkyne to a cis alkene is a **stereoselective reaction,** because only one stereoisomer is formed.

With the Lindlar catalyst, one equivalent of H_2 adds to an alkyne, and the cis alkene product is unreactive to further reduction.

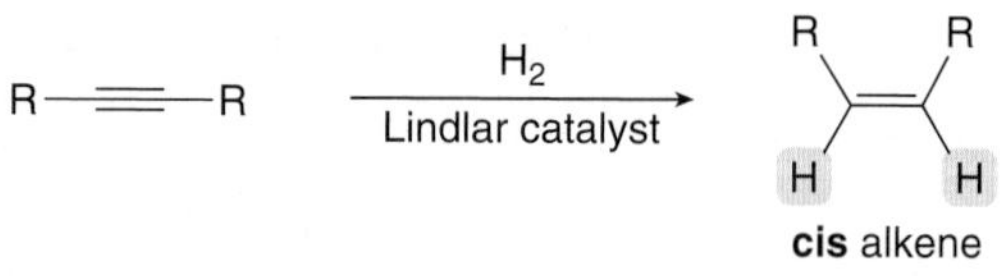

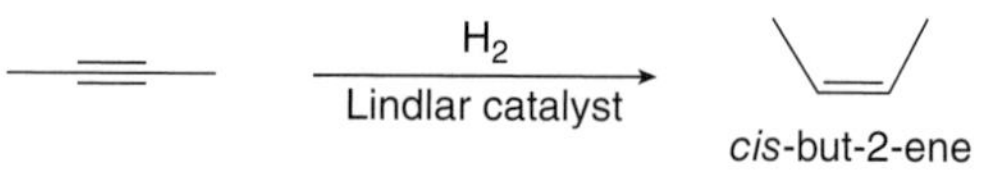

Jasmine flowers are the source of *cis*-jasmone, a perfume component.

Problem 12.10 What is the structure of *cis*-jasmone, a natural product isolated from jasmine flowers, formed by treatment of alkyne **A** with H_2 in the presence of the Lindlar catalyst?

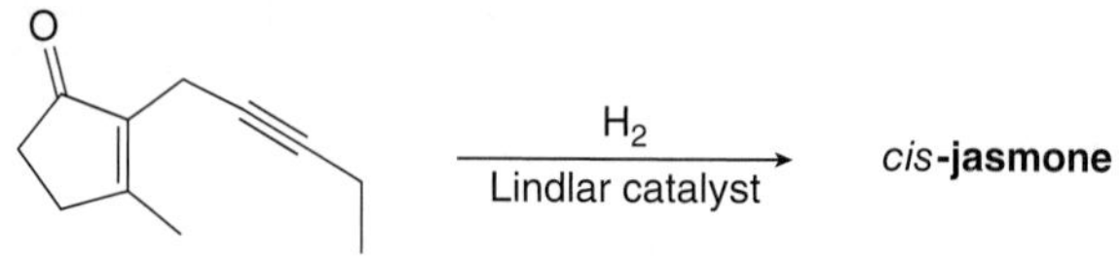

A

Problem 12.11 (a) Draw the structure of a compound of molecular formula C_6H_{10} that reacts with H_2 in the presence of Pd-C but does not react with H_2 in the presence of Lindlar catalyst. (b) Draw the structure of a compound of molecular formula C_6H_{10} that reacts with H_2 when either catalyst is present.

12.5C Reduction of an Alkyne to a Trans Alkene

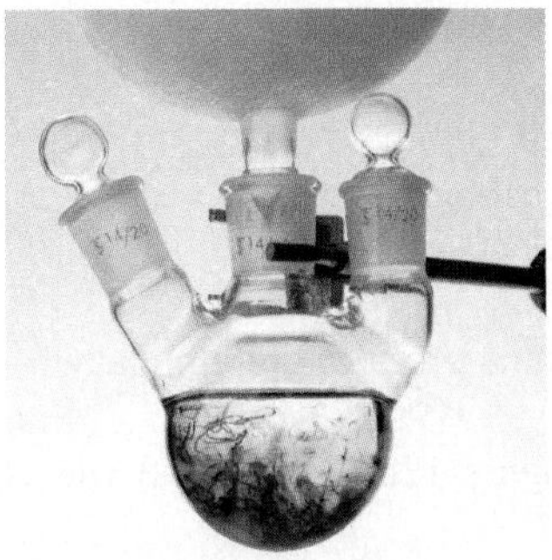

NH_3 has a boiling point of −33 °C, making it a gas at room temperature. To carry out a Na, NH_3 reduction, NH_3 gas is condensed into a flask kept at −78 °C by a cooling bath of solid CO_2 in acetone. When Na is added to the liquid NH_3, a brilliant blue solution is formed.

Although catalytic hydrogenation is a convenient method for preparing cis alkenes from alkynes, it cannot be used to prepare trans alkenes. With a dissolving metal reduction (such as Na in NH_3), however, the elements of H_2 are added in an **anti** fashion to the triple bond, thus forming a **trans alkene.** For example, but-2-yne reacts with Na in NH_3 to form *trans*-but-2-ene.

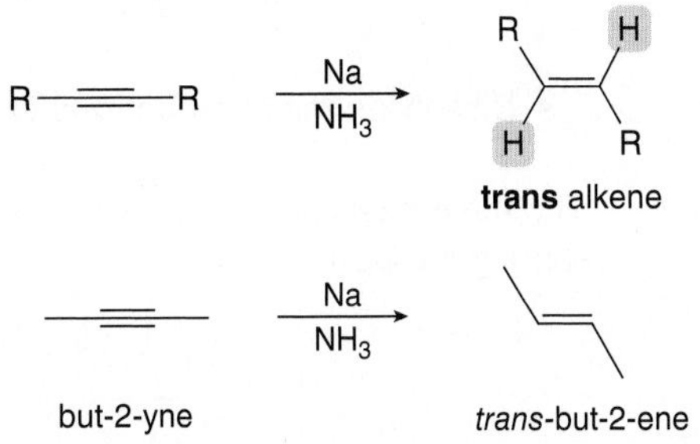

The **mechanism** for the dissolving metal reduction using Na in NH_3 features sequential addition of electrons and protons to the triple bond. Half-headed arrows denoting the movement of a single electron must be used in two steps when Na donates *one* electron. The mechanism can be divided conceptually into two parts, each of which consists of two steps: **addition of an electron followed by protonation of the resulting negative charge,** as shown in Mechanism 12.2.

Mechanism 12.2 Dissolving Metal Reduction of an Alkyne to a Trans Alkene

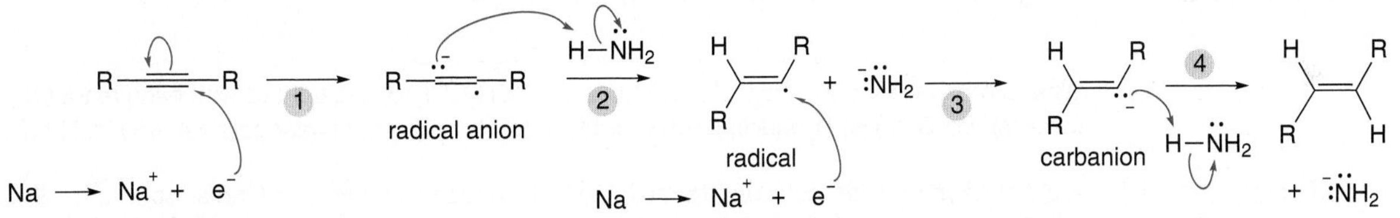

1. Addition of an electron to the triple bond forms a **radical anion,** a species that contains *both* a negative charge *and* an unpaired electron.
2. Protonation of the anion with the solvent NH_3 yields a radical. The net result of the first two steps is the addition of a H atom.
3. Addition of a second electron forms a carbanion.
4. Protonation of the carbanion forms the trans alkene. Steps [3] and [4] add the second H atom to the triple bond.

Although the vinyl carbanion formed in Step [3] could have two different arrangements of its R groups, only the trans alkene is formed from the more stable vinyl carbanion; this carbanion has the larger R groups farther away from each other to avoid steric interactions. Protonation of this anion leads to the more stable trans product.

The larger R groups are farther away from each other.

This more stable vinyl carbanion forms the trans alkene.

Steric interactions between closer R groups destabilize this carbanion.

Dissolving metal reduction of a triple bond with Na in NH_3 is a **stereoselective reaction** because it forms a trans product exclusively.

- **Dissolving metal reductions always form the more stable trans product preferentially.**

The three methods to reduce a triple bond are summarized in Figure 12.5 using hex-3-yne as starting material.

Figure 12.5
Summary: Three methods to reduce a triple bond

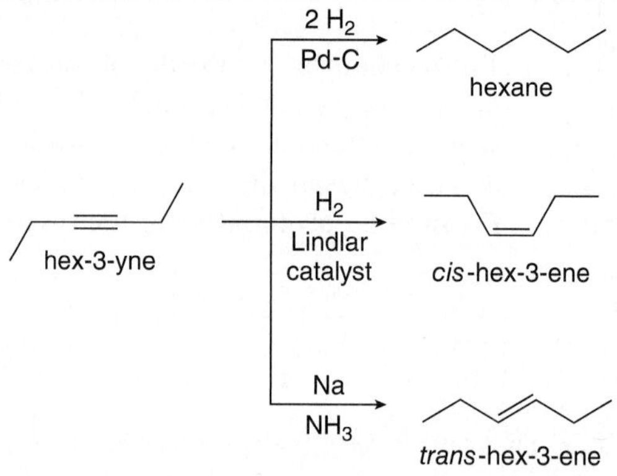

Problem 12.12 What product is formed when $CH_3OCH_2CH_2C{\equiv}CCH_2CH(CH_3)_2$ is treated with each reagent: (a) H_2 (excess), Pd-C; (b) H_2 (1 equiv), Lindlar catalyst; (c) H_2 (excess), Lindlar catalyst; (d) Na, NH_3?

Problem 12.13 A chiral alkyne **A** with molecular formula C_6H_{10} is reduced with H_2 and Lindlar catalyst to **B** having the *R* configuration at its stereogenic center. What are the structures of **A** and **B?**

12.6 The Reduction of Polar C–X σ Bonds

Compounds containing polar C–X σ bonds that react with strong nucleophiles are reduced with metal hydride reagents, most commonly lithium aluminum hydride. Two functional groups possessing both of these characteristics are **alkyl halides** and **epoxides.** Alkyl halides are reduced to alkanes with loss of X^- as the leaving group. Epoxide rings are opened to form alcohols.

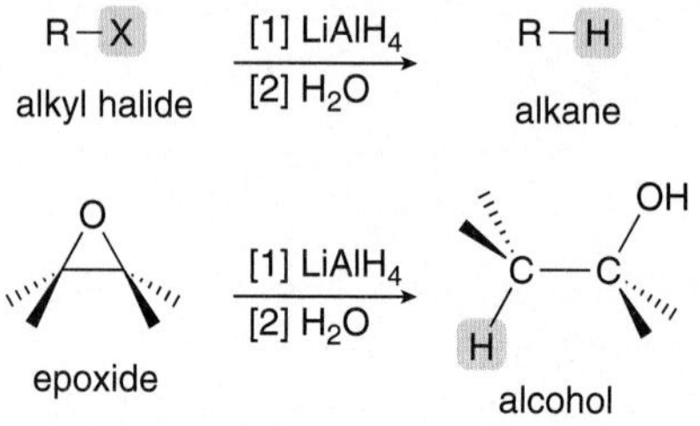

Reduction of these C–X σ bonds is another example of nucleophilic substitution, in which $LiAlH_4$ serves as a source of a hydride nucleophile (H^-). Because H^- is a strong nucleophile, the reaction follows an **S_N2 mechanism,** illustrated for the one-step reduction of an alkyl halide in Mechanism 12.3.

Mechanism 12.3 Reduction of RX with $LiAlH_4$

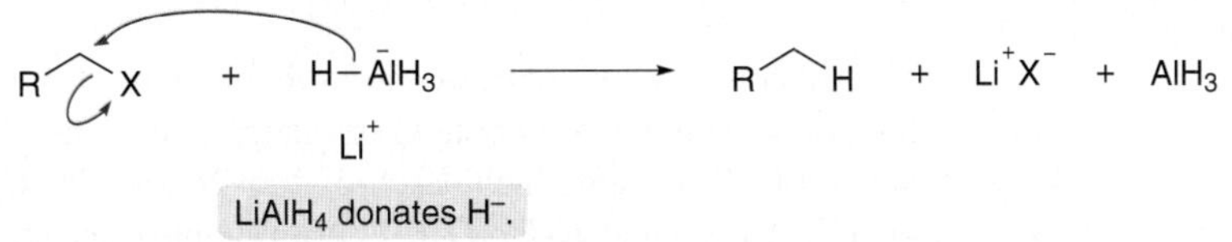
R X
H–AlH3
Li+
R H
Li+X–
AlH3
LiAlH4 donates H–.

- The nucleophile H^- replaces the leaving group X^- in a single step.

Because the reaction follows an S_N2 mechanism:

- **Unhindered CH_3X and 1° alkyl halides are more easily reduced than more substituted 2° and 3° halides.**
- **In unsymmetrical epoxides, nucleophilic attack of H^- (from $LiAlH_4$) occurs at the less substituted carbon atom.**

Examples are shown in Figure 12.6.

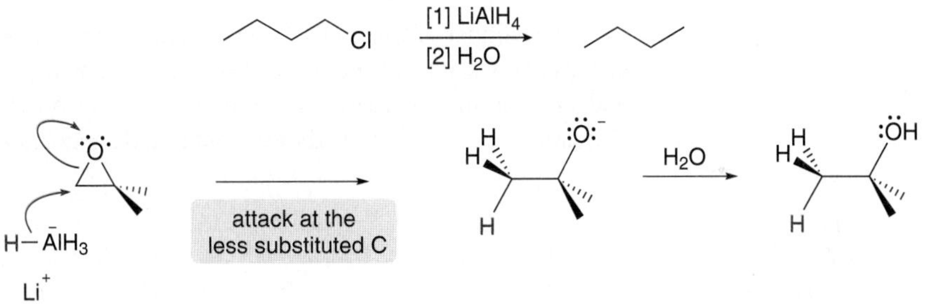
Cl
[1] LiAlH4
[2] H2O
H–AlH3
Li+
attack at the
less substituted C
H2O
:ÖH

Figure 12.6
Examples of reduction of C–X σ bonds with $LiAlH_4$

Problem 12.14 Draw the products of each reaction.

a. 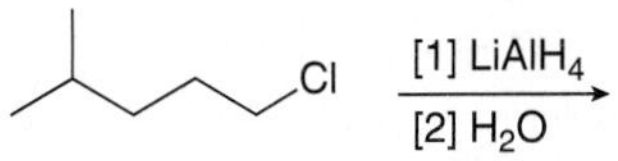
Cl
[1] LiAlH4
[2] H2O

b. 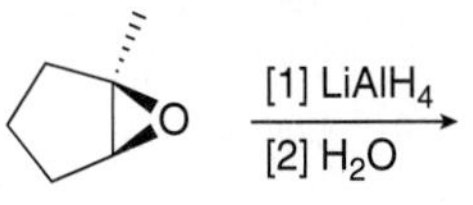
O
[1] LiAlH4
[2] H2O

12.7 Oxidizing Agents

Oxidizing agents fall into two main categories:

- **Reagents that contain an oxygen–oxygen bond**
- **Reagents that contain metal–oxygen bonds**

Oxidizing agents containing an O–O bond include $\mathbf{O_2}$, $\mathbf{O_3}$ (ozone), $\mathbf{H_2O_2}$ (hydrogen peroxide), $\mathbf{(CH_3)_3COOH}$ (*tert*-butyl hydroperoxide), and peroxyacids. **Peroxyacids,** a group of reagents with the general structure $\mathbf{RCO_3H}$**,** have one more O atom than carboxylic acids (RCO_2H). Some peroxyacids are commercially available whereas others are prepared and used without isolation.

Two common peroxyacids are peroxyacetic acid and *meta*-chloroperoxybenzoic acid, abbreviated as **mCPBA.**

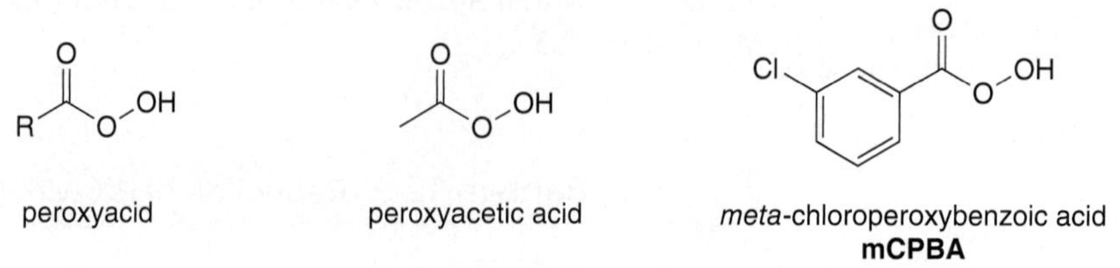

peroxyacid
peroxyacetic acid
meta-chloroperoxybenzoic acid
mCPBA

The most common oxidizing agents with metal–oxygen bonds contain either chromium in the +6 oxidation state (six Cr–O bonds) or manganese in the +7 oxidation state (seven Mn–O bonds). Common Cr^{6+} reagents include chromium(VI) oxide **(CrO_3)** and sodium or potassium dichromate **($Na_2Cr_2O_7$ and $K_2Cr_2O_7$). These reagents are strong oxidants** used in the presence of a strong aqueous acid such as H_2SO_4. **Pyridinium chlorochromate (PCC),** a Cr^{6+} reagent that is soluble in halogenated organic solvents, can be used without strong acid present. This makes it a **more selective Cr^{6+} oxidant,** as described in Section 12.12.

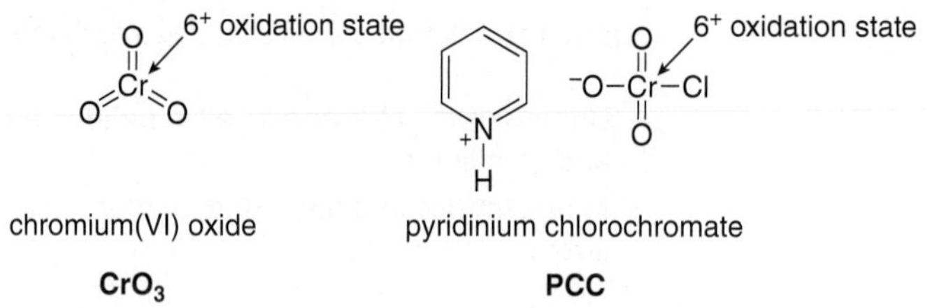

6+ oxidation state
6+ oxidation state
chromium(VI) oxide
CrO3
pyridinium chlorochromate
PCC

The most common Mn^{7+} reagent is **$KMnO_4$** (potassium permanganate), a strong, water-soluble oxidant. Other oxidizing agents that contain metals include **OsO_4** (osmium tetroxide) and **Ag_2O** [silver(I) oxide].

In the remainder of Chapter 12, the oxidation of alkenes, alkynes, and alcohols—three functional groups already introduced in this text—is presented (Figure 12.7). Addition reactions to alkenes and alkynes that increase the number of C–O bonds are described in Sections 12.8–12.11. Oxidation of alcohols to carbonyl compounds appears in Sections 12.12–12.14.

Figure 12.7
Oxidation reactions of alkenes, alkynes, and alcohols

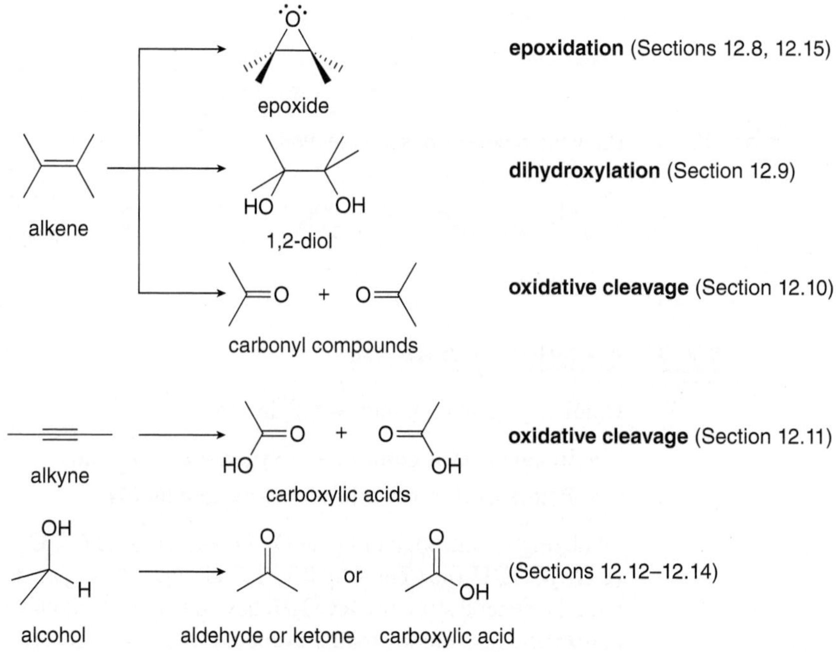

epoxide
epoxidation (Sections 12.8, 12.15)
alkene
1,2-diol
dihydroxylation (Section 12.9)
carbonyl compounds
oxidative cleavage (Section 12.10)
alkyne
carboxylic acids
oxidative cleavage (Section 12.11)
alcohol
aldehyde or ketone
or
carboxylic acid
(Sections 12.12–12.14)

12.8 Epoxidation

Epoxidation is the addition of a single oxygen atom to an alkene to form an **epoxide.**

peroxyacid   epoxide   carboxylic acid

The weak π bond of the alkene is broken and two new C–O σ bonds are formed. Epoxidation is typically carried out with a peroxyacid, resulting in cleavage of the weak O–O bond of the reagent.

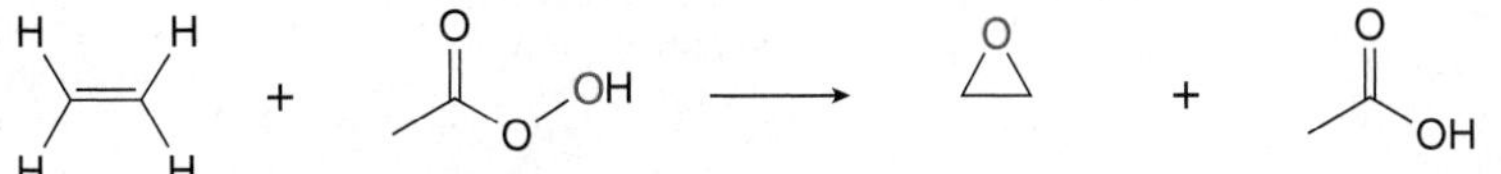

peroxyacetic acid

mCPBA

Epoxidation occurs via the concerted addition of one oxygen atom of the peroxyacid to the π bond as shown in Mechanism 12.4. Epoxidation resembles the formation of the bridged halonium ion in Section 10.13, in that two bonds in a three-membered ring are formed in one step.

Mechanism 12.4 Epoxidation of an Alkene with a Peroxyacid

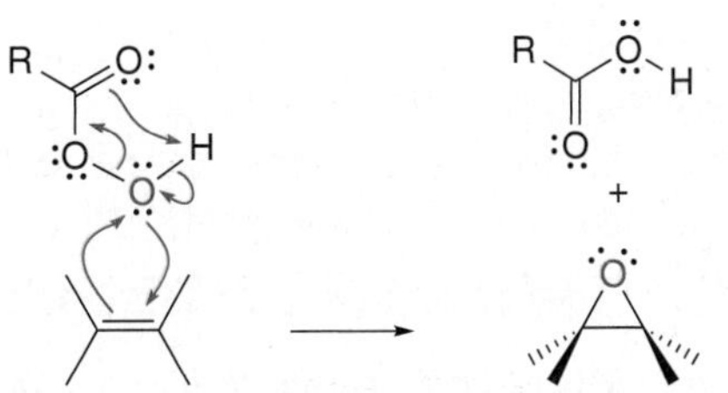

- All bonds are broken and formed in a single step. The two epoxide C–O bonds are formed from one electron pair of the π bond and one lone pair of the peroxyacid. The weak O–O bond is broken.

Problem 12.15 What epoxide is formed when each alkene is treated with mCPBA?

a.  b. 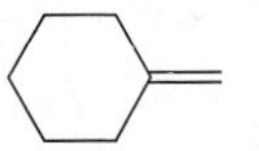 c. 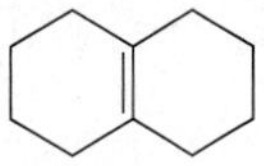

12.8A The Stereochemistry of Epoxidation

Epoxidation occurs via **syn addition** of an O atom from either side of the planar double bond, so that both C–O bonds are formed on the same side. The relative position of substituents in the alkene reactant is **retained** in the epoxide product.

- **A cis alkene gives an epoxide with cis substituents. A trans alkene gives an epoxide with trans substituents.**

Epoxidation is a **stereospecific** reaction because cis and trans alkenes yield different stereoisomers as products, as illustrated in Sample Problem 12.2.

Sample Problem 12.2 Draw the stereoisomers formed when *cis*- and *trans*-but-2-ene are epoxidized with mCPBA.

Solution

To draw each product of epoxidation, add an O atom from either side of the alkene, and keep all substituents in their *original* orientations. The **cis** methyl groups in *cis*-but-2-ene become **cis** substituents in the epoxide. Addition of an O atom from either side of the trigonal planar alkene leads to the same compound—an **achiral meso compound that contains two stereogenic centers,** labeled in blue.

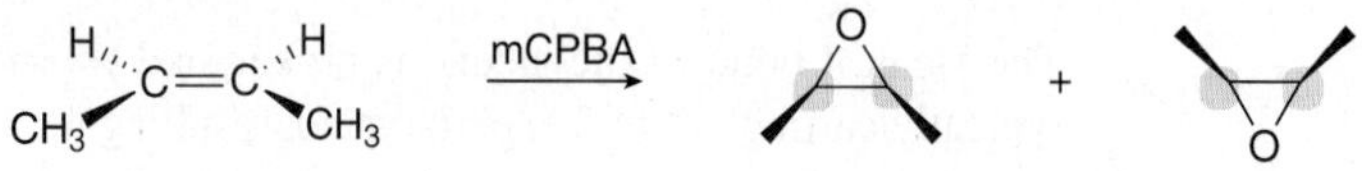

cis CH_3 groups
cis-but-2-ene

cis CH_3 groups
O added from above

cis CH_3 groups
O added from below

Products are identical, an achiral meso compound.

Epoxidation of *cis*- and *trans*-but-2-ene illustrates the general rule about the stereochemistry of reactions: **an achiral starting material gives achiral or racemic products.**

The **trans** methyl groups in *trans*-but-2-ene, become **trans** substituents in the epoxide. Addition of an O atom from either side of the trigonal planar alkene yields an equal mixture of two enantiomers—a **racemic mixture**—with two stereogenic centers labeled in blue.

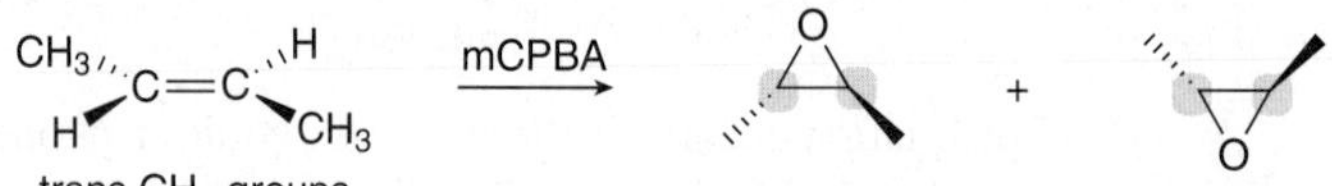

trans CH_3 groups
trans-but-2-ene

trans CH_3 groups
O added from above

trans CH_3 groups
O added from below

Products are enantiomers.

Problem 12.16 Draw all stereoisomers formed when each alkene is treated with mCPBA.

a. 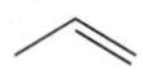 b.  c. 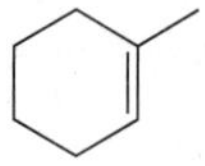

12.8B The Synthesis of Disparlure

Disparlure, the sex pheromone of the female gypsy moth, is synthesized by a stepwise reaction sequence that uses an epoxidation reaction as the final step. Retrosynthetic analysis of disparlure illustrates three key operations:

In 1869, the gypsy moth was introduced into New England in an attempt to develop a silk industry. Some moths escaped into the wild and the population flourished. Mature gypsy moth caterpillars eat an average of one square foot of leaf surface per day, defoliating shade trees and entire forests. Many trees die after a single defoliation.

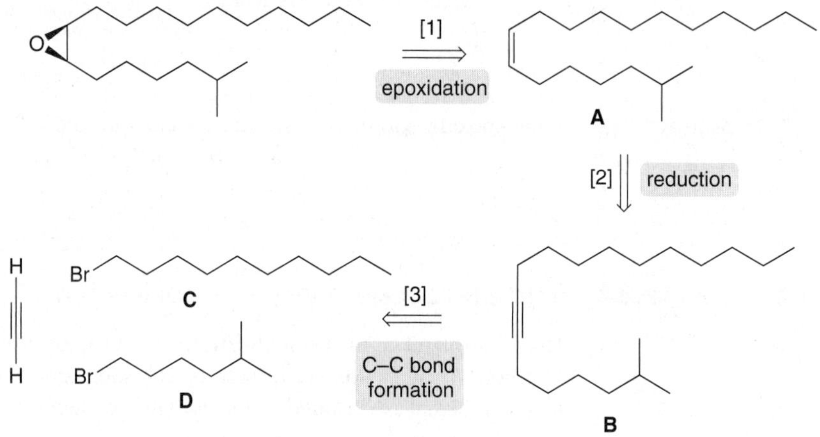

- **Step [1]** The cis epoxide in disparlure is prepared from a cis alkene **A** by epoxidation.
- **Step [2]** **A** is prepared from an internal alkyne **B** by reduction.
- **Step [3]** **B** is prepared from acetylene and two 1° alkyl halides (**C** and **D**) by using S_N2 reactions with acetylide anions.

Figure 12.8 illustrates the synthesis of disparlure beginning with acetylene. The synthesis is conceptually divided into three parts:

- **Part [1]** Acetylene is converted to an internal alkyne **B** by forming two C–C bonds. Each bond is formed by treating an alkyne with base ($NaNH_2$) to form an acetylide anion, which reacts with an alkyl halide (**C** or **D**) in an S_N2 reaction (Section 11.11).
- **Part [2]** The internal alkyne **B** is reduced to a cis alkene **A** by syn addition of H_2 using the Lindlar catalyst (Section 12.5B).
- **Part [3]** The cis alkene **A** is epoxidized to disparlure using a peroxyacid such as mCPBA.

Figure 12.8 The synthesis of disparlure

Part [1] Formation of two C–C bonds using acetylide anions

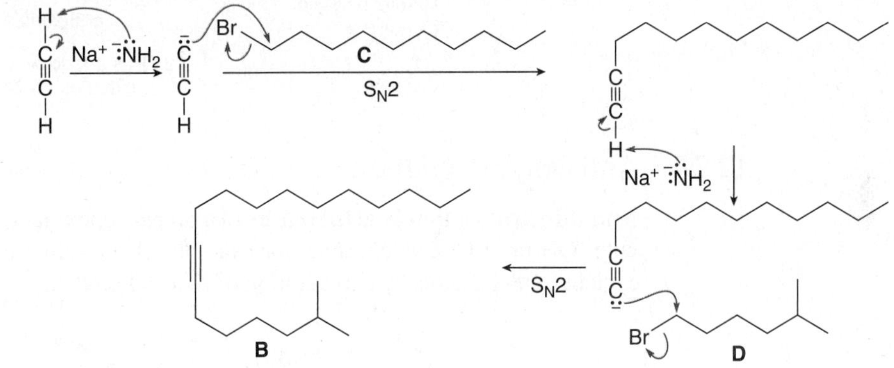

Part [2] Reduction of alkyne **B** to form cis alkene **A**

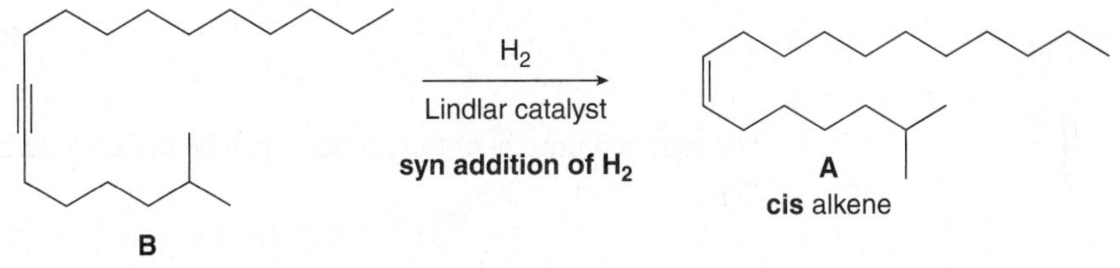

Part [3] Epoxidation of **A** to form disparlure

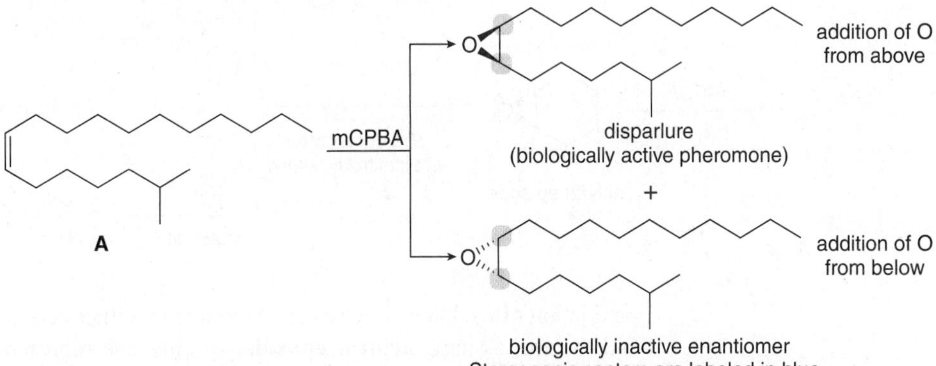

Stereogenic centers are labeled in blue.

- Disparlure has been used to control the spread of the gypsy moth caterpillar, a pest that has periodically devastated forests in the northeastern United States by defoliating many shade and fruit-bearing trees. The active pheromone is placed in a trap containing a poison or sticky substance, and the male moth is lured to the trap by the pheromone. Alternatively, thousands of disparlure-baited traps are placed along the edges of infestation. When the pheromone permeates the air, males are confused and can't locate individual females, so that mating is disrupted. Such a species-specific method presents a new way of controlling an insect population that avoids the widespread use of harmful, nonspecific pesticides.

How to separate a racemic mixture into its component enantiomers is discussed in Section 26.3.

Epoxidation of the cis alkene **A** from two different sides of the double bond affords two cis epoxides in the last step—a racemic mixture of two enantiomers. Thus, half of the product is the desired pheromone disparlure, but the other half is its biologically inactive enantiomer. Separating the desired from the undesired enantiomer is difficult and expensive, because both compounds have identical physical properties. A reaction that affords a chiral epoxide from an achiral precursor without forming a racemic mixture is discussed in Section 12.15.

12.9 Dihydroxylation

Dihydroxylation is the addition of two hydroxy groups to a double bond, forming a **1,2-diol** or **glycol.** Depending on the reagent, the two new OH groups can be added to the opposite sides (**anti** addition) or the same side (**syn** addition) of the double bond.

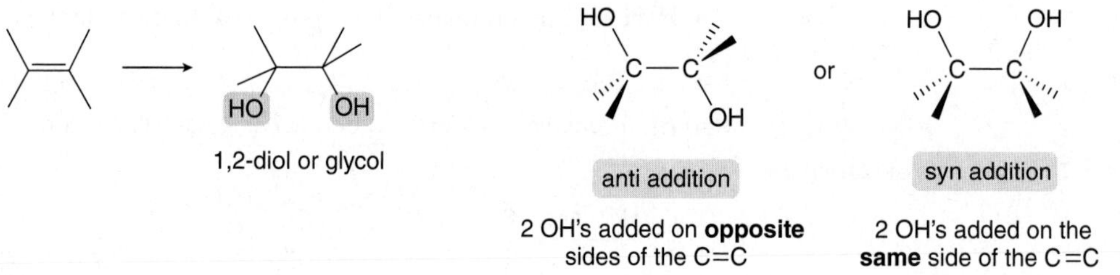
HO
OH
1,2-diol or glycol
HO
OH
anti addition
2 OH's added on opposite sides of the C=C
or
HO
OH
syn addition
2 OH's added on the same side of the C=C

12.9A Anti Dihydroxylation

Anti dihydroxylation is achieved in two steps—epoxidation followed by opening of the ring with ^-OH or H_2O. Cyclohexene, for example, is converted to a racemic mixture of two *trans*-cyclohexane-1,2-diols by anti addition of two OH groups.

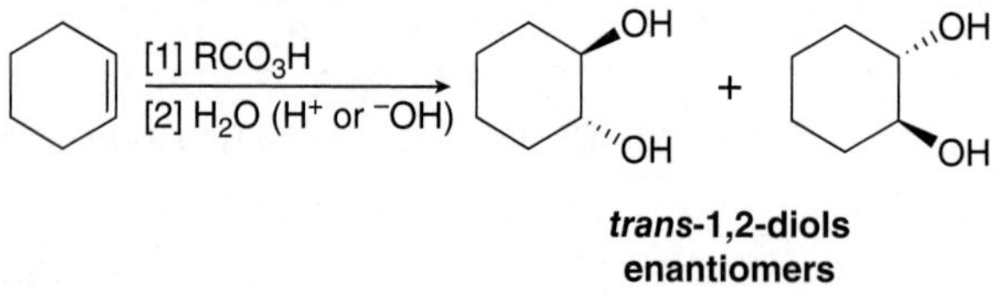
[1] RCO3H
[2] H2O (H+ or −OH)
OH
OH
+
OH
OH
trans-1,2-diols
enantiomers

The stereochemistry of the products can be understood by examining the stereochemistry of each step.

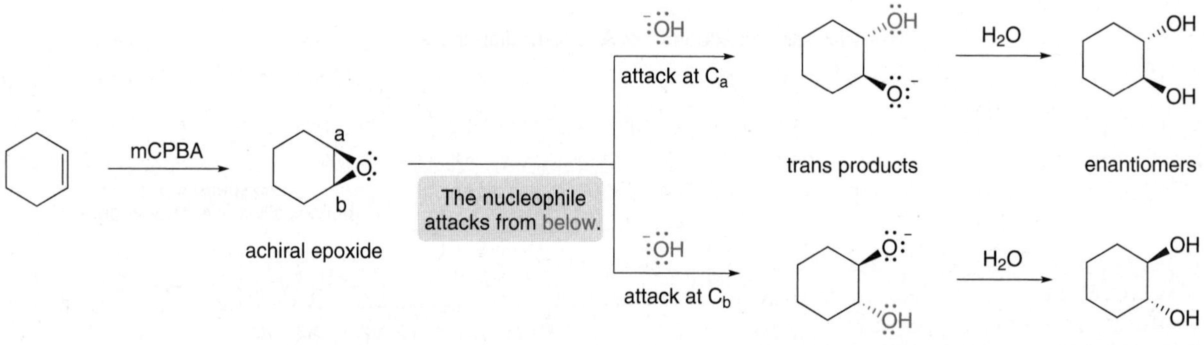
mCPBA
a
b
achiral epoxide
The nucleophile attacks from below.
−OH
attack at Ca
−OH
attack at Cb
trans products
H2O
H2O
OH
OH
enantiomers

Epoxidation of cyclohexene adds an O atom from either above or below the plane of the double bond to form a single **achiral epoxide,** so only one representation is shown. Opening of the epoxide ring then occurs with **backside attack at either C–O bond.** Because the epoxide is drawn above the plane of the six-membered ring, nucleophilic attack occurs from **below** the plane. This reaction is a specific example of the opening of epoxide rings with strong nucleophiles, first presented in Section 9.16.

Because one OH group of the 1,2-diol comes from the epoxide and one OH group comes from the nucleophile (^-OH), the overall result is **anti addition of two OH groups** to an alkene.

Problem 12.17 Draw the products formed when both *cis*- and *trans*-but-2-ene are treated with a peroxyacid followed by ^{-}OH (in H_2O). Explain how these reactions illustrate that anti dihydroxylation is stereospecific.

12.9B Syn Dihydroxylation

Syn dihydroxylation results when an alkene is treated with either $\mathbf{KMnO_4}$ or $\mathbf{OsO_4}$.

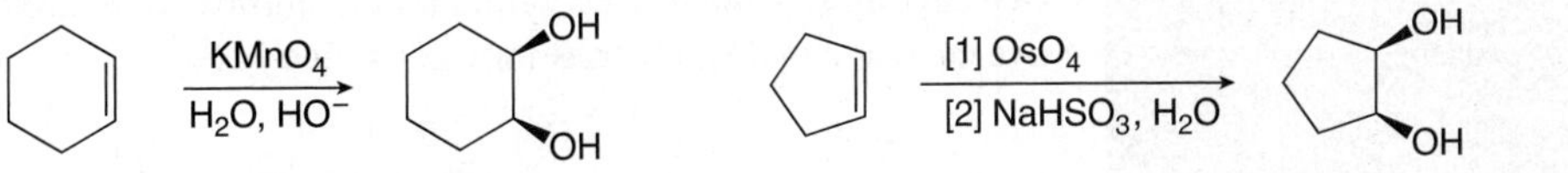

cis-cyclohexane-1,2-diol *cis*-cyclopentane-1,2-diol

Each reagent adds two oxygen atoms to the same side of the double bond—that is, in a **syn** fashion—to yield a cyclic intermediate. Hydrolysis of the cyclic intermediate cleaves the metal–oxygen bonds, forming the *cis*-1,2-diol. With OsO_4, sodium bisulfite ($NaHSO_3$) is also added in the hydrolysis step.

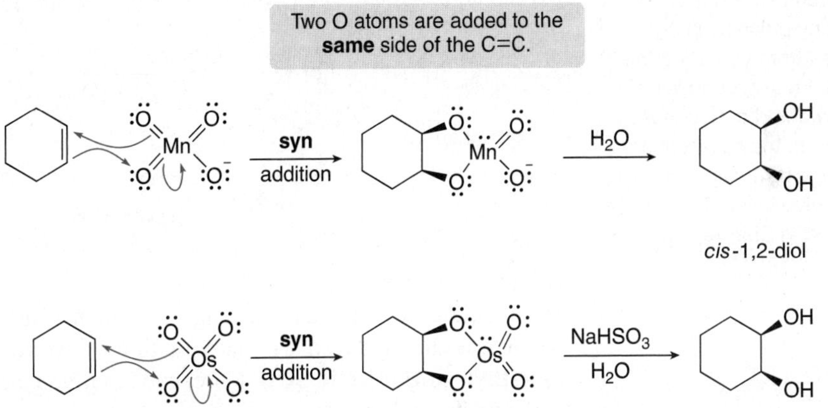

Although $KMnO_4$ is inexpensive and readily available, its use is limited by its insolubility in organic solvents. To prevent further oxidation of the product 1,2-diol, the reaction mixture must be kept basic with added ^{-}OH.

NMO is an **amine oxide.** It is not possible to draw a Lewis structure of an amine oxide having only neutral atoms.

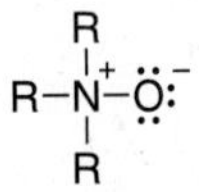

amine oxide

Although OsO_4 is a more selective oxidant than $KMnO_4$ and is soluble in organic solvents, it is toxic and expensive. To overcome these limitations, dihydroxylation can be carried out by using a *catalytic* amount of OsO_4, if the oxidant ***N*-methylmorpholine *N*-oxide (NMO)** is also added.

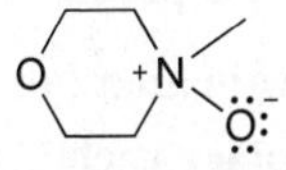

N-methylmorpholine *N*-oxide

NMO

In the catalytic process, dihydroxylation of the double bond converts the Os^{8+} oxidant into an Os^{6+} product, which is then re-oxidized by NMO to Os^{8+}. This Os^{8+} reagent can then be used for dihydroxylation once again, and the catalytic cycle continues.

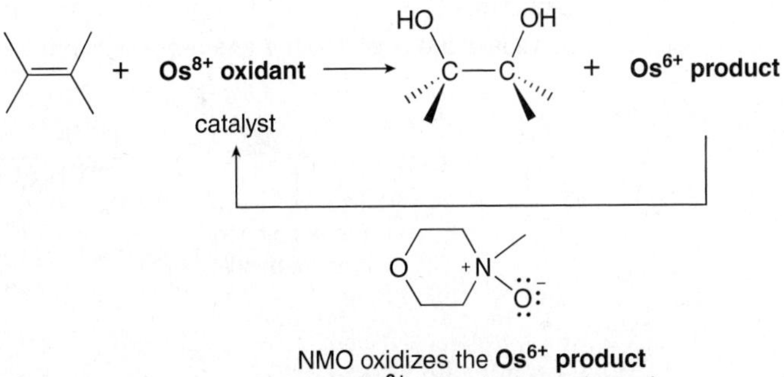

NMO oxidizes the $\mathbf{Os^{6+}}$ **product** back to $\mathbf{Os^{8+}}$ to begin the cycle again.

Problem 12.18 Draw the products formed when both *cis*- and *trans*-but-2-ene are treated with OsO_4, followed by hydrolysis with $NaHSO_3$ + H_2O. Explain how these reactions illustrate that syn dihydroxylation is stereospecific.

12.10 Oxidative Cleavage of Alkenes

Lightning produces O_3 from O_2 during an electrical storm. Moreover, the pungent odor around a heavily used photocopy machine is O_3 produced from O_2 during the process. O_3 at ground level is an unwanted atmospheric pollutant. In the stratosphere, however, it protects us from harmful ultraviolet radiation, as discussed in Chapter 15.

Oxidative cleavage of an alkene breaks both the σ and π bonds of the double bond to form two carbonyl groups. Depending on the number of R groups bonded to the double bond, oxidative cleavage yields either **ketones** or **aldehydes.**

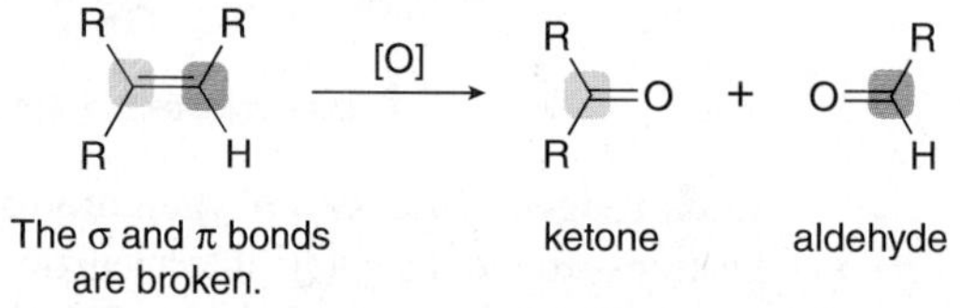

One method of oxidative cleavage relies on a two-step procedure using **ozone (O_3) as the oxidant** in the first step. Cleavage with ozone is called **ozonolysis.**

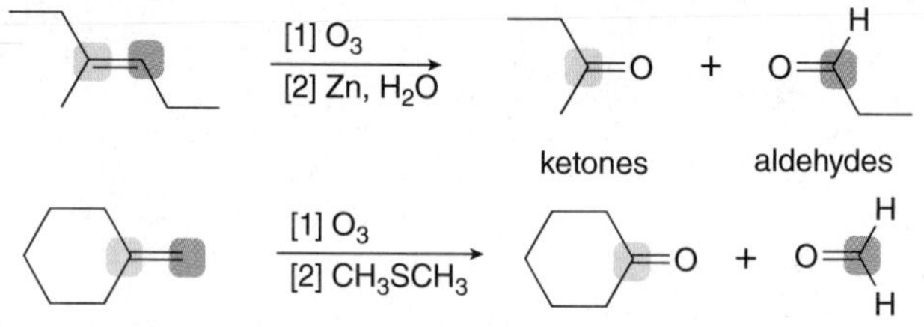

Addition of ozone to the π bond of the alkene forms an unstable intermediate called a **molozonide,** which then rearranges to an **ozonide** by a stepwise process. The unstable ozonide is then reduced without isolation to afford carbonyl compounds. **Zn (in H_2O)** and **dimethyl sulfide (CH_3SCH_3)** are two common reagents used to convert the ozonide to carbonyl compounds.

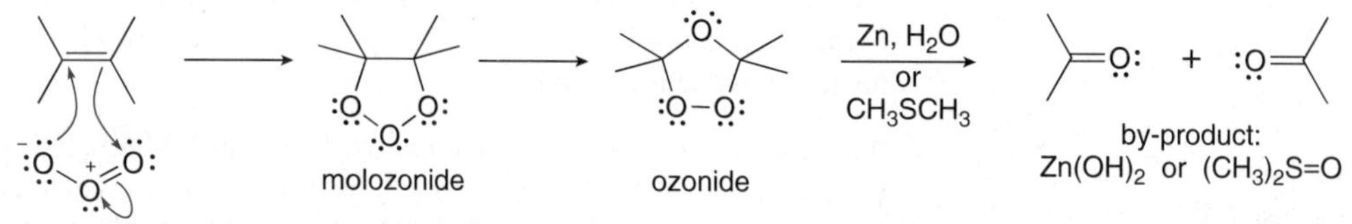

To draw the product of any oxidative cleavage:

- **Locate all π bonds in the molecule.**
- **Replace each C=C by two C=O bonds.**

Sample Problem 12.3 Draw the products when each alkene is treated with O_3 followed by CH_3SCH_3.

a. 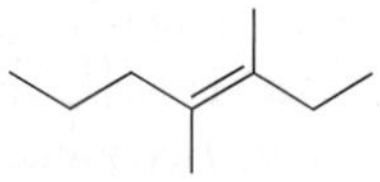 b. 

Solution

a. Cleave the double bond and replace it with two carbonyl groups.

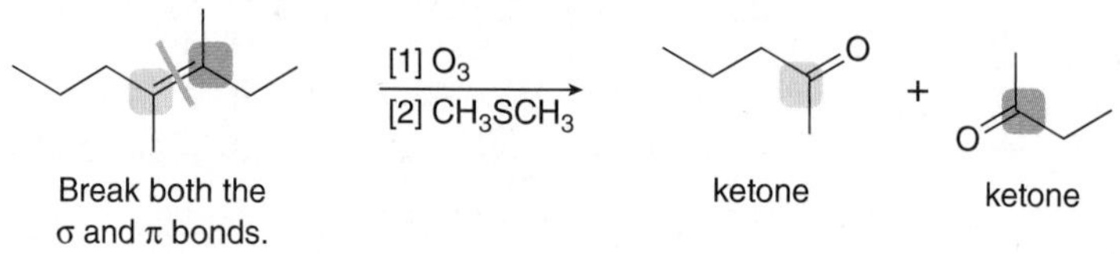

b. For a cycloalkene, oxidative cleavage results in a **single molecule with two carbonyl groups—a dicarbonyl compound.**

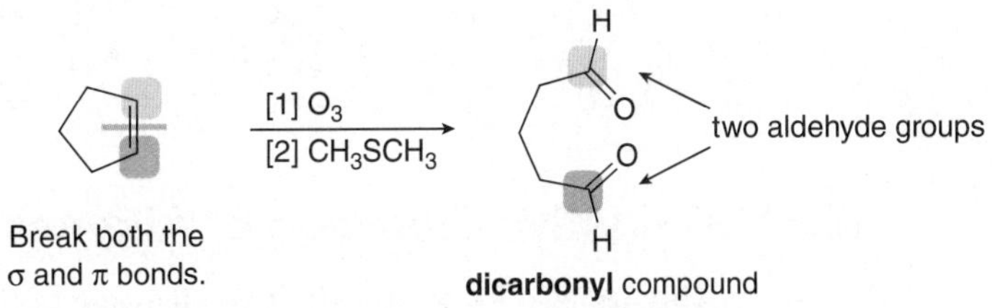
[1] O_3
[2] CH_3SCH_3
two aldehyde groups
Break both the σ and π bonds.
dicarbonyl compound

Problem 12.19 Draw the products formed when each alkene is treated with O_3 followed by Zn, H_2O.

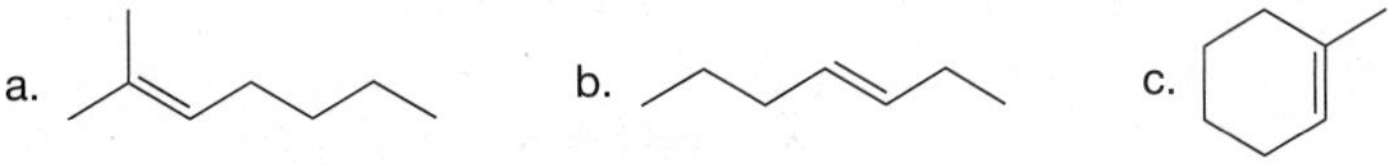
a.
b.
c.

Ozonolysis of dienes (and other polyenes) results in oxidative cleavage of all C=C bonds. The number of carbonyl groups formed in the products is *twice* the number of double bonds in the starting material. The *two* double bonds in limonene are converted to products containing *four* carbonyl groups.

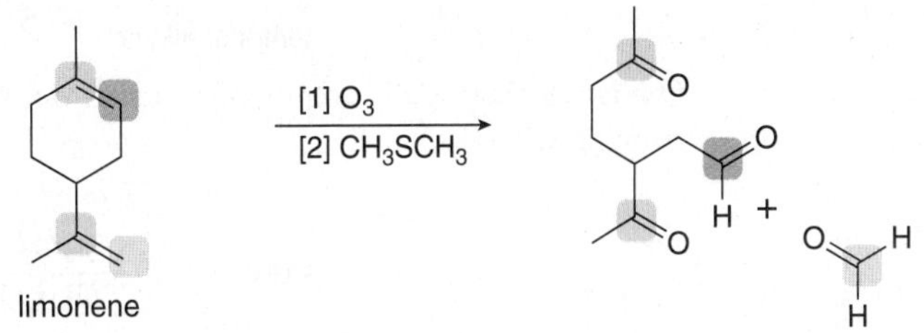
[1] O_3
[2] CH_3SCH_3
limonene

Oxidative cleavage is a valuable tool for structure determination of unknown compounds. The ability to determine what alkene gives rise to a particular set of oxidative cleavage products is thus a useful skill, illustrated in Sample Problem 12.4.

Sample Problem 12.4 What alkene forms the following products after reaction with O_3 followed by CH_3SCH_3?

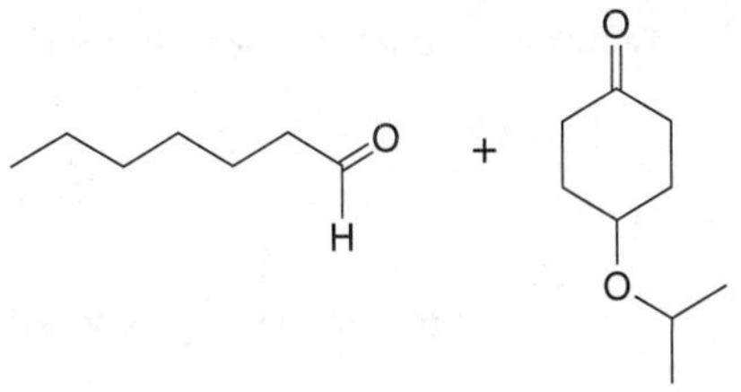

Solution

To draw the starting material, ignore the O atoms in the carbonyl groups and join the carbonyl carbons together by a C=C.

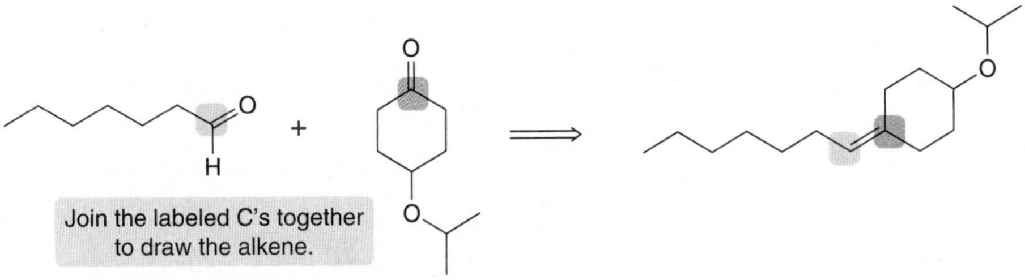
Join the labeled C's together to draw the alkene.

Problem 12.20 What alkene yields each set of oxidative cleavage products?

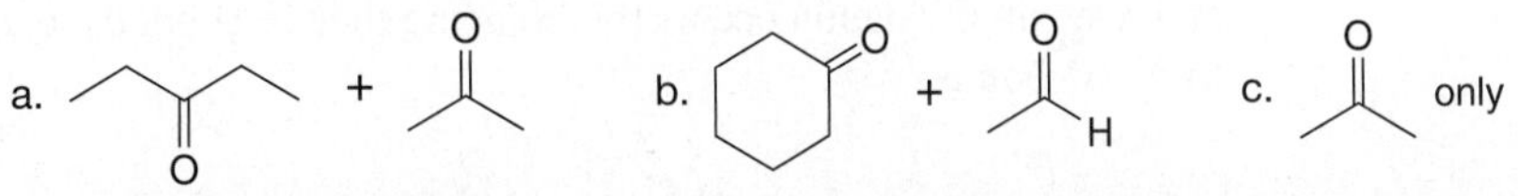
a.
b.
c.
only

Problem 12.21 Draw the products formed when each diene is treated with O_3 followed by CH_3SCH_3.

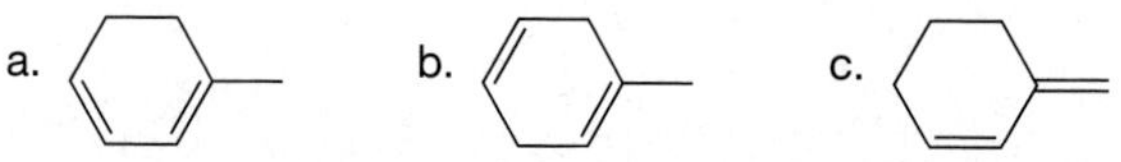

12.11 Oxidative Cleavage of Alkynes

Alkynes also undergo oxidative cleavage of the σ bond and both π bonds of the triple bond. Internal alkynes are oxidized to **carboxylic acids (RCOOH),** whereas terminal alkynes afford carboxylic acids and $\mathbf{CO_2}$ from the *sp* hybridized C–H bond.

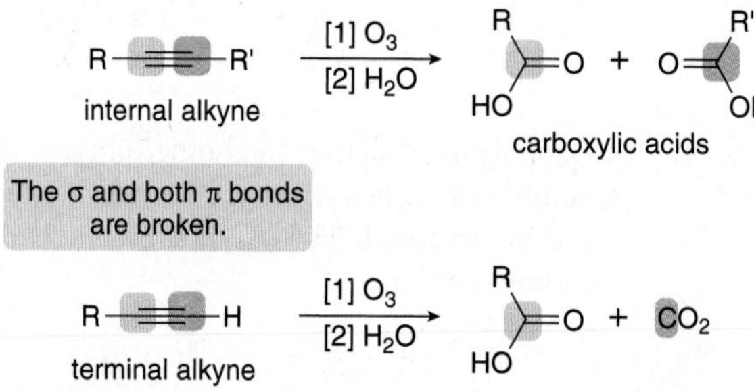

Oxidative cleavage is commonly carried out with O_3, followed by cleavage of the intermediate ozonide with H_2O.

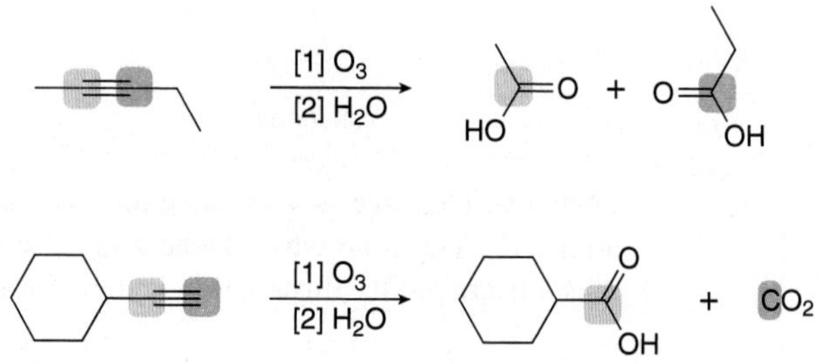

Problem 12.22 Draw the products formed when each alkyne is treated with O_3 followed by H_2O.

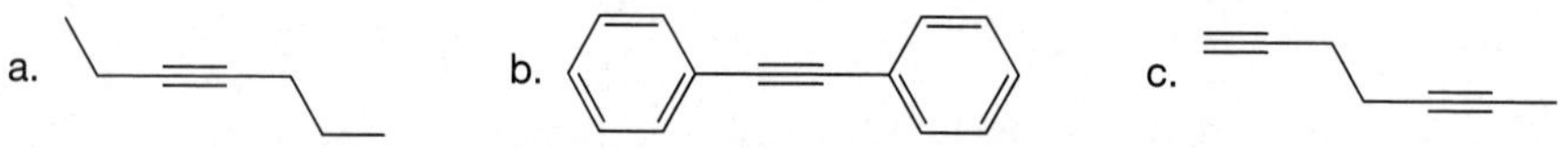

Problem 12.23 What alkyne (or diyne) yields each set of oxidative cleavage products?

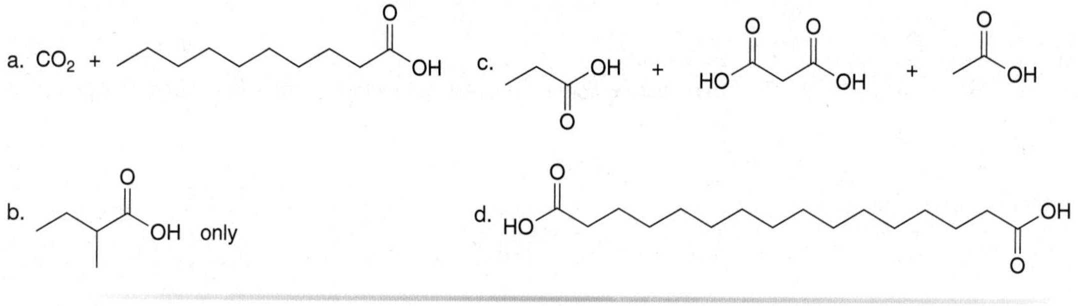

12.12 Oxidation of Alcohols

Alcohols are oxidized to a variety of carbonyl compounds, depending on the type of alcohol and reagent. Oxidation occurs by replacing the C–H bonds *on the carbon bearing the OH group* by C–O bonds.

- **1° Alcohols are oxidized to either aldehydes or carboxylic acids by replacing either one or two C–H bonds by C–O bonds.**

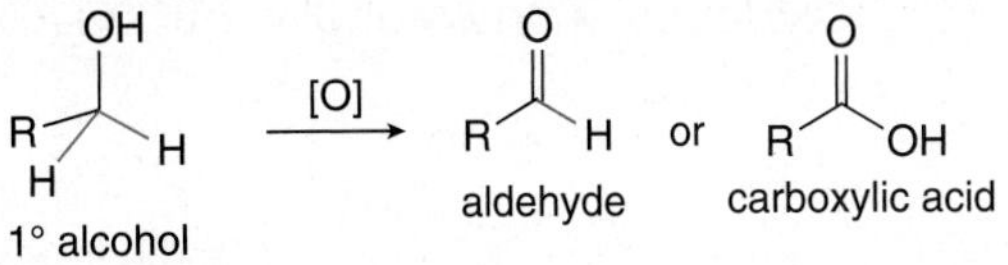

1° alcohol

2 C–H bonds

- **2° Alcohols are oxidized to ketones by replacing the one C–H bond by a C–O bond.**

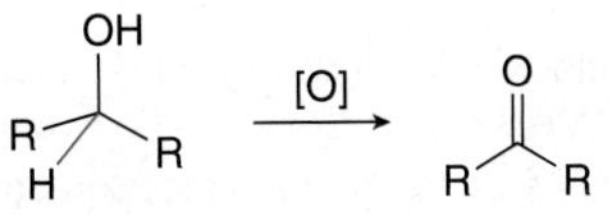

2° alcohol

1 C–H bond

ketone

- **3° Alcohols have no H atoms on the carbon with the OH group, so they are not easily oxidized.**

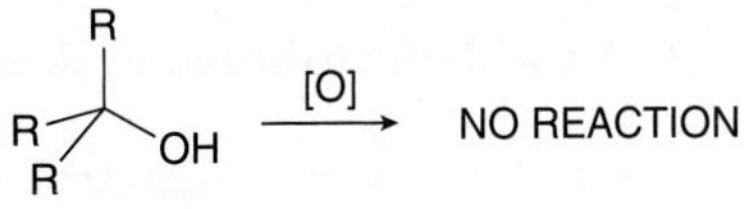

3° alcohol

Alcohol oxidations often occur by a pathway that involves bonding a leaving group Z to the oxygen, where Z is typically a metal in a high oxidation state. Elimination with a base then forms a C=O and a metal in a lower oxidation state.

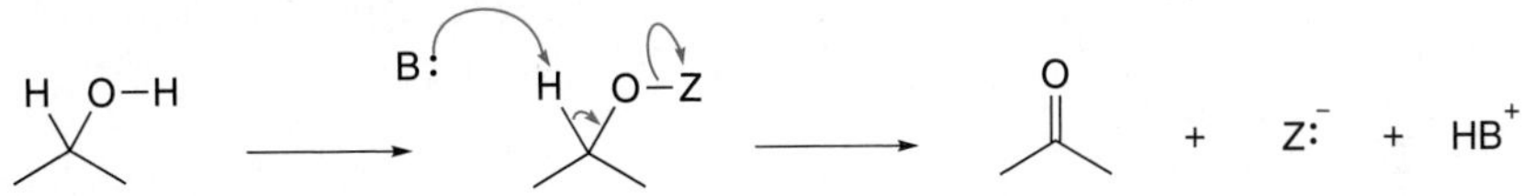

alcohol

carbonyl compound

The oxidation of alcohols to carbonyl compounds is typically carried out with Cr^{6+} oxidants, which are reduced to Cr^{3+} products.

- **CrO_3, $Na_2Cr_2O_7$, and $K_2Cr_2O_7$** are **strong, nonselective oxidants** used in aqueous acid ($H_2SO_4 + H_2O$).
- **PCC** (Section 12.7) is soluble in CH_2Cl_2 (dichloromethane), and can be used without strong acid present, making it a **more selective, milder oxidant.**

12.12A Oxidation of 2° Alcohols

Any of the Cr^{6+} oxidants effectively oxidizes 2° alcohols to ketones.

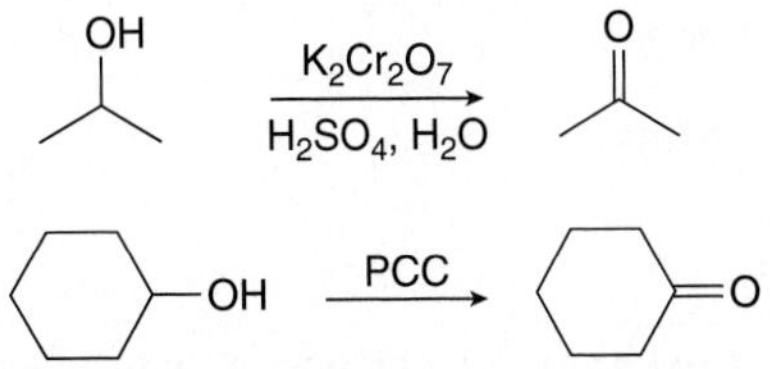

2° alcohol

ketone

The mechanism for alcohol oxidation has two key parts: **formation of a chromate ester** and **loss of a proton.** Mechanism 12.5 is drawn for the oxidation of a general 2° alcohol with CrO_3.

Mechanism 12.5 Oxidation of an Alcohol with CrO_3

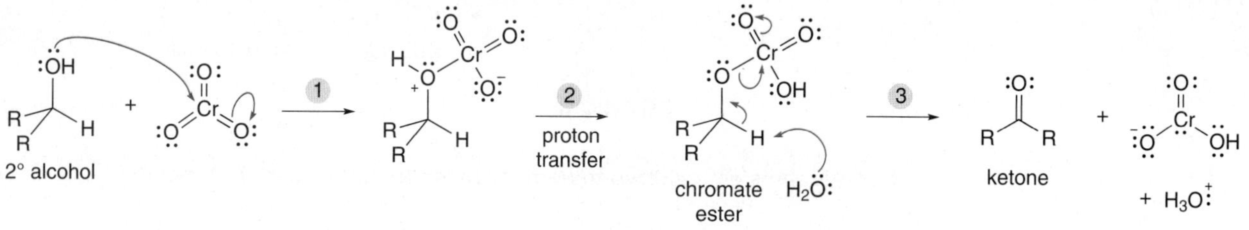
2° alcohol
1
2
proton transfer
chromate ester
3
ketone
+ H_3O^+

1 – 2 Nucleophilic attack of the alcohol on the electrophilic metal (Cr^{6+} oxidation state) followed by proton transfer forms a chromate ester.

3 A base removes a proton and the electron pair in the C–H bond forms the new π bond of the C=O. Carbon is oxidized because the number of C–O bonds increases, and Cr^{6+} is reduced to Cr^{4+}.

These three steps convert the Cr^{6+} oxidant to a Cr^{4+} product, which is then further reduced to a Cr^{3+} product by a series of steps.

12.12B Oxidation of 1° Alcohols

1° Alcohols are oxidized to either aldehydes or carboxylic acids, depending on the reagent.

- **1° Alcohols are oxidized to aldehydes (RCHO) under mild reaction conditions—using PCC in CH_2Cl_2.**
- **1° Alcohols are oxidized to carboxylic acids (RCOOH) under harsher reaction conditions: $Na_2Cr_2O_7$, $K_2Cr_2O_7$, or CrO_3 in the presence of H_2O and H_2SO_4.**

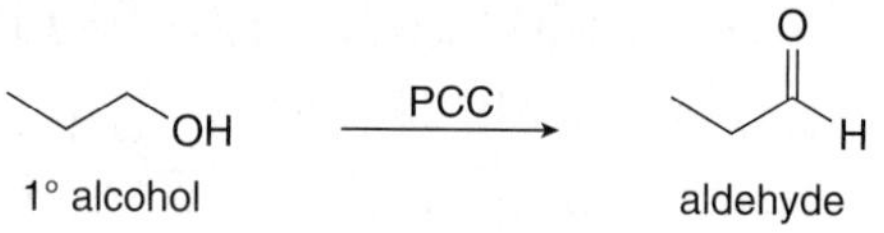
PCC
1° alcohol
aldehyde

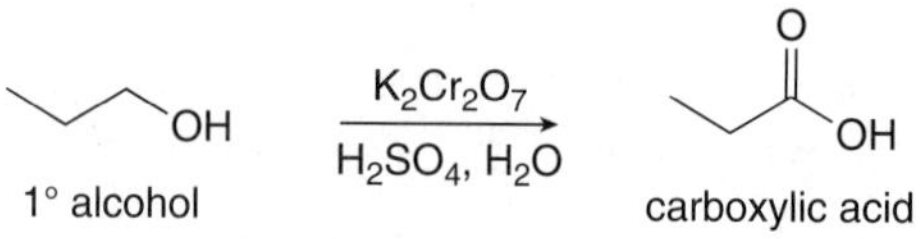
$K_2Cr_2O_7$
H_2SO_4, H_2O
1° alcohol
carboxylic acid

The mechanism for the oxidation of 1° alcohols to aldehydes parallels the oxidation of 2° alcohols to ketones detailed in Section 12.12A. Oxidation of a 1° alcohol to a carboxylic acid requires three operations: **oxidation first to the aldehyde, reaction with water,** and then further **oxidation to the carboxylic acid,** as shown in Mechanism 12.6.

Mechanism 12.6 Oxidation of a 1° Alcohol to a Carboxylic Acid

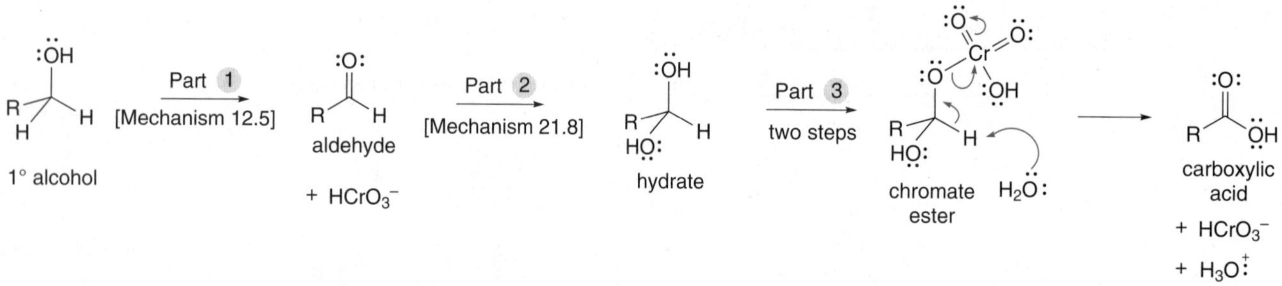
1° alcohol
Part 1
[Mechanism 12.5]
aldehyde
+ $HCrO_3^-$
Part 2
[Mechanism 21.8]
hydrate
Part 3
two steps
chromate ester
carboxylic acid
+ $HCrO_3^-$
+ H_3O^+

Part 1 The 1° alcohol is oxidized to an aldehyde by the three-step sequence in Mechanism 12.5.

Part 2 Water adds to the C=O to form a hydrate, a compound with two OH groups bonded to the same carbon, by a mechanism discussed in Section 21.13.

Part 3 Oxidation of the C–H bond of the hydrate follows Mechanism 12.5—formation of a chromate ester and loss of a proton.

Cr^{6+} oxidations are characterized by a color change, as the **red-orange Cr^{6+} reagent** is reduced to **green Cr^{3+}.** The first devices used to measure blood alcohol content in individuals suspected of "driving under the influence" made use of this color change. Oxidation of CH_3CH_2OH, the 1° alcohol in alcoholic beverages, with orange $K_2Cr_2O_7$ forms CH_3COOH and green Cr^{3+}.

OH
+
K2Cr2O7
red-orange
O
OH
+
Cr3+
green
ethanol
"alcohol"
acetic acid

Blood alcohol level can be determined by having an individual blow into a tube containing $K_2Cr_2O_7$, H_2SO_4, and an inert solid. The alcohol in the exhaled breath is oxidized by the Cr^{6+} reagent, which turns green in the tube (Figure 12.9). The higher the concentration of CH_3CH_2OH in the breath, the more Cr^{6+} is reduced, and the farther the green Cr^{3+} color extends down the length of the sample tube. This value is then correlated with blood alcohol content to determine if an individual has surpassed the legal blood alcohol limit.

Figure 12.9 Blood alcohol screening

a. Schematic of an alcohol testing device

b. Consumer product

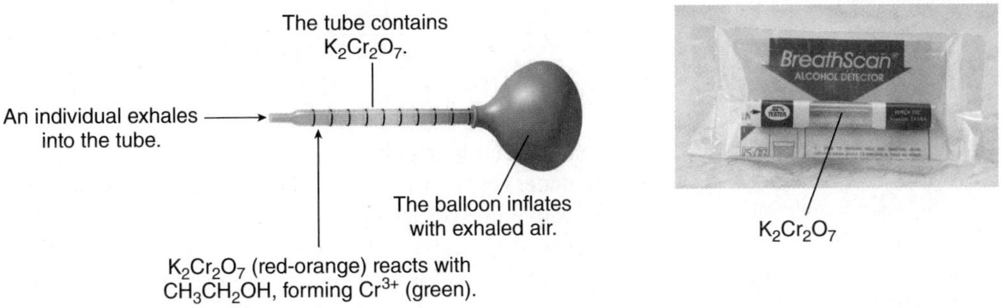
The tube contains
K2Cr2O7.
An individual exhales
into the tube.
The balloon inflates
with exhaled air.
K2Cr2O7 (red-orange) reacts with
CH3CH2OH, forming Cr3+ (green).
BreathScan
ALCOHOL DETECTOR
K2Cr2O7

- The oxidation of CH_3CH_2OH with $K_2Cr_2O_7$ to form CH_3COOH and Cr^{3+} was the first available method for the routine testing of alcohol concentration in exhaled air. Some consumer products for alcohol screening are still based on this technology.

Problem 12.24 Draw the organic products in each of the following reactions.

a. 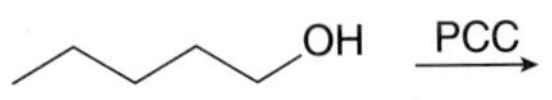
OH
PCC

c. 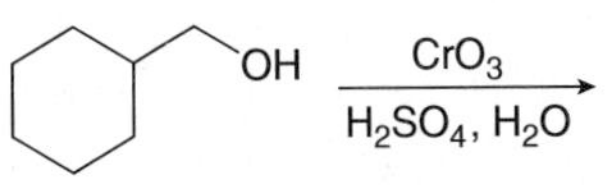
OH
CrO3
H2SO4, H2O

b. 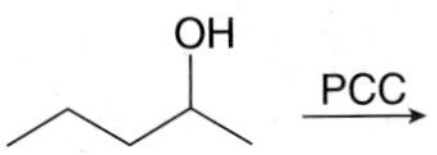
OH
PCC

d. 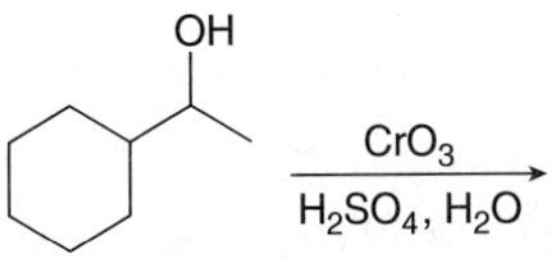
OH
CrO3
H2SO4, H2O

12.13 Green Chemistry

Several new methods of oxidation are based on green chemistry. ***Green chemistry* is the use of environmentally benign methods to synthesize compounds.** Its purpose is to use safer reagents and less solvent, and develop reactions that form fewer by-products and generate less waste.

Green polymer synthesis using starting materials derived from renewable resources (rather than petroleum) is discussed in Chapter 31.

Since many oxidation methods use toxic reagents (such as OsO_4 and O_3) and corrosive acids (such as H_2SO_4), or they generate carcinogenic by-products (such as Cr^{3+}), alternative reactions have been developed. One method uses a polymer-supported Cr^{6+} reagent—$HCrO_4^-$–Amberlyst A-26 resin—that avoids the use of strong acid, and forms a Cr^{3+} by-product that can be easily removed from the product by filtration.

The Amberlyst A-26 resin consists of a complex hydrocarbon network with cationic ammonium ion appendages that serve as counterions to the anionic chromium oxidant, $HCrO_4^-$. Heating the insoluble polymeric reagent with an alcohol results in oxidation to a carbonyl compound, with formation of an insoluble Cr^{3+} by-product. Not only can the metal by-product be removed by filtration without added solvent, it can also be regenerated and reused in a subsequent reaction.

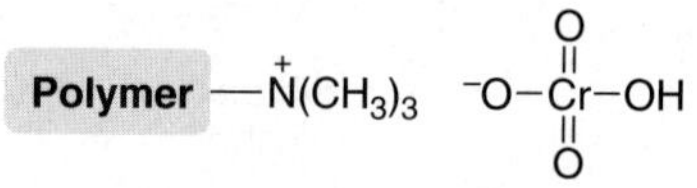

Amberlyst A-26 resin Cr^{6+} oxidant

With $HCrO_4^-$–Amberlyst A-26 resin, 1° alcohols are oxidized to aldehydes and 2° alcohols are oxidized to ketones.

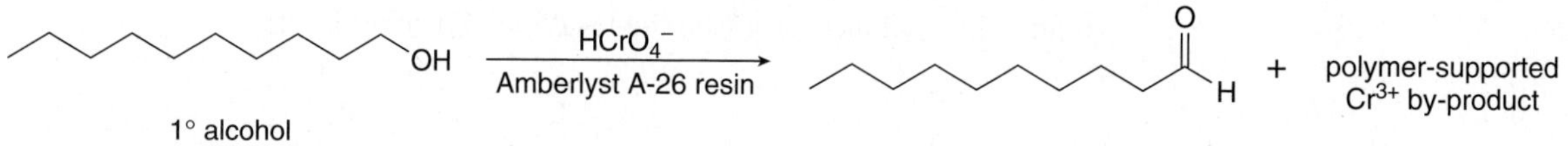

1° alcohol

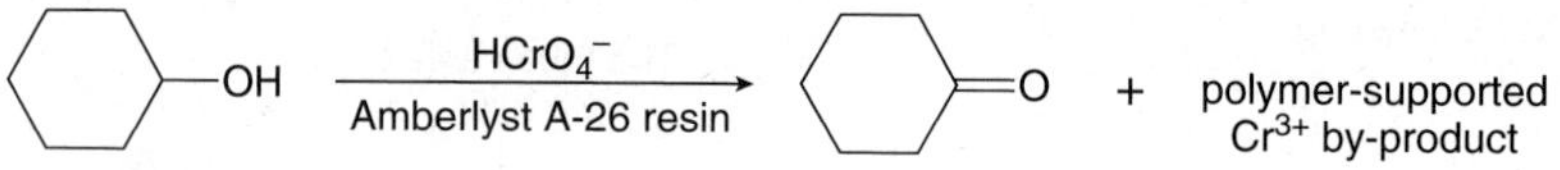

2° alcohol

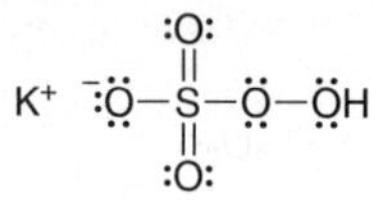

potassium peroxymonosulfate

Many other green approaches to oxidation that avoid the generation of metal by-products entirely are also under active investigation. For example, potassium peroxymonosulfate, $KHSO_5$, is a sulfate derivative of hydrogen peroxide, sold as a triple salt (2 $KHSO_5$•$KHSO_4$•K_2SO_4) under the trade name of Oxone. Oxone oxidizes a variety of substrates without the presence of a heavy metal like chromium or manganese, and in some cases, oxidation reactions can be carried out in water or aqueous solutions. The weak oxygen–oxygen bond of the reagent is cleaved during oxidation, and a sulfate salt (K_2SO_4) is formed as by-product. Two examples of oxidations of alcohols are shown.

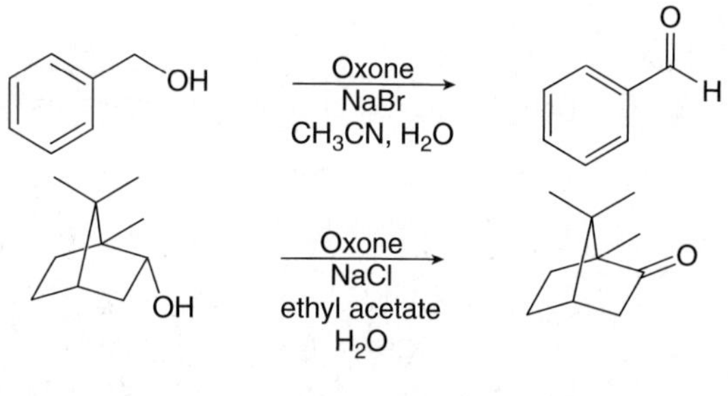

Problem 12.25 What carbonyl compound is formed when each alcohol is treated with $HCrO_4^-$–Amberlyst A-26 resin?

a. 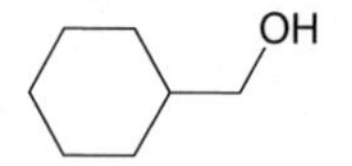
b. 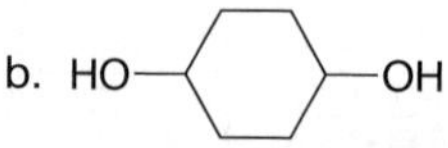
c. 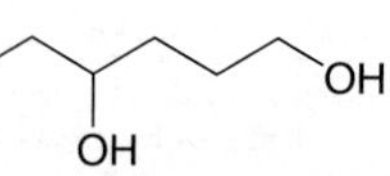

Problem 12.26 Sodium hypochlorite (NaOCl, the oxidant in household bleach) in aqueous CH_3COOH is also touted as a "green" oxidizing agent. For example, oxidation of $(CH_3)_2CHOH$ with NaOCl forms $(CH_3)_2C{=}O$ along with NaCl and H_2O. (a) What advantages and/or disadvantages does this method have over oxidation with $HCrO_4^-$–Amberlyst A-26 resin? (b) What advantages and/or disadvantages does this method have over oxidation with CrO_3, H_2SO_4, H_2O?

12.14 Biological Oxidation

Many reactions in biological systems involve oxidation or reduction. Instead of using Cr^{6+} reagents for oxidation, cells use two organic compounds—a high molecular weight **enzyme** and a simpler **coenzyme** that serves as the oxidizing agent.

The coenzyme often used to oxidize alcohols in biological systems is **nicotinamide adenine dinucleotide,** abbreviated as **NAD^+.** Although the structure is complex, only a portion of the molecule, labeled in red, participates in redox reactions.

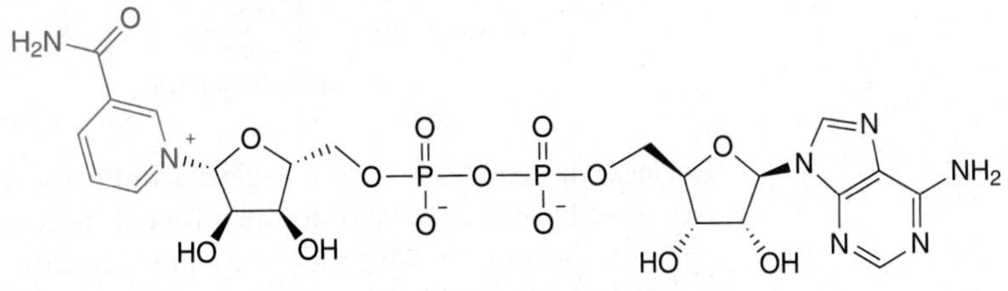

nicotinamide adenine dinucleotide
NAD^+

Biological oxidation of an alcohol occurs by transferring a hydride, a hydrogen atom with two electrons, from the alcohol to NAD^+ to form a carbonyl group. In the process, NAD^+ is reduced to nicotinamide adenine dinucleotide (reduced form), abbreviated as **NADH.** NADH is a biological reducing agent that converts carbonyl compounds to alcohols, as discussed in Section 20.6.

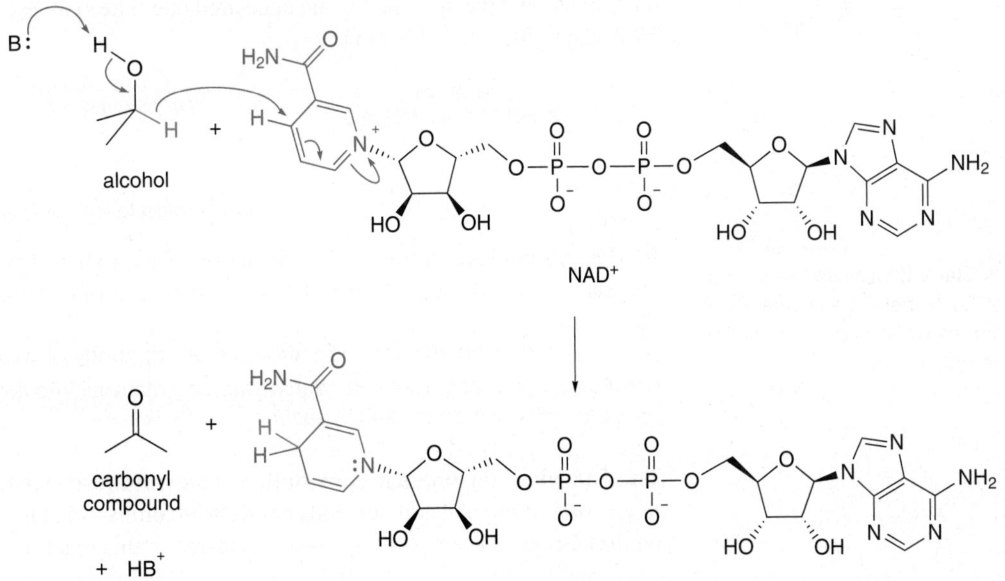

nicotinamide adenine dinucleotide
(reduced form)
NADH

For example, when CH_3CH_2OH (ethanol) is ingested, it is oxidized in the liver by NAD^+ to CH_3CHO (acetaldehyde), and then to CH_3COO^- (acetate anion, the conjugate base of acetic acid). Acetate is the starting material for the synthesis of fatty acids and cholesterol. Both oxidations are catalyzed by a dehydrogenase enzyme.

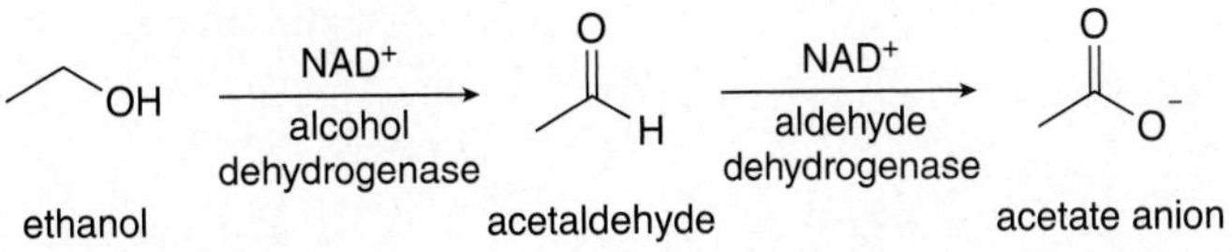

If more ethanol is ingested than can be metabolized in a given time, the concentration of acetaldehyde builds up. This toxic compound is responsible for the feelings associated with a hangover.

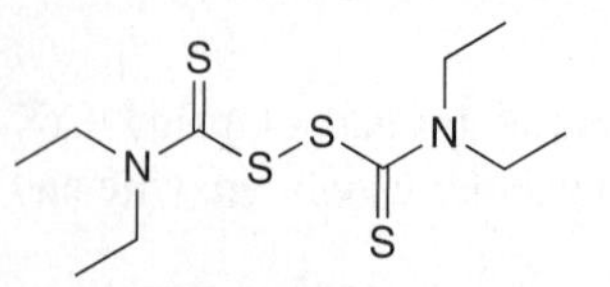

Antabuse

Antabuse, a drug given to alcoholics to prevent them from consuming alcoholic beverages, acts by interfering with the normal oxidation of ethanol. Antabuse inhibits the oxidation of acetaldehyde to the acetate anion. Because the first step in ethanol metabolism occurs but the second does not, the concentration of acetaldehyde rises, causing an individual to become violently ill.

Like ethanol, methanol is oxidized by the same enzymes to give an aldehyde and an acid: formaldehyde and formic acid. These oxidation products are extremely toxic because they cannot be used by the body. As a result, the pH of the blood decreases, and blindness and death can follow.

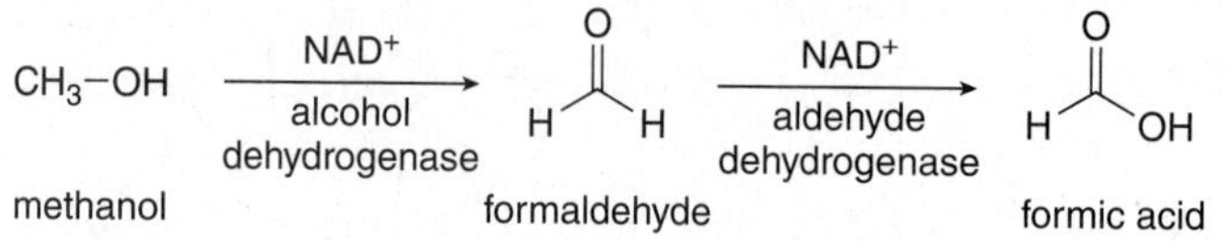

Because the enzymes have a higher affinity for ethanol than methanol, methanol poisoning is treated by giving ethanol to the afflicted individual. With both methanol and ethanol in the patient's system, the enzymes react more readily with ethanol, allowing the methanol to be excreted unchanged without the formation of methanol's toxic oxidation products.

12.15 Sharpless Epoxidation

In all of the reactions discussed so far, an **achiral starting material has reacted with an achiral reagent to give either an achiral product or a racemic mixture of two enantiomers.** If you are trying to make a chiral product, this means that only half of the product mixture is the desired enantiomer and the other half is the undesired one. The synthesis of disparlure, outlined in Figure 12.8, exemplifies this dilemma.

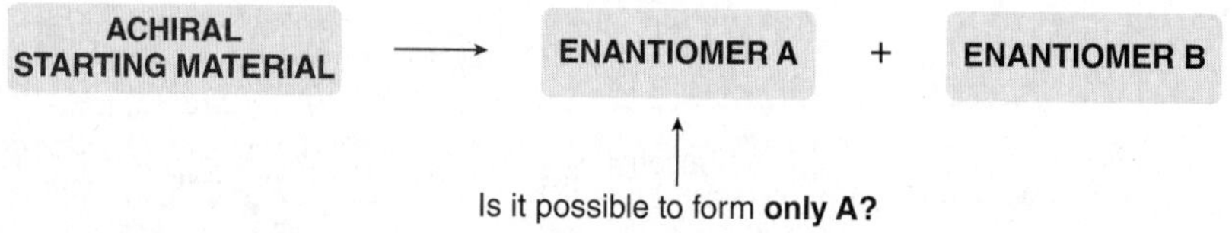

K. Barry Sharpless shared the 2001 Nobel Prize in Chemistry for his work on chiral oxidation reactions.

K. Barry Sharpless, currently at The Scripps Research Institute, reasoned that using a chiral reagent might make it possible to favor the formation of one enantiomer over the other.

- An *enantioselective* reaction affords predominantly or exclusively one enantiomer.
- A reaction that converts an achiral starting material into predominantly one enantiomer is also called an *asymmetric reaction.*

The Sharpless asymmetric epoxidation is an enantioselective reaction that oxidizes alkenes to epoxides. Only the double bonds of **allylic alcohols**—that is, alcohols having a hydroxy group on the carbon adjacent to a C=C—are oxidized in this reaction.

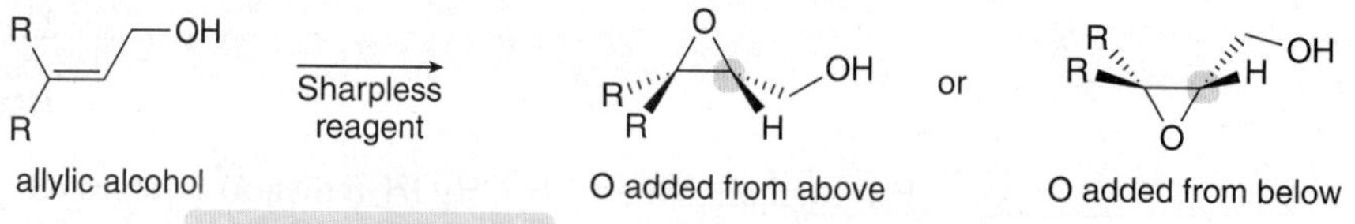

Sharpless reagent
$(CH_3)_3C–OOH$
$Ti[OCH(CH_3)_2]_4$
(+)- or (–)-diethyl tartrate

- With Sharpless reagent, one enantiomer is favored.
- The new stereogenic center is labeled in blue.

The **Sharpless reagent** consists of three components: *tert*-butyl hydroperoxide, **$(CH_3)_3COOH$;** a titanium catalyst—usually titanium(IV) isopropoxide, **$Ti[OCH(CH_3)_2]_4$;** and **diethyl tartrate (DET).** There are two different chiral diethyl tartrate isomers, labeled as (+)-DET or (–)-DET to indicate the direction in which they rotate polarized light.

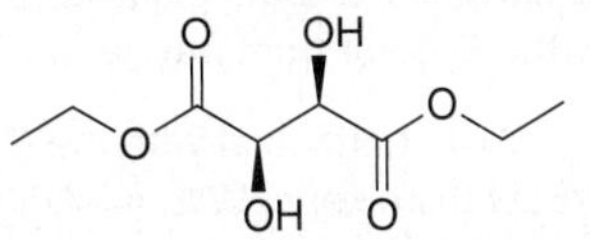

(+)-(*R*,*R*)-diethyl tartrate

(+)-DET

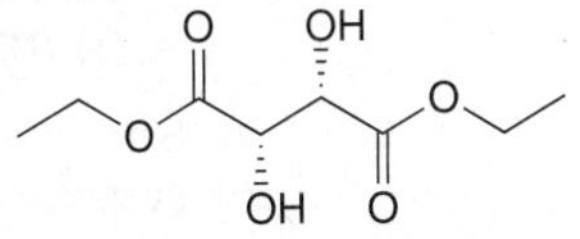

(–)-(*S*,*S*)-diethyl tartrate

(–)-DET

(+)-DET is prepared from (+)-(*R*,*R*)-tartaric acid [$HO_2CCH(OH)CH(OH)CO_2H$], a naturally occurring carboxylic acid found in grapes and sold as a by-product of the wine industry.

Enantiomeric excess = ***ee*** = % of one enantiomer – % of the other enantiomer.

The identity of the DET isomer determines which enantiomer is the major product obtained in the epoxidation of an allylic alcohol with the Sharpless reagent.

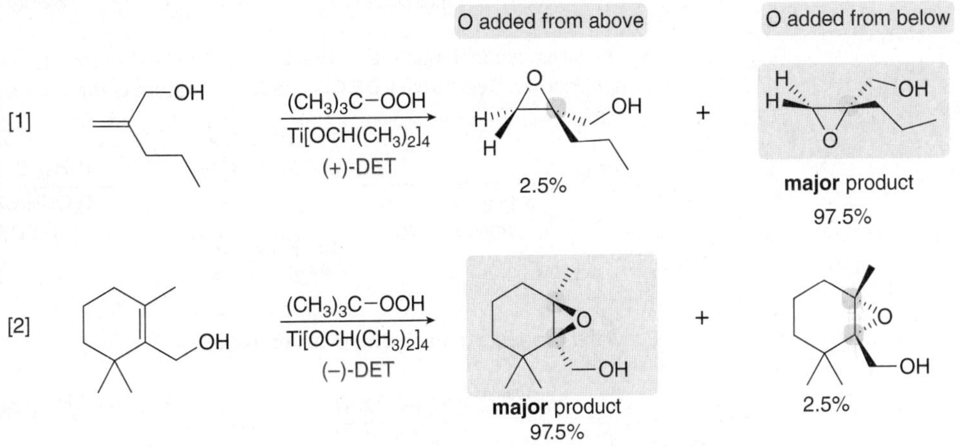

Stereogenic centers are labeled in blue.

The degree of enantioselectivity of a reaction is measured by its enantiomeric excess (***ee***) (Section 5.12D). Reactions [1] and [2] are highly enantioselective because each has an enantiomeric excess of 95% (97.5% of the major enantiomer – 2.5% of the minor enantiomer).

To determine which enantiomer is formed for a given isomer of DET, draw the allylic alcohol in a plane, with the **C=C horizontal and the OH group in the upper right corner;** then:

- **Epoxidation with (–)-DET adds an oxygen atom from above the plane.**
- **Epoxidation with (+)-DET adds an oxygen atom from below the plane.**

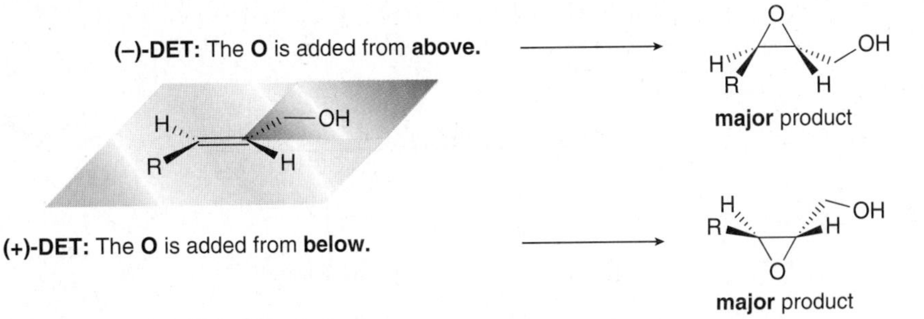

Sample Problem 12.5 Predict the major product in each epoxidation.

a.

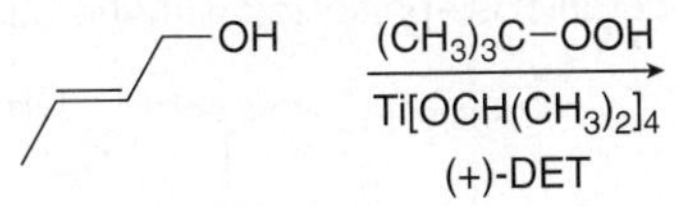

b.

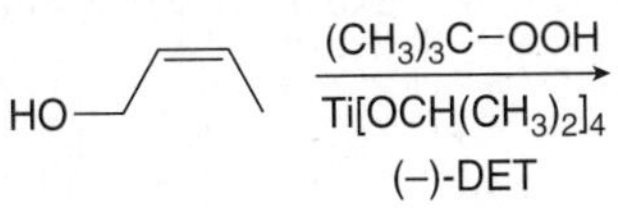

Solution

To draw an epoxidation product:

- Draw the allylic alcohol with the **C=C horizontal and the OH group in the upper right corner of the alkene.** Re-draw the alkene if necessary.
- **(+)-DET** adds the O atom from **below,** and **(–)-DET** adds the O atom from **above.**

a. Since the C=C is drawn horizontal with the OH group in the upper right corner, it is not necessary to re-draw the alkene. With **(+)-DET,** the O atom is added from **below.**

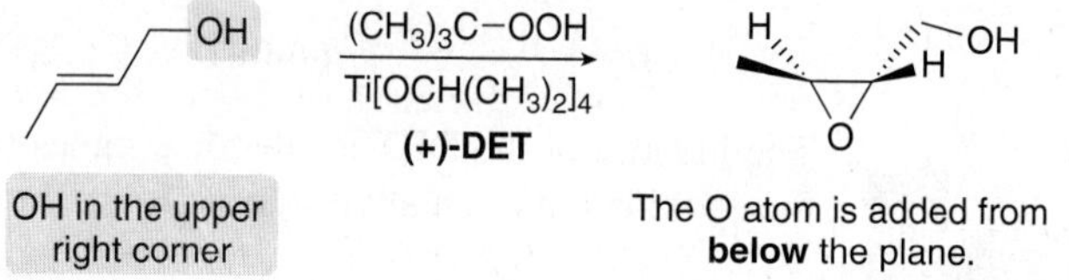

OH
$(CH_3)_3C–OOH$
$Ti[OCH(CH_3)_2]_4$
(+)-DET
H
H
OH
O
OH in the upper right corner
The O atom is added from below the plane.

b. The allylic alcohol must be re-drawn with the C=C horizontal and the OH group in the upper right corner. Because **(–)-DET** is used, the O atom is then added from **above.**

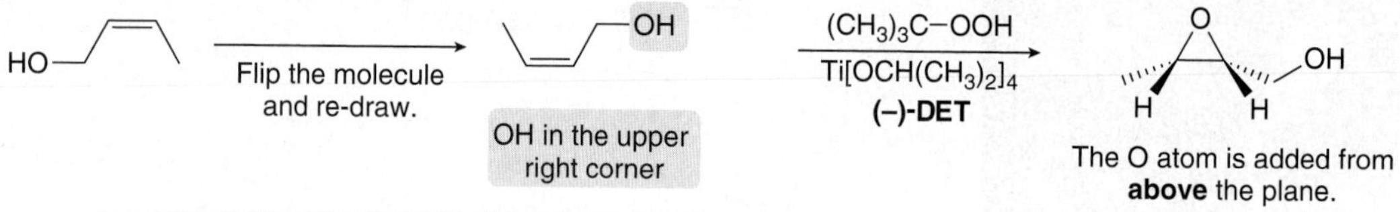

HO
Flip the molecule and re-draw.
OH
OH in the upper right corner
$(CH_3)_3C–OOH$
$Ti[OCH(CH_3)_2]_4$
(–)-DET
O
OH
H
H
The O atom is added from above the plane.

Problem 12.27 Draw the products of each Sharpless epoxidation.

a. 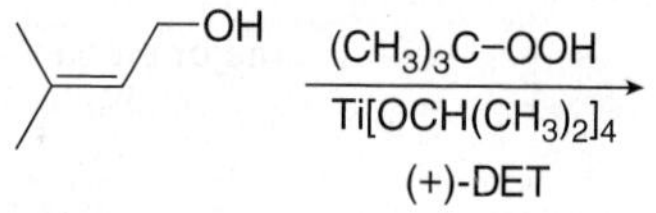

OH
$(CH_3)_3C–OOH$
$Ti[OCH(CH_3)_2]_4$
(+)-DET

b. 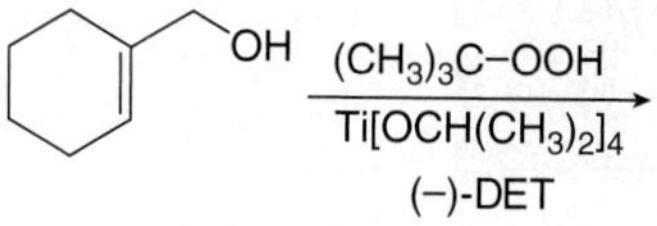

OH
$(CH_3)_3C–OOH$
$Ti[OCH(CH_3)_2]_4$
(–)-DET

The Sharpless epoxidation has been used to synthesize many chiral natural products, including two insect pheromones—(+)-α-multistriatin and (–)-frontalin, as shown in Figure 12.10.

Figure 12.10 The synthesis of chiral insect pheromones using asymmetric epoxidation

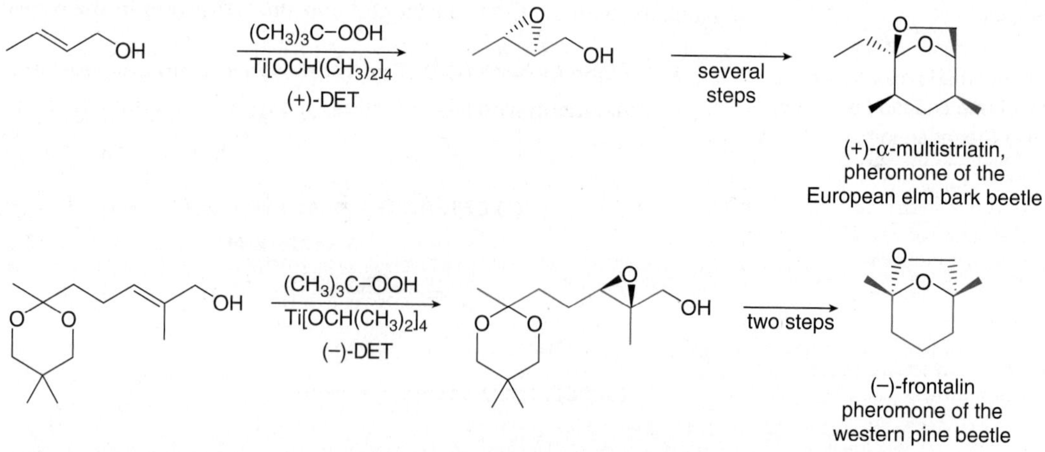

OH
$(CH_3)_3C–OOH$
$Ti[OCH(CH_3)_2]_4$
(+)-DET
O
OH
several steps
O
O
(+)-α-multistriatin, pheromone of the European elm bark beetle
OH
$(CH_3)_3C–OOH$
$Ti[OCH(CH_3)_2]_4$
(–)-DET
O
OH
two steps
(–)-frontalin pheromone of the western pine beetle

- The bonds in the products that originate from the epoxide intermediate are indicated in red.

Problem 12.28 Explain why only one C=C of geraniol is epoxidized with the Sharpless reagent.

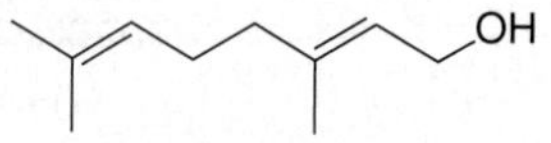

OH

geraniol

KEY CONCEPTS

Oxidation and Reduction

Summary: Terms That Describe Reaction Selectivity

- A **regioselective reaction** forms predominantly or exclusively one constitutional isomer (8.5).
- A **stereoselective reaction** forms predominantly or exclusively one stereoisomer (8.5).
- An **enantioselective reaction** forms predominantly or exclusively one enantiomer (12.15).

Definitions of Oxidation and Reduction (12.1)

Oxidation reactions result in:

- an increase in the number of C–Z bonds, *or*
- a decrease in the number of C–H bonds

Reduction reactions result in:

- a decrease in the number of C–Z bonds, *or*
- an increase in the number of C–H bonds

Reduction Reactions

[1] Reduction of alkenes—Catalytic hydrogenation (12.3)

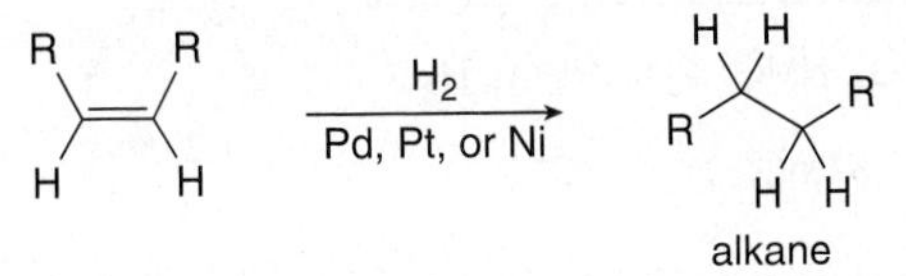

H_2
Pd, Pt, or Ni
alkane

- **Syn addition** of H_2 occurs.
- Increasing alkyl substitution on the C=C decreases the rate of reaction.

[2] Reduction of alkynes

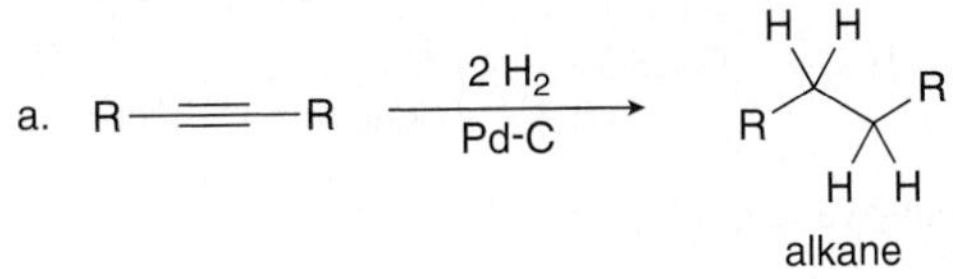

a.
2 H_2
Pd-C
alkane

- Two equivalents of H_2 are added and four new C–H bonds are formed (12.5A).

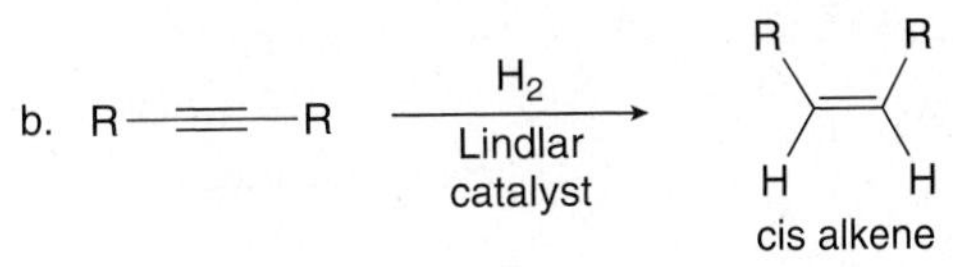

b.
H_2
Lindlar catalyst
cis alkene

- **Syn addition** of H_2 occurs, forming a **cis** alkene (12.5B).
- The Lindlar catalyst is deactivated; reaction stops after one equivalent of H_2 has added.

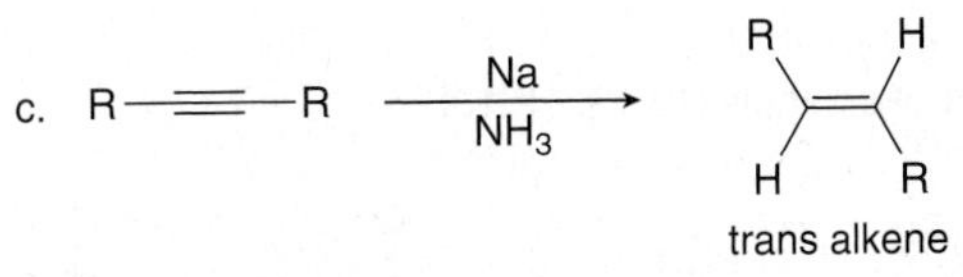

c.
Na
NH_3
trans alkene

- **Anti addition** of H_2 occurs, forming a **trans** alkene (12.5C).

[3] Reduction of alkyl halides (12.6)

$$\text{R–X} \xrightarrow[\text{[2] } H_2O]{\text{[1] } LiAlH_4} \underset{\text{alkane}}{\text{R–H}}$$

- The reaction follows an S_N2 mechanism.
- CH_3X and RCH_2X react faster than a more substituted RX.

[4] Reduction of epoxides (12.6)

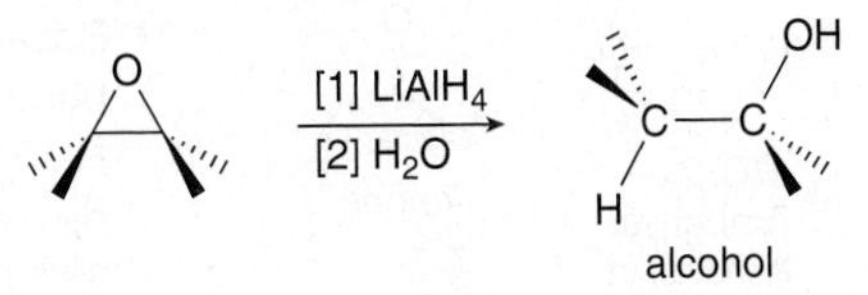

[1] $LiAlH_4$
[2] H_2O
alcohol

- The reaction follows an S_N2 mechanism.
- In unsymmetrical epoxides, H^- (from $LiAlH_4$) attacks at the less substituted carbon.

Oxidation Reactions

[1] Oxidation of alkenes

a. Epoxidation (12.8)

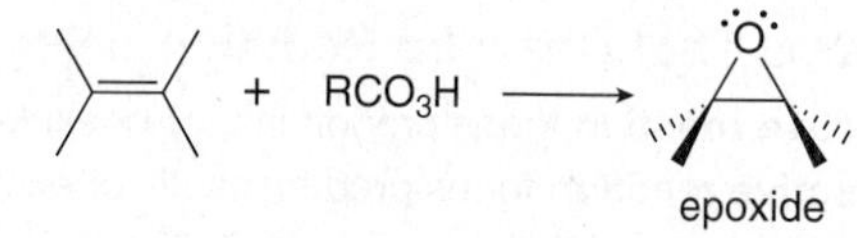
+ RCO3H
epoxide

- The mechanism has **one step.**
- **Syn addition** of an O atom occurs.
- The reaction is stereospecific.

b. Anti dihydroxylation (12.9A)

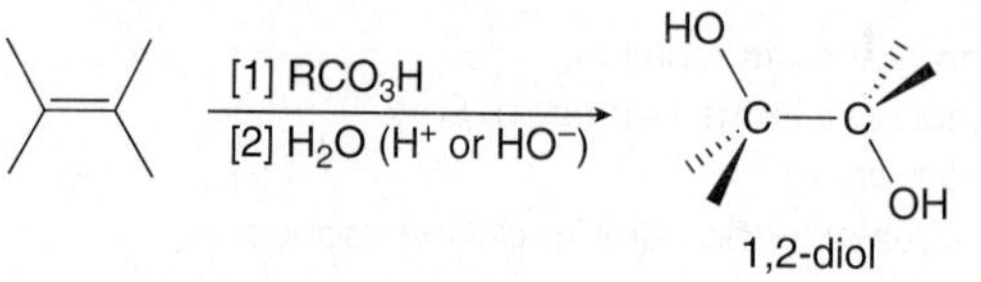
[1] RCO3H
[2] H2O (H+ or HO−)
HO
OH
1,2-diol

- Opening of an epoxide ring intermediate with ^{-}OH or H_2O forms a 1,2-diol with two OH groups added in an **anti** fashion.

c. Syn dihydroxylation (12.9B)

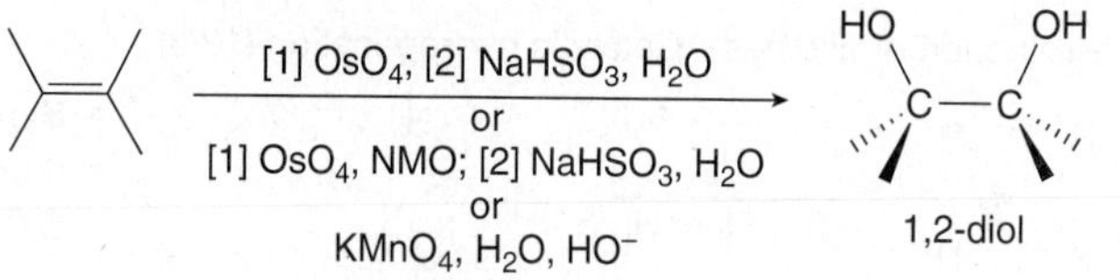
[1] OsO4; [2] NaHSO3, H2O
or
[1] OsO4, NMO; [2] NaHSO3, H2O
or
KMnO4, H2O, HO−
HO
OH
1,2-diol

- Each reagent adds two new C–O bonds to the C=C in a **syn** fashion.

d. Oxidative cleavage (12.10)

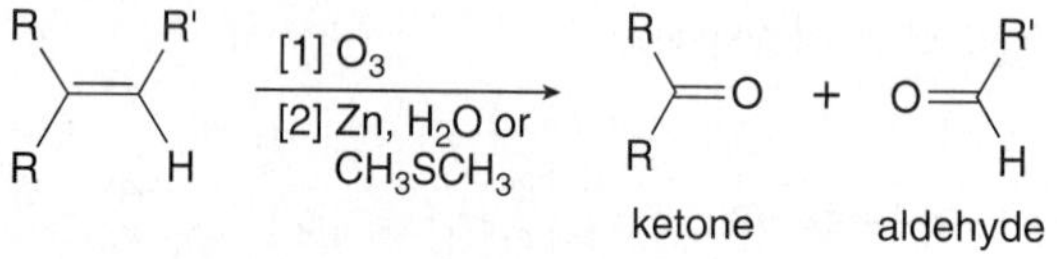
[1] O3
[2] Zn, H2O or CH3SCH3
ketone
aldehyde

- Both the σ and π bonds of the alkene are cleaved to form two carbonyl groups.

[2] Oxidative cleavage of alkynes (12.11)

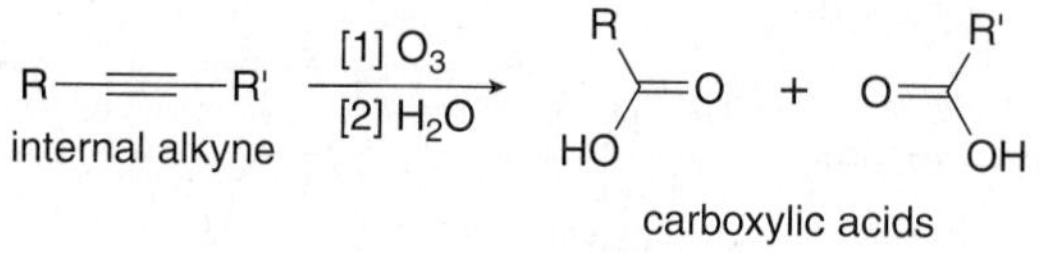
internal alkyne
[1] O3
[2] H2O
carboxylic acids

- The σ bond and both π bonds of the alkyne are cleaved.

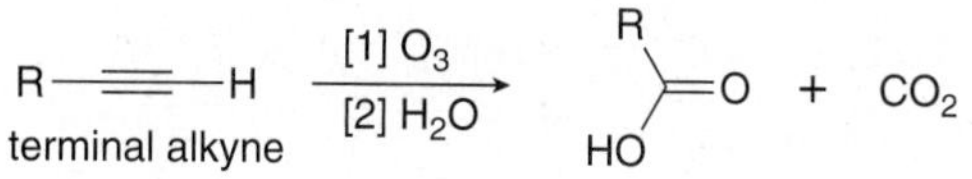
terminal alkyne
[1] O3
[2] H2O
+ CO2

[3] Oxidation of alcohols (12.12, 12.13)

a.

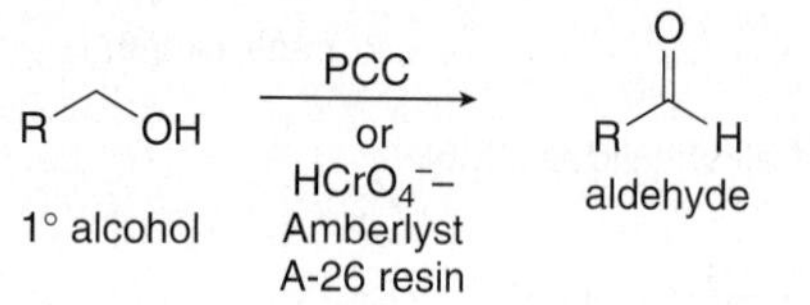
1° alcohol
PCC
or
HCrO4−–Amberlyst A-26 resin
aldehyde

- Oxidation of a 1° alcohol with PCC or $HCrO_4^-$–Amberlyst A-26 resin stops at the aldehyde stage. Only one C–H bond is replaced by a C–O bond.

b.

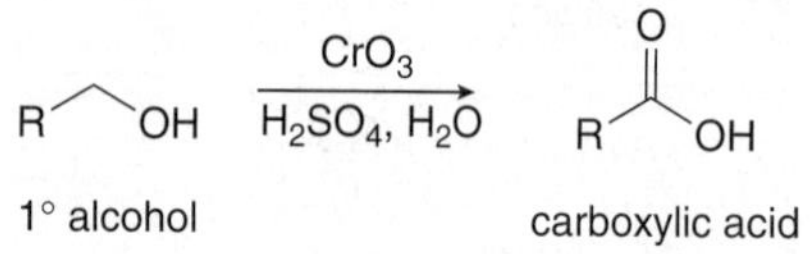
1° alcohol
CrO3
H2SO4, H2O
carboxylic acid

- Oxidation of a 1° alcohol under harsher reaction conditions—CrO_3 (or $Na_2Cr_2O_7$ or $K_2Cr_2O_7$) + H_2O + H_2SO_4—leads to a RCOOH. Two C–H bonds are replaced by two C–O bonds.

c.

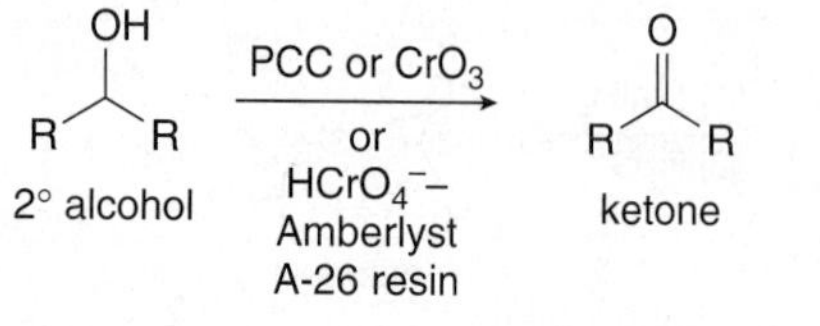
2° alcohol
PCC or CrO3
or
HCrO4−–Amberlyst A-26 resin
ketone

- Because a 2° alcohol has only one C–H bond on the carbon bearing the OH group, all Cr^{6+} reagents—PCC, CrO_3, $Na_2Cr_2O_7$, $K_2Cr_2O_7$, or $HCrO_4^-$ (Amberlyst A-26 resin)—oxidize a 2° alcohol to a ketone.

[4] Asymmetric epoxidation of allylic alcohols (12.15)

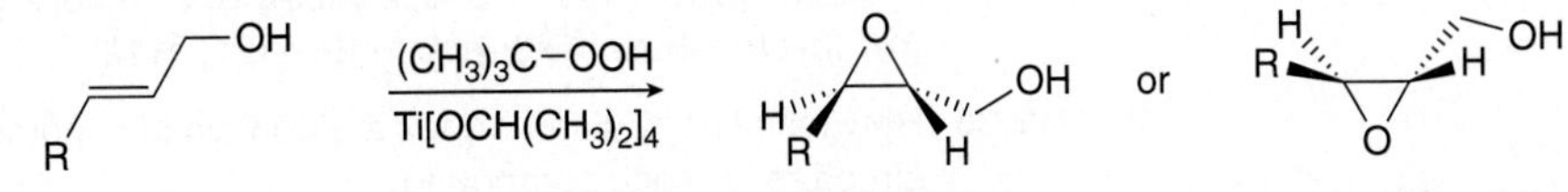

with (−)-DET with **(+)-DET**

PROBLEMS

Problems Using Three-Dimensional Models

12.29 Draw the products formed when **A** is treated with each reagent: (a) H_2 + Pd-C; (b) mCPBA; (c) PCC; (d) CrO_3, H_2SO_4, H_2O; (e) Sharpless reagent with (+)-DET.

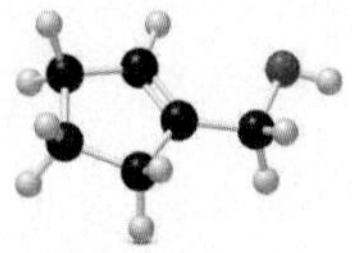

A

12.30 Draw the products formed when the following diene is treated with O_3 followed by CH_3SCH_3.

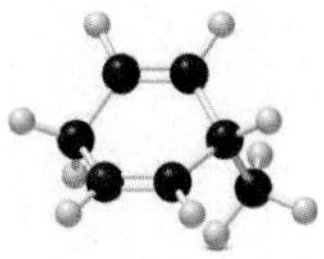

12.31 Devise a synthesis of the following compound from acetylene and organic compounds containing two or fewer carbons. You may use any other required reagents.

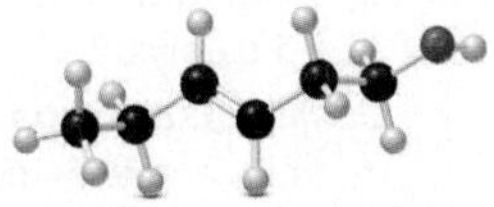

Classifying Reactions as Oxidation or Reduction

12.32 Label each reaction as oxidation, reduction, or neither.

a. 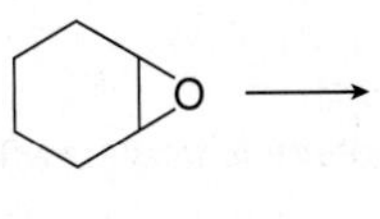

c. 

b. 

d. 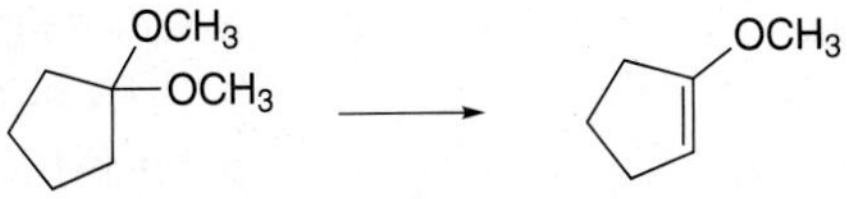

Hydrogenation

12.33 Draw the organic products formed when each alkene is treated with H_2, Pd-C. Indicate the three-dimensional structure of all stereoisomers formed.

a. 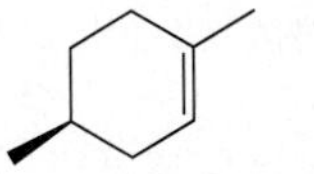
b. 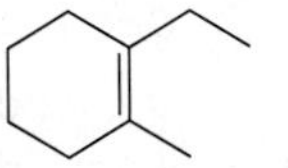
c. 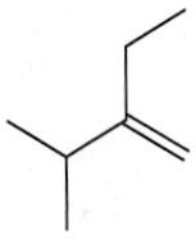

12.34 Match each alkene to its heat of hydrogenation.
Alkenes: 3-methylbut-1-ene, 2-methylbut-1-ene, 2-methylbut-2-ene
$\Delta H°$ (hydrogenation) kJ/mol: −119, −127, −112

12.35 How many rings and π bonds are contained in compounds **A–C?** Draw one possible structure for each compound.

a. Compound **A** has molecular formula C_5H_8 and is hydrogenated to a compound having molecular formula C_5H_{10}.
b. Compound **B** has molecular formula $C_{10}H_{16}$ and is hydrogenated to a compound having molecular formula $C_{10}H_{18}$.
c. Compound **C** has molecular formula C_8H_8 and is hydrogenated to a compound having molecular formula C_8H_{16}.

12.36 For alkenes **A, B,** and **C:** (a) Rank **A, B,** and **C** in order of increasing heat of hydrogenation; (b) rank **A, B,** and **C** in order of increasing rate of reaction with H_2, Pd-C; (c) draw the products formed when each alkene is treated with ozone, followed by Zn, H_2O.

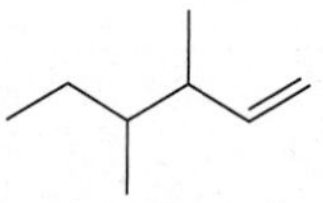
A

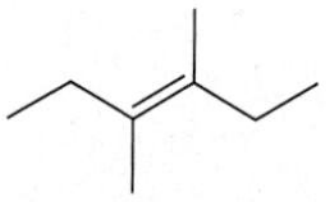
B

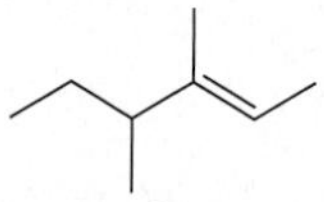
C

12.37 Stearidonic acid ($C_{18}H_{28}O_2$) is an unsaturated fatty acid obtained from oils isolated from hemp and blackcurrant (see also Problem 10.11).

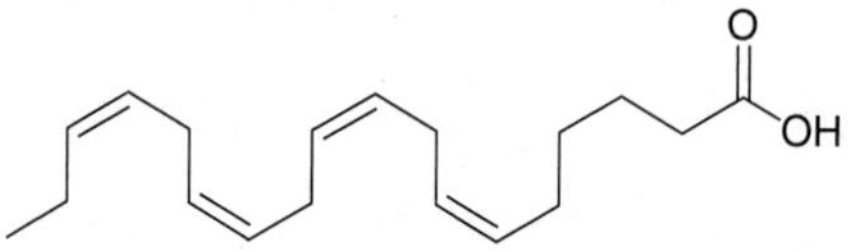
stearidonic acid

a. What fatty acid is formed when stearidonic acid is hydrogenated with excess H_2 and a Pd catalyst?
b. What fatty acids are formed when stearidonic acid is hydrogenated with one equivalent of H_2 and a Pd catalyst?
c. Draw the structure of a possible product formed when stearidonic acid is hydrogenated with one equivalent of H_2 and a Pd catalyst, and one double bond is isomerized to a trans isomer.
d. How do the melting points of the following fatty acids compare: stearidonic acid; one of the products formed in part (b); the product drawn in part (c)?

Reactions—General

12.38 Draw the organic products formed when cyclopentene is treated with each reagent. With some reagents, no reaction occurs.

a. H_2 + Pd-C
b. H_2 + Lindlar catalyst
c. Na, NH_3
d. CH_3CO_3H
e. [1] CH_3CO_3H; [2] H_2O, HO^-
f. [1] OsO_4 + NMO; [2] $NaHSO_3$, H_2O
g. $KMnO_4$, H_2O, HO^-
h. [1] $LiAlH_4$; [2] H_2O
i. [1] O_3; [2] CH_3SCH_3
j. $(CH_3)_3COOH$, $Ti[OCH(CH_3)_2]_4$, (−)-DET
k. mCPBA
l. Product in (k); then [1] $LiAlH_4$; [2] H_2O

12.39 Draw the organic products formed when allylic alcohol **A** is treated with each reagent.

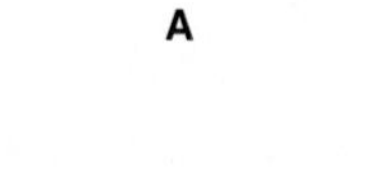
A

a. H_2 + Pd-C
b. mCPBA
c. PCC
d. CrO_3, H_2SO_4, H_2O
e. $(CH_3)_3COOH$, $Ti[OCH(CH_3)_2]_4$, (+)-DET
f. $(CH_3)_3COOH$, $Ti[OCH(CH_3)_2]_4$, (−)-DET
g. [1] PBr_3; [2] $LiAlH_4$; [3] H_2O
h. $HCrO_4^-$–Amberlyst A-26 resin

12.40 Draw the organic products formed in each reaction.

a. 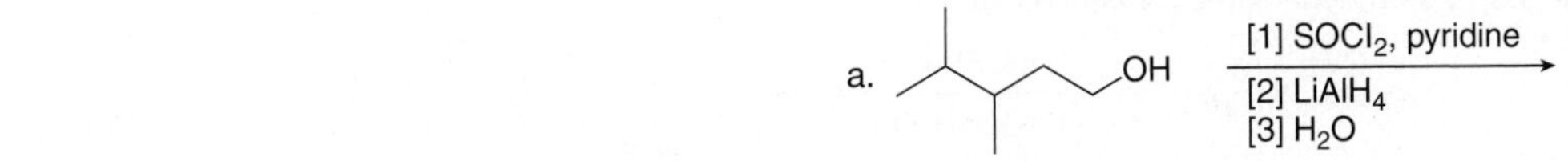
OH
[1] SOCl2, pyridine
[2] LiAlH4
[3] H2O

c. 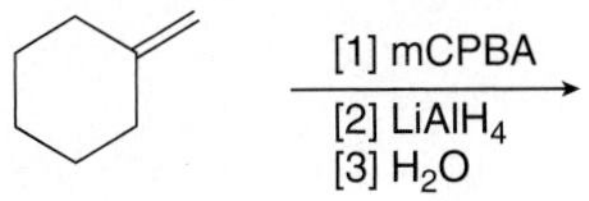
[1] mCPBA
[2] LiAlH4
[3] H2O

b. 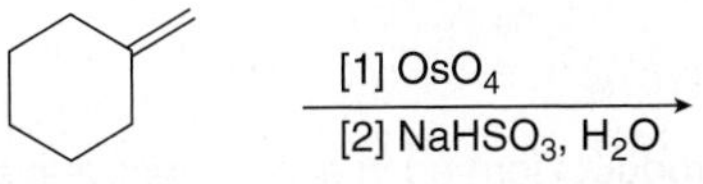
[1] OsO4
[2] NaHSO3, H2O

d. 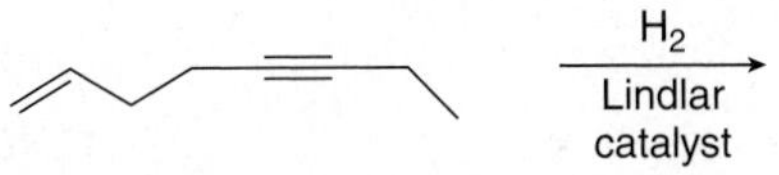
H2
Lindlar
catalyst

12.41 One step in the degradation of fats involves the reaction of (*R*)-glycerol phosphate with NAD^+ in the presence of the enzyme glycerol phosphate dehydrogenase. What products are formed if reaction occurs only at the 2° alcohol?

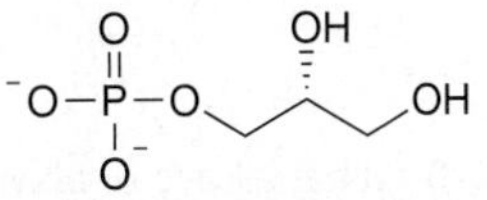
O
OH
O–P–O
OH
O

(*R*)-glycerol phosphate

12.42 Draw the structure of two different epoxides that would yield 2-methylpentan-2-ol [$(CH_3)_2C(OH)CH_2CH_2CH_3$] when reduced with $LiAlH_4$.

12.43 Hydrogenation of alkene **A** with D_2 in the presence of Pd-C affords a single product **B.** Keeping this result in mind, what compound is formed when **A** is treated with each reagent: (a) mCPBA; (b) Br_2, H_2O followed by base? Explain these results.

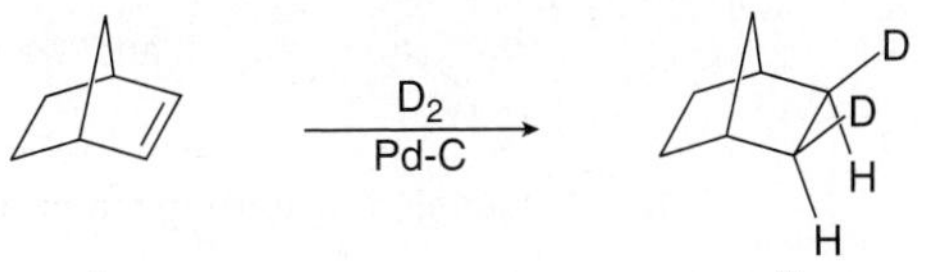
D2
Pd-C
D
D
H
H

A **B**

12.44 What alkene is needed to synthesize each 1,2-diol using [1] OsO_4 followed by $NaHSO_3$ in H_2O; or [2] CH_3CO_3H followed by ^-OH in H_2O?

a. 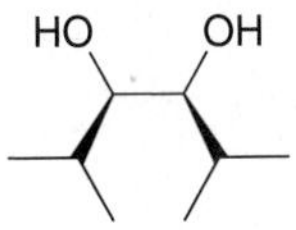
HO
OH

b. 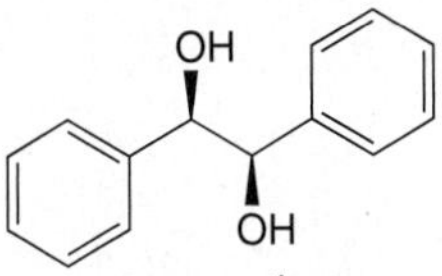
OH
OH

+ enantiomer

c. 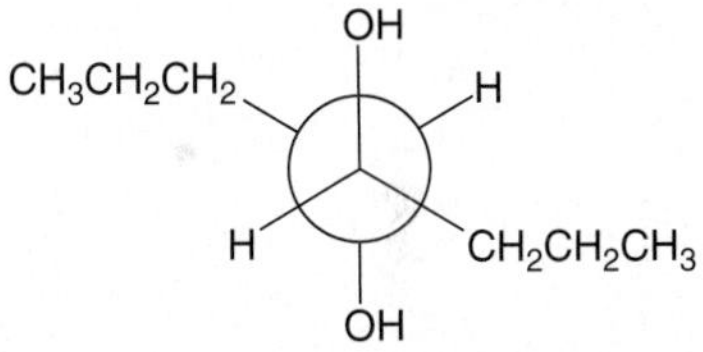
OH
CH3CH2CH2
H
H
CH2CH2CH3
OH

12.45 Draw a stepwise mechanism for the reduction of epoxide **A** to alcohol **B** using $LiAlH_4$. What product would be formed if $LiAlD_4$ were used as reagent? Indicate the stereochemistry of all stereogenic centers in the product using wedges and dashed wedges.

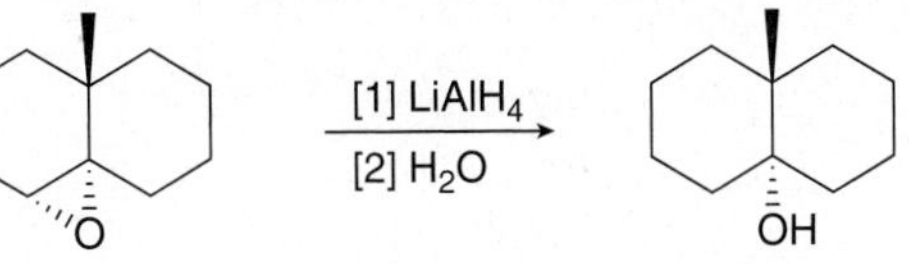
[1] LiAlH4
[2] H2O
O
OH

A **B**

12.46 (a) What product is formed in Step [1] of the following reaction sequence? (b) Draw a mechanism for Step [2] that accounts for the observed stereochemistry. (c) What reaction conditions are necessary to form chiral **A** from prop-2-en-1-ol ($CH_2=CHCH_2OH$)?

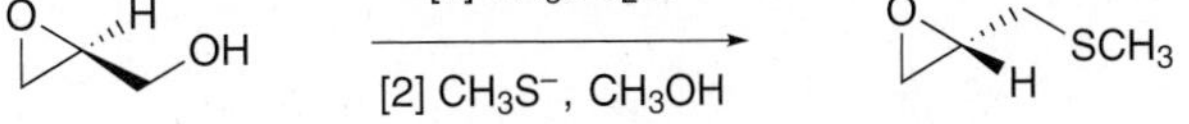
O
H
OH
[1] CH3SO2Cl
[2] CH3S−, CH3OH
O
SCH3
H

A **B**

12.47 Draw the products formed after Steps [1] and [2] in the following three-step sequence. Then draw stepwise mechanisms for each step.

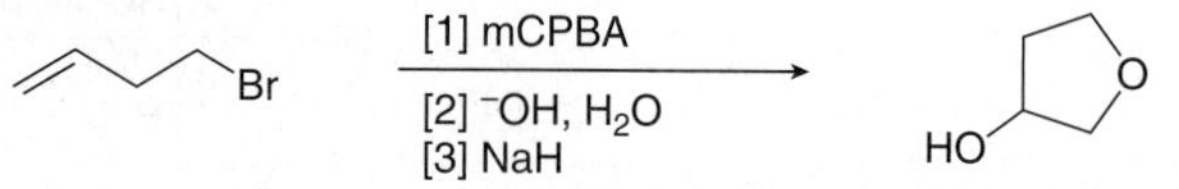
Br
[1] mCPBA
[2] ⁻OH, H₂O
[3] NaH
HO
O

Oxidative Cleavage

12.48 Draw the products formed in each oxidative cleavage.

a. 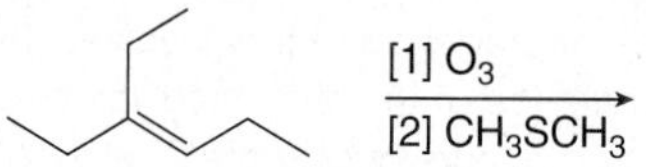
[1] O₃
[2] CH₃SCH₃

c. 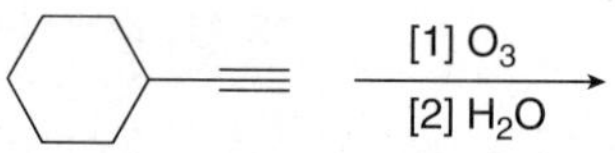
[1] O₃
[2] H₂O

b. 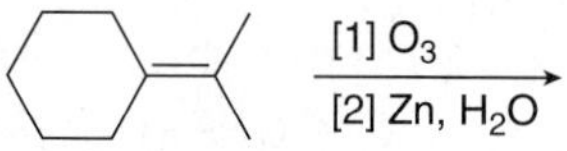
[1] O₃
[2] Zn, H₂O

d. 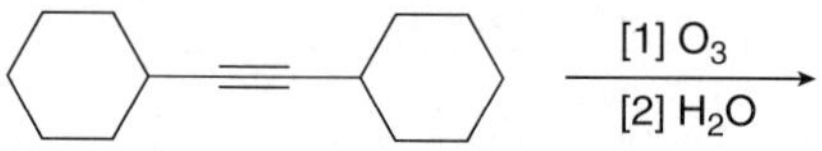
[1] O₃
[2] H₂O

12.49 What alkene or alkyne yields each set of products after oxidative cleavage with ozone?

a. 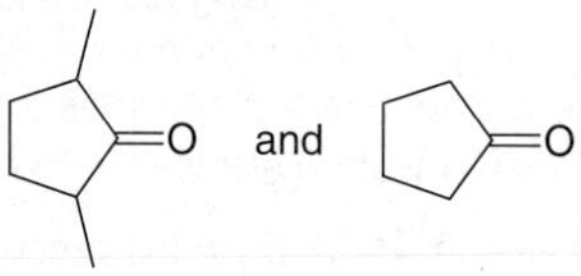
O
and
O

c. 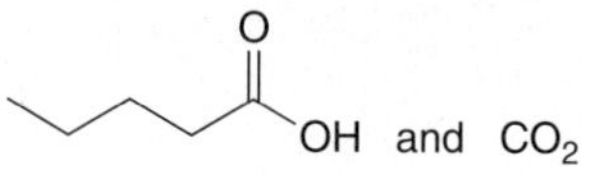
O
OH and CO₂

b. 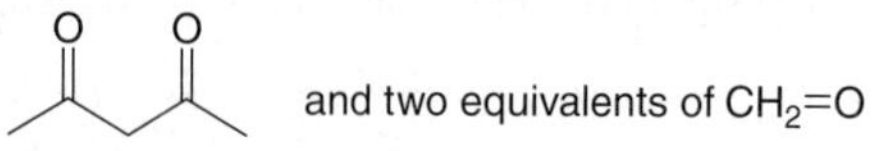
O O
and two equivalents of CH₂=O

d. 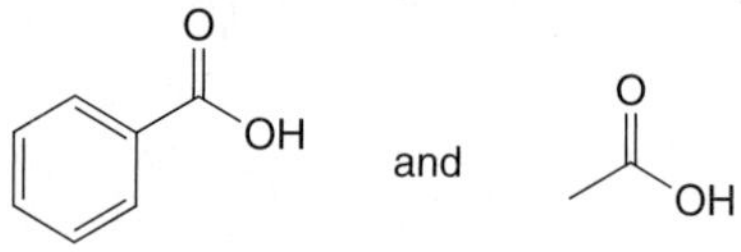
O
OH
and
O
OH

12.50 Identify the starting material in each reaction.

a. 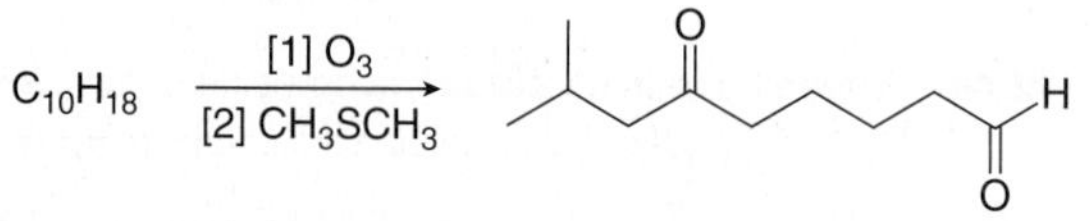
C₁₀H₁₈
[1] O₃
[2] CH₃SCH₃
O
H
O

b. 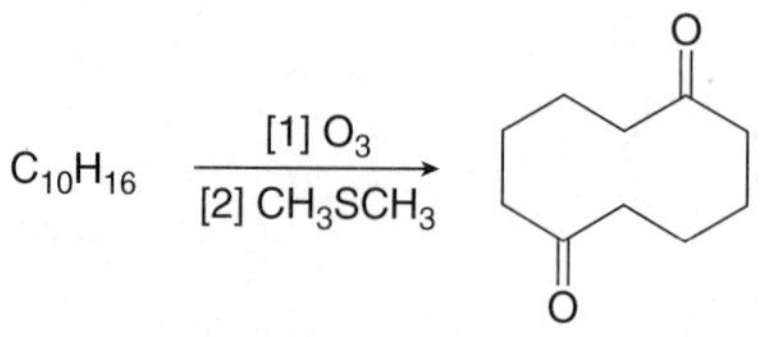
C₁₀H₁₆
[1] O₃
[2] CH₃SCH₃
O
O

12.51 Draw the products formed when each naturally occurring compound is treated with O_3 followed by Zn, H_2O.

a. 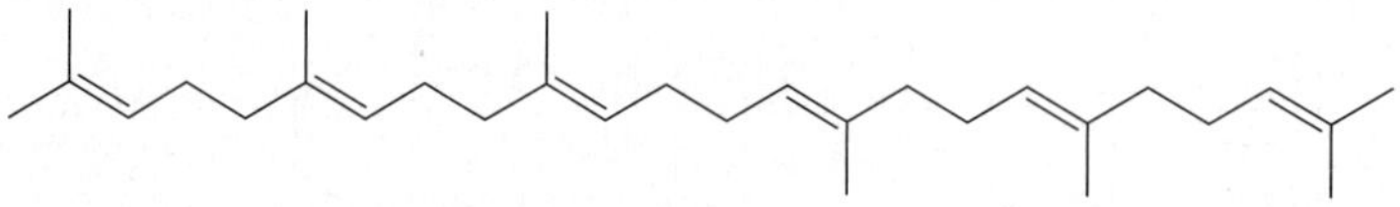

squalene

b. 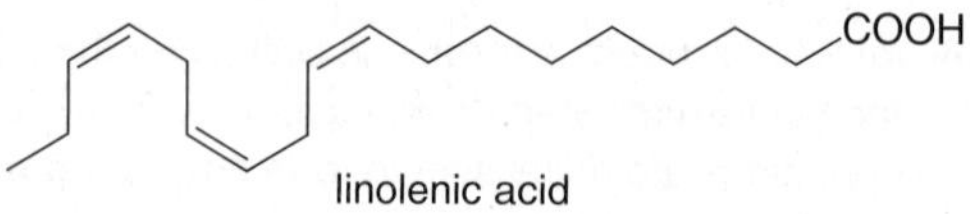
COOH

linolenic acid

c. 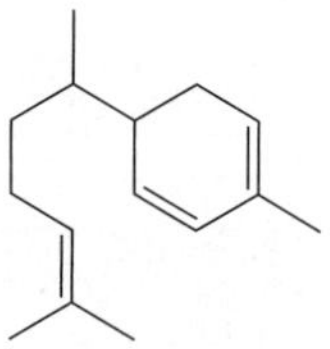

zingiberene

Identifying Compounds from Reactions

12.52 Identify compounds **A, B,** and **C.**

a. Compound **A** has molecular formula C_8H_{12} and reacts with two equivalents of H_2. **A** gives $HCOCH_2CH_2CHO$ as the only product of oxidative cleavage with O_3 followed by CH_3SCH_3.

b. Compound **B** has molecular formula C_6H_{10} and gives $(CH_3)_2CHCH_2CH_2CH_3$ when treated with excess H_2 in the presence of Pd. **B** reacts with $NaNH_2$ and CH_3I to form compound **C** (molecular formula C_7H_{12}).

12.53 Oximene and myrcene, two hydrocarbons isolated from alfalfa that have the molecular formula $C_{10}H_{16}$, both yield 2,6-dimethyloctane when treated with H_2 and a Pd catalyst. Ozonolysis of oximene forms $(CH_3)_2C=O$, $CH_2=O$, $CH_2(CHO)_2$, and CH_3COCHO. Ozonolysis of myrcene yields $(CH_3)_2C=O$, $CH_2=O$ (two equiv), and $HCOCH_2CH_2COCHO$. Identify the structures of oximene and myrcene.

12.54 An unknown compound **A** of molecular formula $C_{10}H_{18}O$ reacts with H_2SO_4 to form two compounds (**B** and **C**) of molecular formula $C_{10}H_{16}$. **B** and **C** both react with H_2 in the presence of Pd-C to form decalin. Ozonolysis of **B** forms **D,** and ozonolysis of **C** forms a diketone **E** of molecular formula $C_{10}H_{16}O_2$. Identify the structures of compounds **A, B, C,** and **E.**

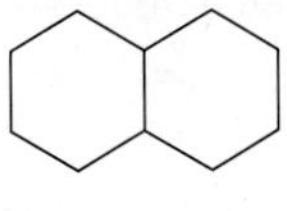

decalin

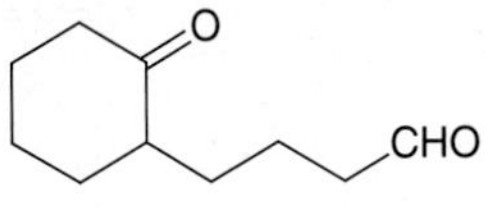
O
CHO

D

12.55 DHA is a fatty acid derived from fish oil and an abundant fatty acid in vertebrate brains. Hydrogenation of DHA forms docosanoic acid [$CH_3(CH_2)_{20}CO_2H$], and ozonolysis forms CH_3CH_2CHO, $CH_2(CHO)_2$ (five equivalents), and $HCOCH_2CH_2CO_2H$. What is the structure of DHA if all double bonds have the *Z* configuration?

12.56 One compound that contributes to the "seashore smell" at beaches in Hawai'i is dictyopterene D', a component of a brown edible seaweed called limu lipoa. Hydrogenation of dictyopterene D' with excess H_2 in the presence of a Pd catalyst forms butylcycloheptane. Ozonolysis with O_3 followed by $(CH_3)_2S$ forms $CH_2(CHO)_2$, $HCOCH_2CH(CHO)_2$, and CH_3CH_2CHO. What are possible structures of dictyopterene D'?

Sharpless Asymmetric Epoxidation

12.57 Draw the product of each asymmetric epoxidation reaction.

a. 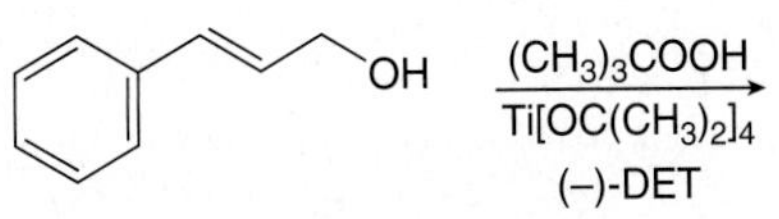
OH
(CH3)3COOH
Ti[OC(CH3)2]4
(–)-DET

b. 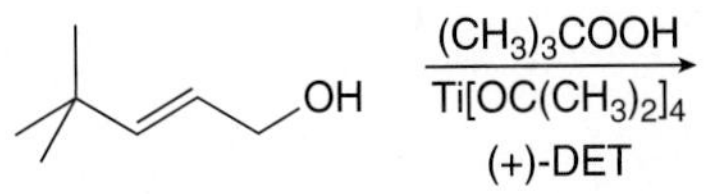
OH
(CH3)3COOH
Ti[OC(CH3)2]4
(+)-DET

12.58 Epoxidation of the following allylic alcohol using the Sharpless reagent with (–)-DET gives two epoxy alcohols in a ratio of 87:13.

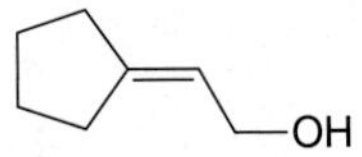
OH

a. Assign structures to the major and minor product.

b. What is the enantiomeric excess in this reaction?

12.59 What allylic alcohol and DET isomer are needed to make each chiral epoxide using a Sharpless asymmetric epoxidation reaction?

a. 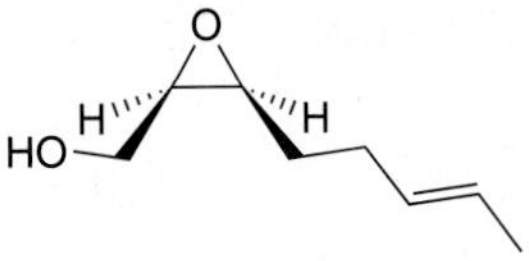
O
H
H
HO

b. 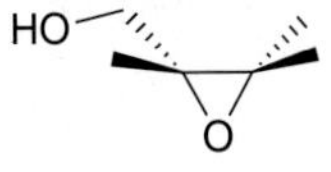
HO
O

c. 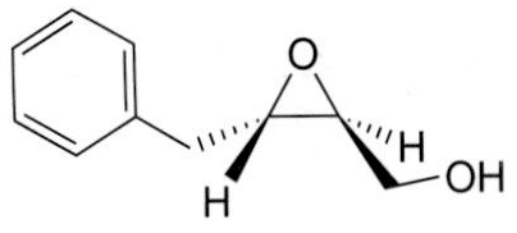
O
H
H
OH

12.60 Identify **A** in the following reaction sequence and draw a mechanism for the conversion of **A** to **B. B** has been converted to (*S,S*)-reboxetine, an antidepressant marketed outside the United States.

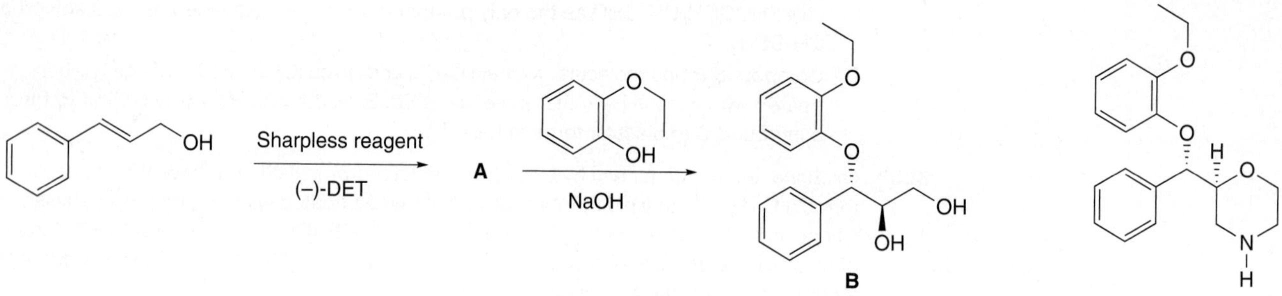

(*S,S*)-reboxetine

Synthesis

12.61 Devise a synthesis of muscalure, the sex pheromone of the common housefly, from acetylene and any other required reagents.

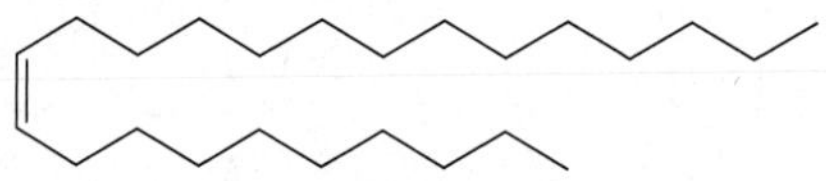

muscalure

12.62 It is sometimes necessary to isomerize a cis alkene to a trans alkene in a synthesis, a process that cannot be accomplished in a single step. Using the reactions you have learned in Chapters 8–12, devise a stepwise method to convert *cis*-but-2-ene to *trans*-but-2-ene.

12.63 Devise a synthesis of each compound from acetylene and any other required reagents.

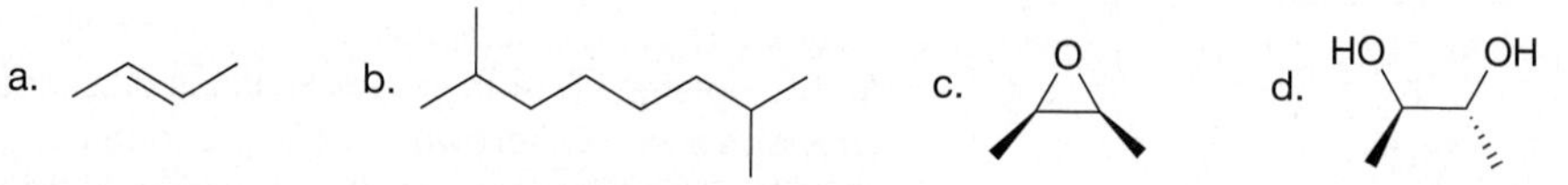

12.64 Devise a synthesis of compound **A** from the given starting materials. You may use any other inorganic reagents or organic alcohols. **A** was used to prepare aliskiren, a drug used to treat hypertension (see also Problem 5.7).

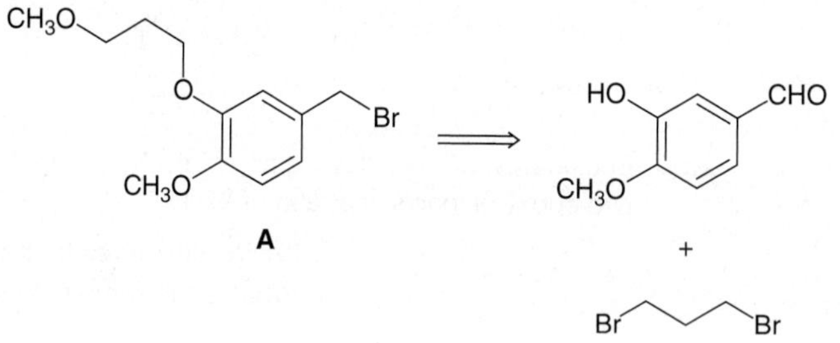

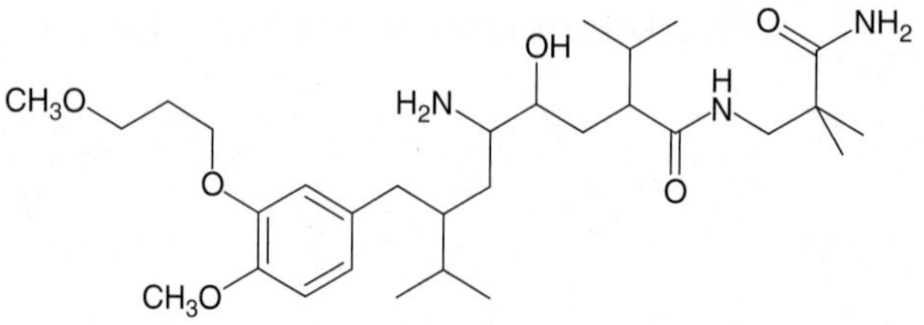

aliskiren

12.65 Devise a synthesis of (*E*)-hex-2-ene from pent-1-ene and any needed organic compounds or inorganic reagents.

12.66 Devise a synthesis of each compound from the indicated starting material, organic compounds containing one or two carbons, and any other required reagents.

a. 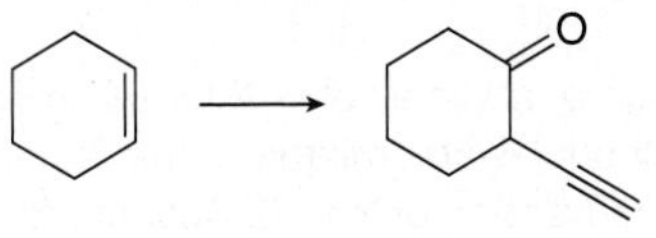

b. 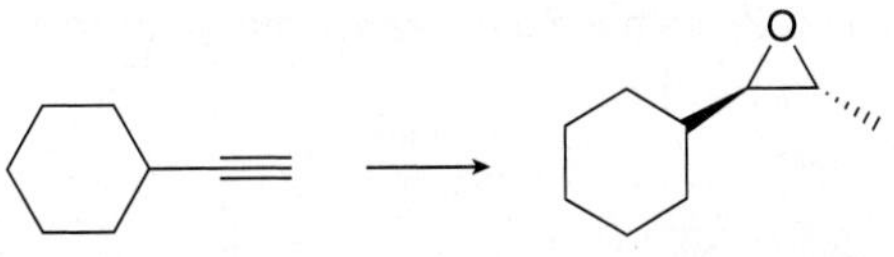

(+ enantiomer)

c. 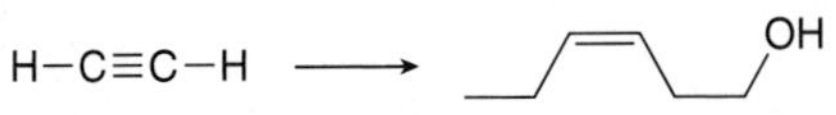

d. 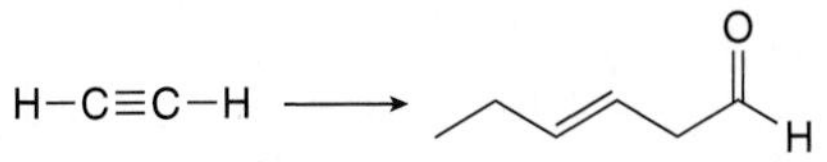

12.67 Devise a synthesis of 1-phenyl-5-methylhexane [$C_6H_5(CH_2)_4CH(CH_3)_2$] from acetylene, alkyl halides, and any required inorganic reagents.

12.68 Devise a synthesis of (3*R*,4*S*)-3,4-dichlorohexane from acetylene and any needed organic compounds or inorganic reagents.

12.69 Devise a synthesis of each compound from CH_3CH_2OH as the only organic starting material; that is, every carbon in the product must come from a molecule of ethanol. You may use any other needed inorganic reagents.

a. 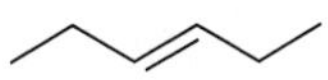

b. 

c. 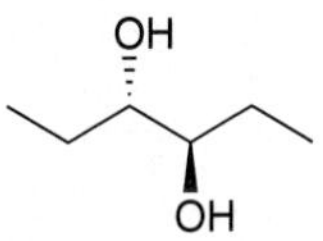

Challenge Problems

12.70 The Birch reduction is a dissolving metal reaction that converts substituted benzenes to cyclohexa-1,4-dienes using Li and liquid ammonia in the presence of an alcohol. Draw a stepwise mechanism for the following Birch reduction.

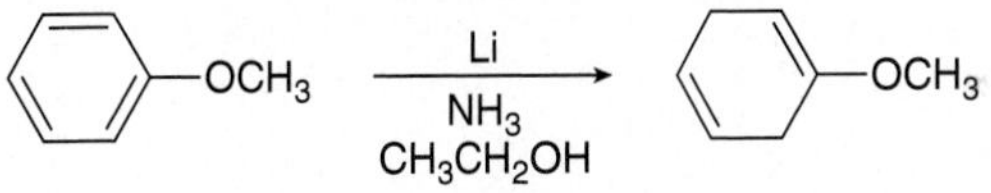

12.71 In the Cr^{6+} oxidation of cyclohexanols, it is generally true that sterically hindered alcohols react faster than unhindered alcohols. Which of the following alcohols should be oxidized more rapidly?

12.72 Draw a stepwise mechanism for the following reaction.

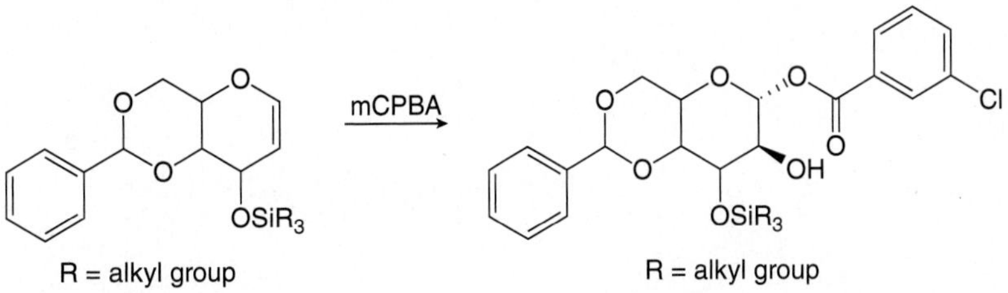

R = alkyl group R = alkyl group

12.73 Dihydroxylation of an alkene can be carried out with H_2O_2 in HCO_2H. In this reaction, *trans*-but-2-ene affords (2*R*,3*S*)-butane-2,3-diol, whereas *cis*-but-2-ene affords a mixture of (2*R*,3*R*)-butane-2,3-diol and (2*S*,3*S*)-butane-2,3-diol. Does dihydroxylation by this method occur with syn or anti addition?

12.74 Draw a stepwise mechanism for the following reaction.

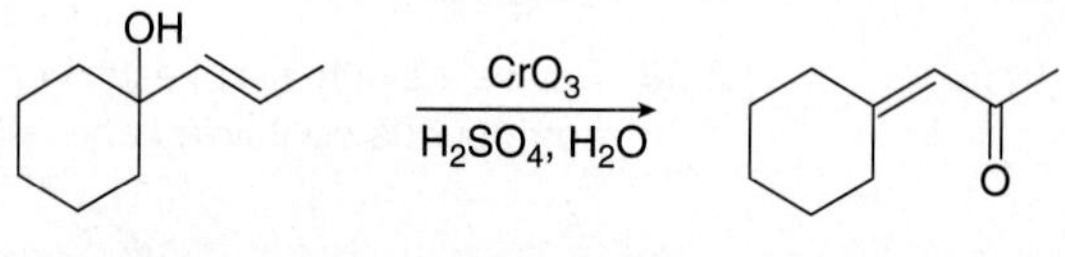
OH
CrO_3
H_2SO_4, H_2O
O

12.75 Sharpless epoxidation of allylic alcohol **X** forms compound **Y.** Treatment of **Y** with NaOH and C_6H_5SH in an alcohol–water mixture forms **Z.** Identify the structure of **Y** and draw a mechanism for the conversion of **Y** to **Z.** Account for the stereochemistry of the stereogenic centers in **Z. Z** has been used as an intermediate in the synthesis of chiral carbohydrates.

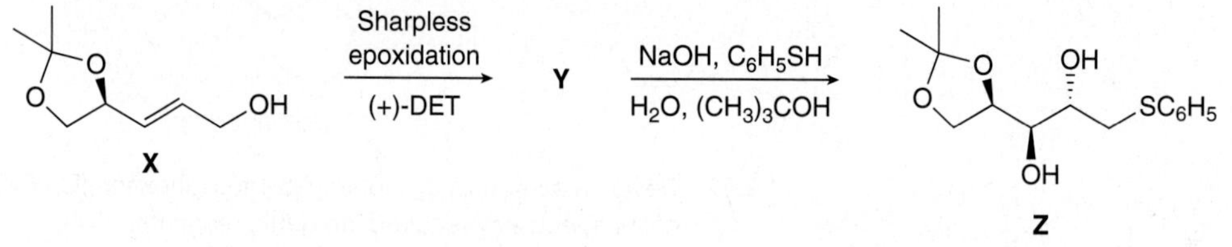
O
O
OH
X
Sharpless
epoxidation
(+)-DET
Y
NaOH, C_6H_5SH
H_2O, $(CH_3)_3COH$
O
O
OH
SC_6H_5
OH
Z

Mass Spectrometry and Infrared Spectroscopy

13

13.1 Mass spectrometry
13.2 Alkyl halides and the M + 2 peak
13.3 Fragmentation
13.4 Other types of mass spectrometry
13.5 Electromagnetic radiation
13.6 Infrared spectroscopy
13.7 IR absorptions
13.8 IR and structure determination

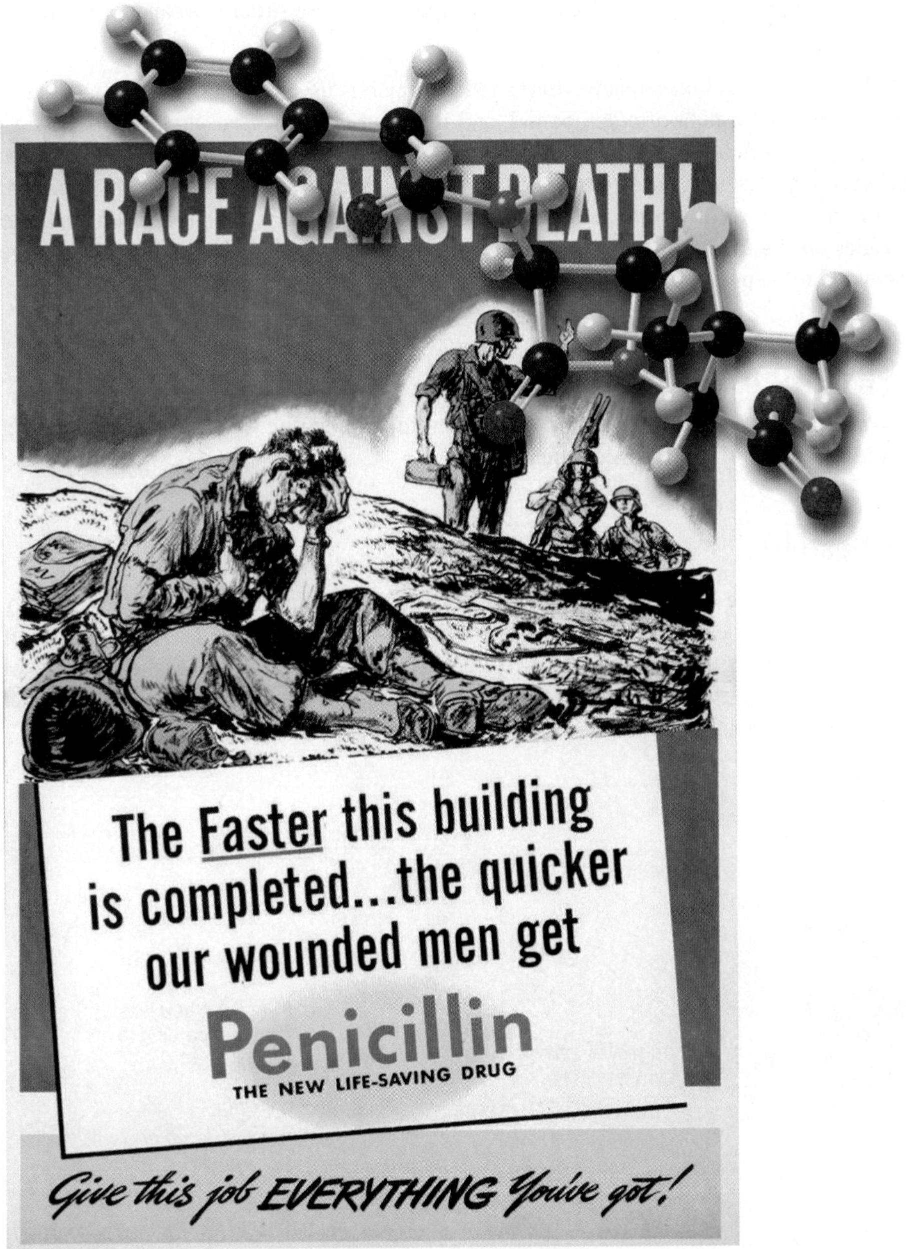

The serendipitous discovery of **penicillin** by Scottish bacteriologist Sir Alexander Fleming in 1928 is considered one of the single most important events in the history of medicine. Penicillin G and related compounds are members of the β-lactam family of antibiotics, all of which contain a strained four-membered amide ring that is responsible for their biological activity. Penicillin was first used to cure a streptococcal infection in 1942, and by 1944 penicillin production was given high priority by the United States government, because it was needed to treat the many injured soldiers in World War II. The unusual structure of penicillin was elucidated by modern instrumental methods in the 1940s. In Chapter 13, we learn about mass spectrometry and infrared spectroscopy, two techniques for characterizing organic compounds like penicillin.

Whether a compound is prepared in the laboratory or isolated from a natural source, a chemist must determine its identity. Seventy years ago, determining the structure of an organic compound involved a series of time-consuming operations: measuring physical properties (melting point, boiling point, solubility, and density), identifying the functional groups using a series of chemical tests, and converting an unknown compound into another compound whose physical and chemical properties were then characterized as well.

Although still a challenging task, structure determination has been greatly simplified by modern instrumental methods. These techniques have both decreased the time needed for compound characterization, and increased the complexity of compounds whose structures can be completely determined.

In Chapter 13 we examine **mass spectrometry (MS),** which is used to determine the molecular weight and molecular formula of a compound, and **infrared (IR) spectroscopy,** a tool used to identify a compound's functional groups. Chapter 14 is devoted to **nuclear magnetic resonance (NMR) spectroscopy,** which is used to identify the carbon–hydrogen framework in a compound, making it the most powerful spectroscopic tool for organic structure analysis. **Each of these methods relies on the interaction of an energy source with a molecule to produce a change that is recorded in a spectrum.**

13.1 Mass Spectrometry

***Mass spectrometry* is a technique used for measuring the molecular weight and determining the molecular formula of an organic molecule.**

13.1A General Features

In the most common type of **mass spectrometer,** a molecule is vaporized and ionized, usually by bombardment with a beam of high-energy electrons, as shown in Figure 13.1. The energy of these electrons is typically about 6400 kJ, or 70 electron volts (eV). Because it takes ~400 kJ of energy to cleave a typical σ bond, 6400 kJ is an enormous amount of energy to come into contact with a molecule. This electron beam ionizes a molecule by causing it to eject an electron.

Figure 13.1
Schematic of a mass spectrometer

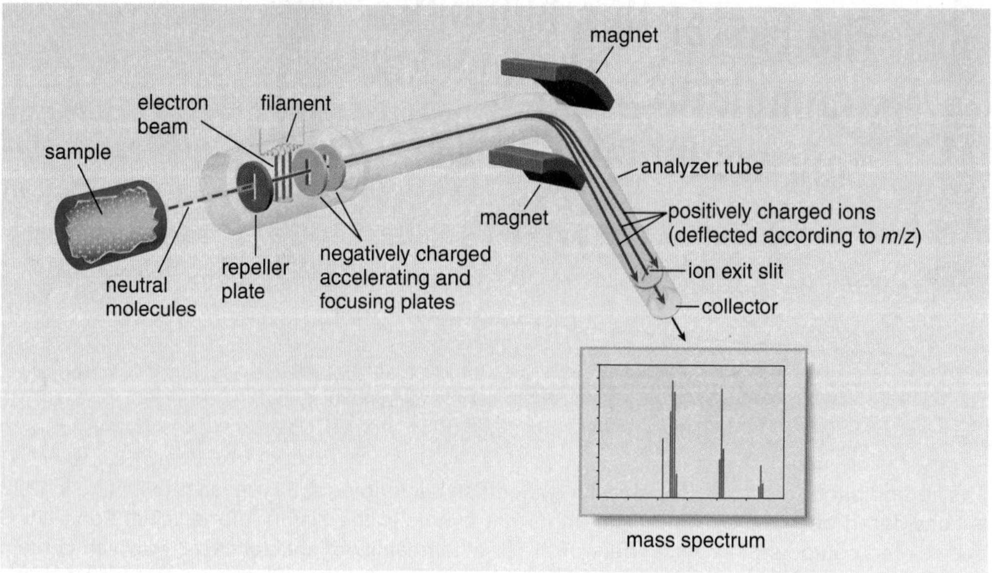

- In a mass spectrometer, a sample is vaporized and bombarded by a beam of electrons to form an unstable radical cation, which then decomposes to smaller fragments. The positively charged ions are accelerated toward a negatively charged plate, and then passed through a curved analyzer tube in a magnetic field, where they are deflected by different amounts depending on their ratio of mass to charge (m/z). A mass spectrum plots the intensity of each ion versus its m/z ratio.

The term **spectroscopy** is usually used for techniques that use electromagnetic radiation as an energy source. Because the energy source in MS is a beam of electrons, the term **mass spectrometry** is used instead.

$$\underset{\text{molecule}}{\mathbf{M}} \xrightarrow[\text{beam of electrons}]{e^-} \underset{\text{radical cation}}{\mathbf{M}^{+\cdot}} + e^-$$

The species formed is a **radical cation,** symbolized $\mathbf{M}^{+\cdot}$**.** It is a radical because it has an unpaired electron, and it is a cation because it has one fewer electron than it started with.

- **The radical cation $M^{+\cdot}$ is called the *molecular ion* or the *parent ion.***

A single electron has a negligible mass, so the **mass of $M^{+\cdot}$ represents the molecular weight** of M. Because the molecular ion $\mathbf{M}^{+\cdot}$ is inherently unstable, it decomposes. Single bonds break to form ***fragments,*** **radicals and cations having a lower molecular weight than the molecular ion.** A mass spectrometer analyzes the masses of cations only. The cations are accelerated in an electric field and deflected in a curved path in a magnetic field, thus sorting the molecular ion and its fragments by their **mass-to-charge (*m/z*) ratio.** Because *z* is almost always +1, *m/z* actually measures the mass (*m*) of the individual ions.

$$\underset{\text{molecule}}{\mathbf{M}} \xrightarrow{e^-} \underset{\text{unstable radical cation}}{\mathbf{M}^{+\cdot}} \longrightarrow \textbf{radicals} + \underset{\text{These fragments are analyzed.}}{\textbf{cations}}$$

- **A *mass spectrum* plots the amount of each cation (its relative abundance) versus its mass.**

The whole-number mass of CH_4 is (1 C × 12 amu) + (4 H × 1 amu) = 16 amu; amu = atomic mass unit.

A mass spectrometer analyzes the masses of *individual* molecules, not the weighted average mass of a group of molecules, so the whole-number masses of the most common individual isotopes must be used to calculate the mass of the molecular ion. Thus, the mass of the molecular ion for CH_4 should be 16. As a result, the mass spectrum of CH_4 shows a line for the molecular ion—the parent peak or **M** peak—at *m/z* = 16.

Mass spectrum of CH_4

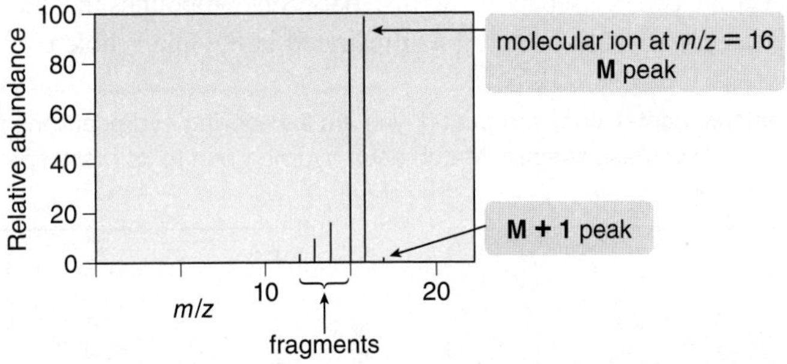

m/z	relative abundance
17	1.2
16	100.0 ← base peak
15	85.9
14	16.1
13	8.1
12	2.8

The tallest peak in a mass spectrum is called the **base peak.** For CH_4, the base peak is also the M peak, although this may *not* always be the case for all organic compounds.

The mass spectrum of CH_4 consists of more peaks than just the M peak. What is responsible for the peaks at *m/z* < 16? Because the molecular ion is unstable, it fragments into other cations and radical cations containing one, two, three, or four fewer hydrogen atoms than methane itself. Thus, the peaks at *m/z* = 15, 14, 13, and 12, are due to these lower molecular weight fragments. The decomposition of a molecular ion into lower molecular weight fragments is called **fragmentation.**

$$CH_4 \xrightarrow{e^-} \underset{\text{mass 16}}{(CH_4)^{+\cdot}} \xrightarrow{-H\cdot} \underset{\text{mass 15}}{CH_3^{+}} \xrightarrow{-H\cdot} \underset{\text{mass 14}}{CH_2^{+\cdot}} \xrightarrow{-H\cdot} \underset{\text{mass 13}}{CH^{+}} \xrightarrow{-H\cdot} \underset{\text{mass 12}}{C^{+\cdot}}$$

molecular ion (mass 16) **fragments** (masses 15–12)

What is responsible for the small peak at m/z = 17 in the mass spectrum of CH_4? Although most carbon atoms have an atomic mass of 12, 1.1% of them have an additional neutron in the nucleus, giving them an atomic mass of 13. When one of these carbon-13 isotopes forms methane, it gives a molecular ion peak at m/z = 17 in the mass spectrum. This peak is called the **M + 1** peak.

These key features—the molecular ion, the base peak, and the M + 1 peak—are illustrated in the mass spectrum of hexane in Figure 13.2.

Figure 13.2
Mass spectrum of hexane ($CH_3CH_2CH_2CH_2CH_2CH_3$)

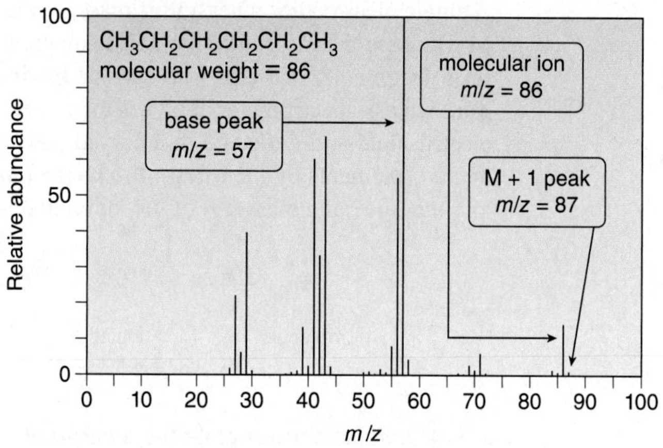

- The molecular ion for hexane (molecular formula C_6H_{14}) is at m/z = 86.
- The base peak (relative abundance = 100) occurs at m/z = 57.
- A small M + 1 peak occurs at m/z = 87.

13.1B Analyzing Unknowns Using the Molecular Ion

Because the **mass of the molecular ion equals the molecular weight of a compound,** a mass spectrum can be used to distinguish between compounds that have similar physical properties but different molecular weights, as illustrated in Sample Problem 13.1.

Sample Problem 13.1 Pentane, pent-1-ene, and pent-1-yne are low-boiling hydrocarbons that have different molecular ions in their mass spectra. Match each hydrocarbon to its mass spectrum.

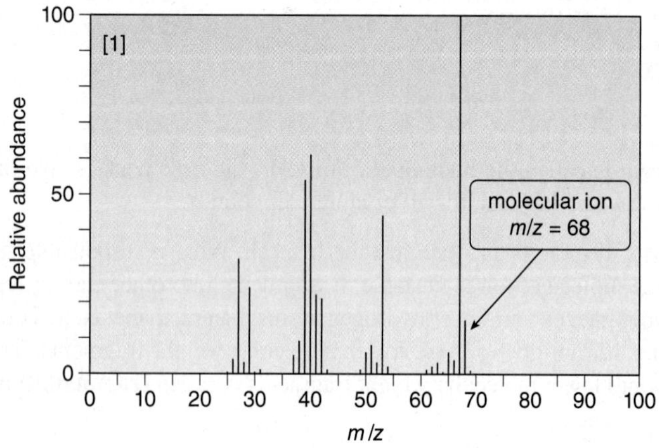

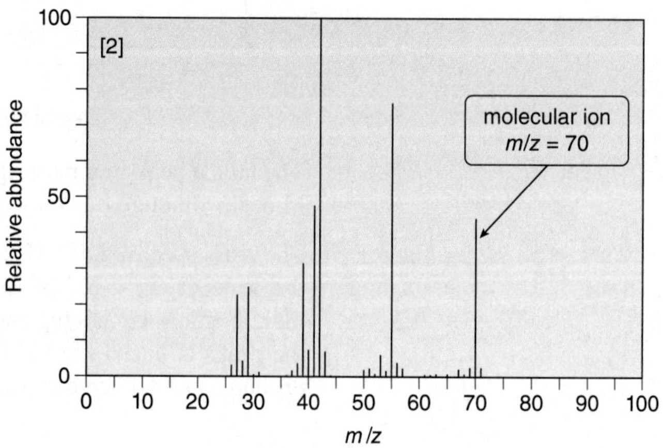

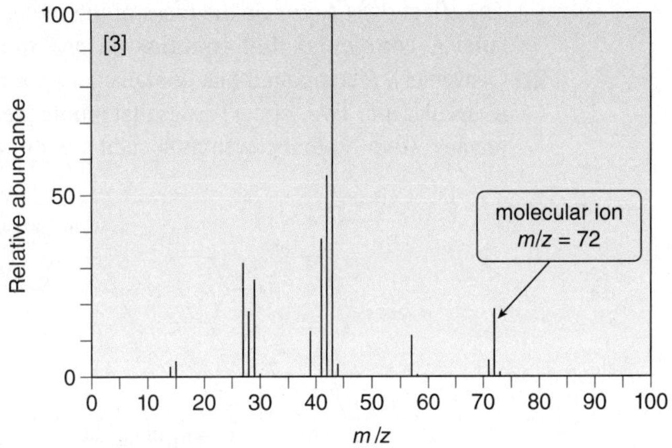

Solution

To solve this problem, first determine the molecular formula and molecular weight of each compound. Then, because the molecular weight of the compound equals the mass of the molecular ion, match the molecular weight to *m/z* for the molecular ion:

Compound	Molecular formula	Molecular weight = *m/z* of molecular ion	Spectrum
 pentane	C_5H_{12}	72	[3]
 pent-1-ene	C_5H_{10}	70	[2]
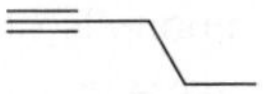 pent-1-yne	C_5H_8	68	[1]

Problem 13.1 What is the mass of the molecular ion formed from compounds having each molecular formula: (a) C_3H_6O; (b) $C_{10}H_{20}$; (c) $C_8H_8O_2$; (d) methamphetamine ($C_{10}H_{15}N$)?

How to use the mass of the molecular ion to propose molecular formulas for an unknown is shown in Sample Problem 13.2. In this process, keep in mind the following useful fact. Hydrocarbons like methane (CH_4) and hexane (C_6H_{14}), as well as compounds that contain only C, H, and O atoms, always have a molecular ion with an *even* mass. An odd molecular ion generally indicates that a compound contains nitrogen.

Sample Problem 13.2 Propose possible molecular formulas for a compound with a molecular ion at $m/z = 86$.

Solution

Because the molecular ion has an **even** mass, the compound likely contains C, H, and possibly O atoms. Begin by determining the molecular formula for a hydrocarbon having a molecular ion at 86. Then, because the mass of an O atom is 16 (the mass of CH_4), replace CH_4 by O to give a molecular formula containing one O atom. Repeat this last step to give possible molecular formulas for compounds with two or more O atoms.

For a molecular ion at $m/z = 86$:

Possible hydrocarbons:

- Divide 86 by 12 (mass of 1 C atom). This gives the maximum number of C's possible.

$$\frac{86}{12} = 7 \text{ C's maximum (remainder} = 2) \longrightarrow C_7H_2$$

- Replace 1 C by 12 H's for another possible molecular formula.

$$C_7H_2 \xrightarrow[+\ 12\ \text{H's}]{-\ 1\ \text{C}} C_6H_{14}$$

Possible compounds with C, H, and O:

- Substitute 1 O for CH_4. (This can't be done for C_7H_2.)

$$C_6H_{14} \xrightarrow[+\ 1\ \text{O}]{-\ CH_4} C_5H_{10}O$$

- Repeat the process.

$$C_5H_{10}O \xrightarrow[+\ 1\ \text{O}]{-\ CH_4} C_4H_6O_2$$

The effect of N atoms on the mass of the molecular ion in a mass spectrum is called the **nitrogen rule: A compound that contains an *odd* number of N atoms gives an odd molecular ion.** Conversely, a compound that contains an *even* number of N atoms (including *zero*) gives an *even* molecular ion. Two "street" drugs that mimic the effects of heroin illustrate this principle: 3-methylfentanyl (two N atoms, even molecular weight) and MPPP (one N atom, odd molecular weight).

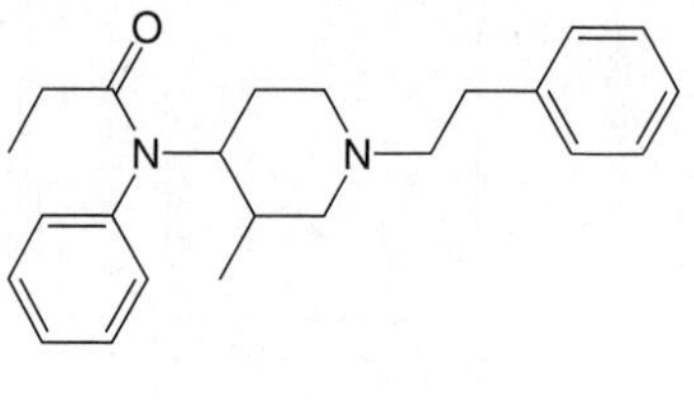

3-methylfentanyl
$C_{23}H_{30}N_2O$
molecular weight = 350

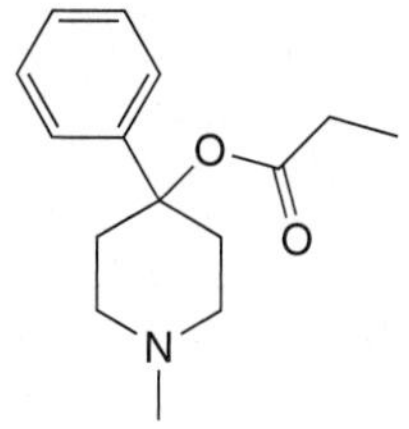

MPPP
(1-methyl-4-phenyl-4-propionoxypiperidine)
$C_{15}H_{21}NO_2$
molecular weight = 247

Problem 13.2 Propose two molecular formulas for each of the following molecular ions: (a) 72; (b) 100; (c) 73.

Problem 13.3 Use the following information to propose a molecular formula for nootkatone, a compound partly responsible for the characteristic odor of grapefruit. Nootkatone contains the elements C, H, and O, has five degrees of unsaturation, and has a molecular ion in its mass spectrum at m/z = 218.

Nootkatone (Problem 13.3) occurs naturally in grapefruits, and has been used for many years as a flavoring in foods and beverages.

13.2 Alkyl Halides and the M + 2 Peak

Most of the elements found in organic compounds, such as carbon, hydrogen, oxygen, nitrogen, sulfur, phosphorus, fluorine, and iodine, have one major isotope. **Chlorine** and **bromine,** on the other hand, have two, giving characteristic patterns to the mass spectra of their compounds.

Chlorine has two common isotopes, ^{35}**Cl** and ^{37}**Cl,** which occur naturally in a 3:1 ratio. Thus, **there are two peaks in a 3:1 ratio for the molecular ion of an alkyl chloride.** The larger peak—the **M** peak—corresponds to the compound containing ^{35}Cl, and the smaller peak—the **M + 2** peak—corresponds to the compound containing ^{37}Cl.

- **When the molecular ion consists of two peaks (M and M + 2) in a 3:1 ratio, a Cl atom is present.**

Sample Problem 13.3 What molecular ions will be present in a mass spectrum of 2-chloropropane, $(CH_3)_2CHCl$?

Solution

Calculate the molecular weight using each of the common isotopes of Cl.

Molecular formula	Mass of molecular ion (m/z)
$C_3H_7{}^{35}Cl$	78 (M peak)
$C_3H_7{}^{37}Cl$	80 (M + 2 peak)

There should be two peaks in a ratio of 3:1, at m/z = 78 and 80, as illustrated in the mass spectrum of 2-chloropropane in Figure 13.3.

Bromine has two common isotopes, ^{79}**Br** and ^{81}**Br,** which occur naturally in a 1:1 ratio. Thus, **there are two peaks in a 1:1 ratio for the molecular ion of an alkyl bromide.** In the mass spectrum of 2-bromopropane (Figure 13.4), for example, there is an M peak at m/z = 122 and an M + 2 peak at m/z = 124.

- **When the molecular ion consists of two peaks (M and M + 2) in a 1:1 ratio, a Br atom is present in the molecule.**

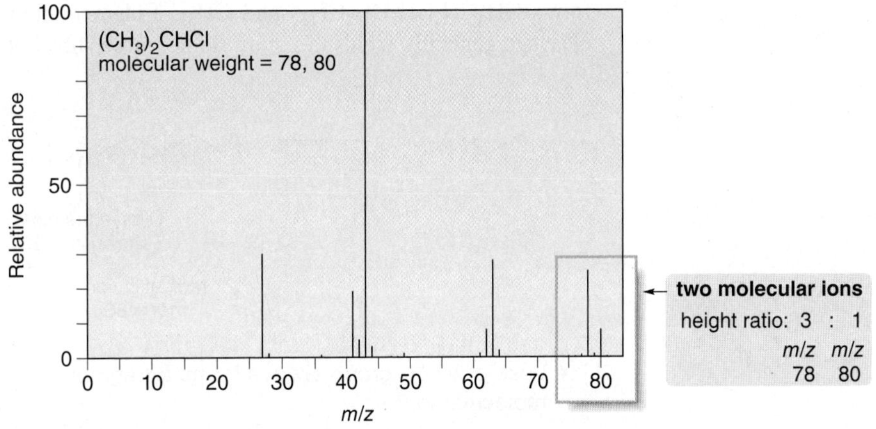

Figure 13.3 Mass spectrum of 2-chloropropane [$(CH_3)_2CHCl$]

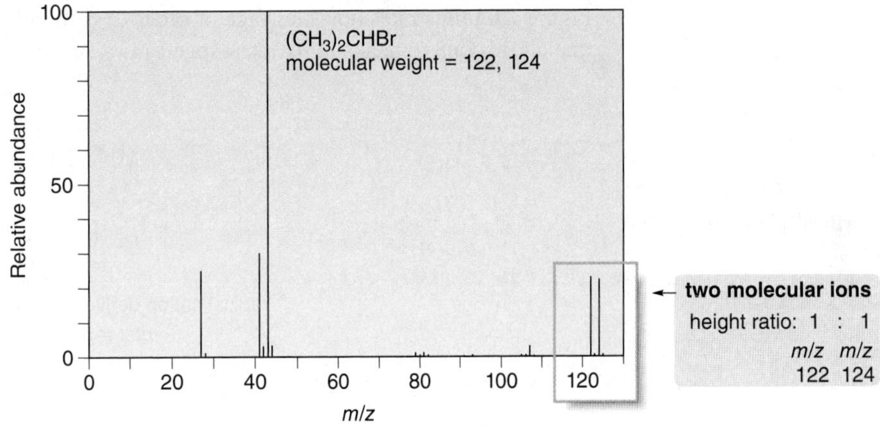

Figure 13.4 Mass spectrum of 2-bromopropane [$(CH_3)_2CHBr$]

Problem 13.4 What molecular ions would you expect for compounds having each of the following molecular formulas: (a) C_4H_9Cl; (b) C_3H_7F; (c) $C_4H_{11}N$; (d) $C_4H_4N_2$?

Problem 13.5 What molecular ions would you expect for the compound depicted in the ball-and-stick model?

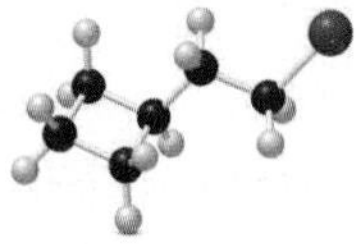

13.3 Fragmentation

While many chemists use a mass spectrum to determine only a compound's molecular weight and molecular formula, additional useful structural information can be obtained from fragmentation patterns. Although each organic compound fragments in a unique way, a particular functional group exhibits common fragmentation patterns.

13.3A General Features of Fragmentation

As an example, consider hexane, whose mass spectrum was shown in Figure 13.2. When hexane is bombarded by an electron beam, it forms a highly unstable radical cation ($m/z = 86$) that can decompose by cleavage of any of the C–C bonds. Thus, cleavage of the terminal C–C bond forms $CH_3CH_2CH_2CH_2CH_2^+$ and $CH_3\bullet$. Fragmentation generates a cation and a radical, and cleavage generally yields the more stable, more substituted carbocation.

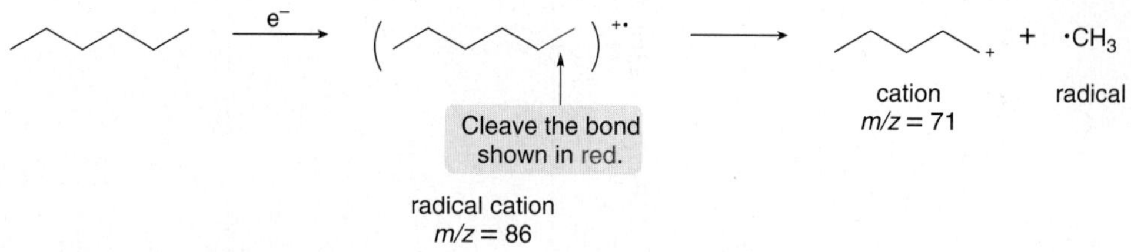

- **Loss of a CH_3 group always forms a fragment with a mass 15 units less than that of the molecular ion.**

As a result, the mass spectrum of hexane shows a peak at $m/z = 71$ due to $CH_3CH_2CH_2CH_2CH_2^+$. Figure 13.5 illustrates how cleavage of other C–C bonds in hexane gives rise to other fragments that correspond to peaks in its mass spectrum.

Figure 13.5
Identifying fragments in the mass spectrum of hexane

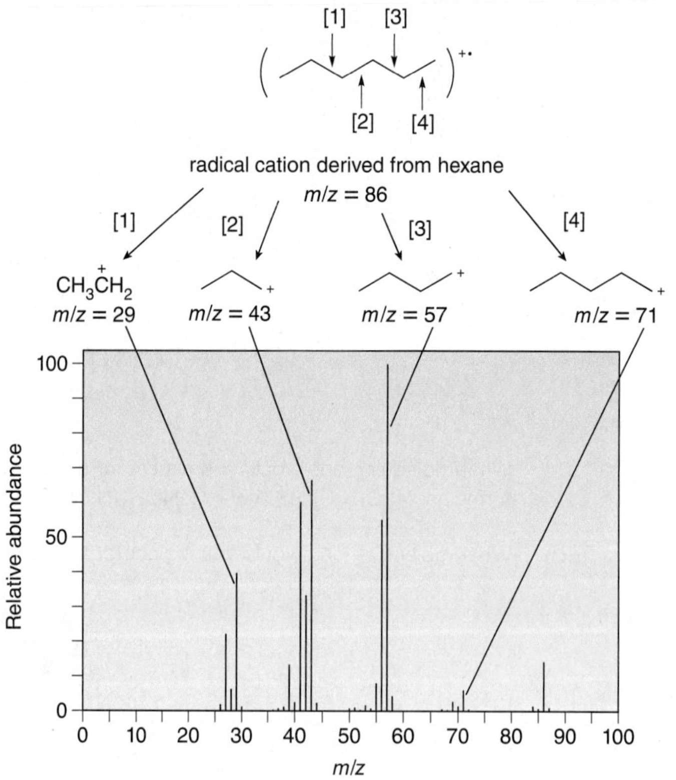

- Cleavage of C–C bonds (labeled [1]–[4]) in hexane forms lower molecular weight fragments that correspond to lines in the mass spectrum. Although the mass spectrum is complex, possible structures can be assigned to some of the fragments, as shown.

Sample Problem 13.4 The mass spectrum of 2,3-dimethylpentane [$(CH_3)_2CHCH(CH_3)CH_2CH_3$] shows fragments at $m/z = 85$ and 71. Propose possible structures for the ions that give rise to these peaks.

Solution

To solve a problem of this sort, first calculate the mass of the molecular ion. Draw out the structure of the compound, break a C–C bond, and calculate the mass of the resulting fragments. Repeat this process on different C–C bonds until fragments of the desired mass-to-charge ratio are formed.

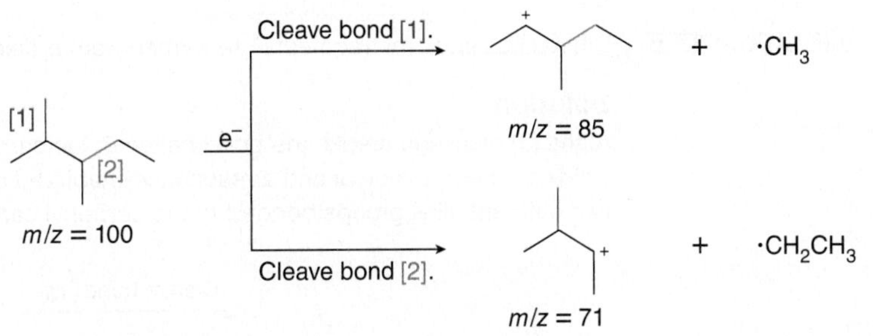

In this example, 2,3-dimethylpentane has a molecular ion at m/z = 100. Cleavage of bond [1] forms a 2° carbocation with m/z = 85 and CH_3•. Cleavage of bond [2] forms another 2° carbocation with m/z = 71 and CH_3CH_2•. Thus, the fragments at m/z = 85 and 71 are possibly due to the two carbocations drawn.

Problem 13.6 The mass spectrum of 2,3-dimethylpentane also shows peaks at m/z = 57 and 43. Propose possible structures for the ions that give rise to these peaks.

Problem 13.7 The base peak in the mass spectrum of 2,2,4-trimethylpentane [$(CH_3)_3CCH_2CH(CH_3)_2$] occurs at m/z = 57. What ion is responsible for this peak and why is this ion the most abundant fragment?

13.3B Fragmentation Patterns of Some Common Functional Groups

Each functional group exhibits characteristic fragmentation patterns that help to analyze a mass spectrum. For example, aldehydes and ketones often undergo the process of **α cleavage, breaking the bond between the carbonyl carbon and the carbon adjacent to it.** Cleavage yields a neutral radical and a resonance-stabilized acylium ion.

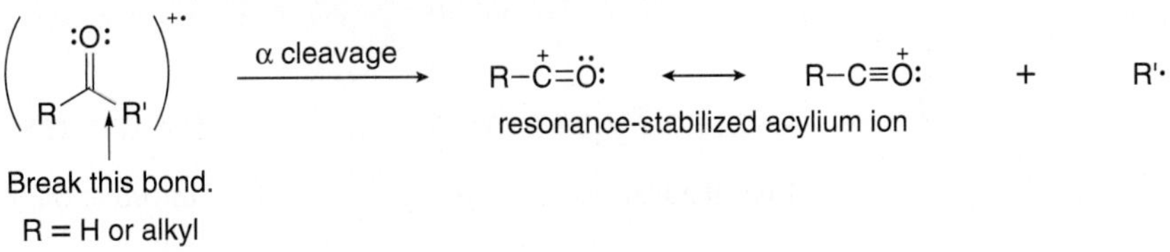

Alcohols undergo fragmentation in two different ways—α cleavage and dehydration. Alpha (α) cleavage occurs by breaking a bond between an alkyl group and the carbon that bears the OH group, resulting in an alkyl radical and a resonance-stabilized carbocation.

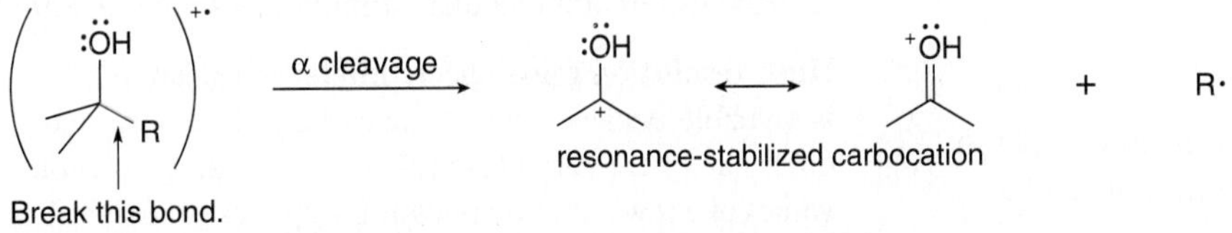

Likewise, alcohols undergo dehydration, the elimination of H_2O, from two adjacent atoms. Unlike fragmentations discussed thus far, dehydration results in the cleavage of two bonds and forms H_2O and the radical cation derived from an alkene.

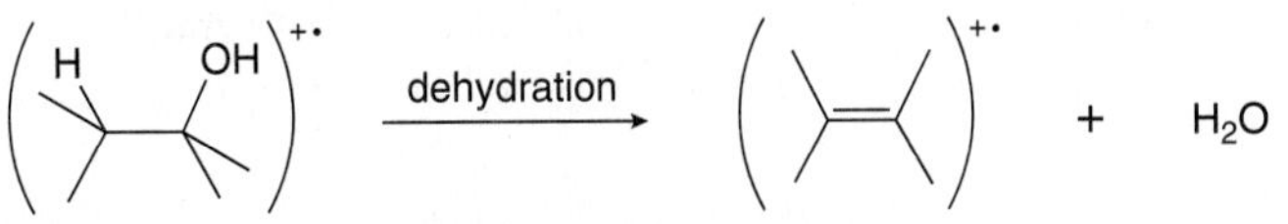

- **Loss of H_2O from an alcohol always forms a fragment with a mass 18 units less than the molecular ion.**

Sample Problem 13.5 What mass spectral fragments are formed from α cleavage of pentan-2-one, $CH_3COCH_2CH_2CH_3$?

Solution

Alpha (α) cleavage breaks the bond between the carbonyl carbon and the carbon adjacent to it, yielding a neutral radical and a resonance-stabilized acylium ion. A ketone like pentan-2-one with two different alkyl groups bonded to the carbonyl carbon has two different pathways for α cleavage.

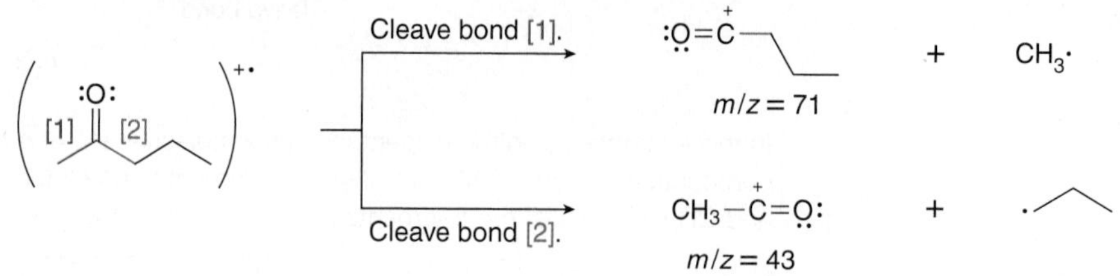

Cleave bond [1].
Cleave bond [2].
[1]
[2]
m/z = 71
m/z = 43
CH3·
CH3–C=O:

As a result, two fragments are formed by α cleavage of pentan-2-one, giving peaks at $m/z = 71$ and 43.

Problem 13.8 (a) What mass spectral fragments are formed by α cleavage of butan-2-ol, $CH_3CH(OH)CH_2CH_3$?
(b) What fragments are formed by dehydration of butan-2-ol?

Problem 13.9 What cations are formed in the mass spectrometer by α cleavage of each of the following compounds?

a. 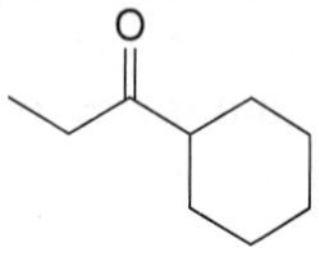

b. 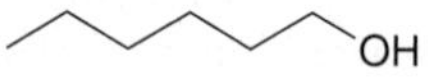

c. 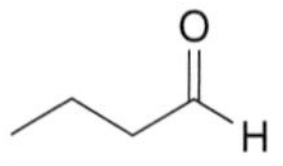

13.4 Other Types of Mass Spectrometry

Recent advances have greatly expanded the information obtained from mass spectrometry.

13.4A High-Resolution Mass Spectrometry

The mass spectra described thus far have been low-resolution spectra; that is, they report *m/z* values to the nearest whole number. As a result, the mass of a given molecular ion can correspond to many different molecular formulas, as shown in Sample Problem 13.2.

High-resolution mass spectrometers measure *m/z* ratios to four (or more) decimal places. This is valuable because except for carbon-12, whose mass is defined as 12.0000, the masses of all other nuclei are very close to—but not exactly—whole numbers. Table 13.1 lists the exact mass values of a few common nuclei. Using these values it is possible to determine the single molecular formula that gives rise to a molecular ion.

Table 13.1
Exact Masses of Some Common Isotopes

Isotope	Mass
^{12}C	12.0000
^{1}H	1.00783
^{16}O	15.9949
^{14}N	14.0031

For example, a compound having a molecular ion at $m/z = 60$ using a low-resolution mass spectrometer could have the following molecular formulas:

Formula	Exact mass
C_3H_8O	60.0575
$C_2H_4O_2$	60.0211
$C_2H_8N_2$	60.0688

If the molecular ion had an exact mass of 60.0578, the compound's molecular formula is C_3H_8O, because its mass is closest to the observed value.

Problem 13.10 The low-resolution mass spectrum of an unknown analgesic **X** had a molecular ion of 151. Possible molecular formulas include $C_7H_5NO_3$, $C_8H_9NO_2$, and $C_{10}H_{17}N$. High-resolution mass spectrometry gave an exact mass of 151.0640. What is the molecular formula of **X?**

13.4B Gas Chromatography–Mass Spectrometry (GC–MS)

Like other controlled substances, the tetrahydrocannabinol from marijuana leaves can be detected in minute amounts by GC–MS.

Two analytical tools—**gas chromatography (GC)** and **mass spectrometry (MS)**—can be combined into a single instrument **(GC–MS)** to analyze mixtures of compounds (Figure 13.6a). The gas chromatograph separates the mixture, and then the mass spectrometer records a spectrum of the individual components.

A gas chromatograph consists of a thin capillary column containing a viscous, high-boiling liquid, all housed in an oven. When a sample is injected into the GC, it is vaporized and swept by an inert gas through the column. The components of the mixture travel through the column at different rates, often separated by boiling point, with lower boiling compounds exiting the column before higher boiling compounds. Each compound then enters the mass spectrometer, where it is ionized to form its molecular ion and lower molecular weight fragments. The GC–MS records a gas chromatogram for the mixture, which plots the amount of each component versus its **retention time**—that is, the time required to travel through the column. Each component of a mixture is characterized by its retention time in the gas chromatogram and its molecular ion in the mass spectrum (Figure 13.6b).

GC–MS is widely used for characterizing mixtures containing environmental pollutants. It is also used to analyze urine and hair samples for the presence of illegal drugs or banned substances thought to improve athletic performance.

Figure 13.6
Compound analysis using GC–MS

a. Schematic of a GC–MS instrument

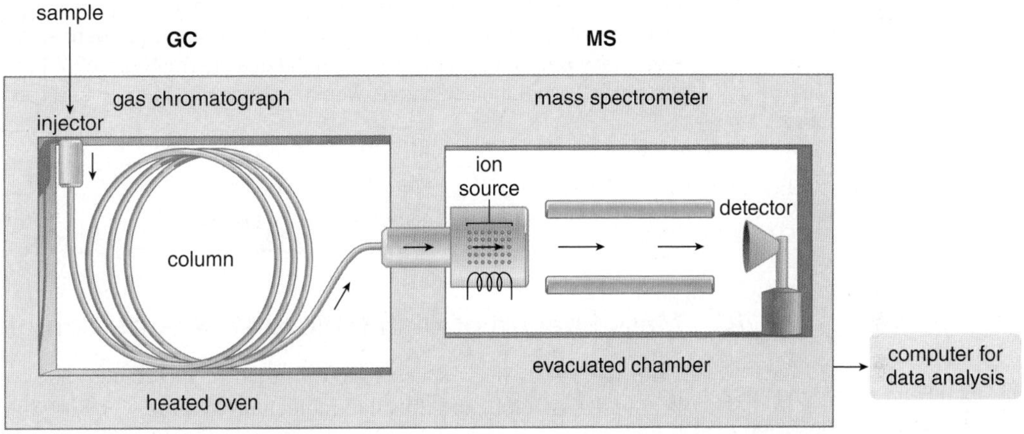

The gas chromatograph separates the mixture into its components.

The mass spectrometer records a spectrum of the individual components.

b. GC trace of a three-component mixture. The mass spectrometer gives a spectrum for each component.

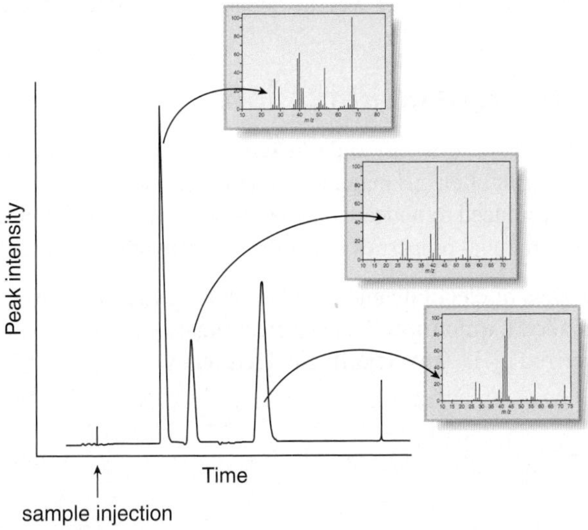

To analyze a urine sample for THC (tetrahydrocannabinol), the principal psychoactive component of marijuana, the organic compounds are extracted from urine, purified, concentrated, and injected into the GC–MS. THC appears as a GC peak with a characteristic retention time (for a given set of experimental parameters), and gives a molecular ion at 314, its molecular weight, as shown in Figure 13.7.

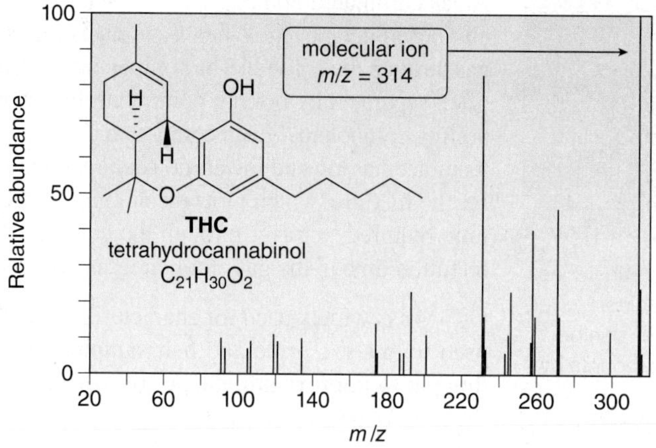

Figure 13.7 Mass spectrum of tetrahydrocannabinol (THC)

Problem 13.11 Benzene, toluene, and *p*-xylene (BTX) are often added to gasoline to boost octane ratings. What would be observed if a mixture of these three compounds were subjected to GC–MS analysis? How many peaks would be present in the gas chromatogram? What would be the relative order of the peaks? What molecular ions would be observed in the mass spectra?

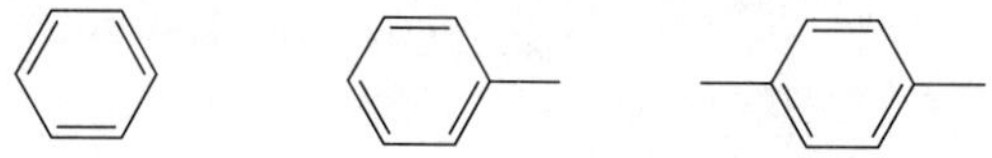

13.4C Mass Spectra of High Molecular Weight Biomolecules

Dr. John Fenn shared the 2002 Nobel Prize in Chemistry for his development of ESI mass spectrometry.

Until the 1980s mass spectra were limited to molecules that could be readily vaporized with heat under vacuum, and thus had molecular weights of < 800. In the last 35 years, new methods have been developed to generate gas phase ions of large molecules, allowing mass spectra to be recorded for large biomolecules such as proteins and carbohydrates. **Electrospray ionization (ESI),** for example, forms ions by creating a fine spray of charged droplets in an electric field. Evaporation of the charged droplets forms gaseous ions that are then analyzed by their *m/z* ratio. ESI and related techniques have extended mass spectrometry into the analysis of nonvolatile compounds with molecular weights greater than 100,000 daltons (atomic mass units).

13.5 Electromagnetic Radiation

Infrared (IR) spectroscopy and **nuclear magnetic resonance (NMR)** spectroscopy (Chapter 14) both use a form of electromagnetic radiation as their energy source. To understand IR and NMR, therefore, you need to understand some of the properties of **electromagnetic radiation**—radiant energy having dual properties of both waves and particles.

The particles of electromagnetic radiation are called **photons,** each having a discrete amount of energy called a **quantum.** Because electromagnetic radiation also has wave properties, it can be characterized by its **wavelength** and **frequency.**

Length units used to report wavelength include:

Unit	Length
meter (m)	1 m
centimeter (cm)	10^{-2} m
micrometer (μm)	10^{-6} m
nanometer (nm)	10^{-9} m
Angstrom (Å)	10^{-10} m

- **Wavelength (λ) is the distance from one point on a wave (e.g., the peak or trough) to the same point on the adjacent wave. A variety of different length units are used for λ, depending on the type of radiation.**
- **Frequency (ν) is the number of waves passing a point per unit time. Frequency is reported in cycles per second (s^{-1}), which is also called hertz (Hz).**

You come into contact with many different kinds of electromagnetic radiation in your daily life. You use visible light to see the words on this page, you may cook with microwaves, and you should use sunscreen to protect your skin from the harmful effects of ultraviolet radiation.

The different forms of electromagnetic radiation make up the **electromagnetic spectrum.** The spectrum is arbitrarily divided into different regions, as shown in Figure 13.8. All electromagnetic radiation travels at the speed of light (c), 3.0×10^8 m/s.

Figure 13.8
The electromagnetic spectrum

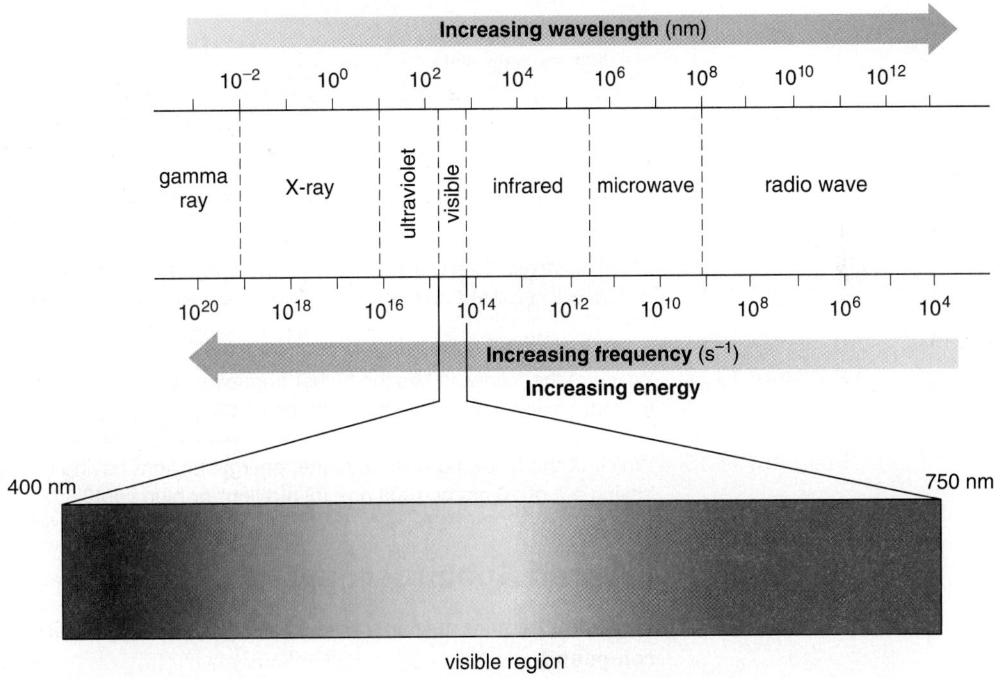

- Visible light occupies only a small region of the electromagnetic spectrum.

The speed of electromagnetic radiation (c) is directly proportional to its wavelength and frequency:

$$c = \lambda\nu$$

The speed of light (c) is a constant, so wavelength and frequency are *inversely* related:

- **$\lambda = c/\nu$: Wavelength increases as frequency decreases.**
- **$\nu = c/\lambda$: Frequency increases as wavelength decreases.**

The energy (E) of a photon is directly proportional to its frequency where h = Planck's constant (6.63×10^{-34} J·s).

$$E = h\nu$$

Frequency and wavelength are *inversely* proportional ($\nu = c/\lambda$), however, so energy and wavelength are *inversely* proportional:

$$E = h\nu = \frac{hc}{\lambda}$$

- **The energy of electromagnetic radiation increases as frequency increases and wavelength decreases.**

When electromagnetic radiation strikes a molecule, some wavelengths—but not all—are absorbed. Only some wavelengths are absorbed because molecules have discrete energy levels. The energies of their electronic, vibrational, and nuclear spin states are *quantized,* not *continuous.*

- **For absorption to occur, the energy of the photon must match the difference between two energy states in a molecule.**

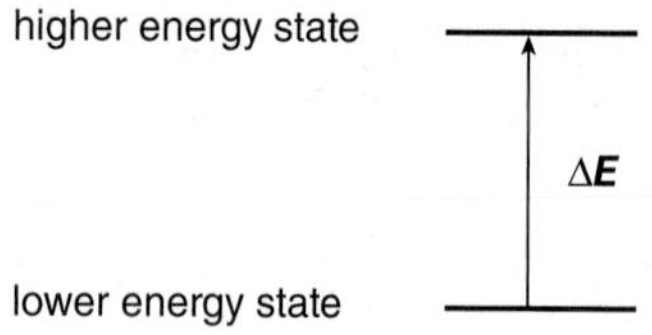

For absorption to occur, the energy of the incident electromagnetic radiation must match ΔE.

ΔE = the energy difference between two states in a molecule

- **The *larger* the energy difference between two states, the *higher* the energy of radiation needed for absorption, the *higher* the frequency, and the *shorter* the wavelength.**

Problem 13.12 Which of the following has the higher frequency: (a) light having a wavelength of 10^2 or 10^4 nm; (b) light having a wavelength of 100 nm or 100 μm; (c) red light or blue light?

Problem 13.13 Which of the following has the higher energy: (a) light having a ν of 10^4 Hz or 10^8 Hz; (b) light having a λ of 10 nm or 1000 nm; (c) red light or blue light?

13.6 Infrared Spectroscopy

Organic chemists use infrared (IR) spectroscopy to identify the functional groups in a compound.

13.6A Background

Using the wavenumber scale results in IR values in a numerical range that is easier to report than the corresponding frequencies given in hertz (4000–400 cm^{-1} compared to 1.2×10^{14}–1.2×10^{15} Hz).

Infrared radiation (λ = 2.5–25 μm) is the energy source in infrared spectroscopy. These are somewhat longer wavelengths than visible light, so they are lower in frequency and lower in energy than visible light. Frequencies in IR spectroscopy are reported using a unit called the **wavenumber** ($\tilde{\nu}$):

$$\tilde{\nu} = \frac{1}{\lambda}$$

Wavenumber is *inversely* proportional to wavelength and reported in reciprocal centimeters (**cm^{-1}**). Wavenumber ($\tilde{\nu}$) is *proportional* to frequency (ν). **Frequency (and therefore energy) increases as the wavenumber increases.** Using the wavenumber scale, IR absorptions occur from **4000 cm^{-1}** to **400 cm^{-1}.**

- **Absorption of IR light causes changes in the vibrational motions of a molecule.**

Covalent bonds are not static. They are more like springs with weights on each end. When two atoms are bonded to each other, the bond stretches back and forth. When three or more atoms are joined together, bonds can also bend. These bond stretching and bending vibrations represent the different vibrational modes available to a molecule.

These vibrations are quantized, so they occur only at specific frequencies, which correspond to the frequency of IR light. When the frequency of IR light matches the frequency of a particular vibrational mode, the IR light is absorbed, causing the amplitude of the particular bond stretch or bond bend to increase.

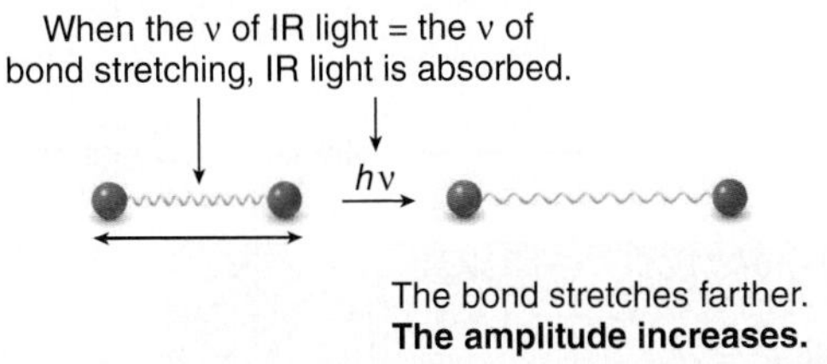

- **Different kinds of bonds vibrate at different frequencies, so they absorb different frequencies of IR light.**
- **IR spectroscopy distinguishes between the different kinds of bonds in a molecule, so it is possible to determine the functional groups present.**

Problem 13.14 Which of the following has higher energy: (a) IR light of 3000 cm^{-1} or 1500 cm^{-1} in wavenumber; (b) IR light having a wavelength of 10 μm or 20 μm?

13.6B Characteristics of an IR Spectrum

In an IR spectrometer, light passes through a sample. Frequencies that match vibrational frequencies are absorbed, and the remaining light is transmitted to a detector. A spectrum plots the amount of transmitted light versus its wavenumber. The IR spectrum of propan-1-ol, $CH_3CH_2CH_2OH$, illustrates several important features of IR spectroscopy.

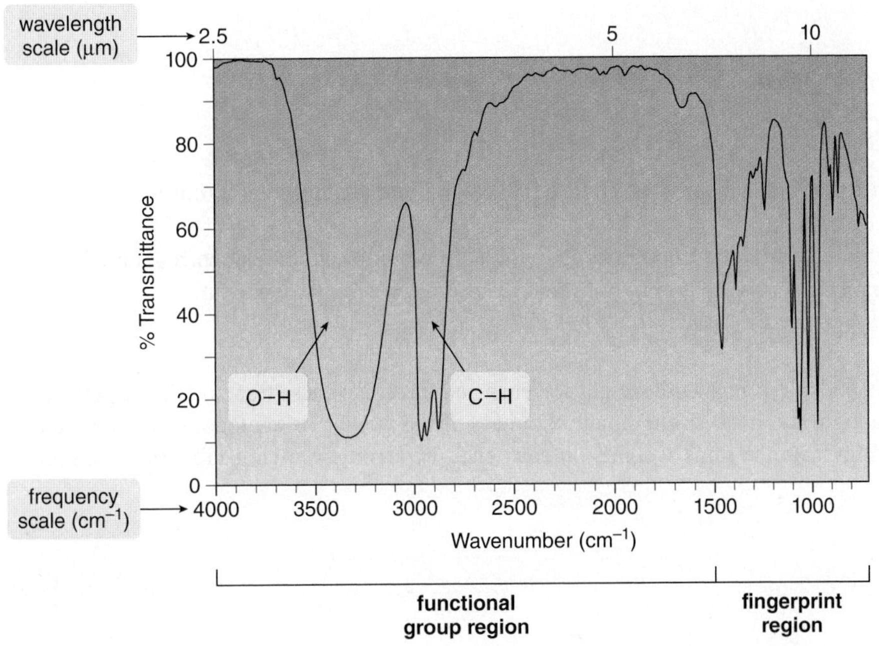

- An IR spectrum has broad lines.
- The absorption peaks go *down* on a page. The *y* axis measures **percent transmittance:** 100% transmittance means that all the light shone on a sample is transmitted and none is absorbed; 0% transmittance means that none of the light shone on a sample is transmitted and all is absorbed. Most absorptions lie between these two extremes.
- **Each peak corresponds to a particular kind of bond, and each bond type (such as O–H and C–H) occurs at a characteristic frequency.**
- IR spectra have both a wavelength and a wavenumber scale on the *x* axis. Wavelengths are recorded in μm (2.5–25). Wavenumber, frequency, and energy *decrease* from left to right. Where a peak occurs is reported in reciprocal centimeters (cm^{-1}).

Conceptually, the IR spectrum is divided into two regions:

- **The functional group region occurs at ≥ 1500 cm^{-1}. Common functional groups give one or two peaks in this region, at a characteristic frequency.**
- **The fingerprint region occurs at < 1500 cm^{-1}. This region often contains a complex set of peaks and is unique for every compound.**

Compare, for example, the IR spectra of 5-methylhexan-2-one (**A**) and ethyl propanoate (**B**) in Figure 13.9. The IR spectra look similar in their functional group regions because both compounds contain a carbonyl group (C=O) and several sp^3 hybridized C–H bonds. Because **A** and **B** are different compounds, however, their fingerprint regions look very different.

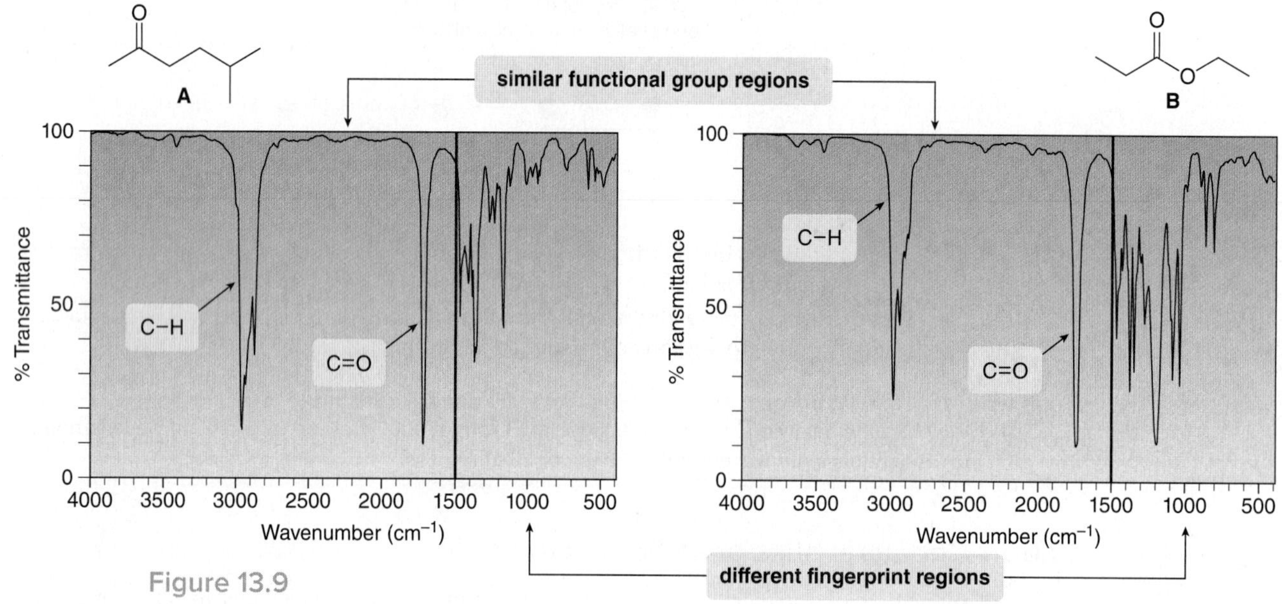

Figure 13.9
Comparing the functional group region and fingerprint region of two compounds

- **A** and **B** show peaks in the same regions for their C=O group and sp^3 hybridized C–H bonds.
- **A** and **B** are different compounds, so their fingerprint regions are quite different.

13.7 IR Absorptions

13.7A Where Particular Bonds Absorb in the IR

Where a particular bond absorbs in the IR depends on **bond strength** and **atom mass.**

- **Bond strength: stronger bonds vibrate at higher frequency, so they absorb at higher $\tilde{\nu}$.**
- **Atom mass: bonds with lighter atoms vibrate at higher frequency, so they absorb at higher $\tilde{\nu}$.**

Thinking of bonds as springs with weights on each end illustrates these trends. The strength of the spring is analogous to bond strength, and the mass of the weights is analogous to atomic mass. For two springs with the same weights on each end, the **stronger spring vibrates at a higher**

frequency. For two springs of the same strength, **springs with lighter weights vibrate at higher frequency** than those with heavier weights. Hooke's law, as shown in Figure 13.10, describes the relationship of frequency to mass and bond strength.

Figure 13.10
Hooke's law: How the frequency of bond vibration depends on atom mass and bond strength

The frequency of bond vibration can be derived from Hooke's law, which describes the motion of a vibrating spring:

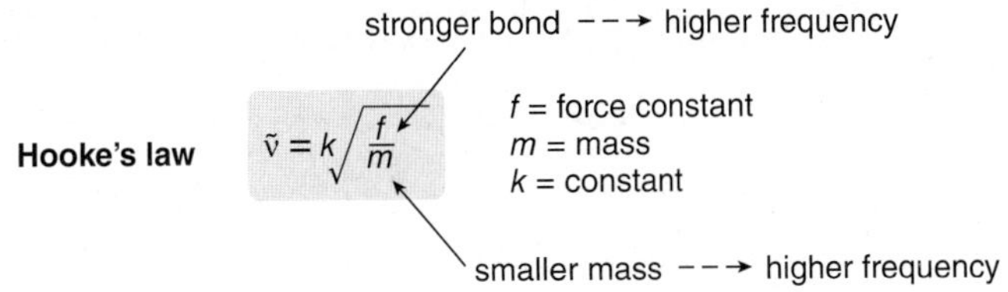

- The force constant (f) is the strength of the bond (or spring). The larger the value of f, the stronger the bond, and the higher the $\tilde{\nu}$ of vibration.
- The mass (m) is the mass of atoms (or weights). The smaller the value of m, the higher the $\tilde{\nu}$ of vibration.

As a result, **bonds absorb in four predictable regions in an IR spectrum.** These four regions, and the bonds that absorb there, are summarized in Figure 13.11. Remembering the information in this figure will help you analyze the spectra of unknown compounds. To help you remember it, keep in mind the following two points:

- **Absorptions for bonds to hydrogen always occur on the left side of the spectrum (the high wavenumber region). H has so little mass that H–Z bonds (where Z = C, O, and N) vibrate at high frequencies.**
- **Bond strength decreases in going from C≡C → C=C → C–C, so the frequency of vibration decreases—that is, the absorptions for these bonds move farther to the right side of the spectrum.**

The functional group region consists of absorptions for single bonds to hydrogen (all H–Z bonds), as well as absorptions for all multiple bonds. Most absorptions in the functional group region are due to bond stretching (rather than bond bending). The fingerprint region consists of absorptions due to all other single bonds (except H–Z bonds), often making it a complex region that is very difficult to analyze.

Figure 13.11
Summary: The four regions of the IR spectrum

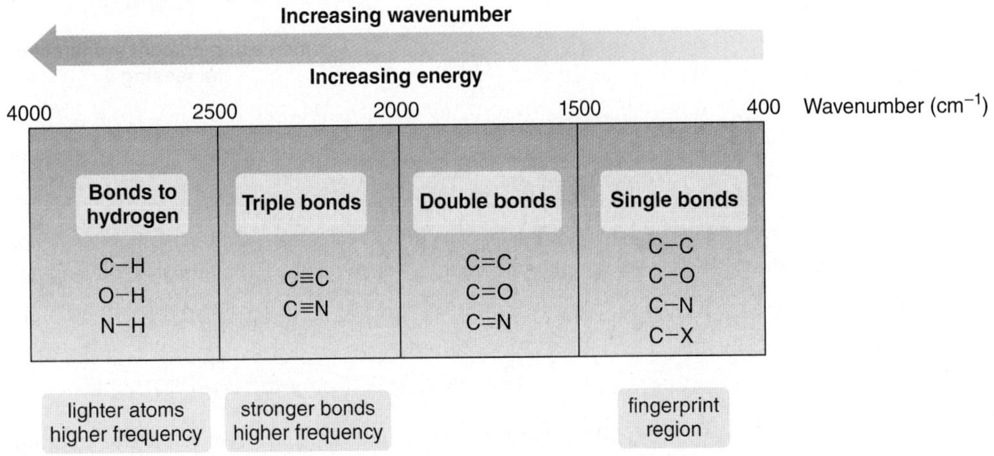

Besides learning the general regions of the IR spectrum, it is useful to learn the specific absorption values for common bonds. Table 13.2 lists the most important IR absorptions in the functional group region. Other details of IR absorptions will be presented in later chapters when new functional groups are introduced. Appendix E contains a detailed list of the characteristic IR absorption frequencies for common bonds.

Table 13.2 Important IR Absorptions

Bond type	Approximate $\tilde{\nu}$ (cm^{-1})	Intensity
O–H	3600–3200	strong, broad
N–H	3500–3200	medium
C–H	~3000	
• C_{sp^3}–H	3000–2850	strong
• C_{sp^2}–H	3150–3000	medium
• C_{sp}–H	3300	medium
C≡C	2250	medium
C≡N	2250	medium
C=O	1800–1650 (often ~1700)	strong
C=C	1650	medium
	1600, 1500	medium

Even subtle differences that affect bond strength affect the frequency of an IR absorption. Recall from Section 1.11 that the strength of a C–H bond increases as the percent *s*-character of the hybrid orbital on the carbon increases; thus:

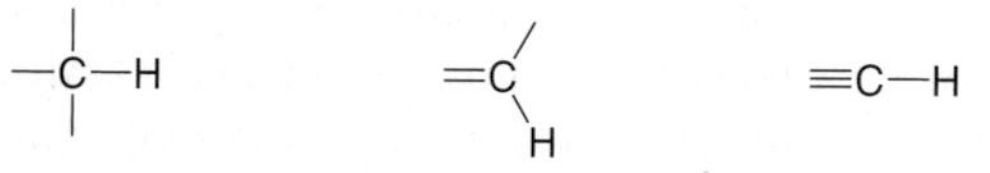

C_{sp^3}–H	C_{sp^2}–H	C_{sp}–H
25% *s*-character	33% *s*-character	50% *s*-character

Increasing percent *s*-character
Increasing $\tilde{\nu}$

- **The higher the percent *s*-character, the stronger the bond and the higher the wavenumber of absorption.**

Problem 13.15 Which highlighted bond in each pair absorbs at higher wavenumber?

a. 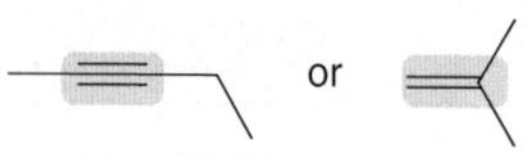

b. 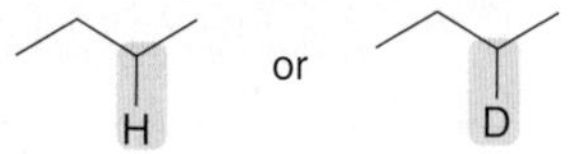

Finally, almost all bonds in a molecule give rise to an absorption peak in an IR spectrum, but a few do not. **For a bond to absorb in the IR, there must be a change in dipole moment during the vibration.** Thus, symmetrical, nonpolar bonds do not absorb in the IR. The carbon–carbon triple bond of but-2-yne, for example, does not have an IR stretching absorption at 2250 cm^{-1} because the C≡C bond is nonpolar and there is no change in dipole moment when the bond stretches along its axis. This type of vibration is said to be **IR inactive.**

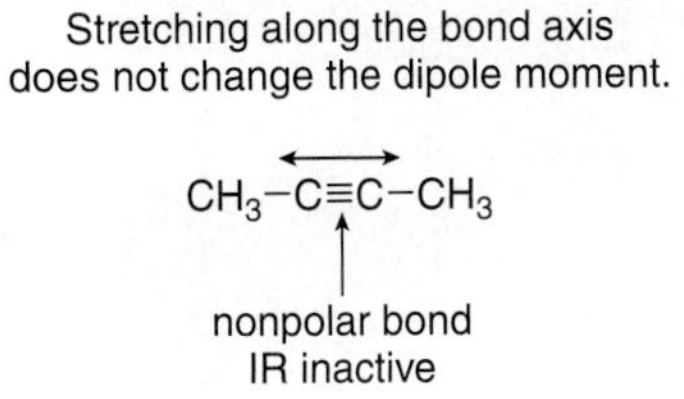

13.7B IR Absorptions in Hydrocarbons

The IR spectra of hexane, hex-1-ene, and hex-1-yne illustrate the important differences that characterize the IR spectra of hydrocarbons above 1500 cm^{-1}. Although all three compounds contain C–C bonds and sp^3 hybridized C–H bonds, the absorption peaks due to C=C and C≡C readily distinguish the alkene and alkyne.

Note, too, that the C–H absorptions in alkanes, alkenes, and alkynes have a characteristic appearance and position. The sp^3 hybridized C–H bonds are often seen as a broad, strong absorption at < 3000 cm^{-1}, whereas sp^2 and sp hybridized C–H bonds absorb at somewhat higher frequency.

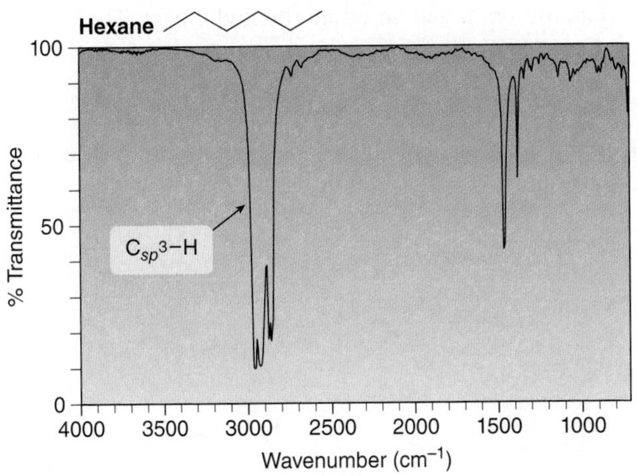

- The **alkane** hexane has only C–C single bonds and sp^3 hybridized C atoms. Therefore, it has only one major absorption above 1500 cm^{-1}, its C_{sp^3}–H absorption at 3000–2850 cm^{-1}.

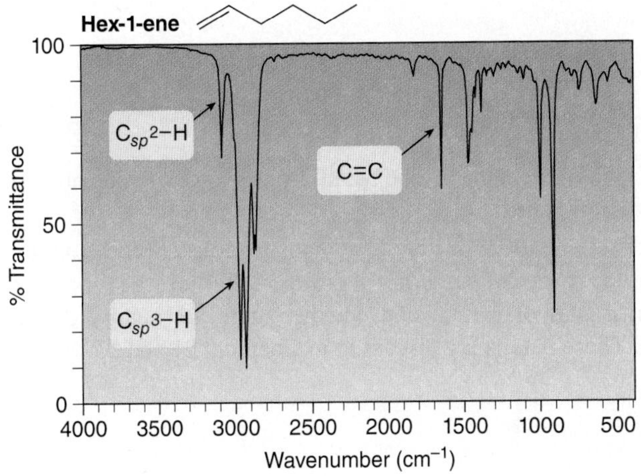

- The **alkene** hex-1-ene has a C=C and C_{sp^2}–H, in addition to its sp^3 hybridized C atoms. Therefore, there are three major absorptions above 1500 cm^{-1}:
  - C_{sp^2}–H at 3150–3000 cm^{-1}
  - C_{sp^3}–H at 3000–2850 cm^{-1}
  - C=C at 1650 cm^{-1}

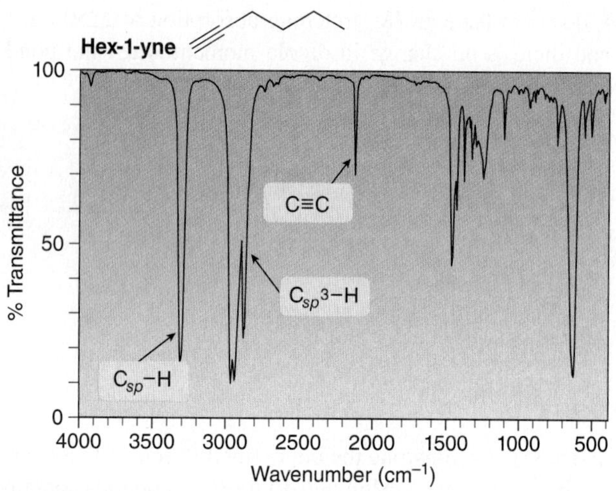

- The **alkyne** hex-1-yne has a C≡C and C_{sp}–H, in addition to its sp^3 hybridized C atoms. Therefore, there are three major absorptions:
  - C_{sp}–H at 3300 cm^{-1}
  - C_{sp^3}–H at 3000–2850 cm^{-1}
  - C≡C at ~2250 cm^{-1}

Problem 13.16 How do the IR spectra of the isomers cyclopentane and pent-1-ene differ?

13.7C IR Absorptions in Oxygen-Containing Compounds

The most important IR absorptions for oxygen-containing compounds occur at **3600–3200 cm^{-1} for an OH group,** and at approximately **1700 cm^{-1} for a C=O,** as illustrated in the IR spectra of an **alcohol** (butan-2-ol), a **ketone** (butan-2-one), and an **ether** (diethyl ether). The peak at ~3000 cm^{-1} in each spectrum is due to C_{sp^3}–H bonds.

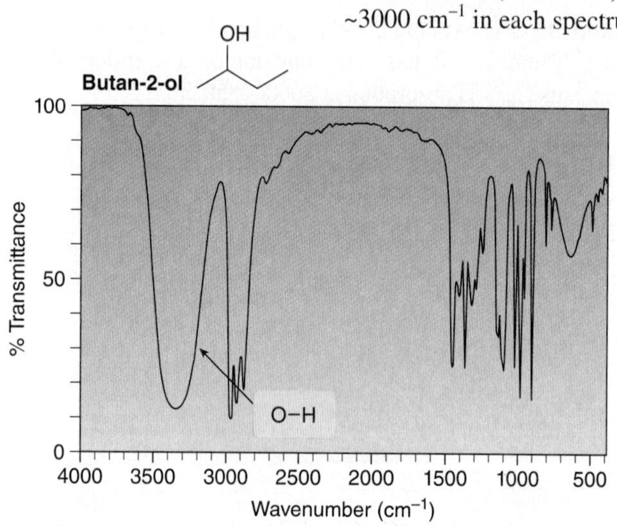

- The **OH** group in the alcohol butan-2-ol shows a strong absorption at 3600–3200 cm^{-1}.

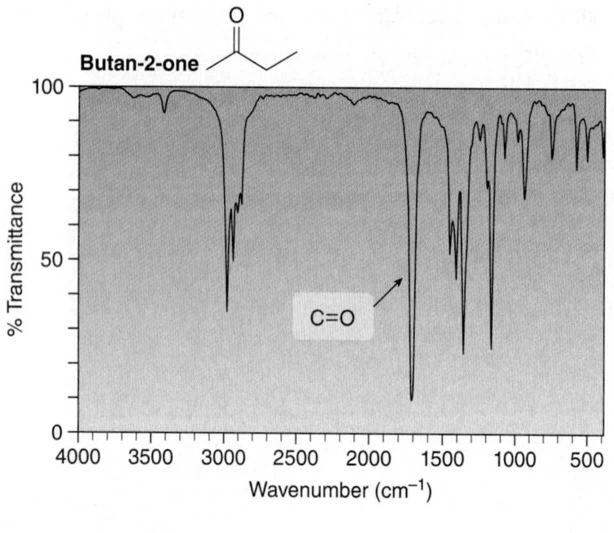

- The **C=O** group in the ketone butan-2-one shows a strong absorption at ~1700 cm^{-1}.
- The exact location of the C=O absorption depends on the particular type of carbonyl group, whether the carbonyl carbon is part of a ring, and whether there are nearby double bonds. These details are discussed in Chapters 21 and 22.

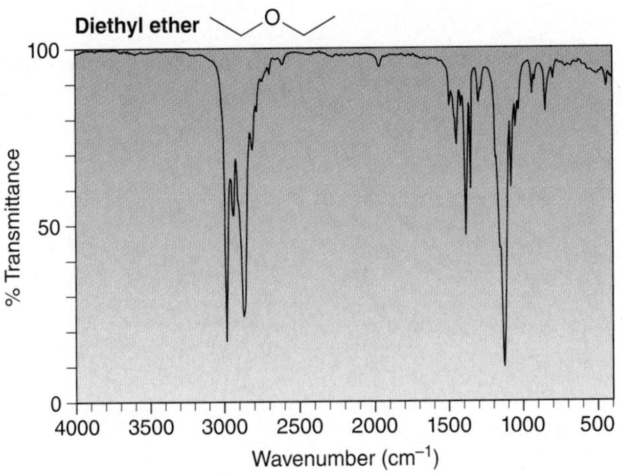

- Diethyl ether has neither an OH group nor a C=O, so its only absorption above 1500 cm^{-1} occurs at ~3000 cm^{-1}, due to sp^3 hybridized C–H bonds. Compounds that contain an oxygen atom but do not show an OH or C=O absorption are **ethers.**

13.7D IR Absorptions in Nitrogen-Containing Compounds

Common functional groups that contain nitrogen atoms are also distinguishable by their IR absorptions above 1500 cm^{-1}, as illustrated by the IR spectra of an **amine** (octylamine), an **amide** (propanamide), and a **nitrile** (octanenitrile). Additional details on the IR spectra of these compounds are given in Chapters 22 and 25.

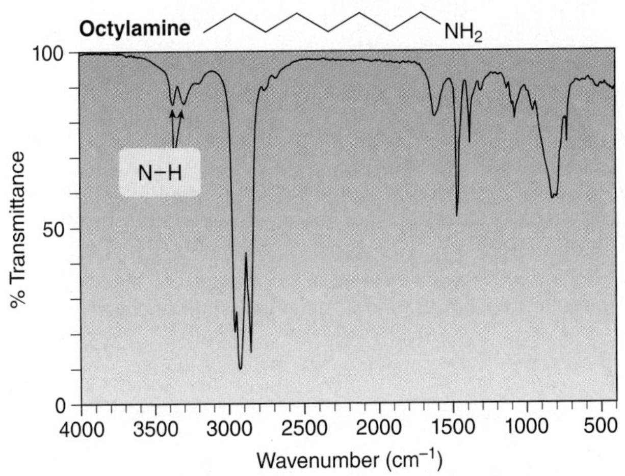

- The **N–H** bonds in the amine octylamine give rise to two weak absorptions at 3300 and 3400 cm^{-1}.

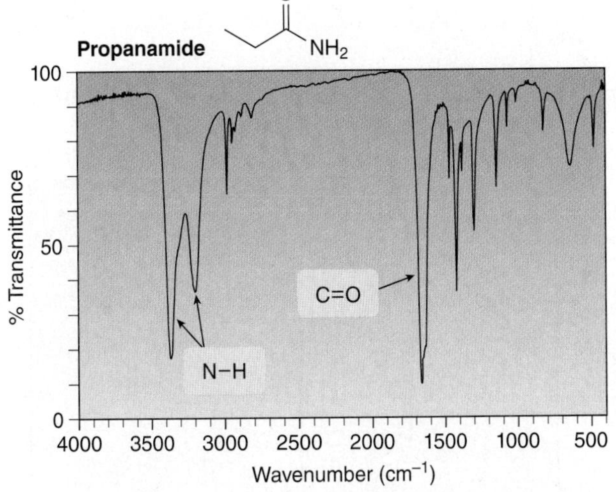

- The **amide** propanamide exhibits absorptions above 1500 cm^{-1} for both its **N–H** and **C=O** groups:
  - N–H (two peaks) at 3200 and 3400 cm^{-1}
  - C=O at 1660 cm^{-1}

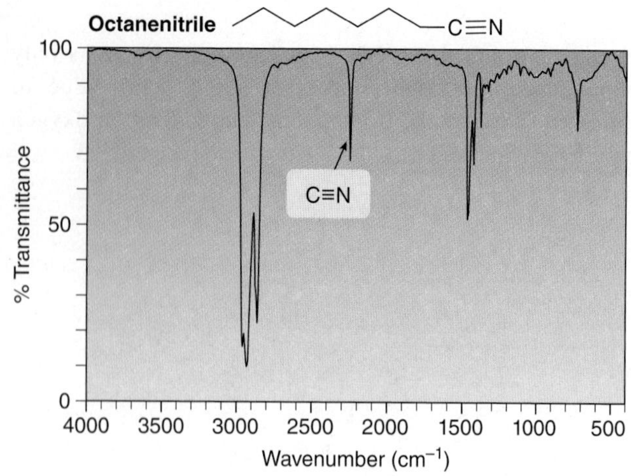

- The **C≡N** group of the **nitrile** octanenitrile absorbs in the triple bond region at ~2250 cm^{-1}.

Sample Problem 13.6 How can the two isomers having molecular formula C_2H_6O be distinguished by IR spectroscopy?

Solution

First, draw the structures of the compounds and then locate the functional groups. One compound is an alcohol and one is an ether.

ethanol

- C–H absorption at ~3000 cm^{-1}
- O–H absorption at 3600–3200 cm^{-1}

dimethyl ether

- C–H absorption at ~3000 cm^{-1} **only**

Although both compounds have sp^3 hybridized C–H bonds, ethanol has an OH group that gives a strong absorption at 3600–3200 cm^{-1}, and dimethyl ether does not. This feature distinguishes the two isomers.

Problem 13.17 How do the three isomers of molecular formula C_3H_6O (**A, B,** and **C**) differ in their IR spectra?

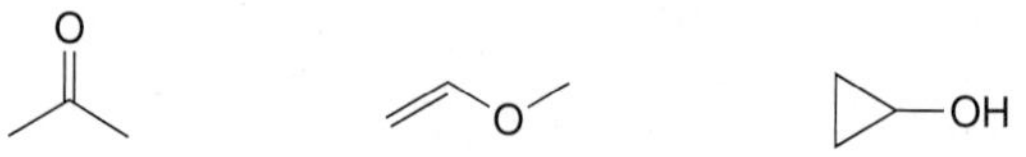

A **B** **C**

Sample Problem 13.7 shows how the region above 1500 cm^{-1} in an IR spectrum can be used for functional group identification.

Sample Problem 13.7 What functional groups are responsible for the absorptions above 1500 cm^{-1} in compounds **A** and **B?**

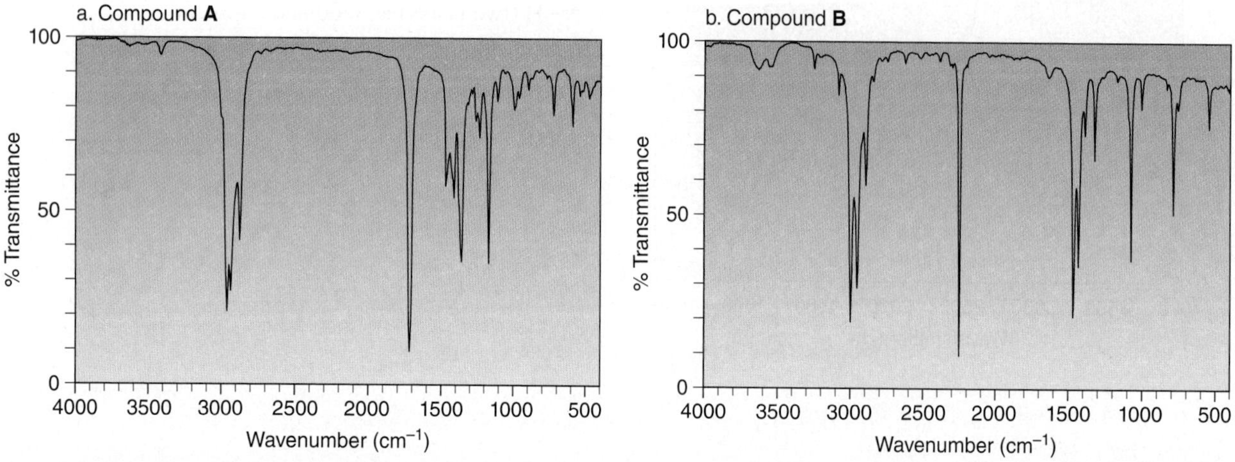

Solution

a. Compound **A** has two major absorptions above 1500 cm^{-1}: The absorption at ~3000 cm^{-1} is due to C–H bonds and the absorption at ~1700 cm^{-1} is due to a C=O group.

b. Compound **B** has two major absorptions above 1500 cm^{-1}: The absorption at ~3000 cm^{-1} is due to C–H bonds and the absorption at ~2250 cm^{-1} is due to a triple bond, either a C≡C or a C≡N. Because there is no absorption due to an *sp* hybridized C–H bond at 3300 cm^{-1}, this IR spectrum *cannot* be due to a terminal alkyne (HC≡CR) but may still be due to an internal alkyne.

Problem 13.18 What functional groups are responsible for the absorptions above 1500 cm^{-1} in the IR spectra for compounds **A** and **B?**

a. Compound **A**

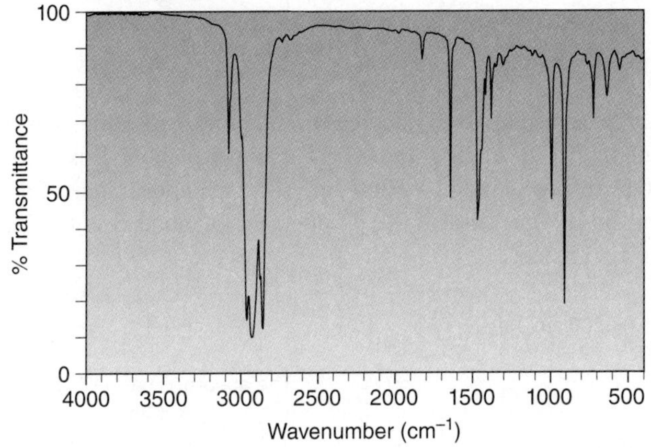

b. Compound **B**

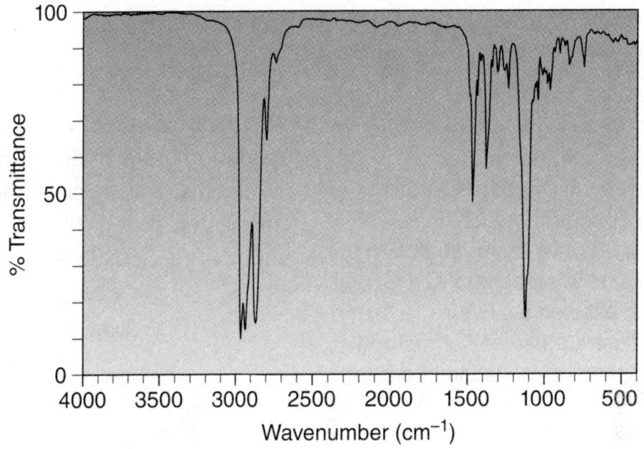

Problem 13.19 What are the major IR absorptions in the functional group region for each compound?

a. 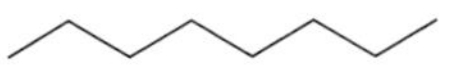

b. 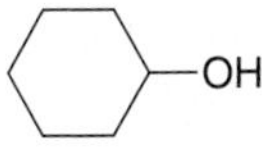

c. 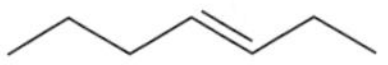

d. 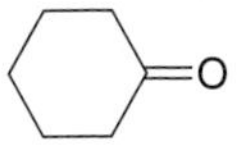

e. 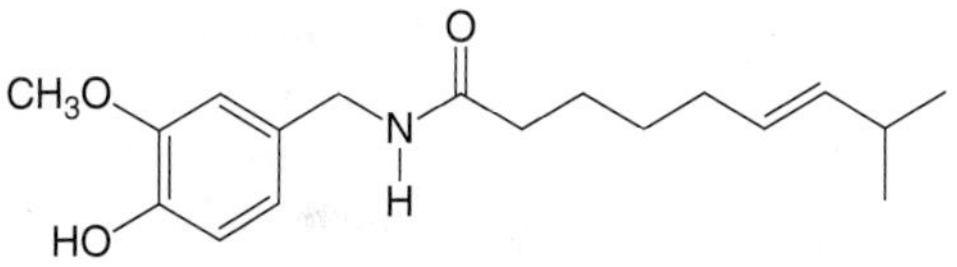

capsaicin
(spicy component of hot peppers)

Problem 13.20 What are the major IR absorptions in the functional group region for oleic acid, a common unsaturated fatty acid (Section 10.6A)?

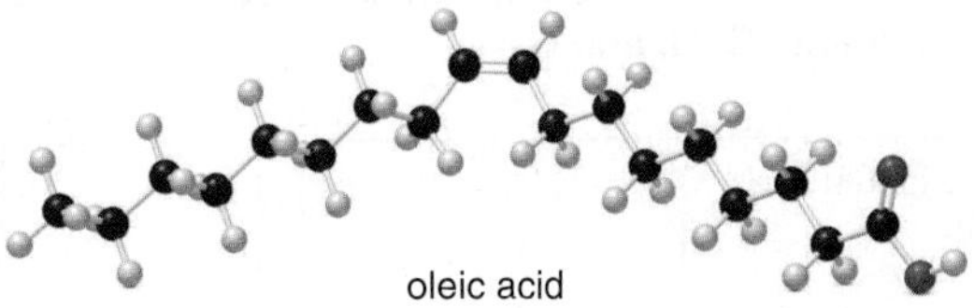

oleic acid

The antibiotic properties of penicillin were discovered when Sir Alexander Fleming noticed that a mold of the genus *Penicillium* inhibited the growth of certain bacteria. Penicillin was used as an antibiotic before its structure was determined conclusively.

13.8 IR and Structure Determination

Since its introduction, IR spectroscopy has proven to be a valuable tool for determining the functional groups in organic molecules.

In the 1940s, IR spectroscopy played a key role in elucidating the structure of the antibiotic penicillin G, the chapter-opening molecule. **β-Lactams,** four-membered rings that contain an amide, have a carbonyl group that absorbs at ~1760 cm^{-1}, a much higher frequency than that observed for most amides and many other carbonyl groups. Because penicillin G had an IR absorption at this frequency, **A** became the leading candidate for the structure of penicillin rather than **B,** a possibility originally considered more likely. Structure **A** was later confirmed by X-ray analysis.

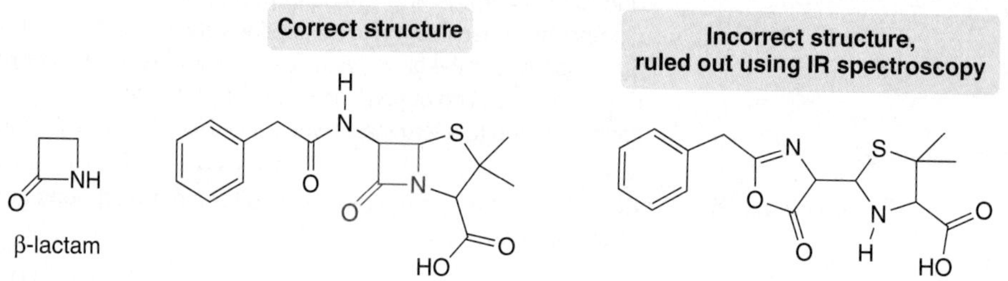

β-lactam

A
penicillin G
(β-lactam in red)

B

New instruments for determining blood alcohol concentration use IR spectroscopy for analyzing the C–H absorption of CH_3CH_2OH in exhaled air. Figure 12.10 illustrated an earlier method based on oxidation chemistry.

IR spectroscopy is often used to determine the outcome of a chemical reaction. For example, oxidation of the hydroxy group in **C** to form the carbonyl group in periplanone B is accompanied by the disappearance of the OH absorption (3600–3200 cm^{-1}) and the appearance of a carbonyl absorption near 1700 cm^{-1} in the IR spectrum of the product. Periplanone B is the sex pheromone of the female American cockroach.

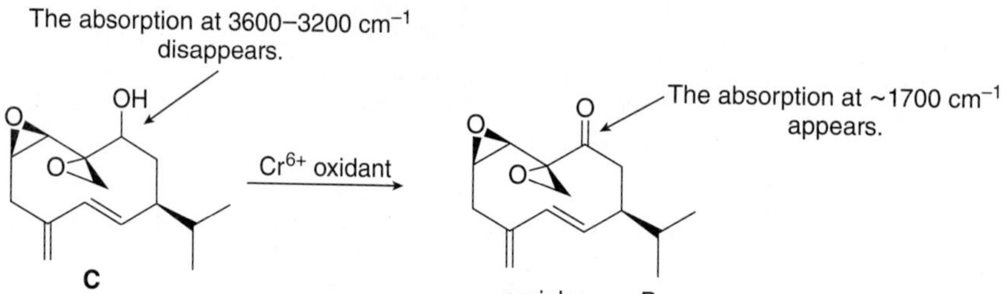

C

periplanone B

The combination of IR and mass spectral data provides key information on the structure of an unknown compound. The mass spectrum reveals the molecular weight of the unknown (and the molecular formula if an exact mass is available), and the IR spectrum helps to identify the important functional groups.

How To Use MS and IR for Structure Determination

Example **What information is obtained from the mass spectrum and IR spectrum of an unknown compound X? Assume X contains the elements C, H, and O.**

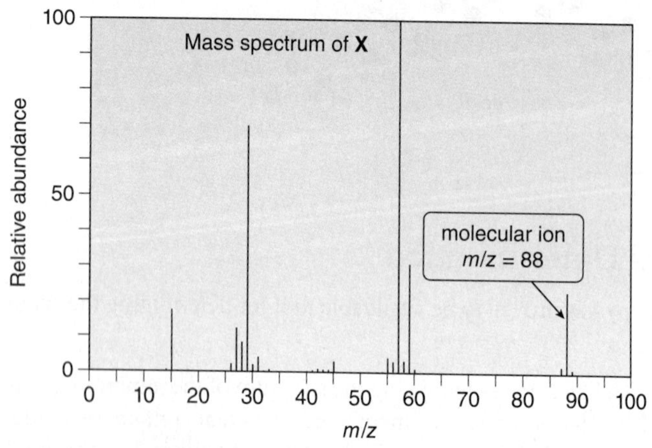

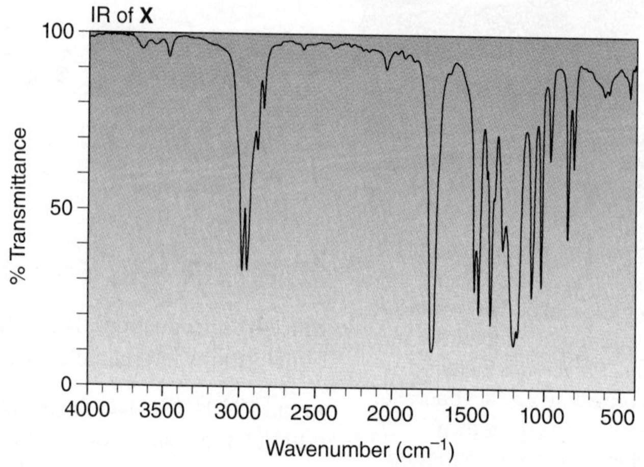

How To, continued . . .

Step [1] Use the molecular ion to determine possible molecular formulas. Use an exact mass (when available) to determine a molecular formula.

- Use the procedure outlined in Sample Problem 13.2 to calculate possible molecular formulas. For a molecular ion at m/z = 88:

$$\frac{88}{12} = 7\text{ C's (maximum) (remainder = 4)} \dashrightarrow C_7H_4 \xrightarrow[+1\text{ O}]{-CH_4} C_6O \xrightarrow[+12\text{ H's}]{-1\text{ C}} C_5H_{12}O \xrightarrow[+1\text{ O}]{-CH_4} C_4H_8O_2 \xrightarrow[+1\text{ O}]{-CH_4} C_3H_4O_3$$

($C_5H_{12}O$, $C_4H_8O_2$, $C_3H_4O_3$: three possible formulas)

- Discounting C_7H_4 (a hydrocarbon) and C_6O (because it contains no H's) gives three possible formulas for **X.**
- If high-resolution mass spectral data are available, the molecular formula can be determined directly. If the molecular ion had an exact mass of 88.0580, the molecular formula of **X** is $\mathbf{C_4H_8O_2}$ (exact mass = 88.0524) rather than $C_5H_{12}O$ (exact mass = 88.0888) or $C_3H_4O_3$ (exact mass = 88.0160).

Step [2] Calculate the number of degrees of unsaturation (Section 10.2).

- For a compound of molecular formula $C_4H_8O_2$, the maximum number of H's = $2n + 2 = 2(4) + 2 = 10$.
- Because the compound contains only 8 H's, it has 10 – 8 = 2 H's fewer than the maximum number.
- Because each degree of unsaturation removes 2 H's, **X** has one degree of unsaturation. **X has one ring or one π bond.**

Step [3] Determine what functional group is present from the IR spectrum.

- The two major absorptions in the IR spectrum above 1500 cm^{-1} are due to sp^3 hybridized C–H bonds (~3000–2850 cm^{-1}) and a C=O group (1740 cm^{-1}). Thus, the one degree of unsaturation in **X** is due to the presence of the **C=O.**

Mass spectrometry and IR spectroscopy give valuable but limited information on the identity of an unknown. Although the mass spectral and IR data reveal that **X** has a molecular formula of $C_4H_8O_2$ and contains a carbonyl group, more data are needed to determine its complete structure. In Chapter 14, we will learn how other spectroscopic data can be used for that purpose.

Problem 13.21 Which of the following possible structures for **X** can be excluded on the basis of its IR spectrum?

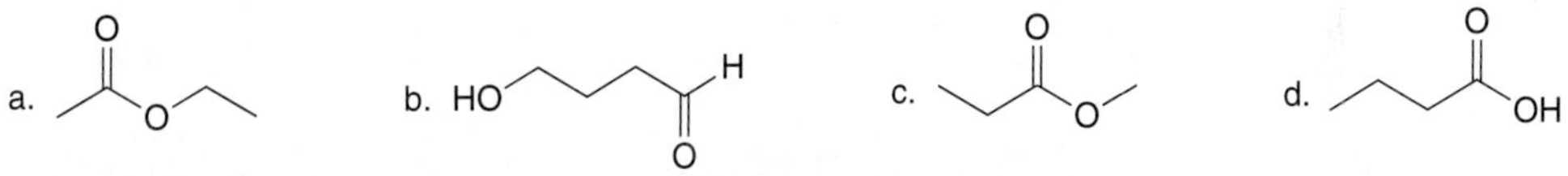

Problem 13.22 Propose structures consistent with each set of data: (a) a hydrocarbon with a molecular ion at m/z = 68 and IR absorptions at 3310, 3000–2850, and 2120 cm^{-1}; (b) a compound containing C, H, and O with a molecular ion at m/z = 60 and IR absorptions at 3600–3200 and 3000–2850 cm^{-1}.

KEY CONCEPTS

Mass Spectrometry and Infrared Spectroscopy

Mass Spectrometry (MS; 13.1–13.4)

- Mass spectrometry measures the molecular weight of a compound (13.1A).
- The mass of the molecular ion (**M**) = the molecular weight of a compound. Except for isotope peaks at M + 1 and M + 2, the molecular ion has the highest mass in a mass spectrum (13.1A).
- The base peak is the tallest peak in a mass spectrum (13.1A).
- A compound with an odd number of N atoms gives an odd molecular ion. A compound with an even number of N atoms (including zero) gives an even molecular ion (13.1B).
- Organic monochlorides show two peaks for the molecular ion (M and M + 2) in a 3:1 ratio (13.2).

- Organic monobromides show two peaks for the molecular ion (M and M + 2) in a 1:1 ratio (13.2).
- The fragmentation of radical cations formed in a mass spectrometer gives lower molecular weight fragments, often characteristic of a functional group (13.3).
- High-resolution mass spectrometry gives the molecular formula of a compound (13.4A).

Electromagnetic Radiation (13.5)

- The wavelength and frequency of electromagnetic radiation are *inversely* related by the following equations: $\lambda = c/\nu$ or $\nu = c/\lambda$ (13.5).
- The energy of a photon is proportional to its frequency; the higher the frequency, the higher the energy: $E = h\nu$ (13.5).

Infrared Spectroscopy (IR; 13.6 and 13.7)

- Infrared spectroscopy identifies functional groups.
- IR absorptions are reported in wavenumbers, $\tilde{\nu} = 1/\lambda$.
- The functional group region from **4000** to **1500 cm^{-1}** is the most useful region of an IR spectrum.
- C–H, O–H, and N–H bonds absorb at high frequency, ≥ 2500 cm^{-1}.
- As bond strength increases, the $\tilde{\nu}$ of absorption increases; thus, triple bonds absorb at higher $\tilde{\nu}$ than double bonds.

C=C ~1650 cm^{-1} → C≡C ~2250 cm^{-1}

Increasing bond strength
Increasing $\tilde{\nu}$

- The higher the percent *s*-character, the stronger the bond, and the higher the $\tilde{\nu}$ of an IR absorption.

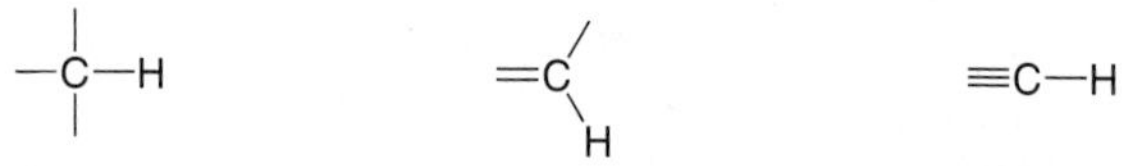

C_{sp^3}–H	C_{sp^2}–H	C_{sp}–H
25% *s*-character	33% *s*-character	50% *s*-character
3000–2850 cm^{-1}	3150–3000 cm^{-1}	3300 cm^{-1}

Increasing percent *s*-character
Increasing $\tilde{\nu}$

PROBLEMS

Problems Using Three-Dimensional Models

13.23 What major IR absorptions are present above 1500 cm^{-1} for each compound?

a. 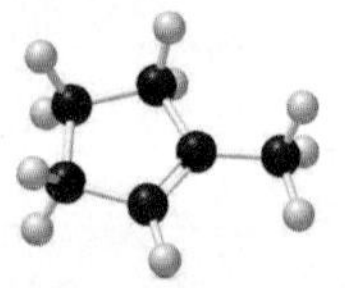 b. 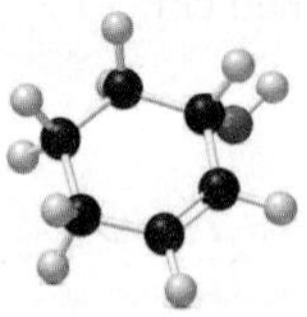

13.24 The mass spectrum of the following compound shows fragments at m/z = 127, 113, and 85. Propose structures for the ions that give rise to these peaks.

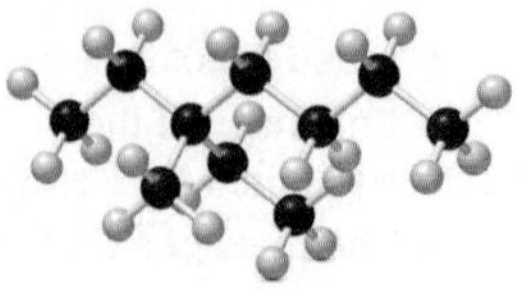

Mass Spectrometry

13.25 What molecular ion is expected for each compound?

a. 

b. 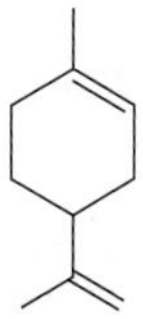

c. 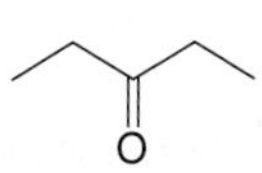

d. 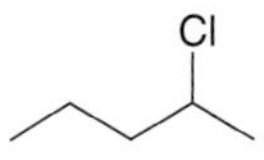

e. 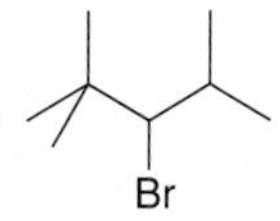

13.26 Which compound gives a molecular ion at *m*/*z* = 122: $C_6H_5CH_2CH_2CH_3$, $C_6H_5COCH_2CH_3$, or $C_6H_5OCH_2CH_3$?

13.27 Propose two molecular formulas for each molecular ion: (a) 102; (b) 98; (c) 119; (d) 74.

13.28 Propose four possible structures for a hydrocarbon with a molecular ion at *m*/*z* = 112.

13.29 What is the molecular formula for α-himachalene, a hydrocarbon obtained from cedar wood, which has four degrees of unsaturation and a molecular ion in its mass spectrum at *m*/*z* = 204?

13.30 Propose a molecular formula for rose oxide, a rose-scented compound isolated from roses and geraniums, which contains the elements of C, H, and O, has two degrees of unsaturation, and has a molecular ion in its mass spectrum at *m*/*z* = 154.

13.31 Match each structure to its mass spectrum.

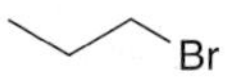

A

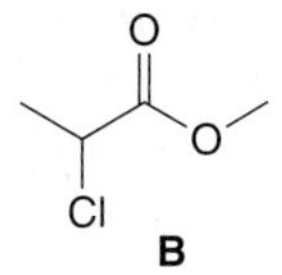

B

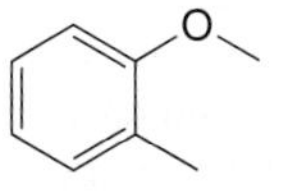

C

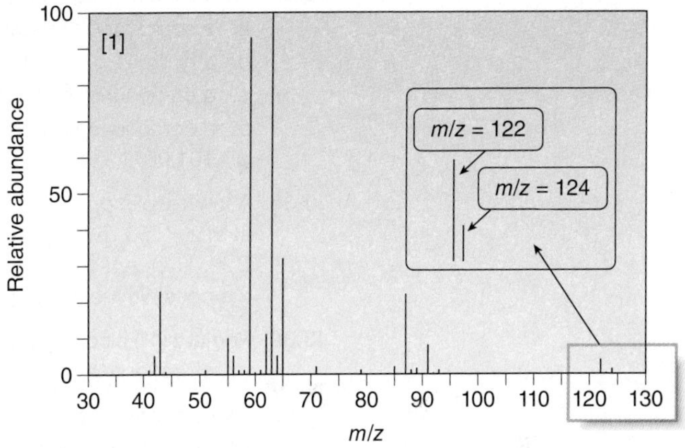

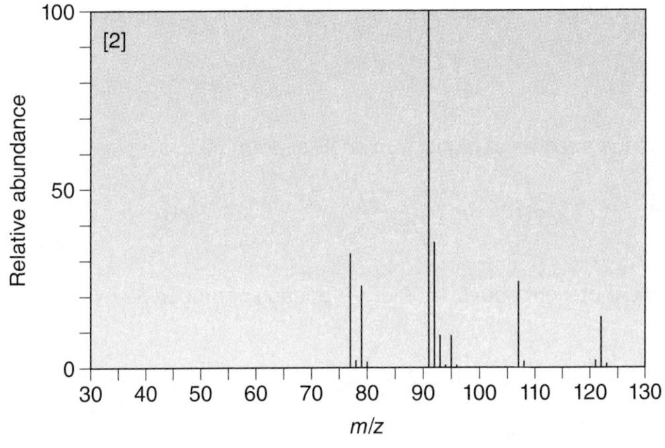

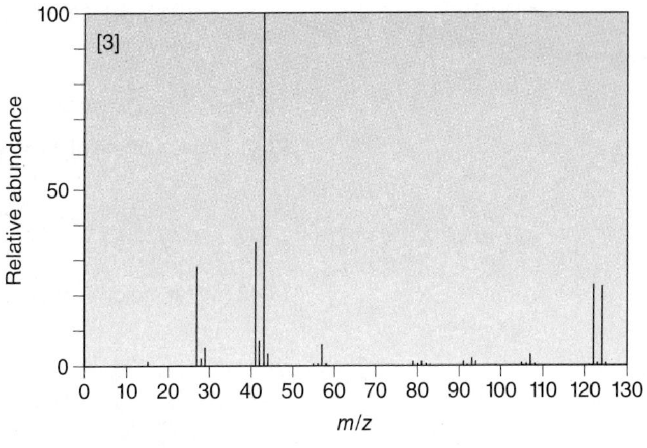

13.32 Propose two possible structures for a hydrocarbon having an exact mass of 96.0939 that forms ethylcyclopentane upon hydrogenation with H_2 and Pd-C.

13.33 What cations are formed in the mass spectrometer by α cleavage of each of the following compounds?

a. 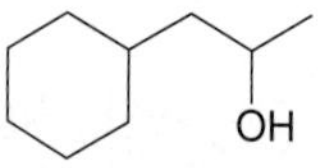 b. 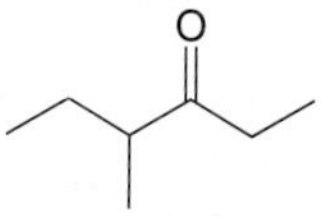 c. 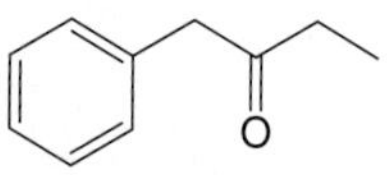

13.34 2,3-Dimethylbutane and 2,2-dimethylbutane have the same molecular ion in the mass spectrum, but only one of these isomers gives a significant fragment at $m/z = 57$. (a) Which isomer shows an intense peak at $m/z = 57$? (b) Propose a structure for the ion that gives rise to this peak. (c) The base peak in the mass spectrum of the other isomer occurs at $m/z = 43$. What ion gives rise to this peak?

13.35 For each compound, assign likely structures to the fragments at each *m/z* value, and explain how each fragment is formed.

a. $C_6H_5CH_2CH_2OH$: peaks at $m/z = 104, 91$

b. $CH_2{=}C(CH_3)CH_2CH_2OH$: peaks at $m/z = 71, 68, 41, 31$

13.36 Suppose you have two bottles, labeled ketone **A** and ketone **B.** You know that one bottle contains $CH_3CO(CH_2)_5CH_3$ and one contains $CH_3CH_2CO(CH_2)_4CH_3$, but you do not know which ketone is in which bottle. Ketone **A** gives a fragment at $m/z = 99$ and ketone **B** gives a fragment at $m/z = 113$. What are the likely structures of ketones **A** and **B** from these fragmentation data?

13.37 Propose a structure consistent with each set of data.

a. a compound that contains a benzene ring and has a molecular ion at $m/z = 107$

b. a hydrocarbon that contains only sp^3 hybridized carbons and a molecular ion at $m/z = 84$

c. a compound that contains a carbonyl group and gives a molecular ion at $m/z = 114$

d. a compound that contains C, H, N, and O and has an exact mass for the molecular ion at 101.0841

13.38 A low-resolution mass spectrum of the neurotransmitter dopamine gave a molecular ion at $m/z = 153$. Two possible molecular formulas for this molecular ion are $C_8H_{11}NO_2$ and $C_7H_{11}N_3O$. A high-resolution mass spectrum provided an exact mass at 153.0680. Which of the possible molecular formulas is the correct one?

13.39 Primary (1°) alcohols often show a peak in their mass spectra at $m/z = 31$. Suggest a structure for this fragment.

13.40 Like alcohols, ethers undergo α cleavage by breaking a carbon–carbon bond between an alkyl group and the carbon bonded to the ether oxygen atom; that is, the red C–C bond in R–CH_2OR' is broken. With this in mind, propose structures for the fragments formed by α cleavage of $(CH_3)_2CHCH_2OCH_2CH_3$. Suggest a reason why an ether fragments by α cleavage.

Infrared Spectroscopy

13.41 Which of the highlighted bonds absorbs at higher $\tilde{\nu}$ in an IR spectrum?

a. 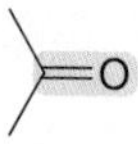 or 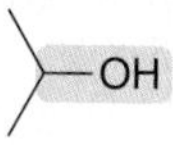 b. 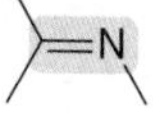 or 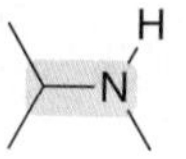 c. 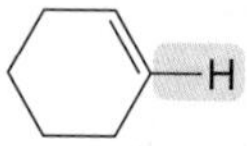 or 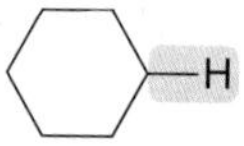

13.42 What major IR absorptions are present above 1500 cm^{-1} for each compound?

a. 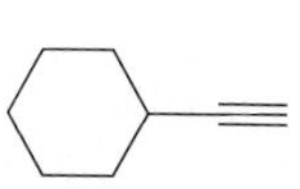 b. 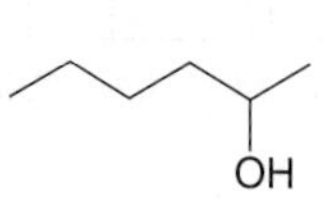 c. 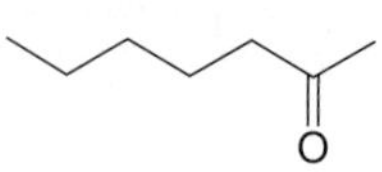 d. 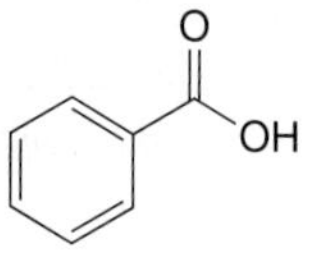

13.43 How would each of the following pairs of compounds differ in their IR spectra?

a.  and 

b. 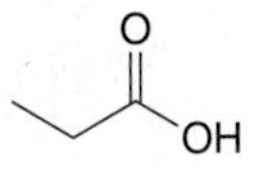 and 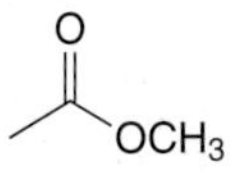

c. 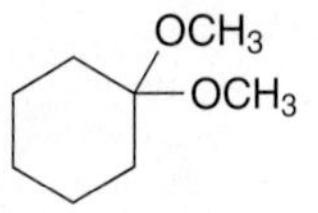 and 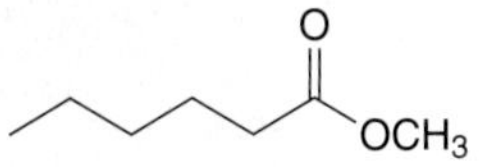

d. 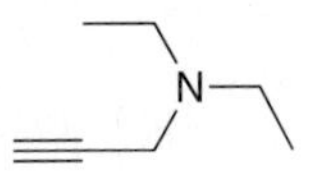 and 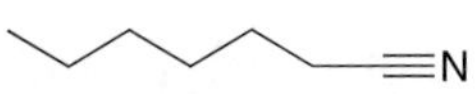

13.44 Morphine, heroin, and oxycodone are three addicting analgesic narcotics. How could IR spectroscopy be used to distinguish these three compounds from each other?

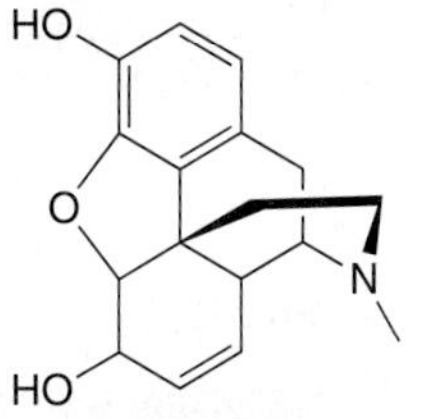

morphine

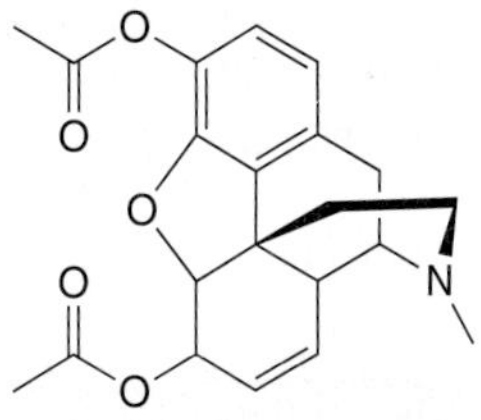

heroin

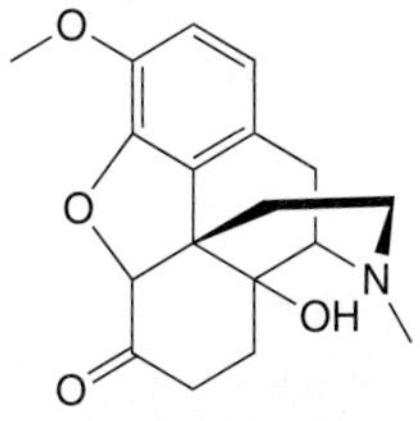

oxycodone

13.45 Reduction of cyclohex-2-enone can yield cyclohexanone, cyclohex-2-enol, or cyclohexanol, depending on the reagent and reaction conditions. How could you use IR spectroscopy to distinguish the three possible products?

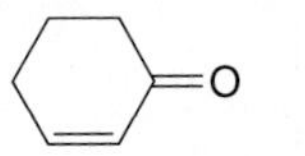

cyclohex-2-enone

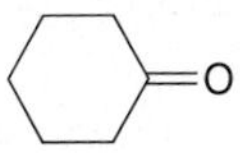

cyclohexanone

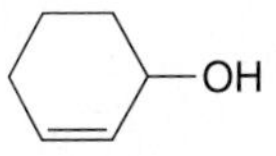

cyclohex-2-enol

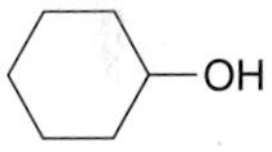

cyclohexanol

13.46 Tell how IR spectroscopy could be used to determine when each reaction is complete.

a. 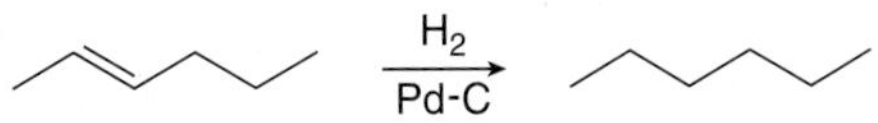

b. 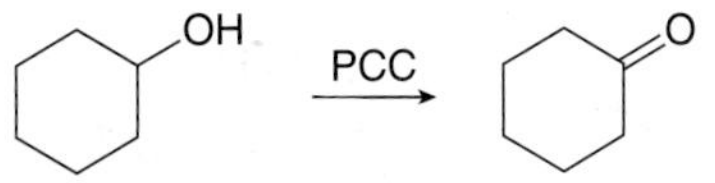

c. 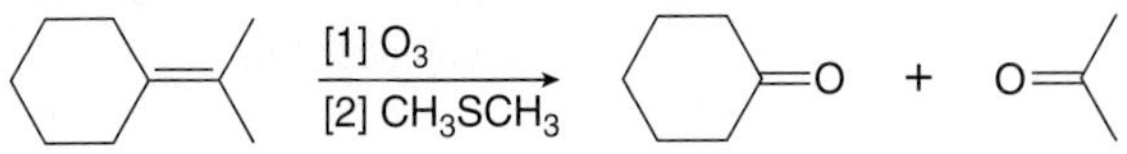

d. 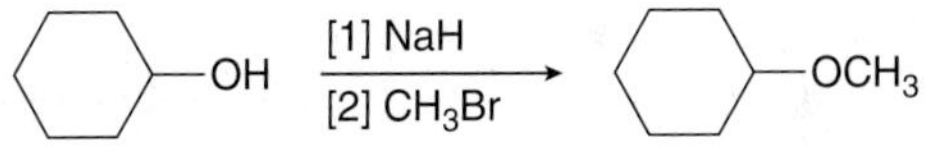

13.47 Match each compound to its IR spectrum.

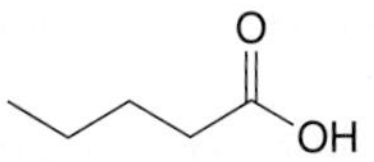

A

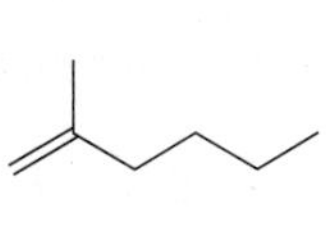
B

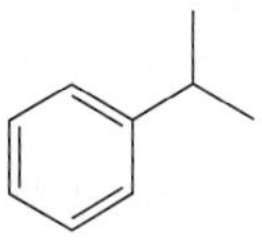
C

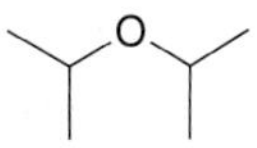

D

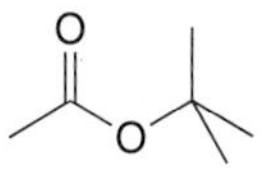

E

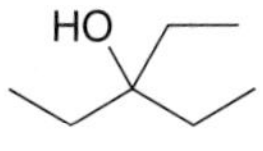

F

Spectrum [1]

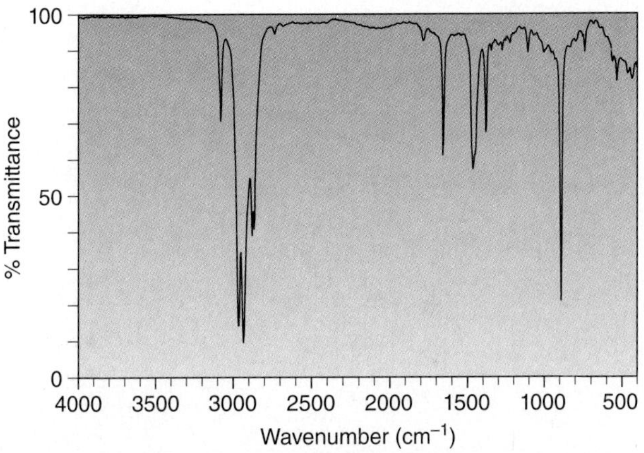

Spectrum [2]

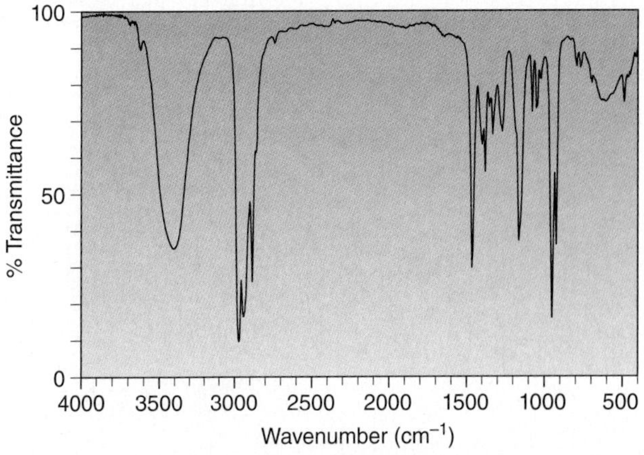

Spectrum [3]

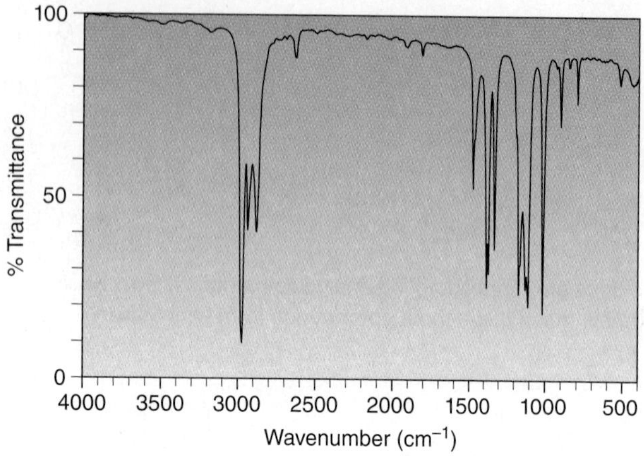

Spectrum [4]

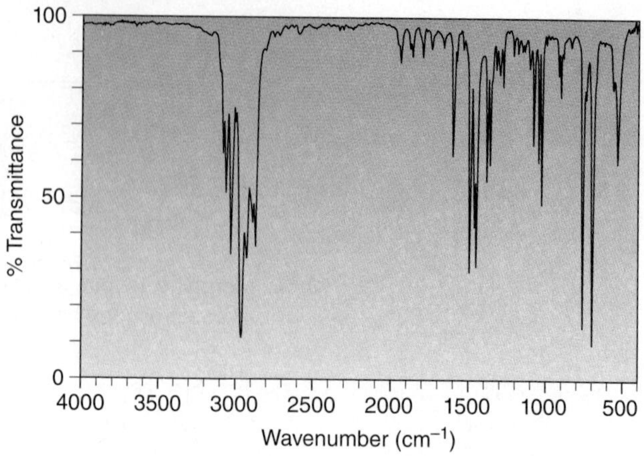

Spectrum [5]

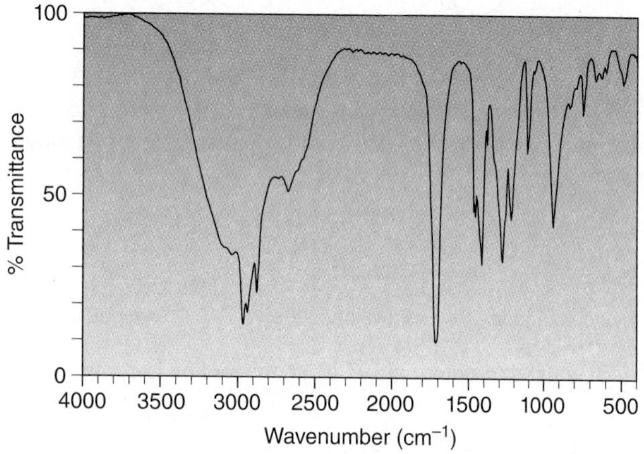

Spectrum [6]

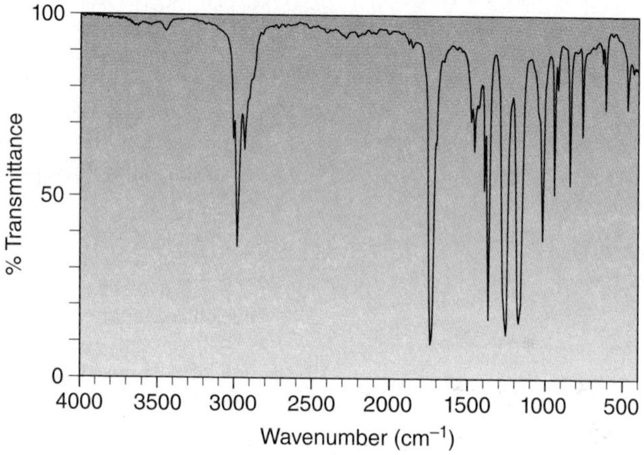

Combined Spectroscopy Problems

13.48 Propose possible structures consistent with each set of data. Assume each compound has an sp^3 hybridized C–H absorption in its IR spectrum, and that other major IR absorptions above 1500 cm^{-1} are listed.

a. a compound having a molecular ion at 72 and an absorption in its IR spectrum at 1725 cm^{-1}

b. a compound having a molecular ion at 55 and an absorption in its IR spectrum at ~2250 cm^{-1}

c. a compound having a molecular ion at 74 and an absorption in its IR spectrum at 3600–3200 cm^{-1}

13.49 A chiral hydrocarbon **X** exhibits a molecular ion at 82 in its mass spectrum. The IR spectrum of **X** shows peaks at 3300, 3000–2850, and 2250 cm^{-1}. Propose a structure for **X.**

13.50 A chiral compound **Y** has a strong absorption at 2970–2840 cm^{-1} in its IR spectrum and gives the following mass spectrum. Propose a structure for **Y.**

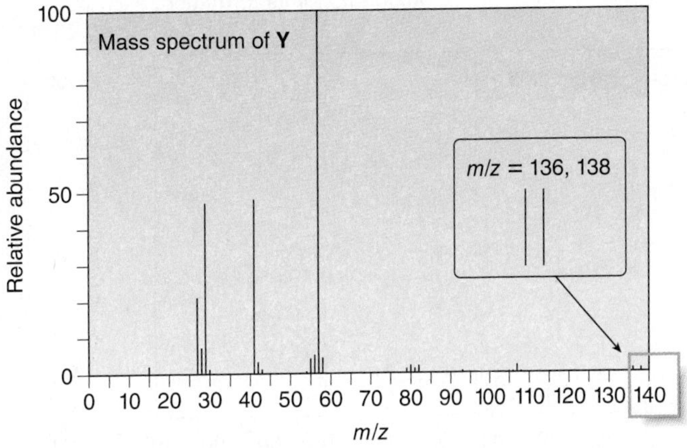

13.51 Treatment of benzoic acid ($C_6H_5CO_2H$) with NaOH followed by 1-iodo-3-methylbutane forms **H. H** has a molecular ion at 192 and IR absorptions at 3064, 3035, 2960–2872, and 1721 cm^{-1}. Propose a structure for **H.**

13.52 Treatment of benzaldehyde (C_6H_5CHO) with Zn(Hg) in aqueous HCl forms a compound **Z** that has a molecular ion at 92 in its mass spectrum. **Z** shows absorptions at 3150–2950, 1605, and 1496 cm^{-1} in its IR spectrum. Give a possible structure for **Z.**

13.53 Reaction of *tert*-butyl pentyl ether [$CH_3CH_2CH_2CH_2CH_2OC(CH_3)_3$] with HBr forms 1-bromopentane ($CH_3CH_2CH_2CH_2CH_2Br$) and compound **B. B** has a molecular ion in its mass spectrum at 56 and gives peaks in its IR spectrum at 3150–3000, 3000–2850, and 1650 cm^{-1}. Propose a structure for **B** and draw a stepwise mechanism that accounts for its formation.

13.54 Reaction of 2-methylpropanoic acid [$(CH_3)_2CHCO_2H$] with $SOCl_2$ followed by 2-methylpropan-1-ol forms **X. X** has a molecular ion at 144 and IR absorptions at 2965, 2940, and 1739 cm^{-1}. Propose a structure for **X.**

13.55 Reaction of pentanoyl chloride ($CH_3CH_2CH_2CH_2COCl$) with lithium dimethyl cuprate [$LiCu(CH_3)_2$] forms a compound **J** that has a molecular ion in its mass spectrum at 100, as well as fragments at m/z = 85, 57, and 43 (base). The IR spectrum of **J** has strong peaks at 2962 and 1718 cm^{-1}. Propose a structure for **J.**

13.56 Benzonitrile (C_6H_5CN) is reduced to two different products depending on the reducing agent used. Treatment with lithium aluminum hydride followed by water forms **K,** which has a molecular ion in its mass spectrum at 107 and the following IR absorptions: 3373, 3290, 3062, 2920, and 1600 cm^{-1}. Treatment with a milder reducing agent forms **L,** which has a molecular ion in its mass spectrum at 106 and the following IR absorptions: 3086, 2820, 2736, 1703, and 1600 cm^{-1}. **L** shows fragments in its mass spectrum at m/z = 105 and 77. Propose structures for **K** and **L** and explain how you arrived at your conclusions.

13.57 Treatment of anisole ($CH_3OC_6H_5$) with Cl_2 and $FeCl_3$ forms **P,** which has peaks in its mass spectrum at m/z = 142 (M), 144 (M + 2), 129, and 127. **P** has absorptions in its IR spectrum at 3096–2837 (several peaks), 1582, and 1494 cm^{-1}. Propose possible structures for **P.**

13.58 Reaction of $BrCH_2CH_2CH_2CH_2NH_2$ with NaH forms compound **W,** which gives the IR and mass spectra shown below. Propose a structure for **W** and draw a stepwise mechanism that accounts for its formation.

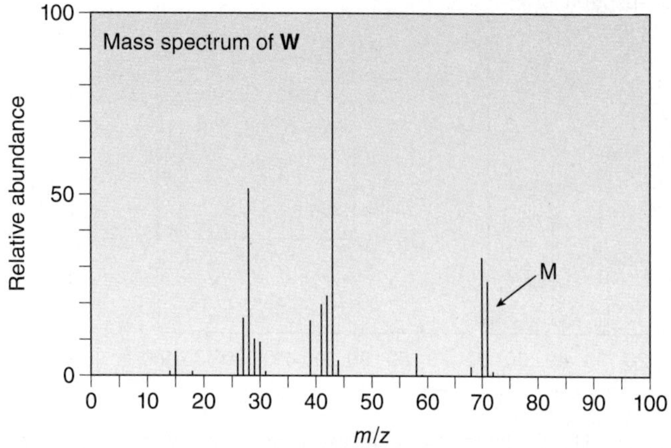

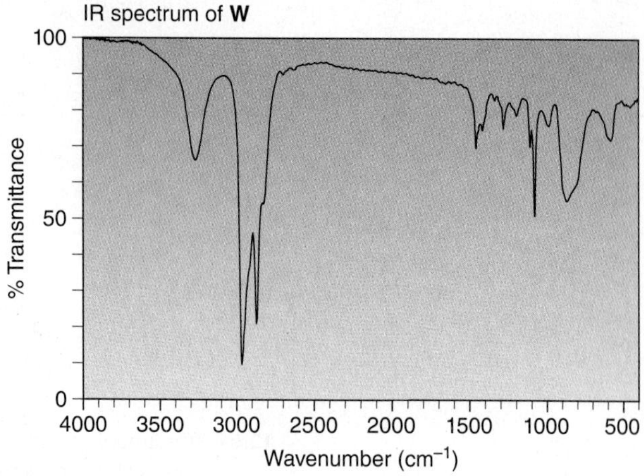

Challenge Problems

13.59 The carbonyl absorption of an amide typically occurs in the 1630–1680 cm^{-1} range, while the carbonyl absorption of an ester occurs at a much higher wavenumber, typically 1735–1745 cm^{-1}. Account for this difference.

13.60 Explain why a carbonyl absorption shifts to lower frequency in an α,β-unsaturated carbonyl compound—a compound having a carbonyl group bonded directly to a carbon–carbon double bond. For example, the carbonyl absorption occurs at 1720 cm^{-1} for cyclohexanone and at 1685 cm^{-1} for cyclohex-2-enone.

cyclohexanone

cyclohex-2-enone (an α,β-unsaturated carbonyl compound)

13.61 Explain why a ketone carbonyl typically absorbs at a lower wavenumber than an aldehyde carbonyl (1715 vs. 1730 cm^{-1}).

13.62 Oxidation of citronellol, a constituent of rose and geranium oils, with PCC in the presence of added $NaOCOCH_3$ forms compound **A. A** has a molecular ion in its mass spectrum at 154 and a strong peak in its IR spectrum at 1730 cm^{-1}, in addition to C–H stretching absorptions. Without added $NaOCOCH_3$, oxidation of citronellol with PCC yields isopulegone, which is then converted to **B** with aqueous base. **B** has a molecular ion at 152, and a peak in its IR spectrum at 1680 cm^{-1} in addition to C–H stretching absorptions.

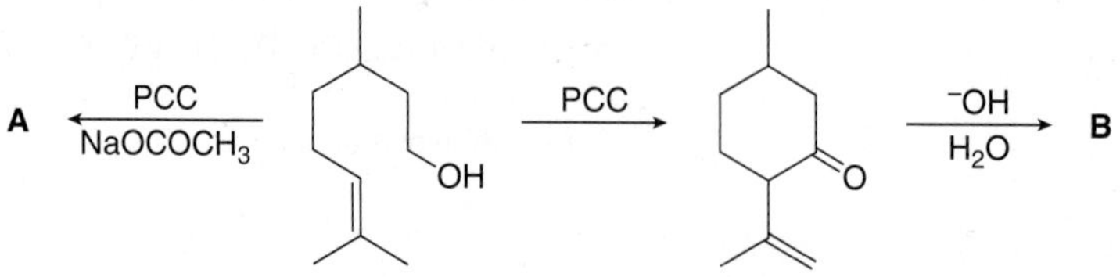

a. Identify the structures of **A** and **B.**

b. Draw a mechanism for the conversion of citronellol to isopulegone.

c. Draw a mechanism for the conversion of isopulegone to **B.**

Nuclear Magnetic Resonance Spectroscopy

14

14.1 An introduction to NMR spectroscopy
14.2 ^{1}H NMR: Number of signals
14.3 ^{1}H NMR: Position of signals
14.4 The chemical shift of protons on sp^2 and *sp* hybridized carbons
14.5 ^{1}H NMR: Intensity of signals
14.6 ^{1}H NMR: Spin–spin splitting
14.7 More complex examples of splitting
14.8 Spin–spin splitting in alkenes
14.9 Other facts about ^{1}H NMR spectroscopy
14.10 Using ^{1}H NMR to identify an unknown
14.11 ^{13}C NMR spectroscopy
14.12 Magnetic resonance imaging (MRI)

Palau'amine is a complex natural product isolated from the sea sponge *Stylotella agminata* collected in the Pacific Ocean near the Republic of Palau. The initial structure proposed for palau'amine in 1993 was revised in 2007 using a variety of modern spectroscopic techniques, including nuclear magnetic resonance spectroscopy. The dense array of functional groups in palau'amine and its antitumor and immunosuppressive properties attracted the attention of dozens of organic chemists, leading to its total synthesis in the laboratory in early 2010. In Chapter 14, we learn how nuclear magnetic resonance spectroscopy plays a key role in structure determination.

In Chapter 14 we continue our study of organic structure determination by learning about **nuclear magnetic resonance (NMR)** spectroscopy. NMR spectroscopy is the most powerful tool for characterizing organic molecules, because it can be used to identify the carbon–hydrogen framework in a compound.

14.1 An Introduction to NMR Spectroscopy

Two common types of NMR spectroscopy are used to characterize organic structure:

- **^{1}H NMR (proton NMR)** is used to determine the number and type of hydrogen atoms in a molecule; and
- **^{13}C NMR (carbon NMR)** is used to determine the type of carbon atoms in a molecule.

Before you can learn how to use NMR spectroscopy to determine the structure of a compound, you need to understand a bit about the physics behind it. Keep in mind, though, that NMR stems from the same basic principle as all other forms of spectroscopy. Energy interacts with a molecule, and absorptions occur only when the incident energy matches the energy difference between two states.

14.1A The Basis of NMR Spectroscopy

The source of energy in NMR is radio waves. Radiation in the radiofrequency region of the electromagnetic spectrum (so-called **RF** radiation) has very long wavelengths, so its corresponding frequency and energy are both low. **When these low-energy radio waves interact with a molecule, they can change the nuclear spins of some elements, including ^{1}H and ^{13}C.**

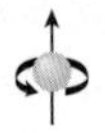

A spinning proton creates a magnetic field.

When a charged particle such as a proton spins on its axis, it creates a magnetic field. For the purpose of this discussion, therefore, a nucleus is a tiny bar magnet, symbolized by ⇵. Normally these nuclear magnets are randomly oriented in space, but in the presence of an external magnetic field, B_0, they are oriented with or against this applied field. More nuclei are oriented *with* the applied field because this arrangement is lower in energy, but the **energy difference between these two states is very small** (< 0.4 J/mol).

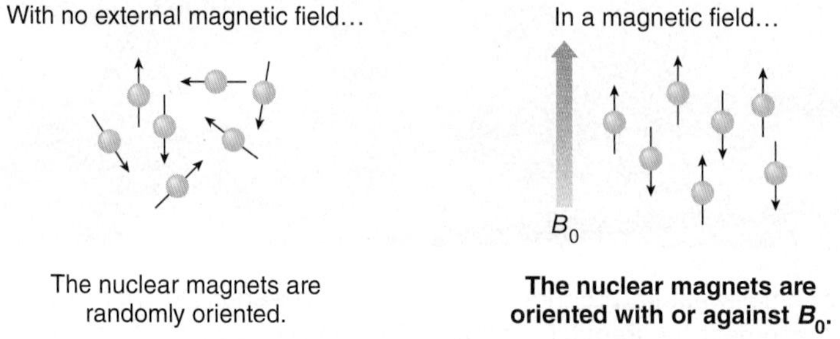

In a magnetic field, there are now two different energy states for a proton:

- **In the *lower energy state* the nucleus is aligned in the *same direction* as B_0.**
- **In the *higher energy state* the nucleus is aligned *opposed* to B_0.**

When an external energy source ($h\nu$) that matches the energy difference (ΔE) between these two states is applied, energy is absorbed, causing the **nucleus to "spin flip" from one orientation to another.** The energy difference between these two nuclear spin states corresponds to the low-frequency radiation in the RF region of the electromagnetic spectrum.

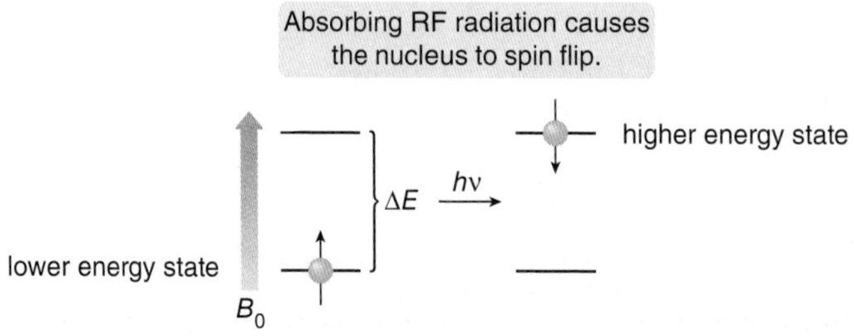

- A nucleus is in *resonance* when it absorbs RF radiation and "spin flips" to a higher energy state.

Thus, two variables characterize NMR:

- **An applied magnetic field, B_0.** Magnetic field strength is measured in tesla (T).
- **The frequency ν of radiation used for resonance,** measured in hertz (Hz) or megahertz (MHz); (1 MHz = 10^6 Hz).

The frequency needed for resonance and the applied magnetic field strength are proportionally related:

$$\nu \propto B_0$$

frequency — applied magnetic field strength

- The stronger the magnetic field, the larger the energy difference between the two nuclear spin states, and the higher the ν needed for resonance.

NMR spectrometers are referred to as 300 MHz instruments, 500 MHz instruments, and so forth, depending on the frequency of RF radiation used for resonance.

Early NMR spectrometers used a magnetic field strength of ~1.4 T, which required RF radiation of 60 MHz for resonance. Modern NMR spectrometers use stronger magnets, thus requiring higher frequencies of RF radiation for resonance. For example, a magnetic field strength of 7.05 T requires a frequency of 300 MHz for a proton to be in resonance. These spectrometers use very powerful magnetic fields to create a small, but measurable energy difference between the two possible spin states. A schematic of an NMR spectrometer is shown in Figure 14.1.

If all protons absorbed at the same frequency in a given magnetic field, the spectra of all compounds would consist of a single absorption, rendering NMR useless for structure determination. Fortunately, however, this is not the case.

Figure 14.1 Schematic of an NMR spectrometer

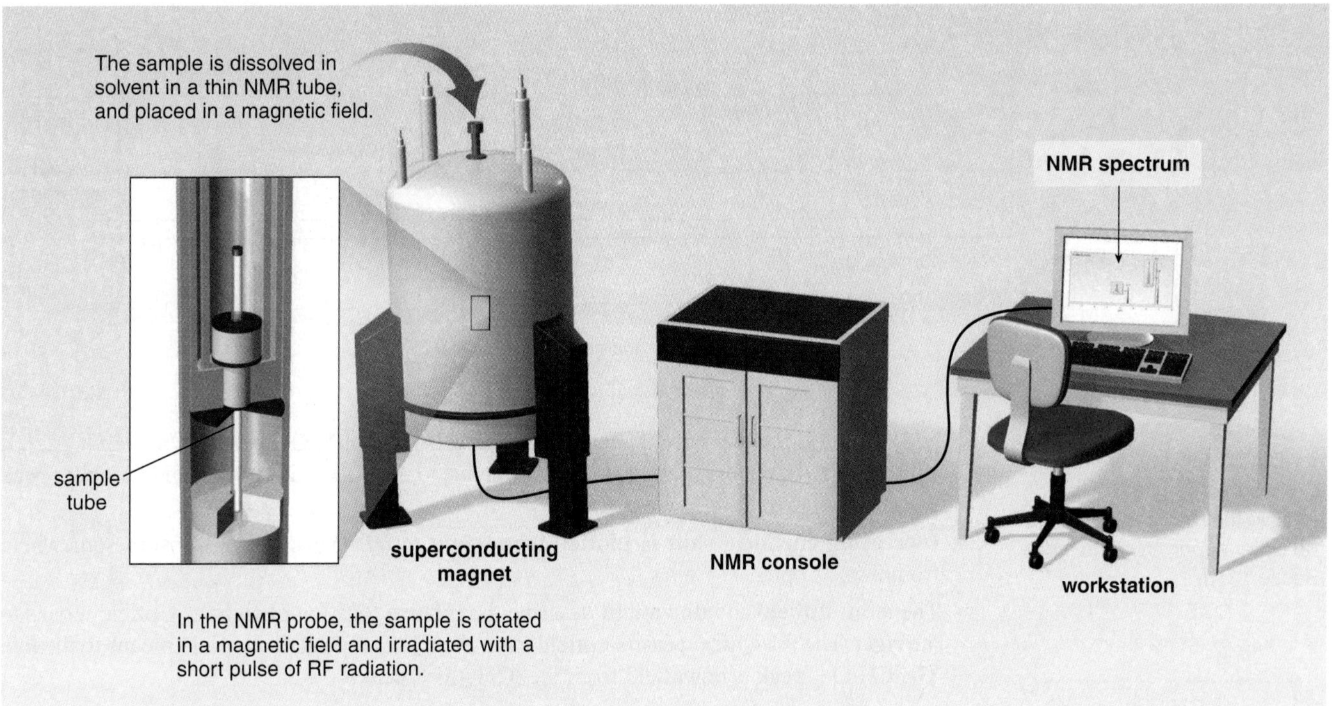

An NMR spectrometer. The sample is dissolved in a solvent, usually $CDCl_3$ (deuterochloroform), and placed in a magnetic field. A radiofrequency generator then irradiates the sample with a short pulse of radiation, causing resonance. When the nuclei fall back to their lower energy state, the detector measures the energy released, and a spectrum is recorded. The superconducting magnets in modern NMR spectrometers have coils that are cooled in liquid helium and conduct electricity with essentially no resistance.

- **All protons do *not* absorb at the same frequency. Protons in different environments absorb at slightly different frequencies, and so they are distinguishable by NMR.**

The frequency at which a particular proton absorbs is determined by its electronic environment, as discussed in Section 14.3. Because electrons are moving charged particles, they create a magnetic field opposed to the applied field B_0, and the size of the magnetic field generated by the electrons around a proton determines where it absorbs. Modern NMR spectrometers use a constant magnetic field strength B_0, and then a narrow range of frequencies is applied to achieve the resonance of all protons.

Only nuclei that contain odd mass numbers (such as ^{1}H, ^{13}C, ^{19}F, and ^{31}P) or odd atomic numbers (such as ^{2}H and ^{14}N) give rise to NMR signals. Because both ^{1}H and ^{13}C, the less abundant isotope of carbon, are NMR active, NMR allows us to map the carbon and hydrogen framework of an organic molecule.

14.1B A ^{1}H NMR Spectrum

An NMR spectrum plots the **intensity of a signal** against its **chemical shift** measured in **parts per million (ppm).** The common scale of chemical shifts is called the **δ (delta) scale.** The proton NMR spectrum of *tert*-butyl methyl ether [$CH_3OC(CH_3)_3$] illustrates several important features:

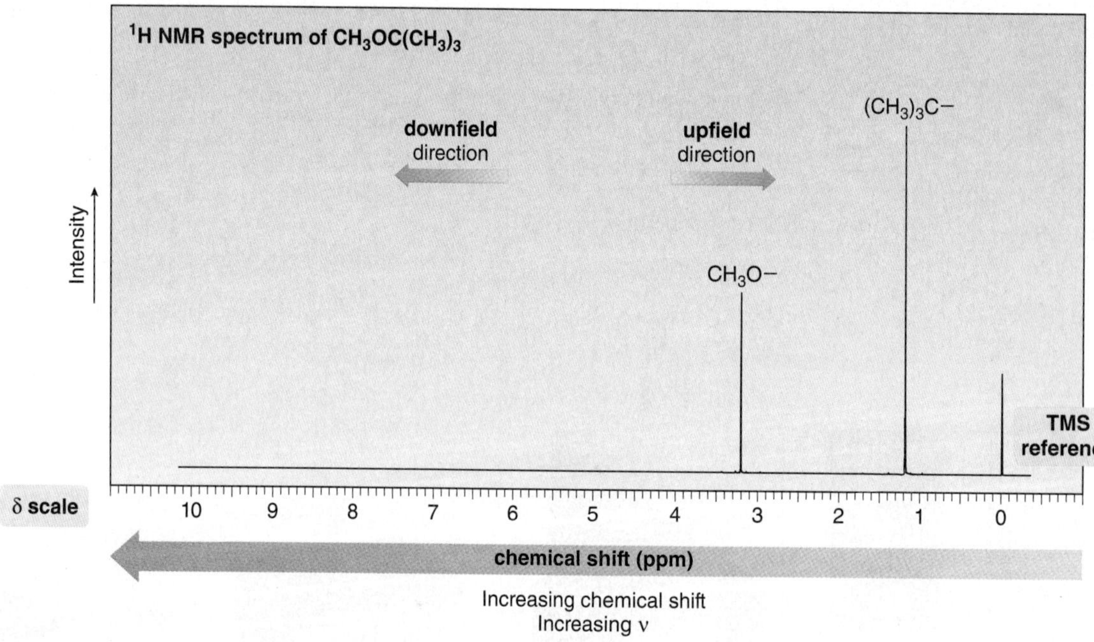

tert-Butyl methyl ether (MTBE) is the high-octane gasoline additive that has contaminated the water supply in some areas (Section 3.4).

- NMR absorptions generally appear as sharp signals. The ^{1}H NMR spectrum of $CH_3OC(CH_3)_3$ consists of two signals: a tall peak at 1.2 ppm due to the $(CH_3)_3C-$ group, and a smaller peak at 3.2 ppm due to the CH_3O- group.
- **Increasing chemical shift is plotted from *right to left.*** Most protons absorb somewhere from 0 to 12 ppm.
- The terms **upfield** and **downfield** describe the relative location of signals. Upfield means to the *right.* The $(CH_3)_3C-$ peak is upfield from the CH_3O- peak. Downfield means to the *left.* The CH_3O- peak is downfield from the $(CH_3)_3C-$ peak.

$(CH_3)_4Si$
tetramethylsilane
TMS

NMR absorptions are measured relative to the position of a reference signal at 0 ppm on the δ scale due to **tetramethylsilane (TMS). TMS** is a volatile and inert compound that gives a single peak upfield from other typical NMR absorptions.

Although chemical shifts are measured relative to the TMS signal at 0 ppm, this reference is often not plotted on a spectrum.

The *positive* direction of the δ scale is *downfield* from TMS. A very small number of absorptions occur upfield from the TMS signal, which is defined as the negative direction of the δ scale. (See Problem 14.74.)

The **chemical shift** on the *x* axis gives the position of an NMR signal, measured in ppm, according to the following equation:

$$\textbf{chemical shift} \text{ (in ppm on the } \delta \text{ scale)} = \frac{\text{observed chemical shift (in Hz) downfield from TMS}}{\nu \text{ of the NMR spectrometer (in MHz)}}$$

A chemical shift gives absorptions as a fraction of the NMR operating frequency, making it independent of the spectrometer used to record a spectrum. Because the frequency of the radiation required for resonance is proportional to the strength of the applied magnetic field, B_0, reporting NMR absorptions in frequency is meaningless unless the value of B_0 is also reported. By reporting the absorption as a fraction of the NMR operating frequency, though, we get units—ppm—that are independent of the spectrometer.

Sample Problem 14.1 Calculate the chemical shift of an absorption that occurs at 1500 Hz downfield from TMS using a 300 MHz NMR spectrometer.

Solution

Use the equation that defines the chemical shift in ppm:

$$\text{chemical shift} = \frac{\text{1500 Hz downfield from TMS}}{\text{300 MHz operating frequency}} = 5 \text{ ppm}$$

Problem 14.1 The ^{1}H NMR spectrum of CH_3OH recorded on a 500 MHz NMR spectrometer consists of two signals, one due to the CH_3 protons at 1715 Hz and one due to the OH proton at 1830 Hz, both measured downfield from TMS. (a) Calculate the chemical shift of each absorption. (b) Do the CH_3 protons absorb upfield or downfield from the OH proton?

Problem 14.2 The ^{1}H NMR spectrum of 1,2-dimethoxyethane ($CH_3OCH_2CH_2OCH_3$) recorded on a 300 MHz NMR spectrometer consists of signals at 1017 Hz and 1065 Hz downfield from TMS. (a) Calculate the chemical shift of each absorption. (b) At what frequency would each absorption occur if the spectrum were recorded on a 500 MHz NMR spectrometer?

Four different features of a ^{1}H NMR spectrum provide information about a compound's structure:

[1] **Number of signals** (Section 14.2)
[2] **Position of signals** (Sections 14.3 and 14.4)
[3] **Intensity of signals** (Section 14.5)
[4] **Spin–spin splitting of signals** (Sections 14.6–14.8)

14.2 ^{1}H NMR: Number of Signals

How many ^{1}H NMR signals does a compound exhibit? The number of NMR signals *equals* the number of different types of protons in a compound.

14.2A General Principles

- **Protons in different environments give different NMR signals. Equivalent protons give the same NMR signal.**

In many compounds, deciding whether two protons are in identical or different environments is intuitive.

Any CH_3 group is different from any CH_2 group, which is different from any CH group in a molecule. Two CH_3 groups may be identical (as in CH_3OCH_3) or different (as in $CH_3OCH_2CH_3$), depending on what each CH_3 group is bonded to.

tert-Butyl methyl ether [$CH_3OC(CH_3)_3$] (Section 14.1) exhibits two NMR signals because it contains two different kinds of protons: one CH_3 group is bonded to $-OC(CH_3)_3$, whereas the other three CH_3 groups are each bonded to the same group, [$-C(CH_3)_2$]OCH_3.

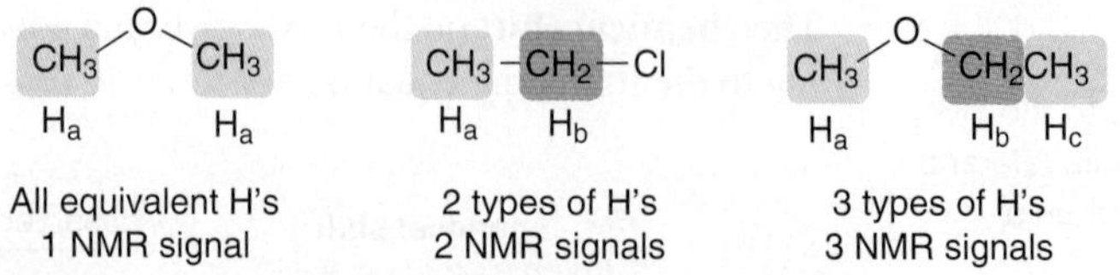
CH3 O CH3
Ha Ha
All equivalent H's
1 NMR signal
CH3–CH2–Cl
Ha Hb
2 types of H's
2 NMR signals
CH3 O CH2CH3
Ha Hb Hc
3 types of H's
3 NMR signals

- **CH_3OCH_3:** Each CH_3 group is bonded to the same group ($-OCH_3$), making both CH_3 groups equivalent.
- **CH_3CH_2Cl:** The protons of the CH_3 group are different from those of the CH_2 group.
- **$CH_3OCH_2CH_3$:** The protons of the CH_2 group are different from those in each CH_3 group. The two CH_3 groups are also different from each other; one CH_3 group is bonded to $-OCH_2CH_3$ and the other is bonded to $-CH_2OCH_3$.

In some cases, it is less obvious by inspection if two protons are equivalent or different. To rigorously determine whether two protons are in identical environments (and therefore give rise to one NMR signal), replace each H atom in question by another atom Z (for example, Z = Cl). **If substitution by Z yields the same compound or enantiomers, the two protons are equivalent,** as shown in Sample Problem 14.2.

Sample Problem 14.2 How many different kinds of H atoms does $CH_3CH_2CH_2CH_2CH_3$ contain?

Solution

In comparing two H atoms, replace each H by Z (for example, Z = Cl), and examine the substitution products that result. The two CH_3 groups are identical because substitution of one H by Cl on each carbon gives the same product, 1-chloropentane.

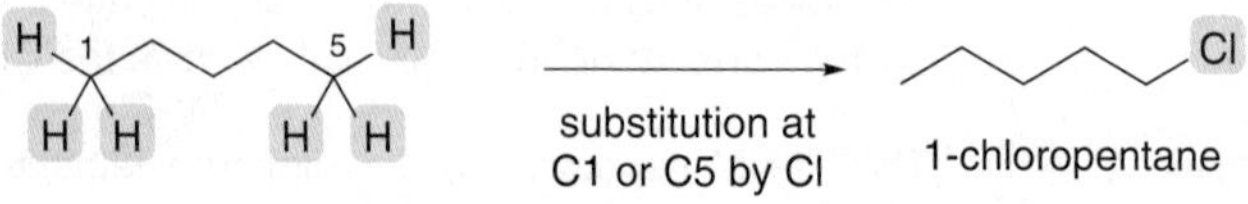
H 1
5 H
H H
H H
substitution at
C1 or C5 by Cl
Cl
1-chloropentane

There are two different types of CH_2 groups. Substitution of Cl for H on C2 or C4 gives the same product, 2-chloropentane, so these H's are identical. Substitution of Cl for H on C3 gives a different product, 3-chloropentane, so this CH_2 group is different from the other two CH_2 groups.

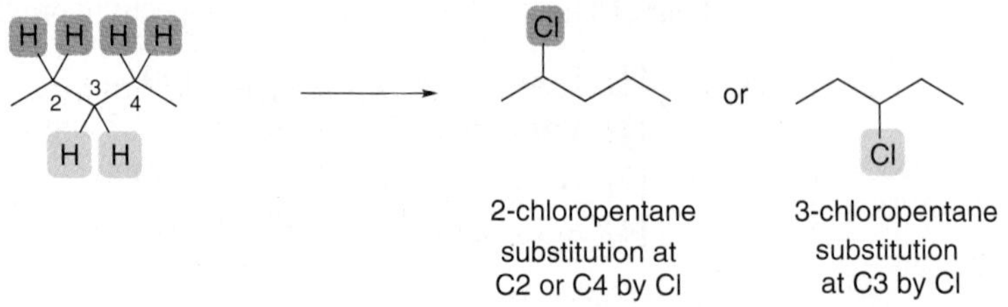
H H H H
2 3 4
H H
Cl
or
Cl
2-chloropentane
substitution at
C2 or C4 by Cl
3-chloropentane
substitution
at C3 by Cl

Thus, $CH_3CH_2CH_2CH_2CH_3$ has three different types of protons and gives three different NMR signals.

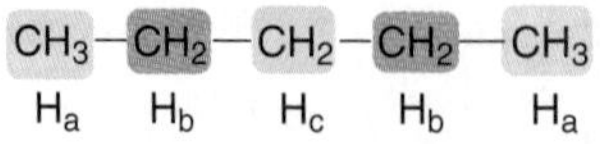
CH3–CH2–CH2–CH2–CH3
Ha Hb Hc Hb Ha

Figure 14.2 gives the number of NMR signals exhibited by four additional molecules. All protons—not just protons bonded to carbon atoms—give rise to NMR signals. Ethanol (CH_3CH_2OH), for example, gives three NMR signals, one of which is due to its OH proton.

Figure 14.2 The number of ^1H NMR signals of some representative organic compounds

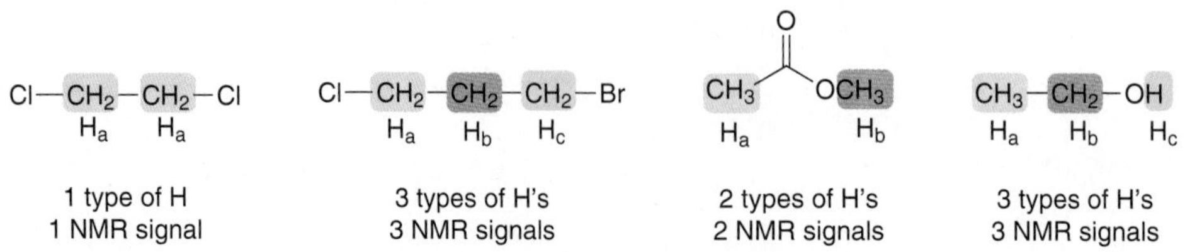
Cl–CH2–CH2–Cl
Ha Ha
1 type of H
1 NMR signal
Cl–CH2–CH2–CH2–Br
Ha Hb Hc
3 types of H's
3 NMR signals
O
CH3 OCH3
Ha Hb
2 types of H's
2 NMR signals
CH3–CH2–OH
Ha Hb Hc
3 types of H's
3 NMR signals

Problem 14.3 How many ^{1}H NMR signals does each compound show?

a. 
c. 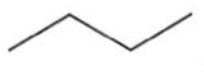
e. 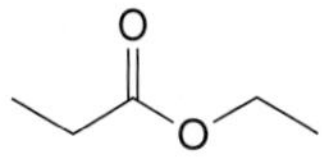
O
O
g. 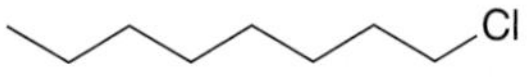
Cl

b. 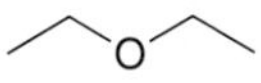
O
d. 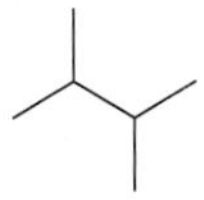
f. 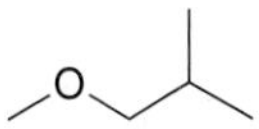
O
h. 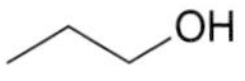
OH

14.2B Determining Equivalent Protons in Alkenes and Cycloalkanes

To determine equivalent protons in cycloalkanes and alkenes that have restricted bond rotation, always draw in all bonds to hydrogen.

Draw 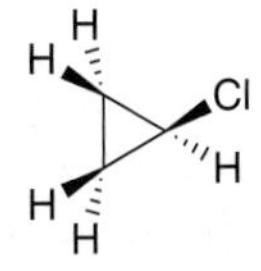
H
H
Cl
H
H
H
NOT 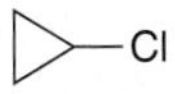
Cl
Draw 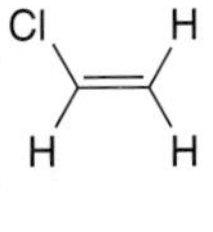
Cl
H
H
H
NOT ClCH=CH$_2$

Then, in comparing two H atoms on a ring or double bond, **two protons are equivalent only if they are cis (or trans) to the same groups,** as illustrated with 1,1-dichloroethylene, 1-bromo-1-chloroethylene, and chloroethylene.

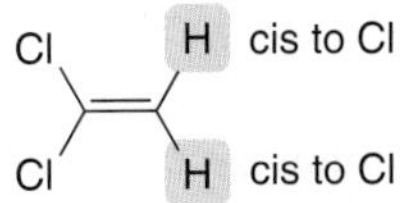
Cl
H cis to Cl
Cl
H cis to Cl
1,1-dichloroethylene

1 type of H
1 NMR signal

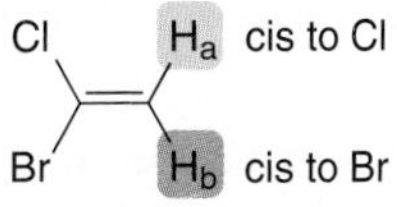
Cl
Ha cis to Cl
Br
Hb cis to Br
1-bromo-1-chloroethylene

2 types of H's
2 NMR signals

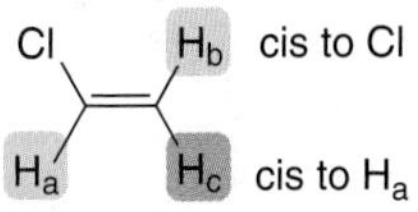
Cl
Hb cis to Cl
Ha
Hc cis to Ha
chloroethylene

3 types of H's
3 NMR signals

- **1,1-Dichloroethylene:** The two H atoms on the C=C are both cis to a Cl atom. Thus, both H atoms are equivalent.
- **1-Bromo-1-chloroethylene:** H_a is cis to a Cl atom and H_b is cis to a Br atom. Thus, H_a and H_b are different, giving rise to two NMR signals.
- **Chloroethylene:** H_a is bonded to the carbon with the Cl atom, making it different from H_b and H_c. Of the remaining two H atoms, H_b is cis to a Cl atom and H_c is cis to a H atom, making them different. All three H atoms in this compound are different.

Proton equivalency in cycloalkanes can be determined similarly.

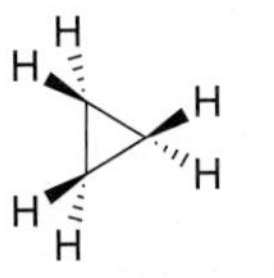
H
H
H
H
H
H
cyclopropane

All H's are equivalent.
1 NMR signal

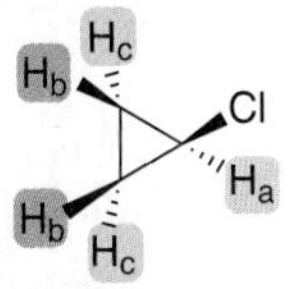
Hc
Hb
Cl
Ha
Hb
Hc
chlorocyclopropane

3 types of H's
3 NMR signals

- **Cyclopropane:** All H atoms are equivalent, so there is only one NMR signal.
- **Chlorocyclopropane:** There are now three kinds of H atoms: H_a is bonded to a carbon bonded to a Cl; both H_b protons are cis to the Cl whereas both H_c protons are cis to another H.

Problem 14.4 How many ^{1}H NMR signals does each dimethylcyclopropane show?

a. 
b. 
c. 

Problem 14.5 How many ^{1}H NMR signals does each compound give?

a. 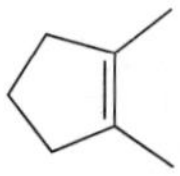

b. 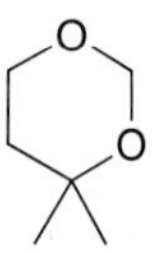

c. 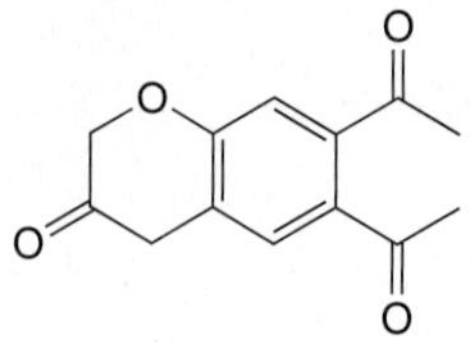

d. 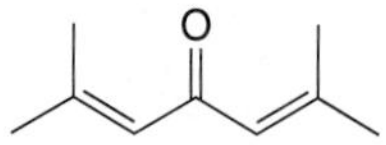

14.2C Homotopic, Enantiotopic, and Diastereotopic Protons

Let's look more closely at the protons of a single sp^3 hybridized CH_2 group to determine whether these two protons are always equivalent to *each other.* Three examples illustrate different outcomes.

$CH_3CH_2CH_3$ has two different types of protons—those of the CH_3 groups and those of the CH_2 group—meaning that the two H atoms of the CH_2 group are *equivalent to each other*. Replacement of each H by Z forms the same product, so they give one NMR signal.

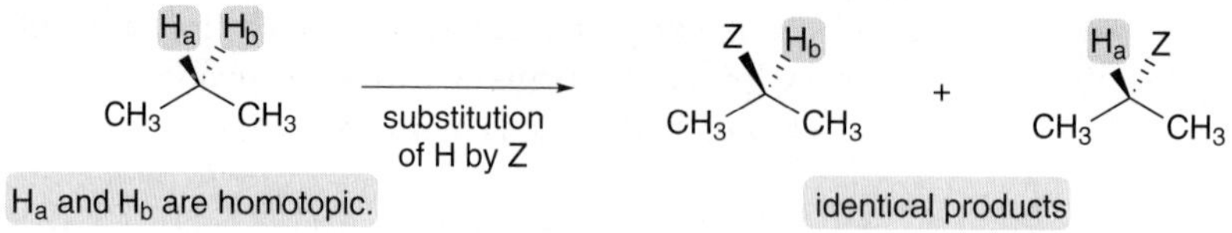

- **When substitution of two H atoms by Z forms the *same* product, these equivalent hydrogens are called *homotopic* protons.**

CH_3CH_2Br has two different types of protons—those of the CH_3 group and those of the CH_2 group—meaning that the two H atoms of the CH_2 group are *equivalent to each other.* Replacement of each H of the CH_2 group by an atom Z creates a new stereogenic center, forming two products that are **enantiomers.**

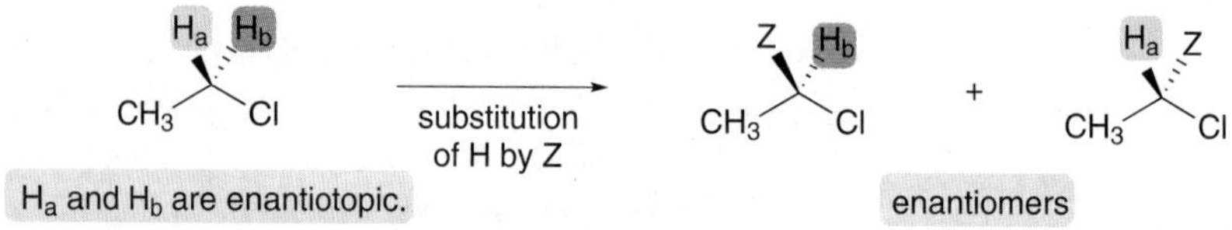

- **When substitution of two H atoms by Z forms *enantiomers*, the two H atoms are equivalent and give a single NMR signal. These two H atoms are called *enantiotopic* protons.**

In contrast, the two H atoms of the CH_2 group in (*R*)-2-chlorobutane, which contains one stereogenic center, are *not* equivalent to each other. Substitution of each H by Z forms two **diastereomers,** and thus, these two H atoms give *different* NMR signals.

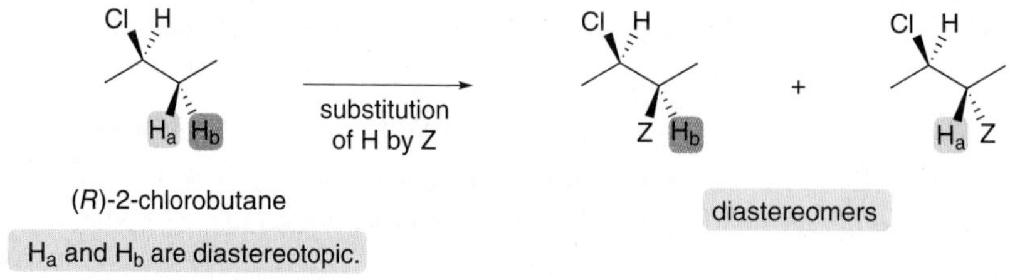

- **When substitution of two H atoms by Z forms *diastereomers*, the two H atoms are *not* equivalent, and give two NMR signals. These two H atoms are called *diastereotopic* protons.**

Sample Problem 14.3 Classify the protons in each labeled CH_2 group as homotopic, enantiotopic, or diastereotopic.

a. 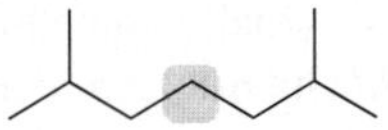 b. 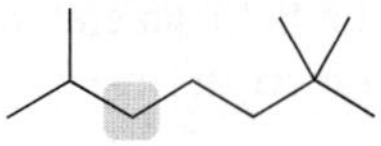 c. 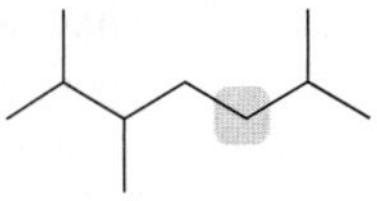

Solution

To determine equivalency in these cases, look for whether the compound has a stereogenic center to begin with and whether a new stereogenic center is formed when H is replaced by Z.

a. The compound is achiral and has no stereogenic center. Replacement of each H on the labeled CH_2 group by Z forms the same product, making them **homotopic.** The H's within the CH_2 group are *equivalent* to each other and give *one* NMR signal.

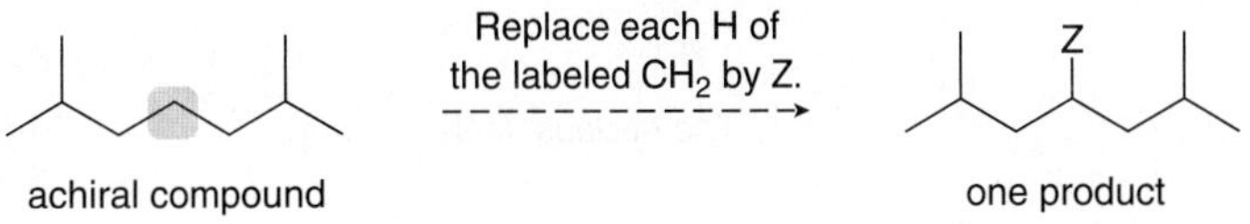

b. The compound is achiral and has no stereogenic center. Because a new stereogenic center is formed on substitution of H by Z, the protons of the CH_2 group are **enantiotopic.** These H's are *equivalent* to each other and give *one* NMR signal.

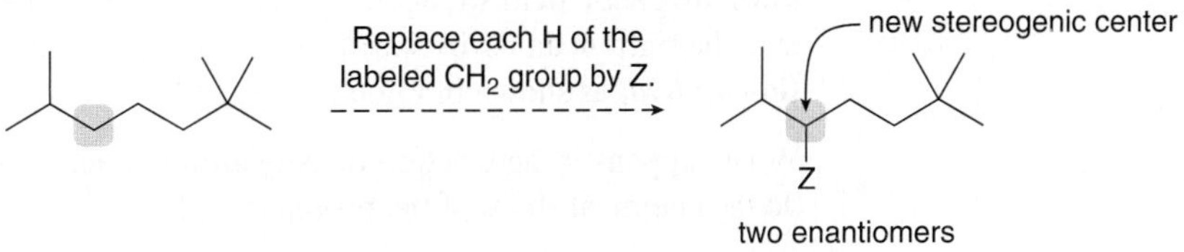

c. The compound has one stereogenic center to begin with. Because a new stereogenic center is formed on substitution of H by Z, the protons are **diastereotopic.** The H's within the CH_2 group are *different* from each other and give *different* NMR signals.

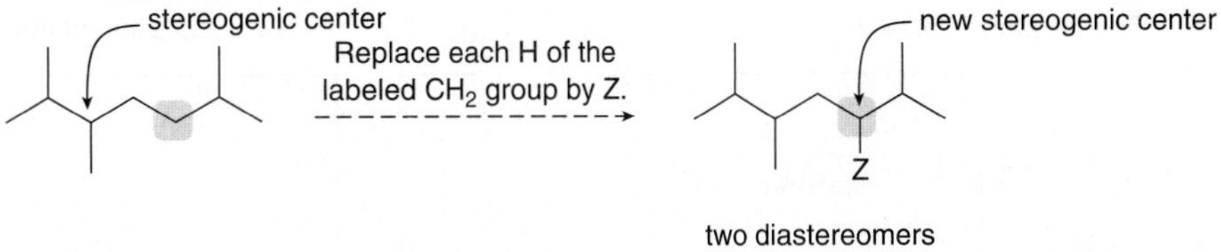

Problem 14.6 Label the protons in each highlighted CH_2 group as enantiotopic, diastereotopic, or homotopic.

a. 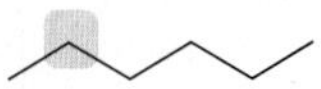 b.  c. 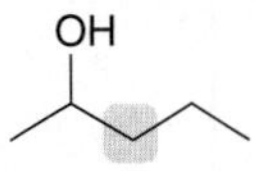

Problem 14.7 How many ¹H NMR signals would you expect for each compound?

a. 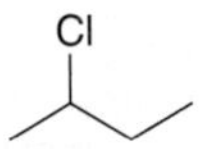

b. 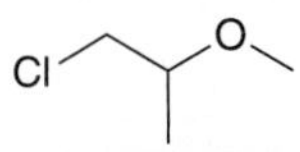

c. 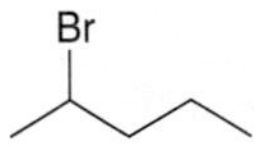

14.3 ¹H NMR: Position of Signals

In the NMR spectrum of *tert*-butyl methyl ether in Section 14.1B, why does the CH_3O- group absorb downfield from the $-C(CH_3)_3$ group?

- **Where a particular proton absorbs depends on its electronic environment.**

14.3A Shielding and Deshielding Effects

To understand how the electronic environment around a nucleus affects its chemical shift, recall that in a magnetic field, an electron creates a small magnetic field that opposes the applied magnetic field, B_0. **Electrons are said to *shield* the nucleus from B_0.**

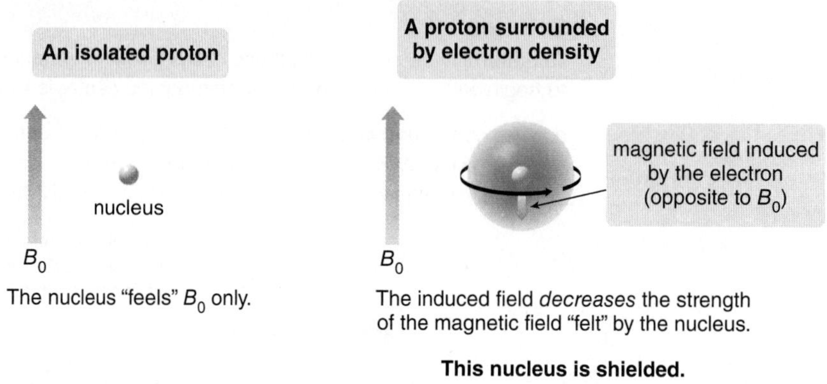

In the vicinity of the nucleus, therefore, the magnetic field generated by the circulating electron *decreases* the external magnetic field that the proton "feels." Because the proton experiences a lower magnetic field strength, it needs a lower frequency to achieve resonance. Lower frequency is to the right in an NMR spectrum, toward lower chemical shift, so **shielding shifts an absorption *upfield,*** as shown in Figure 14.3a.

What happens if the electron density around a nucleus is *decreased,* instead? For example, how do the chemical shifts of the protons in CH_4 and CH_3Cl compare?

Figure 14.3 How chemical shift is affected by electron density around a nucleus

a. **Shielding effects**
- An electron shields the nucleus.
- The absorption shifts *upfield.*

b. **Deshielding effects**
- Decreased electron density deshields a nucleus.
- The absorption shifts *downfield.*

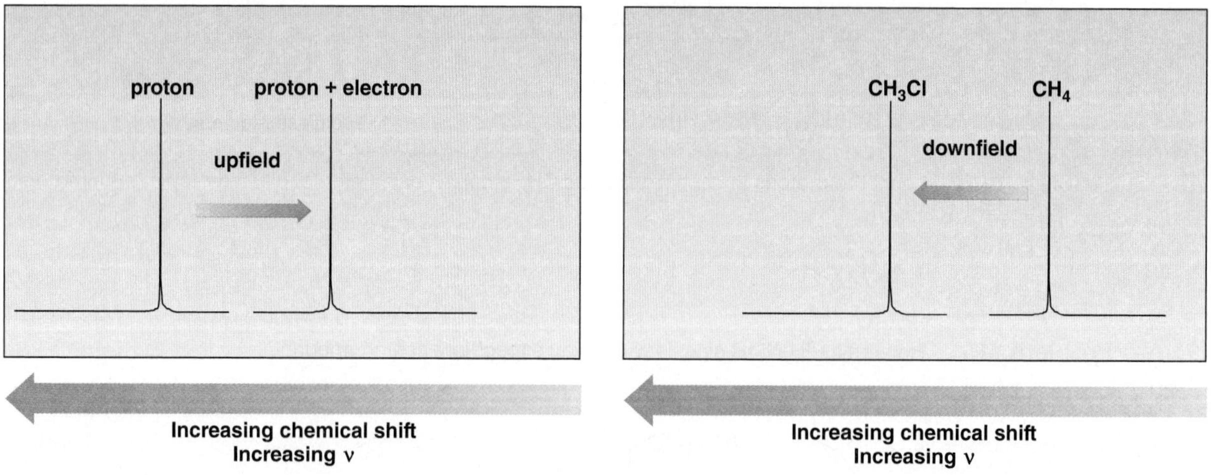

The less shielded the nucleus becomes, the more of the applied magnetic field (B_0) it feels. This *deshielded* nucleus experiences a higher magnetic field strength, so it needs a higher frequency to achieve resonance. Higher frequency is to the *left* in an NMR spectrum, toward higher chemical shift, so **deshielding shifts an absorption downfield,** as shown in Figure 14.3b for CH_3Cl versus CH_4. The electronegative Cl atom withdraws electron density from the carbon and hydrogen atoms in CH_3Cl, thus deshielding them relative to those in CH_4.

Remember the trend:
***D*ecreased electron density *d*eshields a nucleus and an absorption moves *d*ownfield.**

- **Protons near electronegative atoms are deshielded, so they absorb downfield.**

Figure 14.4 summarizes the effects of shielding and deshielding.

These electron density arguments explain the relative position of NMR signals in many compounds.

CH_3-CH_2-Cl (H_a, H_b)

- The H_b protons are **deshielded** because they are closer to the electronegative Cl atom, so they absorb **downfield** from H_a.

$Br-CH_2-CH_2-F$ (H_a, H_b)

- Because F is more electronegative than Br, the H_b protons are more **deshielded** than the H_a protons and absorb farther **downfield.**

$Cl-CH_2-CHCl_2$ (H_a, H_b)

- The larger number of electronegative Cl atoms (two vs. one) **deshields** H_b more than H_a, so it absorbs **downfield** from H_a.

Figure 14.4
Shielding and deshielding effects

a. A shielded nucleus

b. A deshielded nucleus

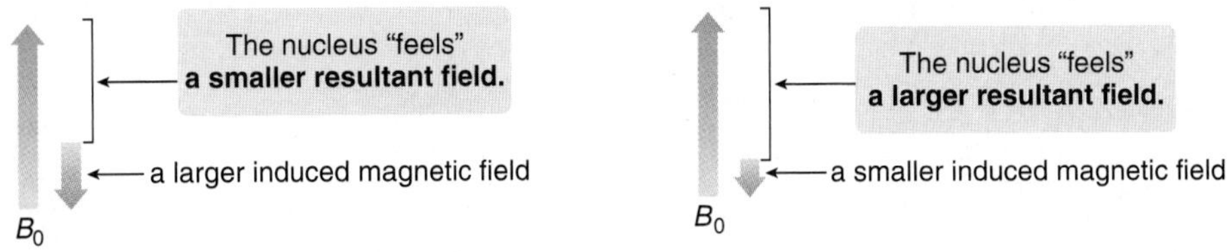

- As the electron density around the nucleus increases, the nucleus feels a smaller resultant magnetic field, so a lower frequency is needed to achieve resonance.
- **The absorption shifts upfield.**

- As the electron density around the nucleus decreases, the nucleus feels a larger resultant magnetic field, so a higher frequency is needed to achieve resonance.
- **The absorption shifts downfield.**

Sample Problem 14.4 Which of the labeled protons in each pair absorbs farther downfield: (a) $CH_3CH_2\mathbf{CH_3}$ or $CH_3O\mathbf{CH_3}$; (b) $CH_3O\mathbf{CH_3}$ or $CH_3S\mathbf{CH_3}$?

Solution

a. The CH_3 group in CH_3OCH_3 is deshielded by the electronegative O atom. **Deshielding shifts the absorption downfield.**
b. Because oxygen is more electronegative than sulfur, the CH_3 group in CH_3OCH_3 is more **deshielded** and absorbs **downfield.**

Problem 14.8 For each compound, which of the protons on the highlighted carbons absorbs farther downfield?

a. 

b. 

c. 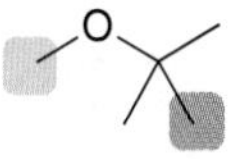

14.3B Chemical Shift Values

Not only is the *relative* position of NMR absorptions predictable, but it is also possible to predict the approximate chemical shift value for a given type of proton.

- **Protons in a given environment absorb in a predictable region in an NMR spectrum.**

A more detailed list of characteristic chemical shift values is found in Appendix F.

Table 14.1 lists the typical chemical shift values for the most common bonds encountered in organic molecules.

Table 14.1 also illustrates that absorptions for a given type of C–H bond occur in a narrow range of chemical shift values, usually 1–2 ppm. For example, all sp^3 hybridized C–H bonds in alkanes and cycloalkanes absorb between 0.9 and 2.0 ppm. By contrast, absorptions due to N–H and O–H protons can occur over a broader range. For example, the OH proton of an alcohol is found anywhere in the 1–5 ppm range. The position of these absorptions is affected by the extent of hydrogen bonding, making it more variable.

Table 14.1 Characteristic Chemical Shifts of Common Types of Protons

Type of proton	Chemical shift (ppm)	Type of proton	Chemical shift (ppm)
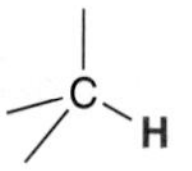	0.9–2	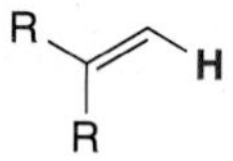	4.5–6
• RCH_3	~0.9	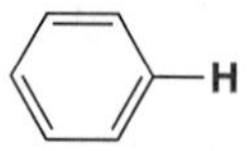	6.5–8
• R_2CH_2	~1.3		
• R_3CH	~1.7		
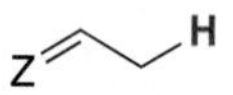 Z = C, O, N	1.5–2.5	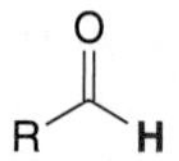	9–10
	~2.5	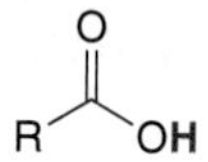	10–12
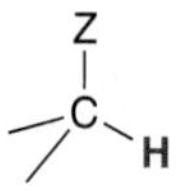 Z = N, O, X	2.5–4	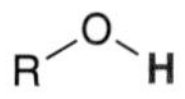 or 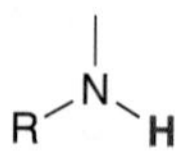	1–5

The chemical shift of a particular type of C–H bond is also affected by the number of R groups bonded to the carbon atom.

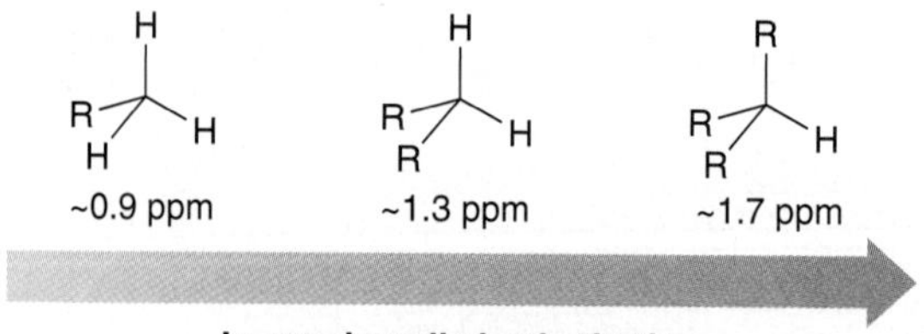

- **The chemical shift of a C–H bond increases with increasing alkyl substitution.**

Problem 14.9 For each compound, first label each different type of proton and then rank the protons in order of increasing chemical shift.

a.  b. 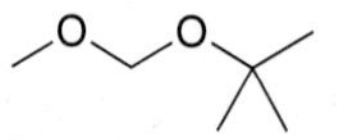 c. 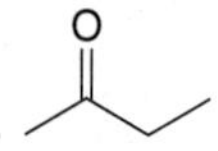

Problem 14.10 Label each statement as True or False.

a. When a nucleus is strongly shielded, the effective field is larger than the applied field and the absorption shifts downfield.
b. When a nucleus is strongly shielded, the effective field is smaller than the applied field and the absorption is shifted upfield.
c. A nucleus that is strongly deshielded requires a lower field strength for resonance.
d. A nucleus that is strongly shielded absorbs at a larger δ value.

14.4 The Chemical Shift of Protons on sp^2 and sp Hybridized Carbons

The chemical shift of protons bonded to benzene rings, C–C double bonds, and C–C triple bonds merits additional comment.

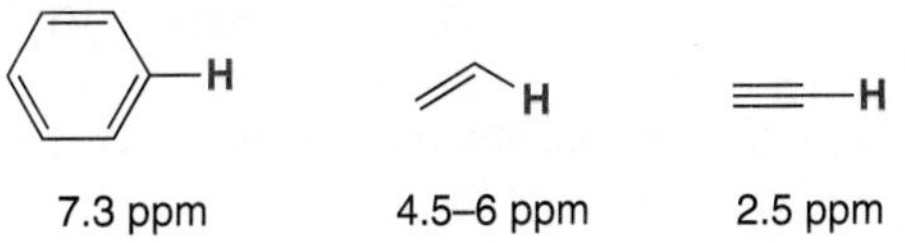

Each of these functional groups contains π bonds with **loosely held π electrons.** When placed in a magnetic field, these π electrons move in a circular path, inducing a new magnetic field. How this induced magnetic field affects the chemical shift of a proton depends on the direction of the induced field *in the vicinity of the absorbing proton.*

Protons on Benzene Rings

In a magnetic field, the six π electrons in **benzene** circulate around the ring, creating a ring current. The magnetic field induced by these moving electrons *reinforces* the applied magnetic field in the vicinity of the protons. The protons thus feel a stronger magnetic field and a higher frequency is needed for resonance, so the **protons are deshielded and the absorption is *downfield.***

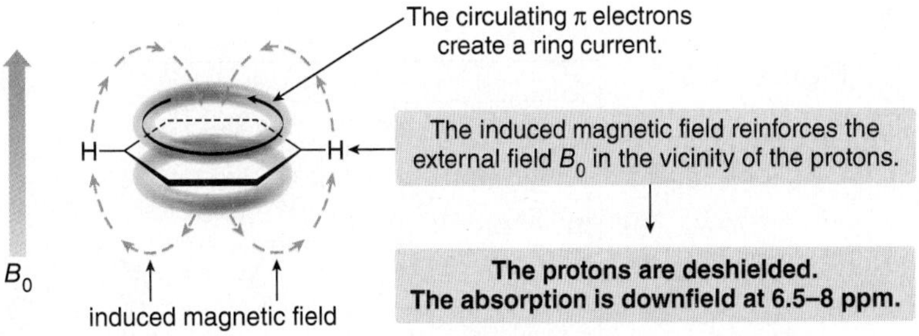

Protons on Carbon–Carbon Double Bonds

A similar phenomenon occurs with protons on carbon–carbon double bonds. In a magnetic field, the loosely held π electrons create a magnetic field that *reinforces* the applied field in the vicinity of the protons. Because the protons now feel a stronger magnetic field, they require a higher frequency for resonance. **The protons are deshielded and the absorption is *downfield.***

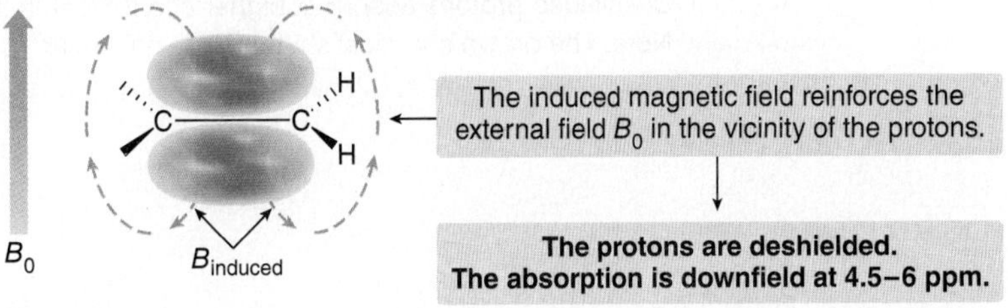

Protons on Carbon–Carbon Triple Bonds

In a magnetic field, the π electrons of a carbon–carbon triple bond are induced to circulate, but in this case the induced magnetic field *opposes* the applied magnetic field (B_0). The proton thus feels a weaker magnetic field, so a lower frequency is needed for resonance. **The nucleus is shielded and the absorption is *upfield*.**

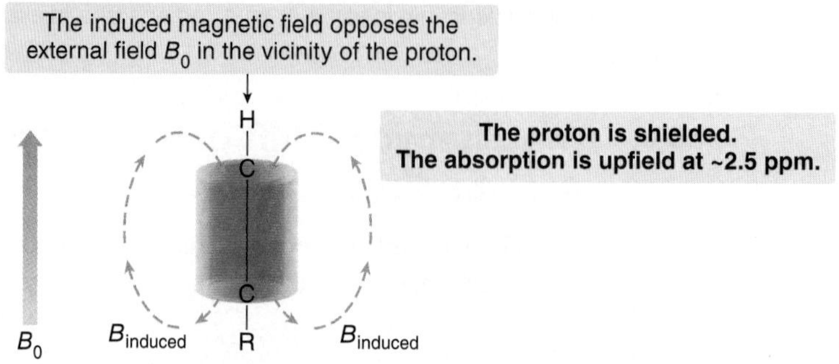

Table 14.2 summarizes the shielding and deshielding effects due to circulating π electrons.

Table 14.2 Effect of π Electrons on Chemical Shift Values

Proton type	Effect	Chemical shift (ppm)
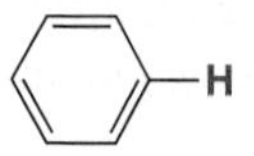	highly deshielded	6.5–8
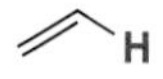	deshielded	4.5–6
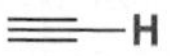	shielded	~2.5

To remember the chemical shifts of some common bond types, it is helpful to think of a ^{1}H NMR spectrum as being divided into six different regions (Figure 14.5).

Figure 14.5
Regions in the ^{1}H NMR spectrum

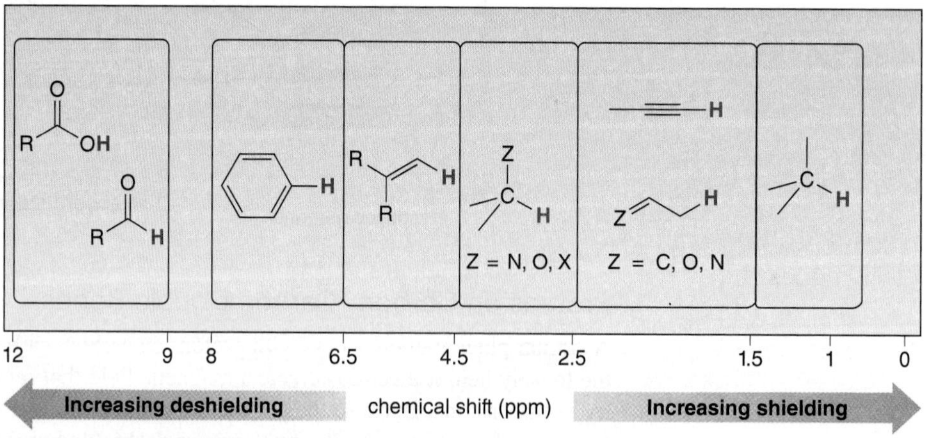

- Shielded protons absorb at lower chemical shift (to the right).
- Deshielded protons absorb at higher chemical shift (to the left).
- Note: The drawn chemical shift scale is not linear.

Sample Problem 14.5 Rank H_a, H_b, and H_c in order of increasing chemical shift.

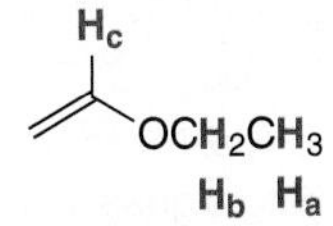

Solution

The H_a protons are bonded to an *sp*^{3} hybridized carbon, so they are shielded and absorb upfield compared to H_b and H_c. Because the H_b protons are deshielded by the electronegative oxygen atom on the C to which they are bonded, they absorb downfield from H_a. The H_c proton is deshielded by two factors. The electronegative O atom withdraws electron density from H_c. Moreover, because H_c is bonded directly to a C=C, the magnetic field induced by the π electrons causes further deshielding. Thus, in order of increasing chemical shift, $H_a < H_b < H_c$.

Problem 14.11 Rank each group of protons in order of increasing chemical shift.

a. 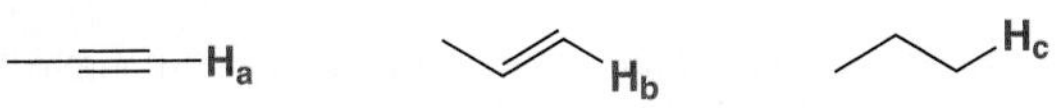

b. 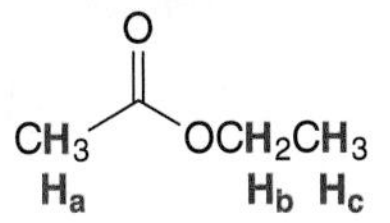

14.5 ^{1}H NMR: Intensity of Signals

The relative intensity of ^{1}H NMR signals also provides information about a compound's structure.

- **The area under an NMR signal is proportional to the number of absorbing protons.**

For example, in the ^{1}H NMR spectrum of $CH_3OC(CH_3)_3$, the ratio of the area under the downfield peak (due to the CH_3O- group) to the upfield peak [due to the $-C(CH_3)_3$ group] is 1:3. An NMR spectrometer automatically integrates the area under the peaks, and prints out a digital display of the *relative* areas of the NMR signals. Older NMR spectrometers print out a stepped curve (an **integral**) on the spectrum. The height of each step is proportional to the area under the peak, which is in turn proportional to the number of absorbing protons.

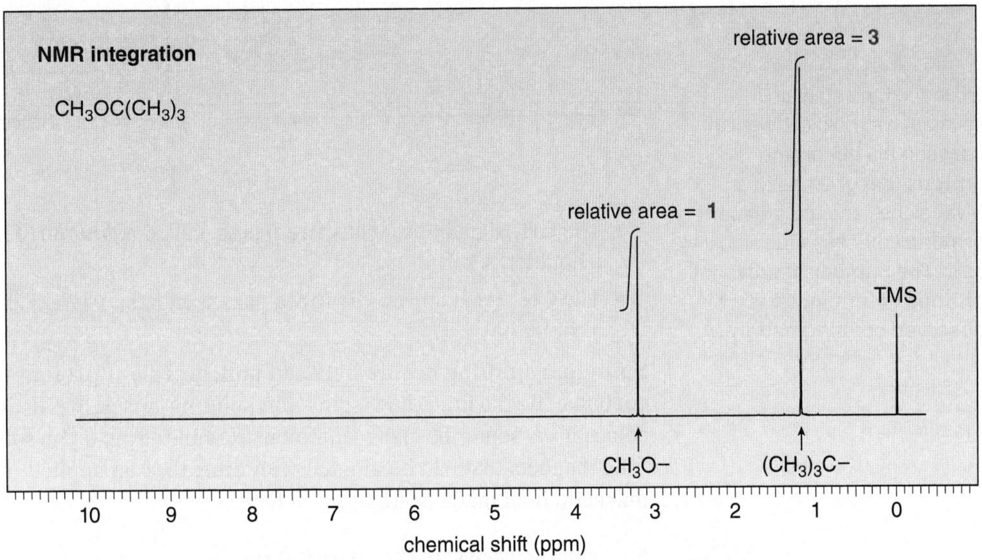

Integrals can be manually measured, but modern NMR spectrometers automatically calculate and plot the value of each integral in arbitrary units. If the heights of two integrals are in a 1:3 ratio, then the ratio of absorbing protons is 1:3, or 2:6, or 3:9, and so forth. This tells the *ratio,* not the absolute number of protons.

Problem 14.12 Which compounds give a ^1H NMR spectrum with two signals in a ratio of 2:3?

a. 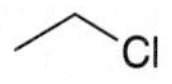 b.  c. 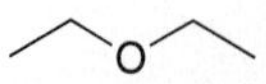 d. 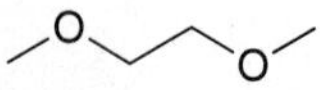

Problem 14.13 Compound **A** exhibits two signals in its ^1H NMR spectrum at 2.64 and 3.69 ppm, and the ratio of the absorbing signals is 2:3. Compound **B** exhibits two signals in its ^1H NMR spectrum at 2.09 and 4.27 ppm, and the ratio of the absorbing signals is 3:2. Which compound corresponds to dimethyl succinate and which compound corresponds to ethylene diacetate?

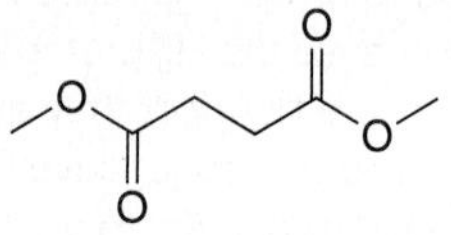

dimethyl succinate

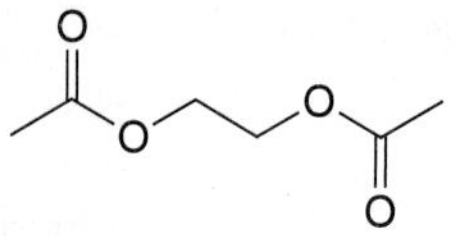

ethylene diacetate

14.6 ^{1}H NMR: Spin–Spin Splitting

The ^1H NMR spectra you have seen up to this point have been limited to one or more single absorptions called **singlets.** In the ^1H NMR spectrum of $BrCH_2CHBr_2$, however, the two signals for the two different kinds of protons are each split into more than one peak. The splitting patterns, the result of **spin–spin splitting,** can be used to determine how many protons reside on the carbon atoms near the absorbing proton.

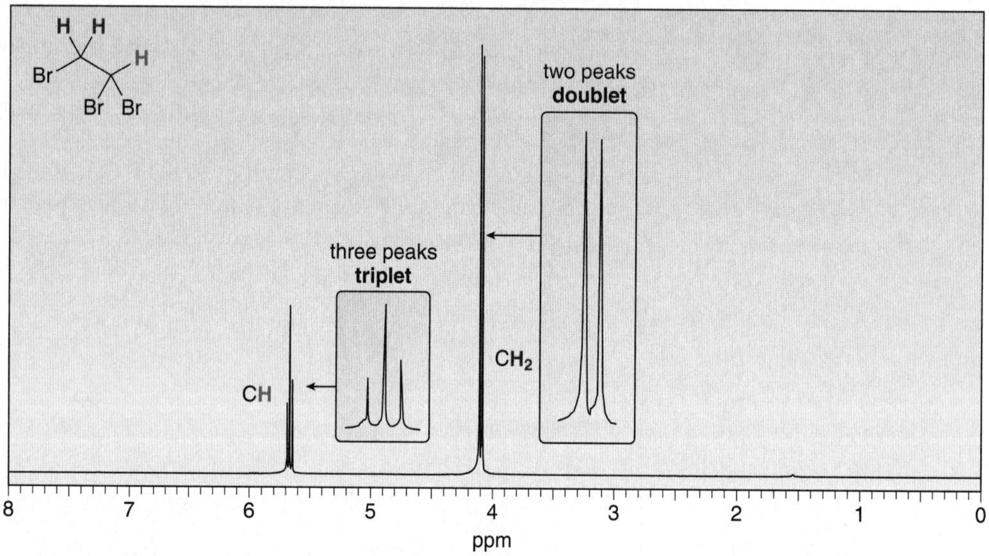

To understand spin–spin splitting, we must distinguish between the **absorbing protons** that give rise to an NMR signal, and the **adjacent protons** that cause the signal to split. **The number of adjacent protons determines the observed splitting pattern.**

- The CH_2 signal appears as **two peaks,** called a ***doublet.*** The relative area under the peaks of a doublet is 1:1.
- The CH signal appears as **three peaks,** called a ***triplet.*** The relative area under the peaks of a triplet is 1:2:1.

Spin–spin splitting occurs between nonequivalent protons on the same carbon or adjacent carbons. To illustrate how spin–spin splitting arises, we'll examine nonequivalent protons on adjacent carbons, the more common example. Spin–spin splitting arises because protons are little magnets that can be aligned with or against an applied magnetic field, and this affects the magnetic field that a nearby proton feels.

14.6A Splitting: How a Doublet Arises

First, let's examine how the doublet due to the CH_2 group in $BrCH_2CHBr_2$ arises. The CH_2 group contains the absorbing protons and the CH group contains the adjacent proton that causes the splitting.

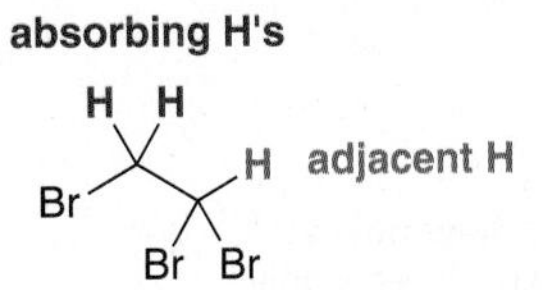
absorbing H's
H H
H adjacent H
Br
Br Br

The adjacent H can be aligned with (↑) or against (↓) B_0.

When placed in an applied magnetic field (B_0), the adjacent proton ($CHBr_2$) can be aligned with (↑) or against (↓) B_0. As a result, the absorbing protons (CH_2Br) feel two slightly different magnetic fields—one slightly larger than B_0 and one slightly smaller than B_0. Because the absorbing protons feel two different magnetic fields, they absorb at two different frequencies in the NMR spectrum, thus splitting a single absorption into a doublet.

How a doublet arises

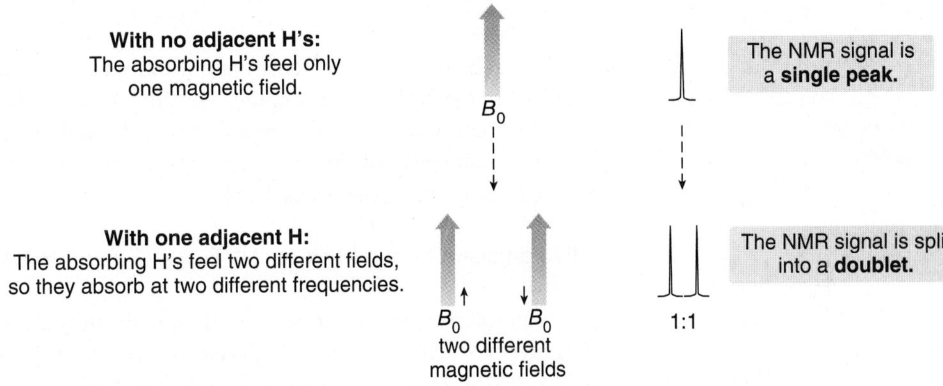
With no adjacent H's:
The absorbing H's feel only one magnetic field.
B_0
The NMR signal is a **single peak.**
With one adjacent H:
The absorbing H's feel two different fields, so they absorb at two different frequencies.
B_0 B_0
two different magnetic fields
1:1
The NMR signal is split into a **doublet.**

Keep in mind the difference between an **NMR signal** and an **NMR peak.** An NMR signal is the entire absorption due to a particular kind of proton. NMR peaks are contained within a signal. A doublet constitutes one signal that is split into two peaks.

- **One adjacent proton splits an NMR signal into a doublet.**

coupling constant, ***J***, in Hz

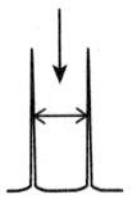

The two peaks of a doublet are approximately equal in area. The area under both peaks—the entire NMR signal—is due to both protons of the CH_2 group of $BrCH_2CHBr_2$.

The frequency difference (measured in Hz) between the two peaks of the doublet is called the **coupling constant,** denoted by ***J*.** Coupling constants are usually in the range of 0–18 Hz, and are **independent of the strength of the applied magnetic field, B_0.**

14.6B Splitting: How a Triplet Arises

Now let's examine how the triplet due to the CH group in $BrCH_2CHBr_2$ arises. The CH group contains the absorbing proton and the CH_2 group contains the adjacent protons (H_a and H_b) that cause the splitting.

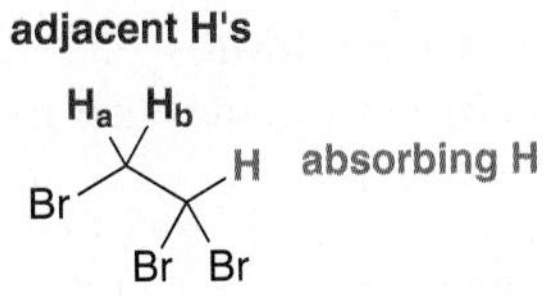
adjacent H's
H_a H_b
H absorbing H
Br
Br Br

H_a and H_b can each be aligned with (↑) or against (↓) B_0.

When placed in an applied magnetic field (B_0), the adjacent protons H_a and H_b can each be aligned with (↑) or against (↓) B_0. As a result, the absorbing proton feels three slightly different magnetic fields—one slightly larger than B_0, one slightly smaller than B_0, and one the same strength as B_0.

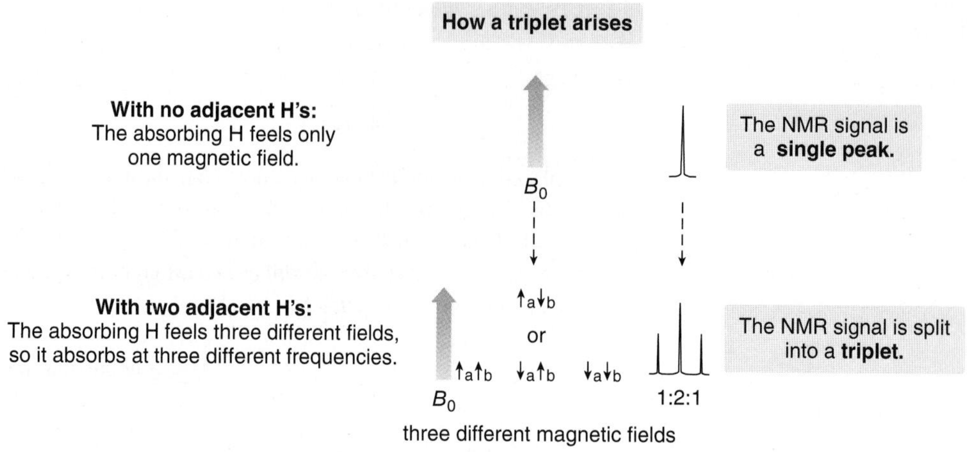

Table 14.3 Names for a Given Number of Peaks in an NMR Signal

Number of peaks	Name
1	singlet
2	doublet
3	triplet
4	quartet
5	quintet
6	sextet
7	septet
> 7	multiplet

Because the absorbing proton feels three different magnetic fields, it absorbs at three different frequencies in the NMR spectrum, thus splitting a single absorption into a triplet. Because there are two different ways to align one proton with B_0 and one proton against B_0—that is, $\uparrow_a\downarrow_b$ and $\downarrow_a\uparrow_b$—the middle peak of the triplet is twice as intense as the two outer peaks, making the ratio of the areas under the three peaks 1:2:1.

- **Two adjacent protons split an NMR signal into a triplet.**

When two protons split each other's NMR signals, they are said to be ***coupled.*** In $BrCH_2CHBr_2$, the CH proton is coupled to the CH_2 protons. The spacing between peaks in a split NMR signal, measured by the *J* value, is *equal* for coupled protons.

14.6C Splitting: The Rules and Examples

Three general rules describe the splitting patterns commonly seen in the ^{1}H NMR spectra of organic compounds.

Rule [1] Equivalent protons don't split each other's signals.

Rule [2] A set of *n* nonequivalent protons splits the signal of a nearby proton into *n* + 1 peaks.

- In $BrCH_2CHBr_2$, for example, *one* adjacent CH proton splits an NMR signal into *two* peaks (a doublet), and *two* adjacent CH_2 protons split an NMR signal into *three* peaks (a triplet). Names for split NMR signals containing two to seven peaks are given in Table 14.3. An NMR signal having more than seven peaks is called a **multiplet.**
- The inside peaks of a split NMR signal are always most intense, with the area under the peaks decreasing from the inner to the outer peaks in a given splitting pattern.

Rule [3] Splitting is observed for nonequivalent protons on the same carbon or adjacent carbons.

If H_a and H_b are not equivalent, splitting is observed in each of the following cases.

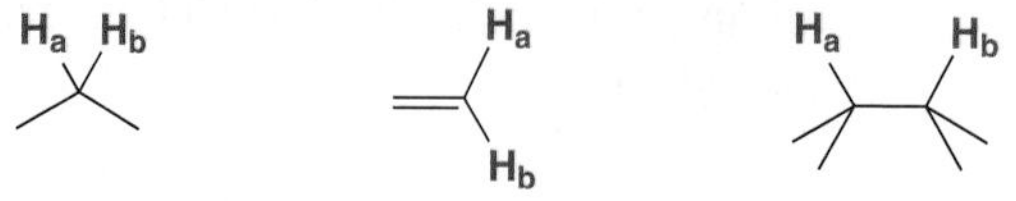

The splitting of an NMR signal reveals the number of nearby nonequivalent protons. It tells nothing about the absorbing proton itself.

Splitting is not generally observed between protons separated by more than three σ bonds. Although H_a and H_b are not equivalent to each other in butan-2-one and ethyl methyl ether, H_a and H_b are separated by four σ bonds, so they are too far away to split each other's NMR signals.

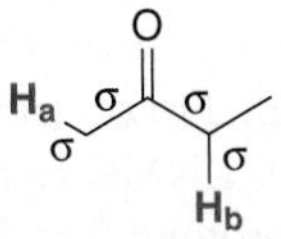

butan-2-one
H_a and H_b are separated by four σ bonds.

no splitting between H_a and H_b

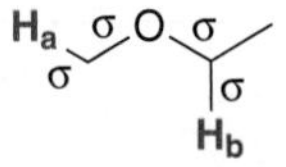

ethyl methyl ether
H_a and H_b are separated by four σ bonds.

no splitting between H_a and H_b

Table 14.4 illustrates common splitting patterns observed for adjacent nonequivalent protons.

Table 14.4 Common Splitting Patterns Observed in ¹H NMR

Example		Pattern	Analysis				
[1]	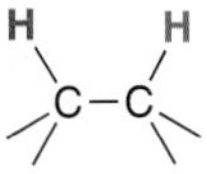	 H H	• H: one adjacent H proton • H: one adjacent H proton	- - -> - - ->	two peaks two peaks	- - -> - - ->	a **doublet** a **doublet**
[2]	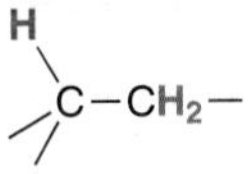	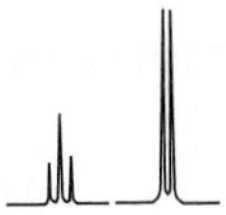 H H	• H: two adjacent H protons • H: one adjacent H proton	- - -> - - ->	three peaks two peaks	- - -> - - ->	a **triplet** a **doublet**
[3]	$—CH_2CH_2—$	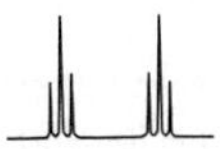 H H	• H: two adjacent H protons • H: two adjacent H protons	- - -> - - ->	three peaks three peaks	- - -> - - ->	a **triplet** a **triplet**
[4]	$—CH_2CH_3$	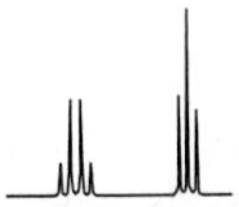 H H	• H: three adjacent H protons • H: two adjacent H protons	- - -> - - ->	four peaks three peaks	- - -> - - ->	a **quartet*** a **triplet**
[5]	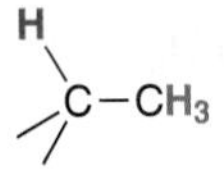	 H H	• H: three adjacent H protons • H: one adjacent H proton	- - -> - - ->	four peaks two peaks	- - -> - - ->	a **quartet*** a **doublet**

*The relative area under the peaks of a quartet is 1:3:3:1.

Predicting splitting is always a two-step process:

- **Determine if two protons are equivalent or different.** Only nonequivalent protons split each other.
- **Determine if two nonequivalent protons are close enough to split each other's signals.** Splitting is observed only for nonequivalent protons on the *same* carbon or *adjacent* carbons.

Several examples of spin–spin splitting in specific compounds illustrate the result of this two-step strategy.

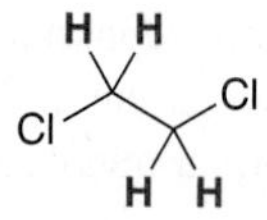

- All protons are equivalent, so there is no splitting and the NMR signal is one singlet.

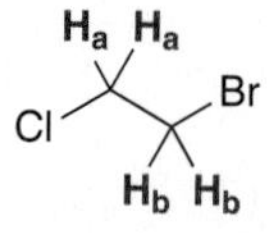

- There are two NMR signals. H_a and H_b are nonequivalent protons bonded to adjacent C atoms, so they are close enough to split each other's NMR signals. The H_a signal is split into a triplet by the two H_b protons. The H_b signal is split into a triplet by the two H_a protons.

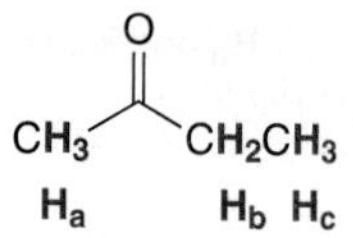

- There are three NMR signals. H_a has no adjacent nonequivalent protons, so its signal is a singlet. The H_b signal is split into a quartet by the three H_c protons. The H_c signal is split into a triplet by the two H_b protons.

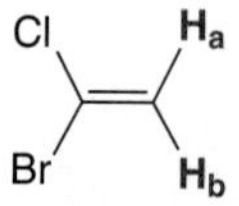

- There are two NMR signals. H_a and H_b are nonequivalent protons on the same carbon, so they are close enough to split each other's NMR signals. The H_a signal is split into a doublet by H_b. The H_b signal is split into a doublet by H_a.

Problem 14.14 Into how many peaks will each proton shown in red be split?

a. 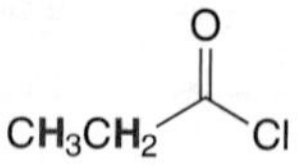 b. 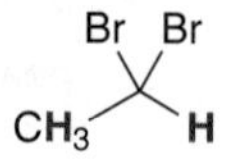 c. 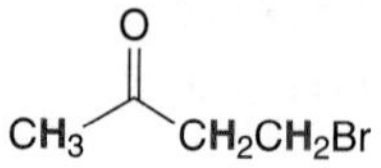 d. 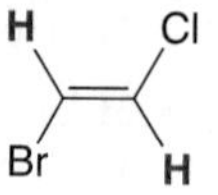 e. 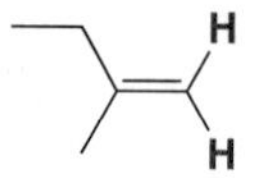 f. 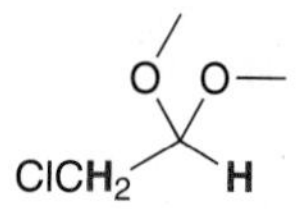

Problem 14.15 For each compound give the number of ^1H NMR signals, and then determine how many peaks are present for each NMR signal.

a. 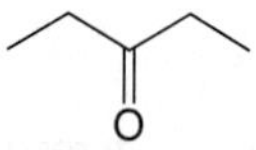 b. 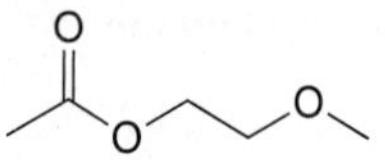 c. 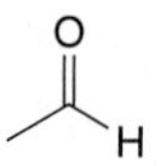 d. 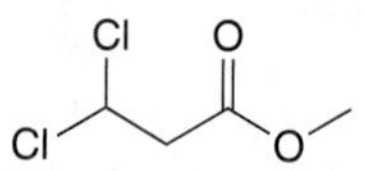

Problem 14.16 Sketch the NMR spectrum of CH_3CH_2Cl, giving the approximate location of each NMR signal.

14.7 More Complex Examples of Splitting

Up to now you have studied examples of spin–spin splitting where the absorbing proton has nearby protons on *one* adjacent carbon only. What happens when the absorbing proton has nonequivalent protons on *two* adjacent carbons? Different outcomes are possible, depending on whether the adjacent nonequivalent protons are *equivalent to* or *different from* each other.

For example, 2-bromopropane [$(CH_3)_2CHBr$] has two types of protons—H_a and H_b—so it exhibits two NMR signals, as shown in Figure 14.6.

Figure 14.6
The ^1H NMR spectrum of 2-bromopropane, $(CH_3)_2CHBr$

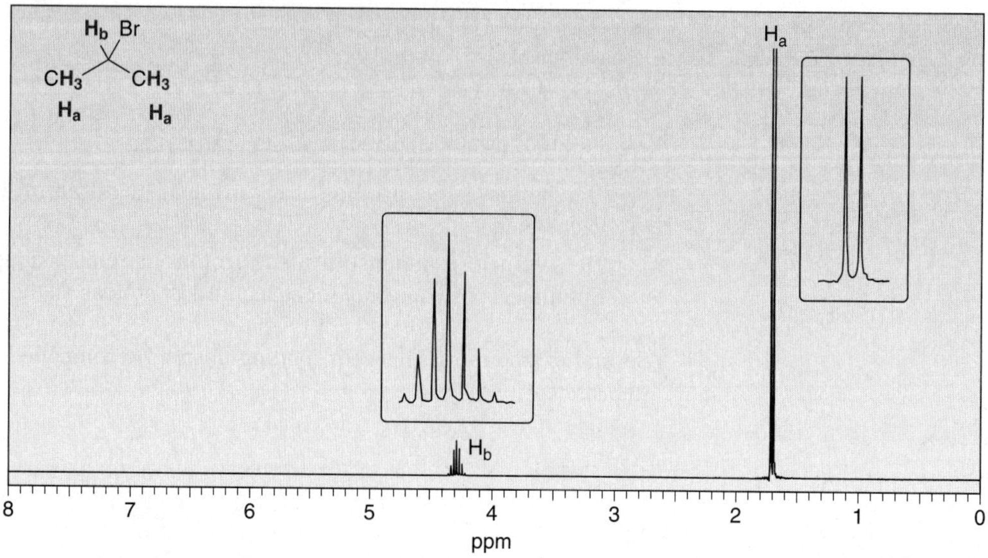

- The H_a protons have only one adjacent nonequivalent proton (H_b), so they are split into two peaks, a **doublet.**
- H_b has three H_a protons on each side. Because the six H_a protons are *equivalent to each other,* the ***n* + 1** rule can be used to determine splitting: 6 + 1 = 7 peaks, a **septet.**

This is a specific example of a general rule:

- **Whenever two (or three) sets of adjacent protons are *equivalent to each other*, use the *n* + 1 rule to determine the splitting pattern.**

A different outcome results when an absorbing proton is flanked by adjacent protons that are *not equivalent to each other.* Consider the splitting pattern expected for the H_b protons in the ^1H NMR spectrum of $CH_3CH_2CH_2Z$. H_b has protons on both adjacent carbons, but since H_a and H_c are *not equivalent to each other,* we cannot merely add them together and use the $n + 1$ rule.

$CH_3CH_2CH_2$—Z
H_a H_b H_c

Instead, to determine the splitting of H_b, we must consider the effect of the H_a protons and the H_c protons *separately.* The three H_a protons split the H_b signal into four peaks, and the two H_c protons split each of these four peaks into three peaks—that is, the NMR signal due to H_b consists of 4 × 3 = **12 peaks.** Figure 14.7 shows a splitting diagram illustrating how these 12 peaks arise.

- **When two sets of adjacent protons are *different from each other* (*n* protons on one adjacent carbon and *m* protons on the other), the number of peaks in an NMR signal = (*n* + 1)(*m* + 1).**

It is possible to see 12 peaks in an NMR spectrum only when the coupling constants between each set of nonequivalent protons—that is, J_{ab} and J_{bc} in this example—are different; in other words, $J_{ab} \neq J_{bc}$. Such is the case with the nonequivalent protons on carbon–carbon double bonds, which is discussed in Section 14.8. **In practice, with flexible alkyl chains it is more common for J_{ab} and J_{bc} to be very similar or identical.** In this case, peaks overlap and many fewer than 12 peaks are observed.

Figure 14.7
A splitting diagram for the H_b protons in $CH_3CH_2CH_2Z$

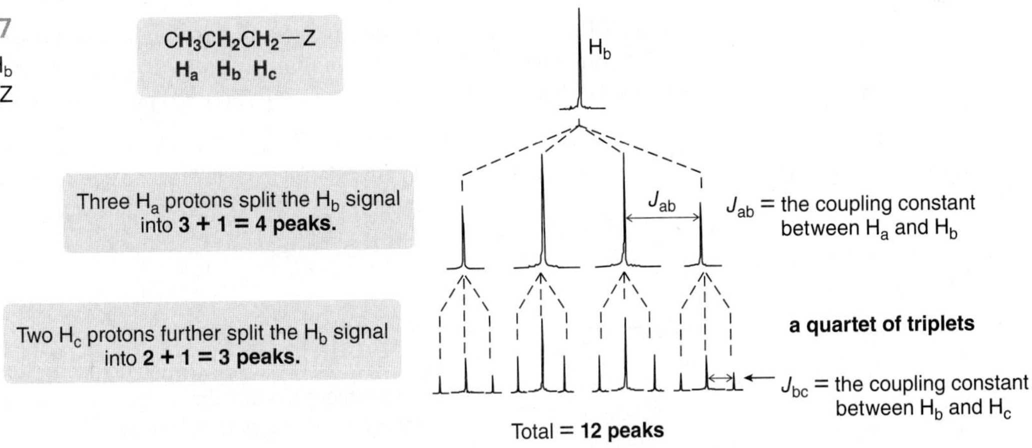

CH3CH2CH2—Z
Ha Hb Hc
Hb
Three Ha protons split the Hb signal into 3 + 1 = 4 peaks.
Jab
Jab = the coupling constant between Ha and Hb
Two Hc protons further split the Hb signal into 2 + 1 = 3 peaks.
a quartet of triplets
Jbc = the coupling constant between Hb and Hc
Total = 12 peaks

- The H_b signal is split into 12 peaks, a quartet of triplets. The number of peaks actually seen for the signal depends on the relative size of the coupling constants, J_{ab} and J_{bc}. When $J_{ab} >> J_{bc}$, as drawn in this diagram, all 12 lines of the pattern are visible. When J_{ab} and J_{bc} are similar in magnitude, peaks overlap and fewer lines are observed.

Figure 14.8
The ^{1}H NMR spectrum of 1-bromopropane, $CH_3CH_2CH_2Br$

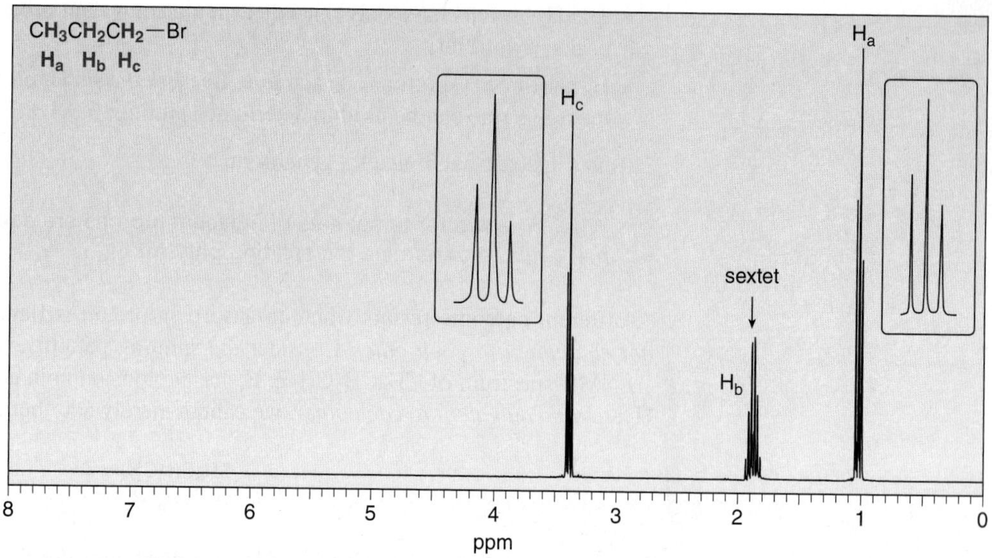

CH3CH2CH2—Br
Ha Hb Hc
Hc
Ha
sextet
Hb
8
7
6
5
4
3
2
1
0
ppm

- H_a and H_c are both triplets.
- The signal for H_b appears as a multiplet of six peaks (a sextet), due to peak overlap; the number of peaks = $n + m + 1 = 3 + 2 + 1 = 6$ peaks.

The ^{1}H NMR spectrum of 1-bromopropane ($CH_3CH_2CH_2Br$) illustrates the result of peak overlap (Figure 14.8).

$$\underset{H_a}{CH_3}\underset{H_b}{CH_2}\underset{H_c}{CH_2}-Br$$

$CH_3CH_2CH_2Br$ has three different types of protons—H_a, H_b, and H_c—so it exhibits three NMR signals. H_a and H_c are each triplets because they are adjacent to two H_b protons. H_b has protons on both adjacent carbons, and H_a and H_c are *not equivalent to each other.* The three H_a protons should split the H_b signal into four peaks, and the two H_c protons should split each of these four peaks into three peaks—that is, the NMR signal due to H_b should once again consist of $4 \times 3 =$ **12 peaks.** However, since $J_{ab} = J_{bc}$ in this case, peak overlap occurs and a multiplet of only six peaks is observed.

In $CH_3CH_2CH_2Br$, **the *n* protons on one adjacent carbon and the *m* protons on the other adjacent carbon split the observed signal into $n + m + 1$ peaks.** In other words, the 3 H_a protons and 2 H_c protons split the NMR signal into $3 + 2 + 1 = 6$ peaks, as shown in the sextet in Figure 14.8.

Sample Problem 14.6 How many peaks are present in the NMR signal of the labeled protons of each compound?

a. 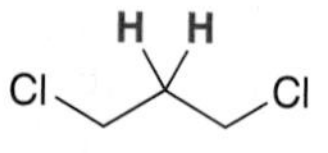

H H
Cl
Cl

b. 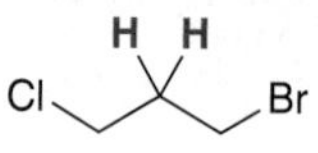

H H
Cl
Br

Solution

a. 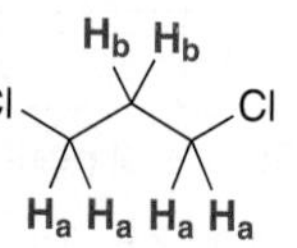

Hb Hb
Cl
Cl
Ha Ha Ha Ha

- H_b has two H_a protons on each adjacent C. Because the four H_a protons are equivalent to each other, the $n + 1$ rule can be used to determine splitting: $4 + 1 =$ **5 peaks,** a quintet.

b. 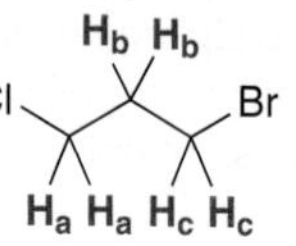

Hb Hb
Cl
Br
Ha Ha Hc Hc

- H_b has two H_a protons on one adjacent C and two H_c protons on the other. Because H_a and H_c are *not equivalent to each other,* the maximum number of peaks for $H_b = (n + 1)(m + 1) = (2 + 1)(2 + 1) =$ **9 peaks.** However, since this molecule has a *flexible alkyl chain,* it is likely that J_{ab} and J_{bc} are very similar, so that peak overlap occurs, and the number of peaks for $H_b = n + m + 1 = 2 + 2 + 1 =$ **5 peaks.**

Problem 14.17 How many peaks are present in the NMR signal of each labeled proton?

a. 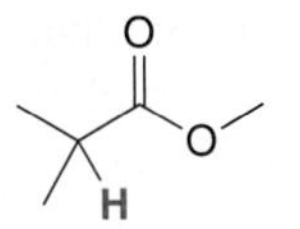 b. 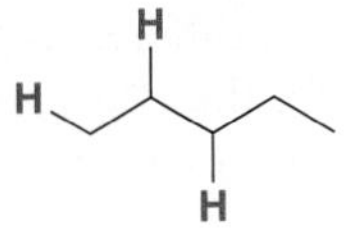 c. 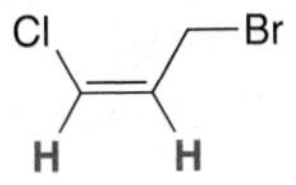 d. 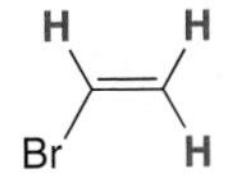

Problem 14.18 Describe the ^1H NMR spectrum of each compound. State how many NMR signals are present, the splitting pattern for each signal, and the approximate chemical shift.

a. 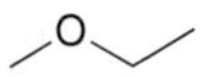 b. 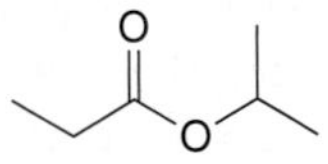 c. 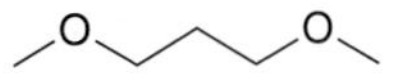 d. 

14.8 Spin–Spin Splitting in Alkenes

Protons on carbon–carbon double bonds often give characteristic splitting patterns. A disubstituted double bond can have two **geminal protons** (on the same carbon atom), two **cis protons,** or two **trans protons.** When these protons are different, each proton splits the NMR signal of the other, so that each proton appears as a doublet. **The magnitude of the coupling constant *J* for these doublets depends on the arrangement of hydrogen atoms.**

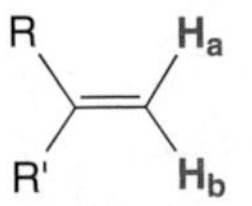		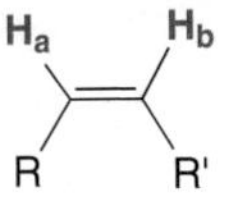		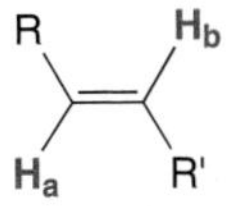
geminal H's		cis H's		trans H's
$J_{geminal}$	<	J_{cis}	<	J_{trans}
0–3 Hz		5–10 Hz		11–18 Hz

Thus, the *E* and *Z* isomers of 3-chloropropenoic acid both exhibit two doublets for the two alkenyl protons, but the coupling constant is larger when the protons are trans compared to when the protons are cis, as shown in Figure 14.9.

Figure 14.9
^1H NMR spectra for the alkenyl protons of (*E*)- and (*Z*)-3-chloropropenoic acid

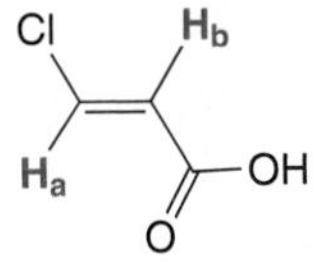

(*E*)-3-chloropropenoic acid

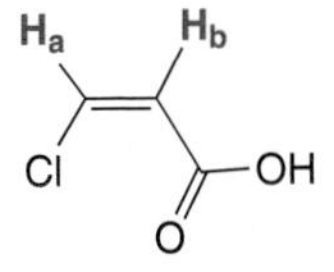

(*Z*)-3-chloropropenoic acid

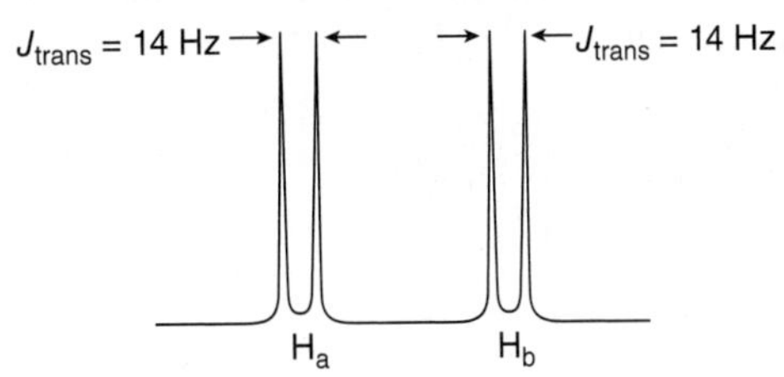

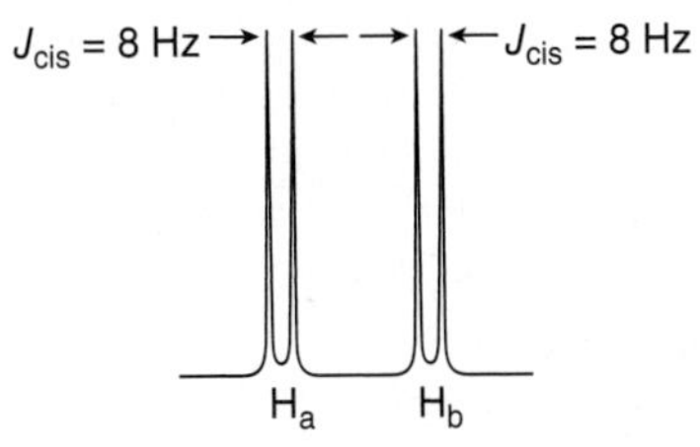

- Although both (*E*)- and (*Z*)-3-chloropropenoic acid show two doublets in their ^1H NMR spectra for their alkenyl protons, $J_{trans} > J_{cis}$.

When a double bond is monosubstituted, there are three nonequivalent protons, and the pattern is more complicated because all three protons are coupled to each other. For example, vinyl acetate (CH_2=CHOCOCH_3) has four different types of protons, three of which are bonded to the double bond. Besides the singlet for the CH_3 group, each proton on the double bond is coupled to two other different protons on the double bond, giving the spectrum in Figure 14.10.

- H_b has two nearby nonequivalent protons that split its signal, the geminal proton H_c and the trans proton H_d. H_d splits the H_b signal into a doublet, and the H_c proton splits the doublet into two doublets. This pattern of four peaks is called a **doublet of doublets.**
- H_c has two nearby nonequivalent protons that split its signal, the geminal proton H_b and the cis proton H_d. H_d splits the H_c signal into a doublet, and the H_b proton splits the doublet into two doublets, forming another **doublet of doublets.**
- H_d has two nearby nonequivalent protons that split its signal, the trans proton H_b and the cis proton H_c. H_b splits the H_d signal into a doublet, and the H_c proton splits the doublet into two doublets, forming another **doublet of doublets.**

Figure 14.10
The ^{1}H NMR spectrum of vinyl acetate (CH_2=CHOCOCH_3)

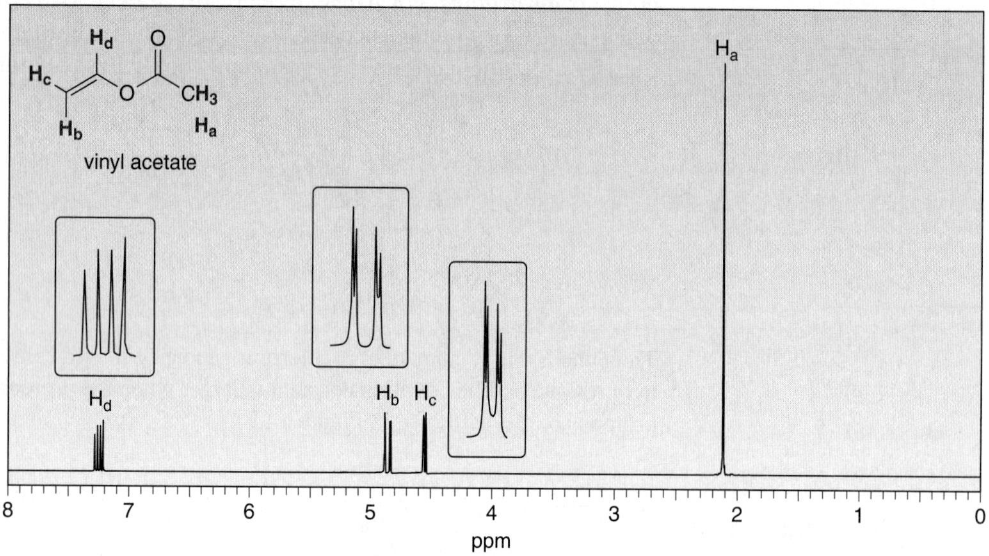

Vinyl acetate is polymerized to poly(vinyl acetate) (Problem 15.26), a polymer used in paints, glues, and adhesives.

Splitting diagrams for the three alkenyl protons in vinyl acetate are drawn in Figure 14.11. Note that each pattern is different in appearance because the magnitude of the coupling constants forming them is different.

Figure 14.11
Splitting diagram for the alkenyl protons in vinyl acetate (CH_2=CHOCOCH_3)

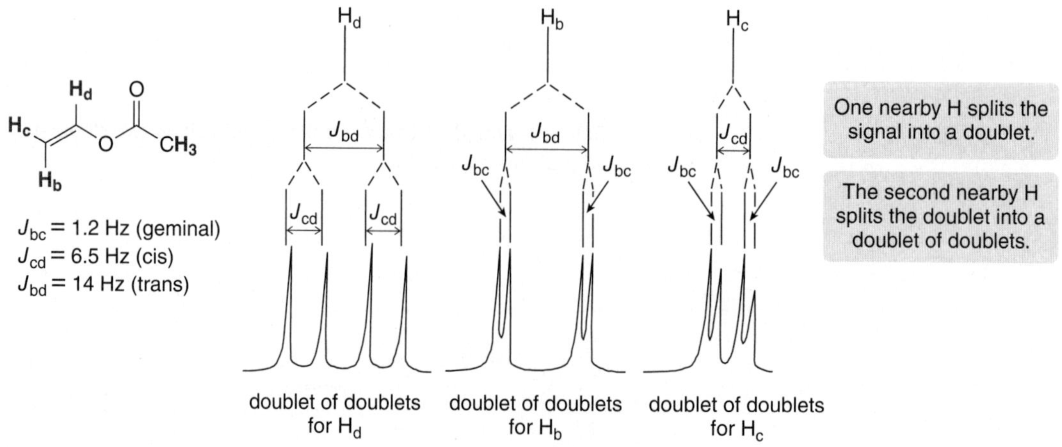

Problem 14.19 Draw a splitting diagram for H_b in *trans*-1,3-dichloropropene, given that J_{ab} = 13.1 Hz and J_{bc} = 7.2 Hz.

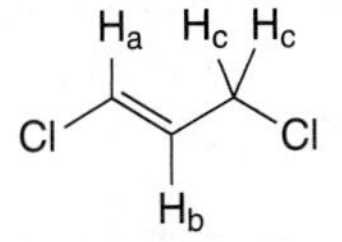

trans-1,3-dichloropropene

Problem 14.20 Identify **A** and **B**, isomers of molecular formula $C_3H_4Cl_2$, from the given ^{1}H NMR data: Compound **A** exhibits signals at 1.75 (doublet, 3 H, *J* = 6.9 Hz) and 5.89 (quartet, 1 H, *J* = 6.9 Hz) ppm. Compound **B** exhibits signals at 4.16 (singlet, 2 H), 5.42 (doublet, 1 H, *J* = 1.9 Hz), and 5.59 (doublet, 1 H, *J* = 1.9 Hz) ppm.

14.9 Other Facts About ^{1}H NMR Spectroscopy

14.9A OH Protons

- **Under usual conditions, an OH proton does not split the NMR signal of adjacent protons.**
- **The signal due to an OH proton is not split by adjacent protons.**

Ethanol (CH_3CH_2OH), for example, has three different types of protons, so there are three signals in its ^{1}H NMR spectrum, as shown in Figure 14.12.

- The H_a signal is split by the two H_b protons into three peaks, a **triplet.**
- The H_b signal is split by only the three H_a protons into four peaks, a **quartet.** The adjacent OH proton does *not* split the signal due to H_b.
- H_c is a **singlet** because OH protons are *not* split by adjacent protons.

Why is a proton bonded to an oxygen atom a singlet in a ^{1}H NMR spectrum? Protons on electronegative elements rapidly **exchange** between molecules in the presence of trace amounts of acid or base. It is as if the CH_2 group in ethanol never "feels" the presence of the OH proton, because the OH proton is rapidly moving from one molecule to another. We therefore see a peak due to the OH proton, but it is a single peak with no splitting. This phenomenon usually occurs with NH and OH protons.

Figure 14.12 The ^{1}H NMR spectrum of ethanol (CH_3CH_2OH)

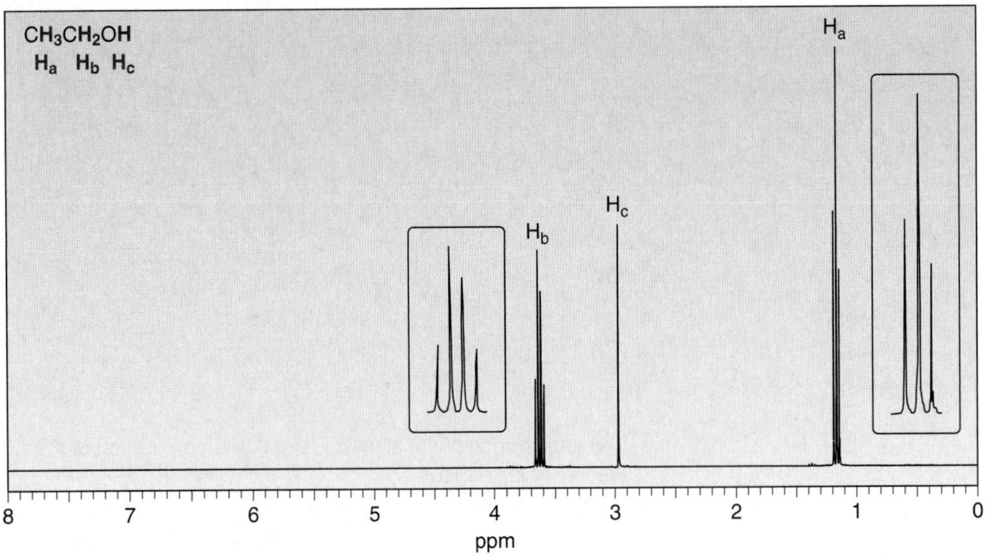

Problem 14.21 How many signals are present in the ^1H NMR spectrum for each molecule? What splitting is observed in each signal?

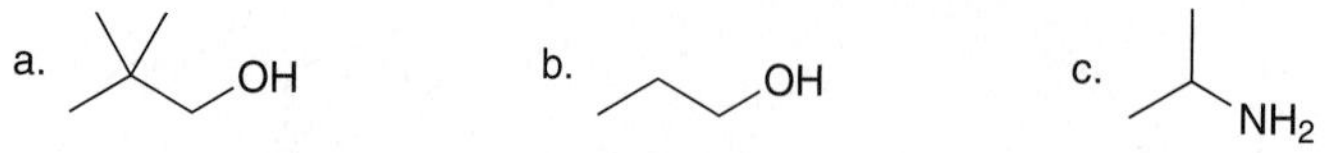

14.9B Cyclohexane Conformations

How does the rotation around carbon–carbon σ bonds and the ring flip of cyclohexane rings affect an NMR spectrum? Because these processes are rapid at room temperature, an NMR spectrum records an **average** of all conformations that interconvert.

Thus, even though each cyclohexane carbon has two different types of hydrogens—one axial and one equatorial—the two chair forms of cyclohexane rapidly interconvert them, and an **NMR spectrum shows a single signal for the average environment** that it "sees."

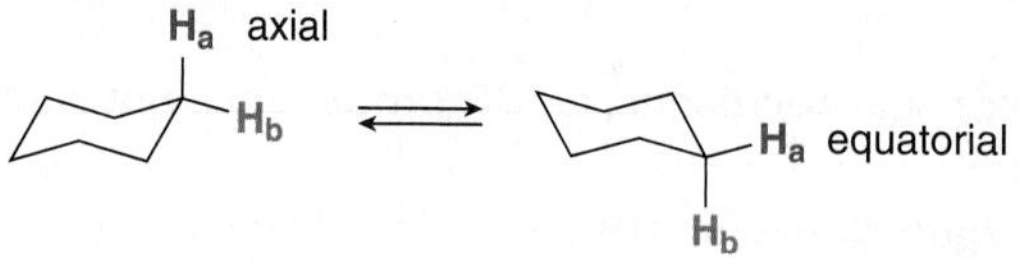

Axial and equatorial H's rapidly interconvert. NMR sees an average environment and shows one signal.

14.9C Protons on Benzene Rings

We will learn more about the spectroscopic absorptions of benzene derivatives in Chapter 17.

Benzene has six equivalent, deshielded protons and exhibits a single peak in its ^1H NMR spectrum at 7.27 ppm. Monosubstituted benzene derivatives—that is, benzene rings with one H atom replaced by another substituent Z—contain five deshielded protons that are no longer all equivalent to each other. The identity of Z determines the appearance of this region of a ^1H NMR spectrum (6.5–8 ppm), as shown in Figure 14.13. We will not analyze the splitting patterns observed for the ring protons of monosubstituted benzenes.

Figure 14.13 The 6.5–8 ppm region of the ^1H NMR spectrum of three benzene derivatives

A monosubstituted benzene ring has three different types of H atoms: H_a, H_b, and H_c.

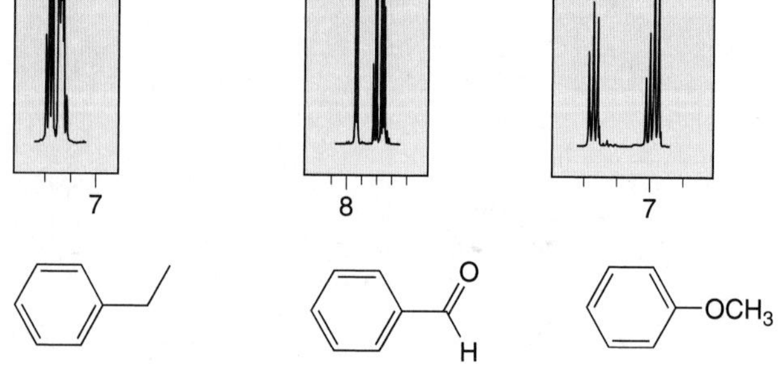

- The appearance of the signals in the 6.5–8 ppm region of the ^1H NMR spectrum depends on the identity of Z in C_6H_5Z.

Problem 14.22 What protons in alcohol **A** give rise to each signal in its ¹H NMR spectrum? Explain all splitting patterns observed for absorptions between 0 and 7 ppm.

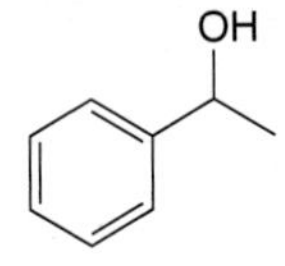

A

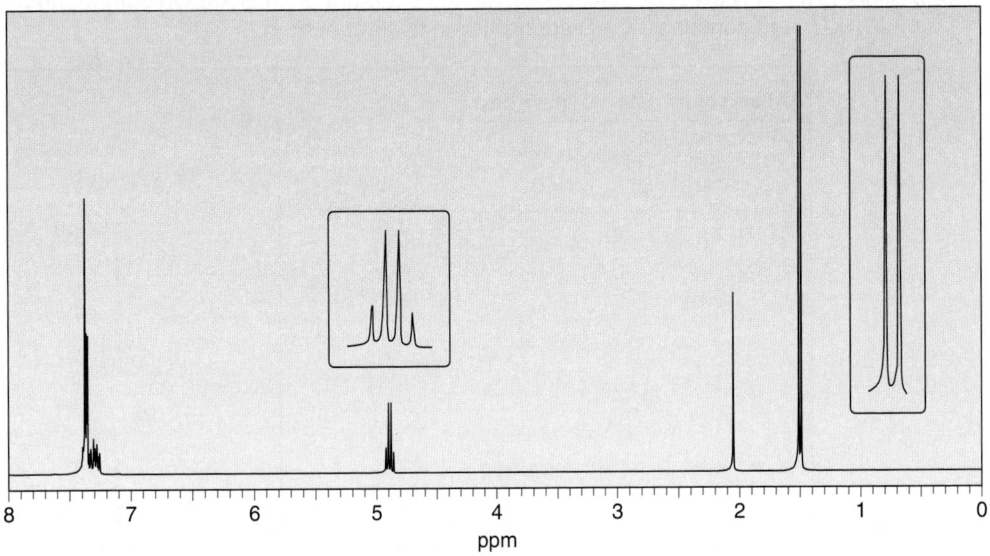

Problem 14.23 How many peaks are observed in the ¹H NMR signal for each proton shown in red in palau'amine, the complex chapter-opening molecule?

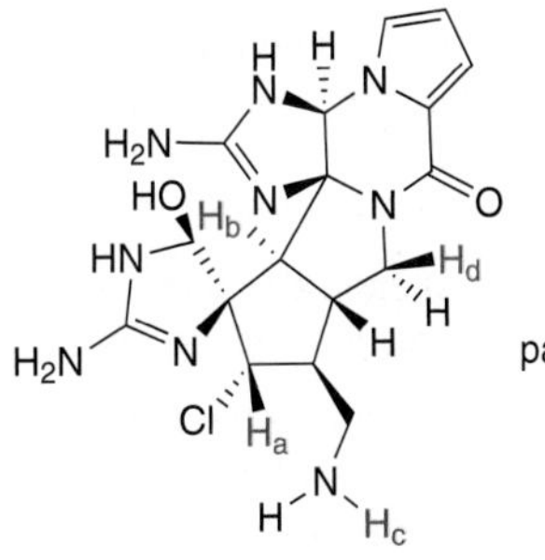

palau'amine

14.10 Using ¹H NMR to Identify an Unknown

Once we know a compound's molecular formula from its mass spectral data and the identity of its functional group from its IR spectrum, we can then use its ¹H NMR spectrum to determine its structure. A suggested procedure is illustrated for compound **X,** whose molecular formula ($C_4H_8O_2$) and functional group (C=O) were determined in Section 13.8.

How To Use ^{1}H NMR Data to Determine a Structure

Example **Using its ^{1}H NMR spectrum, determine the structure of an unknown compound X that has molecular formula $C_4H_8O_2$ and contains a C=O absorption in its IR spectrum.**

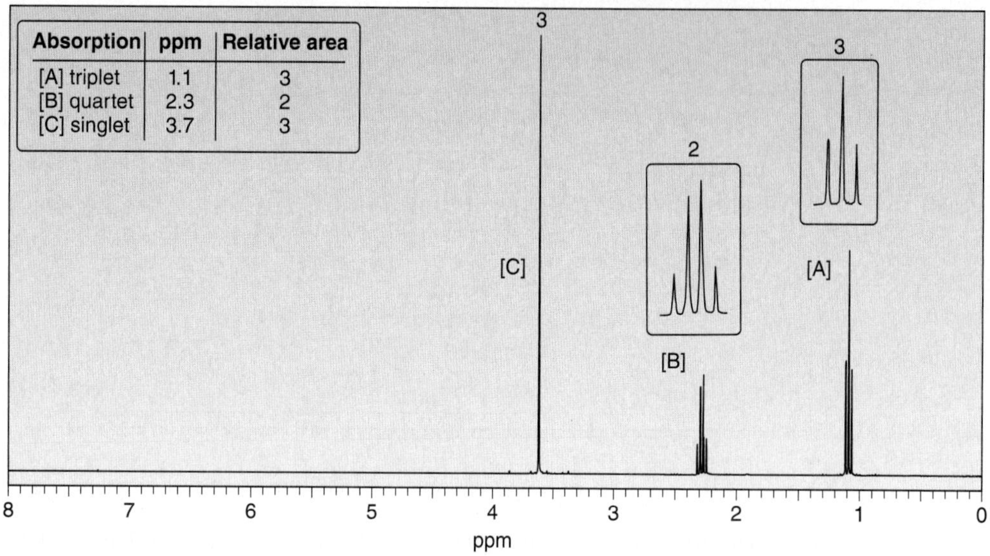

Step [1] **Determine the number of different kinds of protons.**

- The number of NMR signals equals the number of different types of protons.
- This molecule has three NMR signals ([A], [B], and [C]) and therefore **three** types of protons (H_a, H_b, and H_c).

Step [2] **Use the relative area to determine the number of H atoms giving rise to each signal.**

- The relative area (printed on top of each signal) gives the *ratio* of absorbing protons responsible for each signal. In this case, the ratio is 3:2:3 for the signals from left-to-right.
- **When the sum of the relative areas *equals* the number of H's in the molecular formula, the relative area gives the number of absorbing H's responsible for the NMR signal.** In this example, the sum of the relative areas is 3 + 2 + 3 = 8, and the unknown has 8 H's, so the signals are due to 3 H's, 2 H's, and 3 H's from left-to-right in the spectrum.

3 H_a protons signal [A]	2 H_b protons signal [B]	3 H_c protons signal [C]
Three equivalent H's usually means a **CH_3** group.	Two equivalent H's usually means a **CH_2** group.	Three equivalent H's usually means a **CH_3** group.

Step [3] **Use individual splitting patterns to determine what carbon atoms are bonded to each other.**

- Start with the singlets. Signal [C] is due to a CH_3 group with no adjacent nonequivalent H atoms. Possible structures include:

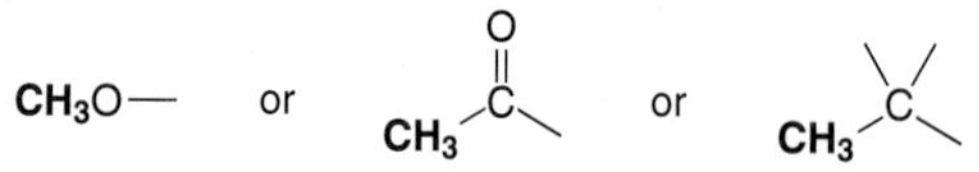

- Because signal [A] is a **triplet,** there must be **2 H's** (CH_2 group) on the adjacent carbon.
- Because signal [B] is a **quartet,** there must be **3 H's** (CH_3 group) on the adjacent carbon.
- This information suggests that **X** has an **ethyl** group ---> CH_3CH_2–.

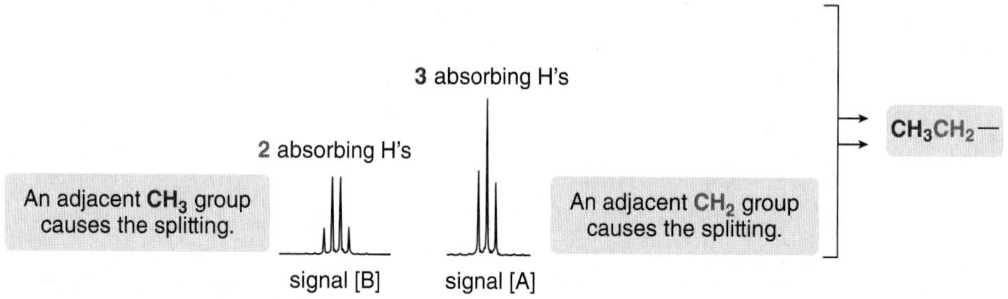

How To, continued . . .

To summarize, **X** contains CH_3-, CH_3CH_2-, and $C{=}O$ (from the IR). Comparing these atoms with the molecular formula shows that one O atom is missing. Because O atoms do not absorb in a ^1H NMR spectrum, their presence can be inferred only by examining the chemical shift of protons near them. O atoms are more electronegative than C, thus deshielding nearby protons, and shifting their absorption downfield.

Step [4] Use chemical shift data to complete the structure.

- Put the structure together in a manner that preserves the splitting data and is consistent with the reported chemical shifts.
- In this example, two isomeric structures (**A** and **B**) are possible for **X** considering the splitting data only:

Structural pieces **Possible structures**

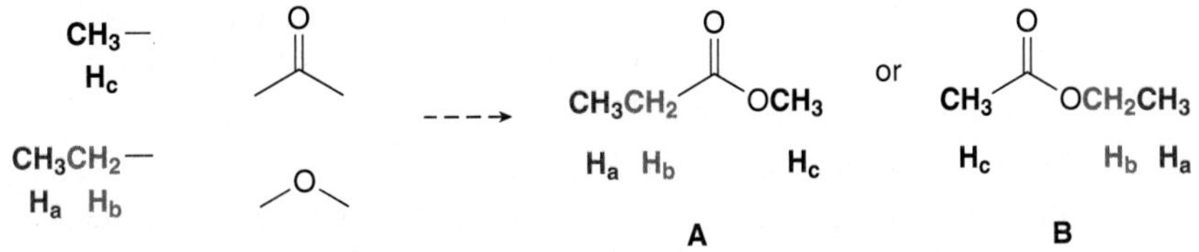
CH3—
Hc
CH3CH2—
Ha Hb
O
O
CH3CH2
OCH3
Ha Hb
Hc
or
CH3
OCH2CH3
Hc
Hb Ha

A **B**

- Chemical shift information distinguishes the two possibilities. The electronegative O atom deshields adjacent H's, shifting them downfield between 3 and 4 ppm. If **A** is the correct structure, the singlet due to the CH_3 group (H_c) should occur downfield, whereas if **B** is the correct structure, the quartet due to the CH_2 group (H_b) should occur downfield.
- Because the NMR of **X** has a singlet (not a quartet) at 3.7, **A is the correct structure.**

Problem 14.24 Propose a structure for a compound of molecular formula $C_7H_{14}O_2$ with an IR absorption at 1740 cm^{-1} and the following ^1H NMR data:

Absorption	ppm	Relative area
singlet	1.2	9
triplet	1.3	3
quartet	4.1	2

Problem 14.25 Propose a structure for a compound of molecular formula C_3H_8O with an IR absorption at 3600–3200 cm^{-1} and the following NMR spectrum:

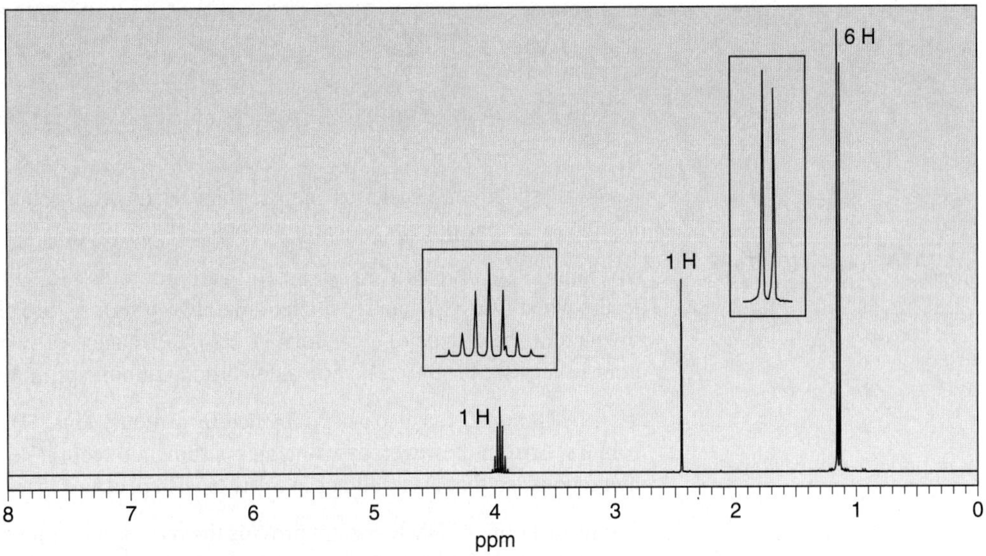
6 H
1 H
1 H
8
7
6
5
4
3
2
1
0
ppm

Problem 14.26 Identify products **A** and **B** from the given ^1H NMR data.

a. Treatment of $CH_2=CHCOCH_3$ with one equivalent of HCl forms compound **A. A** exhibits the following absorptions in its ^1H NMR spectrum: 2.2 (singlet, 3 H), 3.05 (triplet, 2 H), and 3.6 (triplet, 2 H) ppm. What is the structure of **A?**

b. Treatment of acetone [$(CH_3)_2C=O$] with dilute aqueous base forms **B.** Compound **B** exhibits four singlets in its ^1H NMR spectrum at 1.3 (6 H), 2.2 (3 H), 2.5 (2 H), and 3.8 (1 H) ppm. What is the structure of **B?**

14.11 ^{13}C NMR Spectroscopy

^{13}C NMR spectroscopy is also an important tool for organic structure analysis. The physical basis for ^{13}C NMR is the same as for ^1H NMR. When placed in a magnetic field, B_0, ^{13}C nuclei can align themselves with or against B_0. More nuclei are aligned with B_0 because this arrangement is lower in energy, but these nuclei can be made to spin flip against the applied field by applying RF radiation of the appropriate frequency.

^{13}C NMR spectra, like ^1H NMR spectra, plot peak intensity versus chemical shift, using TMS as the reference signal at 0 ppm. ^{13}C occurs in only 1.1% natural abundance, however, so ^{13}C NMR signals are much weaker than ^1H NMR signals. To overcome this limitation, modern spectrometers irradiate samples with many pulses of RF radiation and use mathematical tools to increase signal sensitivity and decrease background noise. The spectrum of acetic acid (CH_3COOH) illustrates the general features of a ^{13}C NMR spectrum.

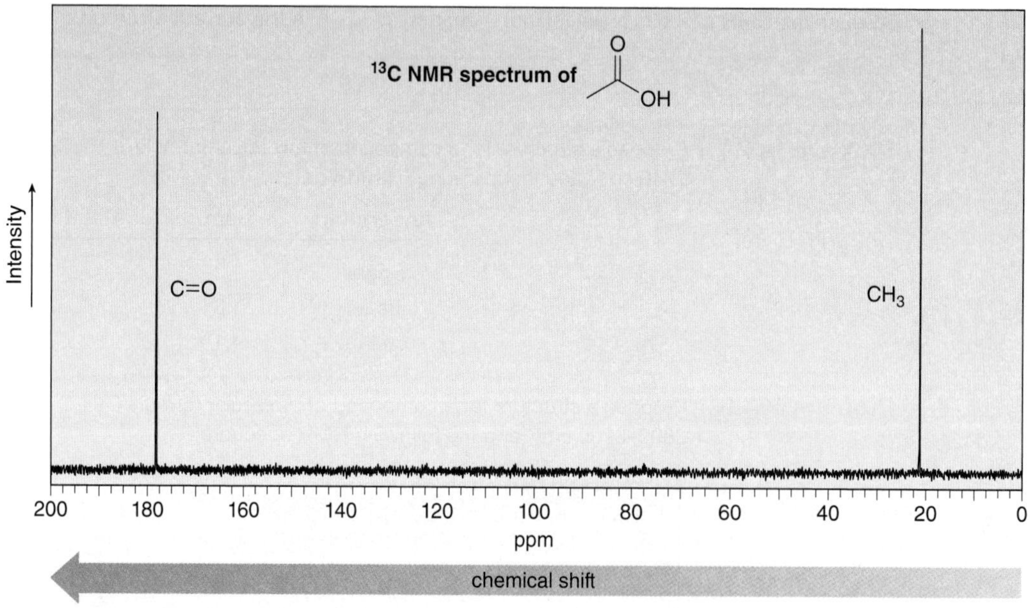

^{13}C NMR spectra are easier to analyze than ^1H spectra because signals are not split. **Each type of carbon atom appears as a single peak.**

Why aren't ^{13}C signals split by nearby carbon atoms? Recall from Section 14.6 that splitting occurs when two NMR active nuclei—like two protons—are close to each other. Because of the low natural abundance of ^{13}C nuclei (1.1%), the chance of two ^{13}C nuclei being bonded to each other is very small (0.01%), and so no carbon–carbon splitting is observed.

A ^{13}C NMR signal can also be split by nearby protons. This $^1H-^{13}C$ splitting is usually eliminated from a spectrum, however, by using an instrumental technique that decouples the proton–carbon interactions, so that every signal in a ^{13}C NMR spectrum is a singlet.

Two features of ^{13}C NMR spectra provide the most structural information: the **number of signals** observed and the **chemical shifts** of those signals.

14.11A ^{13}C NMR: Number of Signals

- **The number of signals in a ^{13}C spectrum gives the number of different types of carbon atoms in a molecule.**

Carbon atoms in the same environment give the same NMR signal, whereas carbons in different environments give different NMR signals. The ^{13}C NMR spectrum of CH_3COOH has two signals because there are two different types of carbon atoms—the C of the CH_3 group and the C of the carbonyl (C=O).

- **Because ^{13}C NMR signals are not split, the number of signals equals the number of lines in the ^{13}C NMR spectrum.**

Thus, the ^{13}C NMR spectra of dimethyl ether, chloroethane, and methyl acetate exhibit one, two, and three lines, respectively, because these compounds contain one, two, and three different types of carbon atoms.

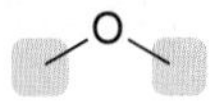

dimethyl ether

1 ^{13}C NMR signal
Both C's are equivalent.

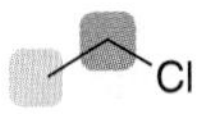

chloroethane

2 ^{13}C NMR signals

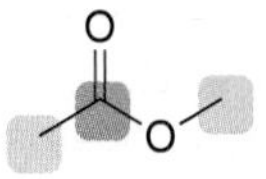

methyl acetate

3 ^{13}C NMR signals

In contrast to what occurs in proton NMR, peak intensity is not proportional to the number of absorbing carbons, so ^{13}C NMR signals are not integrated.

Sample Problem 14.7 How many lines are observed in the ^{13}C NMR spectrum of each compound?

a.  b. 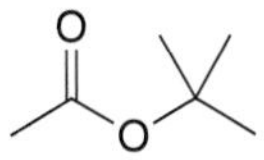

c. 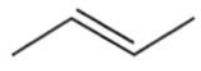

Solution

The number of different types of carbons equals the number of lines in a ^{13}C NMR spectrum.

a. 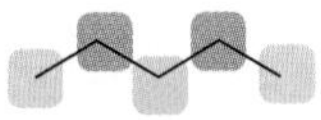

3 types of C's
3 ^{13}C NMR signals

b. 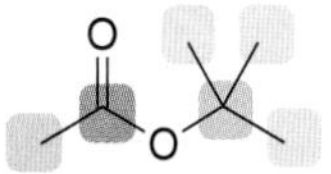

4 types of C's
4 ^{13}C NMR signals

c. 

2 types of C's
2 ^{13}C NMR signals

Problem 14.27 How many lines are observed in the ^{13}C NMR spectrum of each compound?

a. 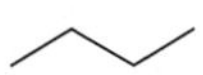 b. 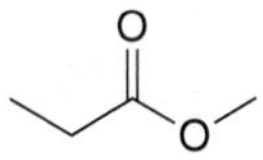

c. 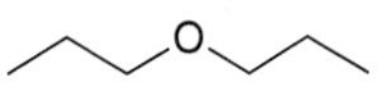

d. 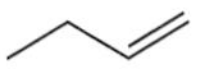

Problem 14.28 Draw all constitutional isomers of molecular formula $C_3H_6Cl_2$.

a. How many signals does each isomer exhibit in its ^{1}H NMR spectrum?
b. How many lines does each isomer exhibit in its ^{13}C NMR spectrum?
c. When only the number of signals in both ^{1}H and ^{13}C NMR spectroscopy is considered, is it possible to distinguish all of these constitutional isomers?

Problem 14.29 Esters of chrysanthemic acid are naturally occurring insecticides. How many lines are present in the ^{13}C NMR spectrum of chrysanthemic acid?

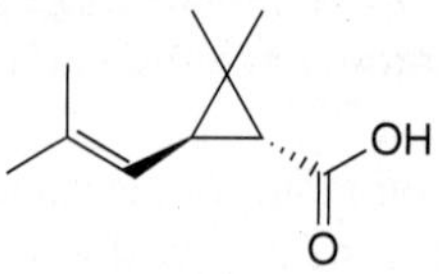

chrysanthemic acid

Esters of chrysanthemic acid are obtained from the flowers of *Chrysanthemum cinerariifolium.* Because they are biodegradable and active against numerous insect species, these esters are widely used insecticides (see also Section 29.4).

14.11B ^{13}C NMR: Position of Signals

In contrast to the small range of chemical shifts in ^{1}H NMR (0–12 ppm usually), ^{13}C NMR absorptions occur over a much broader range, 0–220 ppm. The chemical shifts of carbon atoms in ^{13}C NMR depend on the same effects as the chemical shifts of protons in ^{1}H NMR:

- **The *sp*^{3} hybridized C atoms of alkyl groups are shielded and absorb upfield.**
- **Electronegative elements like halogen, nitrogen, and oxygen shift absorptions downfield.**
- **The *sp*^{2} hybridized C atoms of alkenes and benzene rings absorb downfield.**
- **Carbonyl carbons are highly deshielded, and absorb farther downfield than other carbon types.**

Table 14.5 lists common ^{13}C chemical shift values. The ^{13}C NMR spectra of propan-1-ol ($CH_3CH_2CH_2OH$) and methyl acetate ($CH_3CO_2CH_3$) in Figure 14.14 illustrate these principles.

Table 14.5 Common ^{13}C Chemical Shift Values

Type of carbon	Chemical shift (ppm)	Type of carbon	Chemical shift (ppm)
	5–45	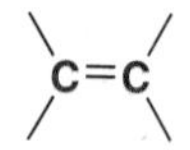	100–140
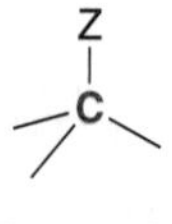 Z = N, O, X	30–80	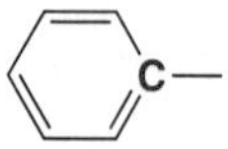	120–150
—C≡C—	65–100	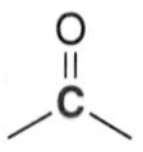	160–210

Problem 14.30 Which of the highlighted carbon atoms in each molecule absorbs farther downfield?

a. 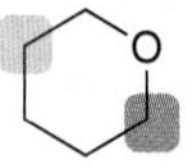 b. 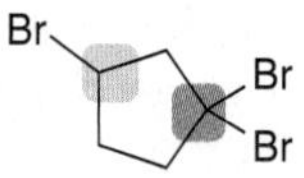 c. 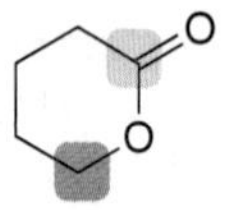 d. 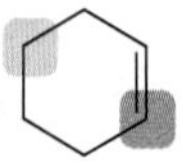

Figure 14.14 Representative ^{13}C NMR spectra

a. Propan-1-ol

b. Methyl acetate

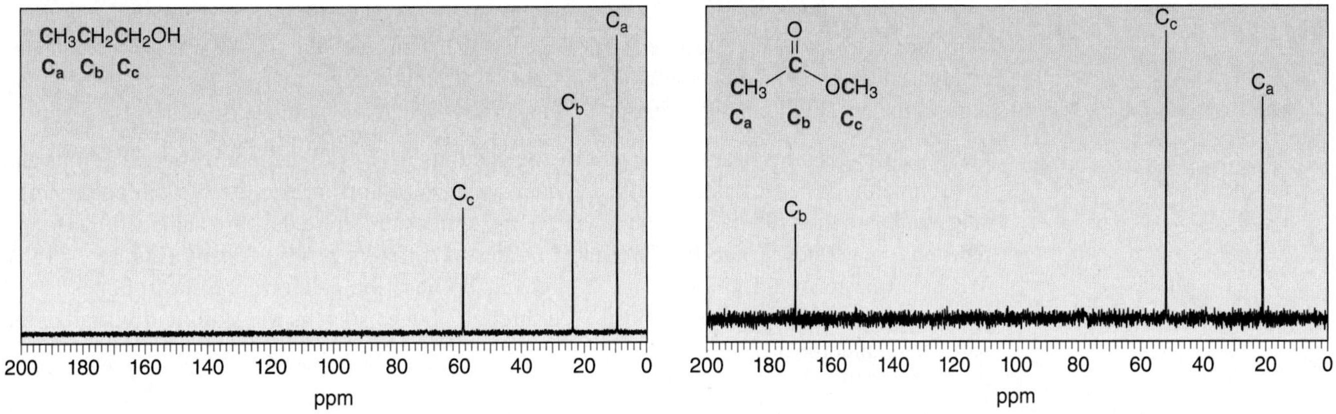

- The three types of C's in propan-1-ol—identified as C_a, C_b, and C_c—give rise to three ^{13}C NMR signals.
- Deshielding increases with increasing proximity to the electronegative O atom, and the absorption shifts downfield; thus, in order of increasing chemical shift: $C_a < C_b < C_c$.

- The three types of C's in methyl acetate—identified as C_a, C_b, and C_c—give rise to three ^{13}C NMR signals.
- The carbonyl carbon (C_b) is highly deshielded, so it absorbs farthest downfield.
- C_a, an *sp*^{3} hybridized C that is not bonded to an O atom, is the most shielded, and so it absorbs farthest upfield.
- Thus, in order of increasing chemical shift: $C_a < C_c < C_b$.

Problem 14.31 Identify the carbon atoms that give rise to each NMR signal.

a.

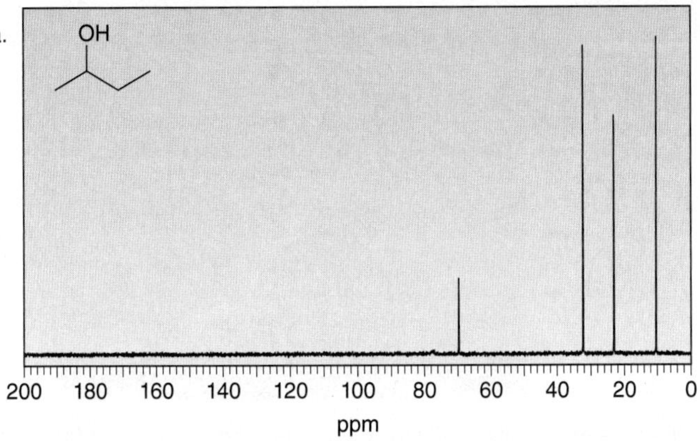

b.

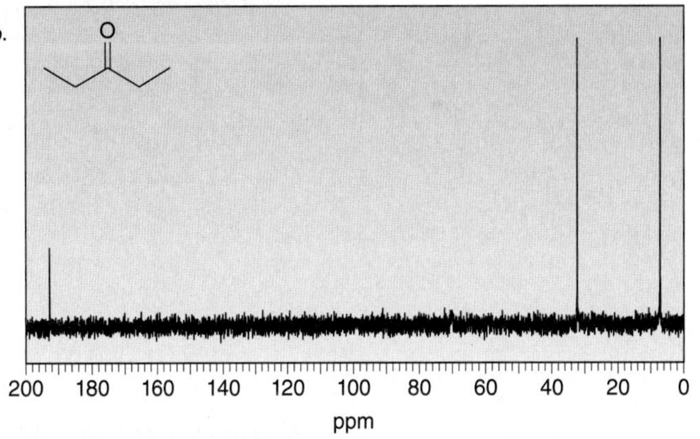

Problem 14.32 A compound of molecular formula $C_4H_8O_2$ shows no IR peaks at 3600–3200 or 1700 cm^{-1}. It exhibits one singlet in its ^{1}H NMR spectrum at 3.69 ppm, and one line in its ^{13}C NMR spectrum at 67 ppm. What is the structure of this unknown?

Problem 14.33 Draw the structure of a compound of molecular formula C_4H_8O that has a signal in its ^{13}C NMR spectrum at > 160 ppm. Then draw the structure of an isomer of molecular formula C_4H_8O that has all of its ^{13}C NMR signals at < 160 ppm.

14.12 Magnetic Resonance Imaging (MRI)

Magnetic resonance imaging (MRI)—NMR spectroscopy in medicine—is a powerful diagnostic technique (Figure 14.15a). The "sample" is the patient, who is placed in a large cavity in a magnetic field, and then irradiated with RF energy. Because RF energy has very low frequency and low energy, the method is safer than X-rays or computed tomography (CT) scans that employ high-frequency, high-energy radiation that is known to damage living cells.

Living tissue contains protons (especially the H atoms in H_2O) in different concentrations and environments. When irradiated with RF energy, these protons are excited to a higher energy spin state, and then fall back to the lower energy spin state. These data are analyzed by a computer that generates a plot that delineates tissues of different proton density (Figure 14.15b). MRIs can be recorded in any plane. Moreover, because the calcium present in bones is not NMR active, an MRI instrument can "see through" bones such as the skull and visualize the soft tissue underneath.

Figure 14.15 Magnetic resonance imaging

a. b.

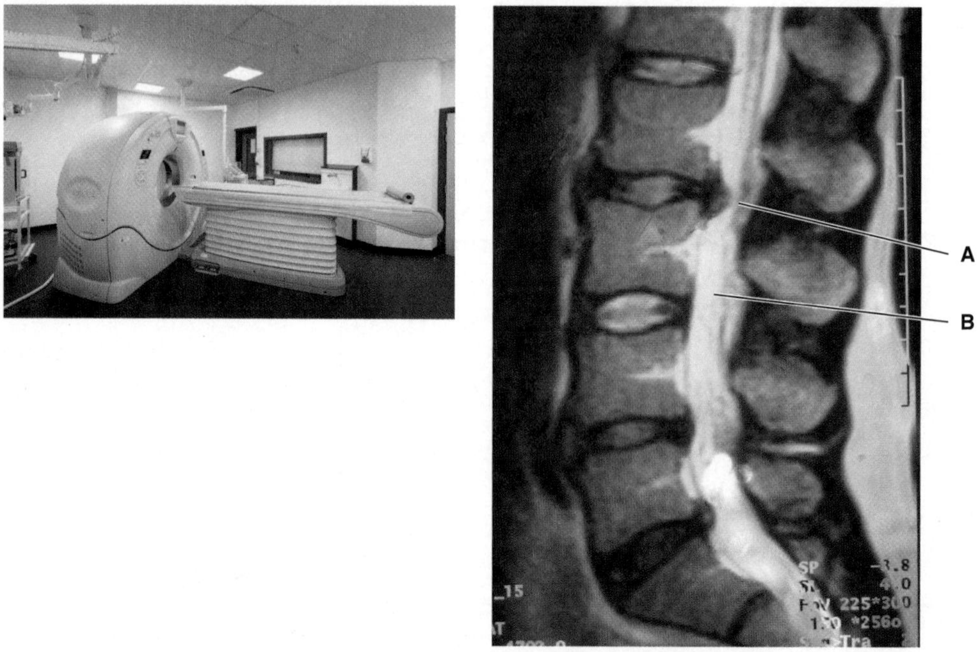

a. An MRI instrument: An MRI instrument is especially useful for visualizing soft tissue. In 2002, 60 million MRI procedures were performed. The 2003 Nobel Prize in Physiology or Medicine was awarded to chemist Paul C. Lauterbur and physicist Sir Peter Mansfield for their contributions in developing magnetic resonance imaging.

b. An MRI image of the lower back: **A** labels spinal cord compression from a herniated disc. **B** labels the spinal cord, which would not be visualized with conventional X-rays.

KEY CONCEPTS

Nuclear Magnetic Resonance Spectroscopy

^{1}H NMR Spectroscopy

[1] The **number of signals** equals the number of different types of protons (14.2).

[2] The **position of a signal** (its chemical shift) is determined by shielding and deshielding effects.
- Shielding shifts an absorption upfield; deshielding shifts an absorption downfield.
- Electronegative atoms withdraw electron density, deshield a nucleus, and shift an absorption downfield (14.3).

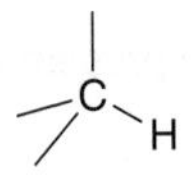

This proton is shielded.
Its absorption is upfield,
0.9–2 ppm.

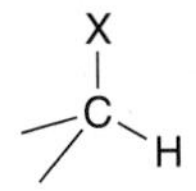

This proton is deshielded.
Its absorption is farther downfield,
2.5–4 ppm.

- Loosely held π electrons can either shield or deshield a nucleus. Protons on benzene rings and double bonds are deshielded and absorb downfield, whereas protons on triple bonds are shielded and absorb upfield (14.4).

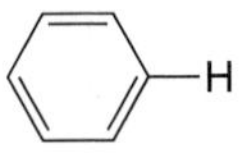

deshielded H
downfield absorption

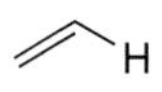

deshielded H
downfield absorption

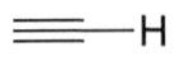

shielded H
upfield absorption

[3] The **area under an NMR signal** is proportional to the number of absorbing protons (14.5).

[4] **Spin–spin splitting** tells about nearby nonequivalent protons (14.6–14.8).
- Equivalent protons do not split each other's signals.
- A set of n nonequivalent protons on the same carbon or adjacent carbons splits an NMR signal into $n + 1$ peaks.
- OH and NH protons do not cause splitting (14.9).
- When an absorbing proton has two sets of nearby nonequivalent protons that are equivalent to each other, use the $n + 1$ rule to determine splitting.
- When an absorbing proton has two sets of nearby nonequivalent protons that are not equivalent to each other, the number of peaks in the NMR signal = $(n + 1)(m + 1)$. In flexible alkyl chains, peak overlap often occurs, resulting in $n + m + 1$ peaks in an NMR signal.

^{13}C NMR Spectroscopy (14.11)

[1] **The number of signals equals the number of different types of carbon atoms.** All signals are **single peaks.**

[2] The relative position of ^{13}C signals is determined by shielding and deshielding effects.
- Carbons that are sp^3 hybridized are shielded and absorb upfield.
- Electronegative elements (N, O, and halogen) shift absorptions downfield.
- The carbons of alkenes and benzene rings absorb downfield.
- Carbonyl carbons are highly deshielded, and absorb farther downfield than other carbon types.

PROBLEMS

Problems Using Three-Dimensional Models

14.34 (a) How many ^{1}H NMR signals does each of the following compounds exhibit? (b) How many ^{13}C NMR signals does each compound exhibit?

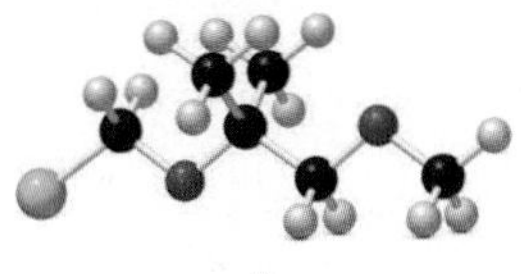

A

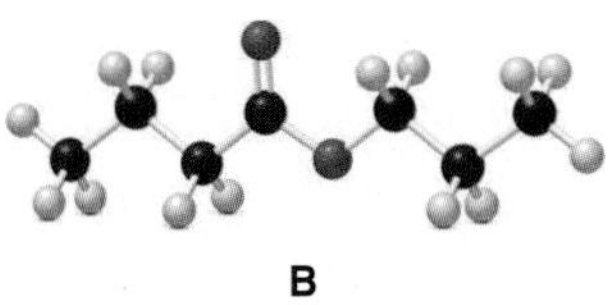

B

14.35 (a) How many ^1H NMR signals does each compound show? (b) Into how many peaks is each signal split?

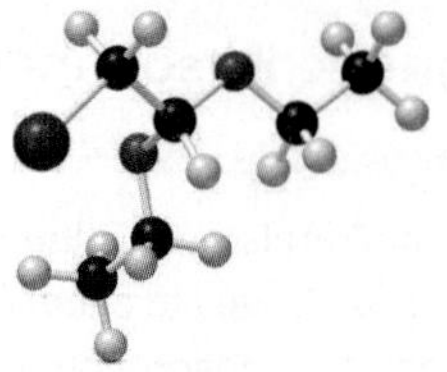

C

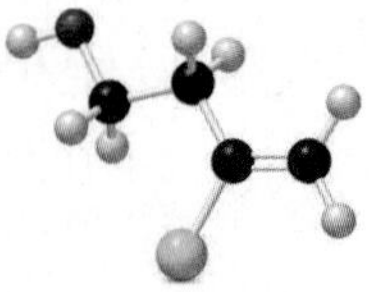

D

^1H NMR Spectroscopy—Determining Equivalent Protons

14.36 How many different types of protons are present in each compound?

a. 

c. 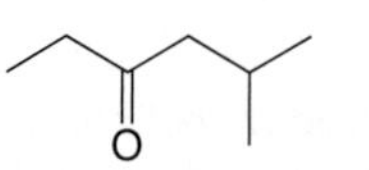

e. 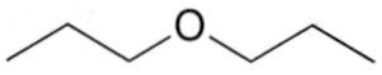

g. 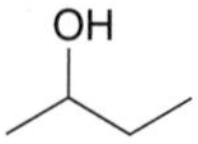

i. 

b. 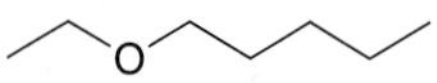

d. 

f. 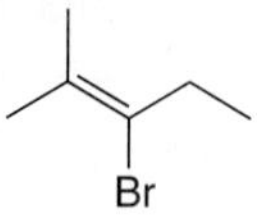

h. 

j. 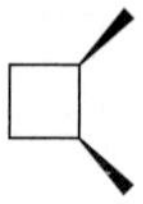

14.37 How many ^1H NMR signals does each natural product exhibit?

a. 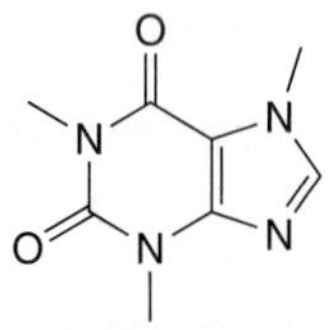

caffeine
(from coffee and tea leaves)

b. 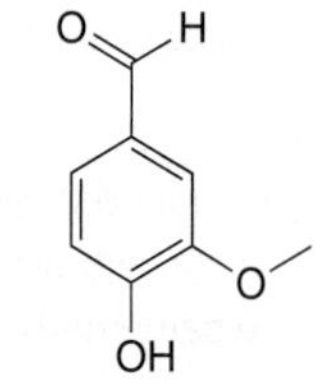

vanillin
(from the vanilla bean)

c. 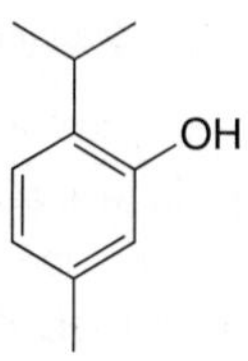

thymol
(from thyme)

d. 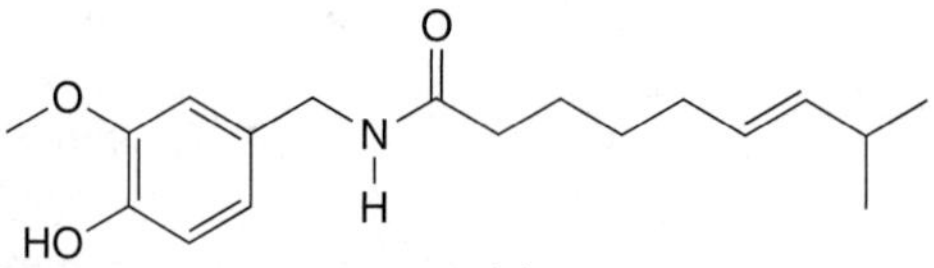

capsaicin
(from hot peppers)

^1H NMR—Chemical Shift and Integration

14.38 Using a 300 MHz NMR instrument:

a. How many Hz downfield from TMS is a signal at 2.5 ppm?

b. If a signal comes at 1200 Hz downfield from TMS, at what ppm does it occur?

c. If two signals are separated by 2 ppm, how many Hz does this correspond to?

14.39 What effect does increasing the operating frequency of a ^1H NMR spectrum have on each value: (a) the chemical shift in δ; (b) the frequency of an absorption in Hz; (c) the magnitude of a coupling constant *J* in Hz?

14.40 Which of the labeled protons in each pair absorbs farther downfield?

a. 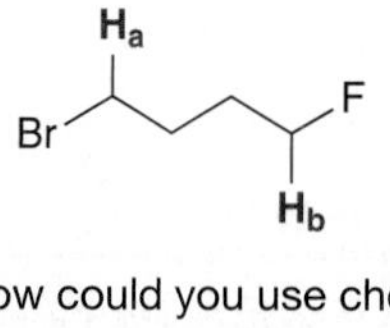

b. 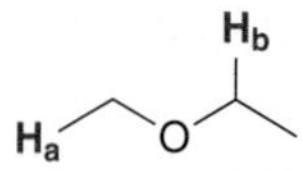

14.41 How could you use chemical shift and integration data in ^1H NMR spectroscopy to distinguish between $CH_3OCH_2CH_2OCH_3$ and $CH_3OCH_2OCH_3$? The ^1H NMR spectrum of each compound contains only singlets.

^{1}H NMR—Splitting

14.42 Into how many peaks will the signal for each of the labeled protons be split?

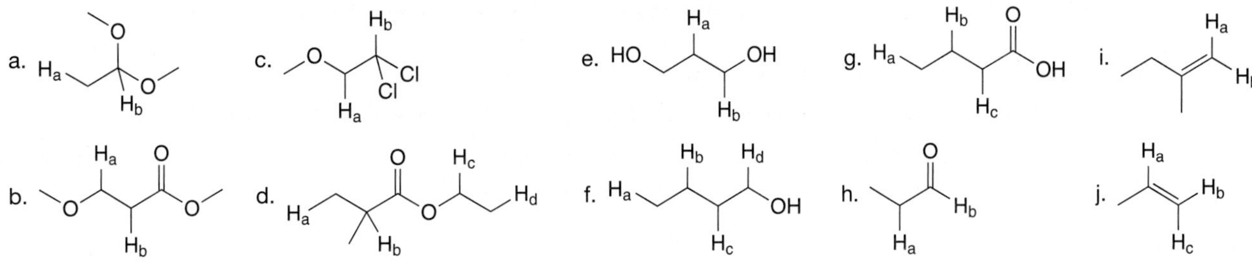

14.43 How can you use ^{1}H NMR spectroscopy to distinguish between $CH_2{=}C(Br)CO_2CH_3$ and methyl (*E*)-3-bromopropenoate, $BrCH{=}CHCO_2CH_3$?

14.44 What splitting pattern is observed for each proton in the following compounds?

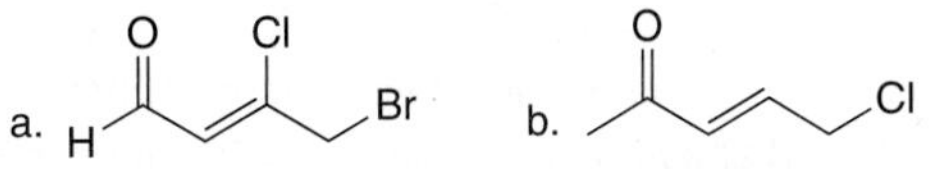

14.45 Label the signals due to H_a, H_b, and H_c in the ^{1}H NMR spectrum of acrylonitrile ($CH_2{=}CHCN$). Draw a splitting diagram for the absorption due to the H_a proton.

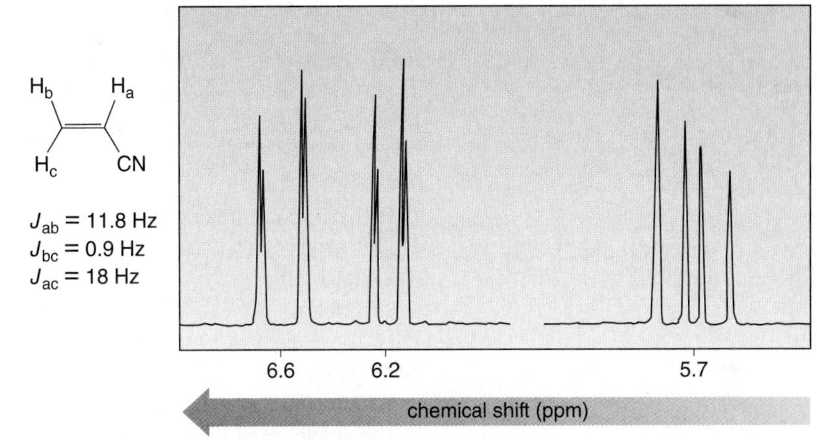

^{13}C NMR

14.46 Draw the four constitutional isomers having molecular formula C_4H_9Br and indicate how many different kinds of carbon atoms each has.

14.47 Explain why the carbonyl carbon of an aldehyde or ketone absorbs farther downfield than the carbonyl carbon of an ester in a ^{13}C NMR spectrum.

14.48 How many ^{13}C NMR signals does each compound exhibit?

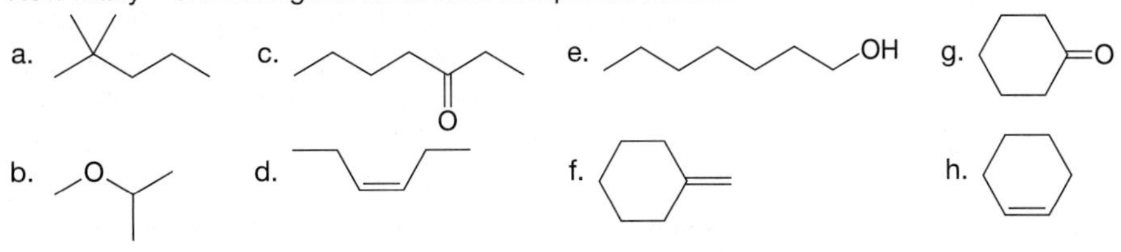

14.49 Rank the highlighted carbon atoms in each compound in order of increasing chemical shift.

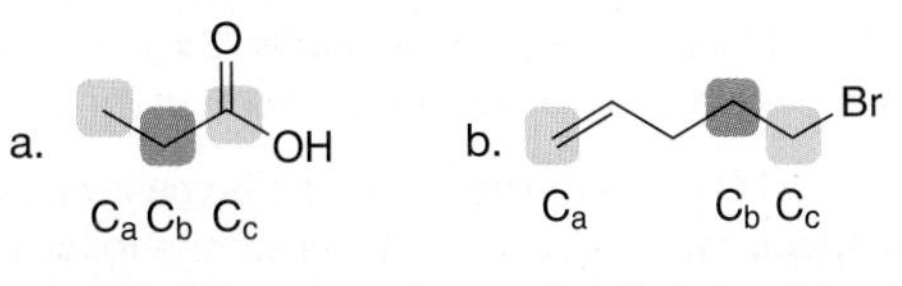

14.50 Identify the carbon atoms that give rise to the signals in the ^{13}C NMR spectrum of each compound.

a. $CH_3CH_2CH_2CH_2OH$; ^{13}C NMR: 14, 19, 35, and 62 ppm
b. $(CH_3)_2CHCHO$; ^{13}C NMR: 16, 41, and 205 ppm
c. $CH_2{=}CHCH(OH)CH_3$; ^{13}C NMR: 23, 69, 113, and 143 ppm

14.51 a. How many signals does dimethyl fumarate ($CH_3O_2CCH{=}CHCO_2CH_3$, with a trans C=C) exhibit in its ^{13}C NMR spectrum?

b. Draw the structure of an isomer of dimethyl fumarate that has each of the following number of signals in its ^{13}C NMR spectrum: [1] three; [2] four; [5] five.

Combined Spectroscopy Problems

Additional spectroscopy problems are located at the end of Chapters 15–23 and 25.

14.52 Answer the following questions about each of the hydroxy ketones: 1-hydroxybutan-2-one (**A**) and 4-hydroxybutan-2-one (**B**).

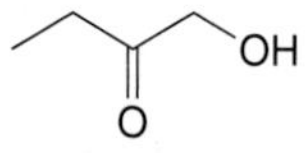

1-hydroxybutan-2-one

A

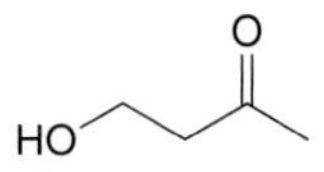

4-hydroxybutan-2-one

B

a. What is the molecular ion in the mass spectrum?

b. What IR absorptions are present in the functional group region?

c. How many lines are observed in the ^{13}C NMR spectrum?

d. How many signals are observed in the ^{1}H NMR spectrum?

e. Give the splitting observed for each type of proton as well as its approximate chemical shift.

14.53 Propose a structure consistent with each set of spectral data:

a. $C_4H_8Br_2$: IR peak at 3000–2850 cm^{-1};
NMR (ppm):
1.87 (singlet, 6 H)
3.86 (singlet, 2 H)

b. $C_3H_6Br_2$: IR peak at 3000–2850 cm^{-1};
NMR (ppm):
2.4 (quintet)
3.5 (triplet)

c. $C_5H_{10}O_2$: IR peak at 1740 cm^{-1};
NMR (ppm):
1.15 (triplet, 3 H) 2.30 (quartet, 2 H)
1.25 (triplet, 3 H) 4.72 (quartet, 2 H)

d. $C_6H_{14}O$: IR peak at 3600–3200 cm^{-1};
NMR (ppm):
0.8 (triplet, 6 H) 1.5 (quartet, 4 H)
1.0 (singlet, 3 H) 1.6 (singlet, 1 H)

e. $C_6H_{14}O$: IR peak at 3000–2850 cm^{-1};
NMR (ppm):
1.10 (doublet, relative area = 6)
3.60 (septet, relative area = 1)

f. C_3H_6O: IR peak at 1730 cm^{-1};
NMR (ppm):
1.11 (triplet)
2.46 (multiplet)
9.79 (triplet)

14.54 Identify the structures of isomers **A** and **B** (molecular formula $C_9H_{10}O$).

Compound **A**: IR peak at 1742 cm^{-1}; ^{1}H NMR data (ppm) at 2.15 (singlet, 3 H), 3.70 (singlet, 2 H), and 7.20 (broad singlet, 5 H).

Compound **B**: IR peak at 1688 cm^{-1}; ^{1}H NMR data (ppm) at 1.22 (triplet, 3 H), 2.98 (quartet, 2 H), and 7.28–7.95 (multiplet, 5 H).

14.55 Reaction of $C_6H_5CH_2CH_2OH$ with CH_3COCl affords compound **W,** which has molecular formula $C_{10}H_{12}O_2$. **W** shows prominent IR absorptions at 3088–2897, 1740, and 1606 cm^{-1}. **W** exhibits the following signals in its ^{1}H NMR spectrum: 2.02 (singlet), 2.91 (triplet), 4.25 (triplet), and 7.20–7.35 (multiplet) ppm. What is the structure of **W?** We will learn about this reaction in Chapter 22.

14.56 Treatment of 2-methylpropanenitrile [$(CH_3)_2CHCN$] with $CH_3CH_2CH_2MgBr$, followed by aqueous acid, affords compound **V,** which has molecular formula $C_7H_{14}O$. **V** has a strong absorption in its IR spectrum at 1713 cm^{-1}, and gives the following ^{1}H NMR data: 0.91 (triplet, 3 H), 1.09 (doublet, 6 H), 1.6 (multiplet, 2 H), 2.43 (triplet, 2 H), and 2.60 (septet, 1 H) ppm. What is the structure of **V?** We will learn about this reaction in Chapter 22.

14.57 Compound **C** has a molecular ion in its mass spectrum at 146 and a prominent absorption in its IR spectrum at 1762 cm^{-1}. **C** shows the following ^{1}H NMR spectral data: 1.47 (doublet, 3 H), 2.07 (singlet, 6 H), and 6.84 (quartet, 1 H) ppm. What is the structure of **C?**

14.58 As we will learn in Chapter 20, reaction of $(CH_3)_2CO$ with $LiC{\equiv}CH$ followed by H_2O affords compound **D,** which has a molecular ion in its mass spectrum at 84 and prominent absorptions in its IR spectrum at 3600–3200, 3303, 2938, and 2120 cm^{-1}. **D** shows the following ^1H NMR spectral data: 1.53 (singlet, 6 H), 2.37 (singlet, 1 H), and 2.43 (singlet, 1 H) ppm. What is the structure of **D?**

14.59 Identify the structures of isomers **E** and **F** (molecular formula $C_4H_8O_2$). Relative areas are given above each signal.

a. **Compound E:** IR absorption at 1743 cm^{-1}

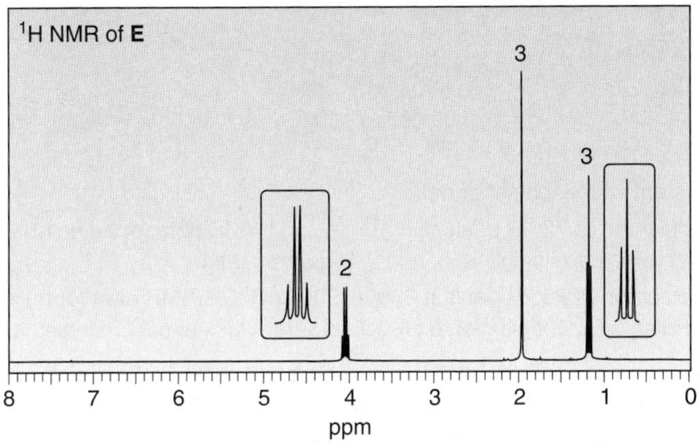

b. **Compound F:** IR absorption at 1730 cm^{-1}

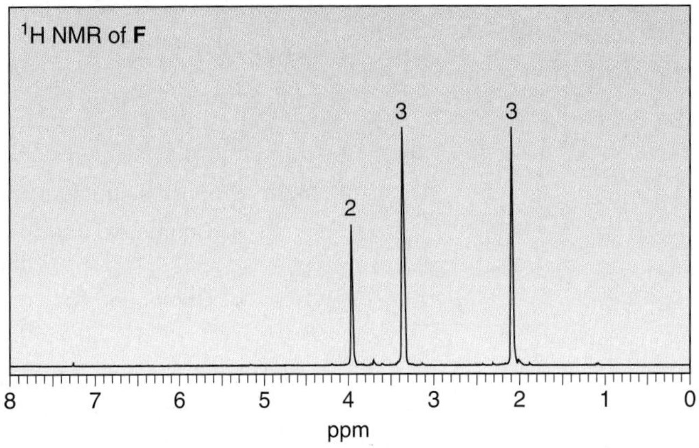

14.60 Identify the structures of isomers **H** and **I** (molecular formula $C_8H_{11}N$).

a. **Compound H:** IR absorptions at 3365, 3284, 3026, 2932, 1603, and 1497 cm^{-1}

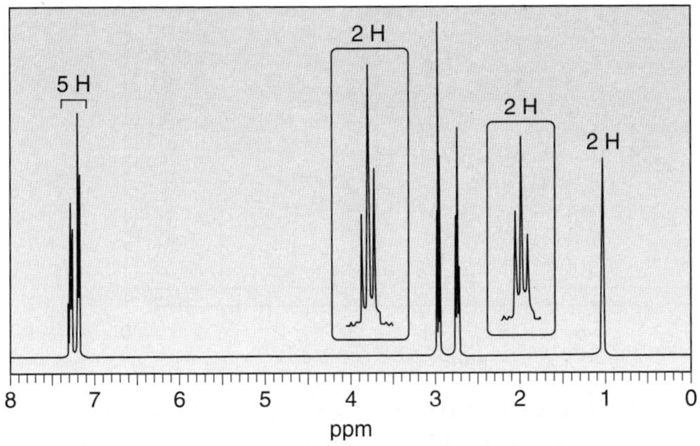

b. **Compound I:** IR absorptions at 3367, 3286, 3027, 2962, 1604, and 1492 cm^{-1}

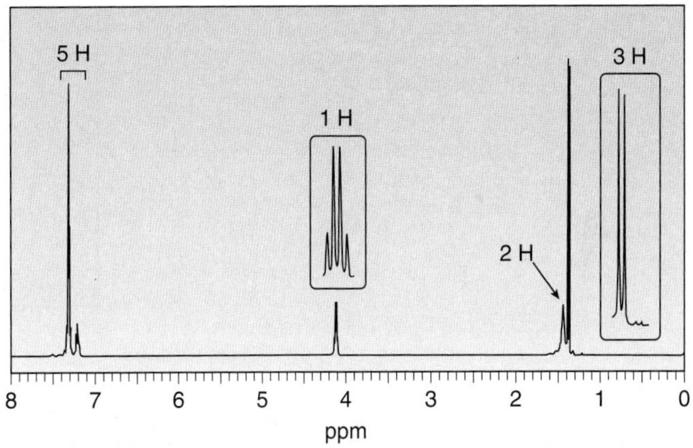

14.61 Propose a structure consistent with each set of data.

a. $C_9H_{10}O_2$: IR absorption at 1718 cm^{-1}

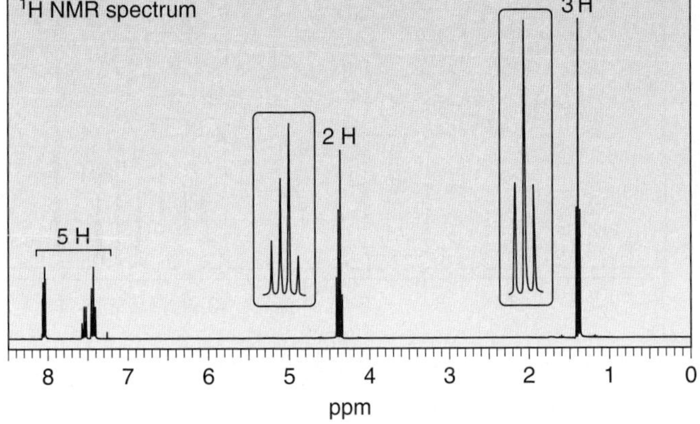

b. C_9H_{12}: IR absorption at 2850–3150 cm^{-1}

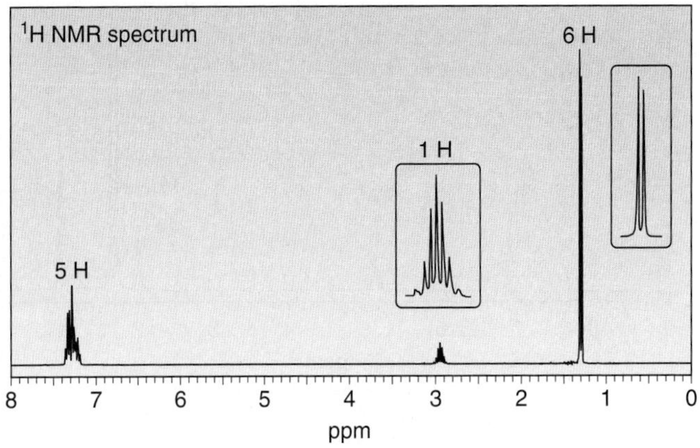

14.62 Reaction of $(CH_3)_3CCHO$ with $(C_6H_5)_3P{=}C(CH_3)OCH_3$, followed by treatment with aqueous acid, affords **R** ($C_7H_{14}O$). **R** has a strong absorption in its IR spectrum at 1717 cm^{-1} and three singlets in its ^1H NMR spectrum at 1.02 (9 H), 2.13 (3 H), and 2.33 (2 H) ppm. What is the structure of **R?** We will learn about this reaction in Chapter 21.

14.63 Reaction of aldehyde **D** with amino alcohol **E** in the presence of NaH forms **F** (molecular formula $C_{11}H_{15}NO_2$). **F** absorbs at 1730 cm^{-1} in its IR spectrum. **F** also shows eight lines in its ^{13}C NMR spectrum, and gives the following ^1H NMR spectrum: 2.32 (singlet, 6 H), 3.05 (triplet, 2 H), 4.20 (triplet, 2 H), 6.97 (doublet, 2 H), 7.82 (doublet, 2 H), and 9.97 (singlet, 1 H) ppm. Propose a structure for **F.** We will learn about this reaction in Chapter 18.

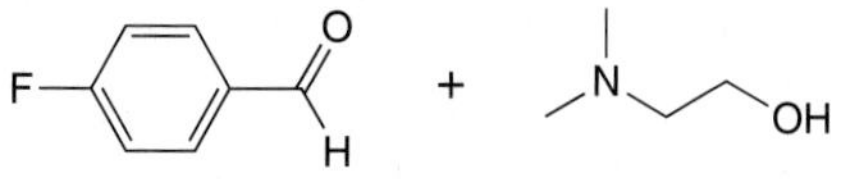

D **E**

14.64 Propose a structure consistent with each set of data.

a. Compound **J:** molecular ion at 72; IR peak at 1710 cm^{-1}; ^1H NMR data (ppm) at 1.0 (triplet, 3 H), 2.1 (singlet, 3 H), and 2.4 (quartet, 2 H)

b. Compound **K:** molecular ion at 88; IR peak at 3600–3200 cm^{-1}; ^1H NMR data (ppm) at 0.9 (triplet, 3 H), 1.2 (singlet, 6 H), 1.5 (quartet, 2 H), and 1.6 (singlet, 1 H)

14.65 In the presence of a small amount of acid, a solution of acetaldehyde (CH_3CHO) in methanol (CH_3OH) was allowed to stand and a new compound **L** was formed. **L** has a molecular ion in its mass spectrum at 90 and IR absorptions at 2992 and 2941 cm^{-1}. **L** shows three signals in its ^{13}C NMR at 19, 52, and 101 ppm. The ^1H NMR spectrum of **L** is given below. What is the structure of **L?**

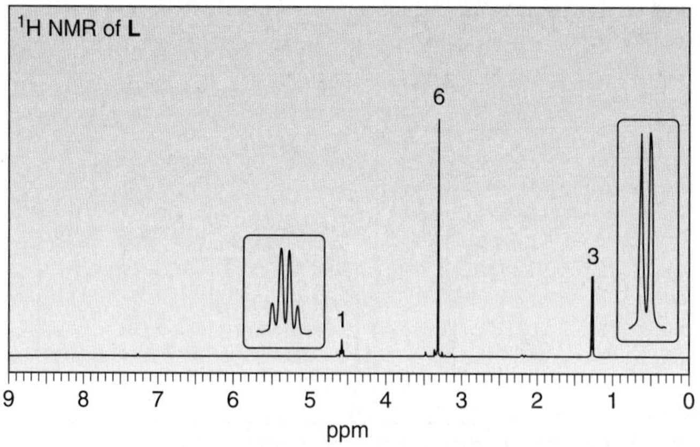

14.66 Treatment of $(CH_3)_2CHCH(OH)CH_2CH_3$ with TsOH affords two products (**M** and **N**) with molecular formula C_6H_{12}. The ^1H NMR spectra of **M** and **N** are given below. Propose structures for **M** and **N** and draw a mechanism to explain their formation.

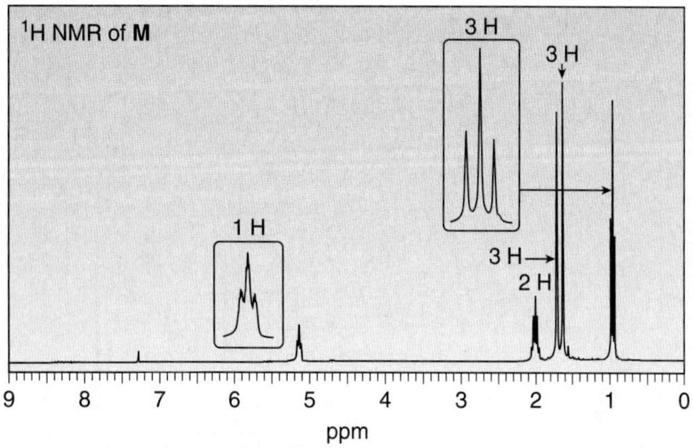

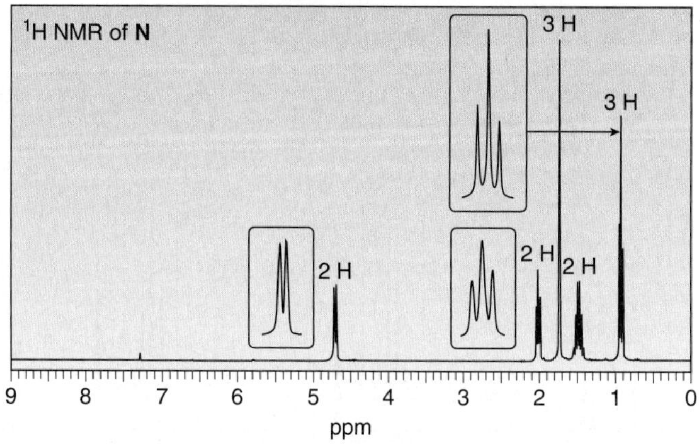

14.67 Compound **O** has molecular formula $C_{10}H_{12}O$ and shows an IR absorption at 1687 cm^{-1}. The ^1H NMR spectrum of **O** is given below. What is the structure of **O?**

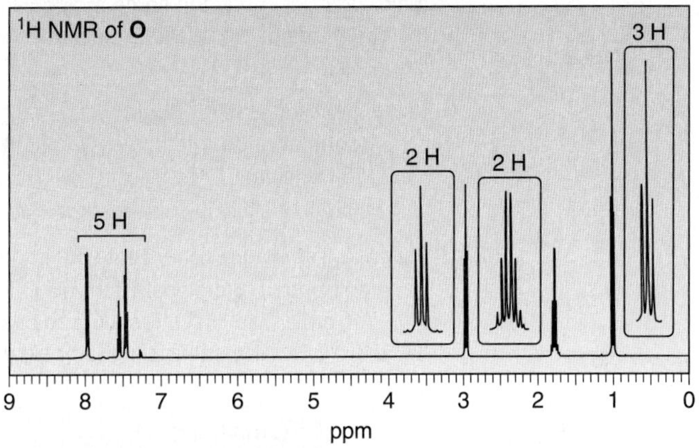

14.68 Compound **P** has molecular formula $C_5H_9ClO_2$. Deduce the structure of **P** from its ^1H and ^{13}C NMR spectra.

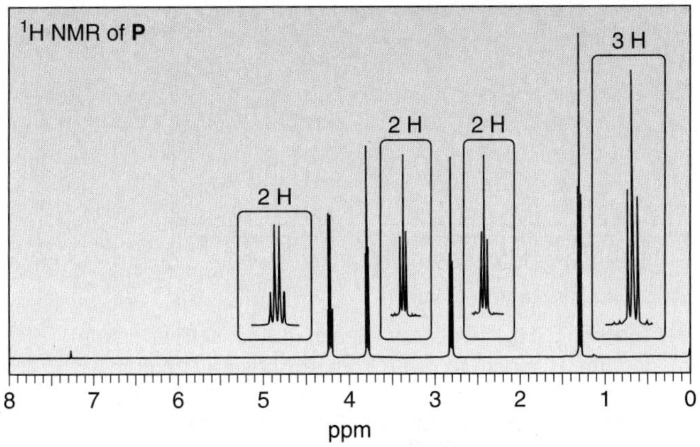

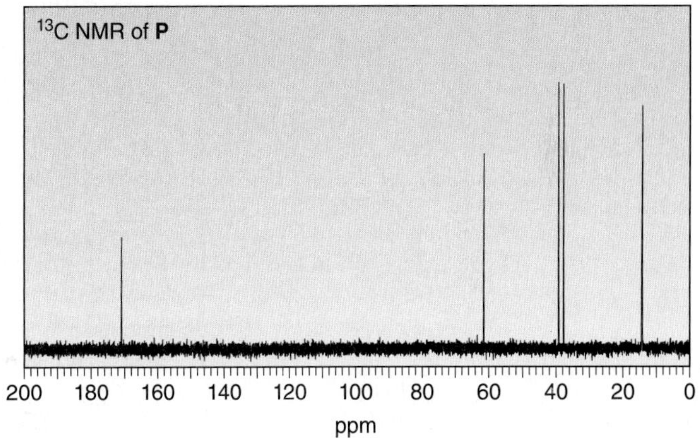

14.69 Treatment of butan-2-one ($CH_3COCH_2CH_3$) with strong base followed by CH_3I forms a compound **Q,** which gives a molecular ion in its mass spectrum at 86. The IR (> 1500 cm^{-1} only) and ^1H NMR spectrum of **Q** are given below. What is the structure of **Q?**

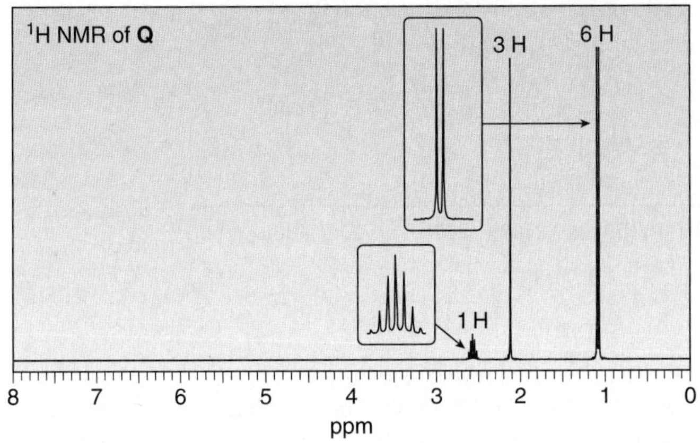

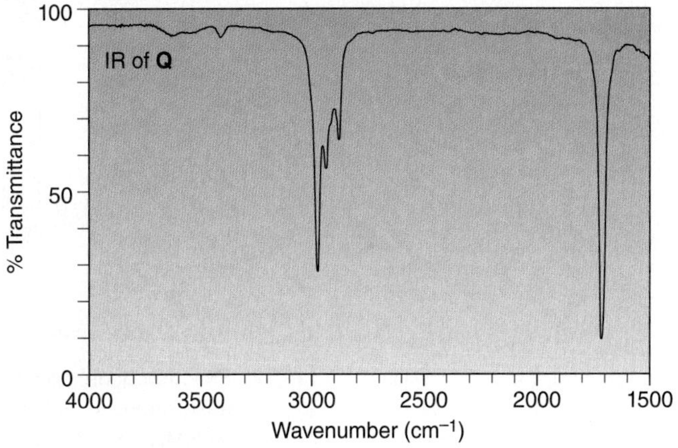

14.70 When 2-bromo-3,3-dimethylbutane is treated with $K^+ \, ^-OC(CH_3)_3$, a single product **T** having molecular formula C_6H_{12} is formed. When 3,3-dimethylbutan-2-ol is treated with H_2SO_4, the major product **U** has the same molecular formula. Given the following ^1H NMR data, what are the structures of **T** and **U?** Explain in detail the splitting patterns observed for the three split signals in **T.**

^1H NMR of **T:** 1.01 (singlet, 9 H), 4.82 (doublet of doublets, 1 H, J = 10, 1.7 Hz), 4.93 (doublet of doublets, 1 H, J = 18, 1.7 Hz), and 5.83 (doublet of doublets, 1 H, J = 18, 10 Hz) ppm

^1H NMR of **U:** 1.60 (singlet) ppm

14.71 Propose a structure consistent with each set of data.

a. A compound **X** (molecular formula $C_6H_{12}O_2$) gives a strong peak in its IR spectrum at 1740 cm^{-1}. The ^1H NMR spectrum of **X** shows only two singlets, including one at 3.5 ppm. The ^{13}C NMR spectrum is given below. Propose a structure for **X.**

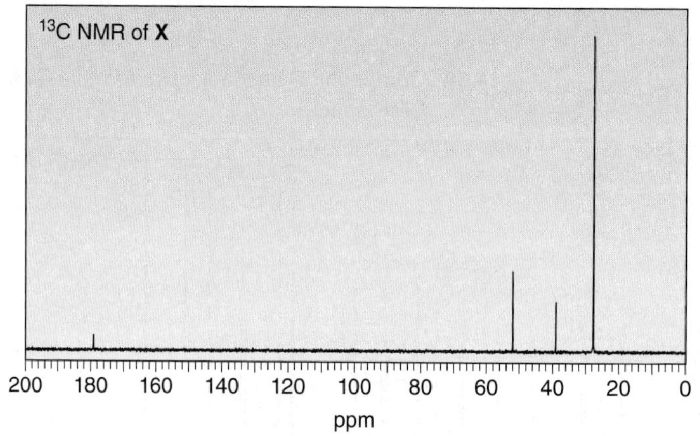

b. A compound **Y** (molecular formula C_6H_{10}) gives four lines in its ^{13}C NMR spectrum (27, 30, 67, and 93 ppm), and the IR spectrum given here. Propose a structure for **Y.**

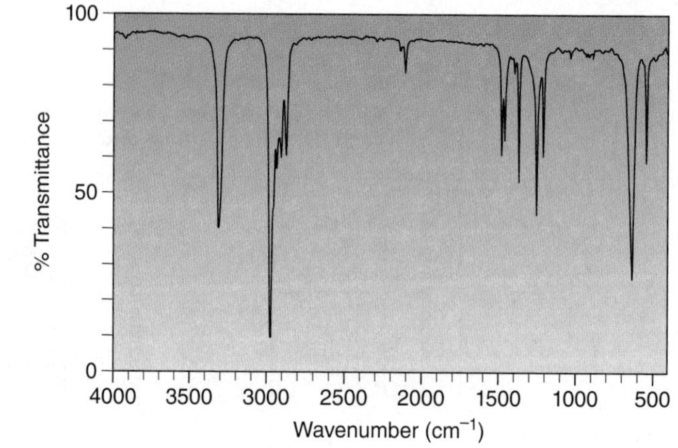

Challenge Problems

14.72 Reaction of unknown **A** with HCl forms chlorohydrin **B** as the major product. **A** shows no absorptions in its IR spectrum at 1700 cm^{-1} or 3600–3200 cm^{-1}, and gives the following ^1H NMR data: 1.4 (doublet, 3 H), 3.0 (quartet of doublets, 1 H), 3.5 (doublet, 1 H), 3.8 (singlet, 3 H), 6.9 (doublet, 2 H), and 7.2 (doublet, 2 H) ppm. (a) Propose a structure for **A,** including stereochemistry. (b) Explain why **B** is the major product in this reaction.

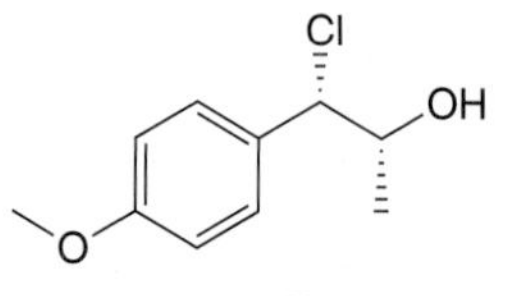

B

14.73 The ^1H NMR spectrum of *N,N*-dimethylformamide shows three singlets at 2.9, 3.0, and 8.0 ppm. Explain why the two CH_3 groups are not equivalent to each other, thus giving rise to two NMR signals.

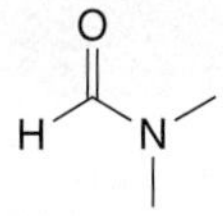

N,N-dimethylformamide

14.74 18-Annulene shows two signals in its ^1H NMR spectrum, one at 8.9 (12 H) and one at –1.8 (6 H) ppm. Using a similar argument to that offered for the chemical shift of benzene protons, explain why both shielded and deshielded values are observed for 18-annulene.

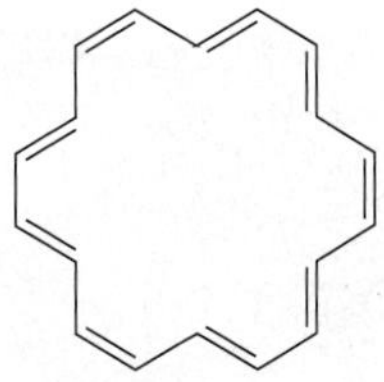

18-annulene

14.75 Explain why the ^{13}C NMR spectrum of 3-methylbutan-2-ol shows five signals.

14.76 Because ^{31}P has an odd mass number, ^{31}P nuclei absorb in the NMR and, in many ways, these nuclei behave similarly to protons in NMR spectroscopy. With this in mind, explain why the ^1H NMR spectrum of methyl dimethylphosphonate, $CH_3PO(OCH_3)_2$, consists of two doublets at 1.5 and 3.7 ppm.

14.77 Cyclohex-2-enone has two protons on its carbon–carbon double bond (labeled H_a and H_b) and two protons on the carbon adjacent to the double bond (labeled H_c). (a) If J_{ab} = 11 Hz and J_{bc} = 4 Hz, sketch the splitting pattern observed for each proton on the sp^2 hybridized carbons. (b) Despite the fact that H_a is located adjacent to an electron-withdrawing C=O, its absorption occurs upfield from the signal due to H_b (6.0 vs. 7.0 ppm). Offer an explanation.

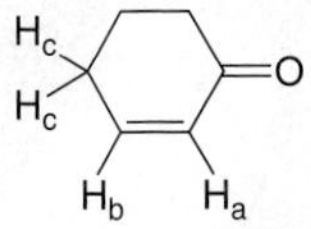

cyclohex-2-enone

15 Radical Reactions

15.1 Introduction
15.2 General features of radical reactions
15.3 Halogenation of alkanes
15.4 The mechanism of halogenation
15.5 Chlorination of other alkanes
15.6 Chlorination versus bromination
15.7 Halogenation as a tool in organic synthesis
15.8 The stereochemistry of halogenation reactions
15.9 Application: The ozone layer and CFCs
15.10 Radical halogenation at an allylic carbon
15.11 Application: Oxidation of unsaturated lipids
15.12 Application: Antioxidants
15.13 Radical addition reactions to double bonds
15.14 Polymers and polymerization

Polystyrene, an inexpensive polymer synthesized from the monomer styrene, $C_6H_5CH{=}CH_2$, is one of the six compounds that account for three-quarters of the synthetic polymers produced in the United States. The polystyrene foam used in packaging materials and drinking cups for hot beverages is called Styrofoam, a trademark of the Dow Chemical Company. Polystyrene is also used to form the housings of small kitchen appliances, televisions, computers, and CD cases. Although recycled polystyrene can be molded into trays and trash cans, the polystyrene used in food packaging and beverage cups is contaminated with food, making it difficult to clean and recycle. In Chapter 15, we learn about the synthesis of polymers like polystyrene.

A small but significant group of reactions involves the homolysis of nonpolar bonds to form highly reactive **radical intermediates.** Although they are unlike other organic reactions, radical transformations are important in many biological and industrial processes. The gases O_2 and NO (nitric oxide) are both radicals. Many oxidation reactions with O_2 involve radical intermediates, and biological processes mediated by NO such as blood clotting and neurotransmission may involve radicals. Many useful industrial products such as Styrofoam and polyethylene are prepared by radical processes.

In Chapter 15 we examine the cleavage of nonpolar bonds by radical reactions.

15.1 Introduction

Radicals were first discussed in Section 6.3.

- A *radical* is a reactive intermediate with a single unpaired electron, formed by homolysis of a covalent bond.

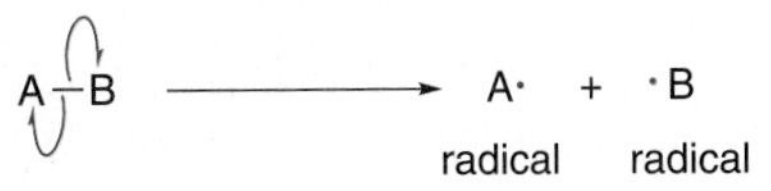

A–B
A·
+
·B
radical
radical

A radical contains an atom that does not have an octet of electrons, making it reactive and unstable. Radical processes involve single electrons, so half-headed arrows are used to show the movement of electrons. **One half-headed arrow is used for each electron.**

Carbon radicals are classified as **primary (1°), secondary (2°),** or **tertiary (3°)** by the number of R groups bonded to the carbon with the unpaired electron. A carbon radical is ***sp*^{2} hybridized** and **trigonal planar,** like *sp*^{2} hybridized carbocations. The unhybridized *p* orbital contains the unpaired electron and extends above and below the trigonal planar carbon.

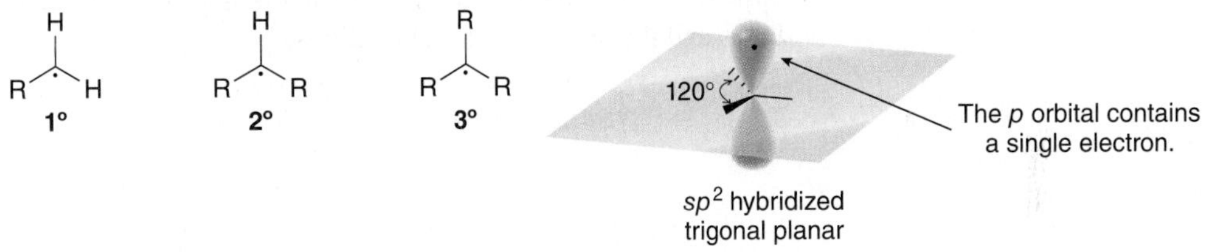

1°
2°
3°
120°
The p orbital contains a single electron.
sp² hybridized
trigonal planar

Bond dissociation energies for the cleavage of C–H bonds are used as a measure of radical stability. For example, two different radicals can be formed by cleavage of the C–H bonds in $CH_3CH_2CH_3$.

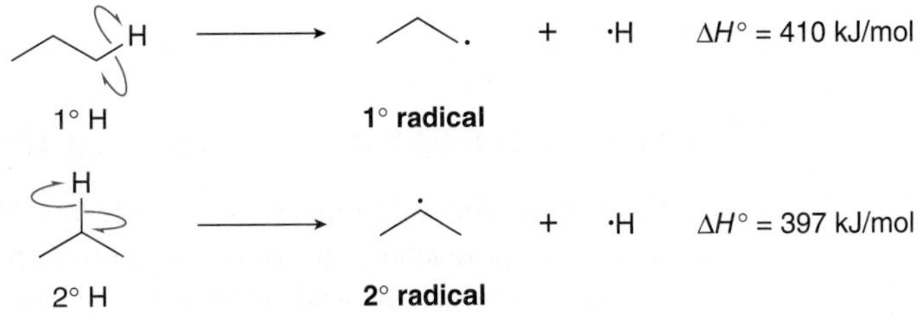

1° H
1° radical
+ ·H
ΔH° = 410 kJ/mol
2° H
2° radical
+ ·H
ΔH° = 397 kJ/mol

Cleavage of the stronger 1° C–H bond to form the 1° radical ($CH_3CH_2CH_2\cdot$) requires more energy than cleavage of the weaker 2° C–H bond to form the 2° radical [$(CH_3)_2CH\cdot$]—410 versus 397 kJ/mol. This makes the 2° radical more stable, because less energy is required for its

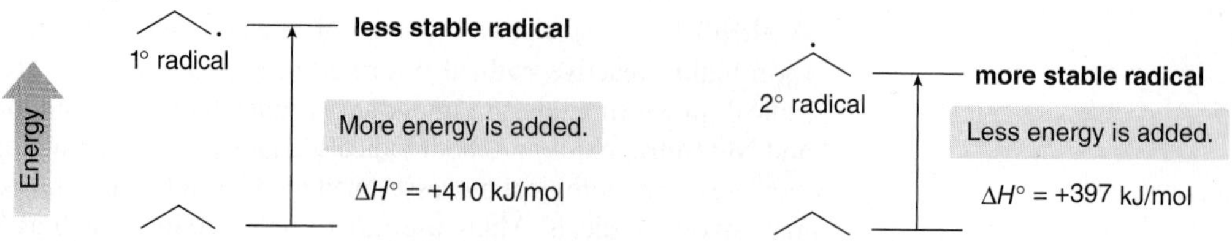

Figure 15.1 The relative stability of 1° and 2° carbon radicals

formation, as illustrated in Figure 15.1. Thus, **cleavage of the weaker bond forms the more stable radical,** a specific example of a general trend.

- **The stability of a radical increases as the number of alkyl groups bonded to the radical carbon increases.**

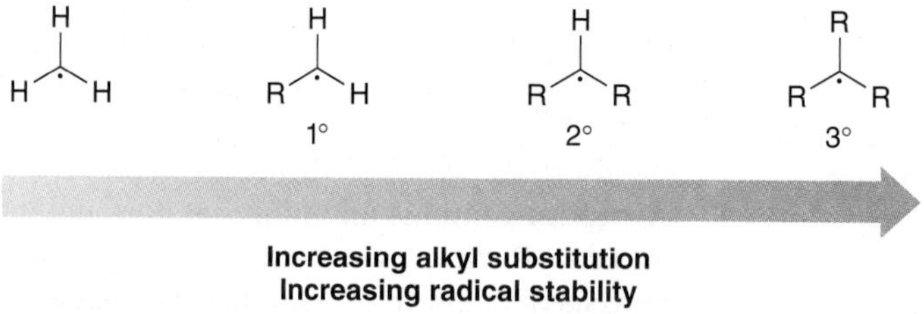

The **lower** the bond dissociation energy for a C–H bond, the **more stable** the resulting carbon radical.

Thus, a 3° radical is more stable than a 2° radical, and a 2° radical is more stable than a 1° radical. Increasing alkyl substitution increases radical stability in the same way it increases carbocation stability. Alkyl groups are more polarizable than hydrogen atoms, so they can more easily donate electron density to the electron-deficient carbon radical, thus increasing stability.

Unlike carbocations, however, **less stable radicals generally do *not* rearrange to more stable radicals.** This difference can be used to distinguish between reactions involving radical intermediates and those involving carbocations.

Problem 15.1 Classify each radical as 1°, 2°, or 3°.

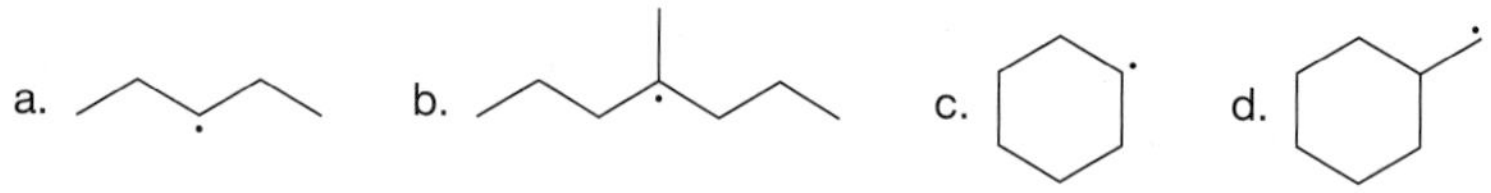

Problem 15.2 Draw the most stable radical that can result from cleavage of a C–H bond in each molecule.

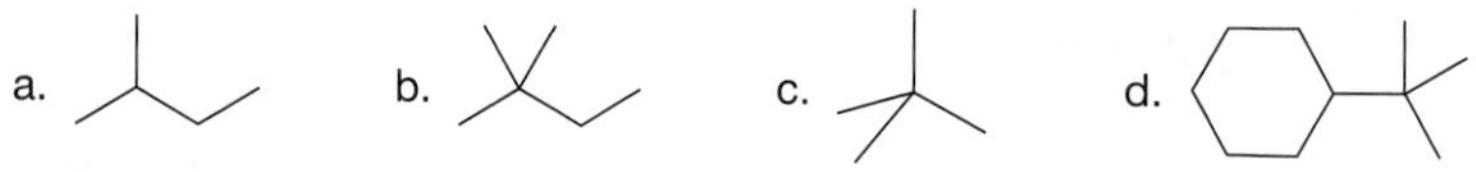

15.2 General Features of Radical Reactions

Radicals are formed from covalent bonds by adding energy in the form of **heat (Δ)** or **light (*h*ν).** Some radical reactions are carried out in the presence of a **radical initiator,** a compound that contains an especially weak bond that serves as a source of radicals. **Peroxides,** compounds with the general structure **RO–OR,** are the most commonly used radical initiators. Heating a peroxide readily causes homolysis of the weak O–O bond, forming two RO· radicals.

15.2A Two Common Reactions of Radicals

Radicals undergo two main types of reactions: **they react with σ bonds,** and **they add to π bonds,** in both cases achieving an octet of electrons.

[1] Reaction of a Radical X· with a C–H Bond

A radical X· abstracts a hydrogen atom from a C–H σ bond to form H–X and a carbon radical. One electron from the C–H bond is used to form the new H–X bond, and the other electron in the C–H bond remains on carbon. The result is that the original radical X· is now surrounded by an octet of electrons, and a new radical is formed.

- One electron in H–X comes from the radical.
- One electron in H–X comes from the C–H bond.

This radical reaction is typically seen with the nonpolar C–H bonds of **alkanes,** which cannot react with polar or ionic electrophiles and nucleophiles.

[2] Reaction of a Radical X· with a C=C

A radical X· also adds to the π bond of a carbon–carbon double bond. One electron from the double bond is used to form a new C–X bond, and the other electron remains on the other carbon originally part of the double bond.

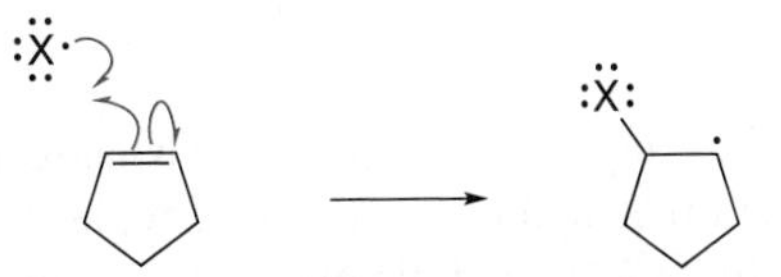

- One electron in C–X comes from the radical.
- One electron in C–X comes from the π bond.

Whenever a radical reacts with a stable single or double bond, a new radical is formed in the products.

Although the electron-rich double bond of an **alkene** reacts with electrophiles by ionic addition mechanisms, it also reacts with radicals because these reactive intermediates are also electron deficient.

15.2B Two Radicals Reacting with Each Other

A radical, once formed, rapidly reacts with whatever is available. Usually that means a stable σ or π bond. Occasionally, however, two radicals come into contact with each other, and they react to form a σ bond.

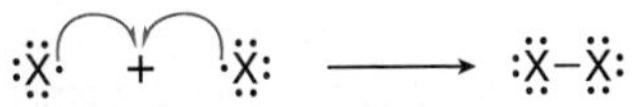

- One electron in X–X comes from each radical.

The reaction of a radical with oxygen, a diradical in its ground state electronic configuration, is another example of two radicals reacting with each other. In this case, the reaction of O_2 with X· forms a new radical, thus preventing X· from reacting with an organic substrate.

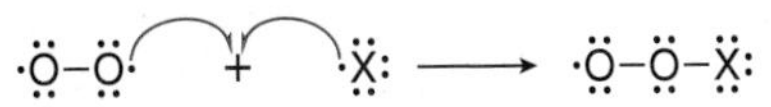

Compounds that prevent radical reactions from occurring are called *radical inhibitors* or *radical scavengers*. Besides O_2, vitamin E and related compounds, discussed in Section 15.12, are radical scavengers, too. The fact that these compounds inhibit a reaction often suggests that the reaction occurs via radical intermediates.

Problem 15.3 Draw the products formed when a chlorine atom (Cl·) reacts with each species.

a.  b. $CH_2=CH_2$ c. :Cl· d. O_2

15.3 Halogenation of Alkanes

In the presence of light or heat, alkanes react with halogens to form alkyl halides. Halogenation is a **radical substitution reaction,** because a halogen atom X replaces a hydrogen via a mechanism that involves radical intermediates.

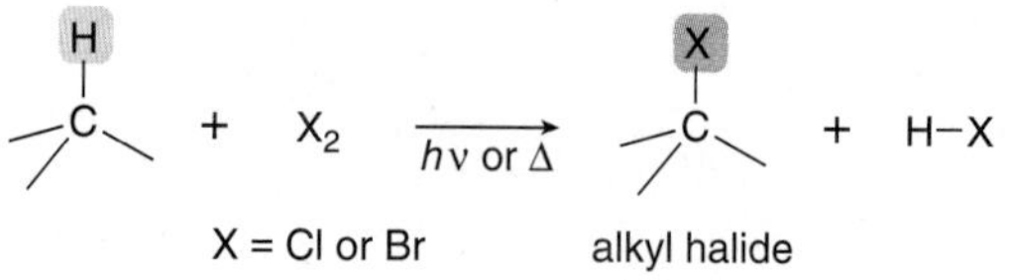
H
X
+ X_2
$h\nu$ or Δ
+ H–X
X = Cl or Br
alkyl halide

Halogenation of alkanes is useful only with Cl_2 and Br_2. Reaction with F_2 is too violent and reaction with I_2 is too slow to be useful. With an alkane that has more than one type of hydrogen atom, a mixture of alkyl halides may result (Reaction [2]).

When asked to draw the products of halogenation of an alkane, **draw the products of monohalogenation only,** unless specifically directed to do otherwise.

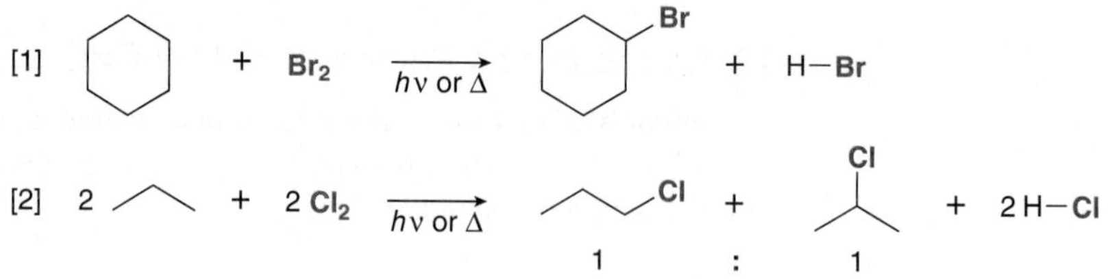
[1]
+ Br_2
$h\nu$ or Δ
Br
+ H–Br
[2] 2
+ 2 Cl_2
$h\nu$ or Δ
Cl
+
Cl
+ 2 H–Cl
1 : 1

In these examples of halogenation, a halogen has replaced a single hydrogen atom on the alkane. Can the other hydrogen atoms be replaced, too? Figure 15.2 shows that when CH_4 is treated with excess Cl_2, all four hydrogen atoms can be successively replaced by Cl to form CCl_4. **Monohalogenation**—the substitution of a single H by X—can be achieved experimentally by adding halogen X_2 to an excess of alkane.

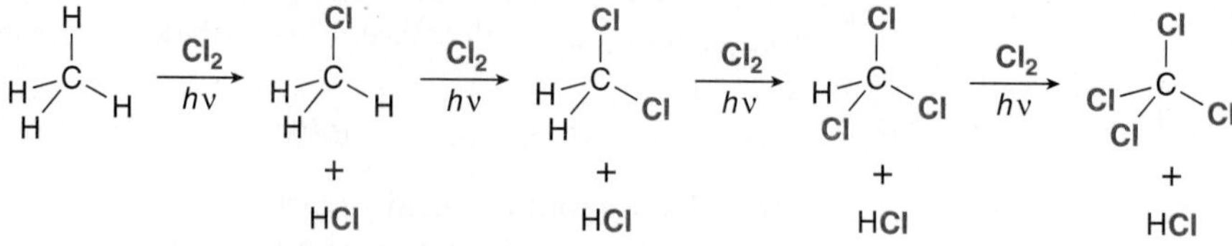
Cl_2
$h\nu$
+
HCl
+
HCl
+
HCl
+
HCl

Figure 15.2
Complete halogenation of CH_4 using excess Cl_2

Sample Problem 15.1 Draw all the constitutional isomers formed by monohalogenation of $(CH_3)_2CHCH_2CH_3$ with Cl_2 and $h\nu$.

Solution

Substitute Cl for H on every carbon, and then check to see if any products are identical. The starting material has five C's, but replacement of one H atom on two C's gives the same product. Thus, **$(CH_3)_2CHCH_2CH_3$ affords four monochloro substitution products.**

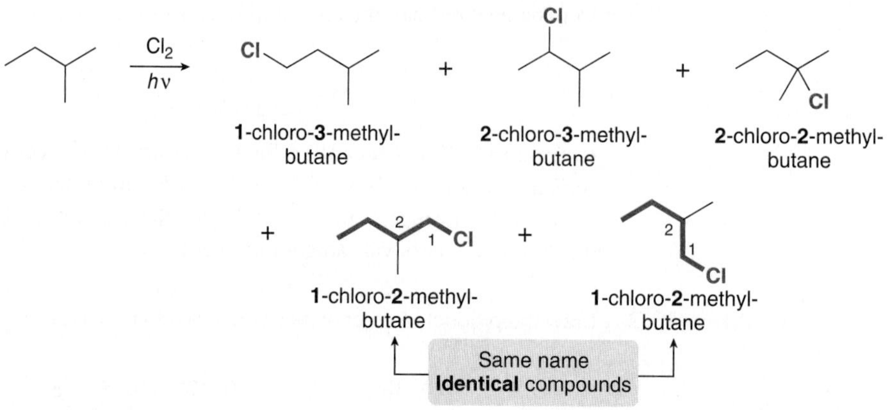
Cl_2
$h\nu$
1-chloro-3-methylbutane
2-chloro-3-methylbutane
2-chloro-2-methylbutane
1-chloro-2-methylbutane
1-chloro-2-methylbutane
Same name
Identical compounds

Problem 15.4 Draw all constitutional isomers formed by monochlorination of each alkane.

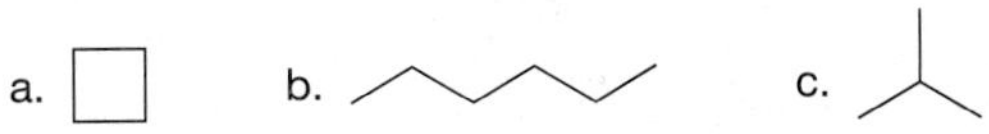

Problem 15.5 Compounds **A** and **B** are isomers having molecular formula C_5H_{12}. Heating **A** with Cl_2 gives a single product of monohalogenation, whereas heating **B** under the same conditions forms three constitutional isomers. What are the structures of **A** and **B?**

15.4 The Mechanism of Halogenation

Unlike nucleophilic substitution, which proceeds by two different mechanisms depending on the starting material and reagent, all halogenation reactions of alkanes—regardless of the halogen and alkane used—proceed by the *same* mechanism. Three facts about halogenation suggest that the mechanism involves radical, not ionic, intermediates.

Fact	Explanation
[1] Light, heat, or added peroxide is necessary for the reaction.	• Light or heat provides the energy needed for homolytic bond cleavage to form radicals. Breaking the weak O–O bond of peroxides initiates radical reactions as well.
[2] O_2 inhibits the reaction.	• The diradical O_2 removes radicals from a reaction mixture, thus preventing reaction.
[3] No rearrangements are observed.	• Radicals do not rearrange.

15.4A The Steps of Radical Halogenation

The chlorination of cyclopentane illustrates the **three distinct parts of radical halogenation** (Mechanism 15.1):

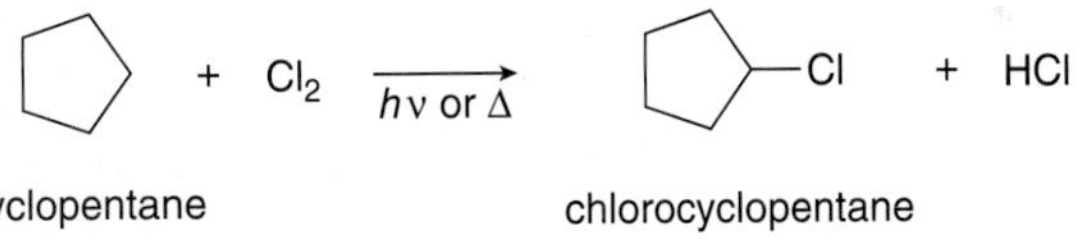

- *Initiation:* **Two radicals are formed by homolysis of a σ bond and this begins the reaction.**
- *Propagation:* **A radical reacts with another reactant to form a new σ bond and another radical.**
- *Termination:* **Two radicals combine to form a stable bond. Removing radicals from the reaction mixture without generating any new radicals stops the reaction.**

Although initiation generates the Cl· radicals needed to begin the reaction, the propagation steps ([2] and [3]) form the two reaction products—chlorocyclopentane and HCl. Once the process has begun, propagation occurs over and over without the need for Step [1] to occur. **A mechanism such as radical halogenation that involves two or more repeating steps is called a *chain mechanism.*** Each propagation step involves a reactive radical abstracting an atom from a stable bond to form a new bond and another radical that continues the chain.

Usually a radical reacts with a stable bond to propagate the chain, but occasionally two radicals combine, and this reaction terminates the chain. Depending on the reaction and the reaction conditions, some radical chain mechanisms can repeat thousands of times before termination occurs.

Mechanism 15.1 Radical Halogenation of Alkanes

Part [1] Initiation

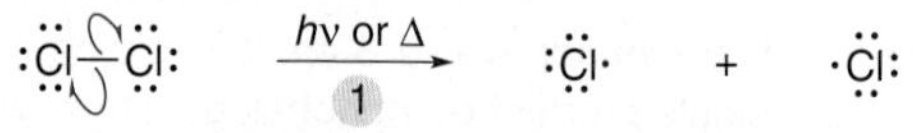

1 Bond cleavage forms two radicals. Homolysis of the weakest bond (Cl–Cl) requires light or heat and forms two chlorine radicals.

Part [2] Propagation

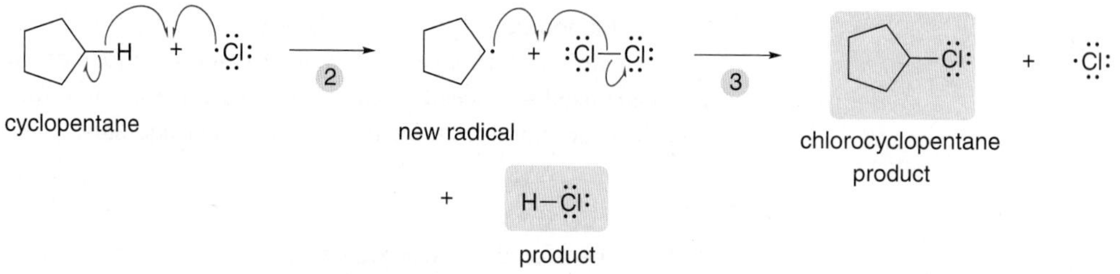

2 The Cl· radical abstracts a hydrogen from cyclopentane to form HCl (a reaction product) and a new carbon radical.

3 The carbon radical abstracts a chlorine atom from Cl_2 to form chlorocyclopentane (a reaction product) and Cl·. Because Cl· is a reactant in Step [2], Steps [2] and [3] can occur repeatedly without additional initiation (Step [1]).

Part [3] Termination

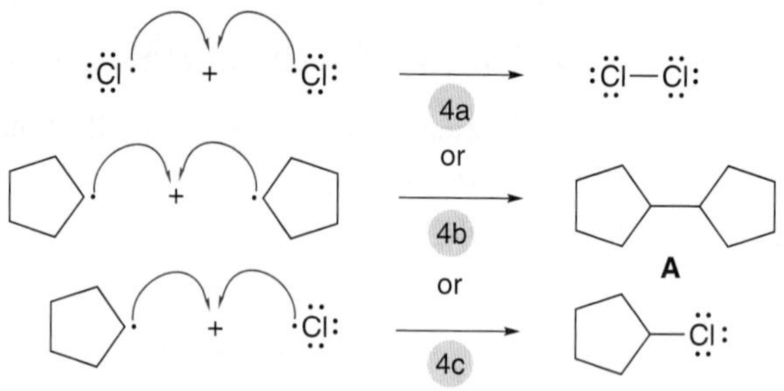

4 Termination of the chain occurs when any two radicals combine to form a bond.

Termination Step [4a] forms Cl_2, a reactant, whereas Step [4c] forms chlorocyclopentane, one of the reaction products. Termination Step [4b] forms **A,** which is neither a reactant nor a desired product. The formation of a small quantity of **A,** however, is evidence that radicals are formed in the reaction.

The most important steps of radical halogenation are those that lead to product formation—the propagation steps—so subsequent discussion of this reaction concentrates on these steps only.

Problem 15.6 Using Mechanism 15.1 as a guide, write the mechanism for the reaction of CH_4 with Br_2 to form CH_3Br and HBr. Classify each step as initiation, propagation, or termination.

15.4B Energy Changes During the Chlorination of Ethane

The chlorination of ethane illustrates how bond dissociation energies (Section 6.4) can be used to calculate $\Delta H°$ in chain propagation.

$$CH_3CH_3 + Cl_2 \xrightarrow{h\nu \text{ or } \Delta} CH_3CH_2Cl + HCl$$

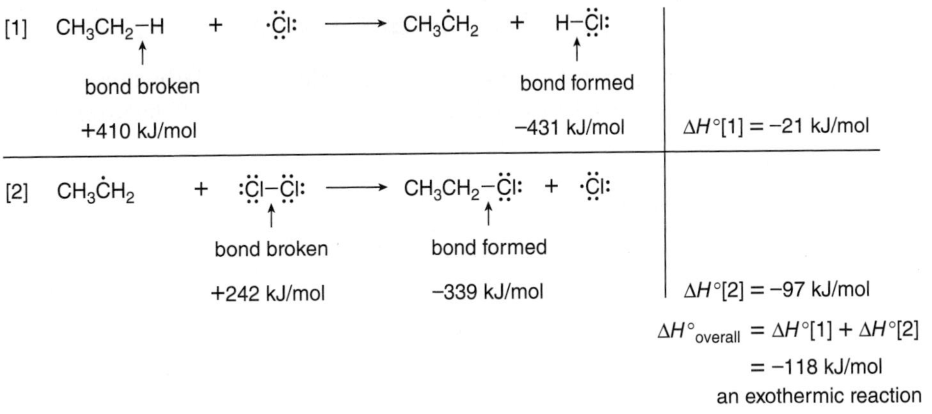

Figure 15.3 Energy changes in the propagation steps during the chlorination of ethane

As shown in Figure 15.3, chain propagation consists of the same two steps drawn in Mechanism 15.1: abstraction of a hydrogen atom to form $CH_3CH_2\cdot$ and HCl, followed by abstraction of a chlorine atom by $CH_3CH_2\cdot$ to form CH_3CH_2Cl and a chlorine radical (Cl·). The $\Delta H°$ for each step is negative, making the overall $\Delta H°$ negative and the reaction exothermic. Because the transition state for the first propagation step is higher in energy than the transition state for the second propagation step, the **first step is rate-determining.** Both of these facts are illustrated in the energy diagram in Figure 15.4.

Problem 15.7 Calculate $\Delta H°$ for the rate-determining step of the reaction of CH_4 with I_2. Explain why this result illustrates that this reaction is extremely slow.

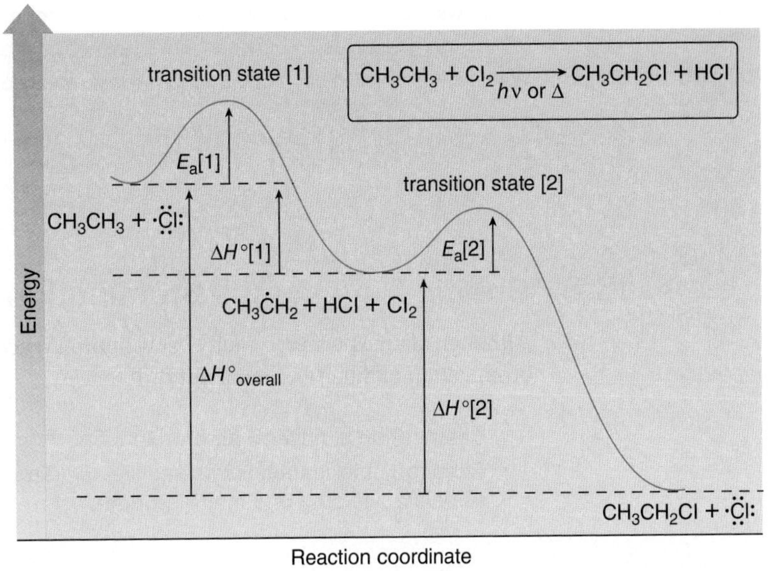

Figure 15.4 Energy diagram for the propagation steps in the chlorination of ethane

- Because radical halogenation consists of two propagation steps, the energy diagram has two energy barriers.
- The first step is rate-determining because its transition state is at higher energy.
- The reaction is exothermic because $\Delta H°_{overall}$ is negative.

15.5 Chlorination of Other Alkanes

Recall from Section 15.3 that the chlorination of $CH_3CH_2CH_3$ affords a 1:1 mixture of $CH_3CH_2CH_2Cl$ (formed by removal of a 1° hydrogen) and $(CH_3)_2CHCl$ (formed by removal of a 2° hydrogen).

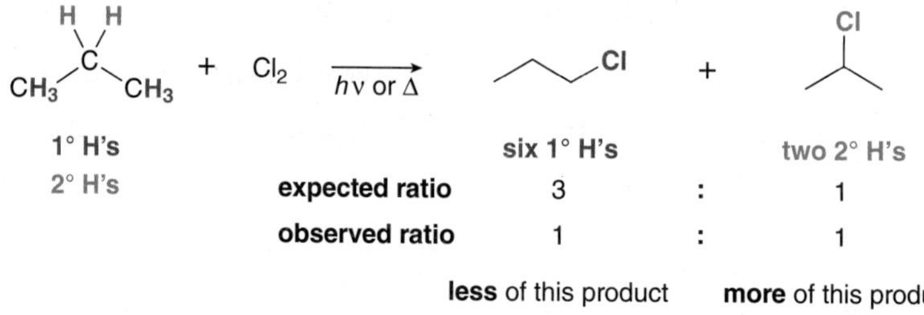
H H
C
CH3
CH3
+
Cl2
hν or Δ
Cl
+
Cl
1° H's
2° H's
six 1° H's
two 2° H's
expected ratio
3 : 1
observed ratio
1 : 1
less of this product
more of this product

$CH_3CH_2CH_3$ has six 1° hydrogen atoms and only two 2° hydrogens, so the expected product ratio of $CH_3CH_2CH_2Cl$ to $(CH_3)_2CHCl$ (assuming all hydrogens are *equally* reactive) is 3:1. Because the observed ratio is 1:1, however, the 2° C–H bonds must be more reactive; that is, **it must be easier to homolytically cleave a 2° C–H bond than a 1° C–H bond.** Recall from Section 15.2 that 2° C–H bonds are weaker than 1° C–H bonds. Thus,

- **The weaker the C–H bond, the more readily the hydrogen atom is removed in radical halogenation.**

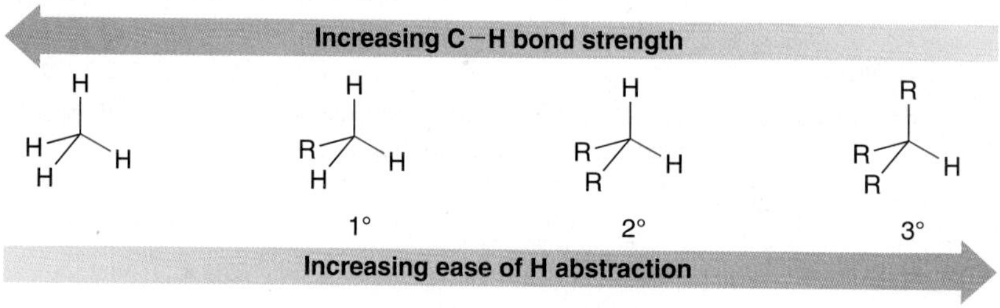
Increasing C–H bond strength
H
H
H
H
H
R
H
H
H
R
R
H
R
R
R
H
1°
2°
3°
Increasing ease of H abstraction

When alkanes react with Cl_2, a mixture of products results, with more product formed by cleavage of the weaker C–H bond than you would expect on statistical grounds.

Problem 15.8 Which C–H bond in each compound is most readily broken during radical halogenation?

a. 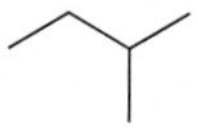 b. 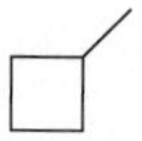 c. 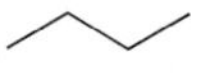

15.6 Chlorination Versus Bromination

Although alkanes undergo radical substitution reactions with both Cl_2 and Br_2, chlorination and bromination exhibit two important differences:

- Chlorination is *faster* than bromination.
- Although chlorination is *unselective,* yielding a mixture of products, bromination is often *selective,* yielding one major product.

For example, propane reacts rapidly with Cl_2 to form a 1:1 mixture of 1° and 2° alkyl chlorides. On the other hand, propane reacts with Br_2 much more slowly and forms 99% $(CH_3)_2CHBr$.

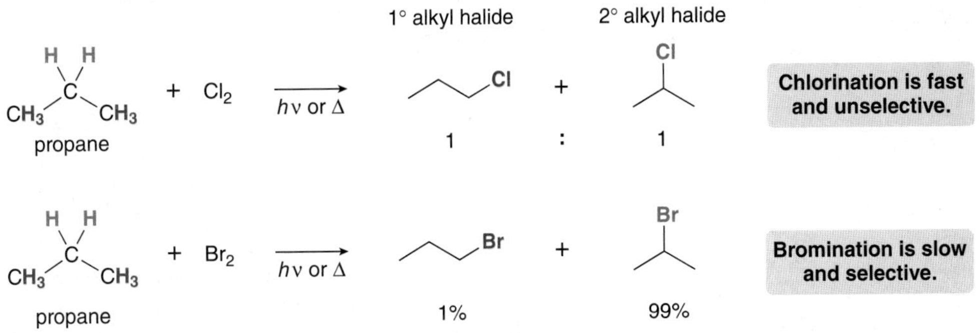

- **In bromination, the major (and sometimes exclusive) product results from cleavage of the *weakest* C–H bond.**

Sample Problem 15.2 Draw the major product formed when 3-ethylpentane is heated with Br_2.

Solution

Keep in mind: **the more substituted the carbon atom, the weaker the C–H bond.** The major bromination product in 3-ethylpentane is formed by cleavage of the sole **3° C–H bond,** its weakest C–H bond.

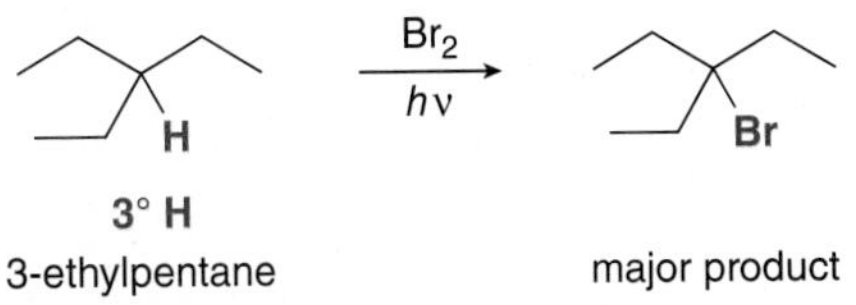

Problem 15.9 Draw the major product formed when each cycloalkane is heated with Br_2.

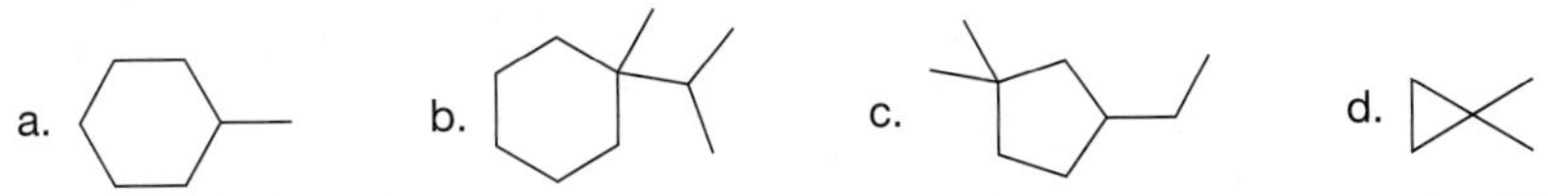

To explain the difference between chlorination and bromination, we return to the Hammond postulate (Section 7.14). The **rate-determining step in halogenation is the abstraction of a hydrogen atom by the halogen radical,** so we must compare these steps for bromination and chlorination. Keep in mind:

- **Transition states in endothermic reactions resemble the products. The more stable product is formed faster.**
- **Transition states in exothermic reactions resemble the starting materials. The relative stability of the products does not greatly affect the relative energy of the transition states, so a mixture of products often results.**

Bromination: $CH_3CH_2CH_3 + Br_2$

A bromine radical can abstract either a 1° or a 2° hydrogen from propane, generating either a 1° radical or a 2° radical. Calculating $\Delta H°$ using bond dissociation energies reveals that both reactions are **endothermic,** but **it takes less energy to form the more stable 2° radical.**

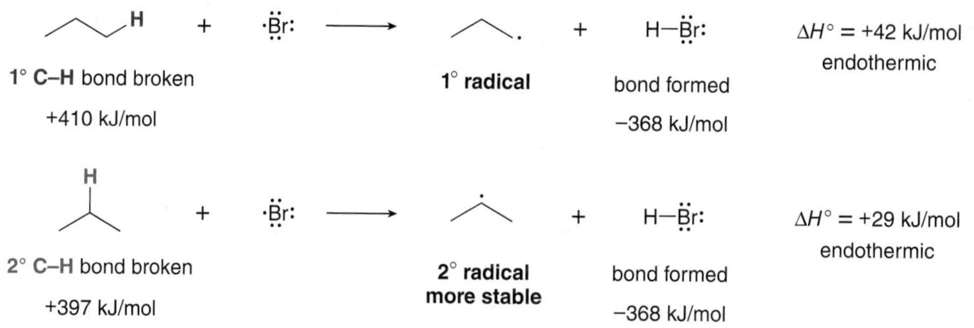

H
+
·Br:
+
H–Br:
ΔH° = +42 kJ/mol
endothermic
1° C–H bond broken
+410 kJ/mol
1° radical
bond formed
−368 kJ/mol
H
+
·Br:
+
H–Br:
ΔH° = +29 kJ/mol
endothermic
2° C–H bond broken
+397 kJ/mol
2° radical
more stable
bond formed
−368 kJ/mol

According to the Hammond postulate, the transition state of an endothermic reaction resembles the products, so the energy of activation to form the more stable 2° radical is lower and it is formed faster, as shown in the energy diagram in Figure 15.5. Because the 2° radical $[(CH_3)_2CH\cdot]$ is converted to 2-bromopropane $[(CH_3)_2CHBr]$ in the second propagation step, this **2° alkyl halide is the major product of bromination.**

- **Conclusion: Because the rate-determining step in bromination is endothermic, the more stable radical is formed faster, and often a single radical halogenation product predominates.**

Figure 15.5
Energy diagram for a selective endothermic reaction

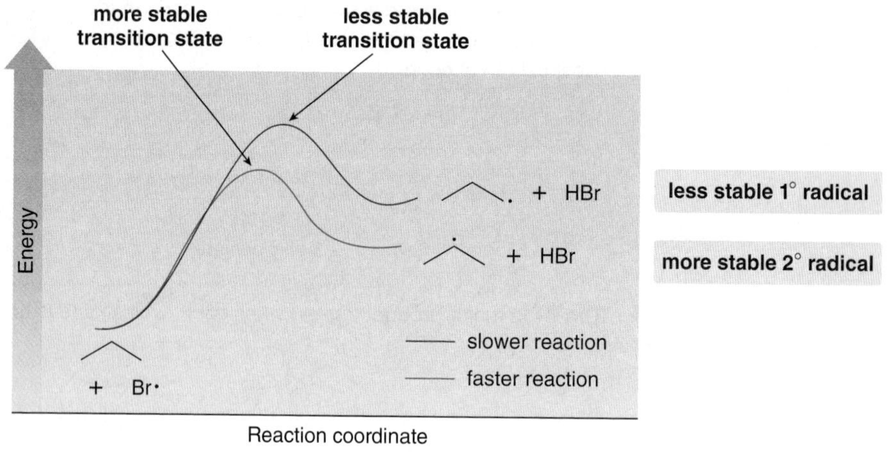

more stable
transition state
less stable
transition state
+ HBr
less stable 1° radical
+ HBr
more stable 2° radical
Energy
slower reaction
faster reaction
+ Br·
Reaction coordinate

- The transition state to form the less stable 1° radical ($CH_3CH_2CH_2\cdot$) is higher in energy than the transition state to form the more stable 2° radical $[(CH_3)_2CH\cdot]$. Thus, **the 2° radical is formed faster.**

Chlorination: $CH_3CH_2CH_3 + Cl_2$

A chlorine radical can also abstract either a 1° or a 2° hydrogen from propane, generating either a 1° radical or a 2° radical. Calculating $\Delta H°$ using bond dissociation energies reveals that both reactions are **exothermic.**

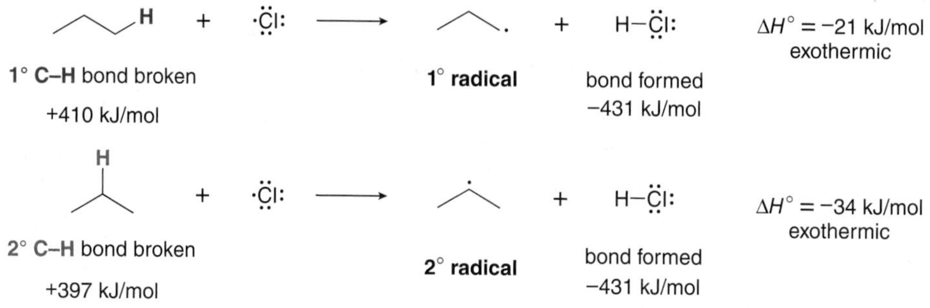

H
+
·Cl:
+
H–Cl:
ΔH° = −21 kJ/mol
exothermic
1° C–H bond broken
+410 kJ/mol
1° radical
bond formed
−431 kJ/mol
H
+
·Cl:
+
H–Cl:
ΔH° = −34 kJ/mol
exothermic
2° C–H bond broken
+397 kJ/mol
2° radical
bond formed
−431 kJ/mol

Figure 15.6
Energy diagram for a nonselective exothermic reaction

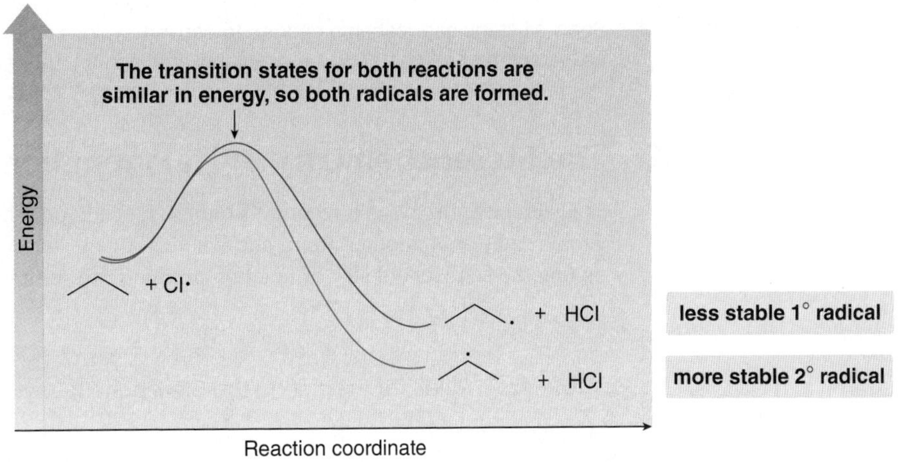

Because chlorination has an *exothermic* rate-determining step, the transition state to form both radicals resembles the same starting material, $CH_3CH_2CH_3$. As a result, the relative stability of the two radicals is much less important and both radicals are formed. An energy diagram for these processes is drawn in Figure 15.6. Because the 1° and 2° radicals are converted to 1-chloropropane ($CH_3CH_2CH_2Cl$) and 2-chloropropane [$(CH_3)_2CHCl$], respectively, in the second propagation step, **both alkyl halides are formed in chlorination.**

- **Conclusion: Because the rate-determining step in chlorination is exothermic, the transition state resembles the starting material, both radicals are formed, and a mixture of products results.**

Problem 15.10 Reaction of $(CH_3)_3CH$ with Cl_2 forms two products: $(CH_3)_2CHCH_2Cl$ (63%) and $(CH_3)_3CCl$ (37%). Why is the major product formed by cleavage of the stronger 1° C–H bond?

15.7 Halogenation as a Tool in Organic Synthesis

Halogenation is a useful tool because it adds a functional group to a previously unfunctionalized molecule, making an **alkyl halide.** These alkyl halides can then be converted to alkenes by elimination, and to alcohols and ethers by nucleophilic substitution.

Sample Problem 15.3 Show how cyclohexane can be converted to cyclohexene by a stepwise sequence.

cyclohexane cyclohexene

Solution

There is no one-step method to convert an alkane to an alkene. A two-step method is needed:

[1] **Radical halogenation** produces an alkyl halide.

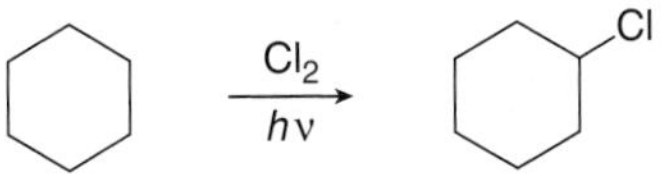

[2] **Elimination of HCl** with a strong base produces cyclohexene.

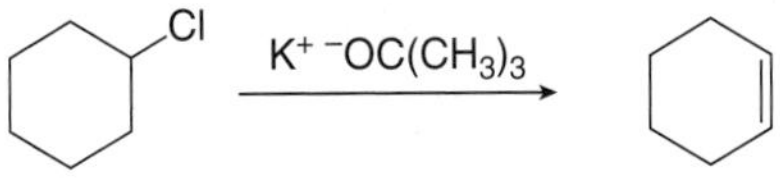

Problem 15.11 Synthesize each compound from $(CH_3)_3CH$.

a. 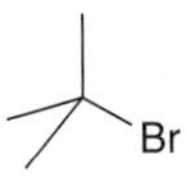

b. 

c. 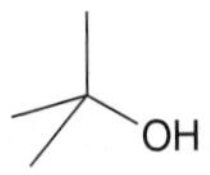

Problem 15.12 Show all steps and reagents needed to convert cyclohexane into each compound: (a) the two enantiomers of *trans*-1,2-dibromocyclohexane; and (b) 1,2-epoxycyclohexane.

15.8 The Stereochemistry of Halogenation Reactions

The stereochemistry of a reaction product depends on whether the reaction occurs at a stereogenic center or at another atom, and whether a new stereogenic center is formed. The rules predicting the stereochemistry of reaction products are summarized in Table 15.1.

Table 15.1 Rules for Predicting the Stereochemistry of Reaction Products

Starting material	Result
Achiral	• An achiral starting material always gives either an achiral or a racemic product.
Chiral	• If a reaction does not occur at a stereogenic center, the configuration at a stereogenic center is retained in the product. • If a reaction occurs at a stereogenic center, we must know the mechanism to predict the stereochemistry of the product.

15.8A Halogenation of an Achiral Starting Material

Halogenation of the **achiral starting material $CH_3CH_2CH_2CH_3$** forms two constitutional isomers by replacement of either a 1° or 2° hydrogen.

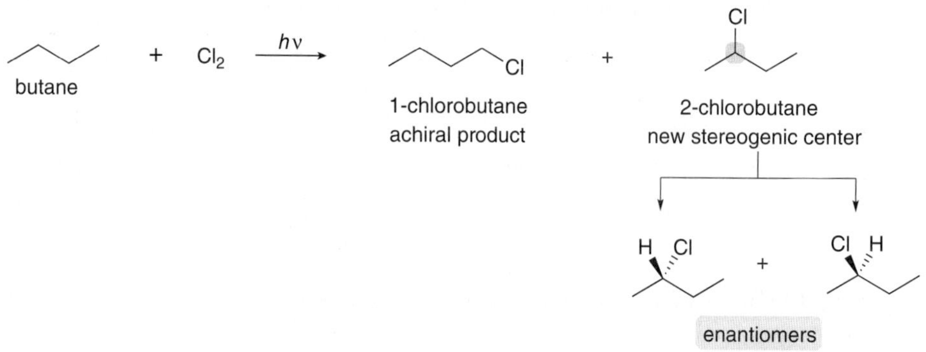

- 1-Chlorobutane ($CH_3CH_2CH_2CH_2Cl$) has no stereogenic center, so it is an **achiral** compound.
- 2-Chlorobutane [$CH_3CH(Cl)CH_2CH_3$] has a new stereogenic center, so an **equal amount of two enantiomers** must form—**a racemic mixture.**

A racemic mixture results when a new stereogenic center is formed because the first propagation step generates a **planar, sp^2 hybridized radical.** Cl_2 then reacts with the planar radical from either the front or back side to form an equal amount of two enantiomers.

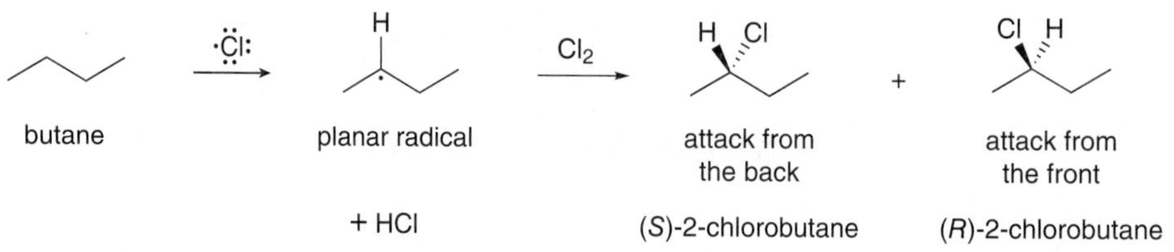

Thus, the achiral starting material butane forms an achiral product (1-chlorobutane) and a racemic mixture of two enantiomers [(*R*)- and (*S*)-2-chlorobutane].

15.8B Halogenation of a Chiral Starting Material

Let's now examine chlorination of the chiral starting material (*R*)-2-bromobutane at C2 and C3.

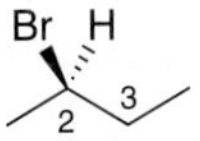
Br H
2
3

(*R*)-2-bromobutane

Chlorination at C2 occurs at the stereogenic center. Abstraction of a hydrogen atom at C2 forms a trigonal planar sp^2 hybridized radical that is now achiral. This achiral radical then reacts with Cl_2 from either side to form a new stereogenic center, resulting in an **equal amount of two enantiomers—a racemic mixture.**

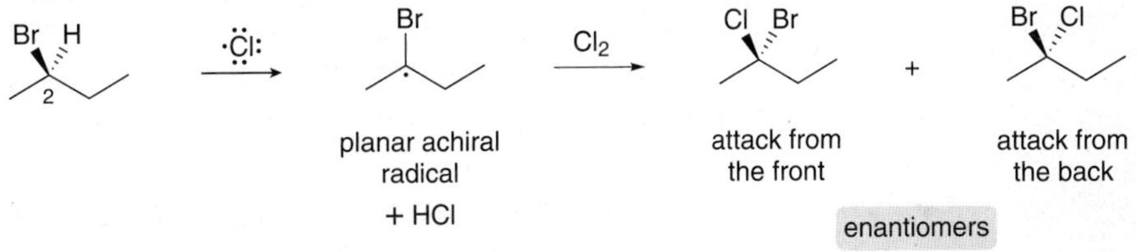
Br H
2
·Cl:
Br
Cl2
Cl Br
+
Br Cl
planar achiral radical
+ HCl
attack from the front
attack from the back
enantiomers

- **Radical halogenation reactions occur with racemization at a stereogenic center.**

Chlorination at C3 does *not* occur at the stereogenic center, but it forms a new stereogenic center. Because no bond is broken to the stereogenic center at C2, **its configuration is retained** during the reaction. Abstraction of a hydrogen atom at C3 forms a trigonal planar sp^2 hybridized radical that still contains this stereogenic center. Reaction of the radical with Cl_2 from either side forms a new stereogenic center, so the products have two stereogenic centers: the configuration at C2 is the same in both compounds, but the configuration at C3 is different, making them **diastereomers.**

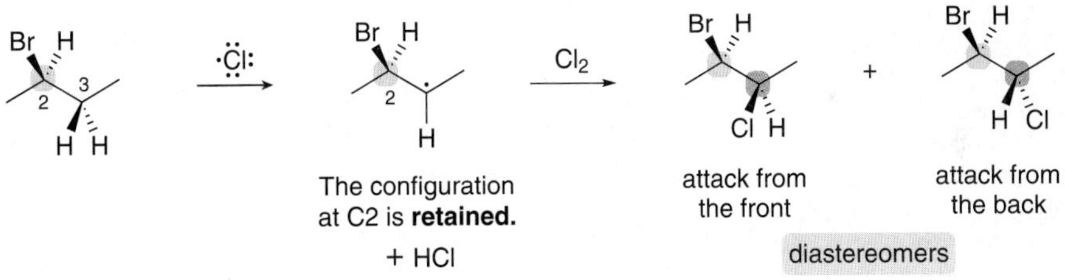
Br H
2
3
H H
·Cl:
Br H
2
H
Cl2
Br H
Cl H
+
Br H
H Cl
The configuration at C2 is retained.
+ HCl
attack from the front
attack from the back
diastereomers

Thus, four isomers are formed by chlorination of (*R*)-2-bromobutane at C2 and C3. Attack at the stereogenic center (C2) gives a product with one stereogenic center, resulting in a mixture of enantiomers. Attack at C3 forms a new stereogenic center, giving a mixture of diastereomers.

Problem 15.13 What products are formed from monochlorination of (*R*)-2-bromobutane at C1 and C4? Assign *R* and *S* designations to each stereogenic center.

Problem 15.14 Draw the monochlorination products formed when each compound is heated with Cl_2. Include the stereochemistry at any stereogenic center.

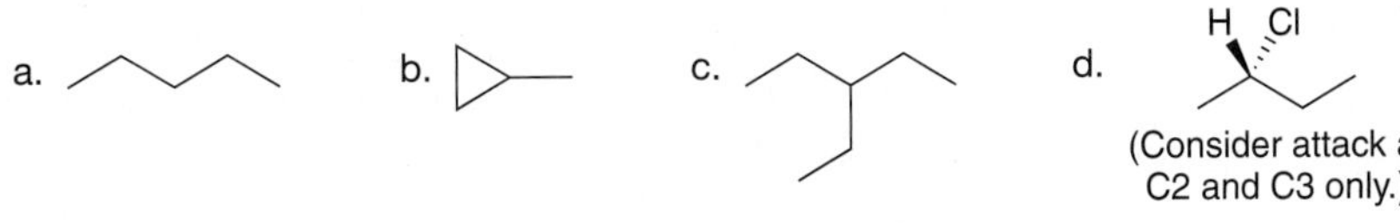
a.
b.
c.
d.
H Cl

(Consider attack at C2 and C3 only.)

15.9 Application: The Ozone Layer and CFCs

The 1995 Nobel Prize in Chemistry was awarded to Mario Molina, Paul Crutzen, and F. Sherwood Rowland for their work in elucidating the interaction of ozone with CFCs.

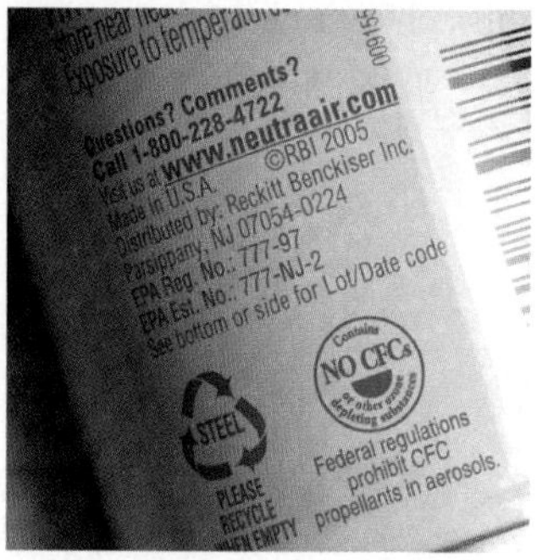

Propane and butane are now used as propellants in spray cans in place of CFCs.

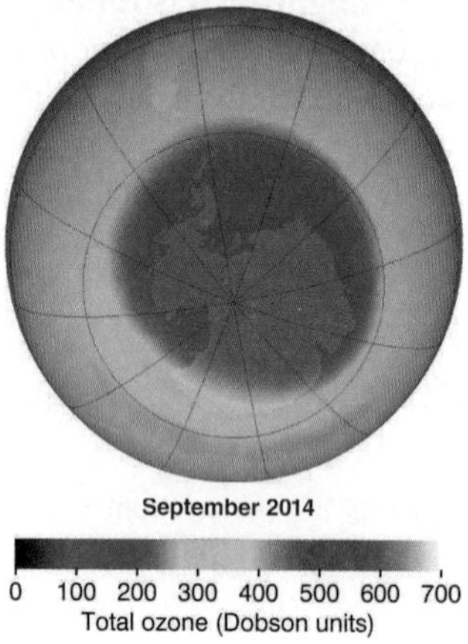

O_3 destruction is most severe in the region of the South Pole, where a large ozone hole is visible with satellite imaging.

Ozone is formed in the upper atmosphere by reaction of oxygen molecules with oxygen atoms. Ozone is also decomposed with sunlight back to these same two species. The overall result of these reactions is to convert high-energy ultraviolet light into heat.

Ozone synthesis: $O_2 + \cdot\ddot{O}\cdot \longrightarrow O_3 + \text{heat}$ (O_3 = ozone)

Ozone decomposition: $O_3 \xrightarrow{h\nu} O_2 + \cdot\ddot{O}\cdot$ (O_3 = ozone)

Ozone is vital to life; it acts like a shield, protecting the earth's surface from destructive ultraviolet radiation. A decrease in ozone concentration in this protective layer would have some immediate consequences, including an increase in the incidence of skin cancer and eye cataracts. Other long-term effects include a reduced immune response, interference with photosynthesis in plants, and harmful effects on the growth of plankton, the mainstay of the ocean food chain.

Current research suggests that **chlorofluorocarbons (CFCs)** are responsible for destroying ozone in the upper atmosphere. **CFCs** are simple halogen-containing organic compounds manufactured under the trade name Freons.

$CFCl_3$
trichlorofluoromethane
CFC 11
Freon 11

CF_2Cl_2
dichlorodifluoromethane
CFC 12
Freon 12

CFCs are inert, odorless, and nontoxic, and they have been used as refrigerants, solvents, and aerosol propellants. Because CFCs are volatile and water insoluble, they readily escape into the upper atmosphere, where they are decomposed by high-energy sunlight to form radicals that destroy ozone by the radical chain mechanism shown in Figure 15.7.

The overall result is that O_3 is consumed as a reactant and O_2 molecules are formed. In this way, a small amount of CFC can destroy a large amount of O_3. These findings led to a ban on the use of CFCs in aerosol propellants in the United States in 1978 and to the phasing out of their use in refrigeration systems.

Newer alternatives to CFCs are **hydrochlorofluorocarbons (HCFCs)** and **hydrofluorocarbons (HFCs)** such as CH_2FCF_3. These compounds have many properties in common with CFCs, but they are largely decomposed by HO· before they reach the stratosphere and therefore they have little impact on stratospheric O_3.

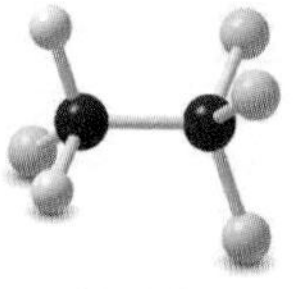

CH_2FCF_3
HFC-134a

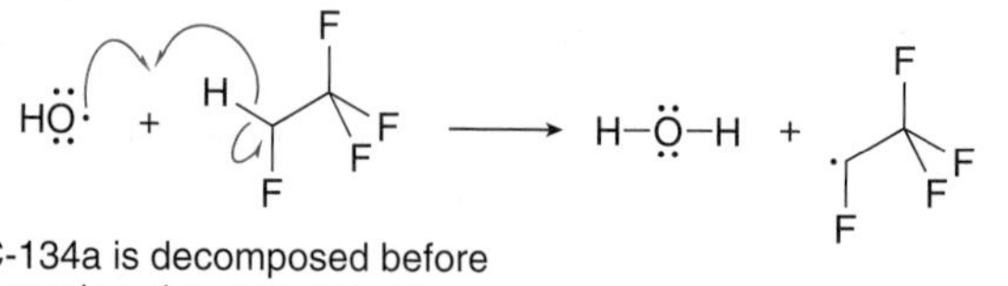

HFC-134a is decomposed before it reaches the stratosphere.

Problem 15.15 Nitric oxide, NO·, is another radical also thought to cause ozone destruction by a similar mechanism. One source of NO· in the stratosphere is supersonic aircraft whose jet engines convert small amounts of N_2 and O_2 to NO·. Write the propagation steps for the reaction of O_3 with NO·.

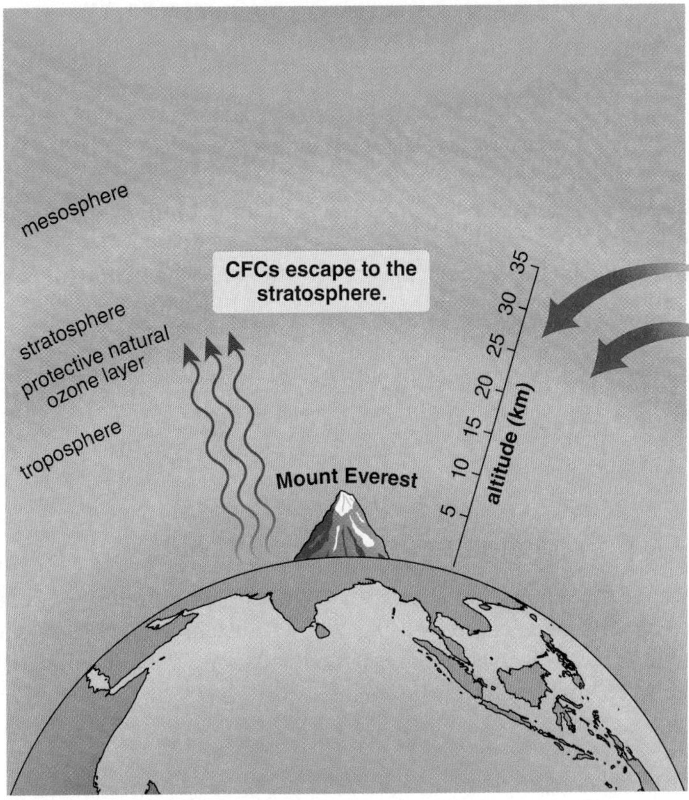

Figure 15.7
CFCs and the destruction of the ozone layer

Initiation: CFCs are decomposed by sunlight to form chlorine radicals.

$$CFCl_3 \xrightarrow{h\nu} \cdot CFCl_2 + \cdot\ddot{Cl}:$$

Propagation: Ozone is destroyed by a chain reaction with radical intermediates.

$$\cdot\ddot{Cl}: + O_3 \longrightarrow :\ddot{Cl}-\ddot{O}\cdot + O_2$$

$$:\ddot{Cl}-\ddot{O}\cdot + \cdot\ddot{O}\cdot \longrightarrow \cdot\ddot{Cl}: + O_2$$

- The chain reaction is initiated by homolysis of a C–Cl bond in $CFCl_3$.
- Propagation consists of two steps. Reaction of Cl· with O_3 forms chlorine monoxide (ClO·), which reacts with oxygen atoms to form O_2 and Cl·.

15.10 Radical Halogenation at an Allylic Carbon

Now let's examine radical halogenation at an ***allylic carbon*—the carbon adjacent to a double bond.** Homolysis of the allylic C–H bond of propene generates the **allyl radical,** which has an unpaired electron on the carbon adjacent to the double bond.

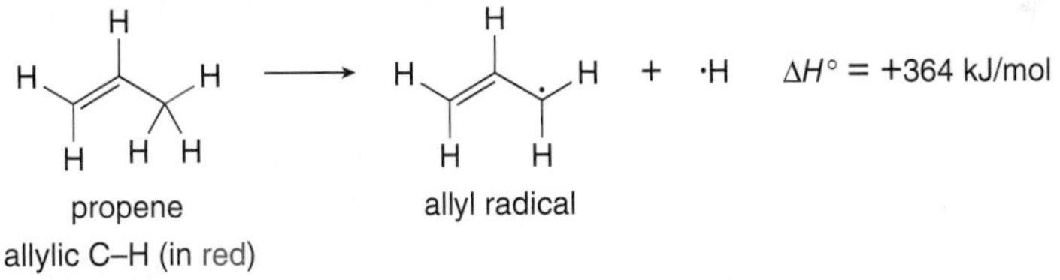

The bond dissociation energy for this process (364 kJ/mol) is even less than that for a 3° C–H bond (381 kJ/mol). Because the weaker the C–H bond, the more stable the resulting radical, an **allyl radical is more stable than a 3° radical,** and the following order of radical stability results:

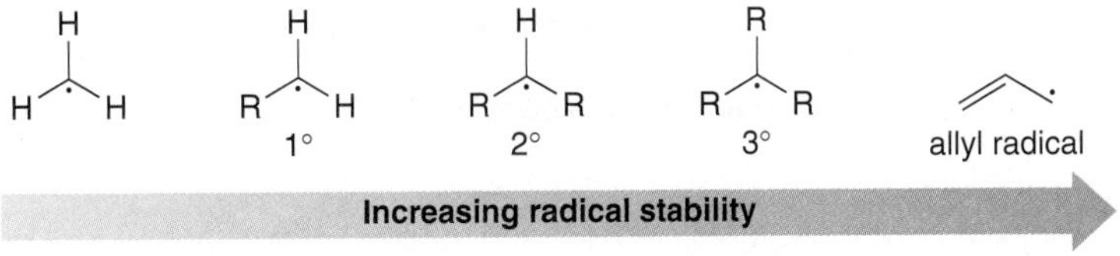

The allyl radical is more stable than other radicals because two resonance structures can be drawn for it.

The position of the atoms and the σ bonds stays the same in drawing resonance structures. Resonance structures differ in the location of only π bonds and nonbonded electrons.

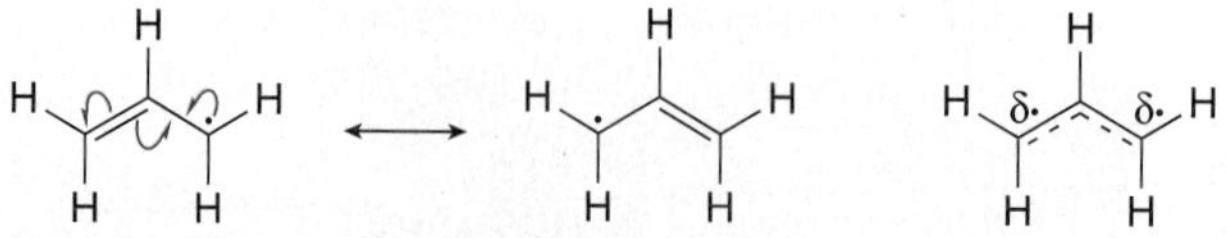

two resonance structures for the allyl radical — hybrid

- **The "true" structure of the allyl radical is a hybrid of the two resonance structures. In the hybrid, the π bond and the unpaired electron are delocalized.**
- **Delocalizing electron density lowers the energy of the hybrid, thus stabilizing the allyl radical.**

Problem 15.16 Draw a second resonance structure for each radical. Then draw the hybrid.

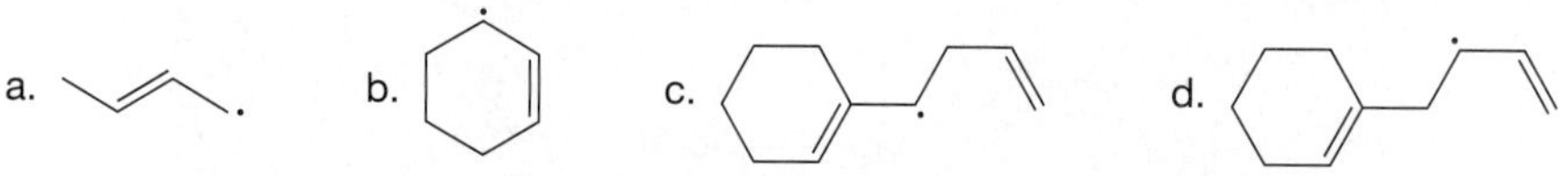

15.10A Selective Bromination at Allylic C–H Bonds

Because allylic C–H bonds are weaker than other sp^3 hybridized C–H bonds, the **allylic carbon can be selectively halogenated** by using *N*-bromosuccinimide (**NBS,** Section 10.15) in the presence of light or peroxides. Under these conditions only the allylic C–H bond in cyclohexene reacts to form an allylic halide.

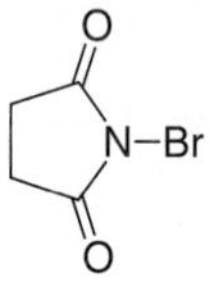

N-bromosuccinimide
NBS

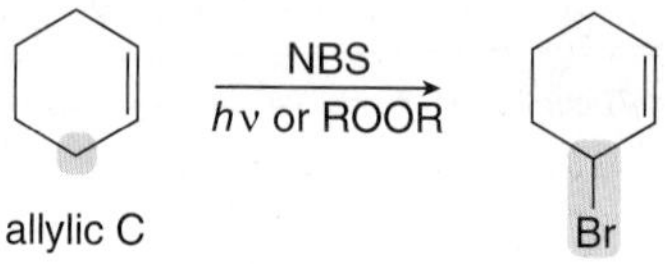

allylic halide

NBS contains a weak N–Br bond that is homolytically cleaved with light to generate a bromine radical, initiating an allylic halogenation reaction. Propagation then consists of the usual two steps of radical halogenation as shown in Mechanism 15.2.

Mechanism 15.2 Allylic Bromination with NBS

Part [1] Initiation

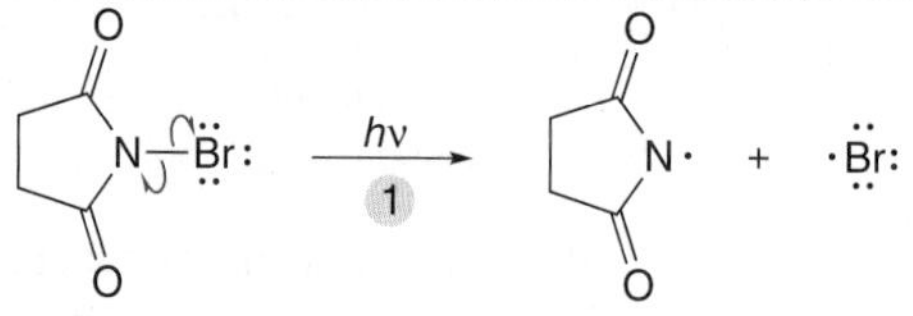

NBS

1 Homolysis of the weak N–Br bond with light energy forms a Br· radical that initiates radical halogenation.

Part [2] Propagation

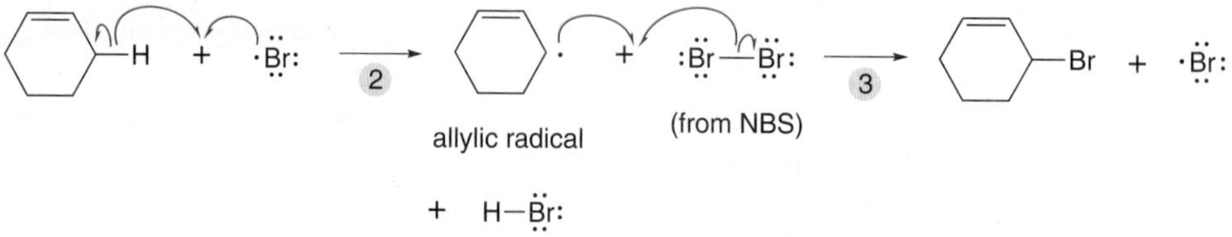

2 The Br· radical abstracts an allylic H to afford an allylic radical. (Only one resonance structure is drawn.)

3 The allylic radical reacts with Br_2 to form the allylic halide. The radical Br· formed in Step [3] can now react in Step [2], so Steps [2] and [3] can repeatedly occur without additional initiation.

Besides acting as a source of Br· to initiate the reaction, NBS generates a low concentration of Br_2 needed in the second chain propagation step (Step [3] of the mechanism). The HBr formed in Step [2] reacts with NBS to form Br_2, which is then used for halogenation in Step [3] of the mechanism.

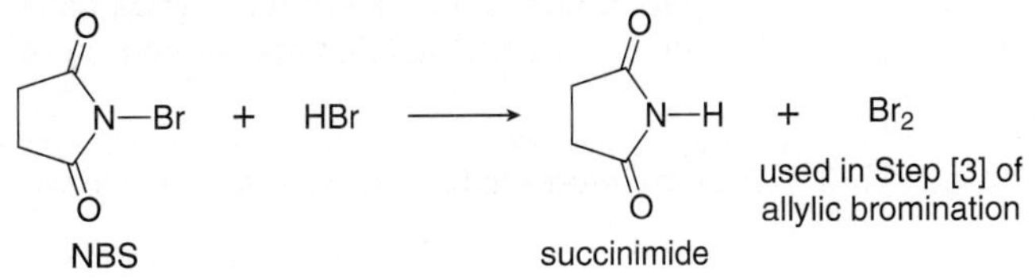

> A **low concentration of Br_2** (from NBS) **favors allylic substitution** (over addition) in part because bromine is needed for only *one* step of the mechanism. When Br_2 adds to a double bond, a low Br_2 concentration would first form a low concentration of bridged bromonium ion (Section 10.13), which must then react with more bromine (in the form of Br^-) in a second step to form a dibromide. **If concentrations of both intermediates—bromonium ion and Br^-—are low, the overall rate of addition is very slow.**

Thus, an alkene with allylic C–H bonds undergoes two different reactions depending on the reaction conditions.

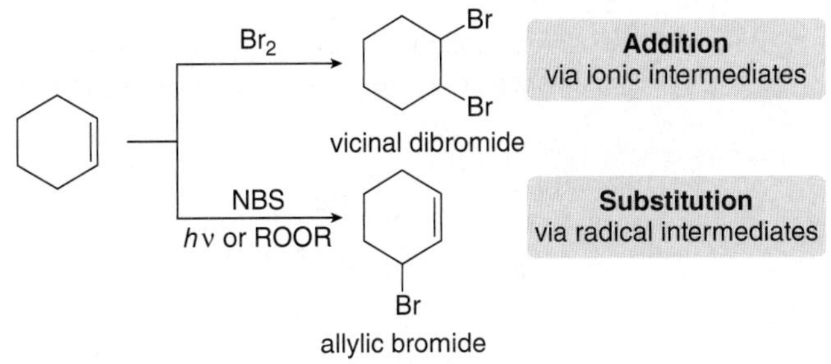

- Treatment of cyclohexene with Br_2 (in an organic solvent like CCl_4) leads to **addition** via **ionic intermediates** (Section 10.13).
- Treatment of cyclohexene with NBS (+ $h\nu$ or ROOR) leads to **allylic substitution,** via **radical intermediates.**

Problem 15.17 Draw the products of each reaction.

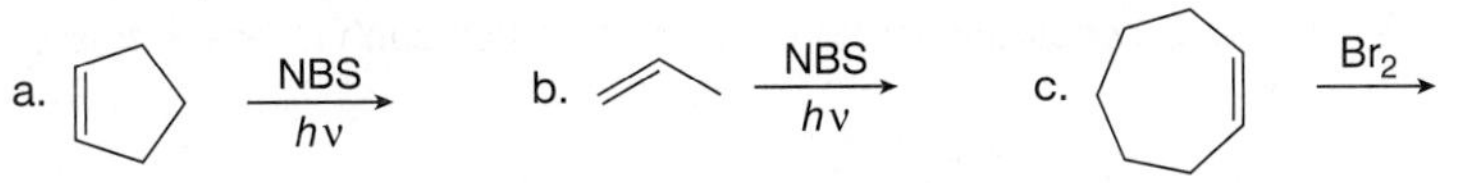

15.10B Product Mixtures in Allylic Halogenation

Halogenation at an allylic carbon often results in a mixture of products. For example, bromination of 3-methylbut-1-ene under radical conditions forms a mixture of 3-bromo-3-methylbut-1-ene and 1-bromo-3-methylbut-2-ene.

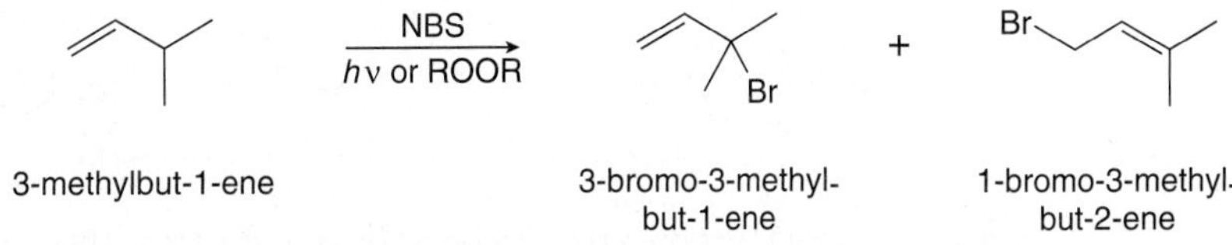

A mixture is obtained because the reaction proceeds by way of a **resonance-stabilized radical.** Abstraction of an allylic hydrogen from the alkene with a Br· radical (from NBS) forms an allylic radical for which **two different Lewis structures** can be drawn.

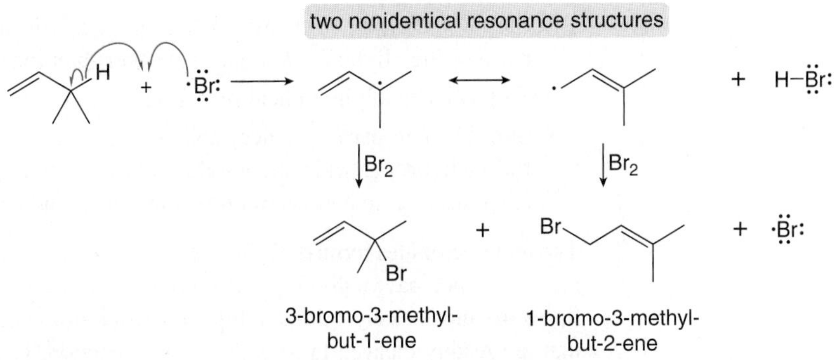

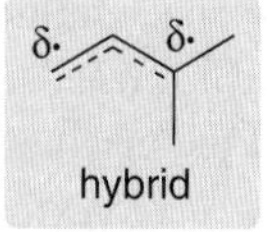

As a result, two different C atoms have partial radical character (indicated by δ·), so that Br_2 reacts at two different sites and two allylic halides are formed.

- **Whenever two different resonance structures can be drawn for an allylic radical, two different allylic halides are formed by radical substitution.**

Sample Problem 15.4 Draw the products formed when **A** is treated with NBS + *hν*.

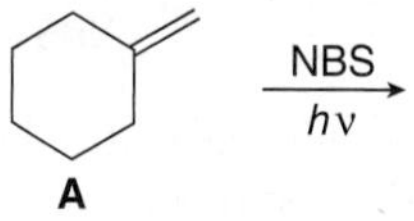

Solution

Hydrogen abstraction at the allylic C forms a resonance-stabilized radical (with two different resonance structures) that reacts with Br_2 to form two constitutional isomers as products.

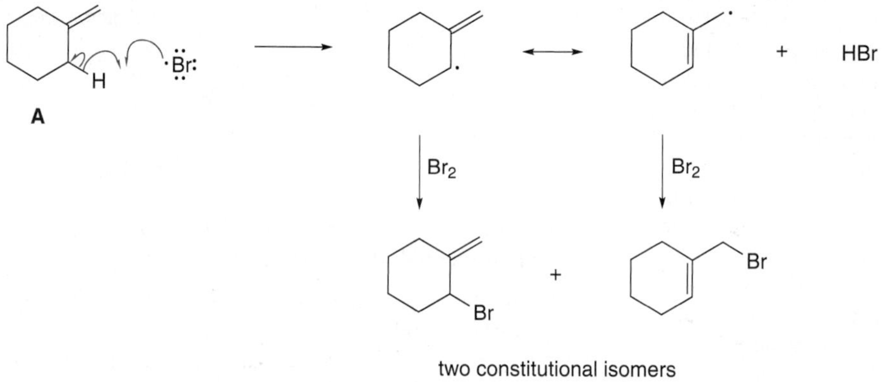

Problem 15.18 Draw all constitutional isomers formed when each alkene is treated with NBS + *hν*.

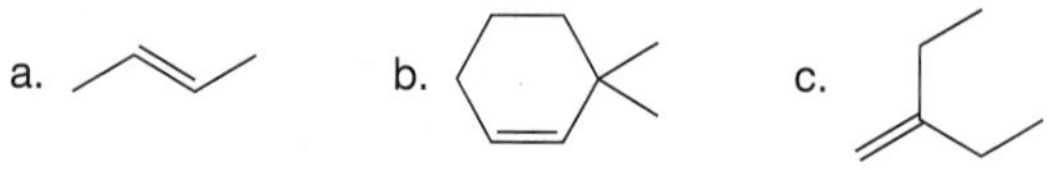

Problem 15.19 Draw the structure of the four allylic halides formed when 3-methylcyclohexene undergoes allylic halogenation with NBS + *hν*.

Problem 15.20 Which compounds can be prepared in good yield by allylic halogenation of an alkene?

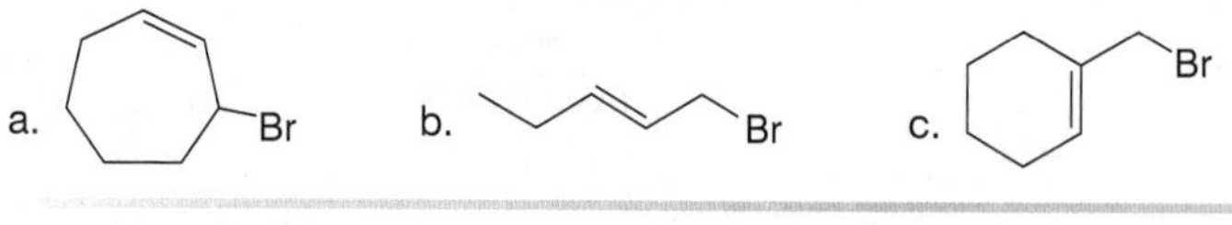

15.11 Application: Oxidation of Unsaturated Lipids

Oils—triacylglycerols having one or more sites of unsaturation in their long carbon chains—are susceptible to oxidation at their allylic carbon atoms. Oxidation occurs by way of a radical chain mechanism, as shown in Figure 15.8.

- **Step [1]** Oxygen in the air abstracts an allylic hydrogen atom to form an allylic radical because the allylic C–H bond is weaker than the other C–H bonds.
- **Step [2]** The allylic radical reacts with another molecule of O_2 to form a peroxy radical.
- **Step [3]** The peroxy radical abstracts an allylic hydrogen from another lipid molecule to form a hydroperoxide and another allylic radical that continues the chain. Steps [2] and [3] can repeat again and again until some other radical terminates the chain.

The hydroperoxides formed by this process are unstable and decompose to other oxidation products, many of which have a disagreeable odor and taste. **This process turns an oil rancid. Unsaturated lipids are more easily oxidized than saturated ones** because they contain weak allylic C–H bonds that are readily cleaved in Step [1] of this reaction, forming resonance-stabilized allylic radicals.

Figure 15.8
The oxidation of unsaturated lipids with O_2

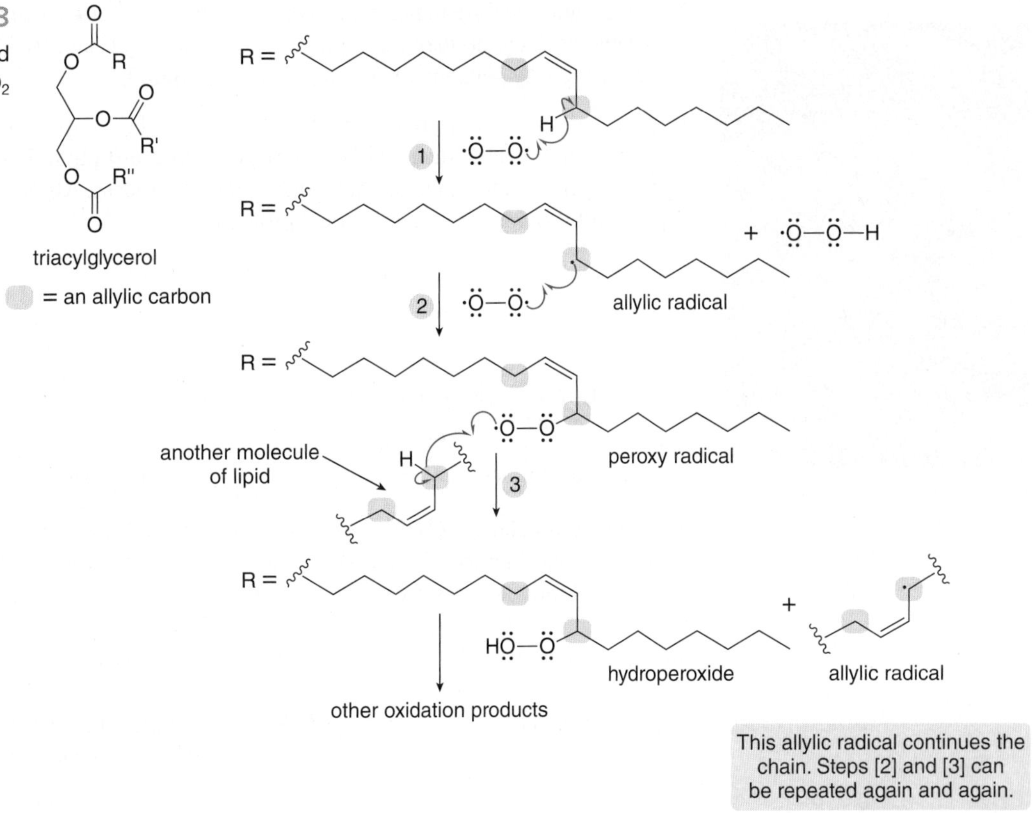

• Oxidation is shown at one allylic carbon only. Reaction at the other labeled allylic carbon is also possible.

Because saturated fats have no double bonds and thus no weak allylic C–H bonds, they are much less susceptible to air oxidation, resulting in increased shelf life of products containing them.

Problem 15.21 Which C–H bond is most readily cleaved in linoleic acid? Draw all possible resonance structures for the resulting radical. Draw all the hydroperoxides formed by reaction of this resonance-stabilized radical with O_2.

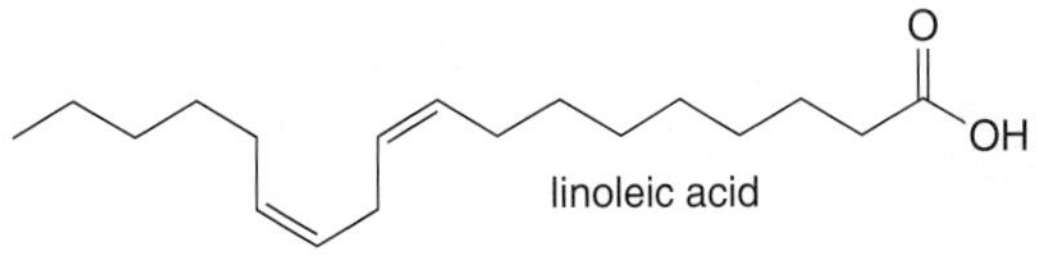
linoleic acid

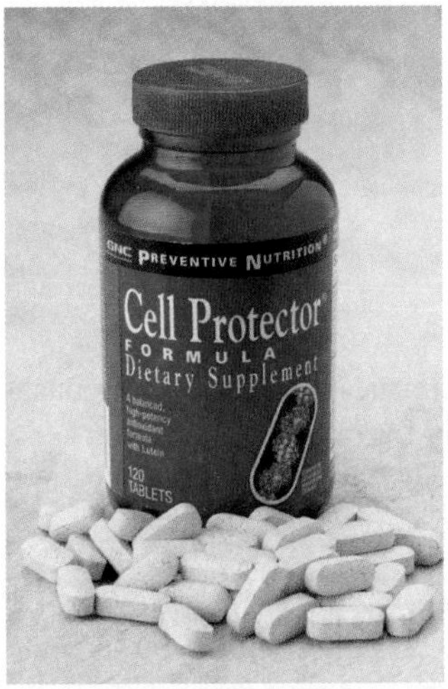

The purported health benefits of antioxidants have made them a popular component in anti-aging formulations.

15.12 Application: Antioxidants

An *antioxidant* is a compound that stops an oxidation reaction from occurring.

- Naturally occurring antioxidants such as **vitamin E** prevent radical reactions that can cause cell damage.
- Synthetic antioxidants such as **BHT**—**b**utylated **h**ydroxy **t**oluene—are added to packaged and prepared foods to prevent oxidation and spoilage.

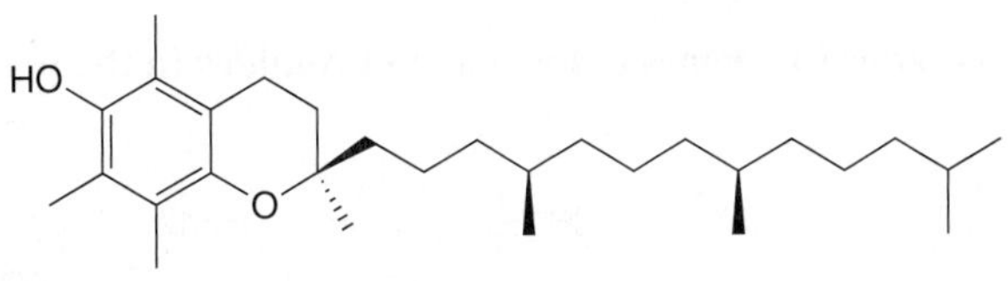
vitamin E

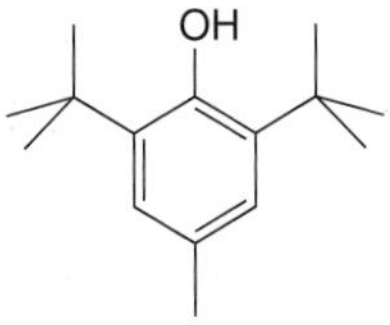
BHT
(butylated hydroxy toluene)

Hazelnuts, almonds, and many other types of nuts are an excellent source of the natural antioxidant vitamin E.

Vitamin E and BHT are radical inhibitors, so they terminate radical chain mechanisms by reacting with radicals. How do they trap radicals? Both vitamin E and BHT use a hydroxy group bonded to a benzene ring—a general structure called a **phenol.**

Radicals (R·) abstract a hydrogen atom from the OH group of an antioxidant, forming a new resonance-stabilized radical. **This new radical does not participate in chain propagation,** but rather terminates the chain and halts the oxidation process. All phenols (including vitamin E and BHT) inhibit oxidation by this radical process.

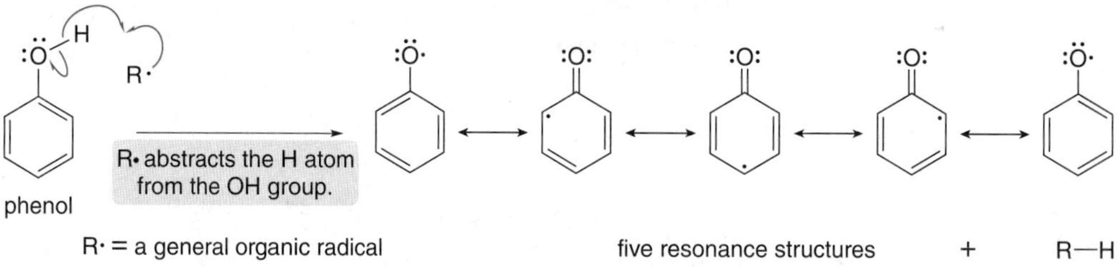

The many nonpolar C–C and C–H bonds of vitamin E make it fat soluble, and thus it dissolves in the nonpolar interior of the cell membrane, where it is thought to inhibit the oxidation of the unsaturated fatty acid residues in the phospholipids. Oxidative damage to lipids in cells via radical mechanisms is thought to play an important role in the aging process. For this reason, many anti-aging formulas with antioxidants like vitamin E are now popular consumer products.

Problem 15.22 Rosmarinic acid is an antioxidant isolated from rosemary. Draw resonance structures for the radical that results from removal of the labeled H atom in rosmarinic acid.

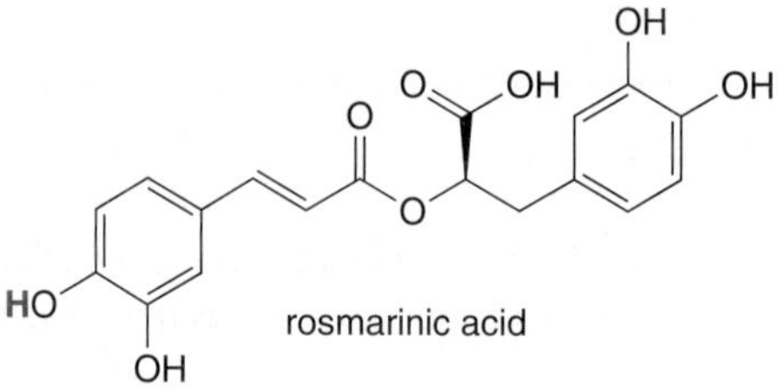

Rosemary extracts contain rosmarinic acid, an antioxidant that helps prevent the oxidation of unsaturated vegetable oils.

15.13 Radical Addition Reactions to Double Bonds

We now turn our attention to the second common reaction of radicals, addition to double bonds. Because an alkene contains an electron-rich, easily broken π bond, it reacts with an electron-deficient radical.

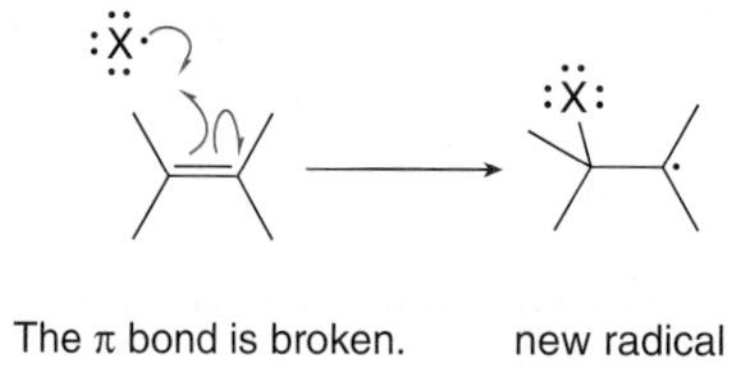

Radicals react with alkenes via a radical chain mechanism that consists of initiation, propagation, and termination steps analogous to those discussed previously for radical substitution.

15.13A Addition of HBr

HBr adds to alkenes to form alkyl bromides in the presence of light, heat, or peroxides.

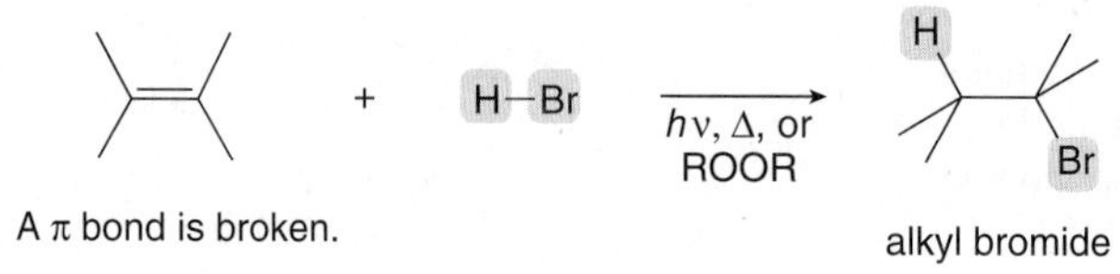

The regioselectivity of addition to an unsymmetrical alkene is *different* from the addition of HBr without added light, heat, or peroxides.

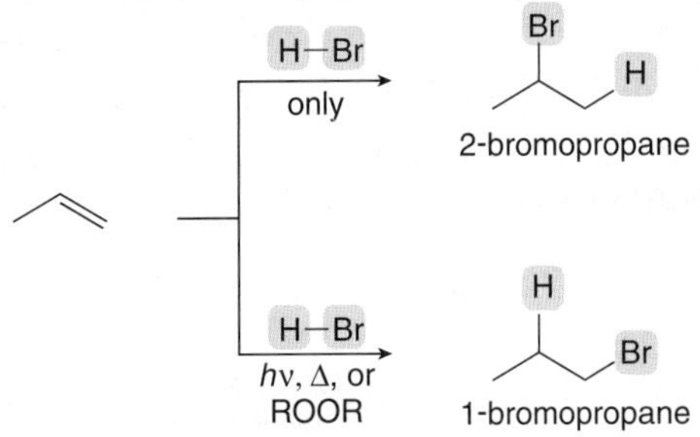

H bonds to the **less substituted** C.

Br bonds to the **less substituted** C.

- HBr addition to propene *without* added light, heat, or peroxides gives 2-bromopropane: the **H atom is added to the less substituted carbon.** This reaction occurs via **carbocation** intermediates (Section 10.10).
- HBr addition to propene *with* added light, heat, or peroxides gives 1-bromopropane: the **Br atom is added to the less substituted carbon.** This reaction occurs via **radical** intermediates.

Problem 15.23 Draw the product(s) formed when each alkene is treated with either [1] HBr alone; or [2] HBr in the presence of peroxides.

15.13B The Mechanism of the Radical Addition of HBr to an Alkene

In the presence of added light, heat, or peroxides, HBr addition to an alkene forms radical intermediates, and like other radical reactions, proceeds by a mechanism with three distinct parts: initiation, propagation, and termination. Mechanism 15.3 is written for the reaction of $CH_3CH{=}CH_2$ with HBr and ROOR to form $CH_3CH_2CH_2Br$.

Mechanism 15.3 Radical Addition of HBr to an Alkene

Part [1] Initiation

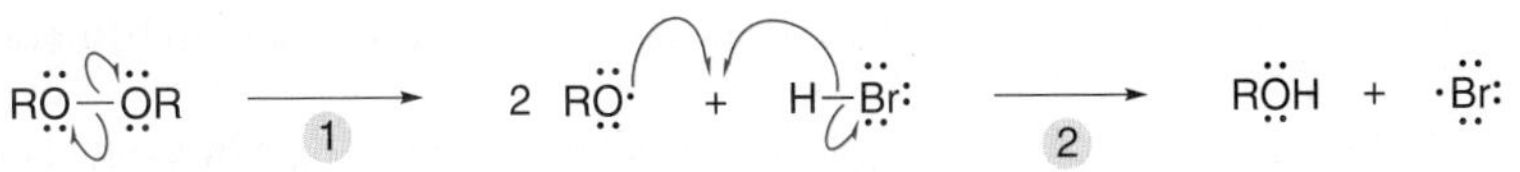

1 – 2 Initiation with ROOR occurs in two steps—homolysis of the weak O–O bond and abstraction of H to form a bromine radical.

Part [2] Propagation

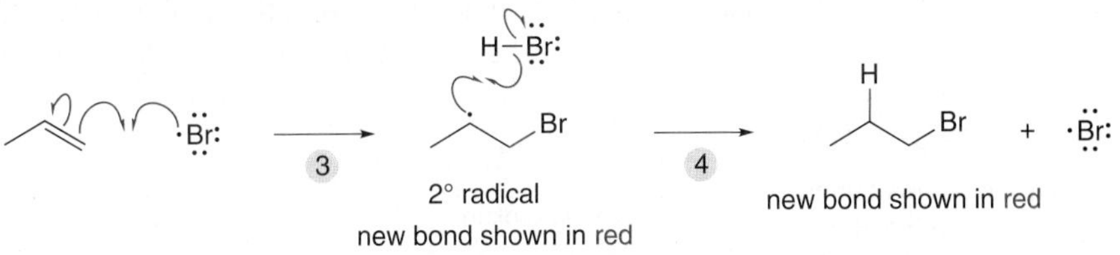

3 Addition of Br· to the terminal carbon forms a 2° radical.

4 Abstraction of H from HBr forms a new C–H bond and a bromine radical, so Steps [3] and [4] can occur repeatedly.

Part [3] Termination

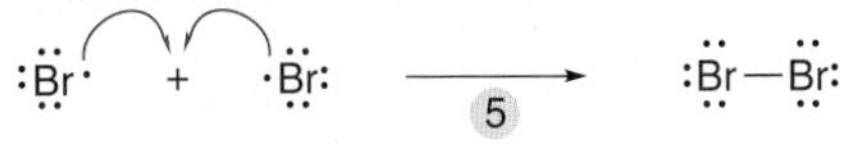

5 Termination of the chain occurs when any two radicals combine to form a bond.

The first propagation step (Step [3] of the mechanism, the addition of Br· to the double bond) is worthy of note. With propene there are two possible paths for this step, depending on which carbon atom of the double bond forms the new bond to bromine. Path [A] forms a less stable 1° radical whereas Path [B] forms a more stable 2° radical. **The more stable 2° radical forms faster,** so Path [B] is preferred.

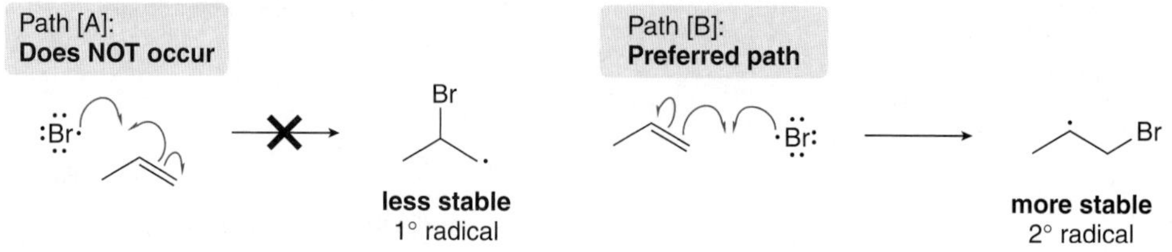

The mechanism also illustrates why the regioselectivity of HBr addition is different depending on the reaction conditions. In both reactions, H and Br add to the double bond, but the *order* of addition depends on the mechanism.

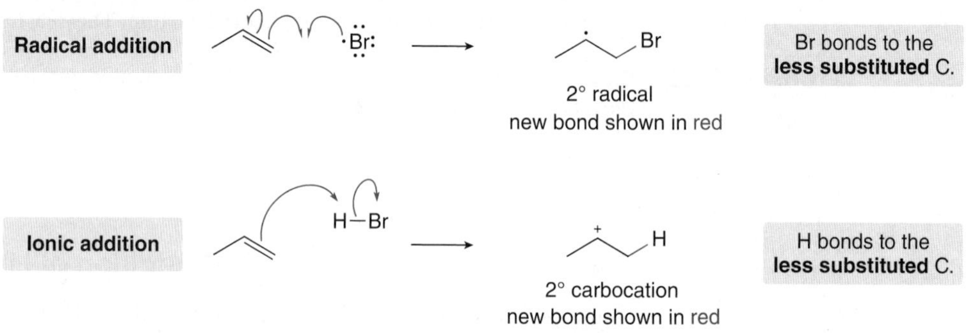

- **In radical addition (HBr with added light, heat, or ROOR), *Br· adds first* to generate the more stable radical.**
- **In ionic addition (HBr alone), *H⁺ adds first* to generate the more stable carbocation.**

Problem 15.24 When HBr adds to $(CH_3)_2C{=}CH_2$ under radical conditions, two radicals are possible products in the first step of chain propagation. Draw the structure of both radicals and indicate which one is formed. Then draw the preferred product from HBr addition under radical conditions.

Problem 15.25 What reagents are needed to convert 1-ethylcyclohexene into (a) 1-bromo-2-ethylcyclohexane; (b) 1-bromo-1-ethylcyclohexane; (c) 1,2-dibromo-1-ethylcyclohexane?

15.13C Energy Changes in the Radical Addition of HBr

The energy changes during propagation in the radical addition of HBr to $CH_2{=}CH_2$ can be calculated from bond dissociation energies, as shown in Figure 15.9.

Figure 15.9 Energy changes during the propagation steps: $CH_2{=}CH_2 + HBr \rightarrow CH_3CH_2Br$

[1] $CH_2{=}CH_2 + \cdot\ddot{Br}: \longrightarrow \dot{C}H_2CH_2{-}Br$

π bond broken	C–Br bond formed	
+267 kJ/mol	–285 kJ/mol	$\Delta H°[1] = -18$ kJ/mol

[2] $\dot{C}H_2CH_2{-}Br + H{-}\ddot{Br}: \longrightarrow H{-}CH_2CH_2Br + \cdot\ddot{Br}:$

bond broken	C–H bond formed	
+368 kJ/mol	–410 kJ/mol	$\Delta H°[2] = -42$ kJ/mol

$\Delta H°_{overall} = \Delta H°[1] + \Delta H°[2]$

$= -60$ kJ/mol

an exothermic reaction

Both propagation steps for the addition of HBr are exothermic, so propagation is exothermic (energetically favorable) overall. For the addition of HCl or HI, however, one of the chain-propagating steps is quite endothermic, and thus too difficult to be part of a repeating chain mechanism. Thus, **HBr adds to alkenes under radical conditions, but HCl and HI do not.**

15.14 Polymers and Polymerization

Polymers*—large molecules made up of repeating units of smaller molecules called *monomers—include such biologically important compounds as proteins and carbohydrates. They also include such industrially important plastics as polyethylene, poly(vinyl chloride) (PVC), and polystyrene.

HDPE (high-density polyethylene) and **LDPE** (low-density polyethylene) are two common types of polyethylene prepared under different reaction conditions and having different physical properties. HDPE is opaque and rigid, and is used in milk containers and water jugs. LDPE is less opaque and more flexible, and is used in plastic bags and electrical insulation. Products containing HDPE and LDPE (and other plastics) are often labeled with a symbol indicating recycling ease: the lower the number, the easier to recycle.

15.14A Synthetic Polymers

Many synthetic polymers—that is, those synthesized in the lab—are among the most widely used organic compounds in modern society. Although some synthetic polymers resemble natural substances, many have different and unusual properties that make them more useful than naturally occurring materials. Soft drink bottles, plastic bags, food wrap, compact discs, Teflon, and Styrofoam are all made of synthetic polymers. In this section we examine polymers derived from alkene monomers. Chapter 31 is devoted to a detailed discussion of the synthesis and properties of several different types of synthetic polymers.

- ***Polymerization*** **is the joining together of monomers to make polymers.**

For example, joining **ethylene monomers** together forms the polymer **polyethylene,** a plastic used in milk containers and sandwich bags.

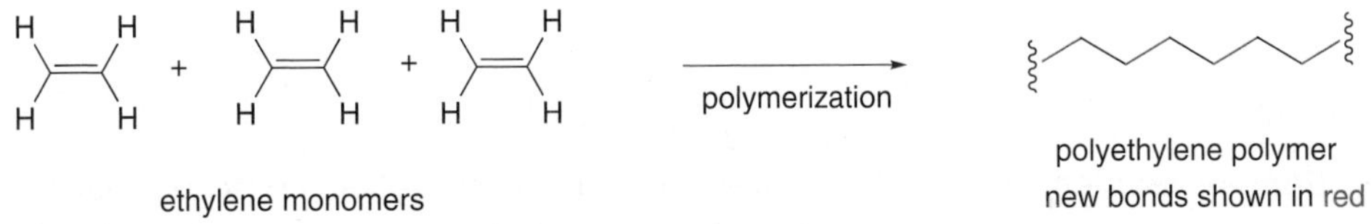

Many ethylene derivatives having the general structure $\mathbf{CH_2{=}CHZ}$ are also used as monomers for polymerization. The identity of Z affects the physical properties of the resulting polymer, making some polymers more suitable for one consumer product (e.g., plastic bags or food wrap) than another (e.g., soft drink bottles or compact discs). Polymerization of $CH_2{=}CHZ$ usually affords polymers with the Z groups on every other carbon atom in the chain. Table 15.2 lists some common monomers and polymers used in medicine or dentistry.

new bonds shown in red

Table 15.2 Common Monomers and Polymers Used in Medicine and Dentistry

Monomer →	Polymer	Consumer product
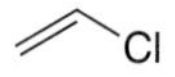 vinyl chloride	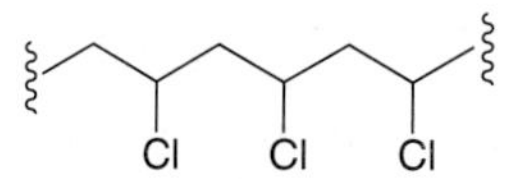 **poly(vinyl chloride) PVC**	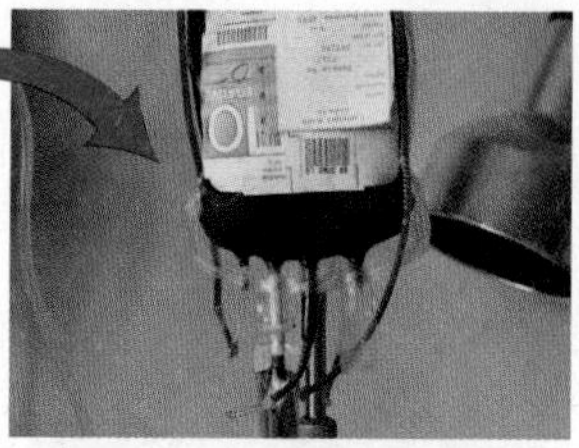 PVC blood bags and tubing
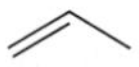 propene	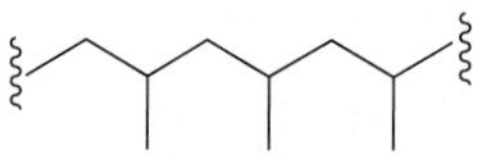 **polypropylene**	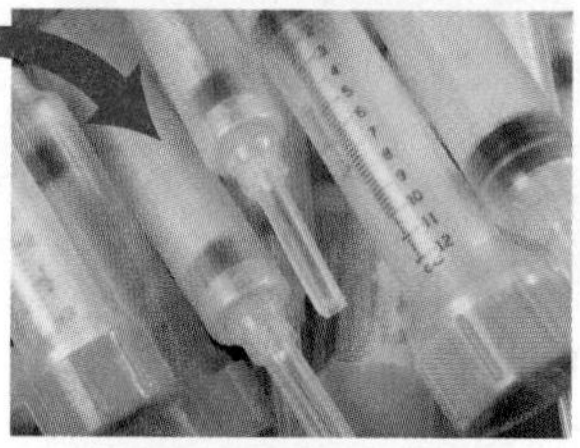 polypropylene syringes
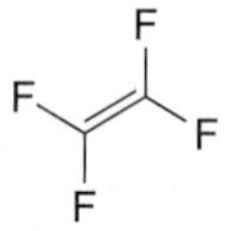 tetrafluoroethylene	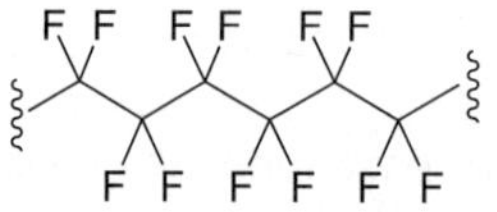 **polytetrafluoroethylene Teflon**	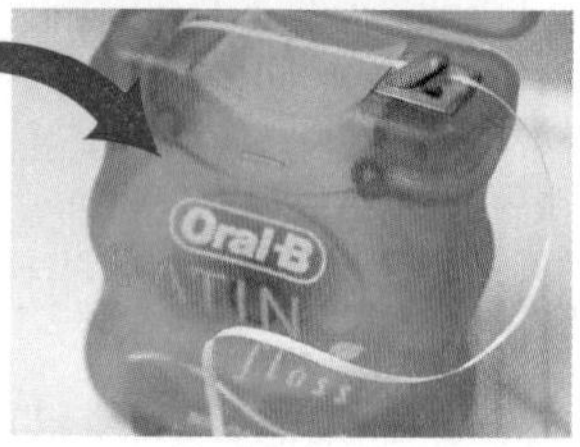 dental floss

Sample Problem 15.5 What polymer is formed when $CH_2=CHCO_2H$ (acrylic acid) is polymerized? The resulting polymer, poly(acrylic acid), is used in disposable diapers because it absorbs 30 times its weight in water.

Solution

Draw three or more alkene monomers, break one bond of each double bond, and join the alkenes together with single bonds. With unsymmetrical alkenes, substituents are bonded to every other carbon.

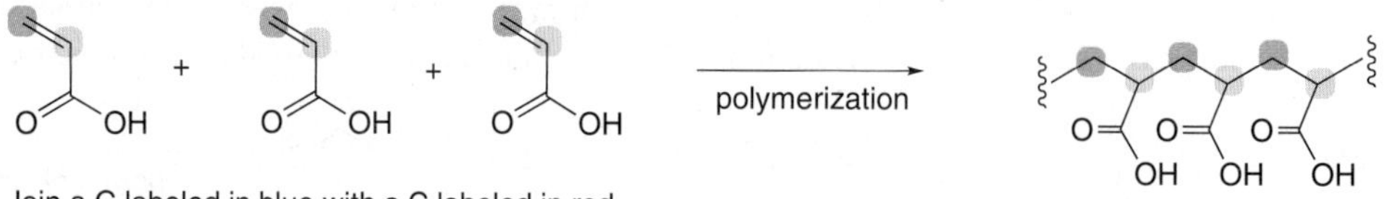

Join a C labeled in blue with a C labeled in red.

poly(acrylic acid)

Problem 15.26 (a) Draw the structure of polystyrene, the chapter-opening molecule, which is formed by polymerizing the monomer styrene, $C_6H_5CH=CH_2$. (b) What monomer is used to form poly(vinyl acetate), a polymer used in paints and adhesives?

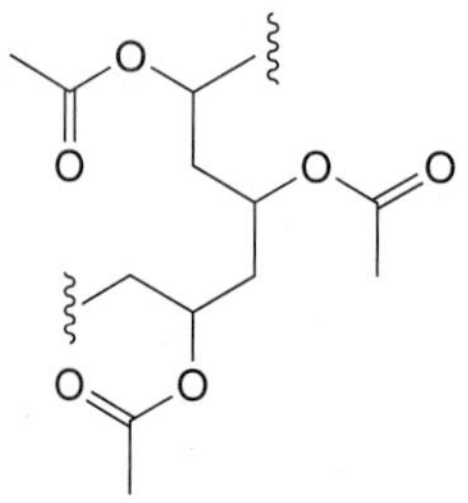

poly(vinyl acetate)

15.14B Radical Polymerization

The alkene monomers used in polymerization are prepared from petroleum.

The polymers described in Section 15.14A are prepared by polymerization of alkene monomers by **adding a radical to a π bond.** The mechanism resembles the radical addition of HBr to an alkene, except that a **carbon radical rather than a bromine atom is added to the double bond.** Mechanism 15.4 is written with the general monomer $CH_2{=}CHZ$, and again has three parts: initiation, propagation, and termination.

Mechanism 15.4 Radical Polymerization of $CH_2{=}CHZ$

Part [1] Initiation

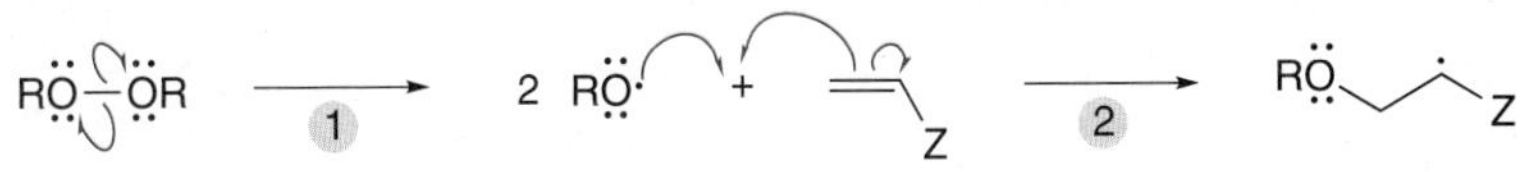

1 – 2 Initiation with ROOR occurs in two steps—homolysis of the weak O–O bond and addition of RO· to the alkene to form a carbon radical.

Part [2] Propagation

3 Chain propagation consists of a single step. The carbon radical adds to another alkene to form a new C–C bond and another carbon radical. Addition forms the radical with the unpaired electron on the atom with the Z substituent.

Part [3] Termination

4 Termination of the chain occurs when any two radicals combine to form a bond.

In radical polymerization, the more substituted radical always adds to the less substituted end of the monomer, a process called **head-to-tail polymerization.**

The **more substituted radical** adds to the **less substituted end** of the double bond.

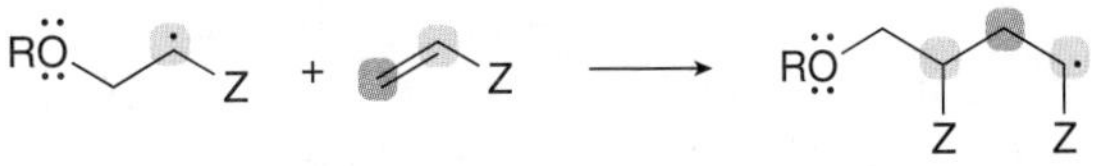

The new radical is always located on the C bonded to Z.

Problem 15.27 Draw the steps of the mechanism that converts vinyl chloride ($CH_2{=}CHCl$) into poly(vinyl chloride).

KEY CONCEPTS

Radical Reactions

General Features of Radicals

- A radical is a reactive intermediate with a single unpaired electron (15.1).
- A carbon radical is sp^2 hybridized and trigonal planar (15.1).
- The stability of a radical increases as the number of C atoms bonded to the radical carbon increases (15.1).
- Allylic radicals are stabilized by resonance, making them more stable than 3° radicals (15.10).

Radical Reactions

[1] Halogenation of alkanes (15.4)

$$\text{R–H} \xrightarrow[h\nu \text{ or } \Delta]{X_2} \text{R–X}$$

X = Cl or Br; R–X = alkyl halide

- The reaction follows a radical chain mechanism.
- The weaker the C – H bond, the more readily the hydrogen is replaced by X.
- Chlorination is faster and less selective than bromination (15.6).
- Radical substitution at a stereogenic center results in racemization (15.8).

[2] Allylic halogenation (15.10)

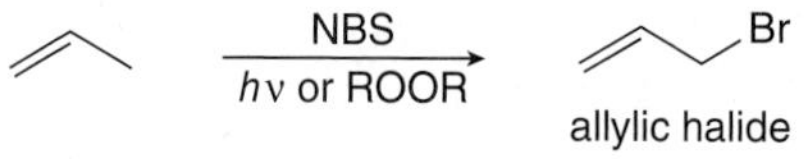
NBS
*h*ν or ROOR
Br
allylic halide

- The reaction follows a radical chain mechanism.

[3] Radical addition of HBr to an alkene (15.13)

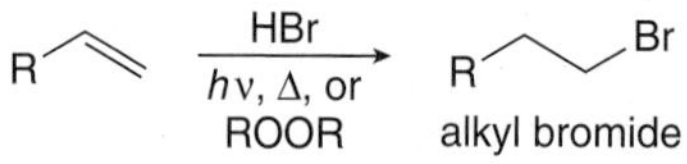
R
HBr
*h*ν, Δ, or ROOR
R
Br
alkyl bromide

- A radical addition mechanism is followed.
- Br bonds to the less substituted carbon atom to form the more substituted, more stable radical.

[4] Radical polymerization of alkenes (15.14)

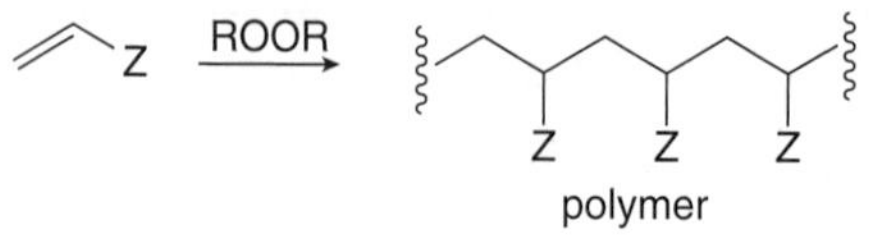
Z
ROOR
Z Z Z
polymer

- A radical addition mechanism is followed.

PROBLEMS

Problems Using Three-Dimensional Models

15.28 (a) Draw all constitutional isomers formed by monochlorination of each alkane with Cl_2 and *h*ν. (b) Draw the major monobromination product formed by heating each alkane with Br_2.

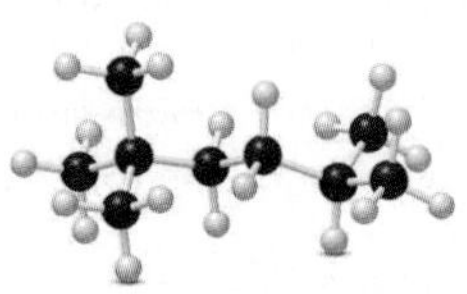

A

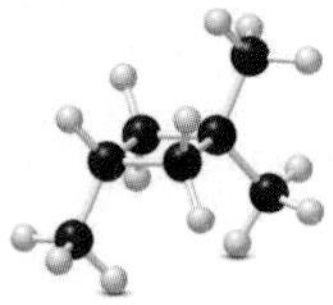

B

15.29 Draw all resonance structures of the radical that results from abstraction of a hydrogen atom from the antioxidant BHA (**b**utylated **h**ydroxy **a**nisole).

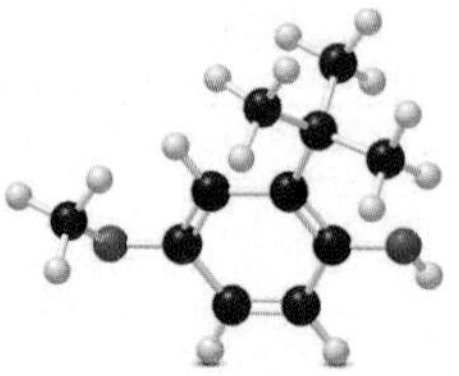

BHA

Radicals and Bond Strength

15.30 With reference to the indicated C – H bonds in 2-methylbutane:

2-methylbutane

a. Rank the C – H bonds in order of increasing bond strength.
b. Draw the radical resulting from cleavage of each C – H bond, and classify it as 1°, 2°, or 3°.
c. Rank the radicals in order of increasing stability.
d. Rank the C – H bonds in order of increasing ease of H abstraction in a radical halogenation reaction.

15.31 Rank the following radicals in order of increasing stability.

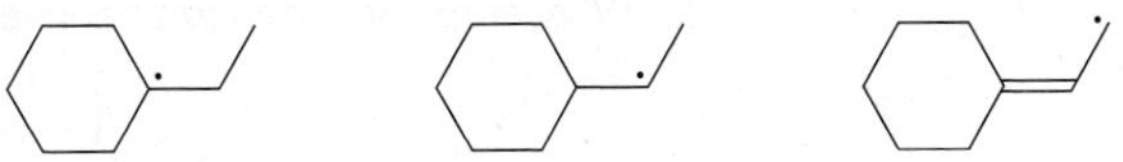

15.32 Why is a benzylic C – H bond (labeled in red) unusually weak?

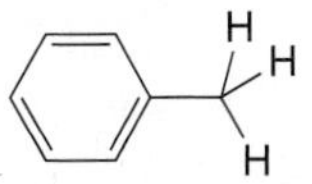

Halogenation of Alkanes

15.33 Rank the indicated hydrogen atoms in order of increasing ease of abstraction in a radical halogenation reaction.

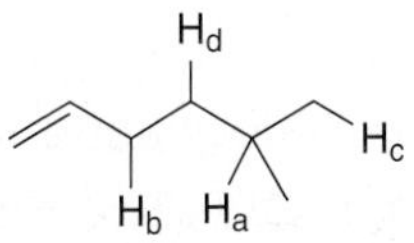

15.34 Draw all constitutional isomers formed by monochlorination of each alkane with Cl_2 and *hν*.

a. 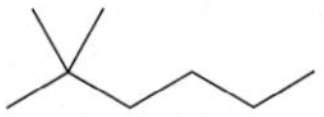 b.  c. 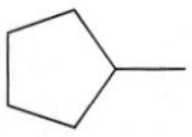

15.35 What is the major monobromination product formed by heating each alkane with Br_2?

a. 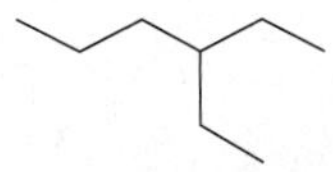 b. 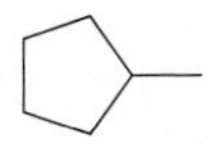 c. 

15.36 Five isomeric alkanes (**A–E**) having the molecular formula C_6H_{14} are each treated with Cl_2 + *hν* to give alkyl halides having molecular formula $C_6H_{13}Cl$. **A** yields five constitutional isomers. **B** yields four constitutional isomers. **C** yields two constitutional isomers. **D** yields three constitutional isomers, two of which possess stereogenic centers. **E** yields three constitutional isomers, only one of which possesses a stereogenic center. Identify the structures of **A–E.**

15.37 What alkane is needed to make each alkyl halide by radical halogenation?

a. 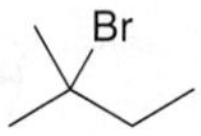 b. 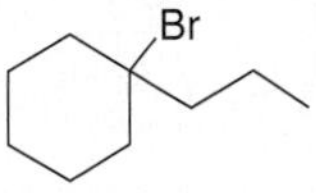 c. 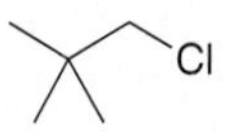

15.38 Which alkyl halides can be prepared in good yield by radical halogenation of an alkane?

a. 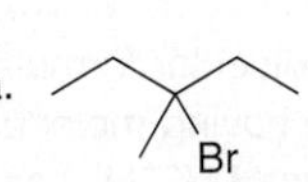 b. 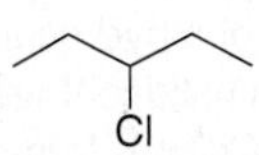 c. 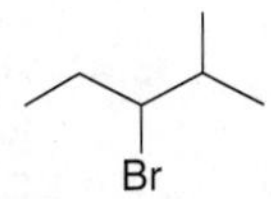

15.39 Draw the products of radical chlorination and bromination of each compound. For which compounds is a single constitutional isomer formed for both reactions? What must be true about the structure of a reactant for both reactions to form a single product?

a. b. 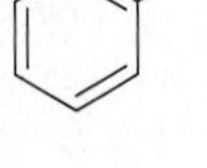 c. 

15.40 Explain why radical bromination of *p*-xylene forms **C** rather than **D.**

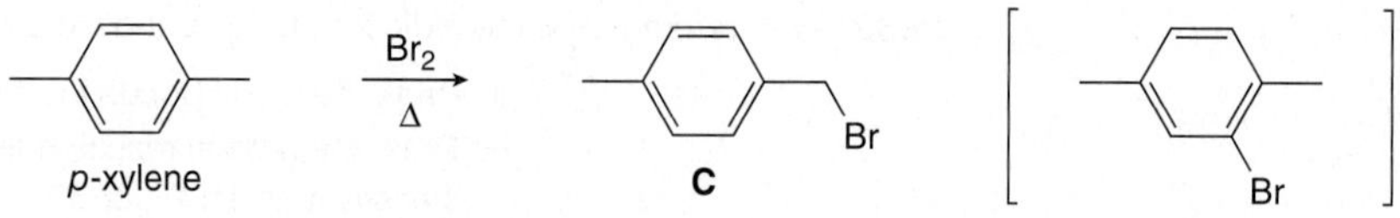

p-xylene **C** **D** NOT formed

15.41 a. What product(s) (excluding stereoisomers) are formed when **Y** is heated with Cl_2?
b. What product(s) (excluding stereoisomers) are formed when **Y** is heated with Br_2?
c. What steps are needed to convert **Y** to the alkene **Z?**

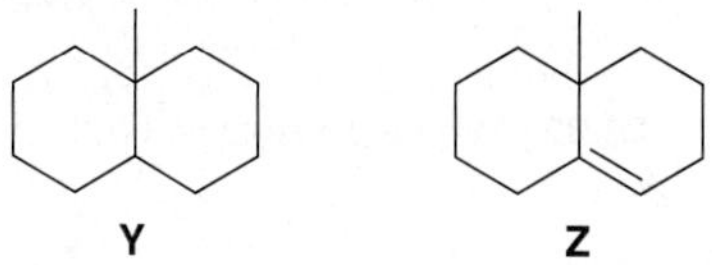

Y **Z**

Resonance

15.42 Draw resonance structures for each radical.

a.  b. 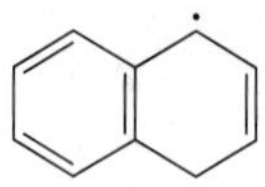 c. 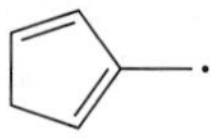

Allylic Halogenation

15.43 Draw the products formed when each alkene is treated with NBS + $h\nu$.

a.  b. 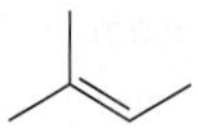 c. 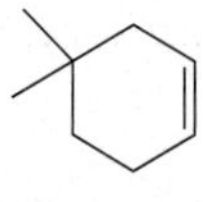

15.44 Draw all constitutional isomers formed when **X** is treated with NBS + $h\nu$.

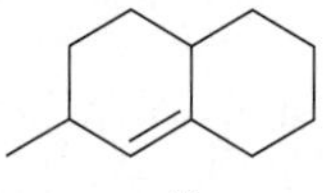

X

Reactions

15.45 Draw the organic products formed in each reaction.

a. 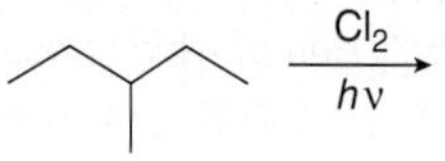

c. 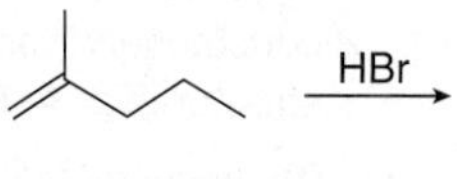

e. 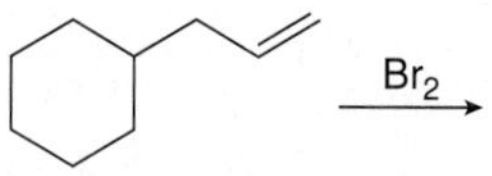

b. 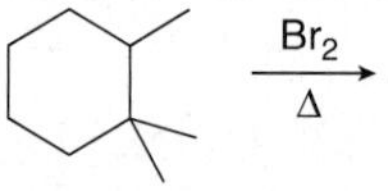

d. 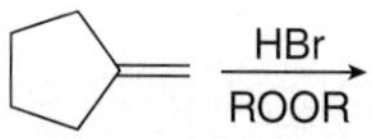

f. 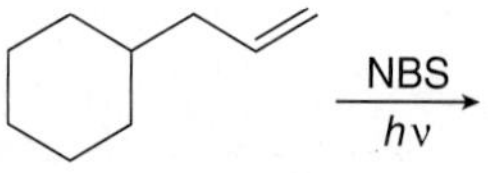

15.46 What reagents are needed to convert cyclopentene into (a) bromocyclopentane; (b) *trans*-1,2-dibromocyclopentane; (c) 3-bromocyclopentene?

15.47 Treatment of a hydrocarbon **A** (molecular formula C_9H_{18}) with Br_2 in the presence of light forms alkyl halides **B** and **C,** both having molecular formula $C_9H_{17}Br$. Reaction of either **B** or **C** with $KOC(CH_3)_3$ forms compound **D** (C_9H_{16}) as the major product. Ozonolysis of **D** forms cyclohexanone and acetone. Identify the structures of **A–D.**

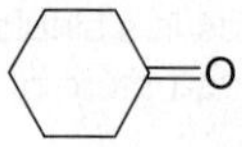

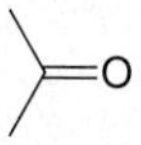

cyclohexanone acetone

Stereochemistry and Reactions

15.48 Draw the products formed in each reaction and include the stereochemistry around any stereogenic centers.

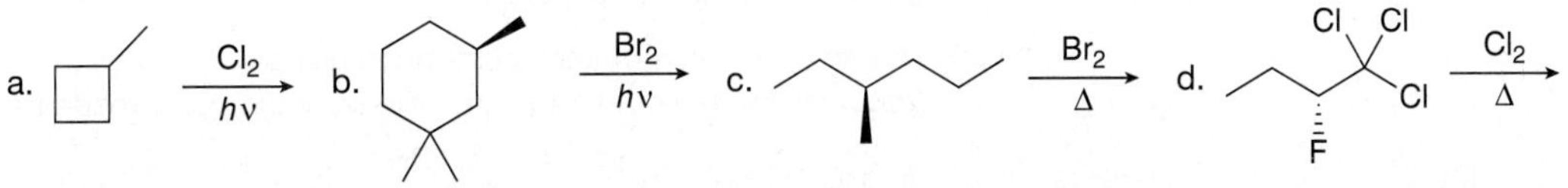

15.49 (a) Draw all stereoisomers of molecular formula $C_5H_{10}Cl_2$ formed when (*R*)-2-chloropentane is heated with Cl_2. (b) Assuming that products having different physical properties can be separated into fractions by some physical method (such as fractional distillation), how many different fractions would be obtained? (c) Which of these fractions would be optically active?

15.50 Draw all the monochlorination products that are formed from (*S*,*S*)-1,2-dimethylcyclopropane.

(*S*,*S*)-1,2-dimethylcyclopropane

15.51 Draw the six products (including stereoisomers) formed when **A** is treated with NBS + *h*ν.

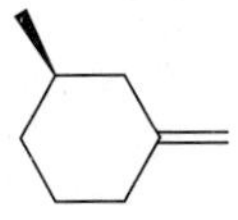

A

15.52 (a) Draw the products (including stereoisomers) formed when 2-methylhex-2-ene is treated with HBr in the presence of peroxides. (b) Draw the products (including stereoisomers) formed when (*S*)-2,4-dimethylhex-2-ene is treated with HBr and peroxides under similar conditions.

Mechanisms

15.53 Consider the following bromination: $(CH_3)_3CH + Br_2 \xrightarrow{\Delta} (CH_3)_3CBr + HBr$.

a. Calculate $\Delta H°$ for this reaction by using the bond dissociation energies in Table 6.2.
b. Draw out a stepwise mechanism for the reaction, including the initiation, propagation, and termination steps.
c. Calculate $\Delta H°$ for each propagation step.
d. Draw an energy diagram for the propagation steps.
e. Draw the structure of the transition state of each propagation step.

15.54 Draw a stepwise mechanism for the following reaction.

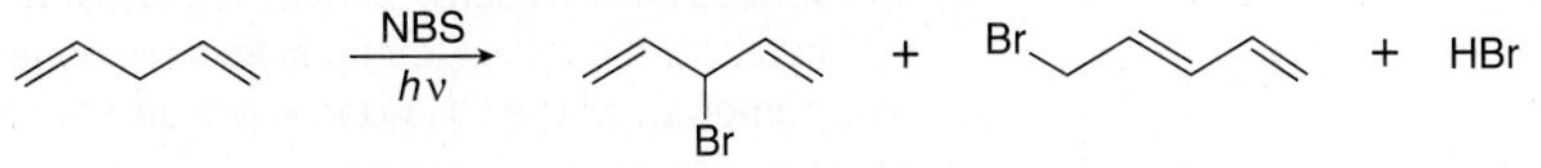

15.55 Although CH_4 reacts with Cl_2 to form CH_3Cl and HCl, the corresponding reaction of CH_4 with I_2 does not occur at an appreciable rate, even though the I–I bond is much weaker than the Cl–Cl bond. Explain why this is so.

15.56 When 3,3-dimethylbut-1-ene is treated with HBr alone, the major product is 2-bromo-2,3-dimethylbutane. When the same alkene is treated with HBr and peroxide, the sole product is 1-bromo-3,3-dimethylbutane. Explain these results by referring to the mechanisms.

Synthesis

15.57 Devise a synthesis of each compound from cyclopentane and any other required organic or inorganic reagents.

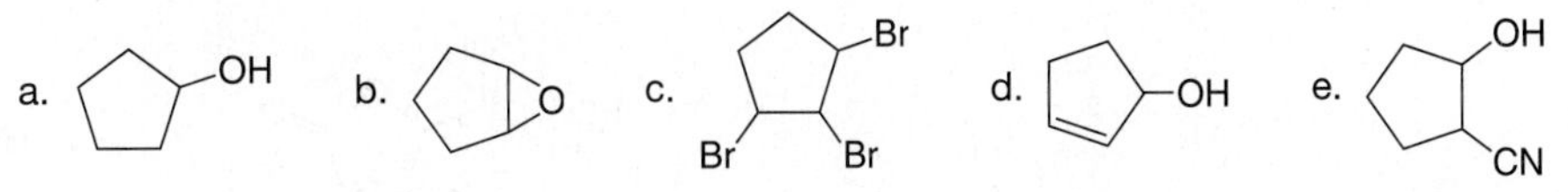

15.58 Devise a synthesis of 1-methylcyclohexene oxide from methylcyclohexane. You may use any other required organic or inorganic reagents.

15.59 Devise a synthesis of $CH_3CH_2CH_2CH_2Br$ from $HC\equiv CH$. You may use any other required organic or inorganic reagents.

15.60 Devise a synthesis of each compound using CH_3CH_3 as the only source of carbon atoms. You may use any other required organic or inorganic reagents.

a. $HC\equiv CH$ b. 

c. 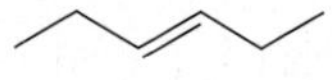
d. 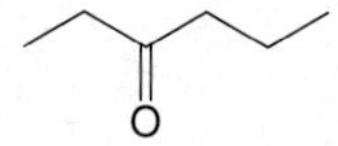

15.61 Devise a synthesis of $OHC(CH_2)_4CHO$ from cyclohexane using any required organic or inorganic reagents.

15.62 Devise a synthesis of hexane-2,3-diol from propane as the only source of carbon atoms. You may use any other required organic or inorganic reagents.

Radical Oxidation Reactions

15.63 As described in Section 9.17, the leukotrienes, important components in the asthmatic response, are synthesized from arachidonic acid via the hydroperoxide 5-HPETE. Write a stepwise mechanism for the conversion of arachidonic acid to 5-HPETE with O_2.

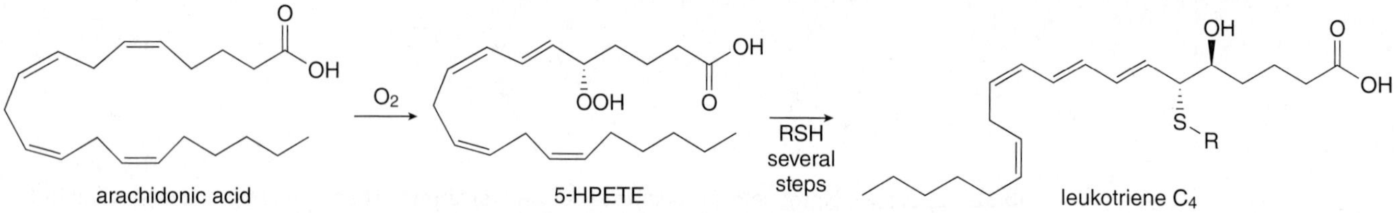

arachidonic acid 5-HPETE leukotriene C_4

15.64 Ethers are oxidized with O_2 to form hydroperoxides that decompose violently when heated. Draw a stepwise mechanism for this reaction.

unstable hydroperoxide

15.65 (a) Ignoring stereoisomers, what two allylic hydroperoxides are formed by the oxidation of hex-1-ene with O_2? (b) Draw a stepwise mechanism that shows how these hydroperoxides are formed.

Antioxidants

15.66 Resveratrol is an antioxidant found in the skin of red grapes. Its anticancer, anti-inflammatory, and various cardiovascular effects are under active investigation. (a) Draw all resonance structures for the radical that results from homolysis of the OH bond shown in red. (b) Explain why homolysis of this OH bond is preferred to homolysis of either OH bond in the other benzene ring.

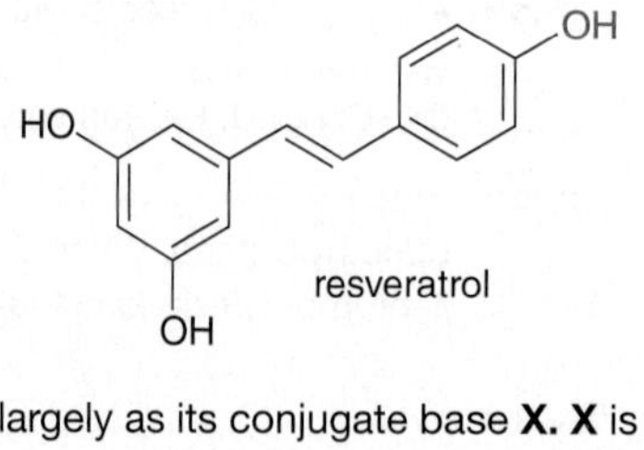

resveratrol

15.67 In cells, vitamin C exists largely as its conjugate base **X. X** is an antioxidant because radicals formed in oxidation processes abstract the labeled H atom, forming a new radical that halts oxidation. Draw the structure of the radical formed by H abstraction, and explain why this H atom is most easily removed.

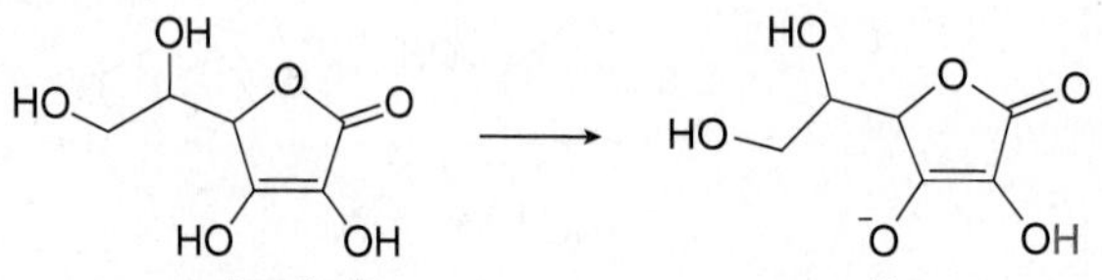

vitamin C **X**

Polymers and Polymerization

15.68 What monomer is needed to form each polymer?

a. 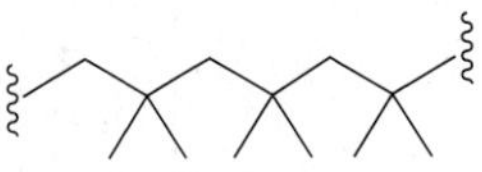

polyisobutylene
(used to make basketballs)

b. 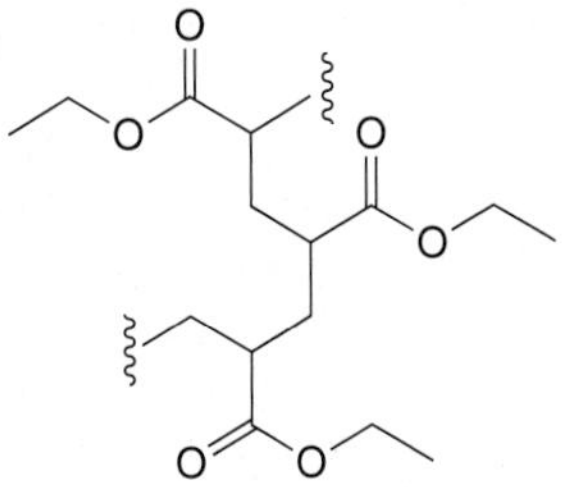

poly(ethyl acrylate)
(used in latex paints)

15.69 (a) Hard contact lenses, which first became popular in the 1960s, were made by polymerizing methyl methacrylate [$CH_2=C(CH_3)CO_2CH_3$] to form poly(methyl methacrylate) (PMMA). Draw the structure of PMMA. (b) More comfortable softer contact lenses introduced in the 1970s were made by polymerizing hydroxyethyl methacrylate [$CH_2=C(CH_3)CO_2CH_2CH_2OH$] to form poly(hydroxyethyl methacrylate) (poly-HEMA). Draw the structure of poly-HEMA. Since neither polymer allows oxygen from the air to pass through to the retina, newer contact lenses that are both comfortable and oxygen-permeable have now been developed.

15.70 Explain why polystyrene is much more readily oxidized by O_2 in the air than polyethylene is. Which H's in polystyrene are most easily abstracted and why?

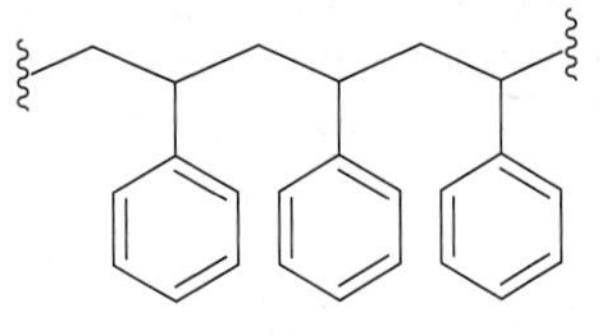

polystyrene

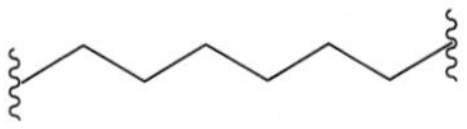

polyethylene

15.71 Draw a stepwise mechanism for the following polymerization reaction.

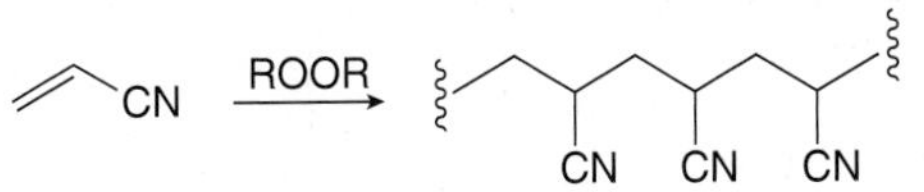

15.72 As we will learn in Chapter 31, styrene derivatives such as **A** can be polymerized by way of cationic rather than radical intermediates. Cationic polymerization is an example of electrophilic addition to an alkene involving carbocations.

A

a. Draw a short segment of the polymer formed by the polymerization of **A.**
b. Why does **A** react faster than styrene ($C_6H_5CH=CH_2$) in a cationic polymerization?

15.73 When two monomers (**X** and **Y**) are polymerized together, a copolymer results. An alternating copolymer is formed when the two monomers **X** and **Y** alternate regularly in the polymer chain. Draw the structure of the alternating copolymer formed when the two monomers, $CH_2=CCl_2$ and $CH_2=CHC_6H_5$, are polymerized together.

Spectroscopy

15.74 **A** and **B,** isomers of molecular formula $C_3H_5Cl_3$, are formed by the radical chlorination of a dihalide **C** of molecular formula $C_3H_6Cl_2$.

a. Identify the structures of **A** and **B** from the following ^1H NMR data:
   Compound **A:** singlet at 2.23 and singlet at 4.04 ppm
   Compound **B:** doublet at 1.69, multiplet at 4.34, and doublet at 5.85 ppm
b. What is the structure of **C?**

15.75 Identify the structure of a minor product formed from the radical chlorination of propane, which has molecular formula $C_3H_6Cl_2$ and exhibits the given ^1H NMR spectrum.

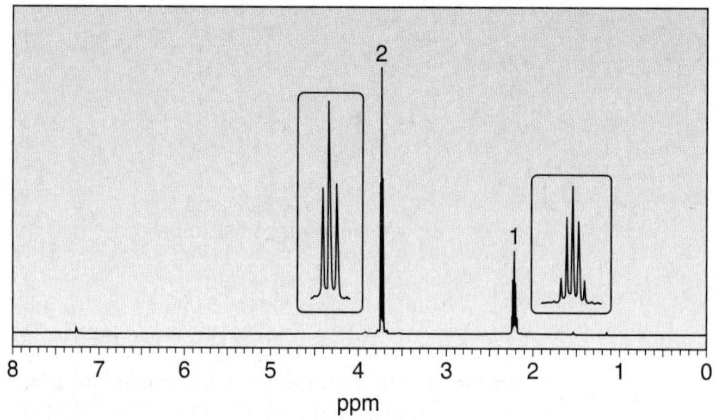

Challenge Problems

15.76 Draw a stepwise mechanism for the following addition reaction to an alkene.

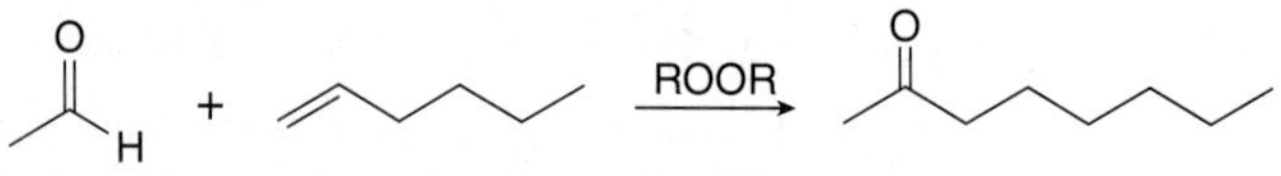

15.77 The triphenylmethyl radical is an unusual persistent radical present in solution in equilibrium with its dimer. For 70 years the dimer was thought to be hexaphenylethane, but in 1970, NMR data showed it to be **A**.

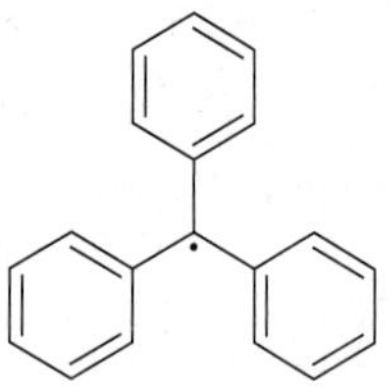
triphenylmethyl radical

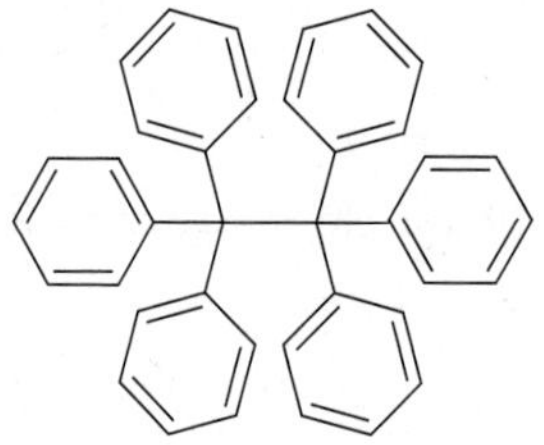
hexaphenylethane

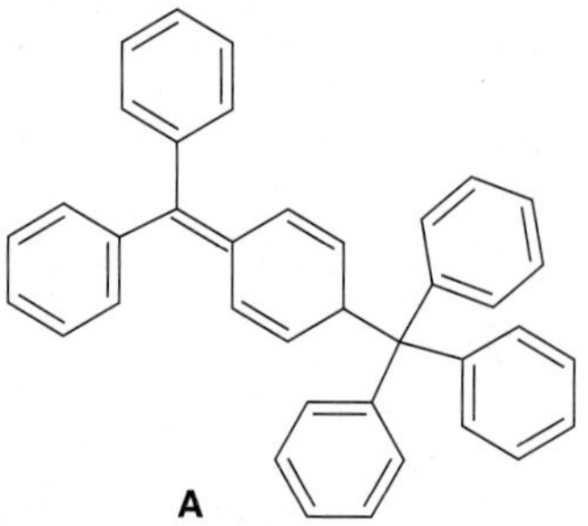
A

a. Why is the triphenylmethyl radical more stable than most other radicals?
b. Use curved arrow notation to show how two triphenylmethyl radicals dimerize to form **A.**
c. Propose a reason for the formation of **A** rather than hexaphenylethane.
d. How could ^1H and ^{13}C NMR spectroscopy be used to distinguish between hexaphenylethane and **A?**

15.78 In the presence of a radical initiator (Z·), tributyltin hydride (R_3SnH, R = $CH_3CH_2CH_2CH_2$) reduces alkyl halides to alkanes: $R'X + R_3SnH \rightarrow R'H + R_3SnX$. The mechanism consists of a radical chain process with an intermediate tin radical:

Initiation: $R_3SnH + Z\cdot \longrightarrow R_3Sn\cdot + HZ$

Propagation:

$R'{-}Br + R_3Sn\cdot \longrightarrow R'\cdot + R_3SnBr$

$R'\cdot + R_3SnH \longrightarrow R'{-}H + R_3Sn\cdot$

This reaction has been employed in many radical cyclization reactions. Draw a stepwise mechanism for the following reaction.

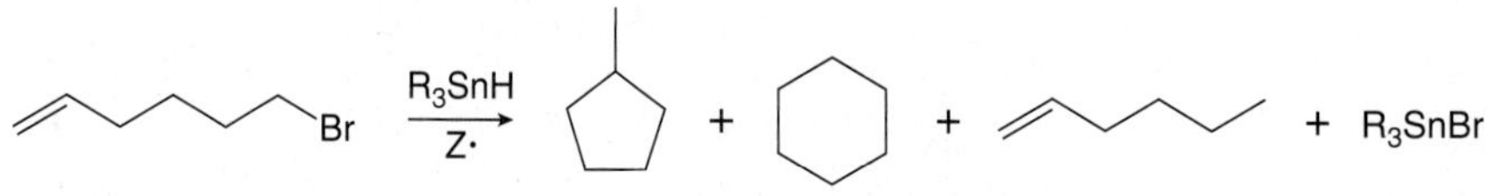

15.79 $PGF_{2\alpha}$ (Section 4.15) is synthesized in cells from arachidonic acid ($C_{20}H_{32}O_2$) using a cyclooxygenase enzyme that catalyzes a multistep radical pathway. Part of this process involves the conversion of radical **A** to PGG_2, an unstable intermediate, which is then transformed to $PGF_{2\alpha}$ and other prostaglandins. Draw a stepwise mechanism for the conversion of **A** to PGG_2. (Hint: The mechanism begins with radical addition to a carbon–carbon double bond to form a resonance-stabilized radical.)

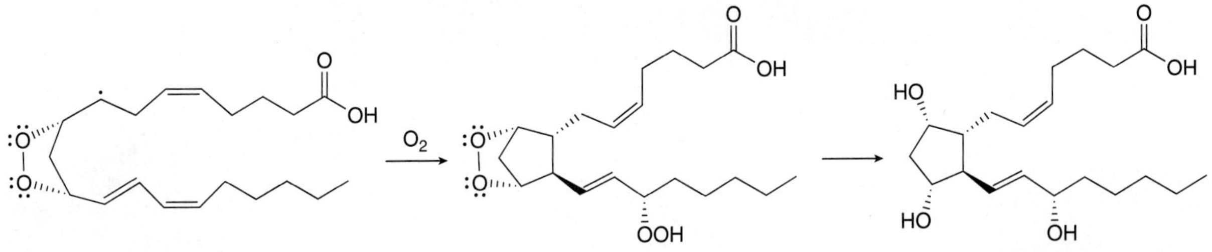

A

PGG_2
unstable intermediate

$PGF_{2\alpha}$

16 Conjugation, Resonance, and Dienes

16.1 Conjugation
16.2 Resonance and allylic carbocations
16.3 Common examples of resonance
16.4 The resonance hybrid
16.5 Electron delocalization, hybridization, and geometry
16.6 Conjugated dienes
16.7 Interesting dienes and polyenes
16.8 The carbon–carbon σ bond length in buta-1,3-diene
16.9 Stability of conjugated dienes
16.10 Electrophilic addition: 1,2- versus 1,4-addition
16.11 Kinetic versus thermodynamic products
16.12 The Diels–Alder reaction
16.13 Specific rules governing the Diels–Alder reaction
16.14 Other facts about the Diels–Alder reaction
16.15 Conjugated dienes and ultraviolet light

Morphine is an analgesic and narcotic isolated from the opium poppy *Papaver somniferum*. Opium has been widely used as a recreational drug and painkilling remedy for centuries, and poppy seed tea, which contains morphine, was used as a folk remedy in parts of England until World War II. A key step in a laboratory synthesis of morphine involves the Diels–Alder reaction, a powerful reaction of conjugated dienes discussed in Chapter 16.

Chapter 16 is the first of three chapters that discuss the chemistry of conjugated molecules—molecules with overlapping *p* orbitals on three or more adjacent atoms. Chapter 16 focuses mainly on acyclic conjugated compounds, whereas Chapters 17 and 18 discuss the chemistry of benzene and related compounds that have a *p* orbital on every atom in a ring.

Much of Chapter 16 is devoted to the properties and reactions of 1,3-dienes. To understand these compounds, however, we must first learn about the consequences of having *p* orbitals on three or more adjacent atoms. Because the ability to draw resonance structures is also central to mastering this material, the key aspects of resonance theory are presented in detail.

16.1 Conjugation

Conjugation **occurs whenever *p* orbitals can overlap on three or more adjacent atoms.** Two common conjugated systems are 1,3-dienes and allylic carbocations.

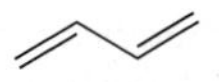

1,3-diene

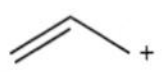

allylic carbocation

16.1A 1,3-Dienes

1,3-Dienes such as buta-1,3-diene contain two carbon–carbon double bonds joined by a single σ bond. Each carbon atom of a 1,3-diene is bonded to three other atoms and has no nonbonded electron pairs, so each carbon atom is sp^2 hybridized and has one *p* orbital containing an electron. **The four *p* orbitals on adjacent atoms make a 1,3-diene a conjugated system.**

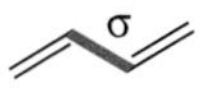

buta-1,3-diene

one σ bond between the double bonds

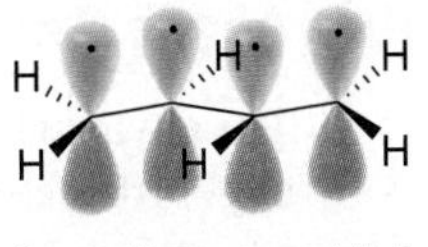

four adjacent *p* orbitals

Each C is sp^2 hybridized and has a *p* orbital containing one electron.

What is special about conjugation? Having three or more *p* orbitals on adjacent atoms allows *p* orbitals to overlap and electrons to delocalize.

The electron density in the two π bonds is delocalized.

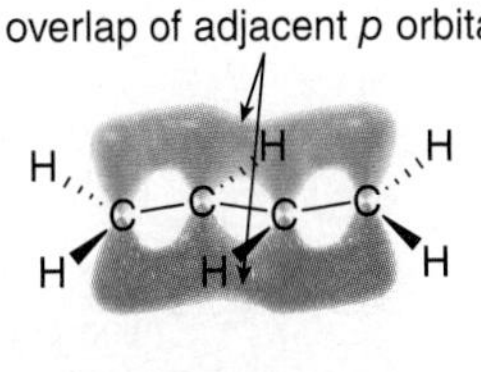

- **When *p* orbitals overlap, the electron density in each of the π bonds is spread out over a larger volume, thus lowering the energy of the molecule and making it more stable.**

Conjugation makes buta-1,3-diene inherently different from penta-1,4-diene, a compound having two double bonds separated by more than one σ bond. The π bonds in penta-1,4-diene are too far apart to be conjugated.

A conjugated diene

buta-1,3-diene

The electrons in the π bonds are **delocalized.**

An isolated diene

penta-1,4-diene

The electrons in the π bonds are **localized.**

Penta-1,4-diene is an **isolated diene.** The electron density in each π bond of an isolated diene is *localized* between two carbon atoms. In buta-1,3-diene, however, the electron density of both π bonds is *delocalized* over the four atoms of the diene. Electrostatic potential maps in Figure 16.1 clearly indicate the difference between these localized and delocalized π bonds.

Figure 16.1 Electrostatic potential plots for a conjugated and an isolated diene

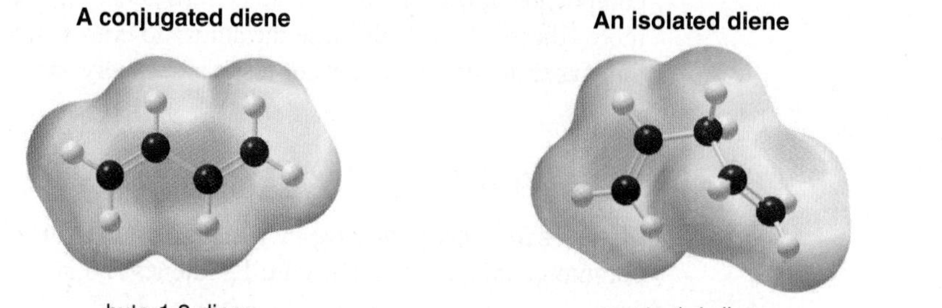

buta-1,3-diene
The red electron-rich region is spread over four adjacent atoms.

penta-1,4-diene
The red electron-rich regions are localized in the π bonds on the two ends of the molecule.

Problem 16.1 Classify each diene as isolated or conjugated.

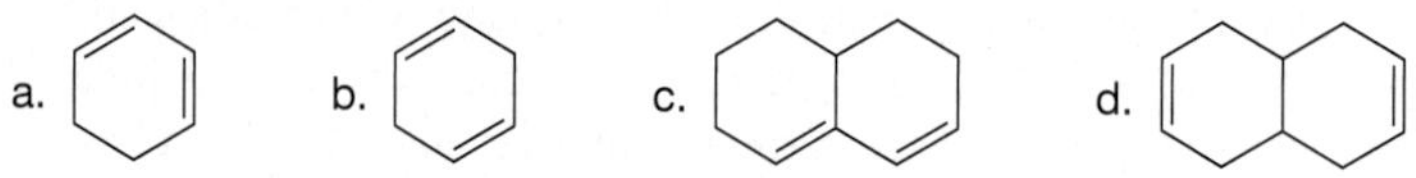

16.1B Allylic Carbocations

The **allyl carbocation** is another example of a conjugated system. The three carbon atoms of the allyl carbocation—the positively charged carbon atom and the two that form the double bond—are sp^2 hybridized and have an unhybridized *p* orbital. The *p* orbitals for the double bond carbons each contain an electron, whereas the *p* orbital for the carbocation is empty.

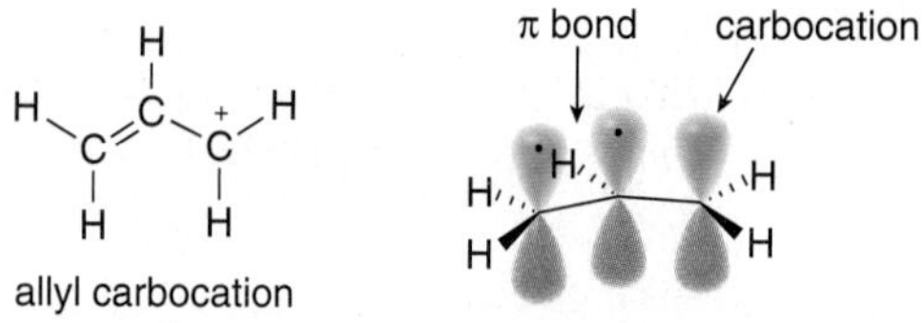

allyl carbocation
Each C is sp^2 hybridized and has a *p* orbital.

three adjacent *p* orbitals

- **Three *p* orbitals on three adjacent atoms, even if one of the *p* orbitals is empty, make the allyl carbocation conjugated.**

Conjugation stabilizes the allyl carbocation because overlap of three adjacent *p* orbitals delocalizes the electron density of the π bond over three atoms.

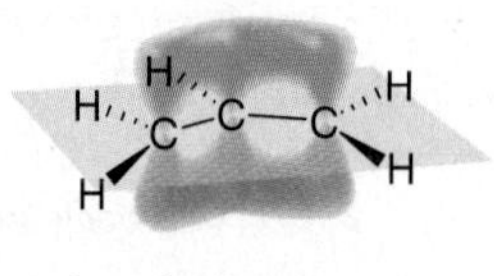

overlap of adjacent *p* orbitals

Problem 16.2 Which of the following species are conjugated?

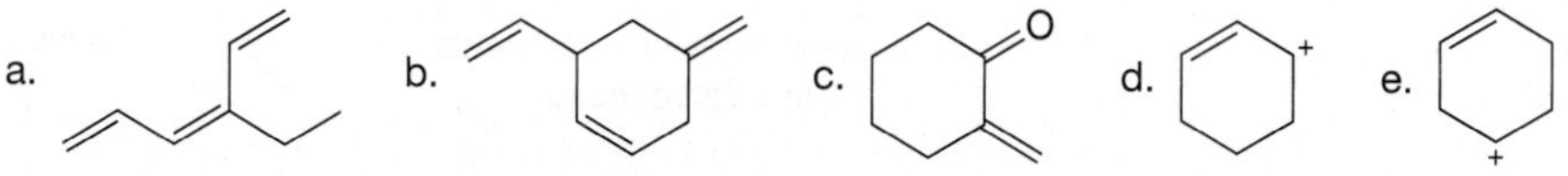

16.2 Resonance and Allylic Carbocations

The word *resonance* is used in two different contexts. In NMR spectroscopy, a nucleus is ***in resonance*** when it absorbs energy, promoting it to a higher energy state. In drawing molecules, there is ***resonance*** when two different Lewis structures can be drawn for the same arrangement of atoms.

Recall from Section 1.6 that resonance structures are two or more different Lewis structures for the same arrangement of atoms. Being able to draw correct resonance structures is crucial to understanding conjugation and the reactions of conjugated dienes.

- **Two resonance structures differ in the placement of π bonds and nonbonded electrons. The placement of atoms and σ bonds stays the same.**

16.2A The Stability of Allylic Carbocations

We have already drawn resonance structures for the acetate anion (Section 2.5C) and the allyl radical (Section 15.10). The **conjugated allyl carbocation** is another example of a species for which two resonance structures can be drawn. Drawing resonance structures for the allyl carbocation is a way to use Lewis structures to illustrate how conjugation delocalizes electrons.

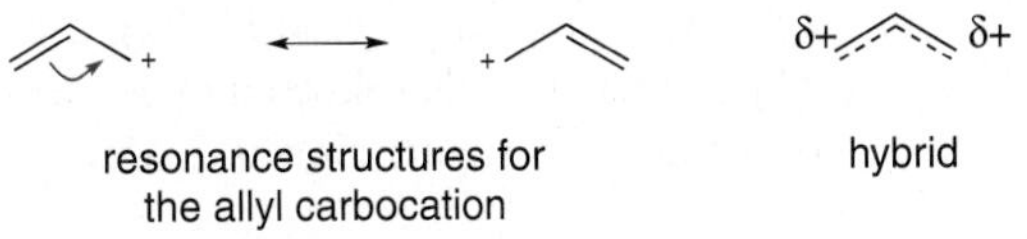

The true structure of the allyl carbocation is a hybrid of the two resonance structures. In the hybrid, the π bond is delocalized over all three atoms. As a result, the positive charge is also delocalized over the two terminal carbons. Delocalizing electron density lowers the energy of the hybrid, thus stabilizing the allyl carbocation and making it more stable than a normal 1° carbocation. Experimental data show that its stability is comparable to a more substituted 2° carbocation.

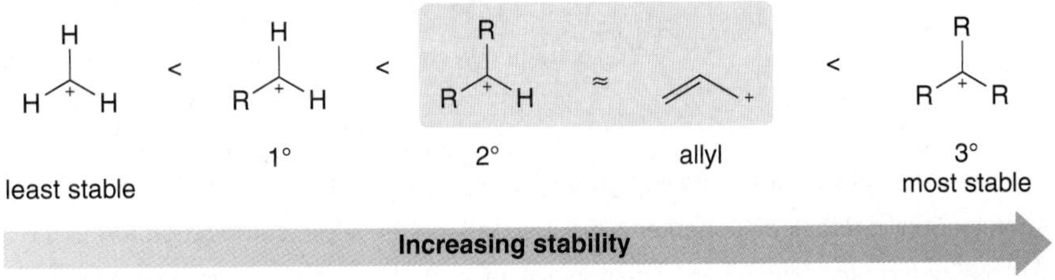

The electrostatic potential maps in Figure 16.2 compare the resonance-stabilized allyl carbocation with $CH_3CH_2CH_2^+$, a localized 1° carbocation. The electron-deficient region—the site of the positive charge—is concentrated on a single carbon atom in the 1° carbocation $CH_3CH_2CH_2^+$. In the allyl carbocation, however, the electron-poor region is spread out on both terminal carbons.

Problem 16.3 Draw a second resonance structure for each carbocation. Then draw the hybrid.

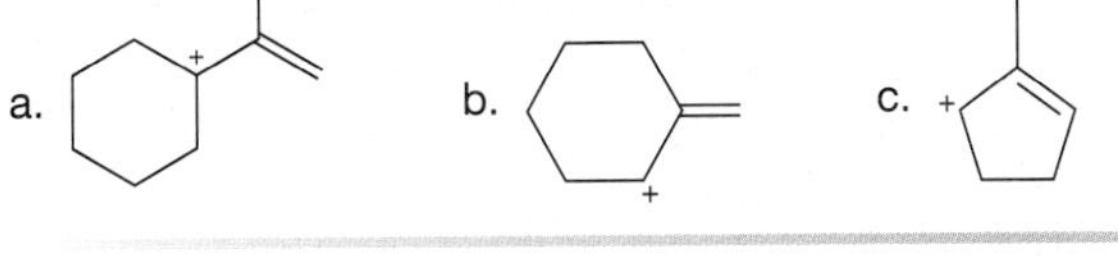

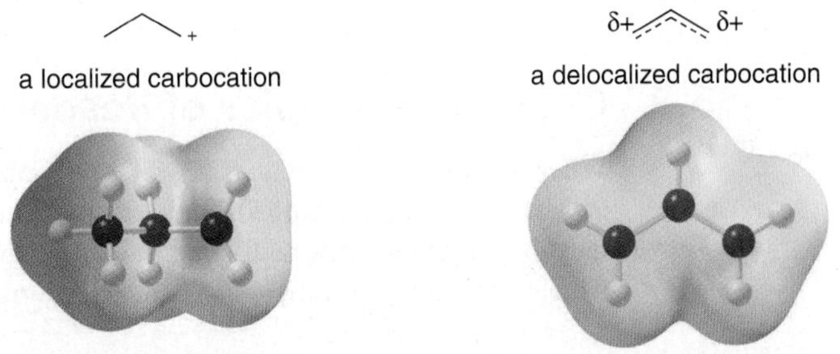

Figure 16.2 Electrostatic potential maps for a localized and a delocalized carbocation

The electron-deficient region (in blue) of a 1° **carbocation** is concentrated on a single carbon atom.

The electron-deficient region (in blue-green) of the **allyl carbocation** is distributed over both terminal carbons.

Problem 16.4 Use resonance theory and the Hammond postulate to explain why 3-chloroprop-1-ene ($CH_2=CHCH_2Cl$) is more reactive than 1-chloropropane ($CH_3CH_2CH_2Cl$) in S_N1 reactions.

16.2B Allylic Carbocations in Biological Reactions

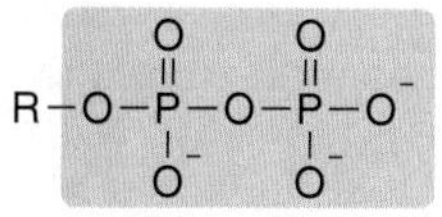

organic diphosphate

R–OPP

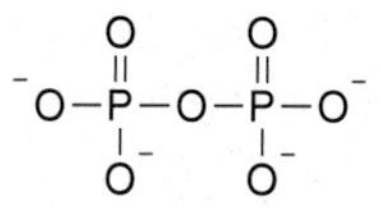

diphosphate leaving group
PP_i

Allylic carbocations formed from diphosphates (Section 7.16) are key intermediates in a variety of biological reactions, including the synthesis of geranyl diphosphate from two five-carbon substrates—dimethylallyl diphosphate and isopentenyl diphosphate. Geranyl diphosphate is the precursor of many lipids that occur in plants and animals.

dimethylallyl diphosphate   isopentenyl diphosphate   geranyl diphosphate

This biological process results in the formation of a new carbon–carbon bond and involves two key steps—loss of a good leaving group (diphosphate, $P_2O_7^{4-}$, abbreviated as PP_i) to form an allylic carbocation, followed by nucleophilic attack with an electron-rich double bond. The steps of the mechanism are shown in Mechanism 16.1.

Mechanism 16.1 Biological Formation of Geranyl Diphosphate

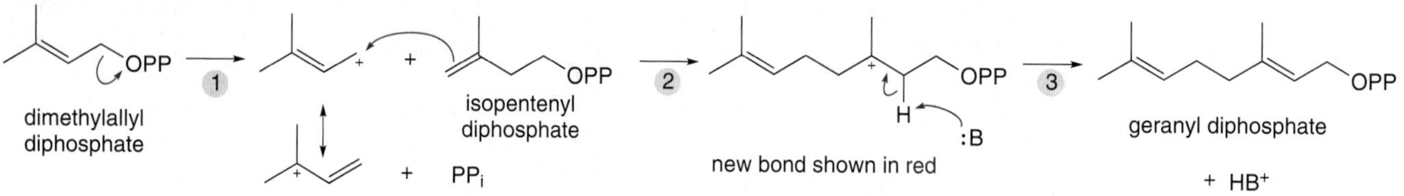

1 Loss of the diphosphate leaving group forms an allylic carbocation.

2 Nucleophilic attack of isopentenyl diphosphate on the allylic carbocation forms the new C–C σ bond.

3 Loss of a proton (shown with the general base, B:) forms geranyl diphosphate.

We will learn more about biological reactions involving allylic carbocations derived from diphosphates in Chapter 28.

Problem 16.5 Farnesyl diphosphate is synthesized from isopentenyl diphosphate and **X** by a pathway similar to Mechanism 16.1. Draw the structure of **X.**

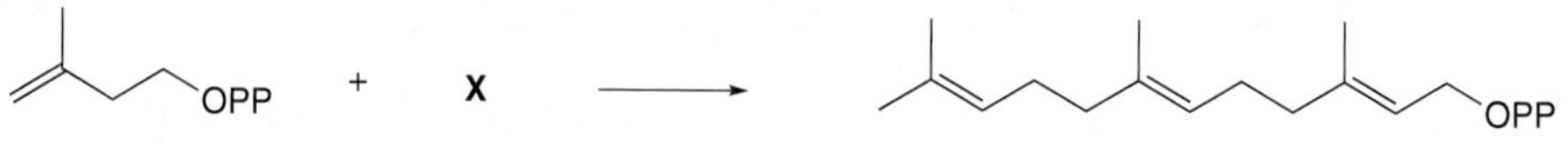

isopentenyl diphosphate   farnesyl diphosphate

16.3 Common Examples of Resonance

When are resonance structures drawn for a molecule or reactive intermediate? Because resonance involves delocalizing π bonds and nonbonded electrons, one or both of these structural features must be present to draw additional resonance forms. There are four common bonding patterns for which more than one Lewis structure can be drawn.

Type [1] The Three Atom "Allyl" System, X=Y–Z*

- **For any group of three atoms having a double bond X=Y and an atom Z that contains a *p* orbital with zero, one, or two electrons, two resonance structures are possible:**

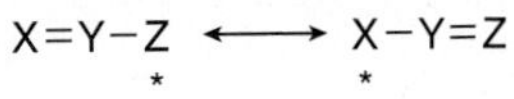
X=Y–Z
*
X–Y=Z
*

The asterisk [*] corresponding to a charge, a radical, or a lone pair.

* = +, –, ·, or ··

This is called **allyl** type resonance because it can be drawn for allylic carbocations, allylic carbanions, and allylic radicals.

X, Y, and Z may all be carbon atoms, as in the case of an allylic carbocation (resonance structures **A** and **B**), or they may be heteroatoms, as in the case of the acetate anion (resonance structures **C** and **D**). The atom Z bonded to the multiple bond can be charged (a net positive or negative charge) or neutral (having zero, one, or two nonbonded electrons). **The two resonance structures differ in the location of the double bond, and either the charge, the radical, or the lone pair, generalized by [*].**

Allylic carbocation **Acetate anion**

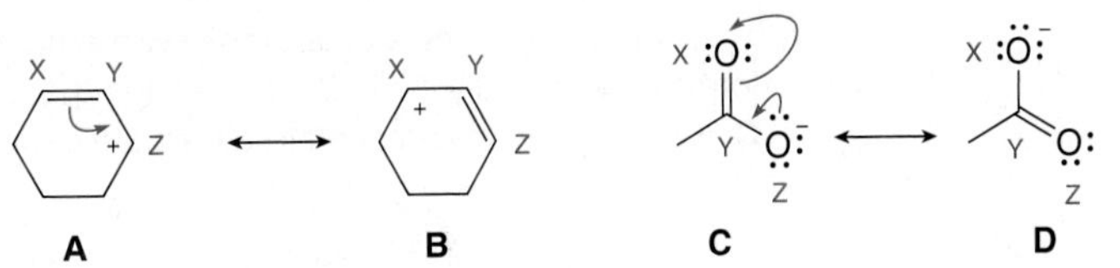
X
Y
Z
X
Y
Z
X
Y
Z
X
Y
Z
A
B
C
D

Type [2] Conjugated Double Bonds

Cyclic, completely conjugated rings like benzene have two resonance structures, drawn by moving the electrons in a cyclic manner around the ring. Three resonance structures can be drawn for conjugated dienes, two of which involve charge separation.

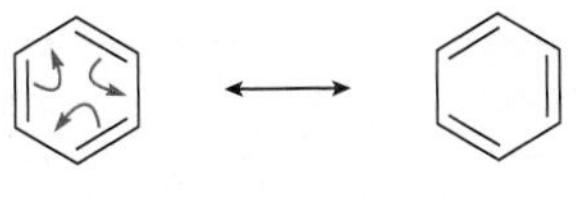

two resonance structures for benzene

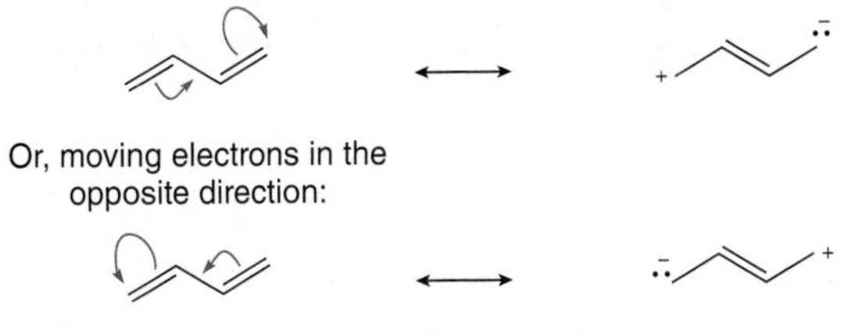
Or, moving electrons in the opposite direction:

three resonance structures for buta-1,3-diene

Type [3] Cations Having a Positive Charge Adjacent to a Lone Pair

- **When a lone pair and a positive charge are located on adjacent atoms, two resonance structures can be drawn.**

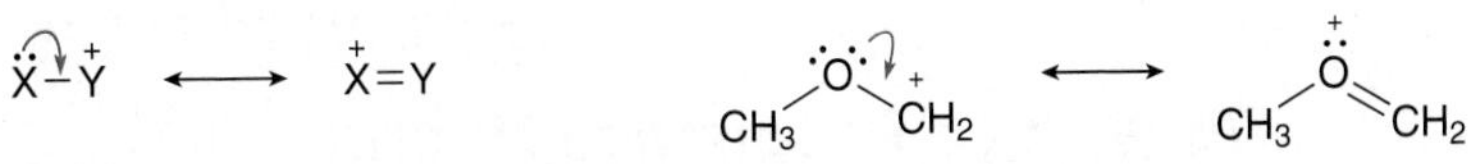
X–Y
X=Y
CH3
CH2
CH3
CH2

The overall charge is the same in both resonance structures. Based on formal charge, a neutral X in one structure must bear a (+) charge in the other.

Type [4] Double Bonds Having One Atom More Electronegative Than the Other

- **For a double bond X=Y in which the electronegativity of Y > X, a second resonance structure can be drawn by moving the π electrons onto Y.**

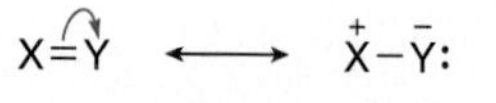

Electronegativity of Y > X.

Sample Problem 16.1 illustrates how to apply these different types of resonance to actual molecules.

Sample Problem 16.1 Draw two more resonance structures for each species.

a.  b. 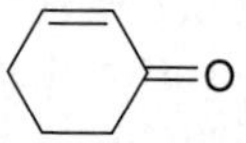

Solution

Mentally breaking a molecule into two- or three-atom units can make it easier to draw additional resonance structures.

a. Think of the top three atoms of the six-membered ring in **A** as an "allyl" unit. Moving the π bond forms a new "allyl" unit in **B,** and moving the π bond in **B** generates a third resonance structure **C.** No new valid resonance structures are generated by moving electrons in **C.**

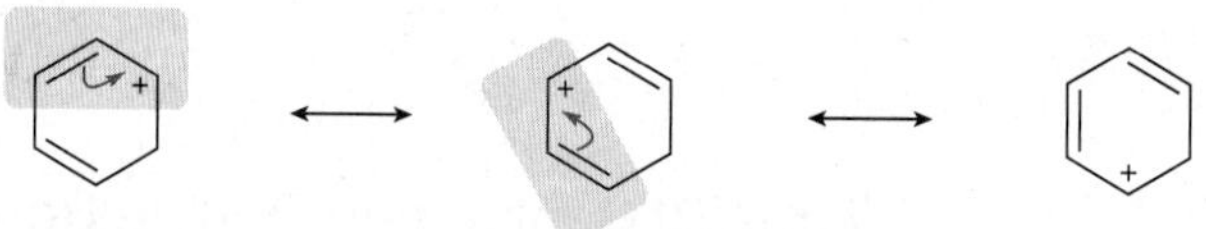

A a three-atom "allyl" unit — **B** another three-atom "allyl" unit — **C**

b. Compound **D** contains a carbonyl group, so moving the electron pair in the double bond to the more electronegative oxygen atom separates the charge and generates structure **E. E** now has a three-atom "allyl" unit, so the remaining π bond can be moved to form structure **F.**

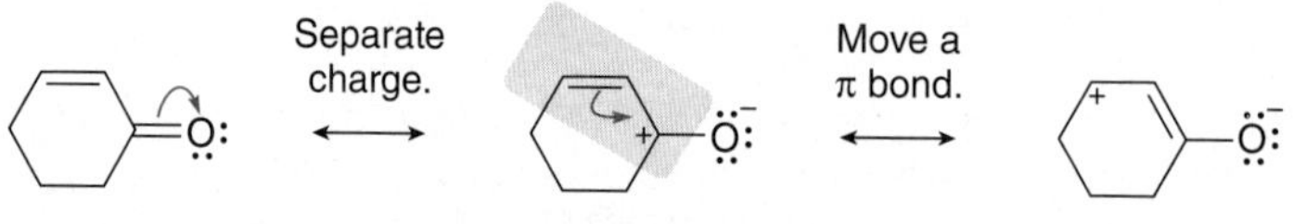
Separate charge.
Move a π bond.

D — **E** a three-atom "allyl" unit — **F**

Problem 16.6 Draw additional resonance structures for each ion.

a. 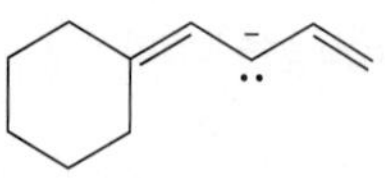 b. 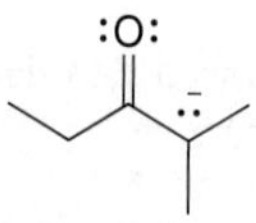 c. 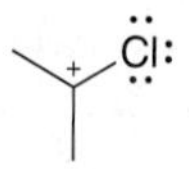 d. 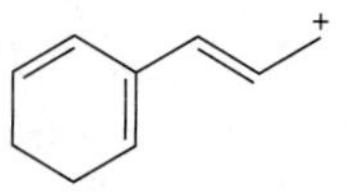

16.4 The Resonance Hybrid

The lower its energy, the more a resonance structure contributes to the overall structure of the hybrid.

Although the resonance hybrid is some combination of all of its valid resonance structures, the **hybrid more closely resembles the best resonance structure.** Recall from Section 1.6C that the best resonance structure is called the **major contributor** to the hybrid, and other resonance structures are called the **minor contributors.** Two identical resonance structures are equal contributors to the hybrid.

Use the following three rules to evaluate the relative energies of two or more valid resonance structures.

Rule [1] **Resonance structures with more bonds and fewer charges are better.**

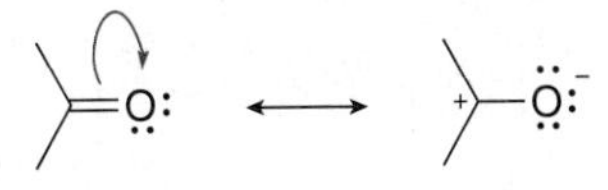

all neutral atoms
one more bond
better resonance structure

charge separation

Rule [2] **Resonance structures in which every atom has an octet are better.**

All second-row elements have an octet.
better resonance structure

In this example, the resonance structure in which all atoms have octets is better, even though it places a (+) charge on a more electronegative O atom.

Rule [3] **Resonance structures that place a negative charge on a more electronegative atom are better.**

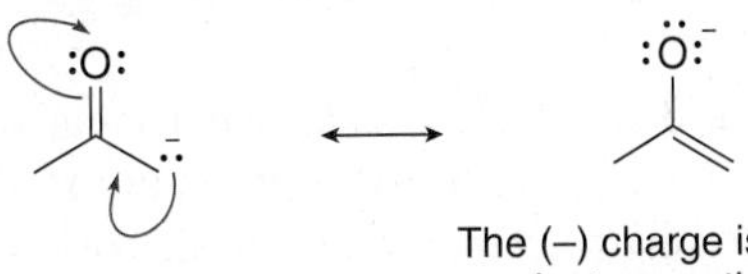

The (–) charge is on the more electronegative O atom.

better resonance structure

Sample Problem 16.2 illustrates how to determine the relative energy of contributing resonance structures and the hybrid.

Sample Problem 16.2 Draw a second resonance structure for carbocation **A,** as well as the hybrid of both resonance structures. Then use Rules [1]–[3] to rank the relative stability of both resonance structures and the hybrid.

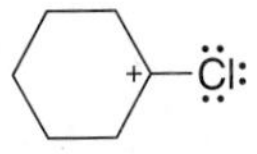

A

Solution

Because **A** contains a positive charge and a lone pair on adjacent atoms, a second resonance structure **B** can be drawn. Because **B** has more bonds and all second-row atoms have octets, **B** is a better resonance structure than **A,** making it the major contributor to the hybrid **C.** Because the hybrid is more stable than either resonance contributor, the order of stability is:

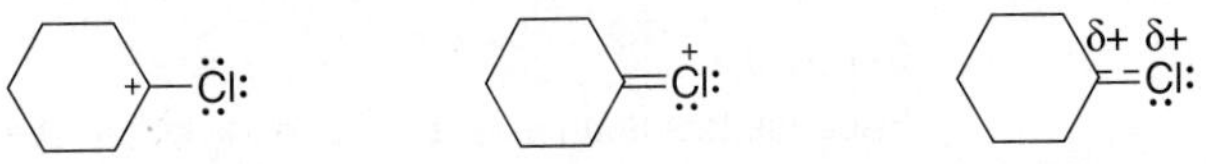

A minor contributor | **B** major contributor | **C** hybrid

Increasing stability →

Problem 16.7 Draw a second resonance structure and the hybrid for each species, and then rank the two resonance structures and the hybrid in order of increasing stability.

a. 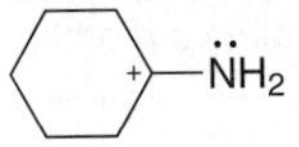
b. 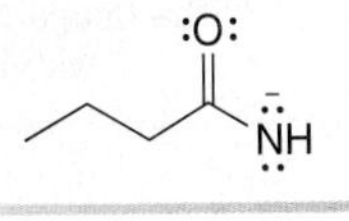
c. 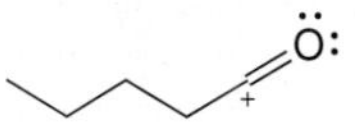
d. 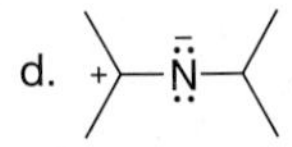

Problem 16.8 Draw all possible resonance structures for the following cation and indicate which structure makes the largest contribution to the resonance hybrid.

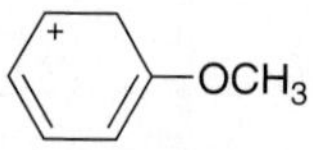

16.5 Electron Delocalization, Hybridization, and Geometry

To delocalize nonbonded electrons or electrons in π bonds, there must be *p* orbitals that can overlap. This may mean that the hybridization of an atom is different than would have been predicted using the rules first outlined in Chapter 1.

For example, there are two Lewis structures (**A** and **B**) for the resonance-stabilized anion $(CH_3COCH_2)^-$.

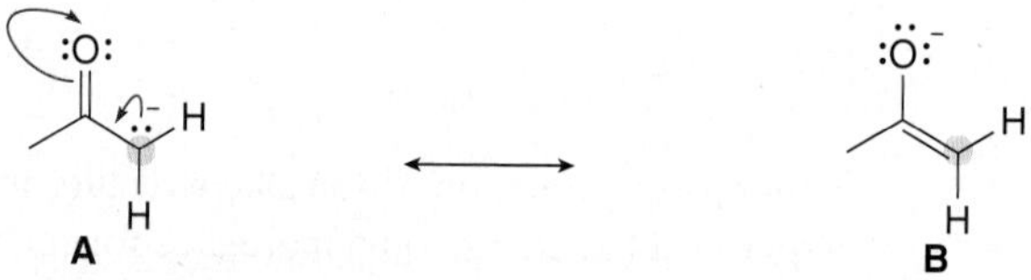

A **B**

The labeled C is surrounded by four groups—three atoms and one nonbonded electron pair. **Is it *sp*³ hybridized?**	The labeled C is surrounded by three groups—three atoms and no nonbonded electron pairs. **Is it *sp*² hybridized?**

Based on structure **A,** the labeled carbon is sp^3 hybridized, with the lone pair of electrons in an sp^3 hybrid orbital. Based on structure **B,** though, it is sp^2 hybridized with the unhybridized *p* orbital forming the π portion of the double bond.

Delocalizing electrons stabilizes a molecule. The electron pair on the carbon atom adjacent to the C=O can only be delocalized, though, if it has a *p* orbital that can overlap with two other *p* orbitals on two adjacent atoms. Thus, the terminal carbon atom is sp^2 hybridized with trigonal planar geometry. **Three adjacent *p* orbitals make the anion conjugated.**

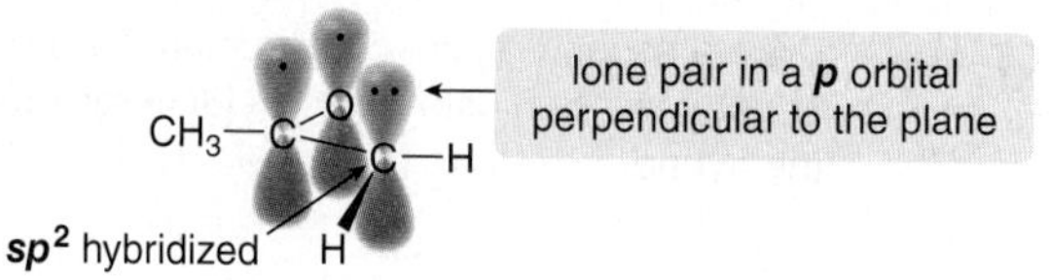

- **In a system X=Y−Z:, Z is generally *sp*² hybridized, and the nonbonded electron pair occupies a *p* orbital to make the system conjugated.**

Sample Problem 16.3 Determine the hybridization around the labeled carbon atom in the following anion.

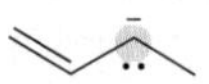

Solution

Because this is an example of an allyl-type system (X=Y−Z*), a second resonance structure can be drawn that "moves" the lone pair and the π bond. To delocalize the lone pair and make the system conjugated, the labeled carbon atom must be sp^2 hybridized with the lone pair occupying a *p* orbital.

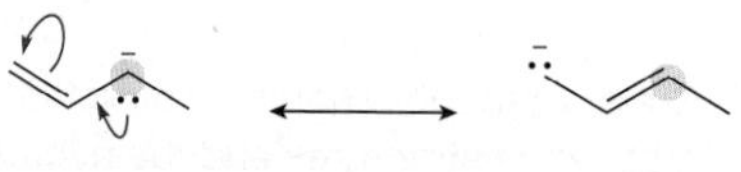

The labeled C atom must be *sp*² hybridized, with the lone pair in a *p* orbital.

Problem 16.9 Determine the hybridization of the labeled atom in each species.

a. 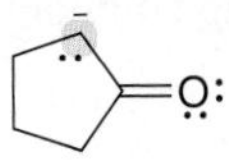 b. 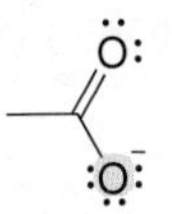 c. 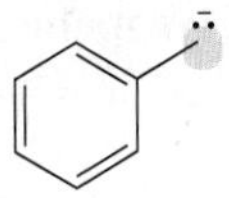

16.6 Conjugated Dienes

Compounds with many π bonds are called **polyenes.**

In the remainder of Chapter 16 we examine **conjugated dienes,** compounds having two double bonds joined by one σ bond. Conjugated dienes are also called **1,3-dienes.** Buta-1,3-diene ($CH_2{=}CH{-}CH{=}CH_2$) is the simplest conjugated diene.

Three stereoisomers are possible for 1,3-dienes with alkyl groups bonded to each end carbon of the diene (RCH=CH−CH=CHR).

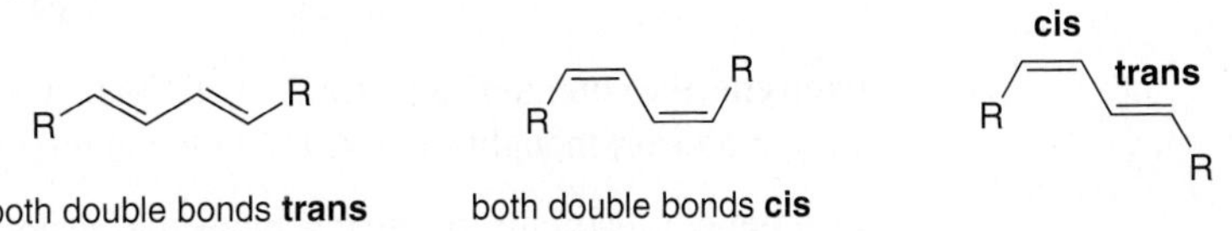
R
R
R
R
cis
trans
R
R

both double bonds **trans**	both double bonds **cis**	
trans,trans-1,3-diene or (*E,E*)-1,3-diene	*cis,cis*-1,3-diene or (*Z,Z*)-1,3-diene	*cis,trans*-1,3-diene or (*Z,E*)-1,3-diene

Two possible conformations result from rotation around the C−C bond that joins the two double bonds.

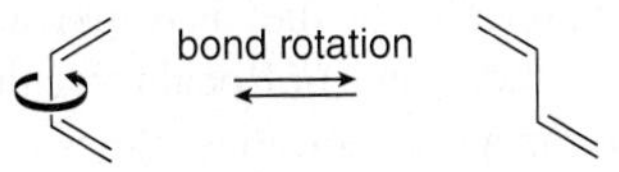
bond rotation

s-cis conformation **s-trans** conformation

- The s-cis conformation has two double bonds on the *same* side of the single bond.
- The s-trans conformation has two double bonds on *opposite* sides of the single bond.

Keep in mind that stereoisomers are discrete molecules, whereas conformations interconvert. Three structures drawn for hexa-2,4-diene illustrate the differences between stereoisomers and conformations in a 1,3-diene:

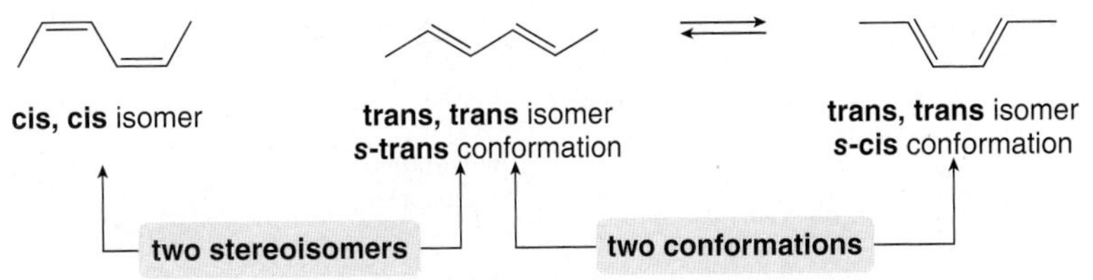
cis, cis isomer
trans, trans isomer
s-trans conformation
trans, trans isomer
s-cis conformation
two stereoisomers
two conformations

Problem 16.10 Draw the structure consistent with each description.

a. (2*E*,4*E*)-octa-2,4-diene in the *s*-trans conformation
b. (3*E*,5*Z*)-nona-3,5-diene in the *s*-cis conformation
c. (3*Z*,5*Z*)-4,5-dimethyldeca-3,5-diene. Draw both the *s*-cis and *s*-trans conformations.

Problem 16.11 Neuroprotectin D1 (NPD1) is synthesized in the body from highly unsaturated essential fatty acids. NPD1 is a potent natural anti-inflammatory agent.

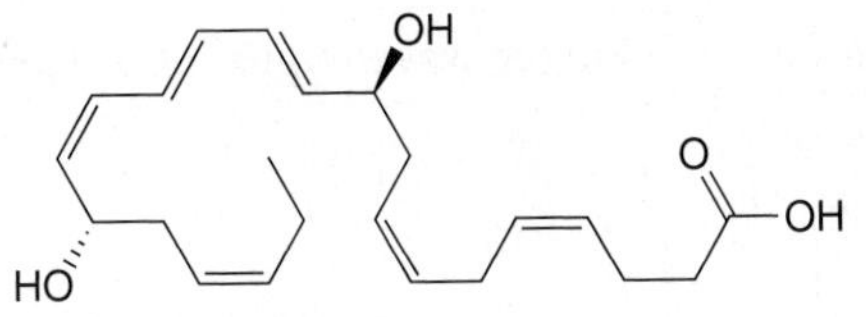
OH
O
OH
HO

NPD1

a. Label each double bond as conjugated or isolated.
b. Label each double bond as *E* or *Z*.
c. For each conjugated system, label the given conformation as *s*-cis or *s*-trans.

16.7 Interesting Dienes and Polyenes

Isoprene is a component of the blue haze seen above forested hillsides, such as Virginia's Blue Ridge Mountains.

Isoprene and **lycopene** are two naturally occurring compounds with conjugated double bonds.

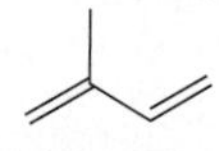

isoprene
(2-methylbuta-1,3-diene)

11 conjugated double bonds shown in red

lycopene

Isoprene, the common name for 2-methylbuta-1,3-diene, is given off by plants as the temperature rises, a process thought to increase a plant's tolerance for heat stress.

Lycopene, a naturally occurring molecule responsible for the red color of tomatoes and other fruits, is an antioxidant like vitamin E. The 11 conjugated double bonds of lycopene cause its red color, a phenomenon discussed in Section 16.15.

Simvastatin and calcitriol are two drugs that contain conjugated double bonds in addition to other functional groups (Figure 16.3). Simvastatin is the generic name of the widely used cholesterol-lowering medicine Zocor. Calcitriol, a biologically active hormone formed from vitamin D_3 obtained in the diet, is responsible for regulating calcium and phosphorus metabolism. Sold under the trade name of Rocaltrol, calcitriol is used to treat patients who are unable to convert vitamin D_3 to the active hormone. Since calcitriol promotes the absorption of calcium ions, it is also used to treat hypocalcemia, the presence of low calcium levels in the blood.

Figure 16.3
Biologically active organic compounds that contain conjugated double bonds

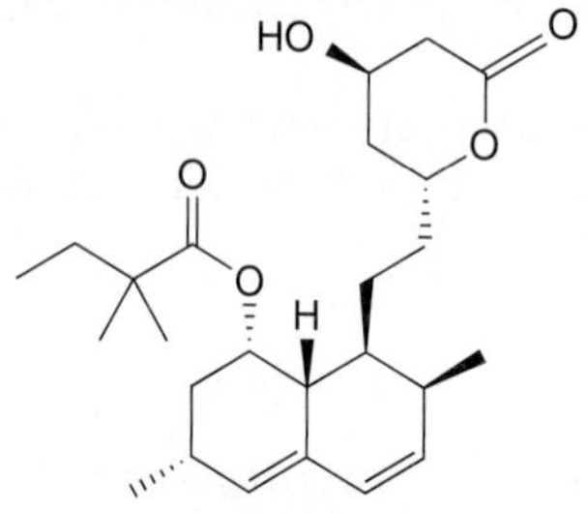

simvastatin
(Zocor)

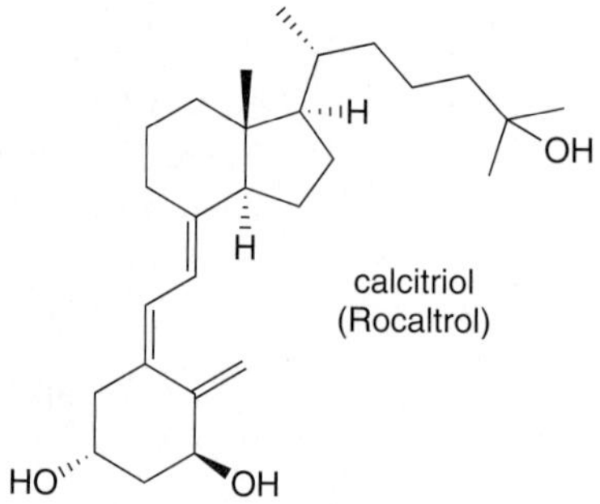

calcitriol
(Rocaltrol)

16.8 The Carbon–Carbon σ Bond Length in Buta-1,3-diene

Four features distinguish conjugated dienes from isolated dienes.

[1] **The C–C single bond joining the two double bonds is unusually short.**

[2] **Conjugated dienes are more stable than similar isolated dienes.**

[3] **Some reactions of conjugated dienes are different than reactions of isolated double bonds.**

[4] **Conjugated dienes absorb longer wavelengths of ultraviolet light.**

The bond length of the carbon–carbon double bonds in buta-1,3-diene is similar to an isolated double bond (134 pm), but the central carbon–carbon single bond is shorter than the C–C bond in ethane (148 pm vs. 153 pm).

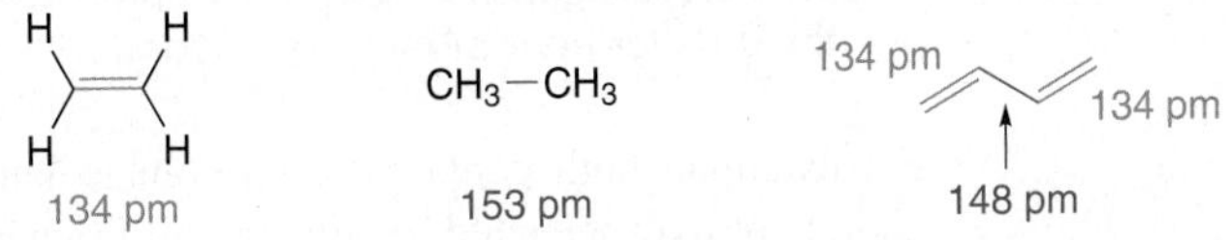

The observed bond distances can be explained by looking at hybridization. Each carbon atom in buta-1,3-diene is sp^2 hybridized, so the central C–C single bond is formed by the overlap of two sp^2 hybridized orbitals, rather than the sp^3 hybridized orbitals used to form the C–C bond in CH_3CH_3.

$CH_3{-}CH_3$	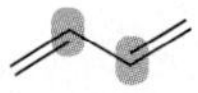
sp^3 C's	sp^2 C's
25% *s*-character	33% *s*-character
lower percent *s*-character **longer bond**	higher percent *s*-character **shorter bond**

Recall from Section 1.11B that increasing percent *s*-character decreases bond length.

- **Based on hybridization, a $C_{sp^2}-C_{sp^2}$ bond should be shorter than a $C_{sp^3}-C_{sp^3}$ bond because it is formed from orbitals having a higher percent *s*-character.**

A resonance argument can also be used to explain the shorter C–C σ bond length of buta-1,3-diene. Buta-1,3-diene can be represented by three resonance structures:

A	**B**	**C**	hybrid
major contributor	minor contributor	minor contributor	The C–C bond (labeled in red) has partial double bond character.

Structures **B** and **C** have charge separation and fewer bonds than **A,** making them less stable resonance structures and only minor contributors to the resonance hybrid. **B** and **C** both contain a double bond between the central carbon atoms, however, so the hybrid must have a partial double bond there. This makes the central C–C bond shorter than a C–C single bond in an alkane.

- **Based on resonance, the central C–C bond in buta-1,3-diene is shorter because it has partial double bond character.**

Problem 16.12 Using hybridization, predict how the bond length of the C–C σ bond in HC≡C–C≡CH should compare with the C–C σ bonds in CH_3CH_3 and $CH_2{=}CH{-}CH{=}CH_2$.

Problem 16.13 Use resonance theory to explain why the labeled C–O bond lengths are equal in the acetate anion.

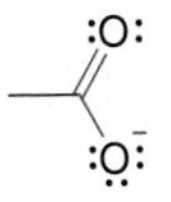

acetate

16.9 Stability of Conjugated Dienes

In Section 12.3 we learned that hydrogen adds to alkenes to form alkanes, and that the heat released in this reaction, the **heat of hydrogenation,** can be used as a measure of alkene stability.

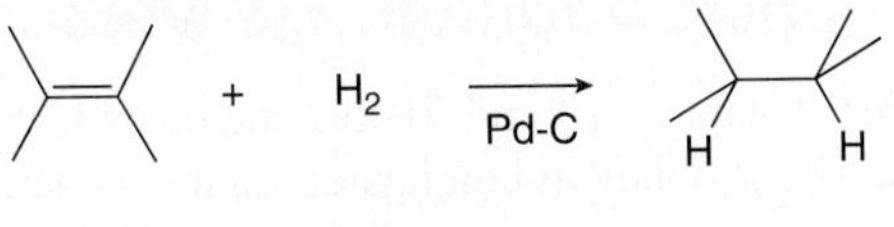

$\Delta H°$ = **heat of hydrogenation**

The relative stability of conjugated and isolated dienes can also be determined by comparing their heats of hydrogenation.

- **When hydrogenation gives the same alkane from two dienes, the more stable diene has the smaller heat of hydrogenation.**

For example, both penta-1,4-diene (an isolated diene) and (*E*)-penta-1,3-diene (a conjugated diene) are hydrogenated to pentane with two equivalents of H_2. Because less energy is released in converting the conjugated diene to pentane, it must be lower in energy (more stable) to begin with. The relative energies of these isomeric pentadienes are illustrated in Figure 16.4.

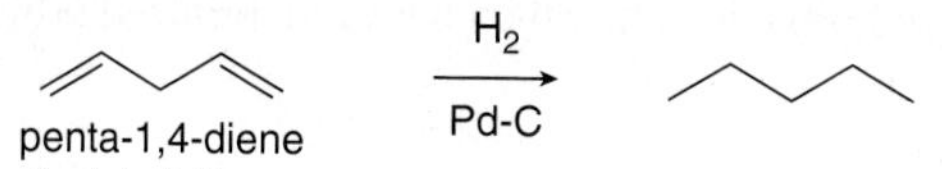

penta-1,4-diene
isolated diene

$\Delta H° = -255$ kJ/mol

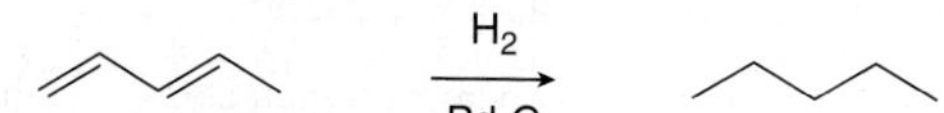

(*E*)-penta-1,3-diene
conjugated diene

more stable starting material

$\Delta H° = -226$ kJ/mol

Less energy is released.

- **A conjugated diene has a smaller heat of hydrogenation and is more stable than a similar isolated diene.**

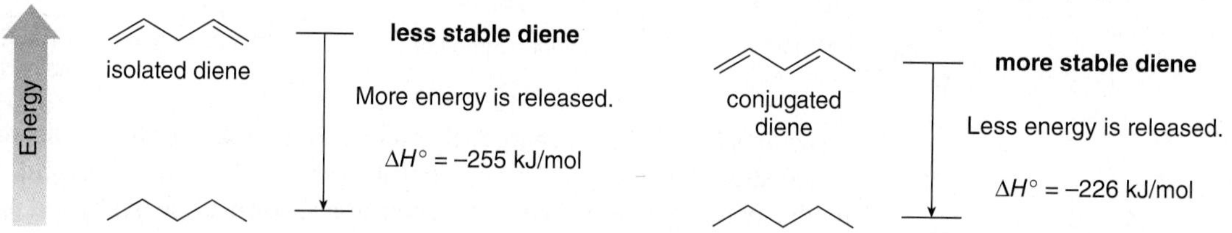

Figure 16.4 Relative energies of an isolated and conjugated diene

In Section 16.1, we learned why a conjugated diene is more stable than an isolated diene. A conjugated diene has overlapping *p* orbitals on four adjacent atoms, so its π electrons are delocalized over four atoms, thus stabilizing the diene. This delocalization cannot occur in an isolated diene, so an isolated diene is less stable than a conjugated diene.

Problem 16.14 Which diene in each pair has the larger heat of hydrogenation?

a. 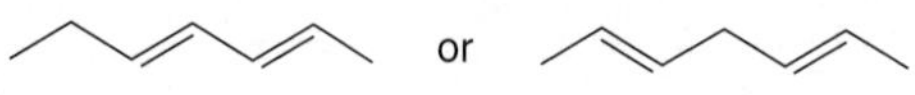

b. 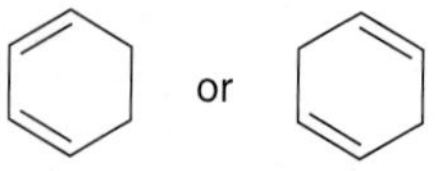

Problem 16.15 Rank the following compounds in order of increasing stability.

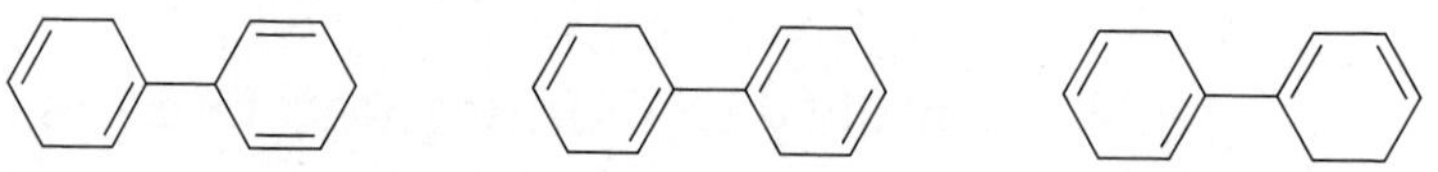

16.10 Electrophilic Addition: 1,2- Versus 1,4-Addition

Recall from Chapters 10 and 11 that the characteristic reaction of compounds with π bonds is **addition.** The π bonds in conjugated dienes undergo addition reactions, too, but they differ in two ways from the addition reactions to isolated double bonds.

- **Electrophilic addition in conjugated dienes gives a mixture of products.**
- **Conjugated dienes undergo a unique addition reaction not seen in alkenes or isolated dienes.**

We learned in Chapter 10 that HX adds to the π bond of alkenes to form alkyl halides.

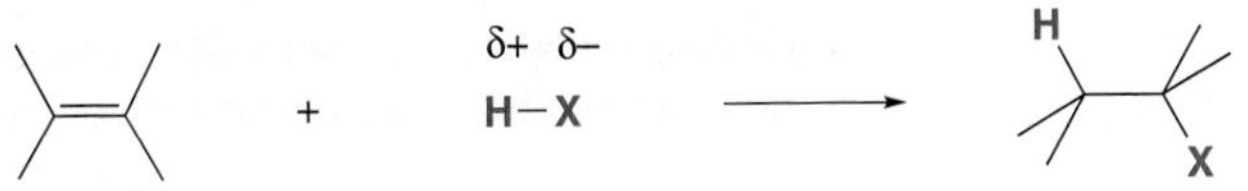

A π bond is broken. alkyl halide

With an **isolated diene,** electrophilic addition of one equivalent of HBr yields *one* product and Markovnikov's rule is followed. The H atom bonds to the less substituted carbon—that is, the carbon atom of the double bond that had more H atoms to begin with.

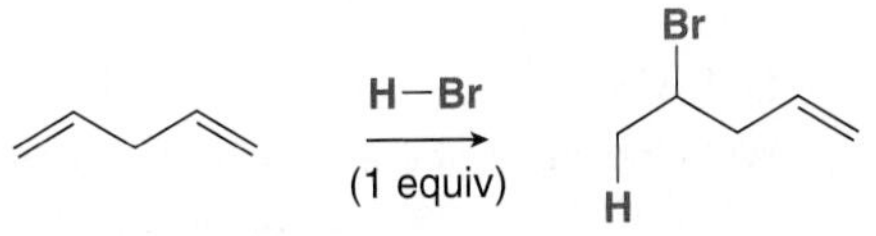

isolated diene

H bonds to the less substituted carbon.

With a **conjugated diene,** electrophilic addition of one equivalent of HBr affords *two* products.

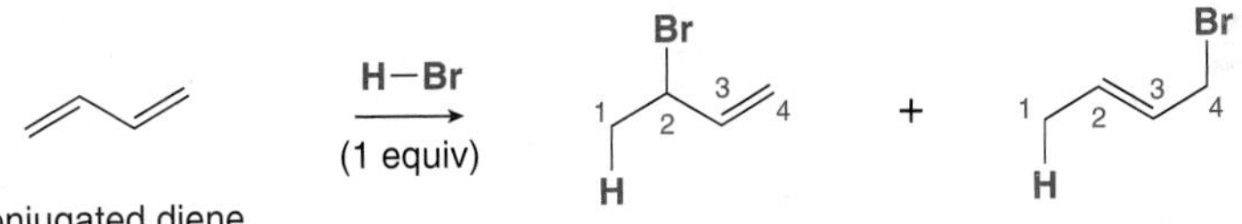

conjugated diene **1,2-product** **1,4-product**

- The **1,2-addition product** results from Markovnikov addition of HBr across two adjacent carbon atoms (C1 and C2) of the diene.
- The **1,4-addition product** results from addition of HBr to the two end carbons (C1 and C4) of the diene. 1,4-Addition is also called **conjugate addition.**

The ends of the 1,3-diene are called C1 and C4 arbitrarily, without regard to IUPAC numbering.

The mechanism of electrophilic addition of HX involves **two steps:** addition of H^+ (from HX) to form a resonance-stabilized carbocation, followed by nucleophilic attack of X^- at either electrophilic end of the carbocation to form two products. Mechanism 16.2 illustrates the reaction of buta-1,3-diene with HBr.

Mechanism 16.2 Electrophilic Addition of HBr to a 1,3-Diene—1,2- and 1,4-Addition

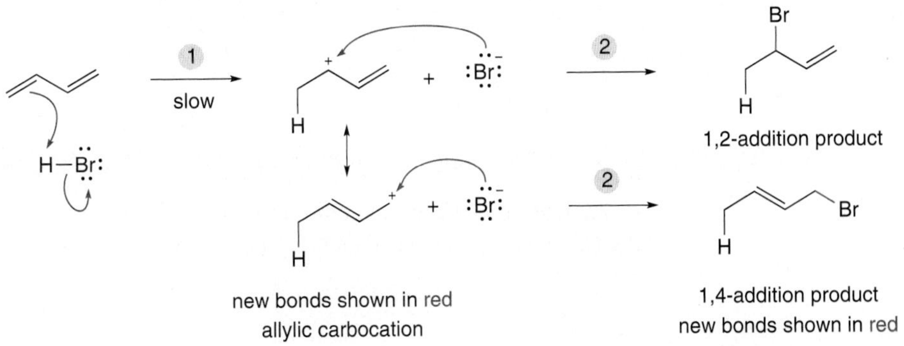

1. H^+ of HBr adds to a terminal carbon of the 1,3-diene to form a resonance-stabilized allylic carbocation.
2. Nucleophilic attack of Br^- occurs at either site of the resonance-stabilized carbocation that bears a (+) charge, forming the 1,2- and 1,4-addition products.

Like the electrophilic addition of HX to an alkene, the addition of HBr to a conjugated diene forms the more stable carbocation in Step [1], the rate-determining step. In this case, however, the carbocation is both 2° and **allylic,** and thus two Lewis structures can be drawn for it. In the second step, nucleophilic attack of Br^- can then occur at two different electrophilic sites, forming two different products.

- **Addition of HX to a conjugated diene forms 1,2- and 1,4-products because of the resonance-stabilized allylic carbocation intermediate.**

Sample Problem 16.4 Draw the products of the following reaction.

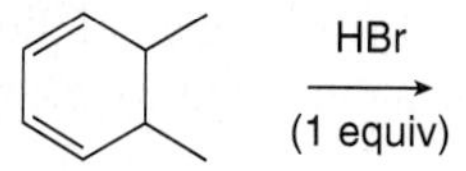

Solution

Write the steps of the mechanism to determine the structure of the products. Addition of H^+ forms the more stable 2° allylic carbocation, for which two resonance structures can be drawn. Nucleophilic attack of Br^- at either end of the allylic carbocation gives two constitutional isomers, formed by 1,2-addition and 1,4-addition to the diene.

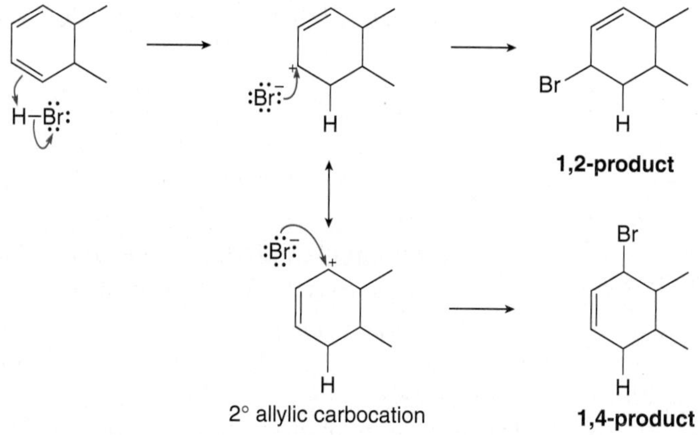

Problem 16.16 Draw the products formed when each diene is treated with one equivalent of HCl.

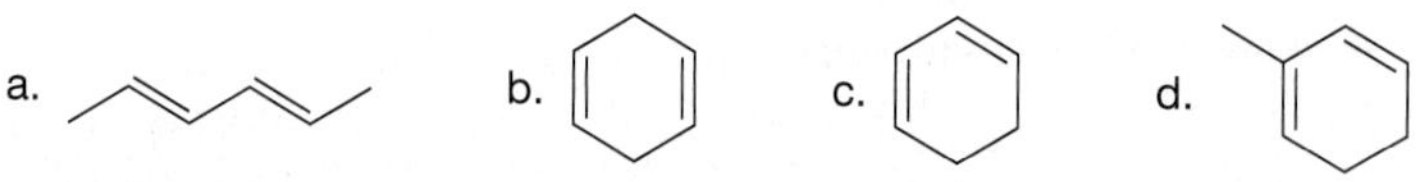

Problem 16.17 Draw a stepwise mechanism for the following reaction.

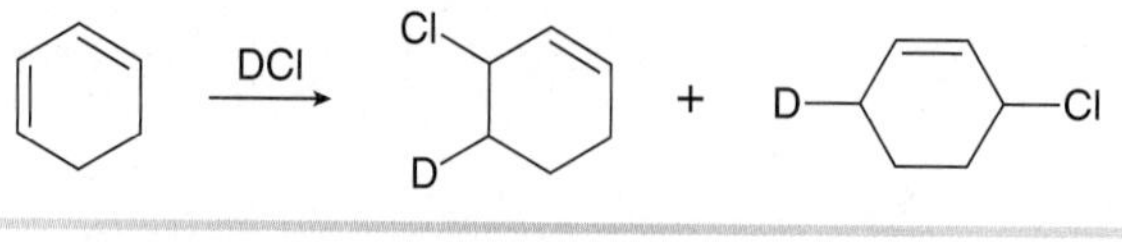

16.11 Kinetic Versus Thermodynamic Products

The amount of 1,2- and 1,4-addition products formed in the electrophilic addition reactions of buta-1,3-diene, a conjugated diene, depends greatly on the reaction conditions.

1,2-product 1,4-product

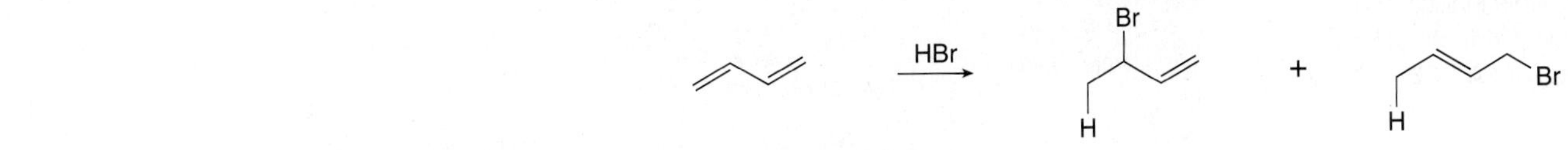

	1,2-product	1,4-product
low temperature (−80 °C)	80%	20%
high temperature (40 °C)	20%	80%

- **At low temperature the major product is formed by 1,2-addition.**
- **At higher temperature the major product is formed by 1,4-addition.**

Moreover, when a mixture containing predominately the 1,2-product is heated, the 1,4-addition product becomes the major product at equilibrium.

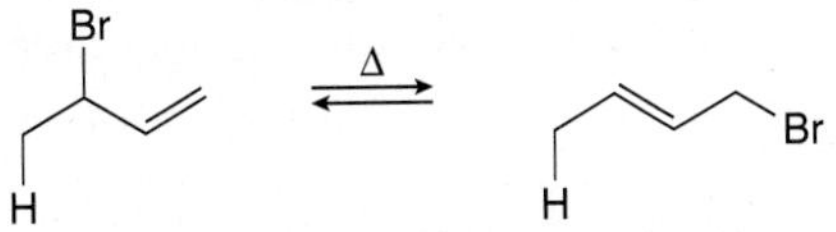

1,2-product — major product at low temperature — **kinetic product**

1,4-product — major product at equilibrium — **thermodynamic product**

- **The 1,2-product is formed faster because it predominates at low temperature. The product that is formed faster is called the *kinetic product.***
- **The 1,4-product must be more stable because it predominates at equilibrium. The product that predominates at equilibrium is called the *thermodynamic product.***

In many of the reactions we have learned thus far, the more stable product is formed faster—that is, the kinetic and thermodynamic products are the same. The electrophilic addition of HBr to buta-1,3-diene is different, in that **the more stable product is formed more slowly**—that is, the kinetic and thermodynamic products are *different.* Why is the more stable product formed more slowly?

To answer this question, recall that the rate of a reaction is determined by its energy of activation (E_a), whereas the amount of product present at equilibrium is determined by its stability (Figure 16.5). When a single starting material **A** forms two different products (**B** and **C**) by two exothermic pathways, the relative height of the energy barriers determines how fast **B** and **C** are formed, whereas the relative energies of **B** and **C** determine the amount of each at equilibrium. In an exothermic reaction, the relative energies of **B** and **C** do not determine the relative energies of activation to form **B** and **C.**

Why, in the addition of HBr to buta-1,3-diene, is the 1,4-product the more stable thermodynamic product? The 1,4-product (1-bromobut-2-ene) is more stable because it has two alkyl groups bonded to the carbon–carbon double bond, whereas the 1,2-product (3-bromobut-1-ene) has only one.

3-bromobut-1-ene
1,2-product
monosubstituted alkene
less stable

1-bromobut-2-ene
1,4-product
disubstituted alkene
more stable
thermodynamic product

Figure 16.5
How kinetic and thermodynamic products form in a reaction: **A** → **B** + **C**

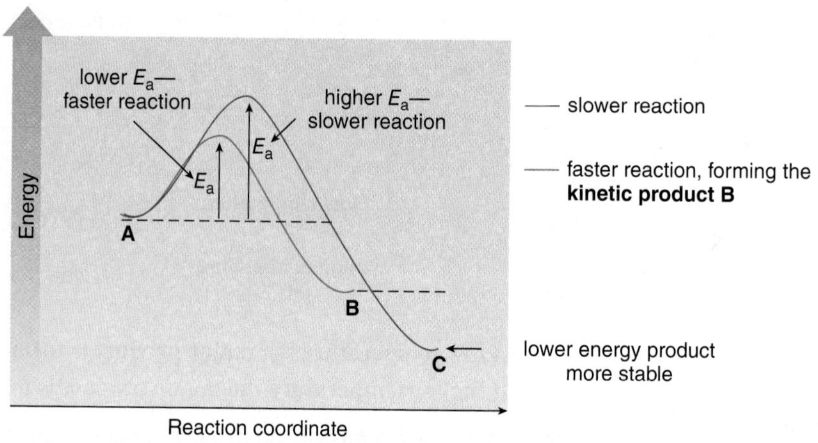

- The conversion of **A** → **B** is a faster reaction because the energy of activation leading to **B** is lower. **B** is the kinetic product.
- Because **C** is lower in energy, **C** is the thermodynamic product.

- **The more substituted alkene—1-bromobut-2-ene in this case—is the thermodynamic product.**

The 1,2-product is the kinetic product because of a **proximity effect.** When H^+ (from HBr) adds to the double bond, Br^- is closer to the adjacent carbon (C2) than it is to C4. Even though the resonance-stabilized carbocation bears a partial positive charge on both C2 and C4, attack at C2 is faster simply because Br^- is closer to this carbon.

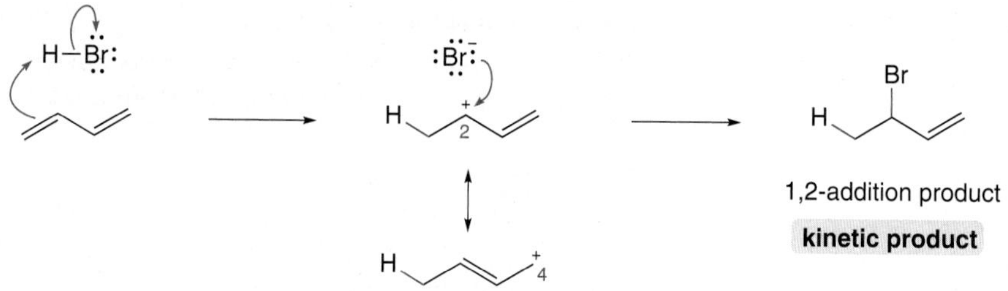

Br^- is closer to C2 than C4.

A **proximity effect** occurs because one species is close to another.

- **The 1,2-product forms faster because of the proximity of Br^- to C2.**

The overall two-step mechanism for addition of HBr to buta-1,3-diene, forming a 1,2-addition product and 1,4-addition product, is illustrated with the energy diagram in Figure 16.6.

Figure 16.6
Energy diagram for the two-step addition of HBr to $CH_2{=}CH{-}CH{=}CH_2$

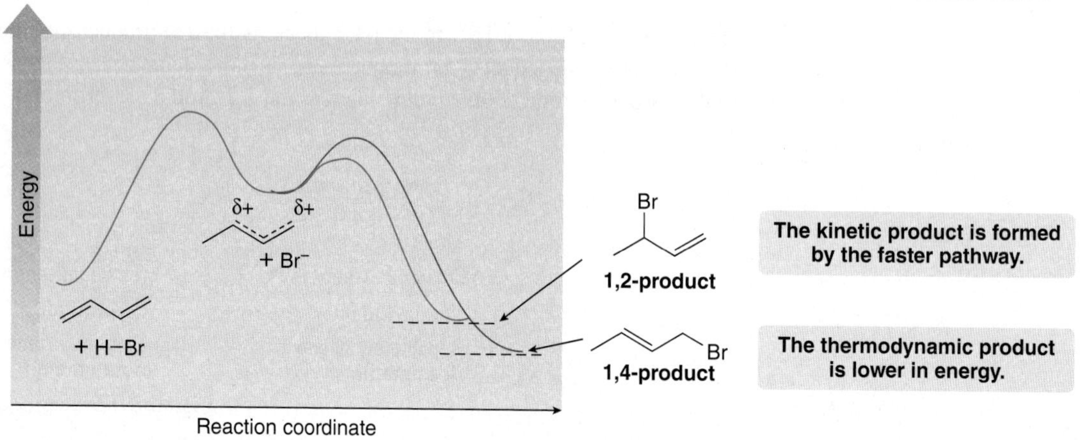

Why is the ratio of products temperature dependent?

- At low temperature, the energy of activation is the more important factor. Because most molecules do not have enough kinetic energy to overcome the higher energy barrier at lower temperature, they react by the faster pathway, forming the kinetic product.
- At higher temperature, most molecules have enough kinetic energy to reach either transition state. The two products are in equilibrium with each other, and the more stable compound—which is lower in energy—becomes the major product.

Problem 16.18 Label each product in the following reaction as a 1,2-product or 1,4-product, and decide which is the kinetic product and which is the thermodynamic product.

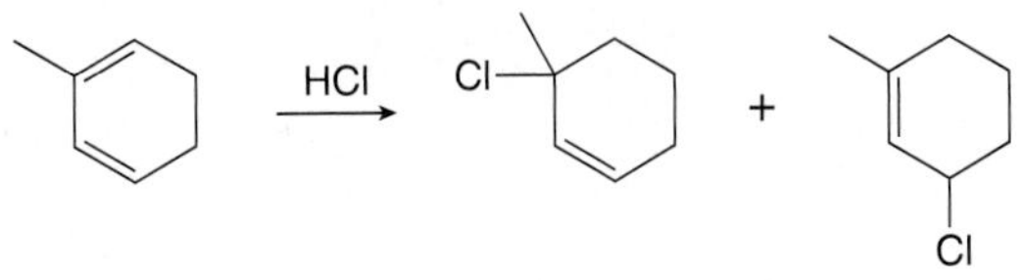

16.12 The Diels–Alder Reaction

Diels and Alder shared the 1950 Nobel Prize in Chemistry for unraveling the intricate details of this remarkable reaction.

The arrows may be drawn in a clockwise or counterclockwise direction to show the flow of electrons in a Diels–Alder reaction.

The **Diels–Alder reaction,** named for German chemists Otto Diels and Kurt Alder, is an addition reaction between a **1,3-diene** and an alkene called a **dienophile,** to form a new six-membered ring.

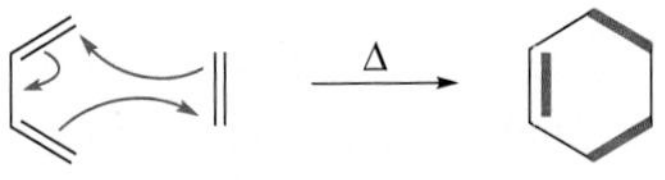

1,3-diene dienophile

Three curved arrows are needed to show the cyclic movement of electron pairs because three π bonds break and two σ bonds and one π bond form. Because each new σ bond is ~100 kJ/mol stronger than a π bond that is broken, a typical Diels–Alder reaction releases ~200 kJ/mol of energy. The following equations illustrate two examples of the Diels–Alder reaction.

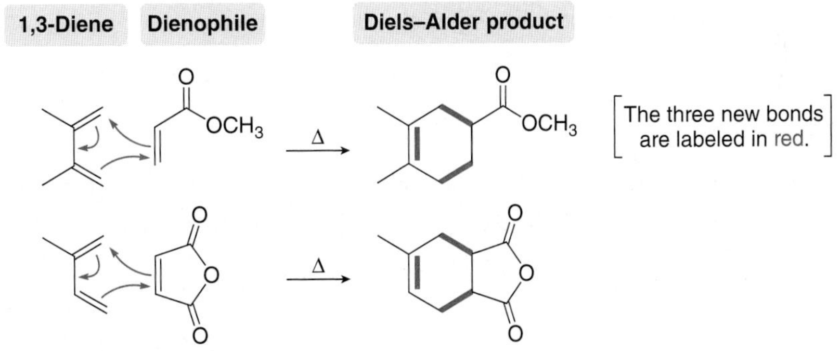

All Diels–Alder reactions have the following features in common:

[1] They are initiated by heat; that is, the Diels–Alder reaction is a *thermal* reaction.

[2] They form new six-membered rings.

[3] Three π bonds break, and two new C – C σ bonds and one new C – C π bond form.

[4] They are concerted; that is, all bonds are broken and formed in a single step.

The Diels–Alder reaction forms new carbon–carbon bonds, so it can be used to synthesize larger, more complex molecules from smaller ones. For example, Figure 16.7 illustrates a Diels–Alder reaction used in the synthesis of tetrodotoxin, a toxin isolated from many different types of puffer fish.

Figure 16.7

Synthesis of a natural product using the Diels–Alder reaction

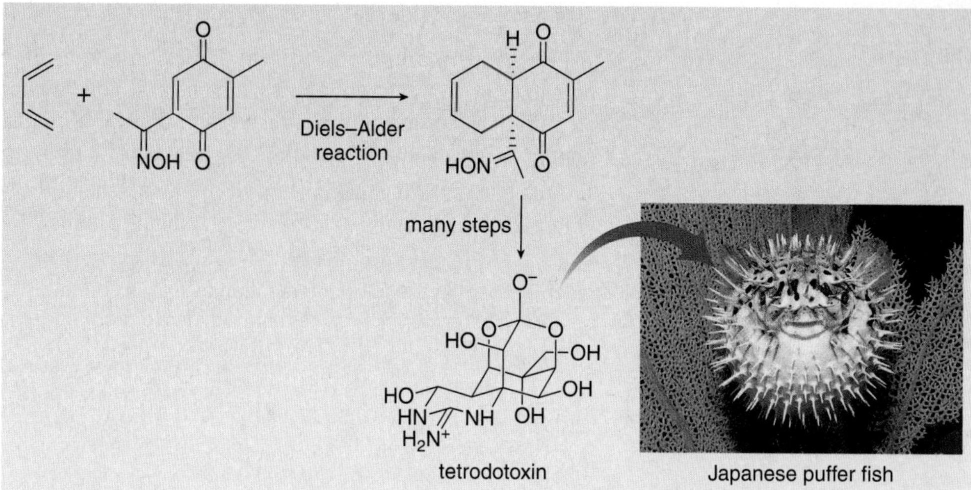

- Tetrodotoxin, a complex natural product containing several six-membered rings joined together, is a poison isolated from the ovaries and liver of the puffer fish, so named because the fish inflates itself into a ball when alarmed. One step in the synthesis of tetrodotoxin involves forming a six-membered ring by a Diels–Alder reaction.

Diels–Alder reactions may seem complicated at first, but they are really less complicated than many of the reactions you have already learned, especially those with multistep mechanisms and carbocation intermediates. The key is to learn how to arrange the starting materials to more easily visualize the structure of the product.

How To Draw the Product of a Diels–Alder Reaction

Example Draw the product of the following Diels–Alder reaction:

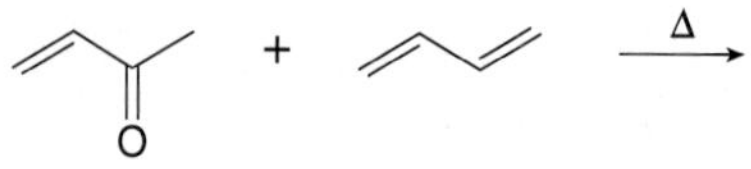

Step [1] **Arrange the 1,3-diene and the dienophile next to each other, with the diene drawn in the *s*-cis conformation.**

- This step is key: **Rotate the diene** so that it is drawn in the **s-cis** conformation, and place the end C's of the diene close to the double bond of the dienophile.

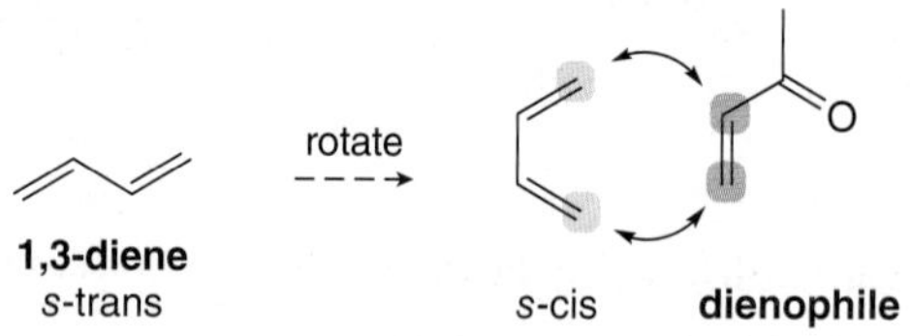

Step [2] **Cleave the three π bonds and use arrows to show where the new bonds will be formed.**

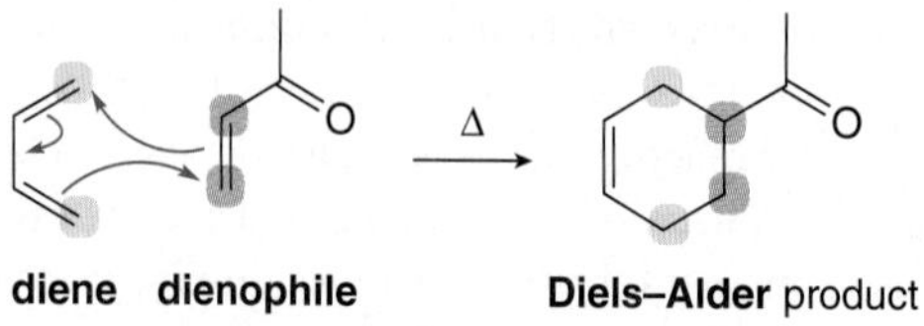

Problem 16.19 Draw the product formed when each diene and dienophile react in a Diels–Alder reaction.

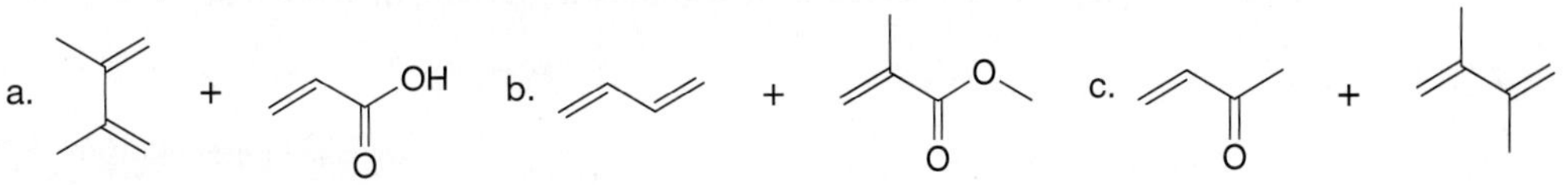

16.13 Specific Rules Governing the Diels–Alder Reaction

Several rules govern the course of the Diels–Alder reaction.

16.13A Diene Reactivity

Rule [1] **The diene can react only when it adopts the s-cis conformation.**

Both ends of the conjugated diene must be close to the π bond of the dienophile for reaction to occur. Thus, an acyclic diene in the *s*-trans conformation must rotate about the central C–C σ bond to form the *s*-cis conformation before reaction can take place.

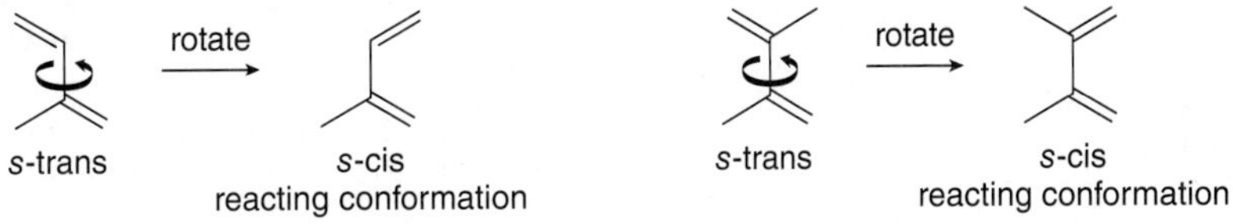

This rotation is prevented in cyclic dienes. As a result:

- **When the two double bonds are constrained in the s-cis conformation, the diene is unusually *reactive*.**
- **When the two double bonds are constrained in the s-trans conformation, the diene is *unreactive*.**

Zingiberene and β-sesquiphellandrene are trienes obtained from ginger root. Ginger is used as a spice in Indian and Chinese cooking. Ginger candy is sometimes used to treat nausea resulting from seasickness.

an *s*-cis 1,3-diene — very reactive

an *s*-trans 1,3-diene — unreactive diene

Problem 16.20 Rank the following dienes in order of increasing reactivity in a Diels–Alder reaction.

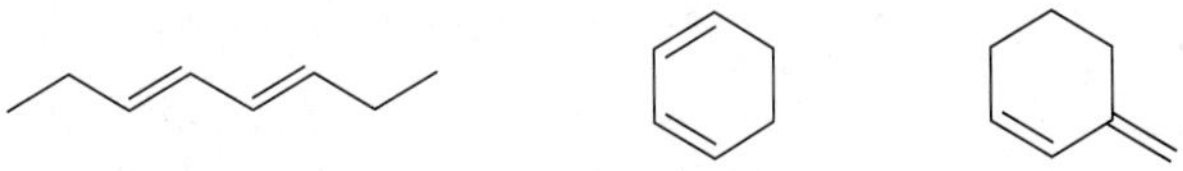

Problem 16.21 Zingiberene and β-sesquiphellandrene, natural products obtained from ginger root, contain conjugated diene units. Which diene reacts faster in the Diels–Alder reaction and why?

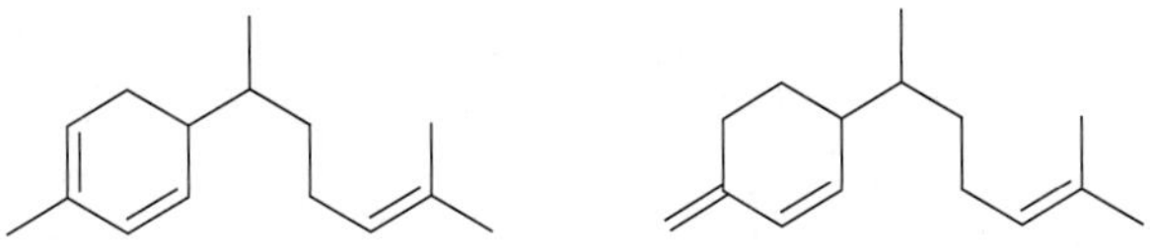

zingiberene β-sesquiphellandrene

16.13B Dienophile Reactivity

Rule [2] **Electron-withdrawing substituents in the dienophile increase the reaction rate.**

In a Diels–Alder reaction, the conjugated diene acts as a nucleophile and the dienophile acts as an electrophile. As a result, electron-withdrawing groups make the dienophile more electrophilic (and, thus, more reactive) by withdrawing electron density from the carbon–carbon double bond. If Z is an electron-withdrawing group, then the reactivity of the dienophile increases as follows:

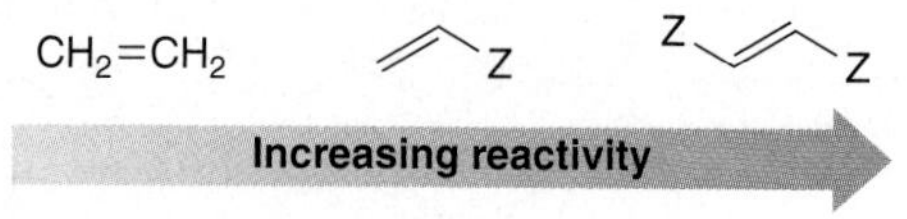

A carbonyl group is an effective electron-withdrawing group because the carbonyl carbon bears a partial positive charge (δ+), which withdraws electron density from the carbon–carbon double bond of the dienophile. Common dienophiles that contain a carbonyl group are shown in Figure 16.8.

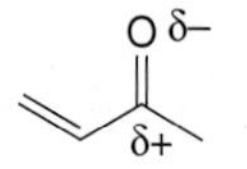

electron-deficient carbonyl carbon

Problem 16.22 Rank the following dienophiles in order of increasing reactivity.

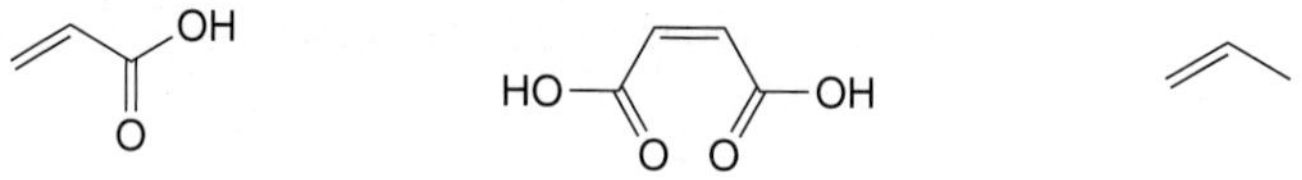

Figure 16.8 Common dienophiles in the Diels–Alder reaction

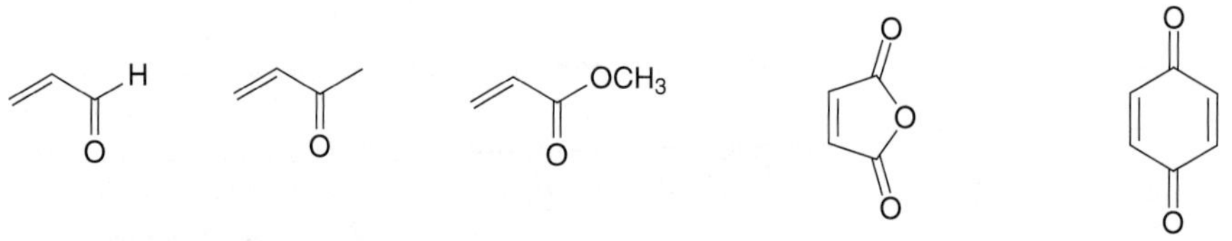

acrolein | methyl vinyl ketone | methyl acrylate | maleic anhydride | benzoquinone

16.13C Stereospecificity

Rule [3] **The stereochemistry of the dienophile is retained in the product.**

- **A cis dienophile forms a cis-substituted cyclohexene.**
- **A trans dienophile forms a trans-substituted cyclohexene.**

The two **cis** CO_2H groups of maleic acid become two **cis** substituents in a Diels–Alder adduct. The CO_2H groups can be drawn both above or both below the plane to afford a single achiral **meso** compound. The **trans dienophile** fumaric acid yields two enantiomers with **trans** CO_2H groups.

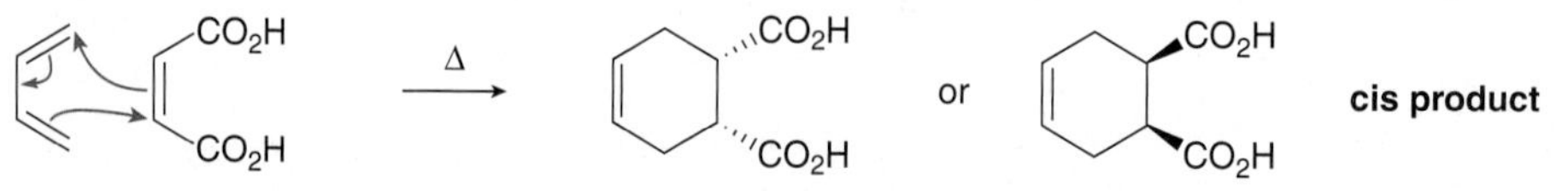

maleic acid
cis dienophile

an achiral meso compound

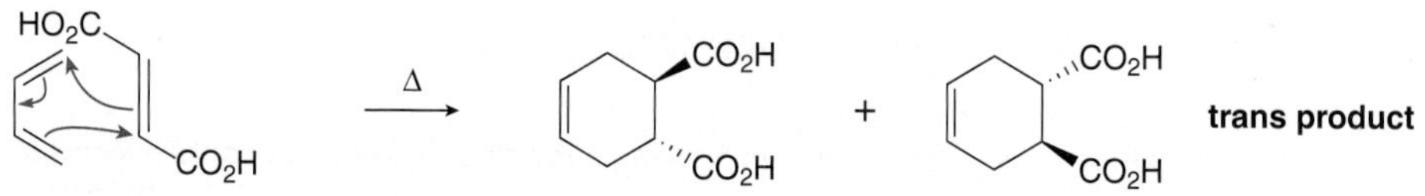

fumaric acid
trans dienophile

enantiomers

A **cyclic dienophile** forms a **bicyclic product.** A bicyclic system in which the two rings share a common C–C bond is called a **fused ring system.** The two H atoms at the ring fusion must be cis, because they were cis in the starting dienophile. A bicyclic system of this sort is said to be **cis-fused.**

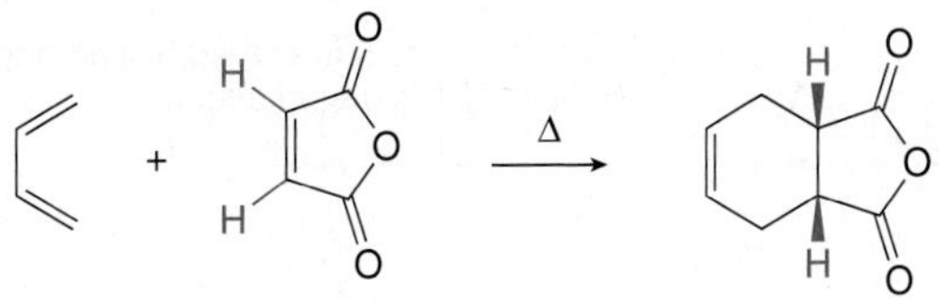

cyclic dienophile
cis H's in the dienophile

bicyclic product
cis H's in the product

Problem 16.23 Draw the products of each Diels–Alder reaction, and indicate the stereochemistry.

a. 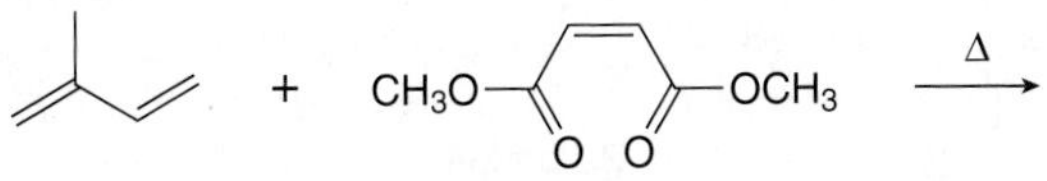

c. 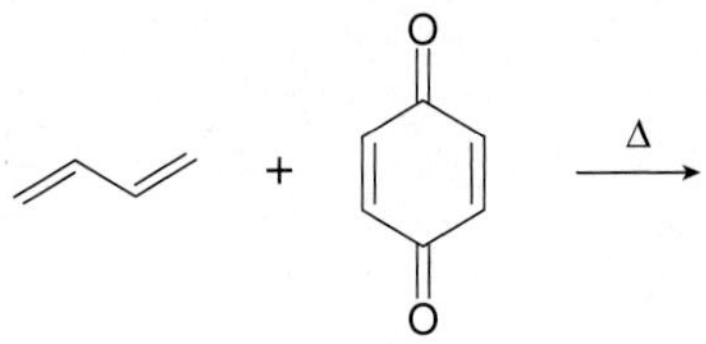

b. 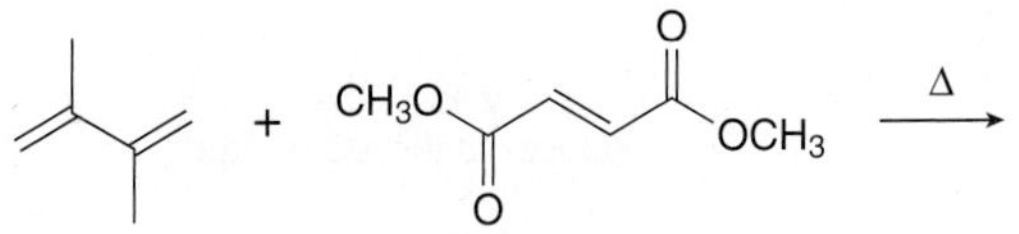

16.13D The Rule of Endo Addition

Rule [4] **When endo and exo products are possible, the endo product is preferred.**

To understand the rule of endo addition, we must first examine Diels–Alder products that result from cyclic 1,3-dienes. When cyclopentadiene reacts with a dienophile such as ethylene, a new six-membered ring forms, and above the ring there is a one atom "bridge," labeled in green. This carbon atom originated as the sp^3 hybridized carbon of the diene that was not involved in the reaction.

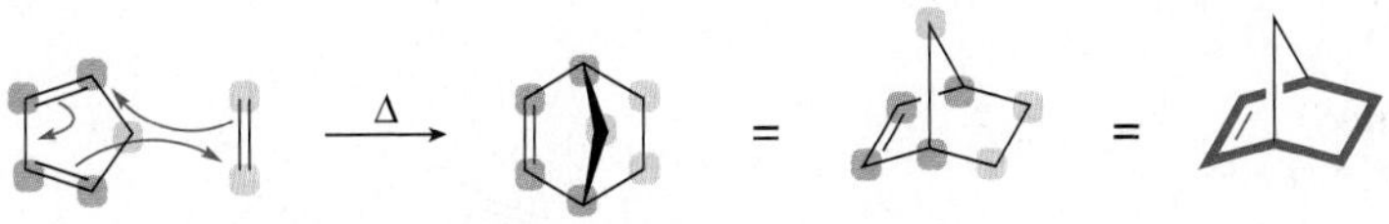

cyclic 1,3-diene

a bridged bicyclic ring system

The product of the Diels–Alder reaction of a cyclic 1,3-diene is bicyclic, but the carbon atoms shared by both rings are *non-adjacent.* Thus, this bicyclic product differs from the fused ring system obtained when the dienophile is cyclic.

- **A bicyclic ring system in which the two rings share non-adjacent carbon atoms is called a *bridged* ring system.**

Figure 16.9
Fused and bridged bicyclic ring systems compared

a. A fused bicyclic system

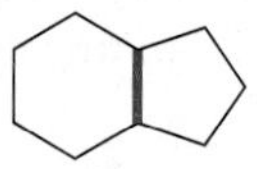

- One bond (in red) is shared by two rings.
- The shared C's are adjacent.

b. A bridged bicyclic system

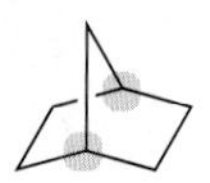

- Two non-adjacent atoms (labeled in blue) are shared by both rings.

Fused and bridged bicyclic ring systems are compared in Figure 16.9.

When cyclopentadiene reacts with a substituted alkene as the dienophile ($CH_2{=}CHZ$), the substituent Z can be oriented in one of two ways in the product. The terms **endo** and **exo** are used to indicate the position of Z.

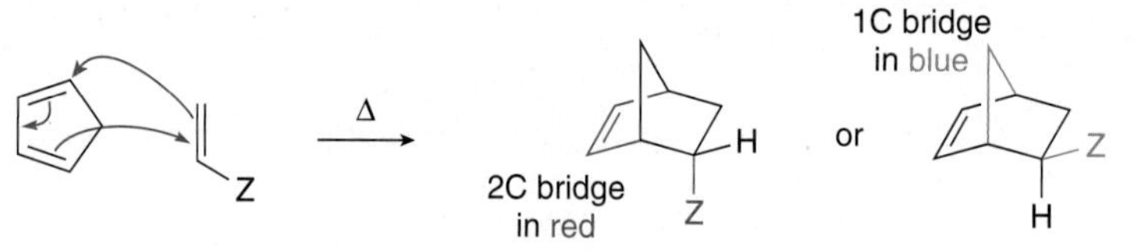

Δ
Z
2C bridge
in red
H
Z
or
1C bridge
in blue
Z
H

Z is exo.
(closer to the 1C bridge)

Z is endo.
(closer to the 2C bridge)

preferred product

- **A substituent on one bridge is *endo* if it is closer to the *longer* bridge that joins the two carbons common to both rings.**
- **A substituent is *exo* if it is closer to the *shorter* bridge that joins the carbons together.**

To help you distinguish endo and exo, remember that e***n***do is u***n***der the newly formed six-membered ring.

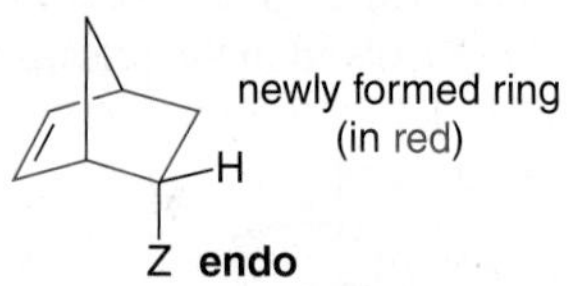

newly formed ring
(in red)
H
Z **endo**

In a Diels–Alder reaction, the **endo** product is preferred, as shown in two examples.

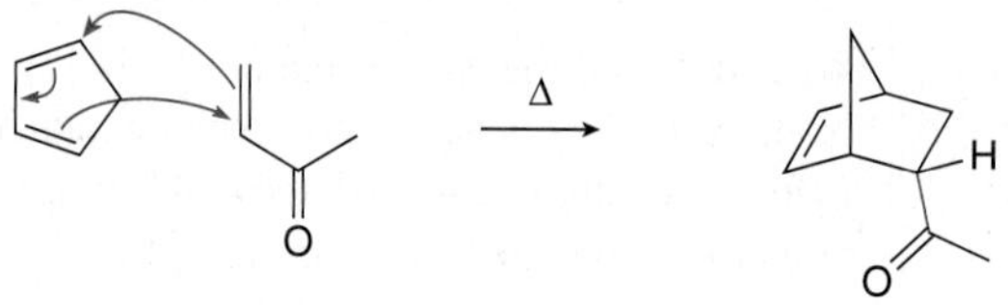

Δ
O
H
O

preferred product
Bonds in red are endo.

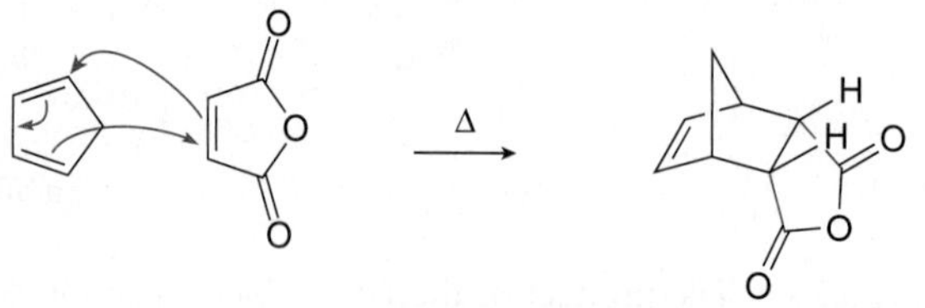

O
O
O
Δ
H
H
O
O
O

preferred product

More details on the Diels–Alder reaction are given in Section 30.4.

The Diels–Alder reaction is concerted, and the reaction occurs with the diene and the dienophile arranged one above the other, as shown in Figure 16.10, not side-by-side. In theory, the substituent Z can be oriented either directly under the diene to form the endo product (Pathway [1] in Figure 16.10) or away from the diene to form the exo product (Pathway [2] in Figure 16.10). In practice, though, the **endo product is the major product.** The transition state leading to the endo product allows more interaction between the electron-rich diene and the electron-withdrawing substituent Z on the dienophile, an energetically favorable arrangement.

Figure 16.10 How endo and exo products are formed in the Diels–Alder reaction

Pathway [1] With Z oriented under the diene, the endo product is formed.

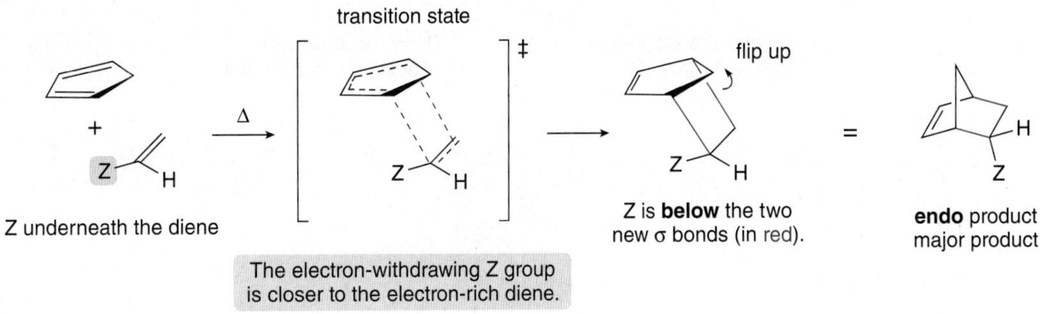
transition state
‡
flip up
Δ
Z
H
Z underneath the diene
Z is **below** the two new σ bonds (in red).
endo product
major product
The electron-withdrawing Z group is closer to the electron-rich diene.

Pathway [2] With Z oriented away from the diene, the exo product is formed.

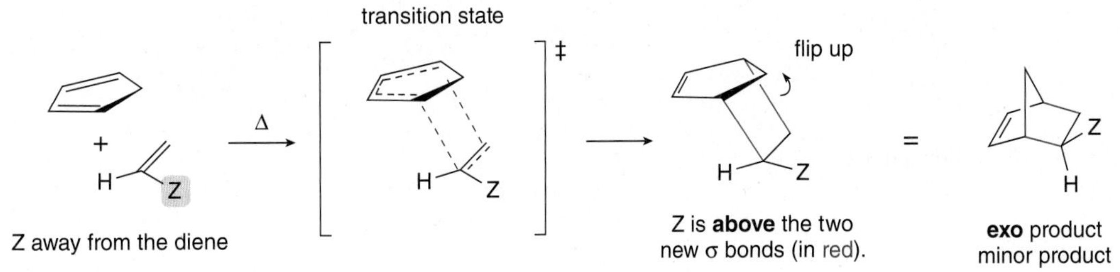
transition state
‡
flip up
Δ
H
Z
Z away from the diene
Z is **above** the two new σ bonds (in red).
exo product
minor product

Problem 16.24 Draw the product of each Diels–Alder reaction.

a. 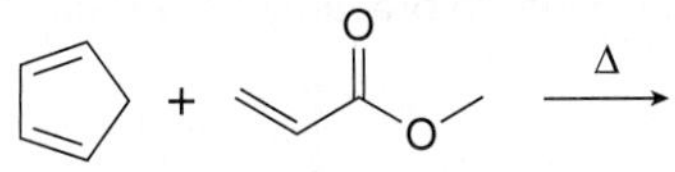

b. 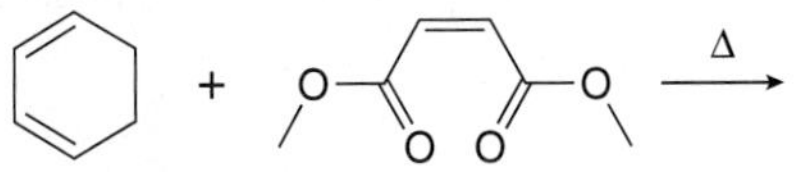

16.14 Other Facts About the Diels–Alder Reaction

16.14A Retrosynthetic Analysis of a Diels–Alder Product

The Diels–Alder reaction is used widely in organic synthesis, so you must be able to look at a compound and determine what conjugated diene and what dienophile were used to make it. To draw the starting materials from a given Diels–Alder adduct:

- **Locate the six-membered ring that contains the C=C.**
- **Draw three arrows around the cyclohexene ring, beginning with the π bond. Each arrow moves two electrons to the adjacent bond, cleaving one π bond and two σ bonds, and forming three π bonds.**
- **Retain the stereochemistry of substituents on the C=C of the dienophile. Cis substituents on the six-membered ring give a cis dienophile.**

Figure 16.11 Finding the diene and dienophile needed for a Diels–Alder reaction

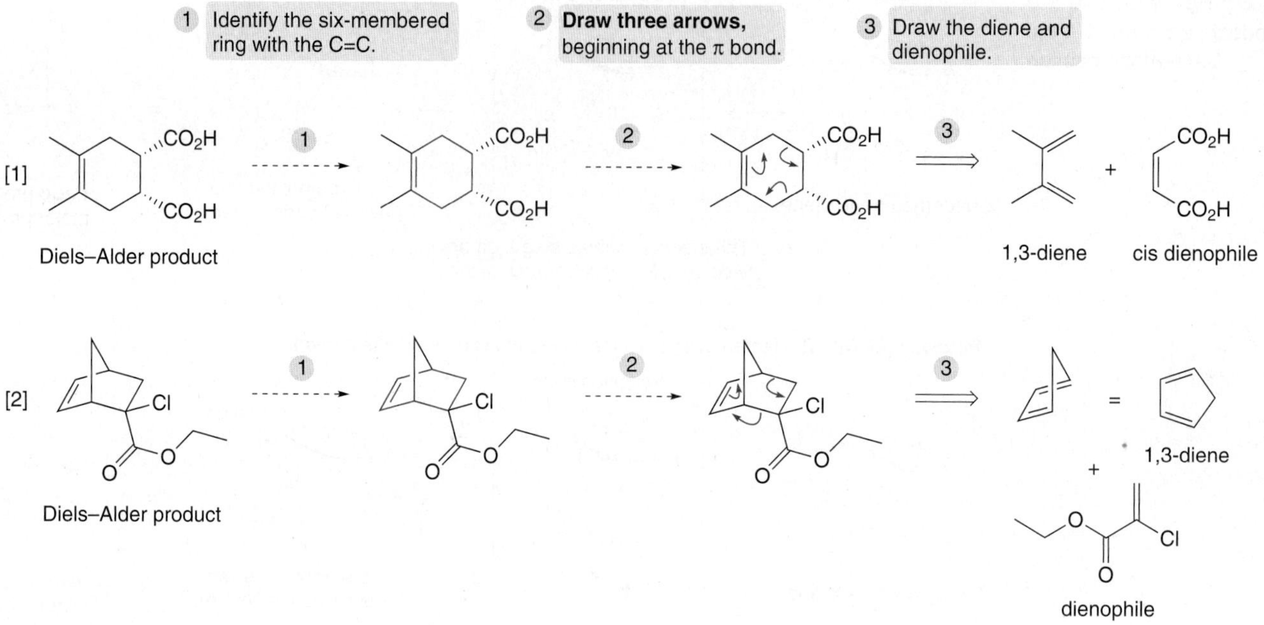
1 Identify the six-membered ring with the C=C.
2 Draw three arrows, beginning at the π bond.
3 Draw the diene and dienophile.
[1]
CO2H
CO2H
Diels–Alder product
1,3-diene
cis dienophile
[2]
Cl
O
Diels–Alder product
1,3-diene
dienophile

This stepwise retrosynthetic analysis gives the 1,3-diene and dienophile needed for any Diels–Alder reaction, as shown in the two examples in Figure 16.11.

Problem 16.25 What diene and dienophile are needed to prepare each product?

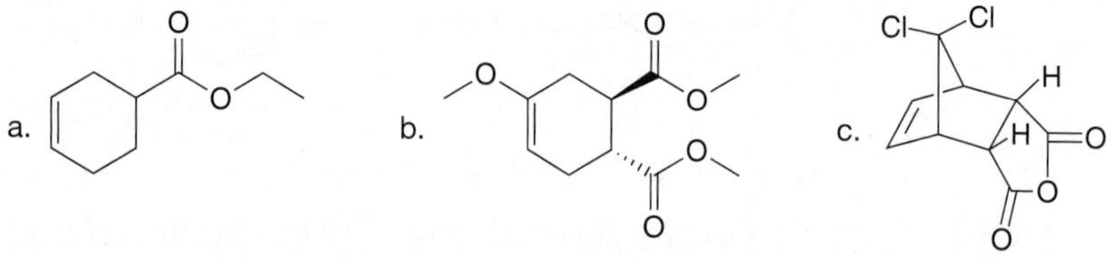
a.
b.
c.
O
Cl
H

16.14B Retro Diels–Alder Reaction

A reactive diene like cyclopenta-1,3-diene readily undergoes a Diels–Alder reaction with *itself;* that is, **cyclopenta-1,3-diene dimerizes because one molecule acts as the diene and another acts as the dienophile.**

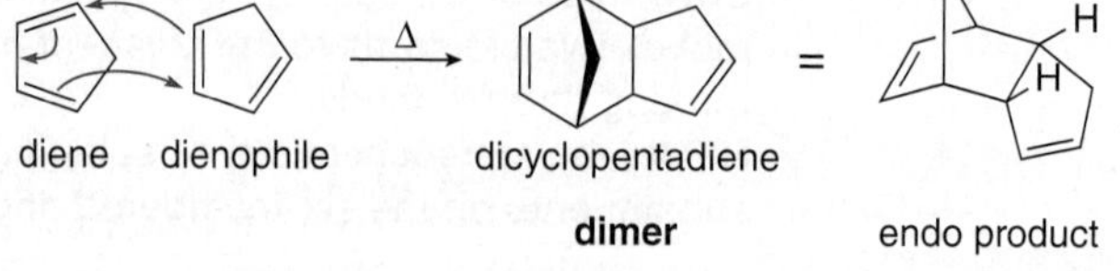
Δ
diene
dienophile
dicyclopentadiene
dimer
H
endo product

The formation of dicyclopentadiene is so rapid that it takes only a few hours at room temperature for cyclopentadiene to completely dimerize. How, then, can cyclopentadiene be used in a Diels–Alder reaction if it really exists as a dimer?

When heated, dicyclopentadiene undergoes a **retro Diels–Alder reaction,** and two molecules of cyclopentadiene are re-formed. If cyclopentadiene is immediately treated with a different dienophile, it reacts to form a new Diels–Alder adduct with this dienophile.

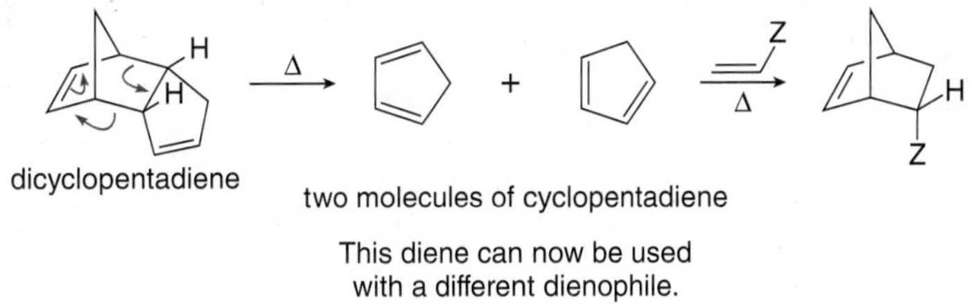

16.14C Application: Diels–Alder Reaction in the Synthesis of Steroids

Recall from Section 4.15 that lipids are water-insoluble biomolecules that have diverse structures.

***Steroids* are tetracyclic lipids containing three six-membered rings and one five-membered ring.** The four rings are designated as **A, B, C,** and **D.**

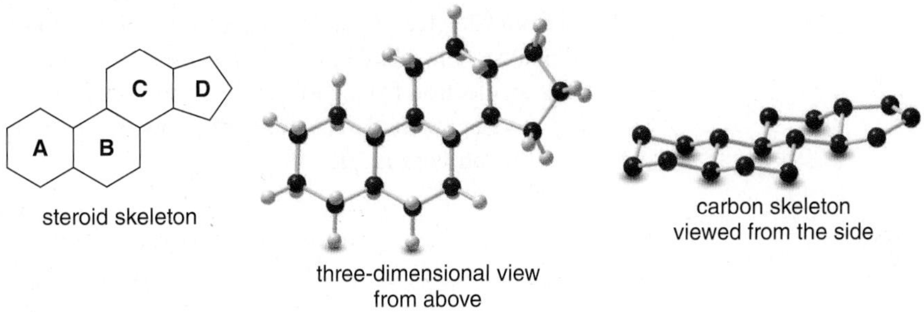

Steroids exhibit a wide range of biological properties, depending on the substitution pattern of functional groups on the rings. They include **cholesterol** (a component of cell membranes that is implicated in cardiovascular disease), **estrone** (a female sex hormone responsible for the regulation of the menstrual cycle), and **cortisone** (a hormone responsible for the control of inflammation and the regulation of carbohydrate metabolism).

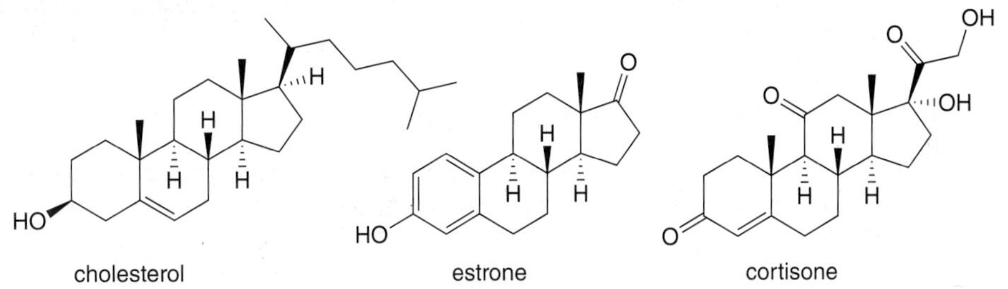

Diels–Alder reactions have been used widely in the laboratory syntheses of steroids. The key Diels–Alder reactions used to prepare the C ring of estrone and the B ring of cortisone are drawn.

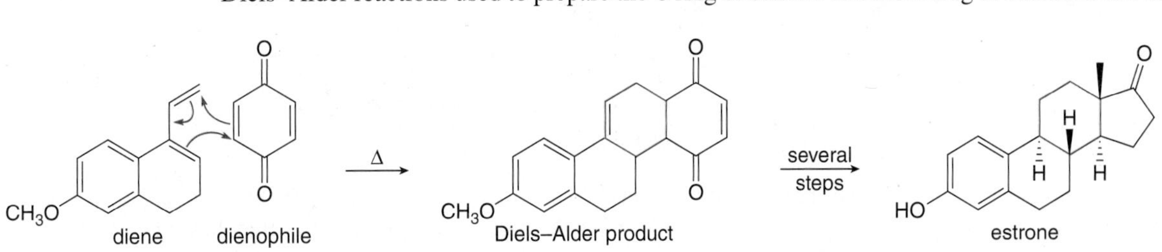

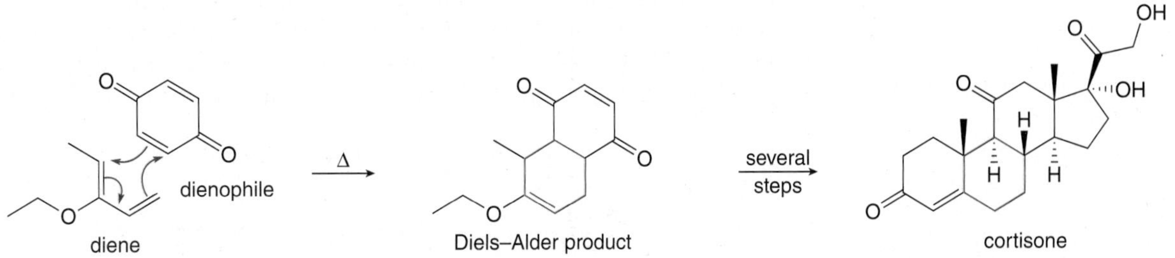

Problem 16.26 Draw the product (**A**) of the following Diels–Alder reaction. **A** was a key intermediate in the synthesis of the addicting pain reliever morphine, the chapter-opening molecule.

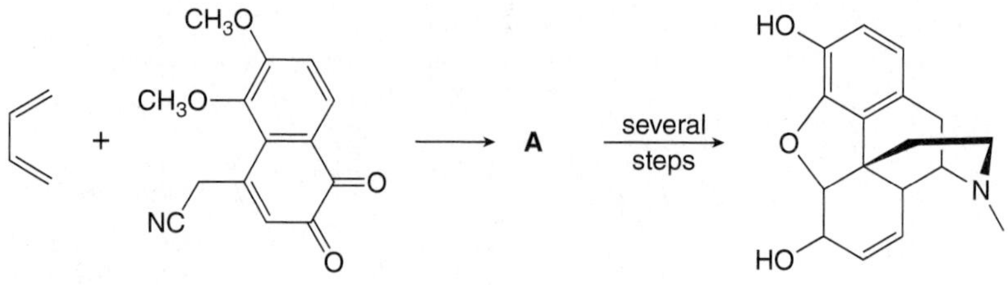

16.15 Conjugated Dienes and Ultraviolet Light

Recall from Chapter 13 that the absorption of infrared light can promote a molecule from a lower vibrational state to a higher one. In a similar fashion, the absorption of ultraviolet (UV) light can promote an electron from a lower electronic state to a higher one. Ultraviolet light has a slightly shorter wavelength (and, thus, higher frequency) than visible light. The most useful region of UV light for this purpose is **200–400 nm.**

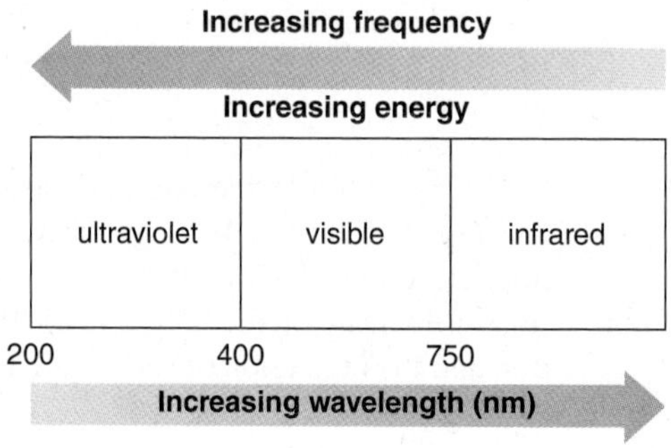

16.15A General Principles

When electrons in a lower energy state (the **ground state**) absorb light having the appropriate energy, an electron is promoted to a higher electronic state (the **excited state**).

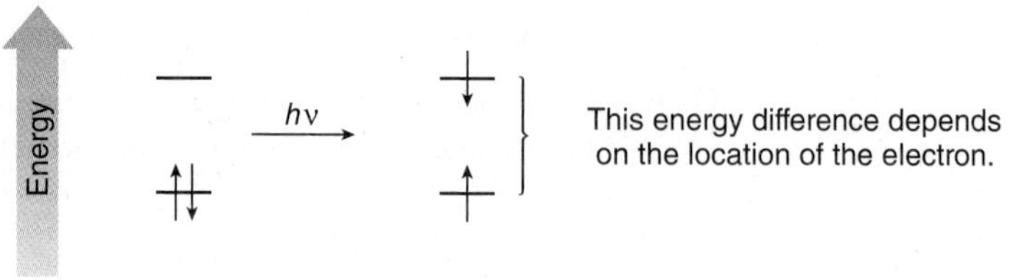

The energy difference between the two states depends on the location of the electron. The promotion of electrons in σ bonds and unconjugated π bonds requires light having a wavelength of < 200 nm; that is, it has a shorter wavelength and higher energy than light in the UV region of the electromagnetic spectrum. With conjugated dienes, however, the energy difference between the ground and excited states decreases, so longer wavelengths of light can be used to promote electrons. The wavelength of UV light absorbed by a compound is often referred to as its λ_{max}. Buta-1,3-diene, for example, absorbs UV light at λ_{max} = 217 nm and cyclohexa-1,3-diene has a λ_{max} of 256 nm.

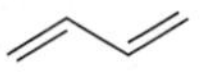

λ_{max} = 217 nm

λ_{max} = 256 nm

- **Conjugated dienes and polyenes absorb light in the UV region of the electromagnetic spectrum (200–400 nm).**

A UV spectrum is a plot of the absorbance of UV light versus wavelength. A spectrum consists of very broad bands, and the maximum absorbance corresponds to the λ_{max}, as shown in the UV spectrum of isoprene in Figure 16.12.

Figure 16.12
UV spectrum of isoprene

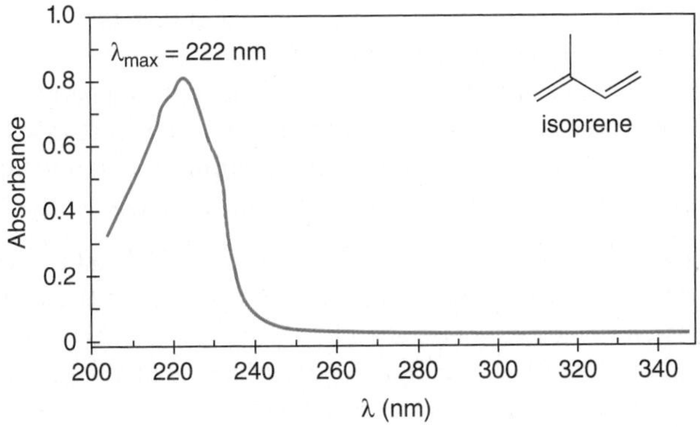
1.0
0.8
0.6
0.4
0.2
0
Absorbance
λmax = 222 nm
isoprene
200
220
240
260
280
300
320
340
λ (nm)

Lycopene is the red pigment found in tomatoes, watermelon, papaya, guava, and pink grapefruit. Lycopene is not destroyed when fruits and vegetables are processed, so tomato juice and ketchup are high in lycopene.

As the number of conjugated π bonds increases, the energy difference between the ground and excited state decreases, shifting the absorption to longer wavelengths.

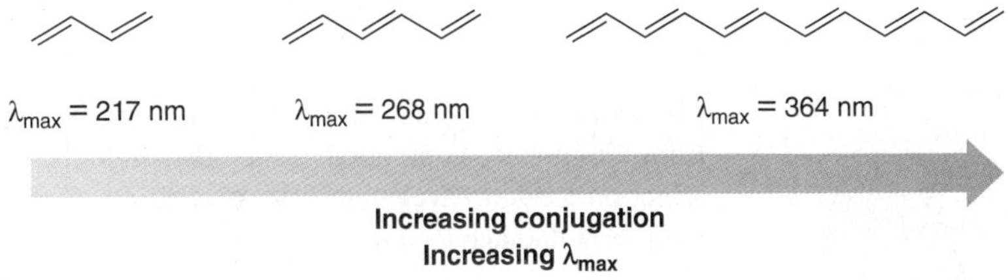
λmax = 217 nm
λmax = 268 nm
λmax = 364 nm
Increasing conjugation
Increasing λmax

With molecules having eight or more conjugated π bonds, the absorption shifts from the UV to the visible region and the compound takes on the color of those wavelengths of visible light it does *not* absorb. For example, lycopene absorbs visible light at λ_{max} = 470 nm, in the blue-green region of the visible spectrum. Because it does not absorb light in the red region, lycopene appears bright red (Figure 16.13).

Figure 16.13 Why lycopene appears red

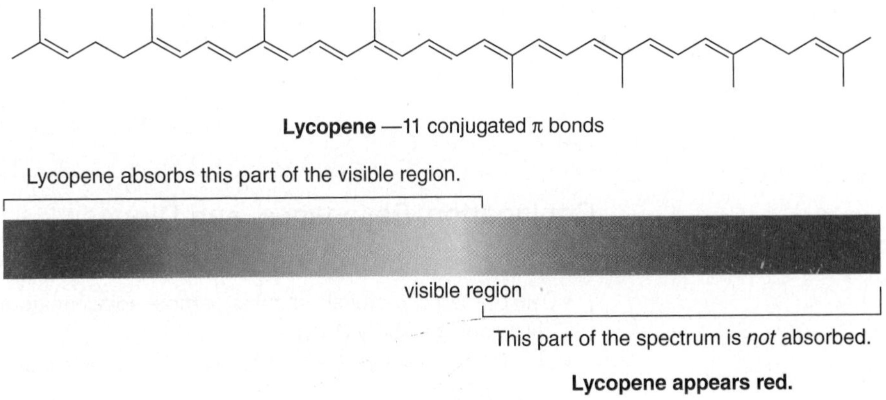
Lycopene —11 conjugated π bonds
Lycopene absorbs this part of the visible region.
visible region
This part of the spectrum is *not* absorbed.
Lycopene appears red.

Problem 16.27 Which compound in each pair absorbs UV light at longer wavelength?

a. 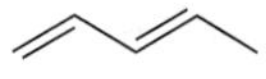 or 

b. 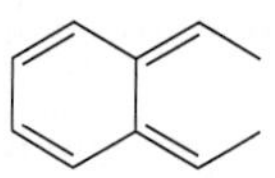 or 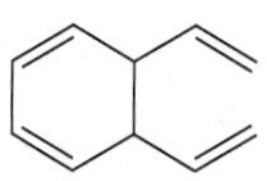

16.15B Sunscreens

Ultraviolet radiation from the sun is high enough in energy to cleave bonds, forming radicals that can prematurely age skin and cause skin cancers. The ultraviolet region is often subdivided, based on the wavelength of UV light: UV-A (320–400 nm), UV-B (290–320 nm), and UV-C (< 290 nm). Fortunately, much of the highest energy UV light (UV-C) is filtered out by the ozone layer, so that only UV light having wavelengths > 290 nm reaches the skin's surface. Much of this UV light is absorbed by **melanin,** the highly conjugated colored pigment in the skin that serves as the body's natural protection against the harmful effects of UV radiation.

Commercial sunscreens are given an **SPF** rating (sun protection factor), according to the amount of sunscreen present. The higher the number, the greater the protection.

Prolonged exposure to the sun can allow more UV radiation to reach your skin than melanin can absorb. A commercial sunscreen can offer added protection, however, because it contains conjugated compounds that absorb UV light, thus shielding your skin (for a time) from the harmful effects of UV radiation. Two sunscreens that have been used for this purpose are *para*-aminobenzoic acid (PABA) and padimate O.

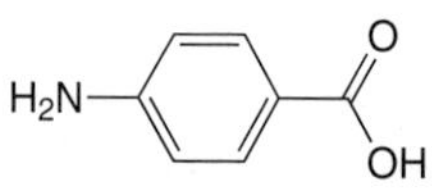

para-aminobenzoic acid (PABA)

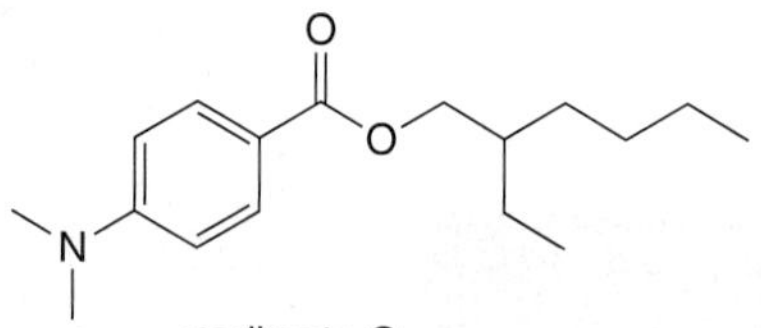

padimate O

Many sunscreens contain more than one component to filter out different regions of the UV spectrum. Conjugated compounds generally shield the skin from UV-B radiation, but often have little effect on longer-wavelength UV-A radiation, which does not burn the skin, but can still cause long-term damage to skin cells.

Problem 16.28 Which of the following compounds might be an ingredient in a commercial sunscreen? Explain why or why not.

a. 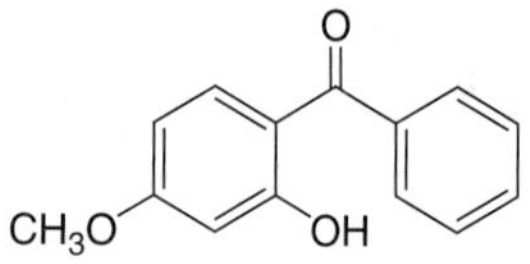

c. 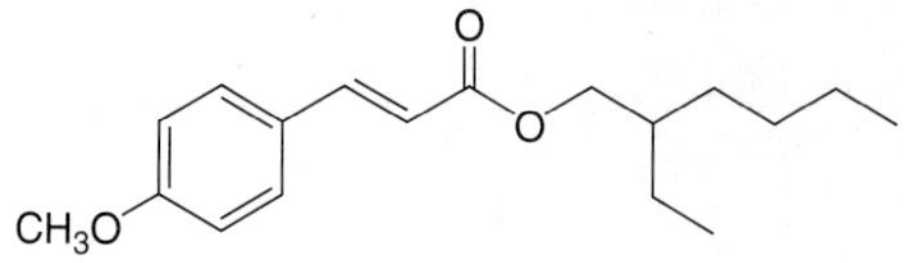

b. 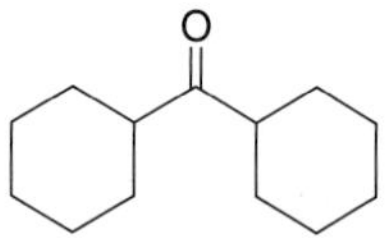

KEY CONCEPTS

Conjugation, Resonance, and Dienes

Conjugation and Delocalization of Electron Density

- The overlap of *p* orbitals on three or more adjacent atoms allows electron density to delocalize, thus adding stability (16.1).
- An allyl carbocation ($CH_2=CHCH_2^+$) is more stable than a 1° carbocation because of *p* orbital overlap (16.2).
- In a system X=Y−Z:, Z is generally sp^2 hybridized to allow the lone pair to occupy a *p* orbital, making the system conjugated (16.5).

Four Common Examples of Resonance (16.3)

[1] The three-atom "allyl" system:

X=Y–Z* ⟷ X*–Y=Z * = +, –, ·, or ··

[2] Conjugated double bonds:

[3] Cations having a positive charge adjacent to a lone pair:

[4] Double bonds involving one atom more electronegative than the other:

 [electronegativity of Y > X]

Rules on Evaluating the Relative "Stability" of Resonance Structures (16.4)

[1] Structures with more bonds and fewer charges are better.

[2] Structures in which every atom has an octet are better.

[3] Structures that place a negative charge on a more electronegative atom are better.

The Unusual Properties of Conjugated Dienes

[1] The C–C σ bond joining the two double bonds is unusually short (16.8).

[2] Conjugated dienes are more stable than the corresponding isolated dienes. $\Delta H°$ of hydrogenation is smaller for a conjugated diene than for an isolated diene converted to the same product (16.9).

[3] The reactions are unusual:
- Electrophilic addition affords products of 1,2-addition and 1,4-addition (16.10, 16.11).
- Conjugated dienes undergo the Diels–Alder reaction, a reaction that does not occur with isolated dienes (16.12–16.14).

[4] Conjugated dienes absorb UV light in the 200–400 nm region. As the number of conjugated π bonds increases, the absorption shifts to longer wavelength (16.15).

Reactions of Conjugated Dienes

[1] Electrophilic addition of HX (X = halogen) (16.10–16.11)

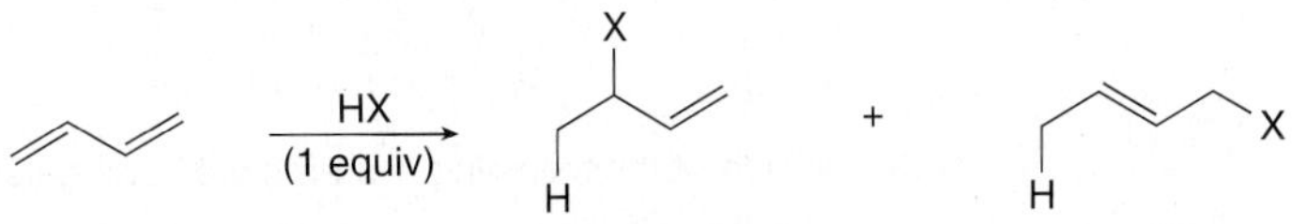

1,2-product **kinetic product** — **1,4-product** **thermodynamic product**

- The mechanism has two steps.
- Markovnikov's rule is followed. Addition of H^+ forms the more stable allylic carbocation.
- The 1,2-product is the kinetic product. When H^+ adds to the double bond, X^- adds to the end of the allylic carbocation to which it is closer (C2 not C4). The kinetic product is formed faster at low temperature.
- The thermodynamic product has the more substituted, more stable double bond. The thermodynamic product predominates at equilibrium. With buta-1,3-diene, the thermodynamic product is the 1,4-product.

[2] Diels–Alder reaction (16.12–16.14)

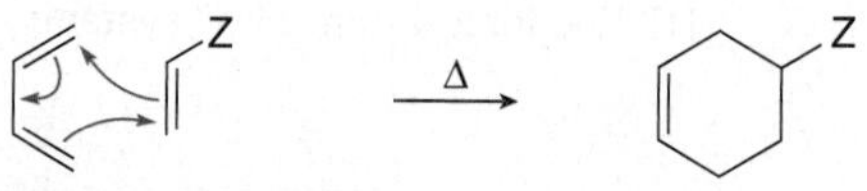

1,3-diene dienophile

- The reaction forms two σ and one π bond in a six-membered ring.
- The reaction is initiated by heat.
- The mechanism is concerted: All bonds are broken and formed in a single step.
- The diene must react in the *s*-cis conformation (16.13A).
- Electron-withdrawing groups in the dienophile increase the reaction rate (16.13B).
- The stereochemistry of the dienophile is retained in the product (16.13C).
- Endo products are preferred (16.13D).

PROBLEMS

Problems Using Three-Dimensional Models

16.29 Name each diene and state whether the ball-and-stick model shows the diene in the *s*-cis or *s*-trans conformation.

a. 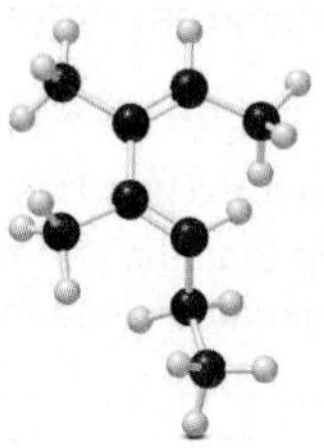 b. 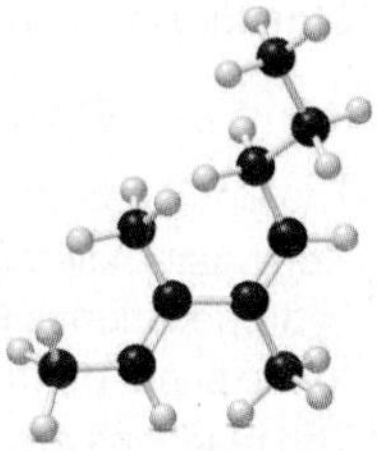

16.30 What diene and dienophile are needed to prepare each compound by a Diels–Alder reaction?

a. 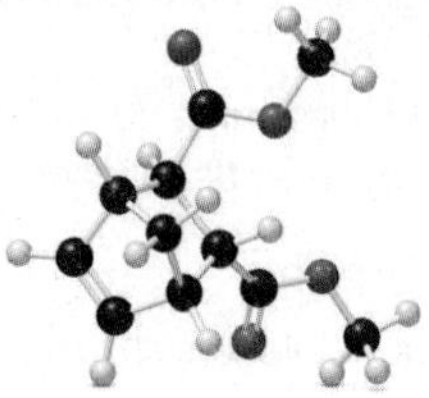 b. 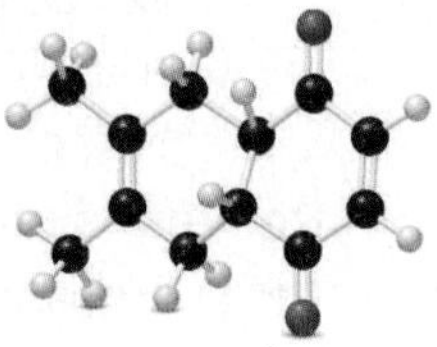

Conjugation

16.31 Which of the following systems are conjugated?

a. 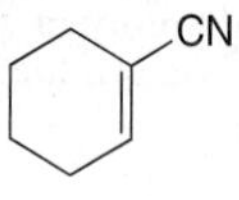 b. 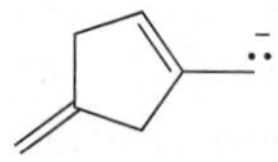 c. 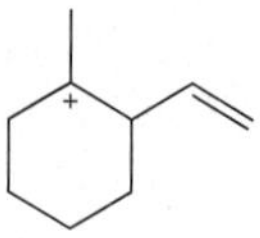 d. 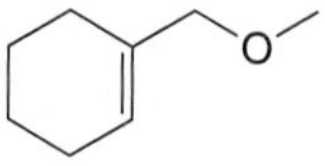

Resonance and Hybridization

16.32 Draw all reasonable resonance structures for each species.

a. 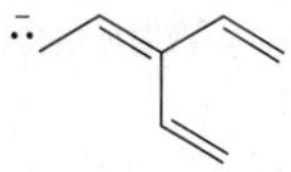 b. 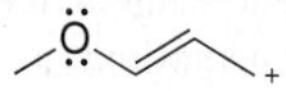 c. 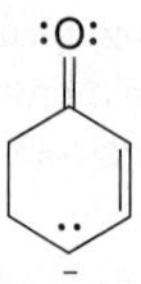 d. 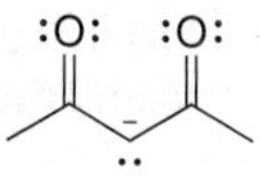 e. 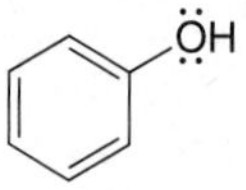 f. 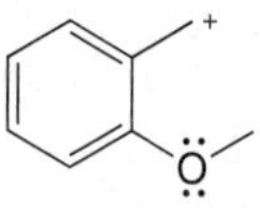

16.33 For which compounds can a second resonance structure be drawn? Draw an additional resonance structure for each resonance-stabilized compound.

a. 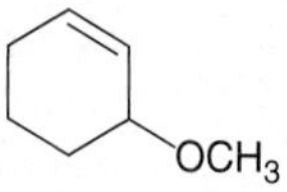 b. 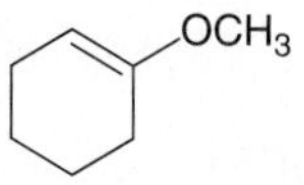 c. 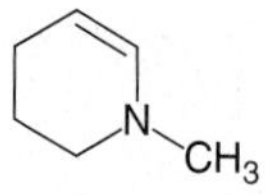 d. 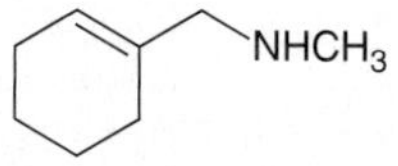

16.34 Draw all reasonable resonance structures for acetic anhydride.

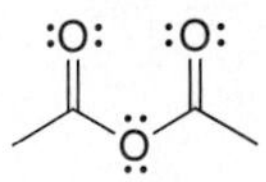

acetic anhydride

16.35 Explain why the cyclopentadienide anion **A** gives only one signal in its ^{13}C NMR spectrum.

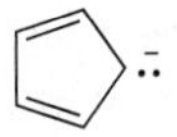 = **A**

16.36 Why is the bond dissociation energy for the C–C bond in ethane much higher than the bond dissociation energy for the labeled C–C bond in but-1-ene?

CH_3—CH_3
ethane
+368 kJ/mol

but-1-ene
+301 kJ/mol

Nomenclature and Stereoisomers in Conjugated Dienes

16.37 Draw the structure of each compound.

a. (*Z*)-penta-1,3-diene in the *s*-trans conformation
b. (2*E*,4*Z*)-1-bromo-3-methylhexa-2,4-diene
c. (2*E*,4*E*,6*E*)-octa-2,4,6-triene
d. (2*E*,4*E*)-3-methylhexa-2,4-diene in the *s*-cis conformation

16.38 Draw all possible stereoisomers of hepta-2,4-diene and label each double bond as *E* or *Z*.

16.39 Label each pair of compounds as stereoisomers or conformations.

a.  and 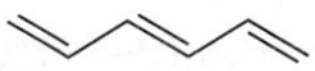

b.  and 

c. 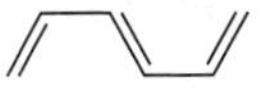 and 

16.40 Rank the following dienes in order of increasing heat of hydrogenation.

Electrophilic Addition

16.41 Draw the products formed when each compound is treated with one equivalent of HBr.

a. 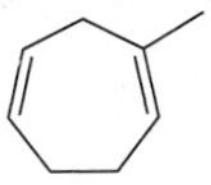

b. 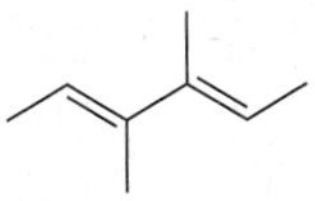

c. 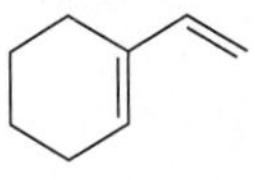

16.42 Ignoring stereoisomers, draw all products that form by addition of HBr to (*E*)-hexa-1,3,5-triene.

16.43 Treatment of alkenes **A** and **B** with HBr gives the same alkyl halide **C.** Draw a mechanism for each reaction, including all reasonable resonance structures for any intermediate.

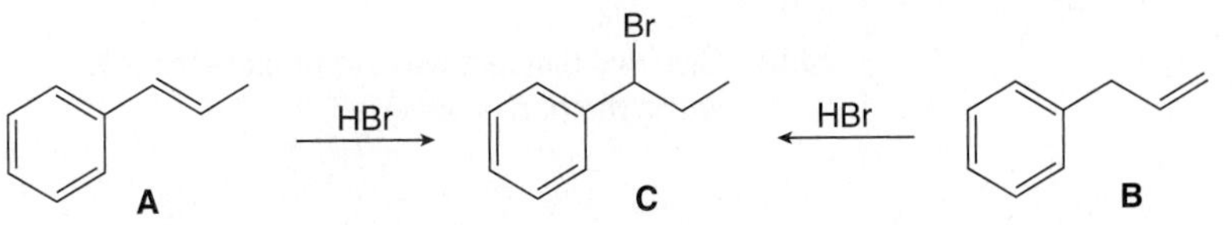

16.44 Draw a stepwise mechanism for the following reaction.

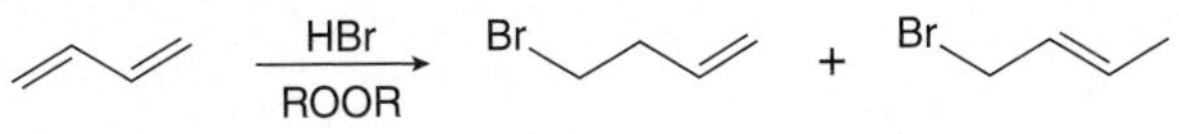

16.45 Addition of HCl to alkene **X** forms two alkyl halides **Y** and **Z.**

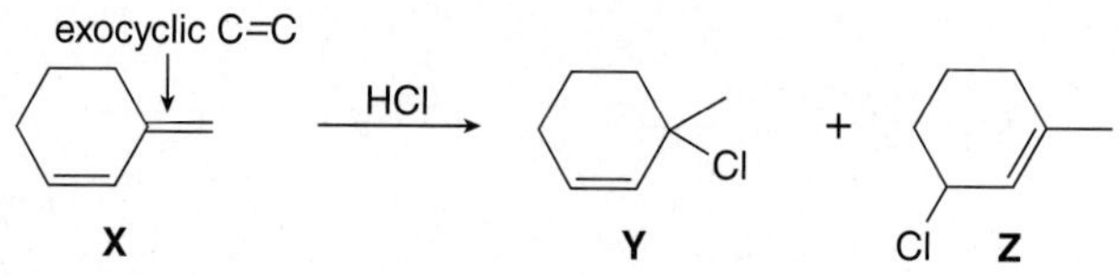

X **Y** **Z**

a. Label **Y** and **Z** as a 1,2-addition product or a 1,4-addition product.
b. Label **Y** and **Z** as the kinetic or thermodynamic product and explain why.
c. Explain why addition of HCl occurs at the indicated C=C (called an exocyclic double bond), rather than the other C=C (called an endocyclic double bond).

16.46 Explain, with reference to the mechanism, why addition of one equivalent of HCl to diene **A** forms only two products of electrophilic addition, even though four constitutional isomers are possible.

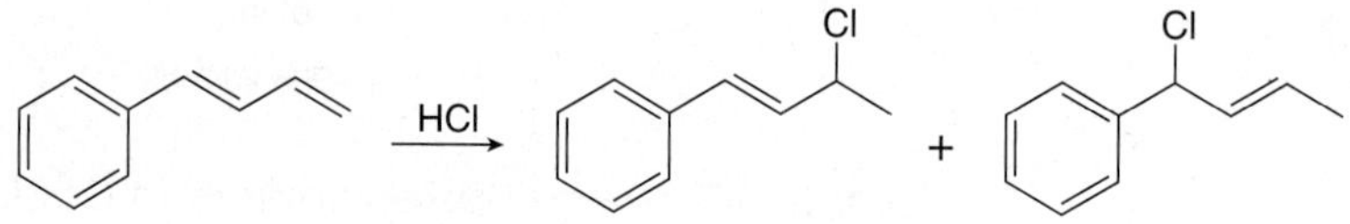

A

16.47 The major product formed by addition of HBr to $(CH_3)_2C=CH-CH=C(CH_3)_2$ is the same at low and high temperature. Draw the structure of the major product and explain why the kinetic and thermodynamic products are the same in this reaction.

Diels–Alder Reaction

16.48 Explain why methyl vinyl ether ($CH_2=CHOCH_3$) is not a reactive dienophile in the Diels–Alder reaction.

16.49 Draw the products of the following Diels–Alder reactions. Indicate stereochemistry where appropriate.

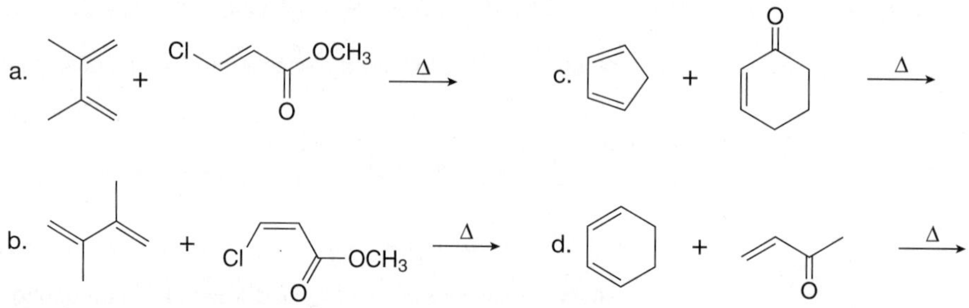

16.50 What diene and dienophile are needed to prepare each Diels–Alder product?

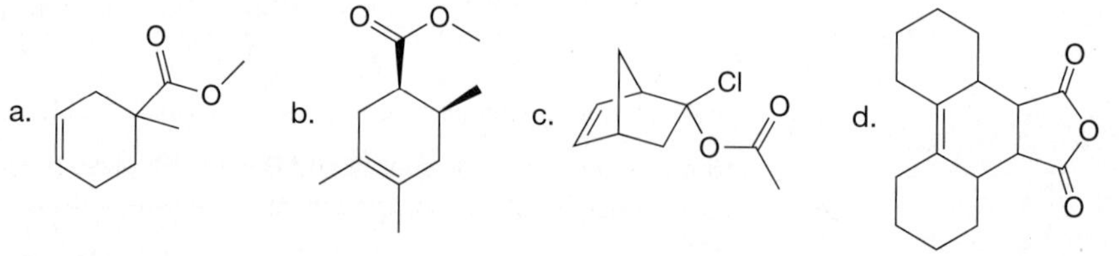

16.51 Give two different ways to prepare the following compound by the Diels–Alder reaction. Explain which method is preferred.

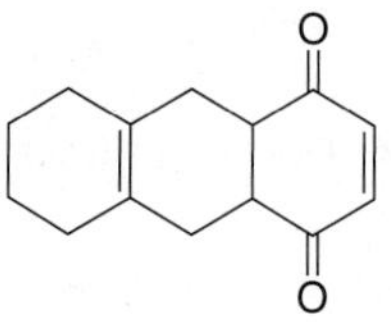

16.52 Compounds containing triple bonds are also Diels–Alder dienophiles. With this in mind, draw the products of each reaction.

a. 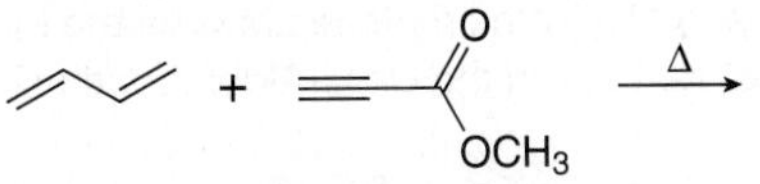
O
OCH3
Δ

b. 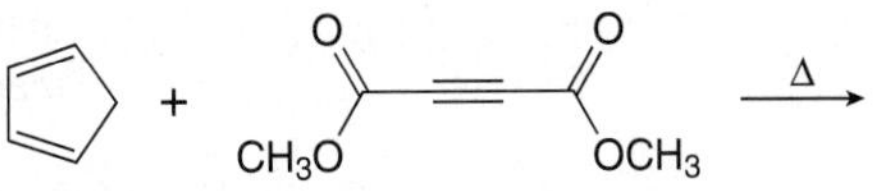
O
O
CH3O
OCH3
Δ

16.53 Diels–Alder reaction of a monosubstituted diene (such as $CH_2=CH-CH=CHOCH_3$) with a monosubstituted dienophile (such as $CH_2=CHCHO$) gives a mixture of products, but the 1,2-disubstituted product often predominates. Draw the resonance hybrid for each reactant and use the charge distribution of the hybrids to explain why the 1,2-disubstituted product is the major product.

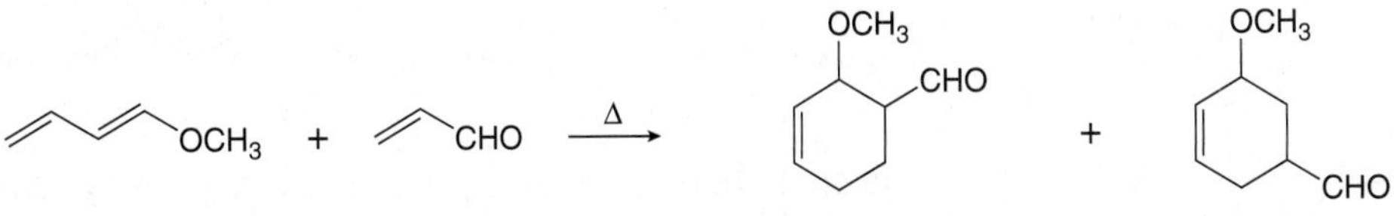
OCH3
CHO
Δ
OCH3
CHO
OCH3
CHO

1,2-disubstituted product **major** 1,3-disubstituted product **minor**

16.54 The following reactions have been used to synthesize dieldrin and aldrin (named for Diels and Alder), two pesticides having a similar story to DDT (Section 7.4). Identify the lettered compounds in this reaction scheme.

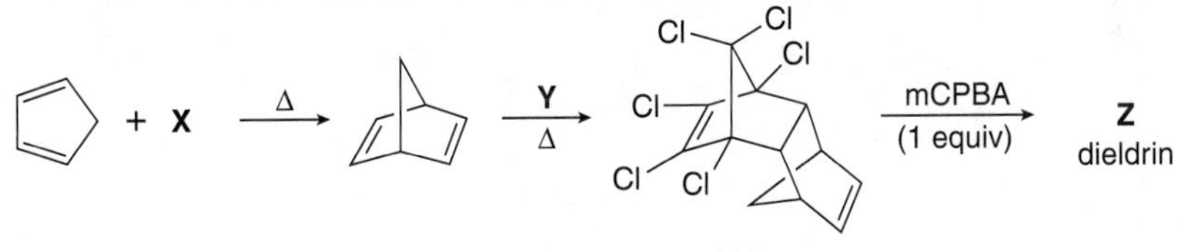
+ X
Δ
Y
Δ
Cl
Cl
Cl
Cl
Cl
Cl
mCPBA
(1 equiv)
Z
dieldrin

aldrin

16.55 Devise a stepwise synthesis of each compound from dicyclopentadiene using a Diels–Alder reaction as one step. You may also use organic compounds having ≤ 4 C's, and any required organic or inorganic reagents.

a. 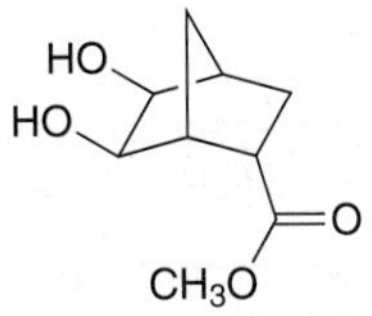
HO
HO
O
CH3O

b. 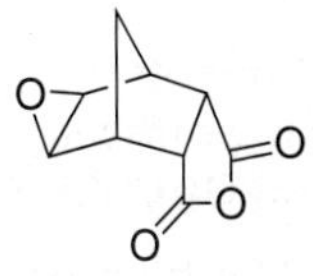
O
O
O
O

c. 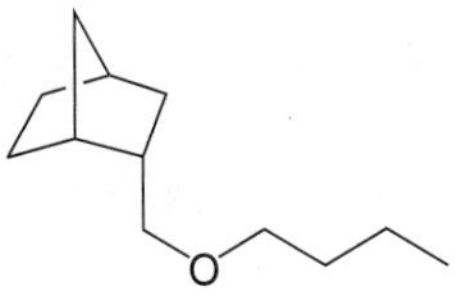
O

16.56 Intramolecular Diels–Alder reactions are possible when a substrate contains both a 1,3-diene and a dienophile, as shown in the following general reaction.

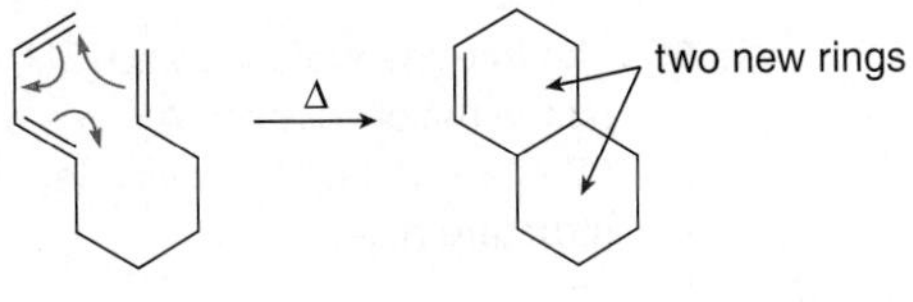
Δ
two new rings

With this in mind, draw the product of each intramolecular Diels–Alder reaction.

a. 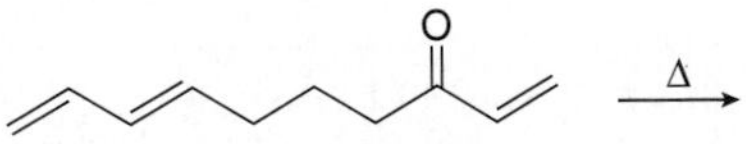
O
Δ

b. 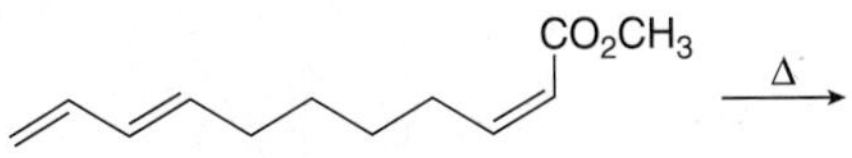
CO2CH3
Δ

16.57 A transannular Diels–Alder reaction is an intramolecular reaction that occurs when the diene and dienophile are contained in one ring, resulting in the formation of a tricyclic ring system. Draw the product formed when the following triene undergoes a transannular Diels–Alder reaction.

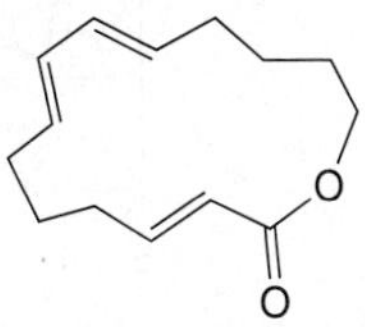
O
O

General Problems

16.58 (a) Which triene has the largest heat of hydrogenation? (b) Which triene has the smallest heat of hydrogenation? (c) Which triene absorbs the longest wavelength of UV light? (d) Which triene is most reactive in the Diels-Alder reaction?

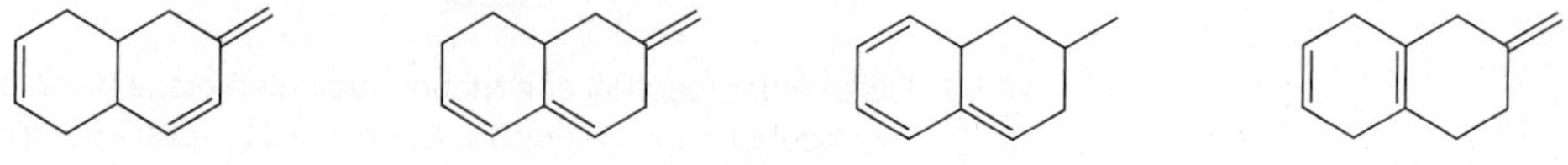

A **B** **C** **D**

16.59 Draw a stepwise mechanism for the following reaction.

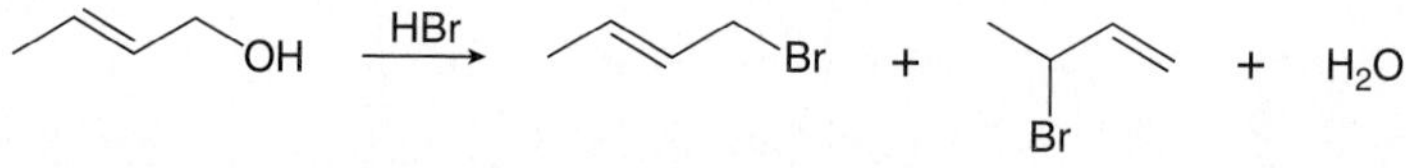

16.60 Draw the products of each reaction. Indicate the stereochemistry of Diels–Alder products.

a. 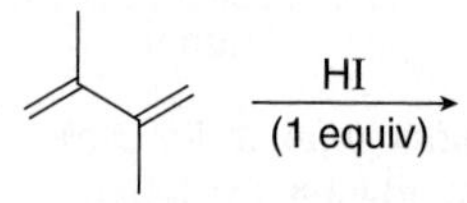

b. 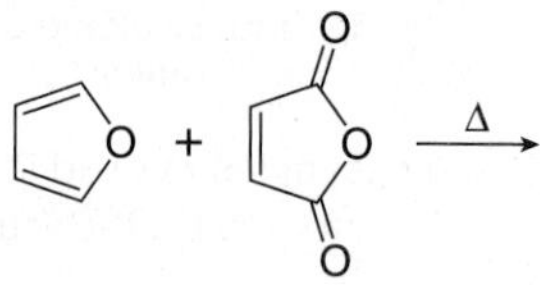

c. 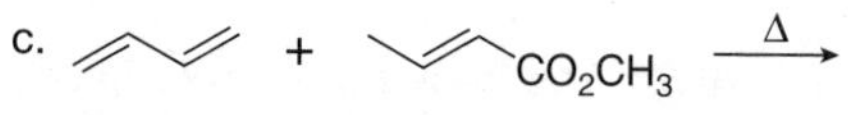

d. 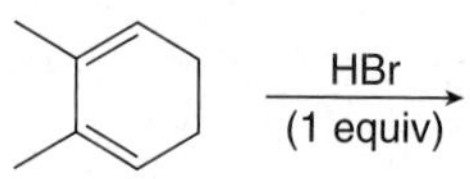

16.61 Draw a stepwise mechanism for the biological conversion of linalyl diphosphate to limonene.

linalyl diphosphate limonene

16.62 Which benzylic halide reacts faster in an S_N1 reaction? Explain.

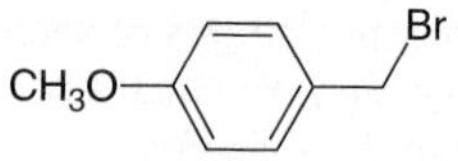

A

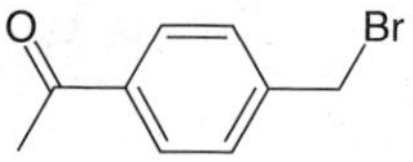

B

16.63 Like alkenes, conjugated dienes can be prepared by elimination reactions. Draw a stepwise mechanism for the acid-catalyzed dehydration of 3-methylbut-2-en-1-ol [$(CH_3)_2C=CHCH_2OH$] to isoprene [$CH_2=C(CH_3)CH=CH_2$].

16.64 (a) Draw the two isomeric dienes formed when $CH_2=CHCH_2CH(Cl)CH(CH_3)_2$ is treated with an alkoxide base. (b) Explain why the major product formed in this reaction does not contain the more highly substituted alkene.

Spectroscopy

16.65 The treatment of isoprene [$CH_2=C(CH_3)CH=CH_2$] with one equivalent of mCPBA forms **A** as the major product. **A** gives a molecular ion at 84 in its mass spectrum, and peaks at 2850–3150 cm^{-1} in its IR spectrum. The ^1H NMR spectrum of **A** is given below. What is the structure of **A**?

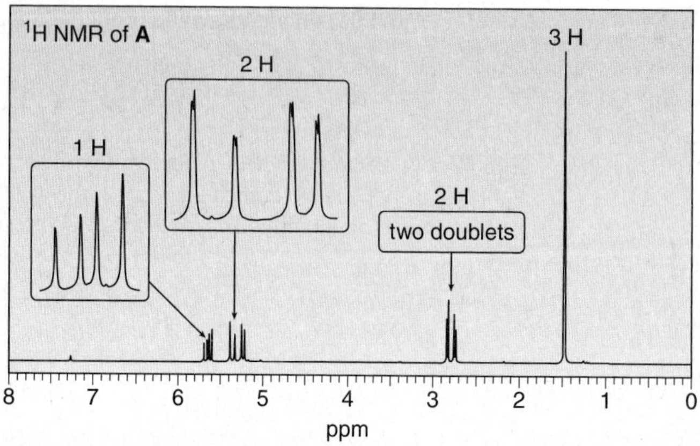

16.66 The treatment of $(CH_3)_2C=CHCH_2Br$ with H_2O forms **B** (molecular formula $C_5H_{10}O$) as one of the products. Determine the structure of **B** from its ^1H NMR and IR spectra.

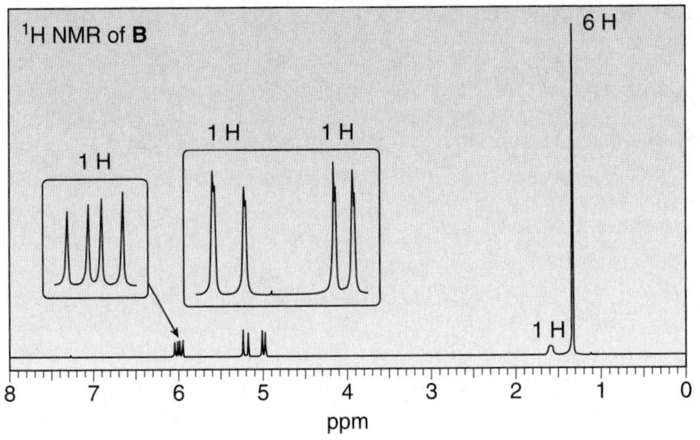

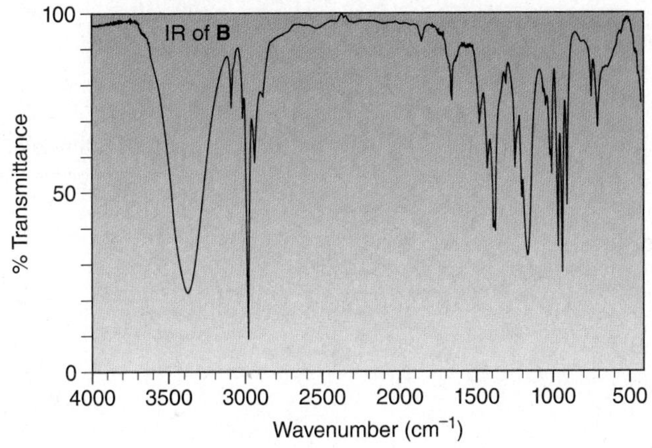

UV Absorption

16.67 Which of the following molecules absorbs a longer wavelength of UV light?

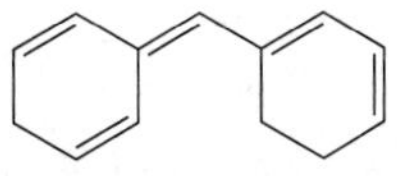

A

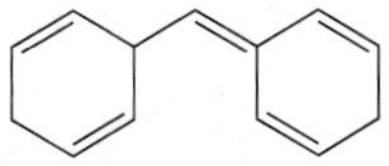

B

16.68 Explain why both **C** and **D** absorb light in the UV region of the electromagnetic spectrum, despite the fact that they are not 1,3-dienes.

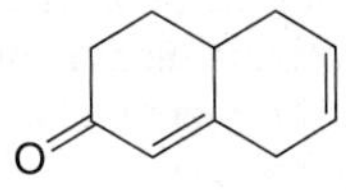

C

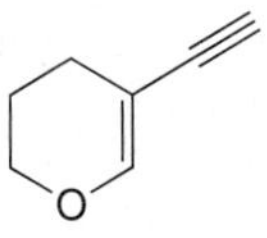

D

16.69 Explain why ferulic acid, a natural product found in rice, oats, and other plants, is both an antioxidant and a sunscreen.

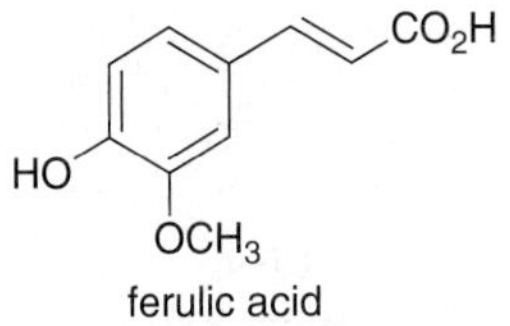

ferulic acid

Challenge Problems

16.70 Addition of HBr to allene ($CH_2=C=CH_2$) forms 2-bromoprop-1-ene rather than 3-bromoprop-1-ene, even though 3-bromoprop-1-ene is formed from an allylic carbocation. Considering the arrangement of orbitals in the allene reactant, explain this result.

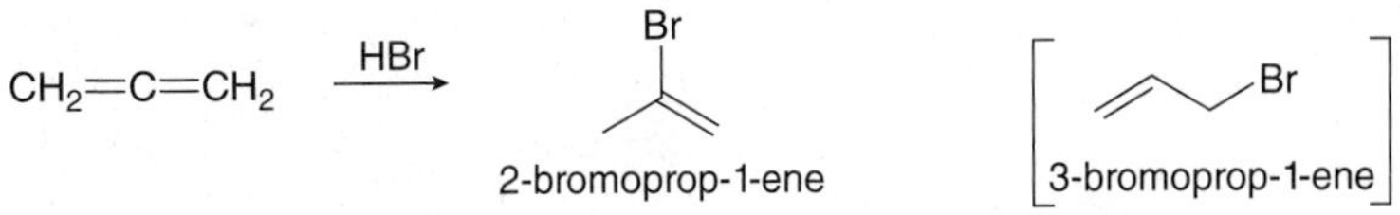

16.71 Determine the hybridization around the N atom in each amine and explain why cyclohexanamine is 10^6 times more basic than aniline.

cyclohexanamine aniline

16.72 Devise a synthesis of **X** from the given starting materials. You may use any organic or inorganic reagents. Account for the stereochemistry observed in **X.**

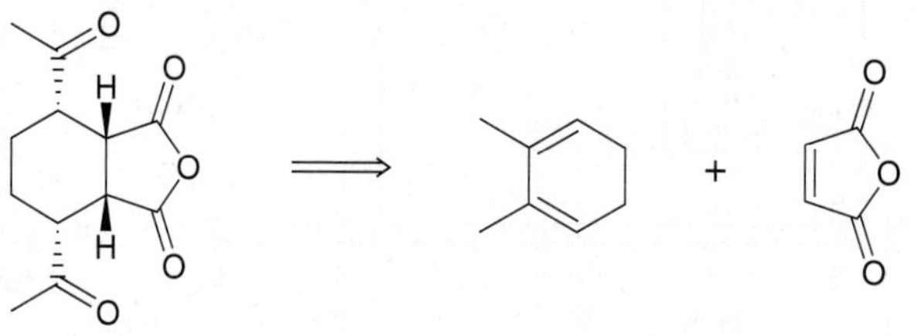

X

16.73 One step in the synthesis of occidentalol, a natural product isolated from the eastern white cedar tree, involved the following reaction. Identify the structure of **A** and show how **A** is converted to **B.**

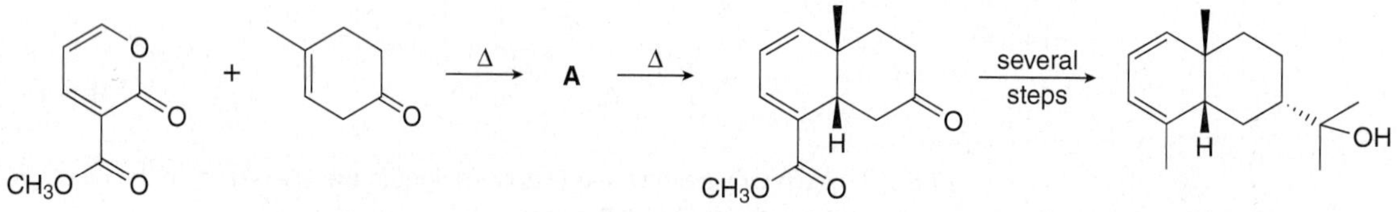

B (–)-occidentalol

16.74 One step in the synthesis of dodecahedrane (Section 4.11) involved reaction of the tetraene **C** with dimethylacetylene dicarboxylate (**D**) to afford two compounds having molecular formula $C_{16}H_{16}O_4$. This reaction has been called a domino Diels–Alder reaction. Identify the two products formed.

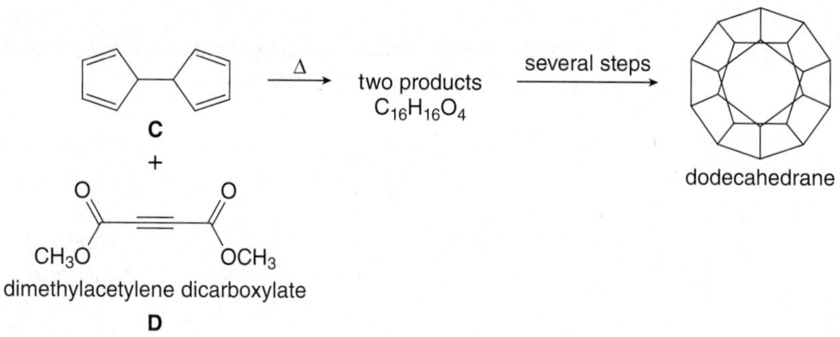

16.75 Devise a stepwise mechanism for the conversion of **M** to **N. N** has been converted in several steps to lysergic acid, a naturally occurring precursor of the hallucinogen LSD (Figure 18.4).

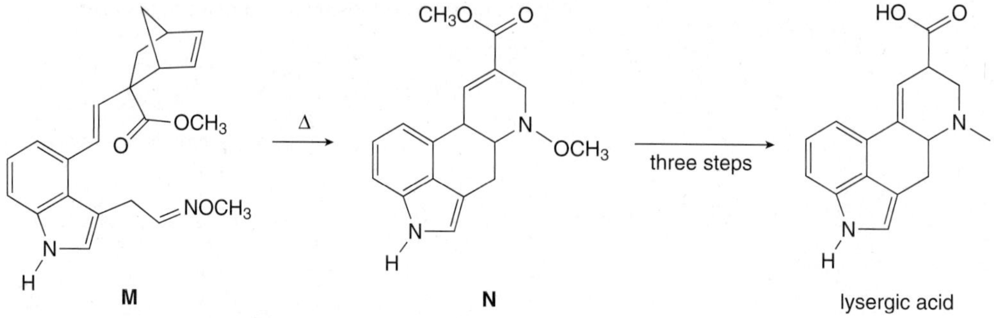

M **N** lysergic acid

Benzene and Aromatic Compounds

17

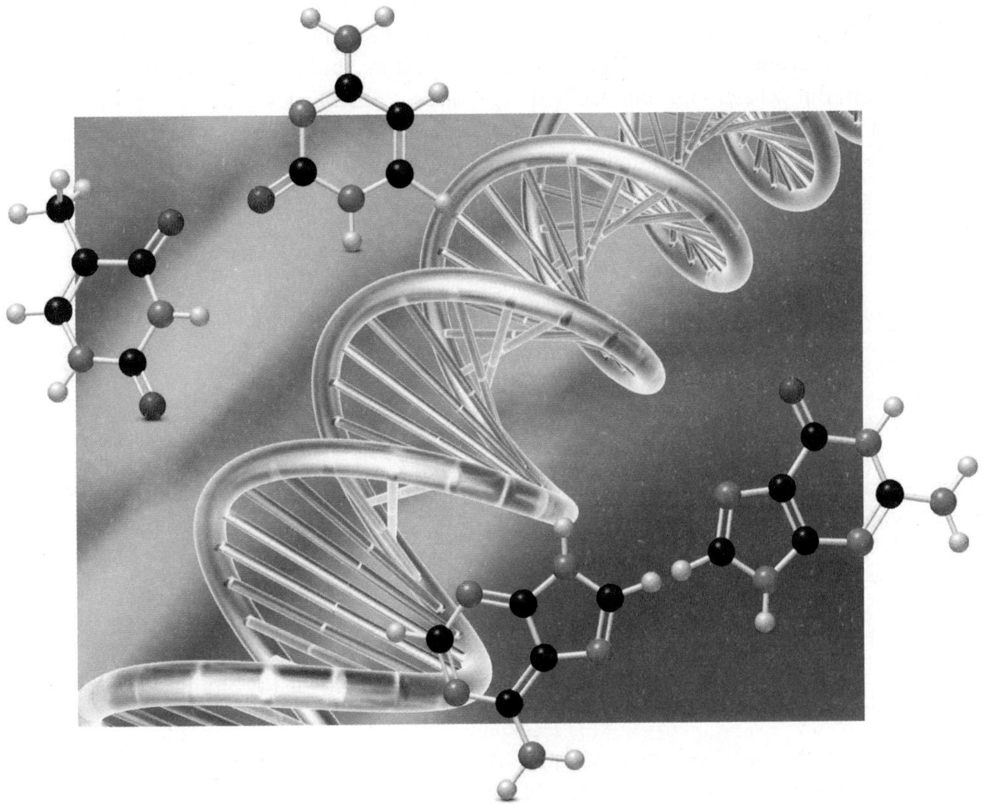

17.1 Background
17.2 The structure of benzene
17.3 Nomenclature of benzene derivatives
17.4 Spectroscopic properties
17.5 Benzene's unusual stability
17.6 The criteria for aromaticity—Hückel's rule
17.7 Examples of aromatic compounds
17.8 Aromatic heterocycles
17.9 What is the basis of Hückel's rule?
17.10 The inscribed polygon method for predicting aromaticity
17.11 Application: Aromatase inhibitors for estrogen-dependent cancer treatment

Each cell in our body contains our genome, the complete set of DNA made up of about three billion aromatic base pairs attached to helical sugar–phosphate backbones. The planar aromatic bases that are stacked together hold encrypted, hereditary, genetic instructions directing our development and cellular processes. Although only two sets of hydrogen-bonded base pairs are geometrically possible, **thymine–adenine** and **cytosine–guanine,** the sequence of bases carries the key to who we are. In Chapter 17, we learn about the unique stability and compactibility of DNA bases and other aromatic compounds.

The hydrocarbons we have examined thus far—including the alkanes, alkenes, and alkynes, as well as the conjugated dienes and polyenes of Chapter 16—have been aliphatic hydrocarbons. In Chapter 17, we continue our study of conjugated systems with **aromatic hydrocarbons.**

We begin with **benzene** and then examine other cyclic, planar, and conjugated ring systems to learn the modern definition of what it means to be aromatic. Then, in Chapter 18, we will learn about the reactions of aromatic compounds, highly unsaturated hydrocarbons that do not undergo addition reactions like other unsaturated compounds. An explanation of this behavior relies on an understanding of the structure of aromatic compounds presented in Chapter 17.

17.1 Background

For 6 C's, the maximum number of H's = $2n + 2$ = $2(6) + 2 = 14$. Because benzene contains only 6 H's, it has $14 - 6 = 8$ H's fewer than the maximum number. This corresponds to 8 H's/2 H's for each degree of unsaturation = **four degrees of unsaturation in benzene.**

Benzene (C_6H_6) is the simplest aromatic hydrocarbon (or arene). Since its isolation by Michael Faraday from the oily residue remaining in the illuminating gas lines in London in 1825, it has been recognized as an unusual compound. Based on the calculation introduced in Section 10.2, **benzene has four degrees of unsaturation, making it a highly unsaturated hydrocarbon.** But, whereas unsaturated hydrocarbons such as alkenes, alkynes, and dienes readily undergo addition reactions, *benzene does not.* For example, bromine adds to ethylene to form a dibromide, but benzene is inert under similar conditions.

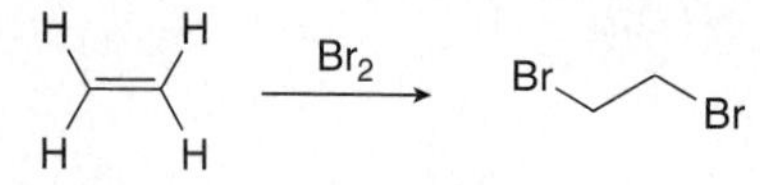

ethylene

addition product

$$C_6H_6 \xrightarrow{Br_2} \textbf{No reaction}$$

benzene

Benzene *does* react with bromine, but only in the presence of $FeBr_3$ (a Lewis acid), and the reaction is a **substitution,** *not* an addition.

$$C_6H_6 \xrightarrow[FeBr_3]{Br_2} C_6H_5Br$$

substitution
Br replaces H

Thus, any structure proposed for benzene must account for its high degree of unsaturation and its lack of reactivity toward electrophilic addition.

In the last half of the nineteenth century, August Kekulé proposed structures that were close to the modern description of benzene. In the Kekulé model, benzene was thought to be a rapidly equilibrating mixture of two compounds, each containing a six-membered ring with three alternating π bonds. These structures are now called **Kekulé structures.** In the Kekulé description, the bond between any two carbon atoms is sometimes a single bond and sometimes a double bond.

Kekulé description:
An equilibrium

Although benzene is still drawn as a six-membered ring with three alternating π bonds, in reality **there is no equilibrium between two different kinds of benzene molecules.** Instead, current descriptions of benzene are based on resonance and electron delocalization due to orbital overlap, as detailed in Section 17.2.

In the nineteenth century, many other compounds having properties similar to those of benzene were isolated from natural sources. Because these compounds possessed strong and characteristic odors, they were called ***aromatic* compounds.** It is their chemical properties, though, not their odor that make these compounds special.

- Aromatic compounds resemble benzene—they are unsaturated compounds that do not undergo the addition reactions characteristic of alkenes.

17.2 The Structure of Benzene

Any structure for benzene must account for the following:

- **Benzene contains a six-membered ring and three additional degrees of unsaturation.**
- **Benzene is planar.**
- **All C–C bond lengths are equal.**

Although the Kekulé structures satisfy the first two criteria, they break down with the third, because having three alternating π bonds would mean that benzene should have three short double bonds alternating with three longer single bonds.

This structure implies that the C–C bonds should have **two different lengths.**

- three longer single bonds in red
- three shorter double bonds in black

Resonance

Some texts draw benzene as a hexagon with an inner circle:

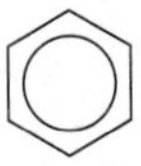

The circle represents the **six π electrons,** distributed over the six atoms of the ring.

Benzene is conjugated, so we must use resonance and orbitals to describe its structure. The resonance description of benzene consists of two equivalent Lewis structures, each with three double bonds that alternate with three single bonds.

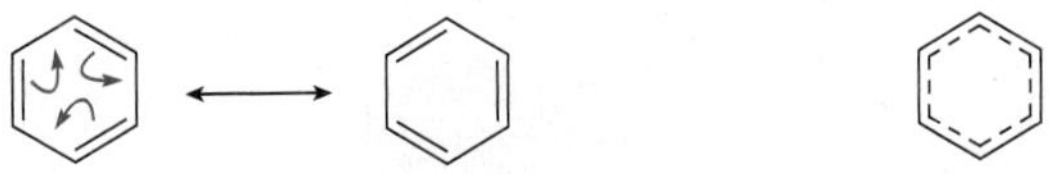

hybrid

The electrons in the π bonds are **delocalized** around the ring.

The resonance description of benzene matches the Kekulé description with one important exception. **The two Kekulé representations are *not* in equilibrium with each other.** Instead, the true structure of benzene is a resonance **hybrid** of the two Lewis structures, with the dashed lines of the hybrid indicating the position of the π bonds.

We will use one of the two Lewis structures and not the hybrid in drawing benzene, because it is easier to keep track of the electron pairs in the π bonds (the π electrons).

- **Because each π bond has two electrons, benzene has six π electrons.**

The resonance hybrid of benzene explains why all C–C bond lengths are the same. Each C–C bond is single in one resonance structure and double in the other, so the actual bond length (139 pm) is intermediate between a carbon–carbon single bond (153 pm) and a carbon–carbon double bond (134 pm).

CH_3-CH_3	$CH_2=CH_2$
153 pm	134 pm

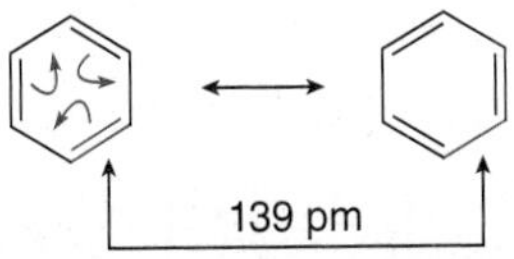

The C–C bonds in benzene are equal and intermediate in length.

Hybridization and Orbitals

Each carbon atom in a benzene ring is surrounded by three atoms and no lone pairs of electrons, making it ***sp*^{2} hybridized and trigonal planar with all bond angles 120°.** Each carbon also has a *p* orbital with one electron that extends above and below the plane of the molecule.

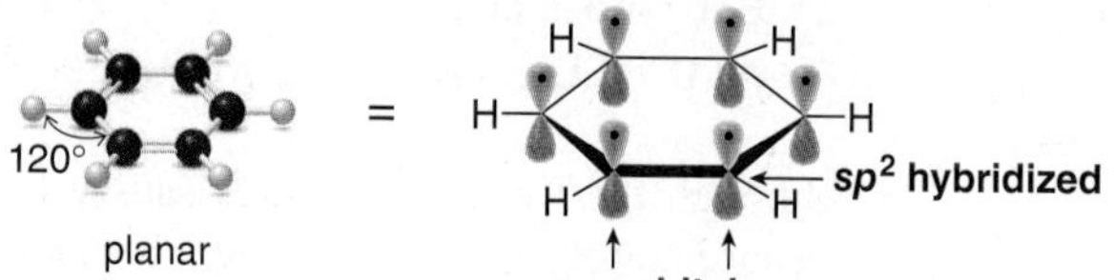

The six adjacent *p* orbitals overlap, delocalizing the six electrons over the six atoms of the ring and making benzene a conjugated molecule. Because each *p* orbital has two lobes, one above and one below the plane of the benzene ring, the overlap of the *p* orbitals creates two "doughnuts" of electron density, as shown in Figure 17.1a. The electrostatic potential plot in Figure 17.1b also shows that the electron-rich region is concentrated above and below the plane of the molecule, where the six π electrons are located.

- **Benzene's six π electrons make it electron rich, so it reacts with electrophiles.**

Figure 17.1 Two views of the electron density in a benzene ring

a. View of the *p* orbital overlap

- Overlap of six adjacent *p* orbitals creates two rings of electron density, one above and one below the plane of the benzene ring.

b. Electrostatic potential plot

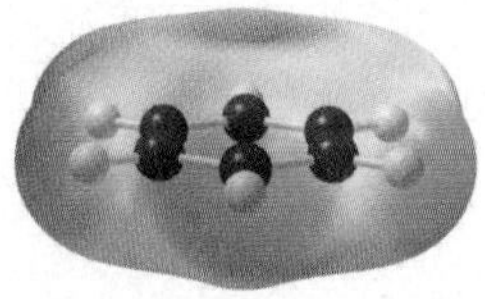

- The electron-rich region (in red) is concentrated above and below the ring carbons, where the six π electrons are located. (The electron-rich region below the plane is hidden from view.)

Problem 17.1 Draw all possible resonance structures for the antihistamine diphenhydramine, the active ingredient in Benadryl.

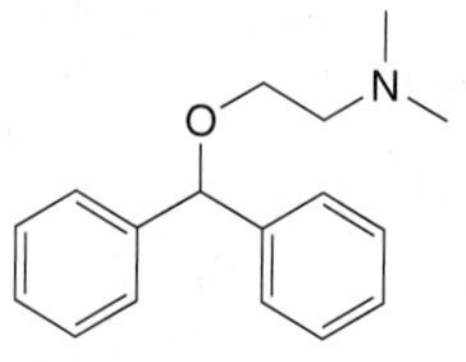

diphenhydramine

Problem 17.2 What orbitals are used to form the labeled bonds in the following molecule? Of the labeled C–C bonds, which is the shortest?

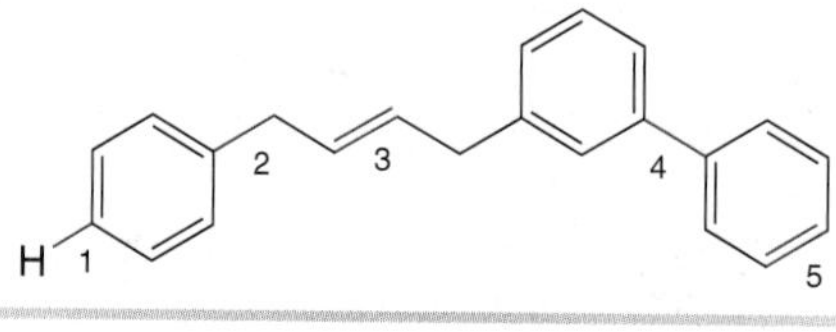

17.3 Nomenclature of Benzene Derivatives

Many organic molecules contain a benzene ring with one or more substituents, so we must learn how to name them. Many common names are recognized by the IUPAC system, however, so this complicates the nomenclature of benzene derivatives somewhat.

17.3A Monosubstituted Benzenes

To name a benzene ring with one substituent, **name the substituent and add the word *benzene.*** Carbon substituents are named as alkyl groups.

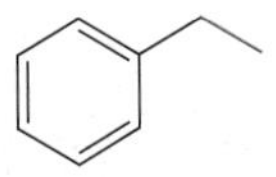
ethylbenzene

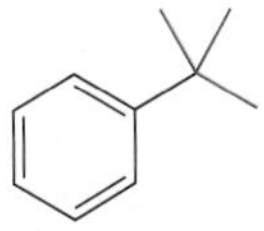
tert-butylbenzene

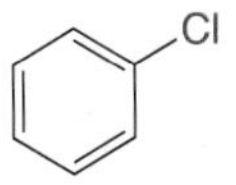
chlorobenzene

Many monosubstituted benzenes, such as those with methyl (CH_3-), hydroxy ($-OH$), and amino ($-NH_2$) groups, have common names that you must learn, too.

Toluene and **benzene** are the simplest aromatic hydrocarbons obtained from petroleum refining.

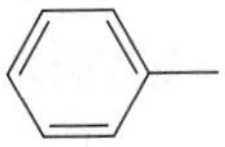
toluene (methylbenzene)

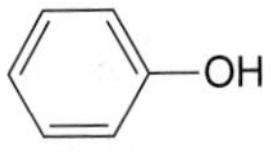
phenol (hydroxybenzene)

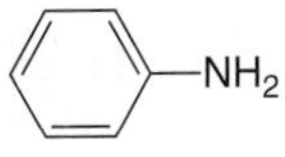
aniline (aminobenzene)

17.3B Disubstituted Benzenes

Novocain (generic name procaine) is a disubstituted aniline derivative that is used as a local anesthetic.

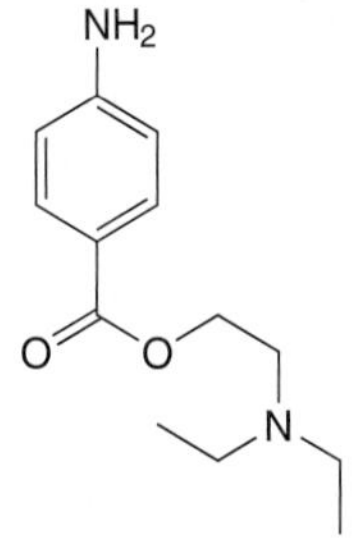

There are three different ways that two groups can be attached to a benzene ring, so a prefix—**ortho, meta,** or **para**—can be used to designate the relative position of the two substituents. Ortho, meta, and para are also abbreviated as ***o*, *m*,** and ***p*,** respectively.

1,2-Disubstituted benzene **ortho** isomer

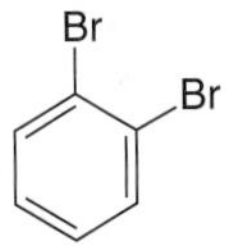
o-dibromobenzene or **1,2**-dibromobenzene

1,3-Disubstituted benzene **meta** isomer

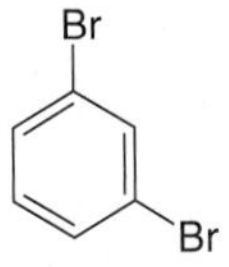
m-dibromobenzene or **1,3**-dibromobenzene

1,4-Disubstituted benzene **para** isomer

p-dibromobenzene or **1,4**-dibromobenzene

If the two groups on the benzene ring are different, **alphabetize the names of the substituents** preceding the word *benzene.* If one of the substituents is part of a **common root,** name the **molecule as a derivative of that monosubstituted benzene.**

Alphabetize two different substituent names:

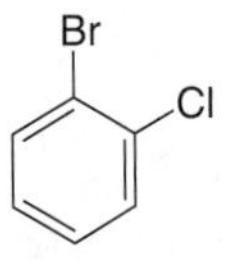
o-bromochloro-benzene

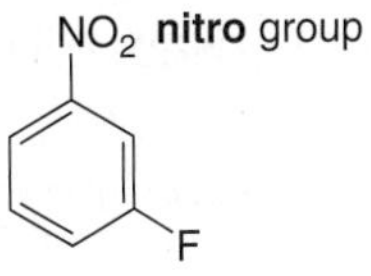

***m*-fluoronitro-benzene**

Use a common root name:

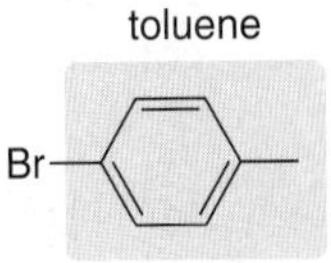

p-bromo**toluene**

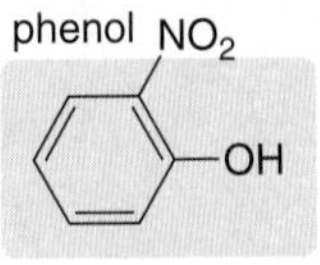

o-nitro**phenol**

17.3C Polysubstituted Benzenes

For three or more substituents on a benzene ring:

[1] Number to give the lowest possible numbers around the ring.

[2] Alphabetize the substituent names.

[3] When substituents are part of common roots, name the molecule as a derivative of that monosubstituted benzene. The substituent that comprises the common root is located at C1.

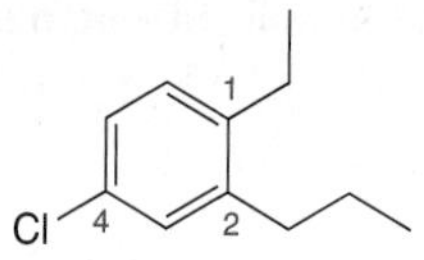

- Assign the lowest set of numbers.
- Alphabetize the names of all the substituents.

4-chloro-1-ethyl-2-propylbenzene

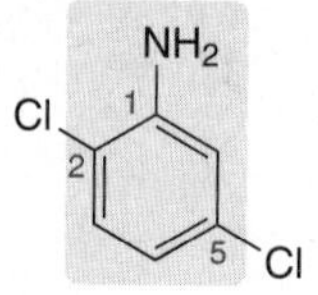

- Name the molecule as a derivative of the common root **aniline.**
- Designate the position of the NH_2 group as "1," and then assign the lowest possible set of numbers to the other substituents.

2,5-dichloroaniline

17.3D Naming Aromatic Rings as Substituents

A benzene substituent (C_6H_5-) is called a **phenyl group,** and it can be abbreviated in a structure as **Ph–.**

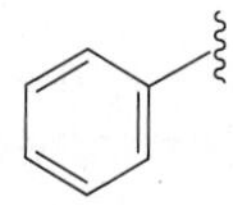

abbreviated as Ph–

phenyl group
C_6H_5-

- **A phenyl group (C_6H_5-) is formed by removing one hydrogen from benzene (C_6H_6).**

Benzene, therefore, can be represented as PhH, and phenol would be PhOH.

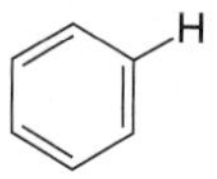

benzene

PhH

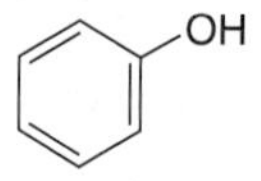

phenol

PhOH

The **benzyl** group contains a benzene ring bonded to a CH_2 group. Thus, a benzyl group and a phenyl group differ by the presence of a CH_2 group.

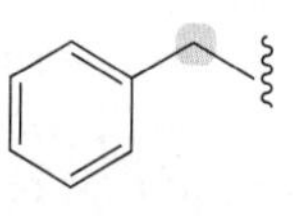

benzyl group
$C_6H_5CH_2-$

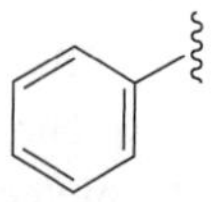

phenyl group
C_6H_5-

Finally, substituents derived from benzene, as well as all other substituted aromatic rings, are collectively called **aryl groups,** abbreviated as Ar–.

Problem 17.3 Give the IUPAC name for each compound.

a. 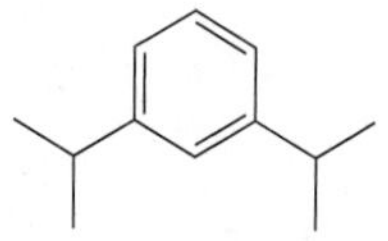

b. 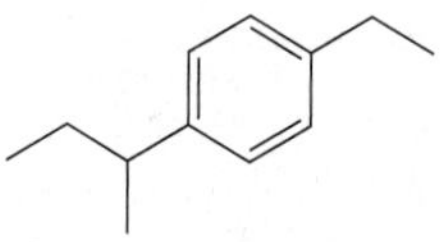

c. 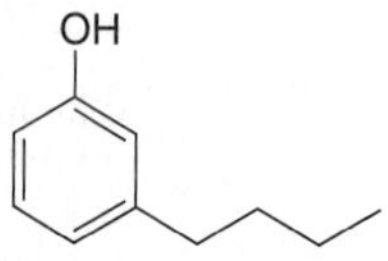

d. 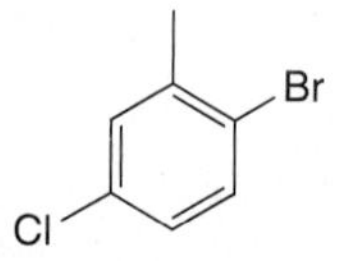

Problem 17.4 Draw the structure corresponding to each name:

a. isobutylbenzene
b. *o*-dichlorobenzene
c. *cis*-1,2-diphenylcyclohexane
d. *m*-bromoaniline
e. 4-chloro-1,2-diethylbenzene
f. 3-*tert*-butyl-2-ethyltoluene

Problem 17.5 What is the structure of propofol, which has the IUPAC name 2,6-diisopropylphenol? Propofol is an intravenous medication used to induce and maintain anesthesia.

17.4 Spectroscopic Properties

The important IR and NMR absorptions of aromatic compounds are summarized in Table 17.1.

Table 17.1 Characteristic Spectroscopic Absorptions of Benzene Derivatives

Type of spectroscopy	Type of C, H	Absorption
IR absorptions	C_{sp^2}–H C=C (arene)	3150–3000 cm^{-1} 1600, 1500 cm^{-1}
^{1}H NMR absorptions	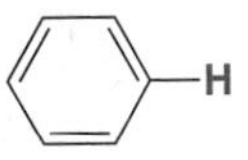 (aryl H)	6.5–8.0 ppm (highly deshielded protons)
	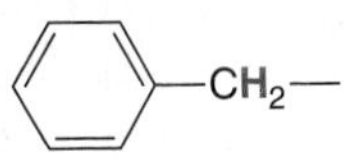 (benzylic H)	1.5–2.5 ppm (somewhat deshielded C_{sp^3}–H)
^{13}C NMR absorption	C_{sp^2} of arenes	120–150 ppm

The absorption at 6.5–8.0 ppm in the ^{1}H NMR spectrum is particularly characteristic of compounds containing benzene rings. **All aromatic compounds have highly deshielded protons due to the ring current effect of the circulating π electrons,** as discussed in Section 14.4. Observing whether a new compound absorbs in this region of a ^{1}H NMR spectrum is one piece of data used to determine if it is aromatic.

^{13}C NMR spectroscopy is used to determine the substitution patterns in disubstituted benzenes, because each line in a spectrum corresponds to a different kind of carbon atom. For example, *o*-, *m*-, and *p*-dibromobenzene each exhibit a different number of lines in its ^{13}C NMR spectrum, as shown in Figure 17.2.

Figure 17.2 ^{13}C NMR absorptions of the three isomeric dibromobenzenes

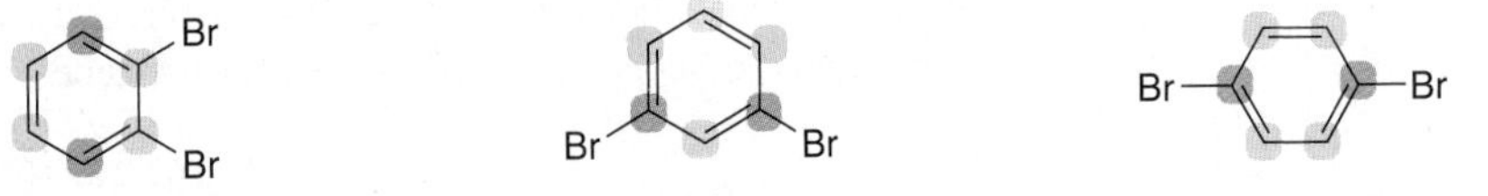

o-dibromobenzene — **three** types of C's, **three** ^{13}C NMR signals

m-dibromobenzene — **four** types of C's, **four** ^{13}C NMR signals

p-dibromobenzene — **two** types of C's, **two** ^{13}C NMR signals

- The number of signals (lines) in the ^{13}C NMR spectrum of a disubstituted benzene with two identical groups indicates whether they are ortho, meta, or para to each other.

Problem 17.6 What is the structure of a compound of molecular formula $C_{10}H_{14}O_2$ that shows a strong IR absorption at 3150–2850 cm^{-1} and gives the following ^{1}H NMR absorptions: 1.4 (triplet, 6 H), 4.0 (quartet, 4 H), and 6.8 (singlet, 4 H) ppm?

Problem 17.7 How many ^{13}C NMR signals does each compound exhibit?

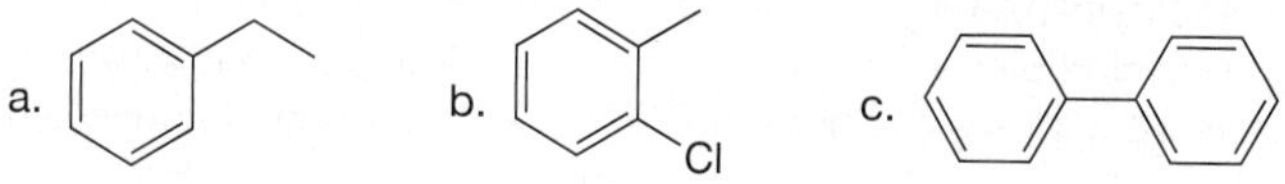

17.5 Benzene's Unusual Stability

Considering benzene as the hybrid of two resonance structures adequately explains its equal C–C bond lengths, but does not account for its unusual stability and lack of reactivity toward addition.

Heats of hydrogenation, which were used in Section 16.9 to show that conjugated dienes are more stable than isolated dienes, can also be used to estimate the stability of benzene. Equations [1]–[3] compare the heats of hydrogenation of cyclohexene, cyclohexa-1,3-diene, and benzene, all of which give cyclohexane when treated with excess hydrogen in the presence of a metal catalyst.

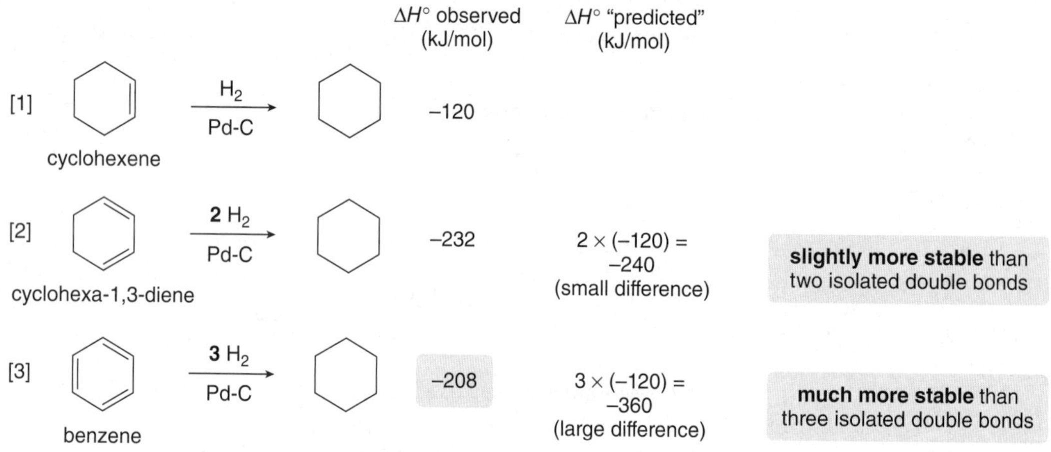

The relative stability of conjugated dienes versus isolated dienes was first discussed in Section 16.9.

The addition of one mole of H_2 to cyclohexene releases –120 kJ/mol of energy (Equation [1]). If each double bond is worth –120 kJ/mol of energy, then the addition of two moles of H_2 to cyclohexa-1,3-diene (Equation [2]) should release 2 × (–120 kJ/mol) = –240 kJ/mol of energy. The observed value, however, is –232 kJ/mol. This is slightly smaller than expected because cyclohexa-1,3-diene is a conjugated diene, and conjugated dienes are more stable than two isolated carbon–carbon double bonds.

The hydrogenations of cyclohexene and cyclohexa-1,3-diene occur readily at room temperature, but benzene can be hydrogenated only under forcing conditions, and even then the reaction is extremely slow. If each double bond is worth –120 kJ/mol of energy, then the addition of three moles of H_2 to benzene should release 3 × (–120 kJ/mol) = –360 kJ/mol of energy. In fact, the observed heat of hydrogenation is only –208 kJ/mol, which is 152 kJ/mol less than predicted and even lower than the observed value for cyclohexa-1,3-diene. Figure 17.3 compares the hypothetical and observed heats of hydrogenation for benzene.

Figure 17.3 A comparison between the observed and hypothetical heats of hydrogenation for benzene

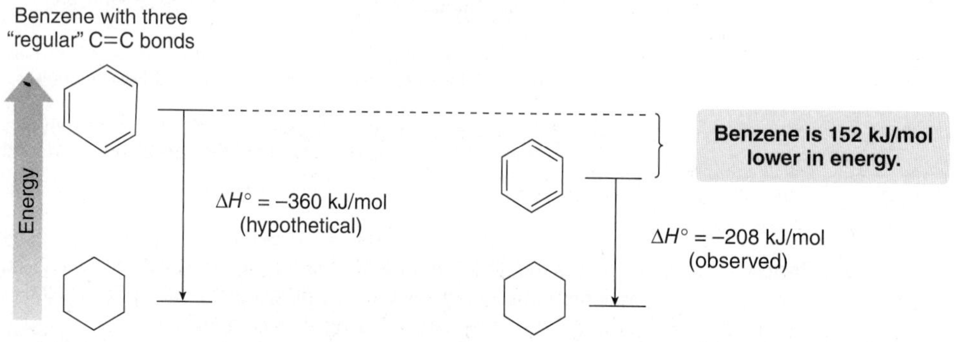

The huge difference between the hypothetical and observed heats of hydrogenation for benzene cannot be explained solely on the basis of resonance and conjugation.

- **The low heat of hydrogenation of benzene means that benzene is especially stable, even more so than the conjugated compounds introduced in Chapter 16. This unusual stability is characteristic of aromatic compounds.**

Benzene's unusual behavior in chemical reactions is not limited to hydrogenation. As mentioned in Section 17.1, **benzene does not undergo addition reactions typical of other highly unsaturated compounds, including conjugated dienes.** Benzene does not react with Br_2 to yield an addition product. Instead, in the presence of a Lewis acid, bromine *substitutes* for a hydrogen atom, thus yielding a product that retains the benzene ring.

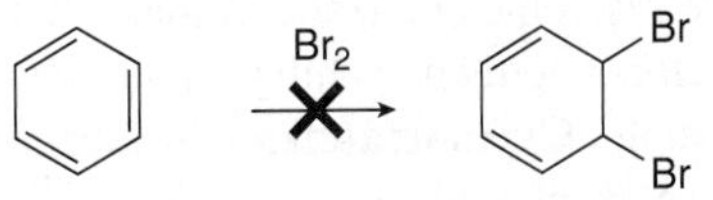

An addition product would no longer contain a benzene ring.

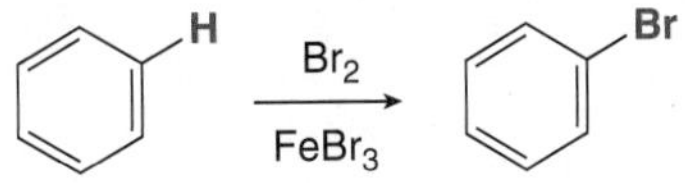

A substitution product still contains a benzene ring.

This behavior is characteristic of aromatic compounds. The structural features that distinguish aromatic compounds from the rest are discussed in Section 17.6.

Problem 17.8 Compounds **A** and **B** are both hydrogenated to methylcyclohexane. Which compound has the larger heat of hydrogenation? Which compound is more stable?

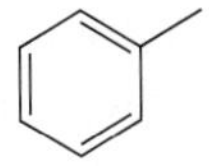

A

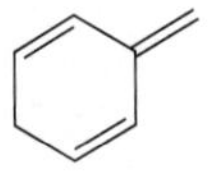

B

17.6 The Criteria for Aromaticity—Hückel's Rule

Four structural criteria must be satisfied for a compound to be aromatic:

- **A molecule must be cyclic, planar, completely conjugated, and contain a particular number of π electrons.**

[1] A molecule must be cyclic.

- **To be aromatic, each *p* orbital must overlap with *p* orbitals on two adjacent atoms.**

The *p* orbitals on all six carbons of benzene continuously overlap, so benzene is aromatic. Hexa-1,3,5-triene has six *p* orbitals, too, but the two on the terminal carbons cannot overlap with each other, so **hexa-1,3,5-triene is not aromatic.**

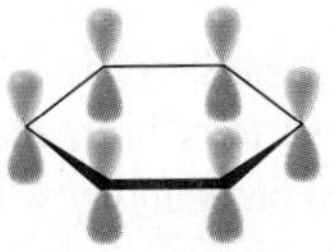

benzene

Every *p* orbital overlaps with two neighboring *p* orbitals.

aromatic

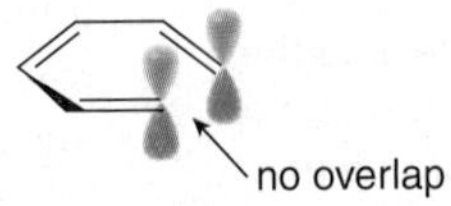

hexa-1,3,5-triene

There can be no overlap between the *p* orbitals on the two terminal C's.

not aromatic

[2] A molecule must be planar.

- **All adjacent *p* orbitals must be aligned so that the π electron density can be delocalized.**

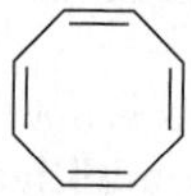

cyclooctatetraene
not aromatic

a tub-shaped, eight-membered ring

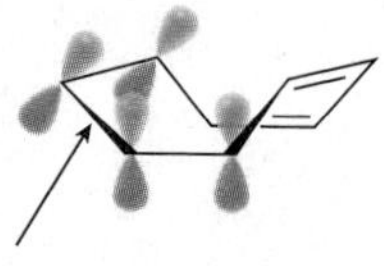
Adjacent *p* orbitals cannot overlap.
Electrons cannot delocalize.

Cyclooctatetraene resembles benzene in that it is a cyclic molecule with alternating double and single bonds. Cyclooctatetraene is tub shaped, however, **not planar,** so overlap between adjacent π bonds is impossible. **Cyclooctatetraene, therefore, is *not* aromatic,** so it undergoes addition reactions like those of other alkenes.

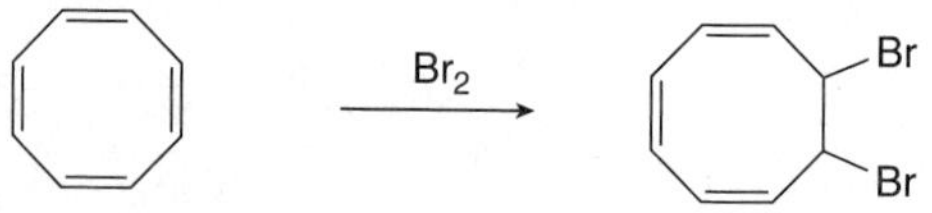
Br_2
Br
Br

cyclooctatetraene **addition** product

[3] A molecule must be completely conjugated.

- **Aromatic compounds must have a *p* orbital on every atom in the ring.**

benzene
a *p* orbital on **every** C
aromatic

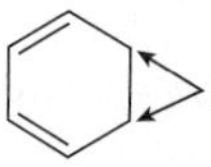
no *p* orbitals

cyclohexa-1,3-diene
not aromatic

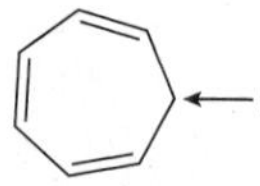
no *p* orbital

cyclohepta-1,3,5-triene
not aromatic

Both cyclohexa-1,3-diene and cyclohepta-1,3,5-triene contain at least one carbon atom that does not have a *p* orbital, and so they are not completely conjugated and therefore ***not* aromatic.**

[4] A molecule must satisfy Hückel's rule, and contain a particular number of π electrons.

Some compounds satisfy the first three criteria for aromaticity, but still they show none of the stability typical of aromatic compounds. For example, **cyclobutadiene** is so highly reactive that it can be prepared only at extremely low temperatures.

cyclobutadiene

a planar, cyclic, completely conjugated molecule that is *not* aromatic

Hückel's rule refers to the number of π electrons, *not* the number of atoms in a particular ring.

It turns out that in addition to being cyclic, planar, and completely conjugated, a compound needs a particular number of π electrons to be aromatic. Erich Hückel first recognized in 1931 that the following criterion, expressed in two parts and now known as **Hückel's rule,** had to be satisfied, as well:

- **An aromatic compound must contain $4n + 2$ π electrons (n = 0, 1, 2, and so forth).**
- **Cyclic, planar, and completely conjugated compounds that contain $4n$ π electrons are especially unstable, and are said to be *antiaromatic*.**

Table 17.2
The Number of π Electrons That Satisfy Hückel's Rule

n	$4n + 2$
0	2
1	6
2	10
3	14
4, etc.	18

Thus, compounds that contain 2, 6, 10, 14, 18, and so forth π electrons are aromatic, as shown in Table 17.2. **Benzene is aromatic and especially stable because it contains 6 π electrons. Cyclobutadiene is antiaromatic and especially unstable because it contains 4 π electrons.**

Benzene
An aromatic compound

$4n + 2$ $= 4(1) + 2 =$
6 π electrons
aromatic

Cyclobutadiene
An antiaromatic compound

$4n$ $= 4(1) =$
4 π electrons
antiaromatic

Considering aromaticity, all compounds can be classified in one of three ways:

[1]	Aromatic	• A cyclic, planar, completely conjugated compound with $4n + 2$ π electrons
[2]	Antiaromatic	• A cyclic, planar, completely conjugated compound with $4n$ π electrons
[3]	Not aromatic or nonaromatic	• A compound that lacks one (or more) of the four requirements to be aromatic or antiaromatic

Note, too, the relationship between each compound type and a similar open-chained molecule having the same number of π electrons.

- An aromatic compound is *more* stable than a similar acyclic compound having the same number of π electrons. Benzene is more stable than hexa-1,3,5-triene.

- An antiaromatic compound is *less* stable than an acyclic compound having the same number of π electrons. Cyclobutadiene is less stable than buta-1,3-diene.

- A compound that is not aromatic is *similar* in stability to an acyclic compound having the same number of π electrons. Cyclohexa-1,3-diene is similar in stability to *cis,cis*-hexa-2,4-diene, so it is not aromatic.

benzene

and

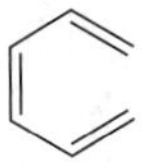
hexa-1,3,5-triene

more stable
aromatic

cyclobutadiene

and

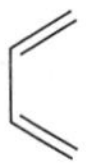
buta-1,3-diene

less stable
antiaromatic

cyclohexa-1,3-diene

and

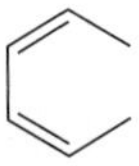
cis,cis-hexa-2,4-diene

similar stability
nonaromatic

^{1}H NMR spectroscopy readily indicates whether a compound is aromatic. The protons on sp^2 hybridized carbons in aromatic hydrocarbons are highly deshielded and absorb at 6.5–8 ppm, whereas unsaturated hydrocarbons that are not aromatic absorb at 4.5–6 ppm, typical of protons bonded to the C=C of an alkene. Thus, benzene absorbs at 7.3 ppm, whereas cyclooctatetraene, which is not aromatic, absorbs farther upfield, at 5.8 ppm for the protons on its sp^2 hybridized carbons.

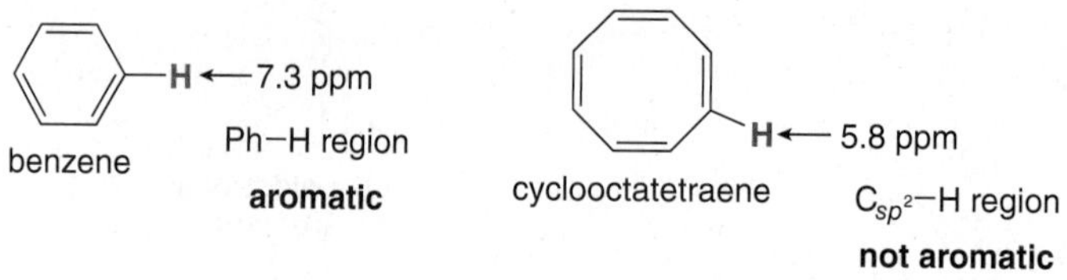
H ← 7.3 ppm
Ph–H region
benzene
aromatic
H ← 5.8 ppm
cyclooctatetraene
C$_{sp^2}$–H region
not aromatic

Many compounds in addition to benzene are aromatic. Several examples are presented in Section 17.7.

Problem 17.9 Estimate where the protons bonded to the sp^2 hybridized carbons will absorb in the ^{1}H NMR spectrum of each compound.

a. b. c.

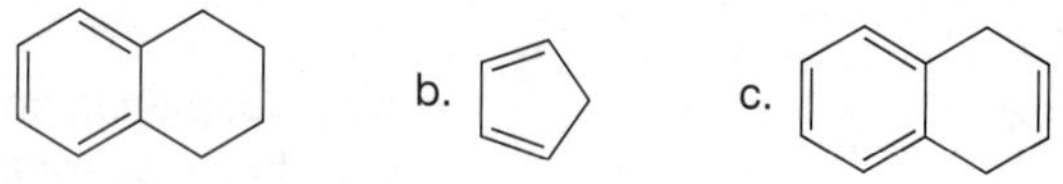

17.7 Examples of Aromatic Compounds

In Section 17.7 we look at many different types of aromatic compounds.

17.7A Aromatic Compounds with a Single Ring

Benzene is the most common aromatic compound having a single ring. **Completely conjugated rings larger than benzene are also aromatic if they are planar and have $4n + 2$ π electrons.**

- **Hydrocarbons containing a single ring with alternating double and single bonds are called *annulenes*.**

To name an annulene, indicate the number of atoms in the ring in brackets and add the word *annulene.* Thus, benzene is [6]-annulene. Both **[14]-annulene** and **[18]-annulene** are cyclic, planar, completely conjugated molecules that follow Hückel's rule, and so they are aromatic.

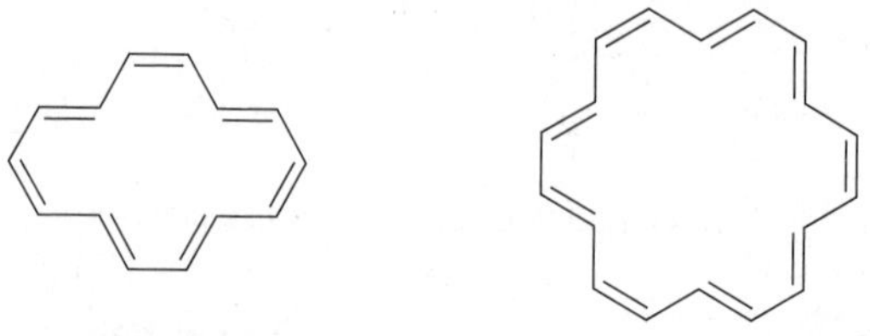

[14]-annulene
$4n + 2 = 4(3) + 2 =$
14 π electrons
aromatic

[18]-annulene
$4n + 2 = 4(4) + 2 =$
18 π electrons
aromatic

[10]-Annulene has 10 π electrons, which satisfies Hückel's rule, but a planar molecule would place the two H atoms inside the ring too close to each other, so the ring puckers to relieve this strain. Because **[10]-annulene is not planar,** the 10 π electrons can't delocalize over the entire ring and it is **not aromatic.**

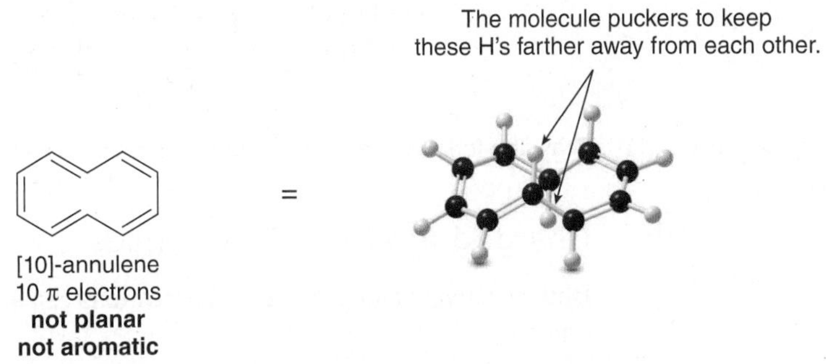

Problem 17.10 Would [16]-, [20]- or [22]-annulene be aromatic if each ring is planar?

17.7B Aromatic Compounds with More Than One Ring

Hückel's rule for determining aromaticity can be applied only to monocyclic systems, but many aromatic compounds containing several benzene rings joined together are also known. Two or more six-membered rings with alternating double and single bonds can be fused together to form **polycyclic aromatic hydrocarbons (PAHs).** Joining two benzene rings together forms **naphthalene,** which is used in mothballs. There are two different ways to join three rings together, forming **anthracene** and **phenanthrene,** and many more complex hydrocarbons are known.

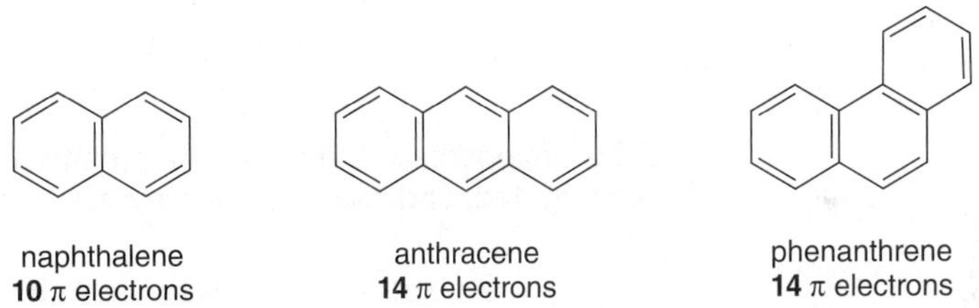

Benzo[*a*]pyrene, a more complicated PAH shown in Figure 17.4, is formed by the incomplete combustion of organic materials. It is found in cigarette smoke, automobile exhaust, and the fumes from charcoal grills. When ingested or inhaled, benzo[*a*]pyrene and other similar PAHs are oxidized to carcinogenic products, as discussed in Section 9.18.

Figure 17.4 Benzo[*a*]pyrene, a common PAH

- Benzo[*a*]pyrene, produced by the incomplete oxidation of organic compounds in tobacco, is found in cigarette smoke.

As the number of fused benzene rings increases, the number of resonance structures increases as well. Although two resonance structures can be drawn for benzene, naphthalene is a hybrid of three resonance structures.

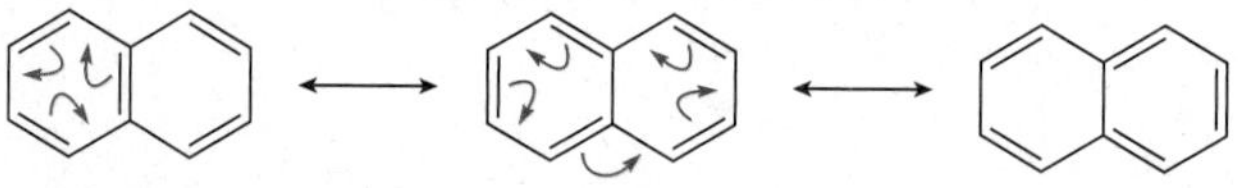

Problem 17.11 Draw the four resonance structures for anthracene.

17.7C Charged Aromatic Compounds

Both negatively and positively charged ions can also be aromatic if they satisfy all the necessary criteria.

Cyclopentadienyl Anion

The **cyclopentadienyl anion** is a cyclic and planar anion with two double bonds and a nonbonded electron pair. The two π bonds contribute four electrons and the lone pair contributes two more, for a total of six. By Hückel's rule, having **six π electrons confers aromaticity.** The **negatively charged carbon atom must be sp^2 hybridized,** and the **nonbonded electron pair must occupy a *p* orbital** for the ring to be completely conjugated.

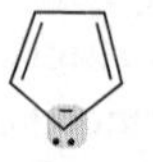

cyclopentadienyl anion
all sp^2 hybridized C's
6 π electrons

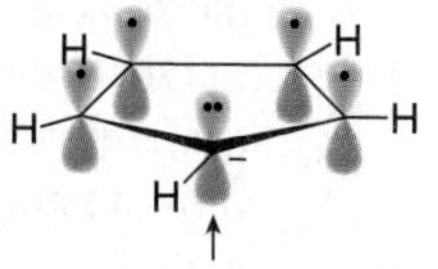

The lone pair resides in a *p* orbital.

- **The cyclopentadienyl anion is aromatic because it is cyclic, planar, completely conjugated, and has six π electrons.**

We can draw **five equivalent resonance structures for the cyclopentadienyl anion,** delocalizing the negative charge over every carbon atom of the ring.

Although five resonance structures can also be drawn for both the **cyclopentadienyl cation** and **radical,** only the cyclopentadienyl anion has six π electrons, a number that satisfies Hückel's rule. The cyclopentadienyl cation has four π electrons, making it antiaromatic and especially unstable. The cyclopentadienyl radical has five π electrons, so it is neither aromatic nor antiaromatic. Having the "right" number of electrons is necessary for a species to be unusually stable by virtue of aromaticity.

cyclopentadienyl anion
- **6** π electrons
- contains $4n + 2$ π electrons

aromatic

cyclopentadienyl cation
- **4** π electrons
- contains $4n$ π electrons

antiaromatic

cyclopentadienyl radical
- **5** π electrons
- does not contain either $4n$ or $4n + 2$ π electrons

nonaromatic

The cyclopentadienyl anion is readily formed from cyclopentadiene by a Brønsted–Lowry acid–base reaction.

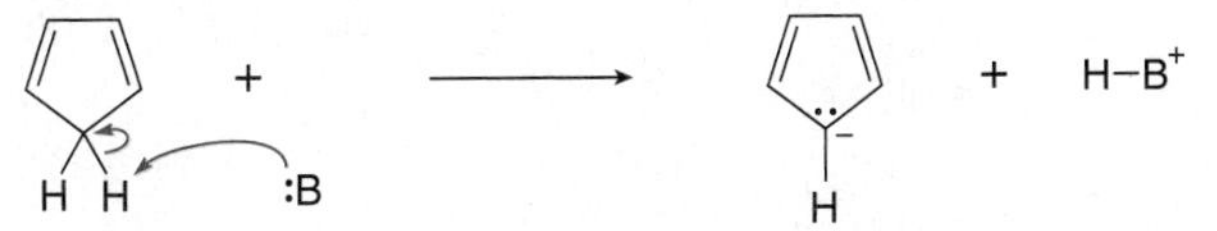

cyclopentadiene
not aromatic
pK_a = 15

cyclopentadienyl anion
aromatic
a stabilized conjugate base

Cyclopentadiene itself is not aromatic because it is not fully conjugated. The cyclopentadienyl anion, however, is aromatic, so it is a very stable base. As such, it makes cyclopentadiene more acidic than other hydrocarbons. In fact, the pK_a of cyclopentadiene is 15, much lower (more acidic) than the pK_a of any C–H bond discussed thus far.

- **Cyclopentadiene is more acidic than many hydrocarbons because its conjugate base is aromatic.**

Problem 17.12 Draw the product formed when cyclohepta-1,3,5-triene (pK_a = 39) is treated with a strong base. Why is its pK_a so much higher than the pK_a of cyclopentadiene?

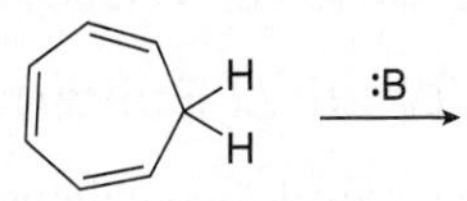

cyclohepta-1,3,5-triene
pK_a = 39

Problem 17.13 Rank the following compounds in order of increasing acidity.

Tropylium Cation

The cyclopentadienyl anion and the tropylium cation both illustrate an important principle: The **number of π electrons determines aromaticity,** not the number of atoms in a ring or the number of *p* orbitals that overlap. The cyclopentadienyl anion and tropylium cation are aromatic because they each have six π electrons.

The **tropylium cation** is a planar carbocation with three double bonds and a positive charge contained in a seven-membered ring. This carbocation is completely conjugated, because the positively charged carbon is sp^2 hybridized and has a vacant *p* orbital that overlaps with the six *p* orbitals from the carbons of the three double bonds. **Because the tropylium cation has three π bonds and no other nonbonded electron pairs, it contains six π electrons,** thereby satisfying Hückel's rule.

tropylium cation
all sp^2 hybridized C's
6 π electrons

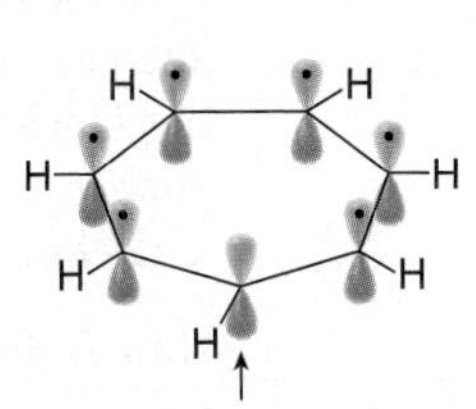

One *p* orbital is vacant.

- **The tropylium cation is aromatic because it is cyclic, planar, completely conjugated, and has six π electrons delocalized over the seven atoms of the ring.**

Problem 17.14 Draw the seven resonance structures for the tropylium cation.

Problem 17.15 Assuming the rings are planar, which ions are aromatic?

a. 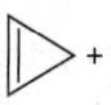 b. 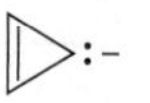 c. 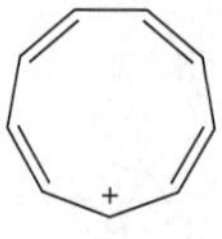 d. 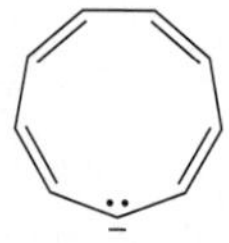

Problem 17.16 Compound **A** exhibits a peak in its ^{1}H NMR spectrum at 7.6 ppm, indicating that it is aromatic. (a) How are the carbon atoms of the triple bonds hybridized? (b) In what type of orbitals are the π electrons of the triple bonds contained? (c) How many π electrons are delocalized around the ring in **A?**

A = 

17.8 Aromatic Heterocycles

17.8A Biological Building Blocks

Recall from Section 9.3 that a **heterocycle** is a ring that contains at least one heteroatom.

Heterocycles containing oxygen, nitrogen, or sulfur—atoms that also have at least one lone pair of electrons—can also be aromatic. With heteroatoms, we must always determine whether the lone pair is localized on the heteroatom or part of the delocalized π system. Two examples of structures frequently found in biomolecules, **pyridine** and **pyrrole,** illustrate these different possibilities.

Pyridine

Pyridine is a heterocycle containing a six-membered ring with three π bonds and one nitrogen atom. Like benzene, two resonance structures (with all neutral atoms) can be drawn.

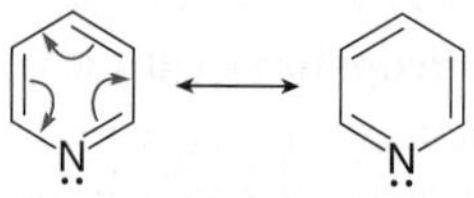

two resonance structures for pyridine
6 π electrons

Pyridine is cyclic, planar, and completely conjugated, because the three single and double bonds alternate around the ring. **Pyridine has six π electrons, two from each π bond, thus satisfying Hückel's rule and making pyridine aromatic.** The nitrogen atom of pyridine also has a nonbonded electron pair, which is localized on the N atom, so it is *not* part of the delocalized π electron system of the aromatic ring.

How is the nitrogen atom of the pyridine ring hybridized? The N atom is surrounded by three groups (two atoms and a lone electron pair), making it ***sp***^{2} **hybridized,** and leaving one unhybridized *p* orbital with one electron that overlaps with adjacent *p* orbitals. The lone pair on N resides in an *sp*^{2} hybrid orbital that is perpendicular to the delocalized π electrons.

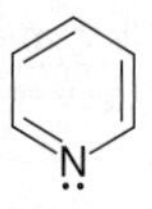

sp^2 hybridized N

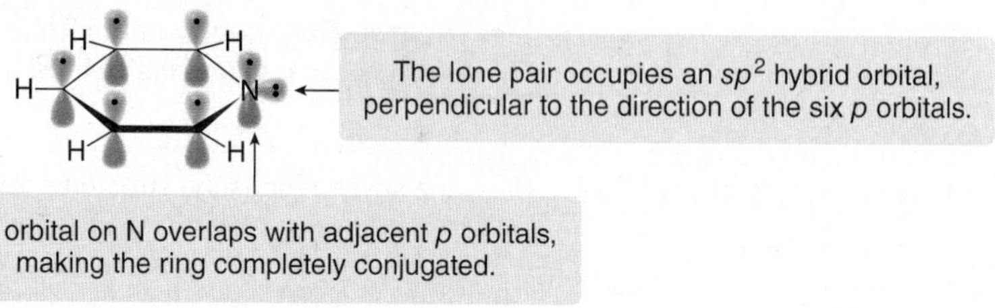

Pyrrole

Pyrrole contains a five-membered ring with two π bonds and one nitrogen atom. The N atom also has a lone pair of electrons.

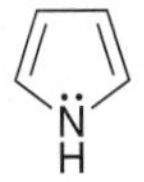

pyrrole

Pyrrole is cyclic and planar, with a total of four π electrons from the two π bonds. Is the nonbonded electron pair localized on N or part of a delocalized π electron system? The lone pair on N is *adjacent* to a double bond. Recall the following general rule from Section 16.5:

- **In a system X=Y–Z:, Z is generally sp^2 hybridized and the lone pair occupies a *p* orbital to make the system conjugated.**

If the lone pair on the N atom occupies a *p* orbital:

- **Pyrrole has a *p* orbital on every adjacent atom, so it is completely conjugated.**
- **Pyrrole has six π electrons—four from the π bonds and two from the lone pair.**

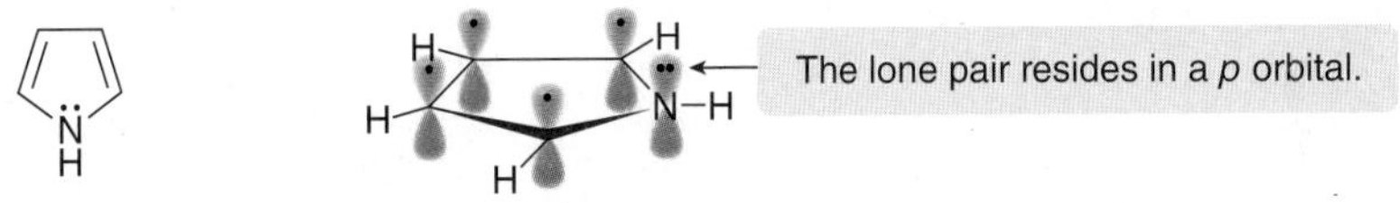

sp^2 hybridized N The ring is completely conjugated with **6 π electrons.**

Because pyrrole is cyclic, planar, completely conjugated, and has $4n + 2$ π electrons, **pyrrole is aromatic. The number of electrons—not the size of the ring—determines whether a compound is aromatic.**

Electrostatic potential maps, shown in Figure 17.5 for pyridine and pyrrole, confirm that the **lone pair in pyridine is localized on N,** whereas the **lone pair in pyrrole is part of the delocalized π system.** Thus, a fundamental difference exists between the N atoms in pyridine and pyrrole.

Figure 17.5 Electrostatic potential maps of pyridine and pyrrole

pyridine

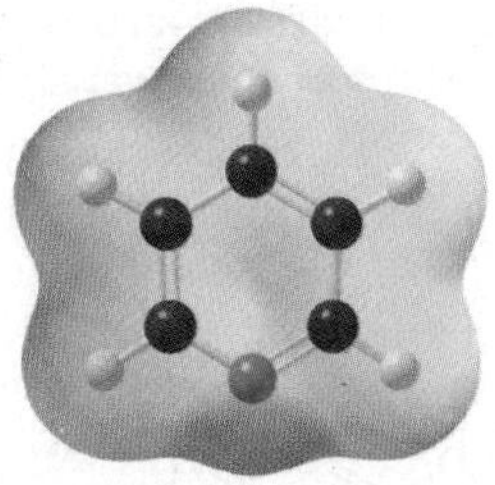

pyrrole

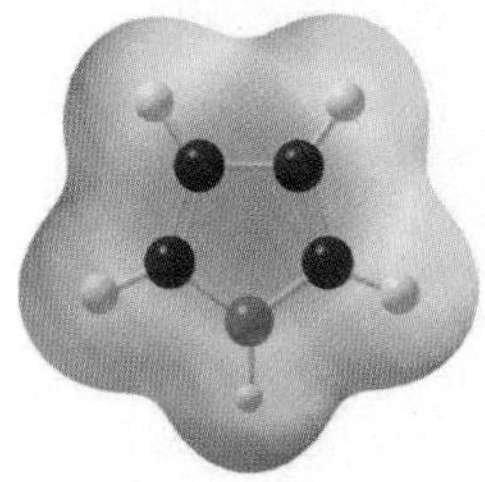

- In pyridine, the nonbonded electron pair is localized on the N atom in an sp^2 hybridized orbital, as shown by the region of high electron density (in red) on N.
- In pyrrole, the nonbonded electron pair is in a *p* orbital and is delocalized over the ring, so the entire ring is electron rich (red).

- **When a heteroatom is already part of a double bond (as in the N of pyridine), its lone pair *cannot* occupy a *p* orbital and so it cannot be delocalized over the ring.**
- **When a heteroatom is *not* part of a double bond (as in the N of pyrrole), its lone pair can be located in a *p* orbital and *delocalized* over a ring to make it aromatic.**

Histamine, a Substituted Imidazole

Imidazole is an aromatic heterocycle with two N atoms, one of which is similar to the N atom of pyridine and one of which is similar to the N atom of pyrrole.

Scombroid fish poisoning, associated with facial flushing, hives, and general itching, is caused by the ingestion of inadequately refrigerated fish, typically mahimahi (pictured) and tuna. Bacteria convert the amino acid histidine (Chapter 26) to histamine, which, when consumed in large amounts, results in this clinical syndrome.

Imidazole has a five-membered ring with two π bonds and two nitrogen atoms, each of which contains a lone pair of electrons. The heterocycle has four π electrons from the two double bonds. The lone pair on the N in red also occupies a *p* orbital, making the heterocycle completely conjugated, and giving it a total of six π electrons. The lone pair on this N atom is thus delocalized over the five-membered ring and the heterocycle is aromatic. The lone pair on the N in blue occupies an sp^2 hybrid orbital perpendicular to the delocalized π electrons.

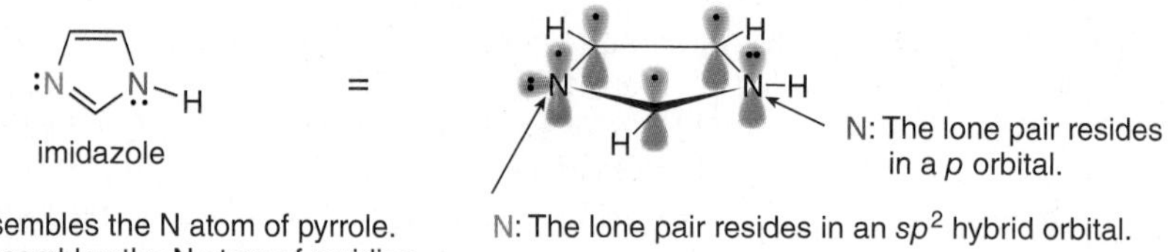

- N (in red) resembles the N atom of pyrrole.
- N (in blue) resembles the N atom of pyridine.

Histamine, a biologically active amine that contains an imidazole, produces a wide range of physiological effects in the body. Excess histamine is responsible for the runny nose and watery eyes symptomatic of hay fever. It also stimulates the overproduction of stomach acid, and contributes to the formation of hives. These effects result from the interaction of histamine with two different cellular receptors. We will learn more about antihistamines and antiulcer drugs, compounds that block the effects of histamine, in Section 25.6.

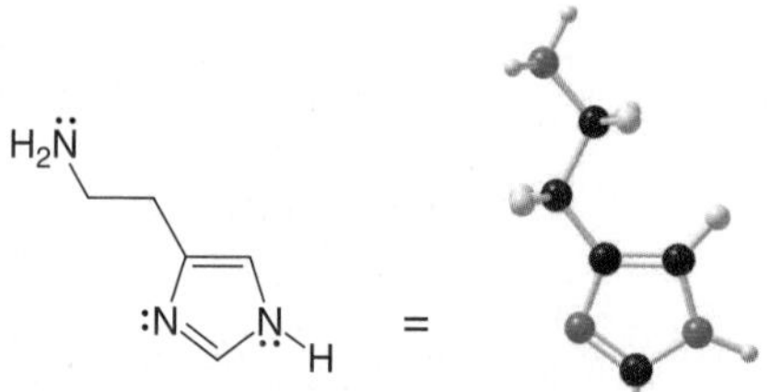

histamine

Quinine is isolated from the bark of the cinchona tree native to the Andes Mountains.

Problem 17.17 Which heterocycles are aromatic?

Problem 17.18 (a) How is each N atom in quinine, an effective antimalarial drug that reduces fever, hybridized? (b) In what type of orbital does the lone pair on each N reside?

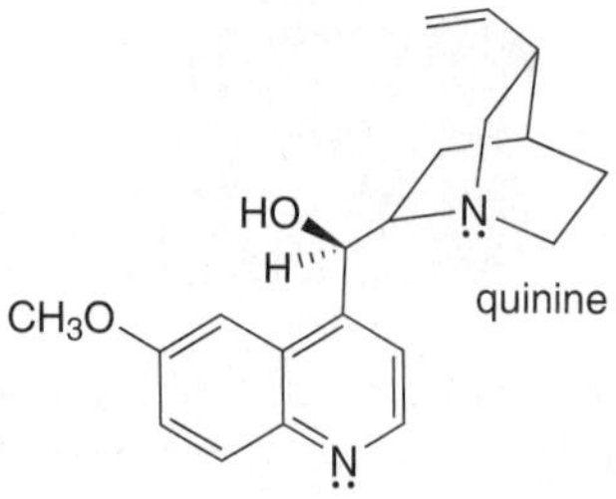

quinine

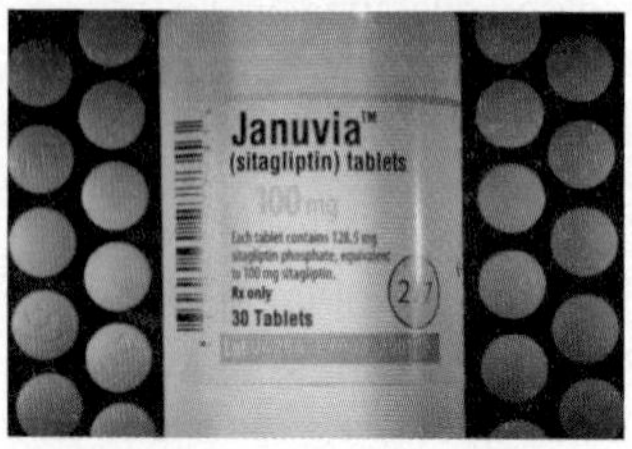

Januvia increases the body's ability to lower blood sugar levels, so it is used alone or in combination with other drugs to treat type 2 diabetes.

Problem 17.19 Januvia, the trade name for sitagliptin, was introduced in 2006 for the treatment of type 2 diabetes. (a) Explain why the five-membered ring in sitagliptin is aromatic. (b) Determine the hybridization of each N atom. (c) In what type of orbital does the lone pair on each N atom reside?

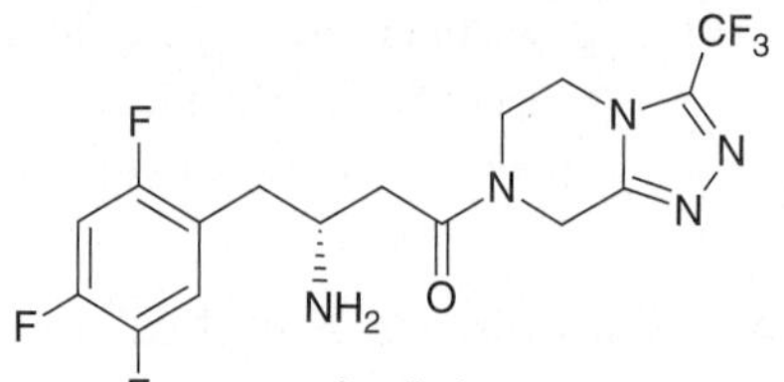

sitagliptin

17.8B Deoxyribonucleic Acid

As we learned in Section 3.9C, deoxyribonucleic acid (**DNA**) has a double helical structure made up of nucleotides, which contain an aromatic nitrogen heterocycle attached to a sugar–phosphate backbone. DNA contains four different bases, derived from the heterocycles **pyrimidine** and **purine.** The π electrons highlighted in red on pyrimidine and purine are delocalized and part of the aromatic system.

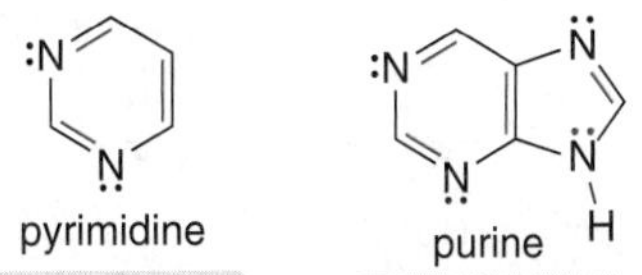

pyrimidine — **6 π electrons**; purine — **10 π electrons**

The lone pair electrons on both N atoms in pyrimidine, like pyridine, reside in sp^2 hybridized orbitals perpendicular to the aromatic π system. Purine, a bicyclic compound that contains one pyrimidine ring fused to an imidazole ring, has a lone pair (highlighted in red) that resides in a *p* orbital and is part of the aromatic system.

Two DNA bases—**cytosine** and **thymine**—are derived from pyrimidine and contain six π electrons, and two bases—**adenine** and **guanine**—are derived from purine and contain 10 π electrons.

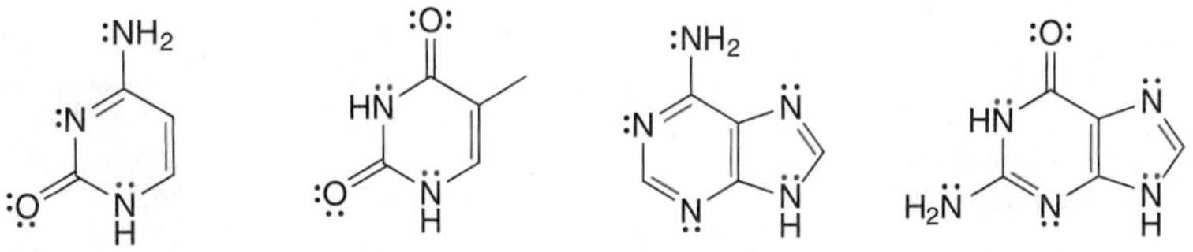

cytosine — **6 π electrons**; thymine — **6 π electrons** (**pyrimidines**)

adenine — **10 π electrons**; guanine — **10 π electrons** (**purines**)

Three of the bases are typically drawn in a form that contains a C=O, so it may not be obvious that each is aromatic. Because these bases contain a lone pair on N that can be delocalized on the C=O, however, a resonance structure can be drawn that clearly shows that six π electrons are delocalized in the ring. Two resonance structures are shown for cytosine.

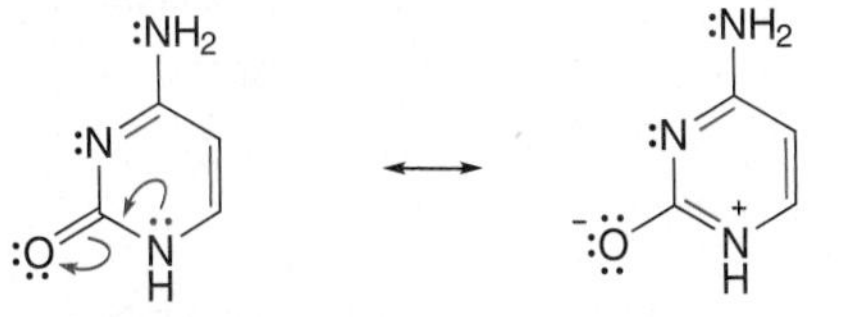

cytosine ⟷ **6 π electrons**

Because DNA holds essential, hereditary information that is critical to our growth and development, it is important that the high molecular weight nucleic acids are stable. Moreover, they must be packaged compactly because, if uncoiled and tied together, all of the DNA in our body would stretch ~10 billion miles.

- **Aromaticity contributes to the stability and packing efficiency of DNA.**

We learned in Section 17.5 that benzene is less reactive than isolated alkenes. Aromatic heterocycles are similarly stable. They are less reactive to covalent modifications, but because they are planar, they are properly aligned to contain noncovalent **hydrogen bonding** and **π stacking**, the interaction of the π electrons of the aromatic rings in close proximity (Section 17.11).

In DNA, hydrogen bonding, the strongest of the noncovalent intermolecular forces (Section 3.3), forms only between two sets of bases, adenine and thymine (A−T), and guanine and cytosine (G−C). The hydrogen bonds create planes of base pairs between the sugar–phosphate backbone (Figure 17.6).

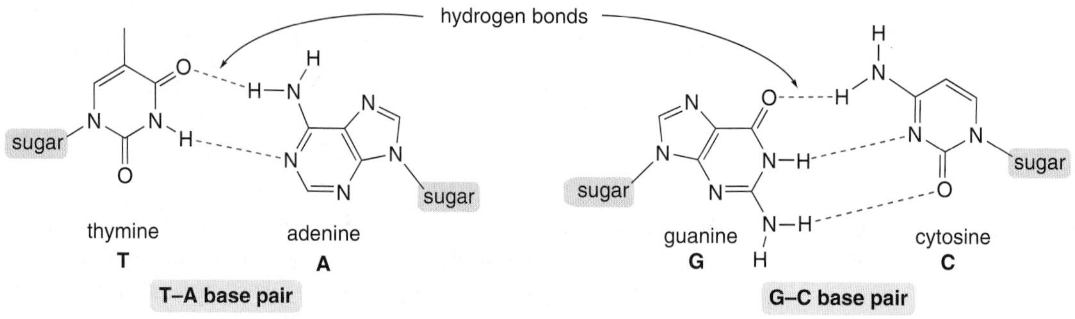

Because of the alignment of the *p* orbitals in the aromatic heterocycles, π stacking also decreases the energy of the closely oriented bases and contributes to the packing efficiency of DNA.

Figure 17.6
Hydrogen bonding and π stacking in DNA

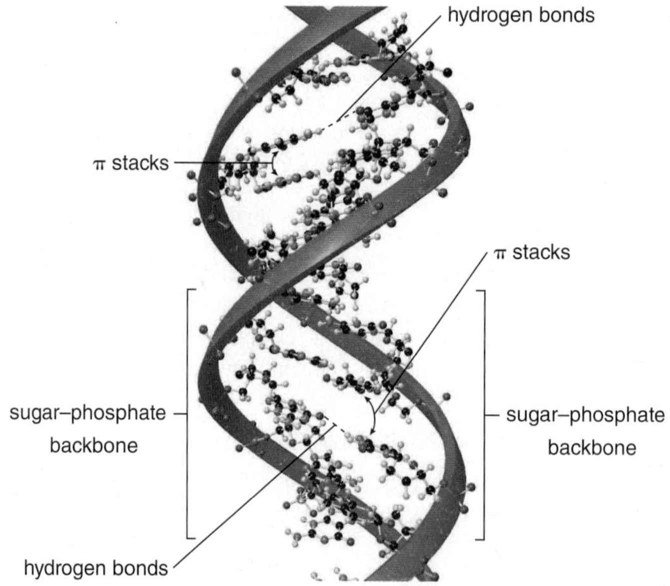

- Aromaticity stabilizes and compacts DNA. The base pairs line up in planes due to the hydrogen bonding between them. The planes of base pairs form parallel π stacks that are perpendicular to the sugar–phosphate backbone.

Problem 17.20 Answer each question for the four bases that comprise DNA. (a) How is each N atom hybridized? (b) In what type of orbital does each lone pair on a N reside?

Problem 17.21 Draw resonance structures for the six-membered thymine and guanine rings that clearly show the π electrons delocalized within the aromatic rings.

17.9 What Is the Basis of Hückel's Rule?

Why does the number of π electrons determine whether a compound is aromatic? Cyclobutadiene is cyclic, planar, and completely conjugated, just like benzene, but why is benzene aromatic and cyclobutadiene antiaromatic?

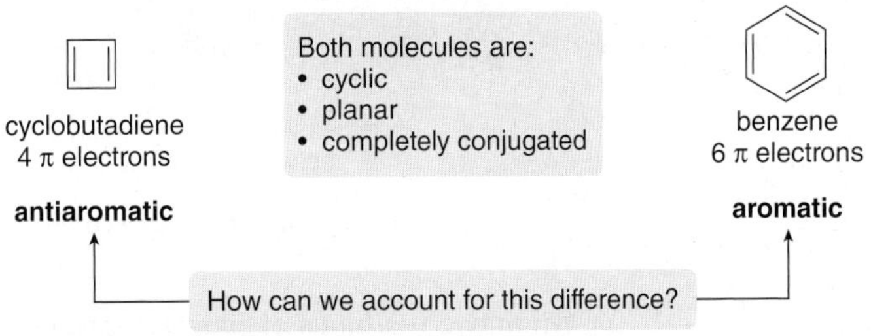

A complete explanation is beyond the scope of an introductory organic chemistry text, but nevertheless, you can better understand the basis of aromaticity by learning more about orbitals and bonding.

17.9A Bonding and Antibonding Orbitals

So far we have used the following basic concepts to describe how bonds are formed:

- **Hydrogen uses its 1*s* orbital to form σ bonds with other elements.**
- **Second-row elements use hybrid orbitals (*sp*, sp^2, or sp^3) to form σ bonds.**
- **Second-row elements use *p* orbitals to form π bonds.**

This description of bonding is called **valence bond theory.** In valence bond theory, a covalent bond is formed by the overlap of two atomic orbitals, and the electron pair in the resulting bond is shared by both atoms. Thus, a carbon–carbon double bond consists of a σ bond, formed by overlap of two sp^2 hybrid orbitals, each containing one electron, and a π bond, formed by overlap of two *p* orbitals, each containing one electron.

This description of bonding works well for most of the organic molecules we have encountered thus far. Unfortunately, it is inadequate for describing systems with many adjacent *p* orbitals that overlap, as there are in aromatic compounds. To more fully explain the bonding in these systems, we must utilize **molecular orbital (MO) theory.**

MO theory describes bonds as the mathematical combination of atomic orbitals that form a new set of orbitals called **molecular orbitals (MOs).** A molecular orbital occupies a region of space *in a molecule* where electrons are likely to be found. When forming molecular orbitals from atomic orbitals, keep in mind:

- **A set of *n* atomic orbitals forms *n* molecular orbitals.**

If *two* atomic orbitals combine, *two* molecular orbitals are formed. This is fundamentally different than valence bond theory. Because aromaticity is based on *p* orbital overlap, what does MO theory predict will happen when two *p* (atomic) orbitals combine?

The two lobes of each *p* orbital are opposite in phase, with a node of electron density at the nucleus. When two *p* orbitals combine, two molecular orbitals should form. The two *p* orbitals can add together constructively—that is, with like phases interacting—or destructively—that is, with opposite phases interacting.

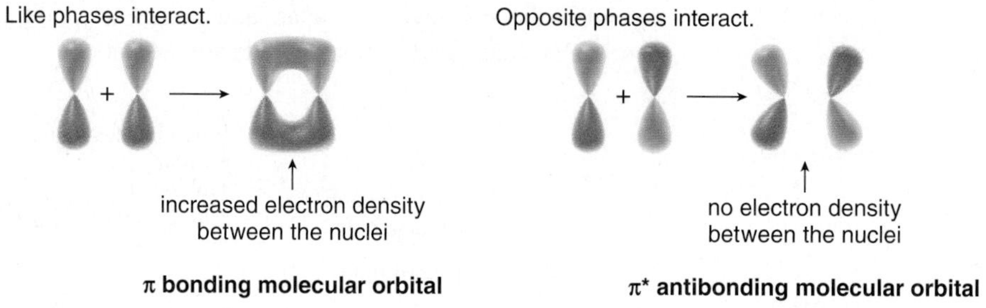

- **When two *p* orbitals of similar phase overlap side-by-side, a π bonding molecular orbital results.**
- **When two *p* orbitals of opposite phase overlap side-by-side, a π* antibonding molecular orbital results.**

A π bonding MO is lower in energy than the two atomic *p* orbitals from which it is formed because a stable bonding interaction results when orbitals of similar phase combine. A bonding interaction holds nuclei together. Similarly, a π* antibonding MO is higher in energy because a destabilizing node results when orbitals of opposite phase combine. A destabilizing interaction pushes nuclei apart.

If two atomic *p* orbitals each have one electron and then combine to form MOs, the two electrons will occupy the lower energy π bonding MO, as shown in Figure 17.7.

Figure 17.7
Combination of two *p* orbitals to form π and π* molecular orbitals

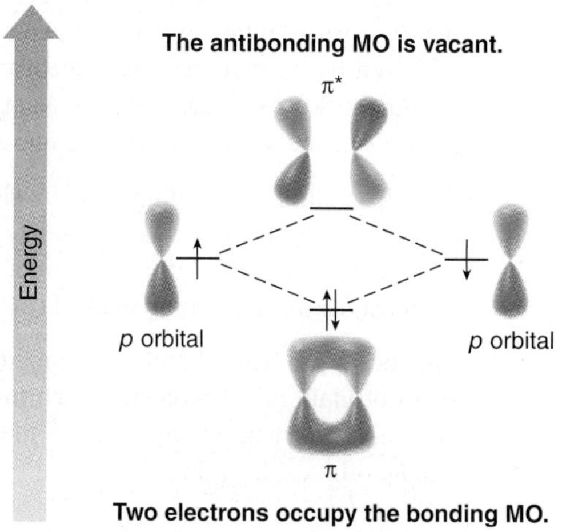

- Two atomic *p* orbitals combine to form two molecular orbitals. The bonding π MO is lower in energy than the two *p* orbitals from which it was formed, and the antibonding π* MO is higher in energy than the two *p* orbitals from which it was formed.
- Two electrons fill the lower energy bonding MO first.

17.9B Molecular Orbitals Formed When More Than Two *p* Orbitals Combine

The molecular orbital description of benzene is much more complex than the two MOs formed in Figure 17.7. Because each of the six carbon atoms of benzene has a *p* orbital, six atomic *p* orbitals combine to form six π molecular orbitals, as shown in Figure 17.8. A description of the exact appearance and energies of these six MOs requires more sophisticated mathematics and

Figure 17.8
How the six p orbitals of benzene overlap to form six molecular orbitals

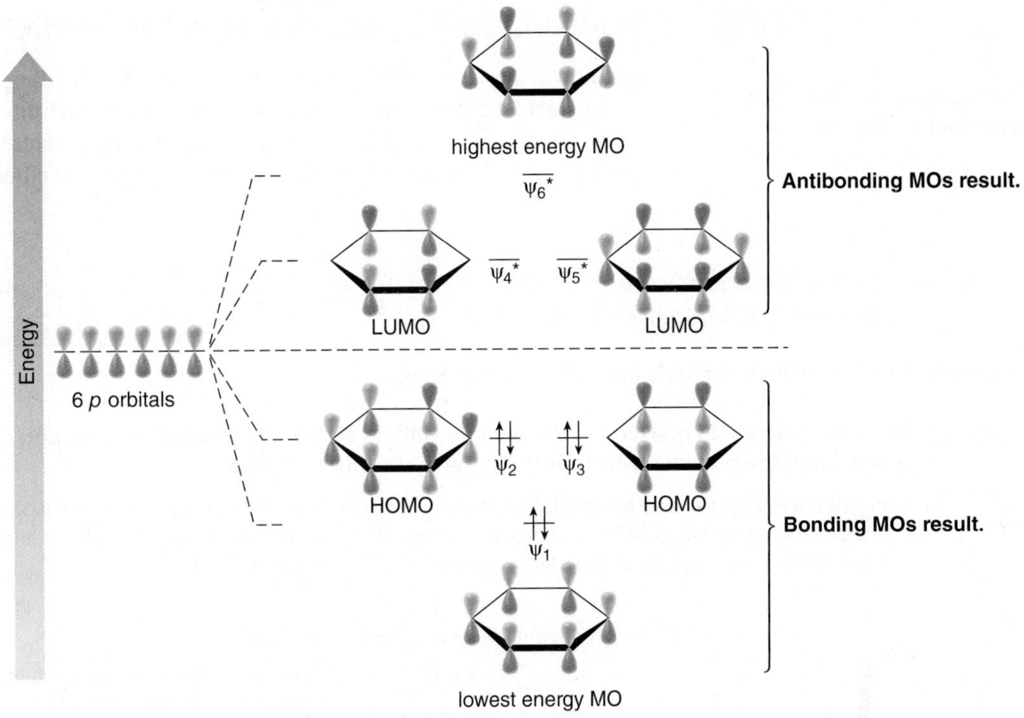

- Depicted in this diagram are the interactions of the six atomic p orbitals of benzene, which form six molecular orbitals. When orbitals of like phase combine, a bonding interaction results. When orbitals of opposite phase combine, a destabilizing node results.

understanding of MO theory than is presented in this text. Nevertheless, note that the six MOs are labeled ψ_1–ψ_6, with ψ_1 being the lowest in energy and ψ_6 the highest.

The most important features of the six benzene MOs are as follows:

- **The larger the number of bonding interactions, the lower in energy the MO.** The lowest energy molecular orbital (ψ_1) has all bonding interactions between the p orbitals.
- **The larger the number of nodes, the higher in energy the MO.** The highest energy MO (ψ_6*) has all nodes between the p orbitals.
- Three MOs are lower in energy than the starting p orbitals, making them bonding MOs (ψ_1, ψ_2, and ψ_3), whereas three MOs are higher in energy than the starting p orbitals, making them antibonding MOs (ψ_4*, ψ_5*, and ψ_6*).
- The two pairs of MOs (ψ_2 and ψ_3; ψ_4* and ψ_5*) with the same energy are called **degenerate orbitals.**
- **The highest energy orbital that contains electrons is called the *highest occupied molecular orbital* (HOMO).** For benzene, the degenerate orbitals ψ_2 and ψ_3 are the HOMOs.
- **The lowest energy orbital that does *not* contain electrons is called the *lowest unoccupied molecular orbital* (LUMO).** For benzene, the degenerate orbitals ψ_4* and ψ_5* are the LUMOs.

To fill the MOs, the six electrons are added, two to an orbital, beginning with the lowest energy orbital. As a result, the six electrons completely fill the bonding MOs, leaving the antibonding MOs empty. This is what gives benzene and other aromatic compounds their special stability, and this is why six π electrons satisfies Hückel's $4n + 2$ rule.

- **All bonding MOs (and HOMOs) are completely filled in aromatic compounds. No π electrons occupy antibonding MOs.**

17.10 The Inscribed Polygon Method for Predicting Aromaticity

An inscribed polygon is also called a **Frost circle.**

To predict whether a compound has π electrons completely filling bonding MOs, we must know how many bonding molecular orbitals and how many π electrons it has. It is possible to predict the relative energies of cyclic, completely conjugated compounds, without sophisticated math (or knowing what the resulting MOs look like) by using the **inscribed polygon method.**

How To Use the Inscribed Polygon Method to Determine the Relative Energies of MOs for Cyclic, Completely Conjugated Compounds

Example Plot the relative energies of the MOs of benzene.

Step [1] Draw the polygon in question inside a circle with its vertices touching the circle and one of the vertices pointing *down*. Mark the points at which the polygon intersects the circle.

- Inscribe a hexagon inside a circle for benzene. The six vertices of the hexagon form six points of intersection, corresponding to the six MOs of benzene. The pattern—a single MO having the lowest energy, two degenerate pairs of MOs, and a single highest energy MO—matches that found in Figure 17.8.

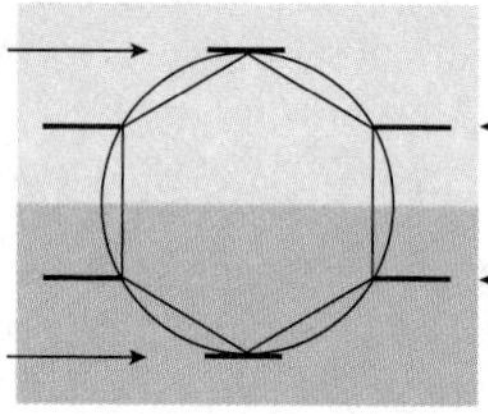

Step [2] Draw a line horizontally through the center of the circle and label MOs as bonding, nonbonding, or antibonding.

- **MOs below this line are bonding,** and lower in energy than the *p* orbitals from which they were formed. Benzene has three bonding MOs.
- **MOs at this line are nonbonding,** and equal in energy to the *p* orbitals from which they were formed. Benzene has no nonbonding MOs.
- **MOs above this line are antibonding,** and higher in energy than the *p* orbitals from which they were formed. Benzene has three antibonding MOs.

Step [3] Add the electrons, beginning with the lowest energy MO.

- **All the bonding MOs (and the HOMOs) are completely filled in aromatic compounds. No π electrons occupy antibonding MOs.**
- Benzene is aromatic because it has six π electrons that completely fill the bonding MOs.

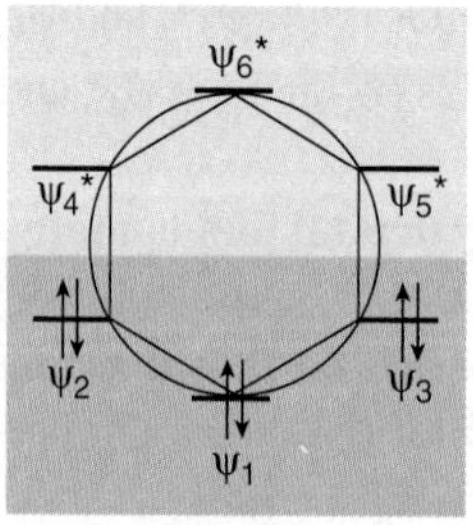

This method works for all monocyclic, completely conjugated hydrocarbons regardless of ring size. Figure 17.9 illustrates MOs for completely conjugated five- and seven-membered rings using this method. The total number of MOs always equals the number of vertices of the polygon. Because both systems have three bonding MOs, each needs six π electrons to fully occupy them, making the cyclopentadienyl anion and the tropylium cation aromatic, as we learned in Section 17.7C.

Figure 17.9 Using the inscribed polygon method for five- and seven-membered rings

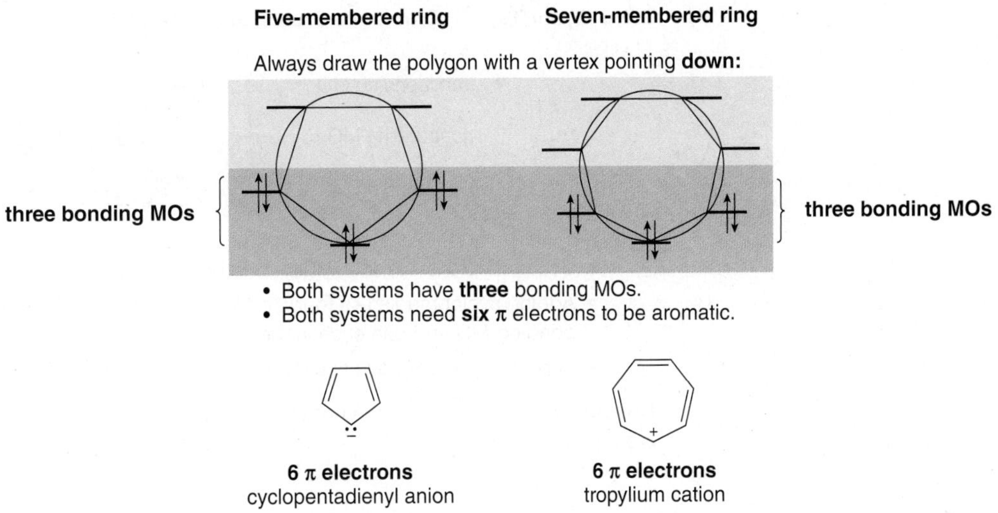

The inscribed polygon method is consistent with Hückel's $4n + 2$ rule; that is, there is always one lowest energy bonding MO that can hold two π electrons and the other bonding MOs come in degenerate pairs that can hold a total of four π electrons. For the compound to be aromatic, these MOs must be completely filled with electrons, so the "magic numbers" for aromaticity fit Hückel's $4n + 2$ rule (Figure 17.10).

Figure 17.10 MO patterns for cyclic, completely conjugated systems

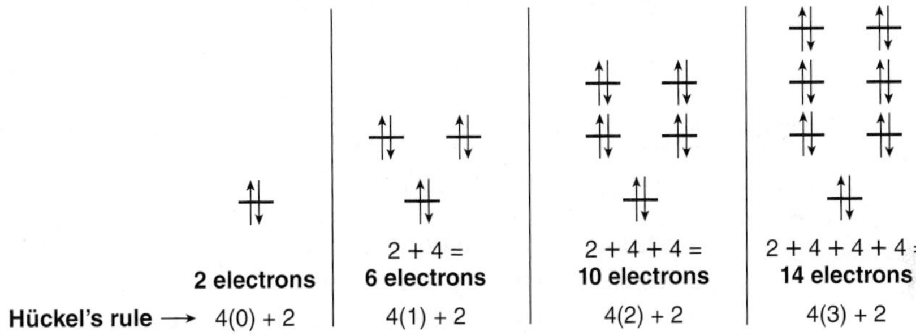

Sample Problem 17.1 Use the inscribed polygon method to show why cyclobutadiene is not aromatic.

cyclobutadiene
4 π electrons

Solution

Cyclobutadiene has four MOs (formed from its four *p* orbitals), to which its four π electrons must be added.

Step [1] Inscribe a square with a vertex down and mark its four points of intersection with the circle.

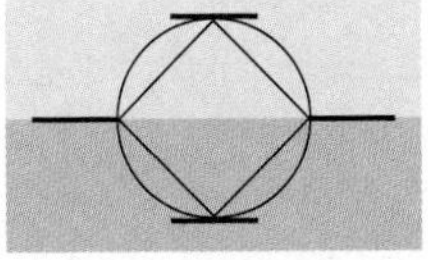

- The four points of intersection correspond to the four MOs of cyclobutadiene.

Steps [2] and [3] Draw a line through the center of the circle, label the MOs, and add the electrons.

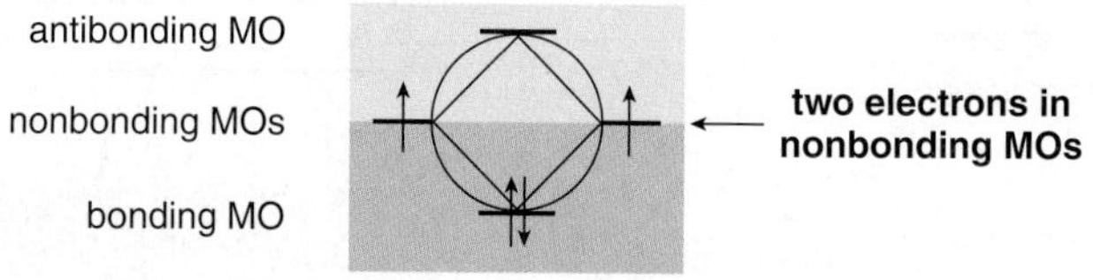

- Cyclobutadiene has four MOs—one bonding, two nonbonding, and one antibonding.
- Adding cyclobutadiene's four π electrons to these orbitals places two in the lowest energy bonding MO and one each in the two nonbonding MOs.
- Separating electrons in two degenerate MOs keeps like charges farther away from each other.

Conclusion: Cyclobutadiene is not aromatic because its HOMOs, two degenerate nonbonding MOs, are not completely filled.

The procedure followed in Sample Problem 17.1 also illustrates why cyclobutadiene is antiaromatic. Having the two unpaired electrons in nonbonding MOs suggests that cyclobutadiene should be a highly unstable diradical. In fact, antiaromatic compounds resemble cyclobutadiene because their HOMOs contain two unpaired electrons, making them especially unstable.

Problem 17.22 Use the inscribed polygon method to show why the following cation is aromatic.

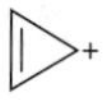

Problem 17.23 Use the inscribed polygon method to show why the cyclopentadienyl cation and radical are not aromatic.

17.11 Application: Aromatase Inhibitors for Estrogen-Dependent Cancer Treatment

We learned in Section 6.11 that amino acids in the active site of an enzyme hold substrates in place electrostatically to undergo chemical transformations. In addition to electrostatic interactions, π stacking is another way in which substrates are shuttled to active sites and held into place. One type of π stacking that is observed in DNA (Section 17.8) is **parallel,** where two aromatic π systems stack on top of each other. Pi stacking may also be **offset,** where only some of the *p* orbitals overlap, or even **perpendicular,** where *p* orbitals overlap by a 90° angle. These interactions tend to be strongest when one or more of the aromatic rings contain heteroatoms. Cation–π stacking is also observed.

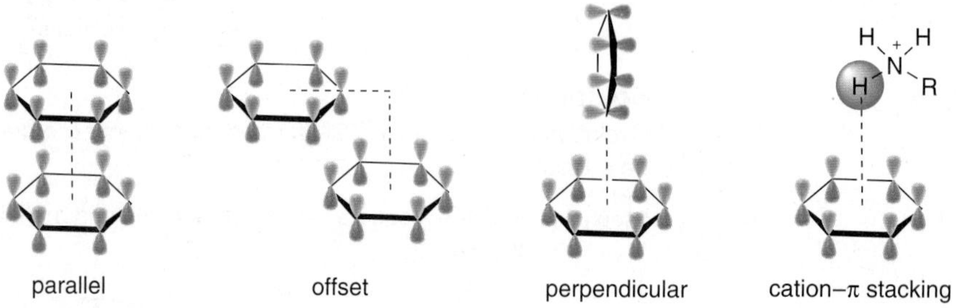

π stacking

Typically the side chains of the aromatic amino acids—**tyrosine, phenylalanine, histidine,** and **tryptophan**—engage in these types of noncovalent interactions.

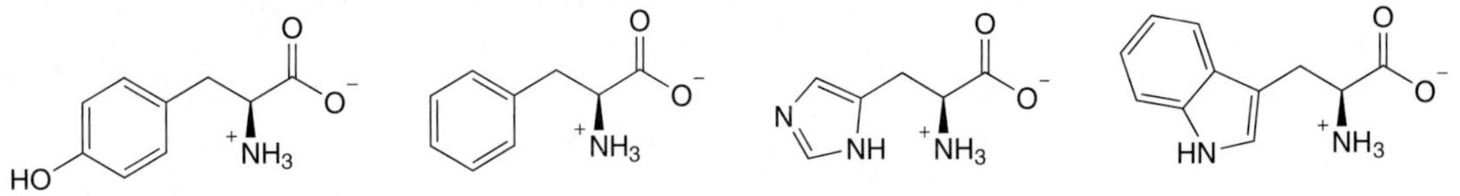

tyrosine phenylalanine histidine tryptophan

The estrogens (female sex hormones), estradiol and estrone, are synthesized from the androgens (male sex hormones), testosterone and androstenedione. Aromatase, an enzyme that forms the aromatic ring on estrogens, uses π stacking to lock the androgen substrate into place. A phenylalanine side chain forms a cation–π stack with a protonated side chain of another amino acid on the enzyme to begin the reaction. Heme, an aromatic compound made up of four pyrrole subunits and 22 π electrons, is tethered into the center of the aromatase active site to catalyze the formation of the phenol ring.

Steroids are composed of three six-membered rings and one five-membered ring joined together as shown in each of these sex hormones. Although functional groups and oxidation states vary among these compounds, the characteristic ring system defines them as steroids. You will learn more about steroids, including female and male sex hormones, in Section 28.8.

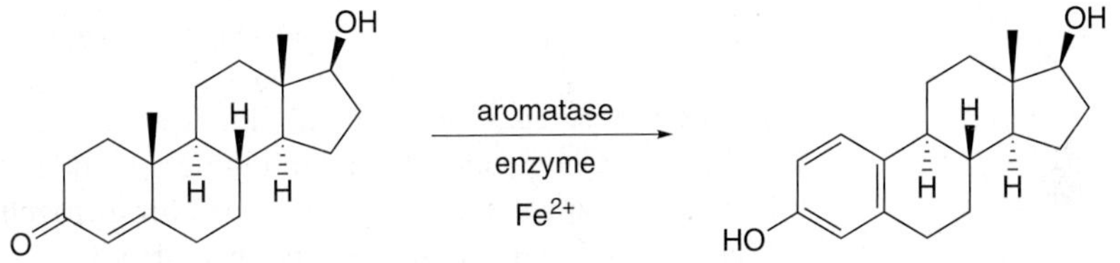
aromatase
enzyme
Fe^{2+}

testosterone estradiol

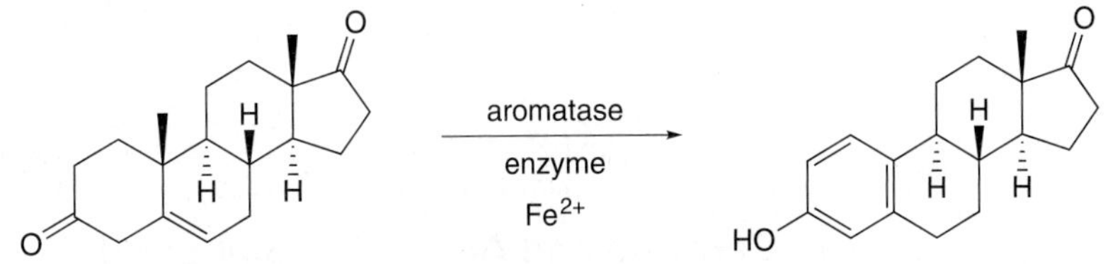
aromatase
enzyme
Fe^{2+}

androstenedione estrone

Figure 17.11 shows both androstenedione and heme in the aromatase active site, where the conversion to estrone occurs.

Figure 17.11 Protein ribbon diagram for aromatase with enlargement of the active site

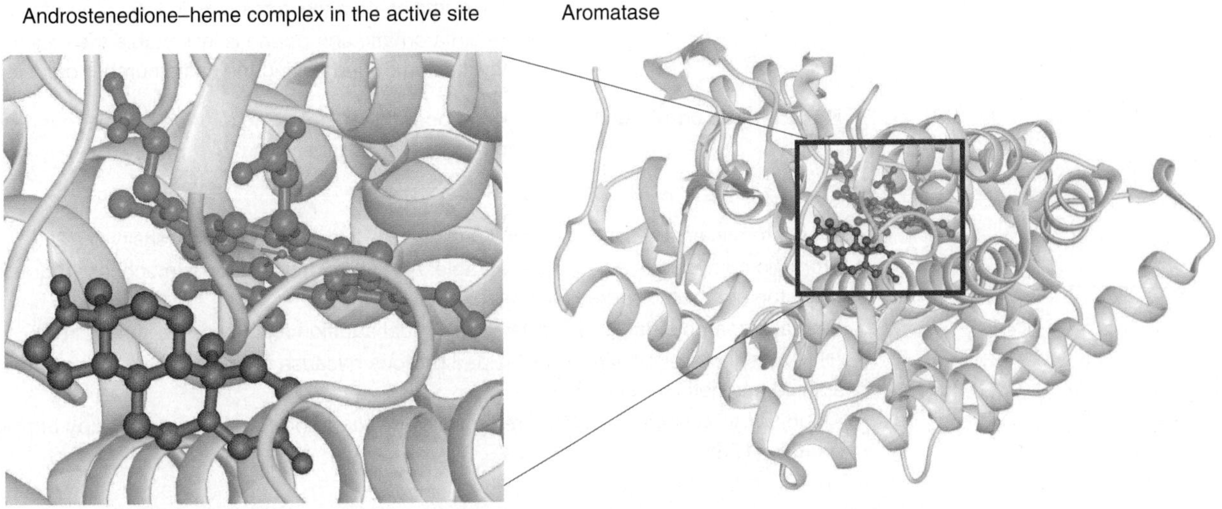
Androstenedione–heme complex in the active site
Aromatase

In the active site of aromatase, androstenedione (gray) is located close to heme (orange) to catalyze the aromatization in the formation of estrone. H atoms are omitted in models.

Ultimately, the estrone synthesized from androstenedione forms a cation–π stack with heme until it is released from the active site.

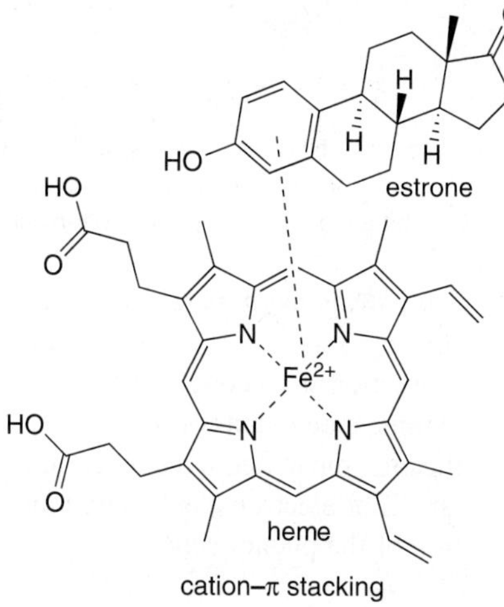

heme–aromatase estrone complex

The aromatase enzyme is the only known enzyme to catalyze the formation of estrogens from androgens. Compounds that inhibit the production of estrogens, so-called aromatase inhibitors, have been and continue to be developed for estrogen-dependent breast cancers (Section 22.18). Many are aromatic compounds because they bind to the active site of the aromatase enzyme, resulting in the decreased production of estrogen, which slows the growth of the cancer.

KEY CONCEPTS

Benzene and Aromatic Compounds

Comparing Aromatic, Antiaromatic, and Nonaromatic Compounds (17.7)

- **Aromatic compound**
  - A cyclic, planar, completely conjugated compound that contains $4n + 2$ π electrons ($n = 0, 1, 2, 3$, and so forth)
  - An aromatic compound is more stable than a similar acyclic compound having the same number of π electrons.
- **Antiaromatic compound**
  - A cyclic, planar, completely conjugated compound that contains $4n$ π electrons ($n = 0, 1, 2, 3$, and so forth)
  - An antiaromatic compound is less stable than a similar acyclic compound having the same number of π electrons.
- **Nonaromatic compound**
  - A compound that lacks one (or more) of the four requirements to be aromatic or antiaromatic

Properties of Aromatic Compounds

- Every atom in an aromatic ring has a *p* orbital to delocalize electron density (17.2).
- Aromatic compounds are unusually stable. $\Delta H°$ for hydrogenation is much less than expected, given the number of degrees of unsaturation (17.5).
- Aromatic compounds do not undergo the usual addition reactions of alkenes (17.5).
- ^{1}H NMR spectra show highly deshielded protons because of ring currents that reinforce the applied magnetic field (17.4).
- All bonding MOs and HOMOs are completely filled and no electrons occupy antibonding orbitals (17.9).

Examples of Aromatic Compounds with Six π Electrons (17.7–17.8)

benzene

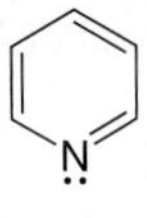
pyridine

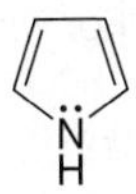
pyrrole

cyclopentadienyl anion

tropylium cation

Examples of Compounds That Are Not Aromatic (17.6)

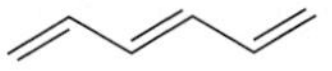
not cyclic

not planar

not completely conjugated

PROBLEMS

Problems Using Three-Dimensional Models

17.24 Name each compound and state how many lines are observed in its ^{13}C NMR spectrum.

a. 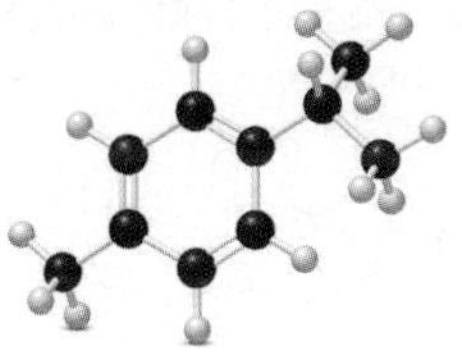

b. 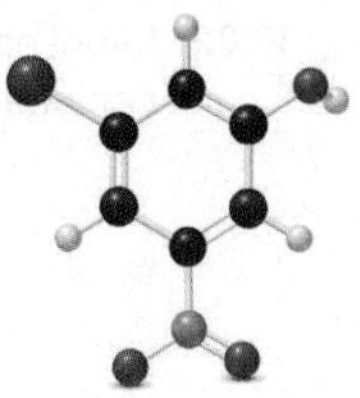

17.25 Classify each compound as aromatic, antiaromatic, or not aromatic.

a. 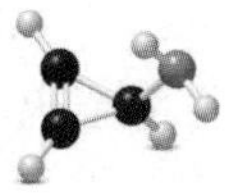

b. 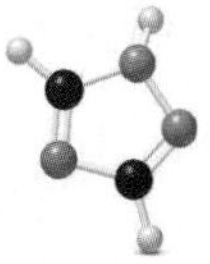

c. 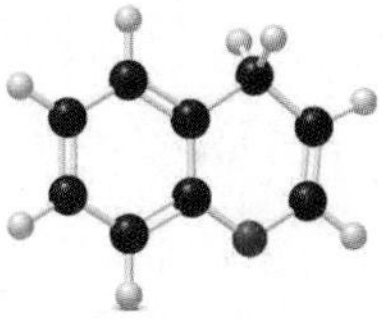

Benzene Structure and Nomenclature

17.26 Draw all aromatic hydrocarbons that have molecular formula C_8H_{10}. For each compound, determine how many isomers of molecular formula C_8H_9Br would be formed if one H atom on the benzene ring were replaced by a Br atom.

17.27 Give the IUPAC name for each compound.

a. 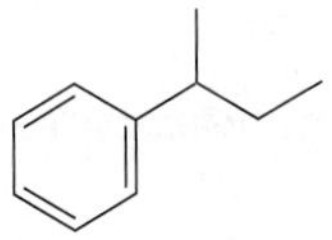

b. 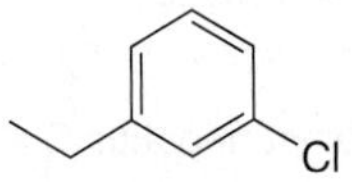

c. 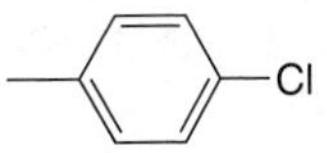

d. 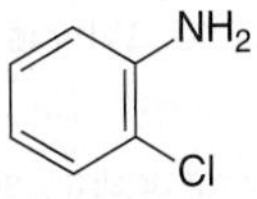

e. 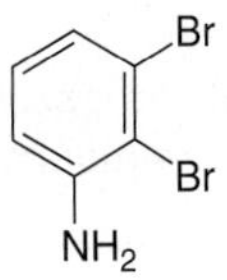

f. 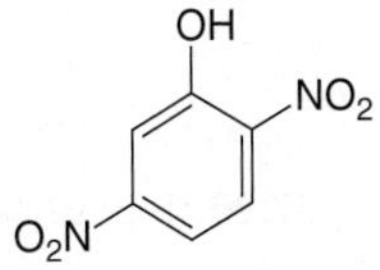

g. 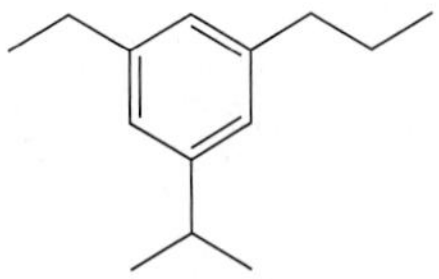

h. 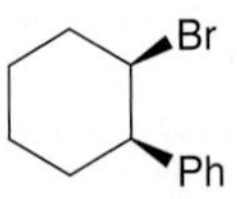

17.28 Draw a structure corresponding to each name.

a. *p*-dichlorobenzene
b. *p*-iodoaniline
c. *o*-bromonitrobenzene
d. 2,6-dimethoxytoluene
e. 2-phenylprop-2-en-1-ol
f. *trans*-1-benzyl-3-phenylcyclopentane

17.29 a. Draw the 14 constitutional isomers of molecular formula C_8H_9Cl that contain a benzene ring.
b. Name all compounds that contain a trisubstituted benzene ring.
c. For which compound(s) are stereoisomers possible? Draw all possible stereoisomers.

Aromaticity

17.30 Which compounds are aromatic? For any compound that is not aromatic, state why this is so.

a. 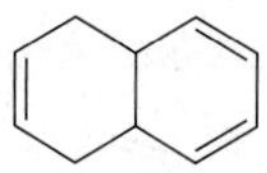
b. 
c. 
d. 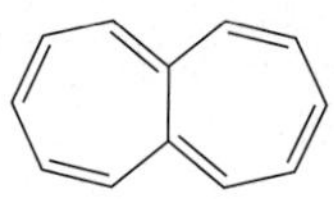

17.31 Which of the following heterocycles are aromatic?

a. 
c. 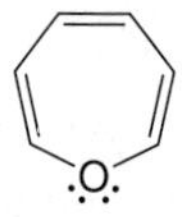
e. 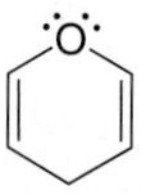
g. 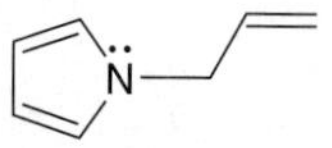
b. 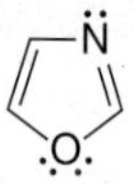
d. 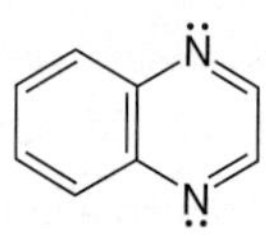
f. 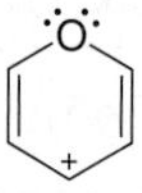
h. 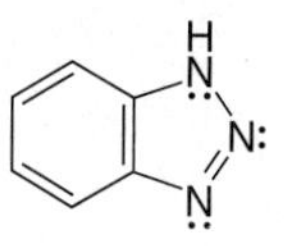

17.32 Label each compound as aromatic, antiaromatic, or not aromatic. Assume all completely conjugated rings are planar.

a. 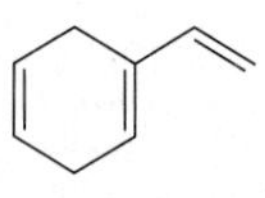
b. 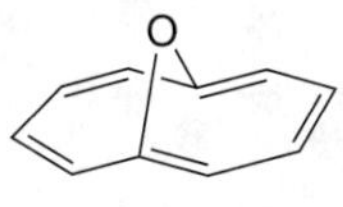
c. 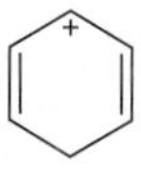
d. 

17.33 Hydrocarbon **A** possesses a significant dipole, even though it is composed of only C–C and C–H bonds. Explain why the dipole arises and use resonance structures to illustrate the direction of the dipole. Which ring is more electron rich?

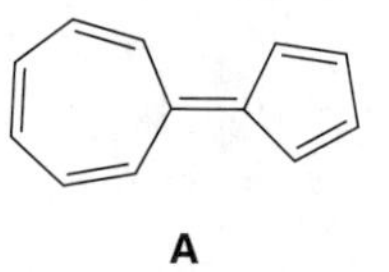

A

17.34 Pentalene, azulene, and heptalene are conjugated hydrocarbons that do not contain a benzene ring. Which hydrocarbons are especially stable or unstable based on the number of π electrons they contain? Explain your choices.

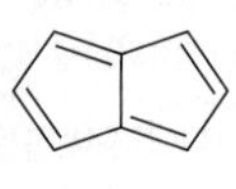
pentalene

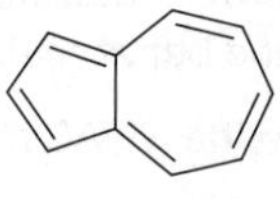
azulene

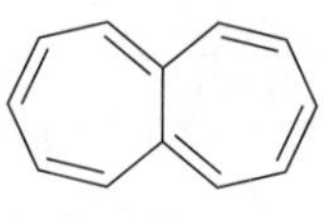
heptalene

17.35 As discussed in Section 17.8B, the purine heterocycle occurs commonly in the structure of DNA.

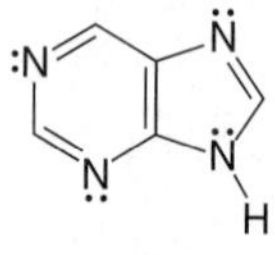
purine

a. How is each N atom hybridized?
b. In what type of orbital does each lone pair on a N atom reside?
c. How many π electrons does purine contain?
d. Why is purine aromatic?

17.36 Ribavirin is an antiviral agent used against hepatitis C and viral pneumonia.

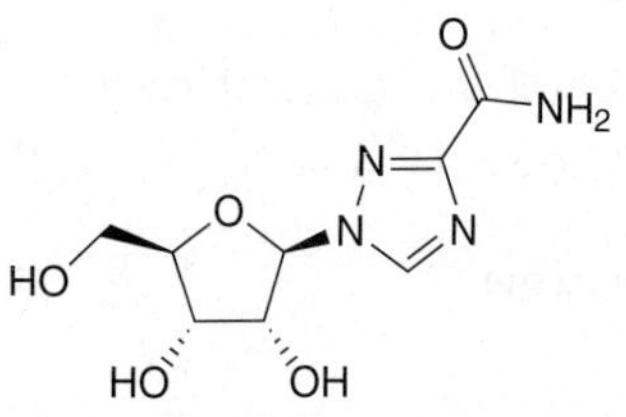
ribavirin

a. Explain why the nitrogen heterocycle is aromatic, and state which electrons are delocalized in the ring.
b. How is each N atom hybridized?
c. In what type of orbital does each lone pair on a N reside?

17.37 AZT was the first drug approved to treat HIV, the virus that causes AIDS. Explain why the six-membered ring of AZT is aromatic.

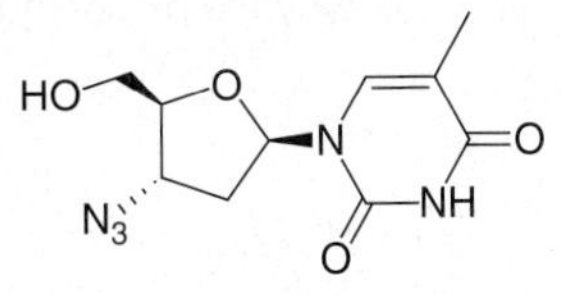

AZT

17.38 Explain the observed rate of reactivity of the following 2° alkyl halides in an S_N1 reaction.

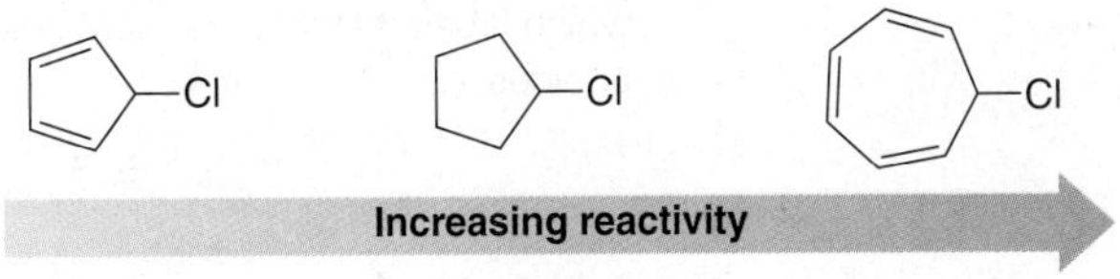

17.39 Draw a stepwise mechanism for the following reaction.

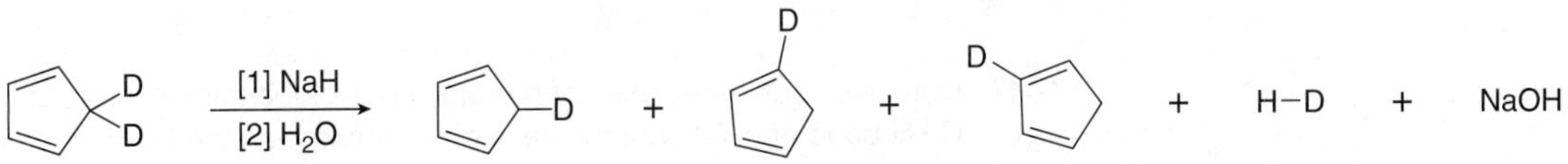

17.40 Explain why α-pyrone reacts with Br_2 to yield a substitution product (like benzene does), rather than an addition product to one of its C=C bonds.

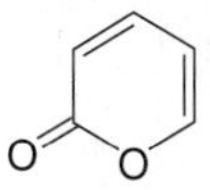

α-pyrone

Resonance

17.41 Draw additional resonance structures for each species.

a. 

cyclopropenyl radical

b. 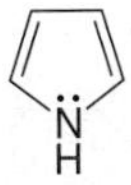

pyrrole

c. 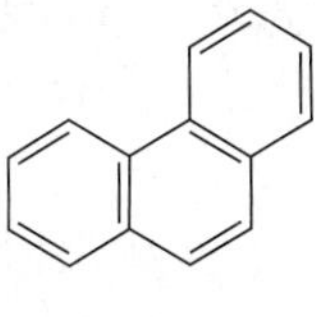

phenanthrene

17.42 The carbon–carbon bond lengths in naphthalene are not equal. Use a resonance argument to explain why bond (a) is shorter than bond (b).

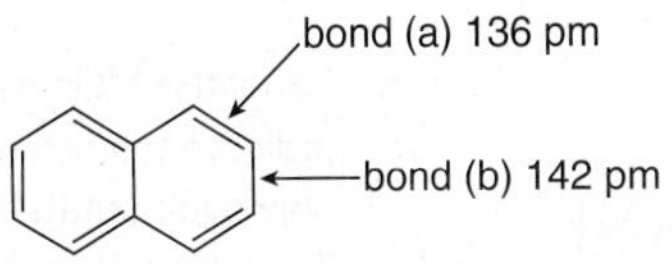

17.43

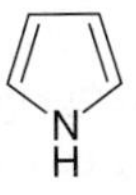

pyrrole

furan

a. Draw all reasonable resonance structures for pyrrole and explain why pyrrole is less resonance stabilized than benzene.
b. Draw all reasonable resonance structures for furan and explain why furan is less resonance stabilized than pyrrole.

Acidity

17.44 Which compound in each pair is the stronger acid?

a.  or 

b.  or 

17.45 Treatment of indene with $NaNH_2$ forms its conjugate base in a Brønsted–Lowry acid–base reaction. Draw all reasonable resonance structures for indene's conjugate base, and explain why the pK_a of indene is lower than the pK_a of most hydrocarbons.

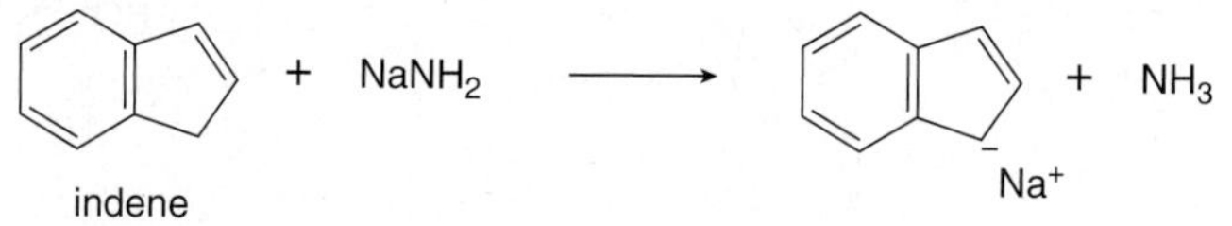

indene
pK_a = 20

17.46 Considering both 5-methylcyclopenta-1,3-diene (**A**) and 7-methylcyclohepta-1,3,5-triene (**B**), which labeled H atom is most acidic? Which labeled H atom is least acidic? Explain your choices.

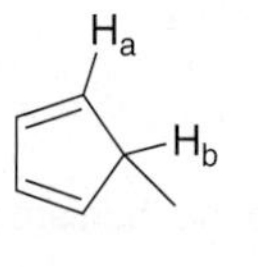

A

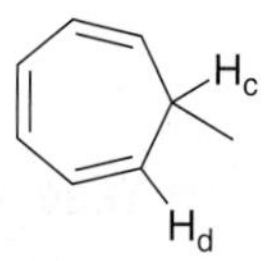

B

17.47 Draw the conjugate bases of pyrrole and cyclopentadiene. Explain why the sp^3 hybridized C–H bond of cyclopentadiene is more acidic than the N–H bond of pyrrole.

17.48 a. Explain why protonation of pyrrole occurs at C2 to form **A,** rather than on the N atom to form **B.**

b. Explain why **A** is more acidic than **C,** the conjugate acid of pyridine.

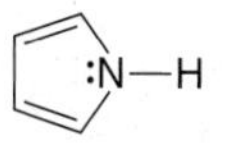

pyrrole

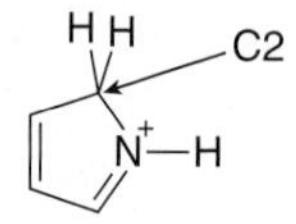

pK_a = 0.4

A

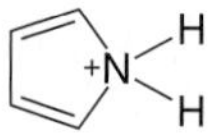

B

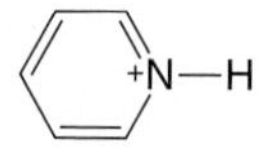

pK_a = 5.3

C

Inscribed Polygon Method

17.49 Use the inscribed polygon method to show the pattern of molecular orbitals in cyclooctatetraene.

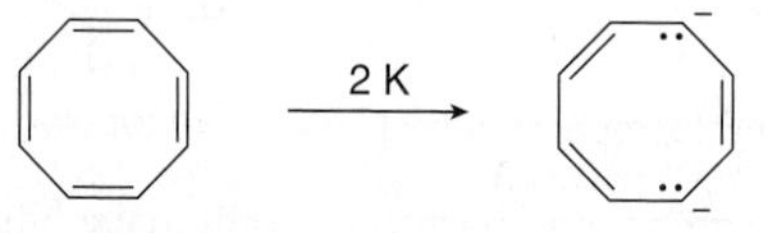

(one resonance structure) + 2 K^+

cyclooctatetraene

dianion of cyclooctatetraene

a. Label the MOs as bonding, antibonding, or nonbonding.

b. Indicate the arrangement of electrons in these orbitals for cyclooctatetraene, and explain why cyclooctatetraene is not aromatic.

c. Treatment of cyclooctatetraene with potassium forms a dianion. How many π electrons does this dianion contain?

d. How are the π electrons in this dianion arranged in the molecular orbitals?

e. Classify the dianion of cyclooctatetraene as aromatic, antiaromatic, or not aromatic, and explain why this is so.

17.50 Use the inscribed polygon method to show the pattern of molecular orbitals in cyclonona-1,3,5,7-tetraene and use it to label its cation, radical, and anion as aromatic, antiaromatic, or not aromatic.

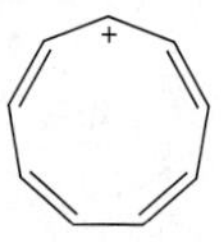

cyclononatetraenyl cation

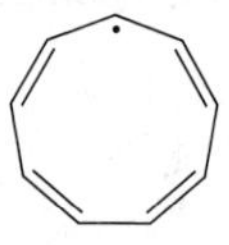

cyclononatetraenyl radical

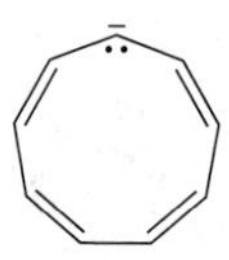

cyclononatetraenyl anion

Spectroscopy

17.51 How many ^{13}C NMR signals does each compound exhibit?

a. 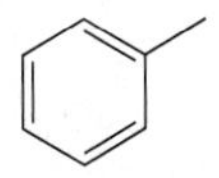

b. 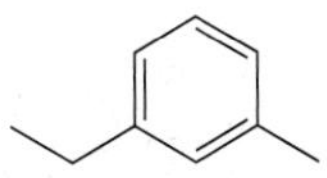

c. 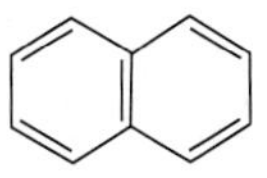

d. 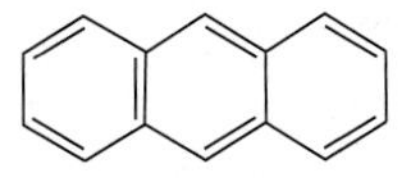

17.52 Which of the diethylbenzene isomers (ortho, meta, or para) corresponds to each set of ^{13}C NMR spectral data?

[A] ^{13}C NMR signals: 16, 29, 125, 127.5, 128.4, and 144 ppm

[B] ^{13}C NMR signals: 15, 26, 126, 128, and 142 ppm

[C] ^{13}C NMR signals: 16, 29, 128, and 141 ppm

17.53 Propose a structure consistent with each set of data.

a. $C_{10}H_{14}$: IR absorptions at 3150–2850, 1600, and 1500 cm^{-1}

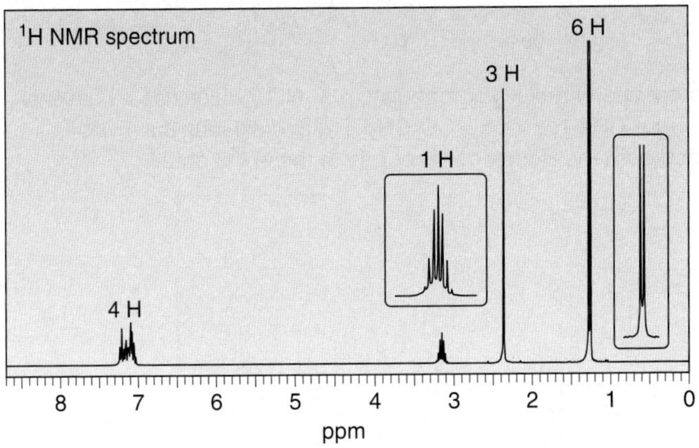

b. C_9H_{12}: ^{13}C NMR signals at 21, 127, and 138 ppm

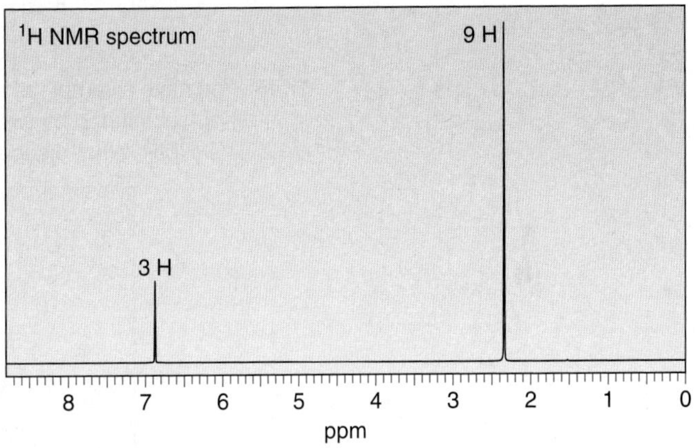

c. C_8H_{10}: IR absorptions at 3108–2875, 1606, and 1496 cm^{-1}

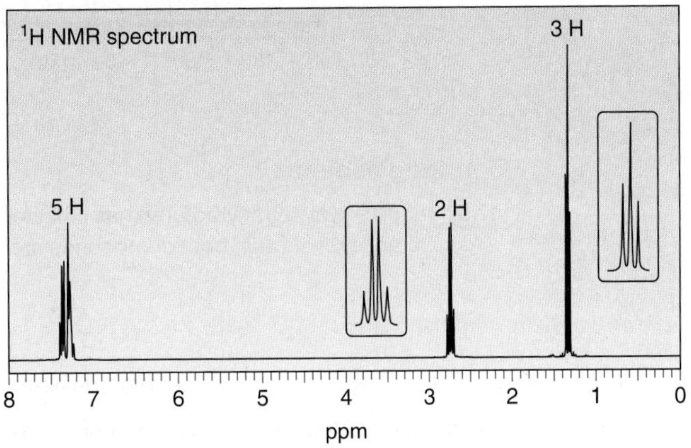

17.54 Propose a structure consistent with each set of data.

a. Compound **A:**

Molecular formula: $C_8H_{10}O$

IR absorption at 3150–2850 cm^{-1}

^1H NMR data: 1.4 (triplet, 3 H), 3.95 (quartet, 2 H), and 6.8–7.3 (multiplet, 5 H) ppm

b. Compound **B:**

Molecular formula: $C_9H_{10}O_2$

IR absorption at 1669 cm^{-1}

^1H NMR data: 2.5 (singlet, 3 H), 3.8 (singlet, 3 H), 6.9 (doublet, 2 H), and 7.9 (doublet, 2 H) ppm

17.55 Thymol (molecular formula $C_{10}H_{14}O$) is the major component of the oil of thyme. Thymol shows IR absorptions at 3500–3200, 3150–2850, 1621, and 1585 cm^{-1}. The ^1H NMR spectrum of thymol is given below. Propose a possible structure for thymol.

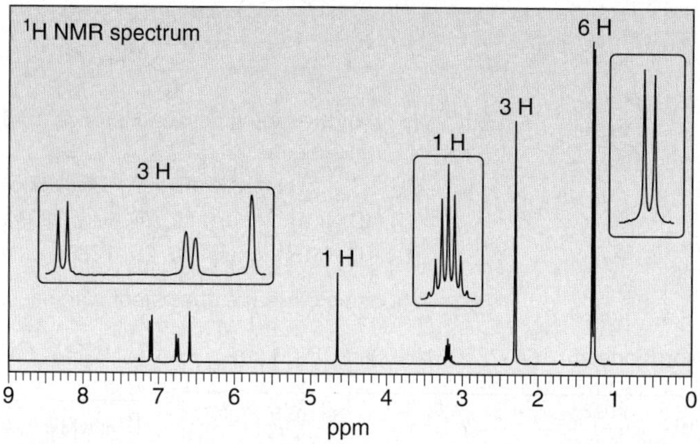

17.56 You have a sample of a compound of molecular formula $C_{11}H_{15}NO_2$, which has a benzene ring substituted by two groups, $(CH_3)_2N-$ and $-CO_2CH_2CH_3$, and exhibits the given ^{13}C NMR. What disubstituted benzene isomer corresponds to these ^{13}C data?

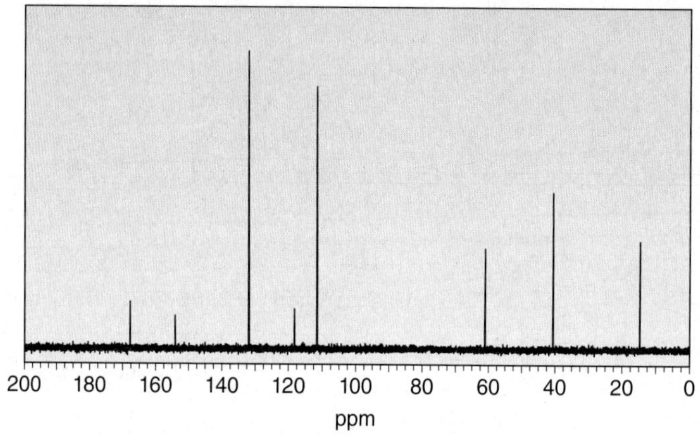

General Problems

17.57 Explain why tetrahydrofuran has a higher boiling point and is much more water soluble than furan, even though both compounds are cyclic ethers containing four carbons.

tetrahydrofuran furan

17.58 Use the four structural criteria of aromaticity to determine if the following compound is aromatic. Explain your answer.

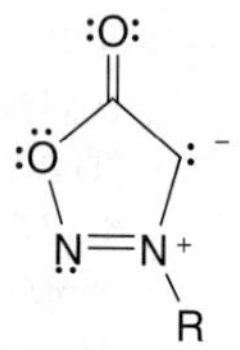

17.59 Thiamine diphosphate is a derivative of vitamin B_1 that is involved in the catalysis of many biological processes. Which heterocycles are aromatic, and which electrons are involved in the aromaticity?

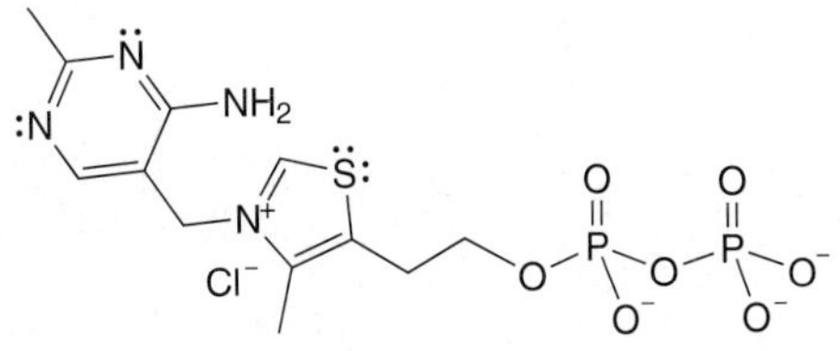

thiamine diphosphate

17.60 Nicotinamide adenine dinucleotide phosphate ($NADP^+$) is a biological oxidant involved in the oxidation of testosterone to estradiol.

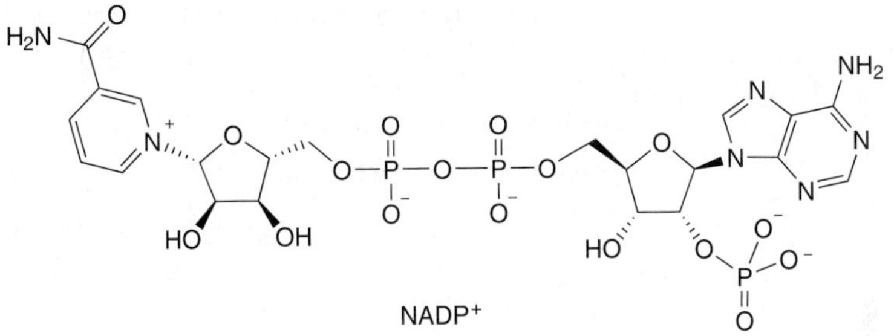

$NADP^+$

a. Which rings in $NADP^+$ are aromatic, and which electrons are involved in the aromaticity?
b. How is each N atom hybridized?
c. In what type of orbital does each lone pair on a N reside?

17.61 Answer the following questions about curcumin, a yellow pigment isolated from turmeric, a tropical perennial in the ginger family and a principal ingredient in curry powder.

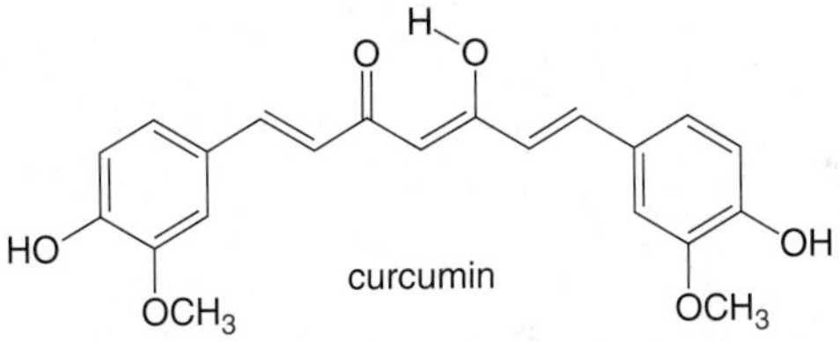

curcumin

a. In Chapter 11 we learned that most enols, compounds that contain a hydroxy group bonded to a C=C, are unstable and tautomerize to carbonyl groups. Draw the keto form of the enol of curcumin, and explain why the enol is more stable than many other enols.
b. Explain why the enol O–H proton is more acidic than an alcohol O–H proton.
c. Why is curcumin colored?
d. Explain why curcumin is an antioxidant.

17.62 Stanozolol is an anabolic steroid that promotes muscle growth. Although stanozolol has been used by athletes and body builders, many physical and psychological problems result from prolonged use and it is banned in competitive sports.

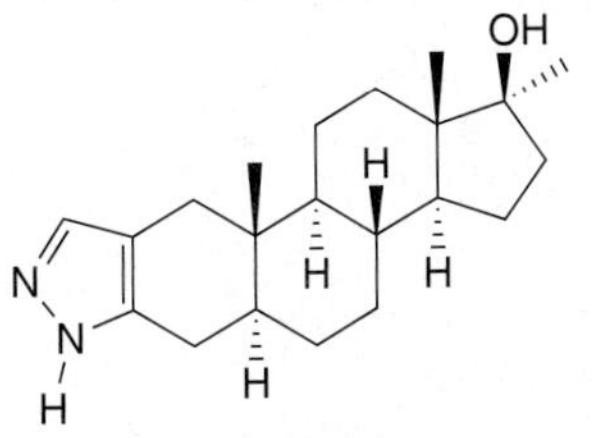

stanozolol

a. Explain why the nitrogen heterocycle—a pyrazole ring—is aromatic.
b. In what type of orbital is the lone pair on each N atom contained?
c. Draw all reasonable resonance structures for stanozolol.
d. Explain why the pK_a of the N–H bond in the pyrazole ring is comparable to the pK_a of the O–H bond, making it considerably more acidic than amines such as CH_3NH_2 (pK_a = 40).

17.63 Tetrahydrofolate, a folic acid derivative, has a key role in the metabolism of amino acids and nucleic acids. Label the aromatic rings, and show which electrons are involved in the aromaticity.

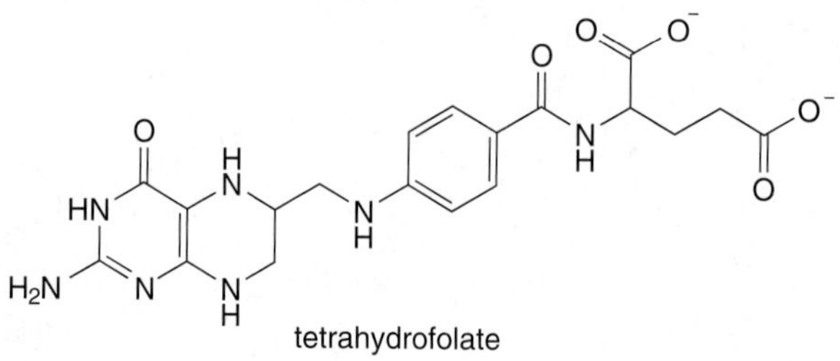

tetrahydrofolate

17.64 Three DNA bases exist primarily as the keto tautomer (Section 11.9) because more hydrogen bonding interactions are possible in this form. However, it was originally thought that the bases existed mainly as the enol tautomers, because they appear to be more traditionally aromatic in this form. Draw the aromatic enol tautomers of the three DNA bases that contain a C= O group.

Challenge Problems

17.65 Explain why **A** is aromatic but **B** is not aromatic.

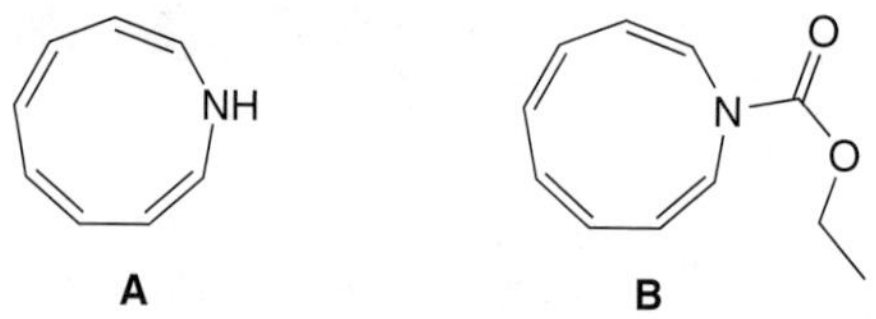

A **B**

17.66 Use the observed ^{1}H NMR data to decide whether **C** and its dianion are aromatic, antiaromatic, or not aromatic. **C** shows NMR signals at –4.25 (6 H) and 8.14–8.67 (10 H) ppm. The dianion of **C** shows NMR signals at –3 (10 H) and 21 (6 H) ppm. Why are the signals shifted upfield (or downfield) to such a large extent?

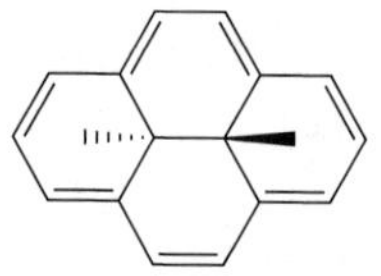

C

17.67 Explain why compound **A** is much more stable than compound **B.**

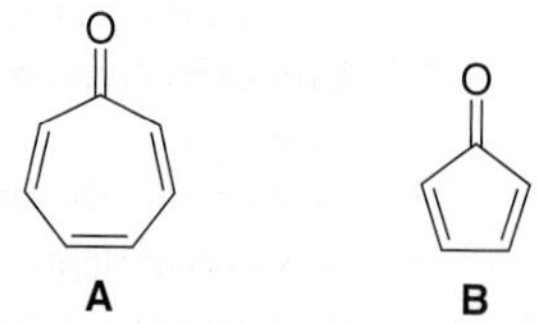

A **B**

17.68 (*R*)-Carvone, the major component of the oil of spearmint, undergoes acid-catalyzed isomerization to carvacrol, a major component of the oil of thyme. Draw a stepwise mechanism and explain why this isomerization occurs.

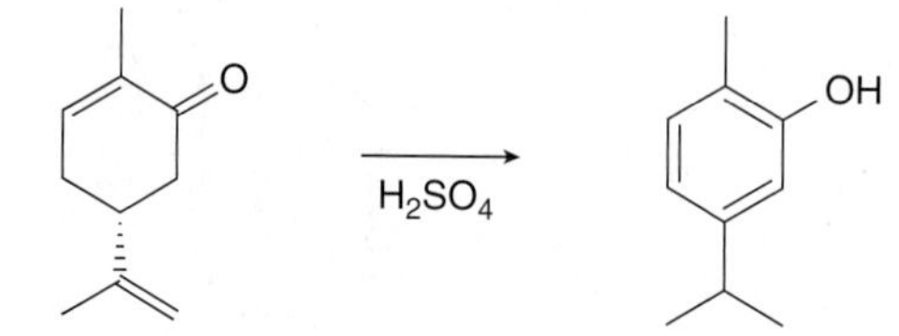

(*R*)-carvone carvacrol

17.69 Explain why triphenylene resembles benzene in that it does not undergo addition reactions with Br_2, but phenanthrene reacts with Br_2 to yield the addition product drawn. (Hint: Draw resonance structures for both triphenylene and phenanthrene, and use them to determine how delocalized each π bond is.)

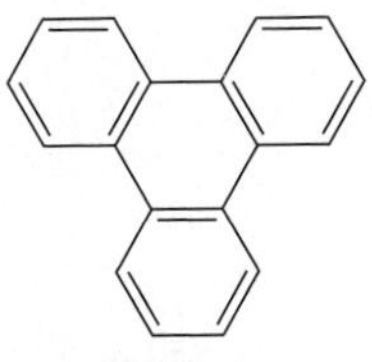

triphenylene

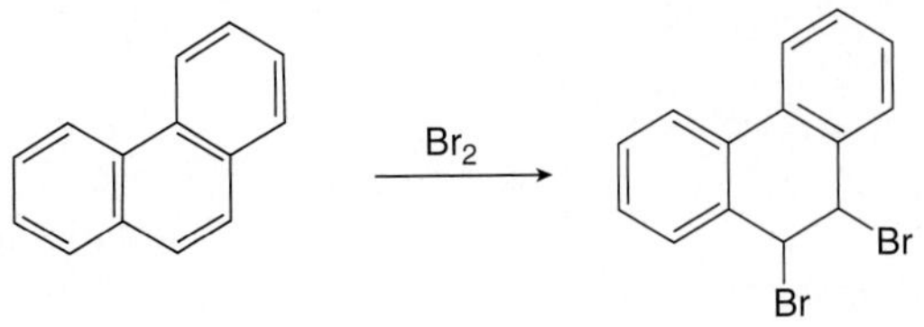
Br2
Br
Br

phenanthrene

17.70 Although benzene itself absorbs at 128 ppm in its ^{13}C NMR spectrum, the carbons of substituted benzenes absorb either upfield or downfield from this value depending on the substituent. Explain the observed values for the carbon ortho to the given substituent in the monosubstituted benzene derivatives **X** and **Y.**

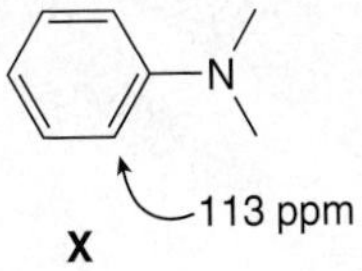
N
113 ppm

X

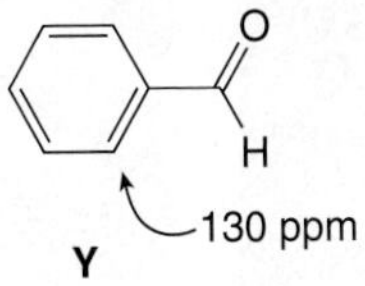
O
H
130 ppm

Y

18 Reactions of Aromatic Compounds

18.1 Electrophilic aromatic substitution
18.2 The general mechanism
18.3 Halogenation
18.4 Nitration and sulfonation
18.5 Friedel–Crafts alkylation and Friedel–Crafts acylation
18.6 Substituted benzenes
18.7 Electrophilic aromatic substitution of substituted benzenes
18.8 Why substituents activate or deactivate a benzene ring
18.9 Orientation effects in substituted benzenes
18.10 Limitations on electrophilic substitution reactions with substituted benzenes
18.11 Disubstituted benzenes
18.12 Synthesis of benzene derivatives
18.13 Nucleophilic aromatic substitution
18.14 Halogenation of alkyl benzenes
18.15 Oxidation and reduction of substituted benzenes
18.16 Multistep synthesis

Vitamin K_1, phylloquinone, is a fat-soluble vitamin that regulates the synthesis of proteins needed for blood to clot. Dietary sources of vitamin K_1 include cauliflower, broccoli, soybeans, leafy greens, and green tea. A severe deficiency of vitamin K_1 leads to excessive and sometimes fatal bleeding because of inadequate blood clotting. Vitamin K_1 is synthesized by a biological Friedel–Crafts reaction, one of the many examples of electrophilic aromatic substitution, a key reaction of aromatic compounds presented in Chapter 18.

Chapter 18 discusses the chemical reactions of benzene and other aromatic compounds. Although aromatic rings are unusually stable, making benzene unreactive in most of the reactions discussed so far, benzene acts as a nucleophile with certain electrophiles, yielding substitution products with an intact aromatic ring.

We begin with the basic features and mechanism of electrophilic aromatic substitution (Sections 18.1–18.5), the most prevalent reaction of benzene. Next, we discuss the electrophilic aromatic substitution of substituted benzenes (Sections 18.6–18.12), and conclude with nucleophilic aromatic substitution and other useful reactions of benzene derivatives (Sections 18.13–18.15). The ability to interconvert resonance structures and evaluate their relative energies is crucial to understanding this material.

18.1 Electrophilic Aromatic Substitution

Based on its structure and properties, what kinds of reactions should benzene undergo? Are any of its bonds particularly weak? Does it have electron-rich or electron-deficient atoms?

- **Benzene has six π electrons delocalized in six *p* orbitals that overlap above and below the plane of the ring. These loosely held π electrons make the benzene ring electron rich, so it reacts with electrophiles.**
- **Because benzene's six π electrons satisfy Hückel's rule, benzene is especially stable. Reactions that keep the aromatic ring intact are therefore favored.**

As a result, **the characteristic reaction of benzene is *electrophilic aromatic substitution*—a hydrogen atom is replaced by an electrophile.**

Electrophilic aromatic substitution

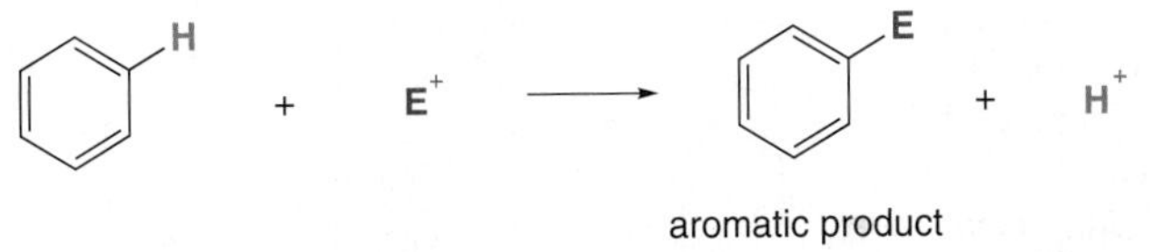
H
+
E+
E
+
H+
aromatic product

As we learned in Section 17.6, benzene does *not* undergo addition reactions like other unsaturated hydrocarbons, because addition would yield a product that is not aromatic. Substitution of a hydrogen, on the other hand, keeps the aromatic ring intact.

Five specific examples of electrophilic aromatic substitution are shown in Figure 18.1. The basic mechanism, discussed in Section 18.2, is the same in all five cases. The reactions differ only in the identity of the electrophile, E^+.

Problem 18.1 Why is benzene less reactive toward electrophiles than an alkene, even though it has more π electrons than an alkene (six versus two)?

Figure 18.1
Five examples of electrophilic aromatic substitution

Reaction	Electrophile
[1] Halogenation—Replacement of H by X (Cl or Br)	

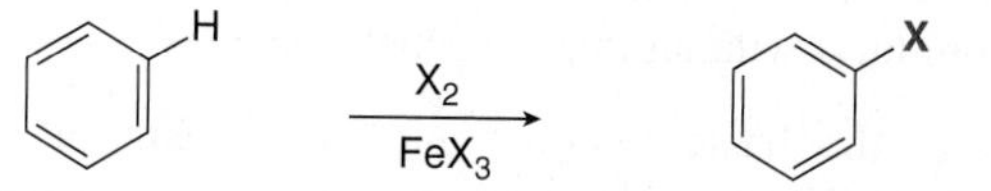
H
X_2
FeX_3
X

X = Cl
X = Br
aryl halide

$E^+ = Cl^+$ or Br^+

[2] Nitration—Replacement of H by NO_2

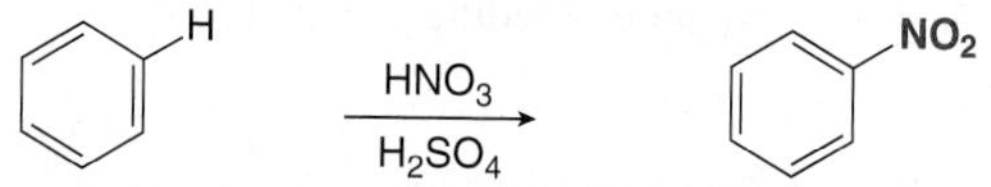
H
HNO_3
H_2SO_4
NO_2

nitrobenzene

$E^+ = \overset{+}{N}O_2$

[3] Sulfonation—Replacement of H by SO_3H

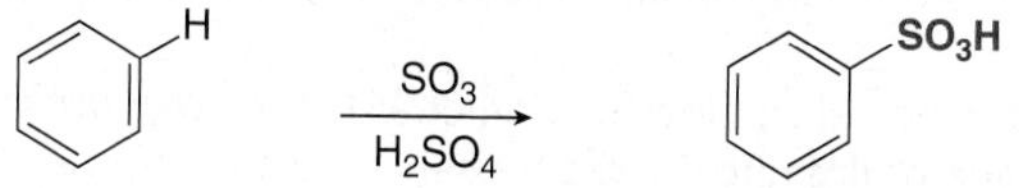
H
SO_3
H_2SO_4
SO_3H

benzenesulfonic acid

$E^+ = \overset{+}{S}O_3H$

[4] Friedel–Crafts alkylation—Replacement of H by R

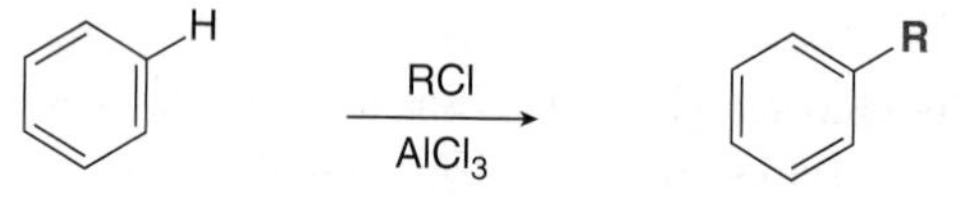
H
RCl
$AlCl_3$
R

alkyl benzene
(arene)

$E^+ = R^+$

Friedel–Crafts alkylation and acylation, named for Charles Friedel and James Crafts who discovered the reactions in the nineteenth century, form new carbon–carbon bonds.

[5] Friedel–Crafts acylation—Replacement of H by RCO

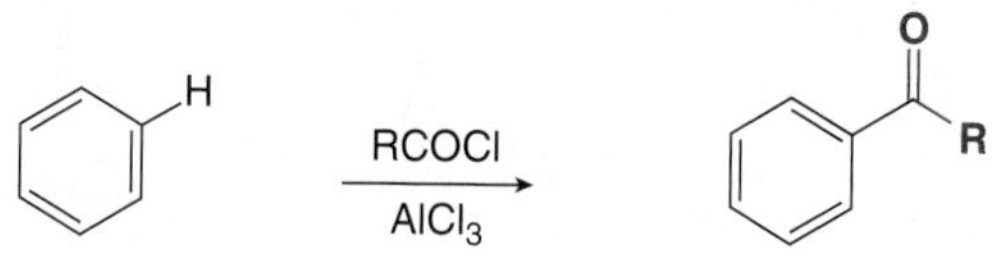
H
RCOCl
$AlCl_3$
O
R

ketone

$E^+ = R-\overset{+}{C}=\ddot{O}:$

18.2 The General Mechanism

No matter what electrophile is used, all electrophilic aromatic substitution reactions occur via a **two-step mechanism:** addition of the electrophile E^+ to form a resonance-stabilized carbocation, followed by deprotonation with base, as shown in Mechanism 18.1.

Mechanism 18.1 General Mechanism—Electrophilic Aromatic Substitution

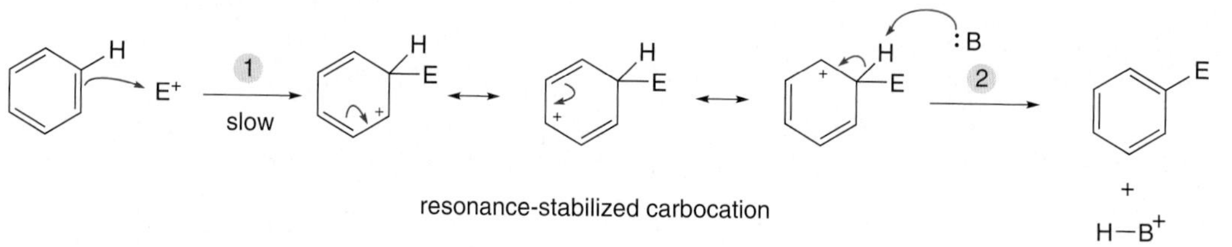
H
E^+
1
slow
H
E
:B
2
E
+
H—B^+

resonance-stabilized carbocation

1 Addition of the electrophile E^+ forms a new C–E bond and a resonance-stabilized carbocation. This step is rate-determining because the aromaticity of the benzene ring is lost.

2 A base removes the proton *on the carbon bonded to the electrophile,* re-forming the aromatic ring. Any resonance structure can be used to draw the product.

The first step in electrophilic aromatic substitution forms a carbocation, for which three resonance structures can be drawn. To help keep track of the location of the positive charge:

- **Always draw in the H atom on the carbon bonded to E. This serves as a reminder that it is the only *sp*^{3} hybridized carbon in the carbocation intermediate.**

- **Notice that the positive charge in a given resonance structure is always located ortho or para to the new C–E bond. In the hybrid, therefore, the charge is delocalized over three atoms of the ring.**

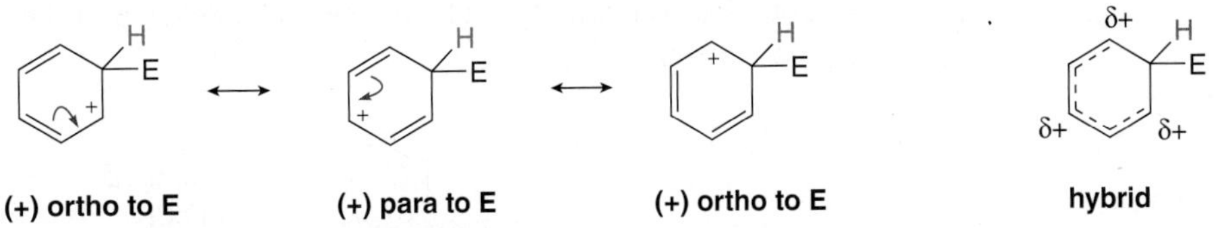

This two-step mechanism for electrophilic aromatic substitution applies to all of the electrophiles in Figure 18.1. **The net result of addition of an electrophile (E^+) followed by elimination of a proton (H^+) is substitution of E for H.**

The energy changes in electrophilic aromatic substitution are shown in Figure 18.2. The mechanism consists of two steps, so the energy diagram has two energy barriers. Because the transition state of the first step is higher in energy, it is rate-determining.

Problem 18.2 In Step [2] of Mechanism 18.1, loss of a proton to form the substitution product was drawn using only one resonance structure. Use curved arrows to show how the other two resonance structures can be converted to the substitution product (PhE) by removal of a proton with :B.

Figure 18.2
Energy diagram for electrophilic aromatic substitution: $PhH + E^+ \rightarrow PhE + H^+$

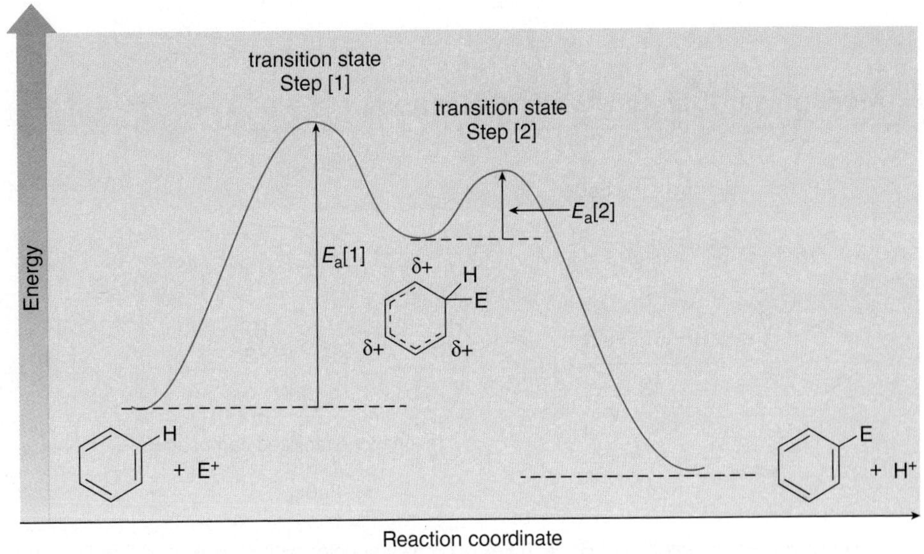

- The mechanism has two steps, so there are two energy barriers.
- Step [1] is rate-determining; its transition state is at higher energy.

18.3 Halogenation

The general mechanism outlined in Mechanism 18.1 can now be applied to each of the five specific examples of electrophilic aromatic substitution shown in Figure 18.1. For each mechanism we must learn how to generate a specific electrophile. This step is *different* with each electrophile. Then, the electrophile reacts with benzene by the two-step process of Mechanism 18.1. These two steps are the *same* for all five reactions.

In **halogenation,** benzene reacts with Cl_2 or Br_2 in the presence of a Lewis acid catalyst, such as $FeCl_3$ or $FeBr_3$, to give the **aryl halides** chlorobenzene or bromobenzene, respectively. Analogous reactions with I_2 and F_2 are not synthetically useful because I_2 is too unreactive and F_2 reacts too violently.

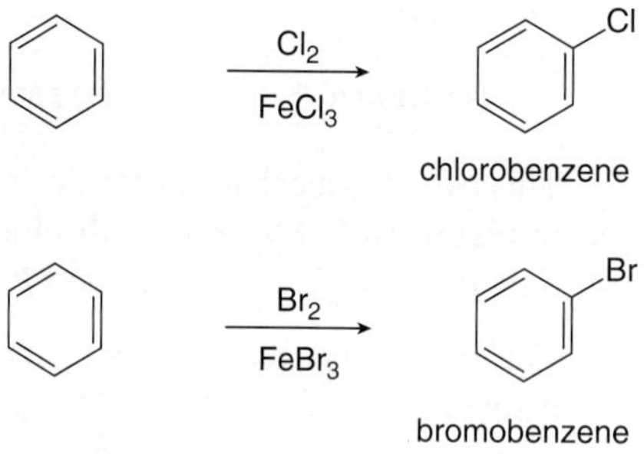

In bromination (Mechanism 18.2), the Lewis acid $FeBr_3$ reacts with Br_2 to form a Lewis acid–base complex that weakens and polarizes the Br–Br bond, making it more electrophilic. This reaction is Step [1] of the mechanism for the bromination of benzene. The remaining two steps follow directly from the general mechanism for electrophilic aromatic substitution: addition of the electrophile (Br^+ in this case) forms a resonance-stabilized carbocation, and loss of a proton regenerates the aromatic ring.

Mechanism 18.2 Bromination of Benzene

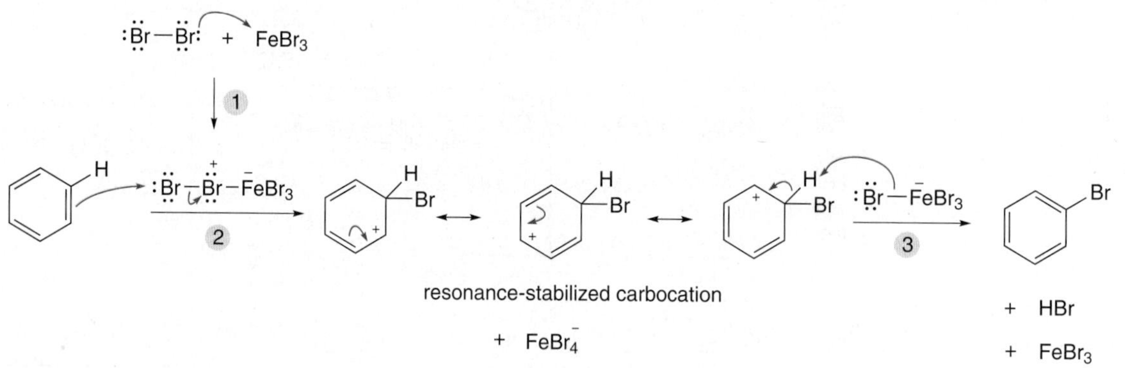

1 Lewis acid–base reaction of Br_2 with $FeBr_3$ forms a species with a weakened Br–Br bond that serves as source of Br^+.

2 Addition of the electrophile forms a new C–Br bond and a resonance-stabilized carbocation.

3 $FeBr_4^-$ removes the proton *on the carbon bonded to the electrophile,* re-forming the aromatic ring. The Lewis acid catalyst $FeBr_3$ is regenerated for another reaction cycle.

Figure 18.3 Examples of biologically active aryl chlorides

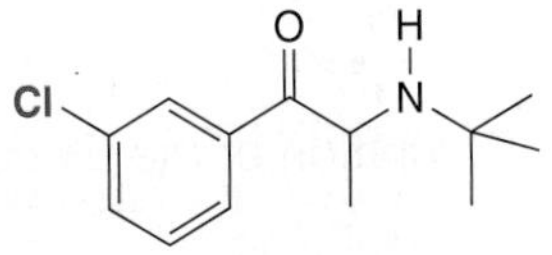

Generic name **bupropion**
Trade names **Wellbutrin, Zyban**
antidepressant,
also used to reduce nicotine cravings

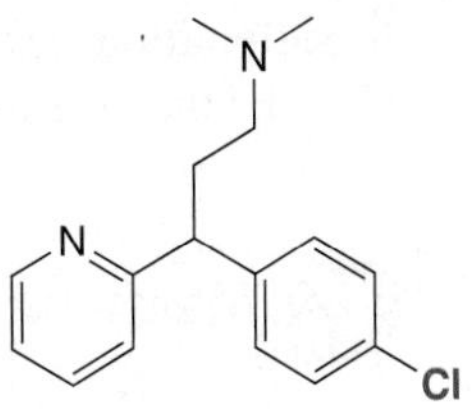

chlorpheniramine
antihistamine

Herbicides were used extensively during the Vietnam War to defoliate dense jungle areas. The concentration of certain herbicide by-products in the soil remains high today.

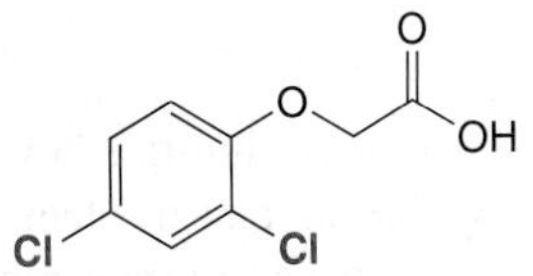

2,4-D
2,4-dichlorophenoxy-acetic acid
herbicide

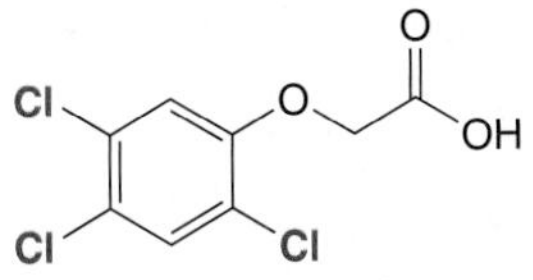

2,4,5-T
2,4,5-trichlorophenoxy-acetic acid
herbicide

the active components in **Agent Orange,** a defoliant used in the Vietnam War

Chlorination proceeds by a similar mechanism. Reactions that introduce a halogen substituent on a benzene ring are widely used, and many halogenated aromatic compounds with a range of biological activity have been synthesized, as shown in Figure 18.3.

Problem 18.3 Draw a detailed mechanism for the chlorination of benzene using Cl_2 and $FeCl_3$.

18.4 Nitration and Sulfonation

Nitration and **sulfonation** of benzene introduce two different functional groups on an aromatic ring. Nitration is an especially useful reaction because a nitro group can then be reduced to an NH_2 group, a common benzene substituent, in a reaction discussed in Section 18.15.

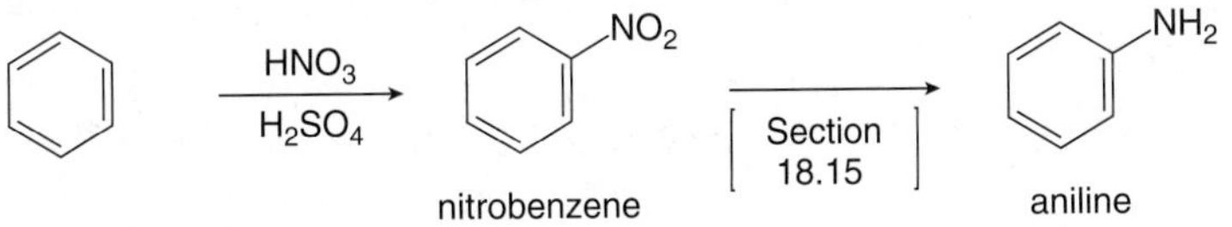

nitrobenzene aniline

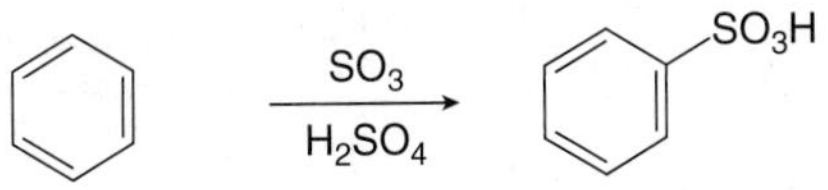

benzenesulfonic acid

Generation of the electrophile in both nitration and sulfonation requires strong acid. In **nitration,** the electrophile is $^+NO_2$ (the **nitronium ion**), formed by protonation of HNO_3 followed by loss of water (Mechanism 18.3).

Mechanism 18.3 Formation of the Nitronium Ion ($^+NO_2$) for Nitration

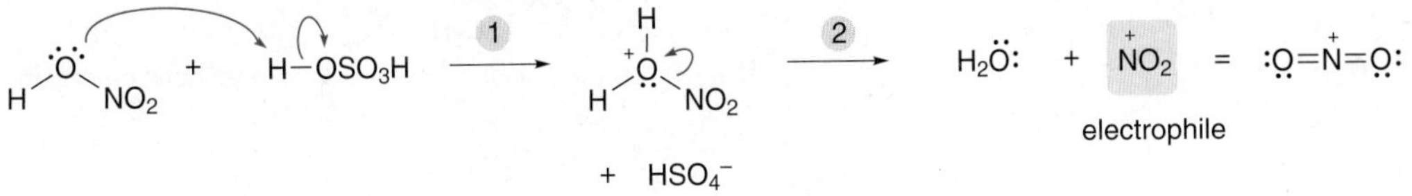

electrophile

In **sulfonation,** protonation of sulfur trioxide, SO_3, forms a positively charged sulfur species ($^+SO_3H$) that acts as an electrophile (Mechanism 18.4).

Mechanism 18.4 Formation of the Electrophile $^+SO_3H$ for Sulfonation

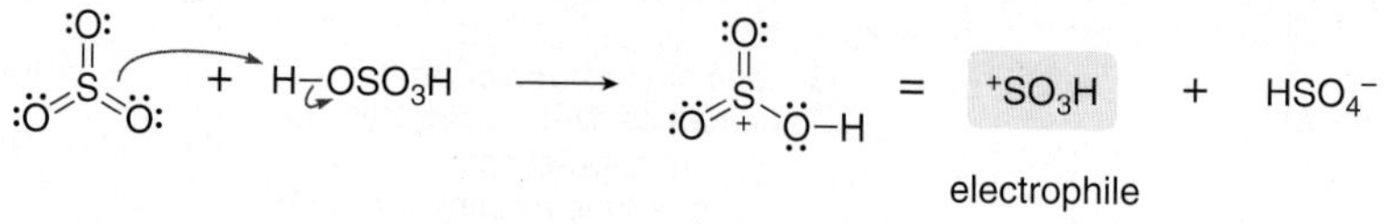
:O:
+ H–OSO3H
:O:
:Ö–H
=
+SO3H
+
HSO4−
electrophile

These steps illustrate how to generate the electrophile E^+ for nitration and sulfonation, the process that begins any mechanism for electrophilic aromatic substitution. To complete either of these mechanisms, you must replace the electrophile E^+ by either $^+NO_2$ or $^+SO_3H$ in the general mechanism (Mechanism 18.1). Thus, **the two-step sequence that replaces H by E is the same regardless of E^+.** This is shown in Sample Problem 18.1 using the reaction of benzene with the nitronium ion.

Sample Problem 18.1 Draw a stepwise mechanism for the nitration of a benzene ring.

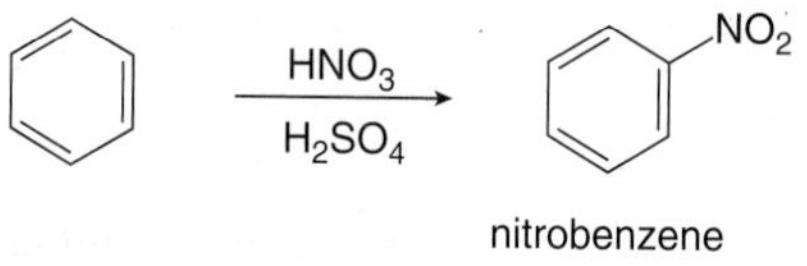
HNO3
H2SO4
NO2
nitrobenzene

Solution

We must first generate the electrophile and then write the two-step mechanism for electrophilic aromatic substitution using it.

Part [1] Generation of the electrophile $^+NO_2$

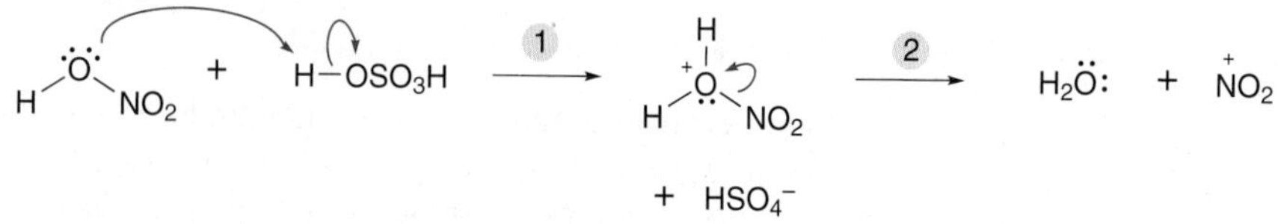
H
NO2
+
H–OSO3H
1
H
H
NO2
2
H2Ö:
+
NO2
+ HSO4−

Part [2] Two-step mechanism for electrophilic aromatic substitution

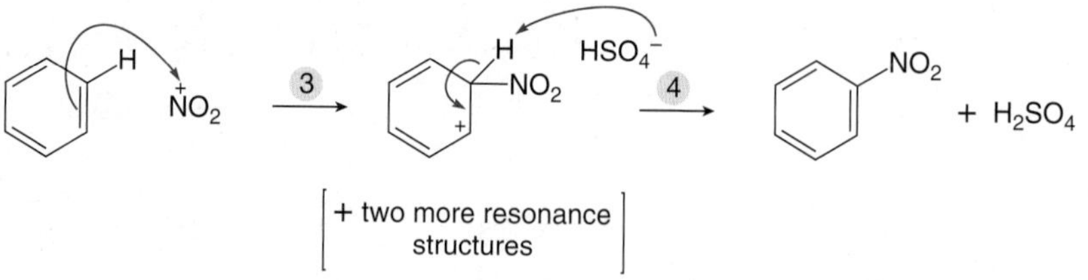
H
NO2
3
H
HSO4−
–NO2
4
NO2
+ H2SO4
+ two more resonance structures

Any species with a lone pair of electrons can be used to remove the proton in the last step. In this case, the mechanism is drawn with HSO_4^-, formed when $^+NO_2$ is generated as the electrophile.

Problem 18.4 Draw a stepwise mechanism for the sulfonation of an alkyl benzene such as **A** to form a substituted benzenesulfonic acid **B.** Treatment of **B** with base forms a sodium salt **C** that can be used as a synthetic detergent to clean away dirt (see Problem 3.22).

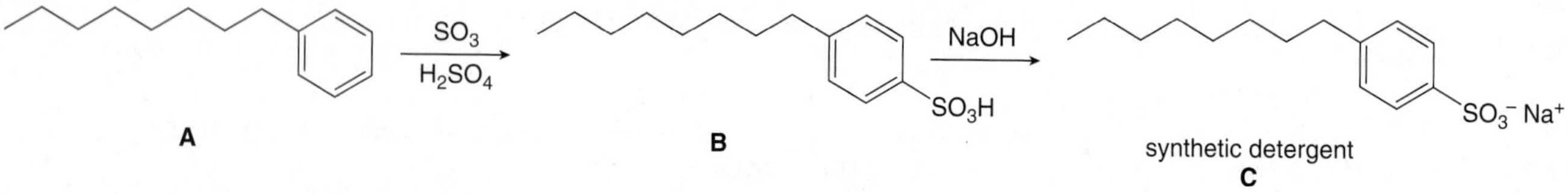
SO3
H2SO4
NaOH
SO3H
SO3− Na+
A
B
synthetic detergent
C

18.5 Friedel–Crafts Alkylation and Friedel–Crafts Acylation

Friedel–Crafts alkylation and **Friedel–Crafts acylation** form new carbon–carbon bonds.

18.5A General Features

In **Friedel–Crafts alkylation,** treatment of benzene with an alkyl halide and a Lewis acid ($AlCl_3$) forms an alkyl benzene. This reaction is an **alkylation** because it results in transfer of an alkyl group from one atom to another (from Cl to benzene).

Friedel–Crafts alkylation

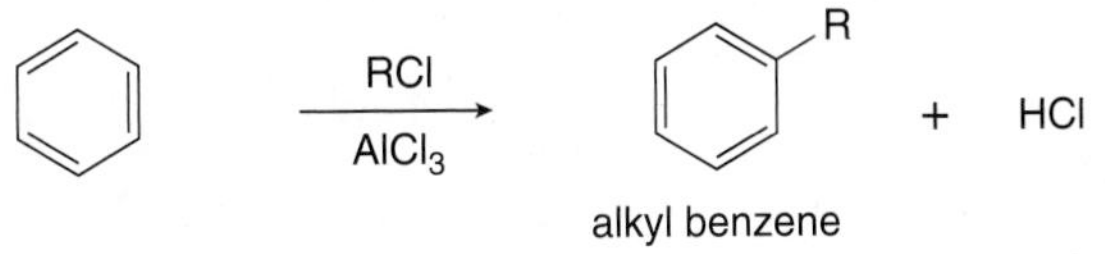

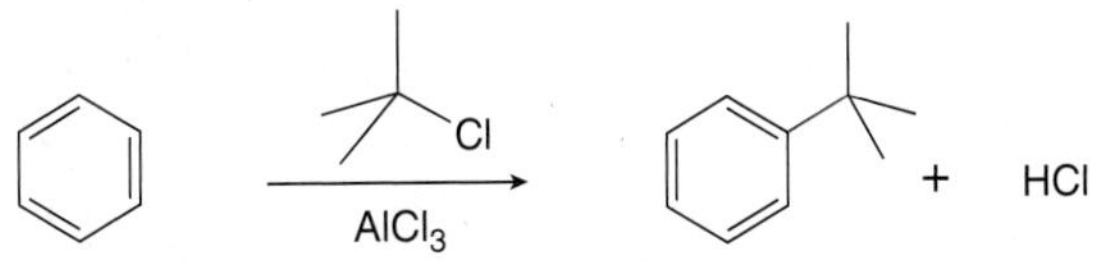

In **Friedel–Crafts acylation,** a benzene ring is treated with an **acid chloride** (RCOCl) and $AlCl_3$ to form a ketone. Because the new group bonded to the benzene ring is called an **acyl group,** the transfer of an acyl group from one atom to another is an **acylation.**

Friedel–Crafts acylation

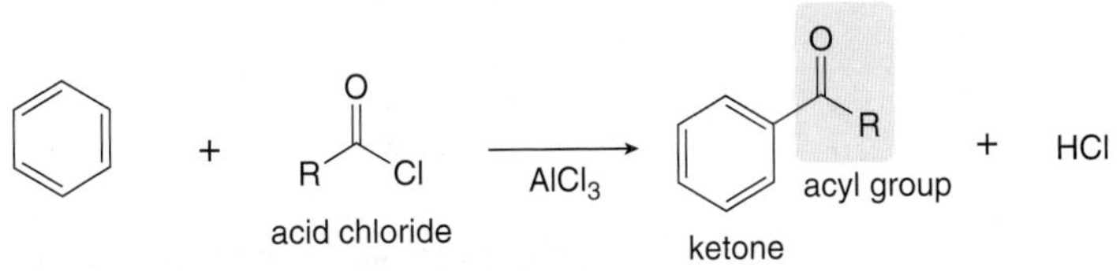

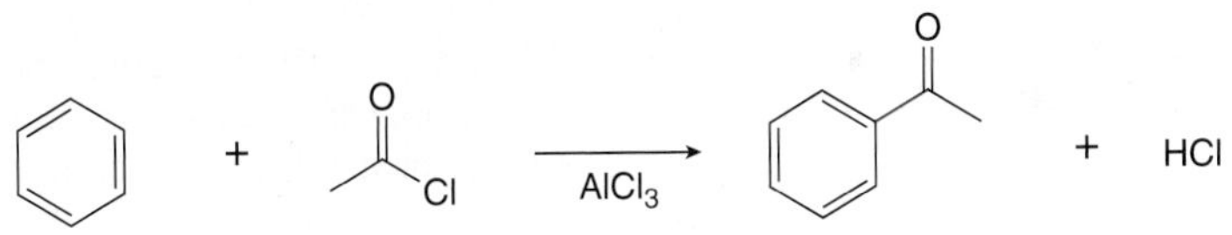

Acid chlorides are also called **acyl chlorides.**

Problem 18.5 What product is formed when benzene is treated with each organic halide in the presence of $AlCl_3$?

a. 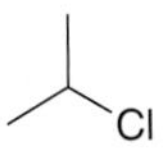

b. 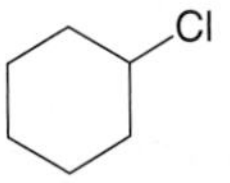

c. 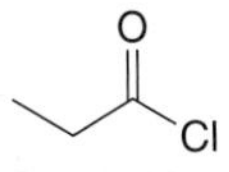

Problem 18.6 What acid chloride would be needed to prepare each of the following ketones from benzene using a Friedel–Crafts acylation?

a. 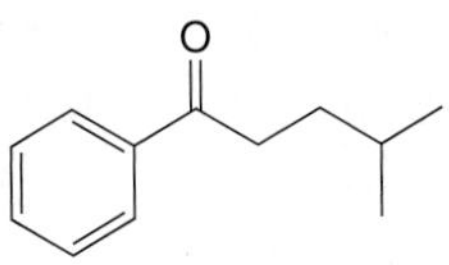

b. 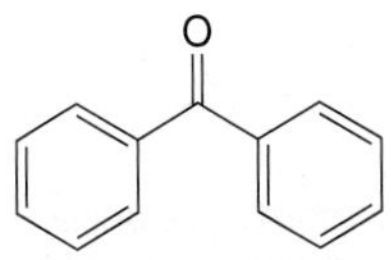

c. 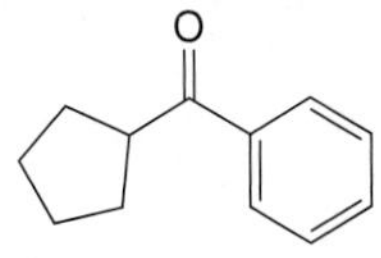

18.5B Mechanism

The mechanisms of alkylation and acylation proceed in a manner analogous to those for halogenation, nitration, and sulfonation. The unique feature in each reaction is how the electrophile is generated.

In **Friedel–Crafts alkylation,** the Lewis acid $AlCl_3$ reacts with the alkyl chloride to form a Lewis acid–base complex, illustrated with CH_3CH_2Cl and $(CH_3)_3CCl$ as alkyl chlorides. The identity of the alkyl chloride determines the exact course of the reaction as shown in Mechanism 18.5.

Mechanism 18.5 Formation of the Electrophile in Friedel–Crafts Alkylation—Two Possibilities

Possibility [1] For CH_3Cl and 1° RCl

Cl:
+
AlCl3
Cl—AlCl3
electrophile
Lewis acid–base complex

Possibility [2] For 2° and 3° RCl

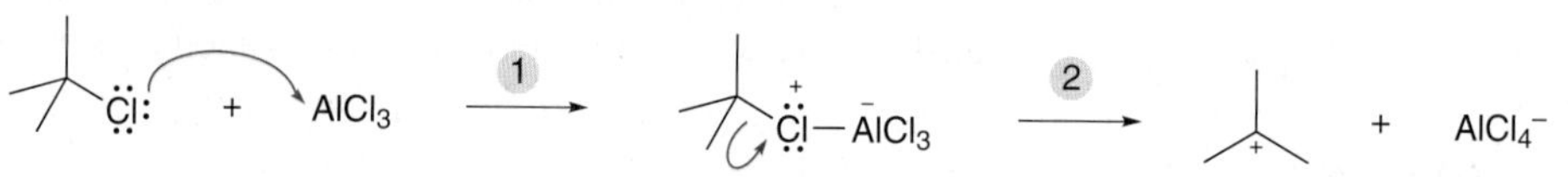

Cl:
+
AlCl3
1
Cl—AlCl3
2
+
AlCl4−
Lewis acid–base complex
electrophile

- **For CH_3Cl and 1° RCl, the Lewis acid–base complex itself serves as the electrophile for electrophilic aromatic substitution.**
- **With 2° and 3° RCl, the Lewis acid–base complex reacts further to give a 2° or 3° carbocation, which serves as the electrophile. Carbocation formation occurs only with 2° and 3° alkyl chlorides, because they afford more stable carbocations.**

In either case, the electrophile goes on to react with benzene in the two-step mechanism characteristic of electrophilic aromatic substitution, illustrated in Mechanism 18.6 using the 3° carbocation, $(CH_3)_3C^+$.

Mechanism 18.6 Friedel–Crafts Alkylation Using a 3° Carbocation

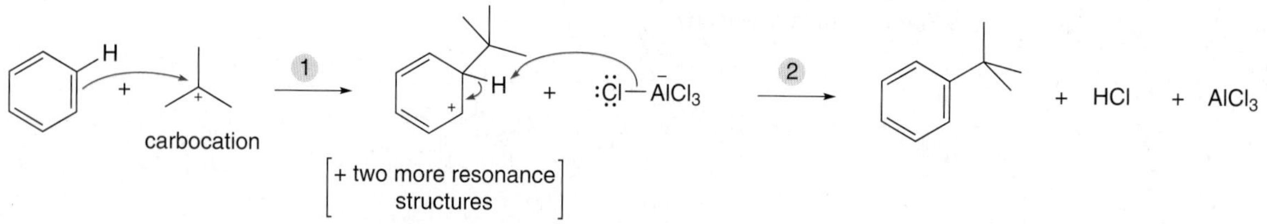

H
+
carbocation
1
H
+
:Cl—AlCl3
[+ two more resonance structures]
2
+ HCl
+ AlCl3

1 Addition of the carbocation electrophile forms a new carbon–carbon bond.

2 $AlCl_4^-$ removes a proton on the carbon bearing the new substituent to re-form the aromatic ring.

In **Friedel–Crafts acylation,** the Lewis acid $AlCl_3$ ionizes the carbon–halogen bond of the acid chloride, thus forming a positively charged carbon electrophile called an **acylium ion,** which is resonance stabilized (Mechanism 18.7). The positively charged carbon atom of the acylium ion then goes on to react with benzene in the two-step mechanism of electrophilic aromatic substitution.

Mechanism 18.7 Formation of the Electrophile in Friedel–Crafts Acylation

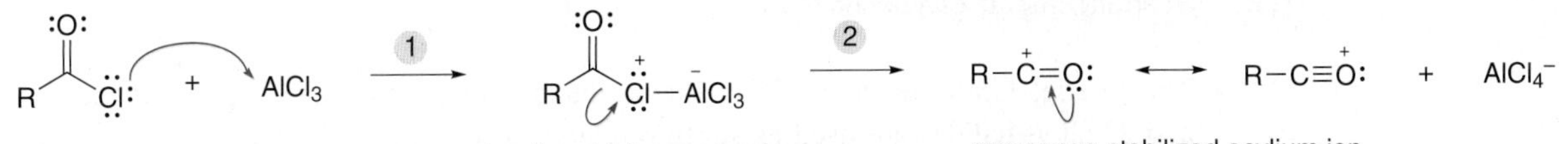

To complete the mechanism for acylation, insert the electrophile into the general mechanism and draw the last two steps, as illustrated in Sample Problem 18.2.

Sample Problem 18.2 Draw a stepwise mechanism for the following Friedel–Crafts acylation.

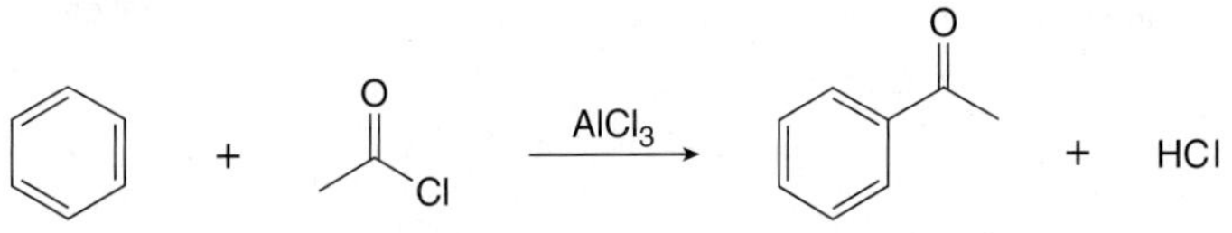

Solution

First generate the acylium ion, and then write the two-step mechanism for electrophilic aromatic substitution using it for the electrophile.

Part [1] Generation of the electrophile $(CH_3CO)^+$

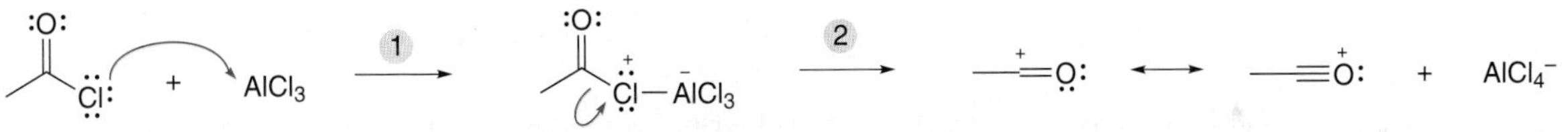

Part [2] Two-step mechanism for electrophilic aromatic substitution

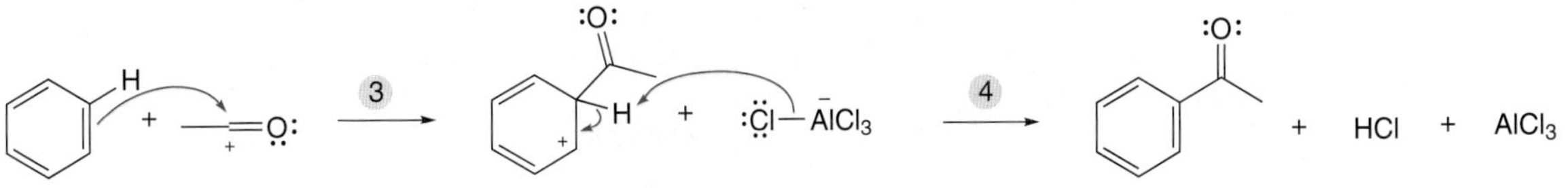

18.5C Other Facts About Friedel–Crafts Alkylation

Three additional facts about Friedel–Crafts alkylations must be kept in mind.

[1] Vinyl halides and aryl halides do *not* react in Friedel–Crafts alkylation.

Most Friedel–Crafts reactions involve carbocation electrophiles. Because the carbocations derived from vinyl halides and aryl halides are highly unstable and do not readily form, these organic halides do *not* undergo Friedel–Crafts alkylation.

vinyl halide unreactive

aryl halide unreactive

Problem 18.7 Which halides are unreactive in a Friedel–Crafts alkylation reaction?

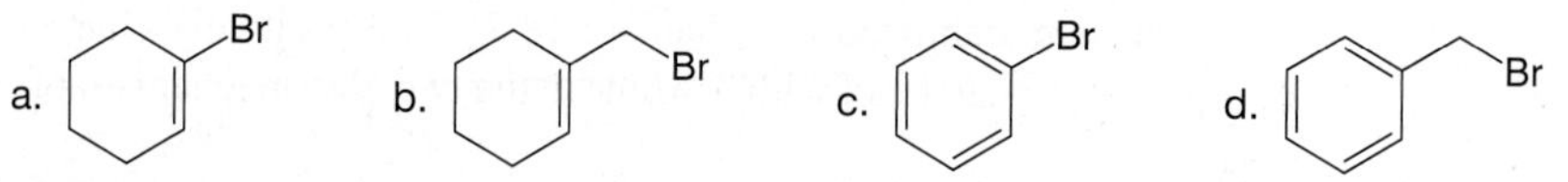
a.
Br
b.
Br
c.
Br
d.
Br

[2] Rearrangements can occur.

The Friedel–Crafts reaction can yield products having rearranged carbon skeletons when 1° and 2° alkyl halides are used as starting materials, as shown in Equations [1] and [2]. In both reactions, the carbon atom bonded to the halogen in the starting material (labeled in blue) is not bonded to the benzene ring in the product, thus indicating that a rearrangement has occurred.

Recall from Section 9.9 that a 1,2-shift converts a less stable carbocation to a more stable carbocation by shift of a hydrogen atom or an alkyl group.

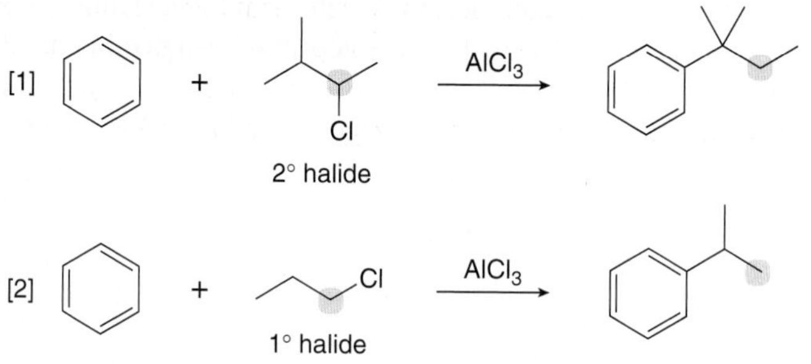
[1]
+
Cl
2° halide
$AlCl_3$
[2]
+
Cl
1° halide
$AlCl_3$

The result in Equation [1] is explained by a carbocation rearrangement involving a 1,2-hydride shift: **the less stable 2° carbocation (formed from the 2° halide) rearranges to a more stable 3° carbocation,** as illustrated in Mechanism 18.8.

Mechanism 18.8 Friedel–Crafts Alkylation Involving Carbocation Rearrangement

Part [1] Formation of a 2° carbocation and rearrangement

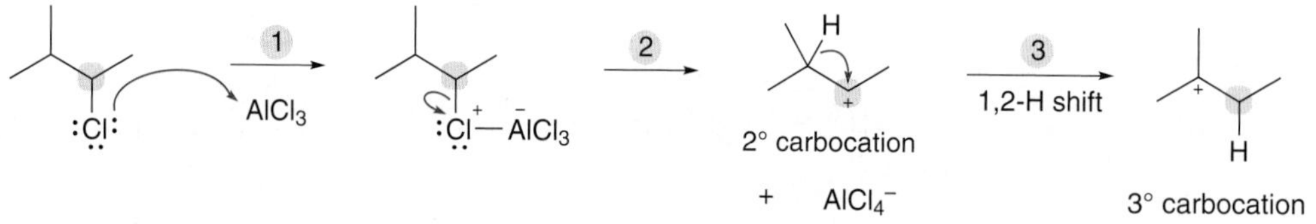
:Cl:
1
$AlCl_3$
:Cl—$AlCl_3$
2
H
2° carbocation
+ $AlCl_4^-$
3
1,2-H shift
H
3° carbocation

1 – 2 Lewis acid–base reaction of the alkyl chloride with $AlCl_3$ and cleavage of the C–Cl bond form a 2° carbocation.

3 1,2-Hydride shift converts a 2° carbocation to a more stable 3° carbocation.

Part [2] Two-step mechanism for electrophilic aromatic substitution

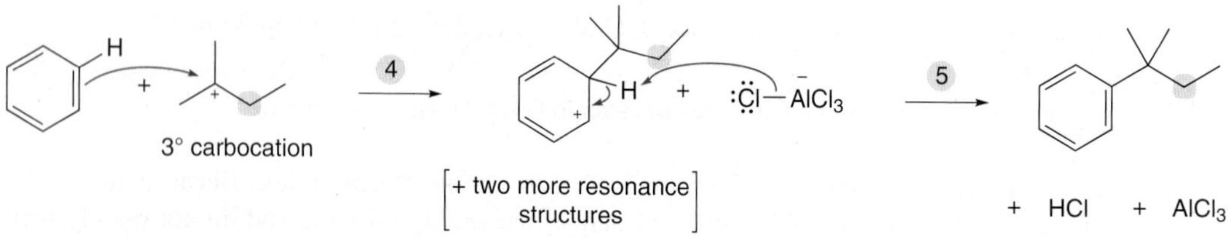
H
+
3° carbocation
4
H
+
:Cl—$AlCl_3$
+ two more resonance structures
5
+ HCl + $AlCl_3$

4 Addition of the 3° carbocation forms a new carbon–carbon bond and a resonance-stabilized carbocation.

5 $AlCl_4^-$ removes a proton on the carbon bearing the new substituent to re-form the aromatic ring.

Rearrangements can occur even when no free carbocation is formed initially. For example, the 1° alkyl chloride in Equation [2] forms a complex with $AlCl_3$, which does *not* decompose to an unstable 1° carbocation, as shown in Mechanism 18.9. Instead, a **1,2-hydride shift** forms a 2° carbocation, which then serves as the electrophile in the two-step mechanism for electrophilic aromatic substitution.

Mechanism 18.9 A Rearrangement Reaction Beginning with a 1° Alkyl Chloride

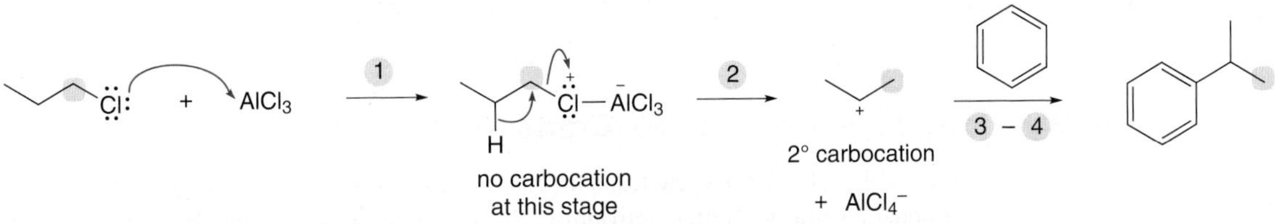

Problem 18.8 Draw a stepwise mechanism for the following reaction.

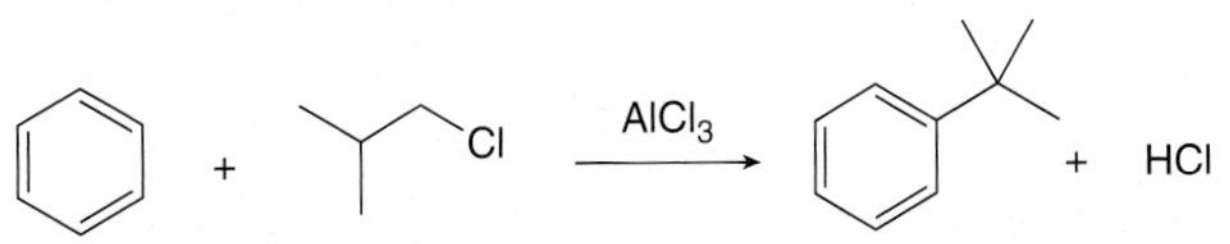

[3] Other functional groups that form carbocations can also be used as starting materials.

Although Friedel–Crafts alkylation works well with alkyl halides, any compound that readily forms a carbocation can be used instead. The two most common alternatives are alkenes and alcohols, both of which afford carbocations in the presence of strong acid.

- **Protonation of an alkene forms a carbocation, which can then serve as an electrophile in a Friedel–Crafts alkylation.**

- **Protonation of an alcohol, followed by loss of water, likewise forms a carbocation.**

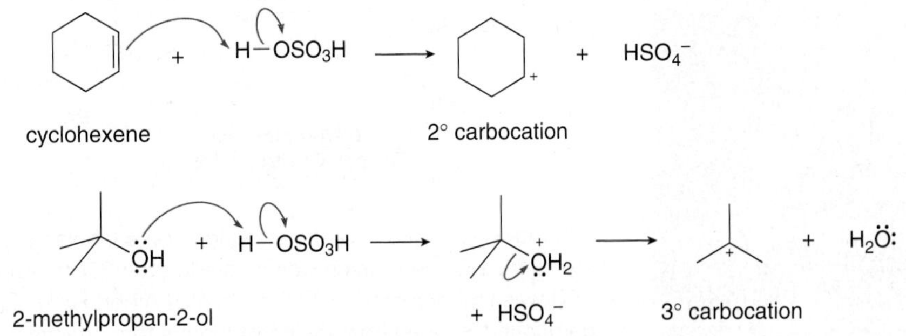

Each carbocation can then go on to react with benzene to form a product of electrophilic aromatic substitution. For example:

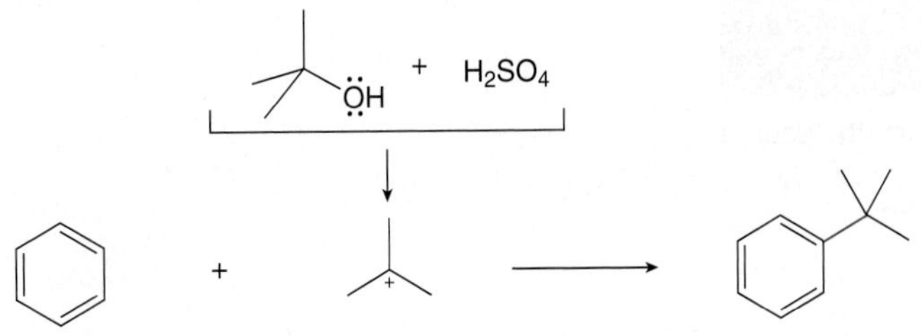

Problem 18.9 Draw the product of each reaction.

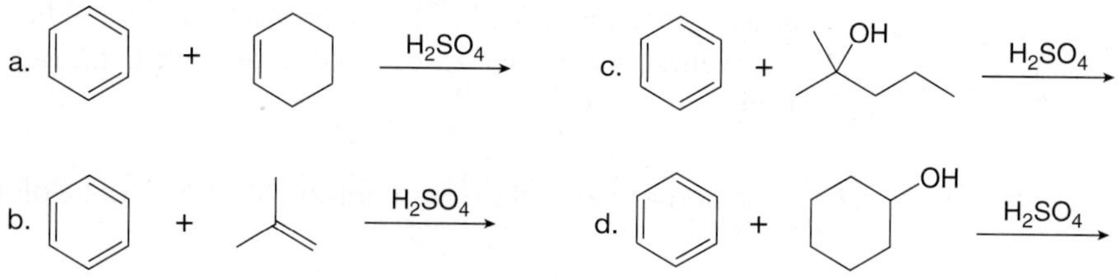

18.5D Intramolecular Friedel–Crafts Reactions

All of the Friedel–Crafts reactions discussed thus far have resulted from intermolecular reaction of a benzene ring with an electrophile. Starting materials that contain both units are capable of **intramolecular reaction,** and this forms a new ring. Treatment of compound **A,** which contains both a benzene ring and an acid chloride, with $AlCl_3$, forms α-tetralone by an intramolecular Friedel–Crafts acylation reaction.

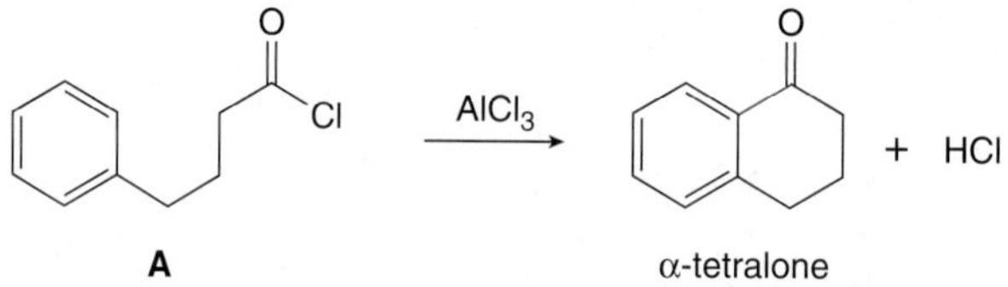

new C–C bond in red

Such an intramolecular Friedel–Crafts acylation was a key step in the synthesis of the hallucinogen LSD, as shown in Figure 18.4.

Figure 18.4 Intramolecular Friedel–Crafts acylation in the synthesis of LSD

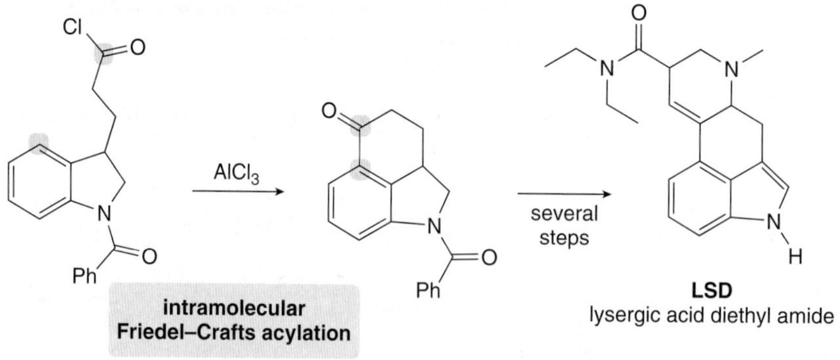

- Intramolecular Friedel–Crafts acylation at the labeled carbons formed a product containing a new six-membered ring, which was converted to LSD in several steps.
- LSD was first prepared by Swiss chemist Albert Hofmann in 1938 from a related organic compound isolated from the ergot fungus that attacks rye and other grains. Ergot has a long history as a dreaded poison, affecting individuals who become ill from eating ergot-contaminated bread. The hallucinogenic effects of LSD were first discovered when Hofmann accidentally absorbed a small amount of the drug through his fingertips.

Ergot-infected grain, the source of lysergic acid

Problem 18.10 Draw a stepwise mechanism for the intramolecular Friedel–Crafts acylation of compound **A** to form **B. B** can be converted in one step to the antidepressant sertraline.

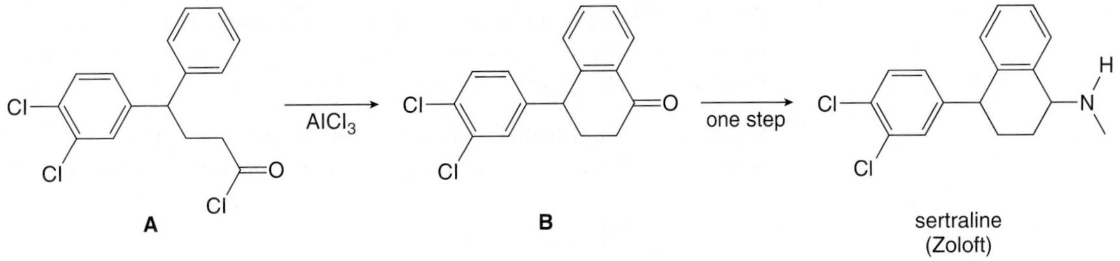

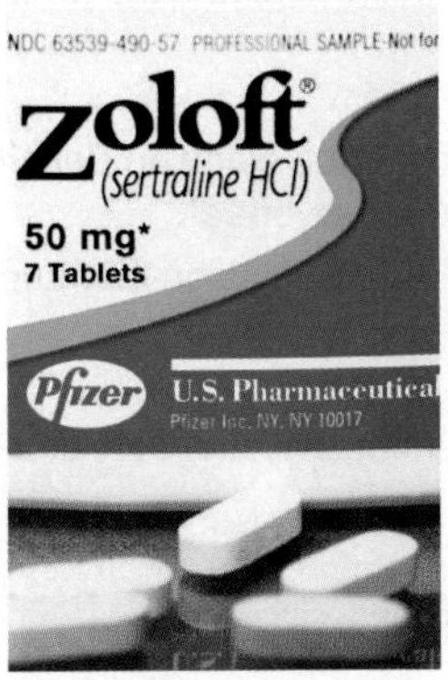

Sertraline (trade name Zoloft) is an effective antidepressant because it increases the concentration of the neurotransmitter serotonin in the brain.

Problem 18.11 Intramolecular reactions are also observed in Friedel–Crafts alkylation. Draw the intramolecular alkylation product formed from each of the following reactants. (Watch out for rearrangements!)

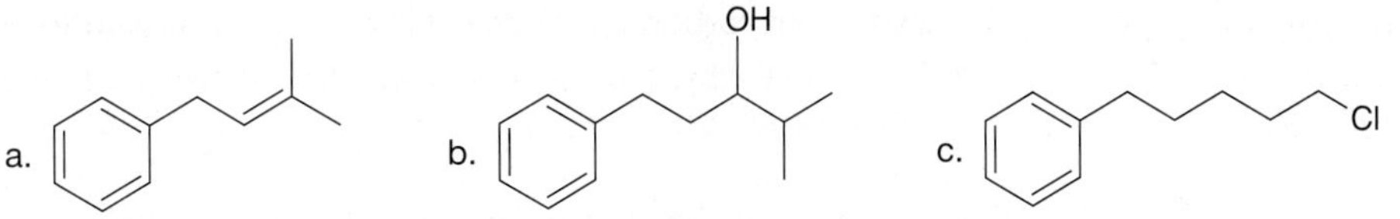

18.5E Biological Friedel–Crafts Reactions

Biological Friedel–Crafts reactions occur as well. As we learned in Section 16.2, allylic diphosphates contain a good leaving group, so they can serve as a source of allylic carbocations. A key step in the biological synthesis of vitamin K_1, the chapter-opening molecule, involves Friedel–Crafts reaction of 1,4-dihydroxynaphthoic acid with phytyl diphosphate to form **X,** which is converted to vitamin K_1 in several steps, as shown in Figure 18.5.

Figure 18.5 Friedel–Crafts reaction in the synthesis of vitamin K_1

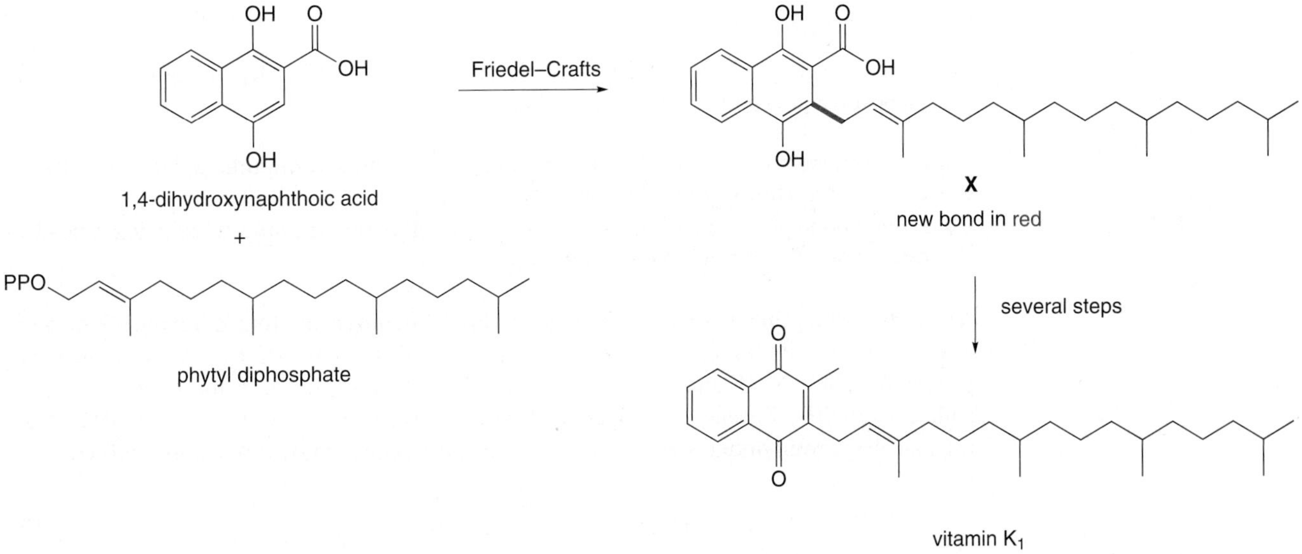

Problem 18.12 (a) Draw resonance structures for the carbocation formed after loss of a leaving group from phytyl diphosphate. (b) Draw the two-step mechanism for Friedel–Crafts alkylation of 1,2-dihydroxynaphthoic acid with this carbocation to form **X.**

18.6 Substituted Benzenes

Many substituted benzene rings undergo electrophilic aromatic substitution. Common substituents include halogens, OH, NH_2, alkyl, and many functional groups that contain a carbonyl. Each substituent either increases or decreases the electron density in the benzene ring, and this affects the course of electrophilic aromatic substitution, as we will learn in Section 18.7.

What makes a substituent on a benzene ring electron donating or electron withdrawing? The answer is **inductive effects** and **resonance effects,** both of which can add or remove electron density.

Inductive Effects

Inductive effects stem from the **electronegativity** of the atoms in the substituent and the **polarizability** of the substituent group.

Inductive and resonance effects were first discussed in Sections 2.5B and 2.5C, respectively.

- **Atoms more electronegative than carbon—including N, O, and X—pull electron density away from carbon and thus exhibit an electron-withdrawing inductive effect.**
- **Polarizable alkyl groups donate electron density, and thus exhibit an electron-donating inductive effect.**

Considering inductive effects *only,* an NH_2 group withdraws electron density and CH_3 donates electron density.

Electron-withdrawing inductive effect

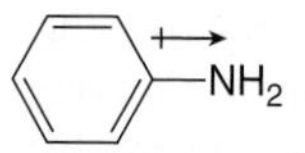

- N is **more electronegative** than C.
- N inductively withdraws electron density.

Electron-donating inductive effect

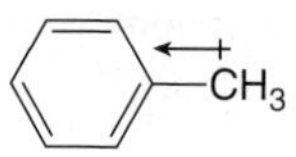

- Alkyl groups are **polarizable,** making them electron-donating groups.

Problem 18.13 Which substituents have an electron-withdrawing and which have an electron-donating inductive effect: (a) $CH_3CH_2CH_2CH_2-$; (b) $Br-$; (c) CH_3CH_2O-?

Resonance Effects

Resonance effects can either donate or withdraw electron density, depending on whether they place a positive or negative charge on the benzene ring.

- **A resonance effect is electron donating when resonance structures place a negative charge on carbons of the benzene ring.**
- **A resonance effect is electron withdrawing when resonance structures place a positive charge on carbons of the benzene ring.**

An electron-donating resonance effect is observed whenever an atom Z having a lone pair of electrons is directly bonded to a benzene ring (general structure—$\mathbf{C_6H_5-Z:}$). Common examples of Z include N, O, and halogen. For example, five resonance structures can be drawn for aniline ($C_6H_5NH_2$). Because three of them place a negative charge on a carbon atom of the benzene ring, an **NH_2 group donates electron density to a benzene ring by a resonance effect.**

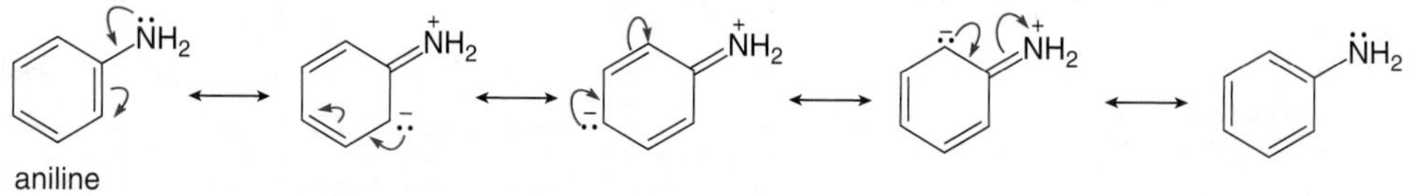

aniline

Three resonance structures place a (–) charge on atoms in the ring.

In contrast, **an electron-withdrawing resonance effect is observed in substituted benzenes having the general structure $\mathbf{C_6H_5-Y{=}Z}$,** where Z is more electronegative than Y. For example, seven resonance structures can be drawn for benzaldehyde (C_6H_5CHO). Because three of them place a positive charge on a carbon atom of the benzene ring, a CHO group withdraws electron density from a benzene ring by a resonance effect.

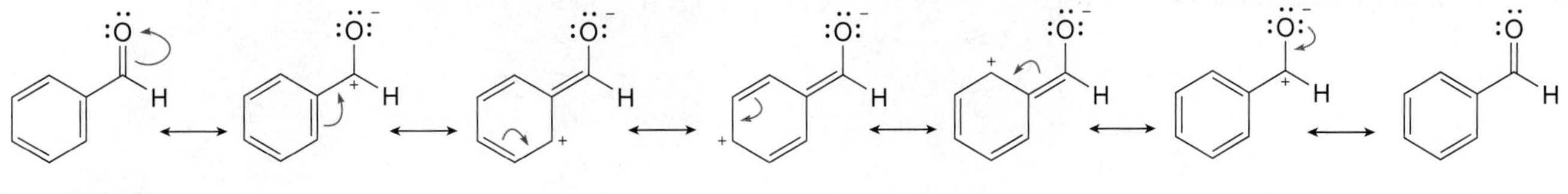

benzaldehyde

Three resonance structures place a (+) charge on atoms in the ring.

Problem 18.14 Draw all resonance structures for each compound and use the resonance structures to determine if the substituent has an electron-donating or electron-withdrawing resonance effect.

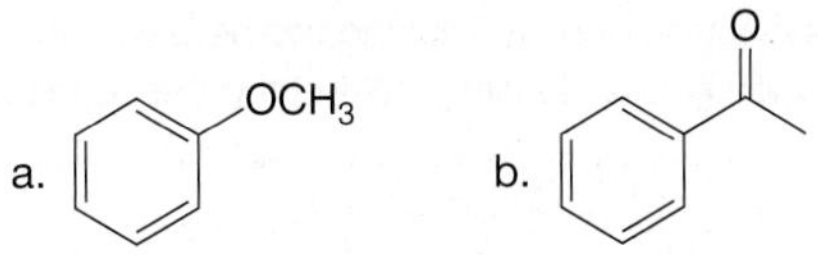

Considering Both Inductive and Resonance Effects

To predict whether a substituted benzene is more or less electron rich than benzene itself, we must consider the **net balance of *both* the inductive and the resonance effects.** Alkyl groups, for instance, donate electrons by an inductive effect, but they have no resonance effect because they lack nonbonded electron pairs or π bonds. As a result,

- **An alkyl group is an electron-donating group and an alkyl benzene is more electron rich than benzene.**

When electronegative atoms, such as N, O, or halogen, are bonded to the benzene ring, they inductively *withdraw* electron density from the ring. All of these groups also have a nonbonded pair of electrons, so they *donate* electron density to the ring by resonance. **The *identity of the element* determines the net balance of these opposing effects.**

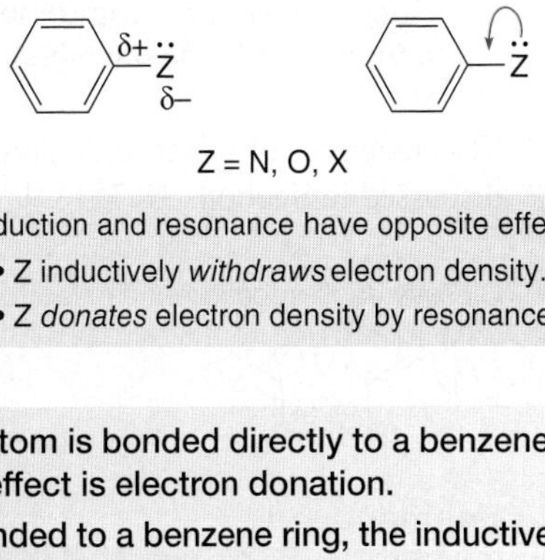

Z = N, O, X

Induction and resonance have opposite effects.
- Z inductively *withdraws* electron density.
- Z *donates* electron density by resonance.

- **When a neutral O or N atom is bonded directly to a benzene ring, the resonance effect dominates and the net effect is electron donation.**
- **When a halogen X is bonded to a benzene ring, the inductive effect dominates and the net effect is electron withdrawal.**

Thus, **NH_2 and OH are electron-donating groups** because the resonance effect predominates, whereas **Cl and Br are electron-withdrawing groups** because the inductive effect predominates.

Finally, the inductive and resonance effects in compounds having the general structure **$C_6H_5-Y=Z$** (with Z more electronegative than Y) are **both electron withdrawing;** in other words, the two effects *reinforce* each other. This is true for benzaldehyde (C_6H_5CHO) and all other compounds that contain a carbonyl group bonded directly to the benzene ring.

Thus, on balance, an **NH_2 group is electron donating,** so the benzene ring of aniline ($C_6H_5NH_2$) has more electron density than benzene. An **aldehyde group (CHO), on the other hand, is**

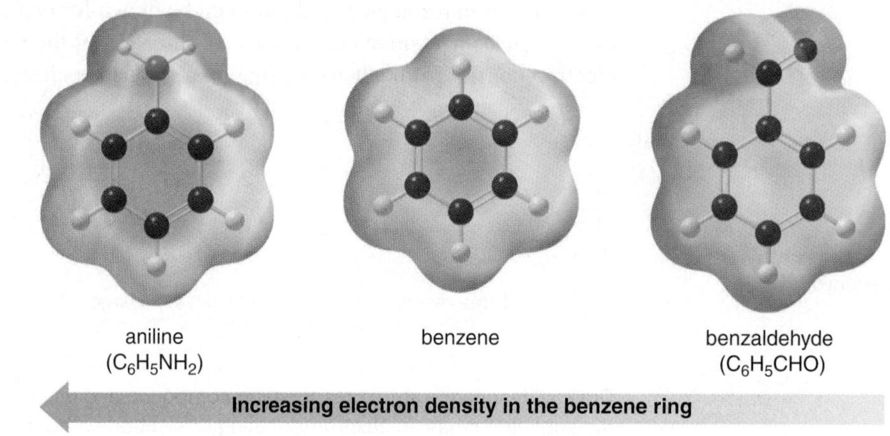
aniline
($C_6H_5NH_2$)
benzene
benzaldehyde
(C_6H_5CHO)
Increasing electron density in the benzene ring

Figure 18.6
The effect of substituents on the electron density in substituted benzenes

- The NH_2 group donates electron density, making the benzene ring more electron rich (redder), whereas the CHO group withdraws electron density, making the benzene ring less electron rich (greener).

electron withdrawing, so the benzene ring of benzaldehyde (C_6H_5CHO) has less electron density than benzene. These effects are illustrated in the electrostatic potential maps in Figure 18.6. These compounds represent examples of the general structural features in electron-donating and electron-withdrawing substituents:

Electron-donating groups **Electron-withdrawing groups**

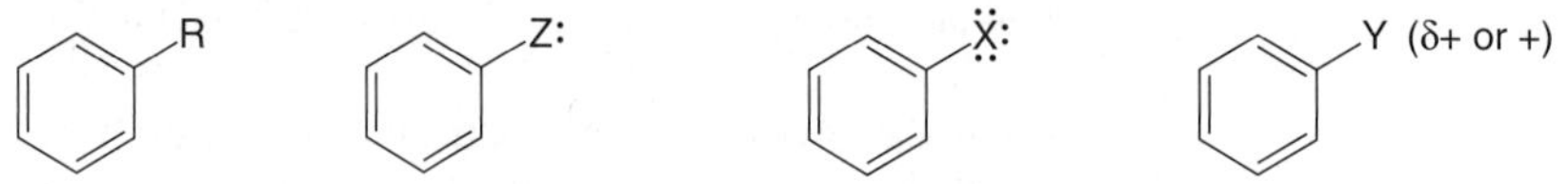
R
Z:
X:
Y (δ+ or +)

R = alkyl Z = N or O X = halogen

- **Common electron-donating groups are alkyl groups or groups with an N or O atom (with a lone pair) bonded to the benzene ring.**

- **Common electron-withdrawing groups are halogens or groups with an atom Y bearing a full or partial positive charge (+ or δ+) bonded to the benzene ring.**

The net effect of electron donation and withdrawal on the reactions of substituted aromatics is discussed in Sections 18.7–18.9.

Sample Problem 18.3

Classify each substituent as electron donating or electron withdrawing.

a. 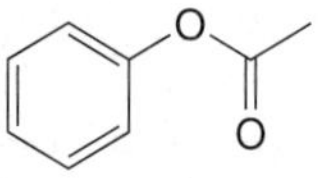 b. 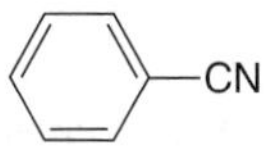
CN

Solution

If necessary, draw out the atoms and bonds of the substituent to clearly see lone pairs and multiple bonds. **Always look at the atom bonded directly to the benzene ring** to determine electron-donating or electron-withdrawing effects. An O or N atom with a lone pair of electrons makes a substituent electron donating. A halogen or an atom with a partial positive charge makes a substituent electron withdrawing.

a.

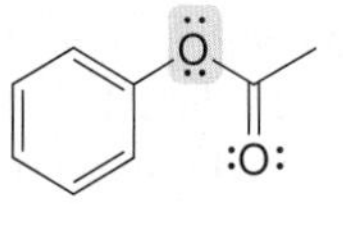
:O:

- An O atom with a lone pair is bonded directly to the benzene ring.

an electron-donating group

b.

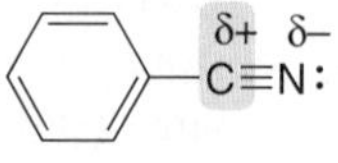
δ+ δ–
C≡N:

- An atom with a partial (+) charge is bonded directly to the benzene ring.

an electron-withdrawing group

Problem 18.15 Classify each substituent as electron donating or electron withdrawing.

a. 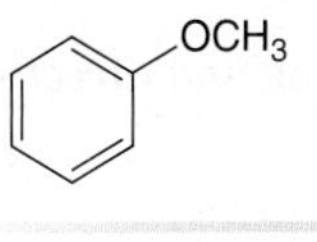

b. 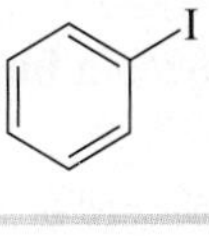

c. 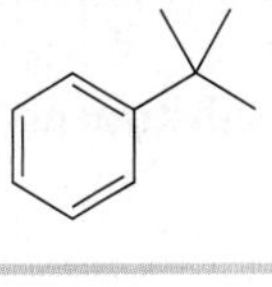

18.7 Electrophilic Aromatic Substitution of Substituted Benzenes

Electrophilic aromatic substitution is a general reaction of *all* aromatic compounds, including polycyclic aromatic hydrocarbons, heterocycles, and substituted benzene derivatives. A substituent affects two aspects of electrophilic aromatic substitution:

- **The rate of reaction:** A substituted benzene reacts faster or slower than benzene itself.
- **The orientation:** The new group is located either ortho, meta, or para to the existing substituent. The identity of the first substituent determines the position of the second substituent.

Toluene ($C_6H_5CH_3$) and nitrobenzene ($C_6H_5NO_2$) illustrate two possible outcomes.

[1] Toluene

Toluene reacts **faster** than benzene in all substitution reactions. Thus, its **electron-donating CH_3 group *activates* the benzene ring** to electrophilic attack. Although three products are possible, compounds with the new group ortho or para to the CH_3 group predominate. The CH_3 group is therefore called an **ortho, para director.**

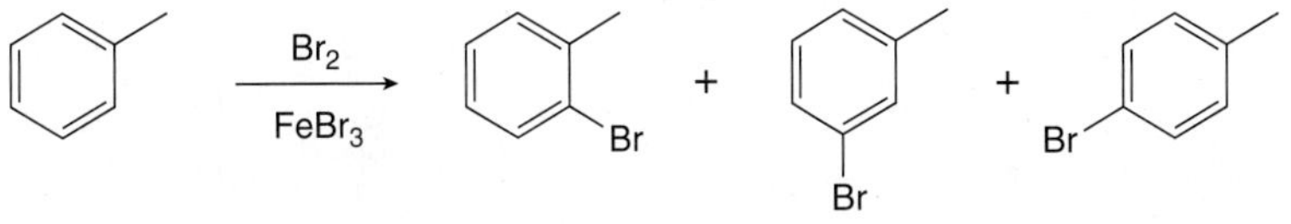

ortho	meta	para
40%	trace	60%

[2] Nitrobenzene

Nitrobenzene reacts **more slowly** than benzene in all substitution reactions. Thus, its **electron-withdrawing NO_2 group *deactivates* the benzene ring** to electrophilic attack. Although three products are possible, the compound with the new group meta to the NO_2 group predominates. The NO_2 group is called a **meta director.**

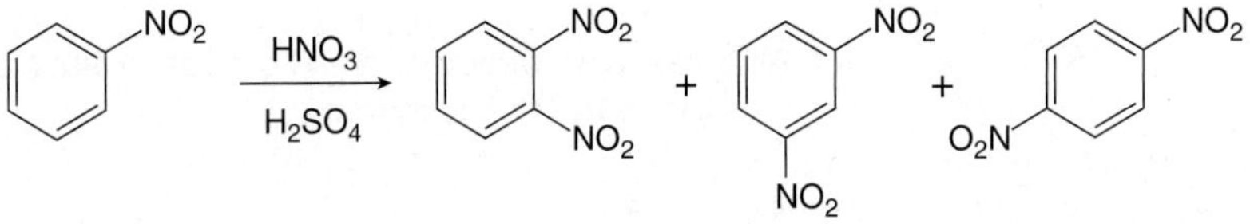

ortho	meta	para
7%	93%	trace

Substituents either activate or deactivate a benzene ring toward electrophiles, and direct selective substitution at specific sites on the ring. **All substituents can be divided into three general types.**

[1] Ortho, para directors and activators

- **Substituents that *activate* a benzene ring and direct substitution ortho and para.**

Increasing activation (↑):

- $-NH_2$, $-NHR$, $-NR_2$
- $-OH$
- $-OR$
- $-NHCOR$
- $-R$

General structure
−R or **−Z:**

[2] Ortho, para deactivators

- **Substituents that *deactivate* a benzene ring and direct substitution ortho and para.**

−F: −Cl: −Br: −I:

[3] Meta directors

- **Substituents that direct substitution meta.**
- **All meta directors *deactivate* the ring.**

Increasing deactivation (↓):

- $-CHO$
- $-COR$
- $-CO_2R$
- $-CO_2H$
- $-CN$
- $-SO_3H$
- $-NO_2$
- $-\overset{+}{N}R_3$

General structure
−Y (δ+ or +)

To learn these lists: **Keep in mind that the halogens are in a class by themselves.** Then learn the general structures for each type of substituent.

- All ortho, para directors are R groups or have a nonbonded electron pair on the atom bonded to the benzene ring.

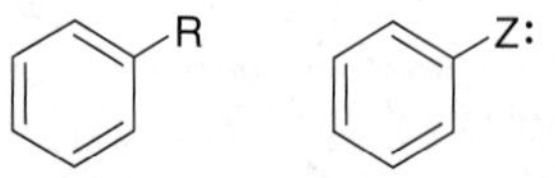

Z = N or O --→ The ring is **activated.**
Z = halogen --→ The ring is **deactivated.**

- All meta directors have a full or partial positive charge on the atom bonded to the benzene ring.

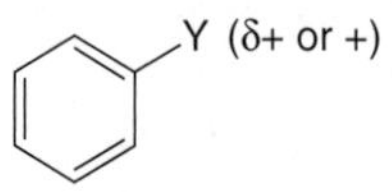

Sample Problem 18.4 shows how this information can be used to predict the products of electrophilic aromatic substitution reactions.

Sample Problem 18.4 Draw the products of each reaction and state whether the reaction is faster or slower than a similar reaction with benzene.

a. 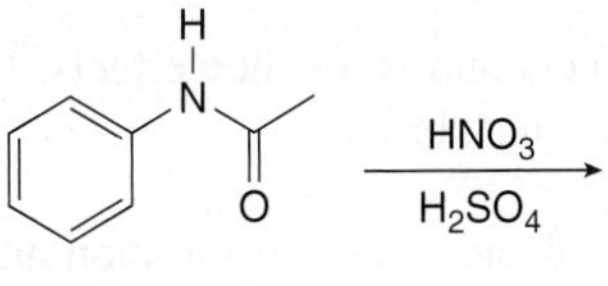
H
N
O
HNO_3
H_2SO_4

b. 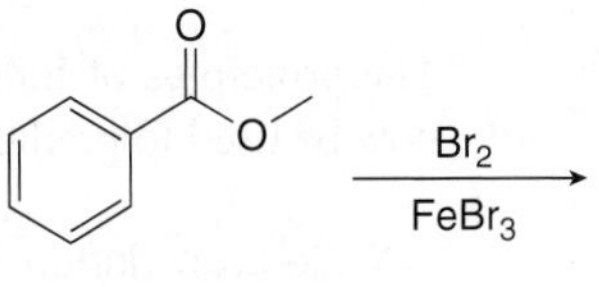
O
O
Br_2
$FeBr_3$

Solution

To draw the products:

- Draw the Lewis structure for the substituent to see if it has a lone pair or partial positive charge on the atom bonded to the benzene ring.
- Classify the substituent—ortho, para activating; ortho, para deactivating; or meta deactivating—and draw the products.

a. 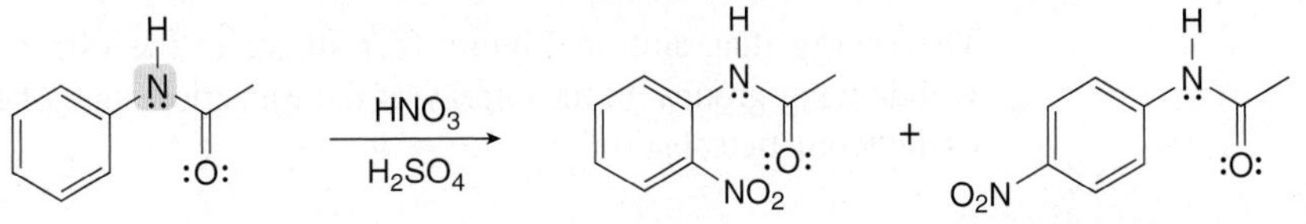
H
N
:O:
HNO_3
H_2SO_4
+
NO_2
O_2N

ortho para

The lone pair on N makes this group an **ortho, para activator. This compound reacts faster than benzene.**

b. 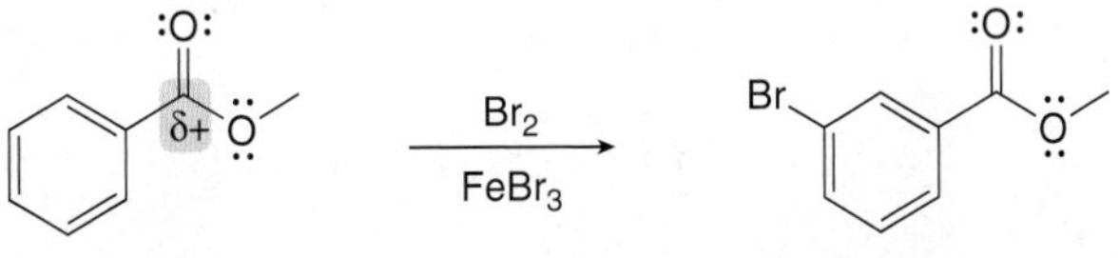
:O:
δ+
O
Br_2
$FeBr_3$
Br

meta

The δ+ on the C bonded to the benzene ring makes the group a **meta deactivator. This compound reacts more slowly than benzene.**

Problem 18.16 Draw the products formed when each compound is treated with HNO_3 and H_2SO_4. State whether the reaction occurs faster or slower than a similar reaction with benzene.

a. 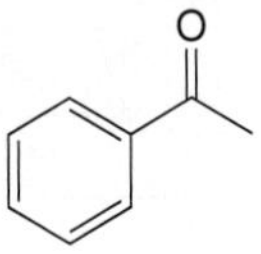
O

b. 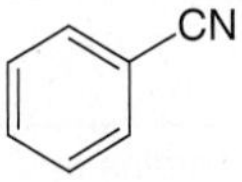
CN

c. 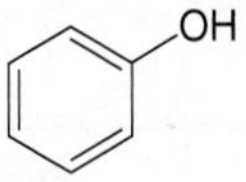
OH

d. 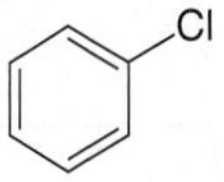
Cl

e. 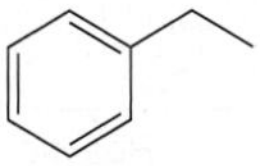

18.8 Why Substituents Activate or Deactivate a Benzene Ring

- **Why do substituents activate or deactivate a benzene ring?**
- **Why are particular orientation effects observed?** Why are some groups ortho, para directors and some groups meta directors?

To understand why some substituents make a benzene ring react *faster* than benzene itself (activators), whereas others make it react *slower* (deactivators), we must evaluate the rate-determining step (the first step) of the mechanism. Recall from Section 18.2 that the first step in electrophilic aromatic substitution is the addition of an electrophile (E^+) to form a resonance-stabilized carbocation. The Hammond postulate (Section 7.14) makes it possible to predict the relative rate of the reaction by looking at the stability of the carbocation intermediate.

- **The more stable the carbocation, the lower in energy the transition state that forms it, and the faster the reaction.**

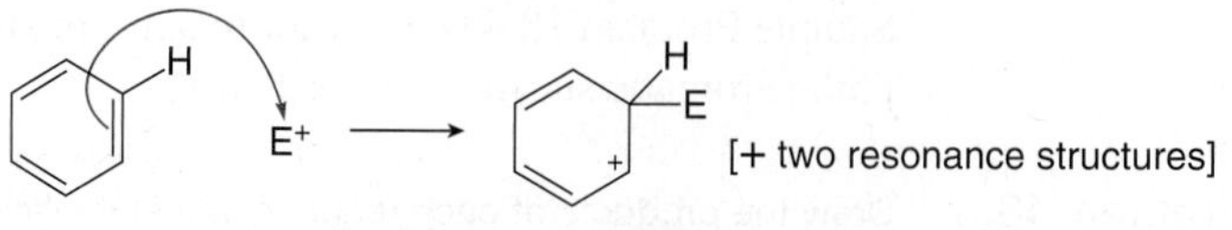

Stabilizing the carbocation makes the reaction faster.

The principles of inductive effects and resonance effects, first introduced in Section 18.6, can now be used to predict carbocation stability.

- **Electron-donating groups stabilize the carbocation and *activate* a benzene ring toward electrophilic attack. All activators are R groups or they have an N or O atom with a lone pair bonded directly to the benzene ring.**

- **Electron-withdrawing groups destabilize the carbocation and *deactivate* a benzene ring toward electrophilic attack. Either all deactivators are halogens, or they have an atom with a full or partial positive charge bonded directly to the benzene ring.**

The energy diagrams in Figure 18.7 illustrate the effect of electron-donating and electron-withdrawing groups on the energy of the transition state of the rate-determining step in electrophilic aromatic substitution.

Figure 18.7 Energy diagrams comparing the rate of electrophilic aromatic substitution of substituted benzenes

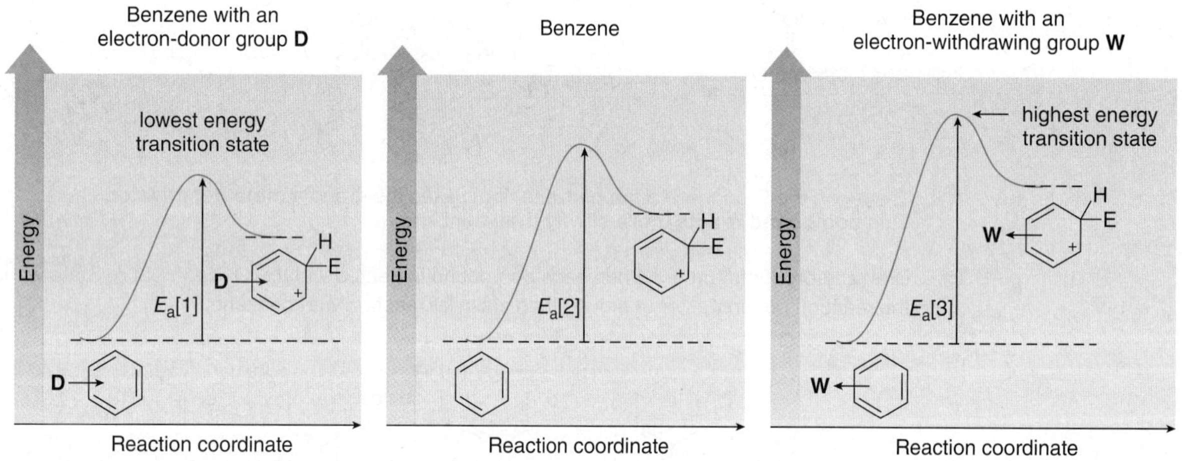

- Electron-donor groups **D** stabilize the carbocation intermediate, lower the energy of the transition state, and increase the rate of reaction.
- Electron-withdrawing groups **W** destabilize the carbocation intermediate, raise the energy of the transition state, and decrease the rate of reaction.

Problem 18.17 Label each compound as more or less reactive than benzene in electrophilic aromatic substitution.

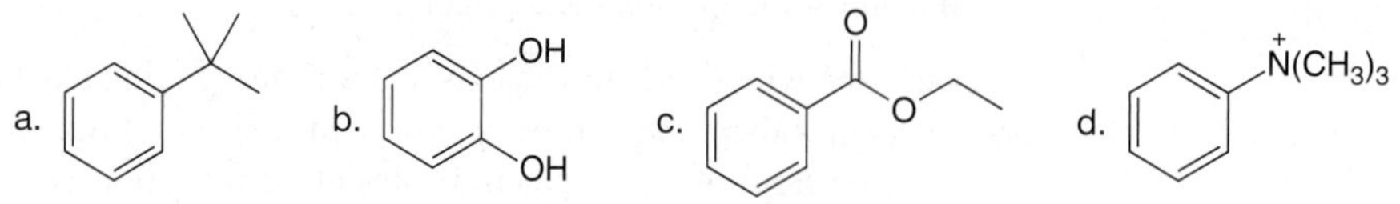

Problem 18.18 Rank the following compounds in order of increasing reactivity in electrophilic aromatic substitution.

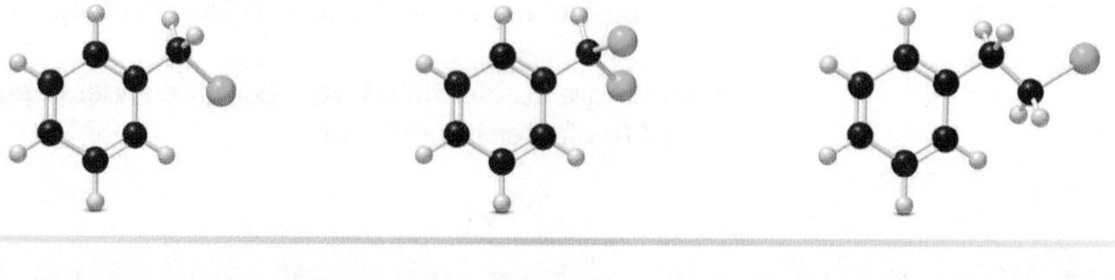

18.9 Orientation Effects in Substituted Benzenes

To understand why particular orientation effects arise, you must keep in mind the general structures for ortho, para directors and for meta directors already given in Section 18.7. There are two general types of ortho, para directors and one general type of meta director:

- **All ortho, para directors are R groups or have a nonbonded electron pair on the atom bonded to the benzene ring.**

- **All meta directors have a full or partial positive charge on the atom bonded to the benzene ring.**

To evaluate the directing effects of a given substituent, we can follow a stepwise procedure.

How To Determine the Directing Effects of a Particular Substituent

Step [1] **Draw all resonance structures for the carbocation formed from attack of an electrophile E^+ at the ortho, meta, and para positions of a substituted benzene (C_6H_5-A).**

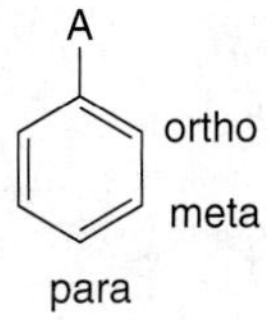

- There are at least three resonance structures for each site of reaction.
- Each resonance structure places a positive charge ortho or para to the new C–E bond.

Step [2] **Evaluate the stability of the intermediate resonance structures. The electrophile attacks at those positions that give the most stable carbocation.**

Sections 18.9A–C show how this two-step procedure can be used to determine the directing effects of the CH_3 group in toluene, the NH_2 group in aniline, and the NO_2 group in nitrobenzene, respectively.

18.9A The CH_3 Group—An ortho, para Director

To determine why a **CH_3 group directs electrophilic aromatic substitution to the ortho and para positions,** first draw all resonance structures that result from electrophilic attack at the ortho, meta, and para positions to the CH_3 group.

Always draw in the H atom at the site of electrophilic attack. This will help you keep track of where the charges go.

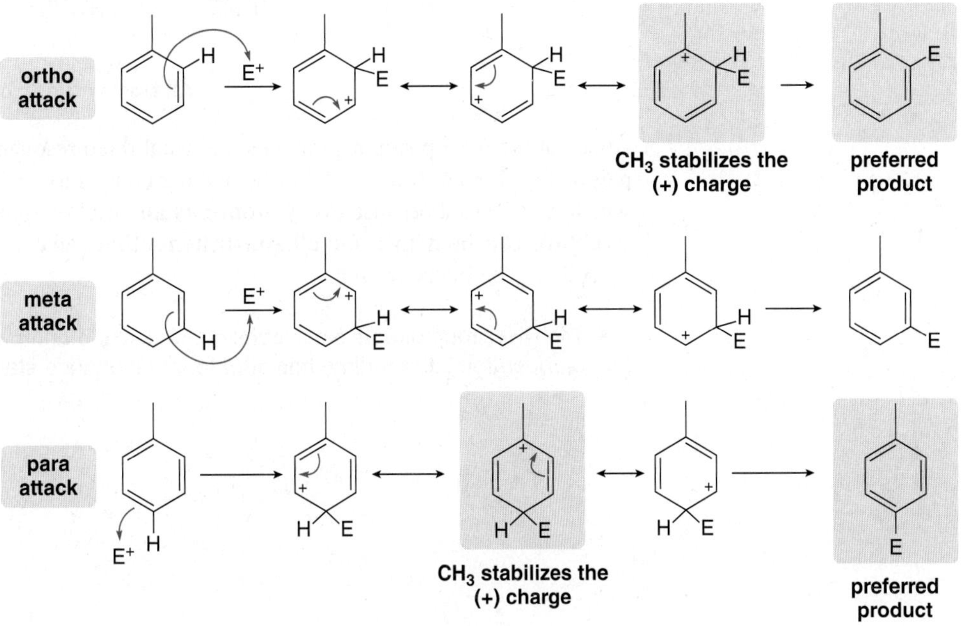

The positive charge in all resonance structures is always ortho or para to the new C–E bond. It is *not* necessarily ortho or para to the CH_3 group.

To evaluate the stability of the resonance structures, determine whether any are especially stable or unstable. In this example, **attack ortho or para to CH_3 generates a resonance structure that places a positive charge on a carbon atom with the CH_3 group.** The electron-donating CH_3 group *stabilizes* the adjacent positive charge. In contrast, attack meta to the CH_3 group does *not* generate any resonance structure stabilized by electron donation. Other alkyl groups are ortho, para directors for the same reason.

- The CH_3 group directs electrophilic attack ortho and para to itself because an electron-donating inductive effect stabilizes the carbocation intermediate.

18.9B The NH_2 Group—An ortho, para Director

To determine why an **amino group (NH_2) directs electrophilic aromatic substitution to the ortho and para positions,** follow the same procedure.

ortho attack

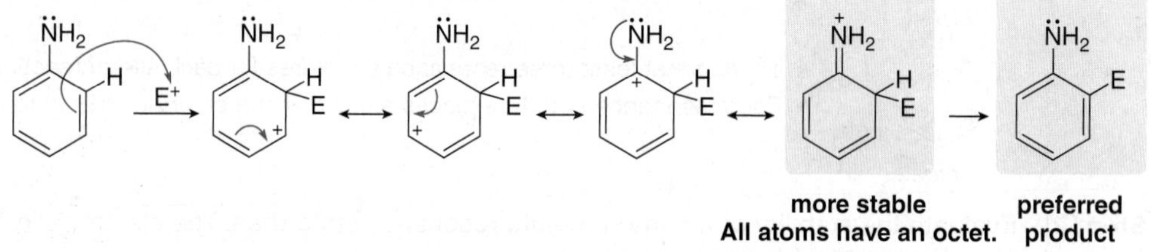

meta attack

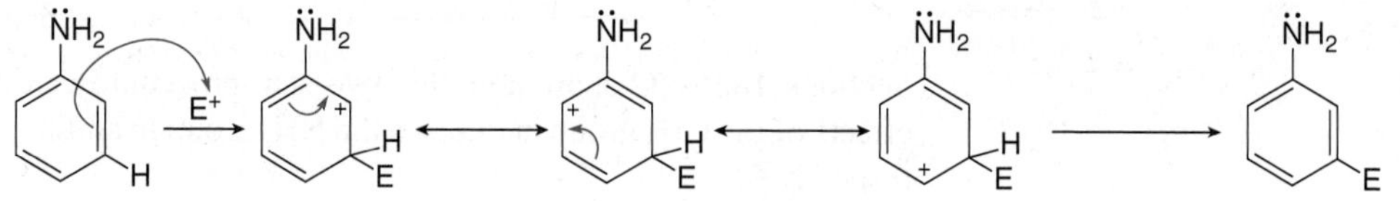

para attack

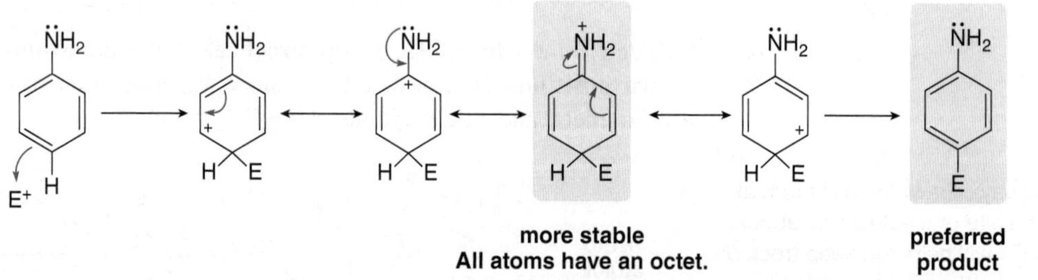

Attack at the meta position generates the usual three resonance structures. Because of the lone pair on the N atom, attack at the ortho and para positions generates a fourth resonance structure, which is stabilized because **every atom has an octet of electrons. This additional resonance structure can be drawn for all substituents that have an N, O, or halogen atom bonded directly to the benzene ring.**

- The NH_2 group directs electrophilic attack ortho and para to itself because the carbocation intermediate has additional resonance stabilization.

18.9C The NO_2 Group—A meta Director

To determine why a **nitro group (NO_2) directs electrophilic aromatic substitution to the meta position,** follow the same procedure.

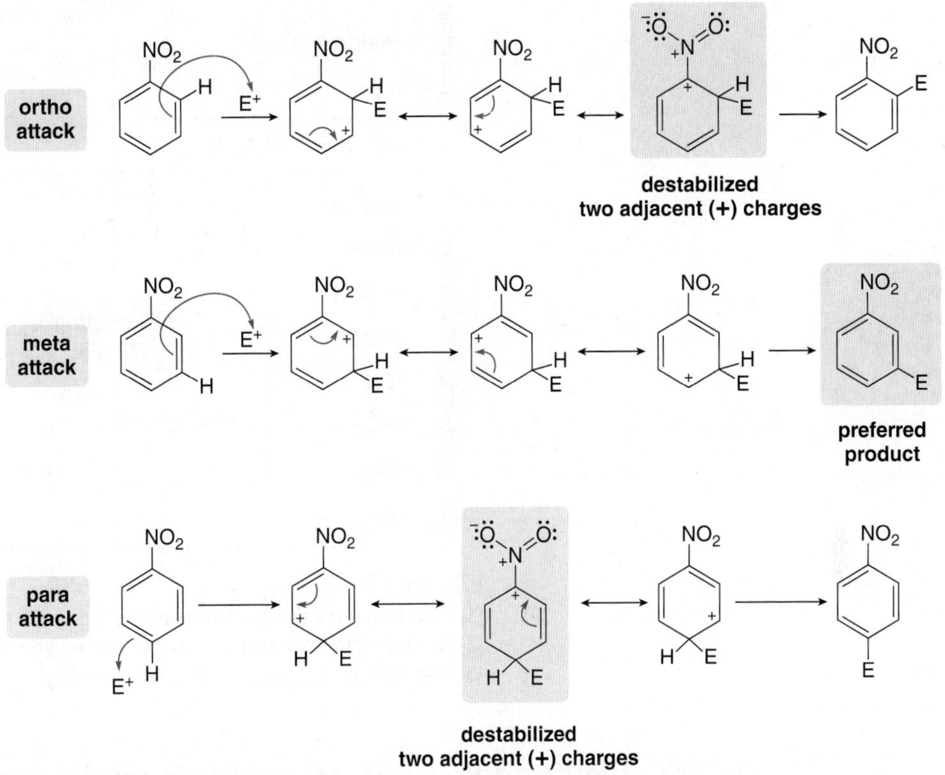

Attack at each position generates three resonance structures. One resonance structure resulting from attack at the ortho and para positions is especially *destabilized,* because it contains a positive charge on two adjacent atoms. Attack at the meta position does not generate any particularly unstable resonance structures.

- **With the NO_2 group (and all meta directors), meta attack occurs because attack at the ortho or para position gives a destabilized carbocation intermediate.**

Problem 18.19 Draw all resonance structures for the carbocation formed by ortho attack of the electrophile $^+NO_2$ on each starting material. Label any resonance structures that are especially stable or unstable.

a. 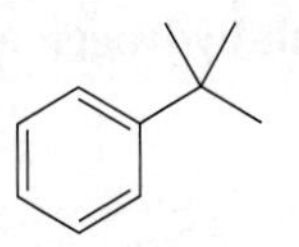

b. 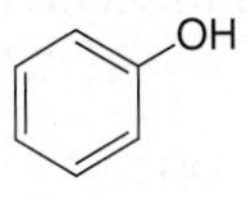

c. 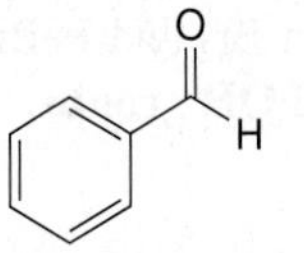

Figure 18.8 summarizes the reactivity and directing effects of the common substituents on benzene rings.

Figure 18.8

The reactivity and directing effects of common substituted benzenes

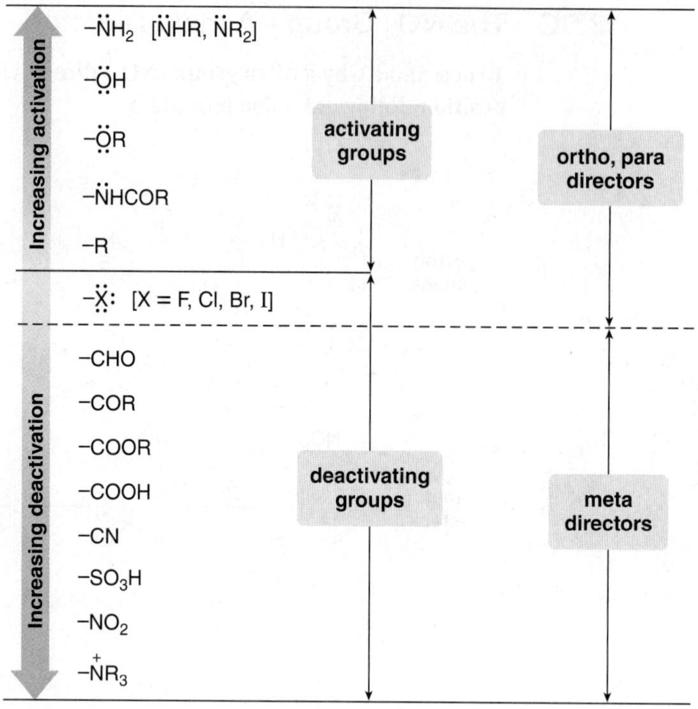

In summary:

[1] All ortho, para directors except the halogens activate the benzene ring.

[2] All meta directors deactivate the benzene ring.

[3] The halogens deactivate the benzene ring.

18.10 Limitations on Electrophilic Substitution Reactions with Substituted Benzenes

Although electrophilic aromatic substitution works well with most substituted benzenes, halogenation and the Friedel–Crafts reactions have some additional limitations that must be kept in mind.

18.10A Halogenation of Activated Benzenes

Considering all electrophilic aromatic substitution reactions, halogenation occurs the most readily. As a result, benzene rings activated by strong electron-donating groups—OH, NH_2, and their alkyl derivatives (OR, NHR, and NR_2)—undergo **polyhalogenation** when treated with X_2 and FeX_3. Aniline ($C_6H_5NH_2$) and phenol (C_6H_5OH) both give a tribromo derivative when treated with Br_2 and $FeBr_3$. **Substitution occurs at all hydrogen atoms ortho and para to the NH_2 and OH groups.**

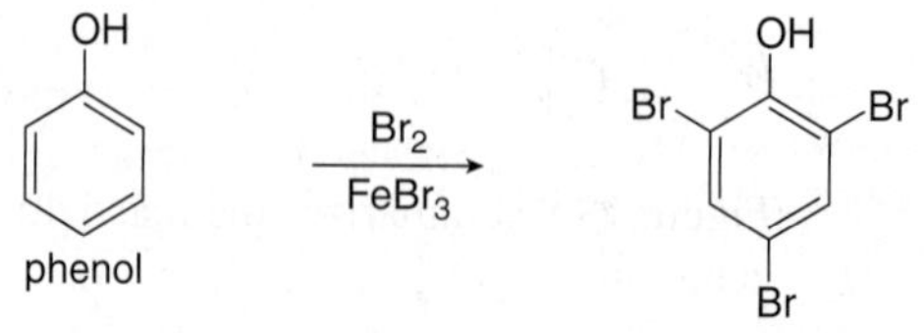

Monosubstitution of H by Br occurs with Br_2 *alone* without added catalyst to form a mixture of ortho and para products.

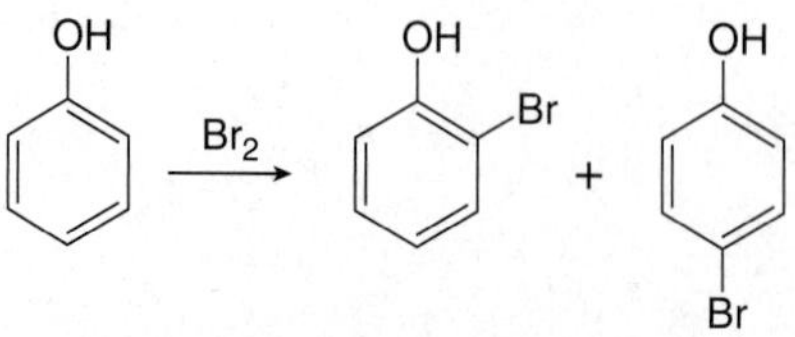

Problem 18.20 Draw the products of each reaction.

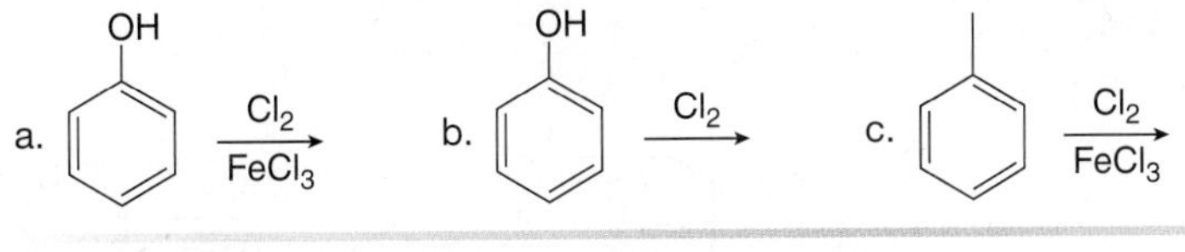

a.
OH
Cl2
FeCl3
b.
OH
Cl2
c.
Cl2
FeCl3

18.10B Limitations in Friedel–Crafts Reactions

Friedel–Crafts reactions are the most difficult electrophilic aromatic substitution reactions to carry out in the laboratory. They do not occur when the benzene ring is substituted with NO_2 (a strong deactivator) or with NH_2, NHR, or NR_2 (strong activators).

A benzene ring deactivated by a strong electron-withdrawing group—that is, any of the meta directors—is not electron rich enough to undergo Friedel–Crafts reactions.

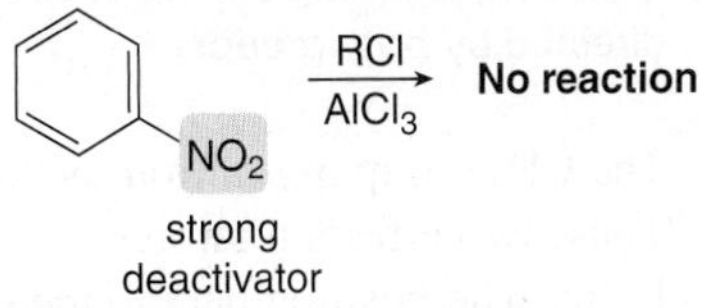

RCl
AlCl3
No reaction
NO2
strong
deactivator

Friedel–Crafts reactions also do not occur with NH_2 groups, which are strong activating groups. NH_2 groups are strong Lewis bases (due to the nonbonded electron pair on N), so they react with $AlCl_3$, the Lewis acid needed for alkylation or acylation. The resulting product contains a positive charge adjacent to the benzene ring, so the ring is now strongly deactivated and therefore unreactive in Friedel–Crafts reactions.

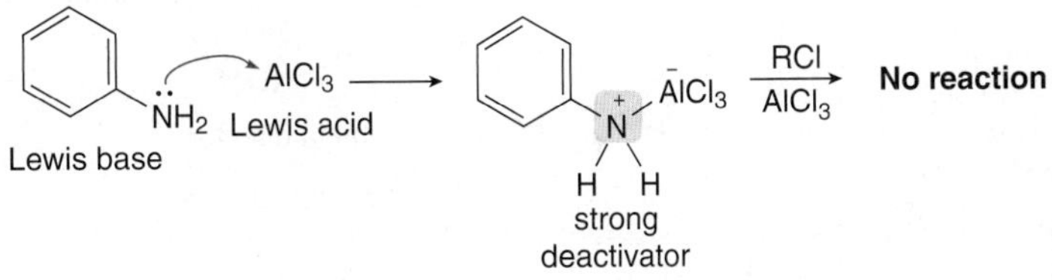

AlCl3
NH2
Lewis acid
Lewis base
AlCl3
N
H H
strong
deactivator
RCl
AlCl3
No reaction

Problem 18.21 Which of the following compounds undergo Friedel–Crafts alkylation with CH_3Cl and $AlCl_3$? Draw the products formed when a reaction occurs.

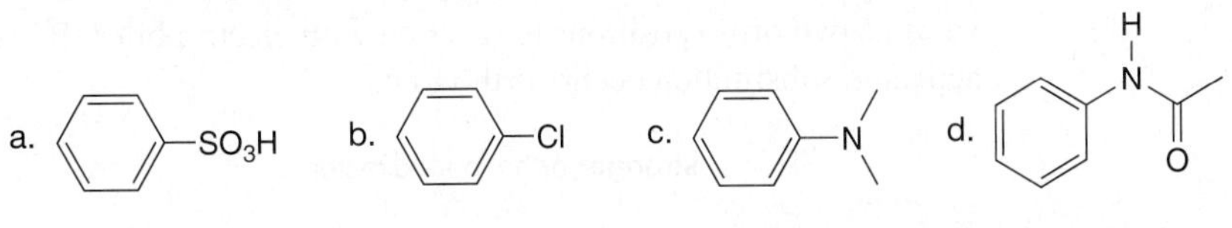

a.
SO3H
b.
Cl
c.
N
d.
H
N
O

Another limitation of the Friedel–Crafts alkylation arises because of **polyalkylation.** Treatment of benzene with an alkyl halide and $AlCl_3$ places an electron-donor R group on the ring. Because R groups activate a ring, the alkylated product (C_6H_5R) is now *more reactive* than benzene itself toward further substitution, and it reacts again with RCl to give products of polyalkylation.

To minimize polyalkylation, a large excess of benzene is used relative to the amount of alkyl halide.

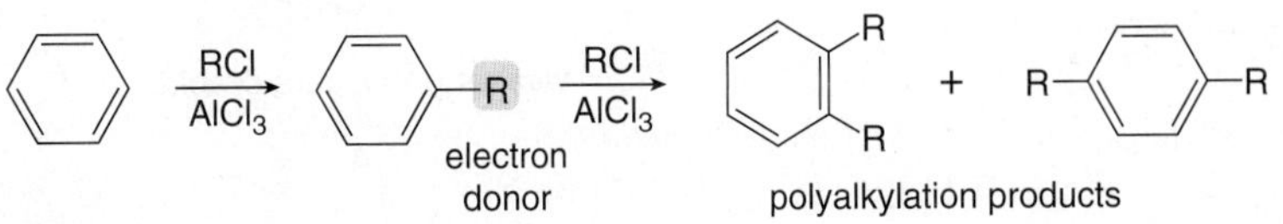

RCl
AlCl3
R
electron
donor
RCl
AlCl3
R
R
+
R
R
polyalkylation products

Polysubstitution does not occur with Friedel–Crafts acylation, because the product now has an electron-withdrawing group that deactivates the ring toward another electrophilic substitution.

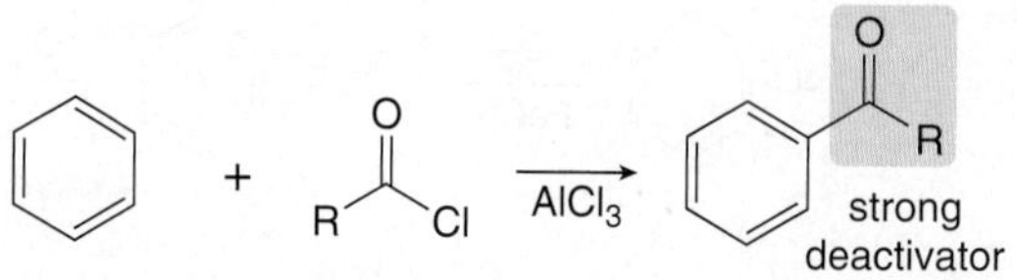

18.11 Disubstituted Benzenes

What happens in electrophilic aromatic substitution when a disubstituted benzene ring is used as starting material? **To predict the products, look at the directing effects of both substituents and then determine the net result,** using the following three guidelines.

Rule [1] **When the directing effects of two groups reinforce, the new substituent is located on the position directed by both groups.**

The CH_3 group in *p*-nitrotoluene is an ortho, para director and the NO_2 group is a meta director. These two effects reinforce each other so that one product is formed on treatment with Br_2 and $FeBr_3$. The position para to the CH_3 group is "blocked" by a nitro group, so no substitution can occur on that carbon.

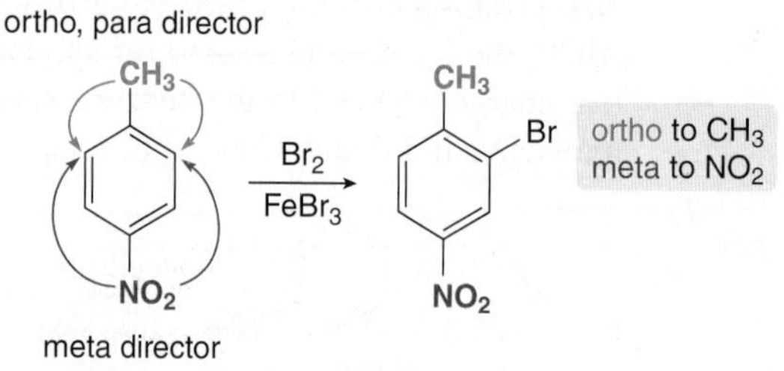

p-nitrotoluene

Rule [2] **If the directing effects of two groups oppose each other, the more powerful activator "wins out."**

In compound **A,** the $NHCOCH_3$ group activates its two ortho positions, and the CH_3 group activates its two ortho positions to reaction with electrophiles. Because the $NHCOCH_3$ is a stronger activator, substitution occurs ortho to it.

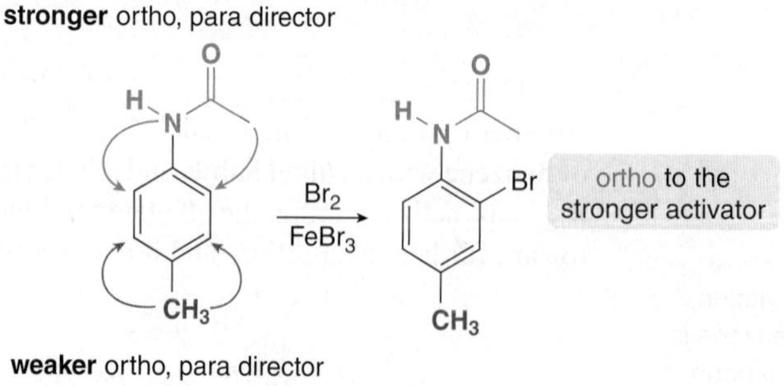

A

Rule [3] **No substitution occurs between two meta substituents because of crowding.**

For example, no substitution occurs at the carbon atom between the two CH_3 groups in *m*-xylene, even though two CH_3 groups activate that position.

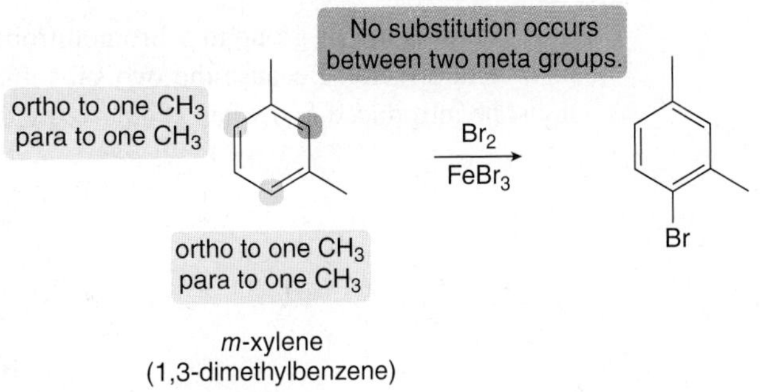

Sample Problem 18.5 Draw the products formed from nitration of each compound.

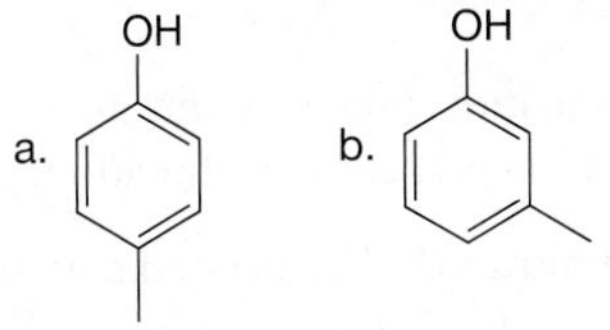

Solution

a. Both the OH and CH_3 groups are ortho, para directors. Because the OH group is a stronger activator, substitution occurs ortho to it.

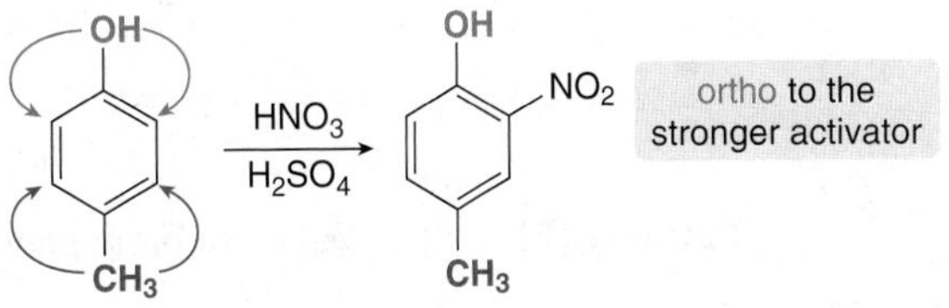

b. Both the OH and CH_3 groups are ortho, para directors whose directing effects reinforce each other in this case. No substitution occurs between the two meta substituents, however, so two products result.

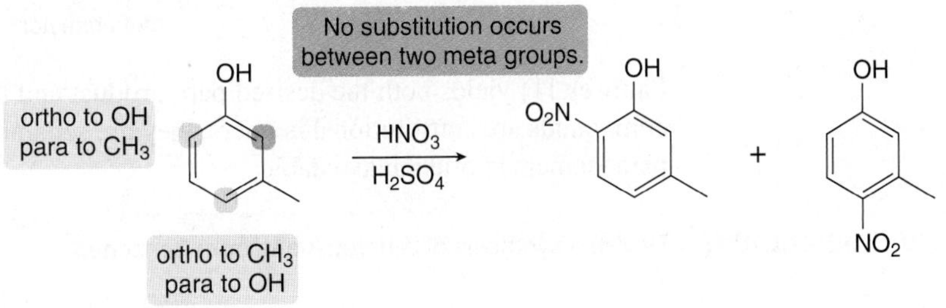

Problem 18.22 Draw the products formed when each compound is treated with HNO_3 and H_2SO_4.

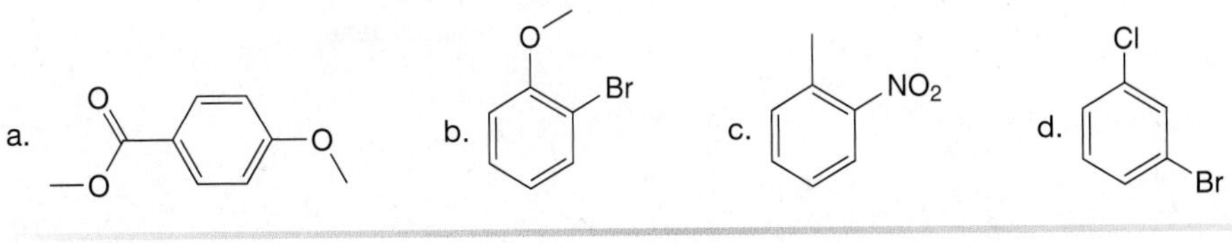

18.12 Synthesis of Benzene Derivatives

To synthesize benzene derivatives with more than one substituent, we must always take into account the directing effects of each substituent. In a disubstituted benzene, for example, **the directing effects indicate which substituent must be added to the ring first.**

For example, the Br group in *p*-bromonitrobenzene is an ortho, para director and the NO_2 group is a meta director. Because the two substituents are para to each other, the ortho, para director must be introduced *first* when synthesizing this compound from benzene.

Br ortho, para director
NO2 meta director

p-bromonitrobenzene

With two para substituents, add the ortho, para director first.

Thus, Pathway [1], in which bromination precedes nitration, yields the desired para product, whereas Pathway [2], in which nitration precedes bromination, yields the undesired meta isomer.

Pathway [1] Bromination before nitration: The desired para product is formed.

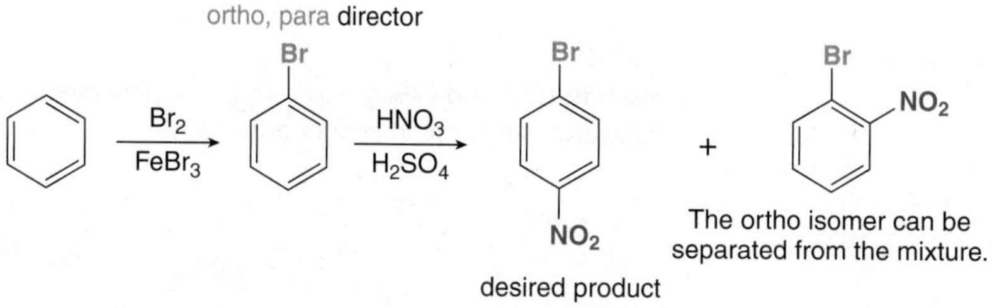
ortho, para director
Br
Br2
FeBr3
HNO3
H2SO4
Br
NO2
desired product
+
Br
NO2
The ortho isomer can be separated from the mixture.

Pathway [2] Nitration before bromination: The undesired meta isomer is formed.

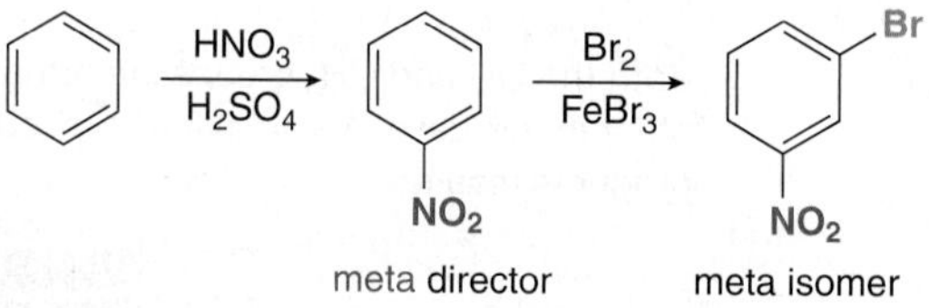
HNO3
H2SO4
NO2
meta director
Br2
FeBr3
Br
NO2
meta isomer

Pathway [1] yields both the desired para product and the undesired ortho isomer. Because these compounds are constitutional isomers, they are separable. Obtaining such a mixture of ortho and para isomers is often unavoidable.

Sample Problem 18.6 Devise a synthesis of *o*-nitrotoluene from benzene.

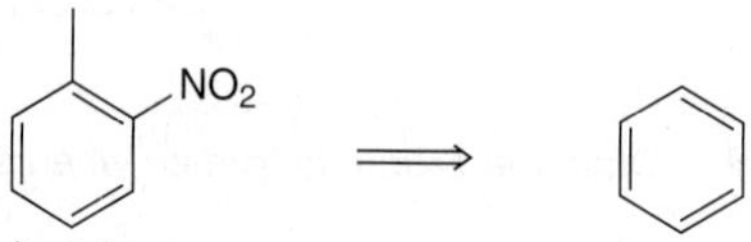
NO2

o-nitrotoluene

Solution

The CH_3 group in *o*-nitrotoluene is an ortho, para director and the NO_2 group is a meta director. Because the two substituents are ortho to each other, the **ortho, para director must be introduced first.** The synthesis thus involves two steps: Friedel–Crafts alkylation followed by nitration.

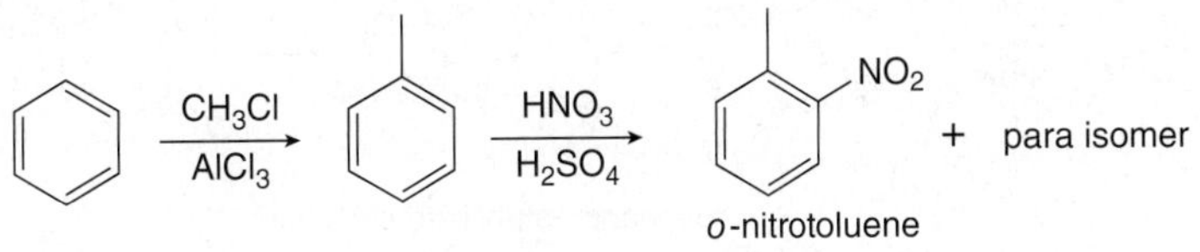

Problem 18.23 Devise a synthesis of each compound from the indicated starting material.

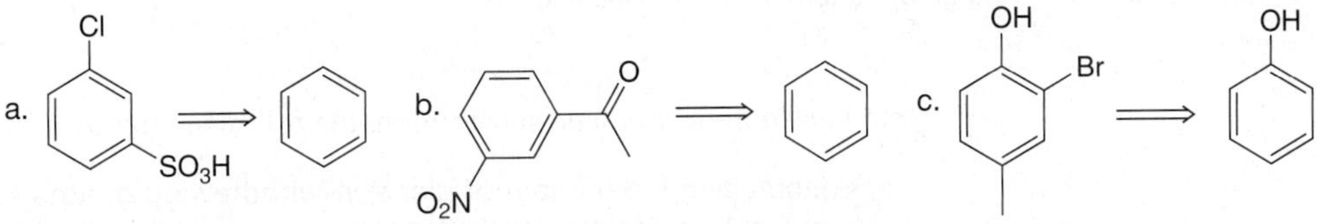

18.13 Nucleophilic Aromatic Substitution

Although most reactions of aromatic compounds occur by way of electrophilic aromatic substitution, aryl halides undergo a limited number of substitution reactions with strong nucleophiles.

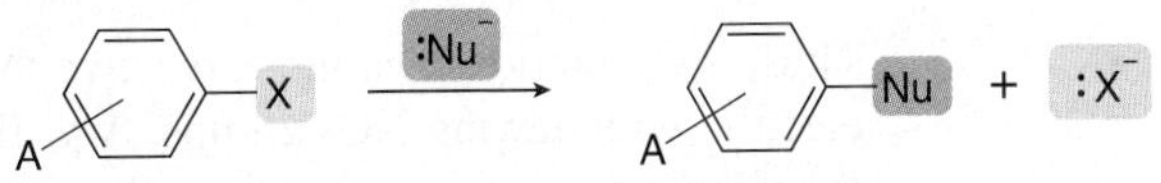

X = F, Cl, Br, I
A = H or electron-withdrawing group

- **Nucleophilic aromatic substitution results in the substitution of a halogen X on a benzene ring by a nucleophile (:Nu⁻).**

As we learned in Section 7.17, these reactions *cannot* occur by an S_N1 or S_N2 mechanism, which take place only at sp^3 hybridized carbons. Instead, two different mechanisms are proposed to explain the results: **addition–elimination** (Section 18.13A) and **elimination–addition** (Section 18.13B).

18.13A Nucleophilic Aromatic Substitution by Addition–Elimination

Aryl halides with strong electron-withdrawing groups (such as NO_2) on the ortho or para positions react with nucleophiles to afford substitution products. Treatment of *p*-chloronitrobenzene with hydroxide (^-OH) affords *p*-nitrophenol by replacement of Cl by OH.

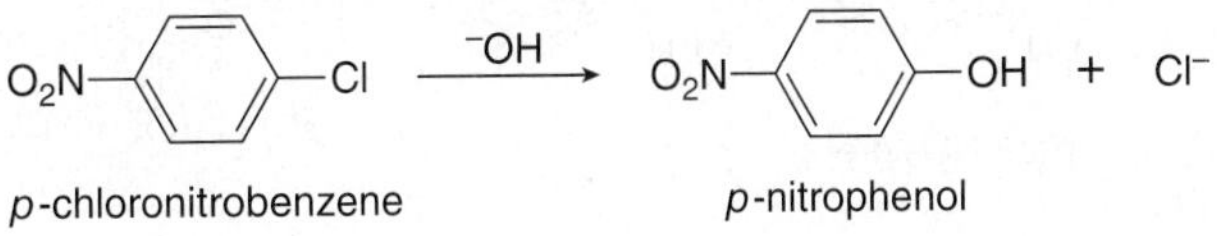

Nucleophilic aromatic substitution occurs with a variety of strong nucleophiles, including ^-OH, ^-OR, $^-NH_2$, ^-SR, and in some cases, neutral nucleophiles such as NH_3 and RNH_2. The mechanism of these reactions has two steps: **addition of the nucleophile** to form a resonance-stabilized carbanion, followed by **elimination of the halogen leaving group.** Mechanism 18.10 is drawn with an aryl chloride containing a general electron-withdrawing group W.

Mechanism 18.10 Nucleophilic Aromatic Substitution by Addition–Elimination

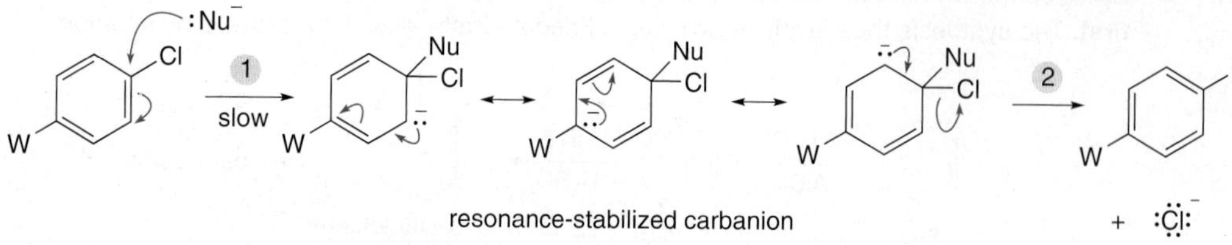

1 Addition of the nucleophile forms a resonance-stabilized carbanion and a new C–Nu bond in the rate-determining step.

2 Loss of the leaving group re-forms the aromatic ring.

In nucleophilic aromatic substitution, the following trends in reactivity are observed.

- **Increasing the number of electron-withdrawing groups increases the reactivity of the aryl halide. Electron-withdrawing groups stabilize the intermediate carbanion, and by the Hammond postulate, lower the energy of the transition state that forms it.**

- **Increasing the electronegativity of the halogen increases the reactivity of the aryl halide. A more electronegative halogen stabilizes the intermediate carbanion by an inductive effect, making aryl fluorides (ArF) much *more* reactive than other aryl halides, which contain less electronegative halogens.**

Thus, aryl chloride **B** is more reactive than *o*-chloronitrobenzene (**A**) because it contains two electron-withdrawing NO_2 groups. Aryl fluoride **C** is more reactive than **B** because **C** contains the more electronegative halogen, fluorine.

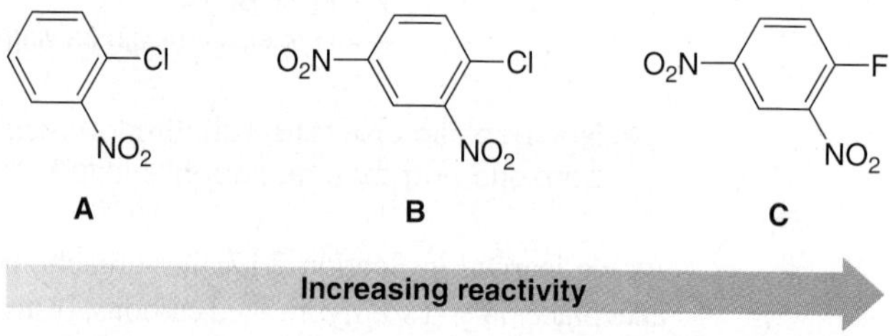

The location of the electron-withdrawing group greatly affects the rate of nucleophilic aromatic substitution. When a nitro group is located ortho or para to the halogen, the negative charge of the intermediate carbanion can be delocalized onto the NO_2 group, thus stabilizing it. With a meta NO_2 group, no such additional delocalization onto the NO_2 group occurs.

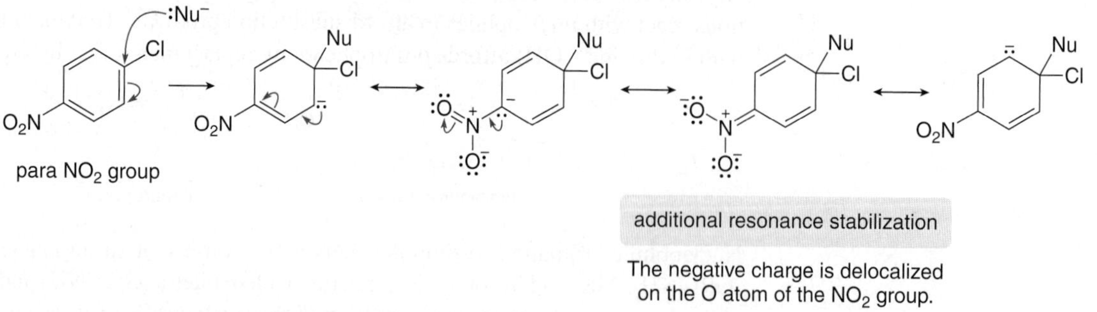

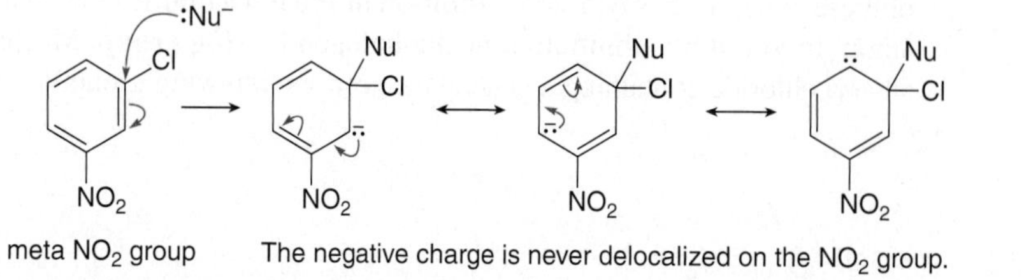

Thus, nucleophilic aromatic substitution by an addition–elimination mechanism occurs only with aryl halides that contain electron-withdrawing substituents at the ortho or para position.

Problem 18.24 Draw the products of each reaction.

a. 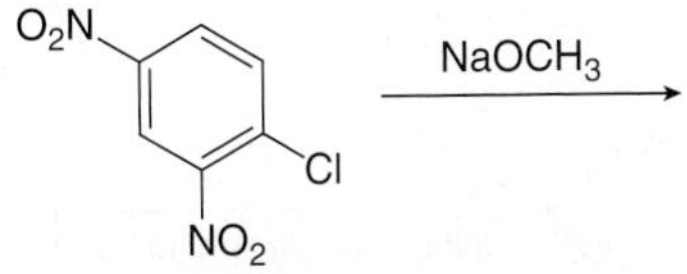
O₂N
NaOCH₃
Cl
NO₂

b. 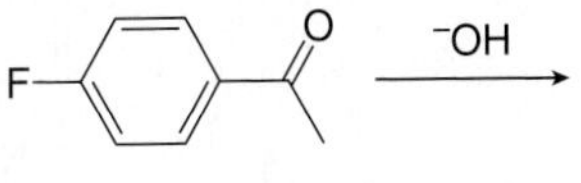
F
O
⁻OH

Problem 18.25 Draw a stepwise mechanism for the following reaction that forms ether **D. D** can be converted to the antidepressant fluoxetine (trade name Prozac) in a single step.

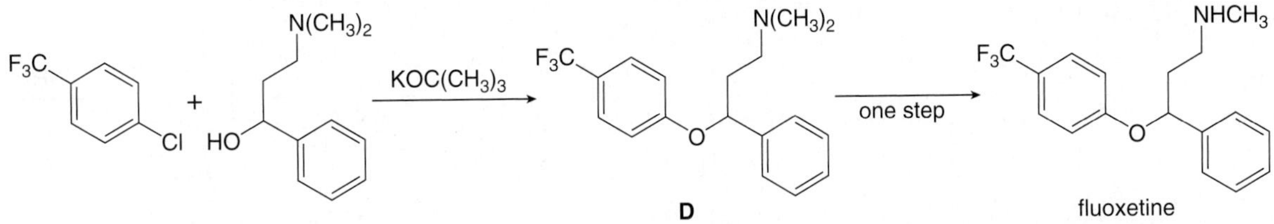
F₃C
Cl
+
HO
N(CH₃)₂
KOC(CH₃)₃
O
one step
NHCH₃

D fluoxetine

18.13B Nucleophilic Aromatic Substitution by Elimination–Addition: Benzyne

Aryl halides that do not contain an electron-withdrawing group generally do not react with nucleophiles. Under extreme reaction conditions, however, nucleophilic aromatic substitution can occur with aryl halides. For example, heating chlorobenzene with NaOH above 300 °C and 170 atmospheres of pressure affords phenol.

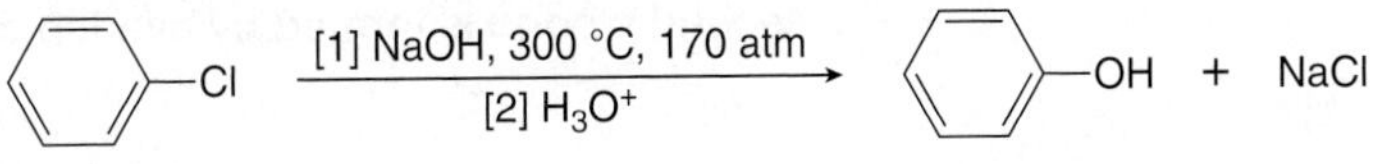
Cl
[1] NaOH, 300 °C, 170 atm
[2] H₃O⁺
OH + NaCl

chlorobenzene phenol

The mechanism proposed to explain this result involves formation of a **benzyne** intermediate (C_6H_4) by elimination–addition. As shown in Mechanism 18.11, benzyne is a highly reactive, unstable intermediate formed by elimination of HX from an aryl halide.

Mechanism 18.11 Nucleophilic Aromatic Substitution by Elimination–Addition: Benzyne

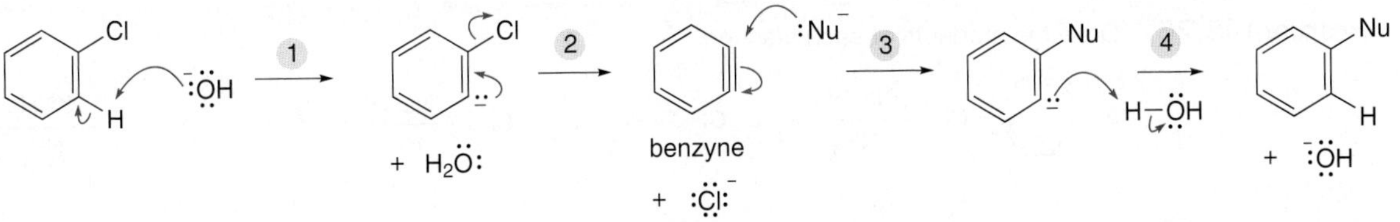
Cl
H
:ÖH
1
+ H₂Ö:
Cl
2
benzyne
+ :Cl:⁻
:Nu⁻
3
Nu
4
H–ÖH
Nu
H
+ :ÖH⁻

1 – 2 Elimination of H and X from two adjacent atoms forms a reactive benzyne intermediate.

3 – 4 Nucleophilic attack and protonation form the substitution product.

Formation of a benzyne intermediate explains why substituted aryl halides form mixtures of products. **Nucleophilic aromatic substitution by an elimination–addition mechanism affords substitution on the carbon directly bonded to the leaving group and the carbon adjacent to it.** As an example, treatment of *p*-chlorotoluene with $NaNH_2$ forms para- and meta-substitution products.

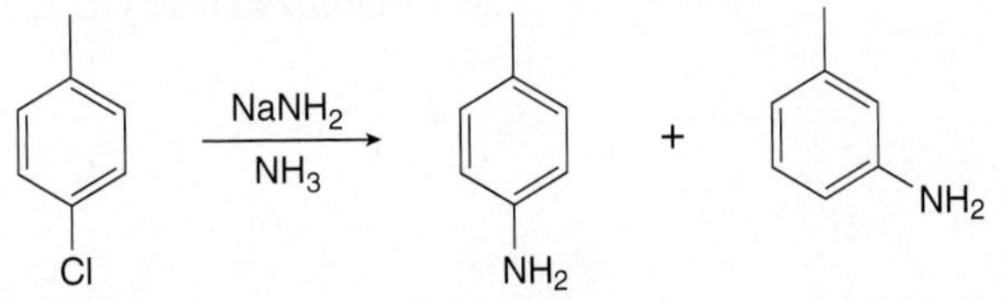
Cl
NaNH₂
NH₃
NH₂
+
NH₂

p-chlorotoluene *p*-methylaniline (*p*-toluidine) *m*-methylaniline (*m*-toluidine)

This result is explained by the fact that nucleophilic attack on the benzyne intermediate may occur at either C3 to form *m*-methylaniline, or C4 to form *p*-methylaniline.

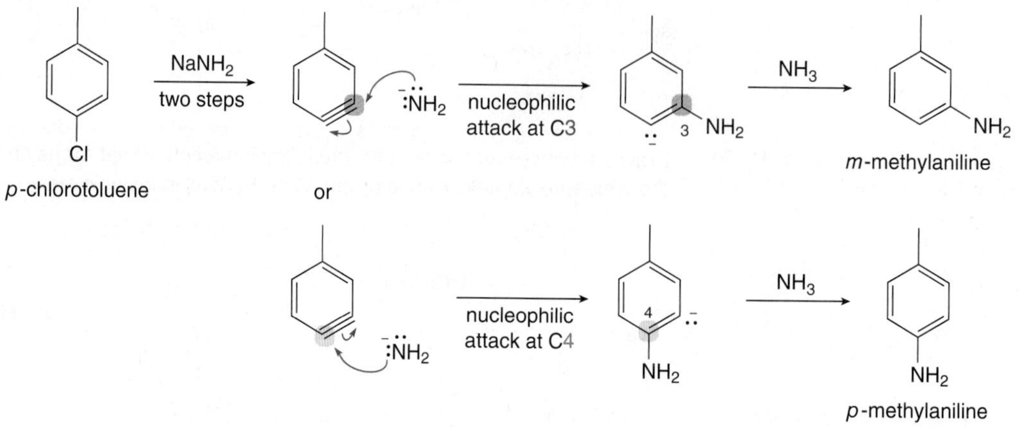

As you might expect, the triple bond in benzyne is unusual. Each carbon of the six-membered ring is sp^2 hybridized, and as a result, the σ bond and two π bonds of the triple bond are formed with the following orbitals.

- **The σ bond is formed by overlap of two sp^2 hybrid orbitals.**
- **One π bond is formed by overlap of two *p* orbitals perpendicular to the plane of the molecule.**
- **The second π bond is formed by overlap of two sp^2 hybrid orbitals.**

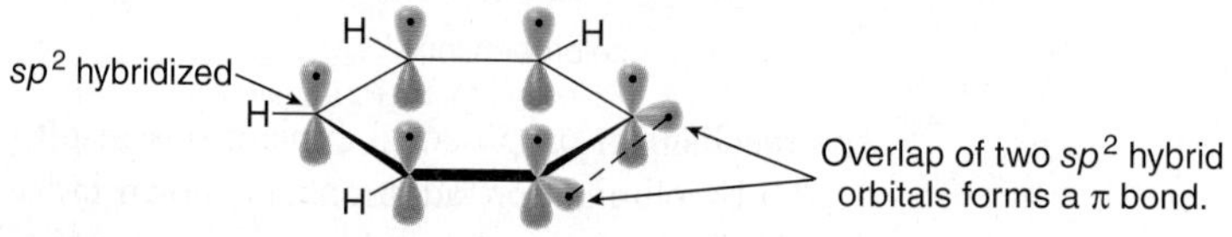

Thus, the second π bond of benzyne differs from all other π bonds seen thus far, because **it is formed by the side-by-side overlap of sp^2 *hybrid orbitals,* not *p* orbitals.** This π bond, located in the plane of the molecule, is extremely weak.

Problem 18.26 Draw the products of each reaction.

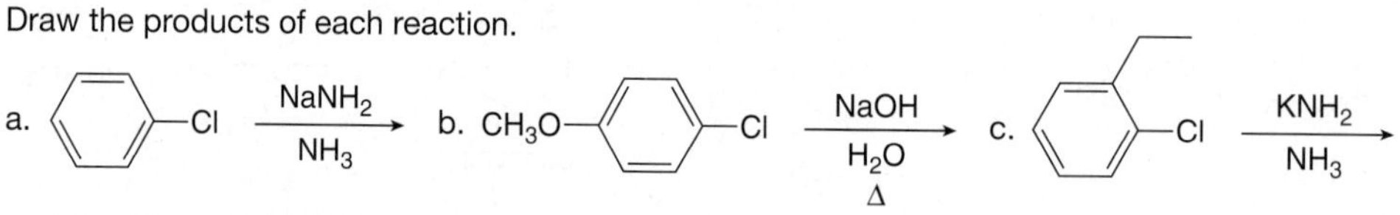

Problem 18.27 Draw all products formed when *m*-chlorotoluene is treated with KNH_2 in NH_3.

18.14 Halogenation of Alkyl Benzenes

Radical halogenation of alkanes was discussed in Chapter 15. The mechanism of radical halogenation at an allylic carbon was given in Section 15.10.

We finish Chapter 18 by learning some additional reactions of substituted benzenes that greatly expand the ability to synthesize benzene derivatives. In Section 18.14 we return to radical halogenation, and in Section 18.15 we examine useful oxidation and reduction reactions.

Benzylic C–H bonds are weaker than most other sp^3 hybridized C–H bonds, because homolysis forms a resonance-stabilized benzylic radical.

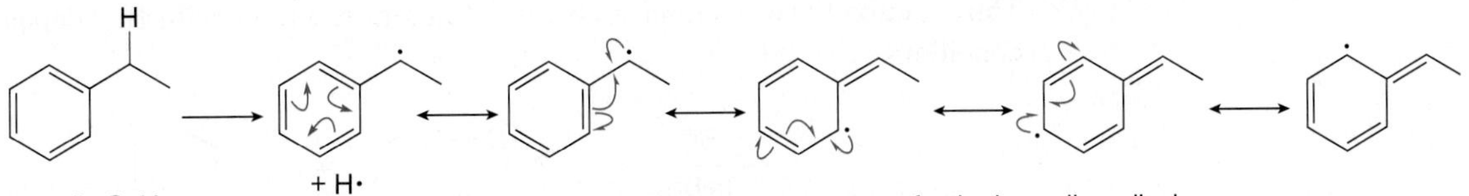
H
+ H·

benzylic C–H in red

five resonance structures for the benzylic radical

The bond dissociation energy for a benzylic C–H bond (356 kJ/mol) is even less than the bond dissociation energy for a 3° C–H bond (381 kJ/mol).

As a result, an alkyl benzene undergoes selective bromination at the weak benzylic C–H bond under radical conditions to form a **benzylic halide.** For example, radical bromination of ethylbenzene using either Br_2 (in the presence of light or heat) or *N*-bromosuccinimide (NBS, in the presence of light or peroxides) forms a benzylic bromide as the sole product.

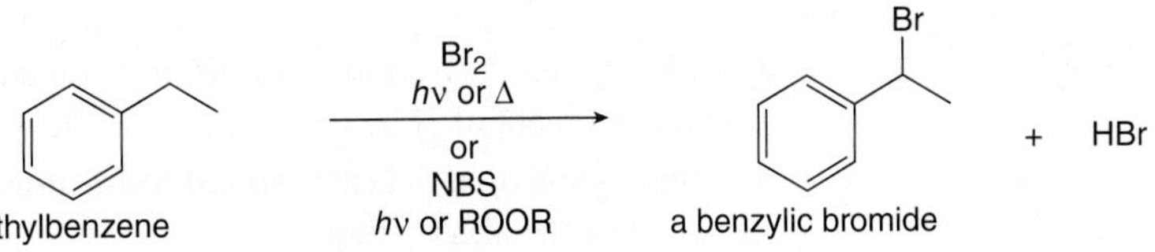
Br2
hν or Δ
or
NBS
hν or ROOR
Br
+ HBr

ethylbenzene

a benzylic bromide

radical conditions

The mechanism for halogenation at the benzylic position resembles other radical halogenation reactions, and so it involves initiation, propagation, and termination. Mechanism 18.12 illustrates the radical bromination of ethylbenzene using Br_2 ($h\nu$ or Δ).

Mechanism 18.12 Benzylic Bromination

Part [1] Initiation

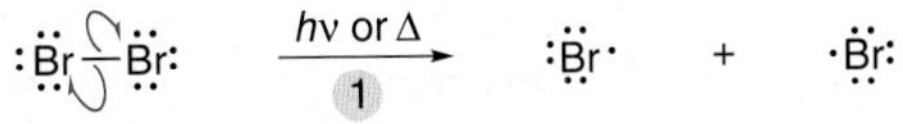
:Br–Br:
hν or Δ
1
:Br·
+
·Br:

1 Homolysis of the Br–Br bond requires light or heat and forms two bromine radicals.

Part [2] Propagation

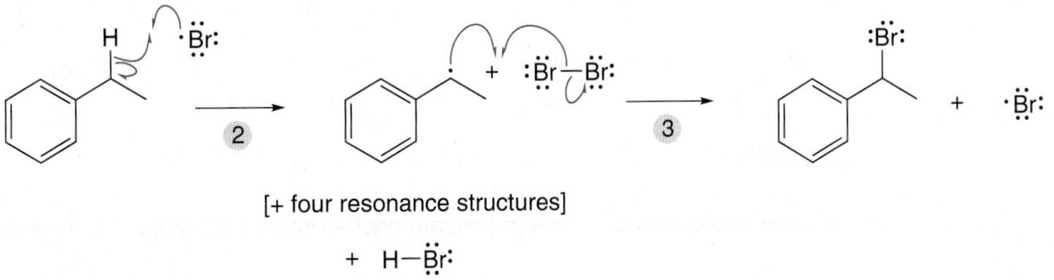
H
·Br:
2
+ :Br–Br:
3
:Br:
+ ·Br:
[+ four resonance structures]
+ H–Br:

2 Abstraction of a benzylic hydrogen by Br· forms the resonance-stabilized benzylic radical.

3 The benzylic radical abstracts a bromine atom to form the benzylic bromide. Because Br· is also formed, Steps [2] and [3] can occur repeatedly without additional initiation.

Part [3] Termination

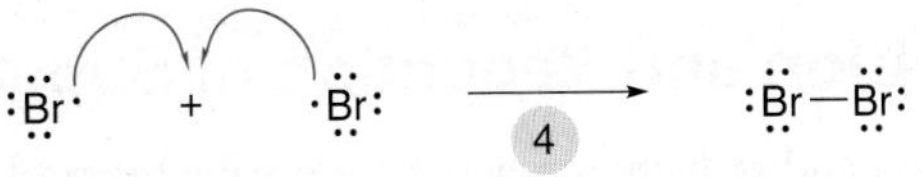
:Br·
+
·Br:
4
:Br–Br:

4 Termination of the chain occurs when any two radicals combine to form a bond.

Thus, an alkyl benzene undergoes two different reactions with Br_2, depending on the reaction conditions.

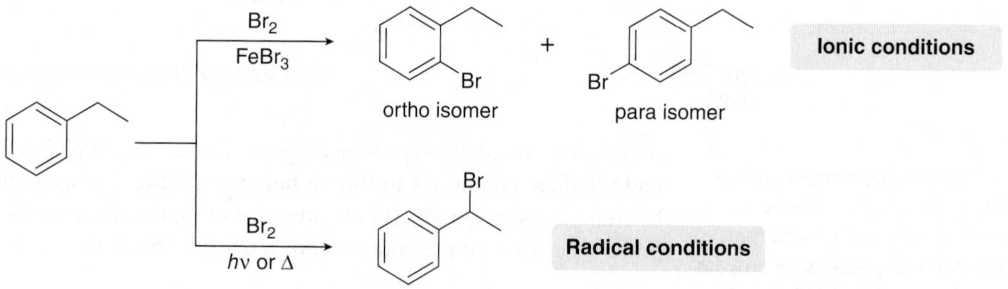
Br2
FeBr3
Br
ortho isomer
+
Br
para isomer
Ionic conditions
Br
Br2
hν or Δ
Radical conditions

- With Br_2 and $FeBr_3$ (**ionic conditions**), electrophilic aromatic substitution occurs, resulting in replacement of H by Br on the aromatic ring to form ortho and para isomers.
- With Br_2 and light or heat (**radical conditions**), substitution of H by Br occurs at the *benzylic* carbon of the alkyl group.

Problem 18.28 Explain why $C_6H_5CH_2CH_2Br$ is not formed during the radical bromination of $C_6H_5CH_2CH_3$.

The radical bromination of alkyl benzenes is a useful reaction because the resulting benzylic halide can serve as starting material for a variety of substitution and elimination reactions, thus making it possible to form many new substituted benzenes. Sample Problem 18.7 illustrates one possibility.

Sample Problem 18.7 Design a synthesis of styrene from ethylbenzene.

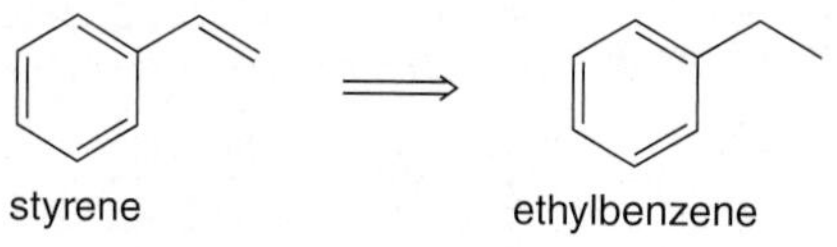
styrene
ethylbenzene

Solution

The double bond can be introduced by a two-step reaction sequence: bromination at the benzylic position under radical conditions, followed by elimination of HBr with strong base to form the π bond.

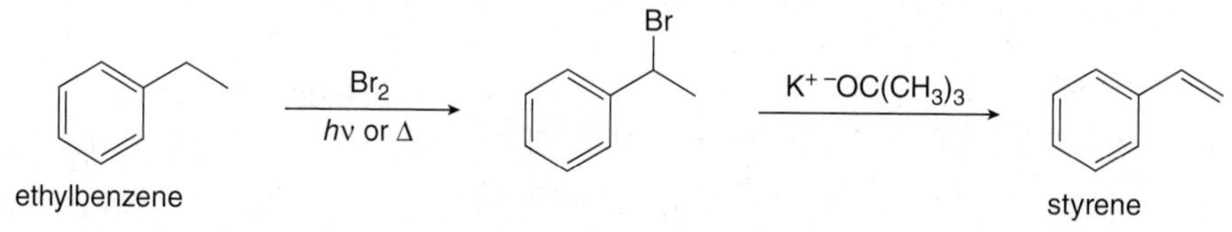
ethylbenzene
Br2
hν or Δ
Br
K+ −OC(CH3)3
styrene

Problem 18.29 How could you use ethylbenzene to prepare each compound? More than one step is required.

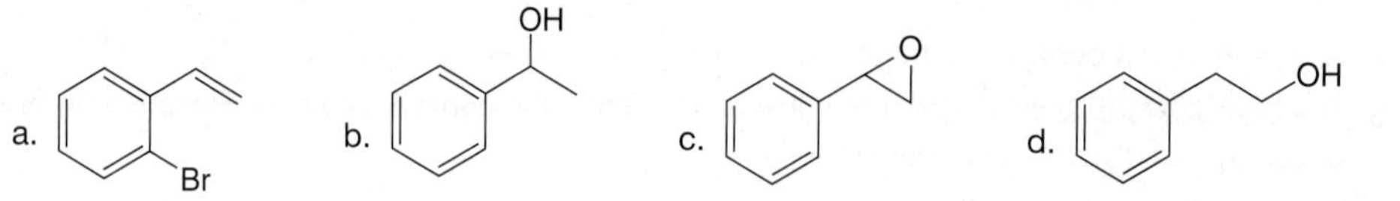
a.
Br
b.
OH
c.
O
d.
OH

18.15 Oxidation and Reduction of Substituted Benzenes

Oxidation and reduction reactions are valuable tools for preparing many other benzene derivatives. Because the mechanisms are complex and do not have general applicability, reagents and reactions are presented only, without reference to the detailed mechanism.

18.15A Oxidation of Alkyl Benzenes

Arenes containing at least one benzylic C–H bond are oxidized with $KMnO_4$ to benzoic acid, a carboxylic acid with the carboxy group (COOH) bonded directly to the benzene ring. With some alkyl benzenes, this also results in the cleavage of carbon–carbon bonds, so the product has fewer carbon atoms than the starting material.

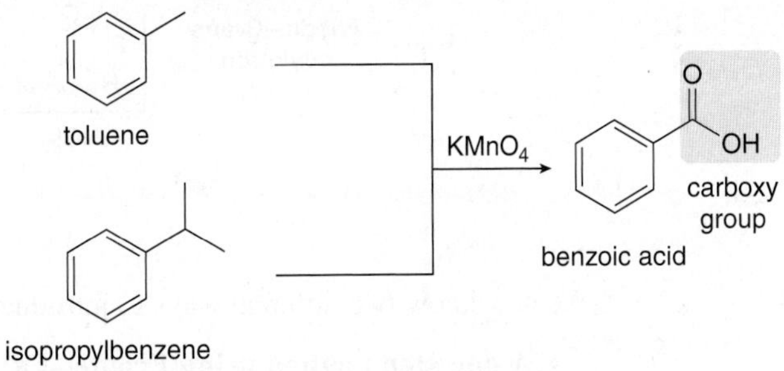
toluene
isopropylbenzene
$KMnO_4$
O
OH
carboxy group
benzoic acid

Substrates with more than one alkyl group are oxidized to dicarboxylic acids. Compounds without a benzylic C–H bond are inert to oxidation.

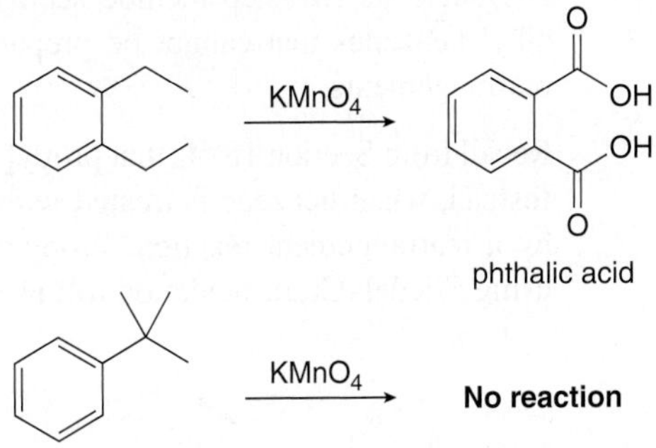
$KMnO_4$
O
OH
OH
O
phthalic acid
$KMnO_4$
No reaction

18.15B Reduction of Aryl Ketones to Alkyl Benzenes

Ketones formed as products in Friedel–Crafts acylation can be reduced to alkyl benzenes by two different methods.

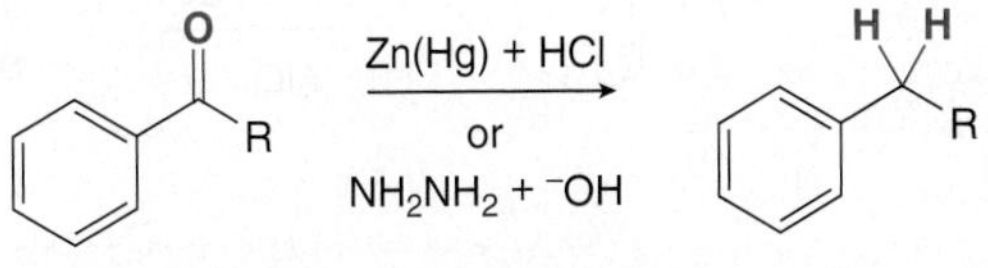
O
R
Zn(Hg) + HCl
or
NH_2NH_2 + ^-OH
H H
R

- The **Clemmensen reduction** uses zinc and mercury in the presence of strong acid.
- The **Wolff–Kishner reduction** uses hydrazine (NH_2NH_2) and strong base (KOH).

Because both C–O bonds in the starting material are converted to C–H bonds in the product, the reduction is difficult and the reaction conditions must be harsh.

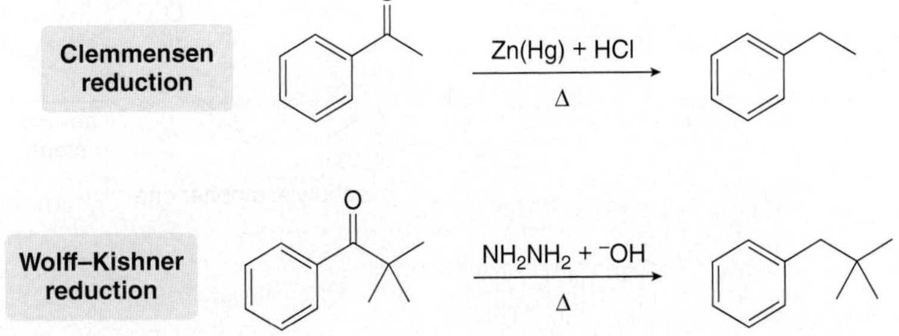
Clemmensen reduction
O
Zn(Hg) + HCl
Δ
Wolff–Kishner reduction
O
NH_2NH_2 + ^-OH
Δ

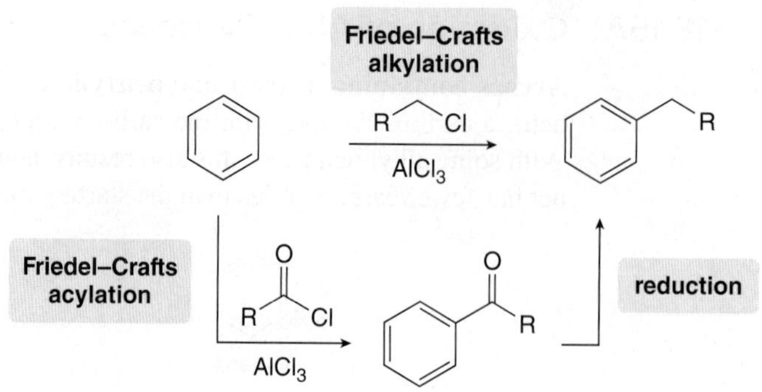

Figure 18.9
Two methods to prepare an alkyl benzene

We now know two different ways to introduce an alkyl group on a benzene ring (Figure 18.9):

- **A one-step method using Friedel–Crafts alkylation**
- **A two-step method using Friedel–Crafts acylation to form a ketone, followed by reduction**

Although the two-step method seems more roundabout, it must be used to synthesize certain alkyl benzenes that cannot be prepared by the one-step Friedel–Crafts alkylation because of rearrangements.

Recall from Section 18.5C that propylbenzene cannot be prepared by a Friedel–Crafts alkylation. Instead, when benzene is treated with 1-chloropropane and $AlCl_3$, isopropylbenzene is formed by a rearrangement reaction. Propylbenzene can be made, however, by a two-step procedure using Friedel–Crafts acylation followed by reduction.

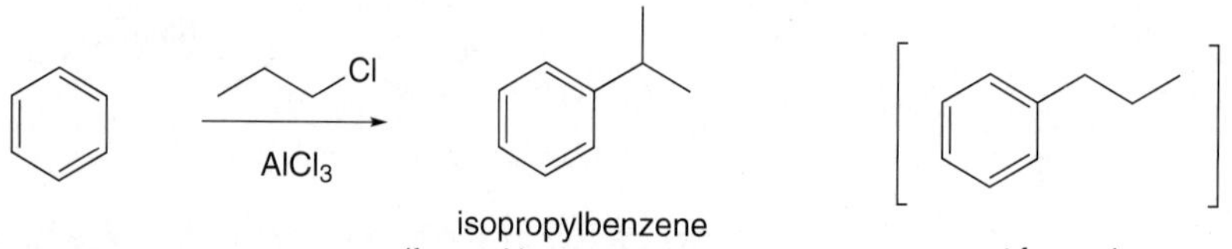

isopropylbenzene
(formed by rearrangemnt)

not formed

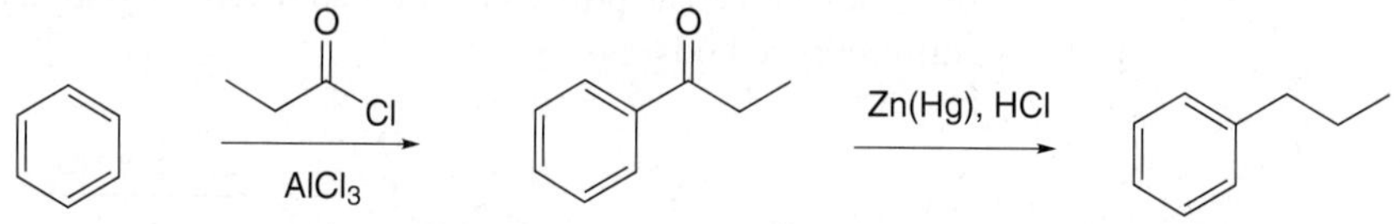

propylbenzene

Problem 18.30 Write out the two-step sequence that converts benzene to each compound.

a. 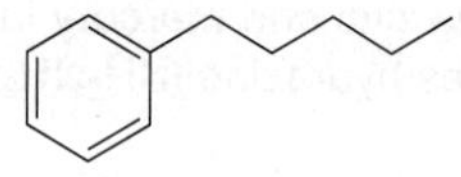 b. 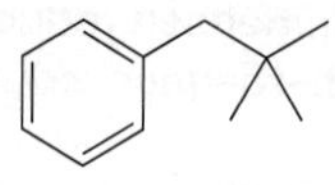

Problem 18.31 What steps are needed to convert benzene into *p*-isobutylacetophenone, a synthetic intermediate used in the synthesis of the anti-inflammatory agent ibuprofen.

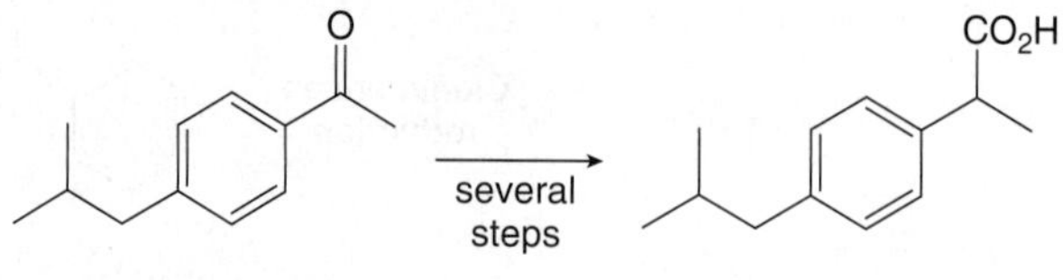

p-isobutylacetophenone ibuprofen

18.15C Reduction of Nitro Groups

A nitro group (NO_2) is easily introduced on a benzene ring by nitration with strong acid (Section 18.4). This process is useful because the nitro group is readily reduced to an amino group (NH_2) under a variety of conditions. The most common methods use H_2 and a catalyst, or a metal (such as Fe or Sn) and a strong acid like HCl.

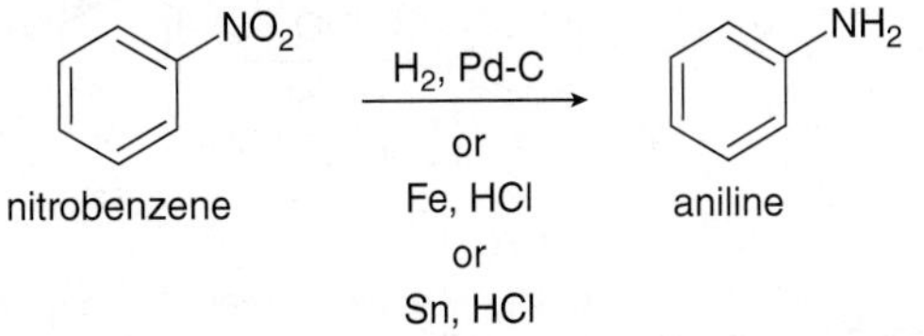

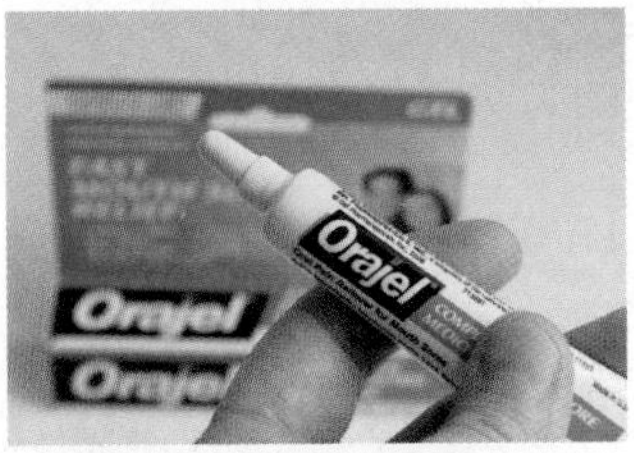

Benzocaine is the active ingredient in the over-the-counter topical anesthetic Orajel.

For example, reduction of ethyl *p*-nitrobenzoate with H_2 and a palladium catalyst forms ethyl *p*-aminobenzoate, a local anesthetic commonly called benzocaine.

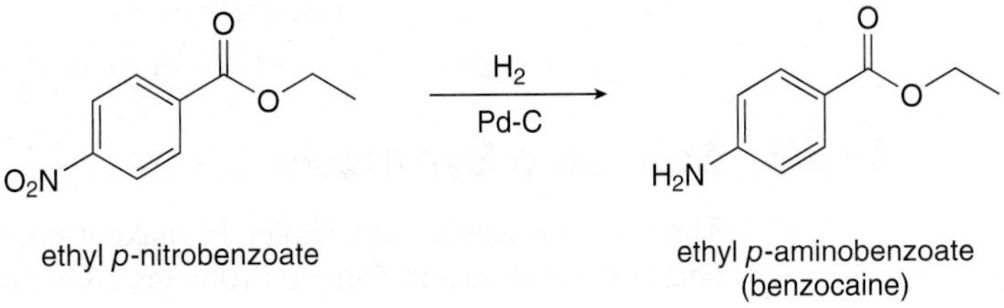

Sample Problem 18.8 illustrates the utility of this process in a short synthesis.

Sample Problem 18.8

Design a synthesis of *m*-bromoaniline from benzene.

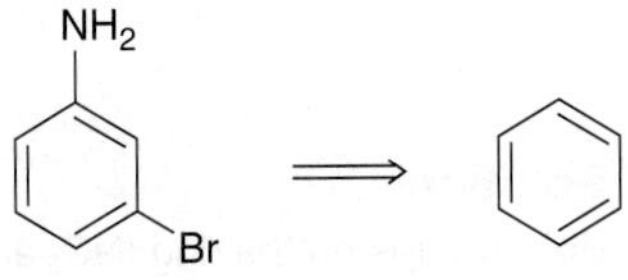

m-bromoaniline

Solution

To devise a retrosynthetic plan, keep in mind:

- The NH_2 group cannot be introduced directly on the ring by electrophilic aromatic substitution. It must be added by a two-step process: nitration followed by reduction.
- Both the Br and NH_2 groups are ortho, para directors, but they are located meta to each other on the ring. However, an NO_2 group (from which an NH_2 group is made) *is* a meta director, and we can use this fact to our advantage.

Retrosynthetic Analysis

Working backwards gives the following **three-step retrosynthetic analysis:**

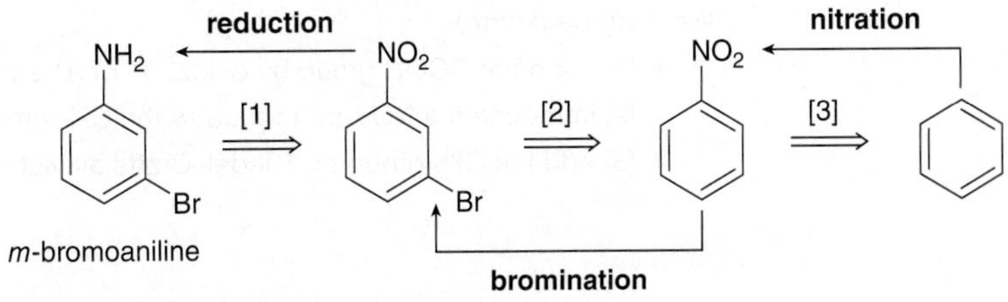

- [1] Form the NH_2 group by reduction of NO_2.
- [2] Introduce the Br group meta to the NO_2 group by halogenation.
- [3] Add the NO_2 group by nitration.

Synthesis

The synthesis involves three steps, and the order is crucial for success. Halogenation (Step [2] of the synthesis) must occur *before* reduction (Step [3]) in order to form the meta substitution product.

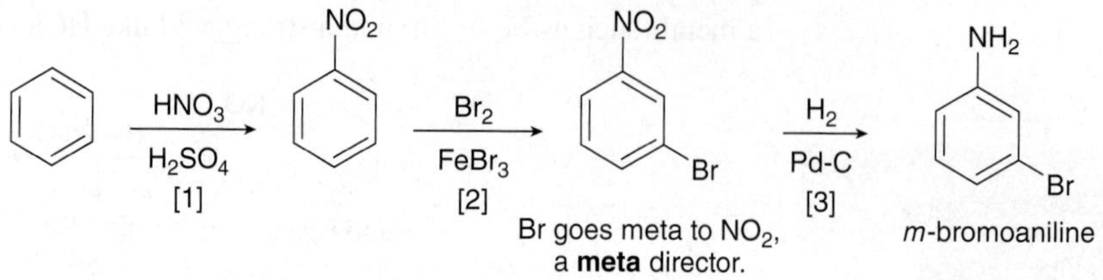

Problem 18.32 Synthesize each compound from benzene.

a. 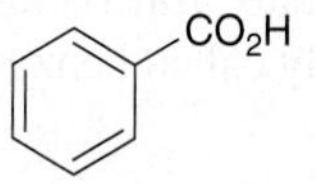 b. 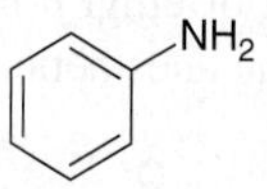 c. 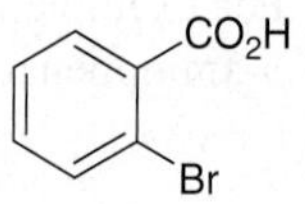

18.16 Multistep Synthesis

The reactions learned in Chapter 18 make it possible to synthesize a wide variety of substituted benzenes, as shown in Sample Problems 18.9–18.11.

Sample Problem 18.9 Synthesize *p*-nitrobenzoic acid from benzene.

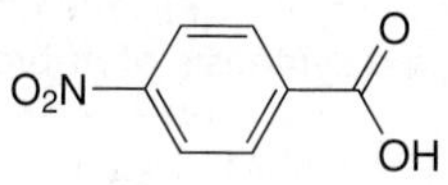

p-nitrobenzoic acid

Solution

Both groups on the ring (NO_2 and COOH) are meta directors. To place these two groups para to each other, remember that the COOH group is prepared by oxidizing an alkyl group, which is an ortho, para director.

Retrosynthetic Analysis

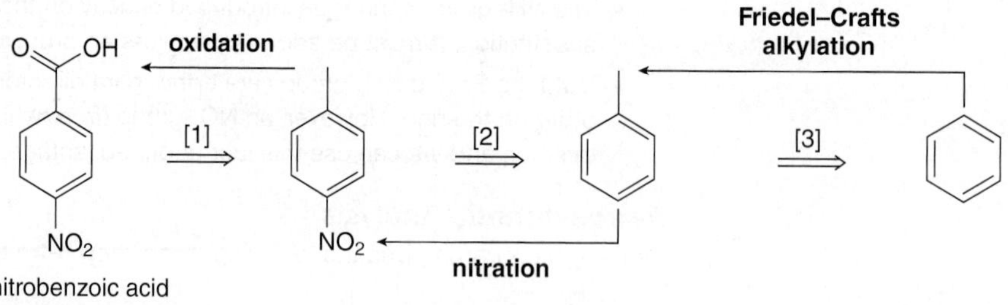

Working backwards:

- [1] Form the COOH group by oxidation of an alkyl group.
- [2] Introduce the NO_2 group para to the CH_3 group (an ortho, para director) by nitration.
- [3] Add the CH_3 group by Friedel–Crafts alkylation.

Synthesis

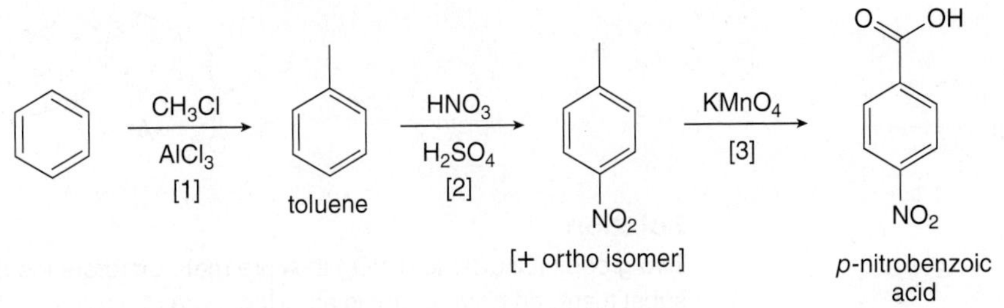

- Friedel–Crafts alkylation with CH_3Cl and $AlCl_3$ forms toluene in Step [1]. Because CH_3 is an ortho, para director, nitration yields the desired para product, which can be separated from its ortho isomer (Step [2]).
- Oxidation with $KMnO_4$ converts the CH_3 group into a COOH group, giving the desired product in Step [3].

Sample Problem 18.10 Synthesize *p*-chlorostyrene from benzene.

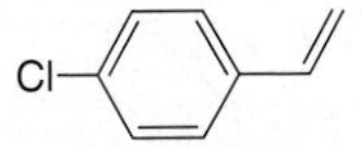

p-chlorostyrene

Solution

Both groups on the ring are ortho, para directors located para to each other. To introduce the double bond in the side chain, we must follow the two-step sequence in Sample Problem 18.7.

Retrosynthetic Analysis

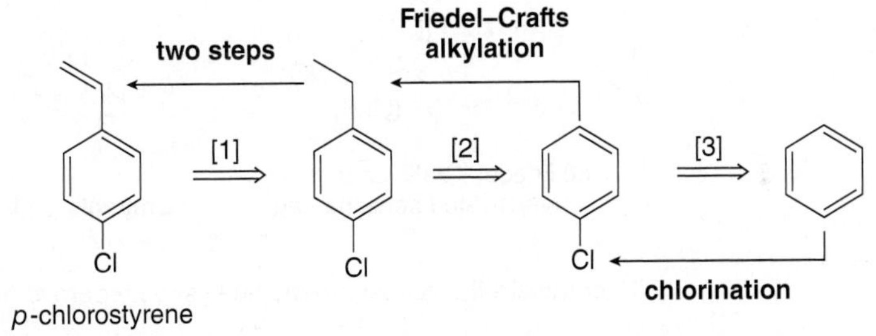

Working backwards:

- [1] Form the double bond by two steps: benzylic halogenation followed by elimination.
- [2] Introduce the CH_3CH_2 group by Friedel–Crafts alkylation.
- [3] Add the Cl atom by chlorination.

Synthesis

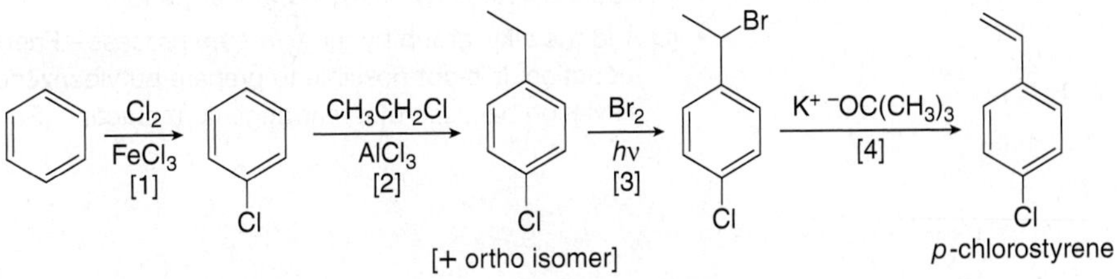

- Chlorination in Step [1] followed by Friedel–Crafts alkylation in Step [2] forms the desired para product, which can be separated from its ortho isomer.
- Benzylic bromination followed by elimination with strong base [$KOC(CH_3)_3$] (Steps [3] and [4]) forms the double bond of the target compound, *p*-chlorostyrene.

Sample Problem 18.11 Synthesize the trisubstituted benzene **A** from benzene.

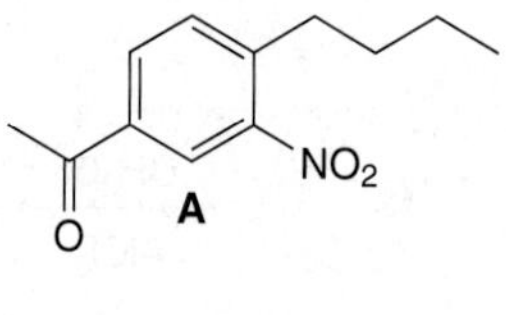

Solution

Two groups (CH_3CO and NO_2) in **A** are meta directors located meta to each other, and the third substituent, an alkyl group, is an ortho, para director.

Retrosynthetic Analysis

With three groups on the benzene ring, begin by determining the possible disubstituted benzenes that are immediate precursors of the target compound, and then eliminate any that cannot be converted to the desired product. For example, three different disubstituted benzenes (**B–D**) can theoretically be precursors to **A.** However, conversion of compounds **B** or **D** to **A** would require a Friedel–Crafts reaction on a deactivated benzene ring, a reaction that does not occur. Thus, only **C** is a feasible precursor of **A.**

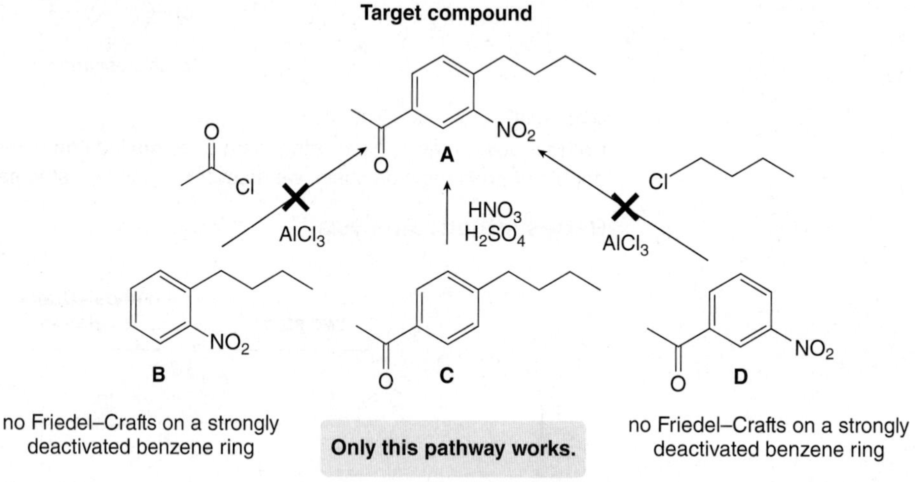

To complete the retrosynthetic analysis, prepare **C** from benzene:

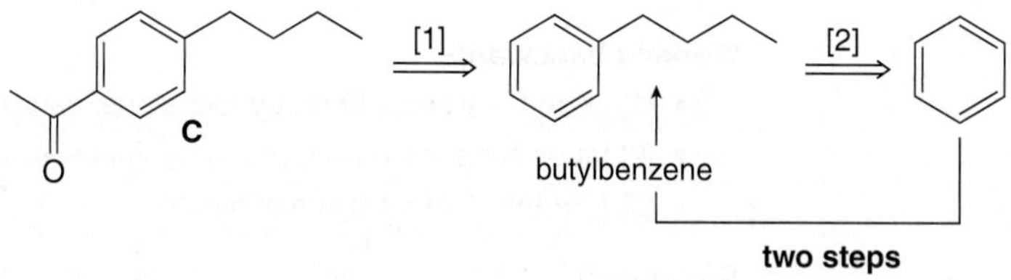

- [1] Add the ketone by Friedel–Crafts acylation.
- [2] Add the alkyl group by the two-step process—Friedel–Crafts acylation followed by reduction. It is not possible to prepare butylbenzene by a one-step Friedel–Crafts alkylation because of a rearrangement reaction (Section 18.15B).

Synthesis

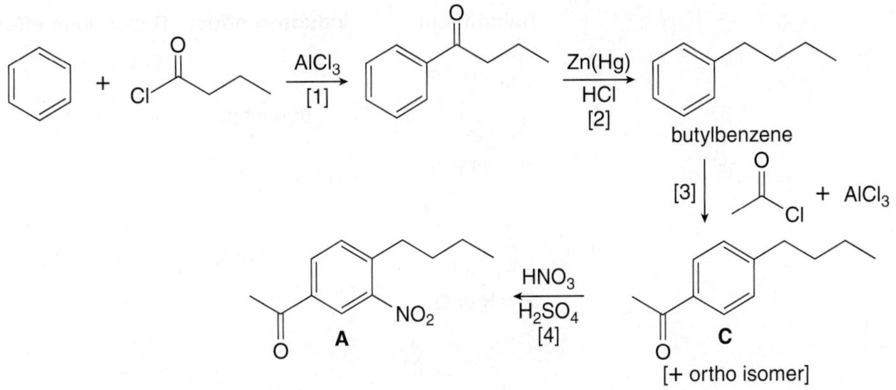
AlCl$_3$
[1]
Zn(Hg)
HCl
[2]
butylbenzene
[3]
+ AlCl$_3$
HNO$_3$
H$_2$SO$_4$
[4]
NO$_2$
A
C
[+ ortho isomer]

- Friedel–Crafts acylation followed by reduction with Zn(Hg), HCl yields butylbenzene (Steps [1]–[2]).
- Friedel–Crafts acylation gives the para product **C,** which can be separated from its ortho isomer (Step [3]).
- Nitration in Step [4] introduces the NO_2 group ortho to the alkyl group (an ortho, para director) and meta to the CH_3CO group (a meta director).

Problem 18.33 Synthesize each compound from benzene.

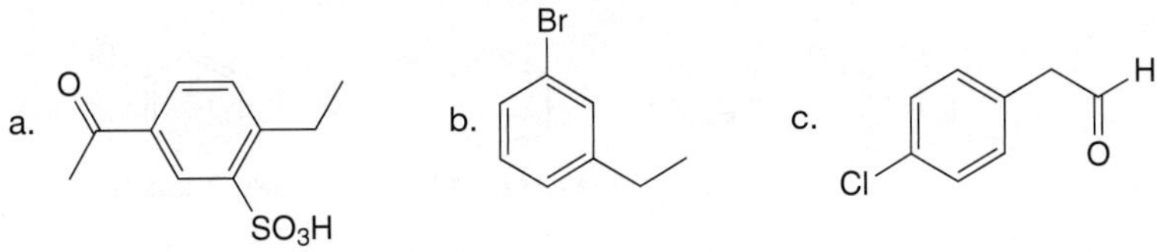
a.
SO$_3$H
b.
Br
c.
Cl
H
O

KEY CONCEPTS

Reactions of Aromatic Compounds

Mechanism of Electrophilic Aromatic Substitution (18.2)

- Electrophilic aromatic substitution follows a two-step mechanism. Reaction of the aromatic ring with an electrophile forms a carbocation, and loss of a proton regenerates the aromatic ring.
- The first step is rate-determining.
- The intermediate carbocation is stabilized by resonance; a minimum of three resonance structures can be drawn. The positive charge is always located ortho or para to the new C – E bond.

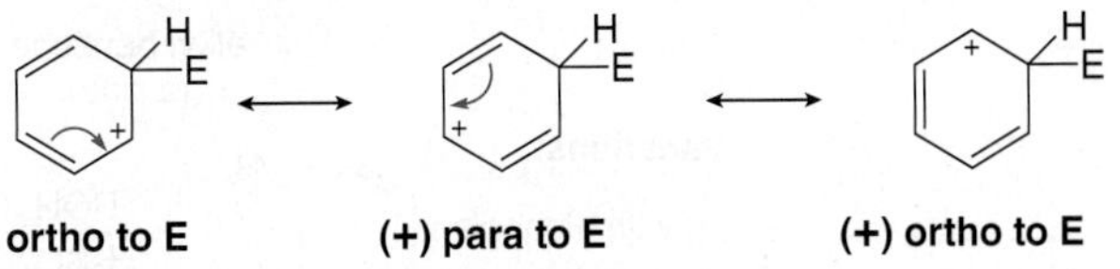
H
E
(+) ortho to E
(+) para to E
(+) ortho to E

Three Rules Describing the Reactivity and Directing Effects of Common Substituents (18.7–18.9)

[1] All ortho, para directors except the halogens activate the benzene ring.
[2] All meta directors deactivate the benzene ring.
[3] The halogens deactivate the benzene ring and direct ortho, para.

Summary of Substituent Effects in Electrophilic Aromatic Substitution (18.6–18.9)

	Substituent	Inductive effect	Resonance effect	Reactivity	Directing effect
[1]	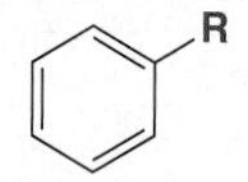R R = alkyl	donating	none	activating	ortho, para
[2]	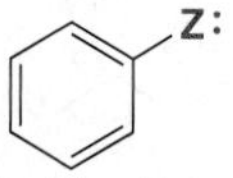Z: Z = N or O	withdrawing	donating	activating	ortho, para
[3]	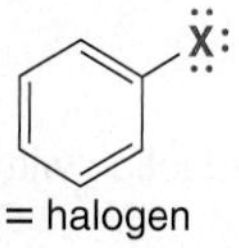X: X = halogen	withdrawing	donating	deactivating	ortho, para
[4]	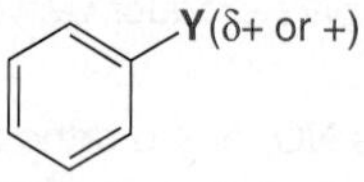Y(δ+ or +)	withdrawing	withdrawing	deactivating	meta

Five Examples of Electrophilic Aromatic Substitution

[1] Halogenation—Replacement of H by Cl or Br (18.3)

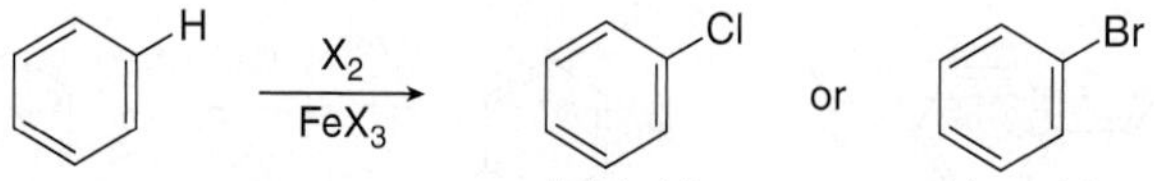
H
X2
FeX3
Cl
or
Br

[X = Cl, Br] aryl chloride aryl bromide

- Polyhalogenation occurs on benzene rings substituted by OH and NH_2 (and related substituents) (18.10A).

[2] Nitration—Replacement of H by NO_2 (18.4)

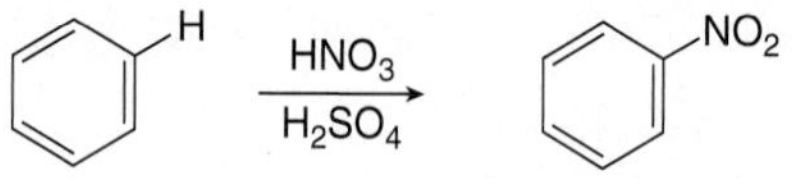
H
HNO3
H2SO4
NO2

nitro compound

[3] Sulfonation—Replacement of H by SO_3H (18.4)

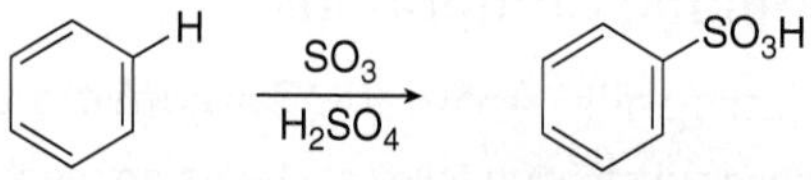
H
SO3
H2SO4
SO3H

benzenesulfonic acid

[4] Friedel–Crafts alkylation—Replacement of H by R (18.5)

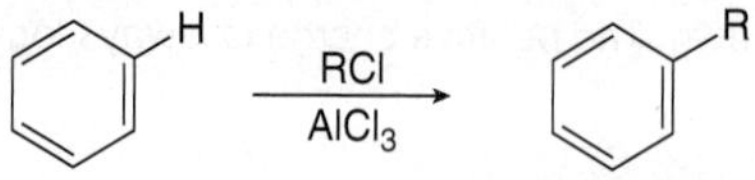
H
RCl
AlCl3
R

alkyl benzene (arene)

Variations:

[1] with alcohols

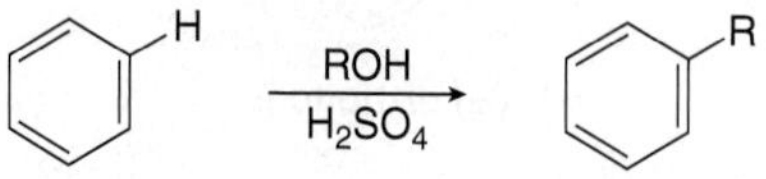
H
ROH
H2SO4
R

[2] with alkenes

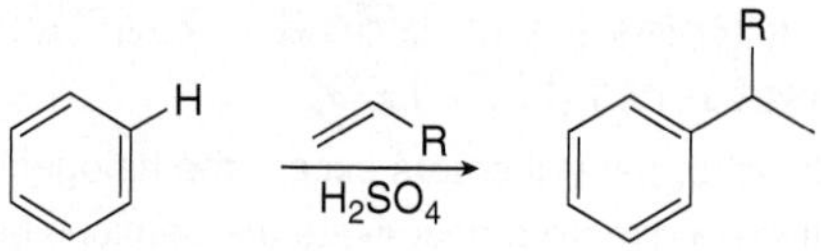
H
R
H2SO4
R

- Rearrangements can occur.
- Vinyl halides and aryl halides are unreactive.
- The reaction does not occur on benzene rings substituted by meta deactivating groups or NH_2 groups (18.10B).
- Polyalkylation can occur.

[5] Friedel–Crafts acylation—Replacement of H by RCO (18.5)

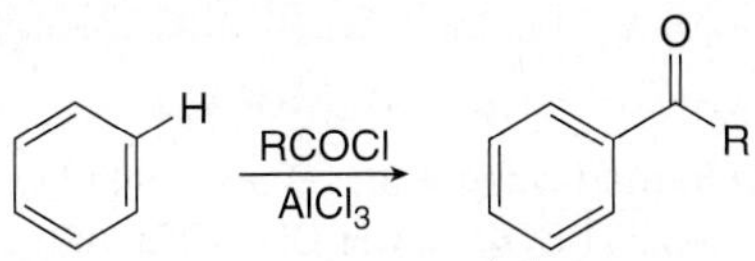

ketone

- The reaction does not occur on benzene rings substituted by meta deactivating groups or NH_2 groups (18.10B).

Nucleophilic Aromatic Substitution (18.13)

[1] Nucleophilic substitution by an addition–elimination mechanism

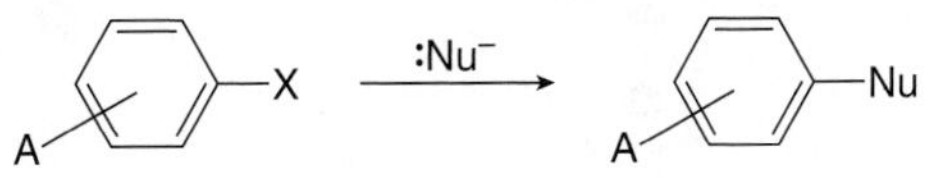

X = F, Cl, Br, I
A = electron-withdrawing group

- The mechanism has two steps.
- Strong electron-withdrawing groups at the ortho or para positions are required.
- Increasing the number of electron-withdrawing groups increases the rate.
- Increasing the electronegativity of the halogen increases the rate.

[2] Nucleophilic substitution by an elimination–addition mechanism

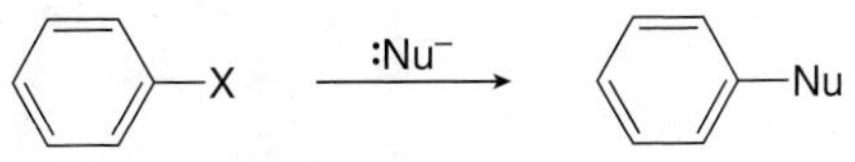

X = halogen

- Reaction conditions are harsh.
- Benzyne is formed as an intermediate.
- Product mixtures may result.

Other Reactions of Benzene Derivatives

[1] Benzylic halogenation (18.14)

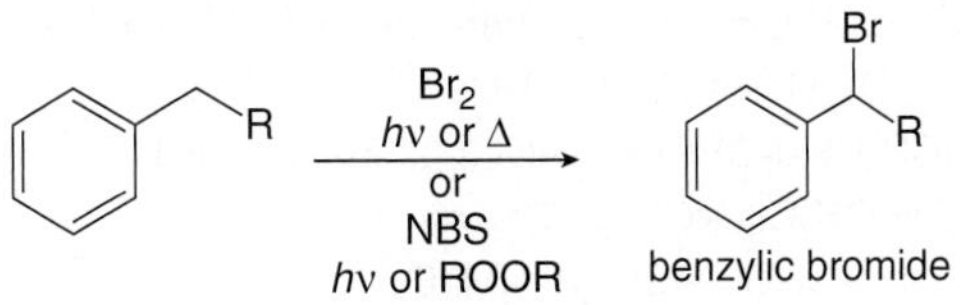

benzylic bromide

[2] Oxidation of alkyl benzenes (18.15A)

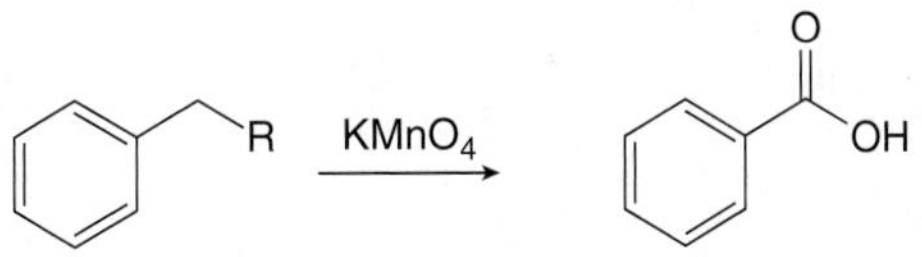

benzoic acid

- A benzylic C – H bond is needed for reaction.

[3] Reduction of ketones to alkyl benzenes (18.15B)

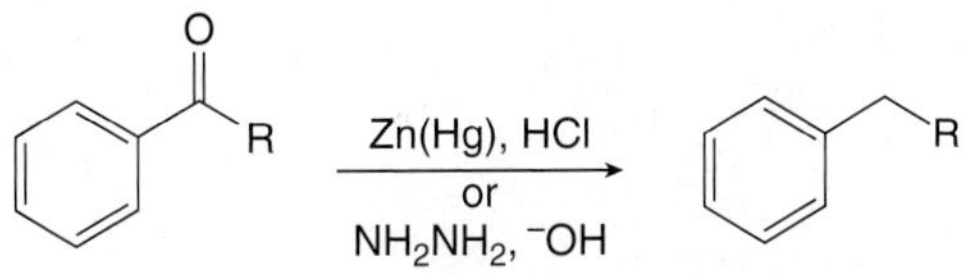

alkyl benzene

[4] Reduction of nitro groups to amino groups (18.15C)

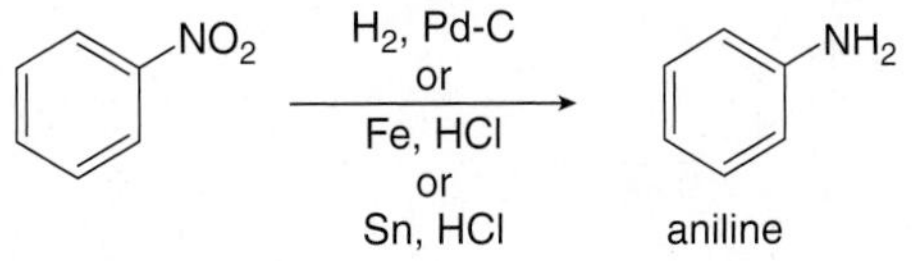

aniline

PROBLEMS

Problems Using Three-Dimensional Models

18.34 Draw the products formed when **A** and **B** are treated with each of the following reagents: (a) Br_2, $FeBr_3$; (b) HNO_3, H_2SO_4; (c) CH_3CH_2COCl, $AlCl_3$.

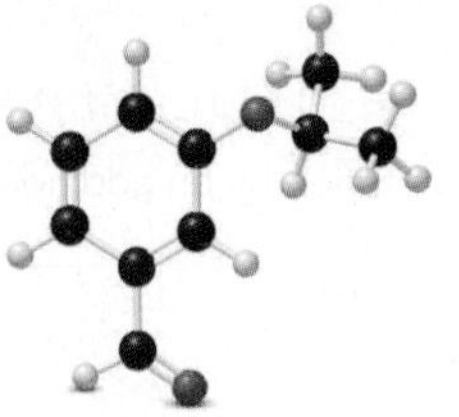

A

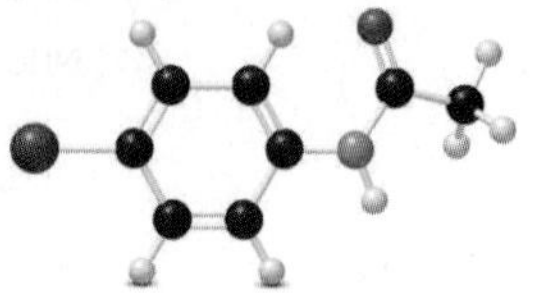

B

18.35 What is the major product formed by an intramolecular Friedel–Crafts acylation of the following compound?

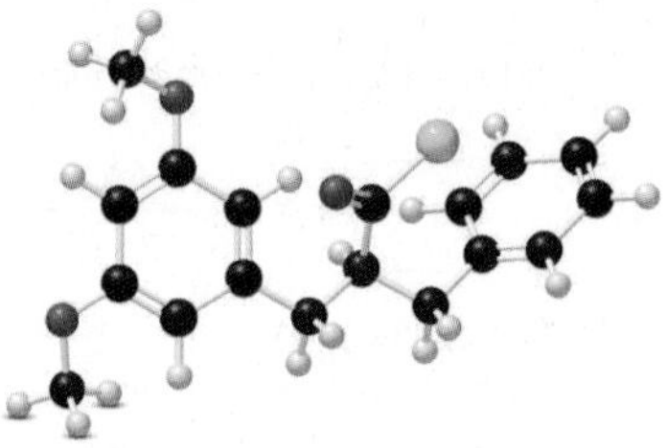

Reactions

18.36 Draw the products formed when phenol (C_6H_5OH) is treated with each set of reagents.

a. [1] HNO_3, H_2SO_4; [2] Sn, HCl
b. [1] $(CH_3CH_2)_2CHCOCl$, $AlCl_3$; [2] Zn(Hg), HCl
c. [1] CH_3CH_2Cl, $AlCl_3$; [2] Br_2, $h\nu$
d. [1] $(CH_3)_2CHCl$, $AlCl_3$; [2] $KMnO_4$

18.37 Draw the products formed when each compound is treated with CH_3CH_2COCl, $AlCl_3$.

a. 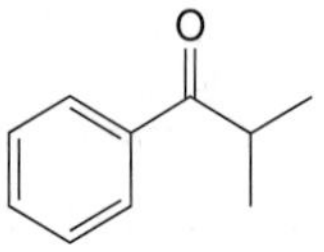

b. 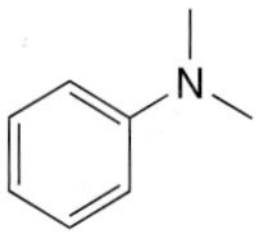

c. 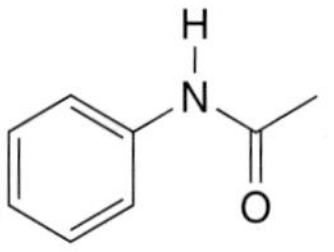

18.38 Draw the products of each reaction.

a. 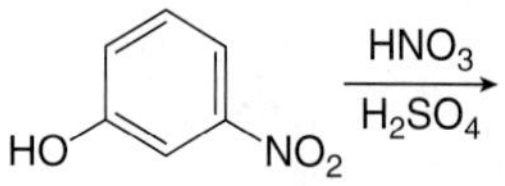

d. 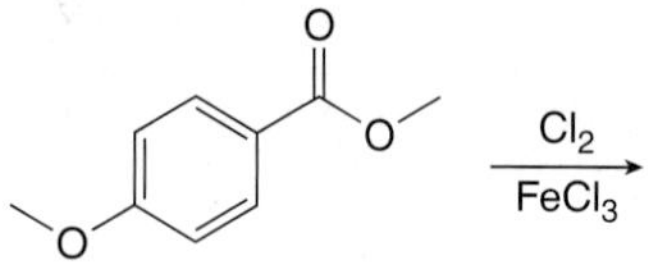

b. 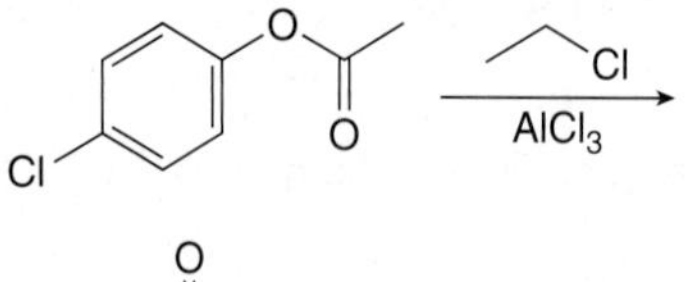

e. 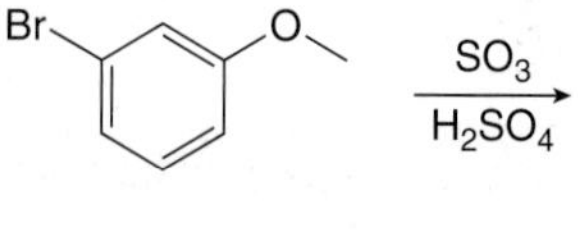

c. 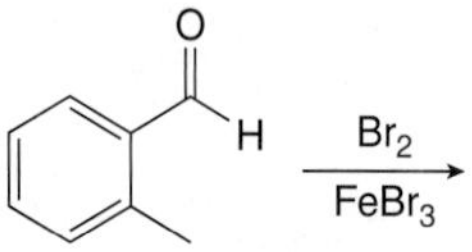

f. 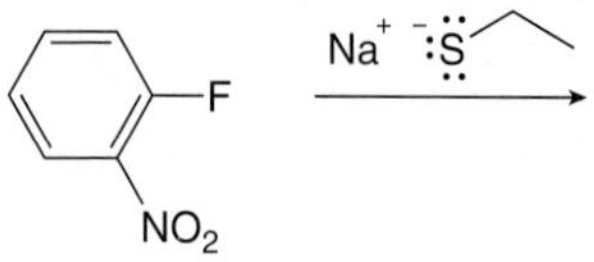

18.39 What products are formed when benzene is treated with each alkyl chloride and $AlCl_3$?

a. 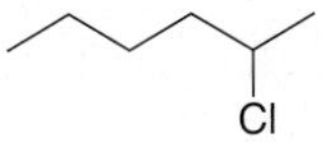
Cl

b. 
Cl

c. 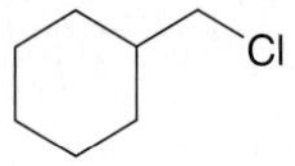
Cl

18.40 Draw the products of each reaction.

a. 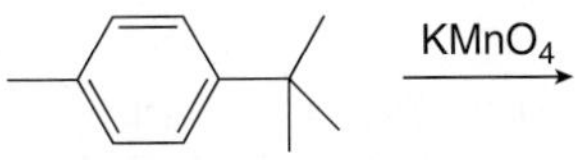
KMnO4

b. 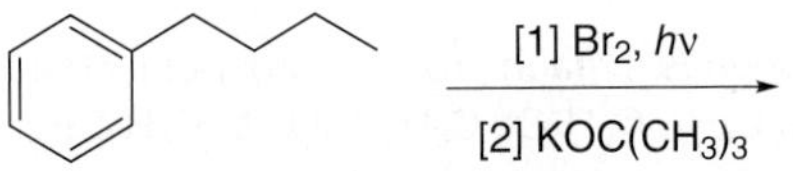
[1] Br2, hν
[2] KOC(CH3)3

c. 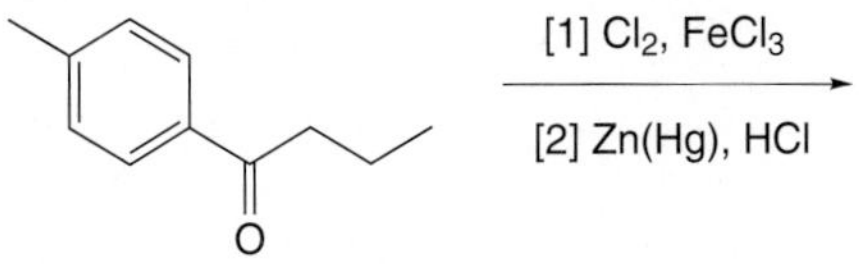
O
[1] Cl2, FeCl3
[2] Zn(Hg), HCl

d. 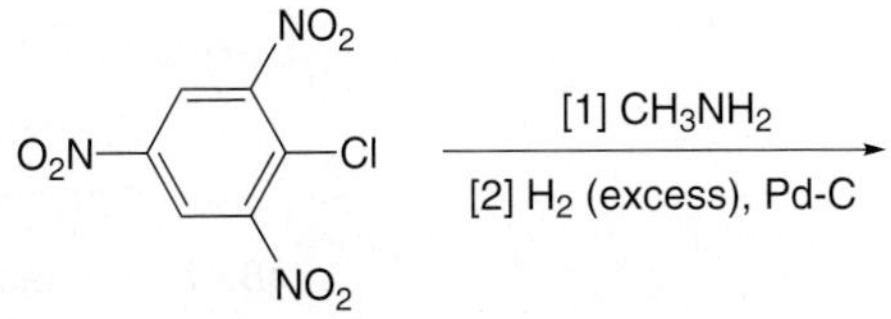
NO2
O2N
Cl
NO2
[1] CH3NH2
[2] H2 (excess), Pd-C

18.41 You have learned two ways to make an alkyl benzene: Friedel–Crafts alkylation, and Friedel–Crafts acylation followed by reduction. Although some alkyl benzenes can be prepared by both methods, it is often true that only one method can be used to prepare a given alkyl benzene. Which method(s) can be used to prepare each of the following compounds from benzene? Show the steps that would be used.

a. 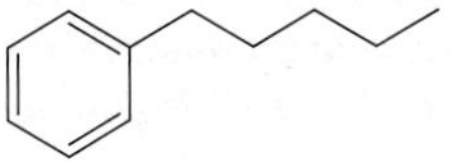

b. 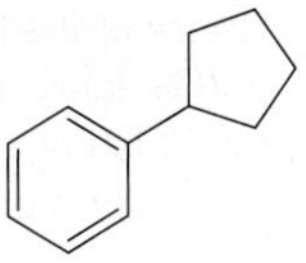

c. 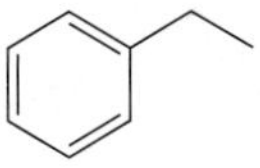

18.42 Draw the structure of **A,** an intermediate in the synthesis of the antipsychotic drug risperidone. Explain why three rings in risperidone are considered aromatic.

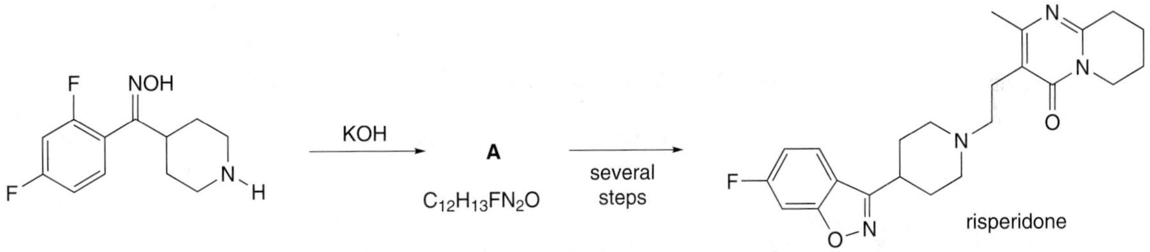
F
NOH
F
N
H
KOH
A
C12H13FN2O
several steps
F
O
N
N
N
N
O
risperidone

18.43 One step in the synthesis of pioglitazone (trade name Actos), a drug used to treat type 2 diabetes, involves the reaction of **A** with **B** in the presence of NaH to afford **C.** What is the structure of **C?**

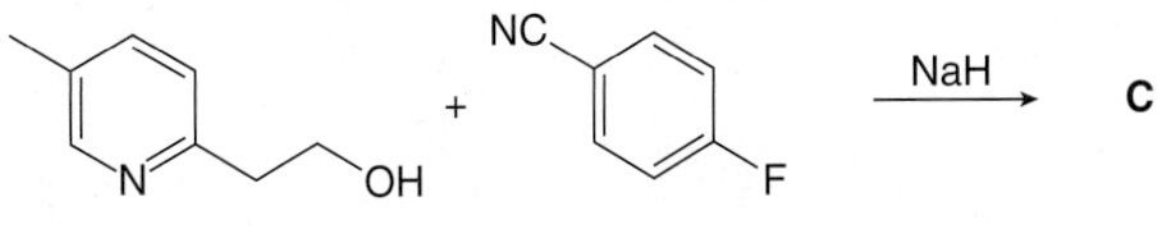
N
OH
+
NC
F
NaH
C

A **B**

18.44 **D** is an intermediate in the synthesis of rosiglitazone (trade name Avandia), a drug used to treat type 2 diabetes. Suggest two different methods to prepare the ether in **D** by substitution reactions.

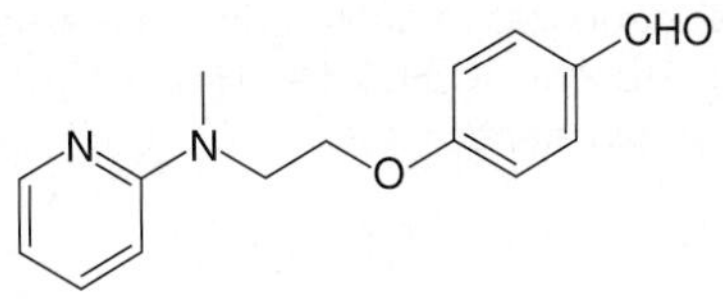
CHO
N
N
O

D

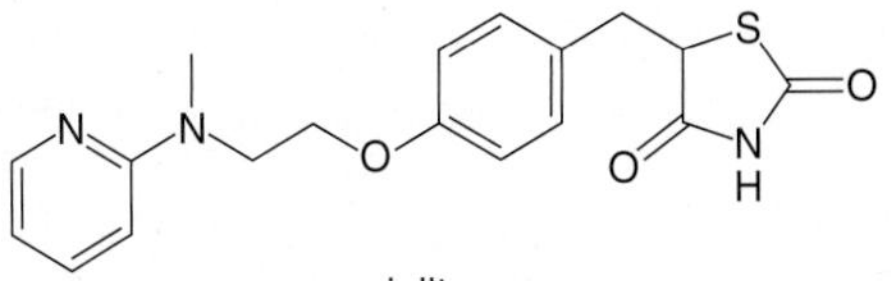
N
N
O
S
O
N
H
O

rosiglitazone

18.45 Explain why each of the following reactions will not form the given product. Then, design a synthesis of **A** from benzene and **B** from phenol (C_6H_5OH).

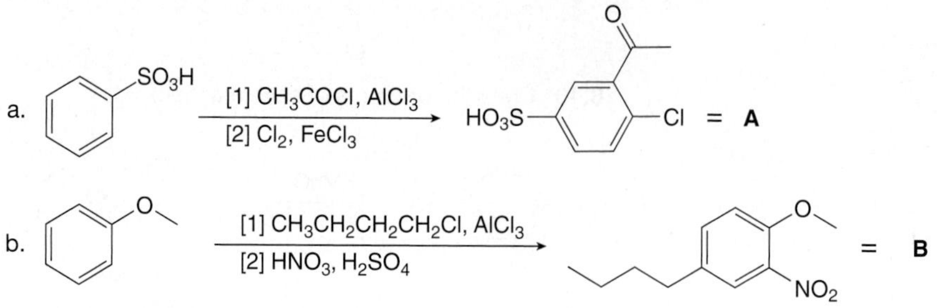

a.
SO3H
[1] CH3COCl, AlCl3
[2] Cl2, FeCl3
HO3S
O
Cl
= A
b.
O
[1] CH3CH2CH2CH2Cl, AlCl3
[2] HNO3, H2SO4
O
NO2
= B

Substituent Effects

18.46 Rank the compounds in each group in order of increasing reactivity in electrophilic aromatic substitution: (a) C_6H_6, C_6H_5Cl, C_6H_5CHO; (b) $C_6H_5CH_3$, $C_6H_5NH_2$, $C_6H_5CH_2NH_2$.

18.47 For each of the following substituted benzenes: [1] C_6H_5Br; [2] C_6H_5CN; [3] $C_6H_5OCOCH_3$:

a. Does the substituent donate or withdraw electron density by an inductive effect?

b. Does the substituent donate or withdraw electron density by a resonance effect?

c. On balance, does the substituent make a benzene ring more or less electron rich than benzene itself?

d. Does the substituent activate or deactivate the benzene ring in electrophilic aromatic substitution?

18.48 Consider the tetracyclic aromatic compound drawn below, with rings labeled as **A, B, C,** and **D.** (a) Which of the four rings is ***most*** reactive in electrophilic aromatic substitution? (b) Which of the four rings is ***least*** reactive in electrophilic aromatic substitution? (c) What are the major product(s) formed when this compound is treated with one equivalent of Br_2?

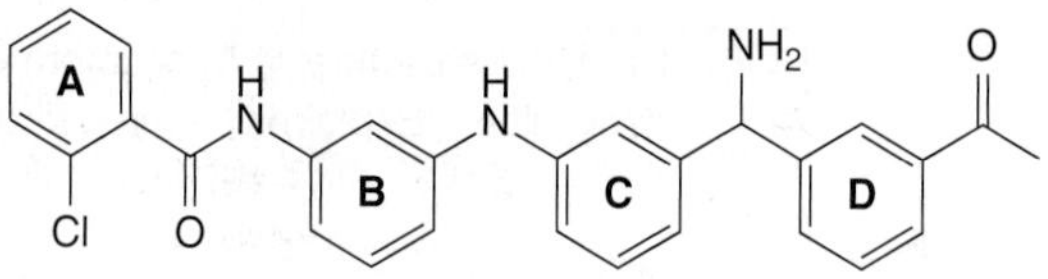

A
Cl
O
H
N
B
H
N
C
NH2
D
O

18.49 For each N-substituted benzene, predict whether the compound reacts faster than, slower than, or at a similar rate to benzene in electrophilic aromatic substitution. Then draw the major product(s) formed when each compound reacts with a general electrophile E^+.

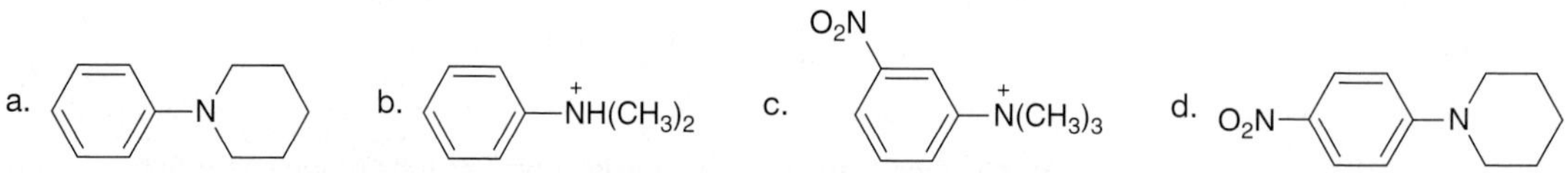

a.
N
b.
+
NH(CH3)2
c.
O2N
+
N(CH3)3
d.
O2N
N

18.50 What is the major product of electrophilic addition of HBr to the following alkene? Explain your choice.

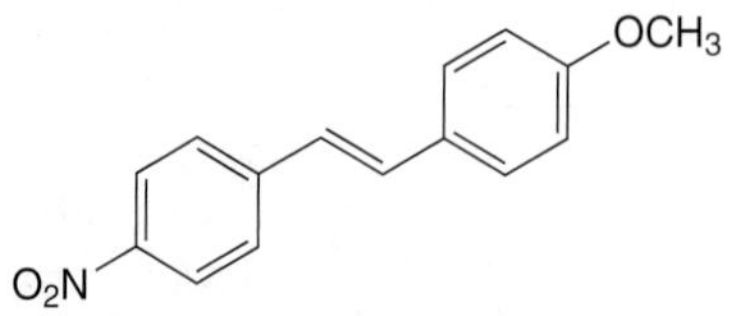

OCH3
O2N

18.51 Using resonance structures, explain why a nitroso group (– NO) is an ortho, para director that deactivates a benzene ring toward electrophilic attack.

18.52 Explain the following observation. Ethyl 3-phenylpropanoate ($C_6H_5CH_2CH_2CO_2CH_2CH_3$) reacts with electrophiles to afford ortho- and para-disubstituted arenes, but ethyl 3-phenylprop-2-enoate ($C_6H_5CH{=}CHCO_2CH_2CH_3$) reacts with electrophiles to afford meta-disubstituted arenes.

18.53 Rank the aryl halides in each group in order of increasing reactivity in nucleophilic aromatic substitution by an addition–elimination mechanism.

a. chlorobenzene, *p*-fluoronitrobenzene, *m*-fluoronitrobenzene
b. 1-fluoro-2,4-dinitrobenzene, 1-fluoro-3,5-dinitrobenzene, 1-fluoro-3,4-dinitrobenzene
c. 1-fluoro-2,4-dinitrobenzene, 4-chloro-3-nitrotoluene, 4-fluoro-3-nitrotoluene

Mechanisms

18.54 Draw a stepwise mechanism for the following reaction.

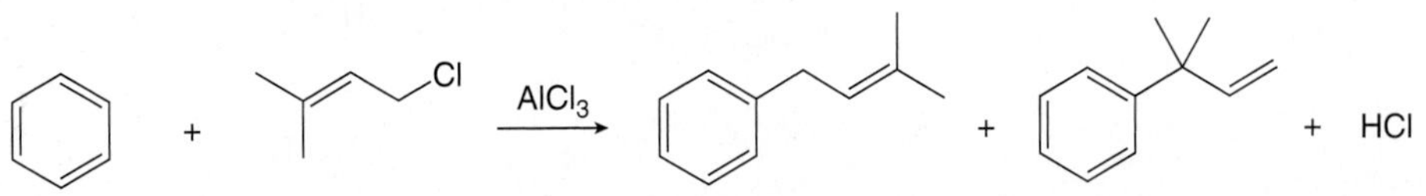

18.55 Draw a stepwise, detailed mechanism for the following intramolecular reaction.

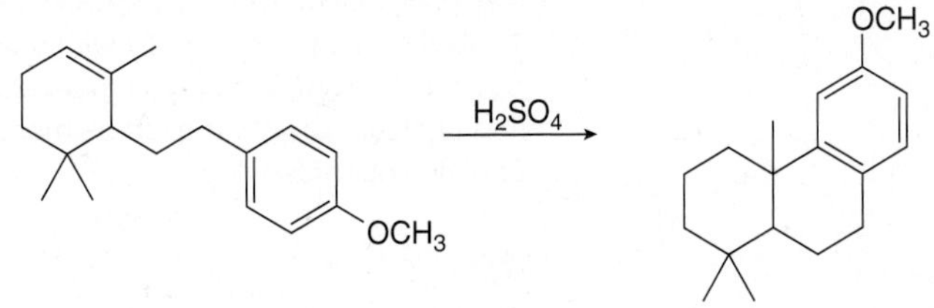

18.56 Draw a stepwise, detailed mechanism for the following reaction.

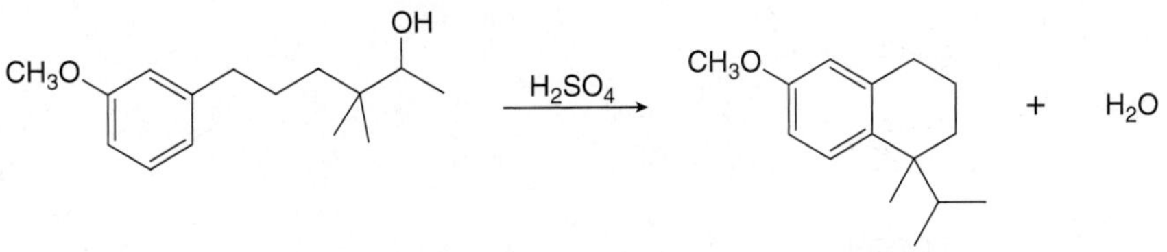

18.57 Friedel–Crafts alkylation of benzene with (*R*)-2-chlorobutane and $AlCl_3$ affords *sec*-butylbenzene.

a. How many stereogenic centers are present in the product?
b. Would you expect the product to exhibit optical activity? Explain, with reference to the mechanism.

18.58 Draw a stepwise mechanism for the following substitution. Explain why 2-chloropyridine reacts faster than chlorobenzene in this type of reaction.

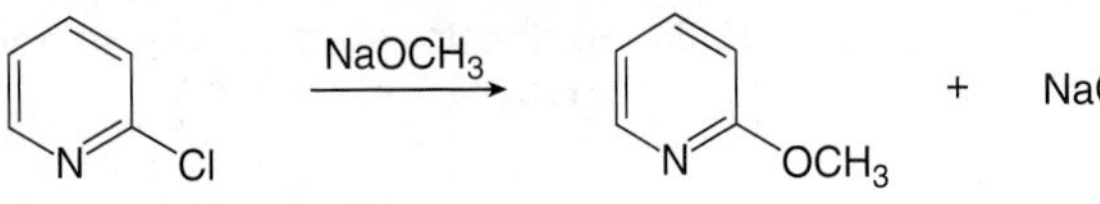

2-chloropyridine

18.59 Draw a stepwise mechanism for the following reaction.

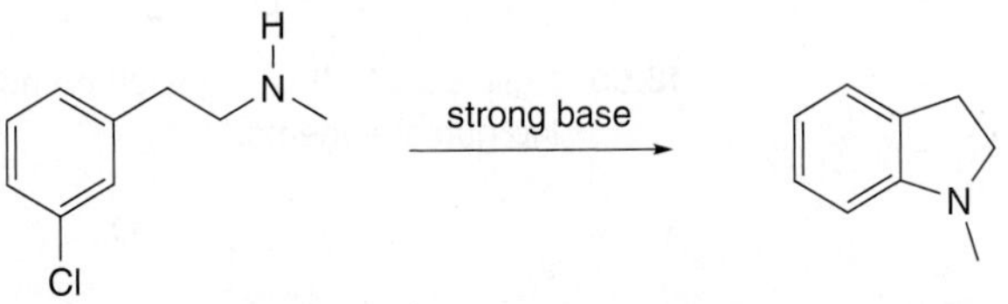

18.60 Although two products (**A** and **B**) are possible when naphthalene undergoes electrophilic aromatic substitution, only **A** is formed. Draw resonance structures for the intermediate carbocation to explain why this is observed.

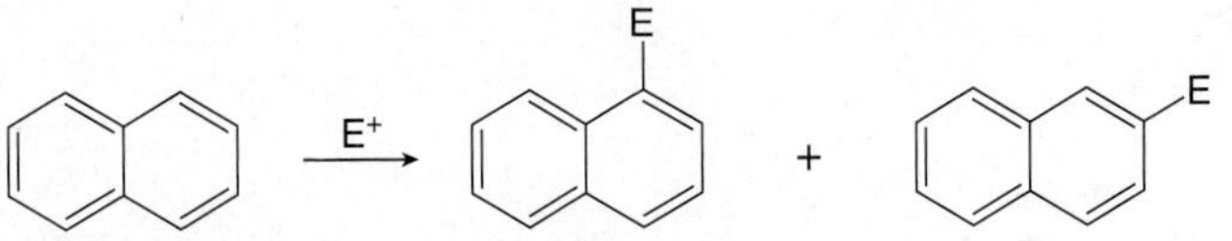

naphthalene

A
This product is formed.

B
This product is *not* formed.

18.61 Draw a stepwise mechanism for the following reaction, which results in the synthesis of bisphenol F (R = H), an additive used in a variety of packaging materials. Bisphenol F is related to BPA (bisphenol A, R = CH_3), a reagent used to harden some plastics, now removed from certain baby products because of its estrogen-like activity that can disrupt endocrine pathways.

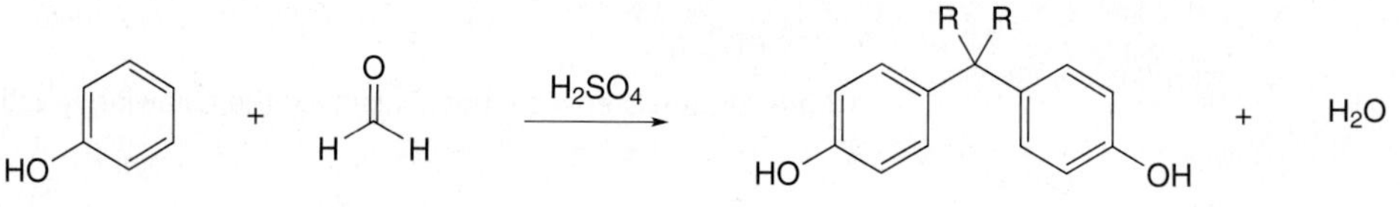

bisphenol F
R = H

18.62 Benzyl bromide ($C_6H_5CH_2Br$) reacts rapidly with CH_3OH to afford benzyl methyl ether ($C_6H_5CH_2OCH_3$). Draw a stepwise mechanism for the reaction, and explain why this 1° alkyl halide reacts rapidly with a weak nucleophile under conditions that favor an S_N1 mechanism. Would you expect the para-substituted benzylic halides $CH_3OC_6H_4CH_2Br$ and $O_2NC_6H_4CH_2Br$ to each be more or less reactive than benzyl bromide in this reaction? Explain your reasoning.

Synthesis

18.63 Synthesize each compound from benzene and any other organic or inorganic reagents.

a. 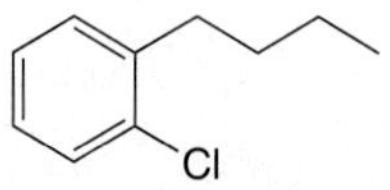

c. 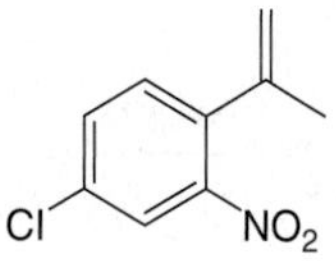

e. 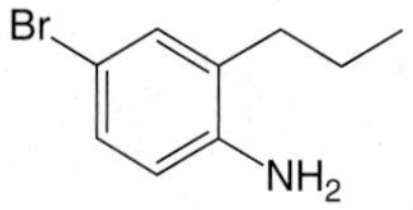

g. 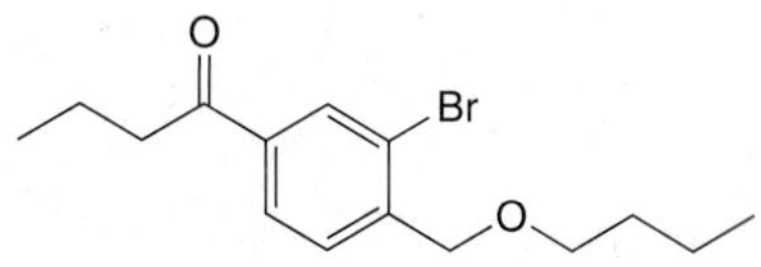

b. 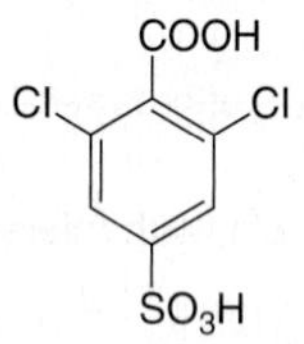

d. 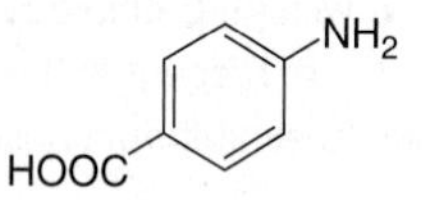

(PABA)
sunscreen component

f. 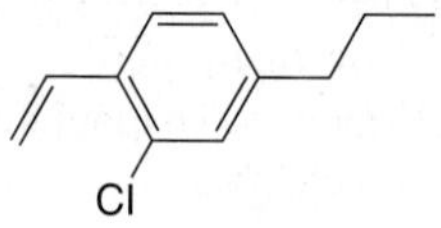

h. 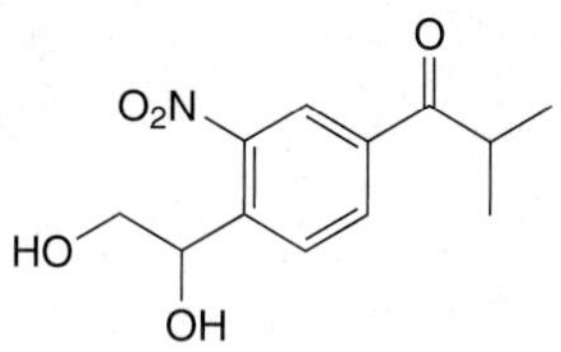

18.64 Synthesize each compound from toluene ($C_6H_5CH_3$) and any other organic or inorganic reagents.

a. 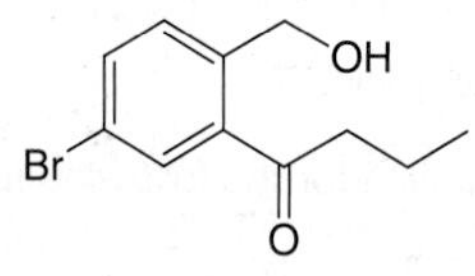

b. 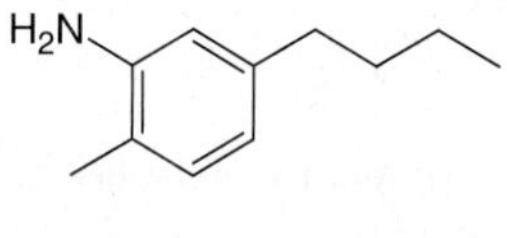

c. 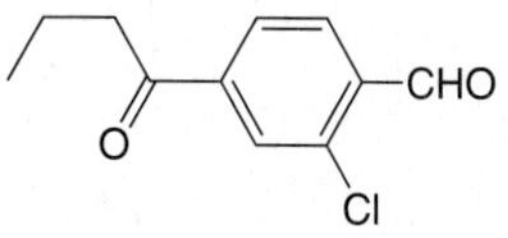

18.65 Devise a synthesis of each compound from phenol (C_6H_5OH) and any other organic or inorganic reagents.

a. 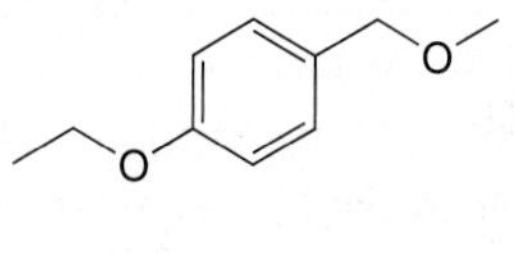

b. 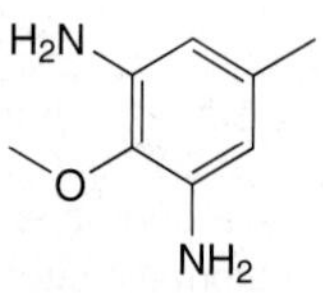

18.66 Use the reactions in this chapter along with those learned in Chapters 11 and 12 to synthesize each compound. You may use benzene, acetylene (HC≡CH), two-carbon alcohols, ethylene oxide, and any inorganic reagents.

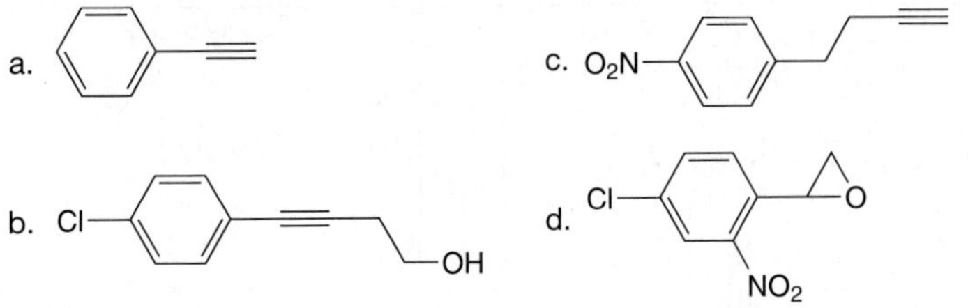

18.67 Ibufenac, a para-disubstituted arene with the structure $HO_2CCH_2C_6H_4CH_2CH(CH_3)_2$, is a much more potent analgesic than aspirin, but it was never sold commercially because it caused liver toxicity in some clinical trials. Devise a synthesis of ibufenac from benzene and organic halides having fewer than five carbons.

18.68 Carboxylic acid **X** is an intermediate in the multistep synthesis of proparacaine, a local anesthetic. Devise a synthesis of **X** from phenol and any needed organic or inorganic reagents.

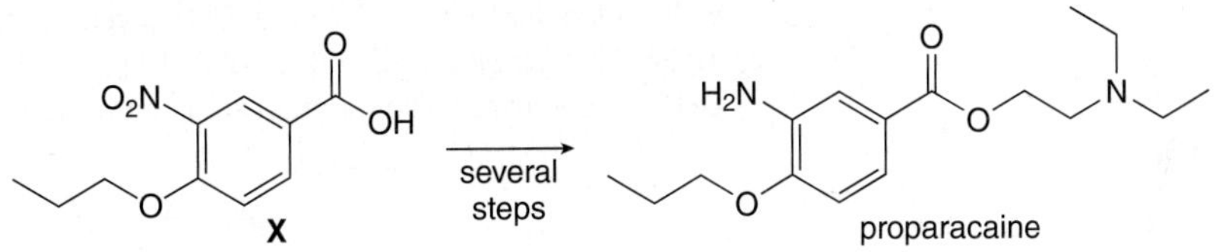

Spectroscopy

18.69 Identify the structures of isomers **A** and **B** (molecular formula C_8H_9Br).

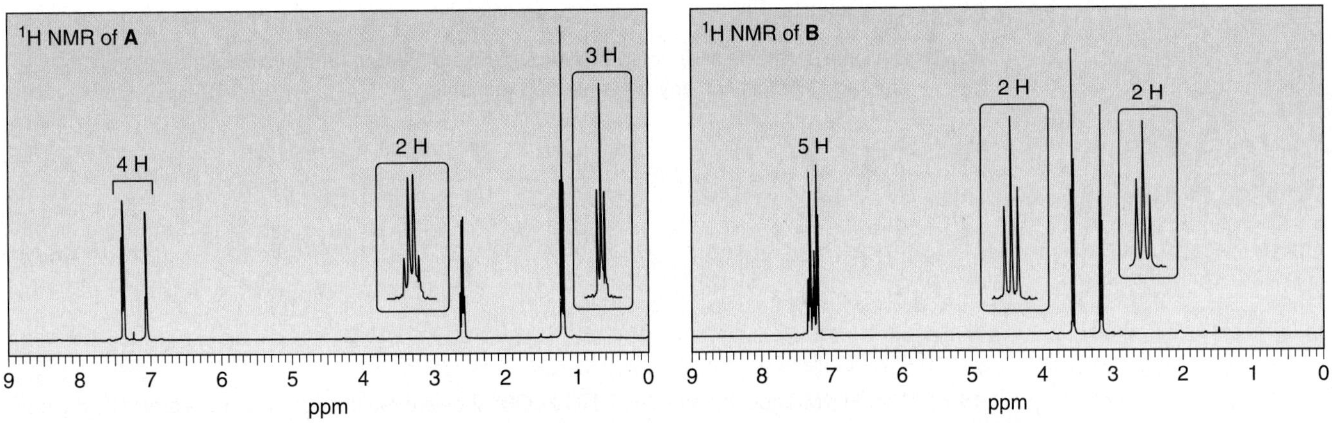

18.70 Propose a structure of compound **C** (molecular formula $C_{10}H_{12}O$) consistent with the following data. **C** is partly responsible for the odor and flavor of raspberries.

Compound **C:** IR absorption at 1717 cm^{-1}

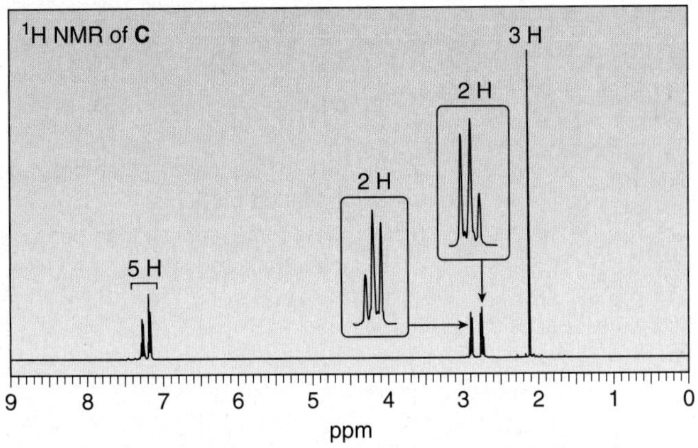

18.71 Compound **X** (molecular formula $C_{10}H_{12}O$) was treated with NH_2NH_2, ^{-}OH to yield compound **Y** (molecular formula $C_{10}H_{14}$). Based on the ^1H NMR spectra of **X** and **Y** given below, what are the structures of **X** and **Y**?

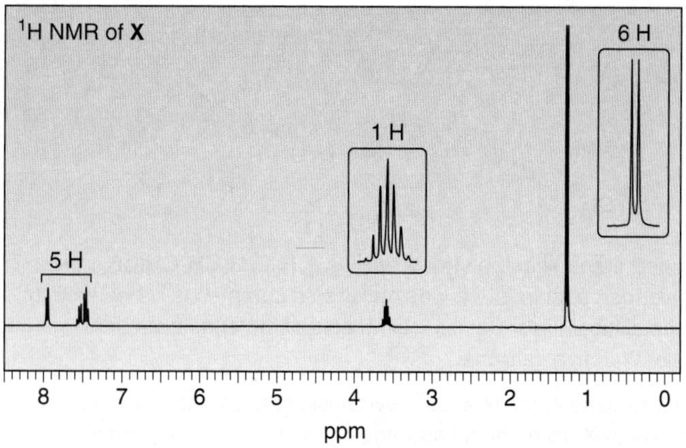

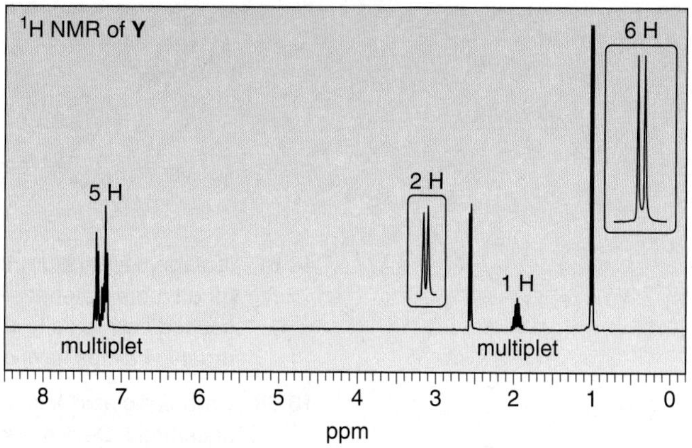

18.72 Reaction of *p*-cresol with two equivalents of 2-methylprop-1-ene affords BHT, a preservative with molecular formula $C_{15}H_{24}O$. BHT gives the following ^1H NMR spectral data: 1.4 (singlet, 18 H), 2.27 (singlet, 3 H), 5.0 (singlet, 1 H), and 7.0 (singlet, 2 H) ppm. What is the structure of BHT? Draw a stepwise mechanism illustrating how it is formed.

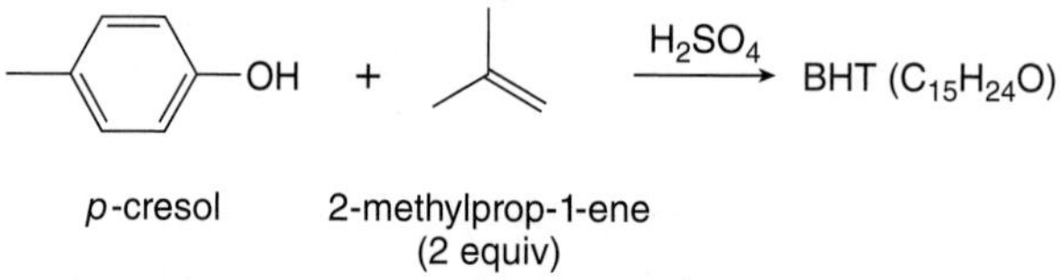

Challenge Problems

18.73 Devise a synthesis of optically active (*S*)-fluoxetine (trade name Prozac) from the given starting materials and any other needed reagents.

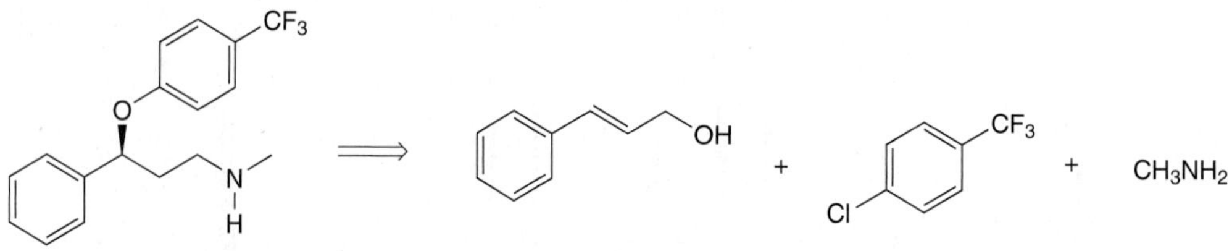

(*S*)-fluoxetine

18.74 The ^1H NMR spectrum of phenol (C_6H_5OH) shows three absorptions in the aromatic region: 6.70 (2 ortho H's), 7.14 (2 meta H's), and 6.80 (1 para H) ppm. Explain why the ortho and para absorptions occur at lower chemical shift than the meta absorption.

18.75 Explain the reactivity and orientation effects observed in each heterocycle.

a. Pyridine is less reactive than benzene in electrophilic aromatic substitution and yields 3-substituted products.

b. Pyrrole is more reactive than benzene in electrophilic aromatic substitution and yields 2-substituted products.

18.76 Draw a stepwise, detailed mechanism for the dienone–phenol rearrangement, a reaction that forms alkyl-substituted phenols from cyclohexadienones.

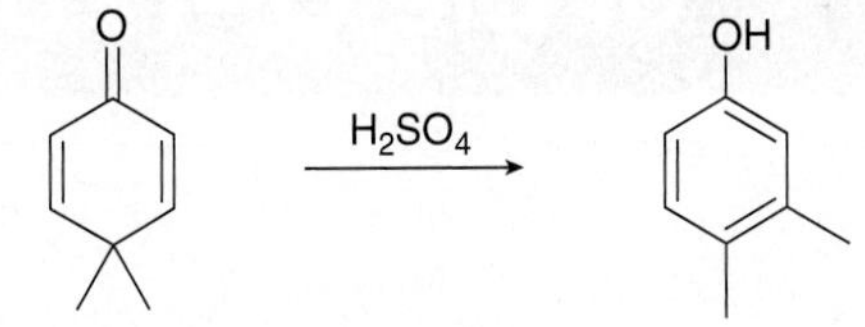

18.77 Draw a stepwise mechanism for the following intramolecular reaction, which is used in the synthesis of the female sex hormone estrone.

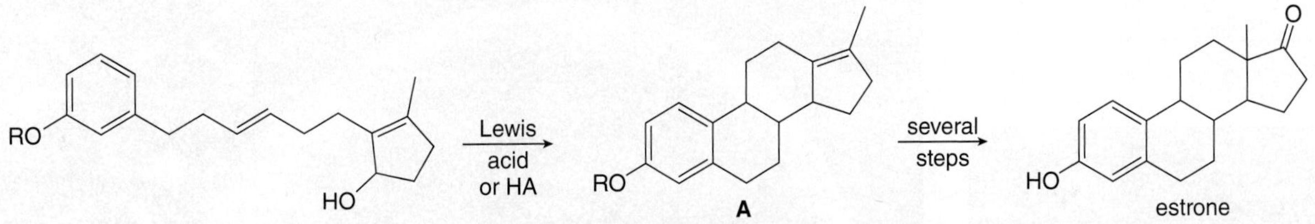

18.78 The bicyclic heterocycles quinoline and indole undergo electrophilic aromatic substitution to give the products shown. (a) Explain why electrophilic substitution occurs on the ring without the N atom for quinoline, but occurs on the ring with the N atom in indole. (b) Explain why electrophilic substitution occurs more readily at C8 than C7 in quinoline. (c) Explain why electrophilic substitution occurs more readily at C3 rather than C2 of indole.

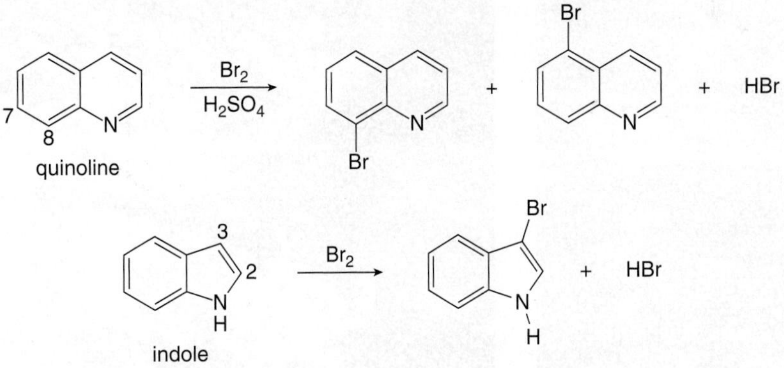

18.79 Devise a stepwise mechanism for the following reaction. The reaction does not take place by direct electrophilic aromatic substitution at C2. (Hint: The mechanism begins with addition of an electrophile at C3.)

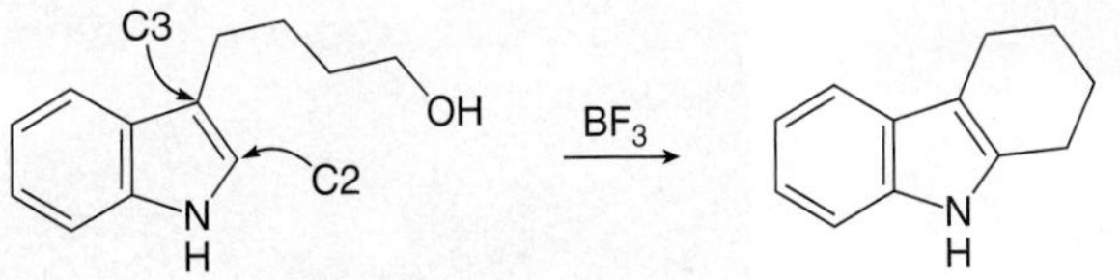

19 Carboxylic Acids and the Acidity of the O–H Bond

19.1 Structure and bonding
19.2 Nomenclature
19.3 Physical properties
19.4 Spectroscopic properties
19.5 Interesting carboxylic acids
19.6 Aspirin, arachidonic acid, and prostaglandins
19.7 Preparation of carboxylic acids
19.8 Reactions of carboxylic acids—General features
19.9 Carboxylic acids—Strong organic Brønsted–Lowry acids
19.10 The Henderson–Hasselbalch equation
19.11 Inductive effects in aliphatic carboxylic acids
19.12 Substituted benzoic acids
19.13 Extraction
19.14 Organic acids containing sulfur and phosphorus
19.15 Amino acids

Lysine is an essential amino acid that is needed for protein synthesis. Because lysine cannot be synthesized by humans and is not stored in the body, it must be ingested on a regular basis. Common food sources of lysine are meat, beans, peas, soy, and peanuts. Although most grains are low in lysine, quinoa is relatively high in lysine content and a good source of essential amino acids for a vegetarian diet. Like other amino acids, lysine contains both a carboxylic acid and an amine base. In Chapter 19 we learn about carboxylic acids and the acid–base properties of amino acids.

Chapter 19 serves as a transition between the preceding discussion of resonance and aromaticity, and the subsequent treatment of carbonyl chemistry. We pause to study the chemistry of the OH group by examining **carboxylic acids (RCO_2H),** and to a lesser extent, **phenols (PhOH)** and **alcohols (ROH).**

In Chapter 19 we concentrate on the acidity of carboxylic acids, and revisit some of the factors that determine acidity, a topic first discussed in Chapter 2. Then, in Chapters 20 and 22 we will learn other reactions of carboxylic acids that occur at the carbonyl group.

19.1 Structure and Bonding

The word **carboxy** (for a COOH group) is derived from ***carb**onyl* (C=O) + hydr***oxy*** (OH).

***Carboxylic acids* are organic compounds containing a carboxy group (COOH).** Although the structure of a carboxylic acid is often abbreviated as **RCOOH** or **RCO_2H,** keep in mind that the central carbon atom of the functional group is doubly bonded to one oxygen atom and singly bonded to another.

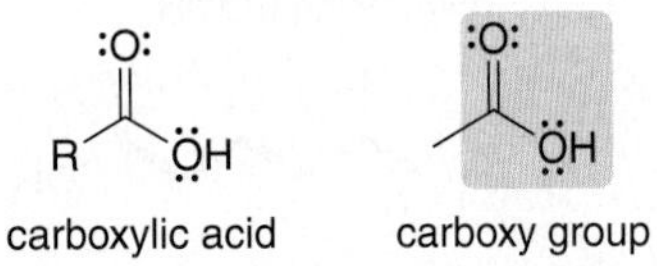

The carbon atom of a carboxy group is surrounded by three groups, making it ***sp*^{2} hybridized** and **trigonal planar,** with bond angles of approximately 120°. The C=O of a carboxylic acid is *shorter* than its C–O.

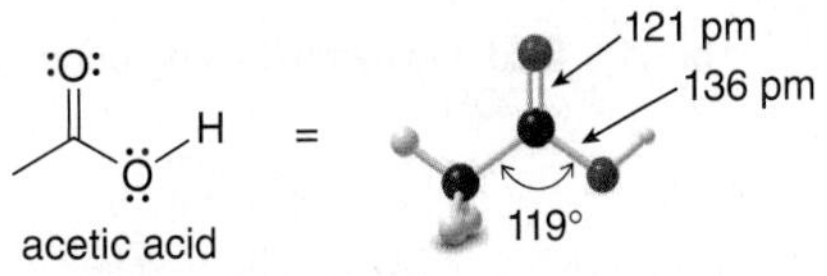

Because oxygen is more electronegative than either carbon or hydrogen, **the C–O and O–H bonds are polar.** The electrostatic potential plot of acetic acid in Figure 19.1 shows that the carbon and hydrogen atoms are electron poor and the oxygen atoms are electron rich.

Figure 19.1
Electrostatic potential plot of acetic acid (CH_3COOH)

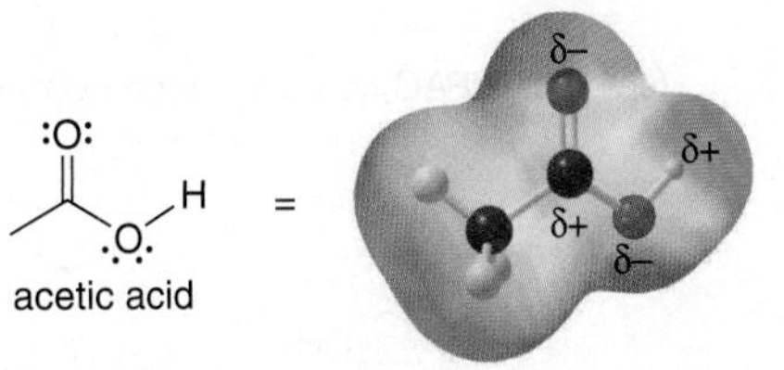

- Acetic acid contains two electron-rich oxygen atoms (in red). Its carbonyl carbon and hydroxy hydrogen are both electron deficient.

19.2 Nomenclature

Both IUPAC and common names are used for carboxylic acids.

19.2A IUPAC System

In IUPAC nomenclature, carboxylic acids are identified by a suffix added to the parent name of the longest chain, and two different endings are used depending on whether the carboxy group is bonded to a chain or ring.

To name a carboxylic acid using the IUPAC system:

[1] If the COOH is bonded to a chain of carbons, find the longest chain containing the COOH group, and change the ***-e*** ending of the parent alkane to the suffix ***-oic acid.*** If the COOH group is bonded to a ring, name the ring and add the words ***carboxylic acid.***

[2] Number the carbon chain or ring to put the **COOH group at C1,** but omit this number from the name. Apply all of the other usual rules of nomenclature.

Sample Problem 19.1 Give the IUPAC name of each compound.

a. 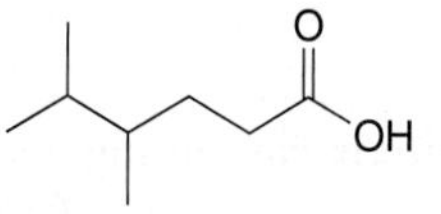

b. 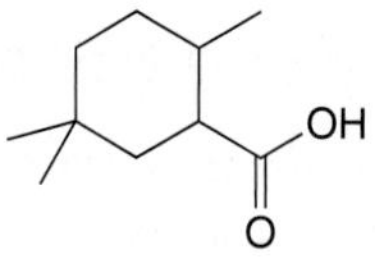

Solution

a. [1] Find and name the longest chain containing COOH:

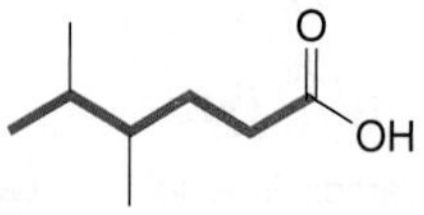

hexane ⟶ hexan*oic acid*
(6 C's)

The COOH contributes one C to the longest chain.

[2] Number and name the substituents:

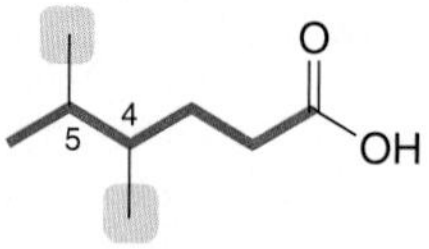

two methyl substituents on C4 and C5

Answer: 4,5-dimethylhexanoic acid

b. [1] Find and name the ring bonded to COOH.

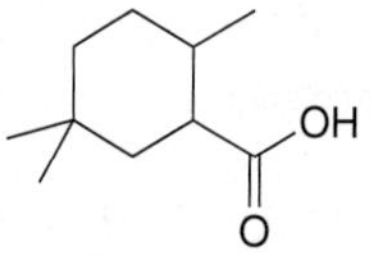

cyclohexane + carboxylic acid
(6 C's)

[2] Number and name the substituents:

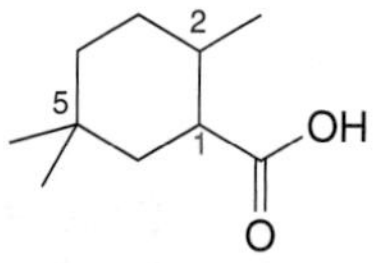

Number to put COOH at C1 and give the second substituent (CH_3) the lower number (C2).

Answer: 2,5,5-trimethylcyclohexanecarboxylic acid

Problem 19.1 Give the IUPAC name for each compound.

a. 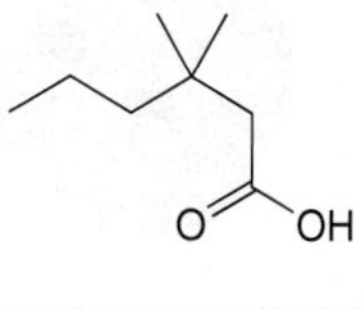

b. 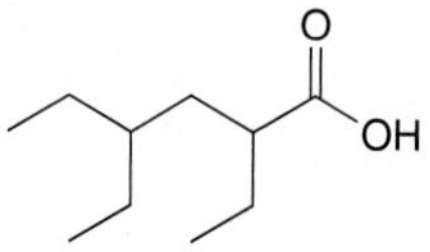

c. 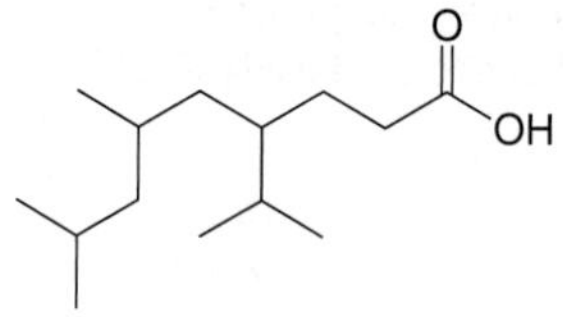

Problem 19.2 Give the structure corresponding to each IUPAC name.

a. 2-bromobutanoic acid
b. 2,3-dimethylpentanoic acid
c. 3,3,4-trimethylheptanoic acid
d. 2-*sec*-butyl-4,4-diethylnonanoic acid
e. 3,4-diethylcyclohexanecarboxylic acid
f. 1-isopropylcyclobutanecarboxylic acid

19.2B Common Names

Most simple carboxylic acids have common names that are more widely used than their IUPAC names.

- **A common name is formed by using a common parent name followed by the suffix *-ic acid.***

Table 19.1 lists common parent names for some simple carboxylic acids. These parent names are used in the nomenclature of many other compounds with carbonyl groups (Chapters 21 and 22).

Table 19.1 Common Names for Some Simple Carboxylic Acids

Number of C atoms	Structure	Parent name	Common name
1	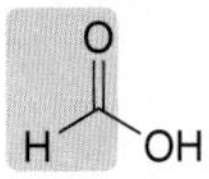	**form-**	formic acid
2	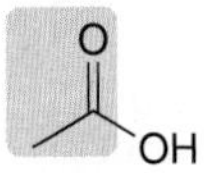	**acet-**	acetic acid
3	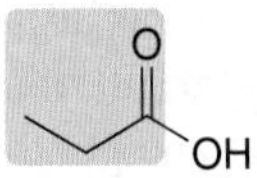	**propion-**	propionic acid
4	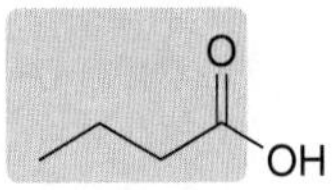	**butyr-**	butyric acid
5	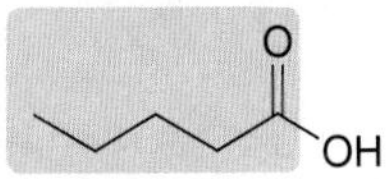	**valer-**	valeric acid
6	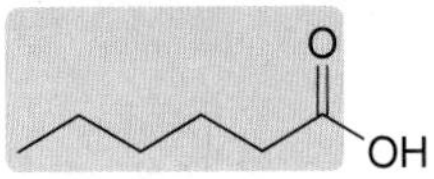	**capro-**	caproic acid
	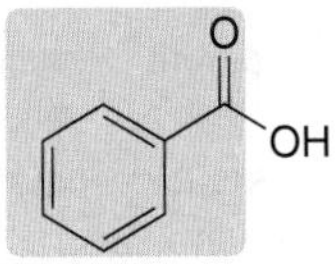	**benzo-**	benzoic acid

Caproic acid, the common name for hexanoic acid, is a foul-smelling carboxylic acid found in the fleshy coat of ginkgo seeds, giving the seeds an unpleasant and even repulsive odor.

Greek letters are used to designate the location of substituents in common names.

- **The carbon adjacent to the COOH is called the α carbon.**
- **The carbon bonded to the α carbon is the β carbon, followed by the γ (gamma) carbon, the δ (delta) carbon, and so forth down the chain. The last carbon in the chain is sometimes called the Ω (omega) carbon.**

The α carbon in the common system is numbered C2 in the IUPAC system.

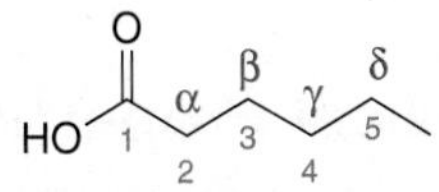

IUPAC numbering begins at the C=O.
Greek lettering begins at the C bonded to the C=O.

Problem 19.3 Draw the structure corresponding to each common name:

a. α-methoxyvaleric acid
b. β-phenylpropionic acid
c. α,β-dimethylcaproic acid
d. α-chloro-β-methylbutyric acid

19.2C Other Nomenclature Facts

Many compounds containing two carboxy groups are also known. In the IUPAC system, **diacids** are named by adding the suffix ***-dioic acid*** to the name of the parent alkane. Many diacids are formed in the citric acid cycle, an enzyme-catalyzed pathway that takes place during the metabolism of carbohydrates, amino acids, and lipids. Three of these diacids, which are most commonly identified by their common names, are shown.

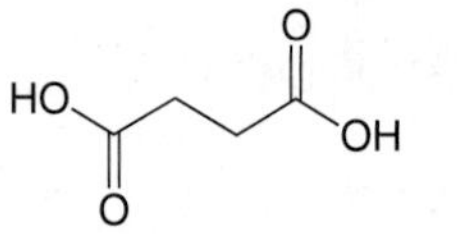

succinic acid
(butanedioic acid)

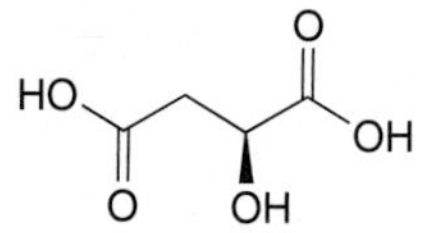

malic acid
[(*S*)-hydroxybutanedioic acid]

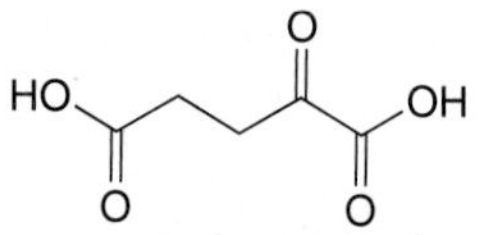

α-ketoglutaric acid
(2-oxopentanedioic acid)

Metal salts of carboxylate anions are formed from carboxylic acids in many reactions in Chapter 19. To name the **metal salt of a carboxylate anion,** change the *-ic acid* ending of the carboxylic acid to the suffix ***-ate*** and put three parts together:

name of the metal cation + parent + suffix

parent: common or IUPAC

suffix: ***-ate***

Two examples are shown in Figure 19.2.

Figure 19.2 Naming the metal salts of carboxylate anions

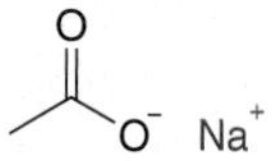

sodium cation

parent + suffix
acet- -ate

sodium acetate

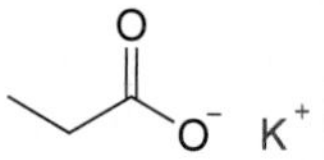

potassium cation

parent + suffix
propano- -ate

potassium propanoate

Problem 19.4 Give the IUPAC name for each metal salt of a carboxylate anion.

a. 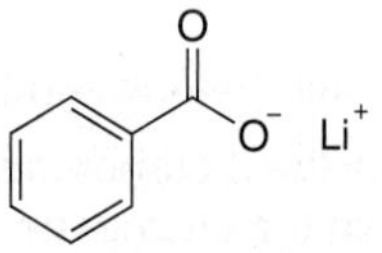

b. 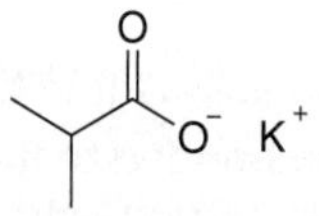

c. 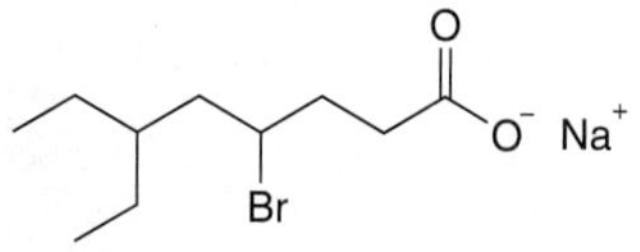

Problem 19.5 Depakote, a drug used to treat seizures and bipolar disorder, consists of a mixture of valproic acid [$(CH_3CH_2CH_2)_2CHCO_2H$] and its sodium salt. Give IUPAC names for each of these compounds.

19.3 Physical Properties

Carboxylic acids exhibit **dipole–dipole** interactions because they have polar C–O and O–H bonds. They also exhibit intermolecular **hydrogen bonding** because they possess a hydrogen atom bonded to an electronegative oxygen atom. Carboxylic acids often exist as **dimers,**

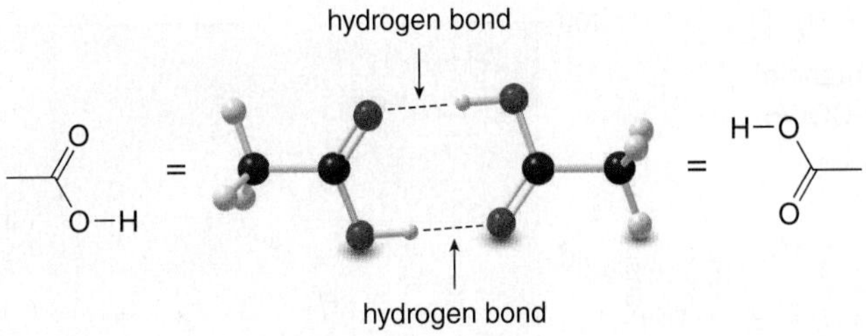

Figure 19.3 Two molecules of acetic acid (CH_3COOH) held together by two hydrogen bonds

held together by *two* intermolecular hydrogen bonds between the carbonyl oxygen atom of one molecule and the OH hydrogen atom of another molecule (Figure 19.3).

How these intermolecular forces affect the physical properties of carboxylic acids is summarized in Table 19.2.

Table 19.2 Physical Properties of Carboxylic Acids

Property	Observation
Boiling point and melting point	• Carboxylic acids have higher boiling points and melting points than other compounds of comparable molecular weight. 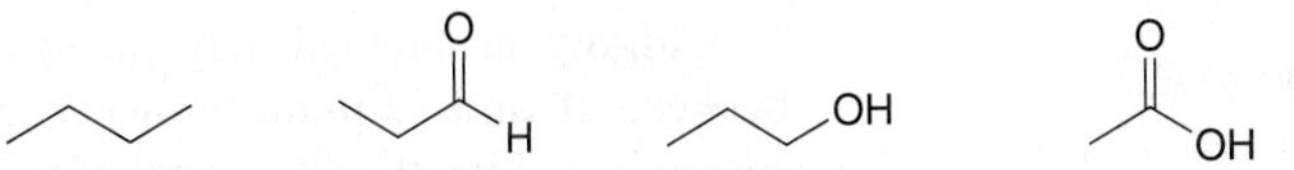 VDW / VDW, DD / VDW, DD, HB / VDW, DD, two HB MW = 58 / MW = 58 / MW = 60 / MW = 60 bp 0 °C / bp 48 °C / bp 97 °C / bp 118 °C  **Increasing strength of intermolecular forces** **Increasing boiling point**
Solubility	• Carboxylic acids are soluble in organic solvents regardless of size. • Carboxylic acids having ≤ 5 C's are water soluble because they can hydrogen bond with H_2O (Section 3.4C). • Carboxylic acids having > 5 C's are water insoluble because the nonpolar alkyl portion is too large to dissolve in the polar H_2O solvent. These "fatty" acids dissolve in a nonpolar fat-like environment but do not dissolve in water.

Key: VDW = van der Waals, DD = dipole–dipole, HB = hydrogen bonding, MW = molecular weight

Problem 19.6 Rank the following compounds in order of increasing boiling point. Which compound is the most water soluble? Which compound is the least water soluble?

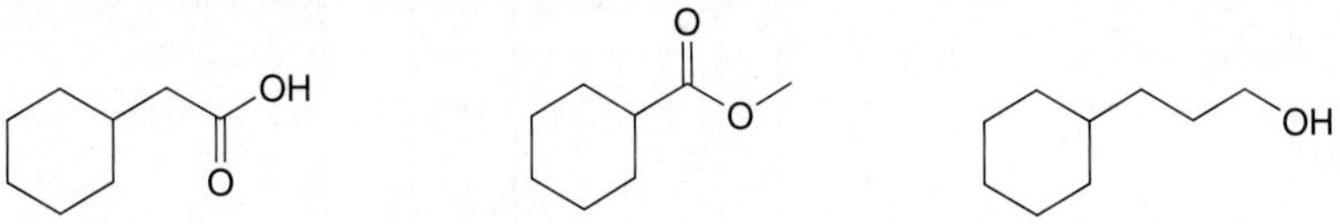

19.4 Spectroscopic Properties

Carboxylic acids have very characteristic IR and NMR absorptions. In the IR, carboxylic acids show two strong absorptions.

- The **C=O group absorbs at about 1710 cm^{-1}**, in the usual region for a carbonyl.
- The **O–H absorption** occurs at **2500–3500 cm^{-1}**. This very broad absorption sometimes obscures the C–H peak at 3000 cm^{-1}.

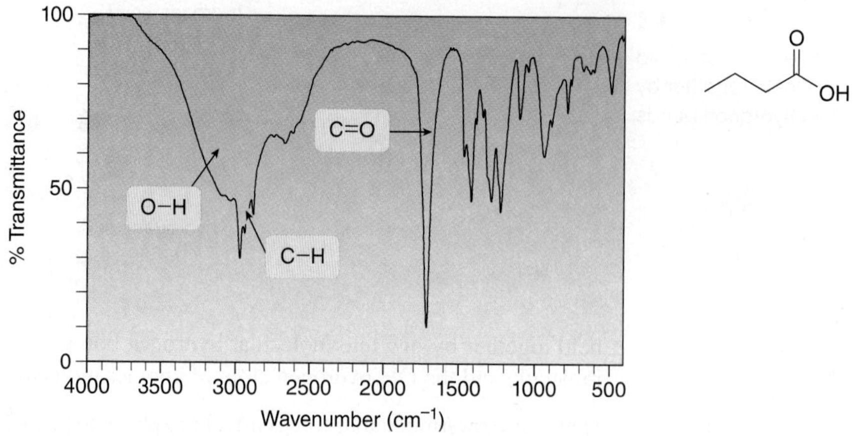

Figure 19.4
The IR spectrum of butanoic acid, $CH_3CH_2CH_2COOH$

- A strong C=O absorption occurs at 1712 cm^{-1}.
- The broad O–H absorption (2500–3500 cm^{-1}) nearly obscures the C–H peak at ~3000 cm^{-1}.

The IR spectrum of butanoic acid in Figure 19.4 illustrates these characteristic peaks.

Carboxylic acids have two noteworthy ^{1}H NMR absorptions and one noteworthy ^{13}C NMR absorption.

- **The highly deshielded OH proton absorbs in the ^{1}H NMR spectrum somewhere between 10 and 12 ppm,** farther *downfield* than all other absorptions of common organic compounds. Like the OH signal of an alcohol, the exact location depends on the degree of hydrogen bonding and the concentration of the sample.
- The protons on the α carbon to the carboxy group are somewhat deshielded, absorbing at 2–2.5 ppm.
- In the ^{13}C NMR spectrum, the carbonyl absorption is highly deshielded, appearing at 170–210 ppm.

Sometimes the OH absorption of a carboxylic acid is very broad, so that it is almost buried in the baseline of the ^{1}H NMR spectrum, making it difficult to see (Figure 19.5).

Figure 19.5 illustrates the ^{1}H and ^{13}C NMR spectra of propanoic acid.

Problem 19.7 Explain how you could use IR spectroscopy to distinguish among the following three compounds.

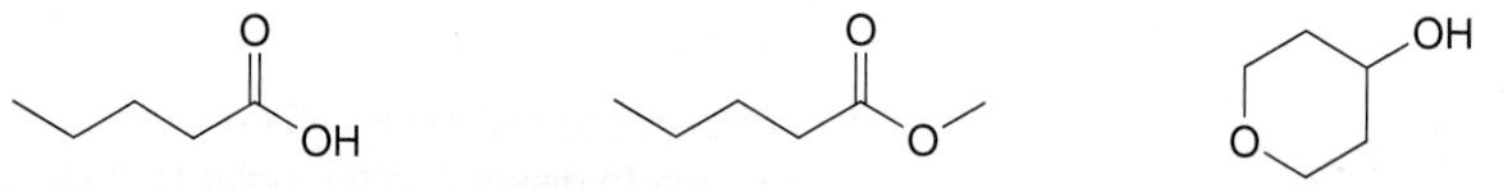

Figure 19.5 The ^{1}H and ^{13}C NMR spectra of propanoic acid

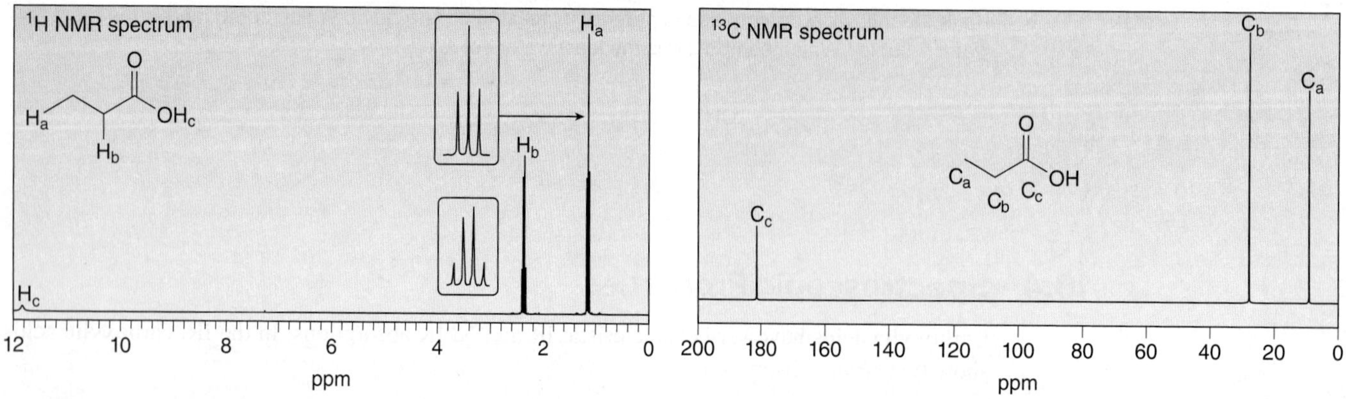

- **^{1}H NMR spectrum:** There are three signals due to three different kinds of H atoms. The H_a and H_b signals are split into a triplet and quartet, respectively. The H_c signal, a singlet, is due to the highly deshielded OH proton.
- **^{13}C NMR spectrum:** There are three signals due to three different kinds of carbon atoms. The carbonyl carbon is highly deshielded.

19.5 Interesting Carboxylic Acids

Several simple carboxylic acids have characteristic odors and flavors.

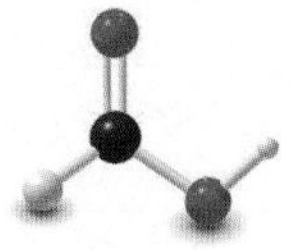
formic acid

- **Formic acid (HCOOH),** a carboxylic acid with an acrid odor and a biting taste, is responsible for the sting of some types of ants. The name is derived from the Latin word *formica,* meaning "ant."

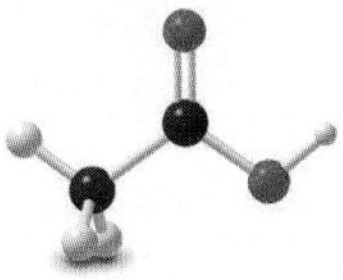
acetic acid

- **Acetic acid (CH_3COOH)** is the sour-tasting component of vinegar. The name comes from the Latin *acetum,* meaning "vinegar." The air oxidation of ethanol to acetic acid is the process that makes "bad" wine taste sour. Acetic acid is an industrial starting material for polymers used in paints and adhesives. Pure acetic acid is often called *glacial* acetic acid, because it freezes just below room temperature (mp = 17 °C), forming white crystals reminiscent of the ice in a glacier.

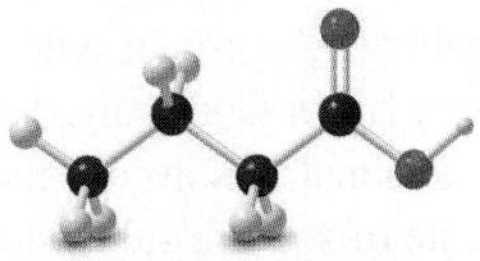
butanoic acid

- **Butanoic acid ($CH_3CH_2CH_2COOH$)** is an oxidation product that contributes to the disagreeable smell of body odor. Its common name, butyric acid, is derived from the Latin word *butyrum,* meaning "butter," because butyric acid gives rancid butter its peculiar odor and taste.

Although oxalic acid is toxic, you would have to eat about nine pounds of spinach at one time to ingest a fatal dose.

Oxalic acid and **lactic acid** are simple carboxylic acids quite prevalent in nature. Oxalic acid occurs naturally in spinach and rhubarb. Lactic acid gives sour milk its distinctive taste.

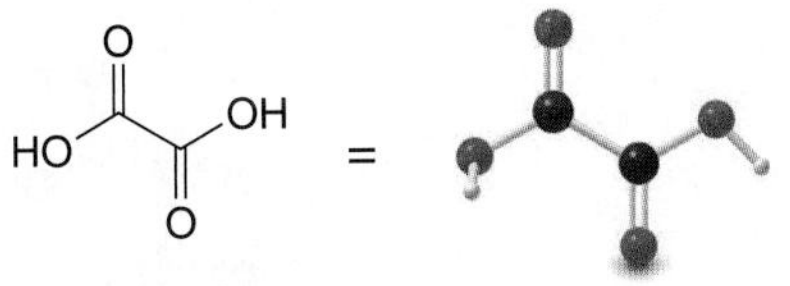
oxalic acid

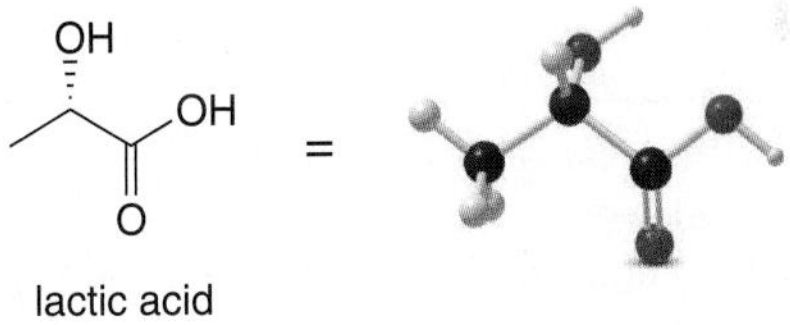
lactic acid

4-Hydroxybutanoic acid, known by its common name γ-hydroxybutyric acid (GHB), is an illegal recreational drug that depresses the central nervous system and results in intoxication. GHB is highly addictive and widely abused, and because its taste can be easily masked in an alcoholic beverage, it has been used as a "date rape" drug.

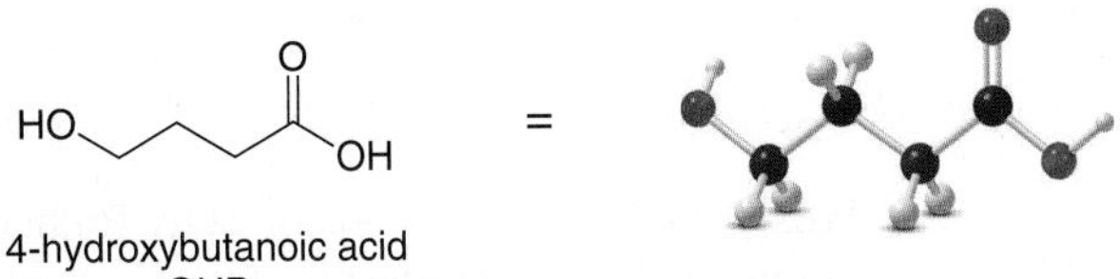
4-hydroxybutanoic acid
GHB

Soaps, the sodium salts of fatty acids, were discussed in Section 3.6.

Salts of carboxylic acids are commonly used as preservatives. Sodium benzoate, a fungal growth inhibitor, is a preservative used in soft drinks, and potassium sorbate is an additive that prolongs the shelf-life of baked goods and other foods.

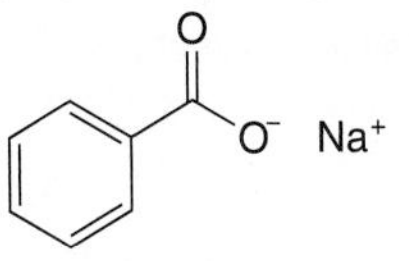
sodium benzoate

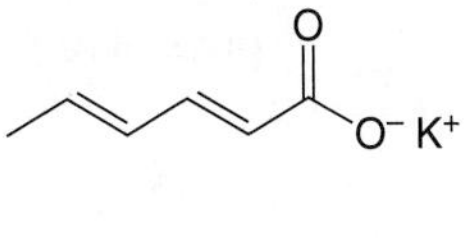
potassium sorbate

19.6 Aspirin, Arachidonic Acid, and Prostaglandins

Recall from Chapter 2 that **aspirin (acetylsalicylic acid)** is a synthetic carboxylic acid, similar in structure to **salicin,** a naturally occurring compound isolated from willow bark, and **salicylic acid,** found in meadowsweet.

The word *aspirin* is derived from the prefix ***a-*** for *acetyl* + ***spir*** from the Latin name *spirea* for the meadowsweet plant.

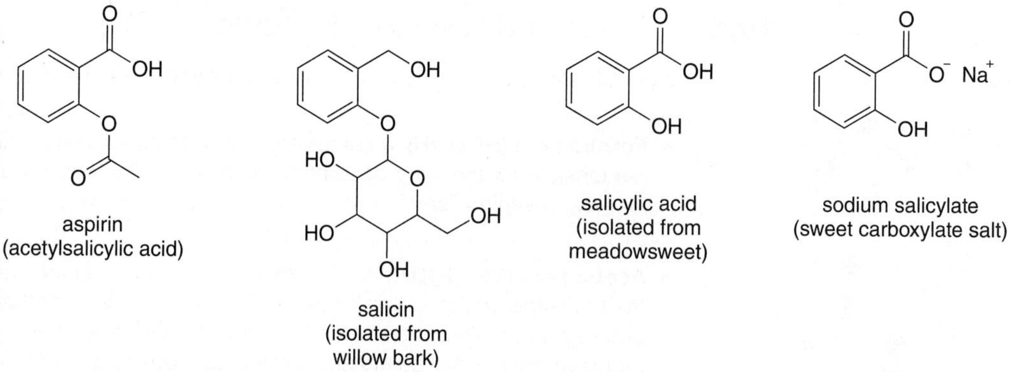

Both salicylic acid and sodium salicylate (its sodium salt) were widely used analgesics in the nineteenth century, but both had undesirable side effects. Salicylic acid irritated the mucous membranes of the mouth and stomach, and sodium salicylate was too sweet for most patients. Aspirin, a synthetic compound, was first sold in 1899 after Felix Hoffmann, a German chemist at Bayer Company, developed a feasible commercial synthesis. Hoffmann's work was motivated by personal reasons; his father suffered from rheumatoid arthritis and was unable to tolerate the sweet taste of sodium salicylate.

How does aspirin relieve pain and reduce inflammation? Aspirin blocks the synthesis of **prostaglandins,** 20-carbon fatty acids with a five-membered ring that are responsible for pain, inflammation, and a wide variety of other biological functions. **PGF$_{2\alpha}$** contains the typical carbon skeleton of a prostaglandin.

Aspirin is the most widely used pain reliever and anti-inflammatory agent in the world, yet its mechanism of action remained unknown until the 1970s. John Vane, Bengt Samuelsson, and Sune Bergström shared the 1982 Nobel Prize in Physiology or Medicine for unraveling the details of its mechanism.

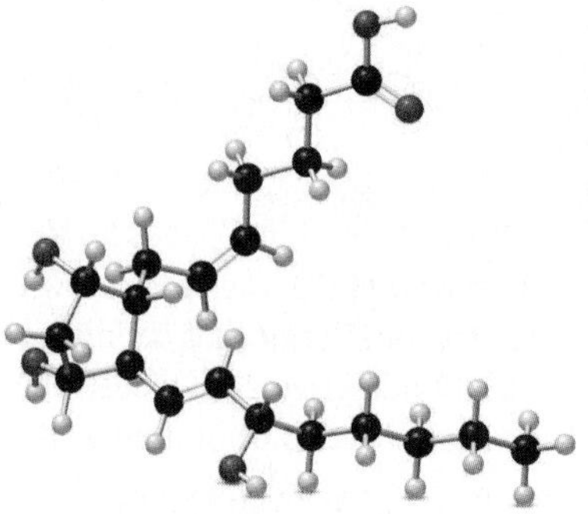

PGF$_{2\alpha}$
a prostaglandin

Prostaglandins are not stored in cells. Rather, they are synthesized from arachidonic acid, a polyunsaturated fatty acid having four cis double bonds. Unlike hormones, which are transported in the bloodstream to their sites of action, prostaglandins act where they are synthesized. Aspirin acts by blocking the synthesis of prostaglandins from arachidonic acid. Aspirin inactivates cyclooxygenase, an enzyme that converts arachidonic acid to PGG$_2$, an unstable precursor of PGF$_{2\alpha}$ and other prostaglandins. **Aspirin lessens pain and decreases inflammation because it prevents the synthesis of prostaglandins, the compounds responsible for both of these physiological responses.**

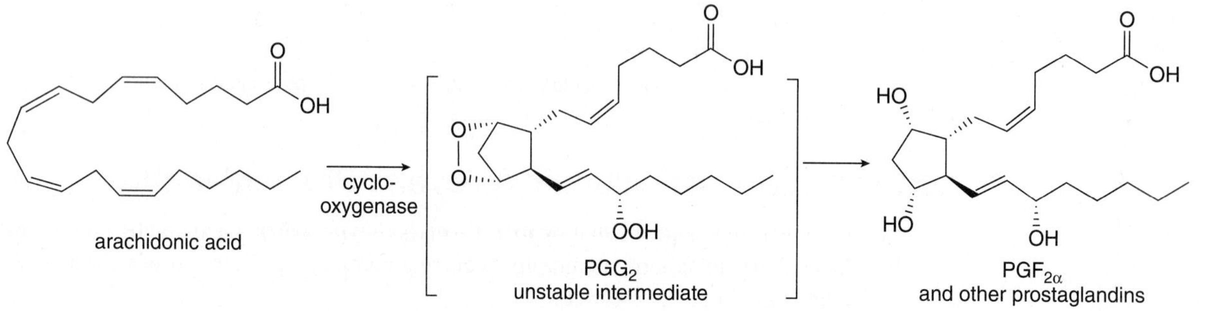

Although prostaglandins have a wide range of biological activity, their inherent instability often limits their usefulness as drugs. Consequently, more stable analogues with useful medicinal properties have been synthesized. For example, latanoprost (trade name Xalatan) and bimatoprost (trade name Lumigan) are prostaglandin analogues used to reduce eye pressure in individuals with glaucoma.

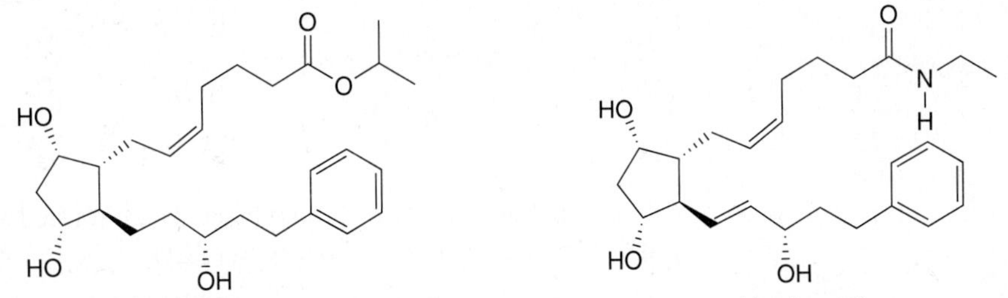

Problem 19.8 How many tetrahedral stereogenic centers does $PGF_{2\alpha}$ contain? Draw its enantiomer. How many of its double bonds can exhibit cis-trans isomerism? Considering both its double bonds and its tetrahedral stereogenic centers, how many stereoisomers are possible for $PGF_{2\alpha}$?

19.7 Preparation of Carboxylic Acids

Our discussion of the reactions involving carboxylic acids begins with a brief list of reactions that synthesize them. In these reactions, the carboxy group is formed in the *product,* and many different functional groups serve as starting materials. Reactions that produce a particular functional group are called **preparations.**

In the remainder of Chapter 19 (and Chapters 20 and 22) we discuss reactions in which a carboxylic acid is a *starting material* that may be converted to a variety of different products. Keep in mind that **reactions of a particular functional group follow a common theme.** For example, alkenes undergo addition reactions. As a result, these reactions are easier to learn than the list of preparations, in which vastly different functional groups undergo a wide variety of reactions to form the same kind of product.

Where have we encountered carboxylic acids as reaction products before? The carbonyl carbon is highly oxidized, because it has three C–O bonds, so **carboxylic acids are typically prepared by oxidation reactions.** Three oxidation methods are summarized below. Two other useful methods to prepare carboxylic acids are presented in Chapter 20.

[1] By oxidation of 1° alcohols (Section 12.12B)

1° Alcohols are converted to carboxylic acids with $Na_2Cr_2O_7$, $K_2Cr_2O_7$, or CrO_3 in the presence of H_2O and H_2SO_4.

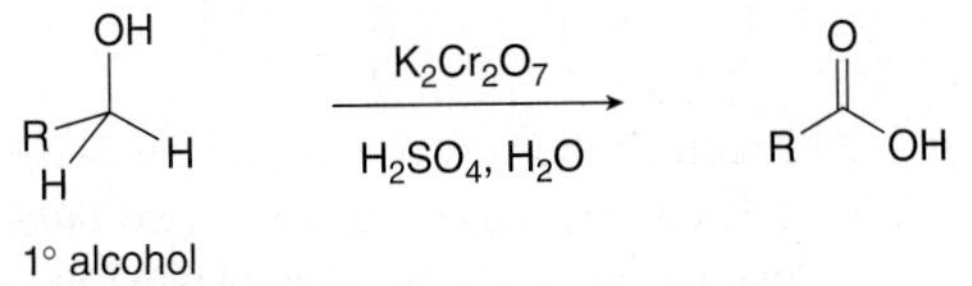

[2] By oxidation of alkyl benzenes (Section 18.15A)

Alkyl benzenes having at least one benzylic C–H bond are oxidized with $KMnO_4$ to benzoic acid.

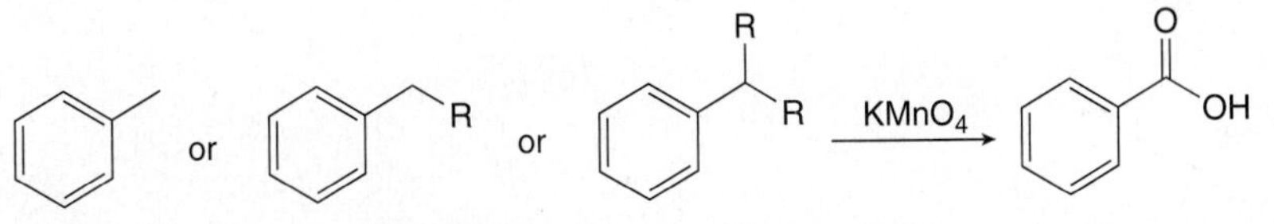

Benzoic acid is *always* the product regardless of the alkyl benzene used as starting material.

[3] By oxidative cleavage of alkynes (Section 12.11)

Both internal and terminal alkynes are oxidatively cleaved with ozone to give carboxylic acids.

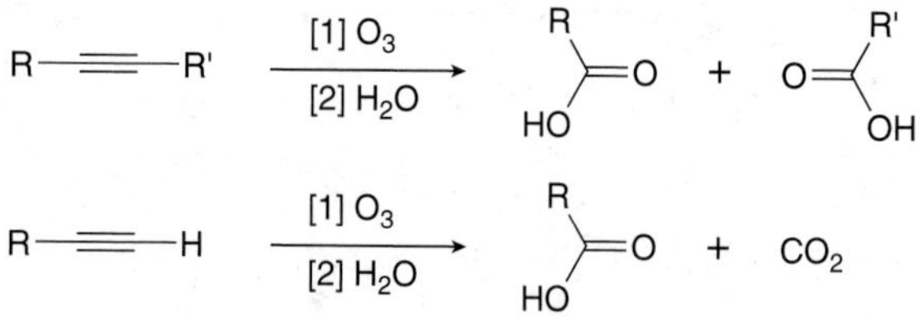

With internal alkynes, two carboxylic acids are formed as products. With terminal alkynes, the *sp* hybridized C–H bond is converted to CO_2.

Problem 19.9 What alcohol can be oxidized to each carboxylic acid?

a. 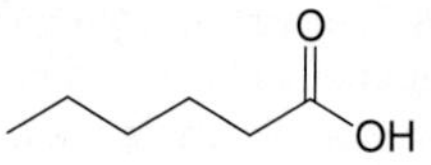

b. 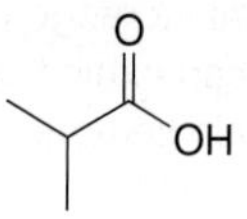

c. 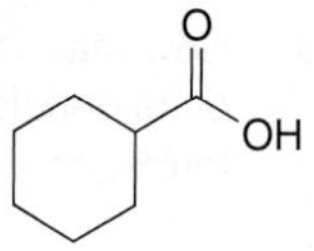

Problem 19.10 Identify **A–D** in the following reactions.

a. **A** 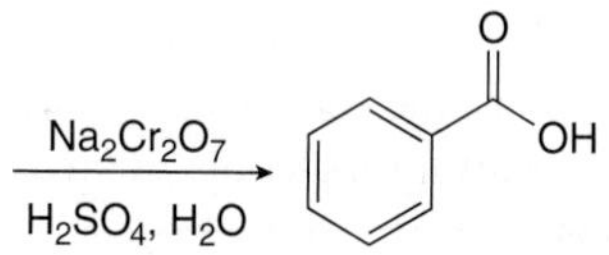

c. **C** 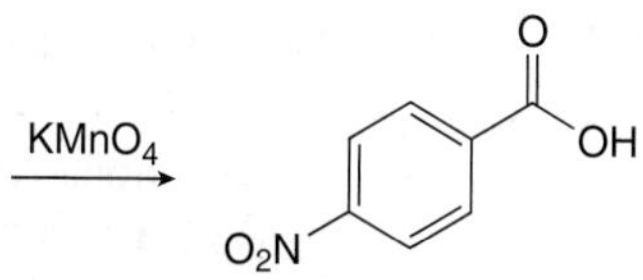

b. **B** 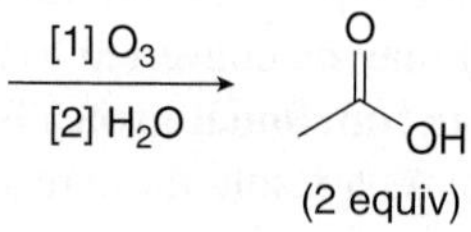

d. **D** 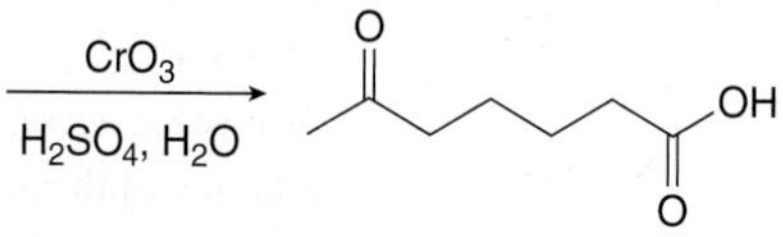

19.8 Reactions of Carboxylic Acids—General Features

The polar C–O and O–H bonds, nonbonded electron pairs on oxygen, and the π bond give a carboxylic acid many reactive sites, complicating its chemistry somewhat. By far, **the most important reactive feature of a carboxylic acid is its polar O–H bond, which is readily cleaved with base.**

- **Carboxylic acids react as Brønsted–Lowry acids—that is, as proton donors.**

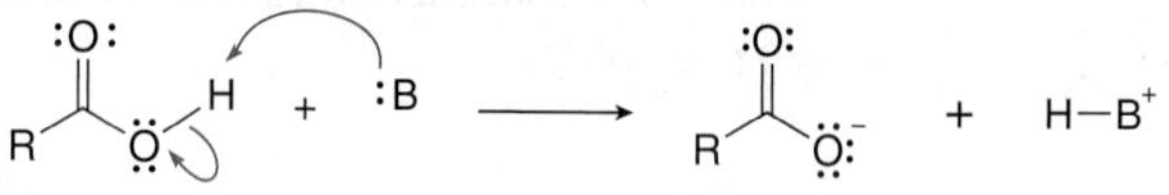

Much of the rest of Chapter 19 is devoted to the acidity of carboxylic acids, as well as some related acid–base reactions. Two other structural features are less important in the reactions of carboxylic acids, but they play a role in the reactions of Chapters 20 and 22.

The nonbonded electron pairs on oxygen create electron-rich sites that can be protonated by strong acids (H–A). Protonation occurs at the carbonyl oxygen because the resulting conjugate acid is resonance stabilized. As a result, **carboxylic acids are weakly basic—they react**

with strong acids by protonation of the carbonyl oxygen. This reaction plays an important role in several mechanisms in Chapter 22.

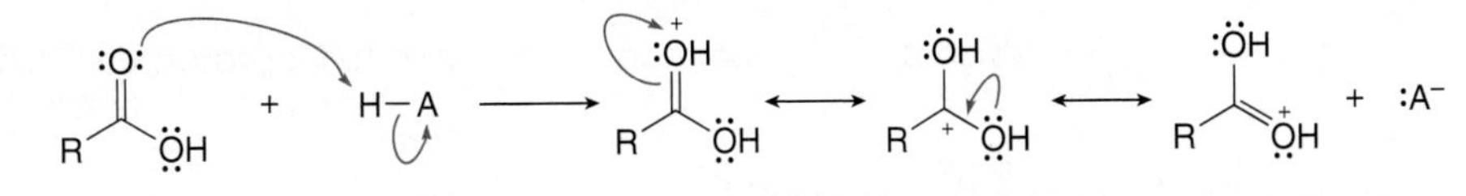

three resonance structures for the conjugate acid

Finally, the polar C−O bonds make the carboxy carbon electrophilic, so carboxylic acids react with nucleophiles. Nucleophilic attack occurs at an sp^2 hybridized carbon atom, so it results in the cleavage of the π bond, as well. This reaction is also discussed in Chapter 22.

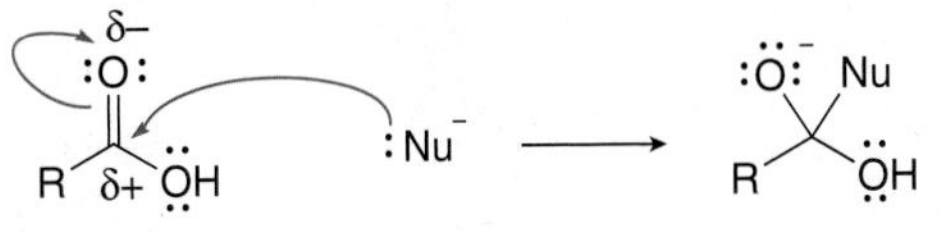

19.9 Carboxylic Acids—Strong Organic Brønsted–Lowry Acids

Carboxylic acids are strong organic acids, and as such, readily react with Brønsted–Lowry bases to form carboxylate anions.

Recall from Section 2.3 that **the lower the pK_a, the stronger the acid.**

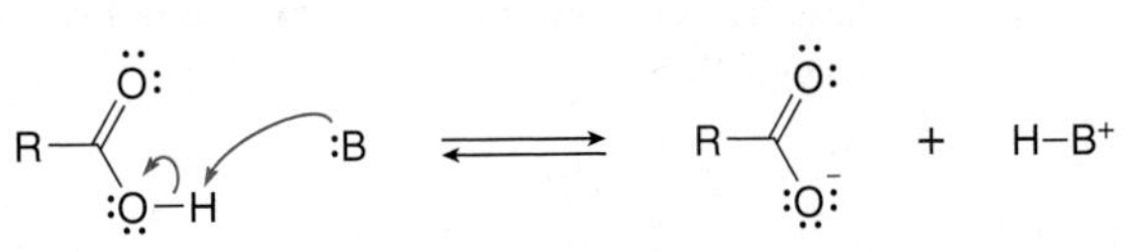

carboxylate anion

What bases are used to deprotonate a carboxylic acid? As we learned in Section 2.3, equilibrium favors the products of an acid–base reaction when the weaker base and acid are formed. Because a weaker acid has a higher pK_a, the following general rule results:

- An acid can be deprotonated by a base that has a conjugate acid with a higher pK_a.

Because the pK_a values of many carboxylic acids are ~5, bases that have conjugate acids with pK_a values higher than 5 are strong enough to deprotonate them. Thus, acetic acid (pK_a = 4.8) and benzoic acid (pK_a = 4.2) can be deprotonated with NaOH and $NaHCO_3$, as shown in the following equations.

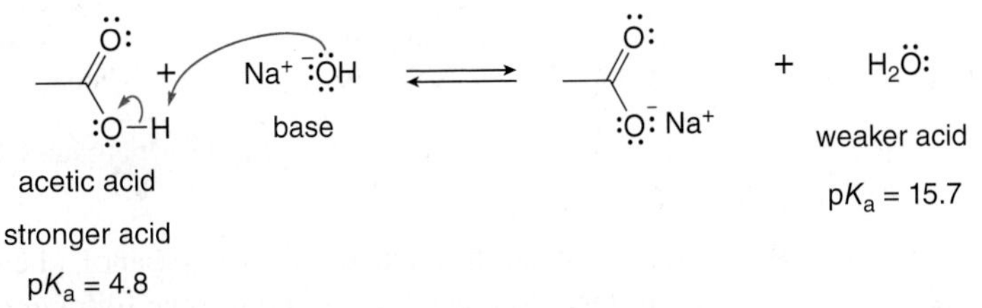

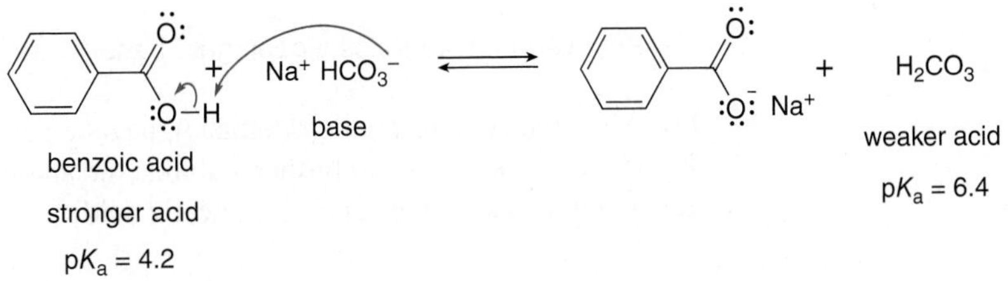

Table 19.3 lists common bases that can be used to deprotonate carboxylic acids. It is noteworthy that even a weak base like $NaHCO_3$ is strong enough to remove a proton from RCOOH.

Table 19.3 Common Bases Used to Deprotonate Carboxylic Acids

Base	Conjugate acid (pK_a)
$Na^+ \ HCO_3^-$	H_2CO_3 (6.4)
NH_3	NH_4^+ (9.4)
Na_2CO_3	HCO_3^- (10.2)
$Na^+ \ ^-OCH_3$	CH_3OH (15.5)
$Na^+ \ ^-OH$	H_2O (15.7)
$Na^+ \ ^-OCH_2CH_3$	CH_3CH_2OH (16)
$Na^+ \ H^-$	H_2 (35)

Increasing basicity (down the table)

Why are carboxylic acids such strong organic acids? Remember that a strong acid has a weak, stabilized conjugate base. **Deprotonation of a carboxylic acid forms a resonance-stabilized conjugate base—a carboxylate anion.** Two equivalent resonance structures can be drawn for acetate (the conjugate base of acetic acid), both of which place a negative charge on an electronegative O atom. In the resonance hybrid, therefore, the negative charge is delocalized over two oxygen atoms.

How resonance affects acidity was first discussed in Section 2.5C.

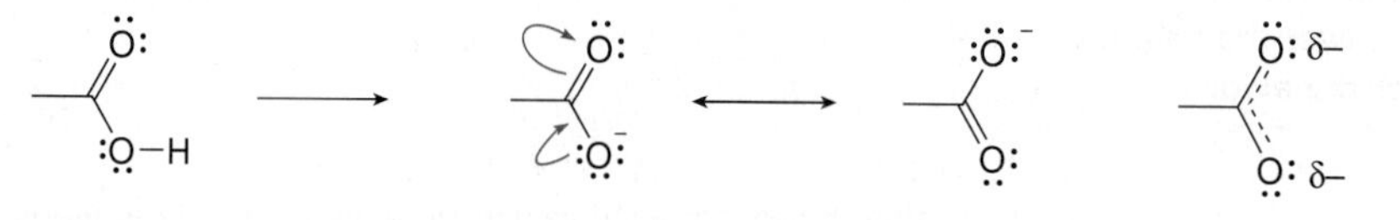

acetic acid — two resonance structures for acetate, the conjugate base — hybrid

Resonance stabilization accounts for why carboxylic acids are more acidic than other compounds with O–H bonds—namely, alcohols and phenols. For example, the pK_a values of ethanol (CH_3CH_2OH) and phenol (C_6H_5OH) are 16 and 10, respectively, both higher than the pK_a of acetic acid (4.8).

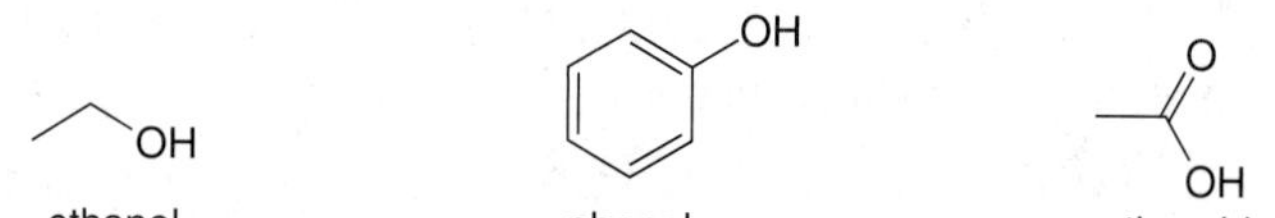

ethanol $pK_a = 16$ — phenol $pK_a = 10$ — acetic acid $pK_a = 4.8$

Increasing acidity

To understand the relative acidity of ethanol, phenol, and acetic acid, we must compare the stability of their conjugate bases and use the following rule:

- **Anything that stabilizes a conjugate base A:$^-$ makes the starting acid H–A more acidic.**

Ethoxide, the conjugate base of ethanol, bears a negative charge on an oxygen atom, but there are no additional factors to further stabilize the anion. Because ethoxide is less stable than acetate, **ethanol is a weaker acid than acetic acid.**

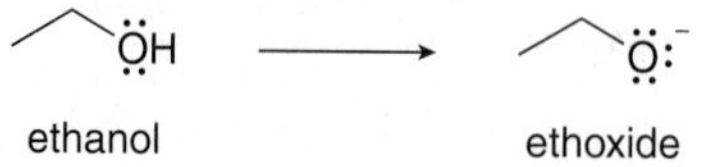

ethanol — ethoxide

Like acetate, **phenoxide** ($C_6H_5O^-$, the conjugate base of phenol) is also resonance stabilized. In the case of phenoxide, however, there are *five* resonance structures that disperse the negative charge over a total of *four* different atoms (three different carbons and the oxygen).

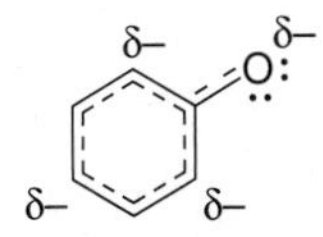

hybrid

The resonance hybrid of phenoxide illustrates that its negative charge is dispersed over four atoms—three C atoms and one O atom.

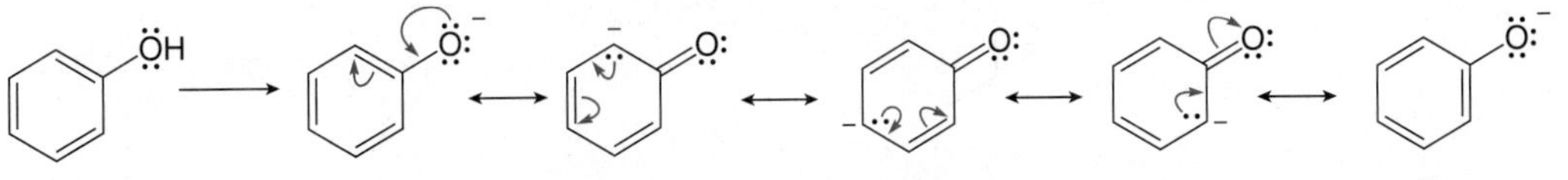

phenol 1 2 3 4 5

Phenoxide is more stable than ethoxide, but less stable than acetate, because acetate has two electronegative oxygen atoms upon which to delocalize the negative charge, whereas phenoxide has only one. Additionally, phenoxide resonance structures **2–4** have the negative charge on a carbon, a less electronegative element than oxygen. As a result, structures **2–4** are less stable than structures **1** and **5,** which have the negative charge on oxygen.

Moreover, resonance structures **1** and **5** have intact aromatic rings, whereas structures **2–4** do not. This, too, makes structures **2–4** less stable than **1** and **5.** Figure 19.6 summarizes this information about phenoxide by displaying the approximate relative energies of its five resonance structures and its hybrid.

Figure 19.6 The relative energies of the five resonance structures for phenoxide and its hybrid

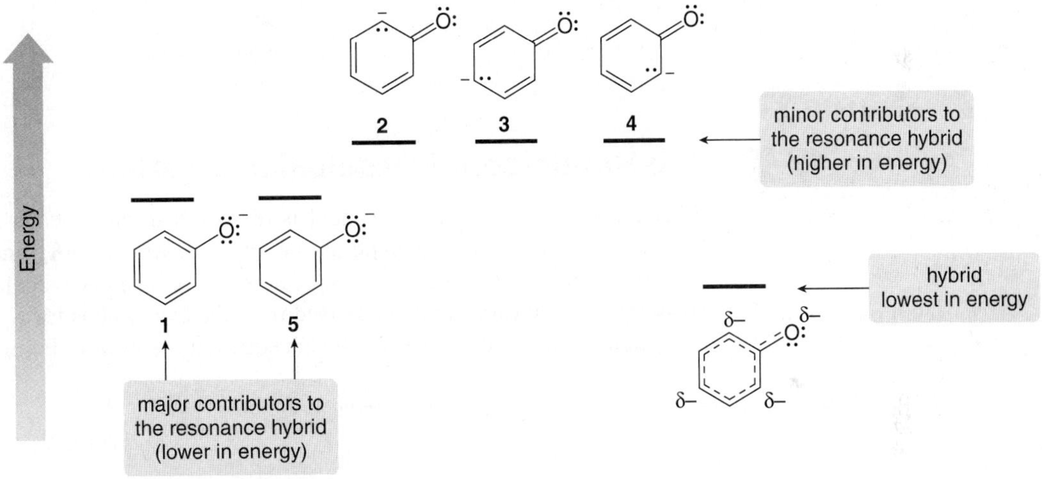

As a result, resonance stabilization of the conjugate base is important in determining acidity, but **the absolute number of resonance structures alone is not what's important.** We must evaluate their relative contributions to predict the relative stability of the conjugate bases.

Keep in mind that although carboxylic acids are strong organic acids, they are still much weaker than strong inorganic acids like HCl and H_2SO_4, which have pK_a values < 0.

- **Because of their O–H bond, RCOOH, ROH, and C_6H_5OH are more acidic than most organic hydrocarbons.**

- **A carboxylic acid is a stronger acid than an alcohol or phenol because its conjugate base is most effectively resonance stabilized.**

Because alcohols and phenols are weaker acids than carboxylic acids, stronger bases are needed to deprotonate them. To deprotonate C_6H_5OH (pK_a = 10), a base whose conjugate acid has a pK_a > 10 is needed. Thus, of the bases listed in Table 19.3, $NaOCH_3$, NaOH, $NaOCH_2CH_3$, and NaH are strong enough. To deprotonate CH_3CH_2OH (pK_a = 16), only NaH is strong enough.

Problem 19.11 Draw the products of each acid–base reaction.

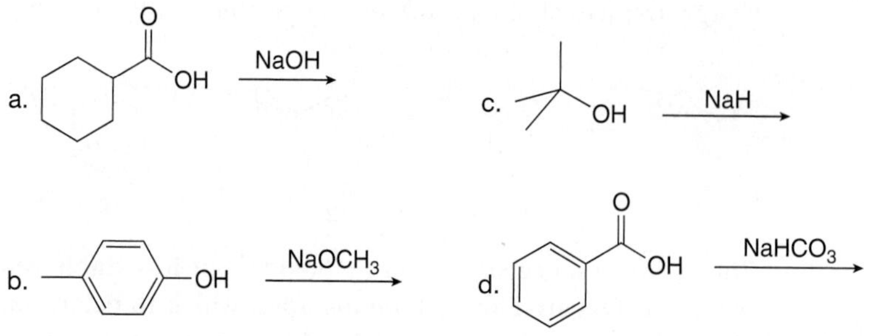
a.
OH
O
NaOH
c.
OH
NaH
b.
OH
NaOCH3
d.
O
OH
NaHCO3

Problem 19.12 Given the pK_a values in Appendix A, which of the following bases are strong enough to deprotonate CH_3COOH: (a) F^-; (b) $(CH_3)_3CO^-$; (c) CH_3^-; (d) $^-NH_2$; (e) Cl^-?

Problem 19.13 Rank the labeled protons (H_a–H_c) in mandelic acid, a naturally occurring carboxylic acid in plums and peaches, in order of increasing acidity. Explain in detail why you chose this order.

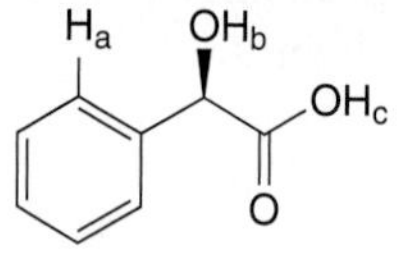
Ha
OHb
OHc
O

mandelic acid

19.10 The Henderson–Hasselbalch Equation

What happens when a particular acid is dissolved in an aqueous solution? **Whether or not the acid will lose a proton depends on two factors—its pK_a and the pH of the solution.** The amount of acid (HA) and its conjugate base ($A:^-$) can be calculated using the **Henderson–Hasselbalch equation,** which is derived from the expressions for K_a and pK_a (Section 2.3). The derivation of the Henderson–Hasselbalch equation is shown in Figure 19.7.

Henderson–Hasselbalch equation

$$pK_a = pH + \log \frac{[HA]}{[A:^-]}$$

Figure 19.7 Derivation of the Henderson–Hasselbalch equation

- Use the definitions of K_a and pK_a to write the equation.

$$pK_a = -\log K_a = -\log \frac{[H_3O^+][A:^-]}{[HA]}$$

- Separate the terms.

$$pK_a = -\log [H_3O^+] - \log \frac{[A:^-]}{[HA]}$$

- Use the definition of pH ($pH = -\log [H_3O^+]$) to simplify.

$$pK_a = pH - \log \frac{[A:^-]}{[HA]}$$

- Invert the terms in the logarithm and change the sign that precedes this term.

$$pK_a = pH + \log \frac{[HA]}{[A:^-]}$$

Henderson–Hasselbalch equation

The Henderson–Hasselbalch equation tells us whether a compound will exist in its acidic form (HA) or as its conjugate base (A:$^-$) at a particular pH.

Recall that a log is an exponent: log $10^2 = 2$; log $10^{-2} = -2$.

- **When the pH of the solution *equals* the pK_a of the acid, the concentration of HA and A:$^-$ must be equal, because log ([HA]/[A:$^-$]) equals zero.**
- **When the pH of the solution is *less* than the pK_a of the acid, the concentration of HA is greater than the concentration of A:$^-$ because log ([HA]/[A:$^-$]) is positive.**
- **When the pH of the solution is *higher* than the pK_a of the acid, the concentration of A:$^-$ is greater than the concentration of HA because log ([HA]/[A:$^-$]) is negative.**

We can summarize these consequences of the Henderson–Hasselbalch equation as follows:

- **An acid exists in its protonated form HA in solutions that are more *acidic* than its pK_a.**
- **An acid exists as its conjugate base A:$^-$ in solutions that are more *basic* than its pK_a.**

In what form does a carboxylic acid like pyruvic acid (pK_a = 2.4), a product of glucose metabolism, exist in the bloodstream, which is buffered to a pH of 7.4? Because the pH of the solution is higher than the pK_a of pyruvic acid, the acid is deprotonated and exists primarily as its conjugate base, pyruvate.

The cellular pH of 7.4 is called **physiological pH.**

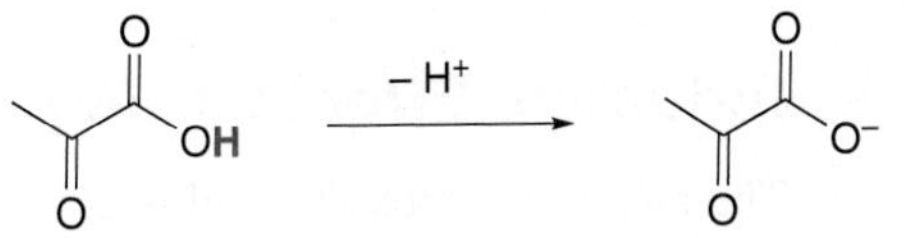

pyruvic acid
pK_a = 2.4

pyruvate
predominant form
at pH = 7.4

Sample Problem 19.2 What is the predominant form of each compound in a solution of pH 5.0: (a) FCH_2CO_2H (pK_a = 2.7); (b) CF_3CH_2OH (pK_a = 12.4)?

Solution

a. Because the pH of the solution (5.0) is more basic than the pK_a of FCH_2CO_2H (2.7), the compound will exist primarily as its conjugate base, $FCH_2CO_2^-$, formed by removal of its most acidic proton.

b. Because the pH of the solution (5.0) is more acidic than the pK_a of CF_3CH_2OH (12.4), the compound will exist primarily as CF_3CH_2OH, the neutral acid that is *not* deprotonated.

Problem 19.14 What form(s) of each compound predominate at the given pH?

a. 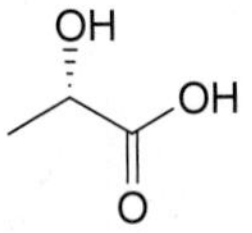

at pH = 7.3

lactic acid
pK_a = 3.9

b. 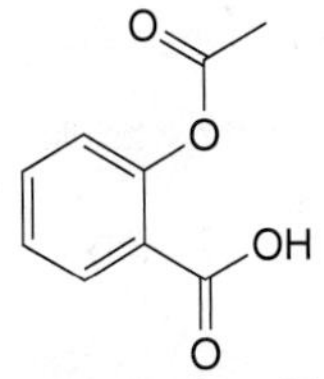

at pH = 3.5

acetylsalicylic acid
pK_a = 3.5

c. 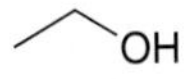

at pH = 2.0

ethanol
pK_a = 16

Some compounds, such as the dicarboxylic acids mentioned in Section 19.2C and the amino acids discussed in Section 19.15, contain two or more functional groups that can lose protons, so two pK_a values are reported. For succinic acid (pK_{a1} = 4.2 and pK_{a2} = 5.6), both pK_a values are less than the physiological pH of 7.4, so both carboxy groups are deprotonated and the predominant species in cells is the dianion succinate.

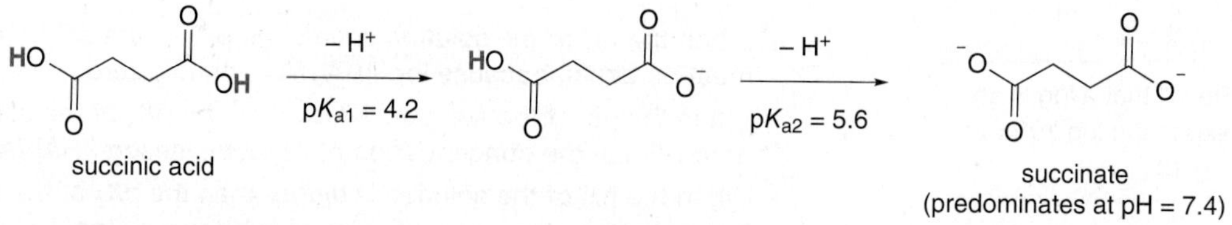
HO
OH
O
O
– H+
pKa1 = 4.2
HO
O
O
– H+
pKa2 = 5.6
succinic acid
succinate
(predominates at pH = 7.4)

Problem 19.15 Citric acid, a metabolic intermediate in the citric acid cycle (Section 19.2C), contains three carboxy groups with pK_a values of 3.1, 4.8, and 6.4. (a) What is the predominant form of citric acid in the stomach (pH = 2)? (b) What is the predominant form of citric acid in the intestines (pH = 8)?

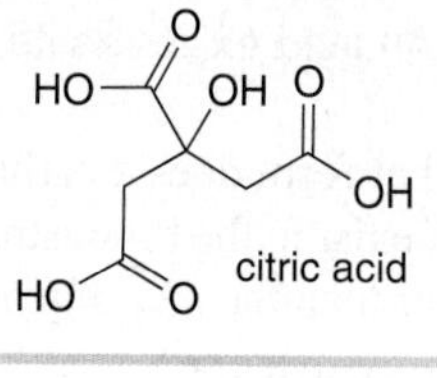
HO
OH
OH
HO
citric acid

19.11 Inductive Effects in Aliphatic Carboxylic Acids

The pK_a of a carboxylic acid is affected by nearby groups that inductively donate or withdraw electron density.

- **Electron-withdrawing groups stabilize a conjugate base, making a carboxylic acid more acidic.**
- **Electron-donating groups destabilize the conjugate base, making a carboxylic acid less acidic.**

The relative acidity of CH_3COOH, $ClCH_2COOH$, and $(CH_3)_3CCOOH$ illustrates these principles in the following equations.

We first learned about inductive effects and acidity in Section 2.5B.

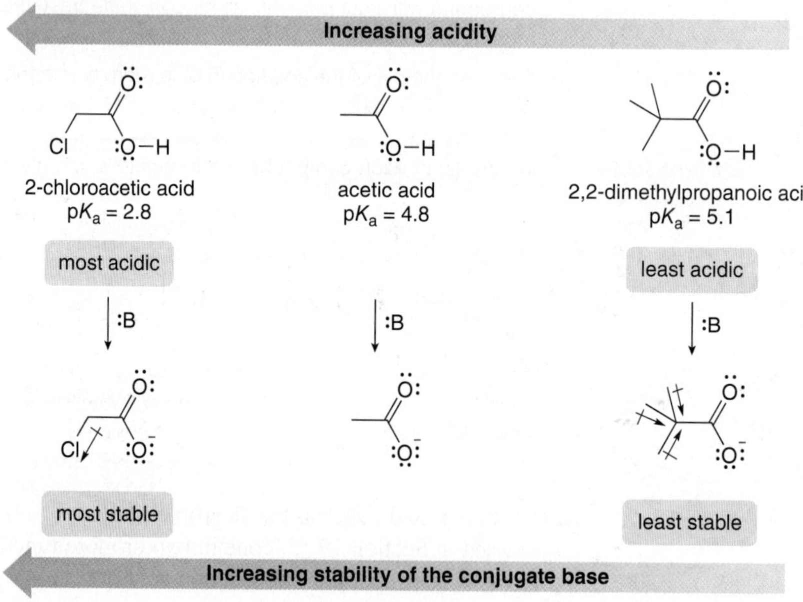
Increasing acidity
2-chloroacetic acid
pKa = 2.8
acetic acid
pKa = 4.8
2,2-dimethylpropanoic acid
pKa = 5.1
most acidic
least acidic
:B
most stable
least stable
Increasing stability of the conjugate base

- $ClCH_2COOH$ is more acidic (pK_a = 2.8) than CH_3COOH (pK_a = 4.8) because its conjugate base is stabilized by the electron-withdrawing inductive effect of the electronegative Cl.
- $(CH_3)_3CCOOH$ is less acidic (pK_a = 5.1) than CH_3COOH because the three polarizable CH_3 groups donate electron density and destabilize the conjugate base.

The number, electronegativity, and location of substituents also affect acidity.

- **The larger the number of electronegative substituents, the stronger the acid.**

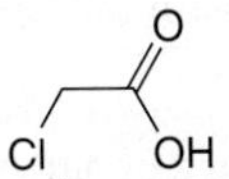
$pK_a = 2.8$

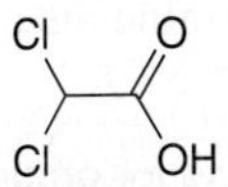
$pK_a = 1.3$

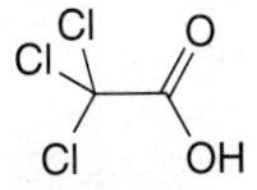
$pK_a = 0.9$

Increasing acidity
Increasing number of electronegative Cl atoms

- **The more electronegative the substituent, the stronger the acid.**

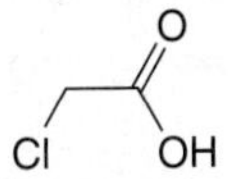
$pK_a = 2.8$

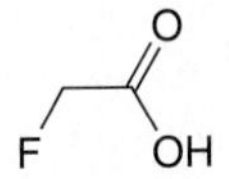
$pK_a = 2.6$

F is more electronegative than Cl.
stronger acid

- **The closer the electron-withdrawing group to the COOH, the stronger the acid.**

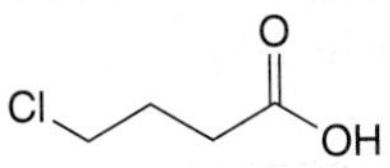
4-chlorobutanoic acid
$pK_a = 4.5$

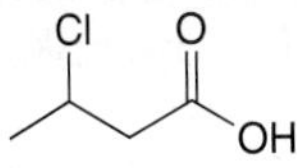
3-chlorobutanoic acid
$pK_a = 4.1$

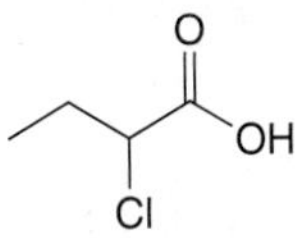
2-chlorobutanoic acid
$pK_a = 2.9$

Increasing acidity
Increasing proximity of Cl to COOH

Problem 19.16 Match each of the following pK_a values (3.2, 4.9, and 0.2) to the appropriate carboxylic acid: (a) CH_3CH_2COOH; (b) CF_3COOH; (c) ICH_2COOH.

Problem 19.17 Rank the compounds in order of increasing acidity.

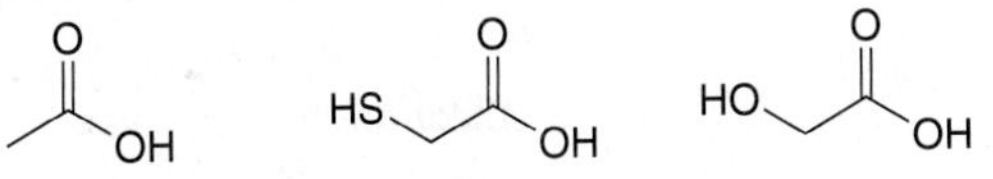

19.12 Substituted Benzoic Acids

Recall from Chapter 18 that substituents on a benzene ring either donate or withdraw electron density, depending on the balance of their inductive and resonance effects. These same effects also determine the acidity of substituted benzoic acids. There are two rules to keep in mind.

Rule [1] **Electron-donor groups destabilize a conjugate base, making an acid less acidic.**

An electron-donor group destabilizes a conjugate base by donating electron density onto a negatively charged carboxylate anion. A benzoic acid substituted by an electron-donor group has a higher pK_a than benzoic acid (pK_a = 4.2).

D = Electron-donor group

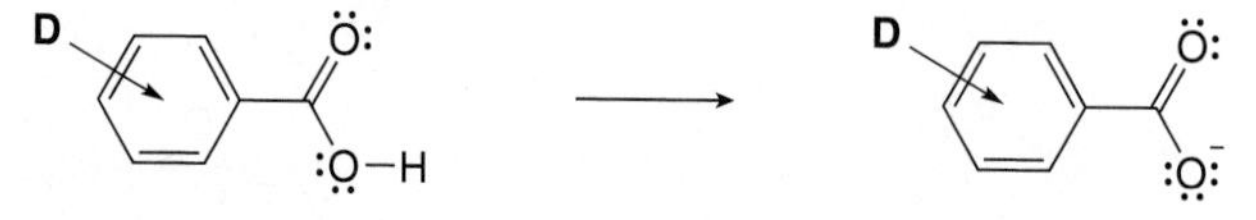

This acid is less acidic than benzoic acid.

pK_a > 4.2

D destabilizes the carboxylate anion.

Rule [2] **Electron-withdrawing groups stabilize a conjugate base, making an acid more acidic.**

An electron-withdrawing group stabilizes a conjugate base by removing electron density from the negatively charged carboxylate anion. A benzoic acid substituted by an electron-withdrawing group has a lower pK_a than benzoic acid (pK_a = 4.2).

W = Electron-withdrawing group

This acid is more acidic than benzoic acid.

pK_a < 4.2

W stabilizes the carboxylate anion.

How do we know which groups are electron donating or electron withdrawing on a benzene ring? We already learned the characteristics of electron-donating and electron-withdrawing groups in Chapter 18, and how they affect the rate of electrophilic aromatic substitution. These principles can now be extended to substituted benzoic acids.

Electron-donor groups D →
- **activate benzene to electrophilic attack**
- **make a benzoic acid less acidic**

Electron-withdrawing groups W →
- **deactivate benzene to electrophilic attack**
- **make a benzoic acid more acidic**

Figure 19.8 illustrates how common electron-donating and electron-withdrawing groups affect both the rate of reaction of a benzene ring toward electrophiles and the acidity of substituted benzoic acids.

Figure 19.8
How common substituents affect the reactivity of a benzene ring toward electrophiles and the acidity of substituted benzoic acids

	Substituent	Effect in electrophilic substitution	Effect on acidity of substituted benzoic acids
electron-donating groups	$-\ddot{N}H_2$ [$\ddot{N}HR$, $\ddot{N}R_2$]	activating groups	These groups make a benzoic acid **less acidic.**
	$-\ddot{O}H$		
	$-\ddot{O}R$		
	$-\ddot{N}HCOR$		
	–R		
electron-withdrawing groups	$-\ddot{X}:$ [X = F, Cl, Br, I]	deactivating groups	These groups make a benzoic acid **more acidic.**
	–CHO		
	–COR		
	–COOR		
	–COOH		
	–CN		
	$-SO_3H$		
	$-NO_2$		
	$-\overset{+}{N}R_3$		

Increasing acidity ↓

- **Groups that donate electron density activate a benzene ring toward electrophilic attack and make a benzoic acid *less* acidic.** Common electron-donating groups are R groups, or groups that have an N or O atom (with a lone pair) bonded to the benzene ring.
- **Groups that withdraw electron density deactivate a benzene ring toward electrophilic attack, and make a benzoic acid *more* acidic.** Common electron-withdrawing groups are the halogens, or groups with an atom Y (with a full or partial positive charge) bonded to the benzene ring.

Problem 19.18 Rank the compounds in each group in order of increasing acidity.

a.
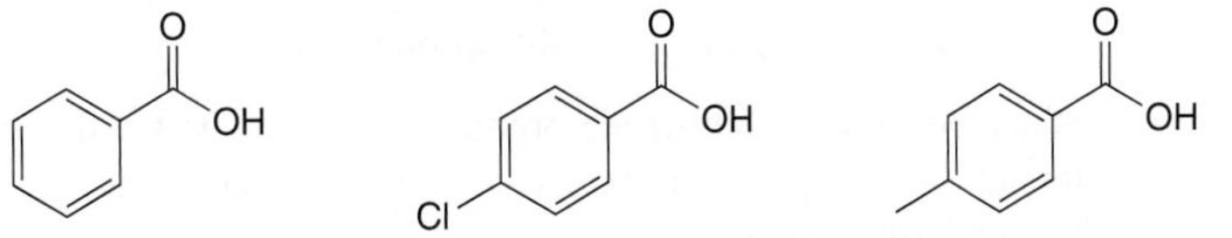

b.
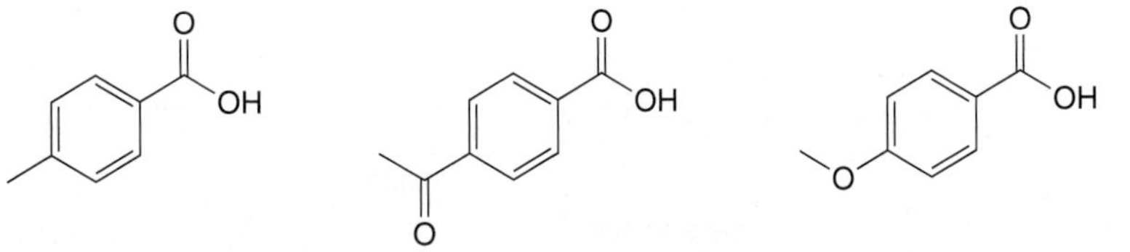

Poison ivy contains the irritant urushiol.

Problem 19.19 Substituted phenols show substituent effects similiar to substituted benzoic acids. Should the pK_a of phenol **A,** one of the naturally occurring phenols called urushiols isolated from poison ivy, be higher or lower than the pK_a of phenol (C_6H_5OH, pK_a = 10)? Explain.

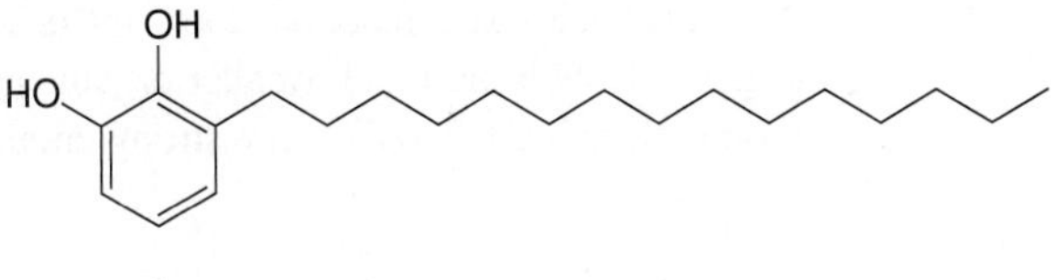

A

19.13 Extraction

An organic chemist in the laboratory must separate and purify mixtures of compounds. One particularly useful technique is **extraction,** which uses solubility differences and acid–base principles to separate and purify compounds.

Extraction has long been and remains the first step in isolating a natural product from its source.

Two solvents are used in extraction: water or an aqueous solution such as 10% $NaHCO_3$ or 10% NaOH; and an organic solvent such as dichloromethane (CH_2Cl_2), diethyl ether, or hexane. **Compounds are separated by their solubility differences in an aqueous and organic solvent.**

An item of glassware called a **separatory funnel** is used for the extraction. When two insoluble liquids are added to the separatory funnel, two layers form, with the less dense liquid on top and the more dense liquid on the bottom.

separatory funnel

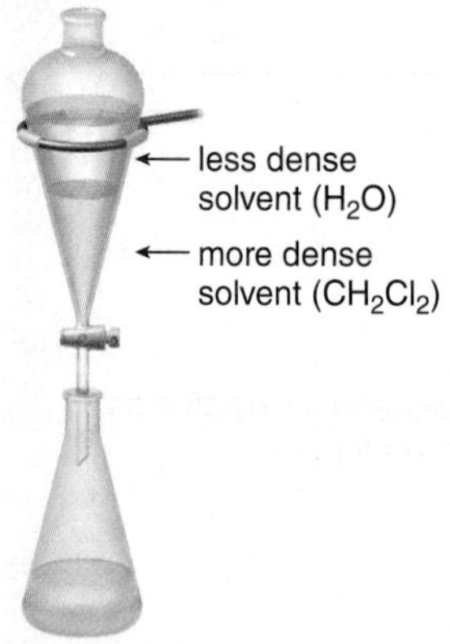

When two insoluble liquids are added to a separatory funnel, two layers are visible, and the less dense liquid forms the upper layer. To separate the layers, the lower layer can be drained from the bottom of the separatory funnel by opening the stopcock. The top layer can then be poured out the top neck of the funnel.

Suppose a mixture of benzoic acid (C_6H_5COOH) and NaCl is added to a separatory funnel containing H_2O and CH_2Cl_2. The benzoic acid would dissolve in the organic layer and the NaCl would dissolve in the water layer. Separating the organic and aqueous layers and placing them in different flasks separates the benzoic acid and NaCl from each other.

How could we separate a mixture of benzoic acid and cyclohexanol? Both compounds are organic, and as a result, both are soluble in an organic solvent such as CH_2Cl_2 and insoluble in water. If a mixture of benzoic acid and cyclohexanol were added to a separatory funnel with CH_2Cl_2 and water, both would dissolve in the CH_2Cl_2 layer, and the two compounds would *not* be separated from each other. Is it possible to use extraction to separate two compounds of this sort that have similar solubility properties?

benzoic acid — cyclohexanol

- insoluble in water
- soluble in CH_2Cl_2

similar solubility properties

- insoluble in water
- soluble in CH_2Cl_2

Recall from Tables 9.1 and 19.2 that alcohols and carboxylic acids having more than five carbons are water insoluble.

If a carboxylic acid is one of the compounds, the answer is *yes,* because we can use acid–base chemistry to change its solubility properties.

When benzoic acid (a strong organic acid) is treated with aqueous NaOH, benzoic acid is deprotonated, forming sodium benzoate. Because sodium benzoate is ionic, it is soluble in water, but insoluble in organic solvents.

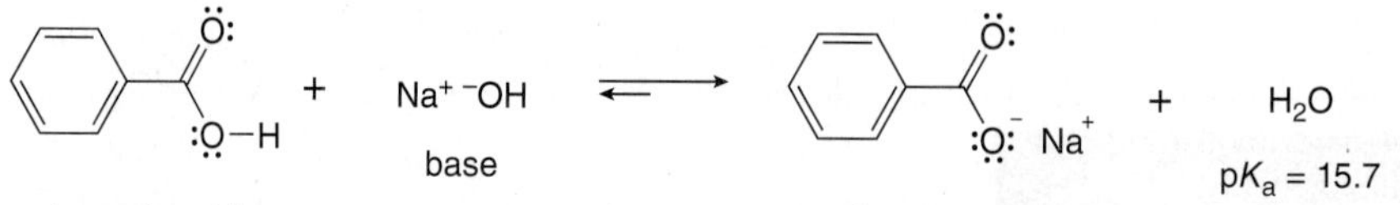

benzoic acid
$pK_a = 4.2$

sodium benzoate

- insoluble in water
- soluble in CH_2Cl_2

different solubility properties

- soluble in water
- insoluble in CH_2Cl_2

A similar acid–base reaction does *not* occur when cyclohexanol is treated with NaOH because organic alcohols are much weaker organic acids, so they can be deprotonated only by a *very strong base* such as NaH. NaOH is not strong enough to form significant amounts of the sodium alkoxide.

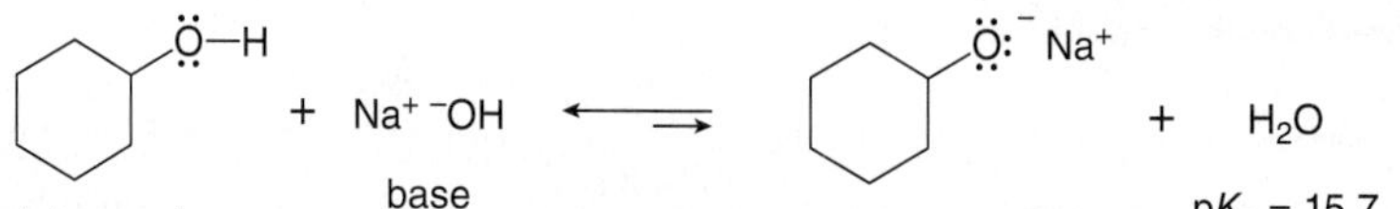

cyclohexanol
pK_a ~ 17

Because equilibrium favors the starting materials, little alkoxide is formed.

Figure 19.9 Separation of benzoic acid and cyclohexanol by an extraction procedure

Step [1] Dissolve benzoic acid and cyclohexanol in CH_2Cl_2.

Step [2] Add 10% NaOH solution to form two layers.

Step [3] Separate the layers.

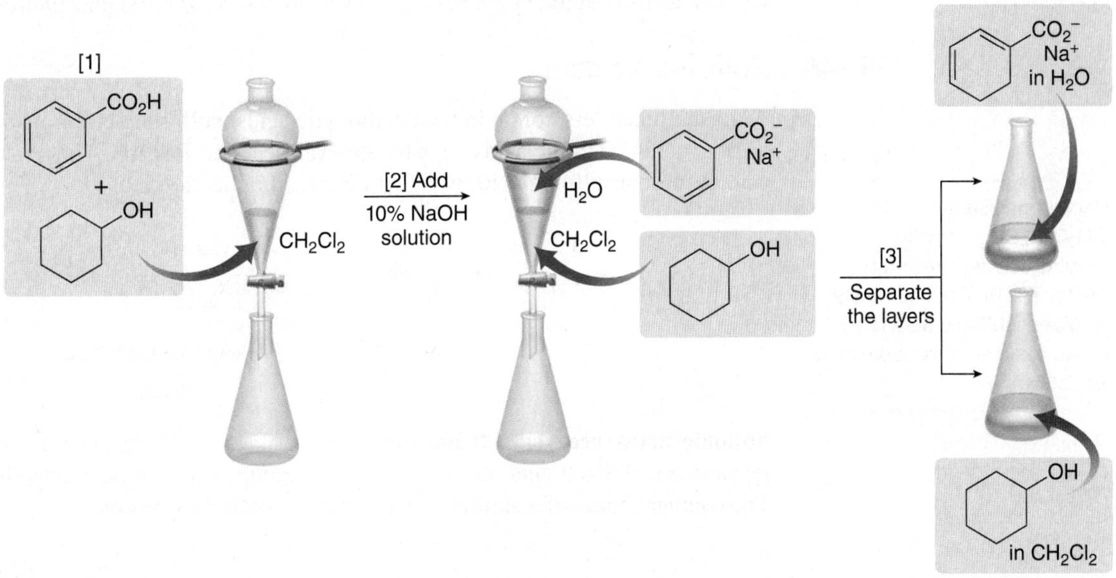

- Both compounds dissolve in the organic solvent CH_2Cl_2.

- Adding 10% aqueous NaOH solution forms two layers. When the two layers are mixed, the NaOH deprotonates $C_6H_5CO_2H$ to form $C_6H_5CO_2^-Na^+$, which dissolves in the aqueous layer.
- The cyclohexanol remains in the CH_2Cl_2 layer.

- Draining the lower layer out the bottom stopcock separates the two layers, and the separation process is complete.
- Cyclohexanol (dissolved in CH_2Cl_2) is in one flask. The sodium salt of benzoic acid, $C_6H_5CO_2^-Na^+$ (dissolved in water) is in another flask.

This difference in acid–base chemistry can be used to separate benzoic acid and cyclohexanol by the stepwise extraction procedure illustrated in Figure 19.9. This extraction scheme relies on two principles:

- **Extraction can separate only compounds having different solubility properties. One compound must dissolve in the aqueous layer and one must dissolve in the organic layer.**

- **A carboxylic acid can be separated from other organic compounds by converting it to a water-soluble carboxylate anion by an acid–base reaction.**

Thus, the water-soluble salt $C_6H_5CO_2^-Na^+$ (derived from $C_6H_5CO_2H$ by an acid–base reaction) can be separated from water-insoluble cyclohexanol by an extraction procedure.

Problem 19.20 Which of the following pairs of compounds can be separated from each other by an extraction procedure?

a. 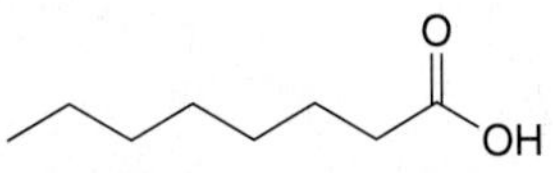

and 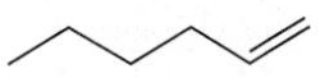

b. 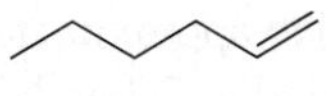
and 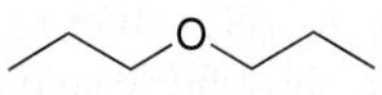

c. 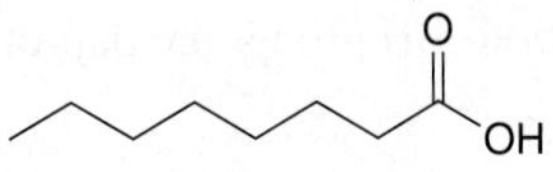

and NaCl

d. NaCl and KCl

19.14 Organic Acids Containing Sulfur and Phosphorus

Organic acids containing hydroxy groups bonded to sulfur and phosphorus are also known.

19.14A Sulfonic Acids

Although much less common than carboxylic acids, **sulfonic acids** constitute a useful group of organic acids. Sulfonic acids have the general structure **RSO_3H.** The most widely used sulfonic acid, ***p*-toluenesulfonic acid,** was first discussed in Section 2.6.

Recall from Section 9.13 that **$CH_3C_6H_4SO_2$–** is called a **tosyl group,** abbreviated by the letters **Ts.** For this reason, *p*-toluenesulfonic acid (also called tosic acid) is abbreviated as **TsOH.**

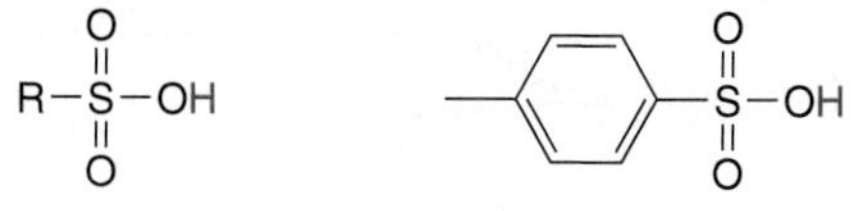

sulfonic acid　　*p*-toluenesulfonic acid

TsOH

Sulfonic acids are very strong acids (pK_a values ≈ –7) because their conjugate bases are resonance stabilized, and all the resonance structures delocalize a negative charge on oxygen. The conjugate base of a sulfonic acid is called a **sulfonate anion.**

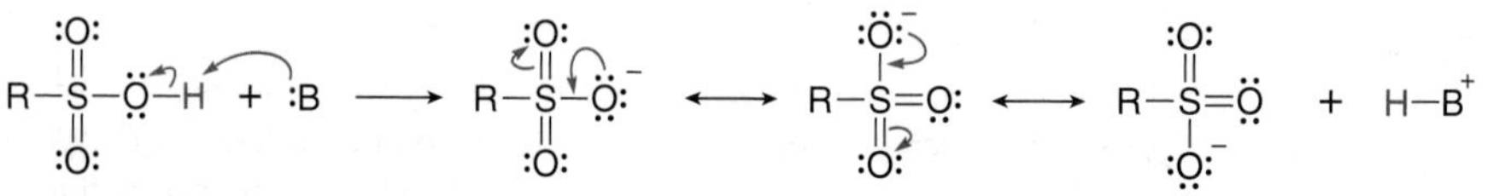

strong acid
pK_a ≈ –7

three resonance structures
All have a negative charge on oxygen.

Because sulfonate anions are such weak bases, they make **good leaving groups** in nucleophilic substitution reactions, as we learned in Section 9.13.

Problem 19.21 Two other commonly used sulfonic acids are methanesulfonic acid (CH_3SO_3H) and trifluoromethanesulfonic acid (CF_3SO_3H). Which has the weaker conjugate base? Which conjugate base is the better leaving group? Which of these acids has the higher pK_a?

19.14B Phosphoric Acid Esters

As we learned in Sections 3.2 and 7.16, several biological compounds are derivatives of phosphoric acid (H_3PO_4) and related compounds. Phosphoric acid itself has three OH groups that can be deprotonated, with pK_a values of 2.1, 6.9, and 12.4, forming $H_2PO_4^-$, HPO_4^{2-}, and PO_4^{3-}, respectively.

All of the phosphoric acid esters in Table 3.4 were drawn as negatively charged species. Now that we have learned about the Henderson–Hasselbalch equation, we can understand why these functional groups are ionized at the physiological pH of 7.4.

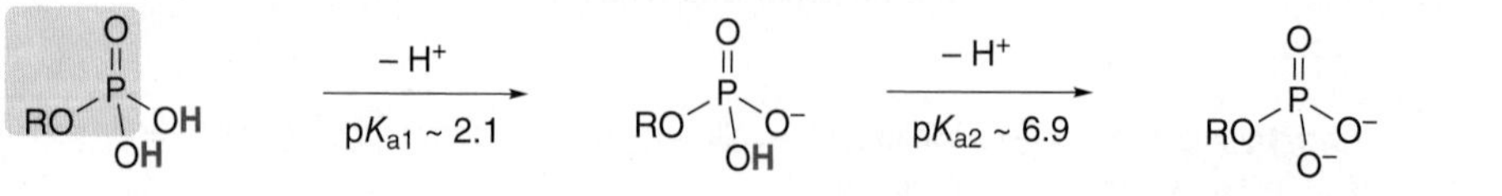

phosphoric acid ester

monophosphate

The dianion predominates at pH = 7.4.

We can use the pK_a values of phosphoric acid to approximate the pK_a values for the two hydroxy protons of a phosphoric acid ester [$ROPO(OH)_2$] as 2.1 and 6.9. Because both pK_a values are *less* than the physiological pH of 7.4, both OH groups are deprotonated and the predominant species at this pH is the dianion of the monophosphate. Thus, glucose 6-phosphate and glyceraldehyde 3-phosphate, two intermediates in glucose metabolism, are monophosphate dianions.

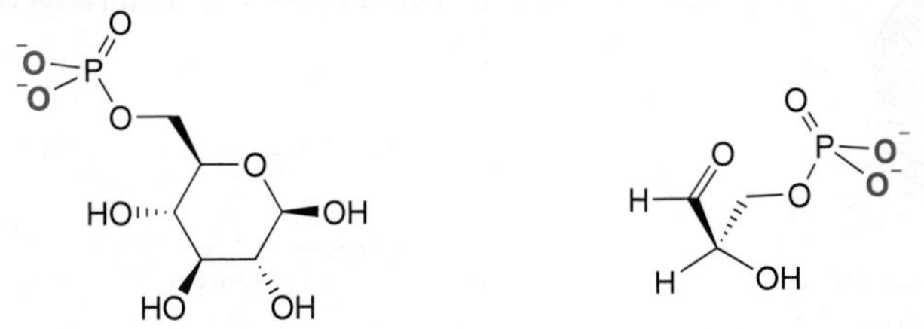

glucose 6-phosphate glyceraldehyde 3-phosphate

Problem 19.22 Use the pK_a values of pyrophosphoric acid (0.9, 2.1, 6.7, and 9.3) to approximate the pK_a values for the nucleotide ADP (adenosine diphosphate), an intermediate formed during metabolism, and draw the predominant form of ADP at physiological pH.

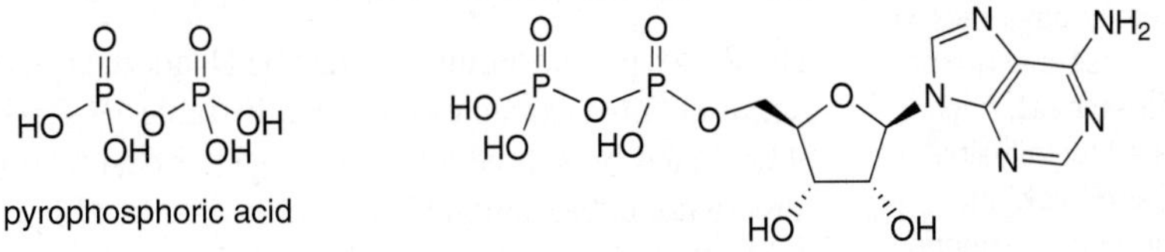

pyrophosphoric acid

adenosine diphosphate
ADP

19.15 Amino Acids

Chapter 26 discusses the synthesis of amino acids and their conversion to proteins.

Amino acids, one of four kinds of small biomolecules that have important biological functions in the cell (Section 3.9), also undergo proton transfer reactions.

19.15A Introduction

Amino acids contain two functional groups—an amino group (NH_2) and a carboxy group (COOH). In most naturally occurring amino acids, the amino group is bonded to the α carbon, so they are called **α-amino acids.** Amino acids are the building blocks of proteins, biomolecules that comprise muscle, hair, fingernails, and many other biological tissues.

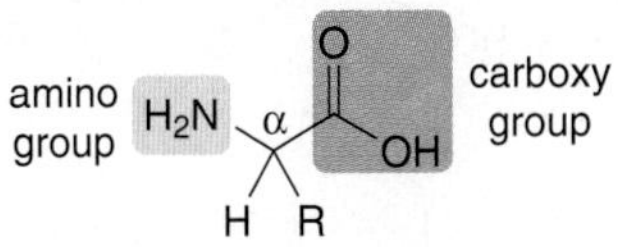

α-amino acid

As we learned in Section 3.9A, an amino acid is both an acid and a base.

- **The NH_2 group has a nonbonded electron pair, making it a base.**
- **The COOH group has an acidic proton, making it an acid.**

Amino acids are never uncharged neutral compounds. They exist as salts, so they have very high melting points and are very soluble in water. Proton transfer from the acidic carboxy group to the basic amino group forms a **zwitterion,** which contains both a positive and a negative charge.

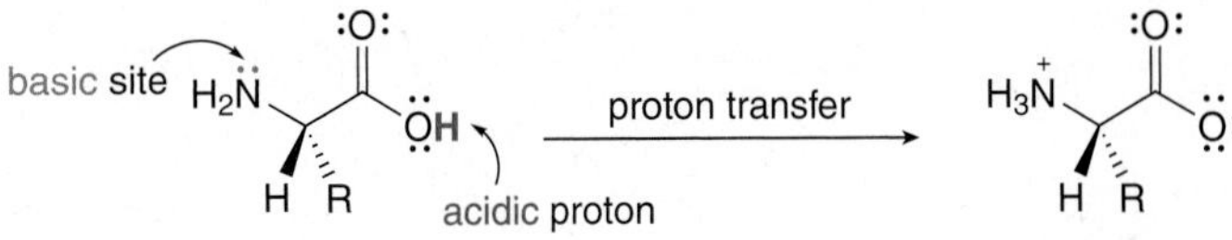

zwitterion

Humans can synthesize only 10 of the 20 amino acids needed for protein synthesis. The remaining 10, called **essential amino acids,** must be obtained from the diet and consumed on a regular, almost daily basis. Vegetarian diets must be carefully balanced to obtain all the essential amino acids. Grains—wheat, rice, and corn—are low in lysine (Figure 26.2 and Problem 19.65), and legumes—beans, peas, and peanuts—are low in methionine, but a combination of these foods provides all the needed amino acids. Thus, a diet of corn tortillas and beans, or rice and tofu, provides all essential amino acids. A peanut butter sandwich on wheat bread does, too.

The 20 amino acids that occur naturally in proteins differ in the identity of the R group bonded to the α carbon. **The simplest amino acid, called glycine, has R = H.** When the R group is any other substituent, **the α carbon is a stereogenic center,** and there are two possible enantiomers.

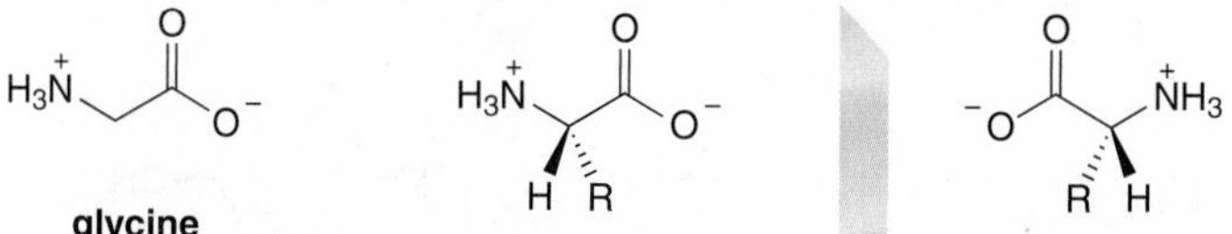

glycine
no stereogenic centers

L amino acid
Only this isomer occurs in proteins.

D amino acid

Amino acids exist in nature as only one of these enantiomers. Except when the R group is CH_2SH, the stereogenic center on the α carbon has the *S* configuration. An older system of nomenclature names the **naturally occurring enantiomer of an amino acid as the L isomer, and its unnatural enantiomer the D isomer.**

The R group of an amino acid can be H, alkyl, aryl, or an alkyl chain containing an N, O, or S atom. Representative examples are listed in Table 19.4. All amino acids have common names, which are abbreviated by a three-letter or one-letter designation. For example, glycine is often written as the three-letter abbreviation **Gly,** or the one-letter abbreviation **G.** These abbreviations are also given in Table 19.4. A complete list of the 20 naturally occurring amino acids is found in Figure 26.2. The pK_a values listed in Table 19.4 are discussed further in Section 19.15B.

Table 19.4 Representative Amino Acids

Amino acid	Name	Abbreviations		pK_a (CO_2H)	pK_a (NH_3^+)
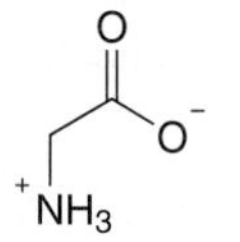	glycine	Gly	G	2.35	9.78
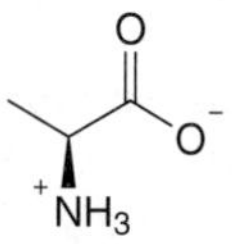	alanine	Ala	A	2.35	9.87
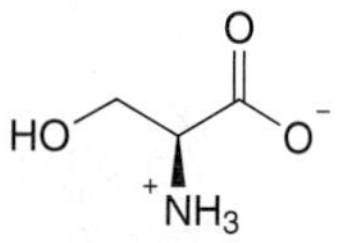	serine	Ser	S	2.21	9.15
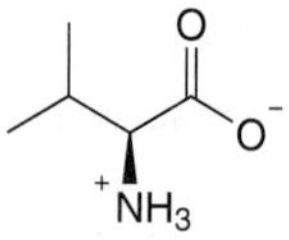	valine	Val	V	2.29	9.72
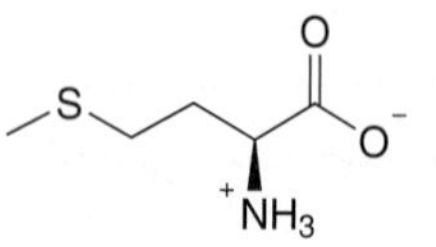	methionine	Met	M	2.28	9.21
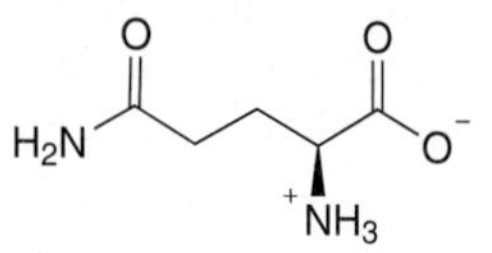	glutamine	Gln	Q	2.17	9.13

Problem 19.23 Draw both enantiomers of the amino acid homocysteine, and label the stereogenic center as *R* or *S*. A high blood level of homocysteine, which is formed from methionine by loss of a methyl group, is considered a risk factor for coronary artery disease.

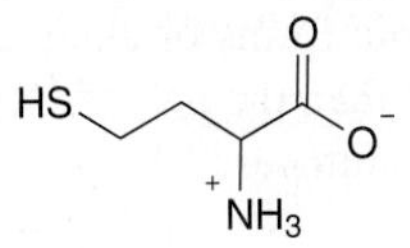

homocysteine

19.15B Acid–Base Properties

Some amino acids contain an additional functional group in the side chain that can be protonated or deprotonated in aqueous solution. The discussion in Section 19.15B is limited to the 13 amino acids that contain only two ionizable functional groups.

Amino acids resemble the dicarboxylic acids discussed in Section 19.10, in that they possess two functional groups—the carboxy group (CO_2H) and the protonated amino group (NH_3^+)— that can lose protons, so two pK_a values are reported. Alanine, [pK_a (CO_2H) = 2.35 and pK_a (NH_3^+) = 9.87] is a representative example from Table 19.4.

- The pK_a of protonated alanine, containing a carboxy group (CO_2H) and a protonated amino group (NH_3^+), is 2.35. Loss of a proton from the CO_2H group yields the zwitterionic form.
- The pK_a of the zwitterion of alanine is 9.87. Loss of a proton from the NH_3^+ group yields an anionic form of alanine, in which both functional groups are now deprotonated.

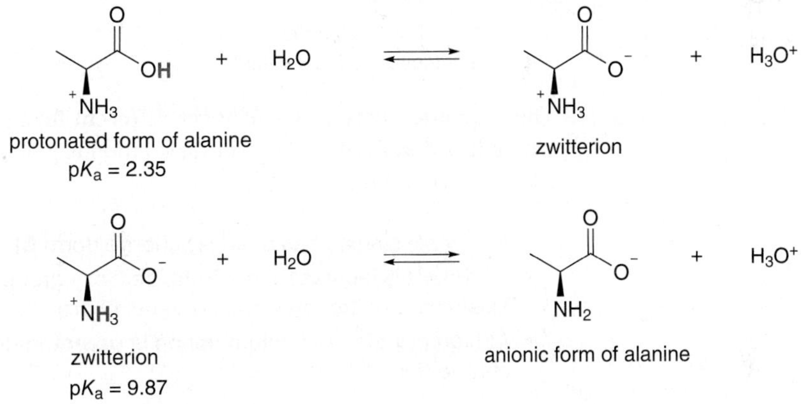

When alanine is dissolved in an aqueous solution, we can use the Henderson–Hasselbalch equation to determine which functional groups retain their protons and which are deprotonated.

- **The ionization of the acidic and basic functional groups of an amino acid depends on the pH of the solution in which it is dissolved.**

To determine the state of the amino acid, we use the pK_a values and consider the effect of pH on each functional group separately.

- **When the pH = 1, the pH of the solution is *less* than both pK_a values, so both functional groups are protonated, and the amino acid has a net +1 charge.**
- **When the pH = 12, the pH of the solution is *higher* than both pK_a values, so both functional groups are deprotonated, and the amino acid has a net −1 charge.**

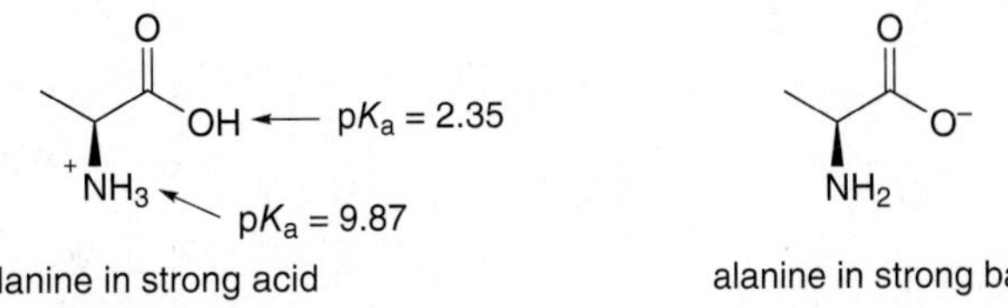

alanine in strong acid
form at pH = 1

alanine in strong base
form at pH = 12

What happens between pH 1 and 12?

The carboxy group: As the pH is gradually increased above 1, the more acidic CO_2H group is deprotonated. When the pH = pK_a (CO_2H) = 2.35, the concentrations of the protonated and zwitterionic forms of alanine are equal. When the pH is increased further and the solution is more basic than the pK_a of CO_2H, the CO_2H group is largely deprotonated, and the zwitterionic form predominates.

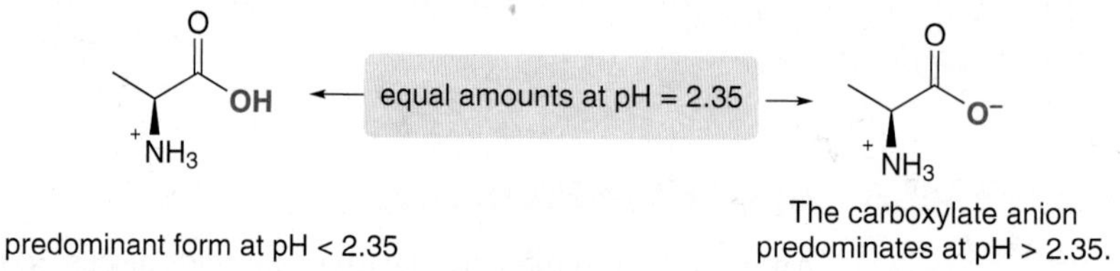
equal amounts at pH = 2.35
predominant form at pH < 2.35
The carboxylate anion
predominates at pH > 2.35.

The NH_3^+ group: When the pH is less than the pK_a of NH_3^+ (< 9.87), the NH_3^+ group remains protonated. When the pH = pK_a (NH_3^+) = 9.87, the concentration of the zwitterionic and anionic forms of alanine are equal. Above pH 9.87, the solution is more basic than the pK_a of NH_3^+, the NH_3^+ group is largely deprotonated, and the anionic form predominates.

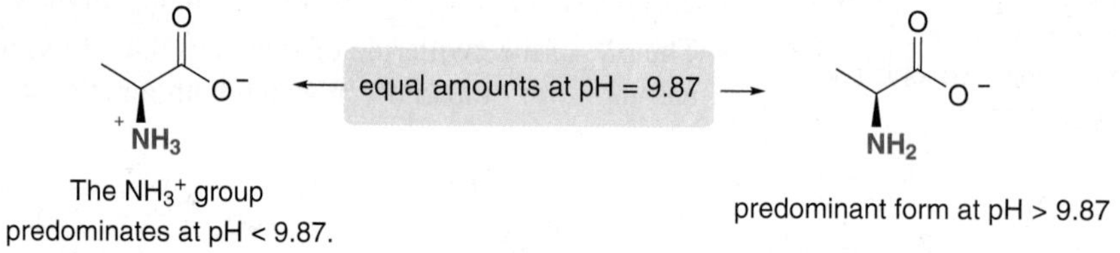
equal amounts at pH = 9.87
The NH3+ group
predominates at pH < 9.87.
predominant form at pH > 9.87

Thus, **alanine exists in one of three different forms depending on the pH of the solution in which it is dissolved.** If the pH of a solution is gradually increased from 1 to 12, the following process occurs (Figure 19.10).

- **At low pH alanine has a net (+) charge (form B).**
- **As the pH is increased to ~6, the carboxy group is deprotonated, and the amino acid exists as a zwitterion with no overall charge (form A).**
- **At high pH, the ammonium cation is deprotonated, and the amino acid has a net (–) charge (form C).**

Problem 19.24 Draw the positively charged, neutral, and negatively charged forms for the amino acid glycine. Which species predominates at pH 11? Which species predominates at pH 1?

Figure 19.10
Summary of the acid–base reactions of alanine

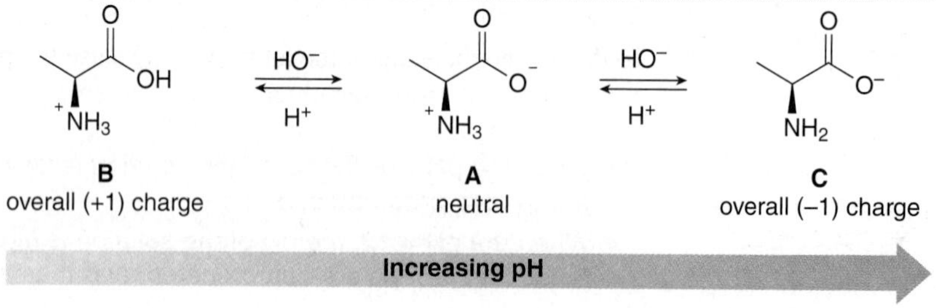
HO−
H+
HO−
H+
B
overall (+1) charge
A
neutral
C
overall (–1) charge
Increasing pH

19.15C Isoelectric Point

As we learned in Section 19.15B, a protonated amino acid has at least two different protons that can be removed, so a pK_a value is reported for each of these protons. Table 26.1 lists these values for all 20 amino acids.

- **The pH at which the amino acid exists primarily in its neutral form is called its *isoelectric point*, abbreviated as p*I*.**

For the amino acids listed in Table 19.4, the isoelectric point is the average of both pK_a values of an amino acid:

$$\textbf{Isoelectric point} = \mathbf{p}I = \frac{pK_a\,(\text{COOH}) + pK_a\,(\text{NH}_3^+)}{2}$$

$$\text{For alanine:} \quad \mathbf{p}I = \frac{2.35 + 9.87}{2} = \begin{matrix} 6.12 \\ \text{p}I\ \text{(alanine)} \end{matrix}$$

Problem 19.25 The pK_a values for the carboxy and ammonium protons of valine are 2.29 and 9.72, respectively. What is the isoelectric point of valine? Draw the structure of valine at its isoelectric point.

KEY CONCEPTS

Carboxylic Acids and the Acidity of the O–H Bond

General Facts

- Carboxylic acids contain a carboxy group (COOH). The central carbon is sp^2 hybridized and trigonal planar (19.1).
- Carboxylic acids are identified by the suffixes *-oic acid, carboxylic acid,* or *-ic acid* (19.2).
- Carboxylic acids are polar compounds that exhibit hydrogen bonding interactions (19.3).

Summary of Spectroscopic Absorptions (19.4)

IR absorptions	C=O	~1710 cm^{-1}
	O–H	3500–2500 cm^{-1} (very broad and strong)
^{1}H NMR absorptions	O–H	10–12 ppm (highly deshielded proton)
	C–H α to COOH	2–2.5 ppm (somewhat deshielded C_{sp^3}–H)
^{13}C NMR absorption	C=O	170–210 ppm (highly deshielded carbon)

General Acid–Base Reaction of Carboxylic Acids (19.9)

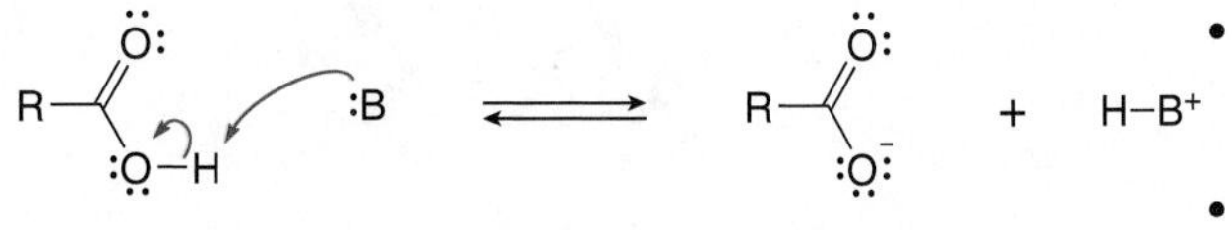

pK_a ≈ 5

carboxylate anion

- Carboxylic acids are especially acidic because carboxylate anions are resonance stabilized.
- For equilibrium to favor the products, the base must have a conjugate acid with a pK_a > 5. Common bases are listed in Table 19.3.

Factors That Affect Acidity

Resonance effects

- A carboxylic acid is more acidic than an alcohol or phenol because its conjugate base is more effectively stabilized by resonance (19.9).

ROH

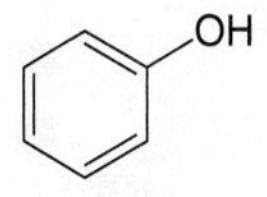

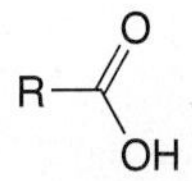

ROH	phenol	RCOOH
pK_a = 16–18	pK_a = 10	pK_a ≈ 5

Increasing acidity →

Inductive effects

- Acidity increases with the presence of electron-withdrawing groups (like the electronegative halogens) and decreases with the presence of electron-donating groups (like polarizable alkyl groups) (19.11).

Substituted benzoic acids (19.12)

- Electron-donor groups **(D)** make a substituted benzoic acid less acidic than benzoic acid.
- Electron-withdrawing groups **(W)** make a substituted benzoic acid more acidic than benzoic acid.

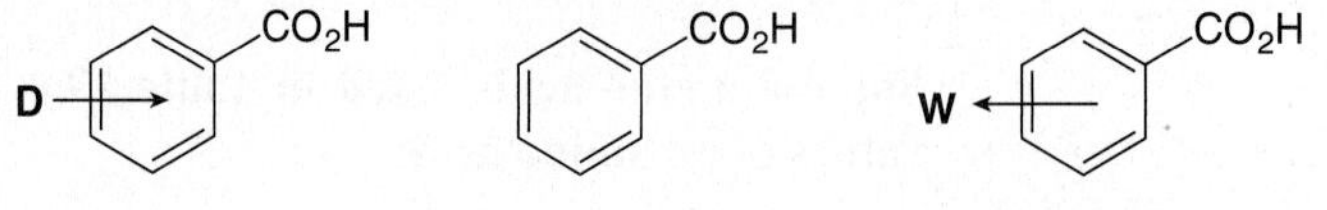

less acidic **higher pK_a** p$K_a > 4.2$	p$K_a = 4.2$	**more acidic** **lower pK_a** p$K_a < 4.2$

Increasing acidity →

Other Facts

- The Henderson–Hasselbalch equation tells whether a compound will exist in its acidic form (HA) or as its conjugate base (A:$^-$) at a particular pH. An acid exists in its protonated form HA in solutions that are more acidic than its pK_a. An acid exists as its conjugate base A:$^-$ in solutions that are more basic than its pK_a (19.10).
- Extraction is a useful technique for separating compounds having different solubility properties. Carboxylic acids can be separated from other organic compounds by extraction, because aqueous base converts a carboxylic acid into a water-soluble carboxylate anion (19.13).
- A sulfonic acid (RSO_3H) is a strong acid because it forms a weak, resonance-stabilized conjugate base on deprotonation (19.14A).
- Phosphoric acid esters [$ROPO(OH)_2$] exist as monophosphate dianions ($ROPO_3^{2-}$) at physiological pH (19.14B), because the pK_a values for the two OH groups are less than 7.4.
- Amino acids have an amino group on the α carbon to the carboxy group [$RCH(NH_2)COOH$]. Amino acids exist as zwitterions at pH ≈ 6. Adding acid forms a species with a net (+1) charge [$RCH(NH_3)COOH]^+$. Adding base forms a species with a net (–1) charge [$RCH(NH_2)COO]^-$ (19.15).

PROBLEMS

Problems Using Three-Dimensional Models

19.26 Answer each question for **A** and **B** depicted in the ball-and-stick models.

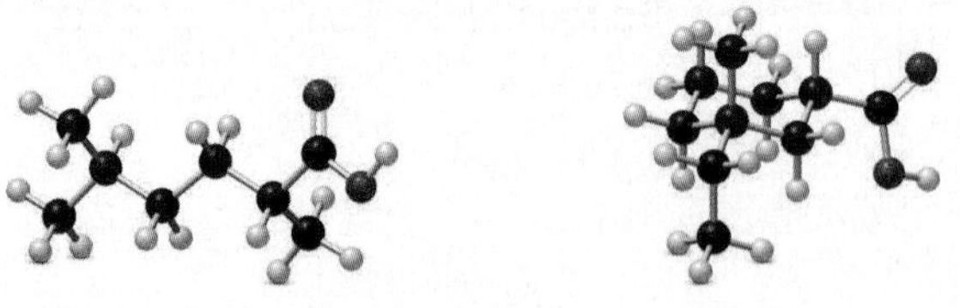

A **B**

a. What is the IUPAC name for each compound?
b. What product is formed when each compound is treated with NaOH?
c. Name the products formed in part (b).
d. Draw the structure of an isomer that is at least 10^5 times less acidic than each compound.

19.27 Rank the carboxylic acids in order of increasing acidity.

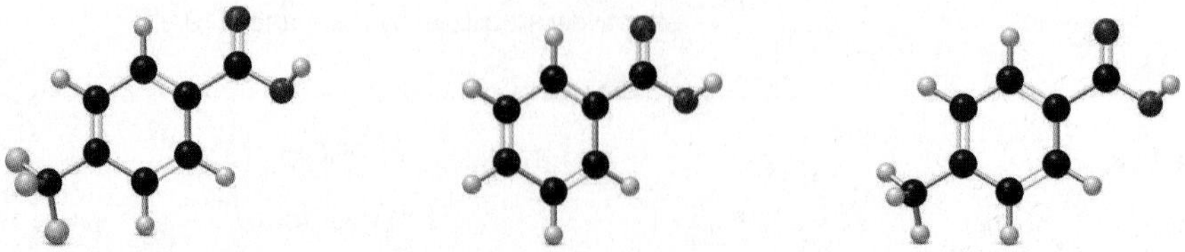

Nomenclature

19.28 Give the IUPAC name for each compound.

a. 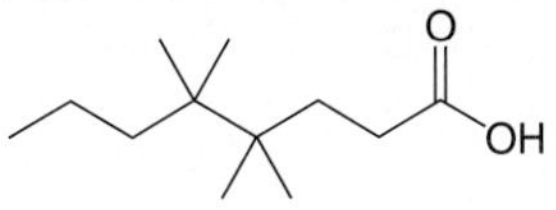
O
OH

d. 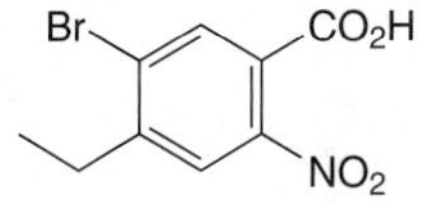
Br
CO_2H
NO_2

b. 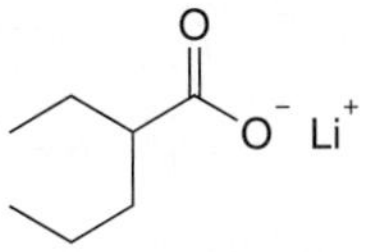
O
O^- Li^+

e. 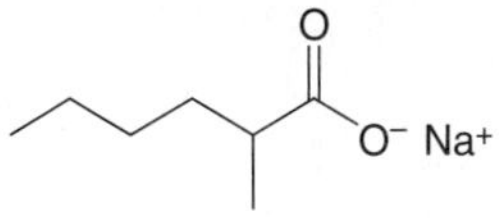
O
O^- Na^+

c. 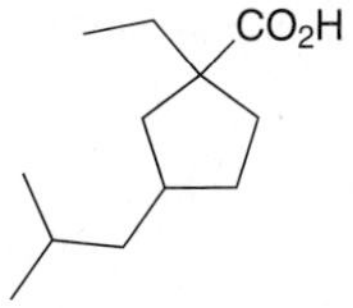
CO_2H

f. 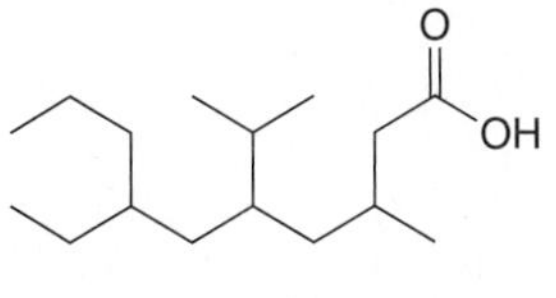
O
OH

19.29 Draw the structure corresponding to each name.

a. 3,3-dimethylpentanoic acid
b. 4-chloro-3-phenylheptanoic acid
c. (*R*)-2-chloropropanoic acid
d. *m*-hydroxybenzoic acid
e. potassium acetate
f. sodium α-bromobutyrate
g. 2,2-dichloropentanedioic acid
h. 4-isopropyl-2-methyloctanedioic acid

Physical Properties

19.30 Rank the following compounds in order of increasing boiling point.

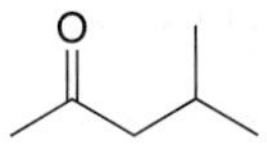
O

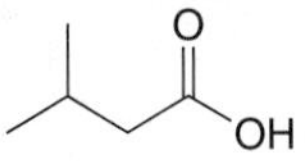
O
OH

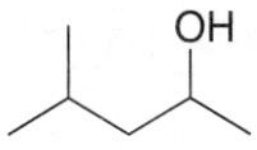
OH

Preparation of Carboxylic Acids

19.31 Draw the organic products formed in each reaction.

a. 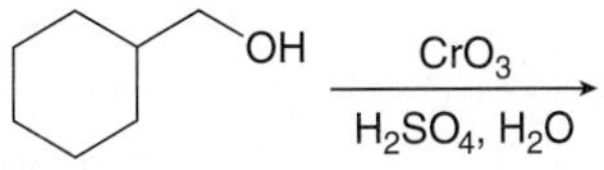
OH
CrO_3
H_2SO_4, H_2O

c. 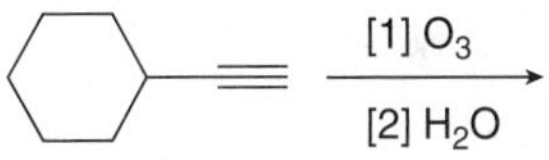
[1] O_3
[2] H_2O

b. 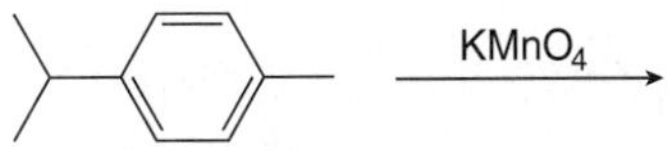
$KMnO_4$

d. 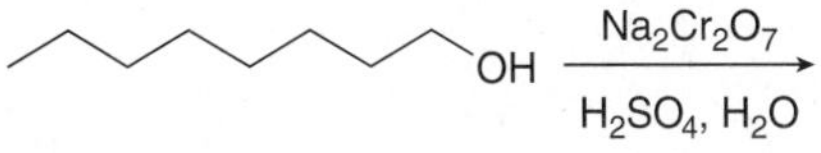
OH
$Na_2Cr_2O_7$
H_2SO_4, H_2O

19.32 Identify the lettered compounds in each reaction sequence.

a. 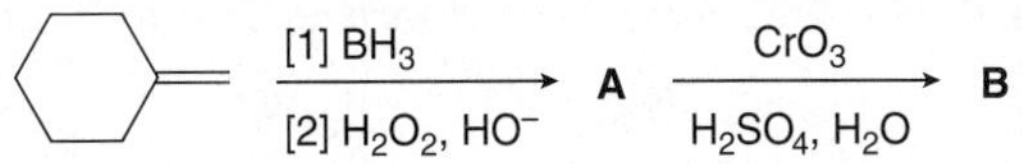
[1] BH_3
[2] H_2O_2, HO^-
A
CrO_3
H_2SO_4, H_2O
B

b. $HC{\equiv}CH \xrightarrow[\text{[2] } CH_3I]{\text{[1] } NaNH_2} \mathbf{C} \xrightarrow[\text{[2] } CH_3CH_2I]{\text{[1] } NaNH_2} \mathbf{D} \xrightarrow[\text{[2] } H_2O]{\text{[1] } O_3} \mathbf{E} + \mathbf{F}$

c. 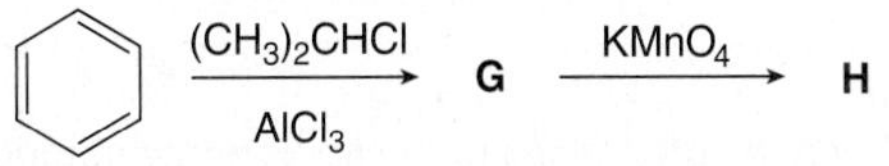
$(CH_3)_2CHCl$
$AlCl_3$
G
$KMnO_4$
H

Acid–Base Reactions; General Questions on Acidity

19.33 Using the pK_a table in Appendix A, determine whether each of the following bases is strong enough to deprotonate the three compounds listed below. Bases: [1] ^{-}OH; [2] $CH_3CH_2^{-}$; [3] $^{-}NH_2$; [4] NH_3; [5] $HC{\equiv}C^{-}$.

a. 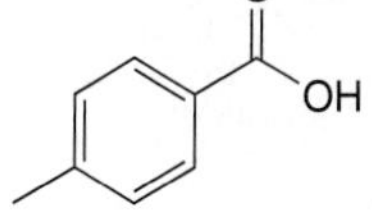

pK_a = 4.3

b. 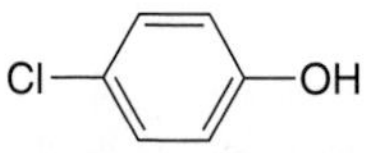

pK_a = 9.4

c. 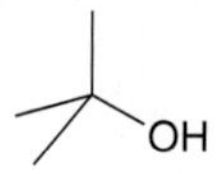

pK_a = 18

19.34 Draw the products of each acid–base reaction, and using the pK_a table in Appendix A, determine if equilibrium favors the reactants or products.

a. 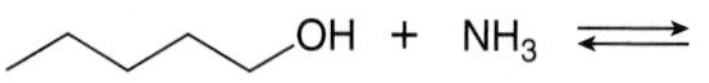

b. 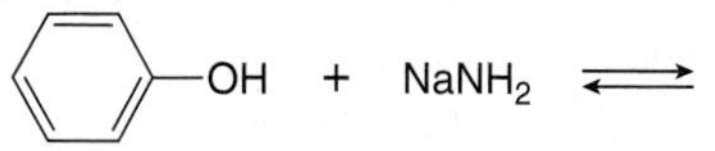

c. 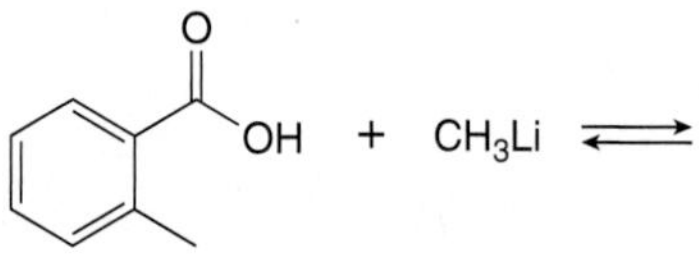

d. 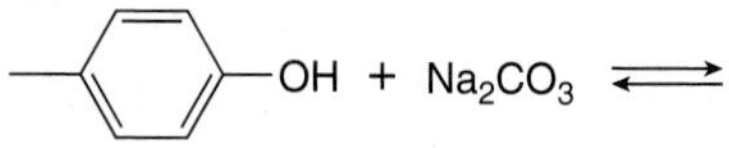

19.35 Which compound in each pair has the lower pK_a? Which compound in each pair has the stronger conjugate base?

a. 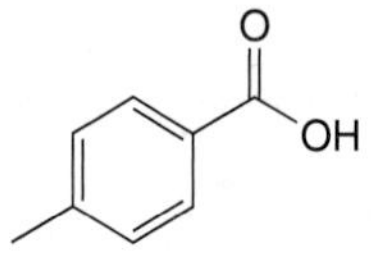 or 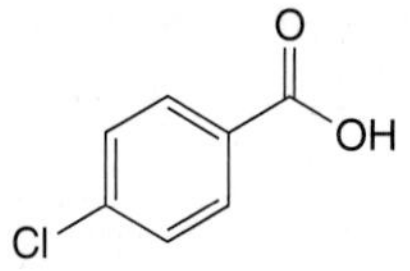

b. 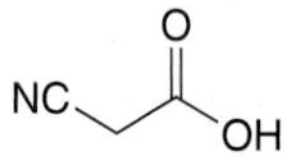 or 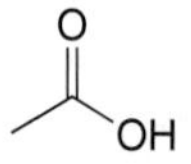

19.36 Rank the compounds in each group in order of increasing acidity.

a. 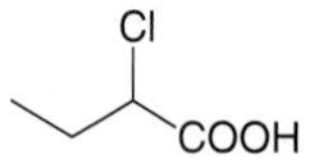 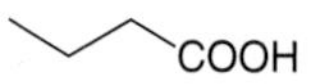 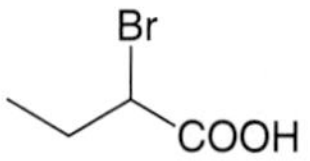

b. 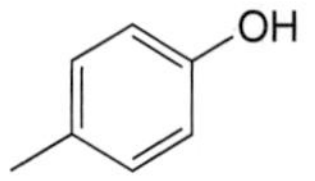 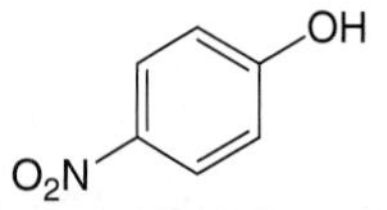 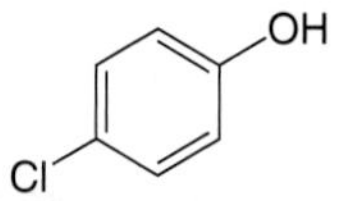

19.37 Rank the compounds in each group in order of increasing basicity.

a. 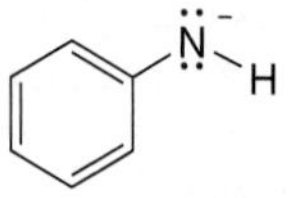 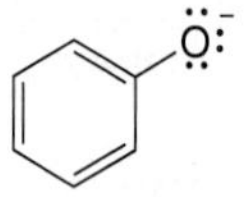 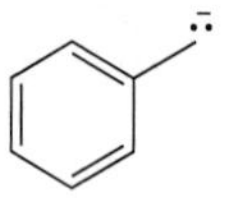

b. 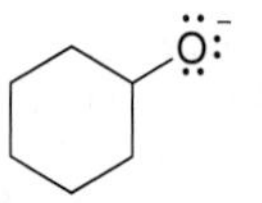 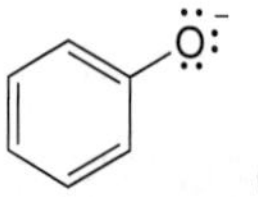 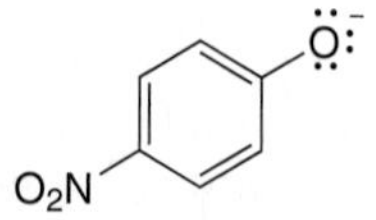

19.38 Match the pK_a values to the appropriate compound. pK_a values: 0.28, 1.24, 2.66, 2.86, and 3.12. Compounds: (a) FCH_2COOH; (b) CF_3COOH; (c) $F_2CHCOOH$; (d) ICH_2COOH; (e) $BrCH_2COOH$.

19.39 Although codeine occurs in low concentration in the opium poppy, most of the codeine used in medicine is prepared from morphine (the principal component of opium) by the following reaction. Explain why selective methylation occurs at only one OH in morphine to give codeine. Codeine is a less potent and less addictive analgesic than morphine.

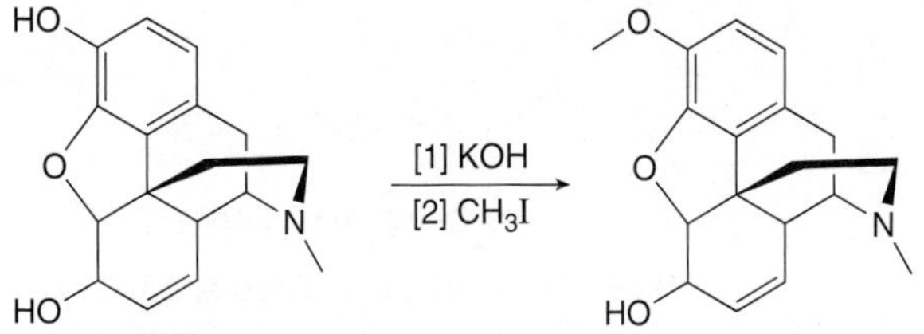

morphine codeine

19.40 Which carboxylic acid has the lower pK_a, pyruvic acid (CH_3COCO_2H) or acetoacetic acid ($CH_3COCH_2CO_2H$)? Explain your choice.

19.41 Explain each statement.

a. The pK_a of *p*-nitrophenol is lower than the pK_a of phenol (7.2 vs. 10).

b. The pK_a of *p*-nitrophenol is lower than the pK_a of *m*-nitrophenol (7.2 vs. 8.3).

19.42 Explain the following statement. Although 2-methoxyacetic acid (CH_3OCH_2COOH) is a stronger acid than acetic acid (CH_3COOH), *p*-methoxybenzoic acid ($CH_3OC_6H_4COOH$) is a weaker acid than benzoic acid (C_6H_5COOH).

19.43 The pK_a of *p*-methylthiophenol ($CH_3SC_6H_4OH$) is 9.53. Is *p*-methylthiophenol more or less reactive in electrophilic aromatic substitution than phenol?

19.44 Rank the following compounds in order of increasing acidity and explain in detail your choice of order.

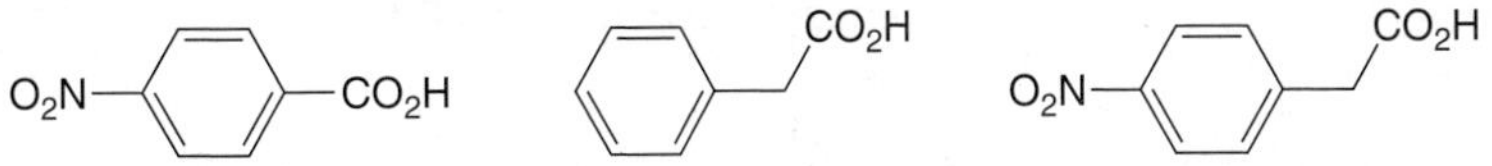

C **D** **E**

19.45 Phthalic acid and isophthalic acid have protons on two carboxy groups that can be removed with base. (a) Explain why the pK_a for loss of the first proton (pK_{a1}) is lower for phthalic acid than isophthalic acid. (b) Explain why the pK_a for loss of the second proton (pK_{a2}) is higher for phthalic acid than isophthalic acid.

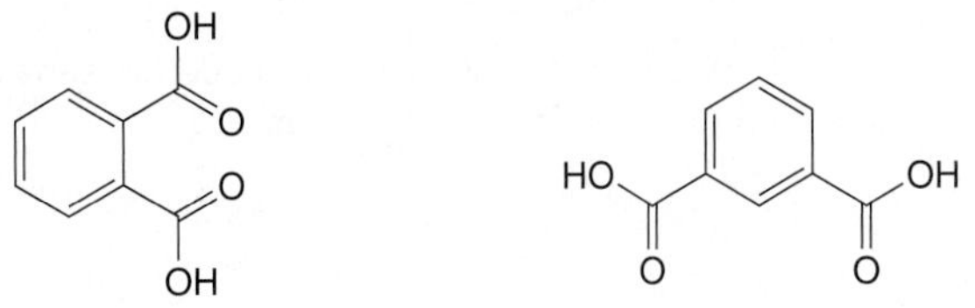

phthalic acid
pK_{a1} = 2.9
pK_{a2} = 5.4

isophthalic acid
pK_{a1} = 3.7
pK_{a2} = 4.6

19.46 Explain the following result. Acetic acid (CH_3COOH), labeled at its OH oxygen with the uncommon ^{18}O isotope (shown in red), was treated with aqueous base, and then the solution was acidified. Two products having the ^{18}O label at different locations were formed.

19.47 Draw all resonance structures of the conjugate bases formed by removal of the labeled protons (H_a, H_b, and H_c) in cyclohexane-1,3-dione and acetanilide. For each compound, rank these protons in order of increasing acidity and explain the order you chose.

a. 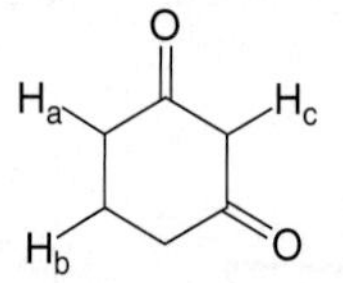

cyclohexane-1,3-dione

b. 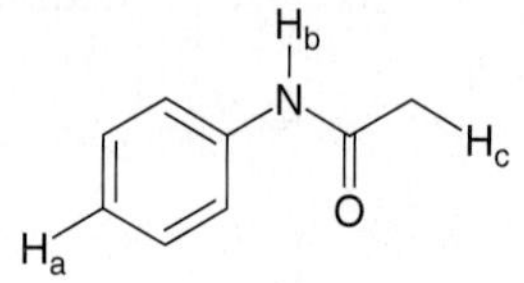

acetanilide

19.48 As we will see in Chapter 23, C–H bonds are sometimes more acidic than O–H bonds. Explain why the pK_a of $CH_2(CHO)_2$ is lower than the pK_a of $HO(CH_2)_3OH$ (9 vs. 16).

19.49 Identify **X** in the following equation, and explain how hexanoic acid (Section 19.2B) is formed by this stepwise reaction sequence.

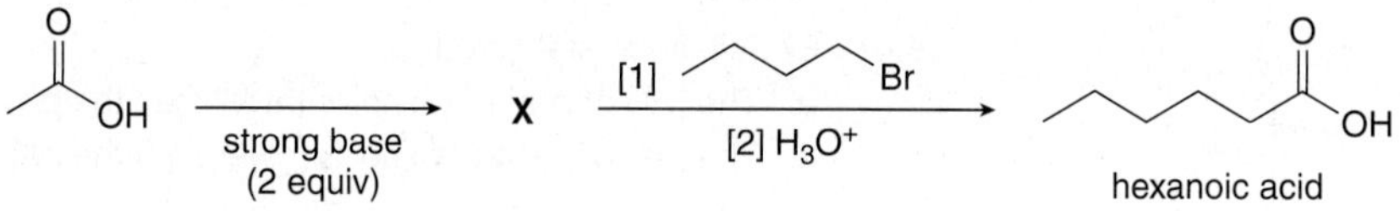

Henderson–Hasselbalch Equation

19.50 Use the Henderson–Hasselbalch equation to determine the ratio of phenol (C_6H_5OH, pK_a = 10.0) to phenoxide ($C_6H_5O^-$) at each of the following pH values: (a) 8; (b) 13; (c) 10; (d) 6.

19.51 The pK_a values for the ionization of the carboxy groups in glutaric acid (**A**) are 4.3 and 5.4. (a) At what pH will there be an equal amount of **A** and **B?** (b) At what pH will there be an equal amount of **B** and **C?**

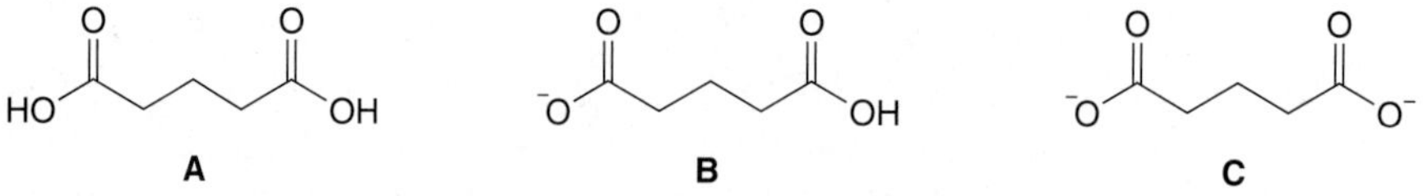

19.52 Consider a carboxylic acid (RCO_2H, pK_a = 4.8) and a protonated amine (RNH_3^+, pK_a = 9). Indicate the predominant form of each compound at the following pH values: (a) 8; (b) 3; (c) 10.

Extraction

19.53 Write out the steps needed to separate hydrocarbon **A** and carboxylic acid **B** by using an extraction procedure.

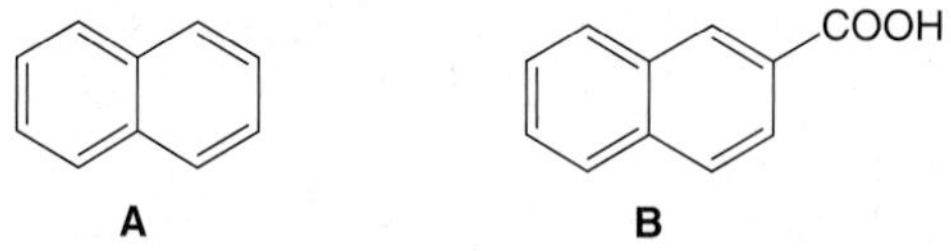

19.54 Because phenol (C_6H_5OH) is less acidic than a carboxylic acid, it can be deprotonated by NaOH but not by the weaker base $NaHCO_3$. Using this information, write out an extraction sequence that can be used to separate C_6H_5OH, benzoic acid, and cyclohexanol. Show what compound is present in each layer at each stage of the process, and if it is present in its neutral or ionic form.

Spectroscopy

19.55 Use the ^1H NMR and IR spectra given below to identify the structures of two isomers (**A** and **B**) having molecular formula $C_4H_8O_2$.

Compound **A:**

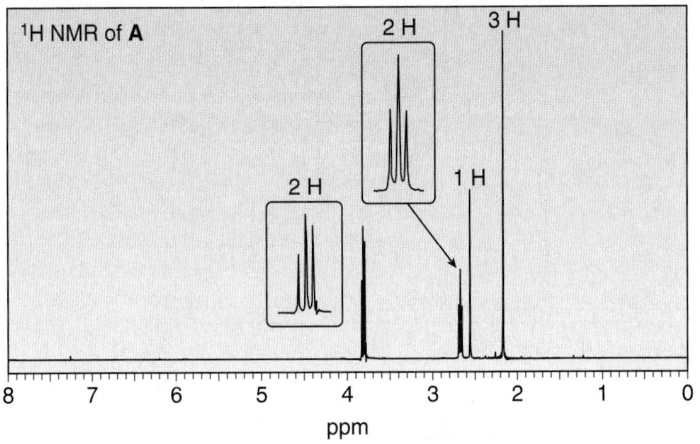

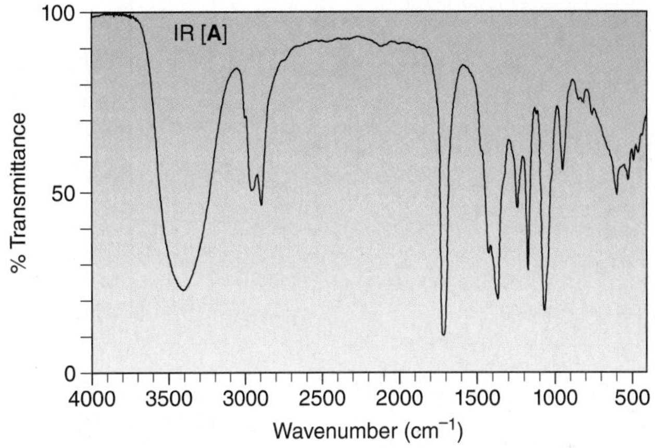

Compound **B:**

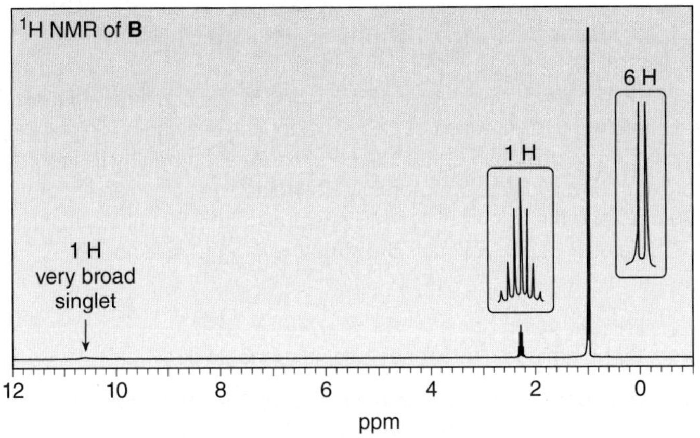

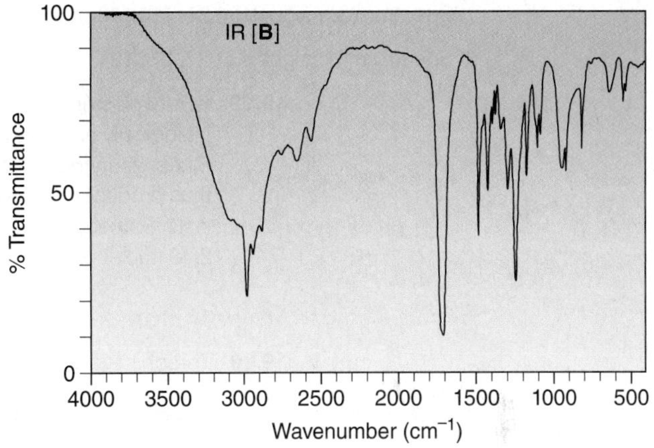

19.56 An unknown compound **C** (molecular formula $C_4H_8O_3$) exhibits IR absorptions at 3600–2500 and 1734 cm^{-1}, as well as the following ^1H NMR spectrum. What is the structure of **C?**

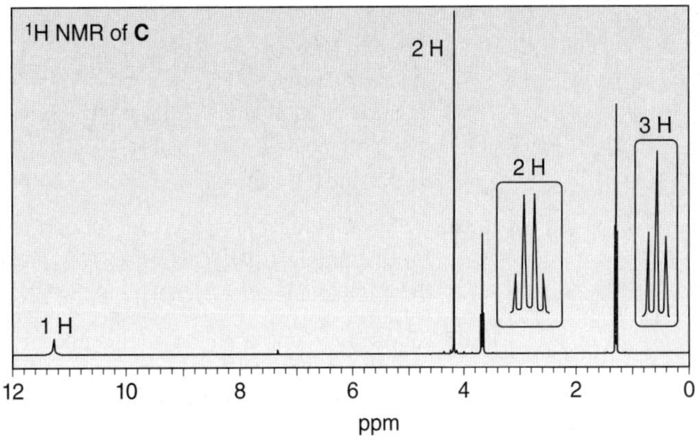

19.57 Propose a structure for **D** (molecular formula $C_9H_9ClO_2$) consistent with the given spectroscopic data.

^{13}C NMR signals at 30, 36, 128, 130, 133, 139, and 179 ppm

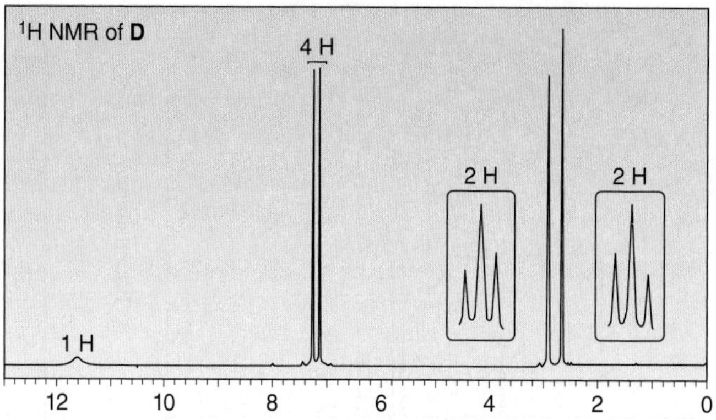

19.58 Match the ^{13}C NMR data to the appropriate structure.

Spectrum [1]: signals at 14, 22, 27, 34, 181 ppm

Spectrum [2]: signals at 27, 39, 186 ppm

Spectrum [3]: signals at 22, 26, 43, 180 ppm

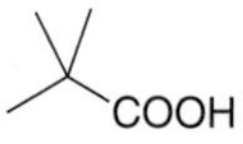

A

B

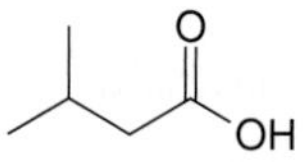

C

19.59 γ-Butyrolactone ($C_4H_6O_2$, GBL) is a biologically inactive compound that is converted to the biologically active recreational drug GHB (Section 19.5) by a lactonase enzyme in the body. Since γ-butyrolactone is more fat soluble than GHB, it is more readily absorbed by tissues and thus produces a faster onset of physiological symptoms. γ-Butyrolactone shows an absorption in its IR spectrum at 1770 cm^{-1} and the following ^1H NMR spectral data: 2.28 (multiplet, 2 H), 2.48 (triplet, 2 H), and 4.35 (triplet, 2 H) ppm. What is the structure of γ-butyrolactone?

Amino Acids

19.60 Threonine is a naturally occurring amino acid that has two stereogenic centers.

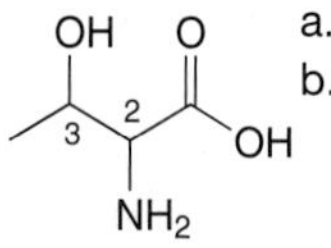

threonine

a. Draw the four possible stereoisomers using wedges and dashes.
b. The naturally occurring amino acid has the 2*S*,3*R* configuration at its two stereogenic centers. Which structure does this correspond to?

19.61 For each amino acid [RCH(NH_2)COOH], draw its neutral, positively charged, and negatively charged forms. Which form predominates at pH = 1, 6, and 11? What is the structure of each amino acid at its isoelectric point?

a. methionine (R = $CH_2CH_2SCH_3$)
b. serine (R = CH_2OH)

19.62 Calculate the isoelectric point for each amino acid.

a. asparagine: pK_a (COOH) = 2.02; pK_a (α-NH_3^+) = 8.80
b. methionine: pK_a (COOH) = 2.28; pK_a (α-NH_3^+) = 9.21

19.63 Consider γ-aminobutyric acid, a neurotransmitter in the brain that contains an amino group on the γ carbon to the carboxy group. (a) If pK_a (CO_2H) = 4.23 and pK_a (NH_3^+) = 10.43, draw the predominant form(s) of the amino acid at each of the following pH values: [1] 1.21; [2] 12.1; [3] 10.43; [4] 4.23. (b) What is the isoelectric point for this amino acid? (c) Explain why the pK_a of the CO_2H group of γ-aminobutyric acid is considerably higher than the pK_a of the CO_2H group of each amino acid in Table 19.4.

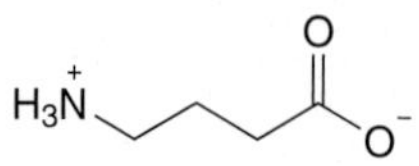

γ-aminobutyric acid

19.64 The amino acid cysteine has three ionizable functional groups with the indicated pK_a values. Draw the predominant form(s) of the amino acid at each of the following pH values: (a) 12; (b) 1; (c) 8; (d) 3.

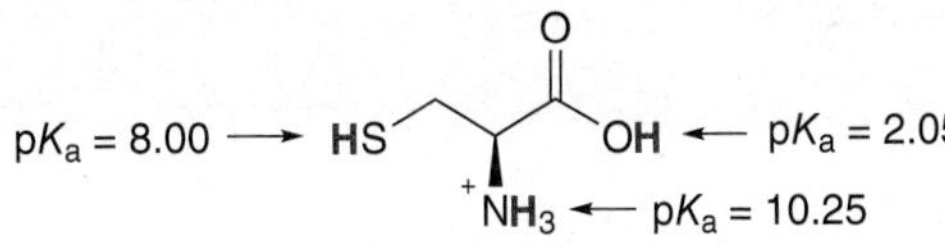
pKa = 8.00
HS
OH
pKa = 2.05
NH3
pKa = 10.25

cysteine

19.65 Lysine and tryptophan are two amino acids that contain an additional N atom in the R group bonded to the α carbon. While lysine is classified as a basic amino acid because it contains an additional basic N atom, tryptophan is classified as a neutral amino acid. Explain why this difference in classification occurs.

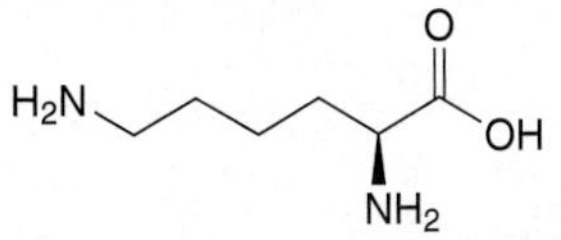

lysine

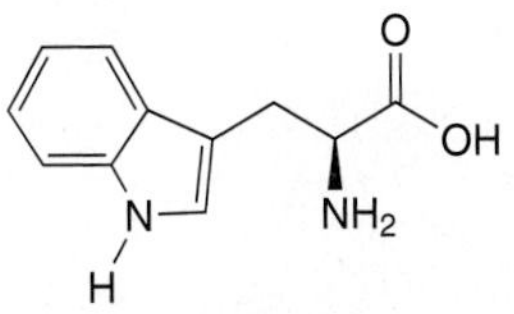

tryptophan

Challenge Problems

19.66 Explain why using one or two equivalents of NaH results in different products in the following reactions.

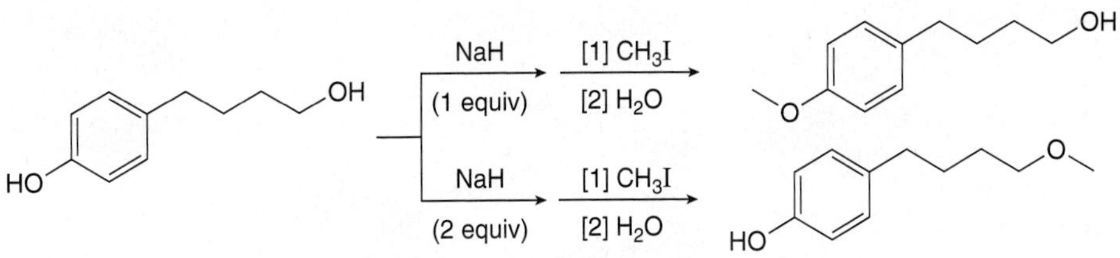
NaH
(1 equiv)
[1] CH3I
[2] H2O
NaH
(2 equiv)
[1] CH3I
[2] H2O

19.67 Although *p*-hydroxybenzoic acid is less acidic than benzoic acid, *o*-hydroxybenzoic acid is slightly more acidic than benzoic acid. Explain this result.

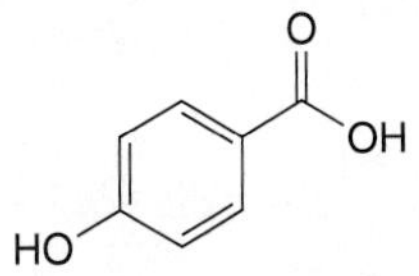

p-hydroxybenzoic acid

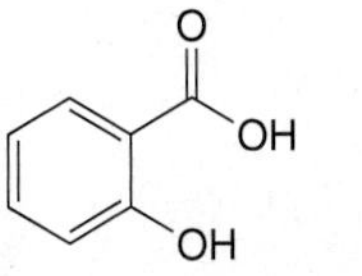

o-hydroxybenzoic acid

19.68 2-Hydroxybutanedioic acid occurs naturally in apples and other fruits. Rank the labeled protons (H_a–H_e) in order of increasing acidity and explain in detail the order you chose.

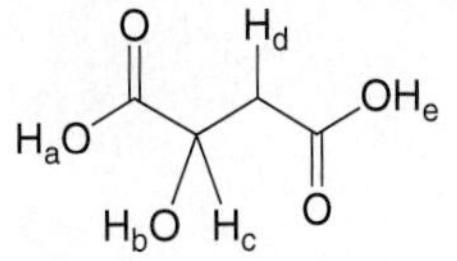

2-hydroxybutanedioic acid

19.69 Although it was initially sold as a rat poison, warfarin is an effective anticoagulant used to prevent blood clots. Label the most acidic proton in warfarin, and explain why its pK_a is comparable to the pK_a of a carboxylic acid.

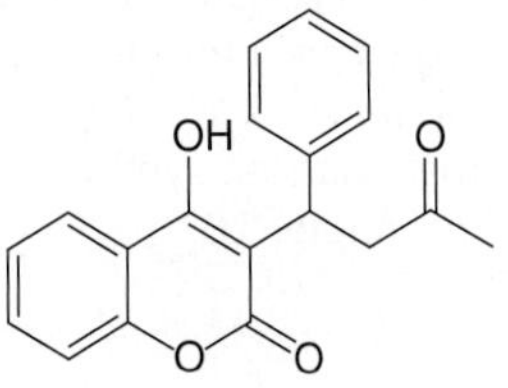

warfarin

20 Introduction to Carbonyl Chemistry; Organometallic Reagents; Oxidation and Reduction

20.1 Introduction
20.2 General reactions of carbonyl compounds
20.3 A preview of oxidation and reduction
20.4 Reduction of aldehydes and ketones
20.5 The stereochemistry of carbonyl reduction
20.6 Enantioselective carbonyl reductions
20.7 Reduction of carboxylic acids and their derivatives
20.8 Oxidation of aldehydes
20.9 Organometallic reagents
20.10 Reaction of organometallic reagents with aldehydes and ketones
20.11 Retrosynthetic analysis of Grignard products
20.12 Protecting groups
20.13 Reaction of organometallic reagents with carboxylic acid derivatives
20.14 Reaction of organometallic reagents with other compounds
20.15 α,β-Unsaturated carbonyl compounds
20.16 Summary—The reactions of organometallic reagents
20.17 Synthesis

Ciguatoxin CTX3C is a potent neurotoxin found in more than 400 species of warm-water fish. Thousands of people contract ciguatera seafood poisoning each year from ingesting tropical reef fish containing this neurotoxin. Interest in providing a practical supply of ciguatoxin CTX3C for biological testing led to its laboratory synthesis in 2001. One step in the multistep synthesis involved a selective reduction, one of the many reactions presented in Chapter 20.

Chapters 20 through 24 of this text discuss carbonyl compounds—aldehydes, ketones, acid halides, esters, amides, and carboxylic acids. **The carbonyl group is perhaps the most important functional group in organic chemistry,** because its electron-deficient carbon and easily broken π bond make it susceptible to a wide variety of useful reactions.

We begin by examining the similarities and differences between two broad classes of carbonyl compounds. We will then spend the remainder of Chapter 20 on reactions that are especially important in organic synthesis. Chapters 21 and 22 present specific reactions that occur at the carbonyl carbon, and Chapters 23 and 24 concentrate on reactions occurring at the α carbon to the carbonyl group.

Although Chapter 20 is "jam-packed" with reactions, most of them follow one of two general pathways, so they can be classified in a well-organized fashion, provided you remember a few basic principles. Keep in mind the following fundamental themes about reactions:

- Nucleophiles attack electrophiles.
- π Bonds are easily broken.
- Bonds to good leaving groups are easily cleaved.

20.1 Introduction

Two broad classes of compounds contain a ***carbonyl group:***

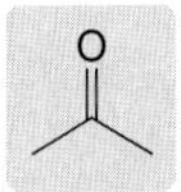

carbonyl group

[1] Compounds that have only carbon and hydrogen atoms bonded to the carbonyl group

- **An aldehyde** has at least one H atom bonded to the carbonyl group.
- **A ketone** has two alkyl or aryl groups bonded to the carbonyl group.

[2] Compounds that contain an electronegative atom bonded to the carbonyl group

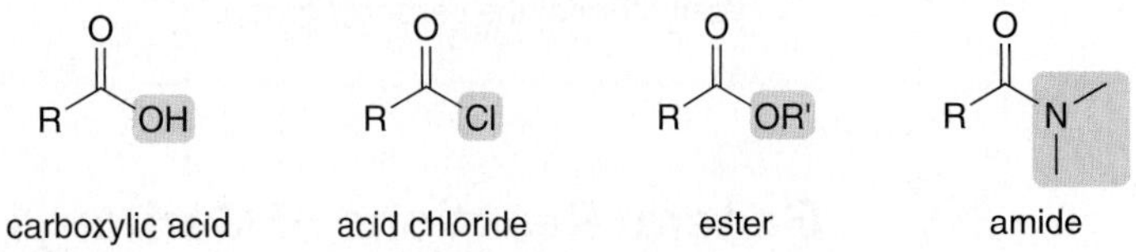

These include **carboxylic acids, acid chlorides, esters,** and **amides,** as well as other similar compounds discussed in Chapter 22. Each of these compounds contains an atom (Cl, O, or N) more electronegative than carbon, capable of acting as a **leaving group.** Acid chlorides, esters, and amides are often called **carboxylic acid derivatives,** because they can be synthesized from carboxylic acids (Chapter 22). Since each compound contains an acyl group (RCO−), they are also called **acyl derivatives.**

- **The presence or absence of a leaving group on the carbonyl carbon determines the type of reactions these compounds undergo (Section 20.2).**

The carbonyl carbon atom is sp^2 hybridized and trigonal planar, and all bond angles are ~120°. The double bond of a carbonyl group consists of one σ bond and one π bond. The π bond is

The aldehyde α-sinensal (Problem 20.1) is the major compound responsible for the orange-like odor of mandarin oil, obtained from the mandarin tree in southern China.

formed by the overlap of two p orbitals, and extends above and below the plane. In these features the carbonyl group resembles the trigonal planar, sp^2 hybridized carbons of a C−C double bond.

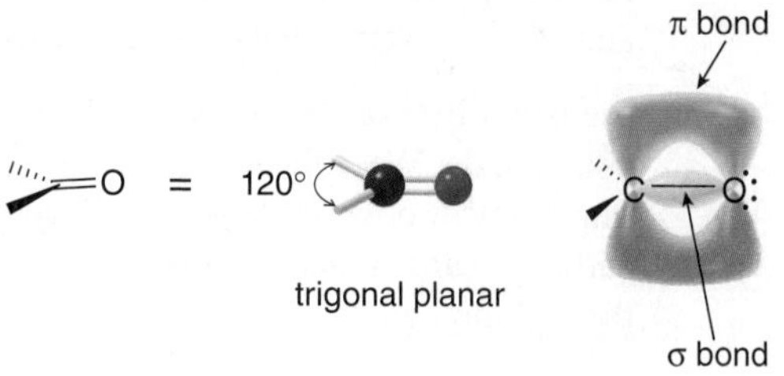

In one important way, though, a C=O and C=C are very different. **The electronegative oxygen atom in the carbonyl group means that the bond is polarized, making the carbonyl carbon electron deficient.** Using a resonance description, the carbonyl group is represented by two resonance structures, with a charge-separated resonance structure a minor contributor to the hybrid. An electrostatic potential plot for formaldehyde, the simplest aldehyde, is shown in Figure 20.1. It clearly indicates the polarized carbonyl group.

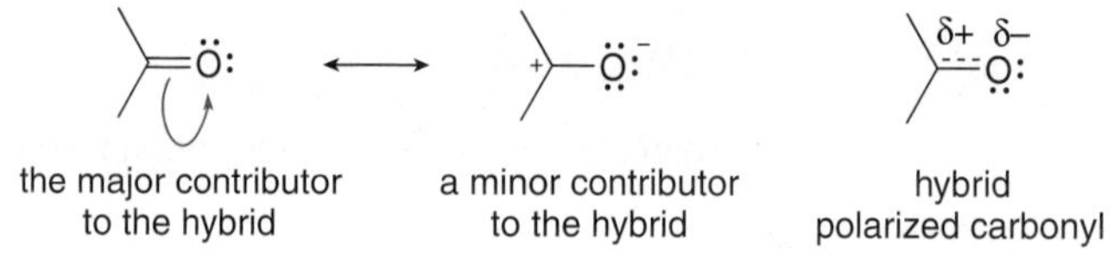

Problem 20.1

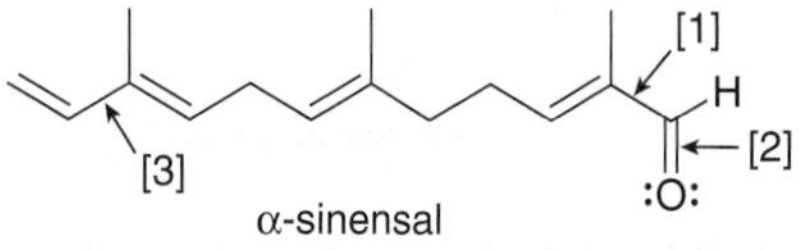

a. What orbitals are used to form the indicated bonds in α-sinensal?

b. In what type of orbitals do the lone pairs on O reside?

Figure 20.1

Electrostatic potential map of formaldehyde, $CH_2=O$

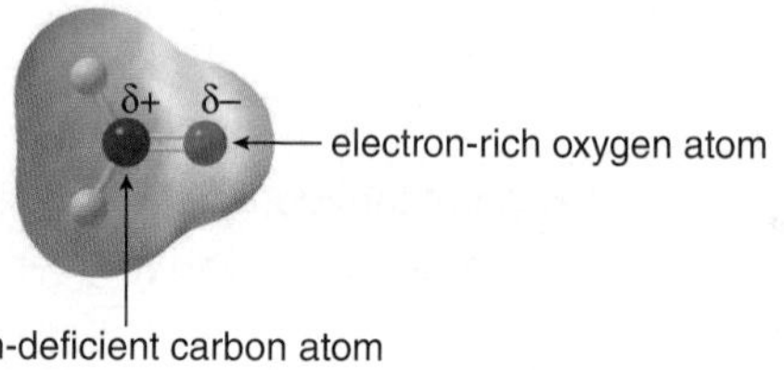

- An electrostatic potential map shows the electron-deficient carbon atom and the electron-rich oxygen atom of the carbonyl group.

20.2 General Reactions of Carbonyl Compounds

With what types of reagents should a carbonyl group react? The electronegative oxygen makes the carbonyl carbon electrophilic, and because it is trigonal planar, a carbonyl carbon is uncrowded. Moreover, a carbonyl group has an easily broken π bond.

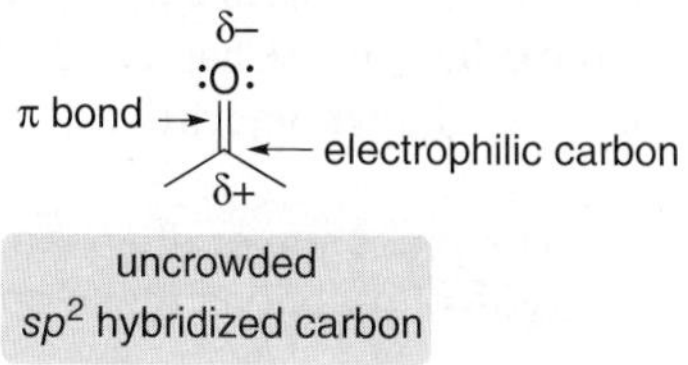

As a result, **carbonyl compounds react with nucleophiles.** The outcome of nucleophilic attack, however, depends on the identity of the carbonyl starting material.

- **Aldehydes and ketones undergo nucleophilic addition.**

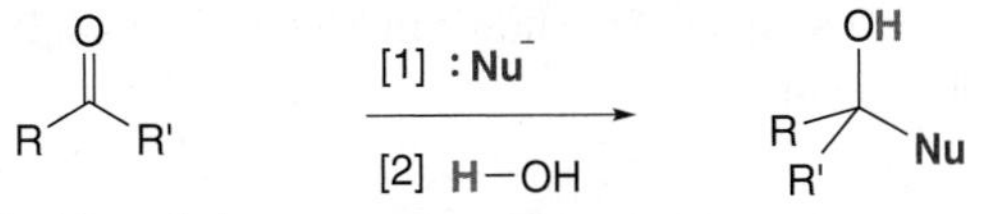
[1] :Nu⁻
[2] H–OH
R' = H or alkyl
nucleophilic addition
H and Nu are added.

- **Carbonyl compounds that contain leaving groups undergo nucleophilic substitution.**

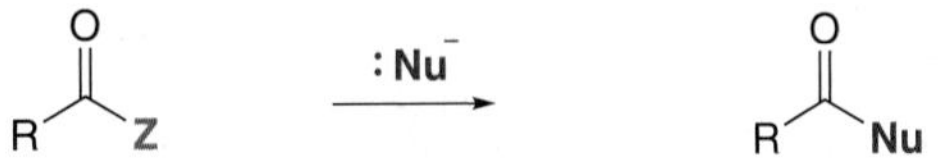
:Nu⁻
nucleophilic substitution
Nu replaces Z.
Z = OH, Cl, OR, NH2

Let's examine each of these general reactions individually.

20.2A Nucleophilic Addition to Aldehydes and Ketones

Aldehydes and ketones react with nucleophiles to form addition products by the two-step process shown in Mechanism 20.1: **nucleophilic attack** followed by **protonation.**

Mechanism 20.1 Nucleophilic Addition—A Two-Step Process

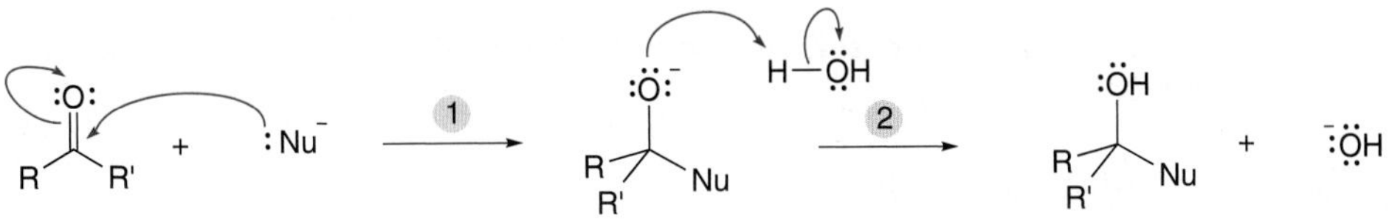

1 The nucleophile attacks the electrophilic carbonyl. The π bond is broken, moving an electron pair out on oxygen and forming an sp^3 hybridized carbon.

2 Protonation of the negatively charged oxygen by H_2O forms the addition product.

More examples of nucleophilic addition to aldehydes and ketones are discussed in Chapter 21.

The net result is that the π bond is broken, two new σ bonds are formed, and the elements of H and Nu are added across the π bond. Nucleophilic addition with two different nucleophiles—**hydride (H:⁻)** and **carbanions (R:⁻)**—is discussed in Chapter 20.

Aldehydes are more reactive than ketones toward nucleophilic attack for both steric and electronic reasons.

aldehyde	**ketone**
• less crowded	• more crowded
• less stable	• more stable
• more reactive	• less reactive

- **The two R groups bonded to the ketone carbonyl group make it more crowded, so nucleophilic attack is more difficult.**
- **The two electron-donor R groups stabilize the partial charge on the carbonyl carbon of a ketone, making it more stable and less reactive.**

20.2B Nucleophilic Substitution of RCOZ (Z = Leaving Group)

Carbonyl compounds with leaving groups react with nucleophiles to form substitution products by the two-step process shown in Mechanism 20.2: **nucleophilic attack,** followed by **loss of the leaving group.**

Mechanism 20.2 Nucleophilic Substitution—A Two-Step Process

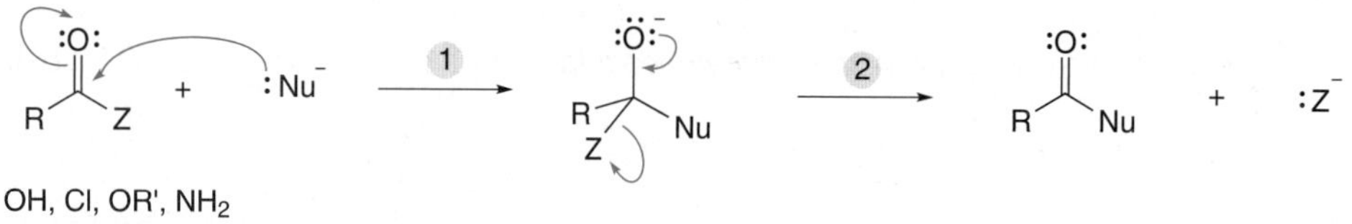

Z = OH, Cl, OR', NH_2

1 The nucleophile attacks the electrophilic carbonyl. The π bond is broken, moving an electron pair out on oxygen and forming an sp^3 hybridized carbon.

2 An electron pair on oxygen re-forms the π bond and Z comes off as a leaving group with the electron pair in the C–Z bond.

The net result is that Nu replaces Z—a nucleophilic substitution reaction. This reaction is often called **nucleophilic *acyl* substitution** to distinguish it from the nucleophilic substitution reactions at sp^3 hybridized carbons discussed in Chapter 7. Nucleophilic substitution with two different nucleophiles—**hydride (H:⁻)** and **carbanions (R:⁻)**—is discussed in Chapter 20. Other nucleophiles are examined in Chapter 22.

Carboxylic acid derivatives differ greatly in their reactivity toward nucleophiles. The order in which they react parallels the leaving group ability of the group Z bonded to the carbonyl carbon.

- **The better the leaving group Z, the more reactive RCOZ is in nucleophilic acyl substitution.**

Recall from Section 7.7 that the weaker the base, the better the leaving group.

Thus, the following trends result:

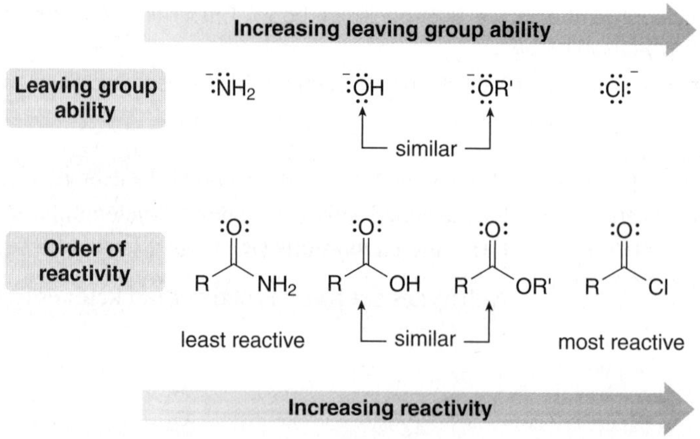

- **Acid chlorides (RCOCl), which have the best leaving group (Cl⁻), are the most reactive carboxylic acid derivatives, and amides ($RCONH_2$), which have the worst leaving group (⁻NH_2), are the least reactive.**
- **Carboxylic acids (RCOOH) and esters (RCOOR'), which have leaving groups of similar basicity (⁻OH and ⁻OR'), fall in the middle.**

Nucleophilic addition and nucleophilic acyl substitution involve the *same* first step—**nucleophilic attack on the electrophilic carbonyl group** to form a tetrahedral intermediate. The difference between them is what then happens to this intermediate. Aldehydes and ketones cannot undergo substitution because they have no leaving group bonded to the newly formed sp^3 hybridized carbon. Nucleophilic substitution with an aldehyde, for example, would form $H:^-$, an extremely strong base and therefore a very poor (and highly unlikely) leaving group.

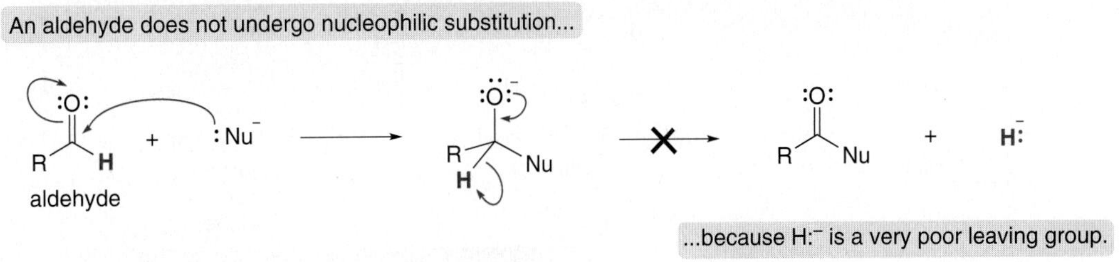

Problem 20.2 Which carbonyl groups in the anticancer drug Taxol (Section 5.5) will undergo nucleophilic addition, and which will undergo nucleophilic substitution?

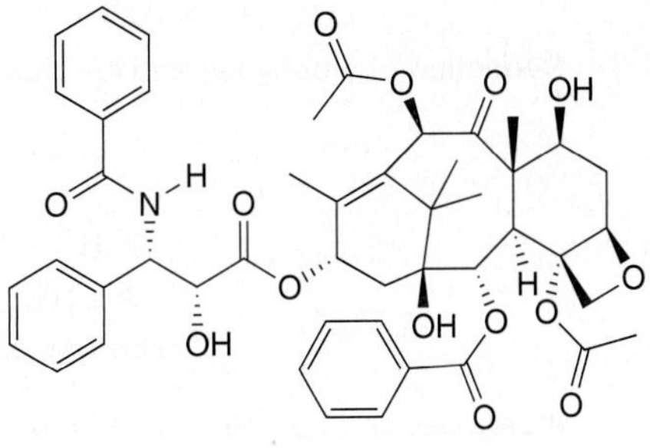

Taxol

Problem 20.3 Which compound in each pair is more reactive toward nucleophilic attack?

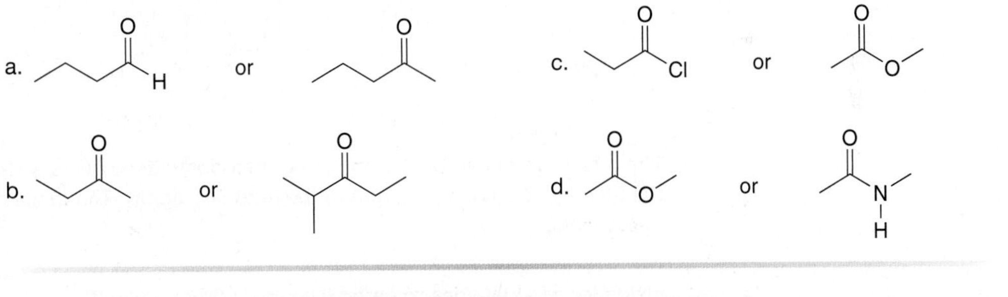

To show how these general principles of nucleophilic substitution and addition apply to carbonyl compounds, we are going to discuss oxidation and reduction reactions, and reactions with organometallic reagents—compounds that contain carbon–metal bonds. We begin with reduction to build on what you learned previously in Chapter 12.

20.3 A Preview of Oxidation and Reduction

Recall the definitions of oxidation and reduction presented in Section 12.1:

- **Oxidation results in an increase in the number of C–Z bonds (usually C–O bonds) or a decrease in the number of C–H bonds.**
- **Reduction results in a decrease in the number of C–Z bonds (usually C–O bonds) or an increase in the number of C–H bonds.**

Carbonyl compounds are either reactants or products in many of these reactions, as illustrated in the accompanying diagram. For example, because aldehydes fall in the middle of this scheme, they can be both oxidized and reduced. Carboxylic acids and their derivatives (RCOZ), on the other hand, are already highly oxidized, so their only useful reaction is reduction.

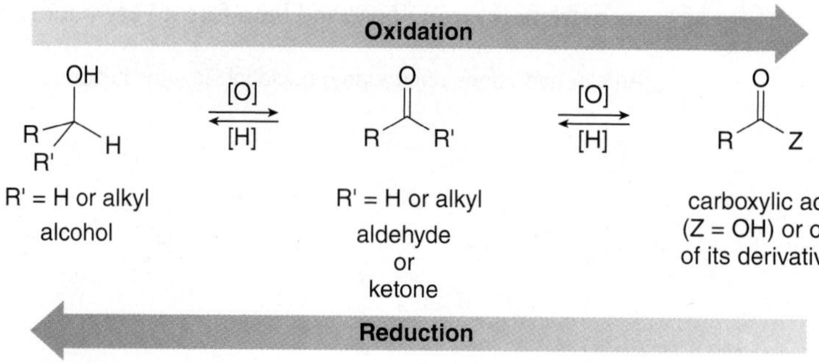

The three most useful oxidation and reduction reactions of carbonyl starting materials can be summarized as follows:

[1] Reduction of aldehydes and ketones to alcohols (Sections 20.4–20.6)

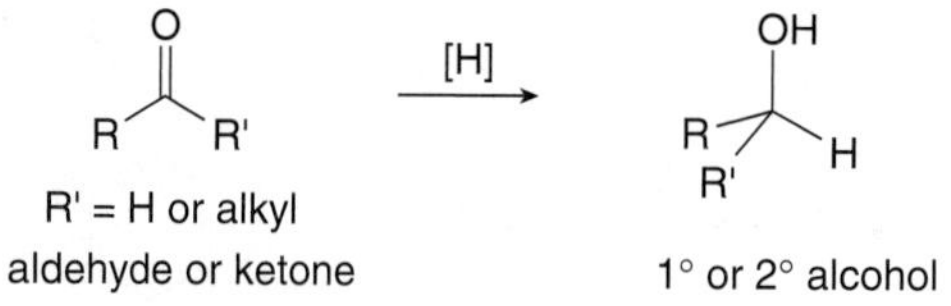

Aldehydes and ketones are reduced to 1° and 2° alcohols, respectively.

[2] Reduction of carboxylic acid derivatives (Section 20.7)

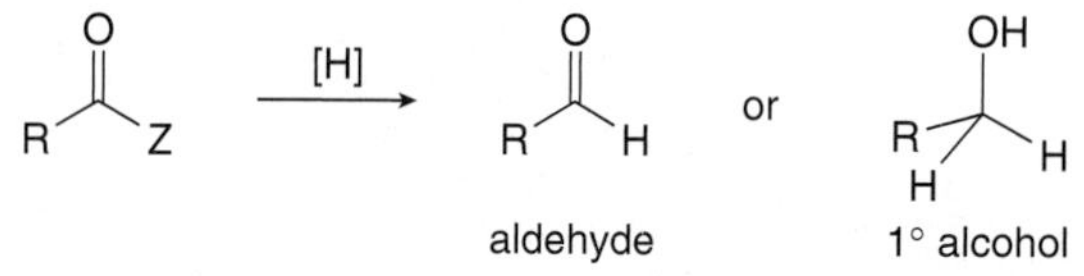

The reduction of carboxylic acids and their derivatives gives a variety of products, depending on the identity of Z and the nature of the reducing agent. The usual products are aldehydes or 1° alcohols.

[3] Oxidation of aldehydes to carboxylic acids (Section 20.8)

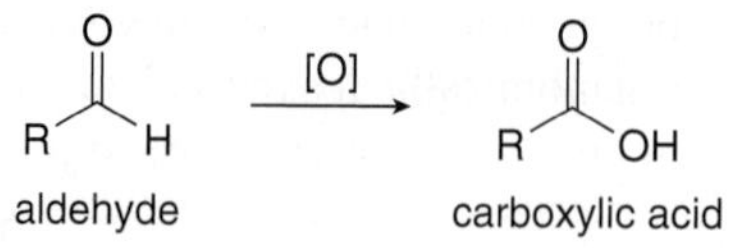

The most useful oxidation reaction of carbonyl compounds is the oxidation of aldehydes to carboxylic acids.

We begin with reduction, because the mechanisms of reduction reactions follow directly from the general mechanisms for nucleophilic addition and substitution.

20.4 Reduction of Aldehydes and Ketones

LiAlH$_4$ and NaBH$_4$ serve as a source of H:$^-$, but there are no free H:$^-$ ions present in reactions with these reagents.

The most useful reagents for reducing aldehydes and ketones are the metal hydride reagents (Section 12.2). The two most common metal hydride reagents are **sodium borohydride (NaBH$_4$)** and **lithium aluminum hydride (LiAlH$_4$).** These reagents contain a polar metal–hydrogen bond that serves as a source of the nucleophile hydride, **H:$^-$.** LiAlH$_4$ is a stronger reducing agent than NaBH$_4$, because the Al–H bond is more polar than the B–H bond.

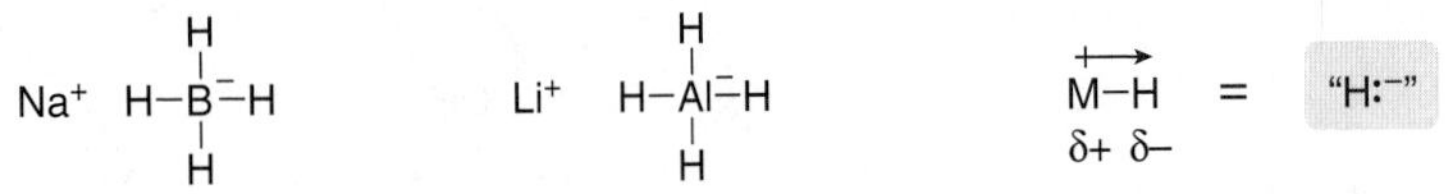

sodium borohydride | lithium aluminum hydride | a polar metal–hydrogen bond

20.4A Reduction with Metal Hydride Reagents

Treating an aldehyde or a ketone with NaBH$_4$ or LiAlH$_4$, followed by water or some other proton source, affords an **alcohol.** This is an addition reaction because **the elements of H$_2$ are added across the π bond,** but it is also a reduction because the product alcohol has fewer C–O bonds than the starting carbonyl compound.

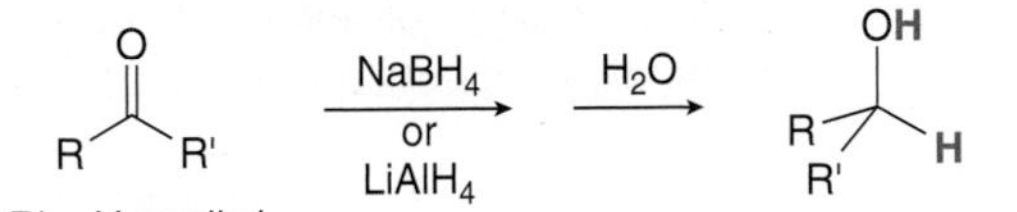

R' = H or alkyl

aldehyde or ketone

1° or 2° alcohol

LiAlH$_4$ reductions must be carried out under anhydrous conditions, because water reacts violently with the reagent. Water is added to the reaction mixture (to serve as a proton source) *after* the reduction with LiAlH$_4$ is complete.

The product of this reduction reaction is a **1° alcohol** when the starting carbonyl compound is an aldehyde, and a **2° alcohol** when it is a ketone.

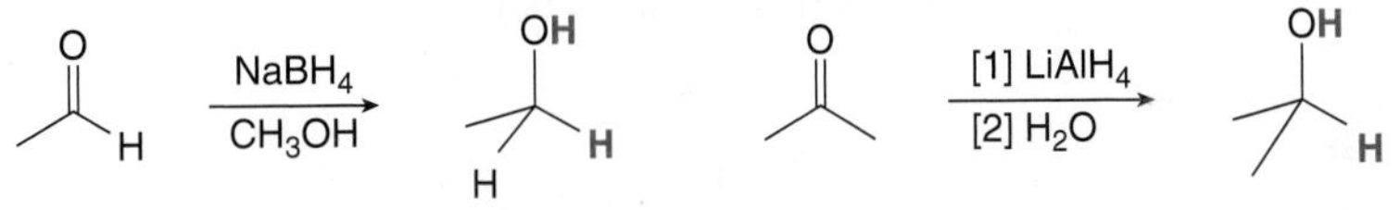

aldehyde | 1° alcohol | ketone | 2° alcohol

NaBH$_4$ selectively reduces aldehydes and ketones in the presence of most other functional groups. Reductions with NaBH$_4$ are typically carried out in CH$_3$OH as solvent. LiAlH$_4$ reduces aldehydes and ketones and many other functional groups as well (Sections 12.6 and 20.7).

Problem 20.4 What alcohol is formed when each compound is treated with NaBH$_4$ in CH$_3$OH?

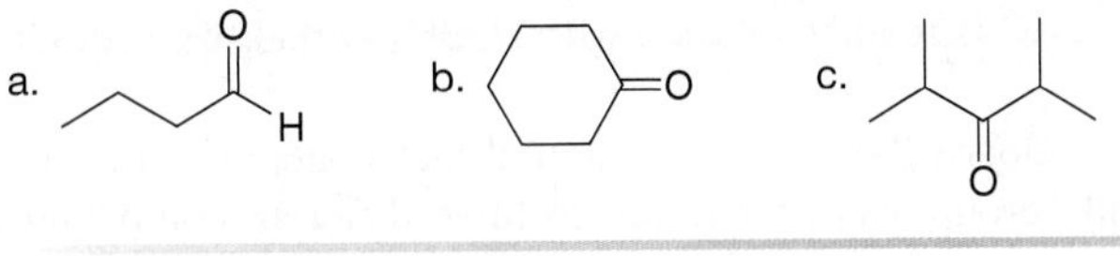

Problem 20.5 What aldehyde or ketone is needed to prepare each alcohol by metal hydride reduction?

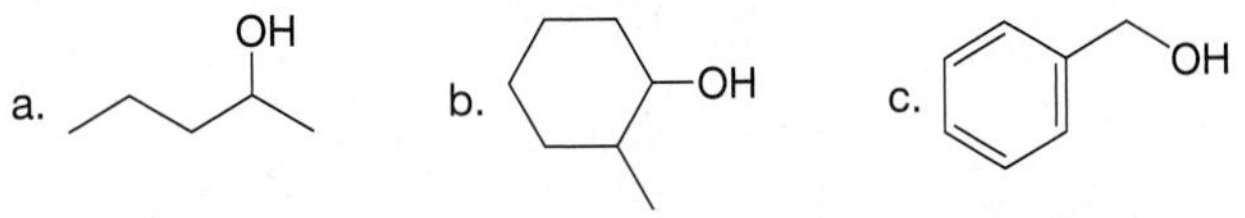

Problem 20.6 Why can't 1-methylcyclohexanol be prepared from a carbonyl compound by reduction?

20.4B The Mechanism of Hydride Reduction

Hydride reduction of aldehydes and ketones occurs via the general mechanism of nucleophilic addition—that is, **nucleophilic attack** followed by **protonation.** Mechanism 20.3 is shown using $LiAlH_4$, but an analogous mechanism can be written for $NaBH_4$.

Mechanism 20.3 $LiAlH_4$ Reduction of RCHO and $R_2C{=}O$

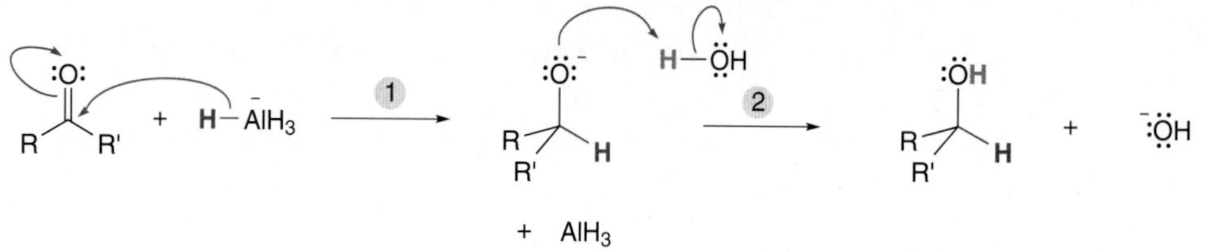

1 The nucleophile (AlH_4^-) donates $H:^-$ to the carbonyl group, breaking the π bond and moving an electron pair out on oxygen. This forms a new C–H bond.

2 Protonation of the negatively charged oxygen by H_2O (or CH_3OH) forms the reduction product with a new O–H bond.

- **The net result of adding $H:^-$ (from $NaBH_4$ or $LiAlH_4$) and H^+ (from H_2O) is the addition of the elements of H_2 to the carbonyl π bond.**

20.4C Catalytic Hydrogenation of Aldehydes and Ketones

Catalytic hydrogenation also reduces aldehydes and ketones to 1° and 2° alcohols, respectively, using H_2 and Pd-C (or another metal catalyst). H_2 adds to the C=O in much the same way that it adds to the C=C of an alkene (Section 12.3). The metal catalyst (Pd-C) provides a surface that binds the carbonyl starting material and H_2, and two H atoms are sequentially transferred with cleavage of the π bond.

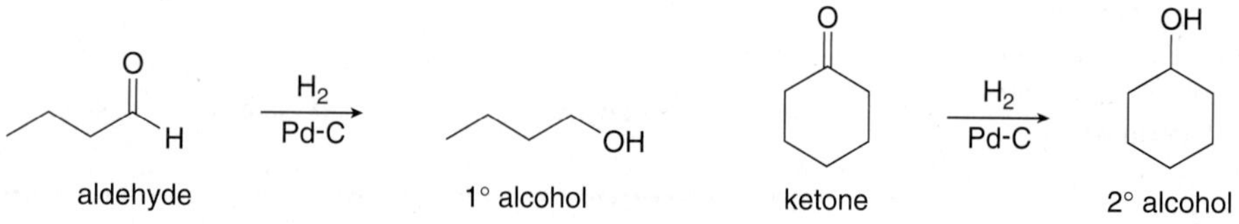

When a compound contains both a carbonyl group and a carbon–carbon double bond, selective reduction of one functional group can be achieved by proper choice of reagent.

- **A C=C is reduced faster than a C=O with H_2 (Pd-C).**
- **A C=O is readily reduced with $NaBH_4$ and $LiAlH_4$, but a C=C is inert.**

Thus, cyclohex-2-enone, a compound that contains both a carbon–carbon double bond and a carbonyl group, can be reduced to three different compounds—an allylic alcohol, a carbonyl compound, or an alcohol—depending on the reagent.

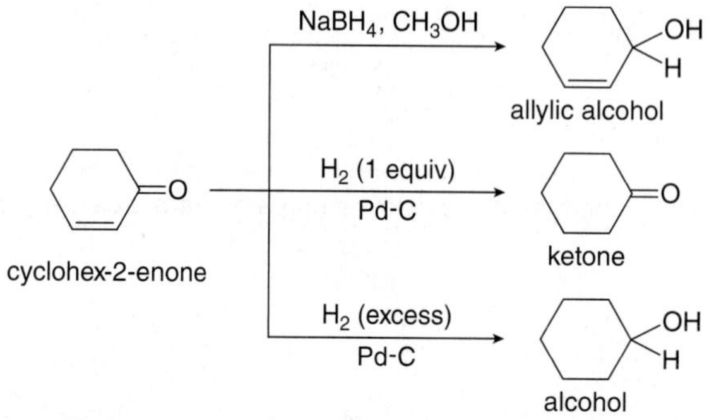

- $NaBH_4$ reduces the C=O selectively to form an allylic alcohol.
- One equivalent of H_2 reduces the C=C selectively to form a ketone.
- Excess H_2 reduces both π bonds to form an alcohol.

Problem 20.7 Draw the products formed when $CH_3COCH_2CH_2CH{=}CH_2$ is treated with each reagent: (a) $LiAlH_4$, then H_2O; (b) $NaBH_4$ in CH_3OH; (c) H_2 (1 equiv), Pd-C; (d) H_2 (excess), Pd-C; (e) $NaBH_4$ (excess) in CH_3OH; (f) $NaBD_4$ in CH_3OH.

The reduction of aldehydes and ketones is a common reaction used in the synthesis of many useful natural products. Two examples are shown in Figure 20.2.

Figure 20.2 $NaBH_4$ reductions used in organic synthesis

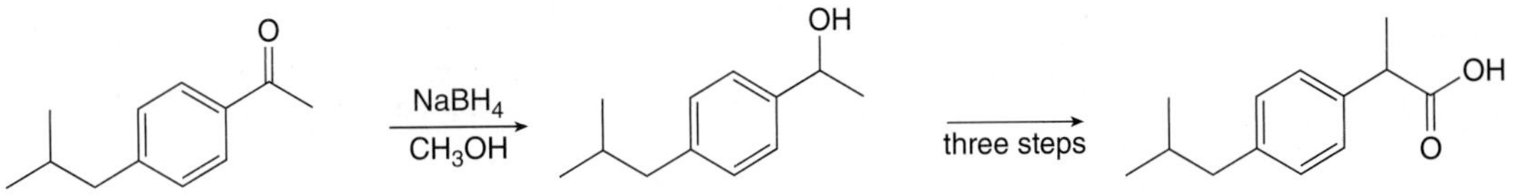

ibuprofen

(anti-inflammatory agent in Motrin and Advil)

The male musk deer, a small antlerless deer found in the mountain regions of China and Tibet, has long been hunted for its musk, a strongly scented liquid used in early medicine and later in perfumery.

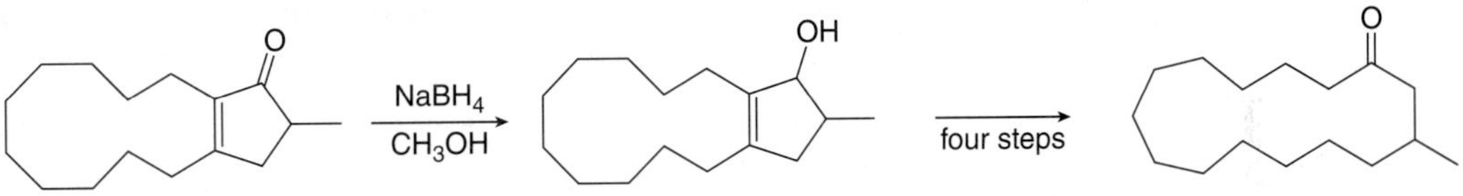

muscone

odor of musk (perfume component)

- **Muscone** is the major compound in musk, one of the oldest known ingredients in perfumes. Musk was originally isolated from the male musk deer, but it can now be prepared synthetically in the laboratory in a variety of ways.

20.5 The Stereochemistry of Carbonyl Reduction

Recall from Section 9.16 that an achiral starting material gives a racemic mixture when a new stereogenic center is formed.

The stereochemistry of carbonyl reduction follows the same principles we have previously learned. Reduction converts a **planar sp^2 hybridized carbonyl carbon to a tetrahedral sp^3 hybridized carbon.** What happens when a new stereogenic center is formed in this process? With an achiral reagent like $NaBH_4$ or $LiAlH_4$, a racemic product is obtained. For example, $NaBH_4$ in CH_3OH solution reduces butan-2-one, an achiral ketone, to butan-2-ol, an alcohol that contains a new stereogenic center. Both enantiomers of butan-2-ol are formed in equal amounts.

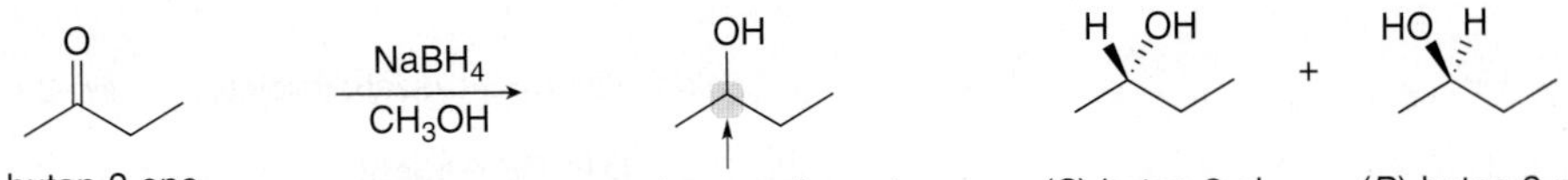

butan-2-one — new stereogenic center — (*S*)-butan-2-ol — (*R*)-butan-2-ol

achiral starting material — butan-2-ol — Two enantiomers are formed.

Why is a racemic mixture formed? Because the carbonyl carbon is sp^2 hybridized and planar, hydride can approach the double bond with equal probability from both sides of the plane, forming two alkoxides, which are **enantiomers** of each other. Protonation of the alkoxides gives an equal amount of two alcohols, which are also **enantiomers.**

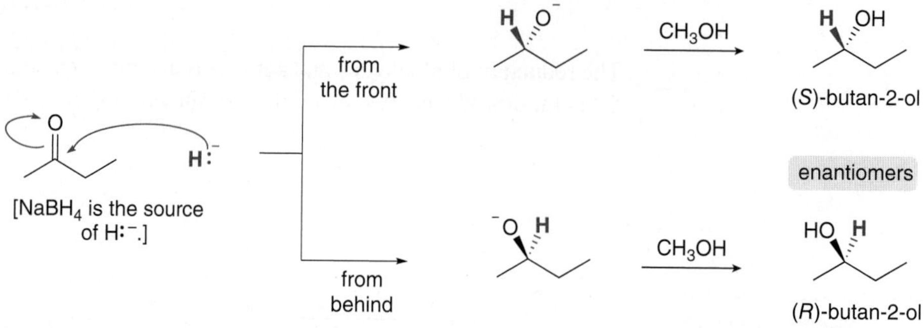

- **Conclusion: Hydride reduction of an achiral ketone with $LiAlH_4$ or $NaBH_4$ gives a racemic mixture of two alcohols when a new stereogenic center is formed.**

Problem 20.8 Draw the products formed (including stereoisomers) when each compound is reduced with $NaBH_4$ in CH_3OH.

a. 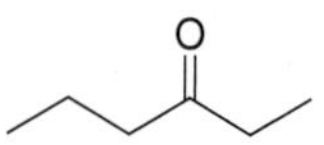 b. 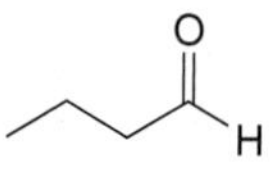 c. 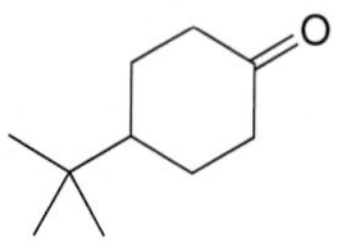

20.6 Enantioselective Carbonyl Reductions

20.6A CBS Reagents

One enantiomer can be formed selectively from the reduction of a carbonyl group, provided a **chiral reducing agent** is used. This strategy is identical to that employed in the Sharpless asymmetric epoxidation reaction (Section 12.15). A reduction that forms one enantiomer predominantly or exclusively is an **enantioselective** or **asymmetric reduction.**

Many different chiral reducing agents have now been prepared for this purpose. One such reagent, formed by reacting borane (**BH_3**) with a heterocycle called an **oxazaborolidine,** has one stereogenic center (and thus two enantiomers).

(*S*)-2-methyl-**CBS**-oxazaborolidine (*R*)-2-methyl-**CBS**-oxazaborolidine

(*S*)-CBS reagent **(*R*)-CBS reagent**

These reagents are called the **(*S*)-CBS reagent** and the **(*R*)-CBS reagent,** named for *C*orey, *B*akshi, and *S*hibata, the chemists who developed these versatile reagents. One B – H bond of BH_3 serves as the source of hydride in this reduction. The stereochemistry of the new stereogenic center in the product is often predictable. For ketones having the general structure C_6H_5COR, draw the starting material with the aryl group on the left side of the carbonyl, as shown with acetophenone. Then, to draw the product, keep in mind:

- **The (*S*)-CBS reagent delivers hydride (H:⁻) from the *front* side of the C=O. This generally affords the *R* alcohol as the major product.**
- **The (*R*)-CBS reagent delivers hydride (H:⁻) from the *back* side of the C=O. This generally affords the *S* alcohol as the major product.**

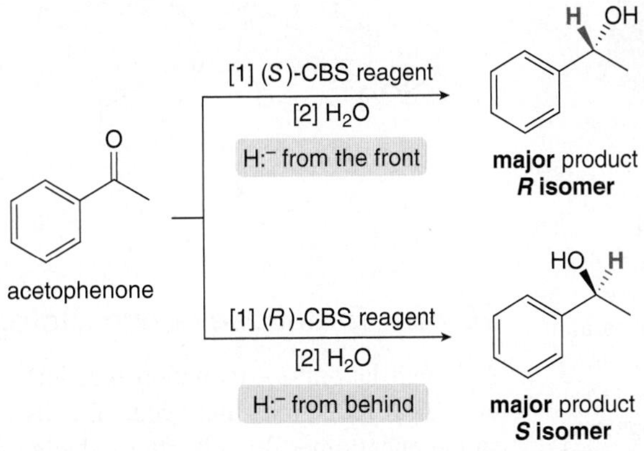
acetophenone
[1] (S)-CBS reagent
[2] H2O
H:⁻ from the front
H OH
major product
R isomer
[1] (R)-CBS reagent
[2] H2O
H:⁻ from behind
HO H
major product
S isomer

These reagents are highly enantioselective. Treatment of propiophenone with the (*S*)-CBS reagent forms the ***R*** alcohol in 97% enantiomeric excess (*ee*). Enantioselective reductions are key steps in the synthesis of several widely used drugs, including salmeterol, a long-acting bronchodilator shown in Figure 20.3. This new technology provides access to single enantiomers of biologically active compounds, often previously available only as a racemic mixture.

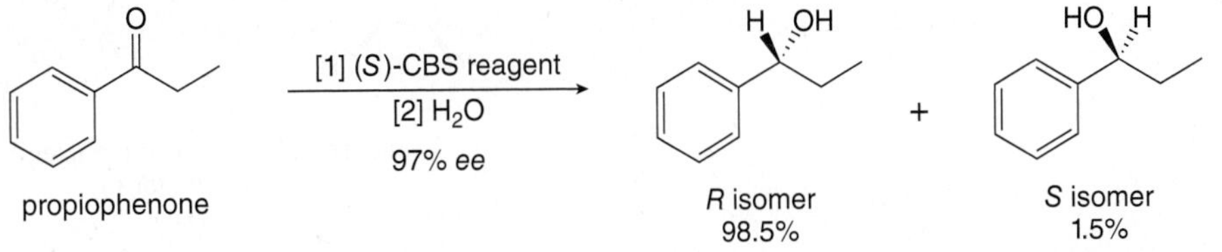
propiophenone
[1] (S)-CBS reagent
[2] H2O
97% ee
H OH
R isomer
98.5%
+
HO H
S isomer
1.5%

Figure 20.3 Enantioselective reduction—A key step in the synthesis of salmeterol

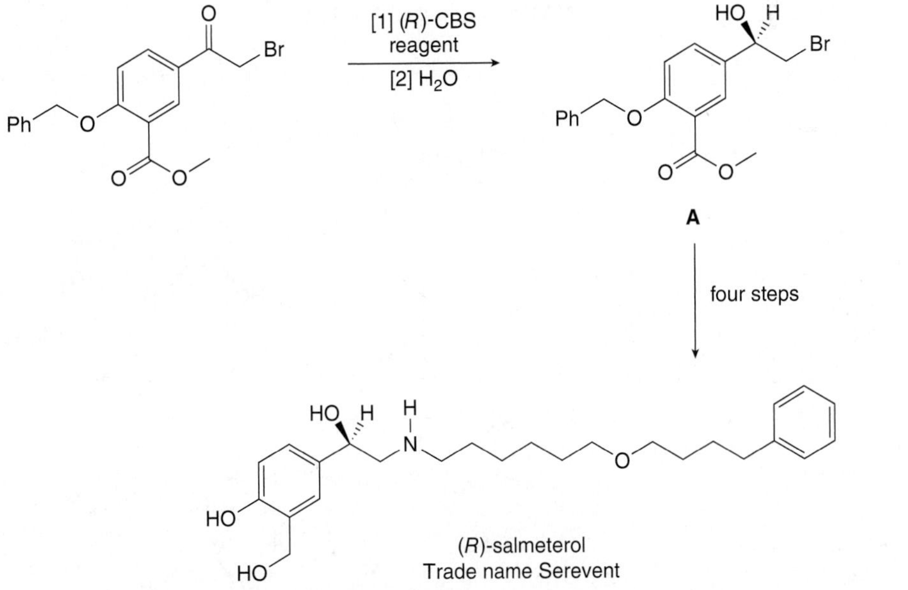
Br
Ph
O
[1] (R)-CBS reagent
[2] H2O
HO H
A
four steps
H
N
(R)-salmeterol
Trade name Serevent

- (*R*)-Salmeterol is a long-acting bronchodilator used for the treatment of asthma.
- In this example, the (*R*)-CBS reagent adds the new H atom from behind, the same result observed with acetophenone and propiophenone. In this case, however, alcohol **A** has the *R* configuration using the rules for assigning priority in Chapter 5.

Ezetimibe is sold as a single medication under the trade name of Zetia, or in combination with simvastatin, another cholesterol-lowering medication, and marketed as Vytorin. These drugs are prescribed for individuals who cannot tolerate or derive no benefit from other cholesterol-lowering medications.

Problem 20.9 What carbonyl compound and CBS reagent are needed to prepare **X,** an intermediate in the synthesis of ezetimibe (trade name Zetia), a drug that lowers cholesterol levels by inhibiting its absorption in the intestines.

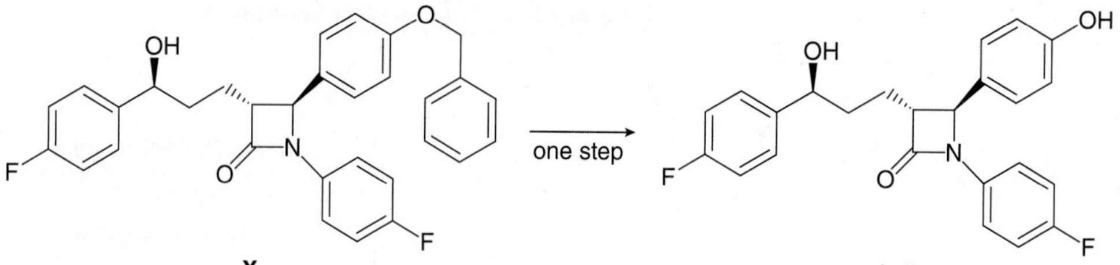

X ezetimibe

20.6B Enantioselective Biological Reduction

Although laboratory reduction reactions often do not proceed with 100% enantioselectivity, biological reductions that occur in cells *always* proceed with complete selectivity, forming a single enantiomer. In cells, the reducing agent is **NADH,** nicotinamide adenine dinucleotide (reduced form), the coenzyme introduced in Section 12.14.

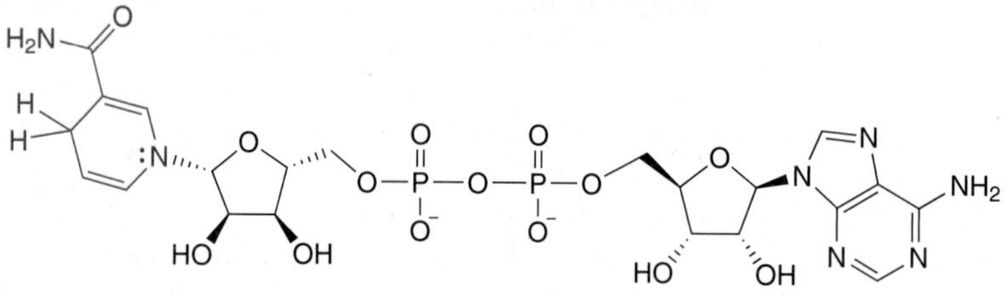

nicotinamide adenine dinucleotide
(reduced form)
NADH

In biological reduction, NADH donates H:$^-$, in much the same way as a metal hydride reagent. Nucleophilic attack of hydride and protonation thus form an alcohol from a carbonyl group, and NADH is converted to NAD$^+$.

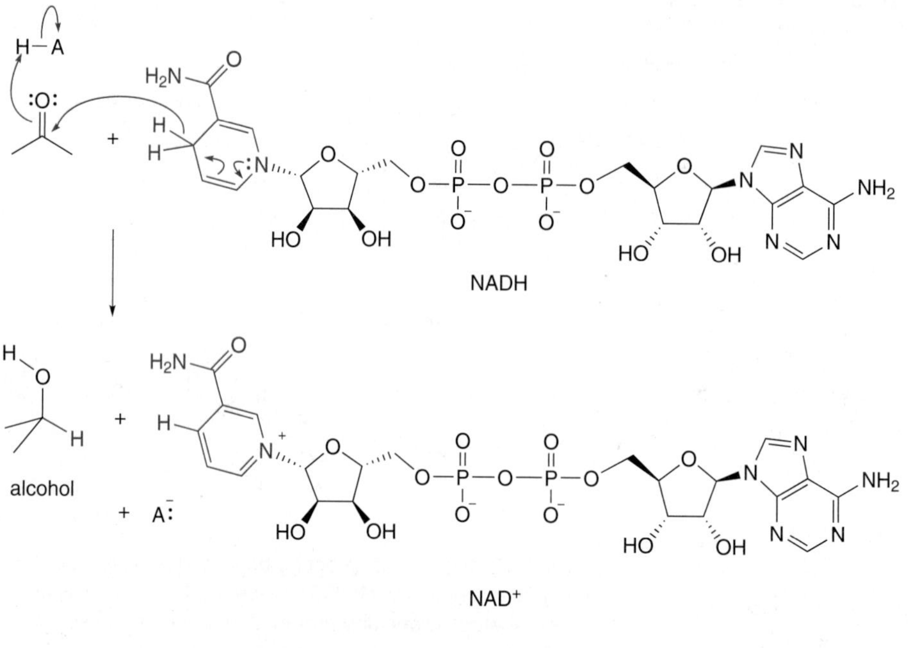

Pyruvic acid is formed during the metabolism of glucose. During periods of strenuous exercise, when there is insufficient oxygen to metabolize pyruvic acid to CO_2, pyruvic acid is reduced to lactic acid. The tired feeling of sore muscles is a result of lactic acid accumulation.

This reaction is completely enantioselective. Reduction of pyruvic acid with NADH catalyzed by lactate dehydrogenase affords a single enantiomer of lactic acid with the *S* configuration. NADH reduces a variety of different carbonyl compounds in biological systems. The configuration of the product (*R* or *S*) depends on the enzyme used to catalyze the process.

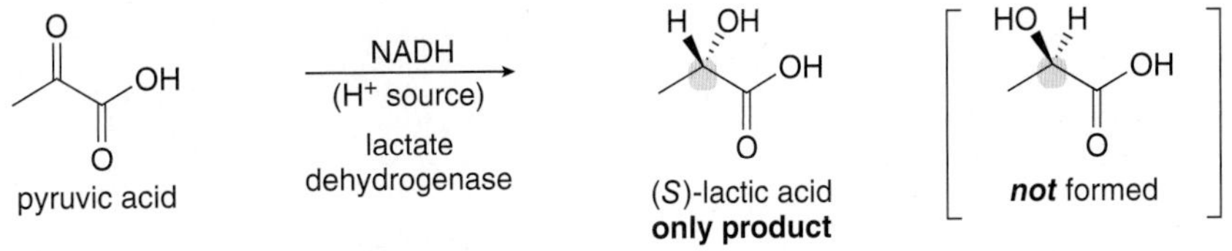
pyruvic acid
NADH
(H⁺ source)
lactate dehydrogenase
(S)-lactic acid
only product
not formed

Niacin can be obtained from foods such as soybeans, which contain it naturally, and from breakfast cereals, which are fortified with it to help people consume their recommended daily allowance of this B vitamin.

As we learned in Section 12.14, **NAD⁺, the oxidized form of NADH, is a biological oxidizing agent** capable of oxidizing alcohols to carbonyl compounds, forming NADH in the process. NAD⁺ is synthesized from the vitamin niacin, which can be obtained from soybeans among other dietary sources.

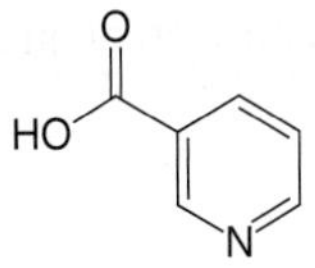

niacin
vitamin B_3

20.7 Reduction of Carboxylic Acids and Their Derivatives

The reduction of carboxylic acids and their derivatives (RCOZ) is complicated because the products obtained depend on the identity of both the leaving group (Z) and the reducing agent. Metal hydride reagents are the most useful reducing reagents. **Lithium aluminum hydride is a strong reducing agent that reacts with *all* carboxylic acid derivatives.** Two other related but more selective reducing agents are also used.

[1] **Diisobutylaluminum hydride, $[(CH_3)_2CHCH_2]_2AlH$,** abbreviated as **DIBAL-H,** has two bulky isobutyl groups, which make this reagent less reactive than $LiAlH_4$.

[2] **Lithium tri-*tert*-butoxyaluminum hydride, $LiAlH[OC(CH_3)_3]_3$,** has three electronegative oxygen atoms bonded to aluminum, which make this reagent less nucleophilic than $LiAlH_4$.

$LiAlH_4$ is a strong, nonselective reducing agent. DIBAL-H and $LiAlH[OC(CH_3)_3]_3$ are milder, more selective reducing agents.

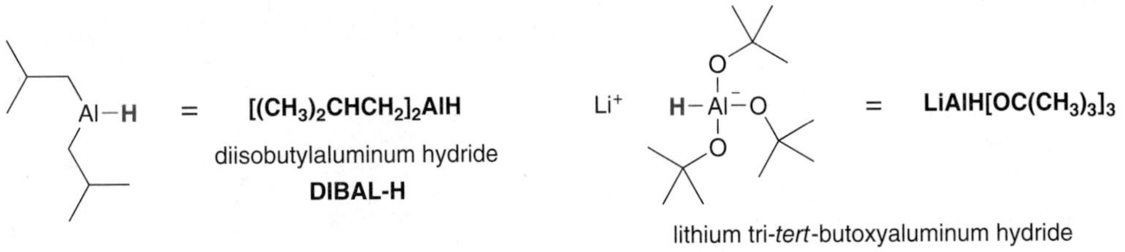
Al–H
= [(CH₃)₂CHCH₂]₂AlH
diisobutylaluminum hydride
DIBAL-H
Li⁺
H–Al–O
= LiAlH[OC(CH₃)₃]₃
lithium tri-tert-butoxyaluminum hydride

In both reagents, the single H atom bonded to Al is donated as $H:^-$ in hydride reductions.

20.7A Reduction of Acid Chlorides and Esters

Acid chlorides and esters can be reduced to either aldehydes or alcohols, depending on the reagent.

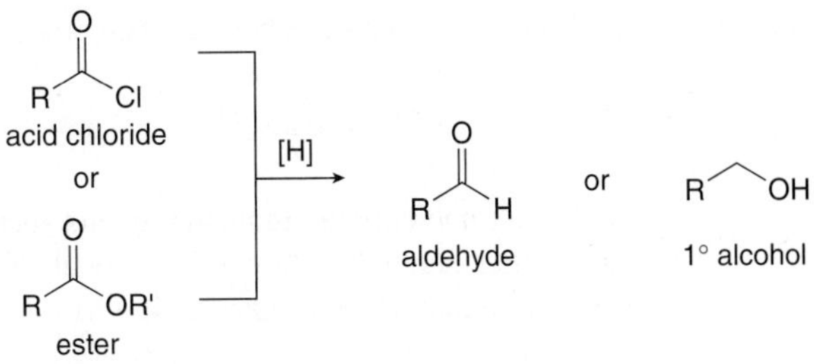
acid chloride
or
ester
[H]
aldehyde
or
1° alcohol

- **$LiAlH_4$ converts RCOCl and RCOOR' to alcohols.**
- **A milder reducing agent (DIBAL-H or $LiAlH[OC(CH_3)_3]_3$) converts RCOCl or RCOOR' to RCHO at low temperatures.**

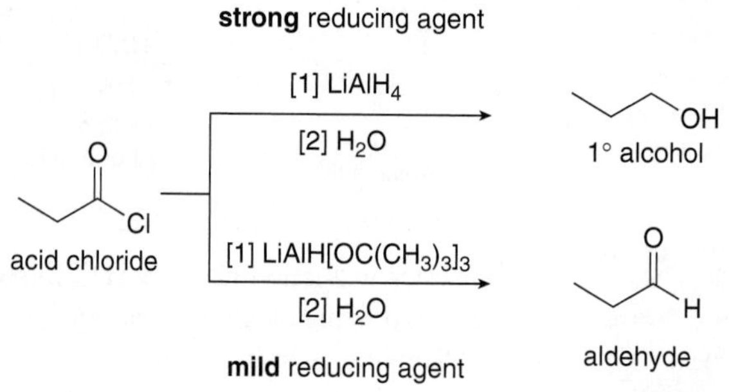

In the reduction of an acid chloride, Cl^- comes off as the leaving group.

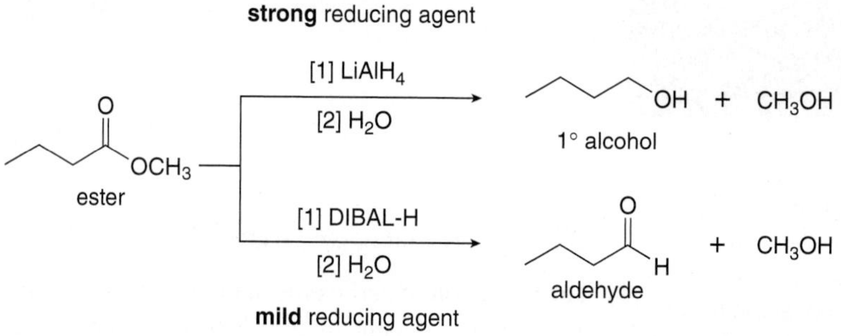

In the reduction of the ester, CH_3O^- comes off as the leaving group, which is then protonated by H_2O to form CH_3OH.

Mechanism 20.4 illustrates why two different products are possible. It can be conceptually divided into two parts: **nucleophilic substitution** to form an aldehyde (Steps [1] and [2]), followed by **nucleophilic addition** to the aldehyde to form an alcohol (Steps [3] and [4]). A general mechanism is drawn using $LiAlH_4$ as reducing agent.

Mechanism 20.4 Reduction of RCOCl and RCOOR' with a Metal Hydride Reagent

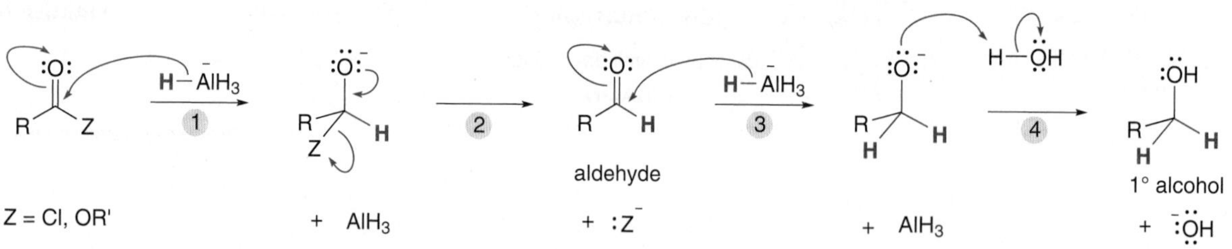

1 Nucleophilic attack of $H:^-$ forms a tetrahedral intermediate with a leaving group Z.

2 The π bond is re-formed and the leaving group Z departs. The overall result of addition of $H:^-$ and elimination of $Z:^-$ is substitution of H for Z.

3 Nucleophilic attack of $H:^-$ forms an alkoxide with no leaving group.

4 Protonation of the alkoxide by H_2O forms the alcohol reduction product. The overall result of Steps [3] and [4] is addition of H_2.

With less nucleophilic reducing agents such as DIBAL-H and $LiAlH[OC(CH_3)_3]_3$, the process stops after reaction with one equivalent of $H:^-$ and the aldehyde is formed as product. With a stronger reducing agent like $LiAlH_4$, two equivalents of $H:^-$ are added and an alcohol is formed.

Problem 20.10 Draw a stepwise mechanism for the following reaction.

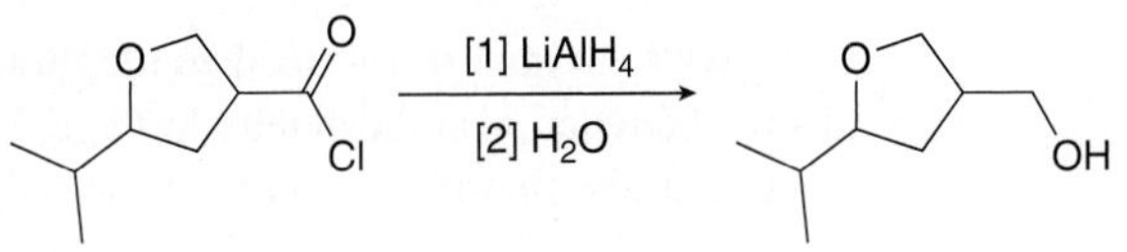

Problem 20.11 Draw the structure of both an acid chloride and an ester that can be used to prepare each compound by reduction.

a. 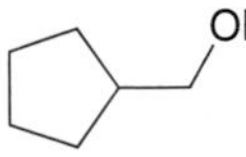

b. 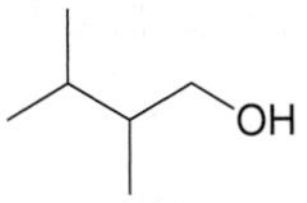

c. 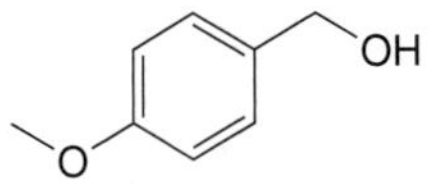

Selective reductions are routinely used in the synthesis of highly complex natural products such as **ciguatoxin CTX3C,** a potent neurotoxin introduced in the chapter opener. One reaction in a synthesis of ciguatoxin CTX3C involved the reduction of an ester to an aldehyde using DIBAL-H, as shown in Figure 20.4.

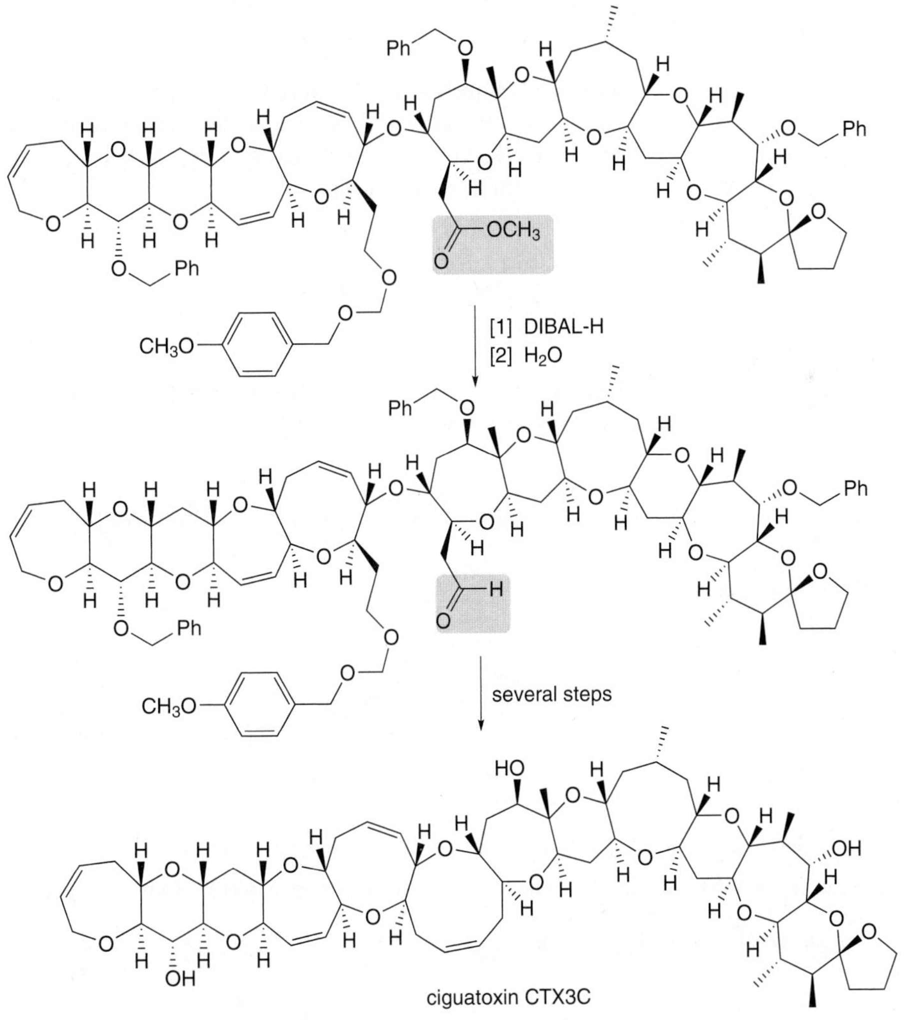

Figure 20.4 The DIBAL-H reduction of an ester to an aldehyde in the synthesis of the marine neurotoxin ciguatoxin CTX3C

- One step in a lengthy synthesis of ciguatoxin CTX3C involved selective reduction of an ester to an aldehyde using DIBAL-H.

20.7B Reduction of Carboxylic Acids and Amides

Carboxylic acids are reduced to alcohols with $LiAlH_4$. $LiAlH_4$ is too strong a reducing agent to stop the reaction at the aldehyde stage, but milder reagents are not strong enough to initiate the reaction in the first place, so this is the only useful reduction reaction of carboxylic acids.

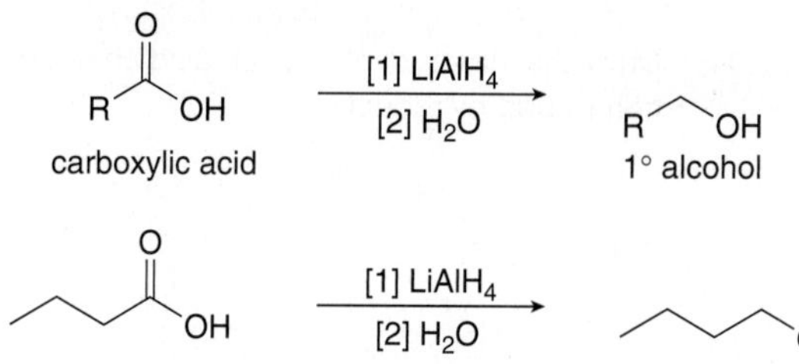

Unlike the $LiAlH_4$ reduction of all other carboxylic acid derivatives, which affords alcohols, the **$LiAlH_4$ reduction of amides forms amines.**

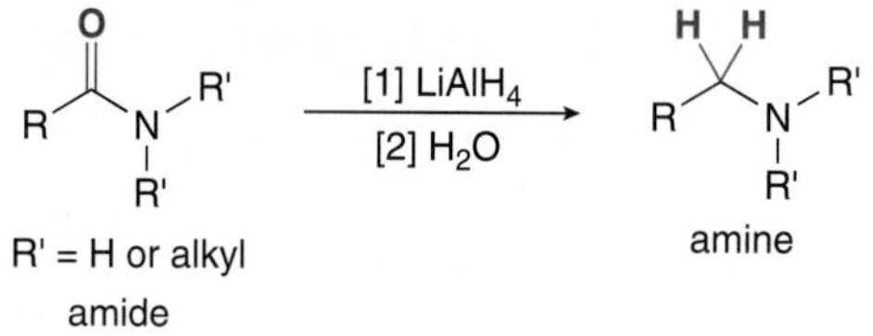

Both C–O bonds are reduced to C–H bonds by $LiAlH_4$, and any H atom or R group bonded to the amide nitrogen atom remains bonded to it in the product. Because $^-NH_2$ (or ^-NHR or $^-NR_2$) is a poorer leaving group than Cl^- or ^-OR, $^-NH_2$ is never lost during reduction, and therefore it forms an amine in the final product.

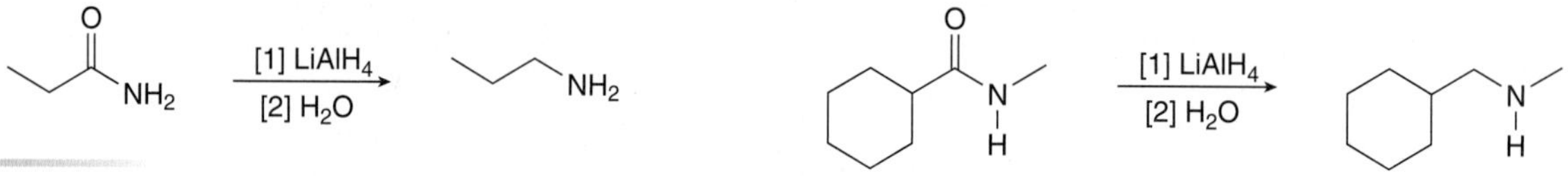

Imines and related compounds are discussed in Chapter 21.

The mechanism, illustrated in Mechanism 20.5 with $RCONH_2$ as starting material, is somewhat different than the previous reductions of carboxylic acid derivatives. Amide reduction proceeds

Mechanism 20.5 Reduction of an Amide to an Amine with $LiAlH_4$

Part [1] Reduction of an amide to an imine

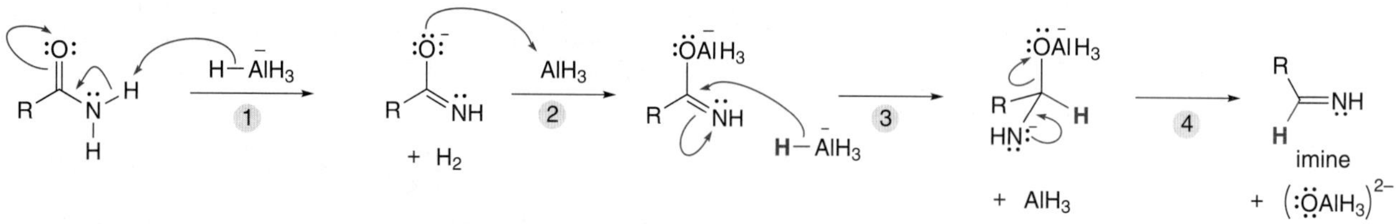

1 – 2 AlH_4^- removes a proton from the amide to form a Lewis base that complexes with AlH_3 in Step [2].

3 – 4 Nucleophilic attack of H:$^-$ and loss of a leaving group, $(OAlH_3)^{2-}$, form an imine.

Part [2] Reduction of an imine to an amine

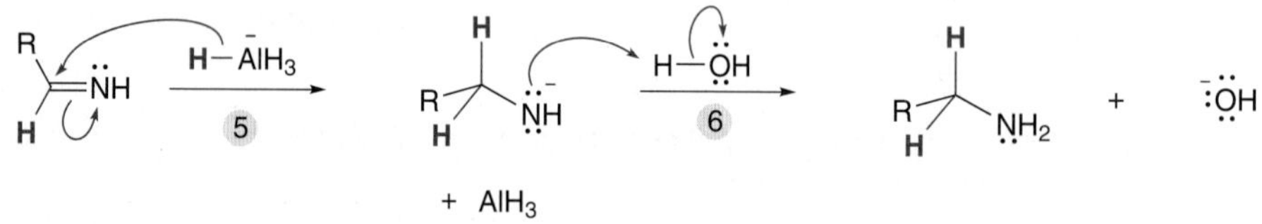

5 – 6 Nucleophilic addition of H:$^-$ and protonation form the amine.

with formation of an intermediate ***imine*****, a compound containing a C–N double bond,** which is then further reduced to an amine.

Problem 20.12 Draw the products formed from $LiAlH_4$ reduction of each compound.

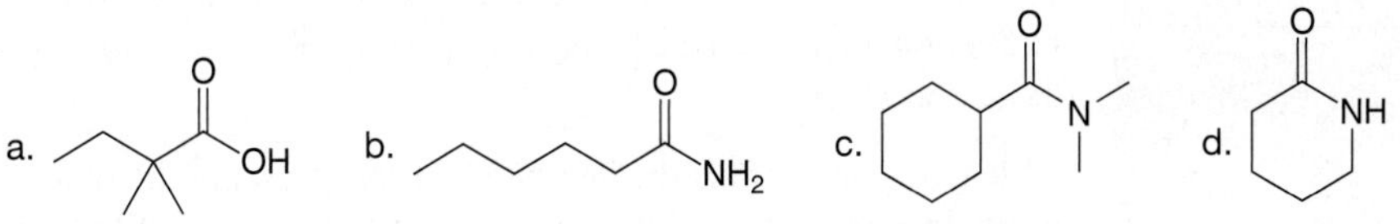

Problem 20.13 What amide(s) will form each of the following amines on treatment with $LiAlH_4$?

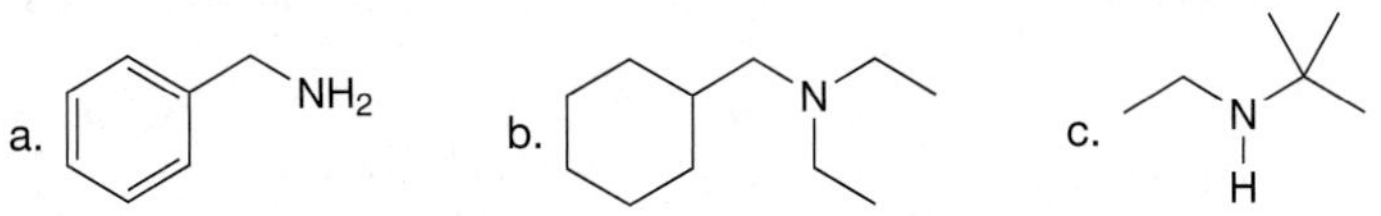

20.7C A Summary of the Reagents for Reduction

The many available metal hydride reagents reduce a wide variety of functional groups. Keep in mind that $LiAlH_4$ is such a strong reducing agent that it *nonselectively* reduces most polar functional groups. All other metal hydride reagents are more selective, and each has its particular reactions that best utilize its reduced reactivity. The reagents and their uses are summarized in Table 20.1.

Problem 20.14 What product is formed when each compound is treated with either $LiAlH_4$ (followed by H_2O), or $NaBH_4$ in CH_3OH?

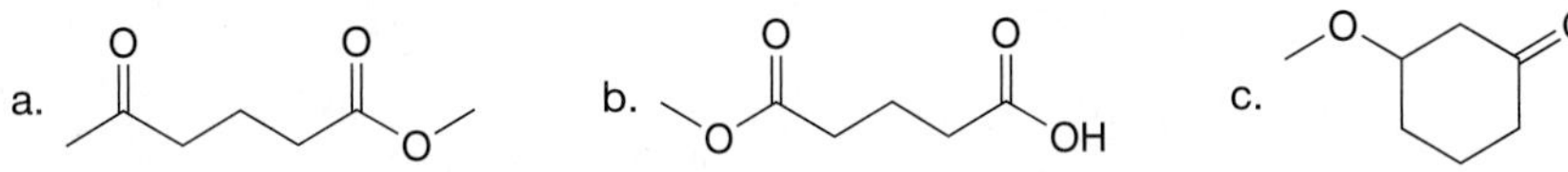

Table 20.1 A Summary of Metal Hydride Reducing Agents

	Reagent	Starting material	→	Product
Strong reagent	$LiAlH_4$	RCHO	→	RCH_2OH
		R_2CO	→	R_2CHOH
		RCOOH	→	RCH_2OH
		RCOOR'	→	RCH_2OH
		RCOCl	→	RCH_2OH
		$RCONH_2$	→	RCH_2NH_2
Milder reagents	$NaBH_4$	RCHO	→	RCH_2OH
		R_2CO	→	R_2CHOH
	$LiAlH[OC(CH_3)_3]_3$	RCOCl	→	RCHO
	DIBAL-H	RCOOR'	→	RCHO

Aldehydes give a positive Tollens test; that is, they react with Ag^+ to form RCOOH and Ag. When the reaction is carried out in a glass flask, a silver mirror is formed on its walls. Other functional groups give a negative Tollens test, because no silver mirror forms.

20.8 Oxidation of Aldehydes

The most common oxidation reaction of carbonyl compounds is the oxidation of **aldehydes to carboxylic acids.** A variety of oxidizing agents can be used, including CrO_3, $Na_2Cr_2O_7$, $K_2Cr_2O_7$, and $KMnO_4$. Cr^{6+} reagents are also used to oxidize 1° and 2° alcohols, as discussed in Section 12.12. Because ketones have no H on the carbonyl carbon, they do not undergo this oxidation reaction.

Aldehydes are oxidized selectively in the presence of other functional groups using **silver(I) oxide in aqueous ammonium hydroxide (Ag_2O in NH_4OH).** This is called **Tollens reagent.** Oxidation with Tollens reagent provides a distinct color change, because the Ag^+ reagent is reduced to silver metal (Ag), which precipitates out of solution.

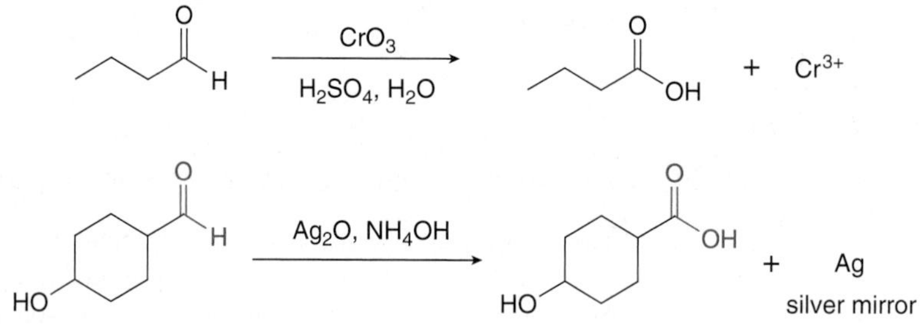

Only the aldehyde is oxidized.

Problem 20.15 What product is formed when each compound is treated with either Ag_2O, NH_4OH or $Na_2Cr_2O_7$, H_2SO_4, H_2O?

a. 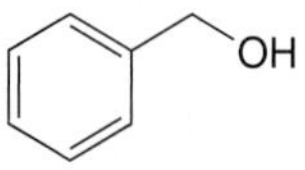

b. 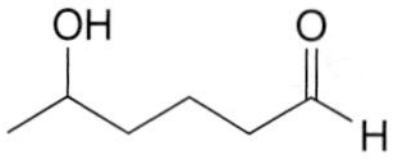

Problem 20.16 Review the oxidation reactions using Cr^{6+} reagents in Section 12.12. Then, draw the product formed when compound **B** is treated with each reagent.

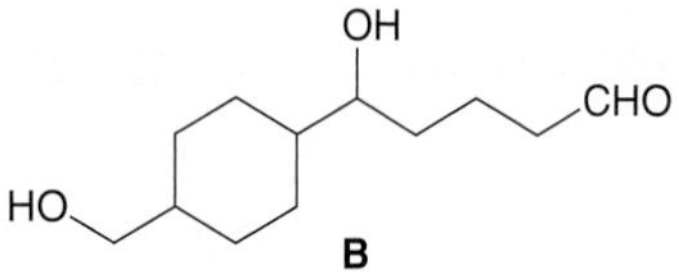

B

a. $NaBH_4$, CH_3OH
b. [1] $LiAlH_4$; [2] H_2O
c. PCC
d. Ag_2O, NH_4OH
e. CrO_3, H_2SO_4, H_2O

20.9 Organometallic Reagents

We will now discuss the reactions of carbonyl compounds with organometallic reagents, another class of nucleophiles.

- *Organometallic reagents* contain a carbon atom bonded to a metal.

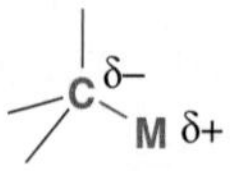

organometallic reagent

= R—M

M = metal

Most common metals:
M = Li, Mg, Cu

Lithium, magnesium, and copper are the most commonly used metals in organometallic reagents, but others (such as Sn, Si, Tl, Al, Ti, and Hg) are known. General structures of the three common organometallic reagents are shown. R can be alkyl, aryl, allyl, benzyl, sp^2 hybridized, and with

M = Li or Mg, *sp* hybridized. Because metals are *more electropositive* (less electronegative) than carbon, they donate electron density toward carbon, so that **carbon bears a partial negative charge.**

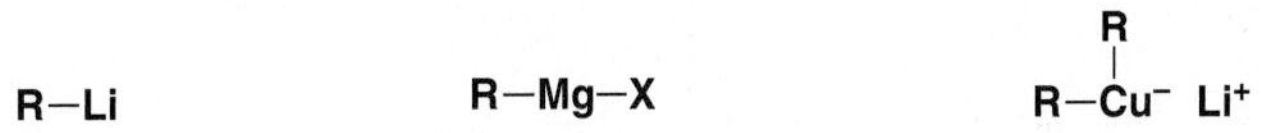
R–Li
R–Mg–X
R
R–Cu⁻ Li⁺

organolithium reagents	organomagnesium reagents or Grignard reagents	organocopper reagents or organocuprates

- **The more polar the carbon–metal bond, the more reactive the organometallic reagent.**

Because both Li and Mg are very electropositive metals, **organolithium (RLi)** and **organomagnesium reagents (RMgX)** contain very polar carbon–metal bonds and are therefore *very reactive* reagents. Organomagnesium reagents are called **Grignard reagents,** after Victor Grignard, who received the Nobel Prize in Chemistry in 1912 for his work with them.

Electronegativity values for carbon and the common metals in R–M reagents are C (2.5), Li (1.0), Mg (1.3), and Cu (1.8).

Organocopper reagents (R_2CuLi), also called **organocuprates,** have a less polar carbon–metal bond and are therefore *less reactive*. Although organocuprates contain two alkyl groups bonded to copper, only one R group is utilized in a reaction.

Regardless of the metal, organometallic reagents are useful synthetically because they react as if they were free carbanions; that is, carbon bears a partial *negative* charge, so the **reagents react as bases and nucleophiles.**

M δ+
δ−
reacts like

carbanion

a base and a nucleophile

20.9A Preparation of Organometallic Reagents

Organolithium and Grignard reagents are typically prepared by reaction of an organic halide with the corresponding metal, as shown in the accompanying equations.

$$R{-}X + 2\,Li \longrightarrow R{-}Li + LiX$$

organolithium reagent

$$CH_3{-}Br + 2\,Li \longrightarrow CH_3{-}Li + LiBr$$

methyllithium

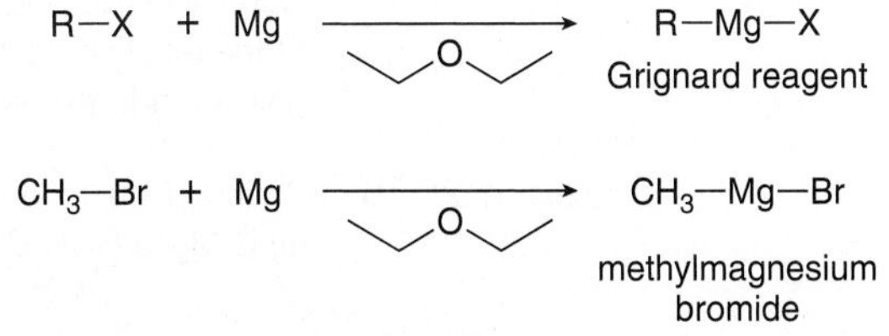
R–X + Mg
R–Mg–X
Grignard reagent
CH3–Br + Mg
CH3–Mg–Br
methylmagnesium bromide

With lithium, the halogen and metal exchange to form the organolithium reagent. With magnesium, the metal inserts in the carbon–halogen bond, forming the Grignard reagent. Grignard reagents are usually prepared in diethyl ether ($CH_3CH_2OCH_2CH_3$) as solvent. It is thought that two ether oxygen atoms complex with the magnesium atom, stabilizing the reagent.

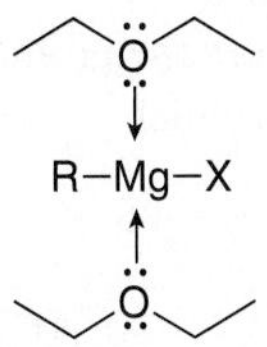
R–Mg–X

Two molecules of diethyl ether complex with the Mg atom of the Grignard reagent.

Organocuprates are prepared from organolithium reagents by reaction with a Cu^+ salt, often CuI.

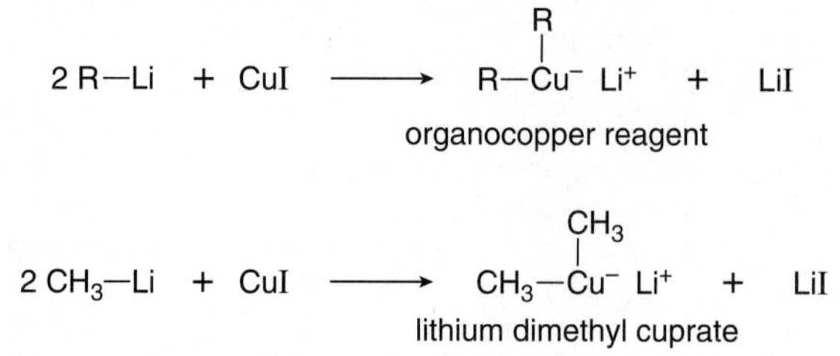
2 R—Li + CuI
R—Cu⁻ Li⁺ + LiI
organocopper reagent
2 CH3—Li + CuI
CH3—Cu⁻ Li⁺ + LiI
lithium dimethyl cuprate

Problem 20.17 Write the step(s) needed to convert CH_3CH_2Br to each reagent: (a) CH_3CH_2Li; (b) CH_3CH_2MgBr; (c) $(CH_3CH_2)_2CuLi$.

20.9B Acetylide Anions

The **acetylide anions** discussed in Chapter 11 are another example of organometallic compounds. These reagents are prepared by an acid–base reaction of an alkyne with a base such as $NaNH_2$ or NaH. We can think of these compounds as **organosodium** reagents. Because sodium is even more electropositive (less electronegative) than lithium, the C–Na bond of these organosodium compounds is best described as **ionic,** rather than polar covalent.

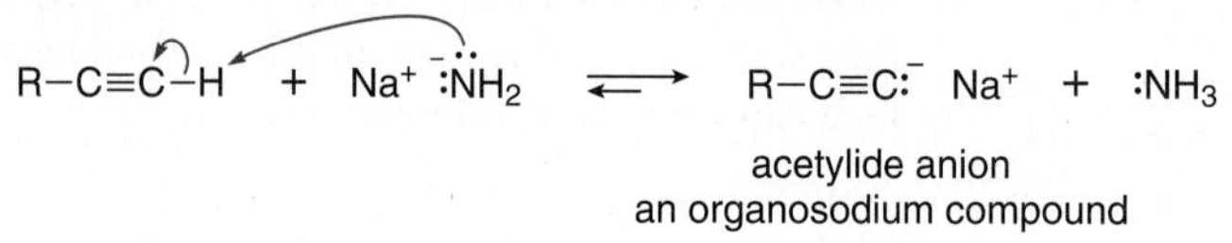
R—C≡C—H + Na⁺ :NH2
R—C≡C:⁻ Na⁺ + :NH3
acetylide anion
an organosodium compound

An acid–base reaction can also be used to prepare *sp* hybridized organolithium compounds. Treatment of a terminal alkyne with CH_3Li affords a lithium acetylide. Equilibrium favors the products because the *sp* hybridized C–H bond of the terminal alkyne is more acidic than the sp^3 hybridized conjugate acid, CH_4, that is formed.

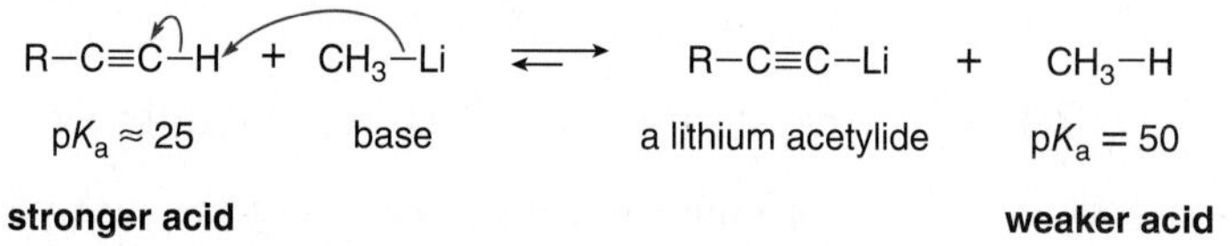
R—C≡C—H + CH3—Li
R—C≡C—Li + CH3—H
pKa ≈ 25
base
a lithium acetylide
pKa = 50
stronger acid
weaker acid

Problem 20.18 Oct-1-yne ($HC{\equiv}CCH_2CH_2CH_2CH_2CH_2CH_3$) reacts rapidly with NaH, forming a gas that bubbles out of the reaction mixture, as one product. Oct-1-yne also reacts rapidly with CH_3MgBr, and a different gas is produced. Write balanced equations for both reactions and identify the gases formed.

Problem 20.19 Which of the following species represent organometallic compounds: (a) $BrMgC{\equiv}CCH_2CH_3$; (b) $NaOCH_2CH_3$; (c) $KOC(CH_3)_3$; (d) PhLi?

20.9C Reaction as a Base

- **Organometallic reagents are strong bases that readily abstract a proton from water to form hydrocarbons.**

The electron pair in the carbon–metal bond is used to form a new bond to the proton. Equilibrium favors the products of this acid–base reaction because H_2O is a much stronger acid than the alkane product.

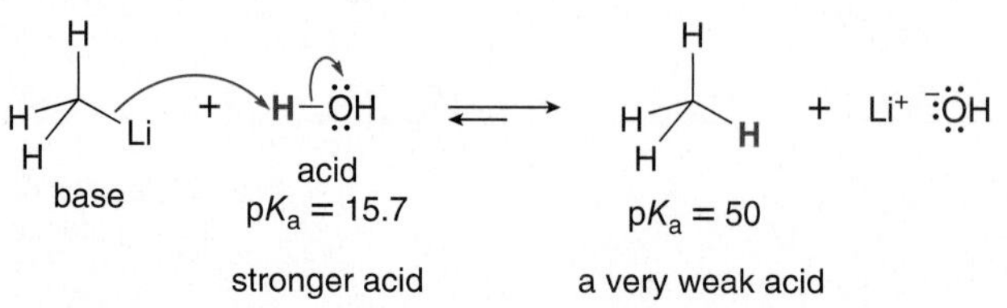
base
H—ÖH
acid
pKa = 15.7
stronger acid
Li⁺ :ÖH
pKa = 50
a very weak acid

Similar reactions occur for the same reason with the O–H proton in alcohols and carboxylic acids, and the N–H protons of amines.

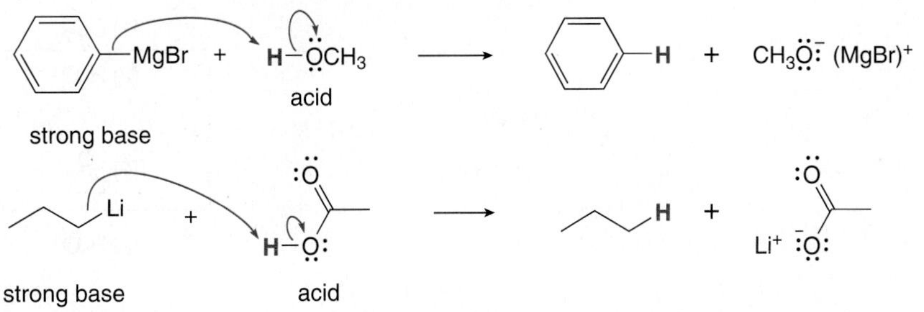

Because organolithium and Grignard reagents are themselves prepared from alkyl halides, a two-step method converts an alkyl halide into an alkane (or another hydrocarbon).

$$\underset{\text{alkyl halide}}{R{-}X} \xrightarrow{M} R{-}M \xrightarrow{H_2O} \underset{\text{alkane}}{R{-}H}$$

Problem 20.20 Draw the product formed when each organometallic reagent is treated with H_2O.

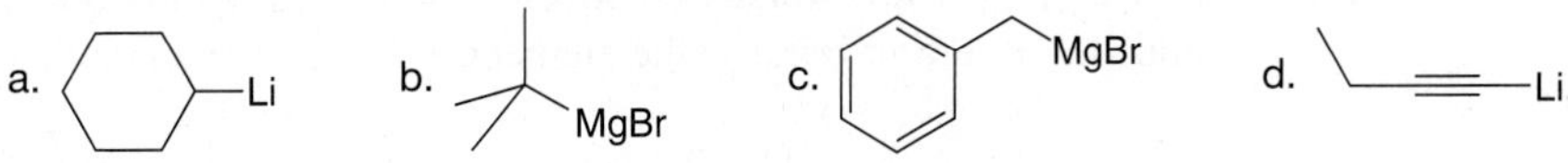

20.9D Reaction as a Nucleophile

Organometallic reagents are also strong nucleophiles that react with electrophilic carbon atoms to form new carbon–carbon bonds. These reactions are very valuable in forming the carbon skeletons of complex organic molecules. The following reactions of organometallic reagents are examined in Sections 20.10, 20.13, and 20.14:

[1] Reaction of R–M with aldehydes and ketones to afford alcohols (Section 20.10)

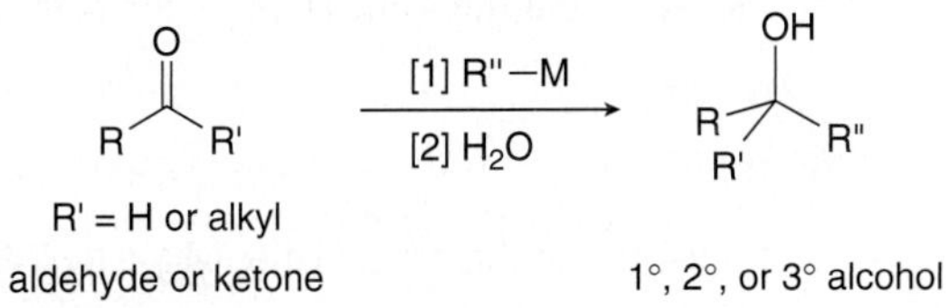

Aldehydes and ketones are converted to 1°, 2°, or 3° alcohols with R"Li or R"MgX.

[2] Reaction of R–M with carboxylic acid derivatives (Section 20.13)

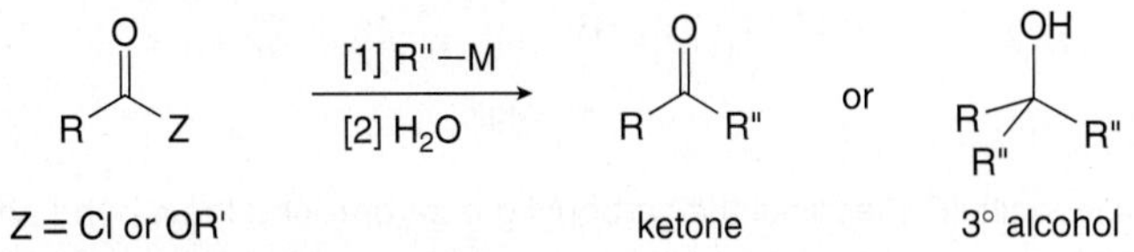

Acid chlorides and esters can be converted to ketones or 3° alcohols with organometallic reagents. The identity of the product depends on the identity of R"–M and the leaving group Z.

[3] Reaction of R–M with other electrophilic functional groups (Section 20.14)

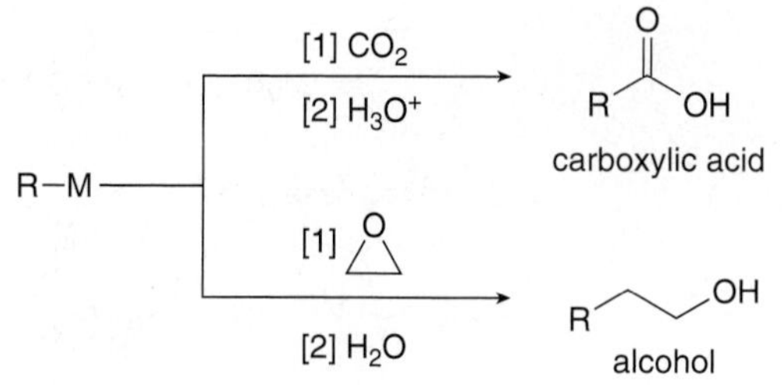

Organometallic reagents also react with CO_2 to form carboxylic acids and with epoxides to form alcohols.

20.10 Reaction of Organometallic Reagents with Aldehydes and Ketones

Treatment of an aldehyde or ketone with either an organolithium or Grignard reagent followed by water forms an alcohol with a new carbon–carbon bond. This reaction is an **addition reaction** because the elements of R" and H are added across the π bond.

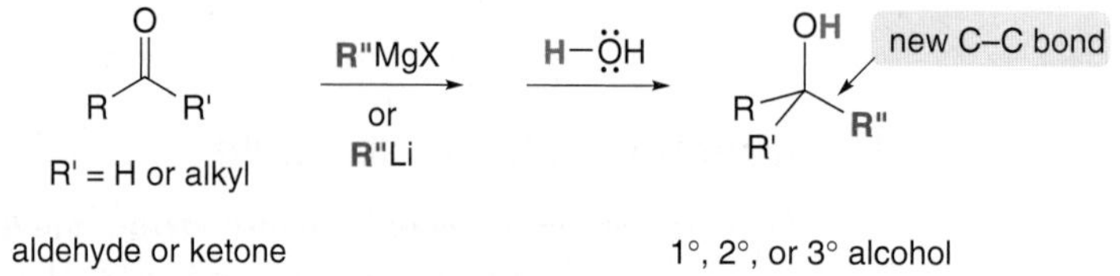

20.10A General Features

This reaction follows the general mechanism for nucleophilic addition (Section 20.2A)—that is, **nucleophilic attack** by a carbanion followed by **protonation.** Mechanism 20.6 is shown using R"MgX, but the same steps occur with organolithium reagents and acetylide anions.

Mechanism 20.6 Nucleophilic Addition of R"MgX to RCHO and $R_2C=O$

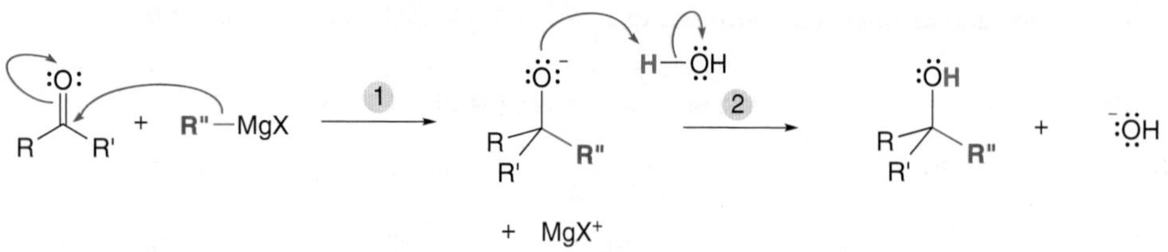

1 The nucleophile $(R")^-$ attacks the carbonyl group, breaking the π bond and yielding an alkoxide. This forms a new carbon–carbon bond.

2 Protonation of the alkoxide by H_2O forms the addition product with a new O–H bond. The overall result is addition of R" and H to the carbonyl group.

This reaction is used to prepare 1°, 2°, and 3° alcohols, depending on the number of alkyl groups bonded to the carbonyl carbon of the aldehyde or ketone.

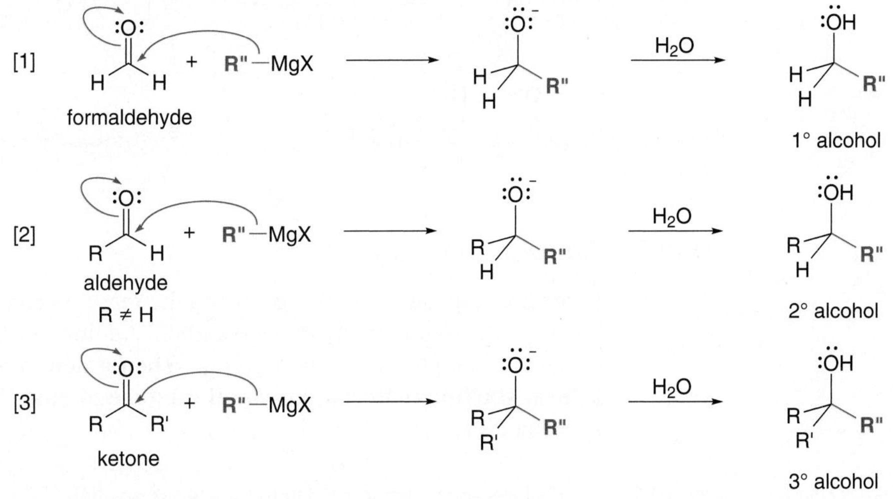

[1] Addition of R"MgX to formaldehyde ($CH_2=O$) forms a 1° alcohol.
[2] Addition of R"MgX to all other aldehydes forms a 2° alcohol.
[3] Addition of R"MgX to ketones forms a 3° alcohol.

Each reaction results in addition of one new alkyl group to the carbonyl carbon, and forms one new carbon–carbon bond. The reaction is general for all organolithium and Grignard reagents, and works for acetylide anions as well, as illustrated in Equations [1]–[3].

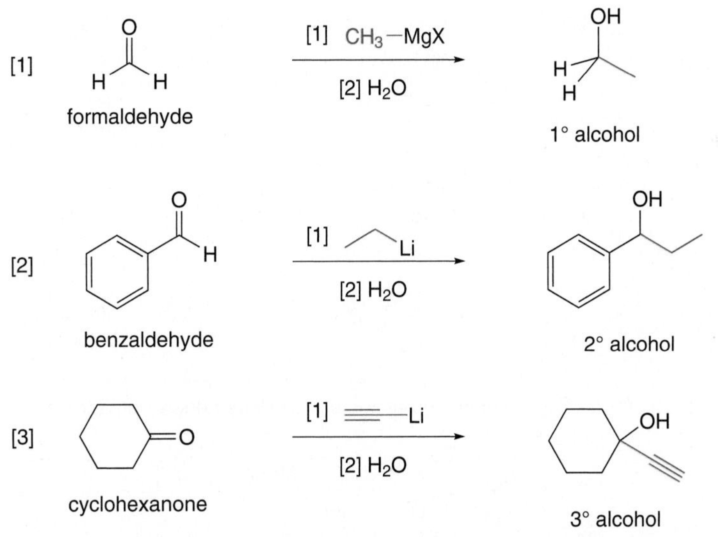

Because organometallic reagents are strong bases that rapidly react with H_2O (Section 20.9C), the addition of the new alkyl group must be carried out under anhydrous conditions to prevent traces of water from reacting with the reagent, thus reducing the yield of the desired alcohol. Water is added after the addition to protonate the alkoxide.

Problem 20.21 Draw the product of each reaction.

a. 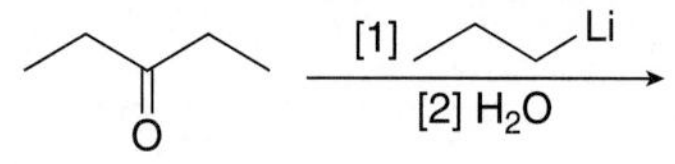
[1] Li
O
[2] H_2O

c. 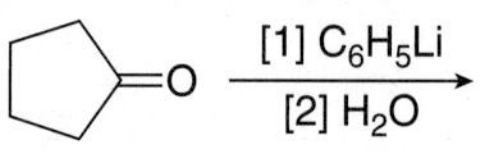
O
[1] C_6H_5Li
[2] H_2O

b. 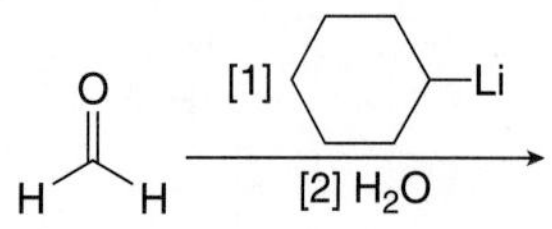
O
H
H
[1]
Li
[2] H_2O

d. 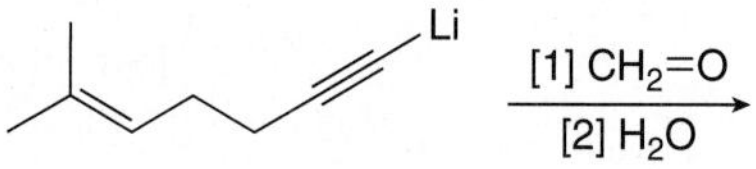
Li
[1] CH_2=O
[2] H_2O

20.10B Stereochemistry

Like reduction, addition of organometallic reagents converts an sp^2 hybridized carbonyl carbon to a tetrahedral sp^3 hybridized carbon. Addition of R–M always occurs from both sides of the trigonal planar carbonyl group. **When a new stereogenic center is formed from an achiral starting material, an equal mixture of enantiomers results,** as shown in Sample Problem 20.1.

Sample Problem 20.1 Draw all stereoisomers formed in the following reaction.

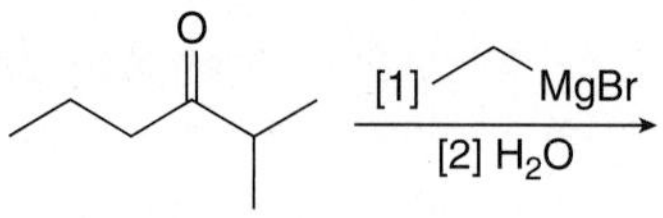
O
[1] MgBr
[2] H_2O

Solution

The Grignard reagent adds from both sides of the trigonal planar carbonyl group, forming two alkoxides, each containing a new stereogenic center labeled in blue. Protonation with water yields **an equal amount of two enantiomers—a racemic mixture.**

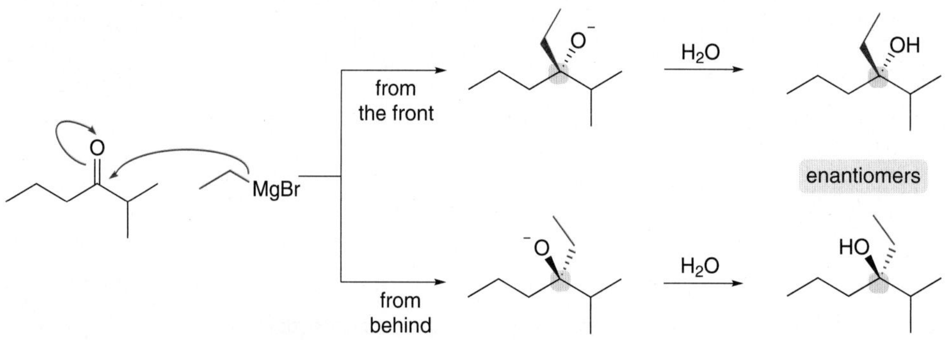
O
MgBr
from
the front
from
behind
O^-
^-O
H_2O
H_2O
OH
HO
enantiomers

Problem 20.22 Draw the products (including stereochemistry) of the following reactions.

a. 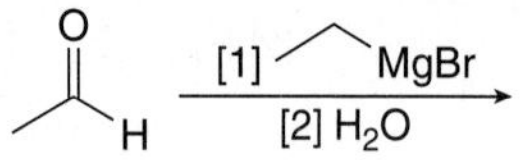
O
H
[1] MgBr
[2] H_2O

b. 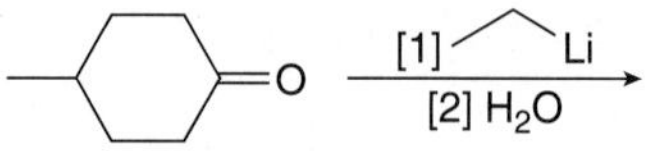
O
[1] Li
[2] H_2O

20.10C Applications in Synthesis

Many syntheses of useful compounds utilize the nucleophilic addition of a Grignard or organolithium reagent to form carbon–carbon bonds. For example, a key step in the synthesis of ethynylestradiol (Section 11.4), an oral contraceptive component, is the addition of lithium acetylide to a ketone, as shown in Figure 20.5.

Figure 20.5
The synthesis of ethynylestradiol

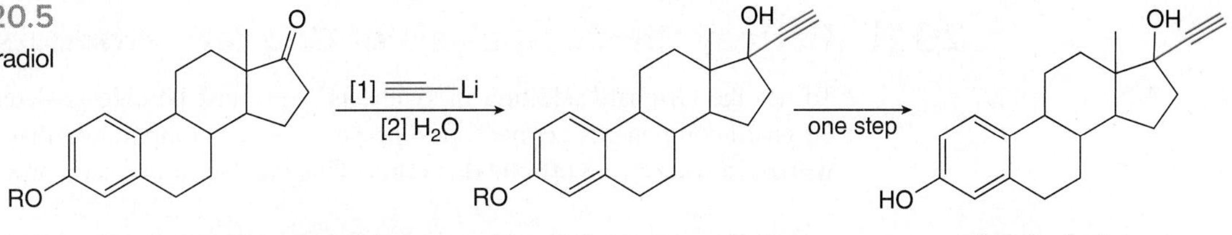

The synthesis of C_{18} juvenile hormone, a member of a group of structurally related molecules that regulate the complex life cycle of an insect, is another example. The last steps of the synthesis are outlined in Figure 20.6.

Figure 20.6
C_{18} juvenile hormone

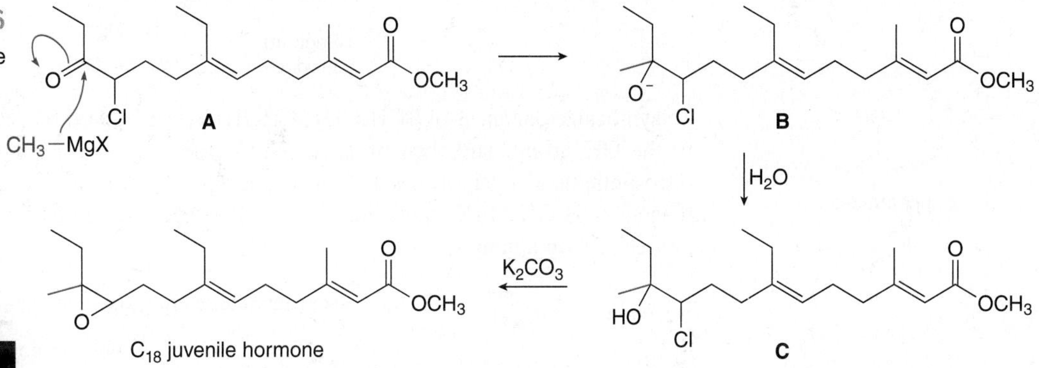

- Addition of CH_3MgX to ketone **A** gives an alkoxide, **B,** which is protonated with H_2O to form 3° alcohol **C.** Although the ester group (–$COOCH_3$) can also react with the Grignard reagent (Section 20.13), it is less reactive than the ketone carbonyl. Thus, with control of reaction conditions, nucleophilic addition occurs selectively at the ketone.
- Treatment of halohydrin **C** with K_2CO_3 forms the C_{18} juvenile hormone in one step. Conversion of a halohydrin to an epoxide was discussed in Section 9.6.

Juvenile hormones regulate the life cycle of the cecropia moth.

Juvenile hormones maintain the juvenile stage of an insect until it is ready for adulthood. This property has been exploited to control mosquitoes and other insects infecting livestock and crops. Although juvenile hormone itself is too unstable in light and too expensive to synthesize for use in controlling insect populations, related compounds, called **juvenile hormone *mimics,*** have been used effectively. Application of these synthetic hormones to an egg or larva of an insect prevents maturation. With no sexually mature adults to propagate the next generation, the insect population is reduced. The best-known example of a synthetic juvenile hormone is called **methoprene,** sold under such trade names as Altocid, Precor, and Diacon. Methoprene is used in cattle salt blocks to control hornflies, in stored tobacco to control pests, and on dogs and cats to control fleas.

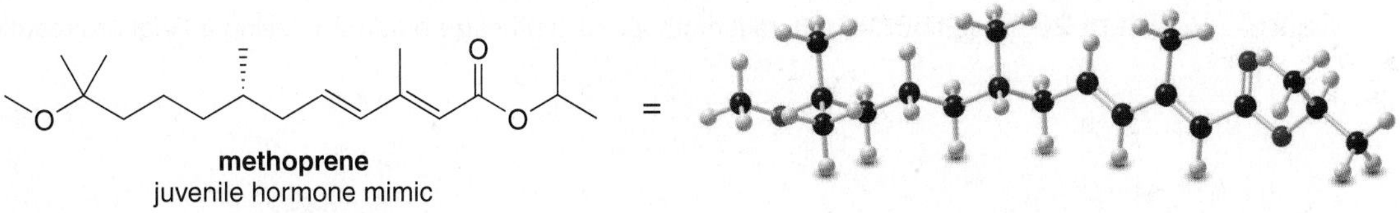

methoprene
juvenile hormone mimic

20.11 Retrosynthetic Analysis of Grignard Products

To use the Grignard addition in synthesis, you must be able to determine what carbonyl and Grignard components are needed to prepare a given compound—that is, **you must work backwards, in the retrosynthetic direction.** This involves a two-step process:

> **Step [1]** Find the carbon bonded to the OH group in the product.
>
> **Step [2]** Break the molecule into two components: One alkyl group bonded to the carbon with the OH group comes from the organometallic reagent. The rest of the molecule comes from the carbonyl component.

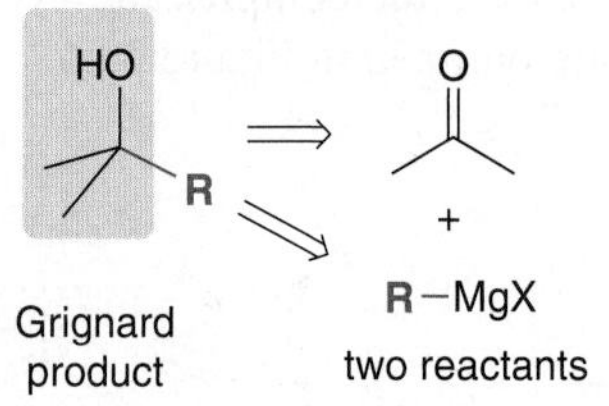

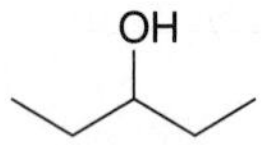
pentan-3-ol

To synthesize pentan-3-ol [$(CH_3CH_2)_2CHOH$] by a Grignard reaction, locate the carbon bonded to the OH group, and then break the molecule into two components at this carbon. Thus, retrosynthetic analysis shows that one of the ethyl groups on this carbon comes from a Grignard reagent (CH_3CH_2MgX), and the rest of the molecule comes from the carbonyl component, a three-carbon aldehyde.

Retrosynthetic analysis

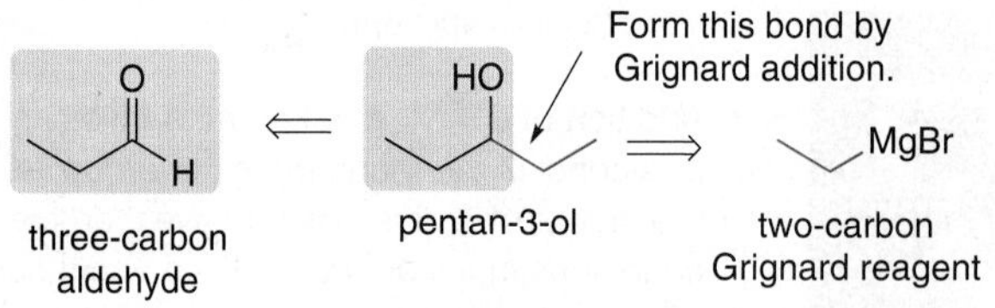

Then, writing the reaction in the synthetic direction—that is, from starting material to product—shows whether the analysis is correct. In this example, a three-carbon aldehyde reacts with CH_3CH_2MgBr to form an alkoxide, which can then be protonated by H_2O to form pentan-3-ol, the desired alcohol.

In the synthetic direction:

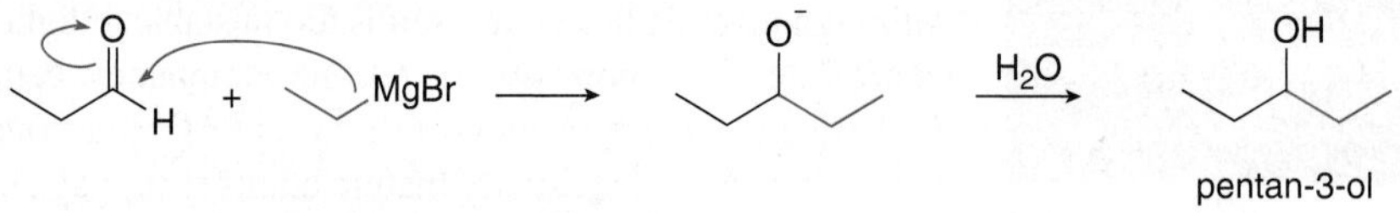

There is often more than one way to synthesize a 2° alcohol by Grignard addition, as shown in Sample Problem 20.2.

Sample Problem 20.2 Show two different methods to synthesize butan-2-ol using a Grignard reaction.

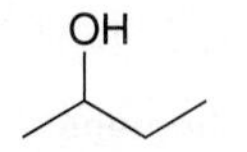

butan-2-ol

Solution

Because butan-2-ol has two different alkyl groups bonded to the carbon bearing the OH group, there are two different ways to form a new carbon–carbon bond by Grignard addition.

Possibility [1] Use CH_3MgX and a three-carbon aldehyde.

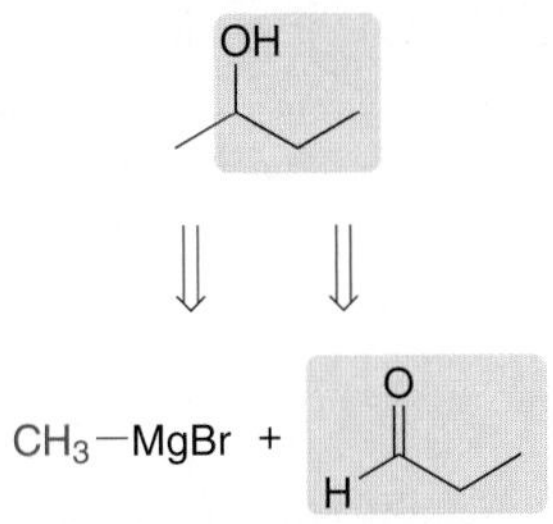

Possibility [2] Use CH_3CH_2MgX and a two-carbon aldehyde.

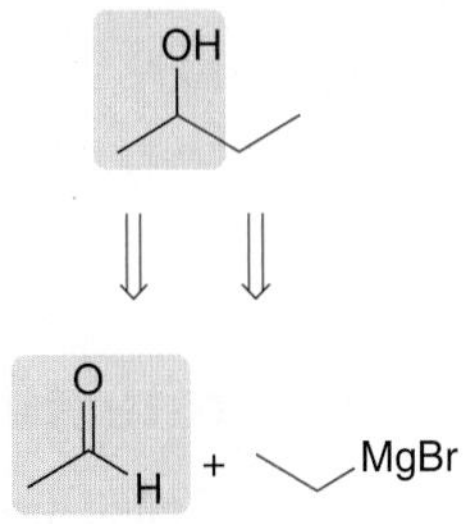

Both methods give the desired product, butan-2-ol, as can be seen by writing the reactions from starting material to product.

Possibility [1]

Possibility [2]

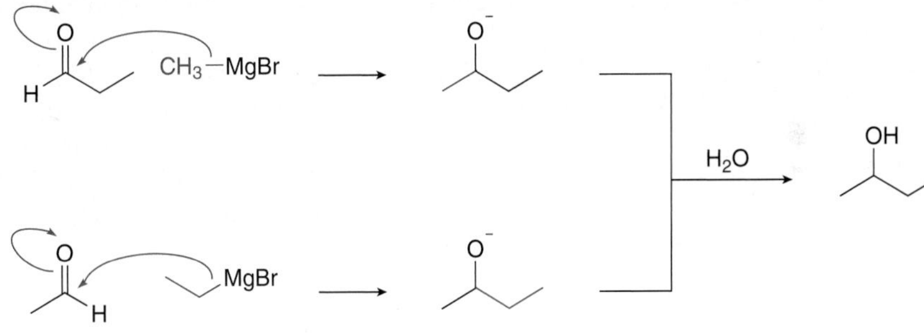

Problem 20.23 What Grignard reagent and carbonyl compound are needed to prepare each alcohol? As shown in part (d), 3° alcohols with three different R groups on the carbon bonded to the OH group can be prepared by three different Grignard reactions.

a. 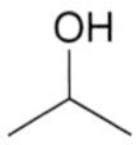

b. 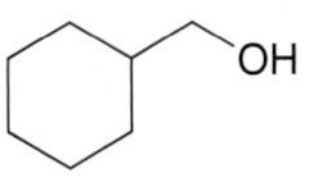

c. 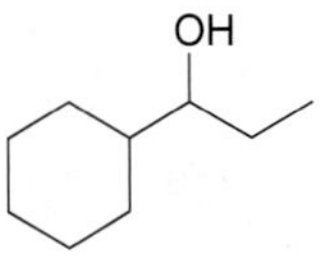

(two methods)

d. 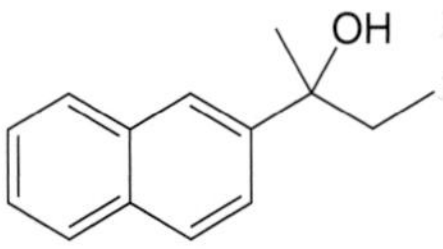

(three methods)

Problem 20.24 Linalool (the Chapter 9 opening molecule) and lavandulol are two of the major components of lavender oil. (a) What organolithium reagent and carbonyl compound can be used to make each alcohol? (b) How might lavandulol be formed by reduction of a carbonyl compound? (c) Why can't linalool be prepared by a similar pathway?

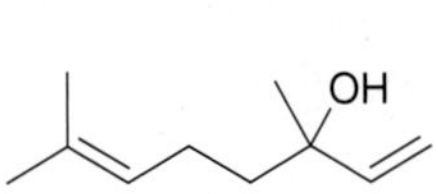

linalool
(three methods)

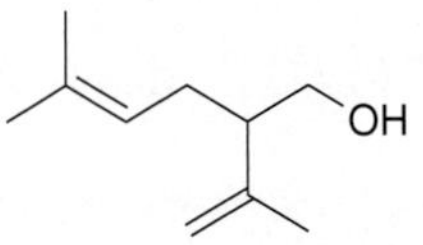

lavandulol

Problem 20.25 What Grignard reagent and carbonyl compound can be used to prepare the antidepressant venlafaxine (trade name Effexor)?

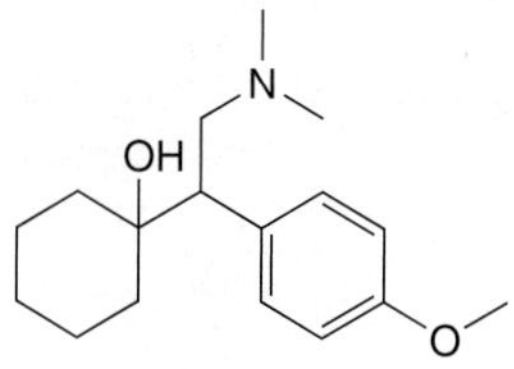

venlafaxine

20.12 Protecting Groups

Although the addition of organometallic reagents to carbonyls is a very versatile reaction, it cannot be used with molecules that contain both a carbonyl group and N–H or O–H bonds.

Rapid acid–base reactions occur between organometallic reagents and all of the following functional groups: ROH, RCOOH, RNH_2, R_2NH, $RCONH_2$, RCONHR, and RSH.

- **Carbonyl compounds that also contain N–H or O–H bonds undergo an acid–base reaction with organometallic reagents, not nucleophilic addition.**

Suppose, for example, that you wanted to add methylmagnesium chloride (CH_3MgCl) to the carbonyl group of 5-hydroxypentan-2-one to form a diol. Nucleophilic addition will *not* occur with this substrate. Instead, **because Grignard reagents are strong bases and proton transfer reactions are fast, CH_3MgCl removes the O–H proton before nucleophilic addition takes place.** The stronger acid and base react to form the weaker conjugate acid and conjugate base, as we learned in Section 20.9C.

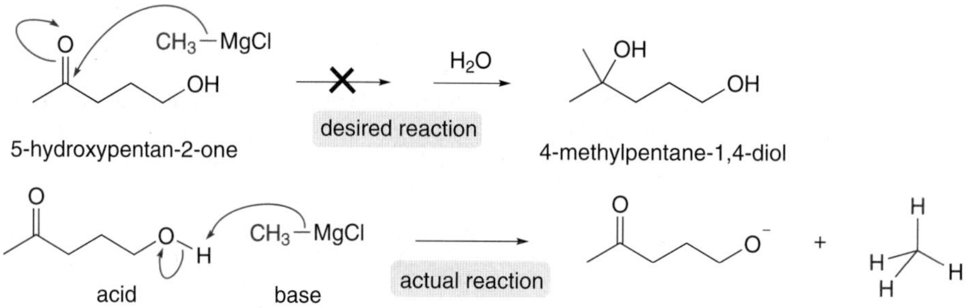

Solving this problem requires a three-step strategy:

Step [1] **Convert the OH group into another functional group that does not interfere with the desired reaction. This new blocking group is called a protecting group, and the reaction that creates it is called *protection.***

Step [2] **Carry out the desired reaction.**

Step [3] **Remove the protecting group. This reaction is called *deprotection.***

Application of the general strategy to the Grignard addition of CH_3MgCl to 5-hydroxypentan-2-one is illustrated in Figure 20.7.

Figure 20.7
General strategy for using a protecting group

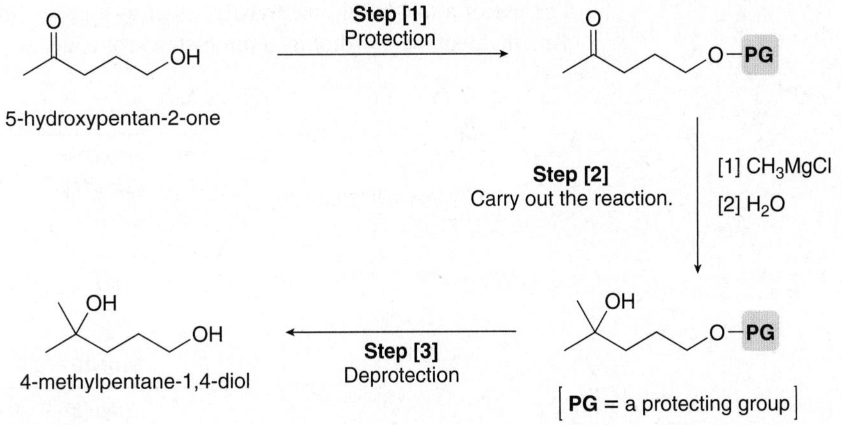

Step [1]
Protection
O-PG
5-hydroxypentan-2-one
Step [2]
Carry out the reaction.
[1] CH3MgCl
[2] H2O
Step [3]
Deprotection
4-methylpentane-1,4-diol
[PG = a protecting group]

- In Step [1], the OH proton in 5-hydroxypentan-2-one is replaced with a protecting group, written as **PG.** Because the product of Step [1] no longer has an OH proton, it can now undergo nucleophilic addition.
- In Step [2], CH_3MgCl adds to the carbonyl group to yield a 3° alcohol after protonation with water.
- Removal of the protecting group in Step [3] forms the desired product, 4-methylpentane-1,4-diol.

A common OH protecting group is a **silyl ether.** A silyl ether has a new O−Si bond in place of the O−H bond of the alcohol. The most widely used silyl ether protecting group is the ***tert*-butyldimethylsilyl ether,** abbreviated as **TBDMS.**

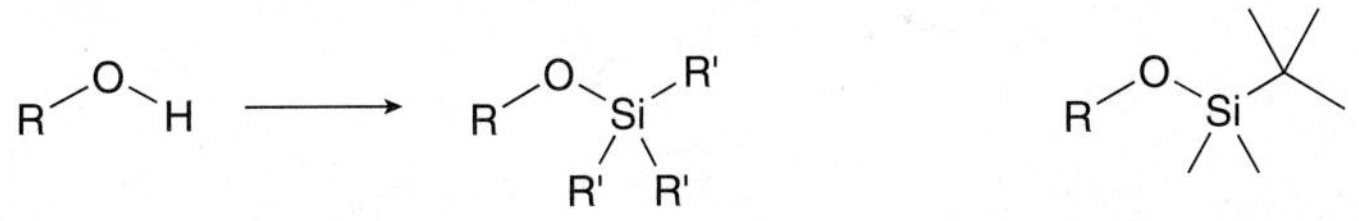

silyl ether

tert-butyldimethylsilyl ether

RO−**TBDMS**

tert-Butyldimethylsilyl ethers are prepared from alcohols by reaction with *tert*-butyldimethylsilyl chloride and an amine base, usually imidazole.

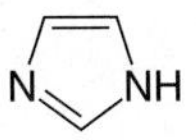

imidazole

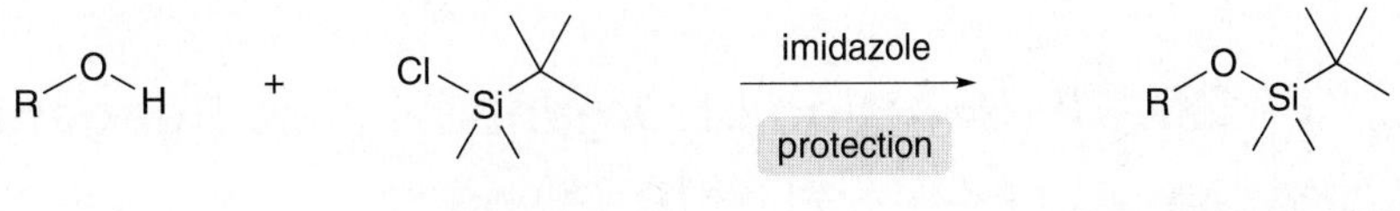

imidazole
protection

tert-butyldimethylsilyl chloride

TBDMS−Cl

tert-butyldimethylsilyl ether

RO−TBDMS

The silyl ether is typically removed with a fluoride salt, usually **tetrabutylammonium fluoride** $(CH_3CH_2CH_2CH_2)_4N^+ F^-$, drawn as $Bu_4N^+F^-$ (Bu = butyl).

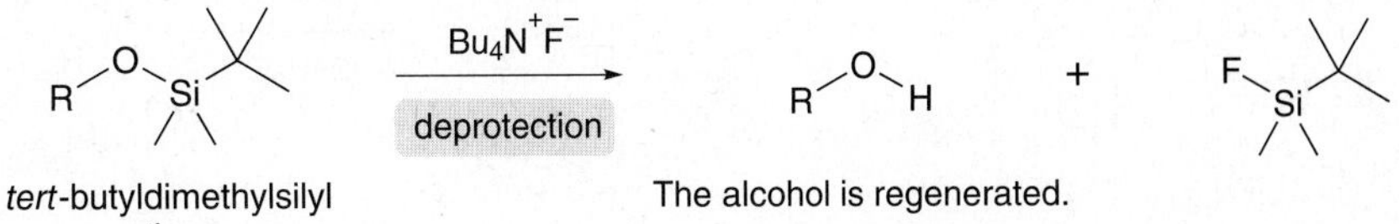

Bu4N+F−
deprotection

tert-butyldimethylsilyl ether

The alcohol is regenerated.

The use of a *tert*-butyldimethylsilyl ether as a protecting group makes possible the synthesis of 4-methylpentane-1,4-diol by a three-step sequence.

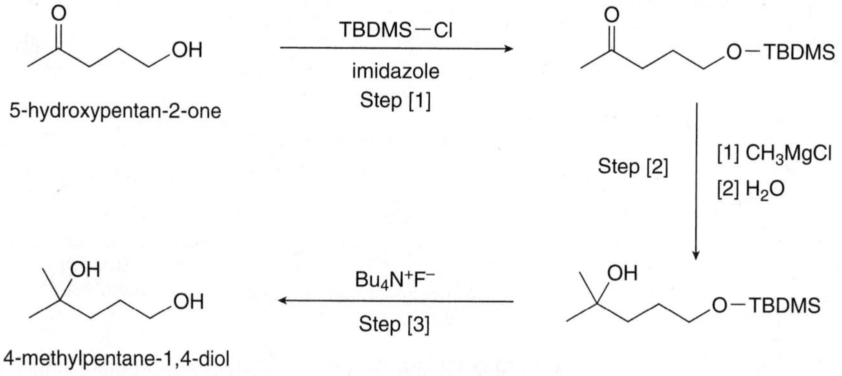

- **Step [1] Protect the OH group** as a *tert*-butyldimethylsilyl ether by reaction with *tert*-butyldimethylsilyl chloride and imidazole.
- **Step [2] Carry out nucleophilic addition** by using CH_3MgCl, followed by protonation.
- **Step [3] Remove the protecting group** with tetrabutylammonium fluoride to form the desired addition product.

Protecting groups block interfering functional groups, and in this way, a wider variety of reactions can take place with a particular substrate. For more on protecting groups, see the discussion of acetals in Section 21.15.

Problem 20.26 Using protecting groups, show how estrone can be converted to ethynylestradiol, a widely used oral contraceptive.

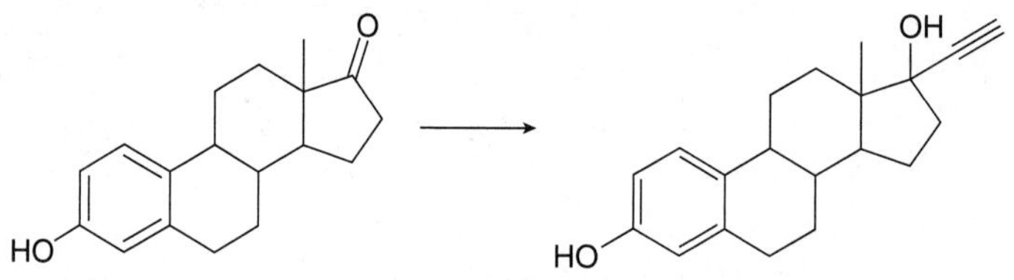

20.13 Reaction of Organometallic Reagents with Carboxylic Acid Derivatives

Organometallic reagents react with carboxylic acid derivatives (RCOZ) to form two different products, depending on the identity of both the leaving group Z and the reagent R–M. The most useful reactions are carried out with esters and acid chlorides, forming either **ketones** or **3° alcohols.**

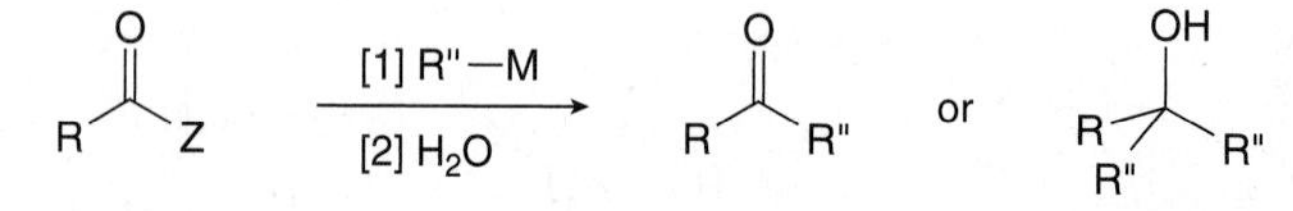

- **Keep in mind that RLi and RMgX are very reactive reagents, whereas R_2CuLi is much less reactive. This reactivity difference makes selective reactions possible.**

20.13A Reaction of RLi and RMgX with Esters and Acid Chlorides

Both esters and acid chlorides form 3° alcohols when treated with two equivalents of either Grignard or organolithium reagents. Two new carbon–carbon bonds are formed in the product.

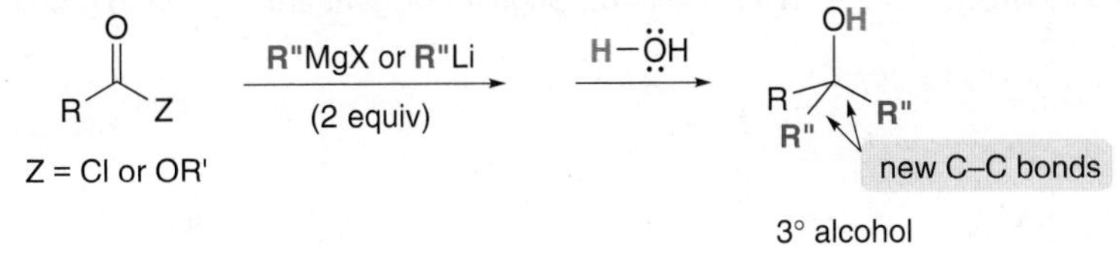
R"MgX or R"Li
(2 equiv)
H–ÖH
Z = Cl or OR'
new C–C bonds
3° alcohol

Two examples using Grignard reagents are shown.

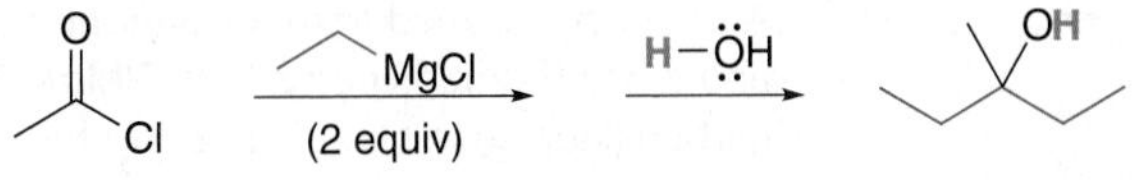
MgCl
(2 equiv)
H–ÖH

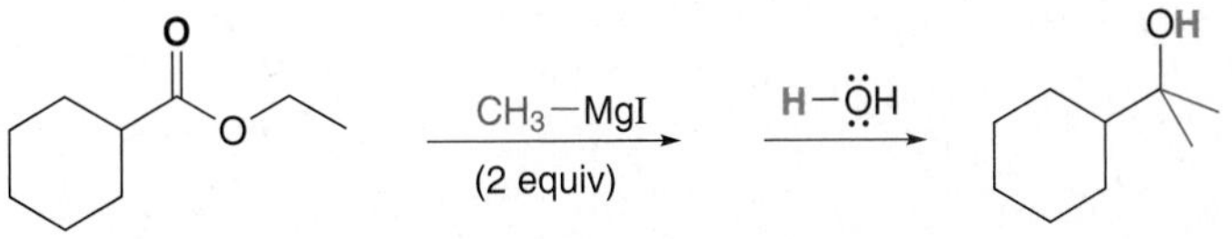
CH3–MgI
(2 equiv)
H–ÖH

Problem 20.27 Draw the product formed when each compound is treated with two equivalents of $CH_3CH_2CH_2CH_2MgBr$ followed by H_2O.

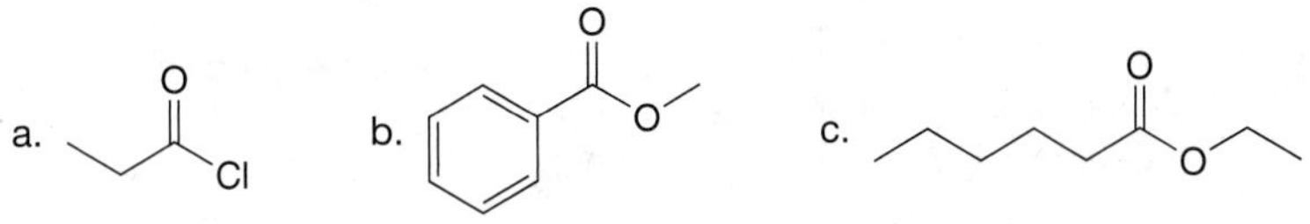
a.
b.
c.

The mechanism for this addition reaction resembles the mechanism for the metal hydride reduction of acid chlorides and esters discussed in Section 20.7A. The mechanism is conceptually divided into two parts: **nucleophilic substitution** to form a ketone (Steps [1] and [2]), followed by **nucleophilic addition** to form a 3° alcohol (Steps [3] and [4]), as shown in Mechanism 20.7.

Mechanism 20.7 Reaction of R"MgX or R"Li with RCOCl and RCOOR'

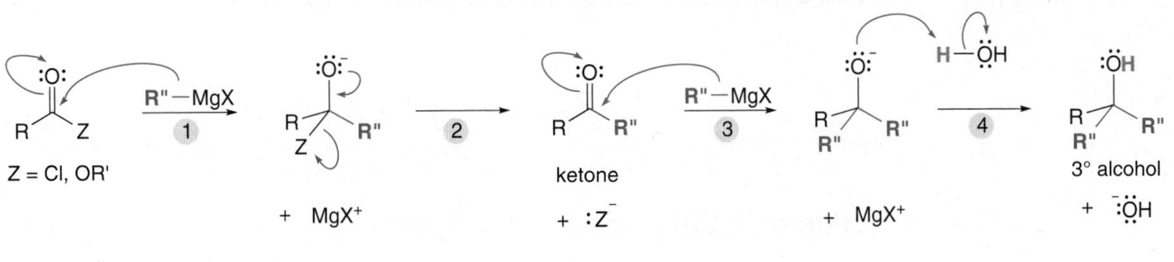
R"–MgX
1
2
R"–MgX
3
4
H–ÖH
Z = Cl, OR'
ketone
3° alcohol
+ MgX+
+ :Z−
+ MgX+
+ :ÖH−

1 Nucleophilic attack of $(R")^-$ forms a tetrahedral intermediate with a leaving group Z.

2 The π bond is re-formed and the leaving group Z departs to form a ketone. The overall result of addition of $(R")^-$ and elimination of $Z:^-$ is substitution of R" for Z.

3 Nucleophilic attack of $(R")^-$ forms an alkoxide with no leaving group.

4 Protonation of the alkoxide by H_2O forms a 3° alcohol.

Organolithium and Grignard reagents afford 3° alcohols when they react with esters and acid chlorides. As soon as the ketone forms by addition of one equivalent of reagent to RCOZ (Steps [1] and [2] of the mechanism), it reacts with a second equivalent of reagent to form the 3° alcohol.

This reaction is more limited than the Grignard addition to aldehydes and ketones, because only 3° alcohols having two identical alkyl groups can be prepared. Nonetheless, it is still a valuable reaction because it forms two new carbon–carbon bonds.

Sample Problem 20.3 What ester and Grignard reagent are needed to prepare the following alcohol?

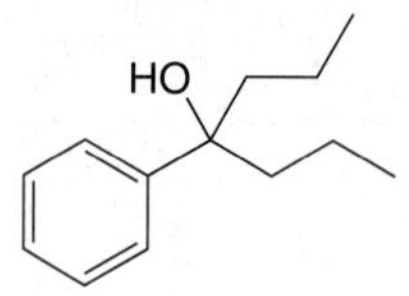

Solution

A 3° alcohol formed from an ester and Grignard reagent must have **two identical R groups,** and these R groups come from RMgX. The remainder of the molecule comes from the ester. The carbon (labeled in blue) bonded to the OH group comes from the carbonyl carbon.

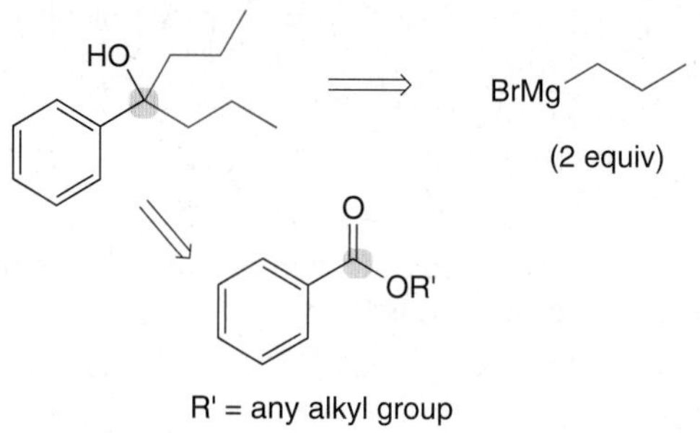

Checking in the synthetic direction:

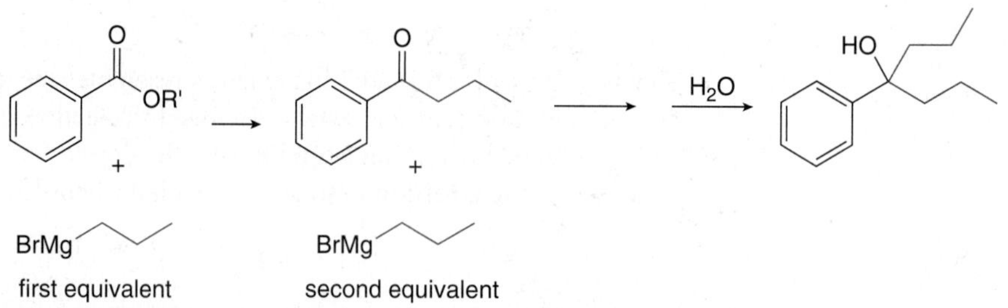

Problem 20.28 What ester and Grignard reagent are needed to prepare each alcohol?

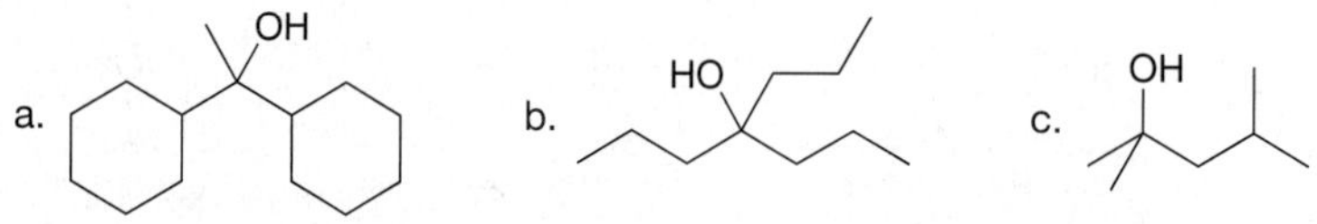

20.13B Reaction of R_2CuLi with Acid Chlorides

To form a ketone from a carboxylic acid derivative, a less reactive organometallic reagent—namely an **organocuprate**—is needed. **Acid chlorides, which have the best leaving group (Cl^-) of the carboxylic acid derivatives, react with R'_2CuLi, to give a ketone as product.** Esters, which contain a poorer leaving group (^-OR), do *not* react with R'_2CuLi.

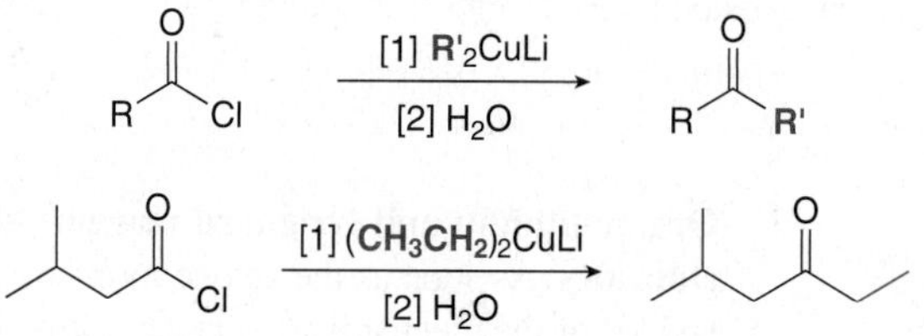

This reaction results in nucleophilic substitution of an alkyl group R' for the leaving group Cl, forming one new carbon–carbon bond.

Problem 20.29 What organocuprate reagent is needed to convert CH_3CH_2COCl to each ketone?

a. 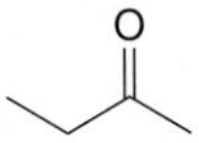 b. 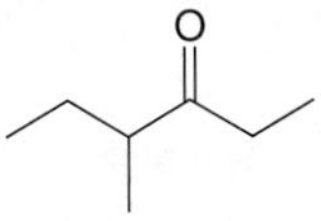 c. 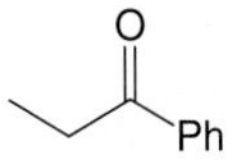

Problem 20.30 What reagent is needed to convert $(CH_3)_2CHCH_2COCl$ into each compound?

a. 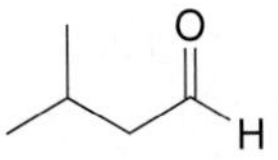 b. 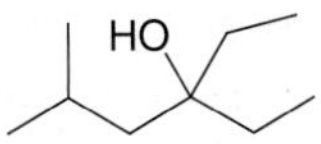 c. 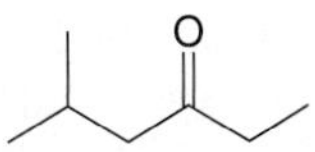 d. 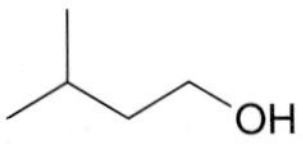

A ketone with two different R groups bonded to the carbonyl carbon can be made by two different methods, as illustrated in Sample Problem 20.4.

Sample Problem 20.4 Show two different ways to prepare pentan-2-one from an acid chloride and an organocuprate reagent.

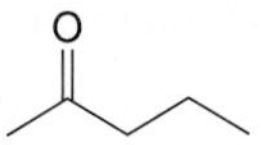

pentan-2-one

Solution

In each case, one alkyl group comes from the organocuprate and one comes from the acid chloride.

Possibility [1] Use $(CH_3)_2CuLi$ and a four-carbon acid chloride.

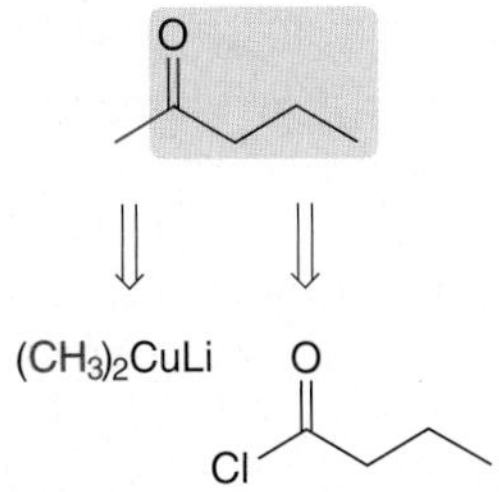

Possibility [2] Use $(CH_3CH_2CH_2)_2CuLi$ and a two-carbon acid chloride.

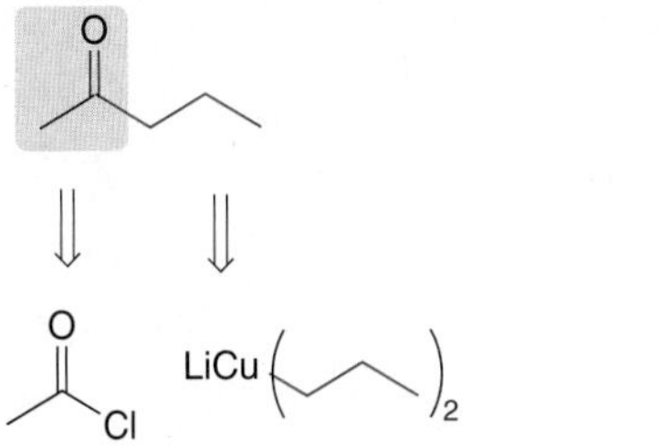

Problem 20.31 Draw two different ways to prepare each ketone from an acid chloride and an organocuprate reagent.

a. 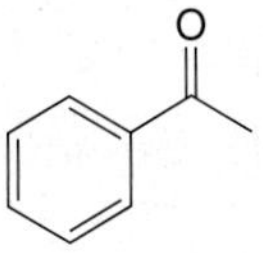 b. 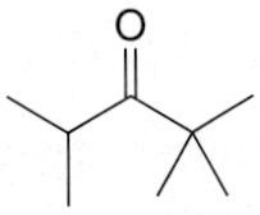

20.14 Reaction of Organometallic Reagents with Other Compounds

Because organometallic reagents are strong nucleophiles, they react with many other electrophiles in addition to carbonyl groups. Because these reactions always lead to the formation of new carbon–carbon bonds, they are also valuable in organic synthesis. In Section 20.14, we examine the reactions of organometallic reagents with **carbon dioxide** and **epoxides.**

20.14A Reaction of Grignard Reagents with Carbon Dioxide

Grignard reagents react with CO_2 to give carboxylic acids after protonation with aqueous acid. This reaction, called **carboxylation,** forms a carboxylic acid with one more carbon atom than the Grignard reagent from which it is prepared.

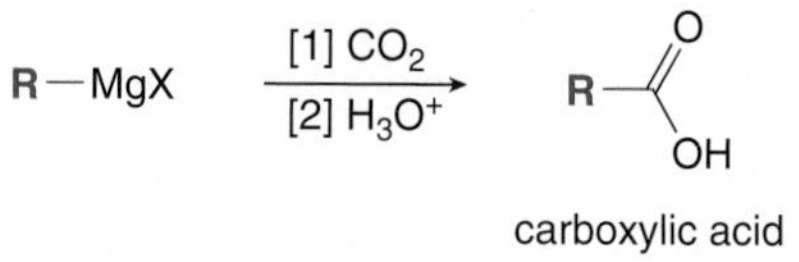

Because Grignard reagents are made from alkyl or aryl halides, an organic halide can be converted to a carboxylic acid having one more carbon atom by a two-step reaction sequence: **formation of a Grignard reagent,** followed by **reaction with CO_2.**

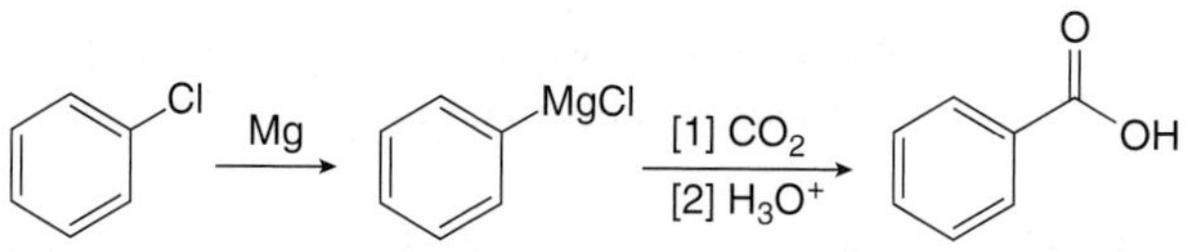

The mechanism resembles earlier reactions of nucleophilic Grignard reagents with carbonyl groups, as shown in Mechanism 20.8.

Mechanism 20.8 Carboxylation—Reaction of RMgX with CO_2

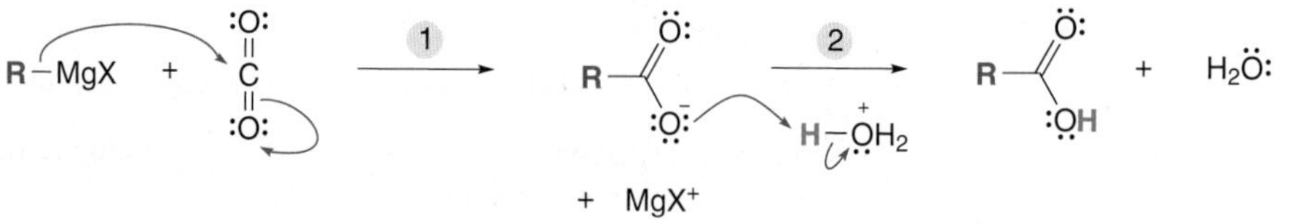

1 The nucleophilic Grignard reagent attacks the electrophilic carbon of CO_2, cleaving the π bond and forming a new carbon–carbon bond.

2 Protonation of the carboxylate anion with aqueous acid forms the carboxylic acid.

Problem 20.32 What carboxylic acid is formed from each alkyl halide on treatment with [1] Mg; [2] CO_2; [3] H_3O^+?

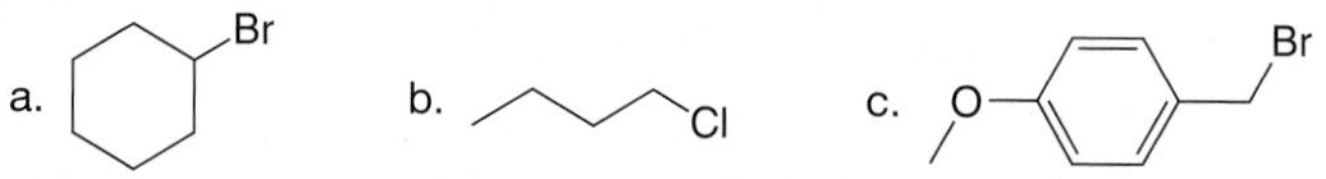

20.14B Reaction of Organometallic Reagents with Epoxides

Like other strong nucleophiles, **organometallic reagents**—RLi, RMgX, and R_2CuLi—**open epoxide rings to form alcohols.**

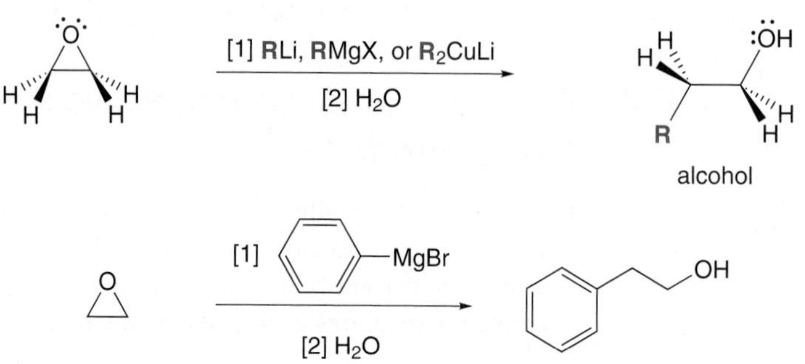

The opening of epoxide rings with negatively charged nucleophiles was discussed in Section 9.16A.

The reaction follows the same two-step process as the opening of epoxide rings with other negatively charged nucleophiles—that is, **nucleophilic attack from the back side of the epoxide ring, followed by protonation of the resulting alkoxide.** In unsymmetrical epoxides, nucleophilic attack occurs at the less substituted carbon atom.

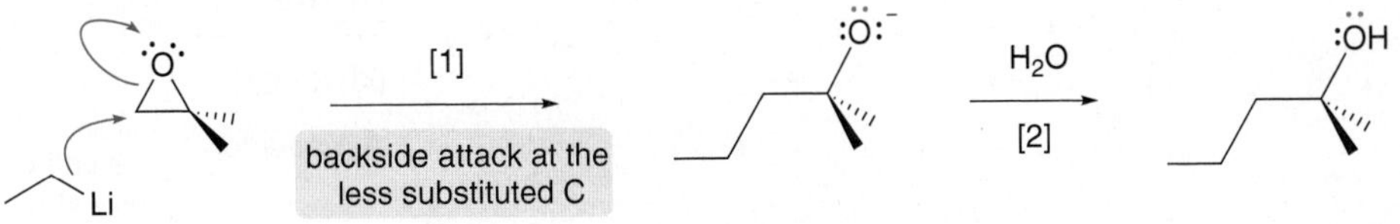

Problem 20.33 What epoxide is needed to convert CH_3CH_2MgBr to each of the following alcohols, after quenching with water?

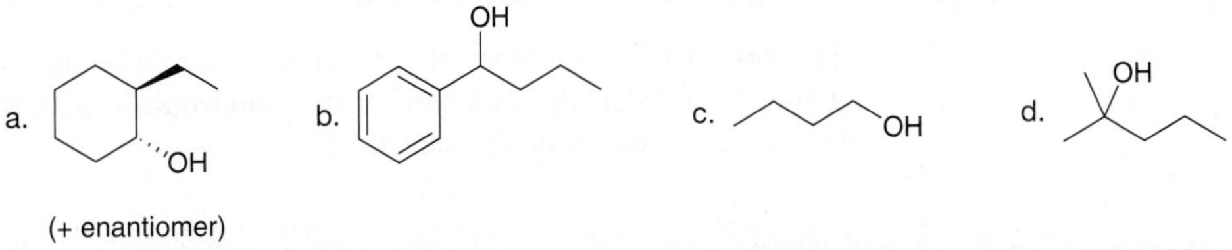

20.15 α,β-Unsaturated Carbonyl Compounds

α,β-Unsaturated carbonyl compounds are conjugated molecules containing a carbonyl group and a carbon–carbon double bond, separated by a single σ bond.

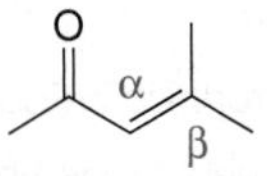

α,β-unsaturated carbonyl compound

Both functional groups of α,β-unsaturated carbonyl compounds have π bonds, but individually, they react with very different kinds of reagents. Carbon–carbon double bonds react with electrophiles (Chapter 10) and carbonyl groups react with nucleophiles (Section 20.2). What happens, then, when these two functional groups having opposite reactivity are in close proximity?

Because the two π bonds are conjugated, the electron density in an α,β-unsaturated carbonyl compound is delocalized over four atoms. Three resonance structures show that the carbonyl carbon and the β carbon bear a partial positive charge. This means that **α,β-unsaturated carbonyl compounds can react with nucleophiles at two different sites.**

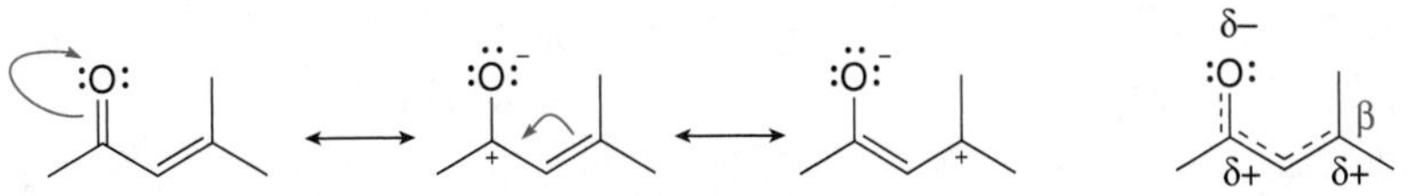

three resonance structures for an α,β-unsaturated carbonyl compound

hybrid
two electrophilic sites

- **Addition of a nucleophile to the carbonyl carbon, called 1,2-addition, adds the elements of H and Nu across the C=O, forming an allylic alcohol.**

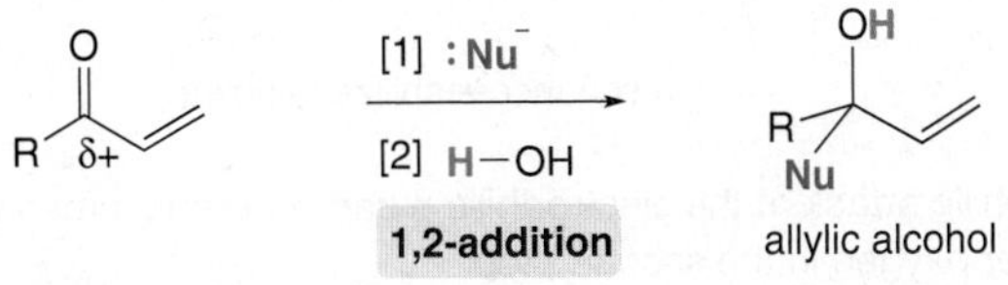

- **Addition of a nucleophile to the β carbon, called 1,4-addition or conjugate addition, forms a carbonyl compound.**

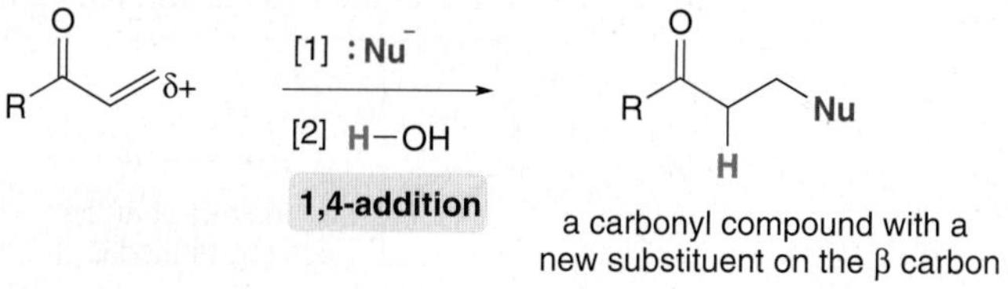
[1] :Nu⁻
[2] H–OH
1,4-addition
a carbonyl compound with a new substituent on the β carbon

Both 1,2- and 1,4-addition result in nucleophilic addition of the elements of H and Nu.

20.15A The Mechanisms for 1,2-Addition and 1,4-Addition

The steps for the mechanism of 1,2-addition are exactly the same as those for the nucleophilic addition to an aldehyde or ketone—that is, **nucleophilic attack,** followed by **protonation** (Section 20.2A), as shown in Mechanism 20.9.

Mechanism 20.9 1,2-Addition to an α,β-Unsaturated Carbonyl Compound

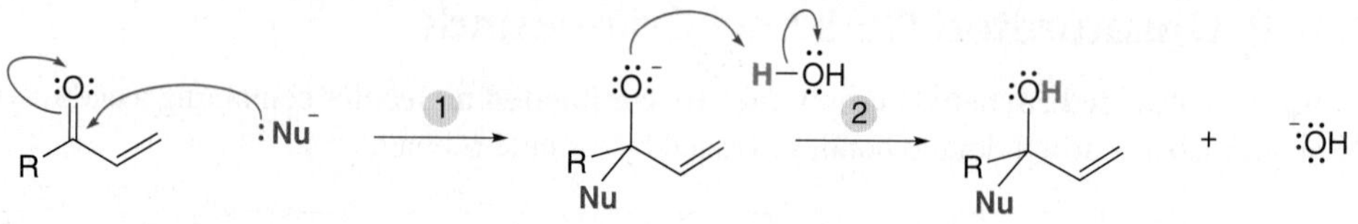

1 The nucleophile attacks the electrophilic carbonyl. The π bond is broken, moving an electron pair out on oxygen.

2 Protonation of the negatively charged oxygen by H_2O forms the addition product. H and Nu are added to the carbonyl group.

The mechanism for 1,4-addition also begins with nucleophilic attack, and then protonation and tautomerization add the elements of H and Nu to the α and β carbons of the carbonyl compound, as shown in Mechanism 20.10.

Mechanism 20.10 1,4-Addition to an α,β-Unsaturated Carbonyl Compound

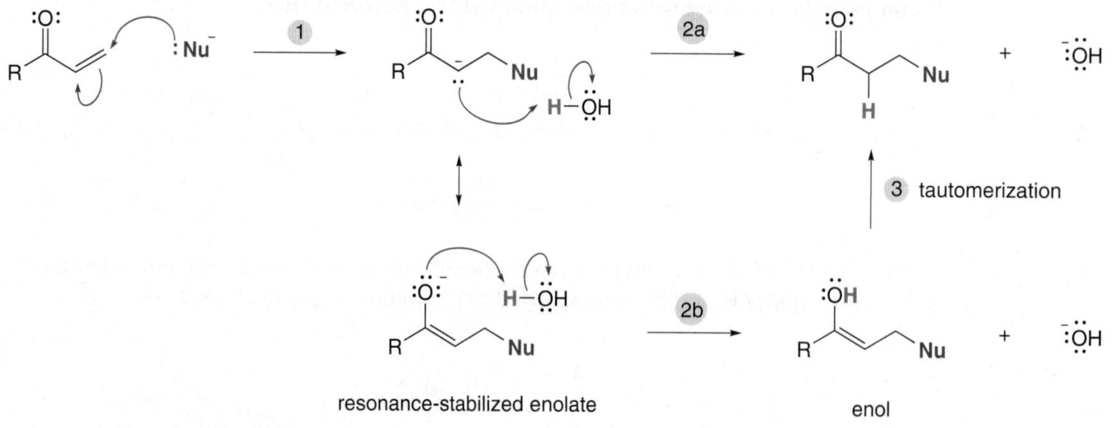
3 tautomerization
resonance-stabilized enolate
enol

1 Nucleophilic attack at the electrophilic β carbon forms a resonance-stabilized enolate anion, which can react on either carbon or oxygen in the second step.

2a Protonation of the carbon end of the enolate forms the 1,4-addition product directly.

2b–3 Protonation of the oxygen end of the enolate forms an enol, which undergoes tautomerization by the two-step process described in Section 11.9. This forms the same 1,4-addition product that results from protonation on carbon.

20.15B Reaction of α,β-Unsaturated Carbonyl Compounds with Organometallic Reagents

The identity of the metal in an organometallic reagent determines whether it reacts with an α,β-unsaturated aldehyde or ketone by 1,2-addition or 1,4-addition.

Why is conjugate addition also called 1,4-addition? If the atoms of the enol are numbered beginning with the O atom, then the elements of H and Nu are bonded to atoms "1" and "4," respectively.

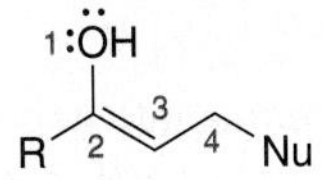

- **Organolithium and Grignard reagents form 1,2-addition products.**

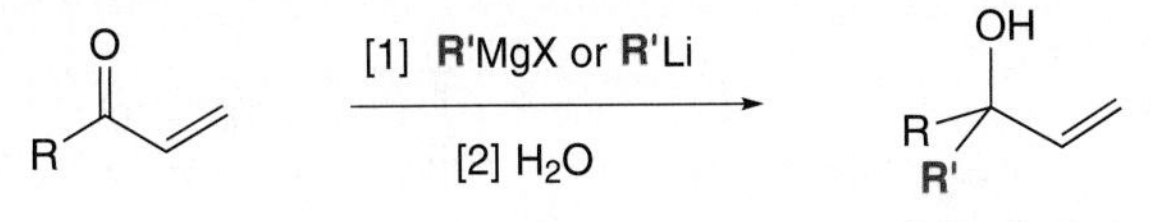

allylic alcohol

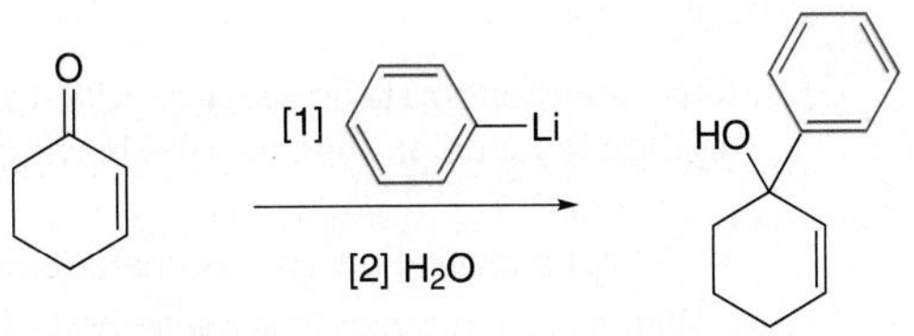

- **Organocuprate reagents form 1,4-addition products.**

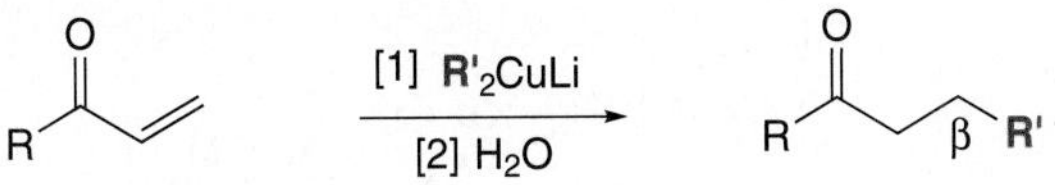

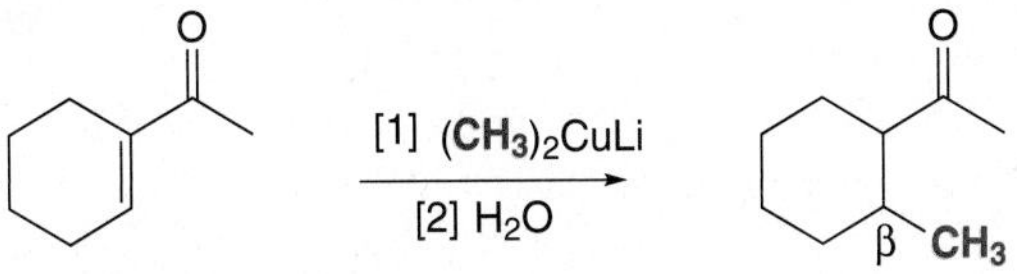

Sample Problem 20.5 Draw the products of each reaction.

a. 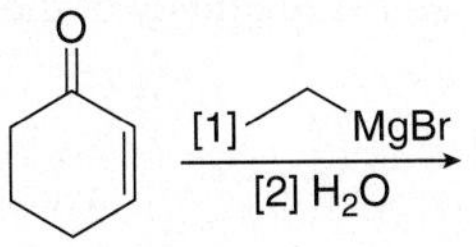

b. 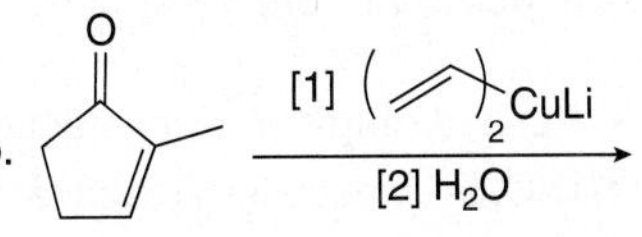

Solution

The characteristic reaction of α,β-unsaturated carbonyl compounds is nucleophilic addition. The reagent determines the mode of addition (1,2- or 1,4-).

a. **Grignard reagents undergo 1,2-addition.** CH_3CH_2MgBr adds a new CH_3CH_2 group at the carbonyl carbon.

b. **Organocuprate reagents undergo 1,4-addition.** The cuprate reagent adds a new vinyl group ($CH_2{=}CH$) at the β carbon.

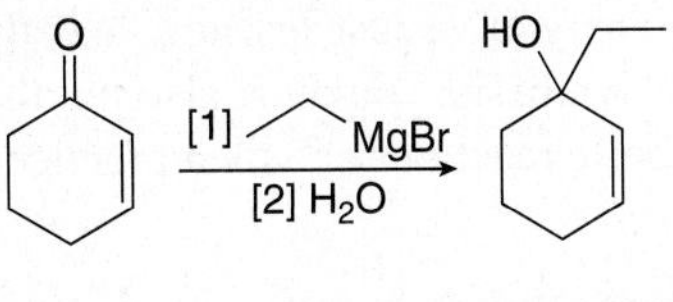

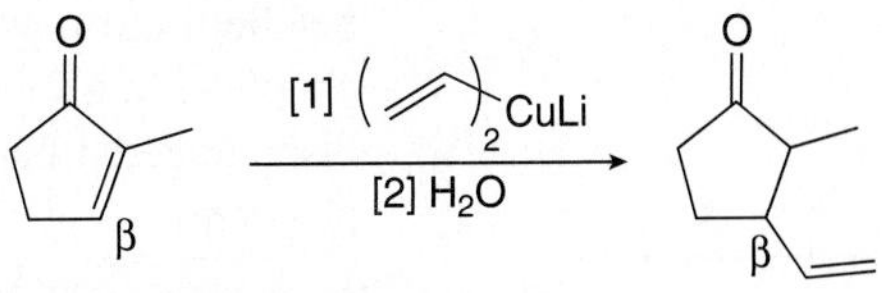

Problem 20.34 Draw the product when each compound is treated with either $(CH_3)_2CuLi$, followed by H_2O, or $HC{\equiv}CLi$, followed by H_2O.

a. 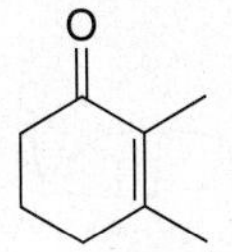

b. 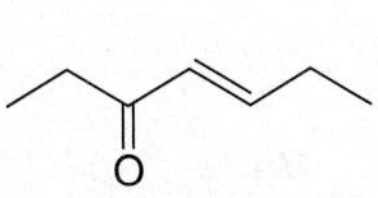

c. 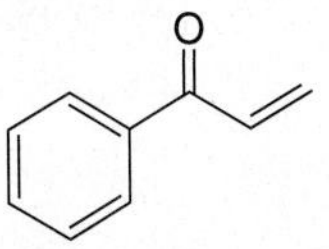

20.16 Summary—The Reactions of Organometallic Reagents

We have now seen many different reactions of organometallic reagents with a variety of functional groups, and you may have some difficulty keeping them all straight. Rather than memorizing them all, keep in mind the following three concepts:

[1] Organometallic reagents (R – M) attack electrophilic carbon atoms, especially the carbonyl carbon.

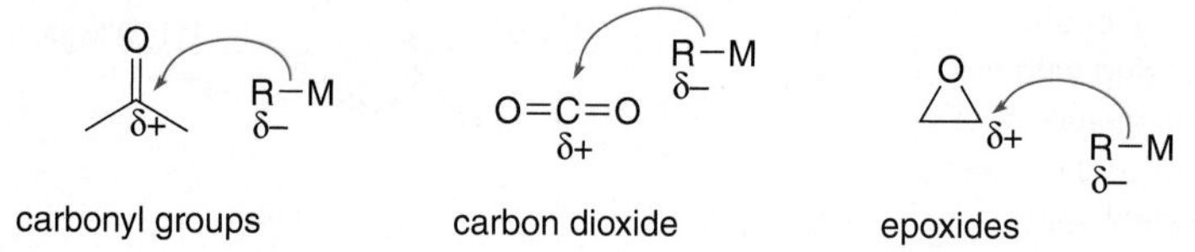

[2] After an organometallic reagent adds to a carbonyl group, the fate of the intermediate depends on the presence or absence of a leaving group.

- Without a leaving group, the characteristic reaction is *nucleophilic addition.*
- With a leaving group, it is *nucleophilic substitution.*

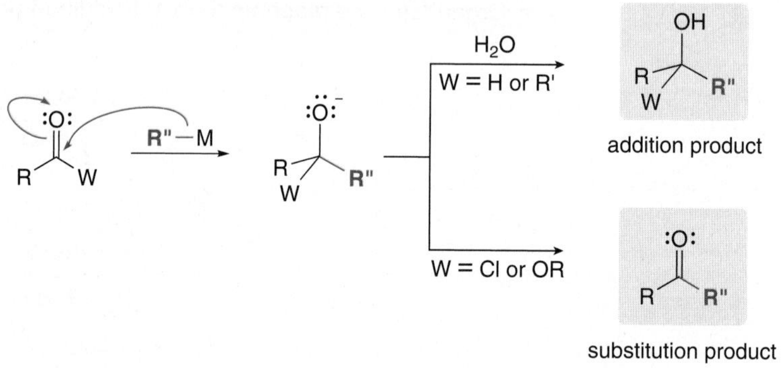

[3] The polarity of the R – M bond determines the reactivity of the reagents.

- RLi and RMgX are very reactive reagents.
- R_2CuLi is much less reactive.

20.17 Synthesis

The reactions learned in Chapter 20 have proven extremely useful in organic synthesis. Oxidation and reduction reactions interconvert two functional groups that differ in oxidation state. Organometallic reagents form new carbon–carbon bonds.

Synthesis is perhaps the most difficult aspect of organic chemistry. It requires you to remember both the new reactions you've just learned, and the ones you've encountered in previous chapters. In a successful synthesis, you must also put these reactions in a logical order. Don't be discouraged. Learn the basic reactions and then practice them over and over again with synthesis problems.

In Sample Problems 20.6–20.8 that follow, keep in mind that the products formed by the reactions of Chapter 20 can themselves be transformed into many other functional groups. For example, hexan-2-ol, the product of Grignard addition of butylmagnesium chloride to acetaldehyde, can be transformed into a variety of other compounds, as shown in Figure 20.8.

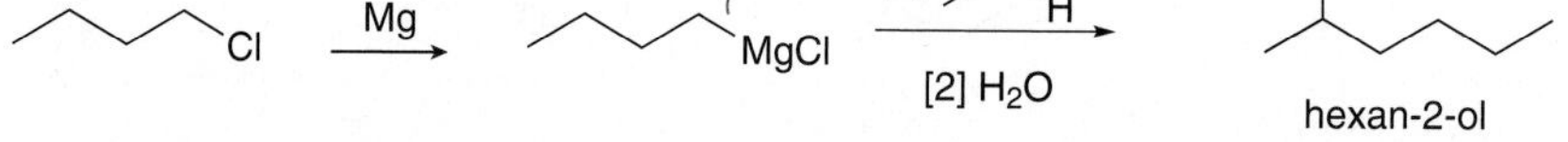

Figure 20.8 Conversion of hexan-2-ol into other compounds

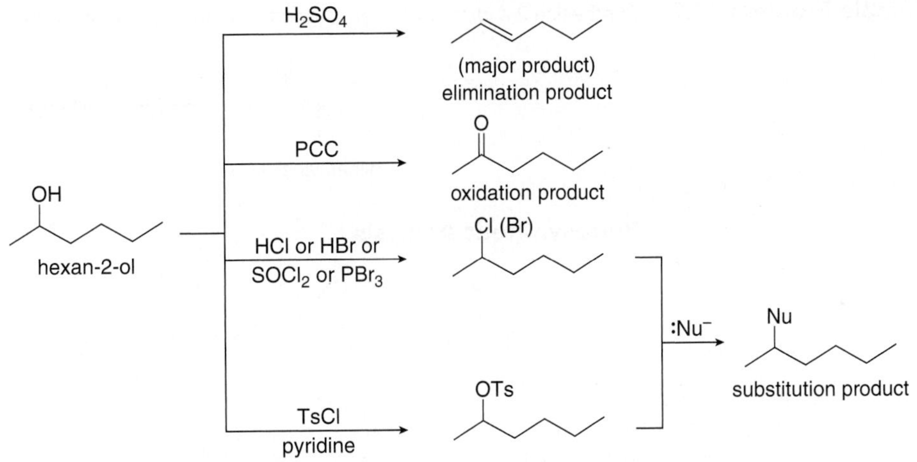
H_2SO_4
(major product)
elimination product
PCC
oxidation product
OH
hexan-2-ol
HCl or HBr or
$SOCl_2$ or PBr_3
Cl (Br)
:Nu^-
Nu
substitution product
OTs
TsCl
pyridine

Before proceeding with Sample Problems 20.6–20.8, you should review the stepwise strategy for designing a synthesis found in Section 11.12.

Sample Problem 20.6 Synthesize 1-methylcyclohexene from cyclohexanone and any organic alcohol.

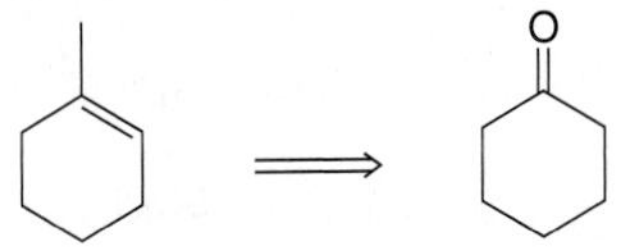

1-methylcyclohexene cyclohexanone

Retrosynthetic Analysis

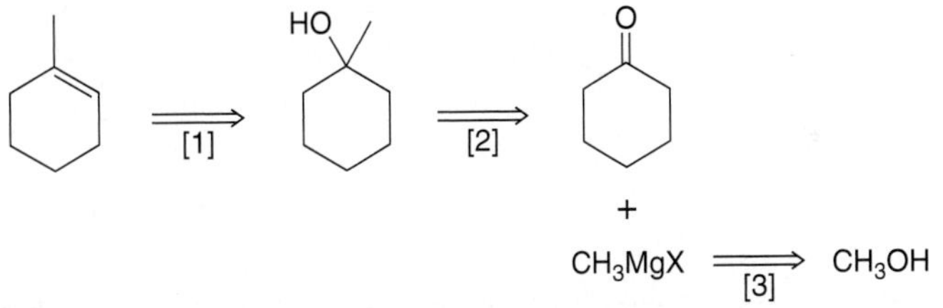
HO
[1]
[2]
+
CH_3MgX
[3]
CH_3OH

Thinking backwards:

- [1] Form the double bond by dehydration of an alcohol.
- [2] Make the 3° alcohol by Grignard addition of CH_3MgX.
- [3] Prepare the Grignard reagent stepwise from an alcohol.

Synthesis

Four steps are needed:

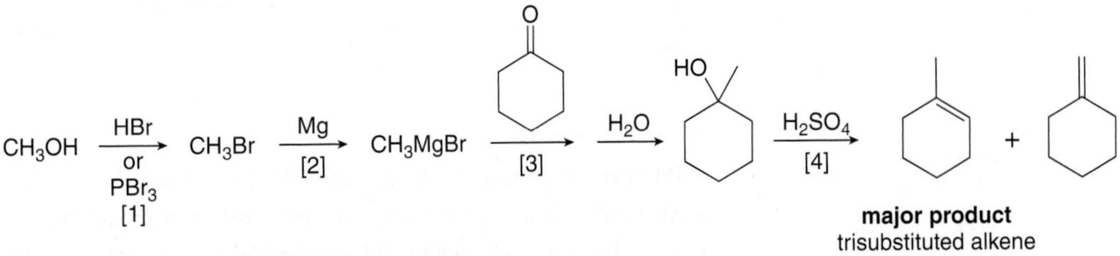
CH_3OH
HBr
or
PBr_3
[1]
CH_3Br
Mg
[2]
CH_3MgBr
[3]
H_2O
HO
H_2SO_4
[4]
+
major product
trisubstituted alkene

- Conversion of CH_3OH to the Grignard reagent CH_3MgBr requires two steps: formation of an alkyl halide (Step [1]), followed by reaction with Mg (Step [2]).
- Addition of CH_3MgBr to cyclohexanone followed by protonation forms an alcohol in Step [3].
- Acid-catalyzed elimination of water in Step [4] forms a mixture of alkenes, with the desired trisubstituted alkene as the major product.

Sample Problem 20.7 Synthesize 2,4-dimethylhexan-3-one from four-carbon alcohols.

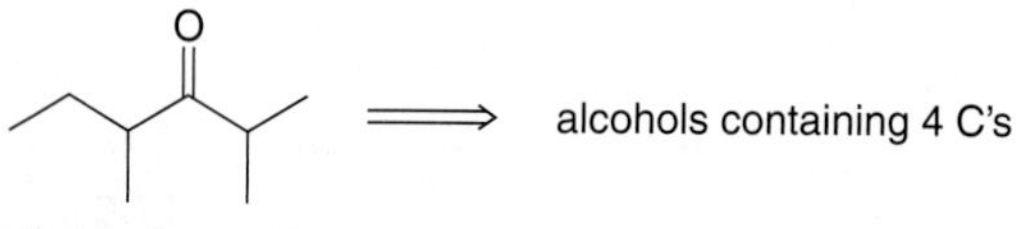

2,4-dimethylhexan-3-one

Retrosynthetic Analysis

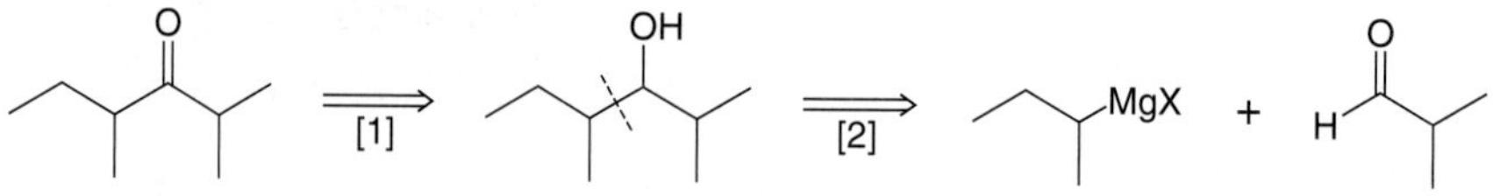

Synthesize each of these components.

Thinking backwards:

- [1] Form the ketone by oxidation of a 2° alcohol.
- [2] Make the 2° alcohol by Grignard addition to an aldehyde. Both of these compounds have 4 C's, and each must be synthesized from an alcohol.

Synthesis

First, make both components needed for the Grignard reaction.

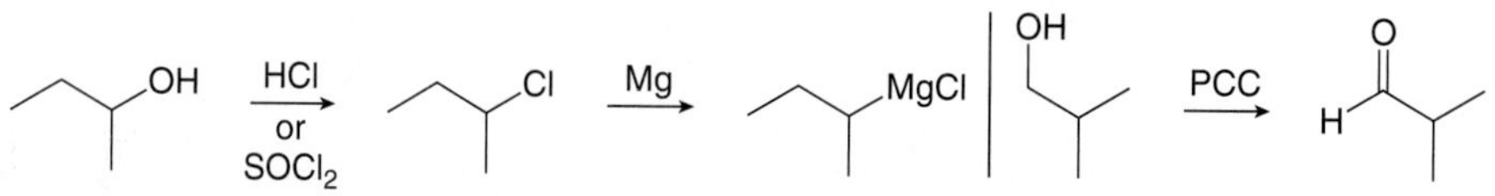

Then, complete the synthesis with Grignard addition, followed by oxidation of the alcohol to the ketone.

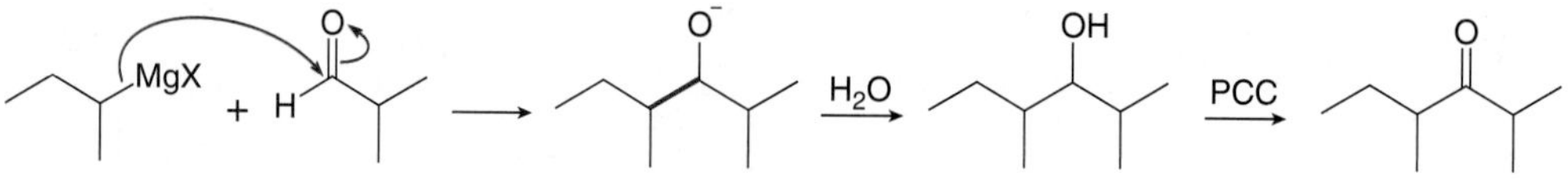

new C–C bond in red

Sample Problem 20.8 Synthesize isopropylcyclopentane from alcohols having ≤ 5 C's.

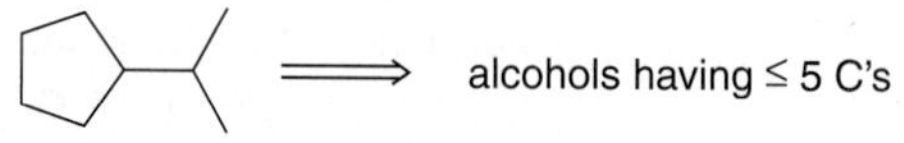

isopropylcyclopentane

Retrosynthetic Analysis

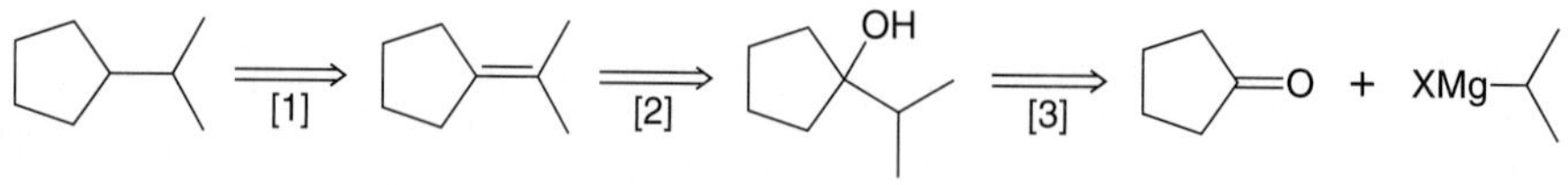

Thinking backwards:

- [1] Form the alkane by hydrogenation of an alkene.
- [2] Introduce the double bond by dehydration of an alcohol.
- [3] Form the 3° alcohol by Grignard addition to a ketone. Both components of the Grignard reaction must then be synthesized.

Synthesis

First make both components needed for the Grignard reaction.

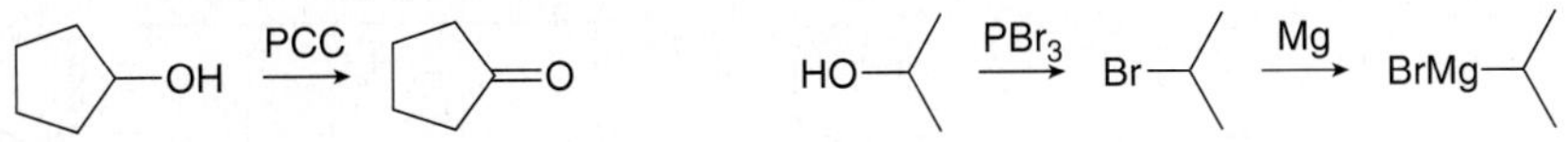
OH
PCC
O
HO
PBr3
Br
Mg
BrMg

Complete the synthesis with Grignard addition, dehydration, and hydrogenation.

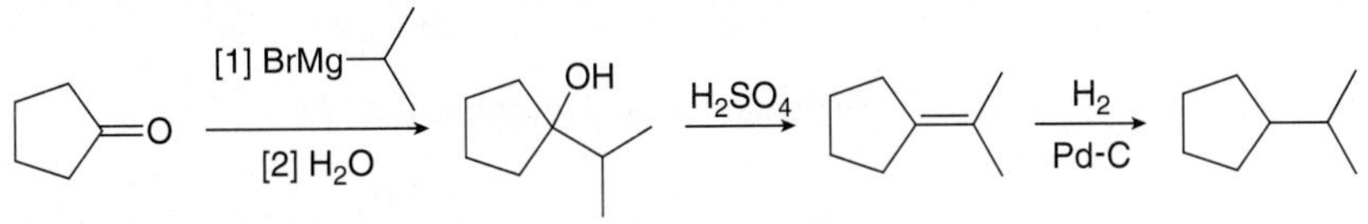
O
[1] BrMg
[2] H2O
OH
H2SO4
H2
Pd-C

major product
tetrasubstituted
double bond

Problem 20.35 Synthesize each compound from cyclohexanol, ethanol, and any other needed reagents.

a. 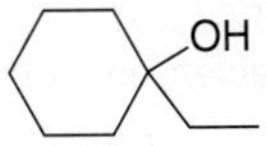 b. 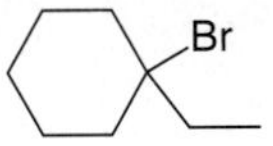 c. 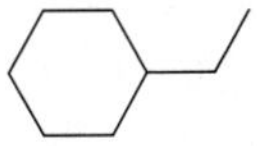 d. 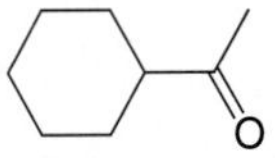 e. 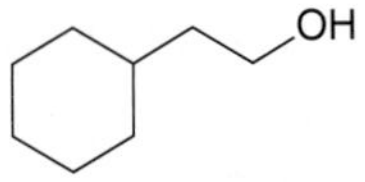

KEY CONCEPTS

Introduction to Carbonyl Chemistry; Organometallic Reagents; Oxidation and Reduction

Reduction Reactions

[1] Reduction of aldehydes and ketones to 1° and 2° alcohols (20.4)

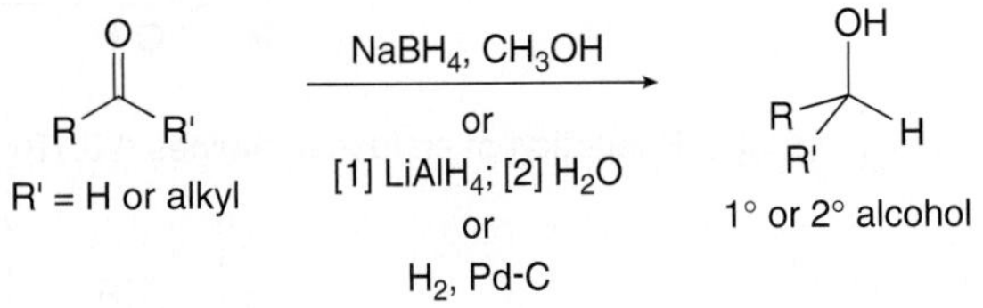
O
R
R'
R' = H or alkyl
NaBH4, CH3OH
or
[1] LiAlH4; [2] H2O
or
H2, Pd-C
OH
R
H
R'
1° or 2° alcohol

[2] Reduction of α,β-unsaturated aldehydes and ketones (20.4C)

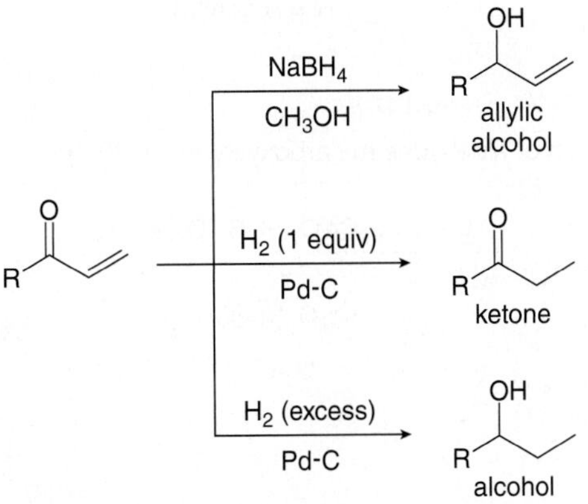
O
R
NaBH4
CH3OH
OH
R
allylic
alcohol
H2 (1 equiv)
Pd-C
O
R
ketone
H2 (excess)
Pd-C
OH
R
alcohol

- reduction of the C=O only
- reduction of the C=C only
- reduction of both π bonds

[3] Enantioselective ketone reduction (20.6)

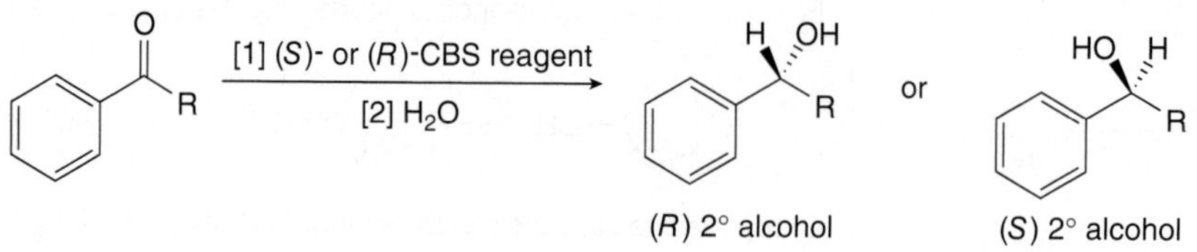
[1] (S)- or (R)-CBS reagent
[2] H2O
or
(R) 2° alcohol
(S) 2° alcohol

- A single enantiomer is formed.

[4] Reduction of acid chlorides (20.7A)

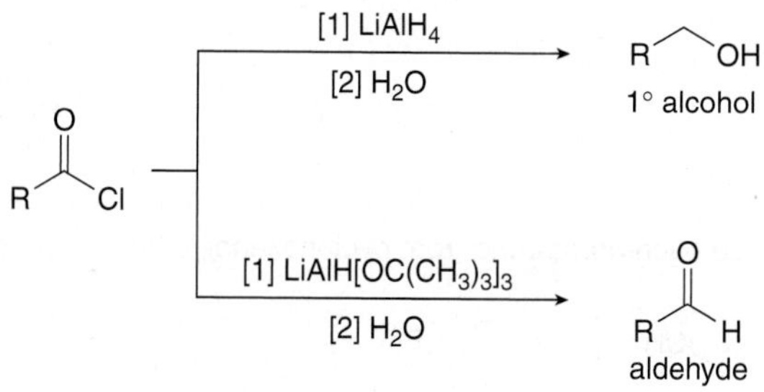
[1] LiAlH4
[2] H2O
1° alcohol
[1] LiAlH[OC(CH3)3]3
[2] H2O
aldehyde

- $LiAlH_4$, a strong reducing agent, reduces an acid chloride to an alcohol.
- With $LiAlH[OC(CH_3)_3]_3$, a milder reducing agent, reduction stops at the aldehyde stage.

[5] Reduction of esters (20.7A)

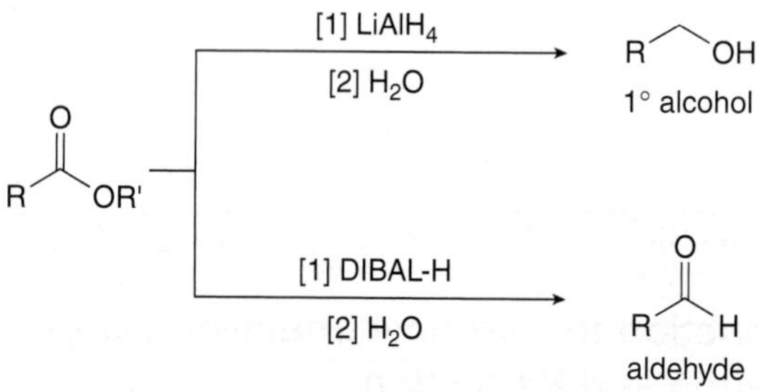
[1] LiAlH4
[2] H2O
1° alcohol
[1] DIBAL-H
[2] H2O
aldehyde

- $LiAlH_4$, a strong reducing agent, reduces an ester to an alcohol.
- With DIBAL-H, a milder reducing agent, reduction stops at the aldehyde stage.

[6] Reduction of carboxylic acids to 1° alcohols (20.7B)

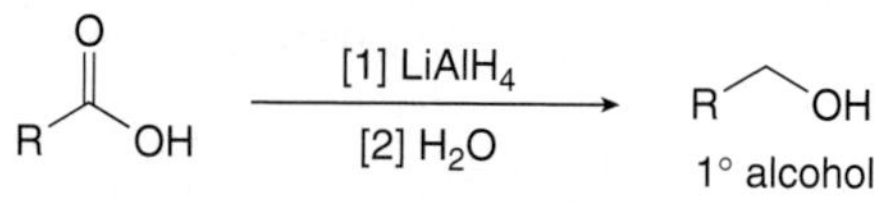
[1] LiAlH4
[2] H2O
1° alcohol

[7] Reduction of amides to amines (20.7B)

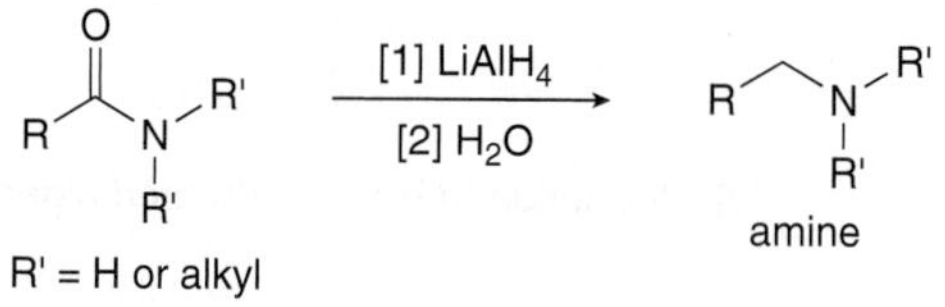
[1] LiAlH4
[2] H2O
R' = H or alkyl
amine

Oxidation Reactions

Oxidation of aldehydes to carboxylic acids (20.8)

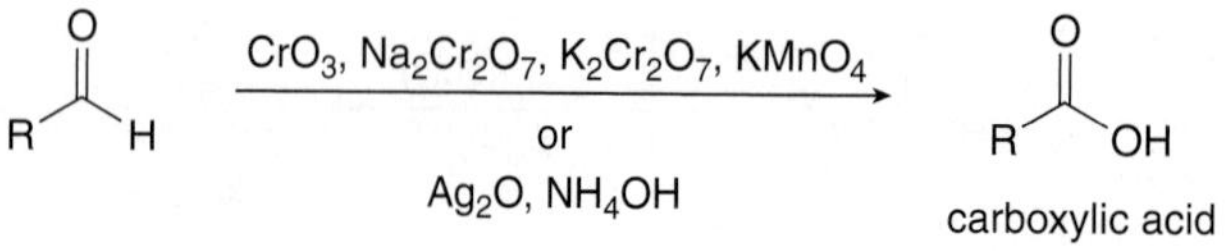
CrO3, Na2Cr2O7, K2Cr2O7, KMnO4
or
Ag2O, NH4OH
carboxylic acid

- All Cr^{6+} reagents except PCC oxidize RCHO to RCOOH.
- Tollens reagent (Ag_2O + NH_4OH) oxidizes RCHO only. Primary (1°) and 2° alcohols do not react with Tollens reagent.

Preparation of Organometallic Reagents (20.9)

[1] Organolithium reagents: $R{-}X + 2\,Li \longrightarrow R{-}Li + LiX$

[2] Grignard reagents:

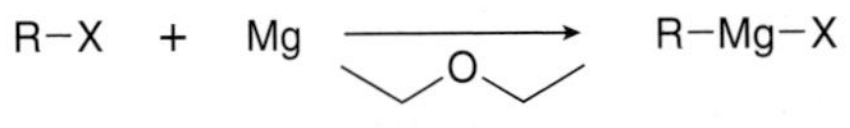

R−X + Mg
R−Mg−X
O

[3] Organocuprate reagents:

$R{-}X + 2\,Li \longrightarrow R{-}Li + LiX$

$2\,R{-}Li + CuI \longrightarrow R_2Cu^- \; Li^+ + LiI$

[4] Lithium and sodium acetylides:

$R{-}C{\equiv}C{-}H \xrightarrow{Na^+ \; ^-NH_2} R{-}C{\equiv}C^- \; Na^+ + NH_3$

a sodium acetylide

$R{-}C{\equiv}C{-}H \xrightarrow{R'{-}Li} R{-}C{\equiv}C{-}Li + R'{-}H$

a lithium acetylide

Reactions with Organometallic Reagents

[1] Reaction as a base (20.9C)

R−M + H−ÖR ⟶ R−H + M⁺ ⁻:Ö−R

- RM = RLi, RMgX, R_2CuLi
- This acid–base reaction occurs with H_2O, ROH, RNH_2, R_2NH, RSH, RCOOH, $RCONH_2$, and RCONHR.

[2] Reaction with aldehydes and ketones to form 1°, 2°, and 3° alcohols (20.10)

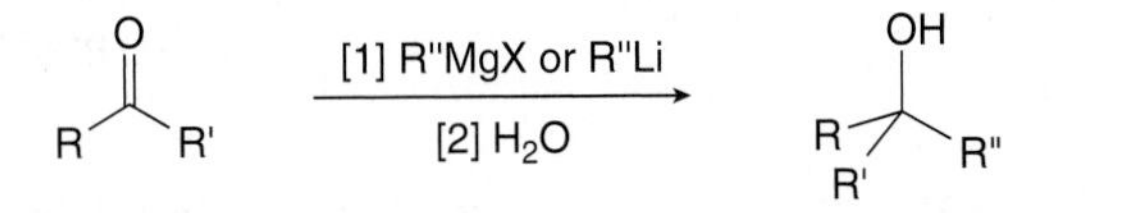

O
R
R'
[1] R"MgX or R"Li
[2] H₂O
OH
R
R'
R"

R' = H or alkyl

1°, 2°, or 3° alcohol

[3] Reaction with esters to form 3° alcohols (20.13A)

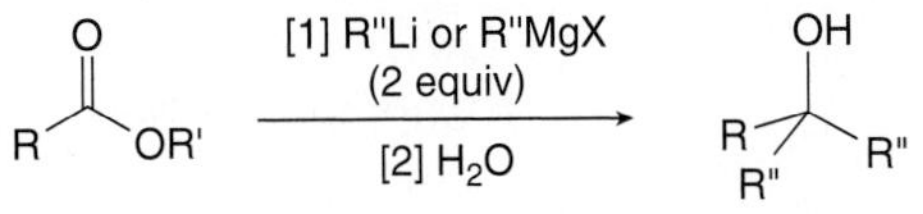

O
R
OR'
[1] R"Li or R"MgX
(2 equiv)
[2] H₂O
OH
R
R"
R"

3° alcohol

[4] Reaction with acid chlorides (20.13B)

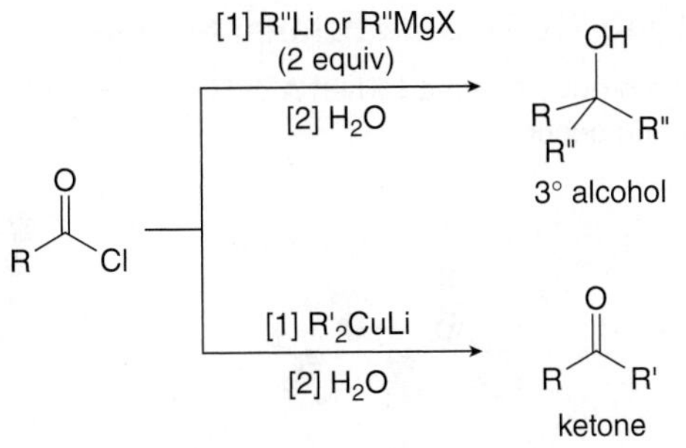

[1] R"Li or R"MgX
(2 equiv)
[2] H₂O
OH
R
R"
R"
3° alcohol
O
R
Cl
[1] R'₂CuLi
[2] H₂O
O
R
R'
ketone

- More reactive organometallic reagents—R"Li and R"MgX—add two equivalents of R" to an acid chloride to form a 3° alcohol with two identical R" groups.
- Less reactive organometallic reagents—R'_2CuLi—add only one equivalent of R' to an acid chloride to form a ketone.

[5] Reaction with carbon dioxide—Carboxylation (20.14A)

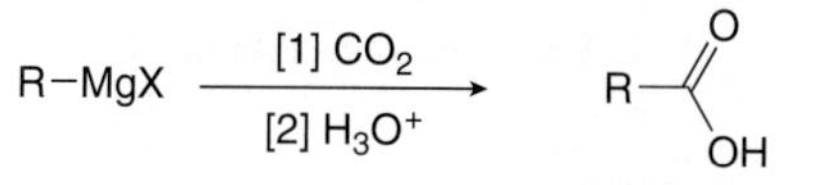
R−MgX
[1] CO_2
[2] H_3O^+
R
O
OH
carboxylic acid

[6] Reaction with epoxides (20.14B)

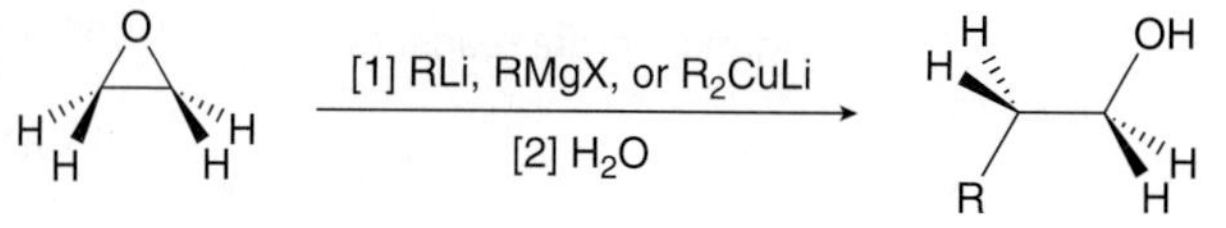
O
H
H
H
H
[1] RLi, RMgX, or R_2CuLi
[2] H_2O
H
H
OH
R
H
H
alcohol

[7] Reaction with α,β-unsaturated aldehydes and ketones (20.15B)

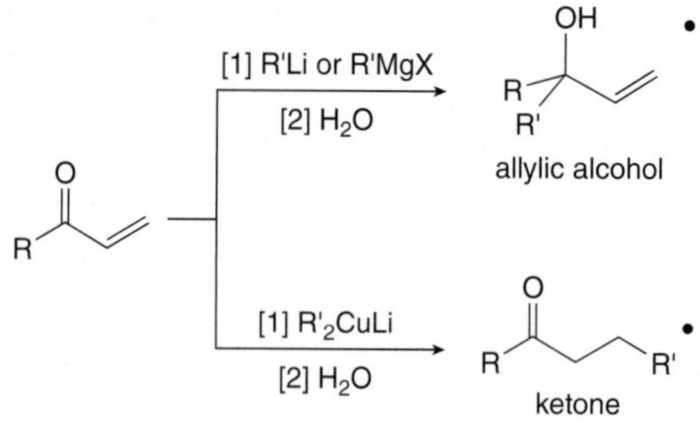
O
R
[1] R'Li or R'MgX
[2] H_2O
OH
R
R'
allylic alcohol
[1] R'_2CuLi
[2] H_2O
O
R
R'
ketone

- More reactive organometallic reagents—R'Li and R'MgX—react with α,β-unsaturated carbonyls by 1,2-addition.
- Less reactive organometallic reagents—R'_2CuLi—react with α,β-unsaturated carbonyls by 1,4-addition.

Protecting Groups (20.12)

[1] Protecting an alcohol as a *tert*-butyldimethylsilyl ether

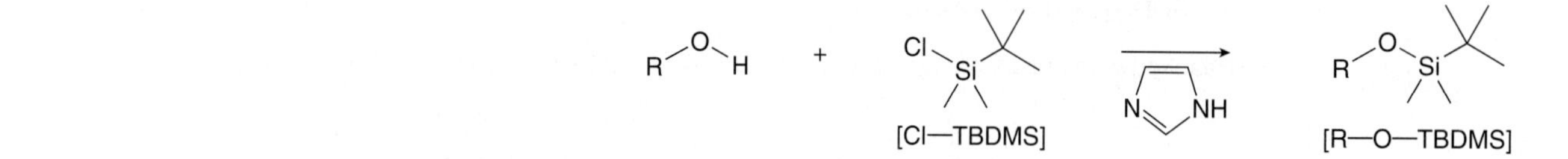
R
O
H
+
Cl
Si
[Cl—TBDMS]
N
NH
R
O
Si
[R—O—TBDMS]
tert-butyldimethylsilyl ether

[2] Deprotecting a *tert*-butyldimethylsilyl ether to re-form an alcohol

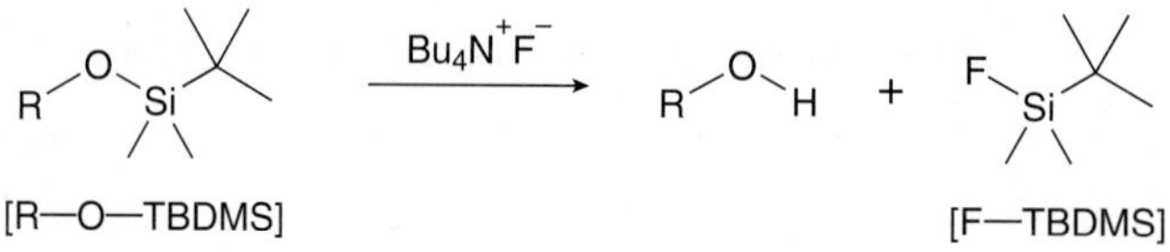
R
O
Si
[R—O—TBDMS]
$Bu_4N^+F^-$
R
O
H
+
F
Si
[F—TBDMS]

PROBLEMS

Problems Using Three-Dimensional Models

20.36 Draw the products formed when **A** or **B** is treated with each reagent. In some cases, no reaction occurs.

A **B**

a. $NaBH_4$, CH_3OH
b. [1] $LiAlH_4$; [2] H_2O
c. [1] CH_3MgBr (excess); [2] H_2O
d. [1] C_6H_5Li (excess); [2] H_2O
e. $Na_2Cr_2O_7$, H_2SO_4, H_2O

20.37 Devise a synthesis of each alcohol from organic alcohols having one or two carbons and any required reagents.

a. 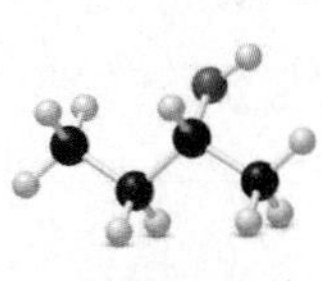 b.  c.  d. 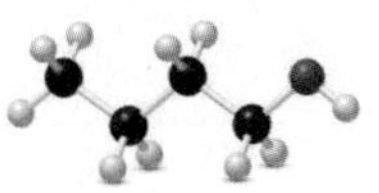

(+ enantiomer)

Reactions and Reagents

20.38 Draw the product formed when pentanal ($CH_3CH_2CH_2CH_2CHO$) is treated with each reagent. With some reagents, no reaction occurs.

a. $NaBH_4$, CH_3OH
b. [1] $LiAlH_4$; [2] H_2O
c. H_2, Pd-C
d. PCC
e. $Na_2Cr_2O_7$, H_2SO_4, H_2O
f. Ag_2O, NH_4OH
g. [1] CH_3MgBr; [2] H_2O
h. [1] C_6H_5Li; [2] H_2O
i. [1] $(CH_3)_2CuLi$; [2] H_2O
j. [1] $HC{\equiv}CNa$; [2] H_2O
k. [1] $CH_3C{\equiv}CLi$; [2] H_2O
l. The product in (a), then TBDMS–Cl, imidazole

20.39 Draw the product formed when $(CH_3CH_2CH_2CH_2)_2CuLi$ is treated with each compound. In some cases, no reaction occurs.

a. 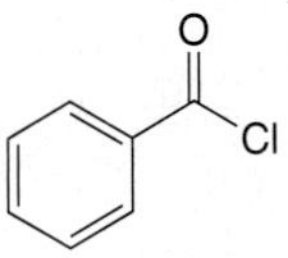

b. 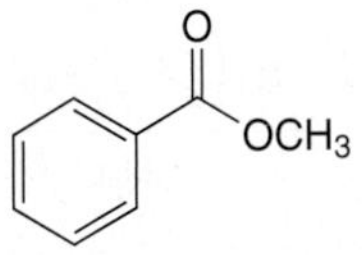

c. 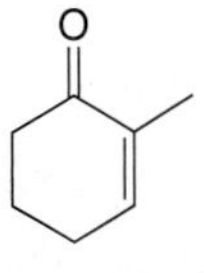

, then H_2O

d. 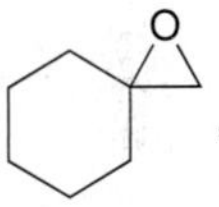

, then H_2O

20.40 The stereochemistry of the products of reduction depends on the reagent used, as you learned in Sections 20.5 and 20.6. With this in mind, how would you convert 3,3-dimethylbutan-2-one [$CH_3COC(CH_3)_3$] to: (a) racemic 3,3-dimethylbutan-2-ol [$CH_3CH(OH)C(CH_3)_3$]; (b) only (*R*)-3,3-dimethylbutan-2-ol; (c) only (*S*)-3,3-dimethylbutan-2-ol?

20.41 Draw the product formed when the α,β-unsaturated ketone **A** is treated with each reagent.

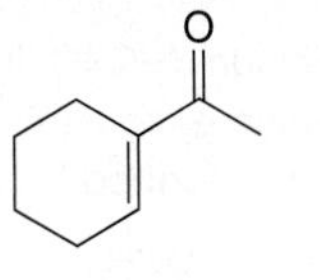

A

a. $NaBH_4$, CH_3OH
b. H_2 (1 equiv), Pd-C
c. H_2 (excess), Pd-C
d. [1] CH_3Li; [2] H_2O
e. [1] CH_3CH_2MgBr; [2] H_2O
f. [1] $(CH_2{=}CH)_2CuLi$; [2] H_2O

20.42 Draw the products of each reduction reaction.

a. 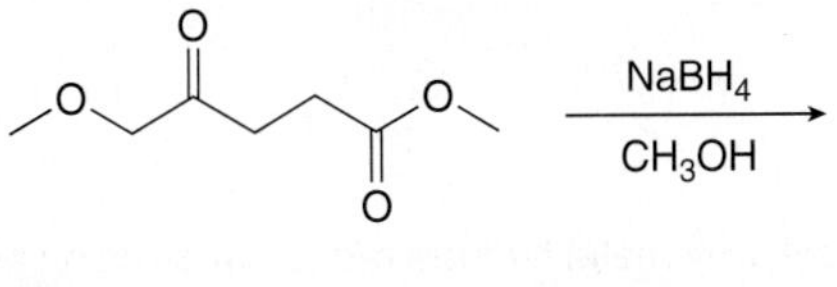

b. 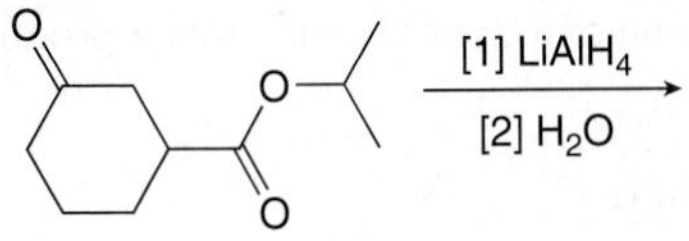

c. 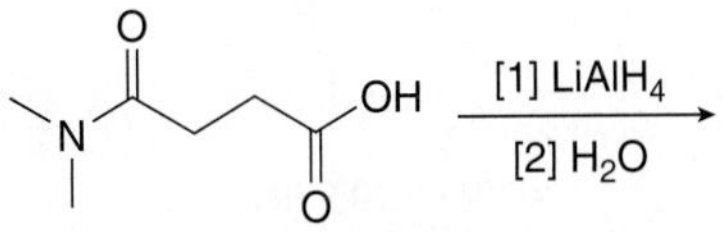

d. 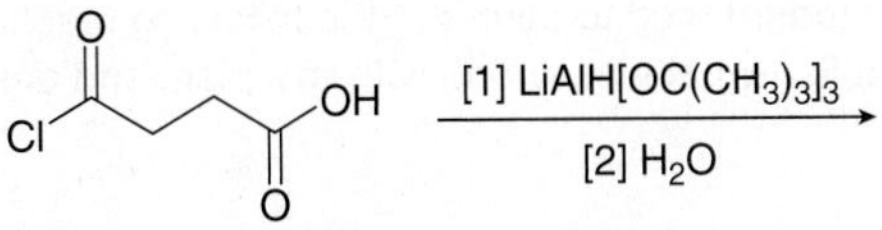

20.43 Draw the products of the following reactions with organometallic reagents.

a. 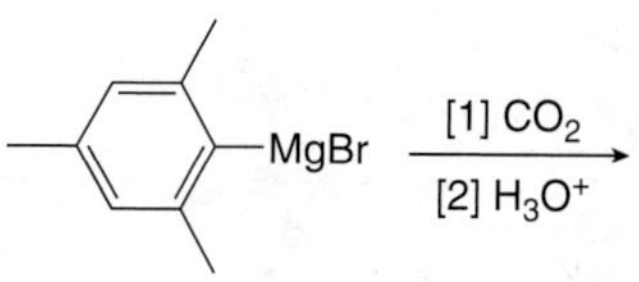

d. 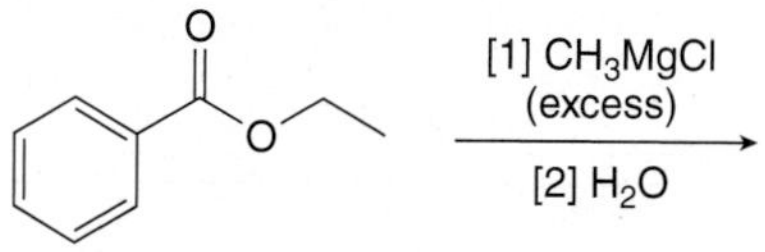

b. 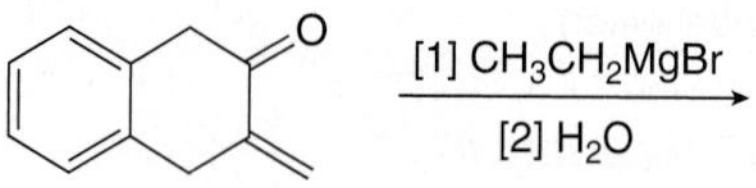

e. 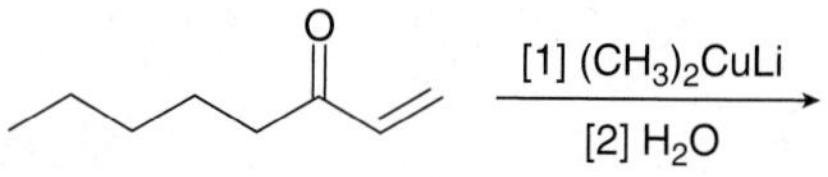

c. 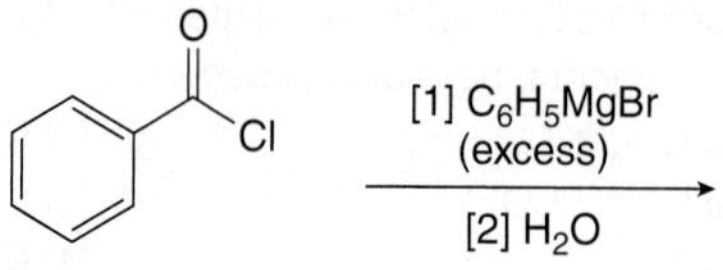

f. 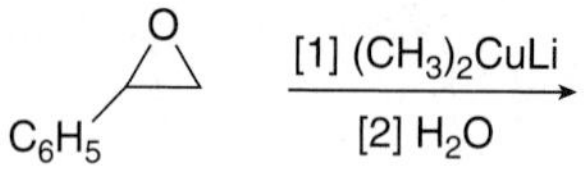

20.44 Draw all stereoisomers formed in each reaction.

a. 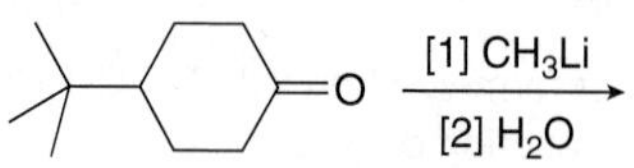

c. 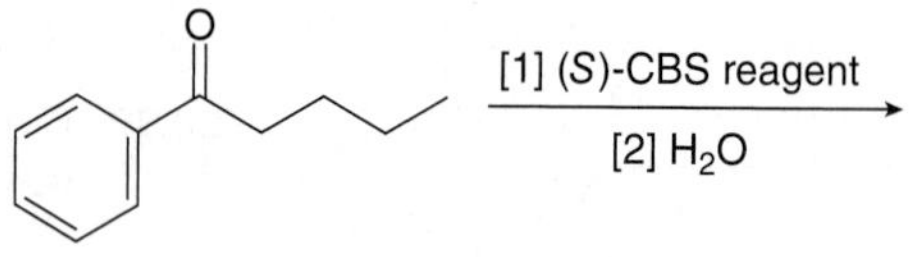

b. 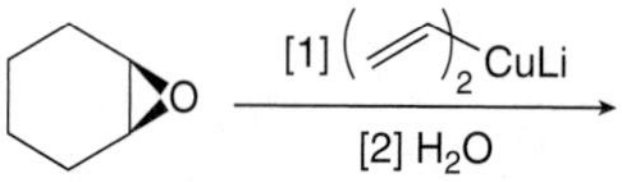

d. 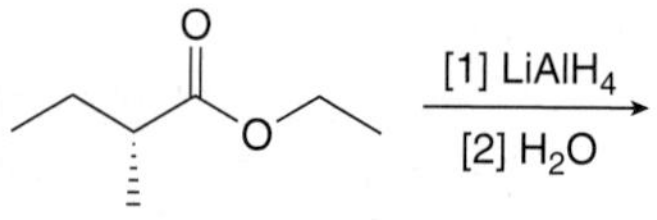

20.45 Which carbon is most electrophilic in **Y?** Explain your choice.

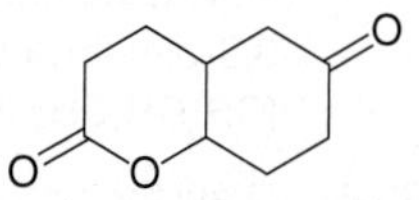

Y

20.46 Treatment of ketone **A** with ethynyllithium (HC≡CLi) followed by D_3O^+ afforded a compound **B** of molecular formula $C_{12}H_{13}DO_3$, which gave an IR absorption at approximately 1715 cm^{-1}. What is the structure of **B** and how is it formed?

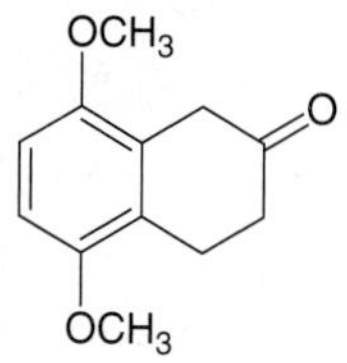

A

20.47 Explain why metal hydride reduction gives an endo alcohol as the major product in one reaction and an exo alcohol as the major product in the other reaction.

endo OH group

exo OH group

20.48 A student tried to carry out the following reaction sequence, but none of diol **A** was formed. Explain what was wrong with this plan, and design a successful stepwise synthesis of **A.**

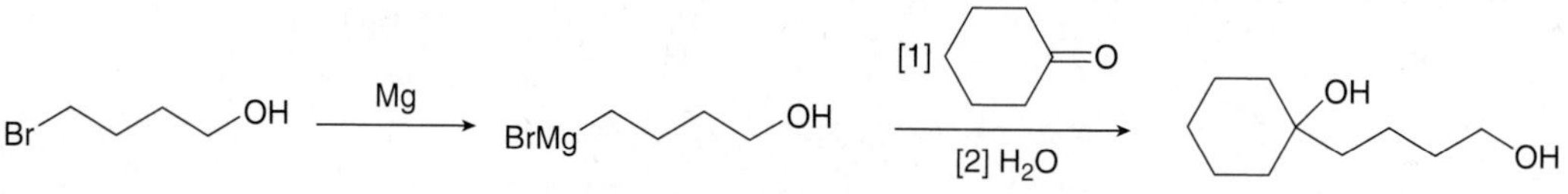

A

20.49 Identify the lettered compounds in the following reaction scheme. Compounds **F, G,** and **K** are isomers of molecular formula $C_{13}H_{18}O$. How could ^1H NMR spectroscopy distinguish these three compounds from each other?

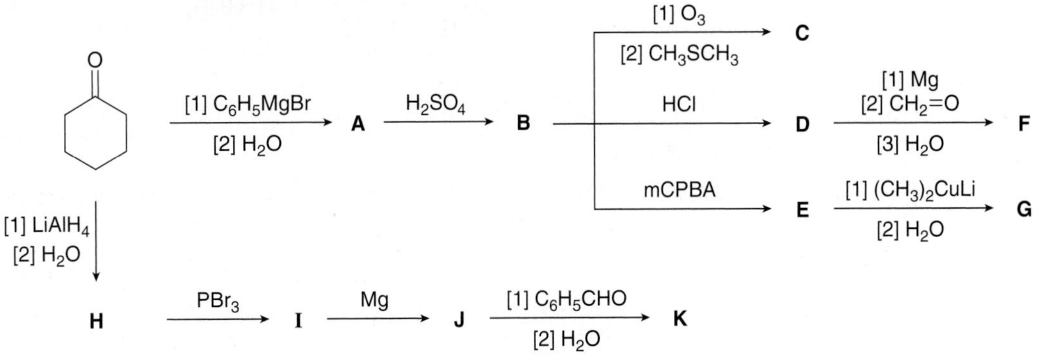
[1] C_6H_5MgBr
[2] H_2O
A
H_2SO_4
B
[1] O_3
[2] CH_3SCH_3
C
HCl
D
[1] Mg
[2] $CH_2{=}O$
[3] H_2O
F
mCPBA
E
[1] $(CH_3)_2CuLi$
[2] H_2O
G
[1] $LiAlH_4$
[2] H_2O
H
PBr_3
I
Mg
J
[1] C_6H_5CHO
[2] H_2O
K

20.50 Fill in the lettered products (**A–D**) in the synthesis of (*R*)-isoproterenol, a drug that increases heart rate and dilates lung passages.

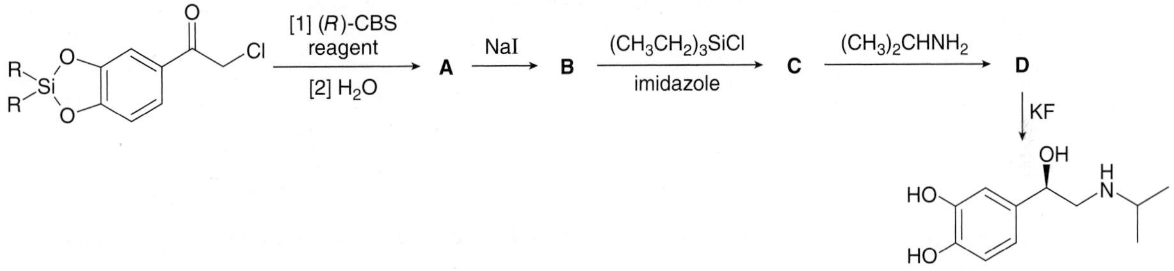
[1] (*R*)-CBS reagent
[2] H_2O
A
NaI
B
$(CH_3CH_2)_3SiCl$
imidazole
C
$(CH_3)_2CHNH_2$
D
KF

(*R*)-isoproterenol
trade name Isuprel

Mechanism

20.51 Draw a stepwise mechanism for the following reaction. Your mechanism must show how both organic products are formed.

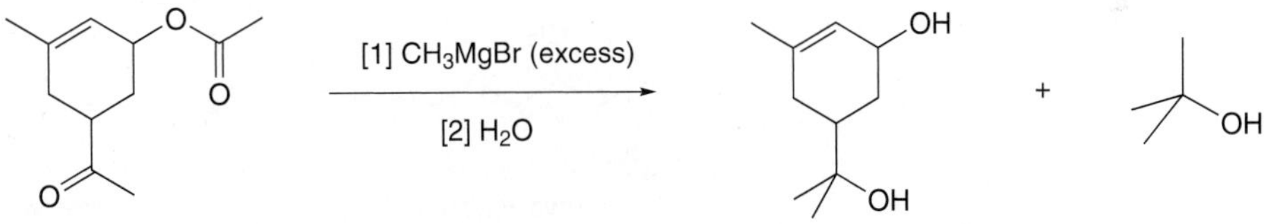
[1] CH_3MgBr (excess)
[2] H_2O

20.52 Draw a stepwise mechanism for the following reaction.

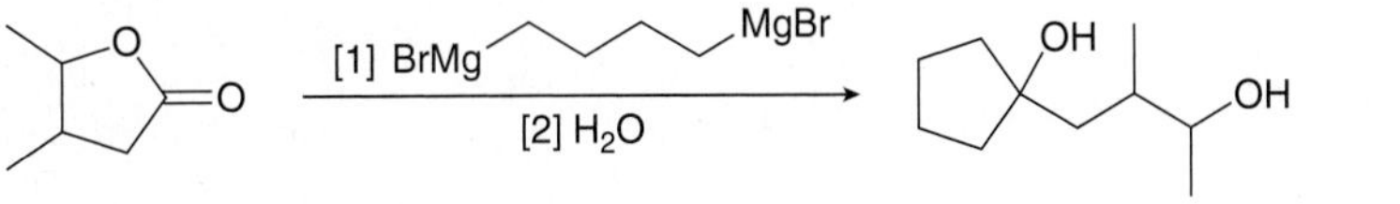
[1] BrMg
MgBr
[2] H_2O

20.53 Tertiary alcohols can be formed by the reaction of dimethyl carbonate $[(CH_3O)_2C{=}O]$ with excess Grignard reagent. Draw a stepwise mechanism for the following transformation.

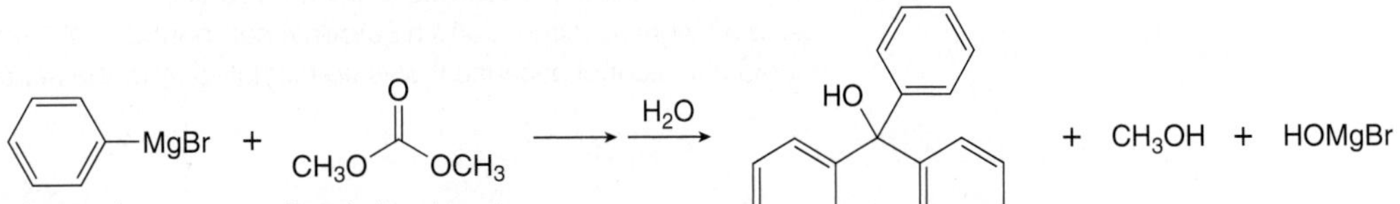
MgBr
+
H_2O
+ CH_3OH + HOMgBr

(excess) dimethyl carbonate

20.54 Draw a stepwise mechanism for the following reduction.

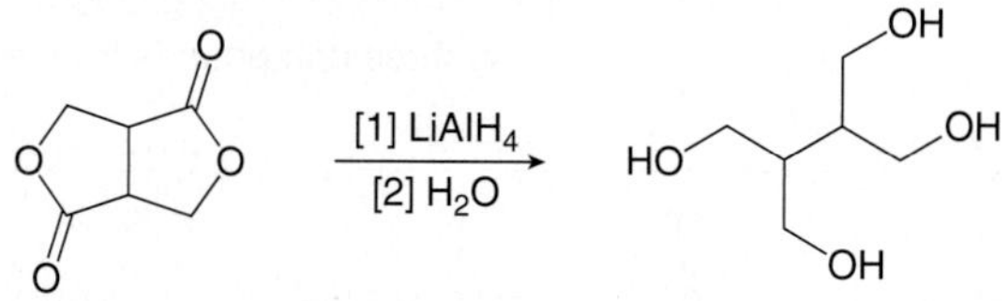

Synthesis

20.55 What Grignard reagent and aldehyde (or ketone) are needed to prepare each alcohol? Show all possible routes.

a. 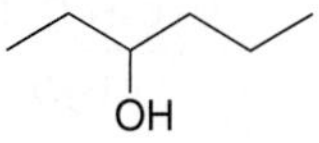

b. 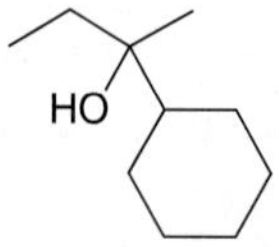

c. 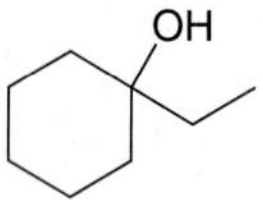

20.56 Procyclidine is a drug that has been used to treat the uncontrolled body movements associated with Parkinson's disease. Draw three different methods to prepare procyclidine using a Grignard reagent.

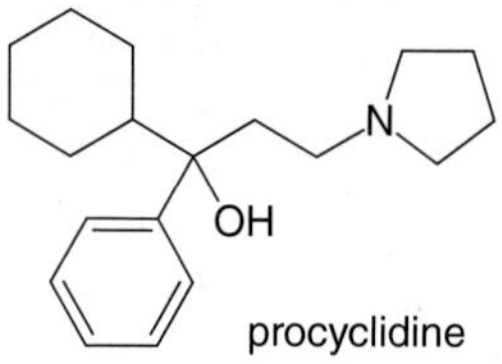

procyclidine

20.57 What ester and Grignard reagent are needed to synthesize each alcohol?

a. 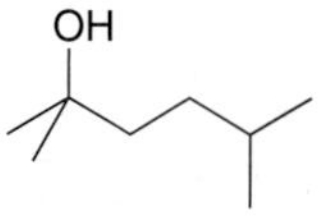

b. 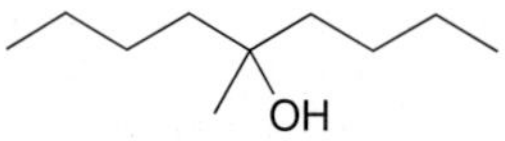

20.58 What organolithium reagent and carbonyl compound can be used to prepare each of the following compounds? You may use aldehydes, ketones, or esters as carbonyl starting materials.

a. 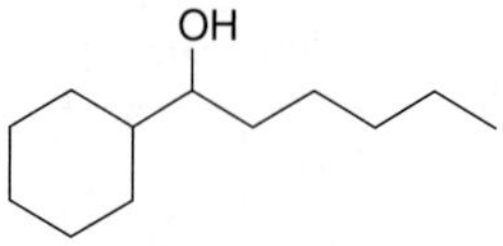

(two ways)

b. 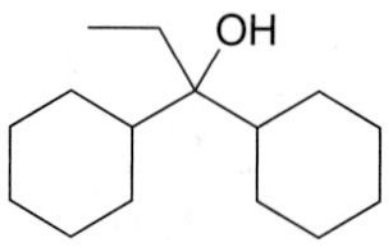

(three ways)

20.59 What epoxide and organometallic reagent are needed to synthesize each alcohol?

a. 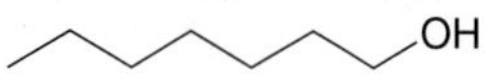

b. 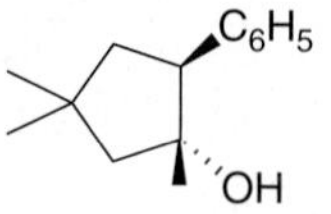

c. 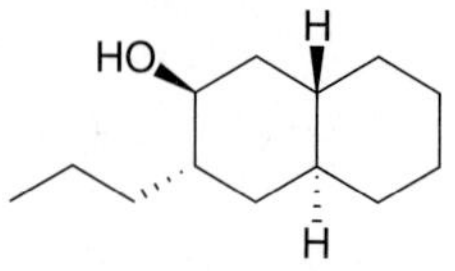

20.60 Propose at least three methods to convert $C_6H_5CH_2CH_2Br$ to $C_6H_5CH_2CH_3$.

20.61 Propose two different methods to synthesize oct-1-en-3-ol [$CH_3(CH_2)_4CH(OH)CH{=}CH_2$] using a Grignard reagent and a carbonyl compound. Oct-1-en-3-ol is commonly called matsutake alcohol because it was first isolated from the Japanese matsutake mushroom.

20.62 Synthesize each compound from cyclohexanol using any other organic or inorganic compounds.

a. 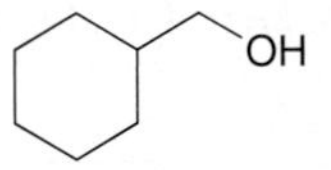 b. 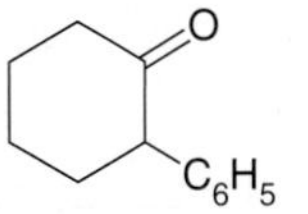 c. 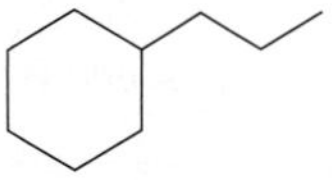 d. 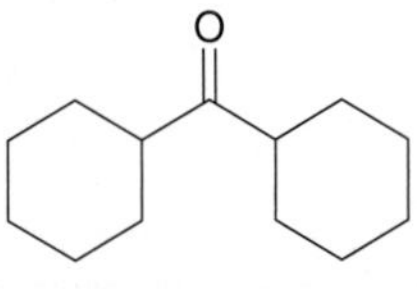

(Each cyclohexane ring must come from cyclohexanol.)

20.63 Convert propan-2-ol [$(CH_3)_2CHOH$] into each compound. You may use any other organic or inorganic compounds.

a. 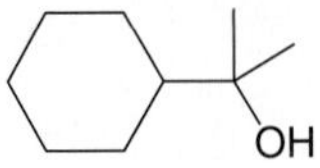 b. 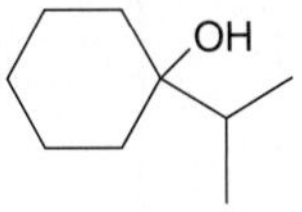

20.64 Convert benzene into each compound. You may also use any inorganic reagents and organic alcohols having three carbons or fewer. One step of the synthesis must use a Grignard reagent.

a. 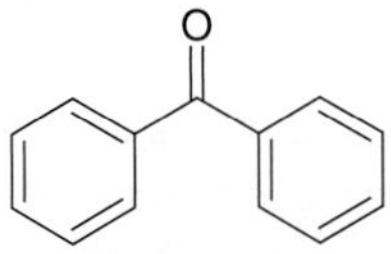 b. 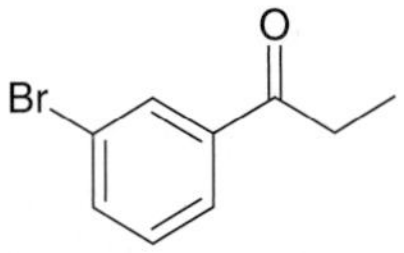

20.65 Design a synthesis of each compound from alcohols having four or fewer carbons as the only organic starting materials. You may use any other inorganic reagents you choose.

a. 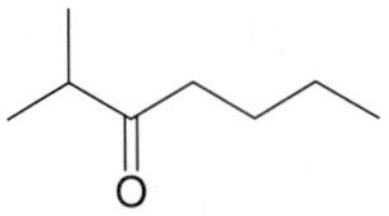 b. 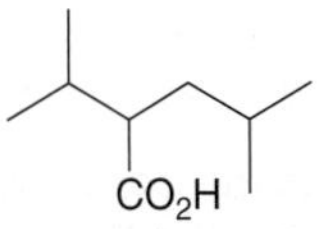 c. 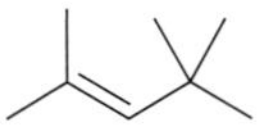

20.66 Synthesize each compound from the given starting material. You may use any other required inorganic reagents.

a. 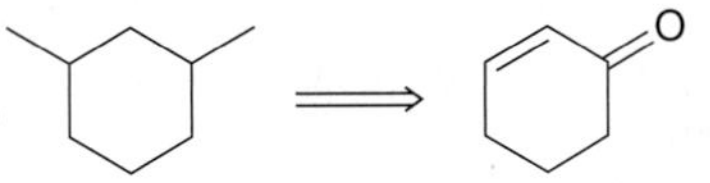

+ organic halides

b. 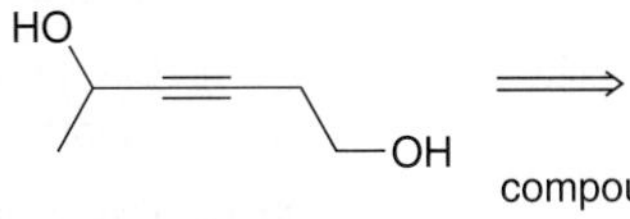 HC≡CH + compounds having ≤ 2 C's

20.67 Devise a synthesis of the given alcohol from benzene, organic alcohols having four or fewer carbons, and any needed inorganic reagents.

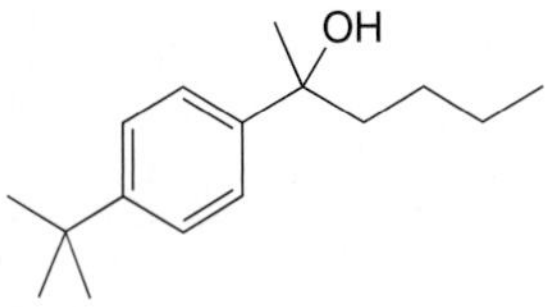

20.68 Devise a synthesis of (*E*)-tetradec-11-enal, a sex pheromone of the spruce budworm, a pest that destroys fir and spruce forests, from acetylene, $Br(CH_2)_{10}OH$, and any needed organic compounds or inorganic reagents.

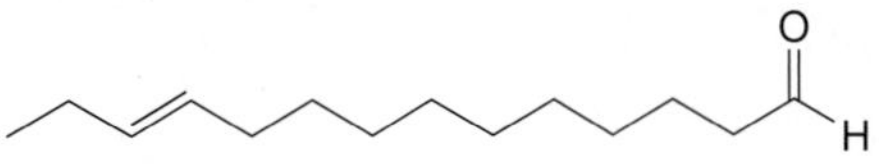

(*E*)-tetradec-11-enal

Spectroscopy

20.69 An unknown compound **A** (molecular formula $C_7H_{14}O$) was treated with $NaBH_4$ in CH_3OH to form compound **B** (molecular formula $C_7H_{16}O$). Compound **A** has a strong absorption in its IR spectrum at 1716 cm^{-1}. Compound **B** has a strong absorption in its IR spectrum at 3600–3200 cm^{-1}. The ^1H NMR spectra of **A** and **B** are given. What are the structures of **A** and **B?**

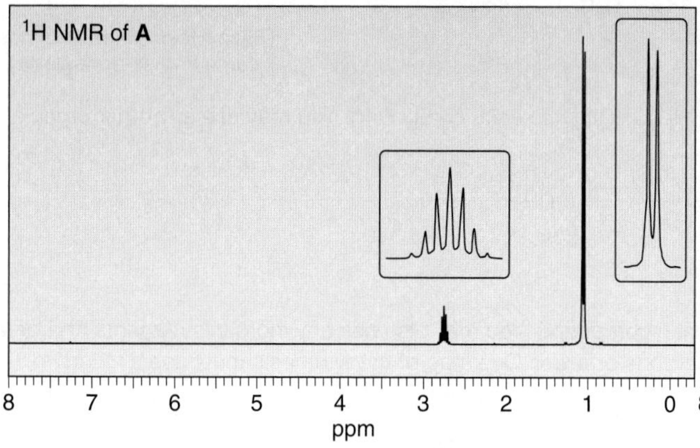

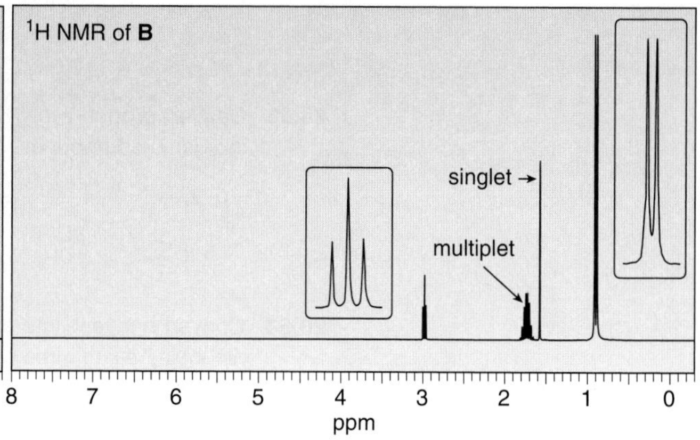

20.70 Treatment of compound **C** (molecular formula C_4H_8O) with C_6H_5MgBr, followed by H_2O, affords compound **D** (molecular formula $C_{10}H_{14}O$). Compound **D** has a strong peak in its IR spectrum at 3600–3200 cm^{-1}. The ^1H NMR spectral data of **C** and **D** are given. What are the structures of **C** and **D?**

Compound **C** signals at 1.3 (singlet, 6 H) and 2.4 (singlet, 2 H) ppm

Compound **D** signals at 1.2 (singlet, 6 H), 1.6 (singlet, 1 H), 2.7 (singlet, 2 H), and 7.2 (multiplet, 5 H) ppm

20.71 Treatment of compound **E** (molecular formula $C_4H_8O_2$) with excess CH_3CH_2MgBr yields compound **F** (molecular formula $C_6H_{14}O$) after protonation with H_2O. **E** shows a strong absorption in its IR spectrum at 1743 cm^{-1}. **F** shows a strong IR absorption at 3600–3200 cm^{-1}. The ^1H NMR spectral data of **E** and **F** are given. What are the structures of **E** and **F?**

Compound **E** signals at 1.2 (triplet, 3 H), 2.0 (singlet, 3 H), and 4.1 (quartet, 2 H) ppm

Compound **F** signals at 0.9 (triplet, 6 H), 1.1 (singlet, 3 H), 1.5 (quartet, 4 H), and 1.55 (singlet, 1 H) ppm

20.72 Reaction of butanenitrile ($CH_3CH_2CH_2CN$) with methylmagnesium bromide (CH_3MgBr), followed by treatment with aqueous acid, forms compound **G. G** has a molecular ion in its mass spectrum at m/z = 86 and a base peak at m/z = 43. **G** exhibits a strong absorption in its IR spectrum at 1721 cm^{-1} and has the ^1H NMR spectrum given below. What is the structure of **G?** We will learn about the details of this reaction in Chapter 22.

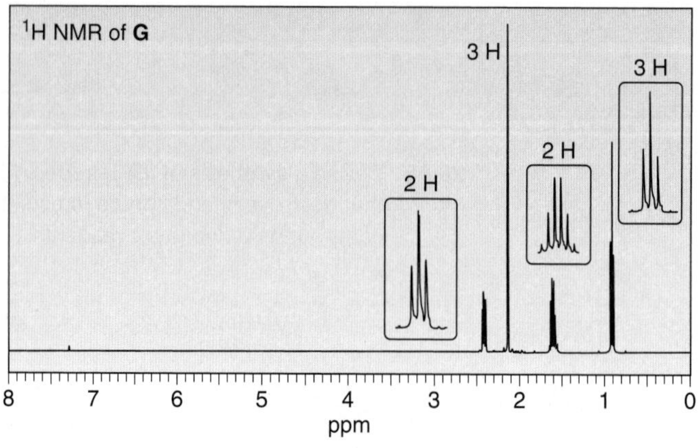

20.73 Treatment of isobutene [$(CH_3)_2C{=}CH_2$] with $(CH_3)_3CLi$ forms a carbanion that reacts with $CH_2{=}O$ to form **H** after water is added to the reaction mixture. **H** has a molecular ion in its mass spectrum at m/z = 86, and shows fragments at 71 and 68. **H** exhibits absorptions in its IR spectrum at 3600–3200 and 1651 cm^{-1}, and has the ^1H NMR spectrum given below. What is the structure of **H?**

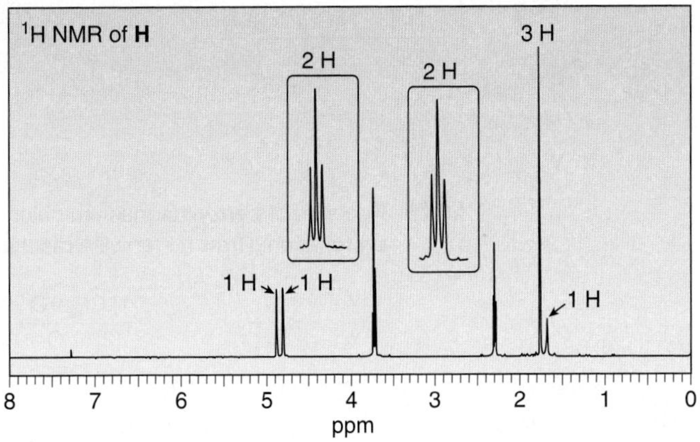

Challenge Problems

20.74 Draw a stepwise mechanism for the following reaction.

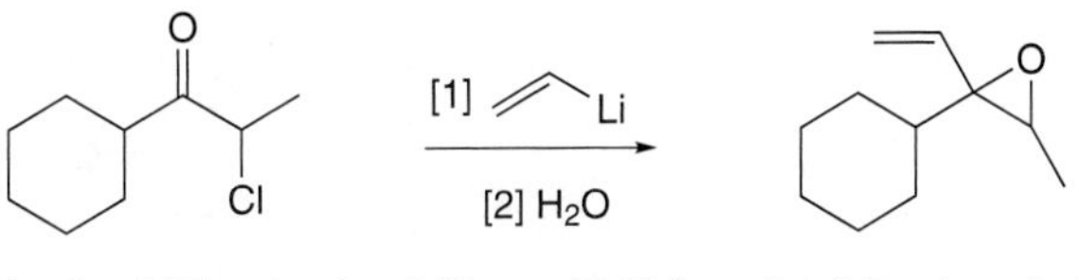

20.75 Design a synthesis of (R)-salmeterol (Figure 20.3) from the following starting materials.

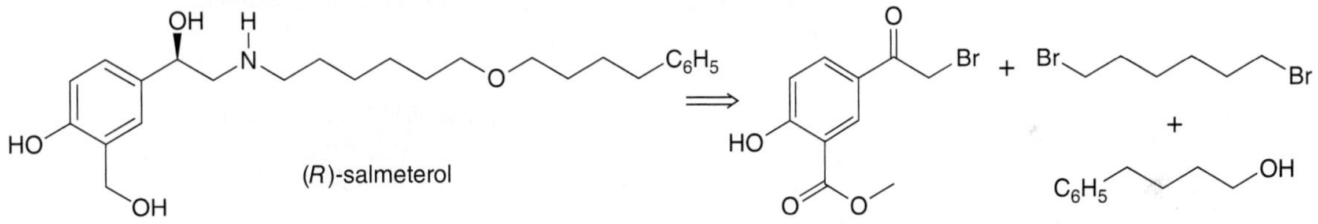

20.76 Lithium tri-*sec*-butylborohydride, also known as L-selectride, is a metal hydride reagent that contains three *sec*-butyl groups bonded to boron. When this reagent is used to reduce cyclic ketones, one stereoisomer often predominates as product. Explain why the reduction of 4-*tert*-butylcyclohexanone with L-selectride forms the cis alcohol as the major product.

$LiBH[CH(CH_3)CH_2CH_3]_3$

lithium tri-*sec*-butylborohydride
L-selectride

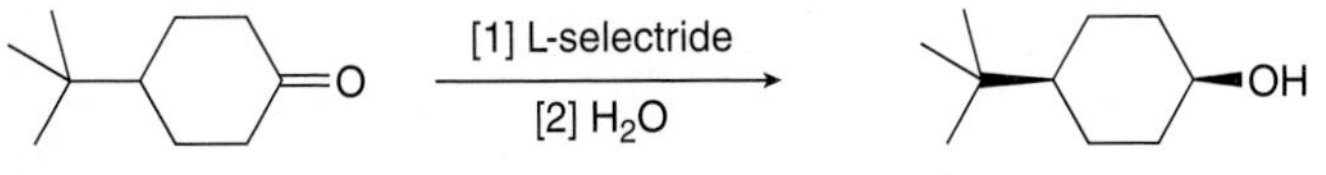

4-*tert*-butylcyclohexanone

cis-4-*tert*-butylcyclohexanol

20.77 Explain why the β carbon of an α,β-unsaturated carbonyl compound absorbs farther downfield in the ^{13}C NMR spectrum than the α carbon, even though the α carbon is closer to the electron-withdrawing carbonyl group. For example, the β carbon of mesityl oxide absorbs at 150.5 ppm, while the α carbon absorbs at 122.5 ppm.

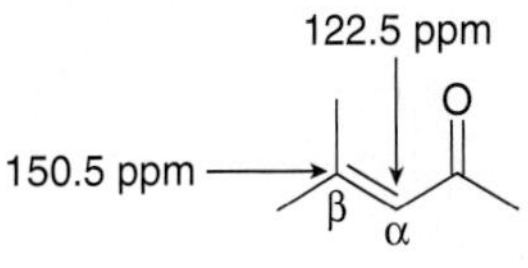

mesityl oxide

20.78 Identify **X** and **Y,** two of the intermediates in a synthesis of the antidepressant venlafaxine (trade name Effexor), in the following reaction scheme. Write a mechanism for the formation of **X** from **W.**

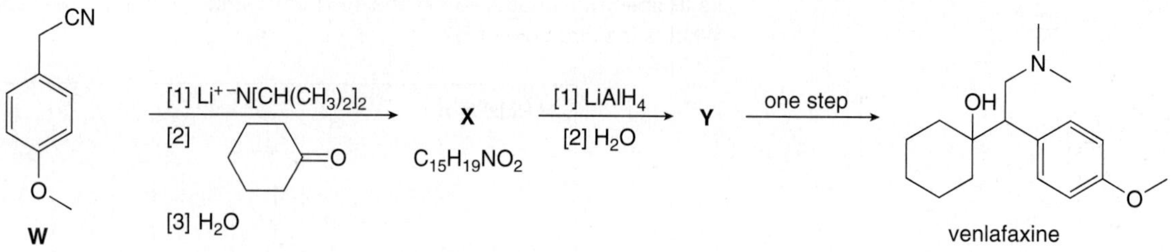

20.79 Reaction of benzylmagnesium chloride with formaldehyde yields alcohols **N** and **P** after protonation. Draw a stepwise mechanism that shows how both products are formed.

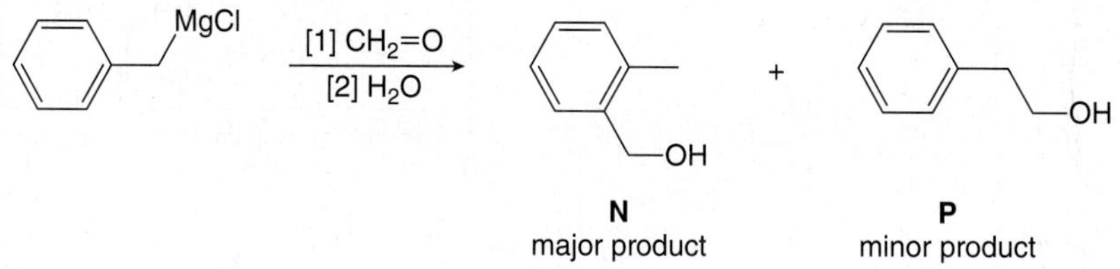

20.80 Draw a stepwise mechanism for the following reaction. (Hint: Conjugate addition can occur with heteroatoms as well as carbon nucleophiles.)

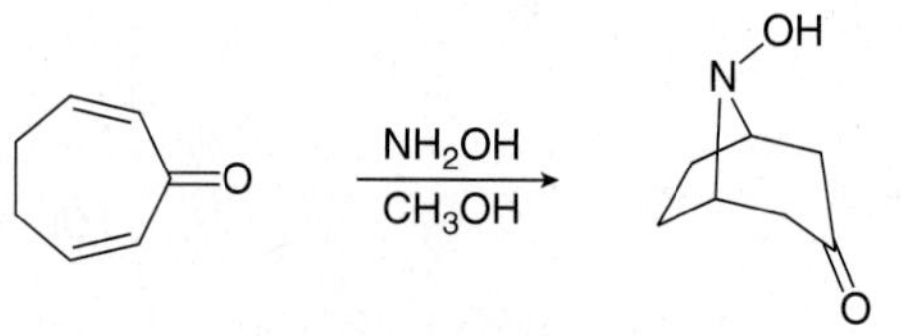

20.81 Draw a stepwise mechanism for the following reaction of a Grignard reagent with a cyclic amide.

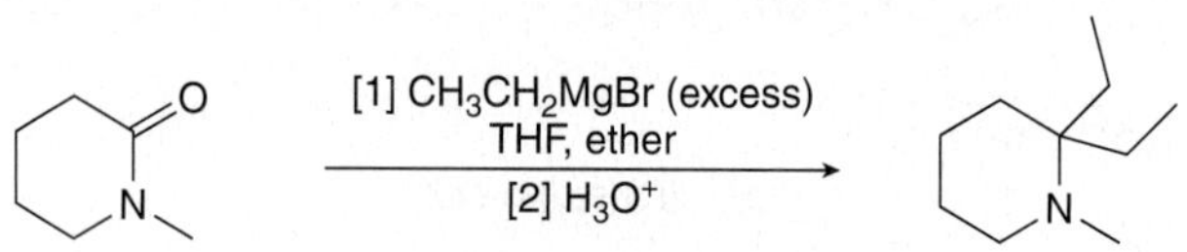

Aldehydes and Ketones—Nucleophilic Addition

21

21.1 Introduction
21.2 Nomenclature
21.3 Physical properties
21.4 Spectroscopic properties
21.5 Interesting aldehydes and ketones
21.6 Preparation of aldehydes and ketones
21.7 Reactions of aldehydes and ketones—General considerations
21.8 Nucleophilic addition of H^- and R^-—A review
21.9 Nucleophilic addition of ^-CN
21.10 The Wittig reaction
21.11 Addition of 1° amines
21.12 Addition of 2° amines
21.13 Addition of H_2O—Hydration
21.14 Addition of alcohols—Acetal formation
21.15 Acetals as protecting groups
21.16 Cyclic hemiacetals
21.17 An introduction to carbohydrates

The natural product **digoxin** has been prescribed since the 1960s for patients with congestive heart failure, a condition that results when fluid builds up in the body because the heart's pumping action is weak. Unlike many commercial medications that are synthesized from simple precursors, digoxin is still obtained by extraction of the leaves of the woolly foxglove plant, which is grown in the Netherlands and shipped to the United States for processing. One thousand kilograms of dried leaves yield one kilogram of digoxin, sold under the trade name of Lanoxin. Digoxin contains three acetal units, which are formed by addition reactions to carbonyl groups. In Chapter 21, we learn about nucleophilic addition, the characteristic reaction of aldehydes and ketones.

In Chapter 21 we continue the study of carbonyl compounds with a detailed look at **aldehydes** and **ketones.** We will first learn about the nomenclature, physical properties, and spectroscopic absorptions that characterize aldehydes and ketones. The remainder of Chapter 21 is devoted to **nucleophilic addition** reactions. Although we have already learned two examples of this reaction in Chapter 20, nucleophilic addition to aldehydes and ketones is a general reaction that occurs with many nucleophiles, forming a wide variety of products.

Every new reaction in Chapter 21 involves nucleophilic addition, so the challenge lies in learning the specific reagents and mechanisms that characterize each reaction.

21.1 Introduction

An aldehyde is often written as **RCHO.** Remember that the **H atom is bonded to the carbon atom,** *not* the oxygen. Likewise, a ketone is written as **RCOR,** or if both alkyl groups are the same, **R_2CO.** Each structure must contain a C=O for every atom to have an octet.

As we learned in Chapter 20, **aldehydes and ketones contain a carbonyl group.** An aldehyde contains at least one H atom bonded to the carbonyl carbon, whereas a ketone has two alkyl or aryl groups bonded to it.

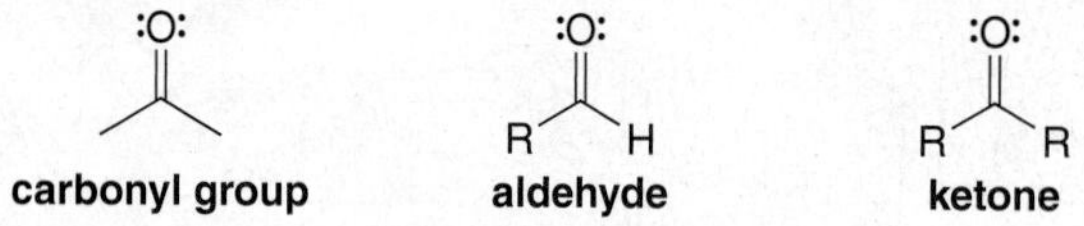

Two structural features determine the chemistry and properties of aldehydes and ketones.

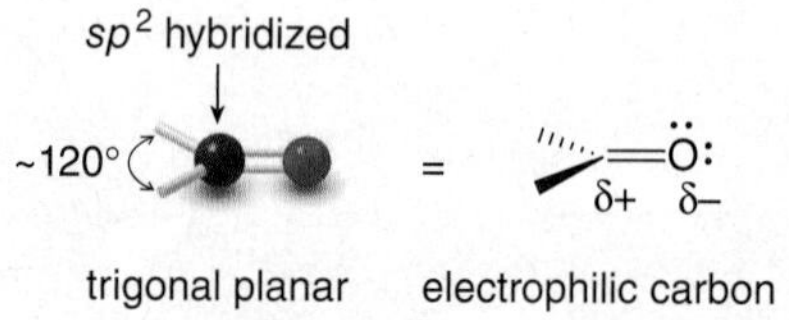

- **The carbonyl group is *sp*^{2} hybridized and trigonal planar, making it relatively uncrowded.**
- **The electronegative oxygen atom polarizes the carbonyl group, making the carbonyl carbon electrophilic.**

As a result, **aldehydes and ketones react with nucleophiles.** The relative reactivity of the carbonyl group is determined by the number of R groups bonded to it. **As the number of R groups around the carbonyl carbon increases, the reactivity of the carbonyl compound decreases,** resulting in the following order of reactivity:

Increasing the number of alkyl groups on the carbonyl carbon decreases reactivity for both steric and electronic reasons, as discussed in Section 20.2B.

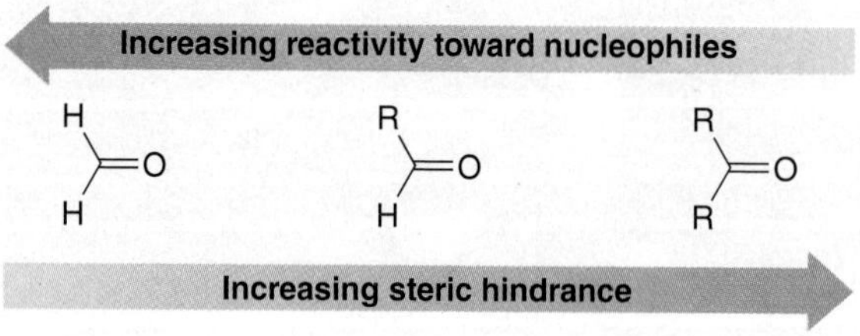

Problem 21.1 Rank the following compounds in order of increasing reactivity toward nucleophilic attack.

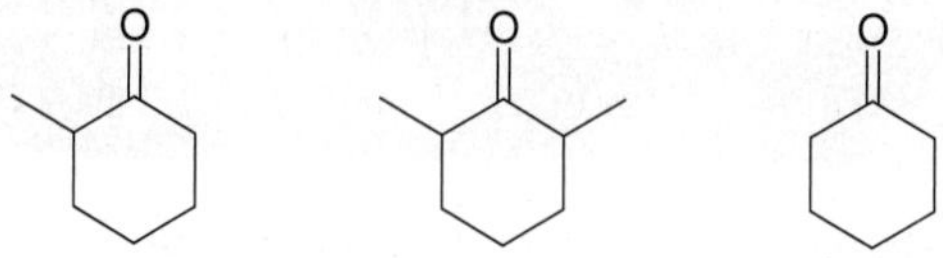

Problem 21.2 Explain why benzaldehyde is less reactive than cyclohexanecarbaldehyde toward nucleophilic attack.

21.2 Nomenclature

Both IUPAC and common names are used for aldehydes and ketones.

21.2A Naming Aldehydes in the IUPAC System

In IUPAC nomenclature, aldehydes are identified by a suffix added to the parent name of the longest chain. Two different suffixes are used, depending on whether the CHO group is bonded to a chain or a ring.

To name an aldehyde using the IUPAC system:

[1] If the CHO is bonded to a chain of carbons, find the longest chain containing the CHO group, and change the ***-e*** ending of the parent alkane to the suffix ***-al.*** If the CHO group is bonded to a ring, name the ring and add the suffix ***-carbaldehyde.***

[2] Number the chain or ring to put the CHO group at C1, but omit this number from the name. Apply all of the other usual rules of nomenclature.

Sample Problem 21.1 Give the IUPAC name for each compound.

a. 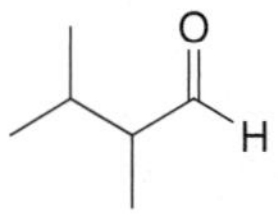

b. 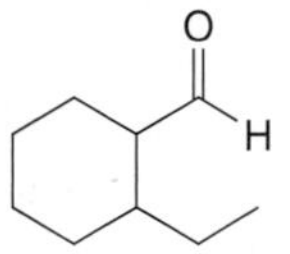

Solution

a. [1] Find and name the longest chain containing the CHO:

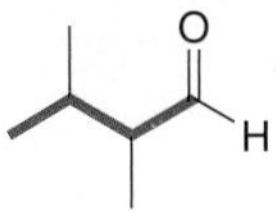

butane ⟶ butan*al*
(4 C's)

[2] Number and name substituents:

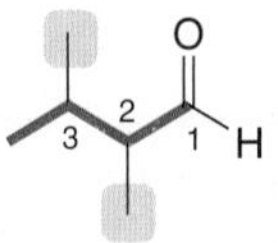

Answer: 2,3-dimethylbutanal

b. [1] Find and name the ring bonded to the CHO group:

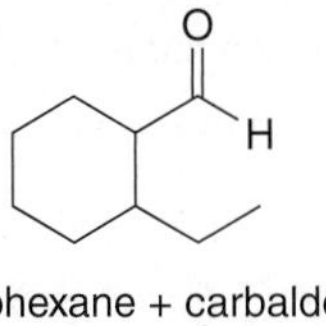

cyclohexane + carbaldehyde
(6 C's)

[2] Number and name substituents:

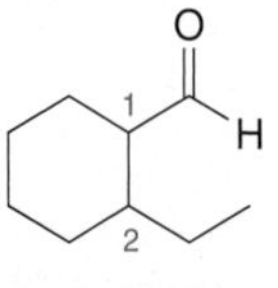

Answer: 2-ethylcyclohexanecarbaldehyde

Problem 21.3 Give the IUPAC name for each aldehyde.

a. 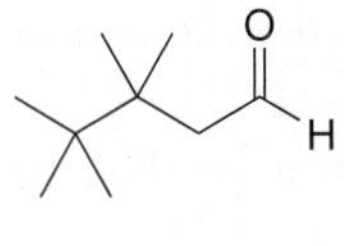

b. 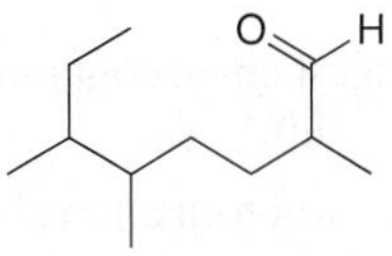

c. 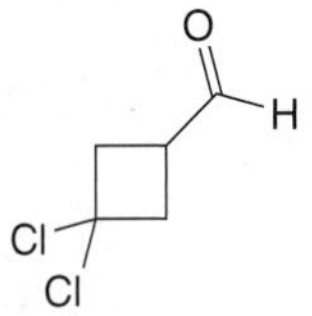

Problem 21.4 Give the structure corresponding to each IUPAC name.

a. 2-isobutyl-3-isopropylhexanal
b. *trans*-3-methylcyclopentanecarbaldehyde
c. 1-methylcyclopropanecarbaldehyde
d. 3,6-diethylnonanal

21.2B Common Names for Aldehydes

Like carboxylic acids, many simple aldehydes have common names that are widely used.

- **A common name for an aldehyde is formed by taking the common parent name and adding the suffix *-aldehyde.***

The common parent names are similar to those used for carboxylic acids, listed in Table 19.1. The common names **formaldehyde, acetaldehyde,** and **benzaldehyde** are virtually always used instead of their IUPAC names.

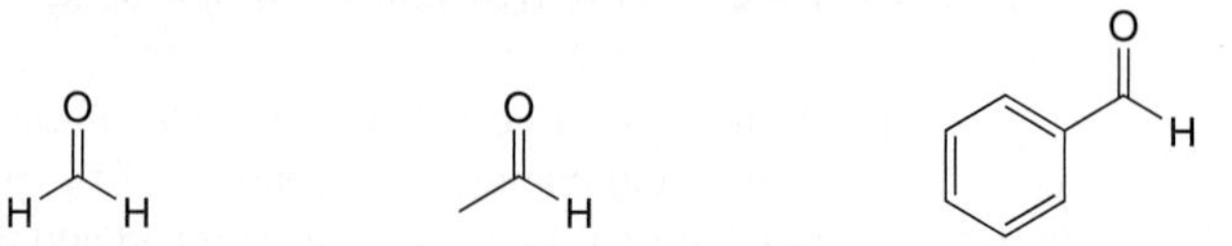

formaldehyde (methanal) — acetaldehyde (ethanal) — benzaldehyde (benzenecarbaldehyde)

(IUPAC names are in parentheses.)

Greek letters are used to designate the location of substituents in common names. **The carbon adjacent to the CHO group is the α carbon,** and so forth down the chain.

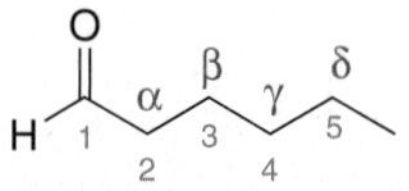

IUPAC numbering begins at the C=O.
Greek lettering begins at the C bonded to the C=O.

Figure 21.1 gives the common and IUPAC names for three aldehydes.

Figure 21.1 Three examples of aldehyde nomenclature

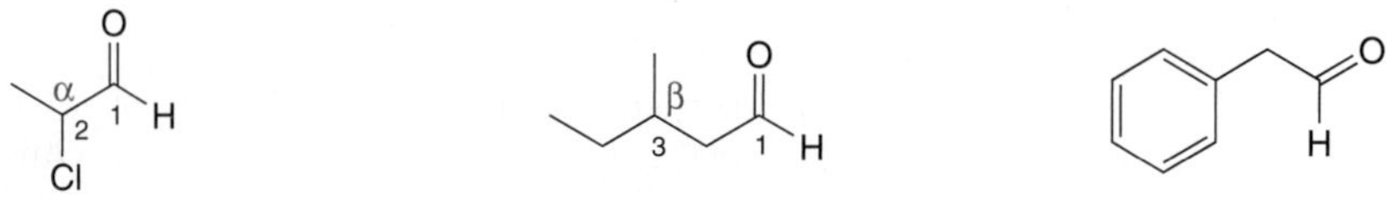

2-chloropropanal (α-chloropropionaldehyde) — 3-methylpentanal (β-methylvaleraldehyde) — phenylethanal (phenylacetaldehyde)

(Common names are in parentheses.)

21.2C Naming Ketones in the IUPAC System

- **In the IUPAC system all ketones are identified by the suffix *-one.***

To name an acyclic ketone using IUPAC rules:

[1] Find the longest chain containing the carbonyl group, and change the ***-e*** ending of the parent alkane to the suffix ***-one.***

[2] Number the carbon chain to give the carbonyl carbon the lower number. Apply all of the other usual rules of nomenclature.

With cyclic ketones, numbering always begins at the carbonyl carbon, but the "1" is usually omitted from the name. The ring is then numbered clockwise or counterclockwise to give the first substituent the lower number.

Sample Problem 21.2 Give the IUPAC name for each ketone.

a. 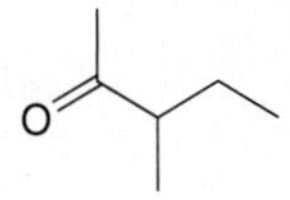 b. 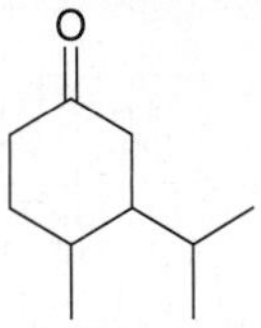

Solution

a. [1] Find and name the longest chain containing the carbonyl group:

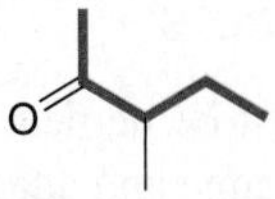

pentane ⟶ pentan*one*
(5 C's)

[2] Number and name substituents:

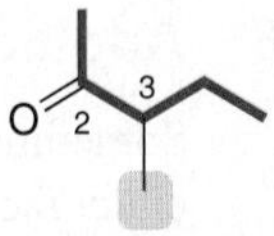

Answer: 3-methylpentan-2-one

b. [1] Name the ring:

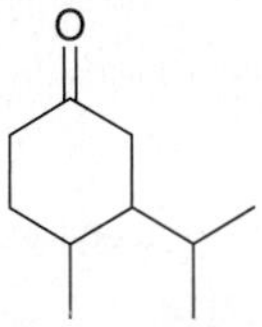

cyclohexane ⟶ cyclohexan*one*
(6 C's)

[2] Number and name substituents:

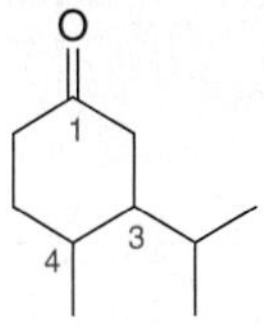

Answer:
3-isopropyl-4-methylcyclohexanone

Problem 21.5 Give the IUPAC name for each ketone.

a. 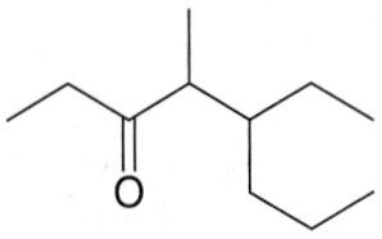 b. 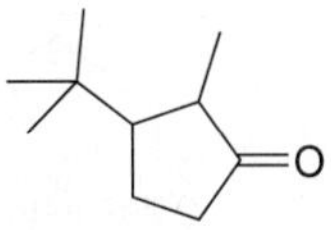 c. 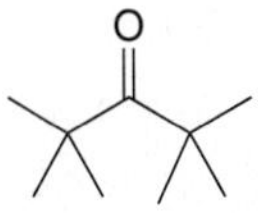

21.2D Common Names for Ketones

Most common names for ketones are formed by **naming both alkyl groups** on the carbonyl carbon, **arranging them alphabetically,** and adding the word ***ketone.*** Using this method, the common name for butan-2-one becomes ethyl methyl ketone.

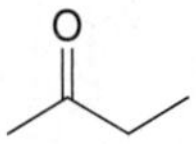

IUPAC name: **butan-2-one**

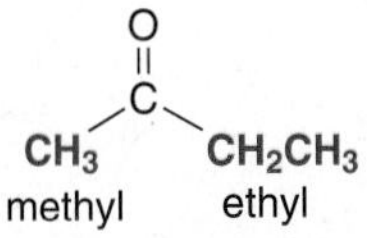

Common name: **ethyl methyl ketone**

Three widely used common names for some simple ketones do not follow this convention:

acetone

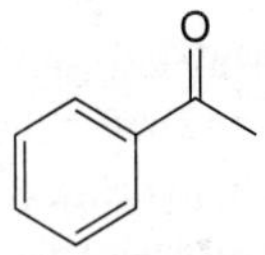

acetophenone

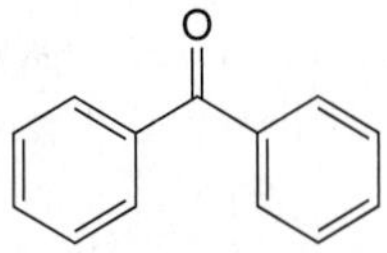

benzophenone

Figure 21.2 gives acceptable names for two ketones.

Figure 21.2 Two examples of ketone nomenclature

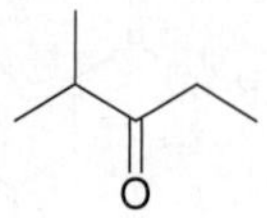

IUPAC name: 2-methylpentan-3-one
Common name: ethyl isopropyl ketone

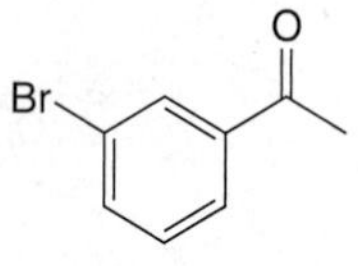

m-bromoacetophenone
or
3-bromoacetophenone

21.2E Additional Nomenclature Facts

Do not confuse a **benzyl** group with a **benzoyl** group.

Sometimes acyl groups (RCO–) must be named as substituents. To name an acyl group, take either the IUPAC or common parent name and add the suffix ***-yl*** or ***-oyl.*** The three most common acyl groups are drawn below.

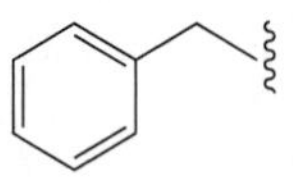

benz*yl* group

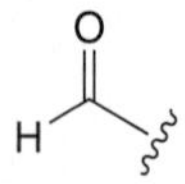

formyl group

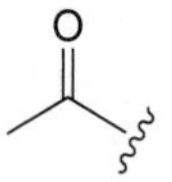

acetyl group

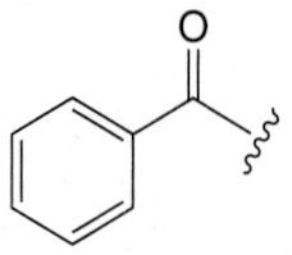

benzoyl group

Compounds containing both a C–C double bond and an aldehyde are named as **enals,** and compounds that contain both a C–C double bond and a ketone are named as **enones.** The chain is numbered to give the carbonyl group the lower number.

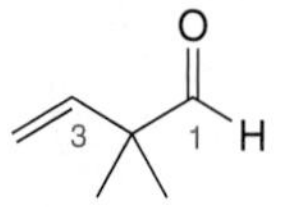

2,2-dimethylbut-3-enal

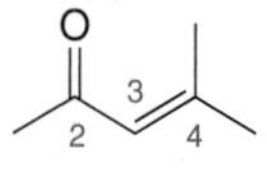

4-methylpent-3-en-2-one

Problem 21.6 Give the structure corresponding to each name: (a) *sec*-butyl ethyl ketone; (b) methyl vinyl ketone; (c) *p*-ethylacetophenone; (d) 3-benzoyl-2-benzylcyclopentanone; (e) 6,6-dimethylcyclohex-2-enone; (f) 3-ethylhex-5-enal.

Problem 21.7 Give the IUPAC name (including any *E,Z* designation) for each unsaturated aldehyde. Neral is obtained from lemongrass, and cucumber aldehyde (Problem 1.30) contributes to the aroma of a fresh mango.

a. 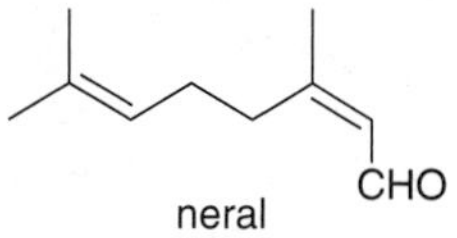

neral

b. 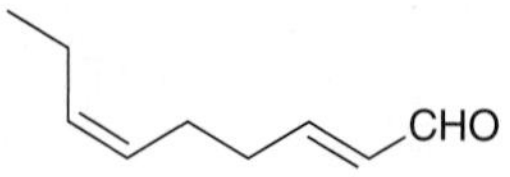

cucumber aldehyde

21.3 Physical Properties

Aldehydes and ketones exhibit dipole–dipole interactions because of their polar carbonyl group. Because they have no O–H bond, two molecules of RCHO or RCOR are incapable of intermolecular hydrogen bonding, making them less polar than alcohols and carboxylic acids. How these intermolecular forces affect the physical properties of aldehydes and ketones is summarized in Table 21.1.

Problem 21.8 The boiling point of butan-2-one (80 °C) is significantly higher than the boiling point of diethyl ether (35 °C), even though both compounds exhibit dipole–dipole interactions and have comparable molecular weights. Offer an explanation.

Table 21.1 Physical Properties of Aldehydes and Ketones

Property	Observation
Boiling point and melting point	• For compounds of comparable molecular weight, bp's and mp's follow the usual trend: The stronger the intermolecular forces, the higher the bp or mp.  VDW, MW = 72, bp 36 °C 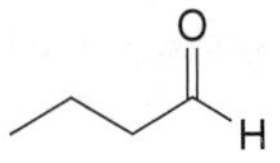 VDW, DD MW = 72, bp 76 °C 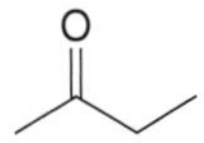 VDW, DD MW = 72, bp 80 °C  VDW, DD, HB, MW = 74, bp 118 °C **Increasing strength of intermolecular forces** **Increasing boiling point**
Solubility	• RCHO and RCOR are soluble in organic solvents regardless of size. • RCHO and RCOR having ≤ 5 C's are H_2O soluble because they can hydrogen bond with H_2O (Section 3.4C). • RCHO and RCOR having > 5 C's are H_2O insoluble because the nonpolar alkyl portion is too large to dissolve in the polar H_2O solvent.

Key: VDW = van der Waals, DD = dipole–dipole, HB = hydrogen bonding, MW = molecular weight

21.4 Spectroscopic Properties

The presence of the carbonyl group in aldehydes and ketones gives them characteristic absorptions in their IR and NMR spectra.

21.4A IR Spectra

Aldehydes and ketones exhibit the following characteristic IR absorptions:

- Like all carbonyl compounds, **aldehydes and ketones give a strong peak at ~1700 cm^{-1} due to the C=O.**
- **The *sp*^{2} hybridized C–H bond of an aldehyde shows one or two peaks at ~2700–2830 cm^{-1}.**

The IR spectrum of propanal in Figure 21.3 illustrates these characteristic peaks.

Figure 21.3
The IR spectrum of propanal, CH_3CH_2CHO

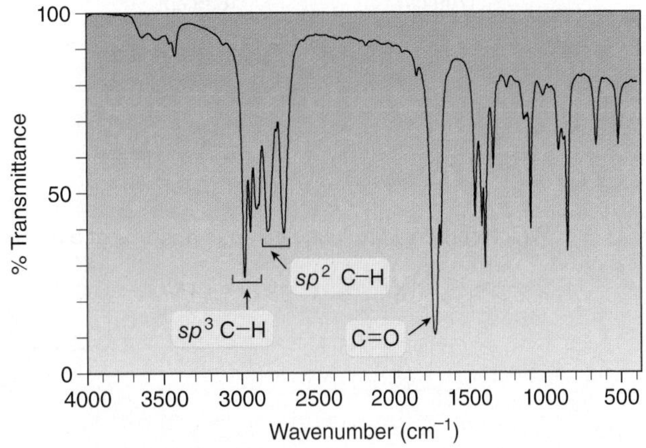

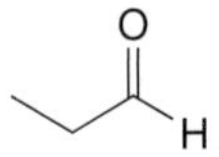
propanal

- A strong C=O occurs at 1739 cm^{-1}.
- The *sp*^{2} C–H of the CHO appears as two peaks at 2813 and 2716 cm^{-1}.

The exact position of the carbonyl absorption often provides additional information about a compound. For example, most aldehydes have a C=O peak around **1730 cm^{-1}**, whereas for ketones, it is typically around **1715 cm^{-1}**. Two other structural features—**ring size** (for cyclic ketones) and **conjugation**—affect the location of the carbonyl absorption in a predictable manner.

[1] The carbonyl absorption of cyclic ketones shifts to higher wavenumber as the size of the ring decreases and the ring strain increases.

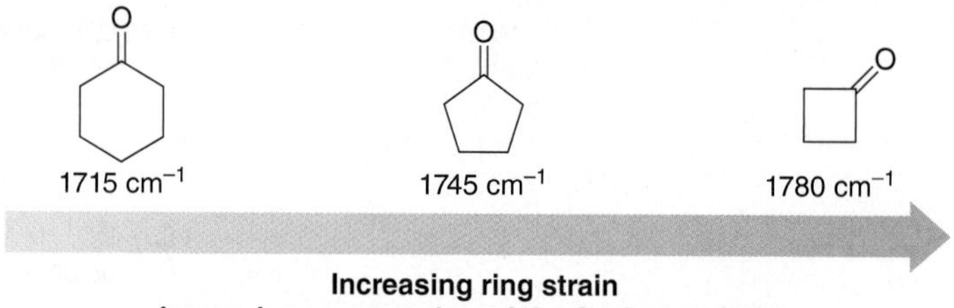
1715 cm⁻¹
1745 cm⁻¹
1780 cm⁻¹
Increasing ring strain
Increasing wavenumber of the C=O absorption

[2] Conjugation of the carbonyl group with a C=C or a benzene ring shifts the absorption to lower wavenumber by ~30 cm^{-1}.

The effect of conjugation on the frequency of the C=O absorption is explained by **resonance.** An α,β-unsaturated carbonyl compound can be written as three resonance structures, two of which place a single bond between the carbon and oxygen atoms of the carbonyl group. Thus, the π bond of the carbonyl group is delocalized, giving the conjugated carbonyl group some single bond character, and making it somewhat **weaker** than an unconjugated C=O. **Weaker bonds absorb at lower frequency (lower wavenumber) in an IR spectrum.**

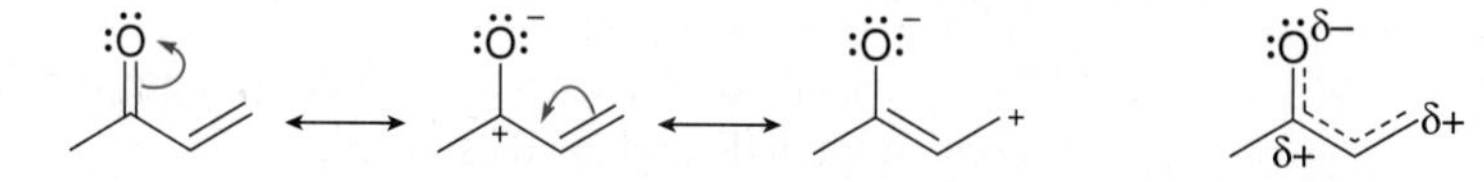
α,β-unsaturated carbonyl group
Two resonance contributors have **a C–O single bond.**
δ–
δ+
δ+
hybrid
The π bonds are delocalized.

Figure 21.4 illustrates the effects of conjugation on the location of the carbonyl absorption in some representative compounds.

Figure 21.4 The effect of conjugation on the carbonyl absorption in an IR spectrum

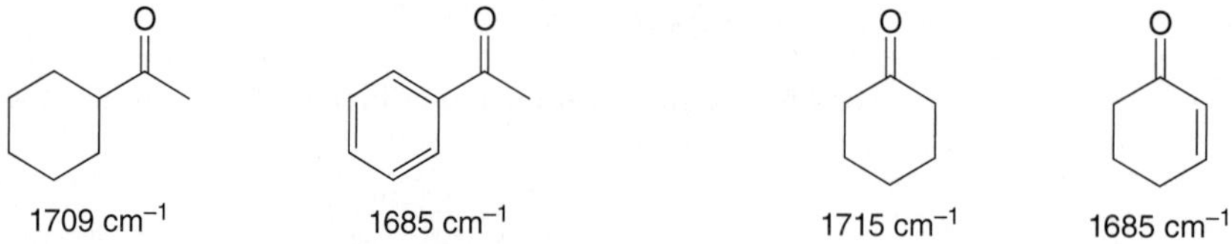
1709 cm⁻¹
1685 cm⁻¹
conjugated C=O
lower wavenumber
1715 cm⁻¹
1685 cm⁻¹
conjugated C=O
lower wavenumber

Problem 21.9 Which carbonyl group in each pair absorbs at a higher frequency?

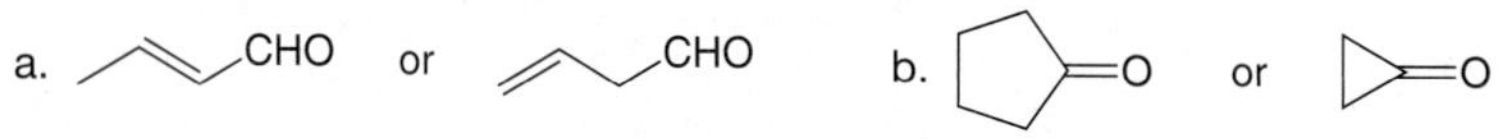
a.
CHO
or
CHO
b.
O
or
O

21.4B NMR Spectra

Aldehydes and ketones exhibit the following characteristic ^1H and ^{13}C NMR absorptions:

- **The *sp*^{2} hybridized C−H proton of an aldehyde is highly deshielded and absorbs far downfield at 9–10 ppm.** Splitting occurs with protons on the α carbon, but the coupling constant is often very small (J = 1–3 Hz).
- **Protons on the α carbon to the carbonyl group absorb at 2–2.5 ppm.** Methyl ketones, for example, give a characteristic singlet at ~2.1 ppm.
- In a ^{13}C NMR spectrum, **the carbonyl carbon is highly deshielded, appearing in the 190–215 ppm region.**

The ^1H and ^{13}C NMR spectra of propanal are illustrated in Figure 21.5.

Figure 21.5 The ^1H and ^{13}C NMR spectra of propanal, CH_3CH_2CHO

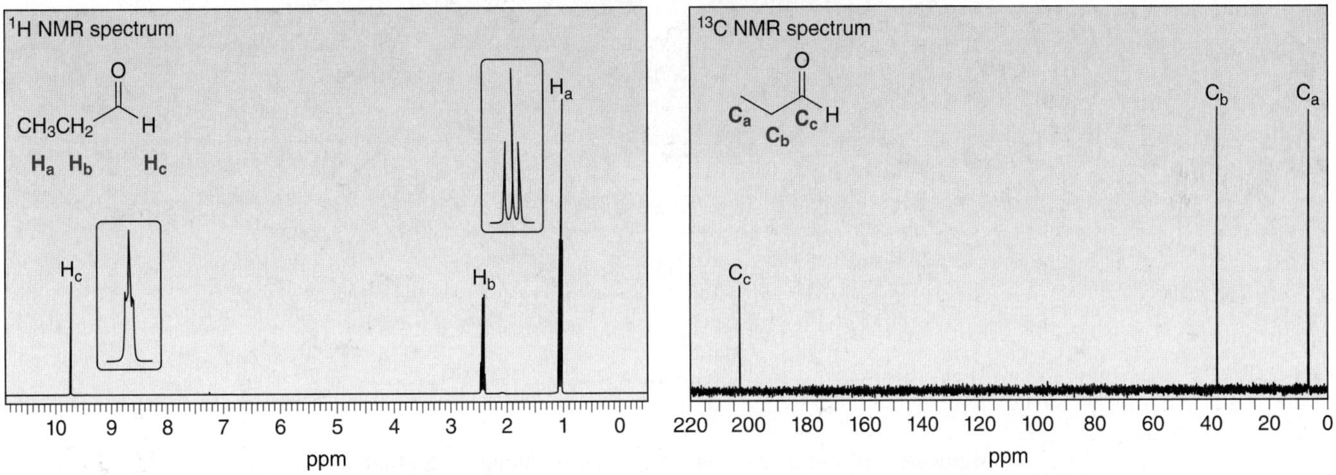

- **^1H NMR:** There are three signals due to the three different kinds of hydrogens, labeled H_a, H_b, and H_c. The deshielded CHO proton occurs downfield at 9.8 ppm. The H_c signal is split into a triplet by the adjacent CH_2 group, but the coupling constant is small.
- **^{13}C NMR:** There are three signals due to the three different kinds of carbons, labeled C_a, C_b, and C_c. The deshielded carbonyl carbon absorbs downfield at 203 ppm.

Problem 21.10 Draw the structure of all constitutional isomers that contain a ketone and have molecular formula $C_5H_{10}O$. Give the IUPAC name for each isomer and state how ^{13}C NMR spectroscopy could be used to distinguish these isomers.

21.5 Interesting Aldehydes and Ketones

formaldehyde
$CH_2{=}O$

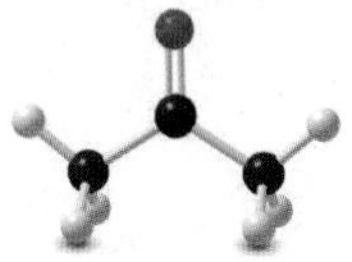

acetone
$(CH_3)_2C{=}O$

Because it is a starting material for the synthesis of many resins and plastics, billions of pounds of **formaldehyde** are produced annually in the United States by the oxidation of methanol (CH_3OH). Formaldehyde is also sold as a 37% aqueous solution called **formalin,** which has been used as a disinfectant, antiseptic, and preservative for biological specimens. Formaldehyde, a product of the incomplete combustion of coal and other fossil fuels, is partly responsible for the irritation caused by smoggy air.

Acetone is an industrial solvent and a starting material in the synthesis of some organic polymers. Acetone is produced in vivo during the breakdown of fatty acids. In diabetes, a common endocrine disease in which normal metabolic processes are altered because of the inadequate secretion of insulin, individuals often have unusually high levels of acetone in their bloodstreams. The characteristic odor of acetone can be detected on the breath of diabetic patients when their disease is poorly controlled.

Many aldehydes with characteristic odors occur in nature, as shown in Figure 21.6.

Figure 21.6 Some naturally occurring aldehydes with strong odors

Many steroid hormones contain a carbonyl along with other functional groups. **Cortisone** and **prednisone** are two anti-inflammatory steroids with closely related structures. Cortisone is secreted by the body's adrenal gland, whereas prednisone is a synthetic analogue used in the treatment of inflammatory diseases such as arthritis and asthma.

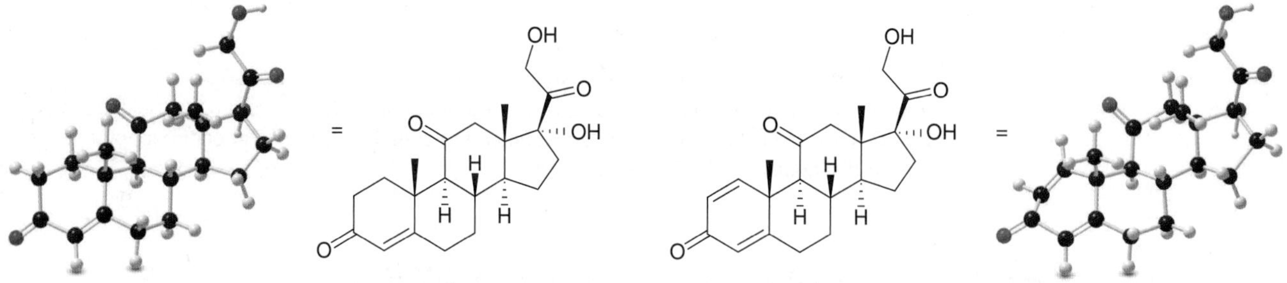

cortisone
(naturally occurring)

prednisone
(synthetic)

21.6 Preparation of Aldehydes and Ketones

Aldehydes and ketones can be prepared by a variety of methods. Because these reactions are needed for many multistep syntheses, Section 21.6 briefly summarizes earlier reactions that synthesize an aldehyde or ketone.

21.6A Common Methods to Synthesize Aldehydes

Aldehydes are prepared from 1° alcohols, esters, acid chlorides, and alkynes.

- **By oxidation of 1° alcohols with PCC**

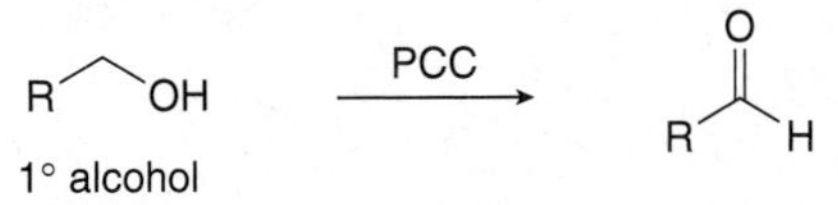
R OH
1° alcohol
PCC
O
R H

(Section 12.12B)

- **By reduction of esters and acid chlorides**

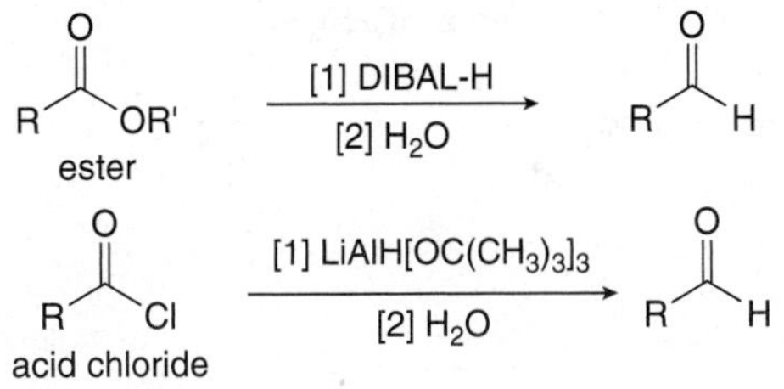
O
R OR'
ester
[1] DIBAL-H
[2] H_2O
O
R H
O
R Cl
acid chloride
[1] $LiAlH[OC(CH_3)_3]_3$
[2] H_2O
O
R H

(Section 20.7A)

- **By hydroboration–oxidation of an alkyne**

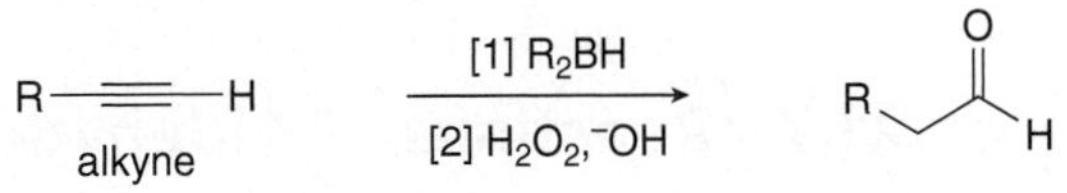
R—≡—H
alkyne
[1] R_2BH
[2] H_2O_2, ^-OH
O
R H

(Section 11.10)

21.6B Common Methods to Synthesize Ketones

Ketones are prepared from 2° alcohols, acid chlorides, and alkynes.

- **By oxidation of 2° alcohols with Cr^{6+} reagents**

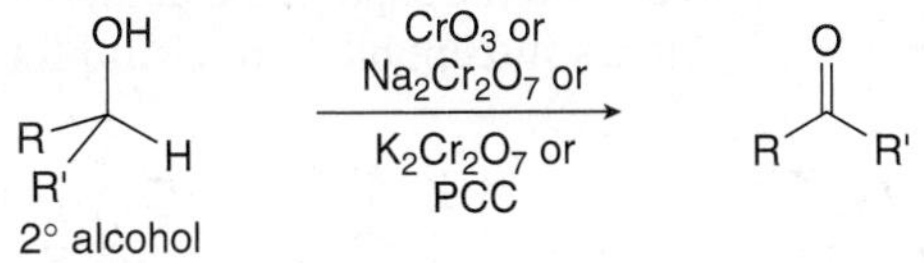
OH
R H
R'
2° alcohol
CrO_3 or
$Na_2Cr_2O_7$ or
$K_2Cr_2O_7$ or
PCC
O
R R'

(Section 12.12A)

- **By reaction of acid chlorides with organocuprates**

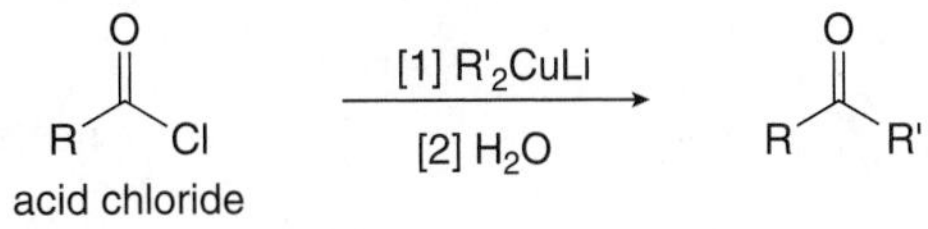
O
R Cl
acid chloride
[1] R'_2CuLi
[2] H_2O
O
R R'

(Section 20.13)

- **By Friedel–Crafts acylation**

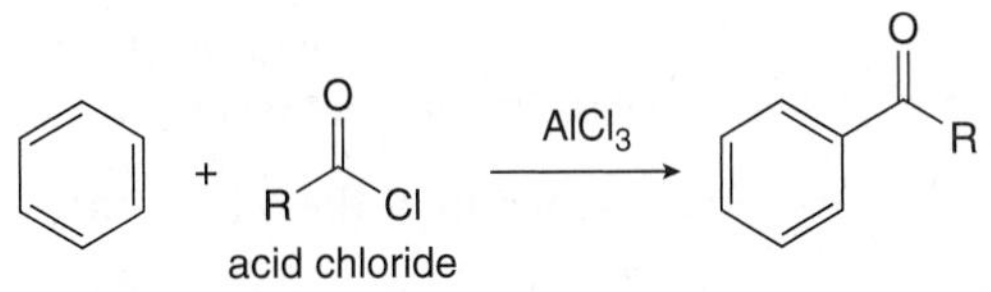
+
O
R Cl
acid chloride
$AlCl_3$
O
R

(Section 18.5)

- **By hydration of an alkyne**

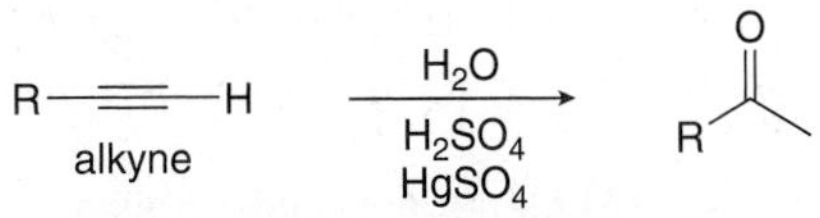
R—≡—H
alkyne
H_2O
H_2SO_4
$HgSO_4$
O
R

(Section 11.9)

Aldehydes and ketones are also both obtained as products of the oxidative cleavage of alkenes (Section 12.10).

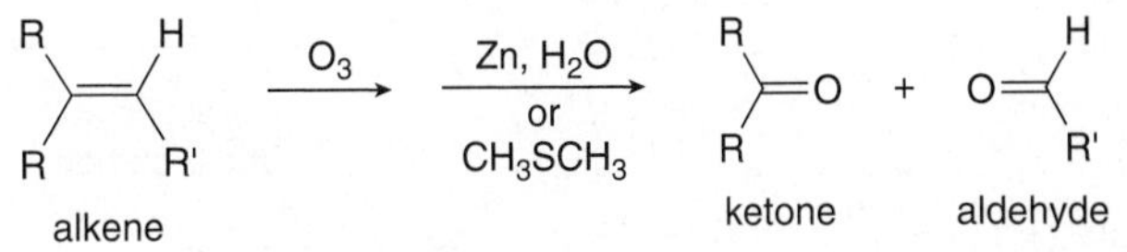
R H
R R'
alkene
O_3
Zn, H_2O
or
CH_3SCH_3
R
R
=O
ketone
+
O=
H
R'
aldehyde

Problem 21.11 What reagents are needed to convert each compound into butanal ($CH_3CH_2CH_2CHO$)?

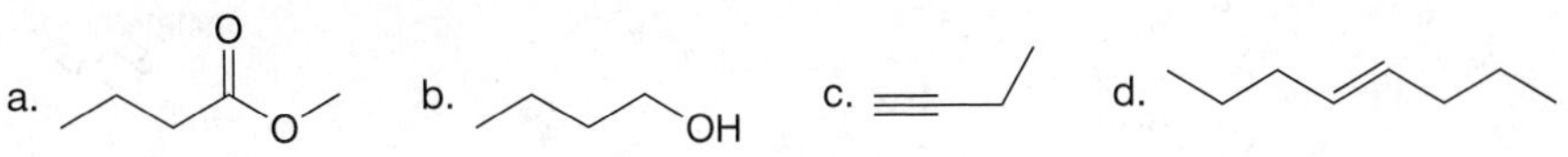
a.
O
O
b.
OH
c.
d.

Problem 21.12 What reagents are needed to convert each compound into acetophenone ($C_6H_5COCH_3$)?

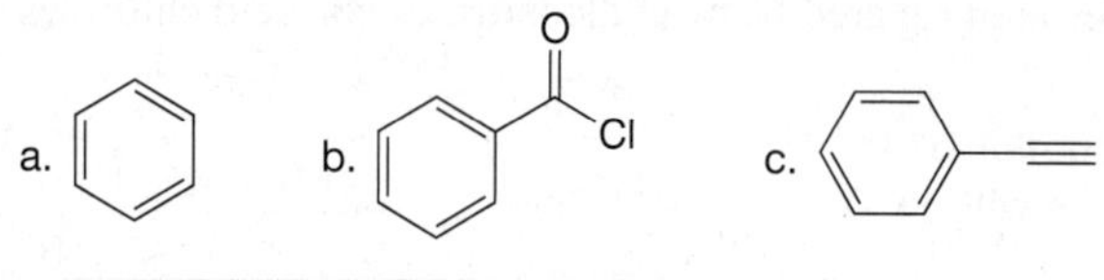

Problem 21.13 What alkene would yield 2,2-dimethoxycyclopentane-1,3-dicarbaldehyde on treatment with O_3 followed by $(CH_3)_2S$?

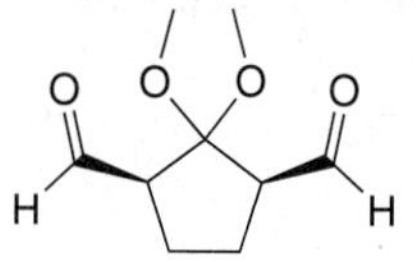

2,2-dimethoxycyclopentane-1,3-dicarbaldehyde

21.7 Reactions of Aldehydes and Ketones—General Considerations

Let's begin our discussion of carbonyl reactions by looking at the two general kinds of reactions that aldehydes and ketones undergo.

[1] Reaction at the carbonyl carbon

Recall from Chapter 20 that the uncrowded, electrophilic carbonyl carbon makes aldehydes and ketones susceptible to **nucleophilic addition** reactions.

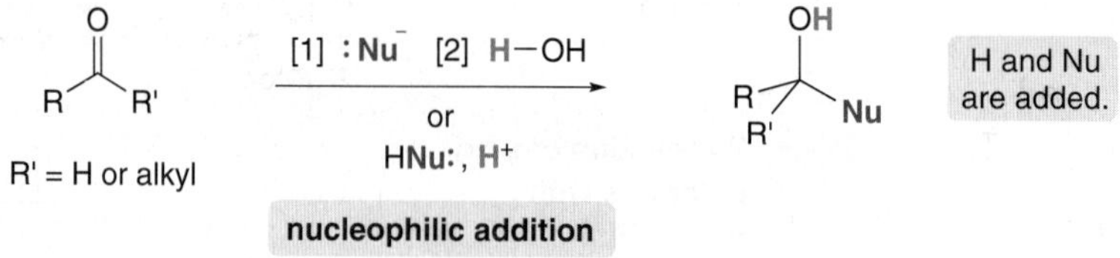

The elements of H and Nu are added to the carbonyl group. In Chapter 20, you learned about this reaction with hydride (H:⁻) and carbanions (R:⁻) as nucleophiles. In Chapter 21, we will discuss similar reactions with other nucleophiles.

[2] Reaction at the α carbon

A second general reaction of aldehydes and ketones involves reaction at the **α carbon.** A C–H bond on the α carbon to a carbonyl group is more acidic than many other C–H bonds, because reaction with base forms a resonance-stabilized enolate anion.

- **Enolates are nucleophiles, and so they react with electrophiles (E^+) to form new bonds on the α carbon.**

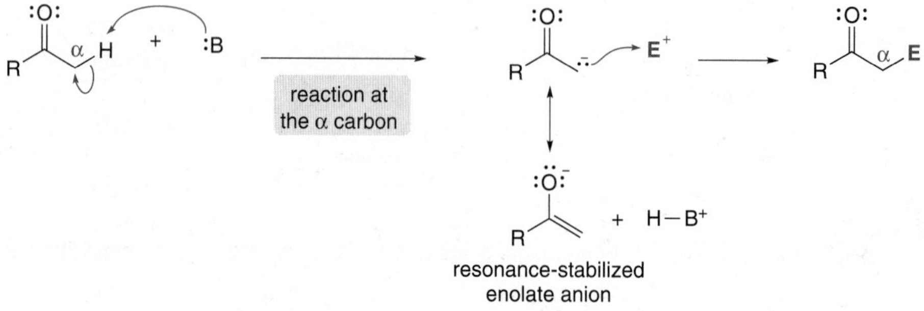

Chapters 23 and 24 are devoted to reactions at the α carbon to a carbonyl group.

- **Aldehydes and ketones react with nucleophiles at the carbonyl carbon.**
- **Aldehydes and ketones form enolates that react with electrophiles at the α carbon.**

21.7A The General Mechanism of Nucleophilic Addition

Two general mechanisms are usually drawn for nucleophilic addition, depending on the nucleophile (negatively charged versus neutral) and the presence or absence of an acid catalyst. With negatively charged nucleophiles, nucleophilic addition follows the two-step process first discussed in Chapter 20—**nucleophilic attack** followed by **protonation,** as shown in Mechanism 21.1.

Mechanism 21.1 General Mechanism—Nucleophilic Addition

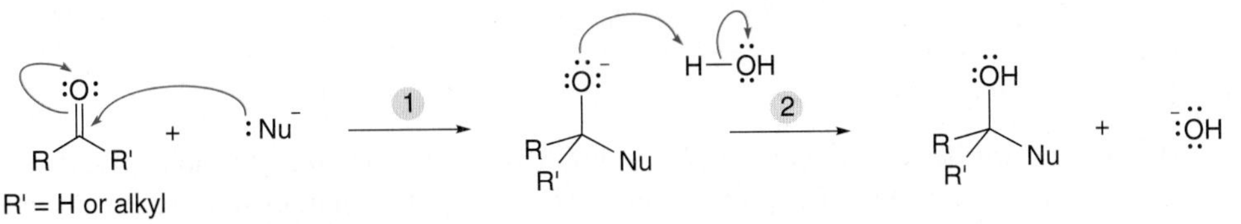

R' = H or alkyl

1 The nucleophile attacks the electrophilic carbonyl. The π bond is broken, moving an electron pair out on oxygen and forming an sp^3 hybridized carbon.

2 Protonation of the negatively charged oxygen by H_2O forms the addition product.

In this mechanism **nucleophilic attack *precedes* protonation.** This process occurs with strong neutral or negatively charged nucleophiles.

With some neutral nucleophiles, however, nucleophilic addition does not occur unless an acid catalyst is added. The general mechanism for this reaction consists of three steps (not two), but the same product results because H and Nu add across the carbonyl π bond. In this mechanism, **protonation *precedes* nucleophilic attack.** Mechanism 21.2 is shown with the neutral nucleophile H–Nu: and a general acid H–A.

Mechanism 21.2 General Mechanism—Acid-Catalyzed Nucleophilic Addition

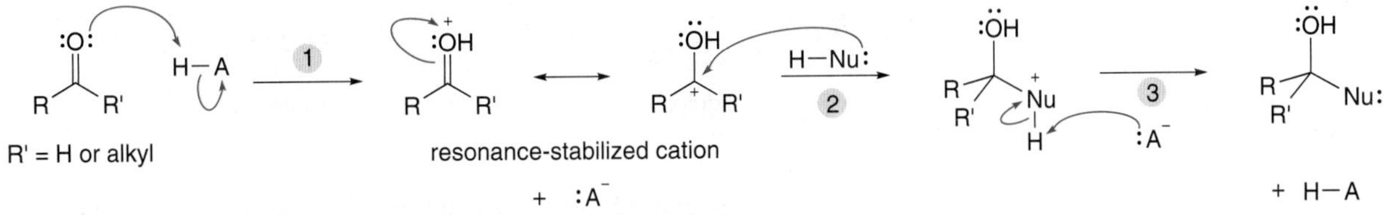

R' = H or alkyl

resonance-stabilized cation

+ :A⁻

+ H–A

1 Protonation of the carbonyl oxygen forms a resonance-stabilized cation.

2 – 3 Nucleophilic attack and deprotonation form the neutral addition product. The overall result is addition of H and Nu to the carbonyl group.

The effect of protonation is to convert a neutral carbonyl group to one having a net positive charge. **This protonated carbonyl group is much more electrophilic,** and much more susceptible to attack by a nucleophile. This step is unnecessary with strong nucleophiles like hydride ($H:^-$) that were used in Chapter 20. With weaker nucleophiles, however, nucleophilic attack does not occur unless the carbonyl group is first protonated.

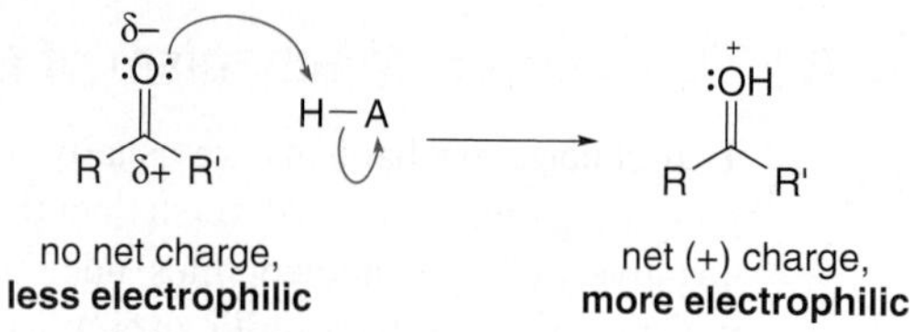

This step is a specific example of a general phenomenon.

- **Any reaction involving a carbonyl group and a strong acid begins with the same first step—protonation of the carbonyl oxygen.**

21.7B The Nucleophile

What nucleophiles add to carbonyl groups? This cannot be predicted solely on the trends in nucleophilicity learned in Chapter 7. Only *some* of the nucleophiles that react well in nucleophilic substitution at sp^3 hybridized carbons give reasonable yields of nucleophilic addition products.

Cl^-, Br^-, and I^- are good nucleophiles in substitution reactions at sp^3 hybridized carbons, but they are ineffective nucleophiles in addition. Addition of Cl^- to a carbonyl group, for example, would cleave the C–O π bond, forming an alkoxide. Because Cl^- is a much weaker base than the alkoxide formed, equilibrium favors the starting materials (the weaker base, Cl^-), *not* the addition product.

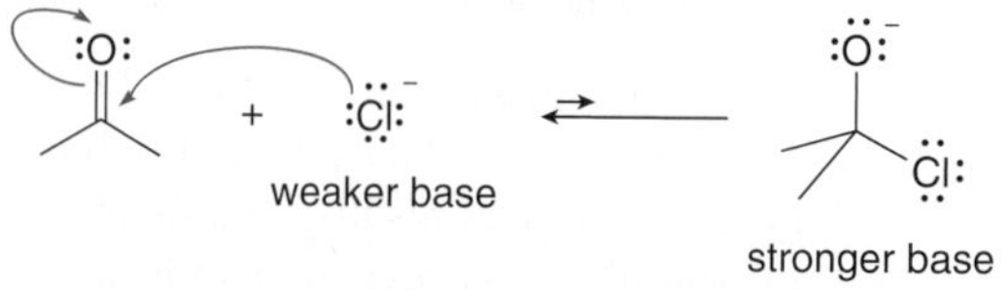

The situation is further complicated because some of the initial nucleophilic addition adducts are unstable and undergo elimination to form a stable product. For example, amines (RNH_2) add to carbonyl groups in the presence of mild acid to form unstable **carbinolamines,** which readily lose water to form **imines.** This addition–elimination sequence replaces a C=O by a C=N. The details of this process are discussed in Section 21.11.

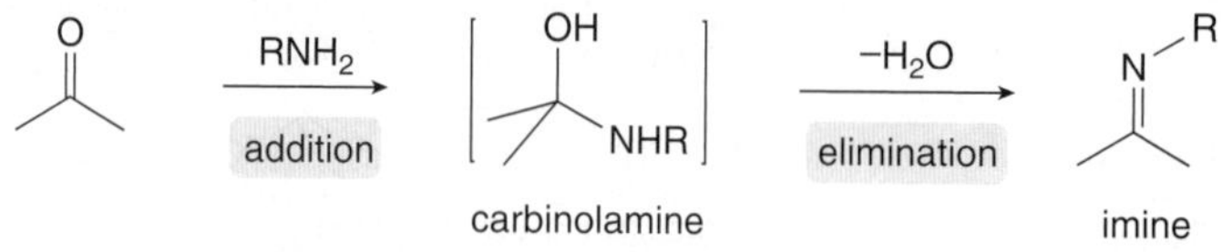

Figure 21.7 lists nucleophiles that add to a carbonyl group, as well as the products obtained from nucleophilic addition using cyclohexanone as a representative ketone. These reactions are discussed in the remaining sections of Chapter 21. In cases in which the initial addition adduct is unstable, it is enclosed within brackets, followed by the final product.

Figure 21.7
Specific examples of nucleophilic addition

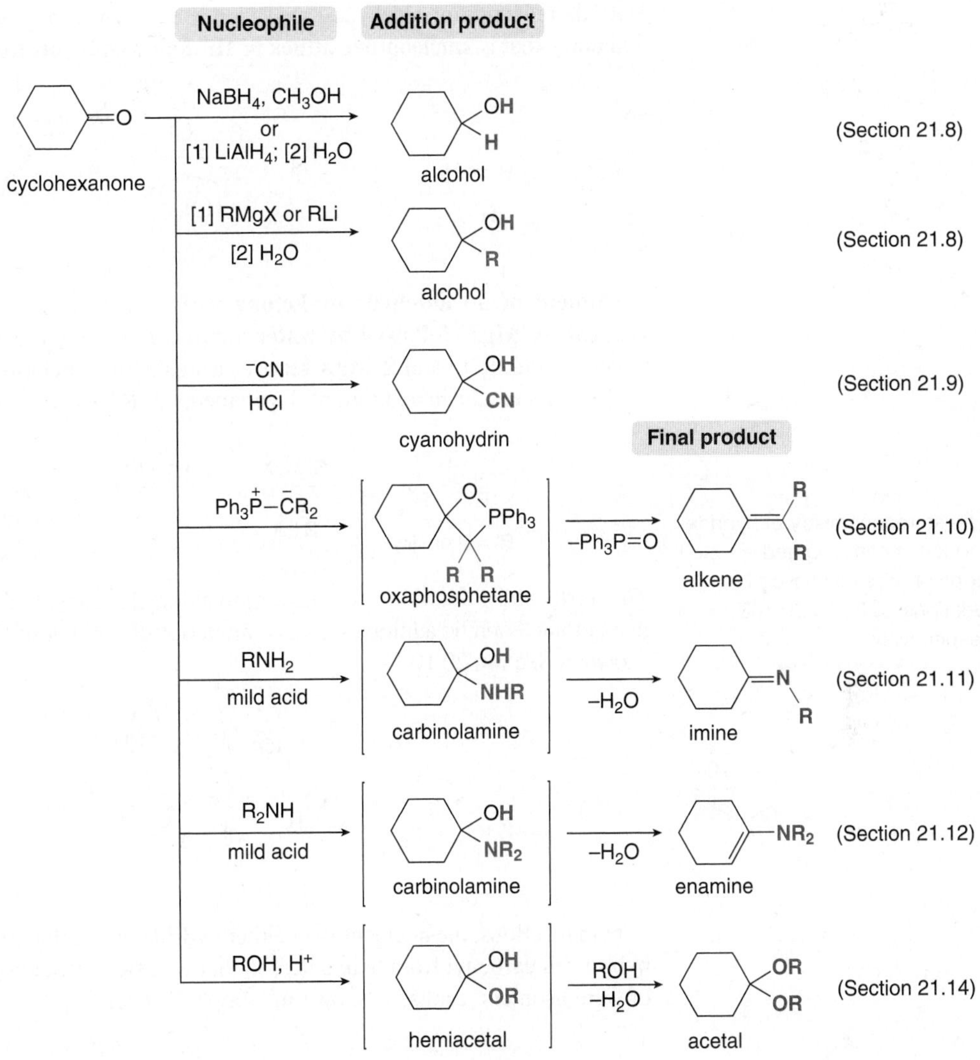

21.8 Nucleophilic Addition of H⁻ and R⁻—A Review

We begin our study of nucleophilic additions to aldehydes and ketones by briefly reviewing nucleophilic addition of hydride and carbanions, two reactions examined in Sections 20.4 and 20.10, respectively.

Treatment of an aldehyde or ketone with either $NaBH_4$ or $LiAlH_4$ followed by protonation forms a 1° or 2° alcohol. $NaBH_4$ and $LiAlH_4$ serve as a source of **hydride, H:⁻—the nucleophile**—and the reaction results in addition of the elements of H_2 across the C–O π bond. Addition of H_2 reduces the carbonyl group to an alcohol.

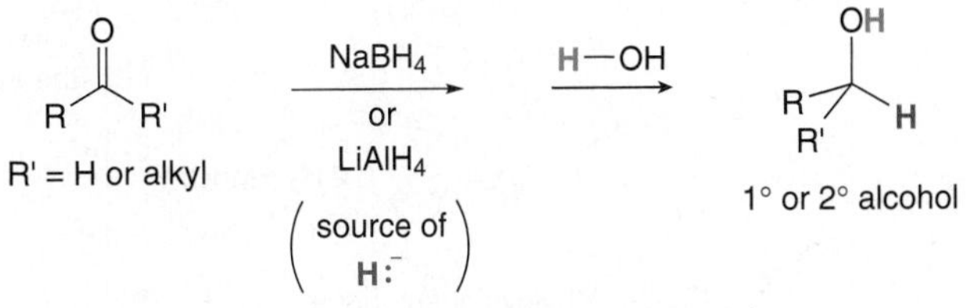

Hydride reduction of aldehydes and ketones occurs via the two-step mechanism of nucleophilic addition—that is, **nucleophilic attack of H:⁻ followed by protonation**—shown in Section 20.4B.

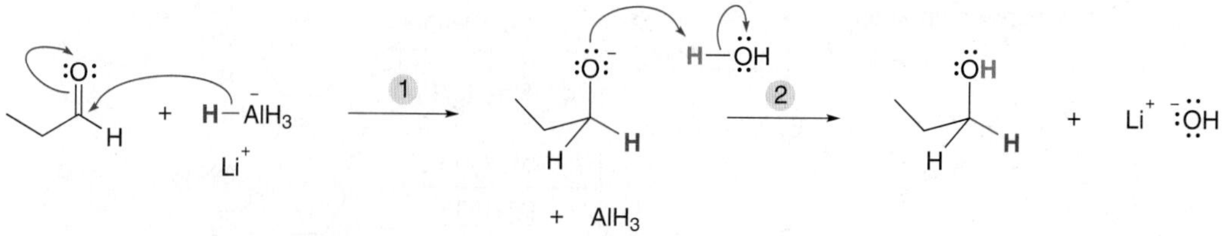

Treatment of an aldehyde or ketone with either an organolithium (R"Li) or Grignard reagent (R"MgX) followed by water forms a 1°, 2°, or 3° alcohol containing a new carbon–carbon bond. R"Li and R"MgX serve as a source of a **carbanion (R")⁻—the nucleophile**—and the reaction results in addition of the elements of R" and H across the C–O π bond.

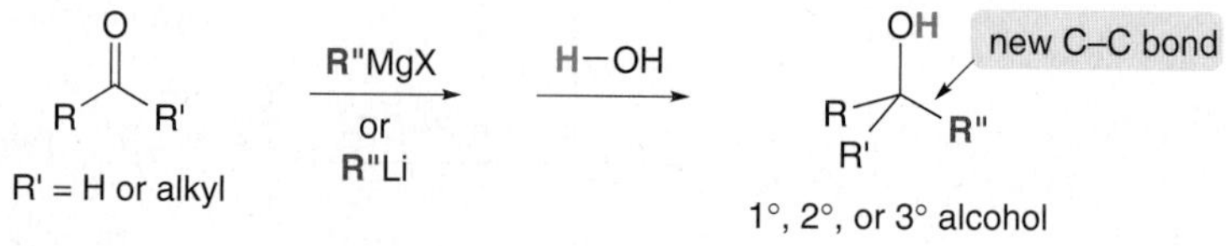

The stereochemistry of hydride reduction and Grignard addition was discussed in Sections 20.5 and 20.10B, respectively.

The nucleophilic addition of carbanions to aldehydes and ketones occurs via the two-step mechanism of nucleophilic addition—that is, **nucleophilic attack of (R")⁻ followed by protonation**—shown in Section 20.10.

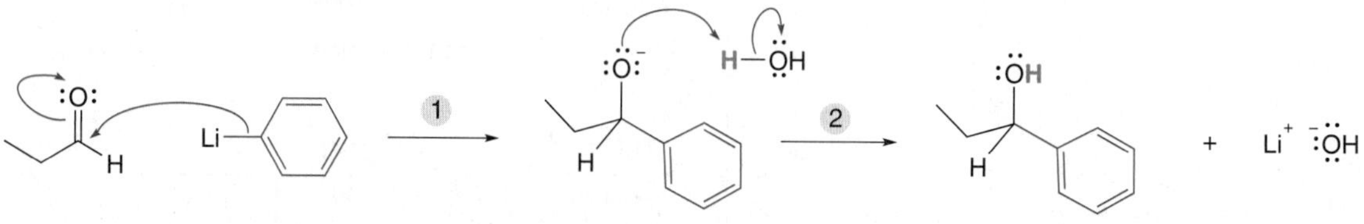

2° alcohol

In both reactions, the nucleophile—either hydride or a carbanion—attacks the trigonal planar sp^2 hybridized carbonyl from both sides, so that when a new stereogenic center is formed, a mixture of stereoisomers results, as shown in Sample Problem 21.3.

Sample Problem 21.3 Draw the products (including the stereochemistry) formed in the following reaction.

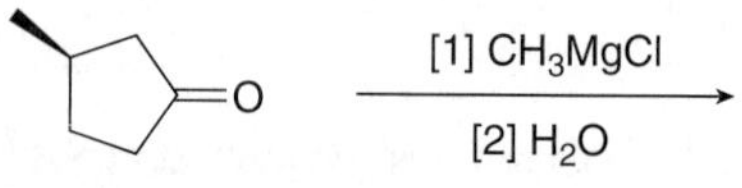

(*R*)-3-methylcyclopentanone

Solution

The Grignard reagent adds CH_3^- from both sides of the trigonal planar carbonyl group, yielding a mixture of 3° alcohols after protonation with water. In this example, the starting ketone and both alcohol products are chiral. The two products, which contain two stereogenic centers, are stereoisomers but not mirror images—that is, they are **diastereomers**.

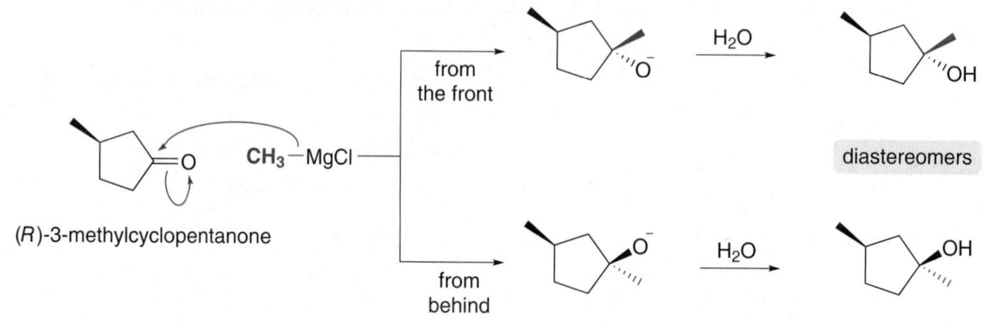

3° alcohols

Problem 21.14 Draw the products of each reaction. Include all stereoisomers formed.

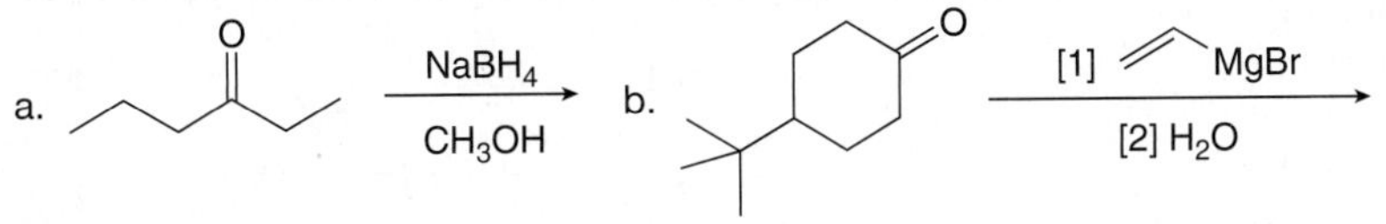
a.
O
NaBH4
CH3OH
b.
O
[1] MgBr
[2] H2O

21.9 Nucleophilic Addition of ⁻CN

Treatment of an aldehyde or ketone with NaCN and a strong acid such as HCl adds the elements of HCN across the carbon–oxygen π bond, forming a **cyanohydrin.**

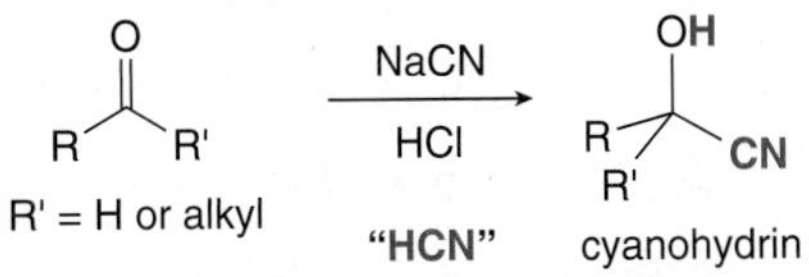
O
R
R'
R' = H or alkyl
NaCN
HCl
"HCN"
OH
R
CN
R'
cyanohydrin

This reaction adds one carbon to the aldehyde or ketone, forming a new carbon–carbon bond.

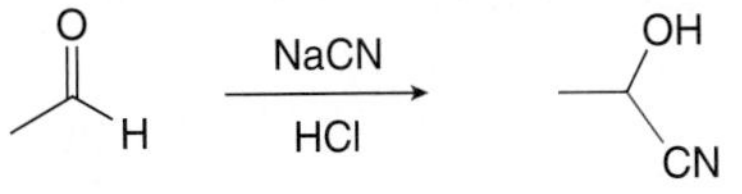
O
H
NaCN
HCl
OH
CN

acetaldehyde cyanohydrin
new C–C bond in red

21.9A The Mechanism

The mechanism of cyanohydrin formation involves the usual two steps of nucleophilic addition: **nucleophilic attack followed by protonation** as shown in Mechanism 21.3.

Mechanism 21.3 Nucleophilic Addition of ⁻CN—Cyanohydrin Formation

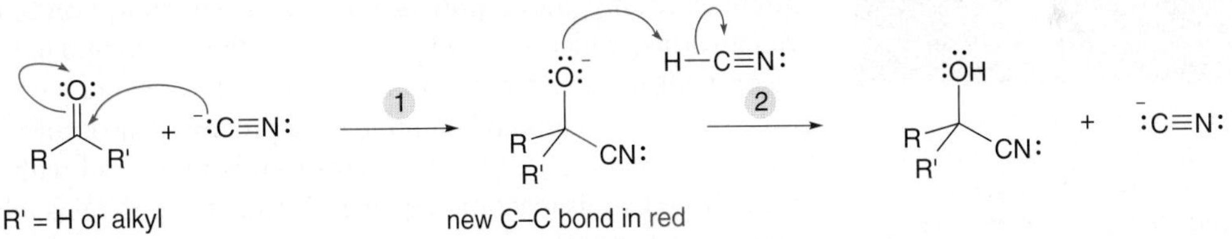
:O:
R
R'
+
⁻:C≡N:
1
:O:⁻
R
R'
CN:
H—C≡N:
2
:OH
R
R'
CN:
+
⁻:C≡N:
R' = H or alkyl
new C–C bond in red

1 Nucleophilic attack of ⁻CN forms a new carbon–carbon bond with cleavage of the C–O π bond.

2 Protonation of the negatively charged oxygen by HCN forms the addition product. The HCN used in this step is formed by the acid–base reaction of ⁻CN with the strong acid, HCl.

This reaction does not occur with HCN alone. The **cyanide anion** makes addition possible because it is a strong nucleophile that attacks the carbonyl group.

Cyanohydrins can be reconverted to carbonyl compounds by treatment with base. This process is just the reverse of the addition of HCN: **deprotonation followed by elimination of ⁻CN.**

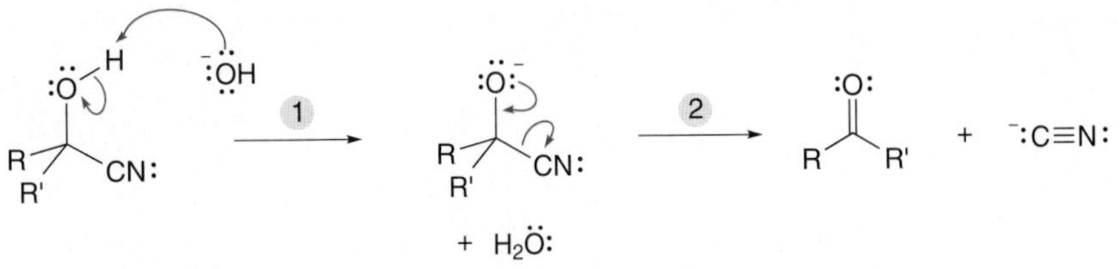
:O
H
R
R'
CN:
⁻:OH
1
:O:⁻
R
R'
CN:
+ H2O:
2
:O:
R
R'
+
⁻:C≡N:

Note the difference between two similar terms. **Hydration** results in *adding* water to a compound. **Hydrolysis** results in *cleaving bonds* with water.

The cyano group (CN) of a cyanohydrin is readily hydrolyzed to a carboxy group (COOH) by heating with aqueous acid or base. **Hydrolysis replaces the three C–N bonds by three C–O bonds.**

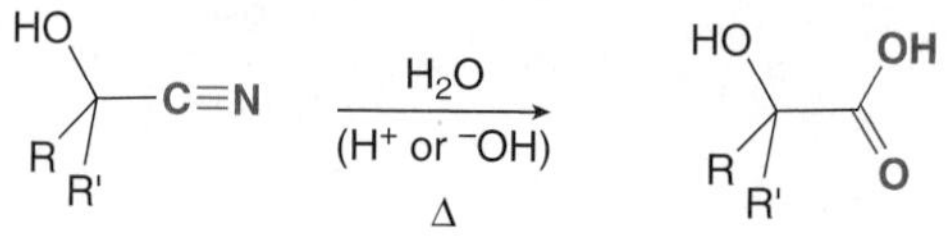

Problem 21.15 Draw the products of each reaction.

a. 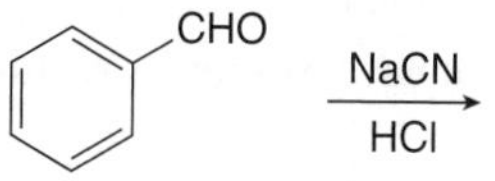

b. 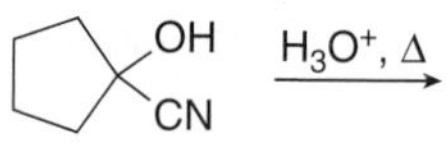

21.9B Application: Naturally Occurring Cyanohydrin Derivatives

Peach and apricot pits are a natural source of the cyanohydrin derivative amygdalin.

Although the cyanohydrin is an uncommon functional group, **linamarin** and **amygdalin** are two naturally occurring cyanohydrin derivatives. Both contain a carbon atom bonded to both an oxygen atom and a cyano group, analogous to a cyanohydrin.

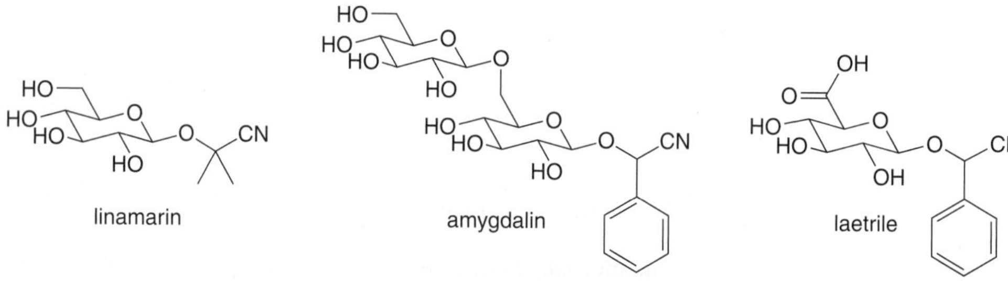

Cassava is a widely grown root crop, first introduced to Africa by Portuguese traders from Brazil in the sixteenth century. The peeled root is eaten after boiling or roasting. If the root is eaten without processing, illness and even death can result from high levels of HCN.

Linamarin is isolated from cassava, a woody shrub grown as a root crop in the humid tropical regions of South America and Africa. **Amygdalin** is present in the seeds and pits of apricots, peaches, and wild cherries. Amygdalin and the related synthetic compound **laetrile** were once touted as anticancer drugs, although their effectiveness is unproven.

Linamarin, amygdalin, and laetrile are toxic compounds because they are metabolized to cyanohydrins, which are hydrolyzed to carbonyl compounds and toxic HCN gas, a cellular poison with a characteristic almond odor. This second step is merely the reconversion of a cyanohydrin to a carbonyl compound, a process that occurs with base in reactions run in the laboratory (Section 21.9A). If cassava root is processed with care, linamarin is enzymatically metabolized by this reaction sequence and the toxic HCN is released before the root is ingested, making it safe to eat.

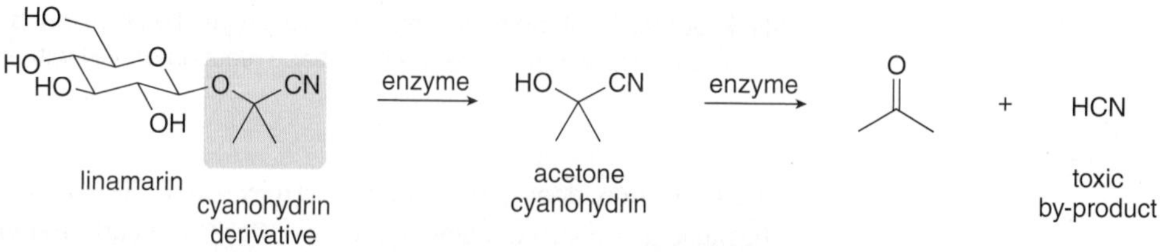

Problem 21.16 What cyanohydrin and carbonyl compound are formed when amygdalin is metabolized in a similar manner to linamarin?

21.10 The Wittig Reaction

The additions of H^-, R^-, and ^-CN all involve the same two steps—nucleophilic attack followed by protonation. Other examples of nucleophilic addition in Chapter 21 are somewhat different. Although they still involve attack of a nucleophile, the initial addition adduct is converted to another product by one or more reactions.

The first reaction in this category is the **Wittig reaction,** named for German chemist Georg Wittig, who was awarded the Nobel Prize in Chemistry in 1979 for its discovery. The Wittig reaction uses a carbon nucleophile, the **Wittig reagent,** to form **alkenes.** When a carbonyl compound is treated with a Wittig reagent, the carbonyl oxygen atom is replaced by the negatively charged alkyl group bonded to the phosphorus—that is, **the C=O is converted to a C=C.**

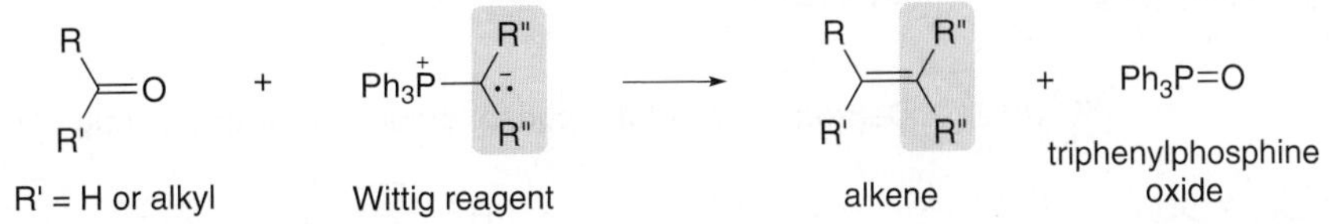

- **A Wittig reaction forms two new carbon–carbon bonds—one new σ bond and one new π bond—as well as a phosphorus by-product, $Ph_3P=O$ (triphenylphosphine oxide).**

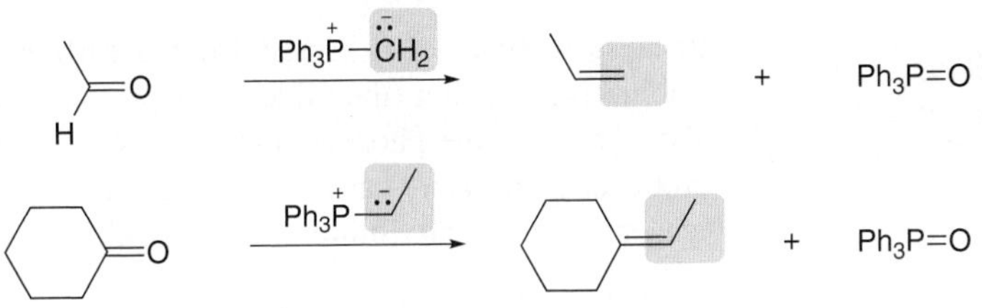

21.10A The Wittig Reagent

A **Wittig reagent** is an **organophosphorus reagent**—a reagent that contains a carbon–phosphorus bond. A typical Wittig reagent has a phosphorus atom bonded to three phenyl groups, plus another alkyl group that bears a negative charge.

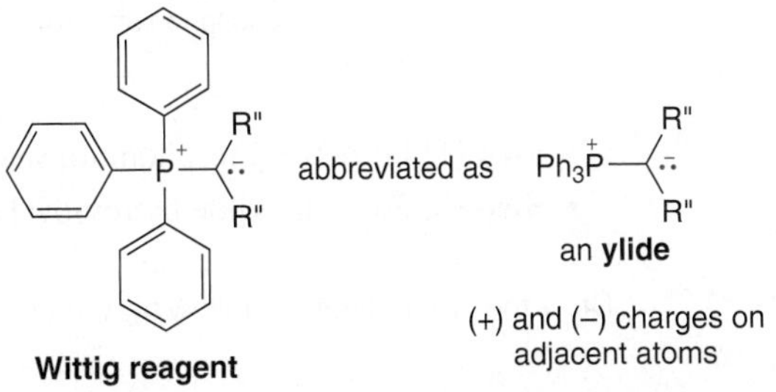

Phosphorus ylides are also called **phosphoranes.**

A Wittig reagent is an ***ylide,* a species that contains two oppositely charged atoms bonded to each other, and both atoms have octets.** In a Wittig reagent, a negatively charged carbon atom is bonded to a positively charged phosphorus atom.

Because phosphorus is a third-row element, it can be surrounded by more than eight electrons. As a result, a second resonance structure can be drawn that places a double bond between carbon and phosphorus. Regardless of which resonance structure is drawn, a **Wittig reagent has no net charge.** In one resonance structure, though, the carbon atom bonded to phosphorus (labeled in blue) bears a net negative charge, so it is nucleophilic.

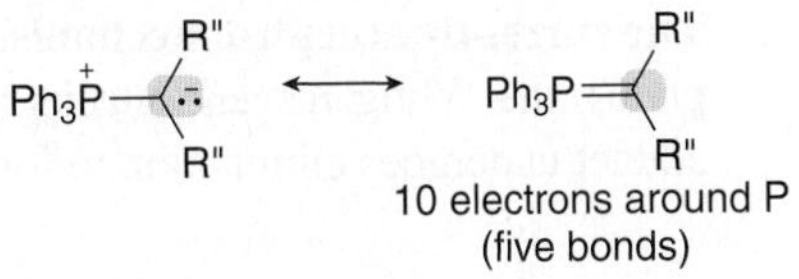

Wittig reagents are synthesized by a two-step procedure.

Step [1] **S_N2 reaction of triphenylphosphine with an alkyl halide forms a phosphonium salt.**

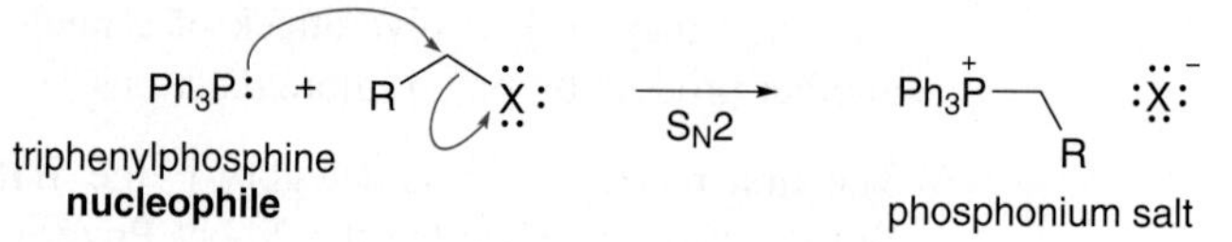
Ph3P: + R X:
SN2
Ph3P R :X:
triphenylphosphine
nucleophile
phosphonium salt

Because phosphorus is located below nitrogen in the periodic table, a neutral phosphorus atom with three bonds also has a lone pair of electrons.

Triphenylphosphine (Ph_3P:), which contains a lone pair of electrons on P, is the nucleophile. Because the reaction follows an S_N2 mechanism, it works best with unhindered CH_3X and 1° alkyl halides (RCH_2X). Secondary alkyl halides (R_2CHX) can also be used, although yields are often lower.

Step [2] **Deprotonation of the phosphonium salt with a strong base (:B) forms the ylide.**

Li

Bu—Li

strong base

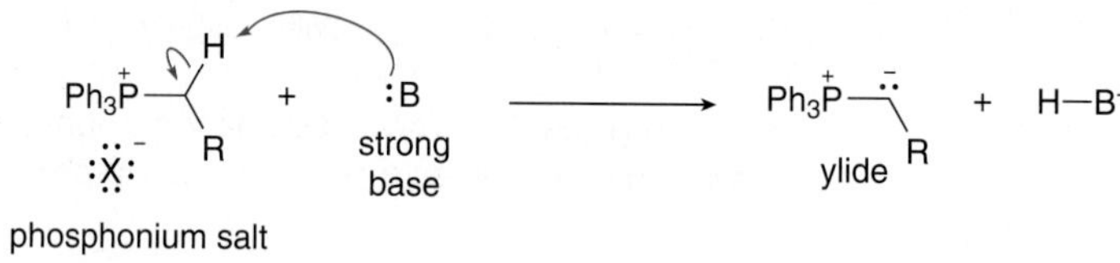
Ph3P H R :X:
+ :B
strong base
phosphonium salt
Ph3P R
ylide
+ H—B+

Section 20.9C discussed the reaction of organometallic reagents as strong bases.

Because removal of a proton from a carbon bonded to phosphorus generates a resonance-stabilized carbanion (the ylide), this proton is somewhat more acidic than other protons on an alkyl group in the phosphonium salt. Very strong bases are still needed, though, to favor the products of this acid–base reaction. Common bases used for this reaction are the organolithium reagents such as butyllithium, $CH_3CH_2CH_2CH_2Li$, abbreviated as BuLi.

To synthesize the Wittig reagent, $Ph_3P{=}CH_2$, use these two steps:

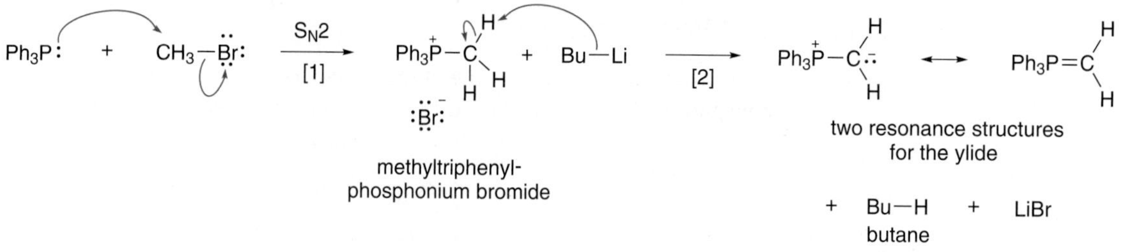
Ph3P: + CH3—Br:
SN2
[1]
Ph3P—C H H H :Br:
+ Bu—Li
[2]
methyltriphenyl-
phosphonium bromide
Ph3P—C H H
Ph3P=C H H
two resonance structures
for the ylide
+ Bu—H + LiBr
butane

- **Step [1]** Form the phosphonium salt by S_N2 reaction of Ph_3P: and CH_3Br.
- **Step [2]** Form the ylide by removal of a proton using BuLi as a strong base.

Problem 21.17 Draw the products of the following Wittig reactions.

a. 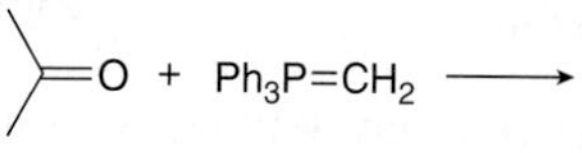
=O + Ph3P=CH2

b. 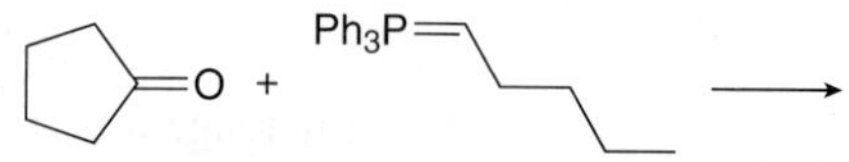
=O +
Ph3P=

Problem 21.18 Outline a synthesis of each Wittig reagent from Ph_3P and an alkyl halide.

a. 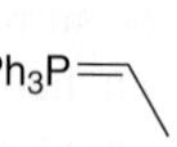
Ph3P=

b. 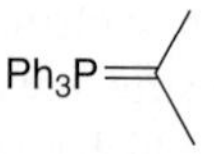
Ph3P=

c. 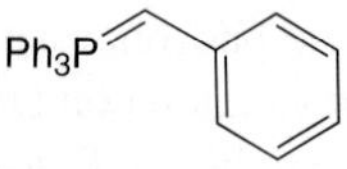
Ph3P

21.10B Mechanism of the Wittig Reaction

The currently accepted mechanism of the Wittig reaction involves two steps. Like other nucleophiles, the Wittig reagent attacks an electrophilic carbonyl carbon, but then the initial addition adduct undergoes elimination to form an alkene. Mechanism 21.4 is drawn using $Ph_3P{=}CH_2$.

Mechanism 21.4 The Wittig Reaction

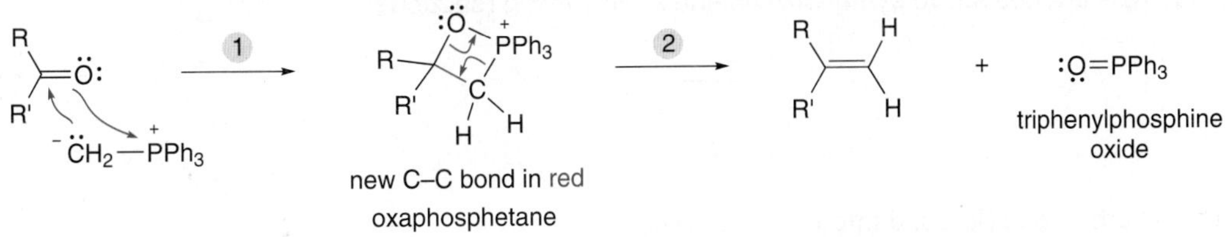

R
R'
:Ö:
CH2—PPh3
1
R
R'
:Ö
PPh3
C
H
H
2
R
R'
H
H
+
:O=PPh3
triphenylphosphine oxide
new C–C bond in red
oxaphosphetane

1 The negatively charged carbon of the ylide attacks the carbonyl carbon as the carbonyl oxygen attacks the positively charged P atom. This step forms two bonds and generates a four-membered ring called an oxaphosphetane.

2 Elimination of triphenylphosphine oxide forms two new π bonds. The formation of the strong P=O provides the driving force for the Wittig reaction.

One limitation of the Wittig reaction is that a mixture of alkene stereoisomers sometimes forms. For example, reaction of propanal (CH_3CH_2CHO) with a Wittig reagent forms the mixture of *E* and *Z* isomers shown.

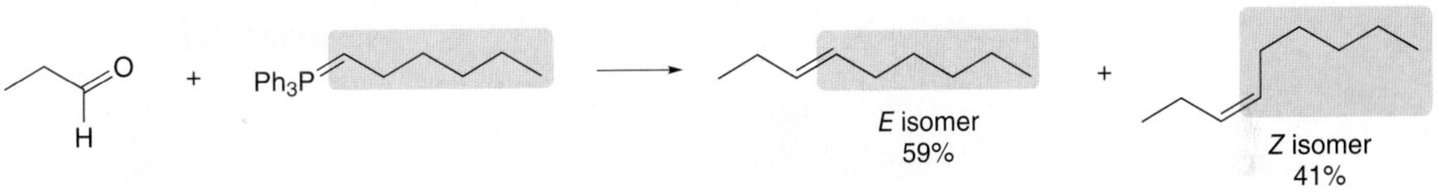

O
H
+
Ph3P
+
E isomer
59%
Z isomer
41%

Because the Wittig reaction forms two carbon–carbon bonds in a single reaction, it has been used to synthesize many natural products, including β-carotene, shown in Figure 21.8.

Figure 21.8
A Wittig reaction used to synthesize β-carotene

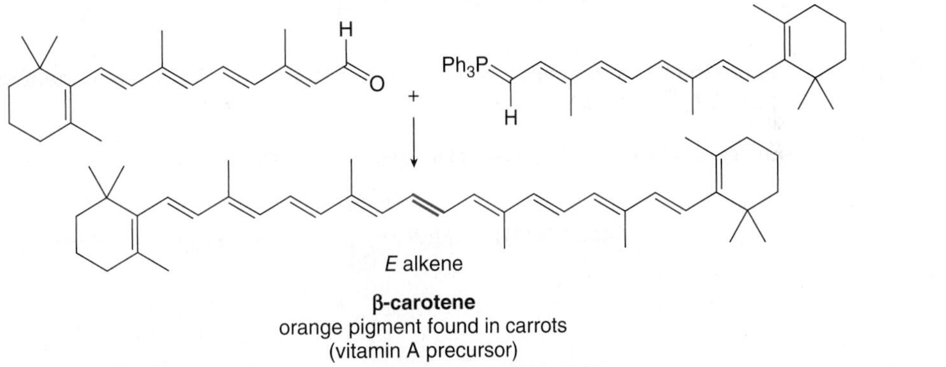

H
O
+
Ph3P
H
E alkene
β-carotene
orange pigment found in carrots
(vitamin A precursor)

- The more stable *E* alkene is the major product in this Wittig reaction.

Problem 21.19 Draw the products (including stereoisomers) formed when benzaldehyde (C_6H_5CHO) is treated with each Wittig reagent.

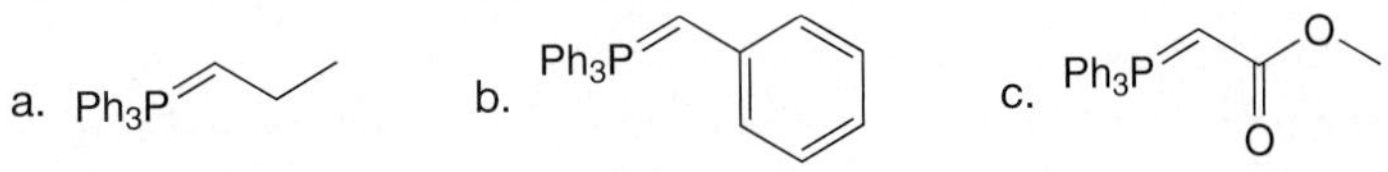

a. Ph3P
b. Ph3P
c. Ph3P
O
O

21.10C Retrosynthetic Analysis

To use the Wittig reaction in synthesis, you must be able to determine what carbonyl compound and Wittig reagent are needed to prepare a given compound—that is, **you must work backwards, in the retrosynthetic direction.** There can be two different Wittig routes to a given alkene, but one is often preferred on steric grounds.

How To Determine the Starting Materials for a Wittig Reaction Using Retrosynthetic Analysis

Example What starting materials are needed to synthesize alkene X by a Wittig reaction?

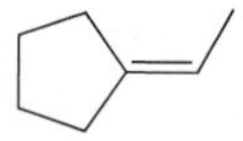

X

Step [1] **Cleave the carbon–carbon double bond into two components.**

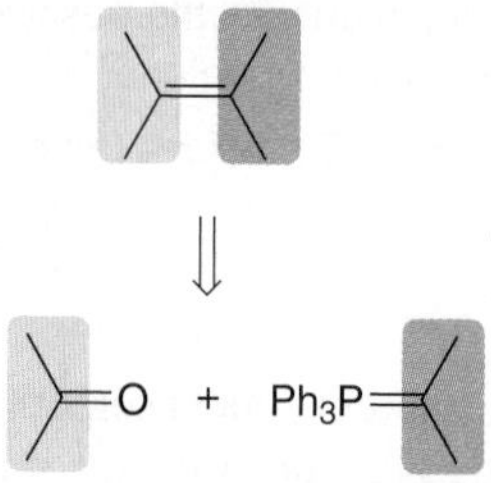
O
+
Ph3P

- Part of the molecule becomes the carbonyl component and the other part becomes the Wittig reagent.

There are usually two routes to a given alkene using a Wittig reaction:

Possibility [1] **Possibility [2]**

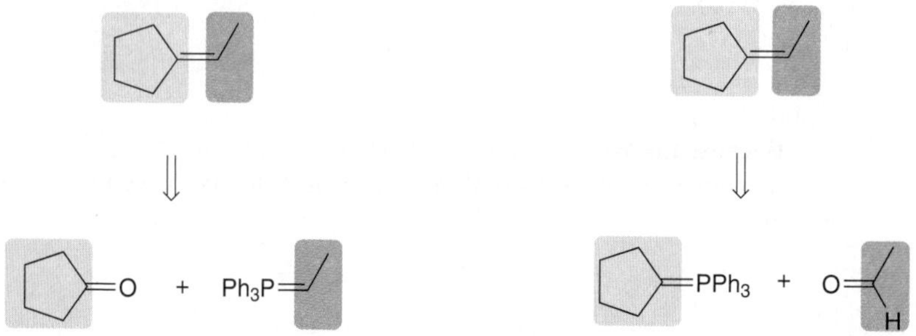
O
+
Ph3P
PPh3
+
O
H

Step [2] **Compare the Wittig reagents. The preferred pathway uses a Wittig reagent derived from an unhindered alkyl halide—CH_3X or RCH_2X.**

Determine what alkyl halide is needed to prepare each Wittig reagent:

Possibility [1]

Possibility [2]

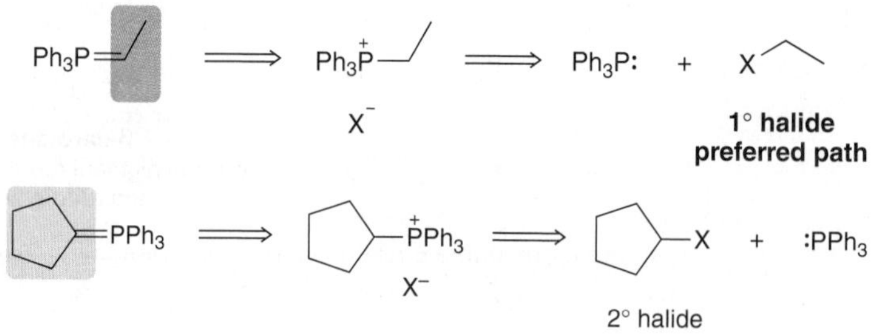
Ph3P
Ph3P
X
Ph3P:
+
X
1° halide
preferred path
PPh3
PPh3
X
X
+
:PPh3
2° halide

Because the synthesis of the Wittig reagent begins with an S_N2 reaction, **the preferred pathway begins with an unhindered methyl halide or 1° alkyl halide.** In this example, retrosynthetic analysis of both Wittig reagents indicates that only one of them ($Ph_3P{=}CHCH_3$) can be synthesized from a 1° alkyl halide, making Possibility [1] the preferred pathway.

Problem 21.20 What starting materials are needed to prepare each alkene by a Wittig reaction? When there are two possible routes, indicate which route, if any, is preferred.

a. 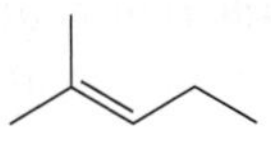

b. 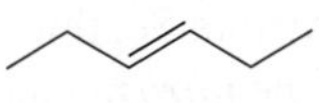

c. 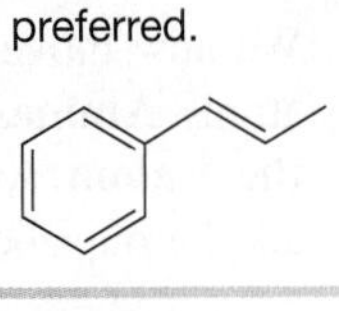

21.10D Comparing Methods of Alkene Synthesis

An advantage in using the Wittig reaction over other elimination methods to synthesize alkenes is that **you always know the location of the double bond.** Whereas other methods of alkene synthesis often give a mixture of constitutional isomers, the Wittig reaction always gives a single constitutional isomer.

For example, two methods can be used to convert cyclohexanone into alkene **B** (methylenecyclohexane): a two-step method consisting of Grignard addition followed by dehydration, or a one-step Wittig reaction.

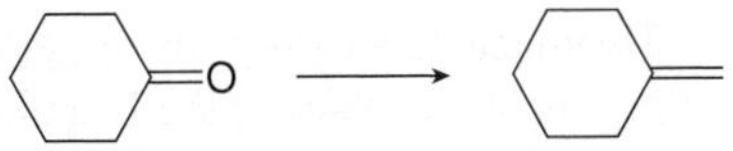

cyclohexanone **B**

Recall from Section 9.8 that the major product formed in acid-catalyzed dehydration of an alcohol is the more substituted alkene.

In a two-step method, treatment of cyclohexanone with CH_3MgBr forms a 3° alcohol after protonation. Dehydration of the alcohol with H_2SO_4 forms a mixture of alkenes, in which the desired disubstituted alkene is the minor product.

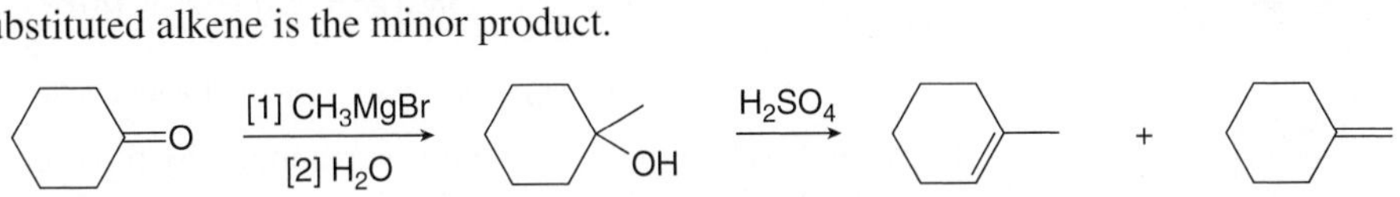
[1] CH_3MgBr
[2] H_2O
OH
H_2SO_4
+
B

cyclohexanone — 3° alcohol — trisubstituted C=C **major product** — disubstituted C=C **minor product**

By contrast, reaction of cyclohexanone with $Ph_3P{=}CH_2$ affords the desired alkene as the only product. The newly formed double bond always joins the carbonyl carbon with the negatively charged carbon of the Wittig reagent. In other words, **the position of the double bond is always unambiguous in the Wittig reaction.** This makes the Wittig reaction an especially attractive method for preparing many alkenes.

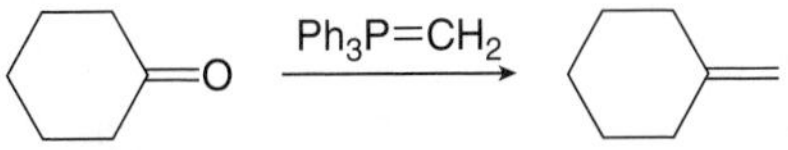
$Ph_3P{=}CH_2$

cyclohexanone **B** only product

Problem 21.21 Show two methods to synthesize each alkene: a one-step method using a Wittig reagent, and a two-step method that forms a carbon–carbon bond with an organometallic reagent in one of the steps.

a. 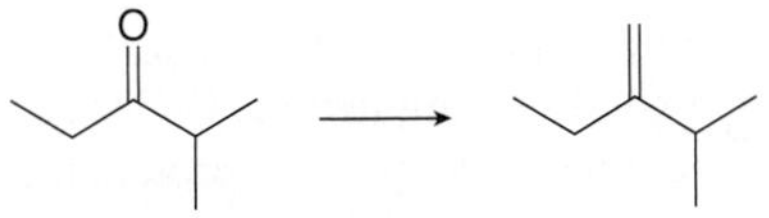

b. 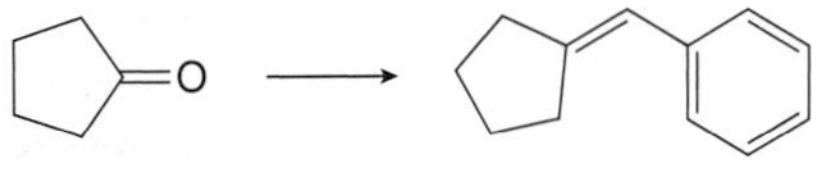

21.11 Addition of 1° Amines

We now move on to the reaction of aldehydes and ketones with nitrogen and oxygen heteroatoms. **Amines are organic nitrogen compounds that contain a nonbonded electron pair on the N atom.** As we learned in Section 3.2, amines are classified as 1°, 2°, or 3° by the number of alkyl groups bonded to the *nitrogen* atom.

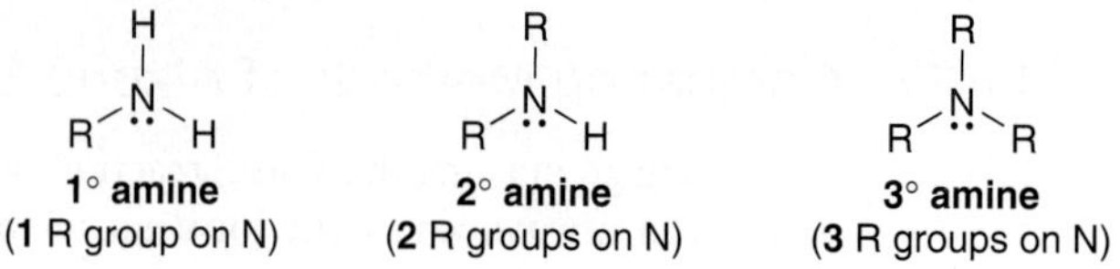

Both 1° and 2° amines react with aldehydes and ketones. We begin by examining the reaction of aldehydes and ketones with 1° amines.

21.11A Formation of Imines

Treatment of an aldehyde or ketone with a 1° amine affords an **imine** (also called a **Schiff base**). Nucleophilic attack of the 1° amine on the carbonyl group forms an unstable **carbinolamine,** which loses water to form an imine. The overall reaction results in **replacement of C=O by C=NR.**

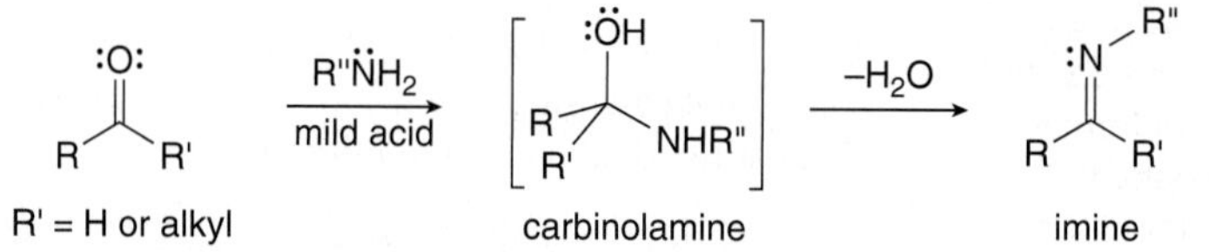

Because the N atom of an imine is surrounded by three groups (two atoms and a lone pair), it is sp^2 hybridized, making the C–N–R" bond angle ~120° (*not* 180°). Imine formation is fastest when the reaction medium is weakly acidic.

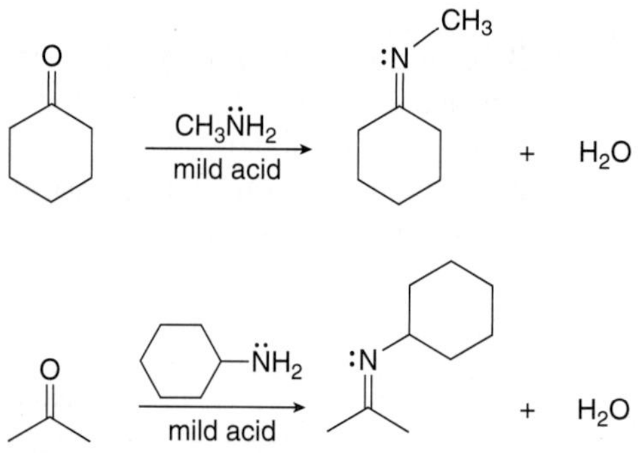

The mechanism of imine formation (Mechanism 21.5) can be divided into two distinct parts: **nucleophilic addition of the 1° amine (Steps [1] and [2]), followed by elimination of H_2O (Steps [3]–[5]).** Each step involves a reversible equilibrium, so that the reaction is driven to completion by removing H_2O.

Imine formation is most rapid at pH 4–5. Mild acid is needed for protonation of the hydroxy group in Step [3] to form a **good leaving group.** Under strongly acidic conditions, the reaction rate decreases because the amine nucleophile is protonated. With no free electron pair, it is no longer a nucleophile, and so nucleophilic addition cannot occur.

Mechanism 21.5 Imine Formation from an Aldehyde or Ketone

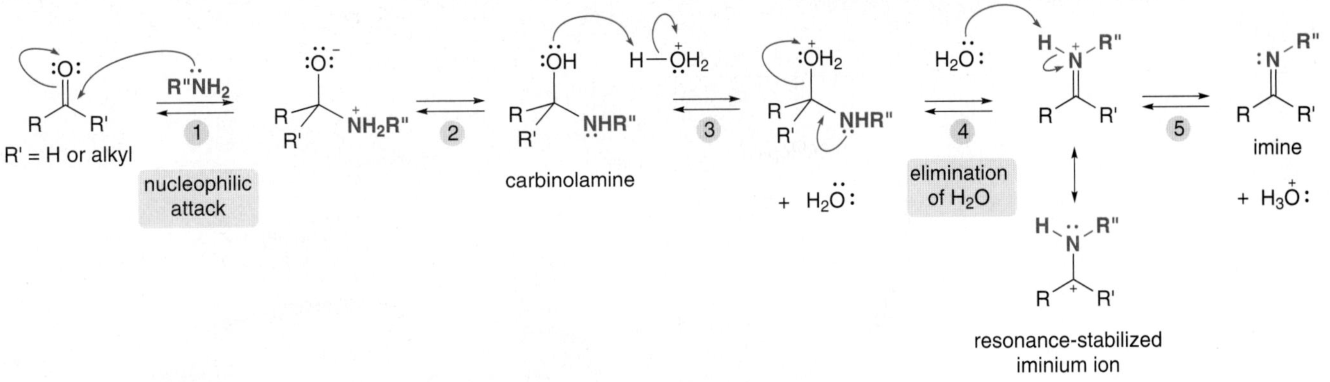

1 - 2 Nucleophilic attack of the amine followed by proton transfer forms the carbinolamine.

3 Protonation of the OH group forms a good leaving group.

4 Loss of H_2O forms a resonance-stabilized iminium ion.

5 Loss of a proton forms the imine.

Problem 21.22 Draw the product formed when $CH_3CH_2CH_2CH_2NH_2$ reacts with each carbonyl compound in the presence of mild acid.

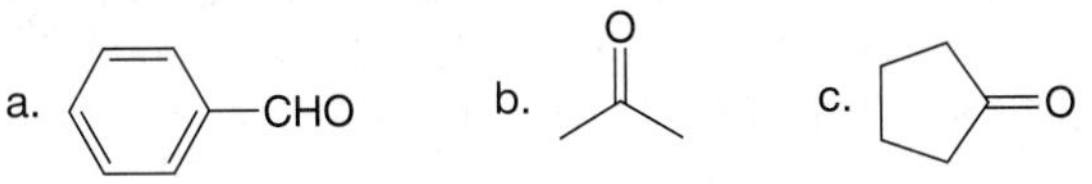

Problem 21.23 What 1° amine and carbonyl compound are needed to prepare each imine?

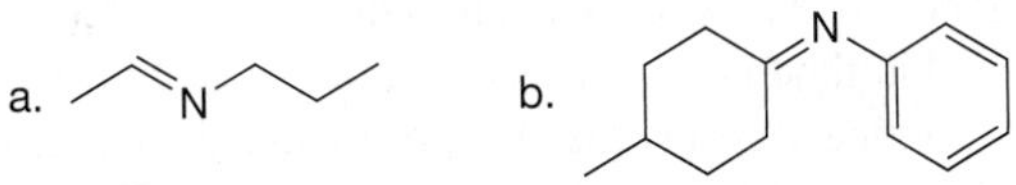

21.11B Application: Retinal, Rhodopsin, and the Chemistry of Vision

Many imines play vital roles in biological systems. A key molecule in the chemistry of vision is the highly conjugated imine **rhodopsin,** which is synthesized in the rod cells of the eye from **11-*cis*-retinal** and a 1° amine in the protein **opsin.**

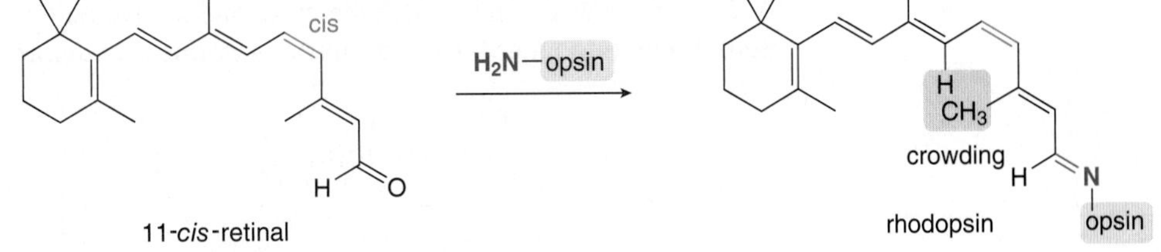

Figure 21.9
The key reaction in the chemistry of vision

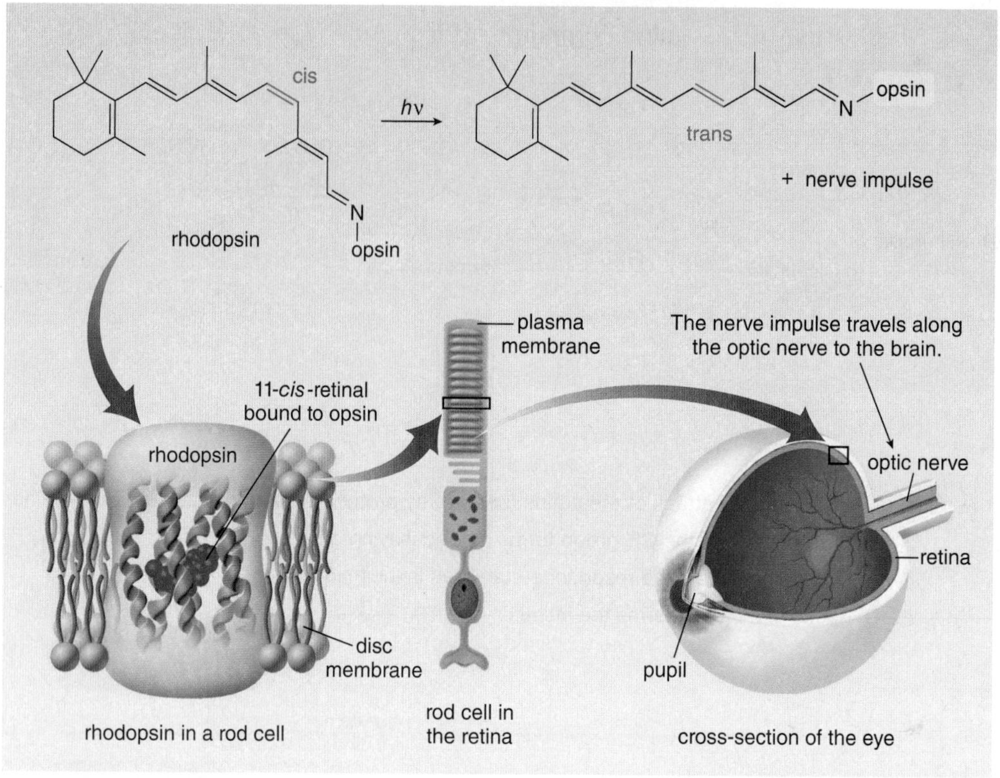

- Rhodopsin is a light-sensitive compound located in the membrane of the rod cells in the retina of the eye. Rhodopsin contains the protein opsin bonded to 11-*cis*-retinal via an imine linkage. When light strikes this molecule, the crowded 11-cis double bond isomerizes to the 11-trans isomer, and a nerve impulse is transmitted to the brain by the optic nerve.

The central role of rhodopsin in the visual process was delineated by Nobel Laureate George Wald of Harvard University.

The complex process of vision centers around this imine derived from retinal (Figure 21.9). The 11-cis double bond in rhodopsin creates crowding in the rather rigid side chain. When light strikes the rod cells of the retina, it is absorbed by the conjugated double bonds of rhodopsin, and the 11-cis double bond is isomerized to the 11-trans arrangement. This isomerization is accompanied by a drastic change in shape in the protein, altering the concentration of Ca^{2+} ions moving across the cell membrane, and sending a nerve impulse to the brain, which is then processed into a visual image.

21.12 Addition of 2° Amines

21.12A Formation of Enamines

A 2° amine reacts with an aldehyde or ketone to give an **enamine.** ***Enamines* have a nitrogen atom bonded to a double bond** (alk*ene* + *amine* = *enamine*).

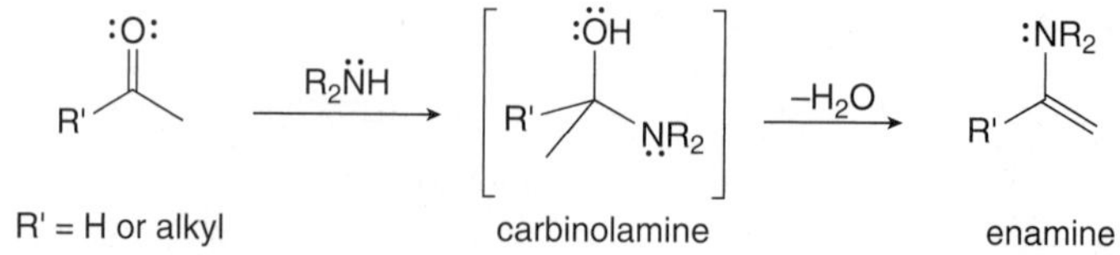

Like imines, enamines are also formed by the addition of a nitrogen nucleophile to a carbonyl group followed by elimination of water. In this case, however, elimination occurs across two adjacent *carbon* atoms to form a new carbon–carbon π bond.

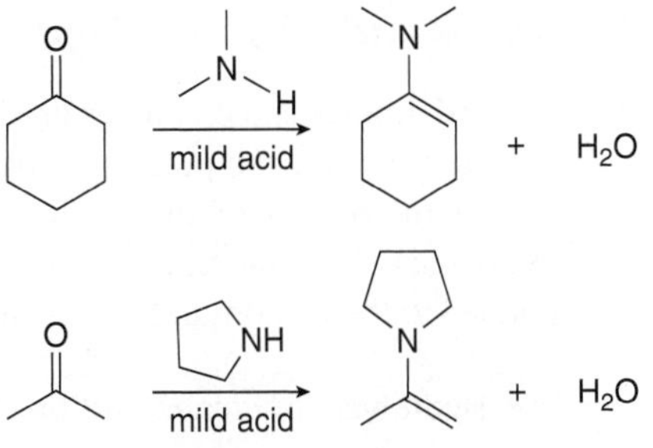

The mechanism for enamine formation (Mechanism 21.6) is identical to the mechanism for imine formation except for the last step, involving formation of the π bond. The mechanism can be divided into two distinct parts: **nucleophilic addition of the 2° amine (Steps [1] and [2]), followed by elimination of H_2O (Steps [3]–[5]).** Each step involves a reversible equilibrium once again, so that the reaction is driven to completion by removing H_2O.

Mechanism 21.6 Enamine Formation from an Aldehyde or Ketone

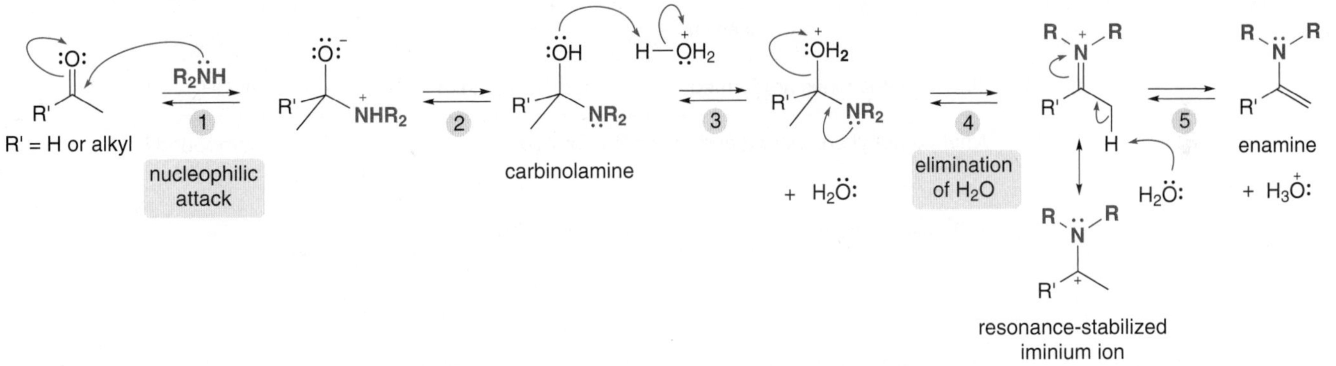

1 - 2 Nucleophilic attack of the amine followed by proton transfer forms the carbinolamine.

3 Protonation of the OH group forms a good leaving group.

4 Loss of H_2O forms a resonance-stabilized iminium ion.

5 Loss of a proton from the adjacent C–H bond forms the enamine.

The mechanisms illustrate why **the reaction of 1° amines with carbonyl compounds forms *imines*, but the reaction with 2° amines forms *enamines*.** In Figure 21.10, the last step of both mechanisms is compared using cyclohexanone as starting material. The position of the double bond depends on which proton is removed in the last step. Removal of an N–H proton forms a C=N, whereas removal of a C–H proton forms a C=C.

Figure 21.10
The formation of imines and enamines compared

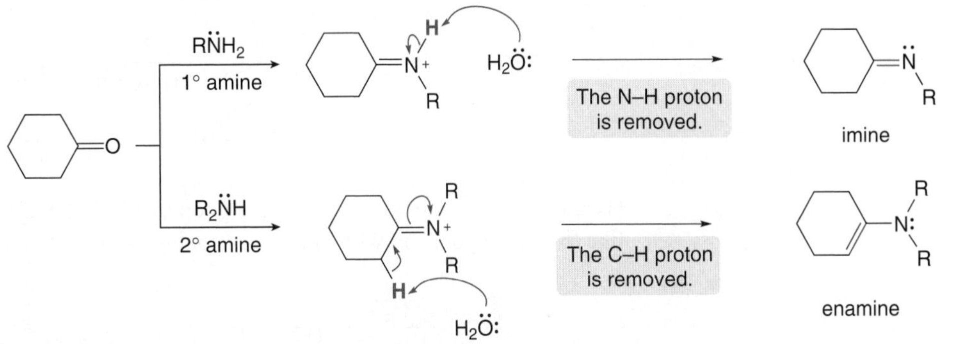

- With a **1° amine,** the intermediate iminium ion still has a proton on the N atom that may be removed to form a C=N.
- With a **2° amine,** the intermediate iminium ion has no proton on the N atom. A proton must be removed from an adjacent C–H bond, and this forms a C=C.

21.12B Imine and Enamine Hydrolysis

Because imines and enamines are formed by a set of reversible reactions, both can be converted back to carbonyl compounds by hydrolysis with mild acid. The mechanism of these reactions is exactly the *reverse* of the mechanism written for the formation of imines and enamines. In the hydrolysis of enamines, the carbonyl carbon in the product comes from the sp^2 hybridized carbon bonded to the N atom in the starting material.

- **Hydrolysis of imines and enamines forms aldehydes and ketones.**

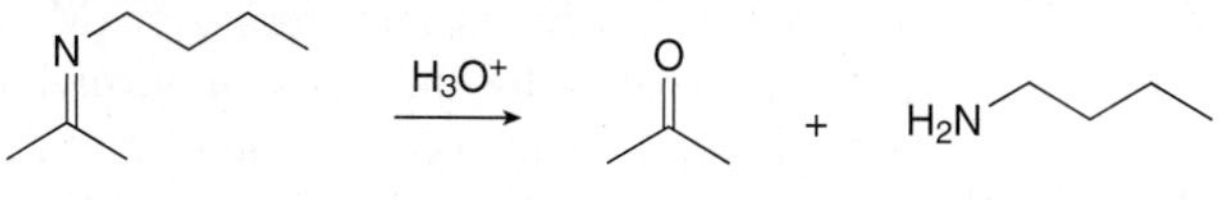

imine

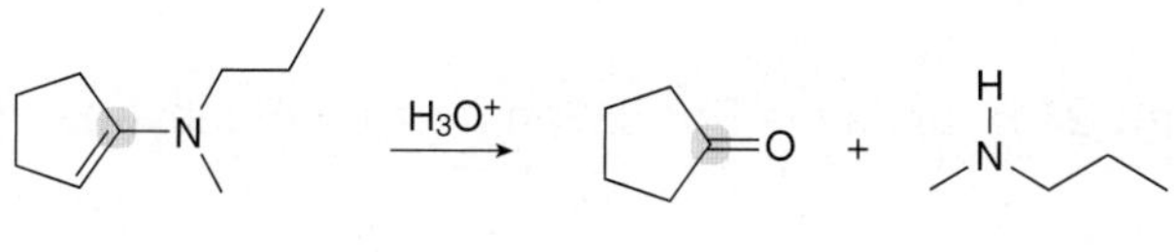

enamine

Problem 21.24 What two enamines are formed when 2-methylcyclohexanone is treated with $(CH_3)_2NH$?

Problem 21.25 What carbonyl compound and amine are formed by the hydrolysis of each compound?

a. 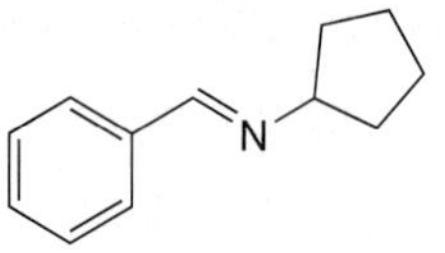

b. 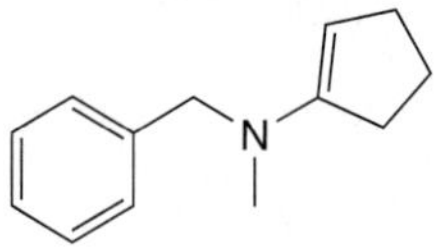

c. 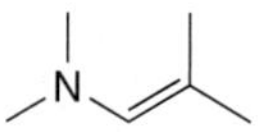

Problem 21.26 Draw a stepwise mechanism for the following imine hydrolysis.

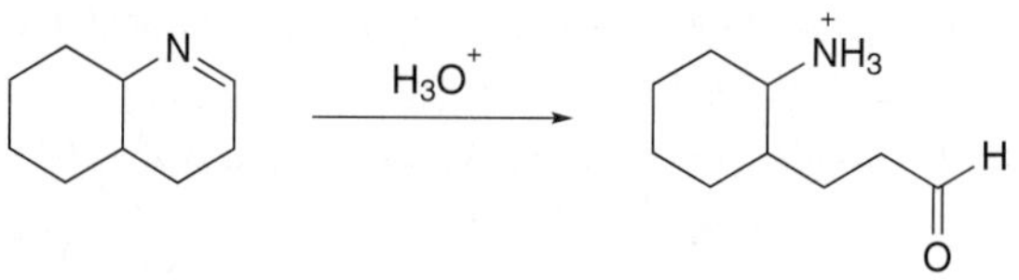

21.13 Addition of H_2O—Hydration

Treatment of a carbonyl compound with H_2O in the presence of an acid or base catalyst **adds the elements of H and OH across the carbon–oxygen π bond,** forming a ***gem*-diol** or **hydrate.**

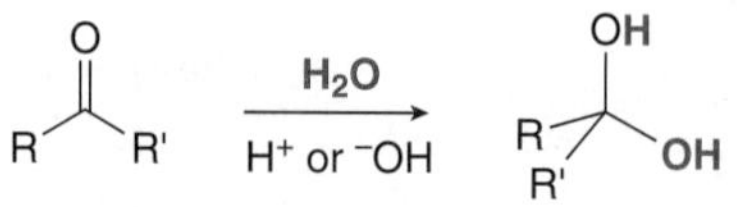

R' = H or alkyl

gem-diol
(hydrate)

Hydration of a carbonyl group gives a good yield of *gem*-diol only with an unhindered aldehyde like formaldehyde, and with aldehydes containing nearby electron-withdrawing groups.

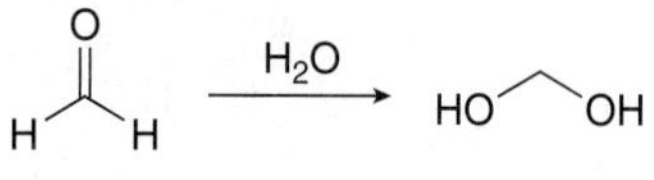

formaldehyde

formaldehyde hydrate

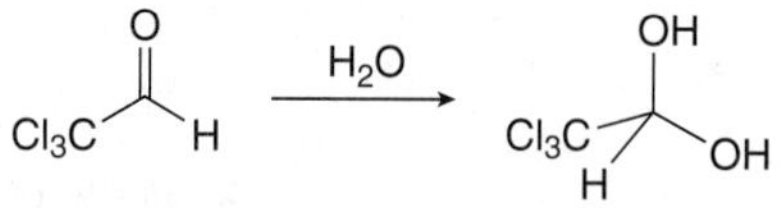

chloral

chloral hydrate

21.13A The Thermodynamics of Hydrate Formation

Whether addition of H_2O to a carbonyl group affords a good yield of the *gem*-diol depends on the relative energies of the starting material and the product. With less stable carbonyl starting materials, equilibrium favors the hydrate product, whereas with more stable carbonyl starting materials, equilibrium favors the carbonyl starting material. Because **alkyl groups stabilize a carbonyl group** (Section 20.2B):

- **Increasing the number of alkyl groups on the carbonyl carbon decreases the amount of hydrate at equilibrium.**

This can be illustrated by comparing the amount of hydrate formed from formaldehyde, acetaldehyde, and acetone.

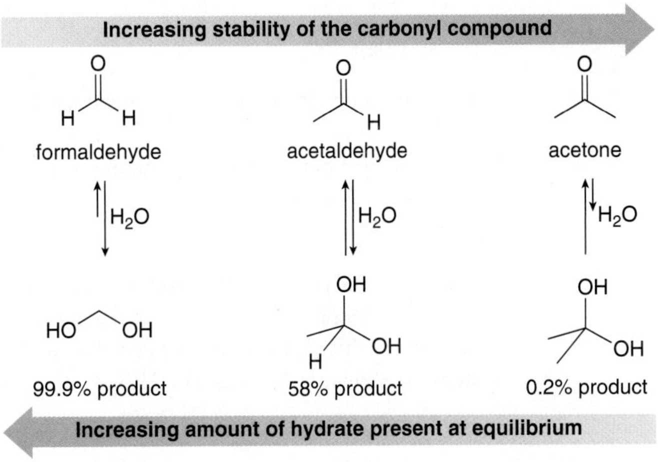

Formaldehyde, the least stable carbonyl compound, forms the largest percentage of hydrate. On the other hand, acetone and other ketones, which have two electron-donor R groups, form < 1% of the hydrate at equilibrium. Other electronic factors come into play as well.

- **Electron-donating groups near the carbonyl carbon stabilize the carbonyl group, decreasing the amount of the hydrate at equilibrium.**
- **Electron-withdrawing groups near the carbonyl carbon destabilize the carbonyl group, increasing the amount of hydrate at equilibrium.**

Chloral hydrate, a sedative sometimes administered to calm a patient prior to a surgical procedure, has also been used for less reputable purposes. Adding it to an alcoholic beverage makes a so-called knock-out drink, causing an individual who drinks it to pass out. Because it is addictive and care must be taken in its administration, it is a controlled substance.

This explains why chloral (trichloroacetaldehyde) forms a large amount of hydrate at equilibrium. Three electron-withdrawing Cl atoms place a partial positive charge on the α carbon to the carbonyl, destabilizing the carbonyl group, and therefore increasing the amount of hydrate at equilibrium.

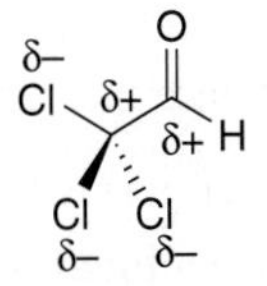

chloral

Adjacent like charges (δ+) *destabilize* the carbonyl and *increase* the amount of hydrate.

Problem 21.27 Which compound in each pair forms the higher percentage of *gem*-diol at equilibrium?

a. 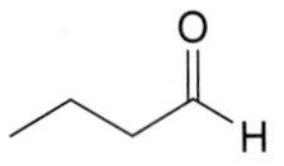 or 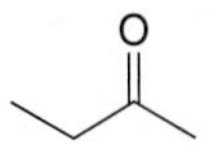

b. 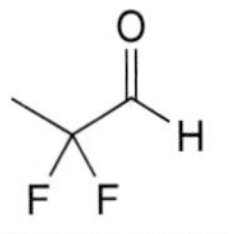 or 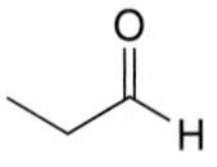

21.13B The Kinetics of Hydrate Formation

Although H_2O itself adds slowly to a carbonyl group, both acid and base catalyze the addition. In base, the nucleophile is ^{-}OH, and the mechanism follows the usual two steps for nucleophilic addition: **nucleophilic attack followed by protonation,** as shown in Mechanism 21.7.

Mechanism 21.7 Base-Catalyzed Addition of H_2O to a Carbonyl Group

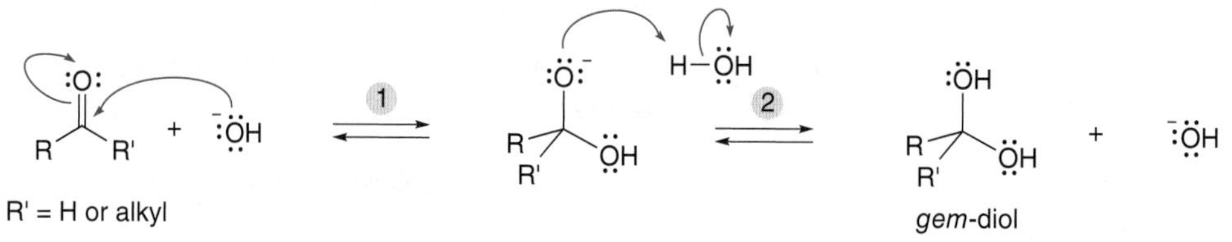

1 The nucleophile (^{-}OH) attacks the carbonyl, breaking the π bond and moving an electron pair out on oxygen.

2 Protonation of the negatively charged oxygen by H_2O forms the hydration product.

The acid-catalyzed addition follows the general mechanism presented in Section 21.7A. For a poorer nucleophile like H_2O to attack a carbonyl group, the **carbonyl must be protonated by acid first; thus, protonation *precedes* nucleophilic attack.** The overall mechanism has three steps, as shown in Mechanism 21.8.

Mechanism 21.8 Acid-Catalyzed Addition of H_2O to a Carbonyl Group

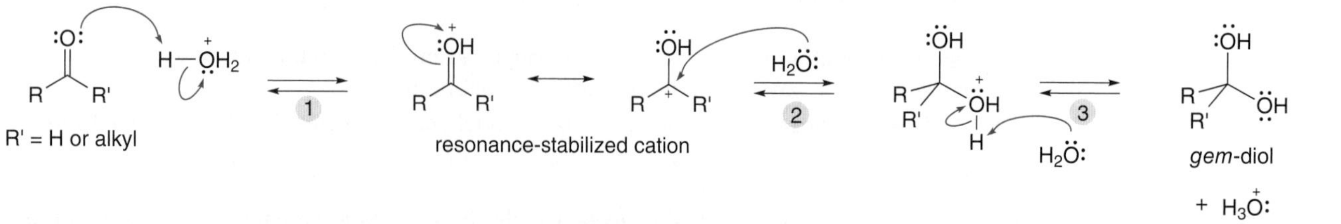

1 Protonation of the carbonyl oxygen forms a resonance-stabilized cation.

2 – 3 Nucleophilic attack and deprotonation form the *gem-diol*. The overall result is addition of H and OH to the carbonyl group.

Acid and base increase the rate of reaction for different reasons.

- **Base converts H_2O into ^{-}OH, a *stronger nucleophile.***
- **Acid protonates the carbonyl group, making it *more electrophilic* toward nucleophilic attack.**

These catalysts increase the rate of the reaction, but they do not affect the equilibrium constant. Starting materials that give a low yield of *gem*-diol do so whether or not a catalyst is present. Because these reactions are reversible, the conversion of *gem*-diols to aldehydes and ketones is also catalyzed by acid and base, and the steps of the mechanism are reversed.

Problem 21.28 Draw a stepwise mechanism for the following reaction.

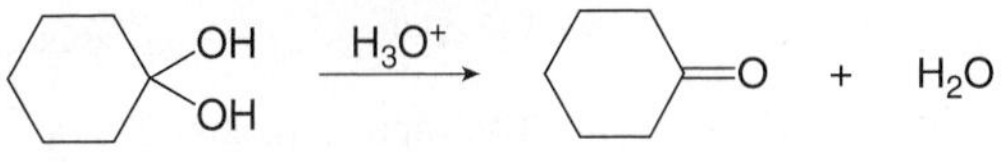
OH
OH
H_3O^+
O
+
H_2O

21.14 Addition of Alcohols—Acetal Formation

Aldehydes and ketones react with *two* equivalents of alcohol to form acetals. In an acetal, the carbonyl carbon from the aldehyde or ketone is now singly bonded to two OR" (alkoxy) groups.

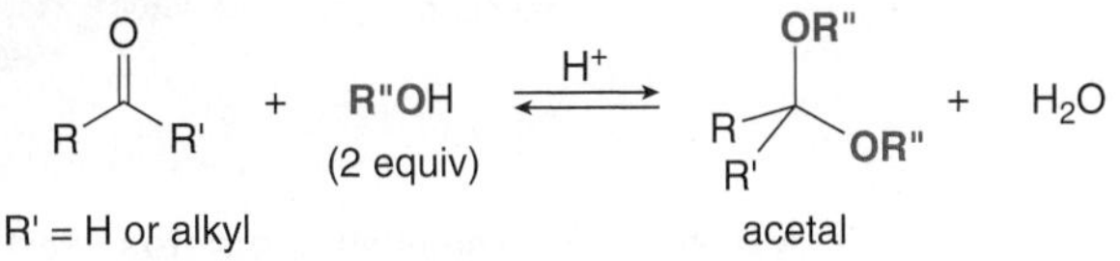
O
R
R'
+
R"OH
(2 equiv)
H^+
OR"
R
R'
OR"
+
H_2O
R' = H or alkyl
acetal

This reaction differs from other additions we have seen thus far, because **two equivalents of alcohol are added to the carbonyl group,** and two new C–O σ bonds are formed. Acetal formation is catalyzed by acids, commonly *p*-toluenesulfonic acid (TsOH).

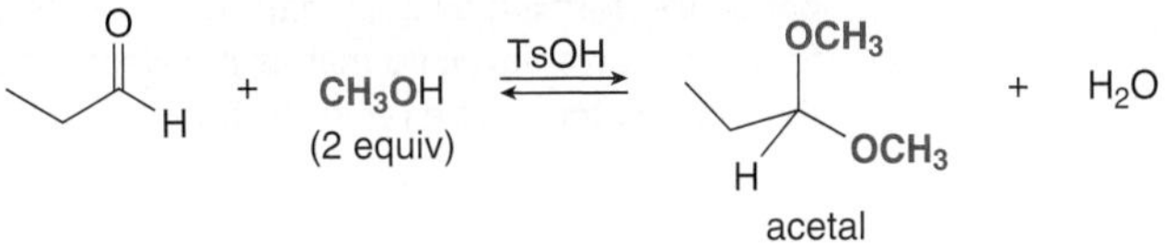
O
H
+
CH_3OH
(2 equiv)
TsOH
OCH_3
H
OCH_3
+
H_2O
acetal

When a diol such as ethylene glycol is used in place of two equivalents of ROH, a cyclic acetal is formed. Both oxygen atoms in the cyclic acetal come from the diol.

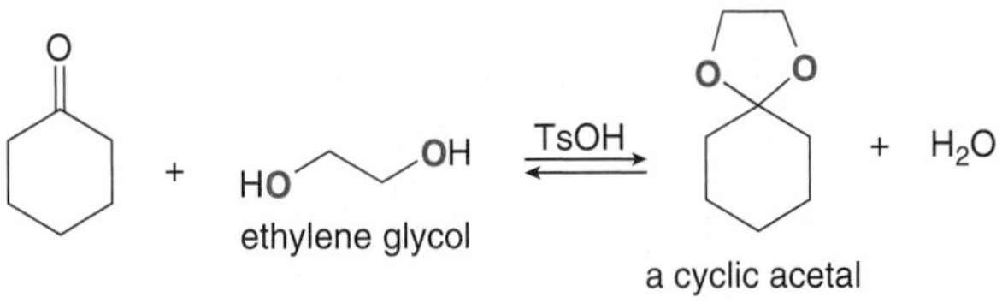
O
+
HO
OH
ethylene glycol
TsOH
O
O
+
H_2O
a cyclic acetal

Like *gem*-diol formation, the synthesis of acetals is reversible, and often the equilibrium favors reactants, not products. In acetal synthesis, however, water is formed as a by-product, so the equilibrium can be driven to the right by removing the water as it is formed. This can be done in a variety of ways in the laboratory. A drying agent can be added that reacts with the water, or more commonly, the water can be distilled from the reaction mixture as it

The term *acetal* refers to any compound derived from an aldehyde or ketone, having two OR groups bonded to a single carbon. The term *ketal* is sometimes used when the starting carbonyl compound is a ketone; that is, the carbon bonded to the alkoxy groups is *not* bonded to a H atom and the general structure is $R_2C(OR')_2$. Because ketals are considered a subclass of acetals in the IUPAC system, we will use the single general term *acetal* for any compound having two OR groups on a carbon atom.

Acetals are *not* ethers, even though both functional groups contain a C–O σ bond. Having two C–O σ bonds on the *same* carbon atom makes an acetal very different from an ether.

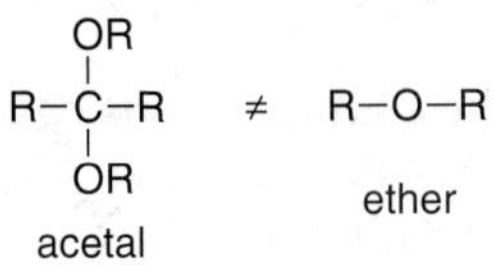
OR
R–C–R
OR
≠
R–O–R
acetal
ether

Figure 21.11
A Dean–Stark trap for removing water

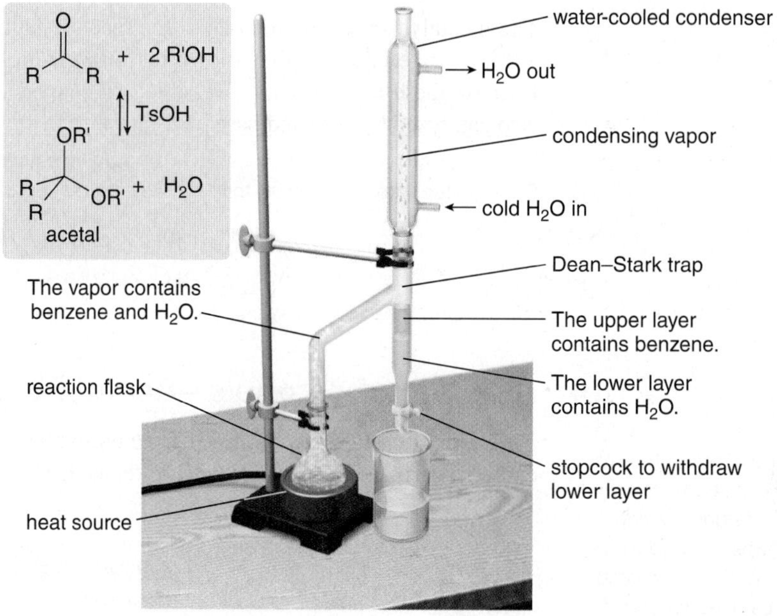

- **A Dean–Stark trap is an apparatus used for removing water from a reaction mixture.** To use a Dean–Stark trap to convert a carbonyl compound to an acetal:

  The carbonyl compound, an alcohol, and an acid are dissolved in benzene. As the mixture is heated, the carbonyl compound is converted to the acetal with water as a by-product. Benzene and water co-distill from the reaction mixture. When the hot vapors reach the cold condenser, they condense, forming a liquid that then collects in the glass tube below. Water, the more dense liquid, forms the lower layer, so that as it collects, it can be drained through the stopcock into a flask. In this way, water can be removed from a reaction mixture, driving the equilibrium.

is formed by using a Dean–Stark trap, as pictured in Figure 21.11. Driving an equilibrium to the right by removing one of the products is an application of Le Châtelier's principle (see Section 9.8).

Problem 21.29 Draw the products of each reaction.

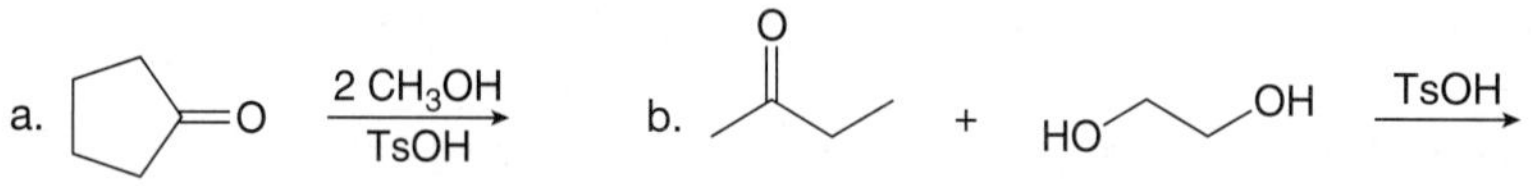

21.14A The Mechanism

The mechanism for acetal formation can be divided into two parts: **the addition of one equivalent of alcohol** to form a **hemiacetal,** followed by the **conversion of the hemiacetal** to the **acetal.** A **hemiacetal** has a carbon atom bonded to one OH group and one OR group.

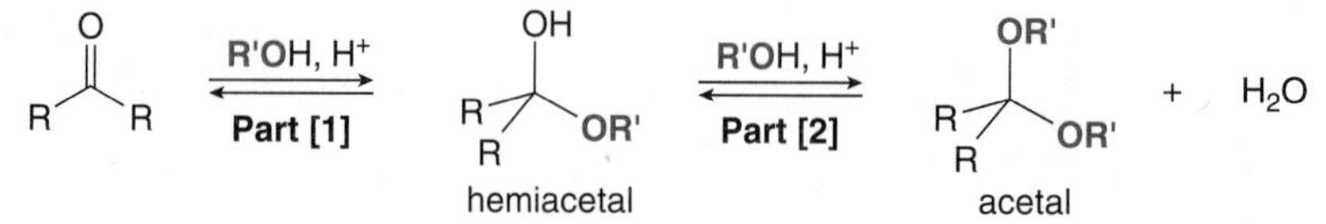

Like *gem*-diols, hemiacetals are often higher in energy than their carbonyl starting materials, making the direction of equilibrium unfavorable for hemiacetal formation. The elimination of H_2O, which can be removed from the reaction mixture to drive the equilibrium to favor product, occurs during the conversion of the hemiacetal to the acetal. This explains why two equivalents of ROH react with a carbonyl compound, forming the acetal as product.

Mechanism 21.9 is written in two parts with a general acid HA.

Mechanism 21.9 Acetal Formation

Part [1] Formation of a hemiacetal

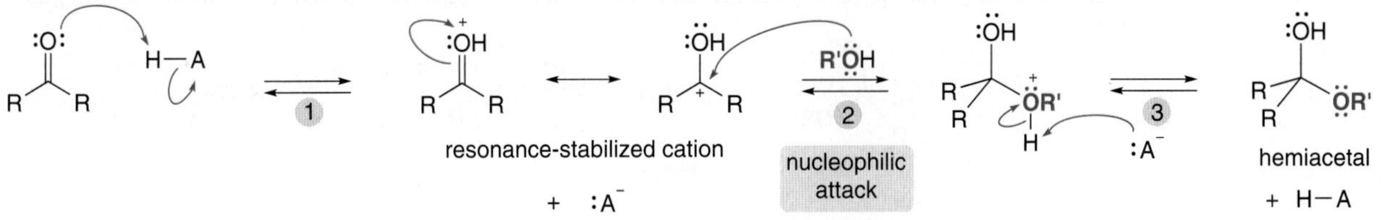

1 Protonation of the carbonyl oxygen forms a resonance-stabilized cation.

2 – 3 Nucleophilic attack by R'OH and deprotonation form the hemiacetal. The overall result is addition of H and OR' to the carbonyl group.

Part [2] Formation of an acetal

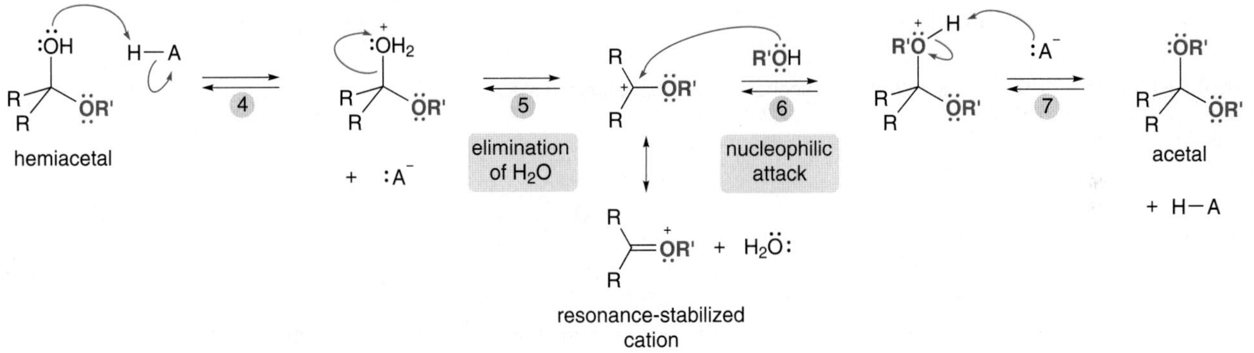

4 Protonation of the OH group of the hemiacetal forms a good leaving group.

5 Loss of H_2O forms a resonance-stabilized cation.

6 – 7 Nucleophilic attack by R'OH followed by loss of a proton forms the acetal. The overall result of Part [2] is the addition of a second OR' group to the carbonyl.

Although this mechanism is lengthy—there are seven steps altogether—there are only three different kinds of reactions: **addition of a nucleophile, elimination of a leaving group,** and **proton transfer.** Steps [2] and [6] involve nucleophilic attack and Step [5] eliminates H_2O. The other four steps in the mechanism shuffle protons from one oxygen atom to another, to make a better leaving group or a more electrophilic carbonyl group.

Problem 21.30 Label each compound as an acetal, a hemiacetal, or an ether.

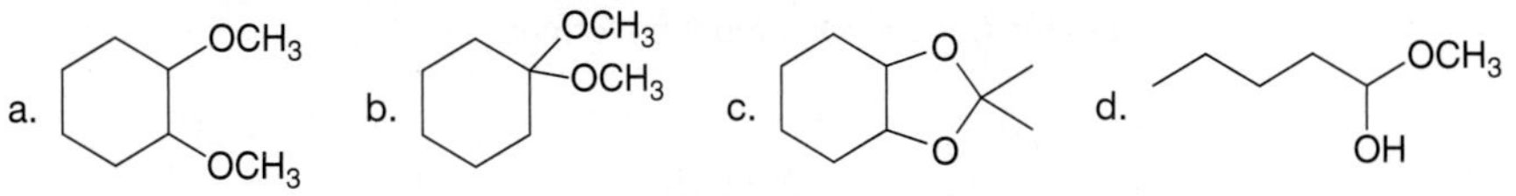

Problem 21.31 Draw a stepwise mechanism for the following reaction.

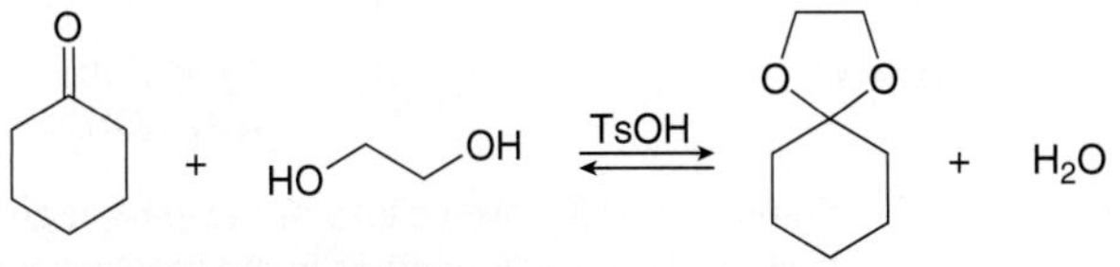

21.14B Hydrolysis of Acetals

Conversion of an aldehyde or ketone to an acetal is a **reversible reaction,** so **an acetal can be hydrolyzed to an aldehyde or ketone by treatment with aqueous acid.** Because this reaction is also an equilibrium process, it is driven to the right by using a large excess of water for hydrolysis.

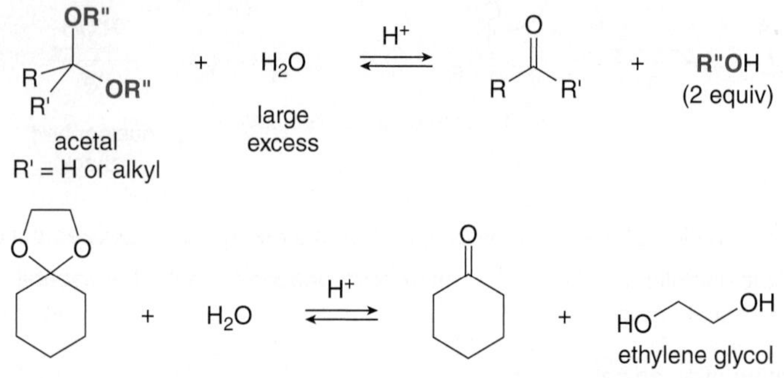

OR"
R
R'
OR"
acetal
R' = H or alkyl
+
H2O
large
excess
H+
O
R
R'
+
R"OH
(2 equiv)
O
O
+
H2O
H+
O
+
HO
OH
ethylene glycol

The mechanism for this reaction is the reverse of acetal synthesis, as illustrated in Sample Problem 21.4.

Sample Problem 21.4 Draw a stepwise mechanism for the following reaction.

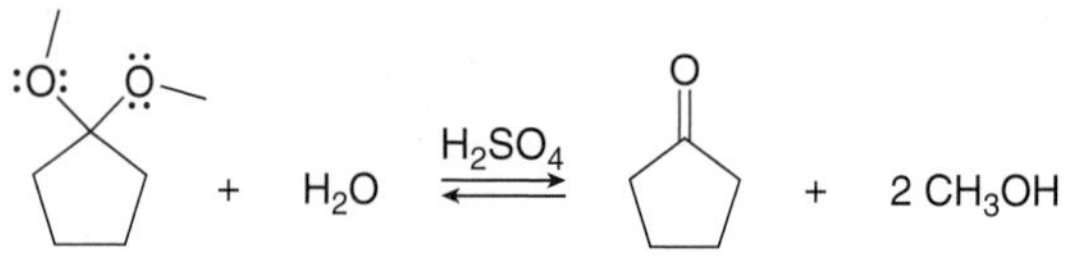

:O:
O
+
H2O
H2SO4
O
+
2 CH3OH

Solution

The mechanism is the reverse of acetal formation and involves two parts—conversion of the acetal to a hemiacetal, followed by conversion of the hemiacetal to the carbonyl compound.

Part [1] Conversion of the acetal to a hemiacetal

To convert this acetal to a hemiacetal, one molecule of CH_3OH must be eliminated and one molecule of H_2O must be added.

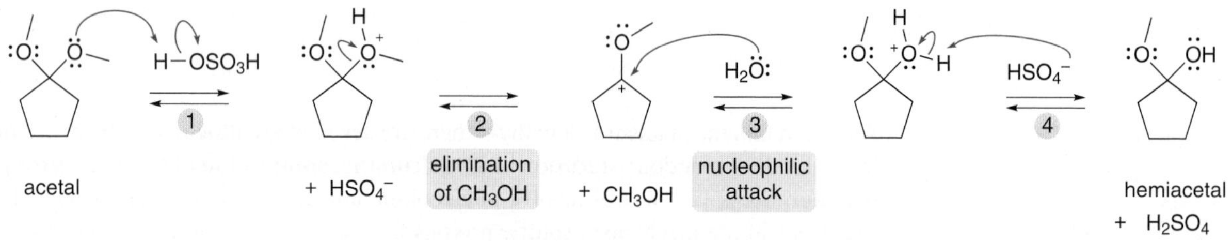

:O:
O
acetal
H–OSO3H
1
:O:
H
O
+ HSO4–
2
elimination
of CH3OH
:O
+ CH3OH
H2O:
3
nucleophilic
attack
:O:
H
O
H
HSO4–
4
:O:
OH
hemiacetal
+ H2SO4

Part [2] Conversion of the hemiacetal to the carbonyl compound

To convert the hemiacetal to a carbonyl compound, one molecule of CH_3OH must be eliminated and the C–O π bond must be formed.

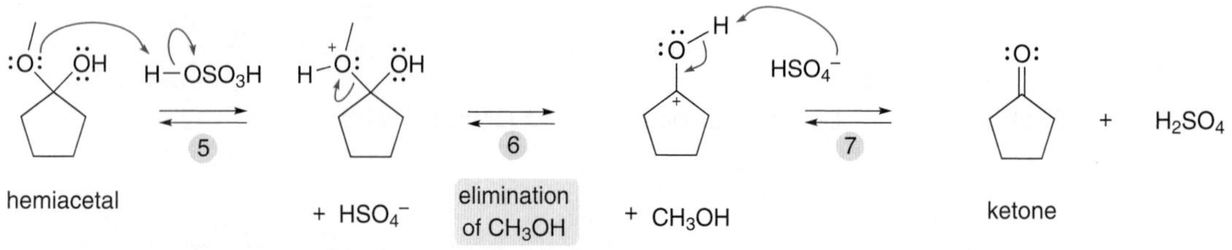

:O:
OH
hemiacetal
H–OSO3H
5
H–O:
OH
+ HSO4–
6
elimination
of CH3OH
:O
H
+ CH3OH
HSO4–
7
:O:
ketone
+
H2SO4

Steps [2] and [6] involve loss of the leaving group (CH_3OH), and Step [3] involves nucleophilic attack of H_2O. The other four steps in the mechanism shuffle protons from one oxygen atom to another. Steps [2] and [6] form resonance-stabilized carbocations, but only one resonance structure is drawn.

Acetal hydrolysis requires a strong acid to make a good leaving group (ROH). In Sample Problem 21.4, H_2SO_4 converts CH_3O^- into CH_3OH, a weak base and neutral leaving group. Acetal hydrolysis does not occur in base.

Problem 21.32 Draw the products of each reaction.

a. 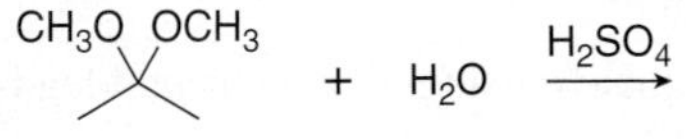
CH3O OCH3
+ H2O
H2SO4

b. 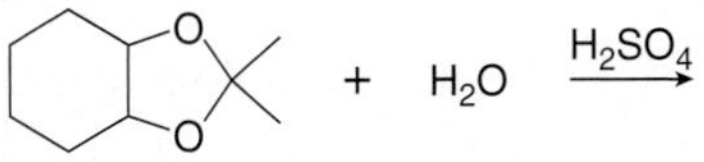
O
O
+ H2O
H2SO4

Problem 21.33 Safrole is a naturally occurring acetal isolated from sassafras plants. Once used as a common food additive in root beer and other beverages, it is now banned because it is carcinogenic. What compounds are formed when safrole is hydrolyzed with aqueous acid?

Sassafras, source of safrole

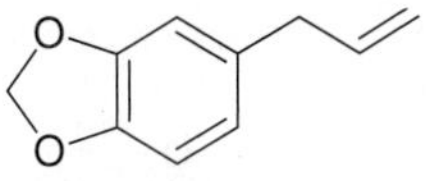
O
O

safrole

Problem 21.34 Identify the acetal in oleandrin, and draw the products formed by acid-catalyzed hydrolysis of the acetal.

Oleandrin, a toxic cardiac glycoside used in some countries for the treatment of congestive heart failure, is obtained from the sap of the oleander plant (*Nerium oleander*), a common ornamental plant in tropical and subtropical regions.

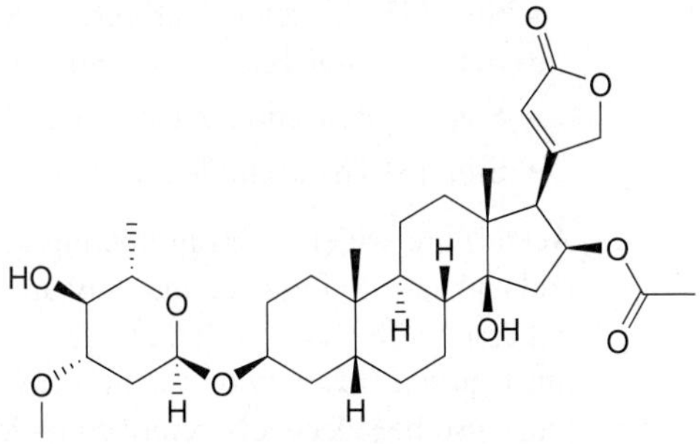
O
O
H
O
HO
O
H
H
OH
O
O
H
O

oleandrin

21.15 Acetals as Protecting Groups

Just as the *tert*-butyldimethylsilyl ethers are used as protecting groups for alcohols (Section 20.12), **acetals are valuable protecting groups for aldehydes and ketones.**

Suppose a starting material **A** contains both a ketone and an ester, and it is necessary to selectively reduce the ester to an alcohol (6-hydroxyhexan-2-one), leaving the ketone untouched. Such a selective reduction is *not* possible in one step. Because ketones are more readily reduced, methyl 5-hydroxyhexanoate is formed instead.

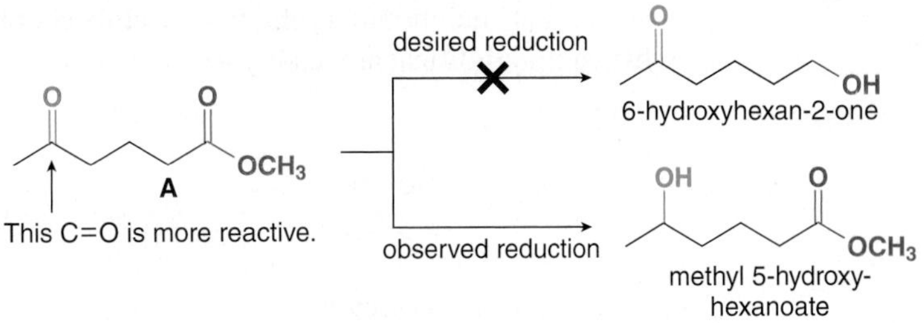
O
O
OCH3
A
This C=O is more reactive.
desired reduction
O
OH
6-hydroxyhexan-2-one
OH
O
OCH3
observed reduction
methyl 5-hydroxy-
hexanoate

To solve this problem we can use a protecting group to block the more reactive ketone carbonyl group. The overall process requires three steps.

[1] Protect the interfering functional group—the ketone carbonyl.
[2] Carry out the desired reaction—reduction.
[3] Remove the protecting group.

The following three-step sequence using a cyclic acetal leads to the desired product.

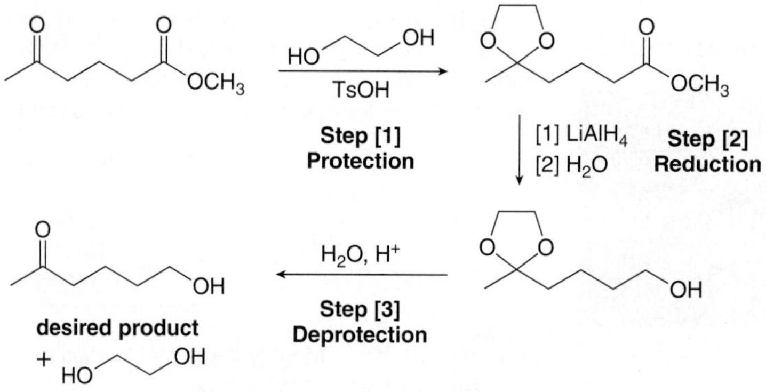

- **Step [1]** The ketone carbonyl is protected as a cyclic acetal by reaction of the starting material with $HOCH_2CH_2OH$ and TsOH.
- **Step [2]** Reduction of the ester is then carried out with $LiAlH_4$, followed by treatment with H_2O.
- **Step [3]** The acetal is then converted back to a ketone carbonyl group with aqueous acid.

Acetals are widely used protecting groups for aldehydes and ketones because they are easy to add and easy to remove, and they are stable to a wide variety of reaction conditions. Acetals do not react with base, oxidizing agents, reducing agents, or nucleophiles. Good protecting groups must survive a variety of reaction conditions that take place at other sites in a molecule, but they must also be selectively removed under mild conditions when needed.

Problem 21.35 How would you use a protecting group to carry out the following transformation?

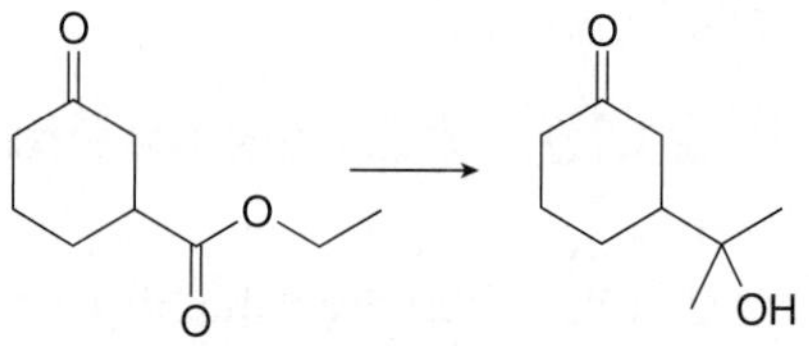

21.16 Cyclic Hemiacetals

Cyclic hemiacetals are also called **lactols.**

Although acyclic hemiacetals are generally unstable and therefore not present in appreciable amounts at equilibrium, **cyclic hemiacetals containing five- and six-membered rings are stable compounds** that are readily isolated.

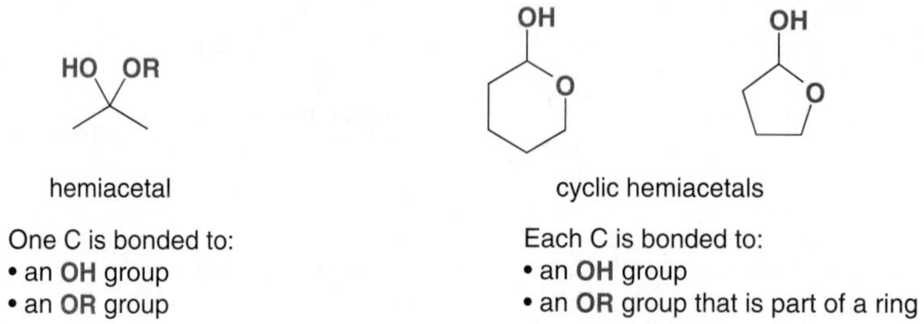

21.16A Forming Cyclic Hemiacetals

All hemiacetals are formed by nucleophilic addition of a hydroxy group to a carbonyl group. In the same way, cyclic hemiacetals are formed by **intramolecular cyclization of hydroxy aldehydes.**

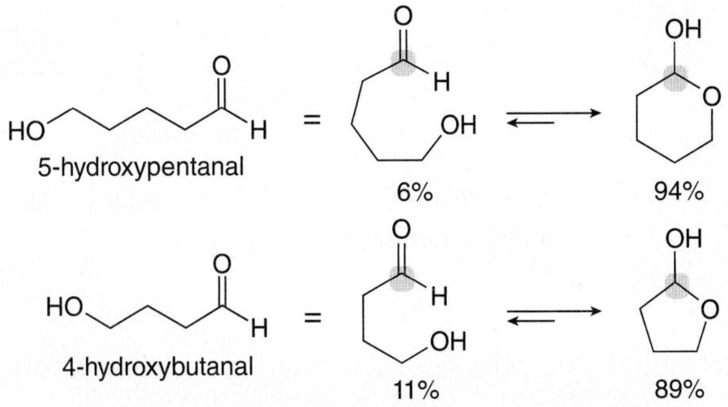

[Equilibrium proportions of each compound are given.]

Such intramolecular reactions to form five- and six-membered rings are faster than the corresponding intermolecular reactions. The two reacting functional groups, in this case OH and C=O, are held in close proximity, increasing the probability of reaction.

Problem 21.36 What lactol (cyclic hemiacetal) is formed from intramolecular cyclization of each hydroxy aldehyde?

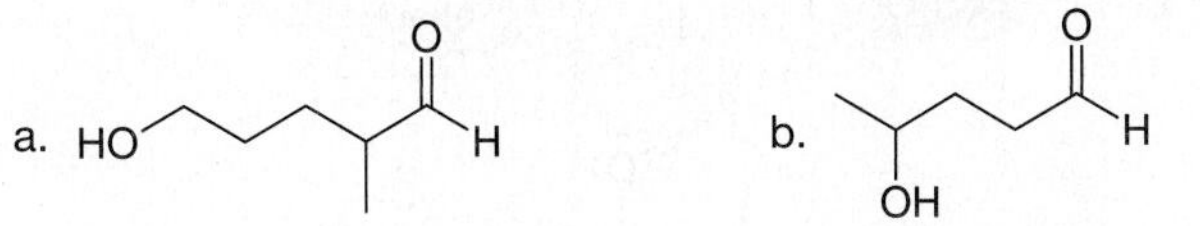

Hemiacetal formation is catalyzed by both acid and base. The acid-catalyzed mechanism is identical to Part [1] of Mechanism 21.9, except that the reaction occurs in an **intramolecular** fashion, as shown for the acid-catalyzed cyclization of 5-hydroxypentanal to form a six-membered cyclic hemiacetal in Mechanism 21.10.

Mechanism 21.10 Acid-Catalyzed Cyclic Hemiacetal Formation

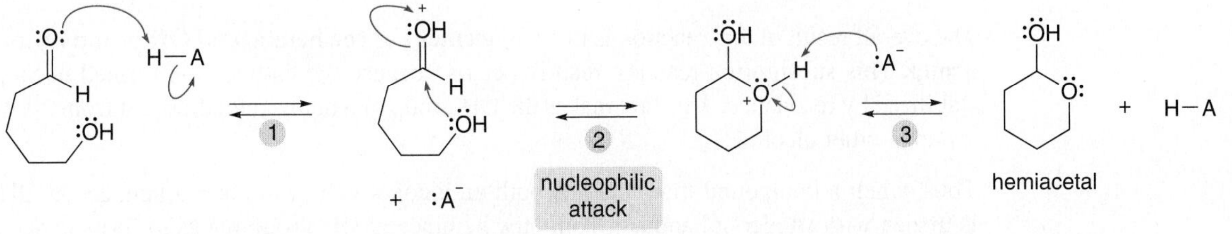

1 – 2 Protonation of the carbonyl oxygen followed by intramolecular nucleophilic attack forms the six-membered ring.

3 Deprotonation forms the neutral cyclic hemiacetal.

Intramolecular cyclization of a hydroxy aldehyde forms a **hemiacetal with a new stereogenic center, so that an equal amount of two enantiomers** results.

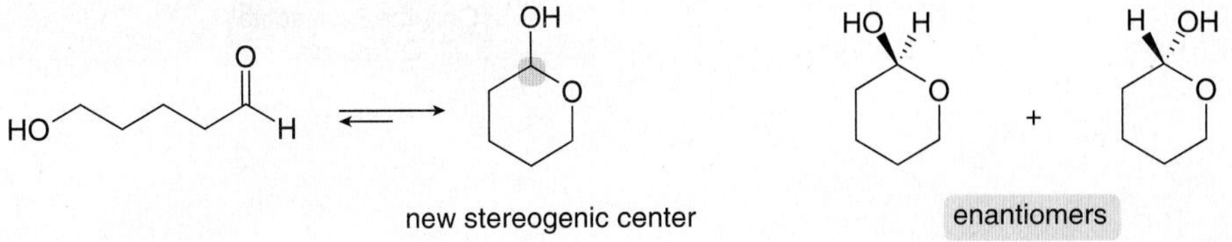

21.16B The Conversion of Hemiacetals to Acetals

Cyclic hemiacetals can be converted to acetals by treatment with an alcohol and acid. This reaction converts the OH group that is part of the hemiacetal to an OR group.

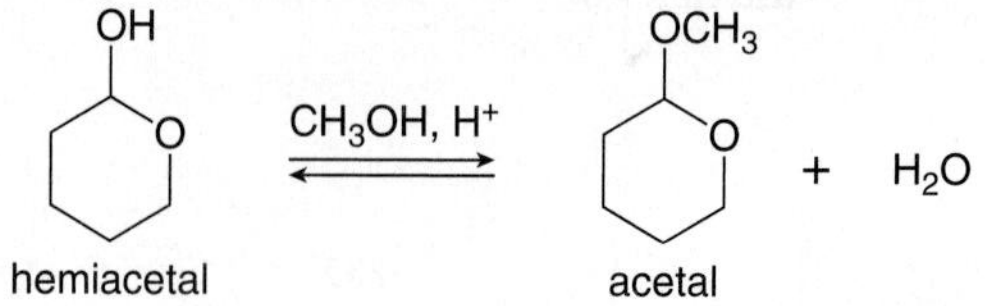

Mechanism 21.11, which is similar to Part [2] of Mechanism 21.9, illustrates the conversion of a cyclic hemiacetal to an acetal.

Mechanism 21.11 A Cyclic Acetal from a Cyclic Hemiacetal

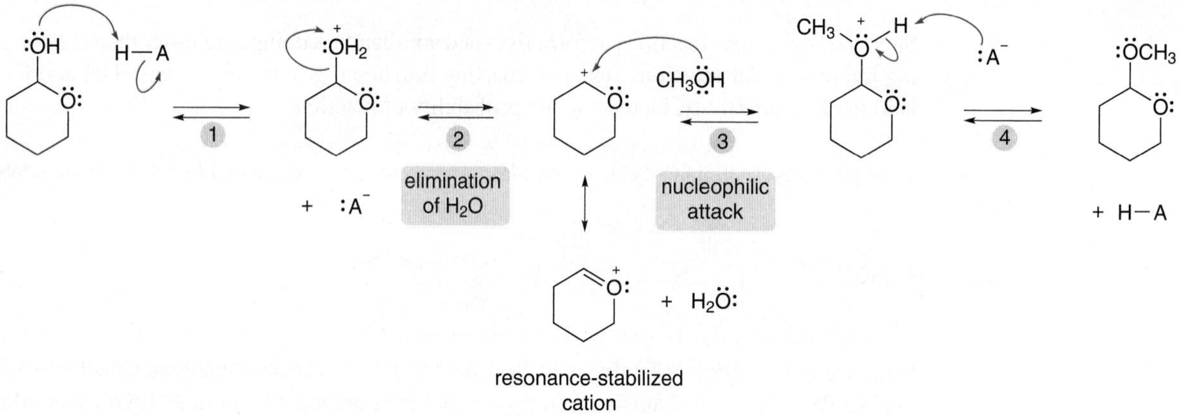

1 Protonation of the OH group of the hemiacetal forms a good leaving group.

2 Loss of H_2O forms a resonance-stabilized cation.

3 - 4 Nucleophilic attack by CH_3OH followed by loss of a proton forms the acetal.

The overall result of this reaction is the **replacement of the hemiacetal OH group by an OCH_3 group.** This substitution reaction readily occurs because the carbocation formed in Step [2] is stabilized by resonance. This fact makes the OH group of a hemiacetal different from the hydroxy group in other alcohols.

Thus, when a compound that contains both an alcohol OH group and a hemiacetal OH group is treated with an alcohol and acid, only the hemiacetal OH group reacts to form an acetal. The alcohol OH group does *not* react.

The conversion of cyclic hemiacetals to acetals is an important reaction in carbohydrate chemistry, as discussed in Chapter 27.

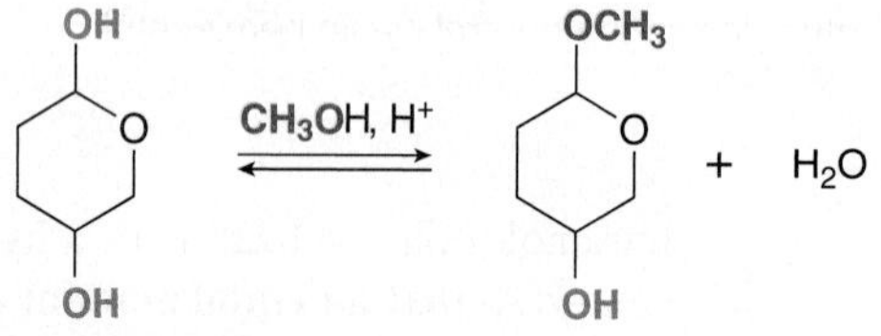

Only the hemiacetal OH reacts.

Problem 21.37 Two naturally occurring compounds that contain stable cyclic hemiacetals and acetals are monensin and digoxin, the chapter-opening molecule. Monensin, a polyether antibiotic produced by *Streptomyces cinamonensis,* is used as an additive in cattle feed. Digoxin is a widely prescribed cardiac drug used to increase the force of heart contractions. Label each acetal, hemiacetal, and ether in both compounds.

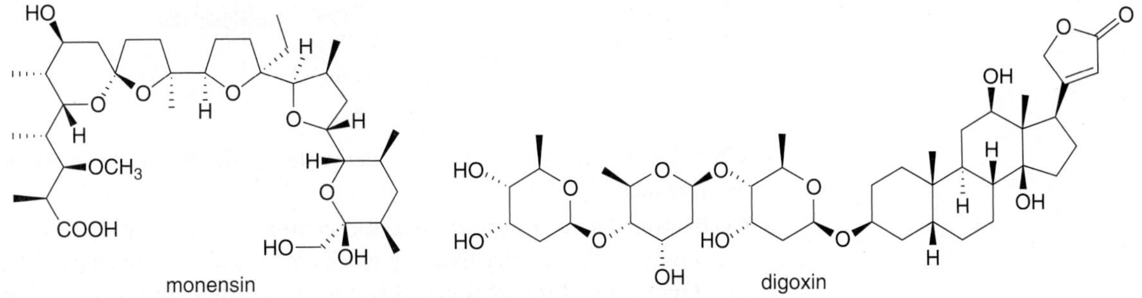
monensin
digoxin

Problem 21.38 Draw the products of each reaction.

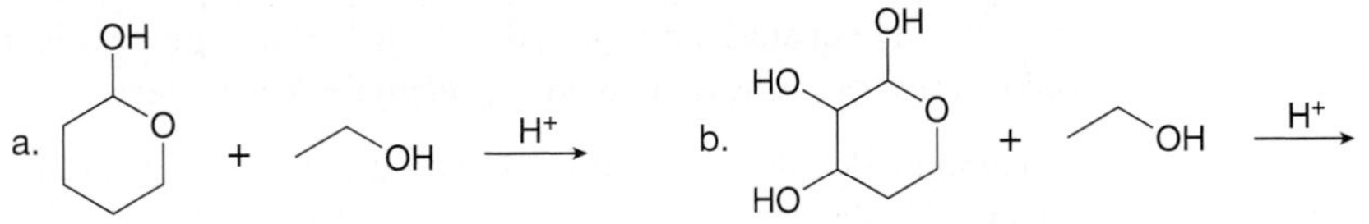

21.17 An Introduction to Carbohydrates

Carbohydrates, commonly referred to as sugars and starches, are polyhydroxy aldehydes and ketones, or compounds that can be hydrolyzed to them. Along with proteins, fatty acids, and nucleotides, they form one of the four main groups of biomolecules responsible for the structure and function of all living cells.

Many carbohydrates contain cyclic acetals or hemiacetals. Examples include **glucose,** the most common simple sugar, and **lactose,** the principal carbohydrate in milk. Hemiacetal carbons are labeled in blue, whereas the acetal carbon is labeled in red.

Glucose is the carbohydrate that is transported in the blood to individual cells. The hormone insulin regulates the level of glucose in the blood. Diabetes is a common disease that results from a deficiency of insulin, resulting in increased glucose levels in the blood and other metabolic abnormalities. Insulin injections control glucose levels.

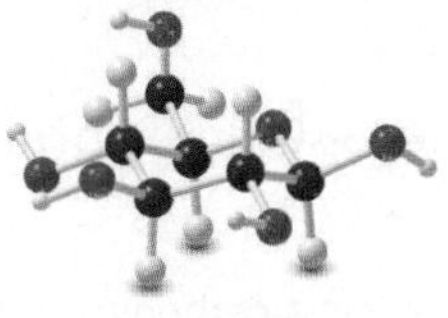

3-D structure

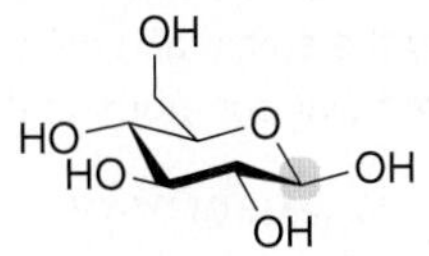

β-D-glucose
(one form of glucose)

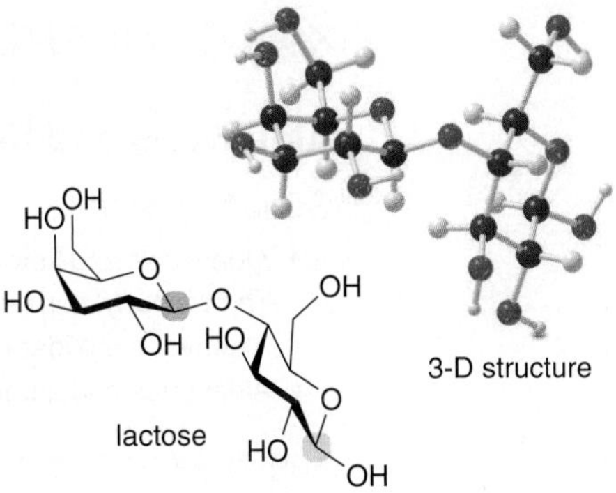
lactose
3-D structure

Hemiacetals in sugars are formed in the same way that other hemiacetals are formed—that is, by **cyclization of hydroxy aldehydes.** Thus, the hemiacetal of glucose is formed by cyclization of an acyclic *poly*hydroxy aldehyde **A,** as shown in the accompanying equation. This process illustrates two important features.

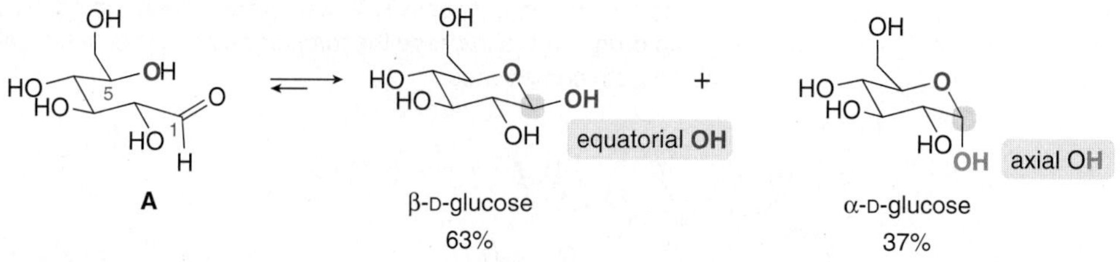

- When the OH group on C5 is the nucleophile, **cyclization yields a six-membered ring,** and this ring size is preferred.
- **Cyclization forms a new stereogenic center** (labeled in blue), exactly analogous to the cyclization of the simpler hydroxy aldehyde (5-hydroxypentanal) in Section 21.16A. **The new OH group of the hemiacetal can occupy either the equatorial or axial position.**

For glucose, this results in two cyclic forms, called **β-D-glucose** (having an equatorial OH group) and **α-D-glucose** (having an axial OH group). Because β-D-glucose has the new OH group in the more roomy equatorial position, this cyclic form of glucose is the major product. At equilibrium, only a trace of the acyclic hydroxy aldehyde **A** is present.

Many more details on this process and other aspects of carbohydrate chemistry are presented in Chapter 27.

Problem 21.39

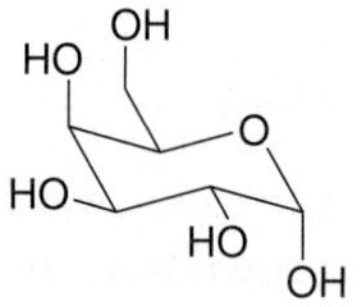

α-D-galactose

a. How many stereogenic centers are present in α-D-galactose?
b. Label the hemiacetal carbon in α-D-galactose.
c. Draw the structure of β-D-galactose.
d. Draw the structure of the polyhydroxy aldehyde that cyclizes to α- and β-D-galactose.
e. From what you learned in Section 21.16B, what product(s) is (are) formed when α-D-galactose is treated with CH_3OH and an acid catalyst?

KEY CONCEPTS

Aldehydes and Ketones—Nucleophilic Addition

General Facts

- Aldehydes and ketones contain a carbonyl group bonded to only H atoms or R groups. The carbonyl carbon is sp^2 hybridized and trigonal planar (21.1).
- Aldehydes are identified by the suffix *-al,* whereas ketones are identified by the suffix *-one* (21.2).
- Aldehydes and ketones are polar compounds that exhibit dipole–dipole interactions (21.3).

Summary of Spectroscopic Absorptions of RCHO and R_2CO (21.4)

IR absorptions	C=O	~1715 cm^{-1} for ketones • Frequency increases with decreasing ring size. ~1730 cm^{-1} for aldehydes • For both RCHO and R_2CO, the frequency decreases with conjugation.
	C_{sp^2}–H of CHO	~2700–2830 cm^{-1} (one or two peaks)
^{1}H NMR absorptions	CHO	9–10 ppm (highly deshielded proton)
	C–H α to C=O	2–2.5 ppm (somewhat deshielded C_{sp^3}–H)
^{13}C NMR absorption	C=O	190–215 ppm

Nucleophilic Addition Reactions

[1] Addition of hydride (H^-) (21.8)

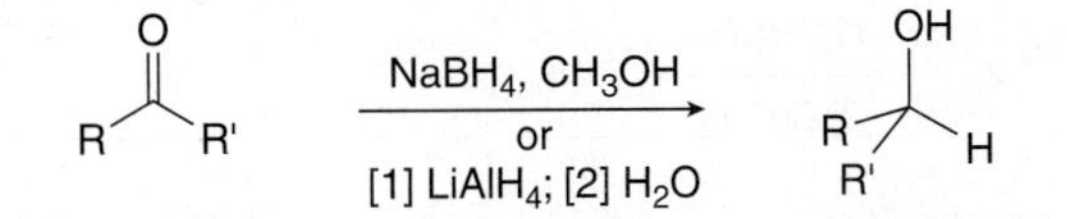
NaBH4, CH3OH
or
[1] LiAlH4; [2] H2O

R' = H or alkyl — 1° or 2° alcohol

- The mechanism has two steps.
- $H:^-$ adds to the planar C=O from both sides.

[2] Addition of organometallic reagents (R^-) (21.8)

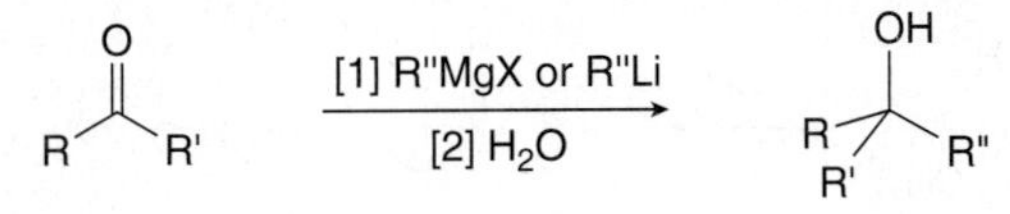
[1] R"MgX or R"Li
[2] H2O

R' = H or alkyl — 1°, 2°, or 3° alcohol

- The mechanism has two steps.
- $(R'')^-$ adds to the planar C=O from both sides.

[3] Addition of cyanide (^-CN) (21.9)

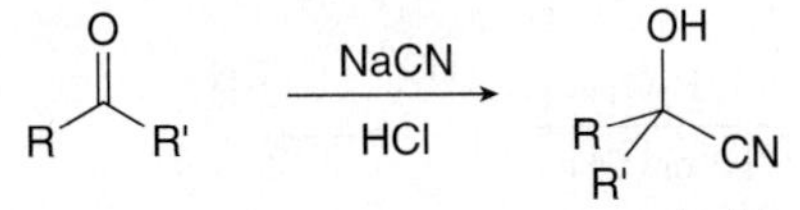
NaCN
HCl

R' = H or alkyl — cyanohydrin

- The mechanism has two steps.
- ^-CN adds to the planar C=O from both sides.

[4] Wittig reaction (21.10)

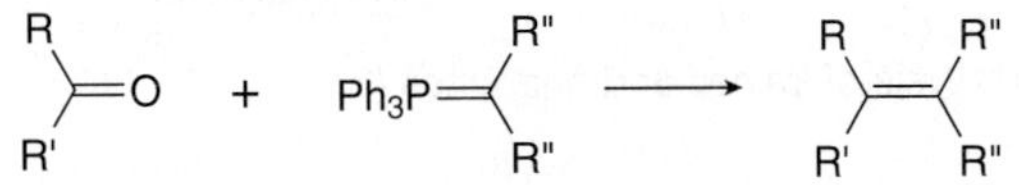

R' = H or alkyl — Wittig reagent — alkene

- The reaction forms a new C–C σ bond and a new C–C π bond.
- $Ph_3P{=}O$ is formed as by-product.

[5] Addition of 1° amines (21.11)

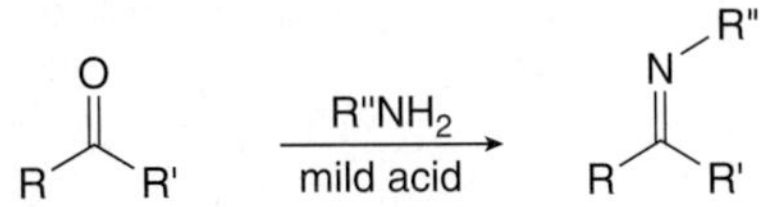
R"NH2
mild acid

R' = H or alkyl — imine

- The reaction is fastest at pH 4–5.
- The intermediate carbinolamine is unstable, and loses H_2O to form the C=N.

[6] Addition of 2° amines (21.12)

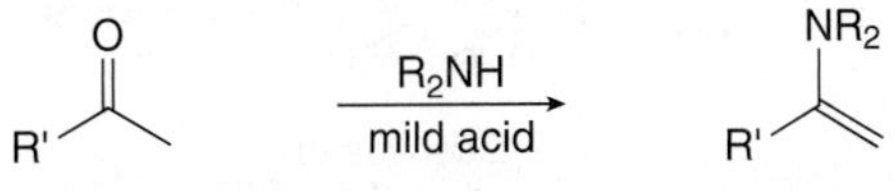
R2NH
mild acid

R' = H or alkyl — enamine

- The reaction is fastest at pH 4–5.
- The intermediate carbinolamine is unstable, and loses H_2O to form the C=C.

[7] Addition of H_2O—Hydration (21.13)

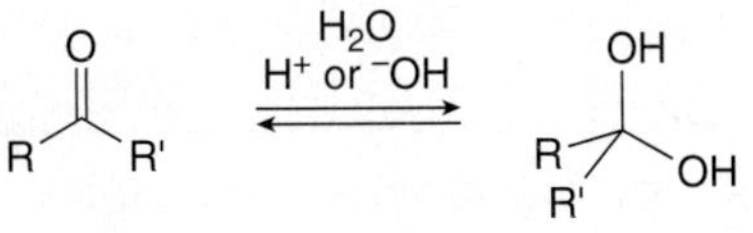
H2O
H+ or −OH

R' = H or alkyl — *gem*-diol

- The reaction is reversible. Equilibrium favors the product only with less stable carbonyl compounds (e.g., H_2CO and Cl_3CCHO).
- The reaction is catalyzed by either H^+ or ^-OH.

[8] Addition of alcohols (21.14)

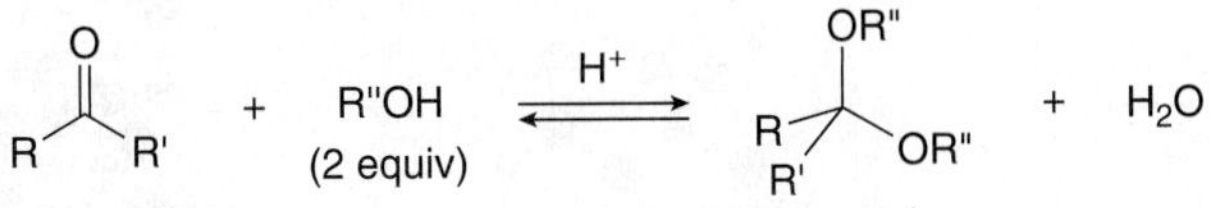
R"OH
(2 equiv)
H+
H2O

R' = H or alkyl — acetal

- The reaction is reversible.
- The reaction is catalyzed by acid.
- Removal of H_2O drives the equilibrium to favor the products.

Other Reactions

[1] Synthesis of Wittig reagents (21.10A)

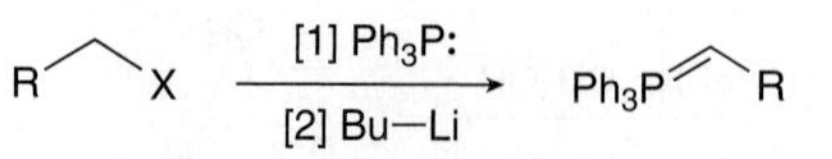
[1] Ph3P:
[2] Bu—Li

- Step [1] is best with CH_3X and RCH_2X because the reaction follows an S_N2 mechanism.
- A strong base is needed for proton removal in Step [2].

[2] Conversion of cyanohydrins to aldehydes and ketones (21.9)

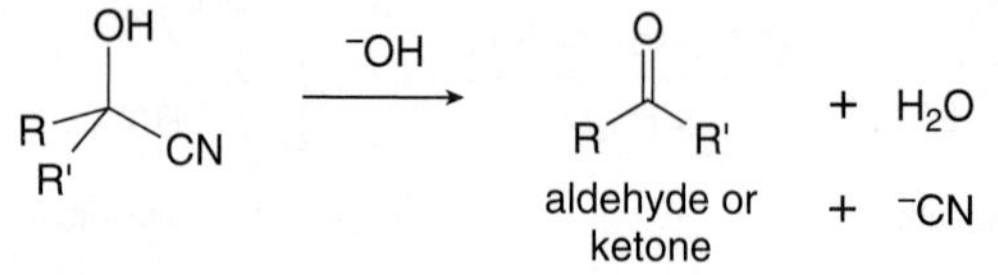
aldehyde or ketone

- This reaction is the reverse of cyanohydrin formation.

[3] Hydrolysis of nitriles (21.9)

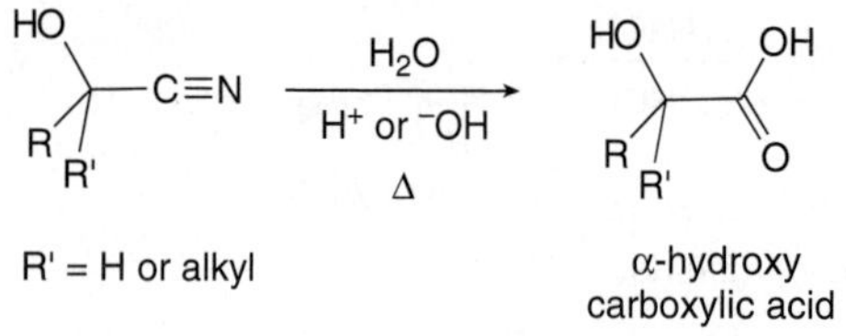
H2O
H+ or −OH
Δ
R' = H or alkyl
α-hydroxy carboxylic acid

[4] Hydrolysis of imines and enamines (21.12)

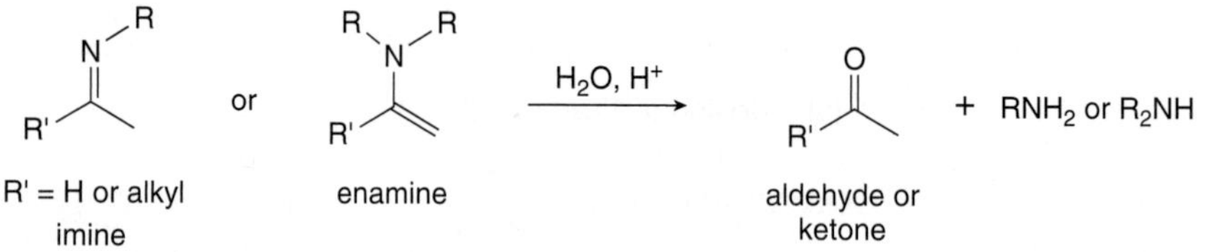
or
H2O, H+
+ RNH2 or R2NH
R' = H or alkyl
imine
enamine
aldehyde or ketone

[5] Hydrolysis of acetals (21.14)

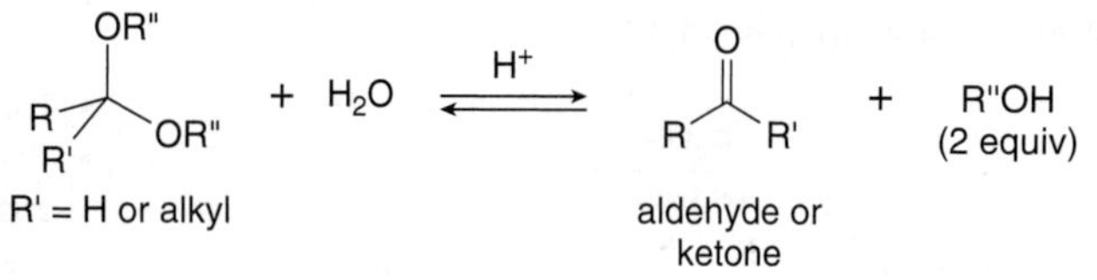
+ H2O
H+
R"OH (2 equiv)
R' = H or alkyl
aldehyde or ketone

- The reaction is acid catalyzed and is the reverse of acetal synthesis.
- A large excess of H_2O drives the equilibrium to favor the products.

PROBLEMS

Problems Using Three-Dimensional Models

21.40 (a) Give the IUPAC name for **A** and **B.** (b) Draw the product formed when **A** or **B** is treated with each reagent: [1] $NaBH_4$, CH_3OH; [2] CH_3MgBr, then H_2O; [3] $Ph_3P{=}CHOCH_3$; [4] $CH_3CH_2CH_2NH_2$, mild acid; [5] $HOCH_2CH_2CH_2OH$, H^+.

A B

21.41 Rank the following compounds in order of increasing reactivity in nucleophilic addition.

21.42 What carbonyl compound and diol are needed to prepare each compound?

a. 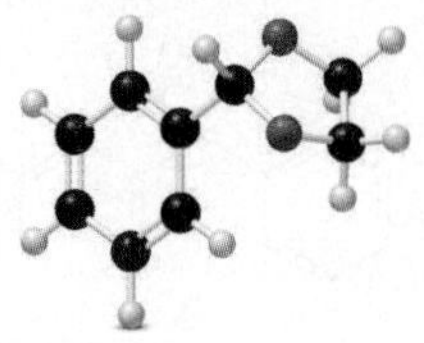

b. 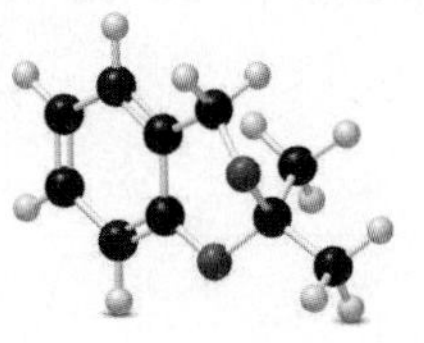

Nomenclature

21.43 Give the IUPAC name for each compound.

a. 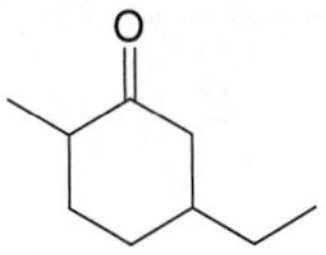

b. 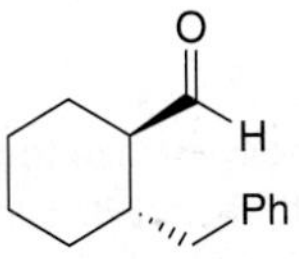

c. 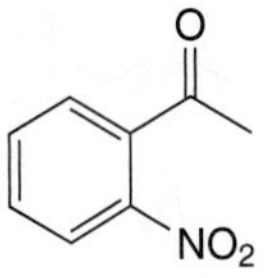

d. 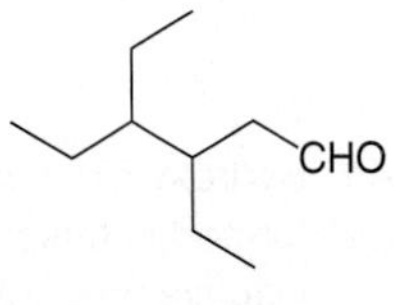

e. 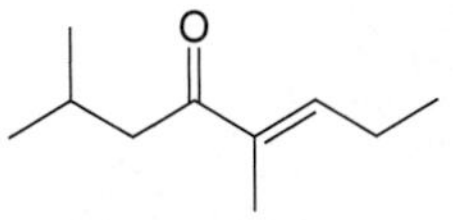

f. 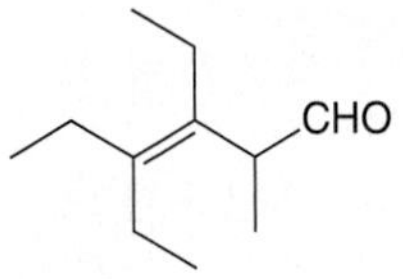

21.44 Give the structure corresponding to each name.

a. 2-methyl-3-phenylbutanal
b. 3,3-dimethylcyclohexanecarbaldehyde
c. 3-benzoylcyclopentanone
d. 2-formylcyclopentanone
e. (*R*)-3-methylheptan-2-one
f. *m*-acetylbenzaldehyde
g. 2-*sec*-butylcyclopent-3-enone
h. 5,6-dimethylcyclohex-1-enecarbaldehyde

Reactions

21.45 Draw the products formed in each reaction sequence.

a. 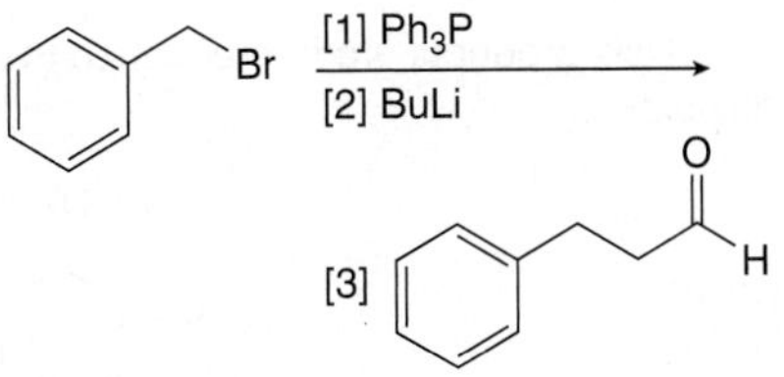

b. 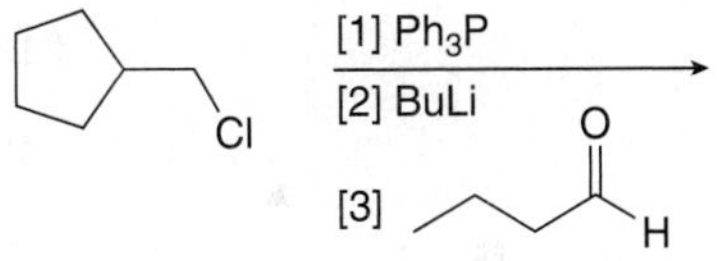

21.46 Draw the products of each reaction.

a. 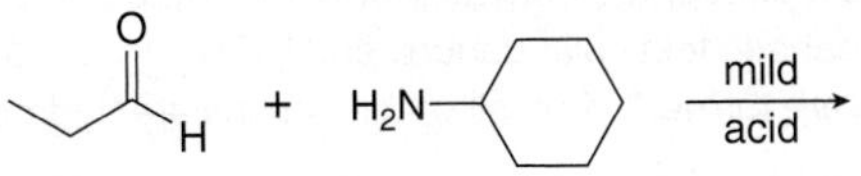

b. 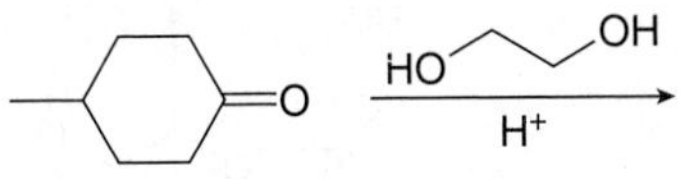

c. 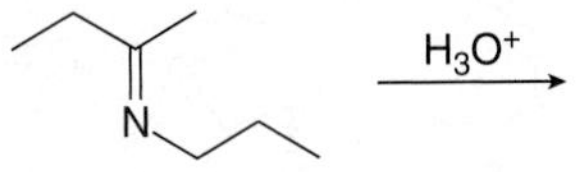

d. 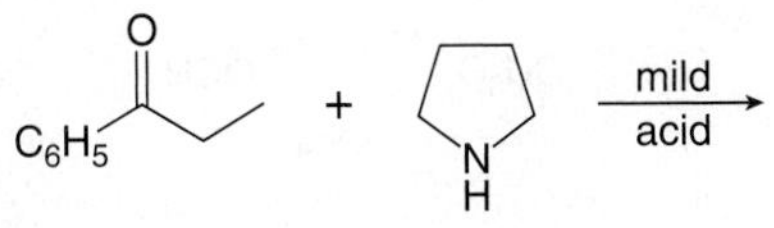

e. 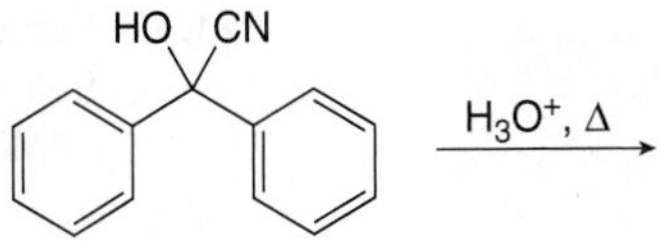

f. 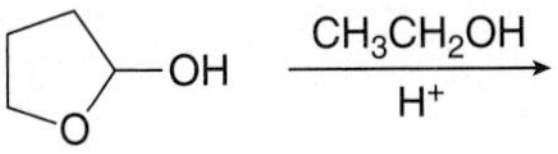

g. 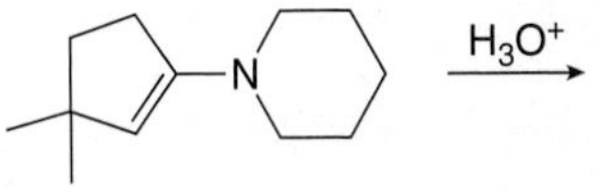

h. 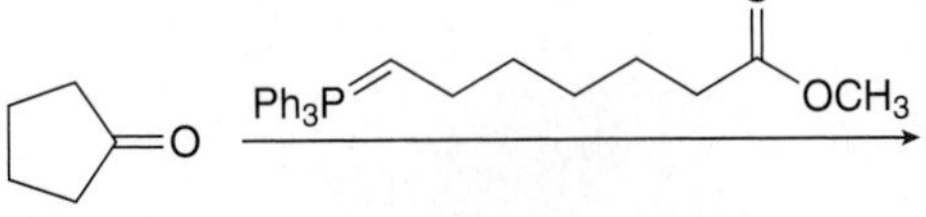

21.47 What carbonyl compound and alcohol are formed by hydrolysis of each acetal?

a. 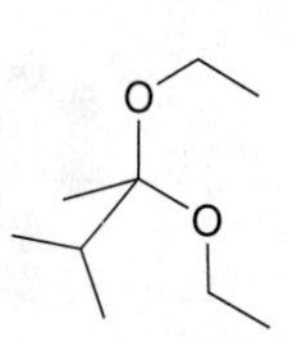

b. 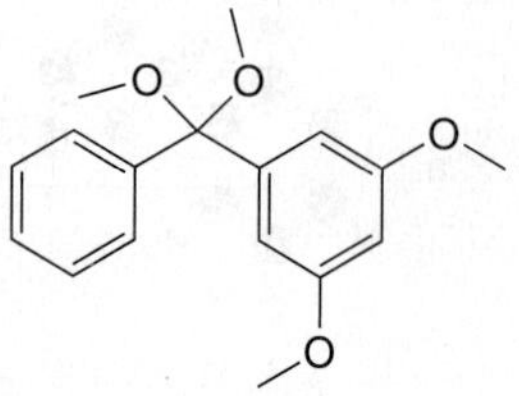

c. 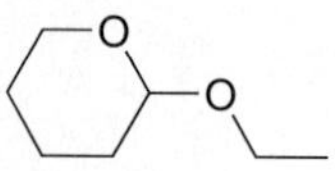

21.48 Draw all stereoisomers formed in each reaction.

a. 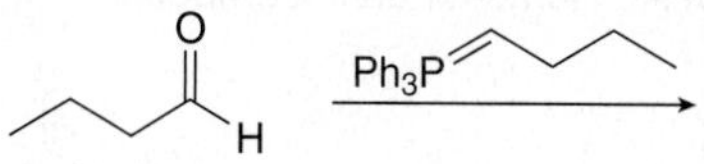
Ph3P

b. 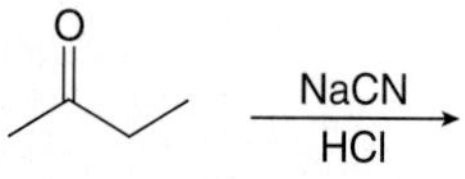
NaCN
HCl

c. 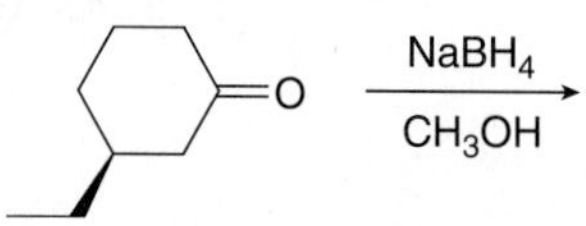
NaBH4
CH3OH

d. 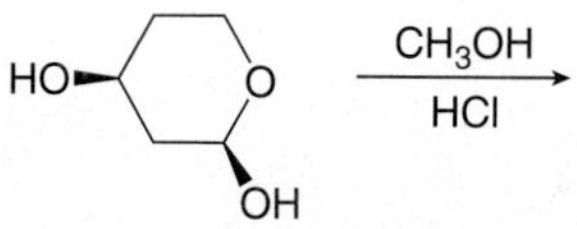
CH3OH
HCl

21.49 Hydroxy aldehydes **A** and **B** readily cyclize to form hemiacetals. Draw the stereoisomers formed in this reaction from both **A** and **B.** Explain why this process gives an optically inactive product mixture from **A,** and an optically active product mixture from **B.**

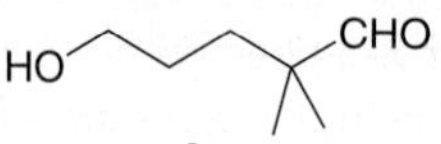

A

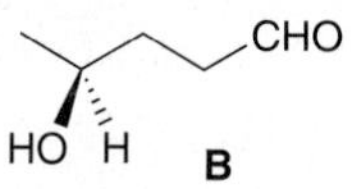

B

21.50 What products are formed when each acetal is hydrolyzed with aqueous acid?

a. 

b. 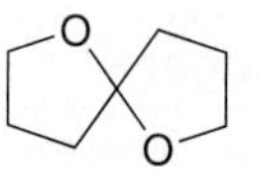

21.51 What hydrolysis products are formed when the following compound is treated with aqueous acid?

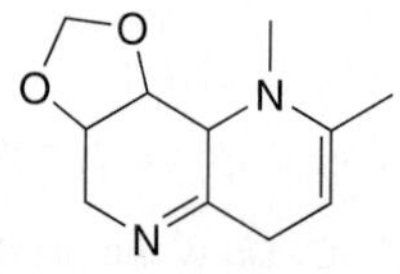

21.52 Etoposide, sold as a phosphate derivative with the trade name of Etopophos, is used for the treatment of lung cancer, testicular cancer, and lymphomas. (a) Locate the acetals in etoposide. (b) What products are formed when all of the acetals are hydrolyzed with aqueous acid?

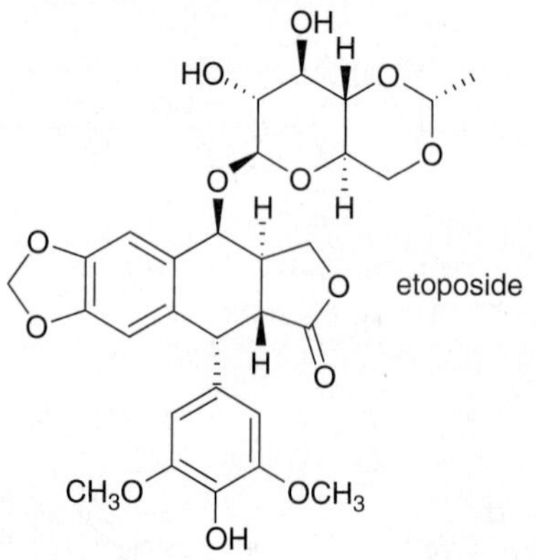

etoposide

Properties of Aldehydes and Ketones

21.53 Which compound forms the higher concentration of hydrate at equilibrium, PhCOCHO or $PhCH_2CHO$? Explain your reasoning.

21.54 Consider the para-substituted aromatic ketones, $NO_2C_6H_4COCH_3$ (*p*-nitroacetophenone) and $CH_3OC_6H_4COCH_3$ (*p*-methoxyacetophenone).
- a. Which carbonyl compound is more stable?
- b. Which compound forms the higher percentage of hydrate at equilibrium?
- c. Which compound exhibits a carbonyl absorption at higher wavenumber in its IR spectrum? Explain your reasoning in each part.

Synthesis

21.55 What Wittig reagent and carbonyl compound are needed to prepare each alkene? When two routes are possible, indicate which route, if any, is preferred.

a. 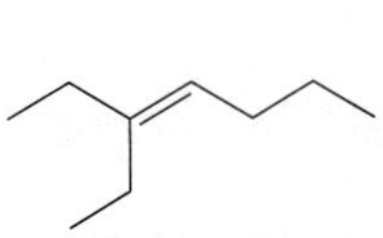

b. 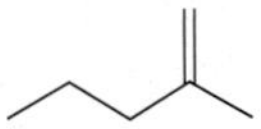

c. 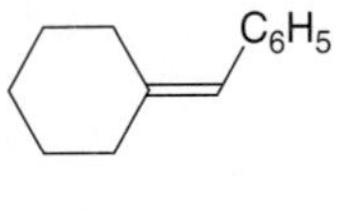

21.56 What carbonyl compound and amine or alcohol are needed to prepare each product?

a. 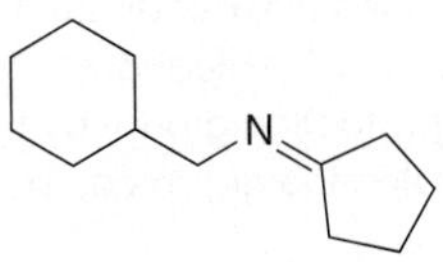

b. 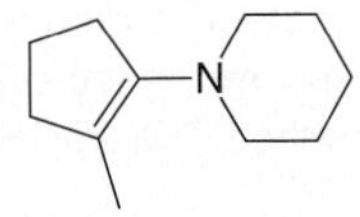

c. 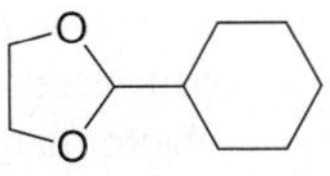

d. 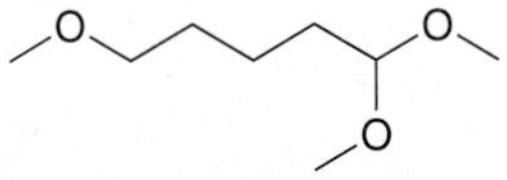

21.57 Show two different methods to carry out the following transformation: a one-step method using a Wittig reagent, and a two-step method using a Grignard reagent. Which route, if any, is preferred?

21.58 Devise a synthesis of each alkene using a Wittig reaction to form the double bond. You may use benzene and organic alcohols having four or fewer carbons as starting materials and any required reagents.

a. 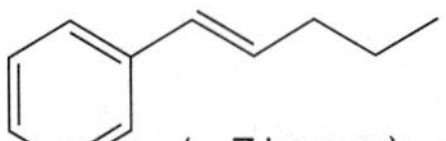
(+ *Z* isomer)

b. 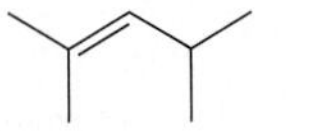

21.59 Devise a synthesis of each compound from cyclohexene and organic alcohols. You may use any other required organic or inorganic reagents.

a. 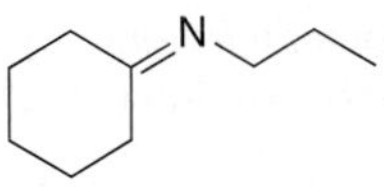

b. 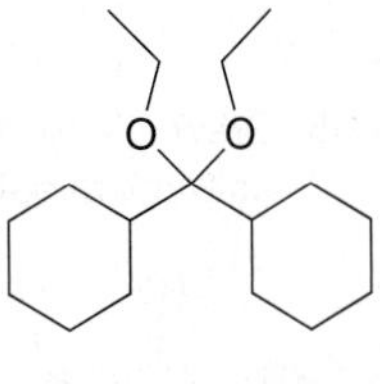

21.60 Devise a synthesis of each compound from the given starting materials. You may also use organic alcohols having four or fewer carbons, and any organic or inorganic reagents.

a. 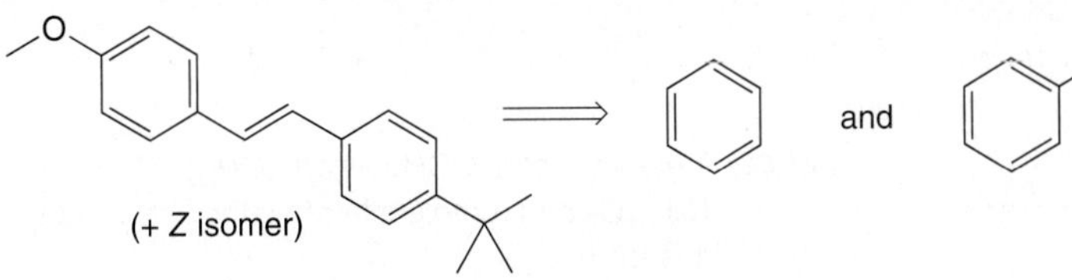

b. 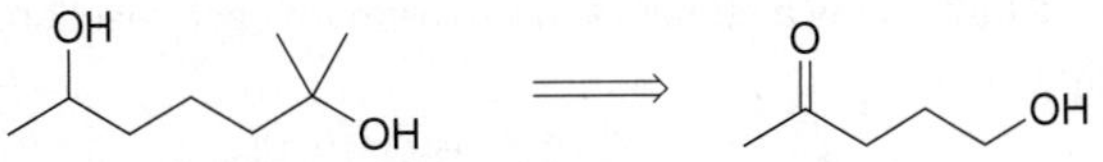

c. 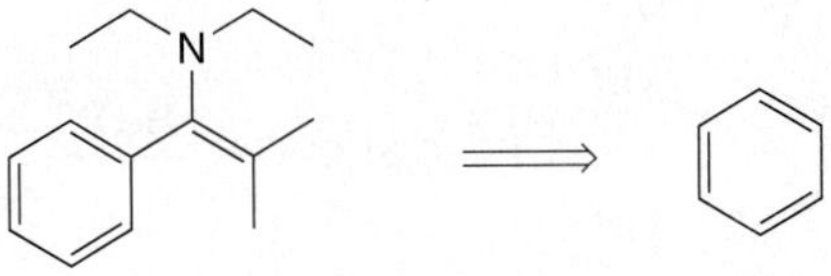

21.61 Devise a synthesis of each compound from ethanol (CH_3CH_2OH) as the only source of carbon atoms. You may use any other organic or inorganic reagents you choose.

a. 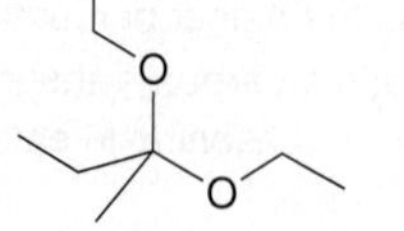 b. 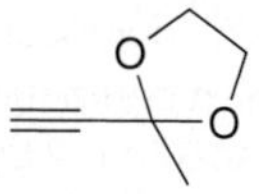

Protecting Groups

21.62 Design a stepwise synthesis to convert cyclopentanone and 4-bromobutanal to hydroxy aldehyde **A.**

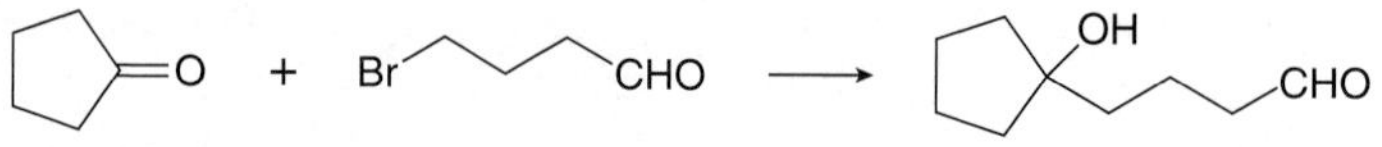

cyclopentanone 4-bromobutanal **A**

21.63 Besides the *tert*-butyldimethylsilyl ethers introduced in Chapter 20, there are many other widely used alcohol protecting groups. For example, an alcohol such as cyclohexanol can be converted to a **meth**o**xy** **m**ethyl ether (a MOM protecting group) by treatment with base and chloromethyl methyl ether, $ClCH_2OCH_3$. The protecting group can be removed by treatment with aqueous acid.

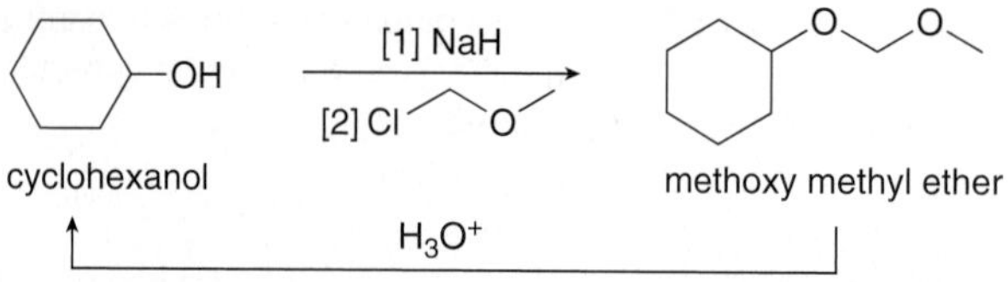

a. Write a stepwise mechanism for the formation of a MOM ether from cyclohexanol.
b. What functional group comprises a MOM ether?
c. Besides cyclohexanol, what other products are formed by aqueous hydrolysis of the MOM ether? Draw a stepwise mechanism that accounts for formation of each product.

Mechanism

21.64 Draw a stepwise mechanism for the following reaction.

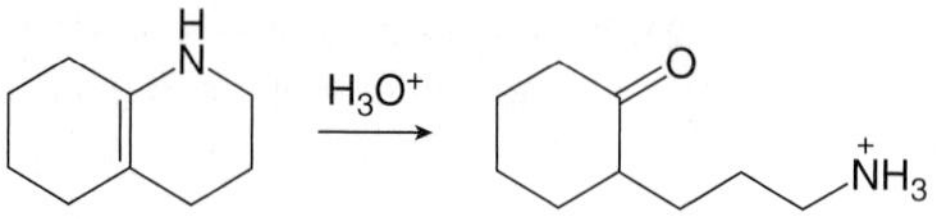

21.65 Draw a stepwise mechanism for the following reaction, a key step in the synthesis of the anti-inflammatory drug celecoxib (trade name Celebrex).

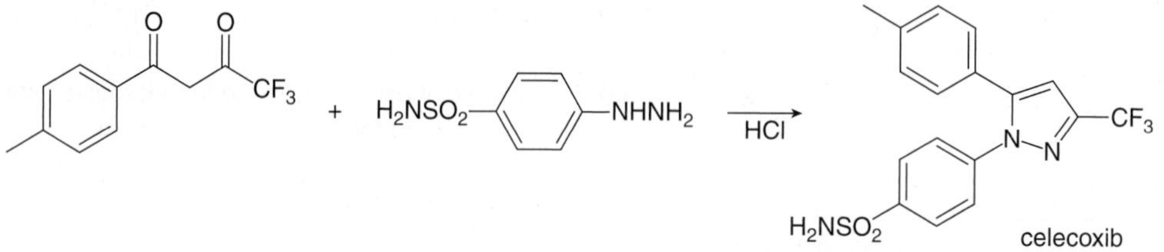

21.66 Treatment of $(HOCH_2CH_2CH_2CH_2)_2CO$ with acid forms a product of molecular formula $C_9H_{16}O_2$ and a molecule of water. Draw the structure of the product and explain how it is formed.

21.67 Draw a stepwise mechanism for each reaction.

a. 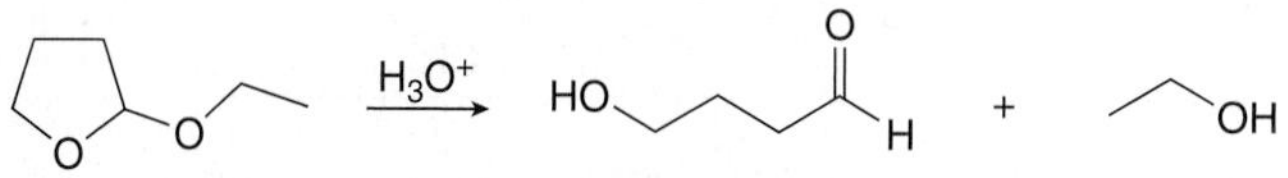

b. 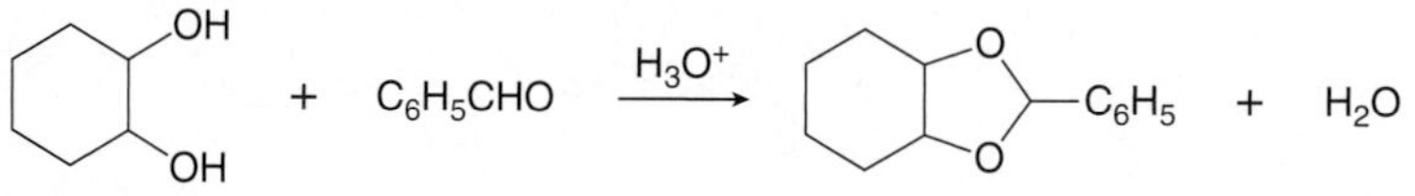

21.68 Draw a stepwise mechanism for the following reaction that converts a dicarbonyl compound to a furan.

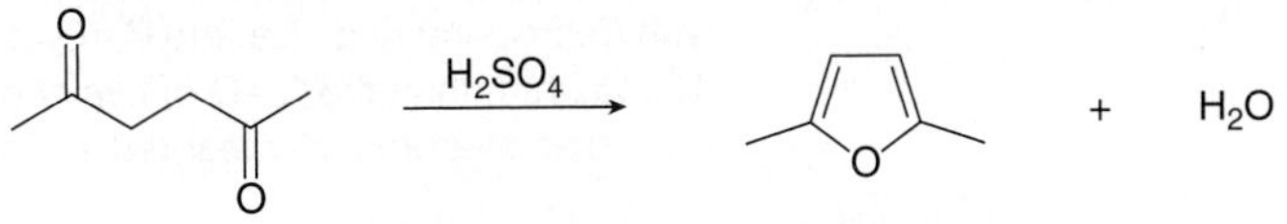

21.69 Draw a stepwise mechanism for the following reaction, a key step in the synthesis of ticlopidine, a drug that inhibits platelet aggregation. Ticlopidine has been used to reduce the risk of stroke in patients who cannot tolerate aspirin.

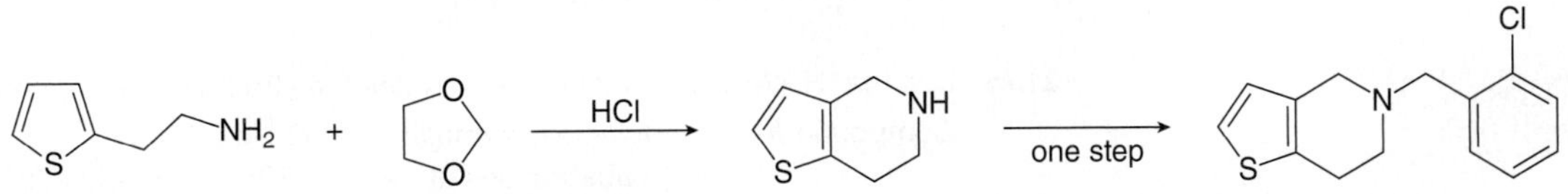

ticlopidine

21.70 Salsolinol is a naturally occurring compound found in bananas, chocolate, and several foods derived from plant sources. Salsolinol is also formed in the body when acetaldehyde, an oxidation product of the ethanol ingested in an alcoholic beverage, reacts with dopamine, a neurotransmitter. Draw a stepwise mechanism for the formation of salsolinol in the following reaction.

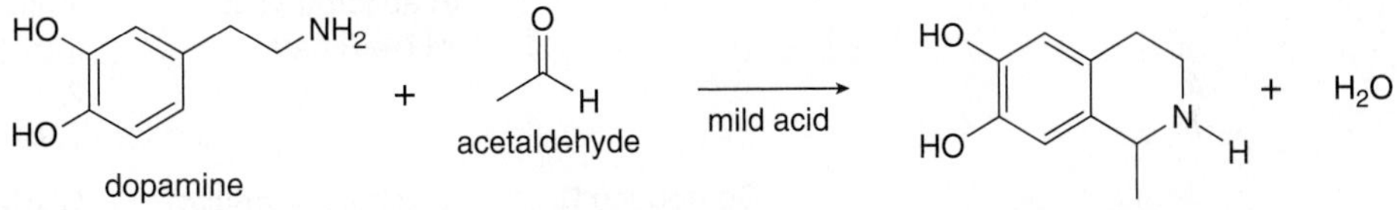

dopamine acetaldehyde salsolinol

21.71 (a) Explain how $NaBH_4$ in CH_3OH can reduce hemiacetal **A** to butane-1,4-diol ($HOCH_2CH_2CH_2CH_2OH$). (b) What product is formed when **A** is treated with $Ph_3P{=}CHCH_2CH(CH_3)_2$? (c) The drug isotretinoin is formed by reaction of **X** and **Y.** What is the structure of isotretinoin? Although isotretinoin (trade name Accutane or Roaccutane) is used for the treatment of severe acne, it is dispensed under strict controls because it also causes birth defects.

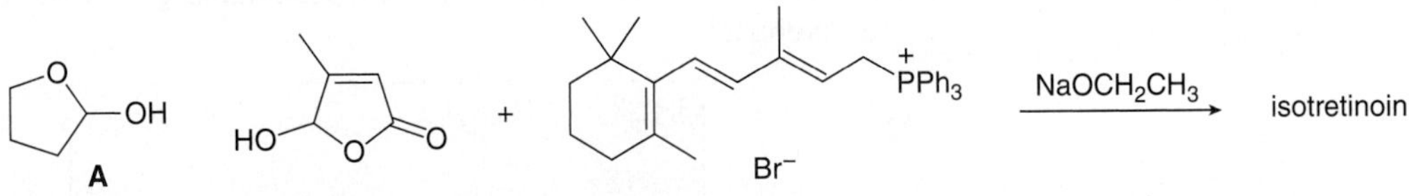

A **X** **Y**

21.72 Reaction of 5,5-dimethoxypentan-2-one with methylmagnesium iodide followed by treatment with aqueous acid forms cyclic hemiacetal **Y.** Draw a stepwise mechanism that illustrates how **Y** is formed.

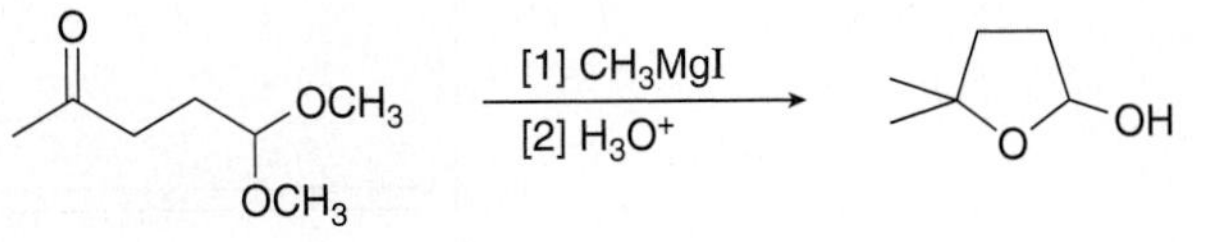

5,5-dimethoxypentan-2-one **Y**

Spectroscopy

21.73 Although the carbonyl absorption of cyclic ketones generally shifts to higher wavenumber with decreasing ring size, the C=O of cyclopropenone absorbs at lower wavenumber in its IR spectrum than the C=O of cyclohex-2-enone. Explain this observation by using the principles of aromaticity learned in Chapter 17.

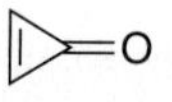

cyclopropenone
(1640 cm^{-1})

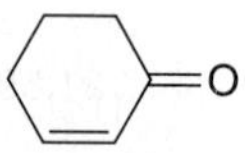

cyclohex-2-enone
(1685 cm^{-1})

21.74 Use the ^1H NMR and IR data to determine the structure of each compound.

Compound **A**	Molecular formula:	$C_5H_{10}O$
	IR absorptions at	1728, 2791, and 2700 cm^{-1}
	^1H NMR data:	1.08 (singlet, 9 H) and 9.48 (singlet, 1 H) ppm
Compound **B**	Molecular formula:	$C_5H_{10}O$
	IR absorption at	1718 cm^{-1}
	^1H NMR data:	1.10 (doublet, 6 H), 2.14 (singlet, 3 H), and 2.58 (septet, 1 H) ppm
Compound **C**	Molecular formula:	$C_{10}H_{12}O$
	IR absorption at	1686 cm^{-1}
	^1H NMR data:	1.21 (triplet, 3 H), 2.39 (singlet, 3 H), 2.95 (quartet, 2 H), 7.24 (doublet, 2 H), and 7.85 (doublet, 2 H) ppm
Compound **D**	Molecular formula:	$C_{10}H_{12}O$
	IR absorption at	1719 cm^{-1}
	^1H NMR data:	1.02 (triplet, 3 H), 2.45 (quartet, 2 H), 3.67 (singlet, 2 H), and 7.06–7.48 (multiplet, 5 H) ppm

21.75 A solution of acetone [$(CH_3)_2C=O$] in ethanol (CH_3CH_2OH) in the presence of a trace of acid was allowed to stand for several days, and a new compound of molecular formula $C_7H_{16}O_2$ was formed. The IR spectrum showed only one major peak in the functional group region around 3000 cm^{-1}, and the ^1H NMR spectrum is given here. What is the structure of the product?

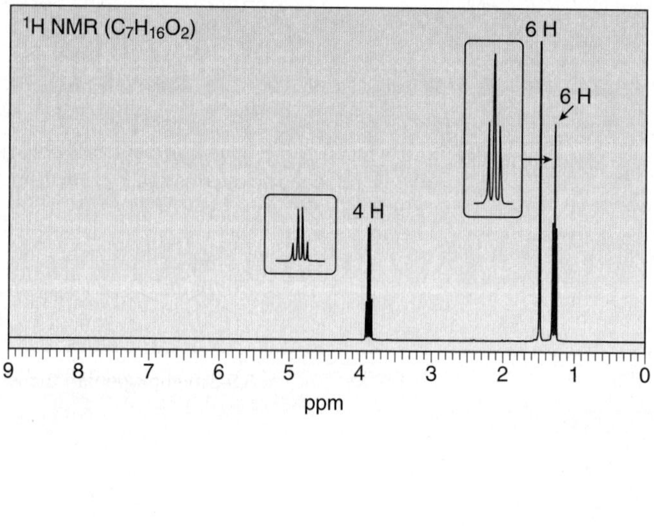

21.76 Compounds **A** and **B** have molecular formula $C_9H_{10}O$. Identify their structures from the ^1H NMR and IR spectra given.

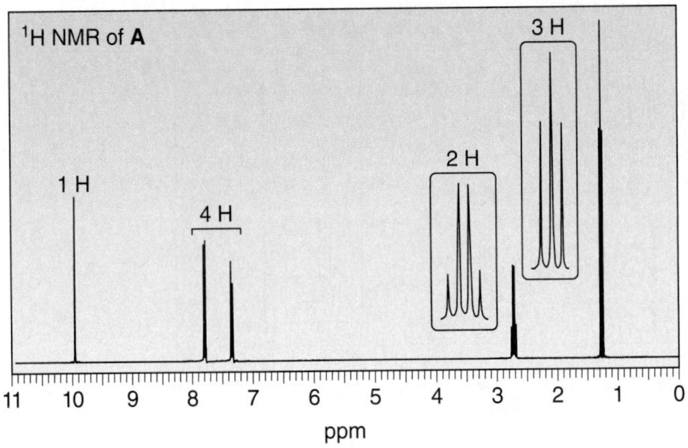

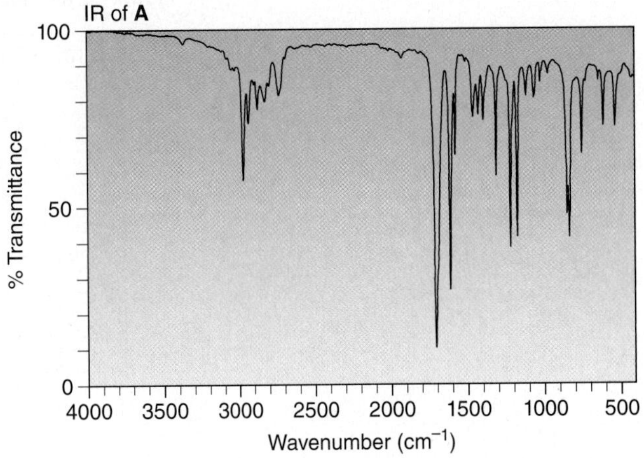

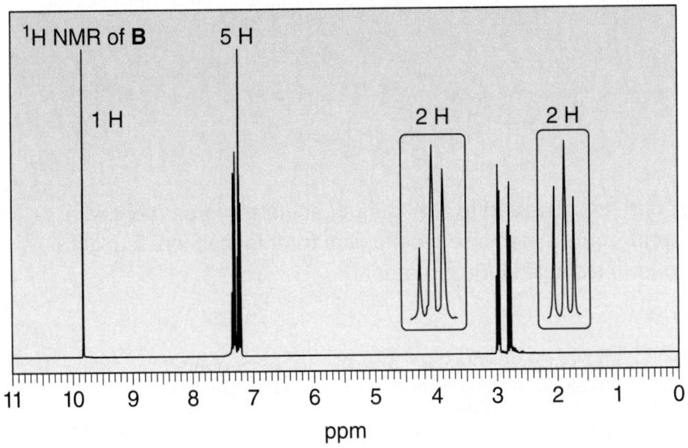

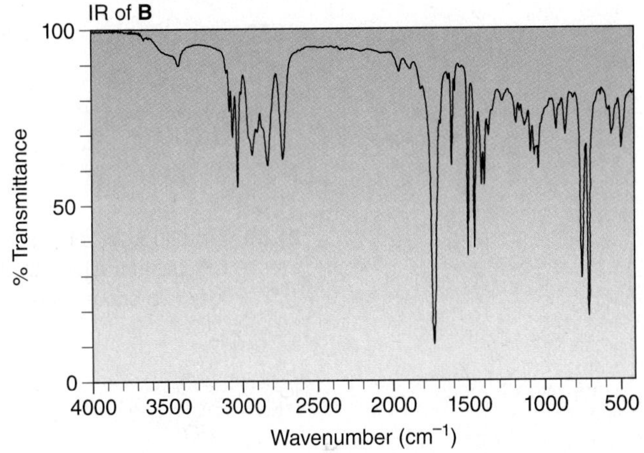

21.77 An unknown compound **C** of molecular formula $C_6H_{12}O_3$ exhibits a strong absorption in its IR spectrum at 1718 cm^{-1} and the given ^1H NMR spectrum. What is the structure of **C?**

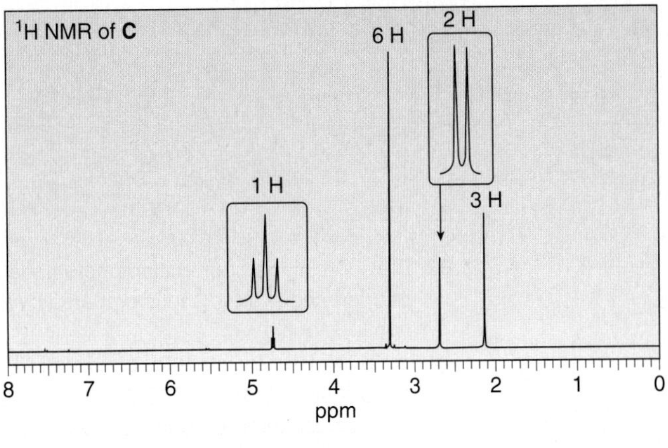

21.78 An unknown compound **D** exhibits a strong absorption in its IR spectrum at 1692 cm^{-1}. The mass spectrum of **D** shows a molecular ion at m/z = 150 and a base peak at 121. The ^1H NMR spectrum of **D** is shown below. What is the structure of **D?**

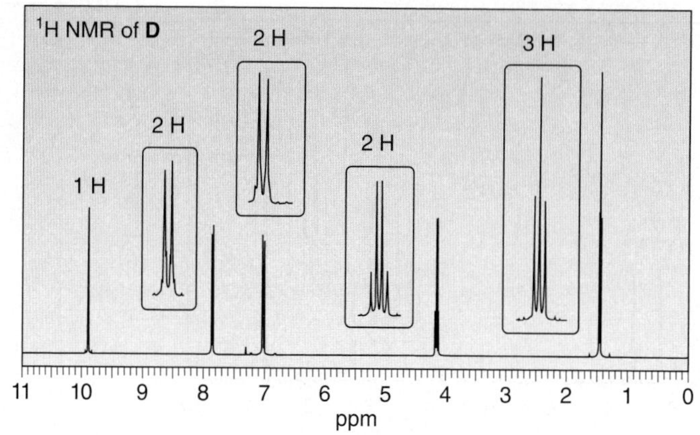

Carbohydrates

21.79 Draw the structure of the acyclic polyhydroxy aldehyde that cyclizes to each hemiacetal.

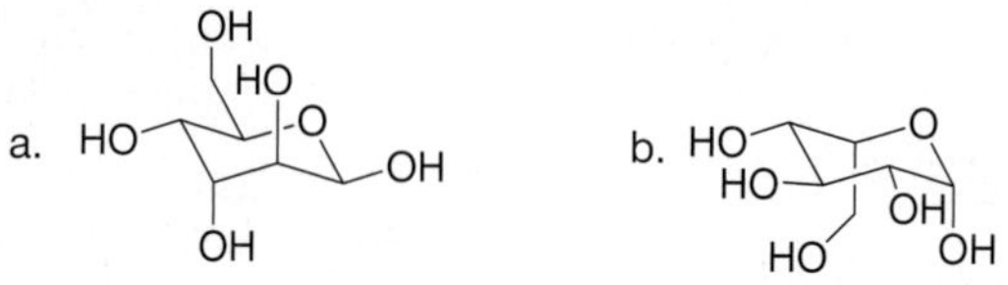

21.80 β-D-Glucose, a hemiacetal, can be converted to a mixture of acetals on treatment with CH_3OH in the presence of acid. Draw a stepwise mechanism for this reaction. Explain why two acetals are formed from a single starting material.

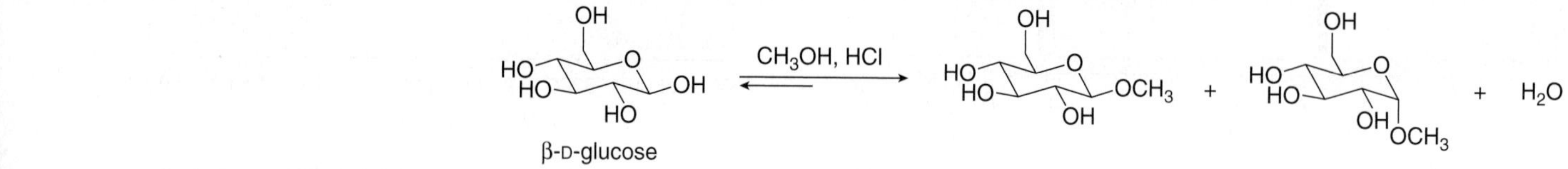

Challenge Problems

21.81 Draw a stepwise mechanism for the following reaction.

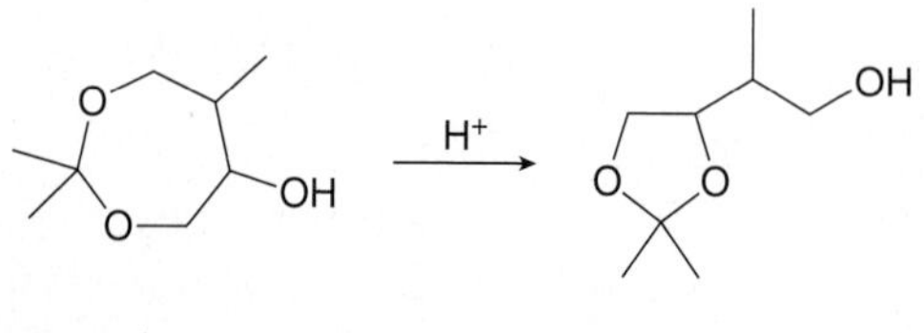

21.82 Brevicomin, the aggregation pheromone of the western pine bark beetle, contains a bicyclic bridged ring system and is prepared by the acid-catalyzed cyclization of 6,7-dihydroxynonan-2-one.

a. Suggest a structure for brevicomin.

b. Devise a synthesis of 6,7-dihydroxynonan-2-one from 6-bromohexan-2-one. You may also use three-carbon alcohols and any required organic or inorganic reagents.

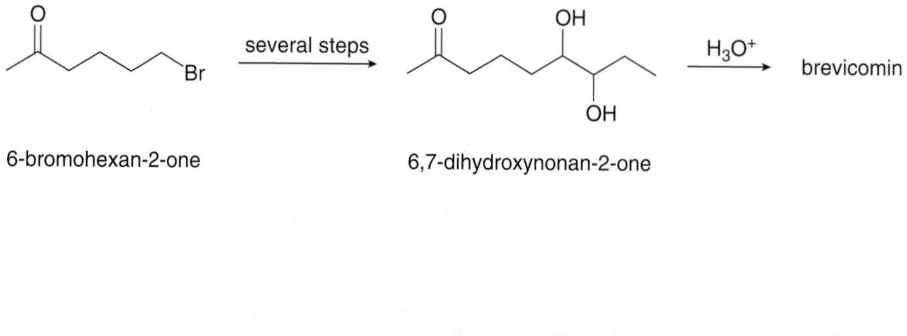

21.83 Draw a stepwise mechanism for the following reaction.

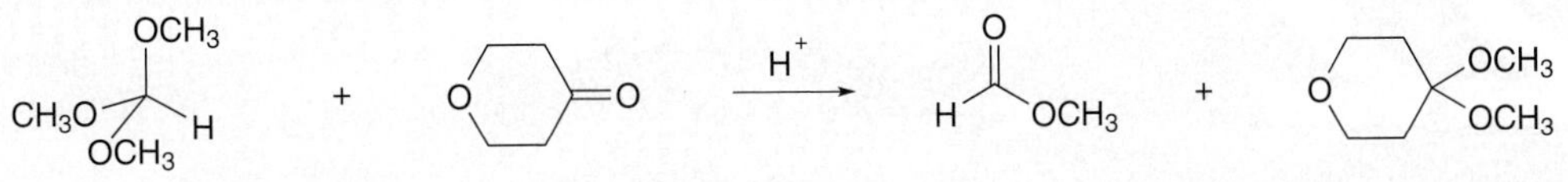

21.84 Maltose is a carbohydrate present in malt, the liquid obtained from barley and other grains. Although maltose has numerous functional groups, its reactions are explained by the same principles we have already encountered.

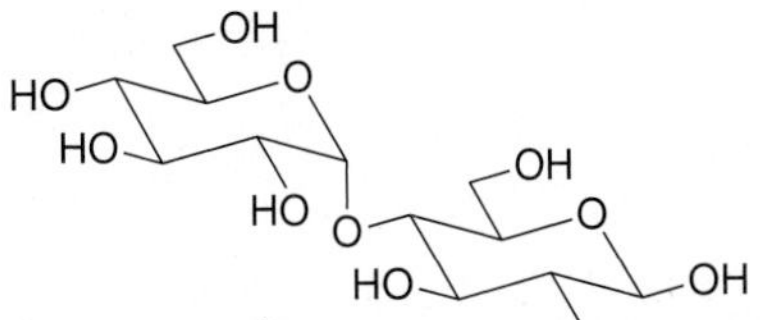

maltose

a. Label the acetal and hemiacetal carbons.

b. What products are formed when maltose is treated with each of the following reagents: [1] H_3O^+; [2] CH_3OH and HCl; [3] excess NaH, then excess CH_3I?

c. Draw the products formed when the compound formed in Reaction [3] of part (b) is treated with aqueous acid.

The reactions in parts (b) and (c) are used to determine structural features of carbohydrates like maltose. We will learn much more about maltose and similar carbohydrates in Chapter 27.

21.85 Identify **R** and **S** in the following reaction sequence, and draw a mechanism for the conversion of **R** to **S** (molecular formula $C_6H_{10}O_3$). **S** was used in the synthesis of darunavir (trade name Prezista), used to treat HIV.

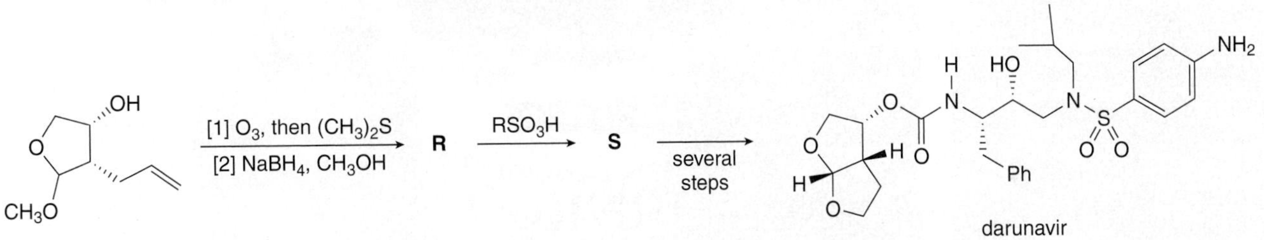
[1] O3, then (CH3)2S
[2] NaBH4, CH3OH
R
RSO3H
S
several steps

darunavir

21.86 Draw a stepwise mechanism for the following reaction, a key step in the synthesis of conivaptan (trade name Vaprisol), a drug used in the treatment of low sodium levels.

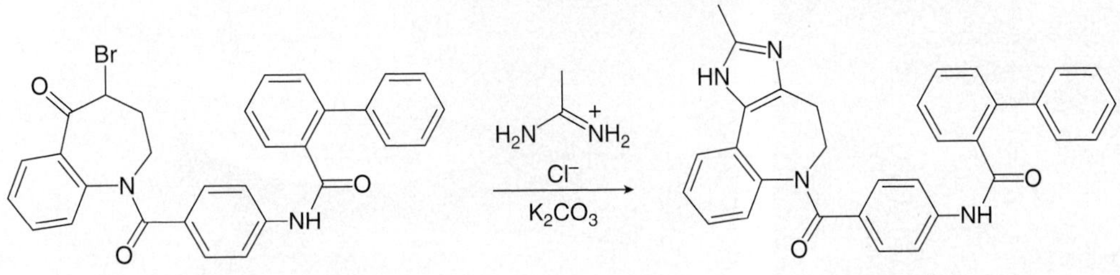
Cl⁻
K2CO3

conivaptan

22 Carboxylic Acids and Their Derivatives—Nucleophilic Acyl Substitution

22.1 Introduction
22.2 Structure and bonding
22.3 Nomenclature
22.4 Physical properties
22.5 Spectroscopic properties
22.6 Interesting esters and amides
22.7 Introduction to nucleophilic acyl substitution
22.8 Reactions of acid chlorides
22.9 Reactions of anhydrides
22.10 Reactions of carboxylic acids
22.11 Reactions of esters
22.12 Application: Lipid hydrolysis
22.13 Reactions of amides
22.14 Application: The mechanism of action of β-lactam antibiotics
22.15 Summary of nucleophilic acyl substitution reactions
22.16 Acyl phosphates—Biological anhydrides
22.17 Reactions of thioesters—Biological acylation reactions
22.18 Nitriles

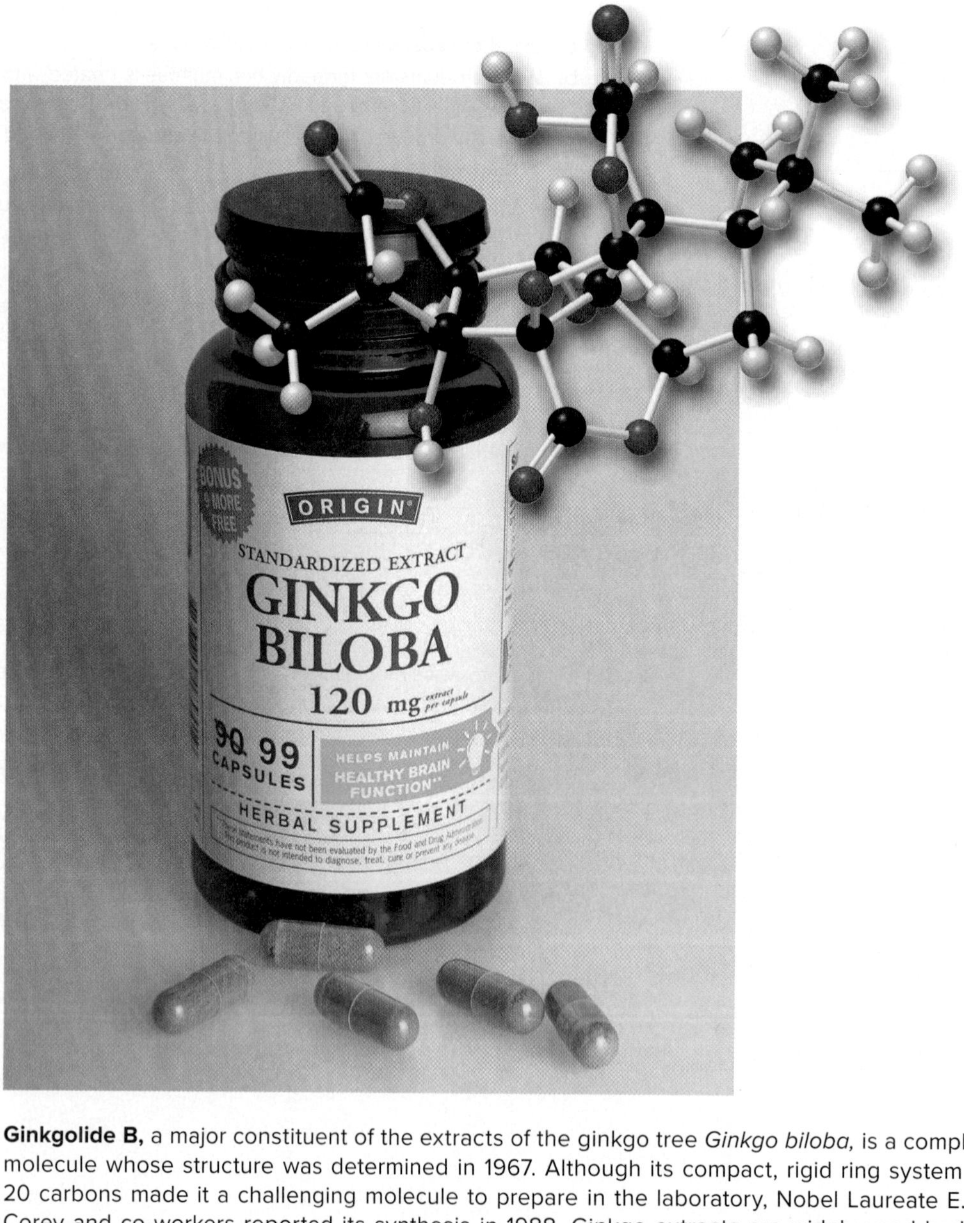

Ginkgolide B, a major constituent of the extracts of the ginkgo tree *Ginkgo biloba,* is a complex molecule whose structure was determined in 1967. Although its compact, rigid ring system of 20 carbons made it a challenging molecule to prepare in the laboratory, Nobel Laureate E. J. Corey and co-workers reported its synthesis in 1988. Ginkgo extracts are widely used herbal supplements, taken to enhance memory and treat dementia. Recent findings of the National Institutes of Health, however, have cast doubt on their efficacy in providing long-term improvement in cognitive function. In Chapter 22 we learn about the reactions of esters, three of the many functional groups in ginkgolide B.

Chapter 22 continues the study of carbonyl compounds with a detailed look at **nucleophilic acyl substitution,** a key reaction of carboxylic acids and their derivatives. Substitution at sp^2 hybridized carbon atoms was introduced in Chapter 20 with reactions involving carbon and hydrogen nucleophiles. In Chapter 22, we learn that nucleophilic acyl substitution is a general reaction that occurs with a variety of heteroatomic nucleophiles. This reaction allows the conversion of one carboxylic acid derivative into another. ***Every* reaction in Chapter 22 that begins with a carbonyl compound or phosphate involves nucleophilic substitution.** Chapter 22 also discusses the properties and chemical reactions of **nitriles,** compounds that contain a carbon–nitrogen triple bond. Nitriles are in the same carbon oxidation state as carboxylic acids, and they undergo reactions that form related products.

22.1 Introduction

Chapter 22 focuses on carbonyl compounds that contain an **acyl group bonded to an electronegative atom.** These include the **carboxylic acids,** as well as carboxylic acid derivatives that can be prepared from them: **acid chlorides, anhydrides, acyl phosphates, esters, thioesters, and amides.**

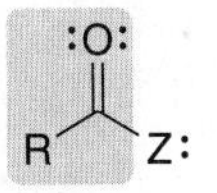

acyl group

Z = an atom more electronegative than carbon

Z = OH

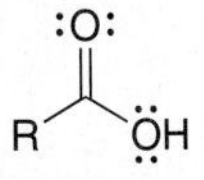

carboxylic acid

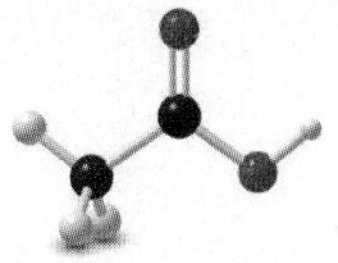

$R = CH_3$
acetic acid

Z = Cl

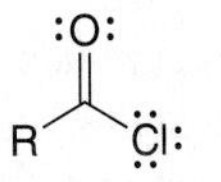

acid chloride

$R = CH_3$
acetyl chloride

Z = OCOR

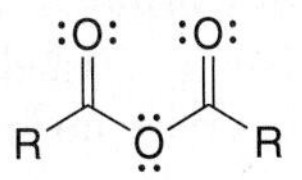

anhydride

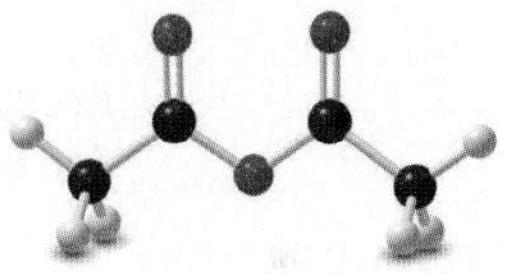

$R = CH_3$
acetic anhydride

$Z = OPO_3^{2-}$

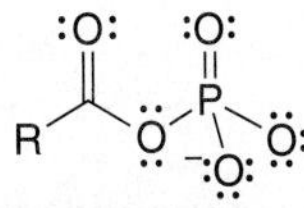

acyl phosphate

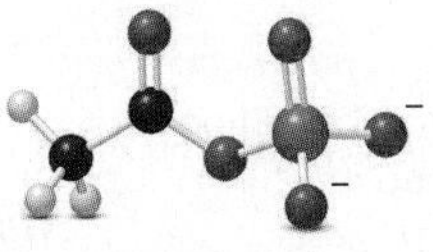

$R = CH_3$
aceytl phosphate

Z = OR'

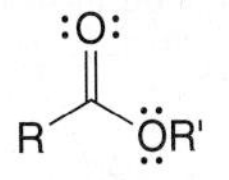

ester

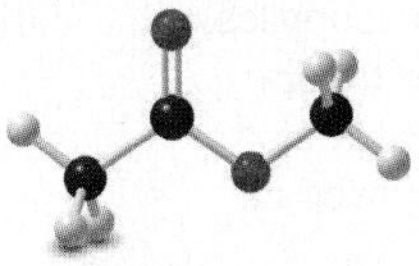

$R = R' = CH_3$
methyl acetate

Z = SR'

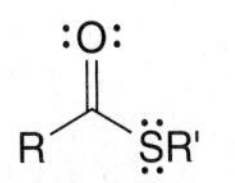

thioester

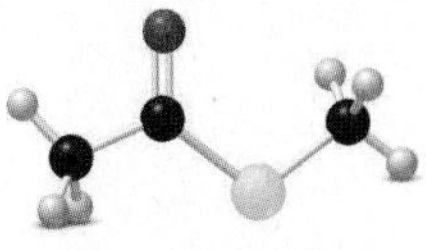

$R = R' = CH_3$
methyl thioacetate

$Z = NR'_2$

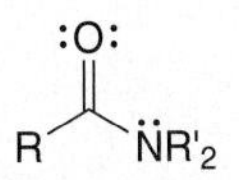

R' = H or alkyl
amide

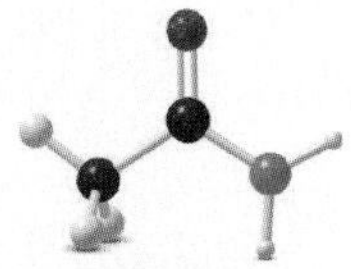

$R = CH_3$, $R' = H$
acetamide

Anhydrides contain two carbonyl groups joined by a single oxygen atom. **Symmetrical anhydrides** have two identical alkyl groups bonded to the carbonyl carbons, and **mixed anhydrides** have two different alkyl groups. **Cyclic anhydrides** are also known. An **acyl phosphate** is a mixed anhydride with a carbonyl group joined to a phosphate by a single oxygen atom. Acyl phosphates, such as glycyl phosphate, are common intermediates in biological pathways.

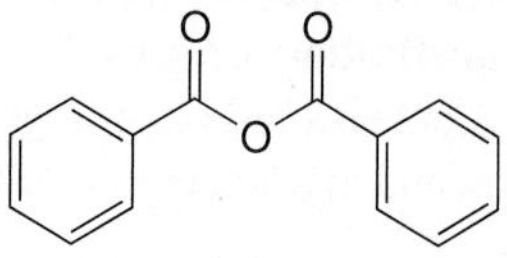

symmetrical anhydride

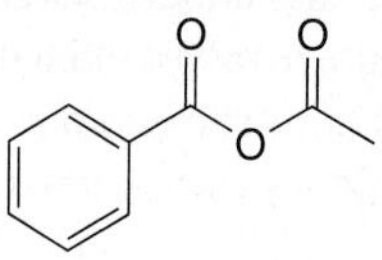

mixed anhydride

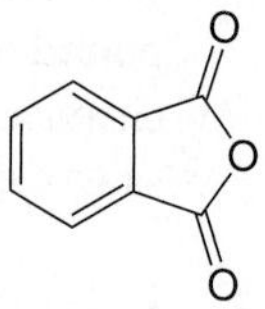

cyclic anhydride

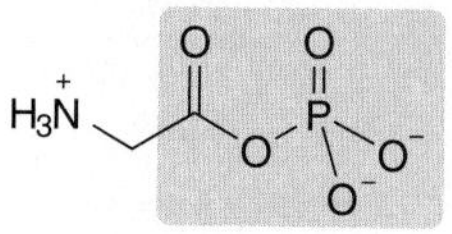

acyl phosphate

glycyl phosphate

As we learned in Section 3.2, **amides** are classified as **1°**, **2°**, or **3°** depending on the number of carbon atoms bonded directly to the *nitrogen* atom.

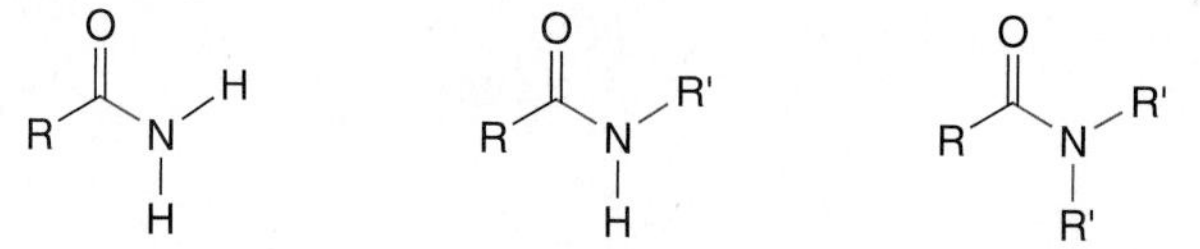

1° amide
1 C—N bond

2° amide
2 C—N bonds

3° amide
3 C—N bonds

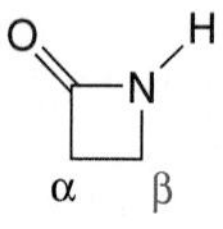

β-lactam

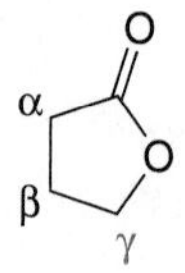

γ-lactone

Cyclic esters and amides are called **lactones** and **lactams,** respectively. The ring size of the heterocycle is indicated by a Greek letter. An amide in a four-membered ring is called a **β-lactam,** because the β carbon to the carbonyl is bonded to the heteroatom. An ester in a five-membered ring is called a **γ-lactone.**

All of these compounds contain an acyl group bonded to an electronegative atom Z that can serve as a **leaving group.** As a result, these compounds undergo **nucleophilic acyl substitution.** Recall from Chapters 20 and 21 that aldehydes and ketones do *not* undergo nucleophilic substitution because they have no leaving group on the carbonyl carbon.

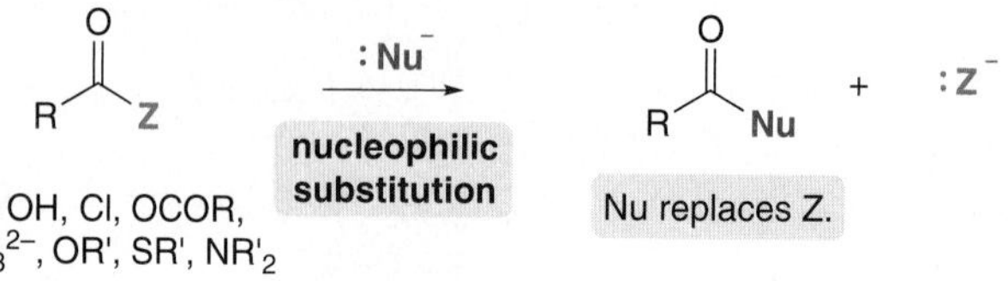

Z = OH, Cl, OCOR, OPO_3^{2-}, OR', SR', NR'_2

Nucleophilic acyl substitution was first discussed in Chapter 20 with R^- and H^- as the nucleophiles. This substitution reaction is general for a variety of nucleophiles, making it possible to form many different substitution products, as discussed in Sections 22.8–22.13.

***Nitriles* are compounds that contain a cyano group, C≡N, bonded to an alkyl group.** Nitriles have no carbonyl group, so they are structurally distinct from carboxylic acids and their derivatives. The carbon atom of the cyano group, however, has the same oxidation state as the carbonyl carbon of carboxylic acid derivatives, so there are certain parallels in their chemistry.

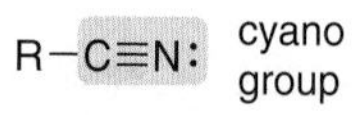

nitrile

Each labeled C has three bonds to a more electronegative element.

Problem 22.1 Oxytocin, sold under the trade name Pitocin, is a naturally occurring hormone used to stimulate uterine contractions and induce labor. Classify each amide in oxytocin as 1°, 2°, or 3°.

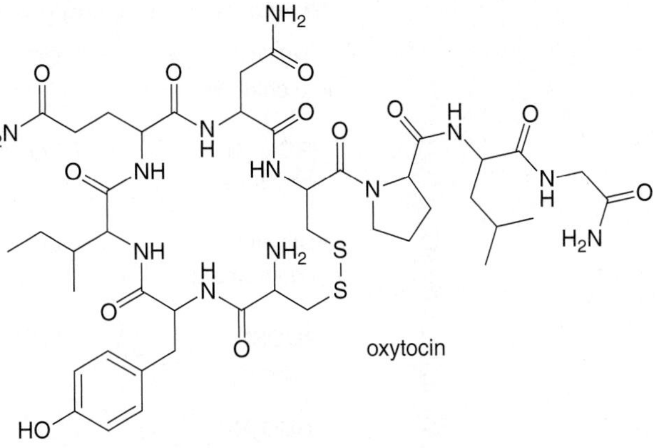

22.2 Structure and Bonding

The two most important features of any carbonyl group, regardless of the other groups bonded to it, are the following:

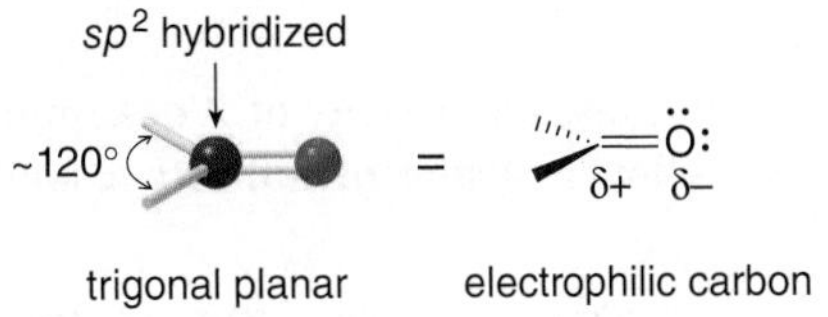

- **The carbonyl carbon is *sp*^{2} hybridized and trigonal planar, making it relatively uncrowded.**
- **The electronegative oxygen atom polarizes the carbonyl group, making the carbonyl carbon electrophilic.**

Because carboxylic acid derivatives (RCOZ) all contain an atom Z with a nonbonded electron pair, three resonance structures can be drawn for RCOZ, compared to just two for aldehydes and ketones (Section 20.1). These three resonance structures stabilize RCOZ by delocalizing electron density. In fact, the more resonance structures **2** and **3** contribute to the resonance hybrid, the more stable RCOZ is.

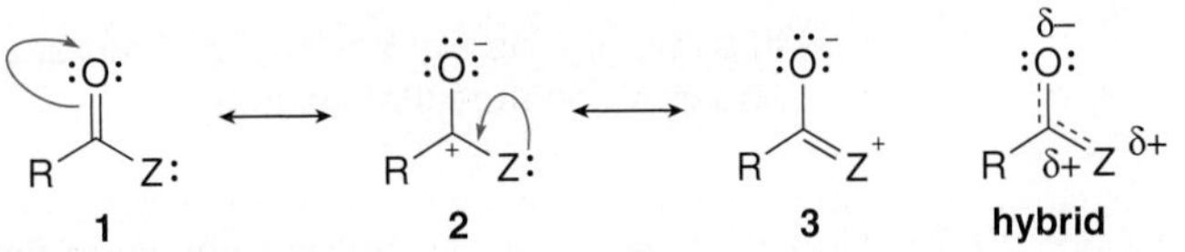

- **The more basic Z is, the more it donates its electron pair, and the more resonance structure 3 contributes to the hybrid.**

To determine the relative basicity of the leaving group Z, we compare the pK_a values of the conjugate acids HZ, given in Table 22.1. The following order of basicity results:

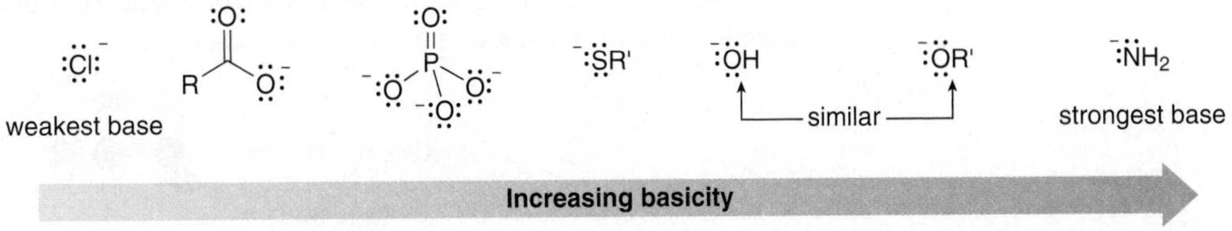

Table 22.1 pK_a Values of the Conjugate Acids (HZ) for Common Z Groups of Acyl Compounds (RCOZ)

Structure	Leaving group (Z^-)	Conjugate acid (HZ)	pK_a
RCOCl acid chloride	Cl^-	HCl	–7
$(RCO)_2O$ anhydride	RCO_2^-	RCO_2H	3–5
$RCO_2PO_3^{2-}$ acyl phosphate	PO_4^{3-}	HPO_4^{2-}	6.9
RCOSR' thioester	$^-SR'$	HSR'	9–11
RCO_2H carboxylic acid	^-OH	H_2O	15.7
RCO_2R' ester	$^-OR'$	R'OH	15.5–18
$RCONR'_2$ amide	$^-NR'_2$	R'_2NH	38–40

Increasing basicity of Z (down) — Increasing acidity of HZ (up)

Because the basicity of Z determines the relative stability of the carboxylic acid derivatives, the following **order of stability** results:

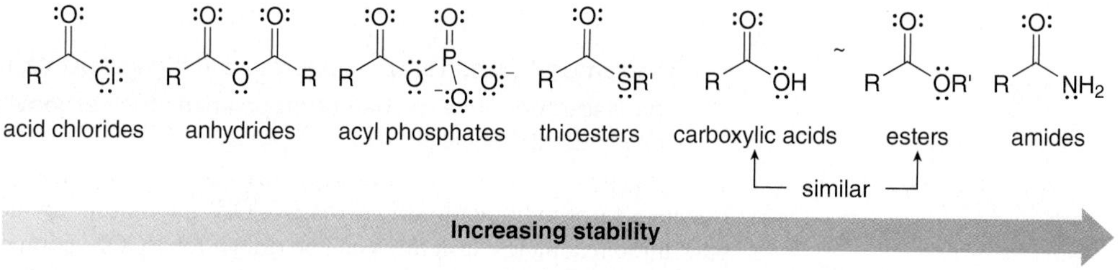
acid chlorides
anhydrides
acyl phosphates
thioesters
carboxylic acids
esters
amides
similar
Increasing stability

Thus, an acid chloride is the least stable carboxylic acid derivative because Cl^- is the weakest base. An amide is the most stable carboxylic acid derivative because $^-NR'_2$ is the strongest base.

- **In summary: As the basicity of Z increases, the stability of RCOZ increases because of added resonance stabilization.**

Problem 22.2 Draw the three possible resonance structures for a thioester and an ester. Then, using the pK_a values in Table 22.1, decide if a thioester is more or less resonance stabilized than an ester.

Problem 22.3 How do the following experimental results support the resonance description of the relative stability of acid chlorides compared to amides? The C–Cl bond lengths in CH_3Cl and CH_3COCl are identical (178 pm), but the C–N bond in $HCONH_2$ is shorter than the C–N bond in CH_3NH_2 (135 pm versus 147 pm).

The structure and bonding in nitriles is very different from the carboxylic acid derivatives, and resembles the carbon–carbon triple bond of alkynes.

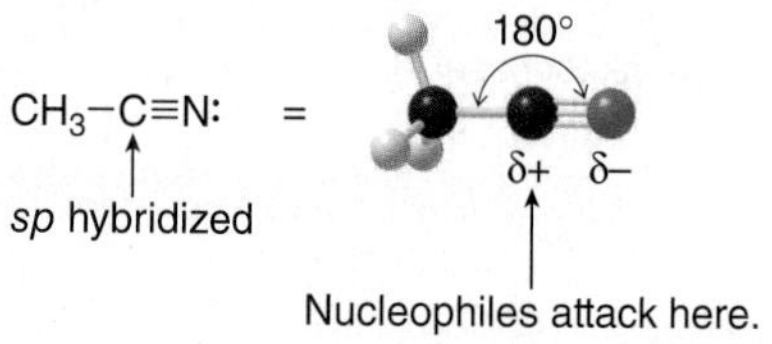
$CH_3-C{\equiv}N:$
=
sp hybridized
180°
δ+ δ–
Nucleophiles attack here.

- **The carbon atom of the C≡N group is *sp* hybridized, making it linear with a bond angle of 180°.**
- **The triple bond consists of one σ and two π bonds.**

Like the carboxylic acid derivatives, **nitriles contain an electrophilic carbon atom,** making them susceptible to nucleophilic attack.

22.3 Nomenclature

The names of carboxylic acid derivatives are formed from the names of the parent carboxylic acids discussed in Section 19.2. Keep in mind that the common names **formic acid, acetic acid,** and **benzoic acid** are virtually always used for the parent acid, so these common parent names are used for their derivatives as well.

22.3A Naming an Acid Chloride (RCOCl) and an Acyl Phosphate ($RCO_2PO_3^{2-}$)

Acid chlorides and acyl phosphates are named by naming the acyl group and adding the word ***chloride*** (for RCOCl) and ***phosphate*** (for $RCO_2PO_3^{2-}$). Two different methods are used.

[1] For acyclic compounds: Change the suffix *-ic acid* of the parent carboxylic acid to the suffix *-yl chloride* for acid chlorides and *-yl phosphate* for acyl phosphates; or

[2] When the functional group is bonded to a ring: Change the suffix *-carboxylic acid* to *-carbonyl chloride* for acid chlorides and *-carbonyl phosphate* for acyl phosphates.

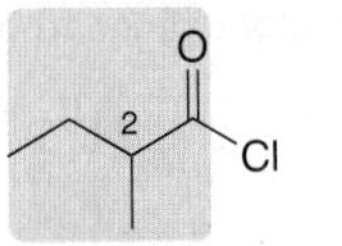

derived from
2-methylbutano*ic acid*

2-methylbutanoyl chloride

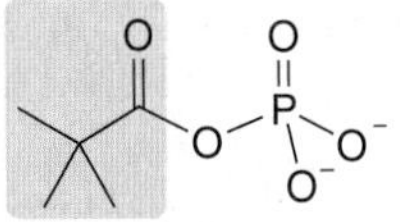

derived from
2,2-dimethylpropano*ic acid*

2,2-dimethylpropanoyl phosphate

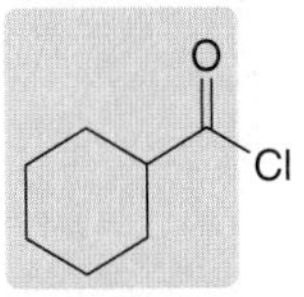

derived from
cyclohexane*carboxylic acid*

cyclohexanecarbonyl chloride

22.3B Naming an Anhydride

The word *anhydride* means "without water." Removing one molecule of water from two molecules of carboxylic acid forms an anhydride.

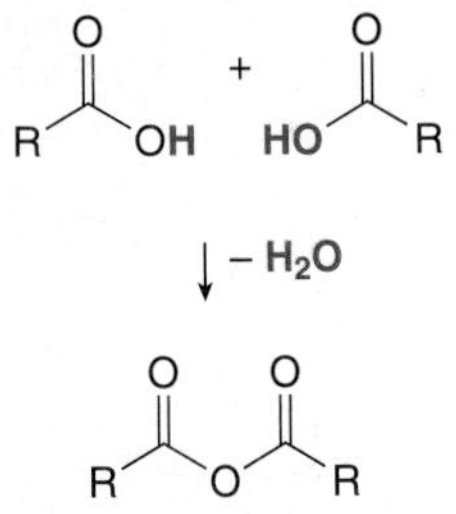

anhydride

Symmetrical anhydrides are named by changing the ***acid*** ending of the parent carboxylic acid to the word ***anhydride.*** **Mixed anhydrides,** which are derived from two different carboxylic acids, are named by alphabetizing the names for both acids and replacing the word ***acid*** by the word ***anhydride.***

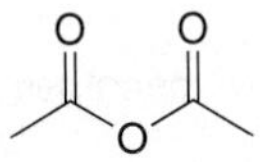

derived from acetic acid

acetic anhydride

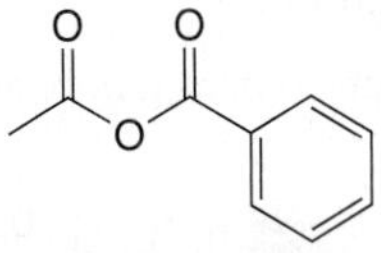

derived from acetic acid
and benzoic acid

acetic benzoic anhydride

Esters are often written as RCOOR', where the alkyl group (R') is written *last.* When an ester is named, however, the R' group appears *first* in the name.

22.3C Naming an Ester (RCOOR') and a Thioester (RCOSR')

Esters and thioesters have two parts to their structures, each of which must be named: an **acyl group (RCO–)** and an **alkyl group** (designated as **R'**) bonded to an oxygen atom in an ester and bonded to a sulfur atom in a thioester.

- **In the IUPAC system, esters are identified by the suffix *-ate.***
- **Thioesters are identified by the suffix *-thioate.***

How To Name an Ester (RCO_2R') Using the IUPAC System

Example **Give a systematic name for each ester:**

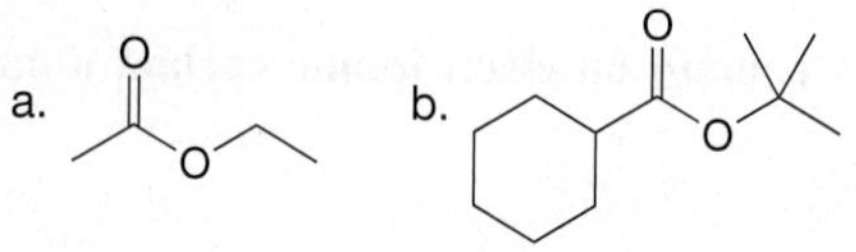

Step [1] **Name the R' group bonded to the oxygen atom as an alkyl group.**

- The name of the alkyl group, ending in the suffix *-yl,* becomes the *first* part of the ester name.

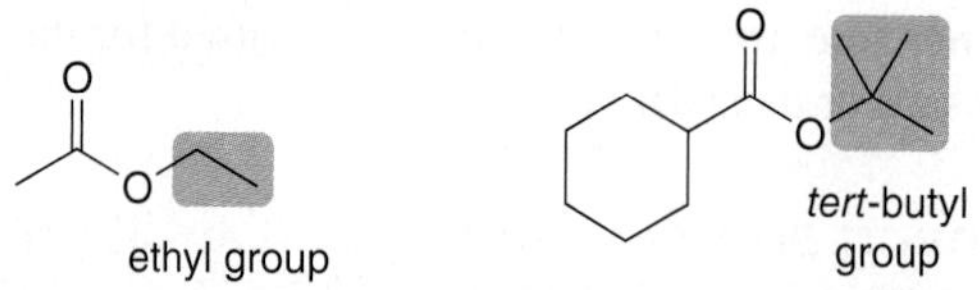

Step [2] **Name the acyl group (RCO–) by changing the *-ic acid* ending of the parent carboxylic acid to the suffix *-ate.***

- The name of the acyl group becomes the *second* part of the name.

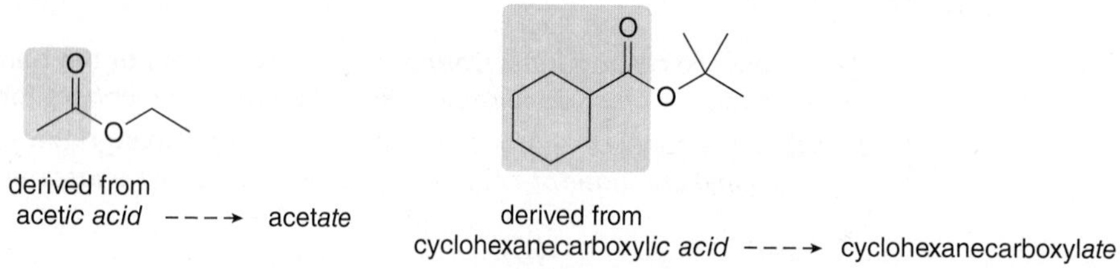

Answer: ethyl acetate **Answer: *tert*-butyl cyclohexanecarboxylate**

How To Name a Thioester (RCOSR') Using the IUPAC System

Example **Give a systematic name for each thioester:**

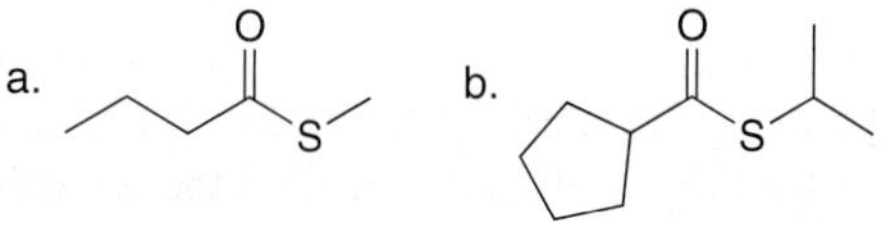

Step [1] **Name the R' group bonded to the sulfur as an alkyl group. Use the prefix "*S*-" preceding the name of the alkyl group.**

- The name of the alkyl group, ending in the suffix *-yl*, becomes the *first* part of the thioester name.

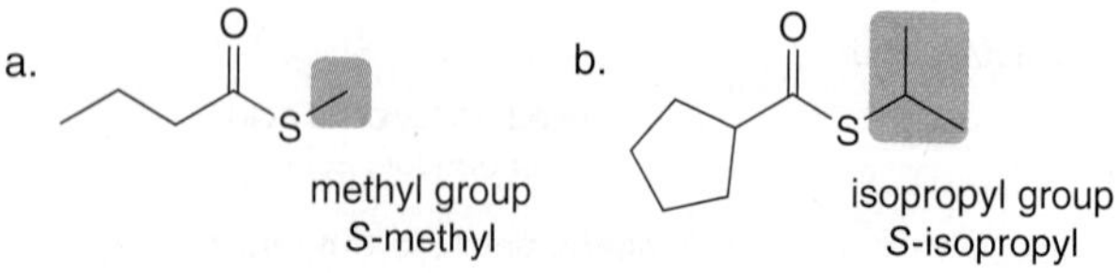

How To, continued . . .

Step [2] **Name the acyl group (RCO–) with the suffix *-thioate*.**

- The name of the acyl group becomes the *second* part of the name.

a. 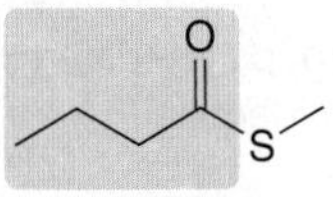

derived from
butan*oic acid* ----→ butan*ethioate*

Answer: *S*-methyl butanethioate

- For acyclic systematic names, change the ***-oic acid*** ending to the suffix ***-ethioate.***

b. 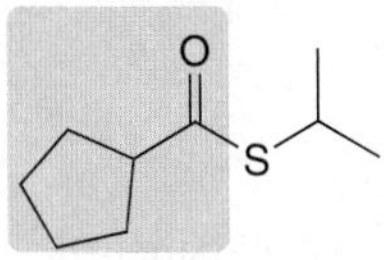

derived from
cyclopentanecarbo*xylic acid* ----→ cyclopentanecarbo*thioate*

Answer: *S*-isopropyl cyclopentanecarbothioate

- For cyclic systematic names, change the ***-xylic acid*** ending to the suffix ***-thioate.***

22.3D Naming an Amide

All 1° amides are named by replacing the ***-ic acid, -oic acid,*** or ***-ylic acid*** ending with the suffix **amide.**

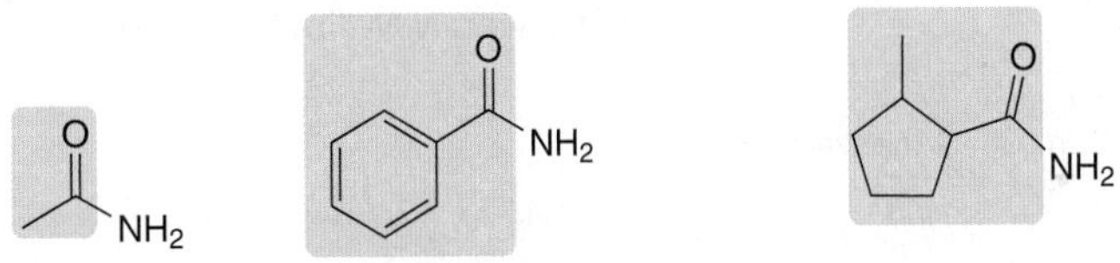

derived from acet*ic acid*	derived from benz*oic acid*	derived from 2-methylcyclopentanecarbox*ylic acid*
acetamide	**benzamide**	**2-methylcyclopentanecarboxamide**

A 2° or 3° amide has two parts to its structure: an **acyl group** that contains the carbonyl group **(RCO−)** and one or two **alkyl groups** bonded to the nitrogen atom.

How To Name a 2° or 3° Amide

Example **Give a systematic name for each amide:**

a. 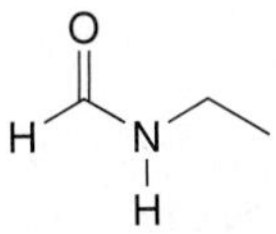 b. 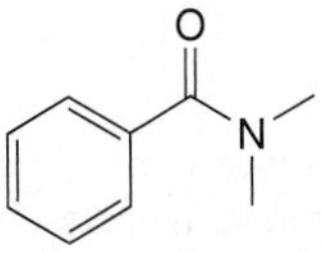

—*Continued*

How To, continued . . .

Step [1] Name the alkyl group (or groups) bonded to the N atom of the amide. Use the prefix "*N*-" preceding the name of each alkyl group.

- The names of the alkyl groups form the first part of each amide name.
- For 3° amides, use the prefix **di-** if the two alkyl groups on N are the same. If the two alkyl groups are different, **alphabetize** their names. One "***N*-**" is needed for each alkyl group, even if both R groups are identical.

a. 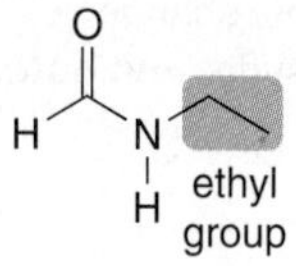

- The compound is a 2° amide with one ethyl group → *N*-ethyl.

b. 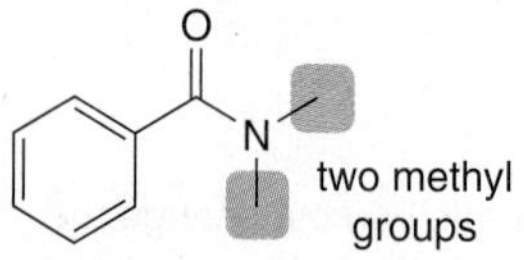

- The compound is a 3° amide with two methyl groups.
- Use the prefix di- and two "*N*-" to begin the name → *N,N*-dimethyl.

Step [2] Name the acyl group (RCO−) with the suffix *-amide.*

a. 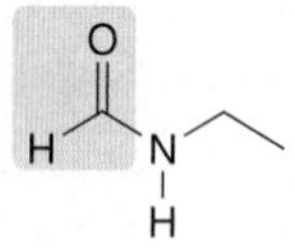

derived from
form*ic acid* - - - - -> form*amide*

- Change the *-ic acid* or *-oic acid* suffix of the parent carboxylic acid to the suffix *-amide.*
- Put the two parts of the name together.
- **Answer: *N*-ethylformamide**

b. 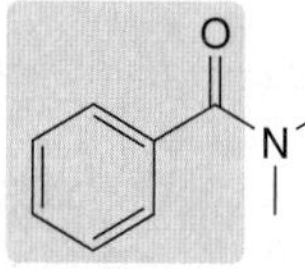

derived from
benzo*ic acid* - - - -> benz*amide*

- Change benzo*ic acid* to benz*amide.*
- Put the two parts of the name together.
- **Answer: *N,N*-dimethylbenzamide**

22.3E Naming a Nitrile

In contrast to the carboxylic acid derivatives, **nitriles are named as alkane derivatives.** To name a nitrile using IUPAC rules:

- **Find the longest chain that contains the CN and add the word *nitrile* to the name of the parent alkane. Number the chain to put CN at C1, but omit this number from the name.**

In naming a nitrile, the CN carbon is one carbon atom of the longest chain. **CH_3CH_2CN is propanenitrile, *not* ethanenitrile.**

Common names for nitriles are derived from the names of the carboxylic acid having the same number of carbon atoms by replacing the ***-ic acid*** ending of the carboxylic acid by the suffix ***-onitrile.***

When CN is named as a substituent, it is called a ***cyano*** group. Figure 22.1 illustrates features of nitrile nomenclature.

Figure 22.1 Summary of nitrile nomenclature

a. IUPAC name for a nitrile

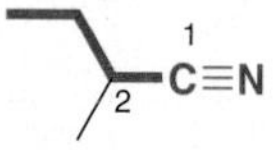

butane + nitrile
(4 C's)

2-methylbutanenitrile

b. Common name for a nitrile

$CH_3-C\equiv N$

derived from
acet*ic acid*

acetonitrile

c. CN as a substituent

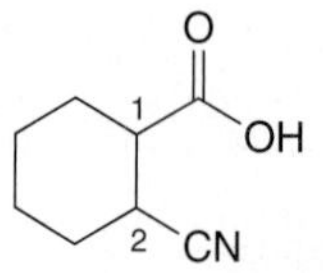

2-cyanocyclohexanecarboxylic acid

Table 22.2 Summary: Nomenclature of Carboxylic Acid Derivatives and Nitriles

Compound	Name ending	Example	Name
acid chloride	**-yl chloride** or **-carbonyl chloride**	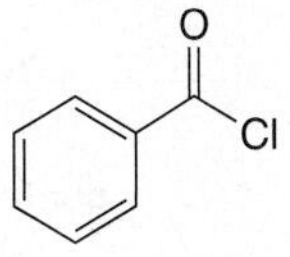	benzoyl chloride
acyl phosphate	**-yl phosphate** or **-carbonyl phosphate**	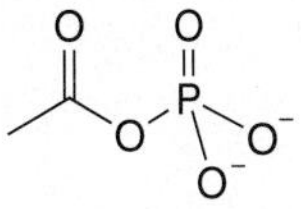	acetyl phosphate
anhydride	**anhydride**	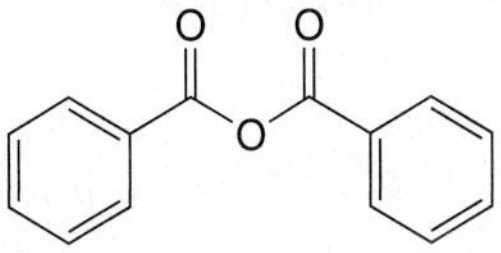	benzoic anhydride
ester	**-ate**	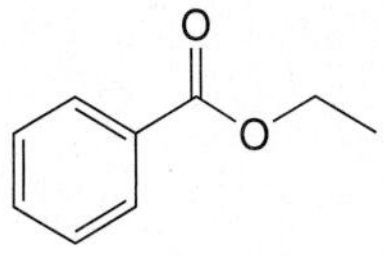	ethyl benzoate
thioester	**-thioate**	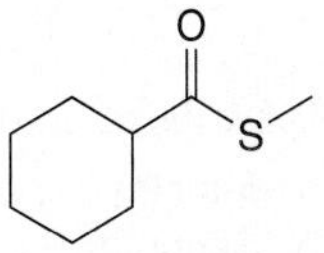	*S*-methyl cyclohexanecarbothioate
amide	**-amide**	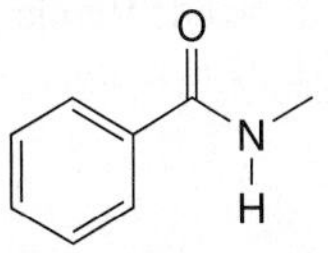	*N*-methylbenzamide
nitrile	**-nitrile** or **-onitrile**	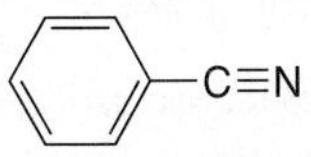	benzonitrile

Table 22.2 summarizes the most important points about the nomenclature of carboxylic acid derivatives and nitriles.

Sample Problem 22.1

Give the IUPAC name for each compound.

a. 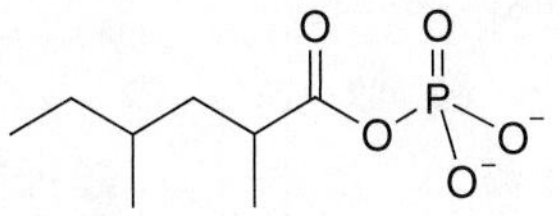

b. 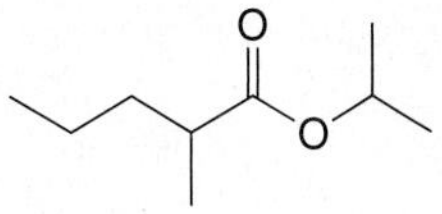

Solution

a. The functional group is an acyl phosphate bonded to a chain of atoms, so the name ends in ***-yl phosphate.***

[1] Find and name the longest chain containing the $CO_2PO_3^{2-}$:

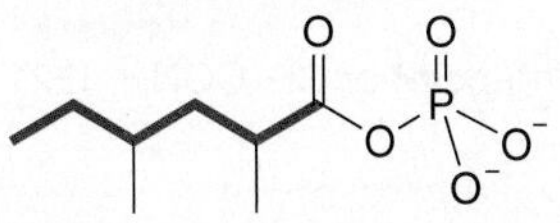

hexano*ic acid* ⟶ **hexano*yl phosphate***
(6 C's)

[2] Number and name the substituents:

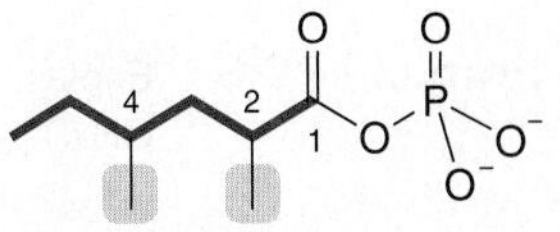

Answer:
2,4-dimethylhexanoyl phosphate

b. The functional group is an ester, so the name ends in ***-ate.***

[1] Find and name the longest chain containing the carbonyl group:

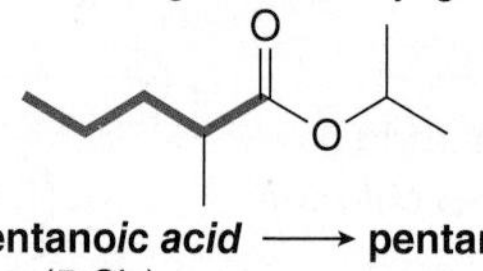

pentan*oic acid* ⟶ **pentan*oate***
(5 C's)

[2] Number and name the substituents:

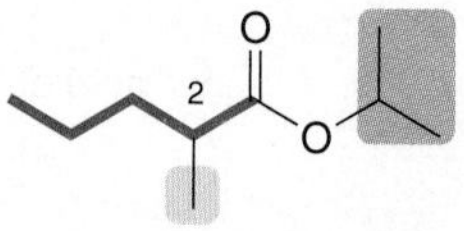

Answer: isopropyl 2-methylpentanoate

The name of the alkyl group on the O atom goes **first** in the name.

Problem 22.4 Give an IUPAC or common name for each compound.

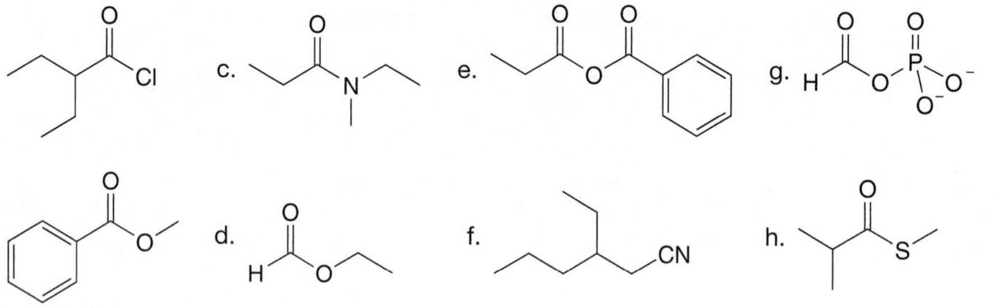

Problem 22.5 Draw the structure corresponding to each name.

a. 5-methylheptanoyl chloride
b. *S*-isopropyl 3-ethylcyclobutanecarbothioate
c. acetic formic anhydride
d. *N*-isobutyl-*N*-methylbutanamide
e. 3-methylpentanenitrile
f. 3-ethyl-2-methylpentanoyl phosphate
g. *sec*-butyl 2-methylhexanoate
h. *N*-ethylhexanamide

22.4 Physical Properties

Because all carbonyl compounds have a polar carbonyl group, they exhibit **dipole–dipole interactions.** Nitriles also have dipole–dipole interactions because they have a polar C≡N group. Because they contain one or two N–H bonds, 1° and 2° amides are capable of intermolecular hydrogen bonding. The N–H bond of one amide intermolecularly hydrogen bonds to the C=O of another amide, as shown using two acetamide molecules (CH_3CONH_2) in Figure 22.2.

Figure 22.2 Intermolecular hydrogen bonding between two CH_3CONH_2 molecules

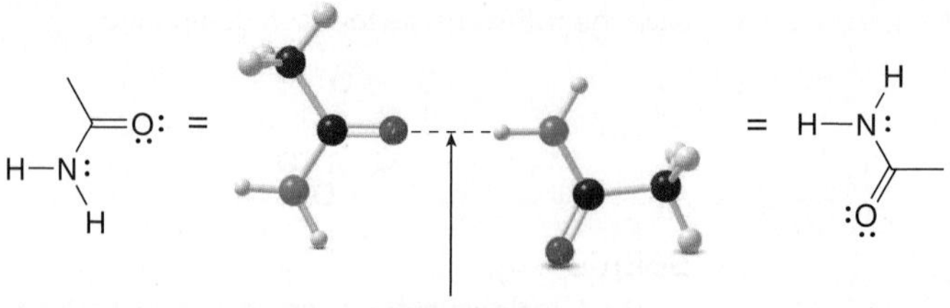

Problem 22.6 Explain why the boiling point of CH_3CONH_2 (221 °C) is significantly higher than the boiling point of CH_3CO_2H (118 °C).

How these factors affect the physical properties of carboxylic acid derivatives is summarized in Table 22.3.

Table 22.3 Physical Properties of Carboxylic Acid Derivatives

Property	Observation
Boiling point and melting point	• Primary (1°) and 2° amides have *higher* boiling points and melting points than compounds of comparable molecular weight. • The boiling points and melting points of other carboxylic acid derivatives are similar to those of other polar compounds of comparable size and shape.  MW = 88, bp 77 °C ~ MW = 92.5, bp 78 °C ~ MW = 90, bp 98 °C < MW = 87, bp 216 °C similar boiling points; higher boiling point, 1° amide
Solubility	• Carboxylic acid derivatives are soluble in organic solvents regardless of size. • Most carboxylic acid derivatives having ≤ 5 C's are H_2O soluble because they can hydrogen bond with H_2O (Section 3.4C). • Carboxylic acid derivatives having > 5 C's are H_2O insoluble because the nonpolar alkyl portion is too large to dissolve in the polar H_2O solvent.

Key: MW = molecular weight

22.5 Spectroscopic Properties

22.5A IR Spectra

The most prominent IR absorptions for carboxylic acid derivatives and nitriles are as follows:

[1] Like all carbonyl compounds, carboxylic acid derivatives have a **strong C=O absorption between 1600 and 1850 cm^{-1}.**

[2] **Primary (1°) and 2° amides** have two additional absorptions due to the N–H bonds:
- one or two N–H stretching peaks at **3200–3400 cm^{-1}**
- an N–H bending absorption at **~1640 cm^{-1}**

[3] **Phosphates** have a **strong P=O absorption between 1100 and 1200 cm^{-1}.**

[4] **Nitriles** have an absorption at **2250 cm^{-1} for the C≡N.**

The exact location of the carbonyl absorption varies with the identity of Z in the carbonyl compound RCOZ. As detailed in Section 22.2, as the basicity of Z increases, resonance stabilization of RCOZ increases, resulting in the following trend:

- **As the carbonyl π bond becomes more delocalized, the C=O absorption generally shifts to lower frequency.**

Thus, the carbonyl group of an acid chloride and anhydride, which are least stabilized by resonance, absorb at higher frequency than the carbonyl group of an amide, which is more stabilized by resonance. Table 22.4 lists specific values for the carbonyl absorptions for some of the carboxylic acid derivatives.

The effects of conjugation and ring size on the location of a carbonyl absorption were first discussed in Section 21.4A.

Conjugation and ring size affect the location of these carbonyl absorptions.

- **Conjugation shifts a carbonyl absorption to lower frequencies.**
- **For cyclic carboxylic acid derivatives, decreasing ring size shifts a carbonyl absorption to higher frequencies.**

Table 22.4 IR Absorptions for the Carbonyl Group of Carboxylic Acid Derivatives

Compound type	Structure (RCOZ)	Carbonyl absorption ($\tilde{\nu}$)
acid chloride	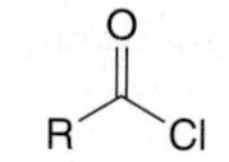	~1800
anhydride	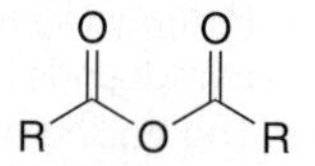	1820 and 1760 (2 peaks)
ester	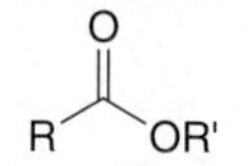	1735–1745
amide	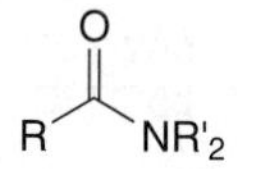 R' = H or alkyl	1630–1680

Increasing basicity of Z (downward arrow) — Increasing $\tilde{\nu}$ of absorption (upward arrow)

Basicity trends do *not* correctly predict the location of the carbonyl absorptions of acyl phosphates and thioesters. Due to the significant electron density of the phosphate and the low electronegativity of the sulfur, the carbonyl π electrons on these compounds are more delocalized and thus shifted to lower frequencies than basicity trends suggest.

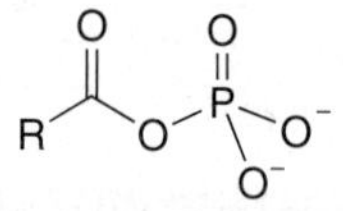

acyl phosphate

1700–1730 cm^{-1}

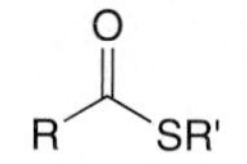

thioester

1690–1720 cm^{-1}

Problem 22.7 How would the compounds in each pair differ in their IR spectra?

a. 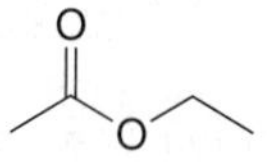 and 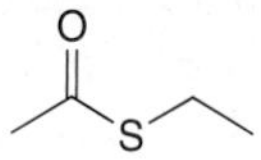

c. 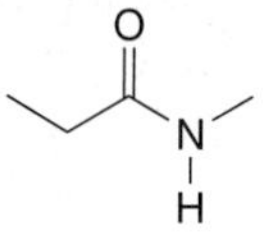 and 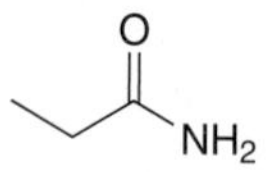

b. 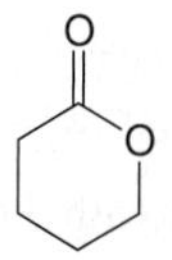 and 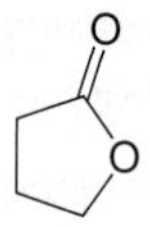

d. 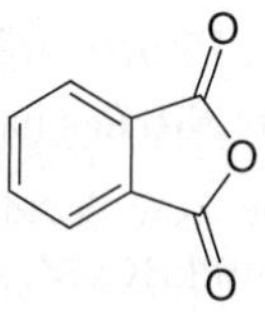 and 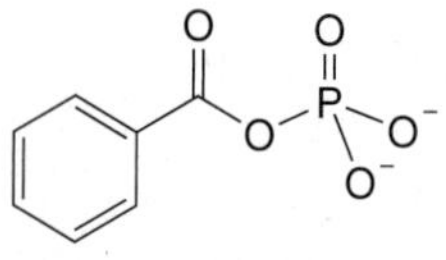

22.5B NMR Spectra

Carboxylic acid derivatives have two characteristic ^{1}H NMR absorptions.

[1] **Protons on the α carbon to the carbonyl absorb at 2–2.5 ppm.**

[2] **The N–H protons of 1° and 2° amides absorb at 7.5–8.5 ppm.**

In their ^{13}C NMR spectra, carboxylic acid derivatives give a highly deshielded peak at 160–180 ppm due to the carbonyl carbon. This is somewhat upfield from the carbonyl absorption of aldehydes and ketones, which occurs at 190–215 ppm.

Nitriles give a peak at 115–120 ppm in their ^{13}C NMR spectrum due to the *sp* hybridized carbon. This is farther downfield than the signal due to the *sp* hybridized carbon of an alkyne, which occurs at 65–100 ppm.

Problem 22.8 Deduce the structures of compounds **A** and **B,** two of the major components of jasmine oil, from the given data.

Compound **A:** $C_9H_{10}O_2$; IR absorptions at 3091–2895 and 1743 cm^{-1}; ^1H NMR signals at 2.06 (singlet, 3 H), 5.08 (singlet, 2 H), and 7.33 (broad singlet, 5 H) ppm.

Compound **B:** $C_{14}H_{12}O_2$; IR absorptions at 3091–2953 and 1718 cm^{-1}; ^1H NMR signals at 5.35 (singlet, 2 H) and 7.26–8.15 (multiplets, 10 H) ppm.

22.6 Interesting Esters and Amides

22.6A Esters

The characteristic odor of many fruits is due to low molecular weight esters.

The coca plant, *Erythroxylon coca,* is the source of the addictive drug cocaine.

Many low molecular weight esters have pleasant and very characteristic odors.

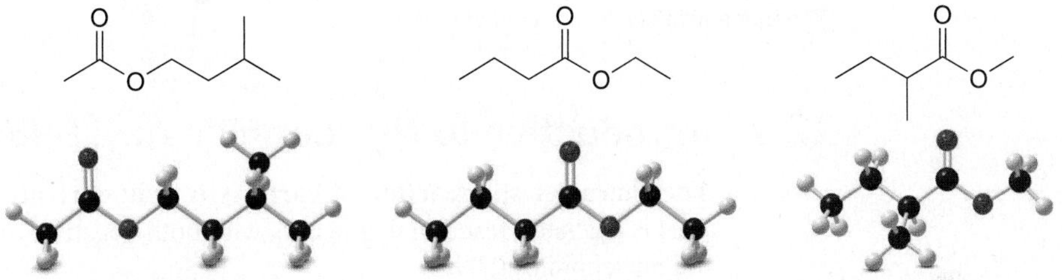

isoamyl acetate
odor of banana

ethyl butanoate
odor of mango

methyl 2-methylbutanoate
odor of pineapple

Several esters, including vitamin C and cocaine, have important biological activities.

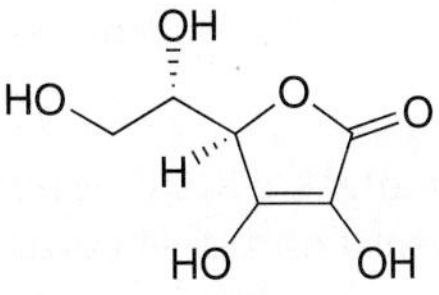

vitamin C

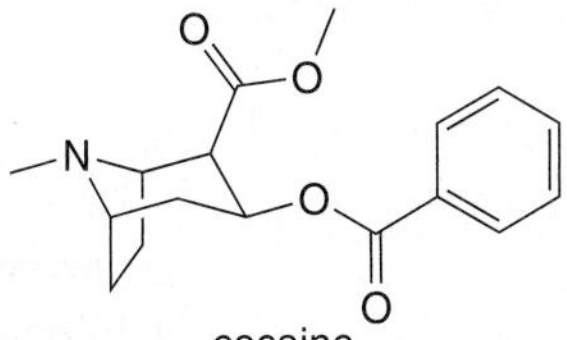

cocaine

- **Vitamin C** (or **ascorbic acid**) is a water-soluble vitamin containing a five-membered lactone that we first discussed in Section 3.5B. Although vitamin C is synthesized in plants, humans do not have the necessary enzymes to make it, and so they must obtain it from their diet.
- **Cocaine** is an addictive stimulant obtained from the leaves of the coca plant. Chewing coca leaves for pleasure has been practiced by the indigenous peoples of South America for over a thousand years, and coca leaves were a very minor ingredient in Coca-Cola for the first 20 years of its production. Cocaine is a widely abused recreational drug, and the possession and use of cocaine is currently illegal in most countries.

22.6B Amides

An important group of naturally occurring amides consists of ***proteins*** **(Section 3.9A), polymers of amino acids joined together by amide linkages.** Proteins differ in the length of the polymer chain, as well as in the identity of the R groups bonded to it. The word *protein* is usually reserved for high molecular weight polymers composed of 40 or more amino acid units, while the designation *peptide* is given to polymers of lower molecular weight.

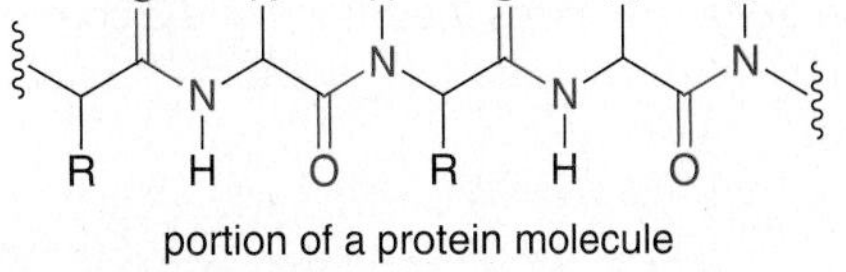

portion of a protein molecule
[Amide bonds are shown in red.]

Peptides and proteins are discussed in detail in Chapter 26.

Proteins and peptides have diverse functions in the cell. They form the structural components of muscle, connective tissue, hair, and nails. They catalyze reactions and transport ions and molecules across cell membranes. **Met-enkephalin,** for example, a peptide with four amide bonds found predominately in nerve tissue cells, relieves pain and acts as an opiate by producing morphine-like effects.

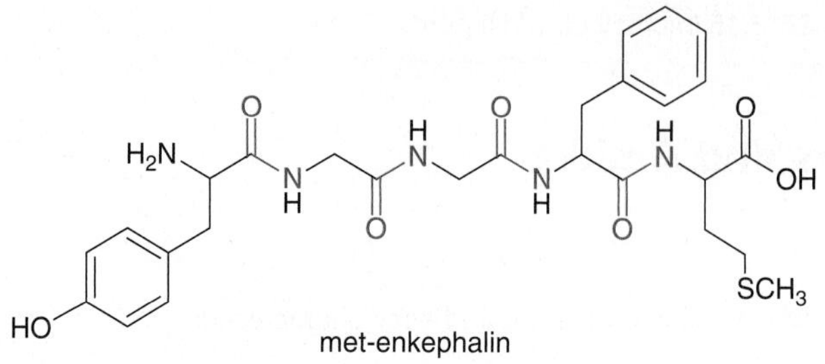

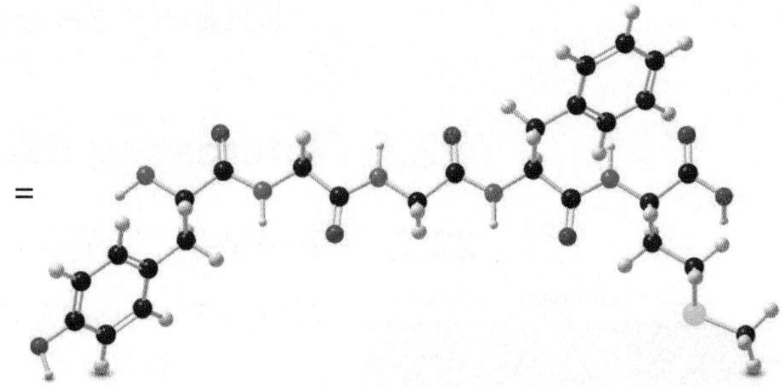

met-enkephalin
[The four amide bonds are shown in red.]

22.7 Introduction to Nucleophilic Acyl Substitution

The characteristic reaction of carboxylic acid derivatives is *nucleophilic acyl substitution.* This is a general reaction that occurs with both negatively charged nucleophiles (Nu:$^-$) and neutral nucleophiles (HNu:).

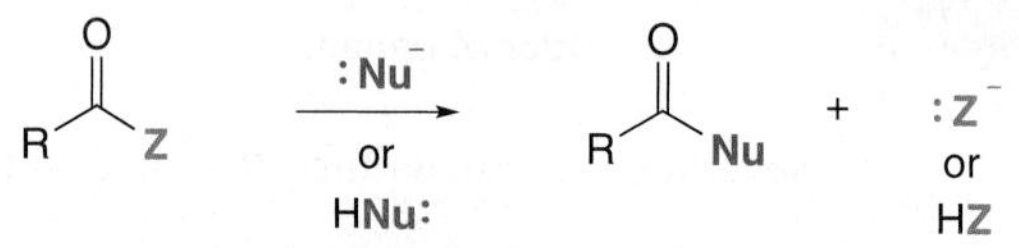

Nu replaces Z.

- **Carboxylic acid derivatives (RCOZ) react with nucleophiles because they contain an electrophilic, unhindered carbonyl carbon.**
- **Substitution occurs, *not* addition, because carboxylic acid derivatives (RCOZ) have a leaving group Z on the carbonyl carbon.**

22.7A The Mechanism

The mechanism for nucleophilic acyl substitution was first presented in Section 20.2.

The general mechanism for nucleophilic acyl substitution is a two-step process: **nucleophilic attack** followed by **loss of the leaving group,** as shown in Mechanism 22.1.

Mechanism 22.1 General Mechanism—Nucleophilic Acyl Substitution

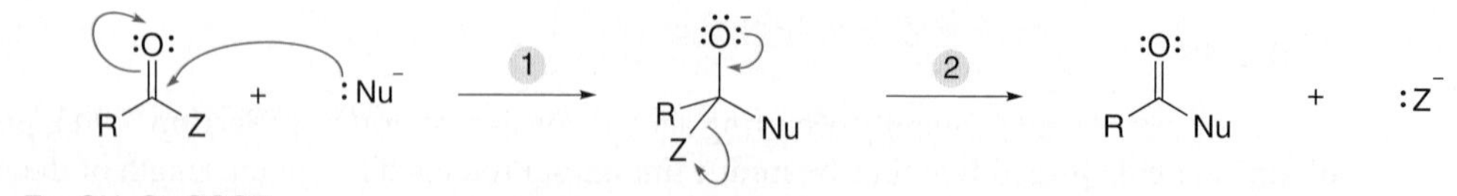

Z = OH, Cl, OCOR, OPO_3^{2-}, OR', SR', NH_2

1 The nucleophile attacks the electrophilic carbonyl group. The π bond is broken, moving an electron pair out on oxygen and forming an sp^3 hybridized carbon.

2 An electron pair on oxygen re-forms the π bond and Z comes off as a leaving group with the electron pair in the C–Z bond.

The overall result of addition of a nucleophile and elimination of a leaving group is substitution of the nucleophile for the leaving group. Recall from Chapter 20 that nucleophilic substitution occurs with carbanions (R^-) and hydride (H^-) as nucleophiles. A variety of oxygen, nitrogen, and sulfur nucleophiles also participate in this reaction.

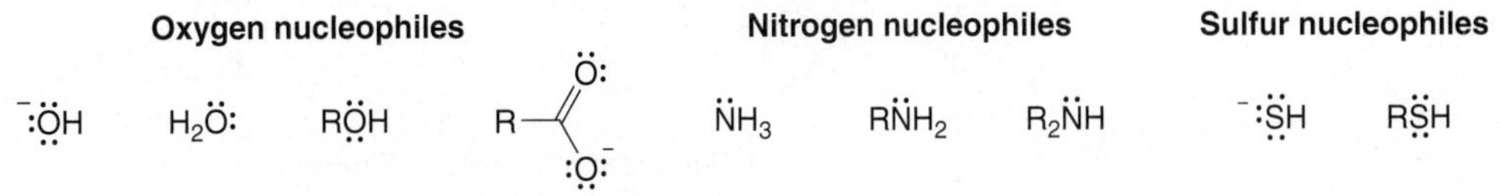

Nucleophilic acyl substitution using heteroatomic nucleophiles results in the conversion of one carboxylic acid derivative into another, as shown in two examples.

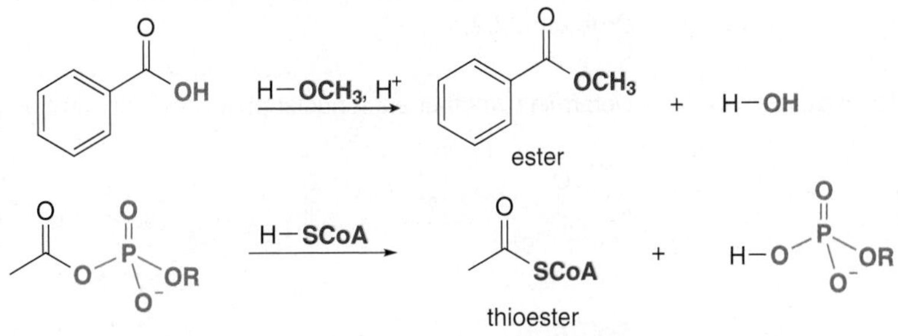

Each reaction results in the replacement of the leaving group by the nucleophile, regardless of the identity of or charge on the nucleophile. To draw any nucleophilic acyl substitution product:

- Find the sp^2 hybridized carbon with the leaving group.
- Identify the nucleophile.
- Substitute the nucleophile for the leaving group. With a neutral nucleophile a proton must be lost to obtain a neutral substitution product.

22.7B Relative Reactivity of Carboxylic Acids and Their Derivatives

As discussed in Section 20.2B, carboxylic acids and their derivatives differ greatly in reactivity toward nucleophiles. The order of reactivity parallels the leaving group ability of the group Z.

- **The better the leaving group, the more reactive RCOZ is in nucleophilic acyl substitution.**

Recall that the **best leaving group is the weakest base.** The relative basicity of the common leaving groups, Z, is given in Table 22.1.

Thus, the following trends result:

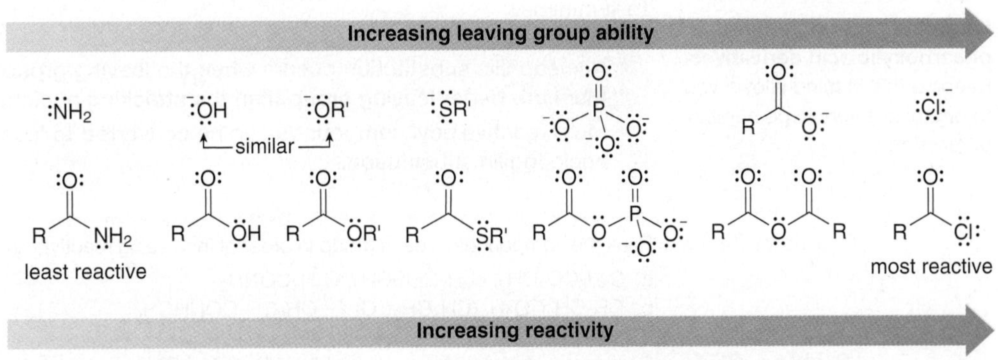

Based on this order of reactivity, **more reactive acyl compounds (acid chlorides, anhydrides, and acyl phosphates) can be converted to less reactive ones (thioesters, carboxylic acids, esters, and amides). The reverse is not usually true.**

To see why this is so, recall that nucleophilic addition to a carbonyl group forms a tetrahedral intermediate with two possible leaving groups, Z^- or $:Nu^-$. The group that is subsequently eliminated is the *better* of the two leaving groups. For a reaction to form a substitution product, therefore, Z^- must be the better leaving group, making the starting material RCOZ a more reactive acyl compound.

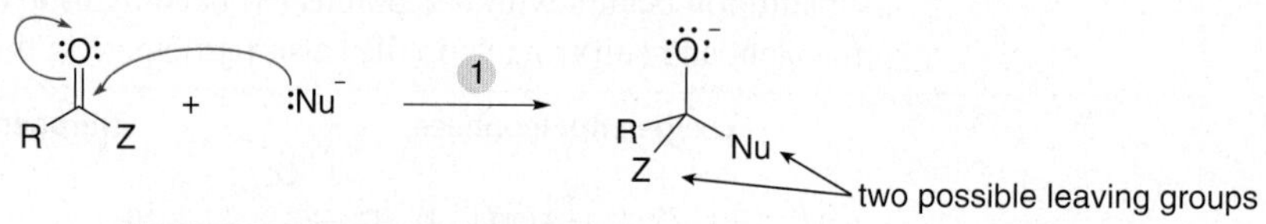

For a reaction to occur, Z must be a better leaving group than Nu.

To evaluate whether a nucleophilic substitution reaction will occur, **compare the leaving group ability of the incoming nucleophile and the departing leaving group,** as shown in Sample Problem 22.2.

Sample Problem 22.2 Determine whether each nucleophilic acyl substitution is likely to occur.

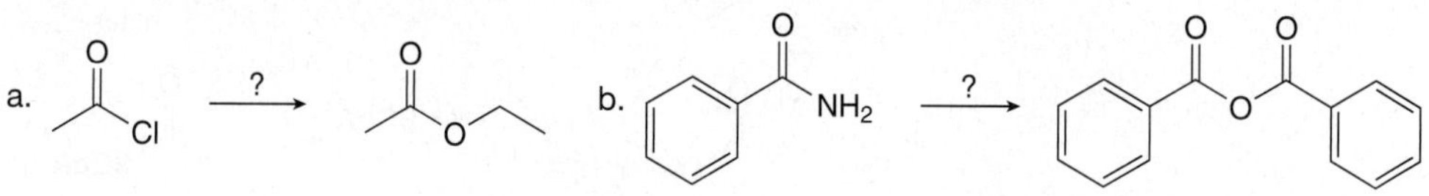

Solution

a. Conversion of CH_3COCl to $CH_3COOCH_2CH_3$ requires the substitution of Cl^- by $^-OCH_2CH_3$. Because Cl^- is a weaker base and therefore a better leaving group than $^-OCH_2CH_3$, **this reaction occurs.**

b. Conversion of $C_6H_5CONH_2$ to $(C_6H_5CO)_2O$ requires the substitution of $^-NH_2$ by $^-OCOC_6H_5$. Because $^-NH_2$ is a stronger base and therefore a poorer leaving group than $^-OCOC_6H_5$, **this reaction does *not* occur.**

Problem 22.9 Without reading ahead in Chapter 22, state whether it should be possible to carry out each of the following nucleophilic substitution reactions.

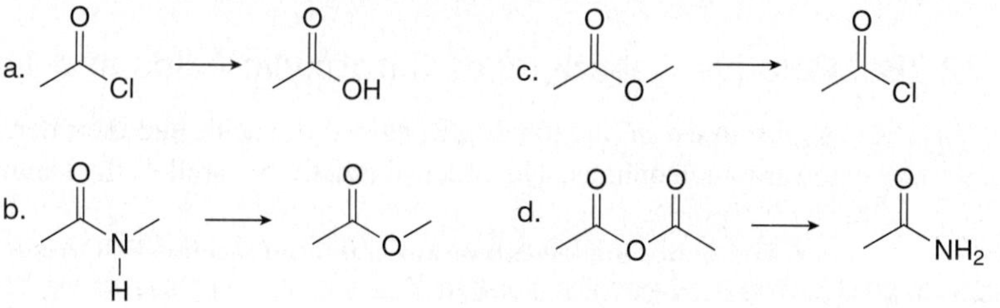

To summarize:

- **Nucleophilic substitution occurs when the leaving group Z^- is a weaker base and therefore better leaving group than the attacking nucleophile $:Nu^-$.**
- **More reactive acyl compounds can be converted to less reactive acyl compounds by nucleophilic substitution.**

Learn the order of reactivity of carboxylic acid derivatives. Keeping this in mind allows you to organize a very large number of reactions.

Problem 22.10 Rank the compounds in each group in order of increasing reactivity in nucleophilic acyl substitution.

a. $C_6H_5CO_2CH_3$, $C_6H_5COSCH_3$, $C_6H_5CONH_2$

b. $CH_3CH_2CO_2H$, $(CH_3CH_2CO)_2O$, $CH_3CH_2CONHCH_3$

Problem 22.11 Explain why trichloroacetic anhydride $[(Cl_3CCO)_2O]$ is more reactive than acetic anhydride $[(CH_3CO)_2O]$ in nucleophilic acyl substitution reactions.

22.7C A Preview of Specific Reactions

Sections 22.8–22.13 are devoted to specific examples of nucleophilic acyl substitution using heteroatoms as nucleophiles. There are a great many reactions, and it is easy to confuse them unless you learn the general order of reactivity of carboxylic acid derivatives. **Keep in mind that every reaction that begins with an acyl starting material involves nucleophilic substitution.**

In this text, all of the nucleophilic substitution reactions are grouped according to the carboxylic acid derivative used as a starting material. We begin with the reactions of acid chlorides, the most reactive acyl compounds, then proceed to less and less reactive carboxylic acid derivatives, ending with amides. Acid chlorides undergo many reactions, because they have the best leaving group of all acyl compounds, whereas amides undergo only one reaction, which must be carried out under harsh reaction conditions, because amides have a poor leaving group.

In general, we will examine nucleophilic acyl substitution with four different nucleophiles, as shown in the following equations.

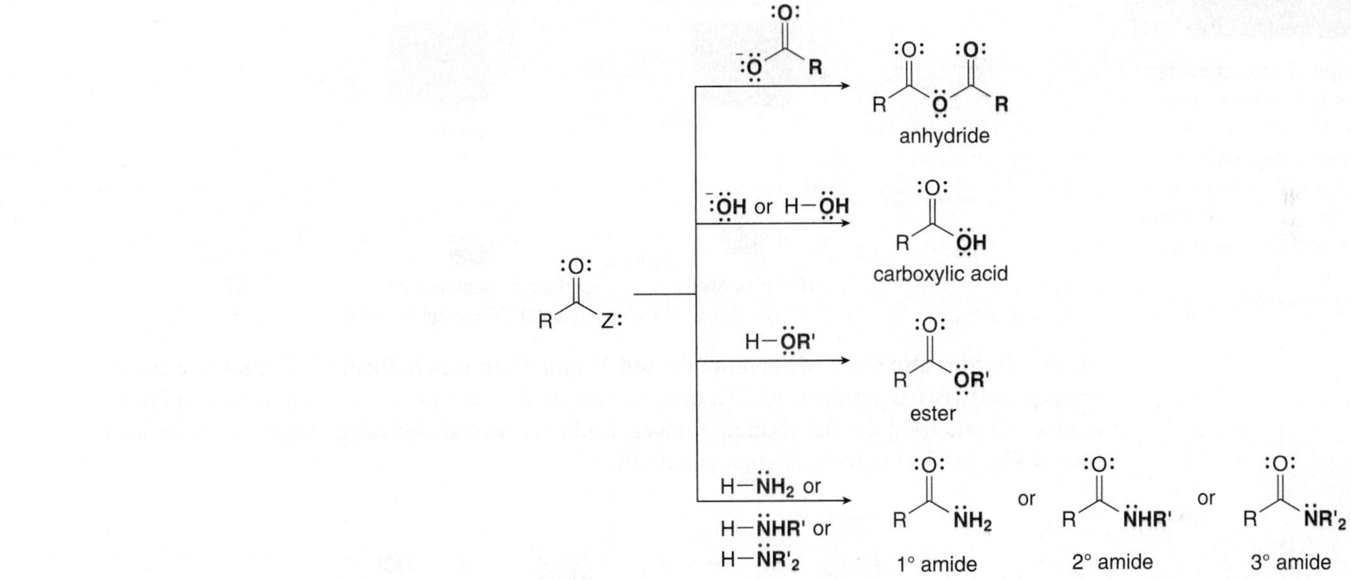

These reactions are used to make anhydrides, carboxylic acids, esters, and amides, but not acid chlorides, from other acyl compounds. Acid chlorides are the most reactive acyl compounds (they have the best leaving group), so they are not easily formed as a product of nucleophilic substitution reactions. They can only be prepared from carboxylic acids using special reagents, as discussed in Section 22.10A.

A fifth nucleophile, a thiol (HSR), is introduced in our discussion of biological acyl phosphates and thioesters.

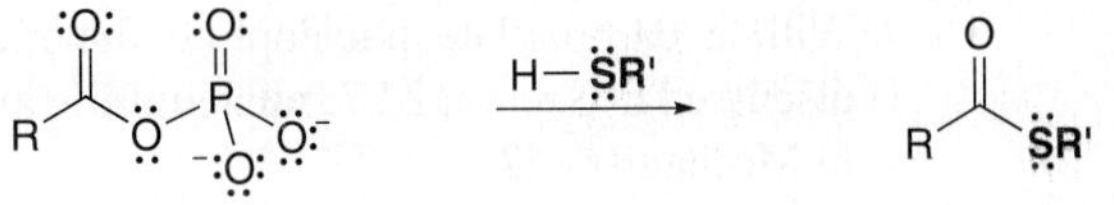

22.8 Reactions of Acid Chlorides

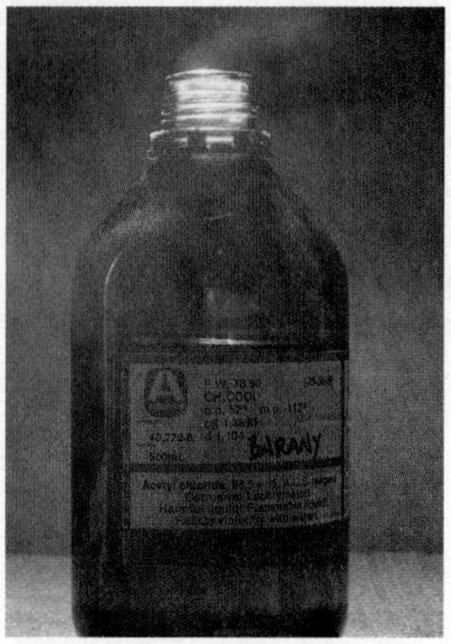

The reaction of acid chlorides with water is rapid. Exposure of an acid chloride to moist air on a humid day leads to some hydrolysis, giving the acid chloride a very acrid odor, due to the HCl formed as a by-product.

Acid chlorides readily react with nucleophiles to form nucleophilic substitution products, with HCl usually formed as a reaction by-product. A weak base like pyridine is added to the reaction mixture to remove this strong acid, forming an ammonium salt.

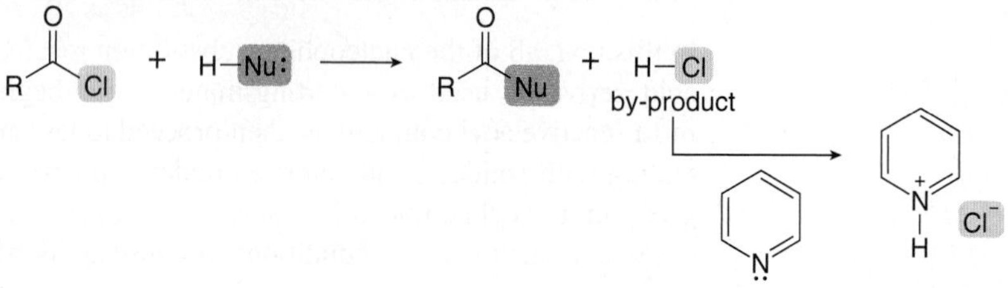

Acid chlorides react with oxygen nucleophiles to form anhydrides, carboxylic acids, and esters.

[1]
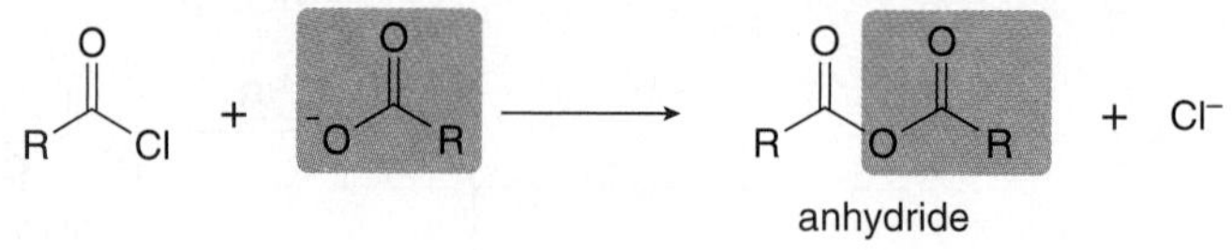

[2]
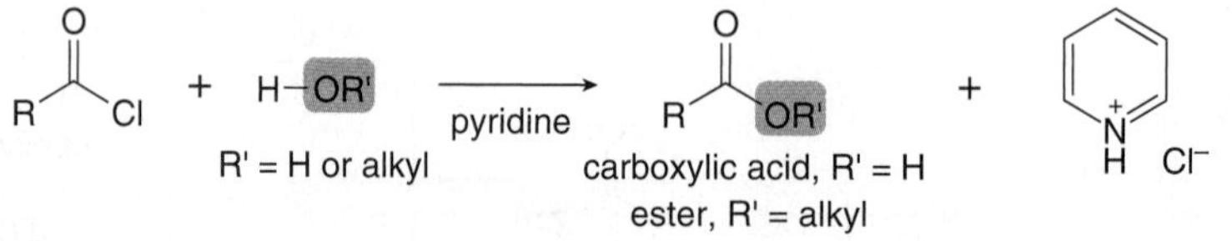

Acid chlorides also react with ammonia and 1° and 2° amines to form 1°, 2°, and 3° amides, respectively. Two equivalents of NH_3 or amine are used. One equivalent acts as a nucleophile to replace Cl and form the substitution product, while the second equivalent reacts as a base with the HCl by-product to form an ammonium salt.

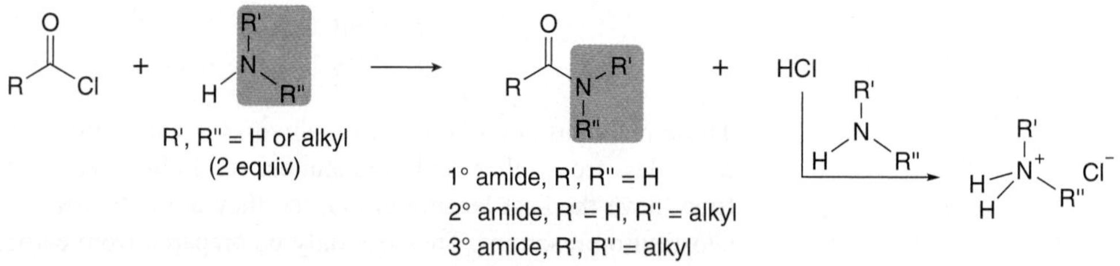

Problem 22.12 Draw the products formed when benzoyl chloride (C_6H_5COCl) is treated with each nucleophile: (a) H_2O, pyridine; (b) CH_3COO^-; (c) NH_3 (excess); (d) $(CH_3)_2NH$ (excess).

With a carboxylate nucleophile the mechanism follows the general, two-step mechanism discussed in Section 22.7: **nucleophilic attack followed by loss of the leaving group,** as shown in Mechanism 22.2.

Mechanism 22.2 Conversion of Acid Chlorides to Anhydrides

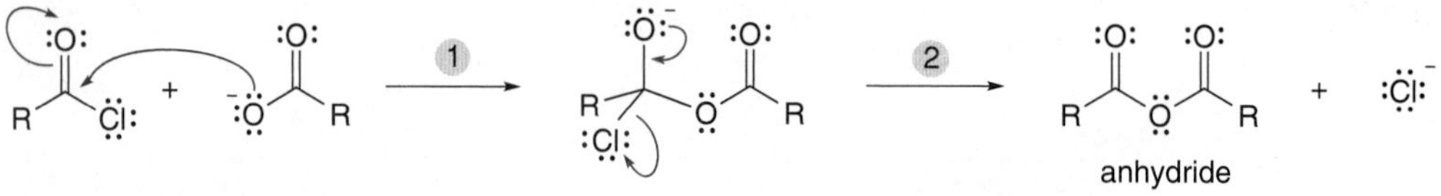

1. The nucleophilic carboxylate anion attacks the carbonyl group, forming an sp^3 hybridized carbon.
2. Elimination of the leaving group (Cl^-) forms the substitution product, an anhydride.

Problem 22.13 Draw a stepwise mechanism for the formation of **A** from an alcohol and acid chloride. **A** was converted in one step to blattellaquinone, the sex pheromone of the female German cockroach *Blattella germanica.*

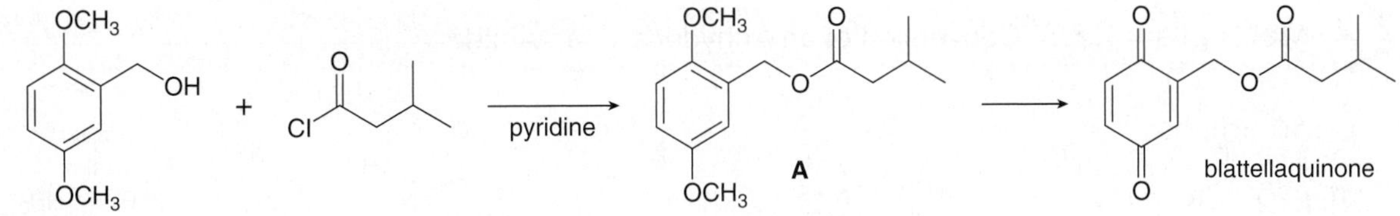

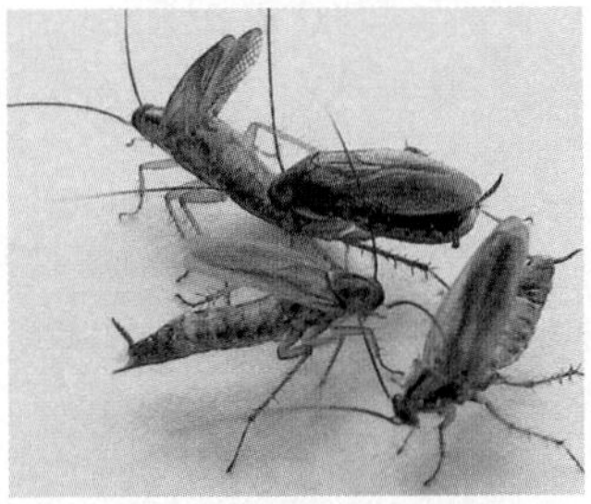

A short laboratory synthesis of blattellaquinone (Problem 22.13), the sex pheromone of the female German cockroach, opens new possibilities for cockroach population control using pheromone-baited traps.

Nucleophilic substitution occurs only when the leaving group is a *weaker* base and therefore a *better* leaving group than the attacking nucleophile.

22.9 Reactions of Anhydrides

Although somewhat less reactive than acid chlorides, anhydrides nonetheless readily react with most nucleophiles to form substitution products. Nucleophilic substitution reactions of anhydrides are no different than the reactions of other carboxylic acid derivatives, even though anhydrides contain two carbonyl groups. **Nucleophilic attack occurs at one carbonyl group, while the second carbonyl becomes part of the leaving group.**

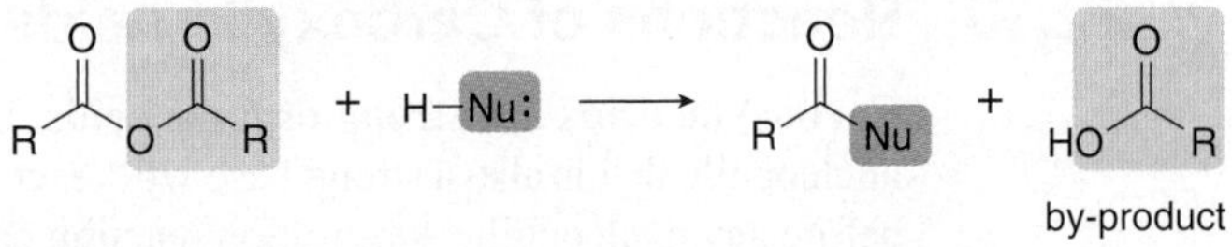

Anhydrides can't be used to make acid chlorides, because $RCOO^-$ is a stronger base and therefore a poorer leaving group than Cl^-. Anhydrides can be used to make all other acyl derivatives, however. Reaction with water and alcohols yields **carboxylic acids** and **esters,** respectively. Reaction with two equivalents of NH_3 or amines forms **1°, 2°,** and **3° amides.** A molecule of carboxylic acid (or a carboxylate salt) is always formed as a by-product.

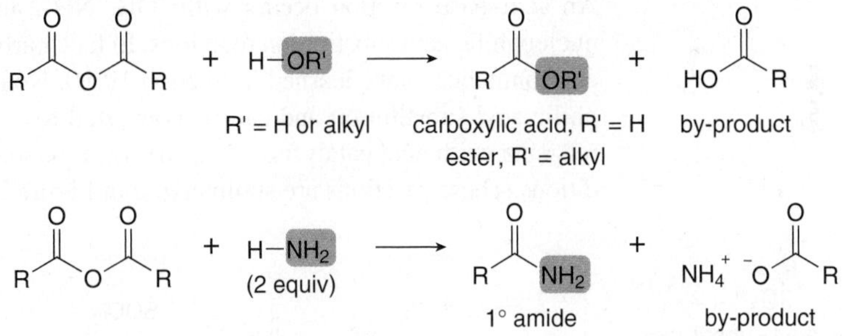

Problem 22.14 Draw the products formed when benzoic anhydride $[(C_6H_5CO)_2O]$ is treated with each nucleophile: (a) H_2O; (b) CH_3OH; (c) NH_3 (excess); (d) $(CH_3)_2NH$ (excess).

The conversion of an anhydride to an amide illustrates the mechanism of nucleophilic acyl substitution with an anhydride as starting material (Mechanism 22.3). Besides the usual steps of **nucleophilic addition** and **elimination of the leaving group,** an additional proton transfer is needed.

Mechanism 22.3 Conversion of an Anhydride to an Amide

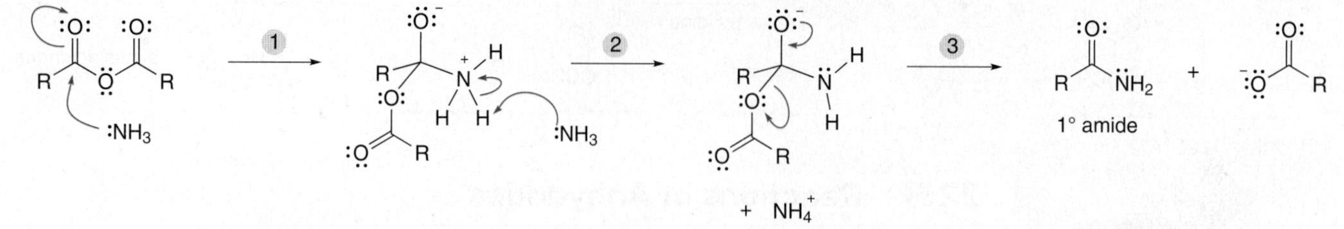

1 The nucleophile (NH_3) attacks the carbonyl, forming an sp^3 hybridized carbon.

2 – 3 Loss of a proton and elimination of the leaving group (RCO_2^-) form the substitution product, a 1° amide.

22.10 Reactions of Carboxylic Acids

Carboxylic acids are strong organic acids. Because acid–base reactions proceed rapidly, any nucleophile that is also a strong base will react with a carboxylic acid by removing a proton ***first,*** before any nucleophilic substitution reaction can take place.

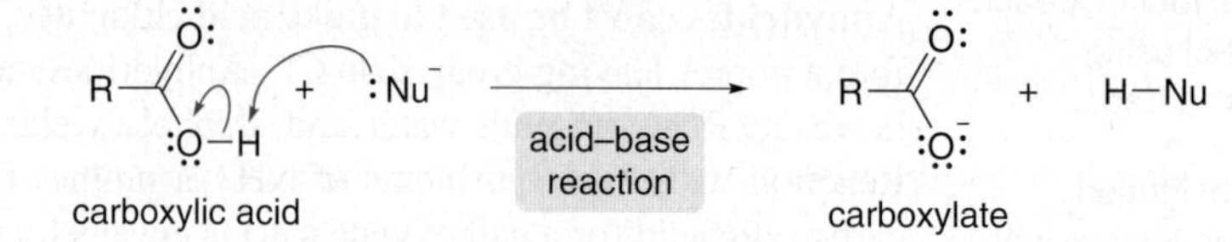

An acid–base reaction occurs with ^-OH, NH_3, and amines, all common nucleophiles used in nucleophilic acyl substitution reactions. In fact, carboxylic acids exist as carboxylates in biological environments, as we learned in Section 19.10. Nonetheless, carboxylic acids do undergo nucleophilic acyl substitution and can be converted to a variety of other acyl derivatives using special reagents, with acid catalysis, using enzymes, or sometimes, by using rather forcing reaction conditions. These reactions are summarized in Figure 22.3 and detailed in Sections 22.10A–22.10D.

Figure 22.3
Nucleophilic acyl substitution reactions of carboxylic acids

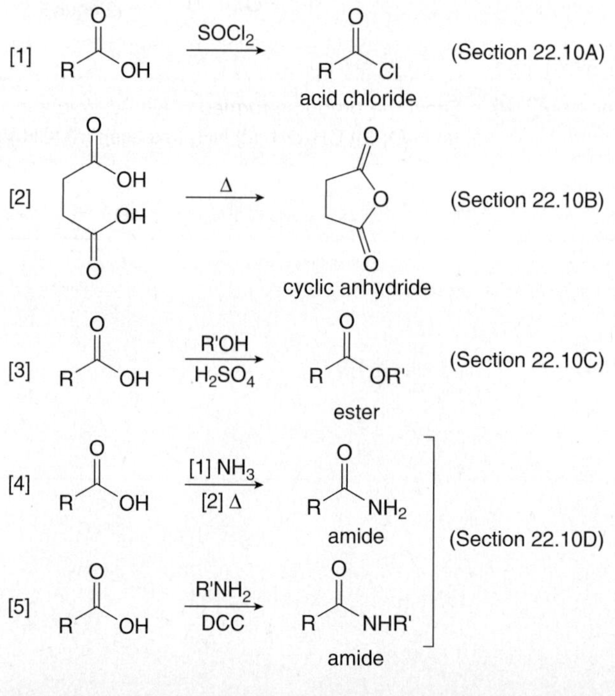

22.10A Conversion of RCOOH to RCOCl

Carboxylic acids can't be converted to acid chlorides by using Cl^- as a nucleophile, because the attacking nucleophile Cl^- is a weaker base than the departing leaving group, ^-OH. But carboxylic acids *can* be converted to acid chlorides using thionyl chloride, $\mathbf{SOCl_2}$**,** a reagent that was introduced in Section 9.12 to convert alcohols to alkyl chlorides.

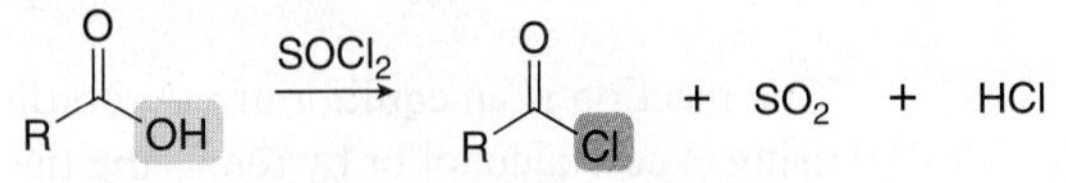

SOCl2
OH
Cl
+ SO2 + HCl

Treatment of benzoic acid with $SOCl_2$ forms benzoyl chloride. This reaction converts a less reactive acyl derivative (a carboxylic acid) into a more reactive one (an acid chloride). This is possible because thionyl chloride converts the OH group of the acid into a better leaving group, and because it provides the nucleophile (Cl^-) to displace the leaving group. The steps in the process are illustrated in Mechanism 22.4.

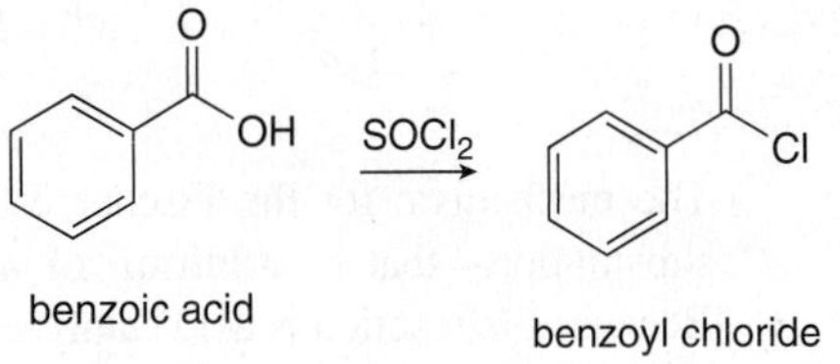

SOCl2
benzoic acid
benzoyl chloride

Mechanism 22.4 Conversion of Carboxylic Acids to Acid Chlorides

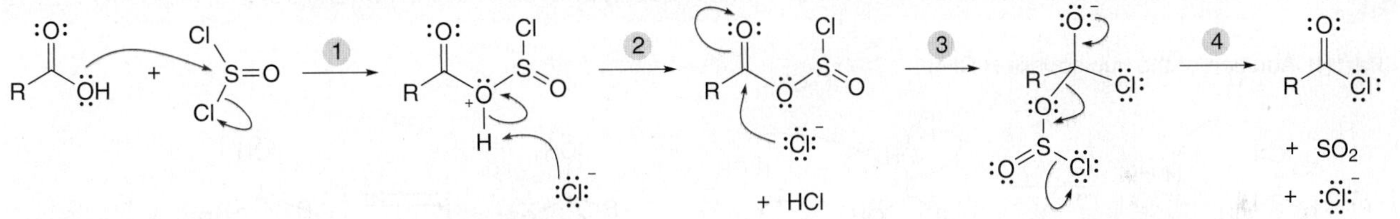

1
2
3
4
+ HCl
+ SO2

1 – 2 Reaction of the carboxylic acid with $SOCl_2$ and loss of a proton convert the OH group to OSOCl, a good leaving group.

3 – 4 Nucleophilic attack of chloride generates a tetrahedral intermediate, and loss of the leaving group (SO_2 and Cl^-) forms the acid chloride.

Problem 22.15 Draw the products of each reaction.

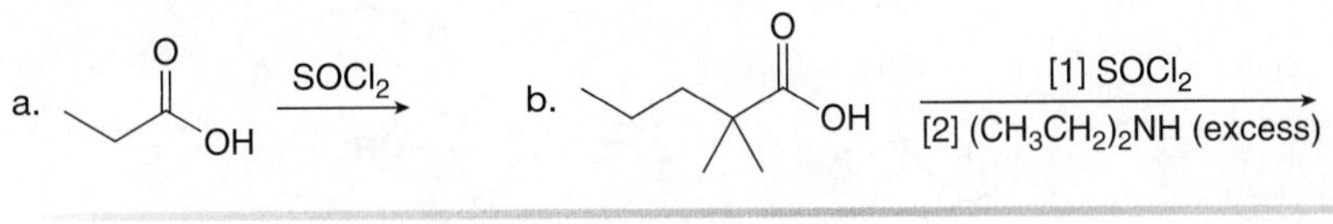

a.
SOCl2
b.
[1] SOCl2
[2] (CH3CH2)2NH (excess)

22.10B Conversion of RCOOH to $(RCO)_2O$

Carboxylic acids cannot be readily converted to anhydrides, but dicarboxylic acids can be converted to cyclic anhydrides by heating to high temperatures. This is a **dehydration** reaction because a water molecule is lost from the diacid.

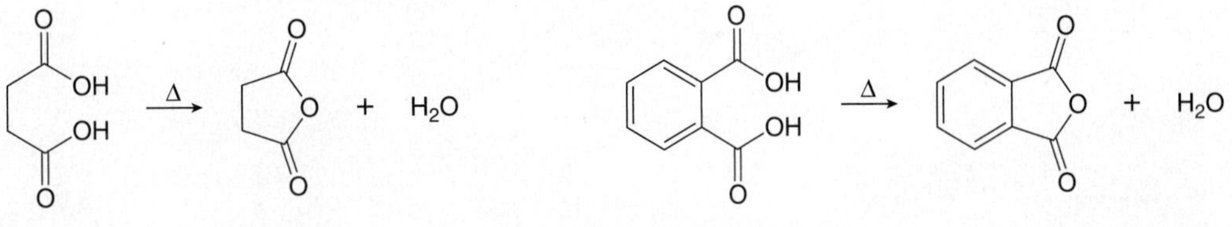

Δ
+ H2O
Δ
+ H2O

22.10C Conversion of RCOOH to RCOOR'

Treatment of a carboxylic acid with an alcohol in the presence of an acid catalyst forms an ester. This reaction is called a **Fischer esterification.**

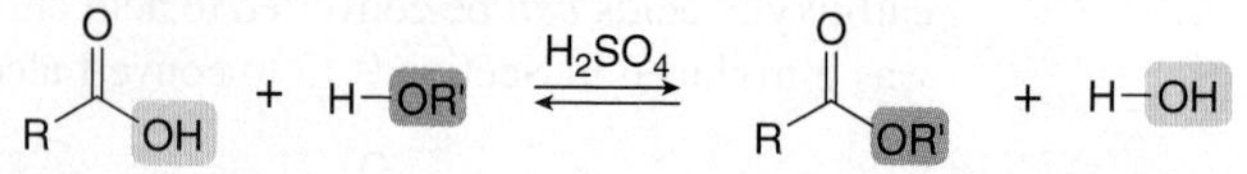

This reaction is an equilibrium. According to Le Châtelier's principle, it is driven to the right by using excess alcohol or by removing the water as it is formed.

Ethyl acetate is a common organic solvent with a characteristic odor. It is used in nail polish remover and model airplane glue.

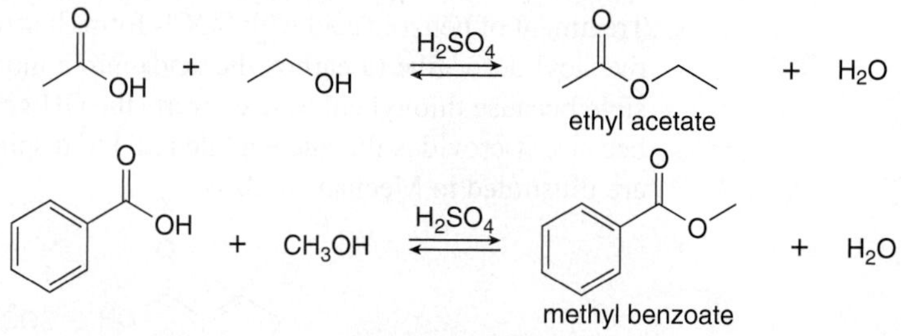

The mechanism for the Fischer esterification involves the usual two steps of nucleophilic acyl substitution—that is, **addition of a nucleophile followed by elimination of a leaving group.** Because the reaction is acid catalyzed, however, there are additional protonation and deprotonation steps. As always, though, the first step of any mechanism with an oxygen-containing starting material and an acid is to **protonate an oxygen atom** as shown with a general acid HA in Mechanism 22.5.

Mechanism 22.5 Fischer Esterification—Acid-Catalyzed Conversion of Carboxylic Acids to Esters

Part [1] Addition of the nucleophile R'OH

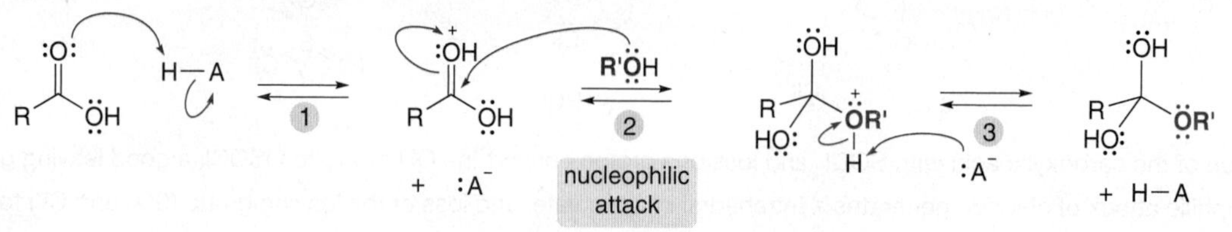

1 Protonation of the carbonyl oxygen makes the carbonyl more electrophilic.

2 – 3 Nucleophilic attack by R'OH forms a tetrahedral intermediate, and deprotonation gives the addition product.

Part [2] Elimination of the leaving group H_2O

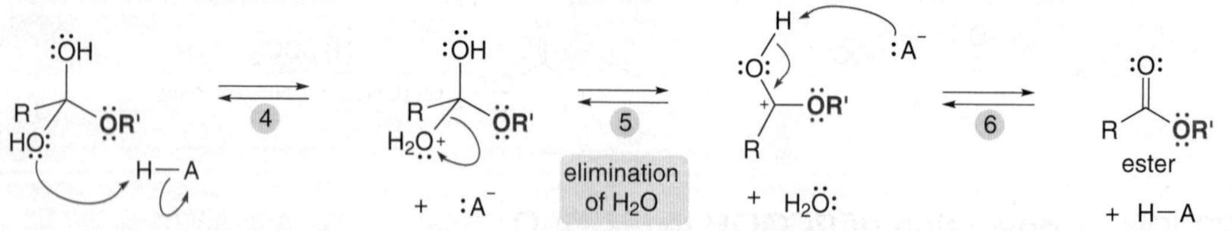

4 Protonation of the OH group forms a good leaving group.

5 – 6 Loss of H_2O and deprotonation give the ester.

Esterification of a carboxylic acid occurs in the presence of acid but not in the presence of base. Base removes a proton from the carboxylic acid, forming an electron-rich carboxylate anion, which does not react with an electron-rich nucleophile.

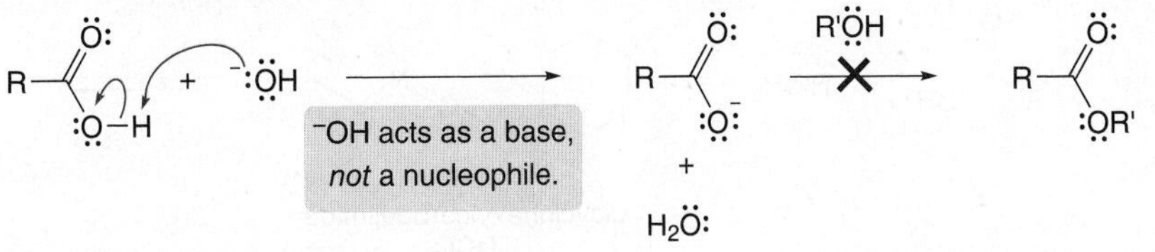

$^{-}$OH acts as a base,
not a nucleophile.

Intramolecular esterification of γ- and δ-hydroxy carboxylic acids forms five- and six-membered lactones.

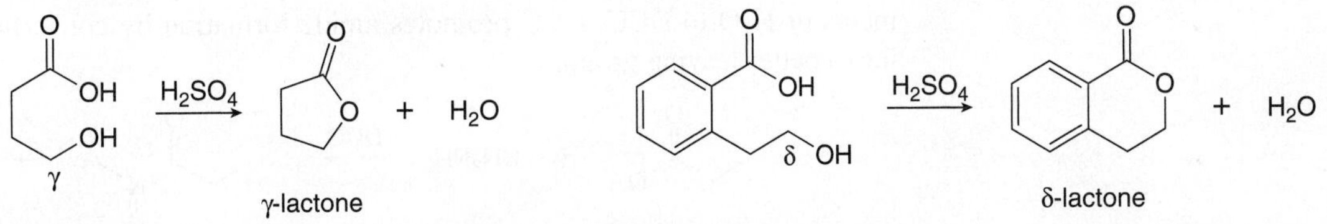

γ-lactone

δ-lactone

Problem 22.16 Draw the products of each reaction.

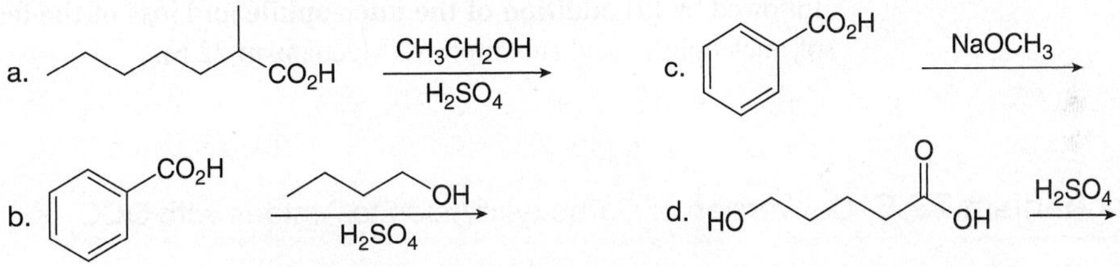

Problem 22.17 Draw the products formed when benzoic acid ($C_6H_5CO_2H$) is treated with CH_3OH having its O atom labeled with ^{18}O ($CH_3{}^{18}OH$). Indicate where the labeled oxygen atom resides in the products.

Problem 22.18 Draw a stepwise mechanism for the following reaction.

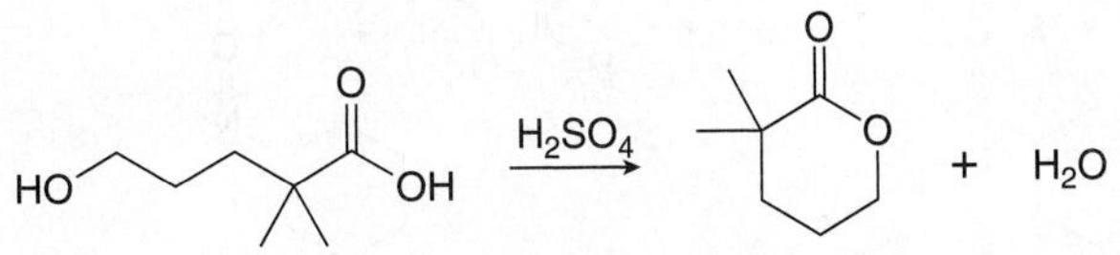

22.10D Conversion of $RCOOH$ to $RCONR'_2$

The direct conversion of a carboxylic acid to an amide with NH_3 or an amine is very difficult, even though a more reactive acyl compound is being transformed into a less reactive one. The problem is that carboxylic acids are strong organic acids and NH_3 and amines are bases, so they undergo an **acid–base reaction to form an ammonium salt** before any nucleophilic substitution occurs.

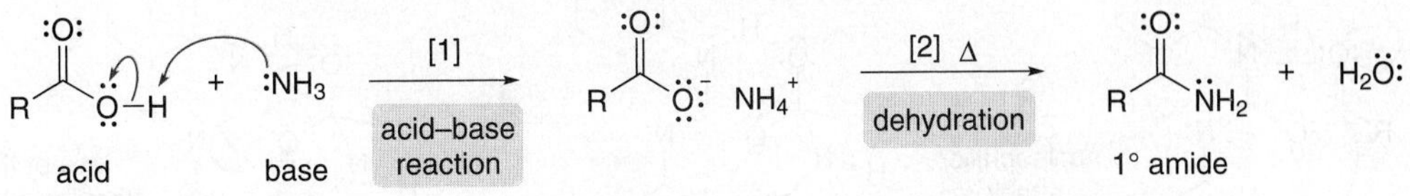

acid
base
[1]
acid–base reaction
[2] Δ
dehydration
1° amide

Heating at high temperature (>100 °C) dehydrates the resulting ammonium salt of the carboxylate anion to form an amide, though the yield can be low.

Amides are much more easily prepared from acid chlorides and anhydrides, as discussed in Sections 22.8 and 22.9.

Therefore, the overall conversion of $RCOOH$ to $RCONH_2$ requires two steps:

[1] **Acid–base reaction of $RCOOH$ with NH_3 to form an ammonium salt**

[2] **Dehydration at high temperature (>100 °C)**

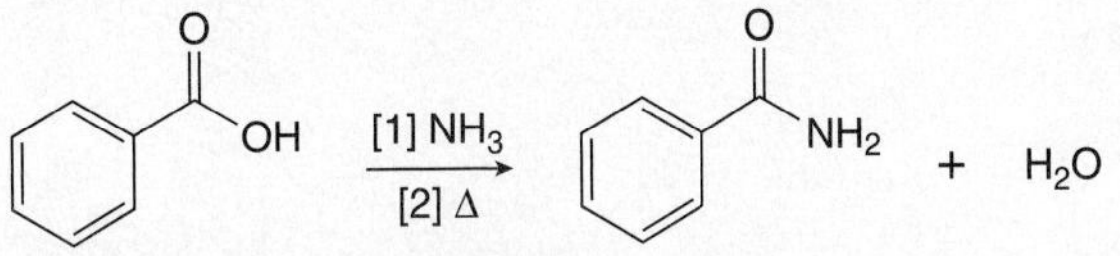

A carboxylic acid and an amine readily react to form an amide in the presence of an additional reagent, **dicyclohexylcarbodiimide (DCC),** which is converted to the by-product dicyclohexylurea in the course of the reaction.

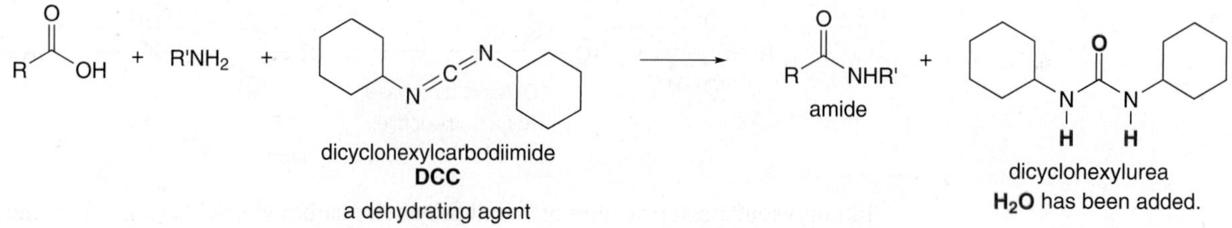

DCC is a dehydrating agent. The dicyclohexylurea by-product is formed by adding the elements of H_2O to DCC. DCC promotes amide formation by converting the carboxy OH group into a better leaving group.

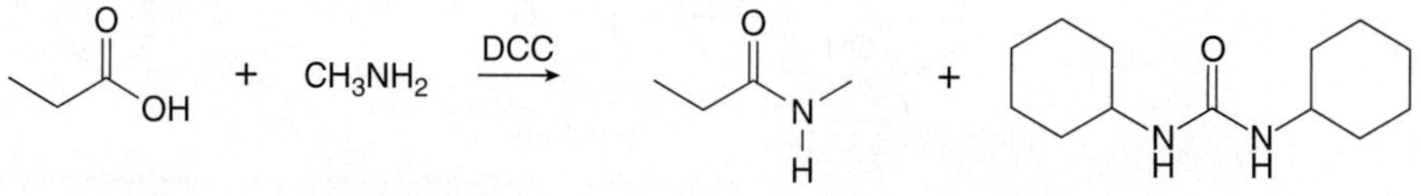

The mechanism consists of two parts: [1] conversion of the OH group into a better leaving group, followed by [2] **addition of the nucleophile and loss of the leaving group** to form the product of nucleophilic acyl substitution (Mechanism 22.6).

Mechanism 22.6 Conversion of Carboxylic Acids to Amides with DCC

Part [1] Conversion of OH into a better leaving group

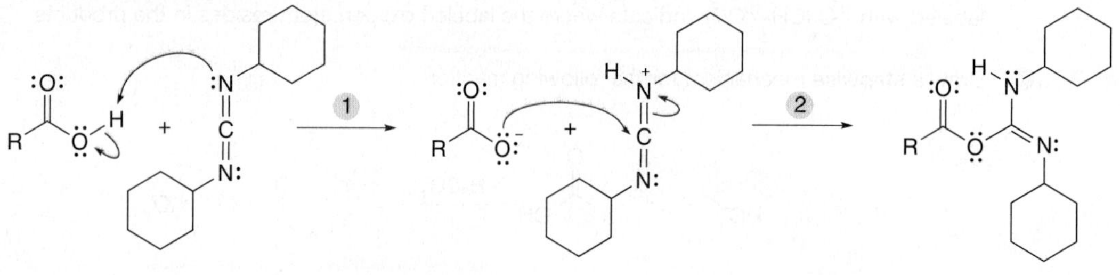

1 Acid–base reaction results in transfer of a proton from the carboxylic acid to DCC.

2 Nucleophilic attack of RCO_2^- on the conjugate acid of DCC forms an addition product. The overall result of Steps [1] and [2] is conversion of OH into a better leaving group.

Part [2] Addition of the nucleophile and loss of the leaving group

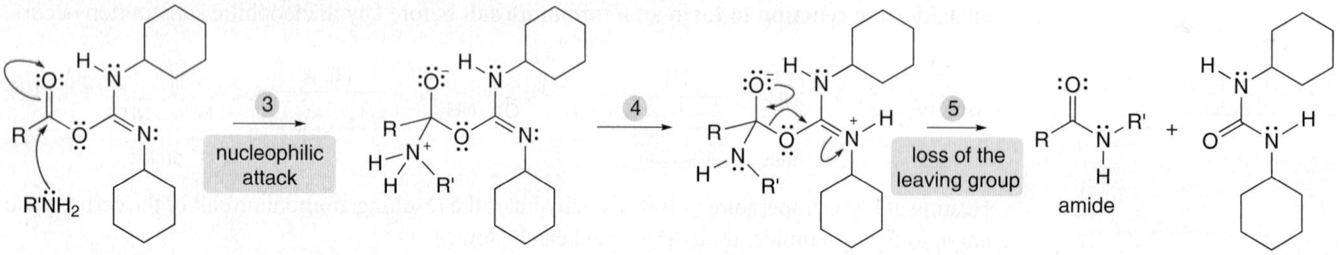

3 Nucleophilic attack of the amine on the activated carboxy group forms a tetrahedral intermediate.

4 – 5 Proton transfer and elimination of dicyclohexylurea as the leaving group form the amide.

The reaction of an acid and an amine with DCC is often used in the laboratory to form the amide bond in peptides, as is discussed in Chapter 26.

Problem 22.19 What product is formed when acetic acid is treated with each reagent: (a) CH_3NH_2; (b) CH_3NH_2, then heat; (c) CH_3NH_2 + DCC?

22.11 Reactions of Esters

Esters can be converted into carboxylic acids and amides.

- **Esters are hydrolyzed with water in the presence of either acid or base to form carboxylic acids or carboxylate anions.**

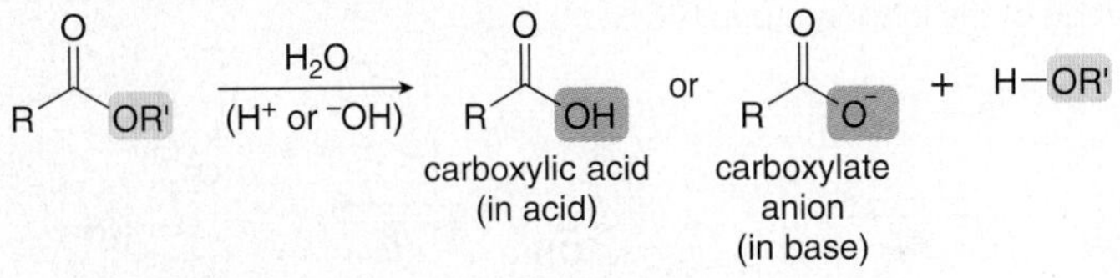

O
R
OR'
H_2O
(H^+ or ^-OH)
O
R
OH
carboxylic acid
(in acid)
or
O
R
O^-
carboxylate
anion
(in base)
+ H−OR'

- **Esters react with NH_3 and amines to form 1°, 2°, or 3° amides.**

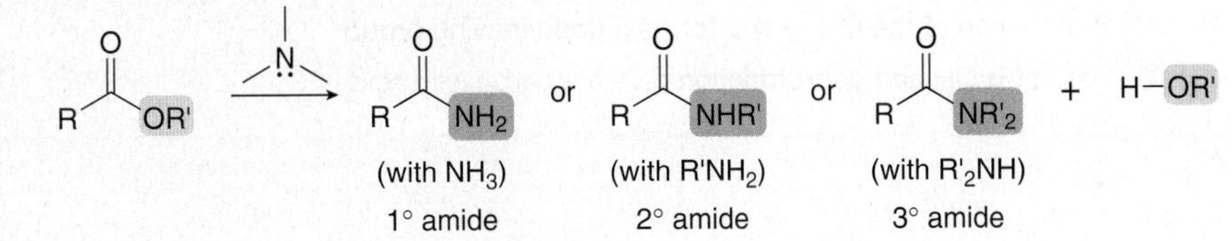

O
R
OR'
N
O
R
NH_2
(with NH_3)
1° amide
or
O
R
NHR'
(with $R'NH_2$)
2° amide
or
O
R
NR'_2
(with R'_2NH)
3° amide
+ H−OR'

22.11A Ester Hydrolysis in Aqueous Acid

The first step in acid-catalyzed ester hydrolysis is **protonation on oxygen,** the same first step of any mechanism involving an oxygen-containing starting material and an acid.

The hydrolysis of esters in aqueous acid is a reversible equilibrium reaction that is driven to the right by using a large excess of water.

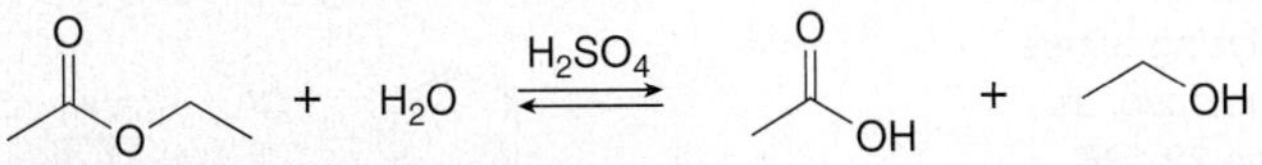

O
O
+ H_2O
H_2SO_4
O
OH
+
OH

The mechanism of ester hydrolysis in acid (shown in Mechanism 22.7) is the reverse of the mechanism of ester synthesis from carboxylic acids (Mechanism 22.5). Thus, the mechanism consists of the addition of the nucleophile and the elimination of the leaving group, the two steps common to all nucleophilic acyl substitutions, as well as several proton transfers, because the reaction is acid-catalyzed.

Mechanism 22.7 Acid-Catalyzed Hydrolysis of an Ester to a Carboxylic Acid

Part [1] Addition of the nucleophile H_2O

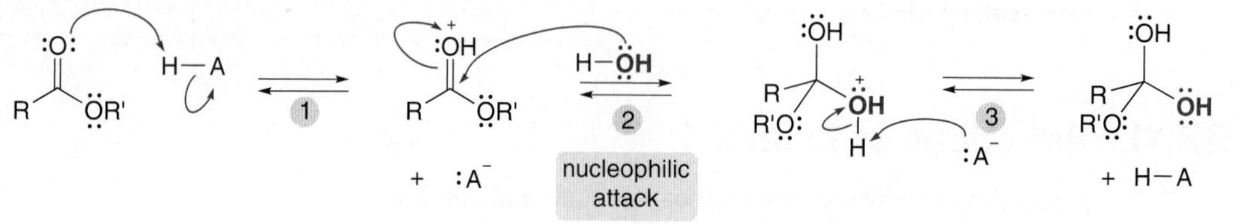

1 Protonation of the carbonyl oxygen makes the carbonyl more electrophilic.

2 – 3 Nucleophilic attack by H_2O forms a tetrahedral intermediate, and deprotonation gives the addition product.

Part [2] Elimination of the leaving group R'OH

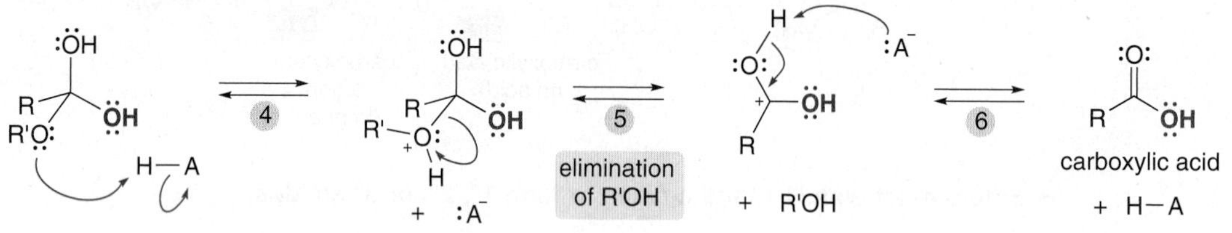

4 Protonation of the OR' group forms a good leaving group.

5 – 6 Loss of R'OH and deprotonation give the carboxylic acid.

22.11B Ester Hydrolysis in Aqueous Base

The word **saponification** comes from the Latin *sapo* meaning **soap.** Soap is prepared by hydrolyzing esters in fats with aqueous base, as explained in Section 22.12B.

Esters are hydrolyzed in aqueous base to form carboxylate anions. Basic hydrolysis of an ester is called **saponification.**

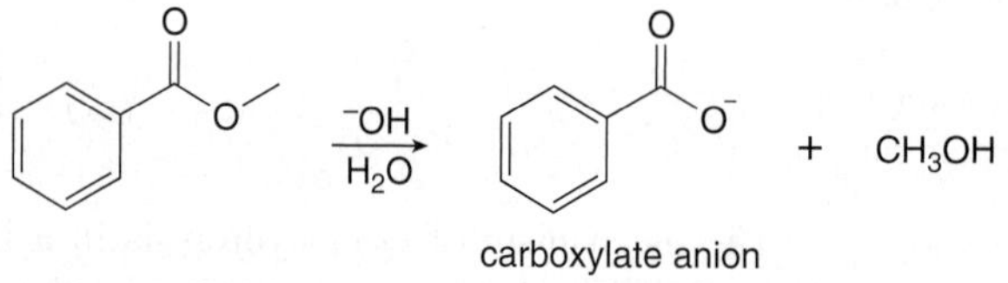

The mechanism for this reaction has the usual two steps of the general mechanism for nucleophilic acyl substitution presented in Section 22.7—**addition of the nucleophile** followed by **loss of a leaving group**—plus an additional step involving proton transfer (Mechanism 22.8).

Mechanism 22.8 Base-Promoted Hydrolysis of an Ester to a Carboxylic Acid

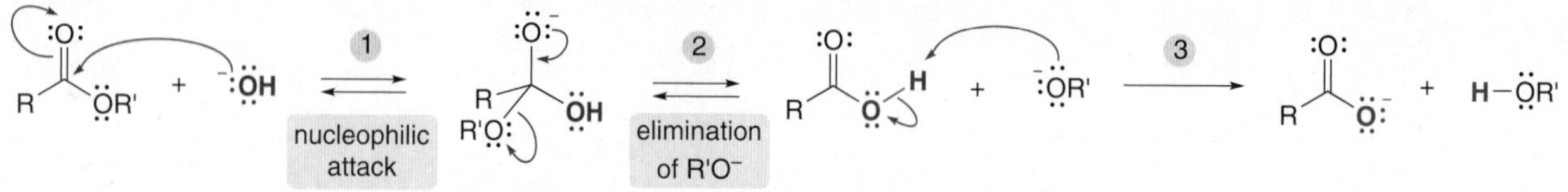

1 – 2 Addition of the nucleophile ($^-$OH) followed by elimination of the leaving group ($^-$OR') form a carboxylic acid. These two steps are reversible.

3 Because the carboxylic acid is a strong organic acid and the leaving group ($^-$OR') is a strong base, an acid–base reaction forms the carboxylate anion.

The carboxylate anion is resonance stabilized, and this drives the equilibrium in its favor. Once the reaction is complete and the carboxylate anion is formed, it can be protonated with strong acid to form the neutral carboxylic acid.

Hydrolysis is base promoted, *not* base catalyzed, because the base (^-OH) is the nucleophile that adds to the ester and forms part of the product. It participates in the reaction and is not regenerated later.

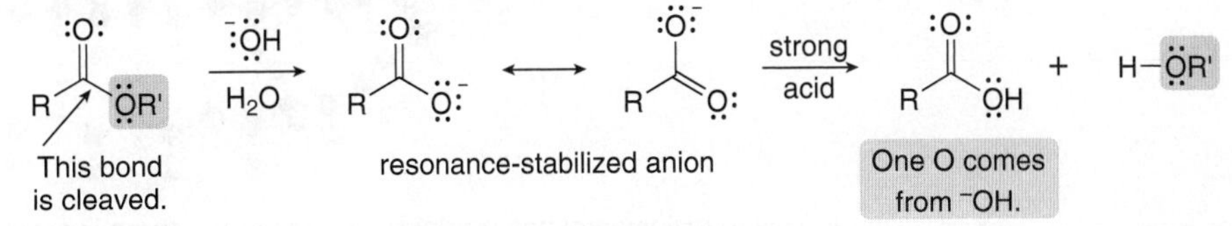

Where do the oxygen atoms in the product come from? The C–OR' bond in the ester is cleaved, so the OR' group becomes the alcohol by-product (R'OH) and one of the oxygens in the carboxylate anion product comes from ^-OH (the nucleophile).

Problem 22.20 Fenofibrate is a cholesterol-lowering medication that is converted to fenofibric acid, the active drug, by hydrolysis during metabolism. What is the structure of fenofibric acid?

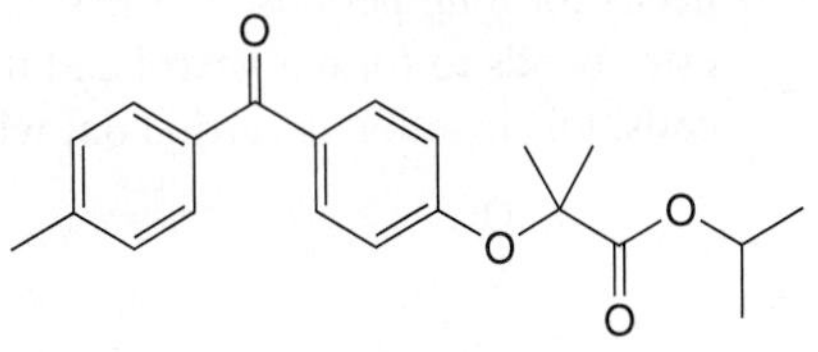

fenofibrate

Problem 22.21 What product is formed when the esters in ginkgolide B, the chapter-opening molecule, are hydrolyzed in aqueous acid? Indicate the stereochemistry of all stereogenic centers.

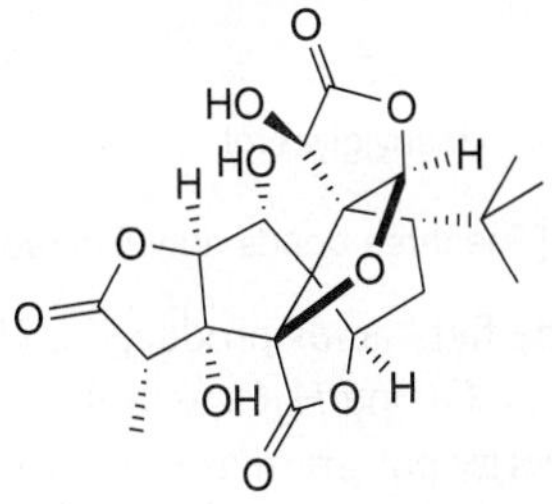

ginkgolide B

22.12 Application: Lipid Hydrolysis

22.12A Olestra—A Synthetic Fat

The most prevalent naturally occurring esters are the **triacylglycerols,** which were first discussed in Section 10.6. **Triacylglycerols are the lipids that comprise animal fats and vegetable oils.**

- **Each triacylglycerol is a triester, containing three long hydrocarbon side chains.**
- **Unsaturated triacylglycerols have one or more double bonds in their long hydrocarbon chains, whereas saturated triacylglycerols have none.**

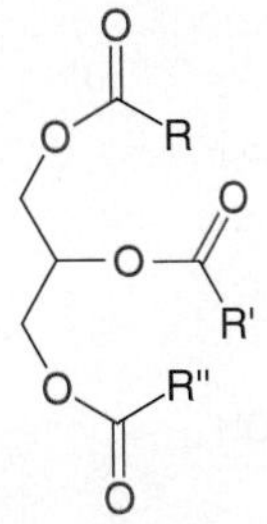

R groups have 11–19 C's.
[Three ester groups are labeled in red.]

triacylglycerol

Figure 22.4
The three-dimensional structure of a saturated triacylglycerol

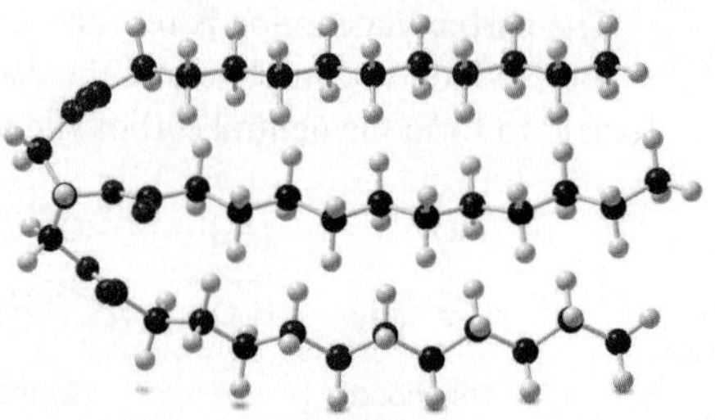

- This triacylglycerol has no double bonds in the three R groups (each with 11 C's) bonded to the ester carbonyls, making it a saturated fat.

Figure 22.4 contains a ball-and-stick model of a saturated fat.

Animals store energy in the form of triacylglycerols, kept in a layer of fat cells below the surface of the skin. This fat serves to insulate the organism, as well as provide energy for its metabolic needs for long periods. The first step in the metabolism of a triacylglycerol is hydrolysis of the ester bonds to form glycerol and three fatty acids. **This reaction is simply ester hydrolysis.** In cells, this reaction is carried out with enzymes called **lipases.**

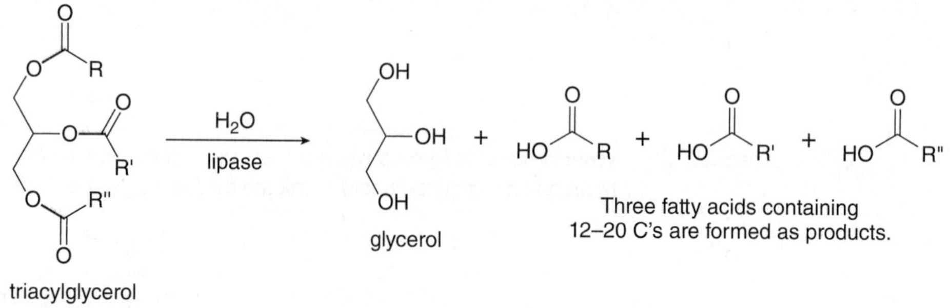

triacylglycerol
H_2O
lipase
glycerol
Three fatty acids containing 12–20 C's are formed as products.

[The three bonds drawn in red are cleaved in hydrolysis.]

The fatty acids produced on hydrolysis are then oxidized in a stepwise fashion, ultimately yielding CO_2 and H_2O, as well as a great deal of energy. Oxidation of fatty acids yields twice as much energy per gram as oxidation of an equivalent weight of carbohydrate.

Diets high in fat content lead to a large amount of stored fat, ultimately causing an individual to be overweight. One attempt to reduce calories in common snack foods has been to substitute "fake fats" such as **olestra** (trade name **Olean**) for triacylglycerols.

sour cream & onion
50% Fewer CALORIES
MADE WITH OLESTRA
Artificially Flavored

Some snack foods contain the "fake fat" olestra, giving them fewer calories for the calorie-conscious consumer.

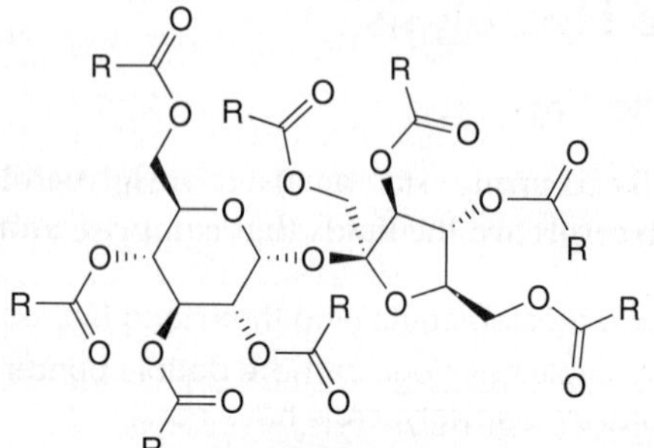

olestra
a polyester of sucrose
a synthetic fat
[R groups contain 11–19 C's.]

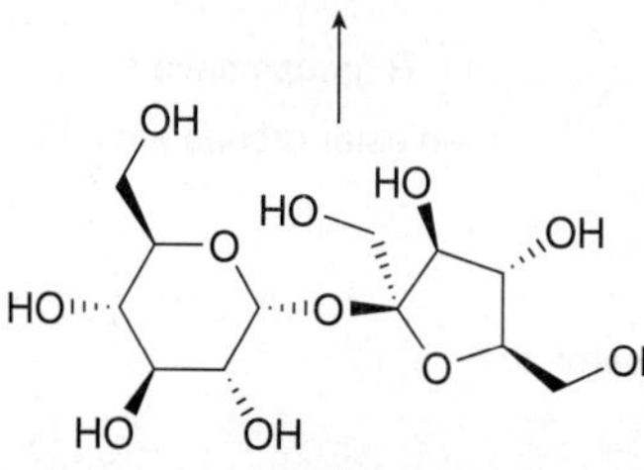

sucrose

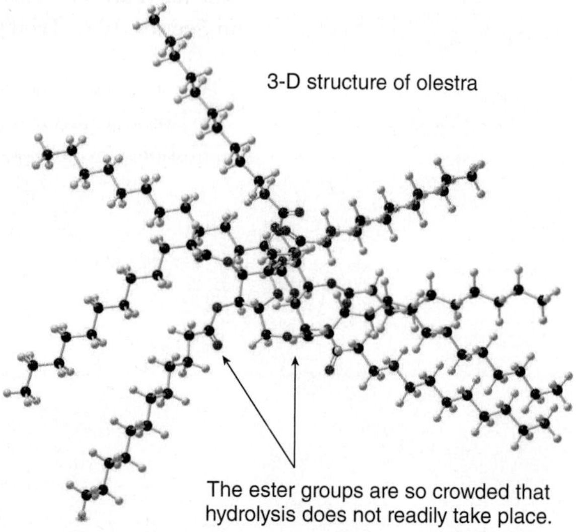

3-D structure of olestra
The ester groups are so crowded that hydrolysis does not readily take place.

Olestra is a polyester formed from long-chain fatty acids and sucrose, the sweet-tasting carbohydrate in table sugar. Naturally occurring triacylglycerols are also polyesters formed from long-chain fatty acids, but olestra has so many ester units clustered together in close proximity that they are too hindered to be hydrolyzed. As a result, olestra is not metabolized. Instead, it passes through the body unchanged, providing no calories to the consumer.

Thus, olestra's many C–C and C–H bonds make it similar in solubility to naturally occurring triacylglycerols, but its three-dimensional structure makes it inert to hydrolysis because of steric hindrance.

Problem 22.22 How would you synthesize olestra from sucrose?

22.12B The Synthesis of Soap

Soap has been previously discussed in Section 3.6.

All soaps are salts of fatty acids. The main difference between soaps is the addition of other ingredients that do not alter their cleaning properties: dyes for color, scents for a pleasing odor, and oils for lubrication. Soaps that float are aerated, so that they are less dense than water.

Soap is prepared by the basic hydrolysis or saponification of a triacylglycerol. Heating an animal fat or vegetable oil with aqueous base hydrolyzes the three esters to form glycerol and sodium salts of three fatty acids. These carboxylate salts are **soaps,** which clean away dirt because of their two structurally different regions. The nonpolar tail dissolves grease and oil and the polar head makes it soluble in water (Figure 3.5). Most triacylglycerols have two or three different R groups in their hydrocarbon chains, so soaps are usually mixtures of two or three different carboxylate salts.

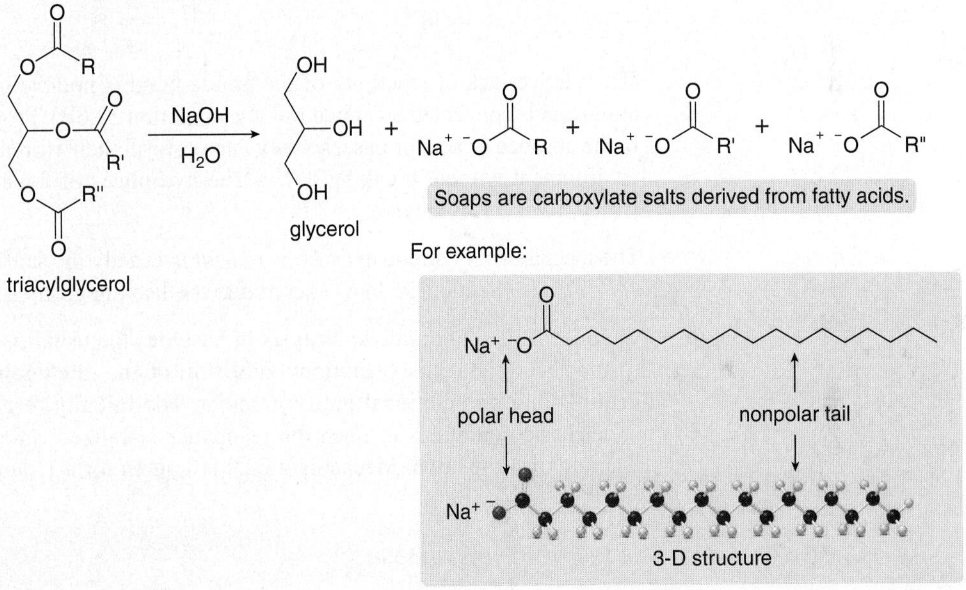

Soaps are typically made from lard (from hogs), tallow (from cattle or sheep), coconut oil, or palm oil. All soaps work in the same way, but have somewhat different properties depending on the lipid source. The length of the carbon chain in the fatty acids and the number of degrees of unsaturation affect the properties of the soap to some extent.

Problem 22.23 What is the composition of the soap prepared by hydrolysis of the following triacylglycerol?

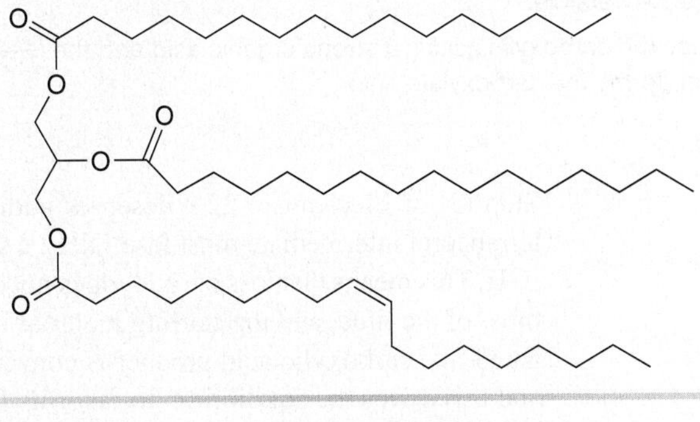

22.13 Reactions of Amides

Because amides have the poorest leaving group of all the carboxylic acid derivatives, they are the least reactive. Under strenuous reaction conditions, **amides are hydrolyzed in acid or base to form carboxylic acids or carboxylate anions.**

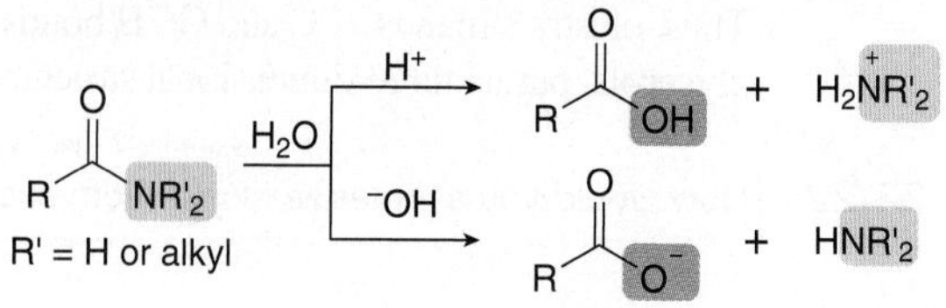

In acid, the amine by-product is protonated as an ammonium ion, whereas in base, a neutral amine is formed.

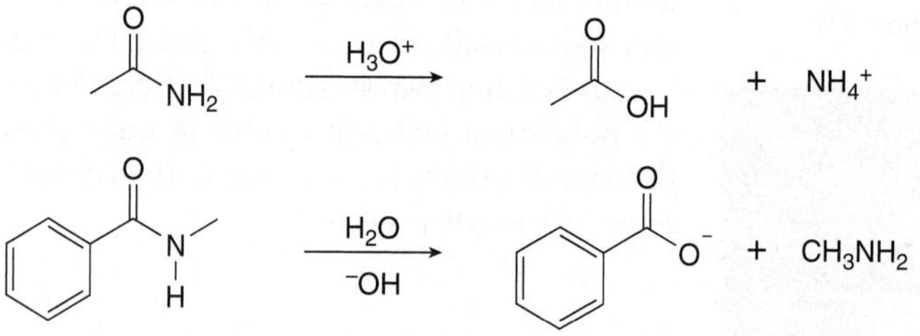

The relative lack of reactivity of the amide bond is notable in proteins, which are polymers of amino acids connected by amide linkages (Section 22.6B). Proteins are stable in aqueous solution in the absence of acid or base, so they can perform their various functions in the aqueous cellular environment without breaking down. The hydrolysis of the amide bonds in proteins requires a variety of specific enzymes.

The mechanism of amide hydrolysis in acid is exactly the same as the mechanism of ester hydrolysis in acid (Section 22.11A) except that the leaving group is different.

The mechanism of amide hydrolysis in base has the usual two steps of the general mechanism for nucleophilic acyl substitution—**addition of the nucleophile** followed by **loss of a leaving group**—plus an additional proton transfer. The initially formed carboxylic acid reacts further under basic conditions to form the resonance-stabilized carboxylate anion, and this drives the reaction to completion. Mechanism 22.9 is written for a 1° amide.

Mechanism 22.9 Amide Hydrolysis in Base

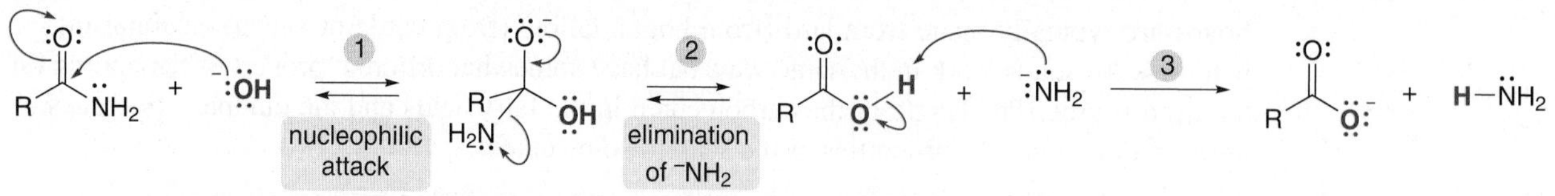

1 - 2 Addition of the nucleophile (^-OH) followed by elimination of the leaving group ($^-NH_2$) form a carboxylic acid. These two steps are reversible.

3 Because the carboxylic acid is a strong organic acid and the leaving group ($^-NH_2$) is a strong base, an acid–base reaction forms the carboxylate anion.

Step [2] of Mechanism 22.9 deserves additional comment. For amide hydrolysis to occur, the tetrahedral intermediate must lose $^-NH_2$, a stronger base and therefore poorer leaving group than ^-OH. This means that loss of $^-NH_2$ does not often happen. Instead, ^-OH is lost as the leaving group most of the time, and the starting material is regenerated. But, when $^-NH_2$ is occasionally eliminated, the carboxylic acid product is converted to a lower energy carboxylate anion in Step [3], and this drives the equilibrium to favor its formation.

Problem 22.24 Draw a stepwise mechanism for the following reaction.

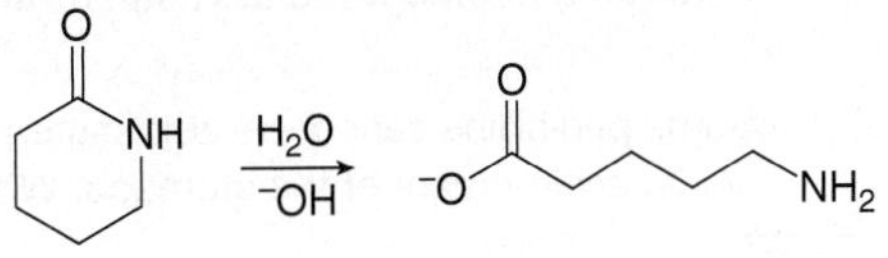

H_2O
^-OH

Problem 22.25 With reference to the structures of acetylsalicylic acid (aspirin, Chapter 2 opening molecule) and acetaminophen (the active ingredient in Tylenol), explain why acetaminophen tablets can be stored in the medicine cabinet for years, but aspirin slowly decomposes over time.

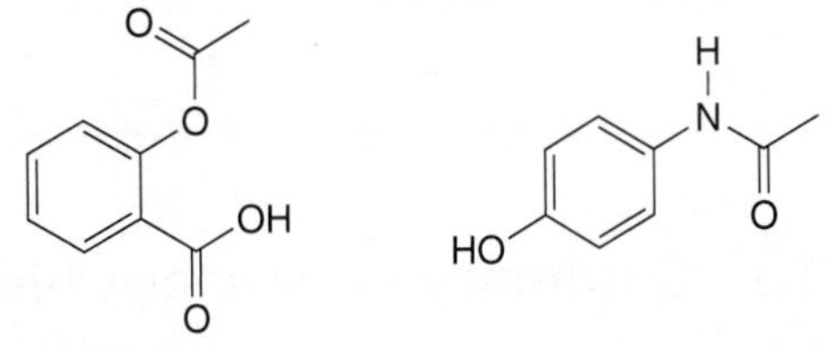

acetylsalicylic acid acetaminophen

22.14 Application: The Mechanism of Action of β-Lactam Antibiotics

Penicillins are a group of structurally related antibiotics, known since the pioneering work of Sir Alexander Fleming led to the discovery of penicillin G in the 1920s. Penicillin and related β-lactams kill bacteria by a nucleophilic acyl substitution reaction. All penicillins have an unreactive amide side chain and a very reactive amide that is part of a β-lactam. The β-lactam is more reactive than other amides because it is part of a strained, four-membered ring that is readily opened with nucleophiles.

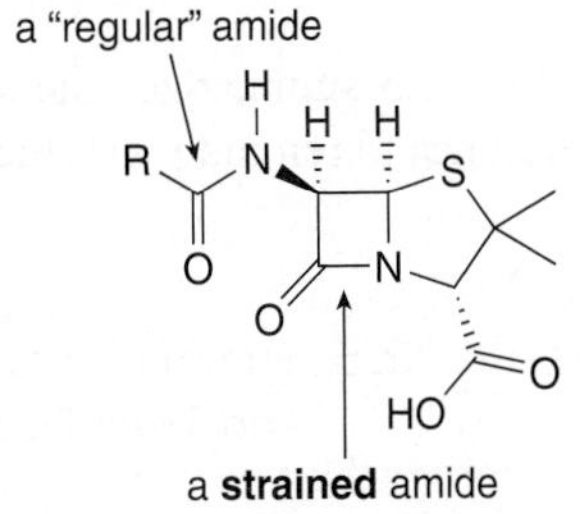

a "regular" amide
a **strained** amide

penicillin

Unlike mammalian cells, bacterial cells are surrounded by a fairly rigid cell wall, which allows the bacterium to live in many different environments. This protective cell wall is composed of carbohydrates linked by peptide chains containing amide linkages, formed using the enzyme **glycopeptide transpeptidase.**

Penicillin interferes with the synthesis of the bacterial cell wall. A nucleophilic OH group of the glycopeptide transpeptidase enzyme cleaves the β-lactam ring of penicillin by a nucleophilic acyl substitution reaction. The opened ring of the penicillin molecule remains covalently bonded to the enzyme, thus deactivating the enzyme, halting cell wall construction, and killing the bacterium. Penicillin has no effect on mammalian cells because they are surrounded by a flexible membrane composed of a lipid bilayer (Chapter 3) and not a cell wall.

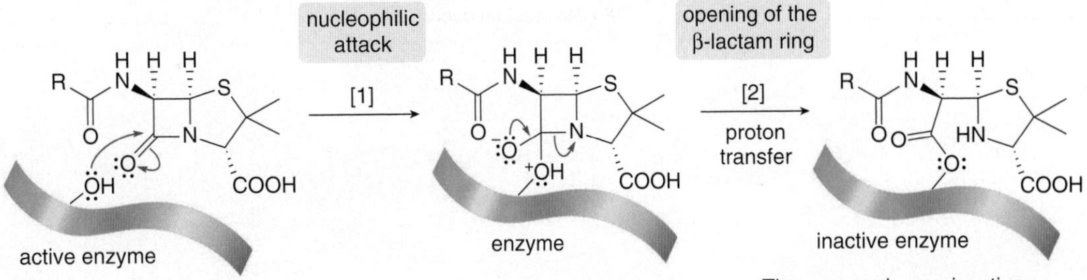

nucleophilic attack
[1]
opening of the β-lactam ring
[2]
proton transfer
active enzyme
enzyme
inactive enzyme

The enzyme is now inactive. Cell wall construction stops.

Thus, penicillin and other β-lactam antibiotics are biologically active precisely because they undergo a nucleophilic acyl substitution reaction with an important bacterial enzyme.

Problem 22.26 Some penicillins cannot be administered orally because their β-lactam is rapidly hydrolyzed by the acidic environment of the stomach. What product is formed in the following hydrolysis reaction?

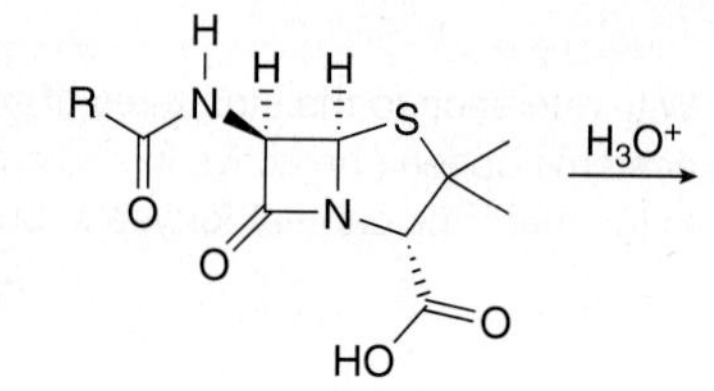

22.15 Summary of Nucleophilic Acyl Substitution Reactions

To help you organize and remember all of the nucleophilic acyl substitution reactions presented thus far, keep in mind the following two principles:

- **The better the leaving group, the more reactive the carboxylic acid derivative.**
- **More reactive acyl compounds can always be converted to less reactive ones. The reverse is not usually true.**

This results in the following order of reactivity:

$RCONR'_2$ RCO_2H ≈ RCO_2R' $(RCO)_2O$ RCOCl

Increasing reactivity →

Table 22.5 summarizes the specific nucleophilic acyl substitution reactions. Use it as a quick reference to remind you which products can be formed from a given starting material.

Table 22.5 Summary of the Nucleophilic Substitution Reactions of Carboxylic Acids and Their Derivatives

Starting material		Product: RCOCl	$(RCO)_2O$	RCO_2H	RCO_2R'	$RCONR'_2$
[1] RCOCl	→	–	✓	✓	✓	✓
[2] $(RCO)_2O$	→	✗	–	✓	✓	✓
[3] RCO_2H	→	✓	✓	–	✓	✓
[4] RCO_2R'	→	✗	✗	✓	–	✓
[5] $RCONR'_2$	→	✗	✗	✓	✗	–

Table key: ✓ = A reaction occurs.
✗ = No reaction occurs.

22.16 Acyl Phosphates—Biological Anhydrides

Because nature uses phosphates to activate carboxylates, both the synthesis and reactions of acyl phosphates are common processes in biological systems.

22.16A The Conversion of $RCOO^-$ to Acyl Phosphates

When carboxylates must be transformed into other functional groups in cells, they are first converted into acyl phosphates.

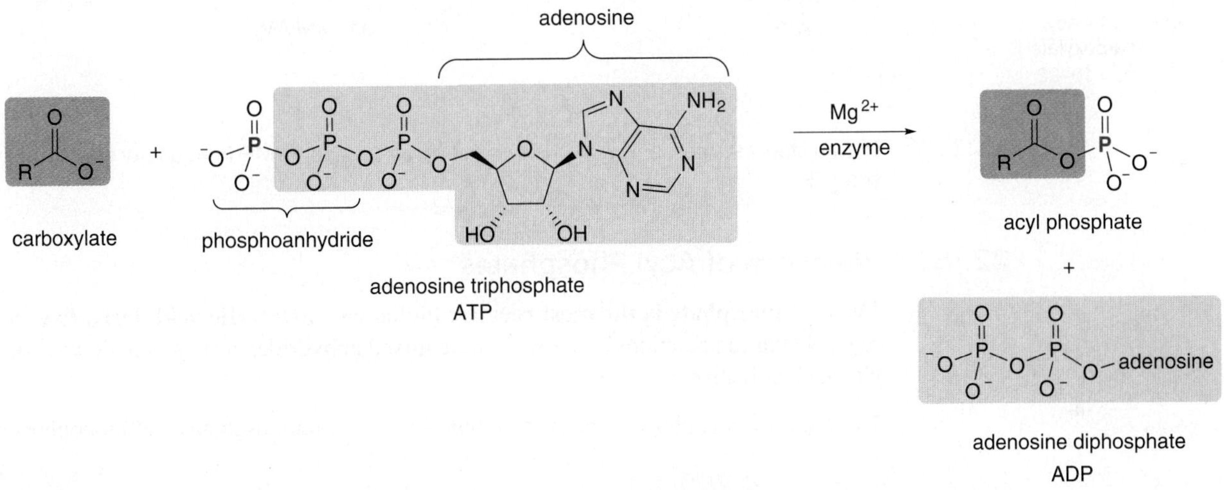

Phosphoric acid anhydrides—also called *phosphoanhydrides*—have two phosphate groups joined together.

In the presence of Mg^{2+} and an enzyme in biological systems, **phosphoric acid anhydrides** react with nucleophiles to form substitution products. Mg^{2+} acts as a Lewis acid to activate a phosphate and increase the electrophilicity of the phosphorus, as shown in the charge-separated resonance structure, which places a full positive charge on phosphorus.

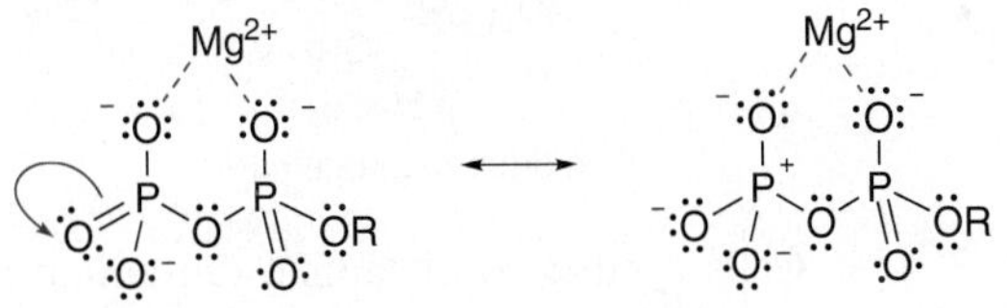

The conversion of a carboxylate into an acyl phosphate is a nucleophilic substitution reaction analogous to nucleophilic acyl substitution, but with a P=O electrophile and a carboxylate nucleophile. Mechanism 22.10 illustrates the steps for this magnesium-activated conversion.

Mechanism 22.10 Biological Conversion of a Carboxylate to an Acyl Phosphate

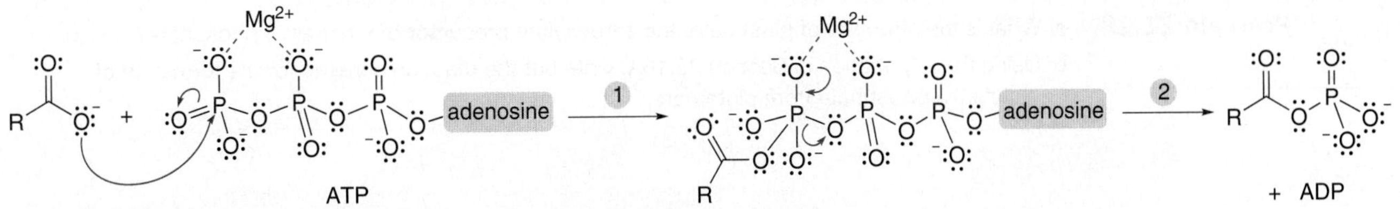

1. The nucleophile (RCO_2^-) attacks the magnesium-activated phosphate.
2. Elimination of the leaving group (ADP) forms the substitution product, an acyl phosphate.

The metabolism of a fatty acid begins with its conversion into a fatty acyl phosphate ester. In this reaction, the carboxylate nucleophile attacks the phosphorus atom that is directly attached to the O–adenosine (labeled in red), displacing diphosphate to form a fatty acyl **adenosine monophosphate** (AMP).

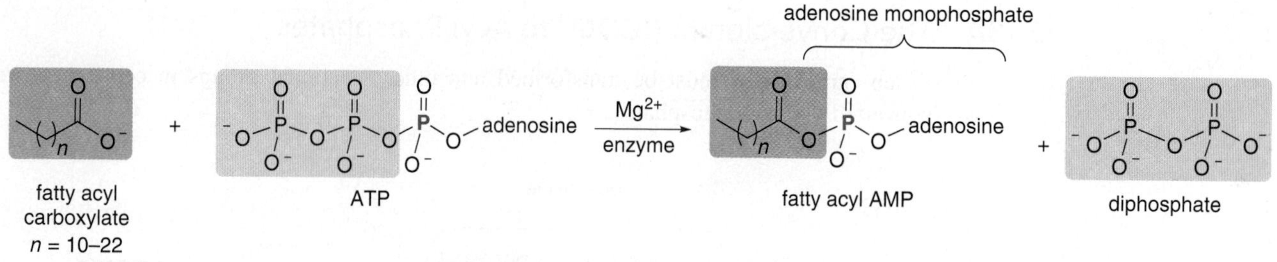

Problem 22.27 Draw a stepwise mechanism for the formation of a fatty acyl AMP from a fatty acyl carboxylate and ATP.

22.16B Reactions of Acyl Phosphates

The acyl phosphate is the most reactive biological carboxylic acid derivative. Nucleophilic acyl substitution reactions occur with these mixed anhydrides just as they do with other carboxylic acid derivatives.

Two types of acyl phosphates are common: simple monophosphates and monophosphate esters.

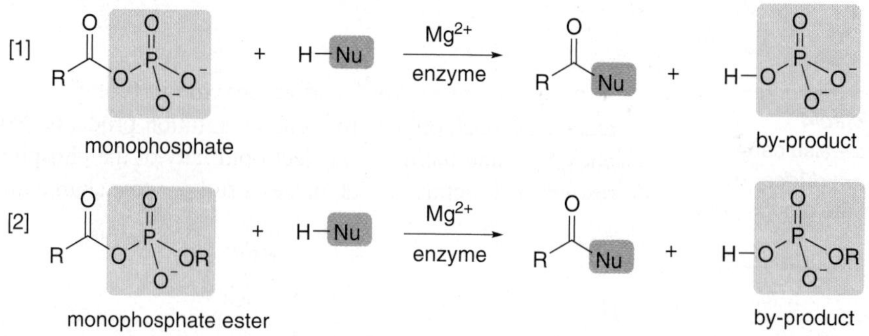

Cells synthesize glutamine (an amino acid) from an acyl phosphate derived from glutamate (another amino acid). In this reaction, ammonia (the nucleophile) displaces the phosphate group on γ-glutamyl phosphate to give glutamine.

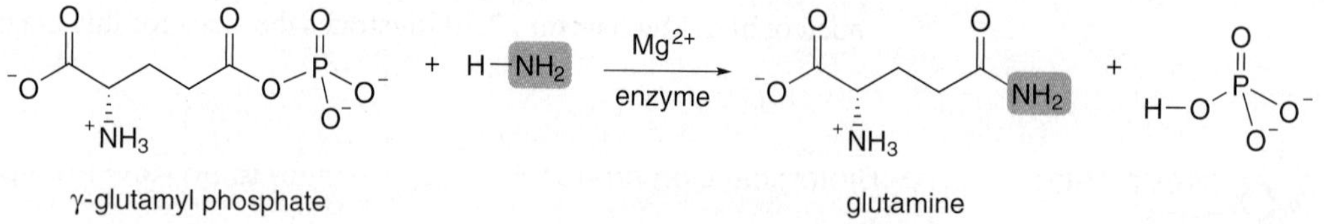

Problem 22.28 a. What is the structure of glutamate, the carboxylate precursor of γ-glutamyl phosphate?

b. Using the information in Section 22.16A, write out the reaction scheme for the formation of γ-glutamyl phosphate from glutamate.

Thiols are also common biological nucleophiles. During fatty acid metabolism, the fatty acyl AMP prepared in Section 22.16A is converted into a thioester, **fatty acyl CoA,** using the thiol in coenzyme A (Section 3.8) as the nucleophile.

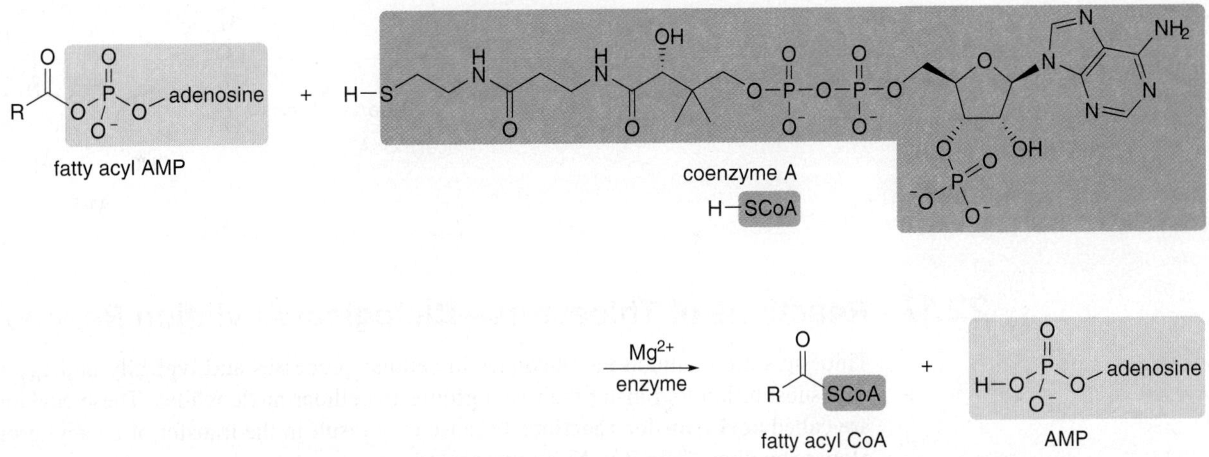

The conversion of an acyl phosphate to a thioester illustrates the mechanism of a biological nucleophilic acyl substitution with an acyl phosphate as starting material (Mechanism 22.11). Mg^{2+} activates the acyl group to increase the electrophilicity of the carbonyl carbon. It is typical in enzyme mechanisms to show the proton transfers as part of the **nucleophilic addition** and **elimination** steps.

Mechanism 22.11 Biological Conversion of an Acyl Phosphate to a Thioester

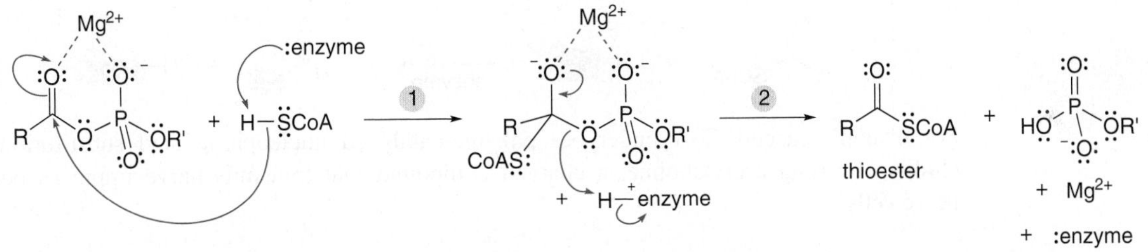

1 The nucleophile (HSCoA) attacks the magnesium-activated carbonyl, forming an sp^3 hybridized carbon.

2 Elimination of the leaving group and protonation form the substitution product, a thioester.

Problem 22.29 Draw the resonance structure of the following compound that shows the electrophilicity of the C=O carbon of the acyl phosphate.

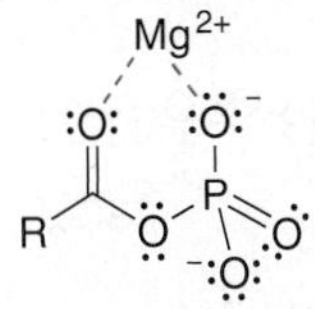

Problem 22.30 Draw the products of the following reaction, one step in the synthesis of the amino acid asparagine.

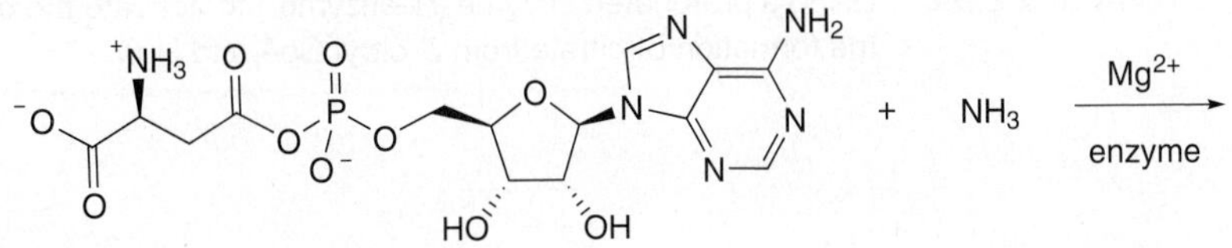

Problem 22.31 Draw a stepwise mechanism for the following reaction. Coenzyme A (HSCoA) reacts with acetyl AMP to give acetyl CoA, a key biomolecule involved in metabolism, and adenosine monophosphate (AMP). You may use "R" groups to simplify the mechanism.

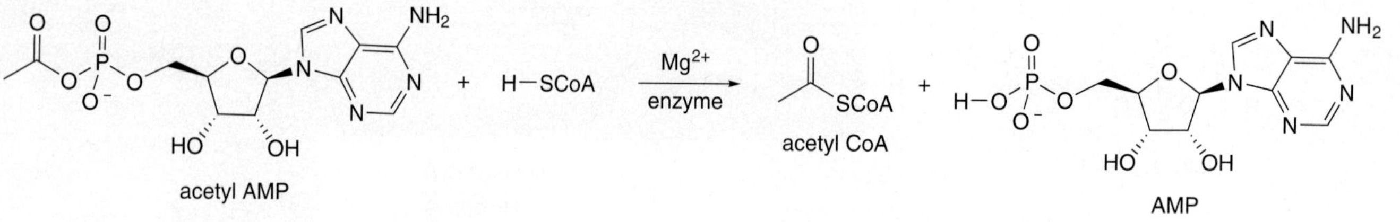

22.17 Reactions of Thioesters—Biological Acylation Reactions

Thioesters are common intermediates in cellular processes and typically undergo nucleophilic acyl substitution to transfer their acyl groups to cellular nucleophiles. These acylation reactions are called **acyl transfer reactions** because they result in the transfer of an acyl group from one atom to another (from Z to Nu in this case).

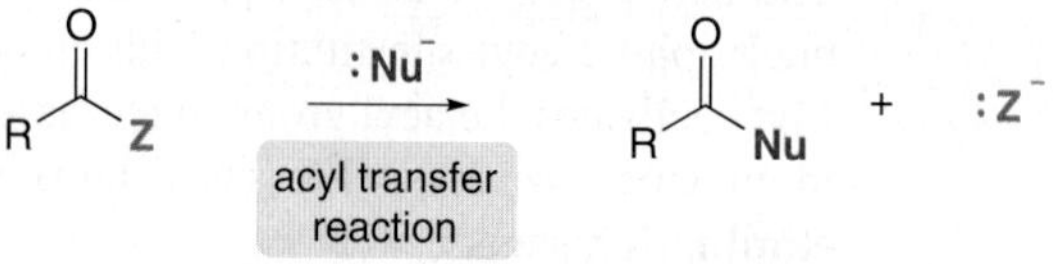

- **A thioester (RCOSR'), like other acyl compounds, undergoes substitution reactions with nucleophiles. With acetyl CoA (Section 22.16), an acetyl group is transferred from SCoA to a nucleophile, Nu.**

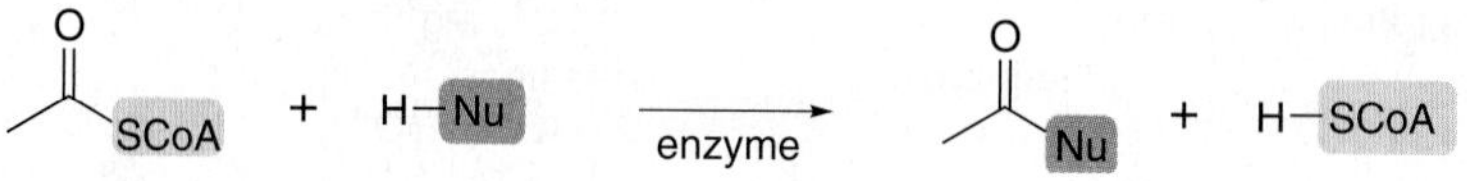

For example, acetyl CoA undergoes enzyme-catalyzed nucleophilic acyl substitution with choline, forming acetylcholine, a charged compound that transmits nerve impulses between nerve cells.

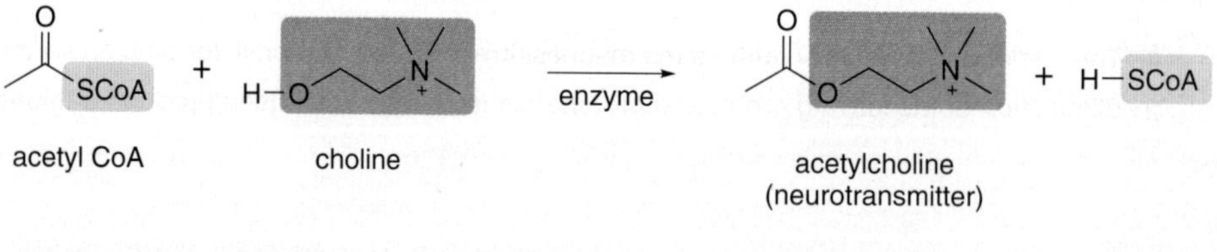

In another example, the acyl group of *S*-citryl CoA is converted to citrate, an intermediate in the citric acid cycle, an energy-generating metabolic cycle. In this reaction, water is the nucleophile that displaces SCoA to ultimately form a carboxylate from a thioester.

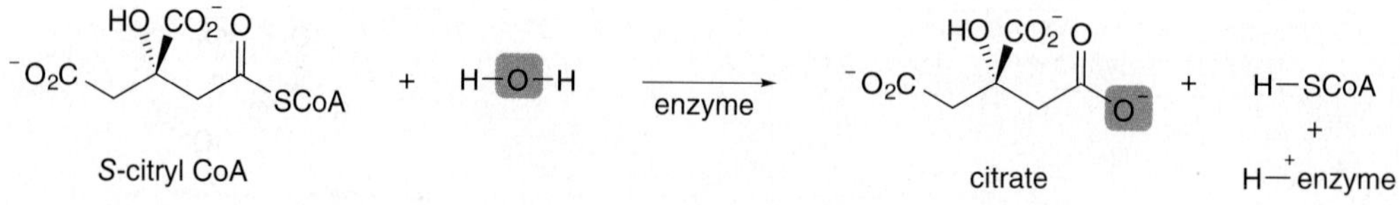

Problem 22.32 Using a protonated enzyme (H–enzyme⁺) to activate the carbonyl, draw a stepwise mechanism for the formation of citrate from *S*-citryl CoA and H_2O.

Many other acyl transfer reactions are important cellular processes. Thioesters of fatty acids react with cholesterol, forming **cholesteryl esters** in an enzyme-catalyzed reaction (Figure 22.5). These esters are the principal form in which cholesterol is stored and transported in the body. Because cholesterol is a lipid, insoluble in the aqueous environment of the blood, it travels through the bloodstream in particles that also contain proteins and phospholipids. These particles are classified by their density.

- **LDL particles** (low density lipoproteins) transport cholesterol from the liver to the tissues.
- **HDL particles** (high density lipoproteins) transport cholesterol from the tissues back to the liver, where it is metabolized or converted to other steroids.

Atherosclerosis is a disease that results from the buildup of fatty deposits on the walls of arteries, forming deposits called **plaque.** Plaque is composed largely of the cholesterol (esterified as an ester) of LDL particles. LDL is often referred to as "bad cholesterol" for this reason. In contrast, HDL particles are called "good cholesterol" because they reduce the amount of cholesterol in the bloodstream by transporting it back to the liver.

Figure 22.5
Cholesteryl esters and lipoprotein particles

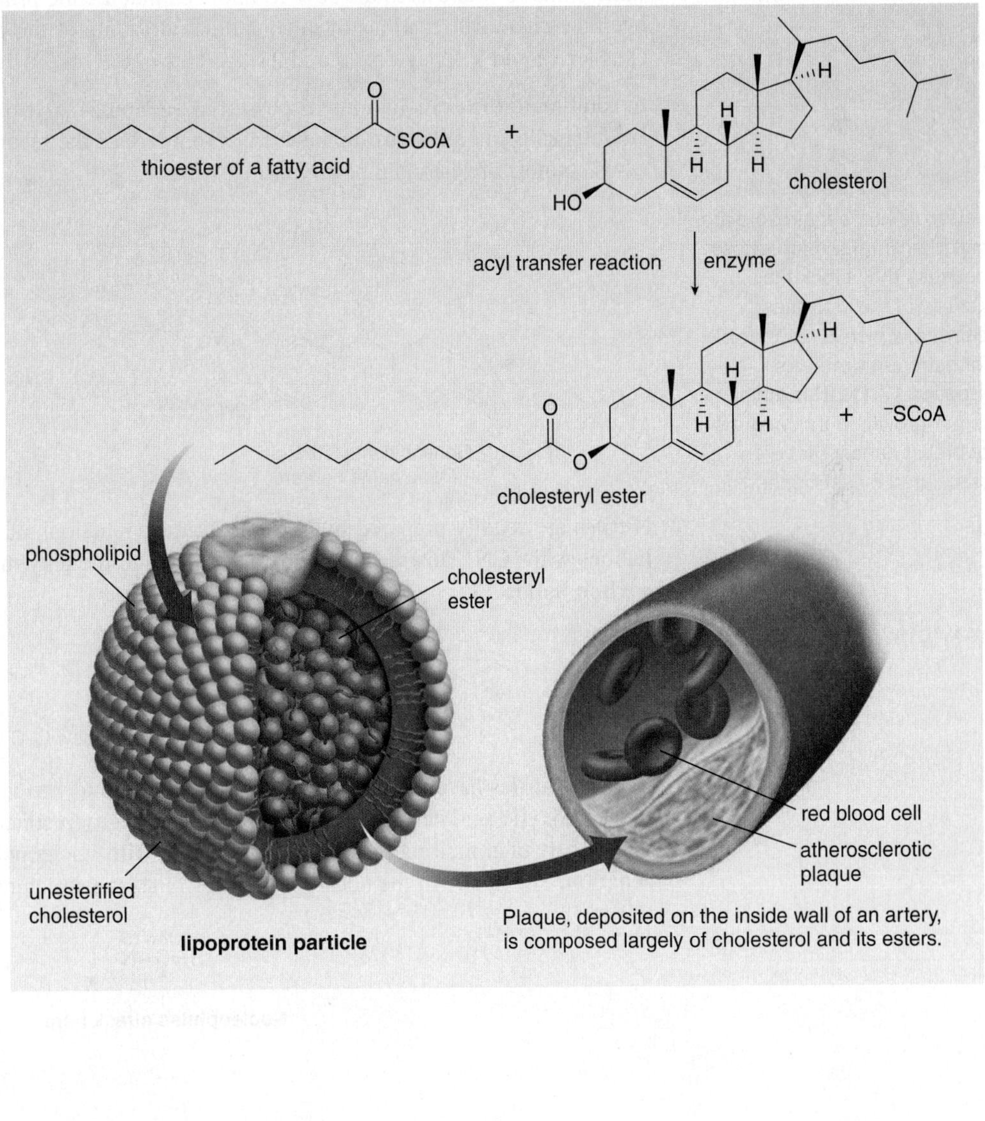

Problem 22.33 Glucosamine is a dietary supplement available in many over-the-counter treatments for osteoarthritis. Reaction of acetyl CoA with glucosamine forms NAG, *N*-acetylglucosamine, the monomer used to form chitin, the carbohydrate that forms the rigid shells of lobsters and crabs. What is the structure of NAG?

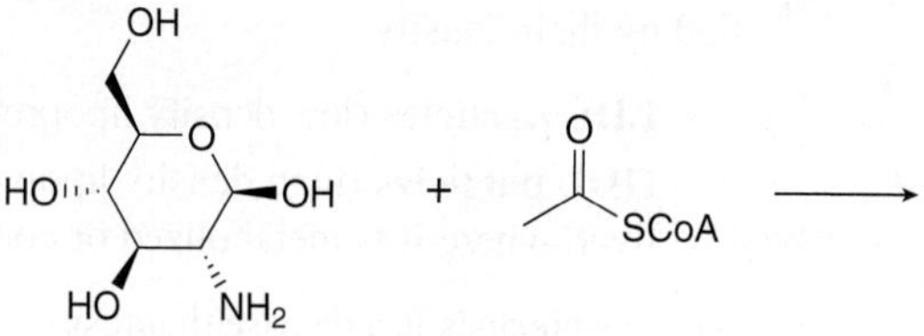

glucosamine

22.18 Nitriles

We end Chapter 22 with the chemistry of **nitriles (RC≡N).** Nitriles have a carbon atom in the same oxidation state as in the acyl compounds that are the principal focus of Chapter 22. Moreover, the chemical reactions of nitriles illustrate some of the concepts first discussed earlier in Chapter 22 and in Chapters 20 and 21.

In addition to the cyanohydrins discussed in Section 21.9, two useful biologically active nitriles are **letrozole** and **anastrozole,** new drugs that reduce the recurrence of breast cancer in women whose tumors are estrogen positive.

Letrozole and anastrozole are called **aromatase inhibitors** because they block the activity of the aromatase enzyme, which is responsible for estrogen synthesis (Section 17.11). This inhibits tumor growth in those forms of breast cancer that are stimulated by estrogen.

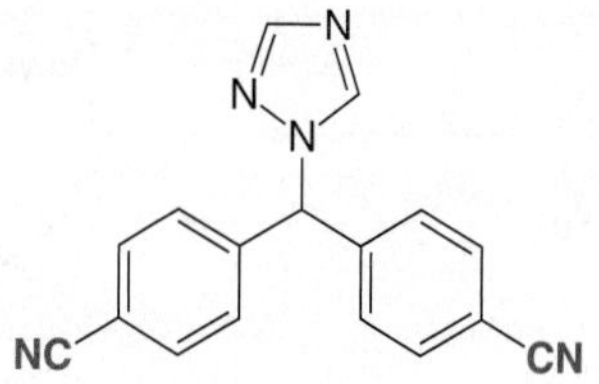

Generic name letrozole
Trade name Femara

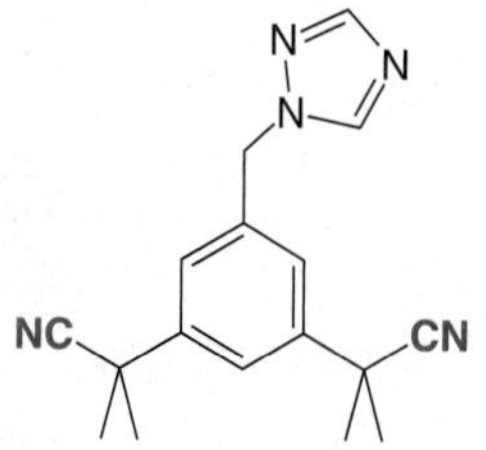

Generic name anastrozole
Trade name Arimidex

Nitriles are readily prepared by S_N2 substitution reactions of unhindered methyl and 1° alkyl halides with $^-$CN. This reaction adds one carbon to the alkyl halide and **forms a new carbon–carbon bond.**

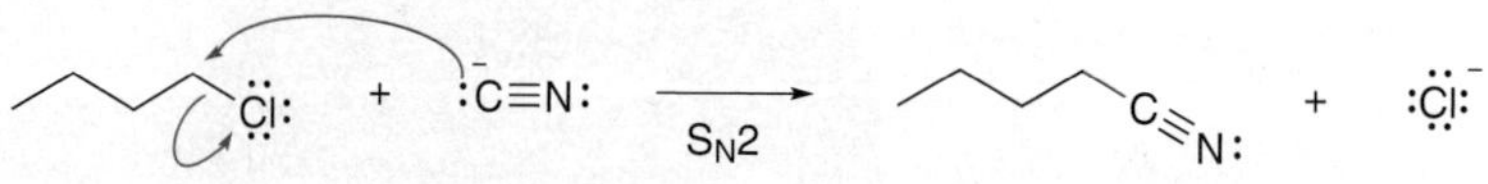

new C–C bond in red

Because nitriles have no leaving group, they do not undergo nucleophilic substitution reactions like carboxylic acid derivatives. Because the cyano group contains an electrophilic carbon atom that is part of a multiple bond, a nitrile reacts with nucleophiles by a **nucleophilic addition reaction.** The nature of the nucleophile determines the structure of the product.

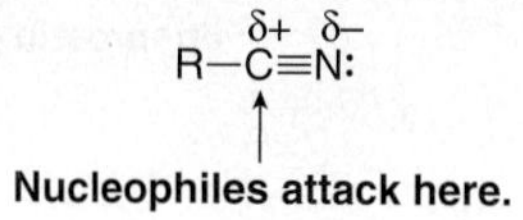

The reactions of nitriles with water, hydride, and organometallic reagents as nucleophiles are as follows:

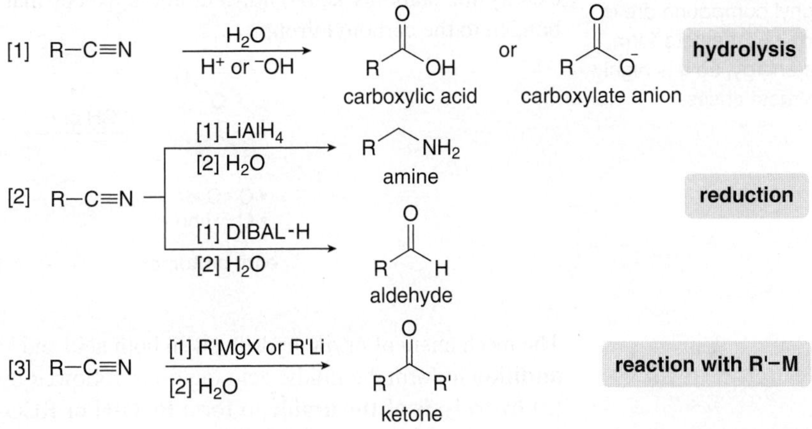
[1] R−C≡N
H_2O
H^+ or ^-OH
carboxylic acid
or
carboxylate anion
hydrolysis
[2] R−C≡N
[1] $LiAlH_4$
[2] H_2O
amine
[1] DIBAL-H
[2] H_2O
aldehyde
reduction
[3] R−C≡N
[1] R'MgX or R'Li
[2] H_2O
ketone
reaction with R'−M

22.18A Hydrolysis of Nitriles

Nitriles are hydrolyzed with water in the presence of acid or base to yield **carboxylic acids** or **carboxylate anions.** In this reaction, the three C−N bonds are replaced by three C−O bonds.

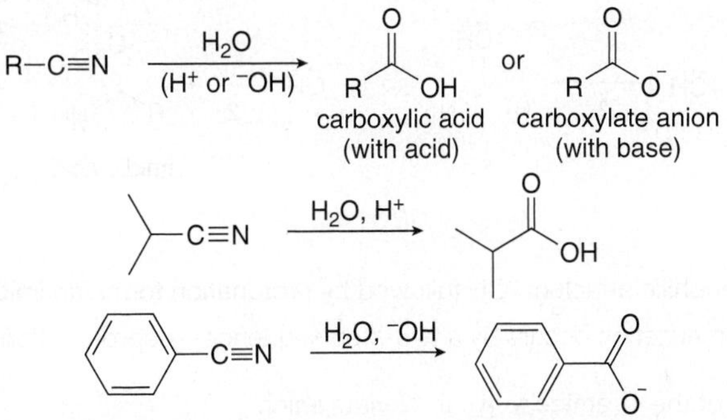
R−C≡N
H_2O
(H^+ or ^-OH)
carboxylic acid
(with acid)
or
carboxylate anion
(with base)
H_2O, H^+
H_2O, ^-OH

The mechanism of this reaction involves the formation of an amide tautomer. Two tautomers can be drawn for any carbonyl compound, and those for a 1° amide are as follows:

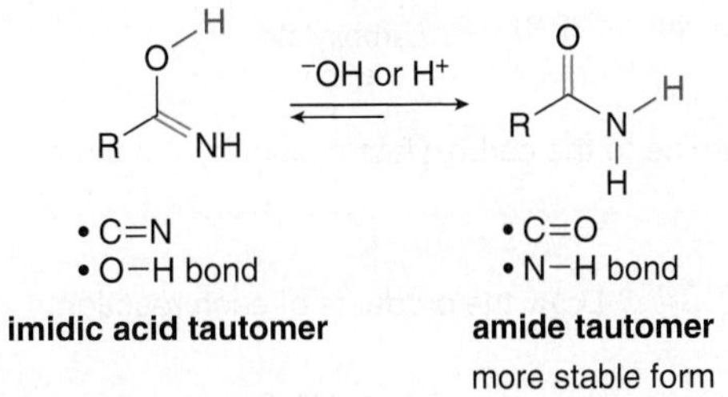
^-OH or H^+
• C=N
• O−H bond
imidic acid tautomer
• C=O
• N−H bond
amide tautomer
more stable form

Recall from Chapter 11 that tautomers are constitutional isomers that differ in the location of a double bond and a proton.

- **The amide form is the more stable tautomer, having a C=O and an N−H bond.**
- **The imidic acid tautomer is the less stable form, having a C=N and an O−H bond.**

Recall from Chapter 11 that the keto and enol tautomers of a carbonyl compound are in equilibrium, but the keto form is lower in energy, so it is highly favored in most cases.

The imidic acid and amide tautomers are interconverted by treating with acid or base, analogous to the keto–enol tautomers of other carbonyl compounds. In fact, the two amide tautomers are exactly the same as keto–enol tautomers except that a nitrogen atom replaces a carbon atom bonded to the carbonyl group.

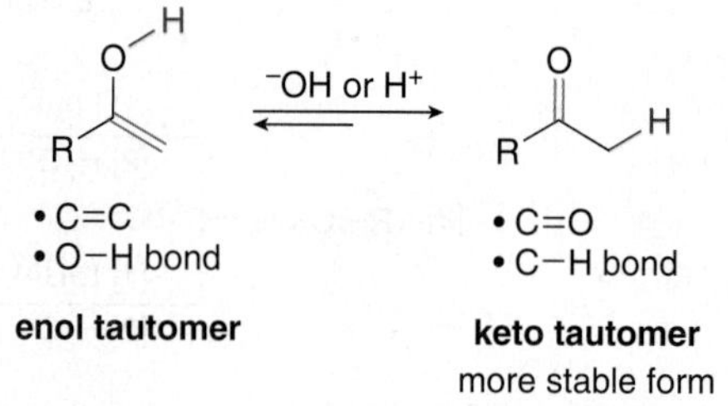

⁻OH or H⁺
• C=C
• O–H bond
enol tautomer
• C=O
• C–H bond
keto tautomer
more stable form

The mechanism of nitrile hydrolysis in both acid and base consists of two parts: [1] **nucleophilic addition** to form the imidic acid tautomer followed by **tautomerization** to form the amide, and [2] **hydrolysis of the amide** to form RCO_2H or RCO_2^-. The mechanism is shown for the basic hydrolysis of RCN to RCO_2^- (Mechanism 22.12).

Mechanism 22.12 Hydrolysis of a Nitrile in Base

Part [1] Conversion of a nitrile to a 1° amide

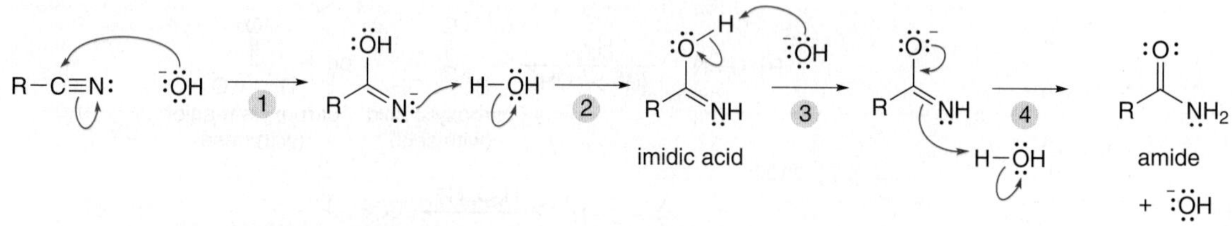

imidic acid
amide

1 – 2 Nucleophilic attack of ⁻OH followed by protonation forms an imidic acid.

3 – 4 Tautomerization occurs by a two-step sequence—deprotonation followed by protonation.

Part [2] Hydrolysis of the 1° amide to a carboxylate anion

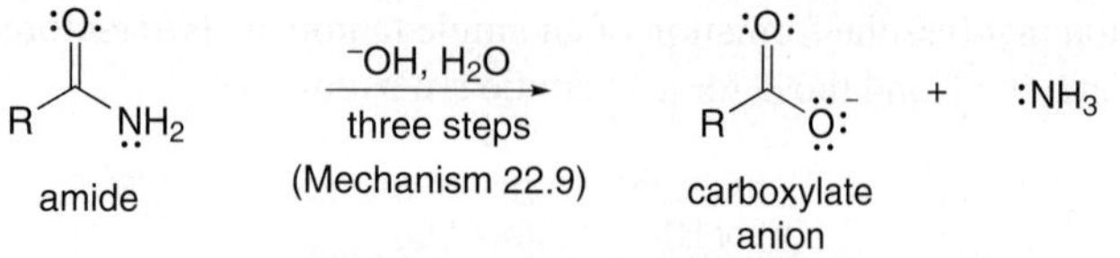

⁻OH, H₂O
three steps
(Mechanism 22.9)
amide
carboxylate anion

Conversion of the amide to the carboxylate occurs by the three-step sequence in Mechanism 22.9.

Problem 22.34 Draw the products of each reaction.

a. 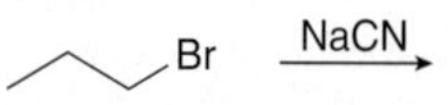

Br
NaCN

c. 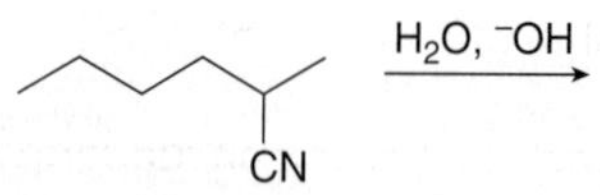

CN
H₂O, ⁻OH

b. 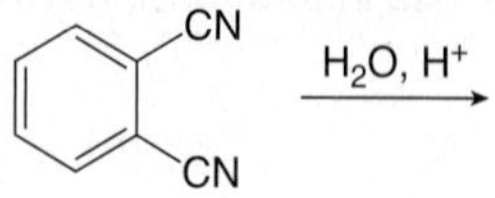

CN
CN
H₂O, H⁺

Problem 22.35 Draw a tautomer of each compound.

a. 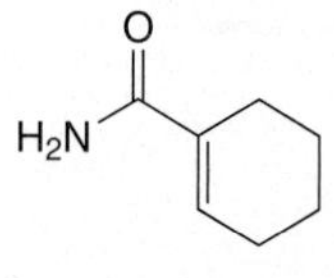

b. 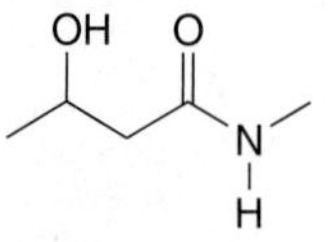

c. 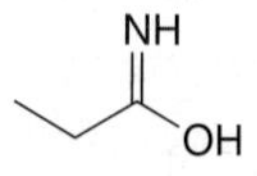

22.18B Reduction of Nitriles

Nitriles are reduced with metal hydride reagents to form either 1° amines or aldehydes, depending on the reducing agent.

- **Treatment of a nitrile with $LiAlH_4$ followed by H_2O adds two equivalents of H_2 across the triple bond, forming a 1° amine.**

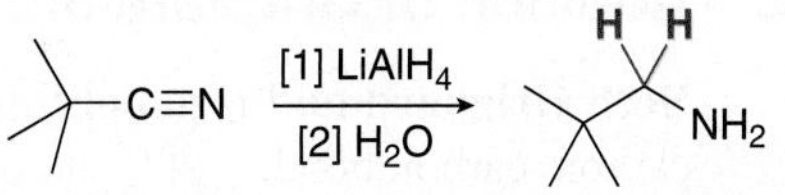

- **Treatment of a nitrile with a milder reducing agent such as DIBAL-H followed by H_2O forms an aldehyde.**

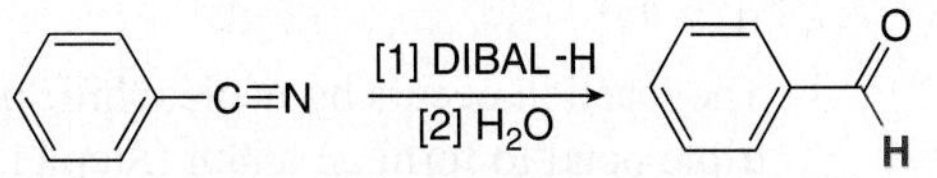

The mechanism of both reactions involves **nucleophilic addition of hydride (H⁻) to the polarized C–N triple bond.** Mechanism 22.13 illustrates that reduction of a nitrile to an amine requires addition of two equivalents of H:⁻ from $LiAlH_4$. It is likely that intermediate nitrogen anions complex with AlH_3 (formed in situ) to facilitate the addition. Protonation of the dianion in Step [4] forms the amine.

Mechanism 22.13 Reduction of a Nitrile with $LiAlH_4$

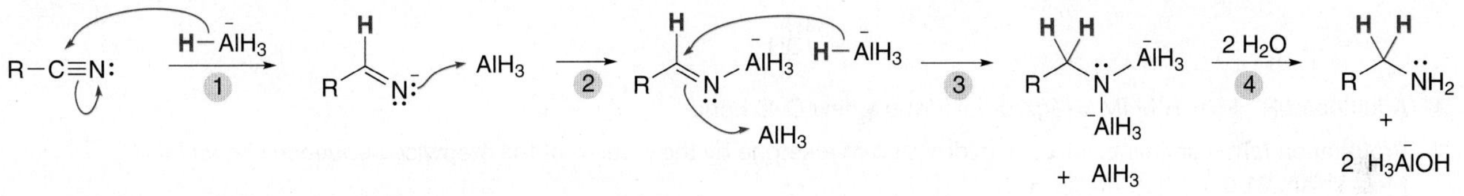

1 - 2 Addition of one equivalent of H:⁻ from $LiAlH_4$ forms an intermediate with one new C–H bond, which complexes with AlH_3.

3 - 4 Nucleophilic attack of a second equivalent of H:⁻ and complexation with AlH_3 form a dianion, which reacts with water to form two new N–H bonds, giving the 1° amine.

With **DIBAL-H,** nucleophilic addition of one equivalent of hydride forms an anion (Step [1]), which is protonated with water to generate an **imine,** as shown in Mechanism 22.14. As described in Section 21.12, imines are hydrolyzed in water to form aldehydes. Mechanism 22.14 is written without complexation of aluminum with the anion formed in Step [1], to emphasize the identity of intermediates formed during reduction.

Mechanism 22.14 Reduction of a Nitrile with DIBAL-H

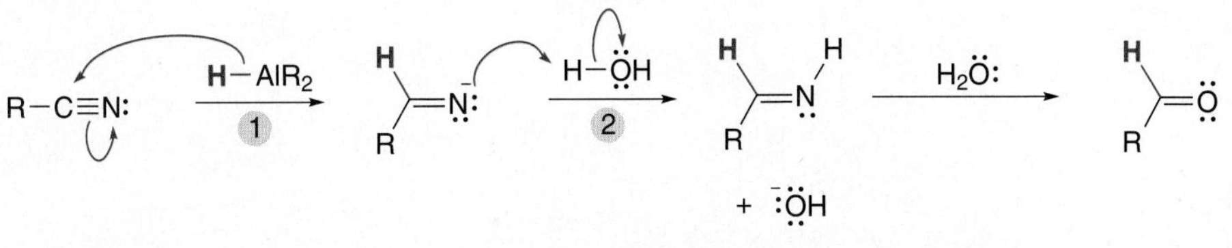

1 Addition of H:⁻ from DIBAL-H (drawn as R_2AlH) forms the new C–H bond.

2 Protonation forms an imine, which is hydrolyzed to an aldehyde by the reverse of the stepwise sequence shown in Mechanism 21.5.

Problem 22.36 Draw the product of each reaction.

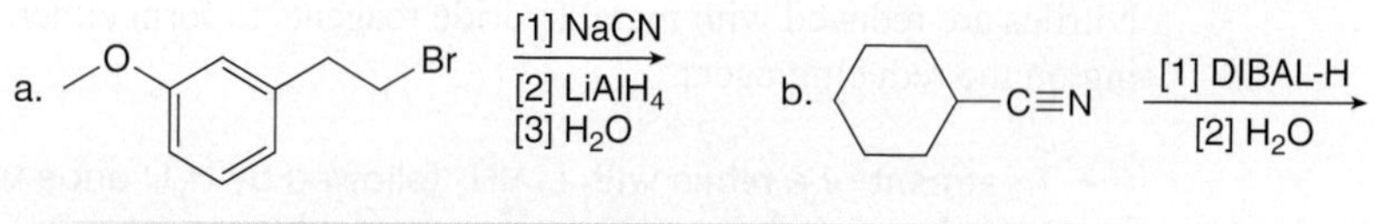
a.
O
Br
[1] NaCN
[2] $LiAlH_4$
[3] H_2O
b.
C≡N
[1] DIBAL-H
[2] H_2O

22.18C Addition of Grignard and Organolithium Reagents to Nitriles

Both Grignard and organolithium reagents react with nitriles to form ketones with a new carbon–carbon bond.

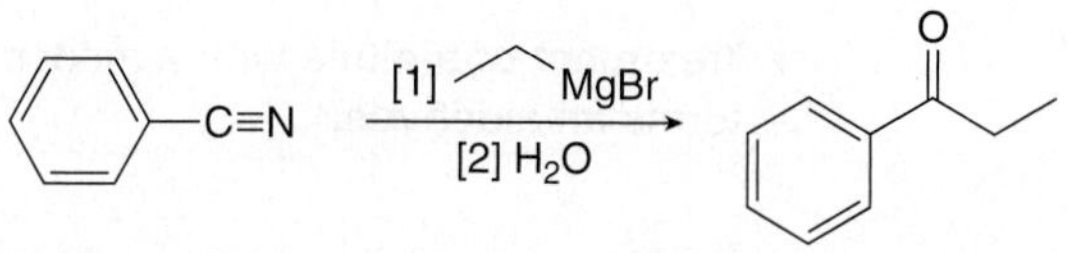
C≡N
[1] MgBr
[2] H_2O
O

The reaction occurs by nucleophilic addition of the organometallic reagent to the polarized C−N triple bond to form an anion (Step [1]), which is protonated with water to form an **imine.** Water then hydrolyzes the imine, replacing the C=N by C=O as described in Section 21.12. The final product is a ketone with a new carbon–carbon bond (Mechanism 22.15).

Mechanism 22.15 Addition of Grignard and Organolithium Reagents (R–M) to Nitriles

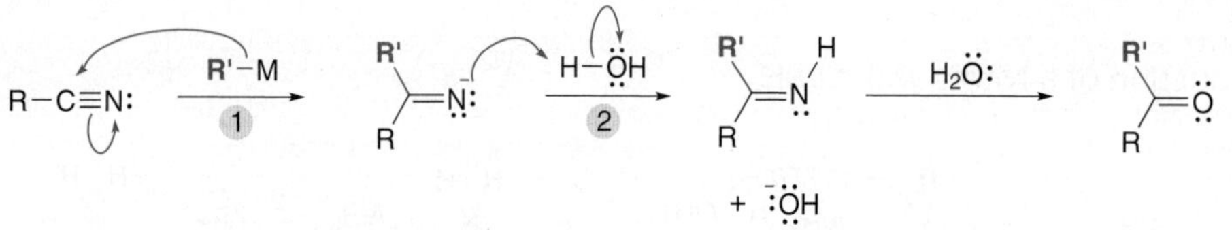
R−C≡N:
R'−M
1
R'
R
N:
H−ÖH
2
R'
H
N
R
+ :ÖH
H_2Ö:
R'
R
Ö

1 Addition of R'⁻ from R'M (M = MgX or Li) forms a new C–C bond.

2 Protonation forms an imine, which is hydrolyzed to a ketone by the reverse of the stepwise sequence shown in Mechanism 21.5.

Problem 22.37 Draw the products of each reaction.

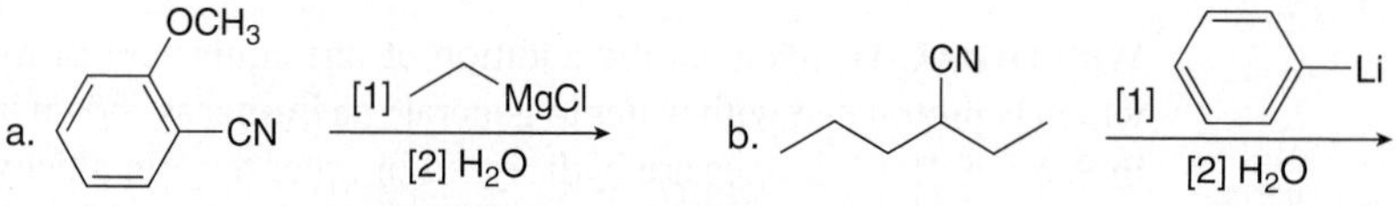
a.
OCH_3
CN
[1] MgCl
[2] H_2O
b.
CN
[1]
Li
[2] H_2O

Problem 22.38 What reagents are needed to convert phenylacetonitrile ($C_6H_5CH_2CN$) to each compound: (a) $C_6H_5CH_2COCH_3$; (b) $C_6H_5CH_2COC(CH_3)_3$; (c) $C_6H_5CH_2CHO$; (d) $C_6H_5CH_2COOH$?

Problem 22.39 Outline two different ways that butan-2-one can be prepared from a nitrile and a Grignard reagent.

KEY CONCEPTS

Carboxylic Acids and Their Derivatives—Nucleophilic Acyl Substitution

Summary of Spectroscopic Absorptions of RCOZ (22.5)

IR absorptions

- All RCOZ compounds have a C=O absorption in the region 1600–1850 cm^{-1}.
  - RCOCl: 1800 cm^{-1}
  - $(RCO)_2O$: 1820 and 1760 cm^{-1} (two peaks)
  - RCO_2R': 1735–1745 cm^{-1}
  - $RCO_2PO_3^{2-}$: 1700–1730 cm^{-1}
  - RCOSR': 1690–1720 cm^{-1}
  - $RCONR'_2$: 1630–1680 cm^{-1}
- Additional amide absorptions occur at 3200–3400 cm^{-1} (N–H stretch) and 1640 cm^{-1} (N–H bending).
- Decreasing the ring size of a cyclic lactone, lactam, or anhydride increases the frequency of the C=O absorption.
- Conjugation shifts the C=O to lower wavenumber.
- Phosphates have a P=O absorption in the region 1100–1200 cm^{-1}.

^1H NMR absorptions

- C–H α to the C=O absorbs at 2–2.5 ppm.
- N–H of an amide absorbs at 7.5–8.5 ppm.

^{13}C NMR absorption

- C=O absorbs at 160–180 ppm.

Summary of Spectroscopic Absorptions of RCN (22.5)

IR absorption

- C≡N absorbs at ~2250 cm^{-1}.

^{13}C NMR absorption

- C≡N absorbs at 115–120 ppm.

Summary: The Relationship between the Basicity of Z^- and the Properties of RCOZ

- **Increasing basicity of the leaving group** (22.2)
- **Increasing resonance stabilization** (22.2)

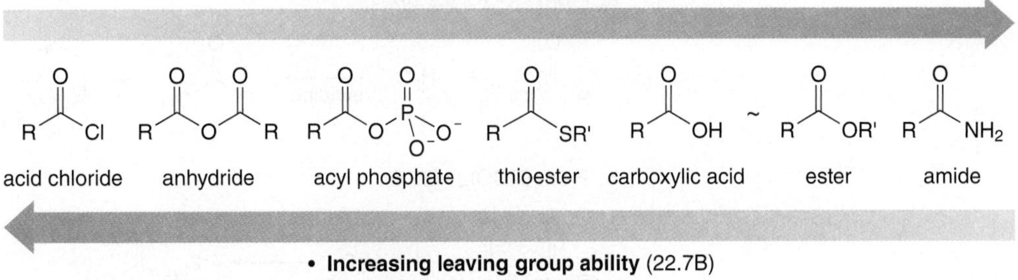
R
Cl
O
acid chloride
anhydride
acyl phosphate
thioester
SR'
carboxylic acid
OH
~
ester
OR'
amide
NH2

- **Increasing leaving group ability** (22.7B)
- **Increasing reactivity** (22.7B)

General Features of Nucleophilic Acyl Substitution

- The characteristic reaction of compounds having the general structure RCOZ is nucleophilic acyl substitution (22.1).
- The mechanism consists of two steps (22.7A):

  [1] Addition of a nucleophile to form a tetrahedral intermediate

  [2] Elimination of a leaving group
- More reactive acyl compounds can be used to prepare less reactive acyl compounds. The reverse is not necessarily true (22.7B).

Nucleophilic Acyl Substitution Reactions

[1] Reactions that produce acid chlorides (RCOCl)

From RCO_2H (22.10A):

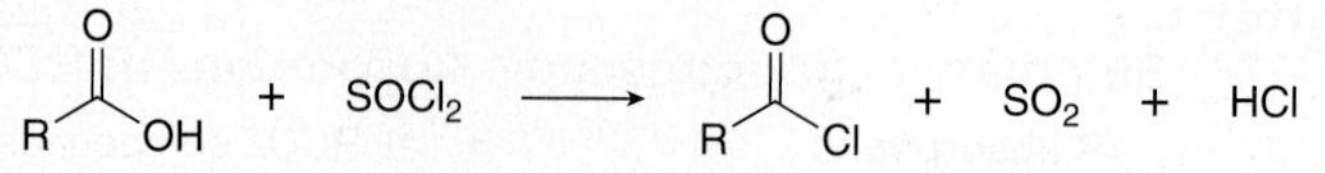

[2] Reactions that produce anhydrides [$(RCO)_2O$]

a. From RCOCl (22.8):

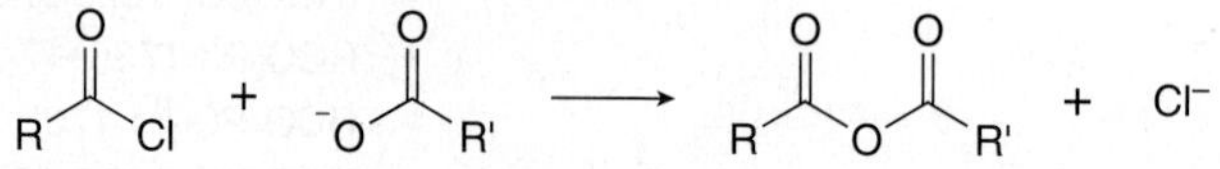

b. From dicarboxylic acids (22.10B):

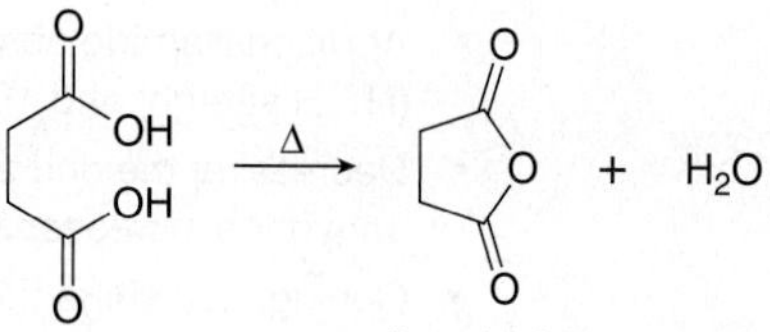

cyclic anhydride

[3] Reactions that produce acyl phosphates ($RCO_2PO_3^{2-}$)

From RCO_2^- (22.16A):

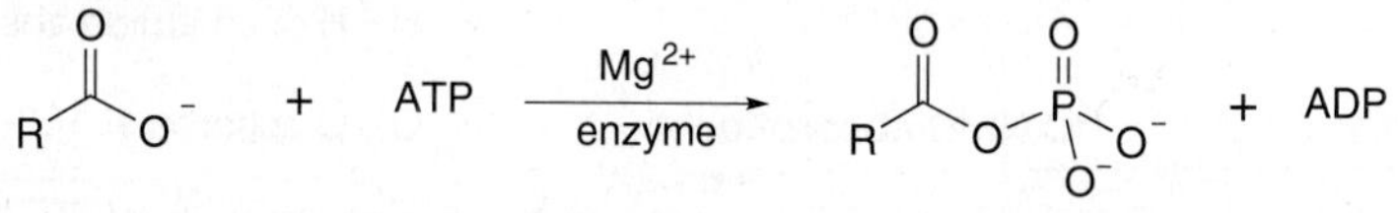

[4] Reactions that produce thioesters (RCOSR')

From $(RCO_2PO_3R)^-$ (22.16B):

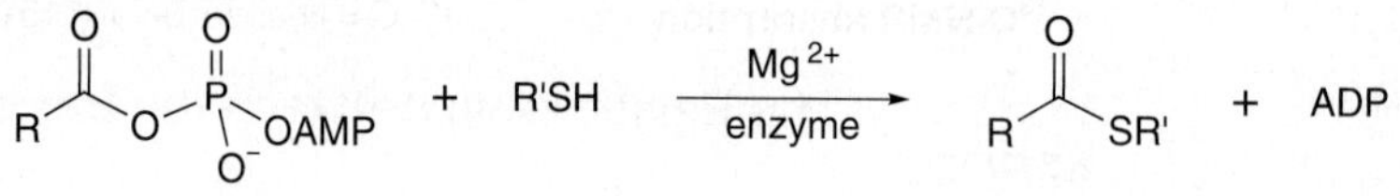

[5] Reactions that produce carboxylic acids (RCOOH) and carboxylates ($RCOO^-$)

a. From RCOCl (22.8):

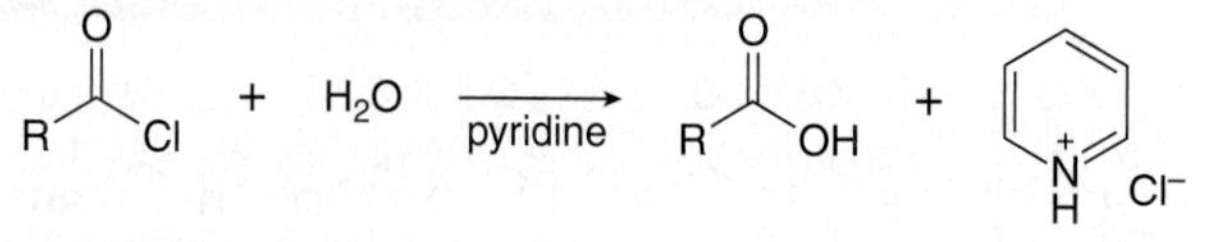

b. From $(RCO)_2O$ (22.9):

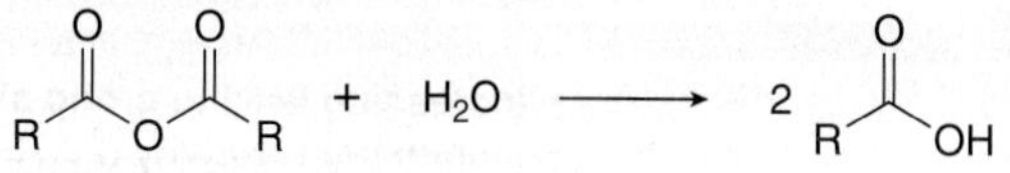

c. From RCOSR' (22.17):

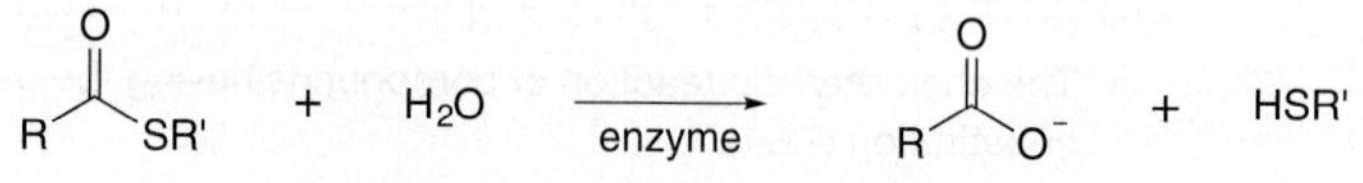

d. From RCO_2R' (22.11):

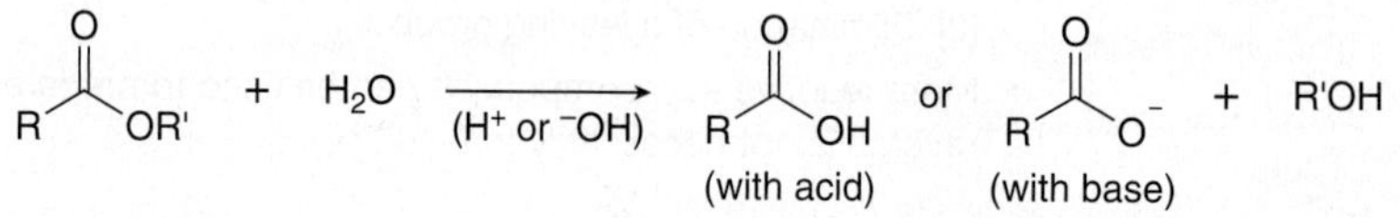
(with acid)
(with base)

e. From $RCONR'_2$ (R' = H or alkyl, 22.13):

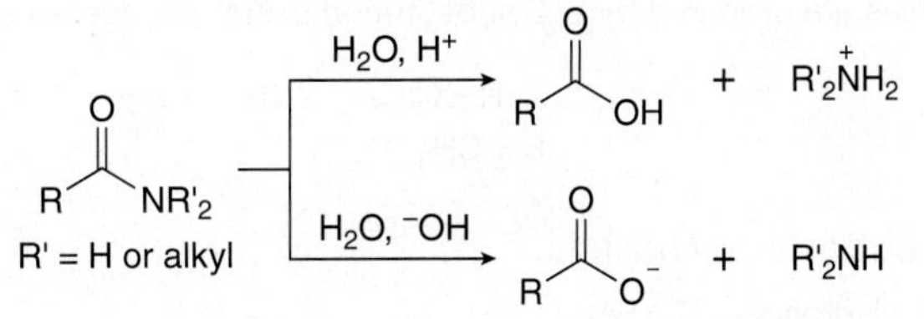

[6] Reactions that produce esters (RCOOR')

a. From RCOCl (22.8):

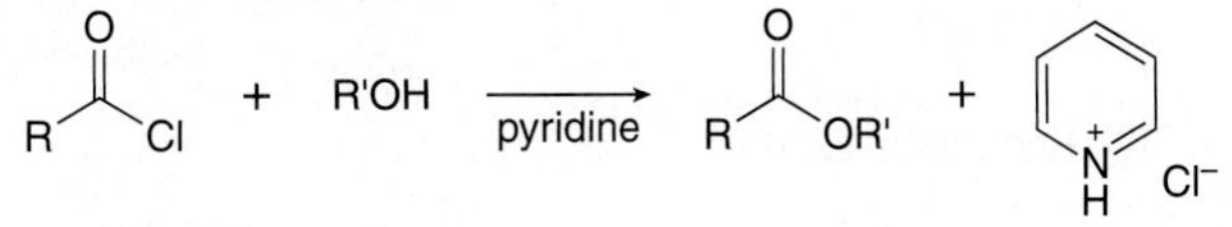

b. From $(RCO)_2O$ (22.9):

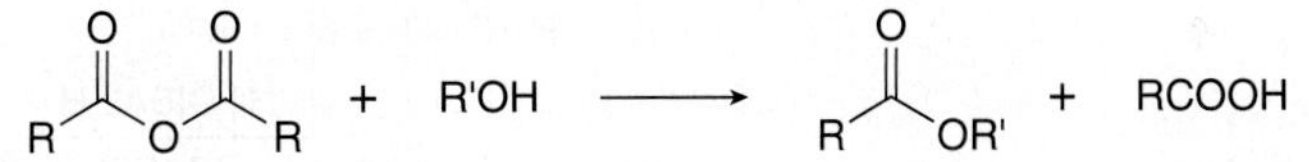

c. From RCOSR' (22.17):

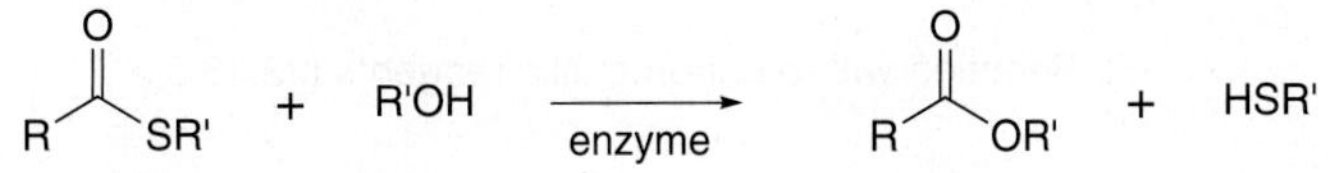

d. From RCO_2H (22.10C):

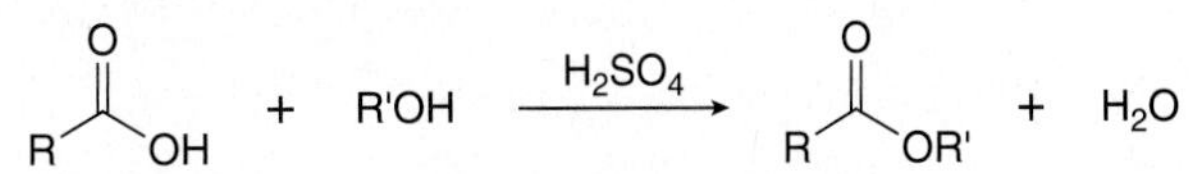

[7] Reactions that produce amides ($RCONH_2$) [The reactions are written with NH_3 as nucleophile to form $RCONH_2$. Similar reactions occur with $R'NH_2$ to form RCONHR', and with R'_2NH to form $RCONR'_2$.]

a. From RCOCl (22.8):

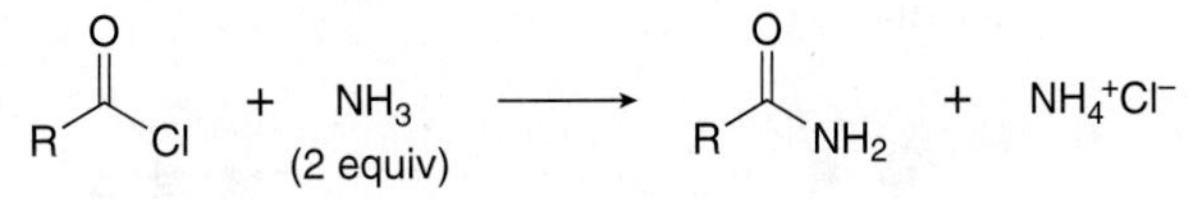

b. From $(RCO)_2O$ (22.9):

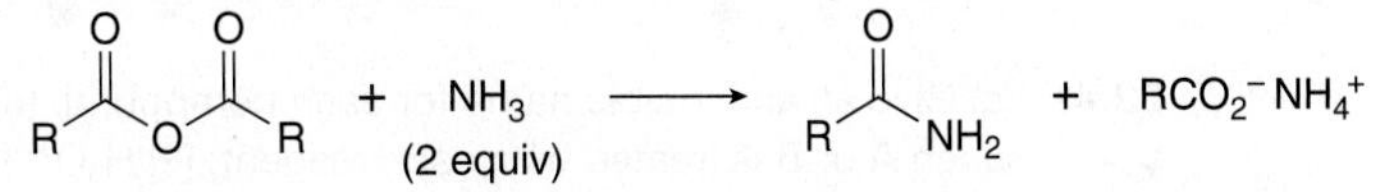

c. From $RCO_2PO_3^{2-}$ (22.16B):

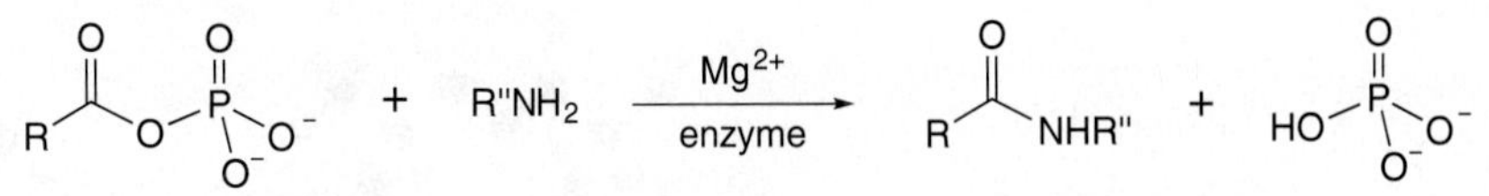

d. From RCO_2H (22.10D):

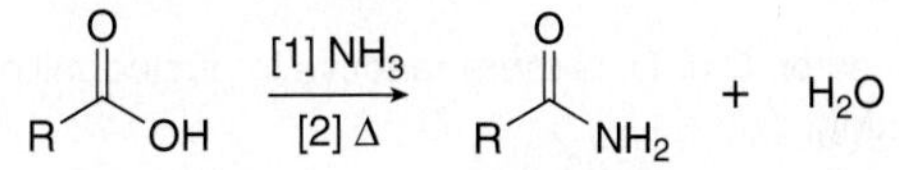

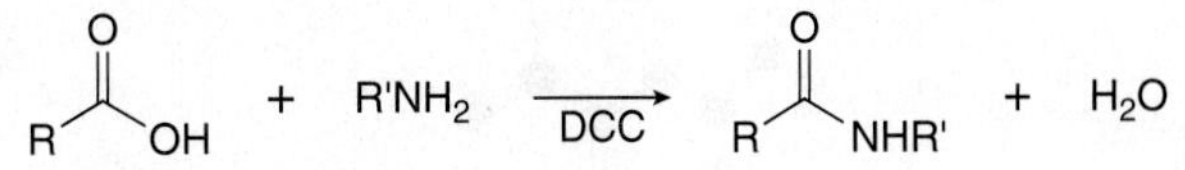

e. From RCO_2R' (22.11):

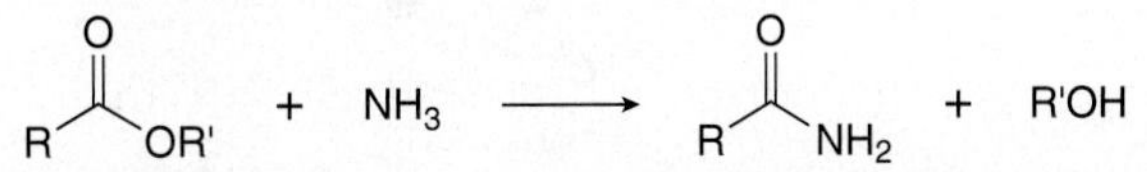

Nitrile Synthesis (22.18)

Nitriles are prepared by S_N2 substitution using unhindered alkyl halides as starting materials.

$$\underset{R = CH_3,\ 1°}{R{-}X} \ + \ {}^{-}CN \xrightarrow[S_N2]{} R{-}C{\equiv}N \ + \ X^{-}$$

Reactions of Nitriles

[1] Hydrolysis (22.18A)

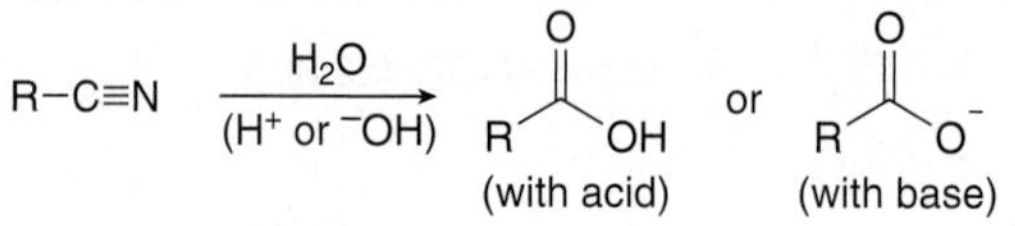
R−C≡N
H₂O
(H⁺ or ⁻OH)
O
R
OH
(with acid)
or
O
R
O⁻
(with base)

[2] Reduction (22.18B)

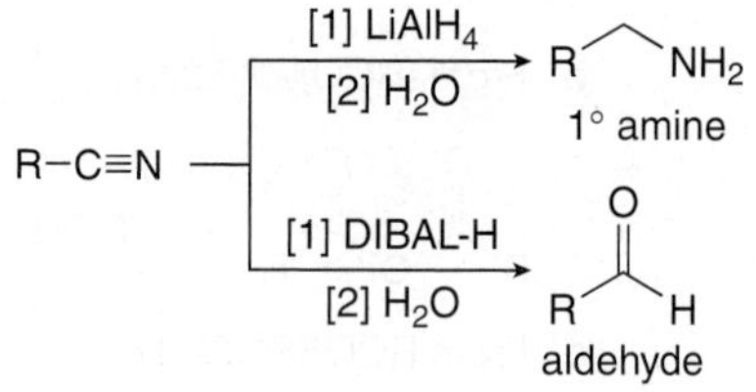
R−C≡N
[1] LiAlH₄
[2] H₂O
R
NH₂
1° amine
[1] DIBAL-H
[2] H₂O
O
R
H
aldehyde

[3] Reaction with organometallic reagents (22.18C)

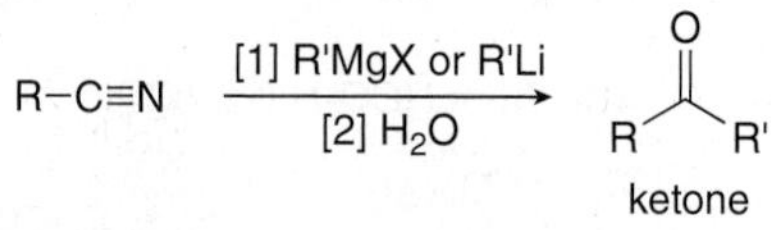
R−C≡N
[1] R'MgX or R'Li
[2] H₂O
O
R
R'
ketone

PROBLEMS

Problems Using Three-Dimensional Models

22.40 Rank the following compounds in order of increasing reactivity in nucleophilic acyl substitution.

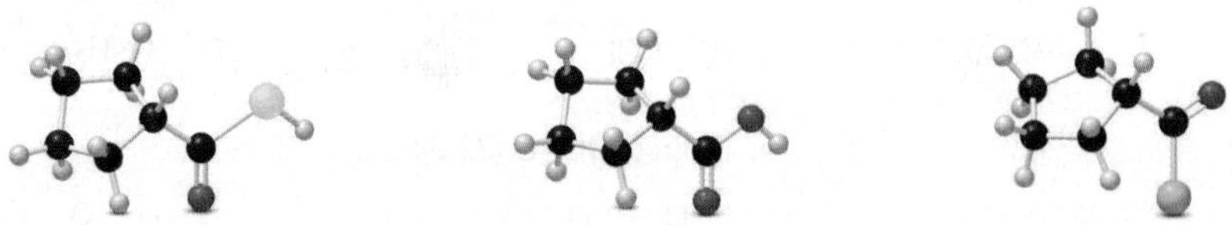

22.41 (a) Give an acceptable name for each compound. (b) Draw the organic products formed when **A** or **B** is treated with each reagent: [1] H_3O^+; [2] ^{-}OH, H_2O; [3] $CH_3CH_2CH_2MgBr$ (excess), then H_2O; [4] $LiAlH_4$, then H_2O.

A **B**

22.42 Which ester, **C** or **D,** is more reactive in nucleophilic acyl substitution? Explain your reasoning.

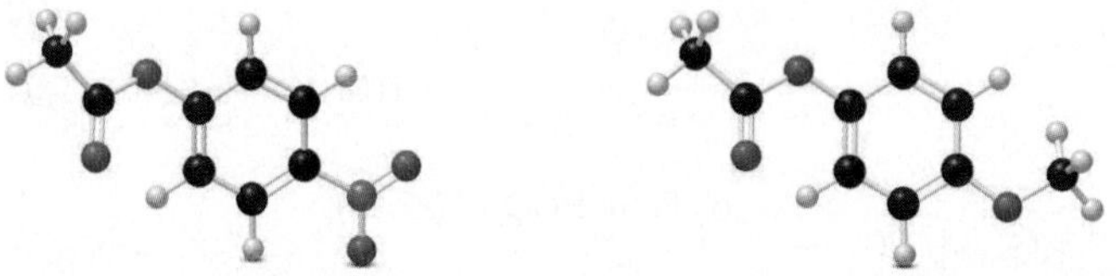

C **D**

Nomenclature

22.43 Give the IUPAC or common name for each compound.

a. 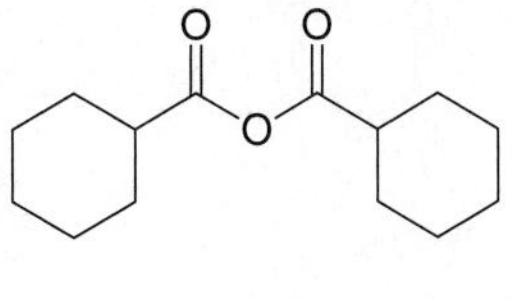

c. 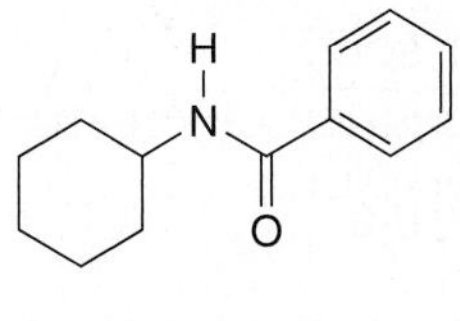

e. 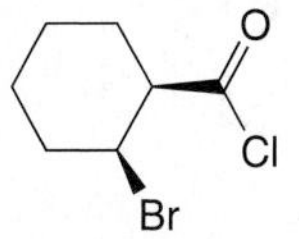

g. 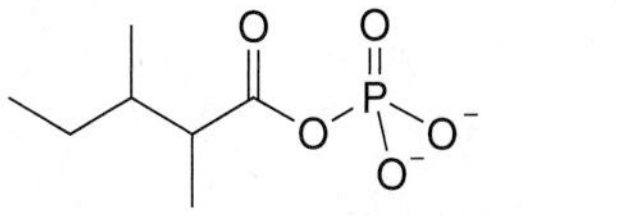

b. 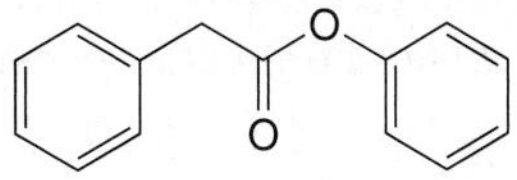

d. 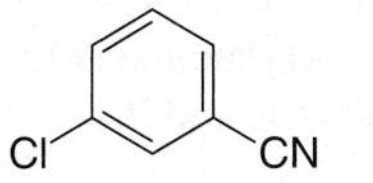

f. 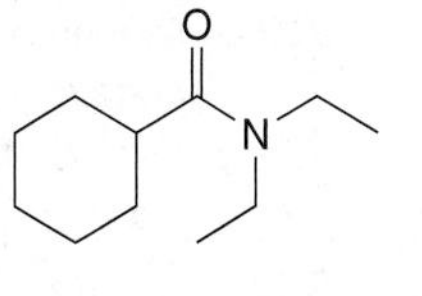

h. 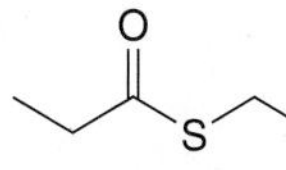

22.44 Give the structure corresponding to each name.

a. cyclohexyl propanoate
b. cyclohexanecarboxamide
c. 4-methylheptanenitrile
d. vinyl acetate
e. benzoic propanoic anhydride
f. 3-methylhexanoyl chloride
g. 3-ethylcyclobutanecarbonyl phosphate
h. *N,N*-dibenzylformamide

Properties of Carboxylic Acid Derivatives

22.45 Explain why CH_3CONH_2 is a stronger acid and a weaker base than $CH_3CH_2NH_2$.

22.46 (a) Propose an explanation for the difference in the frequency of the carbonyl absorptions of phenyl acetate (1765 cm^{-1}) and cyclohexyl acetate (1738 cm^{-1}). (b) Which carbonyl group is more effectively stabilized by resonance? (c) Which ester reacts faster when treated with aqueous base?

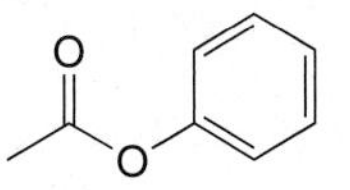

phenyl acetate

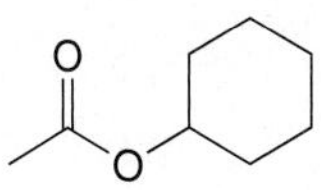

cyclohexyl acetate

Reactions

22.47 Draw the product formed when phenylacetic acid ($C_6H_5CH_2COOH$) is treated with each reagent. With some reagents, no reaction occurs.

a. $NaHCO_3$
b. NaOH
c. $SOCl_2$
d. NaCl
e. NH_3 (1 equiv)
f. NH_3, Δ
g. CH_3OH, H_2SO_4
h. CH_3OH, ^-OH
i. [1] NaOH; [2] CH_3COCl
j. CH_3NH_2, DCC
k. [1] $SOCl_2$; [2] $CH_3CH_2CH_2NH_2$ (excess)
l. [1] $SOCl_2$; [2] $(CH_3)_2CHOH$

22.48 Draw the product formed when phenylacetonitrile ($C_6H_5CH_2CN$) is treated with each reagent.

a. H_3O^+
b. H_2O, ^-OH
c. [1] CH_3MgBr; [2] H_2O
d. [1] CH_3CH_2Li; [2] H_2O
e. [1] DIBAL-H; [2] H_2O
f. [1] $LiAlH_4$; [2] H_2O

22.49 Draw the organic products formed in each reaction.

a. 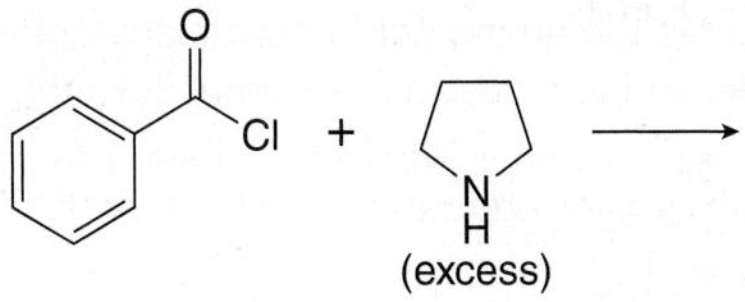

(excess)

c. 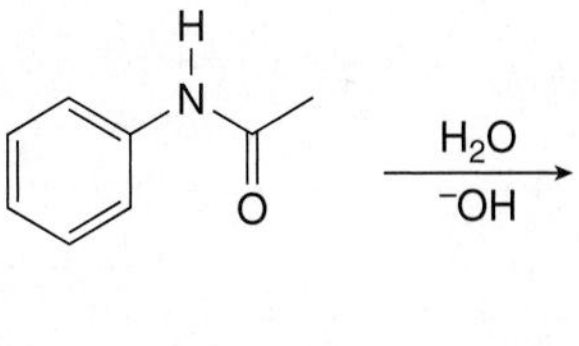

b. 3 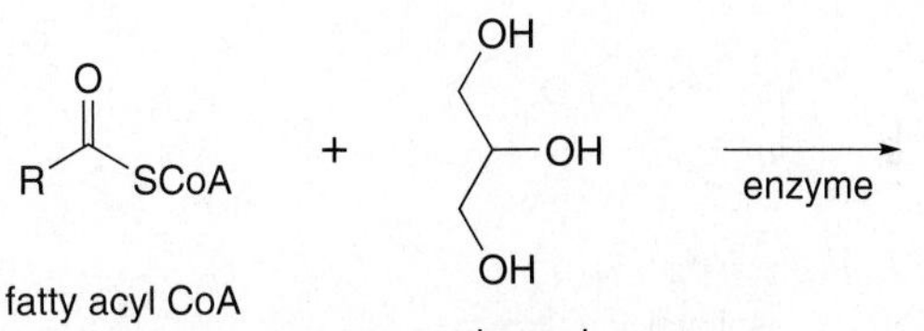

fatty acyl CoA

glycerol

d. 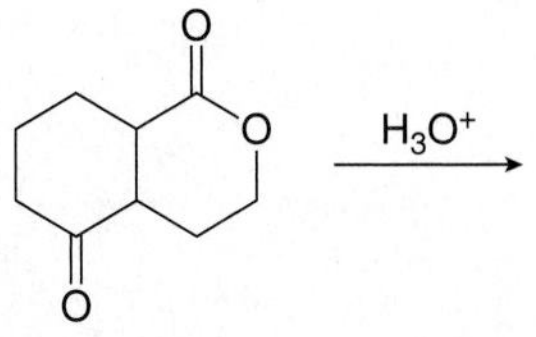

e. 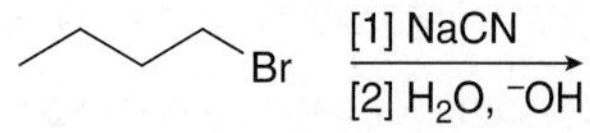
Br
[1] NaCN
[2] H_2O, ^-OH

g. 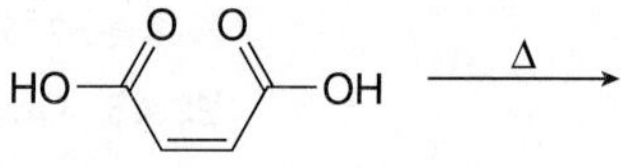
HO
OH
Δ

f. 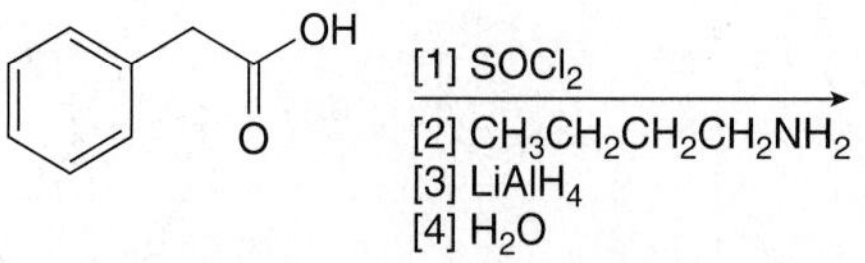
OH
[1] $SOCl_2$
[2] $CH_3CH_2CH_2CH_2NH_2$
[3] $LiAlH_4$
[4] H_2O

h. 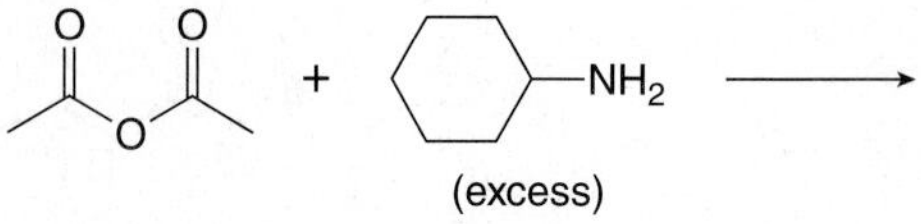
NH_2
(excess)

22.50 Cinnamoylcocaine, a natural product that occurs in coca leaves, can be converted to cocaine by the following reaction sequence. Identify the structure of cinnamoylcocaine, as well as intermediates **X** and **Y.**

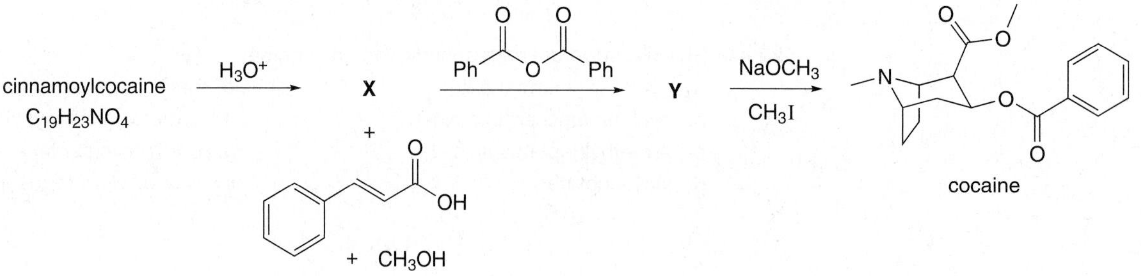
cinnamoylcocaine
$C_{19}H_{23}NO_4$
H_3O^+
X
+
OH
+ CH_3OH
Ph
Ph
Y
$NaOCH_3$
CH_3I
N
cocaine

22.51 What products are formed by hydrolysis of each lactone or lactam with acid?

a. 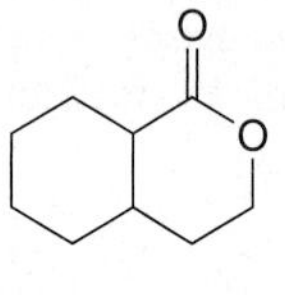

b. 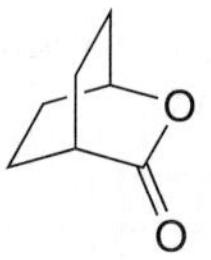

c. 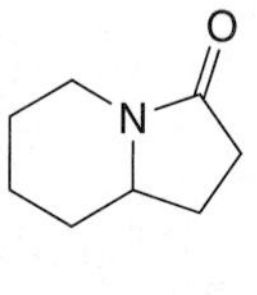

d. 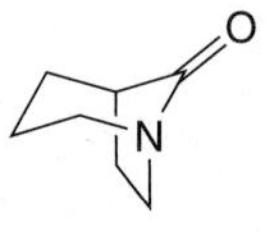

22.52 Draw the products of each reaction and indicate the stereochemistry at any stereogenic centers.

a. 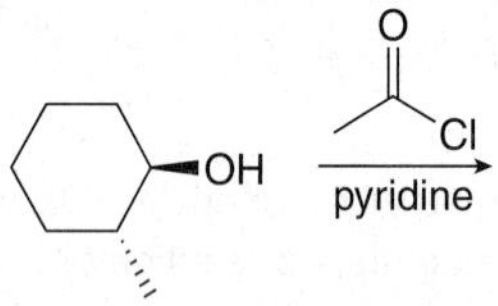
OH
Cl
pyridine

c. 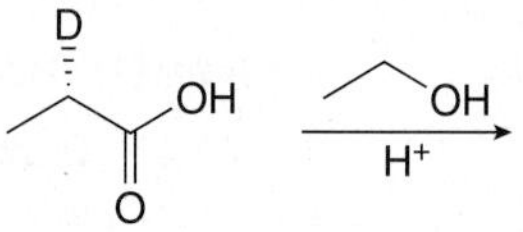
D
OH
OH
H^+

b. 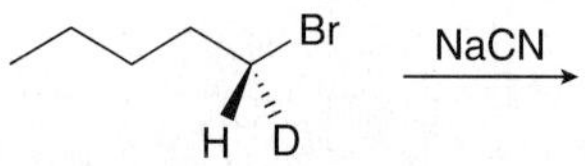
Br
H D
NaCN

d. 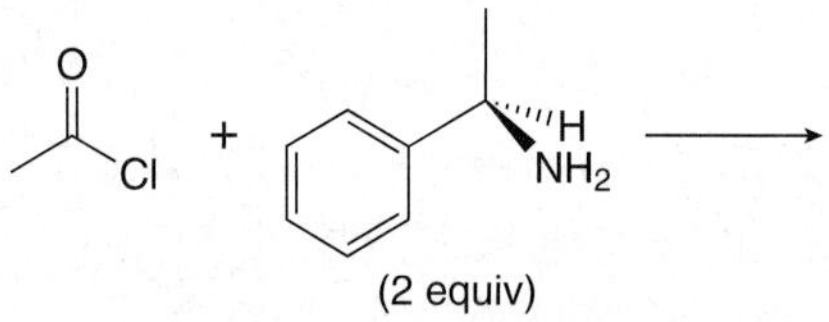
Cl
+
H
NH_2
(2 equiv)

22.53 What products are formed when all of the amide and ester bonds are hydrolyzed in each of the following compounds? **Tamiflu** [part (a)] is the trade name of the antiviral agent oseltamivir, thought to be the most effective agent in treating influenza. **Aspartame** [part (b)] is the artificial sweetener used in Equal and many diet beverages. One of the products of this hydrolysis reaction is the amino acid phenylalanine. Infants afflicted with phenylketonuria cannot metabolize this amino acid, so it accumulates, causing mental retardation. When the affliction is identified early, a diet limiting the consumption of phenylalanine (and compounds like aspartame that are converted to it) can make a normal life possible.

a. 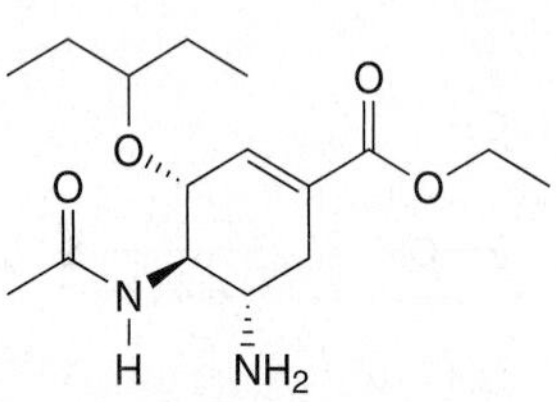
N
H
NH_2

oseltamivir

b. 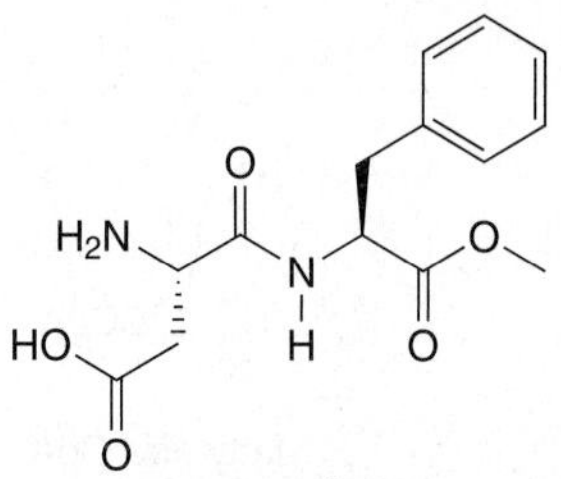
H_2N
N
H
HO

aspartame

22.54 Identify **F** in the following reaction sequence. **F** was converted in several steps to the antidepressant paroxetine (trade name Paxil; see also Problem 9.9).

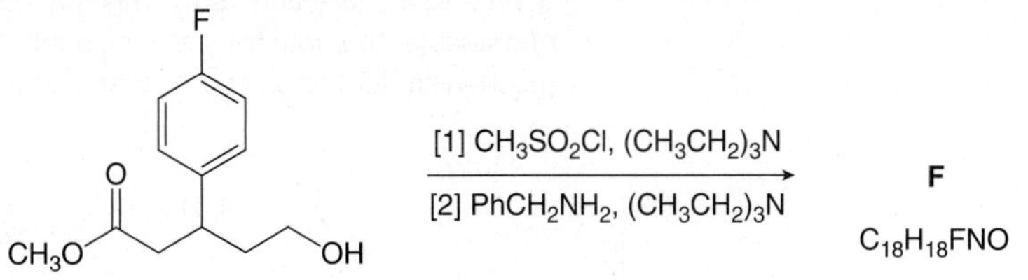
F
O
CH_3O
OH
[1] CH_3SO_2Cl, $(CH_3CH_2)_3N$
[2] $PhCH_2NH_2$, $(CH_3CH_2)_3N$
F
$C_{18}H_{18}FNO$

Mechanism

22.55 Draw a stepwise mechanism for the following reaction.

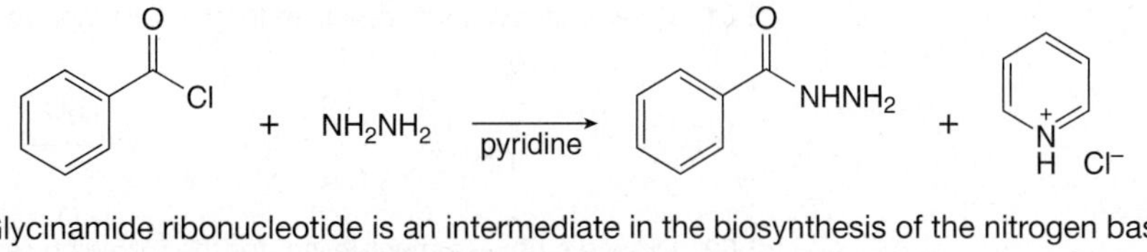
O
Cl
+ NH_2NH_2
pyridine
NHNH$_2$
+
H Cl$^-$

22.56 Glycinamide ribonucleotide is an intermediate in the biosynthesis of the nitrogen bases in DNA. Draw a stepwise mechanism for the formation of glycinamide ribonucleotide from glycyl phosphate and 5-phosphoribosylamine.

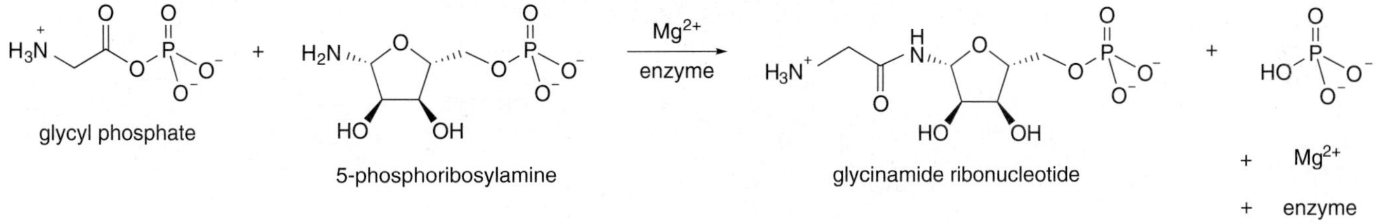
glycyl phosphate
5-phosphoribosylamine
Mg^{2+}
enzyme
glycinamide ribonucleotide
+ Mg^{2+}
+ enzyme

22.57 Using a protonated enzyme (H–enzyme$^+$) to activate the carbonyl, draw a stepwise mechanism for the formation of succinate from succinyl CoA, a step in the citric acid cycle.

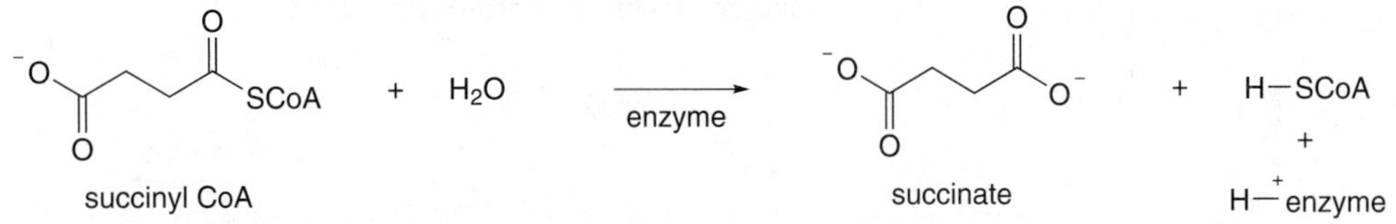
SCoA
+ H_2O
enzyme
succinyl CoA
succinate
+ H–SCoA
+
H–enzyme

22.58 When acetic acid (CH_3COOH) is treated with a trace of acid in water labeled with ^{18}O, the label gradually appears in both oxygen atoms of the carboxylic acid. Draw a mechanism that explains this phenomenon.

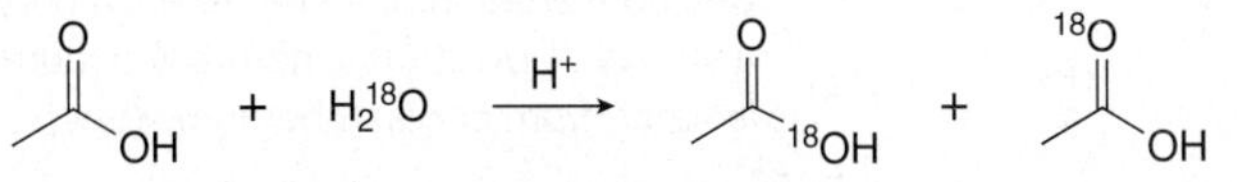
OH
+ $H_2^{18}O$
H^+
^{18}OH
+
^{18}O
OH

22.59 Although γ-butyrolactone (Problem 19.59) is a biologically inactive compound, it is converted in the body to 4-hydroxybutanoic acid (GHB), an addictive and intoxicating recreational drug (Section 19.5). Draw a stepwise mechanism for this conversion in the presence of acid.

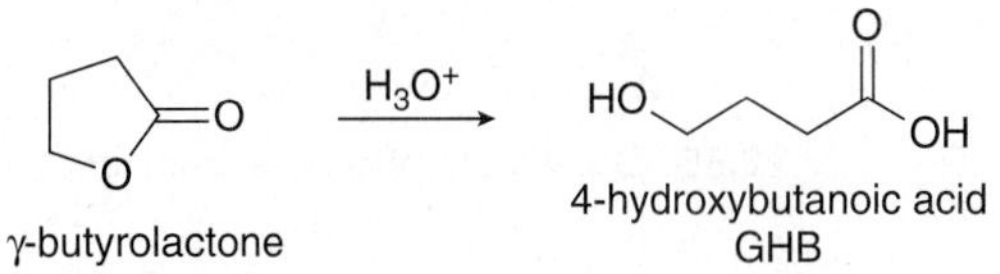
H_3O^+
γ-butyrolactone
4-hydroxybutanoic acid
GHB

22.60 Aspirin is an anti-inflammatory agent because it inhibits the conversion of arachidonic acid to prostaglandins by the transfer of its acetyl group (CH_3CO-) to an OH group at the active site of an enzyme (Section 19.6). This reaction, called transesterification, results in the conversion of one ester to another by a nucleophilic acyl substitution reaction. Draw a stepwise mechanism for the given transesterification.

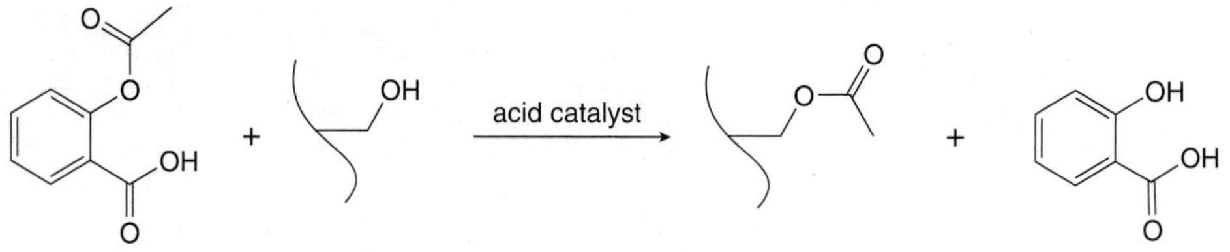
acid catalyst

aspirin　enzyme　inactive enzyme　salicylic acid

22.61 Draw a stepwise mechanism for the following reaction.

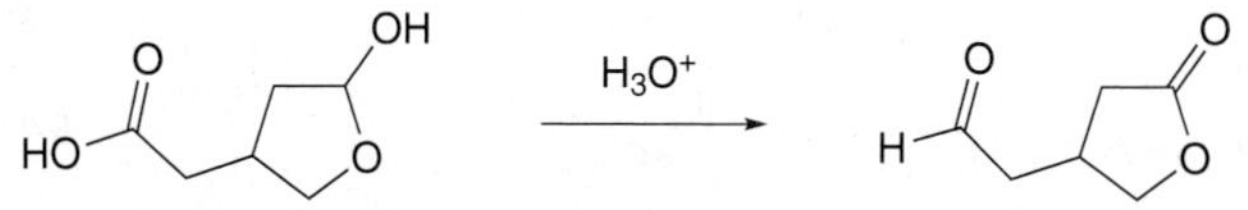
H3O+

22.62 Draw a stepwise mechanism for the following reaction, one step in the synthesis of the cholesterol-lowering drug ezetimibe (Section 20.6).

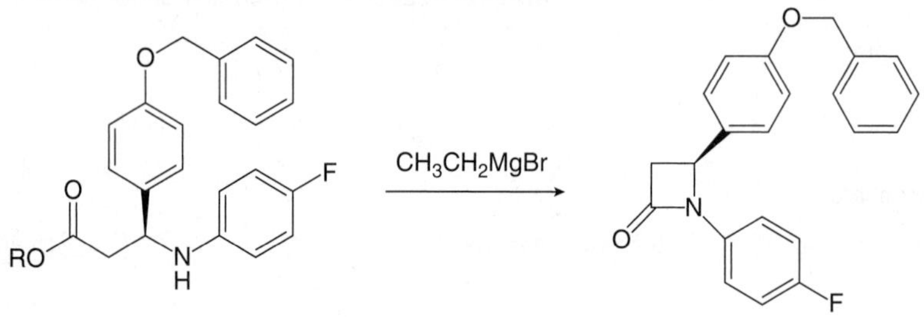
CH3CH2MgBr

22.63 Draw a stepwise mechanism for the conversion of lactone **C** to carboxylic acid **D. C** is a key intermediate in the synthesis of prostaglandins (Section 19.6) by Nobel Laureate E. J. Corey and co-workers at Harvard University.

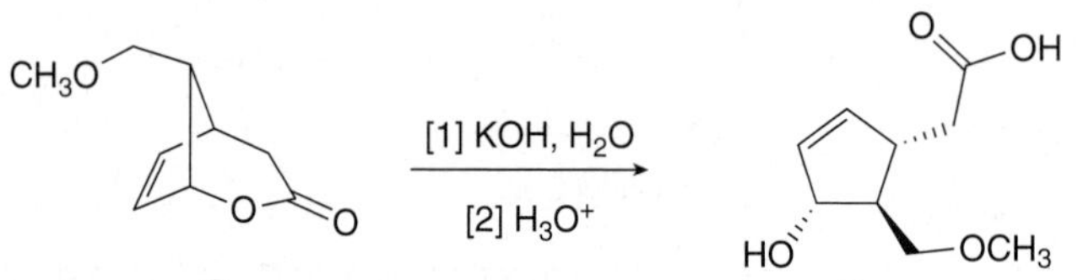
[1] KOH, H2O
[2] H3O+

C　**D**

22.64 Draw a stepwise mechanism for the conversion of lactone **A** to ester **B** using HCl in ethanol. **B** is converted in one step to ethyl chrysanthemate, a useful intermediate in the synthesis of a variety of pyrethrins, naturally occurring insecticides with three-membered rings that are isolated from chrysanthemums (Section 29.4).

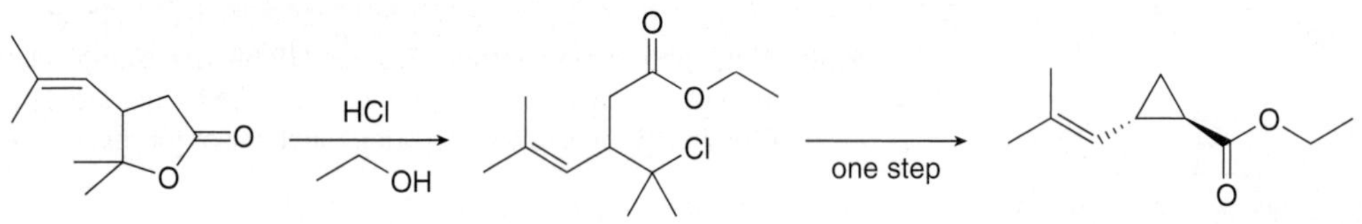
HCl
one step

A　**B**　ethyl chrysanthemate

22.65 Draw a stepwise mechanism for the following reaction.

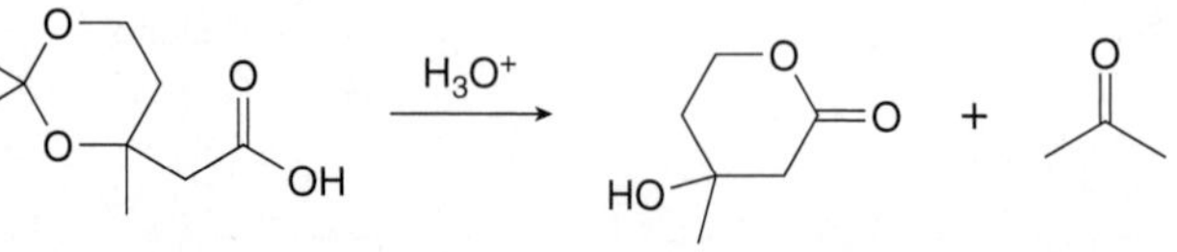
H3O+

22.66 Acid-catalyzed hydrolysis of $HOCH_2CH_2C(CH_3)_2CN$ forms compound **A** ($C_6H_{10}O_2$). **A** shows a strong peak in its IR spectrum at 1770 cm^{-1} and the following signals in its ^1H NMR spectrum: 1.27 (singlet, 6 H), 2.12 (triplet, 2 H), and 4.26 (triplet, 2 H) ppm. Draw the structure for **A** and give a stepwise mechanism that accounts for its formation.

Synthesis

22.67 Write out the steps in the biological conversion of aspartate to asparagine.

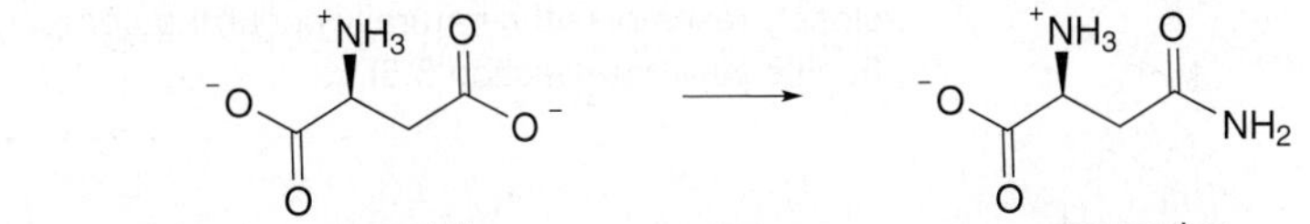

aspartate asparagine

22.68 What carboxylic acid and alcohol are needed to prepare each ester by Fischer esterification?

a. 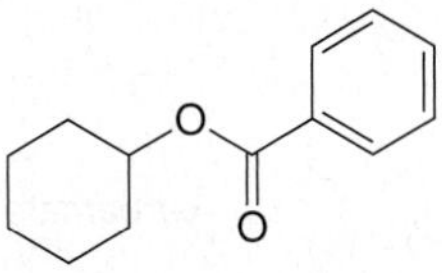
b. 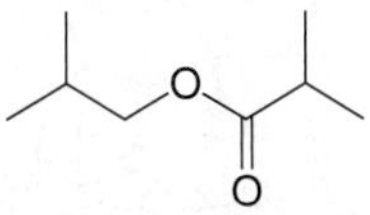

22.69 Devise a synthesis of each compound using 1-bromobutane ($CH_3CH_2CH_2CH_2Br$) as the only organic starting material. You may use any other inorganic reagents.

a. 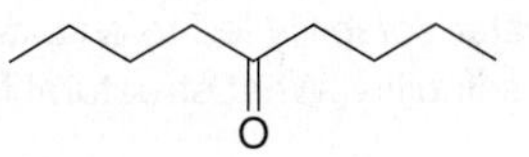
b. 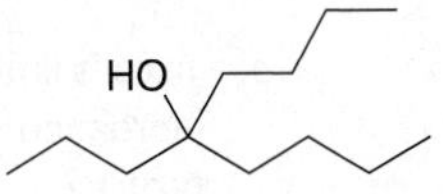

22.70 Two methods convert an alkyl halide into a carboxylic acid having one more carbon atom.

[1] 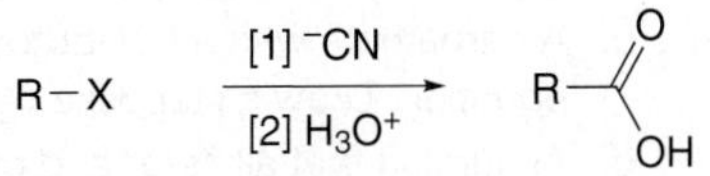 (Section 22.18)

[2] 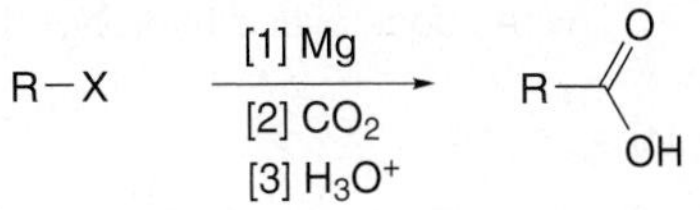 (Section 20.14A)

Depending on the structure of the alkyl halide, one or both of these methods may be employed. For each alkyl halide, write out a stepwise sequence that converts it to a carboxylic acid with one more carbon atom. If both methods work, draw both routes. If one method cannot be used, state why it can't.

a. CH_3Cl
b. 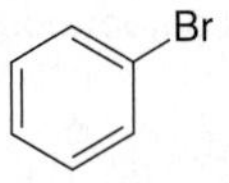
c. 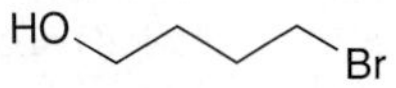

22.71 Devise a synthesis of benzocaine, ethyl *p*-aminobenzoate ($H_2NC_6H_4CO_2CH_2CH_3$), from benzene, organic alcohols, and any needed organic or inorganic reagents. Benzocaine is the active ingredient in the topical anesthetic Orajel (Section 18.15C).

22.72 Devise a synthesis of each analgesic compound from phenol (C_6H_5OH) and any other organic or inorganic reagents.

a. 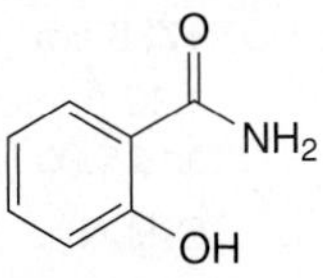
salicylamide

b. 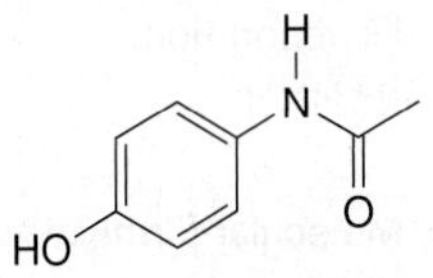
acetaminophen

c. 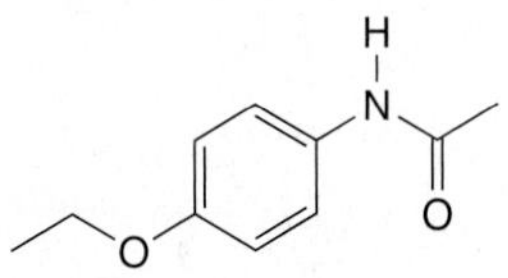
p-acetophenetidin

22.73 Devise a synthesis of each compound from benzene and organic alcohols containing four or fewer carbons. You may also use any required organic or inorganic reagents.

a. 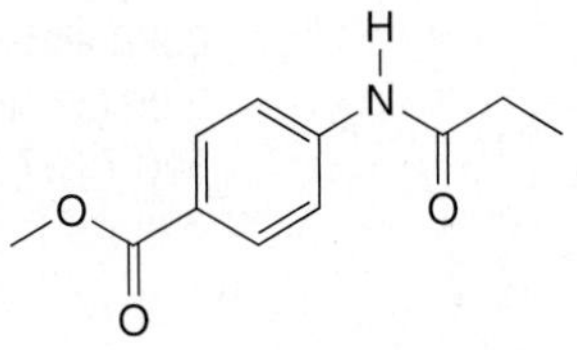
b. 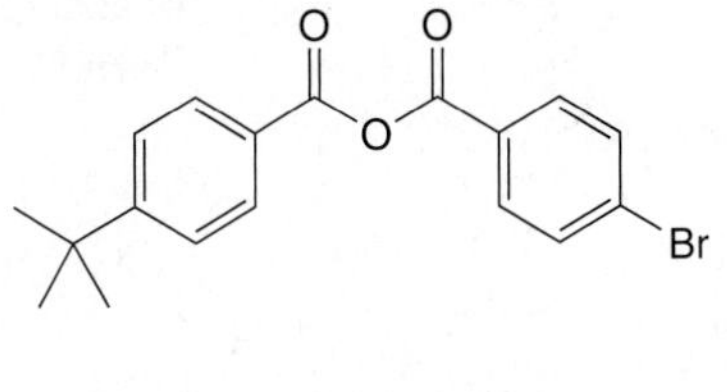

General Problem

22.74 Taxotere is the trade name for the synthetic anticancer drug docetaxel, whose structure closely resembles the naturally occurring compound paclitaxel, which is isolated from the Pacific yew tree (Section 5.5).

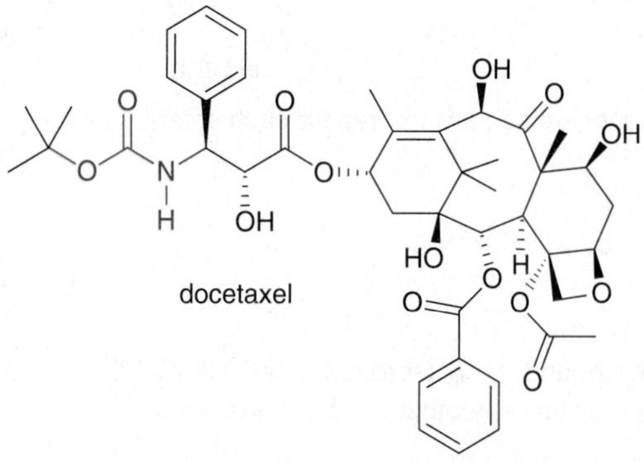

docetaxel

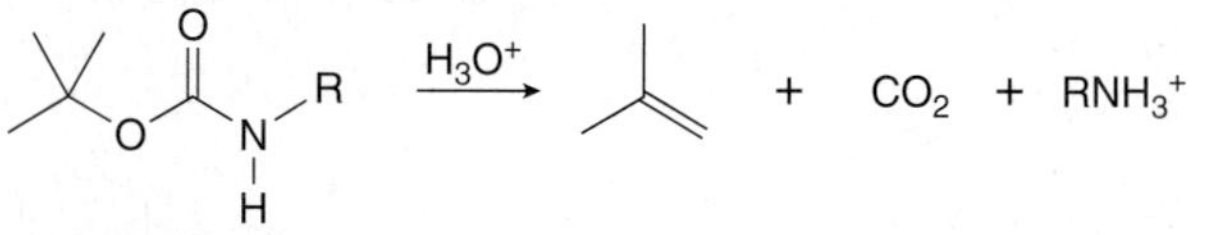

carbamate

a. Taxol's limited water solubility led to an extensive search for related compounds with increased water solubility. What structural features give docetaxel a better water solubility profile?

b. Docetaxel contains a carbamate (labeled in red). Draw three more resonance structures for a carbamate. Rank all four resonance structures in order of increasing stability.

c. A carbamate with a *tert*-butoxy group [$(CH_3)_3CO-$] is hydrolyzed according to the given equation. Draw a stepwise mechanism for this reaction.

d. Assuming that all ester and carbamate bonds are cleaved, draw all products formed when docetaxel is hydrolyzed with aqueous acid.

Spectroscopy

22.75 How can IR spectroscopy be used to distinguish between each pair of isomers?

a. 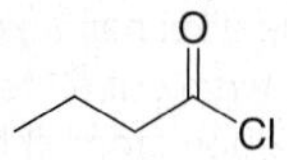

and 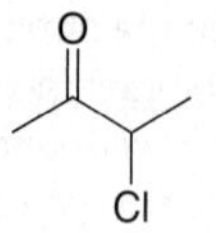

b. 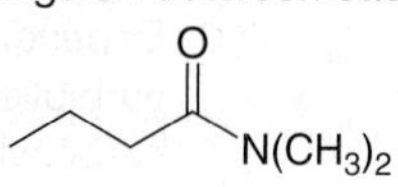

and 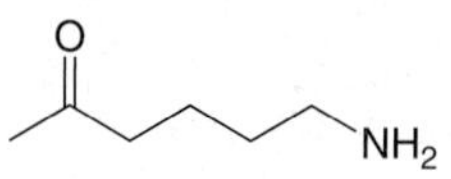

22.76 Rank the following compounds in order of increasing frequency of the C=O absorption in their IR spectra.

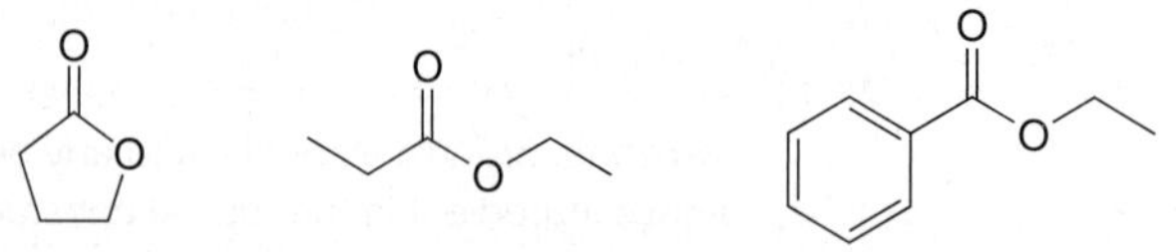

22.77 Identify the structures of each compound from the given data.

a. Molecular formula: $C_6H_{12}O_2$
IR absorption: 1738 cm^{-1}
^1H NMR: 1.12 (triplet, 3 H), 1.23 (doublet, 6 H), 2.28 (quartet, 2 H), and 5.00 (septet, 1 H) ppm

b. Molecular formula: C_4H_7N
IR absorption: 2250 cm^{-1}
^1H NMR: 1.08 (triplet, 3 H), 1.70 (multiplet, 2 H), and 2.34 (triplet, 2 H) ppm

c. Molecular formula: C_8H_9NO
IR absorptions: 3328 and 1639 cm^{-1}
^1H NMR: 2.95 (singlet, 3 H), 6.95 (singlet, 1 H), and 7.3–7.7 (multiplet, 5 H) ppm

22.78 Identify the structures of **A** and **B,** isomers of molecular formula $C_{10}H_{12}O_2$, from their IR data and ^1H NMR spectra.

a. IR absorption for **A** at 1718 cm^{-1}

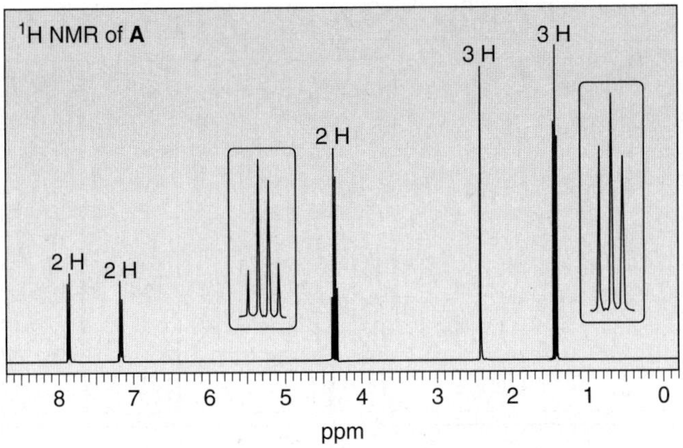

b. IR absorption for **B** at 1740 cm^{-1}

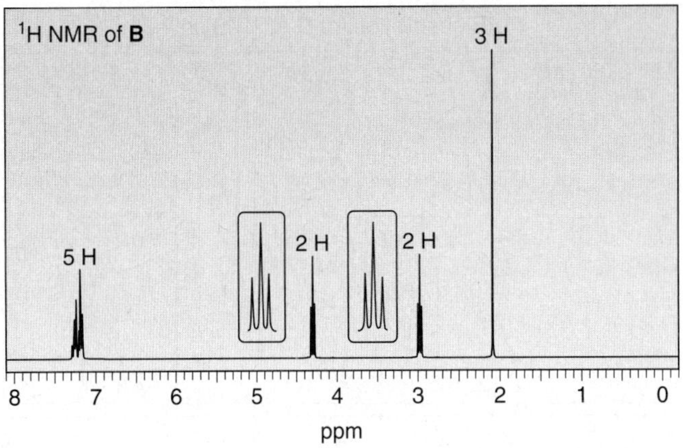

22.79 Phenacetin is an analgesic compound having molecular formula $C_{10}H_{13}NO_2$. Once a common component in over-the-counter pain relievers such as APC (**a**spirin, **p**henacetin, **c**affeine), phenacetin is no longer used because of its liver toxicity. Deduce the structure of phenacetin from its ^1H NMR and IR spectra.

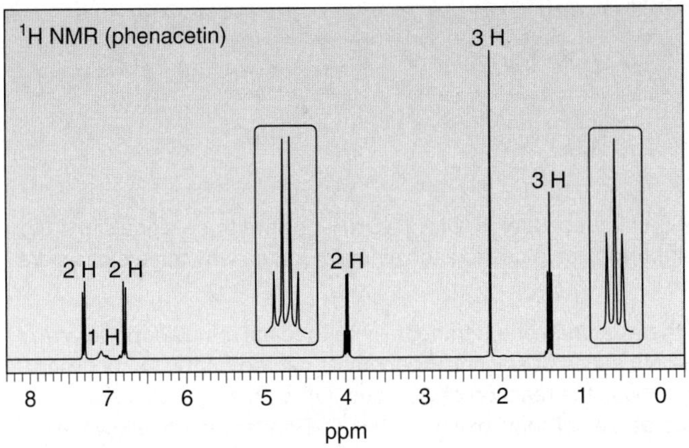

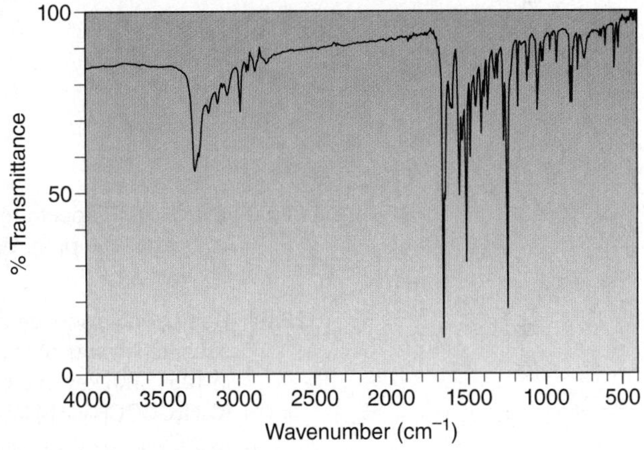

22.80 Identify the structure of compound **C** (molecular formula $C_{11}H_{15}NO_2$), which has an IR absorption at 1699 cm^{-1} and the ^1H NMR spectrum shown below.

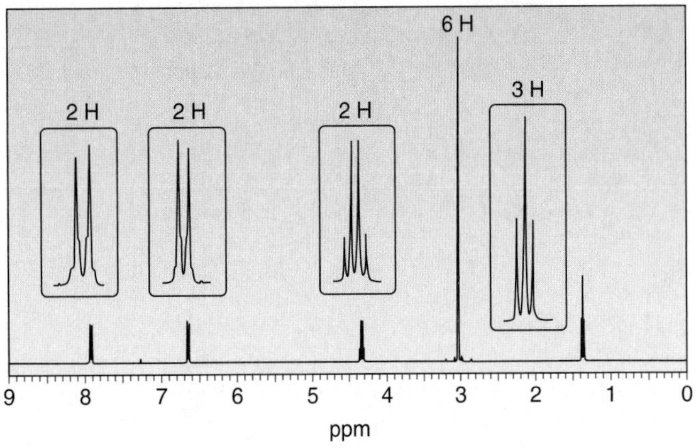

22.81 Identify the structures of **D** and **E,** isomers of molecular formula $C_6H_{12}O_2$, from their IR and ^1H NMR data. Signals at 1.35 and 1.60 ppm in the ^1H NMR spectrum of **D** and 1.90 ppm in the ^1H NMR spectrum of **E** are multiplets.

a. IR absorption for **D** at 1743 cm^{-1}

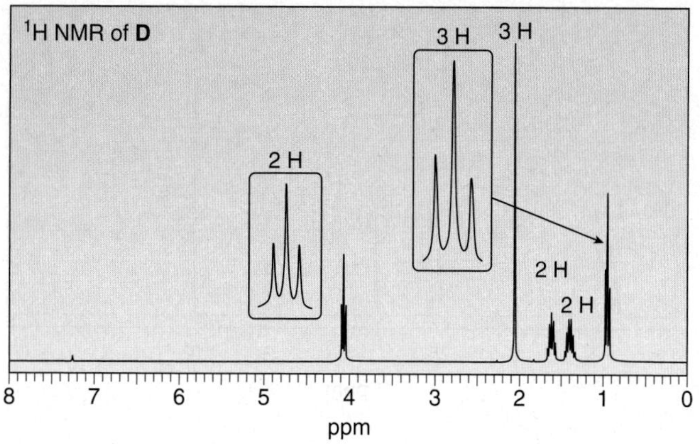

b. IR absorption for **E** at 1746 cm^{-1}

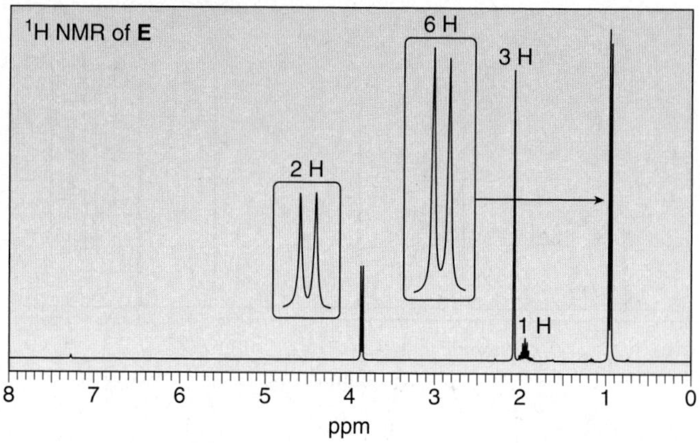

Challenge Problems

22.82 With reference to amides **A** and **B,** the carbonyl of one amide absorbs at a much higher wavenumber in its IR spectrum than the carbonyl of the other amide. Which absorbs at higher wavenumber and why?

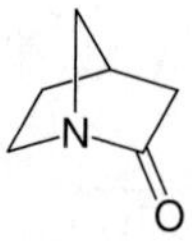

A

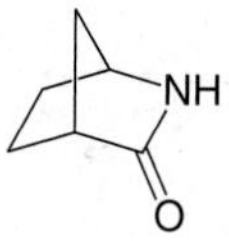

B

22.83 The ^1H NMR spectrum of 2-chloroacetamide ($ClCH_2CONH_2$) shows three signals at 4.02, 7.35, and 7.60 ppm. What protons give rise to each signal? Explain why three signals are observed.

22.84 Compelling evidence for the existence of a tetrahedral intermediate in nucleophilic acyl substitution was obtained in a series of elegant experiments carried out by Myron Bender in 1951. The key experiment was the reaction of aqueous ^-OH with ethyl benzoate ($C_6H_5COOCH_2CH_3$) labeled at the carbonyl oxygen with ^{18}O. Bender did not allow the hydrolysis to go to completion, and then examined the presence of a label in the *recovered starting material.* He found that some of the recovered ethyl benzoate no longer contained a label at the carbonyl oxygen. With reference to the accepted mechanism of nucleophilic acyl substitution, explain how this provides evidence for a tetrahedral intermediate.

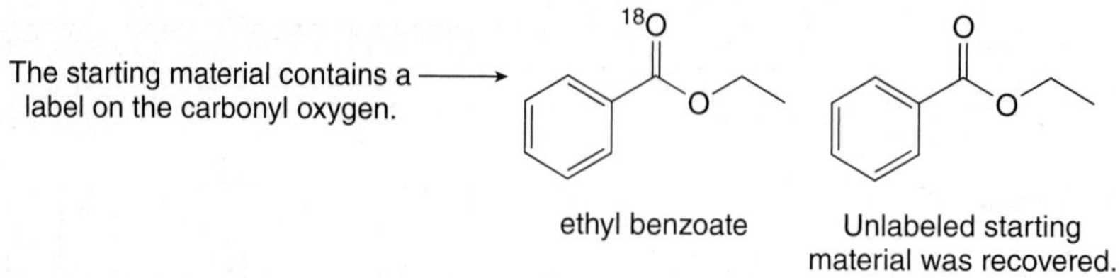

22.85 Draw a stepwise mechanism for the following reactions, two steps in R. B. Woodward's classic synthesis of reserpine in 1958. Reserpine, which is isolated from the extracts of the Indian snakeroot *Rauwolfia serpentina Benth,* was used at one time to manage mild hypertension associated with anxiety.

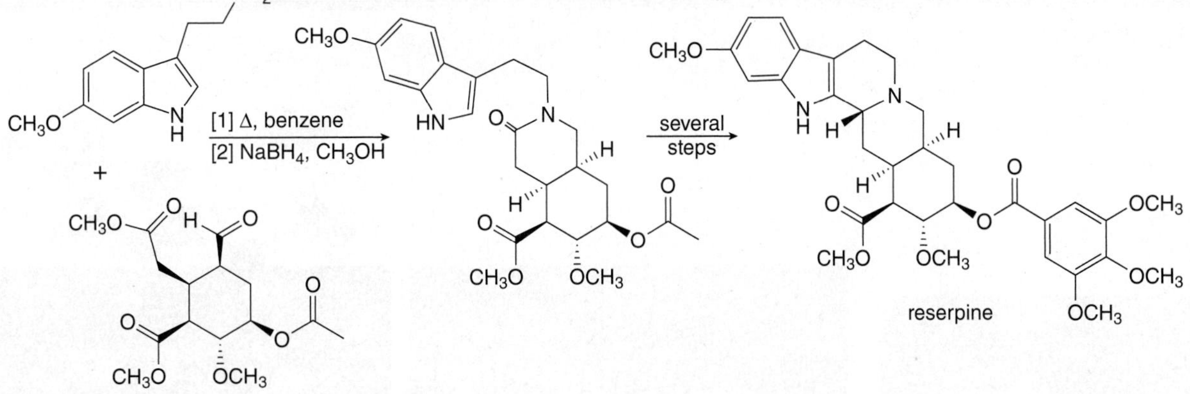

22.86 Draw a stepwise mechanism for the following reaction, the last step in a five-step industrial synthesis of vitamin C that begins with the simple carbohydrate glucose.

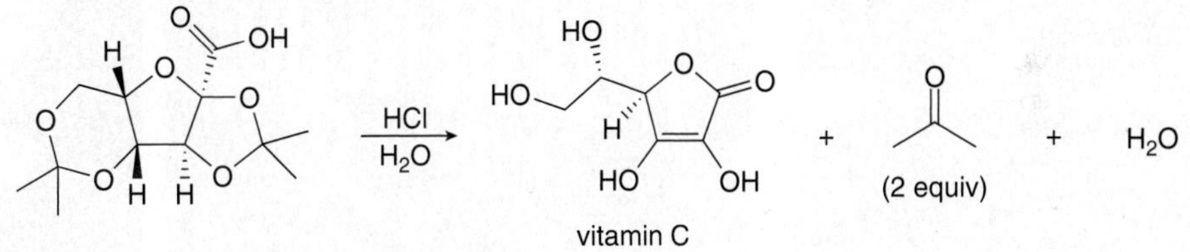

22.87 Draw a stepwise mechanism for the following reaction, a key step in the synthesis of linezolid, an antibacterial agent.

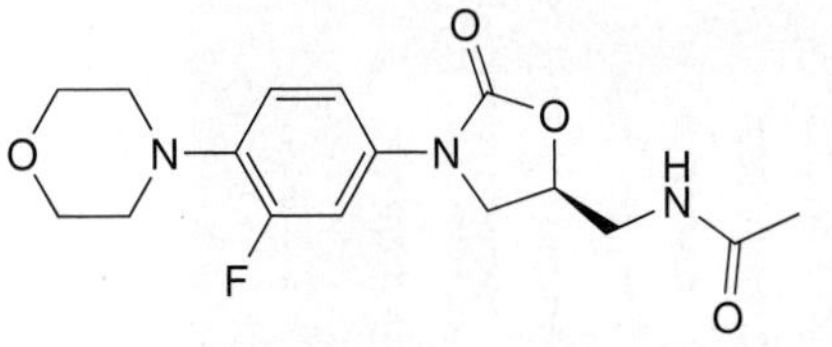
linezolid

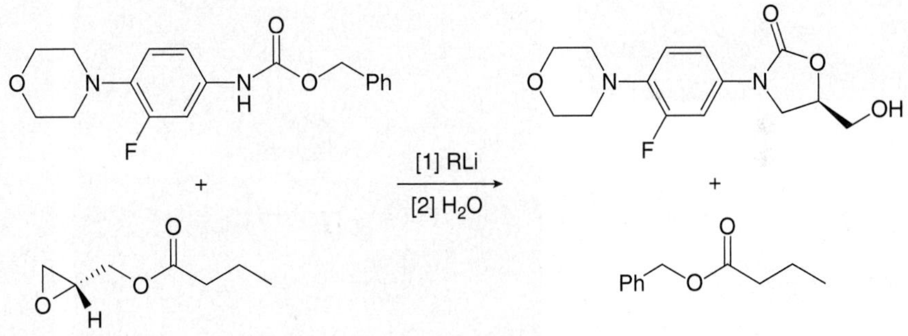

23 Substitution Reactions of Carbonyl Compounds at the α Carbon

23.1 Introduction
23.2 Enols
23.3 Enolates
23.4 Enolates of unsymmetrical carbonyl compounds
23.5 Racemization at the α carbon
23.6 A preview of reactions at the α carbon
23.7 Halogenation at the α carbon
23.8 Direct enolate alkylation
23.9 Malonic ester synthesis
23.10 Acetoacetic ester synthesis

Tamoxifen is a potent anticancer drug used widely in the treatment of breast cancer. Tamoxifen binds to estrogen receptors, and in this way inhibits the growth of breast cancers that are estrogen dependent. One method to synthesize tamoxifen forms a new carbon–carbon bond on the α carbon to a carbonyl group using an intermediate enolate. In Chapter 23 we learn about these and other carbon–carbon bond-forming reactions that occur at the α carbon.

Chapters 23 and 24 focus on reactions that occur at the α carbon to a carbonyl group. These reactions are different from the reactions of Chapters 20–22, all of which involved nucleophilic attack at the electrophilic carbonyl carbon. In reactions at the α carbon, the carbonyl compound serves as a *nucleophile* that reacts with a carbon or halogen electrophile to form a new bond to the α carbon.

Chapter 23 concentrates on **substitution reactions at the α carbon,** whereas Chapter 24 concentrates on reactions between two carbonyl compounds, one of which serves as the nucleophile and one of which is the electrophile. Many of the reactions in Chapter 23 form new carbon–carbon bonds, thus adding to your repertoire of reactions that can be used to synthesize more complex organic molecules from simple precursors. As you will see, the reactions introduced in Chapter 23 have been used to prepare a wide variety of interesting and useful compounds.

23.1 Introduction

Up to now, the discussion of carbonyl compounds has centered on their reactions with nucleophiles at the electrophilic carbonyl carbon. **Two general reactions are observed,** depending on the structure of the carbonyl starting material.

- *Nucleophilic addition* occurs when there is no electronegative atom Z on the carbonyl carbon (as with aldehydes and ketones).

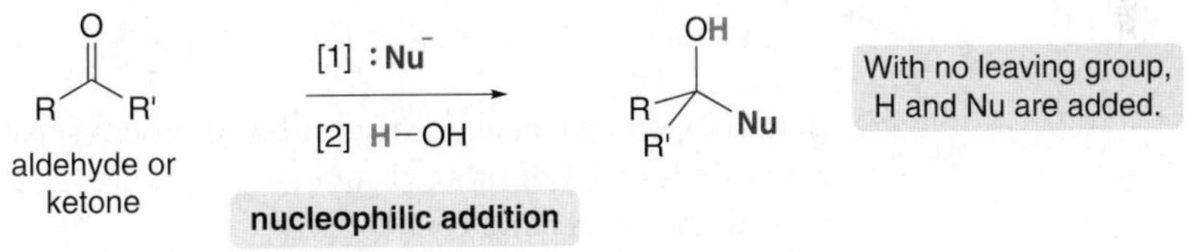

- *Nucleophilic acyl substitution* occurs when there is an electronegative atom Z on the carbonyl carbon (as with carboxylic acids and their derivatives).

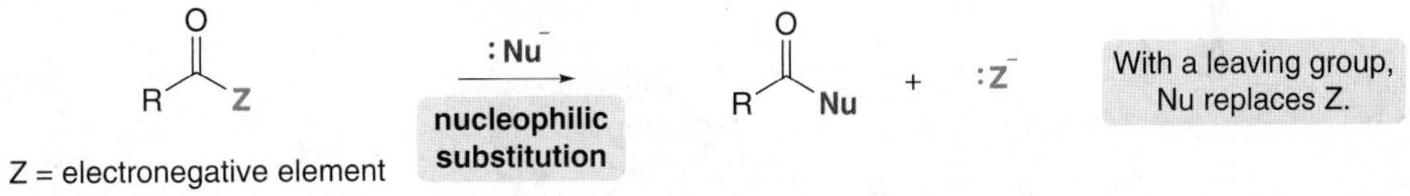

Reactions can also occur at the α carbon to the carbonyl group. These reactions proceed by way of **enols** or **enolates,** two electron-rich intermediates that react with electrophiles, forming a new bond on the α carbon. This reaction results in the **substitution of the electrophile E for hydrogen.**

Hydrogen atoms on the α carbon are called **α hydrogens.**

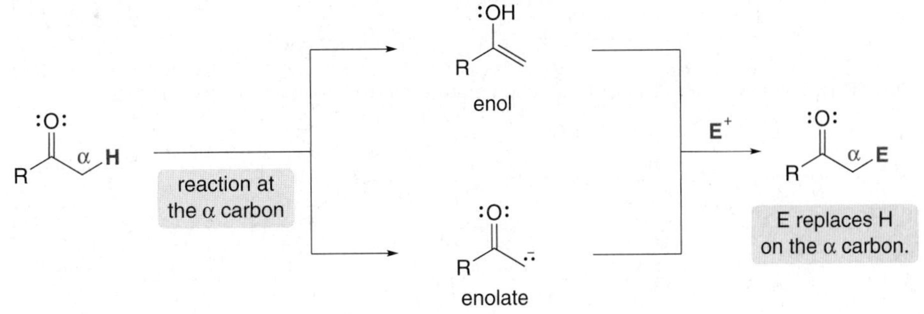

23.2 Enols

Recall from Chapter 11 that **enol and keto forms are tautomers of the carbonyl group that differ in the position of a double bond and a proton.** These constitutional isomers are in equilibrium with each other.

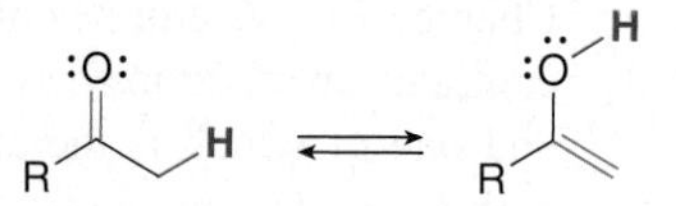

keto form **enol form**

- A keto tautomer has a C=O and an additional C–H bond.
- An enol tautomer has an O–H group bonded to a C=C.

Equilibrium favors the keto form for most carbonyl compounds largely because a C=O is much stronger than a C=C. For simple carbonyl compounds, < 1% of the enol is present at equilibrium. With unsymmetrical ketones, moreover, two different enols are possible, yet they still total < 1%.

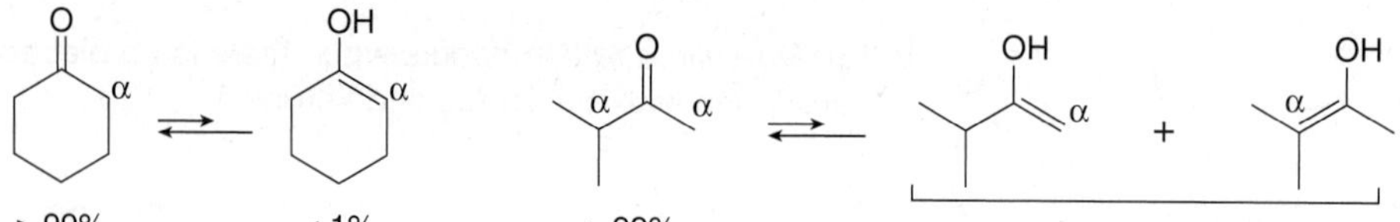

> 99%   < 1%   > 99%   < 1%

With compounds containing two carbonyl groups separated by a single carbon (called β-dicarbonyl compounds or 1,3-dicarbonyl compounds), however, the concentration of the enol form sometimes exceeds the concentration of the keto form.

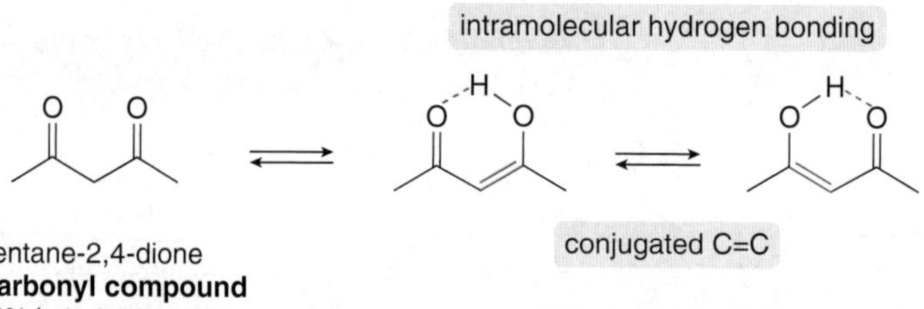

pentane-2,4-dione
β-dicarbonyl compound
24% keto tautomer

76% enol tautomers

Two factors stabilize the enol of β-dicarbonyl compounds: **conjugation** and **intramolecular hydrogen bonding.** The C=C of the enol is conjugated with the carbonyl group, allowing delocalization of the electron density in the π bonds. Moreover, the OH of the enol can hydrogen bond to the oxygen of the nearby carbonyl group. Such intramolecular hydrogen bonds are especially stabilizing when they form a six-membered ring, as in this case.

Sample Problem 23.1 Convert each compound to its enol or keto tautomer.

a. 

b. 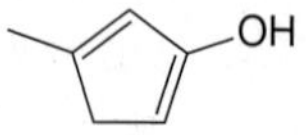

Solution

a. To convert a carbonyl compound to its enol tautomer, draw a double bond between the carbonyl carbon and the α carbon, and change the C=O to C–OH. In this case, both α carbons are identical, so only one enol is possible.

b. To convert an enol to its keto tautomer, change the C–OH to C=O and add a proton to the other end of the C=C.

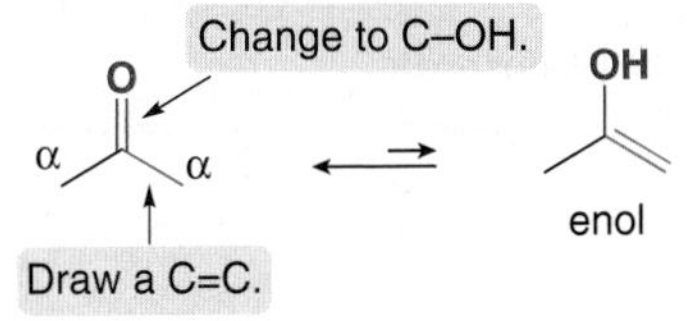

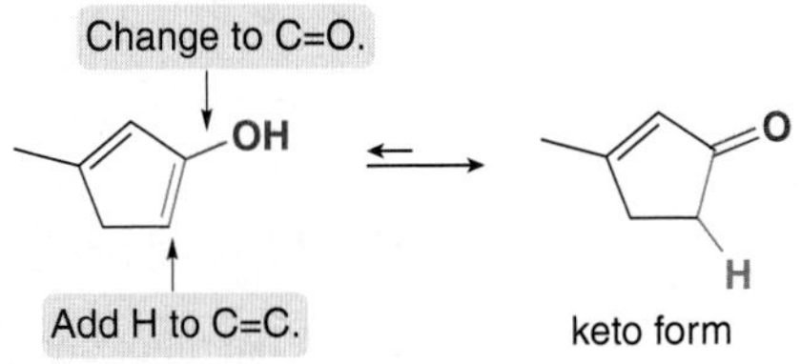

Problem 23.1 Draw the enol or keto tautomer(s) of each compound.

a. 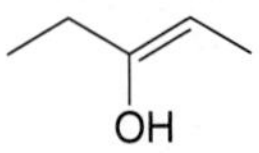

c. 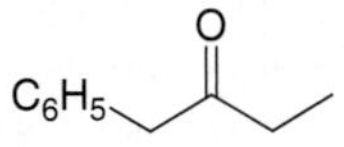

e. 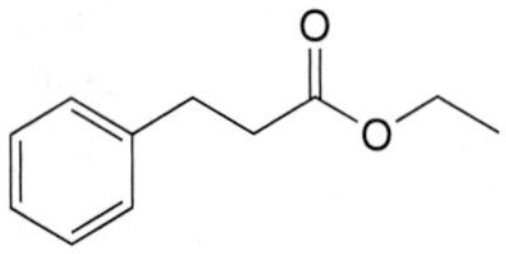

b. 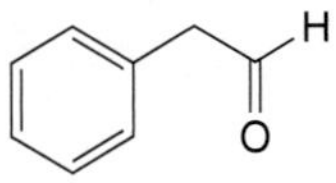

d. 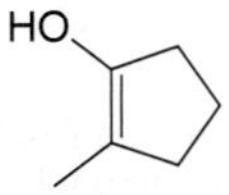

f. 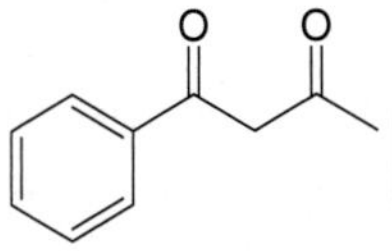 [Draw mono enol tautomers only.]

23.2A The Mechanism of Tautomerization

Tautomerization, the process of converting one tautomer into another, is catalyzed by both acid and base. Tautomerization always requires two steps (**protonation** and **deprotonation**), but the order of these steps depends on whether the reaction takes place in acid or base. In Mechanisms 23.1 and 23.2 for tautomerization, the keto form is converted to the enol form. All of the steps are reversible, though, so they equally apply to the conversion of the enol form to the keto form.

Mechanism 23.1 Tautomerization in Acid

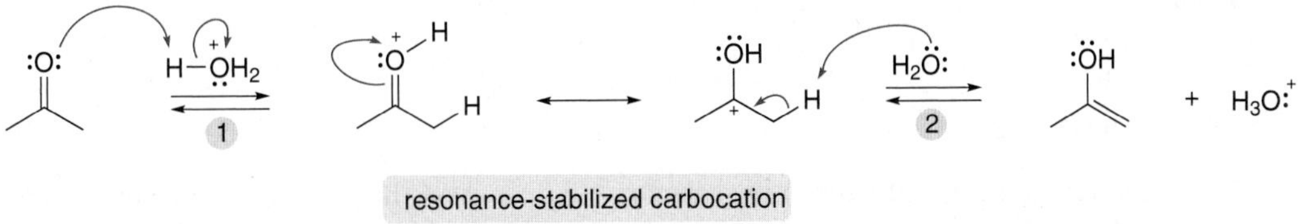

1 With acid, protonation *precedes* deprotonation. Protonation of the carbonyl forms a resonance-stabilized carbocation.

2 Removal of a proton forms the enol.

Mechanism 23.2 Tautomerization in Base

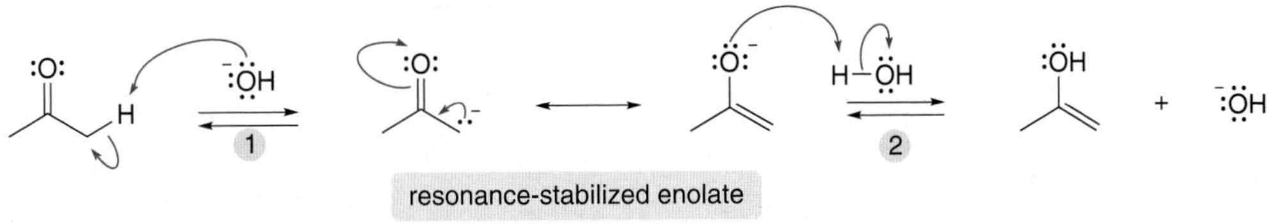

1 With base, deprotonation *precedes* protonation. Removal of a proton on the α carbon forms a resonance-stabilized enolate.

2 Protonation of the enolate forms the enol.

Problem 23.2 During the metabolism of glucose, glyceraldehyde 3-phosphate is converted to dihydroxyacetone phosphate by a process that involves two keto–enol tautomerizations. Draw a stepwise mechanism for this reaction in the presence of acid.

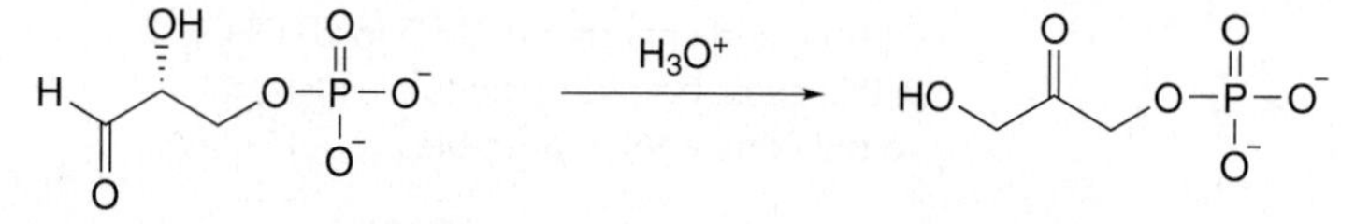

glyceraldehyde 3-phosphate — dihydroxyacetone phosphate

23.2B How Enols React

Like other compounds with carbon–carbon double bonds, **enols are electron rich, so they react as nucleophiles.** Enols are even more electron rich than alkenes, though, because the OH group has a powerful electron-donating resonance effect. A second resonance structure can be drawn for the enol that places a negative charge on one of the carbon atoms. As a result, this carbon atom is especially nucleophilic, and it can react with an electrophile E^+ to form a new bond to carbon. Loss of a proton then forms a neutral product.

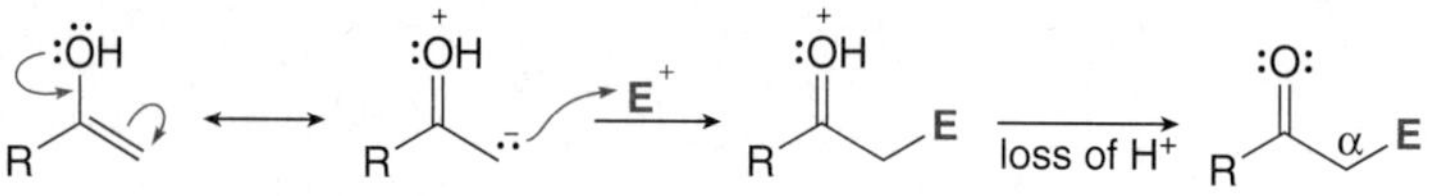

two resonance structures for an enol

- **Reaction of an enol with an electrophile E^+ forms a new C–E bond on the α carbon. The net result is substitution of H by E on the α carbon.**

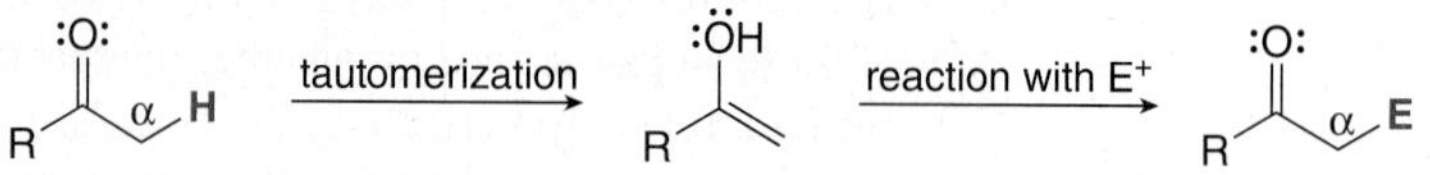

Problem 23.3 When phenylacetaldehyde ($C_6H_5CH_2CHO$) is dissolved in D_2O with added DCl, the hydrogen atoms α to the carbonyl are gradually replaced by deuterium atoms. Write a mechanism for this process that involves enols as intermediates.

23.3 Enolates

Enolates are formed when a base removes a proton on the α carbon to a carbonyl group. A C–H bond on the α carbon is more acidic than many other sp^3 hybridized C–H bonds, because the resulting enolate is resonance stabilized. Moreover, one of the resonance structures is especially stable because it places a negative charge on an electronegative oxygen atom.

Forming enolates from carbonyl compounds was first discussed in Section 21.7.

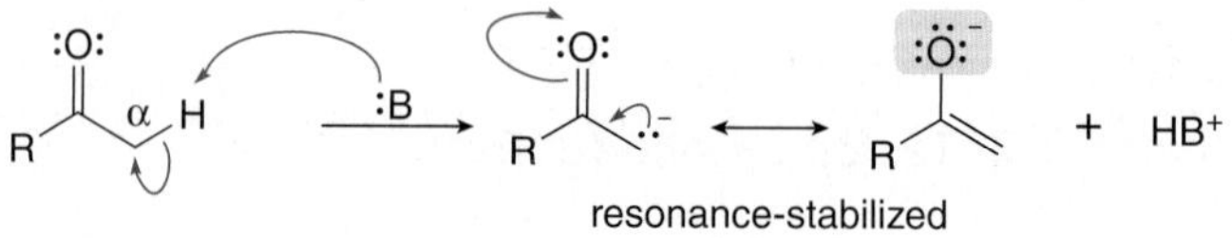

resonance-stabilized enolate anion

Enolates are always formed by removal of a proton on the **α carbon.**

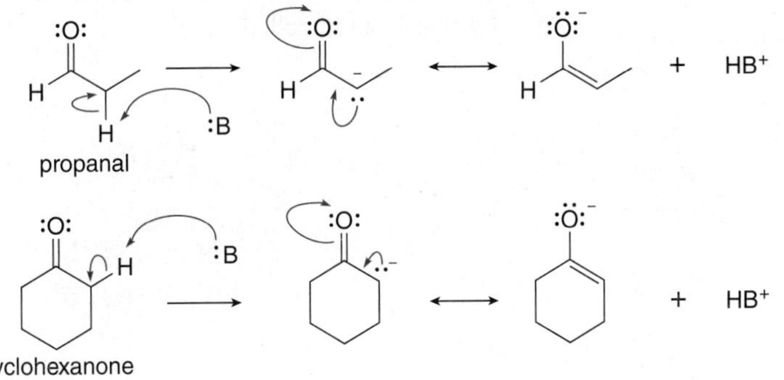

The pK_a of the α hydrogen in an aldehyde or ketone is **~20.** As shown in Table 23.1, this makes it considerably more acidic than the C–H bonds in CH_3CH_3 and $CH_3CH{=}CH_2$. Although C–H bonds α to a carbonyl are more acidic than many other C–H bonds, they are still less acidic than O–H bonds that always place the negative charge of the conjugate base on an electronegative oxygen atom (c.f. CH_3CH_2OH and CH_3COOH in Table 23.1).

Table 23.1 A Comparison of pK_a Values

Compound	pK_a	Conjugate base	Structural features of the conjugate base
CH_3CH_3	50	$CH_3\ddot{C}H_2^-$	• The conjugate base has a (–) charge on C, but is not resonance stabilized.
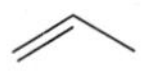	43	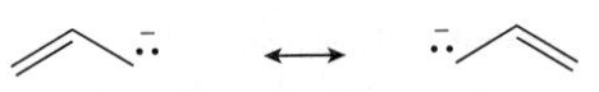	• The conjugate base has a (–) charge on C, and is resonance stabilized.
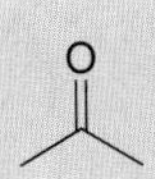	19.2	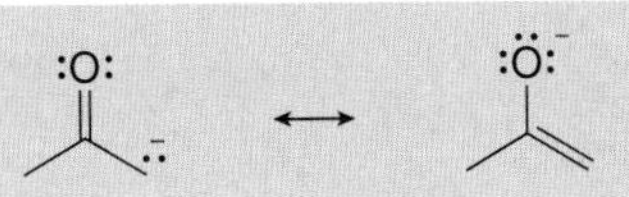	• **The conjugate base has two resonance structures, one of which has a (–) charge on O.**
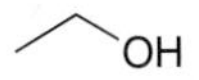	16	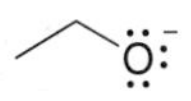	• The conjugate base has a (–) charge on O, but is not resonance stabilized.
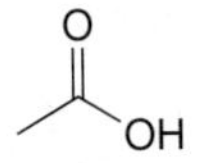	4.8	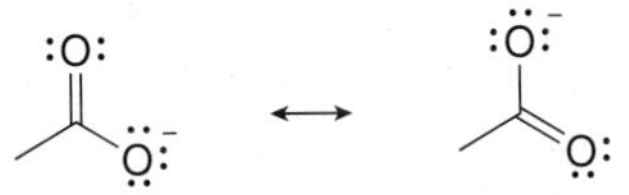	• The conjugate base has two resonance structures, both of which have a (–) charge on O.

Increasing acidity
Increasing stability of the conjugate base

- **Resonance stabilization of the conjugate base increases acidity.**
  - $CH_2{=}CHCH_3$ is more acidic than CH_3CH_3.
  - CH_3COOH is more acidic than CH_3CH_2OH.
- **Placing a negative charge on O in the conjugate base increases acidity.**
  - CH_3CH_2OH is more acidic than CH_3CH_3.
  - CH_3COCH_3 is more acidic than $CH_2{=}CHCH_3$.
  - CH_3COOH (with two O atoms) is more acidic than CH_3COCH_3.

The electrostatic potential plots in Figure 23.1 compare the electron density of the acetone enolate, which is resonance stabilized and delocalized, with that of $(CH_3)_2CHO^-$, an alkoxide that is not resonance stabilized.

Figure 23.1
Electron density in an enolate and an alkoxide

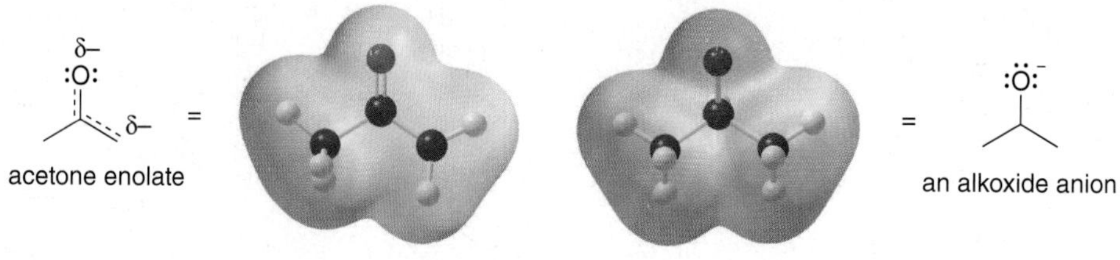

- The acetone enolate is resonance stabilized. The negative charge is delocalized on the oxygen atom (pale red) and the carbon atom (pale green).
- The alkoxide anion is *not* resonance stabilized. The negative charge is concentrated on the oxygen atom only (deep red).

23.3A Examples of Enolates and Related Anions

In addition to enolates from aldehydes and ketones, **enolates from esters and 3° amides can be formed,** although the α hydrogen is somewhat less acidic. **Nitriles** also have acidic protons on the carbon atom adjacent to the cyano group, because the negative charge of the conjugate base is stabilized by delocalization onto an electronegative nitrogen atom.

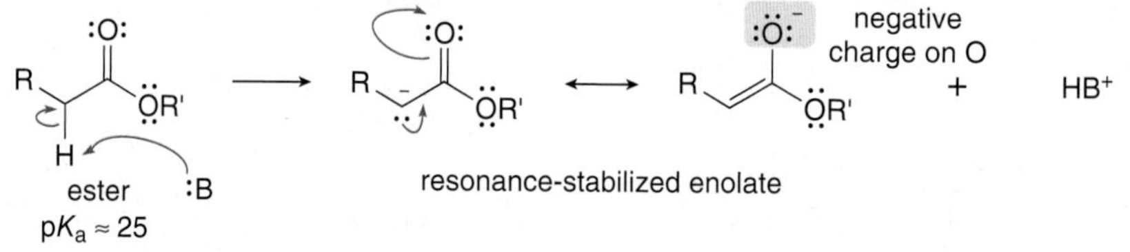

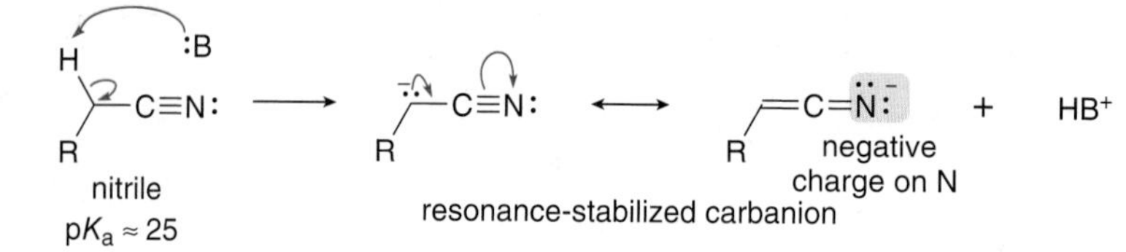

The protons on the carbon between the two carbonyl groups of a β-dicarbonyl compound are especially acidic because resonance delocalizes the negative charge on two different oxygen atoms. Table 23.2 lists pK_a values for β-dicarbonyl compounds as well as other carbonyl compounds and nitriles.

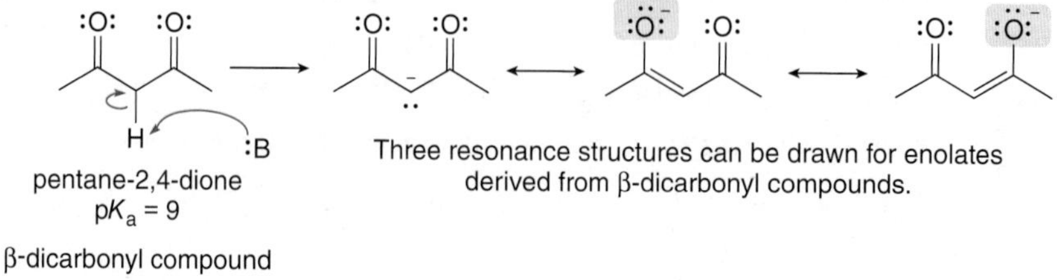

Problem 23.4 Draw additional resonance structures for each anion.

a. 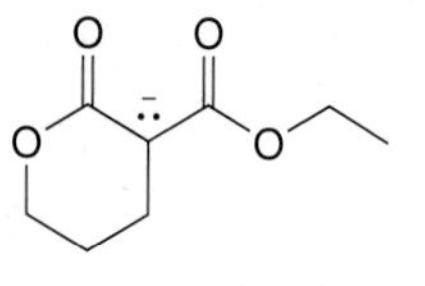

b. 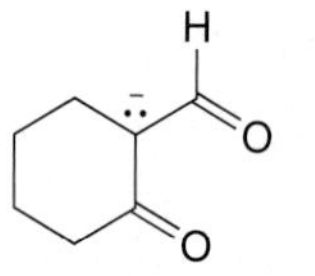

c. 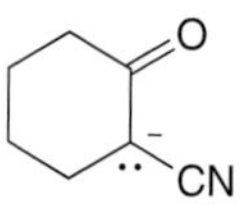

Table 23.2 pK_a Values for Some Carbonyl Compounds and Nitriles

Compound type	Example	pK_a	Compound type	Example	pK_a
[1] Amide	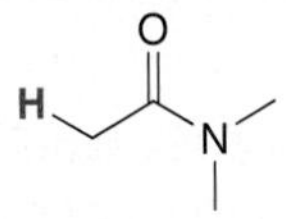	30	[6] 1,3-Diester	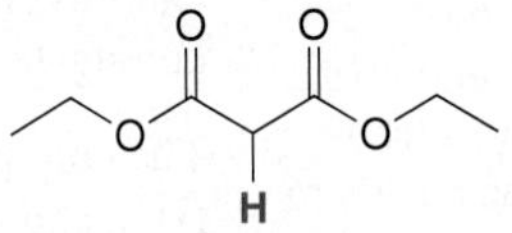	13.3
[2] Nitrile	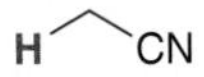	25	[7] 1,3-Dinitrile	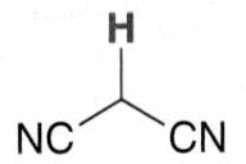	11
[3] Ester	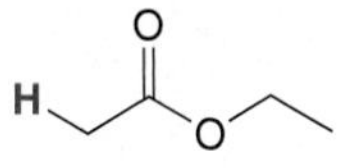	25	[8] β-Keto ester	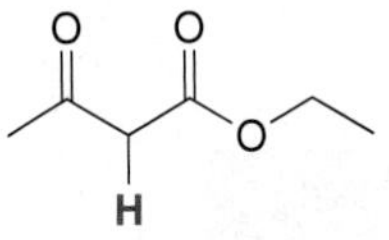	10.7
[4] Ketone	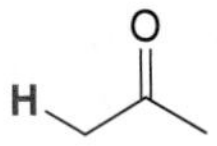	19.2	[9] β-Diketone	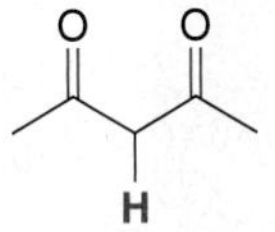	9
[5] Aldehyde	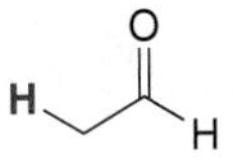	17			

Problem 23.5 Which C–H bonds in the following molecules are acidic because the resulting conjugate base is resonance stabilized?

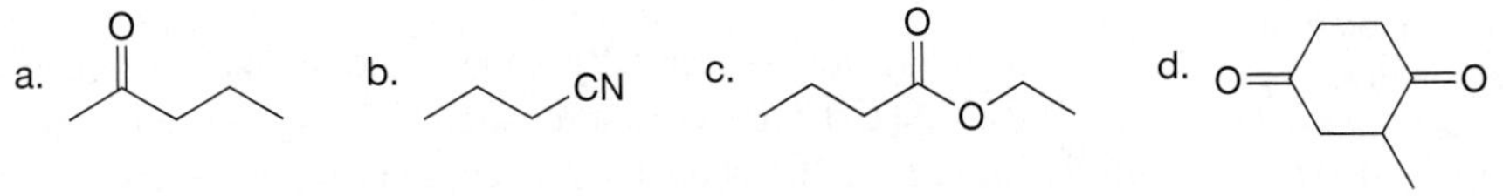

Problem 23.6 Rank the protons in the labeled CH_2 groups in order of increasing acidity, and explain why you chose this order.

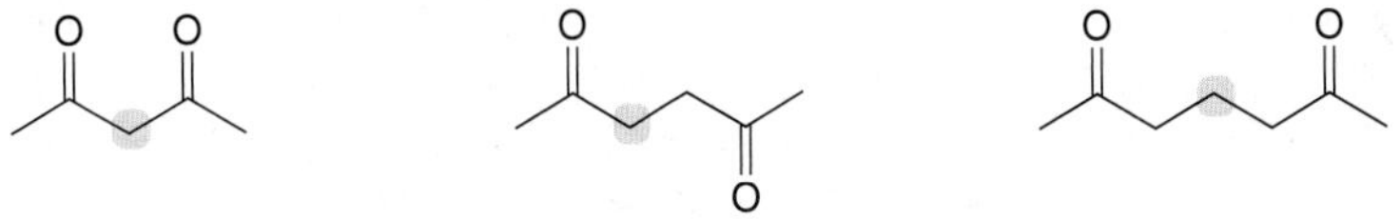

23.3B The Base

The formation of an enolate is an acid–base equilibrium, so the **stronger the base, the more enolate that forms.**

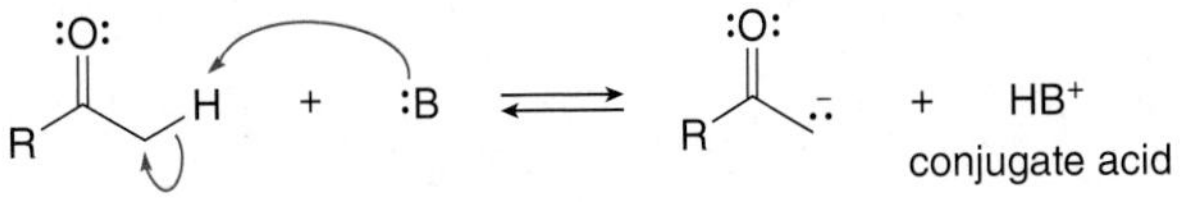

acid
pK_a ≈ 20

Stronger bases drive the equilibrium to the right.

We have now used the term ***amide*** in two different ways—first as a functional group (e.g., the carboxylic acid derivative $RCONH_2$) and now as a base (e.g., $^{-}NH_2$, which can be purchased as a sodium or lithium salt, $NaNH_2$ or $LiNH_2$, respectively). In Chapter 23 we will use dialkylamides, $^{-}NR_2$, in which the two H atoms of $^{-}NH_2$ have been replaced by R groups.

We can predict the extent of an acid–base reaction by comparing the pK_a of the starting acid (the carbonyl compound in this case) with the pK_a of the conjugate acid formed. **The equilibrium favors the side with the weaker acid (the acid with the higher pK_a value).** The pK_a of many carbonyl compounds is ~20, so a significant amount of enolate will form only if the pK_a of the conjugate acid is > 20.

The common bases used to form enolates are hydroxide (^{-}OH), various alkoxides (^{-}OR), hydride (H^{-}), and dialkylamides ($^{-}NR_2$). How much enolate is formed using each of these bases is indicated in Table 23.3.

Table 23.3 Enolate Formation with Various Bases: $RCOCH_3$ (pK_a ≈ 20) + B: → $RCOCH_2^{-}$ + HB^{+}

	Base (B:)	Conjugate acid (HB^{+})	pK_a of HB^{+}	% Enolate
[1]	$Na^{+}\,^{-}OH$	H_2O	15.7	< 1%
[2]	$Na^{+}\,^{-}OCH_2CH_3$	CH_3CH_2OH	16	< 1%
[3]	$K^{+}\,^{-}OC(CH_3)_3$	$(CH_3)_3COH$	18	1–10%
[4]	$Na^{+}H^{-}$	H_2	35	100%
[5]	$Li^{+}\,^{-}N[CH(CH_3)_2]_2$	$HN[CH(CH_3)_2]_2$	40	100%

Enolate formation with LDA is typically carried out at –78 °C, a convenient temperature to maintain in the laboratory because it is the temperature at which dry ice (solid CO_2) sublimes. A low-temperature cooling bath can be made by adding dry ice to acetone until the acetone cools to –78 °C. Immersing a reaction flask in this cooling bath keeps its contents at a constant low temperature.

When the pK_a of the conjugate acid is < 20, as it is for ^{-}OH and all ^{-}OR (entries 1–3), only a small amount of enolate is formed at equilibrium. These bases are more useful in forming enolates when more acidic 1,3-dicarbonyl compounds are used as starting materials. They are also used when both the enolate and the carbonyl starting material are involved in the reaction, as is the case for reactions described in Chapter 24.

To form an enolate in essentially 100% yield, a much stronger base such as lithium diisopropylamide, **$Li^{+}\,^{-}N[CH(CH_3)_2]_2$**, abbreviated as **LDA,** is used (entry 5). **LDA is a strong nonnucleophilic base.** Like the other nonnucleophilic bases (Sections 7.8B and 8.1), its bulky isopropyl groups make the nitrogen atom too hindered to serve as a nucleophile. It is still able, though, to remove a proton in an acid–base reaction.

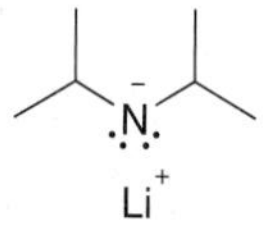

lithium diisopropylamide

LDA

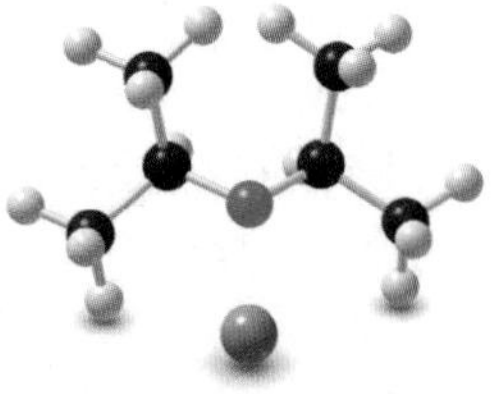

The N atom is too crowded to be a nucleophile.

THF
tetrahydrofuran
a polar aprotic solvent

LDA quickly deprotonates essentially all of the carbonyl starting material, even at –78 °C, to form the enolate product. THF is the typical solvent for these reactions.

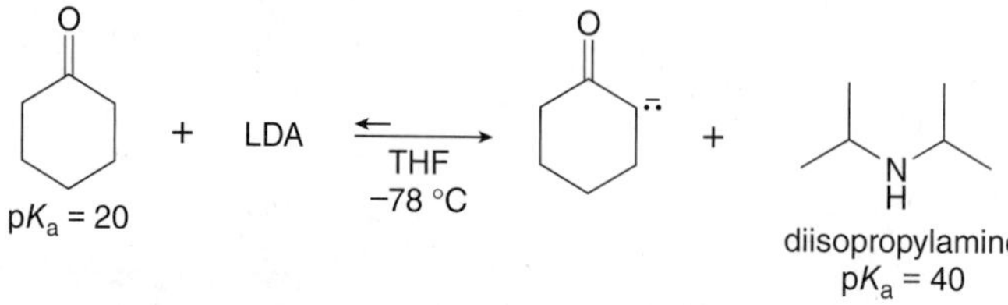

Equilibrium greatly favors the products.
Essentially all of the ketone is converted to enolate.

LDA can be prepared by deprotonating diisopropylamine with an organolithium reagent such as butyllithium, and then used immediately in a reaction.

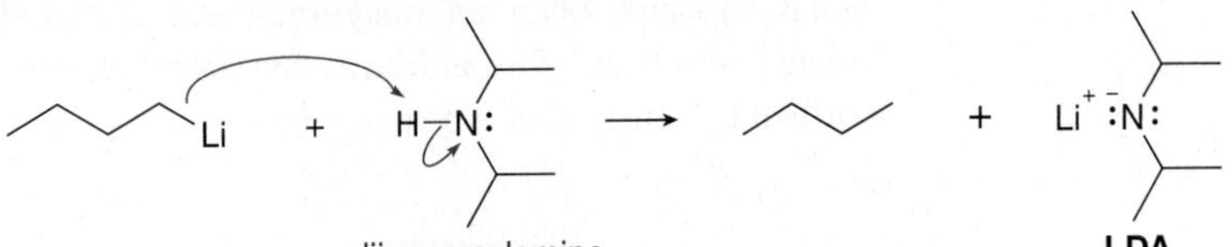

Problem 23.7 Draw the product formed when each starting material is treated with LDA in THF solution at –78 °C.

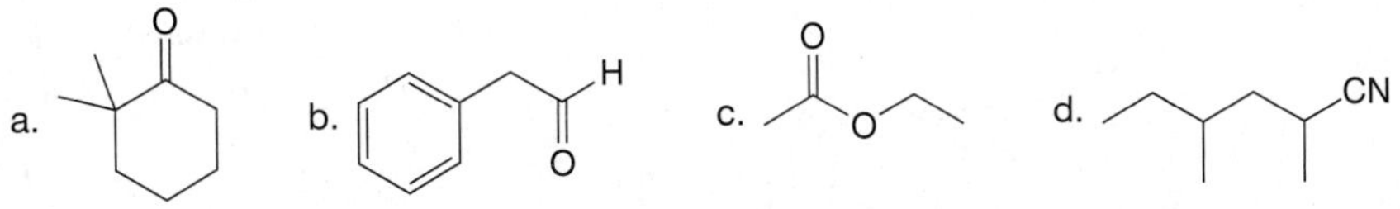

Problem 23.8 As we learned in Chapter 20, organolithium reagents (RLi) are strong bases that readily react with acidic protons. Why aren't organolithium reagents used to generate enolates?

23.3C General Reactions of Enolates

Enolates are nucleophiles, and as such they react with many electrophiles. Because an enolate is resonance stabilized, however, it has two reactive sites—the carbon and oxygen atoms that bear the negative charge. **A nucleophile with two reactive sites is called an *ambident nucleophile.*** In theory, each of these atoms could react with an electrophile to form two different products, one with a new bond to carbon, and one with a new bond to oxygen.

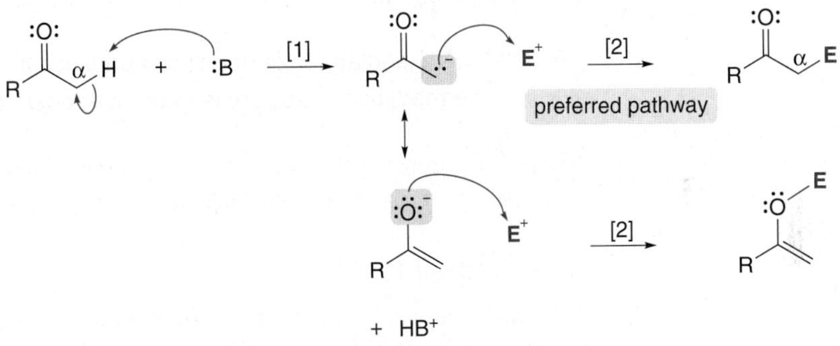

Because enolates usually react at carbon instead of oxygen, the resonance structure that places the negative charge on oxygen will often be omitted in multistep mechanisms.

An enolate usually reacts at the carbon end, however, because this site is more nucleophilic. Thus, **enolates generally react with electrophiles on the α carbon,** so that many reactions in Chapter 23 follow a two-step path:

[1] Reaction of a carbonyl compound with base forms an enolate.
[2] Reaction of the enolate with an electrophile forms a new bond on the α carbon.

23.4 Enolates of Unsymmetrical Carbonyl Compounds

What happens when an unsymmetrical carbonyl compound like 2-methylcyclohexanone is treated with base? **Two enolates are possible,** one formed by removal of a 2° hydrogen, and one formed by removal of a 3° hydrogen.

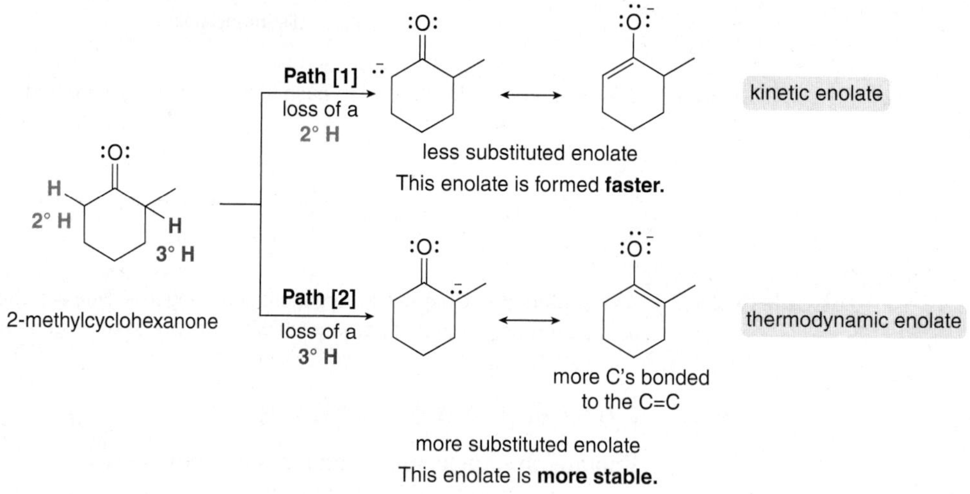

Path [1] occurs *faster* than Path [2] because it results in removal of the less hindered 2° hydrogen, forming an enolate on the less substituted α carbon. Path [2] results in removal of a 3° hydrogen, forming the *more stable* enolate with the more substituted double bond. This enolate predominates at equilibrium.

- The kinetic enolate is formed faster because it is the less substituted enolate.
- The thermodynamic enolate is lower in energy because it is the more substituted enolate.

It is possible to regioselectively form one or the other enolate by the proper use of reaction conditions, because the base, solvent, and reaction temperature all affect the identity of the enolate formed.

Kinetic Enolates

The kinetic enolate forms faster, so mild reaction conditions favor it over slower processes with higher energies of activation. It is the less stable enolate, so it must not be allowed to equilibrate to the more stable thermodynamic enolate. **The kinetic enolate is favored by:**

[1] **A strong nonnucleophilic base.** A strong base ensures that the enolate is formed rapidly. A **bulky base like LDA removes the more accessible proton on the less substituted carbon** much faster than a more hindered proton.

[2] **Polar aprotic solvent.** The solvent must be polar to dissolve the polar starting materials and intermediates. It must be aprotic so that it does not protonate any enolate that is formed. **THF** is both polar and aprotic.

[3] **Low temperature.** The temperature must be low (**–78 °C**) to prevent the kinetic enolate from equilibrating to the thermodynamic enolate.

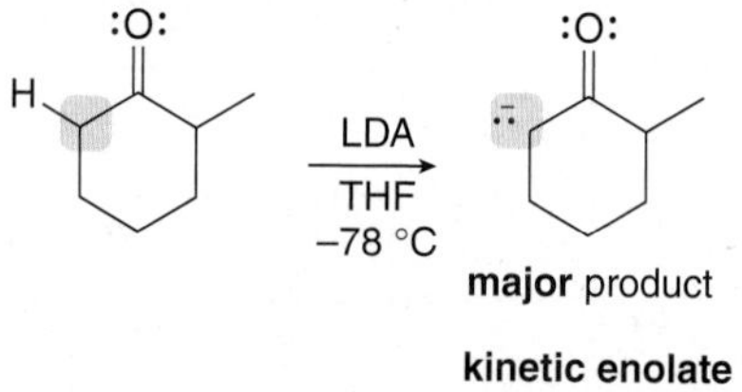

- A kinetic enolate is formed with a strong, nonnucleophilic base (LDA) in a polar aprotic solvent (THF) at low temperature (–78 °C).

Thermodynamic Enolates

A thermodynamic enolate is favored by equilibrating conditions. This is often achieved using a **strong base in a protic solvent.** A strong base yields both enolates, but in a protic solvent, enolates can also be protonated to re-form the carbonyl starting material. At equilibrium, the lower energy intermediate always wins out, so that **the more stable, more substituted enolate is present in higher concentration.** Thus, the **thermodynamic enolate is favored by:**

[1] **A strong base.** $\mathbf{Na^+ \, ^-OCH_2CH_3}$, $\mathbf{K^+ \, ^-OC(CH_3)_3}$, or other alkoxides are common.
[2] **Protic solvent.** $\mathbf{CH_3CH_2OH}$ or other alcohols.
[3] **Room temperature (25 °C).**

To simplify structures, we use abbreviations:
Me = CH_3, so $NaOCH_3$ = NaOMe
Et = CH_2CH_3, so $NaOCH_2CH_3$ = NaOEt
*t*Bu = $C(CH_3)_3$, so $KOC(CH_3)_3$ = KO*t*Bu

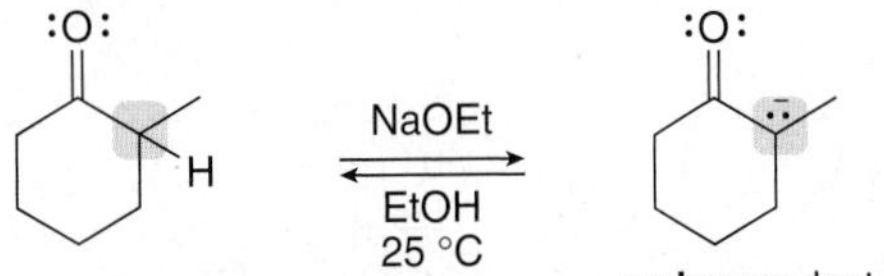

major product

thermodynamic enolate

- **A thermodynamic enolate is formed with a strong base (RO⁻) in a polar protic solvent (ROH) at room temperature.**

Sample Problem 23.2 What is the major enolate formed in each reaction?

a. 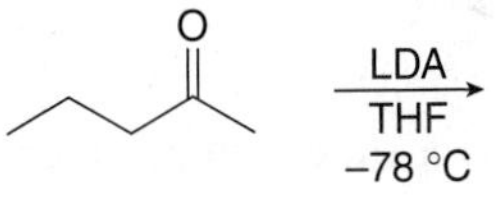

b. 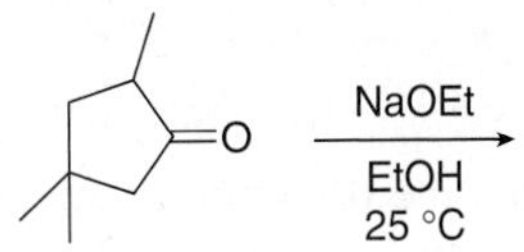

Solution

a. LDA is a strong, nonnucleophilic base that removes a proton on the less substituted α carbon to form the kinetic enolate.

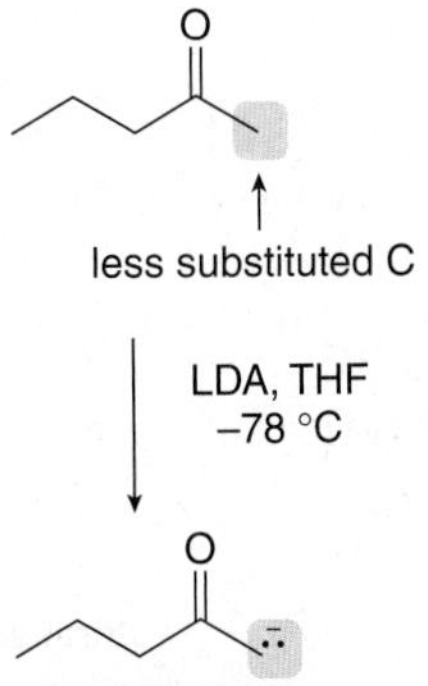

kinetic enolate

b. $NaOCH_2CH_3$ (a strong base) and CH_3CH_2OH (a protic solvent) favor removal of a proton from the more substituted α carbon to form the thermodynamic enolate.

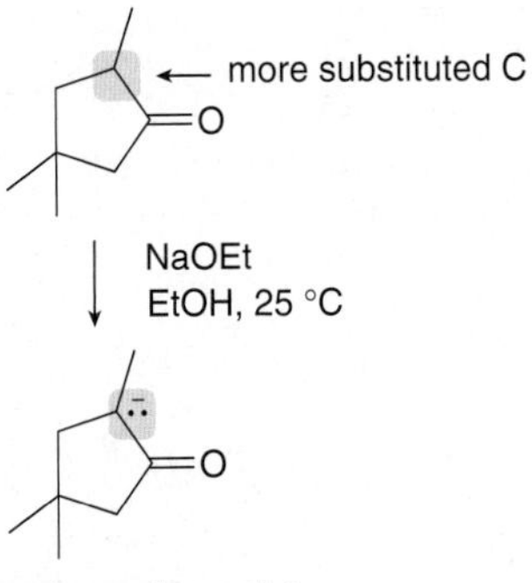

thermodynamic enolate

Problem 23.9 What enolate is formed when each ketone is treated with LDA in THF solution? What enolate is formed when these same ketones are treated with $NaOCH_3$ in CH_3OH solution?

a. 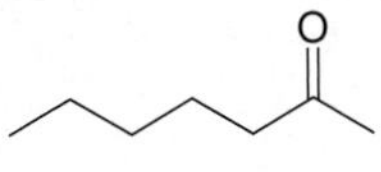

b. 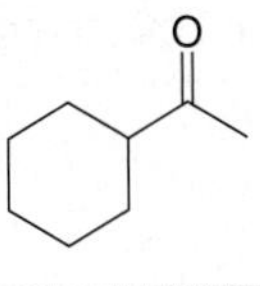

c. 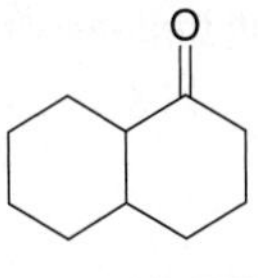

23.5 Racemization at the α Carbon

Recall from Section 16.5 that an enolate can be stabilized by the delocalization of electron density only if it possesses the proper geometry and hybridization.

- **The electron pair on the carbon adjacent to the C=O must occupy a *p* orbital that overlaps with the two other *p* orbitals of the C=O, making an enolate conjugated.**

- **Thus, all three atoms of the enolate are *sp*^{2} hybridized and trigonal planar.**

These bonding features are shown in the acetone enolate in Figure 23.2.

Figure 23.2
The hybridization and geometry of the acetone enolate $(CH_3COCH_2)^-$

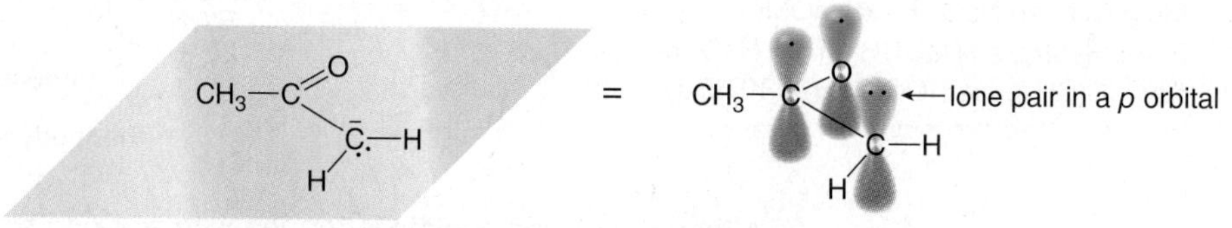

acetone enolate — three adjacent *p* orbitals

- The O atom and both C's of the enolate are *sp*^{2} hybridized and lie in a plane.
- Each atom has a *p* orbital extending above and below the plane; these orbitals overlap to delocalize electron density.

When the α carbon to the carbonyl is a stereogenic center, treatment with aqueous base leads to **racemization** by a two-step process: **deprotonation to form an enolate and protonation to re-form the carbonyl compound.** For example, chiral ketone **A** reacts with aqueous ^-OH to form an achiral enolate having an *sp*^{2} hybridized α carbon. Because the enolate is planar, it can be protonated with H_2O with equal probability from both directions, yielding a racemic mixture of two ketones.

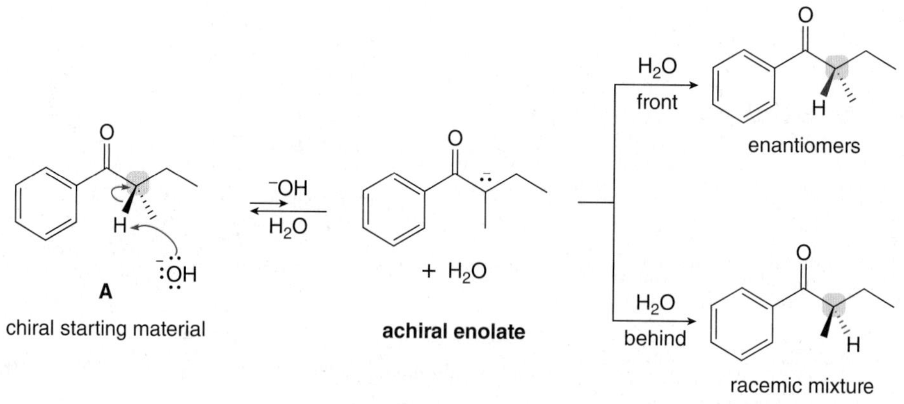

Problem 23.10 Explain each observation: (a) When (*R*)-2-methylcyclohexanone is treated with NaOH in H_2O, the optically active solution gradually loses optical activity. (b) When (*R*)-3-methylcyclohexanone is treated with NaOH in H_2O, the solution remains optically active.

23.6 A Preview of Reactions at the α Carbon

Having learned about the synthesis and properties of enolates, we can now turn our attention to their reactions. Like enols, **enolates are nucleophiles,** but because they are negatively charged, enolates are much more nucleophilic than neutral enols. Consequently, they undergo a wider variety of reactions.

Two general types of reactions of enolates—**substitutions** and **reactions with other carbonyl compounds**—will be discussed in the remainder of Chapter 23 and in Chapter 24. Both reactions form new bonds to the carbon α to the carbonyl.

- **Enolates react with electrophiles to afford substitution products.**

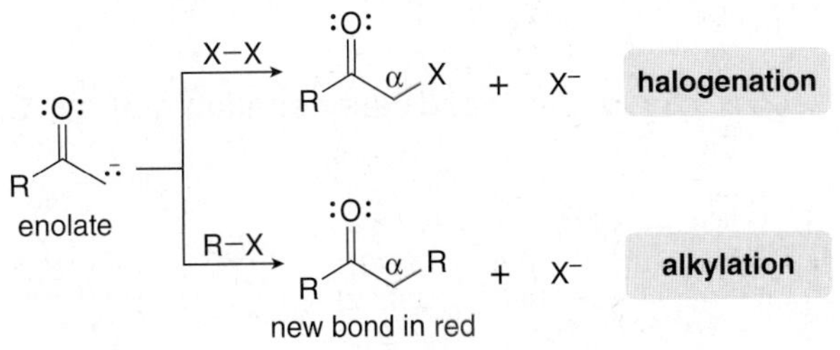

Two different kinds of substitution reactions are examined: **halogenation** with X_2 and **alkylation** with alkyl halides RX. These reactions are detailed in Sections 23.7–23.10.

- **Enolates react with other carbonyl groups at the electrophilic carbonyl carbon.**

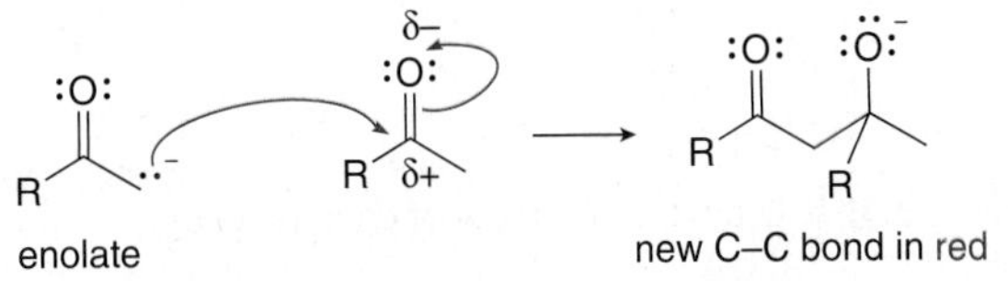

These reactions are more complicated because the initial addition adduct goes on to form different products depending on the structure of the carbonyl group. These reactions form the subject of Chapter 24.

23.7 Halogenation at the α Carbon

The first substitution reaction we examine is **halogenation.** Treatment of a ketone or aldehyde with halogen and either acid or base results in **substitution of X for H on the α carbon,** forming an **α-halo aldehyde or ketone.** Halogenation readily occurs with Cl_2, Br_2, and I_2.

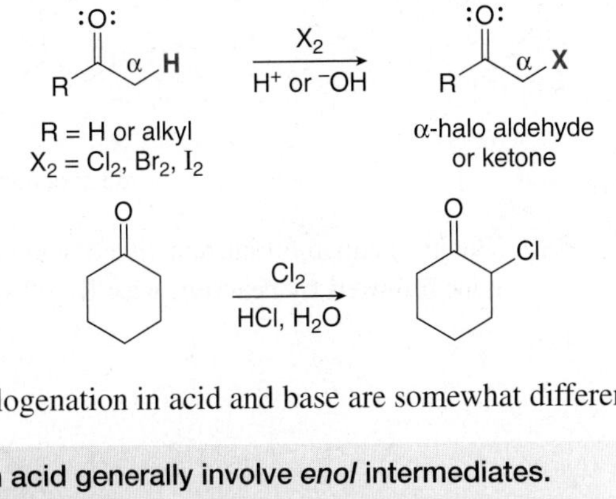

The mechanisms of halogenation in acid and base are somewhat different.

- **Reactions done in acid generally involve *enol* intermediates.**
- **Reactions done in base generally involve *enolate* intermediates.**

23.7A Halogenation in Acid

Halogenation is often carried out by treating a carbonyl compound with a halogen in acetic acid. In this way, acetic acid is both the solvent and the acid catalyst for the reaction.

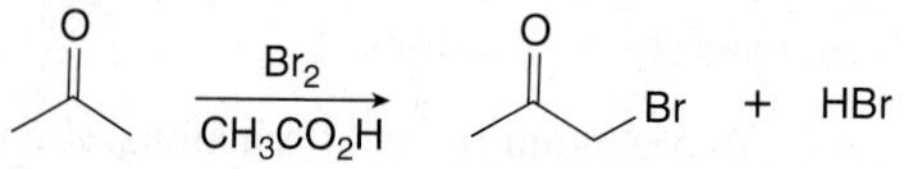

Br_2
CH_3CO_2H
Br
+ HBr

The mechanism of acid-catalyzed halogenation consists of two parts: **tautomerization** of the carbonyl compound to the enol form, and **reaction of the enol with halogen.** Mechanism 23.3 illustrates the reaction of $(CH_3)_2C{=}O$ with Br_2 in CH_3CO_2H.

Mechanism 23.3 Acid-Catalyzed Halogenation at the α Carbon

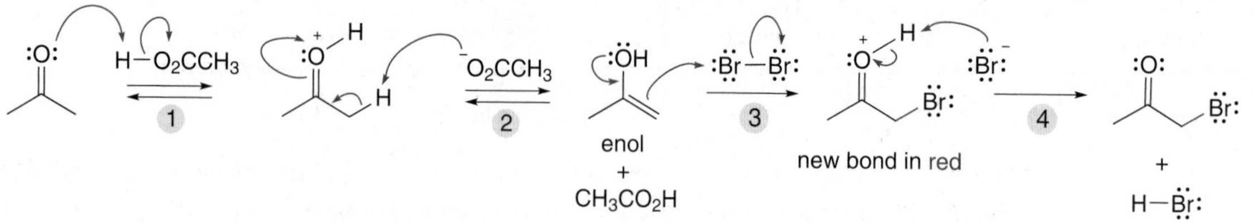

1
2
3
4
enol
+
CH_3CO_2H
new bond in red

1 – 2 The ketone is converted to its enol tautomer by the two-step process of protonation followed by deprotonation.

3 – 4 Addition of the halogen to the enol forms a new bond to Br on the α carbon, and deprotonation yields the substitution product.

Problem 23.11 Draw the products of each reaction.

a. 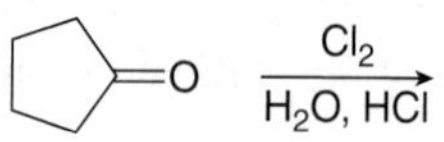

Cl_2
H_2O, HCl

b. 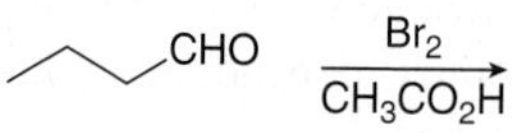

CHO
Br_2
CH_3CO_2H

c. 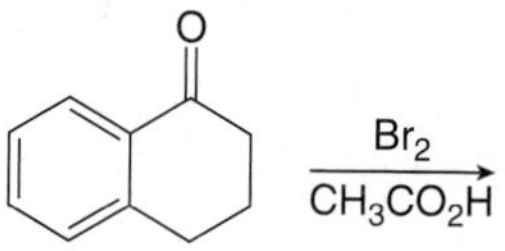

Br_2
CH_3CO_2H

23.7B Halogenation in Base

Halogenation in base is much less useful, because it is often difficult to stop the reaction after addition of just one halogen atom to the α carbon. For example, treatment of propiophenone with Br_2 and aqueous ^-OH yields a dibromo ketone.

Reactions of carbonyl compounds with base invariably involve enolates because the α hydrogens of the carbonyl compound are easily removed.

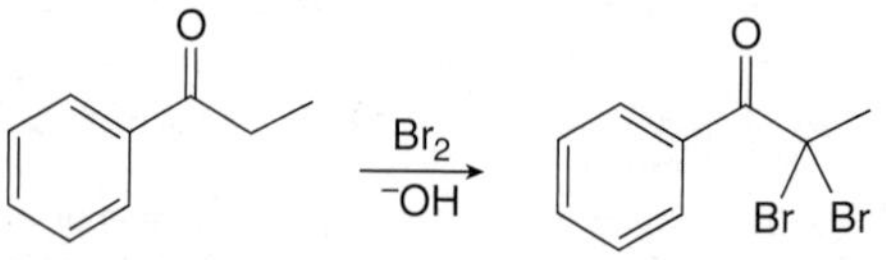

Br_2
^-OH
Br Br

propiophenone

The mechanism for introduction of each Br atom involves the same two steps: **deprotonation with base followed by reaction with Br_2** to form a new C–Br bond, as shown in Mechanism 23.4.

Mechanism 23.4 Halogenation at the α Carbon in Base

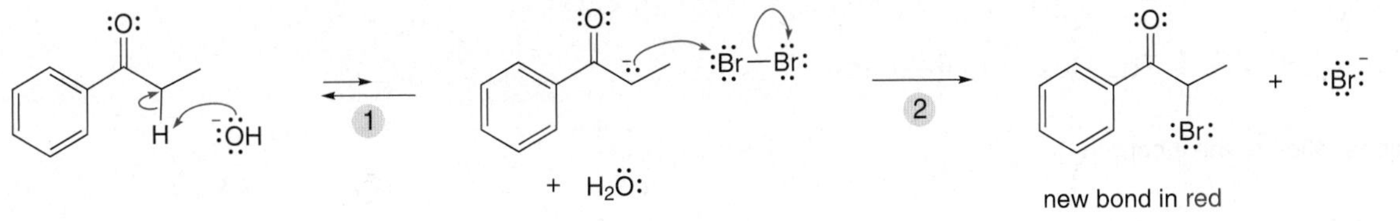

1 Treatment of the ketone with base forms a nucleophilic enolate.

2 Reaction of the enolate with Br_2 forms the substitution product in which one H is replaced by Br on the α carbon.

Only a small amount of the enolate forms at equilibrium using ^{-}OH as base, but the enolate is such a strong nucleophile that it readily reacts with Br_2, thus driving the equilibrium to the right. Then, the same two steps introduce the second Br atom on the α carbon: **deprotonation** followed by **nucleophilic attack.**

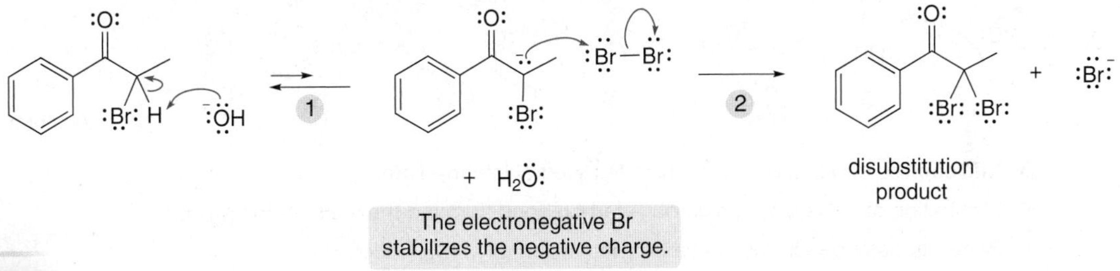

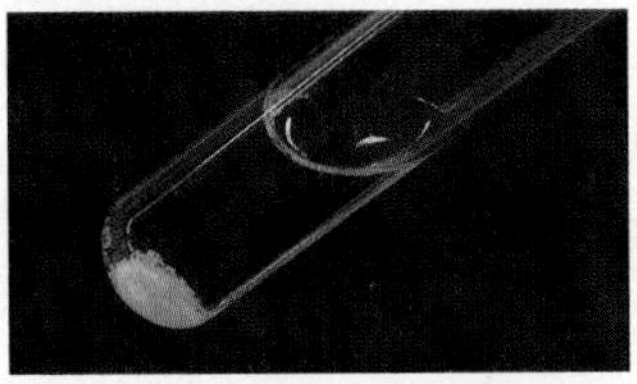

Although all ketones with α hydrogens react with base and I_2, only **methyl** ketones form CHI_3 (iodoform), a pale yellow solid that precipitates from the reaction mixture. This reaction is the basis of the **iodoform test,** once a common chemical method to detect methyl ketones. Methyl ketones give a positive iodoform test (appearance of a yellow solid), whereas other ketones give a negative iodoform test (no change in the reaction mixture).

It is difficult to stop this reaction after the addition of one Br atom because the electron-withdrawing inductive effect of Br stabilizes the second enolate. As a result, the α H of α-bromopropiophenone is more acidic than the α H atoms of propiophenone, making it easier to remove with base.

Halogenation of a methyl ketone with excess halogen, called the **haloform reaction,** results in cleavage of a carbon–carbon σ bond and formation of two products, a carboxylate anion and CHX_3 (commonly called **haloform**).

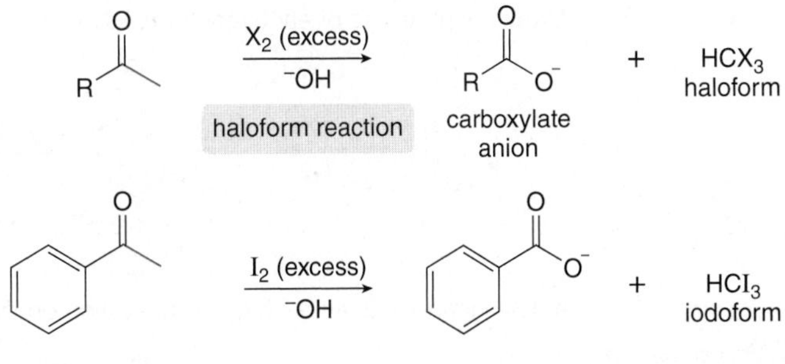

The C–C bond in red is broken.

In the haloform reaction, the three H atoms of the CH_3 group are successively replaced by X to form an intermediate that is oxidatively cleaved with base. Mechanism 23.5 is written with I_2 as halogen, forming CHI_3 (iodoform) as product.

Mechanism 23.5 The Haloform Reaction

Part [1] Halogenation at the α carbon

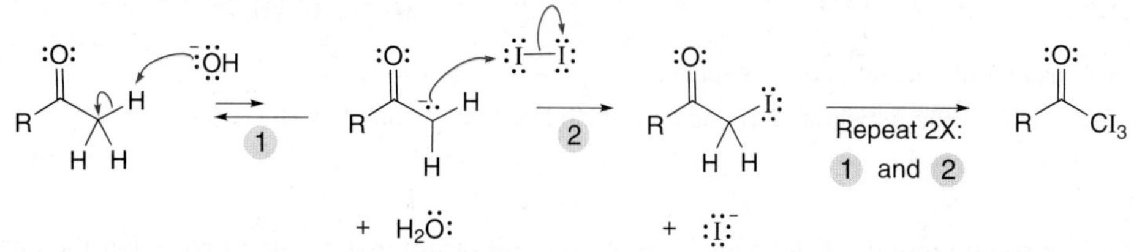

1 – 2 Removal of a proton forms an enolate that reacts with I_2 to yield an α-iodo ketone. Steps [1] and [2] are repeated twice more to form the triiodo substitution product.

Part [2] Oxidative cleavage with ^{-}OH

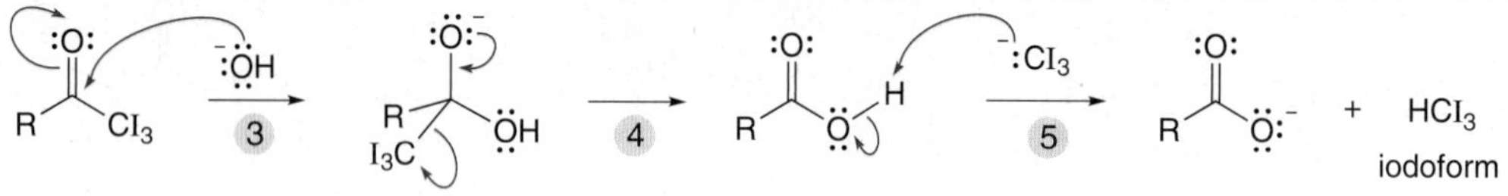

3 Nucleophilic addition of ^{-}OH forms a tetrahedral intermediate.

4 Elimination of $^{-}CI_3$ cleaves a carbon–carbon bond, forming the substitution product.

5 Proton transfer generates the carboxylate anion and HCI_3, iodoform.

Steps [3] and [4] result in a **nucleophilic *substitution*** reaction of a *ketone.* Because ketones normally undergo **nucleophilic *addition,*** this two-step sequence makes the haloform reaction unique. Substitution occurs because the three electronegative halogen atoms make CX_3 (CI_3 in the example) a good leaving group.

Figure 23.3 summarizes the three possible outcomes of halogenation at the α carbon, depending on the substrate and chosen reaction conditions.

Problem 23.12 Draw the products of each reaction. Assume excess halogen is present.

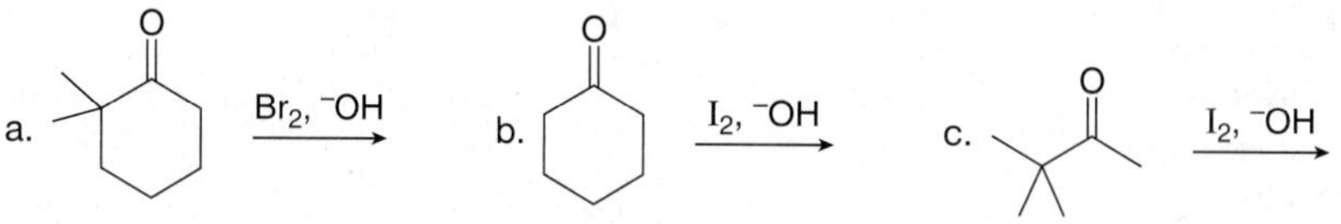

Figure 23.3 Summary: Halogenation reactions at the α carbon to a carbonyl group

a. Halogenation in acid—Monosubstitution on the α carbon

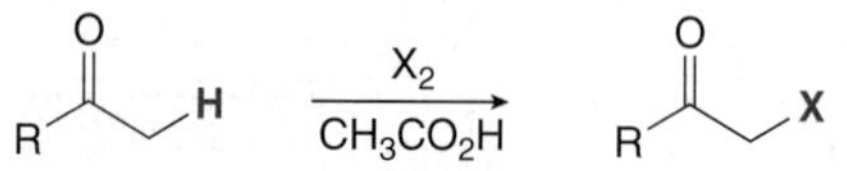

b. General halogenation in base—Polysubstitution on the α carbon

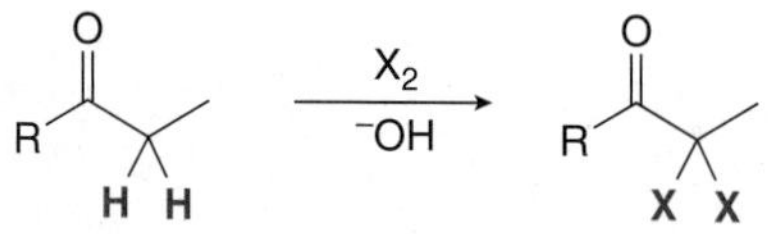

c. Halogenation of *methyl* ketones with excess X_2 and base—Oxidative cleavage

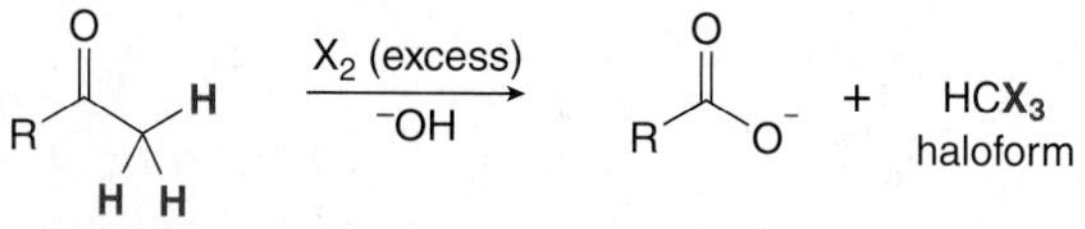

23.7C Reactions of α-Halo Carbonyl Compounds

α-Halo carbonyl compounds undergo two useful reactions—**elimination** with base and **substitution** with nucleophiles.

For example, treatment of 2-bromocyclohexanone with the base Li_2CO_3 in the presence of LiBr in the polar aprotic solvent DMF [$HCON(CH_3)_2$] affords cyclohex-2-enone by elimination of the elements of Br and H from the α and β carbons, respectively. Thus, a two-step method can convert a carbonyl compound such as cyclohexanone into an α,β-unsaturated carbonyl compound such as cyclohex-2-enone.

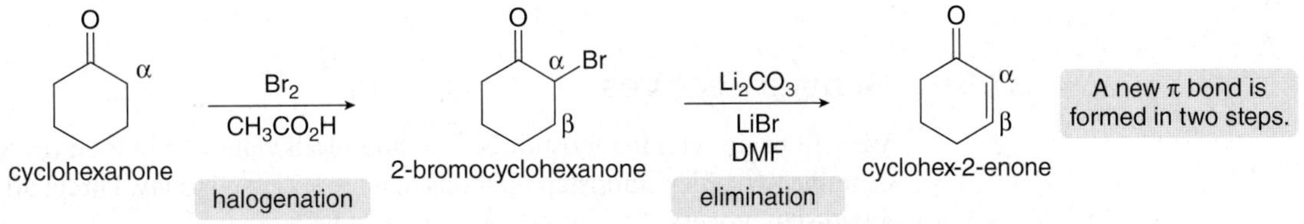

α,β-Unsaturated carbonyl compounds undergo a variety of 1,2- and 1,4-addition reactions as discussed in Section 20.15.

[1] Bromination at the α carbon is accomplished with Br_2 in CH_3CO_2H.
[2] Elimination of Br and H occurs with Li_2CO_3 and LiBr in DMF.

α-Halo carbonyl compounds also react with nucleophiles by S_N2 reactions. For example, reaction of 2-bromocyclohexanone with CH_3NH_2 affords the substitution product **A.** A related intramolecular nucleophilic substitution of an α-halo ketone was a key step in the synthesis of the antimalarial drug quinine, as shown in Figure 23.4.

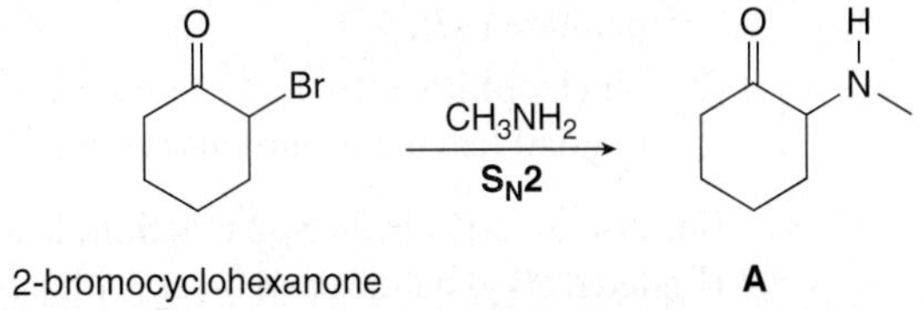

Problem 23.13 Draw the organic products formed when 2-bromopentan-3-one ($CH_3CH_2COCHBrCH_3$) is treated with each reagent: (a) Li_2CO_3, LiBr, DMF; (b) $CH_3CH_2NH_2$; (c) CH_3SH.

Problem 23.14 Identify the product **M** of the following two-step reaction sequence. **M** was converted to the hallucinogen LSD (Figure 18.4) in several steps.

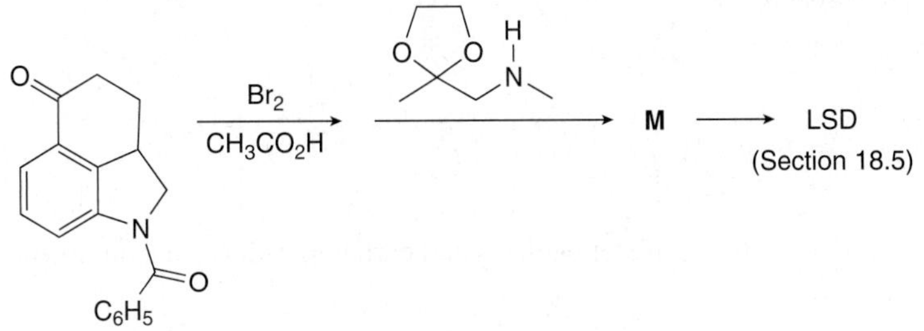

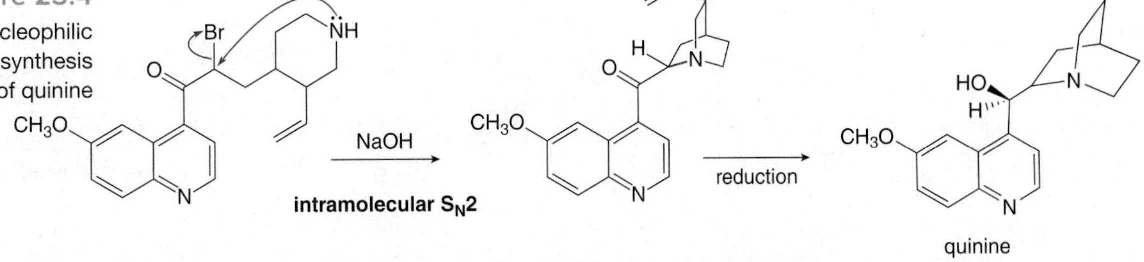

Figure 23.4 Intramolecular nucleophilic substitution in the synthesis of quinine

- Intramolecular S_N2 reaction of a nitrogen nucleophile with an α-halo ketone affords a compound that can be converted to quinine in a single step. The new C–N bond on the α carbon is labeled in red.

23.8 Direct Enolate Alkylation

Treatment of an aldehyde or ketone with base and an alkyl halide (RX) results in ***alkylation*—the substitution of R for H on the α carbon atom.** Alkylation forms a new carbon–carbon bond on the α carbon.

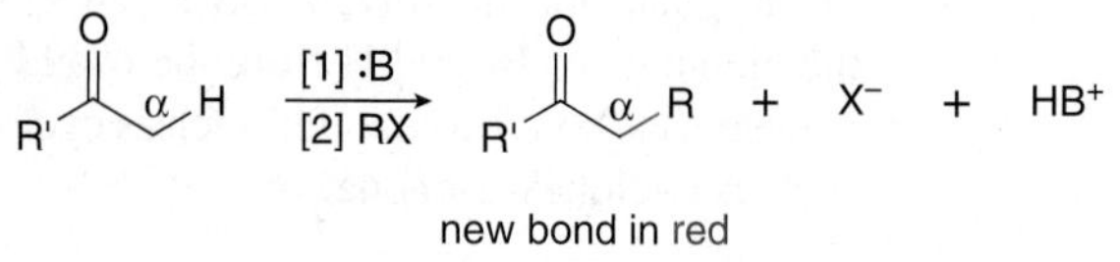

23.8A General Features

We will begin with the most direct method of alkylation, and then (in Sections 23.9 and 23.10) examine two older, multistep methods that are still used today. Direct alkylation is carried out by a two-step process:

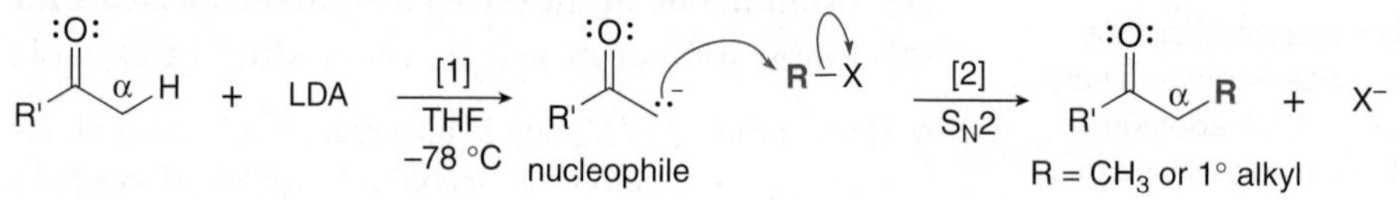

[1] **Deprotonation:** Base removes a proton from the α carbon to generate an enolate. The reaction works best with a strong nonnucleophilic base like LDA in THF solution at low temperature (–78 °C).

[2] **Nucleophilic attack:** The nucleophilic enolate attacks the alkyl halide, displacing the halide (a good leaving group) and forming the alkylation product by an S_N2 reaction.

Because Step [2] is an S_N2 reaction, it works best with unhindered methyl and 1° alkyl halides. Hindered alkyl halides and those with halogens bonded to sp^2 hybridized carbons do not undergo substitution.

R_3CX, $CH_2{=}CHX$, and C_6H_5X do *not* undergo alkylation reactions with enolates, because they are unreactive in S_N2 reactions.

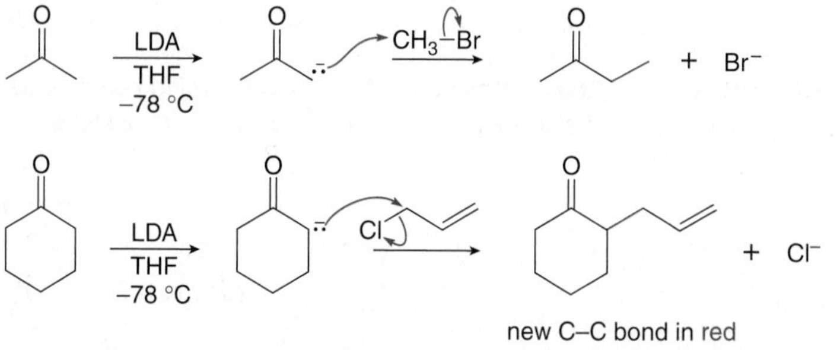

Ester enolates and carbanions derived from nitriles are also alkylated under these conditions.

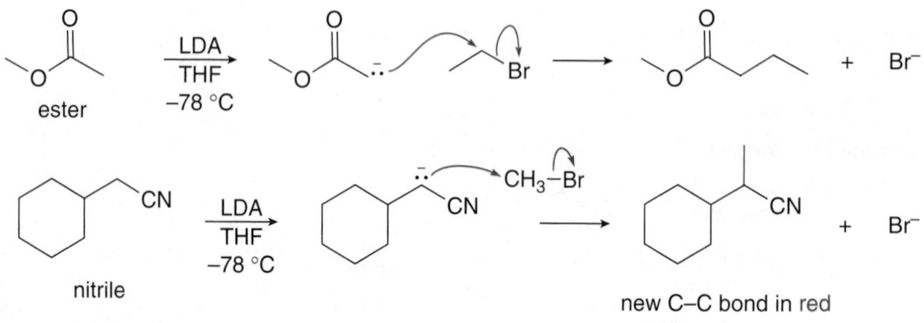

Problem 23.15 What product is formed when each compound is treated first with LDA in THF solution at low temperature, followed by CH_3CH_2I?

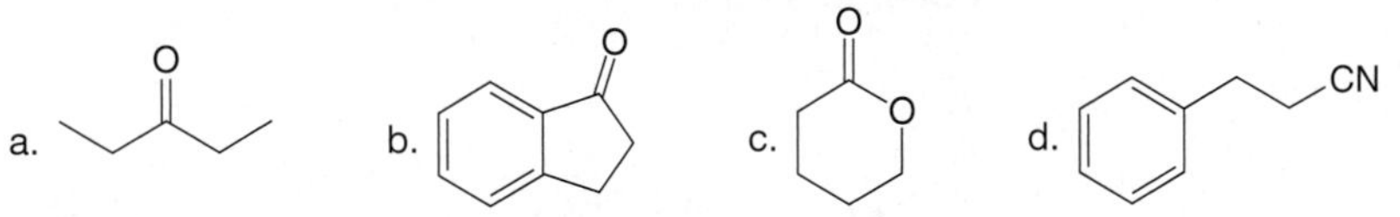

The stereochemistry of enolate alkylation follows the general rule governing the stereochemistry of reactions: **an achiral starting material yields an achiral or racemic product.** For example, when cyclohexanone (an achiral starting material) is converted to 2-ethylcyclohexanone by treatment with base and CH_3CH_2I, a new stereogenic center (labeled in blue) is introduced, and both enantiomers of the product are formed in equal amounts—that is, a **racemic mixture.**

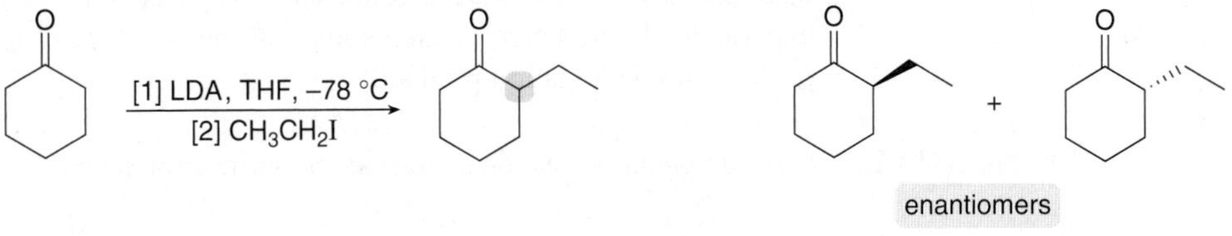

Problem 23.16 Draw the products obtained (including stereochemistry) when each compound is treated with LDA, followed by CH_3I.

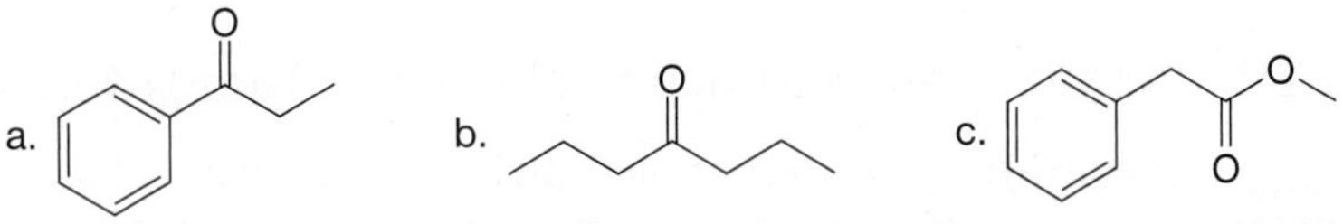

Problem 23.17 The analgesic naproxen can be prepared by a stepwise reaction sequence from ester **A.** Using enolate alkylation in one step, what reagents are needed to convert **A** to naproxen? Write the structure of each intermediate. Explain why a racemic product is formed.

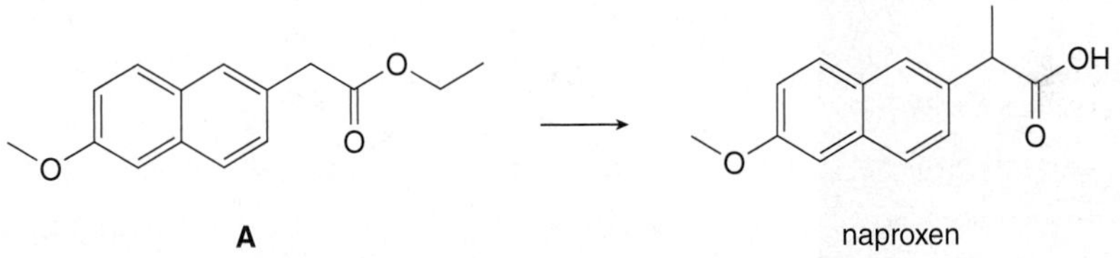

23.8B Alkylation of Unsymmetrical Ketones

An unsymmetrical ketone can be regioselectively alkylated to yield one major product. The strategy depends on the use of the appropriate base, solvent, and temperature to form the kinetic or thermodynamic enolate (Section 23.4), which is then treated with an alkyl halide to form the alkylation product.

For example, 2-methylcyclohexanone can be converted to either 2,6-dimethylcyclohexanone (**A**) or 2,2-dimethylcyclohexanone (**B**) by proper choice of reaction conditions.

- Treatment of 2-methylcyclohexanone with LDA in THF solution at –78 °C gives the less substituted kinetic enolate, which then reacts with CH_3I to form A.

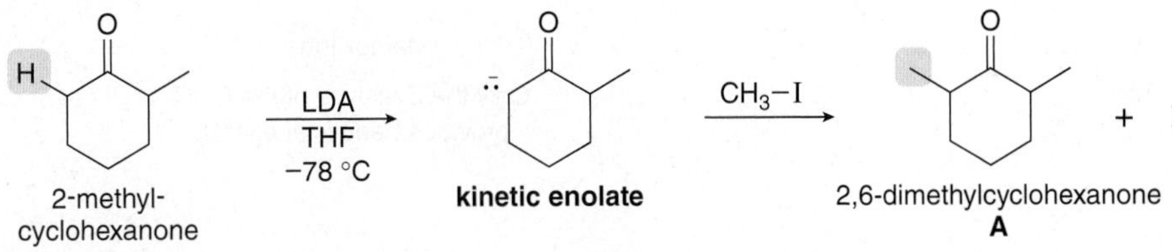

- **Treatment of 2-methylcyclohexanone with $KOC(CH_3)_3$ in $(CH_3)_3COH$ solution at room temperature forms the more substituted thermodynamic enolate, which then reacts with CH_3I to form B.**

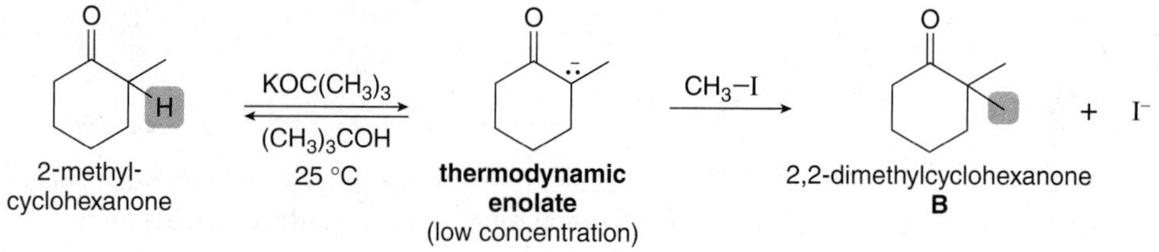

Finally, while enolate alkylation at the less substituted α carbon using LDA is a reliable regioselective reaction, enolate alkylation at the more substituted α carbon with $KOC(CH_3)_3$ may lead to mixtures of products. Regioselectivity depends on the identity of the substrate and the experimental parameters, which sometimes must be carefully monitored to maximize the yield of the desired alkylation product.

Problem 23.18 How can pentan-2-one be converted into each compound?

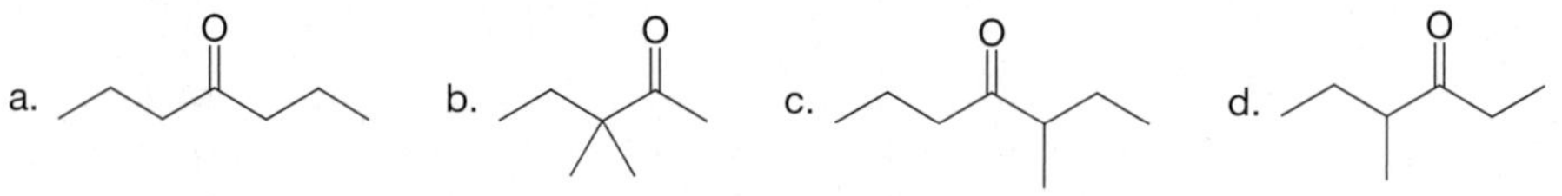

23.8C Application of Enolate Alkylation: Tamoxifen Synthesis

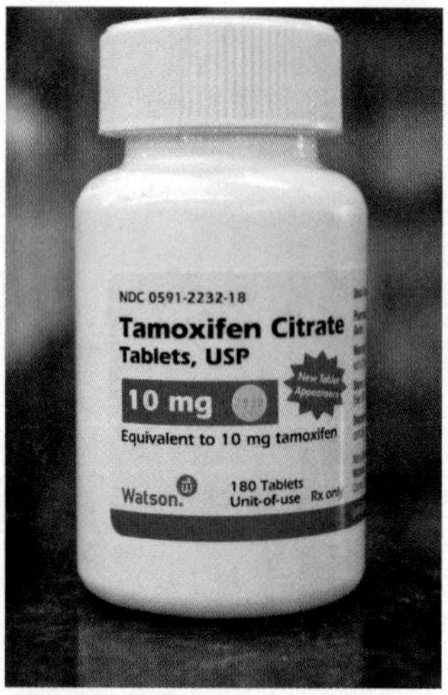

Tamoxifen has been commercially available since the 1970s, sold under the brand name of Nolvadex.

Tamoxifen, the chapter-opening molecule, is a potent anticancer drug that has been used to treat certain forms of breast cancer for many years. One step in the synthesis of tamoxifen involves the treatment of ketone **A** with NaH as base to form an enolate. Alkylation of this enolate with CH_3CH_2I forms **B** in high yield. **B** is converted to tamoxifen in several steps, some of which are reactions you have already learned.

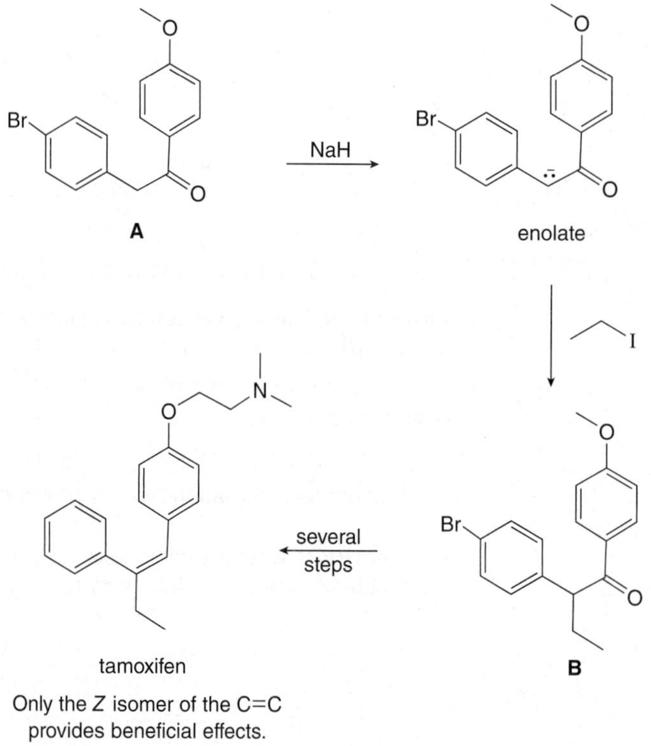

Problem 23.19 Identify **A, B,** and **C,** intermediates in the synthesis of the five-membered ring called an α-methylene-γ-butyrolactone. This heterocyclic ring system is present in some antitumor agents.

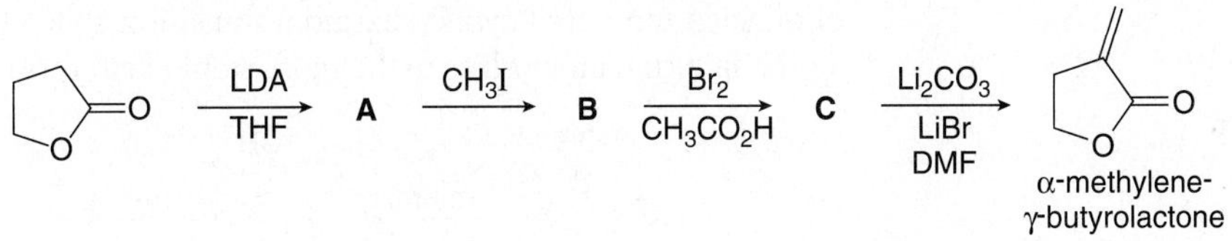

23.9 Malonic Ester Synthesis

Besides the direct method of enolate alkylation discussed in Section 23.8, a new alkyl group can be introduced on the α carbon using the malonic ester synthesis and the acetoacetic ester synthesis.

- **The malonic ester synthesis prepares carboxylic acids** having two general structures:

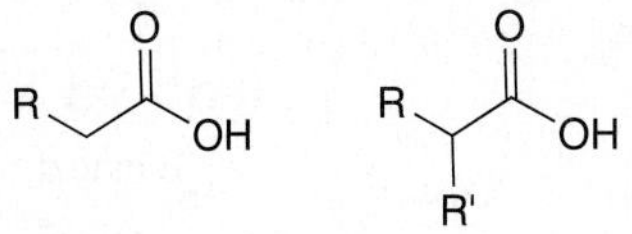

- **The acetoacetic ester synthesis prepares methyl ketones** having two general structures:

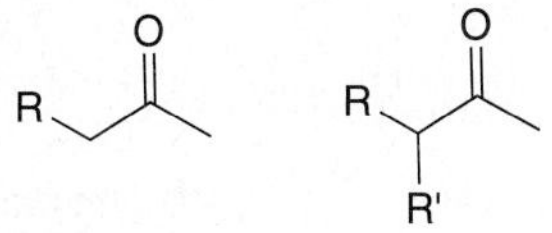

23.9A Background for the Malonic Ester Synthesis

- **The malonic ester synthesis is a stepwise method for converting diethyl malonate into a carboxylic acid having one or two alkyl groups on the α carbon.**

To simplify the structures, the CH_3CH_2 groups of the esters are abbreviated as Et.

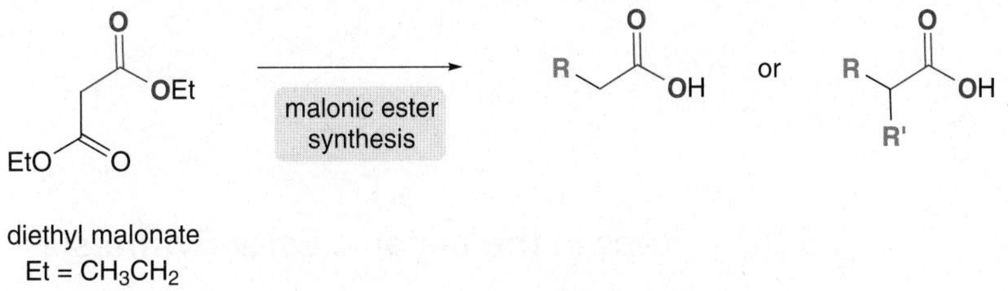

Before writing out the steps in the malonic ester synthesis, recall from Section 22.11 that esters are hydrolyzed by aqueous acid. Thus, heating diethyl malonate with acid and water hydrolyzes both esters to carboxy groups, forming a β-diacid (1,3-diacid).

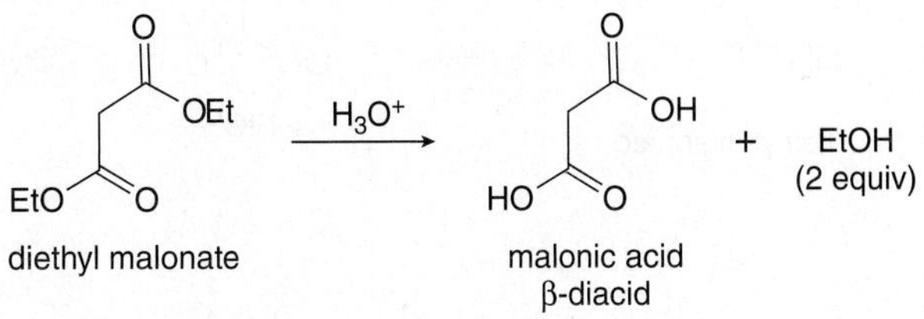

The resulting β-diacids are unstable to heat. They **decarboxylate (lose CO_2),** resulting in cleavage of a carbon–carbon bond and formation of a carboxylic acid. Decarboxylation is not a general reaction of all carboxylic acids. It occurs with β-diacids, however, because CO_2 can be eliminated through a cyclic, six-atom transition state. This forms an enol of a carboxylic acid, which in turn tautomerizes to the more stable keto form.

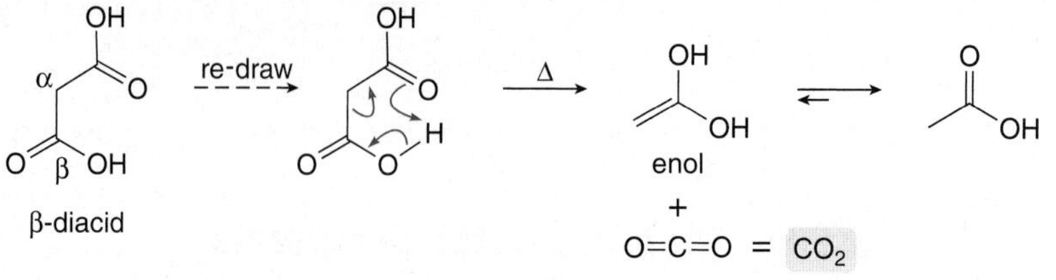

The net result of decarboxylation is cleavage of a carbon–carbon bond on the α carbon, with loss of CO_2.

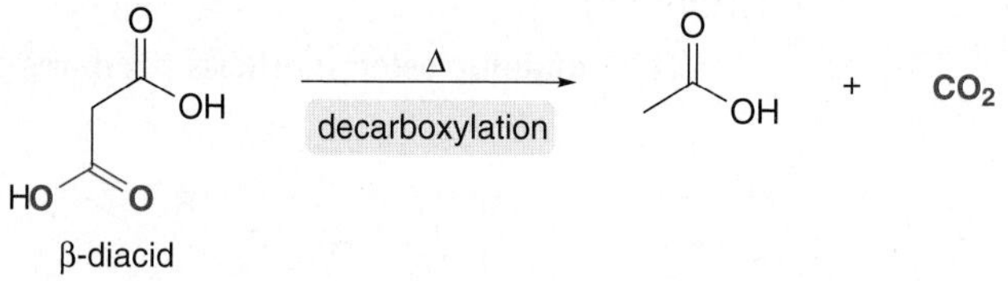

Decarboxylation occurs readily whenever a carboxy group (COOH) is bonded to the α carbon of another carbonyl group. For example, β-keto acids also readily lose CO_2 on heating to form ketones.

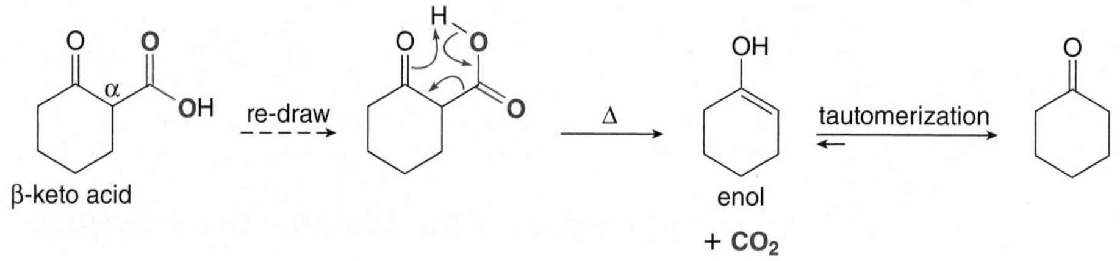

Problem 23.20 Which of the following compounds will readily lose CO_2 when heated?

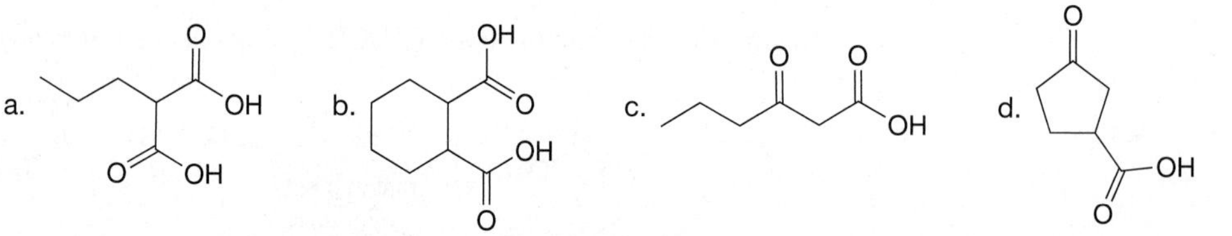

23.9B Steps in the Malonic Ester Synthesis

The malonic ester synthesis converts diethyl malonate to a carboxylic acid in three steps.

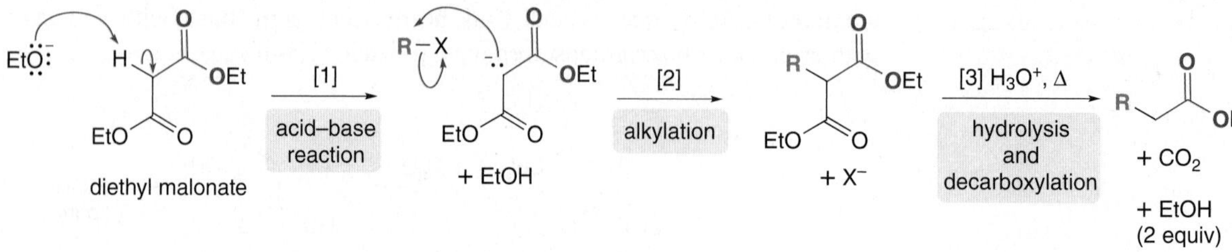

[1] **Deprotonation.** Treatment of diethyl malonate with $^{-}$OEt removes the acidic α proton between the two carbonyl groups. Recall from Section 23.3A that these protons are more acidic than other α protons because three resonance structures can be drawn for the enolate, instead of the usual two. Thus, $^{-}$OEt, rather than the stronger base LDA, can be used for this reaction.

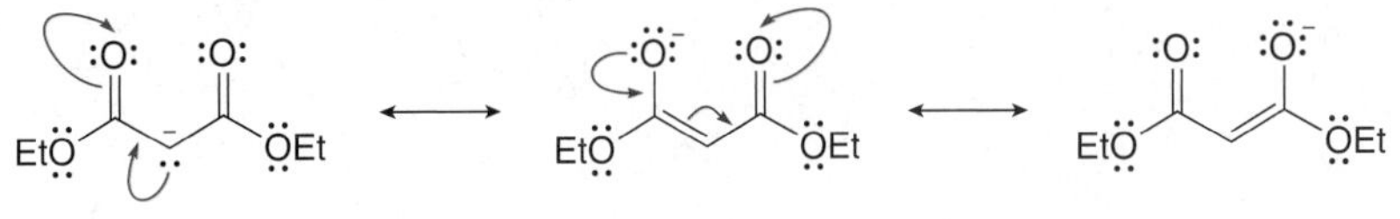

:O:
:O:
EtÖ
ÖEt

three resonance structures for the conjugate base

[2] **Alkylation.** The nucleophilic enolate reacts with an alkyl halide in an **S_N2** reaction to form a substitution product. Because the mechanism is S_N2, the yields are higher when R is CH_3 or a 1° alkyl group.

[3] **Hydrolysis and decarboxylation.** Heating the diester with aqueous acid hydrolyzes the diester to a β-diacid, which loses CO_2 to form a carboxylic acid.

The synthesis of butanoic acid ($CH_3CH_2CH_2COOH$) from diethyl malonate illustrates the basic process:

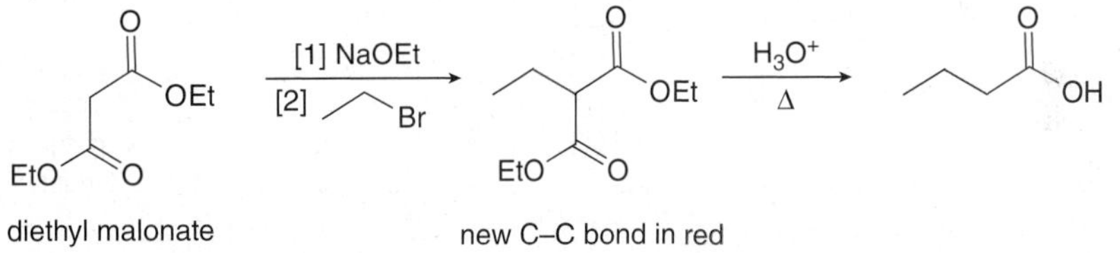

OEt
EtO
O
[1] NaOEt
[2]
Br
H_3O^+
Δ
OH
diethyl malonate
new C–C bond in red

If the first two steps of the reaction sequence are repeated *prior* to hydrolysis and decarboxylation, then a carboxylic acid having *two new alkyl groups* on the α carbon can be synthesized. This is illustrated in the synthesis of 2-benzylbutanoic acid [$CH_3CH_2CH(CH_2C_6H_5)COOH$] from diethyl malonate:

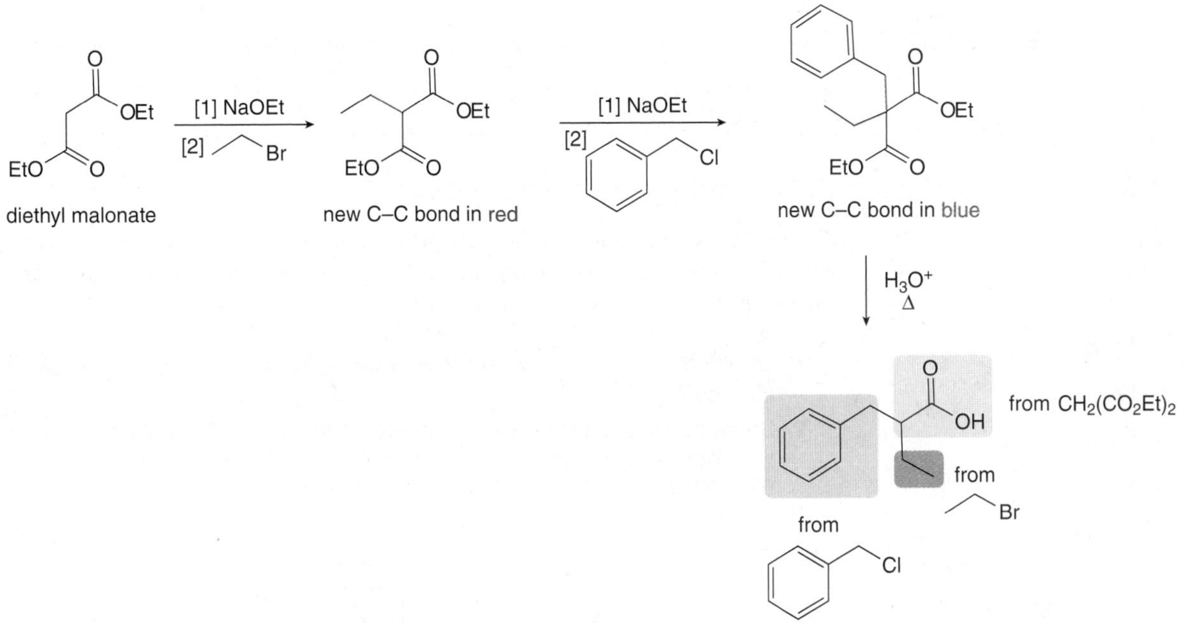

diethyl malonate
[1] NaOEt
[2]
Br
new C–C bond in red
Cl
new C–C bond in blue
H_3O^+
Δ
OH
from $CH_2(CO_2Et)_2$
from
from

An intramolecular malonic ester synthesis can be used to form rings having three to six atoms, provided the appropriate dihalide is used as starting material. For example, cyclopentanecarboxylic acid can be prepared from diethyl malonate and 1,4-dibromobutane ($BrCH_2CH_2CH_2CH_2Br$) by the following sequence of reactions:

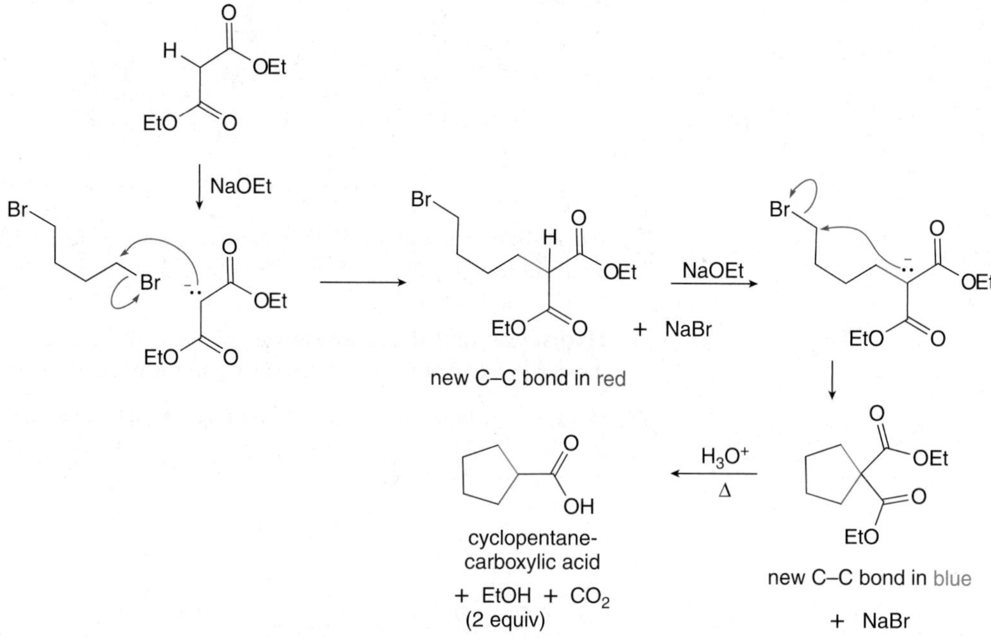
H
OEt
EtO
O
NaOEt
Br
Br
OEt
new C–C bond in red
NaOEt
+ NaBr
H₃O⁺
Δ
OH
cyclopentane-
carboxylic acid
+ EtOH + CO₂
(2 equiv)
new C–C bond in blue
+ NaBr

Problem 23.21 Draw the products of each reaction.

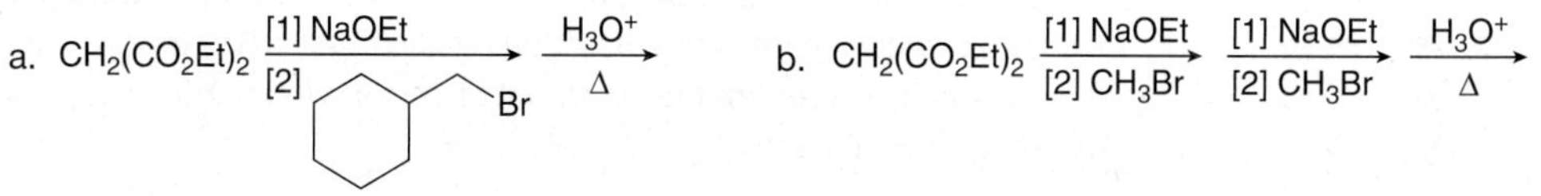
a. CH₂(CO₂Et)₂
[1] NaOEt
[2]
Br
H₃O⁺
Δ
b. CH₂(CO₂Et)₂
[1] NaOEt
[2] CH₃Br
[1] NaOEt
[2] CH₃Br
H₃O⁺
Δ

Problem 23.22 What cyclic product is formed from each dihalide using the malonic ester synthesis?

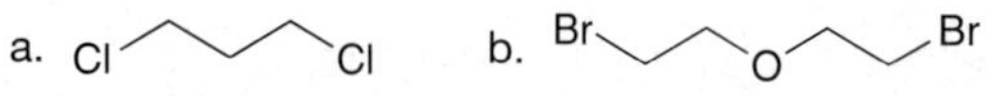
a. Cl
Cl
b. Br
O
Br

23.9C Retrosynthetic Analysis

To use the malonic ester synthesis, you must be able to determine what starting materials are needed to prepare a given compound—that is, you must **work backwards in the retrosynthetic direction.** This involves a two-step process:

[1] Locate the α carbon to the COOH group, and identify all alkyl groups bonded to the α carbon.

[2] Break the molecule into two (or three) components: Each alkyl group bonded to the α carbon comes from an alkyl halide. The remainder of the molecule comes from $CH_2(COOEt)_2$.

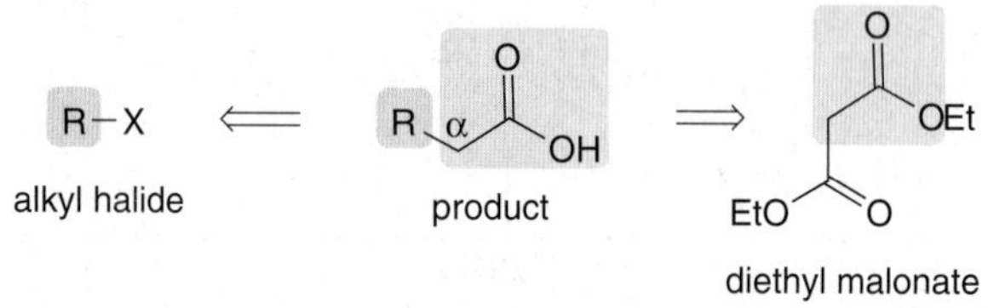
R–X
alkyl halide
R
α
OH
product
O
OEt
EtO
O
diethyl malonate

Sample Problem 23.3 What starting materials are needed to prepare 2-methylhexanoic acid [$CH_3CH_2CH_2CH_2CH(CH_3)COOH$] using a malonic ester synthesis?

Solution

The target molecule has two different alkyl groups bonded to the α carbon, so three components are needed for the synthesis:

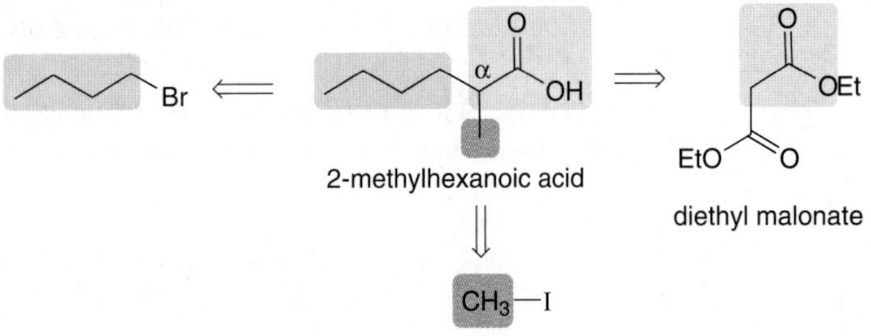

Writing the synthesis in the synthetic direction:

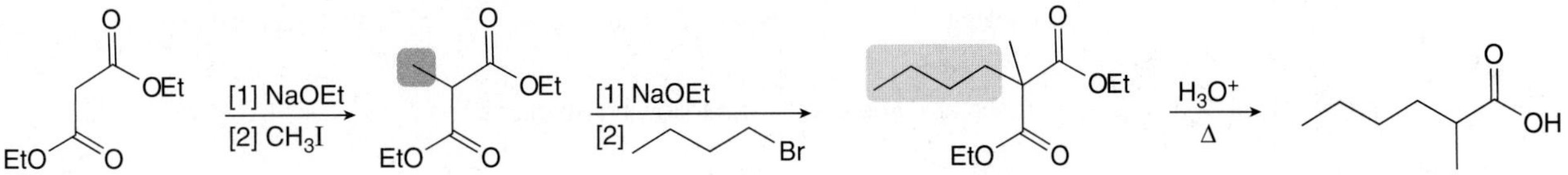

Problem 23.23 What alkyl halides are needed to prepare each carboxylic acid by the malonic ester synthesis?

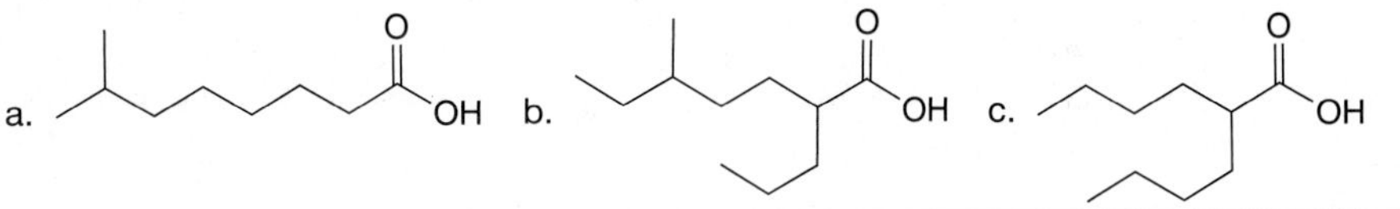

Problem 23.24 Explain why each of the following carboxylic acids cannot be prepared by a malonic ester synthesis.

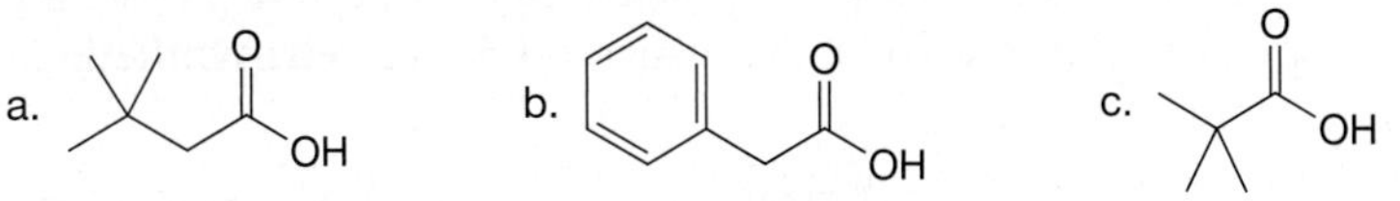

23.10 Acetoacetic Ester Synthesis

- **The acetoacetic ester synthesis is a stepwise method for converting ethyl acetoacetate into a ketone having one or two alkyl groups on the α carbon.**

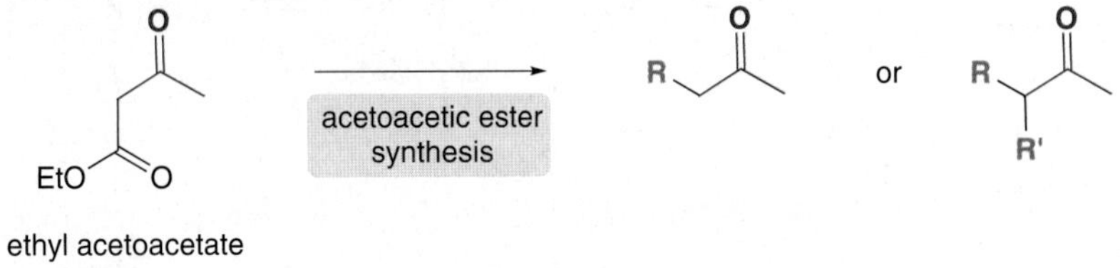

23.10A Steps in the Acetoacetic Ester Synthesis

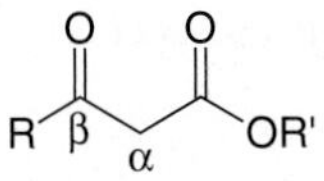

β-keto ester

The steps in the acetoacetic ester synthesis are exactly the same as those in the malonic ester synthesis. Because the starting material, CH_3COCH_2COOEt, is a β-keto ester, the final product is a **ketone,** not a carboxylic acid.

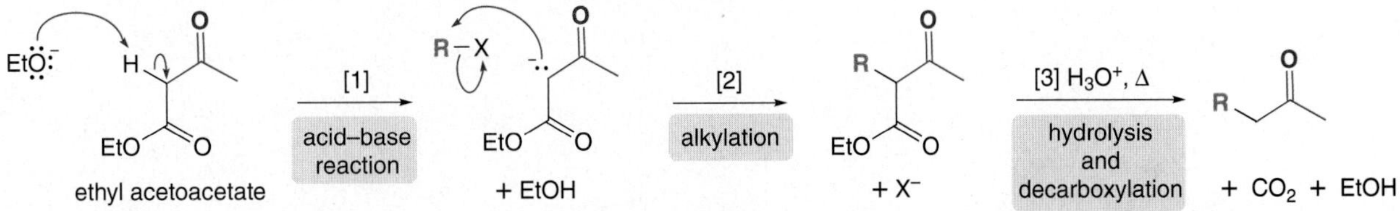

[1] **Deprotonation.** Treatment of ethyl acetoacetate with $^{-}$OEt removes the acidic proton between the two carbonyl groups.

[2] **Alkylation.** The nucleophilic enolate reacts with an alkyl halide (RX) in an S_N2 reaction to form a substitution product. Because the mechanism is S_N2, the yields are higher when R is CH_3 or a 1° alkyl group.

[3] **Hydrolysis and decarboxylation.** Heating the β-keto ester with aqueous acid hydrolyzes the ester to a β-keto acid, which loses CO_2 to form a ketone.

If the first two steps of the reaction sequence are repeated *prior* to hydrolysis and decarboxylation, then a ketone having *two new alkyl groups* on the α carbon can be synthesized.

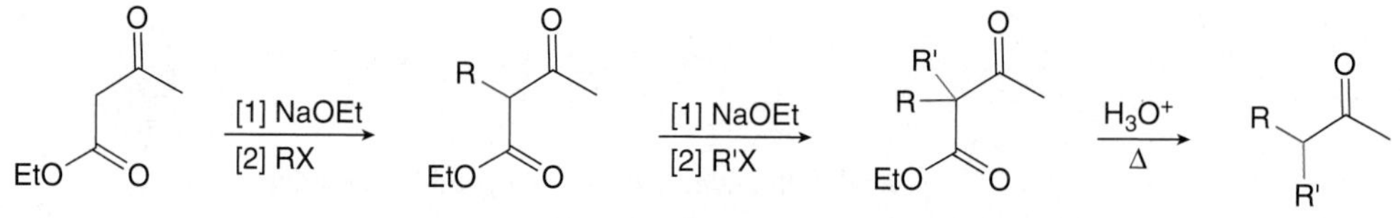

Problem 23.25 What ketones are prepared by the following reactions?

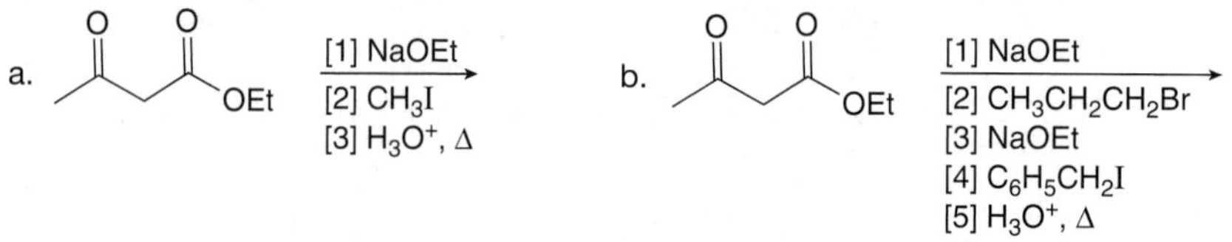

23.10B Retrosynthetic Analysis

To determine what starting materials are needed to prepare a given ketone using the acetoacetic ester synthesis, you must again work in the **retrosynthetic** direction. This involves a two-step process:

[1] Identify the alkyl groups bonded to the α carbon to the carbonyl group.

[2] Break the molecule into two (or three) components: Each alkyl group bonded to the α carbon comes from an alkyl halide. The remainder of the molecule comes from CH_3COCH_2COOEt.

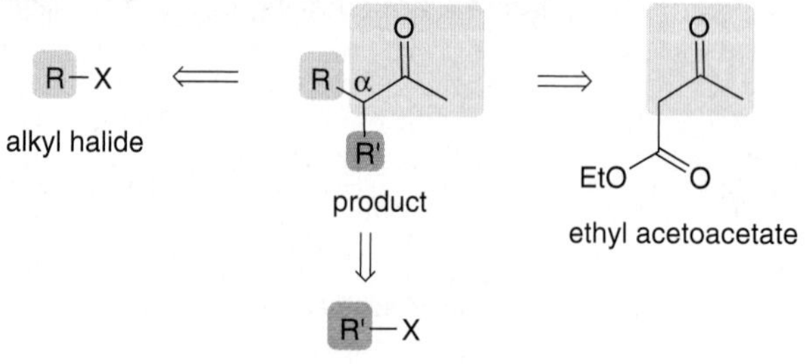

For a ketone with two R groups on the α carbon, three components are needed.

Sample Problem 23.4 What starting materials are needed to synthesize heptan-2-one using the acetoacetic ester synthesis?

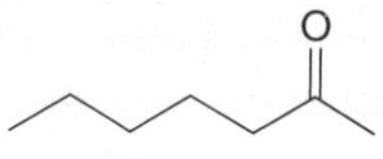

heptan-2-one

Solution

Heptan-2-one has only one alkyl group bonded to the α carbon, so only one alkyl halide is needed in the acetoacetic ester synthesis.

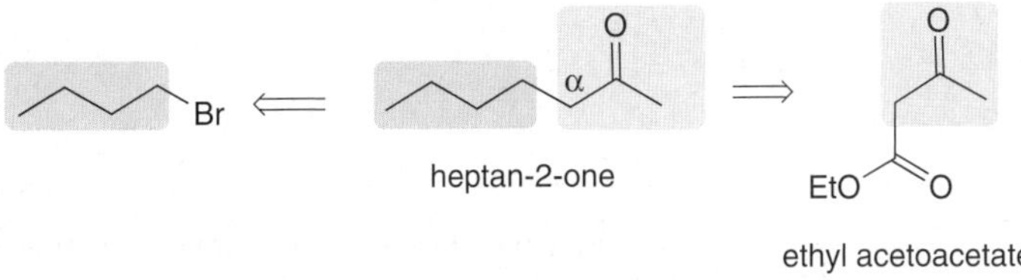

Writing the acetoacetic ester synthesis in the synthetic direction:

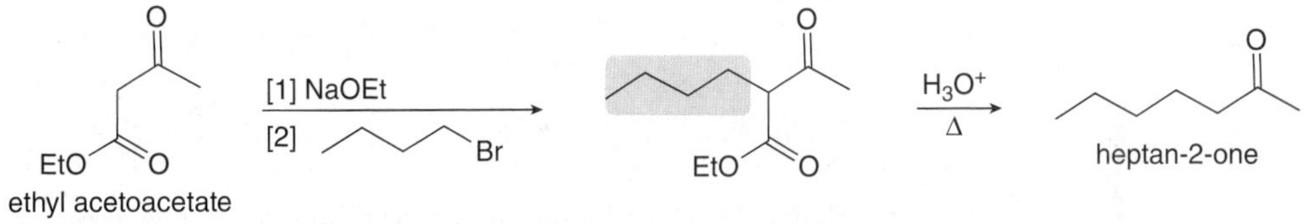

Problem 23.26 What alkyl halides are needed to prepare each ketone using the acetoacetic ester synthesis?

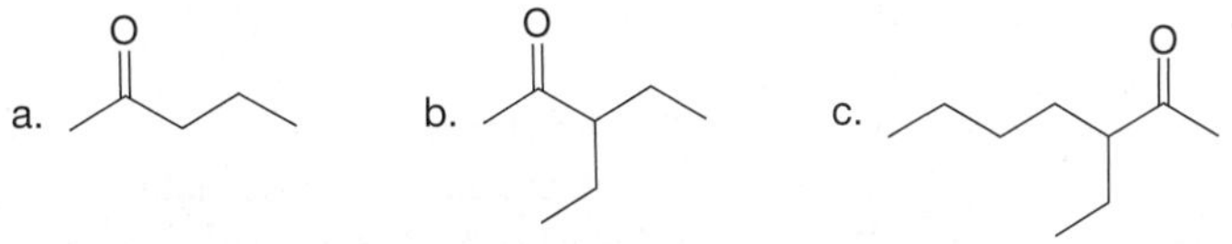

Problem 23.27 Treatment of ethyl acetoacetate with NaOEt (2 equiv) and $BrCH_2CH_2Br$ forms compound **X.** This reaction is the first step in the synthesis of illudin-S, an antitumor substance isolated from the jack-o'-lantern, a poisonous, saffron-colored mushroom. What is the structure of **X?**

The jack-o'-lantern, source of the antitumor agent illudin-S

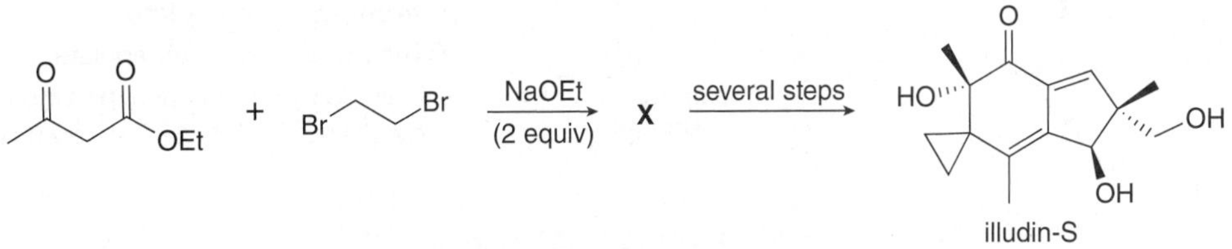

The acetoacetic ester synthesis and direct enolate alkylation are two different methods that prepare similar ketones. Butan-2-one, for example, can be synthesized from acetone by direct enolate alkylation with CH_3I (Method [1]), or by alkylation of ethyl acetoacetate followed by hydrolysis and decarboxylation (Method [2]).

Method [1]
Direct enolate alkylation

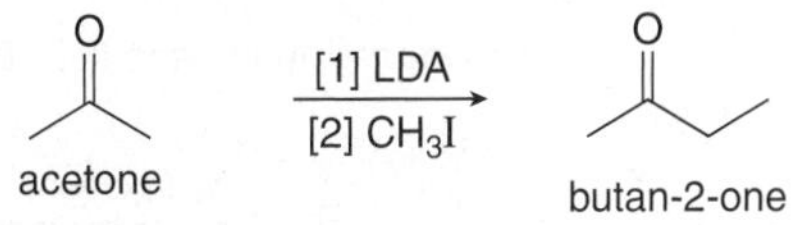

Method [2]
Acetoacetic ester synthesis

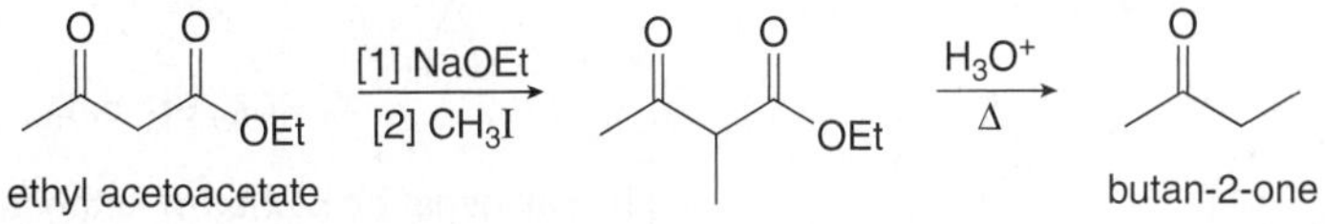

Why would you ever make butan-2-one from ethyl acetoacetate when you could make it in fewer steps from acetone? There are many factors to consider. First of all, synthetic organic chemists like to have a variety of methods to accomplish a single kind of reaction. Sometimes subtle changes in the structure of a starting material make one reaction work better than another.

In the chemical industry, moreover, cost is an important issue. Any reaction needed to make a large quantity of a useful drug or other consumer product must use cheap starting materials. Direct enolate alkylation usually requires a very strong base like LDA to be successful, whereas the acetoacetic ester synthesis utilizes NaOEt. NaOEt can be prepared from cheaper starting materials, and this makes the acetoacetic ester synthesis an attractive method, even though it involves more steps.

Thus, each method has its own advantages and disadvantages, depending on the starting material, the availability of reagents, the cost, and the occurrence of side reactions.

Problem 23.28 Nabumetone is a pain reliever and anti-inflammatory agent sold under the brand name of Relafen.

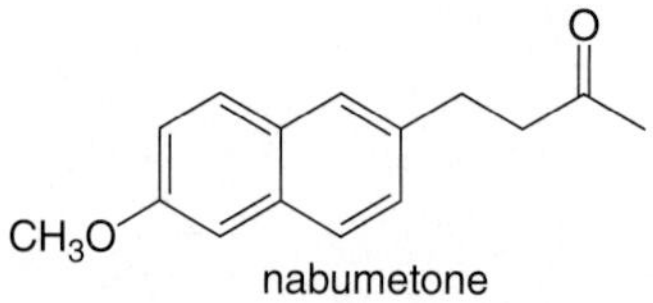
O
CH3O

nabumetone

a. Write out a synthesis of nabumetone from ethyl acetoacetate.
b. What ketone and alkyl halide are needed to synthesize nabumetone by direct enolate alkylation?

KEY CONCEPTS

Substitution Reactions of Carbonyl Compounds at the α Carbon

Kinetic Versus Thermodynamic Enolates (23.4)

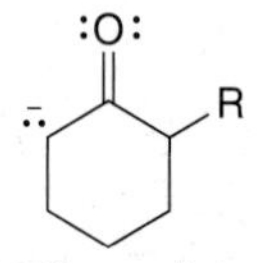
:O:
R

kinetic enolate

Kinetic enolate
- The less substituted enolate
- Favored by strong base, polar aprotic solvent, low temperature: LDA, THF, −78 °C

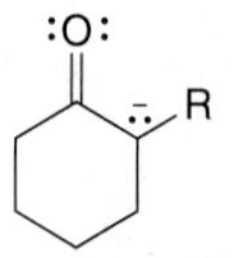
:O:
R

thermodynamic enolate

Thermodynamic enolate
- The more substituted enolate
- Favored by strong base, protic solvent, higher temperature: $NaOCH_2CH_3$, CH_3CH_2OH, room temperature

Halogenation at the α Carbon

[1] Halogenation in acid (23.7A)

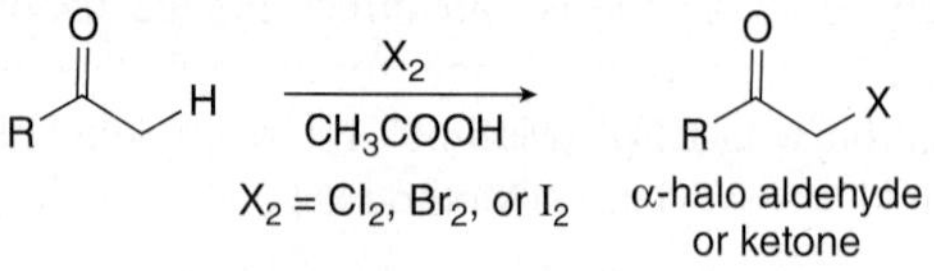
O
R
H
X2
CH3COOH
X2 = Cl2, Br2, or I2
O
R
X
α-halo aldehyde
or ketone

- The reaction occurs via enol intermediates.
- Monosubstitution of X for H occurs on the α carbon.

[2] Halogenation in base (23.7B)

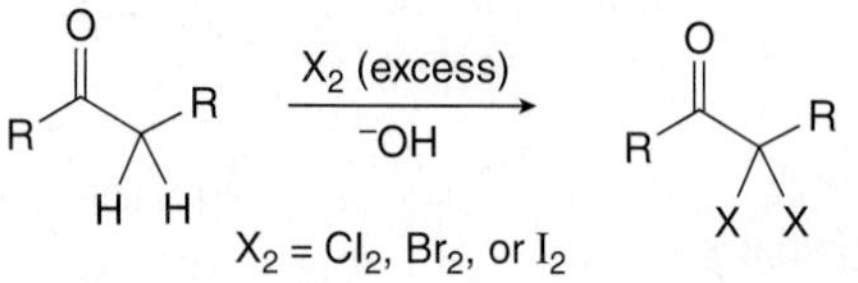
O
R
R
H H
X2 (excess)
−OH
X2 = Cl2, Br2, or I2
O
R
R
X X

- The reaction occurs via enolate intermediates.
- Polysubstitution of X for H occurs on the α carbon.

[3] Halogenation of *methyl* ketones in base—The haloform reaction (23.7B)

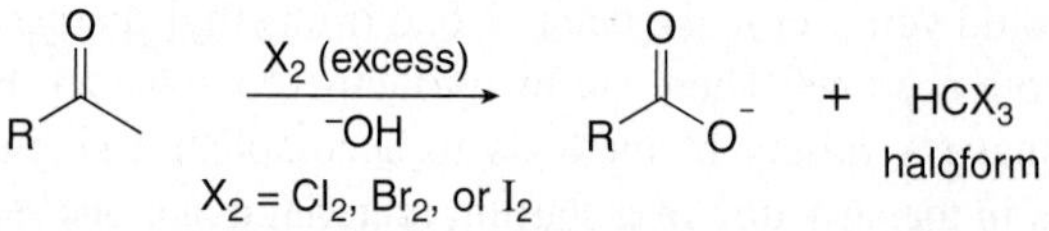
O
R
X2 (excess)
−OH
X2 = Cl2, Br2, or I2
O
R
O−
+
HCX3
haloform

- The reaction occurs with methyl ketones and results in cleavage of a carbon–carbon σ bond.

Reactions of α-Halo Carbonyl Compounds (23.7C)

[1] Elimination to form α,β-unsaturated carbonyl compounds

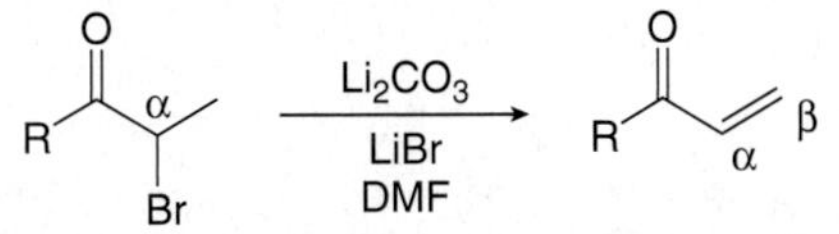

- Elimination of the elements of Br and H forms a new π bond, giving an α,β-unsaturated carbonyl compound.

[2] Nucleophilic substitution

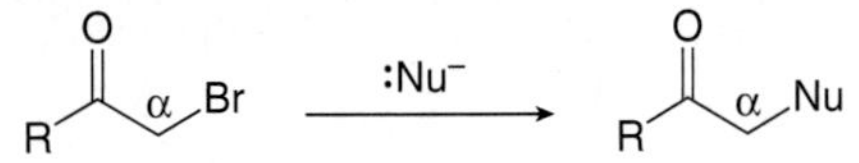

- The reaction follows an S_N2 mechanism, generating an α-substituted carbonyl compound.

Alkylation Reactions at the α Carbon

[1] Direct alkylation at the α carbon (23.8)

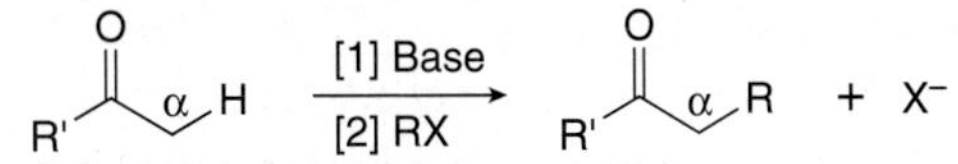

- The reaction forms a new C–C bond to the α carbon.
- LDA is a common base used to form an intermediate enolate.
- The alkylation in Step [2] follows an S_N2 mechanism.

[2] Malonic ester synthesis (23.9)

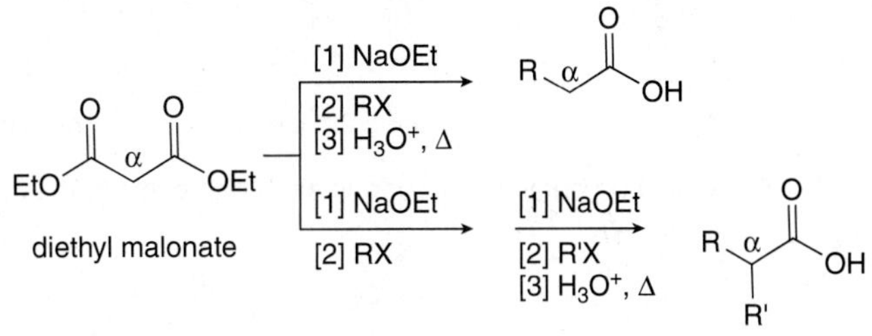

- The reaction is used to prepare carboxylic acids with one or two alkyl groups on the α carbon.
- The alkylation in Step [2] follows an S_N2 mechanism.

[3] Acetoacetic ester synthesis (23.10)

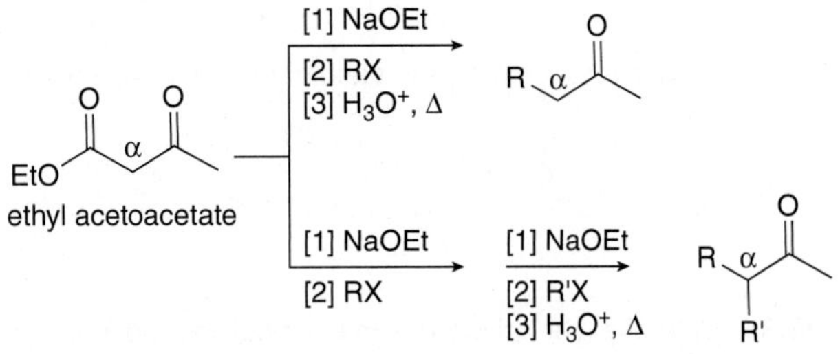

- The reaction is used to prepare ketones with one or two alkyl groups on the α carbon.
- The alkylation in Step [2] follows an S_N2 mechanism.

PROBLEMS

Problems Using Three-Dimensional Models

23.29 Draw enol tautomer(s) for each compound. Ignore stereoisomers.

a. 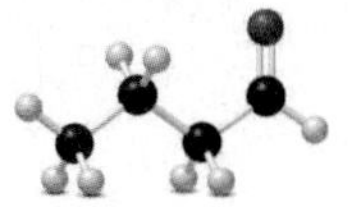

b. 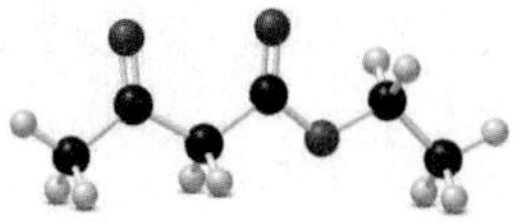

23.30 The cis ketone **A** is isomerized to a trans ketone **B** with aqueous NaOH. A similar isomerization does not occur with ketone **C.** (a) Draw the structure of **B** using a chair cyclohexane. (b) Label the substituents in **C** as cis or trans, and explain the difference in reactivity.

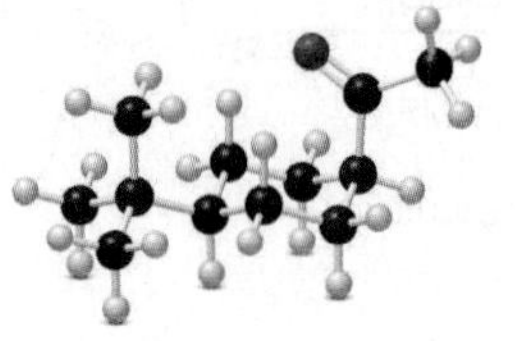

A

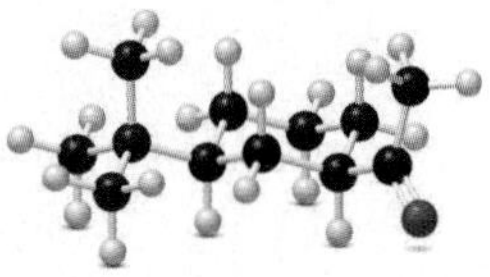

C

Enols, Enolates, and Acidic Protons

23.31 Draw enol tautomer(s) for each compound.

a. 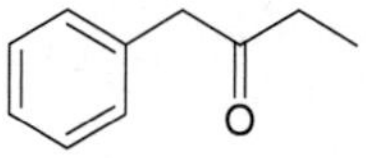 b. 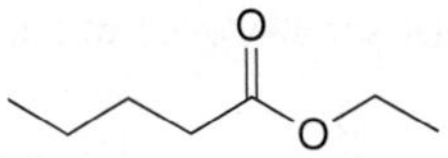 c. 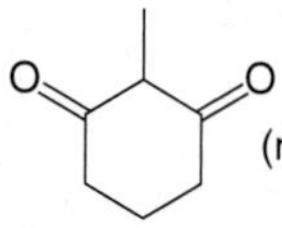 (mono enol form)

23.32 Both pentane-2,4-dione and ethyl acetoacetate have two carbonyl groups separated by a single carbon atom. Although an equilibrium mixture of pentane-2,4-dione tautomers contains 76% of the enol forms, an equilibrium mixture of ethyl acetoacetate tautomers contains only 8% of the enol forms. Suggest a reason for this difference.

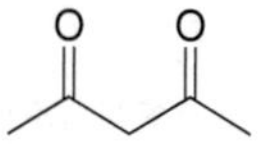

pentane-2,4-dione

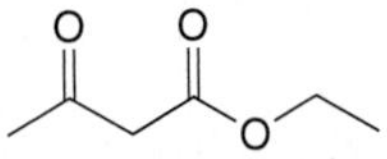

ethyl acetoacetate

23.33 Which carbonyl compound in each pair exhibits the higher percentage of the enol tautomer?

a. 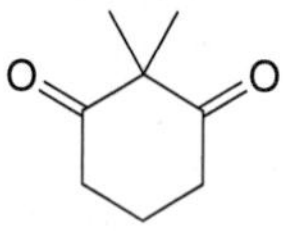 or 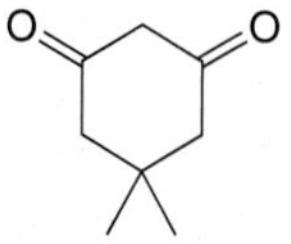 b. 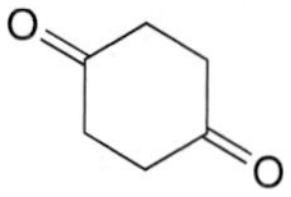 or 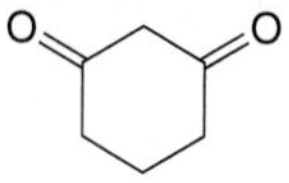

23.34 What hydrogen atoms in each compound have a p$K_a \leq 25$?

a. 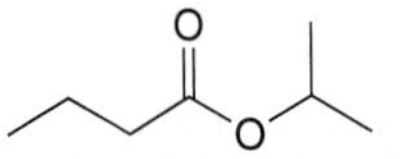 b. 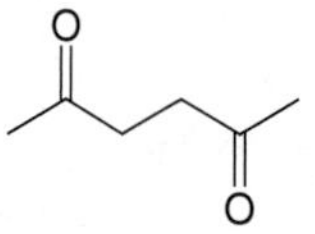 c. 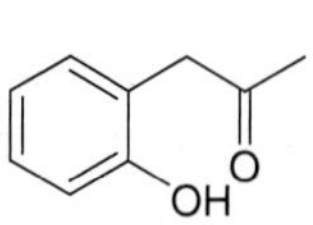

23.35 Rank the labeled protons in each compound in order of increasing acidity.

a. 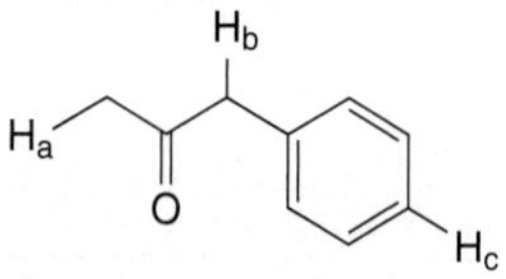

b. 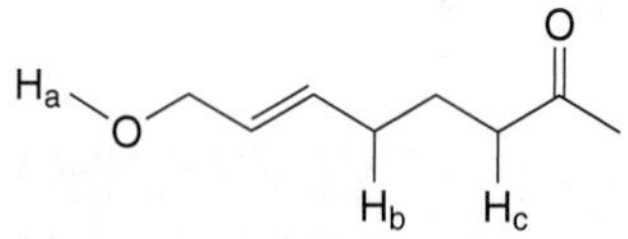

23.36 What is the major enolate (or carbanion) formed when each compound is treated with LDA?

a. 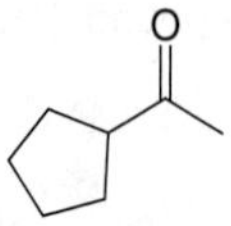 b. 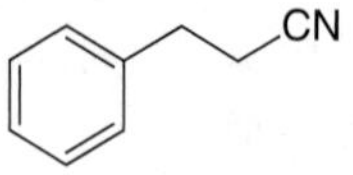

c. 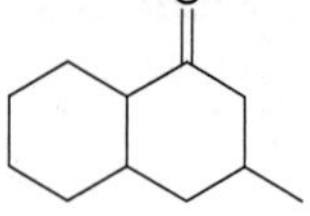 d. 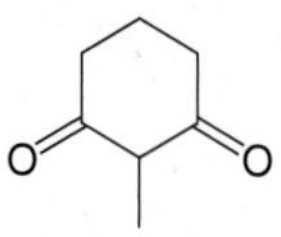

23.37 Why is the pK_a of the H_a protons in 1-acetylcyclohexene higher than the pK_a of the H_b protons?

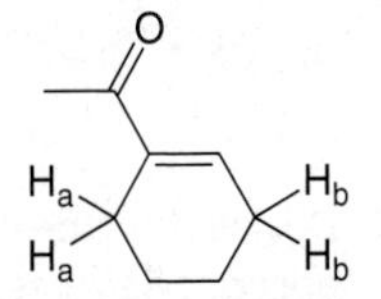
O
Ha
Ha
Hb
Hb

1-acetylcyclohexene

23.38 Acyclovir is an effective antiviral agent used to treat the herpes simplex virus. (a) Draw the enol form of acyclovir, and explain why it is aromatic. (b) Why is acyclovir typically drawn in its keto form, despite the fact that its enol is aromatic?

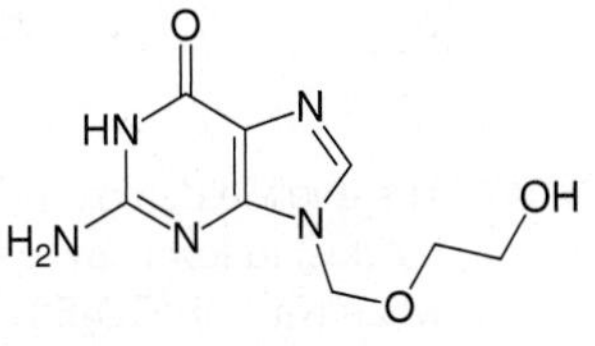
O
HN
N
H2N
N
N
OH
O

acyclovir

23.39 Explain why pentane-2,4-dione forms two different alkylation products (**A** or **B**) when the number of equivalents of base is increased from one to two.

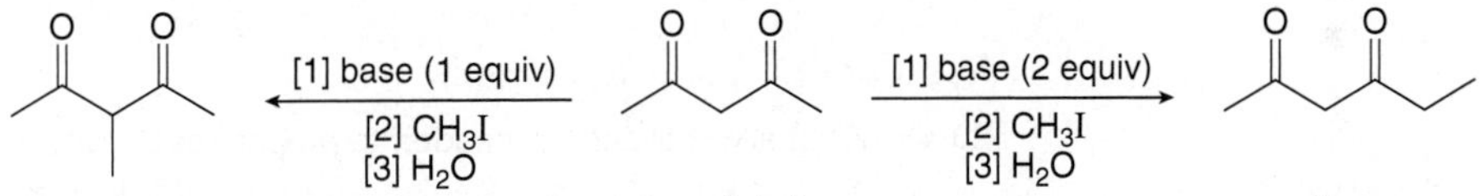
[1] base (1 equiv)
[2] CH3I
[3] H2O
[1] base (2 equiv)
[2] CH3I
[3] H2O

A pentane-2,4-dione **B**

23.40 Vitamin C is a stable enediol. Draw the structure of the two keto tautomers in equilibrium with the enediol and explain why the enediol is more stable than the other tautomers.

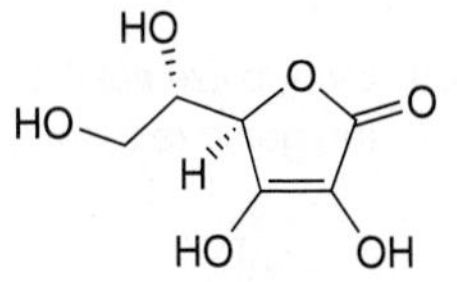
HO
HO
H
O
O
HO
OH

vitamin C
an enediol

Halogenation

23.41 Acid-catalyzed bromination of pentan-2-one ($CH_3COCH_2CH_2CH_3$) forms two products: $BrCH_2COCH_2CH_2CH_3$ (**A**) and $CH_3COCH(Br)CH_2CH_3$ (**B**). Explain why the major product is **B,** with the Br atom on the more substituted side of the carbonyl group.

23.42 Draw a stepwise mechanism for the following reaction.

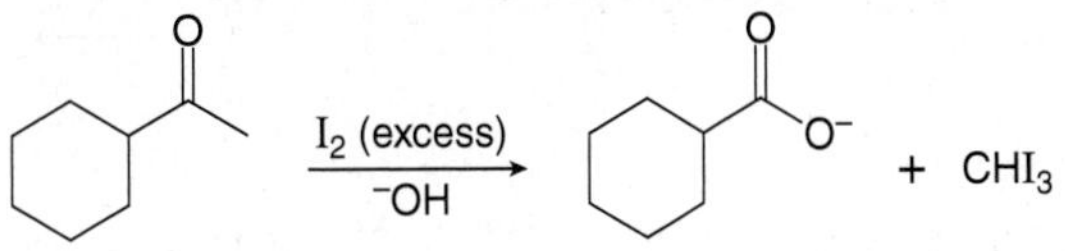
I2 (excess)
−OH
O−
+ CHI3

Malonic Ester Synthesis

23.43 What alkyl halides are needed to prepare each carboxylic acid using the malonic ester synthesis?

a. 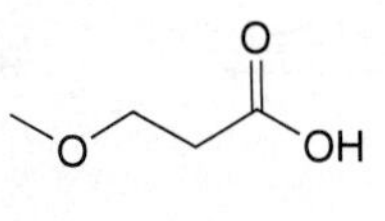
O
O
OH

b. 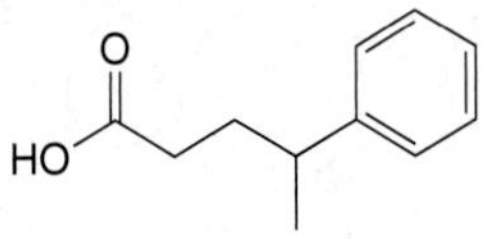
O
HO

c. 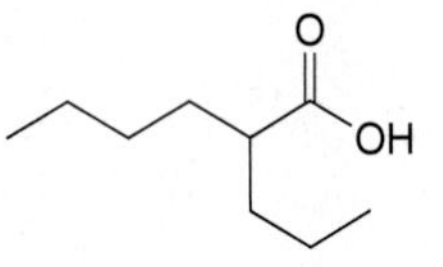
O
OH

23.44 Use the malonic ester synthesis to prepare each carboxylic acid.

a. 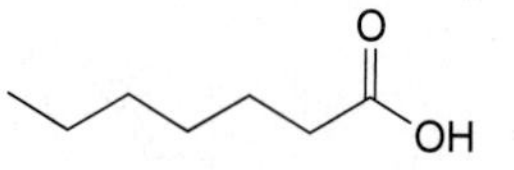

b. 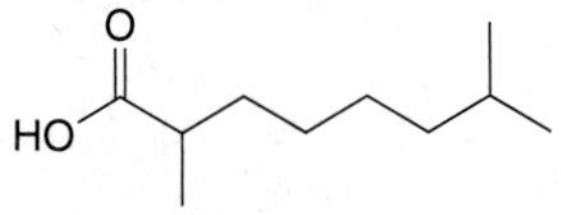

23.45 Devise a synthesis of valproic acid [$(CH_3CH_2CH_2)_2CHCO_2H$], a medicine used to treat epileptic seizures, using the malonic ester synthesis.

23.46 Synthesize each compound from diethyl malonate. You may use any other organic or inorganic reagents.

a. 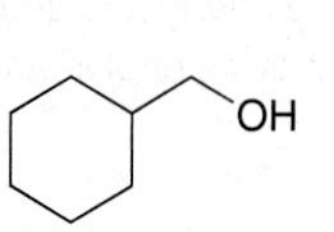

b. 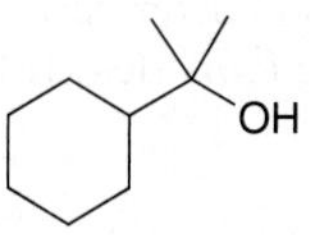

23.47 The enolate derived from diethyl malonate reacts with a variety of electrophiles (not just alkyl halides) to form new carbon–carbon bonds. With this in mind, draw the products formed when $Na^+ \ ^-CH(CO_2Et)_2$ reacts with each electrophile, followed by treatment with H_2O.

a. 

b. 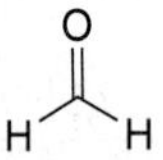

c. 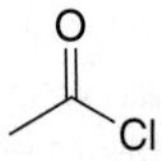

d. 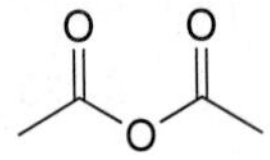

Acetoacetic Ester Synthesis

23.48 What alkyl halides are needed to prepare each ketone using the acetoacetic ester synthesis?

a. 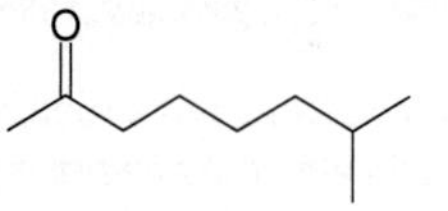

b. 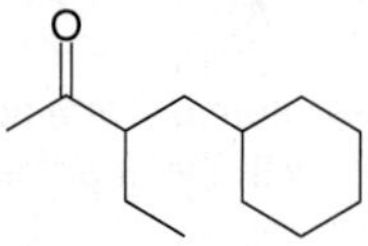

c. 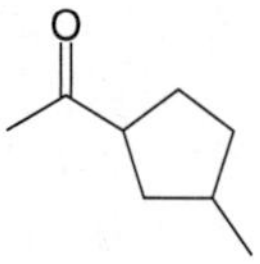

23.49 Synthesize each compound from ethyl acetoacetate. You may use any other organic or inorganic reagents.

a. 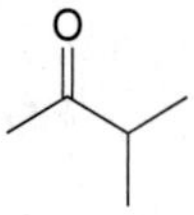

b. 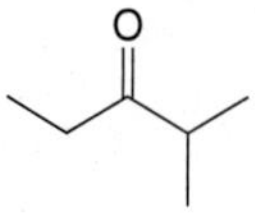

Reactions

23.50 Draw the organic products formed in each reaction.

a. 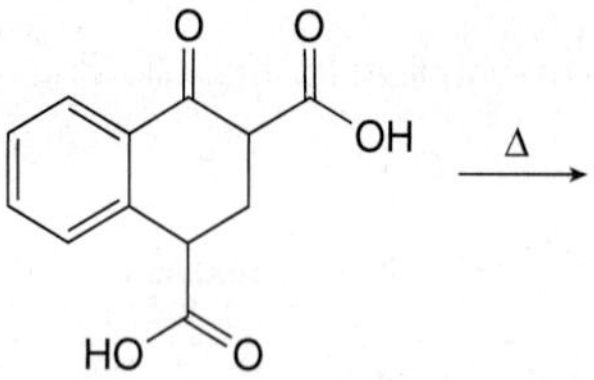
OH
HO
Δ

e. 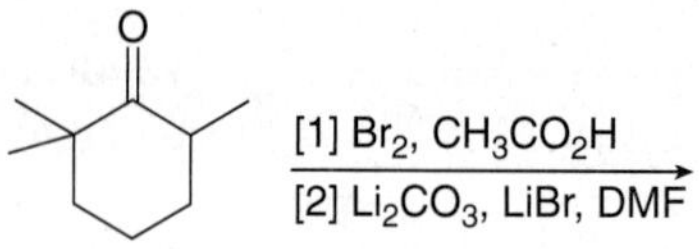
[1] Br2, CH3CO2H
[2] Li2CO3, LiBr, DMF

b. 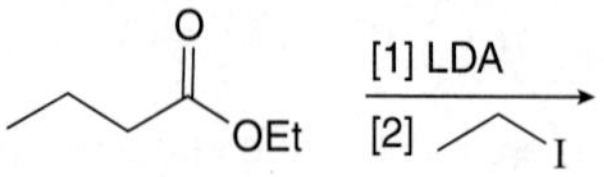
OEt
[1] LDA
[2]

f. 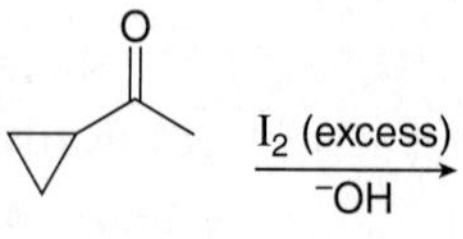
I2 (excess)
−OH

c. 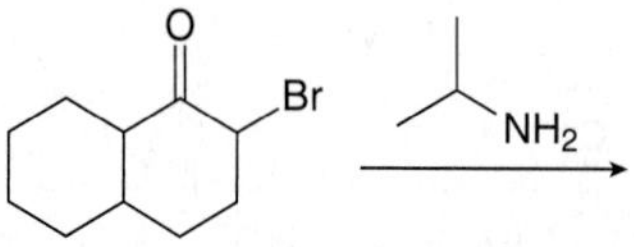
Br
NH2

g. 
Cl
CN
NaH
C6H9N

d. 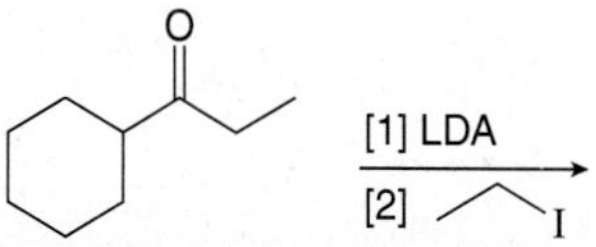
[1] LDA
[2]

h. 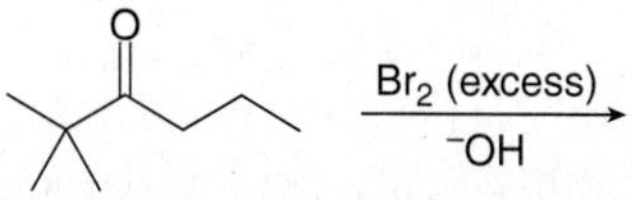
Br2 (excess)
−OH

23.51 Draw the products formed (including stereoisomers) in each reaction.

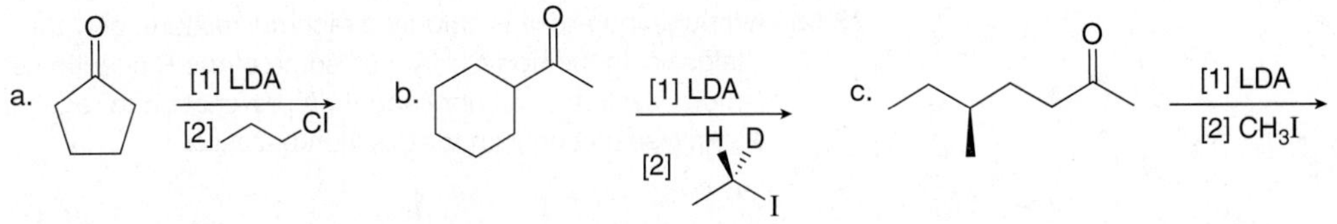
a.
[1] LDA
[2]
Cl
b.
[1] LDA
[2]
H D
I
c.
[1] LDA
[2] CH3I

23.52 a. Identify intermediates **A–C** in the following stepwise conversion of *p*-isobutylbenzaldehyde to the analgesic ibuprofen.

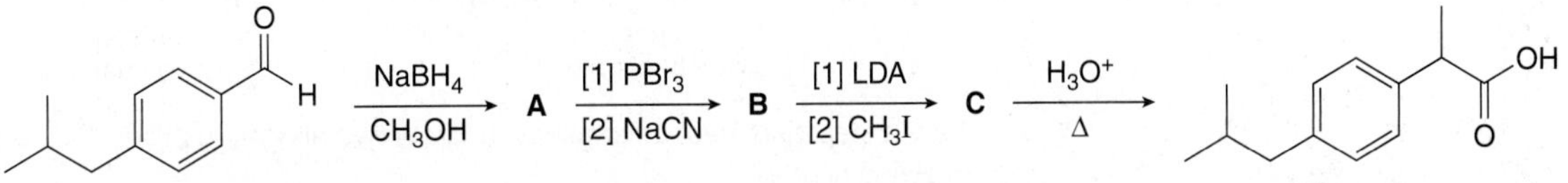
H
NaBH4
CH3OH
A
[1] PBr3
[2] NaCN
B
[1] LDA
[2] CH3I
C
H3O+
Δ
OH

p-isobutylbenzaldehyde

ibuprofen

b. Direct alkylation of **D** by treatment with one equivalent of LDA and CH_3I does not form ibuprofen. Identify the product of this reaction and explain how it is formed.

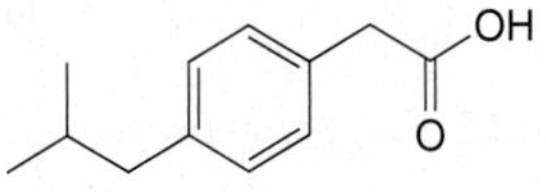
OH

D

23.53 A key step in the synthesis of the narcotic analgesic meperidine (trade name Demerol) is the conversion of phenylacetonitrile to **X.** (a) What is the structure of **X?** (b) What reactions convert **X** to meperidine?

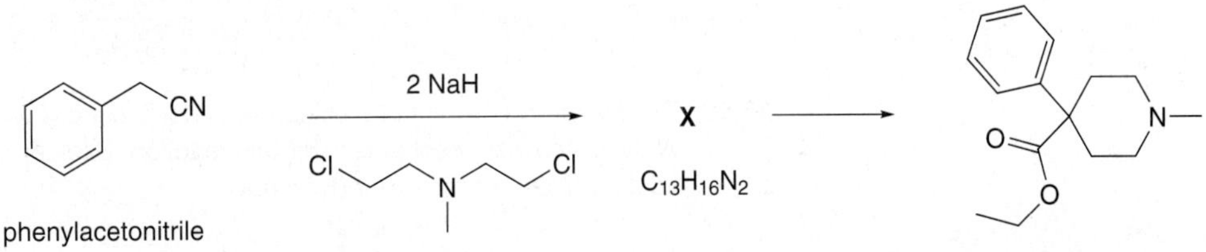
CN
2 NaH
Cl
N
Cl
X
C13H16N2
N

phenylacetonitrile

meperidine

23.54 Clopidogrel is the generic name for Plavix, a drug used to prevent the formation of blood clots in patients who have a history of heart attacks or strokes. A single enantiomer of clopidogrel can be prepared in three steps from the chiral α-hydroxy acid **A.** Identify **B** and **C** in the following reaction sequence, and designate the configuration of the enantiomer formed by this route as *R* or *S.*

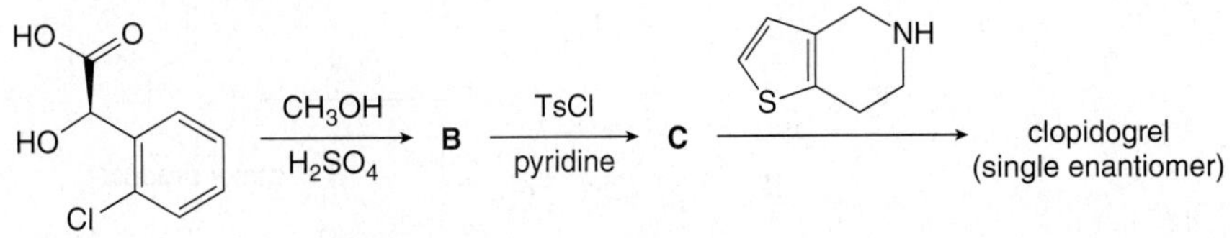
HO
HO
Cl
CH3OH
H2SO4
B
TsCl
pyridine
C
NH
S
clopidogrel
(single enantiomer)

A

23.55 What reaction conditions—base, solvent, and temperature—are needed to convert ketone **A** to either **B** or **C** by an intramolecular alkylation reaction?

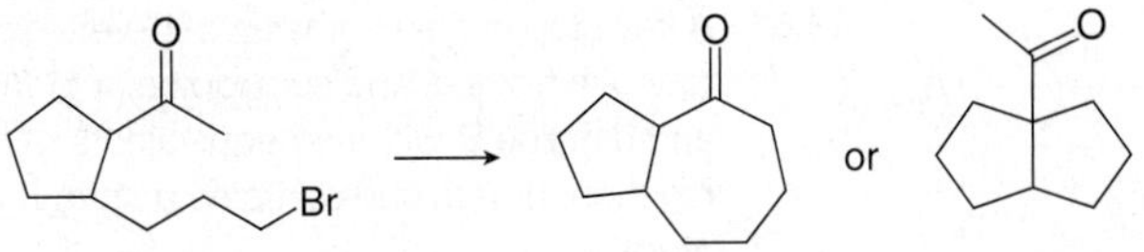
Br
or

A **B** **C**

Mechanism

23.56 Although ibuprofen is sold as a racemic mixture, only the *S* enantiomer acts as an analgesic. In the body, however, some of the *R* enantiomer is converted to the *S* isomer by tautomerization to an enol and then protonation to regenerate the carbonyl compound. Write a stepwise mechanism for this isomerization.

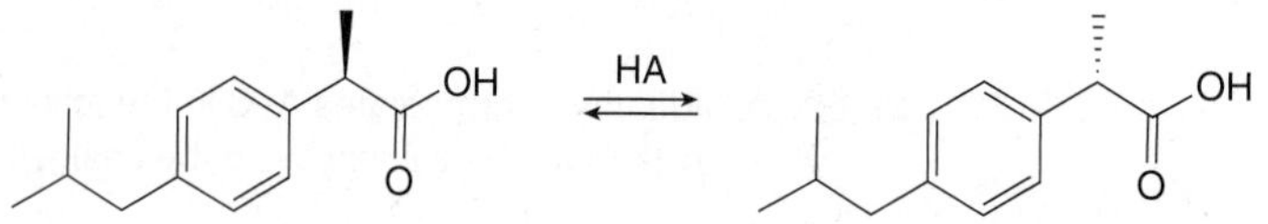
OH
HA
OH
O
O

R isomer
inactive enantiomer

S isomer
active enantiomer

23.57 Draw a stepwise mechanism showing how two alkylation products are formed in the following reaction.

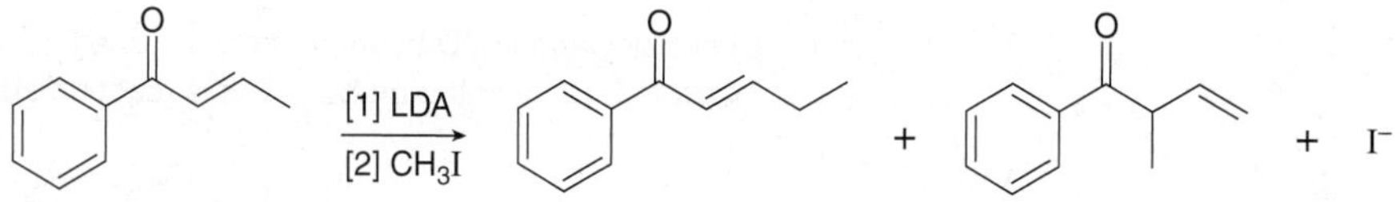
O
[1] LDA
[2] CH3I
O
+
O
+ I⁻

23.58 Draw a stepwise mechanism for the following reaction.

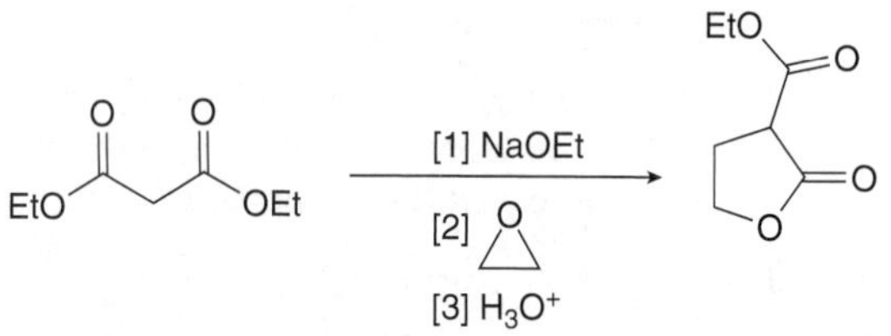
EtO
OEt
[1] NaOEt
[2]
[3] H3O+
EtO
O

23.59 Treatment of α,β-unsaturated carbonyl compound **X** with base forms the diastereomer **Y.** Write a stepwise mechanism for this reaction. Explain why one stereogenic center changes configuration but the other does not.

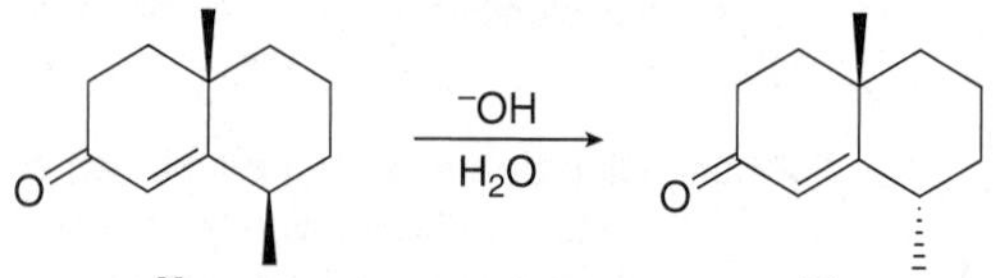
⁻OH
H2O

X **Y**

23.60 Draw stepwise mechanisms illustrating how each product is formed.

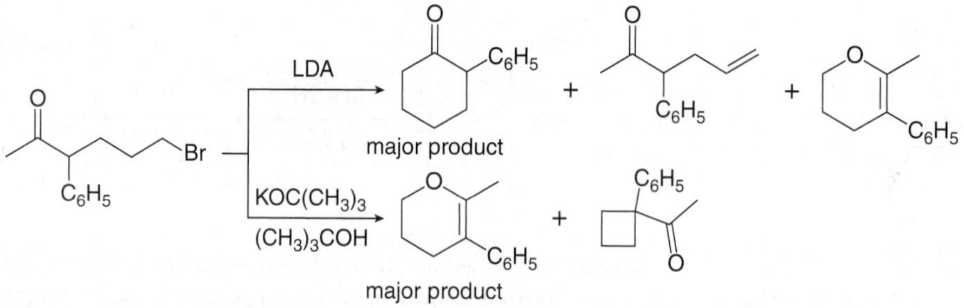
Br
C6H5
LDA
major product
KOC(CH3)3
(CH3)3COH
major product
C6H5

23.61 A key step in the synthesis of β-vetivone, a major constituent of vetiver, a perennial grass found in tropical and subtropical regions of the world, involved the reaction of compound **A** and dihalide **B** with two equivalents of LDA to form **C.** Draw a stepwise mechanism for this reaction. β-Vetivone contains a spiro ring system—that is, two rings that share a single carbon atom.

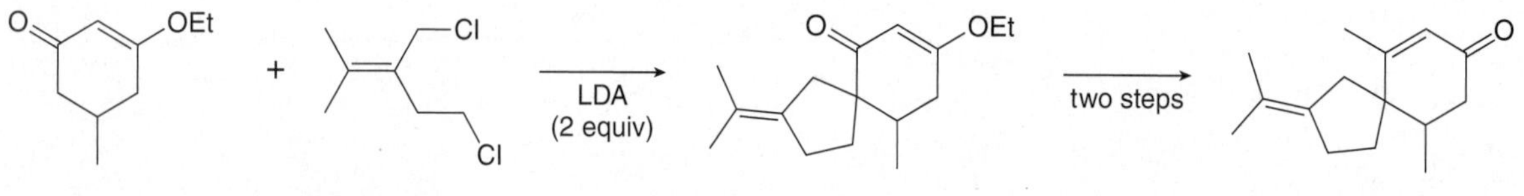
OEt
Cl
Cl
LDA
(2 equiv)
OEt
two steps

A **B** **C** β-vetivone

Synthesis

23.62 (a) Draw two different halo ketones that can form **A** by an intramolecular alkylation reaction.
(b) How can **A** be synthesized by an acetoacetic ester synthesis?

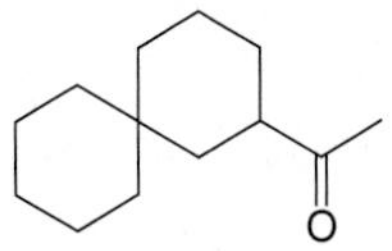

A

23.63 Synthesize each compound from cyclohexanone and organic halides having ≤ 4 C's. You may use any other inorganic reagents.

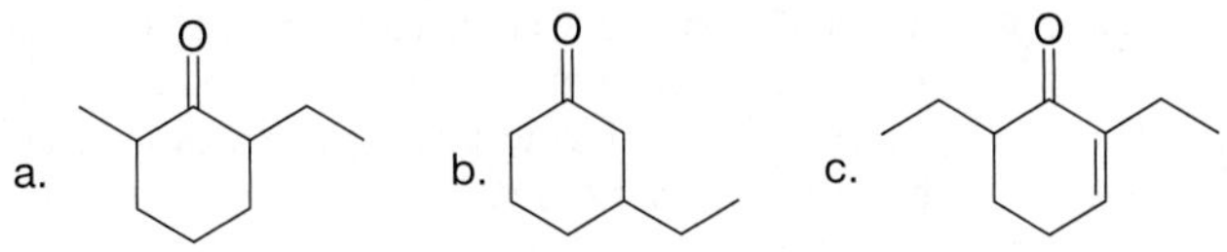

23.64 Bupropion, sold under the trade name of Zyban, is an antidepressant that was approved to aid smoking cessation in 1997. Devise a synthesis of bupropion from benzene, organic compounds that have fewer than five carbons, and any required inorganic reagents.

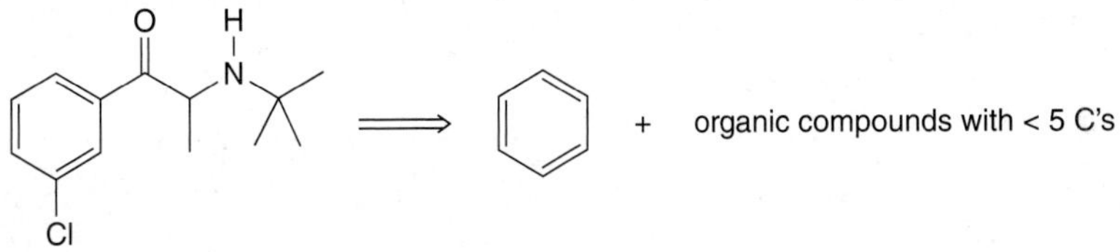

bupropion

23.65 Devise a synthesis of anastrozole, a drug used to reduce the recurrence of breast cancer (Section 22.18), from the given compounds. You may use any other needed organic compounds or inorganic reagents.

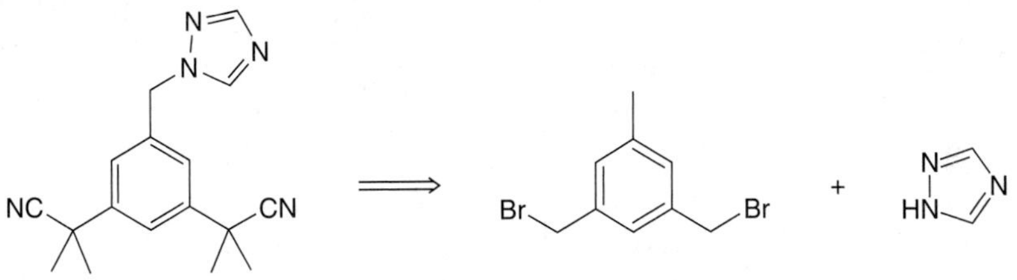

anastrozole

23.66 Synthesize (*Z*)-hept-5-en-2-one from ethyl acetoacetate ($CH_3COCH_2CO_2Et$) and the given starting material. You may also use any other organic compounds or required inorganic reagents.

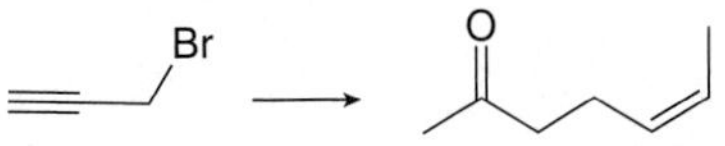

(*Z*)-hept-5-en-2-one

23.67 Treatment of ketone **A** with LDA followed by CH_3CH_2I did not form the desired alkylation product **B.** What product was formed instead? Devise a multistep method to convert **A** to **B,** a synthetic intermediate used to prepare the anticancer drug tamoxifen (Section 23.8C and the chapter-opening molecule).

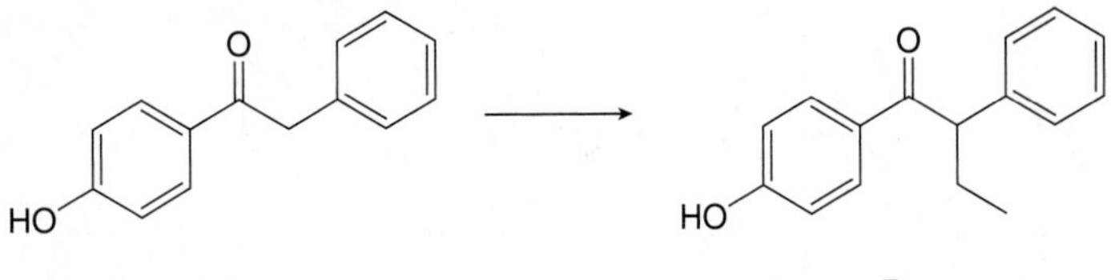

A **B**

23.68 Capsaicin, the spicy component of hot peppers, can be prepared from amine **X** and acid chloride **Y.** Devise a synthesis of **Y** from (*E*)-6-methylhept-4-en-1-ol [$(CH_3)_2CHCH{=}CH(CH_2)_3OH$], $CH_2(CO_2Et)_2$, and any required inorganic reagents.

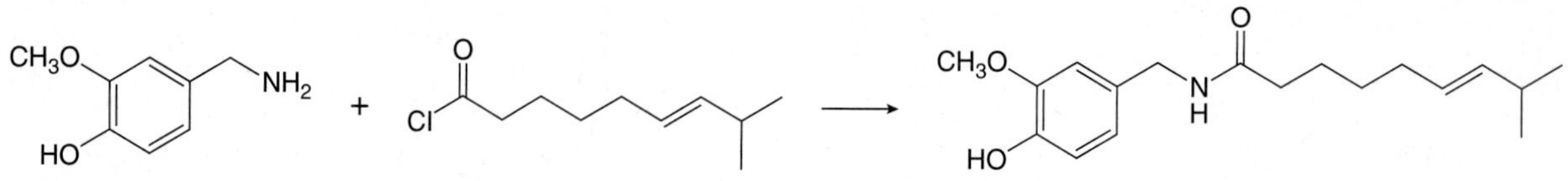

X **Y** capsaicin

Spectroscopy

23.69 Treatment of **W** with CH_3Li, followed by CH_3I, affords compound **Y** ($C_7H_{14}O$) as the major product. **Y** shows a strong absorption in its IR spectrum at 1713 cm^{-1}, and its ^1H NMR spectrum is given below. (a) Propose a structure for **Y.** (b) Draw a stepwise mechanism for the conversion of **W** to **Y.**

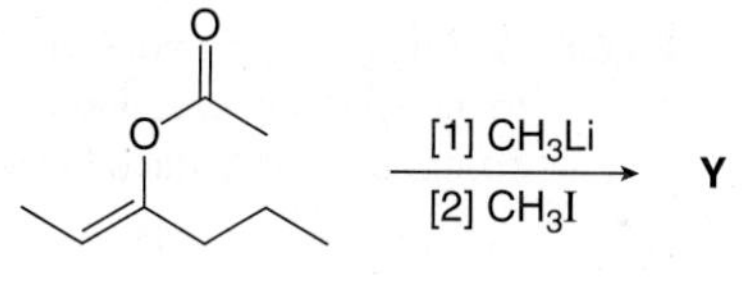

W

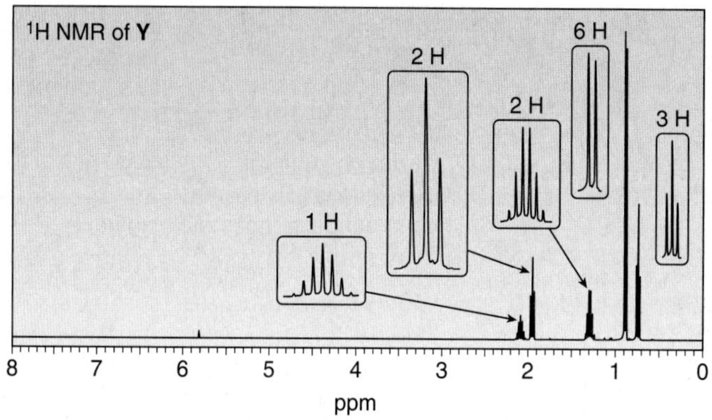

Challenge Problems

23.70 Explain why H_a is much less acidic than H_b. Then draw a mechanism for the following reaction.

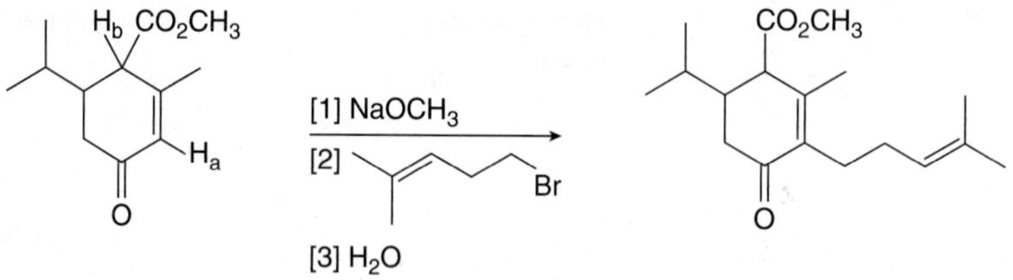

23.71 Devise a stepwise mechanism for the following reaction.

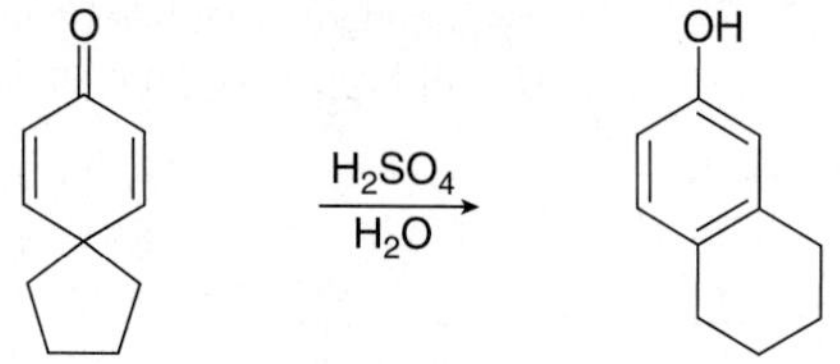

23.72 The last step in the synthesis of β-vetivone (Problem 23.61) involves treatment of **C** with CH_3Li to form an intermediate **X,** which forms β-vetivone with aqueous acid. Identify the structure of **X** and draw a mechanism for converting **X** to β-vetivone.

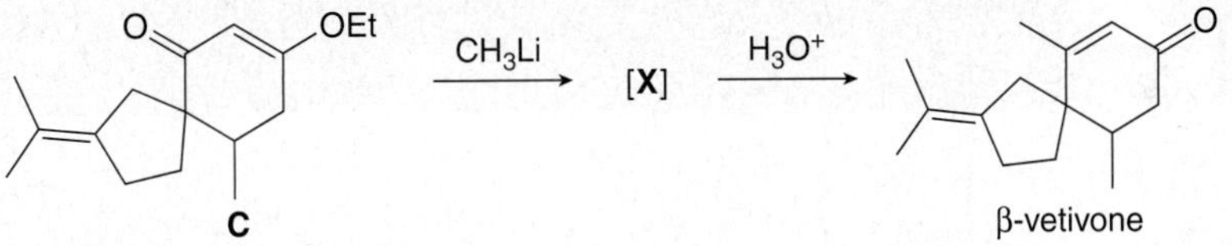

23.73 Keeping in mind the mechanism for the dissolving metal reduction of alkynes to trans alkenes in Chapter 12, write a stepwise mechanism for the following reaction, which involves the conversion of an α,β-unsaturated carbonyl compound to a carbonyl compound with a new alkyl group on the α carbon.

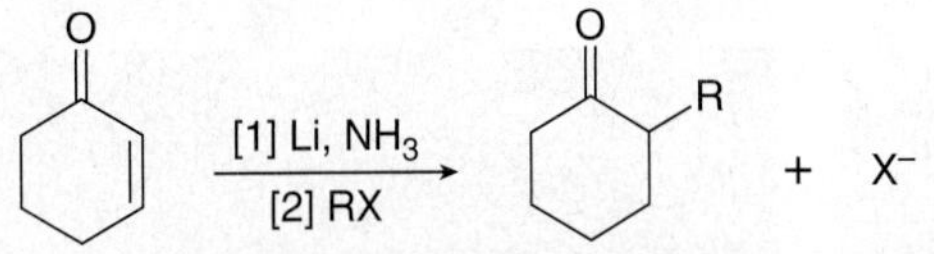

23.74 (–)-Hyoscyamine, an optically active drug used to treat gastrointestinal disorders, is isolated from *Atropa belladonna,* the deadly nightshade plant, by a basic aqueous extraction procedure. If too much base is used during isolation, optically inactive material is isolated. (a) Explain this result by drawing a stepwise mechanism. (b) Explain why littorine, an isomer isolated from the tailflower plant in Australia, can be obtained optically pure regardless of the amount of base used during isolation.

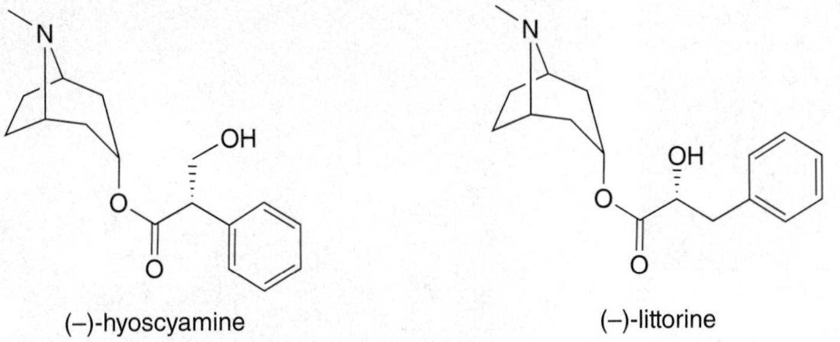

24 Carbonyl Condensation Reactions

24.1 The aldol reaction
24.2 Crossed aldol reactions
24.3 Directed aldol reactions
24.4 Intramolecular aldol reactions
24.5 The Claisen reaction
24.6 The crossed Claisen and related reactions
24.7 The Dieckmann reaction
24.8 Biological carbonyl condensation reactions
24.9 The Michael reaction
24.10 The Robinson annulation

Stearic acid, a saturated fatty acid first discussed in Chapter 10, is a key component of cocoa butter, a mixture of triacylglycerols obtained from the cocoa bean that is used to make chocolate. Because cocoa butter is high in saturated fatty acid content, cocoa butter is a solid at room temperature, so it is classified as a fat. The key carbon–carbon bond-forming reaction that occurs in the biosynthesis of stearic acid and other fatty acids involves the Claisen condensation, a reaction that occurs between the α carbon of one carbonyl component and the carbonyl carbon of another. In Chapter 24, we examine four useful reactions at the α carbon of a carbonyl group that result in carbon–carbon bond formation.

In Chapter 24, we examine **carbonyl condensations**—that is, reactions between two carbonyl compounds—a second type of reaction that occurs at the α carbon of a carbonyl group. Much of what is presented in Chapter 24 applies principles you have already learned. Many of the reactions may look more complicated than those in previous chapters, but they are fundamentally the same. Moreover, a key step in several metabolic pathways involves carbonyl condensations.

Every reaction in Chapter 24 forms a new carbon–carbon bond at the α carbon to a carbonyl group, so these reactions are extremely useful in the synthesis of complex natural products.

24.1 The Aldol Reaction

Chapter 24 concentrates on the second general reaction of enolates—**reaction with other carbonyl compounds.** In these reactions, one carbonyl component serves as the nucleophile and one serves as the electrophile, and a new carbon–carbon bond is formed.

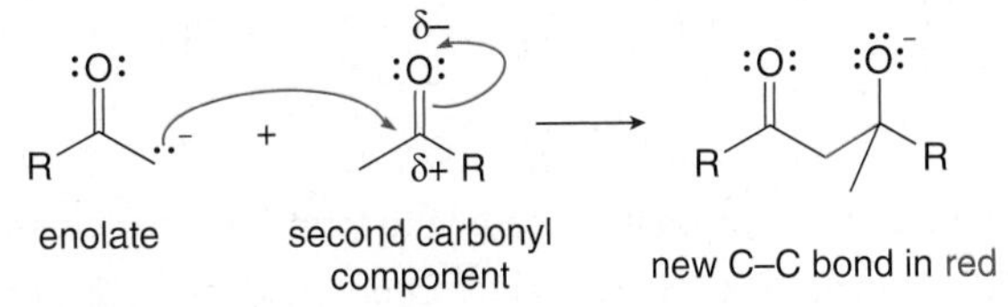

The presence or absence of a leaving group on the electrophilic carbonyl carbon determines the structure of the product. Even though they appear somewhat more complicated, these reactions are often reminiscent of the nucleophilic addition and nucleophilic acyl substitution reactions of Chapters 21 and 22. Four types of reactions are examined:

- **Aldol reaction** (Sections 24.1–24.4)
- **Claisen reaction** (Sections 24.5–24.7)
- **Michael reaction** (Section 24.9)
- **Robinson annulation** (Section 24.10)

24.1A General Features of the Aldol Reaction

In the **aldol reaction,** two molecules of an aldehyde or ketone react with each other in the presence of base to form a **β-hydroxy carbonyl compound.** For example, treatment of acetaldehyde with aqueous ^{-}OH forms 3-hydroxybutanal, a **β-hydroxy aldehyde.**

Many aldol products contain an ***ald***ehyde and an alcoh***ol***—hence the name ***aldol.***

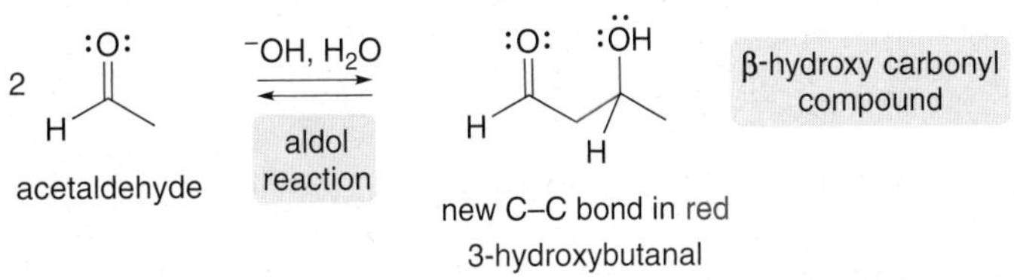

The mechanism of the aldol reaction has **three steps,** as shown in Mechanism 24.1. Carbon–carbon bond formation occurs in Step [2], when the nucleophilic enolate reacts with the electrophilic carbonyl carbon.

Mechanism 24.1 The Aldol Reaction

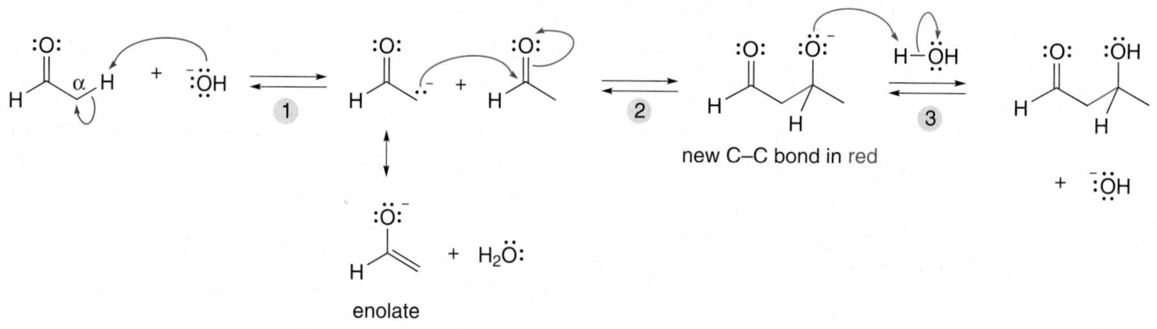

1 The base removes a proton on the α carbon to form a resonance-stabilized enolate.

2 Nucleophilic attack of the enolate on an electrophilic carbonyl in another molecule of aldehyde forms a new C–C bond.

3 Protonation of the alkoxide forms the β-hydroxy aldehyde.

The aldol reaction is a reversible equilibrium, so the position of the equilibrium depends on the base and the carbonyl compound. **⁻OH is the base** typically used in an aldol reaction. Recall from Section 23.3B that only a small amount of enolate forms with ⁻OH. In this case, that's appropriate because the starting aldehyde is needed to react with the enolate in the second step of the mechanism.

Aldol reactions can be carried out with either aldehydes or ketones. With aldehydes, the equilibrium usually favors the products, but with ketones the equilibrium favors the starting materials. There are ways of driving this equilibrium to the right, however, so we will write aldol products whether the substrate is an aldehyde or a ketone.

- **The characteristic reaction of aldehydes and ketones is nucleophilic addition (Section 21.7). An aldol reaction is a nucleophilic addition in which an enolate is the nucleophile. See the comparison in Figure 24.1.**

Figure 24.1
The aldol reaction—An example of nucleophilic addition

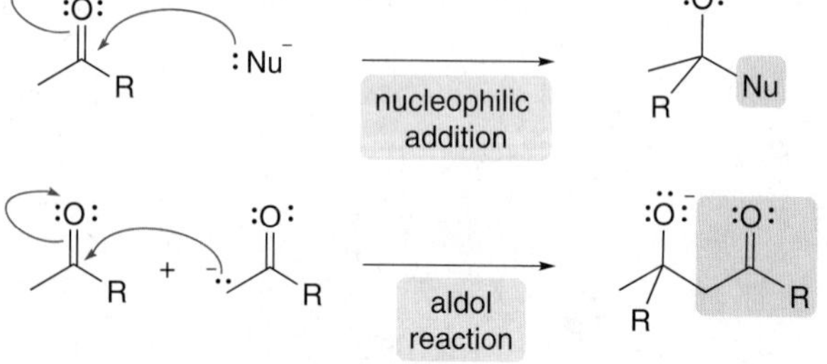

- Aldehydes and ketones react by nucleophilic addition. In an aldol reaction, **an enolate is the nucleophile** that adds to the carbonyl group.

A **second example of an aldol** reaction is shown with propanal as starting material. The two molecules of the aldehyde that participate in the aldol reaction react in opposite ways.

- One molecule of propanal becomes an enolate—an electron-rich *nucleophile.*
- One molecule of propanal serves as the *electrophile* because its carbonyl carbon is electron deficient.

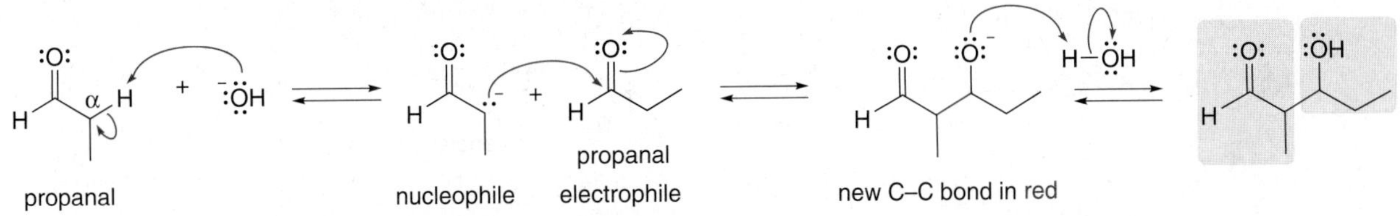

These two examples illustrate the general features of the aldol reaction. **The α carbon of one carbonyl component becomes bonded to the carbonyl carbon of the other component.**

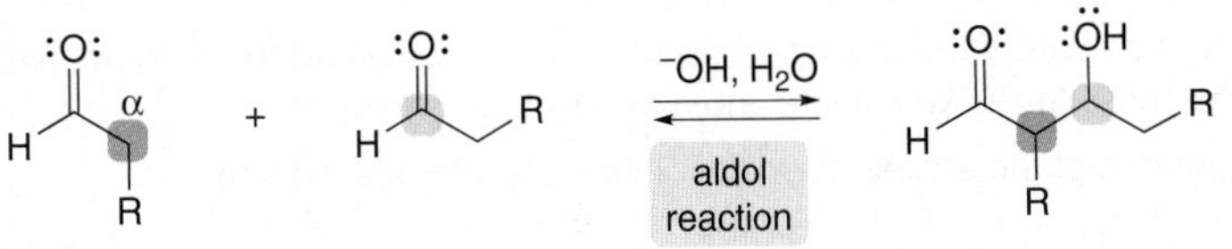

Problem 24.1 Draw the aldol product formed from each compound.

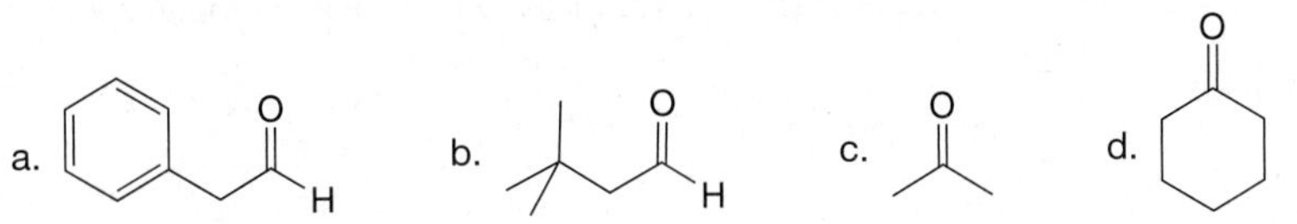

Problem 24.2 Which carbonyl compounds do *not* undergo an aldol reaction when treated with $^{-}$OH in H_2O?

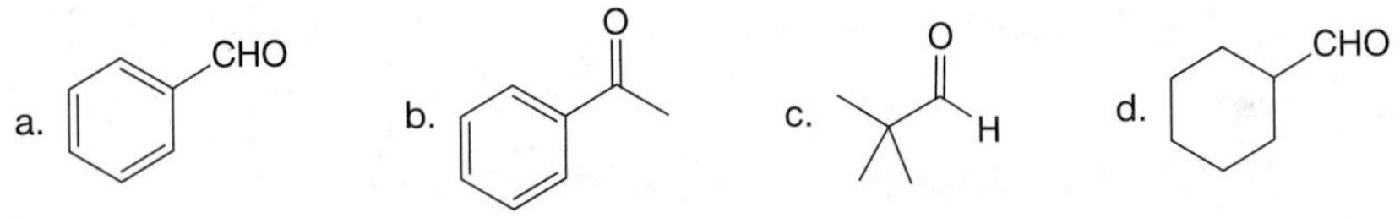

24.1B Retro-Aldol Reaction

Because an aldol reaction is a reversible equilibrium, the β-hydroxy carbonyl products can be re-converted to carbonyl starting materials with heat in the presence of base by a **retro-aldol reaction.** The conversion of 3-hydroxybutanal to acetaldehyde is a retro-aldol reaction, which results in cleavage of the carbon–carbon bond between the α and β carbons.

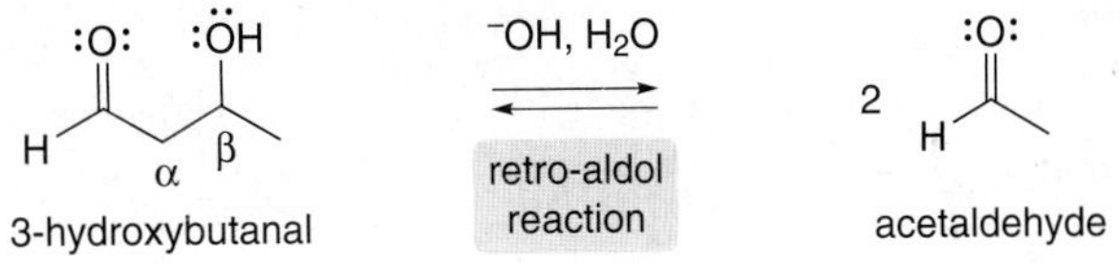

The three-step mechanism of a retro-aldol reaction is just the reverse of an aldol reaction, as shown in Mechanism 24.2. A retro-aldol reaction is a key step in the metabolism of glucose, as we will see in Section 24.8.

Mechanism 24.2 The Retro-Aldol Reaction

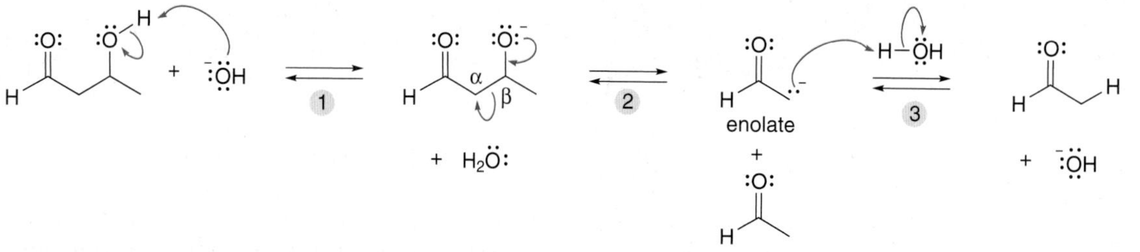

1 The base removes the OH proton to form an alkoxide.

2 An electron pair of the alkoxide is used to form a C=O and the carbon–carbon bond between the α and β carbons is cleaved. This process forms an enolate and a molecule of aldehyde.

3 Protonation of the enolate forms another molecule of aldehyde.

Problem 24.3 What ketone or aldehyde is obtained when each compound is heated in the presence of aqueous base?

a. 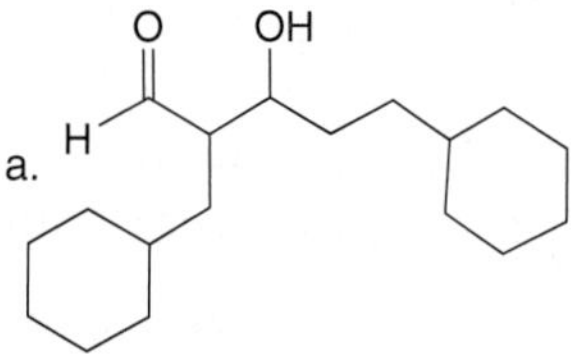

b. 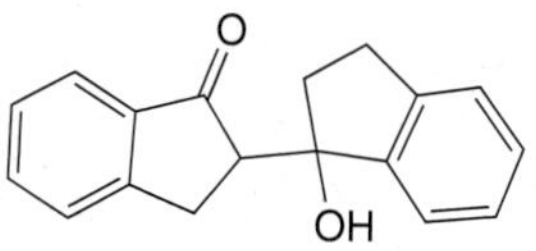

24.1C Dehydration of the Aldol Product

The β-hydroxy carbonyl compounds formed in the aldol reaction dehydrate more readily than other alcohols. In fact, under the basic reaction conditions, the initial aldol product is often not isolated. Instead, **it loses the elements of H_2O from the α and β carbons to form an α,β-unsaturated carbonyl compound, a conjugated product.**

All alcohols—including β-hydroxy carbonyl compounds—dehydrate in the presence of acid. Only β-hydroxy carbonyl compounds dehydrate in the presence of base.

[1] 2 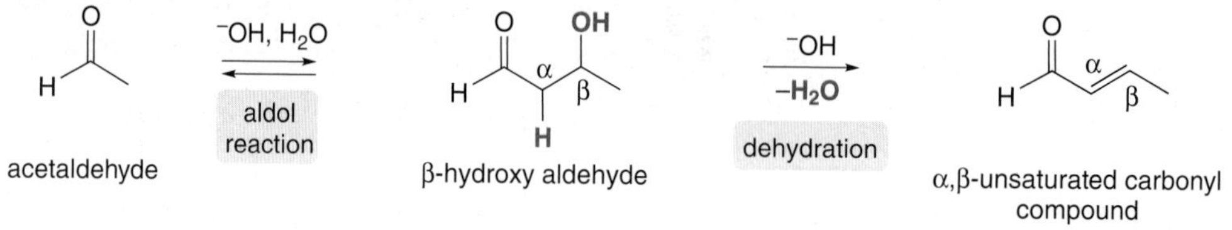

An aldol reaction is often called an **aldol condensation,** because the β-hydroxy carbonyl compound that is initially formed loses H_2O by dehydration. ***A condensation reaction* is one in which a small molecule, in this case H_2O, is eliminated during a reaction.**

[2] 2 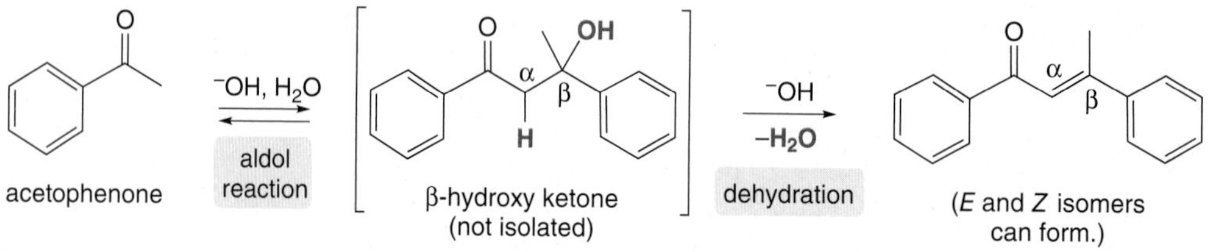

It may or may not be possible to isolate the β-hydroxy carbonyl compound under the conditions of the aldol reaction. When the α,β-unsaturated carbonyl compound is further conjugated with a carbon–carbon double bond or a benzene ring, as in the case of Reaction [2], **elimination of H_2O is spontaneous** and the β-hydroxy carbonyl compound cannot be isolated.

The mechanism of dehydration consists of two steps: **deprotonation followed by loss of $^{-}$OH,** as shown in Mechanism 24.3.

Mechanism 24.3 Dehydration of β-Hydroxy Carbonyl Compounds with Base

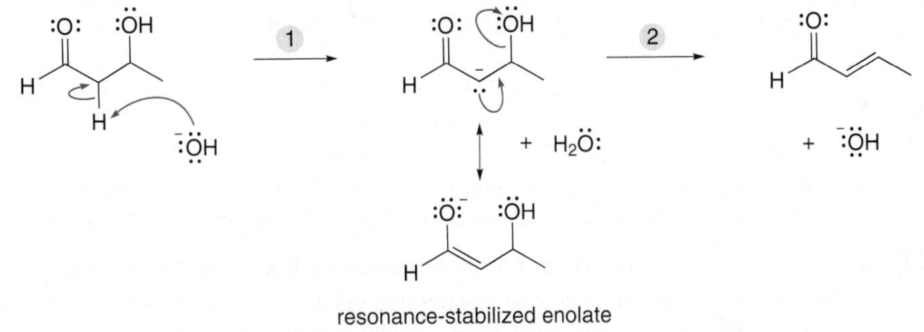

1 The base removes a proton on the α carbon to form a resonance-stabilized enolate.

2 $^{-}$OH is eliminated as the electron pair of the enolate forms the new π bond.

Like E1 elimination, E1cB requires two steps. Unlike E1, though, the intermediate in E1cB is a *carbanion,* not a carbocation. E1cB stands for **Elimination, unimolecular, conjugate base.**

This elimination mechanism, called the **E1cB mechanism,** differs from the two mechanisms of elimination, E1 and E2, which were discussed in Chapter 8. The E1cB mechanism involves two steps, and proceeds by way of an **anionic** intermediate.

Regular alcohols dehydrate only in the presence of acid but not base, because hydroxide is a poor leaving group. When the hydroxy group is β to a carbonyl group, however, loss of H and OH from the α and β carbons forms a **conjugated double bond,** and the stability of the conjugated system makes up for having such a poor leaving group.

Dehydration of the initial β-hydroxy carbonyl compound drives the equilibrium of an aldol reaction to the right, thus favoring product formation. Once the conjugated α,β-unsaturated carbonyl compound forms, it is *not* re-converted to the β-hydroxy carbonyl compound.

The E1cB mechanism is especially common in biological pathways. For example, the dehydration of β-hydroxy thioesters to α,β-unsaturated thioesters, a process that occurs during the biosynthesis of fatty acids, follows an E1cB mechanism.

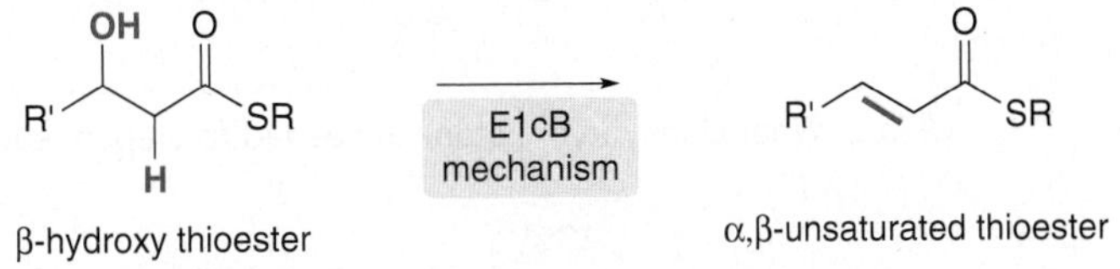

Problem 24.4 What unsaturated carbonyl compound is formed by dehydration of each β-hydroxy carbonyl compound?

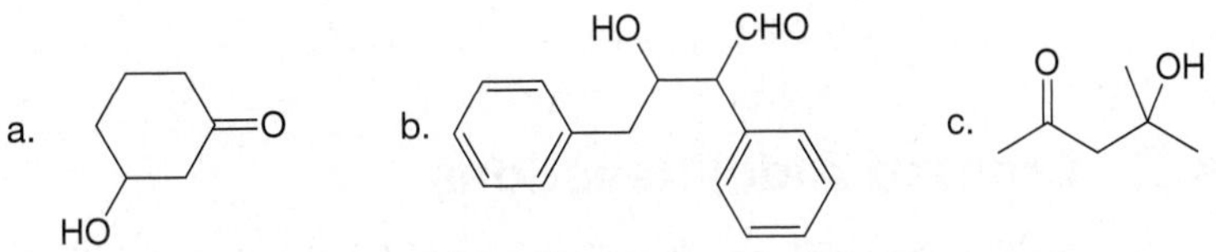

24.1D Retrosynthetic Analysis

To utilize the aldol reaction in synthesis, you must be able to determine which aldehyde or ketone is needed to prepare a particular β-hydroxy carbonyl compound or α,β-unsaturated carbonyl compound—that is, you must be able to **work backwards, in the retrosynthetic direction.**

How To Synthesize a Compound Using the Aldol Reaction

Example What starting material is needed to prepare each compound by an aldol reaction?

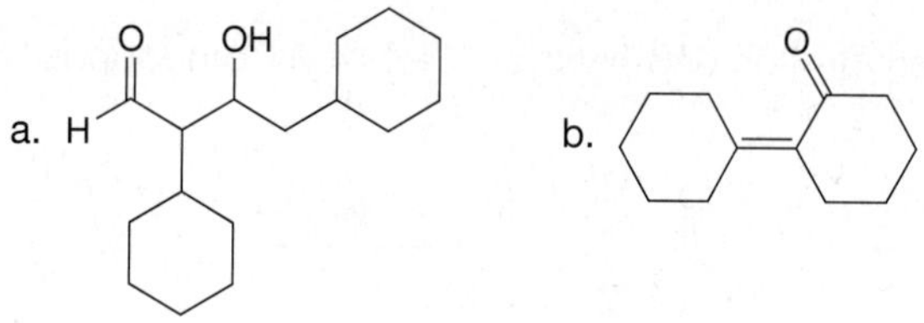

Step [1] **Locate the α and β carbons of the carbonyl group.**

- When a carbonyl group has two different α carbons, choose the side that contains the OH group (in a β-hydroxy carbonyl compound) or is part of the C=C (in an α,β-unsaturated carbonyl compound).

Step [2] **Break the molecule into two components between the α and β carbons.**

- The α carbon and all remaining atoms bonded to it belong to one carbonyl component. The β carbon and all remaining atoms bonded to it belong to the other carbonyl component. Both components are identical in all aldols we have thus far examined.

a. Break the molecule into two halves at the labeled bond.

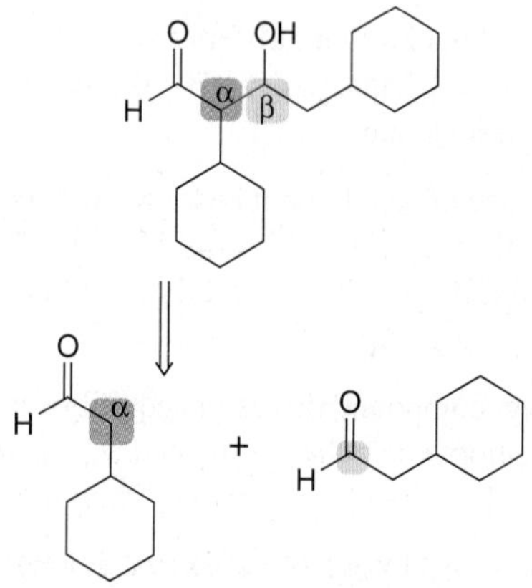

two molecules of the same aldehyde

b. Break the molecule into two halves at the labeled bond.

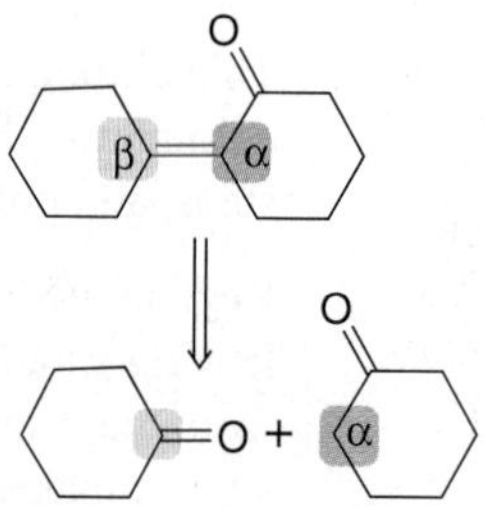

two molecules of cyclohexanone

Problem 24.5 What aldehyde or ketone is needed to prepare each compound by an aldol reaction?

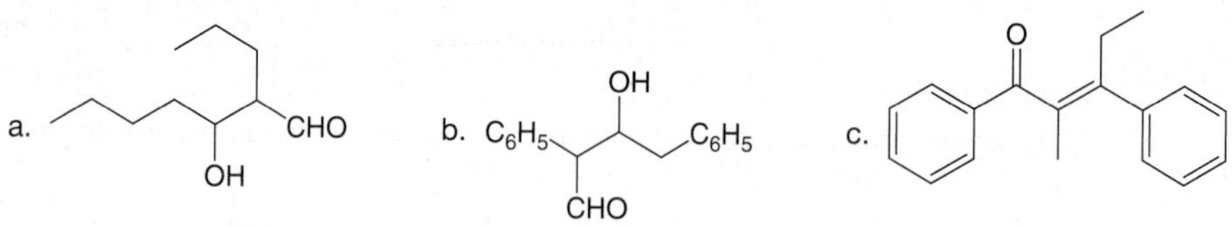

24.2 Crossed Aldol Reactions

In all of the aldol reactions discussed so far, the electrophilic carbonyl and the nucleophilic enolate have originated from the *same* aldehyde or ketone. Sometimes, though, it is possible to carry out an aldol reaction between two *different* carbonyl compounds.

- **An aldol reaction between two different carbonyl compounds is called a *crossed aldol* or *mixed aldol reaction.***

24.2A A Crossed Aldol Reaction with Two Different Aldehydes, Both Having α H Atoms

When two different aldehydes, both having α H atoms, are combined in an aldol reaction, four different β-hydroxy carbonyl compounds are formed. Four products form, not one, because *both* aldehydes can lose an acidic α hydrogen atom and form an enolate in the presence of base. *Both* enolates can then react with *both* carbonyl compounds, as shown for acetaldehyde and propanal in the following reaction scheme.

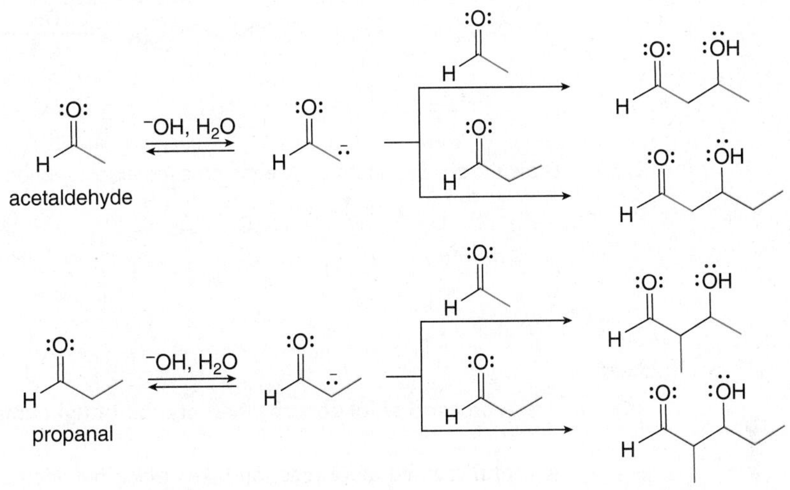

- **Conclusion: When two different aldehydes have α hydrogens, a crossed aldol reaction is *not* synthetically useful.**

24.2B Synthetically Useful Crossed Aldol Reactions

Crossed aldols are synthetically useful in two different situations.

- **A crossed aldol occurs when only one carbonyl component has α H atoms.**

When one carbonyl compound has no α hydrogens, a crossed aldol reaction often leads to one product. Two common carbonyl compounds with no α hydrogens used for this purpose are **formaldehyde ($CH_2{=}O$)** and **benzaldehyde (C_6H_5CHO).**

For example, reaction of C_6H_5CHO (as the electrophile) with either acetaldehyde (CH_3CHO) or acetone [$(CH_3)_2C{=}O$] in the presence of base forms a single α,β-unsaturated carbonyl compound after dehydration.

[1]

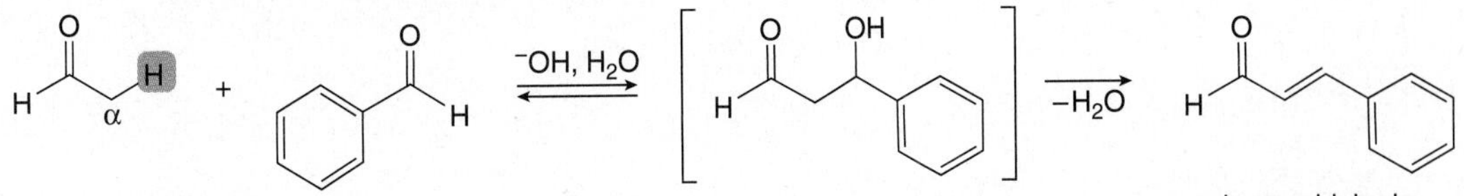

cinnamaldehyde (component of cinnamon)

(+ *Z* isomer)

[2]

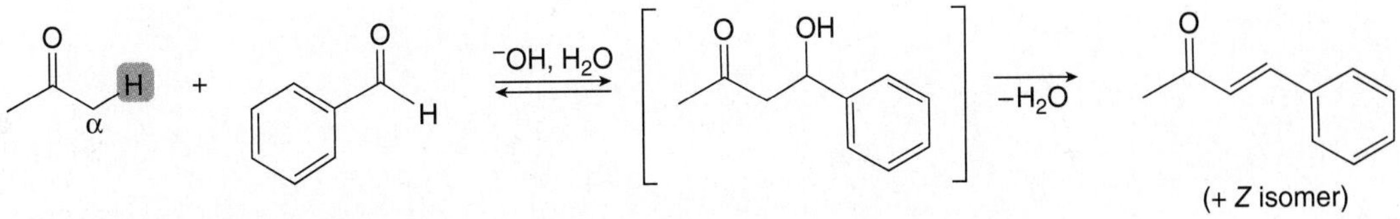

(+ *Z* isomer)

The yield of a single crossed aldol product is increased further if the electrophilic carbonyl component is relatively unhindered (as is the case with most aldehydes), and if it is used in excess.

Problem 24.6 2-Pentylcinnamaldehyde, commonly called flosal, is a perfume ingredient with a jasmine-like odor. Flosal is an α,β-unsaturated aldehyde made by a crossed aldol reaction between benzaldehyde (C_6H_5CHO) and heptanal ($CH_3CH_2CH_2CH_2CH_2CH_2CHO$), followed by dehydration. Draw a stepwise mechanism for the following reaction that prepares flosal.

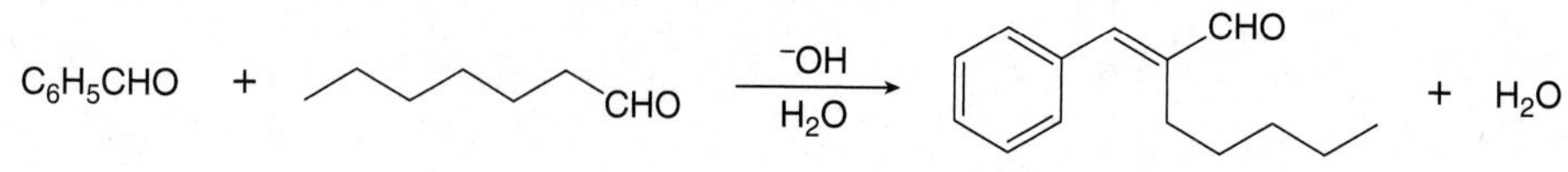

flosal
(perfume component)

Problem 24.7 Draw the products formed in each crossed aldol reaction.

a. 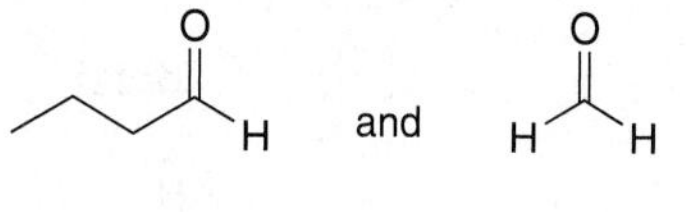

b. 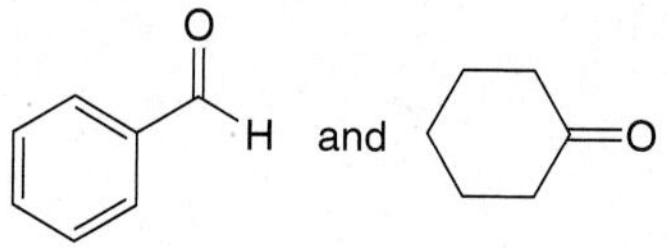

- **A crossed aldol occurs when one carbonyl component has especially acidic α H atoms.**

A useful crossed aldol reaction takes place between an aldehyde or ketone and a β-dicarbonyl (or similar) compound.

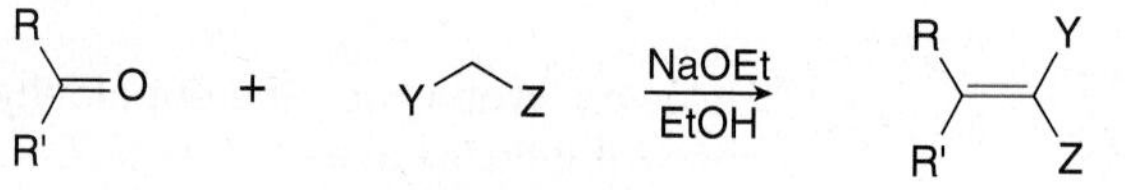

R' = H or alkyl

Y, Z = COOEt, CHO, COR, CN
β-dicarbonyl compound
(and related compounds)

new C–C σ and π bonds in red

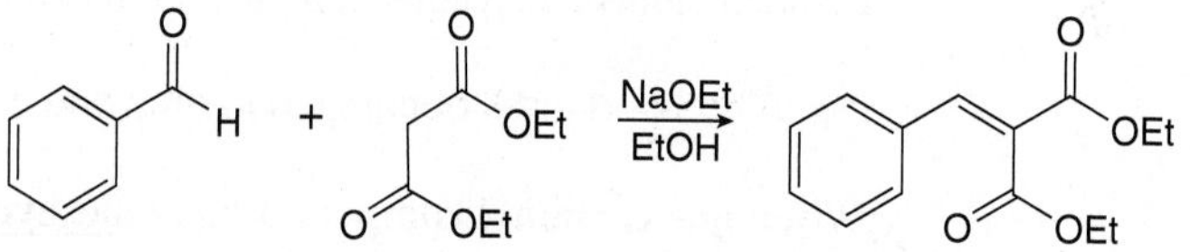

benzaldehyde diethyl malonate

As we learned in Section 23.3, the α hydrogens between two carbonyl groups are especially acidic, and so they are more readily removed than other α H atoms. As a result, **the β-dicarbonyl**

Figure 24.2
Crossed aldol reaction between benzaldehyde and $CH_2(COOEt)_2$

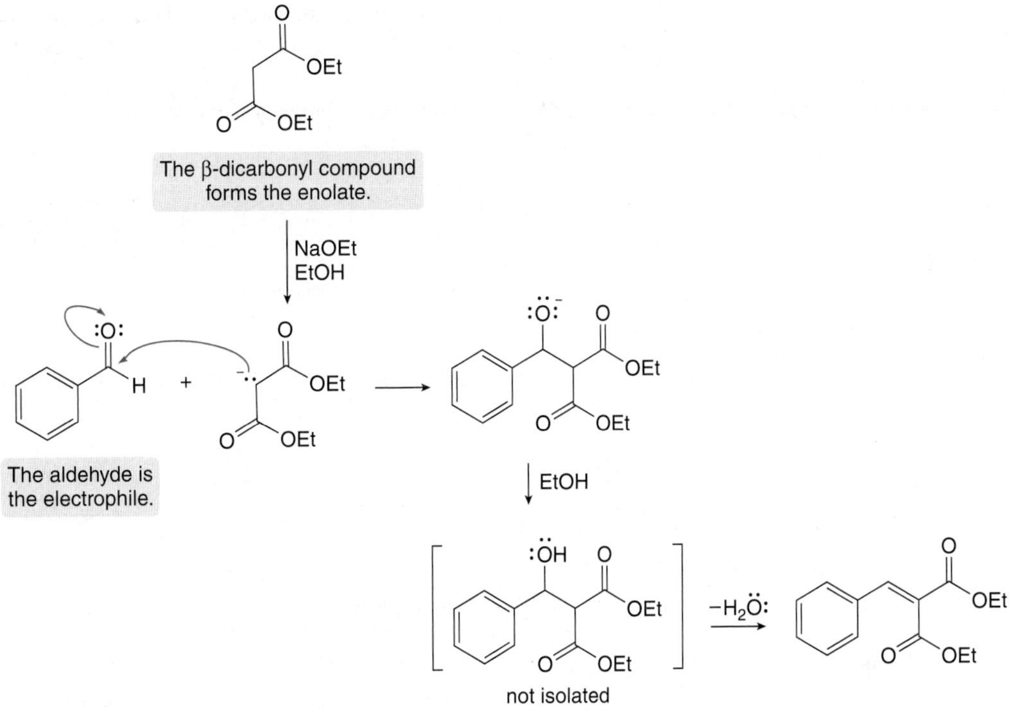

compound always becomes the enolate component of the aldol reaction. Figure 24.2 shows the steps for the crossed aldol reaction between diethyl malonate and benzaldehyde. In this type of crossed aldol reaction, the initial β-hydroxy carbonyl compound *always* loses water to form the highly conjugated product.

β-Dicarbonyl compounds are sometimes called **active methylene compounds** because they are more reactive toward base than other carbonyl compounds. **1,3-Dinitriles** and **α-cyano carbonyl compounds** are also active methylene compounds.

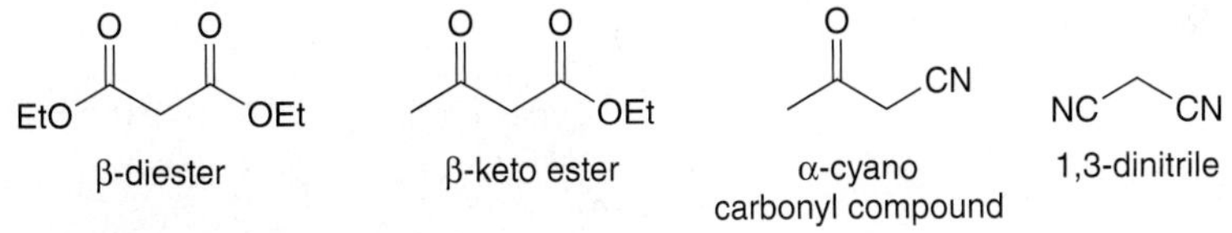

Problem 24.8 Draw the products formed in the crossed aldol reaction of phenylacetaldehyde ($C_6H_5CH_2CHO$) with each compound: (a) $CH_2(COOEt)_2$; (b) $CH_2(COCH_3)_2$; (c) CH_3COCH_2CN.

24.3 Directed Aldol Reactions

A **directed aldol reaction** is a variation of the crossed aldol reaction that clearly defines which carbonyl compound becomes the nucleophilic enolate and which reacts at the electrophilic carbonyl carbon. The strategy of a directed aldol reaction is as follows:

[1] Prepare the enolate of one carbonyl component with LDA.

[2] Add the second carbonyl compound (the electrophile) to this enolate.

Because the steps are done sequentially and a strong nonnucleophilic base is used to form the enolate of one carbonyl component only, a variety of carbonyl substrates can be used in the reaction. Both carbonyl components can have α hydrogens because only one enolate is prepared with LDA. Also, when an unsymmetrical ketone is used, LDA selectively forms the **less substituted, kinetic enolate.**

Sample Problem 24.1 illustrates the steps of a directed aldol reaction between a ketone and an aldehyde, both of which have α hydrogens.

Sample Problem 24.1 Draw the product of the following directed aldol reaction.

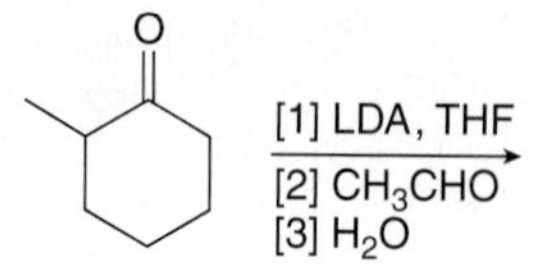

2-methylcyclohexanone

Solution

2-Methylcyclohexanone forms an enolate on the less substituted carbon, which then reacts with the electrophile, CH_3CHO.

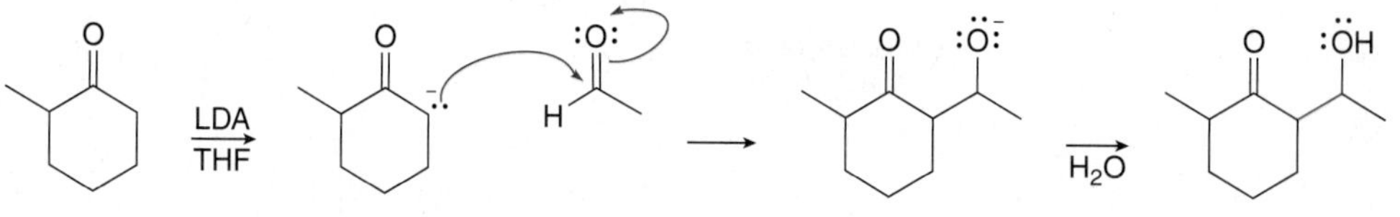

Figure 24.3 illustrates how a directed aldol reaction was used in the synthesis of **periplanone B,** the sex pheromone of the female American cockroach.

Figure 24.3 A directed aldol reaction in the synthesis of periplanone B

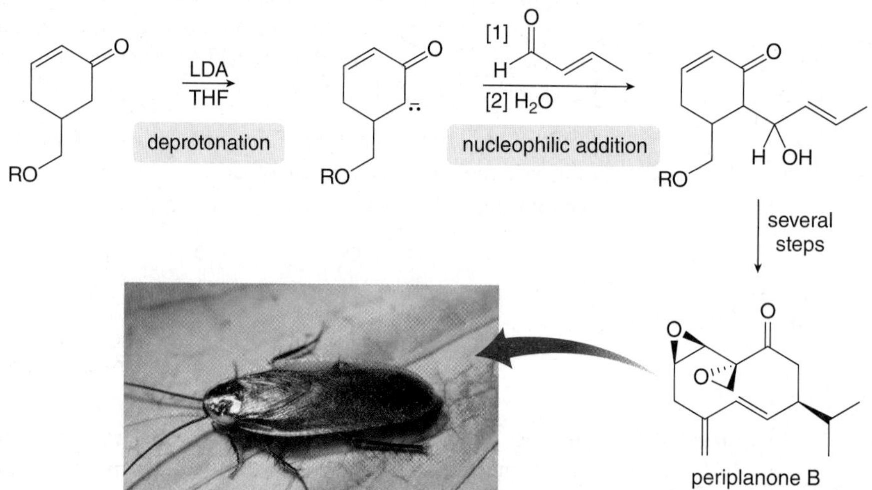

To determine the needed carbonyl components for a directed aldol, follow the same strategy used for a regular aldol reaction in Section 24.1D, as shown in Sample Problem 24.2.

Sample Problem 24.2 What starting materials are needed to prepare *ar*-turmerone using a directed aldol reaction? *ar*-Turmerone is a principal component of the essential oil derived from turmeric root.

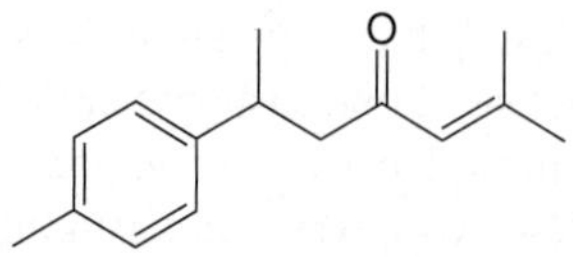

ar-turmerone

The dried and ground root of the turmeric plant, a tropical perennial in the ginger family, is an essential ingredient in curry powder.

Solution

When the desired product is an α,β-unsaturated carbonyl compound, identify the α and β carbons that are part of the C=C, and break the molecule into two components between these carbons.

Break the molecule into two halves.

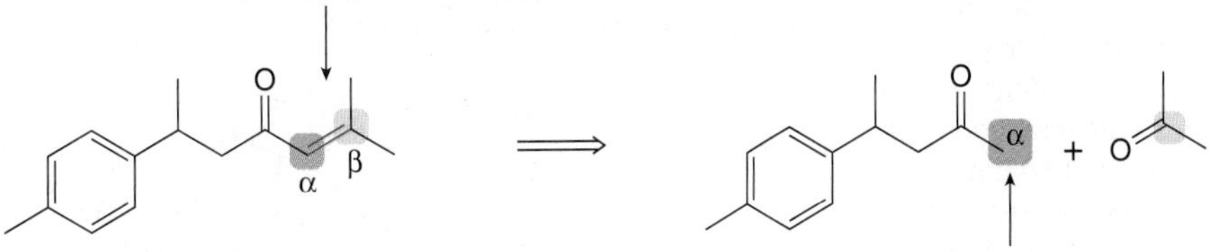

ar-turmerone

Make the enolate here.

Problem 24.9 What carbonyl starting materials are needed to prepare each compound using a directed aldol reaction?

a. 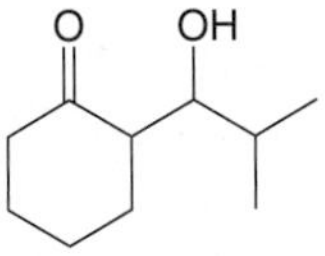

b. 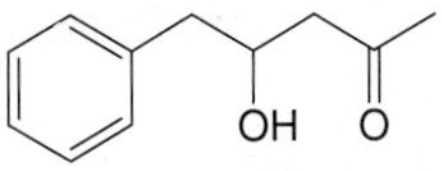

c. 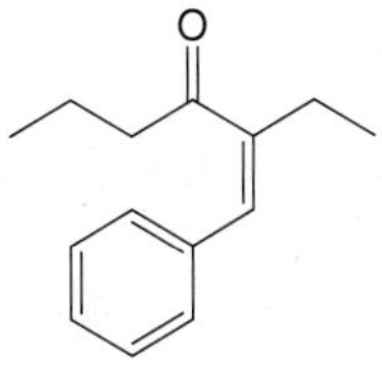

Problem 24.10 Zingerone, a spicy-sweet component of cooked ginger, can be converted to its protected TBDMS ether **A,** as we learned in Chapter 20. How can **A** be converted to gingerol, a compound present in fresh ginger, using a directed aldol reaction as a key step?

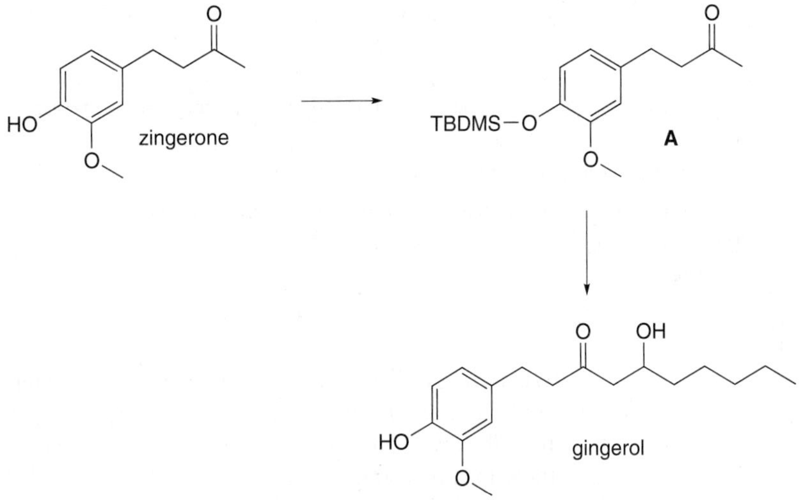

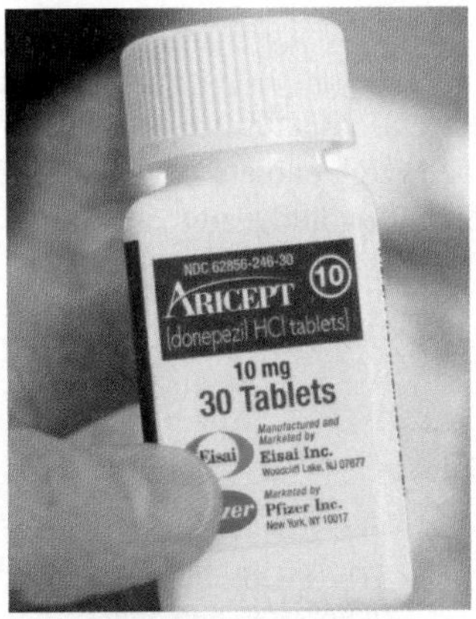

Donepezil (trade name Aricept) is a drug used to improve cognitive function in patients suffering from dementia and Alzheimer's disease.

Problem 24.11 A key step in the synthesis of donepezil (trade name Aricept) is a directed aldol reaction that forms α,β-unsaturated carbonyl compound **X.** What carbonyl starting materials are needed to prepare **X** using a directed aldol reaction? What reagents are needed to convert **X** to donepezil?

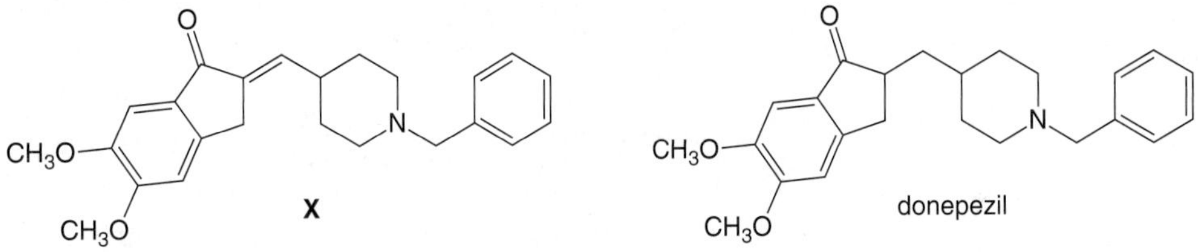

24.4 Intramolecular Aldol Reactions

Hexane-2,5-dione is called a **1,4-dicarbonyl compound** to emphasize the relative positions of its carbonyl groups. 1,4-Dicarbonyl compounds are starting materials for synthesizing **five-membered rings.**

Aldol reactions with dicarbonyl compounds can be used to make five- and six-membered rings. The enolate formed from one carbonyl group is the nucleophile, and the carbonyl carbon of the other carbonyl group is the electrophile. For example, treatment of hexane-2,5-dione with base forms a five-membered ring.

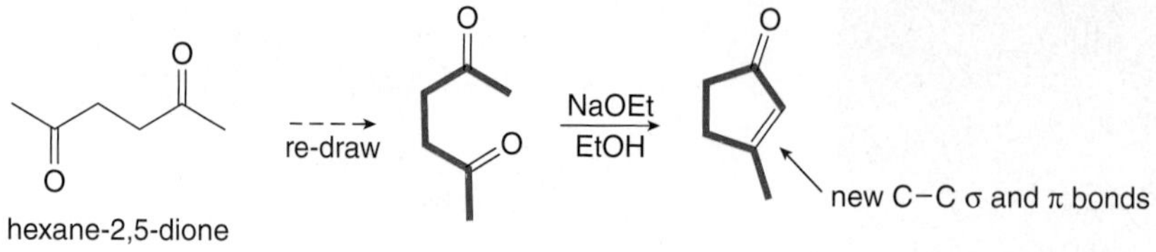

The steps in this process, shown in Mechanism 24.4, are no different from the general mechanisms of the aldol reaction and dehydration described previously in Section 24.1.

Mechanism 24.4 The Intramolecular Aldol Reaction

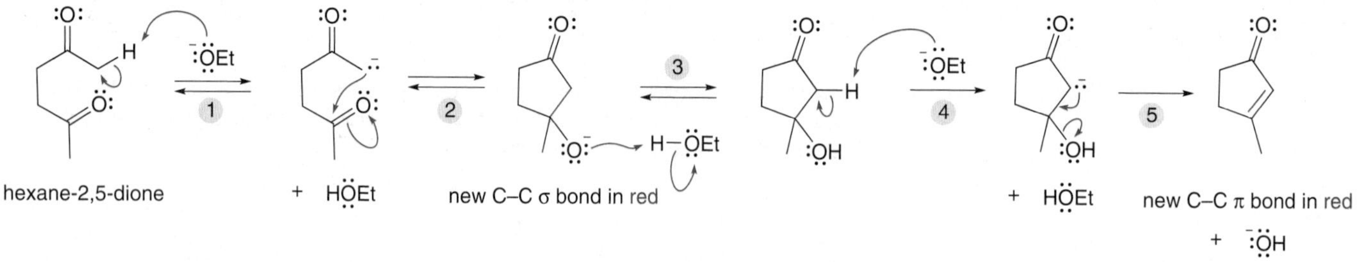

1. The base removes a proton on the α carbon to form a resonance-stabilized enolate.
2. Nucleophilic attack of the enolate on the electrophilic carbonyl *in the same molecule* forms a new C–C σ bond, generating the five-membered ring.
3. Protonation of the alkoxide forms the β-hydroxy carbonyl compound.

4 – 5 Dehydration occurs by the two-step E1cB mechanism—loss of a proton to form an enolate and elimination of ⁻OH to form a π bond.

When hexane-2,5-dione is treated with base in Step [1], two different enolates are possible—enolates **A** and **B,** formed by removal of H_a and H_b, respectively. Although enolate **A** goes on to form the five-membered ring, intramolecular cyclization using enolate **B** would lead to a strained three-membered ring.

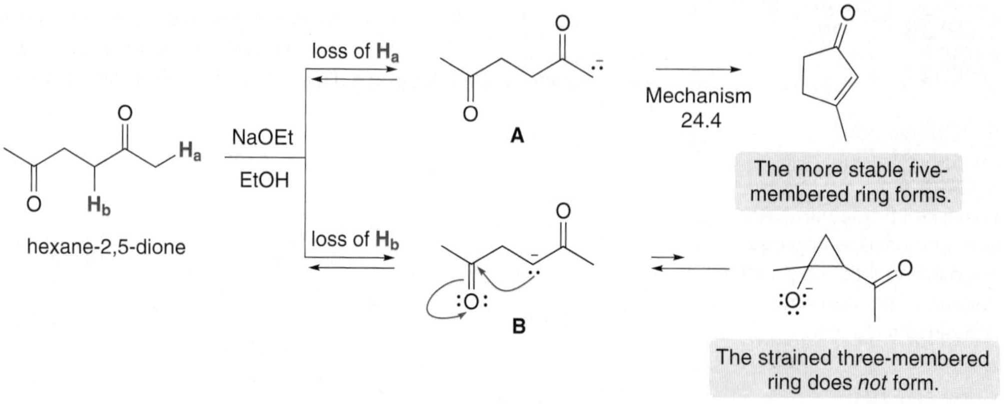

Because the three-membered ring is much higher in energy than the enolate starting material, equilibrium greatly favors the starting materials and the **three-membered ring does not form.** Under the reaction conditions, enolate **B** is re-protonated to form hexane-2,5-dione, because all steps except dehydration are equilibria. **Thus, equilibrium favors formation of the more stable five-membered ring over the much less stable three-membered ring.**

In a similar fashion, six-membered rings can be formed from the intramolecular aldol reaction of **1,5-dicarbonyl compounds.**

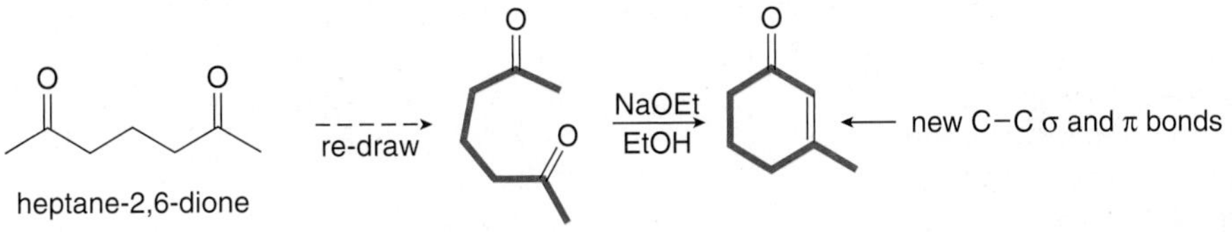

The synthesis of the female sex hormone **progesterone** by W. S. Johnson and co-workers at Stanford University is considered one of the classics in total synthesis. The last six-membered ring needed in the steroid skeleton was prepared by a two-step sequence using an intramolecular aldol reaction, as shown in Figure 24.4.

Figure 24.4 The synthesis of progesterone using an intramolecular aldol reaction

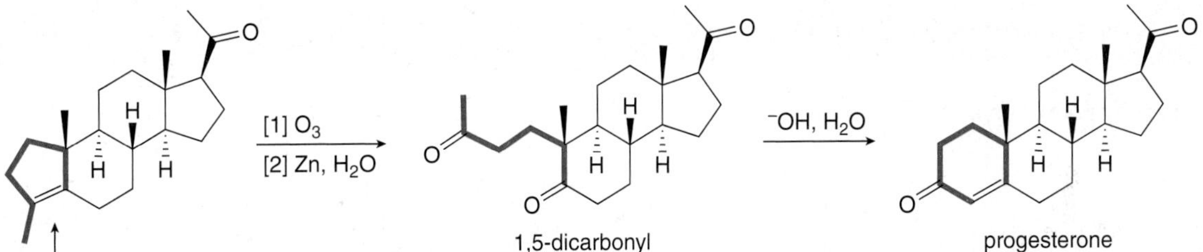

- Oxidative cleavage of the alkene with O_3, followed by Zn, H_2O (Section 12.10), gives the 1,5-dicarbonyl compound.
- Intramolecular aldol reaction of the 1,5-dicarbonyl compound with dilute ^-OH in H_2O solution forms progesterone.
- **This two-step reaction sequence converts a five-membered ring into a six-membered ring.** Reactions that synthesize larger rings from smaller ones are called **ring expansion reactions.**

Problem 24.12 What cyclic product is formed when each 1,5-dicarbonyl compound is treated with aqueous ^-OH?

a. 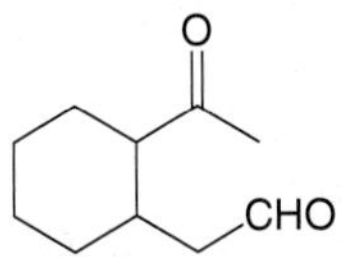

b. 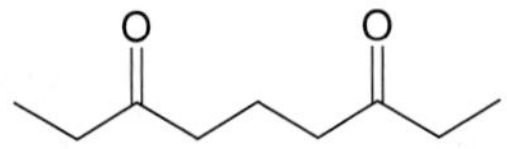

Problem 24.13 Following the two-step reaction sequence depicted in Figure 24.4, write out the steps needed to convert **A** to **B.**

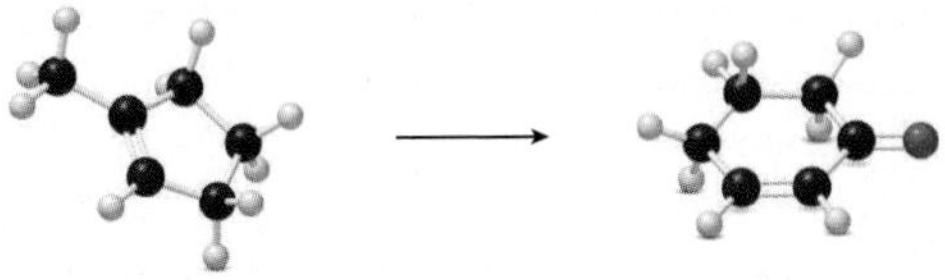

A **B**

24.5 The Claisen Reaction

The **Claisen reaction** is the second general reaction of enolates with other carbonyl compounds. In the Claisen reaction, two molecules of an ester react with each other in the presence of an alkoxide base to form a **β-keto ester.** For example, treatment of ethyl acetate with NaOEt forms ethyl acetoacetate after protonation with aqueous acid.

Unlike the aldol reaction, which is base-catalyzed, a full equivalent of base is needed to deprotonate the β-keto ester formed in Step [3] of the Claisen reaction.

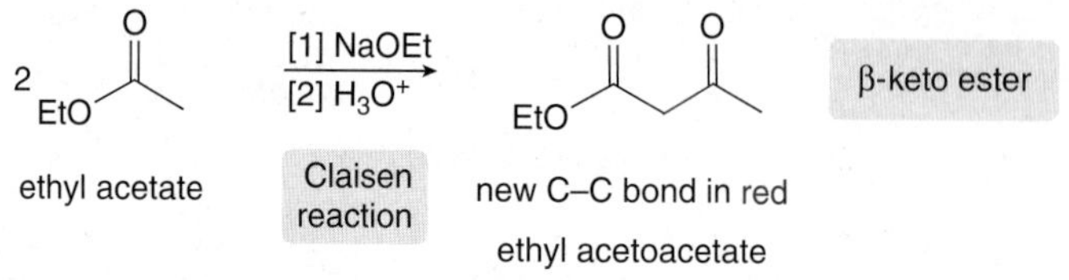

The mechanism for the Claisen reaction (Mechanism 24.5) resembles the mechanism of an aldol reaction in that it involves nucleophilic addition of an enolate to an electrophilic carbonyl group. Because esters have a leaving group on the carbonyl carbon, however, loss of a leaving group occurs to form the product of **substitution, *not* addition.**

Mechanism 24.5 The Claisen Reaction

Part [1] Formation of the β-keto ester

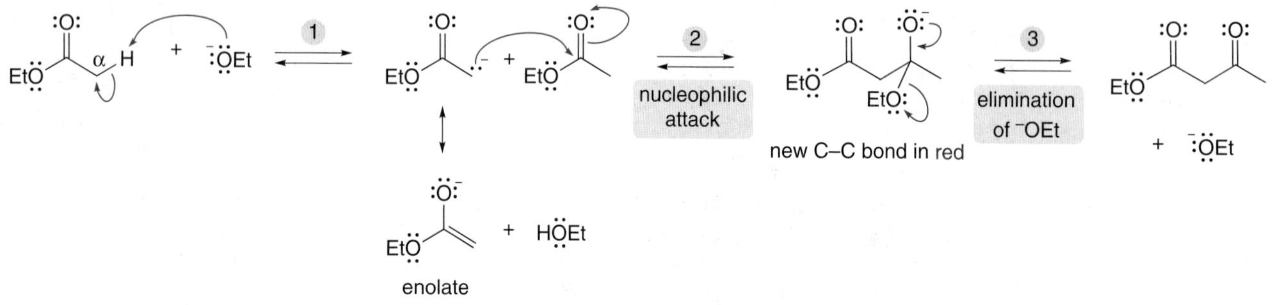

1 The base removes a proton on the α carbon to form a resonance-stabilized enolate.

2 Nucleophilic attack of the enolate on an electrophilic carbonyl in another molecule of ester forms a new C–C bond.

3 Loss of the leaving group (⁻OEt) forms a β-keto ester.

Part [2] Deprotonation and protonation

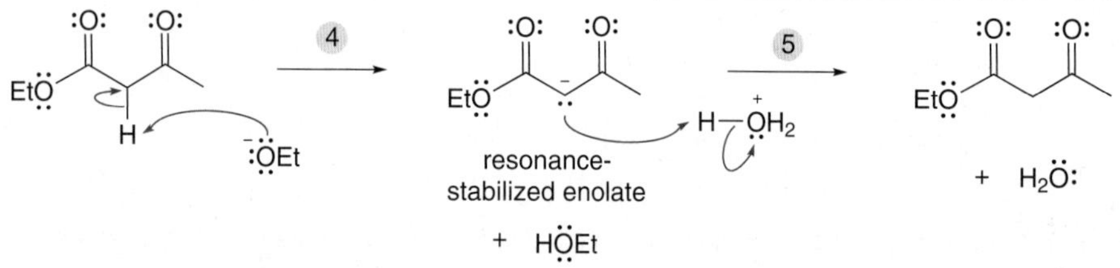

4 Because the β-keto ester formed in Step [3] has especially acidic protons between its two carbonyl groups, the base removes a proton to form a resonance-stabilized enolate.

5 Protonation of the enolate with strong acid re-forms the β-keto ester.

Because the generation of a resonance-stabilized enolate from the product β-keto ester drives the Claisen reaction (Step [4] of Mechanism 24.5), **only esters with two or three hydrogens on the α carbon undergo this reaction;** that is, esters must have the general structure CH_3CO_2R' or RCH_2CO_2R'.

- **Keep in mind: The characteristic reaction of esters is nucleophilic substitution. A Claisen reaction is a nucleophilic substitution in which an enolate is the nucleophile.**

Figure 24.5
The Claisen reaction—An example of nucleophilic substitution

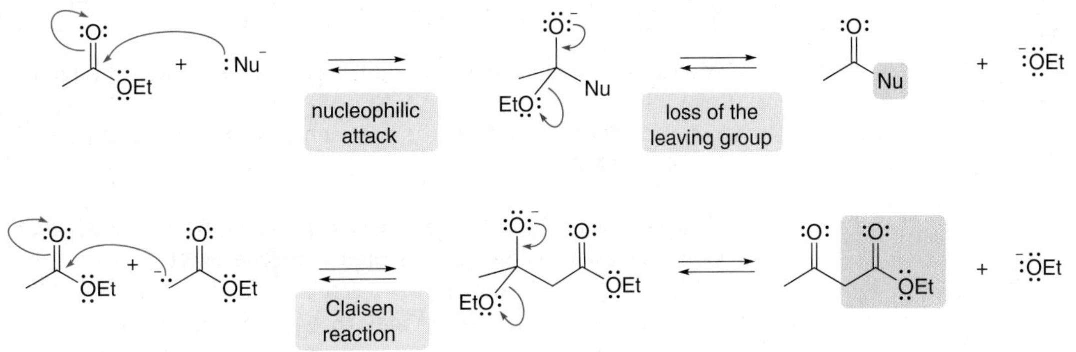

- Esters react by **nucleophilic substitution.** In a Claisen reaction, an enolate is the nucleophile that adds to the carbonyl group.

Figure 24.5 compares the general reaction for nucleophilic substitution of an ester with the Claisen reaction. Sample Problem 24.3 reinforces the basic features of the Claisen reaction.

Sample Problem 24.3 Draw the product of the following Claisen reaction.

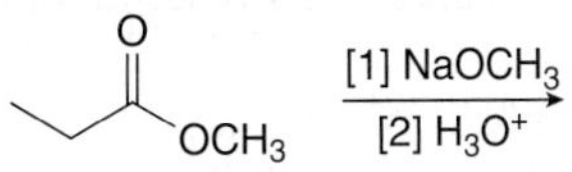

Solution

To draw the product of any Claisen reaction, form a new carbon–carbon bond between the α carbon of one ester and the carbonyl carbon of another ester, with elimination of the leaving group ($^{-}OCH_3$ in this case).

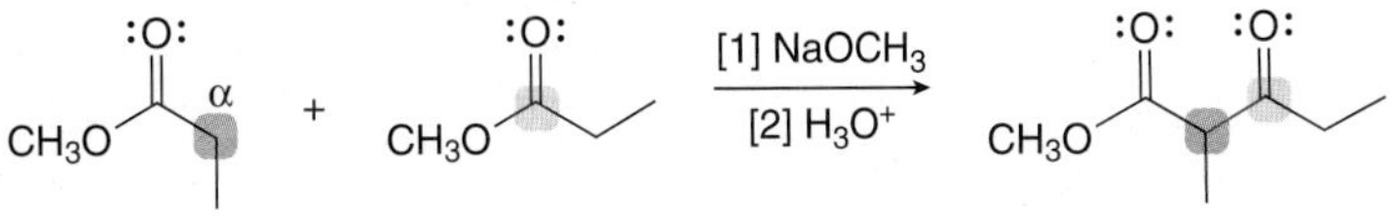

Next, write out the steps of the reaction to verify this product.

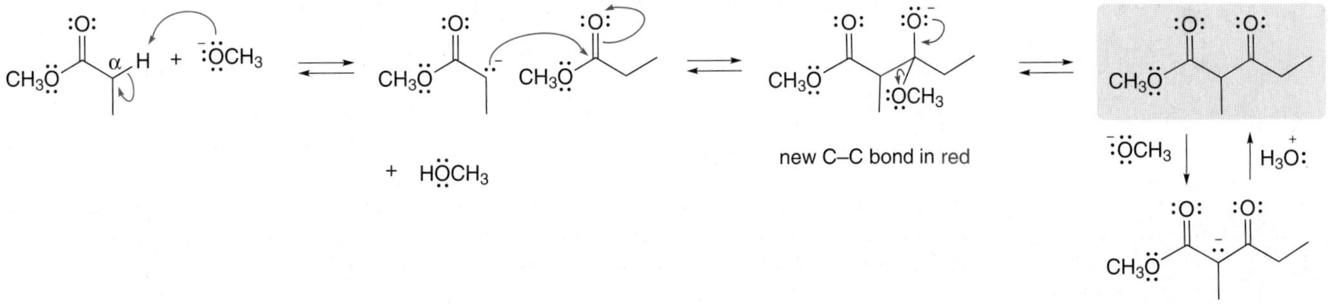

Problem 24.14 What β-keto ester is formed when each ester is used in a Claisen reaction?

a. 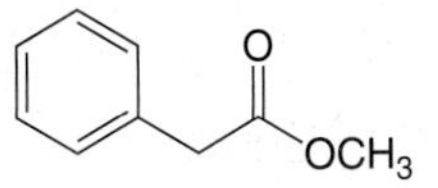

b. 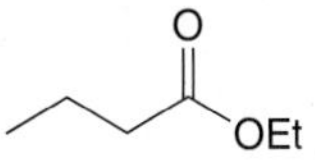

24.6 The Crossed Claisen and Related Reactions

Like the aldol reaction, it is sometimes possible to carry out a Claisen reaction with two different carbonyl components as starting materials.

- **A Claisen reaction between two different carbonyl compounds is called a *crossed Claisen reaction.***

24.6A Two Useful Crossed Claisen Reactions

A crossed Claisen reaction is synthetically useful in two different instances.

- **A crossed Claisen occurs between two different esters when only one has α hydrogens.**

When one ester has no α hydrogens, a crossed Claisen reaction often leads to one product. Common esters with no α H atoms include ethyl formate (**HCO_2Et**) and ethyl benzoate (**$C_6H_5CO_2Et$**). For example, the reaction of ethyl benzoate (as the electrophile) with ethyl acetate (which forms the enolate) in the presence of base forms predominately one β-keto ester.

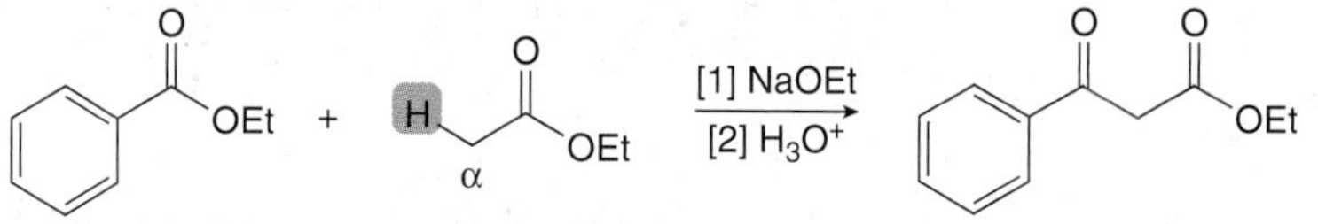

ethyl benzoate — ethyl acetate — Only this ester can form an enolate. — new C–C bond in red — β-keto ester

- **A crossed Claisen occurs between a ketone and an ester.**

The reaction of a ketone and an ester in the presence of base also forms the product of a crossed Claisen reaction. The enolate is generally formed from the ketone component, and the reaction works best when the ester has no α hydrogens. The product of this crossed Claisen reaction is a **β-dicarbonyl compound,** but ***not*** a β-keto ester.

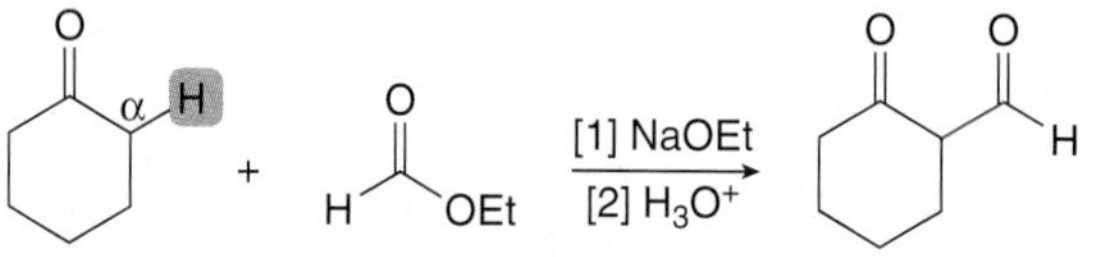

new C–C bond in red

β-dicarbonyl compound

Problem 24.15 What crossed Claisen product is formed from each pair of compounds?

a. 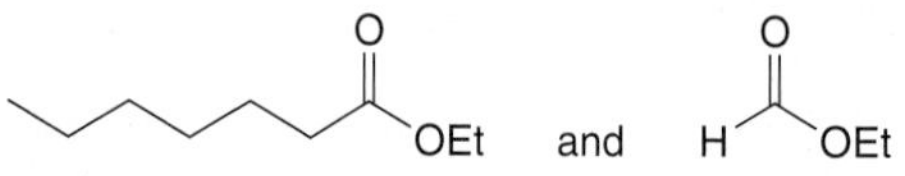

b. 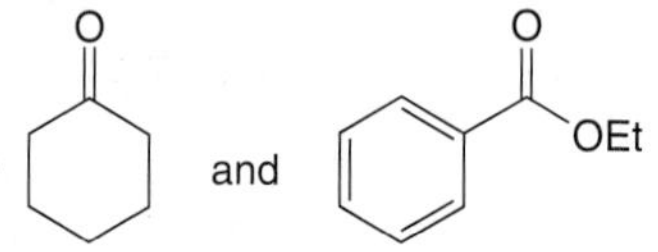

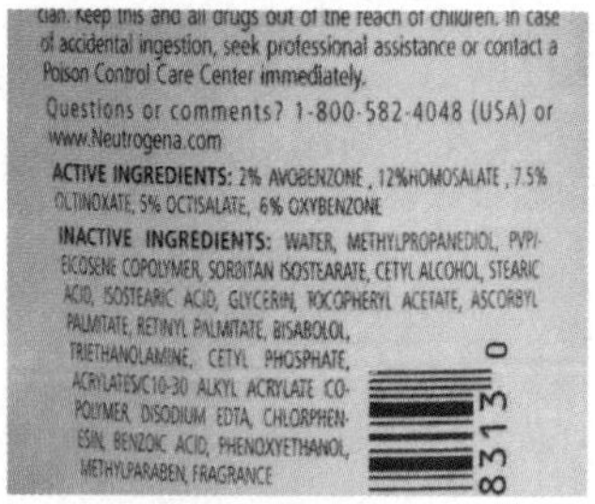

Sunscreen ingredients

Problem 24.16 Avobenzone is a conjugated compound that absorbs ultraviolet light with wavelengths in the 320–400 nm region, so it is a common ingredient in commercial sunscreens. Write out two different crossed Claisen reactions that form avobenzone.

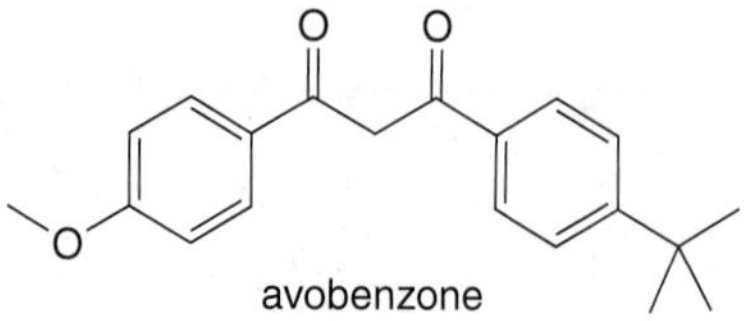

avobenzone

24.6B Other Useful Variations of the Crossed Claisen Reaction

β-Dicarbonyl compounds are also prepared by reacting an enolate with **ethyl chloroformate** and **diethyl carbonate.**

O
Cl OEt
O
EtO OEt

ethyl chloroformate diethyl carbonate

These reactions resemble a Claisen reaction because they involve the same three steps:

[1] Formation of an enolate

[2] Nucleophilic addition to a carbonyl group

[3] Elimination of a leaving group

For example, reaction of an ester enolate with diethyl carbonate yields a β-diester (Reaction [1]), whereas reaction of a ketone enolate with ethyl chloroformate forms a β-keto ester (Reaction [2]). New carbon–carbon bonds are shown in red.

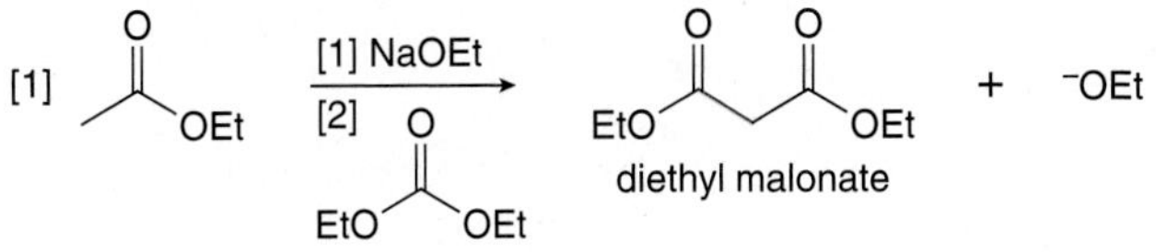
[1]
OEt
[1] NaOEt
[2]
EtO OEt
EtO OEt
diethyl malonate
+ ⁻OEt

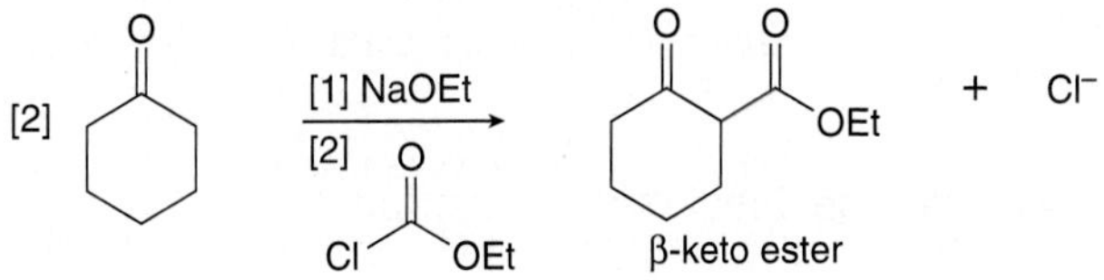
[2]
[1] NaOEt
[2]
Cl OEt
OEt
β-keto ester
+ Cl⁻

Reaction [2] is noteworthy because it provides easy access to **β-keto esters,** which are useful starting materials in the acetoacetic ester synthesis (Section 23.10). In this reaction, Cl^- is eliminated rather than ^-OEt in Step [3], because Cl^- is a better leaving group, as shown in the following steps.

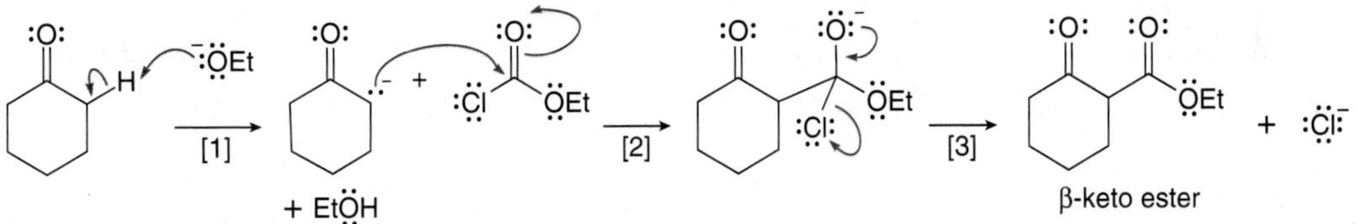
[1]
+ EtÖH
[2]
[3]
β-keto ester

Problem 24.17 Draw the products of each reaction.

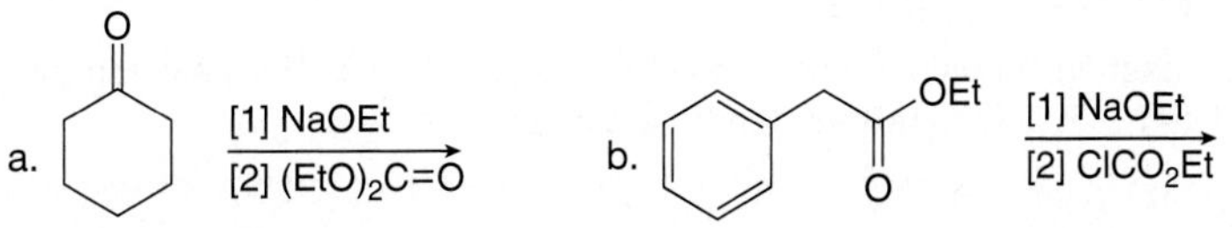
a.
[1] NaOEt
[2] (EtO)₂C=O
b.
OEt
[1] NaOEt
[2] ClCO₂Et

Problem 24.18 Two steps in a synthesis of the analgesic ibuprofen include a carbonyl condensation reaction, followed by an alkylation reaction. Identify intermediates **A** and **B** in the synthesis of ibuprofen.

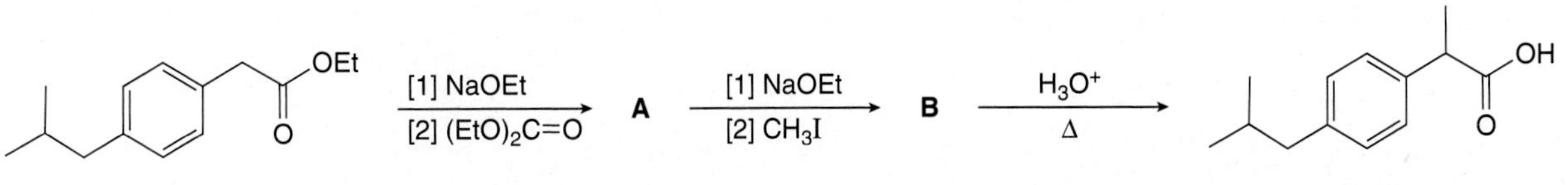
OEt
[1] NaOEt
[2] (EtO)₂C=O
A
[1] NaOEt
[2] CH₃I
B
H₃O⁺
Δ
OH

ibuprofen

24.7 The Dieckmann Reaction

Intramolecular Claisen reactions of diesters form five- and six-membered rings. The enolate of one ester is the nucleophile, and the carbonyl carbon of the other is the electrophile. An intramolecular Claisen reaction is called a **Dieckmann reaction.** Two types of diesters give good yields of cyclic products.

- **1,6-Diesters yield five-membered rings by the Dieckmann reaction.**

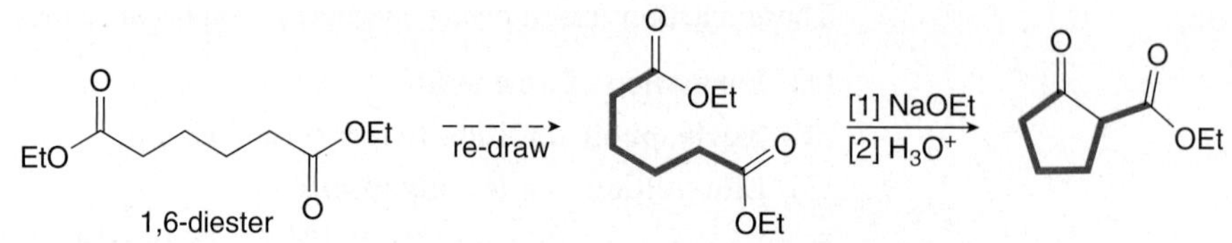

- **1,7-Diesters yield six-membered rings by the Dieckmann reaction.**

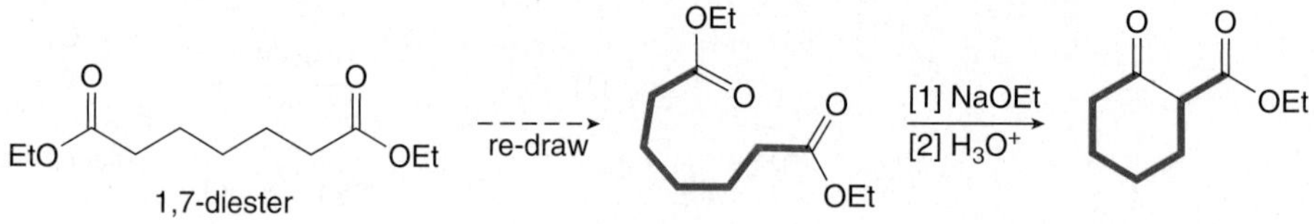

The mechanism of the Dieckmann reaction is exactly the same as the mechanism of an intermolecular Claisen reaction. It is illustrated in Mechanism 24.6 for the formation of a six-membered ring.

Mechanism 24.6 The Dieckmann Reaction

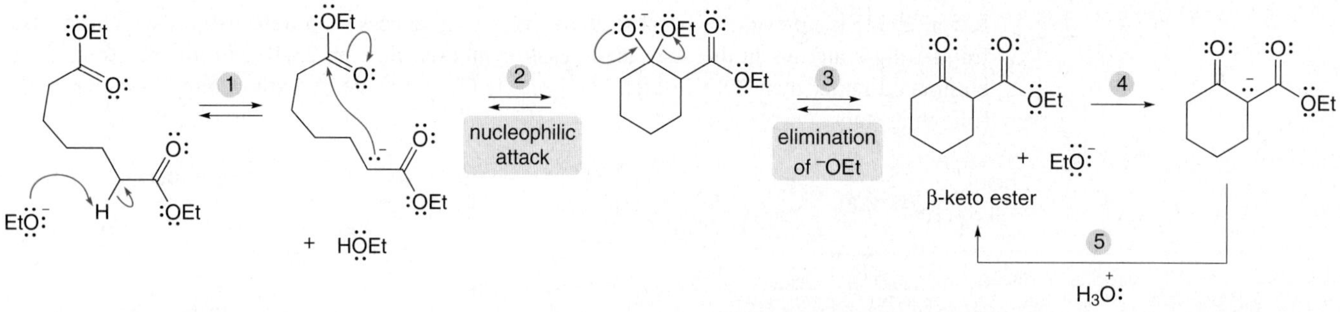

1 – 2 The base removes a proton on the α carbon to form an enolate, which attacks the electrophilic carbonyl of the other ester to form a new C–C bond.

3 Elimination of ⁻OEt forms the β-keto ester.

4 – 5 Under the basic reaction conditions, the proton between the two carbonyls is removed with base to form an enolate, which is protonated with acid to re-form the β-keto ester.

Problem 24.19 What two β-keto esters are formed in the Dieckmann reaction of the following diester?

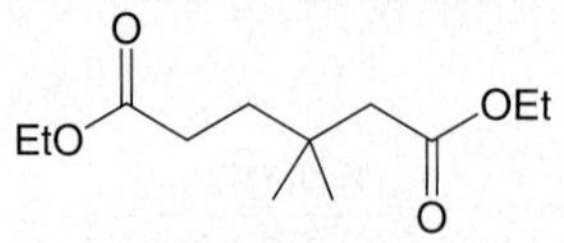

24.8 Biological Carbonyl Condensation Reactions

Aldol, retro-aldol, and Claisen reactions are key steps in several metabolic pathways. In contrast to the stepwise processes illustrated in Mechanisms 24.1, 24.2, and 24.5, the biological reactions, which occur in the presence of enzymes, are often shown with two or more bonds broken or formed at the same time. The acid or base that may be required in a particular reaction, which will be shown with the generic notation HA or :B, respectively, often comes from a functional group located at or near the active site.

24.8A Biological Aldol Reactions

A key step in the biosynthesis of glucose involves the reaction of glyceraldehyde 3-phosphate with dihydroxyacetone phosphate to form fructose 1,6-bisphosphate in the presence of an aldolase enzyme. A new carbon–carbon bond is formed between the aldehyde carbonyl of glyceraldehyde 3-phosphate and the α carbon of dihydroxyacetone phosphate. Fructose 1,6-bisphosphate is drawn as an acyclic ketone in equilibrium with its cyclic hemiacetal (Sections 21.16–21.17).

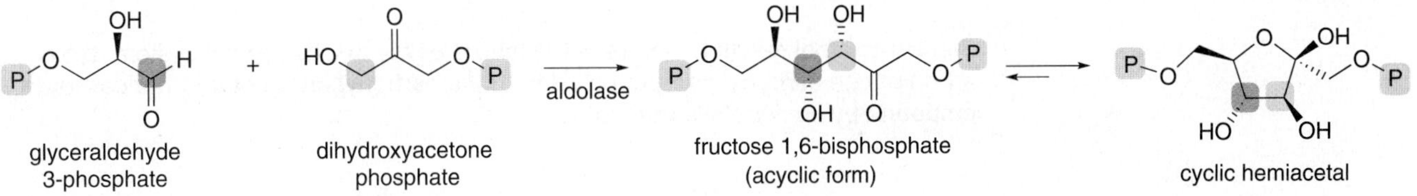

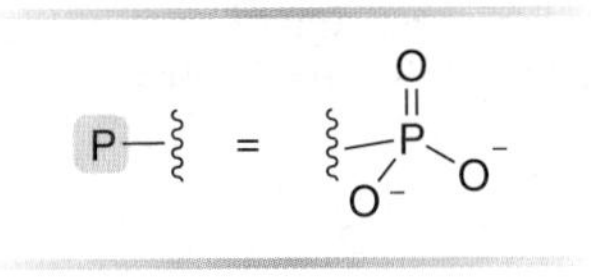

The reaction occurs by way of the tautomerization of dihydroxyacetone phosphate to an enediol, which adds to the carbonyl group of glyceraldehyde 3-phosphate. This enzyme-catalyzed crossed aldol reaction forms a single stereoisomer in the product, even though two new stereogenic centers are generated.

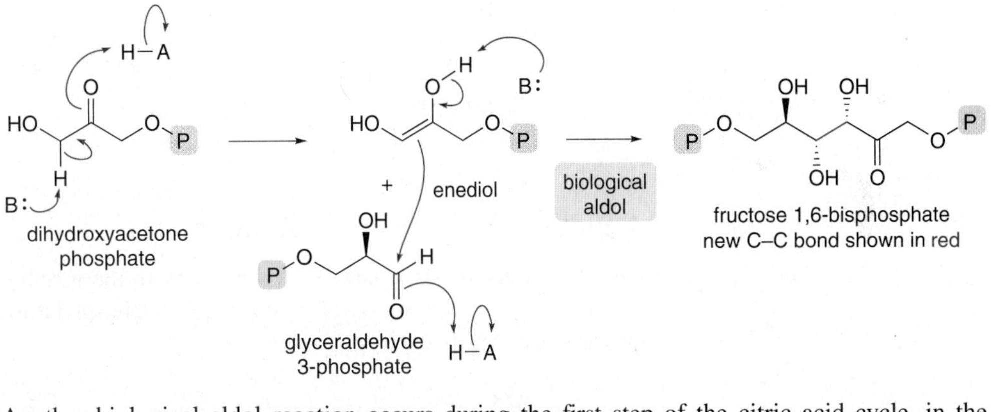

Another biological aldol reaction occurs during the first step of the citric acid cycle, in the reaction of acetyl CoA (Section 22.17) with oxaloacetate to form citrate in the presence of the enzyme citrate synthase. A new carbon–carbon bond is formed between the ketone carbonyl of oxaloacetate and the α carbon of acetyl CoA.

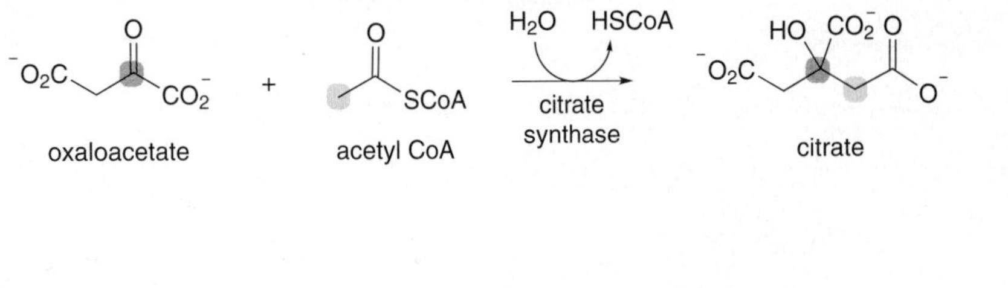

This reaction occurs by way of the tautomerization of the thioester acetyl CoA to an enol-type intermediate, which adds to the carbonyl group of oxaloacetate, forming a new carbon–carbon bond. Aldol condensation first forms the thioester citryl CoA, which is hydrolyzed to citrate.

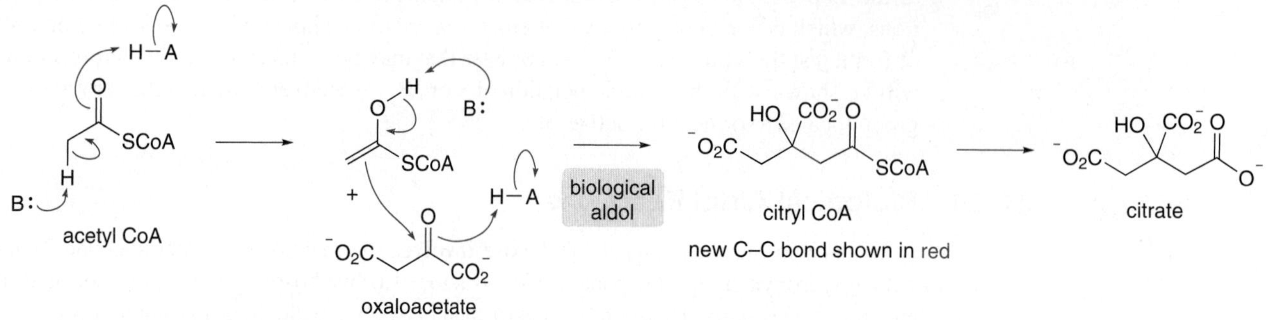

Although these reactions look more complex than the aldol reactions in Section 24.1, they are fundamentally the same.

- **Whether an aldol reaction occurs in the laboratory or in a biological system, the α carbon of one carbonyl compound adds to the carbonyl group of another carbonyl compound by nucleophilic addition.**

Problem 24.20 (a) Identify the carbon atoms in pyruvate and acetyl CoA that are joined together in the given biological aldol reaction to form citramalate. (b) Use curved arrows to show how this reaction occurs.

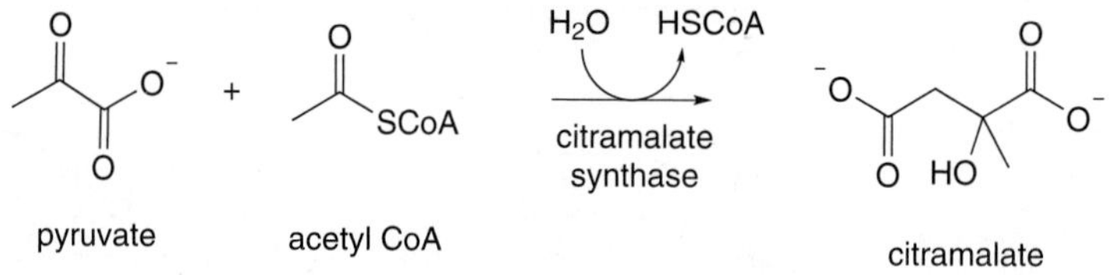

24.8B A Biological Retro-Aldol Reaction

One step in glycolysis, a key 10-step pathway in the metabolism of glucose to CO_2 and H_2O, is the retro-aldol conversion of fructose 1,6-bisphosphate to glyceraldehyde 3-phosphate and dihydroxyacetone phosphate.

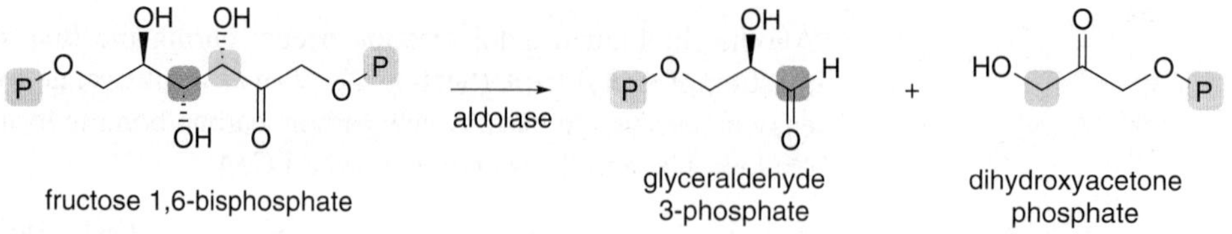

This reaction, catalyzed by an aldolase enzyme, is the reverse of the aldol reaction that prepares fructose 1,6-bisphosphate described in Section 24.8A.

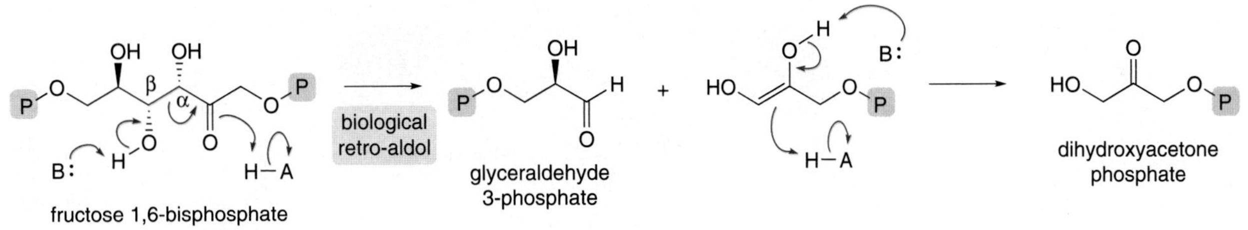
biological
retro-aldol
glyceraldehyde
3-phosphate
dihydroxyacetone
phosphate

fructose 1,6-bisphosphate

The C–C bond between the α and β C's (shown in red) is cleaved.

- **To draw the products of a retro-aldol reaction, always cleave the bond between the α and β carbons to the carbonyl group.**

24.8C Biological Claisen Reactions

Biological Claisen reactions constitute the predominant carbon–carbon bond-forming reactions in the biosynthesis of fatty acids like stearic acid, the chapter-opening molecule. The starting materials in biological systems are *thioesters* (RCOSR'), rather than esters.

The biosynthesis of fatty acids begins with the Claisen condensation of two thioesters (**A** and **B**) to form a four-carbon β-keto thioester with loss of CO_2. A new carbon–carbon bond is formed between the carbonyl carbon of **A** and the α carbon of **B.**

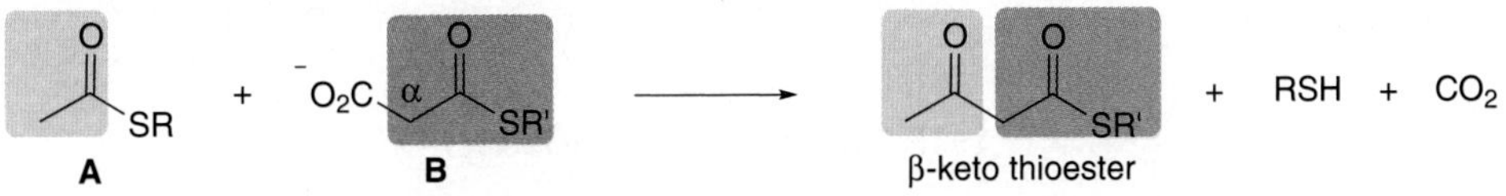
A
B
β-keto thioester
+ RSH + CO2

It is thought that the mechanism of this reaction begins with decarboxylation of **B** to form an enolate (Step [1]), which adds to the carbonyl group of **A** to generate a tetrahedral intermediate (Step [2]). Loss of the leaving group ($^-$SR) then generates the β-keto thioester (Step [3]). Two carbons of **A** and two carbons of **B** are joined to form a four-carbon β-keto thioester.

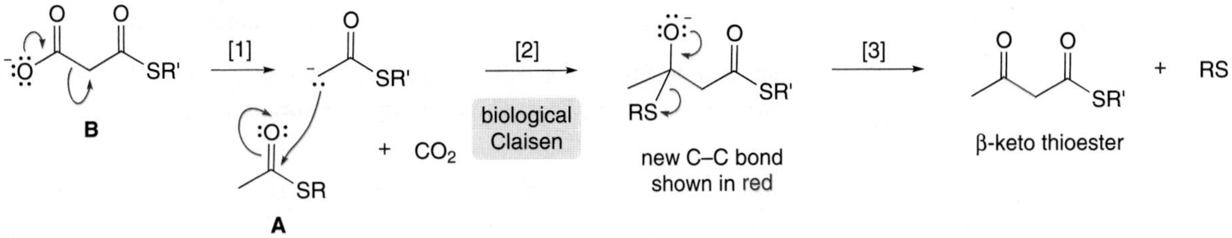
B
[1]
A
+ CO2
[2]
biological
Claisen
new C–C bond
shown in red
[3]
β-keto thioester

In the biosynthesis of fatty acids, the β-keto thioester is converted to a thioester by a three-step process.

three steps
β-keto thioester
thioester
new C–C bond shown in red

The resulting thioester can then undergo another Claisen reaction with **B** to form a β-keto thioester with two more carbons, which can be converted to a six-carbon thioester, **C.**

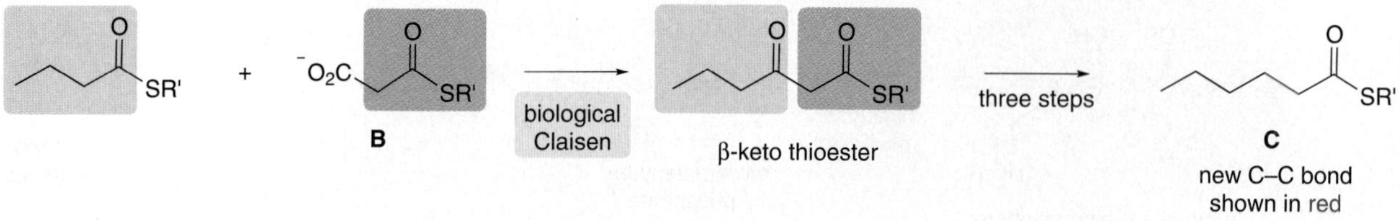

Each time this sequence is carried out, two more carbons are added to the thioester. This process illustrates why most fatty acids have an even number of carbon atoms. In stearic acid, eight carbon–carbon bonds are formed by sequential Claisen reactions that add two carbon units at a time.

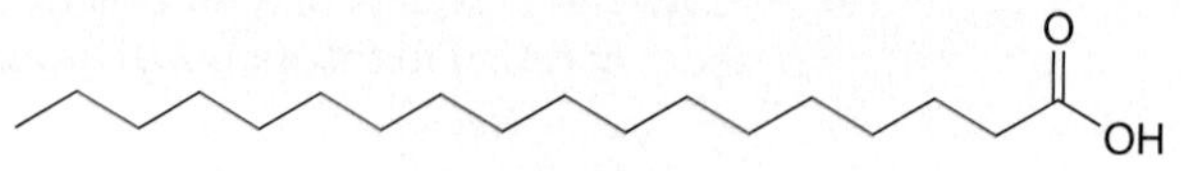

stearic acid
new C–C bonds formed by the
Claisen reaction shown in red

Problem 24.21 Draw the β-keto thioester formed by the biological Claisen reaction of **B** and **X.**

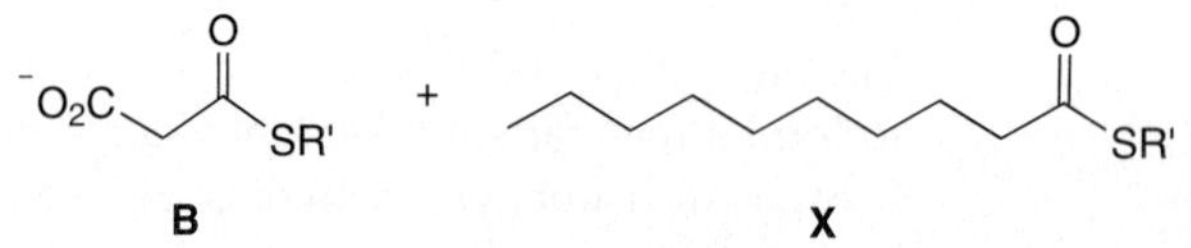

24.9 The Michael Reaction

Like the aldol and Claisen reactions, the **Michael reaction involves two carbonyl components—the enolate of one carbonyl compound and an α,β-unsaturated carbonyl compound.**

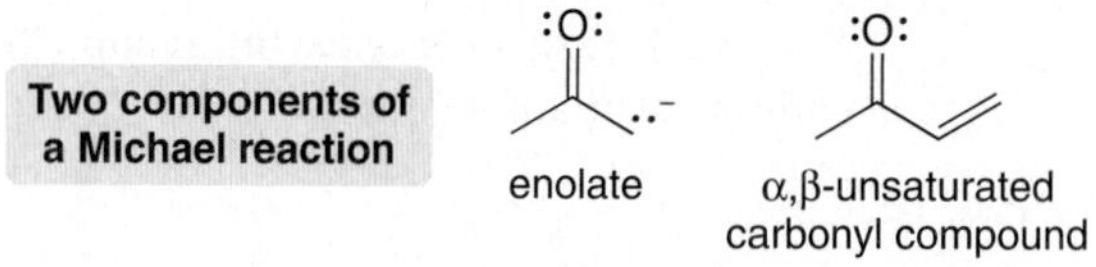

Recall from Section 20.15 that α,β-unsaturated carbonyl compounds are resonance stabilized and have **two electrophilic sites—the carbonyl carbon and the β carbon.**

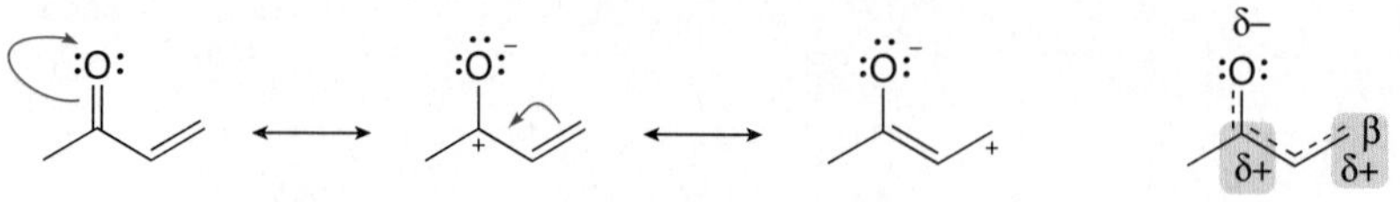

three resonance structures for an
α,β-unsaturated carbonyl compound

hybrid

two electrophilic sites

- **The Michael reaction involves the conjugate addition (1,4-addition) of a resonance-stabilized enolate to the β carbon of an α,β-unsaturated carbonyl system.**

All conjugate additions add the **elements of H and Nu across the α and β carbons.**

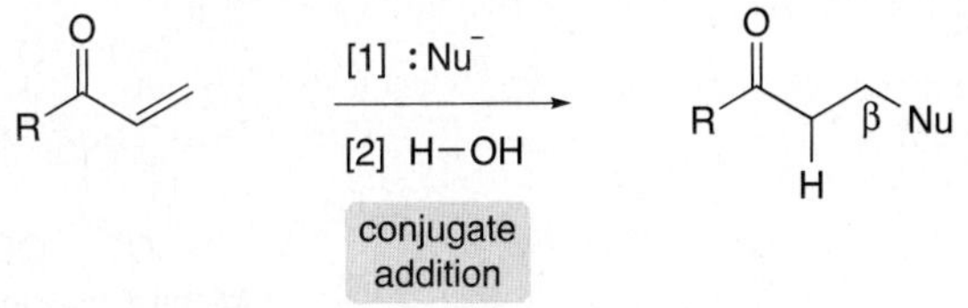

In the Michael reaction, the **nucleophile is an enolate.** Enolates of active methylene compounds are particularly common. The α,β-unsaturated carbonyl component is often called a **Michael acceptor.**

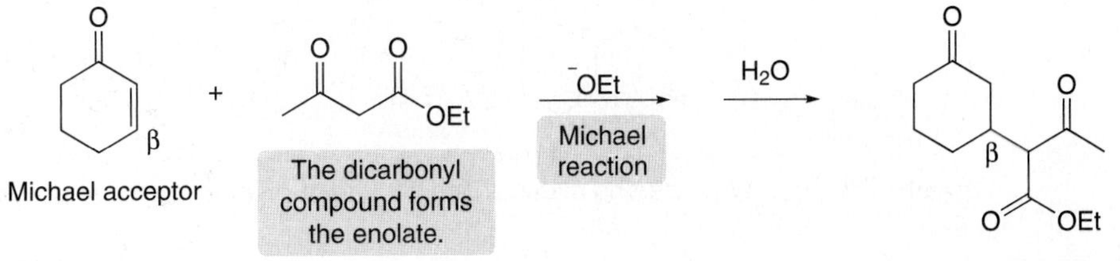

new C–C bond in red

Problem 24.22 Which of the following compounds can serve as Michael acceptors?

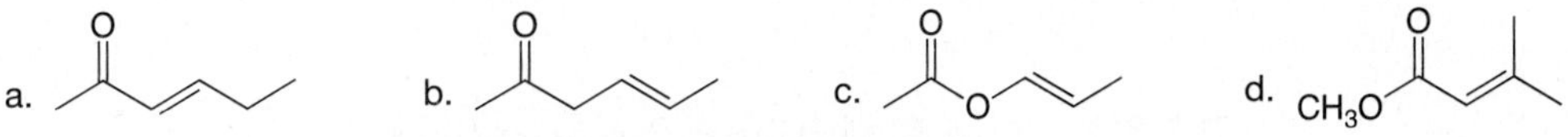

The Michael reaction always forms a new carbon–carbon bond on the β carbon of the Michael acceptor. The key step is nucleophilic addition of the enolate to the β carbon of the Michael acceptor in Step [2], as shown in Mechanism 24.7.

Mechanism 24.7 The Michael Reaction

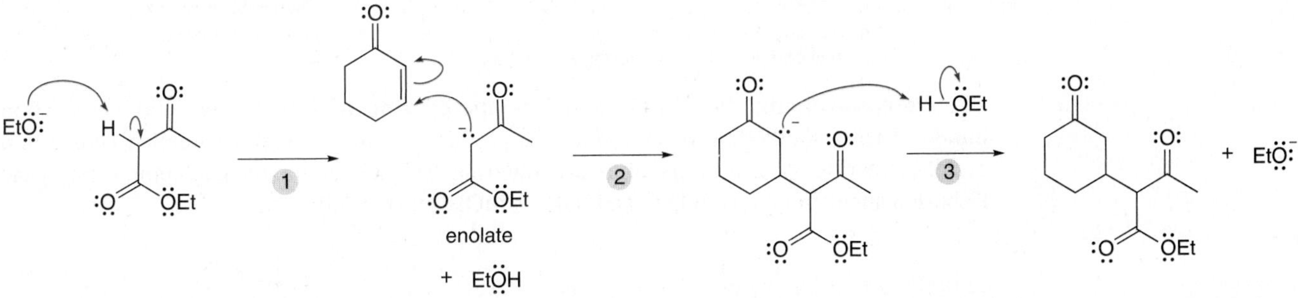

1. The base removes a proton on the carbon between the two carbonyl groups to form an enolate.
2. Nucleophilic addition of the enolate to the β carbon of the α,β-unsaturated carbonyl compound forms a new carbon–carbon bond and another enolate.
3. Protonation of the enolate forms the 1,4-addition product.

When the product of a Michael reaction is also a β-keto ester, it can be hydrolyzed and decarboxylated by heating in aqueous acid, as discussed in Section 23.9. This forms a **1,5-dicarbonyl compound.**

1,5-Dicarbonyl compounds are starting materials for intramolecular aldol reactions, as described in Section 24.4.

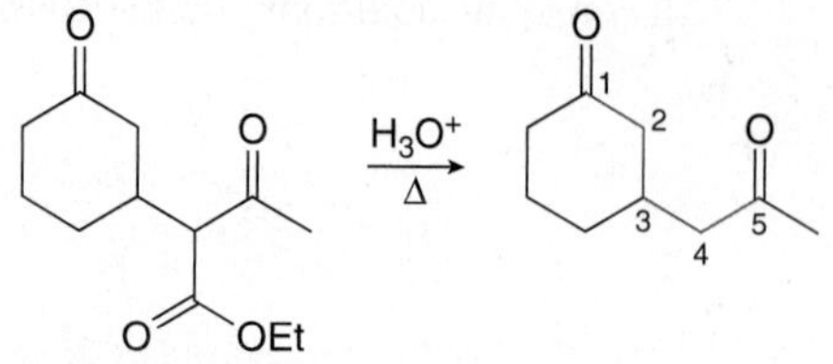

Michael reaction product

1,5-dicarbonyl compound

Problem 24.23 What product is formed when each pair of compounds is treated with NaOEt in ethanol?

a. 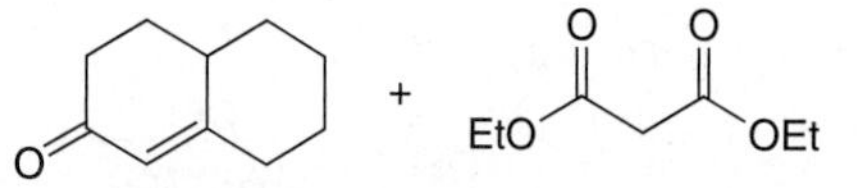

b. 

Problem 24.24 What starting materials are needed to prepare each compound by the Michael reaction?

a. 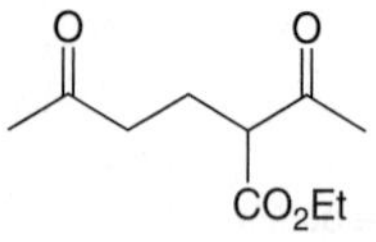

b. 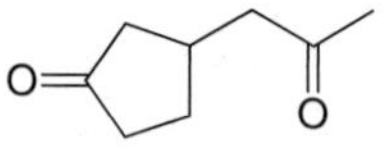

24.10 The Robinson Annulation

The word **annulation** comes from the Greek word *annulus* for ring. The Robinson annulation is named for English chemist Sir Robert Robinson, who was awarded the 1947 Nobel Prize in Chemistry.

The Robinson annulation is a ring-forming reaction that combines a Michael reaction with an intramolecular aldol reaction. Like the other reactions in Chapter 24, it involves enolates and it forms carbon–carbon bonds. The two starting materials for a Robinson annulation are an α,β-unsaturated carbonyl compound and an enolate.

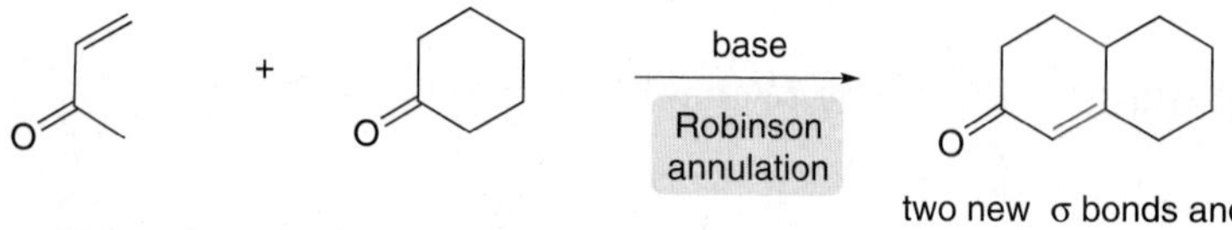

α,β-unsaturated carbonyl compound

carbonyl compound that forms an enolate

two new σ bonds and one new π bond in red

The Robinson annulation forms a six-membered ring and three new carbon–carbon bonds—two σ bonds and one π bond. The product contains an α,β-unsaturated ketone in a cyclohexane ring—that is, a **cyclohex-2-enone** ring. To generate the enolate component of the Robinson annulation, ^-OH in H_2O and ^-OEt in EtOH are typically used.

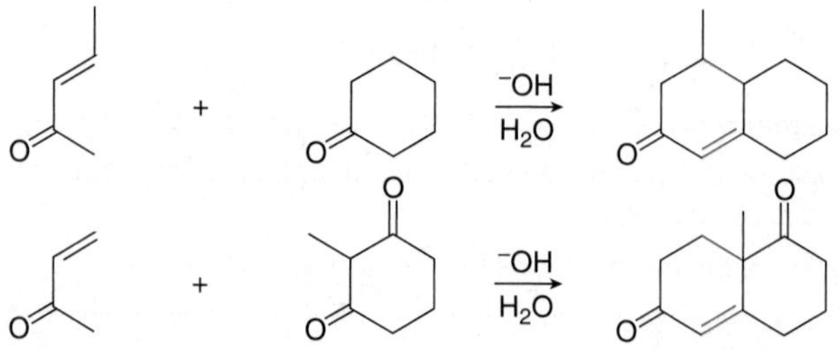

methyl vinyl ketone

2-methylcyclohexane-1,3-dione

new C–C bonds shown in red

The mechanism of the Robinson annulation consists of a **Michael addition** to the α,β-unsaturated carbonyl compound to form a 1,5-dicarbonyl compound, followed by an **intramolecular aldol reaction** to form the six-membered ring. The mechanism is written out in three parts in Mechanism 24.8 for the reaction between methyl vinyl ketone and 2-methylcyclohexane-1,3-dione.

Mechanism 24.8 The Robinson Annulation

Part [1] Michael addition

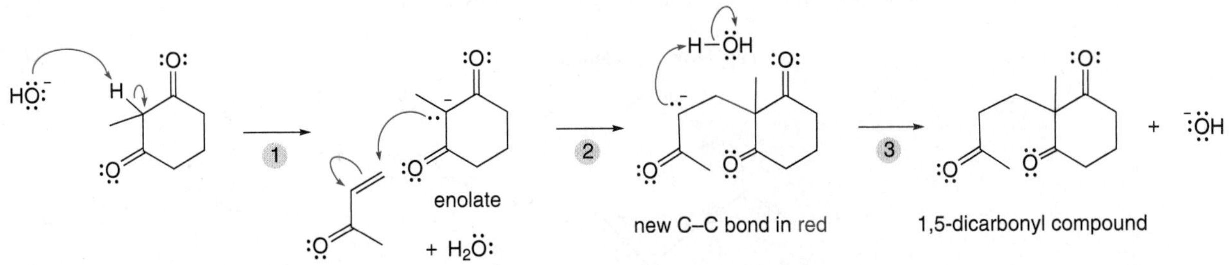

1 - 2 Base removes the most acidic proton—the proton between the two carbonyl groups—to form an enolate. Conjugate addition of the enolate to the α,β-unsaturated carbonyl compound forms a new carbon–carbon bond, generating an enolate.

3 Protonation of the enolate forms a 1,5-dicarbonyl compound.

Part [2] Intramolecular aldol reaction

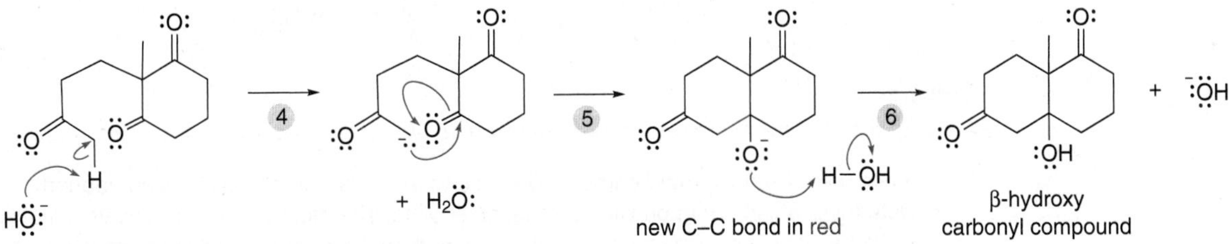

4 - 5 The base removes a proton to form an enolate, which attacks a carbonyl group to form a new C–C σ bond, generating the six-membered ring.

6 Protonation of the alkoxide forms the β-hydroxy carbonyl compound.

Part [3] Dehydration of the β-hydroxy carbonyl compound

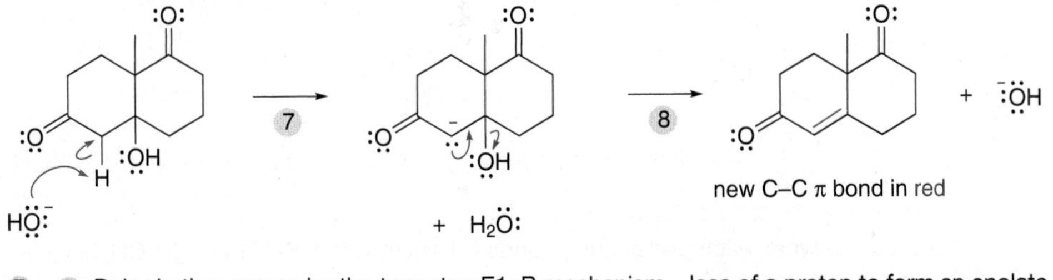

7 - 8 Dehydration occurs by the two-step E1cB mechanism—loss of a proton to form an enolate and elimination of ^{-}OH to form a π bond.

The mechanism begins with the three-step Michael addition that forms the first carbon–carbon σ bond, generating the 1,5-dicarbonyl compound (Part [1]). An intramolecular aldol reaction (Part [2]) forms the second carbon–carbon σ bond, and dehydration of the β-hydroxy ketone (Part [3]) forms the π bond. All of the parts of this mechanism have been discussed in previous sections of Chapter 24. However, the end result of the Robinson annulation—the formation of a cyclohex-2-enone ring—is new.

To draw the product of Robinson annulation without writing out the mechanism each time, **place the α carbon of the compound that becomes the enolate next to the β carbon of the α,β-unsaturated carbonyl compound.** Then, join the appropriate carbons together as shown. If you follow this method of drawing the starting materials, the double bond in the product always ends up in the same position in the six-membered ring.

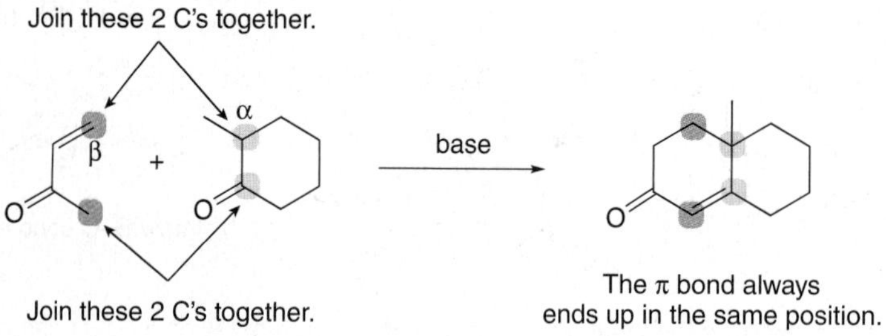

Sample Problem 24.4 Draw the Robinson annulation product formed from the following starting materials.

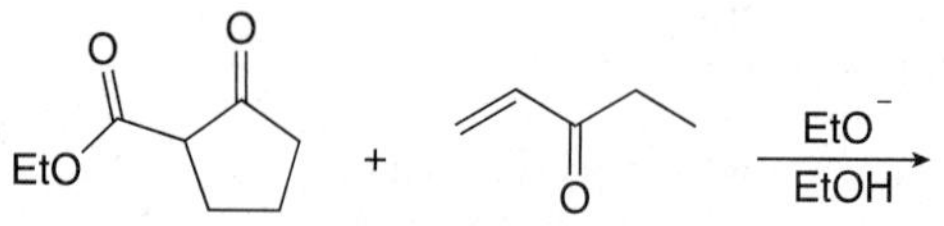

Solution

Arrange the starting materials to put the reactive atoms next to each other. For example:

- Place the α,β-unsaturated carbonyl compound *to the left* of the carbonyl compound.
- Determine which α carbon will become the enolate. The most acidic H is always removed with base first, which in this case is the H (in red) on the α carbon between the two carbonyl groups. This α carbon is drawn adjacent to the β carbon of the α,β-unsaturated carbonyl compound.

Then draw the bonds to form the new six-membered ring.

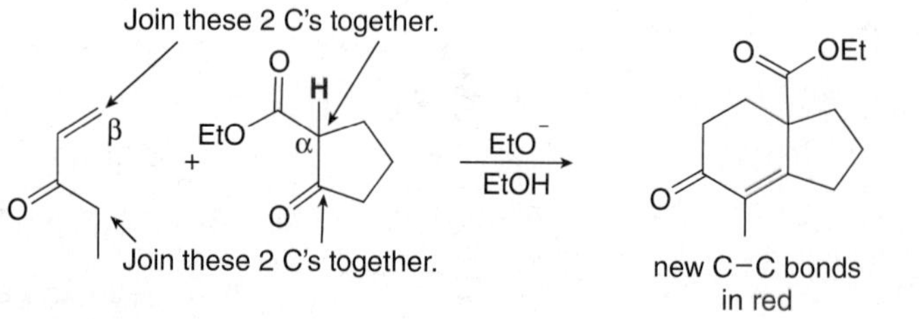

Problem 24.25 Draw the products when each pair of compounds is treated with $CH_3CH_2O^-$, CH_3CH_2OH in a Robinson annulation reaction.

a. 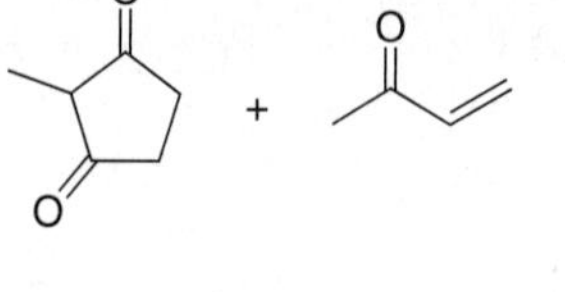

c. 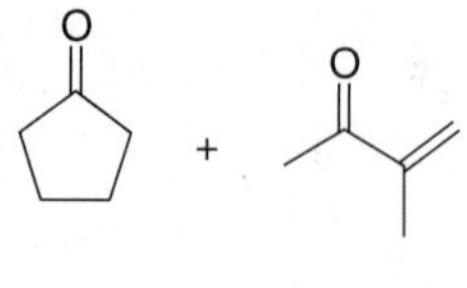

b. 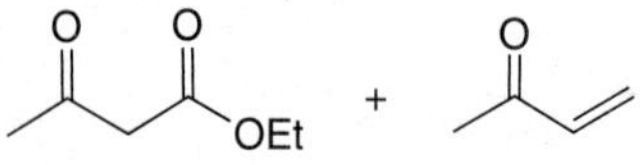

d. 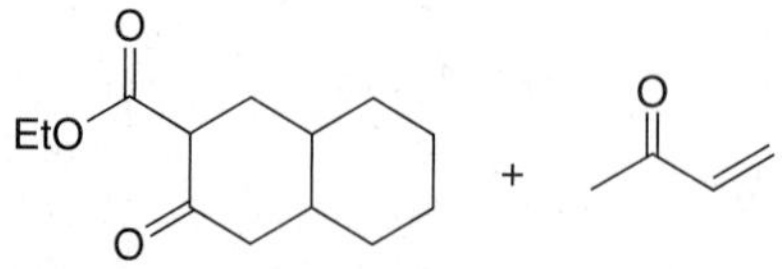

To use the Robinson annulation in synthesis, you must be able to determine what starting materials are needed to prepare a given compound, by working in the retrosynthetic direction.

How To Synthesize a Compound Using the Robinson Annulation

Example **What starting materials are needed to synthesize the following compound using a Robinson annulation?**

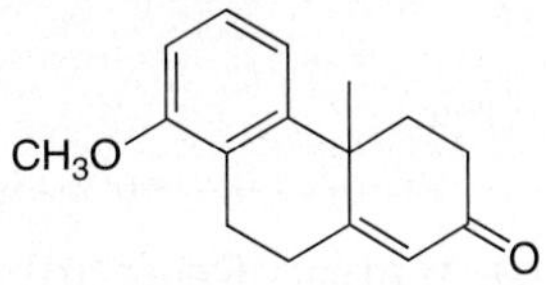
CH3O
O

Step [1] **Locate the cyclohex-2-enone ring and re-draw the target molecule if necessary.**

- To most easily determine the starting materials, always arrange the α,β-unsaturated carbonyl system in the same location. The target compound may have to be flipped or rotated, and you must be careful not to move any bonds to the wrong location during this process.

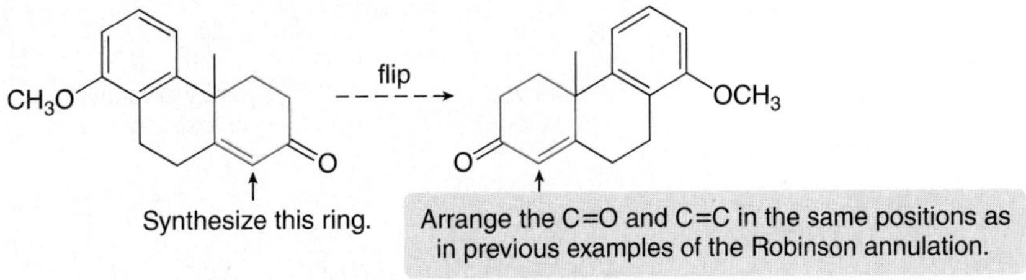
CH3O
flip
OCH3
O
O
Synthesize this ring.
Arrange the C=O and C=C in the same positions as in previous examples of the Robinson annulation.

Step [2] **Break the cyclohex-2-enone ring into two components.**

- Break the C=C. One half becomes the carbonyl group of the enolate component.
- Break the bond between the β carbon and the carbon to which it is bonded.

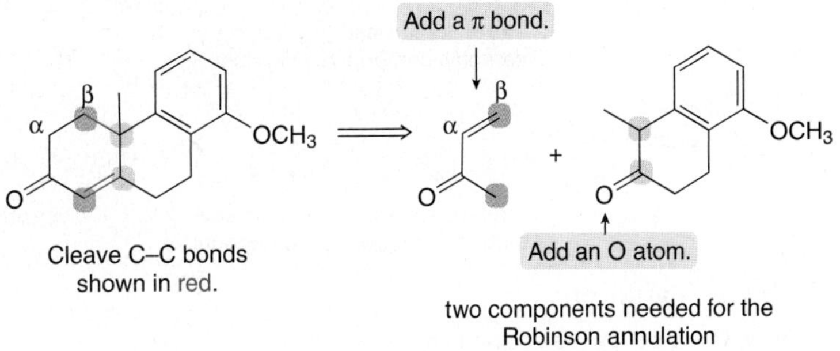
Add a π bond.
α
β
OCH3
α
β
+
OCH3
O
O
O
Add an O atom.
Cleave C–C bonds shown in red.
two components needed for the Robinson annulation

Problem 24.26 Which of the following bicyclic ring systems can be prepared by an intermolecular Robinson annulation?

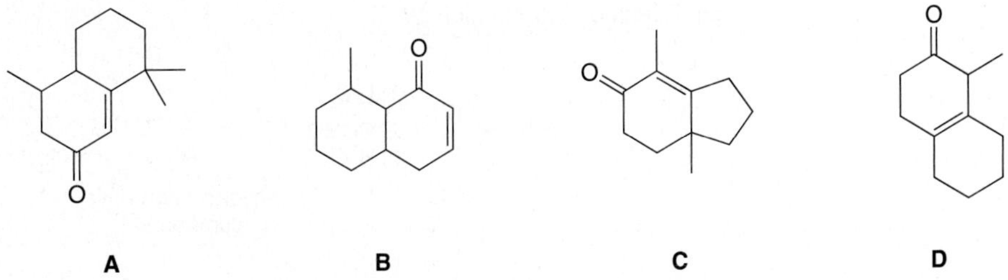
O
O
O
O
A
B
C
D

Problem 24.27 What starting materials are needed to synthesize each compound by a Robinson annulation?

a. 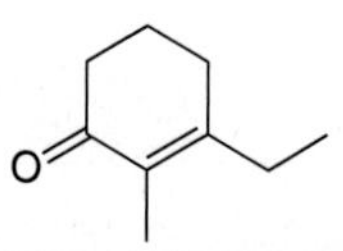
b. 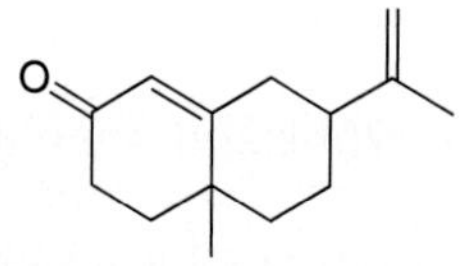
c. 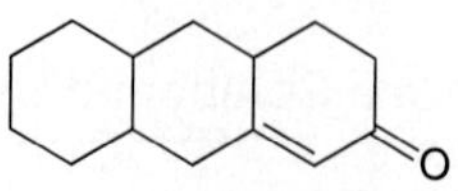

KEY CONCEPTS

Carbonyl Condensation Reactions

The Four Major Carbonyl Condensation Reactions

Reaction type — **Reaction** [New C – C bonds are shown in red.]

[1] Aldol reaction (24.1)

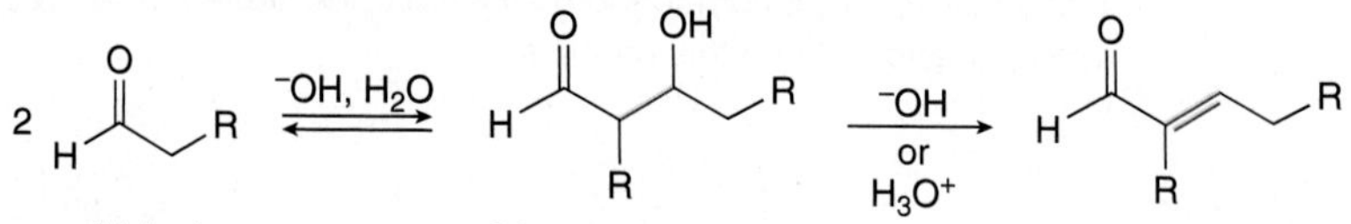

aldehyde (or ketone) — β-hydroxy carbonyl compound — (*E* and *Z*) α,β-unsaturated carbonyl compound

[2] Claisen reaction (24.5)

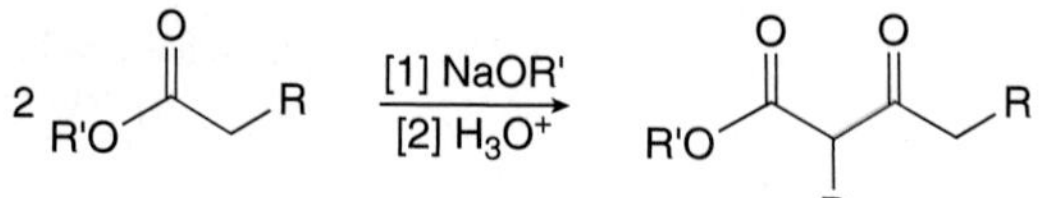

ester — β-keto ester

[3] Michael reaction (24.9)

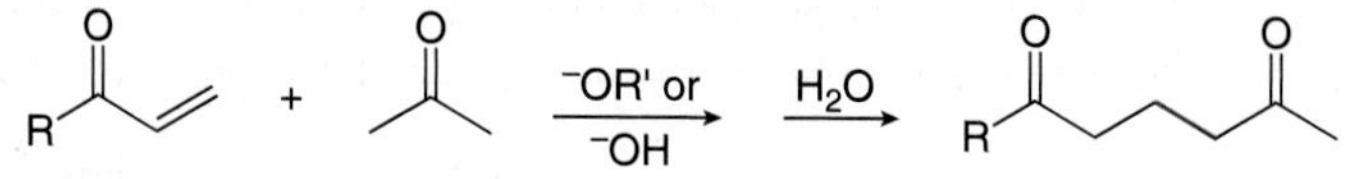

α,β-unsaturated carbonyl compound — carbonyl compound — 1,5-dicarbonyl compound

[4] Robinson annulation (24.10)

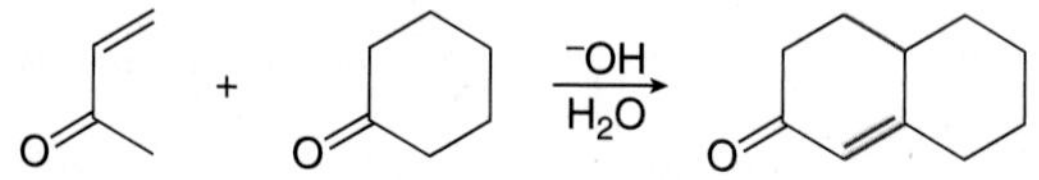

α,β-unsaturated carbonyl compound — carbonyl compound — cyclohex-2-enone

Useful Variations

[New C – C bonds are shown in red.]

[1] Retro-aldol reaction (24.1B)

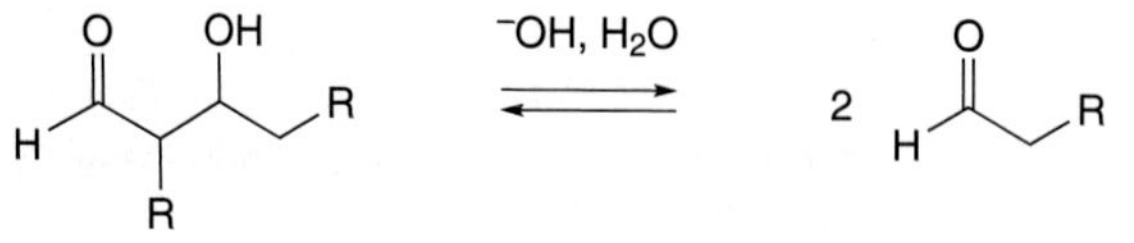

[2] Directed aldol reaction (24.3)

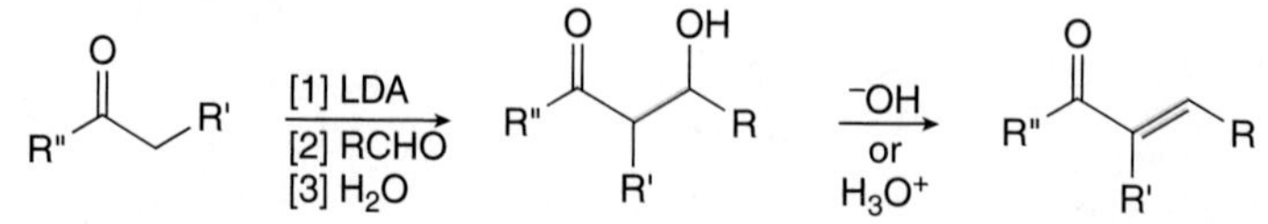

R'' = H or alkyl — β-hydroxy carbonyl compound — (*E* and *Z*) α,β-unsaturated carbonyl compound

[3] Intramolecular aldol reaction (24.4)

a. With 1,4-dicarbonyl compounds:

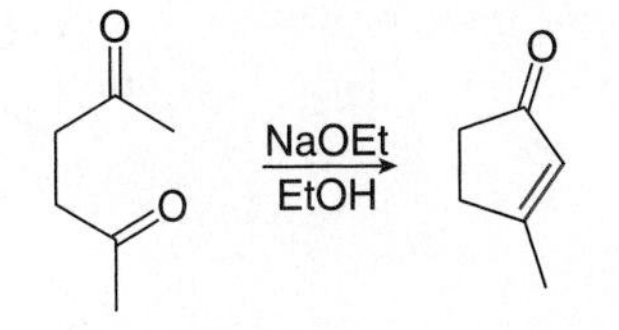

b. With 1,5-dicarbonyl compounds:

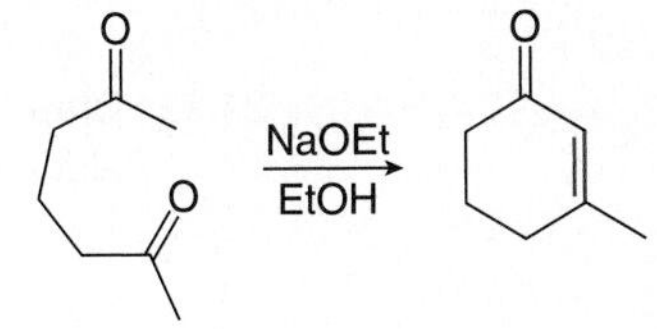

[4] Dieckmann reaction (24.7)

a. With 1,6-diesters:

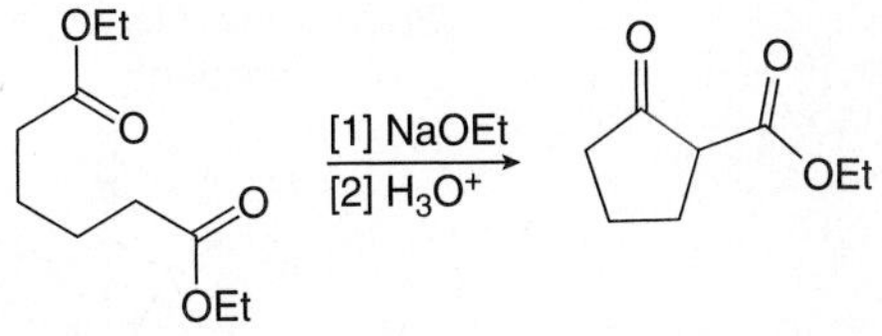

b. With 1,7-diesters:

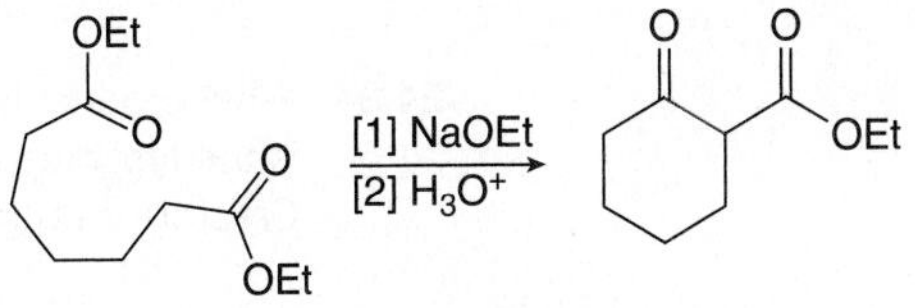

PROBLEMS

Problems Using Three-Dimensional Models

24.28 Draw the aldol product formed from each pair of starting materials using ^{-}OH, H_2O.

a. 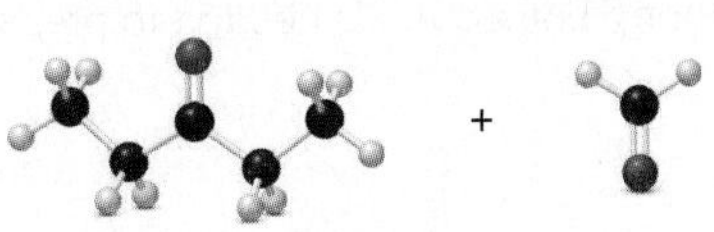

b. 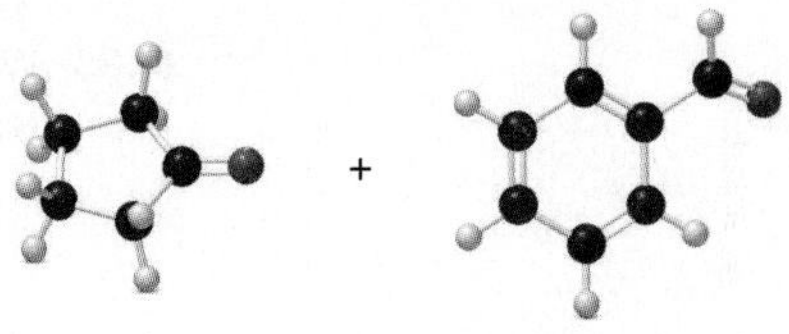

24.29 What steps are needed to convert **A** to **B**?

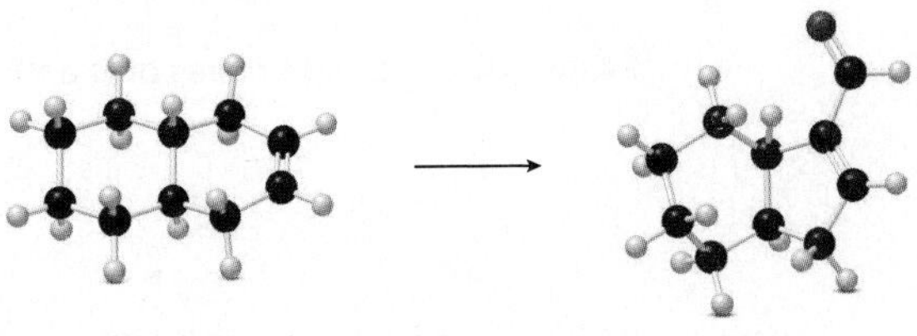

A **B**

The Aldol Reaction

24.30 Draw the product formed from an aldol reaction with the given starting material(s) using ^{-}OH, H_2O.

a. 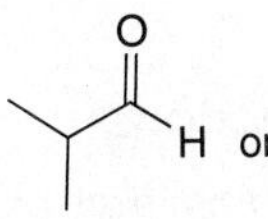 only

b. 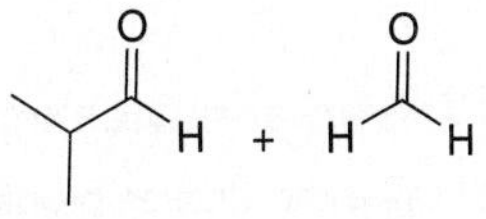

c. 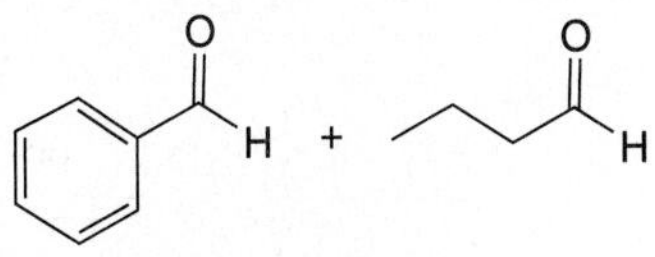

24.31 Draw the product formed in each directed aldol reaction.

a. 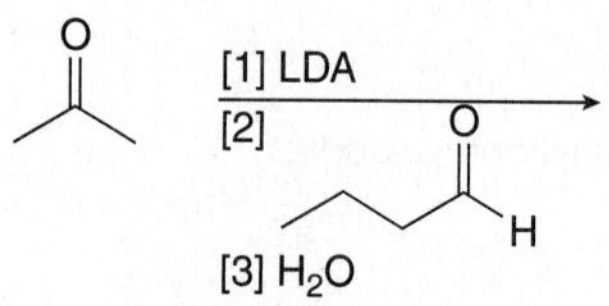

b. 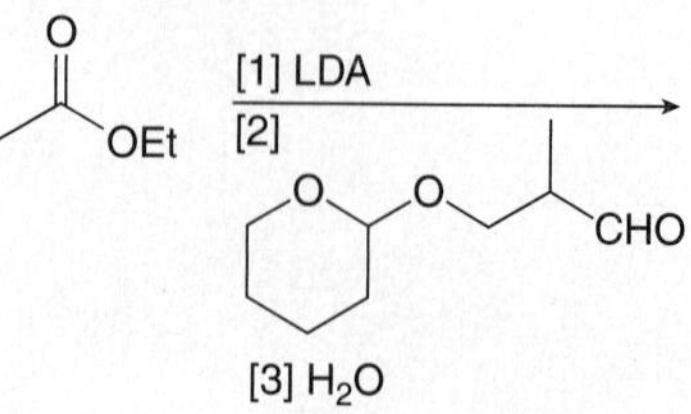

24.32 Draw the product formed when each dicarbonyl compound undergoes an intramolecular aldol reaction followed by dehydration.

a. 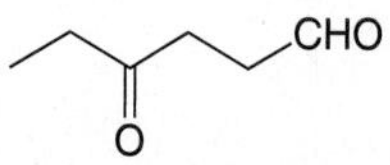

b. 

c. 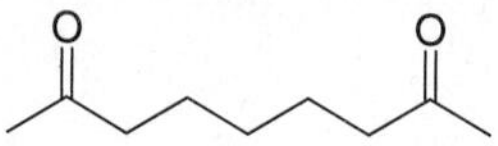

24.33 What starting materials are needed to synthesize each compound using an aldol or similar reaction?

a. 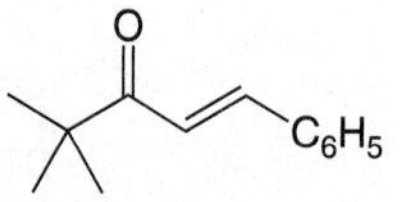

b. 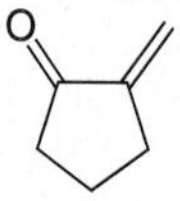

c. 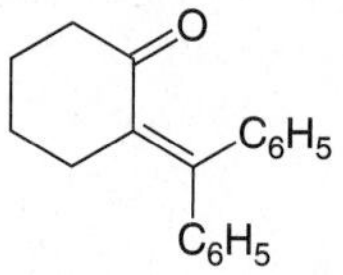

d. 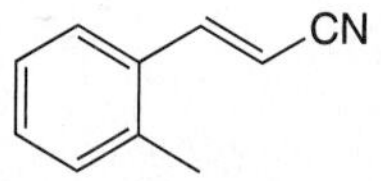

24.34 What product is formed when a solution of **A** and **B** is treated with mild base? This reaction is the first step in the synthesis of rosuvastatin (sold as a calcium salt under the trade name Crestor), a drug used to treat patients with high cholesterol.

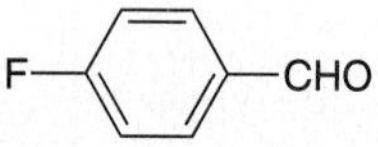

A

+

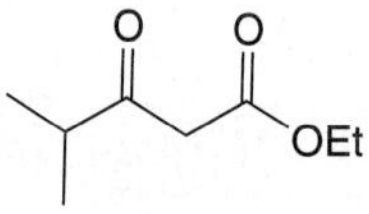

B

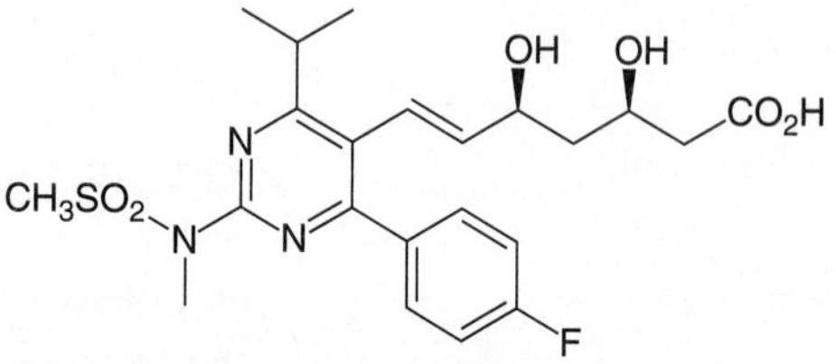

rosuvastatin

24.35 What dicarbonyl compound is needed to prepare each compound by an intramolecular aldol reaction?

a. 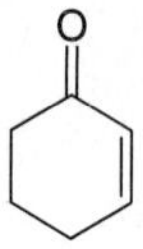

b. 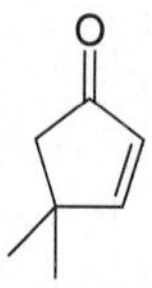

c. 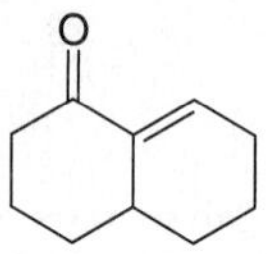

d. 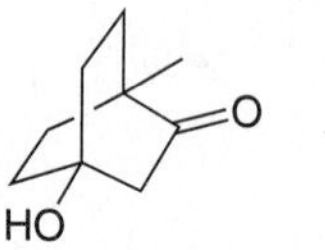

24.36 Identify the structures of **C** and **D** in the following reaction sequence.

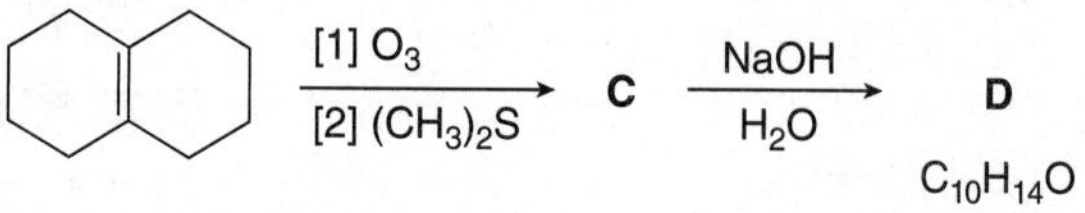

24.37 Explain why ketone **K** undergoes aldol reactions but ketone **J** does not.

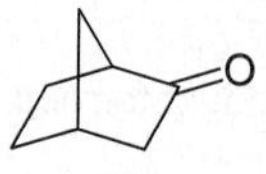

K

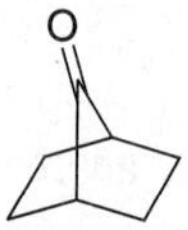

J

The Claisen and Dieckmann Reactions

24.38 Draw the Claisen product formed from each ester.

a. 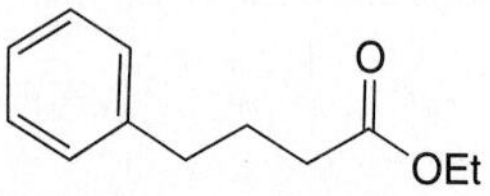

b. 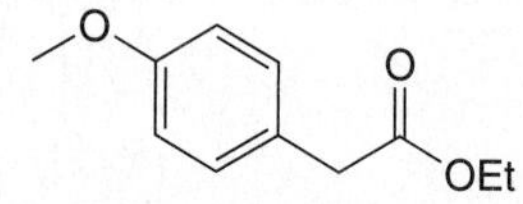

24.39 Draw the product formed from a Claisen reaction with the given starting materials using ^{-}OEt, EtOH.

a. 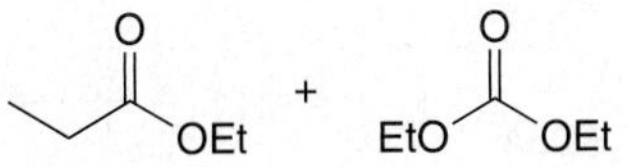
O
OEt
+
O
EtO
OEt

b. 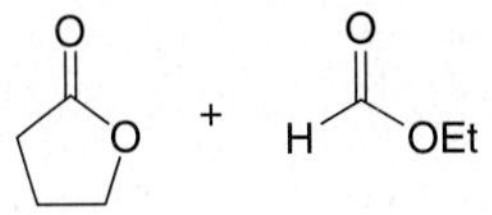
O
O
+
O
H
OEt

24.40 What starting materials are needed to synthesize each compound by a crossed Claisen reaction?

a. 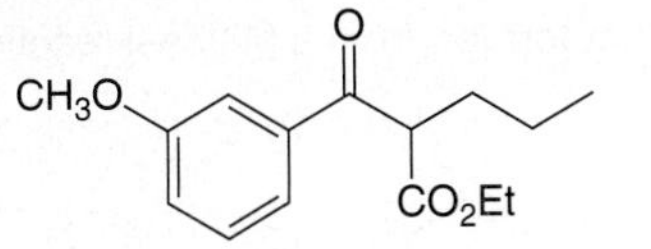
CH3O
O
CO2Et

c. 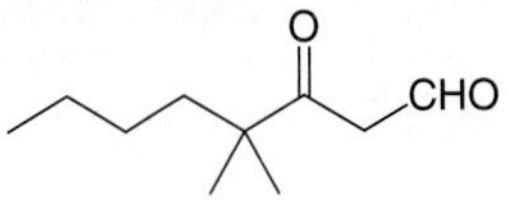
O
CHO

b. 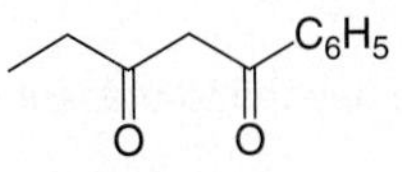
C6H5
O
O

d. 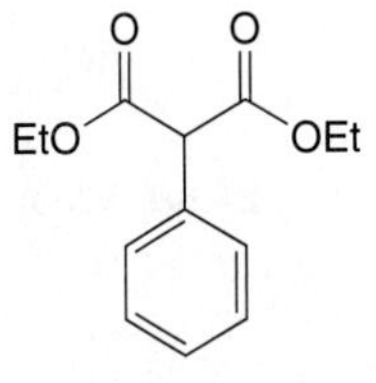
O
O
EtO
OEt

24.41 Even though **B** contains three ester groups, a single Dieckmann product results when **B** is treated with $NaOCH_3$ in CH_3OH, followed by H_3O^+. Draw the structure and explain why it is the only product formed.

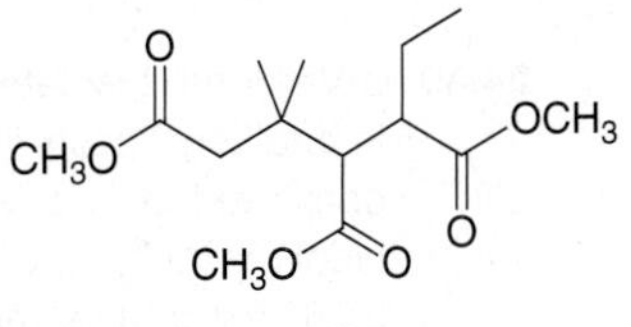
O
CH3O
OCH3
O
CH3O
O

B

Biological Carbonyl Condensation Reactions

24.42 What product is formed when 2-oxoisovalerate reacts with acetyl CoA in a biological aldol reaction?

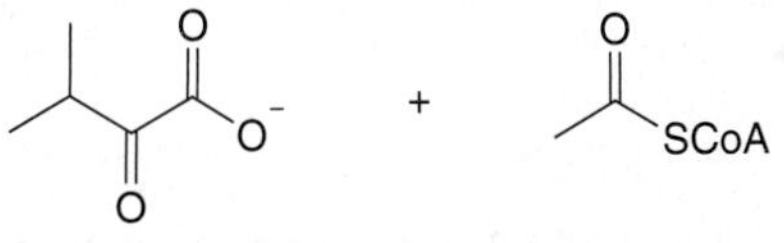
O
O⁻
O
+
O
SCoA

2-oxoisovalerate acetyl CoA

24.43 What products would be formed by a retro-aldol reaction of **X?**

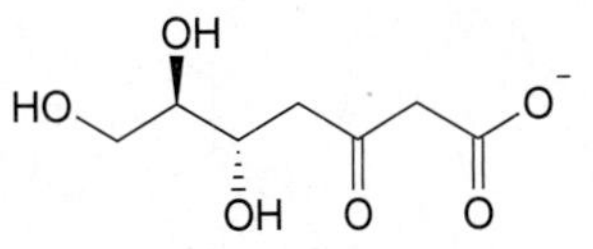
OH
HO
O⁻
OH
O
O

X

24.44 Propose a mechanism for the following reaction, one step in the metabolism of the amino acid leucine.

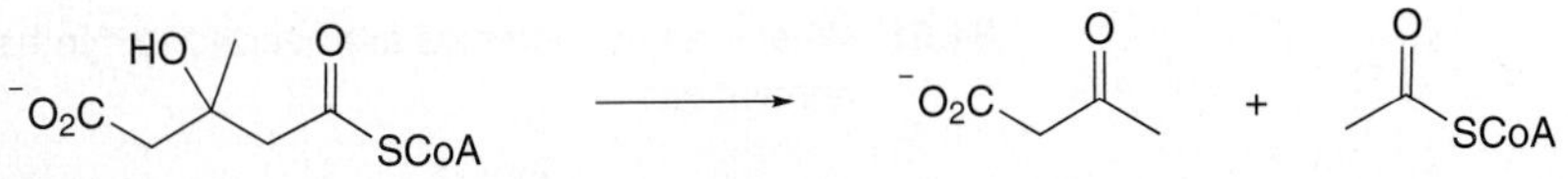
HO
O
⁻O2C
SCoA
⁻O2C
O
+
O
SCoA

24.45 Draw a stepwise mechanism for the conversion of two molecules of acetyl CoA to acetoacetyl CoA.

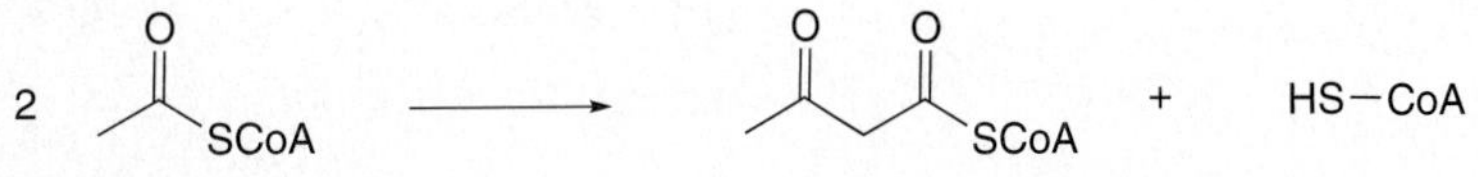
2
O
SCoA
O
O
SCoA
+
HS—CoA

acetyl CoA acetoacetyl CoA

24.46 One step in the metabolism of fatty acids involves the following retro-Claisen reaction. Draw a stepwise mechanism for this process.

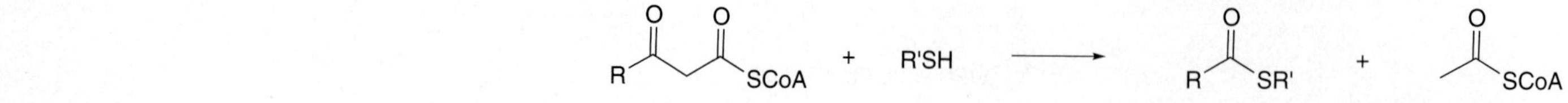

acetyl CoA

Michael Reaction

24.47 Draw the product formed from a Michael reaction with the given starting materials using $^{-}$OEt, EtOH.

a. 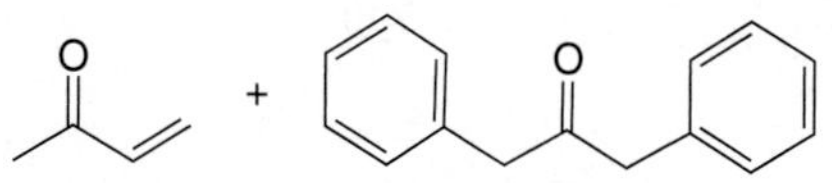

b. 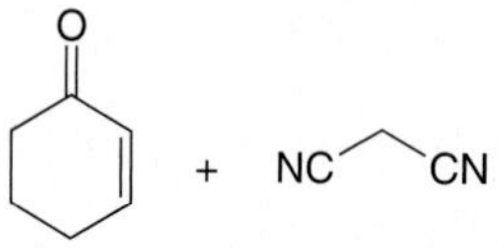

24.48 What starting materials are needed to prepare each compound using a Michael reaction?

a. 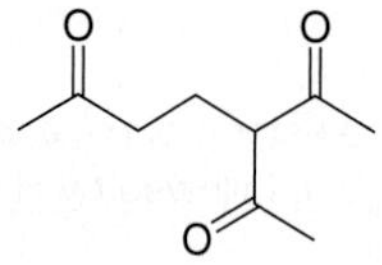

b. 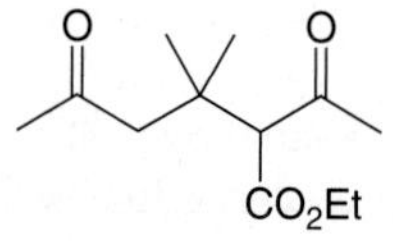

c. 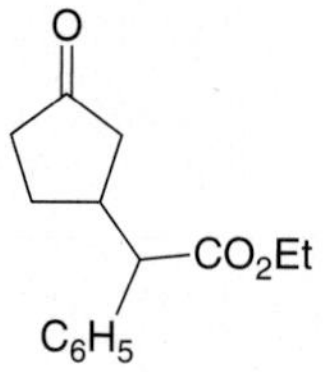

24.49 β-Vetivone is isolated from vetiver, a perennial grass that yields a variety of compounds used in traditional eastern medicine, pest control, and fragrance. In one synthesis, ketone **A** is converted to β-vetivone by a two-step process: Michael reaction, followed by intramolecular aldol reaction. (a) What Michael acceptor is needed for the conjugate addition? (See Problem 23.61 for another method to form the bicyclic ring system of β-vetivone.) (b) Draw a stepwise mechanism for the aldol reaction, which forms the six-membered ring.

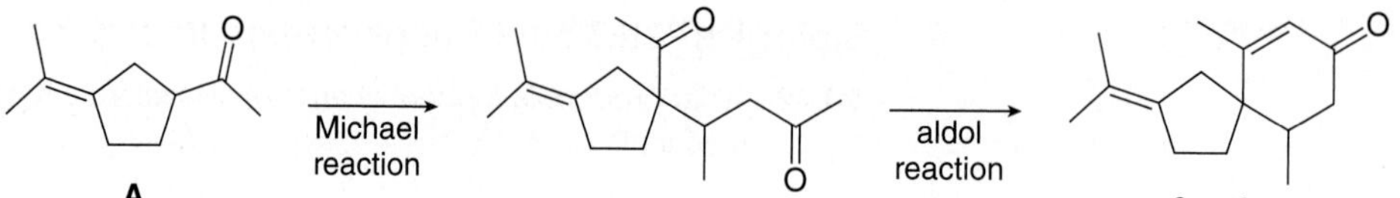

A

β-vetivone

Robinson Annulation

24.50 Draw the product of each Robinson annulation from the given starting materials using $^{-}$OH in H_2O solution.

a. 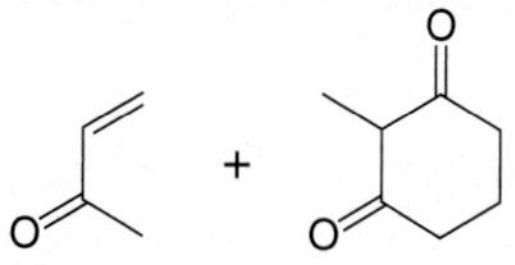

c. 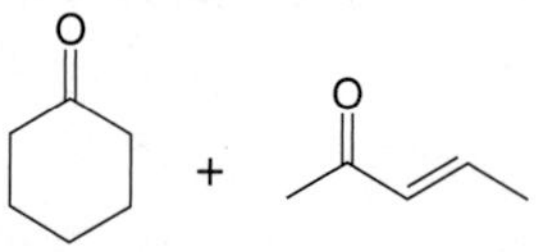

b. 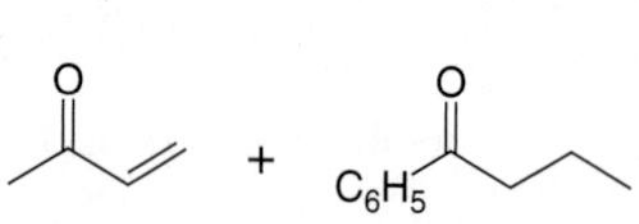

d. 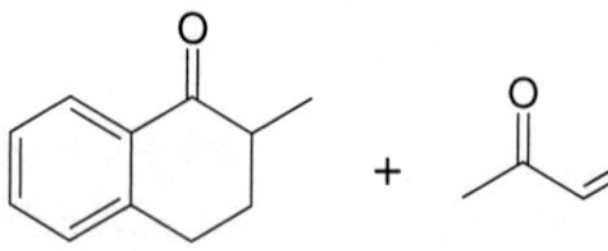

24.51 What starting materials are needed to synthesize each compound using a Robinson annulation?

a. 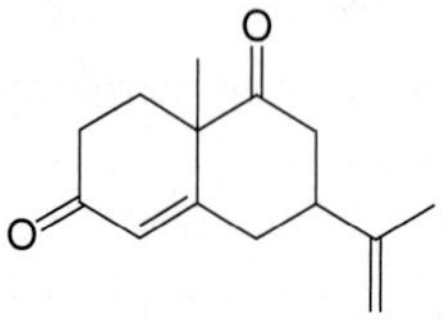

b. 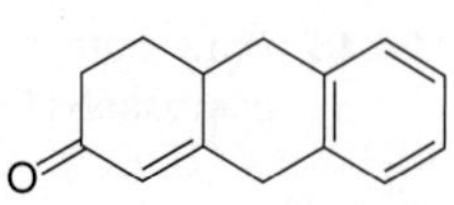

c. 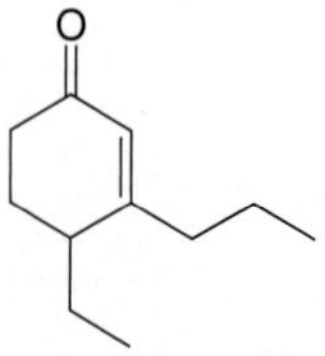

Reactions

24.52 Draw the organic products formed in each reaction.

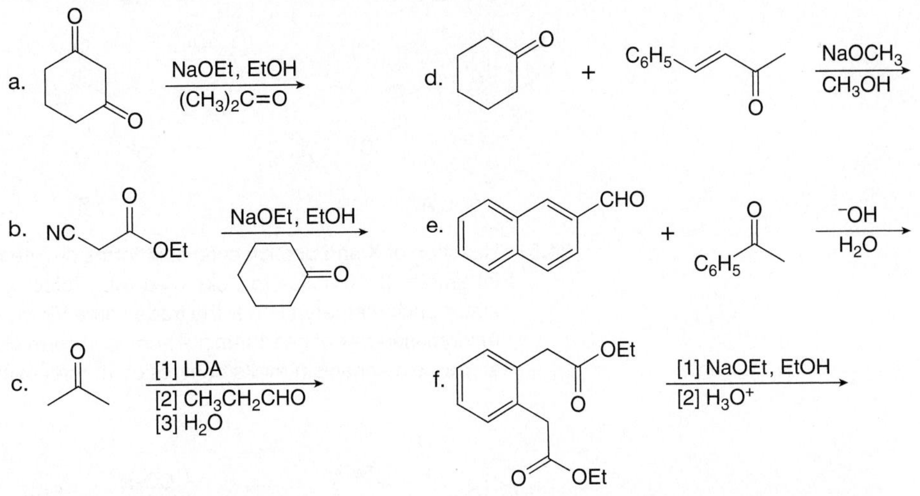
a.
NaOEt, EtOH
(CH3)2C=O
d.
C6H5
NaOCH3
CH3OH
b. NC
OEt
NaOEt, EtOH
e.
CHO
C6H5
-OH
H2O
c.
[1] LDA
[2] CH3CH2CHO
[3] H2O
f.
OEt
[1] NaOEt, EtOH
[2] H3O+

24.53 Identify compounds **A** and **B,** two synthetic intermediates in the 1979 synthesis of the plant growth hormone gibberellic acid by Corey and Smith. Gibberellic acid induces cell division and elongation, thus making plants tall and leaves large.

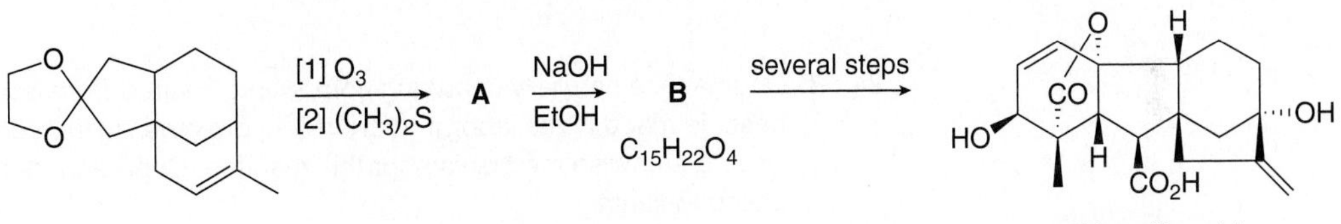
[1] O3
[2] (CH3)2S
A
NaOH
EtOH
B
C15H22O4
several steps
HO
CO
CO2H
OH
gibberellic acid

Mechanisms

24.54 In theory, the intramolecular aldol reaction of 6-oxoheptanal could yield the three compounds shown. It turns out, though, that 1-acetylcyclopentene is by far the major product. Why are the other two compounds formed in only minor amounts? Draw a stepwise mechanism to show how all three products are formed.

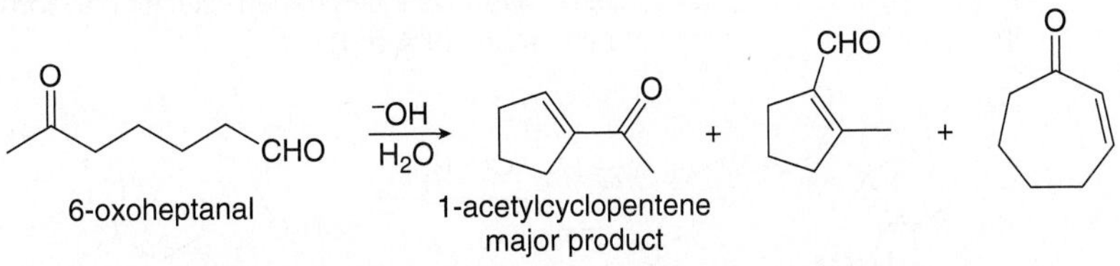
CHO
-OH
H2O
CHO
6-oxoheptanal
1-acetylcyclopentene
major product

24.55 Draw a stepwise mechanism for the following Robinson annulation. This reaction was a key step in a synthesis of the steroid cortisone by R. B. Woodward and co-workers at Harvard University in 1951.

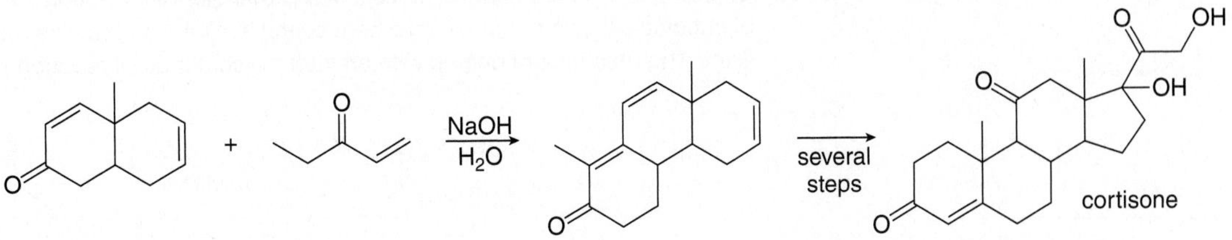
NaOH
H2O
several
steps
OH
OH
cortisone

24.56 One step in the synthesis of sitagliptin (Problem 17.19, a drug used to treat type 2 diabetes) involves reaction of the mixed anhydride **A** with **B** to form **C.** Draw a stepwise mechanism for this reaction.

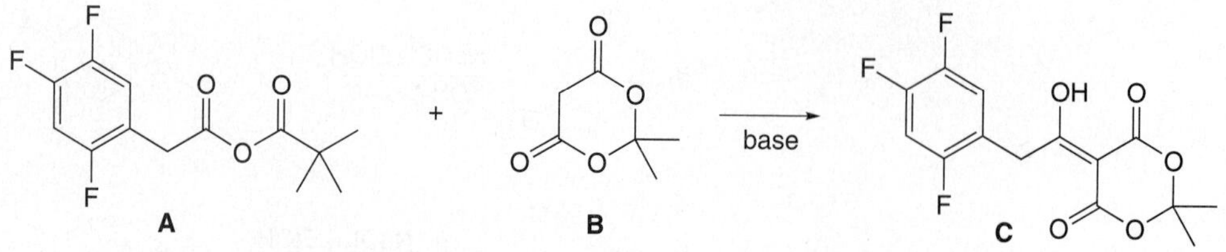

24.57 Reaction of **X** and phenylacetic acid forms an intermediate **Y,** which undergoes an intramolecular reaction to yield rofecoxib. Rofecoxib is a nonsteroidal anti-inflammatory agent once marketed under the trade name Vioxx, now withdrawn from the market because of increased risk of heart attacks from long-term use in some patients. Identify **Y** and draw a stepwise mechanism for its conversion to rofecoxib.

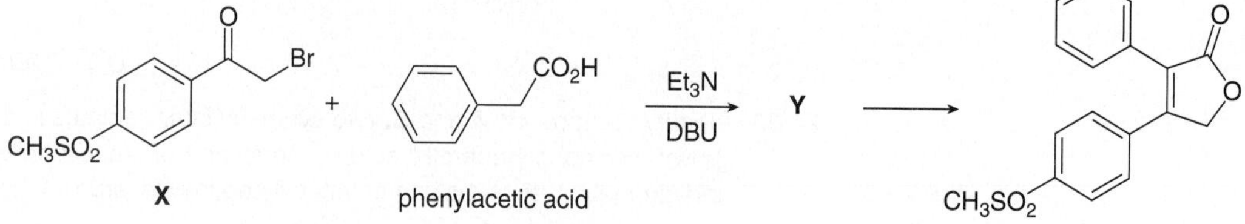

24.58 Coumarin, a naturally occurring compound isolated from lavender, sweet clover, and tonka bean, is made in the laboratory from *o*-hydroxybenzaldehyde by the reaction depicted below. Draw a stepwise mechanism for this reaction. Coumarin derivatives are useful synthetic anticoagulants.

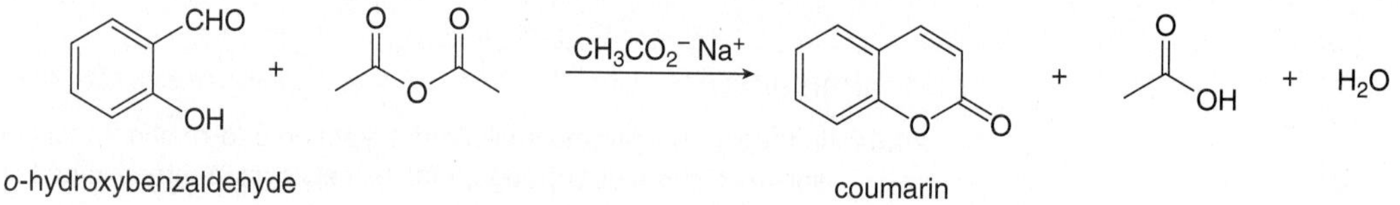

24.59 When **A** is treated with aqueous ^-OH, the major product is compound **B,** which undergoes ester hydrolysis and decarboxylation to form **C.** Draw a stepwise mechanism for the conversion of **A** to **B.**

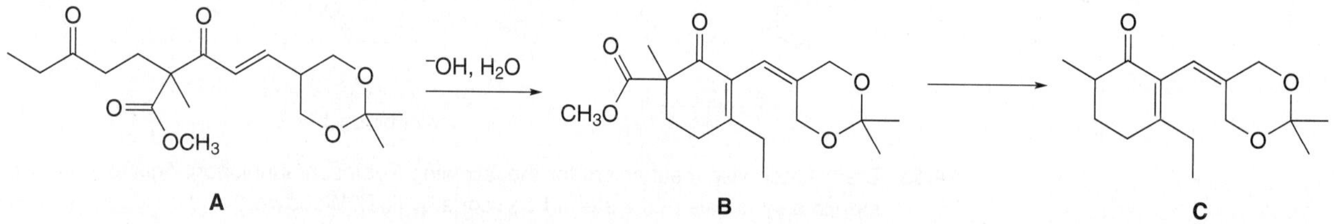

24.60 One step in a recent short synthesis of a prostaglandin (Section 19.6) involves the conversion of succinaldehyde to the bicyclic hemiacetal **X.** Draw a stepwise mechanism for this process. (Hint: The mechanism begins with an intermolecular aldol reaction.)

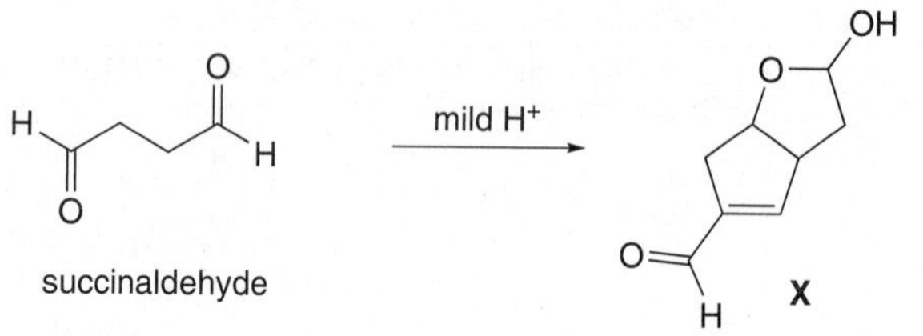

24.61 (a) Draw a stepwise mechanism for the reaction of ethyl hexa-2,4-dienoate with diethyl oxalate in the presence of base. (b) How does your mechanism explain why a new carbon–carbon bond forms on C6? (c) Why is this reaction an example of a crossed Claisen reaction?

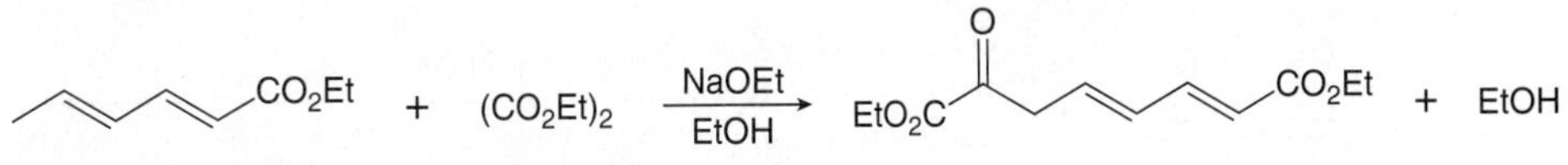

ethyl hexa-2,4-dienoate diethyl oxalate

Synthesis

24.62 Devise a synthesis of each compound from cyclopentanone, benzene, and organic alcohols having ≤ 3 C's. You may also use any required organic or inorganic reagents.

a. 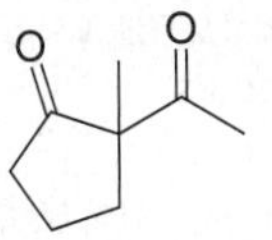

b. 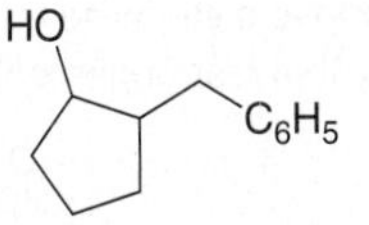

c. 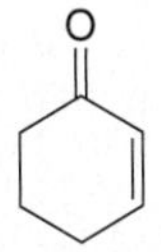

24.63 Devise a synthesis of each compound from $CH_3CH_2CH_2CO_2Et$, benzene, and alcohols having ≤ 2 C's. You may also use any required organic or inorganic reagents.

a. 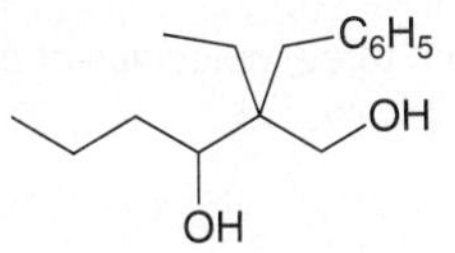

b. 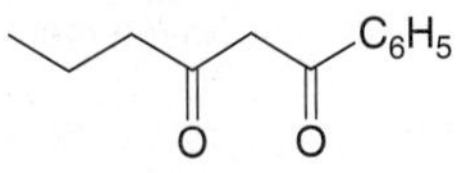

24.64 Devise a synthesis of 2-methylcyclopentanone from cyclohexene. You may also use any required reagents.

24.65 Octinoxate is an unsaturated ester used as an active ingredient in sunscreens. (a) What carbonyl compounds are needed to synthesize this compound using a condensation reaction? (b) Devise a synthesis of octinoxate from the given organic starting materials and any other needed reagents.

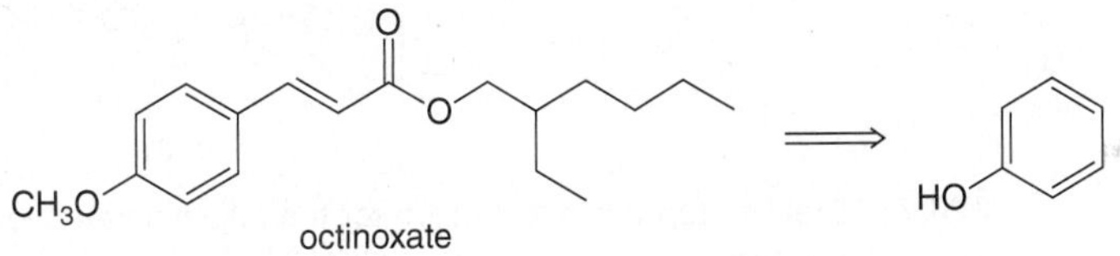

+ alcohols with < 5 C's

octinoxate

General Problem

24.66 Answer the following questions about 2-acetylcyclopentanone.

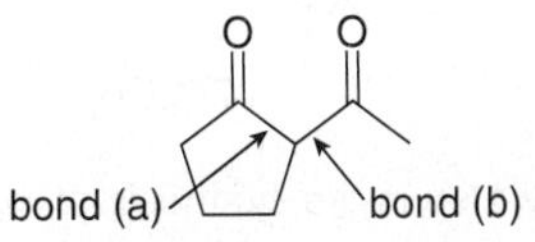

2-acetylcyclopentanone

a. What starting materials are needed to form 2-acetylcyclopentanone by a Claisen reaction that forms bond (a)?

b. What starting materials are needed to form 2-acetylcyclopentanone by a Claisen reaction that forms bond (b)?

c. What product is formed when 2-acetylcyclopentanone is treated with $NaOCH_2CH_3$, followed by CH_3I?

d. Draw the Robinson annulation product(s) formed by reaction of 2-acetylcyclopentanone with methyl vinyl ketone ($CH_2{=}CHCOCH_3$).

e. Draw the structure of the most stable enol tautomer(s).

Challenge Problems

24.67 Draw a stepwise mechanism for the following reaction, which was used in the synthesis of ezetimibe (Section 20.6), a drug used to treat patients with high cholesterol.

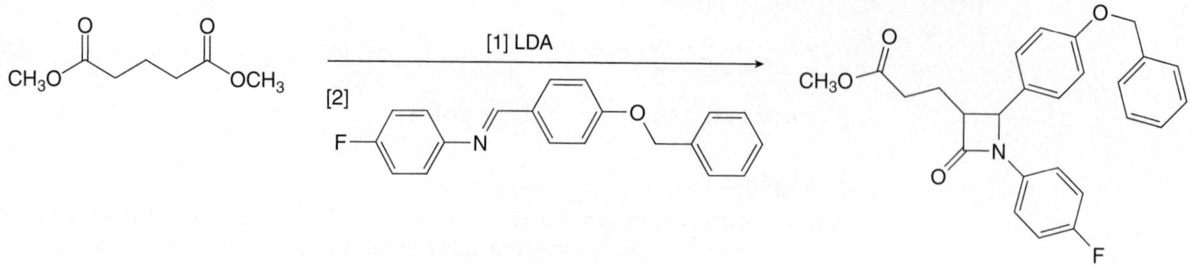

24.68 Propose a stepwise mechanism for the following reaction of a β-keto ester. Suggest a reason why this rearrangement reaction occurs.

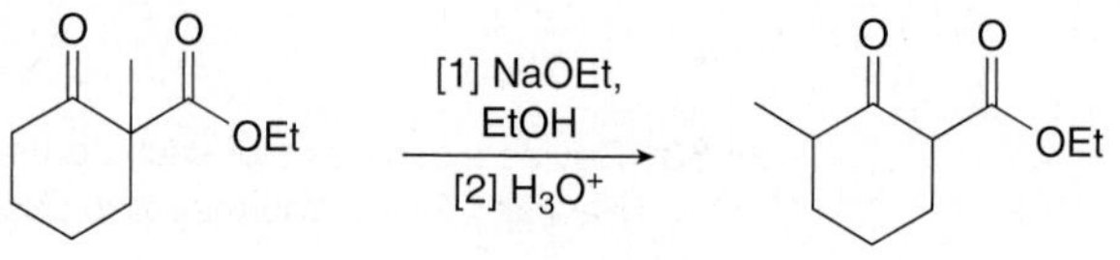

24.69 Isophorone is formed from three molecules of acetone [$(CH_3)_2C{=}O$] in the presence of base. Draw a mechanism for this process.

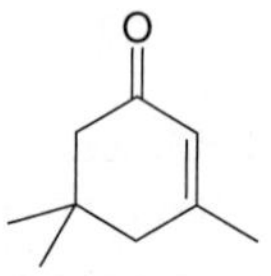

isophorone

24.70 Devise a stepwise mechanism for the following reaction. (Hint: The mechanism begins with the conjugate addition of ^{-}OH.)

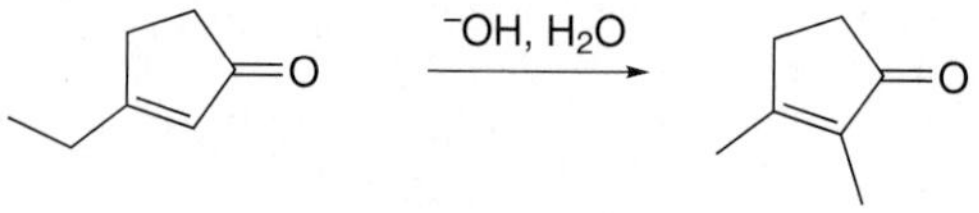

24.71 Draw a stepwise mechanism for the following reaction. (Hint: Two Michael reactions are needed.)

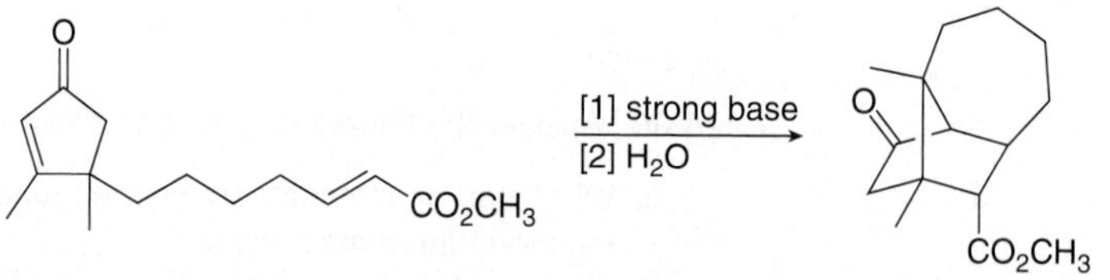

24.72 Draw a stepwise mechanism for the following reaction, one step in the synthesis of the cholesterol-lowering drug pitavastatin, marketed in Japan as a calcium salt under the name Livalo.

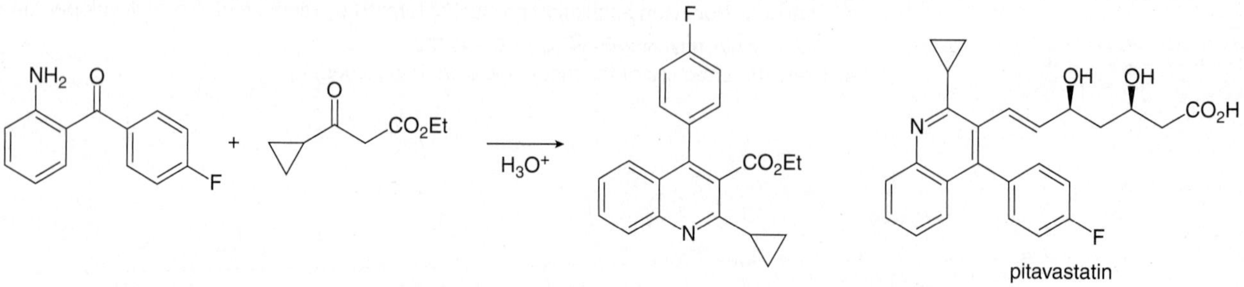

pitavastatin

24.73 Devise a stepwise mechanism for the following reaction, a key step in the synthesis of the antibiotic abyssomicin C. Abyssomicin C was isolated from sediment collected from almost 1000 ft below the surface in the Sea of Japan. (Hint: The mechanism begins with a Dieckmann reaction.)

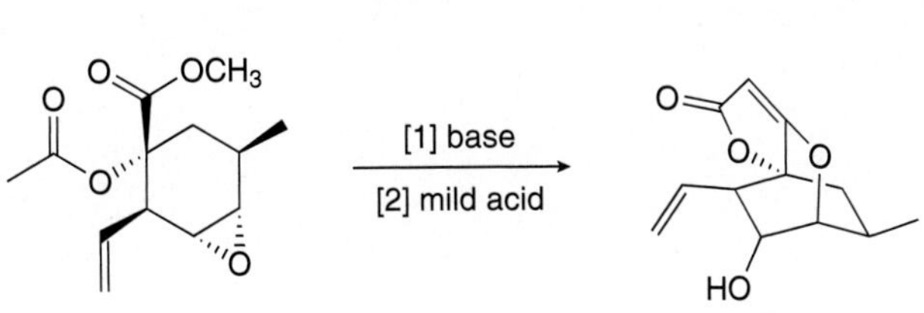

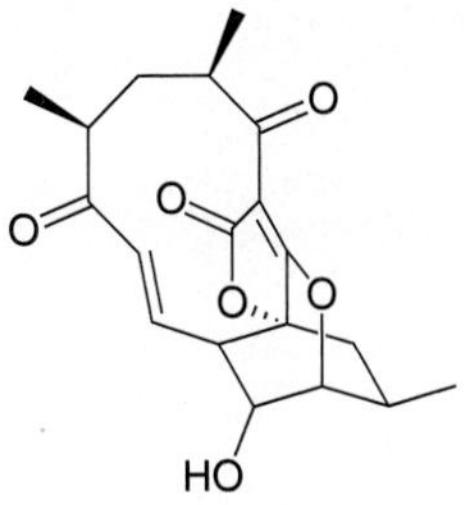

abyssomicin C

25 Amines

25.1 Introduction
25.2 Structure and bonding
25.3 Nomenclature
25.4 Physical properties
25.5 Spectroscopic properties
25.6 Interesting and useful amines
25.7 Preparation of amines
25.8 Reactions of amines—General features
25.9 Amines as bases
25.10 Relative basicity of amines and other compounds
25.11 Amines as nucleophiles
25.12 Hofmann elimination
25.13 Reaction of amines with nitrous acid
25.14 Substitution reactions of aryl diazonium salts
25.15 Coupling reactions of aryl diazonium salts
25.16 Application: Synthetic dyes and sulfa drugs

Scopolamine is a complex amine that occurs in angel's trumpets, ornamental plants with large trumpet-shaped flowers native to South America. Transdermal patches that release a minute dose of scopolamine are used to treat the nausea and vomiting associated with motion sickness. Scopolamine is an **alkaloid,** a naturally occurring amine derived from plant sources. In Chapter 25 we learn about the properties and reactions of amines like scopolamine.

We now leave the chemistry of carbonyl compounds to concentrate on **amines,** organic derivatives of ammonia (NH_3), formed by replacing one or more hydrogen atoms by alkyl or aryl groups. **Amines are stronger bases and better nucleophiles than other neutral organic compounds,** so much of Chapter 25 focuses on these properties.

Like that of alcohols, the chemistry of amines does not fit neatly into one reaction class, and this can make learning the reactions of amines challenging. Many interesting natural products and widely used drugs are amines, so you also need to know how to introduce this functional group into organic molecules.

25.1 Introduction

Amines **are organic nitrogen compounds,** formed by replacing one or more hydrogen atoms of ammonia (NH_3) with alkyl groups. As discussed in Section 3.2, amines are classified as 1°, 2°, or 3° by the number of alkyl groups bonded to the *nitrogen* atom.

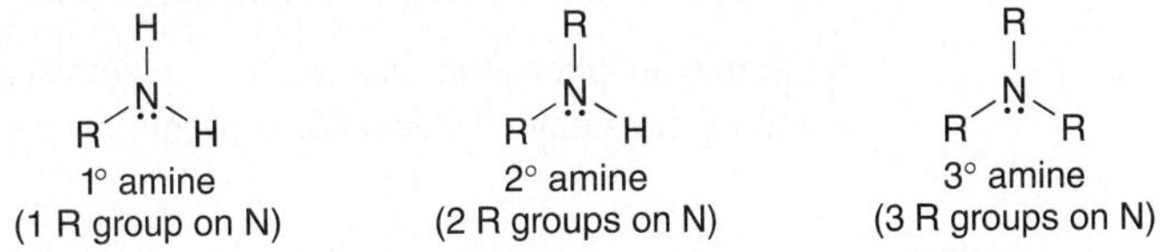
1° amine
(1 R group on N)
2° amine
(2 R groups on N)
3° amine
(3 R groups on N)

Like ammonia, **the amine nitrogen atom has a nonbonded electron pair,** making it both a base and a nucleophile. As a result, amines react with electrophiles to form **ammonium salts**—compounds with a positively charged ammonium ion and an anionic counterion.

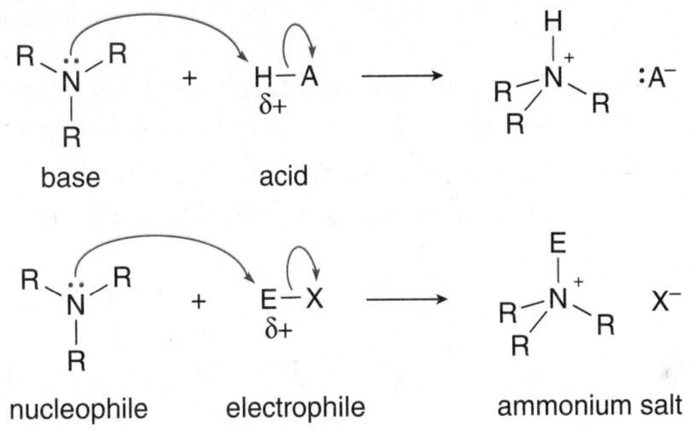
base
acid
nucleophile
electrophile
ammonium salt

- **The chemistry of amines is dominated by the nonbonded electron pair on the nitrogen atom.**

25.2 Structure and Bonding

An amine nitrogen atom is surrounded by three atoms and one nonbonded electron pair, making the N atom ***sp*^{3} hybridized** and **trigonal pyramidal,** with bond angles of approximately 109.5°.

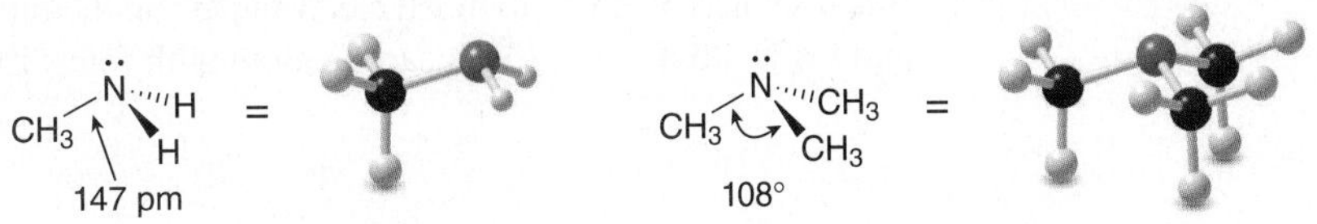
147 pm
108°

Figure 25.1
Electrostatic potential plots of CH_3NH_2 and $(CH_3)_3N$

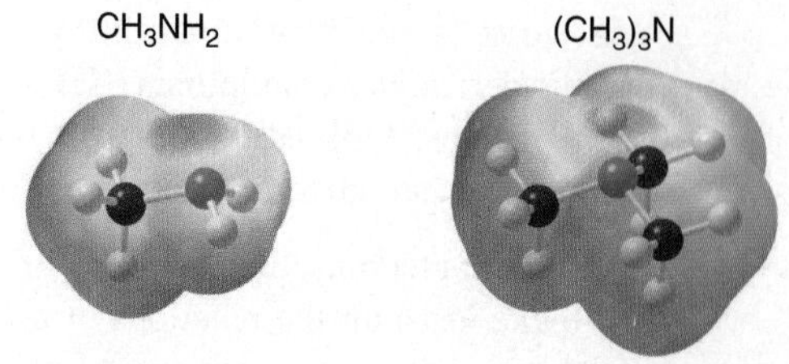

- Both amines clearly show the electron-rich region (in red) at the N atom.

Because nitrogen is much more electronegative than carbon or hydrogen, **the C−N and N−H bonds are all polar,** with the N atom electron rich and the C and H atoms electron poor. The electrostatic potential maps in Figure 25.1 show the polar C−N and N−H bonds in CH_3NH_2 (methylamine) and $(CH_3)_3N$ (trimethylamine).

An amine nitrogen atom bonded to an electron pair and three different alkyl groups is technically a stereogenic center, so two nonsuperimposable trigonal pyramids can be drawn.

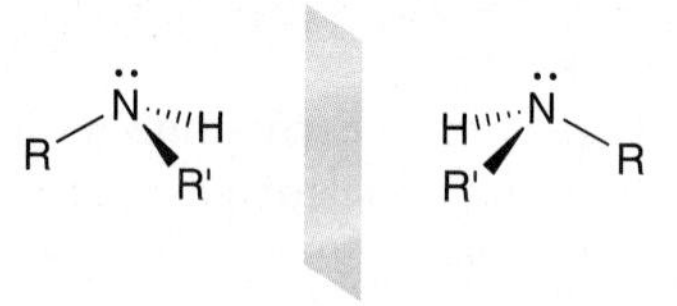

nonsuperimposable mirror images

This does not mean, however, that such an amine exists as two different enantiomers, because one is rapidly converted to the other at room temperature. The amine flips inside out, passing through a trigonal planar (achiral) transition state. **Because the two enantiomers interconvert, we can ignore the chirality of the amine nitrogen.**

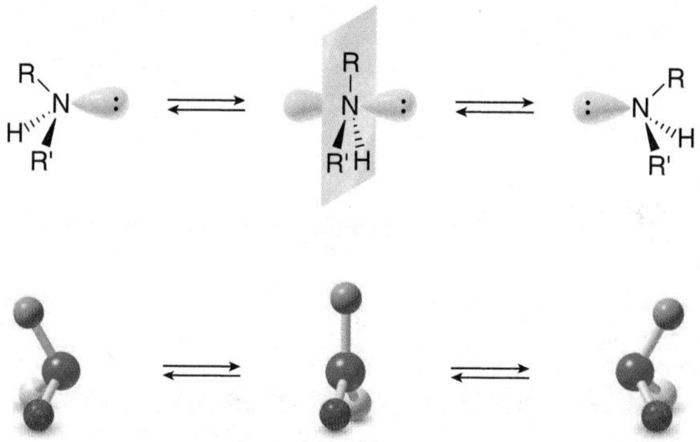

planar transition state

In contrast, the chirality of an ammonium ion with four different groups on N cannot be ignored. Because there is no nonbonded electron pair on the nitrogen atom, interconversion cannot occur, and the N atom is just like a carbon atom with four different groups around it.

chiral ammonium ion

- **The N atom of an ammonium ion is a stereogenic center when N is surrounded by four different groups.**

Problem 25.1 Label the stereogenic centers in each compound.

a. 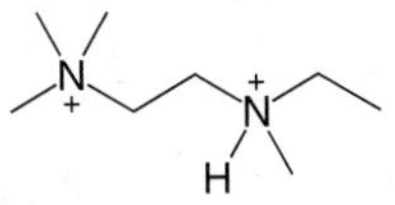

b. 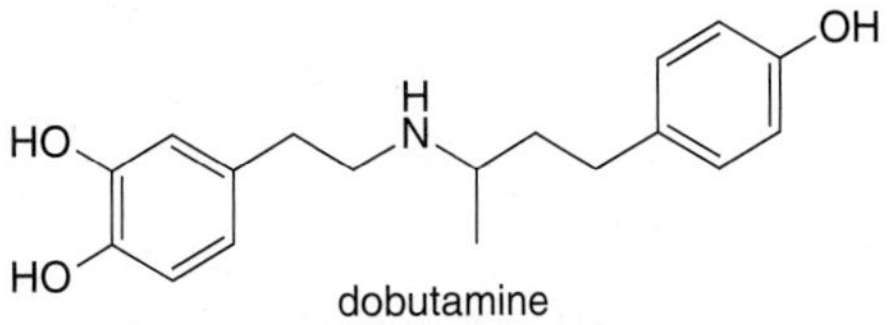

dobutamine
(heart stimulant used in stress tests to measure cardiac fitness)

25.3 Nomenclature

25.3A Primary Amines

Primary amines are named using either systematic or common names.

- To assign the **systematic name,** find the longest continuous carbon chain bonded to the amine nitrogen, and change the **-e** ending of the parent alkane to the suffix ***-amine.*** Then use the usual rules of nomenclature to number the chain and name the substituents.
- To assign a **common name,** name the alkyl group bonded to the nitrogen atom and add the suffix ***-amine,*** forming a single word.

CH_3NH_2

Systematic name: **methanamine**
Common name: **methylamine**

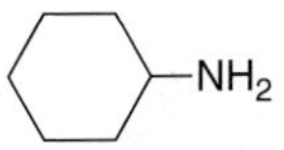

Systematic name: **cyclohexanamine**
Common name: **cyclohexylamine**

25.3B Secondary and Tertiary Amines

Secondary and tertiary amines having identical alkyl groups are named by using the prefix **di-** or **tri-** with the name of the primary amine.

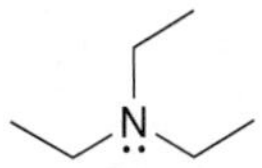

triethylamine

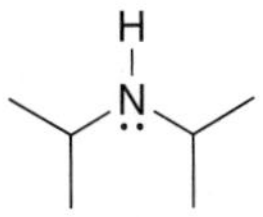

diisopropylamine

Secondary and tertiary amines having more than one kind of alkyl group are named as ***N*-substituted primary amines,** using the following procedure.

How To Name 2° and 3° Amines with Different Alkyl Groups

Example Name the following 2° amine: $(CH_3)_2CHNHCH_3$.

Step [1] Designate the longest alkyl chain (or largest ring) bonded to the N atom as the parent amine and assign a systematic name.

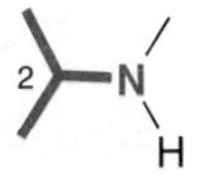

3 C's in the longest chain ----→ propan-2-amine

Step [2] Name the other groups on the N atom as alkyl groups, alphabetize the names, and put the prefix *N*- before the name.

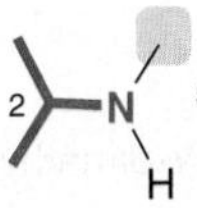

methyl substituent

Answer: ***N*-methylpropan-2-amine**

Sample Problem 25.1 Name each amine.

a. 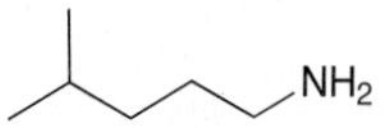

b. 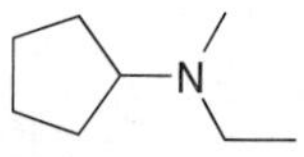

Solution

a. [1] A 1° amine: Find and name the longest chain containing the amine nitrogen.

pentane (**5 C's**) ---→ **pentan*amine***

[2] Number and name the substituents.

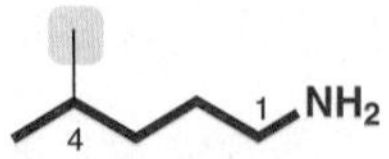

You must use a number to show the location of the NH_2 group.

Answer: 4-methylpentan-1-amine

b. For a 3° amine, one alkyl group on N is the principal R group and the others are substituents.

[1] Name the ring bonded to the N.

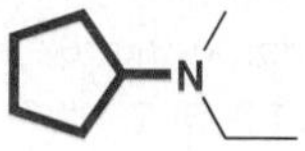

cyclopentanamine

[2] Name the substituents.

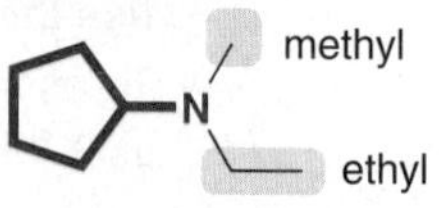

Two N's are needed, one for each alkyl group.

Answer: *N*-ethyl-*N*-methylcyclopentanamine

Problem 25.2 Name each amine.

a. 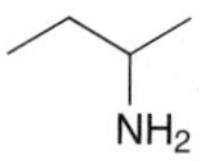

c. 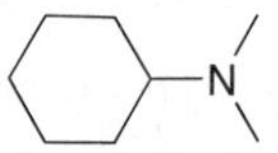

e. 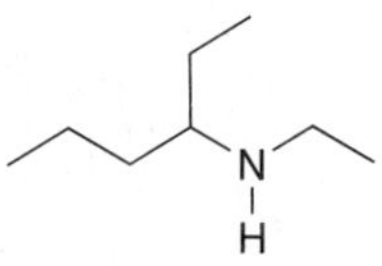

b. 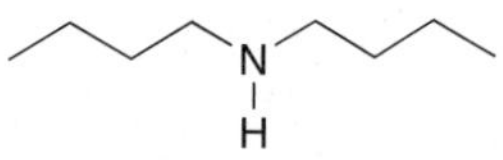

d. 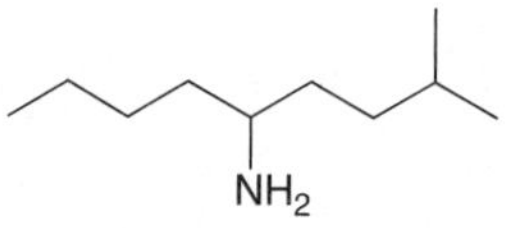

f. 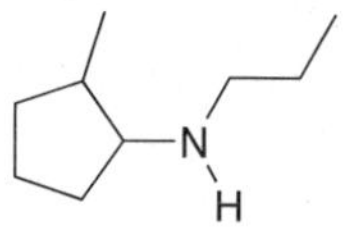

25.3C Aromatic Amines

Aromatic amines are named as derivatives of aniline.

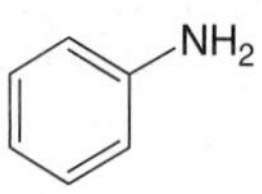

aniline

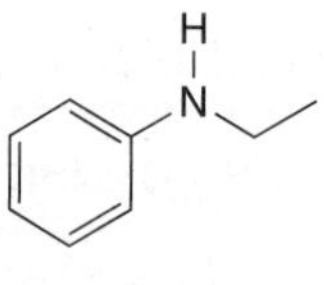

N-ethylaniline

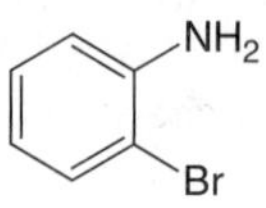

o-bromoaniline

25.3D Miscellaneous Nomenclature Facts

An $\mathbf{NH_2}$ group named as a substituent is called an **amino group.**

There are many different **nitrogen heterocycles,** and each ring type is named differently depending on the number of N atoms in the ring, the ring size, and whether it is aromatic or not. The structures and names of common nitrogen heterocycles are shown in Figure 25.2.

Problem 25.3 Draw a structure corresponding to each name.

a. 2,4-dimethylhexan-3-amine
b. *N*-methylpentan-1-amine
c. *N*-isopropyl-*p*-nitroaniline
d. *N*-methylpiperidine
e. *N*,*N*-dimethylethanamine
f. 2-aminocyclohexanone
g. *N*-methylaniline
h. *m*-ethylaniline

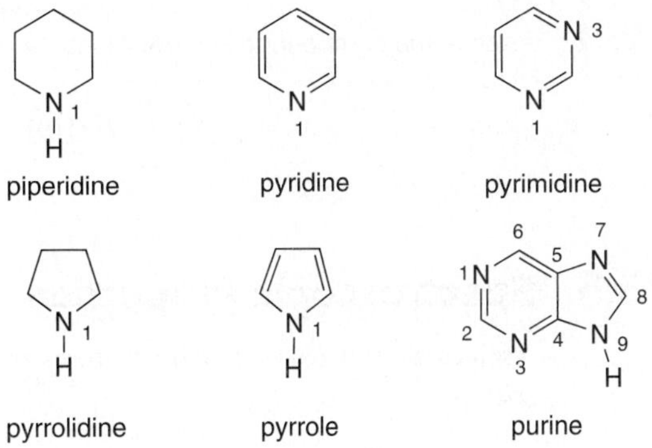
piperidine
pyridine
pyrimidine
pyrrolidine
pyrrole
purine

Figure 25.2 Common nitrogen heterocycles

- Heterocycles with one N atom are numbered to place the N atom at the "1" position.
- Heterocycles with two N atoms are numbered to place one N atom at the "1" position and give the second N atom the lower number.

25.4 Physical Properties

Amines exhibit dipole–dipole interactions because of the polar C–N and N–H bonds. **Primary and secondary amines are also capable of intermolecular hydrogen bonding,** because they contain N–H bonds. Because nitrogen is less electronegative than oxygen, however, intermolecular hydrogen bonds between N and H are *weaker* than those between O and H. How these factors affect the physical properties of amines is summarized in Table 25.1.

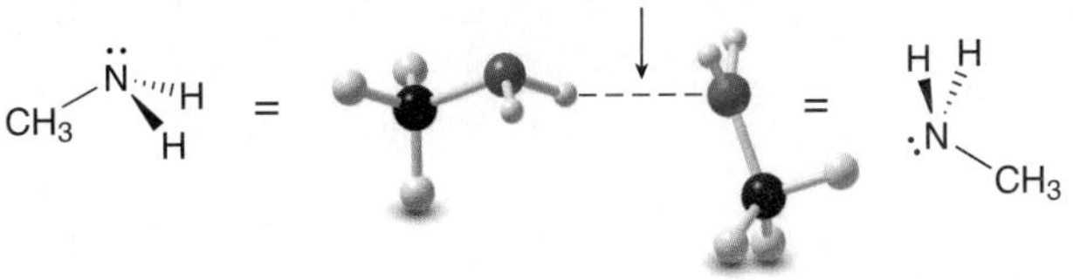
Intermolecular hydrogen bonding
in a 1° amine

Table 25.1 Physical Properties of Amines

Property	Observation
Boiling point and melting point	• Primary (1°) and 2° amines have higher bp's than similar compounds (like ethers) incapable of hydrogen bonding, but lower bp's than alcohols that have stronger intermolecular hydrogen bonds. 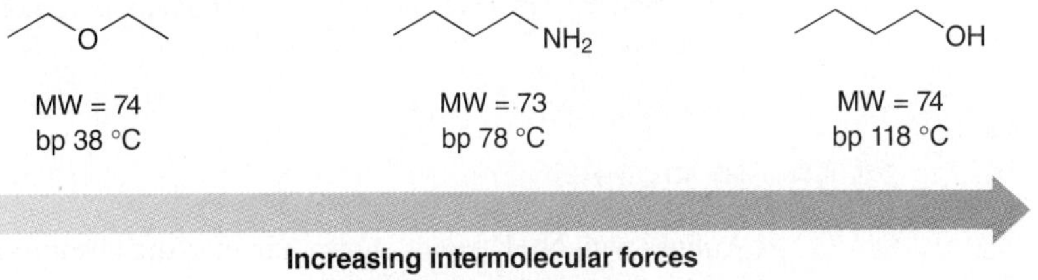 MW = 74, bp 38 °C; MW = 73, bp 78 °C; MW = 74, bp 118 °C **Increasing intermolecular forces** **Increasing boiling point** • Tertiary (3°) amines have lower boiling points than 1° and 2° amines of comparable molecular weight, because they have no N–H bonds and are incapable of hydrogen bonding. 3° amine  2° amine higher bp MW = 73, bp 38 °C, **no N–H bond**; MW = 73, bp 56 °C, **N–H bond**
Solubility	• Amines are soluble in organic solvents regardless of size. • All amines having ≤ 5 C's are H_2O soluble because they can hydrogen bond with H_2O (Section 3.4C). • Amines having > 5 C's are H_2O insoluble because the nonpolar alkyl portion is too large to dissolve in the polar H_2O solvent.

MW = molecular weight

Problem 25.4 Arrange the compounds in order of increasing boiling point.

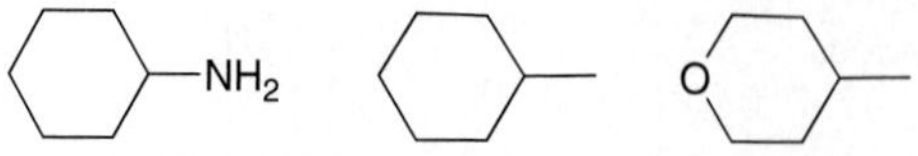

25.5 Spectroscopic Properties

Amines exhibit characteristic features in their mass spectra, IR spectra, and ^{1}H and ^{13}C NMR spectra.

25.5A Mass Spectra

Amines differ from compounds that contain only C, H, and O atoms, which always have a molecular ion with an *even* mass in their mass spectra. This is apparent in the mass spectrum of butan-1-amine, which is shown in Figure 25.3.

The general molecular formula for an amine with one N atom is $C_nH_{2n+3}N$.

- **Amines with an odd number of N atoms give an odd molecular ion in their mass spectra.**

Figure 25.3 Mass spectrum of butan-1-amine

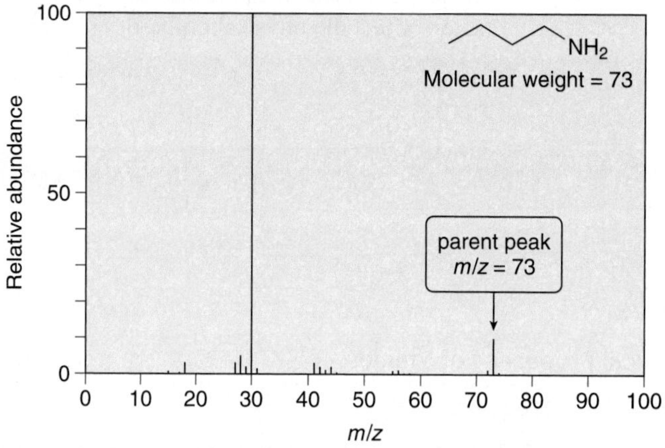

- The molecular ion for $CH_3CH_2CH_2CH_2NH_2$ occurs at $m/z = 73$. This odd mass for a molecular ion is characteristic of an amine with an odd number of N atoms.

25.5B IR Spectra

Amines with N–H bonds show characteristic absorptions in their IR spectra.

- **1° Amines show *two* N–H absorptions at 3300–3500 cm^{-1}.**
- **2° Amines show *one* N–H absorption at 3300–3500 cm^{-1}.**

Because 3° amines have no N–H bonds, they do *not* absorb in this region in their IR spectra. The single bond region (> 2500 cm^{-1}) of the IR spectra for 1°, 2°, and 3° amines illustrates these features in Figure 25.4.

Figure 25.4 The single bond region of the IR spectra for a 1°, 2°, and 3° amine

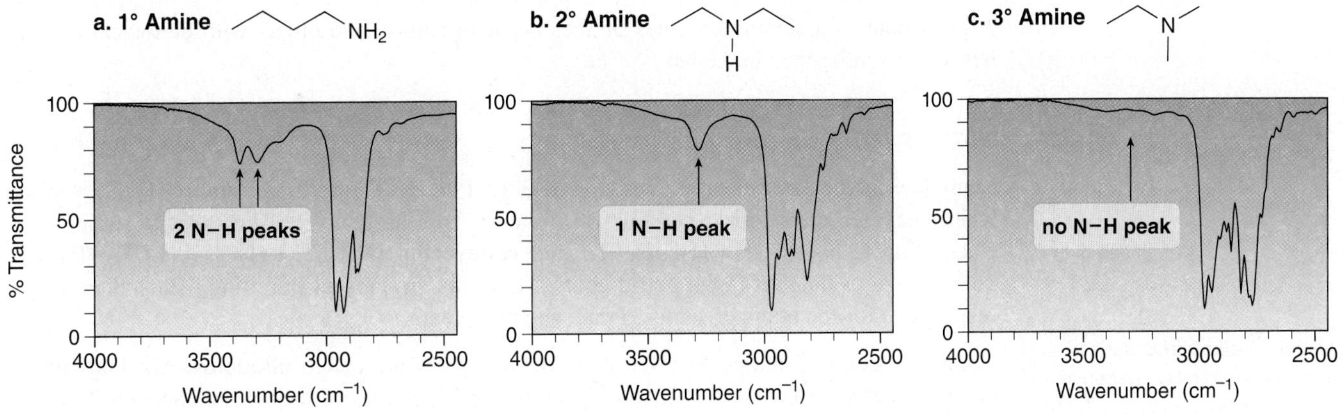

25.5C NMR Spectra

Amines exhibit the following characteristic ^{1}H NMR and ^{13}C NMR absorptions.

- **The NH signal appears between 0.5 and 5.0 ppm.** The exact location depends on the degree of hydrogen bonding and the concentration of the sample.
- The protons on the carbon bonded to the amine nitrogen are deshielded and typically absorb at **2.3–3.0 ppm.**
- In the **^{13}C NMR** spectrum, the carbon bonded to the N atom is deshielded and typically absorbs at **30–50 ppm.**

Like the OH absorption of an alcohol, the **NH absorption is not split by adjacent protons, nor does it cause splitting of adjacent C–H absorptions in a ^{1}H NMR spectrum.** The NH peak of an amine is sometimes somewhat broader than other peaks in the spectrum. The ^{1}H NMR spectrum of *N*-methylaniline is shown in Figure 25.5.

Problem 25.5 What is the structure of an unknown compound with molecular formula $C_6H_{15}N$ that gives the following ^{1}H NMR absorptions: 0.9 (singlet, 1 H), 1.10 (triplet, 3 H), 1.15 (singlet, 9 H), and 2.6 (quartet, 2 H) ppm?

Figure 25.5
The ^{1}H NMR spectrum of *N*-methylaniline

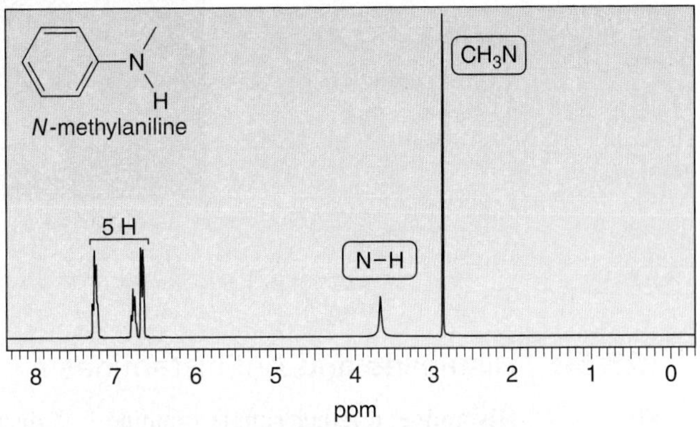

- The CH_3 group appears as a singlet at 2.7 ppm because there is no splitting by the adjacent NH proton.
- The NH proton appears as a broad singlet at 3.6 ppm.
- The five H atoms of the aromatic ring appear as a complex pattern at 6.6–7.2 ppm.

25.6 Interesting and Useful Amines

A great many simple and complex amines occur in nature, and others with biological activity have been synthesized in the lab.

25.6A Simple Amines and Alkaloids

Many low molecular weight amines have *very* foul odors. **Trimethylamine** [$(CH_3)_3N$], formed when enzymes break down certain fish proteins, has the characteristic odor of rotting fish. **Putrescine** ($NH_2CH_2CH_2CH_2CH_2NH_2$) and **cadaverine** ($NH_2CH_2CH_2CH_2CH_2CH_2NH_2$) are both poisonous diamines with putrid odors. They, too, are present in rotting fish, and are partly responsible for the odors of semen, urine, and bad breath.

The word ***alkaloid*** is derived from the word *alkali,* because aqueous solutions of alkaloids are slightly basic.

Naturally occurring amines derived from plant sources are called **alkaloids.** Alkaloids previously encountered in the text include **quinine** (Problem 17.18), **morphine** (Section 2.1), and **cocaine** (Problem 3.55). Three other common alkaloids are **atropine, nicotine,** and **coniine,** illustrated in Figure 25.6.

Figure 25.6
Three common alkaloids—Atropine, nicotine, and coniine

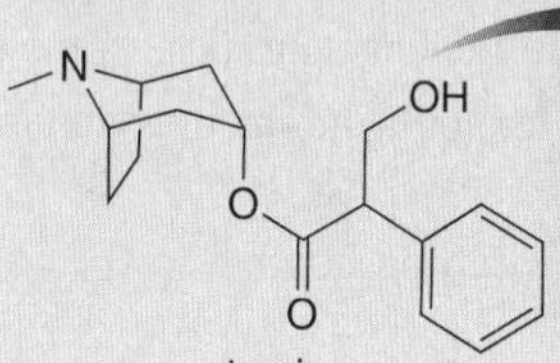
atropine

nightshade

- **Atropine** is an alkaloid isolated from *Atropa belladonna*, the deadly nightshade plant. In the Renaissance, women used the juice of the berries of the nightshade to enlarge the pupils of their eyes for cosmetic reasons. Atropine causes an increase in heart rate, relaxes smooth muscles, and interferes with nerve impulses transmitted by acetylcholine. In higher doses atropine is poisonous, leading to convulsions, coma, and death.

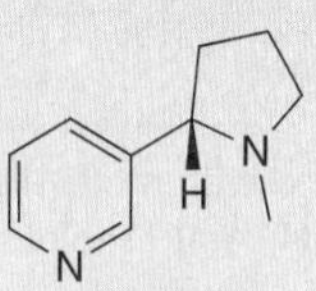
nicotine

tobacco

- **Nicotine** is an addictive and highly toxic compound isolated from tobacco. In small doses it acts as a stimulant, but in large doses it causes depression, nausea, and even death. Nicotine is synthesized in plants as a defense against insect predators, and is used commercially as an insecticide.

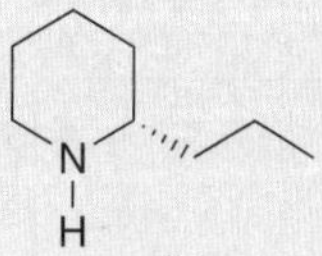
coniine

hemlock

- **Coniine,** a poisonous alkaloid isolated from the seeds, leaves, and roots of hemlock (*Conium maculatum*), has been known since ancient times. Ingestion causes weakness, paralysis, and finally death. The Greek philosopher Socrates was executed by being forced to drink a potion prepared from hemlock in 339 B.C.

25.6B Histamine and Antihistamines

histamine

Histamine, a rather simple triamine first discussed in Section 17.8, is responsible for a wide variety of physiological effects. Histamine is a vasodilator (it dilates capillaries), so it is released at the site of an injury or infection to increase blood flow. It is also responsible for the symptoms of allergies, including a runny nose and watery eyes. In the stomach, histamine stimulates the secretion of acid.

Understanding the central role of histamine in these biochemical processes has helped chemists design drugs to counteract some of its undesirable effects.

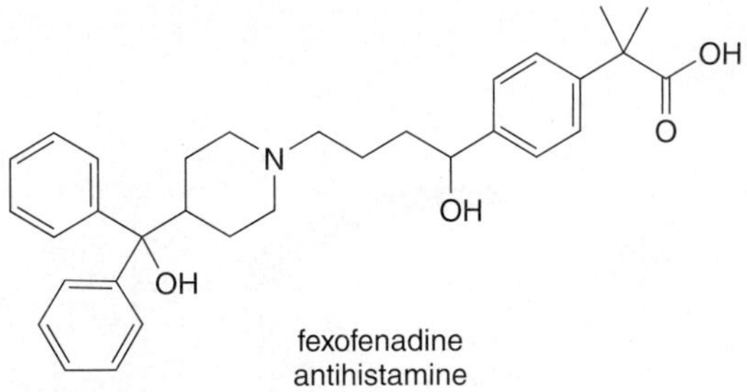

fexofenadine
antihistamine

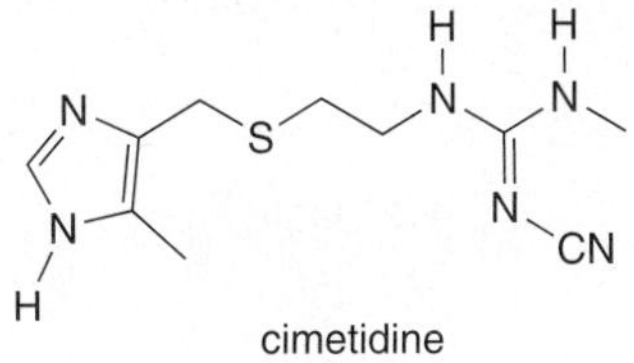

cimetidine
(Tagamet)
antiulcer drug

Antihistamines bind to the same active site of the enzyme that binds histamine in the cell, but they evoke a different response. An antihistamine like **fexofenadine** (trade name Allegra), for example, inhibits vasodilation, so it is used to treat the symptoms of the common cold and allergies. Unlike many antihistamines, fexofenadine does not cause drowsiness because it binds to histamine receptors but does not cross the blood–brain barrier, so it does not affect the central nervous system. **Cimetidine** (trade name Tagamet) is a histamine mimic that blocks the secretion of hydrochloric acid in the stomach, so it is used to treat individuals with ulcers.

25.6C Derivatives of 2-Phenylethanamine

A large number of physiologically active compounds are derived from **2-phenylethanamine, $C_6H_5CH_2CH_2NH_2$.** Some of these compounds are synthesized in cells and needed to maintain healthy mental function. Others are isolated from plant sources or are synthesized in the laboratory and have a profound effect on the brain because they interfere with normal neurochemistry. These compounds include **adrenaline, noradrenaline, methamphetamine,** and **mescaline.** Each contains a benzene ring bonded to a two-carbon unit with a nitrogen atom (shown in red).

adrenaline
(epinephrine)

a hormone secreted in response to stress
(Figure 7.19)

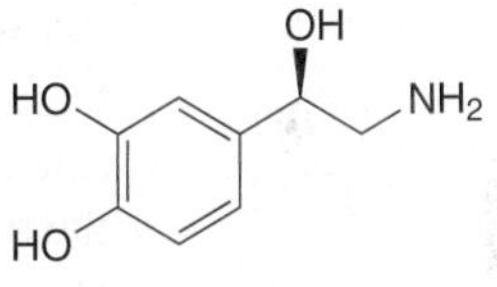

noradrenaline
(norepinephrine)

a neurotransmitter that increases heart rate
and dilates air passages

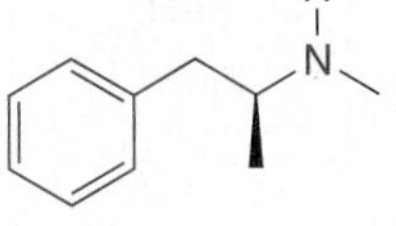

methamphetamine

an addictive stimulant sold as
speed, meth, or crystal meth

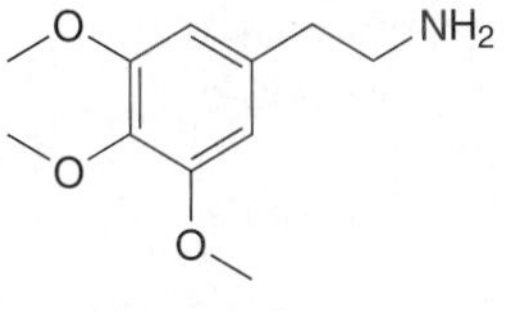

mescaline

a hallucinogen isolated from peyote, a cactus native
to the southwestern United States and Mexico

Another example, **dopamine,** is a neurotransmitter, a chemical messenger released by one nerve cell (neuron), which then binds to a receptor in a neighboring target cell (Figure 25.7). Dopamine affects brain processes that control movement and emotions, so proper dopamine levels are necessary to maintain an individual's mental and physical health. For example, when dopamine-producing neurons die, the level of dopamine drops, resulting in the loss of motor control symptomatic of Parkinson's disease.

Figure 25.7 Dopamine—A neurotransmitter

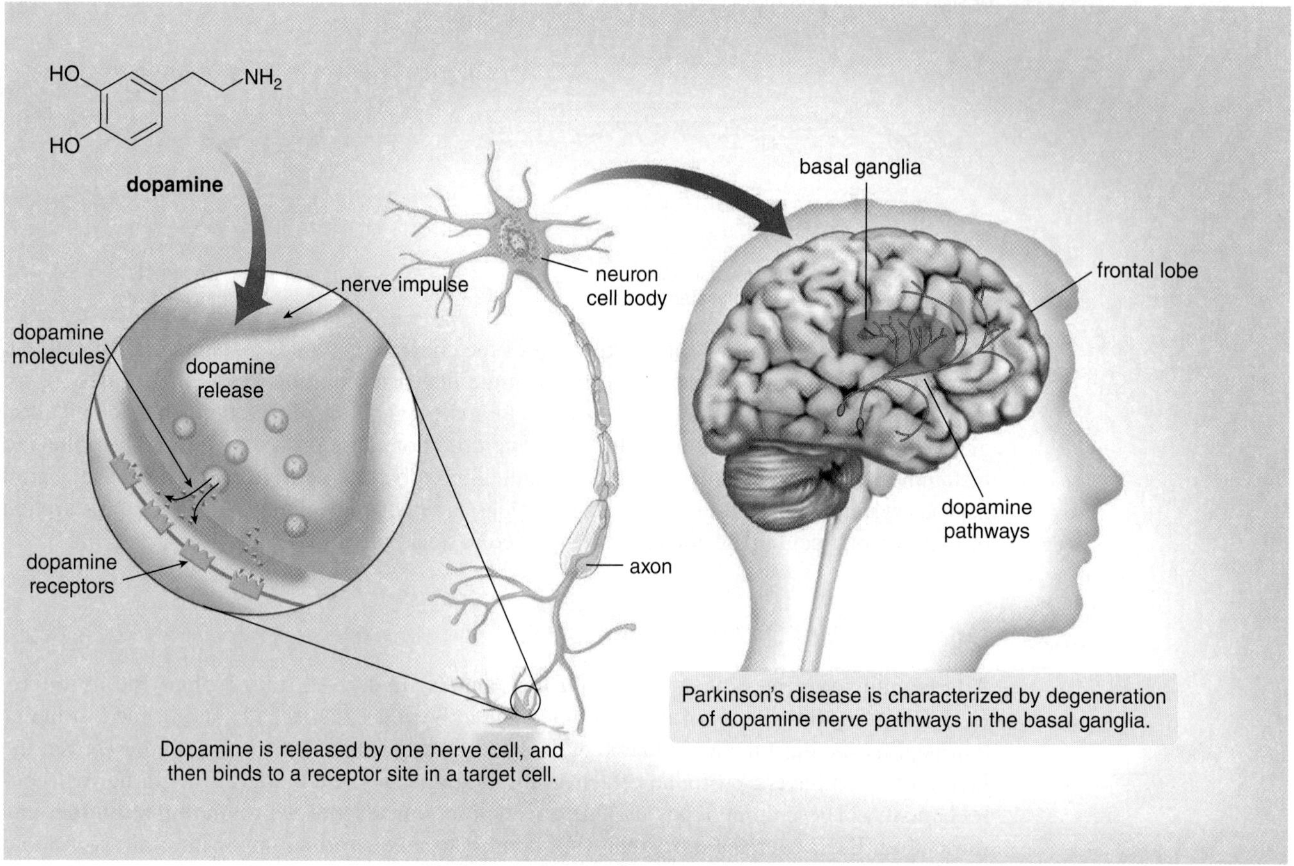
HO
HO
NH2
dopamine
nerve impulse
neuron
cell body
basal ganglia
frontal lobe
dopamine
molecules
dopamine
release
dopamine
receptors
axon
dopamine
pathways
Parkinson's disease is characterized by degeneration of dopamine nerve pathways in the basal ganglia.
Dopamine is released by one nerve cell, and then binds to a receptor site in a target cell.

Serotonin is a neurotransmitter that plays an important role in mood, sleep, perception, and temperature regulation. A deficiency of serotonin causes depression. Understanding the central role of serotonin in determining one's mood has led to the development of a variety of drugs for the treatment of depression. The most widely used antidepressants today are selective serotonin reuptake inhibitors (SSRIs). These drugs act by inhibiting the reuptake of serotonin by the neurons that produce it, thus effectively increasing its concentration. Fluoxetine (trade name Prozac) is a common antidepressant that acts in this way.

Cocaine, amphetamines, and several other addicting drugs increase the level of dopamine in the brain, which results in a pleasurable "high." With time, the brain adapts to increased dopamine levels, so more drug is required for the same sensation.

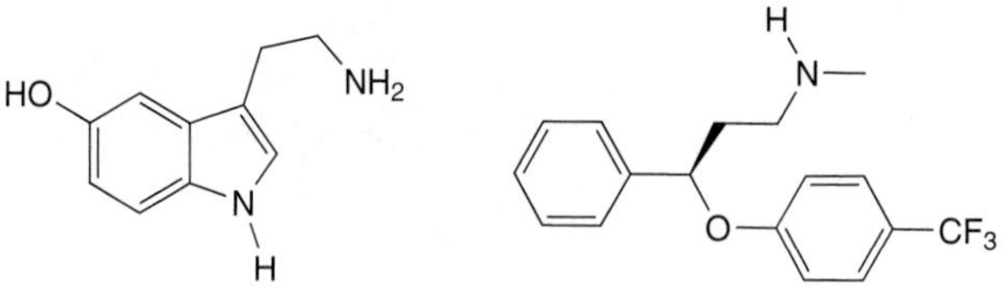
HO
NH2
N
H
H
N
O
CF3

serotonin fluoxetine

Bufo toads from the Amazon jungle are the source of the hallucinogen bufotenin.

Drugs that interfere with the metabolism of serotonin have a profound effect on mental state. For example, bufotenin, isolated from *Bufo* toads from the Amazon jungle, and psilocin, isolated from *Psilocybe* mushrooms, are very similar in structure to serotonin and both cause intense hallucinations.

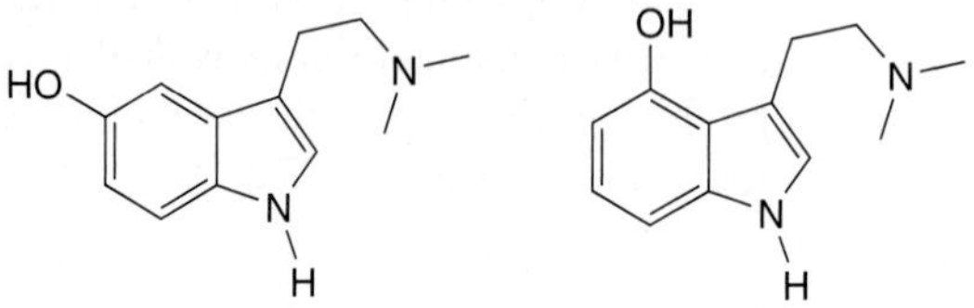
HO
N
N
H
OH
N
N
H

bufotenin psilocin

Problem 25.6 LSD (a hallucinogen) and codeine (a narcotic) are structurally more complex derivatives of 2-phenylethanamine. Identify the atoms of 2-phenylethanamine in each of the following compounds.

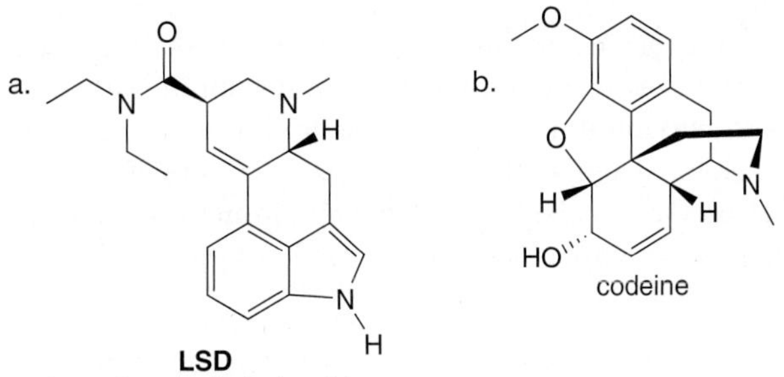

LSD
lysergic acid diethyl amide

25.7 Preparation of Amines

In the preparations of a given functional group, many different starting materials form a common product (amines, in this case).

Three types of reactions are used to prepare an amine:

[1] **Nucleophilic substitution** using nitrogen nucleophiles
[2] **Reduction** of other nitrogen-containing functional groups
[3] **Reductive amination** of aldehydes and ketones

25.7A Nucleophilic Substitution Routes to Amines

Nucleophilic substitution is the key step in two different methods for synthesizing amines: direct nucleophilic substitution and the Gabriel synthesis of 1° amines.

Direct Nucleophilic Substitution

Conceptually, the simplest method to synthesize an amine is by **S_N2 reaction of an alkyl halide with NH_3 or an amine.** The method requires two steps:

[1] **Nucleophilic attack** of the nitrogen nucleophile forms an ammonium salt.
[2] **Removal of a proton** on N forms the amine.

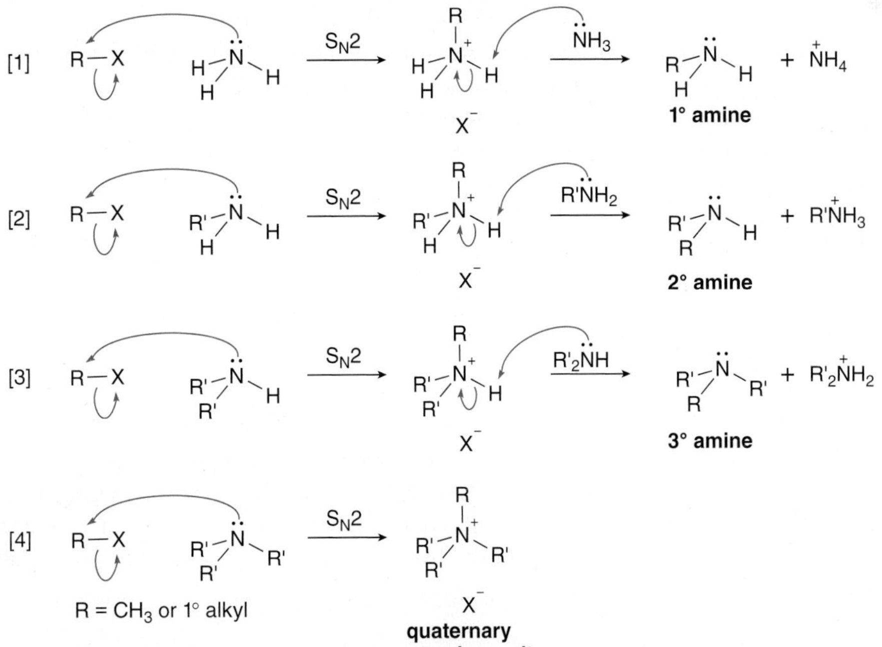

The identity of the nitrogen nucleophile determines the type of amine or ammonium salt formed as product. One new carbon–nitrogen bond is formed in each reaction. Because the reaction follows an S_N2 mechanism, the alkyl halide must be unhindered—that is, CH_3X or RCH_2X.

Although this process seems straightforward, polyalkylation of the nitrogen nucleophile limits its usefulness. **Any amine formed by nucleophilic substitution still has a nonbonded electron pair, making it a nucleophile as well.** It will react with remaining alkyl halide to form a more substituted amine. Because of this, a mixture of 1°, 2°, and 3° amines often results. Only the final product—called a **quaternary ammonium salt** because it has four alkyl groups on N—cannot react further, and so the reaction stops.

As a result, this reaction is most useful for preparing 1° amines by using a very large excess of NH_3 (a relatively inexpensive starting material) and for preparing quaternary ammonium salts by alkylating any nitrogen nucleophile with one or more equivalents of alkyl halide.

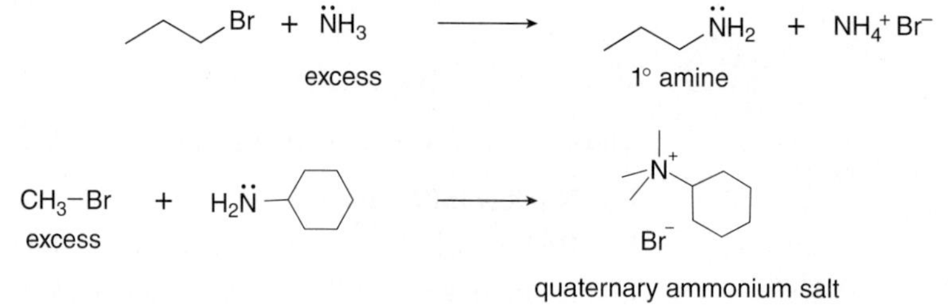

Problem 25.7 Draw the product of each reaction.

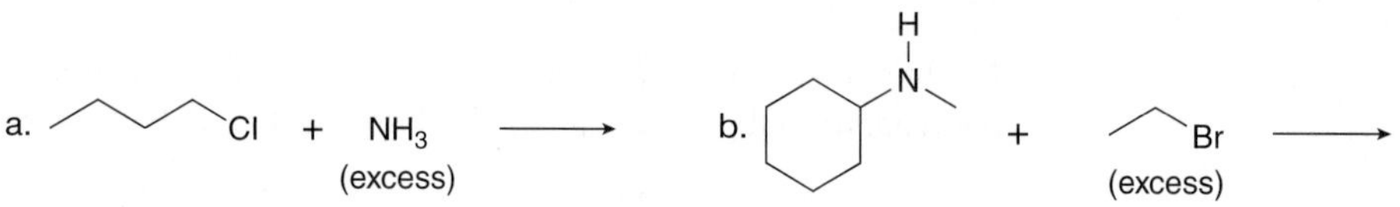

The Gabriel Synthesis of 1° Amines

To avoid polyalkylation, a nitrogen nucleophile can be used that reacts in a single nucleophilic substitution reaction—that is, the reaction forms a product that does not contain a nucleophilic nitrogen atom capable of reacting further.

The **Gabriel synthesis** consists of two steps and uses a resonance-stabilized nitrogen nucleophile to synthesize 1° amines via nucleophilic substitution. The Gabriel synthesis begins with **phthalimide,** one of a group of compounds called **imides.** The **N−H bond of an imide is especially acidic** because the resulting anion is resonance stabilized by the two flanking carbonyl groups.

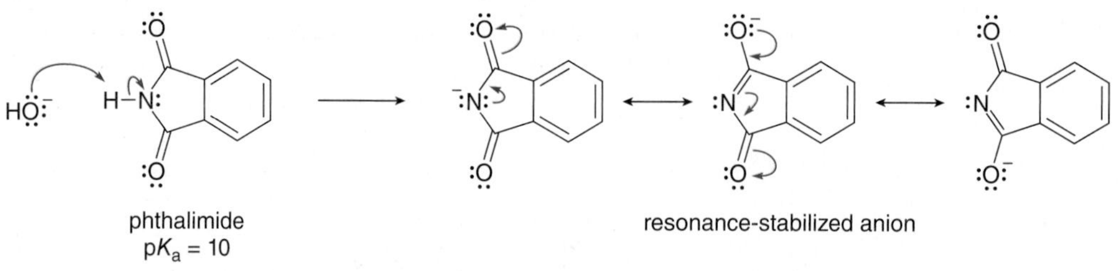

An acid–base reaction forms a nucleophilic anion that can react with an unhindered alkyl halide—that is, CH_3X or RCH_2X—in an S_N2 reaction to form a substitution product. This alkylated imide is then hydrolyzed with aqueous base to give a 1° amine and a dicarboxylate. This reaction is similar to the hydrolysis of amides to afford carboxylate anions and amines, as discussed in

Section 22.13. The overall result of this two-step sequence is **nucleophilic substitution of X by NH₂,** so the Gabriel synthesis can be used to prepare 1° amines only.

Steps in the Gabriel synthesis

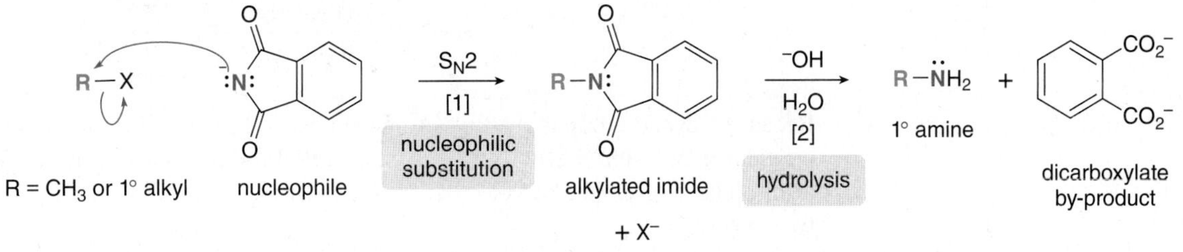
R—X
R = CH3 or 1° alkyl
nucleophile
SN2
[1]
nucleophilic substitution
R—N:
alkylated imide
+ X−
−OH
H2O
[2]
hydrolysis
R—NH2 +
1° amine
CO2−
CO2−
dicarboxylate by-product

- **The Gabriel synthesis converts an alkyl halide into a 1° amine by a two-step process: nucleophilic substitution followed by hydrolysis.**

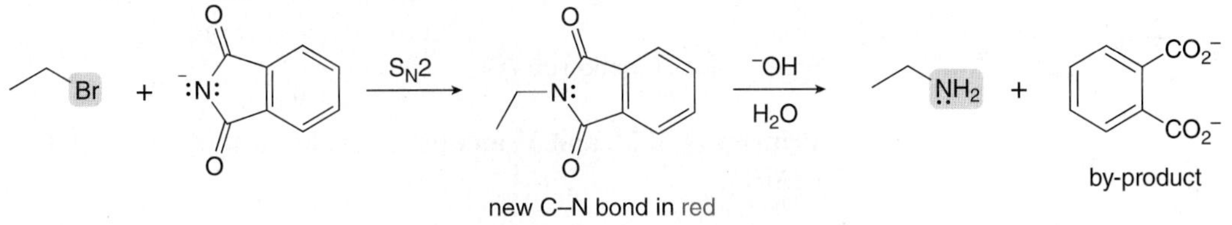
Br
SN2
−OH
H2O
NH2
by-product
new C–N bond in red

Problem 25.8 What alkyl halide is needed to prepare each 1° amine by a Gabriel synthesis?

a.  b. 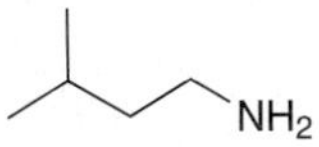 c. 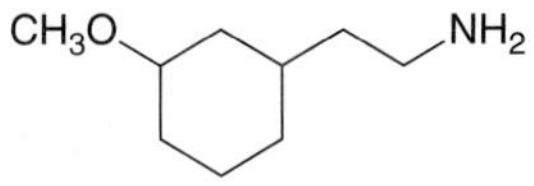

Problem 25.9 Which amines cannot be prepared by a Gabriel synthesis? Explain your choices.

a. 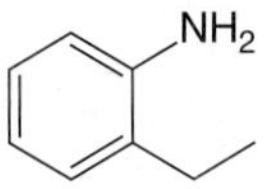 b. 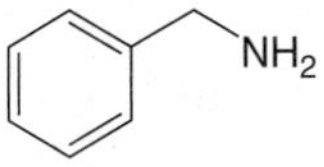 c. 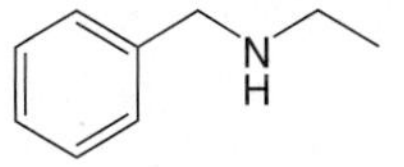 d. 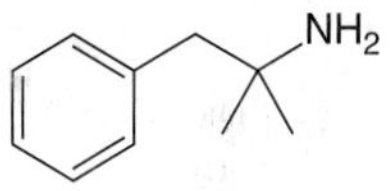

25.7B Reduction of Other Functional Groups That Contain Nitrogen

Amines can be prepared by reduction of nitro compounds, nitriles, and amides. Because the details of these reactions have been discussed previously, they are presented here in summary form only.

[1] From nitro compounds (Section 18.15C)

Nitro groups are reduced to 1° amines using a variety of reducing agents.

$$R{-}NO_2 \xrightarrow[\text{or } Sn,\ HCl]{H_2,\ Pd\text{-}C \text{ or } Fe,\ HCl} R{-}NH_2 \quad (1° \text{ amine})$$

[2] From nitriles (Section 22.18B)

Nitriles are reduced to 1° amines with $LiAlH_4$.

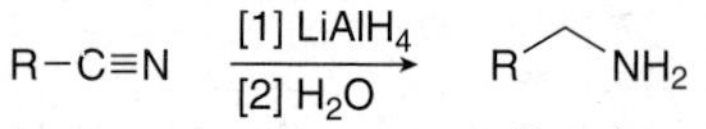

1° amine

Because a cyano group is readily introduced by S_N2 substitution of alkyl halides with ^-CN, this provides **a two-step method to convert an alkyl halide to a 1° amine with one more carbon atom.** The conversion of $(CH_3)_2CHCH_2CH_2Br$ to $(CH_3)_2CHCH_2CH_2CH_2NH_2$ illustrates this two-step sequence.

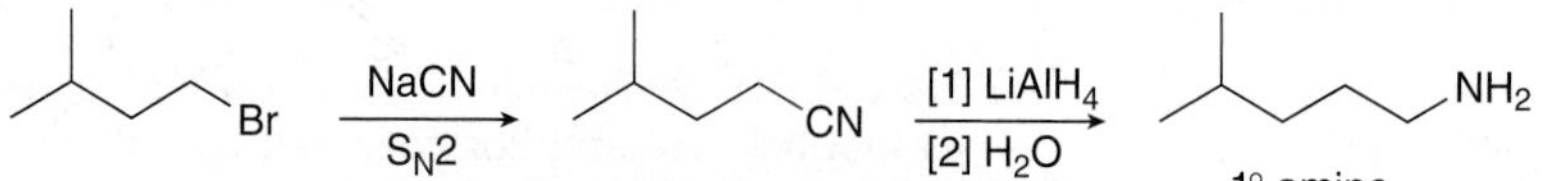

new C–C bond in red

1° amine

[3] From amides (Section 20.7B)

Primary (1°), 2°, and 3° amides are reduced to 1°, 2°, and 3° amines, respectively, by using $LiAlH_4$.

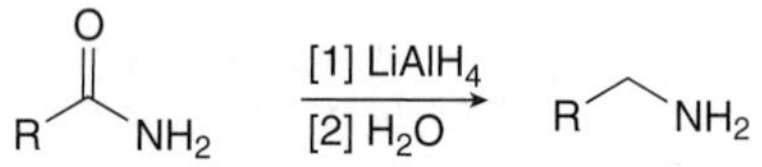

1° amide 1° amine

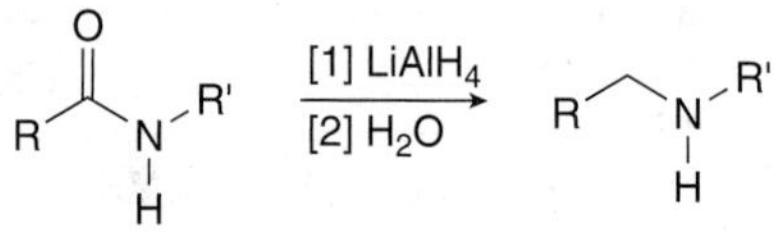

2° amide 2° amine

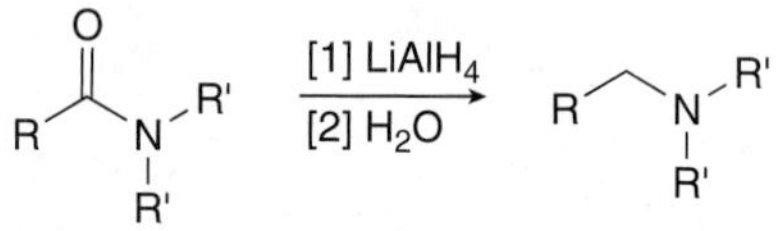

3° amide 3° amine

Problem 25.10 What nitro compound, nitrile, and amide are reduced to each compound?

a. 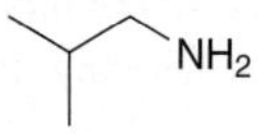 b. 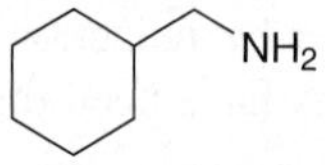 c. 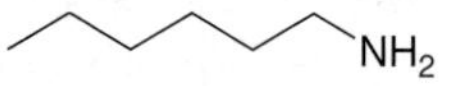

Problem 25.11 What amine is formed by reduction of each amide?

a. 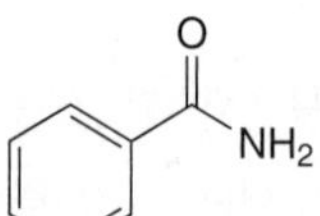 b. 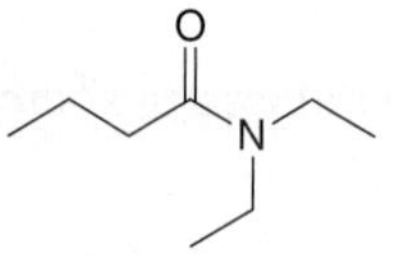 c. 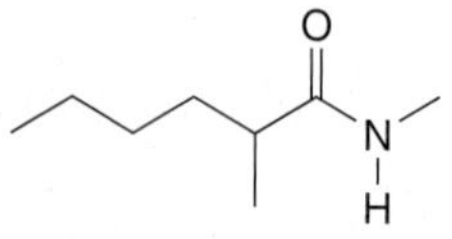

Problem 25.12 Which amines cannot be prepared by reduction of an amide?

a. 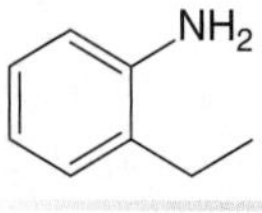 b. 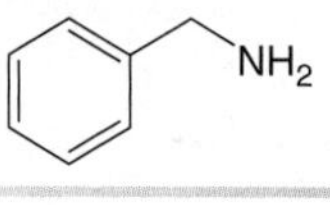 c. 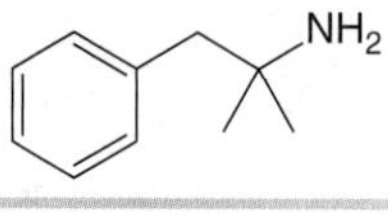 d. 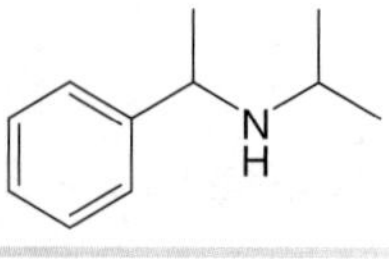

25.7C Reductive Amination of Aldehydes and Ketones

Reductive amination is a two-step method that converts aldehydes and ketones into 1°, 2°, and 3° amines. Let's first examine this method using NH_3 to prepare 1° amines. There are two distinct parts in reductive amination:

[1] **Nucleophilic attack of NH_3 on the carbonyl group forms an imine** (Section 21.11A), which is not isolated; then,

[2] **Reduction of the imine forms an amine** (Section 20.7B).

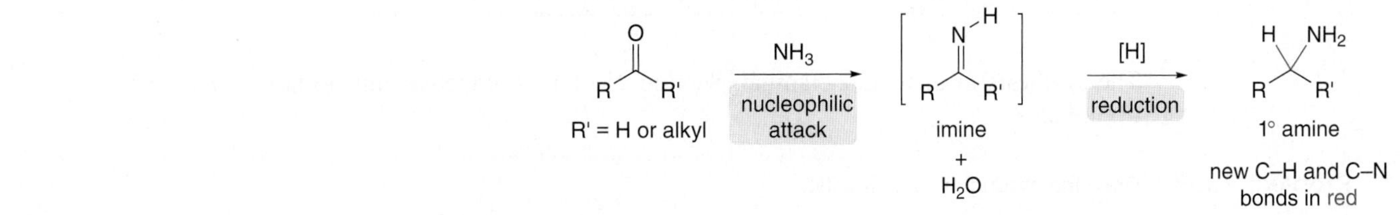

- **Reductive amination replaces a C=O by a C–H and C–N bond.**

The most effective reducing agent for this reaction is sodium cyanoborohydride ($NaBH_3CN$). This hydride reagent is a derivative of sodium borohydride ($NaBH_4$), formed by replacing one H atom by CN.

$NaBH_3CN$
sodium cyanoborohydride

Reductive amination combines two reactions we have already learned in a different way. Two examples are shown. The second reaction is noteworthy because the product is **amphetamine,** a potent central nervous system stimulant.

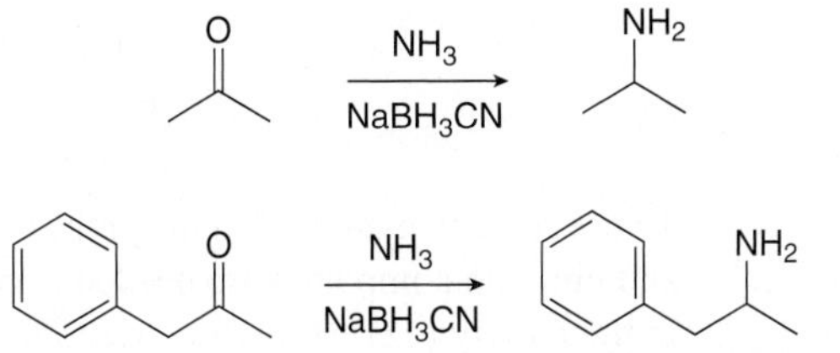

amphetamine
a powerful stimulant

With a 1° or 2° amine as starting material, reductive amination is used to prepare 2° and 3° amines, respectively. Note the result: Reductive amination uses an aldehyde or ketone to replace one H atom on a nitrogen atom by an alkyl group, making a more substituted amine.

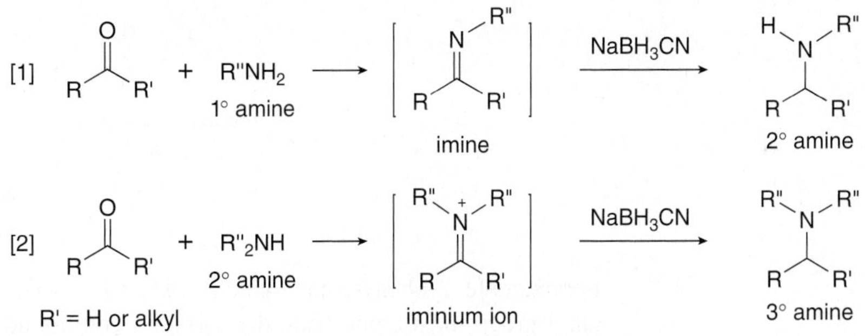

Figure 25.8
Synthesis of methamphetamine by reductive amination

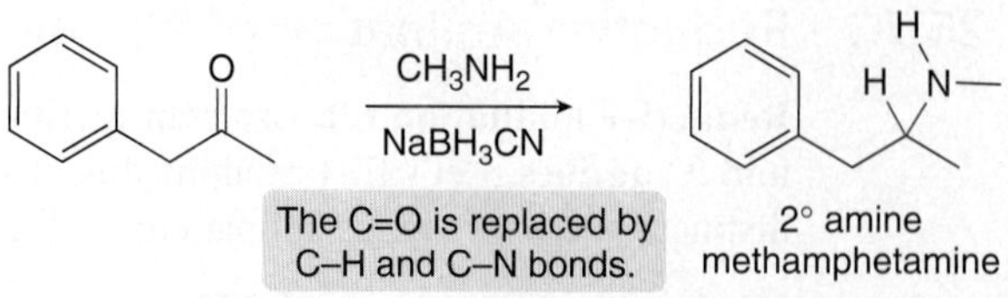
O
CH_3NH_2
$NaBH_3CN$
H
H N—
The C=O is replaced by C–H and C–N bonds.
2° amine
methamphetamine

- In reductive amination, one of the H atoms bonded to N is replaced by an alkyl group. As a result, a 1° amine is converted to a 2° amine and a 2° amine is converted to a 3° amine. In this reaction, CH_3NH_2 (a 1° amine) is converted to methamphetamine (a 2° amine).

The synthesis of methamphetamine (Section 25.6C) by reductive amination is illustrated in Figure 25.8.

Problem 25.13 Draw the product of each reaction.

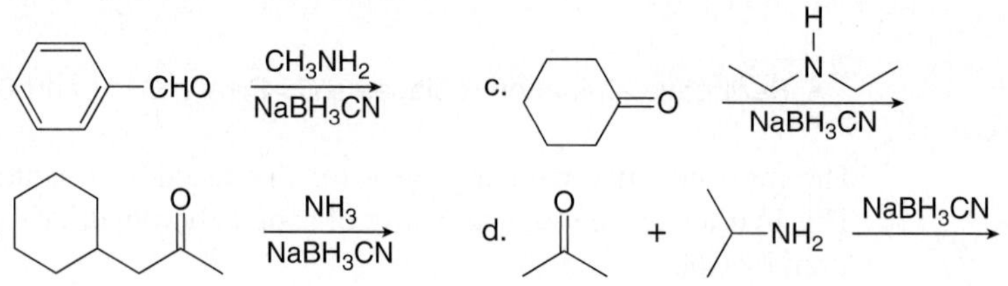
a.
CHO
CH_3NH_2
$NaBH_3CN$
b.
O
NH_3
$NaBH_3CN$
c.
O
H
N
$NaBH_3CN$
d.
O
+
NH_2
$NaBH_3CN$

Problem 25.14 Enalapril, a drug used to treat hypertension, is prepared from compounds **D** and **E** by reductive amination. What is the structure of enalapril?

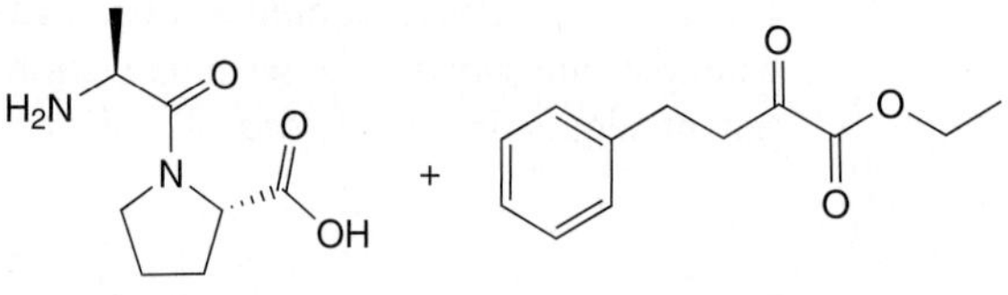
H_2N
O
N
O
OH
+
O
O
O

D **E**

To use reductive amination in synthesis, you must be able to determine what aldehyde or ketone and nitrogen compound are needed to prepare a given amine—that is, you must work backwards in the retrosynthetic direction. Keep in mind the following two points:

- **One alkyl group on N comes from the carbonyl compound.**
- **The remainder of the molecule comes from NH_3 or an amine.**

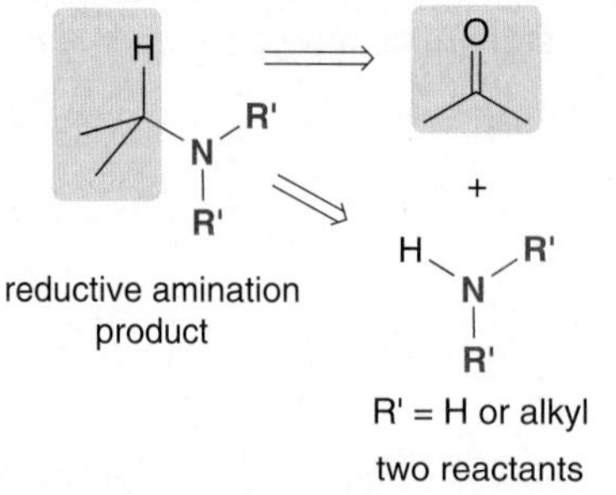
H
R'
N
R'
O
+
H
R'
N
R'
reductive amination product
R' = H or alkyl
two reactants

For example, 2-phenylethanamine is a 1° amine, so it has only one alkyl group bonded to N. This alkyl group must come from the carbonyl compound, and the rest of the molecule then comes from the nitrogen component. **For a 1° amine, the nitrogen component must be NH_3.**

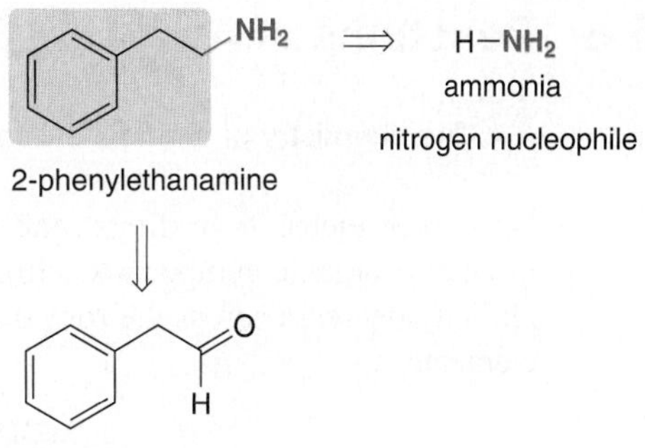

carbonyl component

There is usually more than one way to use reductive amination to synthesize 2° and 3° amines, as shown in Sample Problem 25.2 for a 2° amine.

Sample Problem 25.2 What aldehyde or ketone and nitrogen component are needed to synthesize *N*-ethylcyclohexanamine by a reductive amination reaction?

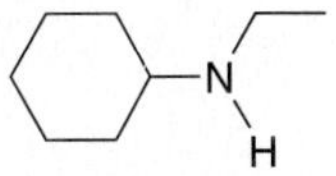

N-ethylcyclohexanamine

Solution

Because *N*-ethylcyclohexanamine has two different alkyl groups bonded to the N atom, either R group can come from the carbonyl component and there are two different ways to form a C–N bond by reductive amination.

Possibility [1] Use $CH_3CH_2NH_2$ and cyclohexanone.

Possibility [2] Use cyclohexanamine and an aldehyde.

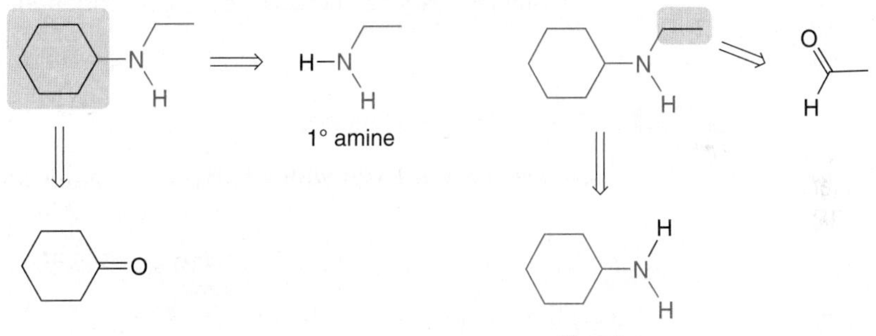

Because reductive amination adds one R group to a nitrogen atom, both routes to form the 2° amine begin with a 1° amine.

Problem 25.15 What starting materials are needed to prepare each drug using reductive amination? Give all possible pairs of compounds when more than one route is possible.

a. 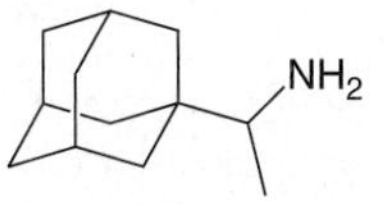

rimantadine
antiviral used to treat influenza

b. 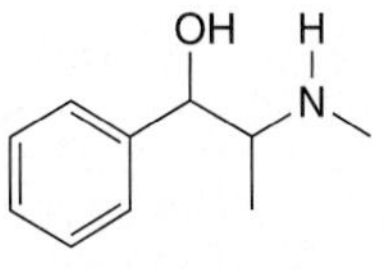

pseudoephedrine
nasal decongestant

Problem 25.16 (a) Explain why phentermine [$PhCH_2C(CH_3)_2NH_2$] can't be made by a reductive amination reaction. (b) Give a systematic name for phentermine, one of the components of the banned diet drug fen–phen.

25.8 Reactions of Amines—General Features

- **The chemistry of amines is dominated by the lone pair of electrons on nitrogen.**

Only three elements in the second row of the periodic table have nonbonded electron pairs in neutral organic compounds: nitrogen, oxygen, and fluorine. Because basicity and nucleophilicity decrease across the row, **nitrogen is the most basic and most nucleophilic** of these elements.

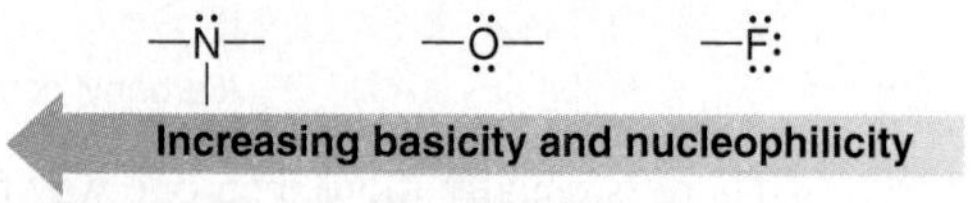
Increasing basicity and nucleophilicity

- **Amines are stronger bases and nucleophiles than other neutral organic compounds.**

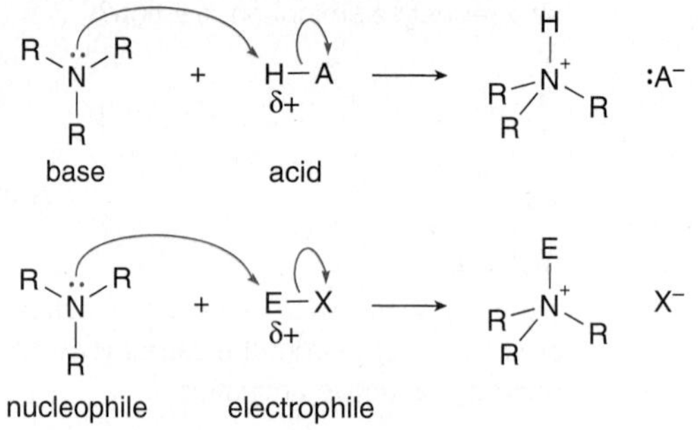
base
acid
nucleophile
electrophile

- **Amines react as bases with compounds that contain acidic protons.**
- **Amines react as nucleophiles with compounds that contain electrophilic carbons.**

25.9 Amines as Bases

Amines react as bases with a variety of organic and inorganic acids.

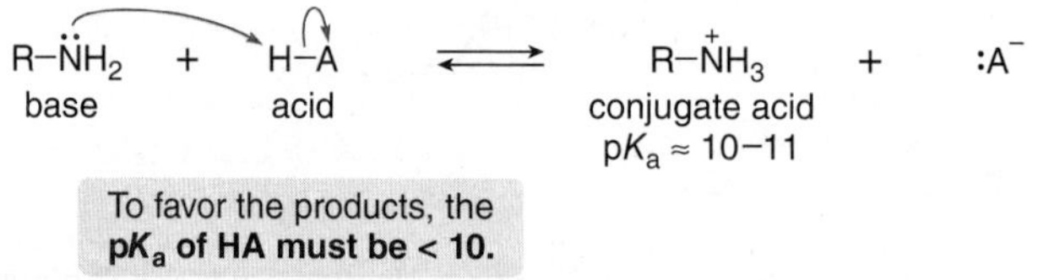
base
acid
conjugate acid
pKa ≈ 10–11
To favor the products, the
pKa of HA must be < 10.

What acids can be used to protonate an amine? Equilibrium favors the products of an acid–base reaction when the weaker acid and base are formed. Because the pK_a of many protonated amines is 10–11, the **pK_a of the starting acid must be less than 10** for equilibrium to favor the products. Amines are thus readily protonated by strong inorganic acids like HCl and H_2SO_4, and by carboxylic acids as well.

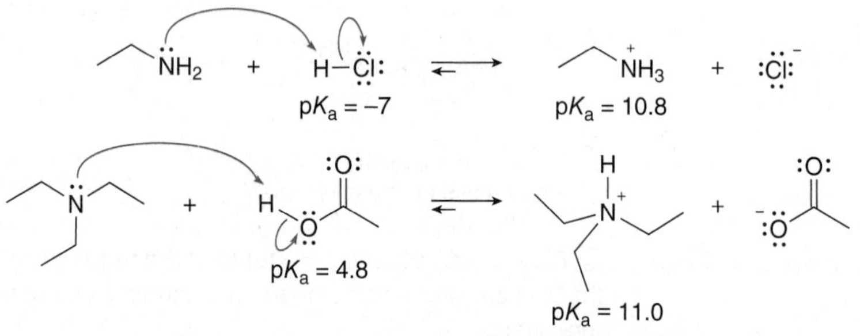
pKa = –7
pKa = 10.8
pKa = 4.8
pKa = 11.0

Equilibrium favors the products.

The principles used in an extraction procedure were detailed in Section 19.13.

Because amines are protonated by aqueous acid, they can be separated from other organic compounds by extraction using a separatory funnel. **Extraction separates compounds based on solubility differences.** When an amine is protonated by aqueous acid, its solubility properties change.

For example, when cyclohexanamine is treated with aqueous HCl, it is protonated, forming an ammonium salt. Because the ammonium salt is ionic, it is soluble in water, but insoluble in organic solvents. A similar acid–base reaction does not occur with other organic compounds like alcohols, which are much less basic.

cyclohexanamine
- insoluble in H_2O
- soluble in CH_2Cl_2

cyclohexanammonium chloride
- soluble in H_2O
- insoluble in CH_2Cl_2

This difference in acid–base chemistry can be used to separate cyclohexanamine and cyclohexanol by the stepwise extraction procedure illustrated in Figure 25.9.

Figure 25.9 Separation of cyclohexanamine and cyclohexanol by an extraction procedure

Step [1] Dissolve cyclohexanamine and cyclohexanol in CH_2Cl_2.

Step [2] Add 10% HCl solution to form two layers.

Step [3] Separate the layers.

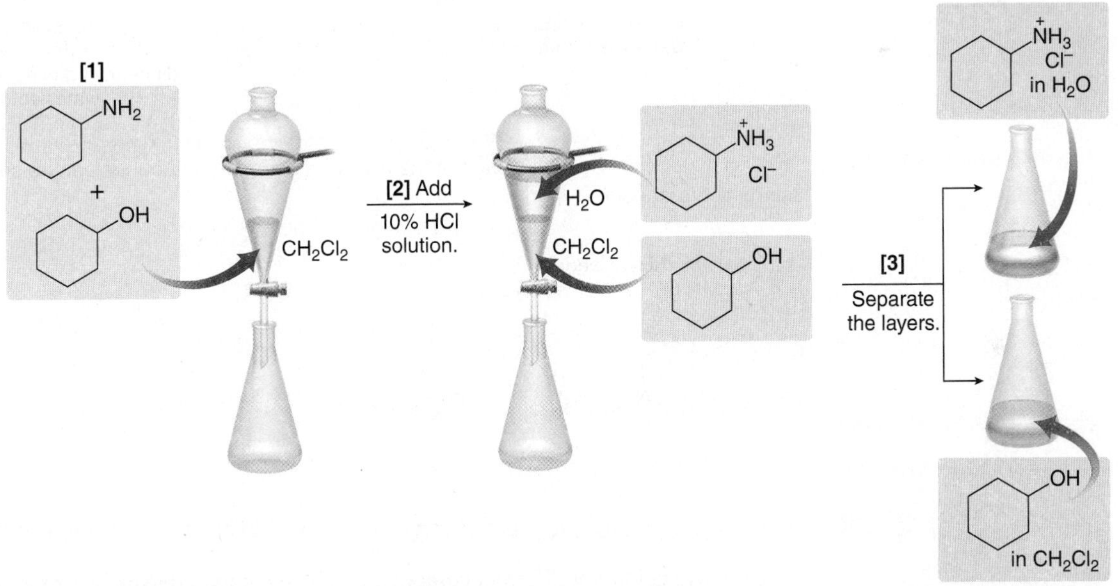

- Both compounds dissolve in the organic solvent CH_2Cl_2.

- Adding 10% aqueous HCl solution forms two layers. When the two layers are mixed, the HCl protonates the amine (RNH_2) to form $RNH_3^+Cl^-$, which dissolves in the aqueous layer.
- The cyclohexanol remains in the CH_2Cl_2 layer.

- Draining the lower layer out the bottom stopcock separates the two layers.
- Cyclohexanol (dissolved in CH_2Cl_2) is in one flask. The ammonium salt, $RNH_3^+Cl^-$ (dissolved in water), is in another flask.

- **An amine can be separated from other organic compounds by converting it to a water-soluble ammonium salt by an acid–base reaction.**

Thus, the water-soluble salt $C_6H_{11}NH_3^+Cl^-$ (obtained by protonation of $C_6H_{11}NH_2$) can be separated from water-insoluble cyclohexanol by an aqueous extraction procedure.

Problem 25.17 Draw the products of each acid–base reaction. Indicate whether equilibrium favors the reactants or products.

a. 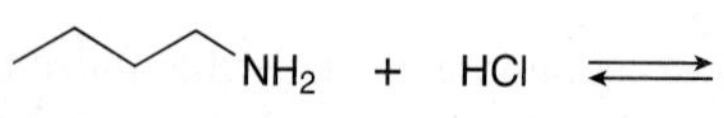

c. 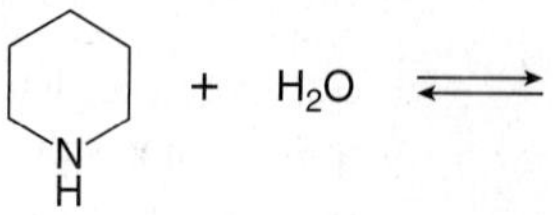

b. 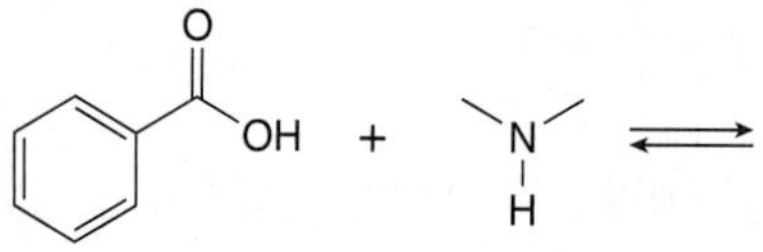

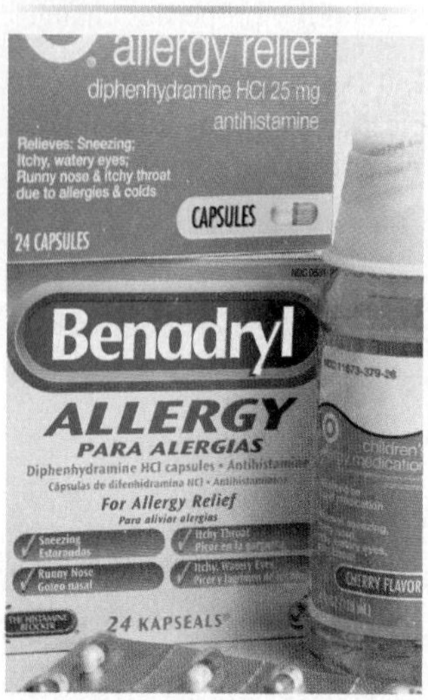

Many antihistamines and decongestants are sold as their ammonium salts.

Many water-insoluble amines with useful medicinal properties are sold as their water-soluble ammonium salts, which are more easily transported through the body in the aqueous medium of the blood. Benadryl, formed by treating diphenhydramine with HCl, is an over-the-counter antihistamine that is used to relieve the itch and irritation of skin rashes and hives.

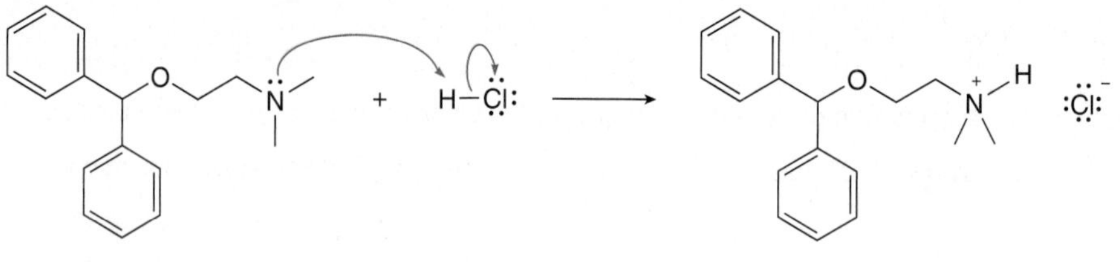

diphenhydramine
water insoluble

ammonium salt
Benadryl
(diphenhydramine hydrochloride)
water soluble

Problem 25.18 Write out steps to show how each of the following pairs of compounds can be separated by an extraction procedure.

a. 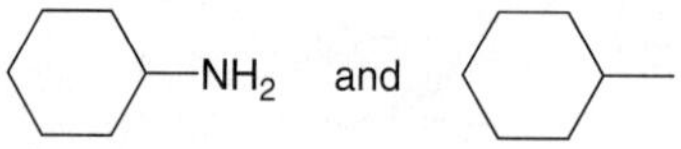

b. 

25.10 Relative Basicity of Amines and Other Compounds

The relative acidity of different compounds can be compared using their pK_a values. The relative *basicity* of different compounds (such as amines) can be compared using the pK_a values of their *conjugate acids.*

- **The weaker the conjugate acid, the higher its pK_a and the stronger the base.**

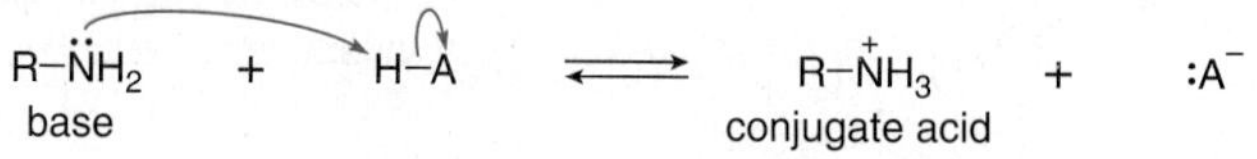

stronger base ⟵ The *weaker* the acid, the *higher* the pK_a.

To compare the basicity of two compounds, keep in mind the following:

- **Any factor that increases the electron density on the N atom increases an amine's basicity.**
- **Any factor that decreases the electron density on N decreases an amine's basicity.**

25.10A Comparing an Amine and NH_3

Because alkyl groups are electron donating, they increase the electron density on nitrogen, which makes an amine like $CH_3CH_2NH_2$ more basic than NH_3. In fact, the pK_a of $CH_3CH_2NH_3^+$ is higher than the pK_a of NH_4^+, so **$CH_3CH_2NH_2$ is a stronger base than NH_3.**

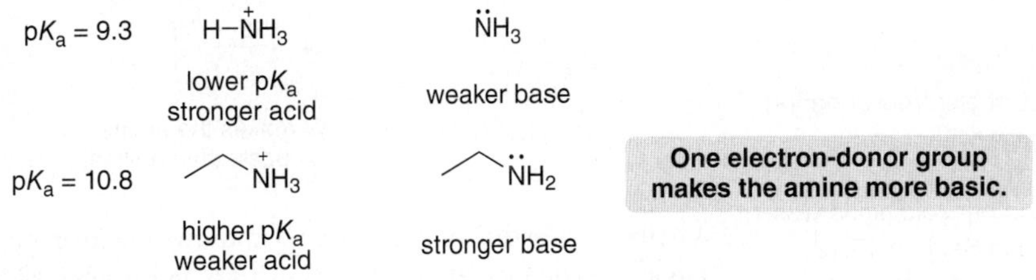

pKa = 9.3
H–NH3
lower pKa
stronger acid
NH3
weaker base
pKa = 10.8
NH3
higher pKa
weaker acid
NH2
stronger base
One electron-donor group makes the amine more basic.

The relative basicity of 1°, 2°, and 3° amines depends on additional factors, and will not be considered in this text.

- **Primary (1°), 2°, and 3° alkylamines are more basic than NH_3 because of the electron-donating inductive effect of the R groups.**

Problem 25.19 Which compound in each pair is more basic: (a) $(CH_3)_2NH$ and NH_3; (b) $CH_3CH_2NH_2$ and $ClCH_2CH_2NH_2$?

25.10B Comparing an Alkylamine and an Arylamine

To compare an alkylamine ($CH_3CH_2NH_2$) and an arylamine ($C_6H_5NH_2$, aniline), we must look at the availability of the nonbonded electron pair on N. With $CH_3CH_2NH_2$, the electron pair is localized on the N atom. With an arylamine, however, the electron pair is now delocalized on the benzene ring. This *decreases* the electron density on N, and makes $C_6H_5NH_2$ less basic than $CH_3CH_2NH_2$.

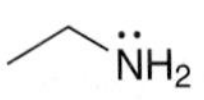

NH2

The electron pair is localized on N.

stronger base

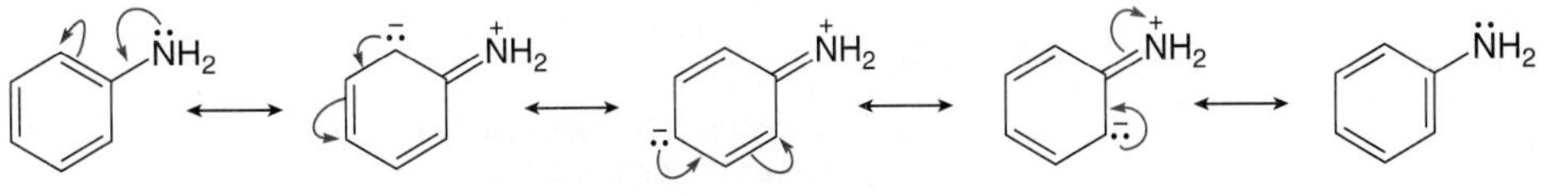

NH2
NH2
NH2
NH2
NH2

The electron pair is delocalized on the benzene ring.

weaker base

The pK_a values support this reasoning. Because the pK_a of $CH_3CH_2NH_3^+$ is higher than the pK_a of $C_6H_5NH_3^+$ (10.8 vs. 4.6), **$CH_3CH_2NH_2$ is a stronger base than $C_6H_5NH_2$.**

- **Arylamines are less basic than alkylamines because the electron pair on N is delocalized.**

Substituted anilines are more or less basic than aniline depending on the nature of the substituent.

- **Electron-donor groups add electron density to the benzene ring, making the arylamine more basic than aniline.**

D = electron-donor group

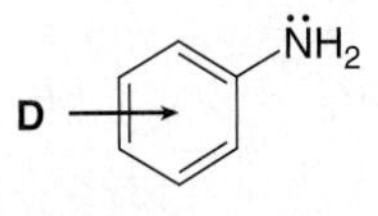

NH2
D

D makes the amine more basic than aniline.

D
–NH_2
–OH
–OR
–NHCOR
–R

- **Electron-withdrawing groups remove electron density from the benzene ring, making the arylamine less basic than aniline.**

W = electron-withdrawing group

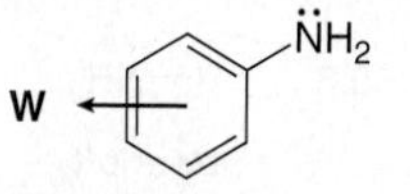

W makes the amine less basic than aniline.

W	
$-X$	$-CN$
$-CHO$	$-SO_3H$
$-COR$	
$-COOR$	$-NO_2$
$-COOH$	$-NR_3^+$

The effect of electron-donating and electron-withdrawing groups on the acidity of substituted benzoic acids was discussed in Section 19.12.

Whether a substituent donates or withdraws electron density depends on the balance of its inductive and resonance effects (Section 18.6 and Figure 18.8).

Sample Problem 25.3 Rank the following compounds in order of increasing basicity.

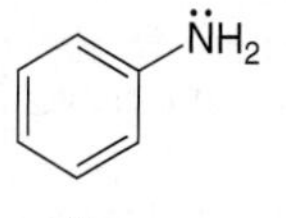

aniline

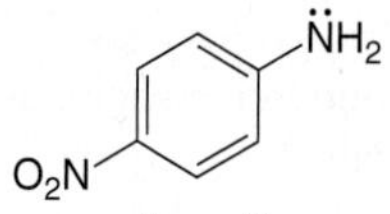

p-nitroaniline

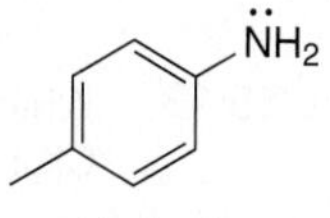

p-methylaniline (*p*-toluidine)

Solution

***p*-Nitroaniline: NO_2** is an electron-withdrawing group, making the amine **less basic** than aniline.

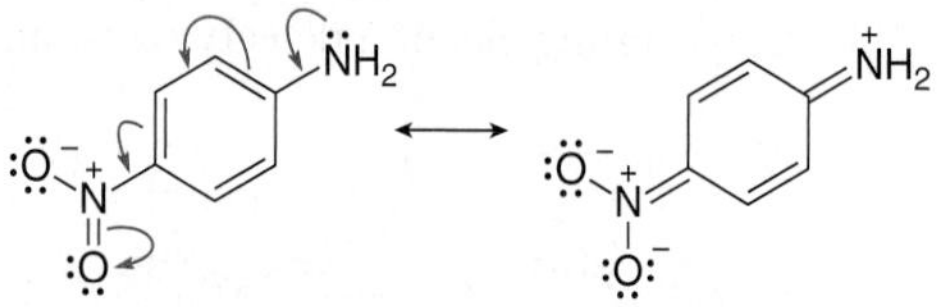

The lone pair on N is delocalized on the O atom, decreasing the basicity of the amine.

***p*-Methylaniline:** CH_3 has an electron-donating inductive effect, making the amine **more basic** than aniline.

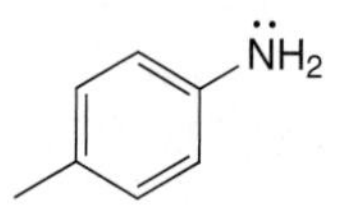

CH_3 inductively donates electron density, increasing the basicity of the amine.

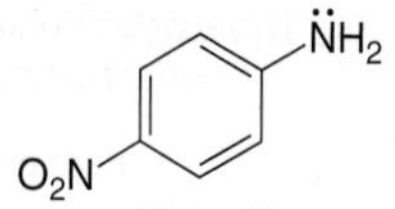

p-nitroaniline

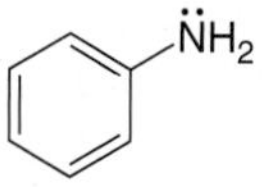

aniline

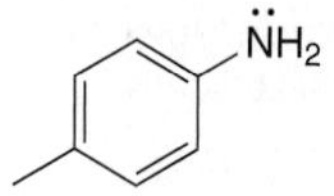

p-methylaniline (*p*-toluidine)

Increasing basicity →

The electrostatic potential plots in Figure 25.10 demonstrate that the electron density of the nitrogen atoms in these anilines increases in the order shown.

Problem 25.20 Rank the compounds in each group in order of increasing basicity.

a. 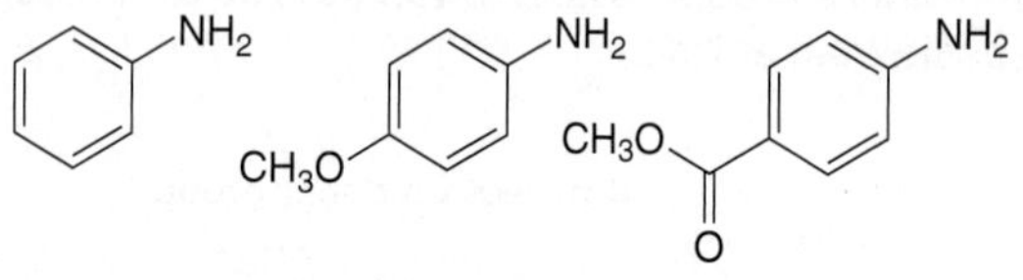

b. 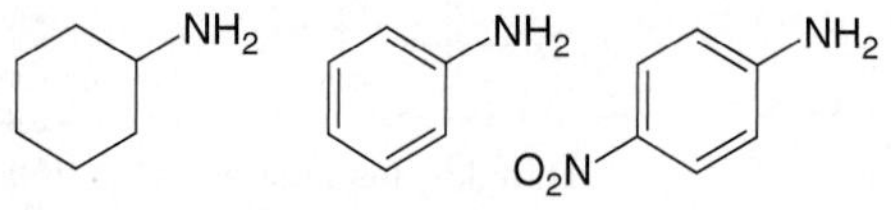

Figure 25.10
Electrostatic potential plots of substituted anilines

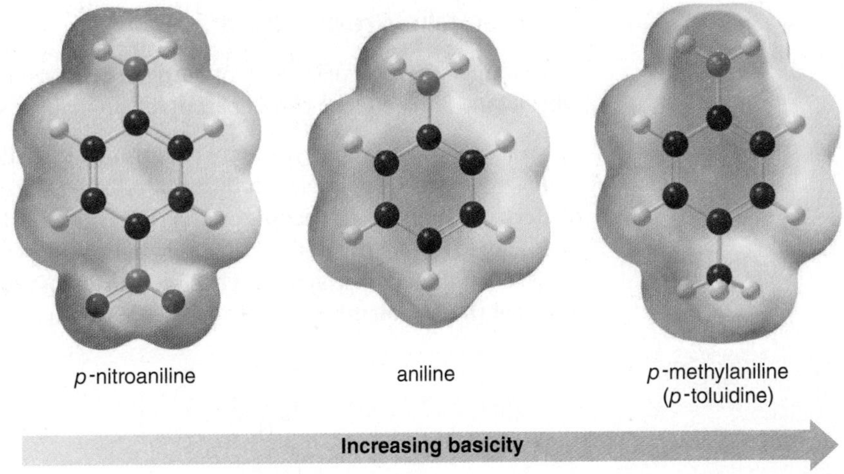

The NH_2 group gets more electron rich as the para substituent changes from $NO_2 \rightarrow H \rightarrow CH_3$. This is indicated by the color change around NH_2 (from green to yellow to red) in the electrostatic potential plot.

25.10C Comparing an Alkylamine and an Amide

To compare the basicity of an alkylamine (RNH_2) and an amide ($RCONH_2$), we must once again compare the availability of the nonbonded electron pair on nitrogen. With RNH_2, the electron pair is localized on the N atom. With an amide, however, the electron pair is delocalized on the carbonyl oxygen by resonance. This *decreases* the electron density on N, making **an amide much less basic than an alkylamine.**

The electron pair on N is delocalized on O by resonance.

- **Amides are much less basic than amines because the electron pair on N is delocalized.**

In fact, amides are not much more basic than any carbonyl compound. When an amide is treated with acid, **protonation occurs at the carbonyl oxygen, *not* the nitrogen,** because the resulting cation is resonance stabilized.

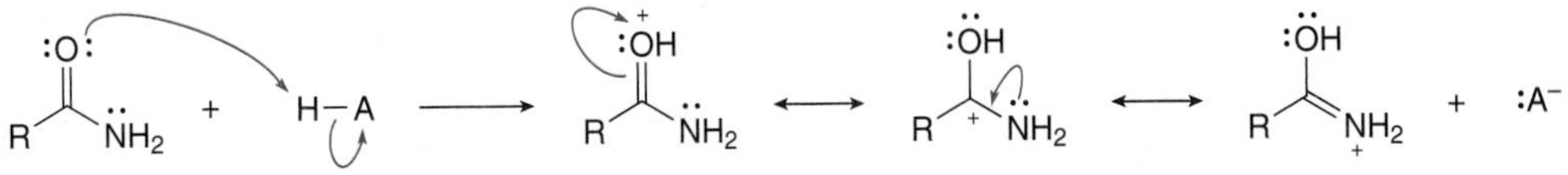

three resonance structures for the conjugate acid

Problem 25.21 Rank the following compounds in order of increasing basicity.

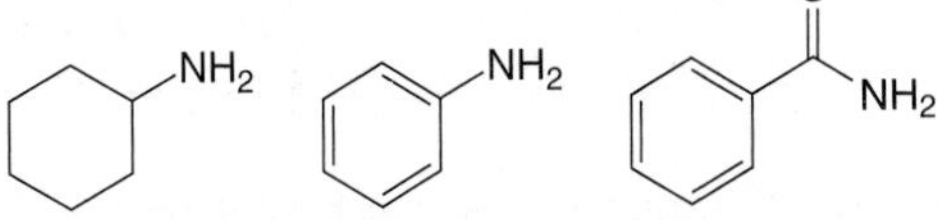

25.10D Heterocyclic Aromatic Amines

To determine the relative basicity of nitrogen heterocycles that are also aromatic, you must know whether the nitrogen lone pair is part of the aromatic π system.

For example, pyridine and pyrrole are both aromatic, but the nonbonded electron pair on the N atom in these compounds is located in different orbitals. Recall from Section 17.8A that the **lone pair of electrons in pyridine occupies an sp^2 hybridized orbital,** perpendicular to the π bonds of the molecule, so it is *not* part of the aromatic system, whereas that of pyrrole resides in a *p* orbital, making it part of the aromatic system. **The lone pair on pyrrole, therefore, is delocalized on all of the atoms of the five-membered ring,** making pyrrole a much weaker base than pyridine.

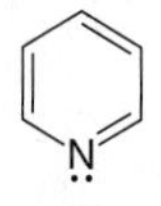
pyridine

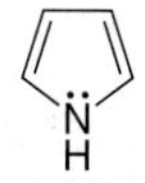
pyrrole

The lone pair resides in an sp^2 hybrid orbital.

The lone pair resides in a *p* orbital and is delocalized in the ring.

Protonation of pyrrole occurs at a ring carbon, not the N atom, as noted in Problem 17.48.

As a result, the pK_a of the conjugate acid of pyrrole is much less than that for the conjugate acid of pyridine.

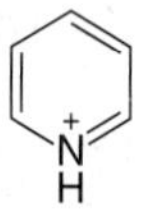
higher pK_a
weaker acid

pK_a = 5.3

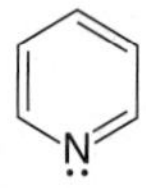
pyridine

stronger base

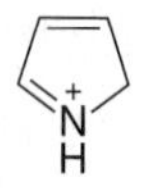
lower pK_a
stronger acid

pK_a = 0.4

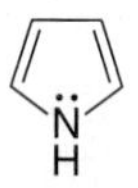
pyrrole

weaker base

- **Pyrrole is much less basic than pyridine because its lone pair of electrons is part of the aromatic π system.**

25.10E Hybridization Effects

The effect of hybridization on the acidity of an H–A bond was first discussed in Section 2.5D.

The hybridization of the orbital that contains an amine's lone pair also affects its basicity. This is illustrated by comparing the basicity of **piperidine** and **pyridine,** two nitrogen heterocycles. The lone pair in piperidine resides in an sp^3 hybrid orbital that has 25% *s*-character. The lone pair in pyridine resides in an sp^2 hybrid orbital that has 33% *s*-character.

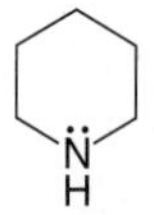
piperidine

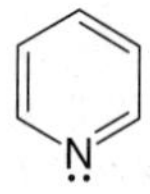
pyridine

The lone pair is in an sp^3 hybrid orbital.

The lone pair is in an sp^2 hybrid orbital.

- **The higher the percent *s*-character of the orbital containing the lone pair, the more tightly the lone pair is held, and the weaker the base.**

Pyridine is a weaker base than piperidine because its nonbonded pair of electrons resides in an sp^2 hybrid orbital. Although pyridine is an aromatic amine, its lone pair is *not* part of the delocalized π system, so its basicity is determined by the hybridization of its N atom. As a result, the pK_a value for the conjugate acid of pyridine is much lower than that for the conjugate acid of piperidine, making pyridine the weaker base.

lower pK_a stronger acid	pyridine weaker base	higher pK_a weaker acid	piperidine stronger base
pK_a = 5.3		pK_a = 11.1	

Problem 25.22 Which nitrogen atom in each compound is more basic?

a. 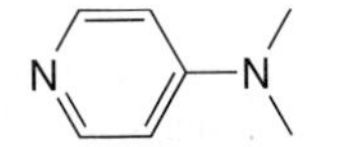

DMAP
4-(*N*,*N*-dimethylamino)pyridine

b. 

nicotine

25.10F Summary

Acid–base chemistry is central to many processes in organic chemistry, so it has been a constant theme throughout this text. Tables 25.2 and 25.3 organize and summarize the acid–base principles discussed in Section 25.10. The principles in these tables can be used to determine the most basic site in a molecule that has more than one nitrogen atom, as shown in Sample Problem 25.4.

Table 25.2 Factors That Determine Amine Basicity

Factor	Example
[1] **Inductive effects:** Electron-donating groups bonded to N increase basicity.	• RNH_2, R_2NH, and R_3N are more basic than NH_3.
[2] **Resonance effects:** Delocalizing the lone pair on N decreases basicity.	• Arylamines ($C_6H_5NH_2$) are less basic than alkylamines (RNH_2). • Amides ($RCONH_2$) are much less basic than amines (RNH_2).
[3] **Aromaticity:** Having the lone pair on N as part of the aromatic π system decreases basicity.	• Pyrrole is less basic than pyridine.  less basic / more basic
[4] **Hybridization effects:** Increasing the percent *s*-character in the orbital with the lone pair decreases basicity.	• Pyridine is less basic than piperidine.  less basic / more basic

Table 25.3 Table of pK_a Values of Some Representative Organic Nitrogen Compounds

	Compound	pK_a of the conjugate acid
Ammonia	NH_3	9.3
Alkylamines[a]	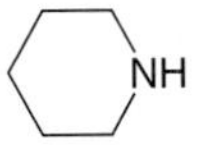	11.1
	$(CH_3CH_2)_2NH$	11.1
	$(CH_3CH_2)_3N$	11.0
	$CH_3CH_2NH_2$	10.8
Arylamines[b]	*p*-$CH_3OC_6H_4NH_2$	5.3
	p-$CH_3C_6H_4NH_2$	5.1
	$C_6H_5NH_2$	4.6
	p-$NO_2C_6H_4NH_2$	1.0
Heterocyclic aromatic amines[c]	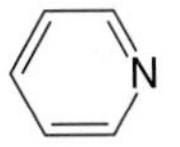	5.3
	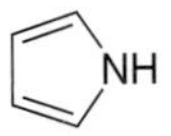	0.4
Amides	$RCONH_2$	−1

[a] Alkylamines have pK_a values of ~10–11.
[b] The pK_a *decreases* as the electron density of the benzene ring *decreases*.
[c] The pK_a depends on whether the lone pair of N is localized or delocalized.

Sample Problem 25.4 Which N atom in chloroquine is the strongest base?

Since 1945 chloroquine has been used to treat malaria, an infectious disease caused by a protozoan parasite that is spread by the *Anopheles* mosquito.

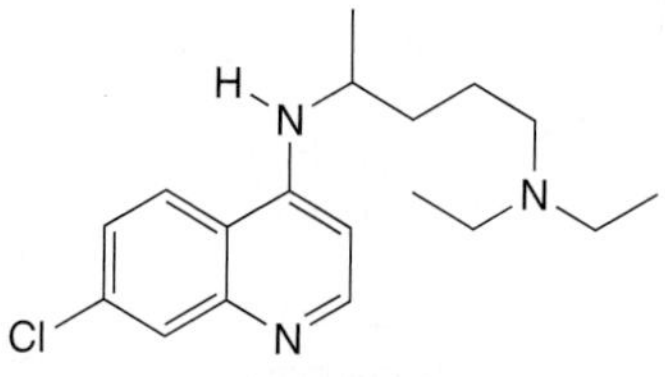

chloroquine

Solution

Examine the nitrogen atoms in chloroquine, labeled in red, blue, and green, and recall that decreasing the electron density on N decreases basicity.

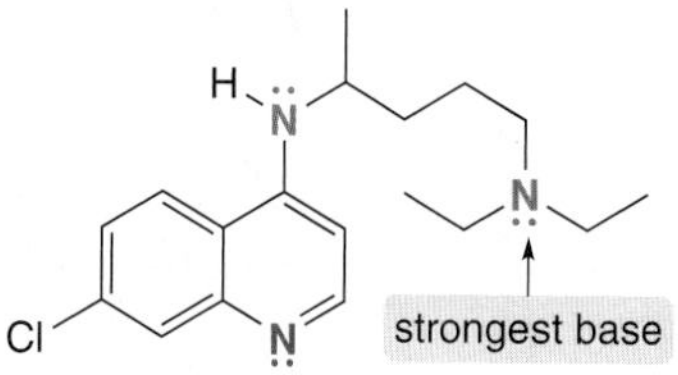

- N is bonded to an aromatic ring, so its lone pair is delocalized in the ring like aniline, decreasing basicity.
- The lone pair is localized on N, but N is sp^2 hybridized. Increasing percent *s*-character decreases basicity.
- N has a localized lone pair and is sp^3 hybridized, making it the most basic site in the molecule.

Problem 25.23 Which N atom in each compound is more basic? What product is formed when each compound is treated with HCl?

Lupin (Problems 25.23 and 25.78) is a genus of flowering plant that is abundant in the Andes of Peru and along the roadside in parts of Alaska.

a. 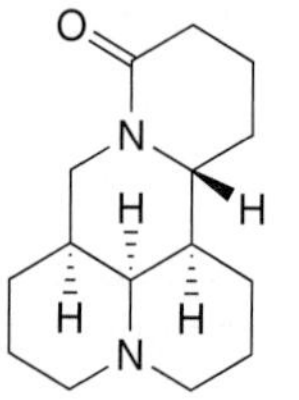

matrine
alkaloid in lupin

b. 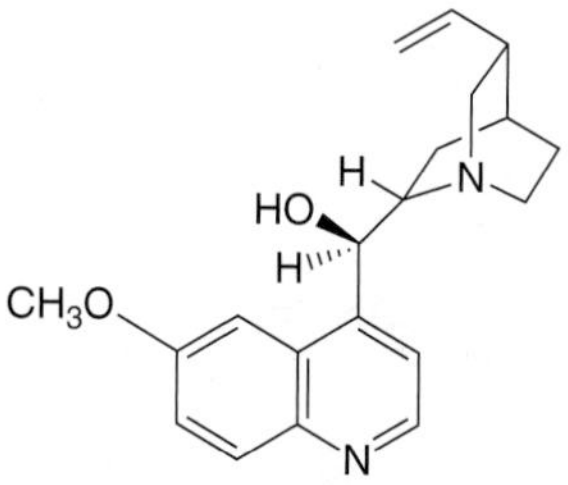

quinine
antimalarial drug

25.11 Amines as Nucleophiles

Amines react as nucleophiles with electrophilic carbon atoms. The details of these reactions have been described in Chapters 21 and 22, so they are summarized here only to emphasize the similar role that the amine nitrogen plays.

- **Amines attack carbonyl groups to form products of nucleophilic addition or substitution.**

The nature of the product depends on the carbonyl electrophile. These reactions are limited to 1° and 2° amines, because only these compounds yield neutral organic products.

[1] Reaction of 1° and 2° amines with aldehydes and ketones (Sections 21.11–21.12)

Aldehydes and ketones react with 1° amines to form **imines** and with 2° amines to form **enamines.** Both reactions involve nucleophilic addition of the amine to the carbonyl group to form a carbinolamine, which then loses water to form the final product.

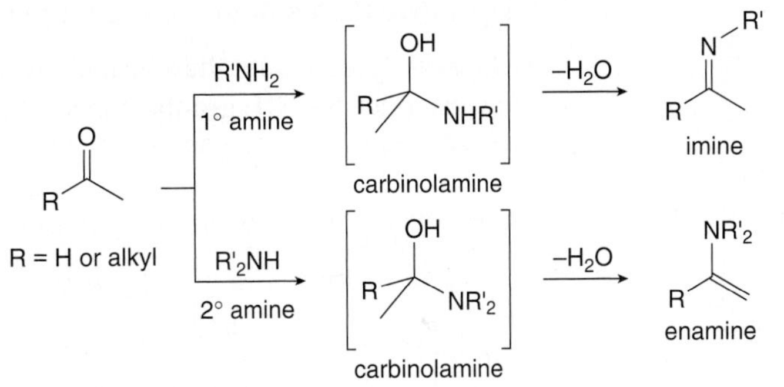
R = H or alkyl
R'NH2
1° amine
R'2NH
2° amine
OH
NHR'
carbinolamine
NR'2
–H2O
imine
enamine

[2] Reaction of NH_3 and 1° and 2° amines with acid chlorides and anhydrides (Sections 22.8–22.9)

Acid chlorides and anhydrides react with NH_3, 1° amines, and 2° amines to form **1°, 2°, and 3° amides,** respectively. These reactions involve attack of the nitrogen nucleophile on the carbonyl group followed by elimination of a leaving group (Cl^- or RCO_2^-). The overall result of this reaction is substitution of the leaving group by the nitrogen nucleophile.

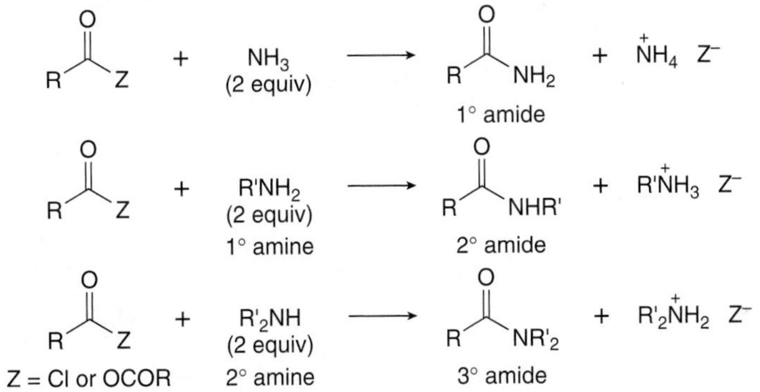
NH3
(2 equiv)
1° amide
R'NH2
(2 equiv)
1° amine
2° amide
R'2NH
(2 equiv)
2° amine
3° amide
Z = Cl or OCOR

Problem 25.24 Draw the products formed when each carbonyl compound reacts with the following amines: [1] $CH_3CH_2CH_2NH_2$; [2] $(CH_3CH_2)_2NH$.

a. 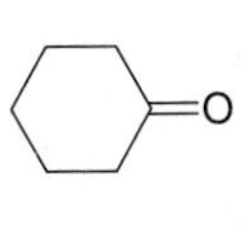 b. 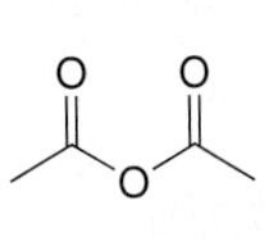 c. 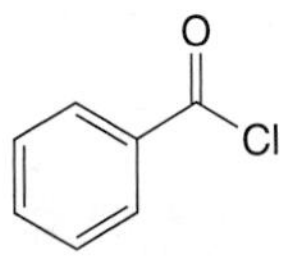

The conversion of amines to amides is useful in the synthesis of substituted anilines. For example, aniline itself does not undergo Friedel–Crafts reactions (Section 18.10B). Instead, its basic lone pair on N reacts with the Lewis acid ($AlCl_3$) to form a deactivated complex that does not undergo further reaction.

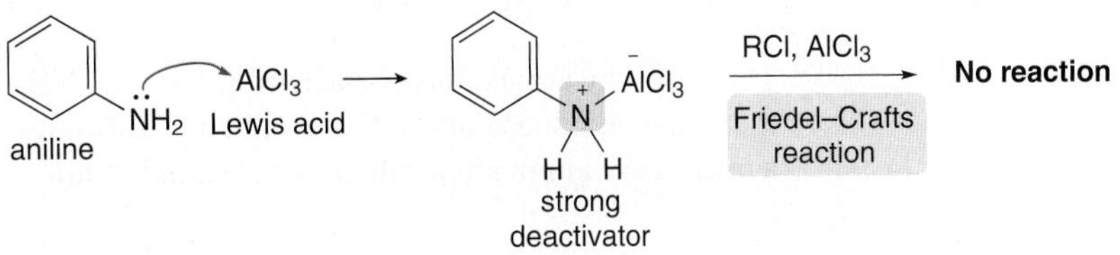
aniline
Lewis acid
strong
deactivator
RCl, AlCl3
Friedel–Crafts
reaction
No reaction

The N atom of an amide, however, is much less basic than the N atom of an amine, so it does not undergo a similar Lewis acid–base reaction with $AlCl_3$. A three-step reaction sequence involving an intermediate amide can thus be used to form the products of the Friedel–Crafts reaction.

[1] **Convert the amine (aniline) into an amide (acetanilide).**

[2] **Carry out the Friedel–Crafts reaction.**

[3] **Hydrolyze the amide** to generate the free amino group.

This three-step procedure is illustrated in Figure 25.11. In this way, **the amide serves as a protecting group for the NH_2 group,** in much the same way that *tert*-butyldimethylsilyl ethers and acetals are used to protect alcohols and carbonyls, respectively (Sections 20.12 and 21.15).

Problem 25.25 Devise a synthesis of each compound from aniline ($C_6H_5NH_2$).

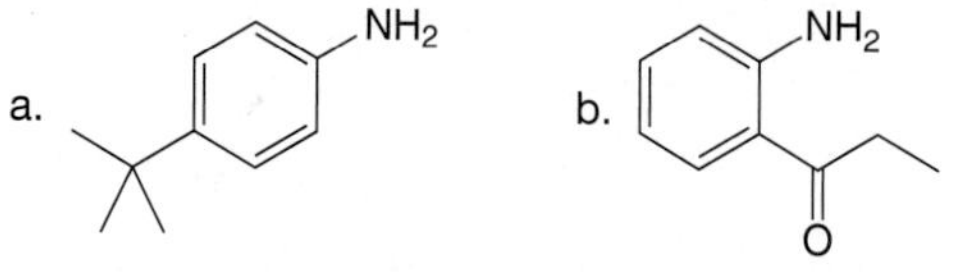

Figure 25.11 An amide as a protecting group for an amine

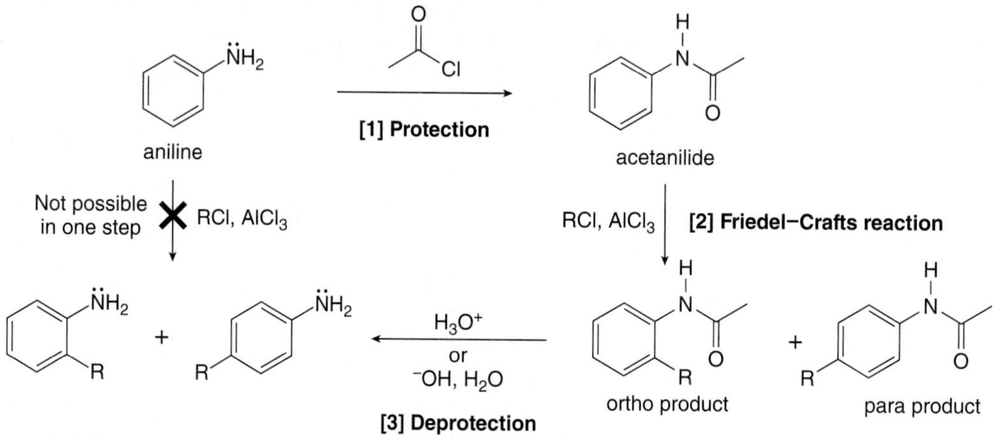

A three-step sequence uses an amide as a protecting group.

[1] Treatment of aniline with acetyl chloride (CH_3COCl) forms an amide (acetanilide).

[2] Acetanilide, having a much less basic N atom compared to aniline, undergoes electrophilic aromatic substitution under Friedel–Crafts conditions, forming a mixture of ortho and para products.

[3] Hydrolysis of the amide forms the Friedel–Crafts substitution products.

25.12 Hofmann Elimination

Amines, like alcohols, contain a poor leaving group. To undergo a β elimination reaction, for example, a 1° amine would need to lose the elements of NH_3 across two adjacent atoms. The leaving group, $^-NH_2$, is such a strong base, however, that this reaction does not occur.

The only way around this obstacle is to convert $^-NH_2$ into a better leaving group. The most common method to accomplish this is called a **Hofmann elimination,** which converts an amine into a quaternary ammonium salt prior to β elimination.

25.12A Details of the Hofmann Elimination

The **Hofmann elimination** converts an amine into an alkene.

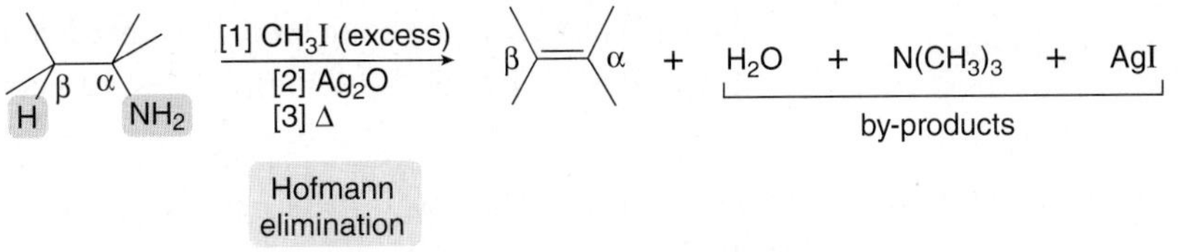

The Hofmann elimination consists of three steps, as shown for the conversion of propan-1-amine to propene.

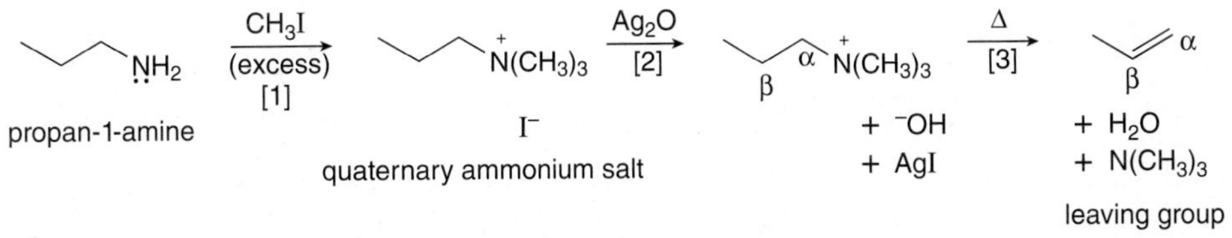

- In Step [1], the amine reacts as a nucleophile in an S_N2 reaction with excess CH_3I to form a quaternary ammonium salt. **The $N(CH_3)_3$ group thus formed is a much better leaving group than $^-NH_2$.**
- Step [2] converts one ammonium salt into another one with a different anion. The silver(I) oxide, Ag_2O, replaces the I^- anion with ^-OH, a strong base.
- When the ammonium salt is heated in Step [3], **^-OH removes a proton from the β carbon atom,** forming the new π bond of the alkene. The mechanism of elimination is **E2,** so:

> - All bonds are broken and formed in a single step.
> - Elimination occurs through an anti periplanar geometry—that is, H and $N(CH_3)_3$ are oriented on opposite sides of the molecule.

The general E2 mechanism for the Hofmann elimination is shown in Mechanism 25.1.

Mechanism 25.1 The E2 Mechanism for the Hofmann Elimination

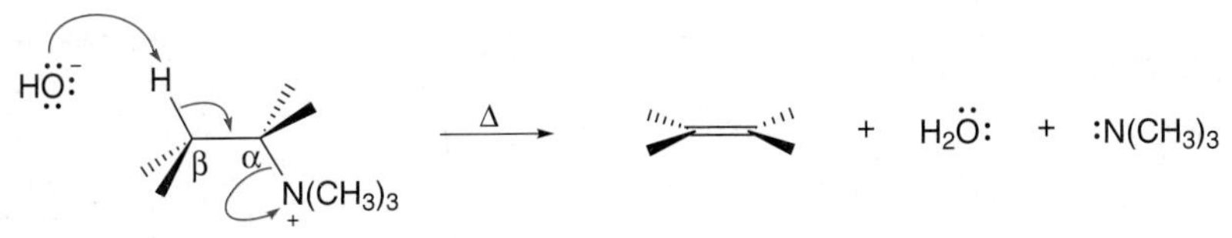

Elimination occurs with an anti periplanar arrangement of H and $N(CH_3)_3$. Base removes a proton on the β carbon, the electron pair in the C–H bond forms the π bond, and $N(CH_3)_3$ comes off as the leaving group.

All Hofmann elimination reactions result in the formation of a new π bond between the α and β carbon atoms, as shown for cyclohexanamine and 2-phenylethanamine.

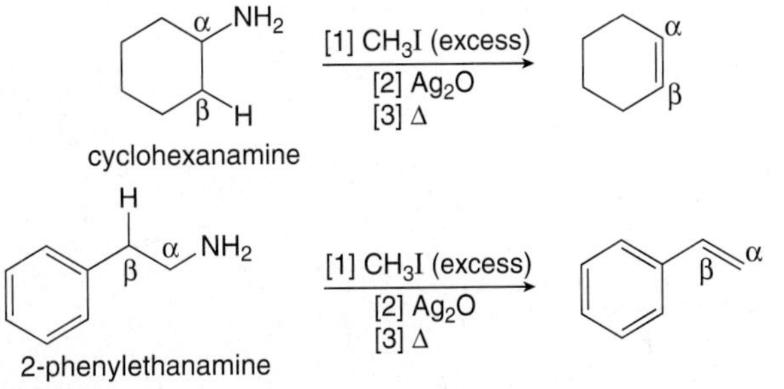

To help remember the reagents needed for the steps of the Hofmann elimination, keep in mind what happens in each step.

- **Step [1]** makes a good leaving group by forming a quaternary ammonium salt.
- **Step [2]** provides the strong base, ^{-}OH, needed for elimination.
- **Step [3]** is the E2 elimination that forms the new π bond.

Problem 25.26 Draw the product formed by treating each compound with excess CH_3I, followed by Ag_2O, and then heat.

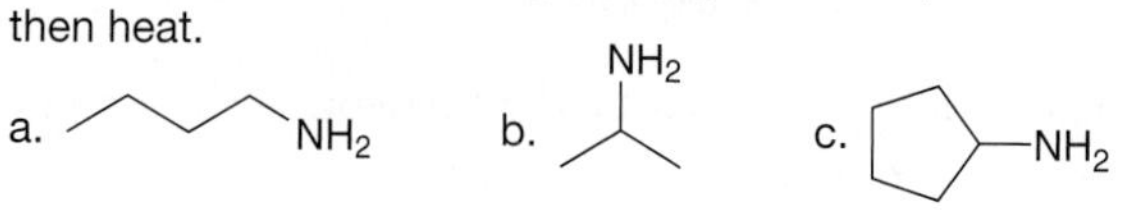
a.
NH2
b.
NH2
c.
NH2

25.12B Regioselectivity of the Hofmann Elimination

There is one major difference between a Hofmann elimination and other E2 eliminations.

- **When constitutional isomers are possible, the major alkene has the *less* substituted double bond in a Hofmann elimination.**

For example, Hofmann elimination of the elements of H and $N(CH_3)_3$ from 2-methylcyclopentanamine, which has two different β carbons (labeled $β_1$ and $β_2$), yields two constitutional isomers: the disubstituted alkene **A** (the major product) and the trisubstituted alkene **B** (the minor product).

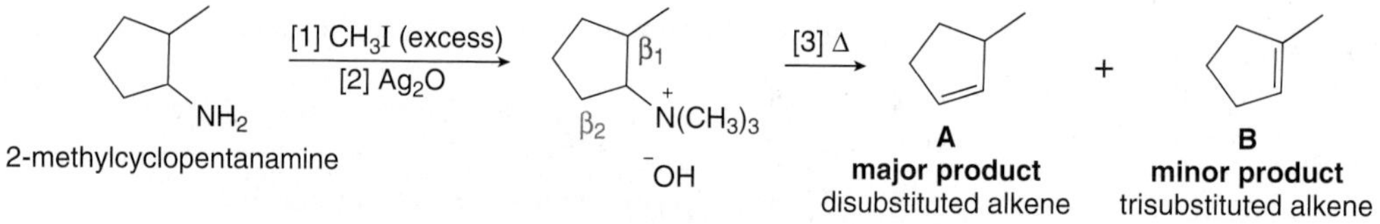
[1] CH3I (excess)
[2] Ag2O
NH2
2-methylcyclopentanamine
β1
β2
N(CH3)3
OH
[3] Δ
A
major product
disubstituted alkene
B
minor product
trisubstituted alkene

This regioselectivity distinguishes a Hofmann elimination from other E2 eliminations, which form the more substituted double bond by the Zaitsev rule (Section 8.5). This result is sometimes explained by the size of the leaving group, $N(CH_3)_3$. **In a Hofmann elimination, the base removes a proton from the less substituted, more accessible β carbon atom, because of the bulky leaving group on the nearby α carbon.**

Sample Problem 25.5 Draw the major product formed from Hofmann elimination of the following amine.

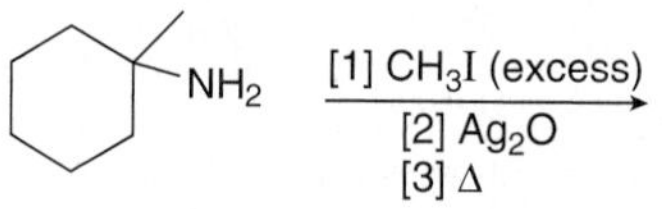
NH2
[1] CH3I (excess)
[2] Ag2O
[3] Δ

Solution

The amine has three β carbons but two of them are identical, so two alkenes are possible. Draw elimination products by forming alkenes having a C=C between the α and β carbons. The major product has the **less substituted double bond**—that is, the alkene with the C=C between the α and $β_1$ carbons in this example.

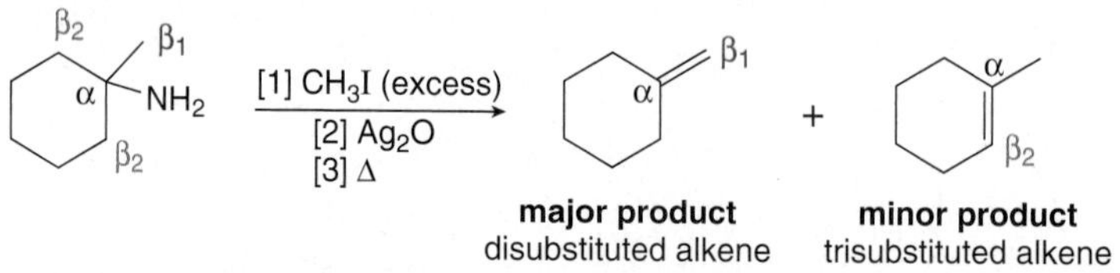
β2
β1
α
NH2
β2
[1] CH3I (excess)
[2] Ag2O
[3] Δ
α
β1
α
β2
major product
disubstituted alkene
minor product
trisubstituted alkene

Figure 25.12
A comparison of E2 elimination reactions using alkyl halides and amines

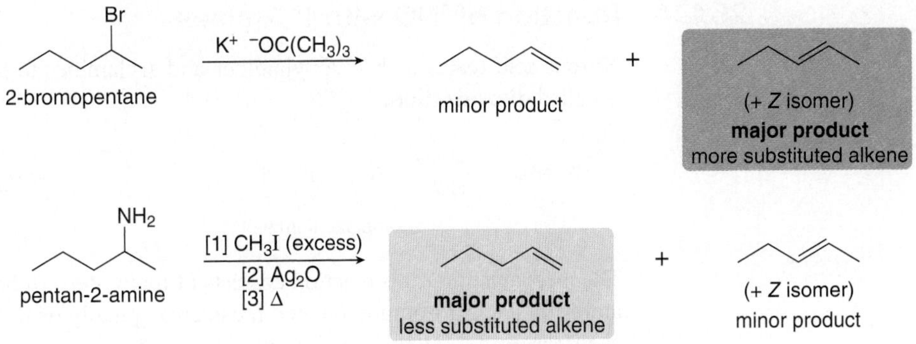

Figure 25.12 contrasts the products formed by E2 elimination reactions using an alkyl halide and an amine as starting materials. Treatment of the alkyl halide (2-bromopentane) with base forms the more substituted alkene as the major product, following the **Zaitsev rule.** In contrast, the three-step Hofmann sequence of an amine (pentan-2-amine) forms the less substituted alkene as major product.

Problem 25.27 Draw the major product formed by treating each amine with excess CH_3I, followed by Ag_2O, and then heat.

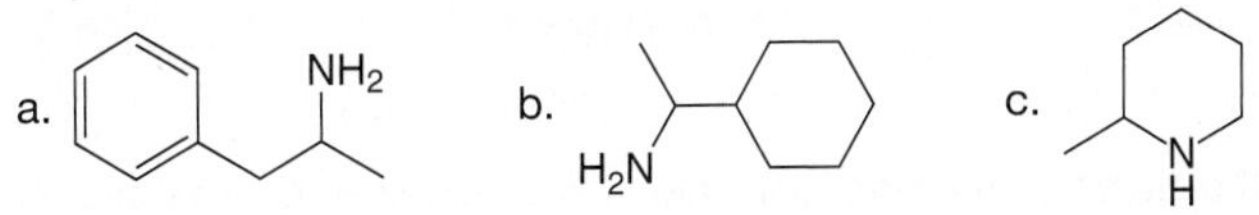

Problem 25.28 Draw the major product formed in each reaction.

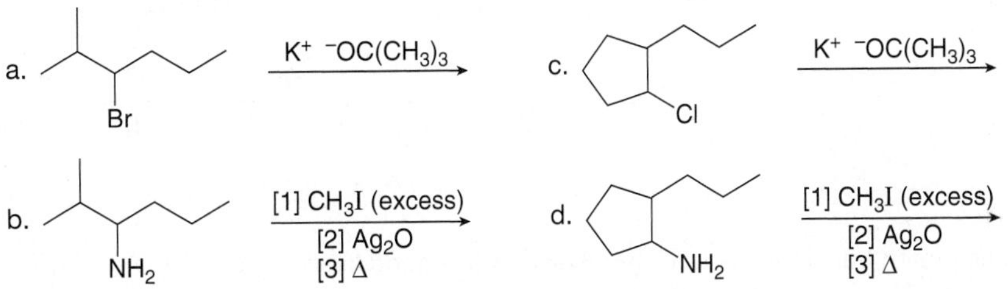

25.13 Reaction of Amines with Nitrous Acid

Nitrous acid, HNO_2, is a weak, unstable acid formed from $NaNO_2$ and a strong acid like HCl.

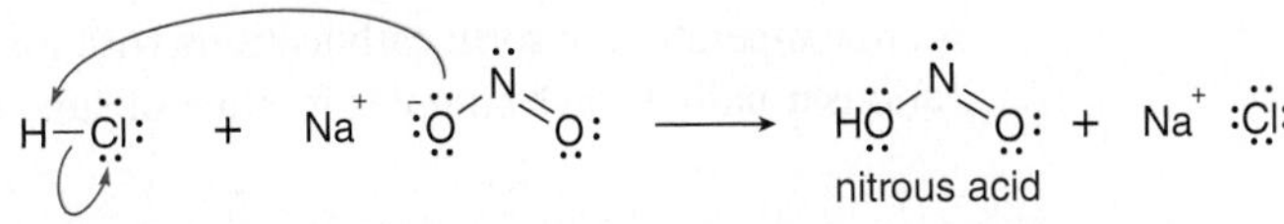

In the presence of acid, nitrous acid decomposes to ^+NO, the **nitrosonium ion.** This electrophile then goes on to react with the nucleophilic nitrogen atom of amines to form **diazonium salts ($RN_2^+Cl^-$)** from 1° amines and ***N*-nitrosamines ($R_2NN{=}O$)** from 2° amines.

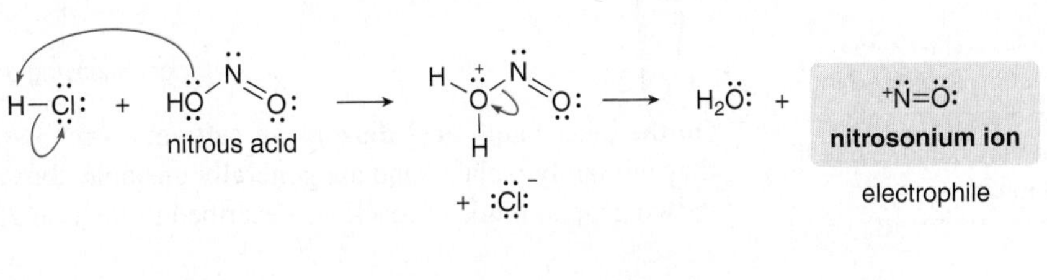

25.13A Reaction of ^+NO with 1° Amines

Nitrous acid reacts with 1° alkylamines and arylamines to form **diazonium salts.** This reaction is called **diazotization.**

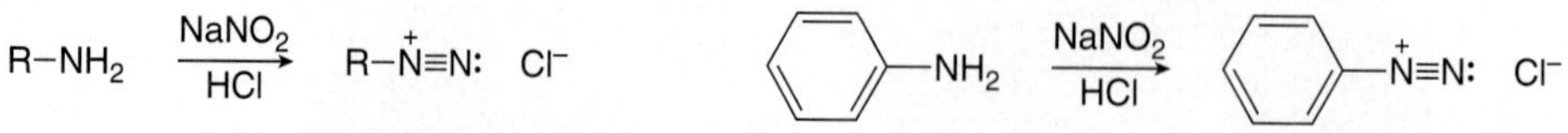

alkyl diazonium salt aryl diazonium salt

The mechanism for this reaction consists of many steps. It begins with nucleophilic attack of the amine on the nitrosonium ion, and it can conceptually be divided into two parts: formation of an ***N*-nitrosamine,** followed by loss of H_2O, as shown in Mechanism 25.2.

Mechanism 25.2 Formation of a Diazonium Salt from a 1° Amine

Part [1] Formation of an *N*-nitrosamine

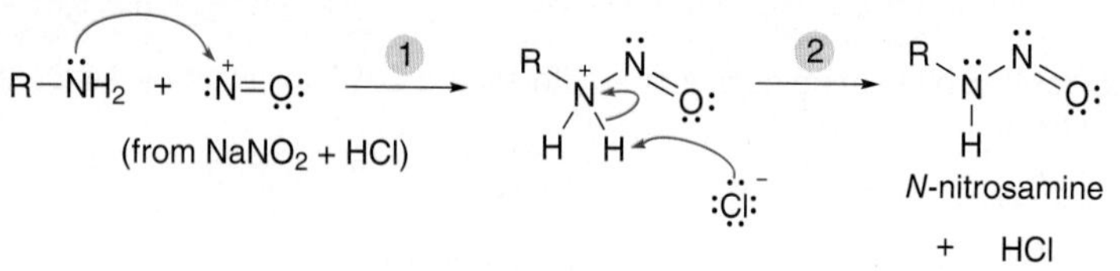
(from NaNO2 + HCl)
N-nitrosamine
+ HCl

1 – 2 Nucleophilic attack of the amine on the nitrosonium ion (^+NO), followed by loss of a proton, forms an *N*-nitrosamine.

Part [2] Formation of a diazonium salt

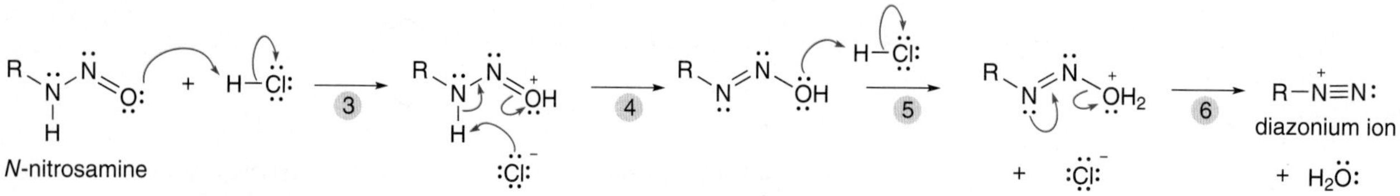
N-nitrosamine
diazonium ion

3 – 5 Three proton transfers form an intermediate with a good leaving group (H_2O).

6 Loss of water forms a diazonium ion (RN_2^+). The diazonium salt formed in this reaction consists of the diazonium ion (RN_2^+) and a chloride anion (Cl^-).

Alkyl diazonium salts are generally not useful compounds. They readily decompose below room temperature to form carbocations with loss of N_2, a very good leaving group. These carbocations usually form a complex mixture of substitution, elimination, and rearrangement products.

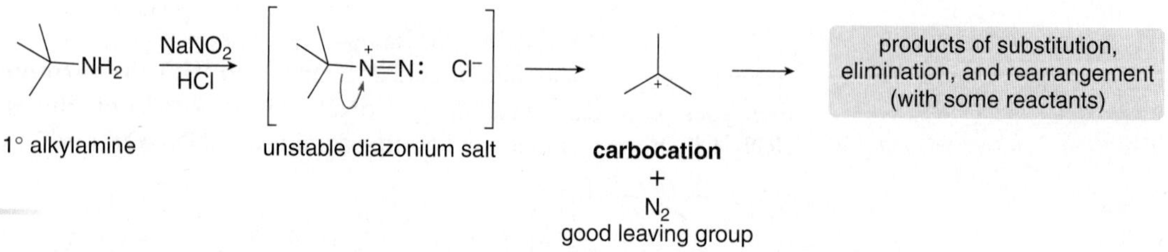
products of substitution, elimination, and rearrangement (with some reactants)
1° alkylamine
unstable diazonium salt
carbocation
+
N2
good leaving group

Care must be exercised in handling diazonium salts, because they can explode if allowed to dry.

On the other hand, **aryl diazonium salts are very useful synthetic intermediates.** Although they are rarely isolated and are generally unstable above 0 °C, they are useful starting materials in two general kinds of reactions described in Section 25.14.

25.13B Reaction of ^{+}NO with 2° Amines

Secondary alkylamines and arylamines react with nitrous acid to form *N*-nitrosamines.

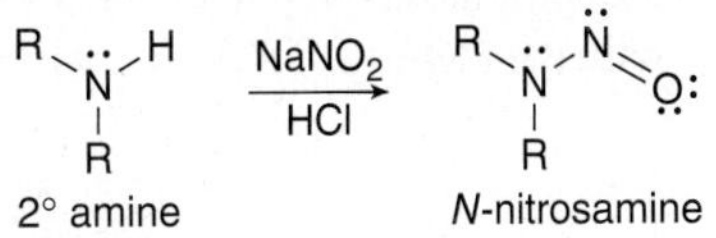

Many *N*-nitrosamines are potent carcinogens found in some food and tobacco smoke. Nitrosamines in food can be formed in the same way they are formed in the laboratory: **reaction of a 2° amine with the nitrosonium ion,** formed from nitrous acid (HNO_2). Mechanism 25.3 is shown for the conversion of dimethylamine [$(CH_3)_2NH$] to *N*-nitrosodimethylamine [$(CH_3)_2NN{=}O$].

Mechanism 25.3 Formation of an *N*-Nitrosamine from a 2° Amine

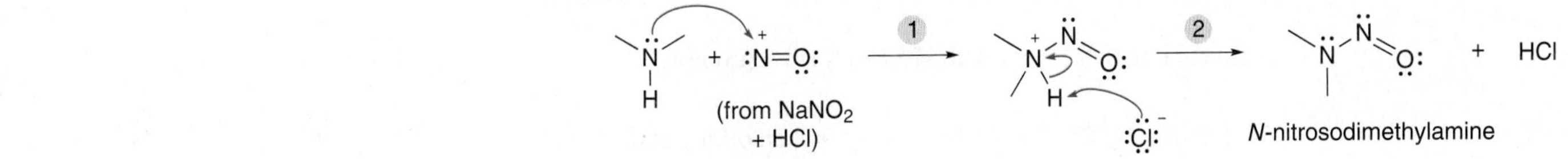

1 – 2 Nucleophilic attack of the amine on ^{+}NO, followed by loss of a proton, forms the *N*-nitrosamine.

Problem 25.29 Draw the product formed when each compound is treated with $NaNO_2$ and HCl.

a. 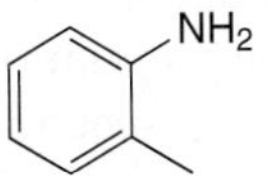 b. 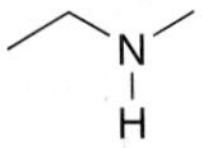 c. 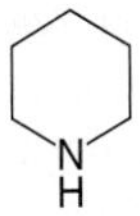 d. 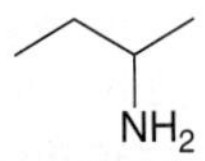

25.14 Substitution Reactions of Aryl Diazonium Salts

Aryl diazonium salts undergo two general reactions.

- **Substitution** of N_2 by an atom or a group of atoms Z forms a variety of substituted benzenes.

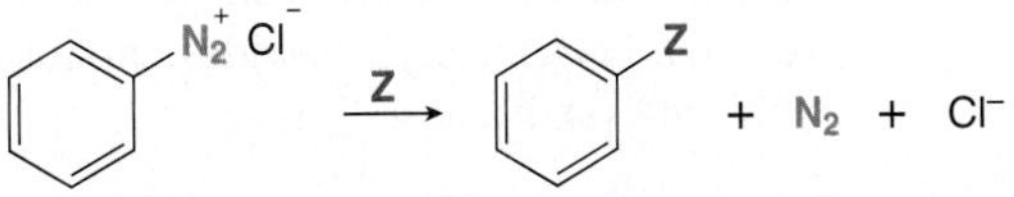

- **Coupling** of a diazonium salt with another benzene derivative forms an **azo compound,** a compound containing a nitrogen–nitrogen double bond.

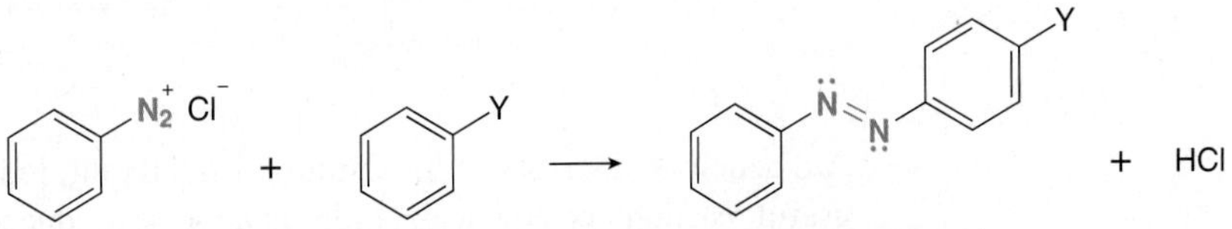

azo compound

Y = NH_2, NHR, NR_2, OH (a strong electron-donor group)

25.14A Specific Substitution Reactions

Aryl diazonium salts react with a variety of reagents to form products in which Z (an atom or group of atoms) replaces N_2, a very good leaving group. The mechanism of these reactions varies with the identity of Z, so we will concentrate on the products of the reactions, not the mechanisms.

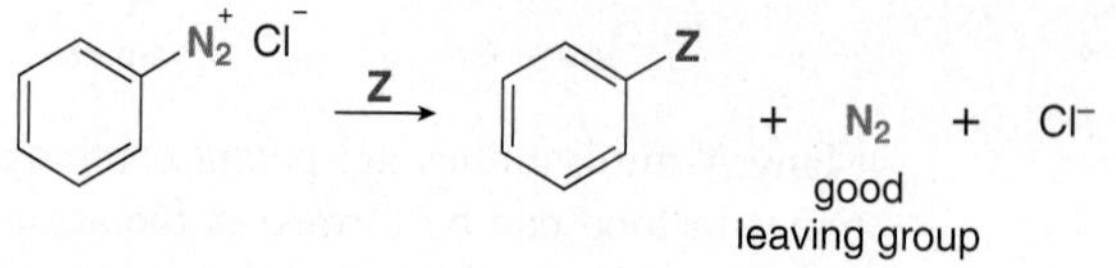

$\overset{+}{N_2}$ $\overset{-}{Cl}$
Z
Z
+ N_2 + Cl^-
good
leaving group

[1] Substitution by OH—Synthesis of phenols

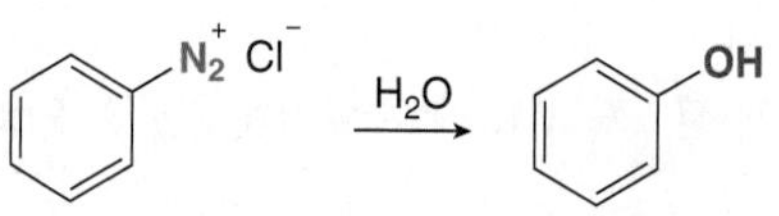

$\overset{+}{N_2}$ $\overset{-}{Cl}$
H_2O
OH

phenol

A diazonium salt reacts with H_2O to form a **phenol.**

[2] Substitution by Cl or Br—Synthesis of aryl chlorides and bromides

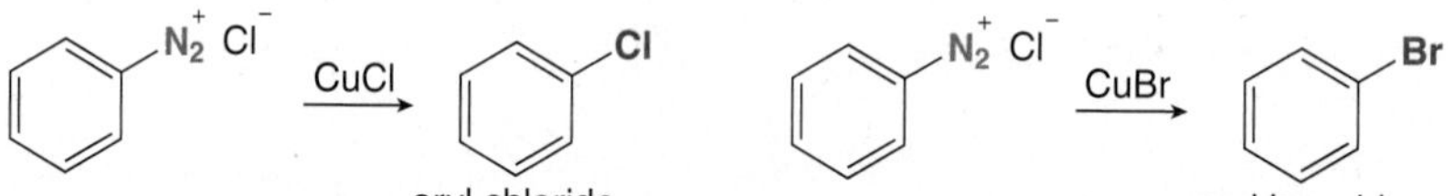

$\overset{+}{N_2}$ $\overset{-}{Cl}$
CuCl
Cl
$\overset{+}{N_2}$ $\overset{-}{Cl}$
CuBr
Br

aryl chloride aryl bromide

A diazonium salt reacts with copper(I) chloride or copper(I) bromide to form an **aryl chloride** or **aryl bromide,** respectively. This is called the **Sandmeyer reaction.** It provides an alternative to direct chlorination and bromination of an aromatic ring using Cl_2 or Br_2 and a Lewis acid catalyst.

[3] Substitution by F—Synthesis of aryl fluorides

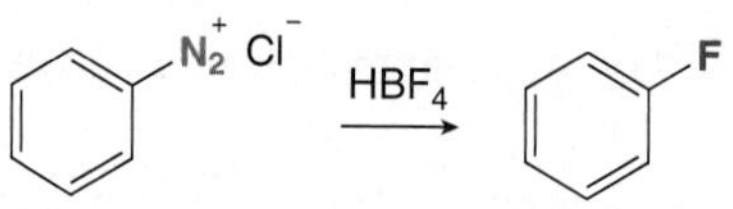

$\overset{+}{N_2}$ $\overset{-}{Cl}$
HBF_4
F

aryl fluoride

A diazonium salt reacts with fluoroboric acid (HBF_4) to form an **aryl fluoride.** This is a useful reaction because aryl fluorides cannot be produced by direct fluorination with F_2 and a Lewis acid catalyst, because F_2 reacts too violently (Section 18.3).

[4] Substitution by I—Synthesis of aryl iodides

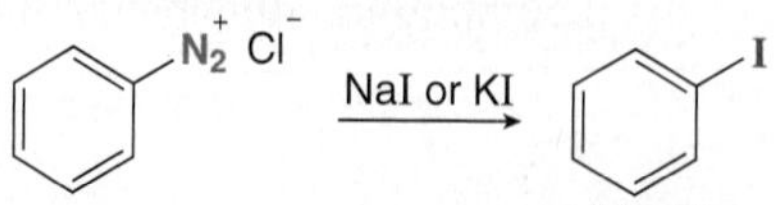

$\overset{+}{N_2}$ $\overset{-}{Cl}$
NaI or KI
I

aryl iodide

A diazonium salt reacts with sodium or potassium iodide to form an aryl iodide. This, too, is a useful reaction because aryl iodides cannot be produced by direct iodination with I_2 and a Lewis acid catalyst, because I_2 reacts too slowly (Section 18.3).

[5] Substitution by CN—Synthesis of benzonitriles

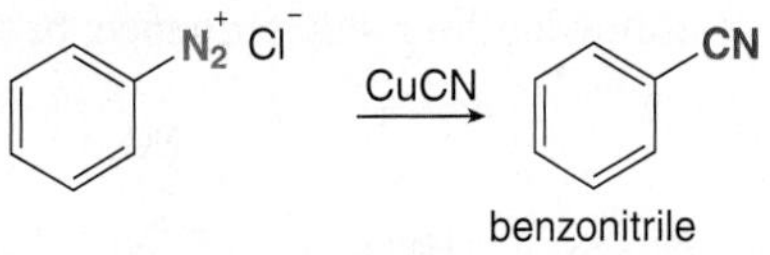

N_2^+ Cl^-
CuCN
CN
benzonitrile

A diazonium salt reacts with copper(I) cyanide to form a benzonitrile. Because a cyano group can be hydrolyzed to a carboxylic acid, reduced to an amine or aldehyde, or converted to a ketone with organometallic reagents, this reaction provides easy access to a wide variety of benzene derivatives using chemistry described in Section 22.18.

[6] Substitution by H—Synthesis of benzene

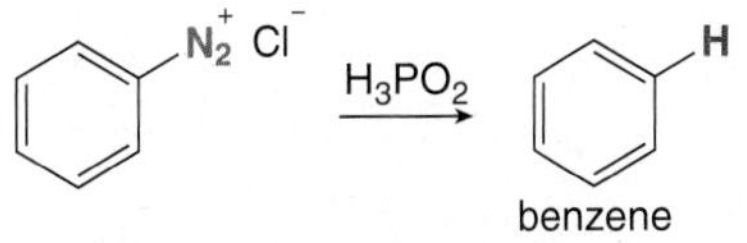

N_2^+ Cl^-
H_3PO_2
H
benzene

A diazonium salt reacts with hypophosphorus acid (H_3PO_2) to form benzene. This reaction has limited utility because it reduces the functionality of the benzene ring by replacing N_2 with a hydrogen atom. Nonetheless, this reaction *is* useful in synthesizing compounds that have substitution patterns that are not available by other means.

For example, it is not possible to synthesize 1,3,5-tribromobenzene from benzene by direct bromination. Because Br is an ortho, para director, bromination with Br_2 and $FeBr_3$ will not add Br substituents meta to each other on the ring.

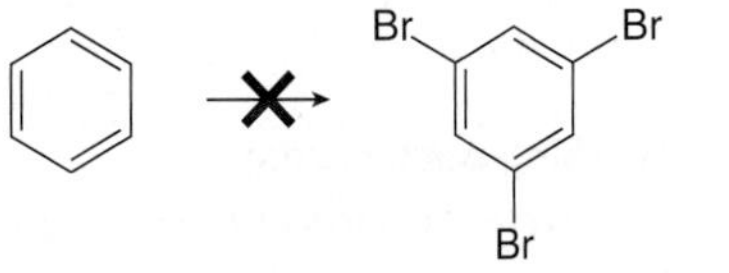

Br
Br
Br
1,3,5-tribromobenzene

It is possible, however, to add three Br atoms meta to each other when aniline is the starting material. Because an NH_2 group is a very powerful ortho, para director, three Br atoms are introduced in a single step on halogenation (Section 18.10A). Then, the NH_2 group can be removed by diazotization and reaction with H_3PO_2.

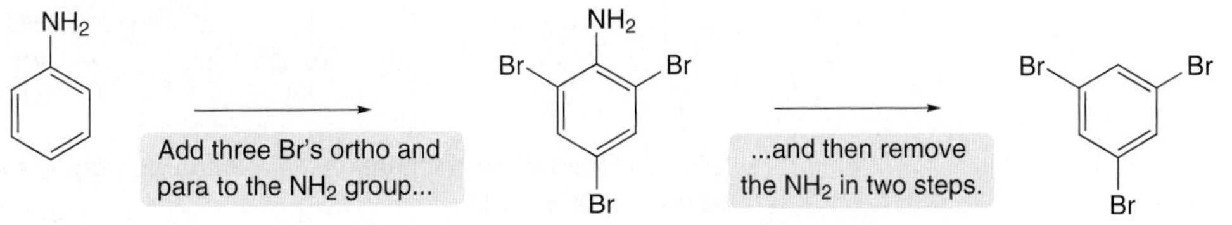

NH_2
Add three Br's ortho and para to the NH_2 group...
NH_2
Br
Br
Br
...and then remove the NH_2 in two steps.
Br
Br
Br

The complete synthesis of 1,3,5-tribromobenzene from benzene is outlined in Figure 25.13.

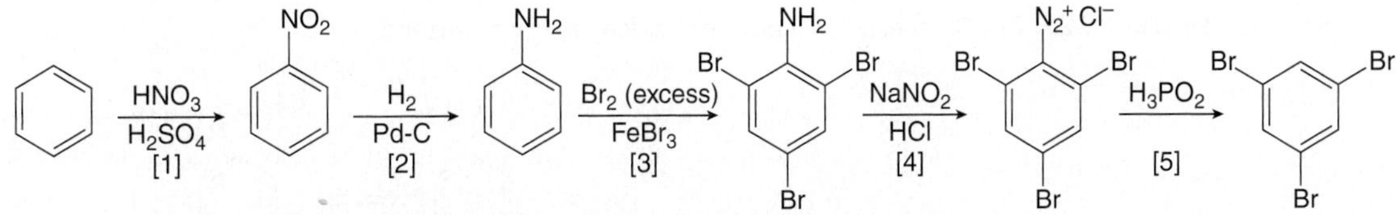

HNO_3
H_2SO_4
[1]
NO_2
H_2
Pd-C
[2]
NH_2
Br_2 (excess)
$FeBr_3$
[3]
NH_2
Br
Br
Br
$NaNO_2$
HCl
[4]
N_2^+ Cl^-
Br
Br
Br
H_3PO_2
[5]
Br
Br
Br

Figure 25.13 The synthesis of 1,3,5-tribromobenzene from benzene

- Nitration followed by reduction forms aniline ($C_6H_5NH_2$) from benzene (Steps [1] and [2]).
- Bromination of aniline yields the tribromo derivative in Step [3].
- The NH_2 group is removed by a two-step process: diazotization with $NaNO_2$ and HCl (Step [4]), followed by substitution of the diazonium ion by H with H_3PO_2.

25.14B Using Diazonium Salts in Synthesis

Diazonium salts provide easy access to many different benzene derivatives. Keep in mind the following four-step sequence, because it will be used to synthesize many substituted benzenes.

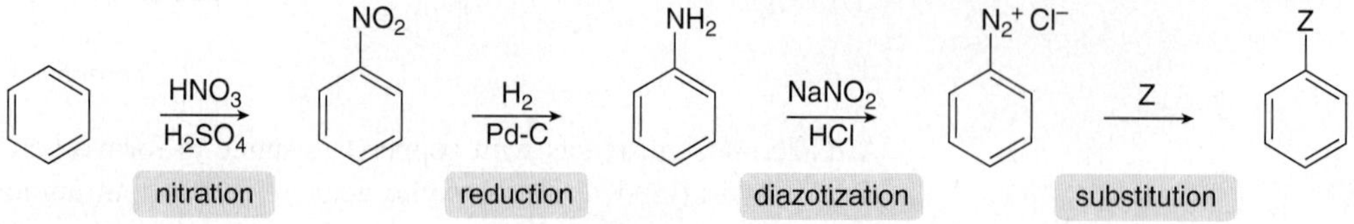
NO2
NH2
N2+ Cl−
Z
HNO3
H2SO4
nitration
H2
Pd-C
reduction
NaNO2
HCl
diazotization
Z
substitution

Sample Problems 25.6 and 25.7 apply these principles to two different multistep syntheses.

Sample Problem 25.6 Synthesize *m*-chlorophenol from benzene.

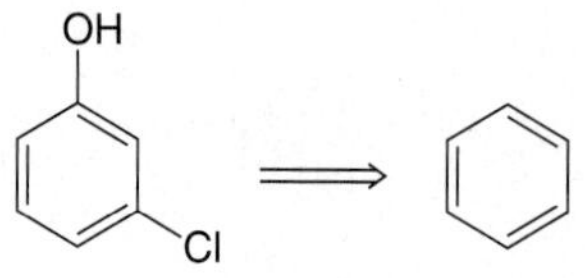
OH
Cl

Solution

Both OH and Cl are ortho, para directors, but they are located meta to each other. The OH group must be formed from a diazonium salt, which can be made from an NO_2 group by a stepwise method.

Retrosynthetic Analysis

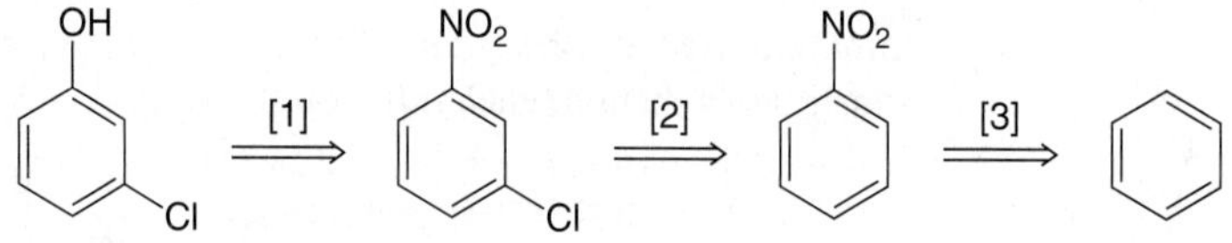
OH
Cl
[1]
NO2
Cl
[2]
NO2
[3]

Working backwards:

- [1] Form the OH group from NO_2 by a three-step procedure using a diazonium salt.
- [2] Introduce Cl meta to NO_2 by halogenation.
- [3] Add the NO_2 group by nitration.

Synthesis

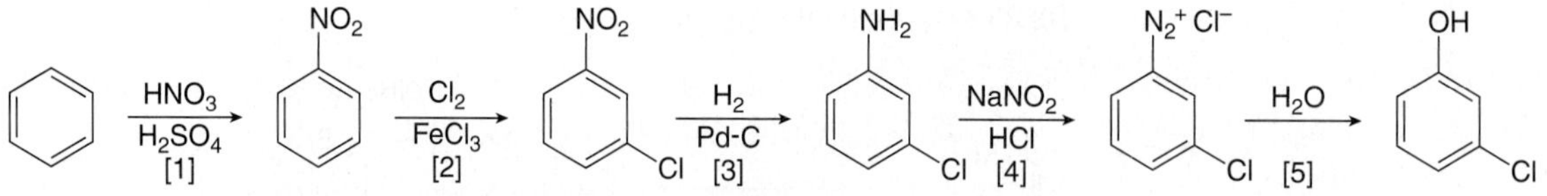
HNO3
H2SO4
[1]
NO2
Cl2
FeCl3
[2]
NO2
Cl
H2
Pd-C
[3]
NH2
Cl
NaNO2
HCl
[4]
N2+ Cl−
Cl
H2O
[5]
OH
Cl

- Nitration followed by chlorination meta to the NO_2 group forms the meta disubstituted benzene (Steps [1]–[2]).
- Reduction of the nitro group (Step [3]) followed by diazotization forms the diazonium salt in Step [4], which is then converted to the desired phenol by treatment with H_2O (Step [5]).

Sample Problem 25.7 Synthesize *p*-bromobenzaldehyde from benzene.

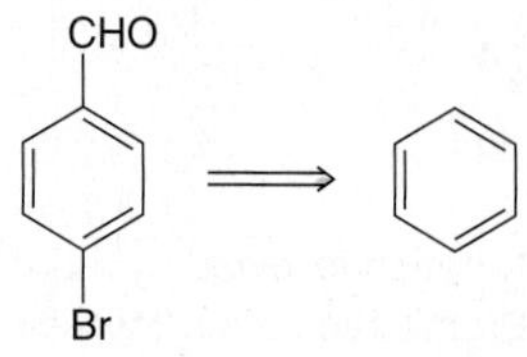
CHO
Br

Solution

Because the two groups are located para to each other and Br is an ortho, para director, Br should be added to the ring first. To add the CHO group, recall that it can be formed from CN by reduction.

Retrosynthetic Analysis

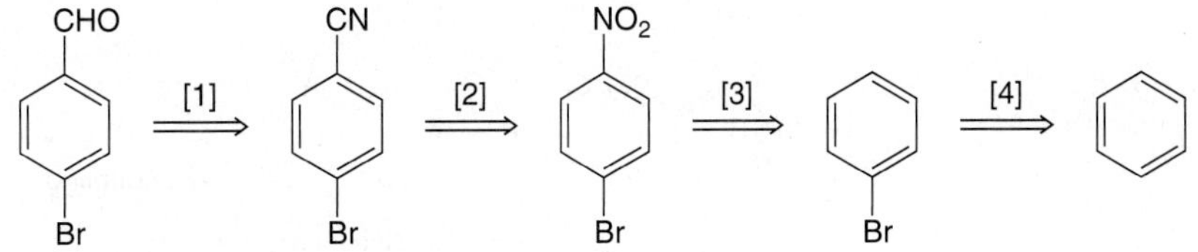
CHO
CN
NO_2
[1]
[2]
[3]
[4]
Br
Br
Br
Br

Working backwards:

- [1] Form the CHO group by reduction of CN.
- [2] Prepare the CN group from an NO_2 group by a three-step sequence using a diazonium salt.
- [3] Introduce the NO_2 group by nitration, para to the Br atom.
- [4] Introduce Br by bromination with Br_2 and $FeBr_3$.

Synthesis

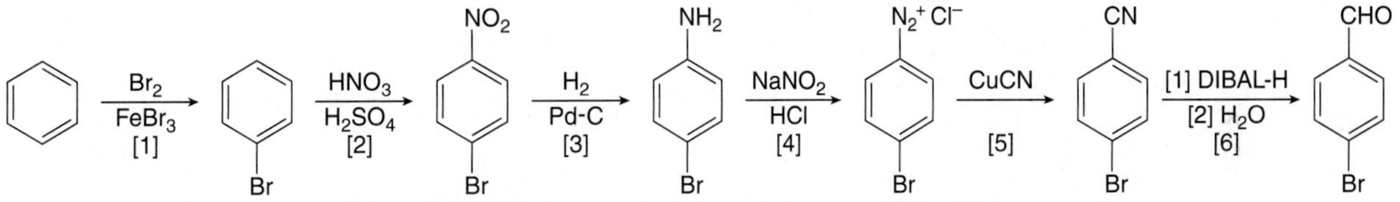
Br_2
$FeBr_3$
[1]
HNO_3
H_2SO_4
[2]
NO_2
H_2
Pd-C
[3]
NH_2
$NaNO_2$
HCl
[4]
N_2^+ Cl^-
CuCN
[5]
CN
[1] DIBAL-H
[2] H_2O
[6]
CHO
Br

- Bromination followed by nitration forms a disubstituted benzene with two para substituents (Steps [1]–[2]), which can be separated from its undesired ortho isomer.
- Reduction of the NO_2 group (Step [3]) followed by diazotization forms the diazonium salt in Step [4], which is converted to a nitrile by reaction with CuCN (Step [5]).
- Reduction of the CN group with DIBAL-H (a mild reducing agent) forms the CHO group and completes the synthesis.

Problem 25.30 Draw the product formed in each reaction.

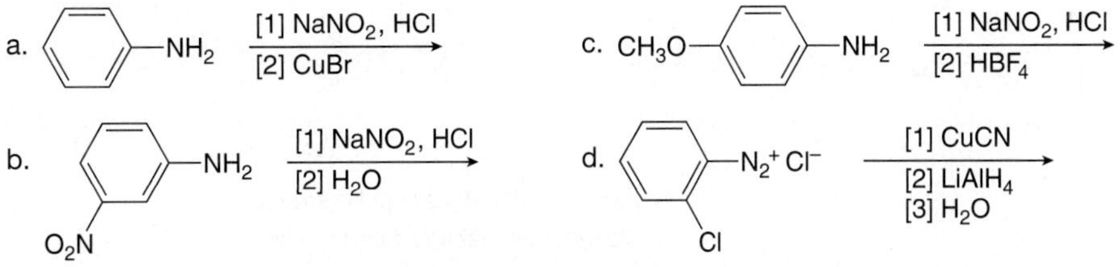
a.
NH_2
[1] $NaNO_2$, HCl
[2] CuBr
c. CH_3O
NH_2
[1] $NaNO_2$, HCl
[2] HBF_4
b.
NH_2
O_2N
[1] $NaNO_2$, HCl
[2] H_2O
d.
N_2^+ Cl^-
Cl
[1] CuCN
[2] $LiAlH_4$
[3] H_2O

Problem 25.31 Devise a synthesis of each compound from benzene.

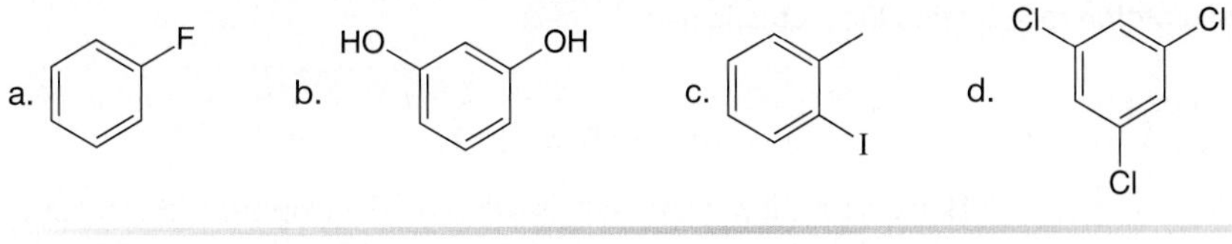
a.
F
b.
HO
OH
c.
I
d.
Cl
Cl
Cl

25.15 Coupling Reactions of Aryl Diazonium Salts

The second general reaction of diazonium salts is **coupling.** When a diazonium salt is treated with an aromatic compound that contains a strong electron-donor group, the two rings join together to form an **azo compound,** a compound with a nitrogen–nitrogen double bond.

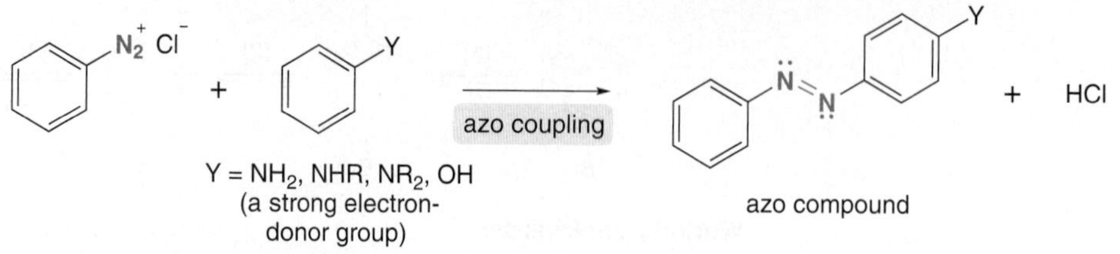

Synthetic dyes are described in more detail in Section 25.16.

Azo compounds are highly conjugated, rendering them colored (Section 16.15). Many of these compounds, such as the azo compound "butter yellow," are synthetic dyes. Butter yellow was once used to color margarine.

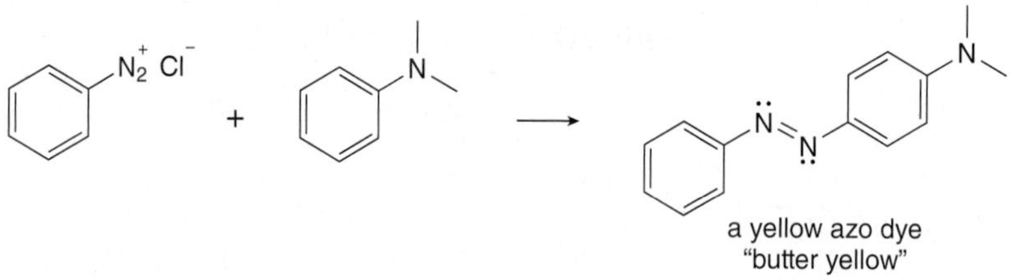

This reaction is another example of **electrophilic aromatic substitution,** with the **diazonium salt acting as the electrophile.** Like all electrophilic substitutions (Section 18.2), the mechanism has two steps: addition of the electrophile (the diazonium ion) to form a resonance-stabilized carbocation, followed by deprotonation, as shown in Mechanism 25.4.

Mechanism 25.4 Azo Coupling

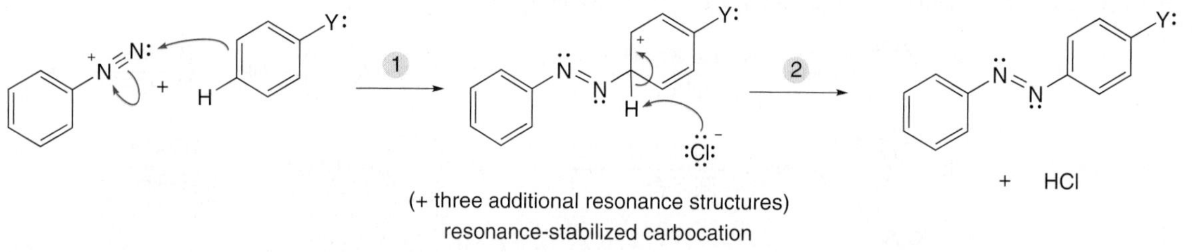

1 The diazonium ion reacts with the benzene ring to form a resonance-stabilized carbocation.

2 Loss of a proton regenerates the aromatic ring.

Because a diazonium salt is weakly electrophilic, the reaction occurs only when the benzene ring has **a strong electron-donor group Y, where Y = NH_2, NHR, NR_2, or OH.** Although these groups activate both the ortho and para positions, para substitution occurs unless the para position already has another substituent present.

To determine what starting materials are needed to synthesize a particular azo compound, always divide the molecule into two components: **one has a benzene ring with a diazonium ion, and one has a benzene ring with a very strong electron-donor group.**

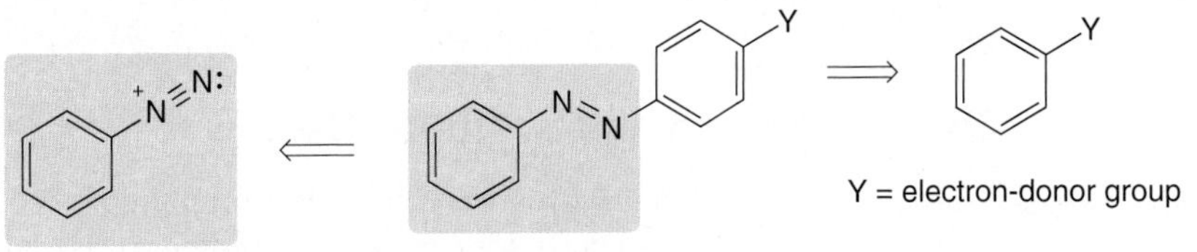
Y
Y
Y = electron-donor group

Sample Problem 25.8 What starting materials are needed to synthesize the following azo compound?

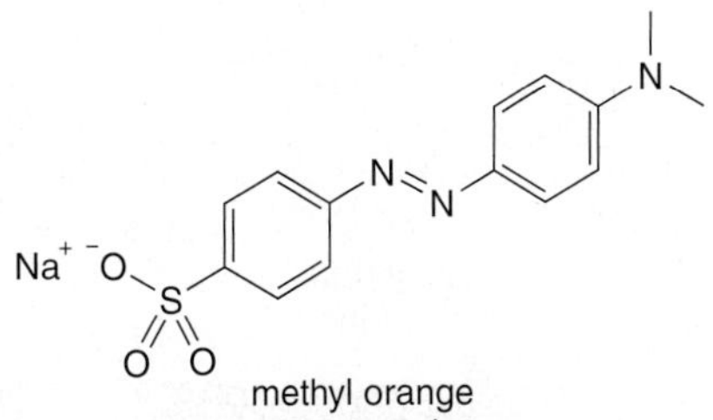

methyl orange
an orange dye

Solution

Both benzene rings in methyl orange have a substituent, but only one group, $N(CH_3)_2$, is a strong electron donor. In determining the two starting materials, the **diazonium ion must be bonded to the ring that is *not* bonded to $N(CH_3)_2$.**

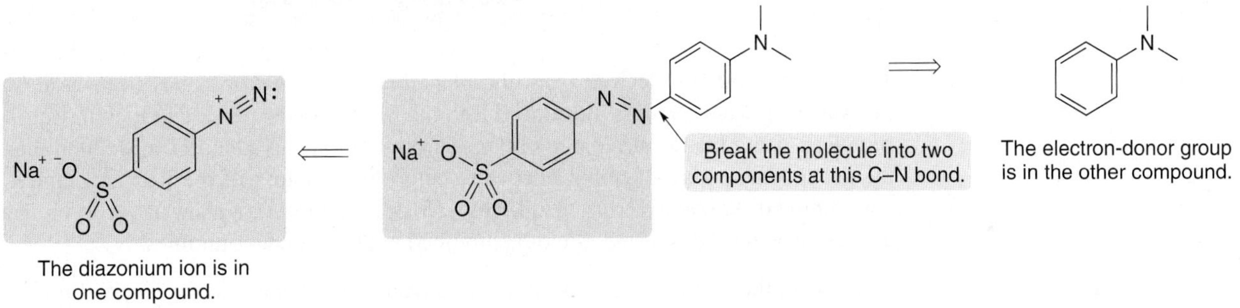
Break the molecule into two components at this C–N bond.
The electron-donor group is in the other compound.
The diazonium ion is in one compound.

Problem 25.32 Draw the product formed when $C_6H_5N_2^+Cl^-$ reacts with each compound.

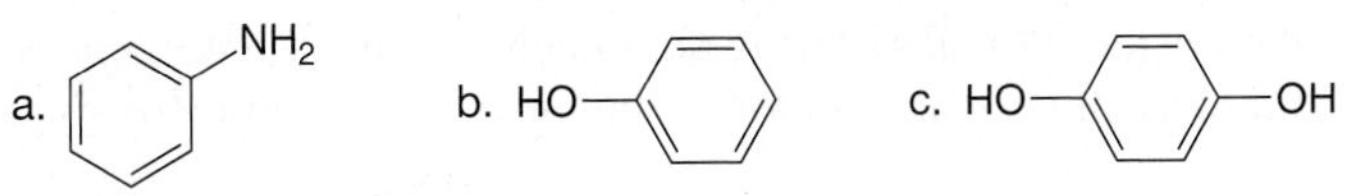
a.
NH_2
b. HO
c. HO
OH

Problem 25.33 What starting materials are needed to synthesize each azo compound?

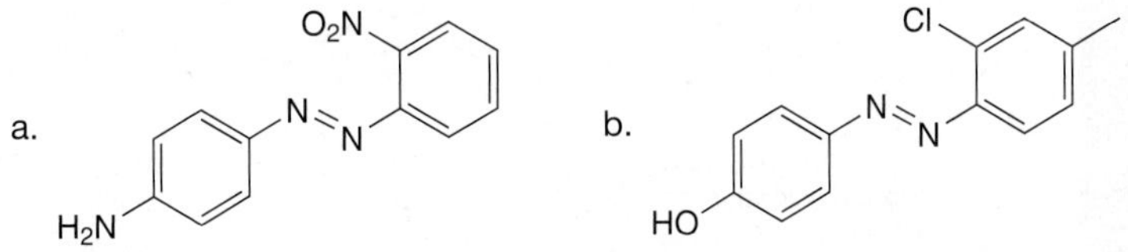
a.
O_2N
H_2N
b.
Cl
HO

25.16 Application: Synthetic Dyes and Sulfa Drugs

Azo compounds have two important applications: as dyes and as sulfa drugs, the first synthetic antibiotics.

25.16A Natural and Synthetic Dyes

Until 1856, all dyes were natural in origin, obtained from plants, animals, or minerals. Three natural dyes known for centuries are **indigo, tyrian purple,** and **alizarin.**

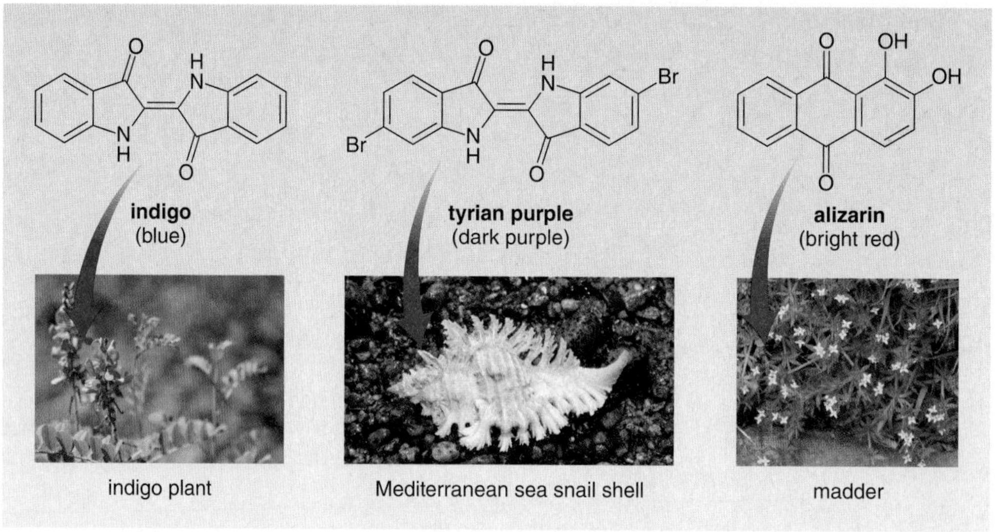

The blue dye **indigo,** derived from the plant *Indigofera tinctoria,* has been used in India for thousands of years. Traders introduced it to the Mediterranean area and then to Europe. **Tyrian purple,** a natural dark purple dye obtained from the mucous gland of a Mediterranean snail of the genus *Murex,* was a symbol of royalty before the collapse of the Roman Empire. **Alizarin,** a bright red dye obtained from madder root (*Rubia tinctorum*), a plant native to India and northeastern Asia, has been found in cloth entombed with Egyptian mummies.

Because all three of these dyes were derived from natural sources, they were difficult to obtain, making them expensive and available only to the privileged. This all changed in 1856 when William Henry Perkin, an 18-year-old student with a makeshift home laboratory, serendipitously prepared a purple dye, which would later be called mauveine, during his failed attempt to synthesize the antimalarial drug quinine. Mauveine is a mixture of two compounds that differ in the presence of only one methyl group on one of the aromatic rings.

A purple shawl dyed with Perkin's mauveine

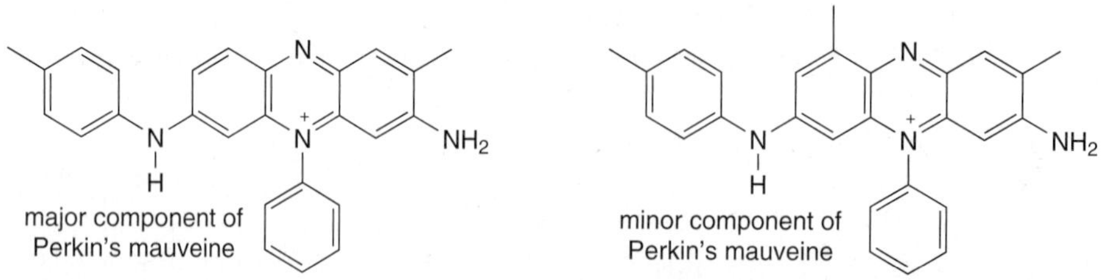

Perkin's discovery marked the beginning of the chemical industry. He patented the dye and went on to build a factory to commercially produce it on a large scale. This event began the surge of research in organic chemistry, not just in the synthesis of dyes, but in the production of perfumes, anesthetics, inks, and drugs as well. Perkin was a wealthy man when he retired at the age of 36 to devote the rest of his life to basic chemical research. The most prestigious award given by the American Chemical Society is named the Perkin Medal in his honor.

Many common synthetic dyes, such as alizarine yellow R, para red, and Congo red, are **azo compounds,** prepared by the diazonium coupling reaction described in Section 25.15.

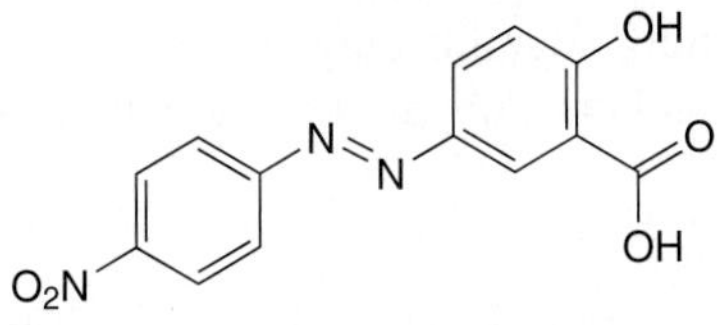

alizarine yellow R

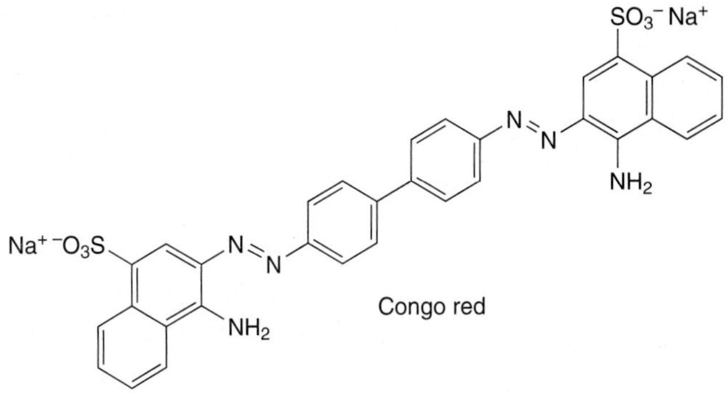

Congo red

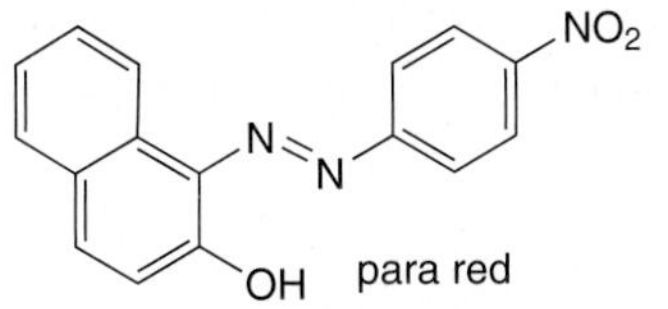

para red

Although natural and synthetic dyes are quite varied in structure, **all of them are colored because they are highly conjugated.** A molecule with eight or more π bonds in conjugation absorbs light in the visible region of the electromagnetic spectrum (Section 16.15A), taking on the color from the visible spectrum that it does *not* absorb.

Problem 25.34 (a) What two components are needed to prepare para red by azo coupling? (b) What two components are needed to prepare alizarine yellow R?

25.16B Sulfa Drugs

Although they may seem quite unrelated, the synthesis of colored dyes led to the development of the first synthetic antibiotics. Much of the early effort in this field was done by the German chemist Paul Ehrlich, who worked with synthetic dyes and used them to stain tissues. This led him on a search for dyes that were lethal to bacteria without affecting other tissue cells, hoping that these dyes could treat bacterial infections. For many years this effort was unsuccessful.

Then, in 1935, Gerhard Domagk, a German physician working for a dye manufacturer, first used a synthetic dye as a drug to kill bacteria. His daughter had contracted a streptococcal infection, and as she neared death, he gave her **prontosil,** an azo dye that inhibited the growth of certain bacteria in mice. His daughter recovered, and the modern era of synthetic antibiotics was initiated. For his pioneering work, Domagk was awarded the Nobel Prize in Physiology or Medicine in 1939.

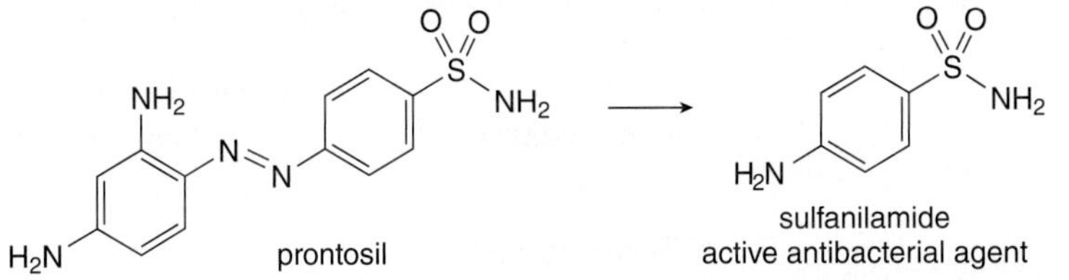

Prontosil and other sulfur-containing antibiotics are collectively called **sulfa drugs.** Prontosil is not the active agent itself. In cells, it is metabolized to **sulfanilamide,** the active drug. To understand how sulfanilamide functions as an antibacterial agent we must examine **folic acid,** which microorganisms synthesize from *p*-aminobenzoic acid.

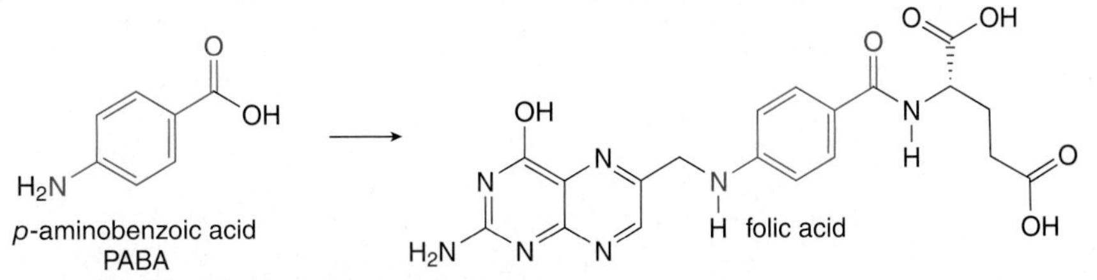

Sulfanilamide and *p*-aminobenzoic acid are similar in size and shape and have related functional groups. Thus, when sulfanilamide is administered, bacteria attempt to use it in place of *p*-aminobenzoic acid to synthesize folic acid. Derailing folic acid synthesis means that the bacteria cannot grow and reproduce. Sulfanilamide only affects bacterial cells, though, because humans do not synthesize folic acid, and must obtain it from their diets.

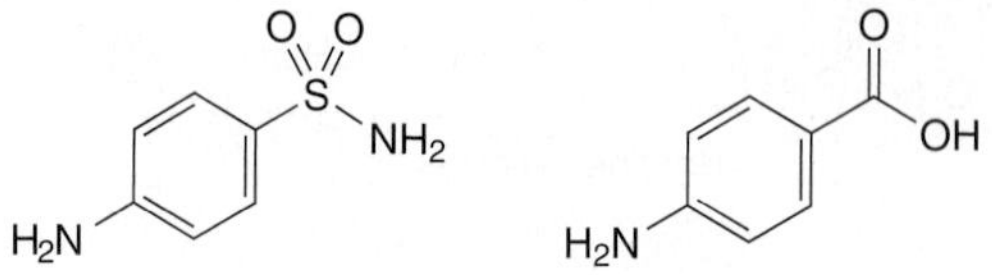

sulfanilamide *p*-aminobenzoic acid

Many other compounds of similar structure have been prepared and are still widely used as antibiotics. The structures of two other sulfa drugs are shown in Figure 25.14.

Figure 25.14
Two common sulfa drugs

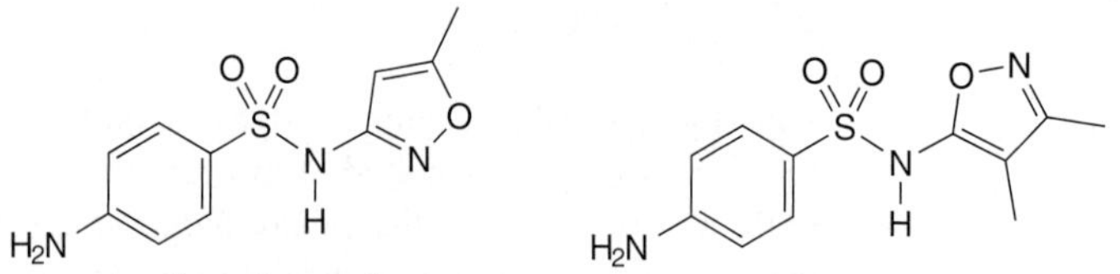

sulfamethoxazole sulfisoxazole

- Sulfamethoxazole is the sulfa drug in Bactrim, and sulfisoxazole is sold as Gantrisin. Both drugs are commonly used in the treatment of ear and urinary tract infections.

KEY CONCEPTS

Amines

General Facts

- Amines are organic nitrogen compounds having the general structure RNH_2, R_2NH, or R_3N, with a lone pair of electrons on N (25.1).
- Amines are named using the suffix *-amine* (25.3).
- All amines have polar C – N bonds. Primary (1°) and 2° amines have polar N – H bonds and are capable of intermolecular hydrogen bonding (25.4).
- The lone pair on N makes amines strong organic bases and nucleophiles (25.8).

Summary of Spectroscopic Absorptions (25.5)

Mass spectra	Molecular ion	Amines with an odd number of N atoms give an odd molecular ion.
IR absorptions	N – H	3300–3500 cm^{-1} (two peaks for RNH_2, one peak for R_2NH)
^{1}H NMR absorptions	NH	0.5–5 ppm (no splitting with adjacent protons)
	CH – N	2.3–3.0 ppm (deshielded C_{sp^3} – H)
^{13}C NMR absorption	C – N	30–50 ppm

Comparing the Basicity of Amines and Other Compounds (25.10)

- Alkylamines (RNH_2, R_2NH, and R_3N) are more basic than NH_3 because of the electron-donating R groups (25.10A).
- Alkylamines (RNH_2) are more basic than arylamines ($C_6H_5NH_2$), which have a delocalized lone pair from the N atom (25.10B).
- Arylamines with electron-donor groups are more basic than arylamines with electron-withdrawing groups (25.10B).
- Alkylamines (RNH_2) are more basic than amides ($RCONH_2$), which have a delocalized lone pair from the N atom (25.10C).
- Aromatic heterocycles with a localized electron pair on N are more basic than those with a delocalized lone pair from the N atom (25.10D).
- Alkylamines with a lone pair in an sp^3 hybrid orbital are more basic than those with a lone pair in an sp^2 hybrid orbital (25.10E).

Preparation of Amines (25.7)

[1] Direct nucleophilic substitution with NH_3 and amines (25.7A)

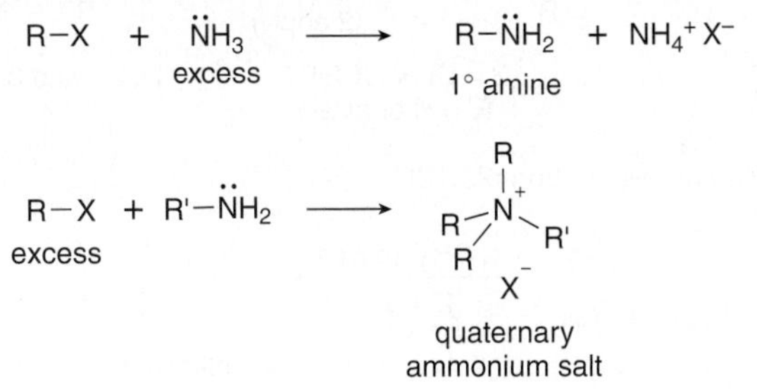
R–X + NH3 → R–NH2 + NH4+ X−
excess
1° amine
R–X + R'–NH2 →
excess
quaternary
ammonium salt

- The mechanism is S_N2.
- The reaction works best for CH_3X or RCH_2X.
- The reaction works best to prepare 1° amines and quaternary ammonium salts.

[2] Gabriel synthesis (25.7A)

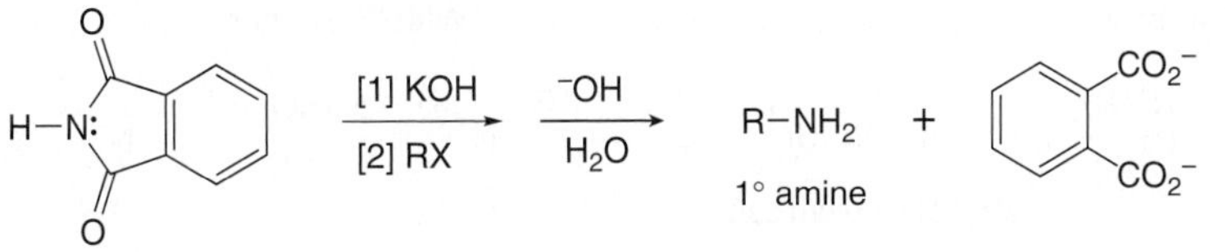
[1] KOH
[2] RX
−OH
H2O
R–NH2
1° amine

- The mechanism is S_N2.
- The reaction works best for CH_3X or RCH_2X.
- Only 1° amines can be prepared.

[3] Reduction methods (25.7B)

a. From nitro compounds

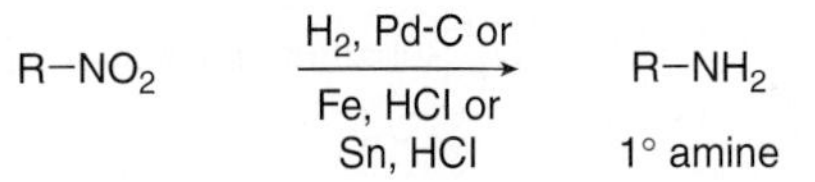
R–NO2
H2, Pd-C or
Fe, HCl or
Sn, HCl
R–NH2
1° amine

b. From nitriles

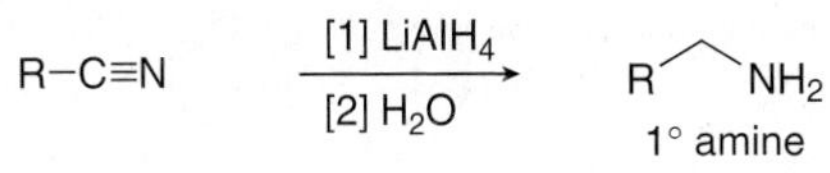
R–C≡N
[1] LiAlH4
[2] H2O
1° amine

c. From amides

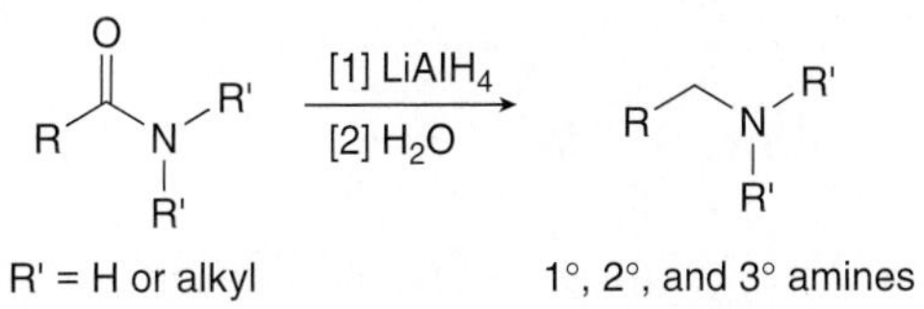
[1] LiAlH4
[2] H2O
R' = H or alkyl
1°, 2°, and 3° amines

[4] Reductive amination (25.7C)

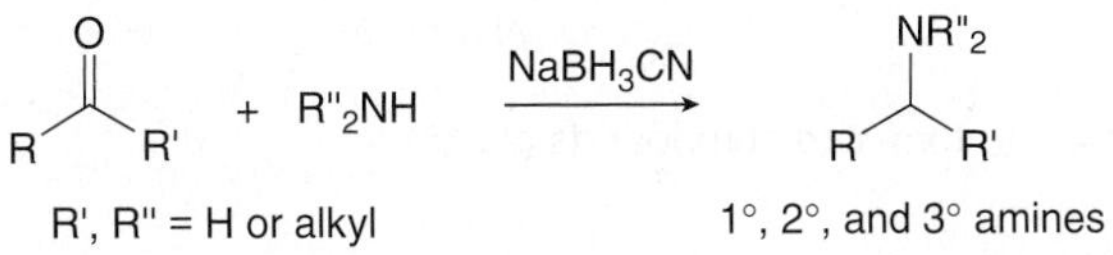
+ R''2NH
NaBH3CN
R', R'' = H or alkyl
1°, 2°, and 3° amines

- Reductive amination adds one alkyl group (from an aldehyde or ketone) to a nitrogen nucleophile.
- Primary (1°), 2°, and 3° amines can be prepared.

Reactions of Amines

[1] Reaction as a base (25.9)

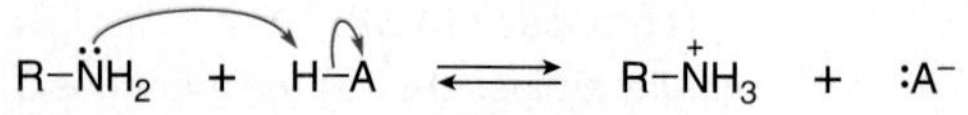
R–N̈H₂ + H–A ⇌ R–N⁺H₃ + :A⁻

[2] Nucleophilic addition to aldehydes and ketones (25.11)

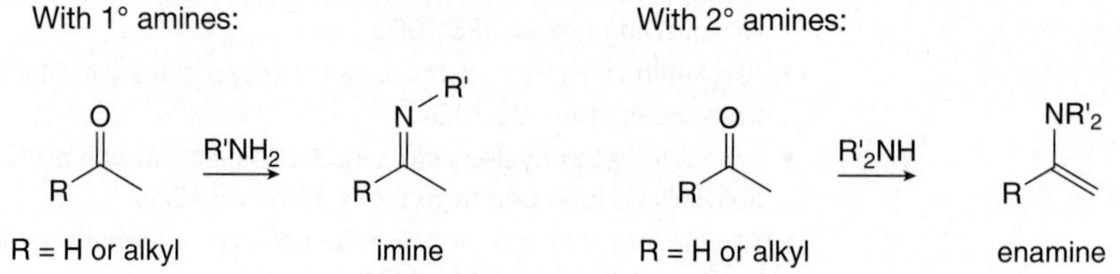
With 1° amines:
R'NH₂
R = H or alkyl
imine
With 2° amines:
R'₂NH
R = H or alkyl
enamine

[3] Nucleophilic substitution with acid chlorides and anhydrides (25.11)

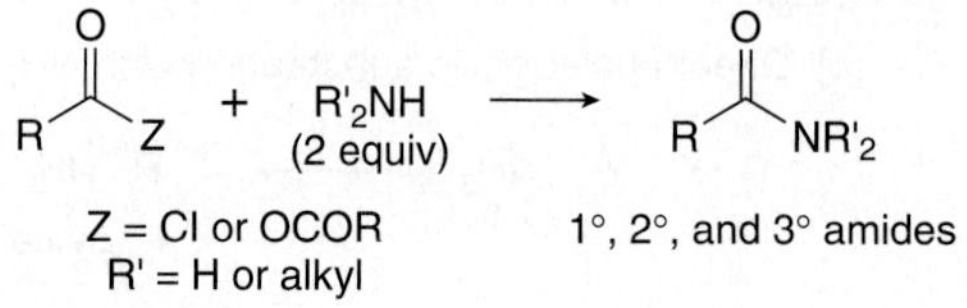
R'₂NH
(2 equiv)
Z = Cl or OCOR
R' = H or alkyl
1°, 2°, and 3° amides

[4] Hofmann elimination (25.12)

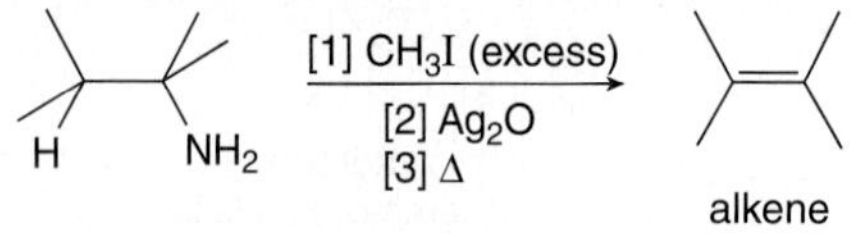
[1] CH₃I (excess)
[2] Ag₂O
[3] Δ
alkene

- The less substituted alkene is the major product.

[5] Reaction with nitrous acid (25.13)

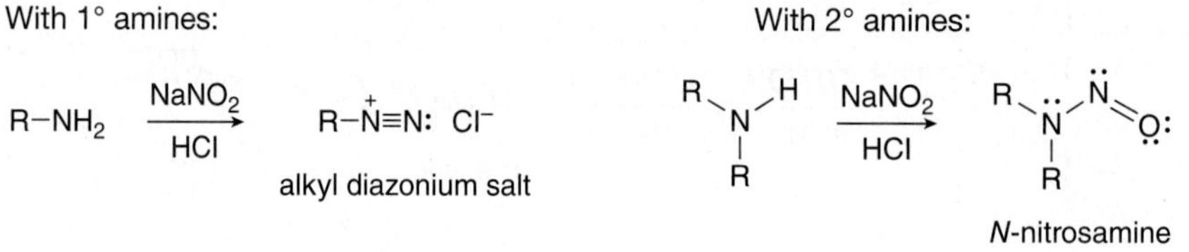
With 1° amines:
R–NH₂
NaNO₂
HCl
R–N⁺≡N: Cl⁻
alkyl diazonium salt
With 2° amines:
NaNO₂
HCl
N-nitrosamine

Reactions of Diazonium Salts

[1] Substitution reactions (25.14)

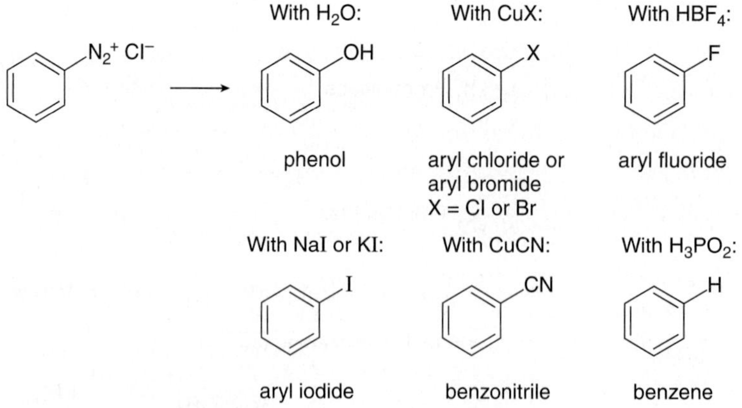
With H₂O:
phenol
With CuX:
aryl chloride or
aryl bromide
X = Cl or Br
With HBF₄:
aryl fluoride
With NaI or KI:
aryl iodide
With CuCN:
benzonitrile
With H₃PO₂:
benzene

[2] Coupling to form azo compounds (25.15)

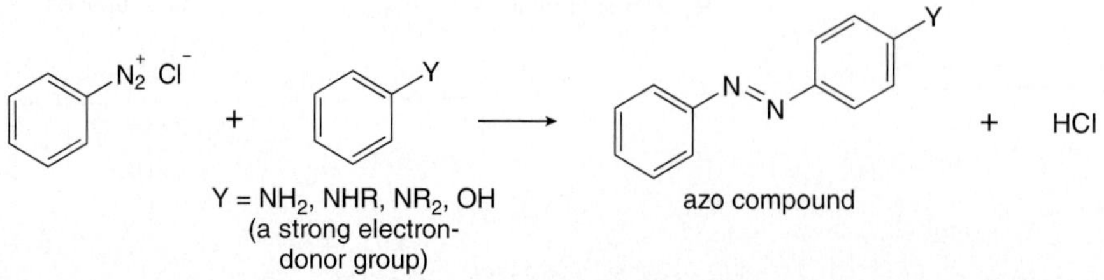
Y = NH₂, NHR, NR₂, OH
(a strong electron-
donor group)
azo compound
+ HCl

PROBLEMS

Problems Using Three-Dimensional Models

25.35 Give a systematic or common name for each compound.

a. 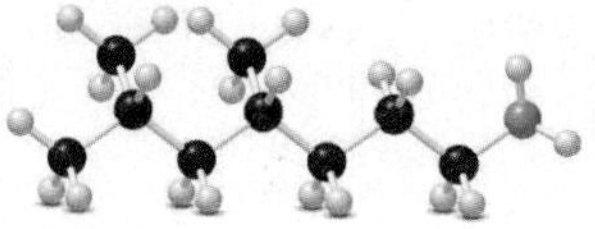

b. 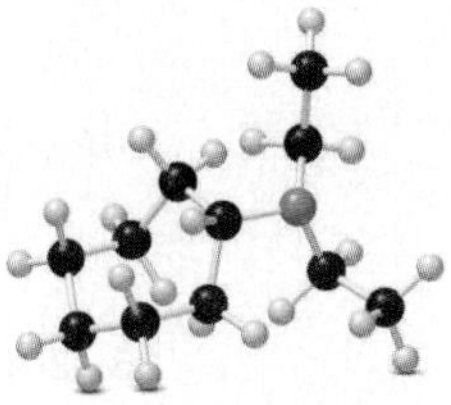

25.36 Which compound is the stronger base?

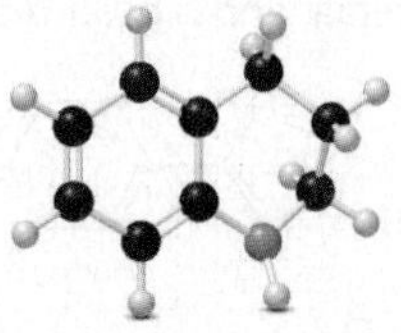

A

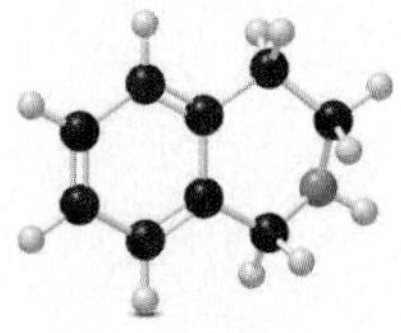

B

25.37 Varenicline (trade name Chantix) is a drug used to help smokers quit their habit. (a) Which N atom in varenicline is most basic? Explain your choice. (b) What product is formed when varenicline is treated with HCl?

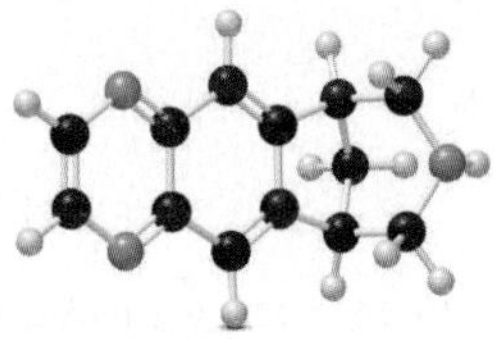

varenicline

Nomenclature

25.38 Give a systematic or common name for each compound.

a. 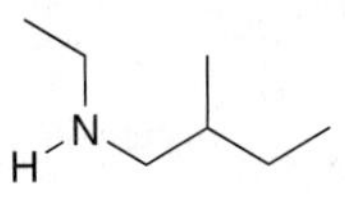

b. 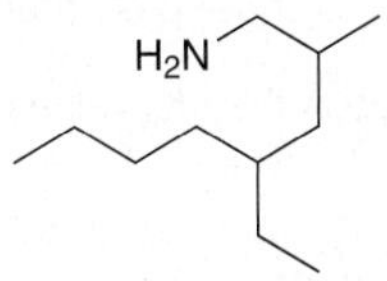

c. 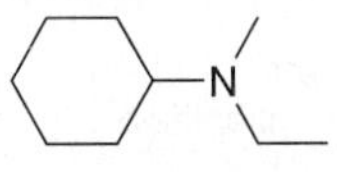

d. 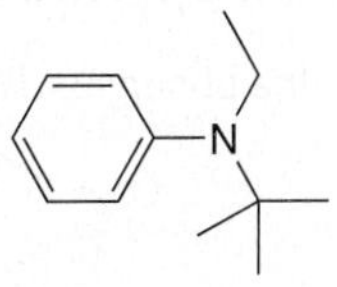

e. 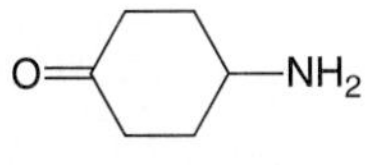

f. 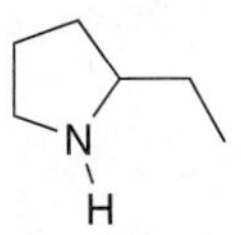

g. 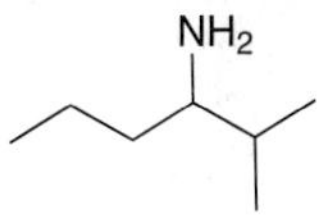

h. 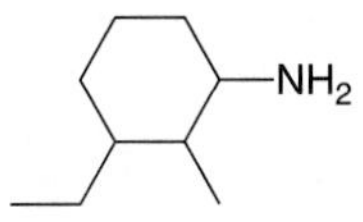

25.39 Draw the structure that corresponds to each name.

a. *N*-isobutylcyclopentanamine
b. tri-*tert*-butylamine
c. *N*,*N*-diisopropylaniline
d. *N*-methylpyrrole
e. *N*-methylcyclopentanamine
f. 3-methylhexan-2-amine
g. 2-*sec*-butylpiperidine
h. (*S*)-heptan-2-amine

Chiral Compounds

25.40 How many stereogenic centers are present in the following tetraalkylammonium salt? Draw all possible stereoisomers.

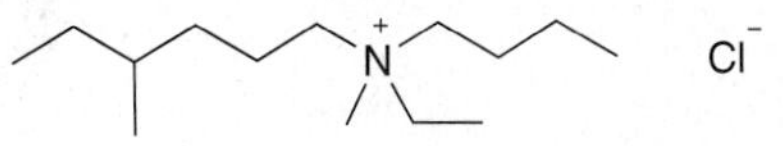

Basicity

25.41 Rank the compounds in each group in order of increasing basicity.

a.

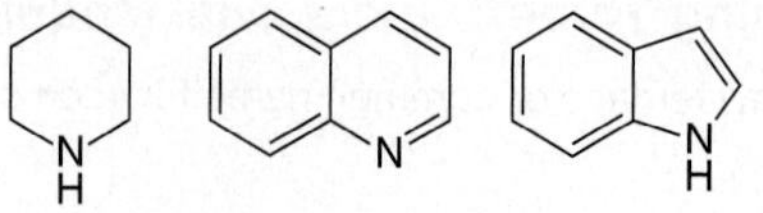

b.

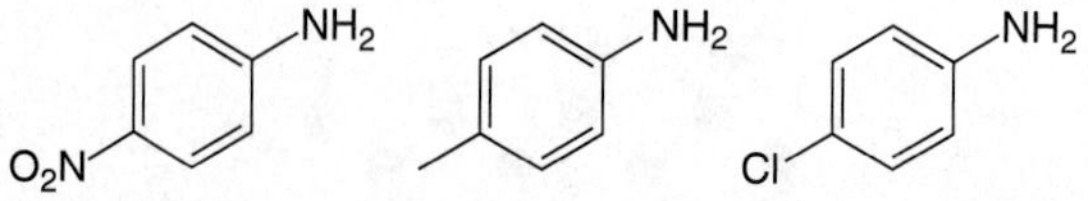

25.42 Decide which N atom in each molecule is most basic and draw the product formed when each compound is treated with CH_3CO_2H. Zolpidem (trade name Ambien) is used to treat insomnia, whereas aripiprazole (trade name Abilify) is used to treat depression, schizophrenia, and bipolar disorders.

a.

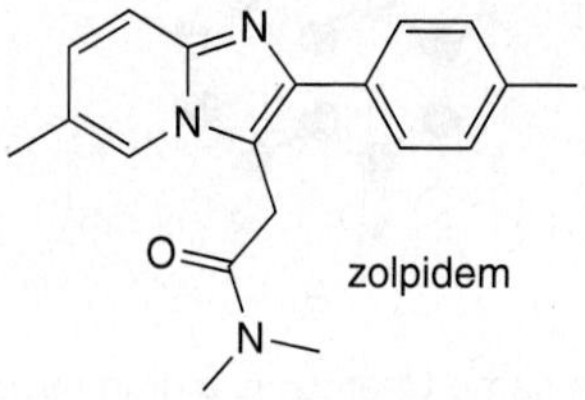

zolpidem

b.

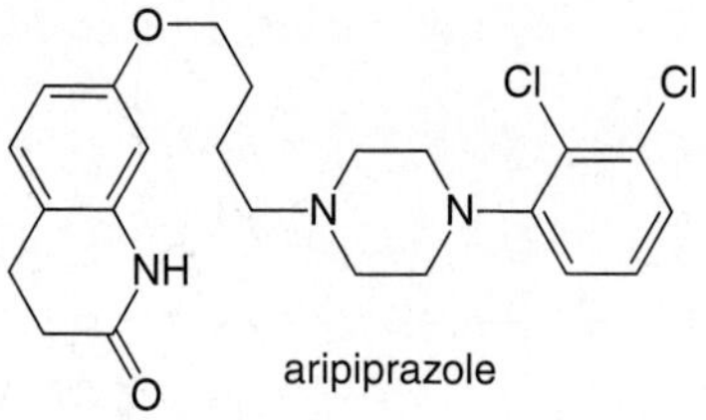

aripiprazole

25.43 Explain why pyrimidine is less basic than pyridine.

pyridine pyrimidine

25.44 Rank the nitrogen atoms in each compound in order of increasing basicity. Isoniazid is a drug used to treat tuberculosis, whereas histamine (Section 25.6B) causes the runny nose and watery eyes associated with allergies.

a.

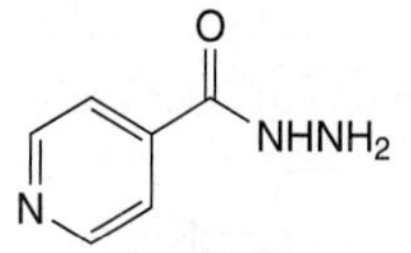

isoniazid

b.

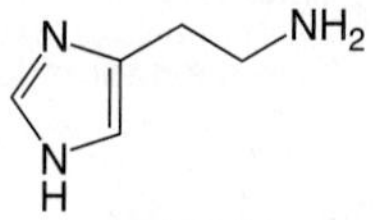

histamine

25.45 Explain why *m*-nitroaniline is a stronger base than *p*-nitroaniline.

25.46 Explain the observed difference in the pK_a values of the conjugate acids of amines **A** and **B.**

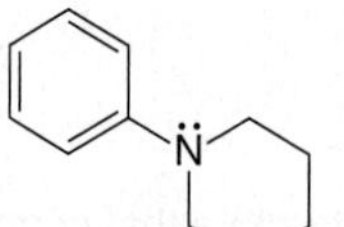

A
$pK_a = 5.2$

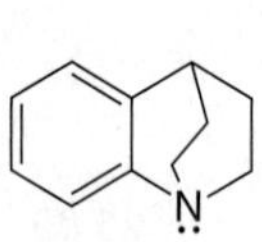

B
$pK_a = 7.29$

25.47 Why is pyrrole more acidic than pyrrolidine?

pyrrole
$pK_a = 23$

pyrrolidine
$pK_a = 44$

Preparation of Amines

25.48 What amide(s) can be used to prepare each amine by reduction?

a. 

b. 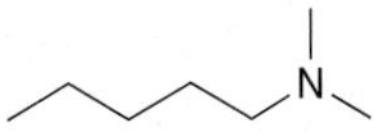

c. 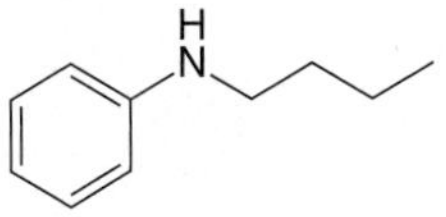

25.49 What carbonyl and nitrogen compounds are needed to make each compound by reductive amination? When more than one set of starting materials is possible, give all possible methods.

a. 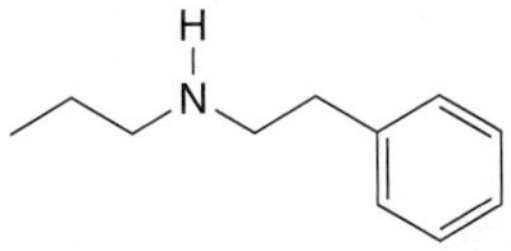

b. 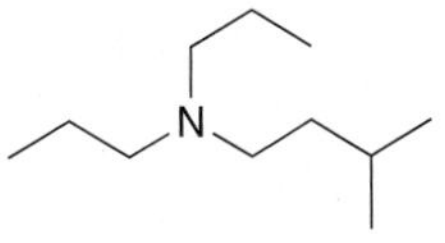

c. 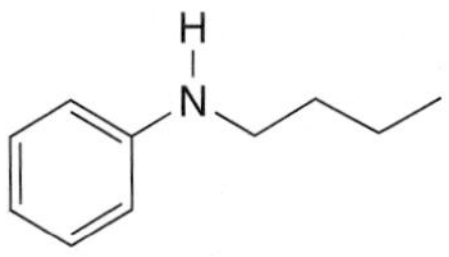

25.50 Draw the product of each reductive amination reaction.

a. 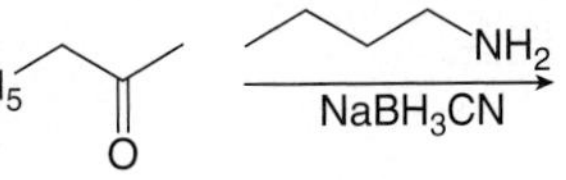

c. 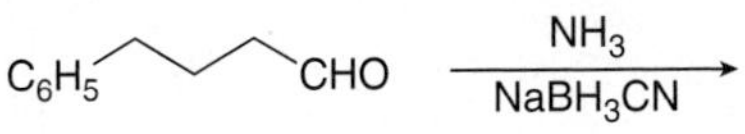

b. 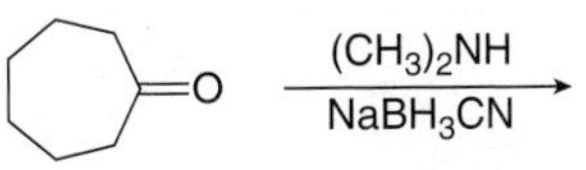

d. 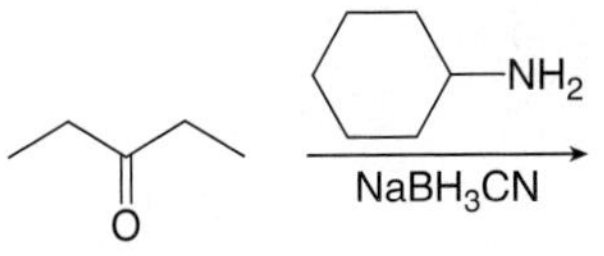

Extraction

25.51 How would you separate toluene ($C_6H_5CH_3$), benzoic acid ($C_6H_5CO_2H$), and aniline ($C_6H_5NH_2$) by an extraction procedure?

Reactions

25.52 Draw the products formed when *p*-methylaniline (*p*-$CH_3C_6H_4NH_2$) is treated with each reagent.

a. HCl
b. CH_3COCl
c. $(CH_3CO)_2O$
d. excess CH_3I
e. $(CH_3)_2C{=}O$
f. CH_3COCl, $AlCl_3$
g. CH_3CO_2H
h. $NaNO_2$, HCl
i. Part (b), then CH_3COCl, $AlCl_3$
j. CH_3CHO, $NaBH_3CN$

25.53 Draw the products formed when each amine is treated with [1] CH_3I (excess); [2] Ag_2O; [3] Δ. Indicate the major product when a mixture results.

a. 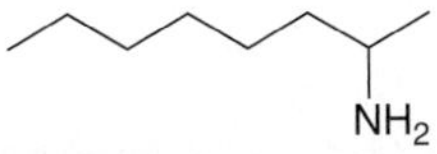

b. 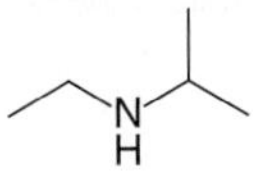

c. 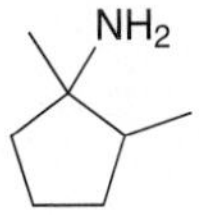

d. 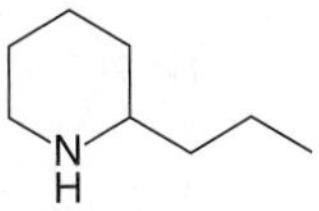

25.54 Answer the following questions about benzphetamine, a habit-forming diet pill sold under the trade name Didrex.

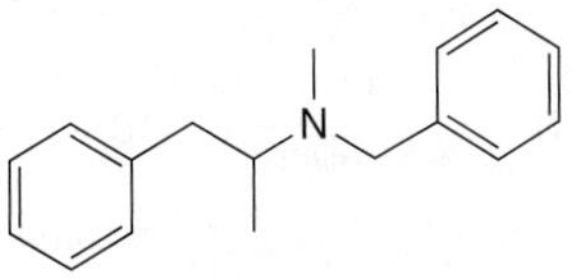

benzphetamine

a. Label the stereogenic center(s).
b. What amide(s) can be reduced to form benzphetamine?
c. What carbonyl compound and amine can be used to make benzphetamine by reductive amination? Draw all possible methods.
d. What products are formed by Hofmann elimination from benzphetamine? Label the major product.

25.55 Draw the organic products formed in each reaction.

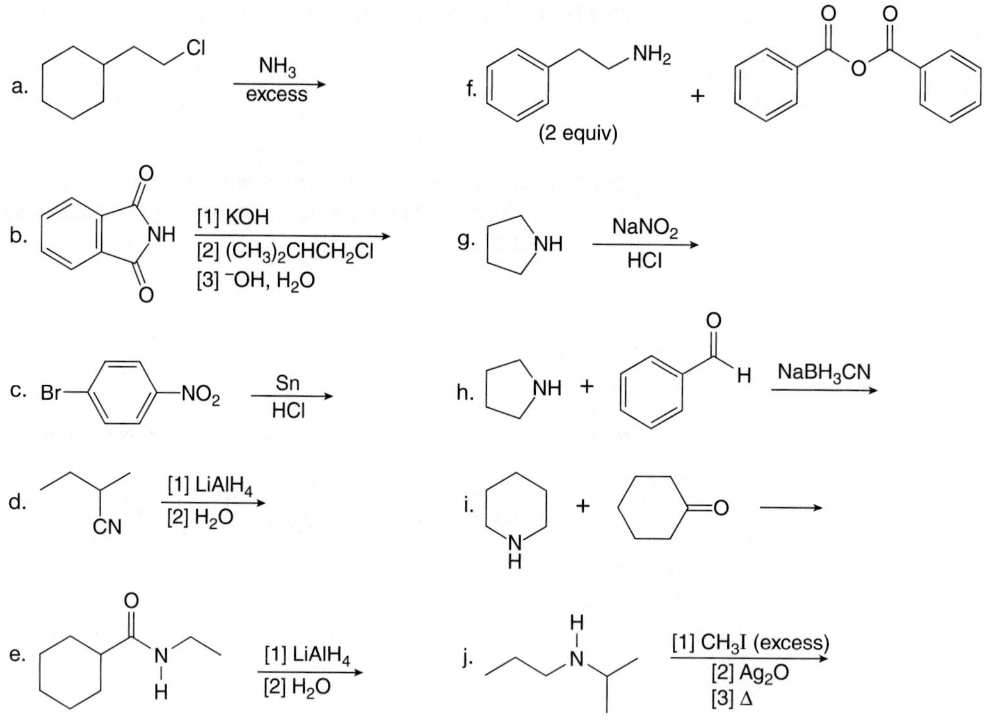
a.
Cl
NH_3
excess
b.
O
NH
O
[1] KOH
[2] $(CH_3)_2CHCH_2Cl$
[3] ^{-}OH, H_2O
c. Br
NO_2
Sn
HCl
d.
CN
[1] $LiAlH_4$
[2] H_2O
e.
O
N
H
[1] $LiAlH_4$
[2] H_2O
f.
NH_2
(2 equiv)
+
O O
O
g.
NH
$NaNO_2$
HCl
h.
NH +
O
H
$NaBH_3CN$
i.
N
H
+
O
j.
H
N
[1] CH_3I (excess)
[2] Ag_2O
[3] Δ

25.56 What is the major Hofmann elimination product formed from each amine?

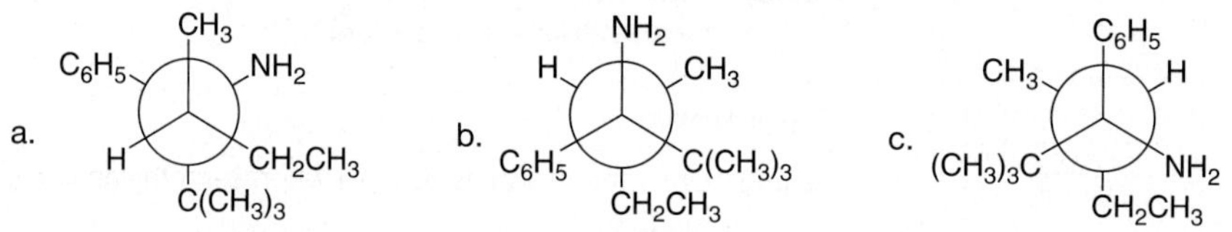
a.
CH_3
C_6H_5
NH_2
H
CH_2CH_3
$C(CH_3)_3$
b.
NH_2
H
CH_3
C_6H_5
$C(CH_3)_3$
CH_2CH_3
c.
C_6H_5
CH_3
H
$(CH_3)_3C$
NH_2
CH_2CH_3

25.57 Identify **A, B,** and **C,** three intermediates in the synthesis of the pain reliever and anesthetic fentanyl.

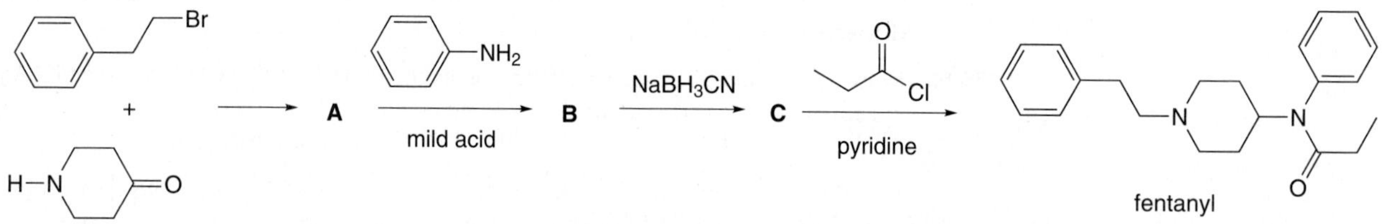
Br
+
H–N
O
A
NH_2
mild acid
B
$NaBH_3CN$
C
O
Cl
pyridine
N
N
O
fentanyl

25.58 Identify the intermediates (**A–C**) in the following reaction sequence, which was used to prepare racemic ofloxacin. One enantiomer of the product, levofloxacin, is an antibiotic used to treat severe bacterial infections that have not responded to other drugs.

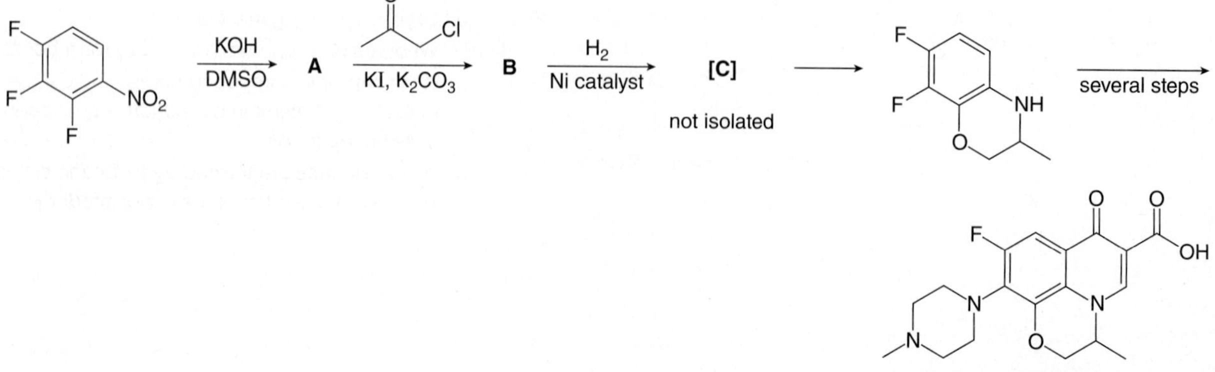
F
F
F
NO_2
KOH
DMSO
A
O
Cl
KI, K_2CO_3
B
H_2
Ni catalyst
[C]
not isolated
F
F
NH
O
several steps
O O
F
OH
N
N
N
O

ofloxacin

25.59 Draw the product formed when **A** is treated with each series of reagents.

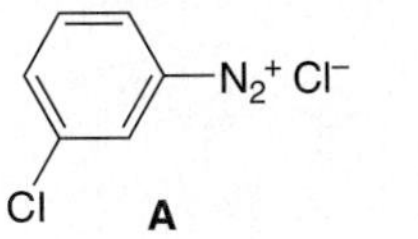

a. [1] H_2O; [2] NaH; [3] CH_3Br
b. [1] CuCN; [2] DIBAL-H; [3] H_2O
c. [1] $C_6H_5NH_2$; [2] CH_3COCl

25.60 A chiral amine **A** having the *R* configuration undergoes Hofmann elimination to form an alkene **B** as the major product. **B** is oxidatively cleaved with ozone, followed by CH_3SCH_3, to form $CH_2{=}O$ and $CH_3CH_2CH_2CHO$. What are the structures of **A** and **B?**

Mechanism

25.61 Draw a stepwise mechanism for each reaction.

a. 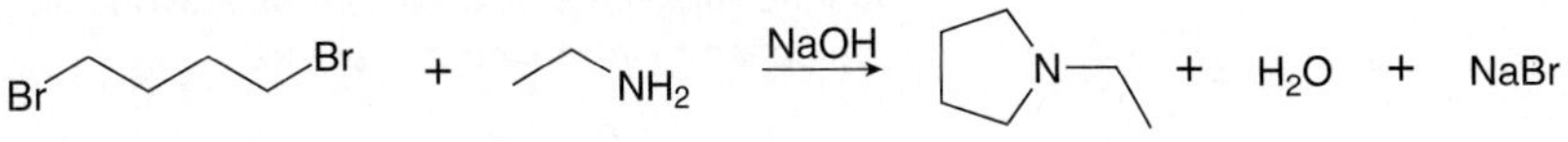
NaOH
+ H2O + NaBr

b. 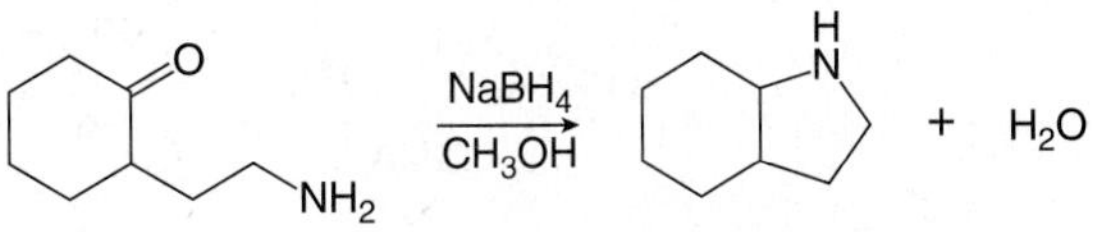
NaBH4
CH3OH
+ H2O

25.62 Draw a stepwise mechanism for the following reaction.

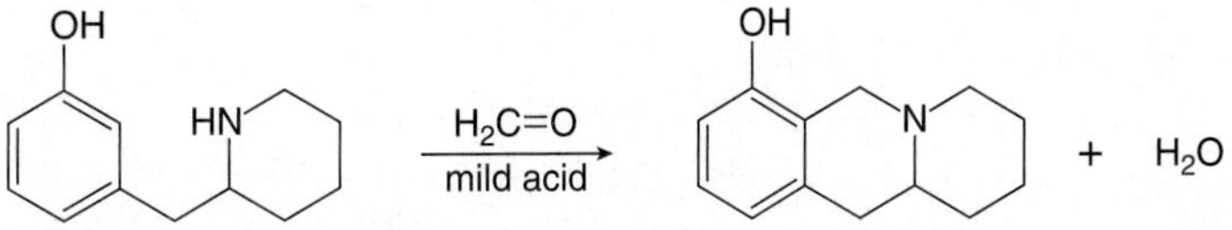
H2C=O
mild acid
+ H2O

25.63 Alkyl diazonium salts decompose to form carbocations, which go on to form products of substitution, elimination, and (sometimes) rearrangement. Keeping this in mind, draw a stepwise mechanism that forms all of the following products.

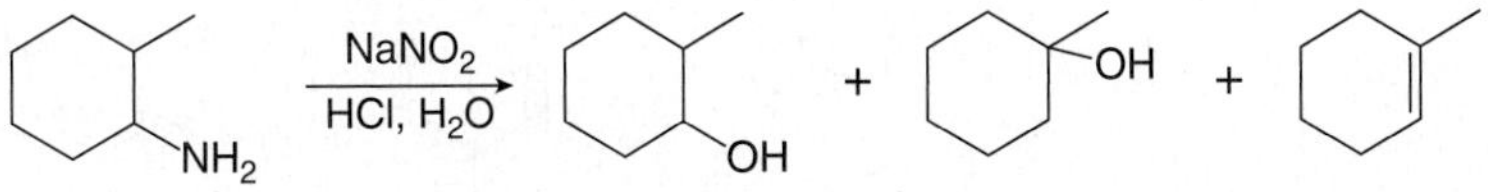
NaNO2
HCl, H2O

25.64 Tertiary (3°) aromatic amines react with $NaNO_2$ and HCl to afford products of electrophilic aromatic substitution. Draw a stepwise mechanism for this nitrosation reaction and explain why it occurs only on benzene rings with strong ortho, para activating groups.

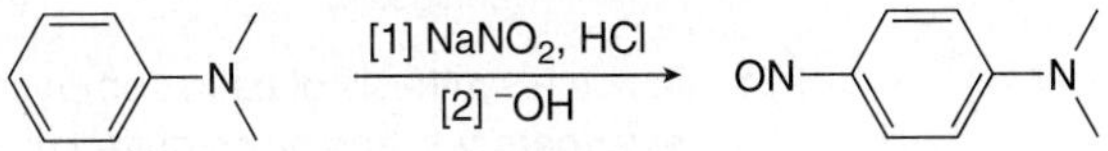
[1] NaNO2, HCl
[2] −OH

Synthesis

25.65 Devise a synthesis of each compound from benzene. You may use any other organic or inorganic reagents.

a. 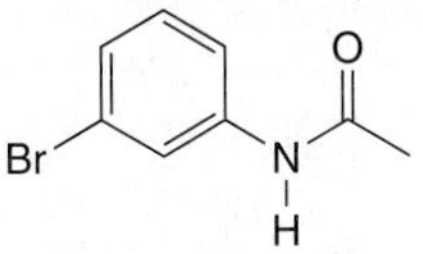

c. 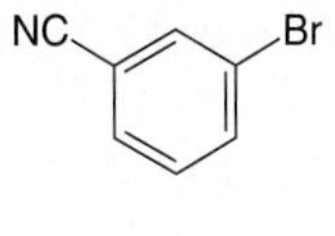

e. 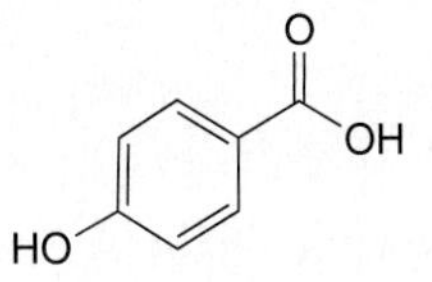

b. 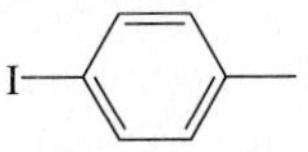

d. 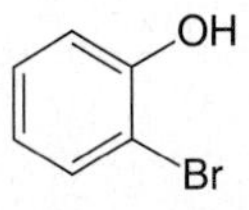

f. 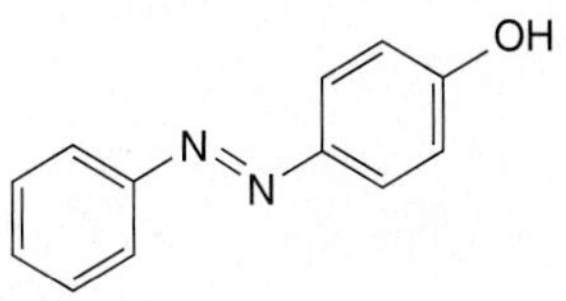

25.66 Devise a synthesis of each compound from aniline ($C_6H_5NH_2$) as starting material.

a. 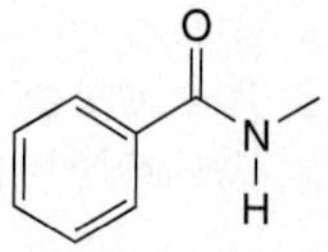 b. 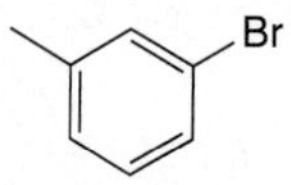 c. 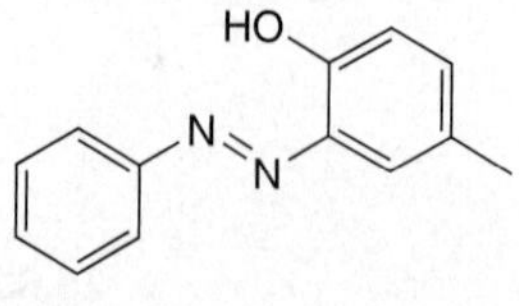

25.67 Devise at least three different methods to prepare *N*-methylbenzylamine ($PhCH_2NHCH_3$) from benzene, any one-carbon organic compounds, and any required reagents.

25.68 Safrole, which is isolated from sassafras (Problem 21.33), can be converted to the illegal stimulant MDMA (3,4-methylenedioxymethamphetamine, "Ecstasy") by a variety of methods. (a) Devise a synthesis that begins with safrole and uses a nucleophilic substitution reaction to introduce the amine. (b) Devise a synthesis that begins with safrole and uses reductive amination to introduce the amine.

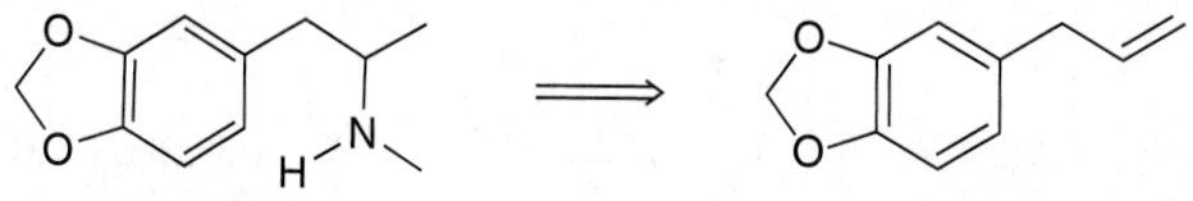

MDMA safrole

25.69 Synthesize each compound from benzene. Use a diazonium salt as one of the synthetic intermediates.

a. 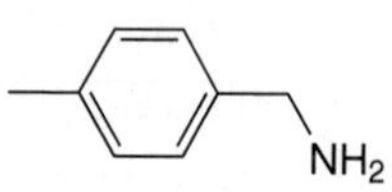 b. 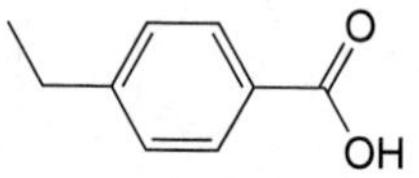 c. 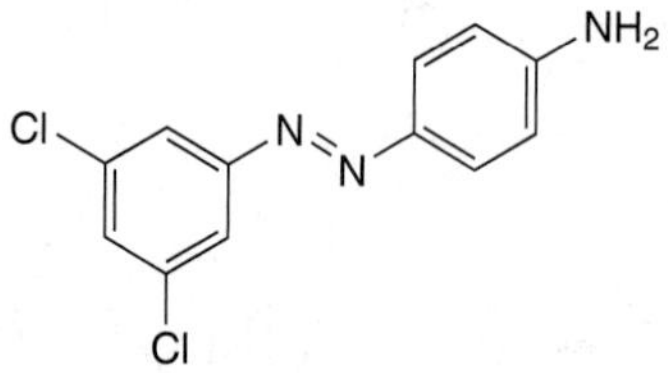

25.70 Devise a synthesis of each biologically active compound from benzene.

a. 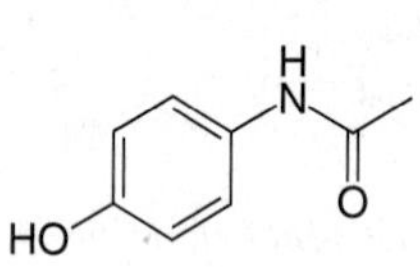

acetaminophen
(analgesic)

b. 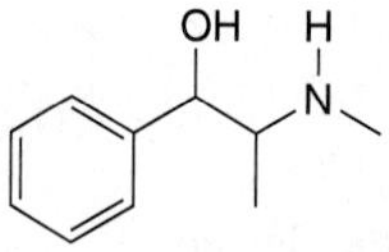

pseudoephedrine
(nasal decongestant)

25.71 Devise a synthesis of each compound from benzene, any organic alcohols having four or fewer carbons, and any required reagents.

a. 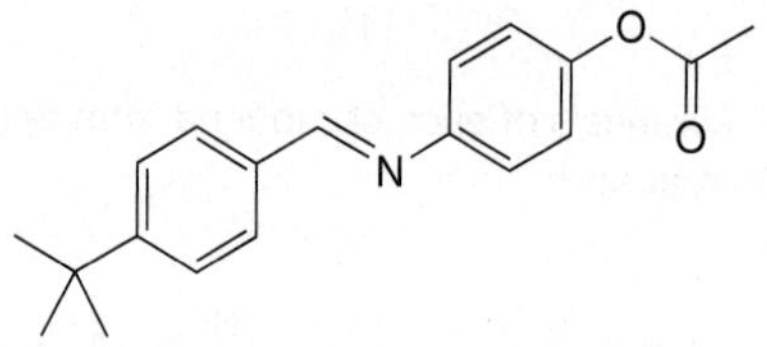

c. 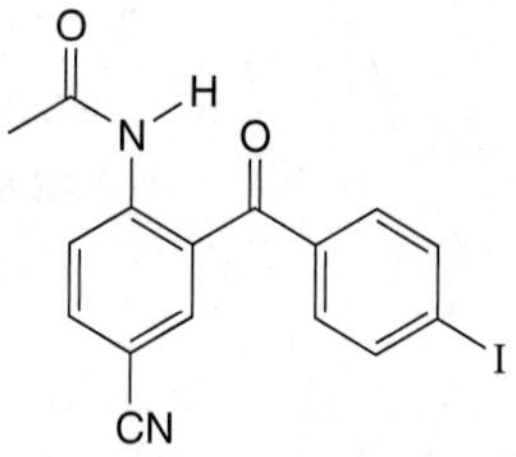

b. 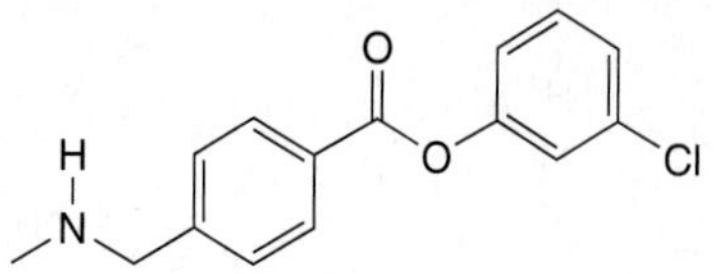

Spectroscopy

25.72 Three isomeric compounds, **A, B,** and **C,** all have molecular formula $C_8H_{11}N$. The ^1H NMR and IR spectral data of **A, B,** and **C** are given below. What are their structures?

Compound **A:** IR peak at 3400 cm^{-1}

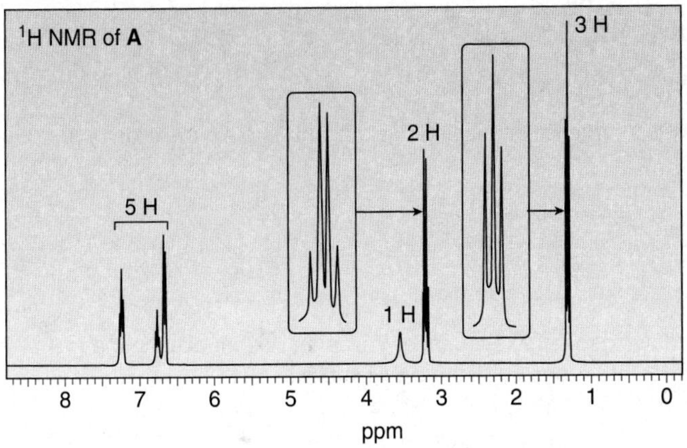

Compound **B:** IR peak at 3310 cm^{-1}

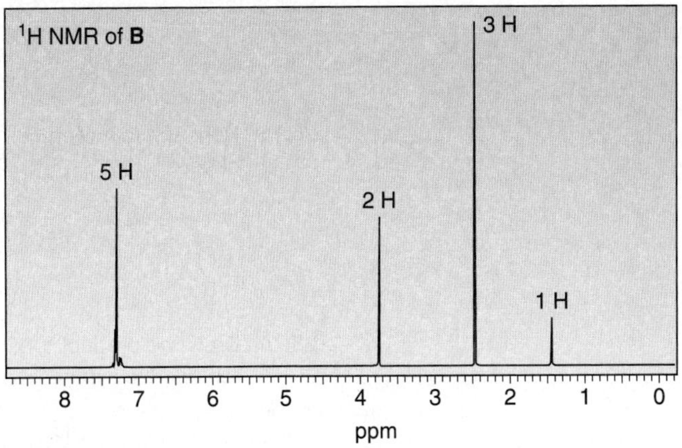

Compound **C:** IR peaks at 3430 and 3350 cm^{-1}

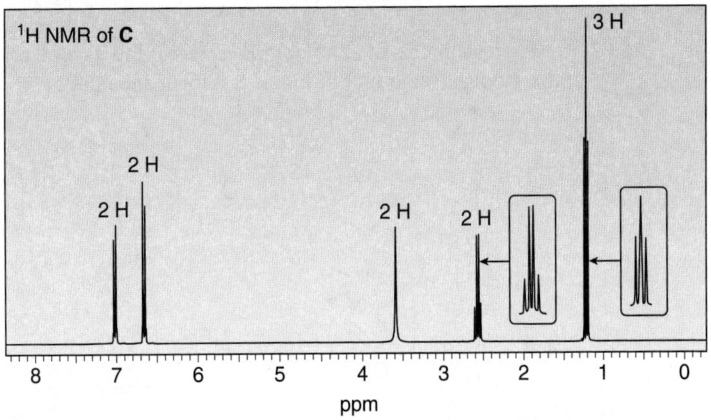

25.73 Treatment of compound **D** with $LiAlH_4$ followed by H_2O forms compound **E. D** shows a molecular ion in its mass spectrum at $m/z = 71$ and IR absorptions at 3600–3200 and 2263 cm^{-1}. **E** shows a molecular ion in its mass spectrum at $m/z = 75$ and IR absorptions at 3636 and 3600–3200 cm^{-1}. Propose structures for **D** and **E** from these data and the given ^1H NMR spectra.

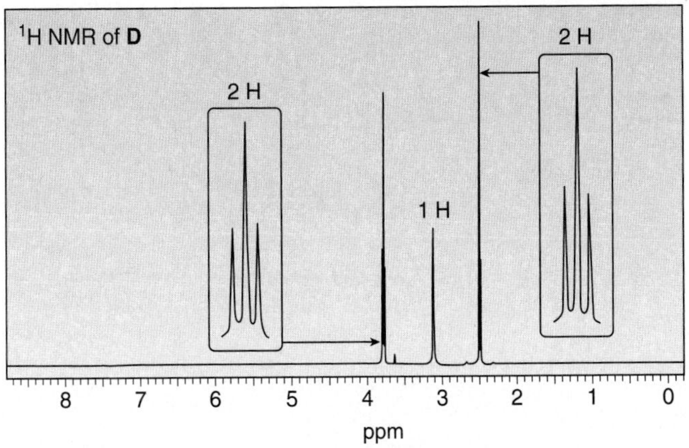

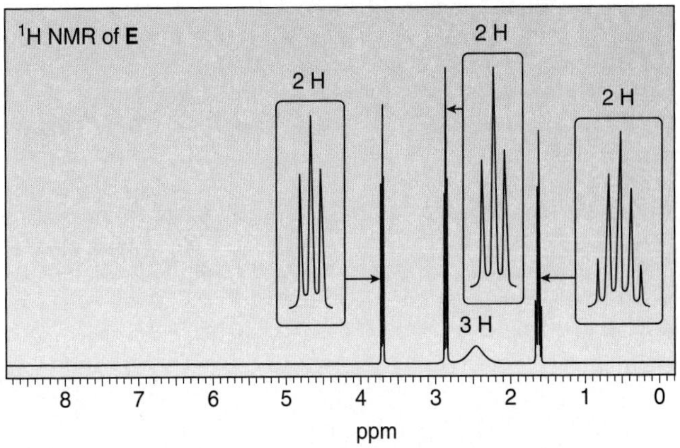

Challenge Problems

25.74 The pK_a of the conjugate acid of guanidine is 13.6, making it one of the strongest neutral organic bases. Offer an explanation.

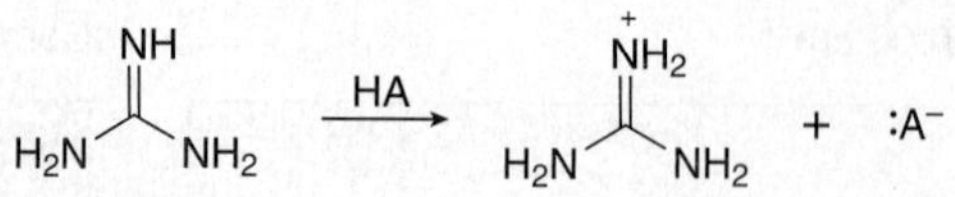

guanidine $\quad$ pK_a = 13.6

25.75 Rank the following compounds in order of increasing basicity and explain the order you chose.

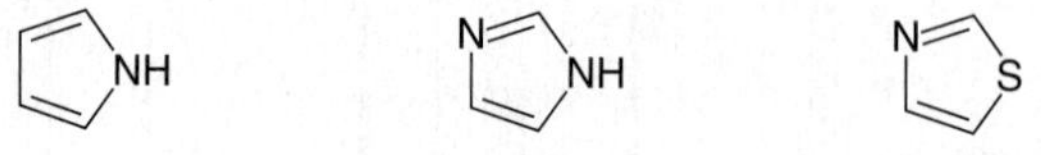

pyrrole $\quad$ imidazole $\quad$ thiazole

25.76 Draw the product **Y** of the following reaction sequence. **Y** was an intermediate in the remarkable synthesis of cyclooctatetraene by Richard Willstatter in 1911.

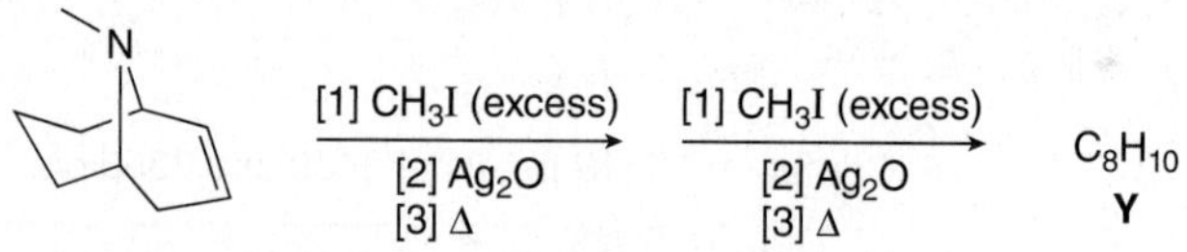

25.77 Devise a synthesis of each compound from the given starting material(s). Albuterol is a bronchodilator and proparacaine is a local anesthetic.

a.

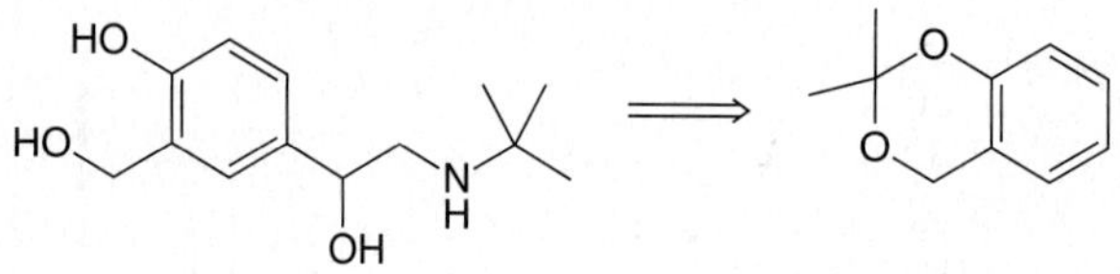

albuterol

b.

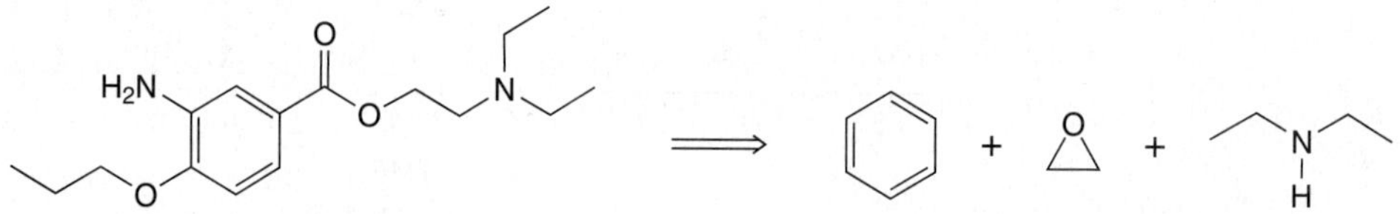

proparacaine

25.78 Heating compound **X** with aqueous formaldehyde forms **Y** ($C_{17}H_{23}NO$), which has been converted to a mixture of lupinine and epilupinine, alkaloids isolated from lupin, a perennial ornamental plant commonly seen on the roadside in parts of Alaska (Section 25.10). Identify **Y** and explain how it is formed.

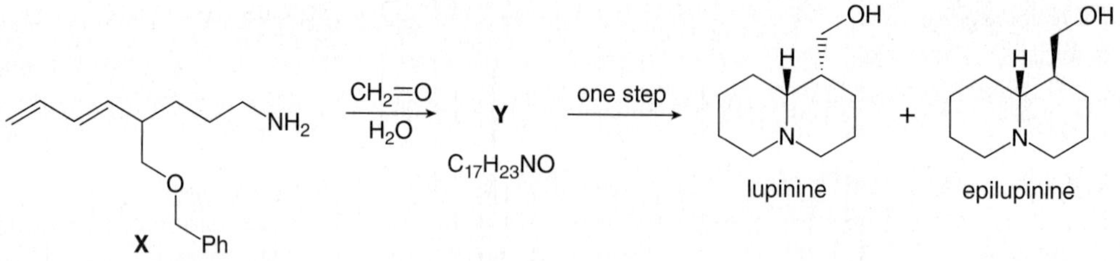

Amino Acids and Proteins

26

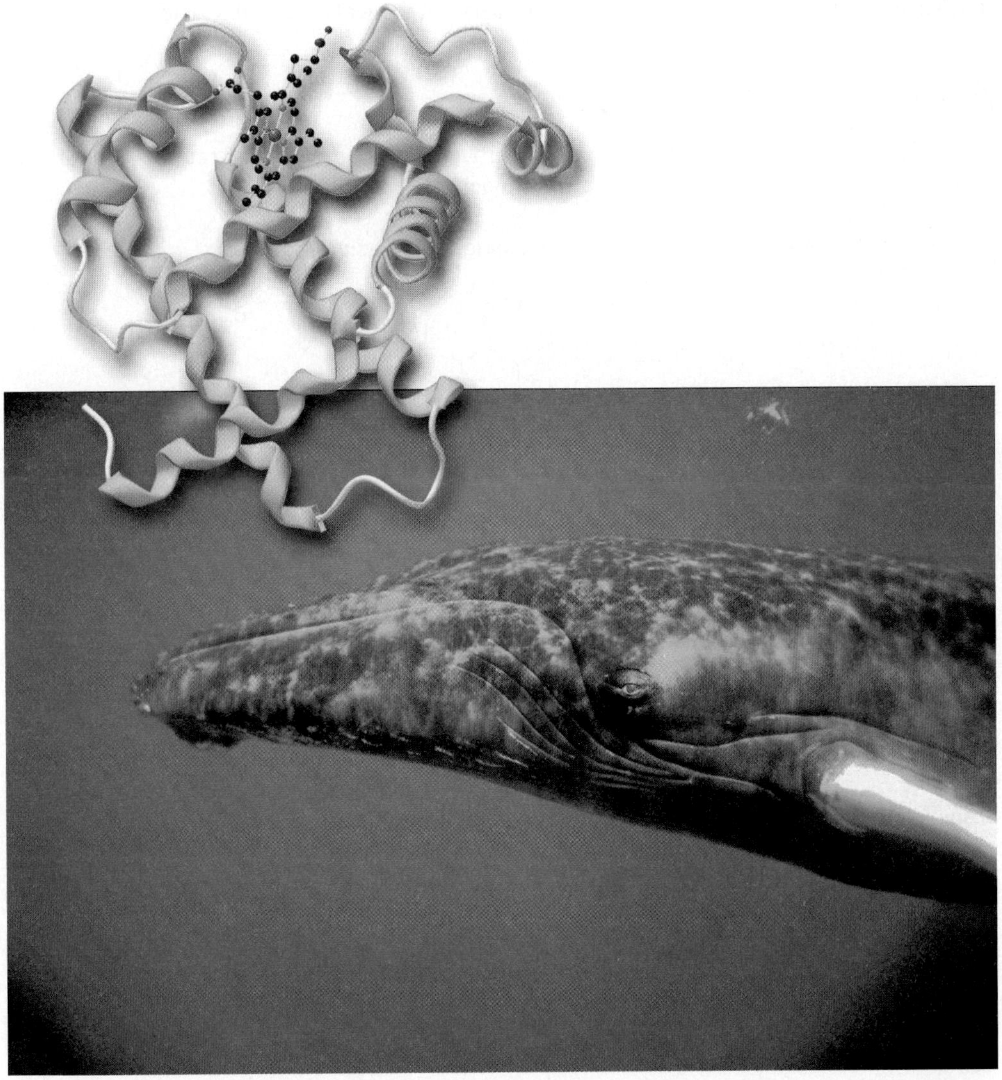

26.1 Amino acids
26.2 Synthesis of amino acids
26.3 Separation of amino acids
26.4 Enantioselective synthesis of amino acids
26.5 Peptides
26.6 Peptide sequencing
26.7 Peptide synthesis
26.8 Automated peptide synthesis
26.9 Protein structure
26.10 Important proteins

Myoglobin is a globular protein that contains 153 amino acids joined together, as well as a non-protein portion called a heme unit. The heme group consists of a large nitrogen heterocycle complexed with the Fe^{2+} cation. The Fe^{2+} cation binds oxygen in the blood and stores it in tissues. Whales have a particularly high myoglobin concentration in their muscles. It serves as an oxygen reservoir for the whale while it is submerged for long periods of time. In Chapter 26, we discuss the properties of proteins and the amino acids from which they are synthesized.

Of the four major groups of biomolecules—lipids, carbohydrates, nucleic acids, and proteins—proteins have the widest array of functions. **Keratin** and **collagen,** for example, are part of a large group of structural proteins that form long insoluble fibers, giving strength and support to tissues. Hair, horns, hooves, and fingernails are all made up of keratin. **Collagen** is found in bone, connective tissue, tendons, and cartilage. **Enzymes** are proteins that catalyze and regulate all aspects of cellular function. **Membrane proteins** transport small organic molecules and ions across cell membranes. **Insulin,** the hormone that regulates blood glucose levels, **fibrinogen** and **thrombin,** which form blood clots, and **hemoglobin,** which transports oxygen from the lungs to tissues, are all proteins.

In Chapter 26 we discuss proteins and their primary components, the amino acids.

26.1 Amino Acids

Amino acids were first discussed in Section 3.9.

Naturally occurring amino acids have an amino group (NH_2) bonded to the α carbon of a carboxy group (COOH), and so they are called **α-amino acids.**

- **All proteins are polyamides formed by joining amino acids together.**

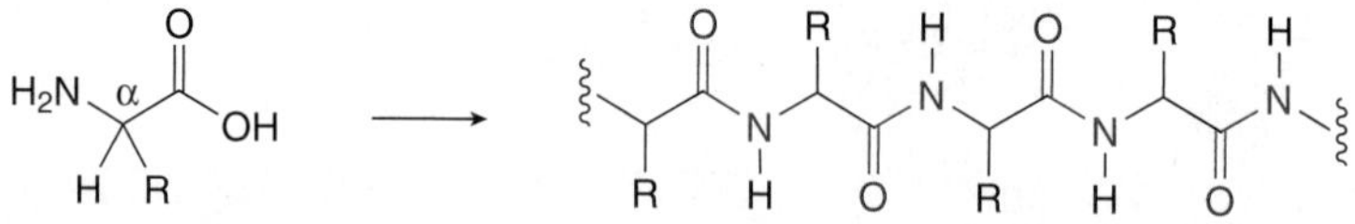

α-amino acid **portion of a protein molecule**

26.1A General Features of α-Amino Acids

The 20 amino acids that occur naturally in proteins differ in the identity of the R group bonded to the α carbon. The R group is called the **side chain** of the amino acid.

The simplest amino acid, called glycine, has R = H. **All other amino acids (R ≠ H) have a stereogenic center on the α carbon.** As is true for monosaccharides (Section 27.2C), the prefixes **D** and **L** are used to designate the configuration at the stereogenic center of amino acids. Common, naturally occurring amino acids are called **L-amino acids.** Their enantiomers, D-amino acids, are rarely found in nature. These general structures are shown in Figure 26.1. According to *R,S* designations, all L-amino acids except cysteine have the ***S* configuration.**

All amino acids have common names. These names can be represented by either a one-letter or a three-letter abbreviation. Figure 26.2 is a listing of the 20 naturally occurring amino acids, together with their abbreviations. Note the variability in the R groups. A side chain can be a simple alkyl group, or it can have additional functional groups such as OH, SH, COOH, or NH_2.

- **Amino acids with an additional COOH group in the side chain are called acidic amino acids.**
- **Those with an additional basic N atom in the side chain are called basic amino acids.**
- **All others are neutral amino acids.**

Figure 26.1 The general features of an α-amino acid

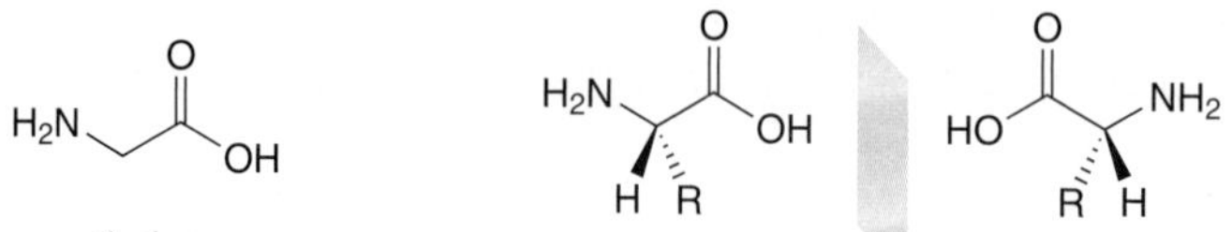

glycine
no stereogenic centers

L-amino acid

D-amino acid

Only this isomer
is common in proteins.

Figure 26.2 The 20 naturally occurring amino acids

Neutral amino acids

Name	Structure	Abbreviations	Name	Structure	Abbreviations
Alanine	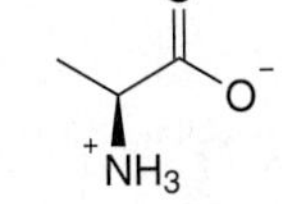	Ala A	Phenylalanine*	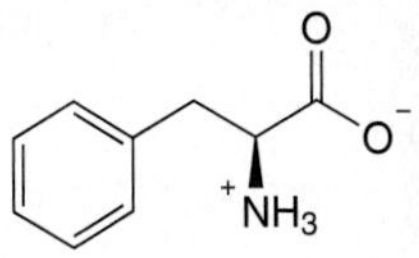	Phe F
Asparagine	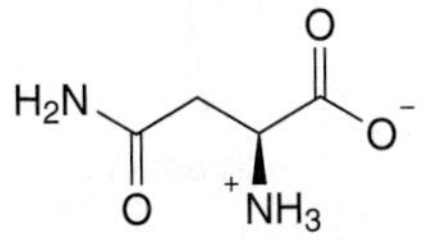	Asn N	Proline	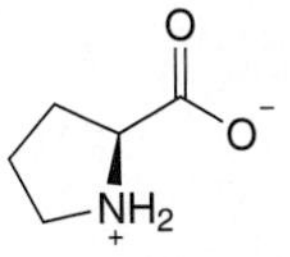	Pro P
Cysteine	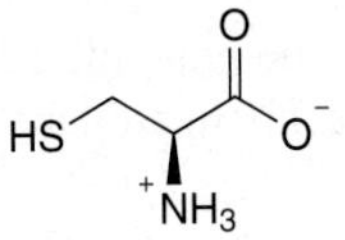	Cys C	Serine	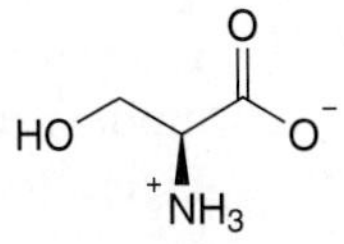	Ser S
Glutamine	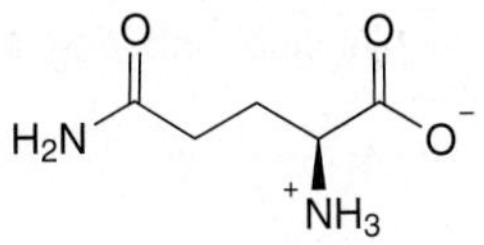	Gln Q	Threonine*	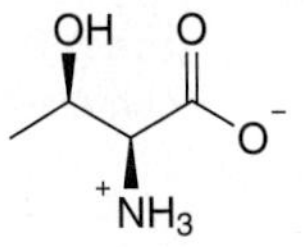	Thr T
Glycine	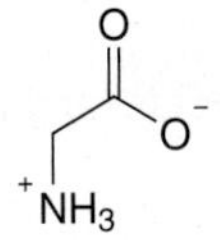	Gly G	Tryptophan*	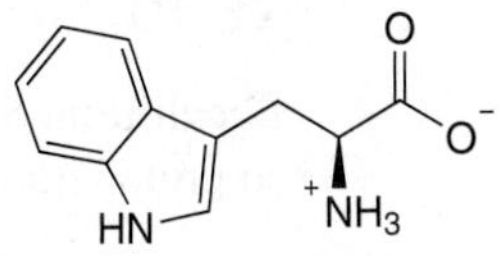	Trp W
Isoleucine*	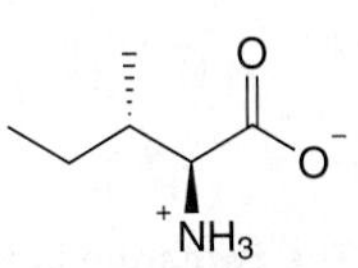	Ile I	Tyrosine	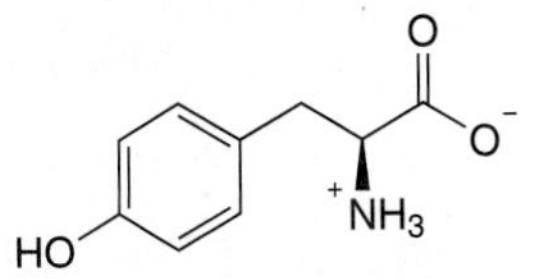	Tyr Y
Leucine*	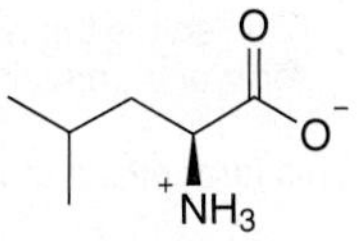	Leu L	Valine*	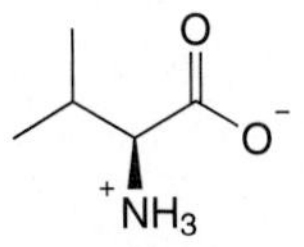	Val V
Methionine*	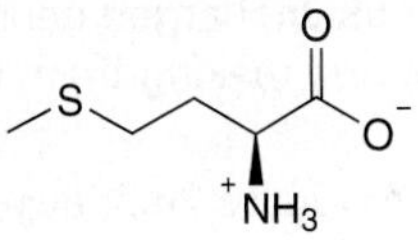	Met M			

Acidic amino acids / **Basic amino acids**

Name	Structure	Abbreviations	Name	Structure	Abbreviations
Aspartic acid	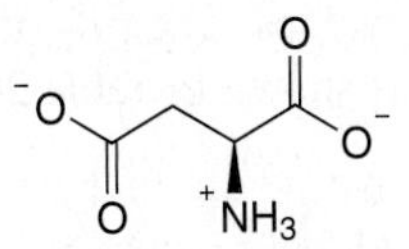	Asp D	Arginine*	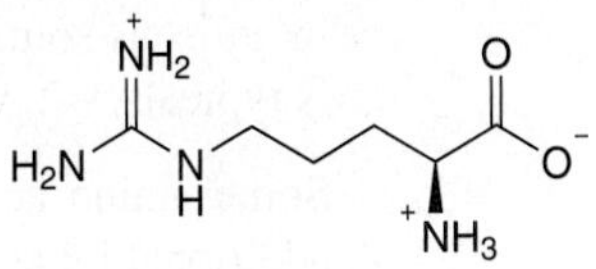	Arg R
Glutamic acid	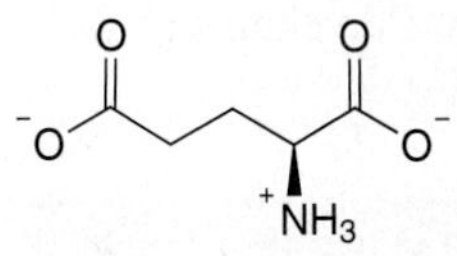	Glu E	Histidine*	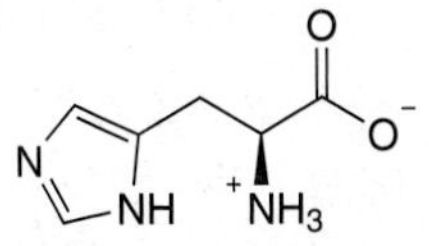	His H
			Lysine*	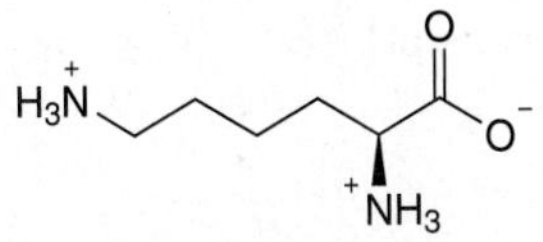	Lys K

Essential amino acids are labeled with an asterisk (*).

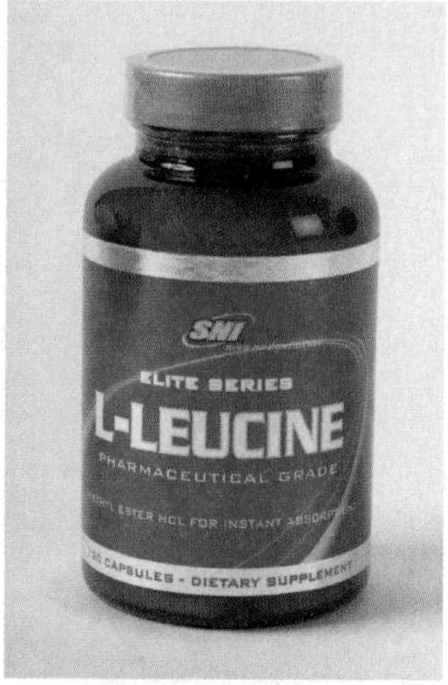

The essential amino acid leucine is sold as a dietary supplement that is used by body builders to help prevent muscle loss and heal muscle tissue after injury.

Look closely at the structures of proline, isoleucine, and threonine.

- **All amino acids are 1° amines except for proline,** which has its N atom in a five-membered ring, making it a **2° amine.**
- **Isoleucine** and **threonine** contain an additional stereogenic center at the β carbon, so there are four possible stereoisomers, only one of which is naturally occurring.

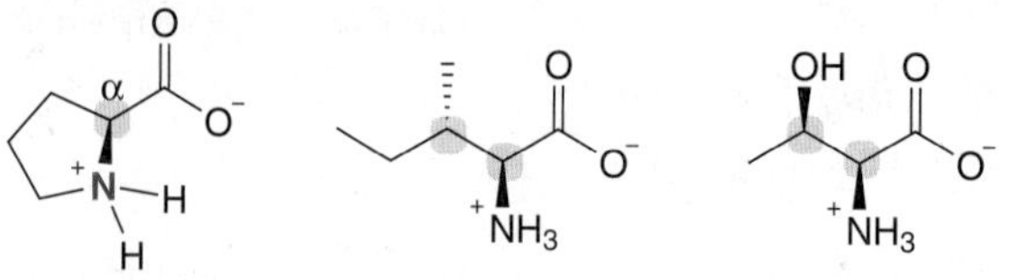

Humans can synthesize only 10 of these 20 amino acids. The remaining 10 are called **essential amino acids** because they must be obtained from the diet. These are labeled with an asterisk in Figure 26.2.

Problem 26.1 Draw the other three stereoisomers of L-isoleucine, and label the stereogenic centers as *R* or *S*.

26.1B Acid–Base Behavior

Recall from Section 19.15B that an amino acid has both an acidic and a basic functional group, so proton transfer forms a salt called a **zwitterion.**

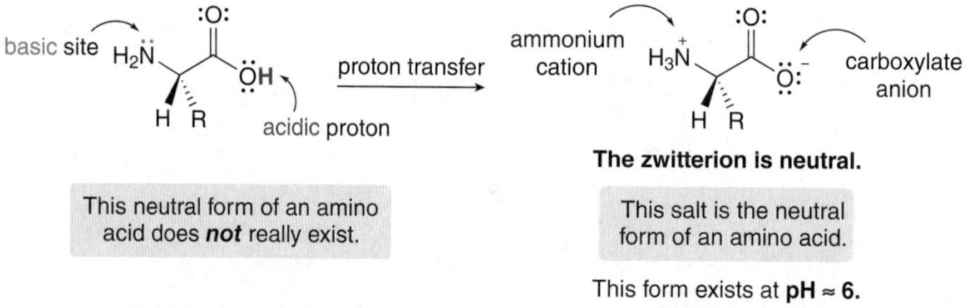

The structures in Figure 26.2 show the charged form of the amino acids at the physiological pH of the blood.

- **Amino acids do not exist to any appreciable extent as uncharged neutral compounds. They exist as salts, giving them high melting points and making them water soluble.**

Amino acids exist in different charged forms, as shown in Figure 26.3, depending on the pH of the aqueous solution in which they are dissolved. For neutral amino acids, the overall charge is +1, 0, or –1. Only at pH ~6 does the zwitterionic form exist.

The $-COOH$ and $-NH_3^+$ groups of an amino acid are ionizable, because they can lose a proton in aqueous solution. As a result, they have different pK_a values. The pK_a of the $-COOH$ group is typically ~2, whereas that of the $-NH_3^+$ group is ~9, as shown in Table 26.1.

Some amino acids, such as aspartic acid and lysine, have acidic or basic side chains. These additional ionizable groups complicate somewhat the acid–base behavior of these amino acids. Table 26.1 lists the pK_a values for these acidic and basic side chains as well.

Figure 26.3 How the charge of a neutral amino acid depends on the pH

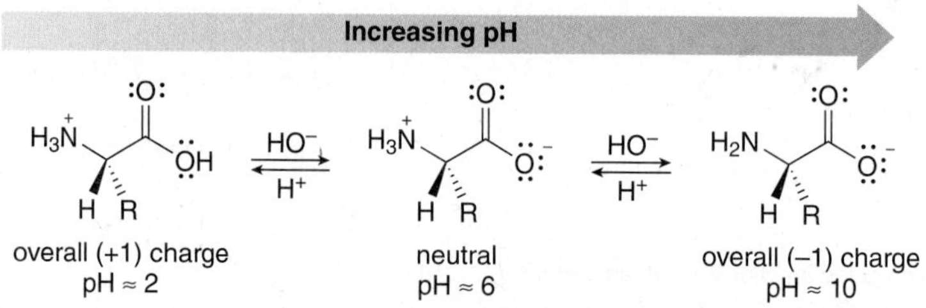

Table 26.1 pK_a Values for the Ionizable Functional Groups of an α-Amino Acid

Amino acid	α-COOH	α-NH_3^+	Side chain	p*I*
Alanine	2.35	9.87	—	6.11
Arginine	2.01	9.04	12.48	10.76
Asparagine	2.02	8.80	—	5.41
Aspartic acid	2.10	9.82	3.86	2.98
Cysteine	2.05	10.25	8.00	5.02
Glutamic acid	2.10	9.47	4.07	3.08
Glutamine	2.17	9.13	—	5.65
Glycine	2.35	9.78	—	6.06
Histidine	1.77	9.18	6.10	7.64
Isoleucine	2.32	9.76	—	6.04
Leucine	2.33	9.74	—	6.04
Lysine	2.18	8.95	10.53	9.74
Methionine	2.28	9.21	—	5.74
Phenylalanine	2.58	9.24	—	5.91
Proline	2.00	10.60	—	6.30
Serine	2.21	9.15	—	5.68
Threonine	2.09	9.10	—	5.60
Tryptophan	2.38	9.39	—	5.88
Tyrosine	2.20	9.11	10.07	5.63
Valine	2.29	9.72	—	6.00

Table 26.1 also lists the isoelectric points (p*I*) for all of the amino acids. Recall from Section 19.15C that the **isoelectric point is the pH at which an amino acid exists primarily in its neutral form,** and that it can be calculated from the average of the pK_a values of the α-COOH and α-NH_3^+ groups (for neutral amino acids only).

Problem 26.2 What form exists at the isoelectric point of each of the following amino acids: (a) valine; (b) leucine; (c) proline; (d) glutamic acid?

Problem 26.3 Explain why the pK_a of the $-NH_3^+$ group of an α-amino acid is lower than the pK_a of the ammonium ion derived from a 1° amine (RNH_3^+). For example, the pK_a of the $-NH_3^+$ group of alanine is 9.87 but the pK_a of $CH_3NH_3^+$ is 10.63.

Problem 26.4 L-Thyroxine, a thyroid hormone and oral medication used to treat thyroid hormone deficiency, is an amino acid that does not exist in proteins. Draw the zwitterionic form of L-thyroxine.

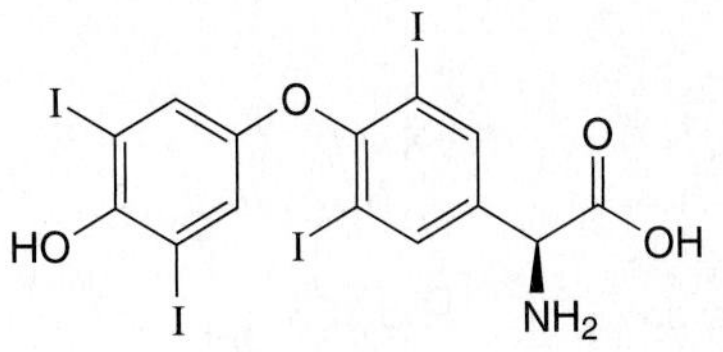

L-thyroxine

26.2 Synthesis of Amino Acids

Amino acids can be prepared in a variety of ways in the laboratory. Three methods are described, each of which is based on reactions learned in previous chapters.

26.2A S_N2 Reaction of α-Halo Acids with NH_3

The most direct way to synthesize an α-amino acid is by **S_N2 reaction of an α-halo carboxylic acid with a large excess of NH_3.**

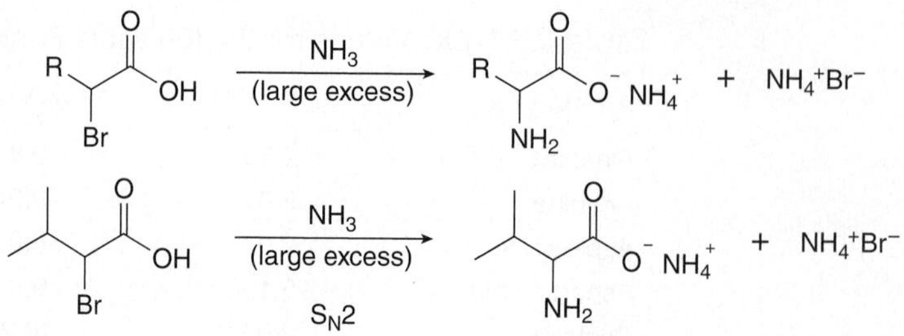

valine

Although the alkylation of ammonia with simple alkyl halides does not generally afford high yields of 1° amines (Section 25.7A), this reaction using α-halo carboxylic acids does form the desired amino acids in good yields. In this case, the amino group in the product is both less basic and more sterically crowded than other 1° amines, so that a single alkylation occurs and the desired amino acid is obtained.

Problem 26.5 What α-halo carbonyl compound is needed to synthesize each amino acid: (a) glycine; (b) isoleucine; (c) phenylalanine?

26.2B Alkylation of a Diethyl Malonate Derivative

The second method for preparing amino acids is based on the malonic ester synthesis. Recall from Section 23.9 that this synthesis converts diethyl malonate to a carboxylic acid with a new alkyl group on its α carbon atom.

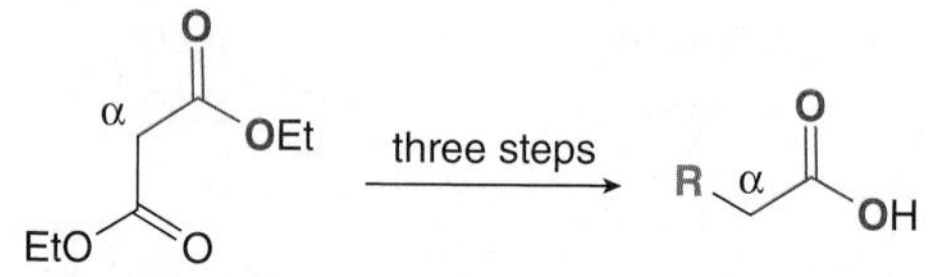

diethyl malonate

This reaction can be adapted to the synthesis of α-amino acids by using a commercially available derivative of diethyl malonate as starting material. This compound, **diethyl acetamidomalonate,** has a nitrogen atom on the α carbon, which ultimately becomes the NH_2 group on the α carbon of the amino acid.

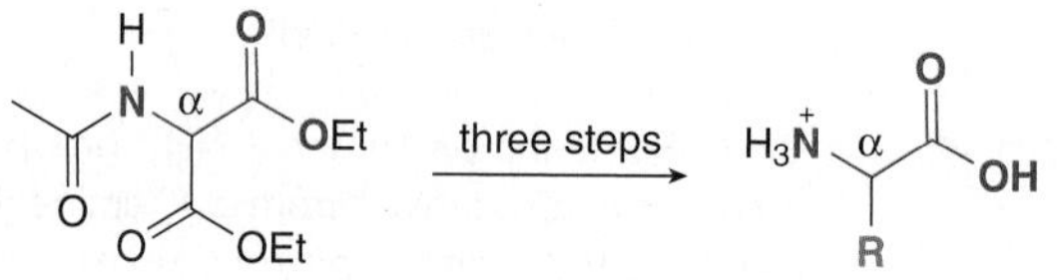

diethyl acetamidomalonate

The malonic ester synthesis consists of three steps, and so does this variation to prepare an amino acid.

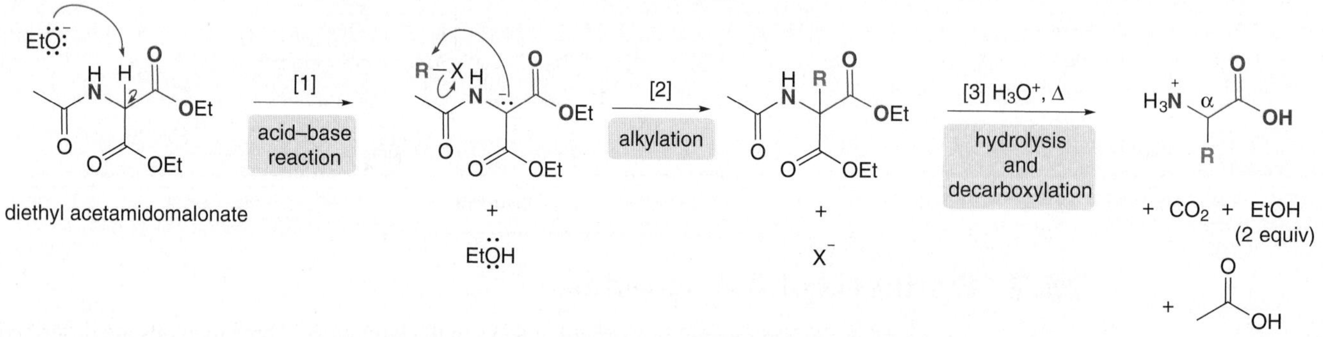

[1] **Deprotonation** of diethyl acetamidomalonate with NaOEt forms an enolate by removal of the acidic proton between the two carbonyl groups.

[2] **Alkylation** of the enolate with an unhindered alkyl halide (usually CH_3X or RCH_2X) forms a substitution product with a new R group on the α carbon.

[3] Heating the alkylation product with aqueous acid results in **hydrolysis** of both esters and the amide, followed by **decarboxylation** to form the amino acid.

Phenylalanine, for example, can be synthesized as follows:

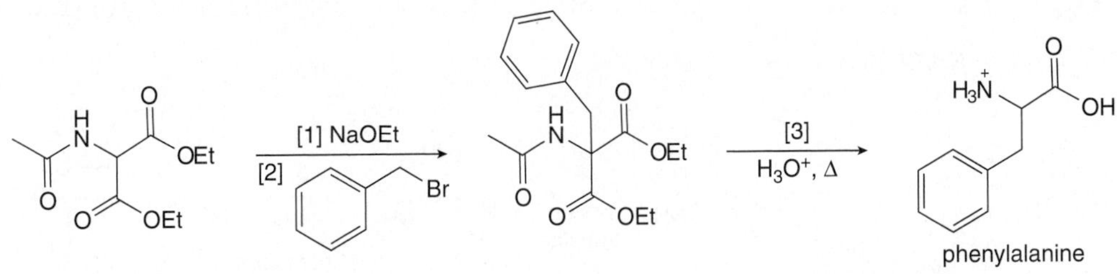
[1] NaOEt
[2]
Br
[3]
H3O+, Δ
phenylalanine

The charge on the amino acid product (+1, –1, or 0) depends on the reaction conditions. Phenylalanine bears a net positive charge because the last step in its synthesis uses strong acid.

Problem 26.6 The enolate derived from diethyl acetamidomalonate is treated with each of the following alkyl halides. After hydrolysis and decarboxylation, what amino acid is formed: (a) CH_3I; (b) $(CH_3)_2CHCH_2Cl$; (c) $CH_3CH_2CH(CH_3)Br$?

Problem 26.7 What amino acid is formed when $CH_3CONHCH(CO_2Et)_2$ is treated with the following series of reagents: [1] NaOEt; [2] $CH_2{=}O$; [3] H_3O^+, Δ?

26.2C Strecker Synthesis

The third method, the **Strecker amino acid synthesis,** converts an aldehyde into an amino acid by a two-step sequence that adds one carbon atom to the aldehyde carbonyl. Treating an aldehyde with NH_4Cl and NaCN first forms an **α-amino nitrile,** which can then be hydrolyzed in aqueous acid to an amino acid.

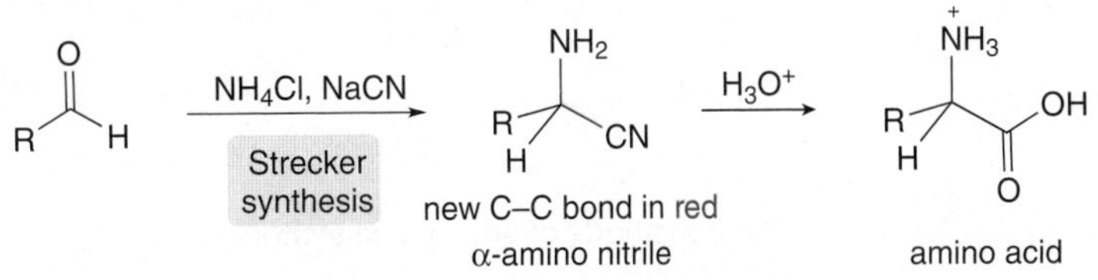
NH4Cl, NaCN
Strecker synthesis
new C–C bond in red
α-amino nitrile
H3O+
amino acid

The Strecker synthesis of alanine, for example, is as follows:

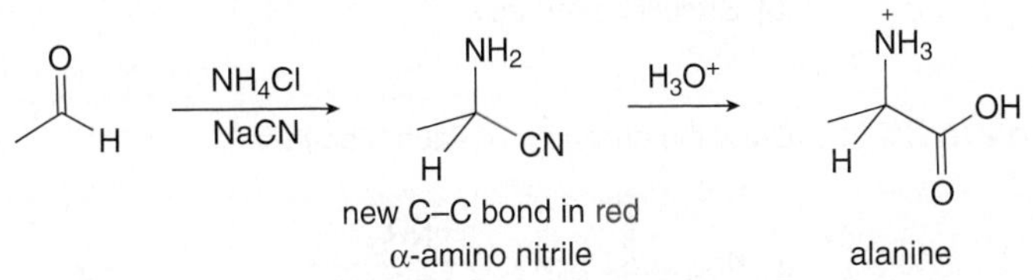
NH4Cl
NaCN
H3O+
new C–C bond in red
α-amino nitrile
alanine

Mechanism 26.1 for the formation of the α-amino nitrile from an aldehyde (the first step in the Strecker synthesis) consists of **nucleophilic addition of NH_3** to form an imine, followed by **addition of cyanide** to the C=N bond. Both parts are related to earlier mechanisms involving imines (Section 21.11) and cyanohydrins (Section 21.9).

The details of the second step of the Strecker synthesis, the hydrolysis of a nitrile (RCN) to a carboxylic acid (RCOOH), have already been presented in Section 22.18A.

Figure 26.4 shows how the amino acid methionine can be prepared by all three methods in Section 26.2.

Problem 26.8 What aldehyde is needed to synthesize each amino acid by the Strecker synthesis: (a) valine; (b) leucine; (c) phenylalanine?

Mechanism 26.1 Formation of an α-Amino Nitrile

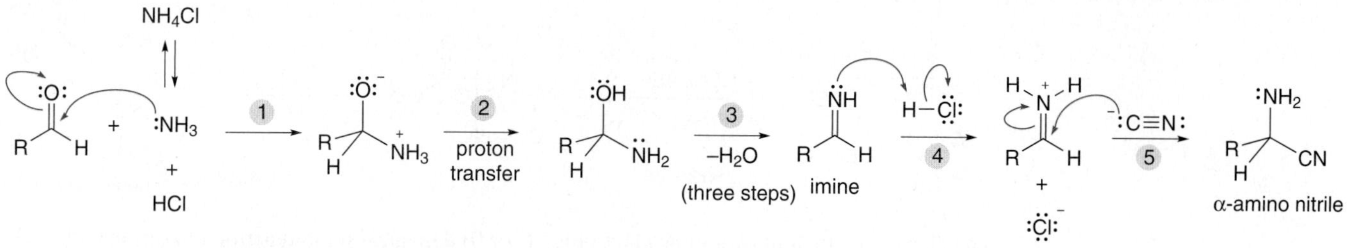
NH4Cl
NH3
HCl
proton transfer
–H2O
(three steps)
imine
α-amino nitrile

1 – 3 Nucleophilic attack of NH_3 followed by proton transfer and loss of H_2O forms an imine. Loss of H_2O occurs by the same three-step process outlined in Mechanism 21.5.

4 – 5 Protonation of the imine followed by nucleophilic attack of ^{-}CN gives the α-amino nitrile.

Figure 26.4
The synthesis of methionine by three different methods

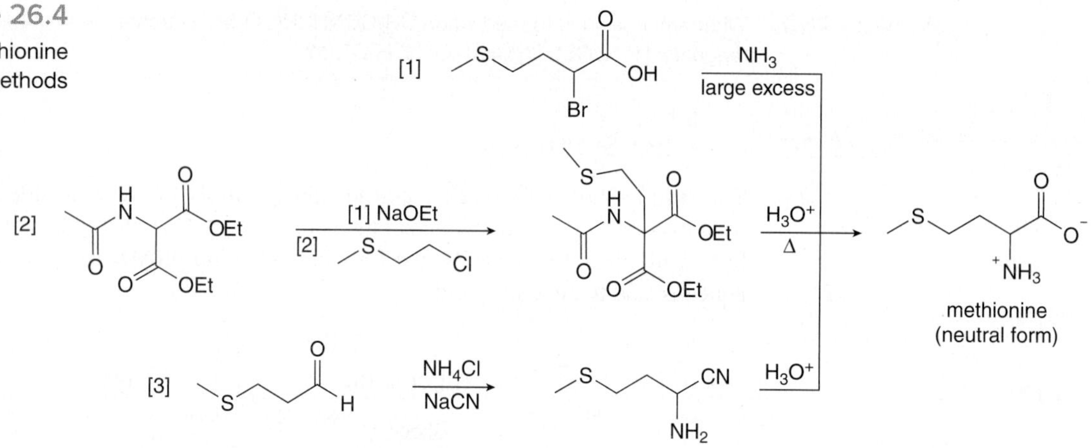
[1]
NH3
large excess
[2]
[1] NaOEt
[2]
H3O+
Δ
methionine
(neutral form)
[3]
NH4Cl
NaCN
H3O+

Three methods of amino acid synthesis:

[1] S_N2 reaction using an α-halo carboxylic acid
[2] Alkylation of diethyl acetamidomalonate
[3] Strecker synthesis

Problem 26.9 Draw the products of each reaction.

a. 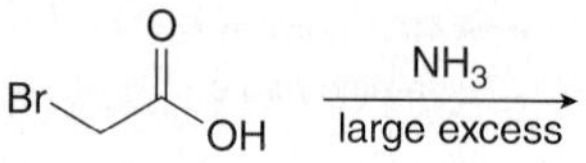
NH3
large excess

c. 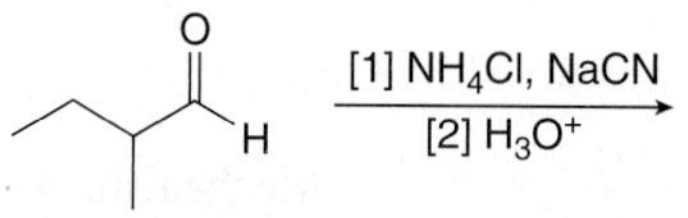
[1] NH4Cl, NaCN
[2] H3O+

b. 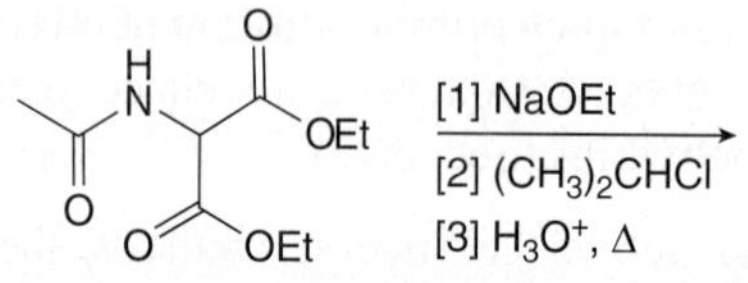
[1] NaOEt
[2] (CH3)2CHCl
[3] H3O+, Δ

d. 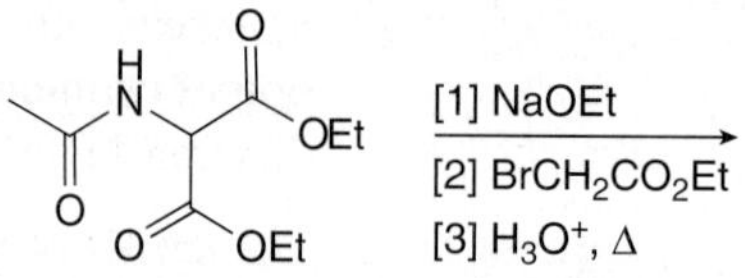
[1] NaOEt
[2] BrCH2CO2Et
[3] H3O+, Δ

26.3 Separation of Amino Acids

No matter which of the preceding methods is used to synthesize an amino acid, all three yield a racemic mixture. Naturally occurring amino acids exist as a single enantiomer, however, so the two enantiomers obtained must be separated if they are to be used in biological applications. This is not an easy task. Two enantiomers have the same physical properties, so they cannot be separated by common physical methods, such as distillation or chromatography. Moreover, they react in the same way with achiral reagents, so they cannot be separated by chemical reactions either.

Nonetheless, strategies have been devised to separate two enantiomers using physical separation techniques and chemical reactions. We examine two different strategies in Section 26.3. Then, in Section 26.4, we will discuss a method that affords optically active amino acids without the need for separation.

- The separation of a racemic mixture into its component enantiomers is called *resolution.* Thus, a racemic mixture is *resolved* into its component enantiomers.

26.3A Resolution of Amino Acids

The oldest, and perhaps still the most widely used method to separate enantiomers, exploits the following fact: **enantiomers have the *same* physical properties, but diastereomers have *different* physical properties.** Thus, a racemic mixture can be resolved using the following general strategy.

[1] **Convert a pair of enantiomers into a pair of diastereomers,** which are now separable because they have different melting points and boiling points.

[2] **Separate the diastereomers.**

[3] **Re-convert each diastereomer into the original enantiomer,** now separated from the other.

This general three-step process is illustrated in Figure 26.5.

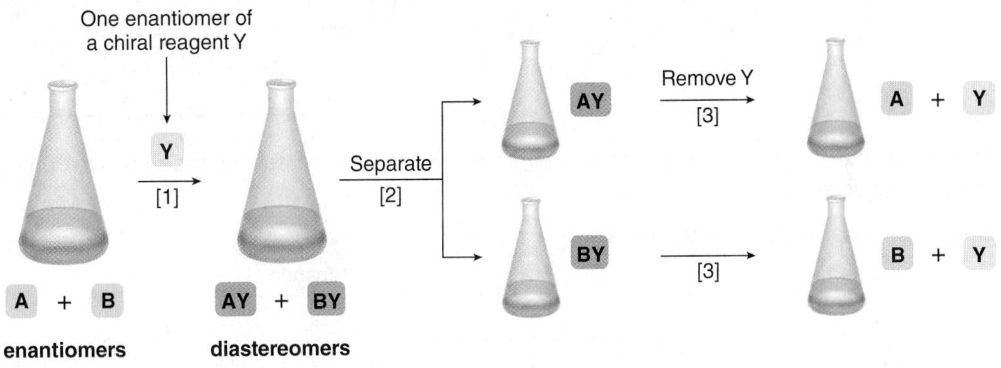

Figure 26.5 Resolution of a racemic mixture by converting it to a mixture of diastereomers

Enantiomers A and B can be separated by reaction with a single enantiomer of a chiral reagent, Y. The process of resolution requires three steps:

[1] Reaction of enantiomers **A** and **B** with **Y** forms two diastereomers, **AY** and **BY.**

[2] Diastereomers **AY** and **BY** have different physical properties, so they can be separated by physical methods such as fractional distillation or crystallization.

[3] **AY** and **BY** are then re-converted to **A** and **B** by a chemical reaction. The two enantiomers **A** and **B** are now separated from each other, and resolution is complete.

To resolve a racemic mixture of amino acids such as (*R*)- and (*S*)-alanine, the racemate is first treated with acetic anhydride to form ***N*-acetyl amino acids.** Each of these amides contains one stereogenic center and they are still enantiomers, so they are *still inseparable.*

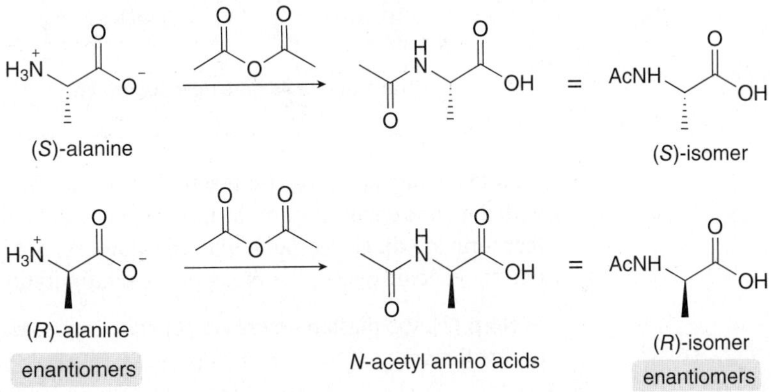

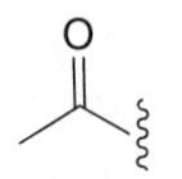

acetyl
Ac

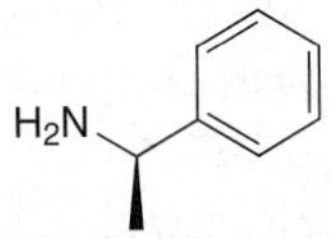

(*R*)-α-methylbenzylamine

a resolving agent

Both enantiomers of *N*-acetyl alanine have a free carboxy group that can react with an amine in an acid–base reaction. **If a chiral amine is used, such as (*R*)-α-methylbenzylamine, the two**

salts formed are diastereomers, *not* enantiomers. Diastereomers can be physically separated from each other, so the compound that converts enantiomers into diastereomers is called a **resolving agent.** Either enantiomer of the resolving agent can be used.

How To Use (*R*)-α-Methylbenzylamine to Resolve a Racemic Mixture of Amino Acids

Step [1] React both enantiomers of an *N*-acetyl amino acid with the *R* isomer of the chiral amine.

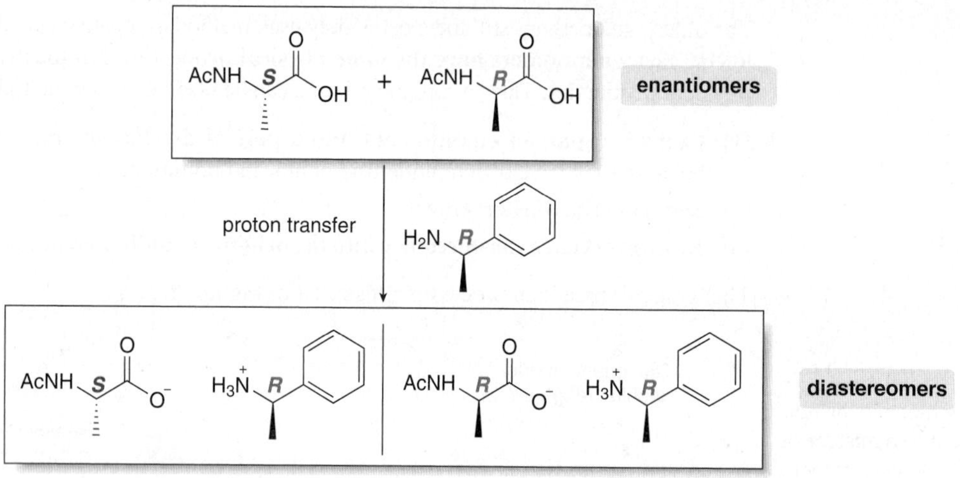

These salts have the *same* configuration around one stereogenic center, but the *opposite* configuration about the other stereogenic center.

Step [2] Separate the diastereomers.

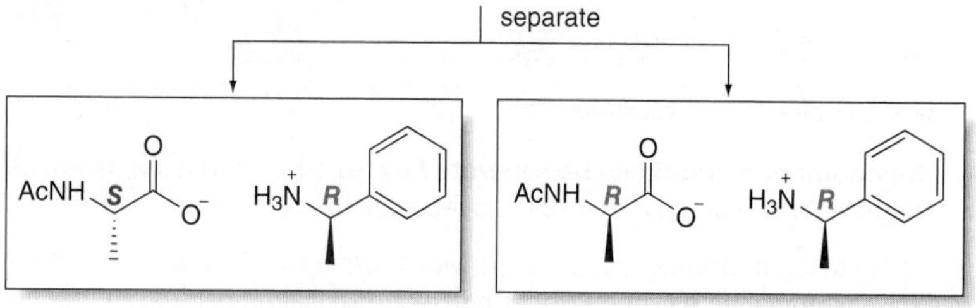

Step [3] Regenerate the amino acid by hydrolysis of the amide.

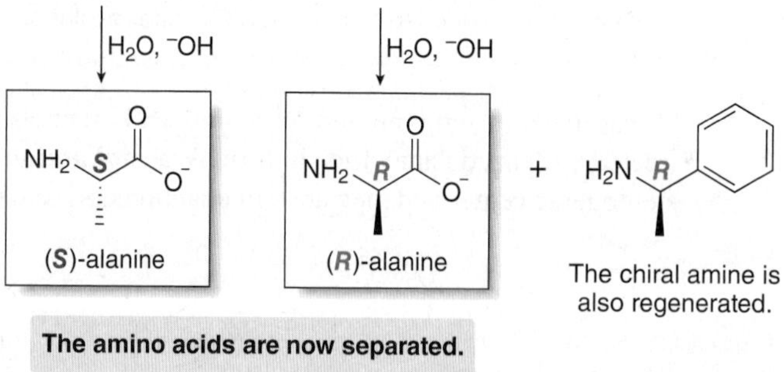

Step [1] is just an acid–base reaction in which the racemic mixture of *N*-acetyl alanines reacts with the same enantiomer of the resolving agent, in this case (*R*)-α-methylbenzylamine. The salts that form are **diastereomers, *not* enantiomers,** because they have the same configuration about one stereogenic center, but the opposite configuration about the other stereogenic center.

In **Step [2],** the diastereomers are separated by some physical technique, such as crystallization or distillation.

In **Step [3],** the amides can be hydrolyzed with aqueous base to regenerate the amino acids. The amino acids are now separated from each other. The optical activity of the amino acids can be measured and compared to their known rotations to determine the purity of each enantiomer.

Problem 26.10 Which of the following amines can be used to resolve a racemic mixture of amino acids?

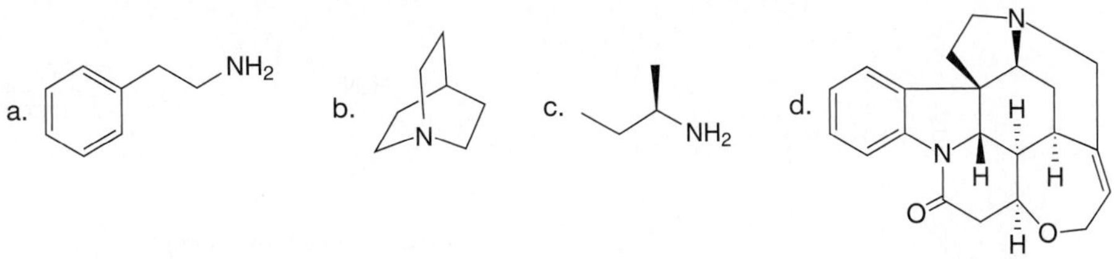

strychnine
(a powerful poison)

Problem 26.11 Write out a stepwise sequence that shows how a racemic mixture of leucine enantiomers can be resolved into optically active amino acids using (*R*)-α-methylbenzylamine.

26.3B Kinetic Resolution of Amino Acids Using Enzymes

A second strategy used to separate amino acids is based on the fact that two enantiomers react differently with chiral reagents. An **enzyme** is typically used as the chiral reagent.

To illustrate this strategy, we begin again with the two enantiomers of *N*-acetyl alanine, which were prepared by treating a racemic mixture of (*R*)- and (*S*)-alanine with acetic anhydride (Section 26.3A). Enzymes called **acylases** hydrolyze amide bonds, such as those found in *N*-acetyl alanine, but only for amides of L-amino acids. Thus, when a racemic mixture of *N*-acetyl alanines is treated with an acylase, only the amide of L-alanine (the *S* stereoisomer) is hydrolyzed to generate L-alanine, whereas the amide of D-alanine (the *R* stereoisomer) is untouched. The reaction mixture now consists of one amino acid and one *N*-acetyl amino acid. Because they have different functional groups with different physical properties, they can be physically separated.

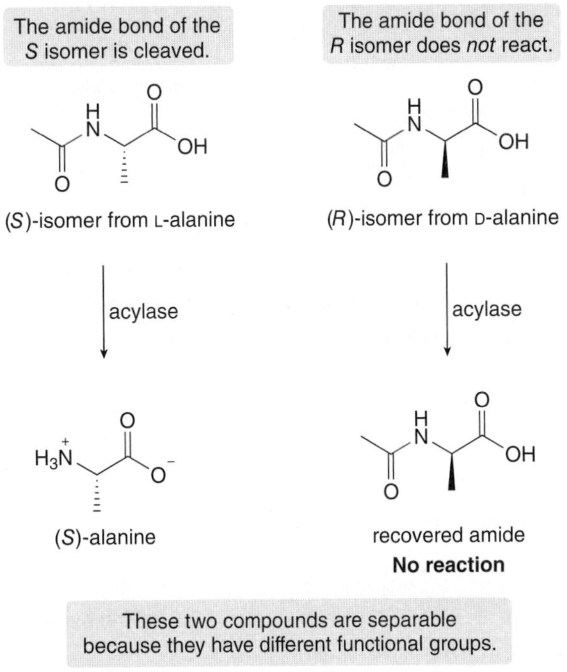

- **Separation of two enantiomers by a chemical reaction that selectively occurs for only one of the enantiomers is called *kinetic resolution*.**

Problem 26.12 Draw the organic products formed in the following reaction.

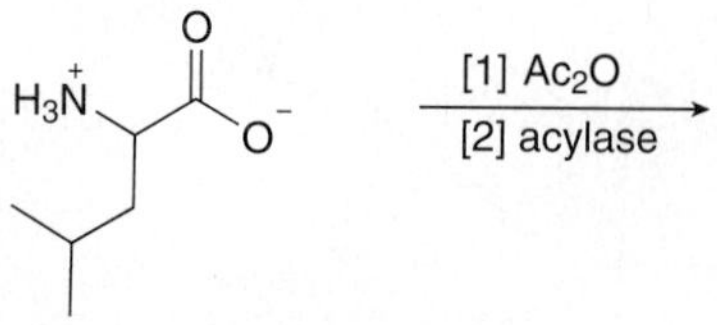

(mixture of enantiomers)

26.4 Enantioselective Synthesis of Amino Acids

Although the two methods introduced in Section 26.3 for resolving racemic mixtures of amino acids make enantiomerically pure amino acids available for further research, half of the reaction product is useless because it has the undesired configuration. Moreover, each of these procedures is costly and time-consuming.

If we use a chiral reagent to synthesize an amino acid, however, it is possible to favor the formation of the desired enantiomer over the other, without having to resort to a resolution. For example, single enantiomers of amino acids have been prepared by using **enantioselective (or asymmetric) hydrogenation reactions.** The success of this approach depends on finding a chiral catalyst, in much the same way that a chiral catalyst is used for the Sharpless asymmetric epoxidation (Section 12.15).

The necessary starting material is an alkene. Addition of H_2 to the double bond forms an *N*-acetyl amino acid with a new stereogenic center on the α carbon to the carboxy group. With proper choice of a chiral catalyst, the naturally occurring *S* configuration can be obtained as product.

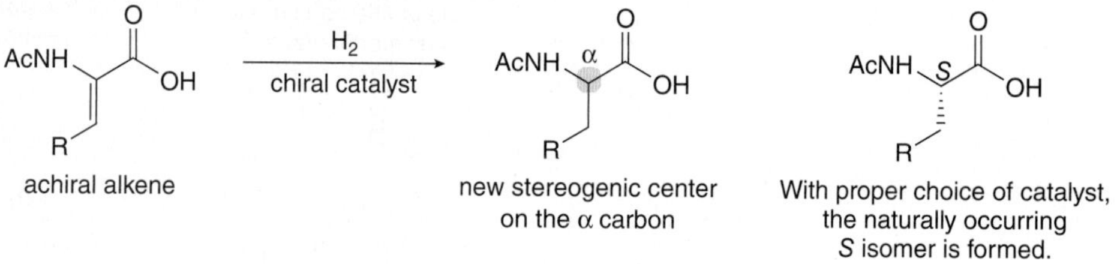

Several chiral catalysts with complex structures have now been developed for this purpose. Many contain **rhodium** as the metal, complexed to a chiral molecule containing one or more phosphorus atoms. One example, abbreviated simply as **Rh***, is drawn below.

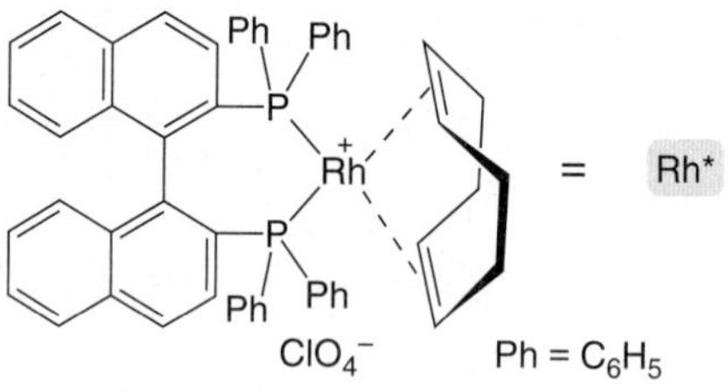

chiral hydrogenation catalyst

Figure 26.6
The structure of BINAP

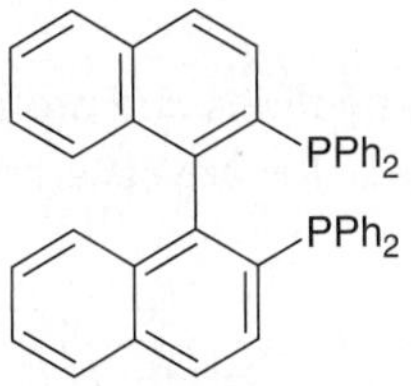

2,2'-bis(diphenylphosphino)-1,1'-binaphthyl

BINAP

3-D model of BINAP

- The two naphthalene rings are oriented at right angles to each other, creating a rigid shape that makes the molecule chiral.

Ryoji Noyori shared the 2001 Nobel Prize in Chemistry for developing methods for asymmetric hydrogenation reactions using the chiral BINAP catalyst.

This catalyst is synthesized from a rhodium salt and a phosphorus compound, 2,2'-bis(diphenylphosphino)-1,1'-binaphthyl **(BINAP).** It is the BINAP moiety (Figure 26.6) that makes the catalyst chiral.

BINAP is one of a small number of molecules that is chiral even though it has no tetrahedral stereogenic centers. Its shape makes it a chiral molecule. The two naphthalene rings of the BINAP molecule are oriented at almost 90° to each other to minimize steric interactions between the hydrogen atoms on adjacent rings. This rigid three-dimensional shape makes BINAP non-superimposable on its mirror image, and thus it is a chiral compound.

Enantioselective hydrogenation can be used to synthesize a single stereoisomer of phenylalanine. Treating achiral alkene **A** with H_2 and the chiral rhodium catalyst Rh* forms the *S* isomer of *N*-acetyl phenylalanine in 100% *ee*. Hydrolysis of the acetyl group on nitrogen then yields a single enantiomer of phenylalanine.

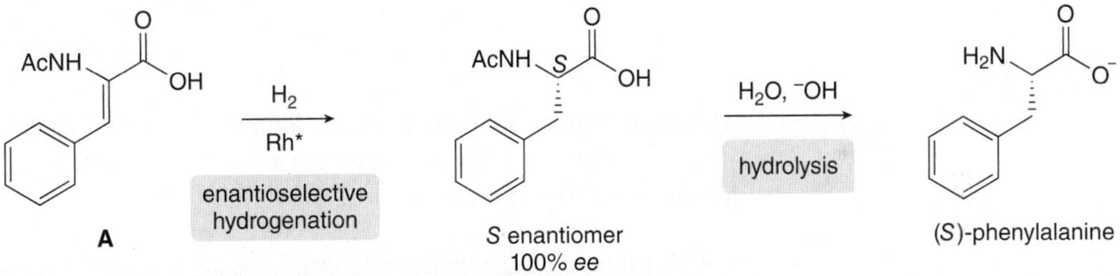

Problem 26.13 What alkene is needed to synthesize each amino acid by an enantioselective hydrogenation reaction using H_2 and Rh*: (a) alanine; (b) leucine; (c) glutamine?

26.5 Peptides

When amino acids are joined together by amide bonds, they form larger molecules called **peptides** and **proteins.**

- A *dipeptide* has two amino acids joined together by *one* amide bond.
- A *tripeptide* has three amino acids joined together by *two* amide bonds.

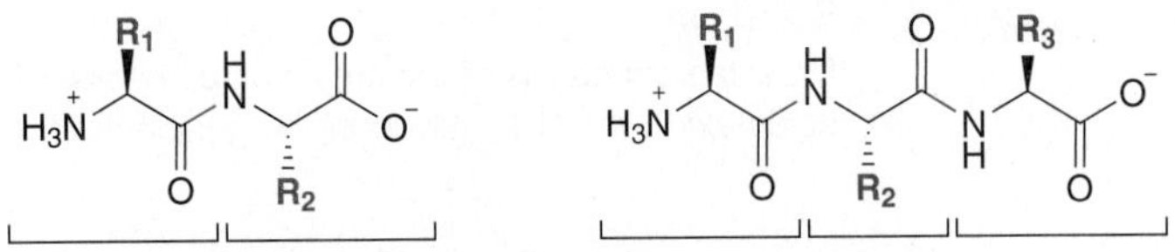

dipeptide
Two amino acids joined together.

tripeptide
Three amino acids joined together.

[Amide bonds are drawn in red.]

Polypeptides and **proteins** both have many amino acids joined together in long linear chains, but the term **protein** is usually reserved for polymers of more than 40 amino acids.

- The amide bonds in peptides and proteins are called *peptide bonds.*
- The individual amino acids are called *amino acid residues.*

26.5A Simple Peptides

To form a dipeptide, the amino group of one amino acid forms an amide bond with the carboxy group of another amino acid. Because each amino acid has both an amino group and a carboxy group, **two different dipeptides can be formed.** This is illustrated with alanine and cysteine.

[1] **The COO^- group of alanine can combine with the NH_3^+ group of cysteine.**

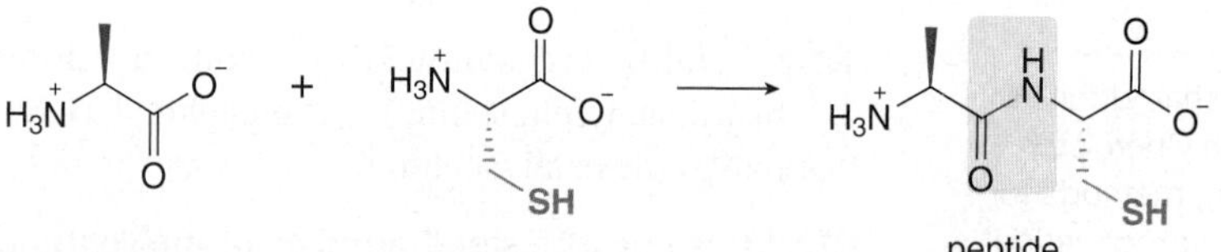
alanine
cysteine
peptide bond
Ala–Cys

[2] **The COO^- group of cysteine can combine with the NH_3^+ group of alanine.**

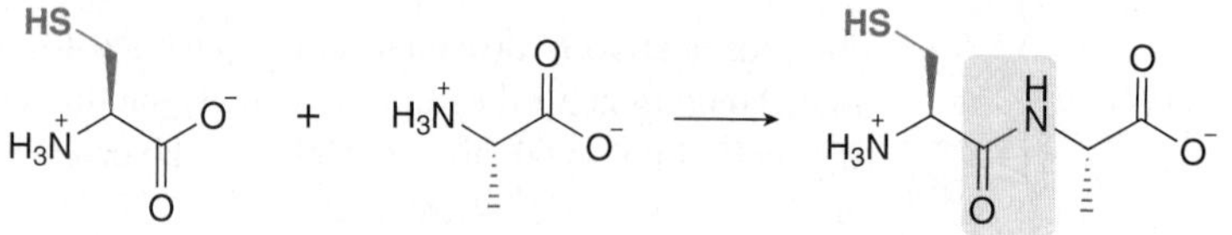
cysteine
alanine
peptide bond
Cys–Ala

These compounds are **constitutional isomers** of each other. Both have a free amino group (protonated as NH_3^+) at one end of their chains and a free carboxy group (deprotonated as a carboxylate anion, COO^-) at the other.

- The amino acid with the free amino group is called the *N-terminal amino acid.*
- The amino acid with the free carboxy group is called the *C-terminal amino acid.*

By convention, **the N-terminal amino acid is always written at the left end of the chain and the C-terminal amino acid at the right.** The peptide can be abbreviated by writing the one- or three-letter symbols for the amino acids in the chain from the N-terminal to the C-terminal end. Thus, Ala–Cys has alanine at the N-terminal end and cysteine at the C-terminal end, whereas Cys–Ala has cysteine at the N-terminal end and alanine at the C-terminal end. Sample Problem 26.1 shows how this convention applies to a tripeptide.

Sample Problem 26.1 Draw the structure of the following tripeptide, and label its N-terminal and C-terminal amino acids: Ala–Gly–Ser.

Solution

Draw the structures of the amino acids in order from **left to right,** placing the COO^- of one amino acid *next* to the NH_3^+ group of the adjacent amino acid. Always draw the **NH_3^+ group on the *left*** and the **COO^- group on the *right*.** Then, join adjacent COO^- and NH_3^+ groups together in amide bonds to form the tripeptide.

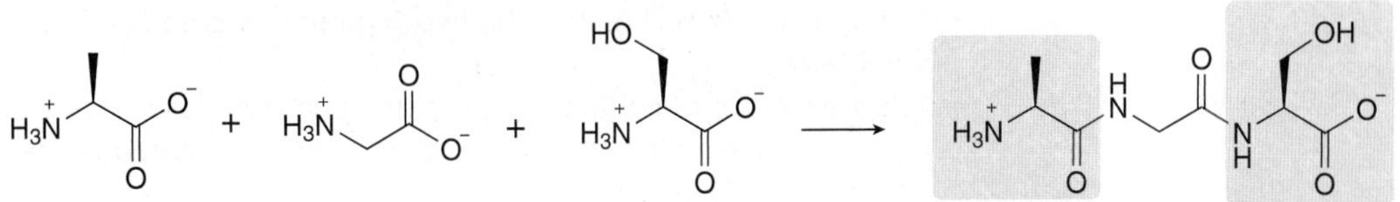

Ala Gly Ser

N-terminal amino acid C-terminal amino acid

tripeptide **Ala–Gly–Ser**
[The new peptide bonds are drawn in red.]

The N-terminal amino acid is **alanine,** and the C-terminal amino acid is **serine.**

The tripeptide in Sample Problem 26.1 has one N-terminal amino acid, one C-terminal amino acid, and two peptide bonds.

- **No matter how many amino acid residues are present, there is only *one* N-terminal amino acid and *one* C-terminal amino acid.**
- **For *n* amino acids in the chain, the number of amide bonds is *n* – 1.**

Problem 26.14 Draw the structure of each peptide. Label the N-terminal and C-terminal amino acids and all amide bonds.

a. Val–Glu b. Gly–His–Leu c. M–A–T–T

Problem 26.15 Name each peptide using both the one-letter and the three-letter abbreviations for the names of the component amino acids.

a. 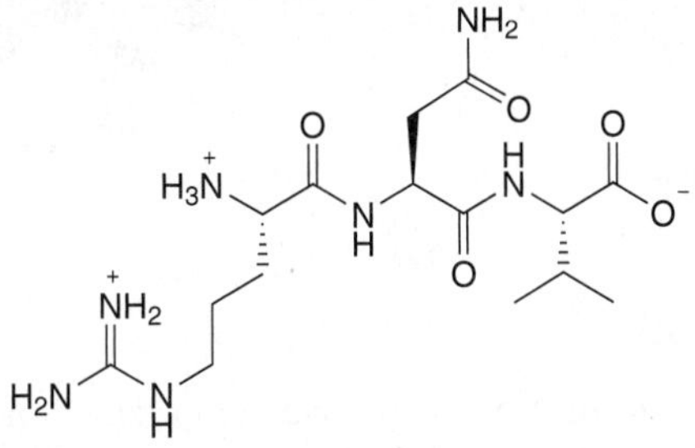

b. 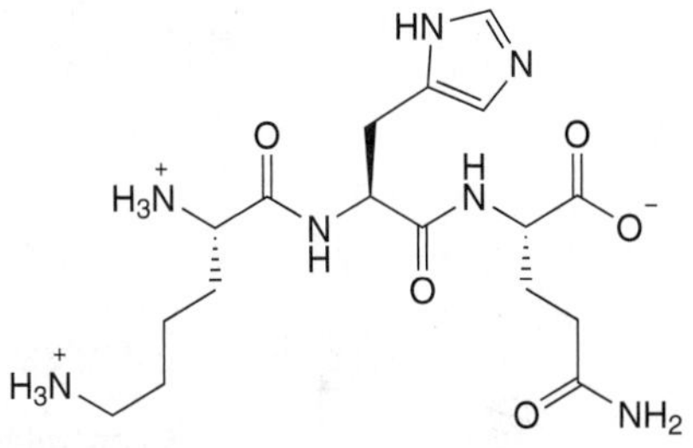

Problem 26.16 How many different tripeptides can be formed from three different amino acids?

26.5B The Peptide Bond

The carbonyl carbon of an amide is ***sp*² hybridized** and has **trigonal planar** geometry. A second resonance structure can be drawn that delocalizes the nonbonded electron pair on the N atom. Amides are more resonance stabilized than other acyl compounds, so the resonance structure having the C=N makes a significant contribution to the hybrid.

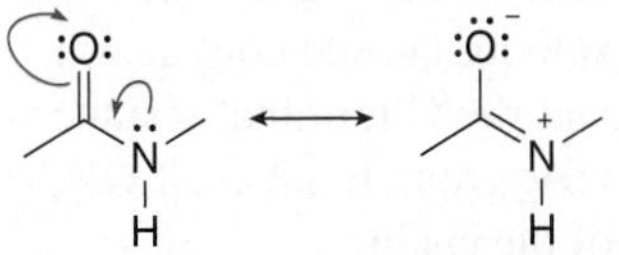

two resonance structures for the peptide bond

Recall from Section 16.6 that buta-1,3-diene can also exist as *s*-cis and *s*-trans conformations. In buta-1,3-diene, the *s*-cis conformation has the two double bonds on the same side of the single bond (dihedral angle = 0°), whereas the *s*-trans conformation has them on opposite sides (dihedral angle = 180°).

Resonance stabilization has important consequences. Rotation about the C–N bond is restricted because it has partial double bond character. As a result, there are two possible conformations.

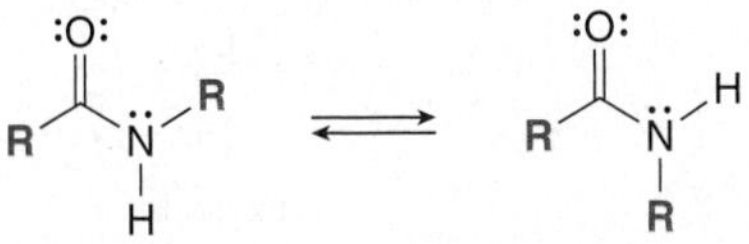

s-trans *s*-cis

- **The *s*-trans conformation has the two R groups oriented on *opposite* sides of the C–N bond.**
- **The *s*-cis conformation has the two R groups oriented on the *same* side of the C–N bond.**
- **The *s*-trans conformation of a peptide bond is typically more stable than the *s*-cis, because the *s*-trans has the two bulky R groups located farther from each other.**

The planar geometry of the peptide bond is analogous to the planar geometry of ethylene (or any other alkene), where the double bond between sp^2 hybridized carbon atoms makes all of the bond angles ~120° and puts all six atoms in the same plane.

A second consequence of resonance stabilization is that **all six atoms involved in the peptide bond lie in the same plane.** All bond angles are ~120°, and the C=O and N–H bonds are oriented 180° from each other.

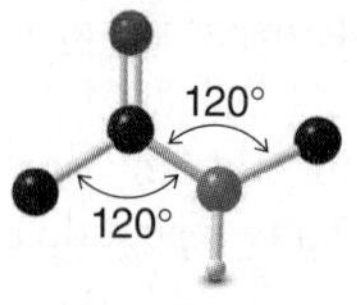
120°
120°

These six atoms lie in a plane.

The structure of a tetrapeptide illustrates the results of these effects in a long peptide chain.

- **The *s*-trans arrangement makes a long chain with a zigzag arrangement.**
- **In each peptide bond, the N–H and C=O bonds lie parallel and at 180° with respect to each other.**

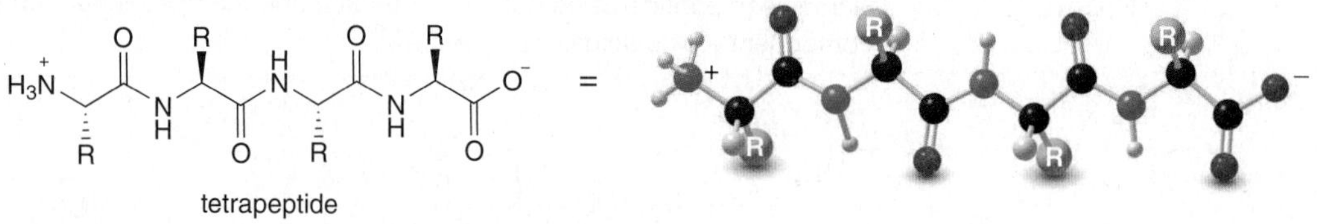

tetrapeptide

26.5C Interesting Peptides

Even relatively simple peptides can have important biological functions. **Bradykinin,** for example, is a peptide hormone composed of nine amino acids. It stimulates smooth muscle contraction, dilates blood vessels, and causes pain. Bradykinin is a component of bee venom.

Arg–Pro–Pro–Gly–Phe–Ser–Pro–Phe–Arg

bradykinin

Oxytocin and **vasopressin** are nonapeptide hormones, too. Their sequences are identical except for two amino acids, yet this is enough to give them very different biological activities. Oxytocin induces labor by stimulating the contraction of uterine muscles, and it stimulates the flow of milk in nursing mothers. Vasopressin, on the other hand, controls blood pressure by regulating smooth muscle contraction. The N-terminal amino acid in both hormones is a cysteine residue, and the C-terminal residue is glycine. Instead of a free carboxy group, both peptides have an NH_2 group in place of OH, so this is indicated with the additional NH_2 group drawn at the end of the chain.

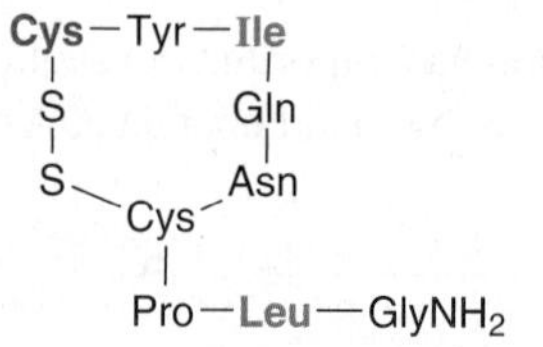
Cys—Tyr—Ile
S
Gln
S
Asn
Cys
Pro—Leu—GlyNH2

oxytocin

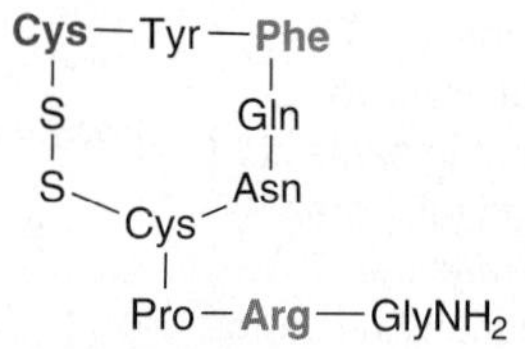
Cys—Tyr—Phe
S
Gln
S
Asn
Cys
Pro—Arg—GlyNH2

vasopressin

- The N-terminal amino acid is labeled in red.
- The amino acids that differ are labeled in blue.

Figure 26.7
Three-dimensional structures of vasopressin and oxytocin

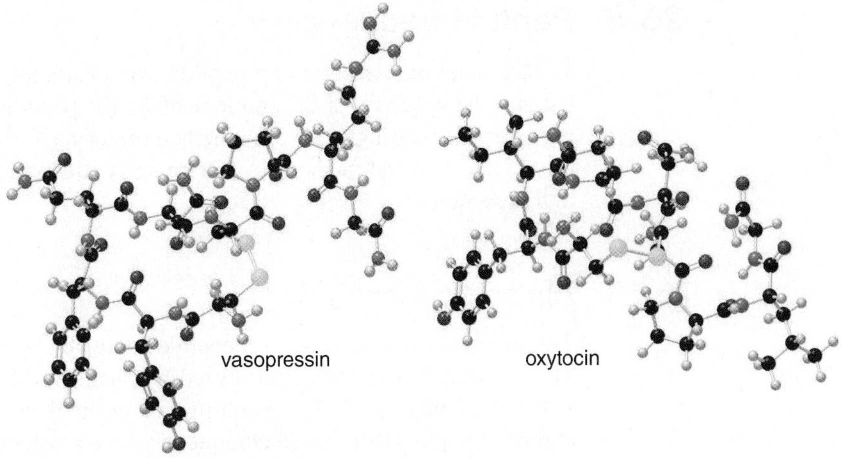

The oxidation of thiols to disulfides was discussed in Section 9.15.

The structure of both peptides includes a **disulfide bond,** a form of covalent bonding in which the –SH groups from two cysteine residues are oxidized to form a sulfur–sulfur bond. In oxytocin and vasopressin, the disulfide bonds make the peptides cyclic. Three-dimensional structures of oxytocin and vasopressin are shown in Figure 26.7.

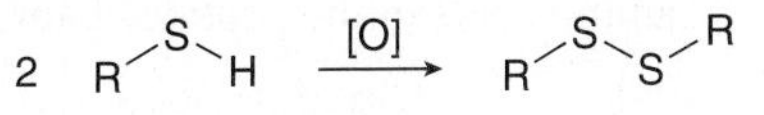

The artificial sweetener **aspartame** (Figure 27.10) is the methyl ester of the dipeptide Asp–Phe. This synthetic peptide is 180 times sweeter (on a gram-for-gram basis) than sucrose (common table sugar). Both of the amino acids in aspartame have the naturally occurring L-configuration. If the D-amino acid is substituted for either Asp or Phe, the resulting compound tastes bitter.

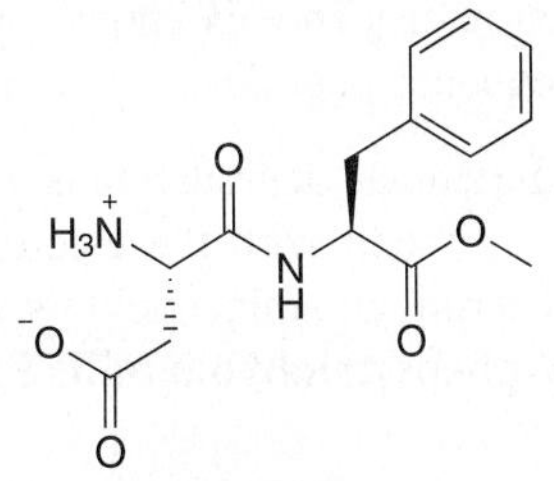

aspartame
the methyl ester of Asp–Phe
a synthetic artificial sweetener

Problem 26.17 Draw the structure of leu-enkephalin, a pentapeptide that acts as an analgesic and opiate, and has the following sequence: Tyr–Gly–Gly–Phe–Leu. (The structure of a related peptide, met-enkephalin, appeared in Section 22.6B.)

Problem 26.18 Glutathione, a powerful antioxidant that destroys harmful oxidizing agents in cells, is composed of glutamic acid, cysteine, and glycine, and has the following structure:

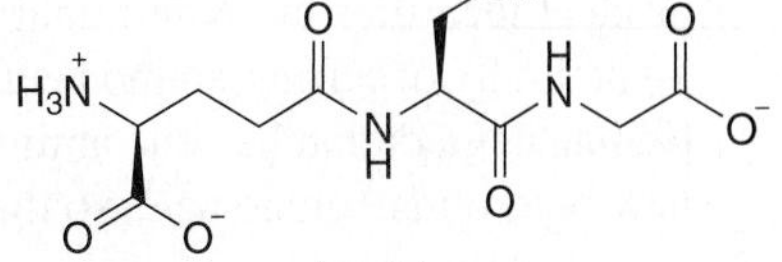

glutathione

a. What product is formed when glutathione reacts with an oxidizing agent?
b. What is unusual about the peptide bond between glutamic acid and cysteine?

26.6 Peptide Sequencing

To determine the structure of a peptide, we must know not only what amino acids comprise it, but also the sequence of the amino acids in the peptide chain. Although mass spectrometry has become an increasingly powerful method for the analysis of high molecular weight proteins (Section 13.4), chemical methods to determine peptide structure are still widely used and presented in this section.

26.6A Amino Acid Analysis

The structure determination of a peptide begins by analyzing the total amino acid composition. The amide bonds are first hydrolyzed by heating with hydrochloric acid for 24 h to form the individual amino acids. The resulting mixture is then separated using high-performance liquid chromatography (HPLC), a technique in which a solution of amino acids is placed on a column and individual amino acids move through the column at characteristic rates, often dependent on polarity.

This process determines both the identity of the individual amino acids and the amount of each present, but it tells nothing about the order of the amino acids in the peptide. For example, complete hydrolysis and HPLC analysis of the tetrapeptide Gly–Gly–Phe–Tyr would indicate the presence of three amino acids—glycine, phenylalanine, and tyrosine—and show that there are twice as many glycine residues as phenylalanine or tyrosine residues. The exact order of the amino acids in the peptide chain must then be determined by additional methods

26.6B Identifying the N-Terminal Amino Acid—The Edman Degradation

To determine the sequence of amino acids in a peptide chain, a variety of procedures are often combined. One especially useful technique is to identify the N-terminal amino acid using the **Edman degradation.** In the Edman degradation, amino acids are cleaved one at a time from the N-terminal end, the identity of the amino acid determined, and the process repeated until the entire sequence is known. Automated sequencers using this methodology are now available to sequence peptides containing up to about 50 amino acids.

The Edman degradation is based on the reaction of the nucleophilic NH_2 group of the N-terminal amino acid with the electrophilic carbon of phenyl isothiocyanate, $C_6H_5N{=}C{=}S$. When the N-terminal amino acid is removed from the peptide chain, two products are formed: **an *N*-phenylthiohydantoin (PTH) and a new peptide with one *fewer* amino acid.**

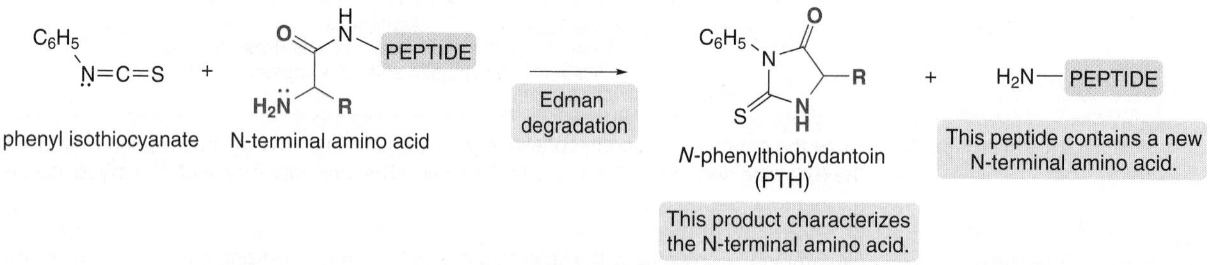

The *N*-phenylthiohydantoin derivative contains the atoms of the N-terminal amino acid. This product identifies the N-terminal amino acid in the peptide because the PTH derivatives of all 20 naturally occurring amino acids are known and characterized. The new peptide formed in the Edman degradation has one amino acid fewer than the original peptide. Moreover, it contains a new N-terminal amino acid, so the process can be repeated.

Mechanism 26.2 illustrates some of the key steps of the Edman degradation. The nucleophilic N-terminal NH_2 group adds to the electrophilic carbon of phenyl isothiocyanate to form an *N*-phenylthiourea, the product of nucleophilic addition (Part [1]). Intramolecular cyclization followed by elimination results in cleavage of the terminal amide bond in Part [2] to form a new peptide with one fewer amino acid. A sulfur heterocycle, called a thiazolinone, is also formed, which rearranges by a multistep pathway to form an *N*-phenylthiohydantoin. The R group in this product identifies the amino acid located at the N-terminal end.

Mechanism 26.2 Edman Degradation

Part [1] Formation of an *N*-phenylthiourea

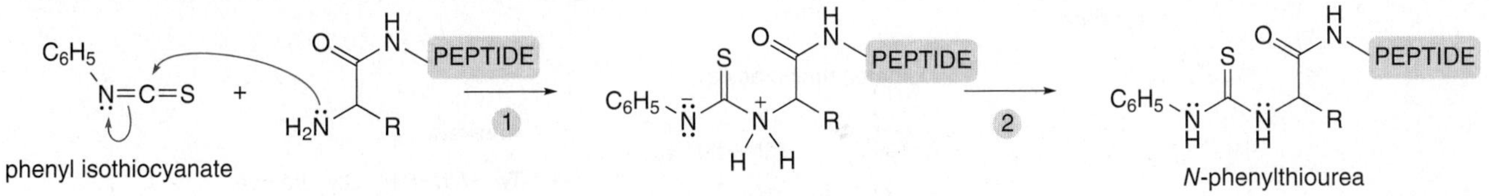

1 – 2 Addition of the amino group of the N-terminal amino acid to phenyl isothiocyanate followed by proton transfer forms an *N*-phenylthiourea.

Part [2] Formation of the N-terminal amino acid and *N*-phenylthiohydantoin (PTH)

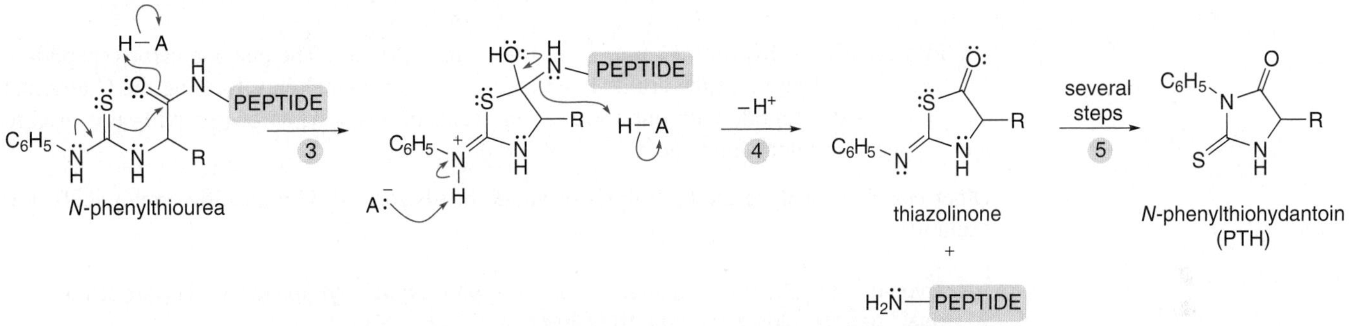

3 Nucleophilic addition of the S atom to the amide carbonyl forms a five-membered ring.

4 Loss of the amino group forms two products—a thiazolinone ring and a peptide chain that contains one fewer amino acid than the original peptide.

5 The thiazolinone rearranges by a multistep pathway to form an *N*-phenylthiohydantoin (PTH) that contains the original amino acid.

In theory a protein of any length can be sequenced using the Edman degradation, but in practice, the accumulation of small quantities of unwanted by-products limits sequencing to proteins having fewer than approximately 50 amino acids.

Problem 26.19 Draw the structure of the *N*-phenylthiohydantoin formed by initial Edman degradation of each peptide: (a) Ala–Gly–Phe–Phe; (b) Val–Ile–Tyr.

26.6C Partial Hydrolysis of a Peptide

Additional structural information can be obtained by cleaving some, but not all, of the amide bonds in a peptide. Partial hydrolysis of a peptide with acid forms smaller fragments in a random fashion. Sequencing these peptides and identifying sites of overlap can be used to determine the sequence of the complete peptide, as shown in Sample Problem 26.2.

Sample Problem 26.2 Give the amino acid sequence of a hexapeptide that contains the amino acids Ala, Val, Ser, Ile, Gly, Tyr, and forms the following fragments when partially hydrolyzed with HCl: Gly–Ile–Val, Ala–Ser–Gly, and Tyr–Ala.

Solution

Looking for points of overlap in the sequences of the smaller fragments shows how the fragments should be pieced together. In this example, the fragment Ala–Ser–Gly contains amino acids common to the two other fragments, thus showing how the three fragments can be joined together.

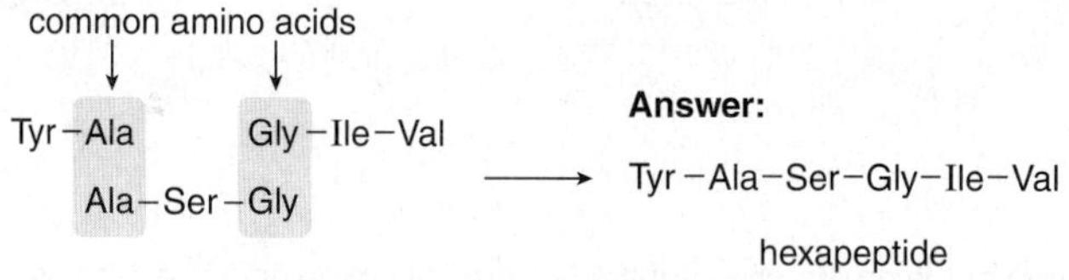

Problem 26.20 Give the amino acid sequence of an octapeptide that contains the amino acids Tyr, Ala, Leu (2 equiv), Cys, Gly, Glu, and Val, and forms the following fragments when partially hydrolyzed with HCl: Val–Cys–Gly–Glu, Ala–Leu–Tyr, and Tyr–Leu–Val–Cys.

Peptides can also be hydrolyzed at specific sites using enzymes. The enzyme carboxypeptidase catalyzes the hydrolysis of the amide bond nearest the C-terminal end, forming the C-terminal amino acid and a peptide with one fewer amino acid. In this way, carboxypeptidase is used to identify the C-terminal amino acid.

Other enzymes catalyze the hydrolysis of amide bonds formed with specific amino acids. For example:

- **Trypsin catalyzes the hydrolysis of amides with a carbonyl group that is part of the basic amino acids arginine and lysine.**
- **Chymotrypsin hydrolyzes amides with carbonyl groups that are part of the aromatic amino acids phenylalanine, tyrosine, and tryptophan.**

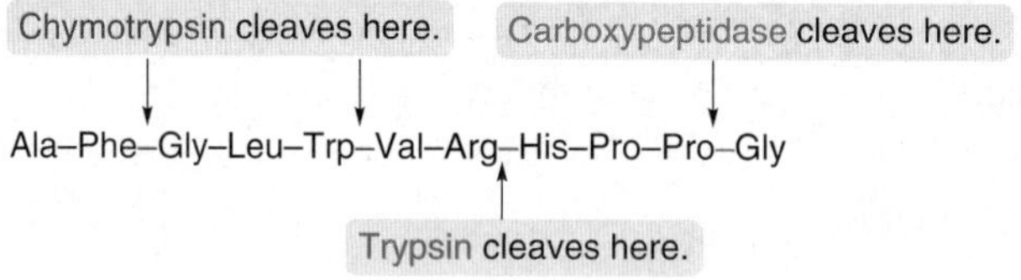

Table 26.2 summarizes these enzyme specificities used in peptide sequencing.

Table 26.2 Cleavage Sites of Specific Enzymes in Peptide Sequencing

Enzyme	Site of cleavage
Carboxypeptidase	Amide bond nearest to the C-terminal amino acid
Chymotrypsin	Amide bond with a carbonyl group from Phe, Tyr, or Trp
Trypsin	Amide bond with a carbonyl group from Arg or Lys

Problem 26.21 (a) What products are formed when each peptide is treated with trypsin? (b) What products are formed when each peptide is treated with chymotrypsin?

[1] Gly–Ala–Phe–Leu–Lys–Ala
[2] Phe–Tyr–Gly–Cys–Arg–Ser
[3] Thr–Pro–Lys–Glu–His–Gly–Phe–Cys–Trp–Val–Val–Phe

Sample Problem 26.3 Deduce the sequence of a pentapeptide that contains the amino acids Ala, Glu, Gly, Ser, and Tyr, from the following experimental data. Edman degradation cleaves Gly from the pentapeptide, and carboxypeptidase forms Ala and a tetrapeptide. Treatment of the pentapeptide with chymotrypsin forms a dipeptide and a tripeptide. Partial hydrolysis forms Gly, Ser, and the tripeptide Tyr–Glu–Ala.

Solution

Use each result to determine the location of an amino acid in the pentapeptide.

Experiment		Result
• Edman degradation identifies the N-terminal amino acid—in this case, Gly.	→	Gly–__–__–__–__
• Carboxypeptidase identifies the C-terminal amino acid (Ala) when it is cleaved from the end of the chain.	→	Gly–__–__–__–Ala
• Chymotrypsin cleaves amide bonds that contain a carbonyl group from an aromatic amino acid—Tyr in this case. Because a dipeptide and tripeptide are obtained after treatment with chymotrypsin, Tyr must be the C-terminal amino acid of either the di- or tripeptide. As a result, Tyr must be either the second or third amino acid in the pentapeptide chain.	→	Gly–Tyr–__–__–Ala or Gly–__–Tyr–__–Ala
• Partial hydrolysis forms the tripeptide Tyr–Glu–Ala. Because Ala is the C-terminal amino acid, this result identifies the last three amino acids in the chain.	→	Gly–__–Tyr–Glu–Ala
• The last amino acid, Ser, must be located at the only remaining position, the second amino acid in the pentapeptide, and the complete sequence is determined.	→	Gly–Ser–Tyr–Glu–Ala

Problem 26.22 Deduce the sequence of a heptapeptide that contains the amino acids Ala, Arg, Glu, Gly, Leu, Phe, and Ser, from the following experimental data. Edman degradation cleaves Leu from the heptapeptide, and carboxypeptidase forms Glu and a hexapeptide. Treatment of the heptapeptide with chymotrypsin forms a hexapeptide and a single amino acid. Treatment of the heptapeptide with trypsin forms a pentapeptide and a dipeptide. Partial hydrolysis forms Glu, Leu, Phe, and the tripeptides Gly–Ala–Ser and Ala–Ser–Arg.

26.7 Peptide Synthesis

The synthesis of a specific dipeptide, such as Ala–Gly from alanine and glycine, is complicated because both amino acids have two functional groups. As a result, four products—namely, Ala–Ala, Ala–Gly, Gly–Gly, and Gly–Ala—are possible.

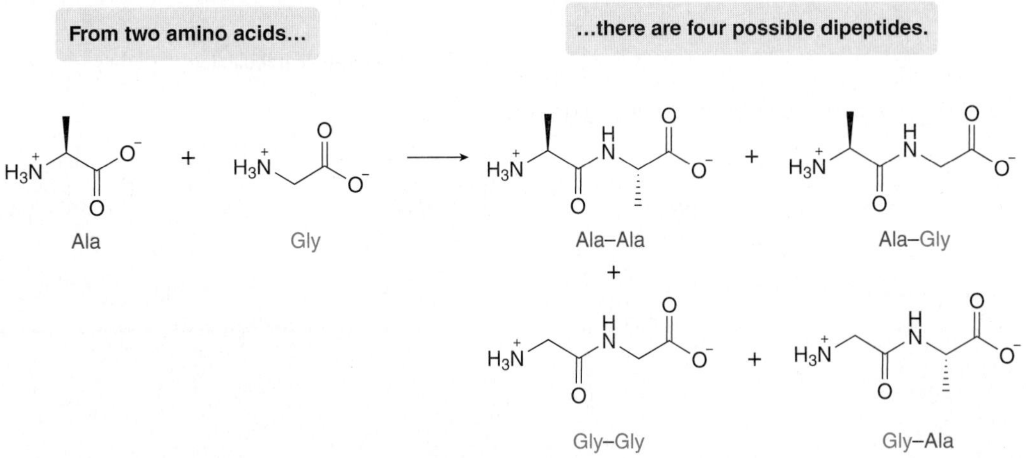
From two amino acids...
...there are four possible dipeptides.
Ala
Gly
Ala–Ala
Ala–Gly
Gly–Gly
Gly–Ala

How do we selectively join the COOH group of alanine with the NH_2 group of glycine?

- **Protect the functional groups that we don't want to react, and then form the amide bond.**

How To Synthesize a Dipeptide from Two Amino Acids

Example

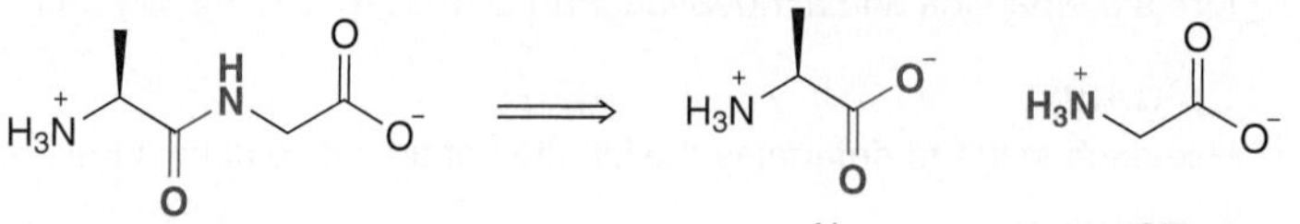

Ala–Gly Ala Gly

Join the functional groups in red.

Step [1] **Protect the NH_2 group of alanine.**

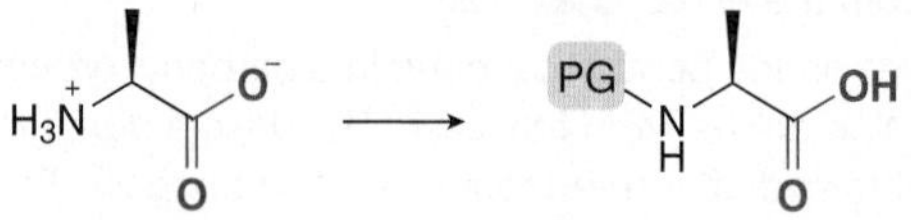

[PG = protecting group]

Ala

- In the neutral amino acid, the NH_2 group exists largely as an ammonium ion, $-NH_3^+$.

Step [2] **Protect the COOH group of glycine.**

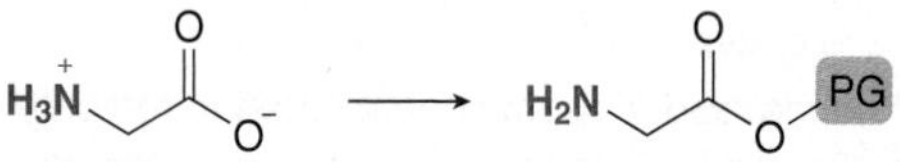

Gly

- In the neutral amino acid, the COOH group exists largely as a carboxylate anion, $-COO^-$.

Step [3] **Form the amide bond with DCC.**

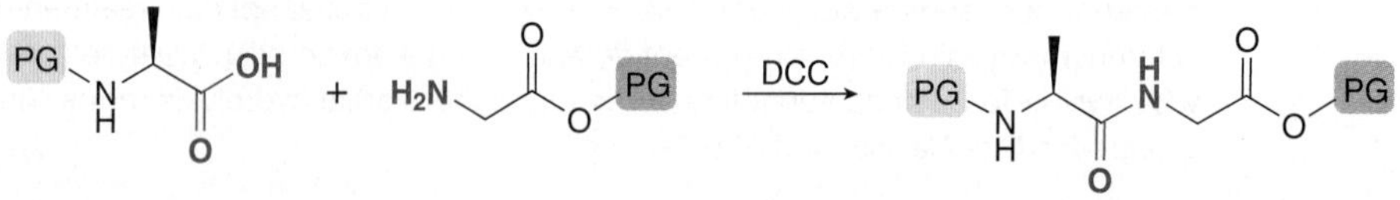

Dicyclohexylcarbodiimide (**DCC**) is a reagent commonly used to form amide bonds (see Section 22.10D). DCC makes the OH group of the carboxylic acid a better leaving group, thus **activating the carboxy group toward nucleophilic attack.**

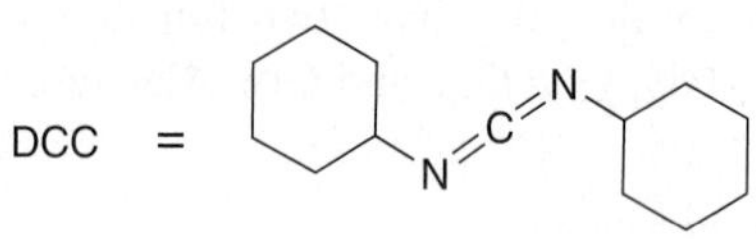

dicyclohexylcarbodiimide

Step [4] **Remove one or both protecting groups.**

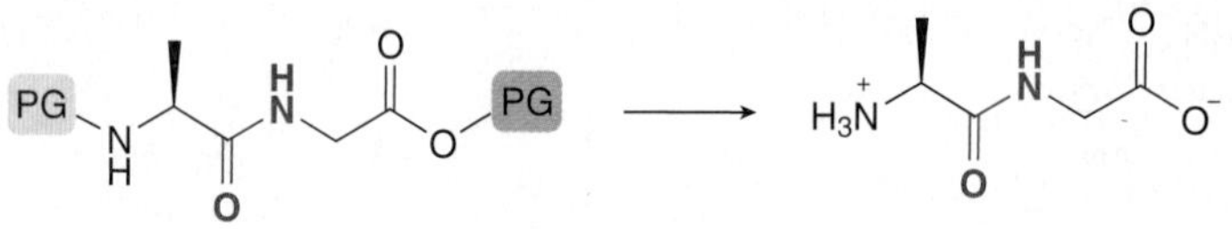

Ala–Gly

Two widely used amino protecting groups convert an amine into a **carbamate,** a functional group having a carbonyl bonded to both an oxygen and a nitrogen atom. Since the N atom of the carbamate is bonded to a carbonyl group, the protected amino group is no longer nucleophilic.

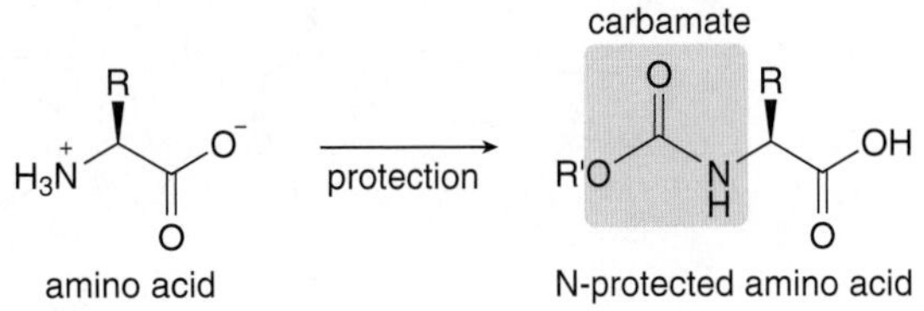

For example, the ***tert*-butoxycarbonyl protecting group,** abbreviated as **Boc,** is formed by reacting the amino acid with di-*tert*-butyl dicarbonate in a nucleophilic acyl substitution reaction.

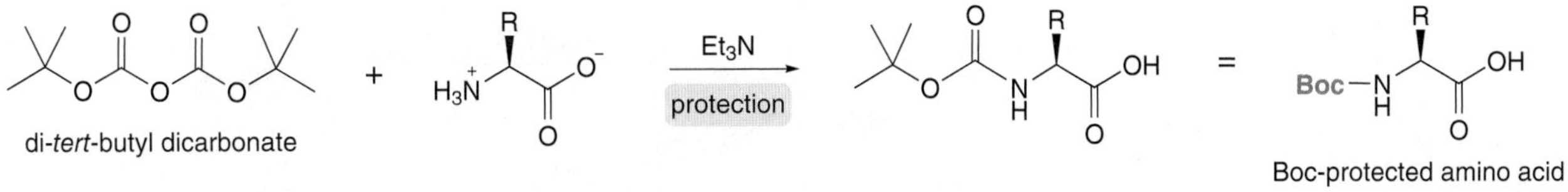

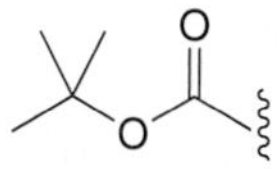

tert-butoxycarbonyl
Boc

To be a useful protecting group, the Boc group must be removed under reaction conditions that do not affect other functional groups in the molecule. It can be removed with an acid such as **trifluoroacetic acid, HCl,** or **HBr.**

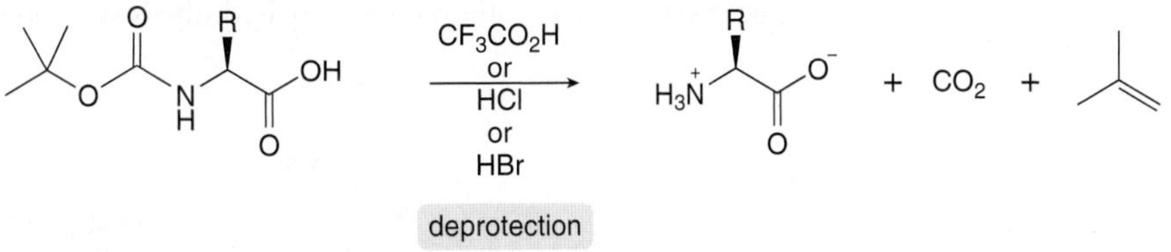

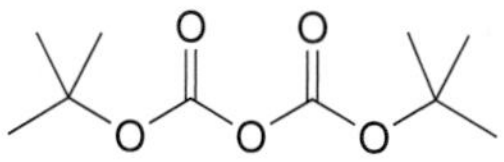

$(Boc)_2O$

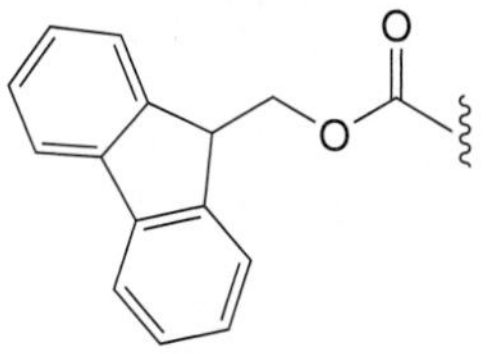

9-fluorenylmethoxycarbonyl
Fmoc

A second amino protecting group, the **9-fluorenylmethoxycarbonyl protecting group,** abbreviated as **Fmoc,** is formed by reacting the amino acid with 9-fluorenylmethyl chloroformate in a nucleophilic acyl substitution reaction.

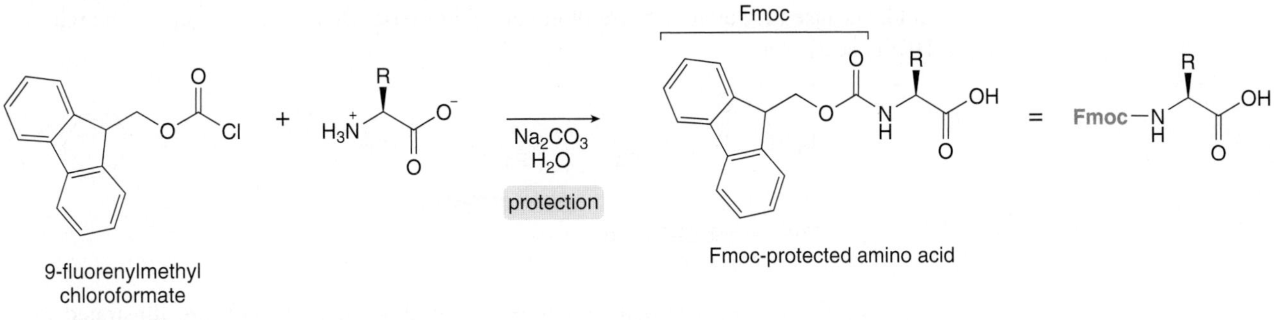

9-fluorenylmethyl
chloroformate
Fmoc–Cl

While the Fmoc protecting group is stable to most acids, it can be removed by treatment with base (NH_3 or an amine).

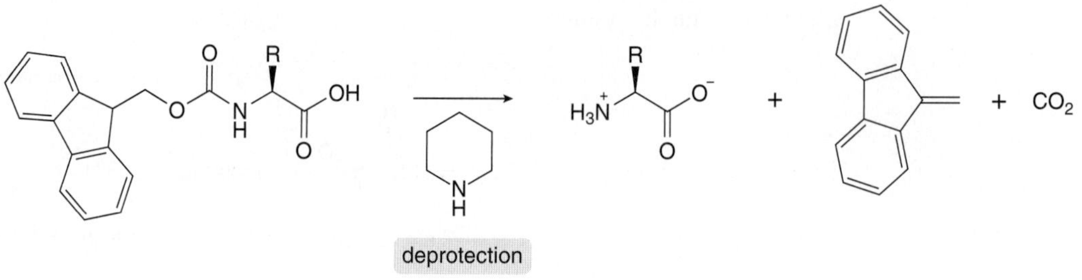

The carboxy group is usually protected as a **methyl** or **benzyl ester** by reaction with an alcohol and an acid.

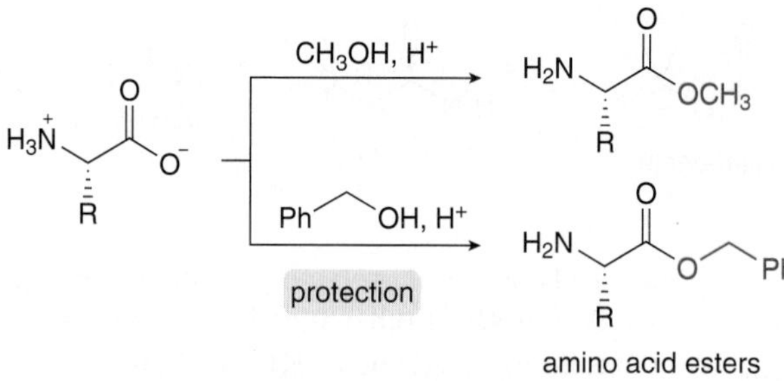

These esters are usually removed by hydrolysis with aqueous base.

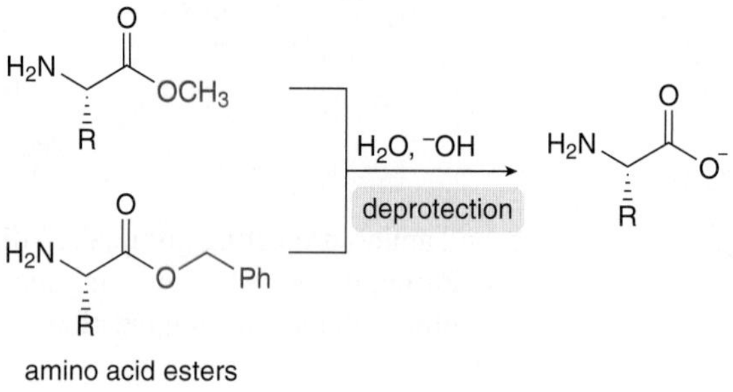

One advantage of using a benzyl ester for protection is that it can also be removed with H_2 in the presence of a Pd catalyst. This process is called **hydrogenolysis.** These conditions are especially mild, because they avoid the use of either acid or base. Benzyl esters can also be removed with HBr in acetic acid.

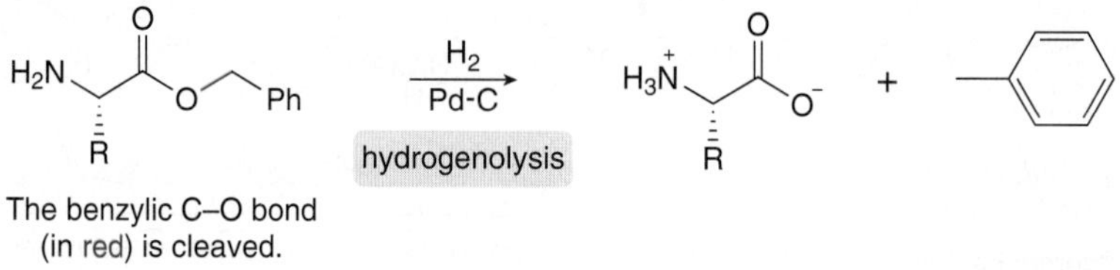

The specific reactions needed to synthesize the dipeptide Ala–Gly are illustrated in Sample Problem 26.4.

Sample Problem 26.4 Draw out the steps in the synthesis of the dipeptide Ala–Gly.

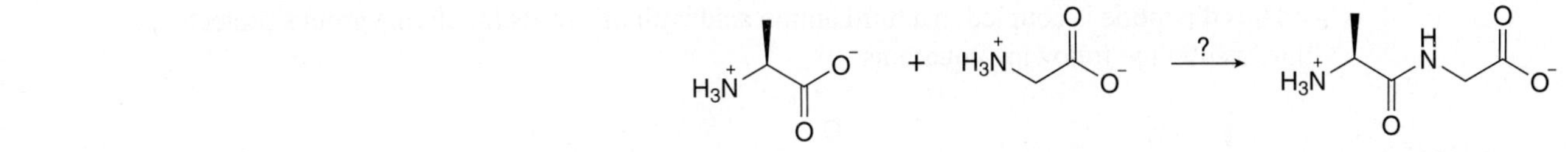

?

Ala Gly Ala–Gly

Solution

Step [1] **Protect the NH_2 group of alanine using a Boc group.**

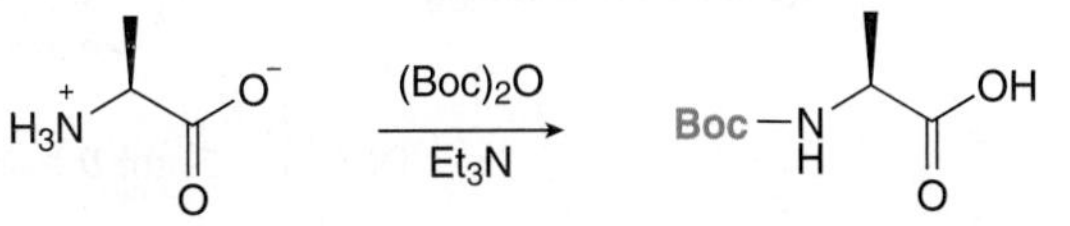

$(Boc)_2O$
Et_3N

Ala Boc–Ala

Step [2] **Protect the COOH group of glycine as a benzyl ester.**

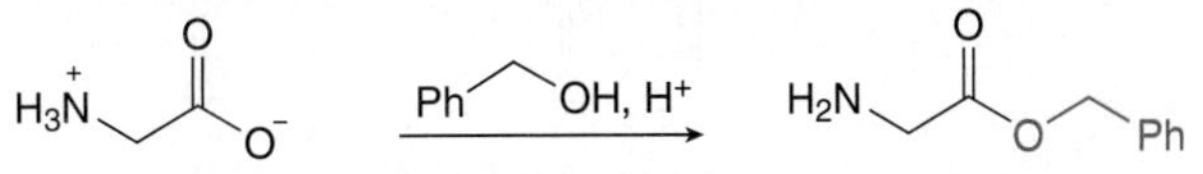

Ph⌒OH, H^+

Gly Gly–OCH_2Ph

Step [3] **Form the amide bond with DCC.**

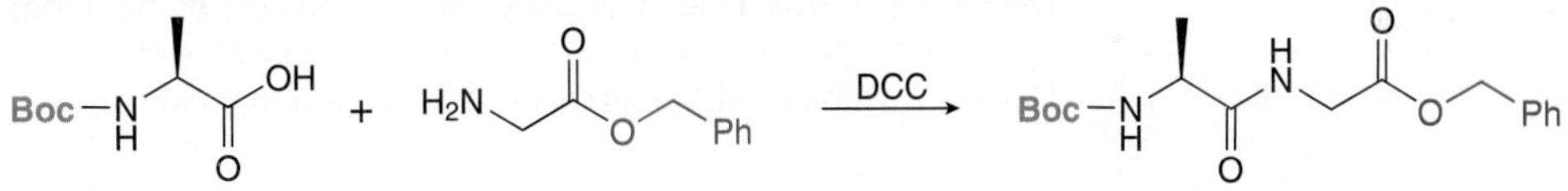

DCC

Boc–Ala Gly–OCH_2Ph Boc–Ala–Gly–OCH_2Ph

Step [4] **Remove one or both protecting groups.**

The protecting groups can be removed in a stepwise fashion, or in a single reaction.

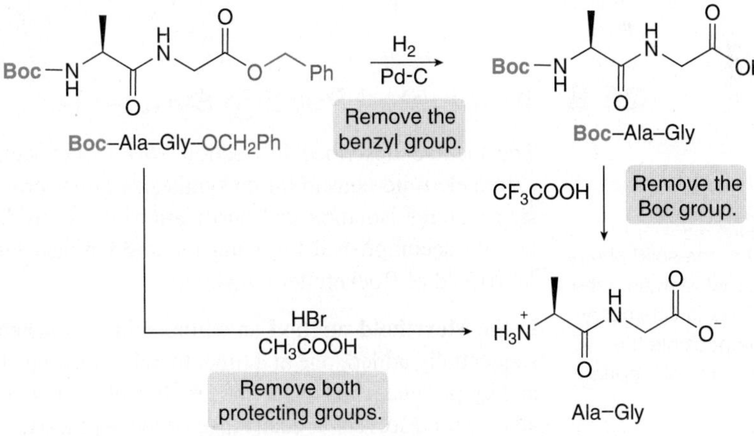

H_2
Pd-C
Remove the benzyl group.
Boc–Ala–Gly–OCH_2Ph
Boc–Ala–Gly
CF_3COOH
Remove the Boc group.
HBr
CH_3COOH
Remove both protecting groups.
Ala–Gly

This method can be applied to the synthesis of tripeptides and even larger polypeptides. After the protected dipeptide is prepared in Step [3], only one of the protecting groups is removed, and this dipeptide is coupled to a third amino acid with one of its functional groups protected, as illustrated in the following equations.

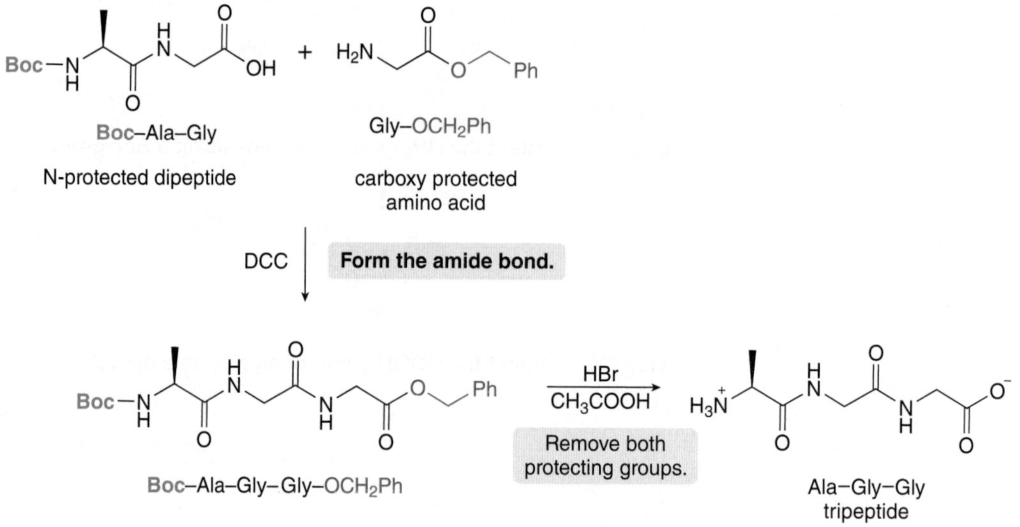

Problem 26.23 Devise a synthesis of each peptide from amino acid starting materials: (a) Leu–Val; (b) Ala–Ile–Gly.

Problem 26.24 Devise a synthesis of the following dipeptide from amino acid starting materials.

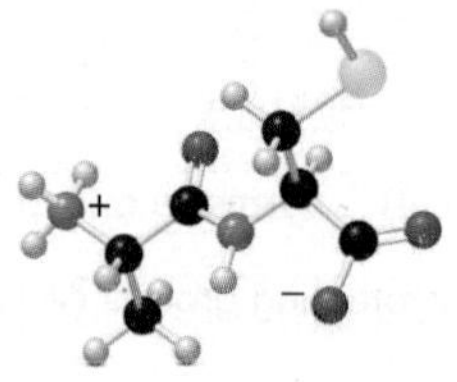

26.8 Automated Peptide Synthesis

The method described in Section 26.7 works well for the synthesis of small peptides. It is extremely time-consuming to synthesize larger proteins by this strategy, however, because each step requires isolation and purification of the product. The synthesis of larger polypeptides is usually accomplished by using the **solid phase technique** originally developed by R. Bruce Merrifield of Rockefeller University.

Development of the solid phase technique earned Merrifield the 1984 Nobel Prize in Chemistry and has made possible the synthesis of many polypeptides and proteins.

In the **Merrifield method** an amino acid is attached to an **insoluble polymer.** Amino acids are sequentially added, one at a time, thereby forming successive peptide bonds. Because impurities and by-products are not attached to the polymer chain, they are removed simply by washing them away with a solvent at each stage of the synthesis.

A commonly used polymer is a **polystyrene derivative** that contains $-CH_2Cl$ groups bonded to some of the benzene rings in the polymer chain. The Cl atoms serve as handles that allow attachment of amino acids to the chain.

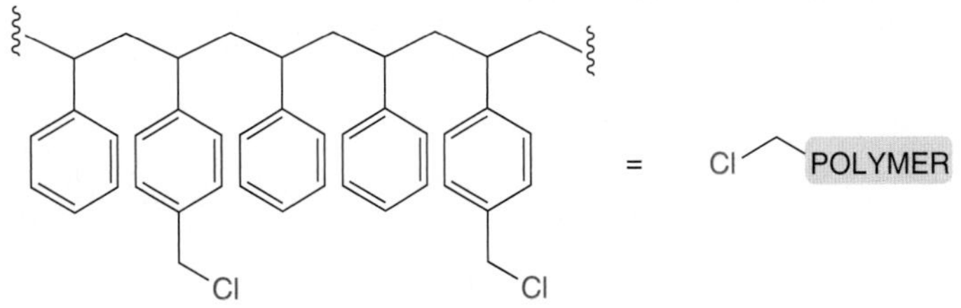

polystyrene derivative with Cl leaving groups

An Fmoc-protected amino acid is attached to the polymer at its carboxy group by an $\mathbf{S_N2}$ reaction.

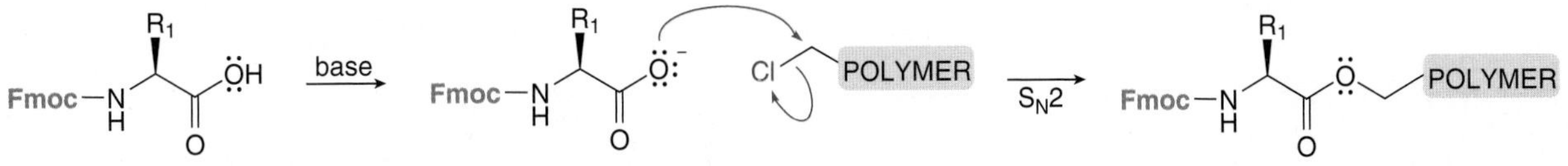

The amino acid is now bound to the insoluble polymer.

Once the first amino acid is bound to the polymer, additional amino acids can be added sequentially. The steps of the solid phase peptide synthesis technique are illustrated in the accompanying scheme. In the last step, HF cleaves the polypeptide chain from the polymer.

How To Synthesize a Peptide Using the Merrifield Solid Phase Technique

Step [1]
Attach an Fmoc-protected amino acid to the polymer.

Step [2]
Remove the protecting group.

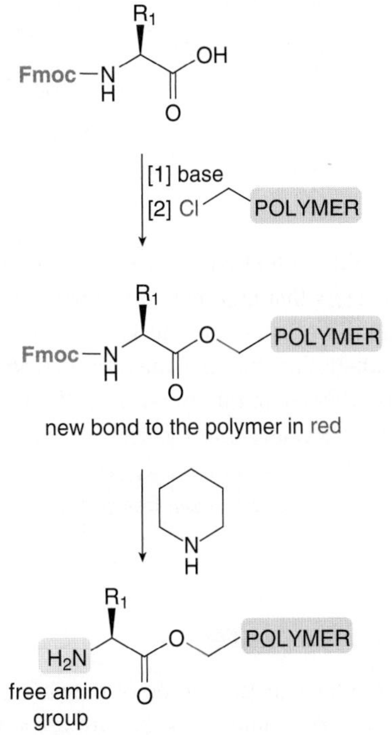

—Continued

How To, continued . . .

Step [3]
Form the amide bond with DCC.

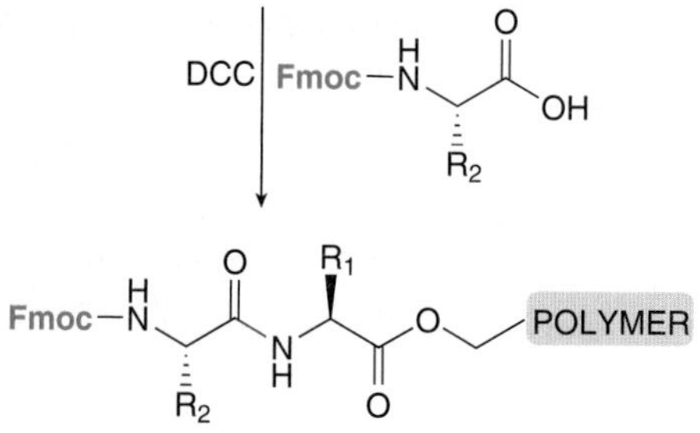

new amide bond in red

Step [4]
Repeat Steps [2] and [3].

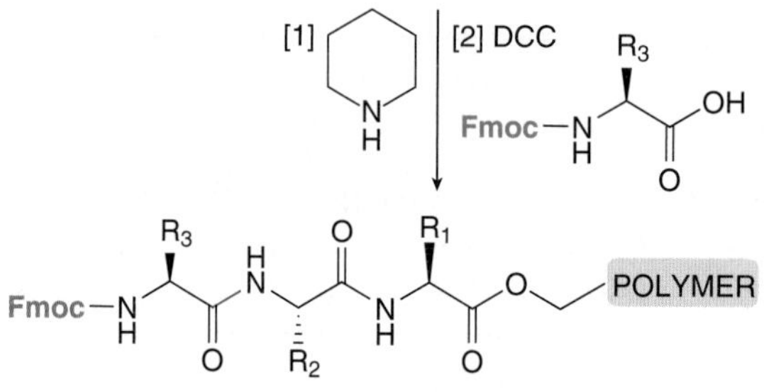

new amide bond in red

Step [5]
Remove the protecting group and detach the peptide from the polymer.

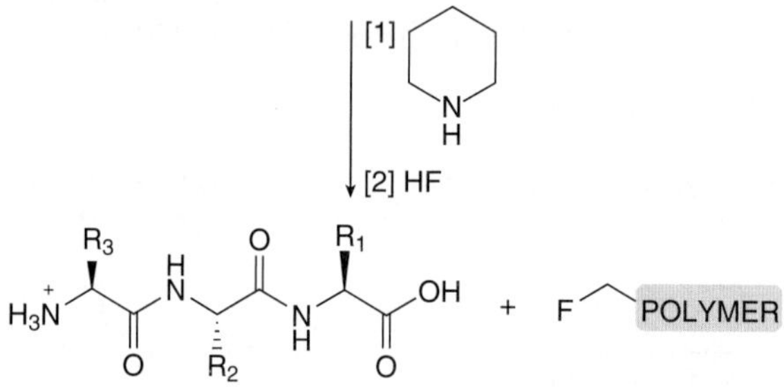

tripeptide

The Merrifield method has now been completely automated, so it is possible to purchase peptide synthesizers that automatically carry out all of the above operations and form polypeptides in high yield in a matter of hours, days, or weeks, depending on the length of the chain of the desired product. For example, the protein ribonuclease, which contains 128 amino acids, has been prepared by this technique in an overall yield of 17%. This remarkable synthesis involved 369 separate reactions, and thus the yield of each individual reaction was > 99%.

Problem 26.25 Outline the steps needed to synthesize the tetrapeptide Ala–Leu–Ile–Gly using the Merrifield technique.

26.9 Protein Structure

Now that you have learned some of the chemistry of amino acids, it's time to study proteins, the large polymers of amino acids that are responsible for so much of the structure and function of all living cells. We begin with a discussion of the **primary, secondary, tertiary, and quaternary structure** of proteins.

26.9A Primary Structure

The *primary structure* of proteins is the particular sequence of amino acids that is joined together by peptide bonds. The most important element of this primary structure is the **amide bond.**

- Rotation around the amide C–N bond is *restricted* because of electron delocalization, and the *s*-trans conformation is the more stable arrangement.
- In each peptide bond, the N–H and C=O bonds are directed 180° from each other.

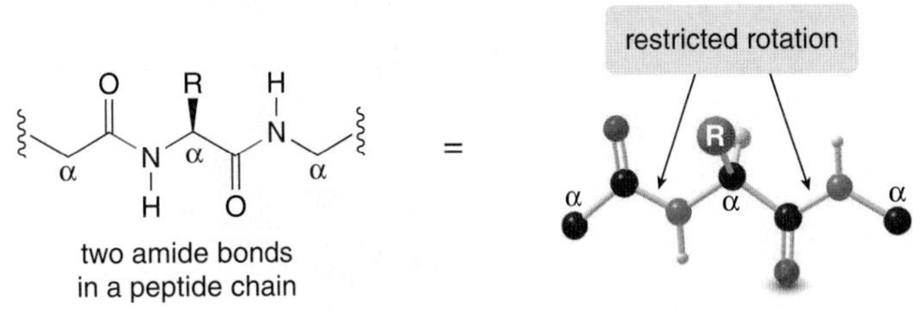

Although rotation about the amide bonds is restricted, **rotation about the other σ bonds in the protein backbone is not.** As a result, the peptide chain can twist and bend into a variety of different arrangements that constitute the secondary structure of the protein.

26.9B Secondary Structure

The three-dimensional conformations of localized regions of a protein are called its secondary structure. These regions arise due to hydrogen bonding between the N–H proton of one amide and C=O oxygen of another. Two arrangements that are particularly stable are called the **α-helix** and the **β-pleated sheet.**

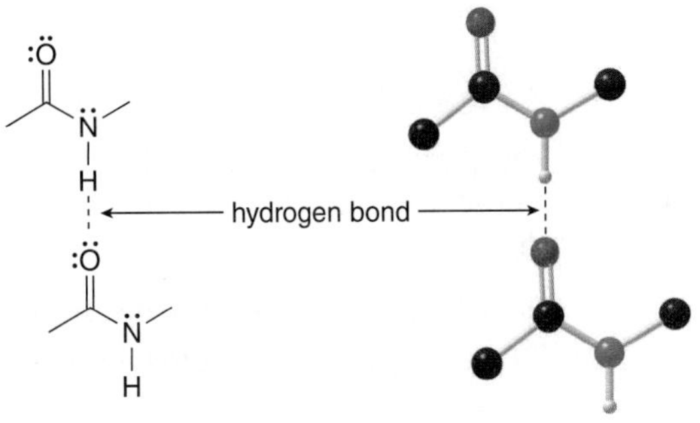

α-Helix

The **α-helix** forms when a peptide chain twists into a right-handed or clockwise spiral, as shown in Figure 26.8. Four important features of the α-helix are as follows:

[1] **Each turn of the helix has 3.6 amino acids.**

[2] **The N–H and C=O bonds point along the axis of the helix.** All C=O bonds point in one direction, and all N–H bonds point in the opposite direction.

[3] **The C=O group of one amino acid is hydrogen bonded to an N–H group four amino acid residues farther along the chain.** Thus, hydrogen bonding occurs between two amino acids ***in the same chain.*** Note, too, that the hydrogen bonds are parallel to the axis of the helix.

[4] The **R groups of the amino acids extend outward** from the core of the helix.

An α-helix can form only if there is rotation about the bonds at the α carbon of the amide carbonyl group, and not all amino acids can do this. For example, proline, the amino acid whose nitrogen atom forms part of a five-membered ring, is more rigid than other amino acids, and its C_α–N bond cannot rotate the necessary amount. Additionally, it has no N–H proton with which to form an intramolecular hydrogen bond to stabilize the helix. Thus, proline cannot be part of an α-helix.

Both the myosin in muscle and α-keratin in hair are proteins composed almost entirely of α-helices.

Figure 26.8
Two different illustrations of the α-helix

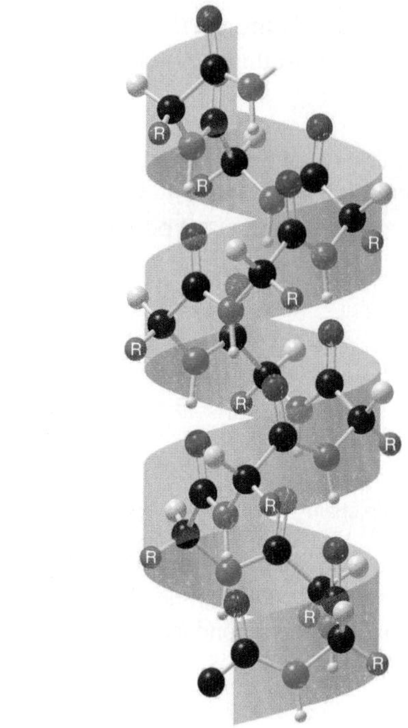

All atoms of the α-helix are drawn in this representation. All C=O bonds are pointing up and all N–H bonds are pointing down.

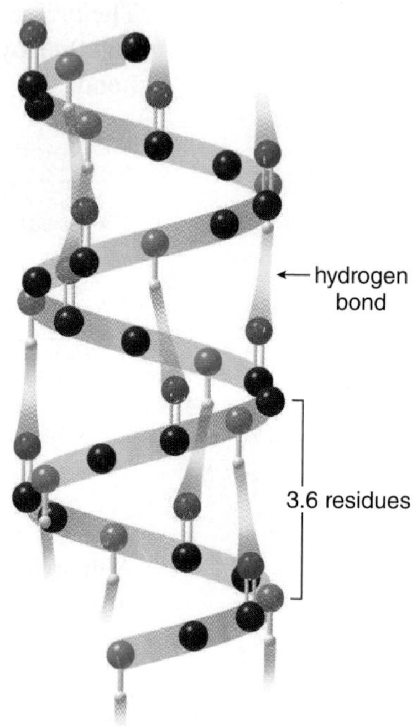

Only the peptide backbone is drawn in this representation. The hydrogen bonds between the C=O and N–H of amino acids four residues away from each other are shown.

β-Pleated Sheet

The **β-pleated sheet** secondary structure forms when two or more peptide chains, called **strands,** line up side-by-side, as shown in Figure 26.9. All β-pleated sheets have the following characteristics:

[1] **The C=O and N–H bonds lie in the plane of the sheet.**

[2] **Hydrogen bonding often occurs between the N–H and C=O groups of nearby amino acid residues.**

[3] The **R groups are oriented above and below the plane** of the sheet, and alternate from one side to the other along a given strand.

Figure 26.9
Three-dimensional structure of the β-pleated sheet

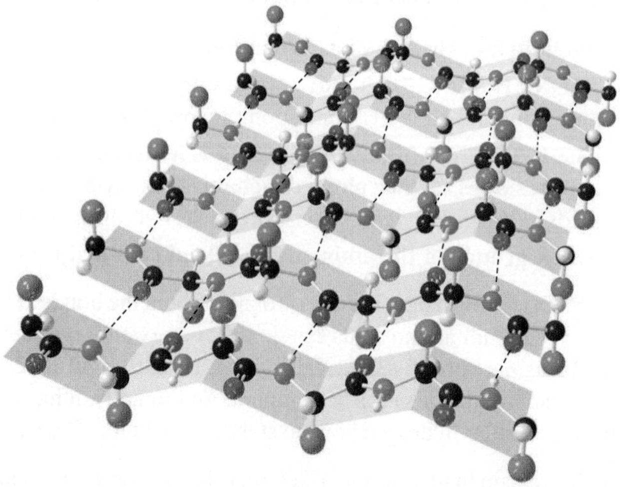

- The β-pleated sheet consists of extended strands of the peptide chains held together by hydrogen bonding. The C=O and N–H bonds lie in the plane of the sheet, and the R groups (shown as orange balls) alternate above and below the plane.

The β-pleated sheet arrangement most commonly occurs with amino acids with small R groups, like alanine and glycine. With larger R groups, steric interactions prevent the chains from getting close together and so the sheet cannot be stabilized by hydrogen bonding.

The peptide strands of β-pleated sheets can actually be oriented in two different ways, as shown in Figure 26.10.

- **In a *parallel* β-pleated sheet, the strands run in the *same* direction from the N- to C-terminal amino acid.**
- **In an *antiparallel* β-pleated sheet, the strands run in the *opposite* direction.**

Figure 26.10 The parallel and antiparallel forms of the β-pleated sheet

Parallel β-pleated sheet

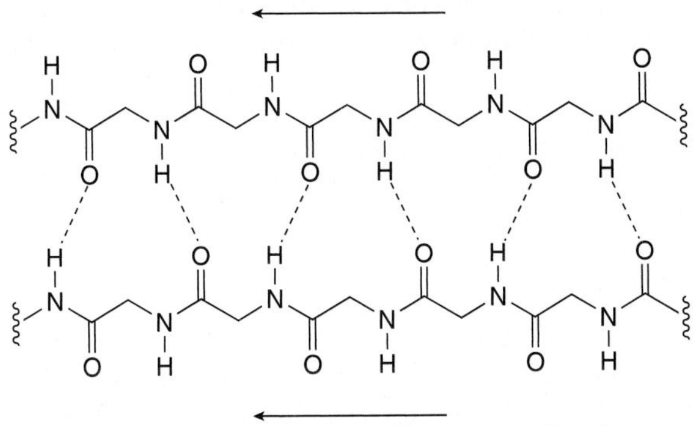

The two peptide chains are arranged in the same direction. Hydrogen bonds occur between N–H and C=O bonds in adjacent chains.

[Note: R groups on the carbon chain are omitted for clarity.]

Antiparallel β-pleated sheet

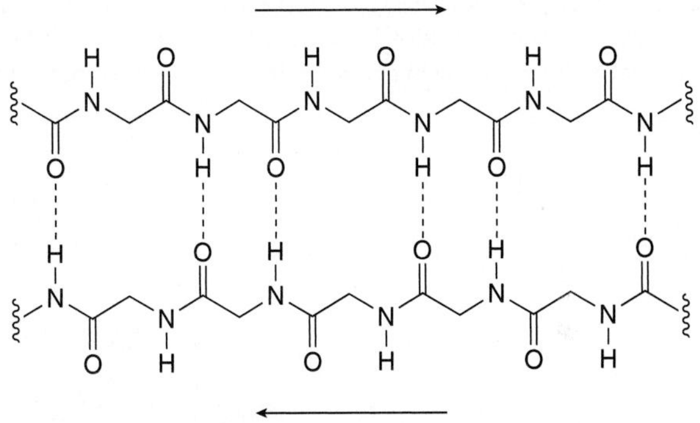

The two peptide chains are arranged in opposite directions. Hydrogen bonding between the N–H and C=O groups still holds the two chains together.

Most proteins have regions of α-helix and β-pleated sheet, in addition to other regions that cannot be characterized by either of these arrangements. Shorthand symbols are often used to indicate regions of a protein that have α-helix or β-pleated sheet. A **flat helical ribbon** is used for the α-helix, and a **flat wide arrow** is used for the β-pleated sheet. These representations are often used in **ribbon diagrams** to illustrate protein structure.

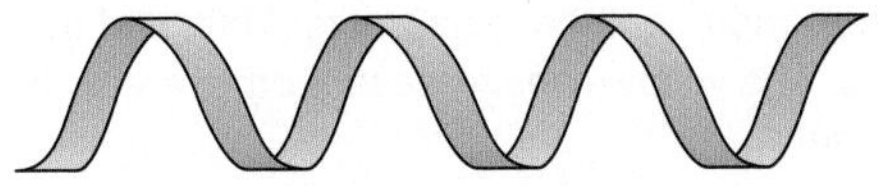

α-helix shorthand

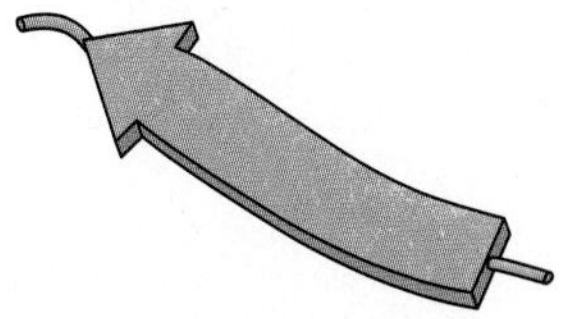

β-pleated sheet shorthand

Proteins are drawn in a variety of ways to illustrate different aspects of their structure. Figure 26.11 illustrates three different representations of the protein lysozyme, an enzyme found in both plants and animals. Lysozyme catalyzes the hydrolysis of bonds in bacterial cell walls, weakening them, often causing the bacteria to burst.

Spider dragline silk is a strong yet elastic protein because it has regions of β-pleated sheet and regions of α-helix (Figure 26.12). α-Helical regions impart elasticity to the silk because the peptide chain is twisted (not fully extended), so it can stretch. β-Pleated sheet regions are almost fully extended, so they can't be stretched further, but their highly ordered three-dimensional structure imparts strength to the silk. Thus, spider silk suits the spider by comprising both types of secondary structure with beneficial properties.

Figure 26.11 Lysozyme

(a) The ball-and-stick model of lysozyme shows the protein backbone with color-coded C, N, O, and S atoms. Individual amino acids are most clearly located using this representation. (b) The space-filling model uses color-coded balls for each atom in the backbone of the enzyme and illustrates how the atoms fill the space they occupy. (c) The ribbon diagram shows regions of α-helix and β-pleated sheet that are not clearly in evidence in the other two representations.

Figure 26.12
Different regions of secondary structure in spider silk

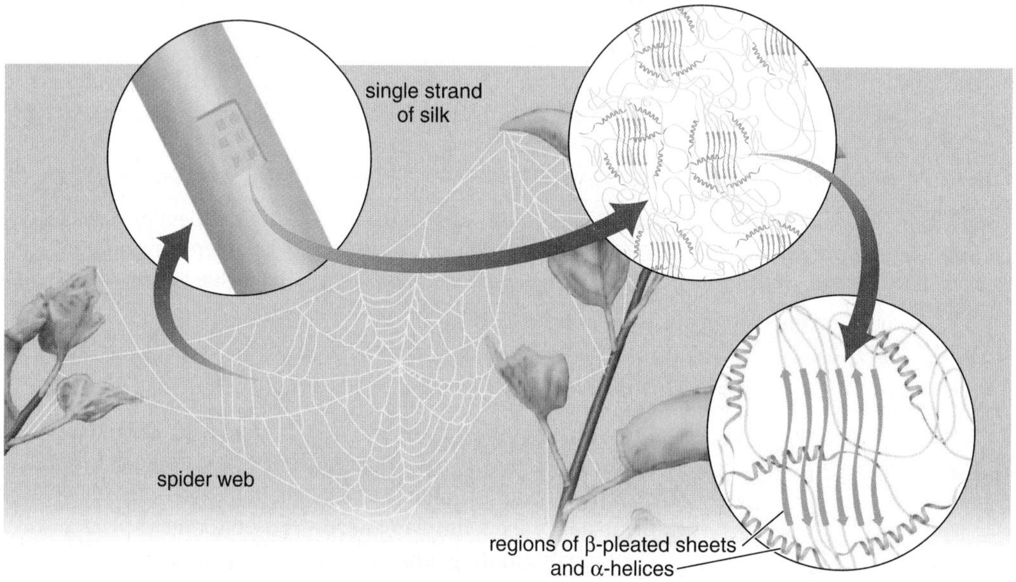

- Spider silk has regions of α-helix and β-pleated sheet that make it both strong and elastic. The green coils represent the α-helical regions, and the purple arrows represent the β-pleated sheet regions. The yellow lines represent other areas of the protein that are neither α-helix nor β-pleated sheet.

Problem 26.26 Consider two molecules of a tetrapeptide composed of only alanine residues. Draw the hydrogen bonding interactions that result when these two peptides adopt a parallel β-pleated sheet arrangement. Answer this same question for the antiparallel β-pleated sheet arrangement.

26.9C Tertiary and Quaternary Structure

The three-dimensional shape adopted by the entire peptide chain is called its tertiary structure. A peptide generally folds into a conformation that maximizes its stability. In the aqueous environment of the cell, proteins often fold in such a way as to place a large number of polar and charged groups on their outer surface, to maximize the dipole–dipole and hydrogen bonding interactions with water. This generally places most of the nonpolar side chains in the interior of the protein, where van der Waals interactions between these hydrophobic groups help stabilize the molecule, too.

In addition, polar functional groups hydrogen bond with each other (not just water), and amino acids with charged side chains like $-COO^-$ and $-NH_3^+$ can stabilize tertiary structure by electrostatic interactions.

Finally, **disulfide bonds are the only covalent bonds that stabilize tertiary structure.** As previously mentioned, these strong bonds form by oxidation of two cysteine residues on either the same polypeptide chain or another polypeptide chain of the same protein.

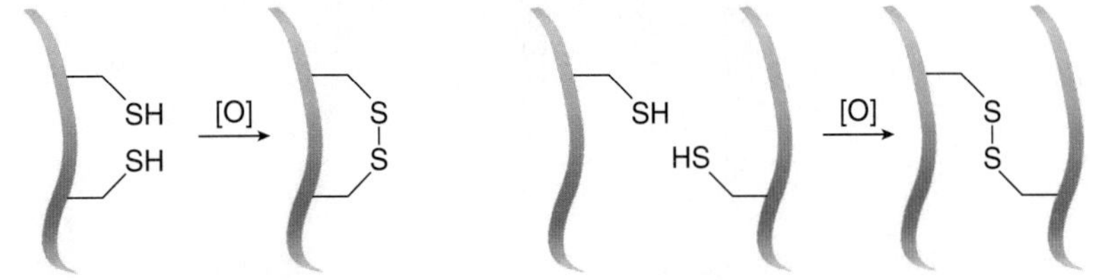
SH
SH
[O]
S
S
SH
HS
[O]
S
S
disulfide bond between two SH groups on the same chain
disulfide bond between two SH groups on different chains

The nonapeptides **oxytocin** and **vasopressin** (Section 26.5C) contain intramolecular disulfide bonds. **Insulin,** on the other hand, consists of two separate polypeptide chains (**A** and **B**) that are covalently linked by two intermolecular disulfide bonds, as shown in Figure 26.13. The **A** chain, which also has an intramolecular disulfide bond, has 21 amino acid residues, whereas the **B** chain has 30.

Figure 26.13
Insulin

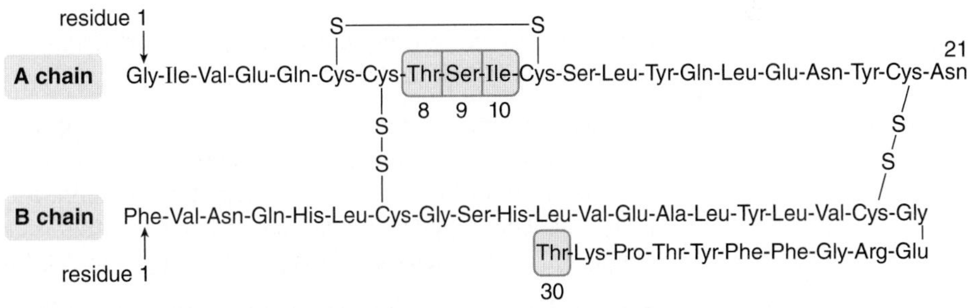
residue 1
A chain
Gly-Ile-Val-Glu-Gln-Cys-Cys-Thr-Ser-Ile-Cys-Ser-Leu-Tyr-Gln-Leu-Glu-Asn-Tyr-Cys-Asn
21
8 9 10
S
S
B chain
Phe-Val-Asn-Gln-His-Leu-Cys-Gly-Ser-His-Leu-Val-Glu-Ala-Leu-Tyr-Leu-Val-Cys-Gly
Thr-Lys-Pro-Thr-Tyr-Phe-Phe-Gly-Arg-Glu
30
residue 1

Insulin is a small protein consisting of two polypeptide chains (designated as the **A** and **B** chains) held together by two disulfide bonds. An additional disulfide bond joins two cysteine residues within the **A** chain.

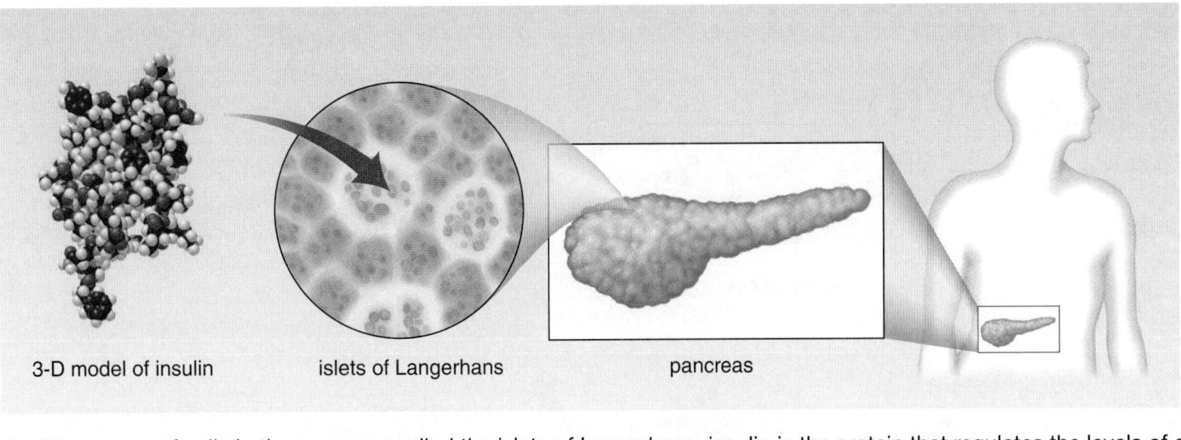
3-D model of insulin
islets of Langerhans
pancreas

Synthesized by groups of cells in the pancreas called the islets of Langerhans, insulin is the protein that regulates the levels of glucose in the blood. Insufficiency of insulin results in diabetes. Many of the abnormalities associated with this disease can be controlled by the injection of insulin. Until the availability of human insulin through genetic engineering techniques, all insulin used by diabetics was obtained from pigs and cattle. The amino acid sequences of these insulin proteins is slightly different from that of human insulin. Pig insulin differs in one amino acid only, whereas bovine insulin has three different amino acids. This is shown in the accompanying table.

	Chain A			Chain B
Position of residue →	**8**	**9**	**10**	**30**
Human insulin	Thr	Ser	Ile	Thr
Pig insulin	Thr	Ser	Ile	Ala
Bovine insulin	Ala	Ser	Val	Ala

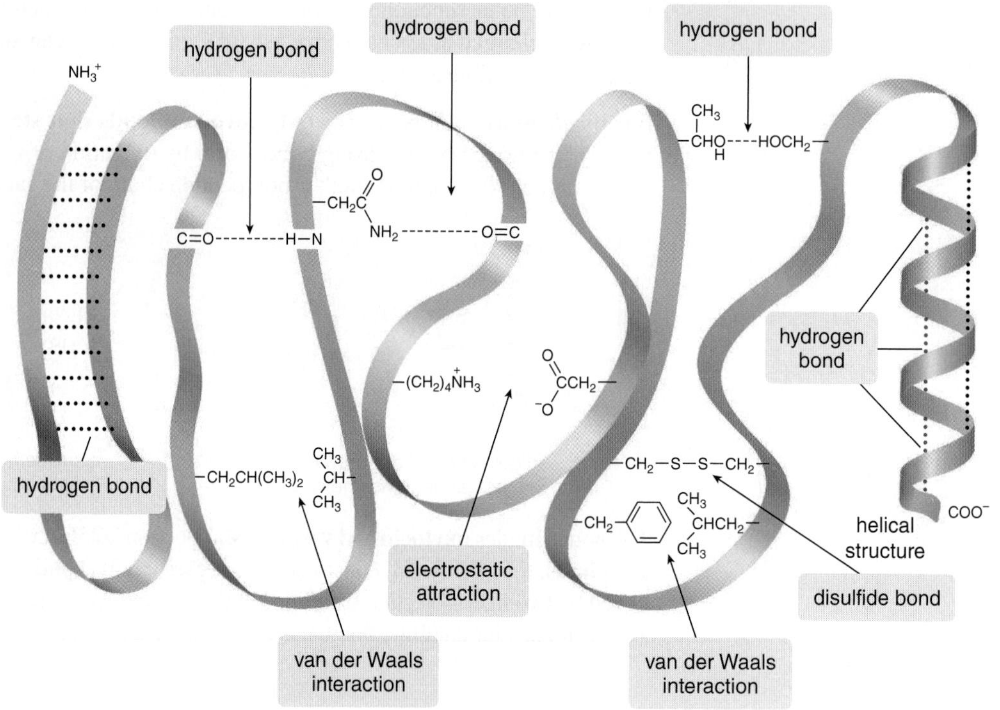

Figure 26.14 The stabilizing interactions in secondary and tertiary protein structure

Figure 26.14 schematically illustrates the many different kinds of intramolecular forces that stabilize the secondary and tertiary structures of polypeptide chains.

The shape adopted when two or more folded polypeptide chains aggregate into one protein complex is called the **quaternary structure** of the protein. Each individual polypeptide chain is called a **subunit** of the overall protein. **Hemoglobin,** for example, consists of two α and two β subunits held together by intermolecular forces in a compact three-dimensional shape. The unique function of hemoglobin is possible only when all four subunits are together.

The four levels of protein structure are summarized in Figure 26.15.

Figure 26.15 The primary, secondary, tertiary, and quaternary structure of proteins

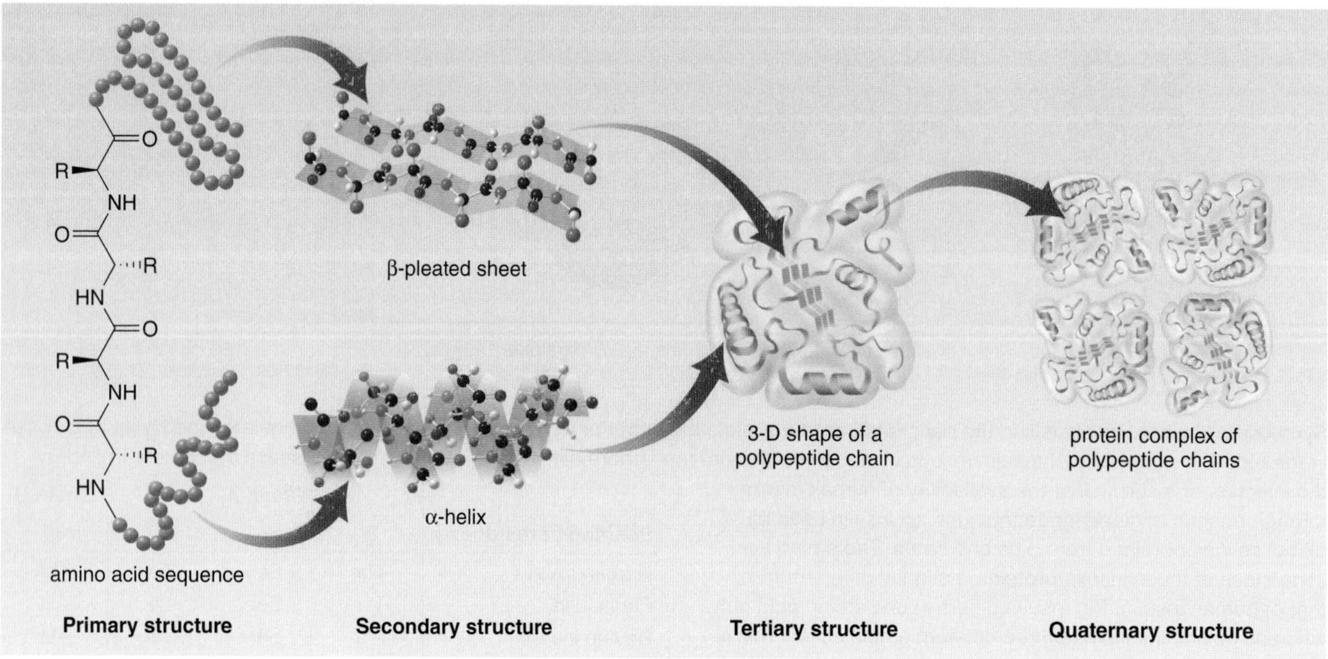

Problem 26.27 What types of stabilizing interactions exist between each of the following pairs of amino acids?

a. Ser and Tyr b. Val and Leu c. two Phe residues

Problem 26.28 The fibroin proteins found in silk fibers consist of large regions of β-pleated sheets stacked one on top of another. (a) Explain why having a glycine at every other residue allows the β-pleated sheets to stack on top of each other. (b) Why are silk fibers insoluble in water?

26.10 Important Proteins

Proteins are generally classified according to their three-dimensional shapes.

- **Fibrous proteins** are composed of long linear polypeptide chains that are bundled together to form rods or sheets. These proteins are insoluble in water and serve structural roles, giving strength and protection to tissues and cells.
- **Globular proteins** are coiled into compact shapes with hydrophilic outer surfaces that make them water soluble. Enzymes and transport proteins are globular to make them soluble in the blood and other aqueous environments in cells.

26.10A α-Keratins

α-Keratins are the proteins found in hair, hooves, nails, skin, and wool. They are composed almost exclusively of long sections of α-helix units, having large numbers of alanine and leucine residues. Because these nonpolar amino acids extend outward from the α-helix, these proteins are very water insoluble. Two α-keratin helices coil around each other, forming a structure called a **supercoil** or **superhelix.** These, in turn, form larger and larger bundles of fibers, ultimately forming a strand of hair, as shown schematically in Figure 26.16.

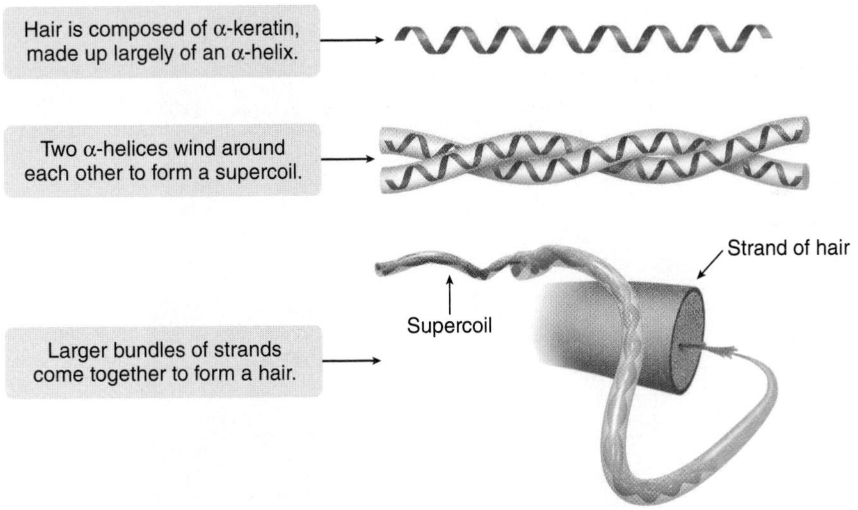
Hair is composed of α-keratin, made up largely of an α-helix.
Two α-helices wind around each other to form a supercoil.
Strand of hair
Supercoil
Larger bundles of strands come together to form a hair.

Figure 26.16 Anatomy of a hair— It begins with α-keratin.

α-Keratins also have a number of cysteine residues, and because of this, disulfide bonds are formed between adjacent helices. The number of disulfide bridges determines the strength of the material. Claws, horns, and fingernails have extensive networks of disulfide bonds, making them extremely hard.

Straight hair can be made curly by cleaving the disulfide bonds in α-keratin, and then rearranging and re-forming them, as shown schematically in Figure 26.17. First, the disulfide bonds in the straight hair are reduced to thiol groups, so the bundles of α-keratin chains are no longer held in their specific "straight" orientation. Then, the hair is wrapped around curlers and treated with an oxidizing agent that converts the thiol groups back to disulfide bonds, now with twists and turns in the keratin backbone. This makes the hair look curly and is the chemical basis for a "permanent."

Figure 26.17
The chemistry of a "permanent"—Making straight hair curly

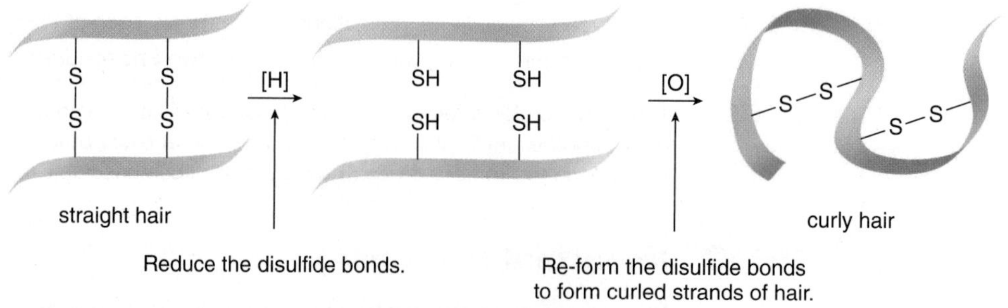
S
S
S
S
[H]
SH
SH
SH
SH
[O]
S–S
S–S
straight hair
curly hair
Reduce the disulfide bonds.
Re-form the disulfide bonds
to form curled strands of hair.

- To make straight hair curly, the disulfide bonds holding the α-helical chains together are cleaved by reduction. This forms free thiol groups (–SH). The hair is turned around curlers and then an oxidizing agent is applied. This re-forms the disulfide bonds in the hair, but between different thiol groups, now giving it a curly appearance.

26.10B Collagen

Collagen, the most abundant protein in vertebrates, is found in connective tissues such as bone, cartilage, tendons, teeth, and blood vessels. Glycine and proline account for a large fraction of its amino acid residues, whereas cysteine accounts for very little. Because of the high proline content, it cannot form a right-handed α-helix. Instead, it forms an elongated left-handed helix, and then three of these helices wind around each other to form a right-handed **superhelix** or **triple helix.** The side chain of glycine is only a hydrogen atom, so the high glycine content allows the collagen superhelices to lie compactly next to each other, thus stabilizing the superhelices via hydrogen bonding. Two views of the collagen superhelix are shown in Figure 26.18.

Figure 26.18
Two different representations for the triple helix of collagen

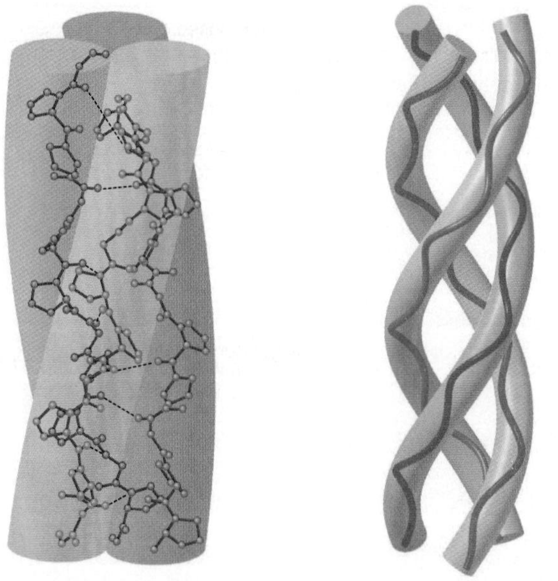

- In collagen, three polypeptide chains having an unusual left-handed helix wind around each other in a right-handed triple helix. The high content of small glycine residues allows the chains to lie close to each other, permitting hydrogen bonding between the chains.

26.10C Hemoglobin and Myoglobin

Hemoglobin and **myoglobin,** two globular proteins, are called **conjugated proteins** because they are composed of a protein unit and a nonprotein molecule called a **prosthetic group.** The prosthetic group in hemoglobin and myoglobin is **heme,** a complex organic compound containing the

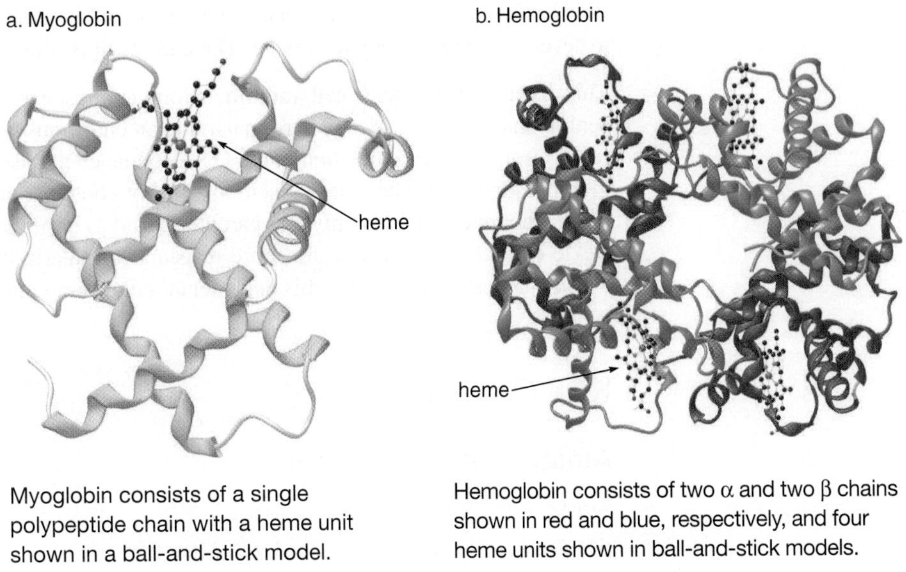
a. Myoglobin
heme
b. Hemoglobin
heme
Myoglobin consists of a single polypeptide chain with a heme unit shown in a ball-and-stick model.
Hemoglobin consists of two α and two β chains shown in red and blue, respectively, and four heme units shown in ball-and-stick models.

Figure 26.19
Protein ribbon diagrams for myoglobin and hemoglobin

Fe^{2+} ion complexed with a nitrogen heterocycle called a **porphyrin.** The Fe^{2+} ion of hemoglobin and myoglobin binds oxygen in the blood. Hemoglobin, which is present in red blood cells, transports oxygen to wherever it is needed in the body, whereas myoglobin stores oxygen in tissues. Ribbon diagrams for myoglobin and hemoglobin are shown in Figure 26.19.

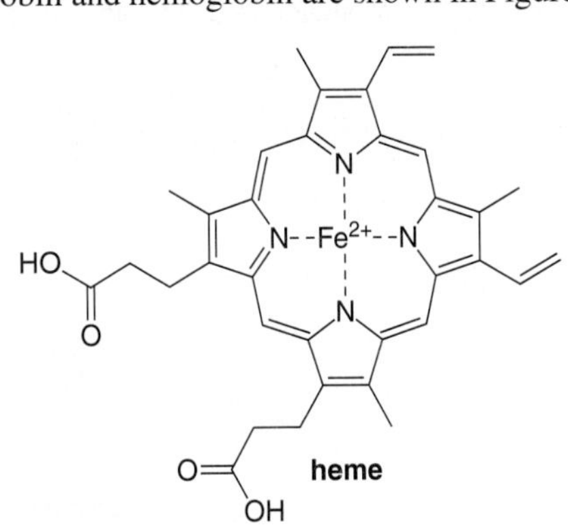
N
N
Fe2+
N
N
HO
O
O
OH

heme

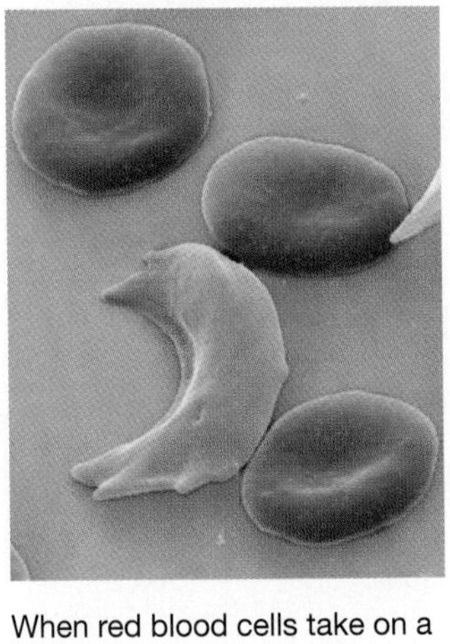

When red blood cells take on a "sickled" shape in persons with sickle cell disease, they occlude capillaries (causing organ injury) and they break easily (leading to profound anemia). This devastating illness results from the change of a single amino acid in hemoglobin. Note the single sickled cell surrounded by three red cells with normal morphology.

Myoglobin, the chapter-opening molecule, has 153 amino acid residues in a single polypeptide chain. It has eight separate α-helical sections that fold back on one another, with the prosthetic heme group held in a cavity inside the polypeptide. Most of the polar residues are found on the outside of the protein so that they can interact with the water solvent. Spaces in the interior of the protein are filled with nonpolar amino acids. Myoglobin gives cardiac muscle its characteristic red color.

Hemoglobin consists of four polypeptide chains (two α subunits and two β subunits), each of which carries a heme unit. Hemoglobin has more nonpolar amino acids than myoglobin. When each subunit is folded, some of these remain on the surface. The van der Waals attraction between these hydrophobic groups is what stabilizes the quaternary structure of the four subunits.

Carbon monoxide is poisonous because it binds to the Fe^{2+} of hemoglobin more strongly than does oxygen. Hemoglobin complexed with CO cannot carry O_2 from the lungs to the tissues. Without O_2 in the tissues for metabolism, cells cannot function, so they die.

The properties of all proteins depend on their three-dimensional shape, and their shape depends on their primary structure—that is, their amino acid sequence. This is particularly well exemplified by comparing normal hemoglobin with **sickle cell hemoglobin,** a mutant variation in which a single amino acid of both β subunits is changed from glutamic acid to valine. The replacement of one acidic amino acid (Glu) with one nonpolar amino acid (Val) changes the shape of hemoglobin, which has profound effects on its function. Deoxygenated red blood cells with sickle cell hemoglobin become elongated and crescent shaped, and they are unusually fragile. As a result, they do not flow easily through capillaries, causing pain and inflammation, and they break open easily, leading to severe anemia and organ damage. The end result is often a painful and premature death.

This disease, called **sickle cell anemia,** is found almost exclusively among people originating from central and western Africa, where malaria is an enormous health problem. Sickle cell hemoglobin results from a genetic mutation in the DNA sequence that is responsible for the synthesis of hemoglobin. Individuals who inherit this mutation from both parents develop sickle cell anemia, whereas those who inherit it from only one parent are said to have the sickle cell trait. They do not develop sickle cell anemia and they are more resistant to malaria than individuals without the mutation. This apparently accounts for this detrimental gene being passed on from generation to generation.

KEY CONCEPTS

Amino Acids and Proteins

Synthesis of Amino Acids (26.2)

[1] From α-halo carboxylic acids by S_N2 reaction

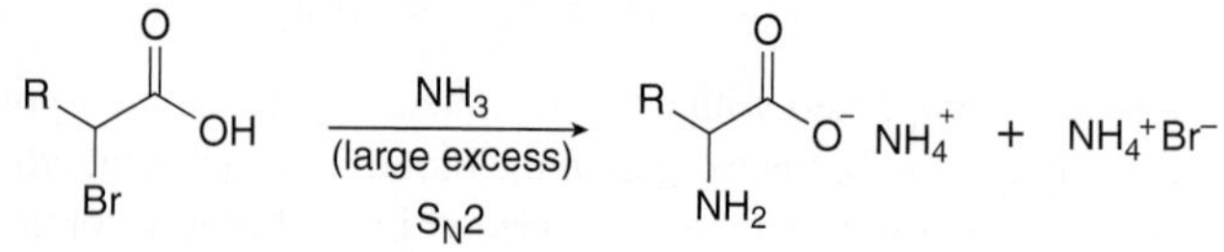

[2] By alkylation of diethyl acetamidomalonate

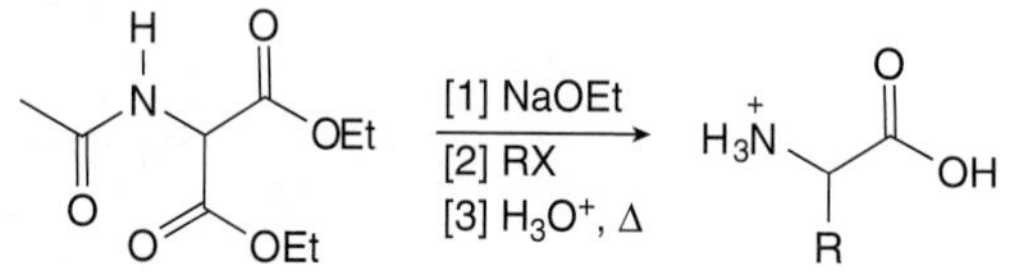

- Alkylation works best with unhindered alkyl halides—that is, CH_3X and RCH_2X.

[3] Strecker synthesis

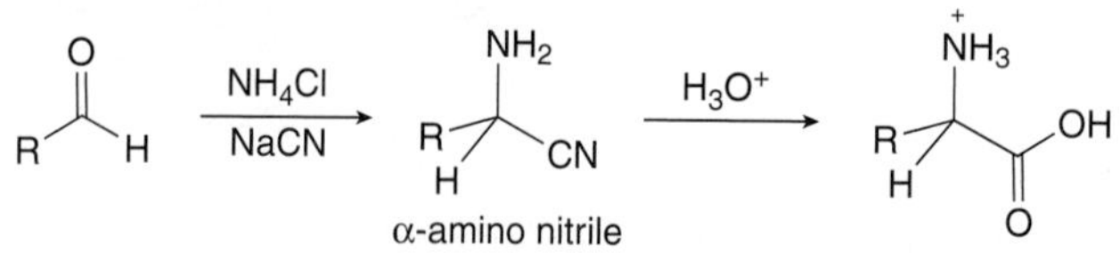

α-amino nitrile

Preparation of Optically Active Amino Acids

[1] Resolution of enantiomers by forming diastereomers (26.3A)

- Convert a racemic mixture of amino acids into a racemic mixture of *N*-acetyl amino acids [(*S*)- and (*R*)-$CH_3CONHCH(R)COOH$].
- Treat the enantiomers with a chiral amine to form a mixture of diastereomers.
- Separate the diastereomers.
- Regenerate the amino acids by protonation of the carboxylate salt and hydrolysis of the *N*-acetyl group.

[2] Kinetic resolution using enzymes (26.3B)

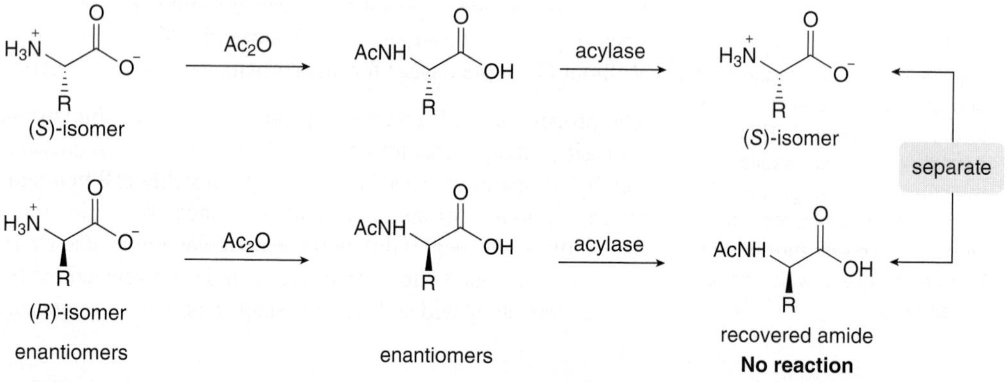

[3] By enantioselective hydrogenation (26.4)

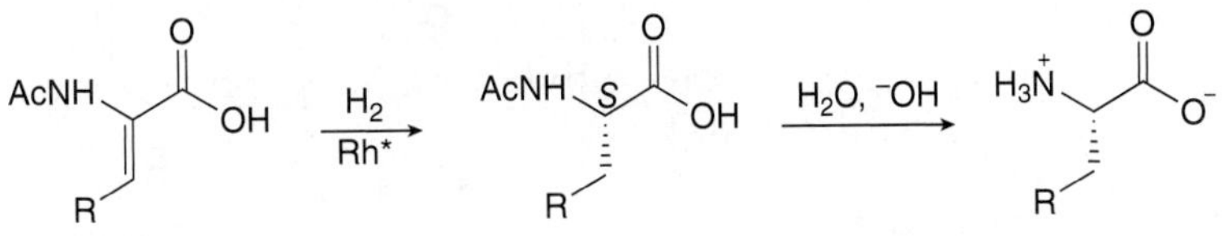

S enantiomer

S amino acid

Rh* = chiral Rh hydrogenation catalyst

Summary of Methods Used for Peptide Sequencing (26.6)

- Complete hydrolysis of all amide bonds in a peptide gives the identity and amount of the individual amino acids.
- Edman degradation identifies the N-terminal amino acid. Repeated Edman degradations can be used to sequence a peptide from the N-terminal end.
- Cleavage with carboxypeptidase identifies the C-terminal amino acid.
- Partial hydrolysis of a peptide forms smaller fragments that can be sequenced. Amino acid sequences common to smaller fragments can be used to determine the sequence of the complete peptide.
- Selective cleavage of a peptide occurs with trypsin and chymotrypsin to identify the location of specific amino acids (Table 26.2).

Adding and Removing Protecting Groups for Amino Acids (26.7)

[1] Protection of an amino group as a Boc derivative

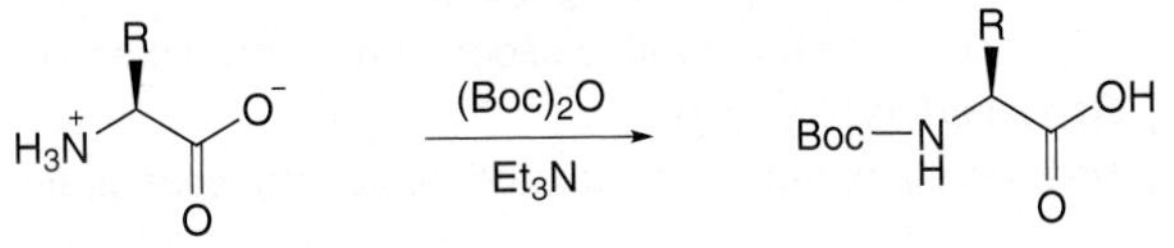

[2] Deprotection of a Boc-protected amino acid

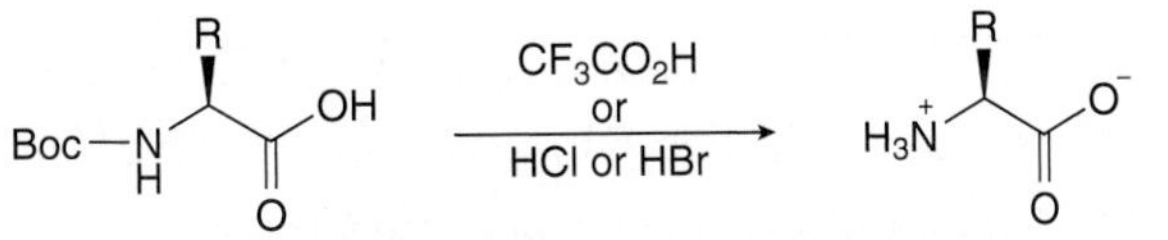

[3] Protection of an amino group as an Fmoc derivative

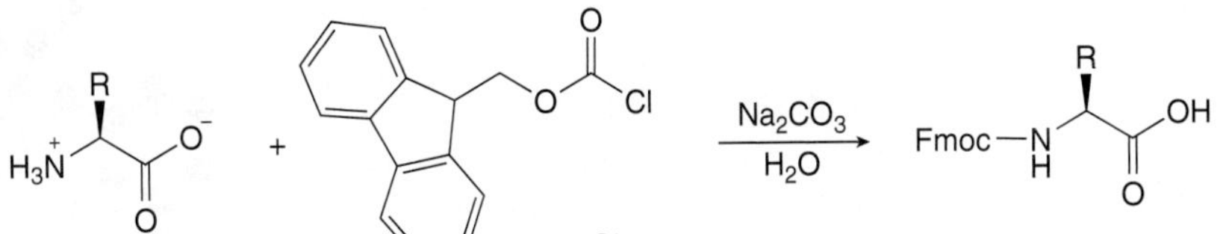

Fmoc–Cl

[4] Deprotection of an Fmoc-protected amino acid

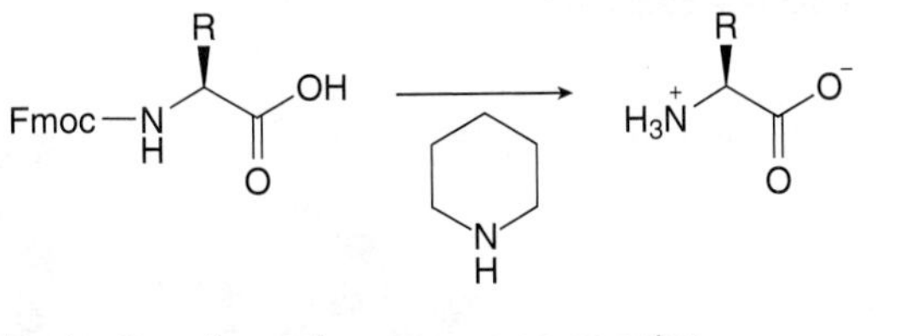

[5] Protection of a carboxy group as an ester

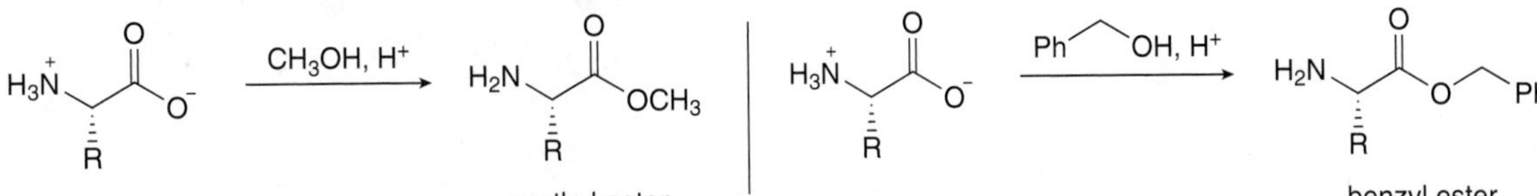

methyl ester

benzyl ester

[6] Deprotection of an ester group

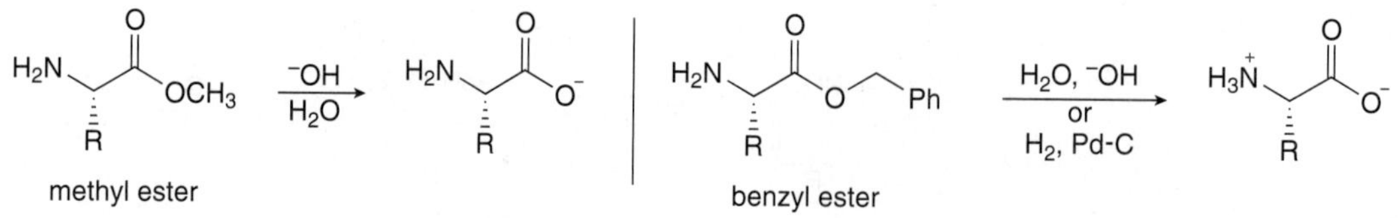
H2N
OCH3
R
methyl ester
⁻OH
H2O
O⁻
benzyl ester
Ph
H2O, ⁻OH
or
H2, Pd-C
H3N

Synthesis of Dipeptides (26.7)

[1] Amide formation with DCC

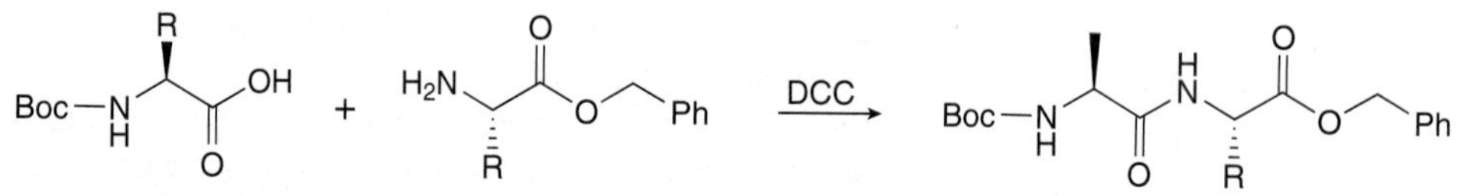
Boc
OH
H2N
Ph
DCC
R

[2] Four steps are needed to synthesize a dipeptide:

a. **Protect** the amino group of one amino acid with a Boc or Fmoc group.
b. **Protect** the carboxy group of the second amino acid as an ester.
c. Form the amide bond with **DCC.**
d. **Remove both protecting groups** in one or two reactions.

Summary of the Merrifield Method of Peptide Synthesis (26.8)

[1] Attach an Fmoc-protected amino acid to a polymer derived from polystyrene.
[2] Remove the Fmoc protecting group.
[3] Form the amide bond with a second Fmoc-protected amino acid by using DCC.
[4] Repeat steps [2] and [3].
[5] Remove the protecting group and detach the peptide from the polymer.

PROBLEMS

Problems Using Three-Dimensional Models

26.29 Draw the product formed when the following amino acid is treated with each reagent: (a) CH_3OH, H^+; (b) CH_3COCl, pyridine; (c) HCl (1 equiv); (d) NaOH (1 equiv); (e) $C_6H_5N{=}C{=}S$.

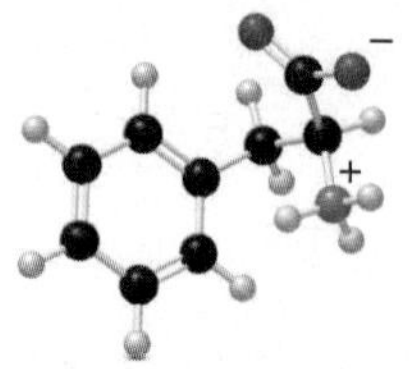

26.30 With reference to the following peptide: (a) Identify the N-terminal and C-terminal amino acids. (b) Name the peptide using one-letter abbreviations. (c) Label all the amide bonds in the peptide backbone.

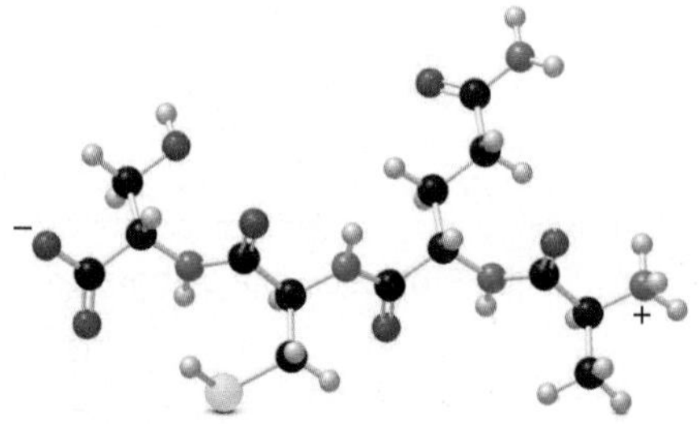

26.31 Devise a synthesis of the following dipeptide from amino acid starting materials.

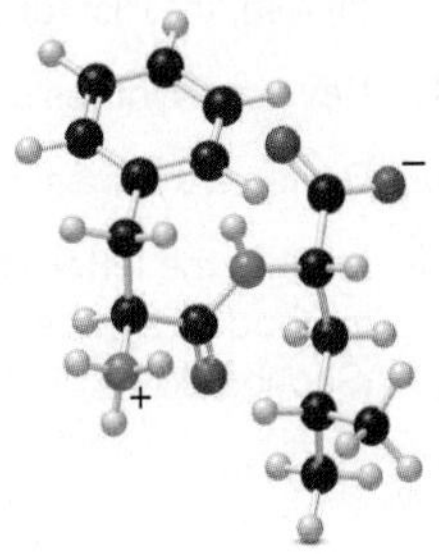

Amino Acids

26.32

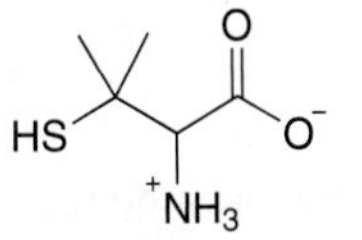
HS
O
O
NH3

penicillamine

a. (*S*)-Penicillamine, an amino acid that does not occur in proteins, is used as a copper chelating agent to treat Wilson's disease, an inherited defect in copper metabolism. (*R*)-Penicillamine is toxic, sometimes causing blindness. Draw the structures of (*R*)- and (*S*)-penicillamine.

b. What disulfide is formed from oxidation of (*S*)-penicillamine?

26.33 Histidine is classified as a basic amino acid because one of the N atoms in its five-membered ring is readily protonated by acid. Which N atom in histidine is protonated and why?

26.34 Tryptophan is not classified as a basic amino acid even though it has a heterocycle containing a nitrogen atom. Why is the N atom in the five-membered ring of tryptophan not readily protonated by acid?

26.35 What is the structure of each amino acid at its isoelectric point: (a) alanine; (b) methionine; (c) aspartic acid; (d) lysine?

26.36 What is the predominant form of each of the following amino acids at pH = 1? What is the overall charge on the amino acid at this pH? (a) threonine; (b) methionine; (c) aspartic acid; (d) arginine

26.37 What is the predominant form of each of the following amino acids at pH = 11? What is the overall charge on the amino acid? (a) valine; (b) proline; (c) glutamic acid; (d) lysine

26.38 a. Draw the structure of the tripeptide A–A–A, and label the two ionizable functional groups.

b. What is the predominant form of A–A–A at pH = 1?

c. The pK_a values for the two ionizable functional groups (3.39 and 8.03) differ considerably from the pK_a values of alanine (2.35 and 9.87; see Table 26.1). Account for the observed pK_a differences.

Synthesis and Reactions of Amino Acids

26.39 Draw the organic products formed in each reaction.

a. 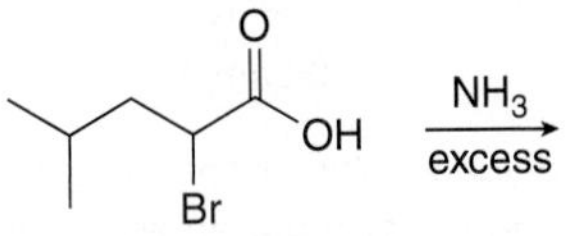
O
OH
Br
NH3
excess

c. 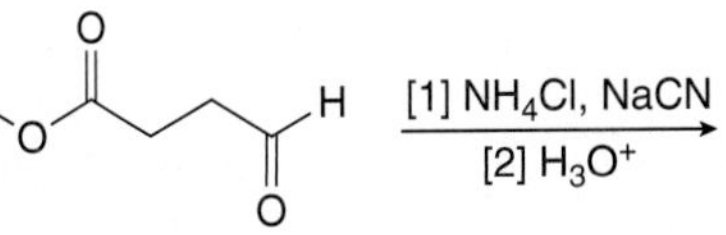
O
O
H
O
[1] NH4Cl, NaCN
[2] H3O+

b. 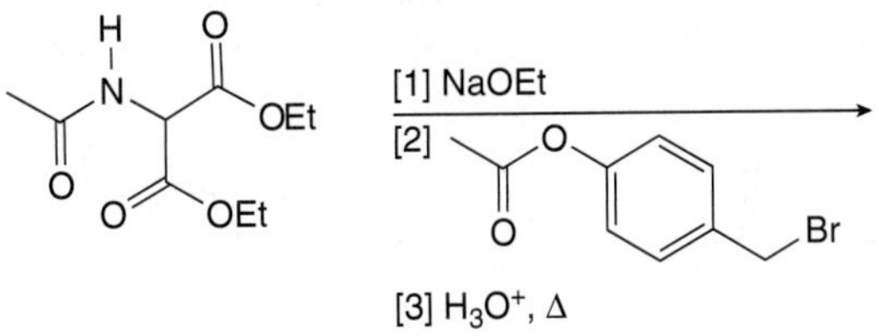
H
N
O
OEt
O
O
OEt
[1] NaOEt
[2]
O
O
Br
[3] H3O+, Δ

d. 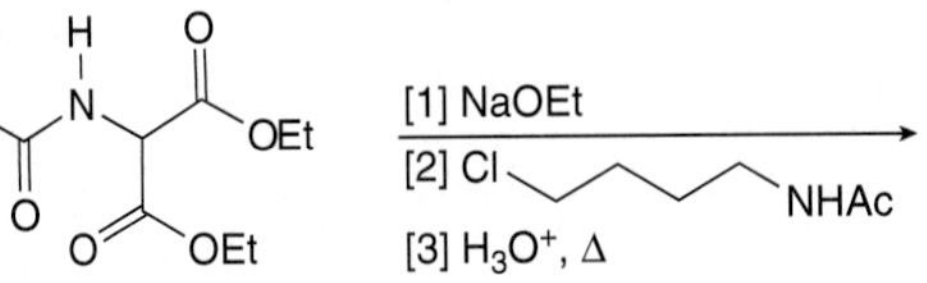
H
N
O
OEt
O
O
OEt
[1] NaOEt
[2] Cl
NHAc
[3] H3O+, Δ

26.40 What alkyl halide is needed to synthesize each amino acid from diethyl acetamidomalonate: (a) Asn; (b) His; (c) Trp?

26.41 Devise a synthesis of threonine from diethyl acetamidomalonate.

26.42 Devise a synthesis of each amino acid from acetaldehyde (CH_3CHO): (a) glycine; (b) alanine.

26.43 Identify the lettered intermediates in the following reaction scheme. This is an alternative method to synthesize amino acids, based on the Gabriel synthesis of 1° amines (Section 25.7A).

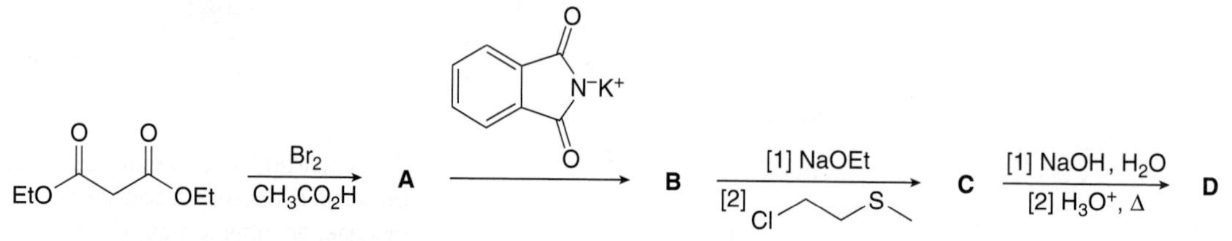

26.44 Glutamic acid is synthesized by the following reaction sequence. Draw a stepwise mechanism for Steps [1]–[3].

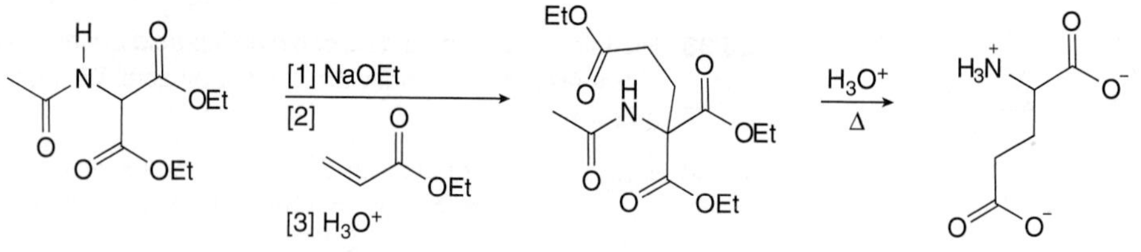

glutamic acid

26.45 Identify **A–E** in the following reaction sequence.

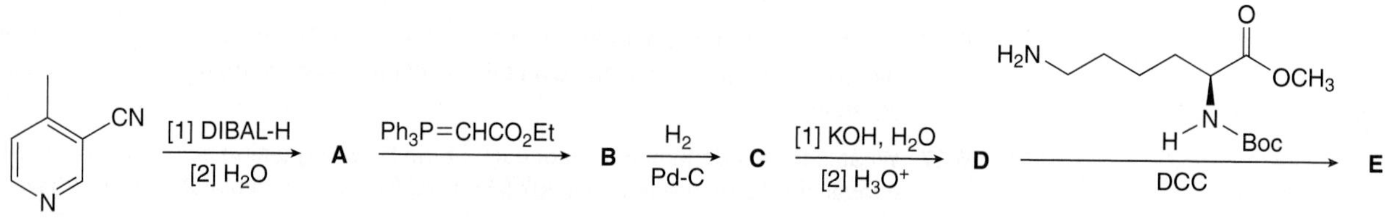

Resolution; The Synthesis of Chiral Amino Acids

26.46 Write out a scheme for the resolution of the two enantiomers of the antiplatelet drug clopidogrel with 10-camphorsulfonic acid.

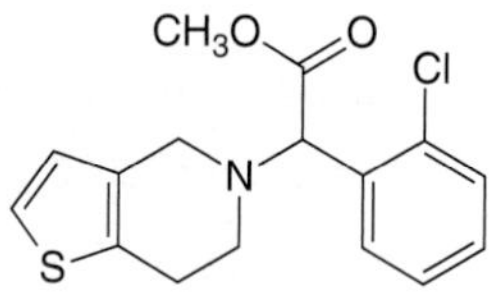

clopidogrel

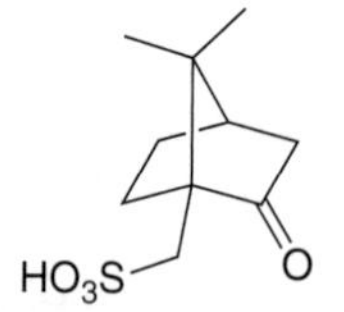

10-camphorsulfonic acid

26.47 Another strategy used to resolve amino acids involves converting the carboxy group to an ester and then using a *chiral carboxylic acid* to carry out an acid–base reaction at the free amino group. Using a racemic mixture of alanine enantiomers and (*R*)-mandelic acid as resolving agent, write out the steps showing how a resolution process would occur.

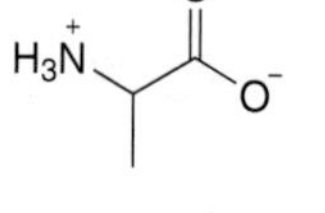

alanine
(two enantiomers)

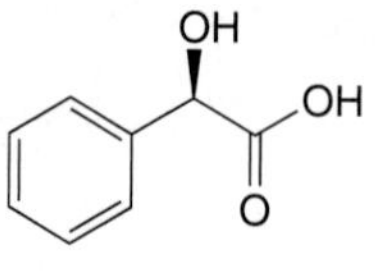

(*R*)-mandelic acid

26.48 Brucine is a poisonous alkaloid obtained from *Strychnos nux vomica,* a tree that grows in India, Sri Lanka, and northern Australia. Write out a resolution scheme similar to the one given in Section 26.3A, which shows how a racemic mixture of phenylalanine can be resolved using brucine.

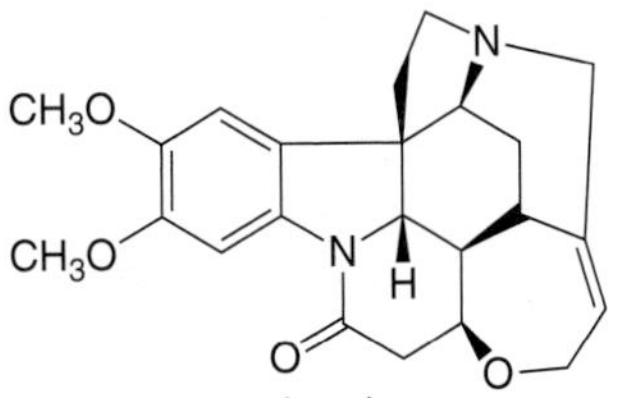

brucine

26.49 Draw the organic products formed in each reaction.

a. 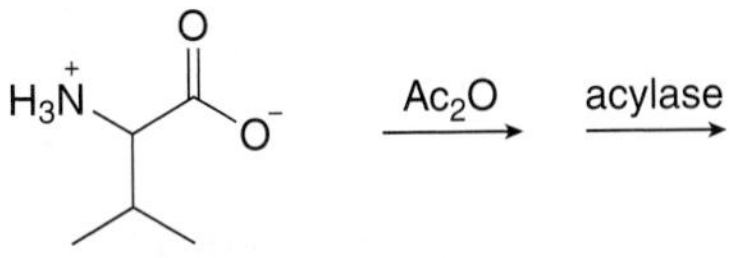

racemic mixture

b. 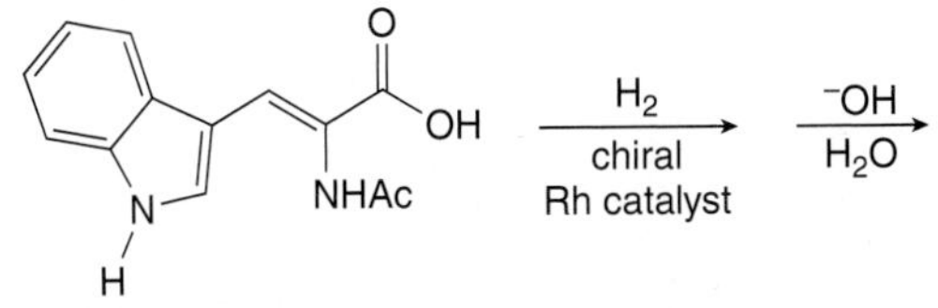

26.50 What steps are needed to convert **A** to L-dopa, an uncommon amino acid that is effective in treating Parkinson's disease? These steps are the key reactions in the first commercial asymmetric synthesis using a chiral transition metal catalyst. This process was developed at Monsanto in 1974.

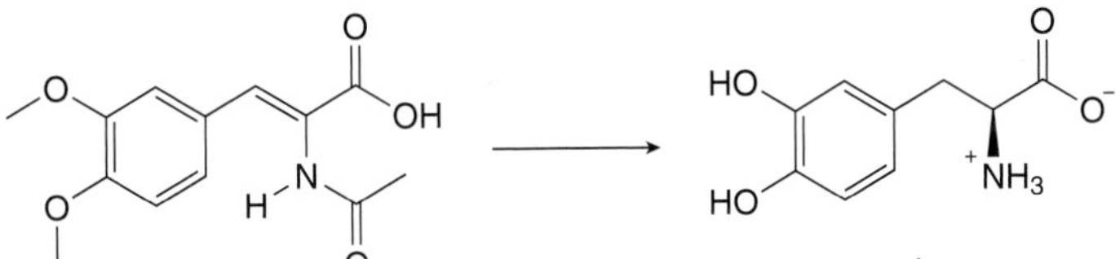

A L-dopa

Peptide Structure and Sequencing

26.51 Draw the structure for each peptide: (a) Phe–Ala; (b) Gly–Gln; (c) Lys–Gly; (d) R–H.

26.52 For the tetrapeptide Asp–Arg–Val–Tyr:

a. Name the peptide using one-letter abbreviations.
b. Draw the structure.
c. Label all amide bonds.
d. Label the N-terminal and C-terminal amino acids.

26.53 Name each peptide using both the three-letter and one-letter abbreviations of the component amino acids.

a. 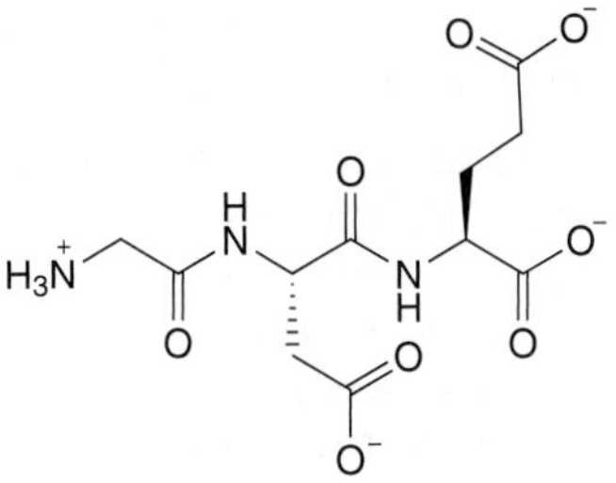

b. 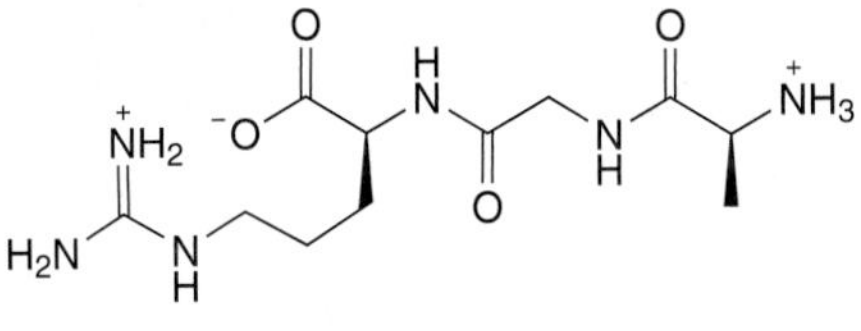

26.54 Gramicidin S, a topical antibiotic produced by the bacterium *Bacillus brevis,* is a cyclic decapeptide formed from five amino acids. Draw the structures of the amino acids that form gramicidin S, and explain why this compound possesses two unusual structural features.

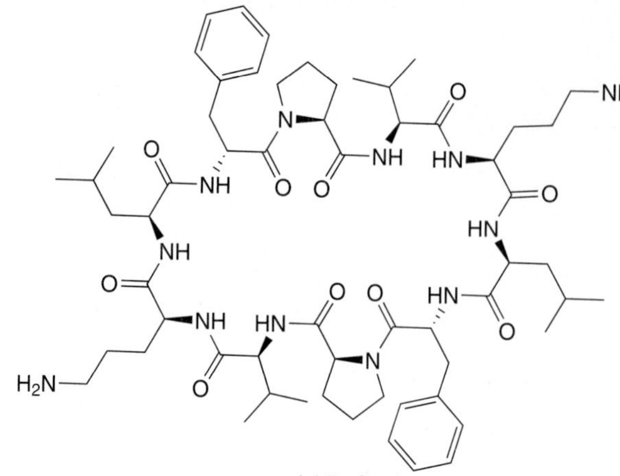

gramicidin S

26.55 Draw the amino acids and peptide fragments formed when the decapeptide A–P–F–L–K–W–S–G–R–G is treated with each reagent or enzyme: (a) chymotrypsin; (b) trypsin; (c) carboxypeptidase; (d) $C_6H_5N{=}C{=}S$.

26.56 Consider the decapeptide angiotensin I.

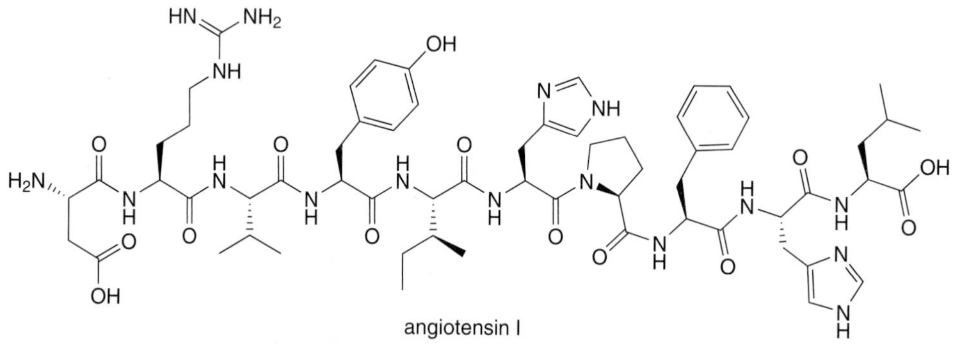

angiotensin I

a. What products are formed when angiotensin I is treated with trypsin?

b. What products are formed when angiotensin I is treated with chymotrypsin?

c. Treatment of angiotensin I with ACE (the angiotensin-converting enzyme) cleaves only the amide bond with the carbonyl group derived from phenylalanine to afford two products. The larger polypeptide is angiotensin II, a hormone that narrows blood vessels and increases blood pressure. Give the amino acid sequence of angiotensin II using three-letter abbreviations. ACE inhibitors are drugs that lower blood pressure by inhibiting the ACE enzyme (Problem 5.15).

26.57 Give the amino acid sequence of each peptide using the fragments obtained by partial hydrolysis of the peptide with acid.

a. a tetrapeptide that contains Ala, Gly, His, and Tyr, which is hydrolyzed to the dipeptides His–Tyr, Gly–Ala, and Ala–His

b. a pentapeptide that contains Glu, Gly, His, Lys, and Phe, which is hydrolyzed to His–Gly–Glu, Gly–Glu–Phe, and Lys–His

26.58 Use the given experimental data to deduce the sequence of an octapeptide that contains the following amino acids: Ala, Gly (2 equiv), His (2 equiv), Ile, Leu, and Phe. Edman degradation cleaves Gly from the octapeptide, and carboxypeptidase forms Leu and a heptapeptide. Partial hydrolysis forms the following fragments: Ile–His–Leu, Gly, Gly–Ala–Phe–His, and Phe–His–Ile.

26.59 An octapeptide contains the following amino acids: Arg, Glu, His, Ile, Leu, Phe, Tyr, and Val. Carboxypeptidase treatment of the octapeptide forms Phe and a heptapeptide. Treatment of the octapeptide with chymotrypsin forms two tetrapeptides, **A** and **B.** Treatment of **A** with trypsin yields two dipeptides, **C** and **D.** Edman degradation cleaves the following amino acids from each peptide: Glu (octapeptide), Glu (**A**), Ile (**B**), Glu (**C**), and Val (**D**). Partial hydrolysis of tetrapeptide **B** forms Ile–Leu in addition to other products. Deduce the structure of the octapeptide and fragments **A–D.**

Peptide Synthesis

26.60 Draw the organic products formed in each reaction.

a. 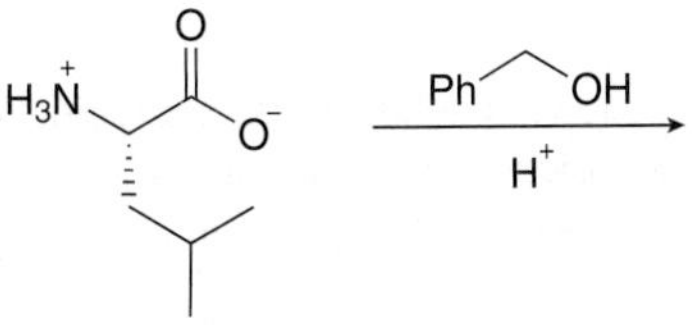

d. 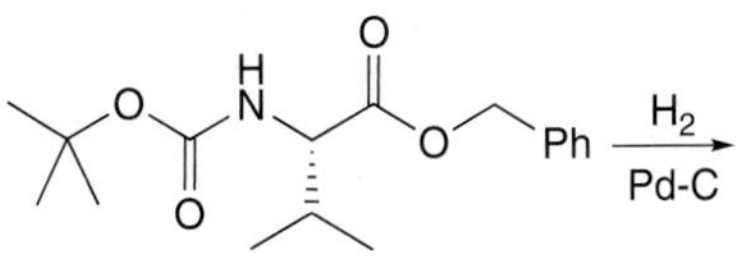

b. 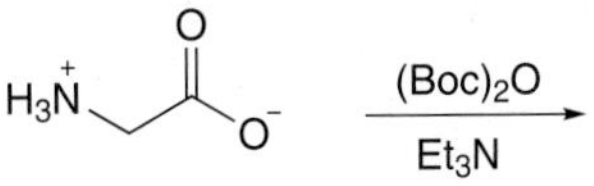

e. product in (d) $\xrightarrow{CF_3COOH}$

c. product in (a) + product in (b) $\xrightarrow{DCC}$

f. 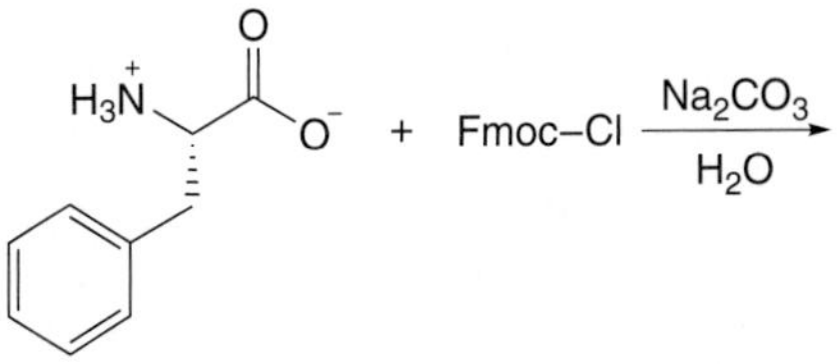

26.61 Draw all the steps in the synthesis of each peptide from individual amino acids: (a) Gly–Ala; (b) Ile–Ala–Phe.

26.62 Write out the steps for the synthesis of each peptide using the Merrifield method: (a) Ala–Leu–Phe–Phe; (b) Phe–Gly–Ala–Ile.

26.63 Besides the Boc and Fmoc protecting groups used in peptide synthesis, amines can also be protected by reaction with benzyl chloroformate ($C_6H_5CH_2OCOCl$). Draw the structure of the product formed by reaction of alanine with benzyl chloroformate.

26.64 Another method to form a peptide bond involves a two-step process:

[1] Conversion of a Boc-protected amino acid to a *p*-nitrophenyl ester.

[2] Reaction of the *p*-nitrophenyl ester with an amino acid ester.

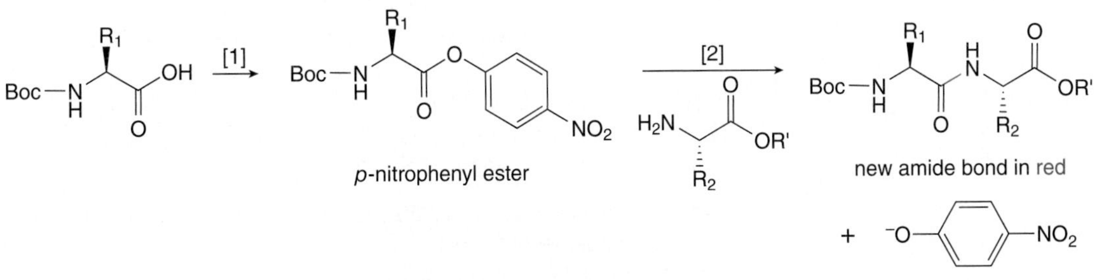
p-nitrophenyl ester
new amide bond in red

a. Why does a *p*-nitrophenyl ester “activate” the carboxy group of the first amino acid to amide formation?

b. Would a *p*-methoxyphenyl ester perform the same function? Why or why not?

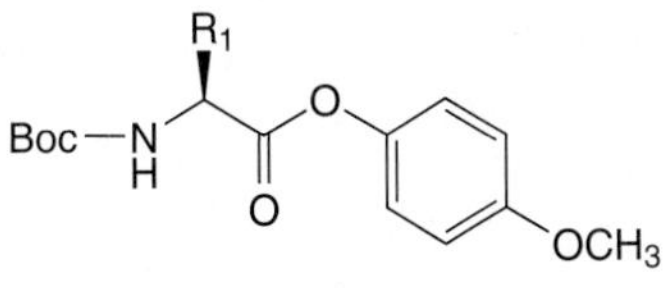

p-methoxyphenyl ester

26.65 Draw the mechanism for the reaction that removes an Fmoc group from an amino acid under the following conditions:

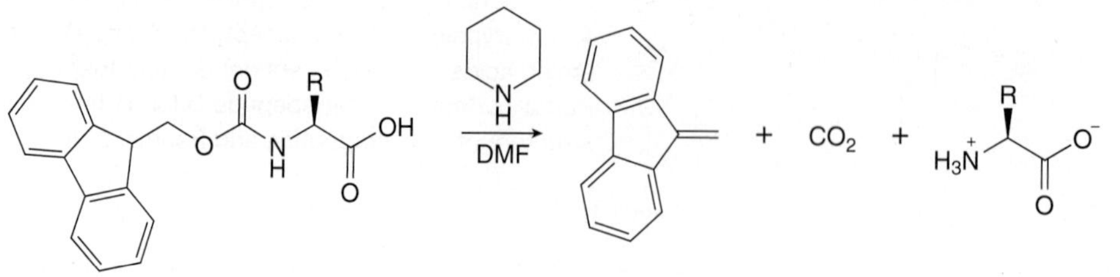

Proteins

26.66 Which of the following amino acids are typically found in the interior of a globular protein, and which are typically found on the surface: (a) phenylalanine; (b) aspartic acid; (c) lysine; (d) isoleucine; (e) arginine; (f) glutamic acid?

26.67 After the peptide chain of collagen has been formed, many of the proline residues are hydroxylated on one of the ring carbon atoms. Why is this process important for the triple helix of collagen?

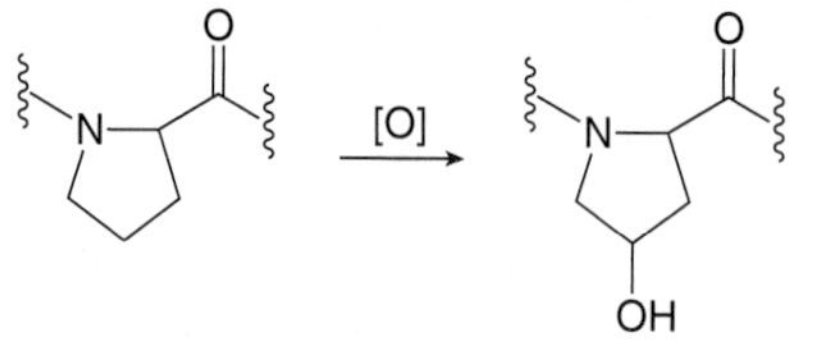

Challenge Problems

26.68 Devise a stepwise synthesis of the tripeptide Val–Leu–Val from 3-methylbutanal [$(CH_3)_2CHCH_2CHO$] as the only organic starting material. You may also use any required inorganic or organic reagents.

26.69 Besides asymmetric hydrogenation (Section 26.4), several other methods are now available for the synthesis of optically active amino acids. How might a reaction like the Strecker synthesis be adapted to the preparation of chiral amino acids?

26.70 The anti-obesity drug orlistat works by irreversibly inhibiting pancreatic lipase, an enzyme responsible for the hydrolysis of triacylglycerols in the intestines, so they are excreted without metabolism. Inhibition occurs by reaction of orlistat with a serine residue of the enzyme, forming a covalently bound, inactive enzyme product. Draw the structure of the product formed during inhibition.

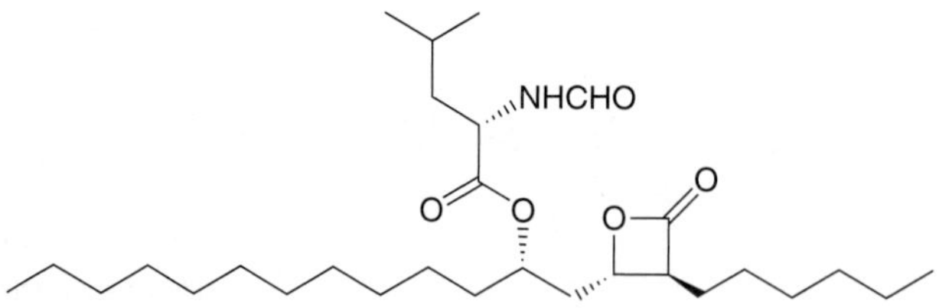

26.71 As shown in Mechanism 26.2, the final steps in the Edman degradation result in rearrangement of a thiazolinone to an *N*-phenylthiohydantoin. Draw a stepwise mechanism for this acid-catalyzed reaction.

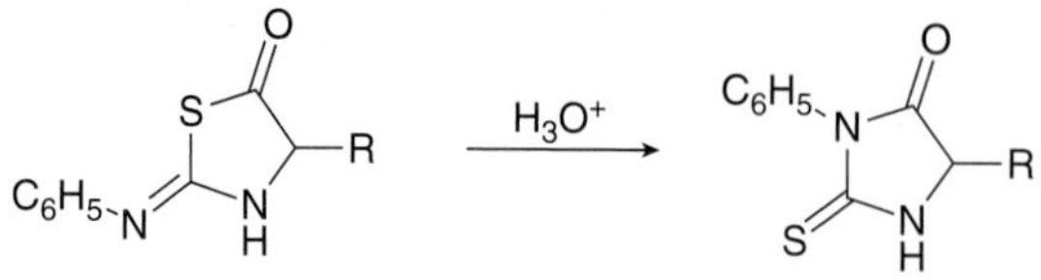

thiazolinone

N-phenylthiohydantoin

Carbohydrates

27

27.1 Introduction
27.2 Monosaccharides
27.3 The family of D-aldoses
27.4 The family of D-ketoses
27.5 Physical properties of monosaccharides
27.6 The cyclic forms of monosaccharides
27.7 Glycosides
27.8 Reactions of monosaccharides at the OH groups
27.9 Reactions at the carbonyl group—Oxidation and reduction
27.10 Reactions at the carbonyl group—Adding or removing one carbon atom
27.11 Disaccharides
27.12 Polysaccharides
27.13 Other important sugars and their derivatives

The leaves, stems, and green spots on potatoes contain the toxin **solanine,** which is produced by the plant as a defense against insects and predators. Solanine is a carbohydrate derivative, formed from a complex amine and three monosaccharide rings, which are joined together by acetals called glycosides. The many hydroxy groups of the carbohydrate portion of solanine increase its water solubility, a useful feature in aqueous biological systems. In Chapter 27, we learn about carbohydrates and carbohydrate derivatives like solanine.

In Chapter 27, we turn our attention to **carbohydrates,** the largest group of organic molecules in nature, comprising approximately 50% of the earth's biomass. Carbohydrates can be simple or complex, having as few as three or as many as thousands of carbon atoms. The glucose metabolized for energy in cells, the sucrose of table sugar, and the cellulose of plant stems and tree trunks are all examples of carbohydrates. Carbohydrates on cell surfaces determine blood type, and carbohydrates form the backbone of DNA, the carrier of all genetic information in the cell. Carbohydrates have many polar functional groups, whose structure and properties can be understood by applying the basic principles of organic chemistry.

27.1 Introduction

Carbohydrates were given their name because molecular formulas of simple carbohydrates could be written as $C_n(H_2O)_n$, making them **hydrates of carbon.**

Carbohydrates such as glucose and cellulose were discussed in Sections 5.1, 6.4, and 21.17.

Carbohydrates, commonly referred to as sugars and starches, are polyhydroxy aldehydes and ketones, or compounds that can be hydrolyzed to them. The cellulose in plant stems and tree trunks and the chitin in the exoskeletons of arthropods and mollusks are both complex carbohydrates. Four examples are shown in Figure 27.1. They include not only glucose and cellulose, but also doxorubicin (an anticancer drug) and 2'-deoxyadenosine 5'-monophosphate (a nucleotide from DNA), both of which have a carbohydrate moiety as part of a larger molecule.

Carbohydrates are storehouses of chemical energy. They are synthesized in green plants and algae by **photosynthesis,** a process that uses the energy from the sun to convert carbon dioxide and water

Figure 27.1
Some examples of carbohydrates

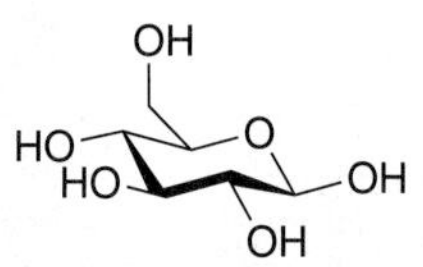

β-D-glucose
most common simple carbohydrate

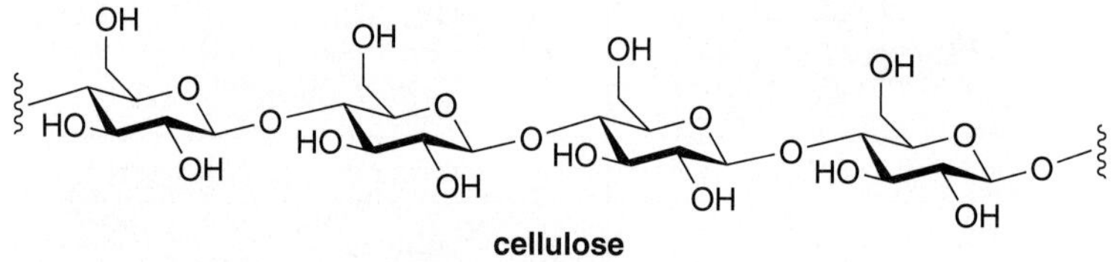

cellulose
main component of wood

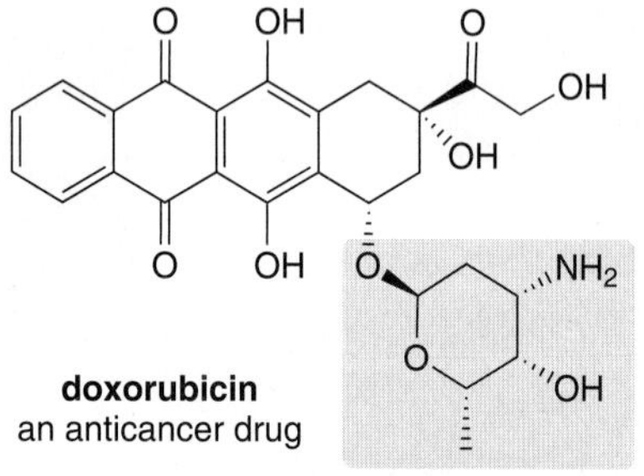

doxorubicin
an anticancer drug

carbohydrate portion

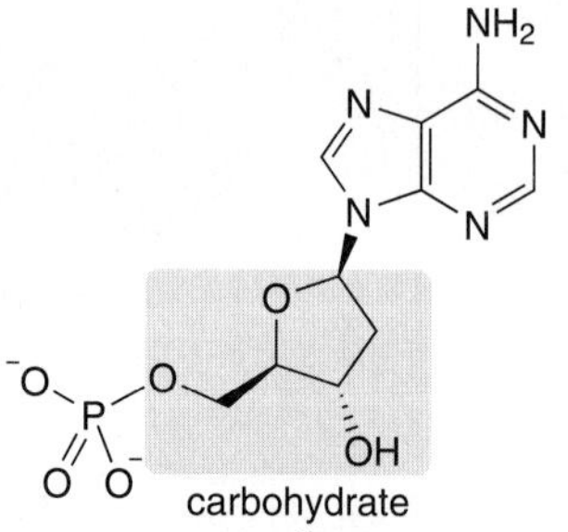

carbohydrate portion

2'-deoxyadenosine 5'-monophosphate
a nucleotide component of DNA

These compounds illustrate the structural diversity of carbohydrates and their derivatives. **Glucose** is the most common simple sugar, whereas **cellulose,** which comprises wood, plant stems, and grass, is the most common carbohydrate in the plant world. **Doxorubicin,** an anticancer drug that has a carbohydrate ring as part of its structure, has been used in the treatment of leukemia, Hodgkin's disease, and cancers of the breast, bladder, and ovaries. **2'-Deoxyadenosine 5'-monophosphate** is one of the four nucleotides that form DNA.

into glucose and oxygen. This energy is released when glucose is metabolized. The oxidation of glucose is a multistep process that forms carbon dioxide, water, and a great deal of energy (Section 6.4).

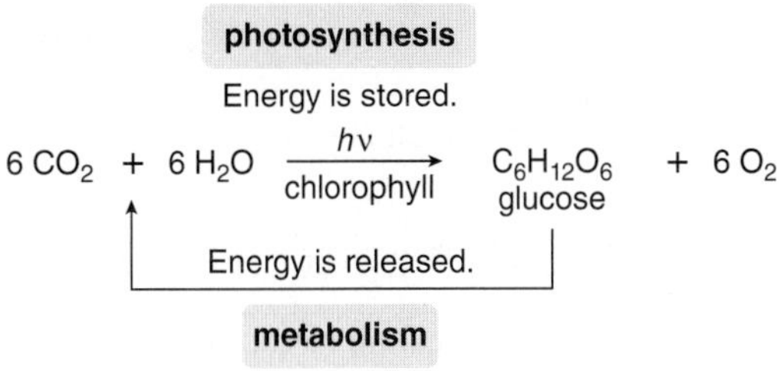

Although the metabolism of lipids provides more energy per gram than the metabolism of carbohydrates, glucose is the preferred source when a burst of energy is needed during exercise. Glucose is water soluble, so it can be quickly and easily transported through the bloodstream to the tissues.

27.2 Monosaccharides

The simplest carbohydrates are called **monosaccharides** or **simple sugars. Monosaccharides have three to seven carbon atoms** in a chain, with a **carbonyl group** at either the terminal carbon (C1) or the carbon adjacent to it (C2). In most carbohydrates, each of the remaining carbon atoms has a **hydroxy group.** Monosaccharides are often drawn vertically, with the carbonyl group at the top. When this convention is used, monosaccharides look different from molecules encountered in prior chapters.

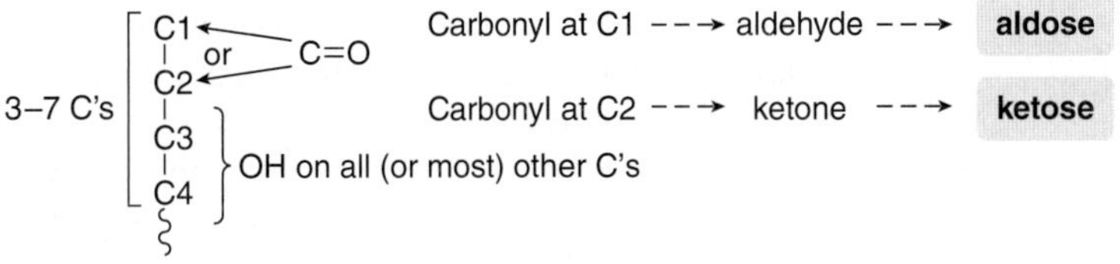

monosaccharide

- **Monosaccharides with an aldehyde carbonyl group at C1 are called aldoses.**
- **Monosaccharides with a ketone carbonyl group at C2 are called ketoses.**

D-Fructose is almost twice as sweet as normal table sugar (sucrose) with about the same number of calories per gram. "Lite" food products use only half as much fructose as sucrose for the same level of sweetness, and so they have fewer calories.

Several examples of simple carbohydrates are shown. D-Glyceraldehyde and dihydroxyacetone have the same molecular formula, so they are **constitutional isomers,** as are D-glucose and D-fructose.

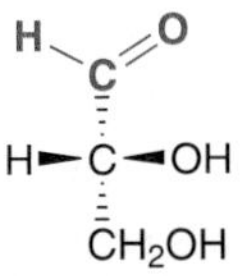

D-glyceraldehyde
an aldehyde
aldose

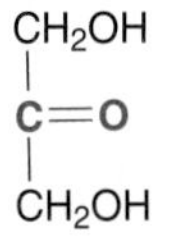

dihydroxyacetone
a ketone
ketose

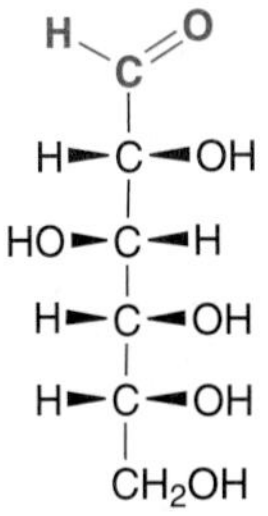

D-glucose
an aldehyde
aldose

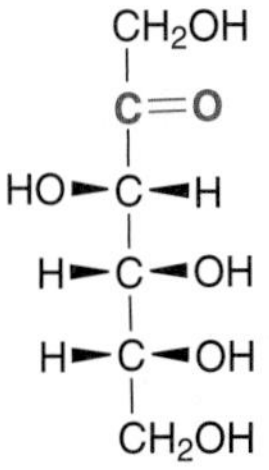

D-fructose
a ketone
ketose

Dihydroxyacetone is the active ingredient in many artificial tanning agents.

All carbohydrates have common names. The simplest aldehyde, glyceraldehyde, and the simplest ketone, dihydroxyacetone, are the only monosaccharides whose names do not end in the suffix ***-ose.*** (The prefix "D-" is explained in Section 27.2C.)

A monosaccharide is called:

- **a triose if it has 3 C's;**
- **a tetrose if it has 4 C's;**
- **a pentose if it has 5 C's;**
- **a hexose if it has 6 C's, and so forth.**

These terms are then combined with the words *aldose* and *ketose* to indicate both the number of carbon atoms in the monosaccharide and whether it contains an aldehyde or ketone. Thus, glyceraldehyde is an aldotriose (three C atoms and an aldehyde), glucose is an aldohexose (six C atoms and an aldehyde), and fructose is a ketohexose (six C atoms and a ketone).

Problem 27.1 Draw the structure of (a) a ketotetrose; (b) an aldopentose; (c) an aldotetrose.

27.2A Fischer Projection Formulas

A striking feature of carbohydrate structure is the presence of stereogenic centers. **All carbohydrates except for dihydroxyacetone contain one or more stereogenic centers.**

The simplest aldehyde, glyceraldehyde, has one stereogenic center, so there are two possible **enantiomers.** Only the enantiomer with the *R* configuration occurs naturally.

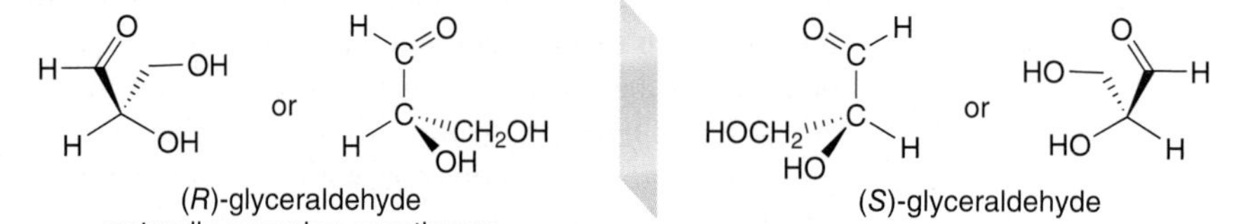

(*R*)-glyceraldehyde
naturally occurring enantiomer

(*S*)-glyceraldehyde

The stereogenic centers in sugars are often depicted following a different convention than is usually seen for other stereogenic centers. Instead of drawing a tetrahedron with two bonds in the plane, one in front of the plane, and one behind it, the **tetrahedron is tipped so that horizontal bonds come forward (drawn on wedges) and vertical bonds go behind (on dashed wedges).** This structure is then abbreviated by a **cross formula,** also called a **Fischer projection formula.** In a Fischer projection formula:

- **A carbon atom is located at the intersection of the two lines of the cross.**
- **The horizontal bonds come forward, on wedges.**
- **The vertical bonds go back, on dashed wedges.**
- **In a carbohydrate, the aldehyde or ketone carbonyl is put at or near the top.**

Carbon atoms that are not stereogenic centers are generally drawn in. Using a Fischer projection formula, (*R*)-glyceraldehyde becomes:

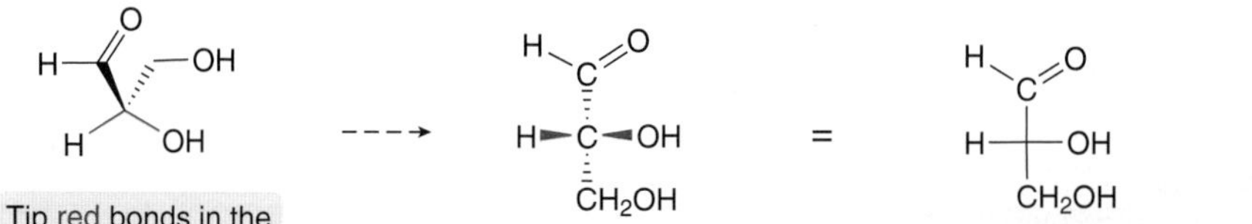

Tip red bonds in the plane forward.

(*R*)-glyceraldehyde

- Horizontal bonds come forward.
- Vertical bonds go back.

Fischer projection formula
(*R*)-glyceraldehyde

Do not rotate a Fischer projection formula in the plane of the page, because you might inadvertently convert a compound into its enantiomer. When using Fischer projections it is usually best to convert them to structures with wedges and dashed wedges, and then manipulate them. Although a Fischer projection formula can be used for the stereogenic center in any compound, it is most commonly used for monosaccharides.

Sample Problem 27.1 Convert each compound to a Fischer projection formula.

a. 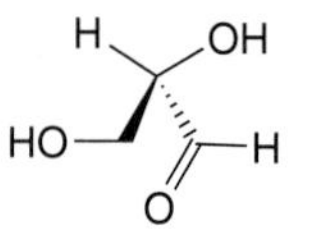

b. 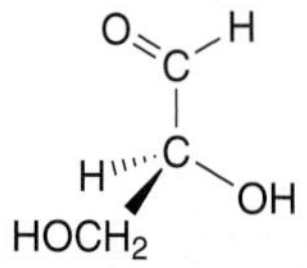

Solution

Rotate and re-draw each molecule to place the horizontal bonds in front of the plane and the vertical bonds behind the plane. Then use a cross to represent the stereogenic center.

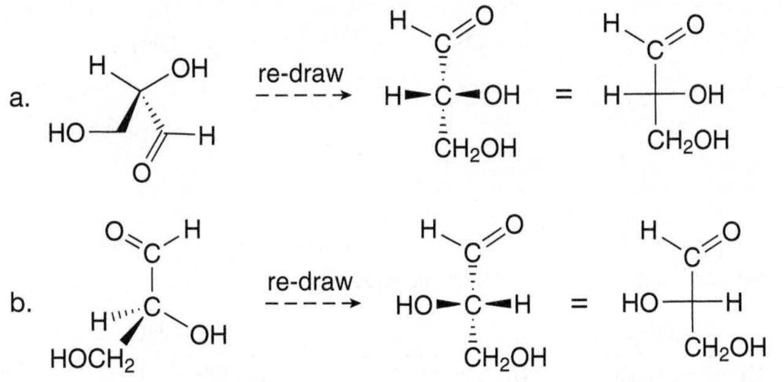

Problem 27.2 Draw each stereogenic center using a Fischer projection formula.

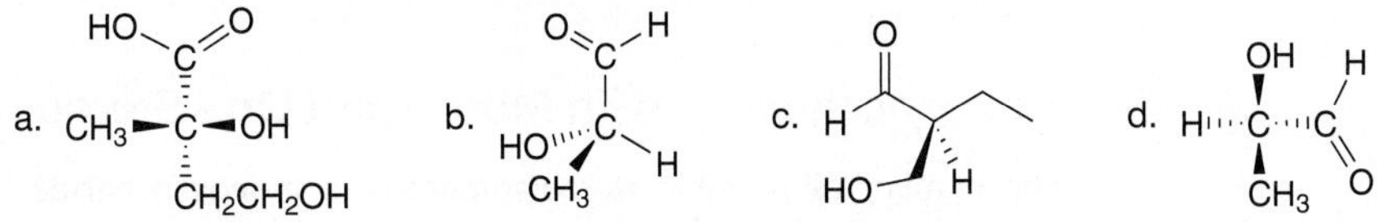

R,S designations can be assigned to any stereogenic center drawn as a Fischer projection formula in the following manner:

[1] Assign priorities (1 → 4) to the four groups bonded to the stereogenic center using the rules detailed in Section 5.6.

[2] When the lowest priority group occupies a vertical bond—that is, it projects *behind* the plane on a dashed wedge—tracing a circle in the clockwise direction (from priority group 1 → 2 → 3) gives the *R* configuration. Tracing a circle in the counterclockwise direction gives the *S* configuration.

[3] When the lowest priority group occupies a horizontal bond—that is, it projects *in front of* the plane on a wedge—reverse the answer obtained in Step [2] to designate the configuration.

Sample Problem 27.2 Re-draw each Fischer projection formula using wedges and dashed wedges for the stereogenic center, and label the center as *R* or *S*.

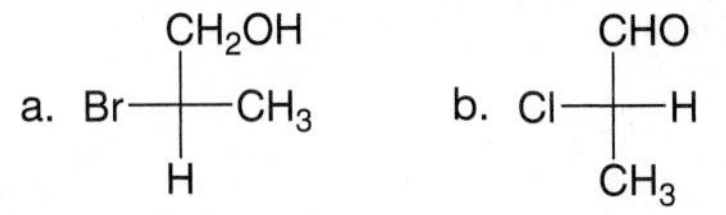

Solution

For each molecule:

[1] Convert the Fischer projection formula to a representation with wedges and dashed wedges.

[2] Assign priorities (Section 5.6).

[3] Determine *R* or *S* in the usual manner. Reverse the answer if priority group [4] is oriented forward (on a wedge).

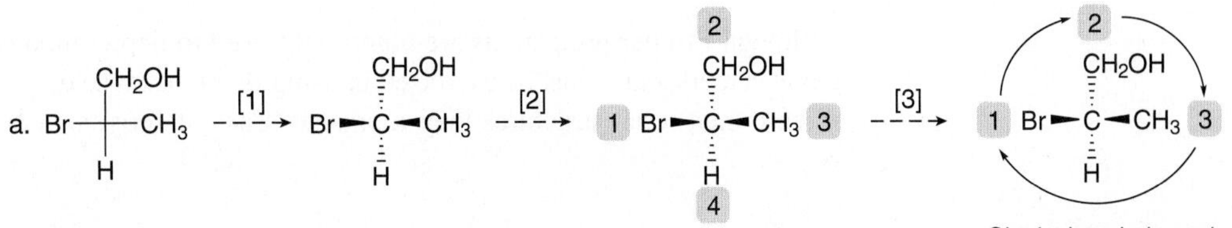

Clockwise circle and group [4] is oriented behind: ***R* configuration**

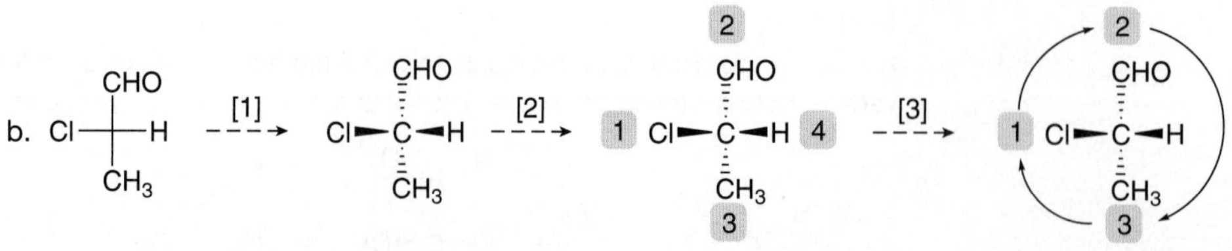

Clockwise circle and
group [4] is oriented forward:
***S* configuration**

Problem 27.3 Label each stereogenic center as *R* or *S*.

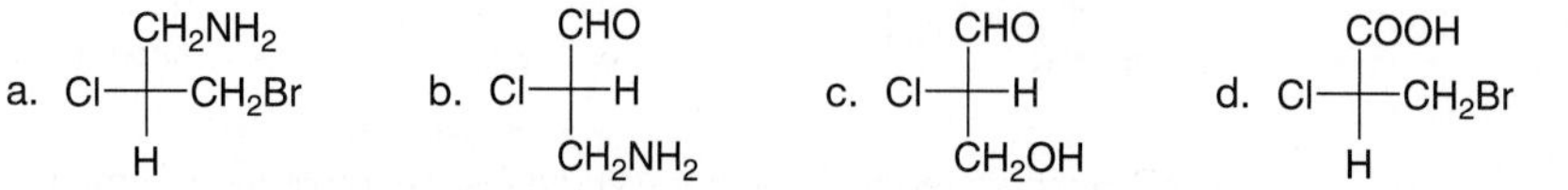

27.2B Monosaccharides with More Than One Stereogenic Center

The number of possible stereoisomers of a monosaccharide increases exponentially with the number of stereogenic centers present. **An aldohexose has four stereogenic centers, and so it has $2^4 = 16$ possible stereoisomers,** or eight pairs of enantiomers.

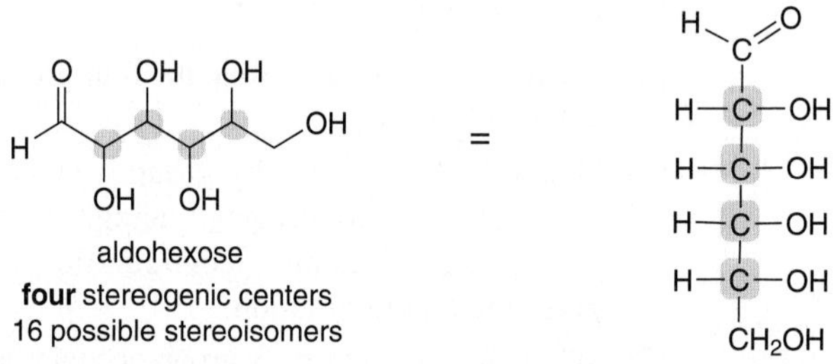

aldohexose
four stereogenic centers
16 possible stereoisomers

vertical representation

Fischer projection formulas are also used for compounds like aldohexoses that contain several stereogenic centers. In this case, the molecule is drawn with a vertical carbon skeleton and the stereogenic centers are stacked one above another. Using this convention, **all horizontal bonds project forward (on wedges).**

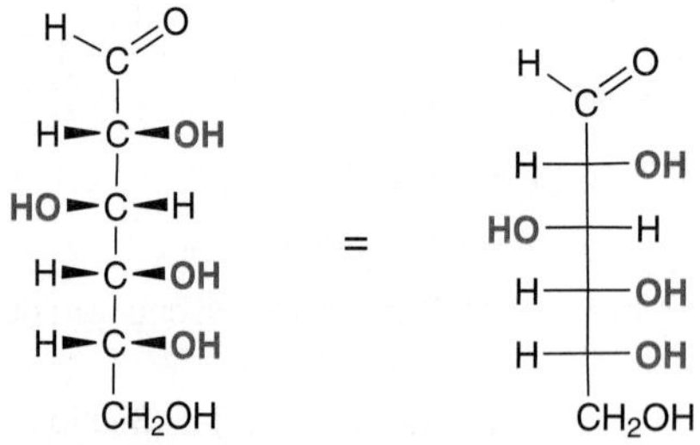

D-glucose
All horizontal bonds are drawn
as **wedges.**

Fischer projection

Although Fischer projections are commonly used to depict monosaccharides with many stereogenic centers, care must be exercised in using them, because they do not give a true picture of the three-dimensional structures they represent. Each stereogenic center is drawn in the less stable

Figure 27.2 A Fischer projection and the 3-D structure of glucose

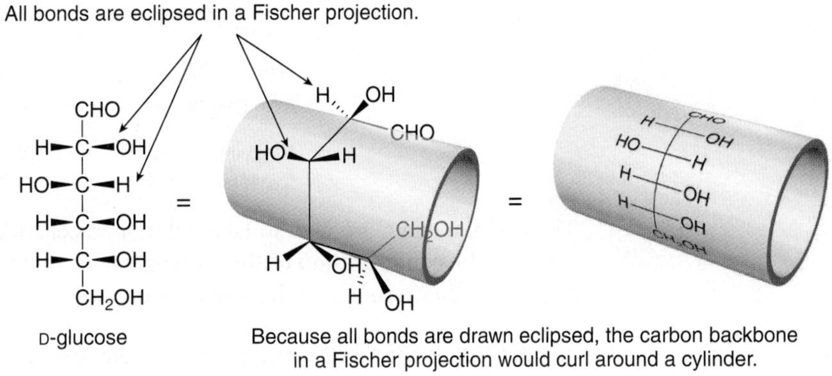

eclipsed conformation, so the Fischer projection of glucose really represents the molecule in a cylindrical conformation, as shown in Figure 27.2.

Sample Problem 27.3 Convert the ball-and-stick model to a Fischer projection.

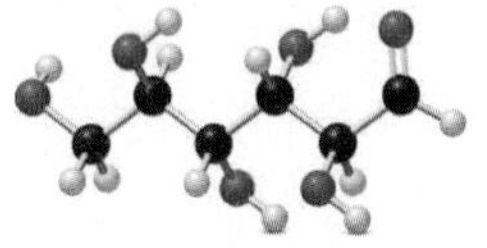

Solution

Since the ball-and-stick model is shown in the more stable staggered conformation, it must be converted to the less stable eclipsed conformation used in a Fischer projection. First, re-draw the model as a skeletal structure (**A**), and rotate it to place the carbonyl group at the top (**B**). To convert the all-staggered form to the all-eclipsed form, rotate about the bonds in **B** to swing two carbons (labeled in red) 180°, forming **C.** Re-draw **C** so that all bonds to H and OH on the four stereogenic centers are drawn on wedges, forming **D.** Groups that are on wedges in **C** (in red) are on the left side of the carbon skeleton in **D,** and groups on dashed wedges in **C** (in blue) are on the right side of the carbon skeleton in **D.** Finally, replace the wedges with crosses to form the Fischer projection.

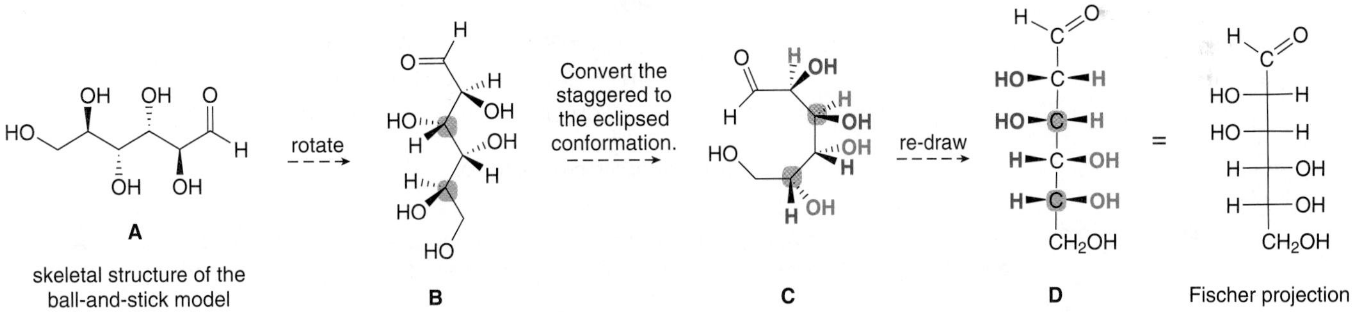

Problem 27.4 Convert the ball-and-stick model to a Fischer projection.

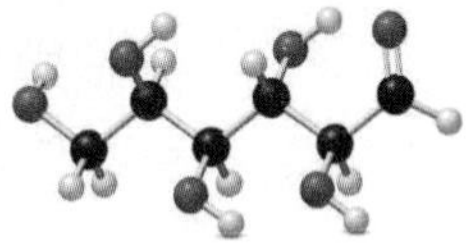

Problem 27.5 Assign *R,S* designations to each stereogenic center in glucose.

27.2C D and L Monosaccharides

Although the prefixes *R* and *S* can be used to designate the configuration of stereogenic centers in monosaccharides, an older system of nomenclature uses the prefixes D- and L-, instead. Naturally occurring glyceraldehyde with the *R* configuration is called the **D-isomer.** Its enantiomer, (*S*)-glyceraldehyde, is called the **L-isomer.**

(*R*)-glyceraldehyde
D-glyceraldehyde

(*S*)-glyceraldehyde
L-glyceraldehyde

The letters D and L are used to label all monosaccharides, even those with multiple stereogenic centers. **The configuration of the stereogenic center *farthest* from the carbonyl group determines whether a monosaccharide is D- or L-.**

- **A D-sugar has the OH group on the stereogenic center farthest from the carbonyl on the right in a Fischer projection (like D-glyceraldehyde).**
- **An L-sugar has the OH group on the stereogenic center farthest from the carbonyl on the left in a Fischer projection (like L-glyceraldehyde).**

The two designations, D and *d,* refer to very different phenomena. The "D" designates the configuration around a stereogenic center in a monosaccharide. The "*d,*" on the other hand, is an abbreviation for "dextrorotatory"; that is, a *d*-compound rotates the plane of polarized light in the clockwise direction. A D-sugar may be dextrorotatory or it may be levorotatory. **There is no direct correlation between D and *d* or L and *l*.**

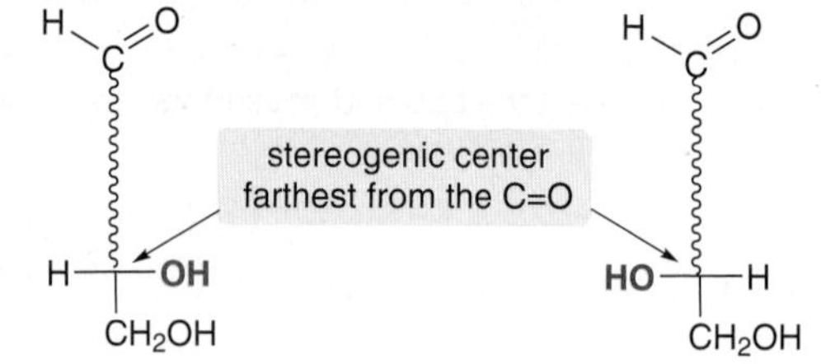

OH on the **right**
D-sugar

OH on the **left**
L-sugar

Glucose and all other naturally occurring sugars are D-sugars. L-Glucose, a compound that does not occur in nature, is the enantiomer of D-glucose. L-Glucose has the opposite configuration at *every* stereogenic center.

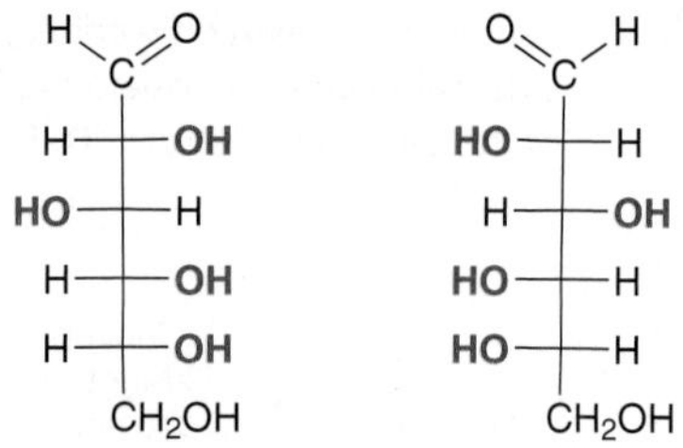

D-glucose
naturally occurring enantiomer

L-glucose

Problem 27.6 (a) Label compounds **A, B,** and **C** as D- or L-sugars. (b) How are compounds **A** and **B** related? **A** and **C? B** and **C?** Choose from enantiomers, diastereomers, or constitutional isomers.

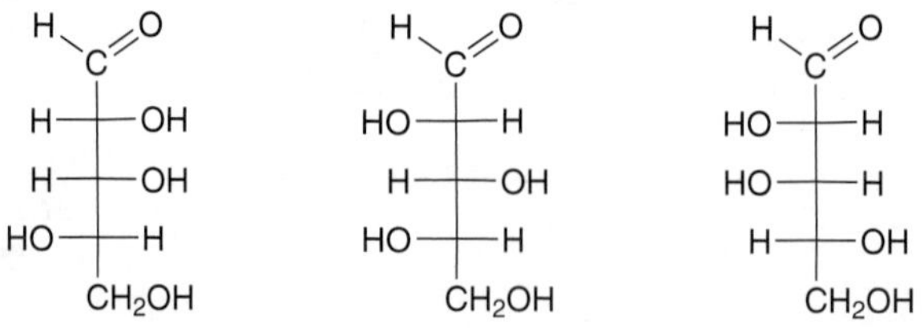

A **B** **C**

27.3 The Family of D-Aldoses

Beginning with D-glyceraldehyde, one may formulate other D-aldoses having four, five, or six carbon atoms by adding carbon atoms (each bonded to H and OH), one at a time, between C1 and C2. **Two D-aldotetroses can be formed from D-glyceraldehyde,** one with the new OH

The common name of each monosaccharide indicates both the number of atoms it contains and the configuration at each of the stereogenic centers. Because the common names are firmly entrenched in the chemical literature, no systematic method has ever been established to name these compounds.

group on the right and one with the new OH group on the left. Their names are D-erythrose and D-threose. They are two diastereomers, each with two stereogenic centers, labeled in blue.

D-erythrose D-threose

Because each aldotetrose has two stereogenic centers, there are 2^2 or four possible stereoisomers. D-Erythrose and D-threose are two of them. The other two are their enantiomers, called L-erythrose and L-threose, respectively. The configuration around each stereogenic center is exactly the opposite in its enantiomer. All four stereoisomers of the aldotetroses are shown in Figure 27.3.

Figure 27.3
The four stereoisomeric aldotetroses

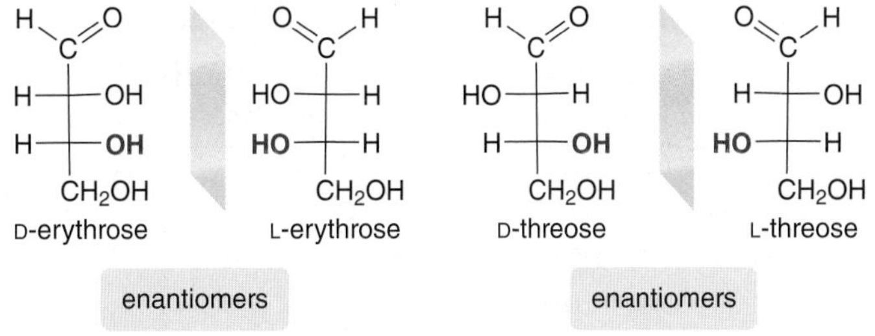
D-erythrose
L-erythrose
D-threose
L-threose
enantiomers
enantiomers

D-Ribose, D-arabinose, and D-xylose are all common aldopentoses in nature. D-Ribose is the carbohydrate component of RNA, the polymer that translates the genetic information of DNA for protein synthesis.

To continue forming the family of D-aldoses, we must add another carbon atom (bonded to H and OH) just below the carbonyl of either tetrose. Because there are two D-aldotetroses to begin with, and there are two ways to place the new OH (right or left), there are now four D-aldopentoses: D-ribose, D-arabinose, D-xylose, and D-lyxose. Each aldopentose now has three stereogenic centers, so there are $2^3 = 8$ possible stereoisomers, or four pairs of enantiomers. The D-enantiomer of each pair is shown in Figure 27.4.

Finally, to form the D-aldohexoses, we must add another carbon atom (bonded to H and OH) just below the carbonyl of all the aldopentoses. Because there are four D-aldopentoses to begin with, and there are two ways to place the new OH (right or left), there are now eight D-aldohexoses. Each aldohexose now has four stereogenic centers, so there are $2^4 = 16$ possible stereoisomers, or eight pairs of enantiomers. Only the D-enantiomer of each pair is shown in Figure 27.4.

The tree of D-aldoses (Figure 27.4) is arranged in pairs of compounds that are bracketed together. Each pair of compounds, such as D-glucose and D-mannose, has the same configuration around all of its stereogenic centers except for one.

Of the D-aldohexoses, only D-glucose and D-galactose are common in nature. **D-Glucose is by far the most abundant of all D-aldoses.** D-Glucose comes from the hydrolysis of starch and cellulose, and D-galactose comes from the hydrolysis of fruit pectins.

- **Two diastereomers that differ in the configuration around only one stereogenic center are called *epimers*.**

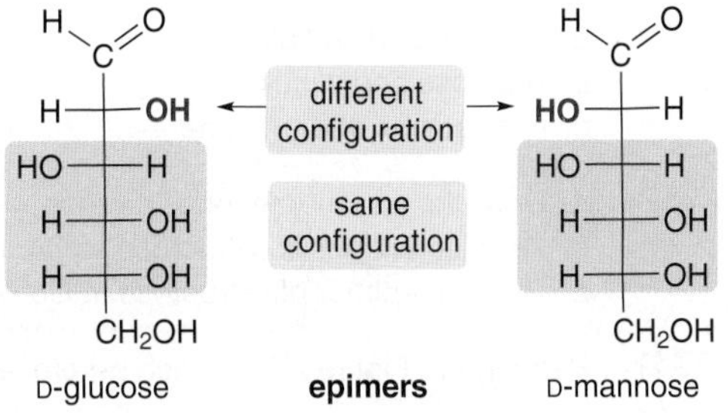
different configuration
same configuration
D-glucose
epimers
D-mannose

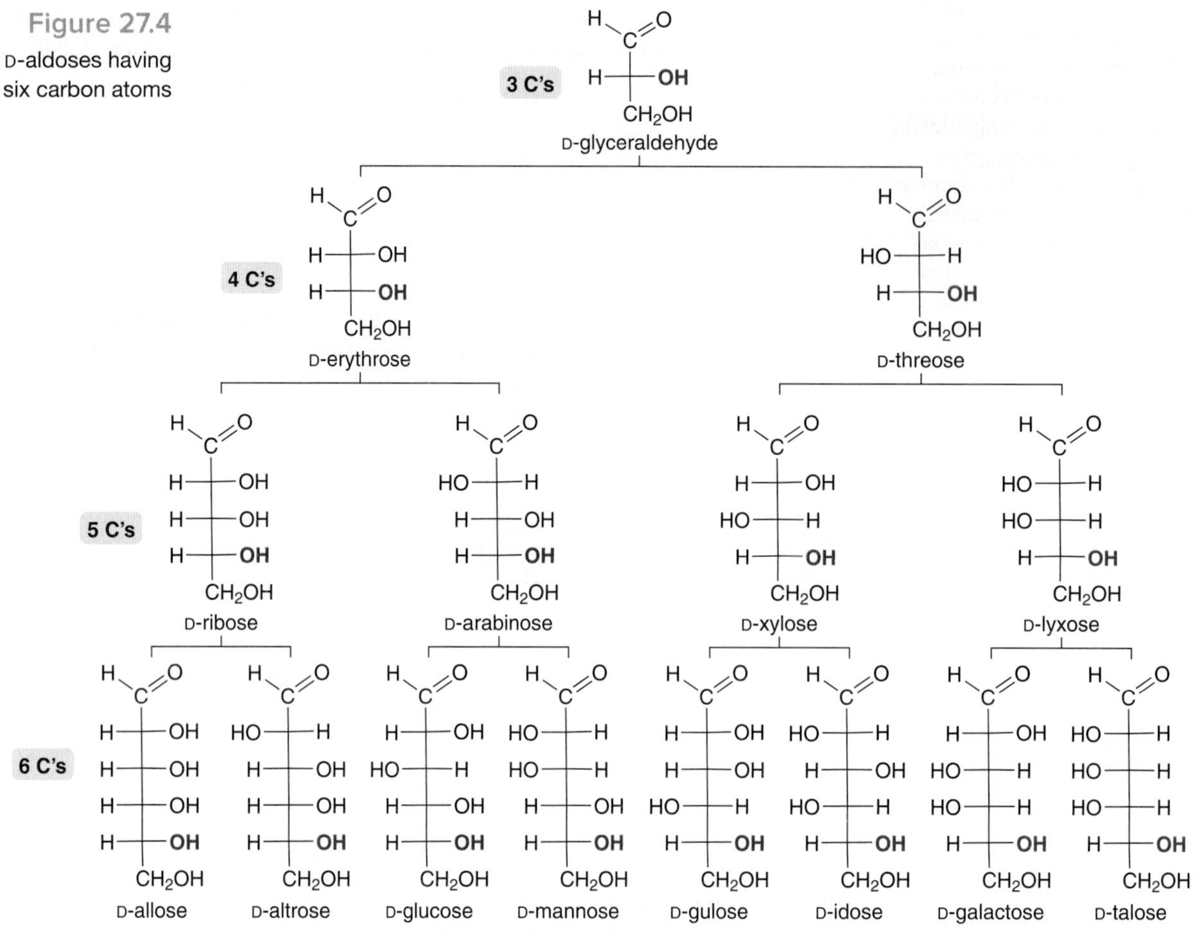
3 C's
D-glyceraldehyde
4 C's
D-erythrose
D-threose
5 C's
D-ribose
D-arabinose
D-xylose
D-lyxose
6 C's
D-allose
D-altrose
D-glucose
D-mannose
D-gulose
D-idose
D-galactose
D-talose

Figure 27.4
The family of D-aldoses having three to six carbon atoms

Problem 27.7 How many different aldoheptoses are there? How many are D-sugars? Draw all D-aldoheptoses having the *R* configuration at C2 and C3.

Problem 27.8 Draw two possible epimers of D-erythrose. Name each of these compounds using Figure 27.4.

27.4 The Family of D-Ketoses

The family of D-ketoses, shown in Figure 27.5, is formed from dihydroxyacetone by adding a new carbon (bonded to H and OH) between C2 and C3. Having a carbonyl group at C2 decreases the number of stereogenic centers in these monosaccharides, so that there are only four D-ketohexoses. The most common naturally occurring ketose is D-fructose.

Problem 27.9 Referring to the structures in Figures 27.4 and 27.5, classify each pair of compounds as enantiomers, epimers, diastereomers but not epimers, or constitutional isomers of each other.

a. D-allose and L-allose
b. D-altrose and D-gulose
c. D-galactose and D-talose
d. D-mannose and D-fructose
e. D-fructose and D-sorbose
f. L-sorbose and L-tagatose

Problem 27.10
a. Draw the enantiomer of D-fructose.
b. Draw an epimer of D-fructose at C4. What is the name of this compound?
c. Draw an epimer of D-fructose at C5. What is the name of this compound?

Problem 27.11 Referring to Figure 27.5, which D-ketohexoses have the *S* configuration at C3?

Figure 27.5
The family of D-ketoses having three to six carbon atoms

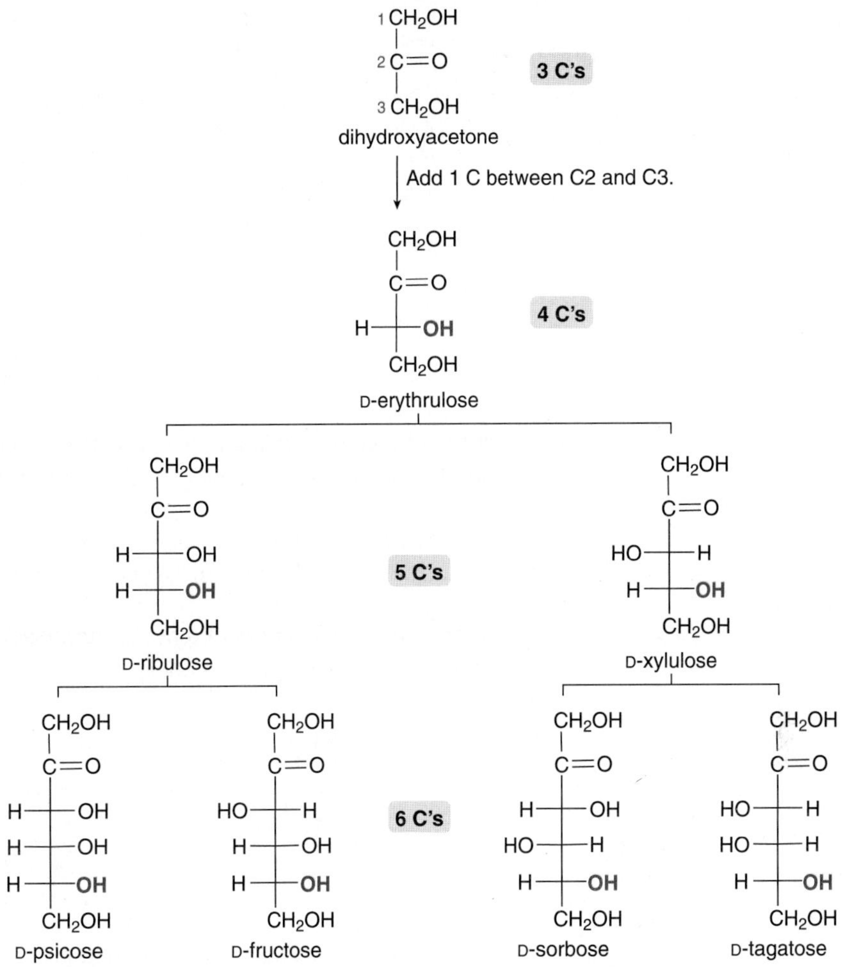

27.5 Physical Properties of Monosaccharides

Monosaccharides have the following physical properties:

- They are all **sweet tasting,** but their relative sweetness varies a great deal.
- They are polar compounds with **high melting points.**
- The presence of so many polar functional groups capable of hydrogen bonding makes them **water soluble.**
- Unlike most other organic compounds, monosaccharides are so polar that they are **insoluble in organic solvents like diethyl ether.**

27.6 The Cyclic Forms of Monosaccharides

Although the monosaccharides in Figures 27.4 and 27.5 are drawn as acyclic carbonyl compounds containing several hydroxy groups, the hydroxy and carbonyl groups of monosaccharides can undergo intramolecular cyclization reactions to form **hemiacetals** having either five or six atoms in the ring. This process was first discussed in Section 21.16.

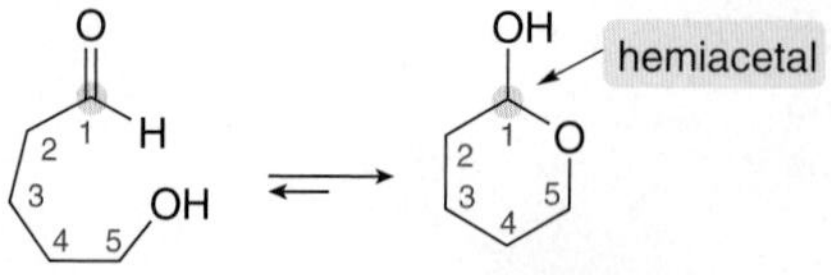

pyranose ring
(a six-membered ring)

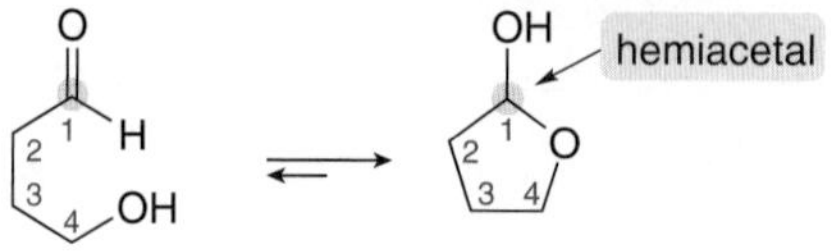

furanose ring
(a five-membered ring)

- **A six-membered ring containing an O atom is called a *pyranose* ring.**
- **A five-membered ring containing an O atom is called a *furanose* ring.**

Cyclization of a hydroxy carbonyl compound always forms a stereogenic center at the hemiacetal carbon, called the **anomeric carbon.** The two hemiacetals are called **anomers.**

- **Anomers are stereoisomers of a cyclic monosaccharide that differ in the position of the OH group at the hemiacetal carbon.**

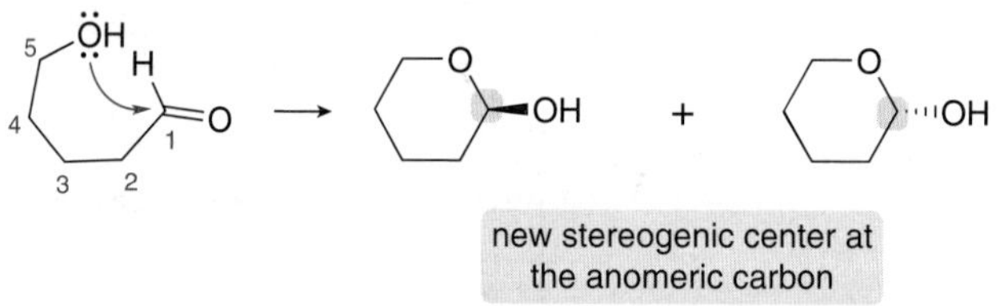

Cyclization forms the more stable ring size in a given molecule. **The most common monosaccharides, the aldohexoses like glucose, typically form a pyranose ring,** so our discussion begins with forming a cyclic hemiacetal from D-glucose.

27.6A Drawing Glucose as a Cyclic Hemiacetal

Which of the five OH groups in glucose is at the right distance from the carbonyl group to form a six-membered ring? The **O atom on the stereogenic center farthest from the carbonyl** (C5) is six atoms from the carbonyl carbon, placing it in the proper position for cyclization to form a pyranose ring.

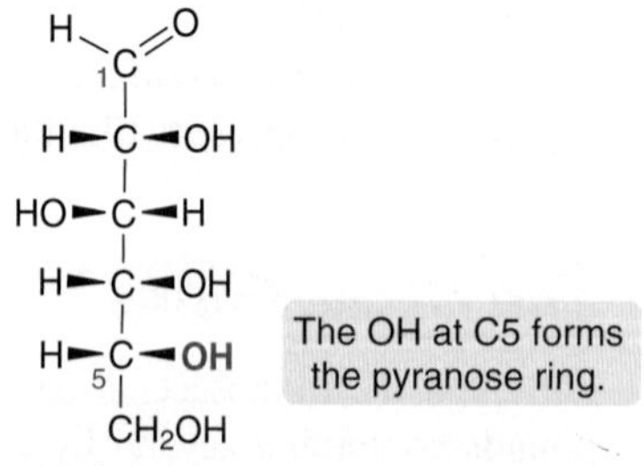

D-glucose

To translate the acyclic form of glucose into a cyclic hemiacetal, we must draw the hydroxy aldehyde in a way that suggests the position of the atoms in the new ring, and then draw the ring. **By convention the O atom in the new pyranose ring is drawn in the upper right-hand corner of the six-membered ring.**

Rotating the groups on the bottom stereogenic center in **A** places all six atoms needed for the ring (including the OH) in a vertical line (**B**). Re-drawing this representation as a Fischer projection makes the structure appear less cluttered (**C**). Twisting this structure and rotating it 90° forms **D.** Structures **A–D** are four different ways of drawing the same acyclic structure of D-glucose.

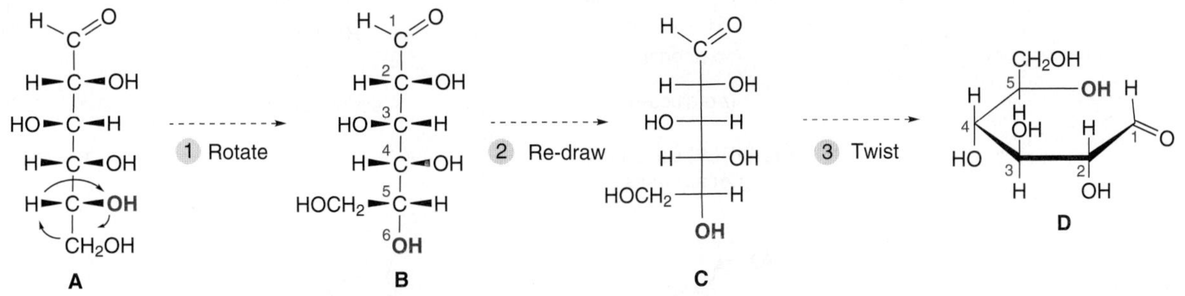

We are now set to draw the cyclic hemiacetal formed by nucleophilic attack of the OH group on C5 on the aldehyde carbonyl. Because cyclization creates a new stereogenic center, there are **two cyclic forms of D-glucose,** an **α anomer** and a **β anomer.** All the original stereogenic centers maintain their configuration in both of the products formed.

- **The α anomer of a D monosaccharide has the OH group drawn *down,* trans to the CH_2OH group at C5. The α anomer of D-glucose is called α-D-glucose, or α-D-glucopyranose (to emphasize the six-membered ring).**
- **The β anomer of a D monosaccharide has the OH group drawn *up,* cis to the CH_2OH group at C5. The β anomer is called β-D-glucose, or β-D-glucopyranose (to emphasize the six-membered ring).**

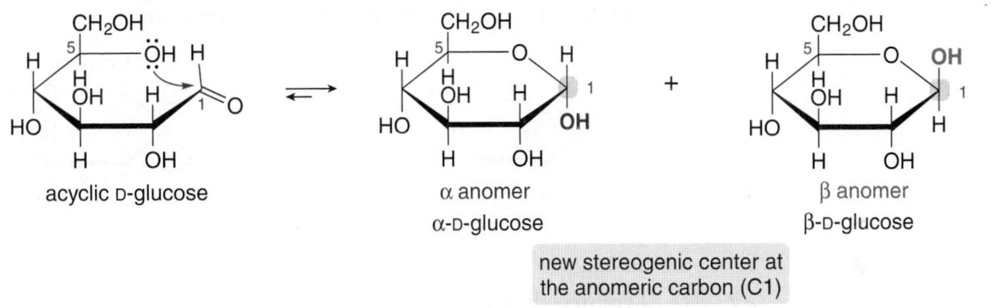

The **α anomer** in any monosaccharide has the anomeric OH group and the CH_2OH group **trans.** The **β anomer** has the anomeric OH group and the CH_2OH group **cis.**

These flat, six-membered rings used to represent the cyclic hemiacetals of glucose and other sugars are called **Haworth projections.** The cyclic forms of glucose now have **five stereogenic centers, the four from the starting hydroxy aldehyde and the new anomeric carbon.** α-D-Glucose and β-D-glucose are **diastereomers,** because only the anomeric carbon has a different configuration.

The mechanism for this transformation is exactly the same as the mechanism that converts a hydroxy aldehyde to a cyclic hemiacetal (Mechanism 21.10). The acyclic aldehyde and two cyclic hemiacetals are all in equilibrium. Each cyclic hemiacetal can be isolated and crystallized separately, but when any one compound is placed in solution, an equilibrium mixture of all three forms results. This process is called **mutarotation.** At equilibrium, the mixture has 37% of the α anomer, 63% of the β anomer, and only trace amounts of the acyclic hydroxy aldehyde (Figure 27.6). Also shown are representations of the three forms of glucose using wedges and dashed wedges.

Figure 27.6
The three forms of glucose

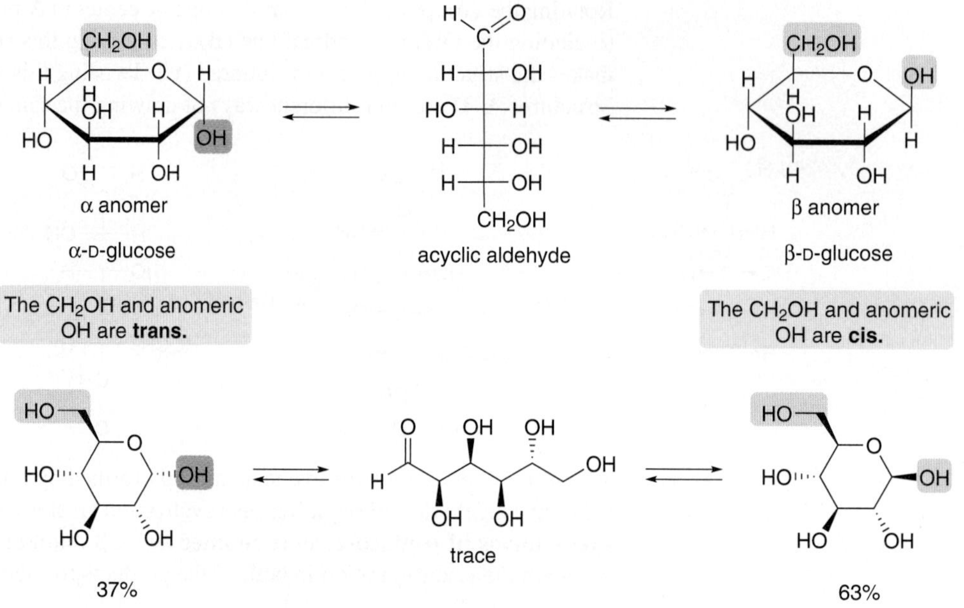
CH2OH
OH
α anomer
α-D-glucose
acyclic aldehyde
β anomer
β-D-glucose
The CH2OH and anomeric OH are **trans.**
The CH2OH and anomeric OH are **cis.**
trace
37%
63%

- Bonds above the ring in a Haworth projection are drawn as wedged bonds.
- Bonds below the ring in a Haworth projection are drawn as dashed wedges.

Problem 27.12 Label each Haworth projection as an α or β anomer and convert the Haworth projection to a six-membered ring with wedges and dashed wedges.

a.
b.

27.6B Haworth Projections

To convert an acyclic monosaccharide to a Haworth projection, follow a stepwise procedure.

How To Draw a Haworth Projection from an Acyclic Aldohexose

Example Convert D-mannose to a Haworth projection.

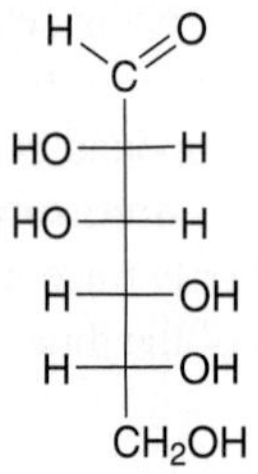
D-mannose

How To, continued . . .

Step [1] **Place the O atom in the upper right corner of a hexagon, and add the CH_2OH group on the first carbon counterclockwise from the O atom.**

- For **D-sugars,** the CH_2OH group is drawn **up.** For **L-sugars,** the CH_2OH group is drawn **down.**

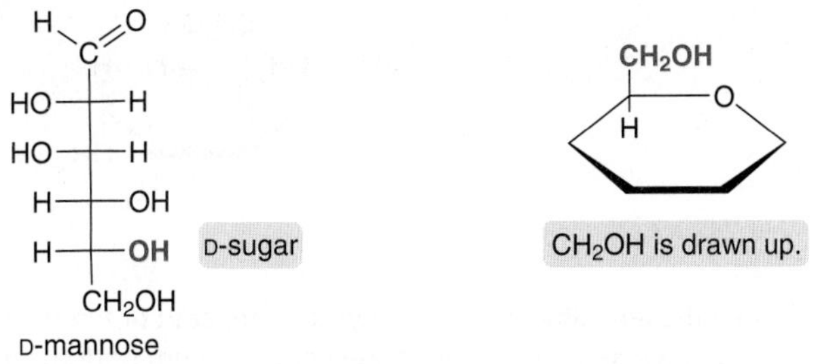

D-sugar
D-mannose
CH₂OH is drawn up.

Step [2] **Place the anomeric carbon on the first carbon clockwise from the O atom.**

- For an α **anomer,** the **OH** is drawn **down** in a D-sugar.
- For a β **anomer,** the **OH** is drawn **up** in a D-sugar.

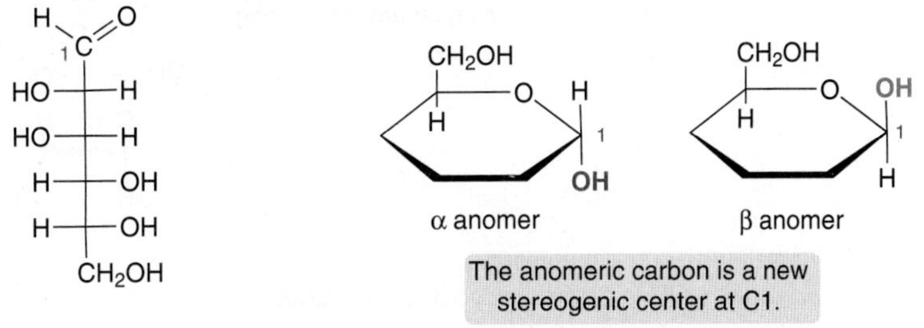

α anomer
β anomer
The anomeric carbon is a new stereogenic center at C1.

- Remember: The carbonyl carbon becomes the anomeric carbon (a new stereogenic center).

Step [3] **Add the substituents at the three remaining stereogenic centers clockwise around the ring.**

- The substituents on the **right side** of the Fischer projection are drawn **down.**
- The substituents on the **left** are drawn **up.**

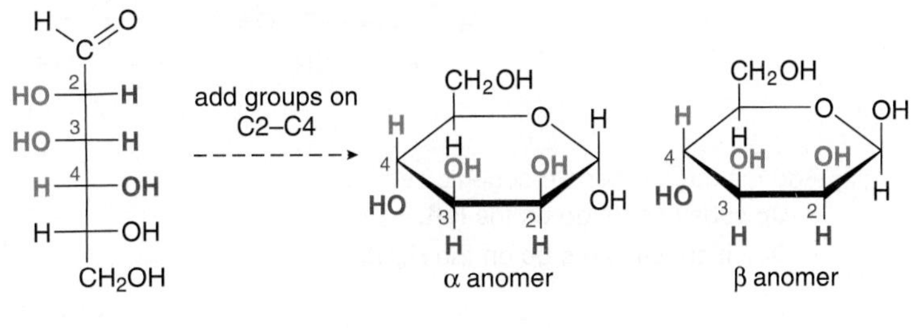

add groups on C2–C4
α anomer
β anomer

Problem 27.13 Convert each aldohexose to the indicated anomer using a Haworth projection.

a. Draw the α anomer of: b. Draw the α anomer of: c. Draw the β anomer of:

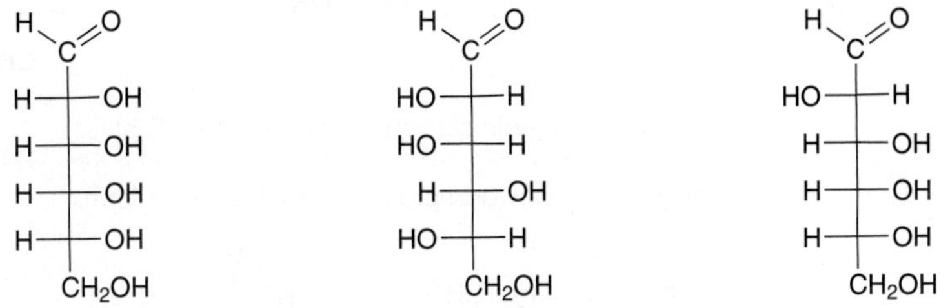

Sample Problem 27.4 shows how to convert a Haworth projection back to the acyclic form of a monosaccharide. It doesn't matter whether the hemiacetal is the α or β anomer, because both anomers give the *same* hydroxy aldehyde.

Sample Problem 27.4 Convert the following Haworth projection to the acyclic form of the aldohexose.

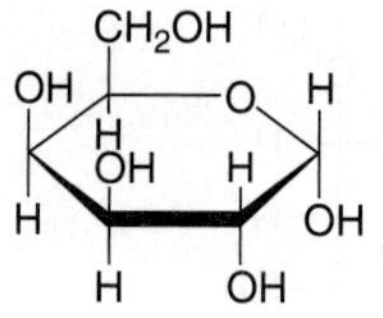

Solution

To convert the substituents to the acyclic form, start at the pyranose O atom, and work in a ***counterclockwise*** fashion around the ring, and from bottom-to-top along the chain.

[1] Draw the carbon skeleton, placing the CHO on the top and the CH_2OH on the bottom.

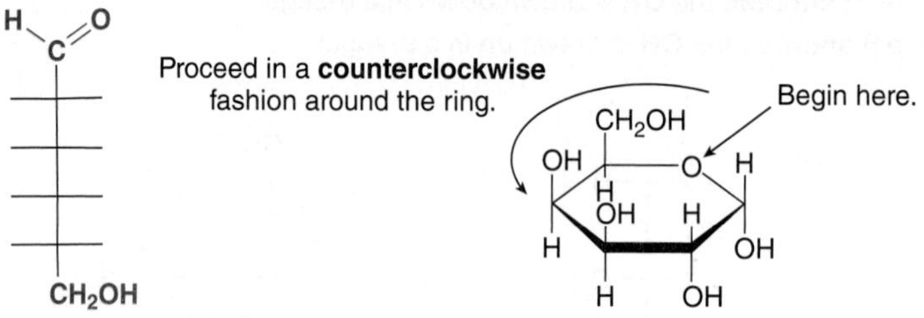

[2] Classify the sugar as D- or L-.

- The CH_2OH is drawn **up,** so it is a **D-sugar.**
- A D-sugar has the OH group on the bottom stereogenic center on the **right.**

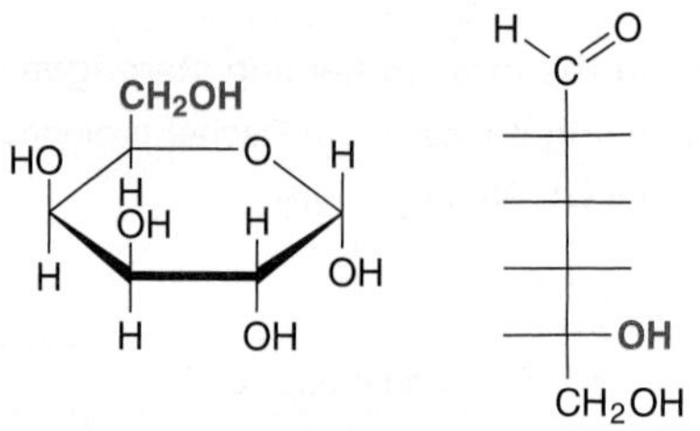

[3] Add the three other stereogenic centers.

- **Up** substituents go on the **left.**
- **Down** substituents go on the **right.**

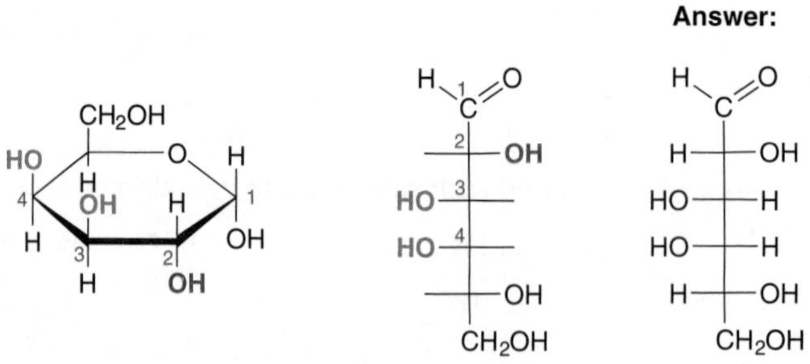

- The anomeric carbon becomes the C=O at C1.

Problem 27.14 Convert each Haworth projection to its acyclic form.

a. 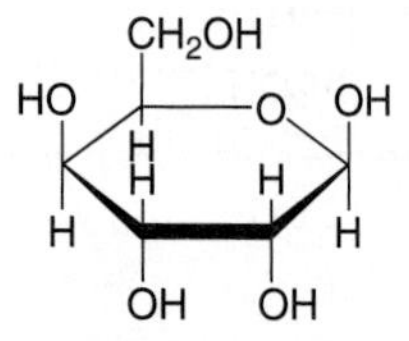

b. 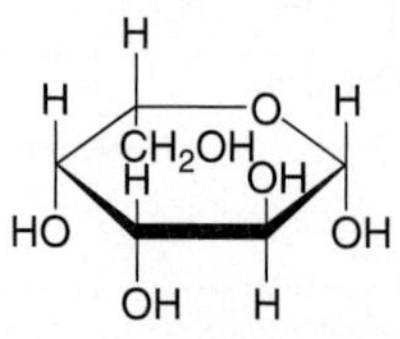

27.6C Three-Dimensional Representations for D-Glucose

Because the chair form of a six-membered ring gives the truest picture of its three-dimensional shape, we must learn to convert Haworth projections into chair forms.

To convert a Haworth projection to a chair form:

- Draw the pyranose ring with the O atom as an "up" atom.
- The "up" substituents in a Haworth projection become the "up" bonds (either axial or equatorial) on a given carbon atom on a puckered six-membered ring.
- The "down" substituents in a Haworth projection become the "down" bonds (either axial or equatorial) on a given carbon atom on a puckered six-membered ring.

As a result, the three-dimensional chair form of β-D-glucose is drawn in the following manner:

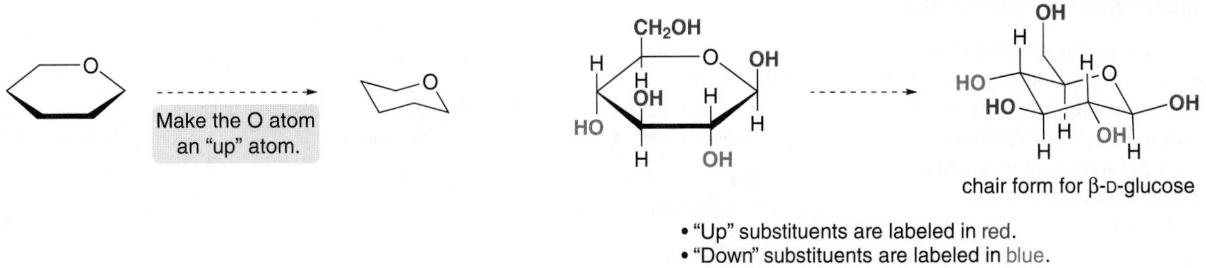
Make the O atom an "up" atom.
chair form for β-D-glucose
• "Up" substituents are labeled in red.
• "Down" substituents are labeled in blue.

Glucose has all substituents larger than a hydrogen atom in the more roomy equatorial positions, making it the most stable and thus most prevalent monosaccharide. The β anomer is the major isomer at equilibrium, moreover, because the hemiacetal OH group is in the equatorial position, too. Figure 27.7 shows both anomers of D-glucose drawn as chair conformations.

Problem 27.15 Convert each Haworth projection in Problem 27.14 to a three-dimensional representation using a chair pyranose ring.

Figure 27.7 Three-dimensional representations for both anomers of D-glucose

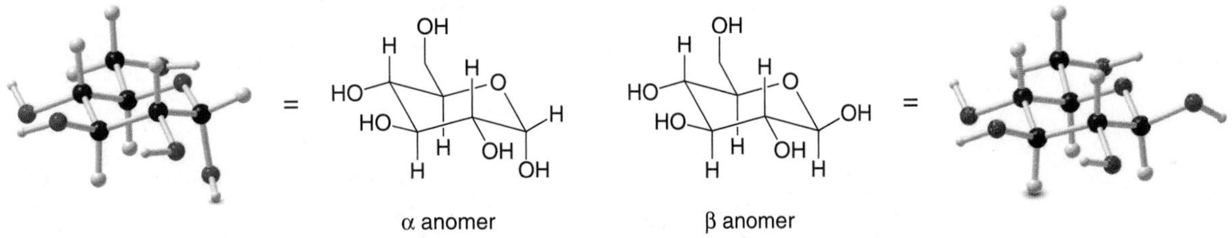
α anomer
β anomer

27.6D Furanoses

Certain monosaccharides—notably aldopentoses and ketohexoses—predominantly form furanose rings, rather than pyranose rings, in solution. The same principles apply to drawing these structures as for drawing pyranose rings, except the ring size is one atom smaller.

- Cyclization always forms a new stereogenic center at the anomeric carbon, so two different anomers are possible. For a D-sugar, the OH group is drawn down in the α anomer and up in the β anomer.
- Use the same drawing conventions for adding substituents to the five-membered ring. With D-sugars, the CH_2OH group is drawn up.

Honey was the first and most popular sweetening agent until it was replaced by sugar (from sugarcane) in modern times. Honey is a mixture consisting largely of D-fructose and D-glucose.

With D-ribose, the OH group used to form the five-membered furanose ring is located on C4. Cyclization yields two anomers at the new stereogenic center, which are called **α-D-ribofuranose** and **β-D-ribofuranose.**

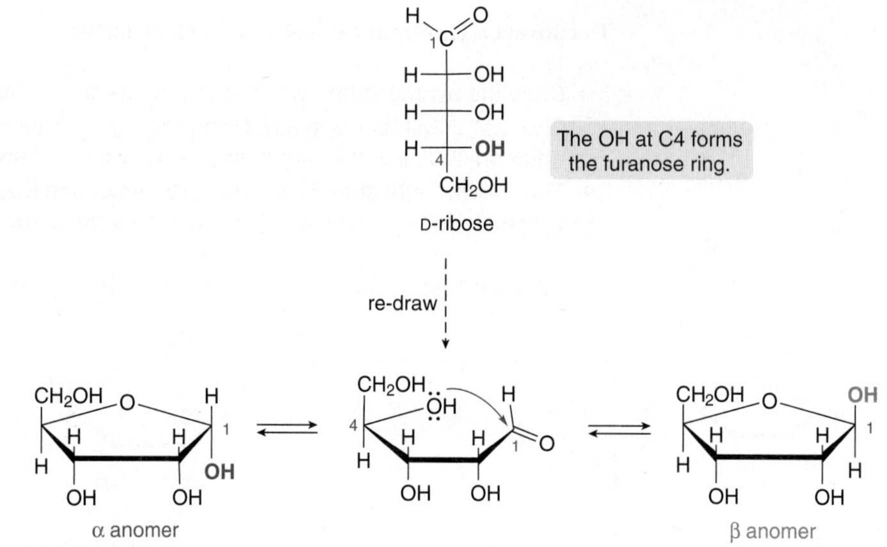

α-D-ribofuranose

β-D-ribofuranose

The same procedure can be used to draw the furanose form of D-fructose, the most common ketohexose. Because the carbonyl group is at C2 (instead of C1, as in the aldoses), the OH group at C5 reacts to form the hemiacetal in the five-membered ring. Two anomers are formed.

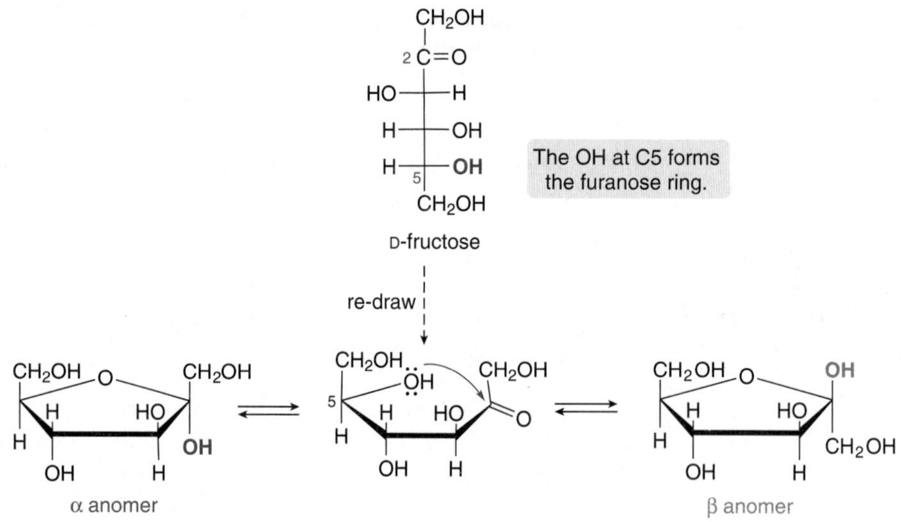

α-D-fructofuranose

β-D-fructofuranose

Problem 27.16 Aldotetroses exist in the furanose form. Draw both anomers of D-erythrose.

27.7 Glycosides

Keep in mind the difference between a hemiacetal and an acetal:

hemiacetal
- one OH group
- one OR group

acetal
- two OR groups

Because monosaccharides exist in solution in an equilibrium between acyclic and cyclic forms, they undergo three types of reactions:

- **Reaction of the hemiacetal**
- **Reaction of the hydroxy groups**
- **Reaction of the carbonyl group**

Even though the acyclic form of a monosaccharide may be present in only trace amounts, the equilibrium can be tipped in its favor by Le Châtelier's principle (Section 9.8). Suppose, for example, that the carbonyl group of the acyclic form reacts with a reagent, thus depleting its equilibrium concentration. The equilibrium will then shift to compensate for the loss, thus producing more of the acyclic form, which can react further.

Note, too, that monosaccharides have two different types of OH groups. Most are "regular" alcohols, and as such, undergo reactions characteristic of alcohols. The anomeric OH group, on the other hand, is part of a hemiacetal, giving it added reactivity.

27.7A Glycoside Formation

Treatment of a monosaccharide with an alcohol and HCl converts the hemiacetal into an acetal called a **glycoside.** For example, treatment of α-D-glucose with CH_3OH and HCl forms two glycosides that are diastereomers at the acetal carbon. The α and β labels are assigned in the same way as anomers: with a D-sugar, an α glycoside has the new OR group (OCH_3 group in this example) down, and a β glycoside has the new OR group up.

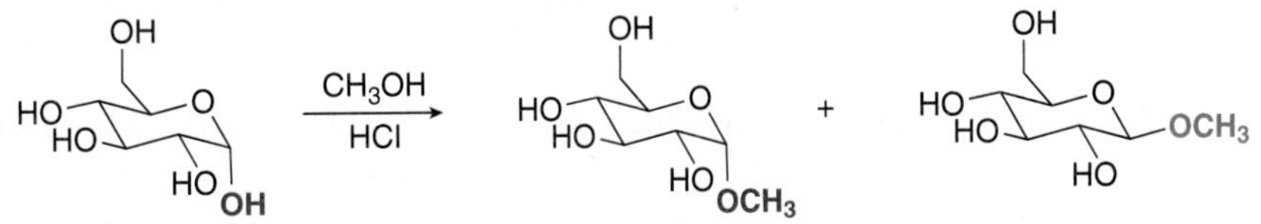

α-D-glucose α glycoside β glycoside

Only the hemiacetal OH reacts.

Mechanism 27.1 explains why a single anomer forms two glycosides. The reaction proceeds by way of a **planar carbocation,** which undergoes nucleophilic attack from two different directions to give a mixture of diastereomers. Because both α- and β-D-glucose form the same planar carbocation, each yields the same mixture of two glycosides.

The mechanism also explains why only the hemiacetal OH group reacts. Protonation of the hemiacetal OH, followed by loss of H_2O, forms a resonance-stabilized carbocation in Step [2]. A resonance-stabilized carbocation is not formed by loss of H_2O from any other OH group.

Unlike cyclic hemiacetals, **glycosides are acetals, and so they do not undergo mutarotation.** When a single glycoside is dissolved in H_2O, it is *not* converted to an equilibrium mixture of α and β glycosides.

- **Glycosides are acetals with an alkoxy group (OR) bonded to the anomeric carbon.**

Problem 27.17 What glycosides are formed when each monosaccharide is treated with CH_3CH_2OH, HCl: (a) β-D-mannose; (b) α-D-gulose; (c) β-D-fructose?

Mechanism 27.1 Glycoside Formation

Part [1] Loss of H_2O from the hemiacetal

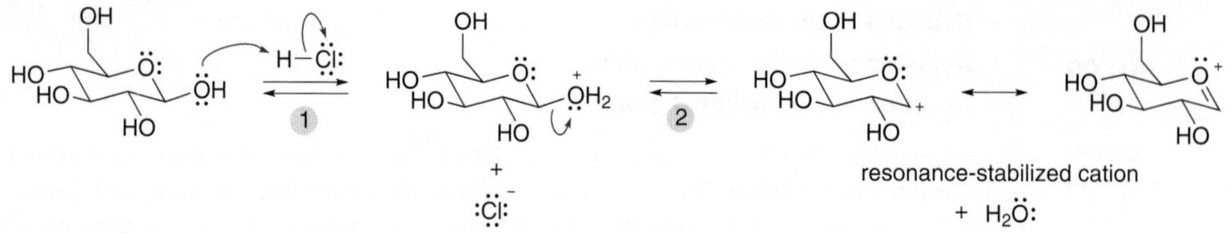

1 – 2 Protonation of the hemiacetal OH followed by loss of H_2O forms a resonance-stabilized carbocation.

Part [2] Formation of the glycosides

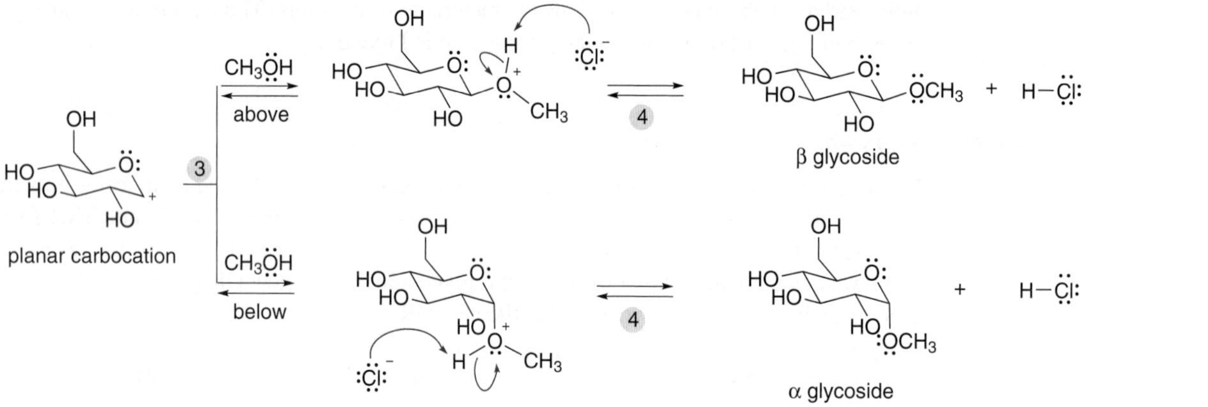

3 – 4 Nucleophilic attack by CH_3OH occurs from both sides of the planar carbocation to yield α and β glycosides after loss of a proton.

27.7B Glycoside Hydrolysis

Because glycosides are acetals, **they are hydrolyzed with acid and water to cyclic hemiacetals and a molecule of alcohol.** A mixture of two anomers is formed from a single glycoside. For example, treatment of methyl α-D-glucopyranoside with aqueous acid forms a mixture of α- and β-D-glucose and methanol.

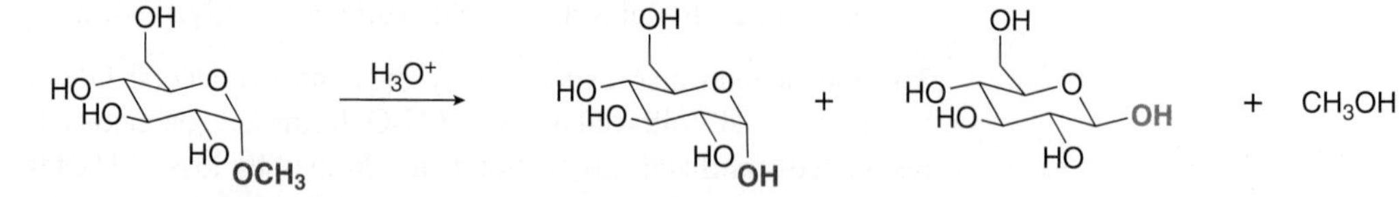

methyl α-D-glucopyranoside α-D-glucose β-D-glucose

The mechanism for glycoside hydrolysis is just the reverse of glycoside formation. It involves two parts: **formation of a planar carbocation,** followed by **nucleophilic attack of H_2O** to form anomeric hemiacetals, as shown in Mechanism 27.2.

Problem 27.18 Draw a stepwise mechanism for the following reaction.

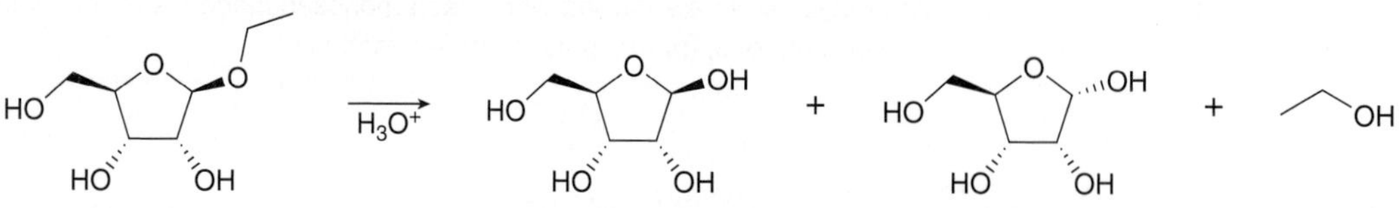

Mechanism 27.2 Glycoside Hydrolysis

Part [1] Loss of CH_3OH from the glycoside

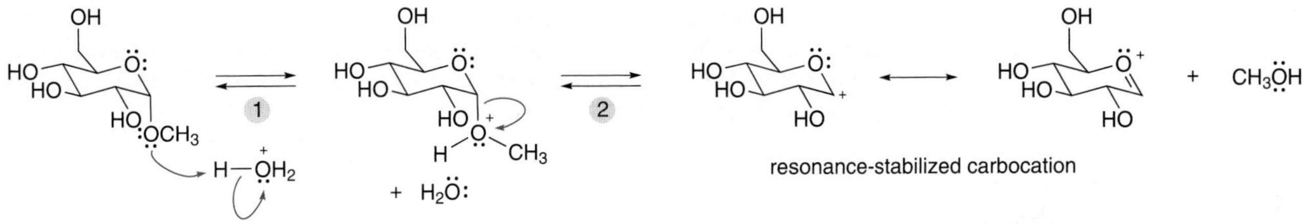

1 – 2 Protonation of the acetal OCH_3 followed by loss of CH_3OH forms a resonance-stabilized carbocation.

Part [2] Formation of the hemiacetals

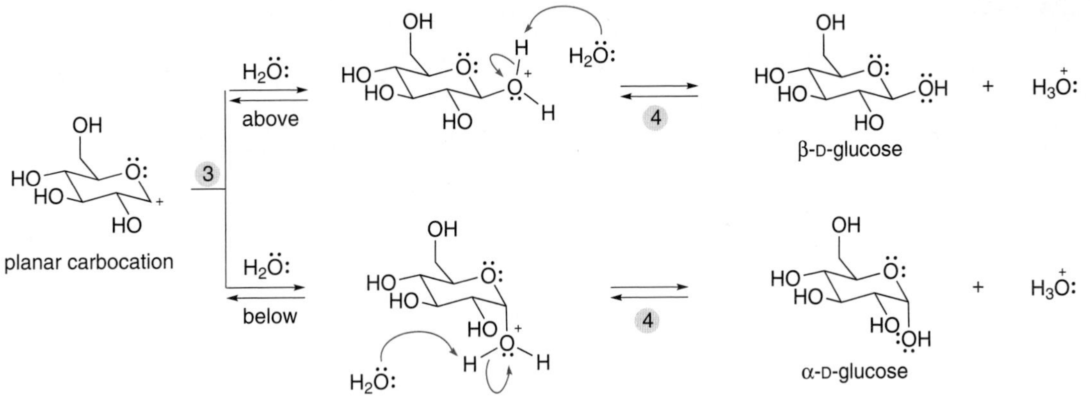

3 – 4 Nucleophilic attack by H_2O occurs from both sides of the planar carbocation to yield α and β anomers after loss of a proton.

27.7C Naturally Occurring Glycosides

The berries of the black nightshade plant (*Solanum nigrum*) are a source of the poisonous alkaloid solanine.

Salicin and **solanine** are two naturally occurring compounds that contain glycoside bonds as part of their structure. Salicin is an analgesic isolated from willow bark, and solanine, the chapter-opening molecule, is a poisonous compound produced in the leaves, stem, and green spots on the skin of potatoes. Solanine is also isolated from the berries of the deadly nightshade plant. It is believed that the role of the sugar rings in both salicin and solanine is to increase their water solubility.

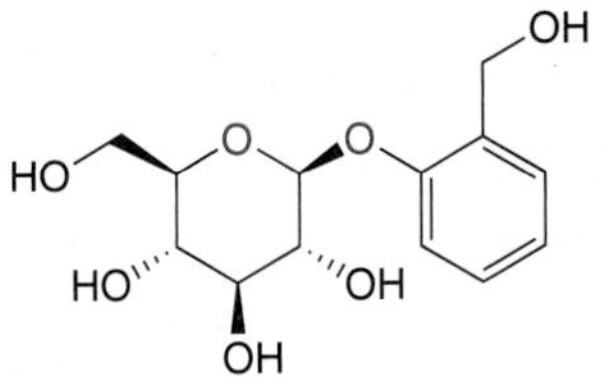

salicin

[The O atoms that are part of the glycoside bonds are drawn in red.]

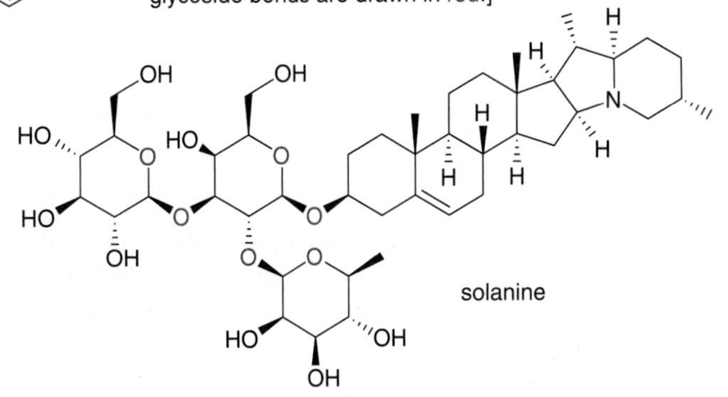

solanine

Glycosides are common in nature. All disaccharides and polysaccharides are formed by joining monosaccharides together with glycosidic linkages. These compounds are discussed in detail beginning in Section 27.11.

Problem 27.19 (a) Label all the O atoms that are part of a glycoside in rebaudioside A. Rebaudioside A, marketed under the trade name Truvia, is a sweet glycoside obtained from the stevia plant, which has been used for centuries in Paraguay to sweeten foods. (b) The alcohol or phenol formed from the hydrolysis of a glycoside is called an **aglycon.** What aglycon and monosaccharides are formed by the hydrolysis of rebaudioside A?

Rebaudioside A, a naturally occurring glycoside about 400 times sweeter than table sugar, is obtained from the leaves of the stevia plant, a shrub native to Central and South America.

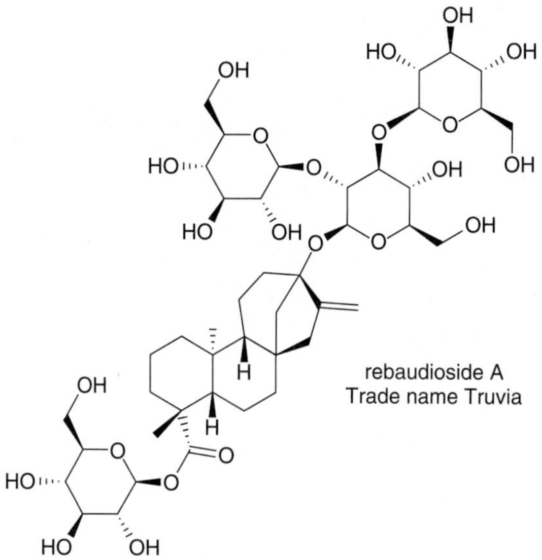

rebaudioside A
Trade name Truvia

27.8 Reactions of Monosaccharides at the OH Groups

Because monosaccharides contain OH groups, they undergo reactions typical of alcohols—that is, they are converted to **ethers** and **esters.** Because the cyclic hemiacetal form of a monosaccharide contains an OH group, this form of a monosaccharide must be drawn as the starting material for any reaction that occurs at an OH group.

All OH groups of a cyclic monosaccharide are converted to ethers by treatment with base and an alkyl halide. For example, α-D-glucose reacts with silver(I) oxide (Ag_2O, a base) and excess CH_3I to form a pentamethyl ether.

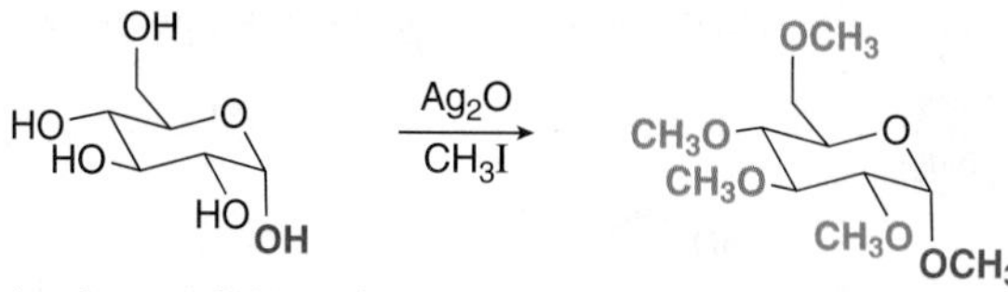

hemiacetal OH in red

pentamethyl ether

- The acetal OCH_3 group is in red.
- Ether OCH_3 groups are in blue.

Ag_2O removes a proton from each alcohol, forming an alkoxide (RO^-), which then reacts with CH_3I in an S_N2 reaction. Because no C–O bonds are broken, the configuration of all substituents in the starting material is **retained,** forming a single product.

The product contains two different types of ether bonds. There are four "regular" ethers formed from the "regular" hydroxyls. The new ether from the hemiacetal is now part of an **acetal**—that is, a **glycoside.**

The four ether bonds that are *not* part of the acetal do not react with any reagents except strong acids like HBr and HI (Section 9.14). **The acetal ether, on the other hand, is hydrolyzed with aqueous acid** (Section 27.7B). Aqueous hydrolysis of a single glycoside (like the pentamethyl ether of α-D-glucose) yields both anomers of the product monosaccharide.

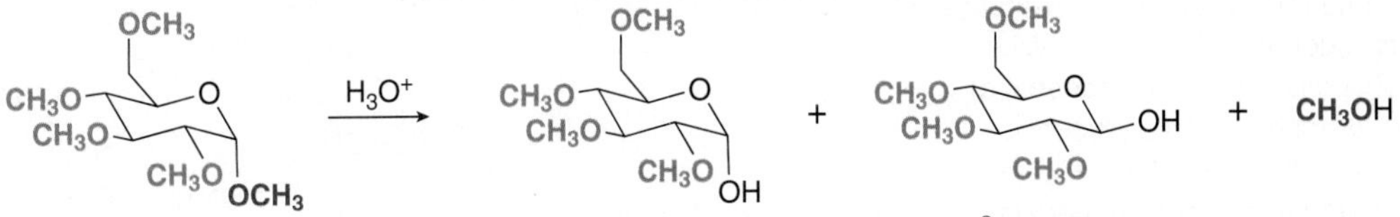

α anomer β anomer

The OH groups of monosaccharides can also be converted to esters. For example, treatment of β-D-glucose with either acetic anhydride or acetyl chloride in the presence of pyridine (a base) converts all OH groups into acetate esters.

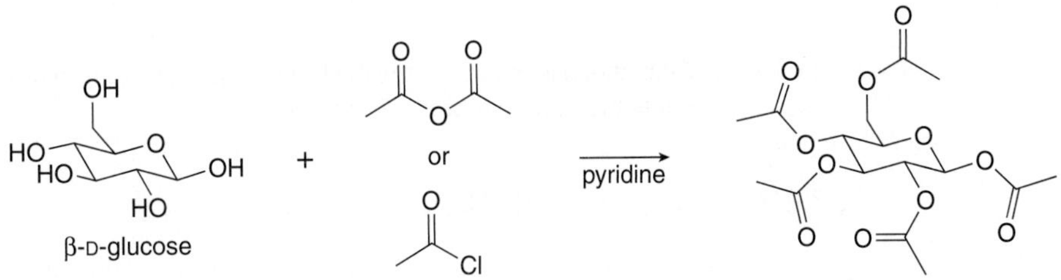

β-D-glucose

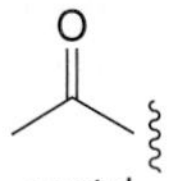

acetyl
Ac

Since it is cumbersome and tedious to draw in all the atoms of the esters, the abbreviation **Ac** is used for the acetyl group, **$CH_3C{=}O$.** The esterification of β-D-glucose can then be written as follows:

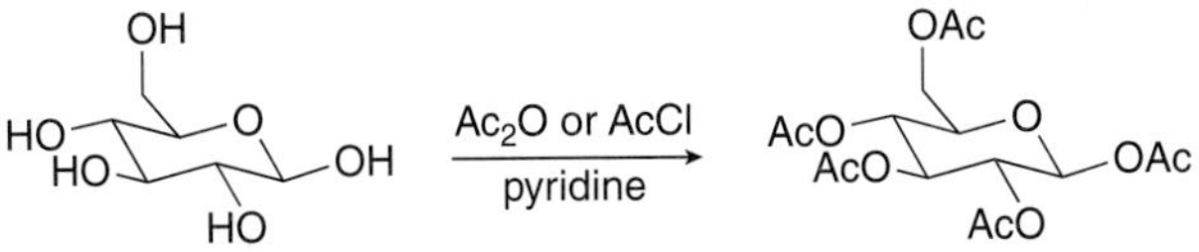

β-D-glucose

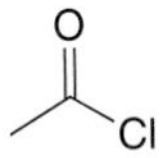

AcCl

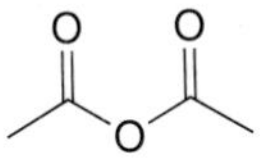

Ac_2O

Monosaccharides are so polar that they are insoluble in common organic solvents, making them difficult to isolate and use in organic reactions. Monosaccharide derivatives that have five ether or ester groups in place of the OH groups, however, are readily soluble in organic solvents.

Problem 27.20 Draw the products formed when β-D-galactose is treated with each reagent.

a. $Ag_2O + CH_3I$
b. $NaH + C_6H_5CH_2Cl$
c. The product in (b), then H_3O^+
d. Ac_2O + pyridine
e. C_6H_5COCl + pyridine
f. The product in (c), then C_6H_5COCl + pyridine

27.9 Reactions at the Carbonyl Group—Oxidation and Reduction

Oxidation and reduction reactions occur at the carbonyl group of monosaccharides, so they all begin with the monosaccharide drawn in the acyclic form. We will confine our discussion to aldoses as starting materials.

27.9A Reduction of the Carbonyl Group

Glucitol occurs naturally in some fruits and berries. It is sometimes used as a substitute for sucrose (table sugar). With six polar OH groups capable of hydrogen bonding, glucitol is readily hydrated. It is used as an additive to prevent certain foods from drying out.

Like other aldehydes, the **carbonyl group of an aldose is reduced to a 1° alcohol using $NaBH_4$.** This alcohol is called an **alditol.** For example, reduction of D-glucose with $NaBH_4$ in CH_3OH yields glucitol (also called sorbitol).

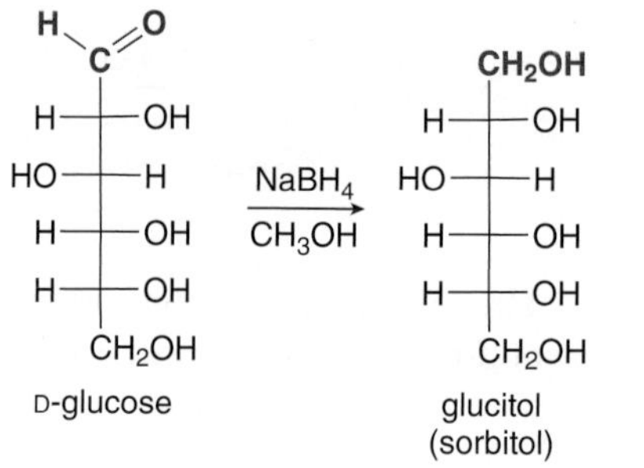
NaBH4
CH3OH
D-glucose
glucitol
(sorbitol)

Problem 27.21 A 2-ketohexose is reduced with $NaBH_4$ in CH_3OH to form a mixture of D-galactitol and D-talitol. What is the structure of the 2-ketohexose?

27.9B Oxidation of Aldoses

Aldoses contain 1° and 2° alcohols and an aldehyde, all of which are oxidizable functional groups. Two different types of oxidation reactions are particularly useful—oxidation of the aldehyde to a carboxylic acid (an **aldonic acid**) and oxidation of both the aldehyde and the 1° alcohol to a diacid (an **aldaric acid**).

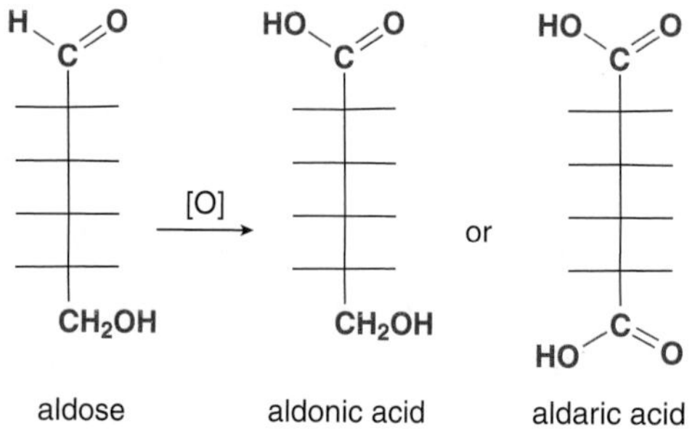
[O]
or
aldose
aldonic acid
aldaric acid

[1] Oxidation of the aldehyde to a carboxylic acid

The aldehyde carbonyl is the most easily oxidized functional group in an aldose, and so a variety of reagents oxidize it to a carboxy group, forming an **aldonic acid.**

Three reagents used for this process produce a characteristic color change because the oxidizing agent is reduced to a colored product that is easily visible. As described in Section 20.8, **Tollens reagent** oxidizes aldehydes to carboxylic acids using Ag_2O in NH_4OH, and forms a mirror of Ag as a by-product. **Benedict's** and **Fehling's reagents** use a blue Cu^{2+} salt as an oxidizing agent, which is reduced to Cu_2O, a brick-red solid. Unfortunately, none of these reagents gives a high yield of aldonic acid. When the aldonic acid is needed to carry on to other reactions, **$Br_2 + H_2O$** is used as the oxidizing agent.

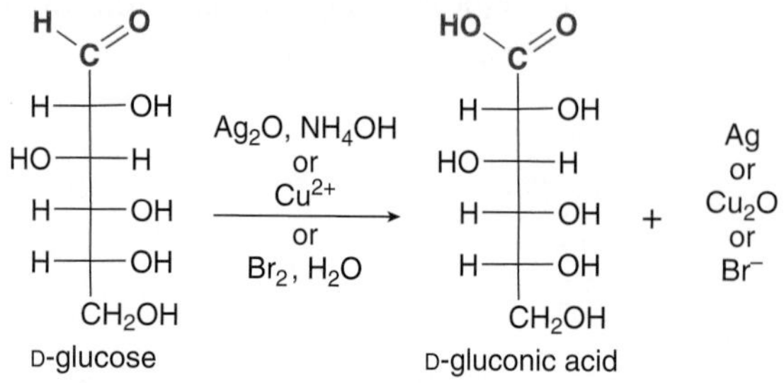
Ag2O, NH4OH
or
Cu2+
or
Br2, H2O
Ag
or
Cu2O
or
Br−
D-glucose
D-gluconic acid

- **Any carbohydrate that exists as a *hemiacetal* is in equilibrium with a small amount of acyclic aldehyde, so it is oxidized to an aldonic acid.**
- **Glycosides are acetals, not hemiacetals, so they are *not* oxidized to aldonic acids.**

Carbohydrates that can be oxidized with Tollens, Benedict's, or Fehling's reagent are called **reducing sugars.** Those that do not react with these reagents are called **nonreducing sugars.** Figure 27.8 shows examples of reducing and nonreducing sugars.

Figure 27.8
Examples of reducing and nonreducing sugars

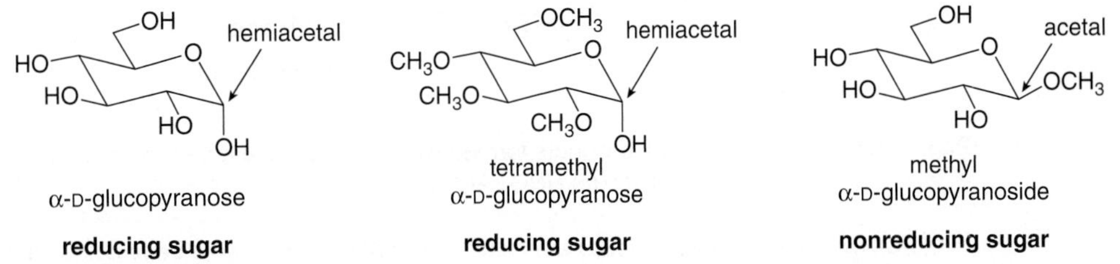
hemiacetal
α-D-glucopyranose
reducing sugar
hemiacetal
tetramethyl α-D-glucopyranose
reducing sugar
acetal
methyl α-D-glucopyranoside
nonreducing sugar

- Carbohydrates containing a hemiacetal are in equilibrium with an acyclic aldehyde, making them reducing sugars.
- Glycosides are acetals, so they are *not* in equilibrium with any acyclic aldehyde, making them nonreducing sugars.

Problem 27.22 Classify each compound as a reducing or nonreducing sugar.

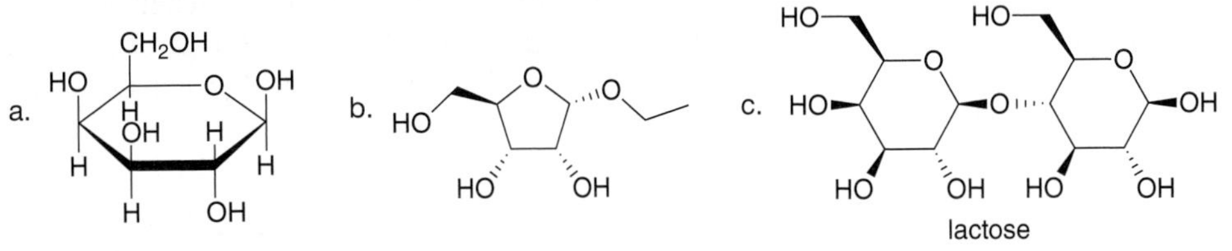
a.
b.
c.
lactose

[2] Oxidation of both the aldehyde and 1° alcohol to a diacid

Both the aldehyde and 1° alcohol of an aldose are oxidized to carboxy groups by treatment with warm nitric acid, forming an **aldaric acid.** Under these conditions, D-glucose is converted to D-glucaric acid.

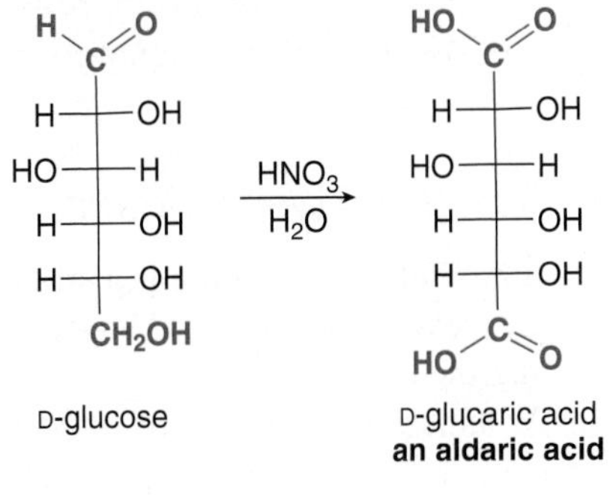
HNO_3
H_2O
D-glucose
D-glucaric acid
an aldaric acid

Because aldaric acids have identical functional groups on both terminal carbons, some aldaric acids contain a plane of symmetry, making them achiral molecules. For example, oxidation of D-allose forms an achiral, optically inactive aldaric acid. This contrasts with D-glucaric acid formed from glucose, which has no plane of symmetry, and is thus still optically active.

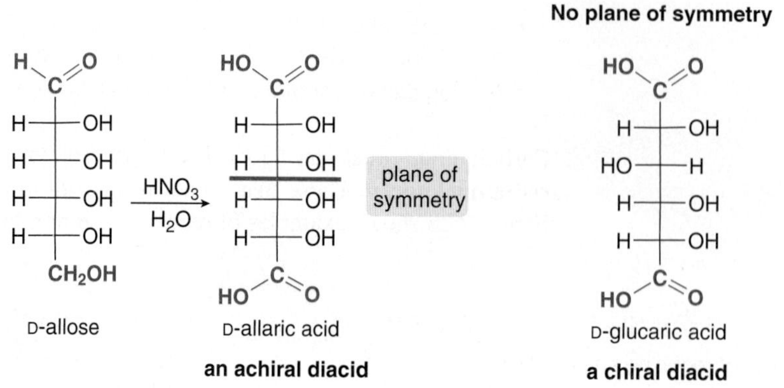
No plane of symmetry
HNO_3
H_2O
plane of symmetry
D-allose
D-allaric acid
an achiral diacid
D-glucaric acid
a chiral diacid

Problem 27.23 Draw the products formed when D-arabinose is treated with each reagent: (a) Ag_2O, NH_4OH; (b) Br_2, H_2O; (c) HNO_3, H_2O.

Problem 27.24 Which aldoses are oxidized to optically inactive aldaric acids: (a) D-erythrose; (b) D-lyxose; (c) D-galactose?

27.10 Reactions at the Carbonyl Group—Adding or Removing One Carbon Atom

Two common procedures in carbohydrate chemistry result in adding or removing one carbon atom from the skeleton of an aldose. The **Wohl degradation** shortens an aldose chain by one carbon, whereas the **Kiliani–Fischer synthesis** lengthens it by one. Both reactions involve cyanohydrins as intermediates. Recall from Section 21.9 that cyanohydrins are formed from aldehydes by addition of the elements of HCN. Cyanohydrins can also be re-converted to carbonyl compounds by treatment with base.

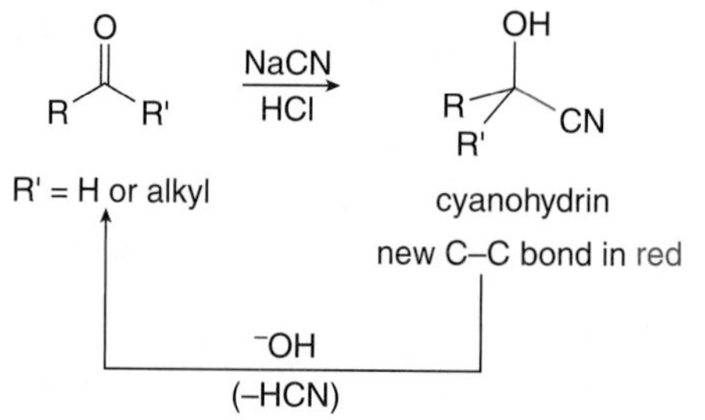
NaCN
HCl
R' = H or alkyl
cyanohydrin
new C–C bond in red
⁻OH
(–HCN)

- **Forming a cyanohydrin adds one carbon to a carbonyl group.**
- **Re-converting a cyanohydrin to a carbonyl compound removes one carbon.**

27.10A The Wohl Degradation

The Wohl degradation is a stepwise procedure that shortens the length of an aldose chain by cleavage of the C1–C2 bond. As a result, an aldohexose is converted to an aldopentose having the same configuration at its bottom three stereogenic centers (C3–C5). For example, the Wohl degradation converts D-glucose into D-arabinose.

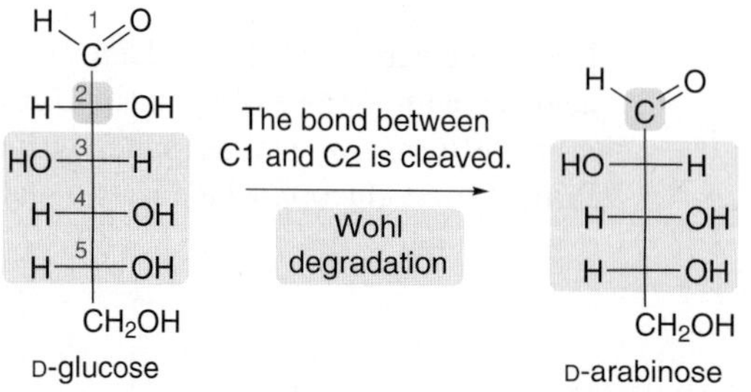
The bond between C1 and C2 is cleaved.
Wohl degradation
D-glucose
D-arabinose

The Wohl degradation consists of three steps, illustrated here beginning with D-glucose.

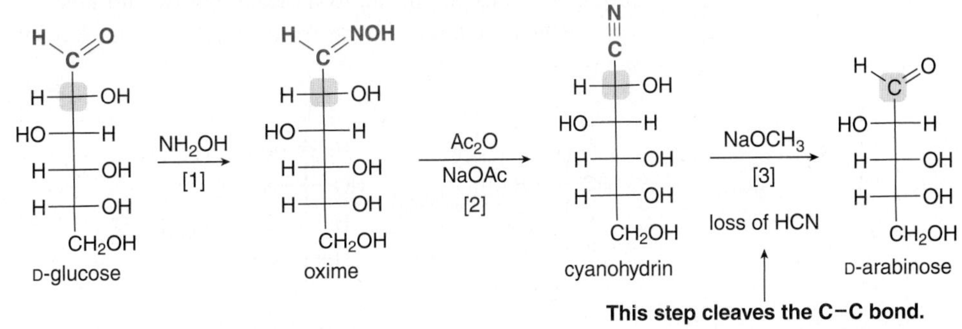
NH2OH
[1]
Ac2O
NaOAc
[2]
NaOCH3
[3]
loss of HCN
D-glucose
oxime
cyanohydrin
D-arabinose
This step cleaves the C–C bond.

[1] Treatment of D-glucose with hydroxylamine (NH_2OH) forms an **oxime** by nucleophilic addition. This reaction is analogous to the formation of imines discussed in Section 21.11.

[2] Dehydration of the oxime to a nitrile occurs with acetic anhydride (Ac_2O) and sodium acetate (NaOAc). The nitrile product is a cyanohydrin.

[3] **Treatment of the cyanohydrin with base results in loss of the elements of HCN to form an aldehyde having one fewer carbon.**

The Wohl degradation converts a stereogenic center at C2 in the original aldose to an sp^2 hybridized C=O. As a result, a pair of aldoses that are epimeric at C2, such as D-galactose and D-talose, yield the same aldose (D-lyxose, in this case) upon Wohl degradation.

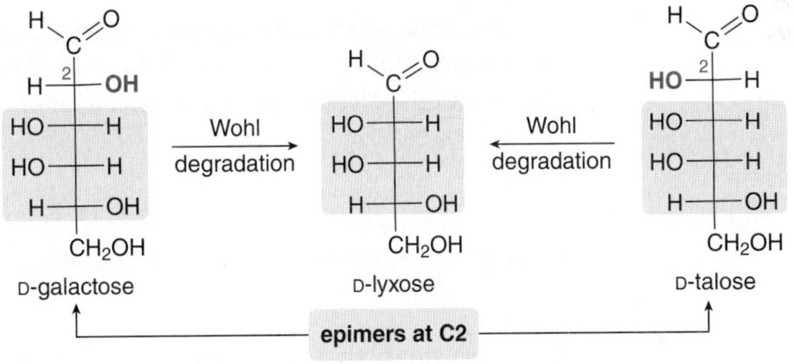
Wohl
degradation
Wohl
degradation
D-galactose
D-lyxose
D-talose
epimers at C2

Problem 27.25 What two aldoses yield D-xylose on Wohl degradation?

27.10B The Kiliani–Fischer Synthesis

The Kiliani–Fischer synthesis lengthens a carbohydrate chain by adding one carbon to the aldehyde end of an aldose, thus forming a new stereogenic center at C2 of the product. The product consists of epimers that differ only in their configuration about the one new stereogenic center. For example, the Kiliani–Fischer synthesis converts D-arabinose into a mixture of D-glucose and D-mannose.

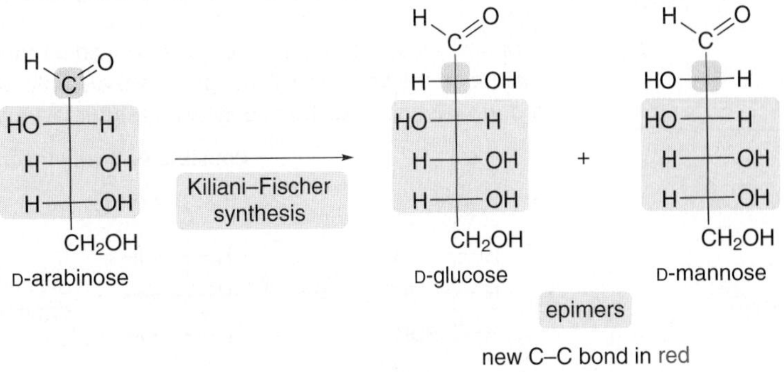
Kiliani–Fischer
synthesis
D-arabinose
D-glucose
D-mannose
epimers
new C–C bond in red

The Kiliani–Fischer synthesis, shown here beginning with D-arabinose, consists of three steps. "Squiggly" lines are meant to indicate that two different stereoisomers are formed at the new stereogenic center. As with the Wohl degradation, the key intermediate is a cyanohydrin.

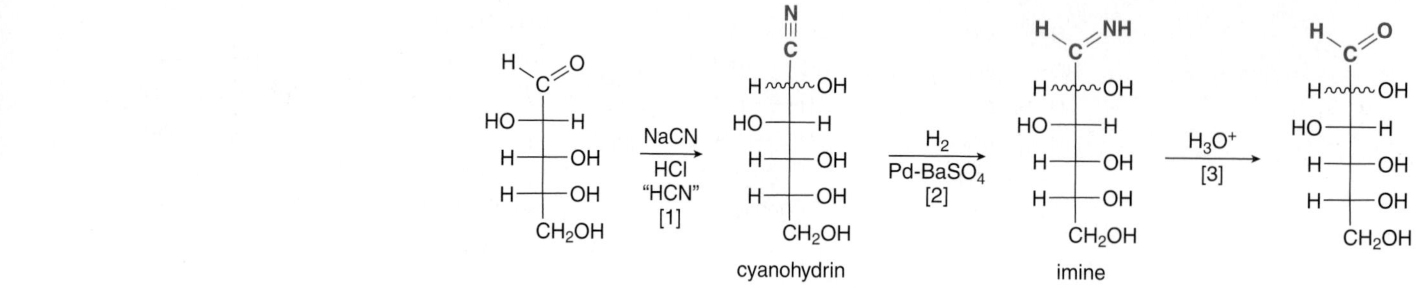

[1] Treating an aldose with NaCN and HCl adds the elements of HCN to the carbonyl group, forming a **cyanohydrin** and a new carbon–carbon bond. Because the sp^2 hybridized carbonyl carbon is converted to an sp^3 hybridized carbon with four different groups, a new stereogenic center is formed in this step.

[2] Reduction of the nitrile with H_2 and Pd-$BaSO_4$, a poisoned Pd catalyst, forms an **imine.**

[3] Hydrolysis of the imine with aqueous acid forms an aldehyde that has one more carbon than the aldose that began the sequence.

Note that the **Wohl degradation and the Kiliani–Fischer synthesis are conceptually opposite transformations.**

- **The Wohl degradation *removes* a carbon atom from the aldehyde end of an aldose. Two aldoses that are epimers at C2 form the same product.**
- **The Kiliani–Fischer synthesis *adds* a carbon to the aldehyde end of an aldose, forming two epimers at C2.**

Problem 27.26 What aldoses are formed when the following aldoses are subjected to the Kiliani–Fischer synthesis: (a) D-threose; (b) D-ribose; (c) D-galactose?

27.10C Determining the Structure of an Unknown Monosaccharide

The reactions in Sections 27.9–27.10 can be used to determine the structure of an unknown monosaccharide, as shown in Sample Problem 27.5.

Sample Problem 27.5 A D-aldopentose **A** is oxidized to an optically inactive aldaric acid with HNO_3. **A** is formed by the Kiliani–Fischer synthesis of a D-aldotetrose **B,** which is also oxidized to an optically inactive aldaric acid with HNO_3. What are the structures of **A** and **B?**

Solution

Use each fact to determine the relative orientation of the OH groups in the D-aldopentose.

Fact [1] A D-aldopentose A is oxidized to an optically *inactive* aldaric acid with HNO_3.

An optically inactive aldaric acid must contain a **plane of symmetry.** Because the OH group on C4 must be on the right for the D-sugar, there are only two ways to arrange the OH groups in a five-carbon D-aldaric acid. Thus, only two structures are possible for **A,** labeled **A'** and **A''.**

Possible optically inactive D-aldaric acids:

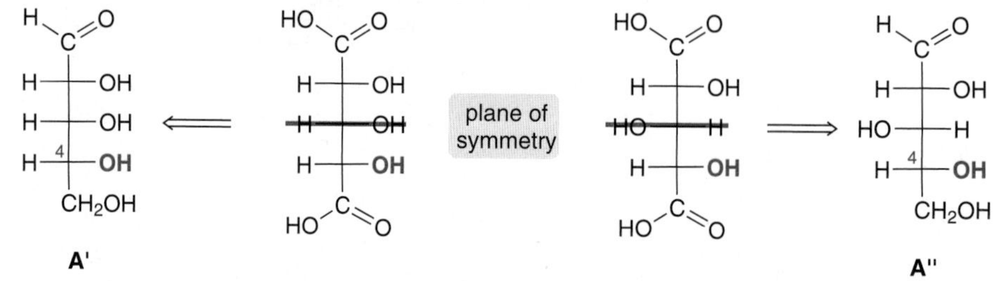

Fact [2] **A is formed by the Kiliani–Fischer synthesis from a D-aldotetrose B.**

A' and **A''** are each prepared from a D-aldotetrose (**B'** and **B''**) that has the same configuration at the bottom two stereogenic centers.

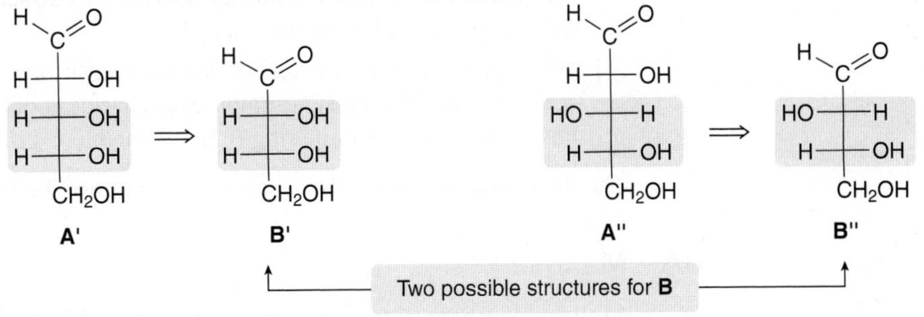

Fact [3] **The D-aldotetrose is oxidized to an optically *inactive* aldaric acid upon treatment with HNO_3.**

Only the aldaric acid from **B'** has a plane of symmetry, making it optically inactive. Thus, **B'** is the correct structure for the D-aldotetrose **B,** and therefore **A'** is the structure of the D-aldopentose **A.**

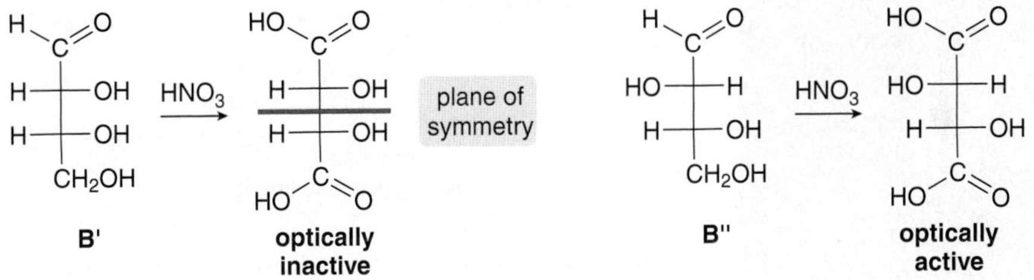

Answer:

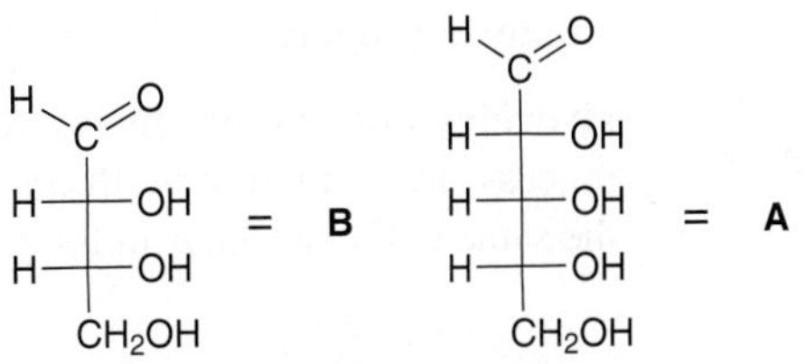

Problem 27.27 D-Aldopentose **A** is oxidized to an optically inactive aldaric acid. On Wohl degradation, **A** forms an aldotetrose **B** that is oxidized to an optically active aldaric acid. What are the structures of **A** and **B?**

Problem 27.28 A D-aldohexose **A** is formed from an aldopentose **B** by the Kiliani–Fischer synthesis. Reduction of **A** with $NaBH_4$ forms an optically inactive alditol. Oxidation of **B** forms an optically active aldaric acid. What are the structures of **A** and **B?**

27.11 Disaccharides

Disaccharides contain two monosaccharides joined together by a glycosidic linkage. The **general features of a disaccharide** include the following:

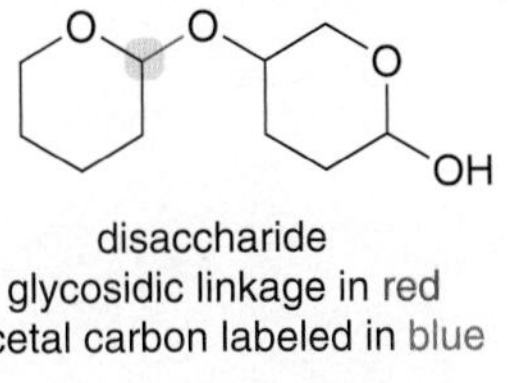
disaccharide
glycosidic linkage in red
acetal carbon labeled in blue

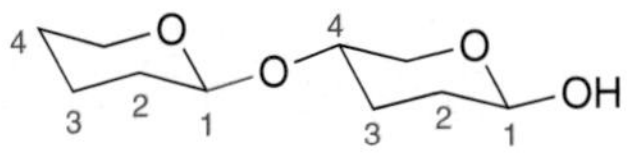
1→4-β-glycosidic linkage

[1] Two monosaccharide rings may be five- or six-membered, but six-membered rings are much more common. The two rings are connected by an O atom that is part of an acetal, called a **glycosidic linkage,** which may be oriented α or β.

[2] The **glycoside is formed from the anomeric carbon of one monosaccharide and any OH group on the other monosaccharide.** All disaccharides have **one acetal,** together with either a hemiacetal or another acetal.

[3] With pyranose rings, the carbon atoms in each ring are numbered beginning with the anomeric carbon. The most common disaccharides contain two monosaccharides in which the hemiacetal carbon of one ring (Cl) is joined to C4 of the other ring.

The three most abundant disaccharides are **maltose, lactose,** and **sucrose.**

27.11A Maltose

Maltose gets its name from malt, the liquid obtained from barley and other cereal grains.

Maltose, a disaccharide formed by the hydrolysis of starch, is found in germinated grains such as barley. Maltose contains two glucose units joined together by a 1→4-α-glycoside bond. Maltose contains one acetal carbon (in red) and one hemiacetal carbon (in blue).

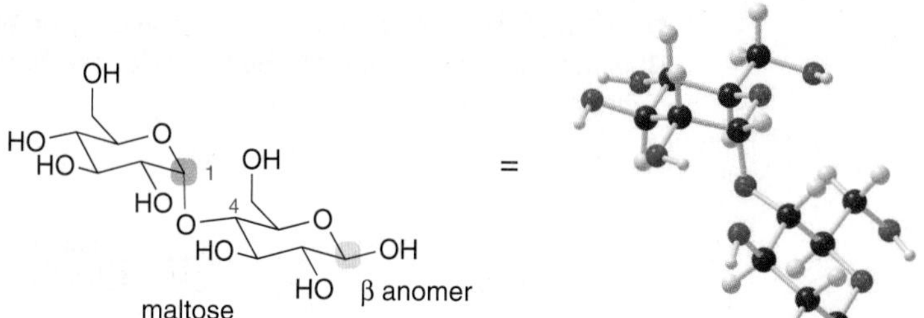

maltose
1→4-α-glycosidic linkage

Because one glucose ring of maltose still contains a hemiacetal, it exists as a mixture of α and β anomers. Only the β anomer is shown. Maltose exhibits two properties of all carbohydrates that contain a hemiacetal: it undergoes **mutarotation,** and it reacts with oxidizing agents, making it a **reducing sugar.**

Hydrolysis of maltose forms two molecules of glucose. The **C1–O** bond is cleaved in this process, and a mixture of glucose anomers forms. The mechanism for this hydrolysis is exactly the same as the mechanism for glycoside hydrolysis in Section 27.7B.

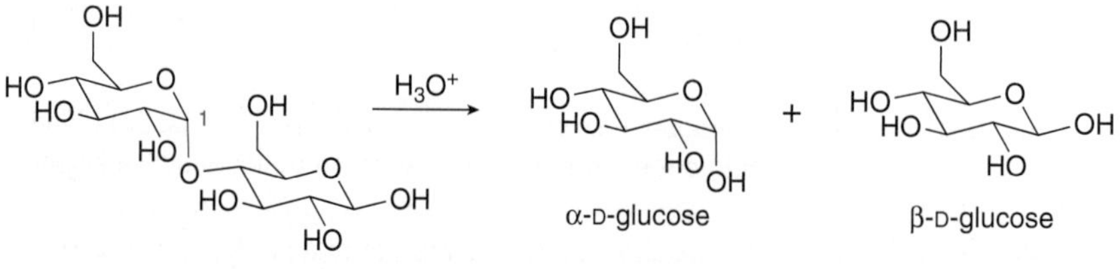

Problem 27.29 Draw the α anomer of maltose. What products are formed on hydrolysis of this form of maltose?

27.11B Lactose

Milk contains the disaccharide lactose.

Lactose is the principal disaccharide found in milk from both humans and cows. Unlike many mono- and disaccharides, lactose is not appreciably sweet. Lactose consists of **one galactose** and **one glucose unit,** joined by a **1→4-β-glycoside bond** from the anomeric carbon of galactose to C4 of glucose.

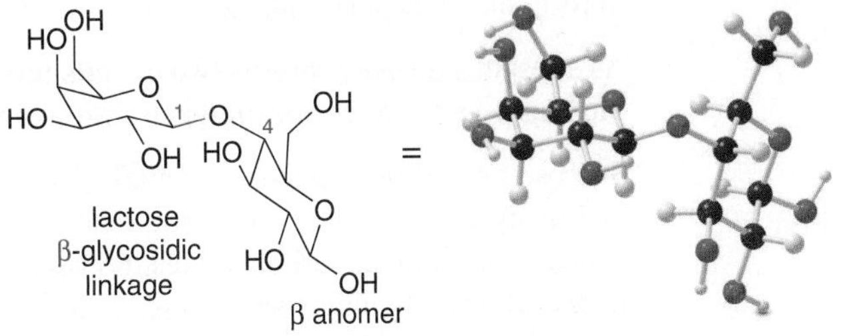

Like maltose, lactose also contains a hemiacetal, so it exists as a mixture of α and β anomers. The β anomer is drawn. Lactose undergoes **mutarotation,** and it reacts with oxidizing agents, making it a **reducing sugar.**

Lactose is digested in the body by first cleaving the 1→4-β-glycoside bond using the enzyme *lactase.* Many individuals, mainly of Asian and African descent, lack adequate amounts of lactase, so they are unable to digest and absorb lactose. This condition, lactose intolerance, is associated with abdominal cramping and recurrent diarrhea when milk and dairy products are ingested.

Problem 27.30 Cellobiose, a disaccharide obtained by the hydrolysis of cellulose, is composed of two glucose units joined together in a 1→4-β-glycoside bond. What is the structure of cellobiose?

27.11C Sucrose

Sucrose, the disaccharide found in sugarcane and used as table sugar (Figure 27.9), is the most common disaccharide in nature. It contains **one glucose unit** and **one fructose unit.**

Figure 27.9 Sucrose

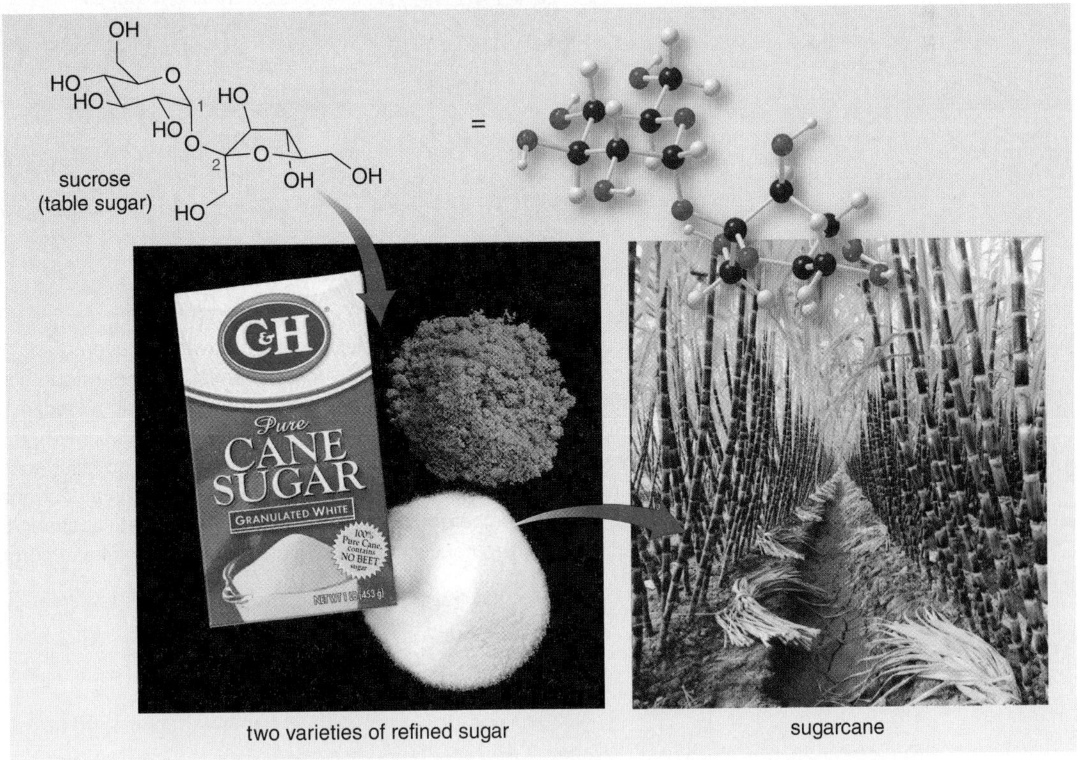

The structure of sucrose has several features that make it different from maltose and lactose. First of all, sucrose contains one six-membered ring (glucose) and one five-membered ring (fructose), whereas both maltose and lactose contain two six-membered rings. In sucrose the six-membered glucose ring is joined by an α-glycosidic bond to C2 of a fructofuranose ring. The numbering in a fructofuranose is different from the numbering in a pyranose ring. The anomeric carbon is now designated as C2, so the anomeric carbons of the glucose and fructose rings are both used to form the glycosidic linkage.

As a result, **sucrose contains two acetals but no hemiacetal.** Sucrose, therefore, is a **nonreducing sugar** and **it does not undergo mutarotation.**

Sucrose's pleasant sweetness has made it a widely used ingredient in baked goods, cereals, bread, and many other products. It is estimated that the average American ingests 100 lb of sucrose annually. Like other carbohydrates, however, sucrose contains many calories. To reduce caloric intake while maintaining sweetness, a variety of artificial sweeteners have been developed. These include sucralose, aspartame, and saccharin (Figure 27.10). These compounds are much sweeter than sucrose so only a small amount of each compound is needed to achieve the same level of perceived sweetness.

Figure 27.10
Artificial sweeteners

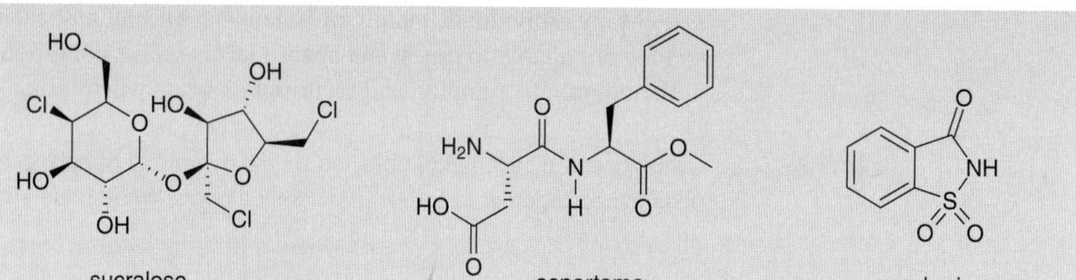

sucralose
(Trade name Splenda)

aspartame
(Trade name Equal)

saccharin
(Trade name Sweet'n Low)

The sweetness of these three artificial sweeteners was discovered accidentally. The sweetness of sucralose was discovered in 1976 when a chemist misunderstood his superior, and so he *tasted* rather than *tested* his compound. Aspartame was discovered in 1965 when a chemist licked his dirty fingers in the lab and tasted its sweetness. Saccharin, the oldest known artificial sweetener, was discovered in 1879 by a chemist who failed to wash his hands after working in the lab. Saccharin was not used extensively until sugar shortages occurred during World War I. Although there were concerns in the 1970s that saccharin causes cancer, there is no proven link between cancer occurrence and saccharin intake at normal levels.

27.12 Polysaccharides

Polysaccharides contain three or more monosaccharides joined together. Three prevalent polysaccharides in nature are **cellulose, starch,** and **glycogen,** each of which consists of repeating glucose units joined by different glycosidic bonds.

27.12A Cellulose

The structure of cellulose was discussed in Section 5.1.

Cellulose is found in the cell walls of nearly all plants, where it gives support and rigidity to wood and plant stems. Cotton is essentially pure cellulose.

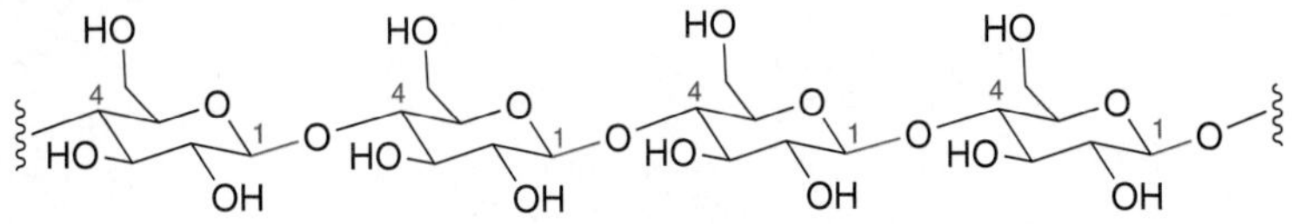

cellulose
1→4-β-glycosidic linkages in red

Ball-and-stick models showing the three-dimensional structures of cellulose and starch were given in Figure 5.2.

Cellulose is an unbranched polymer composed of repeating glucose units joined in a 1→4-β-glycosidic linkage. The β-glycosidic linkage forms long linear chains of cellulose molecules that stack in sheets, creating an extensive three-dimensional array. A network of intermolecular hydrogen bonds between the chains and sheets means that only the few OH groups on the surface are available to hydrogen bond to water, making this very polar compound water insoluble.

Cellulose acetate, a cellulose derivative, is made by treating cellulose with acetic anhydride and sulfuric acid. The resulting product has acetate esters in place of every OH group. Cellulose acetate is spun into fibers that are used for fabrics called *acetates,* which have a deep luster and satin appearance.

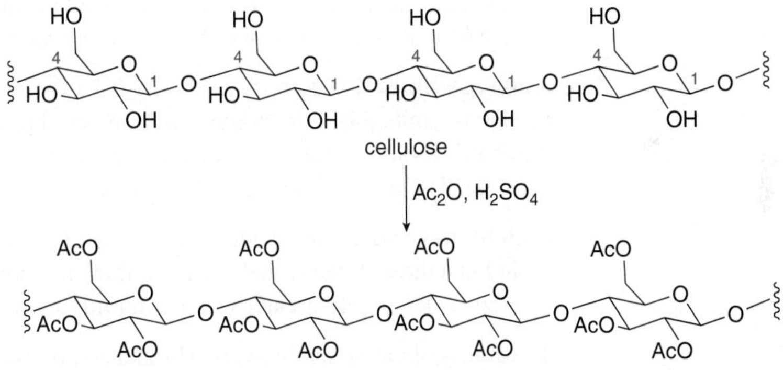
cellulose
Ac_2O, H_2SO_4
cellulose acetate

Cellulose can be hydrolyzed to glucose by cleaving all of the β-glycosidic bonds, yielding both anomers of glucose.

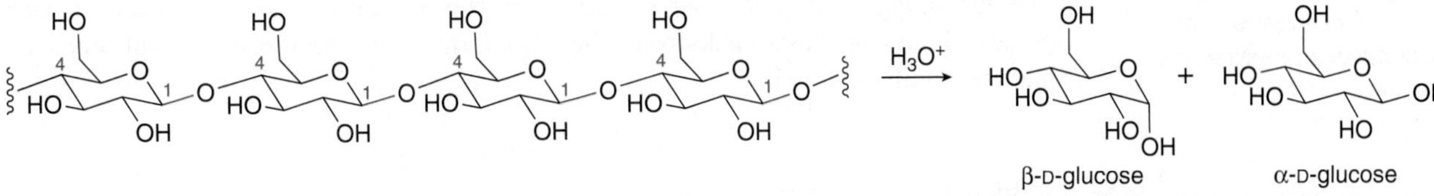
H_3O^+
β-D-glucose
α-D-glucose

A **β-glycosidase** is the general name of an enzyme that hydrolyzes a β-glycosidic linkage.

In cells, the hydrolysis of cellulose is accomplished by an enzyme called a **β-glucosidase,** which cleaves all the β-glycosidic bonds formed from glucose. Humans do not possess this enzyme, and therefore cannot digest cellulose. Ruminant animals, on the other hand, such as cattle, deer, and camels, have bacteria containing a β-glucosidase in their digestive systems, so they can derive nutritional benefit from eating grass and leaves.

27.12B Starch

Starch is the main carbohydrate found in the seeds and roots of plants. Corn, rice, wheat, and potatoes are common foods that contain a great deal of starch.

Starch is a polymer composed of repeating glucose units joined in α-glycosidic linkages. Both starch and cellulose are polymers of glucose, but starch contains α-glycosidic bonds, whereas cellulose contains β-glycosidic bonds. The two common forms of starch are **amylose** and **amylopectin.**

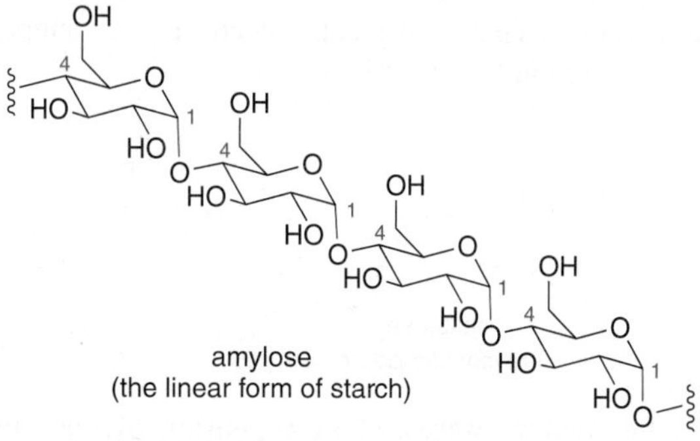
amylose
(the linear form of starch)

1→4-α-glycosidic linkages in red

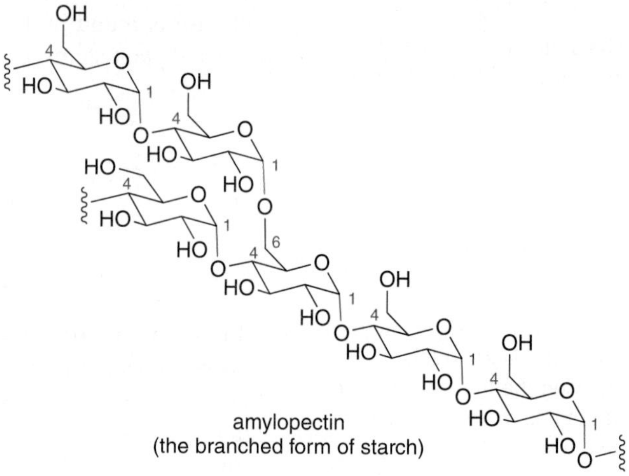
amylopectin
(the branched form of starch)

1→4-α-glycosidic linkages in red
1→6-α-glycosidic linkage in blue

Amylose, which comprises about 20% of starch molecules, has an unbranched skeleton of glucose molecules with **1→4-α-glycosidic bonds.** Because of this linkage, an amylose chain adopts a helical arrangement, giving it a very different three-dimensional shape from the linear chains of cellulose. Amylose was first described in Section 5.1.

Amylopectin, which comprises about 80% of starch molecules, likewise consists of a backbone of glucose units joined in **α-glycosidic bonds,** but it also contains considerable branching along the chain. The linear linkages of amylopectin are formed by **1→4-α-glycosidic bonds,** similar to amylose. The branches are linked to the chain with **1→6-α-glycosidic linkages.**

Both forms of starch are water soluble. Because the OH groups in these starch molecules are not buried in a three-dimensional network, they are more available for hydrogen bonding with water molecules, leading to greater water solubility than cellulose has.

The ability of amylopectin to form branched polymers is a unique feature of carbohydrates. Other types of polymers in the cell, such as the proteins discussed in Chapter 26, occur in nature only as linear molecules.

α-Glycosidase is the general name of an enzyme that hydrolyzes an α-glycosidic linkage.

Both amylose and amylopectin are hydrolyzed to glucose with cleavage of the glycosidic bonds. The human digestive system has the necessary **α-glucosidase** enzymes needed to catalyze this process. Bread and pasta made from wheat flour, rice, and corn tortillas are all sources of starch that are readily digested.

27.12C Glycogen

Glycogen is the major form in which polysaccharides are stored in animals. Glycogen, a polymer of glucose containing **α-glycosidic bonds,** has a branched structure similar to amylopectin, but the branching is much more extensive.

Glycogen is stored principally in the liver and muscle. When glucose is needed for energy in the cell, glucose units are hydrolyzed from the ends of the glycogen polymer, and then further metabolized with the release of energy. Because glycogen has a highly branched structure, there are many glucose units at the ends of the branches that can be cleaved whenever the body needs them.

Problem 27.31 Draw the structure of: (a) a polysaccharide formed by joining D-mannose units in 1→4-β-glycosidic linkages; (b) a polysaccharide formed by joining D-glucose units in 1→6-α-glycosidic linkages. The polysaccharide in (b) is dextran, a component of dental plaque.

27.13 Other Important Sugars and Their Derivatives

Many other examples of simple and complex carbohydrates with useful properties exist in the biological world. In Section 27.13, we examine some carbohydrates that contain nitrogen atoms.

27.13A Amino Sugars and Related Compounds

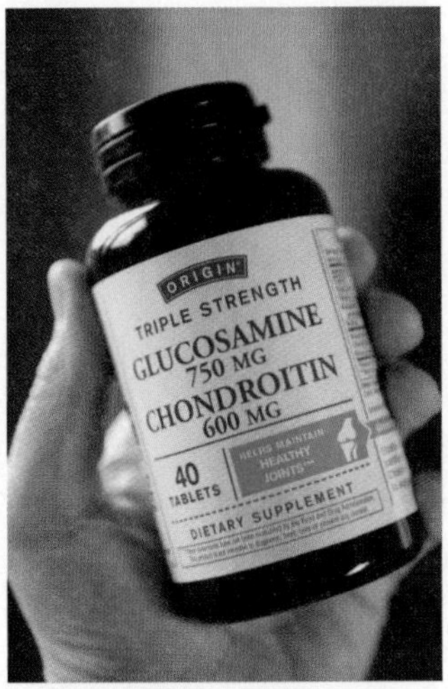

Dietary supplements containing glucosamine are used by individuals suffering from osteoarthritis.

Amino sugars contain an NH_2 group instead of an OH group at a non-anomeric carbon. The most common amino sugar in nature, **D-glucosamine,** is formally derived from D-glucose by replacing the OH at C2 with NH_2. Although it is not classified as a drug, and therefore not regulated by the Food and Drug Administration, glucosamine is available in many over-the-counter treatments for osteoarthritis.

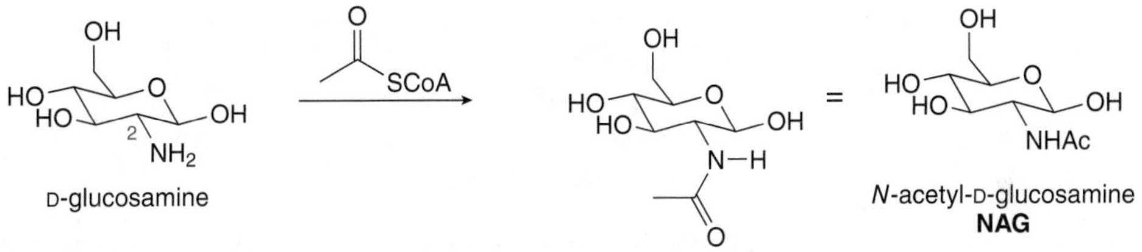

Acetylation of glucosamine with acetyl CoA (Section 22.17) forms ***N*-acetyl-D-glucosamine,** abbreviated as **NAG. Chitin,** the second most abundant carbohydrate polymer, is a polysaccharide formed from NAG units joined together in **1→4-β-glycosidic linkages.** Chitin is identical in structure to cellulose, except that each OH group at C2 is now replaced by $NHCOCH_3$. The exoskeletons of lobsters, crabs, and shrimp are composed of chitin. Like those of cellulose, chitin chains are held together by an extensive network of hydrogen bonds, forming water-insoluble sheets.

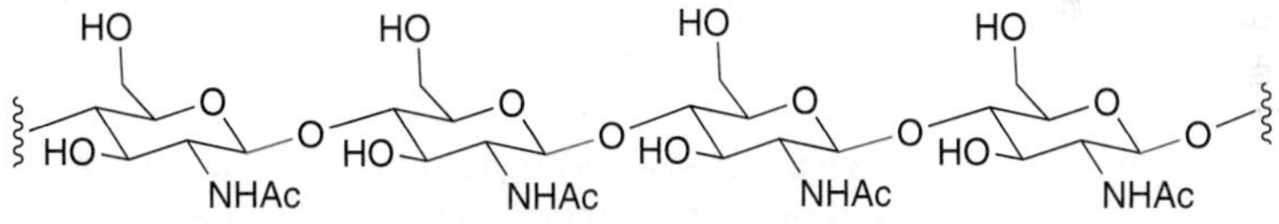

1→4-β-glycosidic linkages in red

chitin

The rigidity of a crab shell is due to chitin, a high molecular weight carbohydrate molecule. Chitin-based coatings have found several commercial applications, such as extending the shelf life of fruits. Processing plants now convert the shells of crabs, lobsters, and shrimp to chitin and various derivatives for use in many consumer products.

Several trisaccharides containing amino sugars are potent antibiotics used in the treatment of certain severe and recurrent bacterial infections. These compounds, such as tobramycin and amikacin, are called **aminoglycoside antibiotics.**

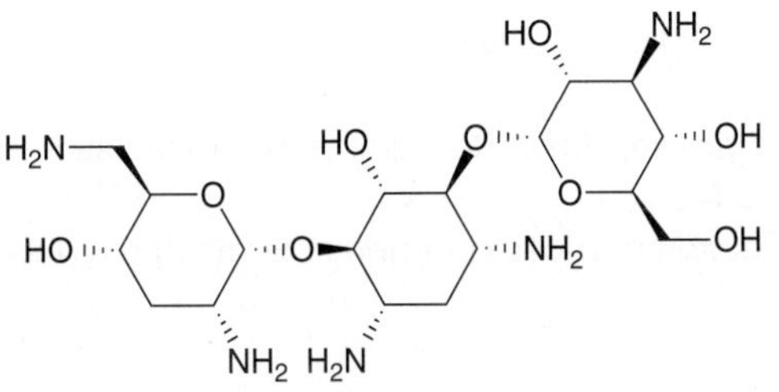

tobramycin

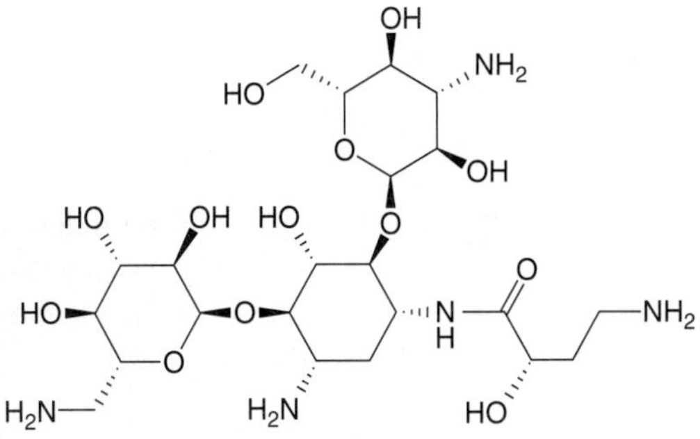

amikacin

Problem 27.32 Treating chitin with H_2O, ^{-}OH hydrolyzes its amide linkages, forming a compound called chitosan. What is the structure of chitosan? Chitosan has been used in shampoos, fibers for sutures, and wound dressings.

27.13B *N*-Glycosides

***N*-Glycosides** are formed when a monosaccharide is reacted with an amine in the presence of mild acid (Reactions [1] and [2]).

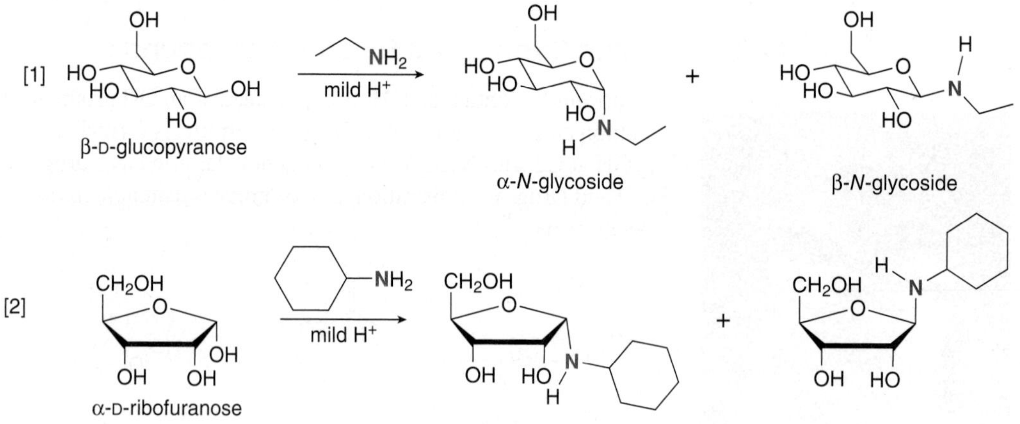
[1]
β-D-glucopyranose
mild H⁺
α-N-glycoside
β-N-glycoside
[2]
α-D-ribofuranose
mild H⁺

The mechanism of *N*-glycoside formation is analogous to the mechanism for glycoside formation, and both anomers of the *N*-glycoside are formed as products.

Problem 27.33 Draw the products of each reaction.

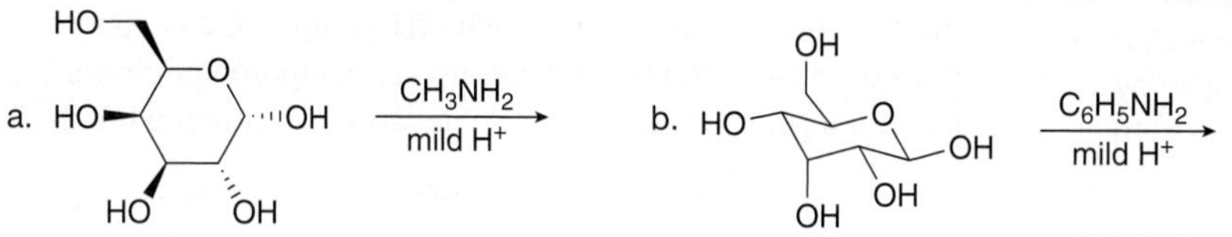
a.
b.
CH₃NH₂
mild H⁺
C₆H₅NH₂
mild H⁺

Problem 27.34 Draw a stepwise mechanism for the conversion of β-D-glucose to both anomers of *N*-ethyl glucopyranoside, the equation written in Reaction [1].

The prefix ***deoxy*** means "without oxygen."

The *N*-glycosides of two sugars, **D-ribose** and **2-deoxy-D-ribose,** are especially noteworthy, because they form the building blocks of RNA and DNA, respectively (Section 3.9C). 2-Deoxyribose is so named because it lacks an OH group at C2 of ribose.

D-ribose 2-deoxy-D-ribose

- **Reaction of D-ribose with certain amine heterocycles forms *N*-glycosides called ribonucleosides.**
- **This same reaction of 2-deoxy-D-ribose forms deoxyribonucleosides.**

An example of a **ribonucleoside** and a **deoxyribonucleoside** are drawn. These *N*-glycosides have the β orientation. Numbering in the sugar ring begins at the anomeric carbon (1'), and proceeds in a clockwise fashion around the ring.

cytidine

a ribonucleoside

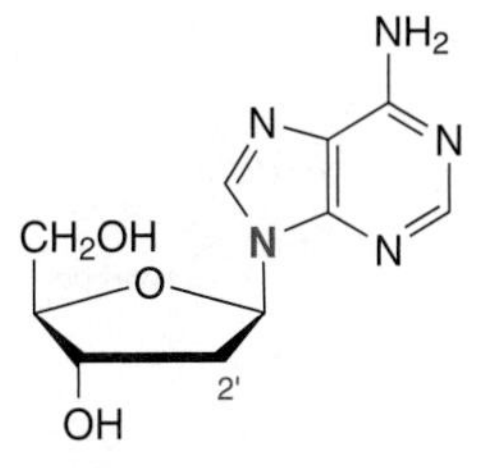

2-deoxyadenosine

a deoxyribonucleoside

Only five common nitrogen heterocycles are used to form these nucleosides (Section 17.8). Three compounds have one ring, and are derived from a nitrogen heterocycle called **pyrimidine.** Two are bicyclic, and are derived from a nitrogen heterocycle called **purine.** These five amines are referred to as ***bases.*** Each base is designated by a one-letter abbreviation, as shown in the names and structures drawn. Uracil (U) occurs only in ribonucleosides and thymine (T) occurs only in deoxyribonucleosides.

- **Each nucleoside has two parts, a sugar and a base, joined together by a β *N*-glycosidic linkage.**

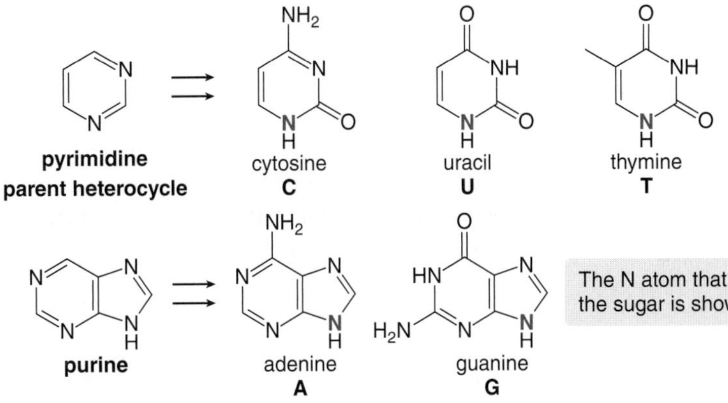
pyrimidine
parent heterocycle
cytosine
C
uracil
U
thymine
T
purine
adenine
A
guanine
G

The N atom that bonds to the sugar is shown in red.

When one OH group of the sugar nucleus is bonded to a phosphate, the derivatives are called **ribonucleotides** and **deoxyribonucleotides.**

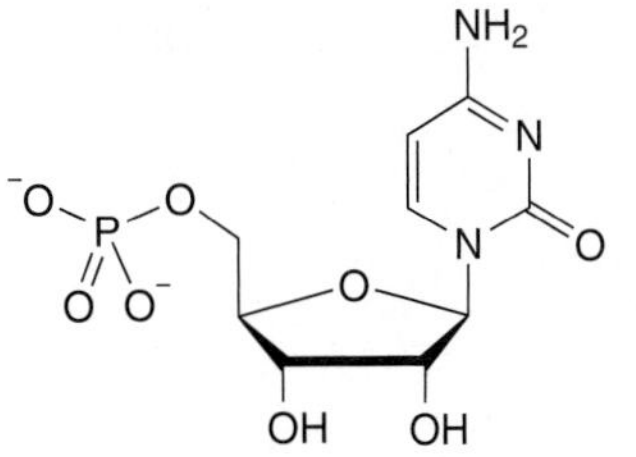

cytidine monophosphate

a ribonucleotide

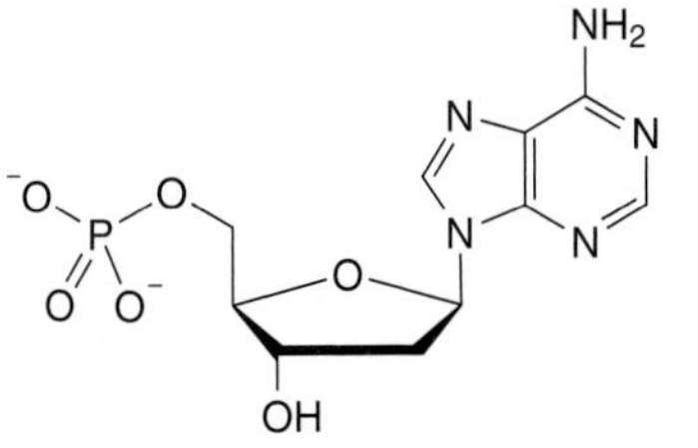

deoxyadenosine monophosphate

a deoxyribonucleotide

- **Ribonucleotides are the building blocks of the polymer ribonucleic acid, or RNA, the messenger molecules that convert genetic information into proteins.**
- **Deoxyribonucleotides are the building blocks of the polymer deoxyribonucleic acid, or DNA, the molecules that are responsible for the storage of all genetic information.**

Figure 27.11 Short segments of RNA and DNA

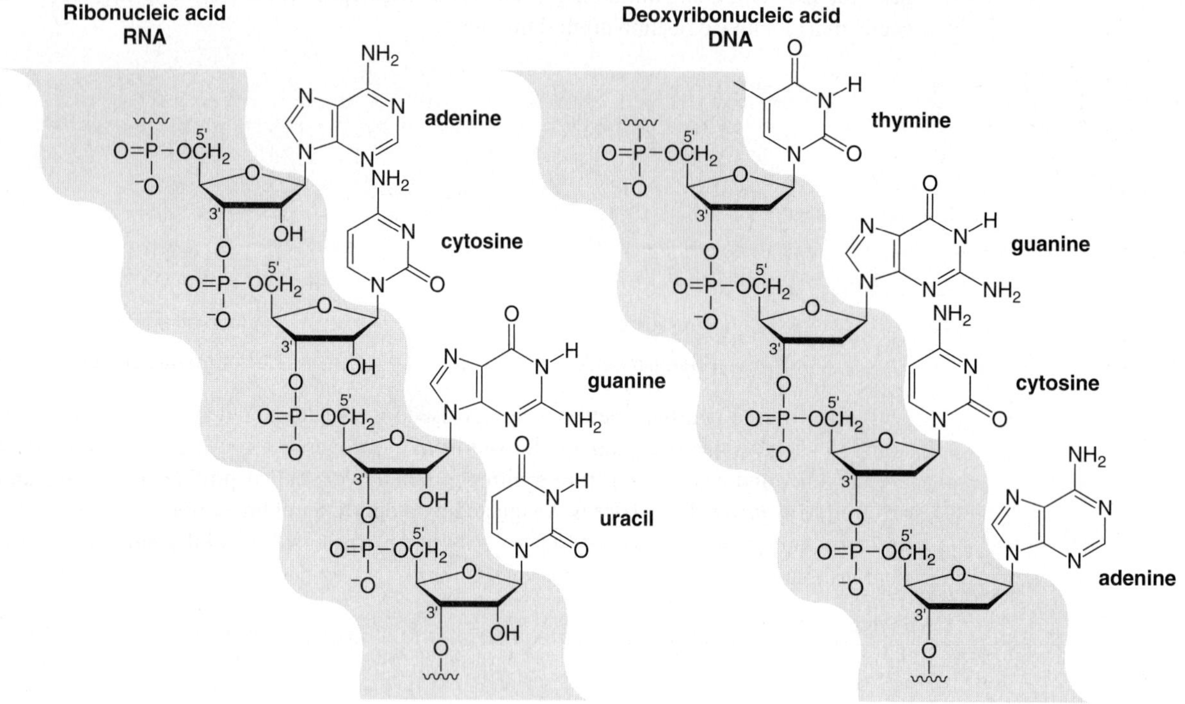

Short segments of both RNA and DNA are shown in Figure 27.11. Note the central role of the sugar moiety in both RNA and DNA. The sugar residues are bonded to two phosphate groups, thus connecting the chain of RNA or DNA together. The sugar residues are also bonded to the nitrogen base via the anomeric carbon.

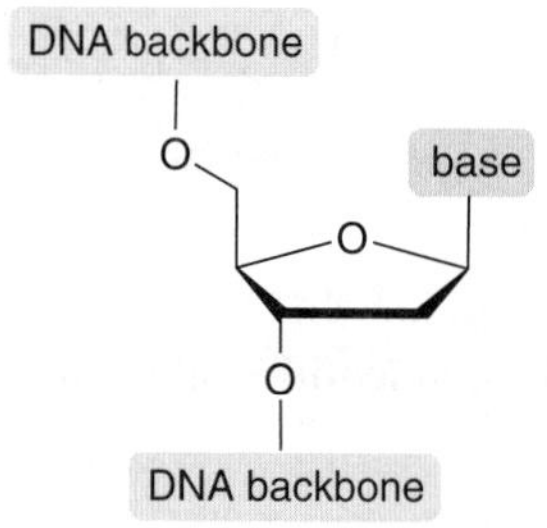

Figure 27.12 DNA—A double helix

DNA double helix

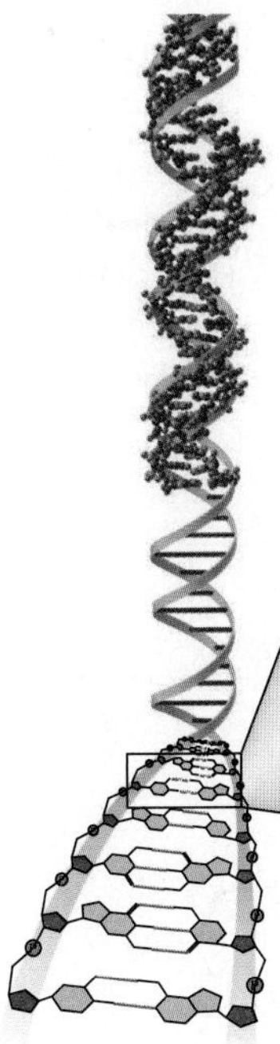

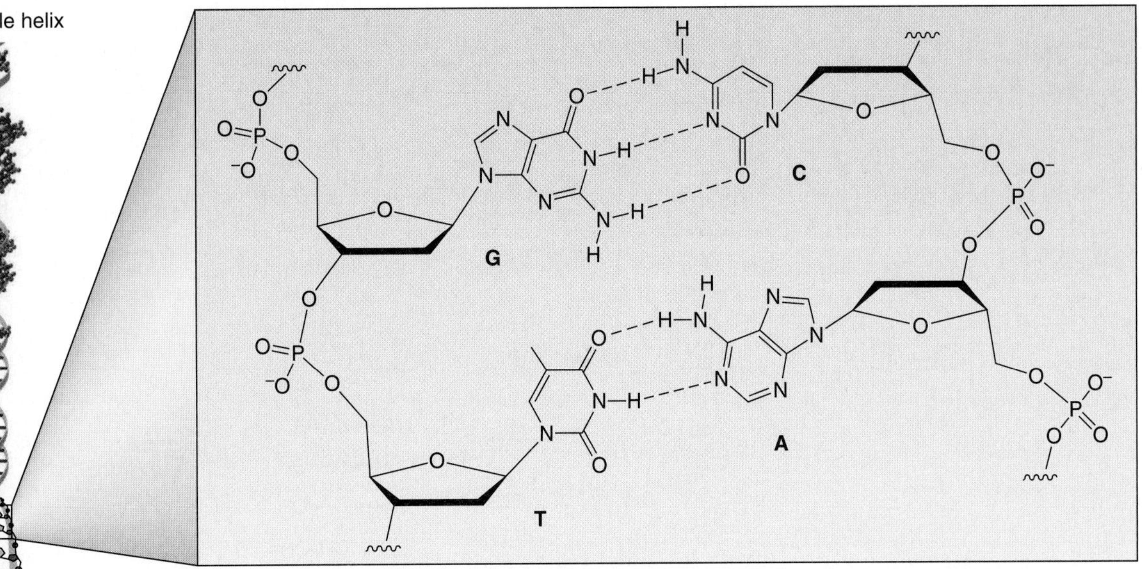

Individual base pairs that hydrogen bond to hold two strands of DNA together

- Two polynucleotide strands form the double helix of DNA. The backbone of each polymer strand is composed of sugar–phosphate residues. Hydrogen bonding of base pairs (A–T and C–G) holds the two strands of DNA together.

DNA is composed of two polynucleotide strands that wind around each other to form a double helix, resembling a spiral ladder, as first described in Section 3.9C. The sides of the ladder are composed of the sugar–phosphate backbone of the polymer and the rungs are composed of the bases, as shown in Figure 27.12.

The nitrogen bases on one strand of DNA hydrogen bond to nitrogen bases on the other strand. A purine base on one strand hydrogen bonds with a pyrimidine base on the other strand. Two types of bases, called **base pairs,** hydrogen bond to each other: adenine hydrogen bonds with thymine (A–T) and cytosine hydrogen bonds with guanine (C–G).

Problem 27.35 Draw the structures of the nucleosides formed from each of the following components: (a) ribose + uracil; (b) 2-deoxyribose + guanine.

Problem 27.36 (a) Why can't two purine bases (A and G) form a base pair and hydrogen bond to each other on two strands of DNA in the double helix? (b) Why is hydrogen bonding between guanine and cytosine more favorable than hydrogen bonding between guanine and thymine?

KEY CONCEPTS

Carbohydrates

Important Terms

- **Aldose** A monosaccharide containing an aldehyde (27.2)
- **Ketose** A monosaccharide containing a ketone (27.2)
- **D-Sugar** A monosaccharide with the OH bonded to the stereogenic center farthest from the carbonyl group drawn on the right in the Fischer projection (27.2C)
- **Epimers** Two diastereomers that differ in configuration around only one stereogenic center (27.3)
- **Anomers** Monosaccharides that differ in configuration at the hemiacetal OH group (27.6)
- **Glycoside** An acetal derived from a monosaccharide hemiacetal (27.7)

Acyclic, Haworth, and 3-D Representations for D-Glucose (27.6)

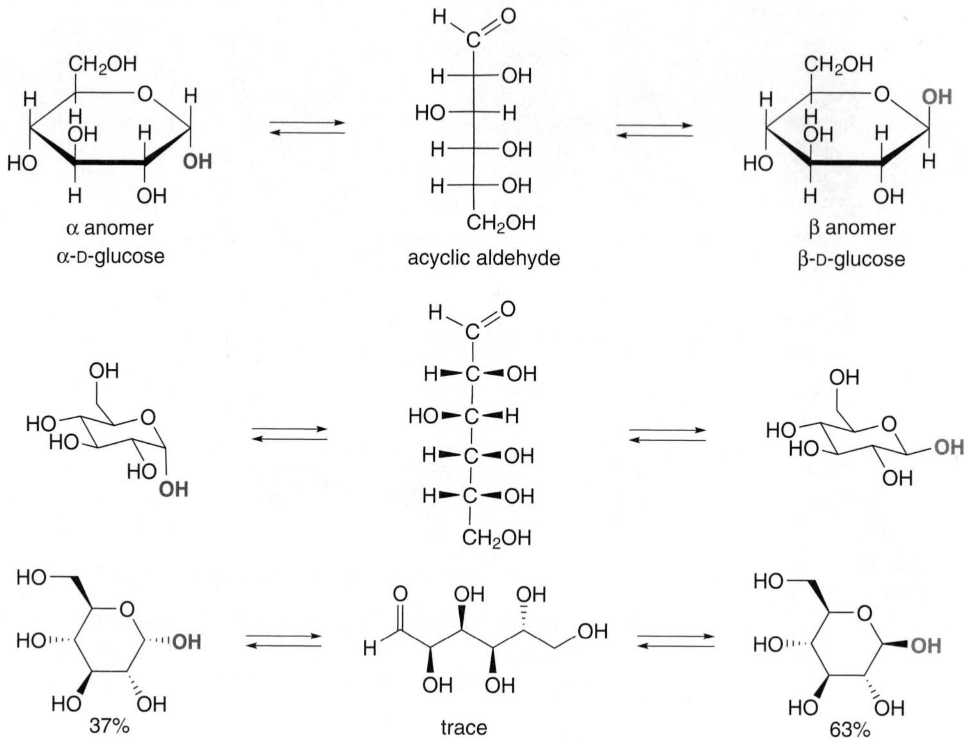
α anomer
α-D-glucose
acyclic aldehyde
β anomer
β-D-glucose
37%
trace
63%

Reactions of Monosaccharides Involving the Hemiacetal

[1] Glycoside formation (27.7A)

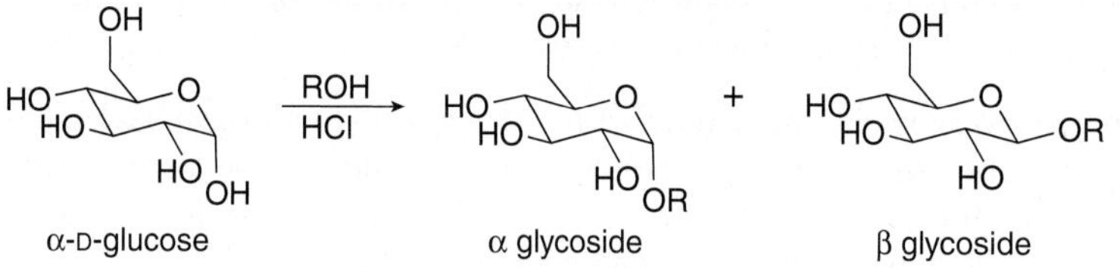
ROH
HCl
α-D-glucose
α glycoside
β glycoside

- Only the hemiacetal OH reacts.
- A mixture of α and β glycosides forms.

[2] Glycoside hydrolysis (27.7B)

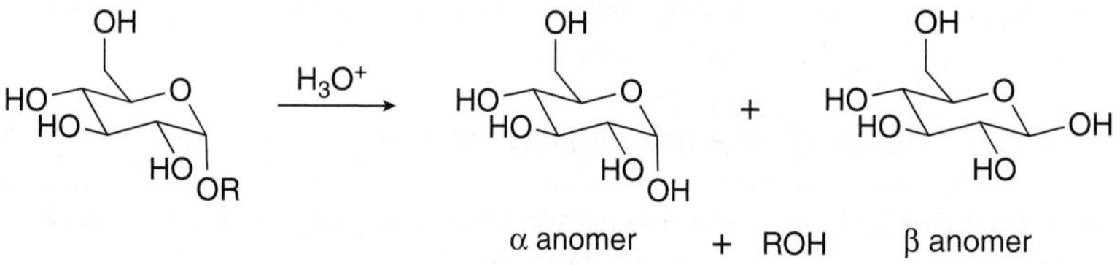
H_3O^+
α anomer
+ ROH
β anomer

- A mixture of α and β anomers forms.

Reactions of Monosaccharides at the OH Groups

[1] Ether formation (27.8)

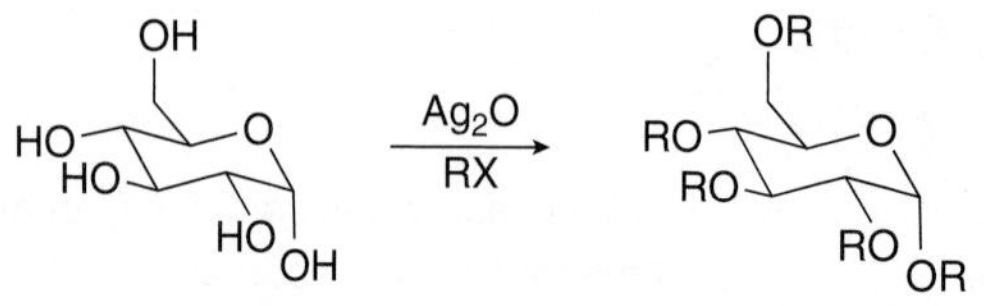
Ag_2O
RX

- All OH groups react.
- The stereochemistry at all stereogenic centers is retained.

[2] Ester formation (27.8)

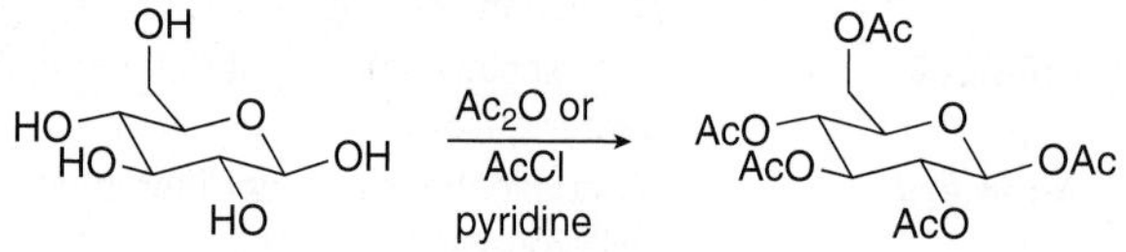
Ac_2O or
AcCl
pyridine

- All OH groups react.
- The stereochemistry at all stereogenic centers is retained.

Reactions of Monosaccharides at the Carbonyl Group

[1] Oxidation of aldoses (27.9B)

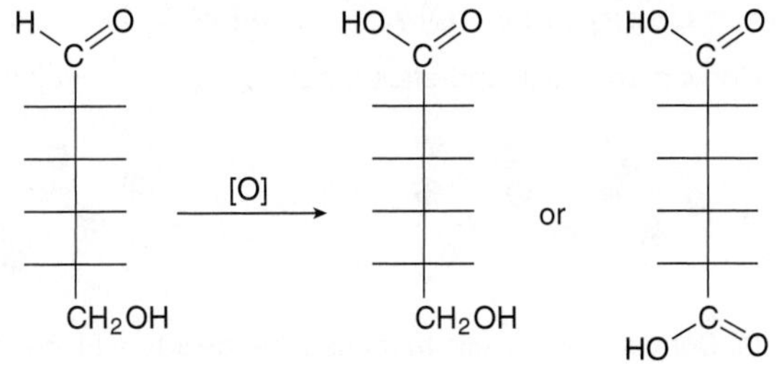
[O]
or

aldose — aldonic acid — aldaric acid

- Aldonic acids are formed using:
  - Ag_2O, NH_4OH
  - Cu^{2+}
  - Br_2, H_2O
- Aldaric acids are formed with HNO_3, H_2O.

[2] Reduction of aldoses to alditols (27.9A)

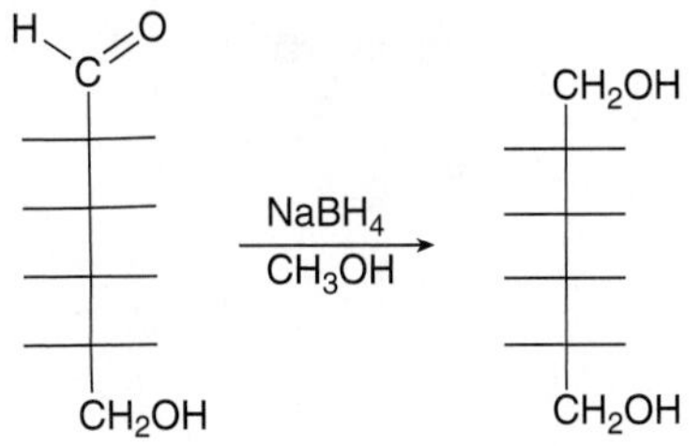
NaBH4
CH3OH

aldose — alditol

[3] Wohl degradation (27.10A)

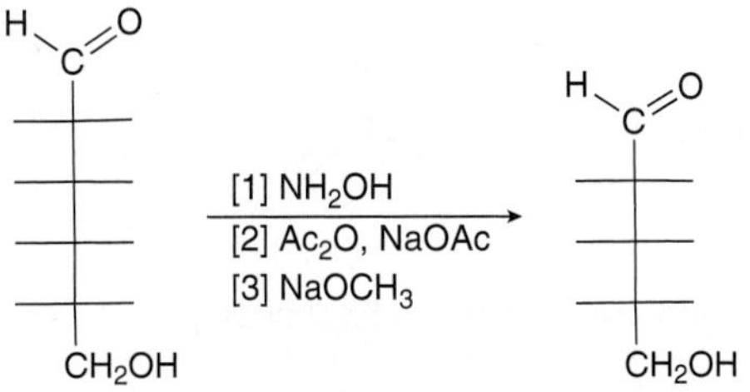
[1] NH2OH
[2] Ac2O, NaOAc
[3] NaOCH3

- The C1 – C2 bond is cleaved to shorten an aldose chain by one carbon.
- The stereochemistry at all other stereogenic centers is retained.
- Two epimers at C2 form the same product.

[4] Kiliani–Fischer synthesis (27.10B)

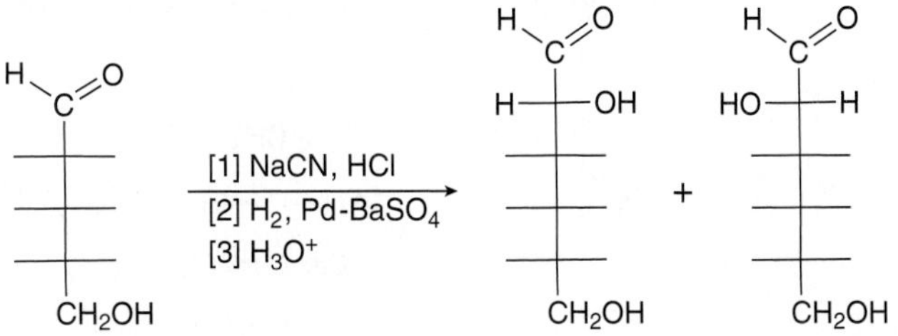
[1] NaCN, HCl
[2] H2, Pd-BaSO4
[3] H3O+

- One carbon is added to the aldehyde end of an aldose.
- Two epimers at C2 are formed.

Other Reactions

[1] Hydrolysis of disaccharides (27.11)

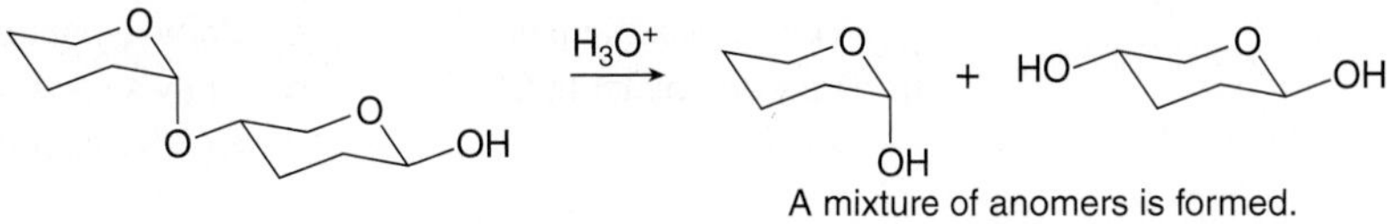
H3O+

A mixture of anomers is formed.

[2] Formation of *N*-glycosides (27.13B)

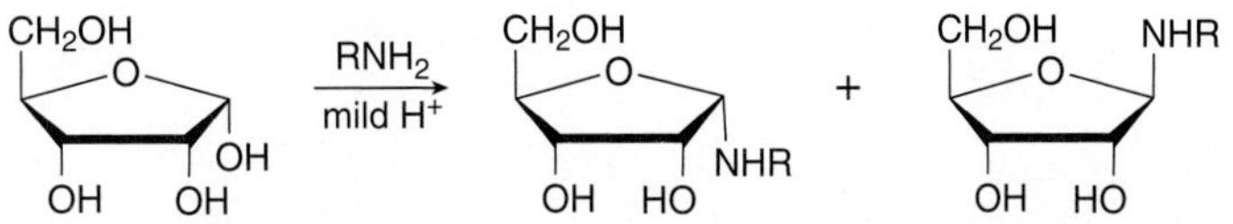
RNH2
mild H+

- Two anomers are formed.

PROBLEMS

Problems Using Three-Dimensional Models

27.37 Convert each ball-and-stick model to a Fischer projection.

a. 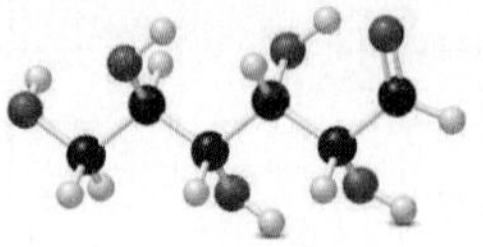

b. 

27.38 (a) Convert each cyclic monosaccharide into a Fischer projection of its acyclic form. (b) Name each monosaccharide. (c) Label the anomer as α or β.

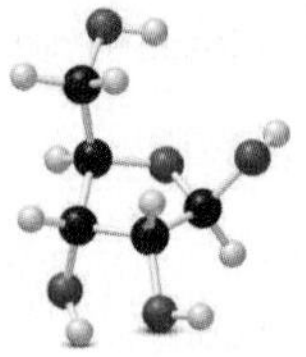

A

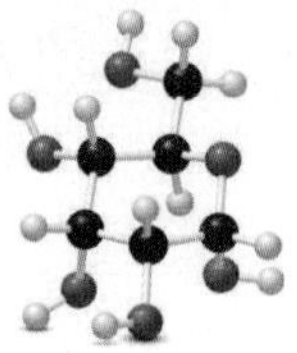

B

Fischer Projections

27.39 Classify each compound as identical to **A** or its enantiomer.

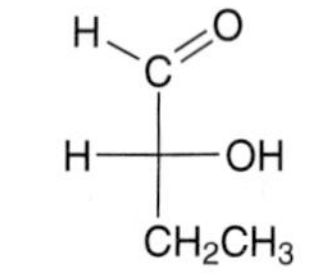

A

a. 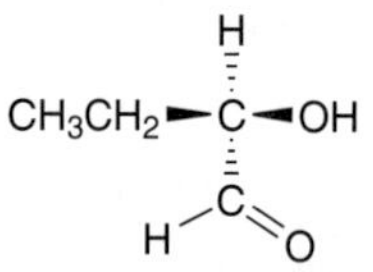

b. 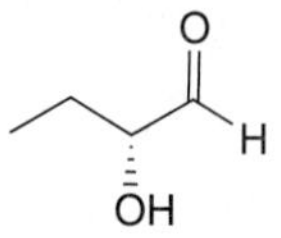

c. 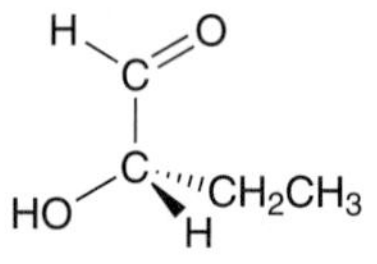

27.40 Convert each compound to a Fischer projection and label each stereogenic center as *R* or *S*.

a. 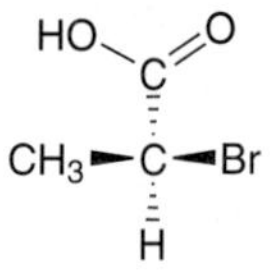

c. 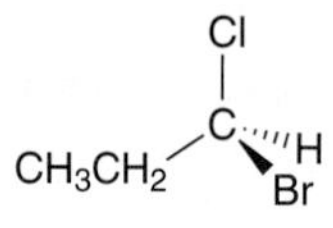

e. 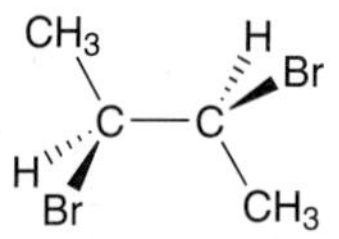

b. 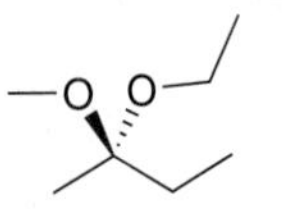

d. 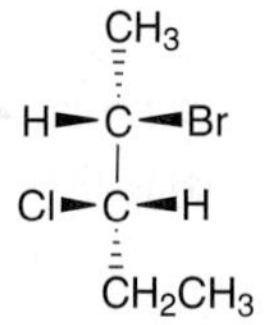

f. 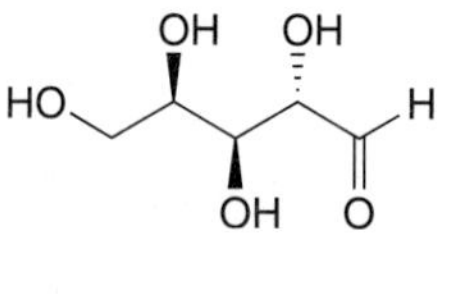

Monosaccharide Structure and Stereochemistry

27.41 For D-arabinose:

a. Draw its enantiomer.

b. Draw an epimer at C3.

c. Draw a diastereomer that is not an epimer.

d. Draw a constitutional isomer that still contains a carbonyl group.

27.42 Consider the following six compounds (**A–F**).

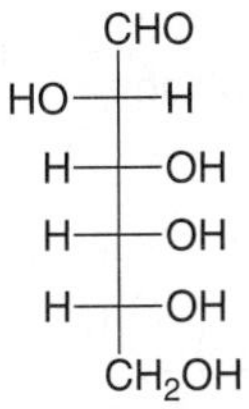

A

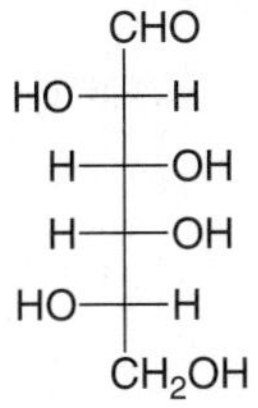

B

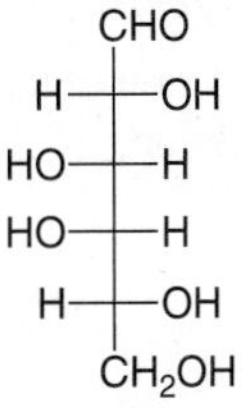

C

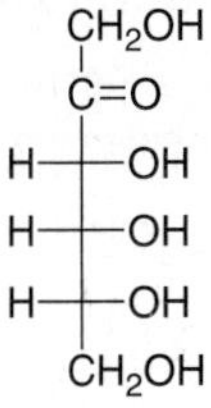

D

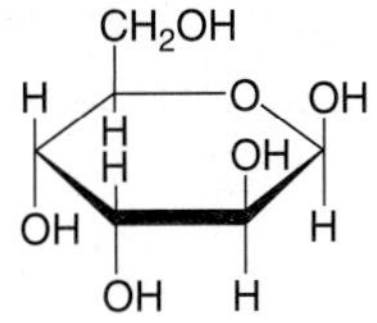

E

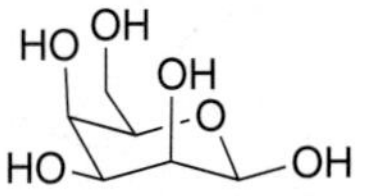

F

How are the two compounds in each pair related? Choose from enantiomers, epimers, diastereomers but not epimers, constitutional isomers, and identical compounds.

a. **A** and **B** b. **A** and **C** c. **B** and **C** d. **A** and **D** e. **E** and **F**

27.43 Draw a Haworth projection for each compound using the structures in Figures 27.4 and 27.5.

a. β-D-talopyranose b. α-D-galactopyranose c. α-D-tagatofuranose

27.44 Draw the β anomer of a monosaccharide epimeric with D-glucose at C2 using a Haworth projection.

27.45 Draw both pyranose anomers of each aldohexose using a three-dimensional representation with a chair pyranose. Label each anomer as α or β.

a.

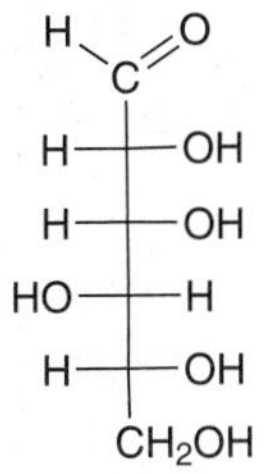

b.

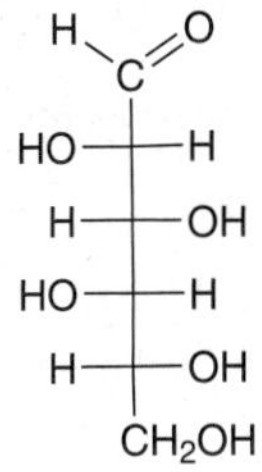

27.46 Convert each cyclic monosaccharide into its acyclic form.

a.

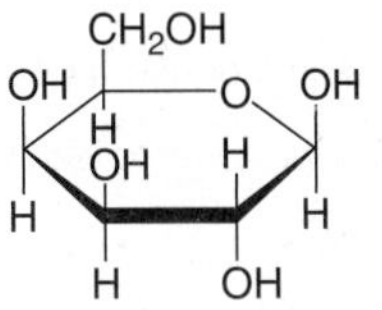

b.

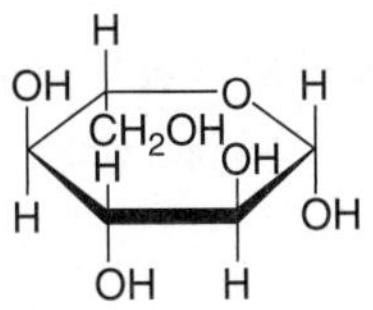

c.

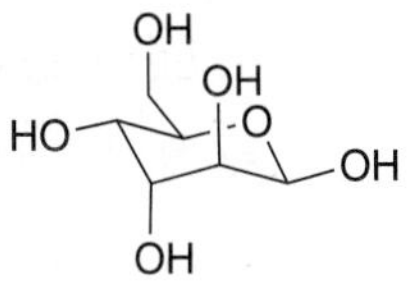

d.

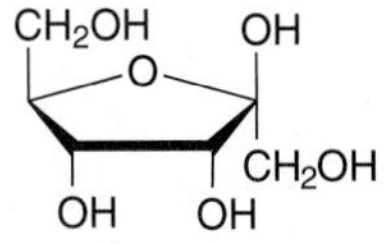

27.47 D-Arabinose can exist in both pyranose and furanose forms.

a. Draw the α and β anomers of D-arabinofuranose.

b. Draw the α and β anomers of D-arabinopyranose.

27.48 The most stable conformation of the pyranose ring of most D-aldohexoses places the largest group, CH_2OH, in the equatorial position. An exception to this is the aldohexose D-idose. Draw the two possible chair conformations of either the α or β anomer of D-idose. Explain why the more stable conformation has the CH_2OH group in the axial position.

Monosaccharide Reactions

27.49 Draw the products formed when α-D-gulose is treated with each reagent.

a. CH_3I, Ag_2O

b. CH_3OH, HCl

c. Ac_2O, pyridine

d. The product in (a), then H_3O^+

e. The product in (b), then Ac_2O, pyridine

f. The product in (d), then $C_6H_5CH_2Cl$, Ag_2O

27.50 Draw the products formed when D-altrose is treated with each reagent.

a. $(CH_3)_2CHOH$, HCl

b. $NaBH_4$, CH_3OH

c. Br_2, H_2O

d. HNO_3, H_2O

e. [1] NH_2OH; [2] $(CH_3CO)_2O$, $NaOCOCH_3$; [3] $NaOCH_3$

f. [1] NaCN, HCl; [2] H_2, Pd-$BaSO_4$; [3] H_3O^+

g. CH_3I, Ag_2O

h. $C_6H_5CH_2NH_2$, mild H^+

27.51 What aglycon and monosaccharides are formed when salicin and solanine (Section 27.7C) are each hydrolyzed with aqueous acid?

27.52 Draw a Fischer projection of the monosaccharide from which each of the following glycosides was prepared.

a. 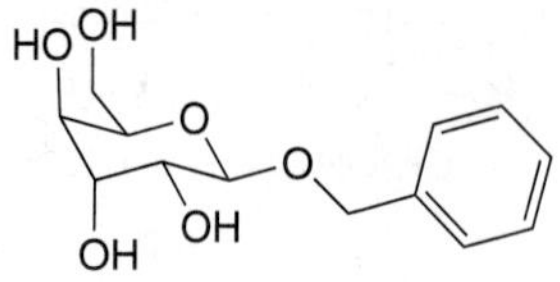

b. 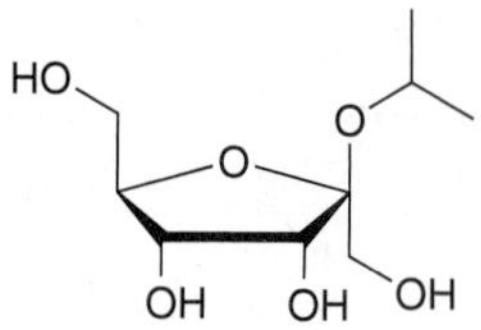

27.53 What two aldohexoses yield D-arabinose upon Wohl degradation?

27.54 What products are formed when each compound is subjected to a Kiliani–Fischer synthesis?

a. 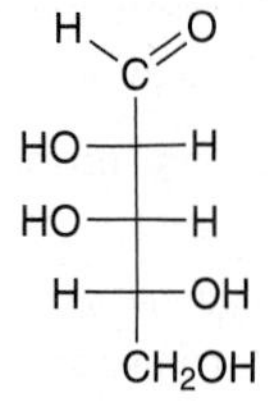

b. 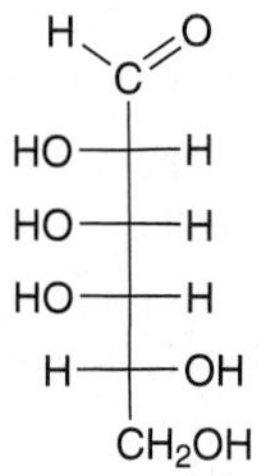

27.55 How would you convert D-glucose into each compound? More than one step is required.

a. 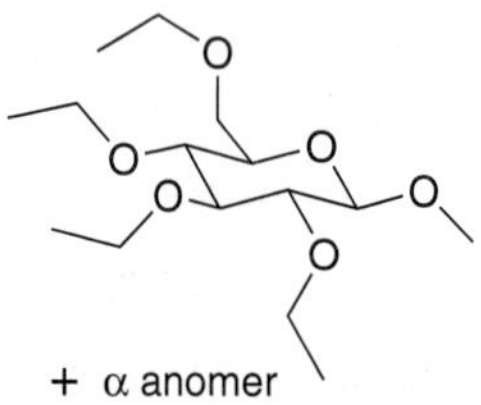

+ α anomer

b. 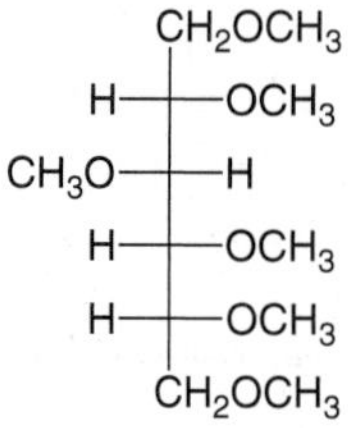

c. 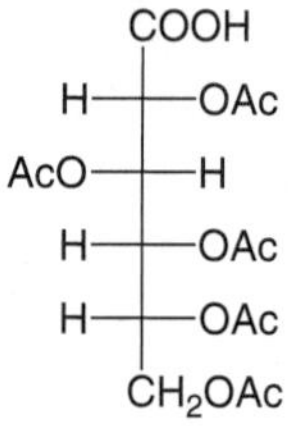

27.56 Which D-aldopentoses are reduced to optically inactive alditols using $NaBH_4$, CH_3OH?

27.57 What products are formed when each compound is treated with aqueous acid?

a. 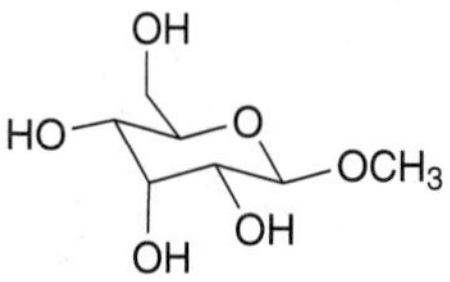

b. 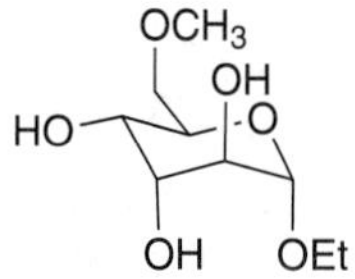

c. 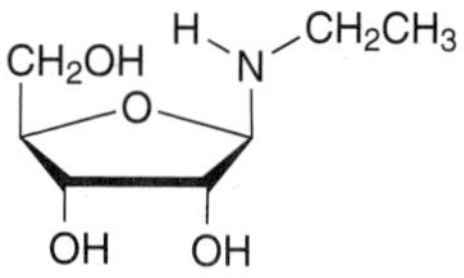

Mechanisms

27.58 Draw a stepwise mechanism for the following reaction.

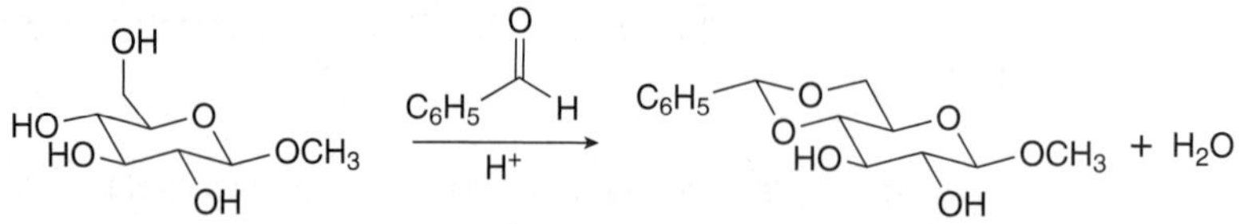

27.59 Draw a stepwise mechanism for the following hydrolysis.

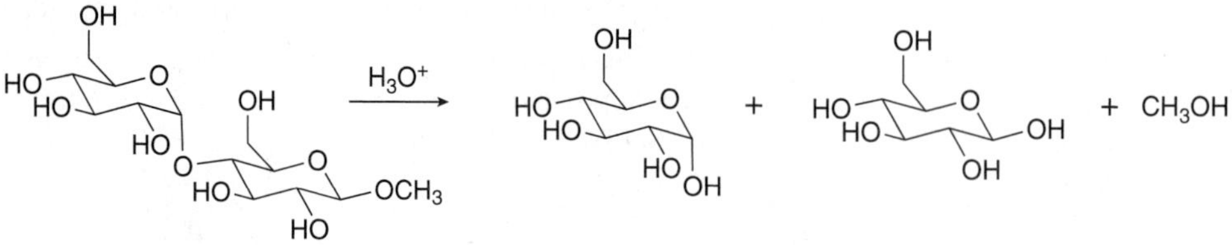

27.60 The following isomerization reaction, drawn using D-glucose as starting material, occurs with all aldohexoses in the presence of base. Draw a stepwise mechanism that illustrates how each compound is formed.

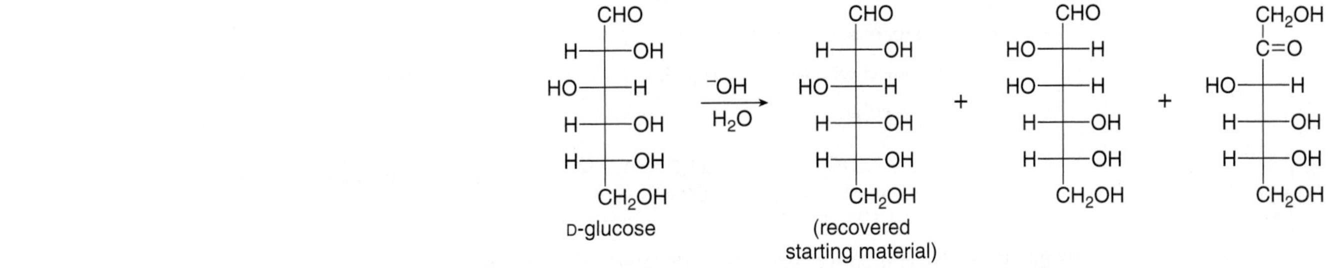
D-glucose
^-OH
H_2O
(recovered starting material)

Identifying Monosaccharides

27.61 Which D-aldopentose is oxidized to an optically active aldaric acid and undergoes the Wohl degradation to yield a D-aldotetrose that is oxidized to an optically active aldaric acid?

27.62 Identify compounds **A–D.** A D-aldopentose **A** is oxidized with HNO_3 to an optically inactive aldaric acid **B. A** undergoes the Kiliani–Fischer synthesis to yield **C** and **D. C** is oxidized to an optically active aldaric acid. **D** is oxidized to an optically inactive aldaric acid.

27.63 A D-aldopentose **A** is reduced to an optically active alditol. Upon Kiliani–Fischer synthesis, **A** is converted to two D-aldohexoses, **B** and **C. B** is oxidized to an optically inactive aldaric acid. **C** is oxidized to an optically active aldaric acid. What are the structures of **A–C?**

Disaccharides and Polysaccharides

27.64 Draw the structure of a disaccharide formed from two mannose units joined by a 1→4-α-glycosidic linkage.

27.65 a. Identify the glycosidic linkage in disaccharide **C,** classify the glycosidic bond as α or β, and use numbers to designate its location.

b. Identify the lettered compounds in the following reaction.

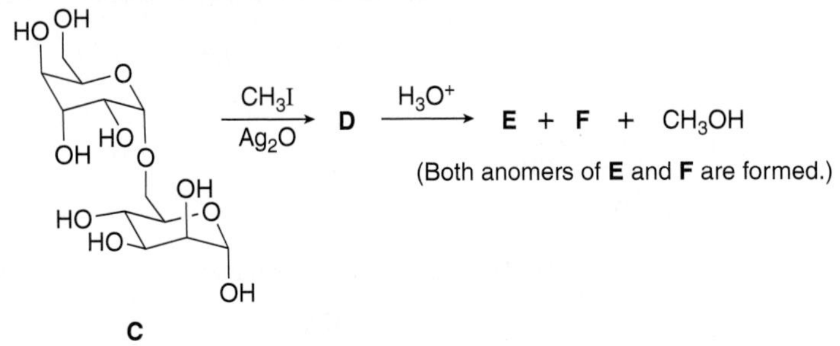
CH_3I
Ag_2O
D
H_3O^+
E + F + CH_3OH
(Both anomers of E and F are formed.)

C

27.66 Consider the tetrasaccharide stachyose drawn below. Stachyose is found in white jasmine, soybeans, and lentils. Because humans cannot digest it, its consumption causes flatulence.

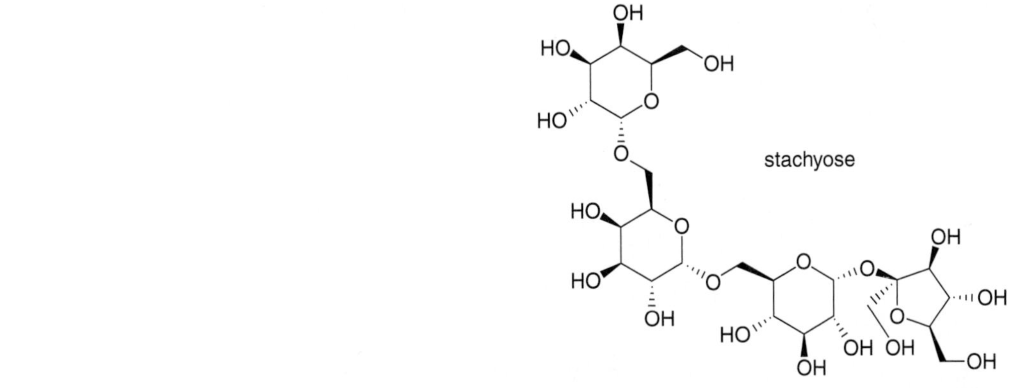
stachyose

a. Label all glycoside bonds.
b. Classify each glycosidic linkage as α or β and use numbers to designate its location between two rings (e.g., 1→4-β).
c. What products are formed when stachyose is hydrolyzed with H_3O^+?
d. Is stachyose a reducing sugar?
e. What product is formed when stachyose is treated with excess CH_3I, Ag_2O?
f. What products are formed when the product in (e) is treated with H_3O^+?

27.67 Deduce the structure of the disaccharide isomaltose from the following data.

[1] Hydrolysis yields D-glucose exclusively.

[2] Isomaltose is cleaved with α-glycosidase enzymes.

[3] Isomaltose is a reducing sugar.

[4] Methylation with excess CH_3I, Ag_2O and then hydrolysis with H_3O^+ forms two products:

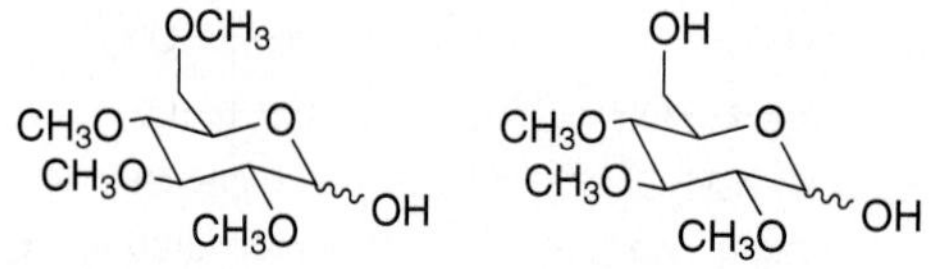

(Both anomers are present.)

27.68 Draw the structure of each of the following compounds:

a. a polysaccharide formed by joining D-glucosamine in 1→6-α-glycosidic linkages

b. a disaccharide formed by joining D-mannose and D-glucose in a 1→4-β-glycosidic linkage using mannose's anomeric carbon

c. an α-*N*-glycoside formed from D-arabinose and $C_6H_5CH_2NH_2$

d. a ribonucleoside formed from D-ribose and thymine

Challenge Problems

27.69 (a) Draw the more stable chair form of fucose, an essential monosaccharide needed in the diet and a component of carbohydrates on mammalian and plant cell surfaces. (b) Classify fucose as a D- or L-monosaccharide. (c) What two structural features are unusual in fucose?

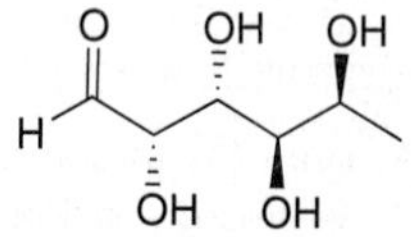

fucose

27.70 As we have seen in Chapter 27, monosaccharides can be drawn in a variety of ways, and in truth, often a mixture of cyclic compounds is present in a solution. Identify each monosaccharide, including its proper D,L designation, drawn in a less-than-typical fashion.

a. 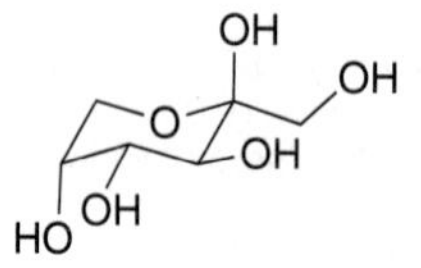

b. 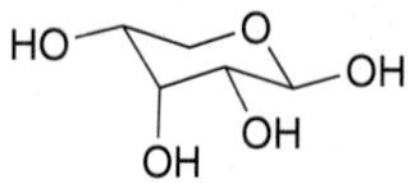

c. 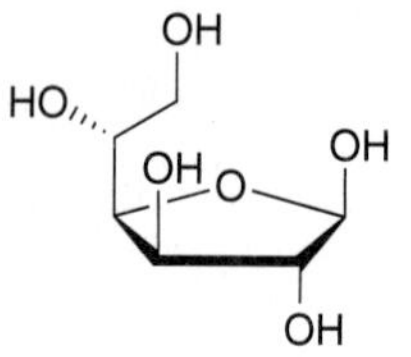

d. 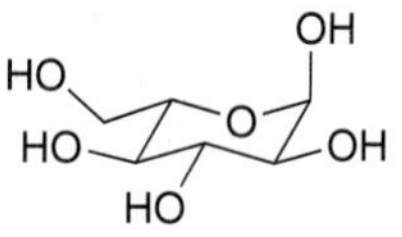

27.71 Draw a stepwise mechanism for the following reaction.

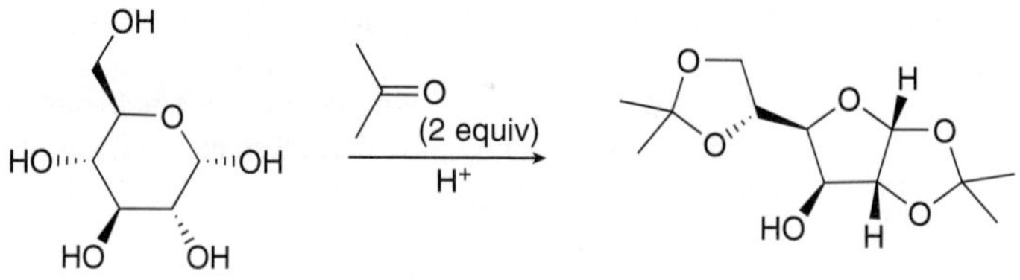

Lipids

28

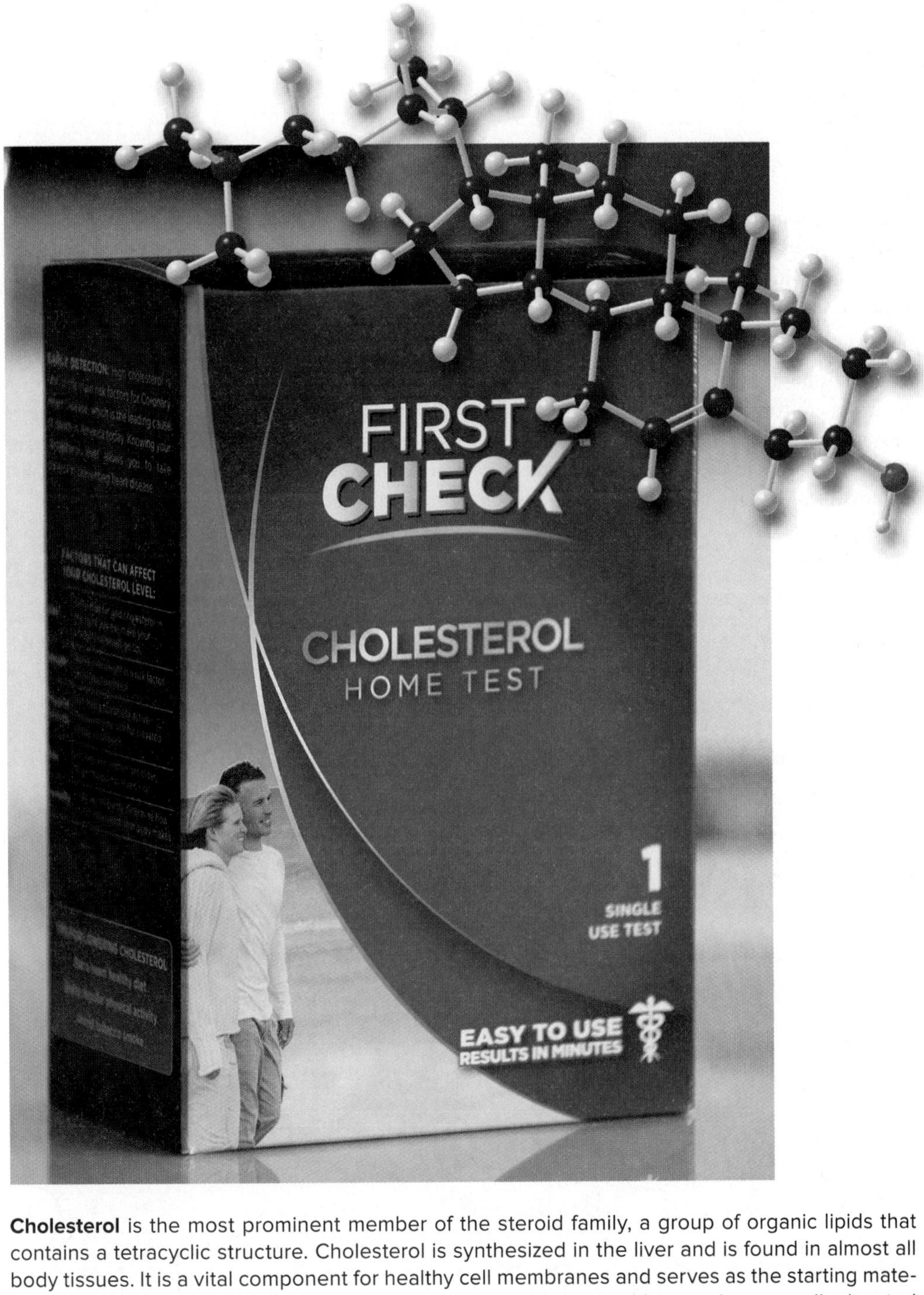

28.1 Introduction
28.2 Waxes
28.3 Triacylglycerols
28.4 Phospholipids
28.5 Fat-soluble vitamins
28.6 Eicosanoids
28.7 Terpenes
28.8 Steroids

Cholesterol is the most prominent member of the steroid family, a group of organic lipids that contains a tetracyclic structure. Cholesterol is synthesized in the liver and is found in almost all body tissues. It is a vital component for healthy cell membranes and serves as the starting material for the synthesis of all other steroids. But, as the general public now knows well, elevated cholesterol levels can lead to coronary artery disease. For this reason, consumer products are now labeled with their cholesterol content. In Chapter 28, we learn about the properties of cholesterol and other lipids.

We conclude the discussion of the organic molecules in biological systems by turning our attention to **lipids,** biomolecules that are soluble in organic solvents. Unlike the carbohydrates in Chapter 27 and the amino acids and proteins in Chapter 26, lipids contain many carbon–carbon and carbon–hydrogen bonds and few functional groups.

Since lipids are the biomolecules that most closely resemble the hydrocarbons we studied in Chapters 4 and 10, we have already learned many facts that directly explain their properties. Since there is no one functional group that is present in all lipids, however, the chemistry of lipids draws upon knowledge learned in many prior chapters.

28.1 Introduction

The word *lipid* comes from the Greek word ***lipos*** for "fat."

- *Lipids* are biomolecules that are soluble in organic solvents.

Lipids are unique among organic molecules because their identity is defined on the basis of a *physical property* and not by the presence of a particular functional group. Because of this, lipids come in a wide variety of structures and they have many different functions in the cell. Three examples are given in Figure 28.1.

The large number of **carbon–carbon and carbon–hydrogen σ bonds in lipids makes them very soluble in organic solvents and insoluble in water.** Monosaccharides (from which carbohydrates are formed) and amino acids (from which proteins are formed), on the other hand, are very polar, so they tend to be water soluble. Because lipids share many properties with hydrocarbons, several features of lipid structure and properties have already been discussed. Table 28.1 summarizes sections of the text where aspects of lipid chemistry were covered previously.

Table 28.1 Summary of Lipid Chemistry Discussed Prior to Chapter 28

Topic	Section	Topic	Section
• Vitamin A	3.5	• Lipid oxidation	15.11
• Soap	3.6	• Vitamin E	15.12
• Phospholipids, the cell membrane	3.7	• Steroid synthesis	16.14
• Triacylglycerols	3.9D	• Prostaglandins	19.6
• Lipids—Part 1	4.15	• Lipid hydrolysis	22.12A
• Leukotrienes	9.17	• Soap	22.12B
• Fats and oils	10.6	• Cholesteryl esters	22.17
• Oral contraceptives	11.4	• Fatty acid biosynthesis	24.8C
• Hydrogenation of oils	12.4		

Figure 28.1 Three examples of lipids

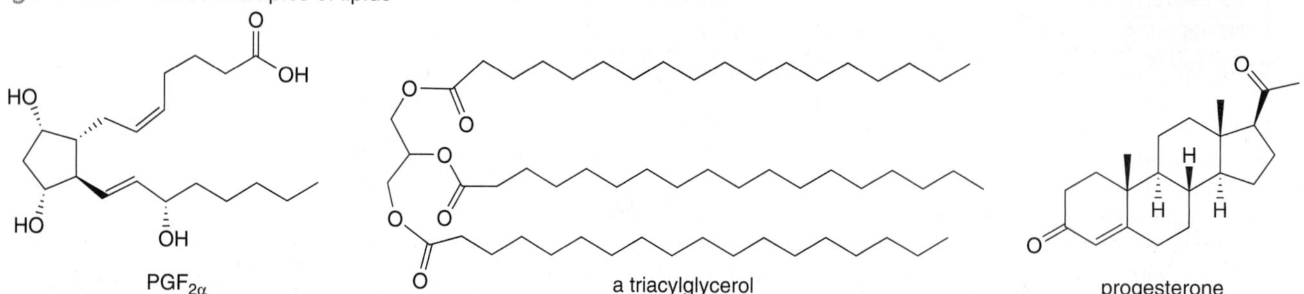

$PGF_{2\alpha}$
a prostaglandin

a triacylglycerol

progesterone
a steroid

- All lipids have many C–C and C–H bonds, but there is no one functional group common to all lipids.

Lipids can be categorized as hydrolyzable or nonhydrolyzable.

[1] ***Hydrolyzable lipids* can be cleaved into smaller molecules by hydrolysis with water.** Most hydrolyzable lipids contain an ester unit. We will examine three subgroups: waxes, triacylglycerols, and phospholipids.

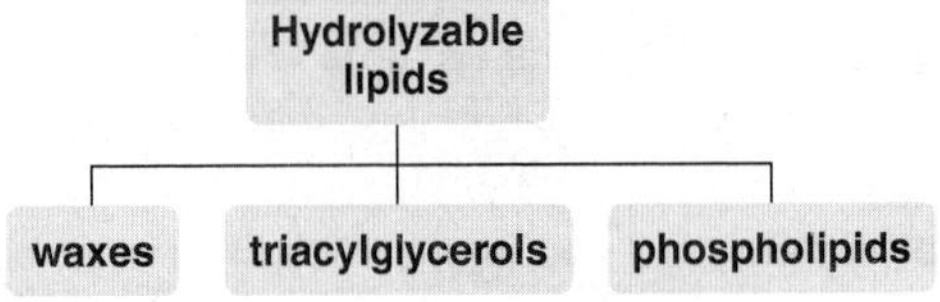

[2] ***Nonhydrolyzable lipids* cannot be cleaved into smaller units by aqueous hydrolysis.** Nonhydrolyzable lipids tend to be more varied in structure. We will examine four different types: fat-soluble vitamins, eicosanoids, terpenes, and steroids.

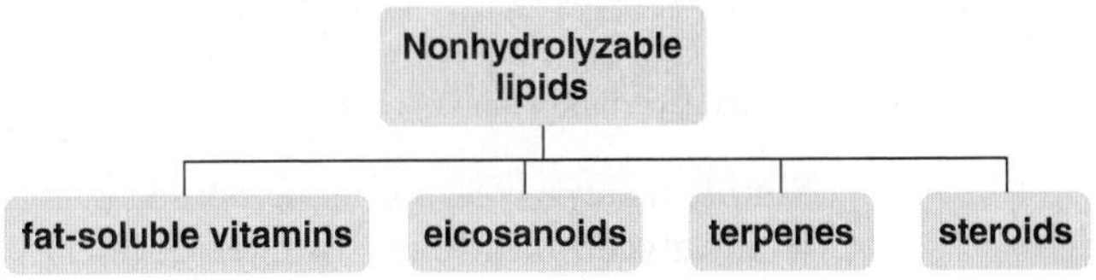

28.2 Waxes

Water beads up on the surface of a leaf because of the leaf's waxy coating.

The seeds of the jojoba plant grown in the southwestern United States are rich in waxes used in cosmetics and personal care products.

Waxes are the simplest hydrolyzable lipids. **Waxes are esters (RCOOR') formed from a high molecular weight alcohol (R'OH) and a fatty acid (RCOOH).**

Because of their long hydrocarbon chains, **waxes are very hydrophobic.** They form a protective coating on the feathers of birds to make them water repellent, and on leaves to prevent water evaporation. **Lanolin,** a wax composed of a complex mixture of high molecular weight esters, coats the wool fibers of sheep. **Spermaceti wax,** isolated from the heads of sperm whales, is largely $CH_3(CH_2)_{14}COO(CH_2)_{15}CH_3$. The three-dimensional structure of this compound shows how small the ester group is compared to the long hydrocarbon chains.

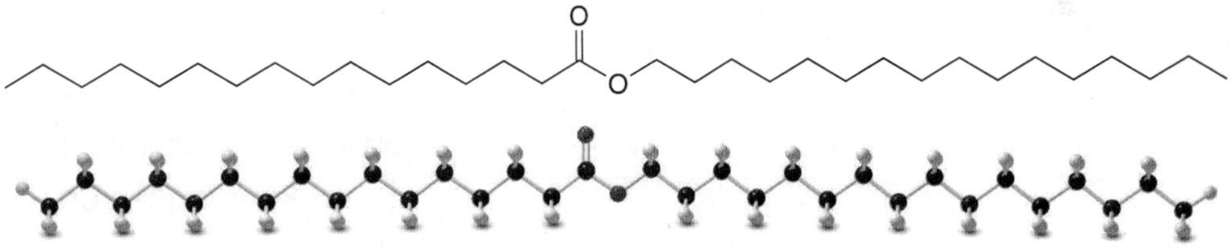

spermaceti wax (from sperm whales)

Problem 28.1 One component of jojoba oil is a wax formed from eicosenoic acid [$CH_3(CH_2)_7CH=CH(CH_2)_9CO_2H$] and $CH_3(CH_2)_7CH=CH(CH_2)_8OH$. Draw the structure of the wax, including the cis geometry of both carbon–carbon double bonds.

28.3 Triacylglycerols

Triacylglycerols, or triglycerides, are the most abundant lipids, and for this reason we have already discussed many of their properties in earlier sections of this text.

- **Triacylglycerols are triesters that produce glycerol and three molecules of fatty acid upon hydrolysis.**

Line structures of stearic, oleic, linoleic, and linolenic acids can be found in Table 10.2. Ball-and-stick models of these fatty acids are shown in Figure 10.6.

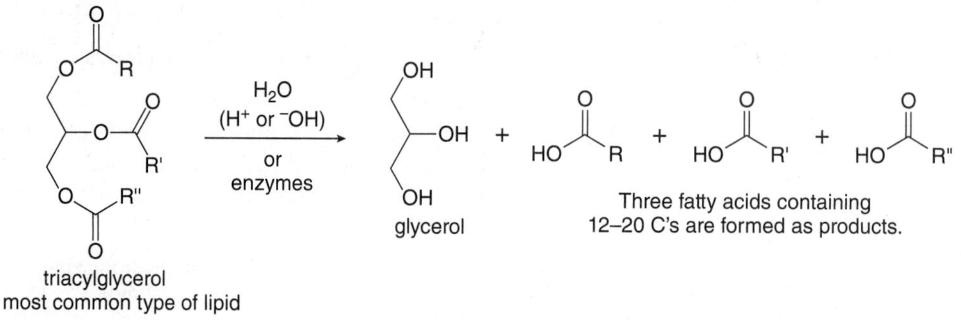

Simple triacylglycerols are composed of three identical fatty acid side chains, whereas **mixed triacylglycerols** have two or three different fatty acids. Table 28.2 lists the most common fatty acids used to form triacylglycerols.

Table 28.2 The Most Common Fatty Acids in Triacylglycerols

Number of C atoms	Number of C=C bonds	Structure	Name	Mp (°C)
		Saturated fatty acids		
12	0	$CH_3(CH_2)_{10}CO_2H$	lauric acid	44
14	0	$CH_3(CH_2)_{12}CO_2H$	myristic acid	58
16	0	$CH_3(CH_2)_{14}CO_2H$	palmitic acid	63
18	0	$CH_3(CH_2)_{16}CO_2H$	stearic acid	69
20	0	$CH_3(CH_2)_{18}CO_2H$	arachidic acid	77
		Unsaturated fatty acids		
16	1	$CH_3(CH_2)_5CH=CH(CH_2)_7CO_2H$	palmitoleic acid	1
18	1	$CH_3(CH_2)_7CH=CH(CH_2)_7CO_2H$	oleic acid	4
18	2	$CH_3(CH_2)_4(CH=CHCH_2)_2(CH_2)_6CO_2H$	linoleic acid	–5
18	3	$CH_3CH_2(CH=CHCH_2)_3(CH_2)_6CO_2H$	linolenic acid	–11
20	4	$CH_3(CH_2)_4(CH=CHCH_2)_4(CH_2)_2CO_2H$	arachidonic acid	–49

What are the characteristics of these fatty acids?

The most common saturated fatty acids are palmitic and stearic acids. The most common unsaturated fatty acid is oleic acid.

- **All fatty acid chains are unbranched, but they may be saturated or unsaturated.**
- **Naturally occurring fatty acids have an even number of carbon atoms.**
- **Double bonds in naturally occurring fatty acids generally have the *Z* configuration.**
- **The melting point of a fatty acid depends on the degree of unsaturation.**

Linoleic and linolenic acids are called **essential fatty acids** because we cannot synthesize them and must acquire them in our diets.

Fats and oils are triacylglycerols; that is, they are triesters of glycerol and these fatty acids.

- **Fats have higher melting points, making them solids at room temperature.**
- **Oils have lower melting points, making them liquids at room temperature.**

Figure 28.2 Three-dimensional structures of a saturated and unsaturated triacylglycerol

a. A saturated triacylglycerol

b. An unsaturated triacylglycerol

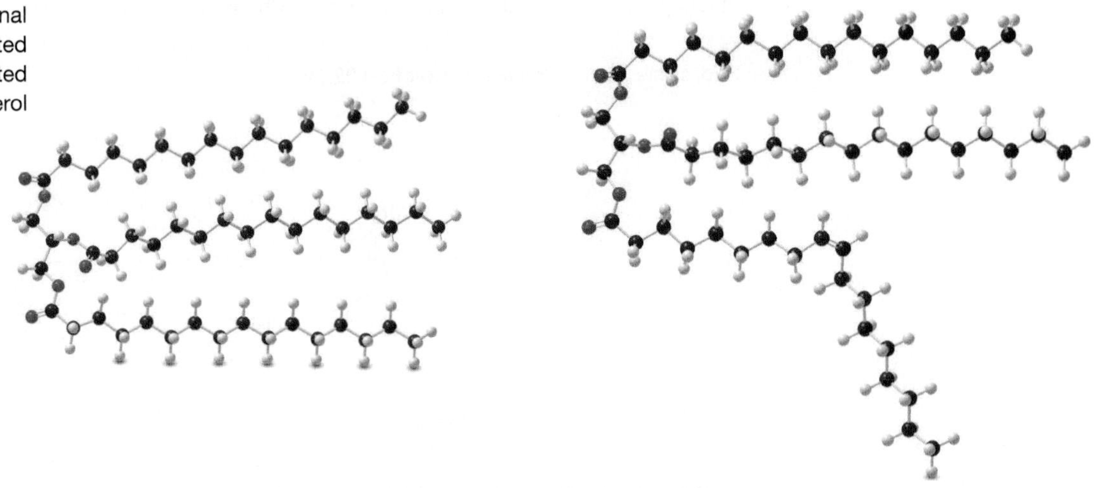

- Three saturated side chains lie parallel to each other, making a compact lipid.
- One *Z* double bond in a fatty acid side chain produces a twist so the lipid is no longer so compact.

This melting point difference correlates with the number of degrees of unsaturation present in the fatty acid side chains. **As the number of double bonds *increases,* the melting point *decreases,* as it does for the constituent fatty acids as well.**

Three-dimensional structures of a saturated and unsaturated triacylglycerol are shown in Figure 28.2. With no double bonds, the three side chains of the saturated lipid lie parallel to each other, making it possible for this compound to pack relatively efficiently in a crystalline lattice, thus leading to a high melting point. In the unsaturated lipid, however, a single *Z* double bond places a kink in the side chain, making it more difficult to pack efficiently in the solid state, thus leading to a lower melting point.

Solid fats have a relatively high percentage of saturated fatty acids and are generally of animal origin. Liquid oils have a higher percentage of unsaturated fatty acids and are generally of vegetable origin. Table 28.3 lists the fatty acid composition of some common fats and oils.

Unlike other vegetable oils, oils from palm and coconut trees are very high in saturated fats. Considerable evidence currently suggests that diets high in saturated fats lead to a greater risk of heart disease. For this reason, the demand for coconut and palm oils has decreased considerably in recent years, and many coconut plantations previously farmed in the South Pacific are no longer in commercial operation.

Table 28.3 Fatty Acid Composition of Some Common Fats and Oils

Source	% Saturated fatty acids	% Oleic acid	% Linoleic acid
beef	49–62	37–43	2–3
milk	37	33	3
coconut	86	7	—
corn	11–16	19–49	34–62
olive	11	84	4
palm	43	40	8
safflower	9	13	78
soybean	15	20	52

Data from *Merck Index,* 10th ed. Rahway, NJ: Merck and Co.; and Wilson, et al., 1967, *Principles of Nutrition,* 2nd ed. New York: Wiley.

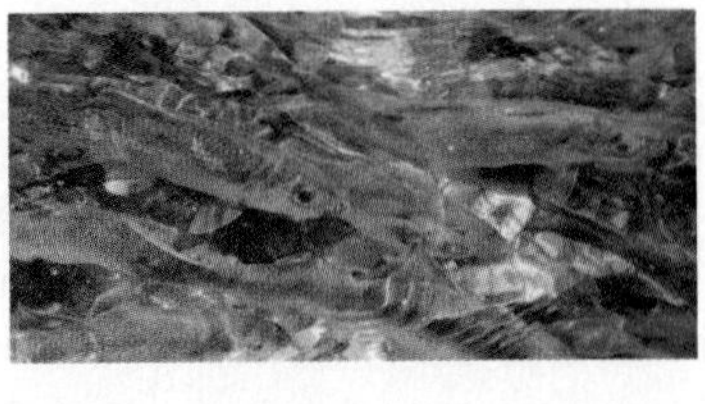

Fish oils, such as cod liver and herring oils, are very rich in polyunsaturated triacylglycerols. These triacylglycerols pack so poorly that they have very low melting points; thus, they remain liquids even in the cold water inhabited by these fish.

The hydrolysis, hydrogenation, and oxidation of triacylglycerols—reactions originally discussed in Chapters 12, 15, and 22—are summarized here for your reference.

[1] Hydrolysis of triacylglycerols (Section 22.12A)

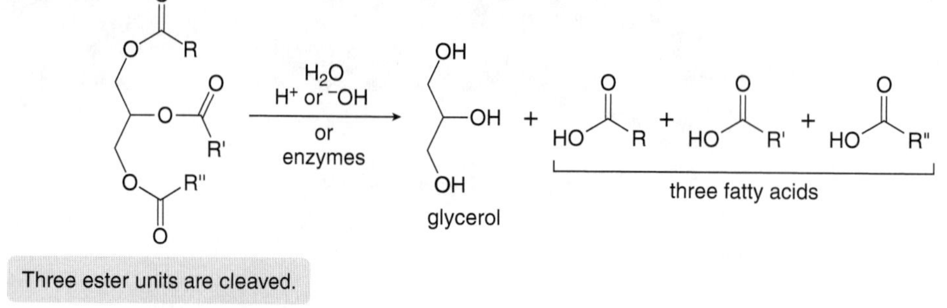
H2O
H+ or -OH
or
enzymes
glycerol
three fatty acids
Three ester units are cleaved.

Hydrolysis of a triacylglycerol with water in the presence of either acid, base, or an enzyme yields glycerol and three fatty acids. This cleavage reaction follows the same mechanism as any other ester hydrolysis (Section 22.11). This reaction is the first step in triacylglycerol metabolism.

[2] Hydrogenation of unsaturated fatty acids (Section 12.4)

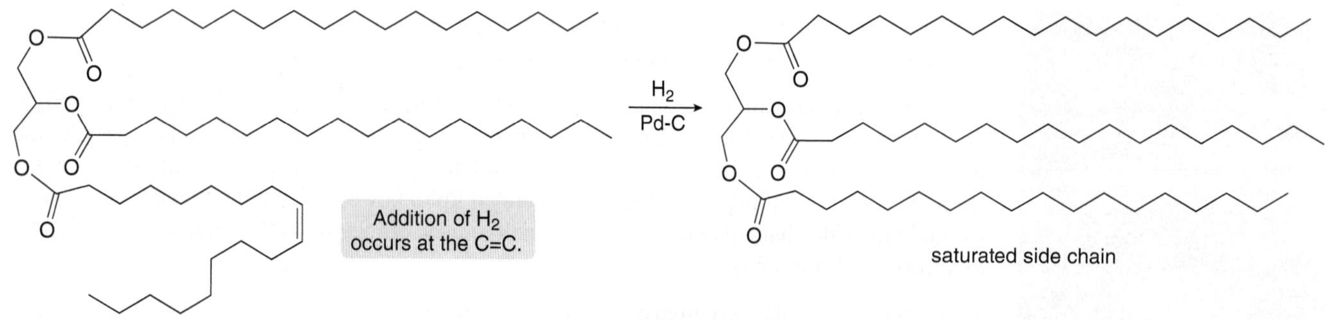
H2
Pd-C
Addition of H2
occurs at the C=C.
saturated side chain

The double bonds of an unsaturated fatty acid can be hydrogenated by using H_2 in the presence of a transition metal catalyst. Hydrogenation converts a liquid oil to a solid fat. This process, sometimes called **hardening,** is used to prepare margarine from vegetable oils.

[3] Oxidation of unsaturated fatty acids (Section 15.11)

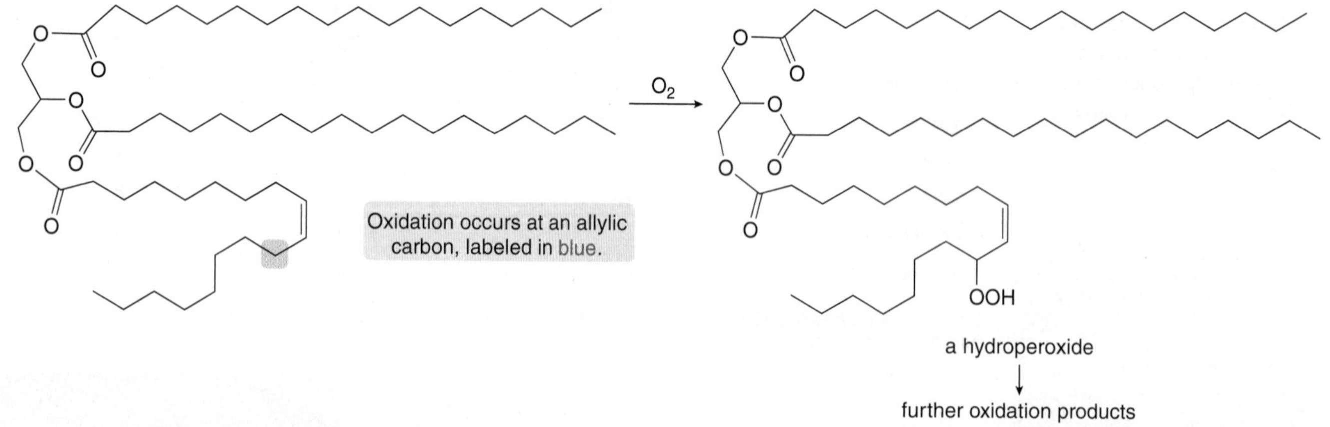
O2
Oxidation occurs at an allylic
carbon, labeled in blue.
OOH
a hydroperoxide
further oxidation products

Allylic C–H bonds are weaker than other C–H bonds and are thus susceptible to oxidation with molecular oxygen by a radical process. The hydroperoxide formed by this process is unstable, and it undergoes further oxidation to products that often have a disagreeable odor. This oxidation process turns an oil rancid.

In the cell, the principal function of triacylglycerols is energy storage. Complete metabolism of a triacylglycerol yields CO_2 and H_2O, and a great deal of energy. This overall reaction is reminiscent of the combustion of alkanes in fossil fuels, a process that also yields CO_2 and H_2O and provides energy to heat homes and power automobiles (Section 4.14B). Fundamentally both processes convert C–C and C–H bonds to C–O bonds, a highly exothermic reaction.

The average body fat content of men and women is ~20% and ~25%, respectively. (For elite athletes, however, the averages are more like <10% for men and <15% for women.) This stored fat can fill the body's energy needs for two or three months.

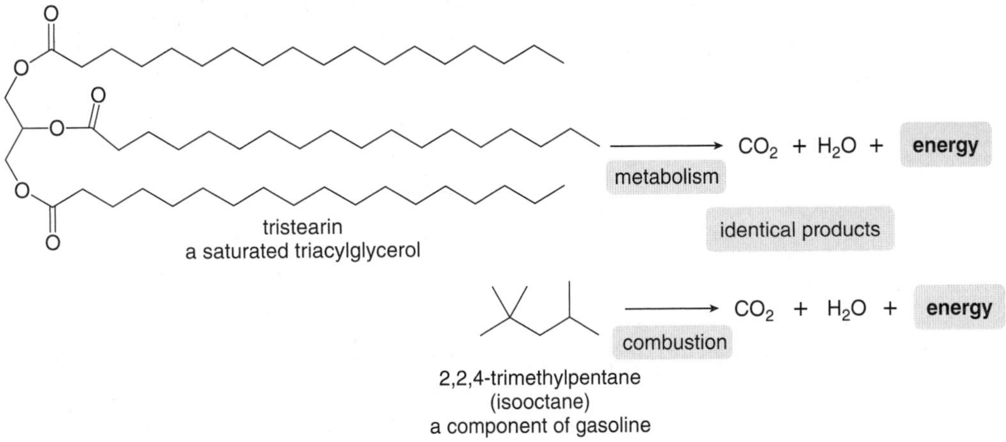

Carbohydrates provide an energy boost, but only for the short term, such as during strenuous exercise. Our long-term energy needs are met by triacylglycerols, because they store ~38 kJ/g, whereas carbohydrates and proteins store only ~16 kJ/g.

Because triacylglycerols release heat on combustion, they can in principle be used as fuels for vehicles. In fact, coconut oil was used as a fuel during both World War I and World War II, when gasoline and diesel supplies ran short. Since coconut oil is more viscous than petroleum products and freezes at 24 °C, engines must be modified to use it and it can't be used in cold climates. Nonetheless, a limited number of trucks and boats can now use vegetable oils, sometimes blended with diesel, as a fuel source. When the price of crude oil is high, the use of these **biofuels** becomes economically attractive.

Problem 28.2 How would you expect the melting point of eicosapentaenoic acid [$CH_3CH_2(CH{=}CHCH_2)_5(CH_2)_2COOH$] to compare with the melting points of the fatty acids listed in Table 28.2?

Problem 28.3 Draw the products formed when triacylglycerol **A** is treated with each reagent. Rank compounds **A, B,** and **C** in order of increasing melting point.

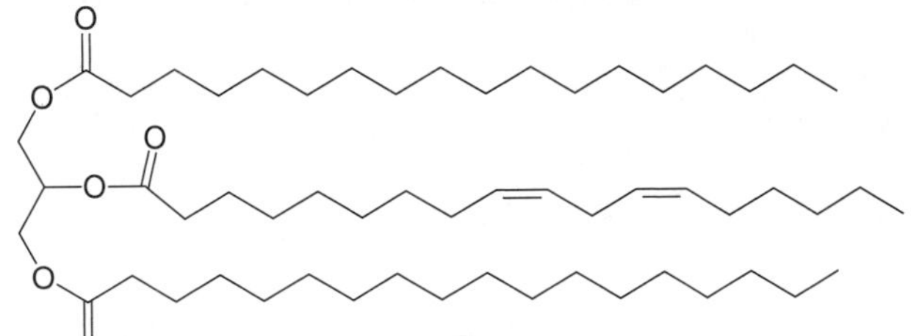

A

a. H_2O, H^+
b. H_2 (excess), Pd-C → **B**
c. H_2 (1 equiv), Pd-C → **C**

The cocoa butter in chocolate is rich in triacylglycerols derived from at least one molecule of oleic acid.

Problem 28.4 The main fatty acid component of the triacylglycerols in coconut oil is lauric acid, $CH_3(CH_2)_{10}COOH$. Explain why coconut oil is a liquid at room temperature even though it contains a large fraction of this saturated fatty acid.

Problem 28.5 Unlike many fats and oils, the cocoa butter used to make chocolate is remarkably uniform in composition. All triacylglycerols contain oleic acid esterified to the 2° OH group of glycerol, and either palmitic acid or stearic acid esterified to the 1° OH groups. Draw the structures of two possible triacylglycerols that compose cocoa butter.

28.4 Phospholipids

Phospholipids are hydrolyzable lipids that contain a phosphorus atom. There are two common types of phospholipids: **phosphoacylglycerols** and **sphingomyelins.** Both classes are found almost exclusively in the cell membranes of plants and animals, as discussed in Section 3.7.

Phospholipids are organic derivatives of phosphoric acid, formed by replacing two of the H atoms by R groups. This type of functional group is called a **phosphodiester,** or a **phosphoric acid diester** (Section 3.2D). These compounds are phosphorus analogues of carboxylic esters. In cells, the remaining OH group on phosphorus loses its proton, giving the phosphodiester a net negative charge.

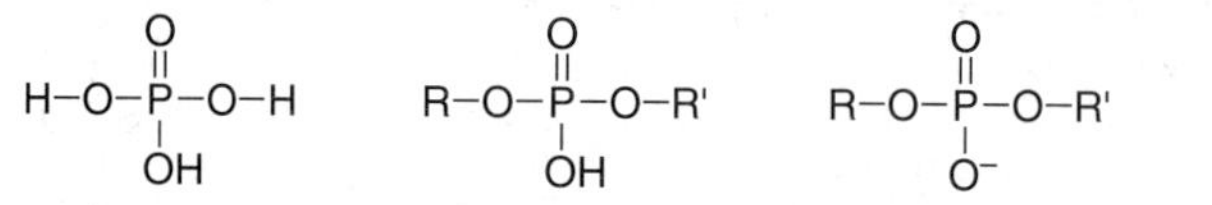

phosphoric acid H_3PO_4 | phosphodiester | This form exists in cells.

28.4A Phosphoacylglycerols

Phosphoacylglycerols (or phosphoglycerides) are the second most abundant type of lipid. They form the principal lipid component of most cell membranes. Their structure resembles the triacylglycerols of Section 28.3 with one important difference. In phosphoacylglycerols, only two of the hydroxy groups of glycerol are esterified with fatty acids. The third OH group is part of a phosphodiester, which is also bonded to another low molecular weight alcohol.

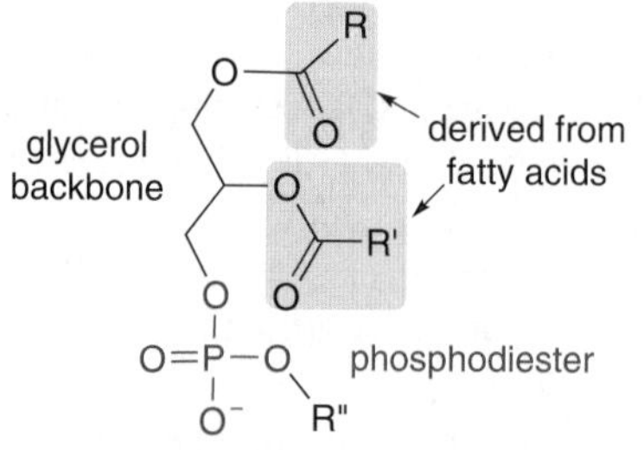

phosphoacylglycerol

There are two prominent types of phosphoacylglycerols. They differ in the identity of the R" group in the phosphodiester.

- When R" = $CH_2CH_2NH_3^+$, the phosphoacylglycerol is called a phosphatidylethanolamine or cephalin.
- When R" = $CH_2CH_2N(CH_3)_3^+$, the phosphoacylglycerol is called a phosphatidylcholine or lecithin.

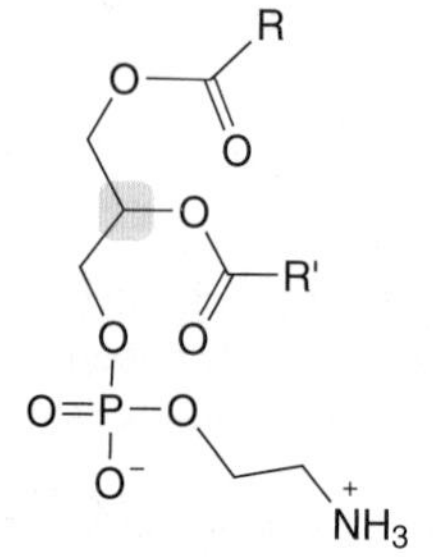

phosphatidylethanolamine or cephalin

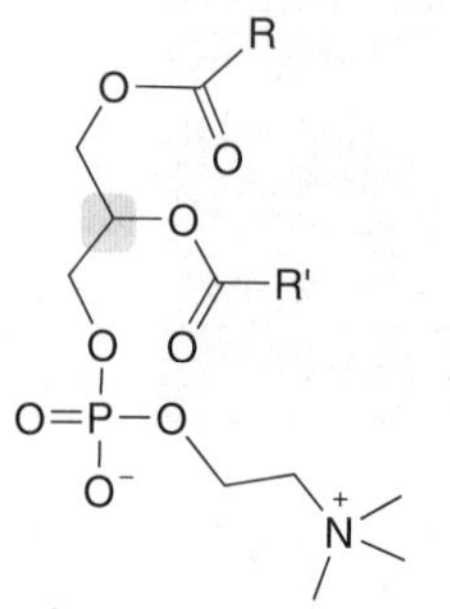

phosphatidylcholine or lecithin

The middle carbon of the glycerol backbone of all of these compounds is a stereogenic center (labeled in blue), usually with the *R* configuration.

The phosphorus side chain of a phosphoacylglycerol makes it different from a triacylglycerol. The two fatty acid side chains form two nonpolar "tails" that lie parallel to each other, while the phosphodiester end of the molecule is a charged or polar "head." A three-dimensional structure of a phosphoacylglycerol is shown in Figure 28.3.

Figure 28.3
Three-dimensional structure of a phosphoacylglycerol

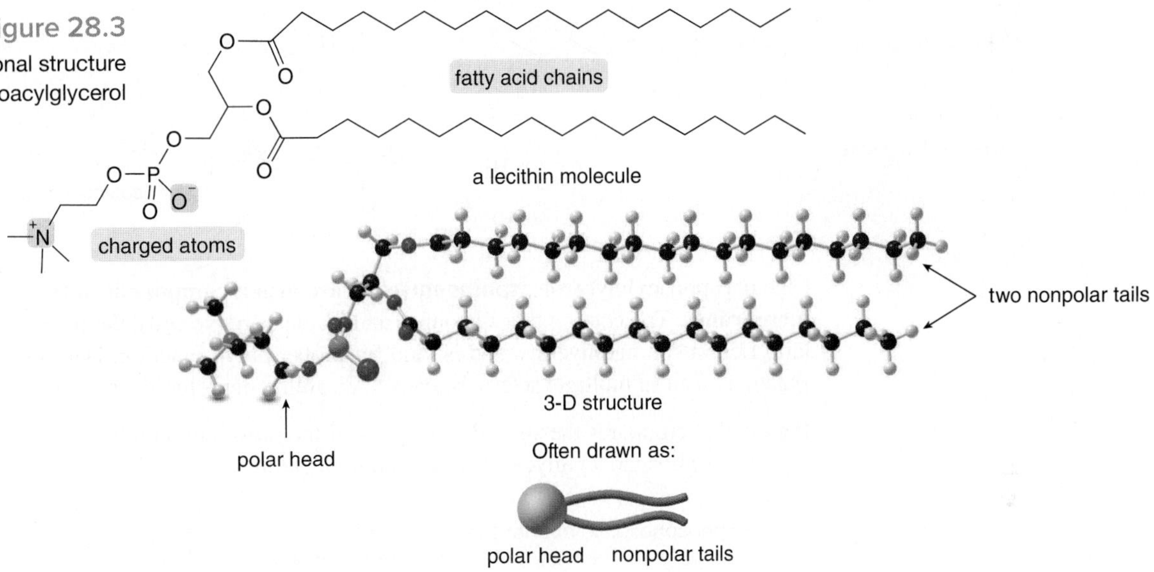

- A phosphoacylglycerol has two distinct regions: two nonpolar tails due to the long-chain fatty acids, and a very polar head from the charged phosphodiester.

As discussed in Section 3.7, when these phospholipids are mixed with water, they assemble in an arrangement called a **lipid bilayer.** The ionic heads of the phospholipid are oriented on the outside and the nonpolar tails on the inside. The identity of the fatty acids in the phospholipid determines the rigidity of this bilayer. When the fatty acids are saturated, they pack well in the interior of the lipid bilayer, and the membrane is quite rigid. When there are many unsaturated fatty acids, the nonpolar tails cannot pack as well and the bilayer is more fluid. Thus, important characteristics of this lipid bilayer are determined by the three-dimensional structure of the molecules that comprise it.

Cell membranes are composed of these lipid bilayers (see Figure 3.7). Proteins and cholesterol are embedded in the membranes as well, but the phospholipid bilayer forms the main fabric of the insoluble barrier that protects the cell.

Problem 28.6 Draw the structure of a lecithin containing oleic acid and palmitic acid as the fatty acid side chains.

Problem 28.7 Phosphoacylglycerols should remind you of soaps (Section 3.6). In what ways are these compounds similar?

28.4B Sphingomyelins

Sphingomyelins, the second major class of phospholipids, are derivatives of the amino alcohol **sphingosine,** in much the same way that triacylglycerols and phosphoacylglycerols are derivatives of glycerol. Other notable features of a sphingomyelin include:

- **A phosphodiester at C1.**
- **An amide formed with a fatty acid at C2.**

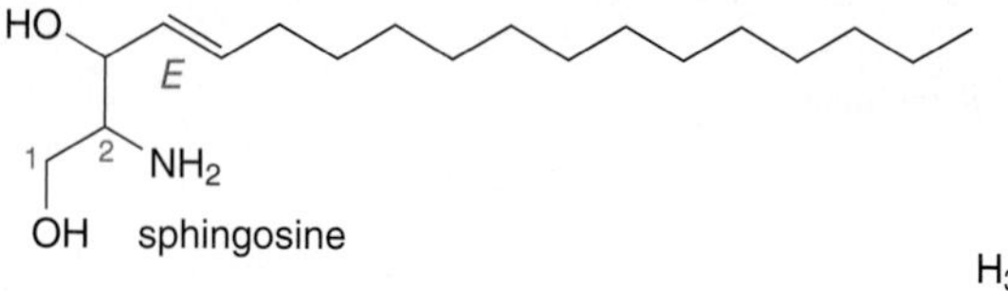

sphingosine

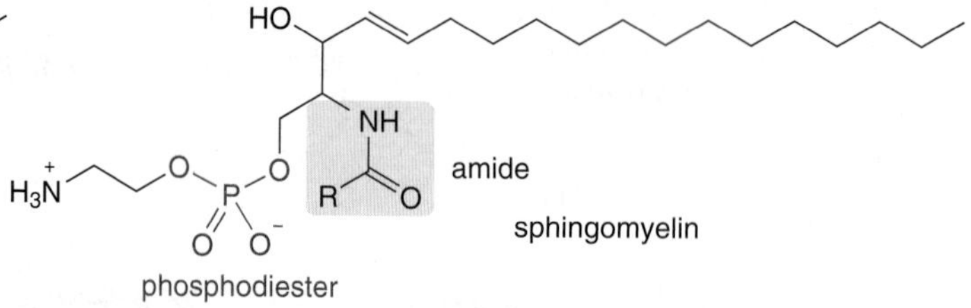

sphingomyelin

Like phosphoacylglycerols, **sphingomyelins are also a component of the lipid bilayer of cell membranes.** The coating that surrounds and insulates nerve cells, the **myelin sheath,** is particularly rich in sphingomyelins, and is vital for proper nerve function. Deterioration of the myelin sheath as seen in multiple sclerosis leads to disabling neurological problems.

Figure 28.4 compares the structural features of the most common hydrolyzable lipids: a triacylglycerol, a phosphoacylglycerol, and a sphingomyelin.

Problem 28.8 Why are phospholipids, but not triacylglycerols, found in cell membranes?

Figure 28.4 A comparison of a triacylglycerol, a phosphoacylglycerol, and a sphingomyelin

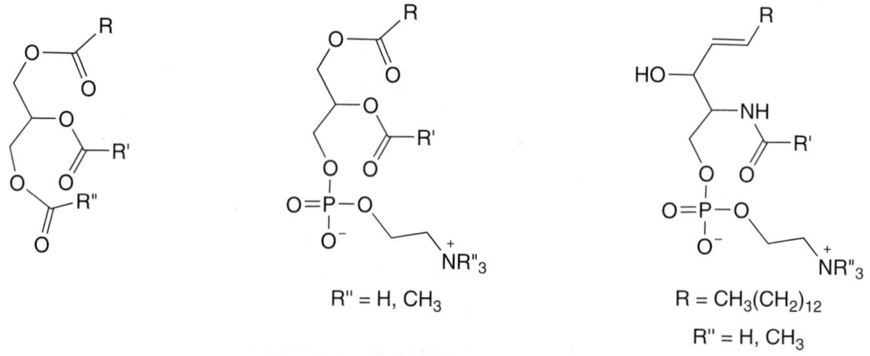

- A triacylglycerol has three nonpolar side chains.
- The three OH groups of glycerol are esterified with three fatty acids.

- A phosphoacylglycerol has two nonpolar side chain tails and one ionic head.
- Two OH groups of glycerol are esterified with fatty acids.
- A phosphodiester is located on a terminal carbon.

- A sphingomyelin has two nonpolar side chain tails and one ionic head.
- A sphingomyelin is formed from sphingosine, not glycerol. One of the nonpolar tails is an amide.
- A phosphodiester is located on a terminal carbon.

28.5 Fat-Soluble Vitamins

Vitamins are organic compounds required in small quantities for normal metabolism (Section 3.5). Because our cells cannot synthesize these compounds, they must be obtained in the diet. Vitamins can be categorized as fat soluble or water soluble. **The fat-soluble vitamins are lipids.**

The four fat-soluble vitamins—**A, D, E,** and **K**—are found in fruits and vegetables, fish, liver, and dairy products. Although fat-soluble vitamins must be obtained from the diet, they do not have to be ingested every day. Excess vitamins are stored in fat cells, and then used when needed. Figure 28.5 shows the structure of these vitamins and summarizes their functions.

Figure 28.5 The fat-soluble vitamins

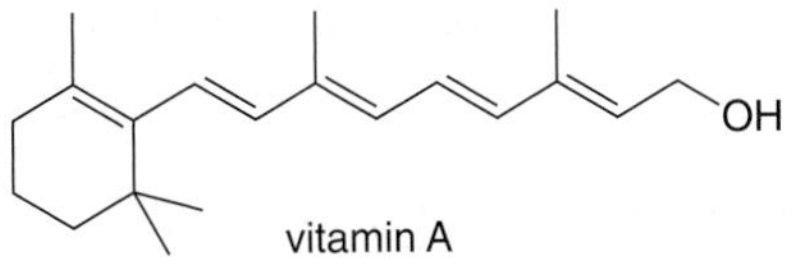

vitamin A

- **Vitamin A** (retinol, Section 3.5) is obtained from fish liver oils and dairy products, and is synthesized from β-carotene, the orange pigment in carrots.
- In the body, vitamin A is converted to 11-*cis*-retinal, the light-sensitive compound responsible for vision in all vertebrates (Section 21.11B). It is also needed for healthy mucous membranes.
- A deficiency of vitamin A causes night blindness, as well as dry eyes and skin.

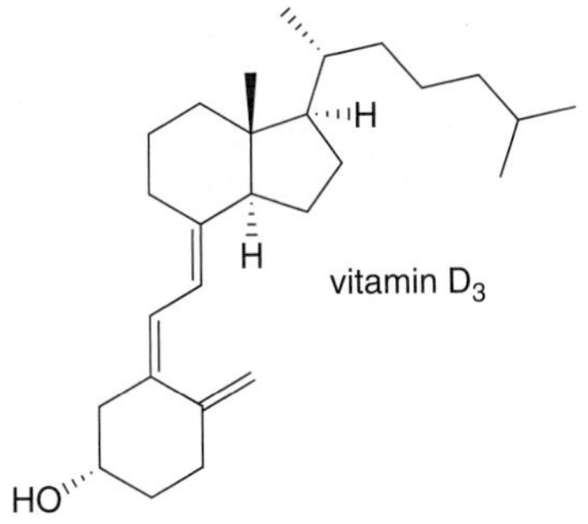

vitamin D_3

- **Vitamin D_3** is the most abundant of the D vitamins. Strictly speaking, it is not a vitamin because it can be synthesized in the body from cholesterol. Nevertheless, it is classified as such, and many foods (particularly milk) are fortified with vitamin D_3 so that we get enough of this vital nutrient.
- Vitamin D helps regulate both calcium and phosphorus metabolism.
- A deficiency of vitamin D causes rickets, a bone disease characterized by knock-knees, spinal curvature, and other deformities.

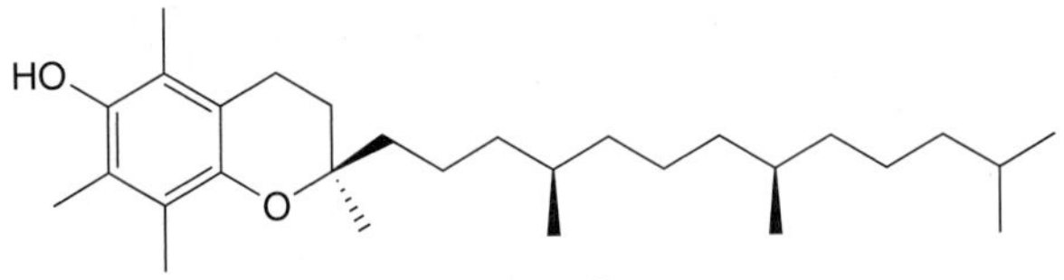

vitamin E
(α-tocopherol)

- The term **vitamin E** refers to a group of structurally similar compounds, the most potent being α-tocopherol (Section 15.12).
- Vitamin E is an antioxidant, so it protects unsaturated side chains in fatty acids from oxidation.
- A deficiency of vitamin E causes numerous neurologic problems.

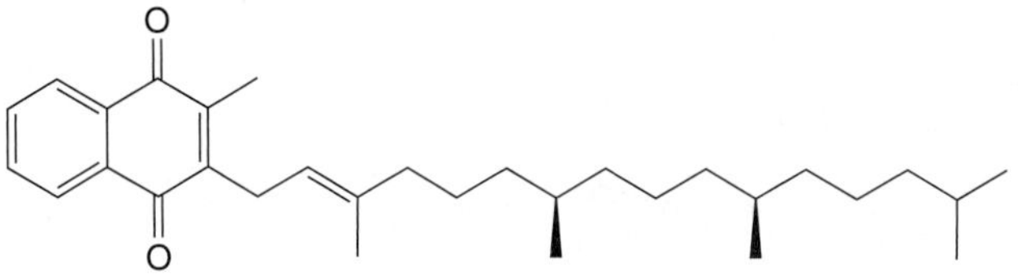

vitamin K

- **Vitamin K** (phylloquinone) regulates the synthesis of prothrombin and other proteins needed for blood to clot.
- A deficiency of vitamin K leads to excessive and sometimes fatal bleeding because of inadequate blood clotting.

Figure 28.6
Electrostatic potential plots of vitamins A and E

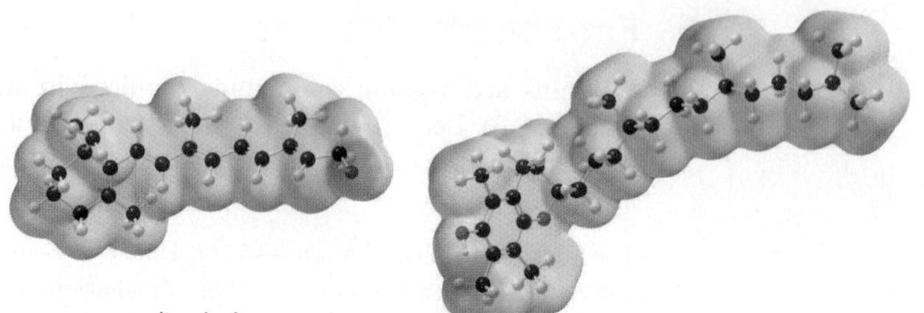

vitamin A vitamin E

- The electron density is distributed fairly evenly among the carbon atoms of these vitamins due to their many nonpolar C–C and C–H bonds.

Electrostatic potential plots of vitamins A and E (Figure 28.6) show that the electron density is virtually uniform in these compounds. The large regions of nonpolar C–C and C–H bonds tend to obscure small dipoles that occur in the one or two polar bonds, making these vitamins nonpolar and hydrophobic.

Problem 28.9 Explain why regularly ingesting a large excess of a fat-soluble vitamin can lead to severe health problems, whereas ingesting a large excess of a water-soluble vitamin often causes no major health problems.

28.6 Eicosanoids

The word ***eicosanoid*** is derived from the Greek word *eikosi,* meaning **20.**

The **eicosanoids** are a group of biologically active compounds containing 20 carbon atoms derived from arachidonic acid. The **prostaglandins** (Section 19.6) and the **leukotrienes** (Section 9.17) are two types of eicosanoids. Two others are the **thromboxanes** and **prostacyclins.**

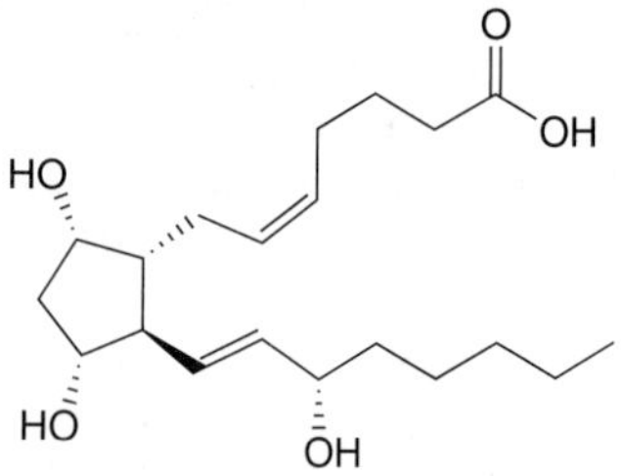

$PGF_{2\alpha}$
a prostaglandin

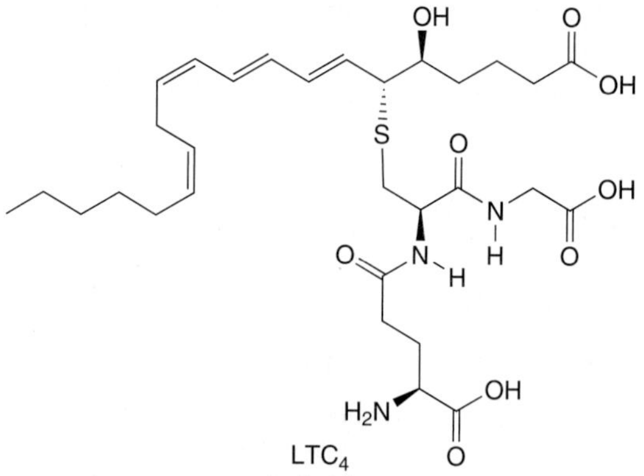

LTC_4
a leukotriene

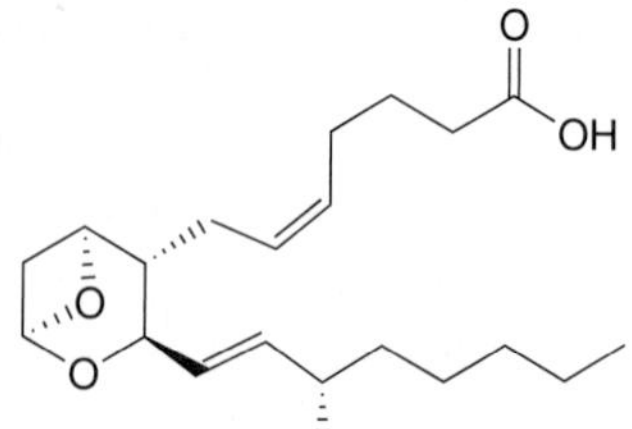

TXA_2
a thromboxane

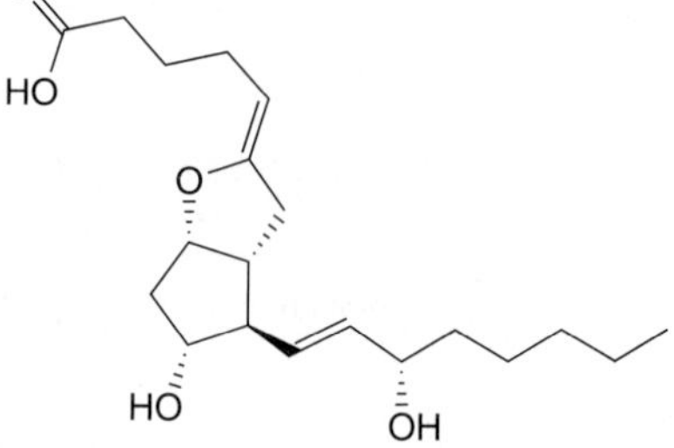

PGI_2
a prostacyclin

All eicosanoids are very potent compounds present in low concentration in cells. They are **local mediators,** meaning that they perform their function in the environment in which they are synthesized. This distinguishes them from **hormones,** which are first synthesized and then transported in the bloodstream to their site of action. Eicosanoids are not stored in cells; rather, they are synthesized from arachidonic acid in response to an external stimulus.

Figure 28.7

The conversion of arachidonic acid to prostaglandins, thromboxanes, prostacyclins, and leukotrienes

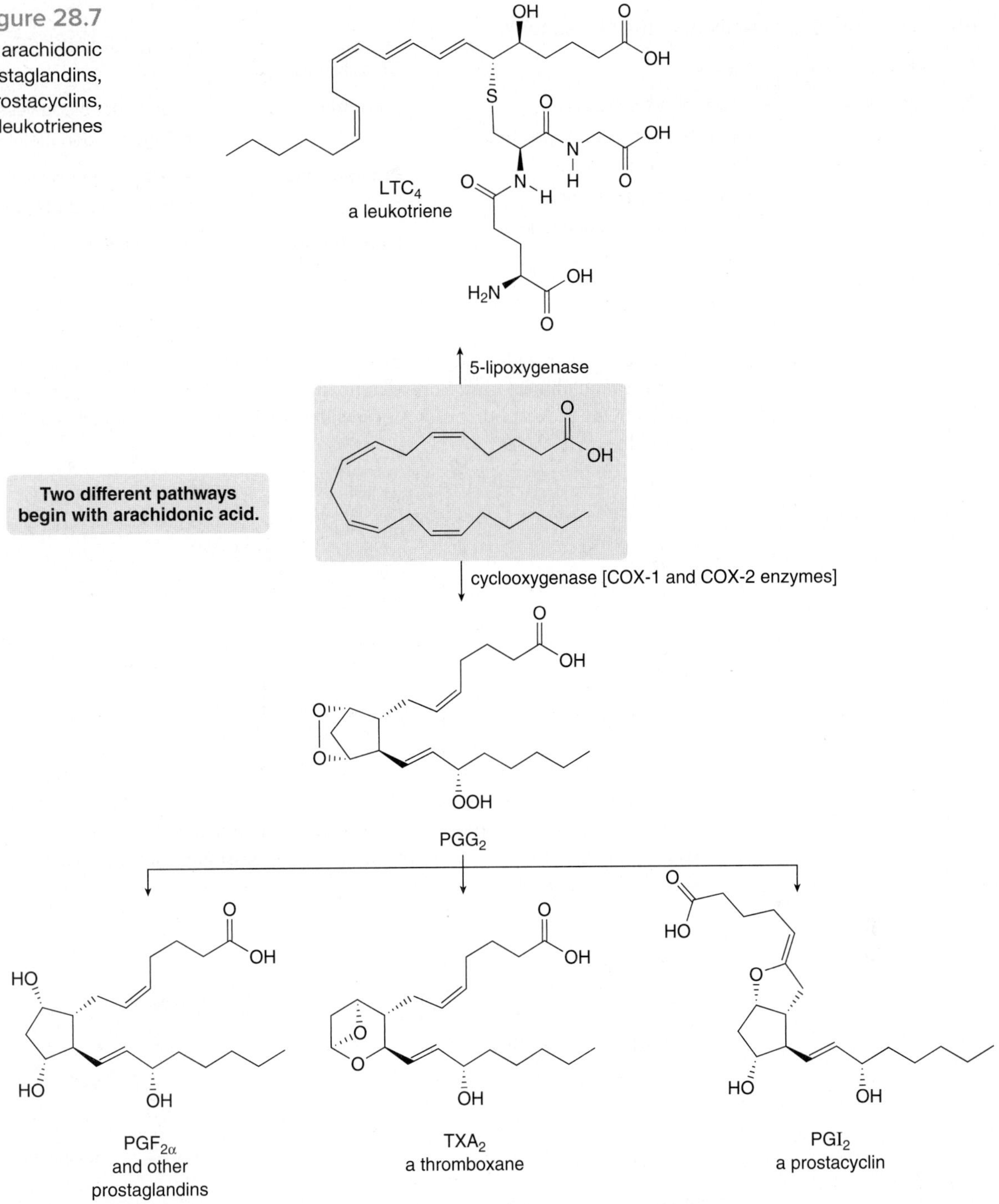

LTC$_4$
a leukotriene
5-lipoxygenase
Two different pathways begin with arachidonic acid.
cyclooxygenase [COX-1 and COX-2 enzymes]
PGG$_2$
PGF$_{2\alpha}$
and other
prostaglandins
TXA$_2$
a thromboxane
PGI$_2$
a prostacyclin

Other details of the biosynthesis of leukotrienes and prostaglandins were given in Sections 9.17 and 19.6, respectively.

The synthesis of prostaglandins, thromboxanes, and prostacyclins begins with the oxidation of arachidonic acid with O_2 by a **cyclooxygenase** enzyme, which forms an unstable cyclic intermediate, PGG_2. PGG_2 is then converted via different pathways to these three classes of compounds. Leukotrienes are formed by a different pathway, using an enzyme called a **lipoxygenase.** These four paths for arachidonic acid are summarized in Figure 28.7.

Each eicosanoid is associated with specific types of biological activity (Table 28.4). In some cases, the effects oppose one another. For example, thromboxanes are vasoconstrictors that trigger blood platelet aggregation, whereas prostacyclins are vasodilators that inhibit platelet aggregation. The levels of these two eicosanoids must be in the right balance for cells to function properly.

Table 28.4 Biological Activity of the Eicosanoids

Eicosanoid	Effect	Eicosanoid	Effect
Prostaglandins	• Lower blood pressure • Inhibit blood platelet aggregation • Control inflammation • Lower gastric secretions • Stimulate uterine contractions • Relax smooth muscles of the uterus	**Thromboxanes**	• Constrict blood vessels • Trigger blood platelet aggregation
		Prostacyclins	• Dilate blood vessels • Inhibit blood platelet aggregation
		Leukotrienes	• Constrict smooth muscle, especially in the lungs

Because of their wide range of biological functions, prostaglandins and their analogues have found several clinical uses. For example, **dinoprostone,** the generic name for **PGE_2,** is administered to relax the smooth muscles of the uterus when labor is induced, and to terminate pregnancies in the early stages.

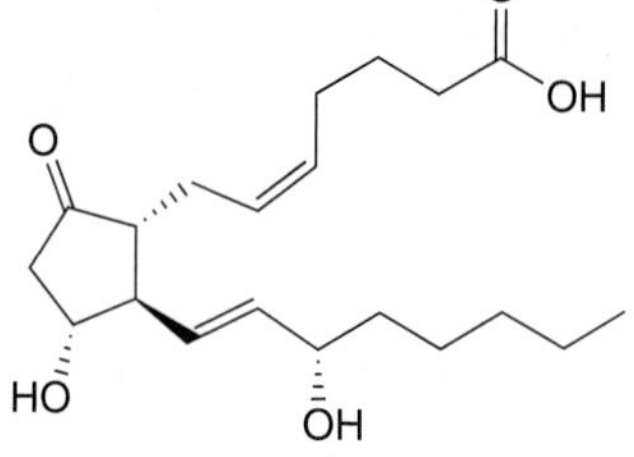

PGE_2
(dinoprostone)

Because prostaglandins themselves are unstable in the body, often having half-lives of only minutes, more stable analogues have been developed that retain their important biological activity longer. Misoprostol is a prostaglandin analogue administered to prevent gastric ulcers in patients who are at high risk of developing them, and unoprostone isopropyl is used to decrease eye pressure in glaucoma patients.

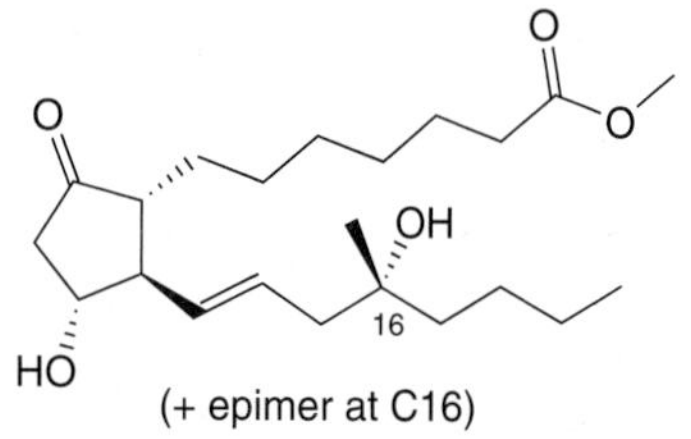

(+ epimer at C16)

misoprostol

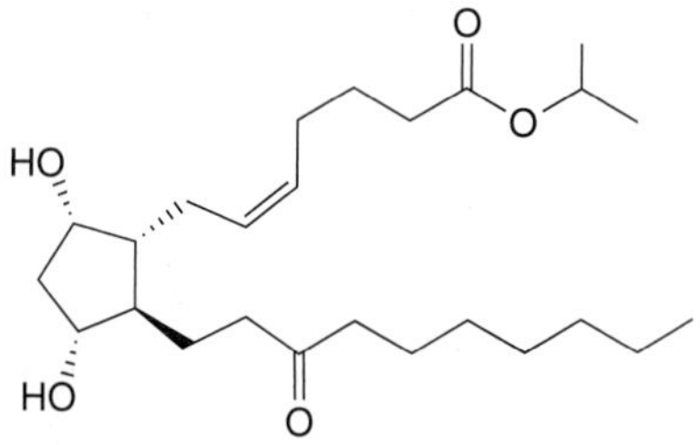

unoprostone isopropyl

Studying the biosynthesis of eicosanoids has led to other discoveries as well. For example, aspirin and other nonsteroidal anti-inflammatory drugs (**NSAIDs**) inactivate the cyclooxygenase enzyme needed for prostaglandin synthesis. In this way, NSAIDs block the synthesis of the prostaglandins that cause inflammation (Section 19.6).

More recently, it has been discovered that two *different* cyclooxygenase enzymes, called **COX-1** and **COX-2,** are responsible for prostaglandin synthesis. COX-1 is involved with the usual production of prostaglandins, but COX-2 is responsible for the synthesis of additional prostaglandins in inflammatory diseases like arthritis. **NSAIDs like aspirin and ibuprofen inactivate both the COX-1 and COX-2 enzymes.** This activity also results in an increase in gastric secretions, making an individual more susceptible to ulcer formation.

A group of anti-inflammatory drugs that block only the COX-2 enzyme was developed in the 1990s. These drugs—**rofecoxib, valdecoxib,** and **celecoxib**—do not cause an increase in gastric secretions, and thus were touted as especially effective NSAIDs for patients with arthritis, who need daily doses of these medications. Unfortunately, both rofecoxib and valdecoxib have now been removed from the market, because their use has been associated with an increased risk of heart attack and stroke.

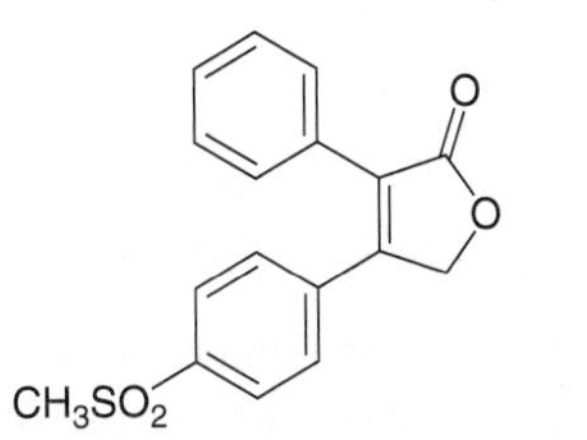

Generic name rofecoxib
Trade name Vioxx

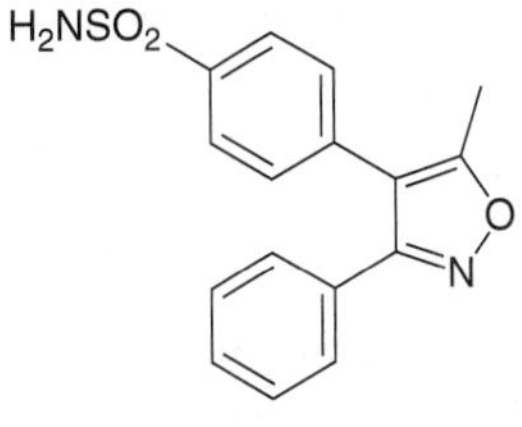

Generic name valdecoxib
Trade name Bextra

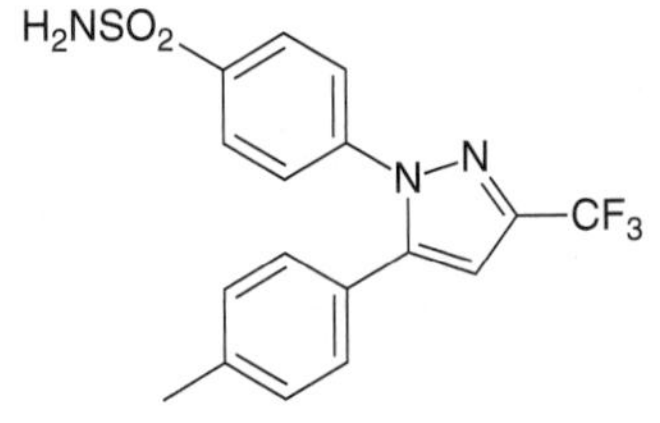

Generic name celecoxib
Trade name Celebrex

The discovery of drugs that block prostaglandin synthesis illustrates how basic research in organic chemistry can lead to important practical applications. Elucidating the structure and biosynthesis of prostaglandins began as a project in basic research. It has now resulted in a number of applications that benefit many individuals with various illnesses.

28.7 Terpenes

Terpenes **are lipids composed of repeating five-carbon units called isoprene units.** An **isoprene unit** has five carbons: four in a row, with a one-carbon branch on a middle carbon.

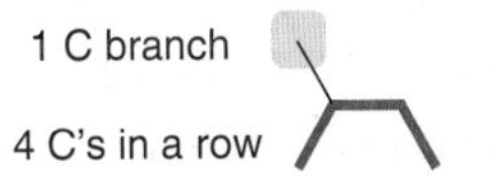

isoprene unit

Terpenes are hydrocarbons that may be acyclic or have one or more rings. The term ***terpenoid*** is used for compounds that contain isoprene units as well as an oxygen heteroatom. Many **essential oils,** a group of compounds isolated from plant sources by distillation, are terpenes and terpenoids. Examples include myrcene from bayberry and menthol from peppermint.

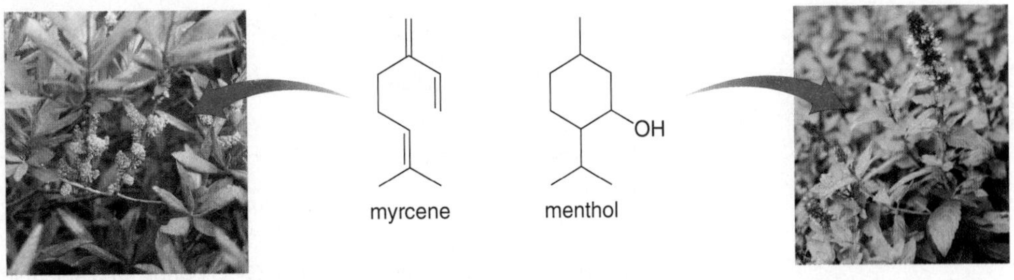

bayberry plant

peppermint plant

28.7A Locating Isoprene Units

How do we identify the isoprene units in these molecules? Start at one end of the molecule near a branch point. Then **look for a four-carbon chain with a one-carbon branch.** This forms one isoprene unit. Continue along the chain or around the ring until all the carbons are part of an isoprene unit. Keep in mind the following:

- An isoprene unit may be composed of C – C σ bonds only, or there may be π bonds at any position.
- Isoprene units are always connected by one or more carbon–carbon bonds.
- Each carbon atom is part of one isoprene unit only.
- Every isoprene unit has five carbon atoms. Heteroatoms may be present but their presence is ignored in locating isoprene units.

Myrcene and menthol, for example, each have 10 carbon atoms, so they are composed of two isoprene units.

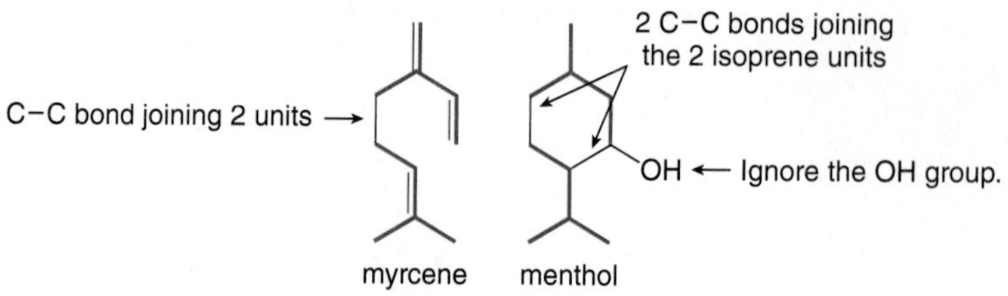
2 C–C bonds joining
the 2 isoprene units
C–C bond joining 2 units
OH
Ignore the OH group.
myrcene
menthol

Terpenes and terpenoids are classified by the number of isoprene units they contain. A ***monoterpene*** **(or *monoterpenoid*) contains 10 carbons** and has two isoprene units, a ***sesquiterpene*** **(or *sesquiterpenoid*) contains 15 carbons** and has three isoprene units, and so forth. The different terpene classes are summarized in Table 28.5.

Several examples, with the isoprene units labeled in red, are given in Figure 28.8.

Table 28.5 Classes of Terpenes and Terpenoids

Name	Number of C atoms	Number of isoprene units
Monoterpene (Monoterpenoid)	10	2
Sesquiterpene (Sesquiterpenoid)	15	3
Diterpene (Diterpenoid)	20	4
Sesterterpene (Sesterterpenoid)	25	5
Triterpene (Triterpenoid)	30	6
Tetraterpene (Tetraterpenoid)	40	8

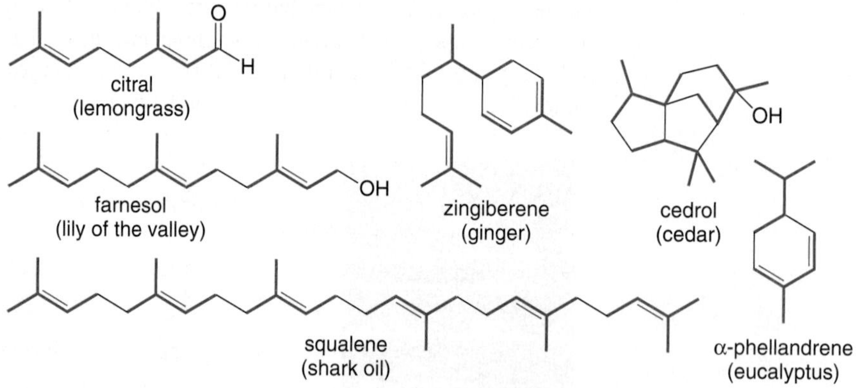
O
H
citral
(lemongrass)
OH
farnesol
(lily of the valley)
zingiberene
(ginger)
OH
cedrol
(cedar)
squalene
(shark oil)
α-phellandrene
(eucalyptus)

Figure 28.8 Examples of some common terpenes and terpenoids

- Isoprene units are labeled in red, with C–C bonds (in black) joining two units.
- The source of each terpene or terpenoid is given in parentheses.

Problem 28.10 Locate the isoprene units in each compound.

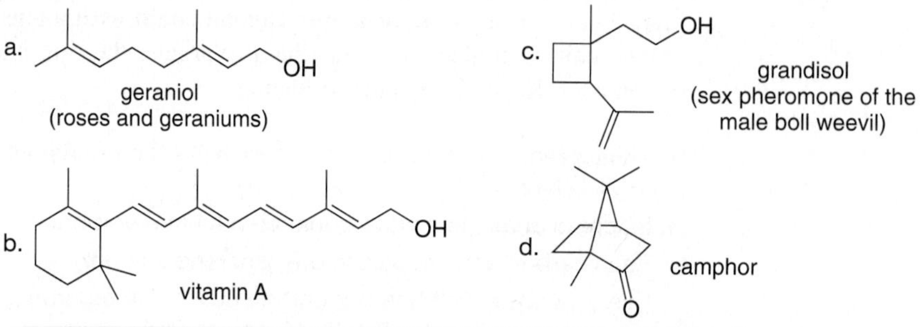
a.
OH
geraniol
(roses and geraniums)
b.
OH
vitamin A
c.
OH
grandisol
(sex pheromone of the
male boll weevil)
d.
camphor
O

Amber, fossilized resin that oozed from trees long ago, contains biformene, as well as many other terpenoids called labdanoids.

Problem 28.11 Locate the isoprene units in biformene, a component of amber, and classify biformene as a monoterpene, sesquiterpene, etc.

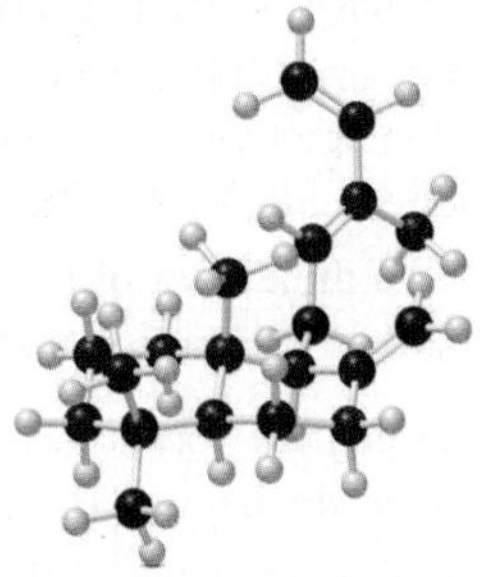

biformene

28.7B The Biosynthesis of Terpenes and Terpenoids

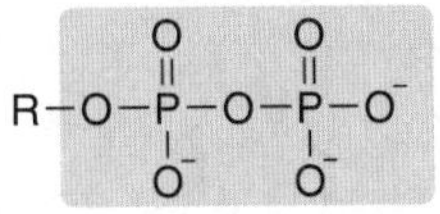

organic diphosphate

R–OPP

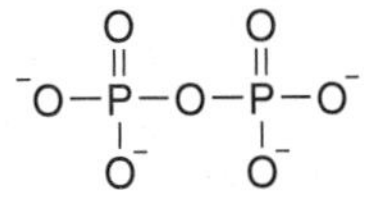

diphosphate leaving group PP_i

Terpene and terpenoid biosynthesis is an excellent example of how syntheses in nature occur with high efficiency. There are two ways this is accomplished.

[1] The same reaction is used over and over again to prepare progressively more complex compounds.

[2] Key intermediates along the way serve as the starting materials for a wide variety of other compounds.

All terpenes and terpenoids are synthesized from dimethylallyl diphosphate and isopentenyl diphosphate. Both of these five-carbon compounds are organic diphosphates (Section 16.2B) with a good leaving group (diphosphate, $P_2O_7^{4-}$, PP_i).

dimethylallyl diphosphate isopentenyl diphosphate

The overall strategy of biosynthesis from dimethylallyl diphosphate and isopentenyl diphosphate is summarized in Figure 28.9.

Figure 28.9 An outline of terpene and terpenoid biosynthesis

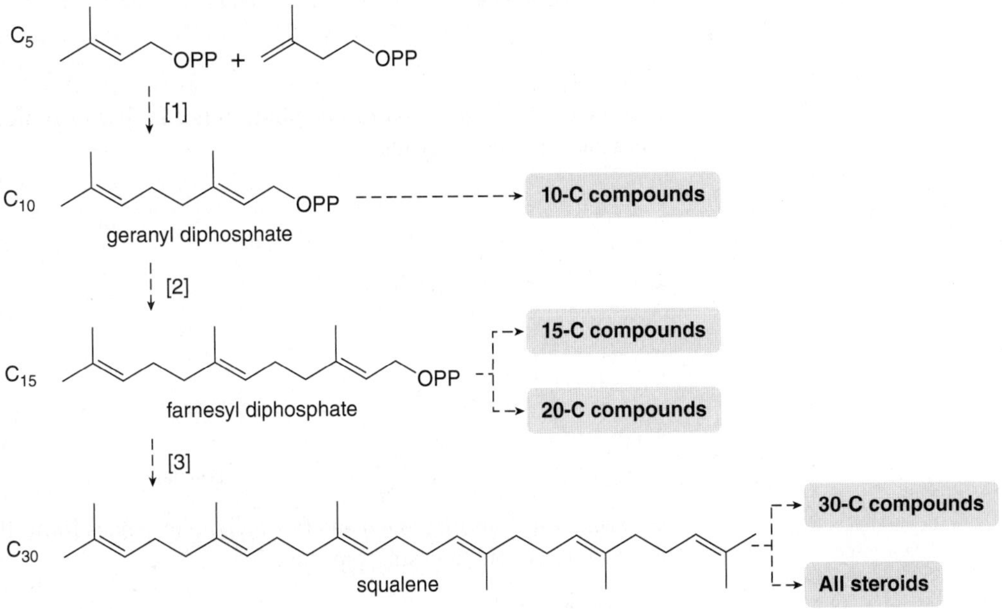

There are three basic parts:

[1] The two C_5 diphosphates are converted to **geranyl diphosphate, a C_{10} monoterpenoid.** Geranyl diphosphate is the starting material for all other monoterpenes and monoterpenoids.

[2] Geranyl diphosphate is converted to **farnesyl diphosphate, a C_{15} sesquiterpenoid,** by addition of a five-carbon unit. Farnesyl diphosphate is the starting material for all sesquiterpenes, diterpenes, and related terpenoids.

[3] Two molecules of farnesyl diphosphate are converted to **squalene, a C_{30} triterpene.** Squalene is the starting material for all triterpenes and steroids.

The biological formation of geranyl diphosphate from the two five-carbon diphosphates was shown in Mechanism 16.1. The biological conversion of geranyl diphosphate to farnesyl diphosphate involves a similar pathway, as shown in Mechanism 28.1.

Mechanism 28.1 Biological Formation of Farnesyl Diphosphate

Part [1] Formation of a new carbon–carbon σ bond

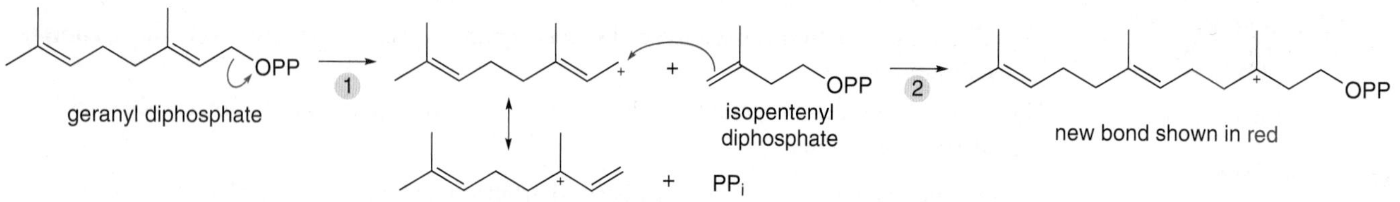

1 Loss of the diphosphate leaving group forms a resonance-stabilized carbocation.

2 Nucleophilic attack of isopentenyl diphosphate on the allylic carbocation forms the new C–C σ bond.

Part [2] Formation of a π bond by loss of a proton

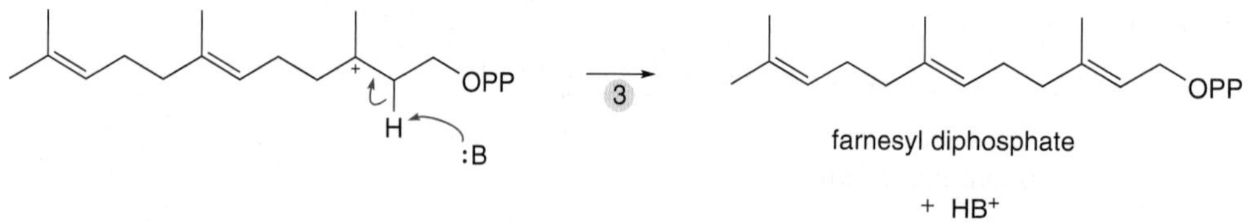

3 Loss of a proton (shown with the general base B:) forms a new π bond and farnesyl diphosphate.

Two molecules of farnesyl diphosphate react to form squalene, from which all other triterpenes and steroids are synthesized.

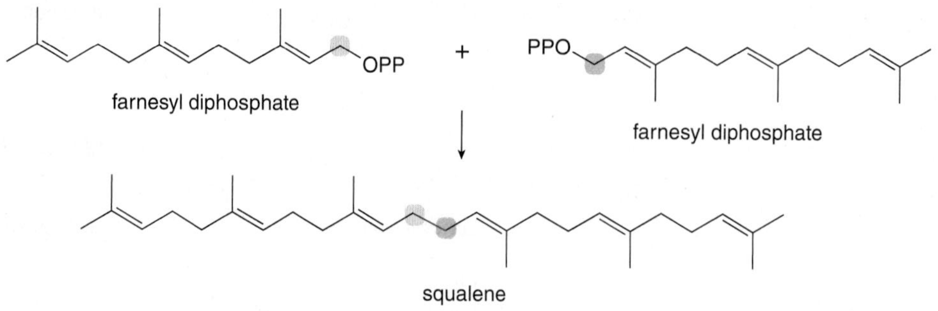

Aqueous hydrolysis of geranyl and farnesyl diphosphates forms the monoterpenoid geraniol and the sesquiterpenoid farnesol, respectively.

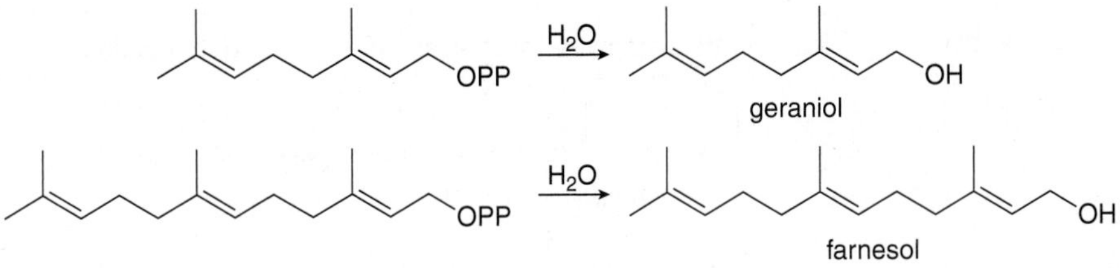

All other terpenes and terpenoids are biologically derived from geranyl and farnesyl diphosphates by a series of reactions. Cyclic compounds are formed by intramolecular reactions involving nucleophilic attack of π bonds on intermediate carbocations. To form some cyclic compounds, the *E* double bond in geranyl diphosphate must first isomerize to an isomeric diphosphate with a *Z* double bond, neryl diphosphate, by the process illustrated in Mechanism 28.2. Isomerization forms a substrate with a leaving group and nucleophilic double bond in close proximity, so that an intramolecular reaction can occur.

Mechanism 28.2 Isomerization of Geranyl Diphosphate to Neryl Diphosphate

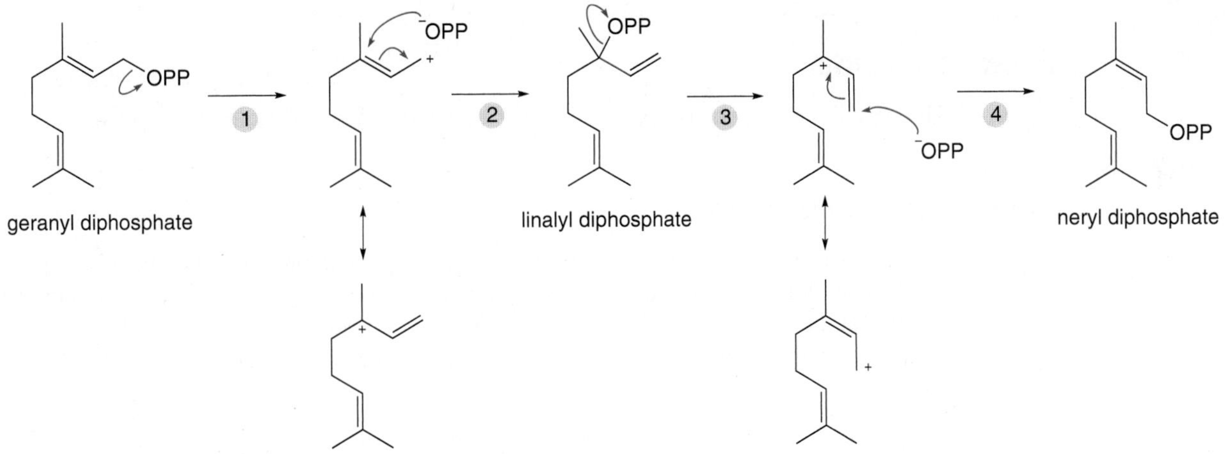

1 – 2 Loss of diphosphate forms a resonance-stabilized carbocation, which reacts with the diphosphate anion to form linalyl diphosphate.

3 Bond rotation of the single bond shown in red and loss of diphosphate forms a resonance-stabilized carbocation.

4 Nucleophilic attack with diphosphate forms neryl diphosphate, which has the leaving group and the double bond at the other end of the chain in close proximity for intramolecular cyclization.

In the synthesis of α-terpineol or limonene, for example, geranyl diphosphate isomerizes to form neryl diphosphate (Step [1] in the following reaction sequence). Neryl diphosphate then cyclizes to a 3° carbocation by intramolecular attack (Steps [2]–[3]). Nucleophilic attack of water on this carbocation yields the monoterpenoid α-terpineol (Step [4]) or loss of a proton yields the monoterpene limonene (Step [5]).

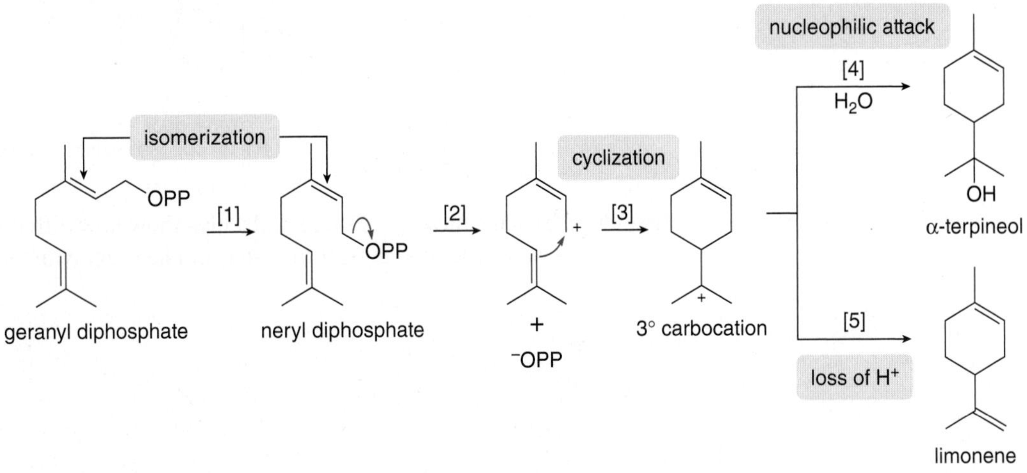

Problem 28.12 Write a stepwise mechanism for the following reaction.

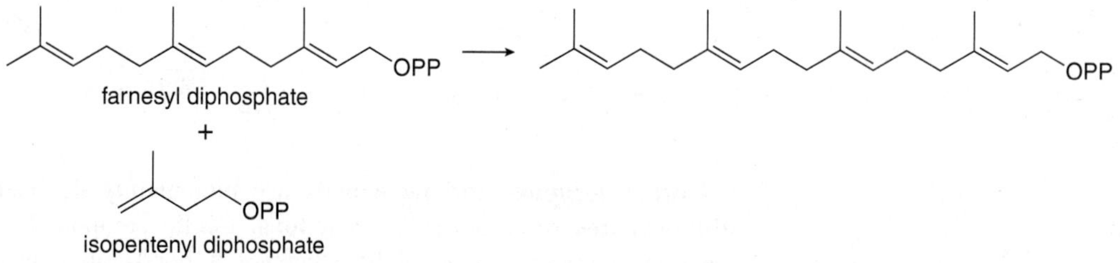

Problem 28.13 Draw a stepwise mechanism for the conversion of geranyl diphosphate to α-terpinene.

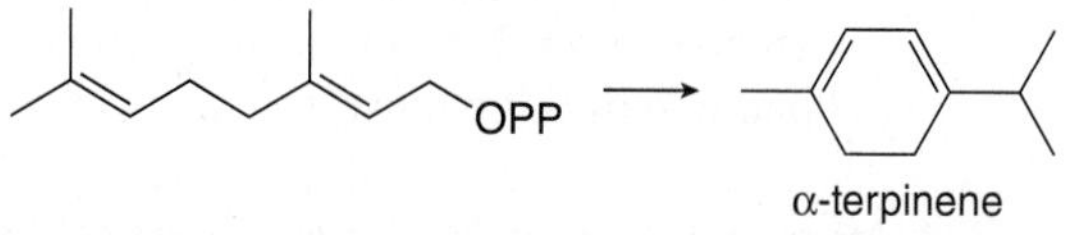

28.8 Steroids

The steroids are a group of tetracyclic lipids, many of which are biologically active.

28.8A Steroid Structure

Steroids are composed of three six-membered rings and one five-membered ring, joined together as drawn. Many steroids also contain two methyl groups, called **angular methyl groups,** at the two ring junctions indicated. The steroid rings are lettered **A, B, C,** and **D,** and the 17 ring carbons are numbered as shown. The two angular methyl groups are numbered C18 and C19.

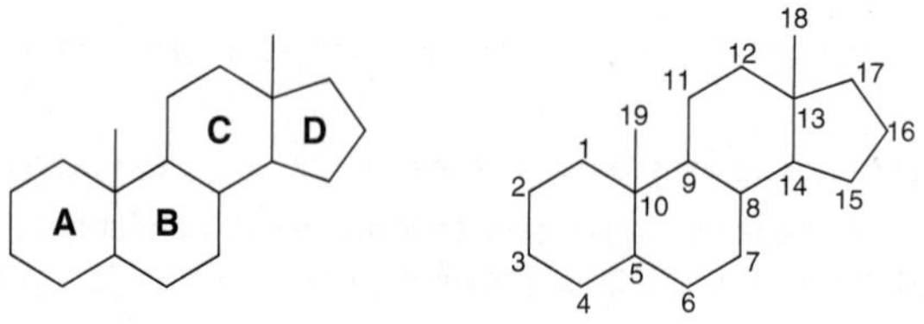

Whenever two rings are fused together, the substituents at the ring fusion can be arranged cis or trans. To see more easily why this is true, consider **decalin,** which consists of two six-membered rings fused together. ***trans*-Decalin** has the two hydrogen atoms at the ring fusion on opposite sides, whereas ***cis*-decalin** has them on the same side.

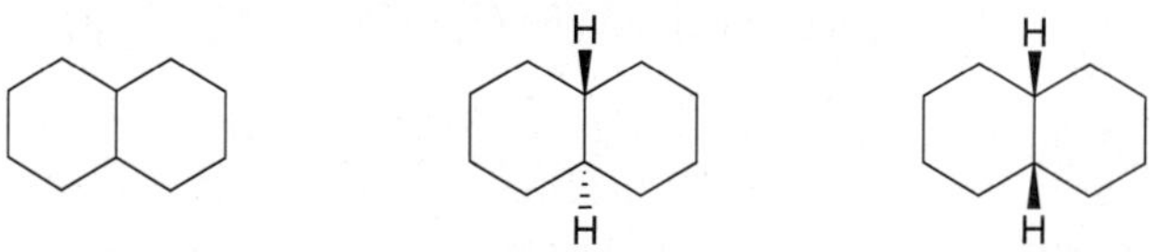

decalin
Two six-membered rings share a C–C bond.

***trans*-decalin**
2 H's on opposite sides

***cis*-decalin**
2 H's on the same side

Three-dimensional structures of these molecules show how different these two possible arrangements actually are. The two rings of *trans*-decalin lie roughly in the same plane, whereas the two rings of *cis*-decalin are almost perpendicular to each other. **The trans arrangement is lower in energy and therefore more stable.**

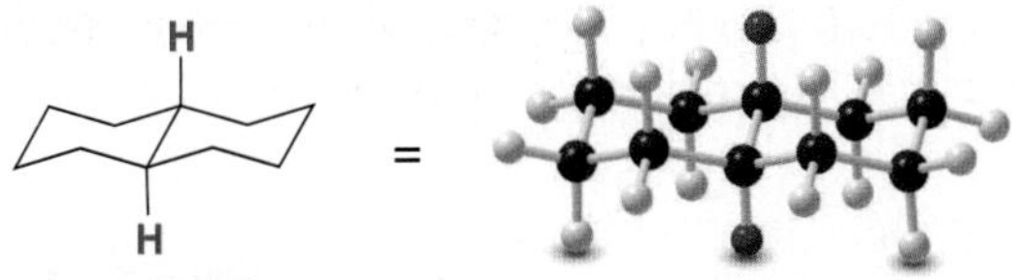

trans-decalin
2 H's on opposite sides

The 2 H's at the ring fusion are shown in red.

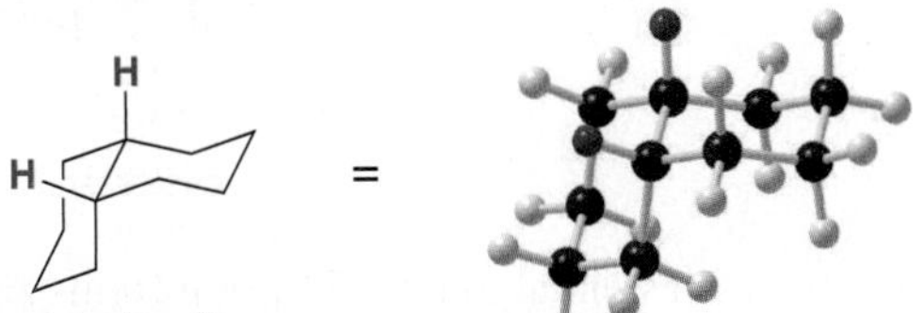

cis-decalin
2 H's on the same side

In steroids, each ring fusion could theoretically have the cis or trans configuration, but, by far the most common arrangement is all trans. Because of this, **all four rings of the steroid skeleton lie in the same plane,** and the ring system is fairly rigid. The two angular methyl groups are oriented perpendicular to the plane of the molecule. These methyl groups make one side of the steroid skeleton significantly more hindered than the other, as shown in Figure 28.10.

Figure 28.10
The three-dimensional structure of the steroid nucleus

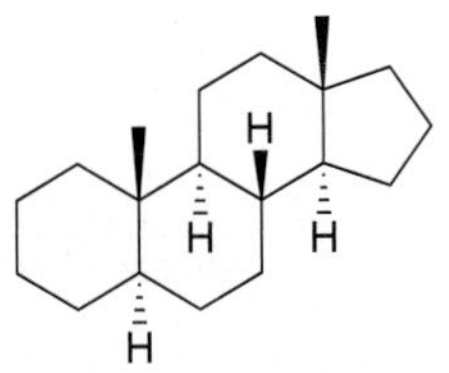

All rings are trans fused.

- The four steroid rings occupy approximately the same plane.
- The 2 CH_3 groups project above the plane of the molecule.

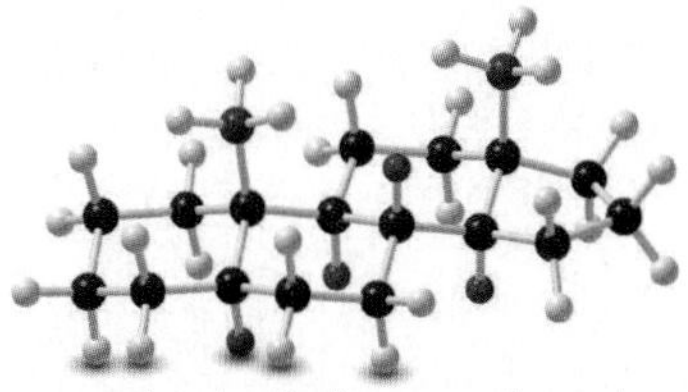

Atoms at the ring fusions are shown in red.

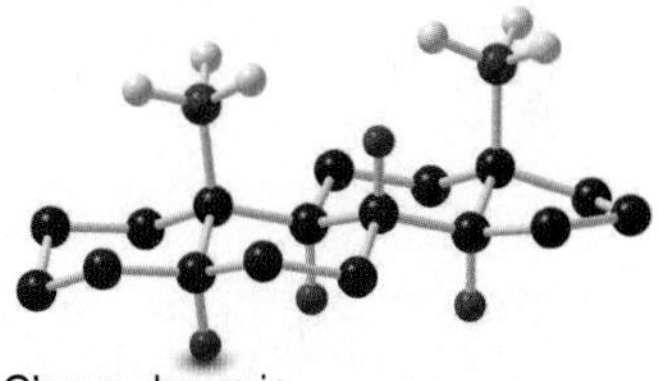

- All C's are drawn in.
- H's and CH_3's at the ring fusions are drawn in.
- All other H's are omitted.

Although steroids have the same fused-ring arrangement of carbon atoms, they differ in the identity and location of the substituents attached to that skeleton.

Problem 28.14 (a) Draw a skeletal structure of the anabolic steroid 4-androstene-3,17-dione, also called "andro," from the following description. Andro contains the tetracyclic steroid skeleton with carbonyl groups at C3 and C17, a double bond between C4 and C5, and methyl groups bonded to C10 and C13.
(b) Add wedges and dashed wedges for all stereogenic centers with the following information: the configuration at C10 is *R*, the configuration at C13 is *S*, and all substituents at ring fusions are trans to each other.

28.8B Cholesterol

Cholesterol has also been discussed in Sections 3.4C and 4.15. The role of cholesterol in plaque formation and atherosclerosis was discussed in Section 22.17.

Cholesterol, the chapter-opening molecule, has the tetracyclic carbon skeleton characteristic of steroids. It also has eight stereogenic carbons (seven on rings and one on a side chain), so there are $2^8 = 256$ possible stereoisomers. In nature, however, only the following stereoisomer exists:

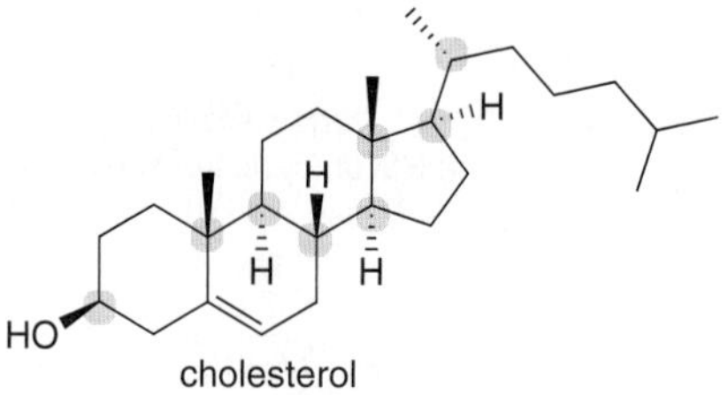

cholesterol

Cholesterol is essential to life because it forms an important component of cell membranes and is the starting material for the synthesis of all other steroids. Humans do not have to ingest cholesterol, because it is synthesized in the liver and then transported to other tissues through the bloodstream. Because cholesterol has only one polar OH group and many nonpolar C–C and C–H bonds, it is insoluble in water (and, thus, in the aqueous medium of the blood).

Konrad Bloch and Feodor Lynen shared the 1964 Nobel Prize in Physiology or Medicine for unraveling the complex transformation of squalene to cholesterol.

Cholesterol is synthesized in the body from squalene, a C_{30} triterpene that is itself prepared from smaller terpenes, as discussed in Section 28.7B. Because the biosynthesis of all terpenes begins with acetyl CoA, every one of the 27 carbon atoms of cholesterol comes from the same two-carbon precursor. The major steps in the conversion of squalene to cholesterol are given in Figure 28.11.

Figure 28.11
The biosynthesis of cholesterol

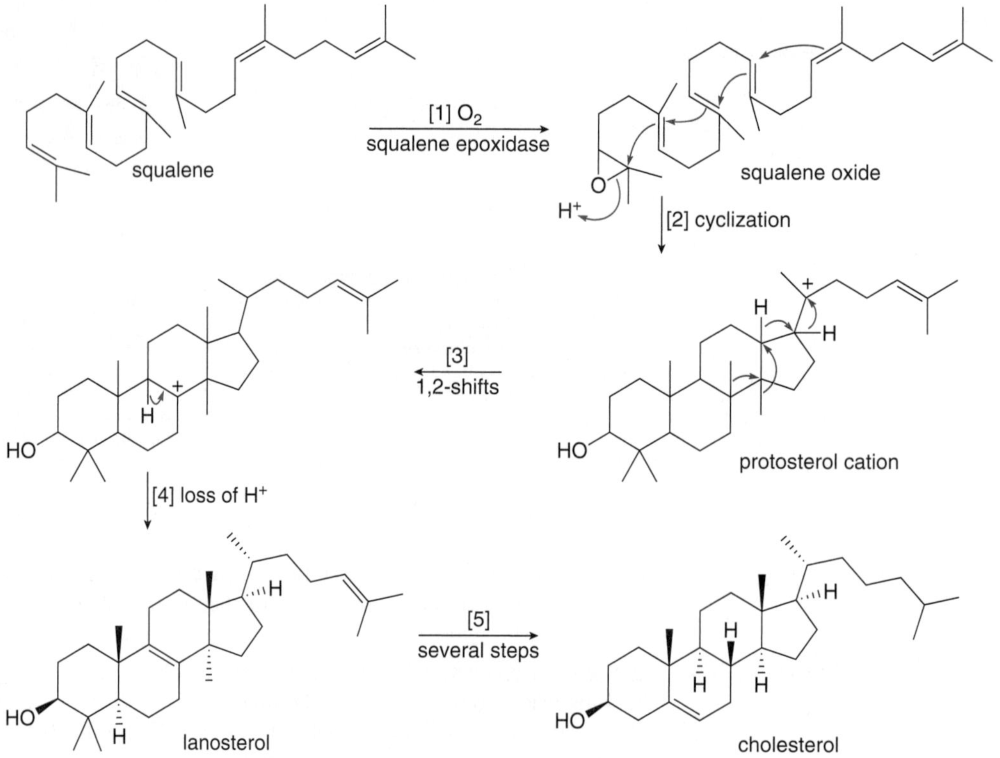

The conversion of squalene to cholesterol consists of five different parts:

[1] **Epoxidation** of squalene with an enzyme, squalene epoxidase, gives squalene oxide, which contains a single epoxide on one of the six double bonds.

[2] **Cyclization** of squalene oxide yields a carbocation, called the protosterol cation. This reaction results in the formation of four new C–C bonds and the tetracyclic ring system.

[3] **The protosterol carbocation rearranges** by a series of 1,2-shifts of either a hydrogen or methyl group to form another 3° carbocation.

[4] **Loss of a proton** gives an alkene called **lanosterol.** Although lanosterol has seven stereogenic centers, a single stereoisomer is formed.

[5] Lanosterol is then converted to cholesterol by a multistep process that results in removal of three methyl groups.

Several drugs called statins are now available to reduce the level of cholesterol in the bloodstream. These compounds act by blocking the biosynthesis of cholesterol at its very early stages. Two examples include atorvastatin (Lipitor) and simvastatin (Zocor), whose structures appear in Figure 28.12.

Figure 28.12
Two cholesterol-lowering drugs

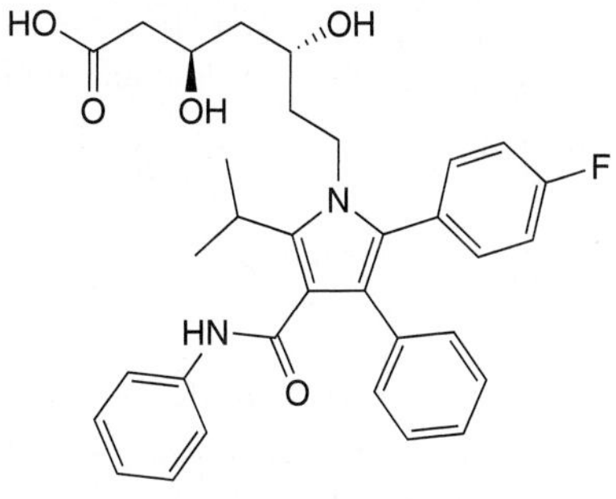

Generic name atorvastatin
Trade name Lipitor

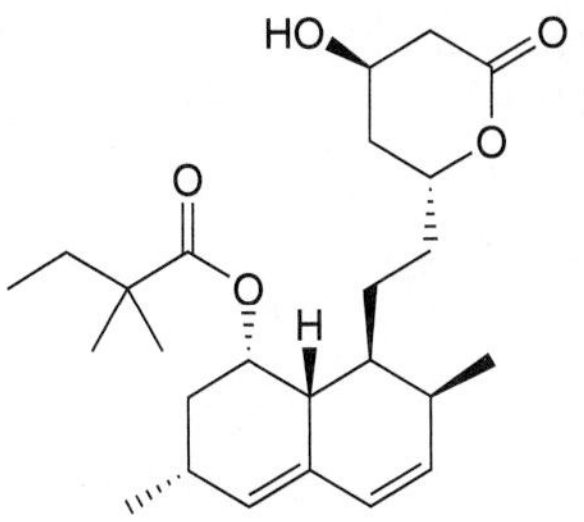

Generic name simvastatin
Trade name Zocor

Problem 28.15 Draw the enantiomer and any two diastereomers of cholesterol. Does the OH group of cholesterol occupy an axial or equatorial position?

Problem 28.16 Treatment of cholesterol with mCPBA results in formation of a single epoxide **A,** with the stereochemistry drawn. Why isn't the isomeric epoxide **B** formed to any extent?

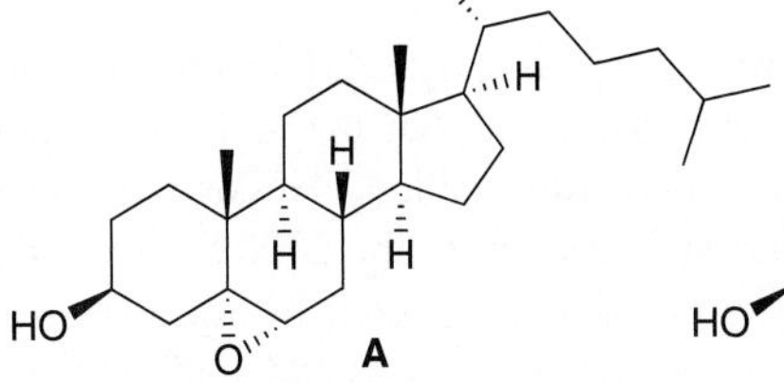

A

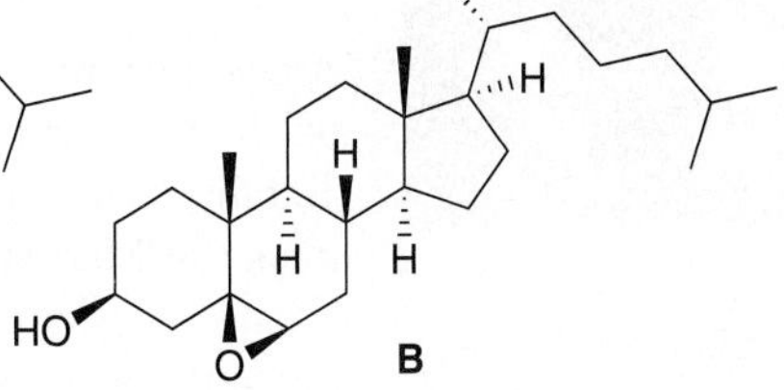

B

28.8C Other Steroids

Many other important steroids are hormones secreted by the endocrine glands. Two classes are the **sex hormones** and the **adrenal cortical steroids.**

There are two types of female sex hormones, **estrogens** and **progestins.** The male sex hormones are called **androgens.** The most important members of each hormone type are given in Table 28.6. The biosynthesis of estrogens from androgens is discussed in Section 17.11.

Table 28.6 The Female and Male Sex Hormones

Structure	Properties
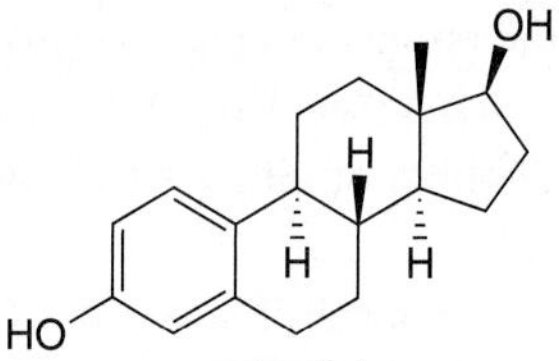 estradiol 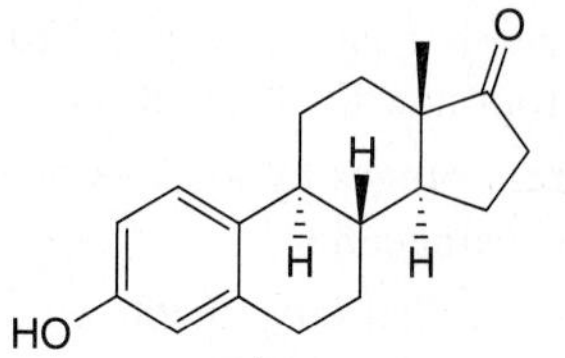 estrone	• **Estradiol** and **estrone** are estrogens synthesized in the ovaries. They control the development of secondary sex characteristics in females and regulate the menstrual cycle.
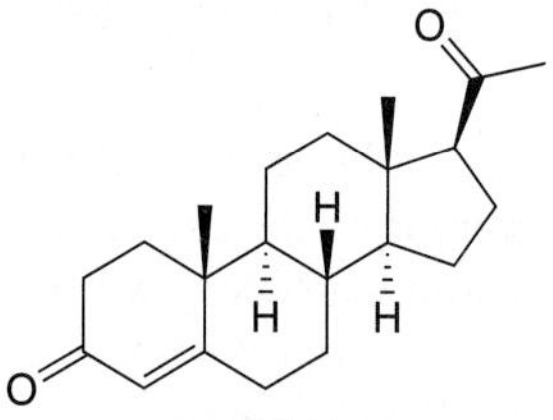 progesterone	• **Progesterone** is often called the "pregnancy hormone." It is responsible for the preparation of the uterus for implantation of a fertilized egg.
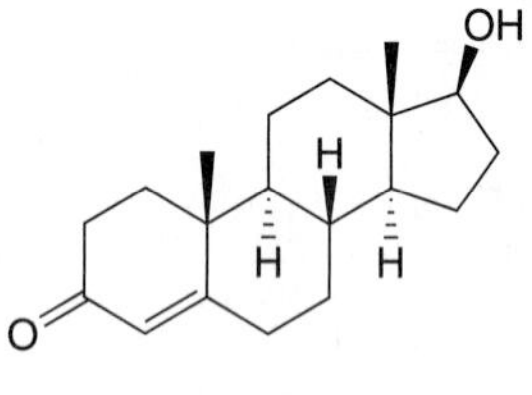 testosterone 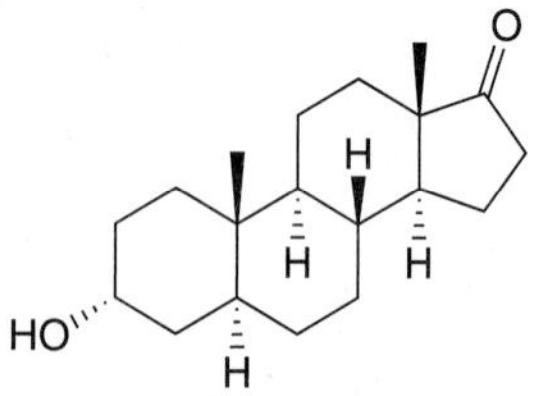 androsterone	• **Testosterone** and **androsterone** are androgens synthesized in the testes. They control the development of secondary sex characteristics in males.

Synthetic analogues of these steroids have found important uses, such as ethynylestradiol and norethindrone in oral contraceptives, mentioned in Section 11.4.

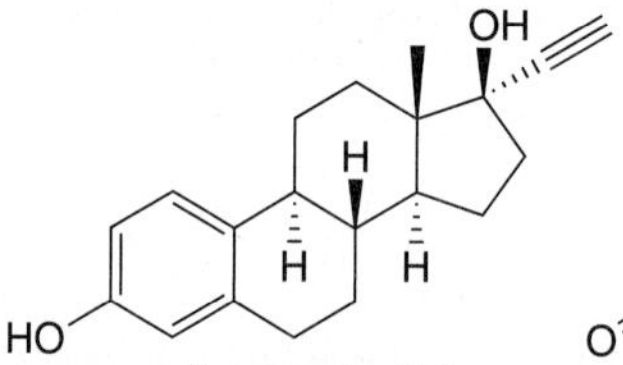
ethynylestradiol

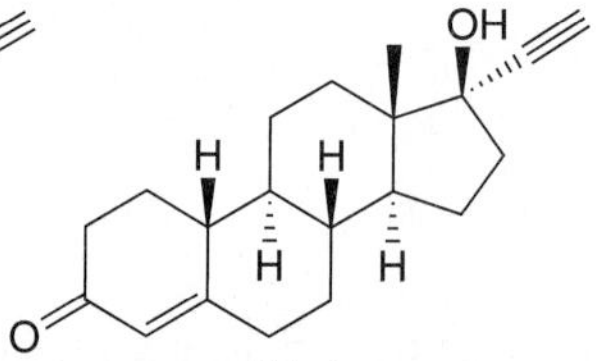
norethindrone

Some body builders use anabolic steroids to increase muscle mass. Long-term or excessive use can cause many health problems, including high blood pressure, liver damage, and cardiovascular disease.

Synthetic androgen analogues, called **anabolic steroids,** promote muscle growth. They were first developed to help individuals whose muscles had atrophied from lack of use following surgery. They have since come to be used by athletes and body builders, although their use is not permitted in competitive sports. Many physical and psychological problems result from their prolonged use.

Anabolic steroids, such as stanozolol, nandrolone, and tetrahydrogestrinone, have the same effect on the body as testosterone, but they are more stable, so they are not metabolized as quickly. Tetrahydrogestrinone (also called THG or The Clear), the performance-enhancing drug used by track star Marion Jones during the 2000 Sydney Olympics, was considered a "designer steroid" because it was initially undetected in urine tests for doping. After its chemical structure and properties were determined, it was added to the list of banned anabolic steroids in 2004.

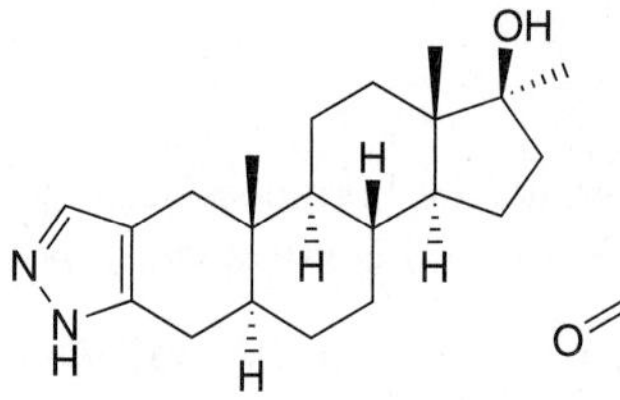
stanozolol

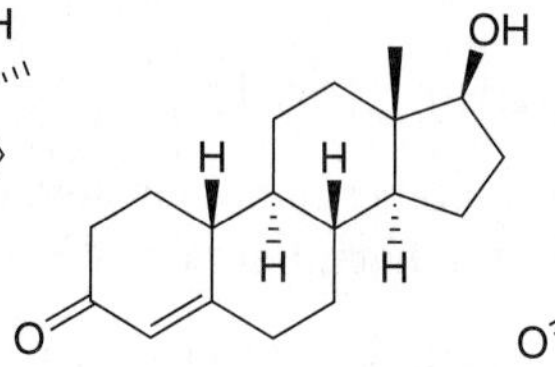
nandrolone

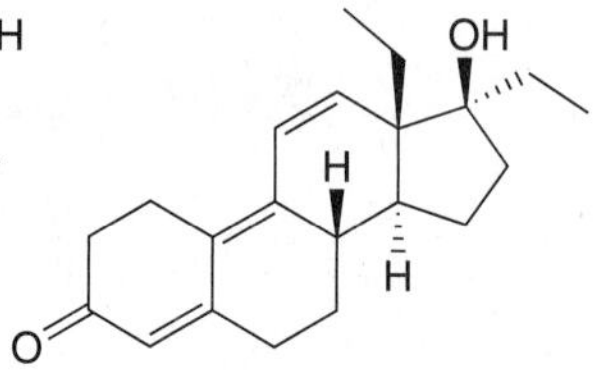
tetrahydrogestrinone

A second group of steroid hormones includes the **adrenal cortical steroids.** Three examples of these hormones are **cortisone, cortisol,** and **aldosterone.** All of these compounds are synthesized in the outer layer of the adrenal gland. Cortisone and cortisol serve as anti-inflammatory agents, and they also regulate carbohydrate metabolism. Aldosterone regulates blood pressure and volume by controlling the concentration of Na^+ and K^+ in body fluids.

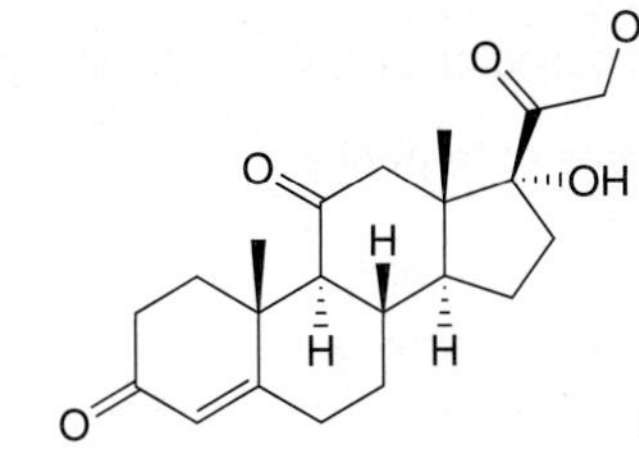

cortisone

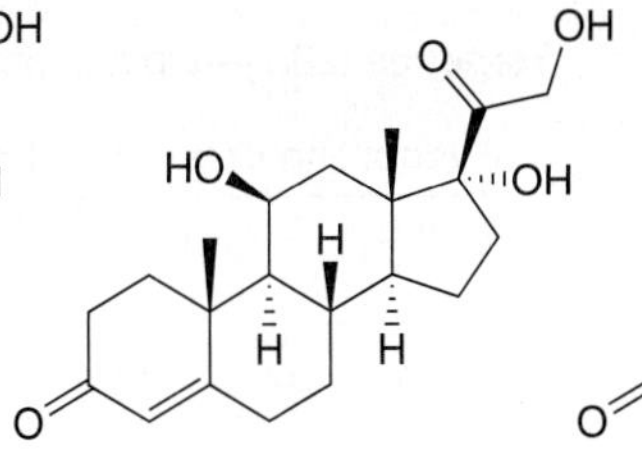

cortisol

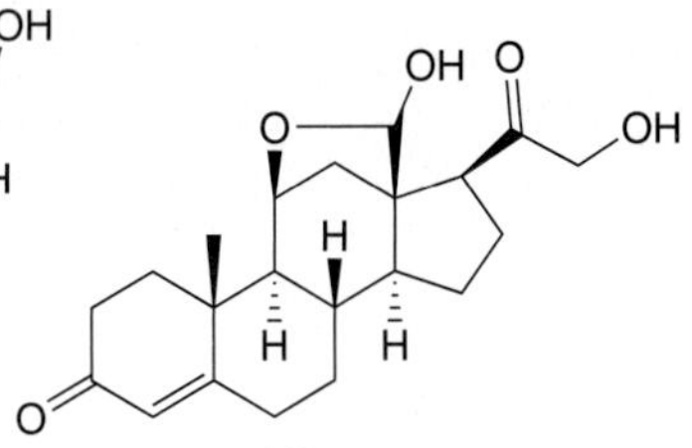

aldosterone

KEY CONCEPTS

Lipids

Hydrolyzable Lipids

[1] **Waxes** (28.2)—Esters formed from a long-chain alcohol and a long-chain carboxylic acid.

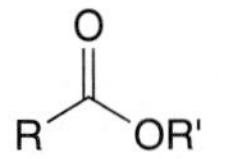

R, R' = long chains of C's

[2] **Triacylglycerols** (28.3)—Triesters of glycerol with three fatty acids.

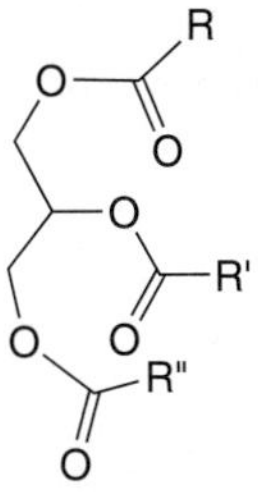

R, R', R'' = alkyl groups with 11–19 C's

[3] **Phospholipids** (28.4)

a. Phosphatidylethanolamine (cephalin)

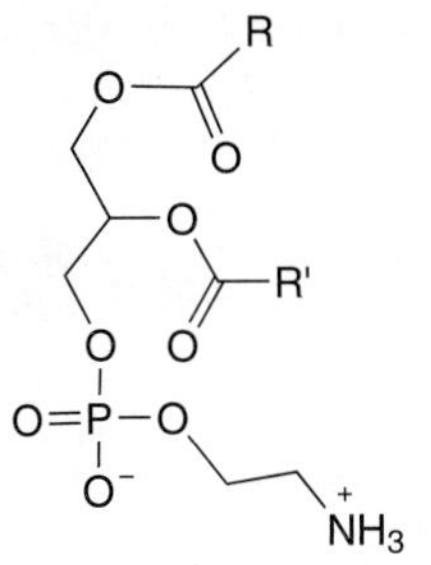

R, R' = long carbon chain

b. Phosphatidylcholine (lecithin)

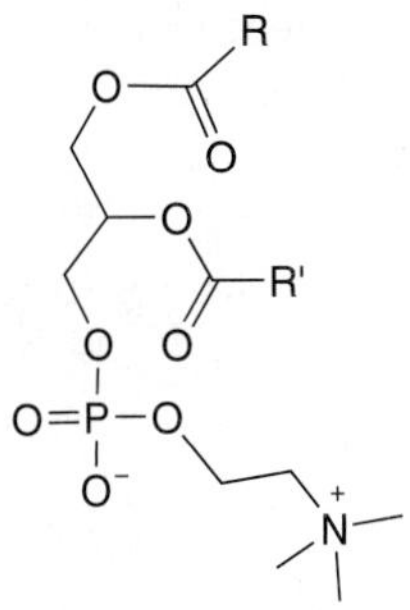

R, R' = long carbon chain

c. Sphingomyelin

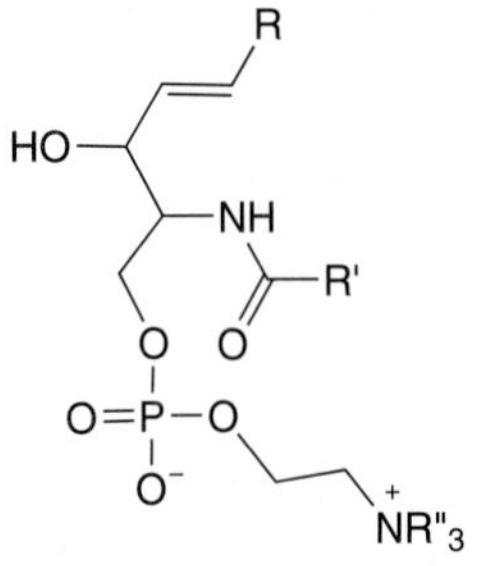

$R = CH_3(CH_2)_{12}$
R' = long carbon chain
R'' = H or CH_3

Nonhydrolyzable Lipids

[1] **Fat-soluble vitamins** (28.5)—Vitamins A, D, E, and K.

[2] **Eicosanoids** (28.6)—Compounds containing 20 C's derived from arachidonic acid. There are four types: prostaglandins, thromboxanes, prostacyclins, and leukotrienes.

[3] **Terpenes** (28.7)—Lipids composed of repeating 5 C units called isoprene units.

Isoprene unit	Types of terpenes			
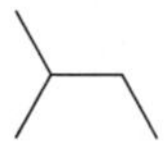	[1] monoterpene	10 C's	[4] sesterterpene	25 C's
	[2] sesquiterpene	15 C's	[5] triterpene	30 C's
	[3] diterpene	20 C's	[6] tetraterpene	40 C's

[4] **Steroids** (28.8)—Tetracyclic lipids composed of three six-membered and one five-membered ring.

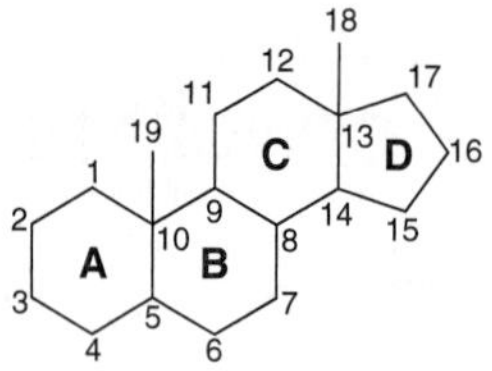

PROBLEMS

Problems Using Three-Dimensional Models

28.17 Locate the isoprene units in each compound.

a. 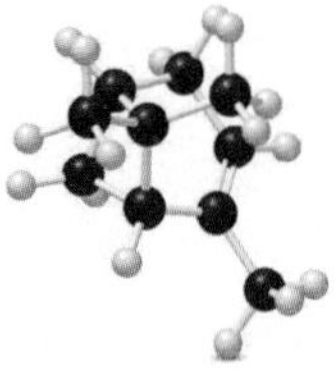

α-pinene

b. 

humulene

28.18 Convert each ball-and-stick model to a skeletal structure that clearly shows the stereochemistry at the ring fusion of these decalin derivatives.

a. 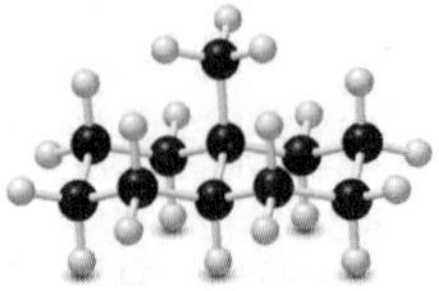

b. 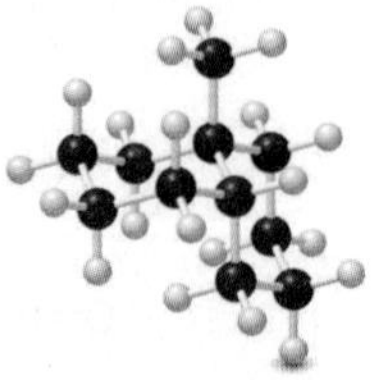

28.19 Convert the ball-and-stick model of androsterone to (a) a skeletal structure using wedges and dashed wedges around all stereogenic centers; and (b) a three-dimensional representation using chair cyclohexane rings.

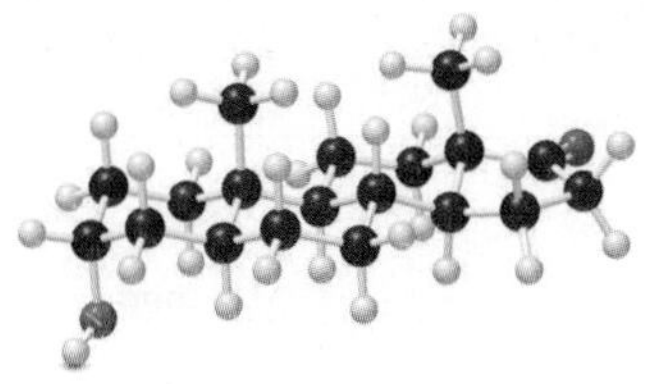

androsterone

Waxes, Triacylglycerols, and Phospholipids

28.20 One component of lanolin, the wax that coats sheep's wool, is derived from cholesterol and stearic acid. Draw its structure, including the correct stereochemistry at all stereogenic centers.

28.21 Draw all possible constitutional isomers of a triacylglycerol formed from one mole each of palmitic, oleic, and linoleic acids. Locate the tetrahedral stereogenic centers in each constitutional isomer.

28.22 What is the structure of an optically inactive triacylglycerol that yields two moles of oleic acid and one mole of palmitic acid when hydrolyzed in aqueous acid?

28.23 Triacylglycerol **L** yields compound **M** when treated with excess H_2, Pd-C. Ozonolysis of **L** ([1] O_3; [2] $(CH_3)_2S$) affords compounds **N – P.** What is the structure of **L?**

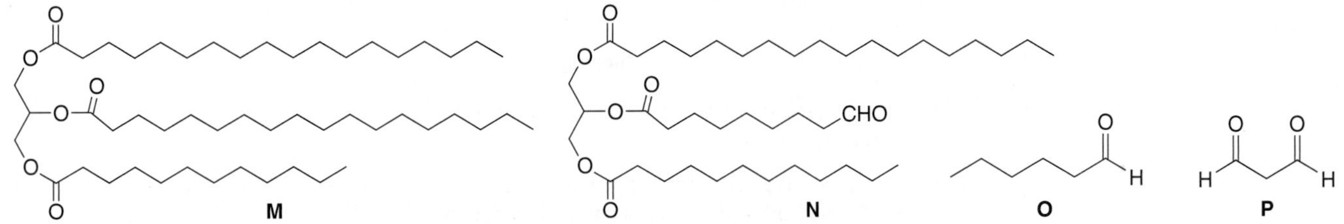

M **N** **O** **P**

28.24 Draw the structure of the following phospholipids:

a. a cephalin formed from two molecules of stearic acid

b. a sphingomyelin formed from palmitic acid

Prostaglandins

28.25 A difficult problem in the synthesis of $PGF_{2\alpha}$ is the introduction of the OH group at C15 in the desired configuration.

a. Label this stereogenic center as *R* or *S*.

b. A well-known synthesis of $PGF_{2\alpha}$ involves reaction of **A** with $Zn(BH_4)_2$, a metal hydride reagent similar in reactivity to $NaBH_4$, to form two isomeric products, **B** and **C.** Draw their structures and indicate their stereochemical relationship.

c. Suggest a reagent to convert **A** to the single stereoisomer **X.**

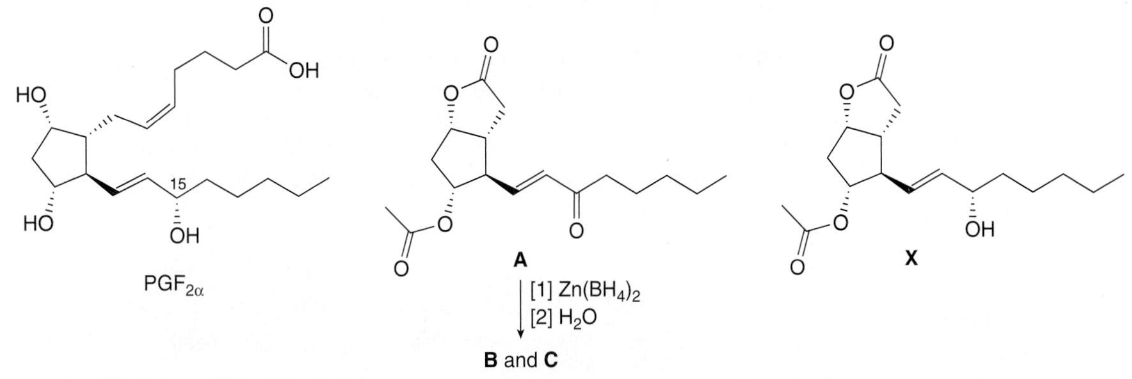

$PGF_{2\alpha}$

Terpenes and Terpenoids

28.26 Locate the isoprene units in each compound.

a. 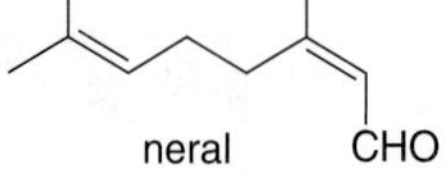

neral

e. 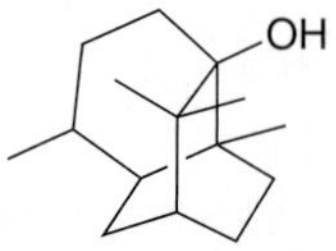

patchouli alcohol

g. 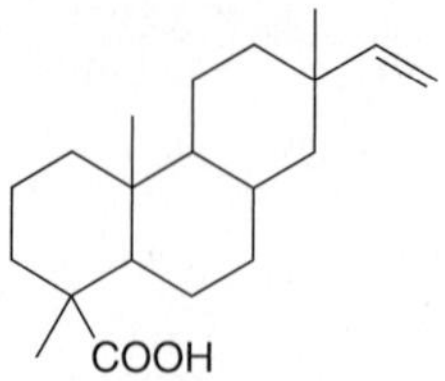

dextropimaric acid

b. 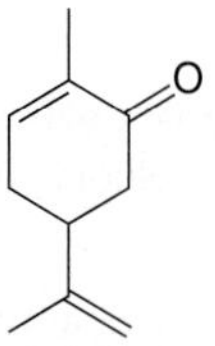

carvone

f. 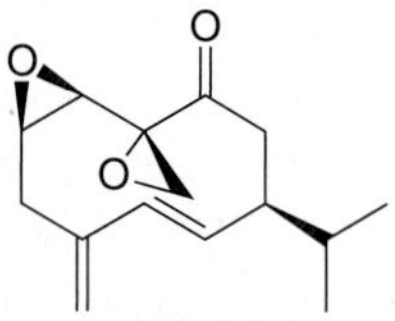

periplanone B

h. 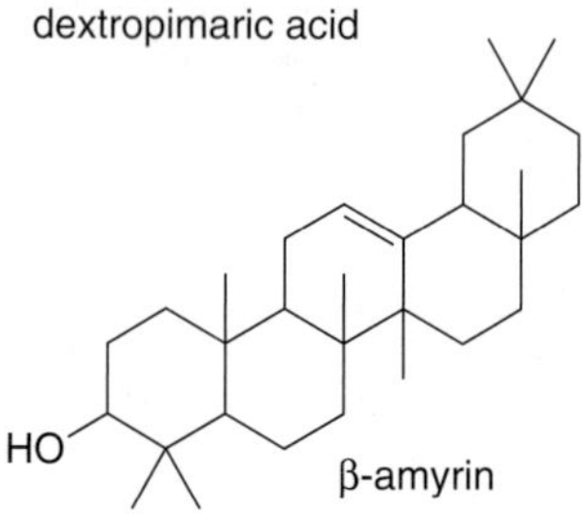

β-amyrin

c. 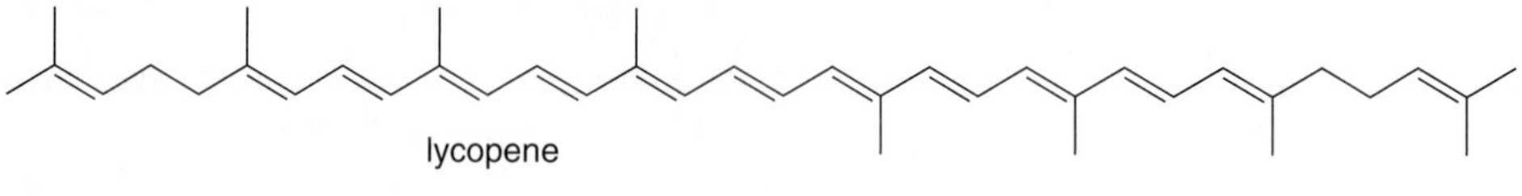

lycopene

d. 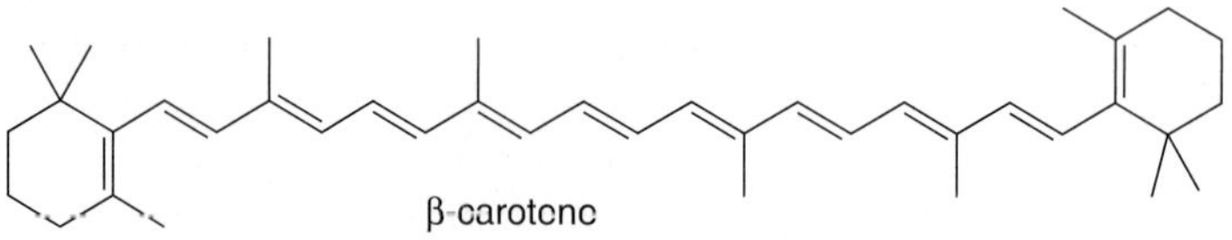

β-carotene

28.27 Classify each terpene and terpenoid in Problem 28.26 (e.g., as a monoterpene, sesquiterpene, etc.).

28.28 Crocin, which occurs naturally in crocus and gardenia flowers, is primarily responsible for the color of saffron. (a) What lipid and monosaccharides are formed by the hydrolysis of crocin? (b) Classify the lipid as a monoterpenoid, diperpenoid, etc., and locate the isoprene units.

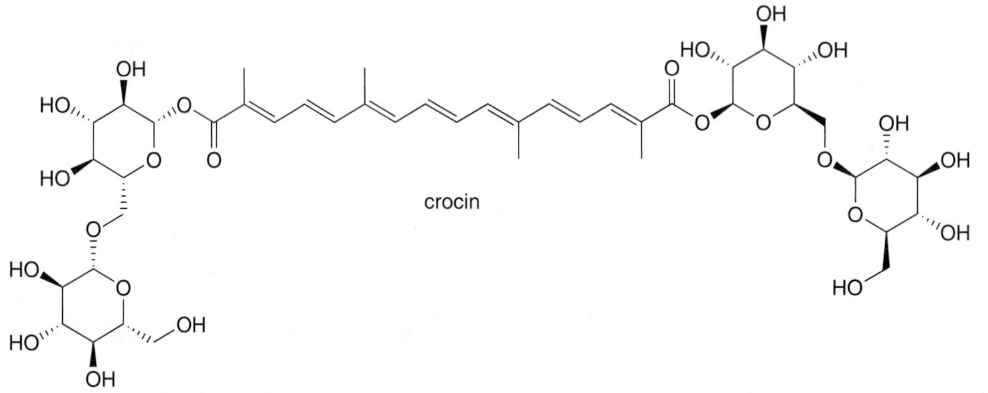

crocin

28.29 An isoprene unit can be thought of as having a head and a tail. The "head" of the isoprene unit is located at the end of the chain nearest the branch point, and the "tail" is located at the end of the carbon chain farthest from the branch point. Most isoprene units are connected in a "head-to-tail" fashion, as illustrated. For both lycopene (Problem 28.26), and squalene (Figure 28.9), decide which isoprene units are connected in a head-to-tail fashion and which are not.

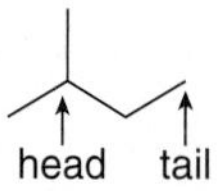

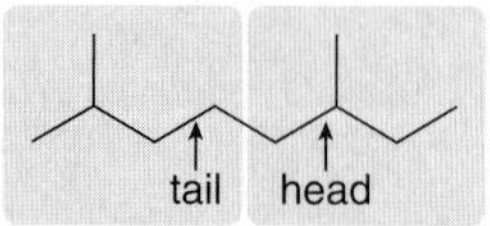

These two isoprene units are connected in a head-to-tail fashion.

28.30 Draw a stepwise mechanism for the conversion of neryl diphosphate to α-pinene. α-Pinene is a component of pine oil and rosemary oil.

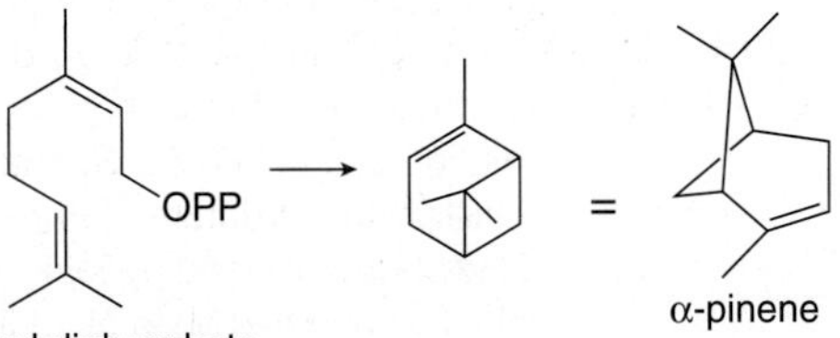

neryl diphosphate α-pinene

28.31 Flexibilene is a terpene isolated from *Sinularia flexibilis,* a soft coral found in the Indian Ocean. Draw a stepwise mechanism for the formation of flexibilene from farnesyl diphosphate and isopentenyl diphosphate. What is unusual about the cyclization that forms the 15-membered ring of flexibilene?

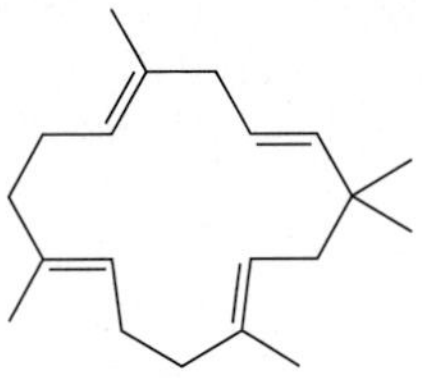

flexibilene

28.32 The biosynthesis of lanosterol from squalene has intrigued chemists since its discovery. It is now possible, for example, to synthesize polycyclic compounds from acyclic or monocyclic precursors by reactions that form several C–C bonds in a single reaction mixture.

a. Draw a stepwise mechanism for the following reaction.

b. Show how **X** can be converted to 16,17-dehydroprogesterone. (Hint: See Figure 24.4 for a related conversion.)

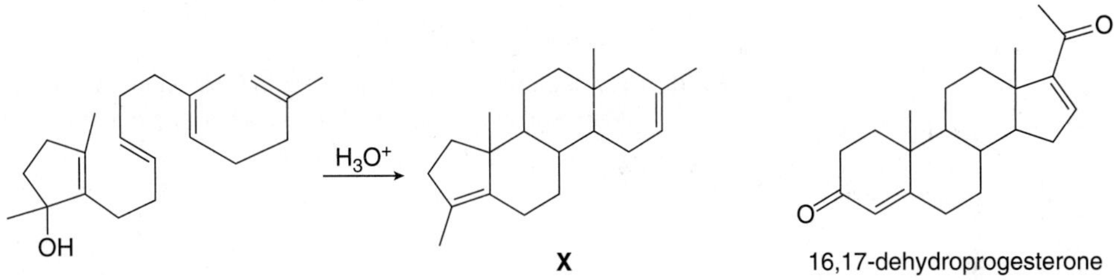

X 16,17-dehydroprogesterone

Steroids

28.33 Draw three-dimensional structures for each alcohol. Label the OH groups as occupying axial or equatorial positions.

a. 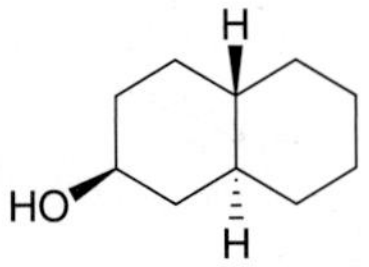
b. 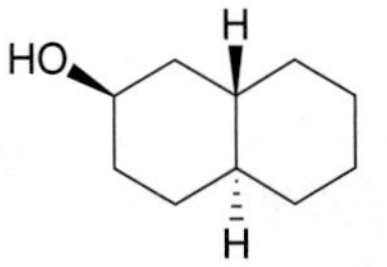
c. 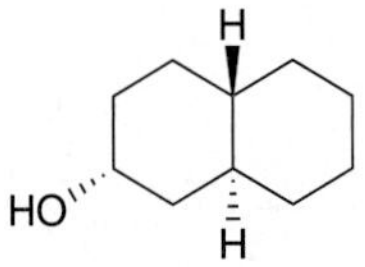
d. 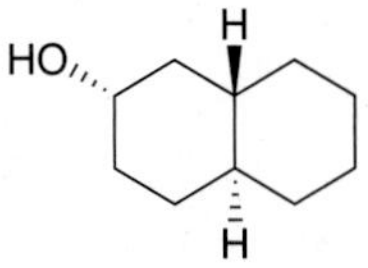

28.34 Axial alcohols are oxidized faster than equatorial alcohols by PCC and other Cr^{6+} oxidants. Which OH group in each compound is oxidized faster?

a. 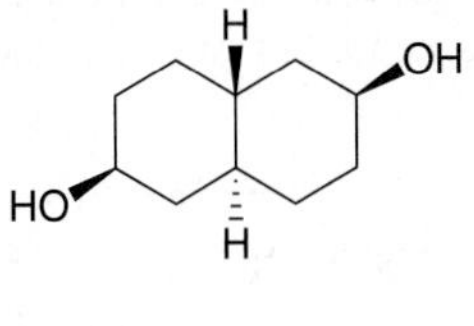
b. 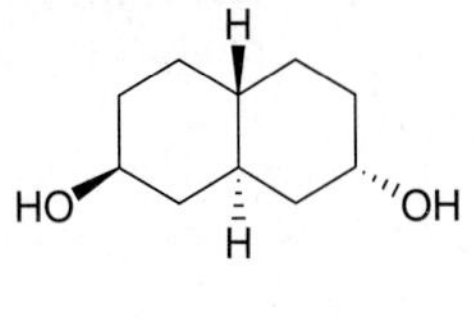

28.35 (a) Draw a skeletal structure of the anabolic steroid methenolone from the following description. Methenolone contains the tetracyclic steroid skeleton with a carbonyl group at C3, a hydroxyl at C17, a double bond between C1 and C2, and methyl groups bonded to C1, C10, and C13. (b) Add wedges and dashed wedges for all stereogenic centers with the following information: the configuration at C10 is *R*, the configuration at C13 is *S*, the configuration at C17 is *S*, and all substituents at ring fusions are trans to each other. (c) Draw the structure of Primobolan, the product formed when methenolone is treated with $CH_3(CH_2)_5COCl$ and pyridine. Primobolan is an anabolic steroid that can be taken orally or by injection and has been used illegally by well-known Major League Baseball players.

28.36 Betamethasone is a synthetic anti-inflammatory steroid used as a topical cream for itching. Betamethasone is derived from cortisol, with the following structural additions: a C=C between C1 and C2, a fluorine at C9, and a methyl group at C16. The configuration at C9 is *R*, and the configuration at C16 is *S*. Draw the structure of betamethasone.

28.37 a. Draw a three-dimensional structure for the following steroid.
b. What is the structure of the single stereoisomer formed by reduction of this ketone with H_2, Pd-C? Explain why only one stereoisomer is formed.

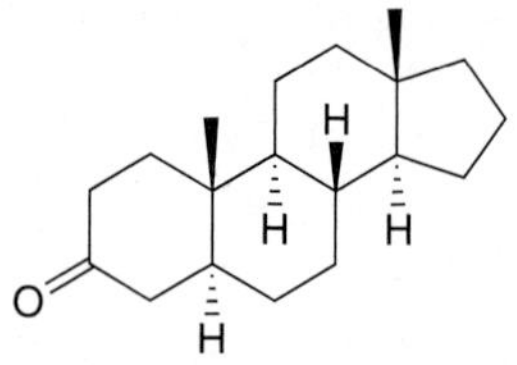

28.38 Draw the products formed when cholesterol is treated with each reagent. Indicate the stereochemistry around any stereogenic centers in the product.

a. CH_3COCl
b. H_2, Pd-C
c. PCC
d. oleic acid, H^+
e. [1] BH_3•THF; [2] H_2O_2, ^{-}OH

Challenge Problems

28.39 Draw a stepwise mechanism for the following conversion, which forms camphene. Camphene is a component of camphor and citronella oils.

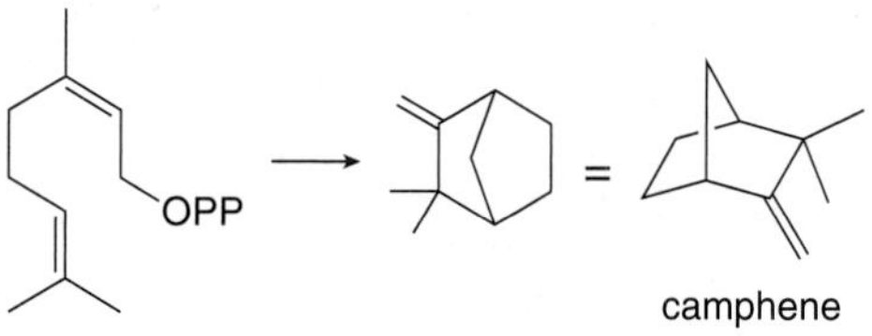

camphene

28.40 Draw a stepwise mechanism for the following reaction.

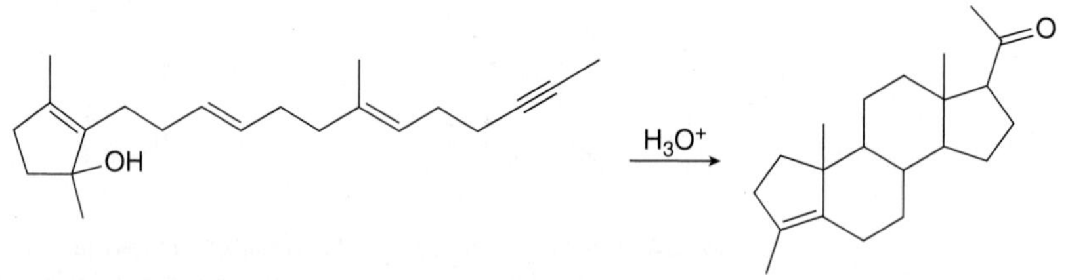

28.41 Farnesyl diphosphate is cyclized to sesquiterpene **A,** which is then converted to the bicyclic product epi-aristolochene. Write a stepwise mechanism for both reactions.

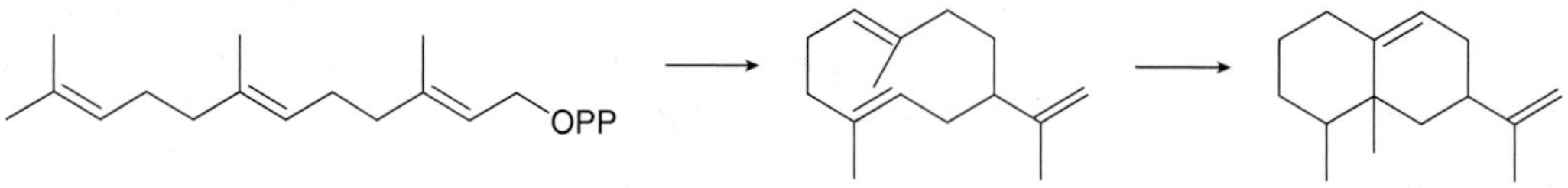

farnesyl diphosphate **A** epi-aristolochene

Carbon–Carbon Bond-Forming Reactions in Organic Synthesis

29

29.1 Coupling reactions of organocuprate reagents
29.2 Suzuki reaction
29.3 Heck reaction
29.4 Carbenes and cyclopropane synthesis
29.5 Simmons–Smith reaction
29.6 Metathesis

Ingenol mebutate is an ester derived from ingenol, a natural product obtained from the sap of *Euphorbia peplus*, a type of milkweed native to Europe, northern Africa, and western Asia. Because ingenol derivatives exhibited useful biological activity and isolation from the natural source did not provide easy access to the material, scientists developed an efficient laboratory synthesis. A gel formulation of ingenol mebutate (trade name Picato) has been approved for the treatment of actinic keratosis, a skin condition resulting from over-exposure to the sun that may result in squamous cell carcinoma, a form of skin cancer. In Chapter 29 we learn about carbon–carbon bond-forming reactions that prepare complex compounds like ingenol.

To form the carbon skeletons of complex molecules, organic chemists need an extensive repertoire of carbon–carbon bond-forming reactions. In Chapter 20, for example, we learned about the reactions of organometallic reagents—organolithium reagents, Grignard reagents, and organocuprates—with carbonyl substrates. In Chapters 23 and 24, we studied the reactions of nucleophilic enolates that form new carbon–carbon bonds.

Chapter 29 presents more carbon–carbon bond-forming reactions that are especially useful tools in organic synthesis. While previous chapters have concentrated on the reactions of one or two functional groups, the reactions in this chapter utilize a variety of starting materials and conceptually different reactions that form many types of products. All follow one central theme—they form new carbon–carbon bonds under mild conditions, making them versatile synthetic methods.

29.1 Coupling Reactions of Organocuprate Reagents

Several carbon–carbon bond-forming reactions involve the coupling of an organic halide (R'X) with an organometallic reagent or alkene. Three useful reactions are discussed in Sections 29.1–29.3:

[1] Reaction of an organic halide with an organocuprate reagent (Section 29.1)

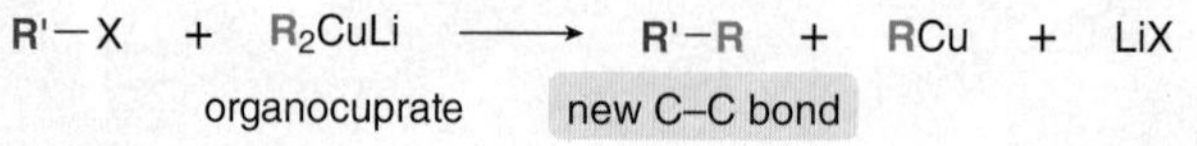
R'−X + R2CuLi → R'−R + RCu + LiX
organocuprate
new C–C bond

[2] Suzuki reaction: Reaction of an organic halide with an organoboron reagent in the presence of a palladium catalyst (Section 29.2)

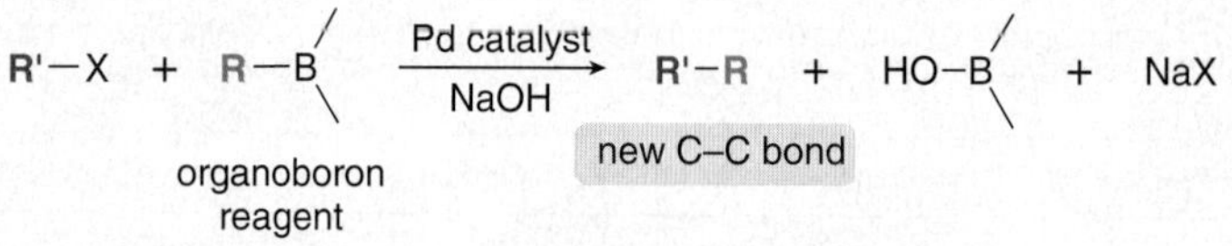
R'−X + R−B
Pd catalyst
NaOH
R'−R + HO−B + NaX
organoboron reagent
new C–C bond

[3] Heck reaction: Reaction of an organic halide with an alkene in the presence of a palladium catalyst (Section 29.3)

A complete list of reactions that form C−C bonds appears in Appendix D.

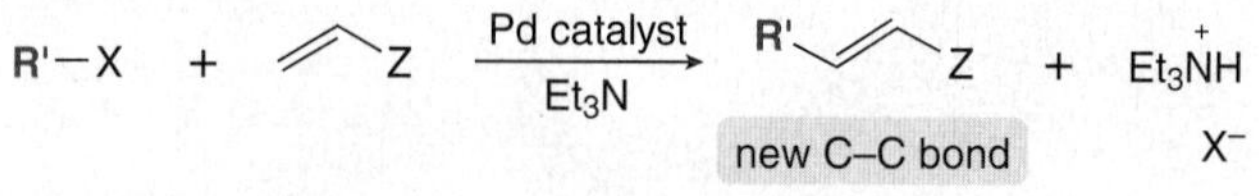
R'−X
Z
Pd catalyst
Et3N
R'
Z
Et3NH
X−
new C–C bond

29.1A General Features of Organocuprate Coupling Reactions

In addition to their reactions with acid chlorides, epoxides, and α,β-unsaturated carbonyl compounds (Sections 20.13–20.15), organocuprate reagents (R_2CuLi) also react with organic halides R'−X to form coupling products R−R' that contain a new C−C bond. Only one R group of the organocuprate is transferred to form the product, while the other becomes part of RCu, a reaction by-product.

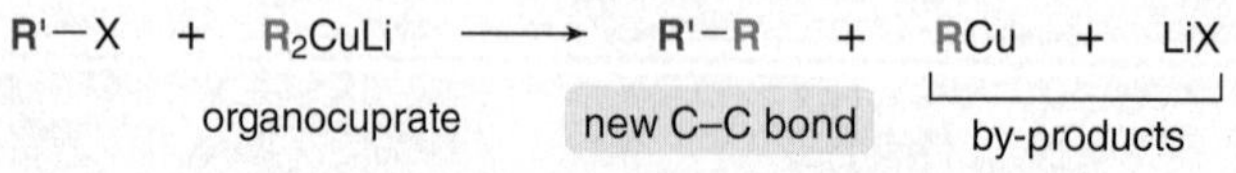
R'−X + R2CuLi → R'−R + RCu + LiX
organocuprate
new C–C bond
by-products

A variety of organic halides can be used, including methyl and 1° alkyl halides, as well as vinyl and aryl halides that contain X bonded to an sp^2 hybridized carbon. Some cyclic 2° alkyl halides give reasonable yields of product, but 3° alkyl halides are too sterically hindered. The halogen X in R'X may be Cl, Br, or I.

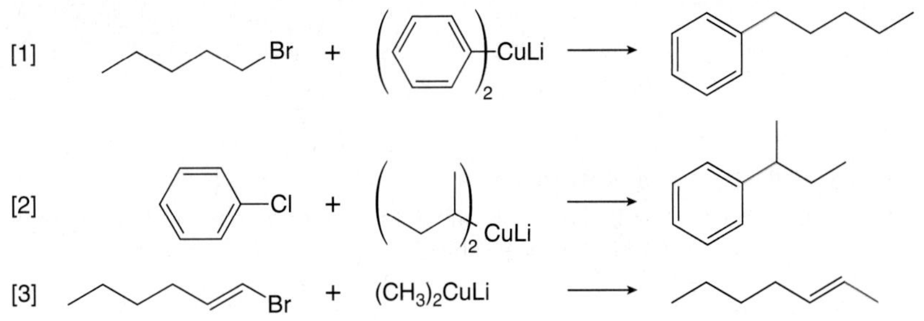
[1]
Br
+
CuLi
2
[2]
Cl
+
CuLi
2
[3]
Br
+
$(CH_3)_2CuLi$

trans-1-bromohex-1-ene *trans*-hept-2-ene

new C–C bonds in red

Coupling reactions with vinyl halides are **stereospecific.** For example, reaction of *trans*-1-bromohex-1-ene with $(CH_3)_2CuLi$ forms *trans*-hept-2-ene as the only stereoisomer (Equation [3]).

Problem 29.1 Draw the product of each coupling reaction.

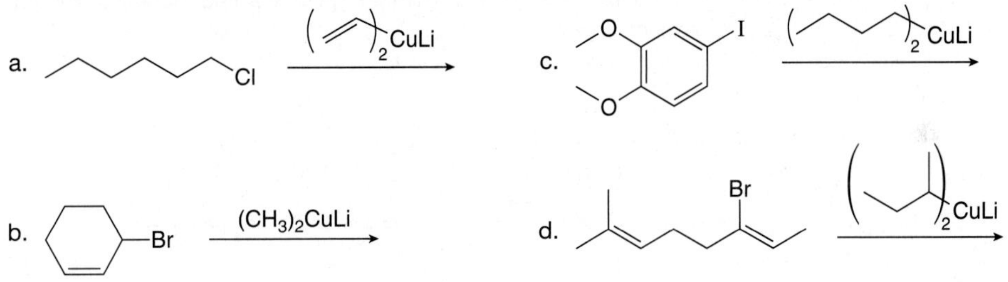
a.
CuLi
2
Cl
c.
O
O
I
CuLi
2
b.
Br
$(CH_3)_2CuLi$
d.
Br
CuLi
2

Problem 29.2 Identify reagents **A** and **B** in the following reaction scheme. This synthetic sequence was used to prepare the C_{18} juvenile hormone (Figure 20.6).

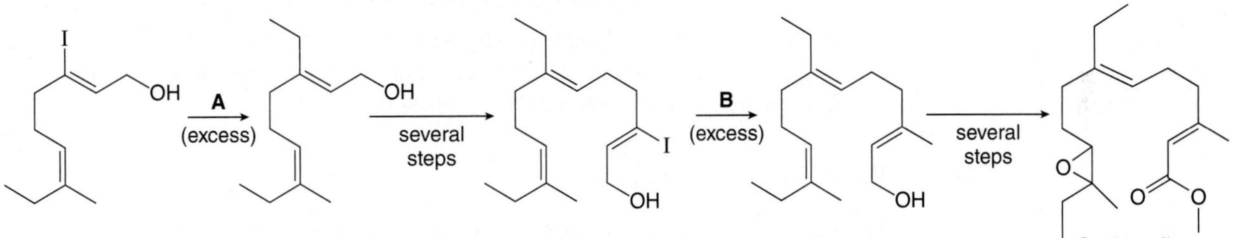
I
OH
A
(excess)
OH
several
steps
I
OH
B
(excess)
OH
several
steps
O
O
O

C_{18} juvenile hormone

29.1B Using Organocuprate Couplings to Synthesize Hydrocarbons

Since organocuprate reagents (R_2CuLi) are prepared in two steps from alkyl halides (RX), this method ultimately converts two organic halides (RX and R'X) into a hydrocarbon R–R' with a new carbon–carbon bond. A hydrocarbon can often be made by two different routes, as shown in Sample Problem 29.1.

$$R{-}X \xrightarrow[\text{(2 equiv)}]{\text{Li}} R{-}Li \xrightarrow[\text{(0.5 equiv)}]{\text{CuI}} R_2CuLi \xrightarrow{R'{-}X} R'{-}R$$

Two organic halides are needed as starting materials.

Sample Problem 29.1 Devise a synthesis of 1-methylcyclohexene from 1-bromocyclohexene and CH_3I.

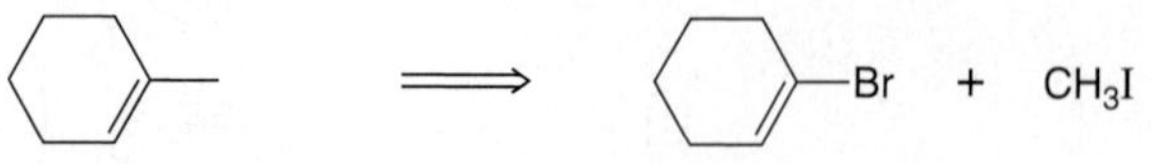
Br
CH3I

1-methylcyclohexene 1-bromocyclohexene

Solution

In this example, either halide can be used to form an organocuprate, which can then be coupled with the second halide.

Possibility [1]

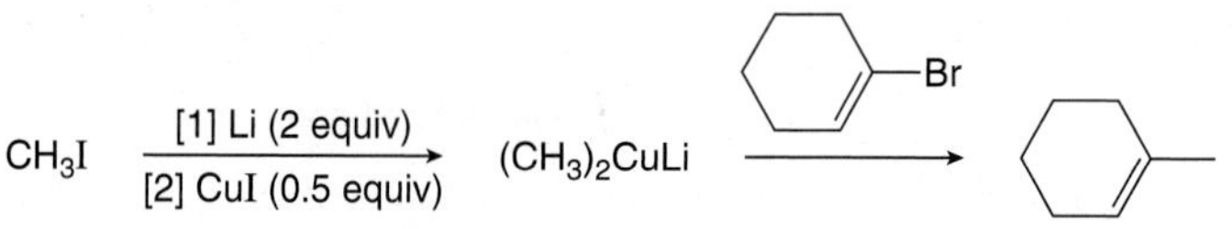
CH3I
[1] Li (2 equiv)
[2] CuI (0.5 equiv)
(CH3)2CuLi
Br

Possibility [2]

Br
[1] Li (2 equiv)
[2] CuI (0.5 equiv)
CuLi
2
CH3I

Problem 29.3 Synthesize each product from the given starting materials using an organocuprate coupling reaction.

a. 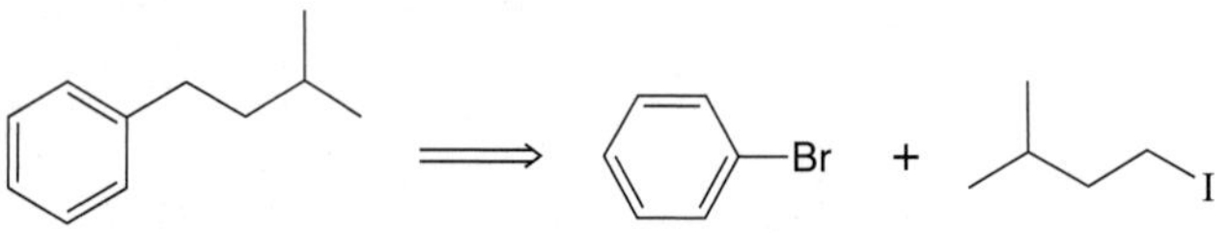
Br
+
I

b. 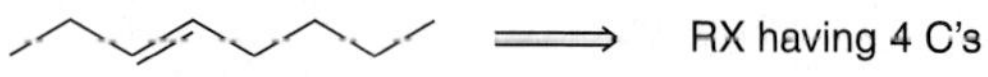
RX having 4 C's

c. 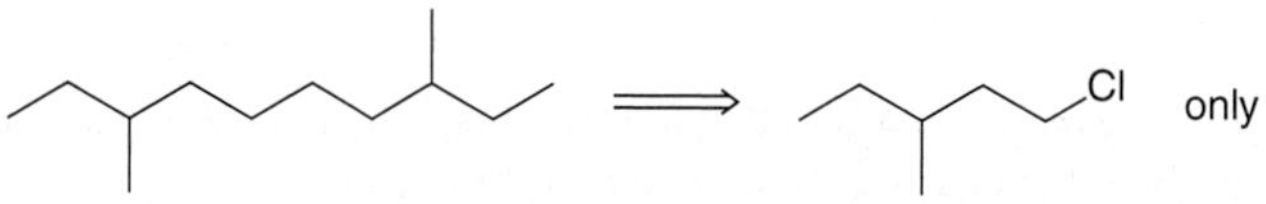
Cl
only

The mechanism of this reaction may vary with the identity of R' in R'−X. Since coupling occurs with organic halides having the halogen X on either an sp^3 or sp^2 hybridized carbon, an S_N2 mechanism cannot explain all the observed results.

29.2 Suzuki Reaction

The **Suzuki reaction** is the first of two reactions that utilize a palladium catalyst and proceed by way of an intermediate organopalladium compound. The second is the Heck reaction (Section 29.3).

29.2A General Features of Reactions with Pd Catalysts

Reactions with palladium compounds share many common features with reactions involving other transition metals. During a reaction, palladium is coordinated to a variety of groups called **ligands,** which donate electron density to (or sometimes withdraw electron density from) the metal. A common electron-donating ligand is a **phosphine,** such as triphenylphosphine, tri(*o*-tolyl)phosphine, or tricyclohexylphosphine.

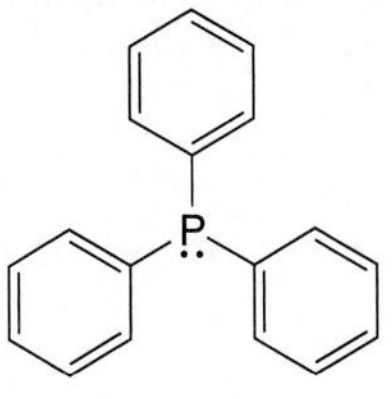

PPh_3
triphenylphosphine

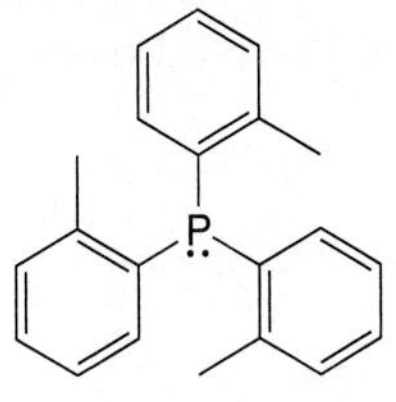

P(*o*-tolyl)$_3$
tri(*o*-tolyl)phosphine
abbreviated as **PAr_3**
Ar = an aryl group

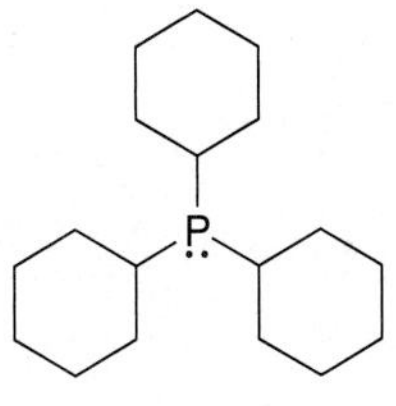

PCy_3
tricyclohexylphosphine

A general ligand bonded to a metal is often designated as **L.** Pd bonded to four ligands is denoted as PdL_4.

Ac is the abbreviation for an acetyl group, **$CH_3C{=}O$,** so OAc (or ^{-}OAc) is the abbreviation for acetate, **$CH_3CO_2^{-}$.**

Organopalladium compounds—compounds that contain a carbon–palladium bond—are generally prepared in situ during the course of a reaction, from another palladium reagent such as $Pd(OAc)_2$ or $Pd(PPh_3)_4$. In most useful reactions only a catalytic amount of palladium reagent is utilized.

Two common processes, called **oxidative addition** and **reductive elimination,** dominate many reactions of palladium compounds.

- *Oxidative addition* is the addition of a reagent (such as RX) to a metal, often increasing the number of groups around the metal by two.

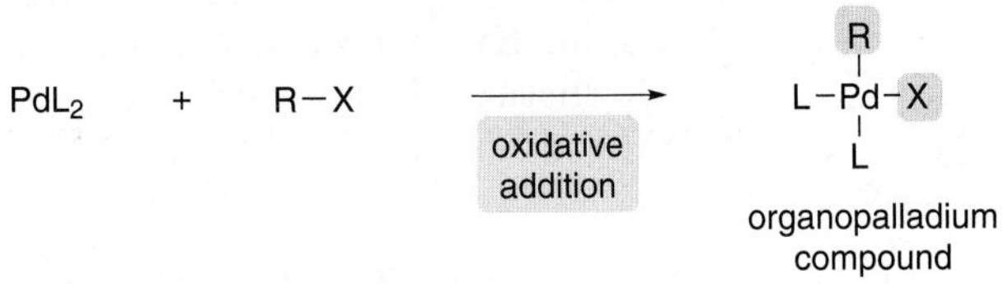
PdL2
+
R–X
oxidative addition
R
L–Pd–X
L
organopalladium compound

- *Reductive elimination* is the elimination of two groups that surround the metal, often forming new C–H or C–C bonds.

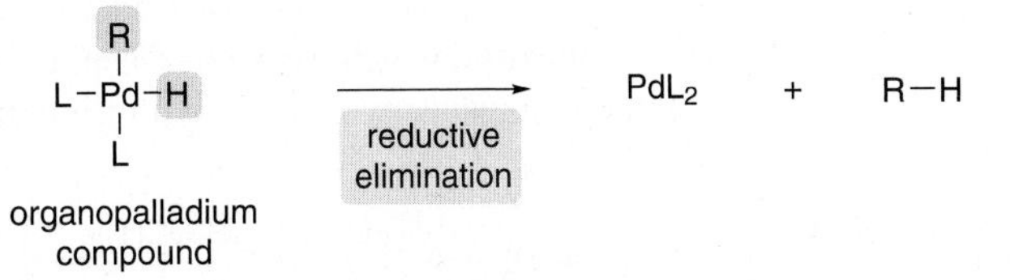
R
L–Pd–H
L
organopalladium compound
reductive elimination
PdL2
+
R–H

Reaction mechanisms with palladium compounds are often multistep. During the course of a reaction, the identity of some groups bonded to Pd will be known with certainty, while the identity of other ligands might not be known. Consequently, only the crucial reacting groups around a metal are usually drawn and the other ligands are not specified.

29.2B Details of the Suzuki Reaction

The **Suzuki reaction** is a palladium-catalyzed coupling of an organic halide (R'X) with an organoborane (RBY_2) to form a product (R–R') with a new C–C bond. $Pd(PPh_3)_4$ is the typical palladium catalyst, and the reaction is carried out in the presence of a base such as NaOH or $NaOCH_2CH_3$.

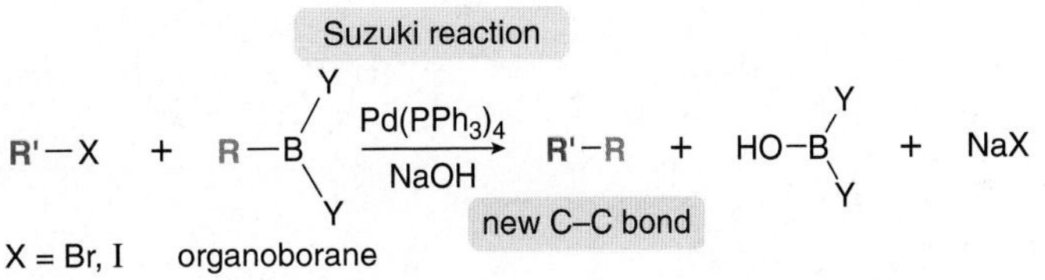
Suzuki reaction
R'–X
+
R–B
Y
Y
Pd(PPh3)4
NaOH
R'–R
+
HO–B
Y
Y
+
NaX
new C–C bond
X = Br, I
organoborane

Vinyl halides and aryl halides, both of which contain a halogen X bonded directly to an sp^2 hybridized carbon, are most often used, and the halogen is usually Br or I. The Suzuki reaction is completely **stereospecific,** as shown in Example [3]; a cis vinyl halide and a trans vinylborane form a *cis,trans*-1,3-diene.

[1]
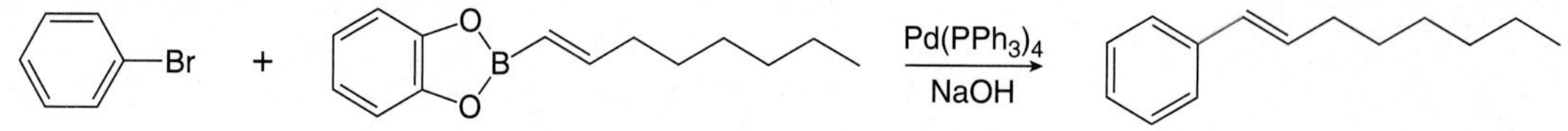
Br
+
O
B
O
Pd(PPh3)4
NaOH

[2]
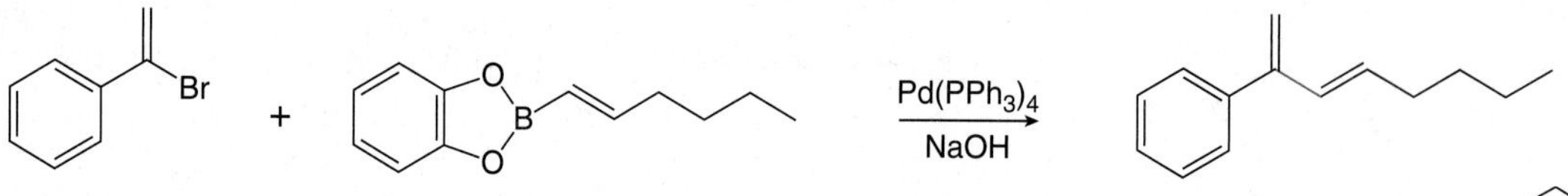
Br
+
O
B
O
Pd(PPh3)4
NaOH

[3]
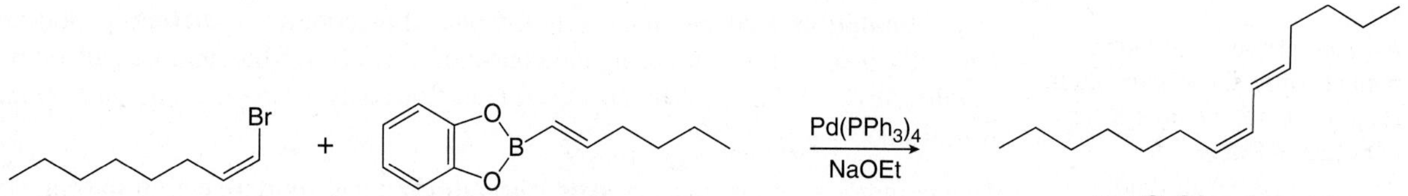
Br
+
O
B
O
Pd(PPh3)4
NaOEt

cis vinyl bromide — trans vinylborane — new C–C bonds in red

The organoboranes used in the Suzuki reaction are prepared from two sources.

- **Vinylboranes,** which have a boron atom bonded to a carbon–carbon double bond, are prepared by hydroboration of an alkyne using catecholborane, a commercially available reagent. **Hydroboration adds the elements of H and B in a syn fashion to form a trans vinylborane.** With terminal alkynes, hydroboration always places the boron atom on the *less substituted* terminal carbon.

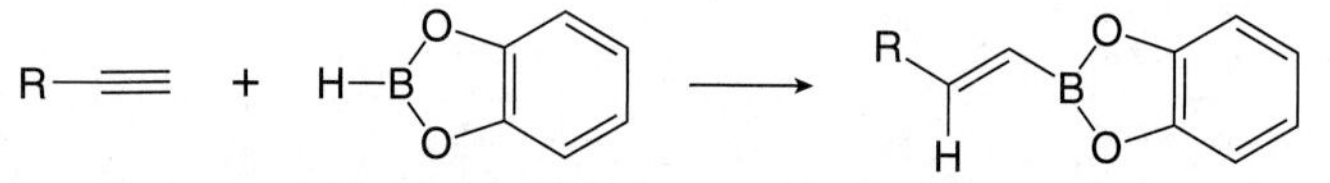
R
+
H–B
O
O
R
B
H
O
O

catecholborane — trans vinylborane

syn addition of H and B

- **Arylboranes,** which have a boron atom bonded to a benzene ring, are prepared from organolithium reagents by reaction with trimethyl borate [$B(OCH_3)_3$].

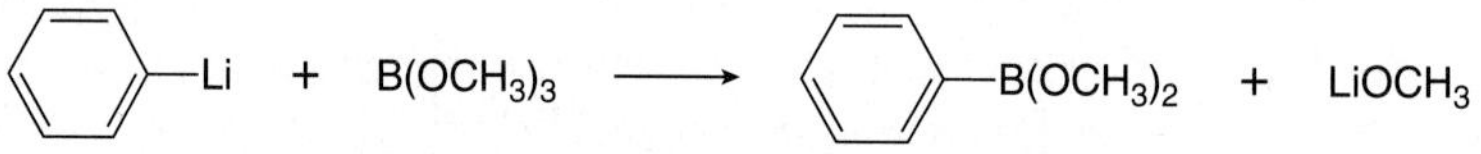
Li
+
B(OCH3)3
B(OCH3)2
+
LiOCH3

trimethyl borate — arylborane

Problem 29.4 Draw the product of each reaction.

a. 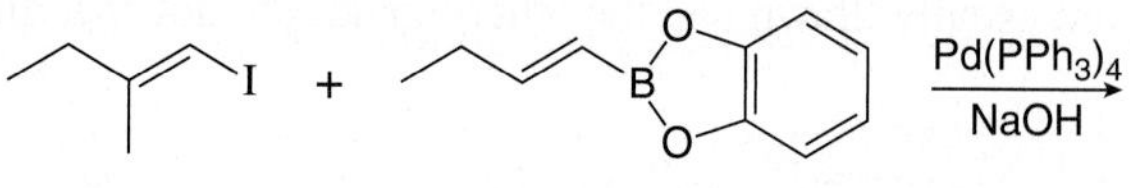
I
+
O
B
O
Pd(PPh3)4
NaOH

b. 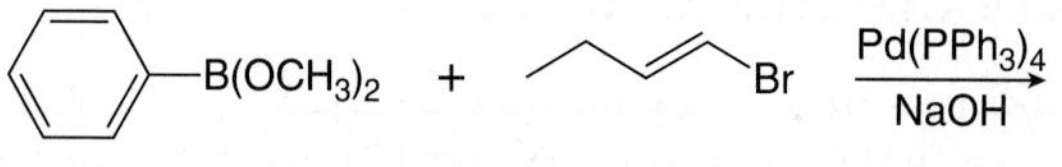
B(OCH3)2
+
Br
Pd(PPh3)4
NaOH

c. 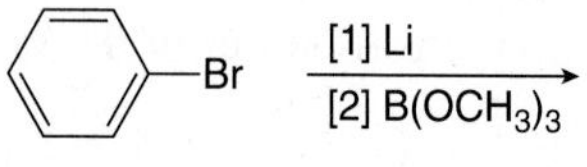
Br
[1] Li
[2] B(OCH3)3

d. 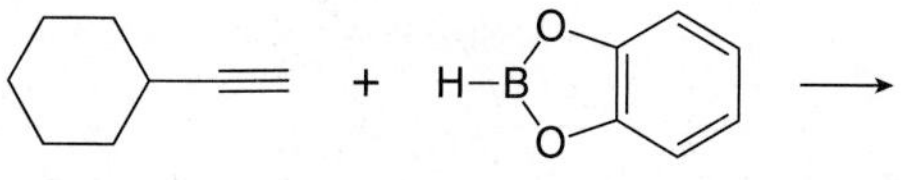
+
H–B
O
O

The structure of bombykol, the sex pheromone of the female silkworm moth *Bombyx mori,* was elucidated in 1959 using 6.4 mg of material obtained from 500,000 silkworm moths.

Problem 29.5 One step in the synthesis of the nonsteroidal anti-inflammatory drug rofecoxib (trade name Vioxx) involves Suzuki coupling of **A** and **B.** What product is formed in this reaction?

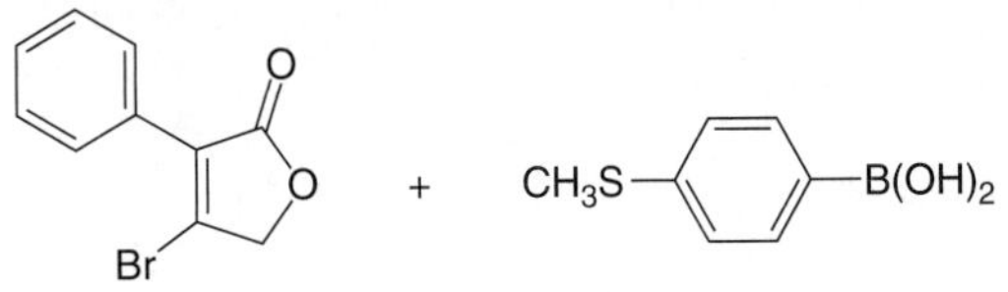

A **B**

The mechanism of the Suzuki reaction consists of oxidative addition of R'–X to the palladium catalyst, transfer of an alkyl group from the organoborane to palladium, and reductive elimination of R–R', forming a new carbon–carbon bond. A general halide R'–X and organoborane $R-BY_2$ are used to illustrate this process in Mechanism 29.1. Since the palladium reagent is regenerated during reductive elimination, only a catalytic amount of palladium is needed.

The Suzuki reaction was a key step in the synthesis of **bombykol,** the sex pheromone of the female silkworm moth, and **humulene,** a lipid isolated from hops, as shown in Figure 29.1. The synthesis of humulene illustrates that an intramolecular Suzuki reaction can form a ring. Sample Problem 29.2 shows how a conjugated diene can be prepared from an alkyne and vinyl halide using a Suzuki reaction.

Mechanism 29.1 Suzuki Reaction

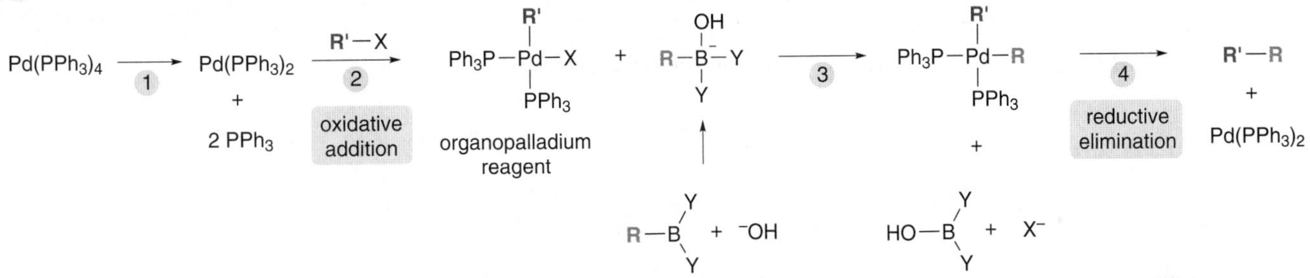

- **1 – 2** Loss of two triphenylphosphine ligands followed by oxidative addition of R'X forms an organopalladium reagent.
- **3** Reaction of the organoborane RBY_2 with ^{-}OH forms a nucleophilic boron intermediate that transfers an alkyl group from boron to palladium.
- **4** Reductive elimination of R'–R forms a new carbon–carbon bond, and the palladium catalyst $Pd(PPh_3)_2$ is regenerated.

Figure 29.1 Synthesis of two natural products using the Suzuki reaction

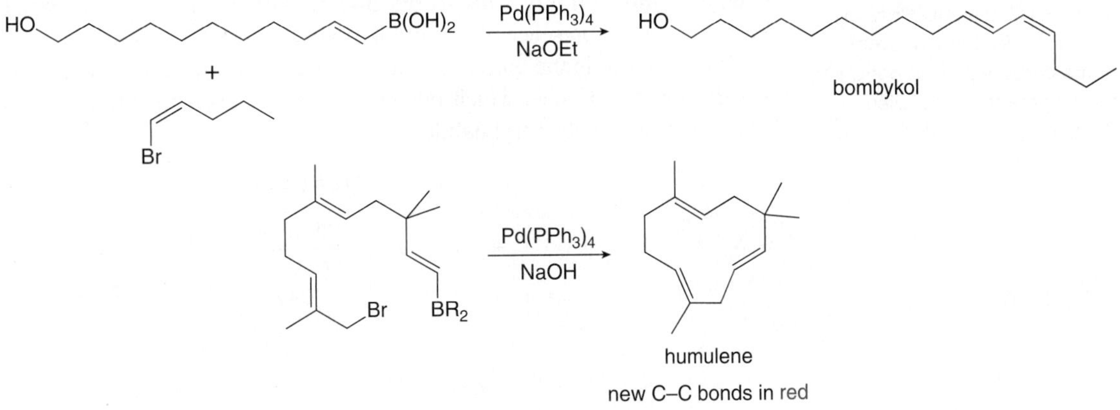

Sample Problem 29.2 Devise a synthesis of (1*Z*,3*E*)-1-phenylocta-1,3-diene from hex-1-yne and (*Z*)-2-bromostyrene using a Suzuki coupling.

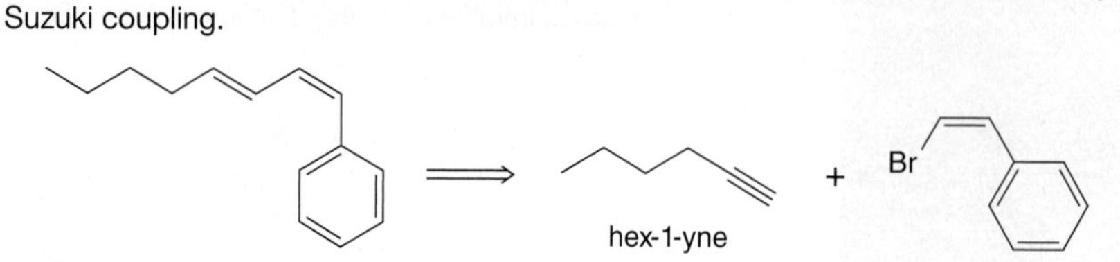
Br
hex-1-yne

(1*Z*,3*E*)-1-phenylocta-1,3-diene (*Z*)-2-bromostyrene

Solution

This synthesis can be accomplished in two steps. Hydroboration of hex-1-yne with catecholborane forms a vinylborane. Coupling of this vinylborane with (*Z*)-2-bromostyrene gives the desired 1,3-diene. The *E* configuration of the vinylborane and the *Z* configuration of the vinyl bromide are both retained in the product.

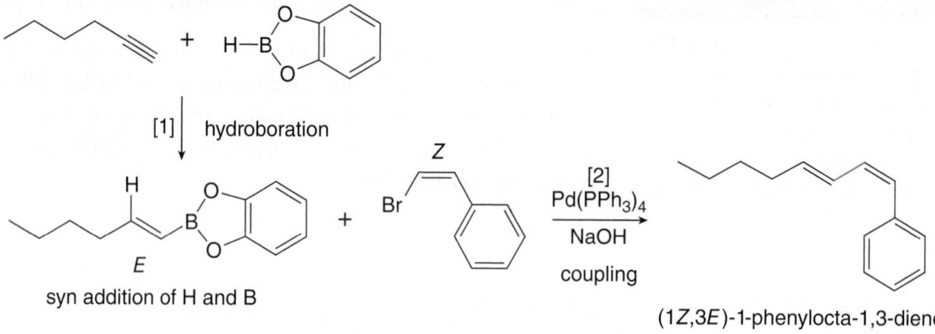
H–B
O
O
[1] hydroboration
H
B
E
syn addition of H and B
Z
Br
[2]
Pd(PPh3)4
NaOH
coupling

(1*Z*,3*E*)-1-phenylocta-1,3-diene

new C–C bond in red

Problem 29.6 Synthesize each compound from the given starting materials.

a. 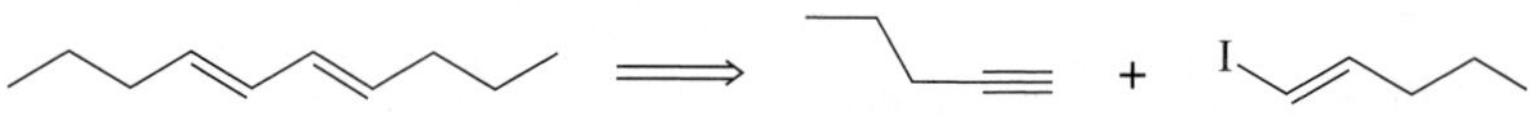
I

b. 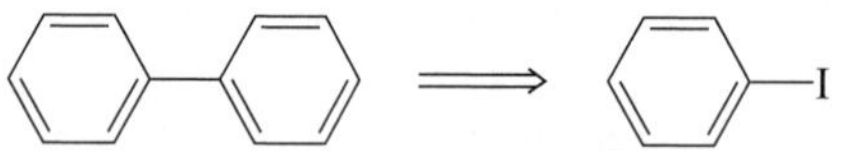
I

c. 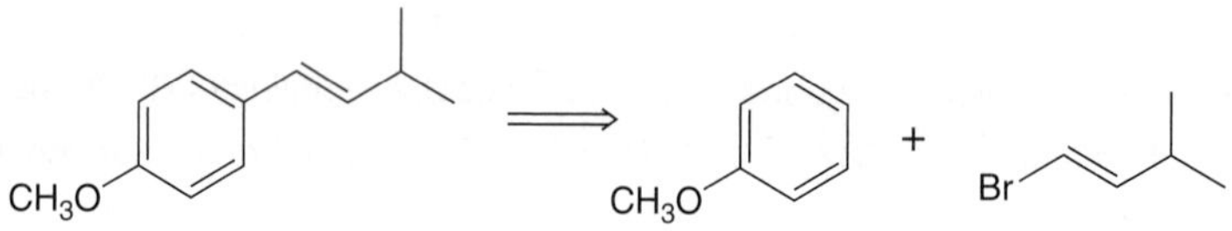
CH3O
CH3O
Br

29.3 Heck Reaction

Richard Heck and Akira Suzuki won the 2010 Nobel Prize in Chemistry for the discovery of the carbon–carbon bond-forming reactions detailed in Sections 29.2 and 29.3.

The **Heck reaction is a palladium-catalyzed coupling of a vinyl or aryl halide with an alkene to form a more highly substituted alkene with a new C–C bond.** Palladium(II) acetate [$Pd(OAc)_2$] in the presence of a triarylphosphine [$P(o\text{-tolyl})_3$] is the typical catalyst, and the reaction is carried out in the presence of a base such as triethylamine (Et_3N). The Heck reaction is a **substitution reaction** in which one H atom of the alkene starting material is replaced by the R' group of the vinyl or aryl halide.

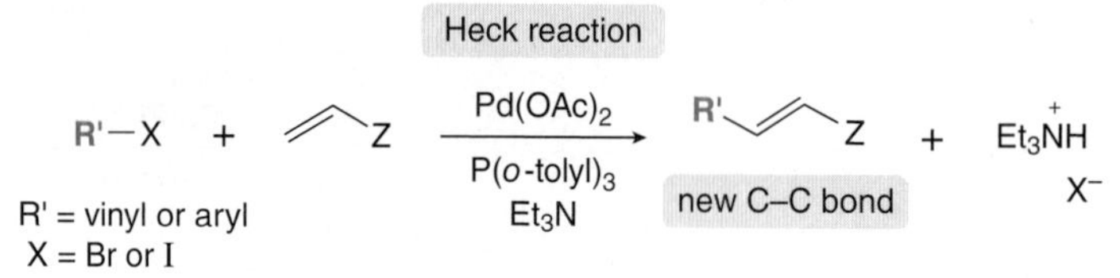
Heck reaction
R'–X
Z
Pd(OAc)2
P(o-tolyl)3
Et3N
R'
Z
new C–C bond
Et3NH+
X–
R' = vinyl or aryl
X = Br or I

The alkene component is typically ethylene or a monosubstituted alkene ($CH_2{=}CHZ$), and the halogen X is usually Br or I. When Z = Ph, COOR, or CN in a monosubstituted alkene, **the new C–C bond is formed on the *less* substituted carbon to afford a trans alkene.** When a vinyl halide is used as the organic halide, the reaction is **stereospecific,** as shown in Example [3]; the trans stereochemistry of the vinyl iodide is retained in the product.

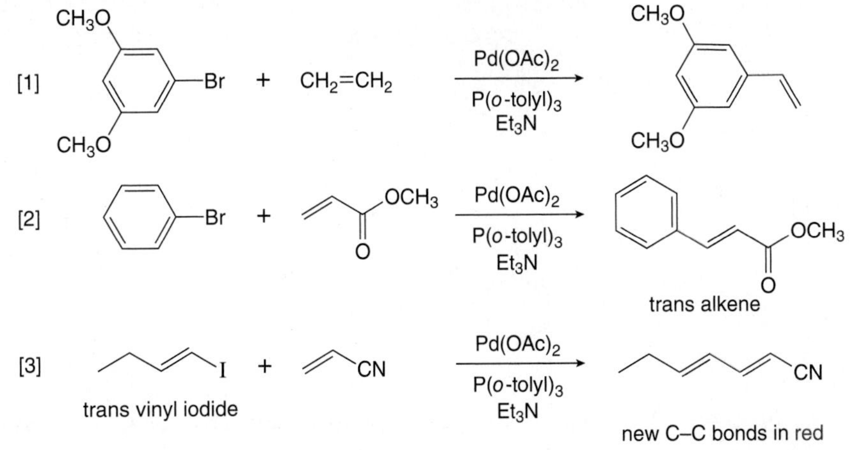

Problem 29.7 Draw the coupling product formed when each pair of compounds is treated with $Pd(OAc)_2$, $P(o\text{-tolyl})_3$, and Et_3N.

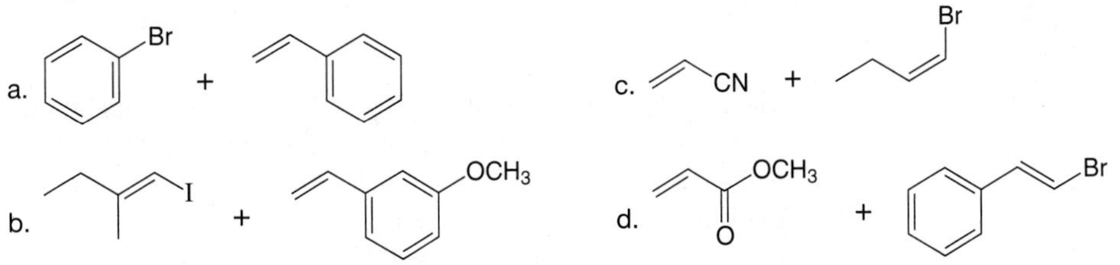

To use the Heck reaction in synthesis, you must determine what alkene and what organic halide are needed to prepare a given compound. To work backwards, locate the double bond with the aryl, COOR, or CN substituent, and break the molecule into two components at the end of the C=C not bonded to one of these substituents. Sample Problem 29.3 illustrates this retrosynthetic analysis.

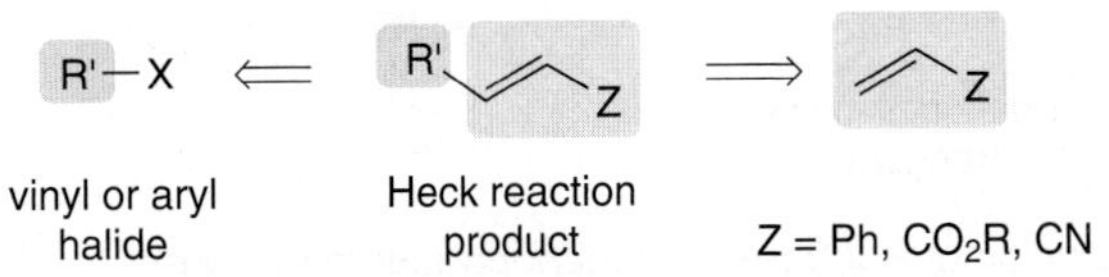

Sample Problem 29.3 What starting materials are needed to prepare each alkene using a Heck reaction?

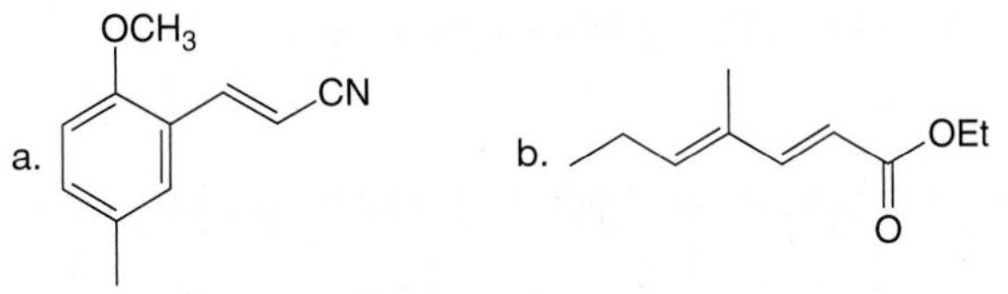

Solution

To prepare an alkene of general formula R'CH=CHZ by the Heck reaction, two starting materials are needed—an alkene (CH_2=CHZ) and a vinyl or aryl halide (R'X).

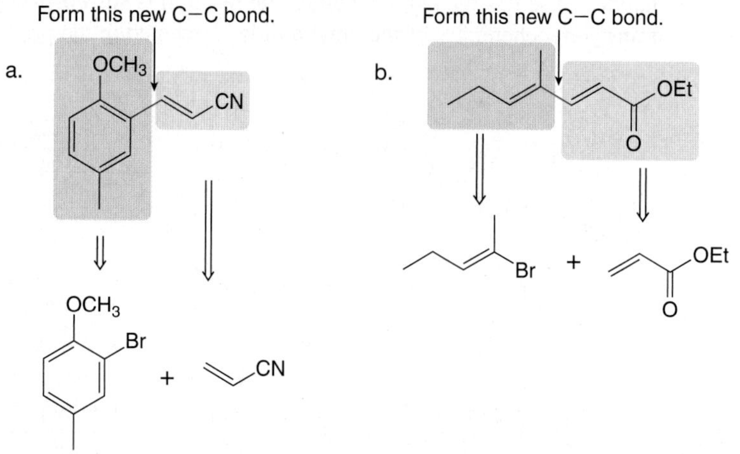

Problem 29.8 What starting materials are needed to prepare each compound using a Heck reaction?

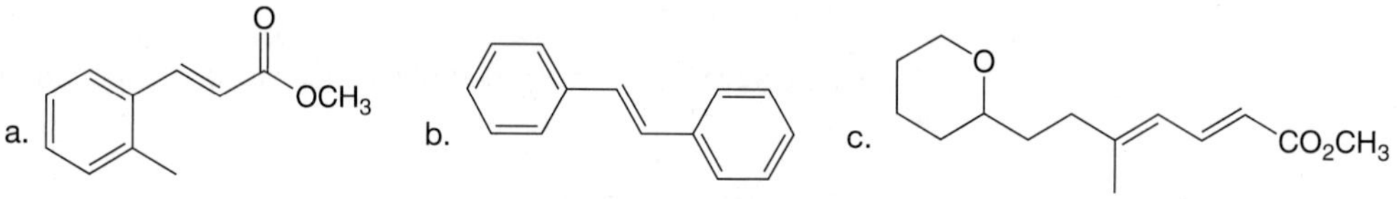

The actual palladium catalyst in the Heck reaction is thought to contain a palladium atom bonded to two tri(*o*-tolyl)phosphine ligands, abbreviated as $Pd(PAr_3)_2$. In this way it resembles the divalent palladium catalyst used in the Suzuki reaction. The mechanism of the Heck reaction consists of oxidative addition of the halide R'X to the palladium catalyst, addition of the resulting organopalladium reagent to the alkene, and two successive eliminations. A general organic halide R'X and alkene CH_2=CHZ are used to illustrate the process in Mechanism 29.2.

Mechanism 29.2 Heck Reaction

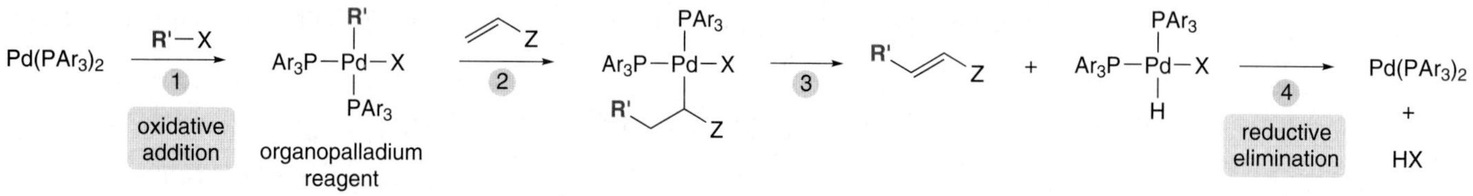

1. Oxidative addition of R'X forms an organopalladium reagent.
2. Addition of R' and Pd to the π bond of CH_2=CHZ places the Pd on the carbon with the Z substituent.
3. Elimination of H and Pd forms the π bond in the reaction product and transfers a hydrogen to Pd.
4. Reductive elimination of HX regenerates the palladium catalyst $Pd(PAr_3)_2$.

29.4 Carbenes and Cyclopropane Synthesis

Another method of carbon–carbon bond formation involves the conversion of alkenes to cyclopropane rings using **carbene** intermediates.

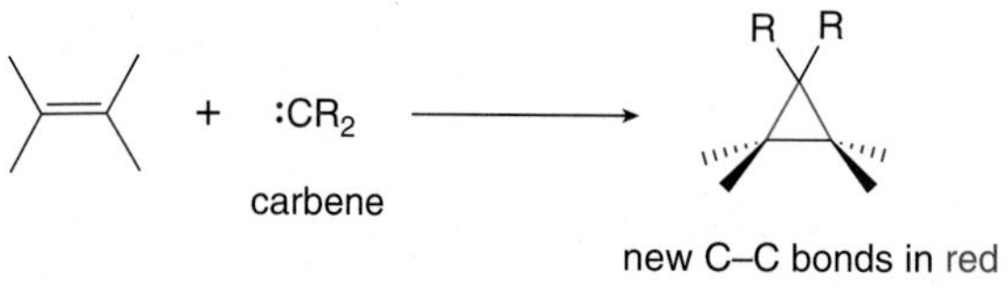

Pyrethrin I and **decamethrin** both contain cyclopropane rings. Pyrethrin I is a naturally occurring biodegradable insecticide obtained from chrysanthemums, whereas **decamethrin** is a more potent synthetic analogue that is widely used as an insecticide in agriculture.

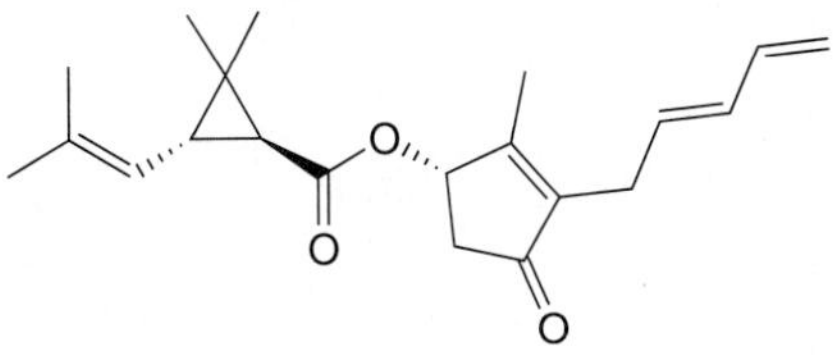

pyrethrin I

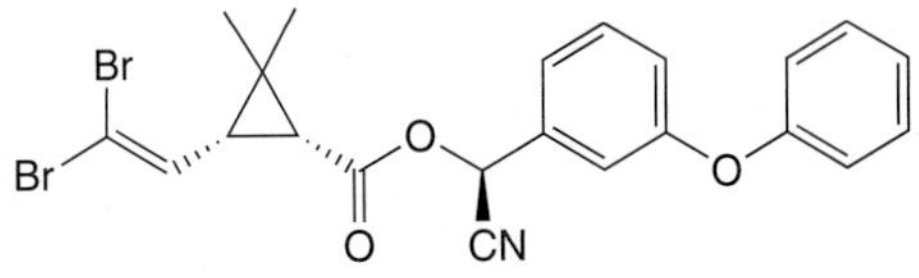

decamethrin

29.4A Carbenes

A *carbene*, R_2C:, is a neutral reactive intermediate that contains a divalent carbon surrounded by six electrons—the lone pair and two each from the two R groups. These three groups make the carbene carbon ***sp*^{2} hybridized,** with a vacant *p* orbital extending above and below the plane containing the C and the two R groups. The lone pair of electrons occupies an *sp*^{2} hybrid orbital.

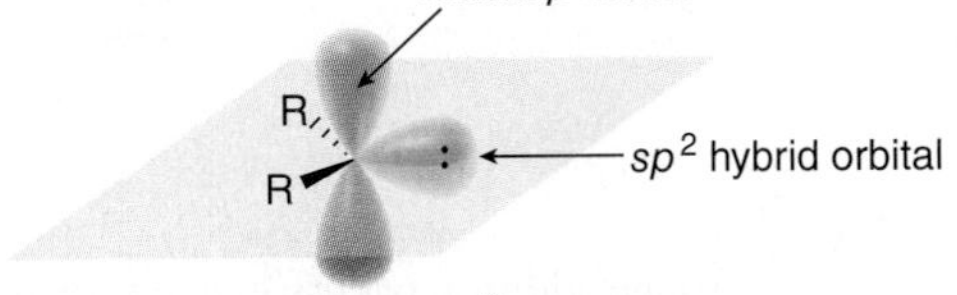

The carbene carbon is *sp*^{2} hybridized.

Carbenes share two features in common with carbocations and carbon radicals.

- **A carbene is highly reactive because carbon does not have an octet of electrons.**
- **A carbene is electron deficient, so it behaves as an electrophile.**

29.4B Preparation and Reactions of Dihalocarbenes

Dihalocarbenes, $:CX_2$, are especially useful reactive intermediates because they are readily prepared from trihalomethanes (CHX_3) by reaction with a strong base. Treatment of chloroform, $CHCl_3$, with $KOC(CH_3)_3$ forms dichlorocarbene, $:CCl_2$.

$$\underset{\text{chloroform}}{CHCl_3} \xrightarrow{KOC(CH_3)_3} \underset{\text{dichlorocarbene}}{:CCl_2} + (CH_3)_3COH + KCl$$

Dichlorocarbene is formed by a two-step process that results in the elimination of the elements of H and Cl from the *same* carbon, as shown in Mechanism 29.3. Loss of two elements from the same carbon is called **α elimination,** to distinguish it from the β eliminations discussed in Chapter 8, in which two elements are lost from *adjacent* carbons.

Mechanism 29.3 Formation of Dichlorocarbene

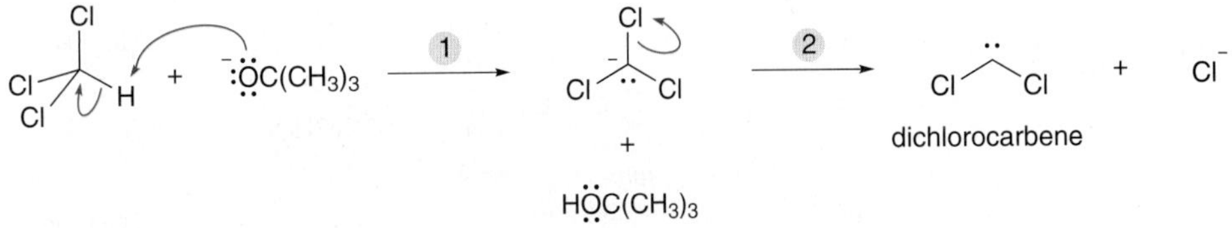

1 Three electronegative Cl atoms acidify the C–H of $CHCl_3$, so it can be removed by strong base to form a carbanion.

2 Elimination of Cl^- forms the carbene.

Since dihalocarbenes are electrophiles, they readily react with double bonds to afford cyclopropanes, forming two new carbon–carbon bonds.

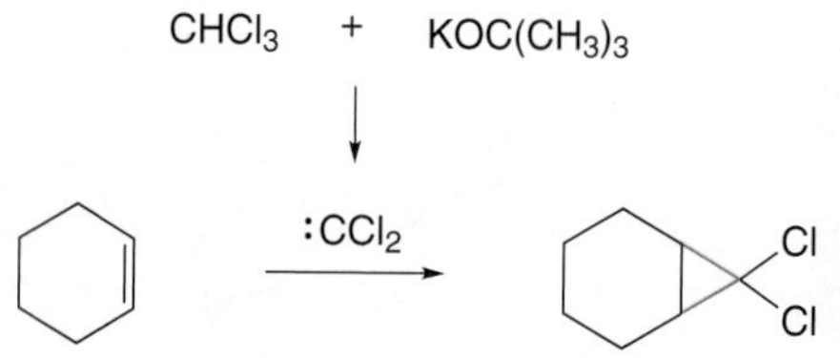

new C–C bonds in red

Cyclopropanation is a concerted reaction, so both C–C bonds are formed in a single step, as shown in Mechanism 29.4.

Mechanism 29.4 Addition of Dichlorocarbene to an Alkene

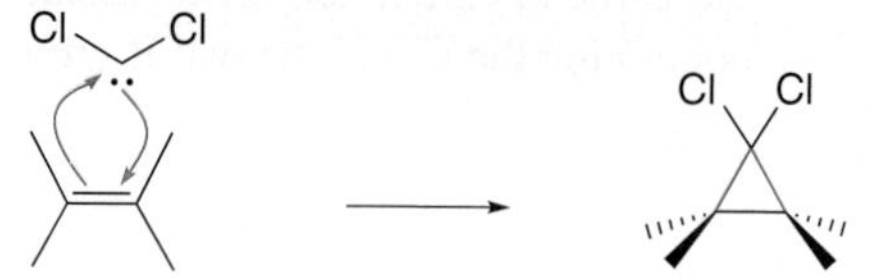

Carbene addition occurs in a **syn** fashion from either side of the planar double bond. The relative position of substituents in the alkene reactant is retained in the cyclopropane product. **Carbene addition is thus a stereospecific reaction,** because cis and trans alkenes yield different stereoisomers as products, as illustrated in Sample Problem 29.4.

Sample Problem 29.4 Draw the products formed when *cis*- and *trans*-but-2-ene are treated with $CHCl_3$ and $KOC(CH_3)_3$.

Solution

To draw each product, add the carbene carbon from either side of the alkene, and keep all substituents in their original orientations. The **cis** methyl groups in *cis*-but-2-ene become **cis** substituents in the cyclopropane. Addition from either side of the alkene yields the same compound—**an achiral meso compound that contains two stereogenic centers**—labeled in blue.

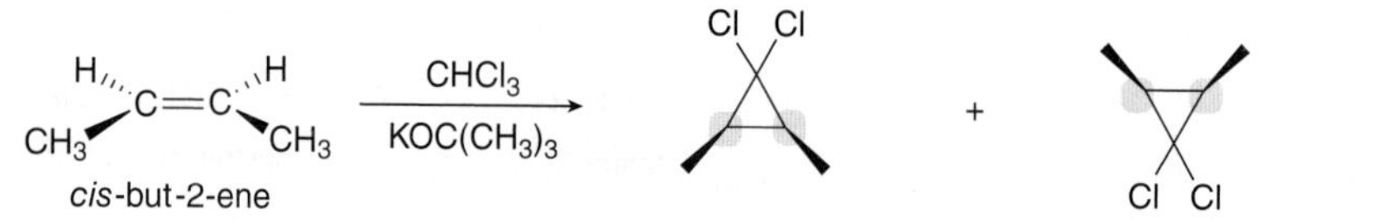

cis-but-2-ene

cis CH_3 groups
$:CCl_2$ added from above

cis CH_3 groups
$:CCl_2$ added from below

Products are identical, an achiral meso compound.

The **trans** methyl groups in *trans*-but-2-ene become **trans** substituents in the cyclopropane. Addition from either side of the alkene yields an equal amount of two enantiomers—**a racemic mixture.**

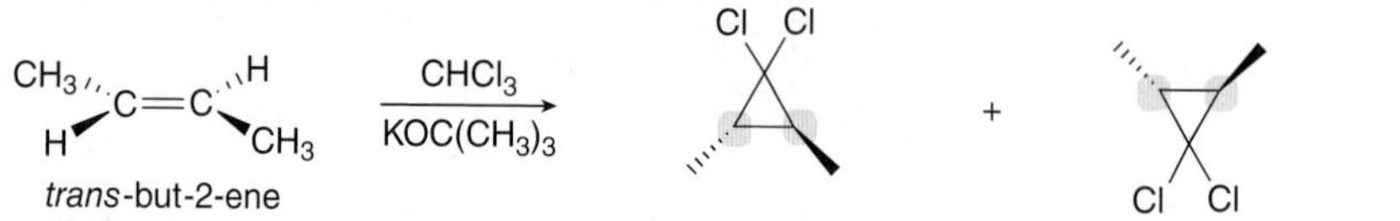

trans-but-2-ene

trans CH_3 groups
$:CCl_2$ added from above

trans CH_3 groups
$:CCl_2$ added from below

Products are enantiomers.

Problem 29.9 Draw all stereoisomers formed when each alkene is treated with $CHCl_3$ and $KOC(CH_3)_3$.

a. 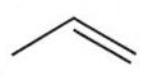 b.  c. 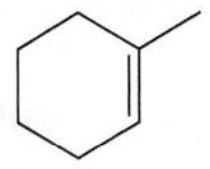

Finally, dihalo cyclopropanes can be converted to dialkyl cyclopropanes by reaction with organocuprates (Section 29.1). For example, cyclohexene can be converted to a bicyclic product having four new C–C bonds by the following two-step sequence: cyclopropanation with dibromocarbene (:CBr_2) and reaction with lithium dimethylcuprate, $LiCu(CH_3)_2$.

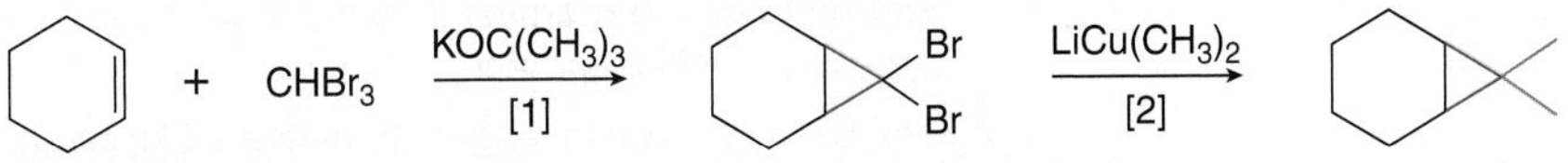

two new C–C bonds in red

two new C–C bonds in blue

Problem 29.10 What reagents are needed to convert 2-methylpropene [$(CH_3)_2C{=}CH_2$] to each compound? More than one step may be required.

a. 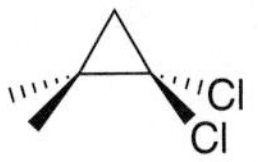

b. 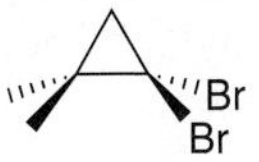

c. 

29.5 Simmons–Smith Reaction

Although the reaction of dihalocarbenes with alkenes gives good yields of halogenated cyclopropanes, this is not usually the case with **methylene, :CH_2,** the simplest carbene. Methylene is readily formed by heating diazomethane, CH_2N_2, which decomposes and loses N_2, but the reaction of :CH_2 with alkenes often affords a complex mixture of products. Thus, this reaction cannot be reliably used for cyclopropane synthesis.

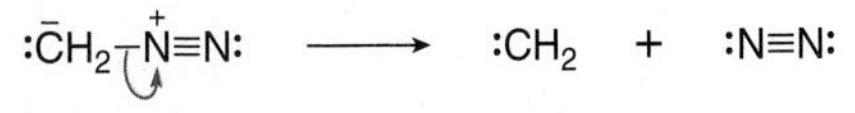

diazomethane methylene

Nonhalogenated cyclopropanes can be prepared by the reaction of an alkene with diiodomethane, CH_2I_2, in the presence of a copper-activated zinc reagent called zinc–copper couple [Zn(Cu)]. This process, the **Simmons–Smith reaction,** is named for H. E. Simmons and R. D. Smith, DuPont chemists who discovered the reaction in 1959.

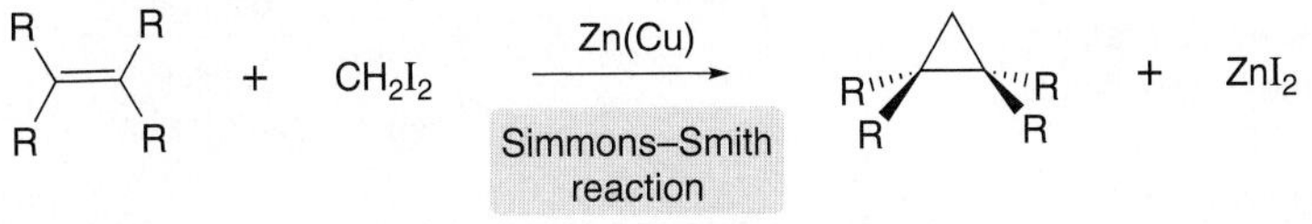

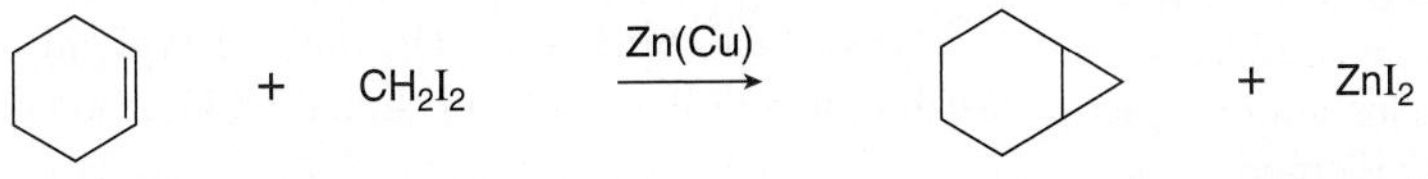

The Simmons–Smith reaction does not involve a free carbene. Rather, the reaction of CH_2I_2 with Zn(Cu) forms (iodomethyl)zinc iodide, which transfers a CH_2 group to an alkene, as shown in Mechanism 29.5.

Mechanism 29.5 Simmons–Smith Reaction

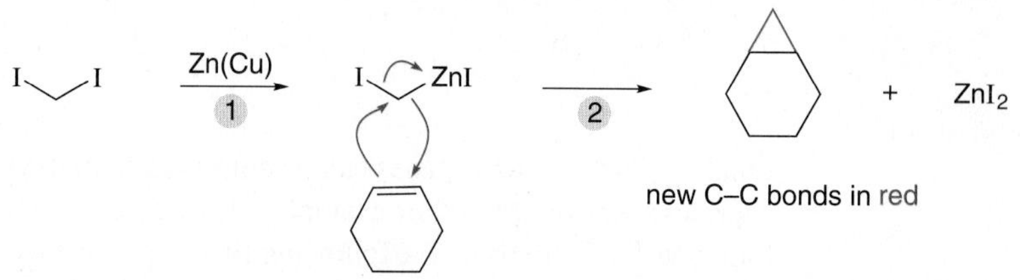

1 Reaction of CH_2I_2 with zinc–copper couple forms ICH_2ZnI [(iodomethyl)zinc iodide], the **Simmons–Smith reagent.** This intermediate is called a *carbenoid,* because the CH_2 does not exist as a free carbene.

2 The Simmons–Smith reagent transfers a CH_2 to an alkene, forming two new C–C bonds.

The Simmons–Smith reaction is stereospecific. The relative position of substituents in the alkene reactant is *retained* in the cyclopropane product, as shown for the conversion of *cis*-hex-3-ene to *cis*-1,2-diethylcyclopropane.

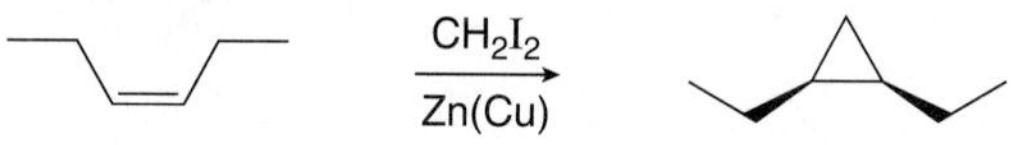

cis-hex-3-ene *cis*-1,2-diethylcyclopropane

Problem 29.11 What product is formed when each alkene is treated with CH_2I_2 and Zn(Cu)?

a. 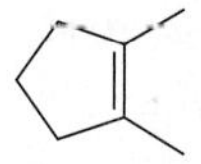 b.  c. 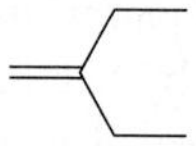

Problem 29.12 What stereoisomers are formed when *trans*-hex-3-ene is treated with CH_2I_2 and Zn(Cu)?

29.6 Metathesis

Recall from Section 10.1 that **olefin** is another name for an **alkene.**

The word *metathesis* is derived from the Greek words *meta* (change) and *thesis* (position). The 2005 Nobel Prize in Chemistry was awarded to Robert Grubbs of the California Institute of Technology, Yves Chauvin of the Institut Français du Pétrole, and Richard Schrock of the Massachusetts Institute of Technology for their work on olefin metathesis.

Alkene metathesis, more commonly called **olefin metathesis,** is a reaction between two alkene molecules that results in the interchange of the carbons of their double bonds. Two σ and two π bonds are broken and two new σ and two new π bonds are formed.

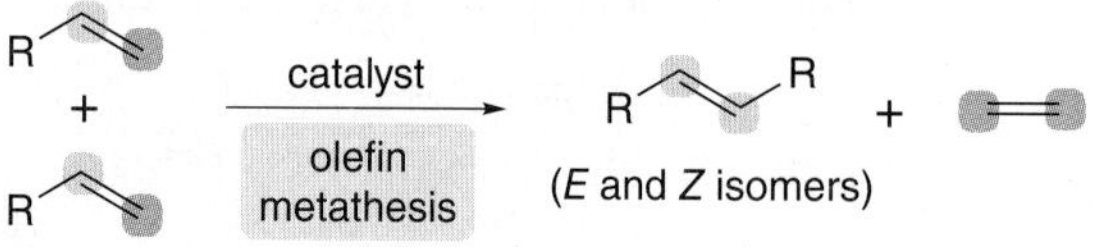

Olefin metathesis occurs in the presence of a complex transition metal catalyst that contains a carbon–metal double bond. The metal is typically ruthenium (Ru), tungsten (W), or molybdenum (Mo). In a widely used catalyst, called **Grubbs catalyst,** the metal is Ru.

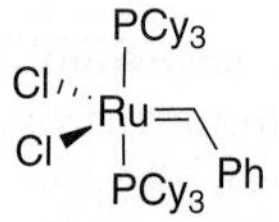

Grubbs catalyst

Olefin metathesis is an equilibrium process and, with many alkene substrates, a mixture of starting material and two or more alkene products is present at equilibrium, making the reaction useless for preparative purposes. With terminal alkenes, however, one metathesis product is $CH_2{=}CH_2$ (a gas), which escapes from the reaction mixture and drives the equilibrium to the right. As a result, monosubstituted alkenes ($RCH{=}CH_2$) and 2,2-disubstituted alkenes ($R_2C{=}CH_2$) are excellent metathesis substrates because high yields of a single alkene product are obtained, as shown in Equations [1] and [2].

[1] 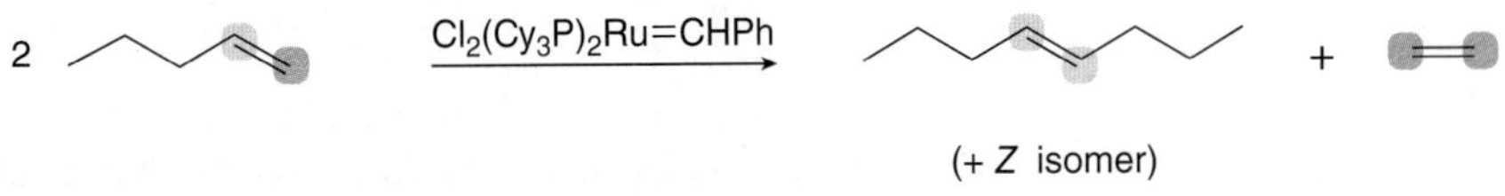
2
$Cl_2(Cy_3P)_2Ru{=}CHPh$
+
(+ *Z* isomer)

[2] 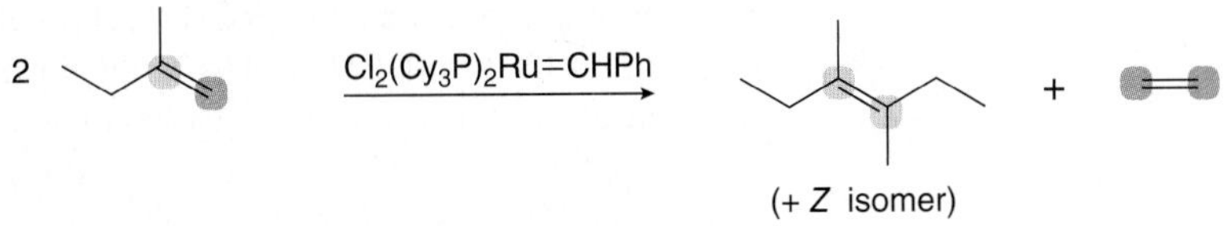
2
$Cl_2(Cy_3P)_2Ru{=}CHPh$
+
(+ *Z* isomer)

To draw the products of any metathesis reaction:

[1] Arrange two molecules of the starting alkene adjacent to each other as in Figure 29.2 where styrene ($PhCH{=}CH_2$) is used as the starting material.

[2] Then, break the double bonds in the starting material and form two new double bonds using carbon atoms that were *not* previously bonded to each other in the starting alkenes.

There are always two ways to arrange the starting alkenes (Pathways [1] and [2] in Figure 29.2). In this example, the two products of the reaction, $PhCH{=}CHPh$ and $CH_2{=}CH_2$, are formed in the first reaction pathway (Pathway [1]), while starting material is re-formed in the second pathway (Pathway [2]). Whenever the starting alkene is regenerated, it can go on to form product when the catalytic cycle is repeated.

Problem 29.13 Draw the products formed when each alkene is treated with Grubbs catalyst.

a. 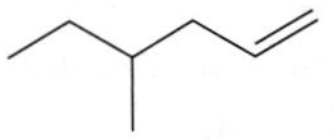

b. 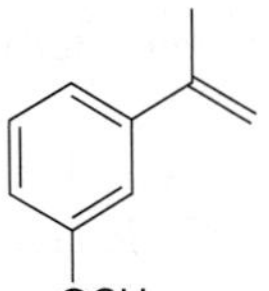
OCH_3

c. 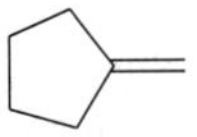

Figure 29.2 Drawing the products of olefin metathesis using styrene ($PhCH{=}CH_2$) as starting material

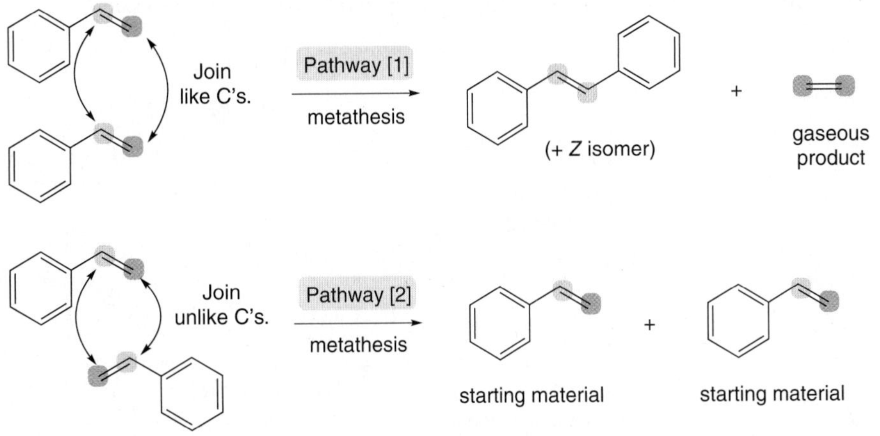
Join
like C's.
Pathway [1]
metathesis
+
(+ *Z* isomer)
gaseous
product
Join
unlike C's.
Pathway [2]
metathesis
+
starting material
starting material

- Overall reaction: $2\ PhCH{=}CH_2 \rightarrow PhCH{=}CHPh + CH_2{=}CH_2$.
- There are always two ways to join the C's of a single alkene together to form metathesis products (Pathways [1] and [2]).
- When like C's of the alkene substrate are joined in the first reaction (Pathway [1]), $PhCH{=}CHPh$ (in a cis and trans mixture) and $CH_2{=}CH_2$ are formed. Because $CH_2{=}CH_2$, escapes as a gas from the reaction mixture, only $PhCH{=}CHPh$ is isolated as product.
- When unlike C's of $PhCH{=}CH_2$ are joined in the second reaction (Pathway [2]), starting material is formed, which can re-enter the catalytic cycle to form product by the first pathway.
- In this way, a single constitutional isomer, $PhCH{=}CHPh$, is isolated.

Problem 29.14 What products are formed when *cis*-pent-2-ene undergoes metathesis? Use this reaction to explain why metathesis of a 1,2-disubstituted alkene (RCH=CHR') is generally not a practical method for alkene synthesis.

The mechanism for olefin metathesis is complex and involves **metal–carbene intermediates—intermediates that contain a metal–carbon double bond.** The mechanism is drawn for the reaction of a terminal alkene ($RCH{=}CH_2$) with Grubbs catalyst, abbreviated as **Ru=CHPh,** to form RCH=CHR and $CH_2{=}CH_2$. To begin metathesis, Grubbs catalyst reacts with the alkene substrate to form two new metal–carbenes **A** and **B** by a two-step process: addition of Ru=CHPh to the alkene to yield two different metallocyclobutanes (Step [1]), followed by elimination to form **A** and **B** (Steps [2a] and [2b]). The alkene by-products formed in this process (RCH=CHPh and $PhCH{=}CH_2$) are present in only a small amount since Grubbs reagent is used catalytically.

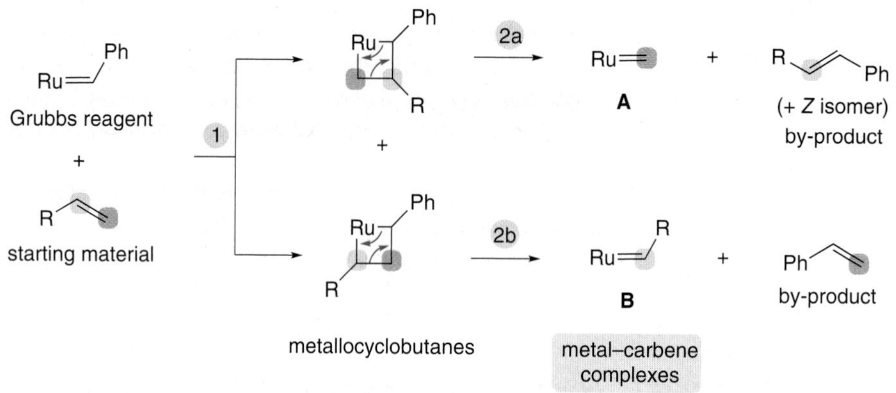

Each of these metal–carbene intermediates **A** and **B** then reacts with more starting alkene to form metathesis products, as shown in Mechanism 29.6. This mechanism is often written in a circle to emphasize the catalytic cycle. The mechanism demonstrates how two molecules of $RCH{=}CH_2$ are converted to RCH=CHR and $CH_2{=}CH_2$. The mechanism can be written beginning with reagent **A** or **B,** and all steps are equilibria.

Mechanism 29.6 Olefin Metathesis: 2 $RCH{=}CH_2 \rightarrow RCH{=}CHR + CH_2{=}CH_2$

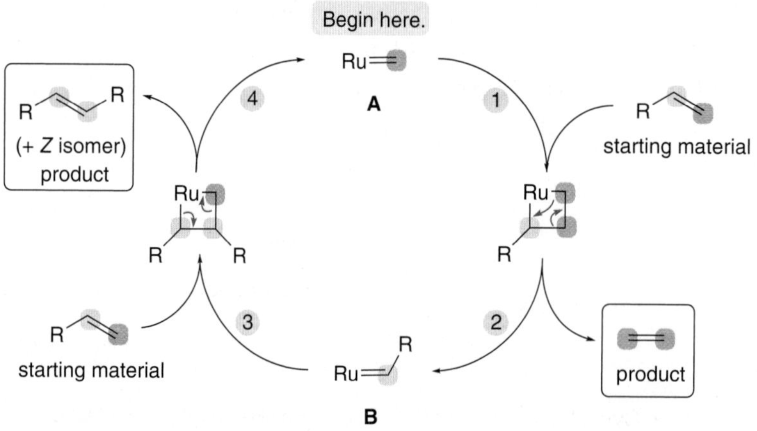

1 Reaction of $Ru{=}CH_2$ (**A**) with $RCH{=}CH_2$ forms a metallocyclobutane. Ru can bond to either the more or less substituted end of the alkene, but product is formed only when Ru bonds to the *more* substituted end, as shown.

2 Elimination forms one metathesis product, $CH_2{=}CH_2$, and metal–carbene complex **B.**

3 Reaction of **B** with $RCH{=}CH_2$ forms a metallocyclobutane. Ru can bond to either the more or less substituted end of the alkene, but product is formed only when Ru bonds to the *less* substituted end, as shown.

4 Elimination forms the other metathesis product, RCH=CHR, and metal–carbene complex **A.** The catalyst is regenerated and the cycle begins again.

A metathesis reaction that forms a ring is called **ring-closing metathesis (RCM).**

When a diene is used as starting material, ring closure occurs.

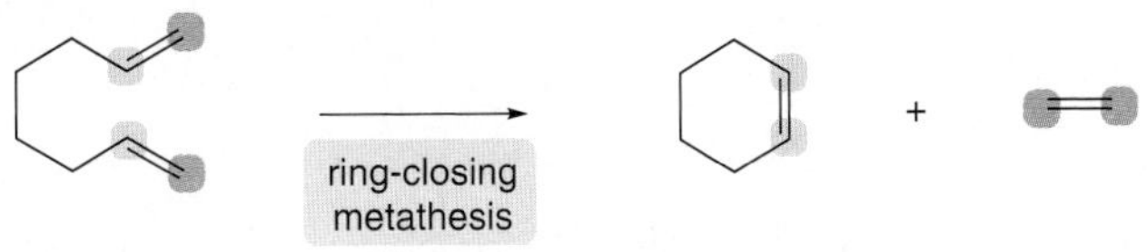

These reactions are typically run in very dilute solution, so that the two reactive ends of the *same* molecule have a higher probability of finding each other for reaction than two functional groups in *different* molecules. These high-dilution conditions thus favor ***intra*molecular** rather than *inter*molecular metathesis. Two examples are shown.

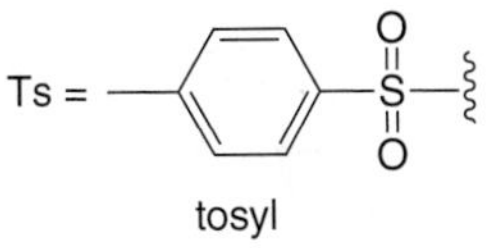

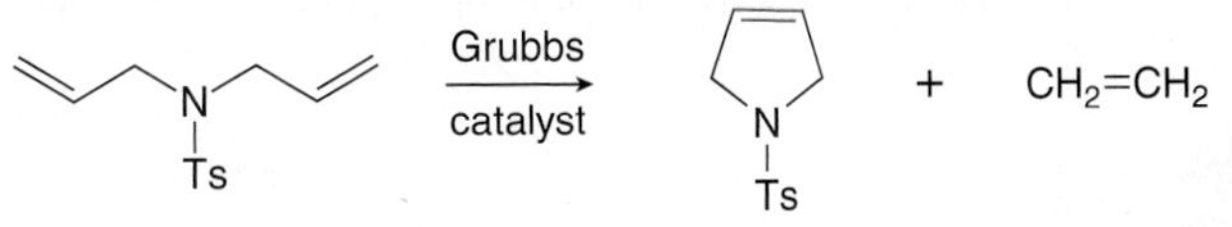

new C=C in red

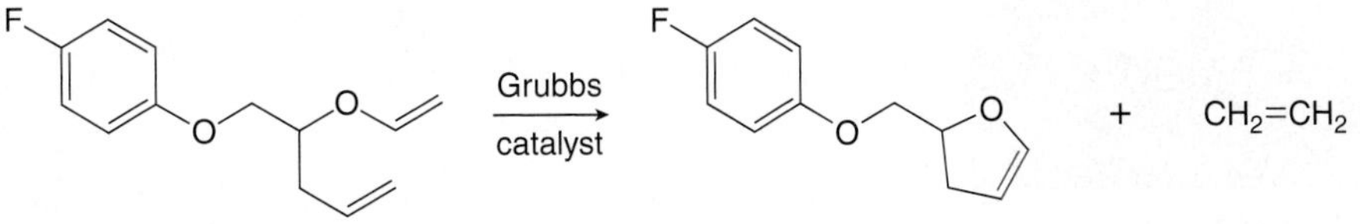

new C=C in red

Because metathesis catalysts are compatible with the presence of many functional groups (such as OH, OR, and C=O) and because virtually any ring size can be prepared, metathesis has been used to prepare many complex natural products (Figure 29.3).

Figure 29.3
Ring-closing metathesis in the synthesis of epothilone A and Sch38516

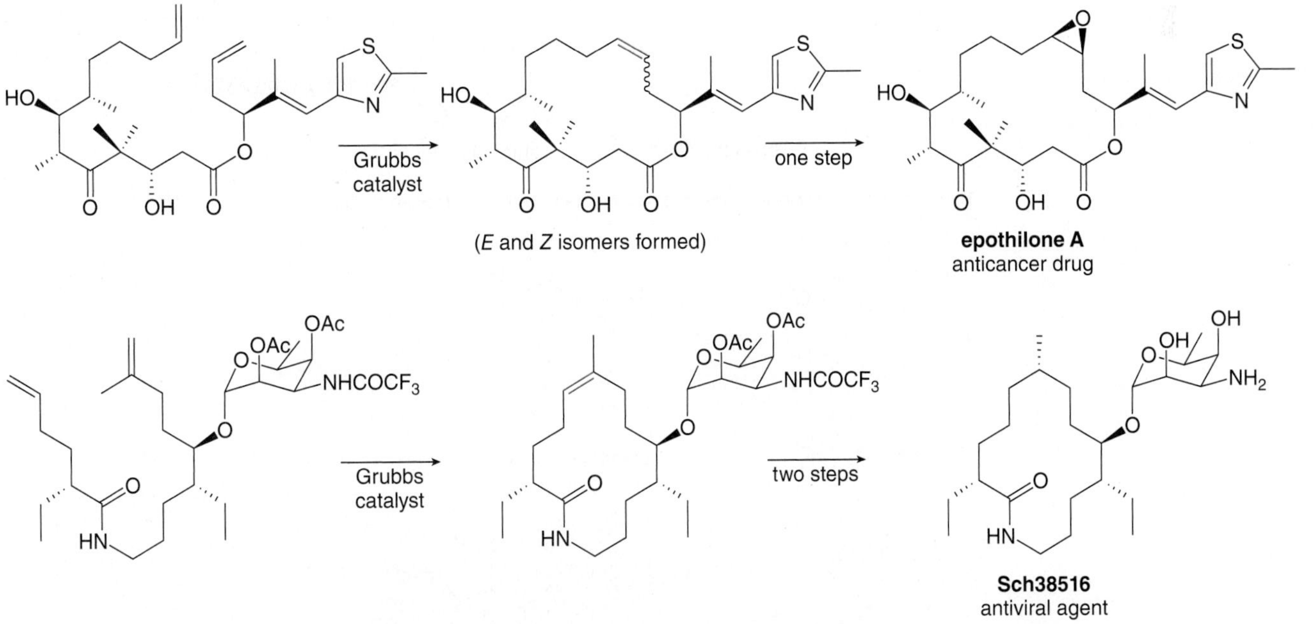

- **Epothilone A,** a promising anticancer agent, was first isolated from soil bacteria collected from the banks of the Zambezi River in South Africa.
- **Sch38516** is an antiviral agent active against influenza A.
- The new C–C bonds formed during metathesis are indicated in red. In both metathesis reactions, $CH_2=CH_2$ is also formed.

Problem 29.15 Draw the product formed from ring-closing metathesis of each compound.

a. 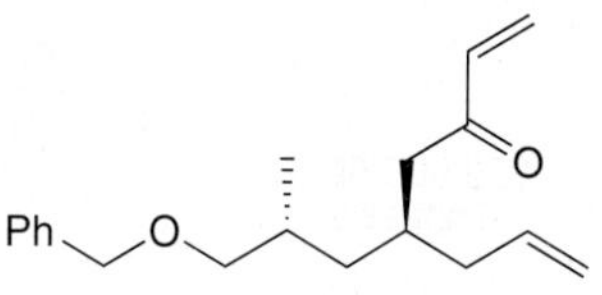
Ph
O
O

b. 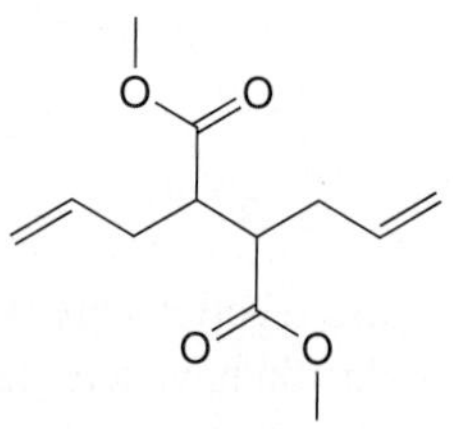
O
O
O
O

Problem 29.16 What product is formed by ring-closing metathesis of compound **V,** a key intermediate in the synthesis of ingenol, a natural product mentioned in the chapter opener?

Ingenol is isolated from the milky liquid obtained from *Euphorbia ingens,* a large cactus commonly called the candelabra tree, which is native to dry areas in southern Africa.

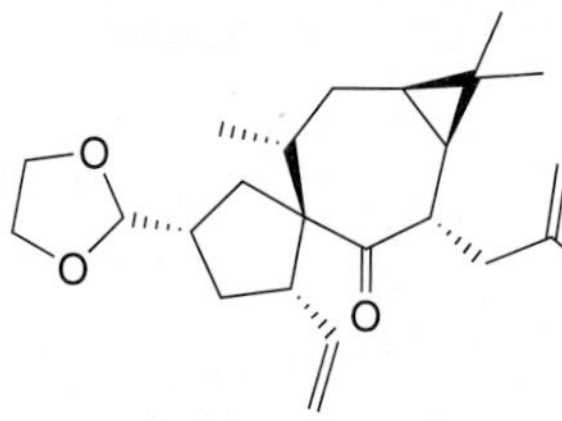
O
O
O
OR

V

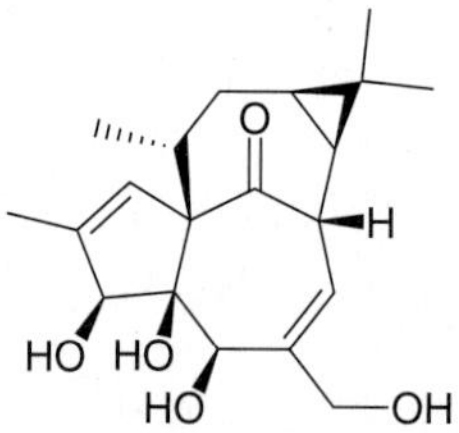
O
H
HO
HO
HO
OH

ingenol

Sample Problem 29.5 What starting material is needed to synthesize each compound by a ring-closing metathesis reaction?

a. 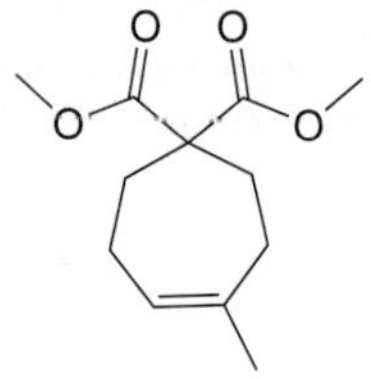
O
O
O
O

b. 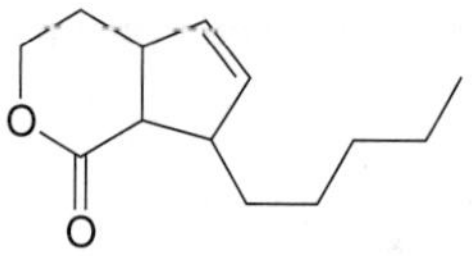
O
O

Solution

To work in the retrosynthetic direction, cleave the C=C in the product, and bond each carbon of the original alkene to a CH_2 group using a double bond.

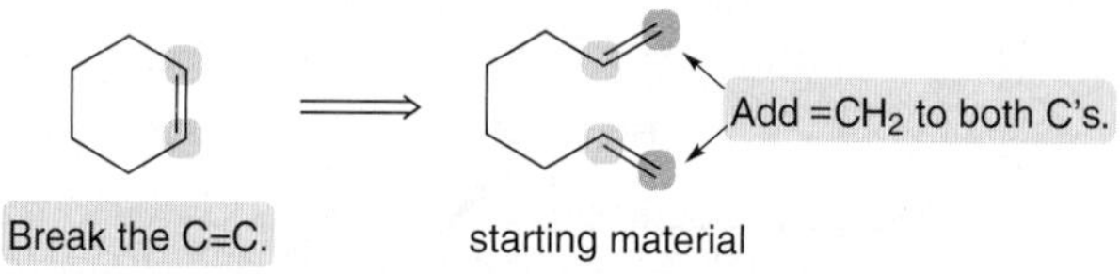
Add =CH2 to both C's.
Break the C=C.
starting material

The resulting compound has a carbon chain with two terminal alkenes.

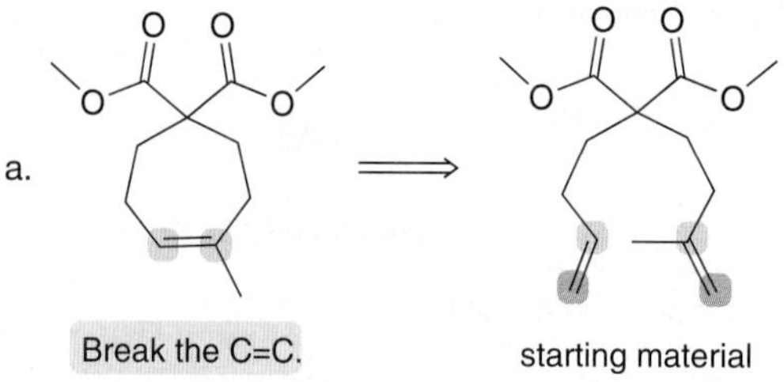
a.
O
O
O
O
Break the C=C.
starting material

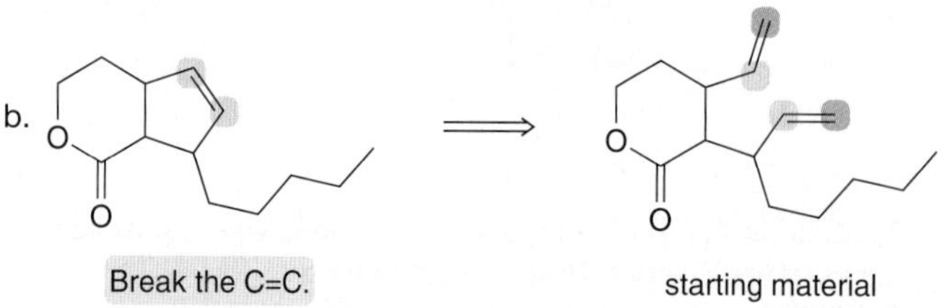
b.
O
O
Break the C=C.
starting material

Problem 29.17 What starting material is needed to synthesize each compound by a ring-closing metathesis reaction?

a. 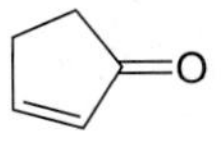

b. 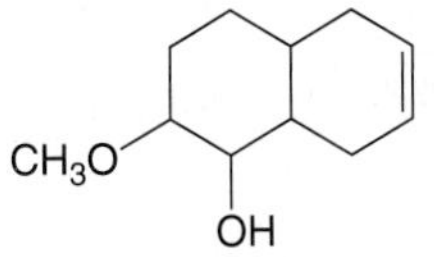
CH3O
OH

c. 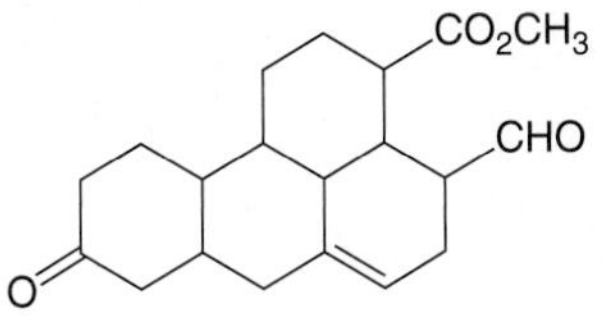
CO2CH3
CHO
O

KEY CONCEPTS

Carbon–Carbon Bond-Forming Reactions in Organic Synthesis

Coupling Reactions

[1] Coupling reactions of organocuprate reagents (29.1)

$$R'{-}X + R_2CuLi \longrightarrow R'{-}R + RCu + LiX$$

X = Cl, Br, I

- R'X can be CH_3X, RCH_2X, 2° cyclic halides, vinyl halides, and aryl halides.
- X may be Cl, Br, or I.
- With vinyl halides, coupling is stereospecific.

[2] Suzuki reaction (29.2)

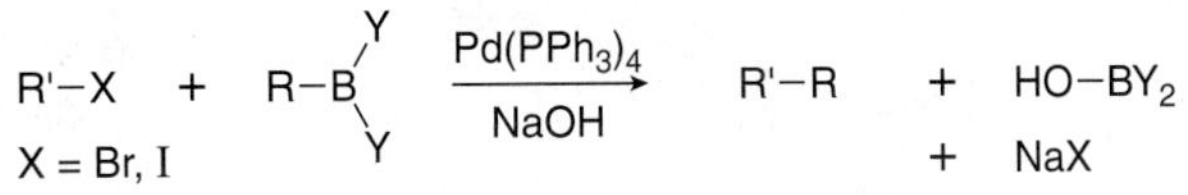
R'–X + R–B(Y)(Y)
Pd(PPh3)4
NaOH
R'–R + HO–BY2
+ NaX
X = Br, I

- R'X is most often a vinyl halide or aryl halide.
- With vinyl halides, coupling is stereospecific.

[3] Heck reaction (29.3)

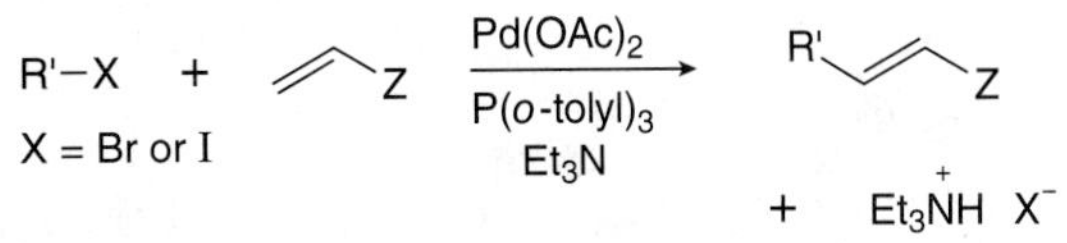
R'–X +
Z
X = Br or I
Pd(OAc)2
P(o-tolyl)3
Et3N
R'
Z
+ Et3NH X⁻

- R'X is a vinyl halide or aryl halide.
- Z = H, Ph, COOR, or CN.
- With vinyl halides, coupling is stereospecific.
- The reaction forms trans alkenes.

Cyclopropane Synthesis

[1] Addition of dihalocarbenes to alkenes (29.4)

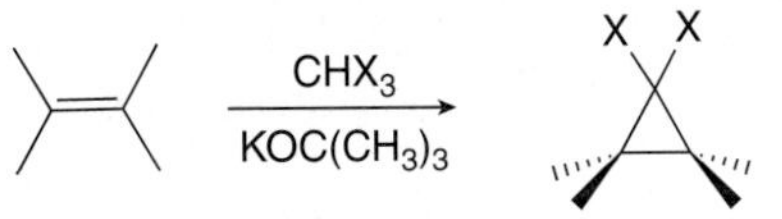
CHX3
KOC(CH3)3
X X

- The reaction occurs with syn addition.
- The position of substituents in the alkene is retained in the cyclopropane.

[2] Simmons–Smith reaction (29.5)

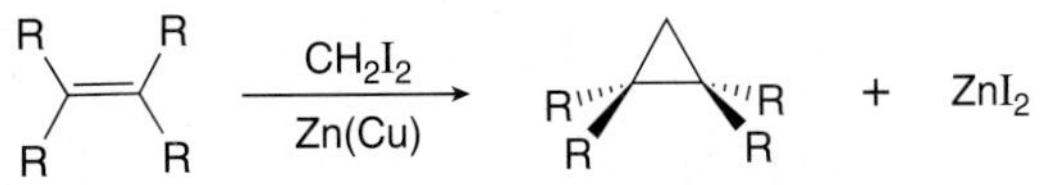
R R
R R
CH2I2
Zn(Cu)
R R
R R
+ ZnI2

- The reaction occurs with syn addition.
- The position of substituents in the alkene is retained in the cyclopropane.

Metathesis (29.6)

[1] Intermolecular reaction

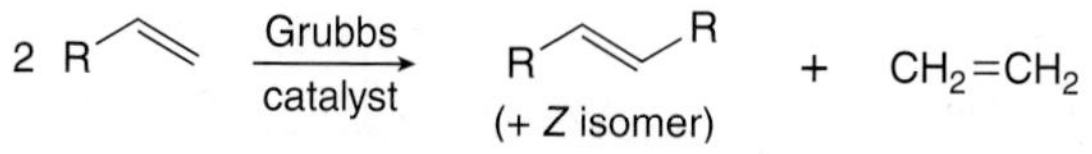
2
Grubbs
catalyst
(+ Z isomer)
+ $CH_2{=}CH_2$

- Metathesis works best when $CH_2{=}CH_2$, a gas that escapes from the reaction mixture, is one of the products formed.

[2] Intramolecular reaction

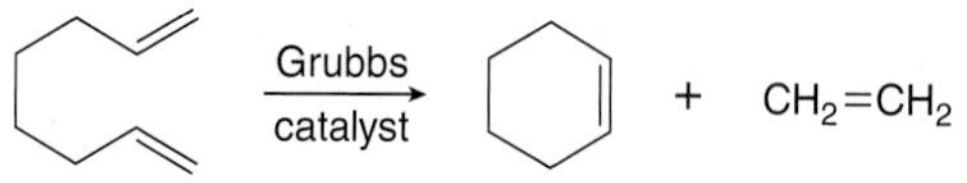
Grubbs
catalyst
+ $CH_2{=}CH_2$

diene

- Ring-closing metathesis forms rings of any size from diene starting materials.

PROBLEMS

Problems Using Three-Dimensional Models

29.18 In addition to organic halides, alkyl tosylates (R'OTs, Section 9.13) also react with organocuprates (R_2CuLi) to form coupling products R–R'. When 2° alkyl tosylates are used as starting materials (R_2CHOTs), inversion of the configuration at a stereogenic center results. Keeping this in mind, draw the product formed when each compound is treated with $(CH_3)_2CuLi$.

a. 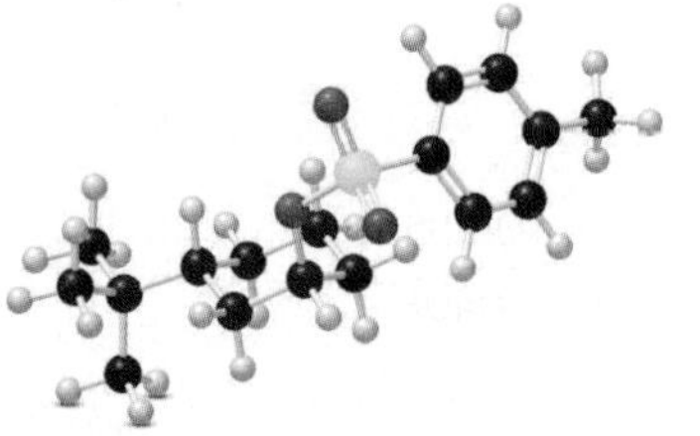

b. 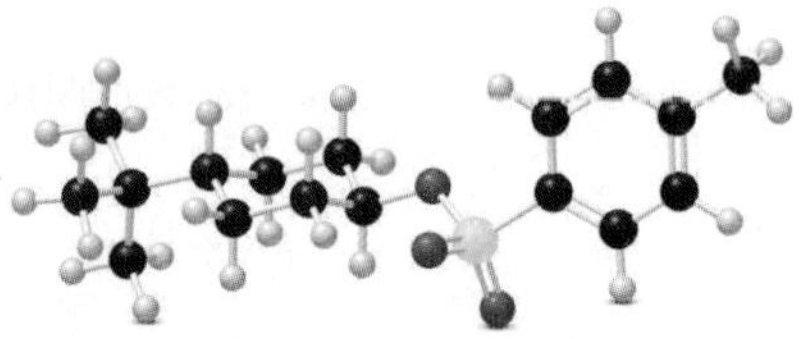

29.19 What product is formed by ring-closing metathesis of each compound?

a. 

b. 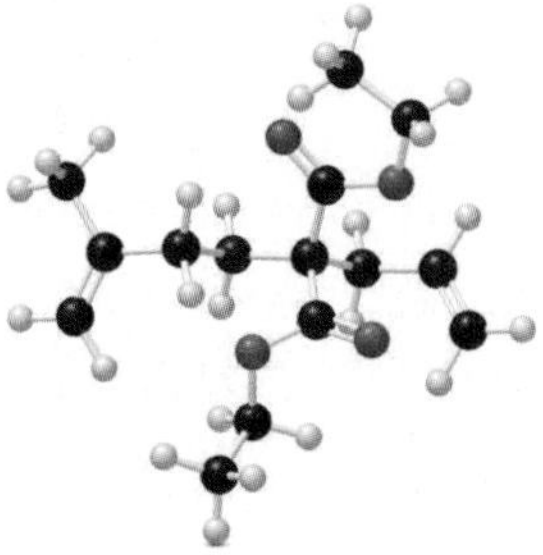

Coupling Reactions

29.20 Draw the products formed in each reaction.

a. 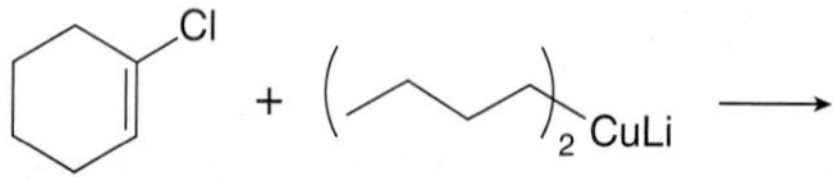
Cl
+
CuLi

b. 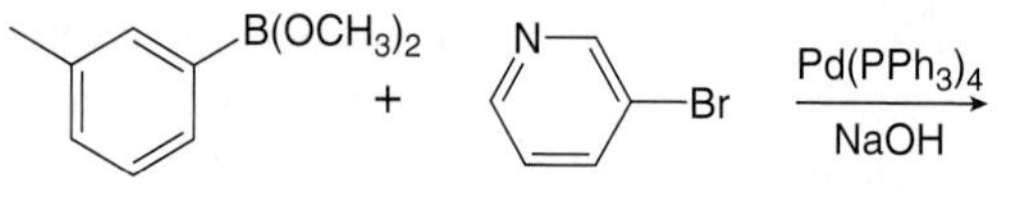
$B(OCH_3)_2$
+
Br
$Pd(PPh_3)_4$
NaOH

c. 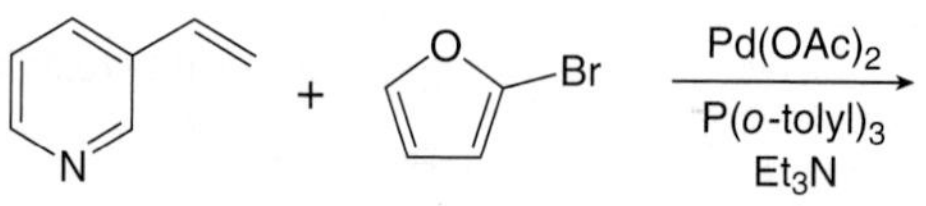
+
Br
$Pd(OAc)_2$
$P(o\text{-tolyl})_3$
Et_3N

d. 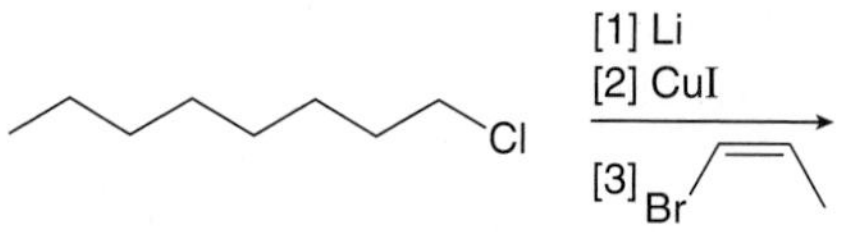
Cl
[1] Li
[2] CuI
[3]
Br

e. 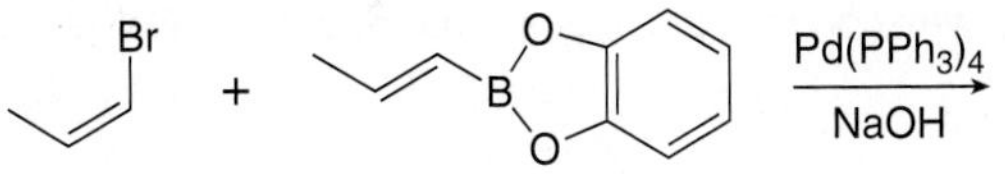
Br
+
B
O
O
Pd(PPh3)4
NaOH

g. 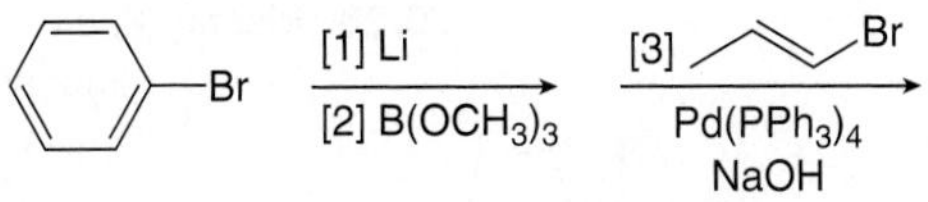
Br
[1] Li
[2] B(OCH3)3
[3]
Br
Pd(PPh3)4
NaOH

f. 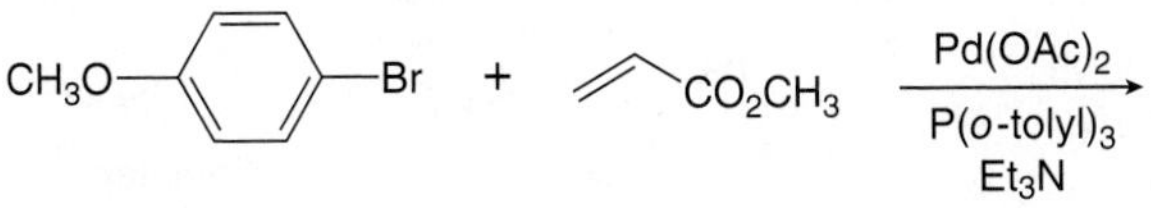
CH3O
Br
+
CO2CH3
Pd(OAc)2
P(o-tolyl)3
Et3N

h. 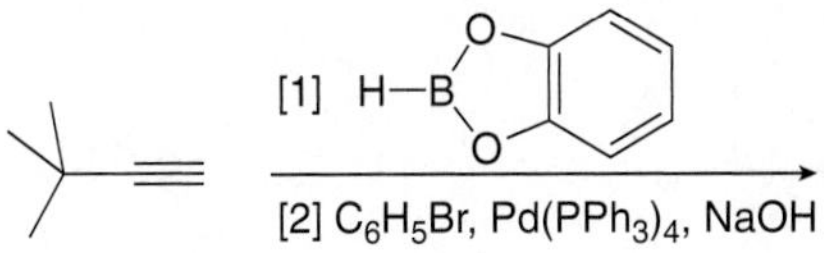
[1] H—B
O
O
[2] C6H5Br, Pd(PPh3)4, NaOH

29.21 What organic halide is needed to convert lithium divinylcuprate [$(CH_2{=}CH)_2CuLi$] to each compound?

a. 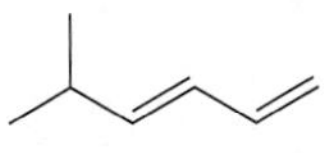

b. 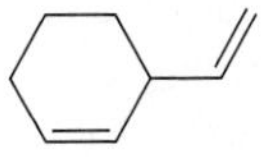

c. 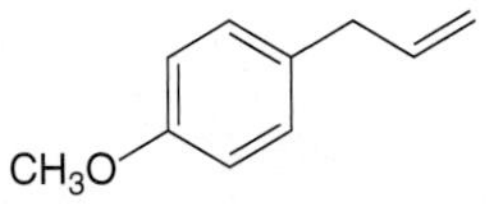
CH3O

29.22 How can you convert ethynylcyclohexane to dienes **A–C** using a Suzuki reaction? You may use any other organic compounds and inorganic reagents. Is it possible to synthesize diene **D** using a Suzuki reaction? Explain why or why not.

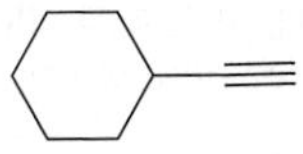

ethynylcyclohexane

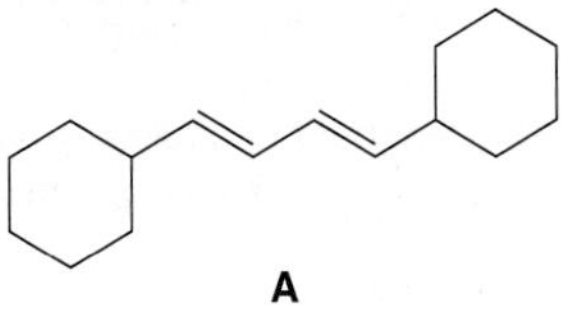

A

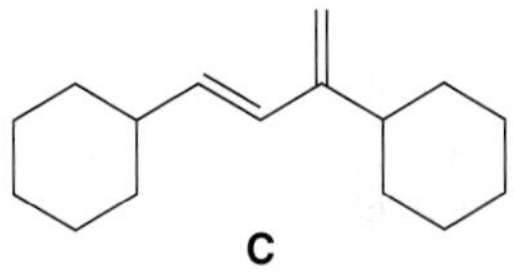

C

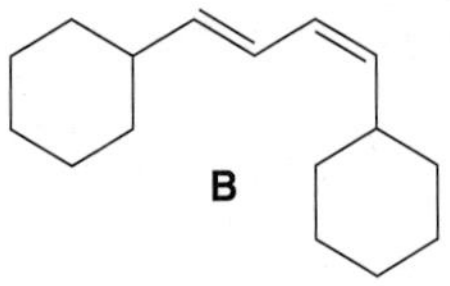

B

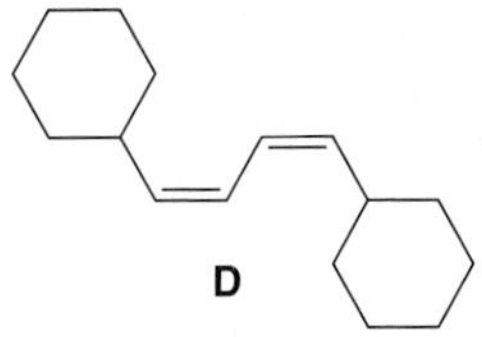

D

29.23 What compound is needed to convert styrene ($C_6H_5CH{=}CH_2$) to each product using a Heck reaction?

a. 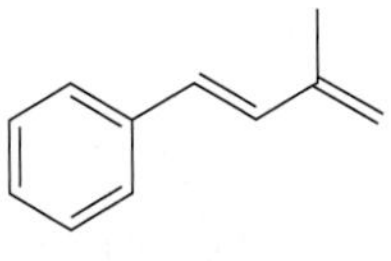

b. 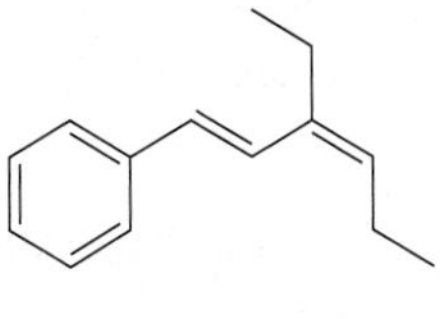

c. 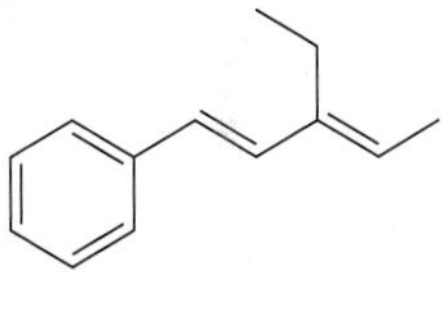

29.24 What steps are needed to convert but-1-ene ($CH_3CH_2CH{=}CH_2$) to octane [$CH_3(CH_2)_6CH_3$] using a coupling reaction with an organocuprate reagent? All carbon atoms in octane must come from but-1-ene.

29.25 What product is formed in the Suzuki coupling of **A** and **B?** This reaction was a key step in the synthesis of losartan, a drug used to treat hypertension.

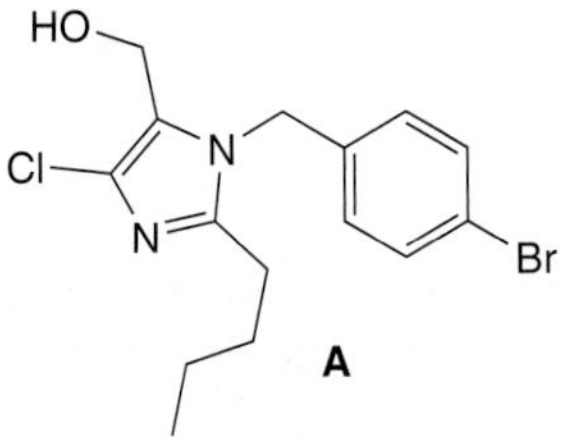
HO
Cl
N
N
Br

A

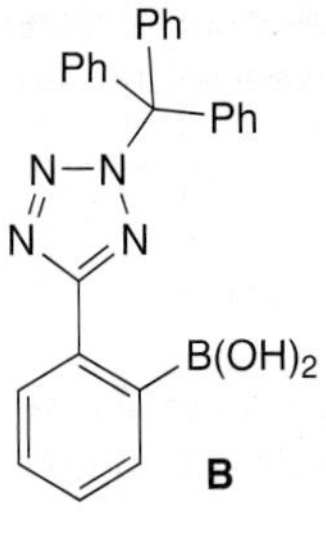
Ph
Ph
Ph
N—N
N
N
B(OH)2

B

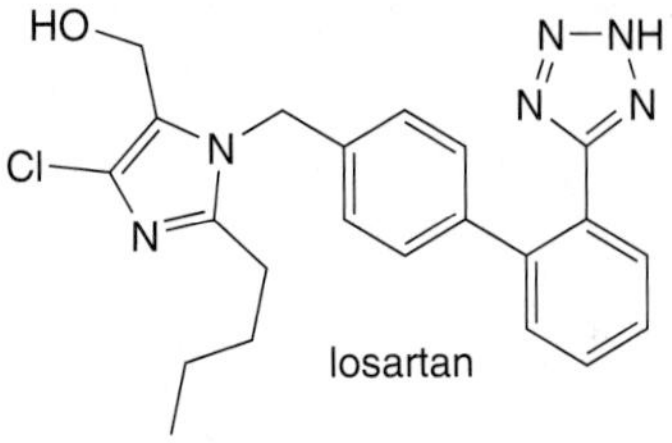
HO
Cl
N
N
N—NH
N
N

losartan

29.26 Identify **X,** an intermediate that was converted to eletriptan (trade name Relpax), a drug used to treat migraines.

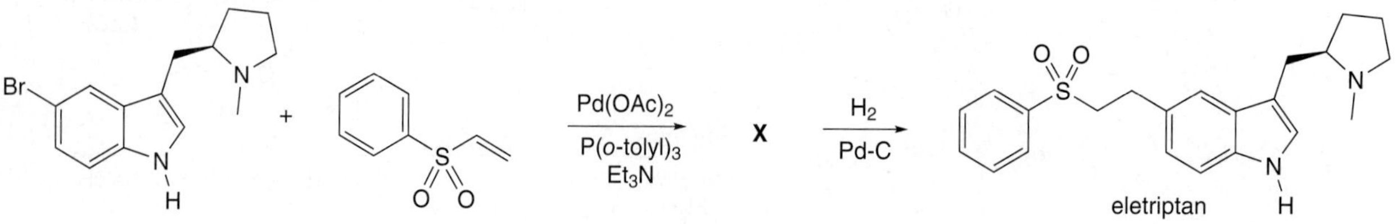

Cyclopropanes

29.27 Draw the products (including stereoisomers) formed in each reaction.

a. 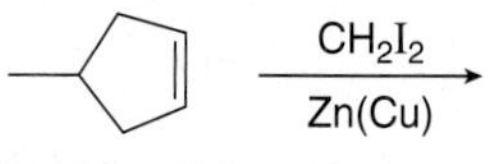

c. 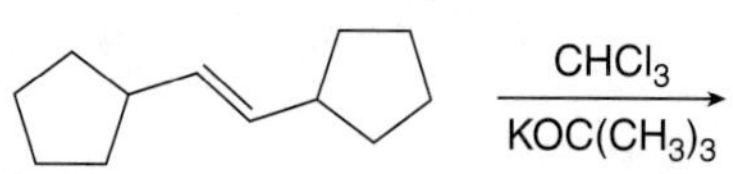

b. 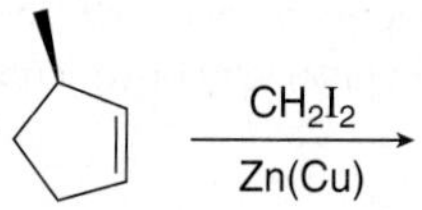

d. 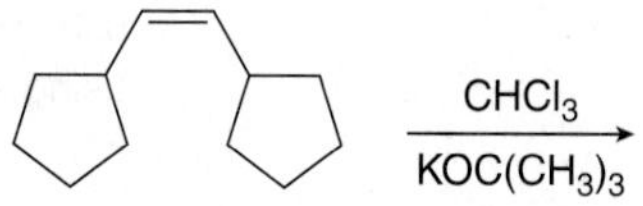

29.28 Treatment of cyclohexene with $C_6H_5CHI_2$ and Zn(Cu) forms two stereoisomers of molecular formula $C_{13}H_{16}$. Draw their structures and explain why two compounds are formed.

Metathesis

29.29 What ring-closing metathesis product is formed when each substrate is treated with Grubbs catalyst under high-dilution conditions?

a. 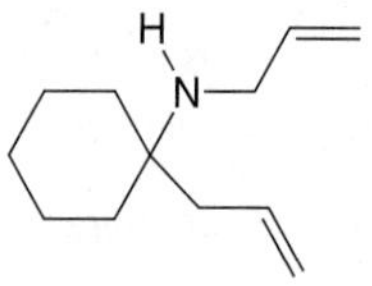

b. 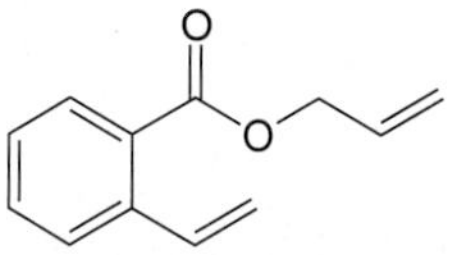

c. 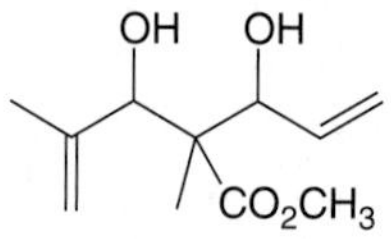

29.30 What starting material is needed to prepare each compound by a ring-closing metathesis reaction?

a. 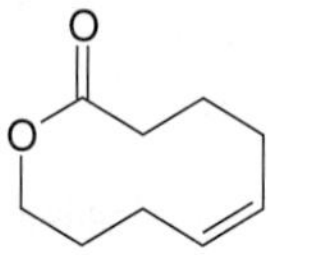

b. 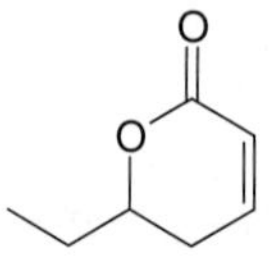

c. 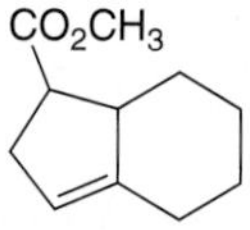

29.31 Metathesis reactions can be carried out with two *different* alkene substrates in one reaction mixture. Depending on the substitution pattern around the C=C, the reaction may lead to one major product or a mixture of many products. For each pair of alkene substrates, draw all metathesis products formed. (Disregard any starting materials that may also be present at equilibrium.) With reference to the three examples, discuss when alkene metathesis with two different alkenes is a synthetically useful reaction.

a. 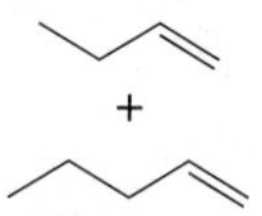

b. 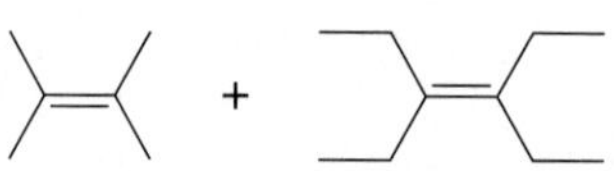

c. 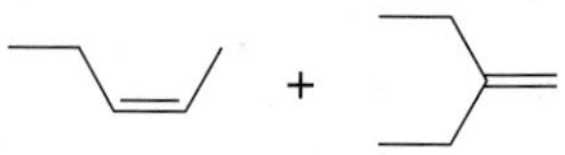

29.32 Draw the structure of the two products of molecular formula $C_{15}H_{26}O_2$ formed when **M** is treated with Grubbs catalyst under high-dilution conditions.

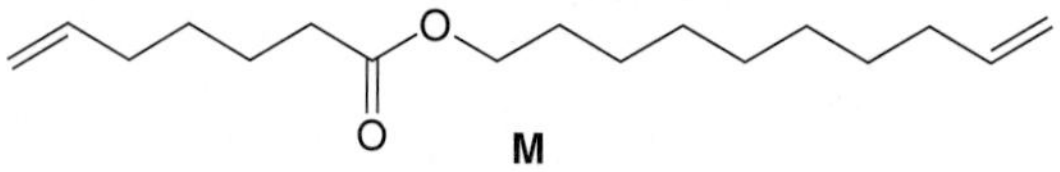

M

29.33 When certain cycloalkenes are used in metathesis reactions, **ring-opening metathesis polymerization (ROMP)** occurs to form a high molecular weight polymer, as shown with cyclopentene as the starting material. The reaction is driven to completion by relief of strain in the cycloalkene.

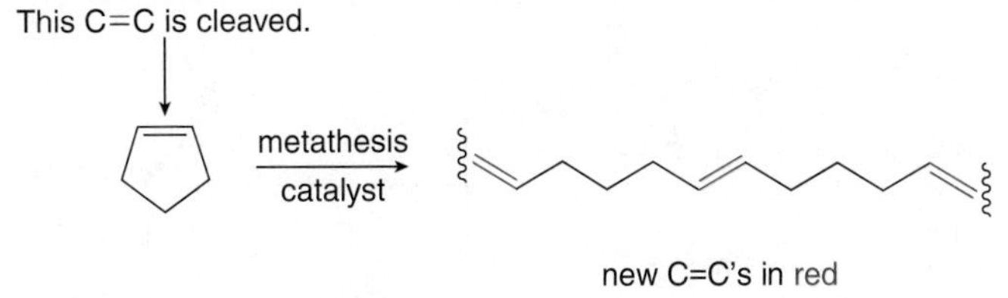

What products are formed by ring-opening metathesis polymerization of each alkene?

a. 

b. 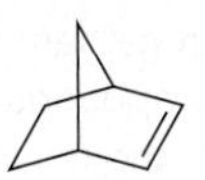

c. 

General Reactions

29.34 Draw the products formed in each reaction.

a. 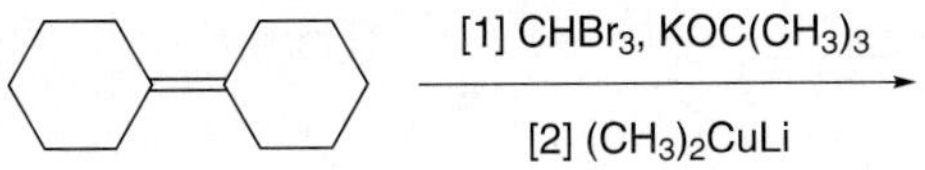

b. 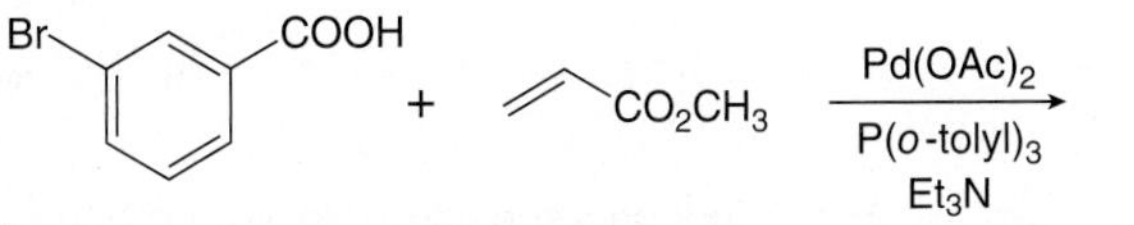

c. 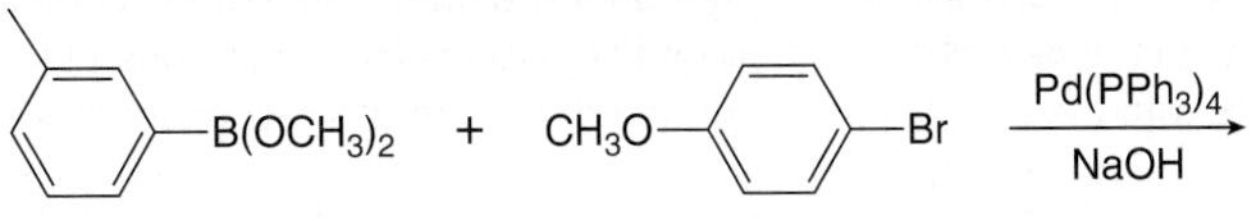

d. 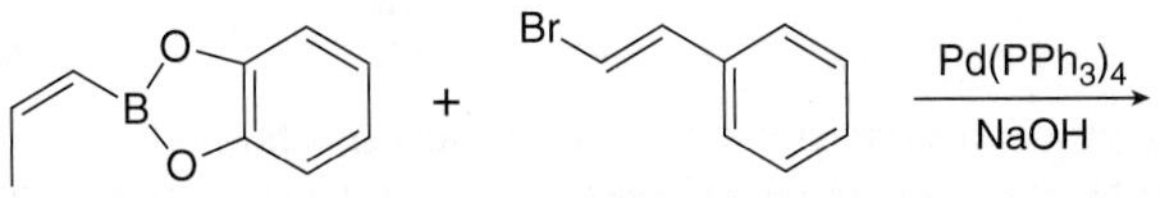

e. 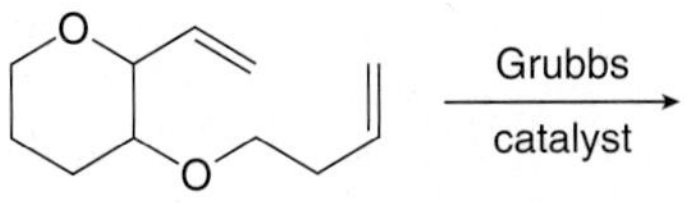

f. 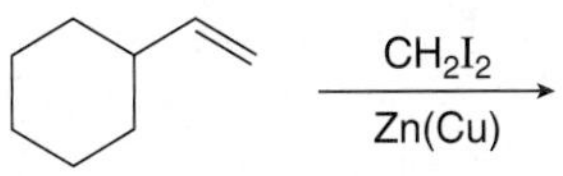

g. 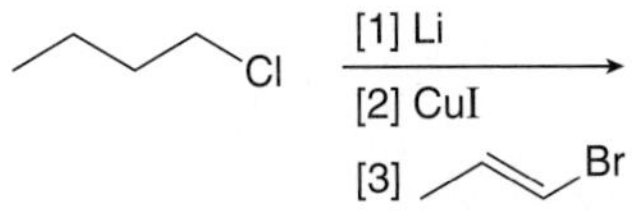

h. 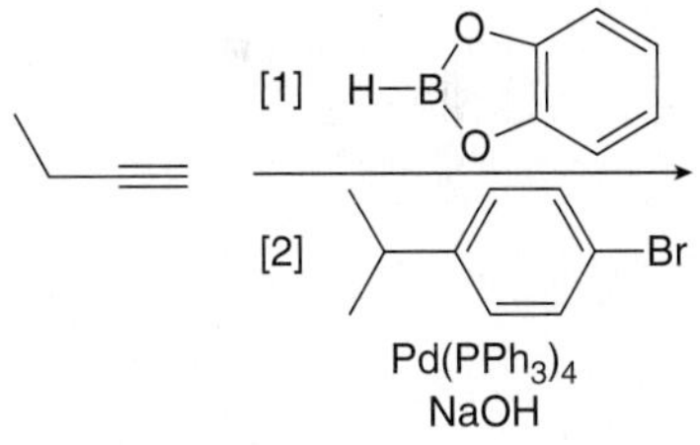

Mechanisms

29.35 In addition to using CHX_3 and base to synthesize dihalocarbenes (Section 29.4), dichlorocarbene (:CCl_2) can be prepared by heating sodium trichloroacetate. Draw a stepwise mechanism for this reaction.

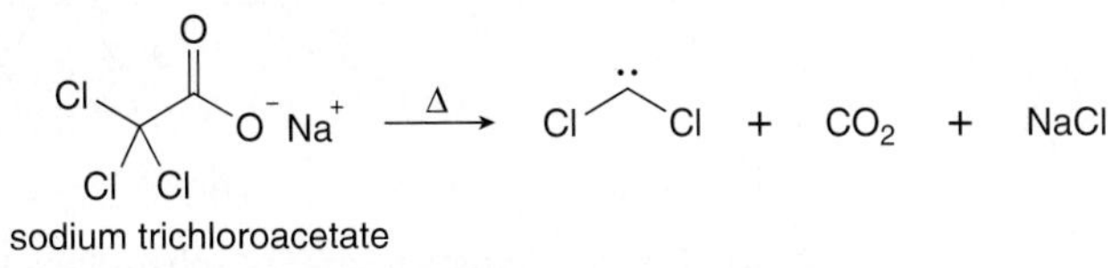

29.36 Draw a stepwise mechanism for the following reaction.

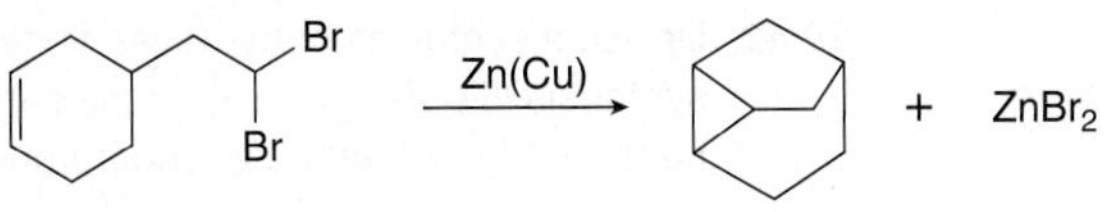

29.37 Sulfur ylides, like the phosphorus ylides of Chapter 21, are useful intermediates in organic synthesis. Methyl *trans*-chrysanthemate, an intermediate in the synthesis of the insecticide pyrethrin I (Section 29.4), can be prepared from diene **A** and a sulfur ylide. Draw a stepwise mechanism for this reaction.

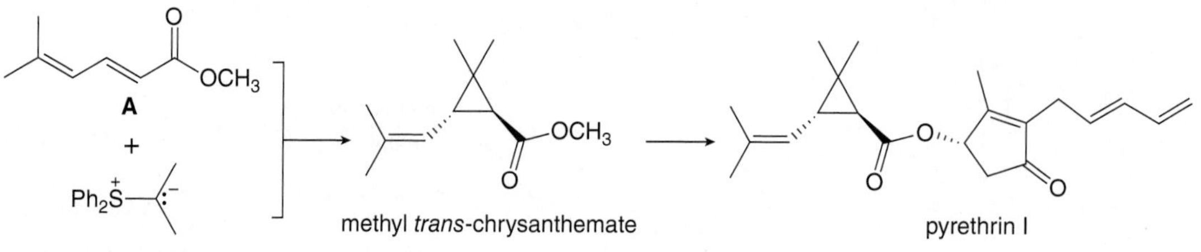

29.38 Although diazomethane (CH_2N_2) is often not a useful reagent for preparing cyclopropanes, other diazo compounds give good yields of more complex cyclopropanes. Draw a stepwise mechanism for the conversion of diazo compound **A** to **B,** an intermediate in the synthesis of sirenin, the sperm attractant produced by the female gametes of the water mold *Allomyces.*

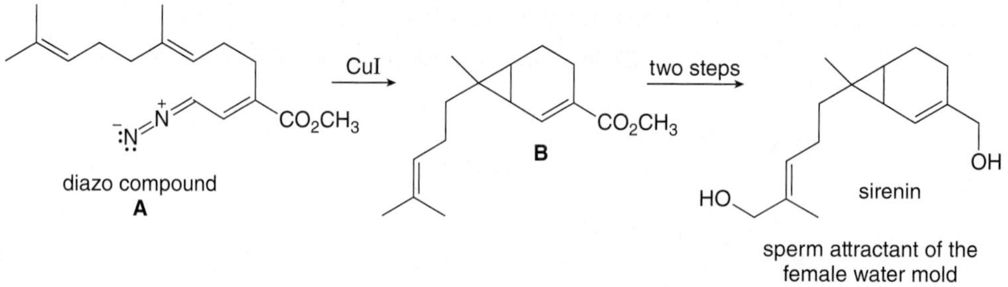

29.39 The reaction of cyclohexene with iodobenzene under Heck conditions forms **E,** a coupling product with the new phenyl group on the allylic carbon, but none of the "expected" coupling product **F** with the phenyl group bonded directly to the carbon–carbon double bond.

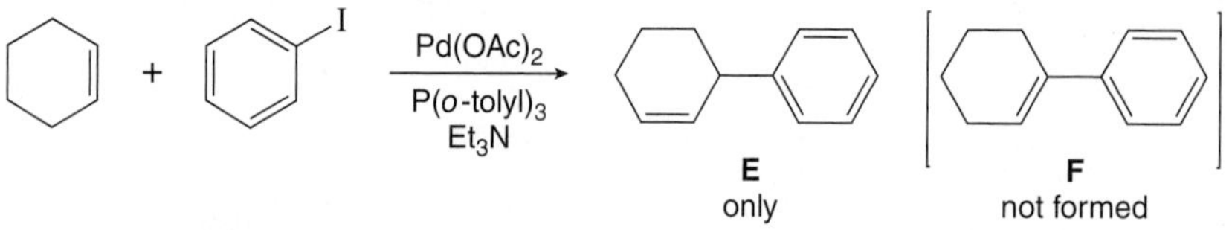

a. Draw a stepwise mechanism that illustrates how **E** is formed.

b. Step [2] in Mechanism 29.2 proceeds with syn addition of Pd and R' to the double bond. What does the formation of **E** suggest about the stereochemistry of the elimination reaction depicted in Step [3] of Mechanism 29.2?

Synthesis

29.40 Devise a synthesis of diene **A** from (*Z*)-2-bromostyrene as the only organic starting material. Use a Suzuki reaction in one step of the synthesis.

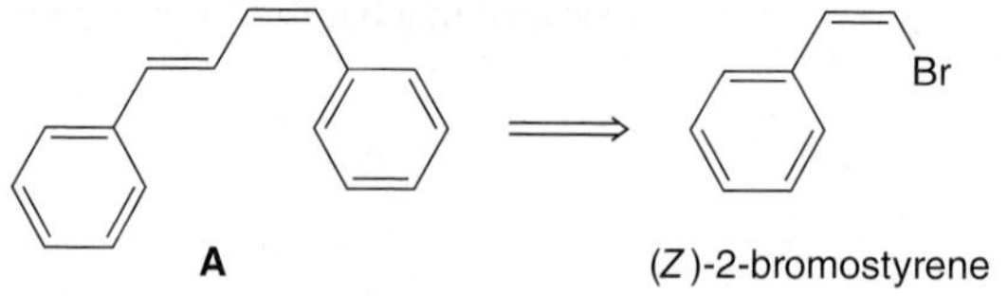

29.41 Devise a synthesis of (*E*)-1-phenylhex-1-ene ($CH_3CH_2CH_2CH_2CH=CHPh$) using hydrocarbons having ≤ 6 C's and a Suzuki reaction as one of the steps.

29.42 Devise a synthesis of the given trans vinylborane, which can be used for bombykol synthesis (Figure 29.1). All of the carbon atoms in the vinylborane must come from acetylene, nonane-1,9-diol, and catecholborane.

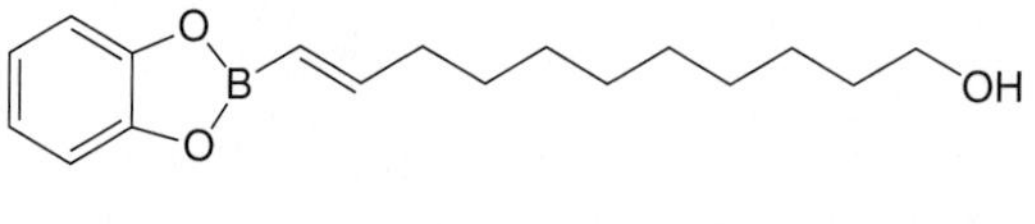

29.43 Devise a synthesis of each compound using a Heck reaction as one step. You may use benzene, $CH_2=CHCO_2Et$, organic alcohols having two carbons or fewer, and any required inorganic reagents.

a. 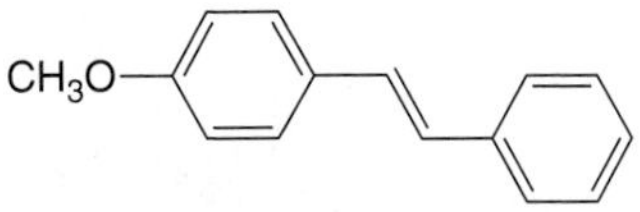

b. 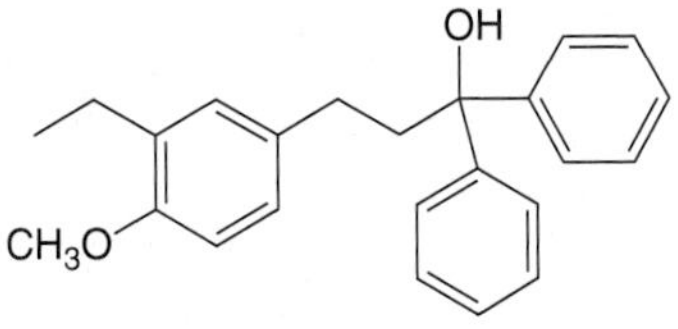

29.44 Devise a synthesis of each compound from cyclohexene and any required organic or inorganic reagents.

a. 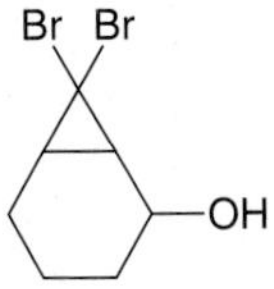

b. 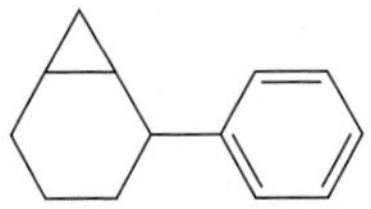

29.45 Devise a synthesis of each compound from benzene. You may also use any organic compounds having four carbons or fewer, and any required inorganic reagents.

a. 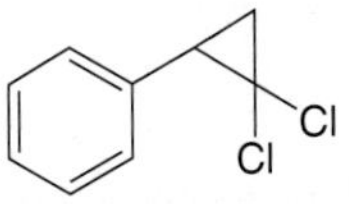

b. 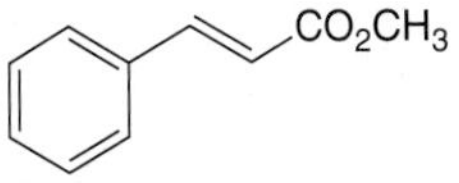

29.46 Devise a synthesis of each substituted cyclopropane. Use acetylene (HC≡CH) as a starting material in part (a), and cyclohexanone as a starting material in part (b). You may use any other organic compounds and any needed reagents.

a. 

+ enantiomer

b. 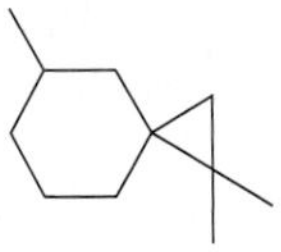

29.47 Biaryls, compounds containing two aromatic rings joined by a C–C bond, can often be efficiently made by two different Suzuki couplings; that is, either aromatic ring can be used to form the organoborane needed for coupling. In some cases, however, only one route is possible. With this in mind, synthesize each of the following biaryls using benzene as the starting material for each aromatic ring. When more than one route is possible, draw both of them. You may use any required organic or inorganic reagents.

a. 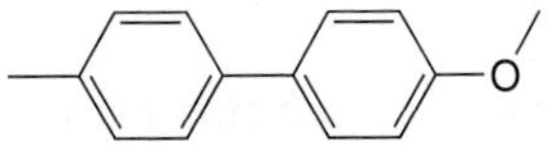

b. 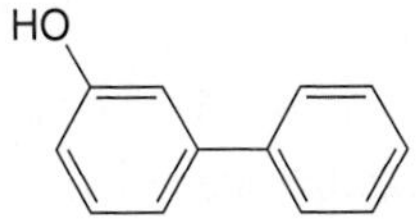

c. 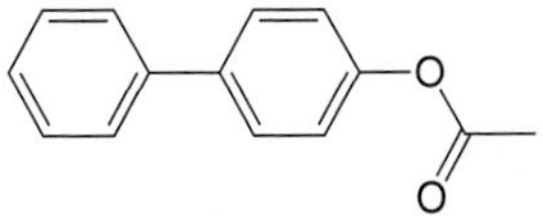

29.48 Draw the product formed from the ring-closing metathesis of each compound. Then, devise a synthesis of each metathesis starting material using any of the following compounds: $CH_2(CO_2Et)_2$, alcohols with less than five carbons, and any needed organic and inorganic reagents.

a. 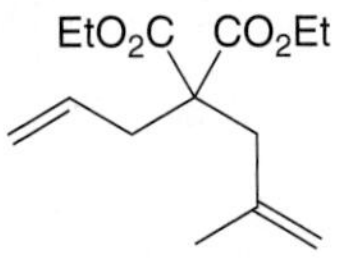

b. 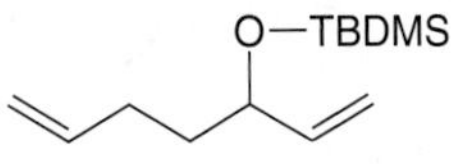

29.49 Draw the product formed from the ring-closing metathesis of each compound. Then, devise a synthesis of each metathesis starting material from benzene, alcohols with less than five carbons, and any needed organic and inorganic reagents.

a. 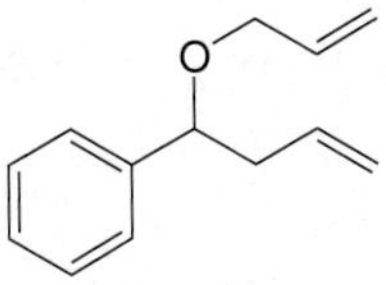

b. 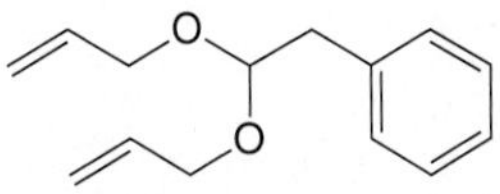

29.50 What reagents are needed to carry out transformations [1]–[3] in the synthesis of aldehyde **A?** **A** can be converted to the antitumor agent maytansine in several steps.

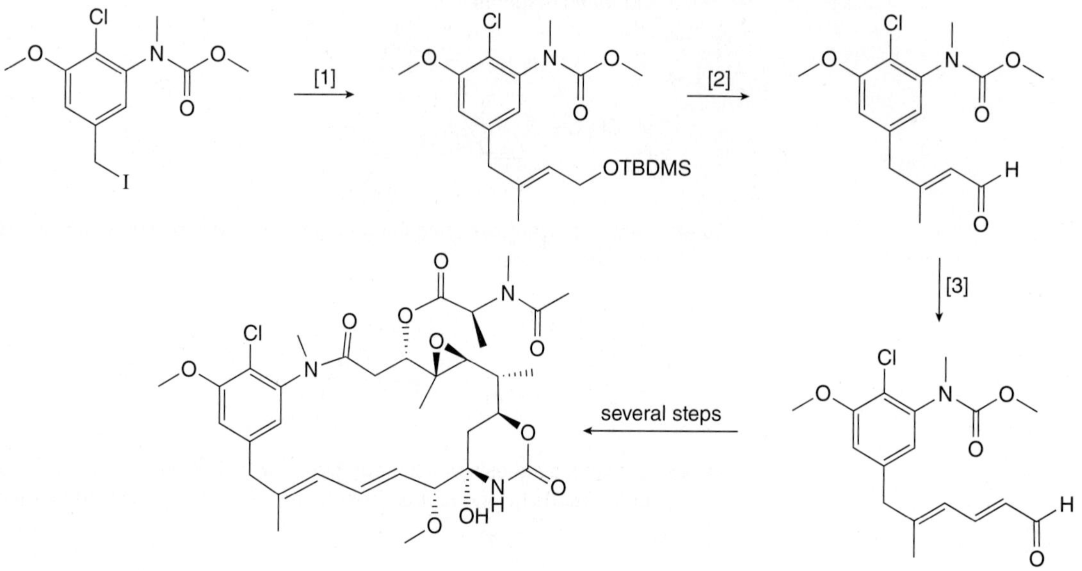

maytansine

A

29.51 Devise a synthesis of each of the following compounds. Besides inorganic reagents, you may use hydrocarbons and halides having ≤ 6 C's, and $CH_2=CHCOOCH_3$ as starting materials. Each synthesis must use at least one of the carbon–carbon bond-forming reactions in this chapter.

a. 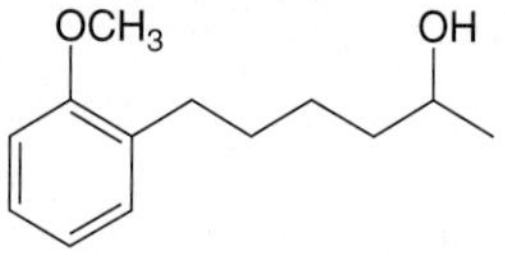

(two enantiomers)

c. 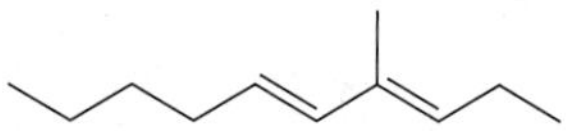

b. 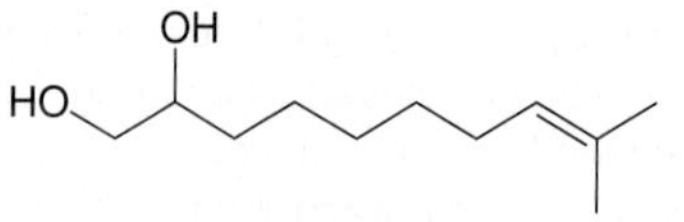

(two enantiomers)

d. 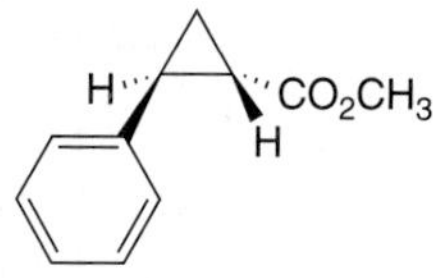

(+ enantiomer)

Challenge Problems

29.52 Many variations of ring-closing metathesis have now been reported. Tandem ring-opening–ring-closing metathesis can occur with cyclic alkenes that contain two additional carbon–carbon double bonds. In this reaction, the cycloalkene is cleaved, and two new rings are formed. [1] What compounds are formed in this tandem reaction with the following substrates? [2] Devise a synthesis of the substrate in part (b) that uses a Diels–Alder reaction with diethyl maleate as the dienophile.

a. 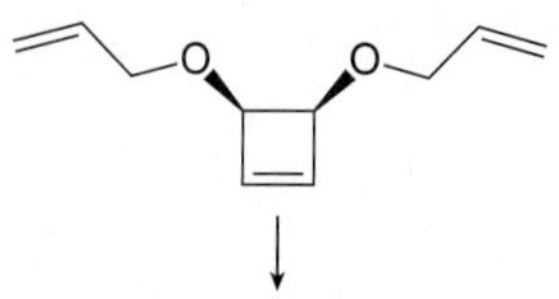

$C_8H_{10}O_2$

b. 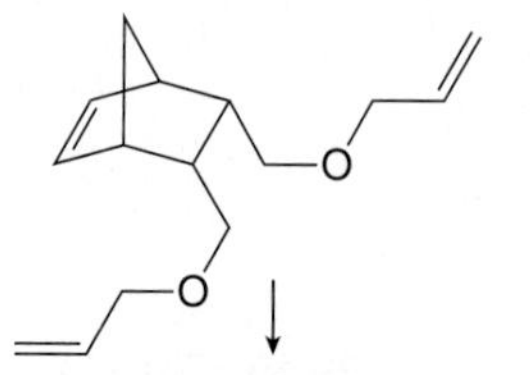

$C_{13}H_{18}O_2$

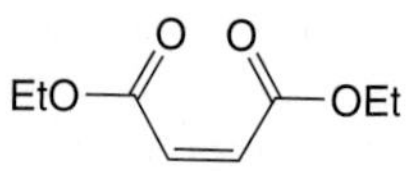

diethyl maleate

29.53 The following conversion, carried out in the presence of Grubbs catalyst and ethylene gas, involves a cascade of metathesis reactions. Draw a reaction sequence that illustrates how the reactant is converted to the product **Z**, and indicate where each labeled atom in the reactant ends up in **Z**.

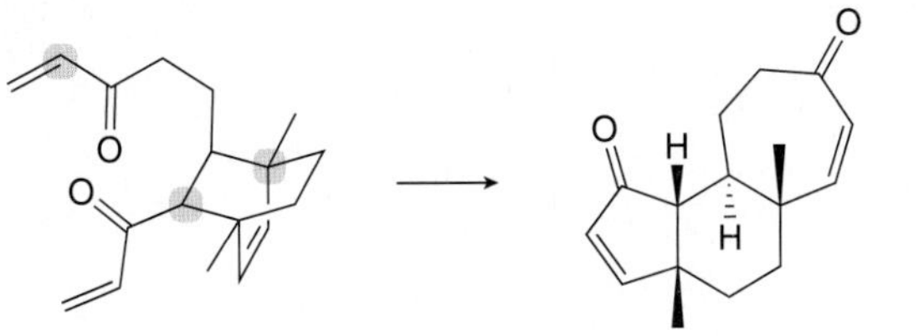

Z

29.54 Suzuki coupling of aryl iodide **A** and vinylborane **B** affords compound **C**, which is converted to **D** in the presence of aqueous acid. Identify compounds **C** and **D** and draw a stepwise mechanism for the conversion of **C** to **D**.

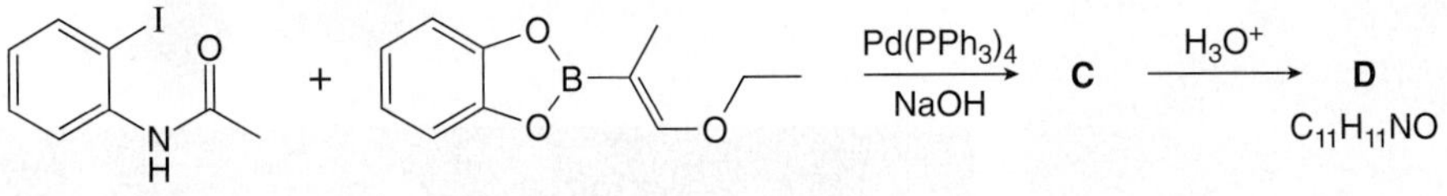

A **B**

29.55 Dimethyl cyclopropanes can be prepared by the reaction of an α,β-unsaturated carbonyl compound **X** with two equivalents of a Wittig reagent **Y**. Draw a stepwise mechanism for this reaction.

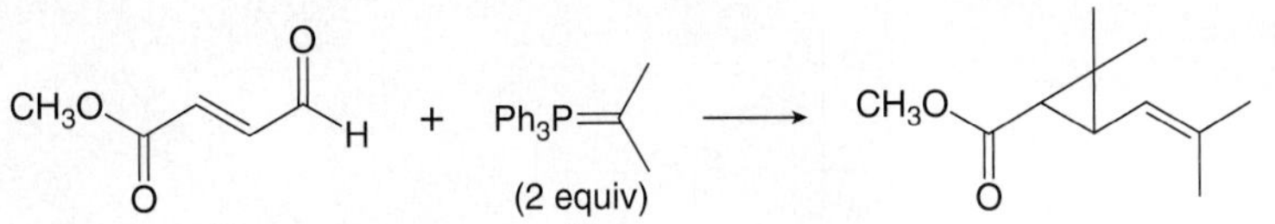

X **Y**

29.56 Dienynes undergo metathesis to afford fused bicyclic ring systems. (a) Explain how **A** is converted to **B**. (b) Keeping this reaction in mind, draw the two products formed by dienyne metathesis of **C**.

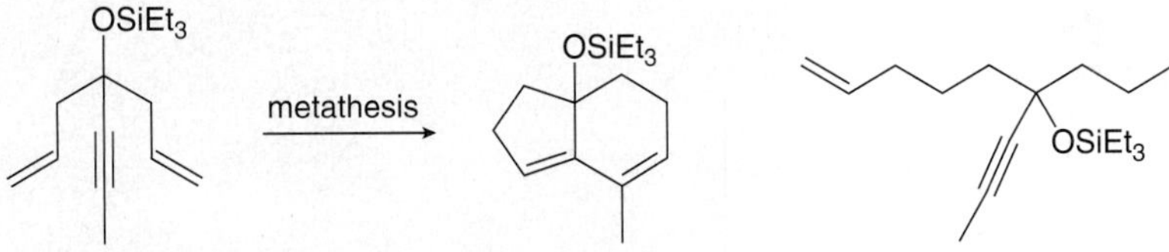

A **B** **C**

30 Pericyclic Reactions

30.1 Types of pericyclic reactions
30.2 Molecular orbitals
30.3 Electrocyclic reactions
30.4 Cycloaddition reactions
30.5 Sigmatropic rearrangements
30.6 Summary of rules for pericyclic reactions

Periplanone B, an unusual diepoxide with a 10-membered ring, is a potent sex pheromone of the female American cockroach. Although periplanone B was isolated in 1952, its structure was not determined until 1976 using 200 µg of material obtained from more than 75,000 female cockroaches. This structure was confirmed by synthesis in 1979, and several subsequent syntheses have been reported. Key steps in an elegant 1984 synthesis of periplanone B involve pericyclic reactions, a group of powerful, stereospecific reactions discussed in Chapter 30.

Many of the reactions thus far encountered in our study of organic chemistry occur by way of reactive intermediates—cations, anions, and radicals. For example, the S_N1 reaction in Chapter 7 and electrophilic aromatic substitutions in Chapter 18 involve carbocations, while the aldol and Claisen reactions in Chapter 24 occur via enolate anions. Other reactions, such as the halogenation of alkanes and the polymerization of alkenes discussed in Chapter 15, take place via radical intermediates.

In Chapter 30, we learn about a small but versatile group of reactions, **pericyclic reactions,** which occurs in a concerted process—all bonds are broken and formed in a single step—with a cyclic transition state. The Diels–Alder reaction in Chapter 16 is an example of one type of pericyclic reaction. Pericyclic reactions involve π bonds, and they are governed by a set of rules that allows us to predict the identity and stereochemistry of the products formed. Consequently, pericyclic reactions are valuable tools for synthesizing organic molecules.

30.1 Types of Pericyclic Reactions

Although most organic reactions take place by way of ionic or radical intermediates, a number of useful reactions occur in one-step processes that do *not* form reactive intermediates.

- **A pericyclic reaction is a concerted reaction that proceeds through a cyclic transition state.**

Stereospecific reactions were first discussed in Chapter 10.

Pericyclic reactions require light or heat and are completely stereospecific; that is, a particular stereoisomer of the reactant forms a particular stereoisomer of the product. There are three categories of pericyclic reactions: **electrocyclic reactions, cycloadditions,** and **sigmatropic rearrangements.**

An **electrocyclic reaction** is a reversible reaction that can involve ring closure or ring opening of one molecule of reactant to form one molecule of product.

- **An electrocyclic ring closure is an intramolecular reaction that forms a cyclic product containing one more σ bond and one fewer π bond than the reactant.**

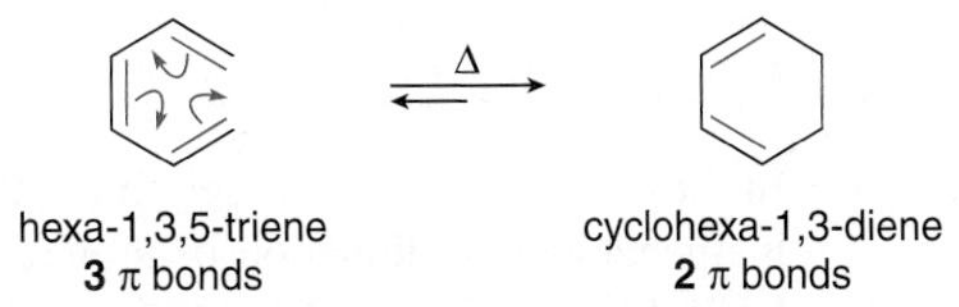
Δ
hexa-1,3,5-triene
3 π bonds
cyclohexa-1,3-diene
2 π bonds

- **An electrocyclic ring opening is a reaction in which a σ bond of a cyclic reactant is cleaved to form a conjugated product with one more π bond.**

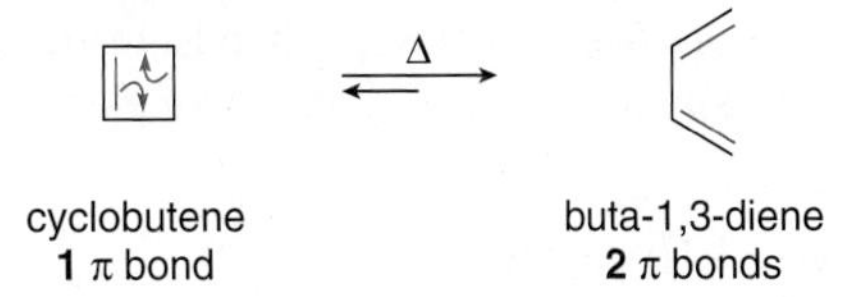
Δ
cyclobutene
1 π bond
buta-1,3-diene
2 π bonds

Cycloaddition reactions form a ring. The Diels–Alder reaction in Chapter 16 is one example of a cycloaddition.

- **A cycloaddition is a reaction between two compounds with π bonds to form a cyclic product with two new σ bonds.**

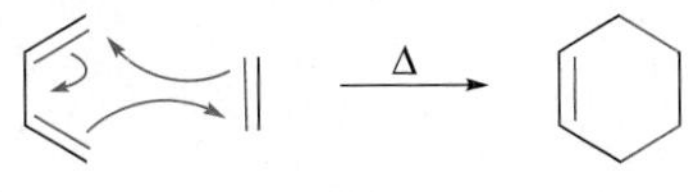
Δ

In contrast to electrocyclic reactions and cycloadditions, in which the number of π bonds differs in the reactants and products, the number of π bonds does *not* change in a **sigmatropic rearrangement.**

- **A sigmatropic rearrangement is a reaction in which a σ bond is broken in the reactant, the π bonds rearrange, and a σ bond is formed in the product.**

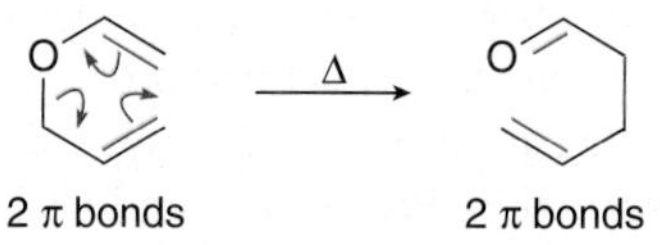

Hoffmann and Fukui received the 1981 Nobel Prize in Chemistry for developing theories that explain the course of pericyclic reactions.

Two features determine the course of the reactions: the number of π bonds involved and whether the reaction occurs in the presence of heat (thermal conditions) or light (photochemical conditions). These reactions follow a set of rules based on orbitals and symmetry first proposed by R. B. Woodward and Roald Hoffmann in 1965, and derived from theory described by Kenichi Fukui in 1954.

To understand pericyclic reactions, we must review and expand upon what we learned about the molecular orbitals of systems with π bonds in Chapter 17.

Problem 30.1 Classify each reaction as an electrocyclic reaction, a cycloaddition, or a sigmatropic rearrangement. Label the σ bonds that are broken or formed in each reaction.

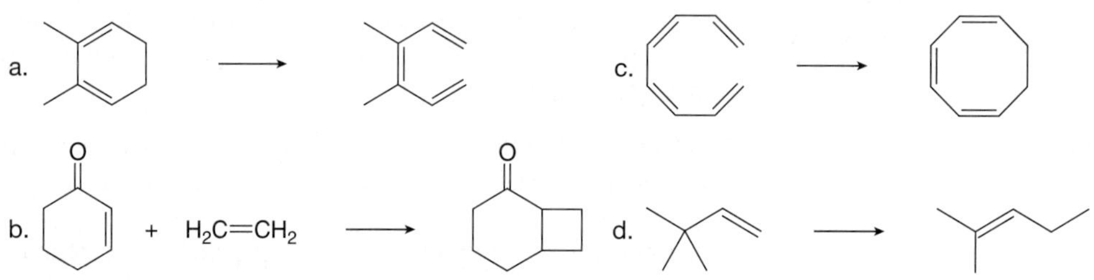

30.2 Molecular Orbitals

In Section 17.9 we learned that molecular orbital (MO) theory describes bonds as the mathematical combination of atomic orbitals that forms a new set of orbitals called **molecular orbitals (MOs). The number of atomic orbitals used *equals* the number of molecular orbitals formed.**

Because pericyclic reactions involve π bonds, let's examine the molecular orbitals that result from *p* orbital overlap in ethylene, buta-1,3-diene, and hexa-1,3,5-triene, molecules that contain one, two, and three π bonds, respectively. Keep in mind that the two lobes of a *p* orbital are opposite in phase, with a node of electron density at the nucleus.

30.2A Ethylene

The π bond in ethylene ($CH_2{=}CH_2$) is formed by side-by-side overlap of two *p* orbitals on adjacent carbons. Two *p* orbitals can combine in two different ways. As shown in Figure 30.1, when two *p* orbitals of similar phase overlap, a π bonding molecular orbital (designated as ψ_1) results. Two electrons occupy this lower-energy bonding molecular orbital. When two *p* orbitals of opposite phase combine, a π* antibonding molecular orbital (designated as ψ_2*) results. A destabilizing node between the orbitals occurs when two orbitals of opposite phase combine.

Figure 30.1
The π and π* molecular orbitals of ethylene

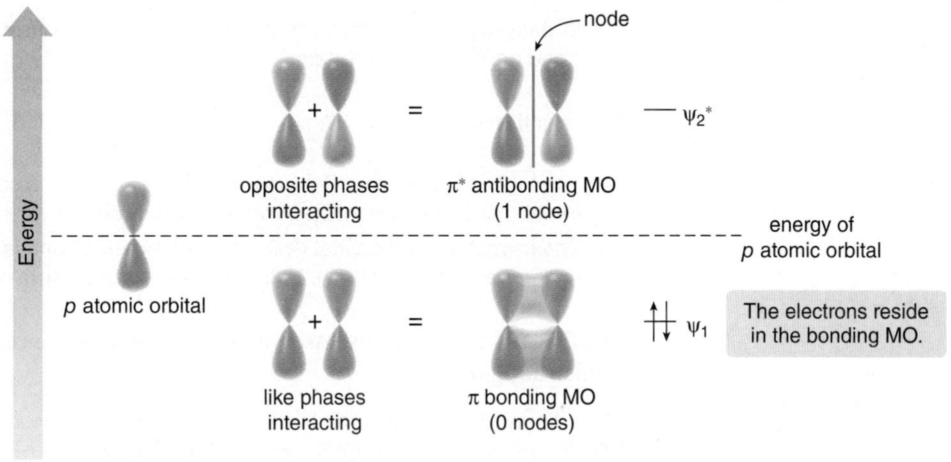

30.2B Buta-1,3-diene

The two π bonds of buta-1,3-diene ($CH_2{=}CH{-}CH{=}CH_2$) are formed by overlap of four *p* orbitals on four adjacent carbons. As shown in Figure 30.2, four *p* orbitals can combine in four different ways to form four molecular orbitals designated as ψ_1–ψ_4. Two are bonding molecular orbitals (ψ_1 and ψ_2), and two are antibonding molecular orbitals (ψ_3* and ψ_4*). The two bonding MOs are lower in energy than the *p* orbitals from which they are formed, whereas the two antibonding MOs are higher in energy than the *p* orbitals from which they are formed. **As the number of bonding interactions decreases and the number of nodes increases, the energy of the molecular orbital increases.**

Figure 30.2
The four π molecular orbitals of buta-1,3-diene

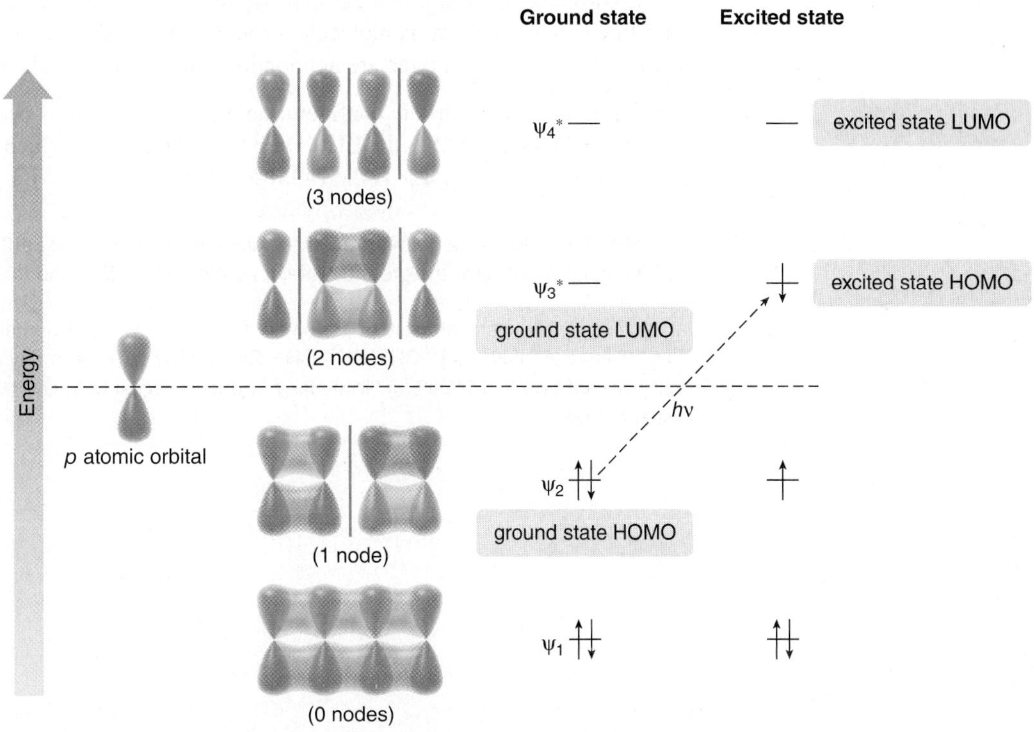

- The two lowest energy molecular orbitals, ψ_1 and ψ_2, are bonding MOs.
- The two highest energy molecular orbitals, ψ_3* and ψ_4*, are antibonding MOs.

- **In the ground-state electronic arrangement, the four π electrons occupy the two bonding molecular orbitals.**

Also recall from Section 17.9:

- **The highest energy orbital that contains electrons is called the highest occupied molecular orbital (HOMO). In the ground state of buta-1,3-diene, ψ_2 is the HOMO.**
- **The lowest energy orbital that contains no electrons is called the lowest unoccupied molecular orbital (LUMO). In the ground state of buta-1,3-diene, ψ_3* is the LUMO.**

The thermal reactions discussed in Section 30.3 utilize reactants in their ground state electronic configuration.

When buta-1,3-diene absorbs light of appropriate energy, an electron is promoted from ψ_2 (the HOMO) to ψ_3* (the LUMO) to form a higher energy electronic configuration, the **excited state.** In the excited state, the HOMO is now ψ_3*. **In the photochemical reactions in Section 30.3, the reactant is in its excited state.** As a result, the HOMO is ψ_3* and the LUMO is ψ_4* for buta-1,3-diene.

All conjugated dienes can be described by a set of molecular orbitals that are similar to those drawn in Figure 30.2 for buta-1,3-diene.

Problem 30.2 For each molecular orbital in Figure 30.2, count the number of bonding interactions (interactions between adjacent orbitals of similar phase) and the number of nodes. (a) How do these two values compare for a bonding molecular orbital? (b) How do these two values compare for an antibonding molecular orbital?

30.2C Hexa-1,3,5-triene

The three π bonds of hexa-1,3,5-triene ($CH_2{=}CH{-}CH{=}CH{-}CH{=}CH_2$) are formed by overlap of six *p* orbitals on six adjacent carbons. As shown in Figure 30.3, six *p* orbitals can combine in six different ways to form six molecular orbitals designated as ψ_1–ψ_6. Three are bonding molecular orbitals (ψ_1–ψ_3), and three are antibonding molecular orbitals (ψ_4*–ψ_6*).

In the ground state electronic configuration, the six π electrons occupy the three bonding MOs, ψ_3 is the HOMO, and ψ_4* is the LUMO. In the excited state, which results from promotion of an electron from ψ_3 to ψ_4*, ψ_4* is the HOMO and ψ_5* is the LUMO.

Problem 30.3 (a) Using Figure 30.2 as a guide, draw the molecular orbitals for hexa-2,4-diene. (b) Label the HOMO and the LUMO in the ground state. (c) Label the HOMO and the LUMO in the excited state.

Problem 30.4 (a) How many π molecular orbitals are present in deca-1,3,5,7,9-pentaene ($CH_2{=}CH{-}CH{=}CH{-}CH{=}CH{-}CH{=}CH{-}CH{=}CH_2$)? (b) How many are bonding MOs and how many are antibonding MOs? (c) How many nodes are present in ψ_1? (d) How many nodes are present in ψ_{10}*?

Figure 30.3 The six π molecular orbitals of hexa-1,3,5-triene

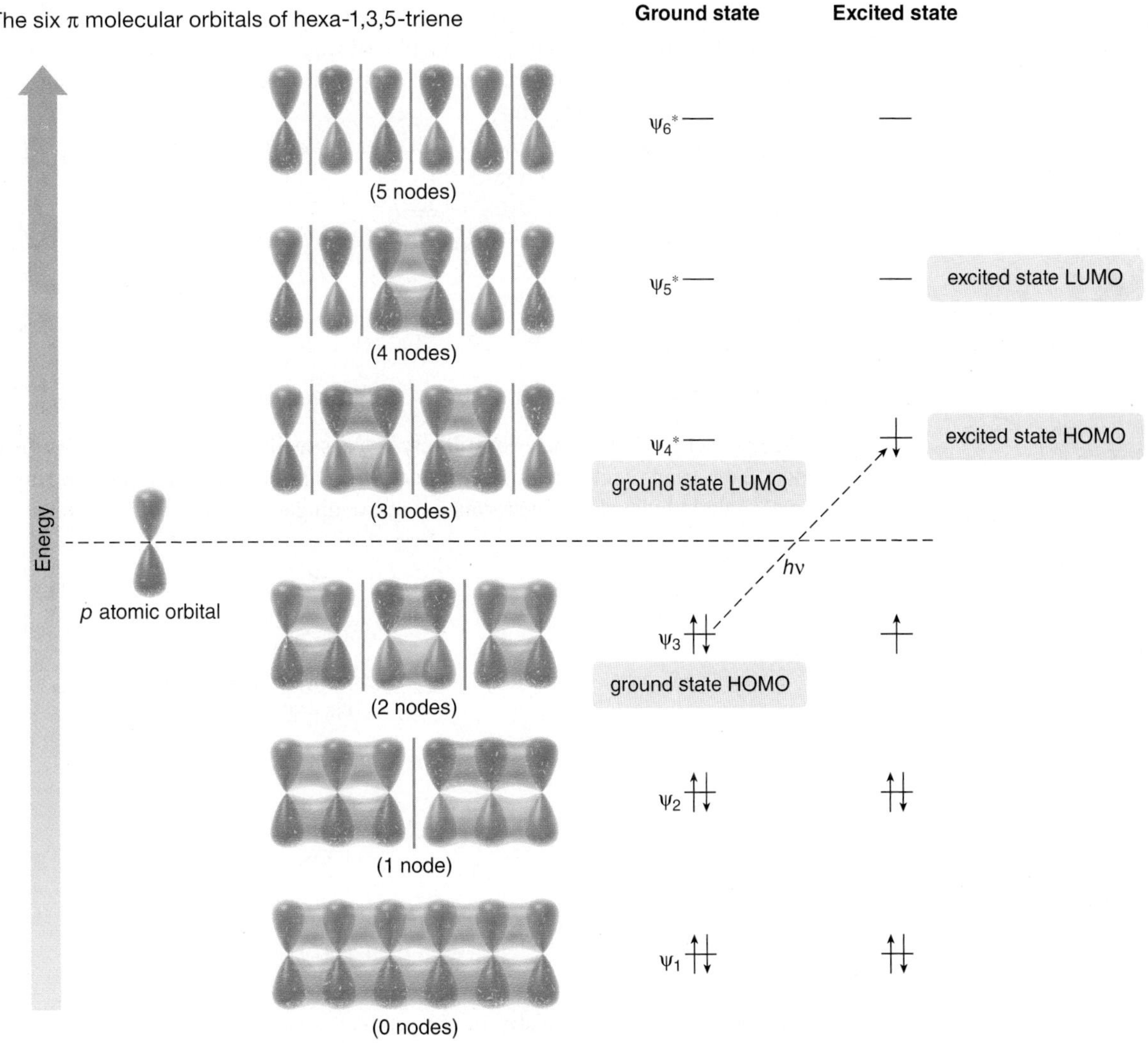

30.3 Electrocyclic Reactions

An electrocyclic reaction is a reversible reaction that involves ring closure of a conjugated polyene to a cycloalkene, or ring opening of a cycloalkene to a conjugated polyene. For example, ring closure of hexa-1,3,5-triene forms cyclohexa-1,3-diene, a product with one more σ bond and one fewer π bond than the reactant. Ring opening of cyclobutene forms buta-1,3-diene, a product with one fewer σ bond and one more π bond than the reactant.

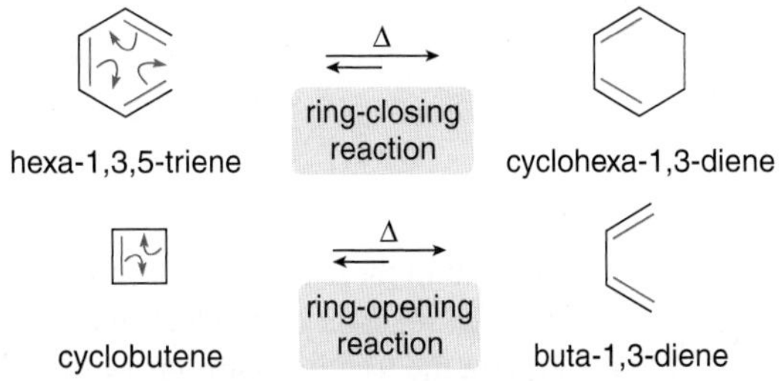

Arrows may be drawn in a clockwise or counterclockwise direction to show the flow of electrons.

- **To draw the product in each reaction, use curved arrows and begin at a π bond. Move the π electrons to an adjacent carbon–carbon bond and continue in a cyclic fashion.**

In a ring-closing reaction, this process forms a new σ bond that now joins the ends of the conjugated polyene. In a ring-opening reaction, this process breaks a σ bond to form a conjugated polyene with one more π bond.

Whether the reactant or product predominates at equilibrium depends on the ring size of the cyclic compound. Generally, a six-membered ring is favored over an acyclic triene at equilibrium. In contrast, an acyclic diene is favored over a strained four-membered ring.

Problem 30.5 Use curved arrows and draw the product of each electrocyclic reaction.

a. 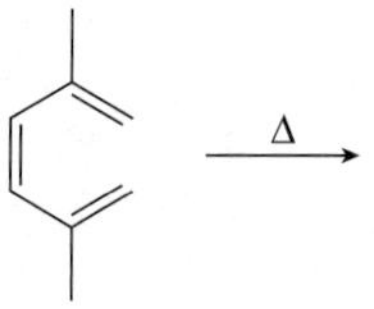

b. 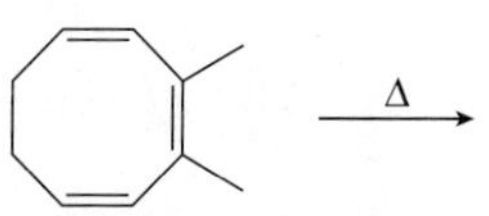

c. 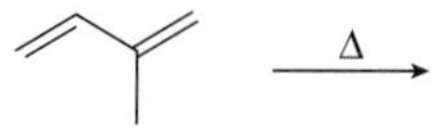

30.3A Stereochemistry and Orbital Symmetry

Electrocyclic reactions are completely stereospecific. For example, ring closure of (2*E*,4*Z*,6*E*)-octa-2,4,6-triene yields a single product with cis methyl groups on the ring. Ring opening of *cis*-3,4-dimethylcyclobutene forms a single conjugated diene with one *Z* alkene and one *E* alkene.

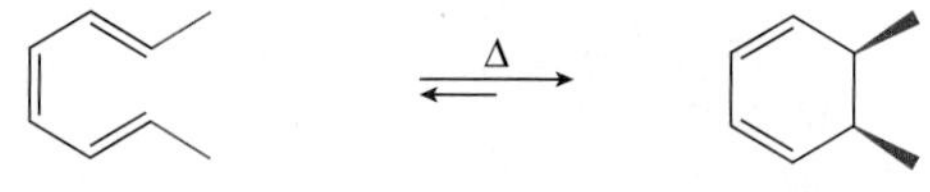

(2*E*,4*Z*,6*E*)-octa-2,4,6-triene — *cis*-5,6-dimethylcyclohexa-1,3-diene, **cis** product only

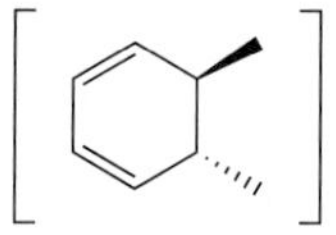

NOT formed

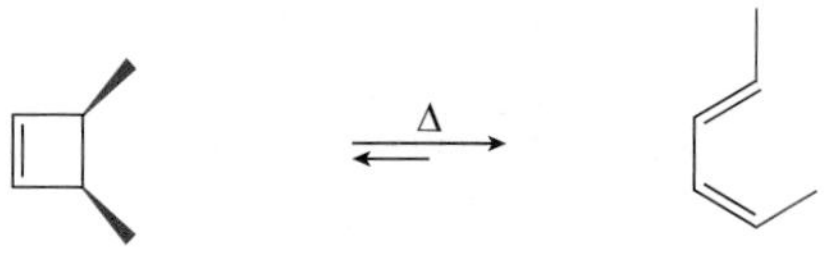

cis-3,4-dimethylcyclobutene — (2*E*,4*Z*)-hexa-2,4-diene, **(2*E*,4*Z*)** diene only

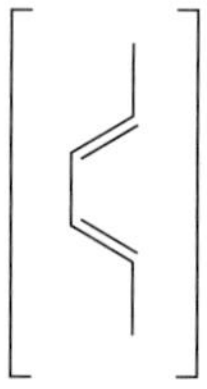

NOT formed

Moreover, the stereochemistry of the product of an electrocyclic reaction depends on whether the reaction is carried out under thermal or photochemical reaction conditions—that is, with heat or light, respectively. Cyclization of (2*E*,4*E*)-hexa-2,4-diene with heat forms a cyclobutene with trans methyl groups, whereas cyclization with light forms a cyclobutene with cis methyl groups.

Electrocyclic ring closure generally forms either an achiral meso compound or a mixture of chiral enantiomers. When enantiomers form, only one enantiomer is drawn in these reactions.

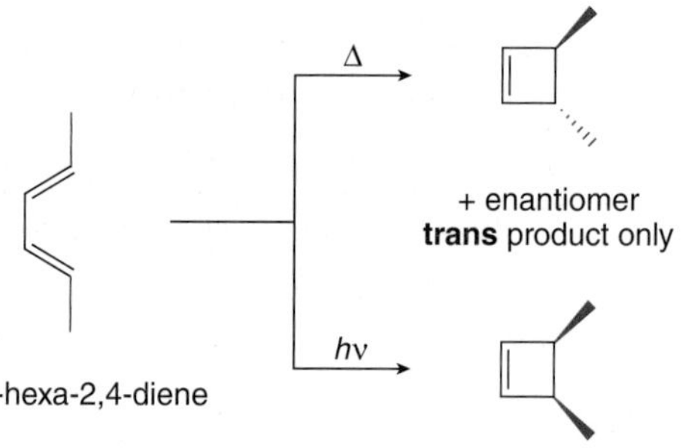

(2*E*,4*E*)-hexa-2,4-diene

cis product only

To understand these results, we must focus on the HOMO of the acyclic conjugated polyene that is either the reactant or product in an electrocyclic reaction. In particular, we must examine the *p* orbitals on the terminal carbons of the HOMO, and determine whether like phases of the orbitals are on the same side or on opposite sides of the molecule.

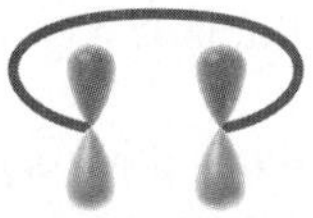

like phases on the same side

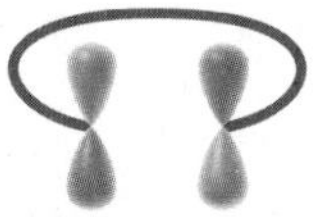

like phases on opposite sides

- An electrocyclic reaction occurs only when like phases of orbitals can overlap to form a bond. Such a reaction is *symmetry allowed.*
- An electrocyclic reaction cannot occur between lobes of opposite phase. Such a reaction is *symmetry forbidden.*

To form a bond, the *p* orbitals on the terminal carbons must rotate so that like phases can interact to form the new σ bond. Two modes of rotation are possible.

- When like phases of the *p* orbitals are on the same side of the molecule, the two orbitals must rotate in opposite directions—one clockwise and one counterclockwise. Rotation in opposite directions is said to be *disrotatory.*

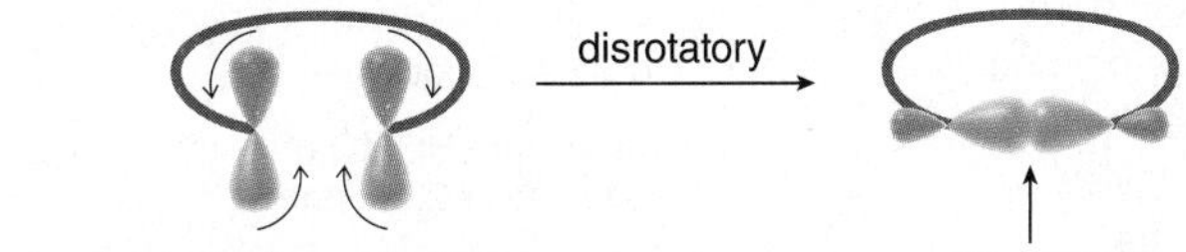

- When like phases of the *p* orbitals are on opposite sides of the molecule, the two orbitals must rotate in the same direction—both clockwise or both counterclockwise. Rotation in the same direction is said to be *conrotatory.*

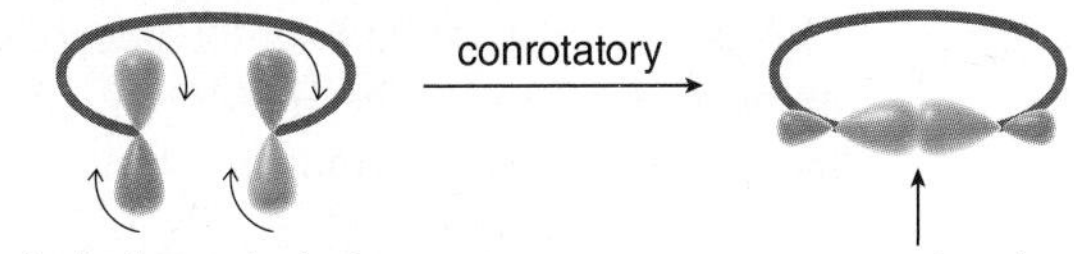

30.3B Thermal Electrocyclic Reactions

To explain the stereochemistry observed in electrocyclic reactions, we must examine the symmetry of the molecular orbital that contains the most loosely held π electrons. In a thermal reaction, we consider the **HOMO of the ground state electronic configuration.** Rotation occurs in a disrotatory or conrotatory fashion so that like phases of the *p* orbitals on the terminal carbons of this molecular orbital combine.

- The number of double bonds in the conjugated polyene determines whether rotation is conrotatory or disrotatory.

Two examples illustrate different outcomes.

Thermal electrocyclic ring closure of (2*E*,4*Z*,6*E*)-octa-2,4,6-triene yields a single product with cis methyl groups on the ring.

Only the *p* orbitals on the terminal carbons of the HOMO are drawn for clarity.

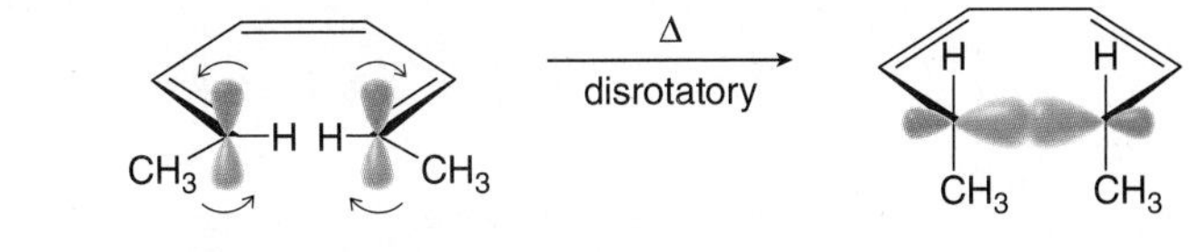

ground state HOMO of (2*E*,4*Z*,6*E*)-octa-2,4,6-triene

cis-5,6-dimethylcyclohexa-1,3-diene

cis product

Cyclization occurs in a disrotatory fashion because the HOMO of a conjugated triene has like phases of the outermost *p* orbitals on the *same* side of the molecule (Figure 30.3). A disrotatory ring closure is symmetry allowed because like phases of the *p* orbitals overlap to form the new σ bond of the ring. In the disrotatory ring closure, both methyl groups are pushed *down* (or *up*), making them *cis* in the product.

This is a specific example of the general process observed for conjugated polyenes with an *odd* number of π bonds. The HOMO of a conjugated polyene with an odd number of π bonds has like phases of the outermost *p* orbitals on the *same* side of the molecule. As a result:

- **Thermal electrocyclic reactions occur in a *disrotatory* fashion for a conjugated polyene with an *odd* number of π bonds.**

In contrast, thermal electrocyclic ring closure of (2*E*,4*E*)-hexa-2,4-diene forms a cyclobutene with trans methyl groups.

The conrotatory ring closure of (2*E*,4*E*)-hexa-2,4-diene is drawn with two clockwise rotations. The conrotatory ring closure could also be drawn with two counterclockwise rotations, leading to the enantiomer of the trans product drawn. Both enantiomers are formed in equal amounts.

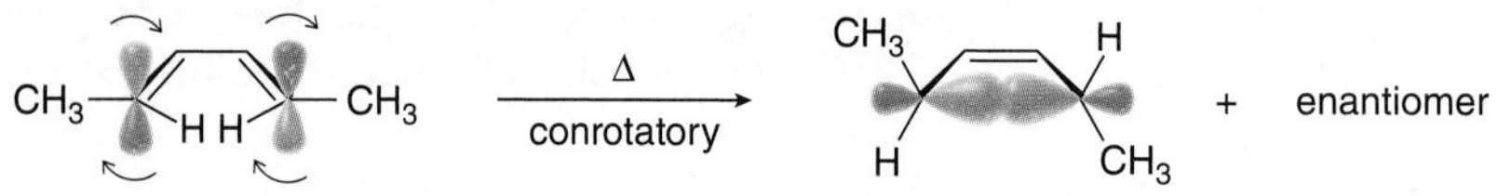

clockwise clockwise

ground state HOMO of (2*E*,4*E*)-hexa-2,4-diene

trans-3,4-dimethylcyclobutene

trans product

Cyclization occurs in a conrotatory fashion because the HOMO of a conjugated diene has like phases of the outermost *p* orbitals on *opposite* sides of the molecule (Figure 30.2). A conrotatory ring closure is symmetry allowed because like phases of the *p* orbitals overlap to form the new σ bond of the ring. In the conrotatory ring closure, one methyl group is pushed *down* and one methyl group is pushed *up,* making them *trans* in the product.

This is a specific example of the general process observed for conjugated polyenes with an *even* number of π bonds. The HOMO of a conjugated polyene with an even number of π bonds has like phases of the outermost *p* orbitals on *opposite* sides of the molecule. As a result:

- **Thermal electrocyclic reactions occur in a *conrotatory* fashion for a conjugated polyene with an *even* number of π bonds.**

Since electrocyclic reactions are reversible, **electrocyclic ring-opening reactions follow the same rules** as electrocyclic ring closures. Thus, thermal ring opening of *cis*-3,4-dimethylcyclobutene—which ring opens to a diene with an *even* number of π bonds—occurs in a conrotatory fashion to form (2*E*,4*Z*)-hexa-2,4-diene as the only product.

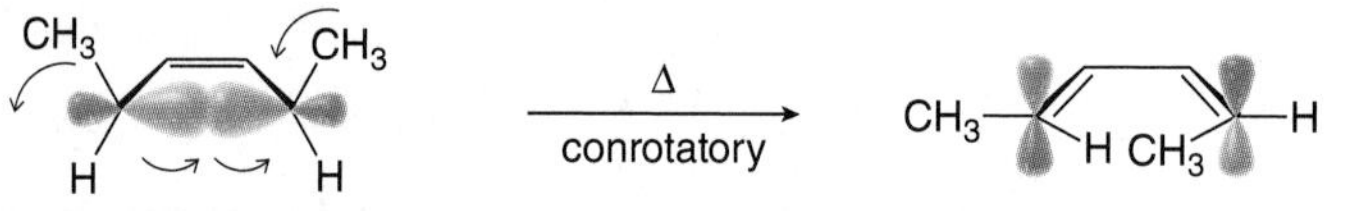

counterclockwise counterclockwise

cis-3,4-dimethylcyclobutene

(2*E*,4*Z*)-hexa-2,4-diene

Sample Problem 30.1 Draw the product of each thermal electrocyclic ring closure.

a. 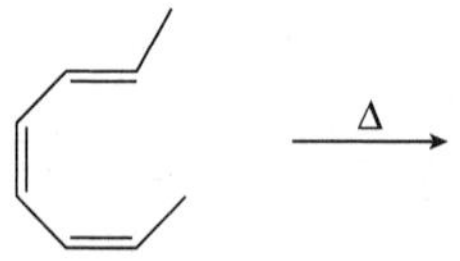

(2*E*,4*Z*,6*Z*)-octa-2,4,6-triene

b. 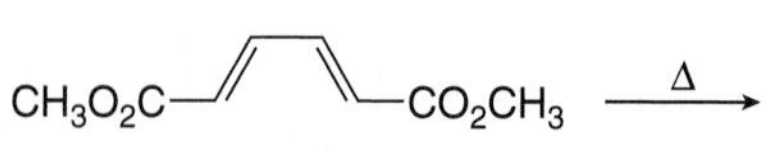

B

Solution

Count the number of π bonds in the conjugated polyene to determine the mode of ring closure in a thermal electrocyclic reaction.

- A conjugated polyene with an **odd** number of π bonds undergoes **disrotatory** cyclization.
- A conjugated polyene with an **even** number of π bonds undergoes **conrotatory** cyclization.

a. (2*E*,4*Z*,6*Z*)-Octa-2,4,6-triene contains three π bonds. The HOMO of a conjugated polyene with an odd number of π bonds has like phases of the outermost *p* orbitals on the *same* side of the molecule, and this results in disrotatory cyclization.

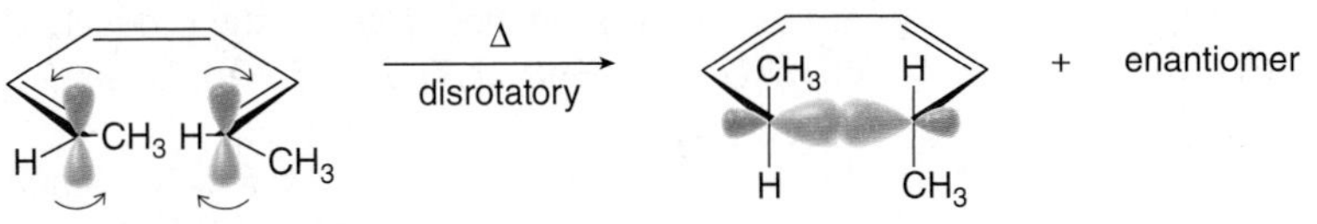

counterclockwise clockwise

(2*E*,4*Z*,6*Z*)-octa-2,4,6-triene

trans-5,6-dimethylcyclohexa-1,3-diene

trans product

b. Diene **B** contains two π bonds. The HOMO of a conjugated polyene with an even number of π bonds has like phases of the outermost *p* orbitals on *opposite* sides of the molecule, and this results in conrotatory cyclization.

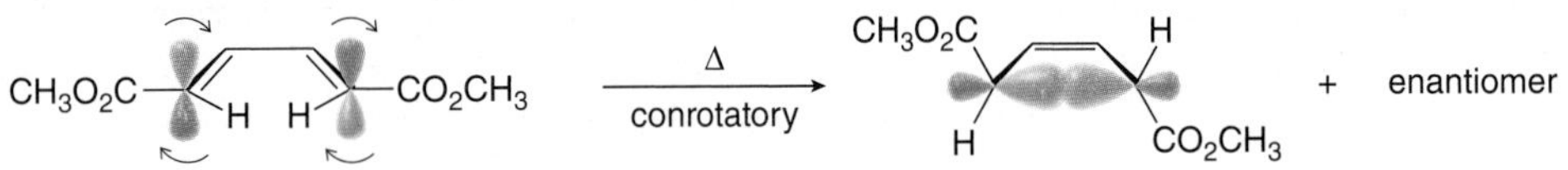

clockwise clockwise

B

trans product

Problem 30.6 What product is formed when each compound undergoes thermal electrocyclic ring opening or ring closure? Label each process as conrotatory or disrotatory, and clearly indicate the stereochemistry around tetrahedral stereogenic centers and double bonds.

a. 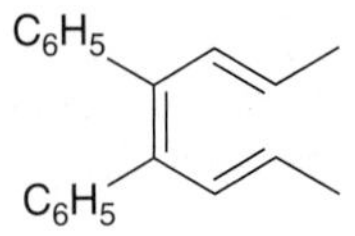

b. 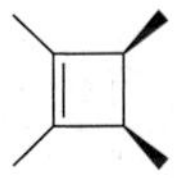

Problem 30.7 What cyclic product is formed when each decatetraene undergoes thermal electrocyclic ring closure?

a. 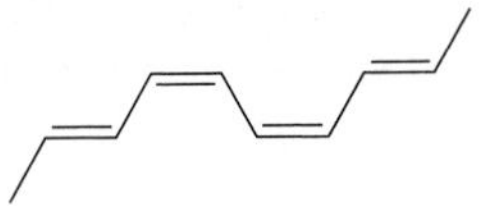

b. 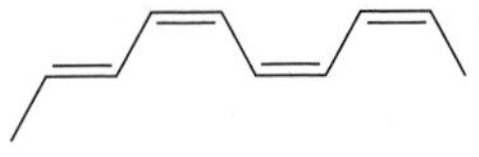

30.3C Photochemical Electrocyclic Reactions

Photochemical electrocyclic reactions follow similar principles as those detailed in thermal reactions with one important difference. **In photochemical reactions, we must consider the orbitals of the HOMO of the *excited* state to determine the course of the reaction.** As a photon is absorbed, an electron in the ground state HOMO is excited to the ground state LUMO. As a result, the excited state HOMO is one energy level higher than before (see Figures 30.2 and 30.3). The excited state HOMO has the *opposite* orientation of the outermost *p* orbitals compared to the HOMO of the ground state. As a result, **the method of ring closure of a photochemical electrocyclic reaction is *opposite* to that of a thermal electrocyclic reaction for the same number of π bonds.**

Photochemical electrocyclic ring closure of (2*E*,4*Z*,6*E*)-octa-2,4,6-triene yields a cyclic product with trans methyl groups on the ring.

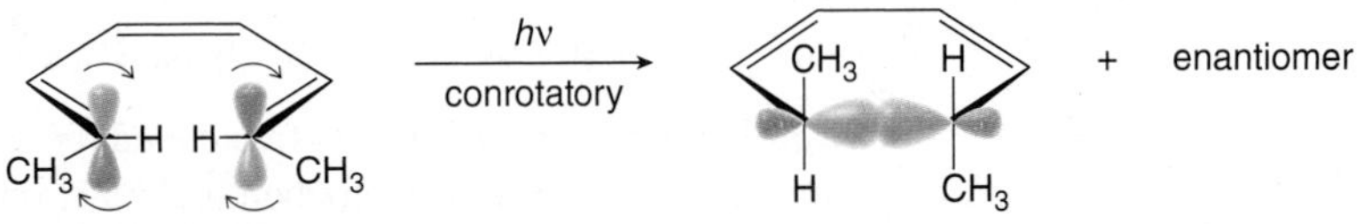

clockwise clockwise

excited state HOMO of (2*E*,4*Z*,6*E*)-octa-2,4,6-triene

trans-5,6-dimethylcyclohexa-1,3-diene

trans product

Cyclization occurs in a conrotatory fashion because the excited state HOMO of a conjugated triene has like phases of the outermost *p* orbitals on the *opposite* sides of the molecule (Figure 30.3). In the conrotatory ring closure, one methyl group is pushed *down* and one methyl group is pushed *up,* making them *trans* in the product. This is a specific example of the general process observed for conjugated polyenes with an *odd* number of π bonds.

- **Photochemical electrocyclic reactions occur in a *conrotatory* fashion for a conjugated polyene with an *odd* number of π bonds.**

Photochemical electrocyclic ring closure of (2*E*,4*E*)-hexa-2,4-diene forms a cyclobutene with cis methyl groups.

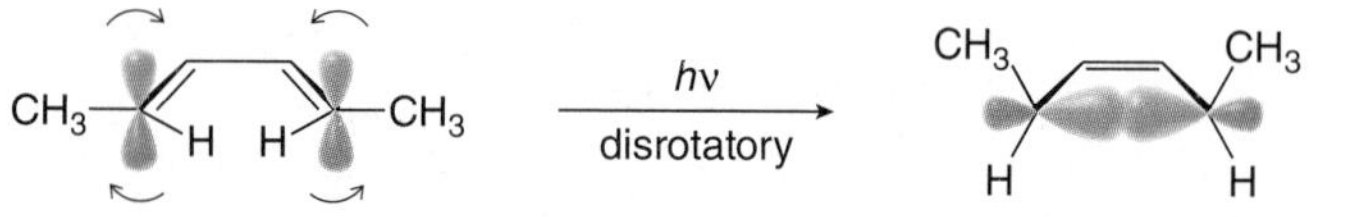

clockwise counterclockwise

excited state HOMO of (2*E*,4*E*)-hexa-2,4-diene

cis-3,4-dimethylcyclobutene
cis product

Cyclization occurs in a disrotatory fashion because the excited state HOMO of a conjugated diene has like phases of the outermost *p* orbitals on the *same* side of the molecule (Figure 30.3). In the disrotatory ring closure, both methyl groups are pushed *down* (or *up*), making them *cis* in the product. This is a specific example of the general process observed for conjugated polyenes with an *even* number of π bonds.

- **Photochemical electrocyclic reactions occur in a *disrotatory* fashion for a conjugated polyene with an *even* number of π bonds.**

Problem 30.8 What product is formed when each compound in Problem 30.6 undergoes photochemical electrocyclic ring opening or ring closure? Label each process as conrotatory or disrotatory, and clearly indicate the stereochemistry around tetrahedral stereogenic centers and double bonds.

Vitamin D (Problem 30.9) regulates calcium absorption, so adequate vitamin D levels are needed for proper bone growth. Vitamin D-fortified milk sold in the United States is produced by exposing milk to ultraviolet light.

Problem 30.9 Vitamin D_3, the most abundant of the D vitamins, is synthesized from 7-dehydrocholesterol, a compound found in milk and fatty fish such as salmon and mackerel. When the skin is exposed to sunlight, a photochemical electrocyclic ring opening forms provitamin D_3, which is then converted to vitamin D_3 by a sigmatropic rearrangement (Section 30.5). Draw the structure of provitamin D_3.

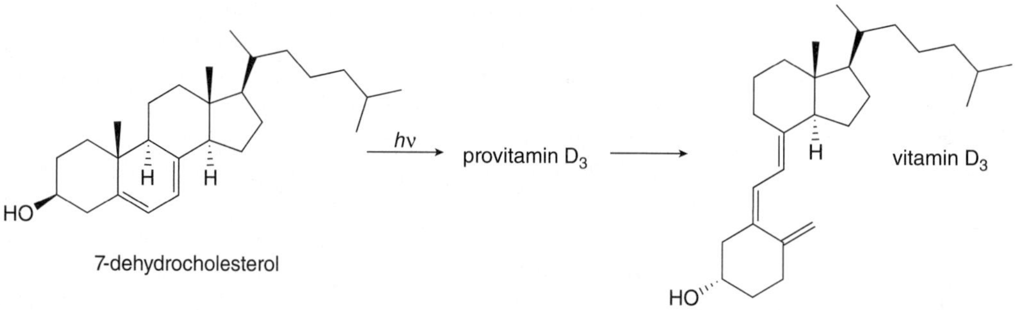

7-dehydrocholesterol

30.3D Summary of Electrocyclic Reactions

Table 30.1 summarizes the rules, often called the **Woodward–Hoffmann rules,** for electrocyclic reactions under thermal or photochemical reaction conditions. The number of π bonds refers to the acyclic conjugated polyene that is either the reactant or product of an electrocyclic reaction.

Table 30.1 Woodward–Hoffmann Rules for Electrocyclic Reactions

Number of π bonds	Thermal reaction	Photochemical reaction
Even	Conrotatory	Disrotatory
Odd	Disrotatory	Conrotatory

Sample Problem 30.2 Identify **A** and **B** in the following reaction sequence. Label each process as conrotatory or disrotatory.

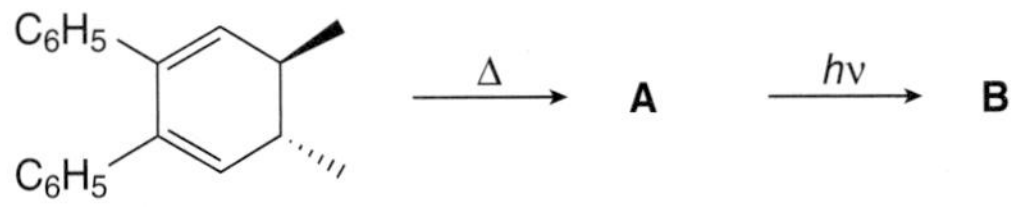

Solution

Ring opening of a cyclohexadiene forms a hexatriene with **three** π bonds. A conjugated polyene with an odd number of π bonds undergoes a thermal electrocyclic reaction in a disrotatory fashion (Table 30.1). The resulting hexatriene (**A**) then undergoes a photochemical electrocyclic reaction in a conrotatory fashion to form a cyclohexadiene with cis methyl groups (**B**).

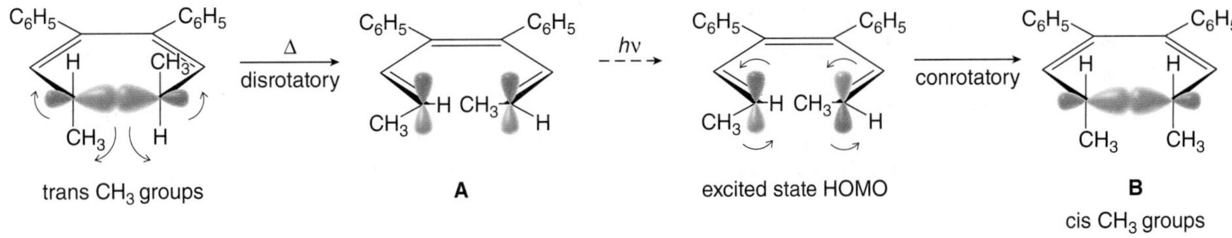

Problem 30.10 Draw the product formed when each triene undergoes electrocyclic reaction under [1] thermal conditions; [2] photochemical conditions.

a. 

b. 

Problem 30.11 What product would be formed by the disrotatory cyclization of the given triene? Would this reaction occur under photochemical or thermal conditions?

30.4 Cycloaddition Reactions

A cycloaddition is a reaction between two compounds with π bonds to form a cyclic product with two new σ bonds. Like electrocyclic reactions, cycloadditions are concerted, stereospecific reactions, and the course of the reaction is determined by the symmetry of the molecular orbitals of the reactants.

Cycloadditions can be initiated by heat (thermal conditions) or light (photochemical conditions). Cycloadditions are identified by the number of π electrons in the two reactants.

The Diels–Alder reaction is a thermal [4 + 2] cycloaddition that occurs between a diene with four π electrons and an alkene (dienophile) with two π electrons (Sections 16.12–16.14).

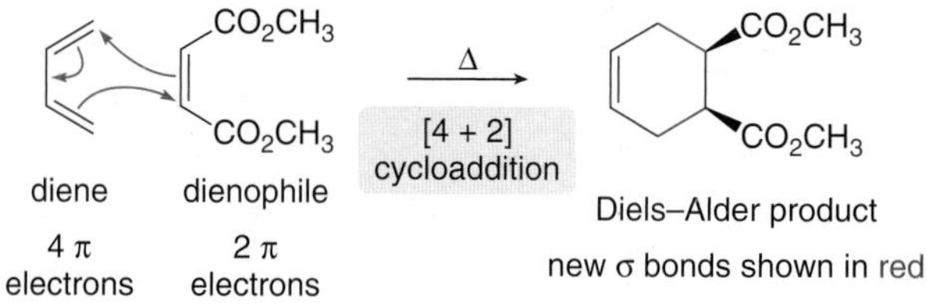

A photochemical [2 + 2] cycloaddition occurs between two alkenes, each with two π electrons, to form a cyclobutane. Thermal [2 + 2] cycloadditions do *not* take place.

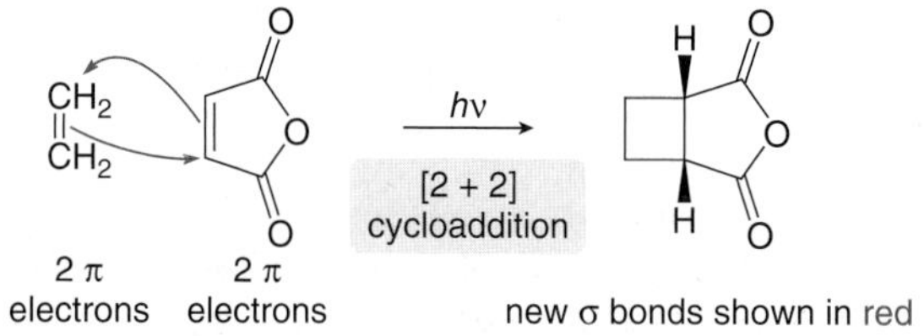

Sample Problem 30.3 What type of cycloaddition is shown in each equation?

a. 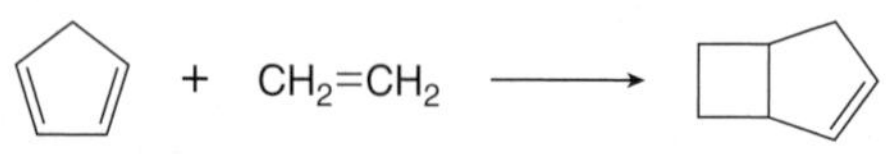

b. 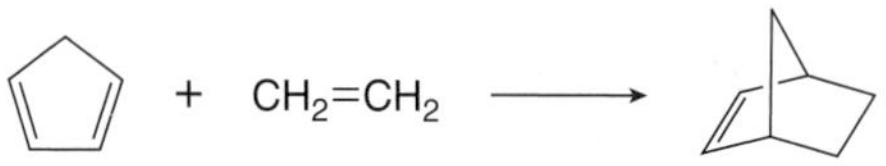

Solution

Count the number of π electrons *involved in each reactant* to classify the cycloaddition.

a. [2 + 2] Cycloaddition

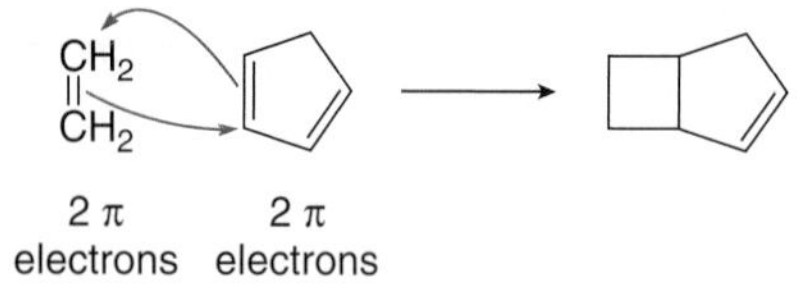

b. [4 + 2] Cycloaddition

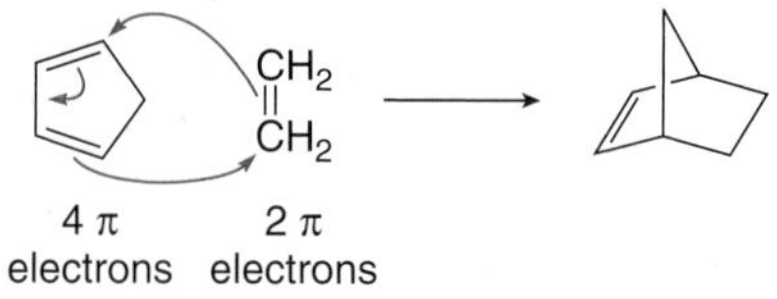

Problem 30.12 Consider cycloheptatrienone and ethylene, and draw a possible product formed from each type of cycloaddition: (a) [2 + 2]; (b) [4 + 2]; (c) [6 + 2].

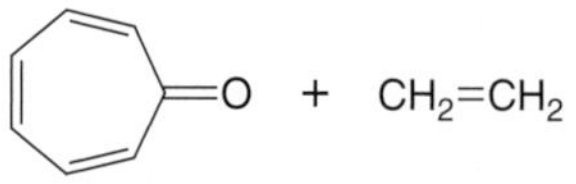

cycloheptatrienone

30.4A Orbital Symmetry and Cycloadditions

To understand cycloaddition reactions, we examine the *p* orbitals of the terminal carbons of both reactants. Bonding can take place only when like phases of both sets of *p* orbitals can combine. Two modes of reaction are possible.

- **A suprafacial cycloaddition occurs when like phases of the *p* orbitals of both reactants are on the *same* side of the π system, so that two bonding interactions result.**

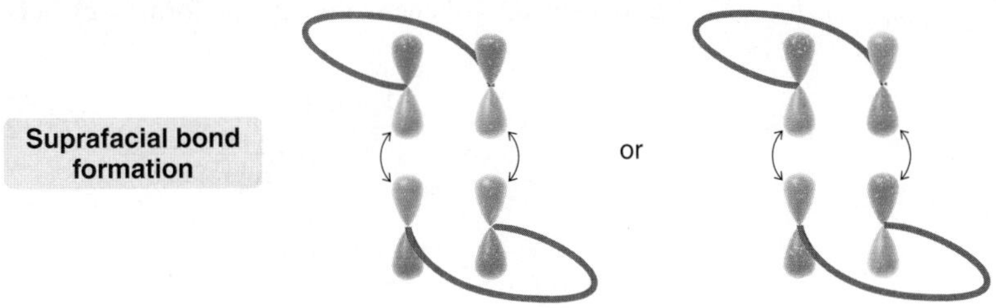

- **An antarafacial cycloaddition occurs when one π system must *twist* to align like phases of the *p* orbitals of the terminal carbons of the reactants.**

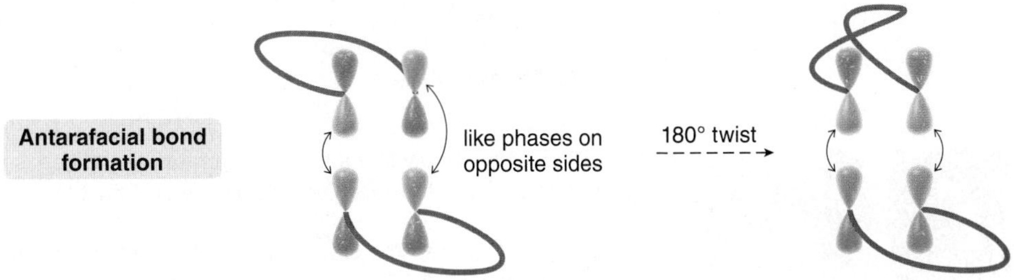

Because of the geometrical constraints of small rings, **cycloadditions that form four- or six-membered rings must take place by suprafacial pathways.**

Since cycloaddition involves the donation of electron density from one reactant to another, one reactant donates its most loosely held electrons—those occupying its HOMO—to a vacant orbital that can accept electrons—the LUMO—of the second reactant. The HOMO of either reactant can be used for analysis.

- **In a cycloaddition we examine the bonding interactions of the HOMO of one component with the LUMO of the second component.**

30.4B [4 + 2] Cycloadditions

To examine the course of a [4 + 2] cycloaddition, let's arbitrarily choose the HOMO of the diene and the LUMO of the alkene, and look at the symmetry of the *p* orbitals on the terminal carbons of both components. Since two bonding interactions result from overlap of the like phases of both sets of *p* orbitals, **a [4 + 2] cycloaddition occurs readily by suprafacial reaction under thermal conditions.**

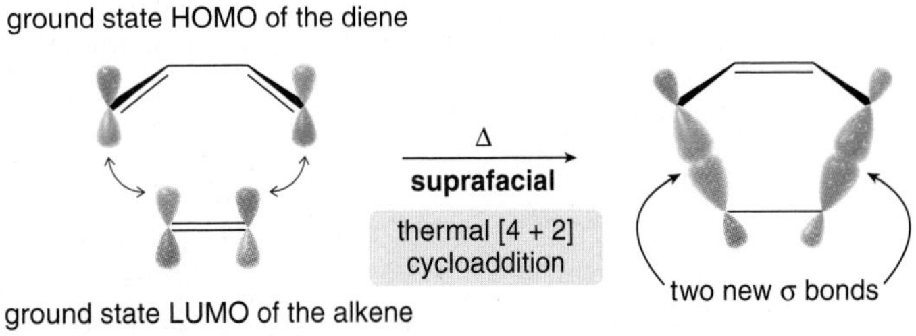

This is a specific example of a general cycloaddition involving an odd number of π bonds (three π bonds total, two from the diene and one from the alkene).

- **Thermal cycloadditions involving an *odd* number of π bonds proceed by a *suprafacial* pathway.**

In Section 16.13, we learned that the stereochemistry of the dienophile is retained in the Diels–Alder product.

Because a Diels–Alder reaction follows a concerted, suprafacial pathway, the **stereochemistry of the diene is retained in the Diels–Alder product.** As a result, reaction of (2*E*,4*E*)-hexa-2,4-diene with ethylene forms a cyclohexene with cis substituents (Reaction [1]), whereas reaction of (2*E*,4*Z*)-hexa-2,4-diene with ethylene forms a cyclohexene with trans substituents (Reaction [2]).

[1]

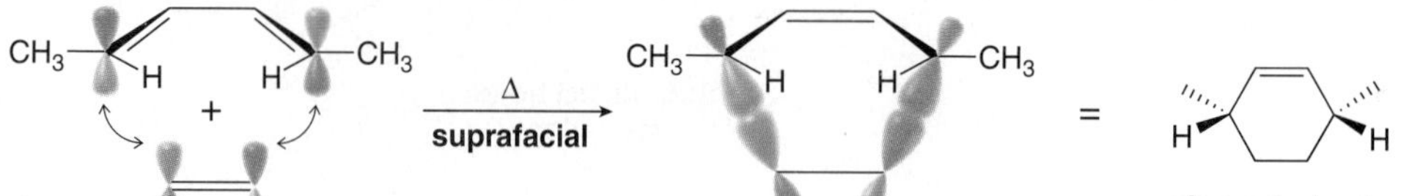

(2*E*,4*E*)-hexa-2,4-diene

cis product only

[2]

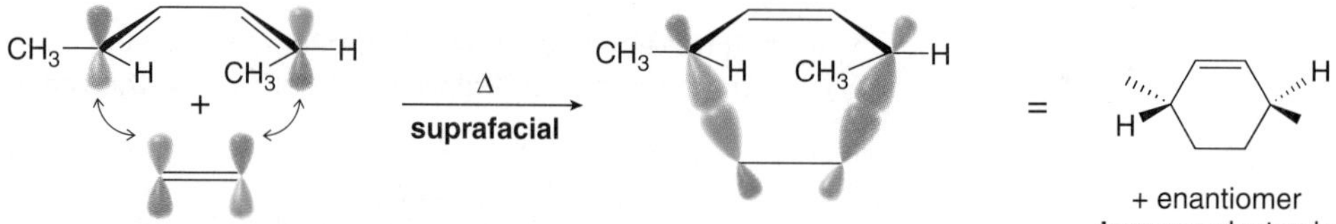

(2*E*,4*Z*)-hexa-2,4-diene

+ enantiomer
trans product only

Reserpine (Problem 30.15) is isolated from the roots of *Rauwolfia serpentina*, a plant used for centuries in traditional Indian medicine for the treatment of mental disorders and venomous bites.

Problem 30.13 Show that a thermal suprafacial addition is symmetry allowed in a [4 + 2] cycloaddition by using the HOMO of the alkene and the LUMO of the diene.

Problem 30.14 Draw the product (including stereochemistry) formed from each pair of reactants in a thermal [4 + 2] cycloaddition reaction.

a.

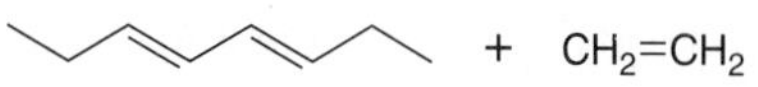

b.

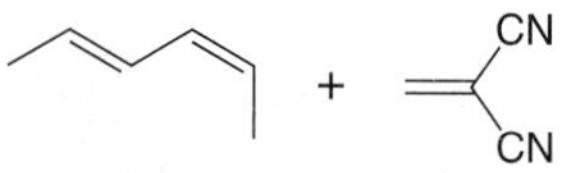

Problem 30.15 (a) Draw the product of the following [4 + 2] cycloaddition, which was carried out in the early stages of the synthesis of the alkaloid reserpine (Problem 22.85). Indicate the stereochemistry at any newly formed stereogenic centers. (b) Draw the *p* orbitals of the alkene and the terminal carbons of the conjugated diene, and show how the orientation of the reactants and orbital overlap lead to the observed stereochemistry.

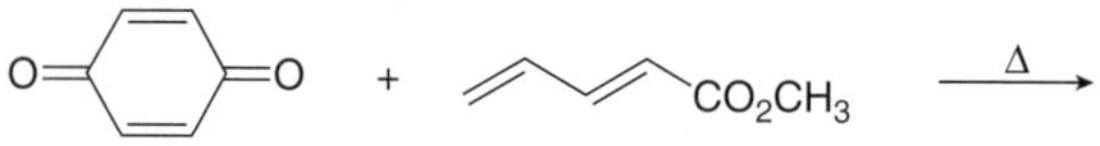

30.4C [2 + 2] Cycloadditions

In contrast to a [4 + 2] cycloaddition, a [2 + 2] cycloaddition does *not* occur under thermal conditions, but *does* take place photochemically. This result is explained by examining the symmetry of the HOMO and LUMO of the alkene reactants.

In a thermal [2 + 2] cycloaddition, like phases of the *p* orbitals on only one set of terminal carbons can overlap. For like phases to overlap on the other terminal carbon, the molecule must twist to allow for an antarafacial pathway. This process *cannot* occur to form small rings.

ground state HOMO of one alkene

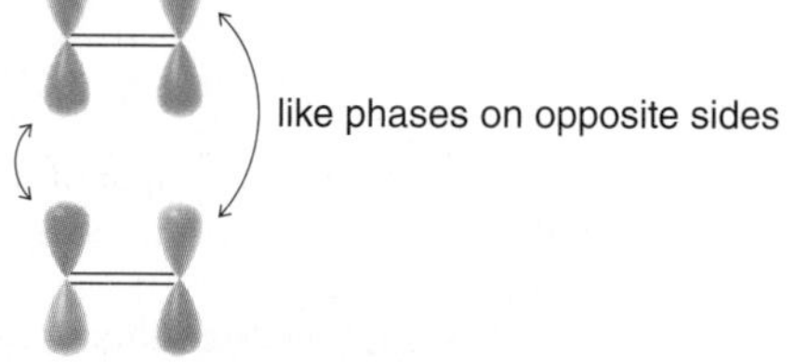

LUMO of the second alkene

In a photochemical [2 + 2] cycloaddition, light energy promotes an electron from the ground state HOMO to form the excited state HOMO (designated as ψ_2^* in Figure 30.1). Interaction of this excited state HOMO with the LUMO of the second alkene then allows for overlap of the like phases of both sets of *p* orbitals. Two bonding interactions result and the reaction occurs by a suprafacial pathway.

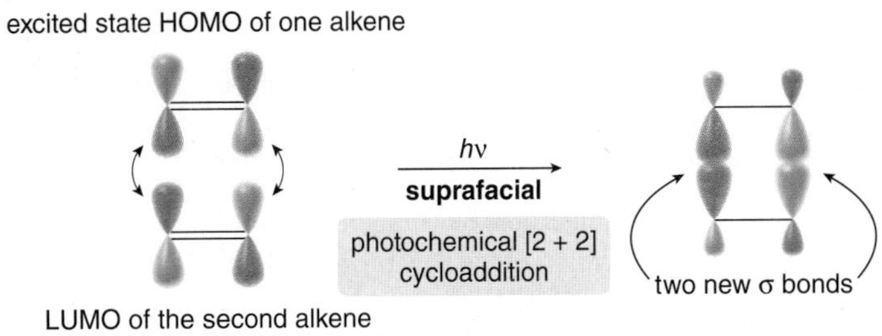

This is a specific example of a general cycloaddition involving an even number of π bonds (two π bonds total, one from each alkene).

- **Photochemical cycloadditions involving an *even* number of π bonds proceed by a *suprafacial* pathway.**

Problem 30.16 Draw the product formed in each cycloaddition.

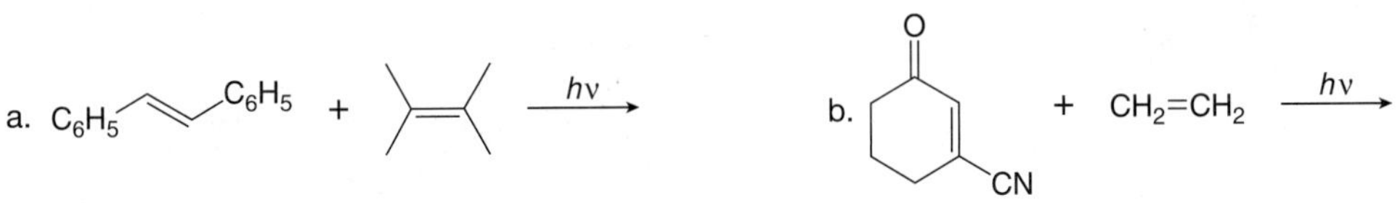

30.4D Summary of Cycloaddition Reactions

Table 30.2 summarizes the Woodward–Hoffmann rules that govern cycloaddition reactions. The number of π bonds refers to the total number of π bonds from both components of the cycloaddition. For a given number of π bonds, the mode of cycloaddition is always *opposite* in thermal and photochemical reactions.

Table 30.2 Woodward–Hoffmann Rules for Cycloaddition Reactions

Number of π bonds	Thermal reaction	Photochemical reaction
Even	Antarafacial	Suprafacial
Odd	Suprafacial	Antarafacial

Problem 30.17 Using the Woodward–Hoffmann rules, predict the stereochemical pathway for each cycloaddition: (a) a [6 + 4] photochemical reaction; (b) an [8 + 2] thermal reaction.

Problem 30.18 Using orbital symmetry, explain why a Diels–Alder reaction does not take place under photochemical reaction conditions.

30.5 Sigmatropic Rearrangements

A sigmatropic rearrangement is an intramolecular pericyclic reaction in which a σ bond is broken in a reactant, the π bonds rearrange, and a new σ bond is formed in the product. In a sigmatropic rearrangement, the number of π bonds in the reactant and product is constant, and the σ bonds broken and formed are **allylic** C−H, C−C, or C−Z bonds (Z = N, O, or S). A sigmatropic rearrangement that results in cleavage and formation of a C−H bond is shown.

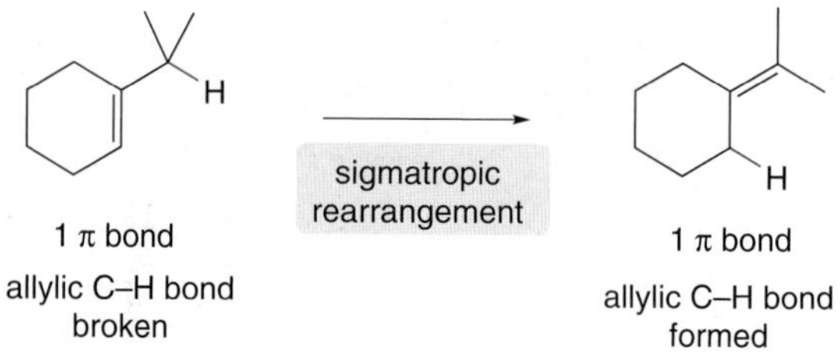

Sigmatropic rearrangements are characterized by a set of numbers in brackets, [*n*,*m*], to indicate the location of the new σ bond relative to the broken σ bond. To designate a sigmatropic rearrangement:

- **Locate the σ bond broken in the reactant and label both atoms in the bond with "1's."**
- **Locate the new σ bond in the product, and count the number of atoms from the broken σ bond to the new σ bond for each fragment.**
- **Place both numbers in brackets, with the lower number first. In a rearrangement involving a C−H bond, the first number is always "1."**

For example, a [3,3] sigmatropic rearrangement converts diene **A** into diene **B** when an allylic C−C bond in **A** is broken and a new allylic C−C bond is formed in **B.**

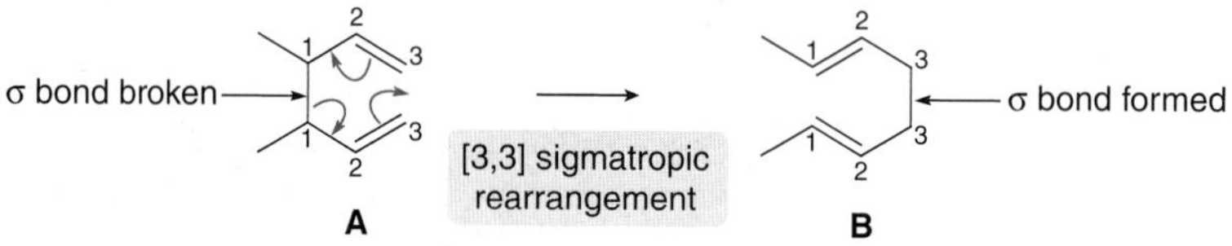

Sample Problem 30.4 What type of sigmatropic rearrangement is illustrated in each equation?

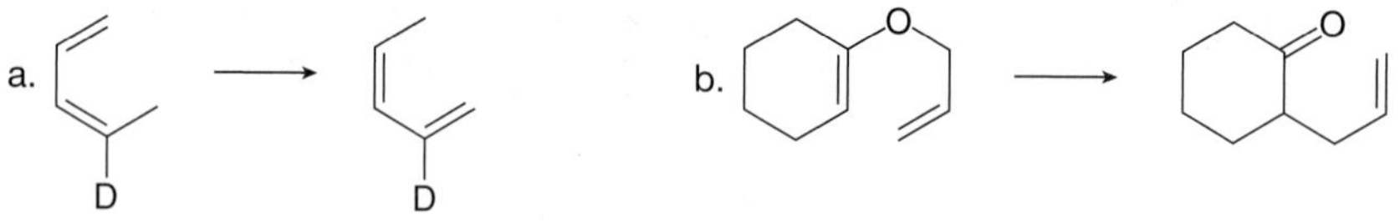

Solution

Locate the atoms in the broken σ bond and label them with 1's. Locate the atoms in the new σ bond, and count the number of atoms from the bond broken to the bond formed. When a C−H bond is broken, the first number in the [*n*,*m*] designation must be 1, because the H atom is bonded to no other atom.

a. A C−H bond is broken on the allylic C and a new C−H bond is formed on C5, so the reaction is a **[1,5] sigmatropic rearrangement.**

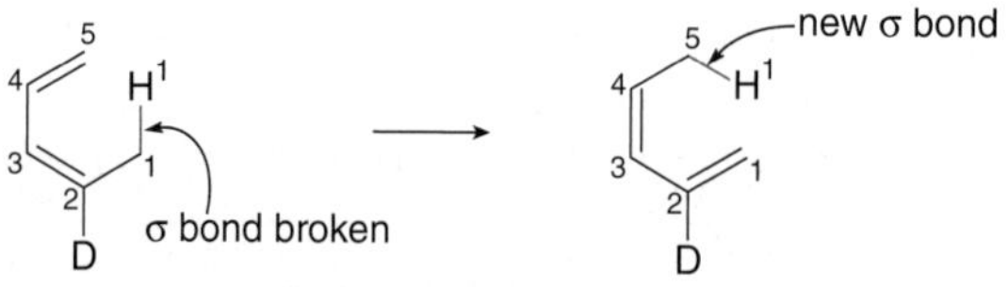

b. The reaction is a **[3,3] sigmatropic rearrangement,** because a C−O σ bond is broken and a new allylic C−C σ bond is formed between carbons that are three atoms removed from the broken bond.

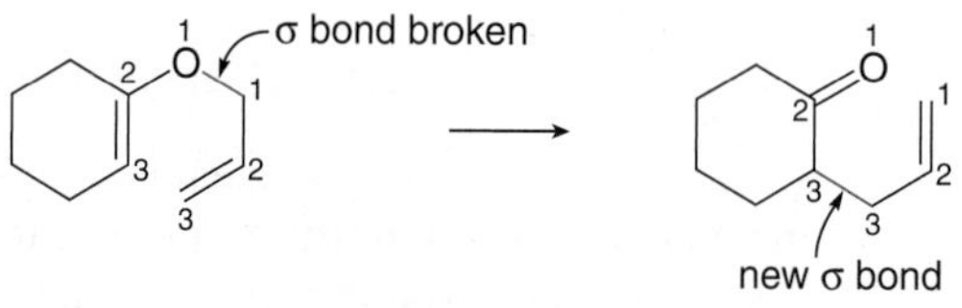

Problem 30.19 What type of sigmatropic rearrangement is illustrated in each equation?

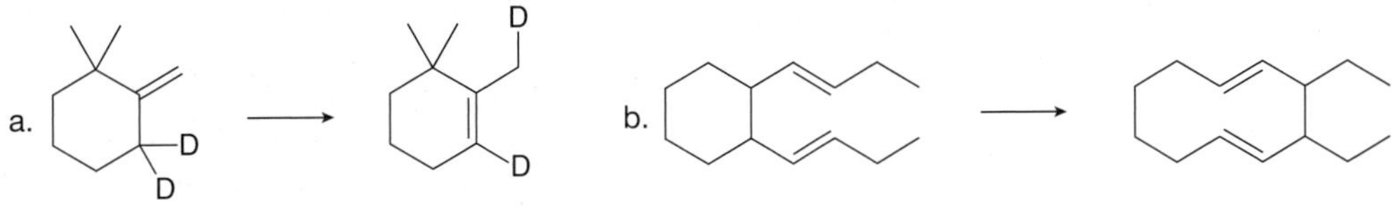

30.5A Sigmatropic Rearrangements and Orbital Symmetry

The stereochemistry of a sigmatropic rearrangement, like that of other pericyclic reactions, is determined by the symmetry of the orbitals involved in the reaction. In sigmatropic rearrangements, we consider the orbitals of the σ bond that is broken and the terminal *p* orbital of the π bond at which the new σ bond forms. Two modes of rearrangement are possible: **suprafacial** and **antarafacial.**

- **In a suprafacial rearrangement, the new σ bond forms on the *same* side of the π system as the broken σ bond.**

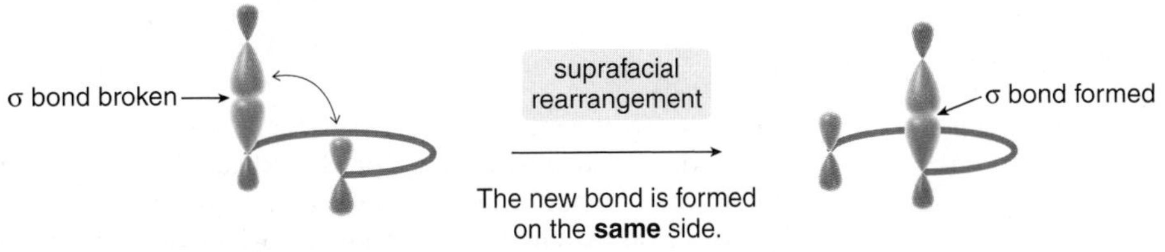

- **In an antarafacial rearrangement, the new σ bond forms on the *opposite* side of the π system as the broken σ bond.**

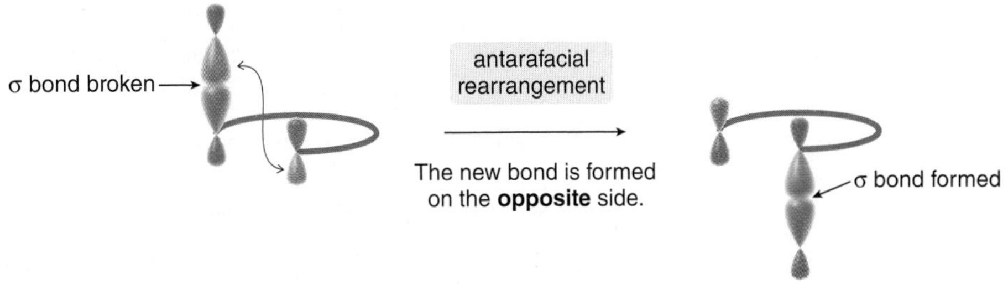

Sigmatropic rearrangements can occur under thermal or photochemical conditions, and follow the same rules observed in cycloaddition reactions. With sigmatropic rearrangements we count the total number of electron pairs in the σ bond that is broken and the π bonds that rearrange (Table 30.3). Because sigmatropic rearrangements involve cyclic transition states and small rings have geometrical constraints, **reactions involving six or fewer atoms must take place by suprafacial pathways.**

Table 30.3 Woodward–Hoffmann Rules for Sigmatropic Rearrangements

Number of electron pairs	Thermal reaction	Photochemical reaction
Even	Antarafacial	Suprafacial
Odd	Suprafacial	Antarafacial

For example, a [1,5] sigmatropic rearrangement of **X** to **Y** involves three electron pairs, one from the σ bond that is broken and two from the π bonds that rearrange.

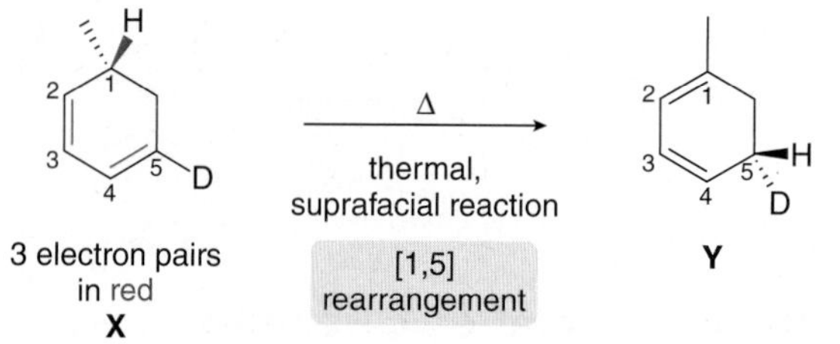

According to Table 30.3, this reaction must occur in a suprafacial mode under thermal conditions and in an antarafacial mode under photochemical conditions. Since this reaction involves only six atoms (including the H atom that migrates), it must take place under thermal conditions in a suprafacial fashion.

Sample Problem 30.5 Classify the following sigmatropic rearrangement and determine whether it takes place readily under thermal or photochemical reaction conditions.

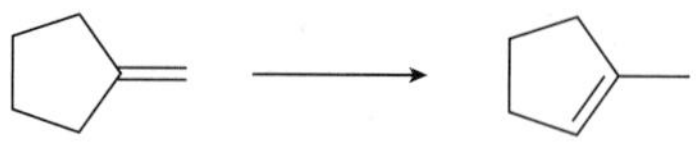

Solution

First, classify the rearrangement as in Sample Problem 30.4: Label the atoms in the broken σ bond with 1's, locate the new σ bond, and count the number of atoms from the bond broken to the bond formed. Then, count the number of electron pairs involved in the reaction, and use Table 30.3 to determine the stereochemical pathway of the reaction. Keep in mind that reactions involving six or fewer atoms must take place by suprafacial pathways.

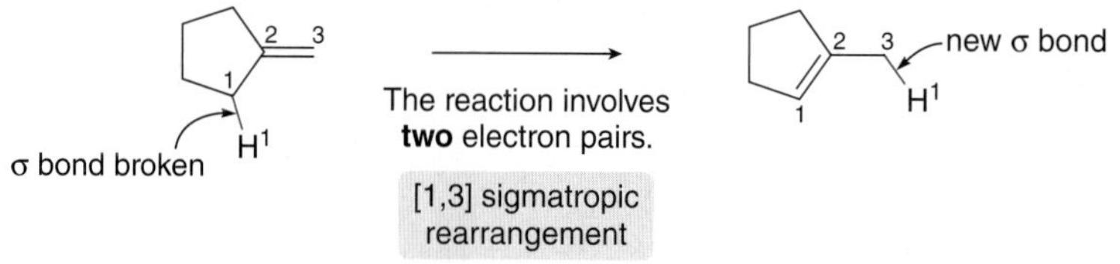

This reaction is a [1,3] sigmatropic rearrangement, involving two electron pairs: the C–H σ bond broken and one π bond. Since the reaction involves four atoms, it must take place via a suprafacial pathway, which occurs under photochemical conditions.

Problem 30.20 (a) What product is formed from the [1,7] sigmatropic rearrangement of a deuterium in the following triene? (b) Does this reaction proceed in a suprafacial or antarafacial manner under thermal conditions? (c) Does this reaction proceed in a suprafacial or antarafacial manner under photochemical conditions?

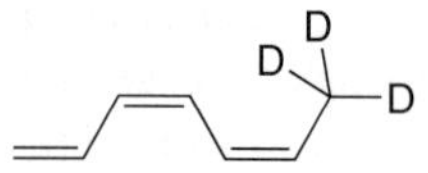

30.5B [3,3] Sigmatropic Rearrangements

Two widely used [3,3] sigmatropic rearrangements in organic synthesis are the **Cope rearrangement** of a 1,5-diene to an isomeric 1,5-diene, and the **Claisen rearrangement** of an unsaturated ether to a γ,δ-unsaturated carbonyl compound.

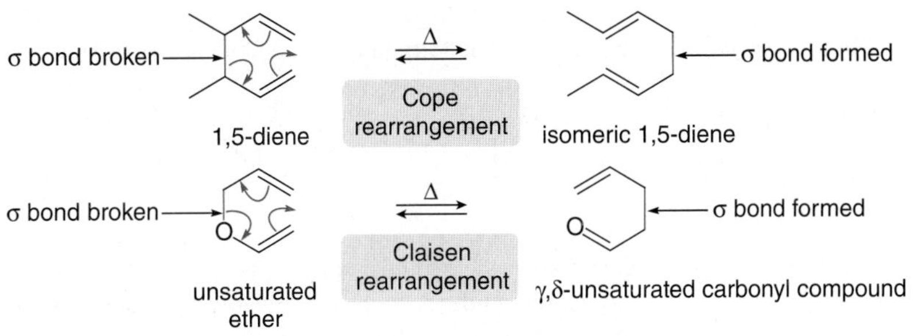

Both reactions involve three electron pairs—two π bonds and one σ bond—and six atoms, and take place readily in a **suprafacial pathway under thermal conditions.**

Cope Rearrangement

Since a Cope rearrangement involves isomeric 1,5-dienes as reactant and product, the more stable diene is favored at equilibrium. Useful Cope rearrangements occur when the reactant 1,5-diene is considerably less stable than the product, as in the case of *cis*-1,2-divinylcyclobutane, which rearranges to cycloocta-1,5-diene with loss of strain from the cyclobutane ring.

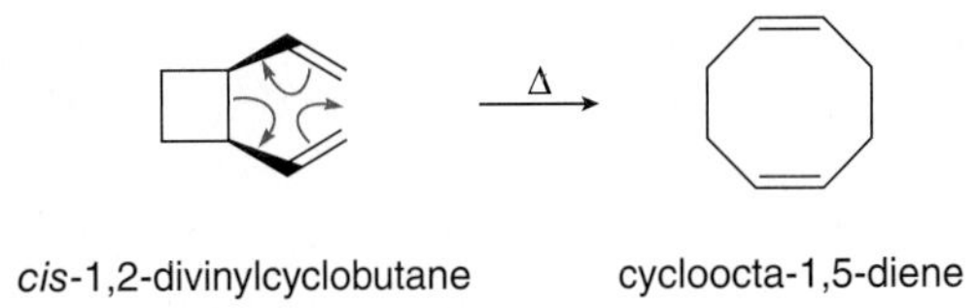

The **oxy-Cope rearrangement** is an especially powerful variation of a Cope rearrangement using an unsaturated alcohol. [3,3] Sigmatropic rearrangement forms an enol initially, which then tautomerizes to form a carbonyl group.

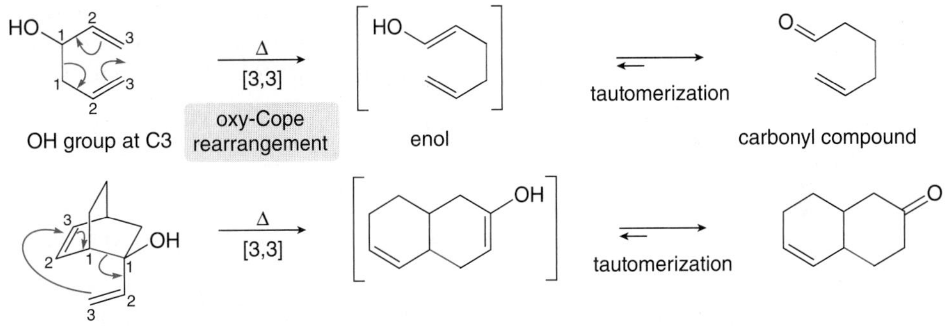

Moreover, ***anionic*** **oxy-Cope rearrangements** often give high yields of rearranged product under very mild reaction conditions. In an anionic oxy-Cope rearrangement, the unsaturated alcohol reactant is first treated with strong base, usually KH in the presence of 18-crown-6 (Section 9.5B), to form an alkoxide. [3,3] Sigmatropic rearrangement then yields a resonance-stabilized enolate, which is protonated to form a carbonyl product.

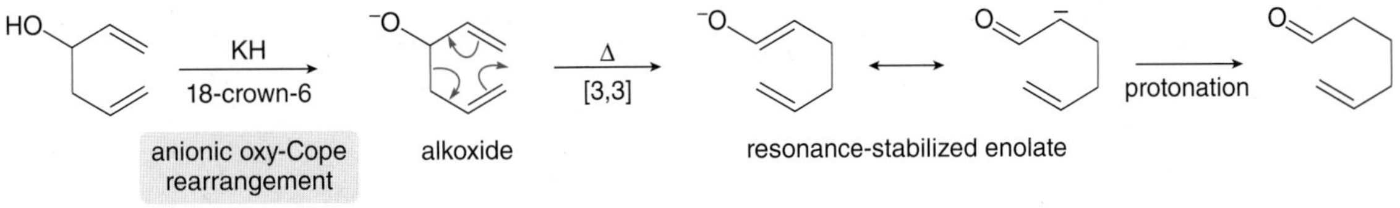

Problem 30.21 What product is formed from the Cope or oxy-Cope rearrangement of each starting material?

a. 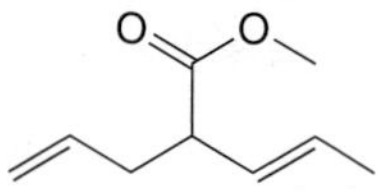 b. 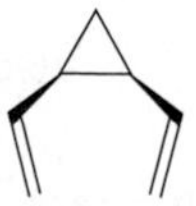 c. 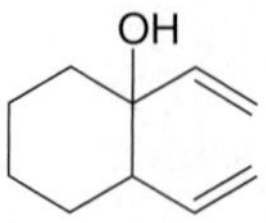

Problem 30.22 One step in the synthesis of periplanone B, the chapter-opening molecule, involved anionic oxy-Cope rearrangement of the following unsaturated alcohol. Draw the product that results after protonation of the intermediate enolate.

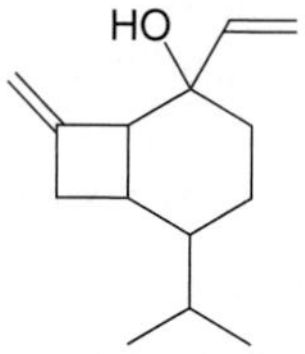

Problem 30.23 What compound forms geranial (Figure 21.6) by a Cope rearrangement?

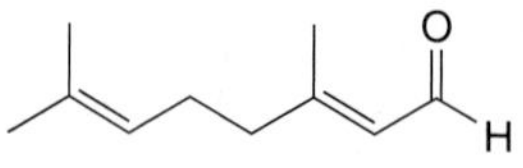

geranial

Claisen Rearrangement

A Claisen rearrangement is a [3,3] sigmatropic rearrangement of an unsaturated ether, either an allyl vinyl ether or an allyl aryl ether. With an allyl vinyl ether, a γ,δ-unsaturated carbonyl compound is formed directly by the concerted rearrangement. With an allyl aryl ether, Claisen rearrangement initially generates a cyclohexadienone intermediate, which tautomerizes to a phenol that contains an allyl group ortho to the OH group.

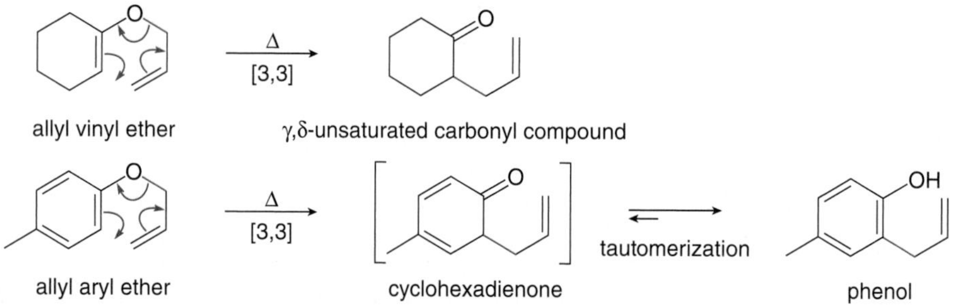

Problem 30.24 What product is formed from the Claisen rearrangement of each starting material?

a. 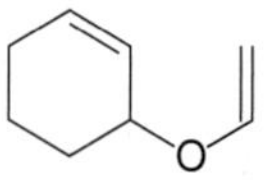 b. 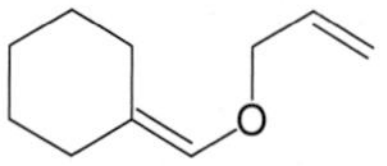 c. 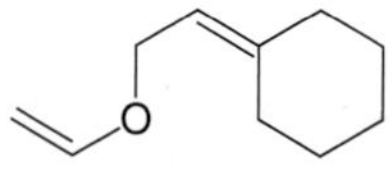

Problem 30.25 (a) What product is formed by the Claisen rearrangement of compound **Z?** (b) Using what you have learned about ring-closing metathesis in Chapter 29, draw the product formed when the product in part (a) is treated with Grubbs catalyst. These two reactions are key steps in the synthesis of garsubellin A, a biologically active natural product that stimulates the synthesis of the neurotransmitter acetylcholine. Compounds of this sort may prove to be useful drugs for the treatment of neurodegenerative diseases such as Alzheimer's disease.

Garsubellin A (Problem 30.25) is isolated from the wood of *Garcinia subelliptica,* a tree grown in Okinawa, Japan.

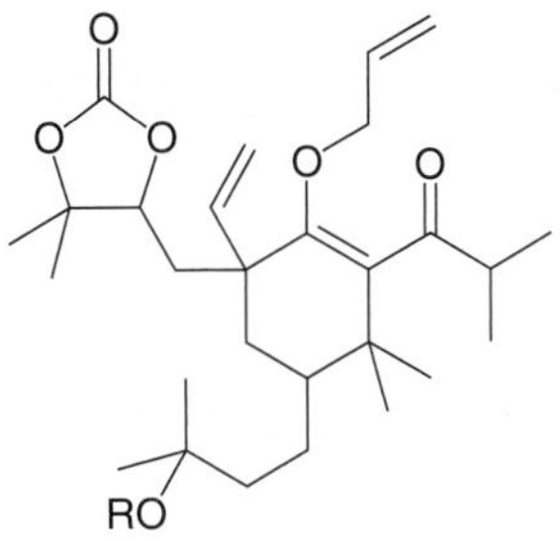

Z

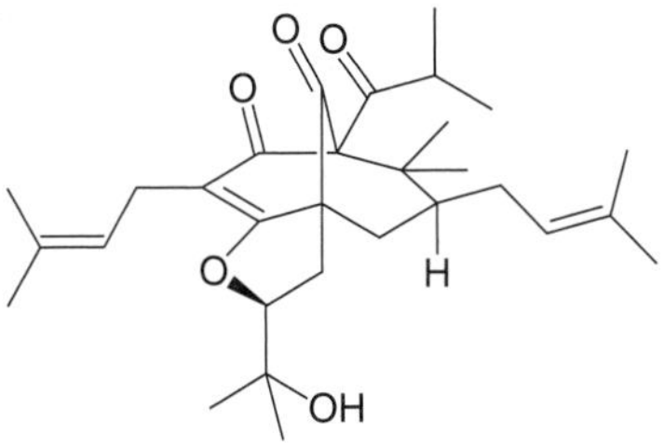

garsubellin A

30.6 Summary of Rules for Pericyclic Reactions

Table 30.4 summarizes the rules that govern pericyclic reactions, and in truth, this table holds a great deal of information. To keep track of this information, it may be helpful to **learn one row in the table only,** and then note the result when one or more conditions change. For example,

- A *thermal* reaction involving an *even* number of electron pairs is *conrotatory* or *antarafacial.*
- If *one* of the reaction conditions changes—either from thermal to photochemical or from an even to an odd number of electron pairs—the stereochemistry of the reaction changes to disrotatory or suprafacial.
- If *both* reaction conditions change—that is, a photochemical reaction with an odd number of electron pairs—the stereochemistry does *not* change.

Table 30.4 Summary of the Stereochemical Rules for Pericyclic Reactions

Reaction conditions	Number of electron pairs	Stereochemistry
Thermal	Even	Conrotatory or antarafacial
	Odd	Disrotatory or suprafacial
Photochemical	Even	Disrotatory or suprafacial
	Odd	Conrotatory or antarafacial

Problem 30.26 Using the Woodward–Hoffmann rules in Table 30.4, predict the stereochemistry of each reaction.

a. a [6 + 4] thermal cycloaddition
b. photochemical electrocyclic ring closure of deca-1,3,5,7,9-pentaene
c. a [4 + 4] photochemical cycloaddition
d. a thermal [5,5] sigmatropic rearrangement

KEY CONCEPTS

Pericyclic Reactions

Electrocyclic Reactions (30.3)

Woodward–Hoffmann rules for electrocyclic reactions

Number of π bonds	Thermal reaction	Photochemical reaction
Even	Conrotatory	Disrotatory
Odd	Disrotatory	Conrotatory

Examples

The stereochemistry of a thermal electrocyclic reaction is opposite to that of a photochemical electrocyclic reaction.

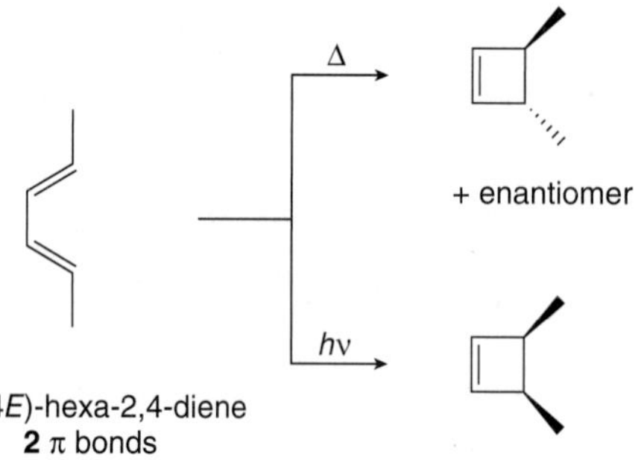

(2*E*,4*E*)-hexa-2,4-diene
2 π bonds

- A thermal electrocyclic reaction with an even number of π bonds occurs in a conrotatory fashion.
- A photochemical electrocyclic reaction with an even number of π bonds occurs in a disrotatory fashion.

Cycloaddition Reactions (30.4)

Woodward–Hoffmann rules for cycloaddition reactions

Number of π bonds	Thermal reaction	Photochemical reaction
Even	Antarafacial	Suprafacial
Odd	Suprafacial	Antarafacial

Examples

[1] A thermal [4 + 2] cycloaddition takes place in a suprafacial fashion with an odd number of π bonds. An antarafacial photochemical [4 + 2] cycloaddition to form a six-membered ring cannot occur, because of the geometrical constraints of forming a six-membered ring.

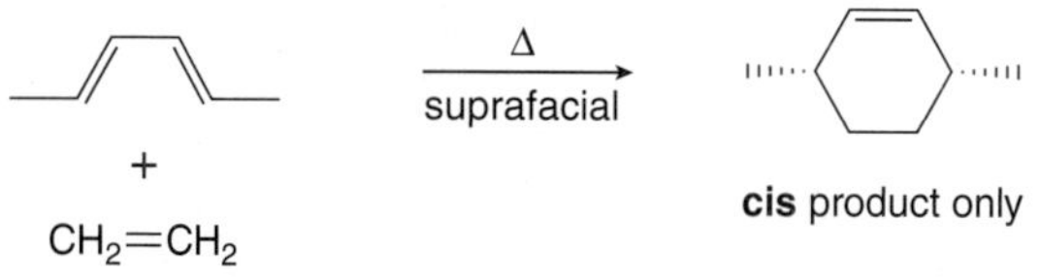

[2] A photochemical [2 + 2] cycloaddition takes place in a suprafacial fashion with an even number of π bonds. An antarafacial thermal [2 + 2] cycloaddition to form a four-membered ring cannot occur, because of the geometrical constraints of forming a four-membered ring.

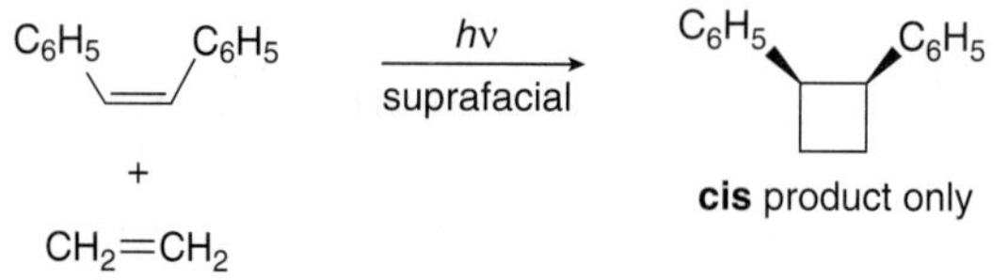

Sigmatropic Rearrangements (30.5)

Woodward–Hoffmann rules for sigmatropic rearrangements

Number of electron pairs	Thermal reaction	Photochemical reaction
Even	Antarafacial	Suprafacial
Odd	Suprafacial	Antarafacial

Examples

[1] A **Cope rearrangement** is a thermal [3,3] sigmatropic rearrangement that converts a 1,5-diene into an isomeric 1,5-diene.

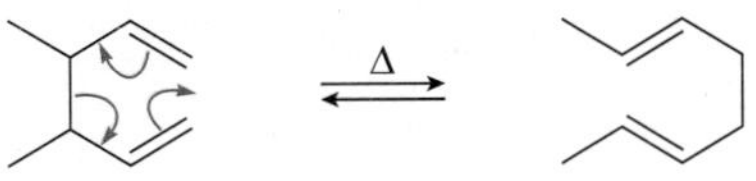

1,5-diene isomeric 1,5-diene

[2] An **oxy-Cope rearrangement** is a thermal [3,3] sigmatropic rearrangement that converts a 1,5-dien-3-ol into a δ,ε-unsaturated carbonyl compound, after tautomerization of an intermediate enol.

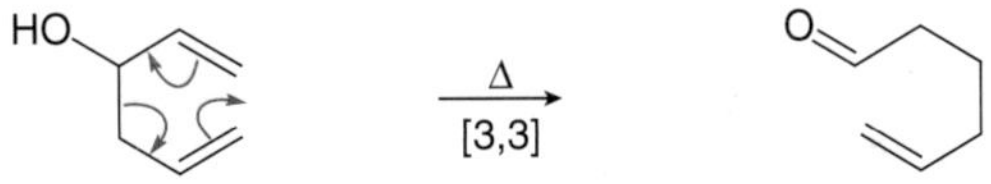

1,5-dien-3-ol δ,ε-unsaturated carbonyl compound

[3] A **Claisen rearrangement** is a thermal [3,3] sigmatropic rearrangement that converts an unsaturated ether into a γ,δ-unsaturated carbonyl compound.

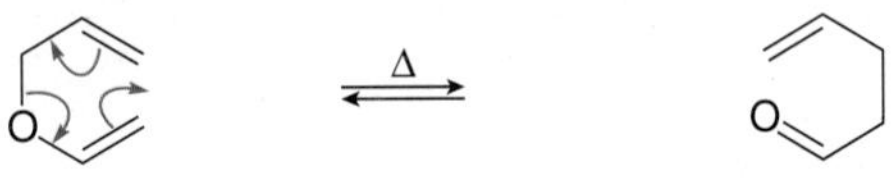

unsaturated ether γ,δ-unsaturated carbonyl compound

PROBLEMS

Problems Using Three-Dimensional Models

30.27 (a) What product is formed when each compound undergoes a thermal electrocyclic ring opening? (b) What product is formed when each compound undergoes a photochemical electrocyclic ring opening?

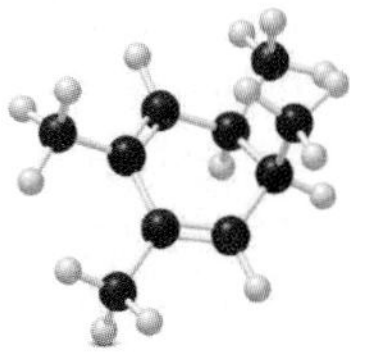

A

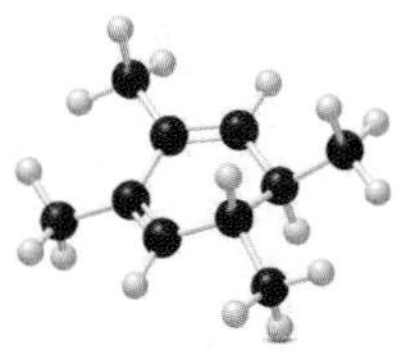

B

30.28 What product is formed by the [3,3] sigmatropic rearrangement of each compound?

a. 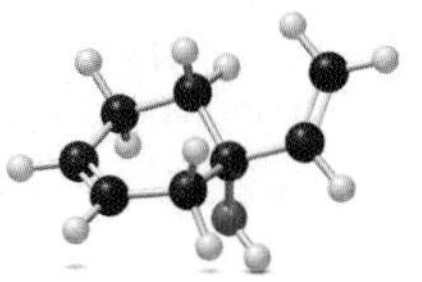

b. 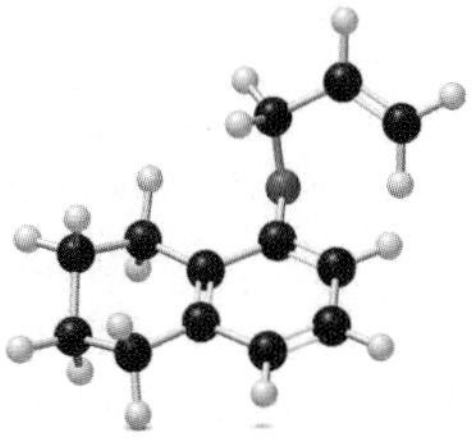

Types of Pericyclic Reactions

30.29 Classify each pericyclic reaction as an electrocyclic reaction, cycloaddition, or sigmatropic rearrangement. Indicate whether the stereochemistry is conrotatory, disrotatory, suprafacial, or antarafacial.

a.

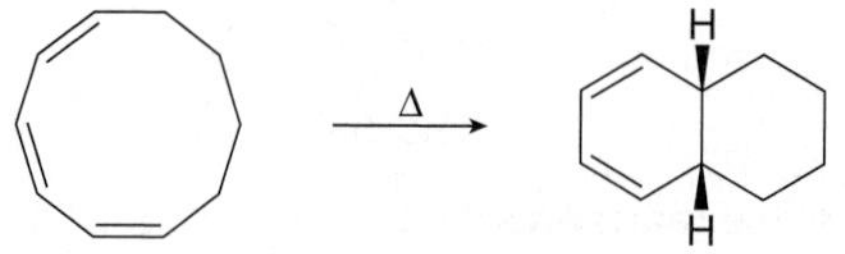

b.

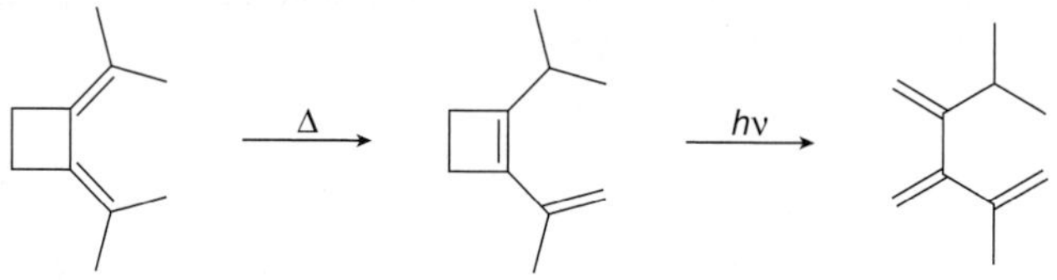

c.

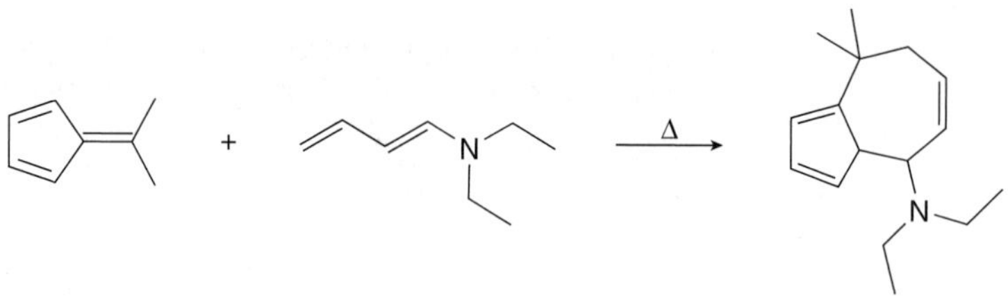

Electrocyclic Reactions

30.30 What product is formed when each compound undergoes thermal electrocyclic ring opening or ring closure? Label each process as conrotatory or disrotatory, and clearly indicate the stereochemistry around tetrahedral stereogenic centers and double bonds.

a.

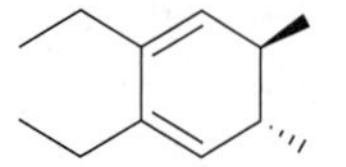

b.

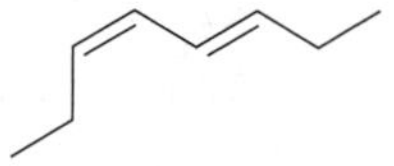

30.31 What product is formed when each compound in Problem 30.30 undergoes photochemical electrocyclic reaction? Label each process as conrotatory or disrotatory, and clearly indicate the stereochemistry around tetrahedral stereogenic centers and double bonds.

30.32 What cyclic product is formed when each decatetraene undergoes photochemical electrocyclic ring closure?

a.

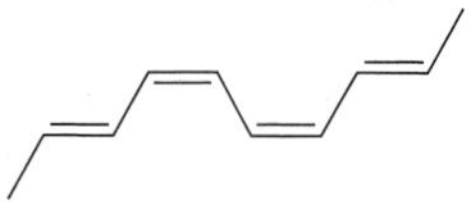

b.

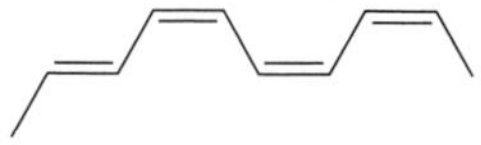

30.33 Draw the product of each electrocyclic reaction.

a. the thermal electrocyclic ring closure of (2*E*,4*Z*,6*Z*)-nona-2,4,6-triene
b. the photochemical electrocyclic ring closure of (2*E*,4*Z*,6*Z*)-nona-2,4,6-triene
c. the thermal electrocyclic ring opening of *cis*-5-ethyl-6-methylcyclohexa-1,3-diene
d. the photochemical electrocyclic ring opening of *trans*-5-ethyl-6-methylcyclohexa-1,3-diene

30.34 Consider the following electrocyclic ring closure. Does the product form by a conrotatory or disrotatory process? Would this reaction occur under photochemical or thermal conditions?

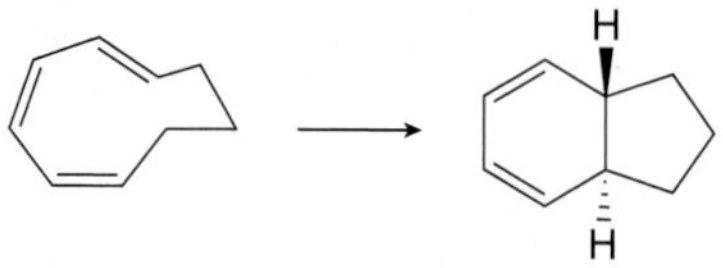

30.35 Draw the product formed when diene **M** undergoes disrotatory cyclization. Indicate the stereochemistry at new sp^3 hybridized carbons. Will the reaction occur under thermal or photochemical conditions?

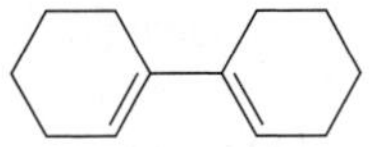

M

30.36 (a) What product is formed when triene **N** undergoes thermal electrocyclic ring closure? (b) What product is formed when triene **N** undergoes photochemical ring closure? (c) Label each process as conrotatory or disrotatory.

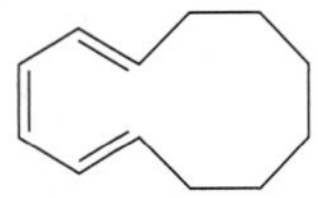

N

30.37 The bicyclic alkene **P** can be prepared by thermal electrocyclic ring closure from cyclodecadiene **Q** or by photochemical electrocyclic ring closure from cyclodecadiene **R.** Draw the structures of **Q** and **R,** and indicate the stereochemistry of the process by which each reaction occurs.

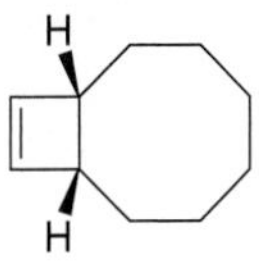

P

Cycloaddition Reactions

30.38 What type of cycloaddition occurs in Reaction [1]? Draw the product of a similar process in Reaction [2]. Would you predict that these reactions occur under thermal or photochemical conditions?

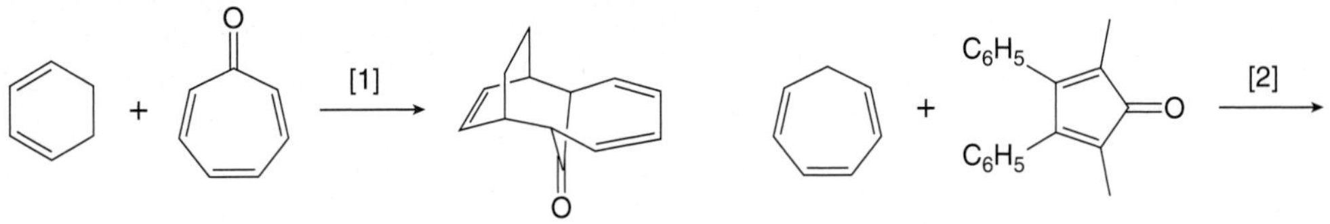

30.39 Draw the product of each Diels–Alder reaction and indicate the stereochemistry at all stereogenic centers.

a. 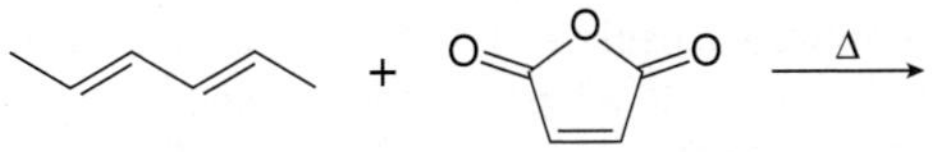

b. 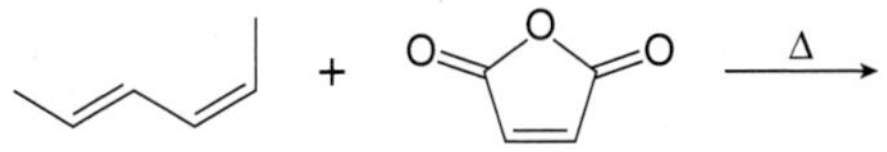

30.40 What cycloaddition products are formed in each reaction? Indicate the stereochemistry of each product.

a. 2 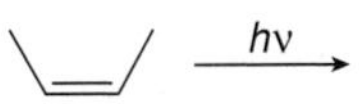

b. 2 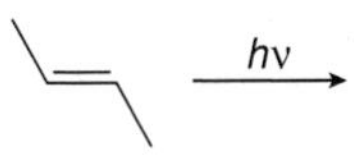

30.41 What starting materials are needed to synthesize each compound by a thermal [4 + 2] cycloaddition?

a. 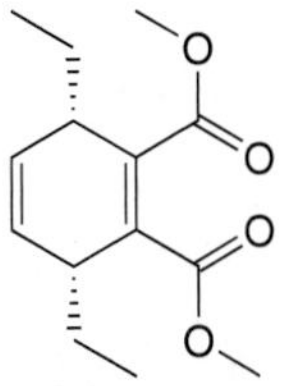

b. 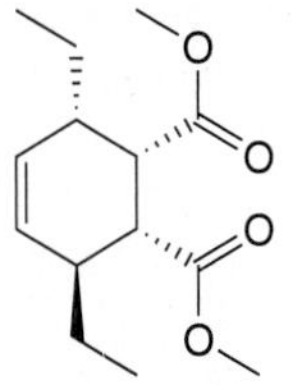

c. 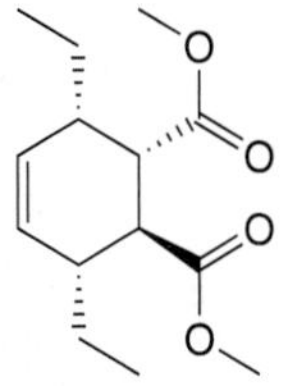

30.42 Explain why heating buta-1,3-diene forms 4-vinylcyclohexene but not cycloocta-1,5-diene.

30.43 How can **X** be prepared from a constitutional isomer by a series of [2 + 2] cycloaddition reactions? Interest in molecules that contain several cyclobutane rings fused together has been fueled by the discovery of pentacycloanammoxic acid methyl ester, a lipid isolated from the membrane of organelles in the bacterium *Candidatus Brocadia anammoxidans.* The role of this unusual natural product is as yet unknown.

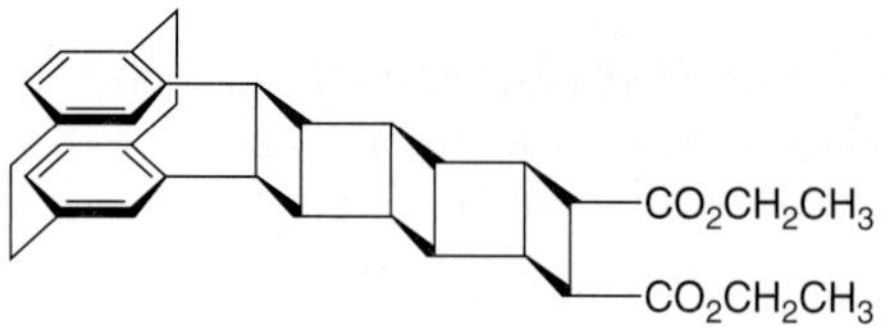

X

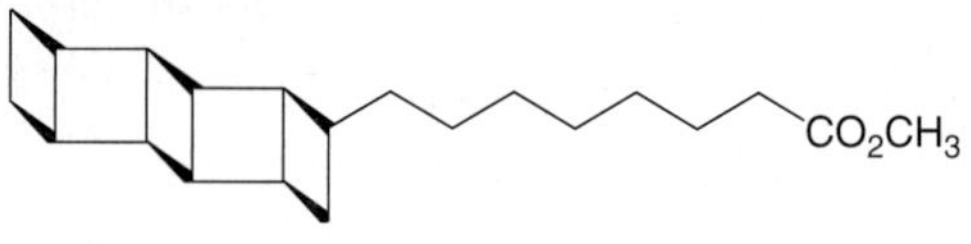

pentacycloanammoxic acid methyl ester

Sigmatropic Rearrangements

30.44 What type of sigmatropic rearrangement is illustrated in each reaction?

a. 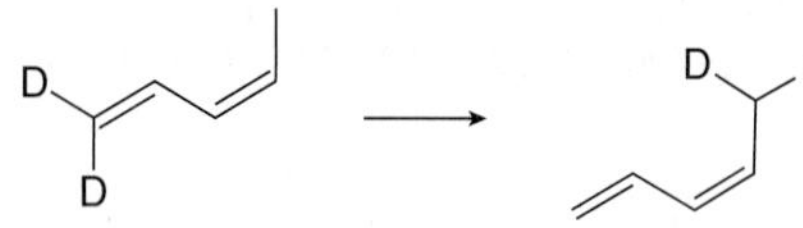

b. 

30.45 Draw the product of the [3,3] sigmatropic rearrangement of each compound.

a. 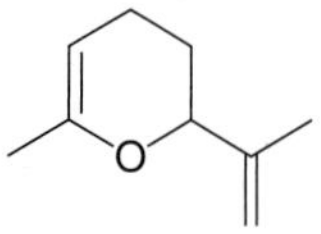

b. 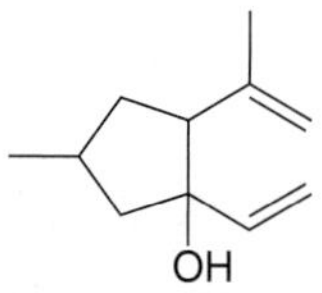

c. 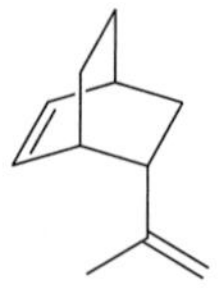

30.46 A solution of 5-methylcyclopenta-1,3-diene rearranges at room temperature to a mixture containing 1-methyl-, 2-methyl-, and 5-methylcyclopenta-1,3-diene. (a) Show how both isomeric products are formed from the starting material by a sigmatropic rearrangement involving a C–H bond. (b) Explain why 2-methylcyclopenta-1,3-diene is not formed directly from 5-methylcyclopenta-1,3-diene by a [1,3] rearrangement.

30.47 What product is formed from the [5,5] sigmatropic rearrangement of the following unsaturated ether?

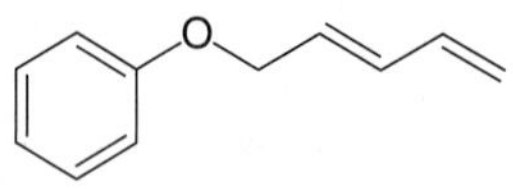

30.48 Heating **A** results in two successive [3,3] sigmatropic rearrangements—Claisen reaction followed by Cope reaction—to afford β-sinensal, a component of mandarin orange oil. What is the structure of β-sinensal?

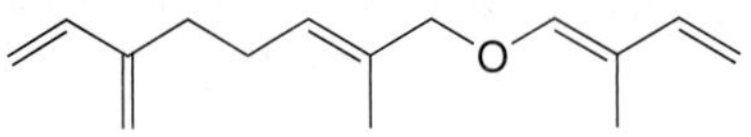

A

30.49 Draw structures for **A, B,** and **C** in the following reaction sequence and identify the process that converts **B** to **C.**

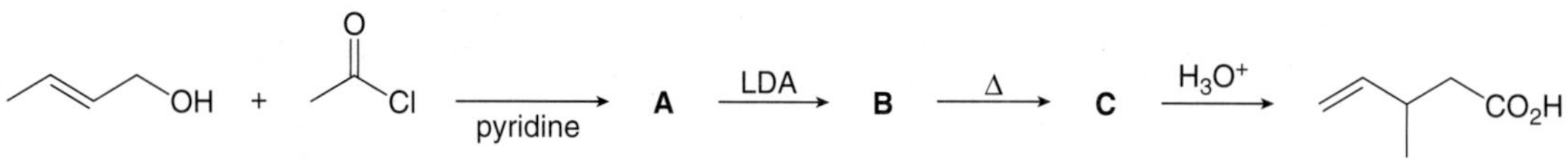

General Pericyclic Reactions

30.50 What type of pericyclic reaction is illustrated in each reaction?

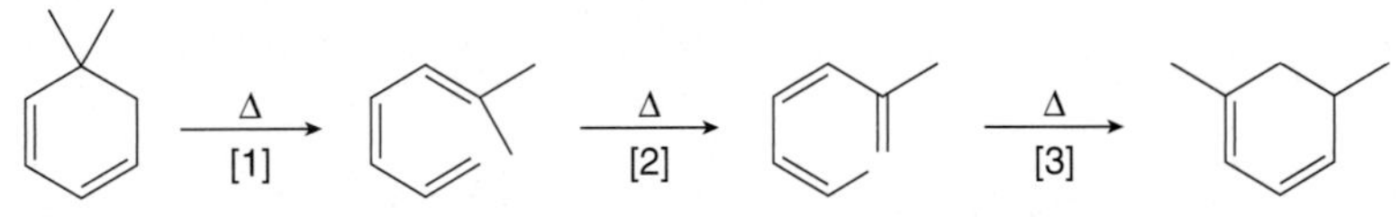

30.51 Draw the product formed (including stereochemistry) in each pericyclic reaction.

a. 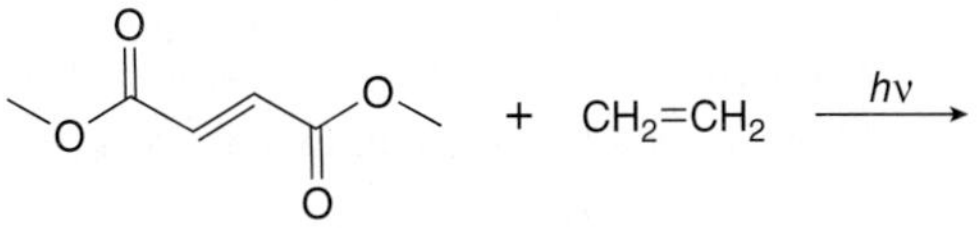

c. 2 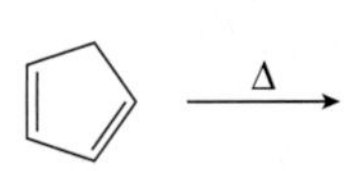

b. 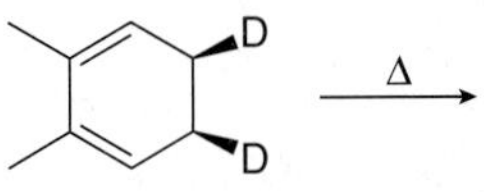

d. 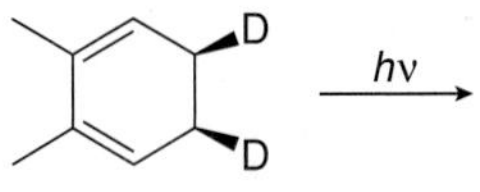

30.52 Draw the products of each reaction.

a. 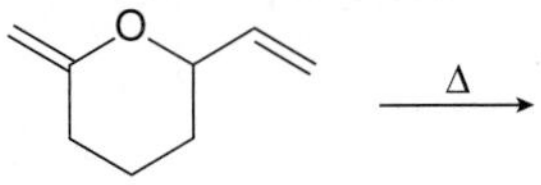

c. 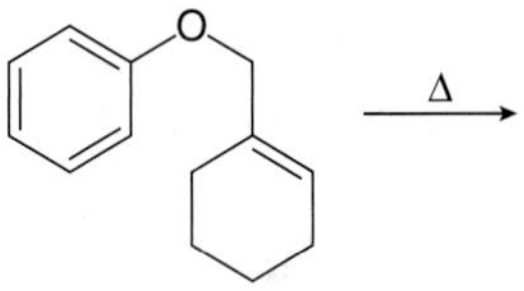

b. 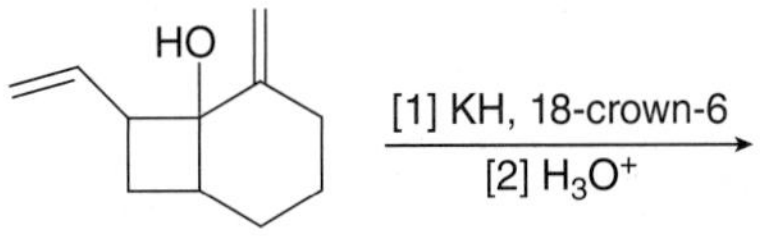

d. 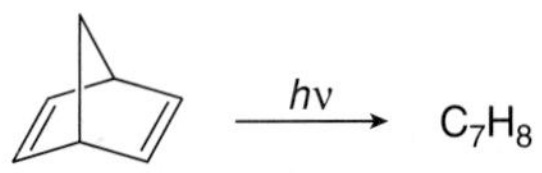

Mechanisms

30.53 When both carbons ortho to the aryl oxygen are not bonded to hydrogen, an allyl aryl ether rearranges to a para-substituted phenol. Draw a stepwise mechanism for the following reaction, which contains two [3,3] sigmatropic rearrangements.

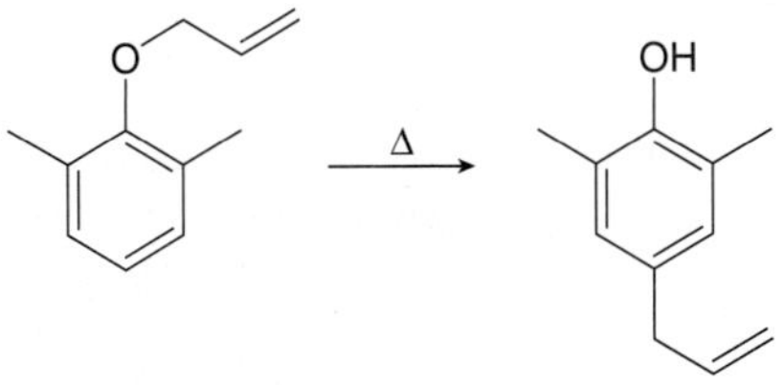

30.54 Draw a stepwise, detailed mechanism for the following reaction.

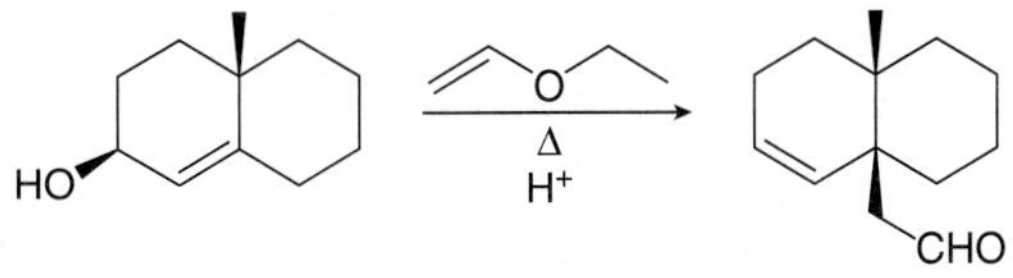

30.55 Show how the following starting material is converted to the given product by a series of two pericyclic reactions. Account for the observed stereochemistry.

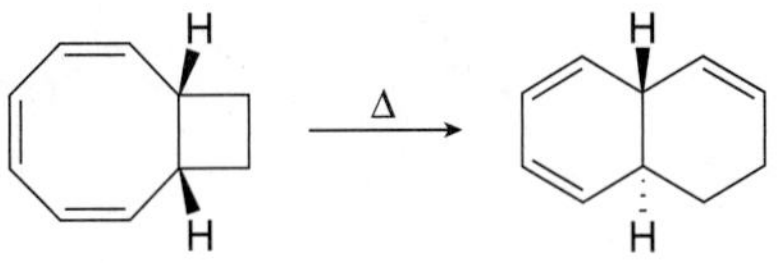

30.56 Use curved arrows to show how **E** is converted to **F** by a two-step reaction sequence consisting of a [1,5] sigmatropic rearrangement followed by a [4 + 2] cycloaddition.

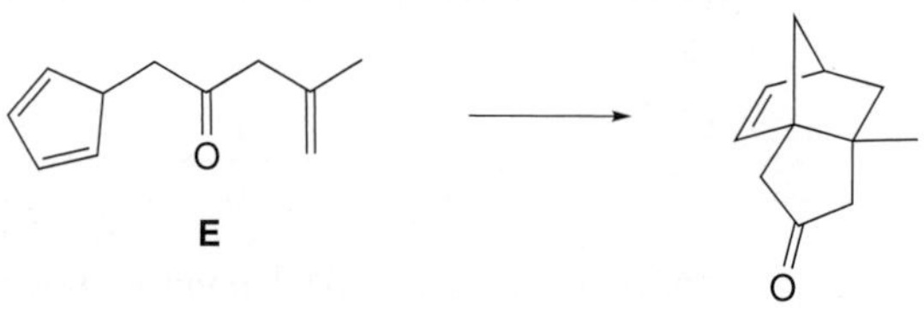

E **F**

30.57 Show how the following starting material is converted to the given product by a series of two pericyclic reactions. Account for the observed stereochemistry.

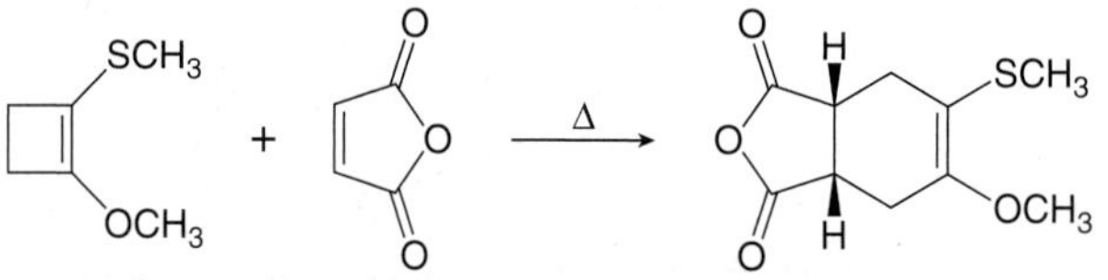

30.58 Draw a stepwise, detailed mechanism for the following reaction.

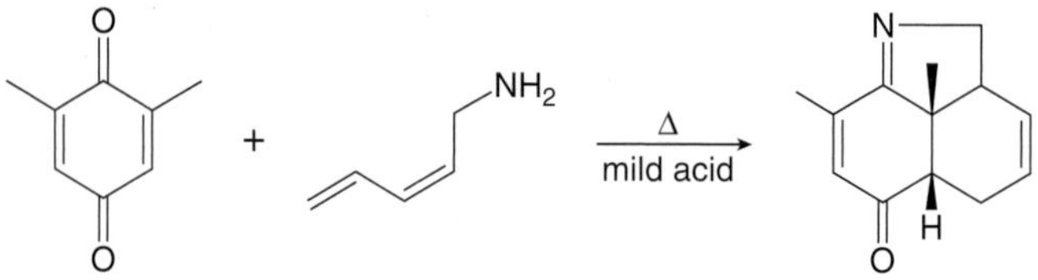

Challenge Problems

30.59 What product is formed by [3,3] sigmatropic rearrangement of the following compound? Clearly indicate the stereochemistry around all tetrahedral stereogenic centers.

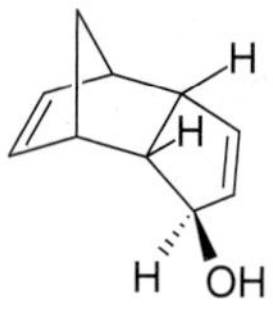

30.60 (a) What is the structure of **C,** which is formed by oxy-Cope rearrangement of **B** with NaOEt? (b) Draw a stepwise mechanism for the conversion of **C** to the bicyclic alcohol **D.**

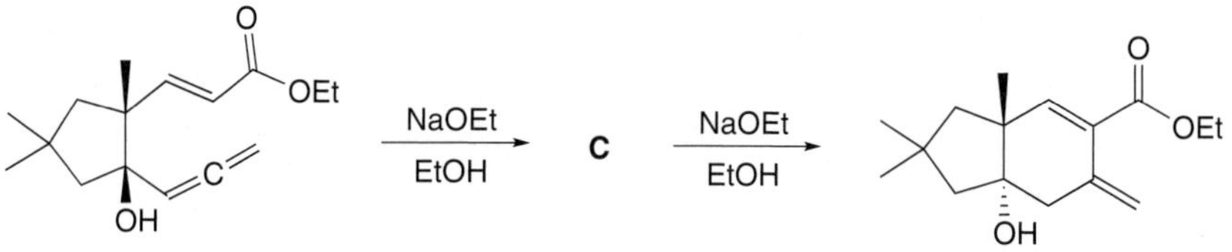

B **D**

30.61 Draw a stepwise mechanism for the Carroll rearrangement, a reaction that prepares a γ,δ-unsaturated carbonyl compound from a β-keto ester and allylic alcohol in the presence of base.

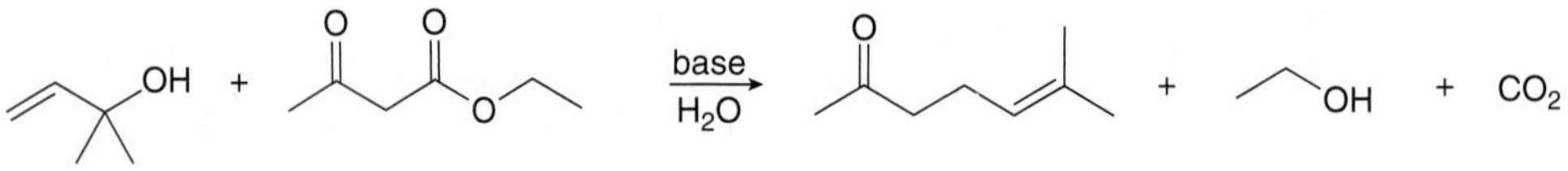

30.62 The endiandric acids comprise a group of unsaturated carboxylic acids isolated from a tree that grows in the rain forests of eastern Australia. The methyl esters of endiandric acids D and E have been prepared from polyene **Y** by a series of two successive electrocyclic reactions: thermal ring closure of the conjugated tetraene followed by ring closure of the resulting conjugated triene. (a) Draw the structures (including stereochemistry) of the methyl esters of endiandric acids D and E. (b) The methyl ester of endiandric acid E undergoes an intramolecular [4 + 2] cycloaddition to form the methyl ester of endiandric acid A. Propose a possible structure for endiandric acid A.

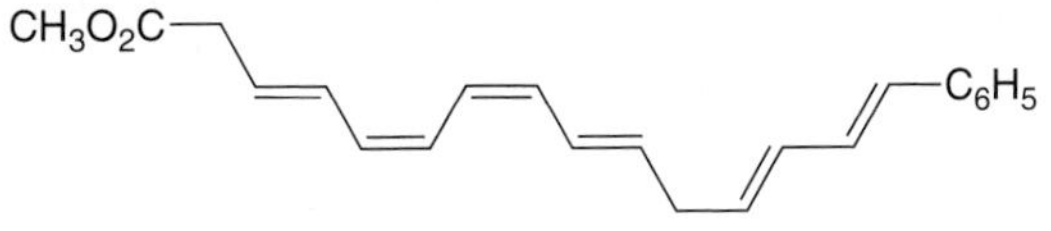

Y

30.63 [4 + 2] Cycloadditions with *o*-quinones such as **B** are often complex because a variety of products are possible.

a. Draw arrows to illustrate how each product is formed when **B** reacts with styrene, and label the "diene" and "dienophile" components.

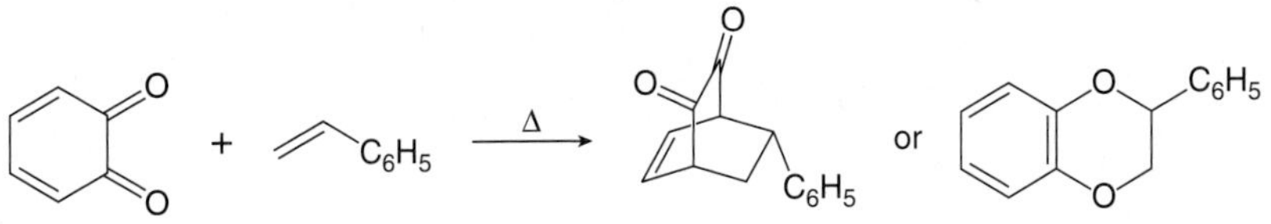

o-quinone **B** styrene

b. Draw arrows to illustrate how each product is formed when **B** reacts with buta-1,3-diene, and label the "diene" and "dienophile" components.

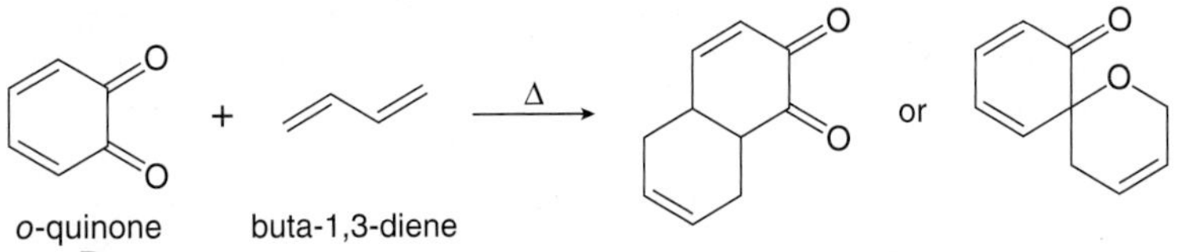

o-quinone **B** buta-1,3-diene

c. *o*-Quinone **C** reacts with diene **D** to form heterocycle **E** by a process that involves a cycloaddition followed by a [3,3] sigmatropic rearrangement. Use curved arrows to illustrate how this two-step sequence occurs. **E** is *not* formed directly from **C** by a Diels–Alder reaction.

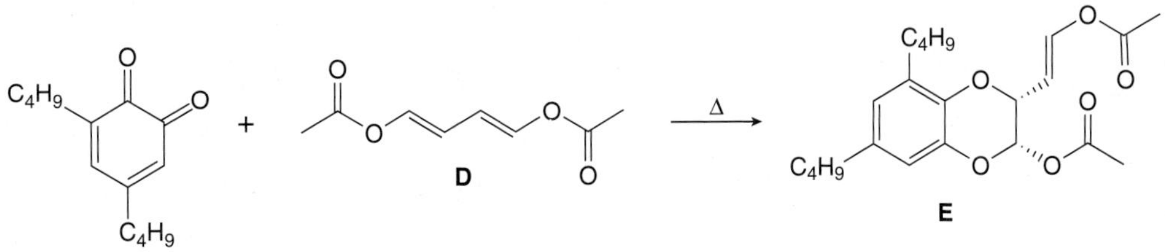

C

APPENDIX A

pK_a Values for Selected Compounds

Compound	pK_a
HI	–10
HBr	–9
H_2SO_4	–9
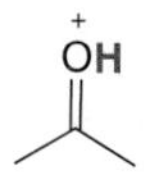	–7.3
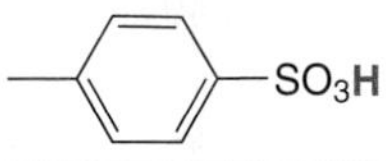	–7
HCl	–7
$[(CH_3)_2OH]^+$	–3.8
$(CH_3OH_2)^+$	–2.5
H_3O^+	–1.7
CH_3SO_3H	–1.2
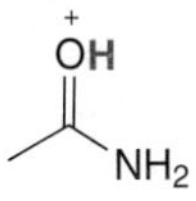	0.0
CF_3CO_2H	0.2
CCl_3CO_2H	0.6
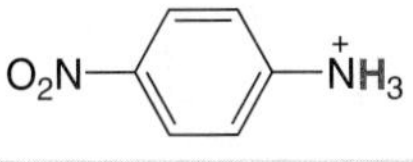	1.0
Cl_2CHCO_2H	1.3
H_3PO_4	2.1
FCH_2CO_2H	2.7
$ClCH_2CO_2H$	2.8
$BrCH_2CO_2H$	2.9
ICH_2CO_2H	3.2
HF	3.2
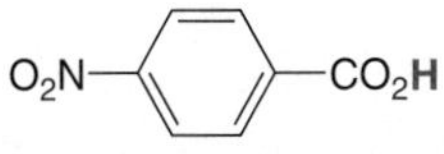	3.4
HCO_2H	3.8

Compound	pK_a
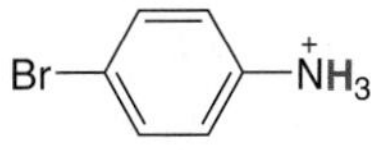	3.9
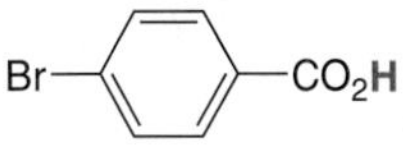	4.0
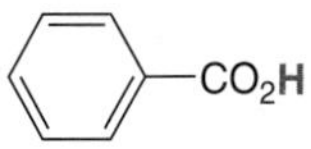	4.2
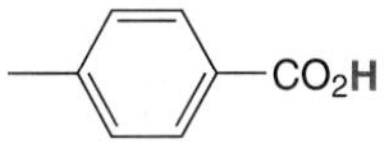	4.3
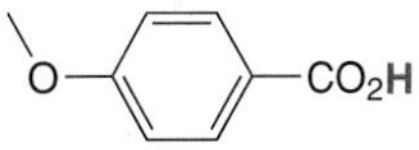	4.5
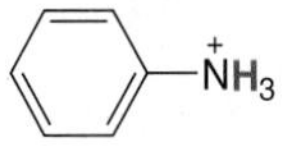	4.6
CH_3CO_2H	4.8
$(CH_3)_3CCO_2H$	5.0
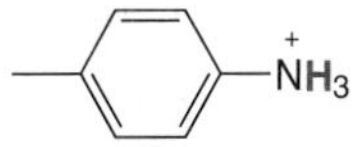	5.1
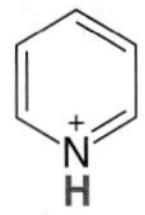	5.3
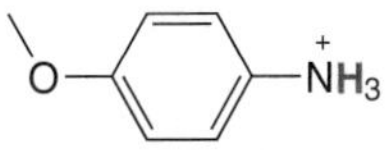	5.3
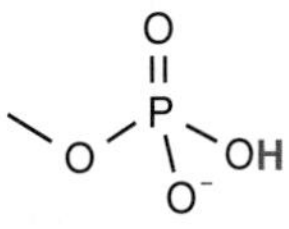	6.3
H_2CO_3	6.4
H_2S	7.0
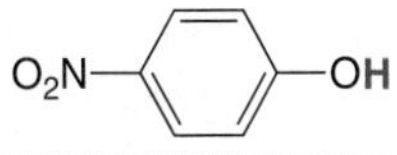	7.1
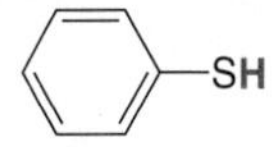	7.8

Compound	pK_a
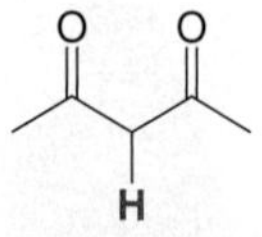	8.9
$HC\equiv N$	9.1
	9.4
NH_4^+	9.4
$H_3\overset{+}{N}CH_2CO_2^-$	9.8
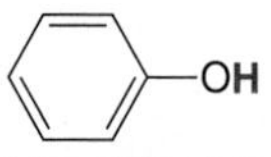	10.0
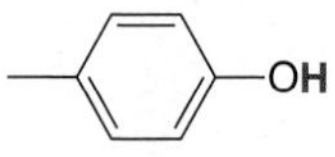	10.2
HCO_3^-	10.2
CH_3NO_2	10.2
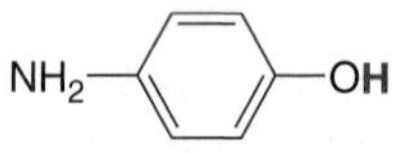	10.3
CH_3SH	10.3
CH_3CH_2SH	10.5
$[(CH_3)_3NH]^+$	10.6
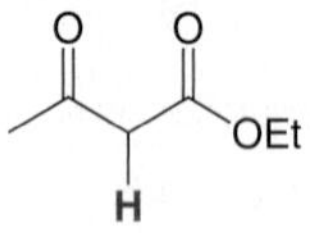	10.7
$(CH_3NH_3)^+$	10.7
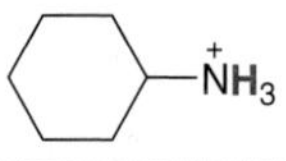	10.7
$[(CH_3)_2NH_2]^+$	10.7
CF_3CH_2OH	12.4
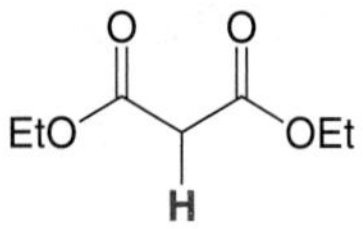	13.3
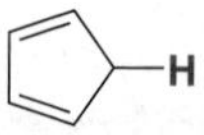	15
CH_3OH	15.5
H_2O	15.7
CH_3CH_2OH	16
CH_3CONH_2	16
CH_3CHO	17
$(CH_3)_3COH$	18
$(CH_3)_2C=O$	19.2
$CH_3CO_2CH_2CH_3$	24.5
$HC\equiv CH$	25
$CH_3C\equiv N$	25
$CHCl_3$	25
$CH_3CON(CH_3)_2$	30
H_2	35
NH_3	38
CH_3NH_2	40
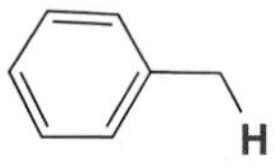	41
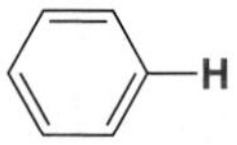	43
$CH_2=CHCH_3$	43
$CH_2=CH_2$	44
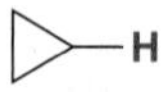	46
CH_4	50
CH_3CH_3	50

APPENDIX

B

Nomenclature

Although the basic principles of nomenclature are presented in the body of this text, additional information is often needed to name many complex organic compounds. Appendix B concentrates on three topics:

- **Naming alkyl substituents that contain branching**
- **Naming polyfunctional compounds**
- **Naming bicyclic compounds**

Naming Alkyl Substituents That Contain Branching

Alkyl groups that contain any number of carbons and no branches are named as described in Section 4.4A: change the *-ane* ending of the parent alkane to the suffix *-yl.* Thus the seven-carbon alkyl group $CH_3CH_2CH_2CH_2CH_2CH_2CH_2-$ is called *heptyl.*

When an alkyl substituent also contains branching, follow a stepwise procedure:

[1] Identify the longest carbon chain of the alkyl group that begins at the point of attachment to the parent. Begin numbering at the point of attachment and use the suffix *-yl* to indicate an alkyl group.

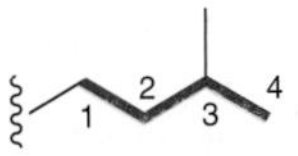

4 C's in the chain ---→ butyl group

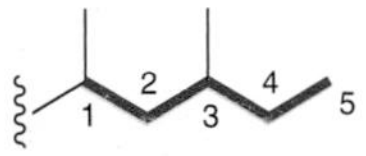

5 C's in the chain ---→ pentyl group

[2] Name all branches off the main alkyl chain and use the numbers from Step [1] to designate their location.

methyl at C3

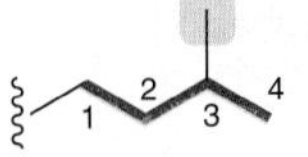

3-methylbutyl

methyls at C1 and C3

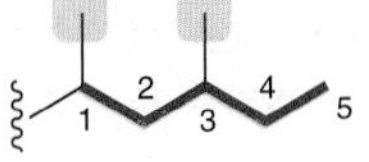

1,3-dimethylpentyl

[3] Set the entire name of the substituent in parentheses, and alphabetize this substituent name by the first letter of the complete name.

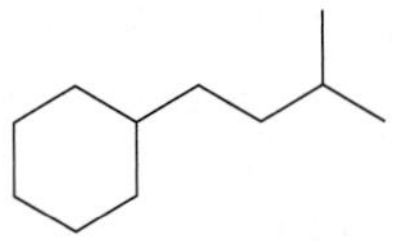

(3-methylbutyl)cyclohexane

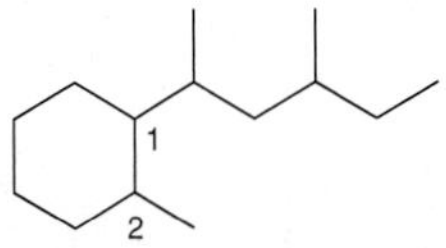

1-(1,3-dimethylpentyl)-2-methylcyclohexane

- Alphabetize the **d** of **d**imethylpentyl before the **m** of **m**ethyl.
- Number the ring to give the lower number to the first substituent alphabetically: place the dimethylpentyl group at C1.

Naming Polyfunctional Compounds

Many organic compounds contain more than one functional group. When one of those functional groups is halo (X−) or alkoxy (RO−), these groups are named as substituents as described in Sections 7.2 and 9.3B. To name other polyfunctional compounds, we must learn which functional group is assigned a higher priority in the rules of nomenclature. Two steps are usually needed:

[1] **Name a compound using the suffix of the highest priority group,** and name other functional groups as *substituents.* Table B.1 lists the common functional groups in order of decreasing priority, as well as the prefixes needed when a functional group must be named as a substituent.

[2] Number the carbon chain to give the lower number to the highest priority functional group that can be named as a suffix, and then follow all other rules of nomenclature. Examples are shown in Figure B.1.

Table B.1 Summary of Functional Group Nomenclature

Increasing priority (upward arrow)

Functional group	Suffix	Substituent name (prefix)
Carboxylic acid	-oic acid	carboxy
Ester	-oate	alkoxycarbonyl
Amide	-amide	amido
Nitrile	-nitrile	cyano
Aldehyde	-al	oxo (= O) or formyl (− CHO)
Ketone	-one	oxo
Alcohol	-ol	hydroxy
Amine	-amine	amino
Alkene	-ene	alkenyl
Alkyne	-yne	alkynyl
Alkane	-ane	alkyl
Ether	—	alkoxy
Halide	—	halo

Figure B.1
Examples of nomenclature of polyfunctional compounds

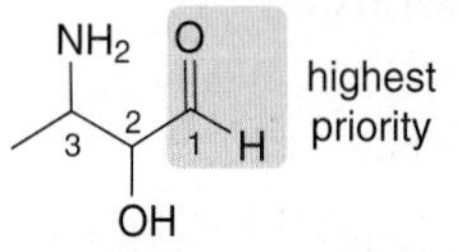

3-amino-2-hydroxybutanal

Name as a derivative of an aldehyde, because CHO is the highest priority functional group.

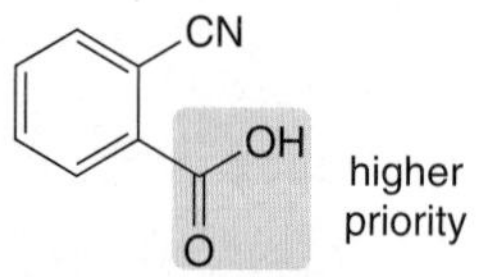

o-cyanobenzoic acid

Name as a derivative of benzoic acid, because COOH is the higher priority functional group.

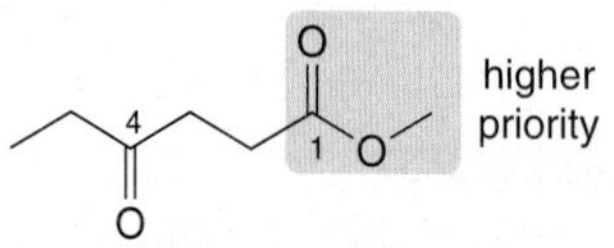

methyl 4-oxohexanoate

Name as a derivative of an ester, because COOR is the higher priority functional group.

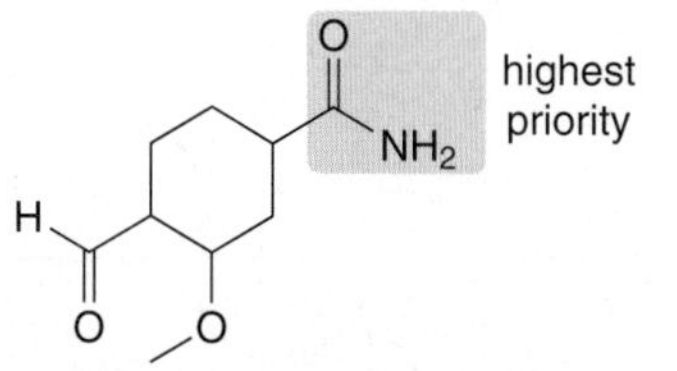

4-formyl-3-methoxycyclohexanecarboxamide

Name as a derivative of an amide, because $CONH_2$ is the highest priority functional group.

Polyfunctional compounds that contain C–C double and triple bonds have characteristic suffixes to identify them, as shown in Table B.2. The higher priority functional group is assigned the lower number.

Table B.2 Naming Polyfunctional Compounds with C–C Double and Triple Bonds

Functional groups	Suffix	Example
C=C and OH	enol	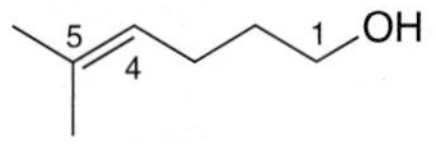 5-methylhex-4-en-1-ol
C=C + C=O (ketone)	enone	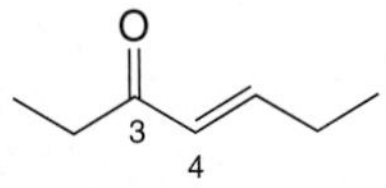 (*E*)-hept-4-en-3-one
C=C + C≡C	enyne	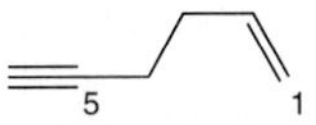 hex-1-en-5-yne

Naming Bicyclic Compounds

Bicyclic ring systems—compounds that contain two rings that share one or two carbon atoms—can be bridged, fused, or spiro.

bridged ring

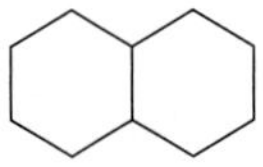

fused ring

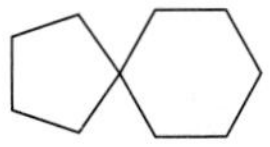

spiro ring

- **A bridged ring system contains two rings that share two non-adjacent carbons.**
- **A fused ring system contains two rings that share a common carbon–carbon bond.**
- **A spiro ring system contains two rings that share one carbon atom.**

Fused and bridged ring systems are named as bicyclo[*x.y.z*]alkanes, where the parent alkane corresponds to the total number of carbons in both rings. The numbers *x, y,* and *z* refer to the number of carbons that join the shared carbons together, written in order of *decreasing* size. For a fused ring system, *z* always equals zero, because the two shared carbons are directly joined together. The shared carbons in a bridged ring system are called the **bridgehead carbons.**

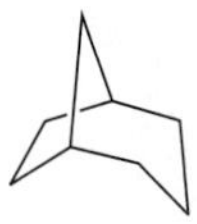

8 C's in the ring system

↓

bicyclooctane

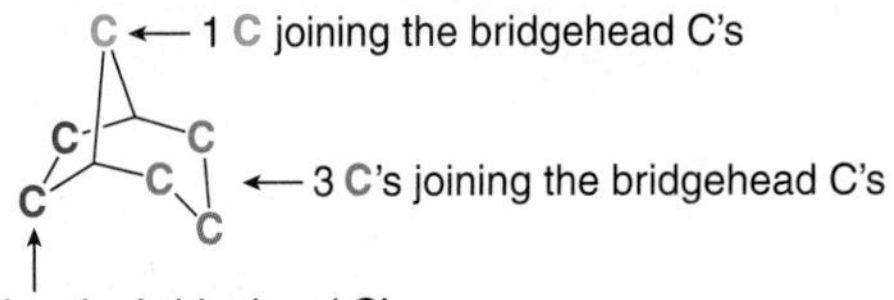
1 C joining the bridgehead C's
3 C's joining the bridgehead C's
2 C's joining the bridgehead C's

Name: bicyclo[3.2.1]octane

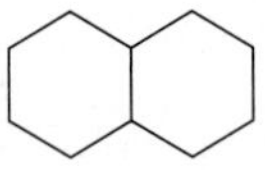

10 C's in the ring system

↓

bicyclodecane

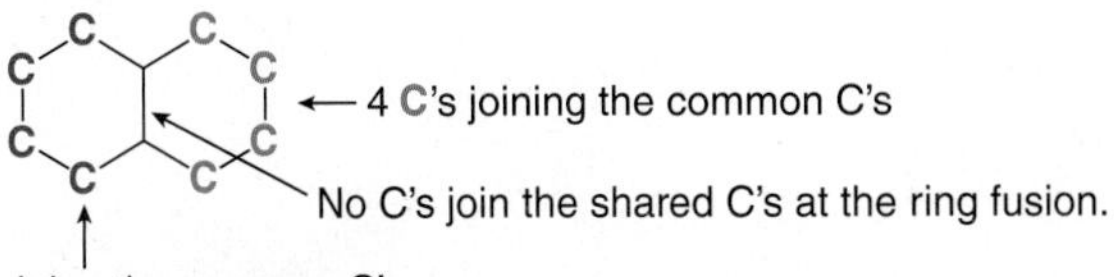
4 C's joining the common C's
No C's join the shared C's at the ring fusion.
4 C's joining the common C's

Name: bicyclo[4.4.0]decane

Rings are numbered beginning at a *shared* carbon, and continuing around the *longest* bridge first, then the next longest, and so forth.

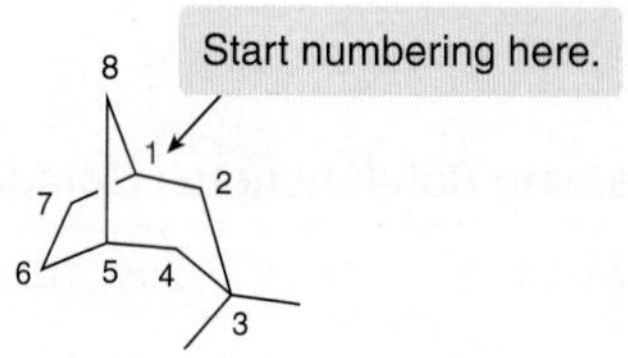

3,3-dimethylbicyclo[3.2.1]octane

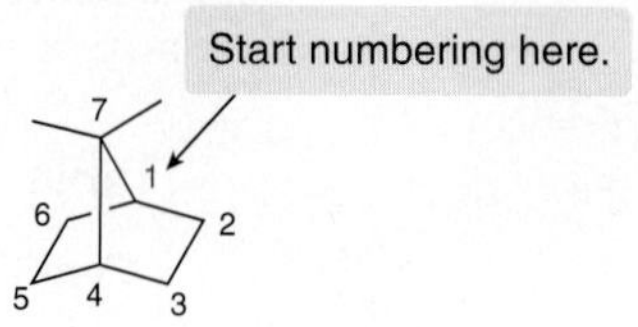

7,7-dimethylbicyclo[2.2.1]heptane

Spiro ring systems are named as spiro[*x.y*]alkanes, where the parent alkane corresponds to the total number of carbons in both rings, and *x* and *y* refer to the number of carbons that join the shared carbon (the spiro carbon), written in order of *increasing* size. When substituents are present, the rings are numbered beginning with a carbon *adjacent* to the spiro carbon in the *smaller* ring.

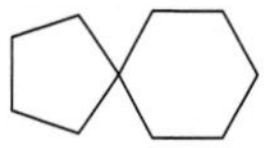

10 C's in the ring system

Name: spiro[4.5]decane

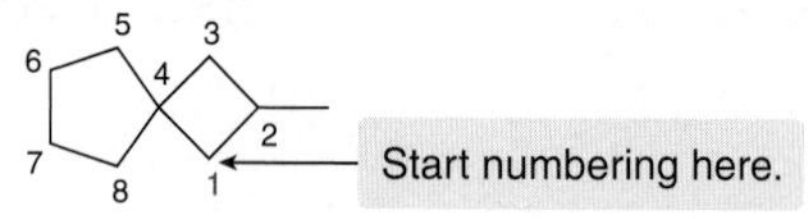

8 C's in the ring system

Name: 2-methylspiro[3.4]octane

APPENDIX C

Bond Dissociation Energies for Some Common Bonds [A–B → A• + •B]

Bond	ΔH° kJ/mol	(kcal/mol)
H–Z bonds		
H–F	569	(136)
H–Cl	431	(103)
H–Br	368	(88)
H–I	297	(71)
H–OH	498	(119)
Z–Z bonds		
H–H	435	(104)
F–F	159	(38)
Cl–Cl	242	(58)
Br–Br	192	(46)
I–I	151	(36)
HO–OH	213	(51)
R–H bonds		
CH_3–H	435	(104)
CH_3CH_2–H	410	(98)
$CH_3CH_2CH_2$–H	410	(98)
$(CH_3)_2CH$–H	397	(95)
$(CH_3)_3C$–H	381	(91)
$CH_2{=}CH$–H	435	(104)
HC≡C–H	523	(125)
$CH_2{=}CHCH_2$–H	364	(87)
C_6H_5–H	460	(110)
$C_6H_5CH_2$–H	356	(85)
R–R bonds		
CH_3–CH_3	368	(88)
CH_3–CH_2CH_3	356	(85)
CH_3–$CH{=}CH_2$	385	(92)
CH_3–C≡CH	489	(117)

Bond	ΔH° kJ/mol	(kcal/mol)
R–X bonds		
CH_3–F	456	(109)
CH_3–Cl	351	(84)
CH_3–Br	293	(70)
CH_3–I	234	(56)
CH_3CH_2–F	448	(107)
CH_3CH_2–Cl	339	(81)
CH_3CH_2–Br	285	(68)
CH_3CH_2–I	222	(53)
$(CH_3)_2CH$–F	444	(106)
$(CH_3)_2CH$–Cl	335	(80)
$(CH_3)_2CH$–Br	285	(68)
$(CH_3)_2CH$–I	222	(53)
$(CH_3)_3C$–F	444	(106)
$(CH_3)_3C$–Cl	331	(79)
$(CH_3)_3C$–Br	272	(65)
$(CH_3)_3C$–I	209	(50)
R–Z bonds		
CH_3–OH	389	(93)
CH_3CH_2–OH	393	(94)
$CH_3CH_2CH_2$–OH	385	(92)
$(CH_3)_2CH$–OH	401	(96)
$(CH_3)_3C$–OH	401	(96)
CH_3–NH_2	331	(79)
CH_3–SH	305	(73)
Other bonds		
$CH_2{=}CH_2$	635	(152)
HC≡CH	837	(200)
O=C=O	535	(128)
O_2	497	(119)

APPENDIX

D Reactions That Form Carbon–Carbon Bonds

Section	Reaction
11.11A	S_N2 reaction of an alkyl halide with an acetylide anion, $^{-}C\equiv CR$
11.11B	Opening of an epoxide ring with an acetylide anion, $^{-}C\equiv CR$
15.14	Radical polymerization of an alkene
16.12	Diels–Alder reaction
18.5	Friedel–Crafts alkylation
18.5	Friedel–Crafts acylation
20.10	Reaction of an aldehyde or ketone with a Grignard or organolithium reagent
20.13A	Reaction of an acid chloride with a Grignard or organolithium reagent
20.13A	Reaction of an ester with a Grignard or organolithium reagent
20.13B	Reaction of an acid chloride with an organocuprate reagent
20.14A	Reaction of a Grignard reagent with CO_2
20.14B	Reaction of an epoxide with an organometallic reagent
20.15	Reaction of an α,β-unsaturated carbonyl compound with an organocuprate reagent
21.9	Cyanohydrin formation
21.10	Wittig reaction to form an alkene
22.18	S_N2 reaction of an alkyl halide with NaCN
22.18C	Reaction of a nitrile with a Grignard or organolithium reagent
23.8	Direct enolate alkylation using LDA and an alkyl halide
23.9	Malonic ester synthesis to form a carboxylic acid
23.10	Acetoacetic ester synthesis to form a ketone
24.1	Aldol reaction to form a β-hydroxy carbonyl compound or an α,β-unsaturated carbonyl compound
24.2	Crossed aldol reaction
24.3	Directed aldol reaction
24.5	Claisen reaction to form a β-keto ester
24.6	Crossed Claisen reaction to form a β-dicarbonyl compound
24.7	Dieckmann reaction to form a five- or six-membered ring
24.9	Michael reaction to form a 1,5-dicarbonyl compound
24.10	Robinson annulation to form a cyclohex-2-enone
25.14	Reaction of a diazonium salt with CuCN
26.2B	Alkylation of diethyl acetamidomalonate to form an amino acid
26.2C	Strecker synthesis of an amino acid
27.10B	Kiliani–Fischer synthesis of an aldose
29.1	Coupling of an organocuprate reagent (R_2CuLi) with an organic halide (R'X)
29.2	The palladium-catalyzed Suzuki reaction of an organic halide with an organoborane
29.3	The palladium-catalyzed Heck reaction of a vinyl or aryl halide with an alkene
29.4	Addition of a dihalocarbene to an alkene to form a cyclopropane
29.5	Simmons–Smith reaction of an alkene with CH_2I_2 and Zn(Cu) to form a cyclopropane
29.6	Olefin metathesis
30.3	Electrocyclic reactions
30.4	Cycloaddition reactions
30.5	Sigmatropic rearrangements
31.2	Chain-growth polymerization
31.4	Polymerization using Ziegler–Natta catalysts

APPENDIX E

Characteristic IR Absorption Frequencies

Bond	Functional group	Wavenumber (cm^{-1})	Comment
O–H			
	• ROH	3600–3200	broad, strong
	• RCO_2H	3500–2500	very broad, strong
N–H			
	• RNH_2	3500–3300	two peaks
	• R_2NH	3500–3300	one peak
	• $RCONH_2$, RCONHR	3400–3200	one or two peaks; N–H bending also observed at 1640 cm^{-1}
C–H			
	• C_{sp}–H	3300	sharp, often strong
	• C_{sp^2}–H	3150–3000	medium
	• C_{sp^3}–H	3000–2850	strong
	• C_{sp^2}–H of RCHO	2830–2700	one or two peaks
C≡C		2250	medium
C≡N		2250	medium
C=O			strong
	• RCOCl	1800	
	• $(RCO)_2O$	1800, 1760	two peaks
	• RCO_2R	1745–1735	increasing $\tilde{\nu}$ with decreasing ring size
	• RCHO	1730	
	• R_2CO	1715	increasing $\tilde{\nu}$ with decreasing ring size
	• $RCO_2PO_3^{2-}$	1700–1730	
	• R_2CO, conjugated	1680	
	• RCO_2H	1710	
	• RCOSR	1690–1720	
	• $RCONH_2$, RCONHR, $RCONR_2$	1680–1630	increasing $\tilde{\nu}$ with decreasing ring size
C=C			
	• Alkene	1650	medium
	• Arene	1600, 1500	medium
C=N		1650	medium

APPENDIX

F Characteristic NMR Absorptions

^{1}H NMR Absorptions

Compound type	Chemical shift (ppm)
Alcohol	
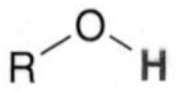	1–5
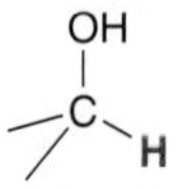	3.4–4.0
Aldehyde	
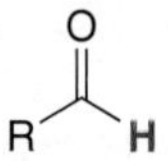	9–10
Alkane	0.9–2.0
RCH_3	~0.9
R_2CH_2	~1.3
R_3CH	~1.7
Alkene	
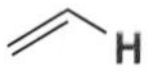 *sp*^{2} C–H	4.5–6.0
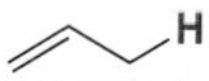 allylic *sp*^{3} C–H	1.5–2.5
Alkyl halide	
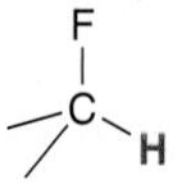	4.0–4.5
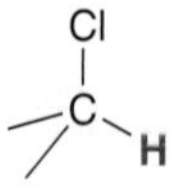	3.0–4.0
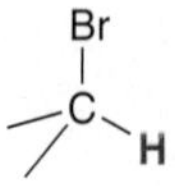	2.7–4.0
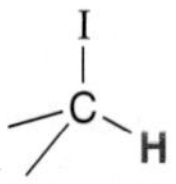	2.2–4.0
Alkyne	
	~2.5

Compound type	Chemical shift (ppm)
Amide	
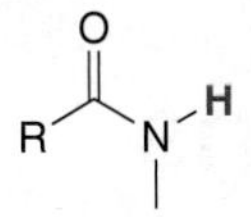	7.5–8.5
Amine	
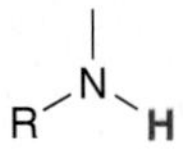	0.5–5.0
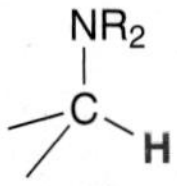	2.3–3.0
Aromatic compound	
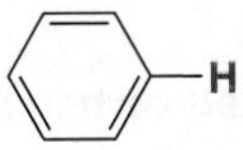 *sp*² C–H	6.5–8
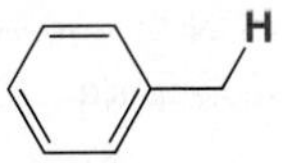 benzylic *sp*³ C–H	1.5–2.5
Carbonyl compound	
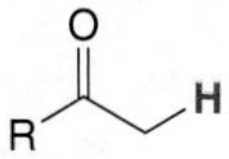 *sp*³ C–H on the α carbon	2.0–2.5
Carboxylic acid	
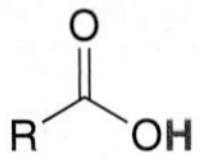	10–12
Ether	
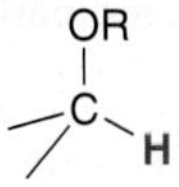	3.4–4.0

^{13}C NMR Absorptions

Carbon type	Structure	Chemical shift (ppm)
Alkyl, *sp*³ hybridized C	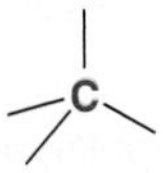	5–45
Alkyl, *sp*³ hybridized C bonded to N, O, or X	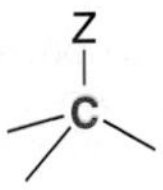 Z = N, O, X	30–80
Alkynyl, *sp* hybridized C	—C≡C—	65–100
Alkenyl, *sp*² hybridized C	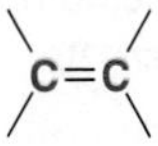	100–140
Aryl, *sp*² hybridized C	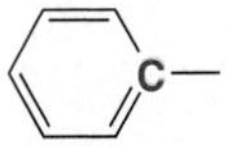	120–150
Carbonyl C	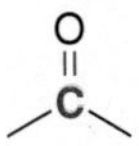	160–210

APPENDIX

G General Types of Organic Reactions

Substitution Reactions

[1] Nucleophilic substitution at an sp^3 hybridized carbon atom

a. Alkyl halides (Chapter 7)

$$R{-}X + :Nu^- \longrightarrow R{-}Nu + X:^-$$

(:Nu⁻ = nucleophile)

b. Alcohols (Section 9.11)

$$R{-}OH + HX \longrightarrow R{-}X + H_2O$$

c. Ethers (Section 9.14)

$$R{-}OR' + HX \longrightarrow R{-}X + R'{-}X + H_2O$$

X = Br or I

d. Epoxides (Section 9.16)

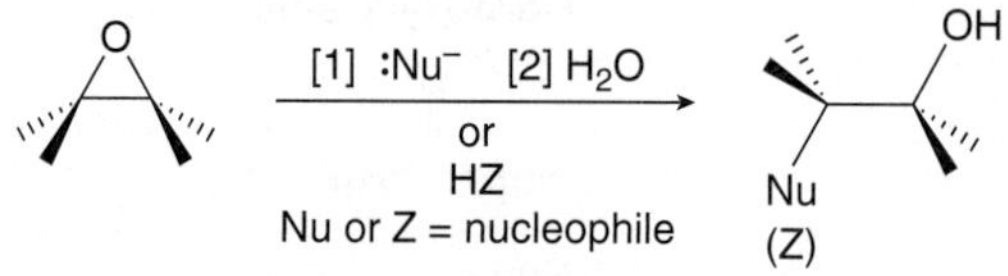

[2] Nucleophilic acyl substitution at an sp^2 hybridized carbon atom

Carboxylic acids and their derivatives (Chapter 22)

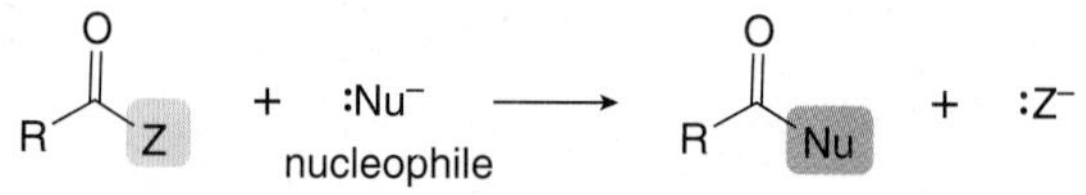

Z = OH, Cl, OCOR, OR', NR'_2

[3] Radical substitution at an sp^3 hybridized C – H bond

Alkanes (Section 15.3)

$$R{-}H + X_2 \xrightarrow{h\nu \text{ or } \Delta} R{-}X + HX$$

[4] Electrophilic aromatic substitution

Aromatic compounds (Chapter 18)

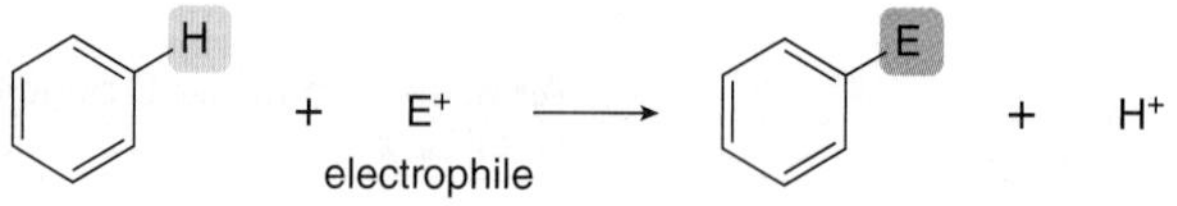

[5] Nucleophilic aromatic substitution

Aromatic compounds (Chapter 18)

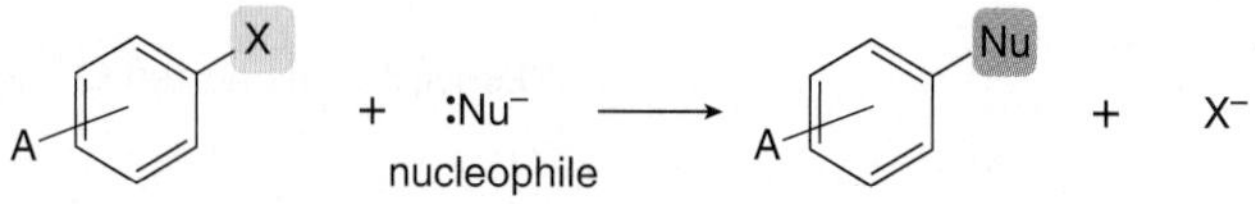

X = F, Cl, Br, I
A = H or electron-withdrawing group

Elimination Reactions

β Elimination at an *sp*^{3} hybridized carbon atom

a. Alkyl halides
(Chapter 8)

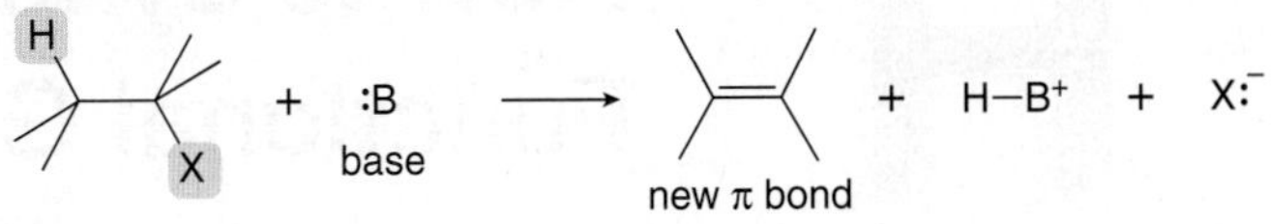

b. Alcohols
(Section 9.8)

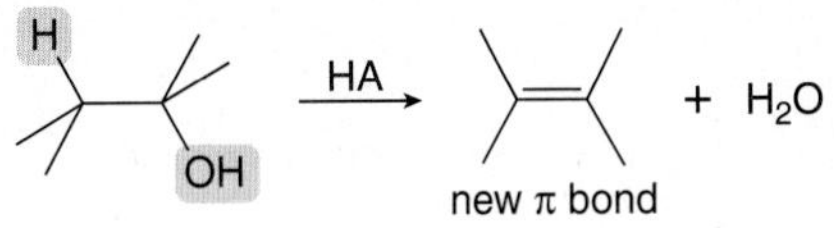

Addition Reactions

[1] Electrophilic addition to carbon–carbon multiple bonds

a. Alkenes
(Chapter 10)

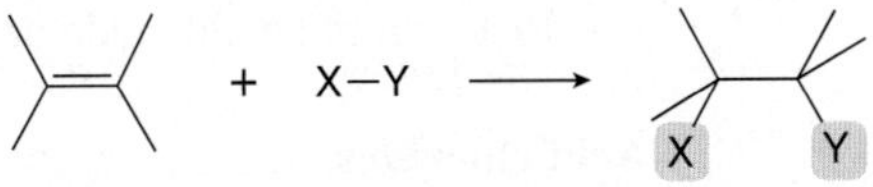

b. Alkynes
(Section 11.6)

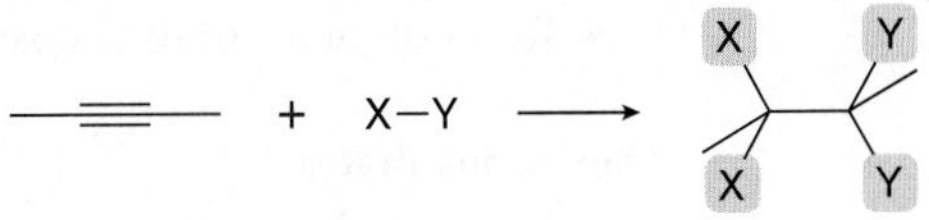

[2] Nucleophilic addition to carbon–oxygen multiple bonds

Aldehydes and ketones
(Chapter 21)

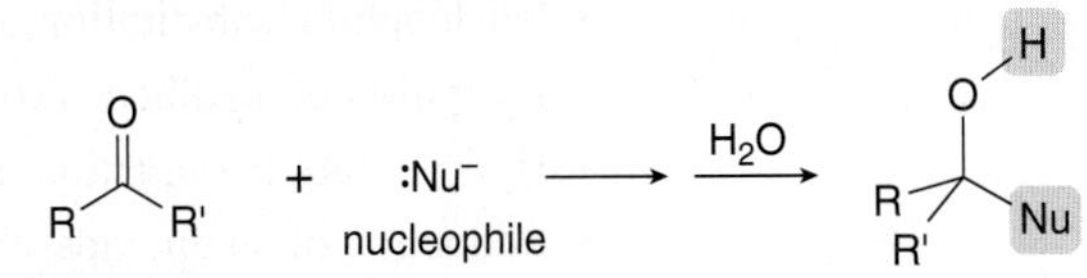

R' = H or alkyl

APPENDIX

H How to Synthesize Particular Functional Groups

Acetals

- Reaction of an aldehyde or ketone with two equivalents of an alcohol (21.14)

Acid chlorides

- Reaction of a carboxylic acid with thionyl chloride (22.10)

Acyl phosphates

- Enzyme-catalyzed reaction of carboxylate anions with ATP (22.16A)

Alcohols

- Nucleophilic substitution of an alkyl halide with ^{-}OH or H_2O (9.6)
- Hydration of an alkene (10.12)
- Hydroboration–oxidation of an alkene (10.16)
- Reduction of an epoxide with $LiAlH_4$ (12.6)
- Reduction of an aldehyde or ketone (20.4)
- Hydrogenation of an α,β-unsaturated carbonyl compound with H_2 + Pd-C (20.4C)
- Enantioselective reduction of an aldehyde or ketone with the chiral CBS reagent (20.6)
- Reduction of an acid chloride with $LiAlH_4$ (20.7)
- Reduction of an ester with $LiAlH_4$ (20.7)
- Reduction of a carboxylic acid with $LiAlH_4$ (20.7)
- Reaction of an aldehyde or ketone with a Grignard or organolithium reagent (20.10)
- Reaction of an acid chloride with a Grignard or organolithium reagent (20.13)
- Reaction of an ester with a Grignard or organolithium reagent (20.13)
- Reaction of an organometallic reagent with an epoxide (20.14B)

Aldehydes

- Hydroboration–oxidation of a terminal alkyne (11.10)
- Oxidative cleavage of an alkene with O_3 followed by Zn or $(CH_3)_2S$ (12.10)
- Oxidation of a 1° alcohol with PCC (12.12)
- Oxidation of a 1° alcohol with $HCrO_4^-$, Amberlyst A-26 resin (12.13)
- Reduction of an acid chloride with $LiAlH[OC(CH_3)_3]_3$ (20.7)
- Reduction of an ester with DIBAL-H (20.7)
- Hydrolysis of an acetal (21.14B)
- Hydrolysis of an imine or enamine (21.12B)
- Reduction of a nitrile (22.18B)

Alkanes

- Catalytic hydrogenation of an alkene with H_2 + Pd-C (12.3)
- Catalytic hydrogenation of an alkyne with two equivalents of H_2 + Pd-C (12.5A)
- Reduction of an alkyl halide with $LiAlH_4$ (12.6)
- Reduction of a ketone to a methylene group (CH_2)—the Wolff–Kishner or Clemmensen reaction (18.15B)
- Protonation of an organometallic reagent with H_2O, ROH, or acid (20.9)
- Coupling of an organocuprate reagent (R_2CuLi) with an alkyl halide, R'X (29.1)
- Simmons–Smith reaction of an alkene with CH_2I_2 and Zn(Cu) to form a cyclopropane (29.5)

Alkenes

- Dehydrohalogenation of an alkyl halide with base (8.3)
- Dehydration of an alcohol with acid (9.8)
- Dehydration of an alcohol using $POCl_3$ and pyridine (9.10)
- β Elimination of an alkyl tosylate with base (9.13)
- Catalytic hydrogenation of an alkyne with H_2 + Lindlar catalyst to form a cis alkene (12.5B)
- Dissolving metal reduction of an alkyne with Na, NH_3 to form a trans alkene (12.5C)
- Wittig reaction (21.10)
- β Elimination of an α-halo carbonyl compound with Li_2CO_3, LiBr, and DMF (23.7C)
- Hofmann elimination of an amine (25.12)
- Coupling of an organocuprate reagent (R_2CuLi) with an organic halide, R'X (29.1)
- The palladium-catalyzed Suzuki reaction of a vinyl or aryl halide with a vinyl- or arylborane (29.2)
- The palladium-catalyzed Heck reaction of a vinyl or aryl halide with an alkene (29.3)
- Olefin metathesis (29.6)

Alkyl halides

- Reaction of an alcohol with HX (9.11)
- Reaction of an alcohol with $SOCl_2$ or PBr_3 (9.12)
- Cleavage of an ether with HBr or HI (9.14)
- Hydrohalogenation of an alkene with HX (10.9)
- Halogenation of an alkene with X_2 (10.13)
- Hydrohalogenation of an alkyne with two equivalents of HX (11.7)
- Halogenation of an alkyne with two equivalents of X_2 (11.8)
- Radical halogenation of an alkane (15.3)
- Radical halogenation at an allylic carbon (15.10)
- Radical addition of HBr to an alkene (15.13)
- Electrophilic addition of HX to a 1,3-diene (16.10)
- Radical halogenation of an alkyl benzene (18.14)
- Halogenation α to a carbonyl group (23.7)
- Addition of a dihalocarbene to an alkene to form a dihalocyclopropane (29.4)

Alkynes

- Dehydrohalogenation of an alkyl dihalide with base (11.5)
- S_N2 reaction of an alkyl halide with an acetylide anion, $^{-}C{\equiv}CR$ (11.11)

Amides

- Reaction of an acid chloride with NH_3 or an amine (22.8)
- Reaction of an anhydride with NH_3 or an amine (22.9)
- Reaction of a carboxylic acid with NH_3 or an amine and DCC (22.10)

- Reaction of an ester with NH_3 or an amine (22.11)
- Enzyme-catalyzed reaction of an acyl phosphate with an amine (22.16B)

Amines

- Nucleophilic aromatic substitution (18.13)
- Reduction of a nitro group (18.15C)
- Reduction of an amide with $LiAlH_4$ (20.7B)
- Reduction of a nitrile (22.18B)
- S_N2 reaction using NH_3 or an amine (25.7A)
- Gabriel synthesis (25.7A)
- Reductive amination of an aldehyde or ketone (25.7C)

Amino acids

- S_N2 reaction of an α-halo carboxylic acid with excess NH_3 (26.2A)
- Alkylation of diethyl acetamidomalonate (26.2B)
- Strecker synthesis (26.2C)
- Enantioselective hydrogenation using a chiral catalyst (26.4)

Anhydrides

- Reaction of an acid chloride with a carboxylate anion (22.8)
- Dehydration of a dicarboxylic acid (22.10)

Aryl halides

- Halogenation of benzene with $X_2 + FeX_3$ (18.3)
- Reaction of a diazonium salt with CuCl, CuBr, HBF_4, NaI, or KI (25.14A)

Carboxylic acids

- Oxidative cleavage of an alkyne with ozone (12.11)
- Oxidation of a 1° alcohol with CrO_3 (or a similar Cr^{6+} reagent), H_2O, H_2SO_4 (12.12B)
- Oxidation of an alkyl benzene with $KMnO_4$ (18.15A)
- Oxidation of an aldehyde (20.8)
- Reaction of a Grignard reagent with CO_2 (20.14A)
- Hydrolysis of a cyanohydrin (21.9)
- Hydrolysis of an acid chloride (22.8)
- Hydrolysis of an anhydride (22.9)
- Hydrolysis of an ester (22.11)
- Hydrolysis of an amide (22.13)
- Enzyme-catalyzed hydrolysis of a thioester (22.17)
- Hydrolysis of a nitrile (22.18A)
- Malonic ester synthesis (23.9)

Cyanohydrins

- Addition of HCN to an aldehyde or ketone (21.9)

1,2-Diols

- Anti dihydroxylation of an alkene with a peroxyacid, followed by ring opening with ^{-}OH or H_2O (12.9A)
- Syn dihydroxylation of an alkene with $KMnO_4$ or OsO_4 (12.9B)

Enamines

- Reaction of an aldehyde or ketone with a 2° amine (21.12)

Epoxides

- Intramolecular S_N2 reaction of a halohydrin using base (9.6)
- Epoxidation of an alkene with mCPBA (12.8)
- Enantioselective epoxidation of an allylic alcohol with the Sharpless reagent (12.15)

Esters

- S_N2 reaction of an alkyl halide with a carboxylate anion, RCO_2^- (7.18)
- Reaction of an acid chloride with an alcohol (22.8)
- Reaction of an anhydride with an alcohol (22.9)
- Fischer esterification of a carboxylic acid with an alcohol (22.10)
- Enzyme-catalyzed reaction of a thioester with an alcohol (22.17)

Ethers

- Williamson ether synthesis—S_N2 reaction of an alkyl halide with an alkoxide, ^-OR (9.6)
- Reaction of an alkyl tosylate with an alkoxide, ^-OR (9.13)
- Addition of an alcohol to an alkene in the presence of acid (10.12)
- Anionic polymerization of epoxides to form polyethers (31.3)

Halohydrins

- Reaction of an epoxide with HX (9.16)
- Addition of X and OH to an alkene (10.15)

Imine

- Reaction of an aldehyde or ketone with a 1° amine (21.11)

Ketones

- Hydration of an alkyne with H_2O, H_2SO_4, and $HgSO_4$ (11.9)
- Oxidative cleavage of an alkene with O_3 followed by Zn or $(CH_3)_2S$ (12.10)
- Oxidation of a 2° alcohol with any Cr^{6+} reagent (12.12, 12.13)
- Friedel–Crafts acylation (18.5)
- Reaction of an acid chloride with an organocuprate reagent (20.13)
- Hydrolysis of an imine or enamine (21.12B)
- Hydrolysis of an acetal (21.14B)
- Reaction of a nitrile with a Grignard or organolithium reagent (22.18C)
- Acetoacetic ester synthesis (23.10)

Nitriles

- S_N2 reaction of an alkyl halide with NaCN (7.18, 22.18)
- Reaction of an aryl diazonium salt with CuCN (25.14A)

Phenols

- Reaction of an aryl diazonium salt with H_2O (25.14A)
- Nucleophilic aromatic substitution (18.13)

Sulfides

- Reaction of an alkyl halide with ^-SR (9.15)

Thioesters

- Enzyme-catalyzed reaction of acyl phosphates with thiols (22.16B)

Thiols

- Reaction of an alkyl halide with ^-SH (9.15)

Glossary

A

Acetal (Section 21.14): A compound having the general structure $R_2C(OR')_2$, where R = H, alkyl, or aryl. Acetals are used as protecting groups for aldehydes and ketones.

Acetoacetic ester synthesis (Section 23.10): A stepwise method that converts ethyl acetoacetate into a ketone having one or two carbons bonded to the α carbon.

Acetylation (Section 22.9): A reaction that transfers an acetyl group (CH_3CO-) from one atom to another.

Acetyl coenzyme A (Section 22.17): A biochemical thioester that acts as an acetylating reagent. Acetyl coenzyme A is often referred to as acetyl CoA.

Acetyl group (Section 21.2E): A substituent having the structure $-COCH_3$.

Acetylide anion (Sections 11.11, 20.9B): An anion formed by treating a terminal alkyne with a strong base. Acetylide anions have the general structure $R-C\equiv C^-$.

Achiral molecule (Section 5.3): A molecule that is superimposable upon its mirror image. An achiral molecule is not chiral.

Acid chloride (Sections 20.1, 22.1): A compound having the general structure RCOCl.

Acidity constant (Section 2.3): A value symbolized by K_a that represents the strength of an acid (HA). The larger the K_a, the stronger the acid.

$$K_a = \frac{[H_3O^+][A:^-]}{[H-A]}$$

Active site (Section 6.11): The region of an enzyme that binds the substrate.

Acyclic alkane (Section 4.1): A compound with the general formula C_nH_{2n+2}. Acyclic alkanes are also called saturated hydrocarbons because they contain the maximum number of hydrogen atoms per carbon.

Acylation (Sections 18.5A, 22.17): A reaction that transfers an acyl group from one atom to another.

Acyl chloride (Section 18.5A): A compound having the general structure RCOCl. Acyl chlorides are also called acid chlorides.

Acyl group (Section 18.5A): A substituent having the general structure RCO–.

Acylium ion (Section 18.5B): A positively charged electrophile having the general structure $(R-C\equiv O)^+$, formed when the Lewis acid $AlCl_3$ ionizes the carbon–halogen bond of an acid chloride.

Acyl phosphate (Sections 22.1, 22.16): A compound having the general structure $RCO_2PO_3^{2-}$.

Acyl transfer reaction (Section 22.17): A reaction that transfers an acyl group from one atom to another.

1,2-Addition (Sections 16.10, 20.15): An addition reaction to a conjugated system that adds groups across two adjacent atoms.

1,4-Addition (Sections 16.10, 20.15): An addition reaction that adds groups to the atoms in the 1 and 4 positions of a conjugated system. 1,4-Addition is also called conjugate addition.

Addition polymer (Section 31.1): A polymer prepared by a chain reaction that adds a monomer to the growing end of a polymer chain. Addition polymers are also called chain-growth polymers.

Addition reaction (Sections 6.2C, 10.8): A reaction in which elements are added to a starting material. In an addition reaction, a π bond is broken and two σ bonds are formed.

Aglycon (Section 27.7C): The alcohol formed from hydrolysis of a glycoside.

Alcohol (Section 9.1): A compound having the general structure ROH. An alcohol contains a hydroxy group (OH group) bonded to an sp^3 hybridized carbon atom.

Aldaric acid (Section 27.9B): The dicarboxylic acid formed by the oxidation of the aldehyde and the primary alcohol of an aldose.

Aldehyde (Section 11.10): A compound having the general structure RCHO, where R = H, alkyl, or aryl.

Alditol (Section 27.9A): A compound formed by the reduction of the aldehyde of an aldose to a primary alcohol.

Aldol condensation (Section 24.1B): An aldol reaction in which the initially formed β-hydroxy carbonyl compound loses water by dehydration.

Aldol reaction (Section 24.1A): A reaction in which two molecules of an aldehyde or ketone react with each other in the presence of base to form a β-hydroxy carbonyl compound.

Aldonic acid (Section 27.9B): A compound formed by the oxidation of the aldehyde of an aldose to a carboxylic acid.

Aldose (Section 27.2): A monosaccharide comprised of a polyhydroxy aldehyde.

Aliphatic (Section 3.2A): A compound or portion of a compound made up of C–C σ and π bonds but not aromatic bonds.

Alkaloid (Section 25.6A): A basic, nitrogen-containing compound isolated from a plant source.

Alkane (Section 4.1): An aliphatic hydrocarbon having only C–C and C–H σ bonds.

Alkene (Section 8.2A): An aliphatic hydrocarbon that contains a carbon–carbon double bond.

Alkoxide (Sections 8.1, 9.6): An anion having the general structure RO^-, formed by deprotonating an alcohol with a base.

Alkoxy group (Section 9.3B): A substituent containing an alkyl group bonded to an oxygen (RO group).

Alkylation (Section 23.8): A reaction that transfers an alkyl group from one atom to another.

Alkyl group (Section 4.4A): A group formed by removing one hydrogen from an alkane. Alkyl groups are named by replacing the suffix *-ane* of the parent alkane with *-yl.*

Alkyl halide (Section 7.1): A compound containing a halogen atom bonded to an sp^3 hybridized carbon atom. Alkyl halides have the general molecular formula $C_nH_{2n+1}X$.

1,2-Alkyl shift (Section 9.9): The rearrangement of a less stable carbocation to a more stable carbocation by the shift of an alkyl group from one carbon atom to an adjacent carbon atom.

Alkyl tosylate (Section 9.13): A compound having the general structure $ROSO_2C_6H_4CH_3$. Alkyl tosylates are also called tosylates and are abbreviated as ROTs.

Alkyne (Section 8.10): An aliphatic hydrocarbon that contains a carbon–carbon triple bond.

Allyl carbocation (Section 16.1B): A carbocation that has a positive charge on the atom adjacent to a carbon–carbon double bond. An allyl carbocation is resonance stabilized.

Allyl group (Section 10.3C): A substituent having the structure $-CH_2-CH{=}CH_2$.

Allylic bromination (Section 15.10A): A radical substitution reaction in which bromine replaces a hydrogen atom on the carbon adjacent to a carbon–carbon double bond.

Allylic carbon (Section 15.10): A carbon atom bonded to a carbon–carbon double bond.

Allylic halide (Section 7.1): A molecule containing a halogen atom bonded to the carbon atom adjacent to a carbon–carbon double bond.

Allyl radical (Section 15.10): A radical that has an unpaired electron on the carbon adjacent to a carbon–carbon double bond. An allyl radical is resonance stabilized.

Alpha (α) carbon (Sections 8.1, 19.2B): In an elimination reaction, the carbon that is bonded to the leaving group. In a carbonyl compound, the carbon that is bonded to the carbonyl carbon.

Ambident nucleophile (Section 23.3C): A nucleophile that has two reactive sites.

Amide (Sections 20.1, 22.1): A compound having the general structure $RCONR'_2$, where R' = H or alkyl.

Amide base (Sections 8.10, 23.3B): A nitrogen-containing base formed by deprotonating an amine or ammonia.

Amine (Sections 21.11, 25.1): A basic organic nitrogen compound having the general structure RNH_2, R_2NH, or R_3N. An amine has a nonbonded pair of electrons on the nitrogen atom.

α-Amino acid (Sections 3.9A, 19.15A, 26.1): A compound having the general structure $RCH(NH_2)COOH$. α-Amino acids are the building blocks of proteins.

Amino acid residue (Section 26.5): The individual amino acids in peptides and proteins.

Amino group (Section 25.3D): A substituent having the structure $-NH_2$.

α-Amino nitrile (Section 26.2C): A compound having the general structure $RCH(NH_2)C{\equiv}N$.

Amino sugar (Section 27.13A): A carbohydrate that contains an NH_2 group instead of a hydroxy group at a non-anomeric carbon.

Ammonium salt (Section 25.1): A compound containing a positively charged nitrogen with four σ bonds; for example, $R_4N^+X^-$.

Angle strain (Section 4.11): An increase in the energy of a molecule resulting when the bond angles of the sp^3 hybridized atoms deviate from the optimum tetrahedral angle of 109.5°.

Angular methyl group (Section 28.8A): A methyl group located at the ring junction of two fused rings of the steroid skeleton.

Anhydride (Section 22.1): A compound having the general structure $(RCO)_2O$.

Aniline (Section 25.3C): A compound having the structure $C_6H_5NH_2$.

Anion (Section 1.2): A negatively charged ion that results from a neutral atom gaining one or more electrons.

Anionic polymerization (Section 31.2C): Chain-growth polymerization of alkenes substituted by electron-withdrawing groups that stabilize intermediate anions.

Annulation (Section 24.10): A reaction that forms a new ring.

Annulene (Section 17.7A): A hydrocarbon containing a single ring with alternating double and single bonds.

α Anomer (Section 27.6): The stereoisomer of a cyclic monosaccharide in which the anomeric OH and the CH_2OH groups are trans. In a D monosaccharide, the hydroxy group on the anomeric carbon is drawn down.

β Anomer (Section 27.6): The stereoisomer of a cyclic monosaccharide in which the anomeric OH and the CH_2OH groups are cis. In a D monosaccharide, the hydroxy group on the anomeric carbon is drawn up.

Anomeric carbon (Section 27.6): The stereogenic center at the hemiacetal carbon of a cyclic monosaccharide.

Antarafacial reaction (Section 30.4): A pericyclic reaction that occurs on opposite sides of the two ends of the π electron system.

Anti addition (Section 10.8): An addition reaction in which the two parts of a reagent are added from opposite sides of a double bond.

Antiaromatic compound (Section 17.6): An organic compound that is cyclic, planar, completely conjugated, and has $4n$ π electrons.

Antibonding molecular orbital (Section 17.9A): A high-energy molecular orbital formed when two atomic orbitals of opposite phase overlap.

Anti conformation (Section 4.10): A staggered conformation in which the two larger groups on adjacent carbon atoms have a dihedral angle of 180°.

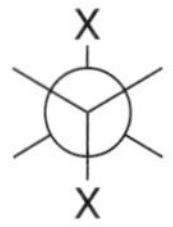

Anti dihydroxylation (Section 12.9A): The addition of two hydroxy groups to opposite faces of a double bond.

Antioxidant (Section 15.12): A compound that stops an oxidation from occurring.

Anti periplanar (Section 8.8A): In an elimination reaction, a geometry where the β hydrogen and the leaving group are on opposite sides of the molecule.

Aromatic compound (Section 17.1): A planar, cyclic organic compound that has p orbitals on all ring atoms and a total of $4n + 2$ π electrons in the orbitals.

Aryl group (Section 17.3D): A substituent formed by removing one hydrogen atom from an aromatic ring.

Aryl halide (Sections 7.1, 18.3): A molecule such as C_6H_5X, containing a halogen atom X bonded to an aromatic ring.

Asymmetric carbon (Section 5.3): A carbon atom that is bonded to four different groups. An asymmetric carbon is also called a stereogenic center, a chiral center, or a chirality center.

Asymmetric reaction (Sections 12.15, 20.6A, 26.4): A reaction that converts an achiral starting material into predominantly one enantiomer.

Atactic polymer (Section 31.4): A polymer having the substituents randomly oriented along the carbon backbone of an elongated polymer chain.

Atomic number (Section 1.1): The number of protons in the nucleus of an element.

Atomic weight (Section 1.1): The weighted average of the mass of all isotopes of a particular element. The atomic weight is reported in atomic mass units (amu).

Axial bonds (Section 4.12A): Bonds located above or below and perpendicular to the plane of the chair conformation of cyclohexane. Three axial bonds point upwards (on the up carbons) and three axial bonds point downwards (on the down carbons).

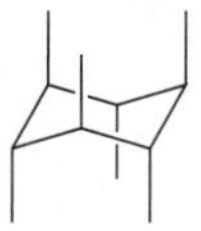

Azo compound (Section 25.15): A compound having the general structure RN=NR'.

B

Backside attack (Section 7.11C): Approach of a nucleophile from the side opposite the leaving group.

Barrier to rotation (Section 4.10): The energy difference between the lowest and highest energy conformations of a molecule.

Base peak (Section 13.1): The peak in the mass spectrum having the greatest abundance value.

Basicity (Section 7.8): A measure of how readily an atom donates its electron pair to a proton.

Benedict's reagent (Section 27.9B): A reagent for oxidizing aldehydes to carboxylic acids using a Cu^{2+} salt, forming brick-red Cu_2O as a side product.

Benzoyl group (Section 21.2E): A substituent having the structure $-COC_6H_5$.

Benzyl group (Section 17.3D): A substituent having the structure $C_6H_5CH_2-$.

Benzylic halide (Sections 7.1, 18.14): A compound such as $C_6H_5CH_2X$, containing a halogen atom X bonded to a carbon that is bonded to a benzene ring.

Benzyne (Section 18.13B): A reactive intermediate formed by elimination of HX from an aryl halide.

Beta (β) carbon (Sections 8.1, 19.2B): In an elimination reaction, the carbon adjacent to the carbon with the leaving group. In a carbonyl compound, the carbon located two carbons from the carbonyl carbon.

Bimolecular reaction (Sections 6.9B, 7.10, 7.11A): A reaction in which the concentration of both reactants affects the reaction rate and both terms appear in the rate equation. In a bimolecular reaction, two reactants are involved in the only step or the rate-determining step.

Biodegradable polymer (Section 31.9B): A polymer that can be degraded by microorganisms naturally present in the environment.

Biomolecule (Section 3.9): An organic compound found in a biological system.

Boat conformation of cyclohexane (Section 4.12B): An unstable conformation adopted by cyclohexane that resembles a boat. The instability of the boat conformation results from torsional strain and steric strain. The boat conformation of cyclohexane is 30 kJ/mol less stable than the chair conformation.

Boiling point (Section 3.4A): The temperature at which molecules in the liquid phase are converted to the gas phase. Molecules with stronger intermolecular forces have higher boiling points. Boiling point is abbreviated as bp.

Bond dissociation energy (Section 6.4): The amount of energy needed to homolytically cleave a covalent bond.

Bonding (Section 1.2): The joining of two atoms in a stable arrangement. Bonding is a favorable process that leads to lowered energy and increased stability.

Bonding molecular orbital (Section 17.9A): A low-energy molecular orbital formed when two atomic orbitals of similar phase overlap.

Bond length (Section 1.7A): The average distance between the centers of two bonded nuclei. Bond lengths are reported in picometers (pm).

Branched-chain alkane (Section 4.1A): An acyclic alkane that has alkyl substituents bonded to the parent carbon chain.

Bridged ring system (Section 16.13D): A bicyclic ring system in which the two rings share non-adjacent carbon atoms.

Bromination (Sections 10.13, 15.6, 18.3): The reaction of a compound with bromine.

Bromohydrin (Section 10.15): A compound having a bromine and a hydroxy group on adjacent carbon atoms.

Brønsted–Lowry acid (Section 2.1): A proton donor, symbolized by HA. A Brønsted–Lowry acid must contain a hydrogen atom.

Brønsted–Lowry base (Section 2.1): A proton acceptor, symbolized by :B. A Brønsted–Lowry base must be able to form a bond to a proton by donating an available electron pair.

C

^{13}C NMR spectroscopy (Section 14.1): A form of nuclear magnetic resonance spectroscopy used to determine the type of carbon atoms in a molecule.

Cahn–Ingold–Prelog system of nomenclature (Section 5.6): The system of designating a stereogenic center as either *R* or *S* according to the arrangement of the four groups attached to the center.

Carbamate (Sections 26.7, 31.6): A functional group containing a carbonyl group bonded to both an oxygen and a nitrogen atom. A carbamate is also called a urethane.

Carbanion (Section 2.5D): An ion with a negative charge on a carbon atom.

Carbene (Section 29.4): A neutral reactive intermediate having the general structure $:CR_2$. A carbene contains a divalent carbon surrounded by six electrons, making it a highly reactive electrophile that adds to C−C double bonds.

Carbinolamine (Section 21.7B): An unstable intermediate having a hydroxy group and an amine group on the same carbon. A carbinolamine is formed during the addition of an amine to a carbonyl group.

Carbocation (Section 7.13C): A positively charged carbon atom. A carbocation is sp^2 hybridized and trigonal planar, and contains a vacant *p* orbital.

Carbohydrate (Sections 3.9B, 21.17, 27.1): A polyhydroxy aldehyde or ketone or a compound that can be hydrolyzed to a polyhydroxy aldehyde or ketone.

Carbonate (Section 31.6D): A compound having the general structure $(RO)_2C=O$.

Carbon backbone (Section 3.1): The C−C and C−H σ bond framework that makes up the skeleton of an organic molecule.

Carbon NMR spectroscopy (Section 14.1): A form of nuclear magnetic resonance spectroscopy used to determine the type of carbon atoms in a molecule.

Carbonyl group (Sections 3.2C, 11.9, 20.1): A functional group that contains a carbon–oxygen double bond (C=O). The polar carbon–oxygen bond makes the carbonyl carbon electrophilic.

Carboxy group (Section 19.1): A functional group having the structure COOH.

Carboxylate anion (Section 19.2C): An anion having the general structure RCO_2^-, formed by deprotonating a carboxylic acid with a Brønsted–Lowry base.
Carboxylation (Section 20.14): The reaction of an organometallic reagent with CO_2 to form a carboxylic acid after protonation.
Carboxylic acid (Section 19.1): A compound having the general structure RCO_2H.
Carboxylic acid derivatives (Section 20.1): Compounds having the general structure RCOZ, which can be synthesized from carboxylic acids. Common carboxylic acid derivatives include acid chlorides, anhydrides, esters, and amides.
Catalyst (Section 6.10): A substance that speeds up the rate of a reaction, but is recovered unchanged at the end of the reaction and does not appear in the product.
Catalytic hydrogenation (Section 12.3): A reduction reaction involving the addition of H_2 to a π bond in the presence of a metal catalyst.
Cation (Section 1.2): A positively charged ion that results from a neutral atom losing one or more electrons.
Cationic polymerization (Section 31.2C): Chain-growth polymerization of alkene monomers involving carbocation intermediates.
CBS reagent (Section 20.6A): A chiral reducing agent formed by reacting an oxazaborolidine with BH_3. CBS reagents predictably give one enantiomer as the major product of ketone reduction.
Cephalin (Section 28.4A): A phosphoacylglycerol in which the phosphodiester alkyl group is $-CH_2CH_2NH_3^+$. Cephalins are also called phosphatidylethanolamines.
Chain-growth polymer (Section 31.1): A polymer prepared by a chain reaction that adds a monomer to the growing end of a polymer chain. Chain-growth polymers are also called addition polymers.
Chain mechanism (Section 15.4A): A reaction mechanism that involves repeating steps.
Chair conformation of cyclohexane (Section 4.12A): A stable conformation adopted by cyclohexane that resembles a chair. The stability of the chair conformation results from the elimination of angle strain (all C−C−C bond angles are 109.5°) and torsional strain (all groups on adjacent carbon atoms are staggered).

Chemical shift (Section 14.1B): The position of an absorption signal on the *x* axis in an NMR spectrum relative to the reference signal of tetramethylsilane.
Chirality center (Section 5.3): A carbon atom bonded to four different groups. A chirality center is also called a chiral center, a stereogenic center, and an asymmetric center.
Chiral molecule (Section 5.3): A molecule that is not superimposable upon its mirror image.
Chlorination (Sections 10.14, 15.5, 18.3): The reaction of a compound with chlorine.
Chlorofluorocarbons (Sections 7.4, 15.9): Synthetic alkyl halides having the general molecular formula CF_xCl_{4-x}. Chlorofluorocarbons, abbreviated as CFCs, were used as refrigerants and aerosol propellants and contribute to the destruction of the ozone layer.
Chlorohydrin (Section 10.15): A compound having a chlorine and a hydroxy group on adjacent carbon atoms.
Chromate ester (Section 12.12A): An intermediate in the chromium-mediated oxidation of an alcohol having the general structure $R-O-CrO_3H$.
***s*-Cis** (Sections 16.6, 26.5B): The conformation of a 1,3-diene that has the two double bonds on the same side of the single bond that joins them.
Cis isomer (Sections 4.13B, 8.2B): An isomer of a ring or double bond that has two groups on the same side of the ring or double bond.
Claisen reaction (Section 24.5): A reaction between two molecules of an ester in the presence of base to form a β-keto ester.
Claisen rearrangement (Section 30.5): A [3,3] sigmatropic rearrangement of an unsaturated ether to a γ,δ-unsaturated carbonyl compound.
α Cleavage (Section 13.3B): A fragmentation in mass spectrometry that results in cleavage of a carbon–carbon bond. With aldehydes and ketones, α cleavage results in breaking the bond between the carbonyl carbon and the carbon adjacent to it. With alcohols, α cleavage occurs by breaking a bond between an alkyl group and the carbon that bears the OH group.
Clemmensen reduction (Section 18.15B): A method to reduce aryl ketones to alkyl benzenes using Zn(Hg) in the presence of a strong acid.
Coenzyme (Section 12.13): A compound that acts with an enzyme to carry out a biochemical process.
Combustion (Section 4.14B): An oxidation–reduction reaction, in which an alkane or other organic compound reacts with oxygen to form CO_2 and H_2O, releasing energy.
Common name (Section 4.6): The name of a molecule that was adopted prior to and therefore does not follow the IUPAC system of nomenclature.
Compound (Section 1.2): The structure that results when two or more elements are joined together in a stable arrangement.
Concerted reaction (Sections 6.3, 7.11B): A reaction in which all bond forming and bond breaking occurs in one step.
Condensation polymer (Section 31.1): A polymer formed when monomers containing two functional groups come together with loss of a small molecule such as water or HCl. Condensation polymers are also called step-growth polymers.
Condensation reaction (Section 24.1B): A reaction in which a small molecule, often water, is eliminated during the reaction process.
Condensed structure (Section 1.8A): A shorthand representation of the structure of a compound in which all atoms are drawn in but bonds and lone pairs are usually omitted. Parentheses are used to denote similar groups bonded to the same atom.
Configuration (Section 5.2): A particular three-dimensional arrangement of atoms.
Conformations (Section 4.9): The different arrangements of atoms that are interconverted by rotation about single bonds.
Conjugate acid (Section 2.2): The compound that results when a base gains a proton in a proton transfer reaction.
Conjugate addition (Sections 16.10, 20.15): An addition reaction that adds groups to the atoms in the 1 and 4 positions of a conjugated system. Conjugate addition is also called 1,4-addition.
Conjugate base (Section 2.2): The compound that results when an acid loses a proton in a proton transfer reaction.
Conjugated diene (Section 16.1A): A compound that contains two carbon–carbon double bonds joined by a single σ bond. Pi (π) electrons are delocalized over both double bonds. Conjugated dienes are also called 1,3-dienes.
Conjugated protein (Section 26.10C): A structure composed of a protein unit and a non-protein molecule.
Conjugation (Section 16.1): The overlap of *p* orbitals on three or more adjacent atoms.

Conrotatory rotation (Section 30.3): Rotation of *p* orbitals in the same direction during electrocyclic ring closure or ring opening.

Constitutional isomers (Sections 1.4, 4.1A, 5.2): Two compounds that have the same molecular formula, but differ in the way the atoms are connected to each other. Constitutional isomers are also called structural isomers.

Coordination polymerization (Section 31.4): A polymerization reaction that uses a homogeneous catalyst that is soluble in the reaction solvents typically used.

Cope rearrangement (Section 30.5): A [3,3] sigmatropic rearrangement of a 1,5-diene to an isomeric 1,5-diene.

Copolymer (Section 31.2D): A polymer prepared by joining two or more different monomers together.

Counterion (Section 2.1): An ion that does not take part in a reaction and is opposite in charge to the ion that does take part in the reaction. A counterion is also called a spectator ion.

Coupled reactions (6.5C): Reactions that are paired together to drive an energetically unfavorable process. The energy released by one reaction provides the energy to drive the other reaction.

Coupling constant (Section 14.6A): The frequency difference, measured in Hz, between the peaks in a split NMR signal.

Coupling reaction (Section 25.15): A reaction that forms a bond between two discrete molecules.

Covalent bond (Section 1.2): A bond that results from the sharing of electrons between two nuclei. A covalent bond is a two-electron bond.

Crossed aldol reaction (Section 24.2): An aldol reaction in which the two reacting carbonyl compounds are different. A crossed aldol reaction is also called a mixed aldol reaction.

Crossed Claisen reaction (Section 24.6): A Claisen reaction in which the two reacting esters are different.

Crown ether (Section 3.7B): A cyclic ether containing multiple oxygen atoms. Crown ethers bind specific cations depending on the size of their central cavity.

Curved arrow notation (Section 1.6A): A convention that shows the movement of an electron pair. The tail of the arrow begins at the electron pair and the head points to where the electron pair moves.

Cyanide anion (Section 21.9A): An anion having the structure $^{-}C{\equiv}N$.

Cyano group (Section 22.1): A functional group consisting of a carbon–nitrogen triple bond ($C{\equiv}N$).

Cyanohydrin (Section 21.9): A compound having the general structure $RCH(OH)C{\equiv}N$. A cyanohydrin results from the addition of HCN across the carbonyl of an aldehyde or a ketone.

Cycloaddition (Section 30.1): A pericyclic reaction between two compounds with π bonds to form a cyclic product with two new σ bonds.

Cycloalkane (Sections 4.1, 4.2): A compound that contains carbons joined in one or more rings. Cycloalkanes with one ring have the general formula C_nH_{2n}.

Cyclopropanation (Section 29.4): An addition reaction to a carbon–carbon double bond that forms a cyclopropane.

D

D-Sugar (Section 27.2C): A sugar with the hydroxy group on the stereogenic center farthest from the carbonyl on the right side in the Fischer projection formula.

Decalin (Section 28.8A): Two fused six-membered rings. *cis*-Decalin has the hydrogen atoms at the ring fusion on the same side of the rings, whereas *trans*-decalin has the hydrogen atoms at the ring fusion on opposite sides of the rings.

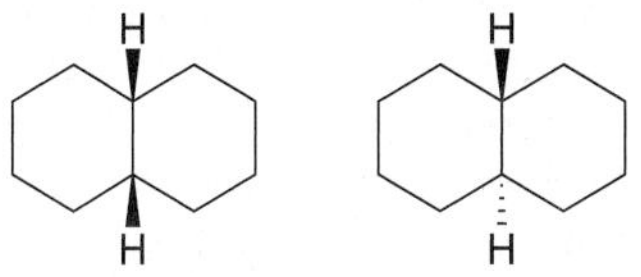

cis trans

Decarboxylation (Section 23.9A): Loss of CO_2 through cleavage of a carbon–carbon bond.

Degenerate orbitals (Section 17.9B): Orbitals (either atomic or molecular) having the same energy.

Degree of unsaturation (Section 10.2): A ring or a π bond in a molecule. The number of degrees of unsaturation compares the number of hydrogens in a compound to that of a saturated hydrocarbon containing the same number of carbons.

Dehydration (Sections 9.8, 22.10B): A reaction that results in the loss of the elements of water from the reaction components.

Dehydrohalogenation (Section 8.1): An elimination reaction in which the elements of hydrogen and halogen are lost from a starting material.

Delta (δ) scale (Section 14.1B): A common scale of chemical shifts used in NMR spectroscopy in which the absorption due to tetramethylsilane (TMS) occurs at zero parts per million.

Deoxy (Section 27.13B): A prefix that means without oxygen.

Deoxyribonucleic acid (Sections 3.9C, 17.8B): The large molecule that stores the genetic information of an organism and transmits that information to the next generation.

Deoxyribonucleoside (Section 27.13B): An *N*-glycoside formed by the reaction of D-2-deoxyribose with certain amine heterocycles.

Deoxyribonucleotide (Section 27.13B): A DNA building block having a deoxyribose and either a purine or pyrimidine base joined together by an *N*-glycosidic linkage, and a phosphate bonded to a hydroxy group of the sugar nucleus.

Deprotection (Section 20.12): A reaction that removes a protecting group, regenerating a functional group.

Deshielding effects (Section 14.3A): An effect in NMR caused by a decrease in electron density, thus increasing the strength of the magnetic field felt by the nucleus. Deshielding shifts an absorption downfield.

Dextrorotatory (Section 5.12A): Rotating plane-polarized light in the clockwise direction. The rotation is labeled *d* or (+).

1,3-Diacid (Section 23.9A): A compound containing two carboxylic acids separated by a single carbon atom. 1,3-Diacids are also called β-diacids.

Dialkylamide (Section 23.3B): An amide base having the general structure R_2N^-.

Diastereomers (Section 5.7): Stereoisomers that are not mirror images of each other. Diastereomers have the same *R,S* designation for at least one stereogenic center and the opposite *R,S* designation for at least one of the other stereogenic centers.

Diastereotopic protons (Section 14.2C): Two hydrogen atoms on the same carbon such that substitution of either hydrogen with a group Z forms diastereomers. The two hydrogen atoms are not equivalent and give two NMR signals.

1,3-Diaxial interaction (Section 4.13A): A steric interaction between two axial substituents of the chair form of cyclohexane. Larger axial substituents create unfavorable 1,3-diaxial interactions, destabilizing a cyclohexane conformation.

Diazonium salt (Section 25.13A): An ionic salt having the general structure $(R-N{\equiv}N)^{+}Cl^{-}$.

Diazotization reaction (Section 25.13A): A reaction that converts 1° alkylamines and arylamines to diazonium salts.

1,3-Dicarbonyl compound (Section 23.2): A compound containing two carbonyl groups separated by a single carbon atom.

1,4-Dicarbonyl compound (Section 24.4): A dicarbonyl compound in which the carbonyl groups are separated by three single bonds. 1,4-Dicarbonyl compounds can undergo intramolecular reactions to form five-membered rings.

1,5-Dicarbonyl compound (Section 24.4): A dicarbonyl compound in which the carbonyl groups are separated by four single bonds. 1,5-Dicarbonyl compounds can undergo intramolecular reactions to form six-membered rings.

Dieckmann reaction (Section 24.7): An intramolecular Claisen reaction of a diester to form a ring, typically a five- or six-membered ring.

Diels–Alder reaction (Section 16.12): An addition reaction between a 1,3-diene and a dienophile to form a cyclohexene ring.

1,3-Diene (Section 16.1A): A compound containing two carbon–carbon double bonds joined by a single σ bond. Pi (π) electrons are delocalized over both double bonds. 1,3-Dienes are also called conjugated dienes.

Dienophile (Section 16.12): The alkene component in a Diels–Alder reaction that reacts with a 1,3-diene.

Dihedral angle (Section 4.9): The angle that separates a bond on one atom from a bond on an adjacent atom.

Dihydroxylation (Section 12.9): Addition of two hydroxy groups to a double bond to form a 1,2-diol.

Diol (Section 9.3A): A compound possessing two hydroxy groups. Diols are also called glycols.

Dipeptide (Section 26.5): Two amino acids joined together by one amide bond.

Diphosphate (Section 7.16): A good leaving group that is often used in biological systems. Diphosphate ($P_2O_7^{4-}$) is abbreviated as PP_i.

Dipole (Section 1.12): A partial separation of electronic charge.

Dipole–dipole interaction (Section 3.3B): An attractive intermolecular interaction between the permanent dipoles of polar molecules. The dipoles of adjacent molecules align so that the partial positive and partial negative charges are in close proximity.

Directed aldol reaction (Section 24.3): A crossed aldol reaction in which the enolate of one carbonyl compound is formed, followed by addition of the second carbonyl compound.

Disaccharide (Sections 3.9B, 27.11): A carbohydrate containing two monosaccharide units joined together by a glycosidic linkage.

Disproportionation (Section 31.2): A method of chain termination in radical polymerization involving the transfer of a hydrogen atom from one polymer radical to another, forming a new C−H bond on one polymer chain and a new double bond on the other.

Disrotatory rotation (Section 30.3): Rotation of *p* orbitals in opposite directions during electrocyclic ring closure or ring opening.

Dissolving metal reduction (Section 12.2): A reduction reaction using alkali metals as a source of electrons and liquid ammonia as a source of protons.

Disubstituted alkene (Section 8.2A): An alkene that has two alkyl groups and two hydrogens bonded to the carbons of the double bond ($R_2C{=}CH_2$ or $RCH{=}CHR$).

Disulfide (Sections 9.15A, 26.5C): A compound having the general structure RSSR', often formed between the side chain of two cysteine residues.

Diterpene (Section 28.7A): A terpene that contains 20 carbons and four isoprene units. A diterpenoid contains at least one oxygen atom as well.

Doublet (Section 14.6): An NMR signal that is split into two peaks of equal area, caused by one nearby nonequivalent proton.

Doublet of doublets (Section 14.8): A splitting pattern of four peaks observed when a signal is split by two different nonequivalent protons.

Downfield shift (Section 14.1B): In an NMR spectrum, a term used to describe the relative location of an absorption signal. A downfield shift means the signal is shifted to the left in the spectrum to higher chemical shift on the δ scale.

E

E1 mechanism (Sections 8.3, 8.6): An elimination mechanism that goes by a two-step process involving a carbocation intermediate. E1 is an abbreviation for "Elimination Unimolecular."

E1cB mechanism (Section 24.1B): A two-step elimination mechanism that goes by a carbanion intermediate. E1cB stands for "Elimination Unimolecular, Conjugate Base."

E2 mechanism (Sections 8.3, 8.4): An elimination mechanism that goes by a one-step concerted process, in which both reactants are involved in the transition state. E2 is an abbreviation for "Elimination Bimolecular."

Eclipsed conformation (Section 4.9): A conformation of a molecule where the bonds on one carbon are directly aligned with the bonds on the adjacent carbon.

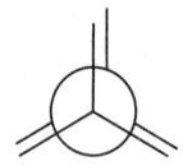

Edman degradation (Section 26.6B): A procedure used in peptide sequencing in which amino acids are cleaved one at a time from the N-terminal end, the identity of the amino acid determined, and the process repeated until the entire sequence is known.

Eicosanoids (Section 28.6): A group of biologically active compounds containing 20 carbon atoms derived from arachidonic acid.

Elastomer (Section 31.5): A polymer that stretches when stressed but then returns to its original shape.

Electrocyclic ring closure (Section 30.1): An intramolecular pericyclic reaction that forms a cyclic product containing one more σ bond and one fewer π bond than the reactant.

Electrocyclic ring-opening reaction (Section 30.1): A pericyclic reaction in which a σ bond of a cyclic reactant is cleaved to form a conjugated product with one more π bond.

Electromagnetic radiation (Section 13.5): Radiant energy having dual properties of both waves and particles. The electromagnetic spectrum contains the complete range of electromagnetic radiation, arbitrarily divided into different regions.

Electron-donating inductive effect (Section 7.13A): An inductive effect in which an electropositive atom or polarizable group donates electron density through σ bonds to another atom.

Electronegativity (Section 1.12): A measure of an atom's attraction for electrons in a bond. Electronegativity indicates how much a particular atom "wants" electrons.

Electron-withdrawing inductive effect (Sections 2.5, 7.13A): An inductive effect in which a nearby electronegative atom pulls electron density toward itself through σ bonds.

Electrophile (Section 2.8): An electron-deficient compound, often symbolized by E^{+}, which can accept a pair of electrons from an electron-rich compound, forming a covalent bond. Lewis acids are electrophiles.

Electrophilic addition reaction (Section 10.9): An addition reaction in which the first step of the mechanism involves addition of the electrophilic end of a reagent to a π bond.

Electrophilic aromatic substitution (Section 18.1): A characteristic reaction of benzene in which a hydrogen atom on the ring is replaced by an electrophile.

Electrospray ionization (Section 13.4C): A method for ionizing large biomolecules in a mass spectrometer. Electrospray ionization is abbreviated as ESI.

Electrostatic potential map (Section 1.12): A color-coded map that illustrates the distribution of electron density in a molecule. Electron-rich regions are indicated in red and electron-deficient regions are indicated in blue. Regions of intermediate electron density are shown in orange, yellow, and green.

α Elimination (Section 29.4): An elimination reaction involving the loss of two elements from the same atom.

β Elimination (Section 8.1): An elimination reaction involving the loss of elements from two adjacent atoms.

Elimination reaction (Sections 6.2B, 8.1): A chemical reaction in which elements of the starting material are "lost" and a π bond is formed.

Enamine (Section 21.12): A compound having an amine nitrogen atom bonded to a carbon–carbon double bond [$R_2C=CH(NR'_2)$].

Enantiomeric excess (Section 5.12D): A measurement of how much one enantiomer is present in excess of the racemic mixture. Enantiomeric excess (*ee*) is also called optical purity; *ee* = % of one enantiomer – % of the other enantiomer.

Enantiomers (Section 5.3): Stereoisomers that are mirror images but are not superimposable upon each other. Enantiomers have the exact opposite *R,S* designation at every stereogenic center.

Enantioselective reaction (Sections 12.15, 20.6A, 26.4): A reaction that affords predominantly or exclusively one enantiomer. Enantioselective reactions are also called asymmetric reactions.

Enantiotopic protons (Section 14.2C): Two hydrogen atoms on the same carbon such that substitution of either hydrogen with a group Z forms enantiomers. The two hydrogen atoms are equivalent and give a single NMR signal.

Endo position (Section 16.13D): A position of a substituent on a bridged bicyclic compound in which the substituent is closer to the longer bridge that joins the two carbons common to both rings.

Endothermic reaction (Section 6.4): A reaction in which the energy of the products is higher than the energy of the reactants. In an endothermic reaction, energy is absorbed and the $\Delta H°$ is a positive value.

Energy diagram (Section 6.7): A schematic representation of the energy changes that take place as reactants are converted to products. An energy diagram indicates how readily a reaction proceeds, how many steps are involved, and how the energies of the reactants, products, and intermediates compare.

Energy of activation (Section 6.7): The energy difference between the transition state and the starting material. The energy of activation, symbolized by E_a, is the minimum amount of energy needed to break bonds in the reactants.

Enolate (Sections 20.15, 23.3): A resonance-stabilized anion formed when a base removes an α hydrogen from the α carbon to a carbonyl group.

Enol tautomer (Sections 9.1, 11.9, 20.15): A compound having a hydroxy group bonded to a carbon–carbon double bond. An enol tautomer [such as $CH_2=C(OH)CH_3$] is in equilibrium with its keto tautomer [$(CH_3)_2C=O$].

Enthalpy change (Section 6.4): The energy absorbed or released in a reaction. Enthalpy change is symbolized by $\Delta H°$ and is also called the heat of reaction.

Entropy (Section 6.6): A measure of the randomness in a system. The more freedom of motion or the more disorder present, the higher the entropy. Entropy is denoted by the symbol $S°$.

Entropy change (Section 6.6): The change in the amount of disorder between reactants and products in a reaction. The entropy change is denoted by the symbol $\Delta S°$. $\Delta S° = S°_{products} - S°_{reactants}$.

Enzyme (Section 6.11): A biochemical catalyst composed of at least one chain of amino acids held together in a very specific three-dimensional shape.

Enzyme–substrate complex (Section 6.11): A structure having a substrate bonded to the active site of an enzyme.

Epoxidation (Section 12.8): Addition of a single oxygen atom to an alkene to form an epoxide.

Epoxide (Section 9.1): A cyclic ether having the oxygen atom as part of a three-membered ring. Epoxides are also called oxiranes.

Epoxy resin (Section 31.6E): A step-growth polymer formed from a fluid prepolymer and a hardener that cross-links polymer chains together.

Equatorial bonds (Section 4.12A): Bonds located in the plane of the chair conformation of cyclohexane (around the equator). Three equatorial bonds point slightly upward (on the down carbons) and three equatorial bonds point slightly downward (on the up carbons).

Equilibrium constant (Section 6.5A): A mathematical expression, denoted by the symbol K_{eq}, which relates the amount of starting material and product at equilibrium. K_{eq} = [products]/[starting materials].

Essential oil (Section 28.7): A class of terpenes isolated from plant sources by distillation.

Ester (Sections 20.1, 22.1): A compound having the general structure RCOOR'.

Esterification (Section 22.10C): A reaction that converts a carboxylic acid or a derivative of a carboxylic acid to an ester.

Ether (Section 9.1): A functional group having the general structure ROR'.

Ethynyl group (Section 11.2): An alkynyl substituent having the structure $-C\equiv C-H$.

Excited state (Sections 1.9B, 16.15A): A high-energy electronic state in which one or more electrons have been promoted to a higher energy orbital by absorption of energy.

Exo position (Section 16.13D): A position of a substituent on a bridged bicyclic compound in which the substituent is closer to the shorter bridge that joins the two carbons common to both rings.

Exothermic reaction (Section 6.4): A reaction in which the energy of the products is lower than the energy of the reactants. In an exothermic reaction, energy is released and the $\Delta H°$ is a negative value.

Extraction (Section 19.13): A laboratory method to separate and purify a mixture of compounds using solubility differences and acid–base principles.

***E,Z* System of nomenclature** (Section 10.3B): A system for unambiguously naming alkene stereoisomers by assigning priorities to the two groups on each carbon of the double bond. The *E* isomer has the two higher priority groups on opposite sides of the double bond, and the *Z* isomer has them on the same side.

F

Fat (Sections 10.6B, 28.3): A triacylglycerol that is solid at room temperature and composed of fatty acid side chains with a high degree of saturation.

Fatty acid (Sections 10.6A, 19.6): A long-chain carboxylic acid having between 12 and 20 carbon atoms.

Fehling's reagent (Section 27.9B): A reagent for oxidizing aldehydes to carboxylic acids using a Cu^{2+} salt as an oxidizing agent, forming brick-red Cu_2O as a by-product.

Fibrous proteins (Section 26.10): Long linear polypeptide chains that are bundled together to form rods or sheets.

Fingerprint region (Section 13.6B): The region in an IR spectrum at $< 1500\ cm^{-1}$. The region often contains a complex set of peaks and is unique for every compound.

First-order rate equation (Sections 6.9B, 7.10): A rate equation in which the reaction rate depends on the concentration of only one reactant.

Fischer esterification (Section 22.10C): An acid-catalyzed esterification reaction between a carboxylic acid and an alcohol to form an ester.

Fischer projection formula (Section 27.2A): A method for representing stereogenic centers with the stereogenic carbon at the intersection of vertical and horizontal lines. Fischer projections are also called cross formulas.

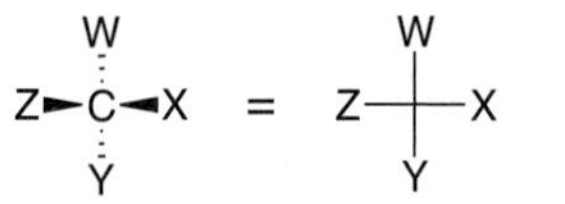

Fishhook (Section 6.3B): A half-headed curved arrow used in a reaction mechanism to denote the movement of a single electron.

Flagpole hydrogens (Section 4.12B): Hydrogens in the boat conformation of cyclohexane that are on either end of the "boat" and are forced into close proximity to each other.

Formal charge (Section 1.3C): The electronic charge assigned to individual atoms in a Lewis structure. The formal charge is calculated by subtracting an atom's unshared electrons and half of its shared electrons from the number of valence electrons that a neutral atom would possess.

Formyl group (Section 21.2E): A substituent having the structure −CHO.

Four-centered transition state (Section 10.16): A transition state that involves four atoms.

Fragment (Section 13.1): Radicals and cations formed by the decomposition of the molecular ion in a mass spectrometer.

Freons (Sections 7.4, 15.9): Chlorofluorocarbons consisting of simple halogen-containing organic compounds that were once commonly used as refrigerants.

Frequency (Section 13.5): The number of waves passing a point per unit time. Frequency is reported in cycles per second (s^{-1}), which is also called hertz (Hz). Frequency is abbreviated with the Greek letter nu (ν).

Friedel–Crafts acylation (Section 18.5A): An electrophilic aromatic substitution reaction in which benzene reacts with an acid chloride in the presence of a Lewis acid to give a ketone.

Friedel–Crafts alkylation (Section 18.5A): An electrophilic aromatic substitution reaction in which benzene reacts with an alkyl halide in the presence of a Lewis acid to give an alkyl benzene.

Frontside attack (Section 7.11C): Approach of a nucleophile from the same side as the leaving group.

Full-headed curved arrow (Section 6.3B): An arrow used in a reaction mechanism to denote the movement of a pair of electrons.

Functional group (Section 3.1): An atom or group of atoms with characteristic chemical and physical properties. The functional group is the reactive part of the molecule.

Functional group interconversion (Section 11.12): A reaction that converts one functional group into another.

Functional group region (Section 13.6B): The region in an IR spectrum at $\geq 1500\ cm^{-1}$. Common functional groups show one or two peaks in this region, at a characteristic frequency.

Furanose (Section 27.6): A cyclic five-membered ring of a monosaccharide containing an oxygen atom.

Fused ring system (Section 16.13D): A bicyclic ring system in which the two rings share one bond and two adjacent atoms.

G

Gabriel synthesis (Section 25.7A): A two-step method that converts an alkyl halide into a primary amine using a nucleophile derived from phthalimide.

Gas chromatography (Section 13.4B): An analytical technique that separates the components of a mixture based on their boiling points and the rate at which their vapors travel through a column.

Gauche conformation (Section 4.10): A staggered conformation in which the two larger groups on adjacent carbon atoms have a dihedral angle of 60°.

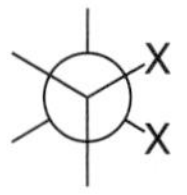

GC–MS (Section 13.4B): An analytical instrument that combines a gas chromatograph (GC) and a mass spectrometer (MS) in sequence.

***gem*-Diol** (Section 21.13): A compound having the general structure $R_2C(OH)_2$. *gem*-Diols are also called hydrates.

Geminal dihalide (Section 8.10): A compound that has two halogen atoms on the same carbon atom.

Gibbs free energy (Section 6.5A): The free energy of a molecule. Gibbs free energy is denoted by the symbol $G°$.

Gibbs free energy change (Section 6.5A): The overall energy difference between reactants and products. The Gibbs free energy change is denoted by the symbol $\Delta G°$. $\Delta G° = G°_{products} - G°_{reactants}$.

Globular proteins (Section 26.10): Polypeptide chains that are coiled into compact shapes with hydrophilic outer surfaces that make them water soluble.

Glycol (Section 9.3A): A compound possessing two hydroxy groups. Glycols are also called diols.

Glycosidase (Section 27.12B): An enzyme that hydrolyzes glycosidic linkages. An α-glycosidase hydrolyzes only α-glycosidic linkages.

Glycoside (Section 27.7A): A monosaccharide with an alkoxy group bonded to the anomeric carbon.

***N*-Glycoside** (Section 27.13B): A monosaccharide containing a nitrogen bonded to the anomeric carbon.

Glycosidic linkage (Section 27.11): An acetal linkage formed between an OH group on one monosaccharide and the anomeric carbon on a second monosaccharide.

Green chemistry (Sections 12.13, 31.8): The use of environmentally benign methods to synthesize compounds.

Grignard reagent (Section 20.9): An organometallic reagent having the general structure RMgX.

Ground state (Sections 1.9B, 16.15A): The lowest energy arrangement of electrons for an atom.

Group number (Section 1.1): The number above a particular column in the periodic table. Group numbers are represented by either an Arabic (1 to 8) or Roman (I to VIII) numeral followed by the letter A or B. The group number of a second-row element is equal to the number of valence electrons in that element.

Grubbs catalyst (Section 29.6): A widely used ruthenium catalyst for olefin metathesis that has the structure $Cl_2(Cy_3P)_2Ru{=}CHPh$.

Guest molecule (Section 9.5B): A small molecule that can bind to a larger host molecule.

H

^{1}H NMR spectroscopy (Section 14.1): A form of nuclear magnetic resonance spectroscopy used to determine the number and type of hydrogen atoms in a molecule. ^{1}H NMR is also called proton NMR spectroscopy.

Half-headed curved arrow (Section 6.3B): An arrow used in a reaction mechanism to denote the movement of a single electron. A half-headed curved arrow is also called a fishhook.

α-Halo aldehyde or ketone (Section 23.7): An aldehyde or ketone with a halogen atom bonded to the α carbon.

Haloform reaction (Section 23.7B): A halogenation reaction of a methyl ketone ($RCOCH_3$) with excess halogen, which results in formation of RCO_2^- and CHX_3 (haloform).

Halogenation (Sections 10.13, 15.3, 18.3): The reaction of a compound with a halogen.

Halohydrin (Sections 9.6, 10.15): A compound that has a hydroxy group and a halogen atom on adjacent carbon atoms.

Halonium ion (Section 10.13): A positively charged halogen atom. A bridged halonium ion contains a three-membered ring and is formed in the addition of a halogen (X_2) to an alkene.

Hammond postulate (Section 7.14): A postulate that states that the transition state of a reaction resembles the structure of the species (reactant or product) to which it is closer in energy.

Haworth projection (Section 27.6A): A representation of the cyclic form of a monosaccharide in which the ring is drawn flat.

Head-to-tail polymerization (Section 15.14B): A mechanism of radical polymerization in which the more substituted radical of the growing polymer chain always adds to the less substituted end of the new monomer.

Heat of hydrogenation (Section 12.3A): The $\Delta H°$ of a catalytic hydrogenation reaction equal to the amount of energy released by hydrogenating a π bond.

Heat of reaction (Section 6.4): The energy absorbed or released in a reaction. Heat of reaction is symbolized by $\Delta H°$ and is also called the change in enthalpy.

Heck reaction (Section 29.3): The palladium-catalyzed coupling of a vinyl or aryl halide with an alkene to form a more highly substituted alkene with a new carbon–carbon bond.

α-Helix (Section 26.9B): A secondary structure of a protein formed when a peptide chain twists into a right-handed or clockwise spiral.

Heme (Section 26.10C): A complex organic compound containing an Fe^{2+} ion coordinated with a porphyrin.

Hemiacetal (Section 21.14A): A compound that contains an alkoxy group and a hydroxy group bonded to the same carbon atom.

Henderson–Hasselbalch equation (Section 19.10): An equation derived from the expressions for pK_a and pH, which tells whether a compound exists in its acidic form or conjugate base at a particular pH.

$$pK_a = pH + \log\frac{[HA]}{[A:^-]}$$

Hertz (Section 13.5): A unit of frequency measuring the number of waves passing a point per second.

Heteroatom (Sections 1.6, 3.1): An atom other than carbon or hydrogen. Common heteroatoms in organic chemistry are nitrogen, oxygen, sulfur, phosphorus, and the halogens.

Heterocycle (Section 9.3B): A cyclic compound containing a heteroatom as part of the ring.

Heterolysis (Section 6.3A): The breaking of a covalent bond by unequally dividing the electrons between the two atoms in the bond. Heterolysis generates charged intermediates. Heterolysis is also called heterolytic cleavage.

Hexose (Section 27.2): A monosaccharide containing six carbons.

Highest occupied molecular orbital (Section 17.9B): The molecular orbital with the highest energy that also contains electrons. The highest occupied molecular orbital is abbreviated as HOMO.

High-resolution mass spectrometer (Section 13.4A): A mass spectrometer that can measure mass-to-charge ratios to four or more decimal places. High-resolution mass spectra are used to determine the molecular formula of a compound.

Hofmann elimination (Section 25.12): An E2 elimination reaction that converts an amine into a quaternary ammonium salt as the leaving group. The Hofmann elimination gives the less substituted alkene as the major product.

Homologous series (Section 4.1B): A group of compounds that differ by only a CH_2 group in the chain.

Homolysis (Section 6.3A): The breaking of a covalent bond by equally dividing the electrons between the two atoms in the bond. Homolysis generates uncharged radical intermediates. Homolysis is also called homolytic cleavage.

Homopolymer (Section 31.2D): A polymer prepared from a single monomer.

Homotopic protons (Section 14.2C): Two equivalent hydrogen atoms such that substitution of either hydrogen with a group Z forms the same product. The two hydrogen atoms give a single NMR signal.

Hooke's law (Section 13.7): A physical law that can be used to calculate the frequency of a bond vibration from the strength of the bond and the masses of the atoms attached to it.

Host–guest complex (Section 9.5B): The complex that is formed when a small guest molecule binds to a larger host molecule.

Host molecule (Section 9.5B): A large molecule that can bind a smaller guest molecule.

Hückel's rule (Section 17.6): A principle that states for a compound to be aromatic, it must be cyclic, planar, completely conjugated, and have $4n + 2$ π electrons.

Hybridization (Section 1.9B): The mathematical combination of two or more atomic orbitals (having different shapes) to form the same number of hybrid orbitals (all having the same shape).

Hybrid orbital (Section 1.9B): A new orbital that results from the mathematical combination of two or more atomic orbitals. The hybrid orbital is intermediate in energy compared to the atomic orbitals that were combined to form it.

Hydrate (Sections 12.12B, 21.13): A compound having the general structure $R_2C(OH)_2$. Hydrates are also called *gem*-diols.

Hydration (Sections 10.12, 21.9A): Addition of the elements of water to a molecule.

Hydride (Section 12.2): A negatively charged hydrogen ion ($H:^-$).

1,2-Hydride shift (Section 9.9): Rearrangement of a less stable carbocation to a more stable carbocation by the shift of a hydrogen atom from one carbon atom to an adjacent carbon atom.

Hydroboration (Section 10.16): The addition of the elements of borane (BH_3) to an alkene or alkyne.

Hydrocarbon (Sections 3.2A, 4.1): A compound made up of only the elements of carbon and hydrogen.

Hydrogen bonding (Section 3.3B): An attractive intermolecular interaction that occurs when a hydrogen atom bonded to an O, N, or F atom is electrostatically attracted to a lone pair of electrons on an O, N, or F atom in another molecule.

Hydrogenolysis (Section 26.7): A reaction that cleaves a σ bond using H_2 in the presence of a metal catalyst.

α Hydrogens (Section 23.1): The hydrogen atoms on the carbon bonded to the carbonyl carbon atom (the α carbon).

Hydrohalogenation (Section 10.9): An electrophilic addition of hydrogen halide (HX) to an alkene or alkyne.

Hydrolysis (Section 21.9A): A cleavage reaction with water.

Hydroperoxide (Section 15.11): An organic compound having the general structure ROOH.

Hydrophilic (Section 3.4C): Attracted to water. The polar portion of a molecule that interacts with polar water molecules is hydrophilic.

Hydrophobic (Section 3.4C): Not attracted to water. The nonpolar portion of a molecule that is not attracted to polar water molecules is hydrophobic.

β-Hydroxy carbonyl compound (Section 24.1A): An organic compound having a hydroxy group on the carbon β to the carbonyl group.

Hydroxy group (Section 9.1): The OH functional group.

Hyperconjugation (Section 7.13B): The overlap of an empty *p* orbital with an adjacent σ bond.

I

Imide (Section 25.7A): A compound having a nitrogen atom between two carbonyl groups.

Imine (Sections 21.7B, 21.11A): A compound with the general structure $R_2C{=}NR'$. Imines are also called Schiff bases.

Iminium ion (Section 21.11A): A resonance-stabilized cation having the general structure $(R_2C{=}NR'_2)^+$, where R' = H or alkyl.

Inductive effect (Sections 2.5B, 7.13A): The pull of electron density through σ bonds caused by electronegativity differences of atoms.

Infrared (IR) spectroscopy (Section 13.6): An analytical technique used to identify the functional groups in a molecule based on their absorption of electromagnetic radiation in the infrared region.

Initiation (Section 15.4A): The initial step in a chain mechanism that forms a reactive intermediate by cleavage of a bond.

Inscribed polygon method (Section 17.10): A method to predict the relative energies of cyclic, completely conjugated compounds to determine which molecular orbitals are filled or empty. The inscribed polygon is also called a Frost circle.

Integration (Section 14.5): The area under an NMR signal that is proportional to the number of absorbing nuclei that give rise to the signal.

Intermolecular forces (Section 3.3): The types of interactions that exist between molecules. Functional groups determine the type and strength of these forces. Intermolecular forces are also called noncovalent interactions or nonbonded interactions.

Internal alkene (Section 10.1): An alkene that has at least one carbon atom bonded to each end of the double bond.

Internal alkyne (Section 11.1): An alkyne that has one carbon atom bonded to each end of the triple bond.

Inversion of configuration (Section 7.11C): The opposite relative stereochemistry of a stereogenic center in the starting material and product of a chemical reaction. In a nucleophilic substitution reaction, inversion results when the nucleophile and leaving group are in the opposite position relative to the three other groups on carbon.

Iodoform test (Section 23.7B): A test for the presence of methyl ketones, indicated by the formation of the yellow precipitate, CHI_3, via the haloform reaction.

Ionic bond (Section 1.2): A bond that results from the transfer of electrons from one element to another. Ionic bonds result from strong electrostatic interactions between ions with opposite charges. The transfer of electrons forms stable salts composed of cations and anions.

Ionophore (Section 3.7B): An organic molecule that can form a complex with cations so they may be transported across a cell membrane. Ionophores have a hydrophobic exterior and a hydrophilic central cavity that complexes the cation.

Isocyanate (Section 31.6C): A compound having the general structure $RN{=}C{=}O$.

Isoelectric point (Sections 19.15C, 26.1A): The pH at which an amino acid exists primarily in its neutral zwitterionic form. Isoelectric point is abbreviated as p*I*.

Isolated diene (Section 16.1A): A compound containing two carbon–carbon double bonds joined by more than one σ bond.

Isomers (Sections 1.4A, 4.1A, 5.1): Two different compounds that have the same molecular formula.

Isoprene unit (Section 28.7): A five-carbon unit with four carbons in a row and a one-carbon branch on one of the middle carbons.

Isotactic polymer (Section 31.4): A polymer having all the substituents on the same side of the carbon backbone of an elongated polymer chain.

Isotope (Section 1.1): Two or more atoms of the same element having the same number of protons in the nucleus but a different number of neutrons. Isotopes have the same atomic number but different mass numbers.

IUPAC system of nomenclature (Section 4.3): A systematic method for naming compounds developed by the International Union of Pure and Applied Chemistry.

K

K_a (Section 2.3): The symbol that represents the acidity constant of an acid HA. The larger the K_a, the stronger the acid.

$$K_a = \frac{[H_3O^+][A{:}^-]}{[H{-}A]}$$

K_{eq} (Section 2.3): The equilibrium constant. K_{eq} = [products]/[starting materials].

Kekulé structures (Section 17.1): Two equilibrating structures for benzene. Each structure contains a six-membered ring and three π bonds alternating with σ bonds around the ring.

Ketal (Section 21.14): A compound having the general structure $R_2C(OR')_2$, where R = alkyl or aryl. Ketals are derived from ketones and constitute a subclass of acetals.

β-Keto ester (Section 23.10): A compound containing a ketone carbonyl on the carbon β to the ester carbonyl group.

Ketone (Section 11.9): A compound with two alkyl groups bonded to the $C{=}O$ carbon atom, having the general structures $R_2C{=}O$ or RCOR'.

Ketose (Section 27.2): A monosaccharide comprised of a polyhydroxy ketone.

Keto tautomer (Section 11.9): A tautomer of a ketone that has a $C{=}O$ and a hydrogen bonded to the α carbon. The keto tautomer is in equilibrium with the enol tautomer.

Kiliani–Fischer synthesis (Section 27.10B): A reaction that lengthens the carbon chain of an aldose by adding one carbon to the carbonyl end.

Kinetic enolate (Section 23.4): The enolate that is formed the fastest—generally the less substituted enolate.

Kinetic product (Section 16.11): In a reaction that can give more than one product, the product that is formed the fastest.

Kinetic resolution (Section 26.3B): The separation of two enantiomers by a chemical reaction that selectively occurs for only one of the enantiomers.

Kinetics (Section 6.5): The study of chemical reaction rates.

L

L-Sugar (Section 27.2C): A sugar with the hydroxy group on the stereogenic center farthest from the carbonyl on the left side in the Fischer projection formula.

Lactam (Section 22.1): A cyclic amide in which the carbonyl carbon–nitrogen σ bond is part of a ring. A β-lactam contains the carbon–nitrogen σ bond in a four-membered ring.

Lactol (Section 21.16): A cyclic hemiacetal.

Lactone (Section 22.1): A cyclic ester in which the carbonyl carbon–oxygen σ bond is part of a ring.

Leaving group (Section 7.6): An atom or group of atoms (Z) that is able to accept the electron density of the C−Z bond during a substitution or elimination reaction.

Leaving group ability (Section 7.7): A measure of how readily a leaving group (Z) can accept the electron density of the C−Z bond during a substitution or elimination reaction.

Le Châtelier's principle (Section 9.8D): The principle that a system at equilibrium will react to counteract any disturbance to the equilibrium.

Lecithin (Section 28.4A): A phosphoacylglycerol in which the phosphodiester alkyl group is $-CH_2CH_2N(CH_3)_3^+$. Lecithins are also called phosphatidylcholines.

Leukotriene (Section 9.16): An unstable and potent biomolecule synthesized in cells by the oxidation of arachidonic acid. Leukotrienes are responsible for biological conditions such as asthma.

Levorotatory (Section 5.12A): Rotating plane-polarized light in the counterclockwise direction. The rotation is labeled *l* or (–).

Lewis acid (Section 2.8): An electron pair acceptor.

Lewis acid–base reaction (Section 2.8): A reaction that results when a Lewis base donates an electron pair to a Lewis acid.

Lewis base (Section 2.8): An electron pair donor.

Lewis structure (Section 1.3): A representation of a molecule that shows the position of covalent bonds and nonbonding electrons. In Lewis structures, unshared electrons are represented by dots and a two-electron covalent bond is represented by a solid line. Lewis structures are also called electron dot structures.

Ligand (Section 29.2A): A group coordinated to a metal, which donates electron density to or sometimes withdraws electron density from the metal.

"Like dissolves like" (Section 3.4C): The principle that compounds dissolve in solvents having similar kinds of intermolecular forces; that is, polar compounds dissolve in polar solvents and nonpolar compounds dissolve in nonpolar solvents.

Lindlar catalyst (Section 12.5B): A catalyst for the hydrogenation of an alkyne to a cis alkene. The Lindlar catalyst is Pd adsorbed onto $CaCO_3$ with lead(II) acetate and quinoline.

Lipid (Sections 3.9D, 4.15, 28.1): A biomolecule with a large number of C−C and C−H σ bonds that is soluble in organic solvents and insoluble in water.

Lone pair of electrons (Section 1.2): A pair of valence electrons that is not shared with another atom in a covalent bond. Lone pairs are also called unshared or nonbonded pairs of electrons.

Lowest unoccupied molecular orbital (Section 17.9B): The molecular orbital with the lowest energy that does not contain electrons. The lowest unoccupied molecular orbital is abbreviated as the LUMO.

M

M peak (Section 13.1): The peak in the mass spectrum that corresponds to the mass of the molecular ion. The M peak is also called the molecular ion peak or the parent peak.

M + 1 peak (Section 13.1): The peak in the mass spectrum that corresponds to the mass of the molecular ion plus one. The M + 1 peak is caused by the presence of isotopes that increase the mass of the molecular ion.

M + 2 peak (Section 13.2): The peak in the mass spectrum that corresponds to the mass of the molecular ion plus two. The M + 2 peak is caused by the presence of isotopes, typically of a chlorine or a bromine atom.

Magnetic resonance imaging (MRI) (Section 14.12): A form of NMR spectroscopy used in medicine.

Malonic ester synthesis (Section 23.9A): A stepwise method that converts diethyl malonate into a carboxylic acid having one or two carbons bonded to the α carbon.

Markovnikov's rule (Section 10.10): The rule that states in the addition of HX to an unsymmetrical alkene, the H atom bonds to the less substituted carbon atom.

Mass number (Section 1.1): The total number of protons and neutrons in the nucleus of a particular atom.

Mass spectrometry (Section 13.1): An analytical technique used for measuring the molecular weight and determining the molecular formula of an organic molecule.

Mass-to-charge ratio (Section 13.1): A ratio of the mass to the charge of a molecular ion or fragment. Mass-to-charge ratio is abbreviated as *m/z*.

Megahertz (Section 14.1A): A unit used for the frequency of the RF radiation in NMR spectroscopy. Megahertz is abbreviated as MHz; 1 MHz = 10^6 Hz.

Melting point (Section 3.4B): The temperature at which molecules in the solid phase are converted to the liquid phase. Molecules with stronger intermolecular forces and higher symmetry have higher melting points. Melting point is abbreviated as mp.

Merrifield method (Section 26.8): A method for synthesizing polypeptides using insoluble polymer supports.

Meso compound (Section 5.8): An achiral compound that contains two or more tetrahedral stereogenic centers.

Meta director (Section 18.7): A substituent on a benzene ring that directs a new group to the meta position during electrophilic aromatic substitution.

Meta isomer (Section 17.3B): A 1,3-disubstituted benzene ring. Meta substitution is abbreviated as *m*-.

Metal hydride reagent (Section 12.2): A reagent containing a polar metal–hydrogen bond that places a partial negative charge on the hydrogen and acts as a source of hydride ions ($H:^-$).

Metathesis (Section 29.6): A reaction between two alkene molecules that results in the interchange of the carbons of their double bonds.

Methylation (Section 7.16): A reaction in which a CH_3 group is transferred from one compound to another.

Methylene group (Sections 4.1B, 10.3C): A CH_2 group bonded to a carbon chain ($-CH_2-$) or part of a double bond ($CH_2=$).

1,2-Methyl shift (Section 9.9): Rearrangement of a less stable carbocation to a more stable carbocation by the shift of a methyl group from one carbon atom to an adjacent carbon atom.

Micelles (Section 3.6): Spherical droplets formed by soap molecules having the ionic heads on the surface and the nonpolar tails packed together in the interior. Grease and oil dissolve in the interior nonpolar region.

Michael acceptor (Section 24.9): The α,β-unsaturated carbonyl compound in a Michael reaction.

Michael reaction (Section 24.9): A reaction in which a resonance-stabilized carbanion (usually an enolate) adds to the β carbon of an α,β-unsaturated carbonyl compound.

Mixed aldol reaction (Section 24.2): An aldol reaction between two different carbonyl compounds. A mixed aldol reaction is also called a crossed aldol reaction.

Mixed anhydride (Section 22.1): An anhydride with two different alkyl groups bonded to the carbonyl carbon atoms.

Molecular ion (Section 13.1): The radical cation having the general structure $M^{+\bullet}$, formed by the removal of an electron from an organic molecule. The molecular ion is also called the parent ion.

Molecular orbital theory (Section 17.9A): A theory that describes bonds as the mathematical combination of atomic orbitals to form a new set of orbitals called molecular orbitals. Molecular orbital theory is also called MO theory.

Molecular recognition (Section 9.5B): The ability of a host molecule to recognize and bind specific guest molecules.

Molecule (Section 1.2): A compound containing two or more atoms bonded together with covalent bonds.

Monomers (Sections 5.1, 15.14): Small organic compounds that can be covalently bonded to each other (polymerized) in a repeating pattern.

Monosaccharide (Sections 3.9B, 27.2): A simple sugar having three to seven carbon atoms.

Monosubstituted alkene (Section 8.2A): An alkene that has one alkyl group and three hydrogens bonded to the carbons of the double bond ($RCH{=}CH_2$).

Monoterpene (Section 28.7A): A terpene that contains 10 carbons and two isoprene units. A monoterpenoid also contains at least one oxygen atom.

Multiplet (Section 14.6C): An NMR signal that is split into more than seven peaks.

Mutarotation (Section 27.6A): The process by which a pure anomer of a monosaccharide equilibrates to a mixture of both anomers when placed in solution.

N

***n* + 1 rule** (Section 14.6C): The rule that an NMR signal for a proton with n nearby nonequivalent protons will be split into $n + 1$ peaks.

Natural product (Section 7.18): A compound isolated from a natural source.

Newman projection (Section 4.9): An end-on representation of the conformation of a molecule. The Newman projection shows the three groups bonded to each carbon atom in a particular C−C bond, as well as the dihedral angle that separates the groups on each carbon.

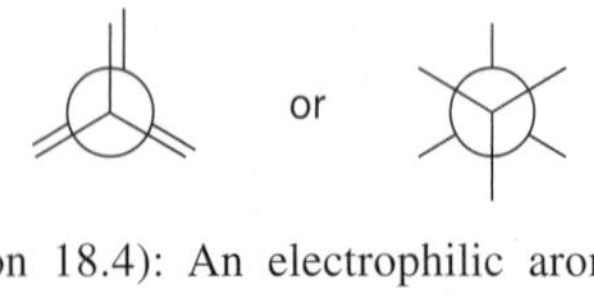

Nitration (Section 18.4): An electrophilic aromatic substitution reaction in which benzene reacts with $^+NO_2$ to give nitrobenzene, $C_6H_5NO_2$.

Nitrile (Sections 22.1, 22.18): A compound having the general structure $RC{\equiv}N$.

Nitronium ion (Section 18.4): An electrophile having the structure $^+NO_2$.

***N*-Nitrosamine** (Sections 7.16, 25.13B): A compound having the general structure $R_2N{-}N{=}O$. Nitrosamines are formed by the reaction of a secondary amine with ^+NO.

Nitrosonium ion (Section 25.13): An electrophile having the structure ^+NO.

NMR peak (Section 14.6A): The individual absorptions in a split NMR signal due to nonequivalent nearby protons.

NMR signal (Section 14.6A): The entire absorption due to a particular kind of proton in an NMR spectrum.

NMR spectrometer (Section 14.1A): An analytical instrument that measures the absorption of RF radiation by certain atomic nuclei when placed in a strong magnetic field.

Nonbonded pair of electrons (Section 1.2): A pair of valence electrons that is not shared with another atom in a covalent bond. Nonbonded electrons are also called unshared or lone pairs of electrons.

Nonbonding molecular orbital (Section 17.10): A molecular orbital having the same energy as the atomic orbitals that formed it.

Nonnucleophilic base (Section 7.8B): A base that is a poor nucleophile due to steric hindrance resulting from the presence of bulky groups.

Nonpolar bond (Section 1.12): A covalent bond in which the electrons are equally shared between the two atoms.

Nonpolar molecule (Section 1.13): A molecule that has no net dipole. A nonpolar molecule has either no polar bonds or multiple polar bonds whose dipoles cancel.

Nonreducing sugar (Section 27.9B): A carbohydrate that cannot be oxidized by Tollens, Benedict's, or Fehling's reagent.

Normal alkane (Section 4.1A): An acyclic alkane that has all of its carbons in a row. A normal alkane is an "*n*-alkane" or a straight-chain alkane.

Nuclear magnetic resonance spectroscopy (Section 14.1): A powerful analytical tool that can help identify the carbon and hydrogen framework of an organic molecule.

Nucleic acid (3.9C): A large molecule composed of nucleotides. DNA is a nucleic acid that stores the genetic information of an organism, and RNA is a nucleic acid responsible for directing protein synthesis.

Nucleophile (Sections 2.8, 7.6): An electron-rich compound, symbolized by $:Nu^-$, which donates a pair of electrons to an electron-deficient compound, forming a covalent bond. Lewis bases are nucleophiles.

Nucleophilic acyl substitution (Sections 20.2B, 22.1): Substitution of a leaving group by a nucleophile at a carbonyl carbon.

Nucleophilic addition (Section 20.2A): Addition of a nucleophile to the electrophilic carbon of a carbonyl group followed by protonation of the oxygen.

Nucleophilic aromatic substitution (Section 18.13): A substitution reaction of an aryl halide with a strong nucleophile.

Nucleophilicity (Section 7.8A): A measure of how readily an atom donates an electron pair to other atoms.

Nucleophilic substitution (Section 7.6): A reaction in which a nucleophile replaces the leaving group in a molecule.

Nucleoside (Section 27.13B): A biomolecule having a sugar and either a purine or pyrimidine base joined together by an *N*-glycosidic linkage.

Nucleotide (Sections 3.9C, 27.13B): A biomolecule having a sugar and either a purine or pyrimidine base joined together by an *N*-glycosidic linkage, and a phosphate bonded to a hydroxy group of the sugar nucleus.

O

Observed rotation (Section 5.12A): The angle that a sample of an optically active compound rotates plane-polarized light. The angle is denoted by the symbol α and is measured in degrees (°).

Octet rule (Section 1.2): The general rule governing the bonding process for second-row elements. Through bonding, second-row elements attain a complete outer shell of eight valence electrons.

Oil (Sections 10.6B, 28.3): A triacylglycerol that is liquid at room temperature and composed of fatty acid side chains with a high degree of unsaturation.

Olefin (Section 10.1): An alkene; a compound possessing a carbon–carbon double bond.

Optically active (Section 5.12A): Able to rotate the plane of plane-polarized light as it passes through a solution of a compound.

Optically inactive (Section 5.12A): Not able to rotate the plane of plane-polarized light as it passes through a solution of a compound.

Optical purity (Section 5.12D): A measurement of how much one enantiomer is present in excess of the racemic mixture. Optical purity is also called enantiomeric excess (*ee*); *ee* = % of one enantiomer – % of the other enantiomer.

Orbital (Section 1.1): A region of space around the nucleus of an atom that is high in electron density. There are four different kinds of orbitals, called *s, p, d,* and *f.*

Order of a rate equation (Section 6.9B): The sum of the exponents of the concentration terms in the rate equation of a reaction.

Organoborane (Section 10.16): A compound that contains a carbon–boron bond. Organoboranes have the general structure RBH_2, R_2BH, or R_3B.

Organocopper reagent (Section 20.9): An organometallic reagent having the general structure R_2CuLi. Organocopper reagents are also called organocuprates.

Organolithium reagent (Section 20.9): An organometallic reagent having the general structure RLi.

Organomagnesium reagent (Section 20.9): An organometallic reagent having the general structure RMgX. Organomagnesium reagents are also called Grignard reagents.

Organometallic reagent (Section 20.9): A reagent that contains a carbon atom bonded to a metal.

Organopalladium compound (Section 29.2): An organometallic compound that contains a carbon–palladium bond.

Organophosphorus reagent (Section 21.10A): A reagent that contains a carbon–phosphorus bond.

Ortho isomer (Section 17.3B): A 1,2-disubstituted benzene ring. Ortho substitution is abbreviated as *o*-.

Ortho, para director (Section 18.7): A substituent on a benzene ring that directs a new group to the ortho and para positions during electrophilic aromatic substitution.

Oxaphosphetane (Section 21.10B): An intermediate in the Wittig reaction consisting of a four-membered ring containing a phosphorus–oxygen bond.

Oxazaborolidine (Section 20.6A): A heterocycle possessing a boron, a nitrogen, and an oxygen. An oxazaborolidine can be used to form a chiral reducing agent.

Oxidation (Sections 4.14A, 12.1): A process that results in a loss of electrons. For organic compounds, oxidation results in an increase in the number of C–Z bonds or a decrease in the number of C–H bonds; Z = an element more electronegative than carbon.

Oxidative addition (Section 29.2A): The addition of a reagent to a metal, often increasing the number of groups around the metal by two.

Oxidative cleavage (Section 12.10): An oxidation reaction that breaks both the σ and π bonds of a multiple bond to form two oxidized products.

Oxime (Section 27.10A): A compound having the general structure $R_2C{=}NOH$.

Oxirane (Section 9.1): A cyclic ether having the oxygen atom as part of a three-membered ring. Oxiranes are also called epoxides.

Oxy-Cope rearrangement (Section 30.5): A [3,3] sigmatropic rearrangement of a 1,5-dien-3-ol to a δ,ε-unsaturated carbonyl compound.

Ozonolysis (Section 12.10): An oxidative cleavage reaction in which a multiple bond reacts with ozone (O_3) as the oxidant.

P

Para isomer (Section 17.3B): A 1,4-disubstituted benzene ring. Para substitution is abbreviated as *p*-.

Parent ion (Section 13.1): The radical cation having the general structure $M^{+\bullet}$, formed by the removal of an electron from an organic molecule. The parent ion is also called the molecular ion.

Parent name (Section 4.4): The portion of the IUPAC name of an organic compound that indicates the number of carbons in the longest continuous chain in the molecule.

Pentose (Section 27.2): A monosaccharide containing five carbons.

Peptide bond (Section 26.5): The amide bond in peptides and proteins.

Peptides (Sections 22.6B, 26.5): Low molecular weight polymers of less than 40 amino acids joined together by amide linkages.

Percent *s*-character (Section 1.11B): The fraction of a hybrid orbital due to the *s* orbital used to form it. As the percent *s*-character increases, a bond becomes shorter and stronger.

Percent transmittance (Section 13.6B): A measure of how much electromagnetic radiation passes through a sample of a compound and how much is absorbed.

Pericyclic reaction (Section 30.1): A concerted reaction that proceeds through a cyclic transition state.

Peroxide (Section 15.2): A reactive organic compound with the general structure ROOR. Peroxides are used as radical initiators by homolysis of the weak O–O bond.

Peroxyacid (Section 12.7): An oxidizing agent having the general structure RCO_3H.

Peroxy radical (Section 15.11): A radical having the general structure ROO·.

Petroleum (Section 4.7): A fossil fuel containing a complex mixture of compounds, primarily hydrocarbons with 1 to 40 carbon atoms.

Phenol (Sections 9.1, 15.12): A compound such as C_6H_5OH, which contains a hydroxy group bonded to a benzene ring.

Phenyl group (Section 17.3D): A group formed by removal of one hydrogen from benzene, abbreviated as C_6H_5- or Ph–.

Pheromone (Section 4.1): A chemical substance used for communication in an animal or insect species.

Phosphate (Section 7.16): A PO_4^{3-} anion.

Phosphate ester (Section 3.2D): A functional group formed by replacing one H atom of phosphoric acid (H_3PO_4) by an alkyl group. In cells, phosphate esters have the formula $ROPO_3^{2-}$.

Phosphatidylcholine (Section 28.4A): A phosphoacylglycerol in which the phosphodiester alkyl group is $-CH_2CH_2N(CH_3)_3^+$. Phosphatidylcholines are also called lecithins.

Phosphatidylethanolamine (Section 28.4A): A phosphoacylglycerol in which the phosphodiester alkyl group is $-CH_2CH_2NH_3^+$. Phosphatidylethanolamines are also called cephalins.

Phosphoacylglycerols (Section 28.4A): A lipid having a glycerol backbone with two of the hydroxy groups esterified with fatty acids and the third hydroxy group as part of a phosphodiester.

Phosphodiester (Sections 3.2D, 28.4): A functional group having the general formula $ROPO_2OR'$ formed by replacing two of the H atoms in phosphoric acid (H_3PO_4) with alkyl groups. In cells, phosphodiesters have a net negative charge [$(RO)_2 PO_2^-$].

Phospholipid (Sections 3.7A, 28.4): A hydrolyzable lipid that contains a phosphorus atom.

Phosphonium salt (Section 21.10A): An organophosphorus reagent with a positively charged phosphorus and a suitable counterion; for example, $R_4P^+X^-$. Phosphonium salts are converted to ylides upon treatment with a strong base.

Phosphorane (Section 21.10A): A phosphorus ylide; for example, $Ph_3P{=}CR_2$.

Phosphoric acid anhydride (Section 22.16): A functional group with two phosphate groups joined together. A phosphoric acid anhydride is also called a phosphoanhydride.

Photon (Section 13.5): A particle of electromagnetic radiation.

Pi (π) bond (Section 1.10B): A bond formed by side-by-side overlap of two *p* orbitals where electron density is not concentrated on the axis joining the two nuclei. Pi (π) bonds are generally weaker than σ bonds.

Pi- (π-) stacking (Sections 17.8B, 17.11): The interaction of the π electrons of aromatic rings held in close proximity.

pK_a (Section 2.3): A logarithmic scale of acid strength. $pK_a = -\log K_a$. The smaller the pK_a, the stronger the acid.

Plane of symmetry (Section 5.3): A mirror plane that cuts a molecule in half, so that one half of the molecule is the mirror reflection of the other half.

Plane-polarized light (Section 5.12A): Light that has an electric vector that oscillates in a single plane. Plane-polarized light, also called polarized light, arises from passing ordinary light through a polarizer.

Plasticizer (Section 31.7): A low molecular weight compound added to a polymer to give it flexibility.

β-Pleated sheet (Section 26.9B): A secondary structure of a protein formed when two or more peptide chains line up side by side.

Poisoned catalyst (Section 12.5B): A hydrogenation catalyst with reduced activity that allows selective reactions to occur. The Lindlar catalyst is a poisoned Pd catalyst that converts alkynes to cis alkenes.

Polar aprotic solvent (Section 7.8C): A polar solvent that is incapable of intermolecular hydrogen bonding because it does not contain an O–H or N–H bond.

Polar bond (Section 1.12): A covalent bond in which the electrons are unequally shared between the two atoms. Unequal sharing of electrons results from bonding between atoms of different electronegativity values, usually with a difference of ≥ 0.5 unit.

Polarimeter (Section 5.12A): An instrument that measures the degree that a compound rotates plane-polarized light.

Polarity (Section 1.12): A characteristic that results from a dipole. The polarity of a bond is indicated by an arrow with the head of the arrow pointing toward the negative end of the dipole and the tail with a perpendicular line through it at the positive end of the dipole. The polarity of a bond can also be indicated by the symbols δ+ and δ–.

Polarizability (Section 3.3B): A measure of how the electron cloud around an atom responds to changes in its electronic environment.

Polar molecule (Section 1.13): A molecule that has a net dipole. A polar molecule has either one polar bond or multiple polar bonds whose dipoles reinforce.

Polar protic solvent (Section 7.8C): A polar solvent that is capable of intermolecular hydrogen bonding because it contains an O–H or N–H bond.

Polyamide (Sections 22.16A, 31.6A): A step-growth polymer that contains many amide bonds. Nylon 6,6 and nylon 6 are polyamides.

Polycarbonate (Section 31.6C): A step-growth polymer that contains many –OC(=O)O– bonds in its backbone, often formed by reaction of $Cl_2C{=}O$ with a diol.

Polycyclic aromatic hydrocarbon (Sections 9.17, 17.7B): An aromatic hydrocarbon containing two or more benzene rings that share carbon–carbon bonds. Polycyclic aromatic hydrocarbons are abbreviated as PAHs.

Polyene (Section 16.7): A compound that contains three or more double bonds.

Polyester (Section 31.6B): A step-growth polymer consisting of many ester bonds between diols and dicarboxylic acids.

Polyether (Sections 9.5B, 31.3): A compound that contains two or more ether linkages.

Polymer (Sections 5.1, 15.14): A large molecule composed of smaller monomer units covalently bonded to each other in a repeating pattern.

Polymerization (Section 15.14A): The chemical process that joins together monomers to make polymers.

Polysaccharide (Sections 3.9B, 27.12): A carbohydrate containing three or more monosaccharide units joined together by glycosidic linkages.

Polyurethane (Section 31.6C): A step-growth polymer that contains many –NHC(=O)O– bonds in its backbone, formed by reaction of a diisocyanate and a diol.

Porphyrin (Section 26.10C): A nitrogen-containing heterocycle that can complex metal ions.

Primary (1°) alcohol (Section 3.2): An alcohol having the general structure RCH_2OH.

Primary (1°) alkyl halide (Section 3.2): An alkyl halide having the general structure RCH_2X.

Primary (1°) amide (Section 3.2): An amide having the general structure $RCONH_2$.

Primary (1°) amine (Section 3.2): An amine having the general structure RNH_2.

Primary (1°) carbocation (Section 7.13): A carbocation having the general structure RCH_2^+.

Primary (1°) carbon (Section 3.2): A carbon atom that is bonded to one other carbon atom.

Primary (1°) hydrogen (Section 3.2): A hydrogen that is bonded to a 1° carbon.

Primary protein structure (Section 26.9A): The particular sequence of amino acids joined together by peptide bonds.

Primary (1°) radical (Section 15.1): A radical having the general structure $RCH_2\cdot$.

Propagation (Section 15.4A): The middle part of a chain mechanism in which one reactive particle is consumed and another is generated. Propagation repeats until a termination step occurs.

Prostaglandin (Section 4.15): A class of lipids containing 20 carbons, a five-membered ring, and a CO_2H group. Prostaglandins possess a wide range of biological activities.

Prosthetic group (Section 26.10C): The non-protein unit of a conjugated protein.

Protecting group (Section 20.12): A blocking group that renders a reactive functional group unreactive, so that it does not interfere with another reaction.

Protection (Section 20.12): The reaction that blocks a reactive functional group with a protecting group.

Proteins (Sections 3.9A, 22.6B, 26.5): High molecular weight polymers of 40 or more amino acids joined together by amide linkages.

Proton (Section 2.1): A positively charged hydrogen ion (H^+).

Proton NMR spectroscopy (Section 14.1): A form of nuclear magnetic resonance spectroscopy used to determine the number and type of hydrogen atoms in a molecule.

Proton transfer reaction (Section 2.2): A Brønsted–Lowry acid–base reaction; a reaction that results in the transfer of a proton from an acid to a base.

Purine (Sections 17.8B, 27.13B): A bicyclic aromatic heterocycle having two nitrogens in each of the rings.

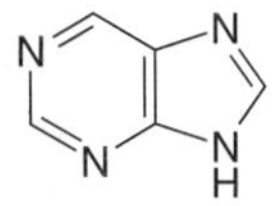

Pyranose (Section 27.6): A cyclic six-membered ring of a monosaccharide containing an oxygen atom.

Pyrimidine (Sections 17.8B, 27.13B): A six-membered aromatic heterocycle having two nitrogens in the ring.

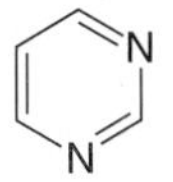

Q

Quantum (Section 13.5): The discrete amount of energy associated with a particle of electromagnetic radiation (i.e., a photon).

Quartet (Section 14.6C): An NMR signal that is split into four peaks having a relative area of 1:3:3:1, caused by three nearby nonequivalent protons.

Quaternary (4°) carbon (Section 3.2): A carbon atom that is bonded to four other carbon atoms.

Quaternary protein structure (Section 26.9C): The shape adopted when two or more folded polypeptide chains aggregate into one protein complex.

Quintet (Section 14.6C): An NMR signal that is split into five peaks caused by four nearby nonequivalent protons.

R

Racemic mixture (Section 5.12B): An equal mixture of two enantiomers. A racemic mixture, also called a racemate, is optically inactive.

Racemization (Section 7.12C): The formation of equal amounts of two enantiomers from an enantiomerically pure starting material.

Radical (Sections 6.3B, 15.1): A reactive intermediate with a single unpaired electron, formed by homolysis of a covalent bond.

Radical anion (Section 12.5C): A reactive intermediate containing both a negative charge and an unpaired electron.

Radical cation (Section 13.1): A species with an unpaired electron and a positive charge, formed in a mass spectrometer by the bombardment of a molecule with an electron beam.

Radical inhibitor (Section 15.2): A compound that prevents radical reactions from occurring. Radical inhibitors are also called radical scavengers.

Radical initiator (Section 15.2): A compound that contains an especially weak bond that serves as a source of radicals.

Radical polymerization (Section 15.14B): A radical chain reaction involving the polymerization of alkene monomers by adding a radical to a π bond.

Radical scavenger (Section 15.2): A compound that prevents radical reactions from occurring. Radical scavengers are also called radical inhibitors.

Rate constant (Section 6.9B): A constant that is a fundamental characteristic of a reaction. The rate constant, symbolized by k, is a complex mathematical term that takes into account the dependence of a reaction rate on temperature and the energy of activation.

Rate-determining step (Section 6.8): In a multistep reaction mechanism, the step with the highest energy transition state.

Rate equation (Section 6.9B): An equation that shows the relationship between the rate of a reaction and the concentration of the reactants. The rate equation depends on the mechanism of the reaction and is also called the rate law.

Reaction coordinate (Section 6.7): The x axis in an energy diagram that represents the progress of a reaction as it proceeds from reactant to product.

Reaction mechanism (Section 6.3): A detailed description of how bonds are broken and formed as a starting material is converted to a product.

Reactive intermediate (Sections 6.3, 10.18): A high-energy unstable intermediate formed during the conversion of a stable starting material to a stable product.

Reciprocal centimeter (Section 13.6A): The unit for wavenumber, which is used to report frequency in IR spectroscopy.

Reducing sugar (Section 27.9B): A carbohydrate that can be oxidized by Tollens, Benedict's, or Fehling's reagent.

Reduction (Sections 4.14A, 12.1): A process that results in the gain of electrons. For organic compounds, reduction results in a decrease in the number of C–Z bonds or an increase in the number of C–H bonds; Z = an element more electronegative than carbon.

Reductive amination (Section 25.7C): A two-step method that converts aldehydes and ketones into amines.

Reductive elimination (Section 29.2A): The elimination of two groups that surround a metal, often forming new carbon–hydrogen or carbon–carbon bonds.

Regioselective reaction (Section 8.5): A reaction that yields predominantly or exclusively one constitutional isomer when more than one constitutional isomer is possible.

Resolution (Section 26.3): The separation of a racemic mixture into its component enantiomers.

Resonance (Section 14.1A): In NMR spectroscopy, when an atomic nucleus absorbs RF radiation and spin flips to a higher energy state.

Resonance hybrid (Sections 1.6C, 16.4): A structure that is a weighted composite of all possible resonance structures. The resonance hybrid shows the delocalization of electron density due to the different locations of electrons in individual resonance structures.

Resonance structures (Sections 1.6, 16.2): Two or more structures of a molecule that differ in the placement of π bonds and nonbonded electrons. The placement of atoms and σ bonds stays the same.

Retention of configuration (Section 7.11C): The same relative stereochemistry of a stereogenic center in the reactant and the product of a chemical reaction.

Retention time (Section 13.4B): The length of time required for a component of a mixture to travel through a chromatography column.

Retro Diels–Alder reaction (Section 16.14B): The reverse of a Diels–Alder reaction in which a cyclohexene is cleaved to give a 1,3-diene and an alkene.

Retrosynthetic analysis (Section 10.18): Working backwards from a product to determine the starting material from which it is made.

RF radiation (Section 14.1A): Radiation in the radiofrequency region of the electromagnetic spectrum, characterized by long wavelength and low frequency and energy.

Ribonucleic acid (Section 3.9C): The large molecule that translates genetic information into proteins needed for cellular functions.

Ribonucleoside (Section 27.13B): An *N*-glycoside formed by the reaction of D-ribose with certain amine heterocycles.

Ribonucleotide (Section 27.13B): An RNA building block having a ribose and either a purine or pyrimidine base joined together by an *N*-glycosidic linkage, and a phosphate bonded to a hydroxy group of the sugar nucleus.

Ring-closing metathesis (Section 29.6): An intramolecular olefin metathesis reaction using a diene starting material, which results in ring closure.

Ring current (Section 14.4): A circulation of π electrons in an aromatic ring caused by the presence of an external magnetic field.

Ring-flipping (Section 4.12B): A stepwise process in which one chair conformation of cyclohexane interconverts with a second chair conformation.

Ring-opening metathesis polymerization (Problem 29.33): An olefin metathesis reaction that forms a high molecular weight polymer from certain cyclic alkenes.

Robinson annulation (Section 24.10): A ring-forming reaction that combines a Michael reaction with an intramolecular aldol reaction to form a cyclohex-2-enone.

R,S **System of nomenclature** (Section 5.6): A system of nomenclature that distinguishes the stereochemistry at a tetrahedral stereogenic center by assigning a priority to each group connected to the stereogenic center. *R* indicates a clockwise orientation of the three highest priority groups, and *S* indicates a counterclockwise orientation of the three highest groups. The system is also called the Cahn–Ingold–Prelog system.

Rule of endo addition (Section 16.13D): The rule that the endo product is preferred in a Diels–Alder reaction.

S

Sandmeyer reaction (Section 25.14A): A reaction between an aryl diazonium salt and a copper(I) halide to form an aryl halide (C_6H_5Cl or C_6H_5Br).

Saponification (Section 22.11B): Basic hydrolysis of an ester to form an alcohol and a carboxylate anion.

Saturated fatty acid (Section 10.6A): A fatty acid having no carbon–carbon double bonds in its long hydrocarbon chain.

Saturated hydrocarbon (Section 4.1): A compound that contains only C–C and C–H σ bonds and no rings, thus having the maximum number of hydrogen atoms per carbon.

Schiff base (Section 21.11A): A compound having the general structure $R_2C{=}NR'$. A Schiff base is also called an imine.

Secondary (2°) alcohol (Section 3.2): An alcohol having the general structure R_2CHOH.

Secondary (2°) alkyl halide (Section 3.2): An alkyl halide having the general structure R_2CHX.

Secondary (2°) amide (Section 3.2): An amide having the general structure RCONHR'.

Secondary (2°) amine (Section 3.2): An amine having the general structure R_2NH.

Secondary (2°) carbocation (Section 7.13): A carbocation having the general structure R_2CH^+.

Secondary (2°) carbon (Section 3.2): A carbon atom that is bonded to two other carbon atoms.

Secondary (2°) hydrogen (Section 3.2): A hydrogen that is attached to a 2° carbon.

Secondary protein structure (Section 26.9B): The three-dimensional conformations of localized regions of a protein.

Secondary (2°) radical (Section 15.1): A radical having the general structure $R_2CH\cdot$.

Second-order rate equation (Sections 6.9B, 7.10): A rate equation in which the reaction rate depends on the concentration of two reactants.

Separatory funnel (Section 19.13): An item of laboratory glassware used for extractions.

Septet (Section 14.6C): An NMR signal that is split into seven peaks caused by six nearby nonequivalent protons.

Sesquiterpene (Section 28.7A): A terpene that contains 15 carbons and three isoprene units. A sesquiterpenoid also contains at least one oxygen atom.

Sesterterpene (Section 28.7A): A terpene that contains 25 carbons and five isoprene units. A sesterterpenoid also contains at least one oxygen atom.

Sextet (Section 14.6C): An NMR signal that is split into six peaks caused by five nearby nonequivalent protons.

Sharpless asymmetric epoxidation (Section 12.15): An enantioselective oxidation reaction that converts the double bond of an allylic alcohol to a predictable enantiomerically enriched epoxide.

Sharpless reagent (Section 12.15): The reagent used in the Sharpless asymmetric epoxidation. The Sharpless reagent consists of *tert*-butyl hydroperoxide, a titanium catalyst, and one enantiomer of diethyl tartrate.

Shielding effects (Section 14.3A): An effect in NMR caused by small induced magnetic fields of electrons in the opposite direction to the applied magnetic field. Shielding decreases the strength of the magnetic field felt by the nucleus and shifts an absorption upfield.

1,2-Shift (Section 9.9): Rearrangement of a less stable carbocation to a more stable carbocation by the shift of a hydrogen atom or an alkyl group from one carbon atom to an adjacent carbon atom.

Sigma (σ) bond (Section 1.9A): A cylindrically symmetrical bond that concentrates the electron density on the axis that joins two nuclei. All single bonds are σ bonds.

Sigmatropic rearrangement (Section 30.1): A pericyclic reaction in which a σ bond is broken in the reactant, the π bonds rearrange, and a σ bond is formed in the product.

Silyl ether (Section 20.12): A common protecting group for an alcohol in which the O–H bond is replaced by an O–Si bond.

Simmons–Smith reaction (Section 29.5): Reaction of an alkene with CH_2I_2 and Zn(Cu) to form a cyclopropane.

Singlet (Section 14.6A): An NMR signal that occurs as a single peak.

Skeletal structure (Section 1.8B): A shorthand representation of the structure of an organic compound in which carbon atoms and the hydrogen atoms bonded to them are omitted. All heteroatoms and the hydrogens bonded to them are drawn in. Carbon atoms are assumed to be at the junction of any two lines or at the end of a line.

S_N1 mechanism (Sections 7.10, 7.12): A nucleophilic substitution mechanism that goes by a two-step process involving a carbocation intermediate. S_N1 is an abbreviation for "Substitution Nucleophilic Unimolecular."

S_N2 mechanism (Sections 7.10, 7.11): A nucleophilic substitution mechanism that goes by a one-step concerted process, where both reactants are involved in the transition state. S_N2 is an abbreviation for "Substitution Nucleophilic Bimolecular."

Soap (Sections 3.6, 22.12B): The carboxylate salts of long-chain fatty acids prepared by the basic hydrolysis or saponification of a triacylglycerol.

Solubility (Section 3.4C): A measure of the extent to which a compound dissolves in a liquid.

Solute (Section 3.4C): The compound that is dissolved in a liquid solvent.

Solvent (Section 3.4C): The liquid component into which the solute is dissolved.

Specific rotation (Section 5.12C): A standardized physical constant for the amount that a chiral compound rotates plane-polarized light. Specific rotation is denoted by the symbol [α] and defined using a specific sample tube length (*l* in dm), concentration (*c* in g/mL), temperature (25 °C), and wavelength (589 nm). $[\alpha] = \alpha/(l \times c)$

Spectator ion (Section 2.1): An ion that does not take part in a reaction and is opposite in charge to the ion that does take part in a reaction. A spectator ion is also called a counterion.

Spectroscopy (Section 13.1): An analytical method using the interaction of electromagnetic radiation with molecules to determine molecular structure.

Sphingomyelin (Section 28.4B): A hydrolyzable phospholipid derived from sphingosine.

Spin flip (Section 14.1A): In NMR spectroscopy, when an atomic nucleus absorbs RF radiation and its magnetic field flips relative to the external magnetic field.

Spin–spin splitting (Section 14.6): Splitting of an NMR signal into peaks caused by nonequivalent protons on the same carbon or adjacent carbons.

Spiro ring system (Problem 23.61, Appendix B): A compound having two rings that share a single carbon atom.

Staggered conformation (Section 4.9): A conformation of a molecule in which the bonds on one carbon bisect the R−C−R bond angle on the adjacent carbon.

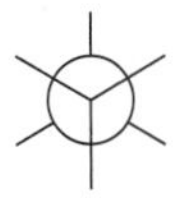

Step-growth polymer (Section 31.1): A polymer formed when monomers containing two functional groups come together with loss of a small molecule such as water or HCl. Step-growth polymers are also called condensation polymers.

Stereochemistry (Sections 4.9, 5.1): The three-dimensional structure of molecules.

Stereogenic center (Section 5.3): A site in a molecule at which the interchange of two groups forms a stereoisomer. A carbon bonded to four different groups is a tetrahedral stereogenic center. A tetrahedral stereogenic center is also called a chirality center, a chiral center, or an asymmetric center.

Stereoisomers (Sections 4.13B, 5.1): Two isomers that differ only in the way the atoms are oriented in space.

Stereoselective reaction (Section 8.5): A reaction that yields predominantly or exclusively one stereoisomer when two or more stereoisomers are possible.

Stereospecific reaction (Section 10.14): A reaction in which each of two stereoisomers of a starting material yields a particular stereoisomer of a product.

Steric hindrance (Section 7.8B): A decrease in reactivity resulting from the presence of bulky groups at the site of a reaction.

Steric strain (Section 4.10): An increase in energy resulting when atoms in a molecule are forced too close to one another.

Steroid (Sections 16.14C, 28.8): A tetracyclic lipid composed of three six-membered rings and one five-membered ring.

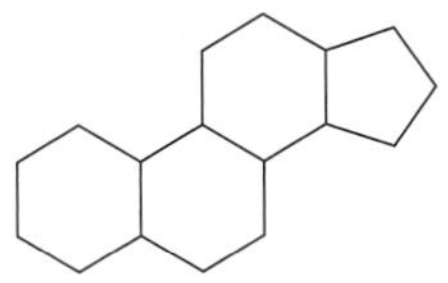

Straight-chain alkane (Section 4.1A): An acyclic alkane that has all of its carbons in a row. Straight-chain alkanes are also called normal alkanes.

Strecker amino acid synthesis (Section 26.2C): A reaction that converts an aldehyde into an α-amino acid by way of an α-amino nitrile.

Structural isomers (Sections 4.1A, 5.2): Two compounds that have the same molecular formula but differ in the way the atoms are connected to each other. Structural isomers are also called constitutional isomers.

Substituent (Section 4.4): A group or branch attached to the longest continuous chain of carbons in an organic molecule.

Substitution reaction (Section 6.2A): A reaction in which an atom or a group of atoms is replaced by another atom or group of atoms. Substitution reactions involve σ bonds: one σ bond breaks and another is formed at the same atom.

Substrate (Section 6.11): An organic molecule that is transformed by the action of an enzyme.

Sulfide (Section 9.15): A compound having the general structure RSR'.

Sulfonate anion (Section 19.14): An anion having the general structure RSO_3^-, formed by deprotonating a sulfonic acid with a Brønsted–Lowry base.

Sulfonation (Section 18.4): An electrophilic aromatic substitution reaction in which benzene reacts with $^+SO_3H$ to give a benzenesulfonic acid, $C_6H_5SO_3H$.

Sulfonic acid (Section 19.14): A compound having the general structure RSO_3H.

Suprafacial reaction (Section 30.4): A pericyclic reaction that occurs on the same side of the two ends of the π electron system.

Suzuki reaction (Section 29.2): The palladium-catalyzed coupling of an organic halide (R'X) with an organoborane (RBY_2) to form a product R−R'.

Symmetrical anhydride (Section 22.1): An anhydride that has two identical alkyl groups bonded to the carbonyl carbon atoms.

Symmetrical ether (Section 9.1): An ether with two identical alkyl groups bonded to the oxygen.

Syn addition (Section 10.8): An addition reaction in which two parts of a reagent are added from the same side of a double bond.

Syn dihydroxylation (Section 12.9B): The addition of two hydroxy groups to the same face of a double bond.

Syndiotactic polymer (Section 31.4): A polymer having the substituents alternating from one side of the backbone of an elongated polymer chain to the other.

Syn periplanar (Section 8.8): In an elimination reaction, a geometry in which the β hydrogen and the leaving group are on the same side of the molecule.

Systematic name (Section 4.3): The name of a molecule indicating the compound's chemical structure. The systematic name is also called the IUPAC name.

T

Target compound (Section 11.12): The final product of a synthetic scheme.

Tautomerization (Sections 11.9, 23.2A): The process of converting one tautomer into another.

Tautomers (Section 11.9): Constitutional isomers that are in equilibrium and differ in the location of a double bond and a hydrogen atom.

Terminal alkene (Section 10.1): An alkene that has the double bond at the end of the carbon chain.

Terminal alkyne (Section 11.1): An alkyne that has the triple bond at the end of the carbon chain.

C-Terminal amino acid (Section 26.5A): The amino acid at the end of a peptide chain with a free carboxy group.

N-Terminal amino acid (Section 26.5A): The amino acid at the end of a peptide chain with a free amino group.

Termination (Section 15.4A): The final step of a chain reaction. In a radical chain mechanism, two radicals combine to form a stable bond.

Terpene (Section 28.7): A hydrocarbon composed of repeating five-carbon isoprene units.

Terpenoid (Section 28.7): A lipid that contains isoprene units as well as at least one oxygen heteroatom.

Tertiary (3°) alcohol (Section 3.2): An alcohol having the general structure R_3COH.

Tertiary (3°) alkyl halide (Section 3.2): An alkyl halide having the general structure R_3CX.

Tertiary (3°) amide (Section 3.2): An amide having the general structure $RCONR'_2$.

Tertiary (3°) amine (Section 3.2): An amine having the general structure R_3N.

Tertiary (3°) carbocation (Section 7.13): A carbocation having the general structure R_3C^+.

Tertiary (3°) carbon (Section 3.2): A carbon atom that is bonded to three other carbon atoms.

Tertiary (3°) hydrogen (Section 3.2): A hydrogen that is attached to a 3° carbon.

Tertiary protein structure (Section 26.9C): The three-dimensional shape adopted by an entire peptide chain.

Tertiary (3°) radical (Section 15.1): A radical having the general structure $R_3C\cdot$.

Tesla (Section 14.1A): A unit used to measure the strength of a magnetic field. Tesla is denoted with the symbol "T."

Tetramethylsilane (Section 14.1B): An internal standard used as a reference in NMR spectroscopy. The tetramethylsilane (TMS) reference peak occurs at 0 ppm on the δ scale.

Tetrasubstituted alkene (Section 8.2A): An alkene that has four alkyl groups and no hydrogens bonded to the carbons of the double bond ($R_2C{=}CR_2$).

Tetraterpene (Section 28.7A): A terpene that contains 40 carbons and eight isoprene units. A tetraterpenoid contains at least one oxygen atom as well.

Tetrose (Section 27.2): A monosaccharide containing four carbons.

Thermodynamic enolate (Section 23.4): The enolate that is lower in energy—generally the more substituted enolate.

Thermodynamic product (Section 16.11): In a reaction that can give more than one product, the product that predominates at equilibrium.

Thermodynamics (Section 6.5): A study of the energy and equilibrium of a chemical reaction.

Thermoplastics (Section 31.7): Polymers that can be melted and then molded into shapes that are retained when the polymer is cooled.

Thermosetting polymer (Section 31.7): A complex network of cross-linked polymer chains that cannot be re-melted to form a liquid phase.

Thioester (Section 22.17): A compound with the general structure RCOSR'.

Thiol (Section 9.15): A compound having the general structure RSH.

Tollens reagent (Sections 20.8, 27.9B): A reagent that oxidizes aldehydes, and consists of silver(I) oxide in aqueous ammonium hydroxide. A Tollens test is used to detect the presence of an aldehyde.

***p*-Toluenesulfonate** (Section 9.13): A very good leaving group having the general structure $CH_3C_6H_4SO_3^-$ and abbreviated as TsO^-. Compounds containing a *p*-toluenesulfonate leaving group are called alkyl tosylates and are abbreviated ROTs.

Torsional energy (Section 4.9): The energy difference between the staggered and eclipsed conformations of a molecule.

Torsional strain (Section 4.9): An increase in the energy of a molecule caused by eclipsing interactions between groups attached to adjacent carbon atoms.

Tosylate (Section 9.13): A very good leaving group having the general structure $CH_3C_6H_4SO_3^-$ and abbreviated as TsO^-.

***s*-Trans** (Sections 16.6, 26.5B): The conformation of a 1,3-diene that has the two double bonds on opposite sides of the single bond that joins them.

Trans diaxial (Section 8.8B): In an elimination reaction of a cyclohexane, a geometry in which the β hydrogen and the leaving group are trans with both in the axial position.

Trans isomer (Sections 4.13B, 8.3B): An isomer of a ring or double bond that has two groups on opposite sides of the ring or double bond.

Transition state (Section 6.7): An unstable energy maximum as a chemical reaction proceeds from reactants to products. The transition state is at the top of an energy "hill" and can never be isolated.

Triacylglycerol (Sections 3.9D, 10.6, 22.12A, 28.3): A lipid consisting of the triester of glycerol with three long-chain fatty acids. Triacylglycerols are the lipids that comprise animal fats and vegetable oils. Triacylglycerols are also called triglycerides.

Triose (Section 27.2): A monosaccharide containing three carbons.

Triphosphate (Section 7.16): A good leaving group used in biological systems. Triphosphate ($P_3O_{10}^{5-}$) is abbreviated as PPP_i.

Triplet (Section 14.6): An NMR signal that is split into three peaks having a relative area of 1:2:1, caused by two nearby nonequivalent protons.

Trisubstituted alkene (Section 8.2A): An alkene that has three alkyl groups and one hydrogen bonded to the carbons of the double bond ($R_2C{=}CHR$).

Triterpene (Section 28.7A): A terpene that contains 30 carbons and six isoprene units. A triterpenoid contains at least one oxygen atom as well.

U

Ultraviolet (UV) light (Section 16.15): Electromagnetic radiation with a wavelength from 200 to 400 nm.

Ultraviolet (UV) spectrum (Section 16.15): A plot of the absorbance of ultraviolet light versus wavelength, often recorded for conjugated systems.

Unimolecular reaction (Sections 6.9B, 7.10, 7.12A): A reaction that has only one reactant involved in the rate-determining step, so the concentration of only one reactant appears in the rate equation.

α,β-Unsaturated carbonyl compound (Section 20.15): A conjugated compound containing a carbonyl group and a carbon–carbon double bond separated by a single σ bond.

Unsaturated fatty acid (Section 10.6A): A fatty acid having one or more carbon–carbon double bonds in its hydrocarbon chain. In natural fatty acids, the double bonds generally have the *Z* configuration.

Unsaturated hydrocarbon (Section 10.2): A hydrocarbon that has fewer than the maximum number of hydrogen atoms per carbon atom. Hydrocarbons with π bonds or rings are unsaturated.

Unsymmetrical ether (Section 9.1): An ether in which the two alkyl groups bonded to the oxygen are different.

Upfield shift (Section 14.1B): In an NMR spectrum, a term used to describe the relative location of an absorption signal. An upfield shift means a signal is shifted to the right in the spectrum to lower chemical shift.

Urethane (Section 31.6C): A compound that contains a carbonyl group bonded to both an OR group and an NHR (or NR_2) group. A urethane is also called a carbamate.

V

Valence bond theory (Section 17.9A): A theory that describes covalent bonding as the overlap of two atomic orbitals with the electron pair in the resulting bond being shared by both atoms.

Valence electrons (Section 1.1): The electrons in the outermost shell of orbitals. Valence electrons determine the properties of a given element. Valence electrons are loosely held and participate in chemical reactions.

van der Waals forces (Section 3.3B): Very weak intermolecular interactions caused by momentary changes in electron density in molecules. The changes in electron density cause temporary dipoles, which are attracted to temporary dipoles in adjacent molecules. Van der Waals forces are also called London forces.

Vicinal dihalide (Section 8.10): A compound that has two halogen atoms on adjacent carbon atoms.

Vinyl group (Section 10.3C): An alkene substituent having the structure $-CH{=}CH_2$.

Vinyl halide (Section 7.1): A molecule containing a halogen atom bonded to the sp^2 hybridized carbon of a carbon–carbon double bond.

Vitamins (Sections 3.5, 28.5): Organic compounds needed in small amounts by biological systems for normal cell function.

VSEPR theory (Section 1.7B): Valence shell electron pair repulsion theory. A theory that determines the three-dimensional shape of a molecule by the number of groups surrounding a central atom. The most stable arrangement keeps the groups as far away from each other as possible.

W

Walden inversion (Section 7.11C): The inversion of a stereogenic center involved in an S_N2 reaction.

Wavelength (Section 13.5): The distance from one point of a wave to the same point on the adjacent wave. Wavelength is abbreviated with the Greek letter lambda (λ).

Wavenumber (Section 13.6A): A unit for the frequency of electromagnetic radiation that is inversely proportional to wavelength. Wavenumber, reported in reciprocal centimeters (cm^{-1}), is used for frequency in IR spectroscopy.

Wax (Sections 4.15, 28.2): A hydrolyzable lipid consisting of an ester formed from a high molecular weight alcohol and a fatty acid.

Williamson ether synthesis (Section 9.6): A method for preparing ethers by reacting an alkoxide (RO^-) with a methyl or primary alkyl halide.

Wittig reaction (Section 21.10): A reaction of a carbonyl group and an organophosphorus reagent that forms an alkene.

Wittig reagent (Section 21.10A): An organophosphorus reagent having the general structure $Ph_3P{=}CR_2$.

Wohl degradation (Section 27.10A): A reaction that shortens the carbon chain of an aldose by removing one carbon from the aldehyde end.

Wolff–Kishner reduction (Section 18.15B): A method to reduce aryl ketones to alkyl benzenes using hydrazine (NH_2NH_2) and strong base (KOH).

Woodward–Hoffmann rules (Section 30.3): A set of rules based on orbital symmetry used to explain the stereochemical course of pericyclic reactions.

Y

Ylide (Section 21.10A): A chemical species that contains two oppositely charged atoms bonded to each other, and both atoms have octets of electrons.

Z

Zaitsev rule (Section 8.5): In a β elimination reaction, a rule that states that the major product is the alkene with the most substituted double bond.

Ziegler–Natta catalysts (Section 31.4): Polymerization catalysts prepared from an organoaluminum compound and a Lewis acid such as $TiCl_4$, which afford polymer chains without significant branching and with controlled stereochemistry.

Zwitterion (Sections 3.9A, 19.15B, 26.1B): A neutral compound that contains both a positive and negative charge.

Credits

Front Matter

Page iii(Smith): © Daniel C. Smith; (Vollmer-Snarr): © Sy Snarr; p. v(pod): © Walter B. McKenzie/Getty Images RF; (aspirin): © Jill Braaten; (scurvy grass): © Brian & Sophia Fuller/Alamy; (oil spill): © Vince Bevan/Alamy; p. vi (tree bark): © George Ostertag/Alamy RF; (honey): © ninikas/Getty Images RF; (nasal spray): © McGraw-Hill Education/Mark Dierker, photographer; (osprey): © Comstock/PunchStock RF; (lavender): © Daniel C. Smith; p.vii(candlenuts): © Inga Spence/Science Source; (birth control): © McGraw-Hill Education, Chris Kerrigan, photographer; (wine): © Royalty Free/Corbis; (penicillin): Source: Library of Congress, LC-USZC4-1986; p. viii(sea sponge): © Daniel C. Smith; (coffee): © Jamie Grill/Getty Images RF; (poppy seeds): © John Foxx/Getty Images RF; (DNA): © Pasieka/Science Photo Library/Getty Images RF; p. ix(broccoli): © Jill Braaten; (quinoa): © Sarka Babicka/Getty Images RF; (fish): © Daniel C. Smith; (foxglove): © Richard Cech, Horizon Herbs, LLC; p. x(ginkgo biloba): © Jill Braaten; (breast cancer): © starry_eyed_girl/Getty Images RF; (cocoa beans): © Deborah Van Kirk/The Image Bank/Getty Images; (trumpet flower): © James Forte/Getty Images RF; p. xi: (whale): © Royalty Free/Corbis; (potatoes): © yuelan/iStock/Getty Images RF; (cholesterol test): © Jill Braaten; (euphorbia peplus): © De Agostini Picture Library/Getty Images; (cockroach): © Nature Picture Library/Alamy; p. xii(PET label): © Stuart Burford/Alamy.

Prologue

Figure 1a: © Comstock/PunchStock RF; 1b: © Corbis Premium/Alamy RF; 1c: © Steven May/Alamy; 1d: © Layne Kennedy/Getty Images; 5(cinchona pubescens): Source: US Geological Survey-Photo by Forest & Kim Starr; (Avloclor): © Chris Howes/Wild Places Photography/Alamy; p. 5 © Gregory G. Dimijian, M.D./Science Source.

Chapter 1

Opener: © Walter B. McKenzie/Getty Images RF; p. 11: © Jill Braaten; p.19(both): © Steve Gschmeissner/Science Source; p. 30: © John Somerville; p. 34(top): © Author's Image/PunchStock RF; (bottom): © McGraw-Hill Education/Elite Images, photographer; p. 40: © Royalty Free/Corbis; p. 41: © Creatas/PunchStock RF; p. 42: © Phillip Spears/Getty Images RF; p. 45: © McGraw-Hill Education/Mark Dierker, photographer; p. 50: © Cristina Pedrazzini/Science Source.

Chapter 2

Opener: © Jill Braaten; p. 63: © Bon Appetit/Alamy; p. 72&74: © McGraw-Hill Education/Jill Braaten, photographer; p. 79: © Wilmer Stratton; 2.7: © Skip Brown/National Geographic/Getty Images.

Chapter 3

Opener: © Brian & Sophia Fuller/Alamy; p. 97&99: © Jill Braaten; p. 102(top): © Andrew Syred/Science Source; (bottom): © Erik Tham/Alamy; p. 111: © Comstock/PunchStock RF; p. 112(top): © McGraw-Hill Education/Mary Reeg, photographer; (bottom): © Digital Vision/Getty Images RF; 3.9: © Shutterstock/smereka RF.

Chapter 4

Opener: © Vince Bevan/Alamy; p. 135(top): © God of Insects; (bottom): © Pixtal/age fotostock RF; p. 138: © Daniel C. Smith; p. 139: © Jill Braaten; 4.5: © Glow Images RF; p. 148: © McGraw-Hill Education/Jill Braaten, photographer; p. 170: © Riorita/Getty Images RF; p. 171: © Paul Nicklen/Getty Images.

Chapter 5

Opener: © George Ostertag/Alamy RF; 5.1(starch): © Keith Weller/USDA; (wheat): © Pixtal/age fotostock RF; (cotton plant): © Royalty Free/Corbis; (cotton fabric): © Daniel C. Smith; p. 184: © Daniel C. Smith; p. 186: © McGraw-Hill Education/Jill Braaten, photographer; p. 187: © Mark W. Skinner; p. 190: © Shallenberger Photography; p. 196: © Robert Brook/Science Source; p. 201: © McGraw-Hill Education/Jill Braaten, photographer; p. 209: © McGraw-Hill Education/Elite Images, photographer; p. 210(caraway seeds): © McGraw-Hill Education/Elite Images, photographer; (spearmint): © DAJ/Getty Images RF; (celery): © Aaron Roeth Photography RF.

Chapter 6

Opener: © ninikas/Getty Images RF; p. 226: © Tim Hill/Alamy; p.241(top): © Royalty Free/Corbis; (bottom): © McGraw-Hill Education/Jill Braaten, photographer.

Chapter 7

Opener: © McGraw-Hill Education/Mark Dierker, photographer; p. 260: © Michael Guiry; p. 263&292: © Jill Braaten; p. 295: © McGraw-Hill Education/Mark Dierker, photographer.

Chapter 8

Opener: © Comstock/PunchStock RF; p. 308: © McGraw-Hill Education/Jill Braaten, photographer; p. 309: © Steven P. Lynch RF; p. 315: Source: Forest & Kim Starr.

Chapter 9

Opener: © Daniel C. Smith; p.341: © Peter J.S. Franks, Scripps Institution of Oceanography; 9.4(left): © Pixtal/age fotostock RF; (right): © Mark Wilson/Getty Images; p. 360: © Henriette Kress, http://www.henriettes-herb.com; p. 373: © Purestock/SuperStock RF; p. 379: © McGraw-Hill Education/Jill Braaten, photographer; p. 381: © McGraw-Hill Education, John Thoeming, photographer.

Chapter 10

Opener: © Inga Spence/Science Source; p. 398: © Daniel C. Smith; 10.5(carrots): © Corbis/RF; (ginger): © Mitch Hrdlicka/Getty Images RF; (oranges): Source: USDA; (apples): © Pixtal/age fotostock RF; p. 401: © Jill Braaten; p. 403: © McGraw-Hill Education/Jill Braaten, photographer; p. 413: © Andrew Lambert/Science Source; p. 423: © Henriette Kress, http://www.henriettes-herb.com.

Chapter 11

Opener: © McGraw-Hill Education, Chris Kerrigan, photographer; 11.3: © Michael Sewell/Getty Images; p. 451: © Michael G. Moye.

Chapter 12

Opener: © Royalty Free/Corbis; p. 470(peanut butter and label): © McGraw-Hill Education/Elite Images, photographer; 12.4(both): © McGraw-Hill Education/Jill Braaten, photographer; p. 472: © Charlotte Björnström / EyeEm/Getty Images RF; p. 473: © 1998 Richard Megna, Fundamental Photographs, NYC; p. 478: Source: USDA APHIS PPQ Archive, USDA APHIS PPQ, Bugwood.org; p. 482: © Balazs Kovacs/Getty Images RF; 12.9: © McGraw-Hill Education/Elite Images, photographer; p. 491: © Image Source RF.

Chapter 13

Opener: Source: Library of Congress, LC-USZC4-1986; p. 508: © MizC/Getty Images RF; p. 513: © Ted Kinsman/Science Source; p. 525: © Andrew McClenaghan/Science Source.

Chapter 14

Opener: © Daniel C. Smith; p. 566: © Brand X Pictures/PunchStock RF; 14.15a: © Peter Noyce PLB/Alamy; 14.15b: © With permission, Daniel C. Smith. Spectra courtesy of the Chemistry Department at Rutgers University.

Chapter 15

Opener: © Jamie Grill/Getty Images RF; p. 592(top): © McGraw-Hill Education/Jill Braaten, photographer; (bottom): Source: NASA Ozone Watch; p. 597: © McGraw-Hill Education/Elite Images, photographer; p. 598(top): © Stockbyte/Corbis RF; (bottom): © Pixtal/SuperStock RF; p. 601: © McGraw-Hill Education/Jill Braaten, photographer; p. 602(top): © Image Source/Getty Images RF; (middle): © Royalty Free/Corbis; (bottom): © Jill Braaten.

Chapter 16

Opener: © John Foxx/Getty Images RF; p. 622: © Franz-Marc Frei/Getty Images; 16.7: © Stephen Frink/Getty Images RF; p. 631: © Alvis Upitis/Getty Images RF; p. 639: © C Squared Studios/Getty Images RF; p. 640: © McGraw-Hill Education/ Jill Braaten, photographer.

Chapter 17

Opener: © Pasieka/Science Photo Library/Getty Images RF; 17.4(left): © C. Zachariasen/PhotoAlto RF; (right): © Photo by Bob Nichols, USDA Natural Resources Conservation Service; p. 666(top): © Daniel C. Smith; (middle): Source: Forest & Kim Starr; (bottom): © Jb Reed/Bloomberg/ Getty Images.

Chapter 18

Opener: © Jill Braaten; p. 691: © Dick Swanson/Time Life Pictures/Getty Images; 18.4: © Rene Dulhoste/Science Source; p. 699: © Bill Aron/Photo Edit, Inc.; p. 723: © McGraw-Hill Education/Jill Braaten, photographer.

Chapter 19

Opener: © Sarka Babicka/Getty Images RF; p. 741: © PicturePartners/iStock/ Getty Images RF; p. 745: © Ryman Cabannes/PhotoCuisine RM/Alamy; p. 746: © Biopix.dx http://www.biopix.dk; p. 757: © Ken Samuelsen/Getty Images RF; p. 762: © FoodCollection RF.

Chapter 20

Opener: © Daniel C. Smith; p. 776: © Gerald D. Carr, PhD; p. 783: © M.K. Ranjitsinh/Science Source; p. 786: © John Kaprielian/Science Source; p. 787: © C Squared Studios/Getty Images RF; p. 792: © The McGraw-Hill Companies, Inc./Charles D. Winters, photographer; p. 799: © George Grall/National Geographic/Getty Images.

Chapter 21

Opener: © Richard Cech, Horizon Herbs, LLC; 21.6: © McGraw-Hill Education/ Jill Braaten, photographer; p. 844(top): © McGraw-Hill Education/Jill Braaten, photographer; (bottom): © Daniel C. Smith; p 861(top): © kj2011/iStock/Getty Images RF; (bottom): © alessandro0770/iStock/Getty Images RF.

Chapter 22

Opener: © Jill Braaten; p. 891(top): © Jill Braaten; (bottom): © blickwinkel/ Alamy; p. 896: © McGraw-Hill Education/Joe Franek, photographer; p. 897: © Coby Schal/NC State University; p. 906&907: © McGraw-Hill Education/Jill Braaten, photographer.

Chapter 23

Opener: © starry_eyed_girl/Getty Images RF; p. 942&949: © McGraw-Hill Education/Joe Franek, photographer; p. 954: © McGraw-Hill Education/Mary Reeg, photographer; p. 961: © AwakenedEye/Getty Images RF.

Chapter 24

Opener: © Deborah Van Kirk/The Image Bank/Getty Images; 24.3: © James H. Robinson/Science Source; p. 983(both): © Jill Braaten; p. 988: © McGraw-Hill Education/Jill Braaten, photographer.

Chapter 25

Opener: © James Forte/Getty Images RF; 25.6(nightshade): © E.A. Janes/Getty Images; (tobacco): © Creatas Images RF; (hemlock): © Steven P. Lynch RF; p. 1020: © Daniel C. Smith; p. 1030: © McGraw-Hill Education/Jill Braaten, photographer; p. 1036(top): © James Gathany/CDC; (bottom): © Daniel C. Smith; p. 1050(indigo): © Mantonature/istock/Getty Images RF; (sea snail): © Andrew J. Martinez/Science Source; (madder): © Bob Gibbons/Alamy; (purple shawl): © Science & Society Picture Library/Getty Images.

Chapter 26

Opener: © Royalty Free/Corbis; p. 1066: © Jill Braaten; p. 1099: © Eye of Science/Science Source.

Chapter 27

Opener: © yuelan/iStock/Getty Images RF; p. 1111(top): © Jill Braaten; (bottom): © McGraw-Hill Education/Elite Images, photographer; p. 1126: © Brand X Pictures/PunchStock RF; p. 1129: © Nature Picture Library/Alamy; p. 1130: © inga spence/Alamy; p. 1138: © Sundell Larsen/Getty Images RF; p. 1139: © Mitch Hrdlicka/Getty Images RF; 27.9(left): © McGraw-Hill Education/Elite Images, photographer; (right): © George Clark/E+/Getty Images RF; 27.10&p.1143(top): © McGraw-Hill Education/Jill Braaten, photographer; (bottom): © Comstock Images/PictureQuest RF.

Chapter 28

Opener: © Jill Braaten; p. 1157(top): © Daniel C. Smith; (bottom): Source: USDA, ARS, National Genetic Resources Program; p. 1159(palm trees): © Royalty-Free/Corbis; (oils): © McGraw-Hill Education/Elite Images, photographer; (fish): © Darlyne A. Murawski/Getty Images; p. 1161(top): © David Crigger/Bristol Herald Courier/AP Images; (bottom): © C Squared Studios/Getty Images RF; p. 1169(left): © Henriette Kress, http://www.henriettes-herb.com; (right): © Mark Turner/Getty Images; p. 1171: © Thomas J. Abercrombie/Getty Images; p. 1178: © Comstock/JupiterImages RF.

Chapter 29

Opener: © De Agostini Picture Library/Getty Images; p. 1191: © Pascal Goetgheluck/Science Source; p. 1195: © McGraw-Hill Education/John Thoeming, photographer; p. 1202: © Ross Warner/Alamy.

Chapter 30

Opener: © Nature Picture Library/Alamy; p. 1222: © McGraw-Hill Education/ Mary Reeg, photographer; p. 1226: © blickwinkel/Alamy; p. 1233: © Marina Khaytarova, TopTropicals.com.

Index

Page numbers followed by f indicate figures; those followed by t indicate tables. An A- before page numbers indicates appendix pages.

A

ABS, 1251
Abyssomicin C, 1009
Acebutolol, 131
Acetaldehyde
  aldol reaction of, 973, 976, 979
  ethanol oxidized to, 489–490
  hydrate formed from, 855
  retro-aldol reaction to form, 975
  structure of, 830
Acetal(s)
  conversion to hemiacetals, 860–861
  cyclic, 857
    in carbohydrates, 865–866
    conversion of cyclic hemiacetal to, 864–865
  ether compared, 857
  glycosides, 1127–1130
  hemiacetals compared to, 1127
  hydrolysis of, 860–861
  as protecting groups, 861–862
  structure of, 1127
  synthesis of
    addition of alcohols to aldehydes and ketones, 857–861
    from hemiacetals, 858–859, 864–865
  terminology of, 857
Acetamide, 23–24, 59, 89
  structure of, 879, 885, 888, 888f
Acetaminophen, 59, 909, 929, 1060
Acetanilide, 770, 1038, 1038f
Acetate
  bond length in, 623
  formation from acetaldehyde oxidation, 489
  as nucleophile, 270–272, 277
  resonance structures of, 74–75, 617, 750
Acetic acid, 745, 879
  acidity of, 74–75, 77, 79, 750, 754
  as Brønsted–Lowry acid, 62f
  deprotonation of, 749
  dimers, 743f
  electrostatic potential plot of, 75f, 739
  in ethyl acetate formation, 244
  glacial, 745
  halogenation of carbonyl compounds in, 948
  melting point of, 107
  NMR spectrum of, 564
  pK_a of, 749
  as polar protic solvent, 267f
  structure of, 741t, 879
Acetic anhydride, 643, 879, 883, 1131
Acetic benzoic anhydride, 883
Acetoacetic ester synthesis, 955, 959–962, 989
Acetoacetyl coenzyme A, 1003
Acetone
  acidity of, 75
  aldol reactions of, 979
  butan-2-one synthesis, 961
  dipoles in adjacent molecules, 102
  hydrate formed from, 855
  odor of, 835
  as polar aprotic solvent, 268, 268f
  solubility of, 109
  structure and bonding of, 44, 606, 831, 835
  surface area and boiling point of, 105, 106f
  uses of, 835
Acetone enolate, 940f, 946f
Acetonitrile, 58
  acidity of, 75
  deprotonation of, 69
  as polar aprotic solvent, 268f
  structure of, 886f
p-Acetophenetidin, 929
Acetophenone, 785
  aldol reaction of, 976
  structure of, 831
N-Acetyl amino acids, 1171–1173
Acetyl chloride, 252, 879, 1131
  structure of, 879, 883
Acetylcholine, 914
Acetyl coenzyme A, 914, 916, 991, 992, 1003, 1176
1-Acetylcyclohexene, 965
2-Acetylcyclopentanone, 1007
1-Acetylcyclopentene, 1005
Acetylene
  acetylide anion formation from, 454–455
  in acid–base reactions, 68–69
  acidity of, 75–76, 76f
  bond angles, 435
  bond dissociation energies of, 435
  bonds in, 42–43, 43f, 44t
  combustion of, 438
  deprotonation of, 454
  disparlure synthesis from, 479, 479f
  electrostatic potential map of, 440, 441f
  functional group in, 93, 94t
  Lewis structure, 16
  molecular shape of, 27
  nomenclature, 437
N-Acetyl-D-glucosamine (NAG), 916, 1143
Acetyl group, 1131
  abbreviation for, 1189
  structure, 832
Acetylide anions
  conversion of terminal alkynes to, 449
  formation by deprotonation of terminal alkynes, 441
  as organometallic compounds, 794
  reactions with alkyl halides, 449–452, 451f
  reactions with epoxides, 452
  retrosynthetic analysis with, 454–455
  synthesis from acetylene, 454–455
  in synthesis of disparlure, 479, 479f
Acetyl phosphate, 879
Acetylsalicylic acid, 61, 745, 753, 909. *See also* Aspirin
Achiral enolate, 946
Achiral epoxides, 480
Achiral intermediate, 414
Achiral meso compound, 199, 201, 415, 416f, 478, 632, 1196
Achiral molecules
  definition of, 184–186, 186f
  halogenation of, 590, 590t
  optical inactivity of, 204
Achiral product, 410
Achiral reactant, 376, 410
Achiral starting material, 478, 490, 590, 783, 798
Acid–base reactions. *See also* Nucleophilic substitution reactions
  amines, 1028–1030
  Brønsted–Lowry, 63–65, 262, 280
  of carboxylic acids, 898–903
  description, 63–65
  Lewis, 82–83, 262
  organometallic reagents and, 794–795
  predicting the outcome of, 68–69
Acid chlorides
  acidity/basicity of, 882, 882t
  boiling point of, 889t
  functional group in, 98t
  IR spectra of, 889, 890t
  leaving group, 778, 806
  melting point of, 889t
  nomenclature of, 883, 887t
  nucleophilic substitution reactions, 778
  odor of, 896
  physical properties of, 889t
  pK_a, 882t
  reactions of, 896–897, 910t
    with ammonia or amines to form amides, 1037
    conversion to amides, 896
    conversion to anhydrides, 896
    conversion to carboxylic acids, 896
    conversion to esters, 896
    Friedel–Crafts acylation reactions, 693
    with organometallic reagents, 795, 805–807
  reactivity, 893–894
  reduction of, 787–788
  solubility of, 889t
  stability of, 882
  structure and bonding of, 775, 879, 882
  synthesis from carboxylic acids, 898f, 899
Acidity
  basicity, inverse relationship with, 67, 67t
  of carbon–hydrogen bonds, 71
  of carboxylic acids, 748, 749–754, 750
  determining the relative acidity of protons, 77–78
  electronegativity and, 70–73
  factors affecting, 70–78
    element effects, 70–72, 77f
    hybridization effects, 75–76, 76f, 77f
    inductive effects, 72–74, 73f, 77f
    resonance effects, 74–75, 75f, 77f
  periodic trends in, 70–72
Acidity constant (K_a), 66
Acid rain, 79
Acid(s)
  amine reactions with, 1028–1030
  amino. *See* Amino acids
  Brønsted–Lowry
    carboxylic acids, 749–752
    description of, 62–63
    reactions of, 63–65
  carboxylic. *See* Carboxylic acid(s)
  as catalysts, 244–245
  common, 78–79
  conjugate, 63–65, 264–266, 264t
  dehydration of alcohols to alkenes, 353–358
  as electron pair acceptors, 62, 65, 81
  in epoxide ring opening reactions, 377–379, 378f
  halogenation in, 948, 950f
  Lewis, 81–83
  origin of word, 62
  as proton donors, 62, 69
  reaction of ethers with strong, 370–371
  strength, 66–78
  sulfonic, 760
Acrolein, 632f
Acrylic acid, 602
Acrylonitrile, 571, 1249f
Actinic keratosis, 1185
Activation energy. *See* Energy of activation (E_a)
Active methylene compounds, 981, 995
Active site, enzyme, 245, 246f
Acyclic alkanes
  branched-chain, 136
  conformations of, 150–157
    butane, 154–157, 154f, 156f
    ethane, 150–153, 152f, 153f
  constitutional isomers, 136–138, 137t
  having more than five carbon atoms, 137–138, 137t
  having one to five carbon atoms, 135–137, 137t
  homologous series, 137, 137t
  molecular formula, 135, 137, 137t
  naming, 141–144, 143f
  normal (*n*) (straight-chain), 136–137, 137t
  structure, 135–136
  torsional and steric strain energies in, 156t
Acyclovir, 965

Acylases, 1173
Acylation
  biological, 914–916
  definition of, 693
  Friedel–Crafts. *See* Friedel–Crafts acylation
Acyl derivatives, 775
Acyl group, 693, 879, 883, 884, 885
  naming, 832
Acylium ion, 511, 695
Acyl phosphate, 100t, 879, 887t, 911–914
Acyl transfer reactions, 914–915, 915f
Addition polymers. *See* Chain-growth polymers
Addition reactions
  in alkenes, 392, 404–426, A-13
    anti addition, 405, 411, 411t, 413, 414, 417
    halogenation, 405f, 413–415, 416f
    halohydrin formation, 405f, 416–418, 418t
    hydration, 405f, 412–413, 423
    hydroboration–oxidation, 405f, 419–423, 420f, 421f, 422t, 423f
    hydrohalogenation, 405–412, 405f, 407f, 409f, 411t
    overview of, 404–405, 405f
    stereochemistry of, 405, 410–412, 411t
    syn addition, 405, 411, 411t, 419, 421
  allylic bromination, 595
  anti additions, 471, 473, 480, 481
  carbene addition, 1196
  conjugate, 995, 997
  in conjugated dienes
    Diels–Alder reaction, 629–638, 630f
    electrophilic, 624–626
    kinetic versus, 626–629
    thermodynamic product, 626–629
  description of, 222
  dihydroxylation, 480–482
  epoxidation, 477–480, 490–492
  exothermic reactions, 406, 406f
  ionic addition, 600
  in nucleophilic aromatic substitution, 715–718
  oxidative, 1189, 1191, 1194
  radical reactions, 595, 598–601
  syn additions, 466, 468, 471, 477, 479, 479f, 480, 481–482, 1196
Adenine, 649, 667, 1145, 1146f
Adenosine diphosphate (ADP), 233–235, 234f, 911
Adenosine monophosphate (AMP), 911
Adenosine triphosphate (ATP), 101, 233–235, 234f, 292, 374
S-Adenosylmethionine (SAM), 292–293, 374
-adiene (suffix), 395
Adipic acid, 1255, 1262, 1262f, 1263
Adrenal cortical steroids, 1177, 1179
Adrenaline
  structure of, 1019
  synthesis of, 293–294, 293f
Advair. *See* Fluticasone
Advil. *See* Ibuprofen
Aerosol propellant, 260, 592
Agent Orange, 691f
Aglycon, 1130
-al (suffix), 829
Alanine
  abbreviation for, 762t
  acid–base reactions of, 763–764, 764f
  in dipeptide synthesis, 1183–1184, 1187
  enantiomers of, 189f
  N-acetyl, 1171–1173
  isoelectric point of, 765, 1167t
  in peptide formation, 1176
  physical characteristics, 206
  pK_a values for, 1167t
  structure of, 1165f
  synthesis of, 1169
Albuterol, 187–188, 379, 379f, 1062
Alcohol dehydrogenase, 489–490
Alcoholic beverages, 346, 463
Alcohols, 339–390
  alkene preparation by acid-catalyzed dehydration, 403
  allylic, epoxidation of, 490–492
  boiling point of, 346t
  bonding in, 341
  classification of, 96, 340
  dehydration reactions of, 353–360
    alkene synthesis in, 353–360
    carbocation rearrangements in, 356–358
    dehydration to alkenes, 353–360, 356f
    E1 mechanism in, 354–355
    E2 mechanism in, 355, 359
    enthalpy change in, 355, 356f
    Le Châtelier's principle in, 355
    in phosphorus oxychloride and pyridine, 359, 360f
    in strong acid, 353–358
  electrostatic potential maps, 341f
  elimination reactions, 351–360, 369, 370f
  ether formation from, 414
  fragmentation pattern in mass spectrum, 511
  Friedel–Crafts alkylation of, 697
  functional group, 93, 96t
  β-hydroxy carbonyl compounds, 973–975
  infrared (IR) spectra of, 522
  interesting examples, 346–347, 347f
  leaving groups of, 352, 354–355, 357–362, 364–369
  melting point of, 346t
  NMR spectra of, 559–560
  nomenclature, 342–343
    common names, 343
    for cyclic alcohols, 342, 342f
    IUPAC system, 342, 343
  oxidation of, 476f, 484–490
  physical properties of, 345–346, 346t
  as polar protic solvents, 267, 267f
  primary (1°), 423
    classification as, 340
    conversion to alkyl halides with HX, 360–362, 365t
    conversion to alkyl halides with phosphorus tribromide, 365, 365t
    conversion to alkyl halides with thionyl chloride, 364, 365t
    dehydration by E2 mechanism, 355
    hydrogen bonding extent, 345
    from nucleophilic addition to aldehydes, 788
    oxidation to carboxylic acids, 485, 486–487, 747
    from reactions of aldehydes with organometallic reagents, 795–797
    from reduction of acid chlorides, 787–788
    from reduction of aldehydes, 469, 780, 781, 782
    from reduction of carboxylic acid derivatives, 780
    from reduction of carboxylic acids, 790
    from reduction of esters, 787–788
    from reduction of monosaccharides, 1132
  products of S_N2 substitution reactions, 294, 295t
  racemic mixture of, 422
  reactions
    conversion to alkyl halides with HX, 360–363, 365t
    conversion to alkyl halides with thionyl chloride and phosphorus tribromide, 364–365, 365t
    conversion to alkyl tosylates, 367
    elimination reactions, 351–360
    general features, 351–352
    nucleophilic substitution reactions, 352, 360–369, 370f
    summary of, 369, 370f
  secondary (2°), 423
    classification as, 340
    conversion to alkyl halides with HX, 360–363, 365t
    conversion to alkyl halides with phosphorus tribromide, 365, 365t
    conversion to alkyl halides with thionyl chloride, 364, 365t
    dehydration by an E1 mechanism, 354–355
    hydrogen bonding extent, 345
    oxidation to ketones, 485–486
    from reactions of aldehydes with organometallic reagents, 795–797
    from reduction of ketones, 469, 780, 781, 782
  solubility, 346t
  structure of, 340, 341
  substitution reactions, 351–352, 360–369, 370f
  synthesis of, 349–350
    from alkyl halides, 349–350
    hydration of alkenes, 412–413
    hydroboration–oxidation of alkenes, 419–423
    by nucleophilic substitution reaction, 349–350
    reactions of organometallic reagents with aldehydes and ketones, 795–797, 842–843
    reactions of organometallic reagents with epoxides, 808–809
    reduction of epoxides, 474–475, 475f
  tertiary (3°)
    classification as, 340
    conversion to alkyl halides with HX, 360–363, 365t
    dehydration by an E1 mechanism, 354–355
    hydrogen bonding extent, 345
    from reaction of esters and acid chlorides with organometallic reagents, 804–806
    from reaction of ketones with organometallic reagents, 795–797
Aldaric acids, 1132, 1133, 1136–1137
  D-aldaric acids, 1136
-aldehyde (suffix), 830
Aldehyde dehydrogenase, 489–490
Aldehyde group, as electron-withdrawing group, 701–702
Aldehydes. *See also* Carbonyl compounds
  aldol reactions of, 973–985
  boiling point of, 833t
  carbohydrates, 865–866
  electrostatic potential map of, 776, 776f
  fragmentation pattern in mass spectrum, 511
  functional group in, 98t
  hydration of, 854–857
    kinetics of hydrate formation, 856–857
    thermodynamics of hydrate formation, 855
  hydroxy
    cyclization of, 863, 865–866
    β-hydroxy, synthesis in aldol reactions, 973–985
  interesting examples, 835–836, 836f
  IR spectra, 833–834, 833f, 834f
  melting point of, 833t
  NMR spectra, 835, 835f
  nomenclature, 829–830, 832
    common names, 830, 830f
    IUPAC system, 829, 830f
  nucleophilic addition reactions, 777, 782, 788, 827–877
    carbanion addition, 842
    cyanide addition, 843–844
    hydride addition, 841–842
    mechanism of, 839–840
    nucleophiles in, 840, 841f
    primary amine addition, 850–852, 853f
    secondary amine addition, 852–854, 853f
  odors of, 836, 836f
  oxidation reactions, 780, 792
    to carboxylic acids, 1132–1133
  physical properties of, 832, 833t
  pK_a, 941t
  polyhydroxy, 865–866
  protecting groups for, 861–862
  reactions of
    with amines, 1037
    at the carbonyl carbon, 838, 839
    halogenation at the α carbon, 947–951, 950f
    with organometallic reagents, 795–797, 842
    at the α carbon, 838–839
  reactivity of, 777
  reduction of, 841–842
    to alcohols, 469, 780, 781, 782
  reductive amination to amines, 1025–1027
  solubility of, 833t
  structure of, 775, 828
  synthesis of, 836–838
    alcohol oxidation, 476f, 485, 486–487
    carboxylic acid reduction, 780

hydroboration–oxidation of alkynes, 448, 837
hydrolysis of acetals, 860–861
hydrolysis of imines and enamines, 854
oxidation of primary alcohols, 837
oxidative cleavage of alkenes, 482–483, 837
reduction of acid chlorides, 787
reduction of acid chlorides and esters, 837
reduction of esters, 787
reduction of nitriles, 919
unreactivity to nucleophilic substitution, 779
Wittig reaction, 845–849
Alder, Kurt, 629
Alditol, 1132
Aldohexoses, 1112
D-aldohexoses, 1117, 1118f
Haworth projections of, 1122–1124
pyranose ring of, 1120
stereogenic centers of, 1114, 1117
Aldol reactions
biological, 991–993
crossed, 979–981
synthetically useful reactions, 979–981, 981f
dehydration of aldol products, 976–977
directed, 981–983, 982f
features of, 973–975, 974f
intramolecular, 984–985, 985f
in Robinson annulation, 996–1000
mechanism of, 974
retro-, 975–976, 992–993
retrosynthetic analysis using, 977–978
Aldonic acid, 1132, 1133
D-Aldopentoses, 1117, 1118f, 1136
Aldoses
D-aldose family, 1116–1118, 1118f
in Kiliani–Fischer synthesis, 1134, 1135–1137
oxidation of, 1132–1134
reduction of, 1132
structure of, 1111–1112
in Wohl degradation, 1134–1135
Aldosterone, 1179
Aldotetroses
D-aldotetroses, 1117, 1117f, 1118f, 1136, 1137
stereoisomers of, 1117, 1117f
Aldrin, 645
Alendronic acid (Fosamax), 19
Aleve. *See* Naproxen
Aliphatic carboxylic acids, 754–755
Aliphatic hydrocarbons, 93, 135. *See also* Alkanes; Alkenes; Alkynes
Aliskiren, 188, 500
Alizarin, 1050
Alizarine yellow R, 1051
Alkaloids, 1010, 1018, 1018f
Alkanes, 134–179
acyclic
branched-chain, 136
conformations of, 150–157, 152f, 153f, 154f, 156f
constitutional isomers, 136–138, 137t
having more than five carbon atoms, 137–138, 137t
having one to five carbon atoms, 135–137, 137t
homologous series, 137, 137t
molecular formula, 135, 137, 137t
normal (*n*) (straight-chain), 136–137, 137t
structure of, 135–136
torsional and steric strain energies in, 156t
boiling point of, 149, 149t
bonds in, 93
combustion of, 168–169
cycloalkanes
molecular formula, 135, 138
naming, 144–146, 146f
structure, 135, 138, 158f
substituted, 162–167, 163f, 166f
definition of, 135
fossil fuels, 147–148
halo, 257
halogenation, 582–591
of achiral starting materials, 590, 590t
bromination, 586–588, 588f
of chiral starting materials, 590t, 591
chlorination, 582–591, 589f
energy changes during, 584–585, 585f
mechanism of, 583–584
in organic synthesis, 589
stereochemistry of, 590–591, 590t
infrared (IR) spectra of, 521
melting point of, 149, 149t
nomenclature, 139–147, 143f, 146f
common names, 147, 147f
IUPAC system, 141–146, 143f
substituents, 140
nonreactivity of, 94
oxidation of, 167–169
physical properties, 149–150, 149t
radical reactions, 581
solubility of, 149, 149t
synthesis of
reduction of alkenes, 464, 464f, 466–469
reduction of alkyl halides, 474, 795
reduction of alkynes, 464f, 471–472, 474f
Alkene oxides, 344, 345
Alkenes, 391–433
addition reactions, 392, 404–426
anti addition, 405, 411, 411t, 413, 414, 417
halogenation, 405f, 413–415, 416f
halohydrin formation, 405f, 416–418, 418t
hydration, 405f, 412–413, 423
hydroboration–oxidation, 405f, 419–423, 420f, 421f, 422t, 423f
hydrohalogenation, 405–412, 405f, 406f, 407f, 409f, 411t
overview of, 404–405, 405f
stereochemistry of, 405, 410–412, 411t
syn addition, 405, 411, 411t, 414, 419, 421
amino acid synthesis from, 1174–1175
boiling point of, 399
bonds in, 93
classification of, 308, 308f
constitutional isomers of, 353
conversion to alkynes, 439–440
conversion to cyclopropanes, 1194–1197
in coupling reactions with organic halides (Heck reaction), 1186, 1192–1194
cycloaddition between, 1224
cyclopropane synthesis in Simmons–Smith reaction, 1197–1198
degrees of unsaturation, calculating, 393–395, 468–469
dehydration of alcohols to, 353–360
diastereomers of, 309
dihydroxylation of, 480–482
disubstituted, 308–311, 308f, 314, 316–317, 320, 324–325
double bonds in, 306–310, 392, 393t
restricted rotation around, 308–309, 309f
electrostatic potential plots, 404, 404f
epoxidation of, 476f, 477–478
fatty acids, 401–403, 402f, 402t
Friedel–Crafts alkylation of, 697
functional group in, 93, 94t
hybrid orbitals in, 307–308, 310
hydrogenation of, 466–471
alkene stability and, 466–467, 467f
cis and trans isomers compared, 466, 467f
degrees of unsaturation, determination of, 468–469
mechanism of catalytic hydrogenation, 467–468
infrared (IR) spectra of, 521
interesting examples, 399, 400f
internal, 392
lipids, 401–403
melting point of, 399, 401, 402t
metathesis, 1198–1203, 1199f, 1201f
molecular formula of, 393
monomers in polymerization, 601–603, 602t
monosubstituted, 308, 308f, 314
NMR spectra of, 541, 557–559, 557f, 558f
nomenclature, 395–398
common names, 398, 398f
IUPAC system, 395–396
for stereoisomers, 396–398
organic synthesis, 425–426
oxidative cleavage of, 476f, 482–484, 837
physical properties of, 399
polymerization of, 1244–1251, 1252–1253
radical reactions, 581, 598–601
reduction of, 464, 464f, 466–469
solubility of, 399
stability of, 310–311, 466–467, 467f
stereoisomers of, 308–310
alkyne reduction and, 471–474
synthesis of, 403–404
amine conversion, 1038–1041
comparison of methods, 849
dehydration of alcohols, 403
dehydrohalogenation of alkyl halides with strong base, 403
by elimination reactions, 306–307, 403–404
by Hofmann elimination reaction, 1038–1041
reduction of alkynes to cis alkenes, 471, 472, 474f
reduction of alkynes to trans alkenes, 471, 473–474, 474f
substituted alkenes in Heck reaction, 1192–1193
Wittig reaction, 845–849
terminal, 392
tetrasubstituted, 308
trisubstituted, 308, 308f, 316–317, 320, 324–325
Alkenols, 396
Alkoxides, 306
electron density plot, 940, 940f
protonation of, 784, 796, 797, 800, 805, 809
Alkoxide salt, in ether synthesis, 350
Alkoxy group, 96t, 344
Alkylamines
basicity of, 1031–1032, 1035t
reaction with nitrous acid, 1042–1043
Alkylation reactions
in acetoacetic ester synthesis, 955, 959–962
definition of, 693
of a diethyl malonate derivative, 1168–1169, 1170f
of enolates, 952–955, 961
general features of, 952–953
tamoxifen synthesis, 954
of unsymmetrical ketones, 953–954
Friedel–Crafts. *See* Friedel–Crafts alkylation
in malonic ester synthesis, 955–959
polyalkylation, 711
Alkyl benzenes
electron density of, 701–702
halogenation of, 718–720
oxidation of, 721, 747
sulfonation of, 692
synthesis of
Friedel–Crafts alkylation, 688f, 693–694, 722, 722f
by reduction of aryl ketones, 721–722
Alkylboranes
formation by hydroboration, 419–423
oxidation of, 421–422
Alkyl bromides
mass spectrum of, 508, 509f
synthesis of, 598–601
Alkyl chlorides
in Friedel–Crafts alkylation reactions, 694, 696–697
mass spectrum of, 508, 509f
Alkyl diazonium salts, 1042, 1054
Alkyl group, 140–143, 257, 261, 275, 280–283, 398, 693, 883–886
in alkyl tosylates, 367
carbon radical stability and, 579–580, 580f
as electron-donating group, 700–702
in ethers, 340, 343–344
nomenclature
in alcohols, 343
for branching molecules, A-3
in ethers, 343–344
polarizability, 700
stabilization of carbonyl group by, 855

- Alkyl halides
  - acetylide anion reactions with, 449–452, 451f
  - alkane formation from, 795
  - allylic halides, 256, 256f
  - benzylic halides, 256, 256f
  - classification of, 256, 328
  - conversion of alcohols to, 360–366
    - with HX, 360–363, 365t
    - importance of, 366
    - with thionyl chloride and phosphorus tribromide, 364–365, 365t
  - conversion to carboxylic acids, 808
  - conversion to ethers of monosaccharides, 1130
  - conversion to primary amines, 1023, 1024
  - coupling reaction with organocuprates, 1186–1187
  - dehydrohalogenation with strong base, 403
  - in direct enolate alkylation, 952–954
  - elimination reactions, 261, 305–338
    - E1 mechanism, 320–321, 327–330
    - E2 mechanism, 311–315, 315t, 322, 324–325, 327–330, 1041, 1041f
  - examples, 256f, 259–260, 259f
  - functional group, 96, 96t
  - mass spectrometry of, 508–509, 509f
  - molecular formula of, 256
  - nomenclature, 257–258, 258f
    - common names, 257
    - IUPAC, 257, 258f
  - nucleophilic substitution reactions, 262–296
    - alcohol synthesis, 349–350
    - alkyl halide structure and, 275, 281, 287, 287f
    - ether synthesis, 349–350
    - to form amines, 1021–1023
    - S_N1 mechanism, 327–331, 328t
    - S_N2 mechanism, 327, 328t, 329–330
  - physical properties, 258, 258t
  - polar carbon–halogen bond, 260–261, 261f
  - primary (1°)
    - acetylide anion reactions with, 449–450
    - classification as, 256
    - E2 elimination reactions, 314, 328–329
    - S_N2 substitution reactions, 275, 287, 287f, 328t, 329
    - synthesis of, 360
  - reaction mechanism, determining type of, 327–331, 328t
  - reduction of, 474–475, 475f
  - secondary (2°)
    - acetylide anion reactions with, 449–450
    - classification as, 256
    - E1 elimination reactions, 319, 328f, 329
    - E2 elimination reactions, 314, 328t, 329
    - S_N1 substitution reactions, 281, 287, 287f, 328t, 329
    - S_N2 substitution reactions, 275, 287, 287f, 328f, 329
    - synthesis of, 360
  - synthesis of
    - halogenation of alkanes, 582, 587–589
    - hydrohalogenation of alkenes, 405–412, 411t
  - tertiary (3°)
    - acetylide anion reaction with, 449–450
    - classification as, 256
    - E1 elimination reactions, 319, 321, 328, 328t
    - E2 elimination reactions, 314, 328, 328t
    - S_N1 substitution reactions, 279, 281, 287, 287f, 321, 328, 328t
    - synthesis of, 360, 363
  - unreactive, 256, 256f
  - vinyl halides, 256, 256f, 294
- 1,2-Alkyl shift, 356
- Alkylthio group, 96t, 374
- Alkyl titanium compound, 1253
- Alkyl tosylates, 367–369, 370f, 403
  - conversion of alcohols to, 367, 369
  - reactions of, 368
- Alkynes, 434–462
  - addition reactions
    - halogenation, 441f, 444–445
    - hydration, 441f, 445–447
    - hydroboration–oxidation, 441f, 447–449
    - hydrohalogenation, 441f, 442–444
    - overview of, 440
  - boiling point of, 437
  - bond dissociation energies of, 435
  - bonds in, 93
  - cycloalkynes, 436
  - degrees of unsaturation, 435
  - deprotonation of, 441–442, 441t
  - examples, 438–439, 438f, 439f
  - functional group in, 93, 94t
  - hydroboration of, 1190, 1192
  - hydroboration–oxidation of, 441f, 447–449
  - infrared (IR) spectra of, 522
  - internal
    - definition of, 435
    - hydration of, 445
    - hydroboration–oxidation of, 448
    - oxidative cleavage of, 484, 748
    - synthesis in acetylide anion reactions with alkyl halides, 449–451
  - melting point of, 437
  - molecular formula of, 435
  - nomenclature, 436–437
  - oxidative cleavage of, 476f, 484, 748
  - physical properties of, 437
  - reduction of, 464, 464f, 471–474, 474f
    - to alkanes, 464f, 471–472, 474f
    - to cis alkenes, 471, 472, 474f
    - to trans alkenes, 471, 473–474, 474f
  - solubility of, 437
  - structure of, 435–436
  - synthesis of, 439–440, 454–455
    - from alkenes, 439–440
    - by dehydrohalogenation, 326–327, 327f
    - from dihalides, 439–440
    - E2 elimination reactions, 306, 326–327, 327f
    - multistep, 454–455
  - terminal
    - definition of, 435
    - hydration of, 445
    - hydroboration–oxidation, 448
    - hydrohalogenation, 442
    - oxidative cleavage of, 484, 748
    - reaction as an acid, 441–442
    - synthesis in acetylide anion reactions with alkyl halides, 449–451
  - triple bond of, 435–436
- D-Allaric acid, 1134
- Allegra. *See* Fexofenadine
- Allenes, 217, 647
- Allicin, 138
- D-Allose, 1118f, 1134
- Allyl aryl ether, 1232
- 2-Allylcycloheptanol, 398f
- Allyl group, 398
- Allylic alcohol
  - epoxidation of, 490–492
  - synthesis from α,β-unsaturated carbonyl compounds, 809
- Allylic carbocations, 626
  - in biological reactions, 616
  - as conjugated system, 613, 614
  - electrostatic potential plot, 615, 615f
  - resonance of, 615–617
  - resonance-stabilized, 626
  - stability, 615
- Allylic carbon
  - oxidation of, 470, 470f
  - radical halogenation at, 593–596
    - bromination, 594–596
    - product mixtures in, 595–596
  - in triacylglycerols, 470, 470f
- Allylic halides, 256, 256f
- Allylic sigma bonds, 1228
- Allylic substitution, 595
- Allyl radical, 593–596, 597f
- Allyl vinyl ether, 1232
- Alpha (α) carbon, 306–307, 741, 744, 761
  - of aldehydes and ketones, 830, 835, 838–839
  - substitution reactions of carbonyl compounds at the α carbon, 934–971
- Alpha (α) cleavage, 511–512
- Alpha (α) elimination reaction, 1195
- Alpha (α) helix, 1191, 1192f, 1193, 1194f, 1196f
- Alpha (α) hydrogens, 935, 939, 948–949, 979–980, 981, 988
- Altocid. *See* Methoprene
- D-Altrose, 1118f
- Aluminum chloride, in Friedel–Crafts alkylations, 693–699, 711
- Alzheimer's disease, 983, 1233
- Amber, 1171
- Amberlyst A-26 resin, 487–488
- Ambident nucleophile, 943
- -amide (suffix), 885, 887t
- Amide bond, 1175–1177, 1191
- Amides
  - for alkyne synthesis by dehydrohalogenation, 327
  - basicity of, 882t, 1032, 1035t
  - boiling point of, 889t
  - classification, 98
  - cyclic, 880
  - functional group in, 98t, 99
  - hydrolysis of, 908
  - infrared (IR) spectra of, 523
  - interesting examples, 891
  - IR spectra of, 889, 890t
  - leaving group, 778
  - melting point of, 889t
  - nitrogen bases, 79, 79f
  - NMR spectra of, 890
  - nomenclature, 885–886, 887t
  - nucleophilic substitution reactions, 778
  - physical properties of, 888–889, 888f, 889t
  - pK_a, 882t, 941t
  - primary (1°), 880, 885, 888, 889t, 890, 896, 897, 903
    - tautomers, 917
  - as a protecting group for an amine, 1038, 1038f
  - proteins, 891–892
  - reactivity, 893–894
  - reduction to amines, 790, 1024
  - secondary (2°), 880, 885–886, 888, 889t, 890, 903
  - solubility of, 889t
  - stability of, 882
  - structure and bonding of, 775, 879, 880
  - synthesis of
    - from acid chlorides, 896, 1037
    - from anhydrides, 897, 1037
    - from carboxylic acids, 898f, 901–903
    - from esters, 903
  - tautomers, 917–918
  - term usage, 942
  - tertiary (3°), 880, 885–886, 896, 897, 903
    - enolate formation from, 940
- Amide tautomer, 917–918
- Amikacin, 1143
- -amine (suffix), 1013
- Amine oxide, 481
- Amines, 295t, 1010–1062
  - aromatic
    - basicity of heterocyclic, 1034, 1035t
    - nomenclature, 1014
  - basicity of, 1011, 1028–1035, 1035t
    - aromaticity effects on, 1034, 1035t
    - hybridization effects on, 1034–1035, 1035t
    - inductive effects on, 1035t
    - pK_a values, 1028, 1030–1031, 1035t
    - relative to other compounds, 1030–1035, 1035t
    - resonance effects of, 1035t
  - boiling point of, 1015t
  - classification of, 97, 1011
  - electrostatic potential plots of, 1012, 1012f
  - functional group, 96t, 97
  - general molecular formula for, 1016
  - infrared (IR) spectra of, 523
  - interesting examples, 1018–1021
    - derivatives of 2-phenylethanamine, 1019–1020, 1020f
    - histamine and antihistamines, 1018–1019
    - simple amines and alkaloids, 1018, 1018f
  - IR spectra of, 1016, 1017f
  - mass spectra of, 1016, 1017f
  - melting point of, 1015t
  - NMR spectra of, 1017, 1017f
  - nomenclature of, 1013–1015
  - as nucleophiles, 1011, 1028, 1036–1038

odors of, 1018
physical properties of, 1015, 1015t
primary (1°)
addition to aldehydes and ketones, 850–852, 852f, 853f
amino acids, 1166
from direct nucleophilic substitution, 1021–1022
Gabriel synthesis, 1022–1023
IR spectra of, 1017f
nomenclature, 1013
reactions with acid chlorides and anhydrides, 1037
reactions with aldehydes and ketones, 1037
reactions with nitrous acid, 1042
from reduction of amides, 1024
from reduction of nitriles, 919, 1024
from reduction of nitro compounds, 1023
structure of, 1011
protecting group for, 1038, 1038f
reactions of
with acid chlorides, 896, 1037
with acids, 1028–1030
with aldehydes and ketones, 1037
with anhydrides, 897–898, 1037
with carboxylic acids, 892–894, 898f
conversion to alkenes, 1038–1041
conversion to amides, 1037–1038
with esters, 903
general features, 1028
Hofmann elimination, 1038–1041
with monosaccharides, 1144
with nitrous acid, 1042–1043
secondary (2°)
addition to aldehydes and ketones, 852–854, 853f
from direct nucleophilic substitution, 1021–1022
IR spectra of, 1016, 1017f
nomenclature, 1013–1014
proline, 1166
reactions with acid chlorides and anhydrides, 1037
reactions with aldehydes and ketones, 1037
reaction with nitrous acid, 1043
from reduction of amides, 1024
structure of, 1011
separation from other organic compounds, 1029, 1029f
solubility of, 1015t
spectroscopic properties of, 1016–1017, 1017f
structure and bonding of, 1011–1013
synthesis of, 1021–1027
direct nucleophilic substitution reactions, 1021–1022
Gabriel synthesis, 1022–1023
reduction of amides, 790, 1024
reduction of nitriles, 1024
reduction of nitro compounds, 1023
reductive amination of aldehydes and ketones, 1025–1027
tertiary (3°)
from direct nucleophilic substitution, 1021–1022
IR spectra of, 1016, 1017f
nomenclature, 1013–1014
from reduction of amides, 1024
structure of, 1011

Amino acids, 119–120, 761–765, 1163–1208. *See also* Peptides; Proteins
abbreviations for, 762t, 1165f
N-acetyl, 1171–1173
acid–base behavior of, 763–764, 764f, 1166–1167, 1166f, 1167t
α-amino acids, 761, 1164, 1164f, 1167t
D-amino acids, 761, 1164, 1164f
L-amino acids, 761, 1164, 1164f
analysis of peptide composition, 1180
enantiomers of, 1164, 1164f
separation of, 1170–1173
essential, 761, 1166
functional groups in, 120
general features of, 1164, 1164f, 1166
isoelectric points, 765
naturally occurring, 1164, 1165f
peptide synthesis from, 1183–1190
pK_a values, 1166, 1167, 1167t
representative examples, 762t
resolution of
by conversion to diastereomers, 1171–1173, 1171f
kinetic, using enzymes, 1173–1174
separation of, 1170–1174
stereogenic center of, 1164, 1166
structure of, 761, 1164, 1164f, 1165f, 1166
synthesis of, 1167–1170
alkylation of a diethyl malonate derivative, 1168–1169, 1170f
enantioselective hydrogenation, 1174–1175
reaction of α-halo acids with ammonia, 1167–1168, 1170f
Strecker synthesis, 1169–1170, 1170f
zwitterion form, 761, 1166
Aminobenzene, 653. *See also* Aniline
p-Aminobenzoic acid (PABA), 640, 929, 1051–1052
Aminoglycoside antibiotics, 1143
Amino group, 96t, 120, 1014, 1033, 1038
as activating group, 711
as electron-donating group, 700–701, 702f, 710
inductive effect of, 700
nitro reduction to, 723
as ortho, para director, 708
resonance effect of, 700
6-Aminohexanoic acid, 1256
α-Amino nitrile, 1169–1170
Amino sugars, 1143
Ammonia
basicity of, 1031
boiling point, 473
in dissolving metal reductions, 465, 473
hybrid orbitals in, 39, 39f
molecular shape, 28
pK_a of, 68
reactions with
acid chlorides, 896, 1037
alkyl halides to form amines, 1021–1022
anhydrides, 897, 1037
carboxylic acids, 901
esters, 903
α-halo acids, 1167–1168

Ammonium cation, of amino acids, 764
Ammonium hydroxide, 792
Ammonium salt, 901
in amine extraction process, 1029, 1029f
enantiomers, 1012
in Hofmann elimination reaction, 1038–1041
synthesis of, 1011, 1022
Amoxicillin, 3, 58, 213
Amphetamine, 81, 85, 213, 1020, 1025
Amygdalin, 216, 844
Amylopectin, 1142
Amylose, 121, 181f, 1142
β-Amyrin, 1182
Anabolic steroids, 1178
Analgesics, 612, 746, 929, 931, 953, 962, 967, 968
Anastrozole, 916, 969
Androgens, 1178
Androsterone, 675, 1178t, 1181
-ane (suffix), 33, 137, 140, 141, 145, 147, 395
Anesthetics, 95, 115, 132, 259f, 348, 723, 735
Angiotensin, 196, 1206
Angle strain, 157–158
in epoxides, 341, 352
Angstrom (Å), 26
Angular methyl groups, 1174–1175
Anhydrides
acidity/basicity of, 882t
boiling point of, 889t
cyclic, 880, 898f, 899
derivation of term, 883
IR spectra of, 890t
melting point of, 889t
mixed, 880, 883
nomenclature, 883, 887t
physical properties of, 889t
pK_a, 882t
reactions of, 897–898, 910t
with ammonia or amines to form amides, 1037
conversion to amides, 897
conversion to carboxylic acids, 897
conversion to esters, 897
reactivity, 893–894
solubility of, 889t
stability of, 882
structure and bonding of, 879–883, 882
symmetrical, 880, 883
synthesis of
from acid chlorides, 896
dehydration of dicarboxylic acids, 890
Aniline, 648, 1037
basicity of, 1031–1032, 1033f
electrophilic aromatic substitution reaction, 691, 708, 710
electrostatic potential plot of, 702f, 1033f
nomenclature and, 654
polyhalogenation of, 710
protection of, 1038, 1038f
resonance structures, 700
structure of, 653, 1014
synthesis of
electrophilic aromatic substitution reaction, 691
by nitrobenzene reduction, 723
Anionic oxy-Cope rearrangements, 1232

Anionic polymerization, 1249–1250, 1249f
of epoxides, 1251–1252
Anions
acetate, 74–75
carbanion, 76, 76f, 79, 225, 225f
conjugated, 620
cyclopentadienyl, 662–663, 672
definition of, 8
enolate, 22
naked, 268
radical, 473
salt formation, 12
solvation by hydrogen bonding, 267
Anisole, 371
Annulation
origin of word, 996
Robinson, 996–1000
Annulenes, 660–661
18-annulene, 577
Anomeric carbon, 1120–1125, 1127, 1138–1140, 1145
Anomers, 1121–1126, 1125f
Antabuse, 490
Antarafacial bond formation, 1225, 1227t
Antarafacial rearrangement, 1229–1230, 1229t
Anthracene, 661
Anti addition, 405, 411, 411t, 413, 414, 417, 471, 473, 480, 481
Antiaromatic compounds, 658, 659, 662
Antibiotics, 2f, 3
aminoglycoside, 1143
amoxicillin, 213
chloramphenicol, 187–188
ionophores, 116–117
β-lactam, 511, 525–526, 909–910
penicillin(s), 511, 525–526, 909–910
sulfa drugs, 1051–1052, 1052f
Antibonding molecular orbital, 669–670, 670f, 671f, 672, 674
Anticancer drugs
anastrozole, 969
aromatase inhibitors, 674–676, 675f, 916
doxorubicin, 1110, 1110f
epothilone A, 1201f
illudin-S, 961
imatinib mesylate, 276
tamoxifen, 934, 954
Anti conformation, 154–156, 154f, 156f, 309f
Antidepressant, 276, 500, 691f, 699, 717, 1020
Antifreeze, 347f
Antihistamines, 652, 691f, 1019, 1030
Anti-inflammatory drugs
aspirin, 61, 80–81, 241, 295, 295f, 745–746, 1168
dexamethasone, 97
fluticasone, 255, 301
ibuprofen, 139, 170, 208, 722, 783f, 989, 1168
nabumetone, 962
naproxen, 86, 209, 953
nonsteroidal anti-inflammatory drugs (NSAIDs), 1168–1169
Antimalarial drugs, 315, 315f, 422–423, 423f, 666, 951, 1036
Antimicrobials. *See* Antibiotics
Antioxidants, 254, 597–598
Anti periplanar geometry, 322–323, 322f

Antiseptics, 263
Antiviral agents, 99, 1201f
Applied magnetic field, 537, 539, 547, 551
Aprotic solvents, polar, 268, 268f, 289, 290t, 313, 315t, 944
D-Arabinose, 1117, 1118f, 1134–1135
Arachidic acid, 1158t
Arachidonic acid, 608, 611
leukotriene synthesis by oxidation of, 380
prostaglandin synthesis from, 746
structure of, 111, 1158t
synthesis of eicosanoids from, 1167, 1167f
Arene, 650, 688f
Arene oxide, 177
Arginine
isoelectric point for, 1167t
pK_a values for, 1167t
structure of, 1165f
Aricept. *See* Donepezil
Arimidex. *See* Anastrozole
Aripiprazole, 1056
Aromatase inhibitors, 674–676, 675f, 916
Aromatic compounds, 649–685
background on, 650
bonding and antibonding orbitals, 669–670
charged, 662–663
electrophilic substitution. *See* Electrophilic aromatic substitution reactions
examples, 660–664
charged compounds, 662–664
heterocycles, 664–668, 665f
multiple rings, 661–662
single ring, 660–661
Hückel's rule, 657–660
basis of, 669–671
molecular orbitals, 669–674, 670f, 671f
nomenclature, 652–655
predicting aromaticity using the inscribed polygon method, 672–674, 673f
reactions of, 686–737
electrophilic aromatic substitution, 687–715
nucleophilic aromatic substitution, 715–718
spectroscopic properties, 655, 655f, 655t
Aromatic hydrocarbons, 93–94, 94t
Aromaticity, effect on basicity of amines, 1034, 1035t
Arrows
coupled reaction, 226t
curved, 21–22, 60, 63, 64, 83, 224, 225–227, 226t, 262, 1217
double-headed, 19, 63–64, 226t
equilibrium, 63–64, 69
half-headed, 473, 579
open, 452
use in describing organic reactions, 224, 225–227, 226t
Artemisinin, 5, 217, 422–423, 423f
Artificial sweeteners, 1140, 1140f, 1179
Arylamines
basicity of, 1031–1032, 1035t
reaction with nitrous acid, 1042–1043
Arylboranes, 1190
Aryl bromides, synthesis from aryl diazonium salts, 1044
Aryl chlorides
examples of biologically active, 691f
reactivity of, 716
synthesis from aryl diazonium salts, 1044
Aryl diazonium salt, 1043–1049
Aryl fluoride, 716
Aryl fluorides, synthesis from aryl diazonium salts, 1044
Aryl groups, 654
Aryl halides, 256, 256f, 294
coupling reactions
with alkenes, 1192–1194
with organoboranes, 1190
with organocuprates, 1186
examples of biologically active, 691f
nucleophilic aromatic substitution, 715–718
reactivity of, 695
synthesis of, 688f, 690–691
Aryl iodides, synthesis from aryl diazonium salts, 1044
Aryl ketones, reduction to alkyl benzenes, 721–722
Ascorbic acid, 90, 91, 112. *See also* Vitamin C
Asparagine
isoelectric point for, 1167t
pK_a values for, 1167t
structure of, 1165f
Aspartame, 926, 1140, 1140f, 1179
Aspartic acid, 1166, 1267
isoelectric point for, 1167t
pK_a values for, 1167t
structure of, 1165f
Aspirin, 58, 61, 80–81, 745–746, 909, 928
crossing of cell membrane, 81
derivation of word, 746
mechanism of action, 241, 746
mode of action, 1168
proton transfer reaction, 80–81
structure, 746
synthesis of, 295, 295f
Asthma, 380
Asymmetric carbon. *See* Stereogenic centers
Asymmetric reaction, 490
Asymmetric reduction, 784
Atactic polymer, 1252–1253
-ate (suffix), 742, 883, 884, 887t
Atenolol, 88
Atherosclerosis, 915, 915f
Atmosphere
carbon dioxide concentration in, 169, 169f
ozone layer destruction, 260
Atomic mass unit (amu), 8
Atomic number, 8
Atomic orbitals, number of, 1214
Atomic weight, 8
Atom mass, infrared (IR) absorption and, 518–519, 519f
Atoms
components of, 8
excited state, 37
ground state, 36–37
Atorvastatin, 1177, 1177f
Atropine, 1018, 1018f
Automated peptide synthesis, 1188–1190
Avobenzone, 34, 130, 988
Avocado, 112
Axial bond, 182
Axial hydrogens, 159–160, 159f, 163f
Aziridines, 390
Azo compound
applications of, 1050–1052
dyes, 1050–1051
sulfa drugs, 1051–1052, 1052f
formation of, 1043, 1048–1049
Azo coupling, 1048–1049
AZT (azidodeoxythymidine), 3, 679
Azulene, 678

B

Backside attack, in nucleophilic substitution reactions, 273–274, 275t, 288
Baekeland, Leo, 1261f
Baeyer, Adolf von, 138
Baeyer strain theory, 157
Bakelite, 1260–1261, 1261f
Ball-and-stick model
of acyclic alkanes, 135–136
of molecular shape, 27–29
Barbituric acid, 138
Barrier to rotation, 156
Base pairs, 649, 667–668, 1147, 1147f
Base peak, in mass spectrum, 505
Base(s)
for alkyne deprotonation, 441, 441t
for alkyne synthesis by dehydrohalogenation, 326–327
amines as, 1028–1030
Brønsted–Lowry
description of, 62–63
reactions of, 63–65
substitution reactions with alkyl halides, 261
common, 79–80, 79f
conjugate, 63–65, 67, 67t
dehydration of β-hydroxy carbonyl compounds with, 976
as electron pair donors, 62, 65, 81
in elimination reactions, 306, 306t, 325t
in E1 elimination reactions, 320, 320t, 321, 325, 325t
in E2 elimination reactions, 313, 313f, 315t, 322, 325, 325t
enolate formation with, 941–943, 942t
halogenation in, 948–949, 950f
hydrides, 79, 79f
Lewis, 81–83
nucleophile as, 261, 265, 277
nitrogen, 79, 79f
nonnucleophilic, 266
nucleoside, 1145, 1147
organometallic reagents, 793, 794–795
oxygen, 79, 79f
as proton acceptors, 62
strong, 79
used to deprotonate carboxylic acids, 749, 750t
weak as good leaving group in nucleophilic substitution reactions, 263–265
Basicity
acidity, inverse relationship with, 67, 67t
of amines, 1011, 1028–1035, 1035t
of carboxylic acid derivatives, 881–882, 882t
of carboxylic acids, 748–749
leaving group ability and, 263–265
nucleophilicity compared to, 265–266
periodic trends in, 70–72
Basketane, 147f
Beeswax, 170, 170f
Benadryl, 1030. *See also* Diphenhydramine
Bender, Myron, 932–933
Benedict's reagent, 1132–1133
Bent molecular shape
alcohols, 341
description of, 29
ethers, 341
polarity and, 49
Benzaldehyde
aldol reactions of, 979, 980, 981, 981f
electron density of, 702, 702f
resonance structures of, 701
secondary alcohol formed from, 797
structure of, 828, 830
Benzamide, structure of, 885
Benzene, 58, 93, 94t, 128, 514
as annulene, 660
background on, 650
in BTX mixture, 522
p-chlorostyrene synthesis from, 725
conjugation of, 658
degrees of unsaturation, 650, 651
electrostatic potential plot of, 652, 652f, 702f
Friedel–Crafts acylation of, 688f, 693–699, 722, 722f, 727
Friedel–Crafts alkylation of, 688f, 693–699, 722, 722f, 724–725
halogenation of, 688f, 690–691, 714
heat of hydrogenation, 656–657, 656f
hybridization and orbitals, 652
lack of addition reactivity, 650, 657, 687
metabolism of, 177
molecular orbitals of, 670–672, 671f
nitration of, 688f, 691–692, 714, 724
p-nitrobenzoic acid synthesis from, 724–725
NMR spectrum of, 547, 548f, 548t, 660
pi (π) electrons of, 651, 652, 659, 664, 687
in polymer synthesis, 1262, 1262f
resonance, 651
stability, 656–657, 659
structure of, 651–652
substitution reactions, 650, 657, 687–699. *See also* Electrophilic aromatic substitution reactions
sulfonation of, 688f, 691–692
synthesis from aryl diazonium salts, 1045
synthesis of *m*-bromoaniline from, 723–724
synthesis of bromonitrobenzene from, 714
synthesis of *o*-nitrotoluene from, 714–715
synthesis of 1,3,5-tribromobenzene from, 1045, 1045f
trisubstituted benzene synthesis from, 726–727
Benzenecarbaldehyde. *See* Benzaldehyde
Benzene derivatives
disubstituted, 712–713
electrophilic aromatic substitution reactions of, 712–713
halogenation of, 712–713

nitration of, 715
synthesis of, 714–715
NMR spectra, 560–561, 560f
nomenclature, 653–655
aromatic rings as substituents, 654
disubstituted, 653
monosubstituted, 653
polysubstituted, 654
spectroscopic properties, 655–656, 655f, 655t
substituted
activating and deactivating substituents, 703–706, 706f
directing effects of substituents, 703–710, 710f
electrophilic aromatic substitution, 703–706, 706f, 710–712
electrostatic potential plot of, 702, 702f
Friedel–Crafts acylation of, 712
Friedel–Crafts alkylation of, 711
halogenation of, 710–711
inductive effects on, 700–702
multistep synthesis of, 724–727
orientation effects on, 707–710
oxidation of, 721
reduction of, 723–724
resonance effects on, 700–702
synthesis of, 715–716, 724–727
with diazonium salts, 1045–1047
trisubstituted, synthesis of, 726–727
Benzene ring
of arylamines, 1031
NMR spectra of, 547, 548t, 560–561, 560f
Benzenesulfonic acid, 688f, 691
Benzo[*a*]pyrene, 381, 661f
Benzocaine, 723, 929
Benzoic acid, 899
acidity of, 756–757, 757f
deprotonation of, 749
extraction of, 758
separation from cyclohexanol by extraction procedure, 758–759, 759f
structure of, 741t
substituted, 756–757, 757f
synthesis of
by arene oxidation, 721
by oxidation of alkyl benzenes, 747
Benzoic anhydride, 887t
Benzonitriles, 887t, 1045
Benzophenone, structure of, 831
Benzoquinone, 632f
Benzoyl chloride, 887t, 899
Benzoyl group, 832
Benzphetamine, 1057
2-Benzylbutanoic acid, 957
Benzyl ester, 1186
Benzyl group, 654, 832
Benzylic bromination, 719–720, 725
Benzylic C–H bond, 719, 721, 747
Benzylic halides, 256, 256f, 719–720
Benzylic radical, 719, 1246
Benzyl salicylate, 1273
Benzyne, 717–718
Bergström, Sune, 746
Beryllium, octet rule exceptions in, 19
Beryllium hydride
hybrid orbitals in, 39
molecular shape of, 27
Beta (β) carbon, 306–307, 316, 323, 741
Beta (β) elimination reactions. *See also* Elimination reactions
of alcohols, 351–360
dehydrohalogenation as, 306
Zaitsev rule, 316–317
Beta (β) lactam family of antibiotics, 511, 525–526
Beta (β) lactam(s), 338, 880, 909–910
Bextra. *See* Valdecoxib
BHA (butylated hydroxy anisole), 604
BHT (butylated hydroxy toluene), 597–598
Biaryls, 1209
Bicyclic compounds, naming, A-5–6
Bicyclic product, 633
Bicyclic ring system, bridged, 633–634, 634f
Biformene, 1171
"Big Six" synthetic polymers, 1264–1265, 1265t
Bile acid, 178
Bile salt, 178
Bimatoprost, 747
Bimolecular elimination. *See* E2 elimination reactions
Bimolecular reactions
elimination reactions, 311
in nucleophilic substitution reactions, 271
organic reactions, 242
Biocatalyst, 1262, 1263, 1263f
Biodegradable polymers, 1266–1267
Biofuels, 1161
Biomolecules, 119–124, 122f, 1110. *See also* Carbohydrates; DNA; Lipids; Proteins
definition of, 119
families of, 119
mass spectra of, 514
Birch reduction, 501
2,2'-Bis(diphenylphosphino)-1,1'-binaphthyl (BINAP), 1175
Bisphenol A (BPA), 1258, 1259f, 1264, 1271
Bisphenol F, 734
Blattellaquinone, 897
Bloch, Konrad, 1176
Blood alcohol content, measurement of, 487, 487f
Blood–brain barrier, 114–115
Boat form, of cyclohexane, 161, 161f
Boc (*tert*-butoxycarbonyl protecting group), 1185, 1187
Boiling point
of alcohols, 346t
of aldehydes and ketones, 833t
of alkanes, 149, 149t
of alkenes, 399
of alkyl halides, 258t
of alkynes, 437
of amines, 1015t
of carboxylic acids, 743t
definition of, 105
of enantiomers, 205, 205t
of epoxides, 346t
of ethers, 346t
intermolecular forces and, 105–106
polarizability and, 105, 106f
of racemic mixtures, 205, 205t
surface area and, 105, 106f
Bombykol, 1191, 1191f
Bond angles
in alkynes, 435
four groups around an atom, 26–29, 29t, 44t
molecular shape and, 26–29
three groups around an atom, 26, 27, 29t, 44t
two groups around an atom, 26, 27, 29t, 44t
Bond cleavage
heterolytic, 224
homolytic, 224
in nucleophilic substitution reactions, 269–272, 277–278
Bond dissociation energy, 227–230
in alkynes, 435
for benzylic C–H bond, 719
bond length and, 228
bond strength and, 228
carbon radical stability and, 579–580
of common bonds, 228t, A-7
definition of, 227
enthalpy change calculation from, 229
limitations of, 230
periodic table trend in, 228
Bonding (shared) electrons, 13
Bonding molecular orbital, 669–674, 670f, 671f, 673f
Bond length
average for common bonds, 26t
in benzene, 651
bond dissociation energy and, 228
bond strength and, 45–46, 46t
in conjugated dienes, 622–623
percent *s*-character and, 46, 623
periodic trends in, 26
Bond polarity, 48–50, 50f
Bond(s) and bonding, 11–13. *See also specific bond types*
bond angle, 26–29, 29t, 44t
bond length, 26, 26t, 45–46, 46t
breaking and making in organic reactions, 223–227, 225f
covalent, 11, 12–13
definition of, 11
double, 15–16, 21–25, 31, 31f, 42, 42f, 45–46, 46t
energy and entropy, 235–236
ionic, 11–12
Lewis structures, 13–19
in nucleophilic substitution reactions, 269–272, 277–278
octet rule, 11
pi (π), 42–43, 42f, 43f, 44t
polar, 48–49, 50f
rotation, 41, 42
sigma (σ), 36, 38, 41–43, 42f, 43f, 44t
triple, 16, 42–43, 45, 46t
Bond strength
bond dissociation energy and, 228–229
bond length and, 45–46, 46t
infrared (IR) absorption and, 518–520, 518f, 520t
percent *s*-character and, 46
9-Borabicyclo[3.3.1]nonane (9-BBN), 420
Borane, 419–422, 420f, 784
Boron, octet rule exceptions in, 19
Boron trifluoride, 27, 39
BPA (bisphenol A), 1258, 1259f, 1264, 1271
Bradykinin, 1178
Breast cancer, 944
Brevenal, 341
Brevicomin, 876
Bridged bicyclic ring system, 633–634, 634f
Bridged halonium ion, 414–418, 418t, 444
Bridgehead carbons, A-5
Bromination, 723
of alkanes, 586–588, 588f
of alkenes, 598–601
at allylic C–H bonds, 594–596
with NBS, 594–596
product mixtures in, 595–596
selective, 594–595
substitution versus addition, 595
of benzene, 690, 714
benzylic, 719–720, 725
chlorination compared, 586–589
as endothermic reaction, 587–588
as exothermic reaction, 599–600
polybromination, 710
of substituted benzenes, 710
as test for a pi (π) bond, 413
Bromine, isotopes of, 508
m-Bromoacetophenone, 832f
m-Bromoaniline, 723–724
o-Bromoaniline, 1014
p-Bromobenzaldehyde, synthesis of, 1046–1047
Bromobenzene, 371, 690
4-Bromobutanal, 872
1-Bromobutane, 314
2-Bromobutane, 307, 591
1-Bromobut-2-ene, 627–628
3-Bromobut-1-ene, 627
o-Bromochlorobenzene, 653
1-Bromo-1-chloroethylene, NMR spectrum of, 541
Bromocyclohexane, 410
cis-4-Bromocyclohexanol, 389
trans-4-Bromocyclohexanol, 389
2-Bromocyclohexanone, 951
1-Bromocyclohexene, 1188
3-Bromohexane, 187
enantiomers of, 189f
6-Bromohexan-2-one, 876
trans-1-Bromohex-1-ene, 1187
1-Bromohex-2-ene, 300
Bromohydrins, 417
Bromomethane, 350
in nucleophilic substitution reactions, 270–272, 275f
2-Bromo-2-methylbutane, 316
1-Bromo-3-methylbut-2-ene, 595
3-Bromo-3-methylbut-1-ene, 595
cis-1-Bromo-4-methylcyclohexane, 287
1-Bromo-1-methylcyclopentane, 320
2-Bromo-2-methylpropane, 307, 314, 450
Bromonitrobenzene, 714
Bromonium ion, 595
2-Bromopentane, in an E2 elimination reaction, 1041, 1041f
1-Bromo-1-phenylpropane, 304
1-Bromopropane
NMR spectrum of, 556, 556f
synthesis of, 599
2-Bromopropane, 350
mass spectrum of, 508, 509f
NMR spectrum of, 554, 554f
synthesis of, 588, 599
2-Bromoprop-1-ene, 647
α-Bromopropiophenone, 949
(*Z*)-2-Bromostyrene, 1192, 1208
N-Bromosuccinimide (NBS), 417, 594–596
p-Bromotoluene, 653
Bronchodilator, 187, 349, 379, 379f

Brønsted–Lowry acid
carboxylic acids, 749–752
description of, 62–63
reactions of, 63–65
Brønsted–Lowry base
alkyl halide and elimination reactions, 261
description of, 62–63
in elimination reactions with alkyl halides, 306
reactions of, 63–65
Brucine, 1205
BTX mixture, 93, 522
Bufotenin, 1020
Bufo toads, 1020
Bupivacaine, 88
Bupropion, 691f, 969
Buta-1,2-diene, 459
Buta-1,3-diene, 621, 1213, 1217, 1241
conformations of, 1177
conjugation in, 613–614, 614f, 621
electrostatic potential plot for, 614, 614f
HBr addition to, 625–628, 628f
molecular orbitals of, 1215–1216, 1215f
polymerization of, 1251, 1255
resonance structures, 617, 623
sigma (σ) bond length in, 622–623
stability of, 659
ultraviolet absorption, 638
Butanal
boiling point of, 105
intermolecular forces in, 103, 104
melting point of, 107
Butan-1-amine, mass spectrum of, 1016, 1017f
Butane
alkene conversion to, 466
anti conformation, 309f
conformations of, 154–157, 154f, 156f
constitutional isomers of, 136, 137t, 140
eclipsed conformation, 309f
halogenation of, 590
molecular formula of, 137t
as propellant, 592
rotation about carbon–carbon single bonds of, 308, 309f
solubility of, 109
structure of, 136
Butanedioic acid, 742
Butane-2,3-diol, 377
Butanenitrile, synthesis from propan-1-ol, 366
Butanoic acid, 744f, 745
tert-Butanol, as polar protic solvent, 267f
Butan-1-ol
boiling point of, 105
intermolecular forces in, 103, 104
melting point of, 107
naming, 342
Butan-2-ol
(*S*)-butan-2-ol, conversion to its tosylate, 367
enantiomers of, 192–194, 783–784
infrared (IR) spectrum, 522
stereogenic center of, 188
synthesis using Grignard reaction, 800–801
Butan-2-one, 457
boiling point of, 832
infrared (IR) spectrum, 522
reduction of, 783–784
structure of, 831
synthesis of, 961
But-1-ene, 310, 334, 393t
bond dissociation energy, 643
But-2-ene
cis-but-2-ene, 393t, 399, 407, 415, 416f
from alkyne reduction, 472
carbene addition to, 1196
epoxidation of, 478
hydrogenation of, 466–467, 467f
trans-but-2-ene, 393t, 399, 415, 416f, 417
from but-2-yne reduction, 473
carbene addition to, 1196
epoxidation of, 478
hydrogenation of, 466, 467f
stereoisomers of, 308–310, 309f, 334
tert-Butoxide, nucleophilicity of, 266
tert-Butoxycarbonyl protecting group (Boc), 1185, 1187
tert-Butoxy group, 344
Butter, 403, 470
Butter yellow dye, 1048
(*R*)-*sec*-Butylamine, 218
Butylbenzene, 726–727
tert-Butylbenzene, 653
Butylcyclohexane, 146f
tert-Butylcyclohexane, 163f
tert-Butyl cyclohexanecarboxylate, 884
cis-4-*tert*-Butylcyclohexanol, 825
4-*tert*-Butylcyclohexanone, 825
tert-Butylcyclopentane, 145
1-*tert*-Butylcyclopentene, 389
tert-Butyldimethylsilyl ether (TBDMS), 803–804, 818
sec-Butyl ethyl sulfide, 373
Butyl group, 140
sec-butyl group, 140
tert-butyl group, 140, 163
tert-Butyl hydroperoxide, 475, 491
tert-Butyl iodide, 257
Butyllithium, 79, 846, 943
Butylmagnesium chloride, 812
1-*sec*-Butyl-3-methylcyclohexane, 146f
sec-Butyl methyl ether, 343
tert-Butyl methyl ether (MTBE), 109
formation of, 413
as fuel additive, 413, 538
NMR spectrum of, 538, 540, 543
5-*tert*-Butyl-3-methylnonane, 143
But-1-yne, 459, 462
addition reactions of, 440, 441f
hydrohalogenation of, 441f, 442, 443
But-2-yne, 459, 462, 473
hydration of, 446
Butyric acid, structure of, 741t
γ-Butyrolactone, 927

C

Cadaverine, 1018
Caffeine, 58, 115, 129, 570
Cahn–Ingold–Prelog system, 191
Calcitriol, 622, 622f
Calorie (cal), 45
Camphene, 1184
Camphor, 1170
10-Camphorsulfonic acid, 1204
Cancer, 381, 674–676, 675f. *See also* Anticancer drugs; Polycyclic aromatic hydrocarbons
Candlenuts, 391
Capnellene, 451, 451f
Caproic acid, 741, 741t
ε-Caprolactam, 1256, 1266
ε-Caprolactone, 1273
Capsaicin, 4, 56, 525, 570, 970
Caraway, odor of, 210
-carbaldehyde suffix, 829
Carbamate, 930, 1185, 1257
Carbanion(s), 225, 225f
addition to aldehydes and ketones, 842
defined, 76
electrostatic potential plots, 76f
formation in alkyne reduction, 473–474
nucleophilic addition with, 777, 842
nucleophilic substitution with, 778
as strong bases, 79
vinyl, 474
Carbenes
intermediates in cyclopropane synthesis, 1194–1197
metal–carbene intermediates in metathesis, 1200
preparation of dihalocarbenes, 1195–1197
structure of, 1194, 1195
Carbenoid, 1198
Carbidopa, 50, 51
Carbinolamines, 840, 841f, 850, 851, 853, 1037
Carbocation intermediates, 599
Carbocation rearrangements
conversions of alcohols to alkyl halides, 363
in dehydration of alcohols, 356–358
Friedel–Crafts alkylation, 696–697
in hydration reactions, 412
1,2-shifts, 356–358
1,2-hydride shift, 696–697
1,2-methyl shift, 357
in S_N1 reactions, 363
Carbocation(s), 22, 225, 225f
achiral, 411
allylic, 626
in biological reactions, 616
as conjugated system, 613, 614–615
electrostatic potential plot, 615, 615f
resonance of, 615–617
stability, 615
classification, 281
in the E1 mechanism, 318–321, 325t
in electrophilic addition reactions, 407–412
in electrophilic aromatic substitution reactions, 688–690, 705–706, 706f
electrostatic potential maps, 282, 282f
energy diagram for formation of, 286f
formation in Lewis acid–base reactions, 83
Hammond postulate, 408, 409f
as Lewis acid, 407
localized versus delocalized, 615f
in nucleophilic substitution reactions, 270
planar, 278–280, 288, 411, 1127–1128
primary (1°), 281–283
as reactive intermediate, 358
resonance-stabilized, 443–444, 446, 511, 615, 615f, 625–626, 628, 688, 690, 705–706, 716, 937
secondary (2°), 281–283, 694, 696, 697
in S_N1 reactions, 321
in S_N1 substitution reactions, 277–286, 286f, 288–289, 292
tertiary (3°), 281–283, 694, 696, 697
trigonal planar, 278–279
vinyl, 294, 442–443, 446
Carbocation stability, 281–283
of allylic carbocations, 615
conjugation and, 615
in electrophilic aromatic substitution reactions, 696, 705–706, 706f
Hammond postulate and, 283–286
hyperconjugation and, 282–283
inductive effects and, 281
1,2-shifts and, 356
Carbohydrates, 120–121, 121f, 865–866, 1109–1154
amino sugars, 1143
cellulose, 181–182, 181f, 182f, 1110f, 1142
disaccharides, 1137–1140, 1139*f*, 1140*f*
energy in, 1110–1111
examples of, 1110, 1110f
glycosides, 1127–1130
N-glycosides, 1144–1147
molecular formula of, 1110
monosaccharides, 1111–1137. *See also* Monosaccharides
nomenclature, 1111–1112, 1117
polysaccharides, 1141–1143
starch, 181–182, 181f, 182f, 347, 347f, 1142
Carbon
alpha (α), 306–307, 741, 744, 761
of aldehydes and ketones, 830, 835, 838–839
substitution reactions of carbonyl compounds at the α carbon, 934–971
beta (β), 306–307, 316, 322, 323, 741
bonding of, 12, 13f, 16, 17, 18t, 36–38
charged atom, 35
chiral. *See* Stereogenic centers
classification of atoms, 94–95
common nucleophiles, 269t
delta (Δ), 741
electron-deficient, 73, 92, 95, 117–118, 225, 260–261, 261f
electronegativity value, 793
electrophilic, 261
excited state, 37
formal charge of, 18t
gamma (γ), 741
ground state, 36–37
isotopes of, 8, 512t
mass of, 512t
omega, 741
periodic table entry, 9
radicals, 579–580, 580f
tetravalence of, 4, 28, 30, 32–34
valence electrons of, 36
Carbonate, 1258
Carbon backbone (skeleton), 92, 109
Carbon–carbon bonds
bond length and bond strength, 45–46, 46t
in condensed structures, 31, 31f
double. *See also* Double bonds
in alkenes, 307–309. *See also* Alkenes
properties of, 393t
radical reactions, 598–601

restricted rotation around, 308–309, 309f
double bond, 16
NMR spectra of, 547–548, 548f, 548t
nonpolar nature of, 48
reactions forming, 1185–1211, A-8
coupling reactions of organocuprates, 1186–1188
cyclopropane synthesis, 1194–1197
Heck reaction, 1186, 1192–1194
metathesis, 1198–1203, 1199f, 1201f
Simmons–Smith reaction, 1197–1198
Suzuki reaction, 1186, 1188–1192, 1191f
rotation, 41, 42
rotation around, 150–151
Carbon dioxide
atmospheric concentration of, 169, 169f
electrostatic potential plot for, 50f
global climate change, role in, 169
as nonpolar molecule, 49
reaction of Grignard reagents with, 808
release in combustion reactions, 168–169
Carbon–halogen bond, polar, 260–261, 261f
Carbon–hydrogen bonds
acidity of, 71
benzylic, 719, 721
bond length and bond strength, 46
chemical shifts, 546
halogenation of, 593–596
nonpolar nature of, 48
radical reaction with, 581
Carbon monoxide, 1199
Carbon NMR, 536, 564–567
of aldehydes, 835, 835f
of amines, 1017
basis of, 564
of benzene derivatives, 655, 655f, 655t
of carboxylic acid derivatives, 890
of carboxylic acids, 744, 744f
characteristic absorptions, A-11
chemical shifts in, 564, 566, 566t, 567f
examples of spectra, 567f
of ketones, 835
of nitriles, 890
number of signals in, 565–566
position of signals in, 566–567
Carbon–oxygen bond(s)
carbonyl group, 98, 98t
condensed structures containing a double, 31, 31f
in functional groups, 98–100, 98t
polar nature of, 48
Carbon skeleton (backbone), 92, 109
Carbon tetrachloride, 259
as solvent, 108
-carbonyl chloride (suffix), 883, 887t
Carbonyl compounds. *See also specific classes*
classes of, 775
condensation reactions, 972–1009
conversion of ozonides to, 482
electron deficiency of, 776
electrostatic potential map of, 776, 776f
formation by oxidation of alcohols, 484–486
β-hydroxy
synthesis of, in aldol reactions, 973–985
useful transformations of, 989
leaving groups of, 775, 777, 778–779
pK_a, 939, 939t, 940t, 942
reactions at the α carbon, 934–971
acetoacetic ester synthesis, 959–962
enolate alkylation, 947, 952–955
halogenation, 947–951, 950f
malonic ester synthesis, 955–959
racemization, 946, 946f
reactions of
general, 776–779
nucleophilic addition, 777
nucleophilic substitution, 777, 778–779
with organometallic reagents, 795–799, 802–804, 803f
oxidation, 780, 792
reduction, 779–791
reactivity, 828
reduction of, 779–791
biological, 786–787
carboxylic acids and derivatives, 787–791
by catalytic hydrogenation, 782
enantioselective, 784–787, 785f
with metal hydride reagents, 781–783, 783f, 791t
stereochemistry of, 783–784
synthesis by Cope rearrangement, 1231
synthesis by hydroboration–oxidation of alkynes, 447–449
tautomers, 917–918
α,β-unsaturated compounds, 809–811
unsymmetrical, enolates of, 944–945
Carbonyl condensation reactions, 972–1109
aldol reactions, 973–985. *See also* Aldol reactions
biological, 991–994, 1003–1004
Claisen reaction, 986–989, 987f
Dieckmann reaction, 990
Michael reaction, 994–996, 996f
Robinson annulation, 996–1000
Carbonyl groups, 445, 775
conjugated, 834, 834f
as electron-withdrawing group, 632
features of, 881
formation by oxidative cleavage of alkenes, 476f, 482–483
as functional group, 98, 98t
hydrogenation of, 469
IR absorption of, 744, 744f, 833–834, 833f, 834f
of monosaccharides, 1111
nucleophilic addition to, 838–840, 841f, 856
reactions of, 1131–1137
reactivity of, 98
stabilization by alkyl groups, 855
tautomers of, 936–937
Carboxy, derivation of term, 739
Carboxy group, 98t, 120, 739
of benzene derivatives, 721, 724–725
in prostaglandins, 170
Carboxylate anion
of amino acids, 764
conversion of, to acyl phosphates, 911–912
formation of, 749
metal salt of, 742, 742f
synthesis of
hydrolysis of amides, 908
hydrolysis of esters, 903–905
hydrolysis of nitriles, 917–918
Carboxylation, 808
-carboxylic acid (suffix), 883
Carboxylic acid derivatives. *See also* Acid chlorides; Amides; Anhydrides; Esters; Thioesters
acidity/basicity of, 881–882, 882t
boiling point of, 889t
IR spectra of, 889–890, 890t
leaving groups of, 880, 882t, 892–893
melting point of, 889t
NMR spectra of, 890
nomenclature of, 883–888, 887t
nucleophilic acyl substitution reactions of, 778
mechanism of, 892–893
relative reactivity, 893–894, 910
summary of, 910, 910t
physical properties of, 888–889, 889t
reaction of organometallic reagents with, 795, 804–807
reactivity with nucleophiles, 778
reduction of, 780, 787–791
resonance structures of, 881
solubility of, 889t
stability of, 882
structure and bonding of, 775, 879–883
Carboxylic acid(s), 738–773
acidity of, 748, 749–754, 750, 882t, 898
aliphatic, inductive effects in, 754–755
amino acids, 761–765
aspirin as, 745–746
basicity of, 748–749
benzene derivatives, 721
boiling point of, 743t
deprotonation of, 749–750, 750t
diacids, 742, 955–957
as dimers, 742–743, 743f
dipole–dipole interactions of, 742–743
electrostatic potential plot, 739, 739f
functional group in, 98t
hydrogen bonding, 742, 743f
interesting examples, 745
intermolecular forces, 742, 743f, 743t
IR spectra of, 744, 744f
leaving group, 778
melting point of, 743t
NMR spectra of, 744, 744f
nomenclature of, 739–742, 742f
common names, 740–741, 741t
IUPAC system, 740–742, 742f
nucleophilic acyl substitution reactions, 898–903, 898f, 910t
conversion to acid chlorides, 898f, 899
conversion to amides, 898f, 901–903
conversion to cyclic anhydrides, 898f, 899
conversion to esters, 898f, 900
nucleophilic substitution reactions, 778
oxidation of aldehydes to, 1132
physical properties, 742–743, 743t, 889t
pK_a of, 882t
protonation of, 748
reactions of
ester synthesis, 898f, 900
general features, 748–749
nucleophilic acyl substitution reactions, 898–903, 898f, 910t
reduction, 780, 788, 790
reactivity, 893–894
resonance stabilization of, 750–751, 751f
salts of, 746
separation from other organic compounds, 758–759, 759f
solubility of, 743t
spectroscopic properties, 743–744, 744f
stability of, 882
structure and bonding of, 739, 775, 879, 882
substituted benzoic acids, 756–757, 757f
synthesis of, 747–748
conversion of acid chlorides, 896, 897
conversion of anhydrides, 897
hydrolysis of amides, 908
hydrolysis of esters, 903–905
hydrolysis of nitriles, 917–918
malonic ester synthesis, 955–959
oxidation of aldehydes, 780, 792
oxidation of alkyl benzenes, 747
oxidation of alkynes, 748
oxidation of primary (1°) alcohols, 476f, 485, 486–487, 747
oxidative cleavage of alkynes, 476f, 484
reaction of organometallic reagents with carbon dioxide, 808
Carboxypeptidase, 1182, 1182t
Carcinogen, 381
Cardinol, 1264
Carotatoxin, 129
β-Carotene, 111, 400f, 847, 847f, 1182
Carroll rearrangement, 1240
Carson, Rachel, 260
Carvone, 210, 1182
(*R*)-Carvone, 684
Caryophyllene, 333
Cassava, 844
Catalysts, 244–245. *See also specific catalysts*
acids as, 244–245
biocatalyst, 1262, 1263, 1263f
chiral, 1174–1175
definition of, 244
energy of activation lowered by, 244, 245f
enzymes, 245–246, 246f
homogeneous, 1253
hydrogenation, 466
Lewis acid, 362
Lindlar, 472, 474f
metal, 244–245, 466
palladium, 244, 782. *See also* Palladium-catalyzed reactions
Ziegler–Natta, 1253, 1254
Catalytic amount, 244
Catalytic hydrogenation, 466–468
of aldehydes and ketones, 782
Catecholborane, 1190, 1192

Cationic polymerization, 1248–1249, 1249f
Cation(s)
carbocation. *See* Carbocation(s)
counterions, 62
cyclopentadienyl, 662–663
definition of, 8
radical, 505, 509
salt formation, 12
solvation by ion–dipole interactions, 267–268, 288
spectator ions, 62
tropylium, 663–664, 672
Cation π stacking, 674
CBS reagent, 784–786, 785f
Cedrol, 1170f
Celecoxib (Celebrex), 872, 1168–1169
Celery ketone, 210
Cell membrane
cholesterol in, 171, 171f, 1176
functions, 114
phospholipids in, 1162–1164
structure of, 114–115, 132
transport across, 116–117, 117f
Cellobiose, 1139
Cells, 114
Cellulose
hydrolysis of, 181f, 182, 1141
solubility of, 183
structure of, 181–183, 181f, 182f, 1110f, 1141
Cellulose acetate, 1141
Cephalin, 1162
Cetylpyridinium chloride (CPC), 263
CFCs (chlorofluorocarbons), 260, 592, 593f
Chain branching, in polymerization, 1246–1247
Chain-growth polymerization, 1244–1251
anionic, 1249–1250, 1249f, 1251–1252
cationic, 1248–1249, 1249f
chain branching in, 1246–1247
copolymers, 1251
of epoxides, 1251–1252
ionic, 1248–1250, 1249f
radical, 1245–1246, 1246f
Chain-growth polymers, 1243, 1244–1251
Chain mechanism, 583, 592, 601
Chain termination, 1245–1246
Chair conformation
of cyclohexanes, 158–167, 159f
of glucose, 1125
Charge
Brønsted–Lowry acids and bases, 62, 62f, 65
on carbon atom, 35
dipole, 48, 49, 50f
formal, 16–18, 18t
partial, 25, 48
in resonance hybrid, 25
Chauvin, Yves, 1198
Chemical shift
in ^{1}H NMR
equation for, 539
predicting values of, 545–546
protons on benzene rings, 547, 548t
protons on carbon–carbon double bonds, 547, 548t
protons on carbon–carbon triple bonds, 548, 548t
scale, 538
values for common proton types, 546t
in ^{13}C NMR, 564, 566, 566t, 567f
Chiral catalysts, 1174–1175
Chiral drugs, 208–209
Chirality, 184–186
basic principles of, 186f
of naturally occurring objects, 186
in substituted cycloalkanes, 202
test for, 184
Chirality center, 185. *See also* Stereogenic centers
Chiral molecules, 184–186, 186f
definition of, 184, 185
halogenation of, 590t, 591
optical activity of, 204
Chiral reagent, 1173, 1174
Chiral reducing agent, 784
Chitin, 1110, 1143
Chloral, 854, 855
Chloral hydrate, 854, 855
Chloramphenicol, 187–188
Chlorination
of alkanes, 582–591, 585f, 589f
of benzene, 691, 725
bromination compared to, 586–589
of ethane, 584–585, 585f
as exothermic reaction, 585, 585f, 588–589
mechanism of, 583–584
Chlorine
isotopes of, 508
mass spectra of alkyl chlorides, 508, 509f
2-Chloroacetic acid, 754
Chlorobenzene, 653, 690, 717
2-Chlorobuta-1,3-diene, 1255
1-Chlorobutane, synthesis of, 590
2-Chlorobutane
NMR spectrum of, 542
synthesis of, 590
Chlorobutanoic acid, 755
Chlorocyclohexane, chair conformation of, 323–324, 323f
Chlorocyclopentane, 583–584
Chlorocyclopropane, NMR spectrum of, 541
Chloroethane, 57
NMR spectrum of, 565
Chloroethylene, NMR spectrum of, 541
4-Chloro-1-ethyl-2-propylbenzene, 654
Chlorofluorocarbons (CFCs), 3, 260, 592, 593f
Chloroform, 259
dichlorocarbene preparation from, 1195
Chlorohydrin, 416, 418f
Chloromethane, 259f
1-Chloro-2-methylbutane, 582
1-Chloro-3-methylbutane, 582, 593
2-Chloro-2-methylbutane, 582, 593
2-Chloro-3-methylbutane, 582, 593
1-Chloro-2-methylcyclohexane
cis-1-chloro-2-methylcyclohexane, 323
trans-1-chloro-2-methylcyclohexane, 323
1-Chloro-1-methylcyclopropane, 335
2-Chloro-5-methylheptane, 257
1-Chloro-2-methylpropane, 258f
p-Chloronitrobenzene, 715
1-Chloropentane, 540
2-Chloropentane, 540
3-Chloropentane, 540
meta-Chloroperoxybenzoic acid, 476–479
m-Chlorophenol, 1046
2-Chloropropanal, 830f
1-Chloropropane, 408, 589, 722
2-Chloropropane, 408, 589
mass spectrum of, 508, 509f
2-Chloropropene, 442
3-Chloropropenoic acid, NMR spectra of, 557, 557f
α-Chloropropionaldehyde, 830f
2-Chloropyridine, 733
Chloroquine, 1036
p-Chlorostyrene, 725
N-Chlorosuccinimide, 461
p-Chlorotoluene, 717–718
1-Chloro-1,3,3-trimethylcyclohexane, 410
Chlorpheniramine, 691f
Cholesterol, 637, 915, 915f, 1155, 1176–1177
biosynthesis of, 1176–1177, 1176f
in cell membranes, 171, 171f, 1176
cholesterol-lowering drugs, 1177, 1177f
solubility of, 108, 109, 170
specific rotation of, 207
structure of, 170, 170f, 191, 1176, 1176f
synthesis stimulation by saturated fats, 403
trans fats and, 471
Cholesteryl esters, 914–915, 915f
Cholic acid, 178
Choline, 914
Chondrocole A, 260
Chromate ester, 485–486
Chromium
anionic oxidant of, 488
in blood alcohol screening, 487, 487f
in oxidizing agents, 476, 485–488, 792
Chromium(VI) oxide, 476
Chromosome, 122
Chrysanthemic acid, 566
Churchane, 147f
Churchill, Winston, 260
Chymotrypsin, 1182, 1182t
Ciguatoxin (CTX3C), 774, 789, 789f
Cimetidine, 1019
Cinnamaldehyde, 836f, 979
Cinnamoylcocaine, 926
Cis (prefix), 396–397
Cis-fused, 633
Cis geometry
alkenes, 393, 393t, 396–397, 411
cycloalkenes, 393
Cis isomers
of but-2-ene, 309–310, 309f
of cycloalkanes, 164–167
definition of, 164, 309
of disubstituted cycloalkanes, 201
stability, 311
Cis protons, 557, 557f
Citalopram, 214
Citral, 1170f
Citramalate, 992
Citric acid, 53
Citric acid cycle, 914, 991
Citronellal, 836f
Citronellol, 534
Claisen condensation, 972
Claisen reactions, 986–989, 987f
biological, 993–994
crossed, 987–989
intramolecular, 990
Claisen rearrangement, 1231, 1232–1233
Clavulanic acid, 428
Clemmensen reduction, 721
Clopidogrel, 196, 967, 1204
^{13}C NMR. *See* Carbon NMR
Cocaine, 87, 131, 891, 926, 1018, 1020
Cocoa butter, 972, 1161
Coconut oil, 403, 1159, 1161
Codeine, 769, 1021
Coenzyme, 489
NADH, 489, 786
Coenzyme A, 118–119, 913
Coffee, decaffeinated, 259f
Collagen, 91, 1164
structure of, 1198, 1198f
Column, in the periodic table, 9
Combustion
of acetylene, 438
of alkanes, 168–169
definition of, 168
of triacylglycerols, 1161
Common names
of alcohols, 343
of aldehydes, 830, 830f
of alkanes, 147, 147f
of alkenes, 398, 398f
of alkyl halides, 257
of amines, 1013
of benzene derivatives, 652–653
of carbohydrates, 1111, 1117
of carboxylic acid derivatives, 883
of carboxylic acids, 740–741, 741t
of epoxides, 345
of ethers, 343
of ketones, 831, 832f
of nitriles, 886, 886f
Compounds
examples of, 11
formation of, 11
ionic, 11–12
Concentration, reaction rate and, 241–243
Concerted reaction, 223, 312
Condensation polymer. *See* Step-growth polymers
Condensation reactions
aldol. *See* Aldol reactions
carbonyl, 972–1109
definition of, 973, 976
Condensed structures, 30–32, 31f
Configuration
inversion of
in alkyl tosylate reactions, 368, 369
retention of
in chlorination, 591
in conversion of alcohols to alkyl tosylates, 367, 369
in oxidation of alkylboranes, 421
in S_N2 reactions, 273–274, 274f, 275t, 288
of stereoisomers, 183
Conformations
of acyclic alkanes, 150–157
butane, 154–157, 154f, 156f
ethane, 150–153, 152f
anti, 154–156, 154f, 156f
of cycloalkanes, 157–167
disubstituted, 164–167
monosubstituted, 162–164
of cyclohexane(s), 158–167
boat, 161, 161f
chair, 158–167, 159f
disubstituted cyclohexane, 164–167, 166f
monosubstituted cyclohexane, 162–164, 163f
ring-flipping, 161, 161f

definition of, 150
eclipsed, 150–157, 152f, 153f, 154f, 156f, 156t, 309, 322f
free energy change and, 232–233
gauche, 154–155, 154f, 156f
staggered, 150–157, 152f, 153f, 154f, 156f, 322, 322f
Congo red, 1051
Coniine, 1018, 1018f
Conivaptan, 877
Conjugate acid, 63–65, 264–265, 264t
Conjugate addition, 625, 995, 997
Conjugate base, 63–65, 67, 67t
Conjugated dienes, 621–640
addition reactions
Diels–Alder reaction, 629–638, 630f
electrophilic, 624–626
kinetic versus thermodynamic product, 626–629
carbon–carbon bond length in, 622–623
conformation of
s-cis, 621
s-trans, 621
electrostatic potential plot for, 614, 614f
features of, 622
hydrogenation of, 623–624
interesting examples of, 622
stability of, 623–624, 624f, 662
stereoisomers of, 621
synthesis in Suzuki reaction, 1192
ultraviolet (UV) light absorption, 638–640, 639f, 647
Conjugated double bond, 976
Conjugated proteins, 1198
Conjugated systems
allylic carbocation, 613, 614–616
1,3-dienes, 613–614, 614f
Conjugation, 613–648
definition of, 613
β-dicarbonyl compounds, 936
effect on carbonyl group IR absorption, 834, 834f
Hückel's rule and, 658
Conrotatory, 1219–1223, 1223t
Constitutional isomers
of alkanes, 136–138, 137t
of alkenes, 353
definition of, 19
of dipeptides, 1176
in elimination reactions, 307
features of, 183, 208
ketone and aldehyde, 448
of monosaccharides, 1111
stereoisomers compared, 183f
tautomers, 445–447
Contraceptives, 2f. *See also* Oral contraceptives
Cope rearrangements, 1231–1232
Copolymers, 1251, 1256
Copper
electronegativity value of, 793
organocuprate reagents, 793, 794, 1186–1188
Copper(I) cyanide, 1045
Corey, E. J., 380, 878, 928
Corn
ethanol production from, 347, 347f
use in green polymer synthesis, 1263, 1263f
Coronary artery disease, 170, 196, 403
Cortisol, 340, 1179
Cortisone, 836, 1005, 1179
Diels–Alder reactions, 637
Cotton, 1141
Coumarin, 1006
Counterions, 62
Coupled reaction arrows, 226t
Coupled reactions, in metabolism, 233–235, 234f
Coupling constant, 551, 557
Coupling reactions
alkenes with organic halides (Heck reaction), 1186, 1192–1194
of aryl diazonium salts, 1048–1049
of organocuprate reagents, 1186–1188
Covalent bonding
description of, 11, 12–13
in Lewis structures, 13–16
Covalent molecules
boiling points of, 105
intermolecular forces, 101–104, 102f, 103t
melting points of, 107
solubility of, 110t
COX-1, 1167f, 1168
COX-2, 1167f, 1168
CPC (cetylpyridinium chloride), 263
Cracking, 399
Crafts, James, 688
Creatine, 54
p-Cresol, 736, 1273
Crestor, 1002
Crocin, 1182
Crossed aldol reactions, 979–981
synthetically useful reactions, 979–981, 981f
with two different aldehydes, both having α H atoms, 979
useful transformations of aldol products, 989
Crossed Claisen reactions, 987–989
Cross formula, 1112
Crown ethers, 116, 348–349, 348f
Crude oil, 147–148, 148f, 150
Crutzen, Paul, 592
Crystallites, 1260
CTX3C (ciguatoxin), 774, 789, 789f
Cubane, 147f, 158
Cucumber aldehyde, 45, 832
Curcumin, 683
Curved arrow notation, 21–22, 60, 63, 64, 83, 224, 225–227, 226t, 262, 1217
Cyanide
addition to aldehydes and ketones, 843–844
addition to imines, 1169
α-Cyano carbonyl compounds, 981
2-Cyanocyclohexanecarboxylic acid, 886f
Cyano group, 844
of nitriles, 880, 886
Cyanohydrins
in Kiliani–Fischer synthesis, 1134, 1136
naturally occurring derivatives, 844
synthesis by cyanide addition to aldehydes and ketones, 843–844
in Wohl degradation, 1134, 1135
Cyclic anhydrides, 880, 898f, 899
Cyclic compounds
entropy decrease with formation of, 235
stereogenic centers in, 189–191
Cyclization, of hydroxy aldehydes, 863, 865–866
Cyclo- (prefix), 138, 144
Cycloaddition reactions, 1223–1227. *See also* Diels–Alder reaction
[2 + 2] cycloadditions, 1224, 1226–1227
[4 + 2] cycloadditions, 1224, 1225–1226
description of, 1213
orbital symmetry and, 1224–1225
Woodward–Hoffmann rules, 1227, 1227t
Cycloalkanes
conformations of, 157–167
disubstituted, 164–167
monosubstituted, 162–164
disubstituted, 201–202
molecular formula, 135, 138, 393
naming, 144–146, 146f
NMR spectra, 541–542
stereoisomers, 164–167
structure of, 135, 138, 158f
substituted, 162–167, 163f, 166f
Cycloalkenes, 392
cis and trans geometry of, 393
nomenclature, 396, 396f
Cycloalkynes, 436
Cyclobutadiene, 658–659, 669, 673–674
Cyclobutane
conformation, 157
structure of, 138, 158f
Cyclobutene, 1213, 1217, 1218, 1222, 1224
1-Cyclobutylhexane, 146f
Cyclodecane, structure of, 158f
α-Cyclogeraniol, 433
Cycloheptane, structure of, 158f
Cyclohepta-1,3,5-triene, 658, 663
Cycloheptatrienone, 1224
Cyclohexa-1,3-diene, 658, 1213, 1217
heat of hydrogenation, 656
stability of, 659
ultraviolet absorption, 638
Cyclohexadienone, 1232
Cyclohexanamine, 648, 1013
ammonium salt of, 1029, 1029f
in Hofmann elimination reaction, 1039
separation from cyclohexanol, 1029, 1029f
Cyclohexanecarbaldehyde, 828
Cyclohexanecarbonyl chloride, structure of, 883
Cyclohexane-1,2-diols, 480
cis-cyclohexane-1,2-diol, 481
Cyclohexane-1,4-diol, 1271
Cyclohexane-1,3-dione, 770
Cyclohexane(s)
conformations of, 158–167
boat, 161, 161f
chair, 158–167, 159f, 323–324, 323f
disubstituted cyclohexane, 164–167, 166f
free energy change and, 232–233
monosubstituted cyclohexane, 162–164, 163f
ring-flipping, 161, 161f, 166
conversion to cyclohexene, 589
E2 elimination in, 323–325
formation from cyclohexene, 244–245
in mango, 135
NMR spectrum, 560
skeletal structures, 32
structure of, 135, 138, 145
Cyclohexanol, 531, 758, 872
cyclohexane formation from, 426
dehydration of, 359, 425
1,2-dibromocyclohexane synthesis from, 425–426
separation from benzoic acid by extraction procedure, 758–759, 759f
separation of cyclohexanamine from, 1029, 1029f
synthesis from cyclohexene, 412
Cyclohexanone, 531, 534, 813
aldol reactions of, 978
alkylation of, 953
conversion to an alkene, 849
enolate formation from, 939
halogenation of, 951
nucleophilic addition reactions, 840, 841f
structure, 606
tertiary alcohol formed from, 797
Cyclohexene, 697
addition reactions of, 405f, 410, 412, 413
bromination of, 594
1,2-dibromocyclohexane synthesis from, 425
dihydroxylation of, 480
heat of hydrogenation, 656
as Lewis base, 83
reaction with hydrogen, 244–245
stereoisomers, 632
synthesis
cyclohexane conversion, 589
from cyclohexanol, 426
by dehydration of cyclohexanol, 359
as synthetic intermediate, 426
Cyclohex-2-en-1-ol, 389, 531
Cyclohex-3-en-1-ol, 389
Cyclohex-2-enone, 531, 534, 577, 875
formation in Robinson annulation, 996
reduction of, 782
synthesis of, 951
Cyclohexyl acetate, 925
Cyclohexylamine. *See* Cyclohexanamine
Cyclononatetraenyl anion, 680
Cyclononatetraenyl cation, 680
Cyclononatetraenyl radical, 680
Cycloocta-1,5-diene, 1231
Cyclooctane, 209, 209f
Cyclooctatetraene, 658, 680
NMR spectrum of, 660
Cyclooctene, cis and trans isomers of, 393
Cyclooctyne, 436
Cyclooxygenase, 746, 1167, 1167f, 1168
Cyclopentadiene, 633–634, 636, 663
Cyclopentadienyl anion, 662–663, 672
Cyclopentadienyl cation, 662
Cyclopentadienyl radical, 662
Cyclopentane, 583–584
structure of, 138, 158f
Cyclopentanecarboxylic acid, 958
cis-Cyclopentane-1,2-diol, 481
trans-Cyclopentane-1,2-diol, 343
Cyclopentanone, 872
Cyclopentene, chlorination of, 414
Cyclopropane(s)
conformation, 157
dihalo to dialkyl conversion, 1197
NMR spectrum of, 541
structure of, 138

Cyclopropane(s) (*Continued*)
synthesis of
in Simmons–Smith reaction, 1197–1198
using carbene intermediates, 1194–1197
Cyclopropenone, 874
Cyclopropenyl radical, 679
Cysteine
abbreviation for, 762t
functional groups, 97
isoelectric point for, 1167t
in peptide formation, 1176
pK_a values for, 1167t
structure of, 1165f
Cytidine, 1145
Cytidine monophosphate, 1145
Cytosine, 649, 667, 1145, 1146f

D

2,4-D (2,4-dichlorophenoxyacetic acid herbicide), 691f
Darunavir, 877
Darvon, 127. *See also* Propoxyphene
DBN (1,5-diazabicyclo[4.3.0]non-5-ene), 313, 313f, 328
DBU (1,8-diazabicyclo[5.4.0]undec-7-ene), 89, 313, 328
DCC. *See* Dicyclohexylcarbodiimide
d (dextrorotatory) compounds, 205
DDE (dichlorodiphenyldichloroethylene), 305, 307
DDT (dichlorodiphenyltrichloroethane), 4, 241, 260
biological effects of, 305
DDE formed from, 305
degradation of, 307
Dean–Stark trap, 858, 858f
(2*Z*,4*Z*)-2,4-Decadienoic acid, 1262–1263
Decalin, 178, 499, 1174–1175
Decamethrin, 1195
Decane, 134, 144
constitutional isomers of, 137t
molecular formula of, 137t
Decarboxylation, 1168
in acetoacetic ester synthesis, 959–960
of β-acids, 956
of β-keto acids, 956
in malonic ester synthesis, 956–958
Decongestants, 1030, 1060
Dehydrating agent, 902
Dehydration reactions
of alcohols to alkenes, 353–360, 403
alkene synthesis in, 353–360
carbocation rearrangements in, 356–358
E1 mechanism in, 354–355
E2 mechanism in, 355, 359
enthalpy change in, 355, 356f
Le Châtelier's principle in, 355
in phosphorus oxychloride and pyridine, 359, 360f
in strong acid, 353–358
of aldol product, 976–977
defined, 353
dicarboxylic acid conversion to cyclic anhydrides, 899
as β-elimination reaction, 353
fragmentation pattern in mass spectrum, 511
7-Dehydrocholesterol, 1222
Dehydrohalogenation
alkene synthesis by, 306
of alkyl halides with strong base, 403
alkyne synthesis by, 326–327, 327f
bases commonly used in, 306, 306t
drawing products of, 307
E1 mechanism of, 318
E2 mechanism of, 311, 317
16,17-Dehydroprogesterone, 1183
Delocalization, resonance, 75
Delta (δ) carbon, 741
Delta (δ) scale, 538, 539
Dementia, 983
Demerol. *See* Meperidine
Dendrobates histrionicus (poison dart frog), 439, 439f
Deoxy (prefix), 1144
2-Deoxyadenosine, 1145
Deoxyadenosine monophosphate, 1110, 1110f, 1145
Deoxyribonucleic acid (DNA), 120–121, 122–123, 122f, 649, 667–668, 674, 1144, 1145–1147, 1145f, 1146f, 1147f
Deoxyribonucleosides, 1144–1145
Deoxyribonucleotides, 1145
2-Deoxyribose, 122, 1145
Deprotection, 802
Deprotonation
in acetoacetic ester synthesis, 960
of acetylene, 454
of alkyne, 441–442, 441t, 446–447
of carboxylic acids, 749–750, 750t
of diethyl acetamidomalonate, 1168
in direct enolate alkylation, 952
in glycoside formation, 1128
in halogenation at the α carbon of carbonyl compounds, 948–950
in malonic ester synthesis, 957
racemization at the α carbon of carbonyl compounds, 946
in tautomerization, 937
Deshielding effects, in NMR, 544–545, 544f, 545f, 548f, 548t
DET (diethyl tartrate), 491–492
Detergents, 114
Deuterium, 8
Deuterochloroform, 537f
Dexamethasone, 97
Dextropimaric acid, 1182
Dextrorotatory, 1116
DHA, 499
Di- (prefix), 142, 343, 886, 1013
Diabetes, 835
Diacids, 742, 955–957
formation from monosaccharide oxidation, 1132, 1133–1134
in nylon synthesis, 1255
Diacon. *See* Methoprene
Dialkylamides, 942
Dialkylborane, 420
Diamine, in nylon synthesis, 1255
1,4-Diaminobenzene, 1256
1,6-Diaminohexane, 1255, 1262, 1262f
Diastereomers, 196–198, 198f, 842
alkenes, 393t, 396–397
of but-2-ene, 309
definition of, 198
disubstituted cycloalkanes, 201–202
epimers, 1117
of glucose, 1121, 1122f
halogenation reactions and, 591
of monosaccharides, 1117
NMR spectrum of, 542–543
nucleophilic substitution reactions and, 288
physical properties of, 207–208, 207f
separation of amino acid enantiomers into, 1171–1172, 1171f
Diastereotopic protons, 542–543
1,3-Diaxial interactions, 163
Diazomethane, 1197
Diazonium salts, 1041–1049
alkyl, 1042
aryl, 1043–1049
coupling reactions of, 1048–1049
explosiveness of, 1042
formation from primary amines, 1042
substitution reactions of, 1043–1047
synthesis of aryl chlorides and bromides, 1044
synthesis of aryl fluorides, 1044
synthesis of aryl iodides, 1044
synthesis of benzene, 1045
synthesis of benzonitriles, 1045
synthesis of phenols, 1044
uses in synthesis, 1046–1047
Diazotization, 1042, 1045
DIBAL-H (diisobutylaluminum hydride), 787–789, 789f, 791t, 919–920
Dibromobenzene
meta-, 653, 655, 655f
ortho-, 653, 655, 655f
para-, 653, 655, 655f
1,4-Dibromobutane, 958
2,2-Dibromobutane, 442–443, 459
2,3-Dibromobutane, 187, 199–200, 200f
Dibromocarbene, 1197
1,2-Dibromocyclohexane, synthesis from cyclohexanol, 425–426
1,3-Dibromocyclopentane, 201–202
2,3-Dibromopentane, stereoisomers of, 197–198, 198f, 200
Di-*tert*-butyl dicarbonate, 1185
Dibutyl phthalate, 1261, 1270
Dicarbonyl compounds, 483
aldol reactions with, 984–985
1,4-dicarbonyl compounds, 984
1,5-dicarbonyl compounds, 985, 985f
in Robinson annulation, 997
β-dicarbonyl compounds, 980–981, 981f
pK_a values, 940, 941t
synthesis in crossed Claisen reactions, 988
tautomers, 936
2,5-Dichloroaniline, 654
4,4'-Dichlorobiphenyl, 109
Dichlorocarbene, 1195
1,2-Dichlorocyclopentane, 414
Dichlorodifluoromethane, 592
Dichlorodiphenyldichloroethylene. *See* DDE
Dichlorodiphenyltrichloroethane. *See* DDT
1,1-Dichloroethylene, NMR spectrum of, 541
Dichloromethane, 259f, 485
as solvent in extraction procedure, 758, 759f
2,4-Dichlorophenoxyacetic acid herbicide (2,4-D), 691f
trans-1,3-Dichloropropene, 559
Dictyopterene, 499
Dicyclohexylcarbodiimide (DCC), 898f, 902, 1184, 1187, 1188
Dicyclohexylurea, 902
Dicyclopentadiene, 636
Dieckmann reaction, 990
Dieldrin, 645
Diels, Otto, 629
Diels–Alder reaction, 252, 612, 629–638
cycloaddition, 1224, 1226
diene reactivity in, 631
dienophile reactivity in, 631–632
drawing product of, 630–631
features of, 629
retro, 636
retrosynthetic analysis of product, 635–636
rule of endo addition, 633–635
rules governing, 631–635
stereospecificity of, 632–633
in steroid synthesis, 637
tetrodotoxin synthesis, 630, 630f
as thermal reaction, 629
Dienes, 395
conjugated, 621–640. *See also* Conjugated dienes
Cope rearrangements, 1230, 1231
cycloaddition, 1224
isolated, 613–614, 614f, 622, 624, 624f, 625, 656
in metathesis, 1201
ozonolysis of, 483
reactivity, 631
stability, 656
1,3-Dienes, 613–614, 614f, 621. *See also* Conjugated dienes
polymerization of, 1255
synthesis in Suzuki reaction, 1192
Dienophiles, 1224
cis, 630, 631–632, 633, 636
cyclic, 631, 633
Diels–Alder reaction, 629–636, 632f
reactivity of, 631–632
stereochemistry, 632–633
trans, 631, 632
Diesel fuel, composition of hydrocarbons in, 148
Diesters
in Dieckmann reaction, 990
1,3-diesters, pK_a of, 941t
1,6-diesters, in Dieckmann reaction, 990
1,7-diesters, in Dieckmann reaction, 990
β-diesters
as active methylene compounds, 989
synthesis in crossed Claisen reaction, 989
Diethyl acetamidomalonate, 1168
Diethylbenzene, 681
Diethyl carbonate, in crossed Claisen reaction, 989
cis-1,2-Diethylcyclopropane, 1198
Diethyl ether, 95, 129, 1119
as anesthetic, 259f, 348
boiling point of, 832
in extraction procedures, 758
Grignard reagent preparation, 793
infrared (IR) spectrum, 523
naming, 343
solubility of, 108, 129
as solvent, 108, 370, 758
Diethyl maleate, 1210
Diethyl malonate, 980, 981f, 989
alkylation of derivatives, 1168–1169, 1170f
conversion to a carboxylic acid, 955–959
Diethyl oxalate, 1007

Diethyl sulfide, 373
Diethyl tartrate (DET), 491–492
Digoxin, 827, 865
Dihalides, 326–327, 327f
  alkyne synthesis from, 439–440
Dihalocarbenes, 1195–1197
Dihedral angle, 151, 152f, 153f, 154, 156f, 322
Dihydroxyacetone, 1111, 1112, 1118, 1119f
Dihydroxyacetone phosphate, 938, 991, 992, 993
1,4-Dihydroxybenzene, 1259
Dihydroxylation, 476f, 480–482
  anti, 480–481
  syn, 480, 481–482
1,4-Dihydroxynapththoic acid, 699, 699f
6,7-Dihydroxynonan-2-one, 876
Diiodomethane, 1197
Diisobutylaluminum hydride (DIBAL-H), 787–789, 789f, 791t, 919–920
Diisocyanate, 1257
Diisopropylamine, 943, 1013
β-Diketone, pK_a of, 941t
Dimers, carboxylic acids as, 742–743, 743f
2,2-Dimethoxycyclopentane-1,3-dicarbaldehyde, 838
1,2-Dimethoxyethane, 539
5,5-Dimethoxypentan-2-one, 873
N,N-Dimethylacetamide, 69
Dimethylacetylene dicarboxylate, 648
Dimethylallyl diphosphate, 616, 1171
4-(*N,N*-Dimethylamino)pyridine, 1035
N,N-Dimethylbenzamide, 886
1,3-Dimethylbenzene, 713
2,3-Dimethylbutanal, 829
3,3-Dimethylbutan-2-ol, 356, 357
2,2-Dimethylbut-3-enal, 832
Dimethyl carbonate, 821
cis-3,4-Dimethylcyclobutene, 1218, 1220, 1222
trans-3,4-Dimethylcyclobutene, 1220
cis-5,6-Dimethylcyclohexa-1,3-diene, 1218–1219
trans-5,6-Dimethylcyclohexa-1,3-diene, 1218–1219
1,2-Dimethylcyclohexane, 146f
1,3-Dimethylcyclohexane, 145
1,4-Dimethylcyclohexane, 164–166, 166f
2,2-Dimethylcyclohexanone, 953, 954
2,6-Dimethylcyclohexanone, 953
1,6-Dimethylcyclohexene, 396f
1,2-Dimethylcyclopentane, 164
  cis isomer, 183f
  trans isomer, 183f
1,3-Dimethylcyclopentane, 208
(*S,S*)-1,2-Dimethylcyclopropane, 607
Dimethylcyclopropanes, 216, 541
Dimethyl ether
  isomers of, 18–19
  NMR spectrum of, 565
Dimethylformamide (DMF)
  boiling point of, 106
  as solvent, 268f, 951
N,N-Dimethylformamide, 577
2,5-Dimethylhept-3-yne, 437f
2,5-Dimethylhexa-2,3-diene, 462
4,5-Dimethylhexanoic acid, 740
2,4-Dimethylhexan-3-one, 814
2,4-Dimethylhexanoyl chloride, 887
2,5-Dimethylhex-3-yne, 462
6,6-Dimethyloct-3-yne, 437
2,2-Dimethyloxirane, 345, 378
2,3-Dimethyloxirane, 377
2,3-Dimethylpentane, 143f
  mass spectrum of, 510–511
2,2-Dimethylpropane (neopentane), 102, 102f, 107, 136
2,2-Dimethylpropanoic acid, 754
Dimethyl succinate, 550
Dimethyl sulfide, conversion of ozonide to carbonyl compounds, 482–484
Dimethyl sulfoxide (DMSO), 19
  aqueous, 417
  as polar aprotic solvent, 268f, 290t
  as solvent for alkyne synthesis, 327, 327f
Dimethyl terephthalate, 1257, 1265
1,3-Dinitriles
  active methylene compounds, 981
  pK_a, 941t
Dinoprostone, 1168
-dioic acid (suffix), 742
-diol (suffix), 343
Diol epoxide, 381
Diols, 343
  1,2-diol formation by dihydroxylation, 476f, 480–482
  gem-diol, 854–857
p-Dioxanone, 1273
Dioxybenzone, 130
Dipeptides
  constitutional isomers of, 1176
  synthesis of, 1175–1176, 1183–1187
Diphenhydramine, 301, 652, 1030
Diphenyl carbonate, 1258, 1264
Diphosphate, 100t, 291, 616
  as leaving group, 616, 1171, 1173
Dipole, 48, 49, 50f
  in alkenes, 399
  temporary, 101–102
Dipole–dipole interactions, 102, 103t
  of aldehydes and ketones, 832
  of alkyl halides, 258
  of carbonyl compounds, 888
  of carboxylic acids, 742
  of nitriles, 888
  in polar aprotic solvents, 268, 289
  in polar protic solvents, 267
Dipole moment, infrared (IR) absorption and, 520
Directed aldol reactions, 981–983, 982f
Disaccharides, 121, 1137–1140
  artificial sweeteners, 1140, 1140f
  general features, 1137–1138
  lactose, 1133, 1139
  maltose, 1138
  sucrose, 1139–1140, 1139f
Discodermolide, 218
Disorder, entropy as measure of, 235–236
Disparlure, synthesis of, 478–480, 479f
Disproportionation, 1246
Disrotatory, 1219–1223, 1223t
Dissolving metal reductions, 465, 473–474
Distillation, of crude petroleum, 148, 148f
trans-1,2-Disubstituted cycloalkanes, 376, 377
Disulfide, 96t
Disulfide bonds, 373
  in peptides, 1179
  in tertiary structure of proteins, 1195, 1195f, 1196f
  in vulcanized rubber, 1254, 1254f
  in α-keratins, 1197, 1198f
Diterpene, 1170t, 1172
cis-1,2-Divinylcyclobutane, 1231
Diynes, 436
DMAP, 1035
DMF. *See* Dimethylformamide
DMSO. *See* Dimethyl sulfoxide
DNA, 120–121, 122–123, 122f, 649, 667–668, 674, 1144, 1145–1147, 1145f, 1146f, 1147f
Docetaxel, 930
Dodecahedrane, 147, 158, 648
Dodecane, 137t
Domagk, Gerhard, 1051
Donepezil, 983
Dopamine, 7, 873, 1019–1020, 1020f
Double bonds. *See also* Alkenes
  in alkenes, 93, 94t, 307–309
  in alkyl halides, 256
  bond length and bond strength, 45–46, 46t
  components of carbon–carbon, 42, 42f
  in condensed structures, 31, 31f
  conjugated, 976
  conjugated dienes, 621
  degrees of unsaturation, calculating, 468–469
  effect on melting point of fatty acids, 399
  in fats and oils, 403
  in functional group, 93, 94t
  hydrogenation of, 469–470, 470f
  Lewis structure, 15–16
  metal–carbon, 1200
  nitrogen–nitrogen, 1048
  NMR spectra of, 547, 548t
  pi (π) bonds, 93, 98
  planar, 411, 414, 420, 422
  properties of, 393t
  radical reactions with, 581, 598–601
  in resonance structure, 21–25
  resonance structures
    conjugated, 617
    major contributor, 618–619, 623
    minor contributors, 618–619, 623
    with one atom more electronegative than the other, 618
  rotation restriction, 42, 308–309, 309f, 393t
  in trans fats, 470–471
  in triacylglycerols, 1158–1160, 1158t
  in unsaturated fatty acids, 391, 401–403
  Z configuration, 401
Double dagger notation, 237
Double-headed arrow, 19, 63–64, 226t
Doublet, NMR spectrum, 550–551, 552t, 553t, 554, 555
Doublet of doublets, NMR spectrum, 558, 558f
Doxorubicin, 97, 1110, 1110f
Dyes
  azo compounds, 1048, 1050–1051
  natural and synthetic, 1050–1051
  sulfa drugs, 1051–1052, 1052f
Dyneema, 1253

E

E (prefix), 397–398
E1cB elimination reactions, 977, 997
E1 elimination reactions, 311, 318–321
  acid-catalyzed dehydration of alcohols, 403
  base in, 320, 320t, 321, 325, 325t
  characteristics of, 320t
  dehydration of secondary and tertiary alcohols, 354–355
  E2 mechanism compared, 318, 321, 325, 325t, 327–331, 328t
  energy diagram for, 318, 319f
  kinetics of, 318, 320t
  leaving group in, 318, 320t
  rate of reaction, 319–320, 320t
  regioselectivity of, 320
  S_N1 substitution reactions compared, 321, 327–331, 328t
  S_N2 substitution reactions compared, 327–331, 328t
  solvent for, 320t
  two-step mechanism, 318–319, 320t
E2 elimination reactions
  acetylide anion reactions with alkyl halides, 449–450
  alkyl halide identity in, 314–315, 315t
  alkyl tosylates reactions, 368
  alkyne synthesis, 326–327, 439–440
  anti periplanar geometry in, 322–323, 322f
  base in, 313, 313f, 315t, 325, 325t
  characteristics of, 315t
  dehydration of alcohols, 355, 359
  dehydrohalogenation of alkyl halides, 403
  E1 mechanism compared, 318, 321, 325, 325t, 327–331, 328t
  energy diagram of, 312, 312f
  in Hofmann elimination, 1039–1041, 1041f
  kinetics of, 311, 315t
  leaving group in, 312–314, 315t
  one-step mechanism, 312–313, 315t
  organic synthesis examples, 315f
  rate of reaction, 313, 314
  regioselectivity of, 316
  S_N1 and S_N2 substitution reactions compared, 327–331, 328t
  solvent for, 313, 315t
  stereochemistry of, 322–325, 322f
  stereoselectivity of, 317
  syn periplanar geometry in, 322, 322f
  trans diaxial geometry in, 323–324, 323f
  Zaitsev rule, 316–317, 324
E-85 fuel, 347, 347f
Eclipsed conformation, 150–157, 152f, 153f, 154f, 156f, 156t, 197, 309f, 322f
Edman degradation, 1180–1181
Effexor. *See* Venlafaxine
Ehrlich, Paul, 1051
Eicosanoids, 1166–1169
  biological activity of, 1166, 1168t
  biosynthesis of, 1167, 1167f, 1168
  derivation of word, 1166
E isomer, 397
Elaidic acid, 429
Elastomer, 1254
Electrocyclic reaction, 1217–1223
  description of, 1213
  photochemical, 1221–1222
  stereochemistry and orbital symmetry, 1218–1219
  thermal, 1219–1221
  Woodward–Hoffmann rules for, 1222, 1223t
Electrocyclic ring closure, 1213, 1217–1222

Electrocyclic ring opening, 1213, 1217–1218, 1220
Electromagnetic radiation, 514–516, 515f
Electromagnetic spectrum, 515, 515f
Electron cloud, 8
Electron-deficient sites/species, 48–49, 50f, 65, 81–83, 92, 95, 117–118, 225, 776
Electron density, delocalization, 74
Electron-donating groups, 284, 700–702, 706, 711, 754–757, 757f
Electron dot representation. *See* Lewis structures
Electronegativity
  acidity trends and, 70–73
  bond polarity and, 48–49
  definition of, 47
  inductive effects on substituted benzenes, 700
  oxidation and, 167
  periodic trends in, 47, 47f
  values for common elements, 47f
Electron pair acceptor, 62, 81
Electron pair donor, 62, 81
Electron pairs
  in Brønsted–Lowry acid–base definition, 62
  Lewis acids and bases, 81–83
  movement of, 63, 224
Electron-poor sites/species, 82
Electron-rich sites/species, 48, 50f, 65, 82, 117–118, 404
Electrons
  bonded (shared), 13
  delocalized, 20, 24, 594, 615, 620, 624, 651, 664–666
  density of, 9, 36, 43, 48–49, 49f, 101
  excited state, 638–639
  ground state, 638–639
  in Lewis structure, 13–19
  lone pairs. *See* Lone pairs
  nonbonded (unshared), 13, 13f
  number of, 8, 11
  in resonance structures, 19–25
  unpaired, 224
  valence, 11–13, 18t
Electron-withdrawing groups, 284, 631–632, 700–702, 706, 711–712, 715–717, 754–757, 757f
Electrophiles, 98
  alkene reaction with, 404
  alkyne reactions with, 440
  carbene, 1195
  carbocations as, 225
  Lewis acid, 82
  meaning of term, 82
  radicals as, 225
  reactivity of, 117–118
Electrophilic addition reactions, A-13
  of conjugated dienes
    1,2- versus 1,4-addition, 624–626
    kinetic versus thermodynamic products, 626–629, 628f
  energy diagram for, 407, 407f
  halogenation of alkynes, 444
  hydration of alkenes, 412–413, 423
  hydrohalogenation of alkenes, 405–412, 405f, 406f, 407f, 409f, 411t
  hydrohalogenation of alkynes, 442–444
  Markovnikov's rule on, 408–410, 409f, 411t
  stereochemistry of, 410–412, 411t
Electrophilic aromatic substitution reactions, 687–715, A-12
  azo coupling, 1048–1049
  energy changes in, 689, 689f
  examples of, 688f
  Friedel–Crafts acylation, 688f, 693–699, 698f, 712, 721–722, 722f, 726–727
  Friedel–Crafts alkylation, 688f, 693–699, 711, 721–722, 722f, 724–725
  halogenation, 688f, 690–691, 710–713
  mechanism of, 688–689
  nitration, 688f, 691–692, 714–715
  of substituted benzenes, 700–715
    disubstituted benzenes, 712–713
    limitations on reactions, 710–712
    rate of reactions, 705, 706, 706f
  sulfonation, 688f, 691–692
Electrophilic carbon, in alkyl halides, 261
Electrospray ionization (ESI), 514
Electrostatic interactions
  in ionic compounds, 101
  between ions, 11–12
Electrostatic potential plot
  of acetic acid, 75f, 739, 739f
  of acetylene, 440, 441f
  of alcohols, 341f
  of alkoxide, 940, 940f
  of amines, 1012, 1012f
  of benzene, 652, 652f
  of carbanions, 76f
  of carbocations, 282, 282f
    localized and delocalized, 615, 615f
  of carbon dioxide, 50f
  description of, 48
  of dienes, 614, 614f
  of enolates, 940, 940f
  of epoxides, 341f
  of ethanol, 73f, 75f
  of ethers, 341f
  of ethylene, 404, 404f
  of formaldehyde, 776, 776f
  of halomethanes, 261f
  of methyl chloride, 49, 49f
  of pyridine and pyrrole, 665, 665f
  of substituted anilines, 1032, 1033f
  of 2,2,2-trifluoroethanol, 73f
  of vitamins A and E, 1166, 1166f
  of water, 50f
Elements, periodic table, 8–10, 9f
Element symbol, 8
Elemicin, 126
Eleostearic acid, 429
Eletriptan, 1206
Elimination reactions, A-13
  of alcohols, 351–360, 369, 370f
  alkene synthesis, 306–307, 403–404
  of alkyl halides, 261, 305–338
  alkyne synthesis, 306, 326–327, 327f, 439–440
  alpha (α), 1195
  beta (β), 306
  bimolecular, 311. *See also* E2 elimination reactions
  description of, 221–222
  elimination, unimolecular, conjugate base (E1cB) mechanism, 977, 997
  features of, 306–307
  Hofmann elimination, 1038–1041
  mechanisms of, 311
    E1. *See* E1 elimination reactions
    E1 and E2 compared, 318, 325, 325t, 327–331, 328t
    E2. *See* E2 elimination reactions
  reductive, 1189, 1191, 1194
  stereochemistry, 322–325
  unimolecular, 311, 318. *See also* E1 elimination reactions
  Zaitsev rule on, 316–317, 320
Enalapril, 1026
Enals, 832
Enamines, 458
  hydrolysis of, 854
  synthesis of
    from aldehydes and ketones, 841f, 853–854, 853f, 1037
Enanthotoxin, 30
Enantiomeric excess, 206–207, 491, 785
Enantiomers
  of amino acids, 762, 1164, 1164f
    separation of, 1170–1174, 1171f
  of an ammonium salt, 1012
  in carbonyl reductions, 783–784
  chemical properties of, 208–210
  chiral drugs, 208–209
  of cyclic compounds, 190
  cyclic hemiacetals, 863
  definition of, 185, 196, 198
  dienophile, 632
  dihydroxylation, 480
  disubstituted cycloalkanes, 202
  drawing, 188–189, 189f
  in electrophilic addition to alkenes, 410–411, 411t
  enantioselective reactions, 490–491
  enantiotopic protons, 542–543
  in enolate, 946, 953
  from epoxidation of alkenes, 477, 480
  in epoxide ring-opening reactions, 376–377
  formation in halogenation of alkenes, 413–415
  halogenation reactions and, 590–591
  in halohydrin formation, 417
  in hydroboration–oxidation reactions, 422
  labeling stereogenic centers with *R* or *S*, 191–196, 193f, 195f, 200–201
  of monosaccharides, 1112–1113, 1115–1117, 1117f
  in nucleophilic substitution reactions, 273, 279–280
  optical purity, 206–207
  in organometallic reagent reactions, 798
  physical properties, 203–208, 205t, 207f
  racemic mixture of, 205, 205t, 410–411
  sense of smell and, 209–210, 209f
  three-dimensional representations for, 189f
Enantioselective reactions, 490–491
  amino acid synthesis, 1174–1175
  biological reduction, 786–787
  carbonyl reductions, 784–787, 785f
Enantiotopic protons, NMR signals of, 542–543
Enclomiphene, 428
Endergonic reaction, 236
Endiandric acids, 1241
Endo addition, 633–635, 635f
Endo product, 633–634, 635f
Endothermic reaction
  bond breakage and, 227–228
  bromination, 587–588
  dehydration of alcohols, 356f
  energy diagram of, 238f
  enthalpy change and, 236
  reaction rate, 284
  transition state in, 284, 284f
-ene (suffix), 395
Energy
  nonrenewable, 148
  of photons, 514, 515–516
  release in bond formation, 225
  stored in triacylglycerols, 1161
  torsional, 153
  units of measurement, 45
Energy barrier, 239–241, 240, 312, 318, 319f
Energy change
  during the chlorination of ethane, 584–585, 585f
  during radical addition of HBr, 600–601, 600f
Energy diagrams, 236–237, 240f
  for carbocation formation, 286f
  for chlorination of ethane, 585f
  for E1 elimination reactions, 318, 319f
  for E2 elimination reactions, 312, 312f
  for electrophilic addition of HBr to buta-1,3-diene, 628, 628f
  for electrophilic additions, 407, 407f
  for electrophilic aromatic substitution, 689, 689f
  for halogenations, 585f, 588f, 589f
  rate of electrophilic aromatic substitution of substituted benzenes, 706, 706f
  for S_N1 substitution reactions, 278, 278f, 285, 286f
  for S_N2 substitution reactions, 272, 272f
  for two-step reaction mechanisms, 239–241, 240f
Energy of activation (E_a)
  catalysts and, 244, 245f
  definition of, 237
  energy barrier height determined by, 237
  in energy diagrams, 237–238, 238f, 241f
  in exothermic reactions, 285, 285f
  rate constant, relationship to, 241
  reaction rate and, 241
  transition state energy effect on, 284
-enol (suffix), A-5
Enolate(s), 22, 838, 935, 938–945
  achiral, 946
  in aldol reactions, 973, 974, 974f, 976, 977, 979, 981, 981f, 982, 983, 985, 989
  alkylation, 952–955, 961
    general features of, 952–953
    tamoxifen synthesis, 954
    of unsymmetrical ketones, 953–954
  in Claisen reactions, 986–989
  electrostatic potential plot of, 940, 940f
  examples, 940
  kinetic, 944, 945, 953, 981
  in Michael reactions, 994–996, 996f
  racemization, 946, 946f, 953
  reactions of, 943, 947, 952–955
  resonance-stabilized, 977, 986, 995

resonance structures, 938
in Robinson annulation, 996–1000
synthesis of, 810
with bases, 941–943, 942t
from carbonyl compounds, 938–939
from esters, 940
from tertiary amides, 940
thermodynamic, 944, 945, 953, 954
of unsymmetrical carbonyl compounds, 944–945
Enol(s), 935–938
hydroxy group of, 340
reaction mechanism, 938
resonance structures, 938
synthesis of, 810
hydration of alkynes, 445–447
hydroboration–oxidation of alkynes, 447–448
tautomers, 936–937
Enol tautomers, 445–447, 918, 936–937
-enone (suffix), A-5
Enones, 832
Entacapone, 59
Enthalpy change ($\Delta H°$), 227–230
calculation from bond dissociation energies, 228–229
in dehydration of alcohols, 355, 356f
E1 elimination reactions, 318, 319f
in energy diagrams, 237, 238f, 240f
free energy change, relation to, 235–236
negative, 227, 235
positive, 227, 235
Entropy ($S°$), 235–236
Entropy change ($\Delta S°$), 235–236
-enyne (suffix), A-5
Enynes, 436
Enzymatic reactions, laboratory reactions versus, 246t
Enzymes, 246f, 1164
active site of, 245, 246f
catalytic action of, 245–246, 246f
in kinetic resolution of amino acids, 1173
in oxidation reactions, 489–490
in peptide sequencing, 1182, 1182t
Enzyme–substrate complex, 245, 246f
Ephedrine, 187, 214
Epi-aristolochene, 1184
Epichlorohydrin, 388, 1258–1259, 1259f
Epilupinine, 1062
Epimers
definition of, 1117
of monosaccharides, 1117, 1135–1136
Epinephrine
structure of, 1019
synthesis of, 293–294, 293f
Eplerenone, 349
Epothilone A, 1201f
Epoxidation, 476f, 477–480
in anti dihydroxylation, 480
Sharpless, 490–492
of squalene, 1176f, 1177
stereochemistry of, 477–478
in synthesis of disparlure, 478–480, 479f
Epoxides, 304
achiral, 480
from alkene oxidation, 476f, 477–480
angle strain, 341, 352
anionic polymerization of, 1251–1252
applications, 379–381
boiling point of, 346t
bonding, 341
electrostatic potential plot of, 341f
interesting examples, 349
leaving group, 352
melting point of, 346t
nomenclature, 344–345
physical properties of, 345–346, 346t
reactions of, 375–379
with acetylide anions, 452
with acids HZ, 377–379, 378f
general features, 352
with organometallic reagents, 808–809
reduction, 474–475, 475f
with strong nucleophiles, 375–377
reduction of, 474–475, 475f
ring opening reactions, 1251–1252
with acids, 377–379
enantiomers in, 376–377
nucleophilic substitution reactions, 375–379
regioselectivity of, 377–378
solubility of, 346t
structure of, 340, 341
synthesis of, 351
from halohydrin, 351
Epoxy (prefix), 344
Epoxyalkane, 344
1,2-Epoxycyclohexane, 344, 376
1,2-Epoxy-2-methylpropane, 344
cis-2,3-Epoxypentane, 344
Epoxy resins, 1258–1259, 1259f
Equal. *See* Aspartame
Equations for organic reactions, writing, 220–221
Equatorial bond, 182
Equatorial hydrogens, 159–160, 159f, 163f
Equilibrium, 231–232, 232f, 232t
acid–base reactions, 68–69
direction of, 241
nucleophilic substitution and, 265
Equilibrium arrow, 63–64, 69, 226t
Equilibrium constant (K_{eq}), 231–232, 232f, 232t, 241
Ergot, 698f
Erlotinib, 456
Erythrose, 215
D-Erythrose, 1117, 1117f, 1118f
L-Erythrose, 1117, 1117f
D-Erythrulose, 1119f
ESI (electrospray ionization), 514
Esmolol, 89
Essential amino acids, 762
Essential fatty acids, 401
Essential oils, 1169
Esterification, 254
Fischer, 900
Esters
acidity/basicity of, 882t
boiling point of, 889t
cholesteryl, 914–915, 915f
chromate, 485–486
cyclic, 880
enolates from, 940
functional group in, 98t
general structure of, 401
interesting examples, 891
IR spectra of, 890t
leaving group, 778
melting point of, 889t
nomenclature of, 883–884, 887t
nucleophilic substitution reactions, 778
odors of, 891
physical properties of, 889t
pK_a, 882t, 941t
reactions of, 903–905, 910t
with ammonia and amines to form amides, 903
carboxylic acid preparation by malonic ester synthesis, 955–959
Claisen reaction, 986–989, 987f
hydrolysis to carboxylic acids, 903–905
lipid hydrolysis, 905–907
methyl ketone preparation by acetoacetic ester synthesis, 955, 959–962
reaction with organometallic reagents, 795, 804–806
reactivity, 893–894
reduction of, 787–789
solubility of, 889t
stability of, 882
structure and bonding of, 775, 879, 882
synthesis of
from acid chlorides, 896
from anhydrides, 897
from carboxylic acids, 898f, 900
from monosaccharides, 1130–1131
thioesters, 914–915
in triacylglycerols, 401
waxes, 1157
Estradiol, 56, 675
analogs of, 438
function in body, 1178t
structure of, 438, 1178t
synthesis of, 315, 315f
Estrogen-dependent cancer treatment, 674–676, 675f
Estrogens, 1178, 1178t
Estrone, 675, 737
function in the body, 1178t
structure of, 804, 1178t
synthesis of
from chlorohydrin, 418f
in Diels–Alder reactions, 637
using a Michael reaction, 996, 996f
Eth- (prefix), 139
Ethambutol, 56, 213, 276, 276f
Ethanal. *See* Acetaldehyde
Ethane
acidity of, 75–76, 76f
bond dissociation energy, 643
bonds in, 40–41, 44t
chlorination of, 584–585, 585f
conformations of, 150–153, 152f, 153f
energy, 153
molecular formula, 135, 137t
nonreactivity of, 92
structure of, 92, 93, 135
Ethanedioic acid, 742
Ethane-1,2-diol, 343
Ethanethiol, 372
Ethanol, 130, 753
acidity of, 72–75, 750
in alcoholic beverages, 346, 463
blood alcohol concentration, 526
boiling point, 372
description, 3
electrostatic potential plot, 73f, 75f
in ethyl acetate formation, 244–245
isomers of, 18–19
NMR spectrum of, 540, 559, 559f
oxidation of, 489–490
pK_a of, 750
as polar protic solvent, 267f
preparation from ethylene, 400f
production by fermentation of carbohydrates, 346–347, 347f
reactivity of, 93
as renewable fuel source, 347, 347f
sodium ethoxide production from, 350
solubility of, 109
structure of, 93
Ethene, 398. *See also* Ethylene
Ethers
acetal compared, 857
boiling point of, 346t
bonding of, 341
Claisen rearrangement, 1232
crown, 116, 348–349, 348f
electrostatic potential maps, 341f
formation by alcohol addition to alkenes, 413
functional group, 96t
infrared (IR) spectra of, 523
interesting examples, 348–349, 348f
leaving group, 352
melting point of, 346t
nomenclature, 343–344
physical properties of, 345–346, 346t
preparation by nucleophilic substitution reaction, 294, 295t
reactions of
alkyl halide formation, 370–371
general features, 352
with strong acid, 370–371
simple and complex, 343–344
solubility of, 346t
structure of, 340, 341
symmetrical, 340, 349
synthesis of
from alkyl halides, 349–350
from monosaccharides, 1130–1131
by nucleophilic substitution reactions, 349–350
Williamson ether synthesis, 349, 350
unsymmetrical, 340, 349
Ethoxide, 74, 750
nucleophilicity of, 266
Ethoxy group, 344
4-Ethoxyoctane, 344
Ethyl acetate, 254
in Claisen reaction, 986–987
formation from acetic acid and ethanol, 244–245
as solvent, 900
structure of, 884
synthesis of, 900
Ethyl acetoacetate, 964
conversion to a ketone, 959–961
synthesis in Claisen reaction, 986
Ethyl *p*-aminobenzoate, 723
N-Ethylaniline, 1014
Ethylbenzene, 653
bromination of, 719
styrene synthesis from, 720
Ethyl benzoate, 887t, 933
in crossed Claisen reactions, 988
Ethyl butanoate, 891
Ethyl chloride, 257

Ethyl chloroformate, in crossed Claisen reaction, 989
Ethyl chrysanthemate, 928
Ethyl α-cyanoacrylate, 1249f
Ethylcyclobutane, 146f
N-Ethylcyclohexanamine, 1027
2-Ethylcyclohexanecarbaldehyde, 829
2-Ethylcyclohexanone, 953
4-Ethyl-3,4-dimethyloctane, 143f
5-Ethyl-2,6-dimethyloctane, 142
Ethylene, 398
  acidity of, 75–76, 76f
  bond dissociation energies, 435
  bonds of, 41–42, 42f, 44t, 392
  bromine addition to, 650
  electrostatic potential plot, 404, 404f
  fruit ripening and, 308
  functional group in, 93, 94t
  in Heck reaction, 1193
  as industrial starting material, 399, 400f
  Lewis structure, 15–16
  molecular orbitals of, 1214, 1215f
  molecular shape of, 27
  monomers in polymerization, 601
  pK_a of, 68
  polymerization of, 1247, 1253
  preparation from petroleum, 399
  structure, 307–308, 1246f
Ethylene diacetate, 550
Ethylene glycol, 860
  acetal formation from, 857
  in antifreeze, 347f
  formation from PET recycling, 1265
  naming, 343
  in polyethylene terephthalate (PET), 1242, 1244f, 1256–1257
  in polyurethane synthesis, 1257
  preparation from ethylene, 400f
  structure of, 347f
  synthesis, 1272
Ethylene oxide, 345
  polymerization of, 1252
  ring opening of, 1252
1-Ethyl-2-fluorocyclopentane, 258f
N-Ethylformamide, 886
Ethyl formate, in crossed Claisen reactions, 988
Ethyl group, 140
2-Ethylhep-1-ene, 395
Ethyl hexa-2,4-dienoate, 1006
Ethyl isopropyl ether, 350
Ethyl isopropyl ketone, 832f
1-Ethyl-3-methylcyclohexane, 145
N-Ethyl-*N*-methylcyclopentanamine, 1014
Ethyl methyl ether, 553
Ethyl methyl ketone, 831
4-Ethyl-5-methyloctane, 143f
Ethyl *p*-nitrobenzoate, 723
3-Ethylpentane, bromination of, 587
Ethyl propyl sulfide, 374
Ethyne, 437. *See also* Acetylene
Ethynylcyclohexane, 1205
Ethynylestradiol, 434, 438, 438f, 798, 799f, 804, 1178
Ethynyl group, 437
1-Ethynyl-2-isopropylcyclohexane, 437f
Ethynyllithium, 820
Etoposide, 870
Euphorbia peplus, 1185
Excited state
  of atom, 37
  of electron, 638–639
Exergonic reaction, 236
Exo product, 633–634, 635f
Exothermic reactions
  addition reaction, 406, 406f
  bond formation and, 227–228, 230
  chlorination, 585, 585f, 588–589
  energy diagram of, 237, 238f
  enthalpy change and, 236
  hydrogenation, 466
  radical addition of HBr to alkene, 599–600
  reaction rate, 285
  transition state in, 284–285, 285f
Extraction, 758–759, 759f
  for separation of amines, 1029, 1029f
Ezetimibe, 211, 786, 928, 1008
E,Z system of nomenclature, 397–398

F

Faraday, Michael, 650
α-Farnesene, 400f
Farnesol, 1170f, 1172, 1173
Farnesyl diphosphate, 616, 1171f, 1172–1174, 1184
Fats, 403, 469–471, 905–907
  body content of, 1161
  degree of unsaturation, 469–470
  fatty acid composition, 1159, 1159t
  melting point of, 403, 1158–1159
  oxidation, 597
  saturated, 403, 597
  trans, 470–471
Fat-soluble vitamins, 1165–1166, 1165f, 1166f
Fatty acids, 124f
  biosynthesis of, 993–994
  characteristics of, 1158
  composition in fats and oils, 1158–1159, 1159t
  essential, 401, 1158
  melting point of, 401, 402t
  omega-3, 402, 403
  oxidation of, 906
  saturated, 401, 470–471, 1158–1159, 1158t, 1159t
  synthesis in lipid hydrolysis, 906
  thioesters of, 914
  trans, 470–471
  triacylglycerol formation and, 124
  in triacylglycerols, 401, 403, 1158–1161, 1158t
  unsaturated, 391, 401, 1158–1160, 1158t, 1159t
    hydrogenation of, 1160
    oxidation of, 1160
Fatty acyl coenzyme A, 913
Feedstock, 1262
Fehling's reagent, 1132–1133
Femara. *See* Letrozole
Fenfluramine, 86
Fenn, Dr. John, 514
Fenofibrate, 905
Fentanyl, 133, 1058
Fermentation, alcohol production by, 346–347, 347f
Ferulic acid, 647
Fexofenadine, 56, 1019
Fibrinogen, 1164
Fibrous proteins, 1197
Fingerprint, of IR spectrum, 517–519, 518f, 519f
First-order kinetics, 243
  of E1 elimination reactions, 318, 320t
  of nucleophilic substitution reactions, 271
  of S_N1 substitution reactions, 277
First-order rate equation, 243, 270–271
Fischer projection formulas, 1112–1116, 1115f
Fishhook notation, 224, 226t
Five-membered rings, synthesis of
  in Dieckmann reaction, 990
  in intramolecular aldol reactions, 984–985
Flagpole hydrogens, 161, 161f
Flat helical ribbon, shorthand for α-helix, 1193
Flat polygons, 201
Flat wide arrow, shorthand for β-pleated sheet, 1193
Fleming, Sir Alexander, 511, 525, 909
Flexibilene, 1183
Flonase. *See* Fluticasone
Flosal, 980
9-Fluorenylmethoxycarbonyl protecting group (Fmoc), 1185, 1189–1190
Fluorine
  polarizability, 102
  valence electrons of, 12
Fluoroboric acid, 1044
Fluoromethane, polarizability and boiling point of, 106f
m-Fluoronitrobenzene, 653
Fluoxetine, 3, 86, 208, 276, 276f, 717, 736, 1020
Fluticasone, 255, 301
Fmoc (9-fluorenylmethoxycarbonyl protecting group), 1185, 1189–1190
Folic acid, 100, 1051–1052
Formal charge, 16–18, 18t
Formaldehyde, 16
  aldol reactions of, 979
  Bakelite preparation from, 1260–1261, 1261f
  electrostatic potential map of, 776, 776f
  formation from methanol oxidation, 490
  hydration of, 854–855
  primary alcohol formed from, 797
  structure of, 830, 835
  uses of, 835
Formaldehyde hydrate, 854
Formalin, 835
Formic acid, 68, 138, 490, 741t, 745
Formyl group, 832
Fosamax (alendronic acid), 19
Fossil fuels, 79, 147–148, 169
Four-centered transition state, 420
Fragmentation, 505, 509–512, 510f
Free energy change ($\Delta G°$), 231–232, 232f, 232t
Freon 11, 260, 592
Freon 12, 592
Frequency
  definition of, 515
  of electromagnetic radiation, 514–516, 515f
  Hooke's law, 519, 519f
  of infrared radiation, 516–517, 519
Friedel, Charles, 688
Friedel–Crafts acylation, 686, 688f, 693–699, 726–727
  alkyl benzene synthesis, 722, 722f
  of benzene, 688f, 722f
  electrophilic aromatic substitution reaction, 688f
  general features of, 693
  intramolecular, 698–699, 698f
  ketone synthesis, 688f, 693, 721
  limitations of, 711–712
  mechanism of, 695
Friedel–Crafts alkylation, 688f, 693–699, 724–725
  of alcohols, 697
  of alkenes, 697
  in alkyl benzene synthesis, 693–694, 722, 722f
  of benzene, 688f, 722f
  carbocation rearrangements, 696–697
  electrophilic aromatic substitution reaction, 688f
  general features of, 693
  intramolecular reactions, 699
  limitations in, 711–712
  mechanism of, 694
  polyalkylation, 711
  unreactive halides, 695–696
Friedel–Crafts reaction of amides, 695, 1038, 1038f
Frontalin, 492, 492f
Frontside attack, in nucleophilic substitution reactions, 273
Frost circle, 672
Fructofuranose
  α-D-fructofuranose, 1126
  β-D-fructofuranose, 1126
Fructofuranose ring, 1140
Fructose
  D-fructose, 1111, 1118, 1119f, 1126
  phosphorylation of, 234
  stereogenic centers, 187
  in sucrose, 1139–1140
Fructose 1,6-bisphosphate, 991, 992, 993
Fructose 6-phosphate, 234
Fucose, 1165
Fuels, renewable, 347, 347f
Fukui, Kenichi, 1214
Full-headed curved arrow, 224, 225–227, 225f, 226t
Fumaric acid, 133, 632
Functional groups, 92–101. *See also specific functional groups*
  compounds containing a C–O group, 98–100, 98t
  compounds containing a C–Z σ bond, 95–97, 96t
  definition of, 92
  fragmentation pattern in mass spectrum, 511–512
  grouping organic compounds by, 93
  hydrocarbons, 93–95, 94t
  importance of, 99
  infrared (IR) spectroscopy and, 516–520, 518f, 519f, 520t
  intermolecular forces and, 101–104
  nomenclature of, A-4–5
  nucleophilic substitutions and, 294, 295t
  reactivity and, 117–118
  solubility and, 108–109
  synthesis of, A-14–17
Funk, Casimir, 111
Funnel, separatory, 758, 759f, 1029, 1029f
Furan, 679, 682
Furandicarboxylic acid, 1257
Furanose rings, 1120, 1125–1126
Fused ring system, 633–634, 634f

G

Gabapentin, 191
Gabriel synthesis of primary (1°) amines, 1022–1023
Galactose, 1139
  D-galactose, 866, 1117, 1118f, 1135
  in lactose, 245, 246f
  Wohl degradation, 1135
β-Galactosidase, 245
Gamma (γ) carbon, 741
Garlic, 138
Garsubellin A, 1233
Gas chromatography–mass spectrometry (GC–MS), 513–514, 513f, 514f
Gasohol, 347, 347f
Gasoline
  additives
    benzene, 93
    BTX mixture, 522
    ethanol, 347
    high-octane, 522, 538
    MTBE, 109
  combustion of, 168–169
  composition of hydrocarbons in, 148, 150
  octane and, 93, 109
  refining, 148f
Gauche conformation, 154–155, 154f, 156f
GC–MS (gas chromatography–mass spectrometry), 513–514, 513f, 514f
Geckos, 102
Geminal dichloride, 439
Geminal dihalide, 326
  alkyne synthesis from, 439–440
  synthesis in hydrohalogenation of alkynes, 442
Geminal protons, 557
Gene, 122
Generic name, 139
Geranial, 836f, 1232
Geraniol, 292, 492, 1170, 1173
Geranyl diphosphate, 292, 616, 1171f, 1172–1173
GHB [γ-hydroxybutyric acid], 746
Gibberellic acid, 1005
Gibbs free energy ($G°$), 231
Ginger, 631
Gingerol, 983
Ginkgo biloba, 741
Ginkgolide B, 878, 905
Glacial acetic acid, 745
Global climate change, 169
Globular proteins, 1197
D-Glucaric acid, 1133–1134
Glucitol, 1132
D-Gluconic acid, 1132
Glucopyranose
  α-D-glucopyranose, 1133f
  β-D-glucopyranose, 1121, 1144
Glucosamine, 916
  D-glucosamine, 1143
Glucose, 219, 865–866
  acyclic aldehyde, 1121, 1122f
  anomers of, 1121, 1122f, 1125, 1125f
  in complex carbohydrates, 120–121
  conversion to esters, 1130–1131
  conversion to ethers, 1130–1131
  diastereomers of, 1121, 1122f
  in disaccharides, 1138–1140
  drawing as cyclic hemiacetal, 1120–1122, 1122f
  energy in, 1111
  Fischer projection of, 1114–1115, 1115f
  D-glucose, 1111, 1114, 1115f, 1116, 1117, 1118f, 1120–1121, 1122f, 1125, 1125f, 1126, 1132–1135, 1164, 1262–1263
    α-D-glucose, 866, 1121, 1122f, 1125f, 1127–1129, 1131, 1138, 1141
    β-D-glucose, 865–866, 876, 1110f, 1121, 1122f, 1125, 1125f, 1127–1129, 1131, 1138, 1141
  L-glucose, 1116
  glycoside formation from, 1127–1128
  in honey, 1126
  from hydrolysis
    of cellulose, 1141
    of glycosides, 1128–1129
    of maltose, 1138
    of starch, 181f, 182, 1142
  from Kiliani–Fischer synthesis, 1135
  in lactose, 245, 246f
  metabolism, 753, 992–993
  oxidation of, 230, 233–234, 235–236, 1132–1133
  reduction of carbonyl group, 1132
  structure, 121, 181f
  in sucrose, 1139–1140
  three-dimensional representations for D-glucose, 1125, 1125f
  Wohl degradation, 1134–1135
Glucosidase
  α-glucosidase, 1142
  β-glucosidase, 1141
Glutamate, 912
Glutamic acid, 773
  abbreviation for, 762t
  isoelectric point for, 1167t
  pK_a values for, 1167t
  structure of, 1165f
Glutamine, 762t
  isoelectric point for, 1167t
  pK_a values for, 1167t
  structure of, 1165f
  synthesis of, 912
γ-Glutamyl phosphate, 912
Glutathione, 99, 1179
Glyceraldehyde
  enantiomers of, 1112–1113, 1115–1116
  D-glyceraldehyde, 1111, 1116, 1118f
  L-glyceraldehyde, 1116
  (*R*)-glyceraldehyde, 1112, 1115–1116
  (*S*)-glyceraldehyde, 1112, 1115–1116
  optical activity, 205
Glyceraldehyde 3-phosphate, 938, 991, 992, 993
Glycerol
  naming, 343
  synthesis in lipid hydrolysis, 906
  in triacylglycerols, 401, 1158
(*R*)-Glycerol phosphate, 497
Glycinamide ribonucleotide, 927
Glycine, 762
  abbreviation for, 762t
  in dipeptide synthesis, 1183–1184, 1187
  isoelectric point for, 1167t
  pK_a values for, 1167t
  structure of, 1164f, 1165f
  zwitterion form, 90
Glycogen, 1142
Glycolic acid, 16, 74
  copolymerization with lactic acid, 1256
Glycols, 343
  formation by dihydroxylation, 480
Glycolysis, 992–993
Glycopeptide transpeptidase, 909
Glycoside(s), 1127–1130
  formation, 1127–1128
  α glycoside, 1127–1128
  β glycoside, 1127–1128
  hydrolysis of, 1128–1129
  naturally occurring, 1129–1130
N-Glycoside(s), 1144–1147, 1149f
  α-*N*-glycoside, 1144
  β-*N*-glycoside, 1144
Glycosidic bond/linkage
  in disaccharides, 1137–1140
  α-glycosidic linkage, 1140, 1142
    1→4-α-glycosidic bond, 1138, 1142
    1→6-α-glycosidic bond, 1142
  β-glycosidic linkage, 1137, 1139, 1143, 1145
    1→4-β-glycosidic bond, 1137, 1139, 1143
  in polysaccharides, 1141–1143
Goodyear, Charles, 1254
Gramicidin S, 1206
Grandisol, 1170
Grapefruit mercaptan, 373
Green chemistry, 487–488
  Amberlyst A-26 resin and, 487–488
  definition of, 487
  green polymer synthesis, 1261–1264, 1262f, 1263f
Grignard, Victor, 793
Grignard products, retrosynthetic analysis of, 800–802
Grignard reagents, 793
  addition to nitriles, 920–921
  organic synthesis with, 812–815
  preparation of, 793
  reaction as base, 795, 802
  reaction with aldehydes and ketones, 796–799, 842
  reaction with carbon dioxide, 808
  reaction with epoxides, 808–809
  reaction with esters and acid chlorides, 805–807
  reaction with α,β-unsaturated carbonyl compounds, 811
Ground state
  of atom, 36–37
  of electron, 638–639
Group, in periodic table, 9
Group number, 9, 10
Grubbs catalyst, 1198–1201, 1201f
Grubbs, Robert, 1198
Guanidine, 1062
Guanine, 649, 667, 1145, 1146f
D-Gulose, 1118f
Gutta-percha, 1254, 1255, 1270
Gypsy moth, 478, 479f

H

Half-headed curved arrow, 224, 225–227, 225f, 226t, 473, 579
Halides. *See also* Alkyl halides; Aryl halides; Hydrogen halides; Vinyl halides
  allylic, 256, 256f, 594
  benzylic, 719–720
  methyl
    S_N2 substitution reactions, 280–281, 287
  organic
    coupling with alkenes (Heck reaction), 1186, 1192–1194
    coupling with organoboron reagents (Suzuki reaction), 1186, 1189–1192
    coupling reaction with organocuprate reagents, 1186–1188
  unreactive, 695
α-Halo aldehyde or ketone, synthesis of, 947–951
Halo alkane, 257
α-Halo carbonyl compounds
  reactions of, 951
  synthesis of, 947–950
α-Halo carboxylic acids, amino acid synthesis from, 1167–1168, 1170f
Halocyclohexanes, E2 elimination in, 323–325, 323f
Haloethanes, 178
Haloform, 949–950, 950f
Haloform reaction, 949–950
Halogenation
  of activated benzenes, 710–711
  of alkanes, 582–591
    of achiral starting material, 590, 590t
    bromination, 586–588, 588f
    of chiral starting material, 590t, 591
    chlorination, 582–591, 589f
    energy changes during, 584–585, 585f
    mechanism of, 583–584
    in organic synthesis, 589
    stereochemistry of, 590–591, 590t
  of alkenes, 405f, 413–415, 416f
  of alkyl benzenes, 718–720
  of alkynes, 441f, 444–445
  of benzene, 688f, 690–691, 714
  of carbonyl compounds, at the α carbon, 947–951, 950f
    in acid, 948, 950f
    in base, 948–949, 950f
  electrophilic aromatic substitution reaction, 688f, 690–691, 710–713
  haloform reaction, 949–950
  monohalogenation, 582
  polyhalogenation, 710
  stereochemistry of, 414–415, 416f
  transition states in, 585, 585f, 587, 589, 589f
Halogens
  in alkyl halides. *See* Alkyl halides
  carbon–halogen bond, polar, 260–261, 261f
  common nucleophiles, 269t
  electronegativity, 716
  as electron-withdrawing groups, 701–702, 706
  halo group, 96t
  polarizable, 413
  X as representation for, 13
Halo group, 96t
Halohydrin
  epoxide synthesis from, 351
  synthesis from alkenes, 405f, 416–418, 418t
    mechanism of, 416–417, 418t

Halohydrin (*Continued*)
regioselectivity of, 417–418, 418t
stereochemistry of, 417–418, 418t
use in organic synthesis, 418
Halolactonization, 432
Halomethanes, electrostatic potential map of, 261f
Halomon, 256
Halonium ion, bridged, 414–418, 418t, 444
Halothane, 259f
Hammond postulate, 283–286, 320, 408, 409f, 443, 588, 716
electrophilic aromatic substitution reaction, 705
features of, 283–285
S_N1 substitution reactions, 283, 285–286
Handedness, 181, 184, 186
Hansen's disease (leprosy), 190
Hardener, 1258–1259, 1259f
Hardening, 470, 1160
Haworth projections, 1121, 1122–1124
converting to chair form, 1125
HCFCs (hydrochlorofluorocarbons), 592
HDL particles (high density lipoproteins), 915
HDPE (high-density polyethylene), 601, 1246, 1253, 1265, 1265t
Heat of hydrogenation, 466–467, 623–624
benzene, 656–657, 656f
Heat of reaction, 227. *See also* Enthalpy change
Heat symbol, 220
Heck reaction, 1186, 1192–1194
Heck, Richard, 1192
Helium, electronic configuration of, 10
Heme, 1163, 1198–1199
Hemiacetal(s)
acetals compared to, 1127
in carbohydrates, 865–866
conversion to acetals, 858–859
conversion to carbonyl compounds, 860
cyclic
conversion to cyclic acetals, 864–865
structure, 862
synthesis of, 863–864
monosaccharides, 1119–1126
structure of, 862
synthesis of, 858–860
Hemibrevetoxin B, 95
Hemlock water dropwort, 30
Hemoglobin, 119, 1164
sickle cell, 1199–1200
structure, 1196, 1198–1199, 1199f
Henderson–Hasselbalch equation, 752–754, 752f
Heptalene, 678
Heptanal, 980
Heptane
constitutional isomers of, 137t
molecular formula, 137t
Heptane-2,6-dione, 985
Heptan-2-one, synthesis of, 961
trans-Hept-2-ene, 1187
Herbicides, 691, 691f
Heroin, 115, 213, 531
Heteroatom(s), 4
in condensed structures, 31
definition of, 23, 92
in functional groups, 92, 95, 117
in resonance structures, 23
in skeletal structures, 32, 35
stereogenic centers, 188
Heterocycle(s), 344
aromatic, 664–668, 665f
nitrogen, 1014, 1015f
Heterocyclic aromatic amines, basicity of, 1034, 1035t
Heterogeneous reaction mixture, 466
Heterolysis (heterolytic cleavage)
carbocation formation, 277, 283–285, 289, 294
description, 224
in E1 elimination reactions, 318
reactive intermediates of, 225, 225f
in two-step reaction mechanism, 239
Hexachloroethane, 209
Hexa-2,4-diene, 621, 659
(2*E*,4*Z*)-Hexa-2,4-diene, 1218–1220, 1222, 1226
Hexa-1,3-diyne, 437f
Hexamethylphosphoramide (HMPA), 268f
Hexane, 134, 413, 474f
constitutional isomers of, 137t
infrared (IR) spectrum, 521
mass spectrum of, 506–507, 506f, 509–510, 510f
molecular formula, 137t
skeletal structures, 32
as solvent, 108
solvent in extraction procedure, 758
Hexane-2,5-dione, aldol reactions of, 984
Hexanoic acid, 740, 741, 770
Hexan-2-ol, conversion to other compounds, 812, 813f
Hexaphenylethane, 610
Hexa-1,3,5-triene, 657, 659, 1213, 1217
molecular orbitals of, 1216, 1217f
Hex-1-ene, 413, 521
Hex-2-ene, 395
Hex-3-ene, 474f
cis-hex-3-ene, 396–397, 1198
trans-hex-3-ene, 397
Hexokinase, 234, 234f
Hex-1-yne, 522, 1192
Hex-3-yne, 474f
HFCs (hydrofluorocarbons), 592
High density lipoproteins (HDLs), 915
High-density polyethylene (HDPE), 601, 1246, 1253, 1265, 1265t
Highest occupied molecular orbital (HOMO), 672, 1216, 1219–1222, 1225–1227
High-performance liquid chromatography (HPLC), 1180
Histamine, 665–666, 1018–1019, 1056
Histidine
isoelectric point for, 1167t
pK_a values for, 1167t
structure of, 1165f
Histrionicotoxin, 127, 439, 439f
HMPA (hexamethylphosphoramide), 268f
^{1}H NMR. *See* Proton NMR
Hoffmann, Felix, 746
Hoffmann, Roald, 1214
Hofmann, Albert, 698f
Hofmann elimination, 1038–1041
HOMO (highest occupied molecular orbital), 672, 1216, 1219–1222, 1225–1227
Homogeneous catalysts, 1253
Homologous series, 137, 137t
Homolysis (homolytic cleavage)
of allylic C–H bonds, 594
bond dissociation energy for, 227–228
chlorination of alkanes, 583, 584, 586
description, 224
entropy increase with, 235
radical formation, 580
reactive intermediates of, 224–225, 225f
Homopolymers, 1251
Homotopic protons, 542–543
Honey, 1126
Honeycomb, 170
Hooke's law, 519, 519f
Hormones. *See also specific hormones*
local mediators distinct from, 1166
peptide, 1178–1179, 1179f
sex, 1177–1178, 1178t
Host–guest complex, 348
Housane, 147f
5-HPETE (5-hydroperoxyeicosatetraenoic acid), 380, 608
Hückel, Erich, 658
Hückel's rule, 657–660
basis of, 669–671
Humulene, 333, 1180, 1191, 1191f
Hybridization, 36–45, 652
conjugated systems and, 619–620
definition of, 37
Hybridization effects
on acidity, 75–76, 76f, 77f
on basicity of amines, 1034–1035, 1035t
Hybrid orbitals, 36–45
percent *s*-character, 46, 76, 310
sp orbitals, 38–39, 39t, 42–43, 43f, 44t
in alkynes, 435, 436, 440, 441
percent *s*-character, 76
sp^2 orbitals, 38–39, 39t, 41–42, 42f, 44t
in alkenes, 307–308, 310
in aromatic heterocycles, 664–665, 665f
in benzene, 652
carbon radical, 579
carboxy group, 739
percent *s*-character, 76
sp^3 orbitals, 37–38, 38f, 39, 39t, 40, 44t
in acyclic alkanes, 135
in alkenes, 310
percent *s*-character, 76
superscripts, meaning of, 39
Hydrates
oxidation of, 486
synthesis from aldehydes and ketones, 854–857
kinetics, 856–857
thermodynamics, 855
Hydration
of aldehydes and ketones, 854–857
kinetics, 856–857
thermodynamics, 855
of alkenes, 405f, 412–413, 423
definition of, 844
hydroboration–oxidation compared, 423
reactions, 441f, 445–447
Hydrazine, in Wolff–Kishner reduction, 721
Hydrides
as base, 79, 79f
metal hydride reagents
reducing agents, 465, 474, 781–783, 787–791, 791t
reduction of aldehydes and ketones, 781–783
reduction of nitriles, 919
nucleophilic addition with, 777, 782
nucleophilic substitution with, 778
reduction of ketones and aldehydes, 781–783, 783f, 791t, 841–842
1,2-Hydride shift, 356–358, 696–697
Hydroboration
of alkenes, 419–421
of alkynes, 1190, 1192
definition of, 419
mechanism of, 419–420
regioselectivity of, 420, 422t
Hydroboration–oxidation reactions, 447–449
of alkenes, 405f, 419–423
mechanism of, 419–421, 422t
regioselectivity of, 420, 422t
stereochemistry of, 421–422, 422t
hydration compared, 423
synthesis of artemisinin, 422–423, 423f
Hydrocarbons
aliphatic, 93, 135
aromatic, 93–94, 94t, 649–685
definition of, 2, 93
functional groups in, 93–95, 94t
infrared (IR) absorptions in, 521–522
oxidation and reduction of, 464, 464f
polycyclic aromatic (PAH), 381
saturated, 135
synthesis in organocuprate couplings, 1187–1188
unsaturated, 393. *See also* Alkenes
Hydrochloric acid
in amine extraction, 1029, 1029f
in epoxide ring opening reactions, 377–378, 378f
as Lewis acid, 83
pK_a of, 70, 78
for removal of Boc protecting group, 1185
Hydrochlorofluorocarbons (HCFCs), 592
Hydrofluorocarbons (HFCs), 592
Hydrogen
addition to alkenes, 466–471
addition to alkynes, 471–473, 474f
alpha (α), 935, 939, 948–949, 979–980, 981, 988
classification, 94–95
covalent bond formation by, 12, 13f
electronic configuration, 10
isotopes of, 8, 512t
mass of, 512t
octet rule exceptions, 19
primary (1°), 140
as reducing agent, 465. *See also* Hydrogenation
secondary (2°), 140
sigma (σ) bond in molecular hydrogen, 36
tertiary (3°), 140
Hydrogenation
of aldehydes and ketones, 782
of alkenes, 466–471
alkene stability and, 466–467, 467f
cis and trans isomers compared, 466, 467f

degrees of unsaturation, determination of, 468–469
mechanism of catalytic hydrogenation, 467–468
alkene stability and, 466–467, 467f
catalytic, 466–468
of conjugated dienes, 623–624
enantioselective synthesis of amino acids, 1174–1175
as exothermic reactions, 466
heat of, 466–467, 623–624, 656–657, 656f
of oils, 469–471, 470f
of unsaturated fatty acids, 1160
Hydrogen bonding
in alcohols, 345
in amides, 888, 888f
in amines, 1015
anion solvation by, 267
in carboxylic acids, 742, 743f
in DNA, 104, 104f, 122f, 123, 668, 1147, 1147f
intramolecular in β-dicarbonyl compounds, 936
in polar protic solvents, 267, 289
in protein secondary structure, 1191–1193, 1192f, 1193f
solubility and, 108–109, 110t
strength of, 103, 103t
in water, 103
Hydrogen bromide, for removal of Boc protecting group, 1185
Hydrogen chloride. *See* Hydrochloric acid
Hydrogen cyanide, 16, 844
Hydrogen fluoride, 13
Hydrogen halides
addition to alkenes, 405–412, 405t, 406t, 407t, 411t
addition to alkynes, 441f, 442–444
alcohol conversion to alkyl halides in, 360–363
reactivity of, 362
Hydrogen ions (H^+), Brønsted–Lowry acid definition and, 62
Hydrogenolysis
of benzyl esters, 1186
in peptide synthesis, 1186–1187
Hydrogen peroxide, as oxidizing agent, 475
Hydrohalogenation
of alkenes, 405–412, 405f
Markovnikov's rule, 408–410, 409f, 411t
mechanism of, 405–408, 406f, 407f, 411t
stereochemistry of, 410–412, 411t, 412f
of alkynes, 441f, 442–444
Markovnikov's rule, 442–444
Hydrolysis
of acetals, 860–861
in acetoacetic ester synthesis, 959, 960
of amides, 908
in amino acid synthesis, 1168
base-promoted, 904
definition of, 844
of esters, 903–905
acid-catalyzed, 903–904
applications of, 905–907
base-promoted, 904
glucose formation by, 1117, 1128–1129, 1138, 1141, 1142
of imines and enamines, 854
of lipids, 905–907, 1157
in malonic ester synthesis, 955–957
of nitriles, 917–918
partial of peptides, 1182–1183
of starch, 347, 347f
of triacylglycerols, 1160
Hydronium ion, formal charge on, 17
Hydroperoxides, 596, 597f, 608, 1160
5-Hydroperoxyeicosatetraenoic acid (5-HPETE), 380, 608
Hydrophilic, 109, 113, 114
Hydrophobic, 109, 113, 114, 170, 1157
Hydroxide (^-OH)
as Brønsted–Lowry base, 79
nucleophile in alcohol synthesis, 350
oxygen bases, 808, 808f
α-Hydroxy acids, 74
Hydroxy aldehyde
cyclization of, 863, 865–866
β-hydroxy, synthesis in aldol reactions, 973–985
o-Hydroxybenzaldehyde, 1006
Hydroxybenzene, 653. *See also* Phenol(s)
o-Hydroxybenzoic acid, 773
p-Hydroxybenzoic acid, 773
3-Hydroxybutanal, 973, 975
4-Hydroxybutanal, 863
2-Hydroxybutanedioic acid, 773
4-Hydroxybutanoic acid, 745, 927
1-Hydroxybutan-2-one, 572
4-Hydroxybutan-2-one, 572
3-Hydroxybutyric acid, 1266
γ-Hydroxybutyric acid (GHB), 745
β-Hydroxy carbonyl compounds
synthesis of, in aldol reactions, 973–985
useful transformations of, 981
3-Hydroxy carboxylic acids, 1266
Hydroxy group, 96t
of alcohols, 340, 352
of enols, 340
of ethanol, 93
hydrophilic nature of, 109
IR absorption, 522, 744, 744f
of monosaccharides, 1111, 1127, 1130–1131
NMR absorption, 744, 744f
OH proton, NMR spectra of, 559–560, 559f
of phenols, 340
6-Hydroxyhexan-2-one, 861
Hydroxylamine, 1135
5-Hydroxypentanal, 863
5-Hydroxypentan-2-one, 802–804, 803f
3-Hydroxyvaleric acid, 1266
(-)-Hyoscyamine, 971
Hyperconjugation, 281, 282–283
Hypertension, 132
Hypophosphorus acid, 1045

I

Ibufenac, 735
Ibuprofen, 127, 967, 968
anti-inflammatory action of, 170, 1168
enantiomers, 208
mode of action, 1168
synthesis of, 722, 783f, 989
systematic name of, 139
-ic acid (suffix), 741, 883, 885
Icosane, 137t
-ide (suffix), 257
D-Idose, 1118f
Iejimalide B, 429
Illudin-S, 961
Imatinib mesylate, 276
Imidazole, 665–666, 803, 804
Imidazolide, 925
Imides, in Gabriel synthesis of primary amines, 1022
Imidic acid tautomer, 917–918
Imines, 458, 791, 1136
hydrolysis of, 854
reduction to amines, 1025
synthesis of
from aldehydes and ketones, 850–852, 852f, 853f, 1037, 1169
from nucleophilic addition reactions, 840, 841f
reduction of nitriles, 919–920
Iminium ion, 851, 853, 853f
Indene, 680
Indigo, 1050
Indinavir, 128
Indole, 737
Inductive effects
on acidity, 72–74, 73f, 77f
in aliphatic carboxylic acids, 754–755
on basicity of amines, 1035t
in carbocation stability, 281
definition of, 73
electron-donating, 700–702
electron-withdrawing, 73, 700–702
on substituted benzenes, 700–702
-ine (suffix), 257
Influenza, 99
Infrared (IR) absorptions/spectra
of aldehydes, 833–834, 834f
of amines, 1016, 1017f
atom mass and, 518–519, 519f
of benzene derivatives, 655t
bond strength and, 518–520, 519f, 520t
of carboxylic acid derivatives, 889–890, 890t
of carboxylic acids, 743–744, 744f
characteristic frequencies, A-9
functional groups and, 516–525, 518f, 519f, 520t
of hydrocarbons, 521–522
important, 520, 520t
of ketones, 833–834, 834f
of nitriles, 889
of nitrogen-containing compounds, 523–524
of oxygen-containing compounds, 522–523
percent *s*-character and, 520
vibration of bonds and, 516, 518–521
wavenumber scale of, 516, 517
Infrared (IR) spectroscopy, 512, 514, 516–527
background on, 516–517
functional group identification, 516–525, 518f, 519f, 520t
IR inactive vibration, 521
isomers distinguished by, 524
structure determination in, 525–527
vibration modes, 517
wavenumber scale, 516, 517
Infrared (IR) spectrum
characteristics of, 517–518, 518f
fingerprint region, 517–519, 518f, 519f
functional group region, 517–519, 518f
Ingenol, 1185, 1202
Ingenol mebutate, 1185
Initiator, 1244, 1245, 1249, 1250
Inorganic compounds, 1
Inscribed polygon method for predicting aromaticity, 672–674, 673f
Insecticide, 260, 1195
Inspra. *See* Eplerenone
Insulin, 119, 865, 1164, 1195, 1195f
Integral, NMR spectrum, 549
Intermolecular forces, 101–104
of aldehydes and ketones, 832, 833t
boiling point and, 105–106, 106f
in carboxylic acids, 742, 743f, 743t
dipole–dipole interactions, 103, 103t
electrostatic interactions, 101
hydrogen bonding, 103, 103t, 1015
ion–ion interactions, 101, 103t
melting point and, 106–107
solubility and, 107–111, 110t
van der Waals forces, 101–102, 102f, 103t
International Union of Pure and Applied Chemistry, 138. *See also* IUPAC nomenclature
Intramolecular cyclization, of hydroxy aldehydes, 863
Intramolecular hydrogen bonding, in β-dicarbonyl compounds, 936
Intramolecular malonic ester synthesis, 958
Intramolecular reactions
aldol reactions, 984–985, 985f
Robinson annulation, 996–1000
electrocyclic ring closure, 1213
Friedel–Crafts reactions, 698–699, 698f
intramolecular Claisen (Dieckmann reaction), 990
sigmatropic rearrangements, 1227–1233
Inversion of configuration
in alcohol reactions with HX, 362
in alkyl tosylate reactions, 368, 369
in S_N2 substitution reactions, 273–274, 274f, 275t, 288
Iodine, polarizability of, 102
Iodoform, 949–950
Iodoform test, 949
Iodomethane, polarizability and boiling point of, 106f
1-Iodo-1-methylcyclohexane, 316
Ion–dipole interactions
cation solvation by, 267–268, 289
description of, 108, 110t
Ionic addition, 600
Ionic bonding, 11–12
Ionic compounds
boiling points of, 105
formation of, 11–12
intermolecular forces, 101, 103t
melting points of, 107
solubility of, 108, 110t
Ionic intermediates, 595
Ionic polymerization, 1248–1250, 1249f
Ion–ion interactions, 101, 103t
Ionophores, 116–117, 117f
Ion(s)
anions
acetate, 74–75
carbanion, 76, 76f, 79, 225, 225f
cyclopentadienyl, 662–663, 672
definition of, 8
enolate, 22
naked, 268
radical, 473
salt formation, 12
solvation by hydrogen bonding, 267

Ion(s) (*Continued*)
bond formation by, 225
cations
carbocation. *See* Carbocation(s)
counterions, 62
cyclopentadienyl, 662
definition of, 8
radical, 505
salt formation, 12
solvation by ion–dipole interactions, 267–268, 288
spectator ions, 62
tropylium, 663–664, 672
counterions, 62
polyatomic, 17
spectator ions, 62
transport across cell membranes, 116–117, 117f
IR spectroscopy. *See* Infrared (IR) spectroscopy
Iso- (prefix), 140
Isoamyl acetate, 891
Isobutane, 136
p-Isobutylacetophenone, 722
p-Isobutylbenzaldehyde, 967
Isobutyl chloride, 258f
Isobutylene, 1249f
Isobutyl group, 140
Isocomene, 432
Isocyanate, 1257
Isoelectric point (p*I*), for amino acids, 765, 1167, 1167t
Isoleucine
isoelectric point for, 1167t
pK_a values for, 1167t
structure of, 1165f, 1166
Isomers
constitutional. *See* Constitutional isomers
definition of, 18, 182, 183
infrared (IR) spectroscopy of, 524
Lewis structures, 18–19
meta, ortho, and para, 714–715
resonance structures compared, 20
stereoisomers. *See* Stereoisomers
types of, 183, 183f
Isoniazid, 1056
Isooctane, 53
combustion of, 168
Isopentane, 136, 147. *See also* 2-Methylbutane
Isopentenyl diphosphate, 118, 616, 1171, 1172, 1174
Isophorone, 1008
Isophthalic acid, 769
Isoprene, 622, 639, 639f, 646, 1254, 1270
Isoprene unit, 1169–1171, 1170t, 1182
Isopropyl alcohol, 343
Isopropylamine, 388
Isopropylbenzene
oxidation to benzoic acid, 721
synthesis from benzene, 722
Isopropylcyclopentane, 814–815
Isopropyl group, 140
3-Isopropyl-4-methylcyclohexanone, 831
Isopropyl 2-methylpentanoate, 888
Isoproterenol, 821
Isopulegone, 534
Isotactic polymer, 1252–1253
Isotopes, 8
of bromine, 508
of carbon, 8, 512t
of chlorine, 508
definition of, 8
hydrogen, 512t
masses of common, 512t
of nitrogen, 10, 512t
of oxygen, 512t
Isotretinoin, 873
IUPAC nomenclature
of alcohols, 342, 342f, 343
of aldehydes, 829, 830f
of alkanes, 141–146, 146f
of alkenes, 395–396
of alkyl halides, 257, 258f
of alkynes, 436–437
of benzene derivatives, 652–655
of carboxylic acids, 740–742, 742f
of cycloalkanes, 144–146, 146f
description of, 138–139
of epoxides, 344–345
of ethers, 343–344
of ketones, 830–831, 832f

J

Jack-o'-lantern mushroom, 961
Januvia. *See* Sitagliptin
cis-Jasmone, 472
Johnson, W. S., 985
Jojoba, 1157
Joule (J), 45
Juvenile hormones, 799, 799f, 1187
mimics of, 799

K

Kavain, 398
Kekulé, August, 650
Kekulé structures, 650, 651
Kelsey, Frances Oldham, 190
Keratins, 1164
α-keratins, 1191, 1197, 1197f, 1198f
Kerosene, composition of hydrocarbons in, 148
Ketal, 857
Ketene, 57
Keto–enol tautomers, 445–447
β-Keto esters
as active methylene compounds, 981
conversion to ketones, 959–960
pK_a, 941t
synthesis of
Claisen reaction, 985f, 986–989, 987f, 993
Michael reaction, 996
α-Ketoglutaric acid, 742
D-Ketohexoses, 1118, 1119f, 1125–1126
Ketones. *See also* Carbonyl compounds
aldol reactions, 973–985
boiling point of, 833t
carbohydrates, 865–866
in crossed Claisen reactions, 988–989
fragmentation pattern in the mass spectrum, 511
functional group in, 98t
hydration of, 854–857
kinetics of hydrate formation, 856–857
thermodynamics of hydrate formation, 855
interesting examples, 835–836
iodoform test of, 949
IR spectra of, 522, 833–834, 834f
melting point of, 833t
NMR spectra of, 835
nomenclature, 830–831, 832
common names, 831, 832f
IUPAC system, 830–831, 832f
nucleophilic addition reactions, 777, 782, 827–877
carbanion addition, 842
cyanide addition, 843–844
hydride addition, 841–842
mechanism of, 839–840
nucleophiles in, 840, 841f
primary amine addition, 850–852, 853f
secondary amine addition, 852–854, 853f
physical properties, 832, 833t
pK_a, 941t
protecting groups for, 861–862
reactions of
aldol reactions, 973–985
with amines, 1037
at the carbonyl carbon, 838, 839
crossed Claisen reactions, 988–989
halogenation at the α carbon, 947–951, 950f
hydration, 854–857
nucleophilic addition reactions, 777, 827–877
with organometallic reagents, 795–799, 799f, 842
at the α carbon, 838–839
reactivity of, 777
reduction of, 841–842
to alcohols, 780, 781, 782
of aryl ketones to alkyl benzenes, 721–722
reductive amination to amines, 1025–1027
to secondary alcohols, 469
solubility of, 833t
structure of, 775, 828
synthesis
from carboxylic acid derivatives, 806–807
synthesis of, 836–838
acetoacetic ester synthesis, 955, 959–962
alcohol oxidation, 485–486
Friedel–Crafts acylation, 688f, 693, 721, 837
hydration of alkynes, 445–447, 837
hydroboration–oxidation of alkyne, 448
hydrolysis of acetals, 860–861
hydrolysis of imines and enamines, 854
oxidation of secondary alcohols, 837
oxidative cleavage of alkenes, 482–483, 837
reactions of acid chlorides with organocuprates, 837
unreactivity to nucleophilic substitution, 779
unsymmetrical, alkylation of, 953–954
Wittig reaction, 845–849
Ketoprofen, 78, 213
Ketoses
D-ketose family, 1118, 1119f
structure, 1111–1112
Keto tautomers, 684, 918, 936–937
β-Keto thioester, 993, 994
Kevlar, 1256, 1270
Kiliani–Fischer synthesis, 1134, 1135–1137
Kinetic enolates, 944, 945, 953, 981
Kinetic product, in electrophilic addition reactions of conjugated dienes, 627–629, 628f
Kinetic resolution, of amino acids using enzymes, 1173–1174
Kinetics
definition of, 231, 241
E1 elimination reactions, 318, 320t
E2 elimination reactions, 311, 315t
first-order, 243
rate equations, 242–244
second-order, 243
S_N1 substitution reactions, 277, 281t
S_N2 substitution reactions, 271, 275t
Kukui nuts, 391

L

β-Lactam family of antibiotics, 511, 525–526
Lactams, 880, 1256
β-Lactam(s), 880, 909–910
Lactase, 245, 246f, 1139
Lactate dehydrogenase, 787
Lactic acid, 32, 745, 753, 787
copolymerization with glycolic acid, 1256
optical activity of, 205
specific rotation of, 207
Lactide, 1263
Lactols, 862
Lactone(s), 880
synthesis by esterification of carboxylic acids, 901
Lactose, 121, 865, 1133, 1139
breakdown of, 245, 246f
Lactose intolerance, 1139
Laetrile, 844
Lanolin, 1157
Lanosterol, 1176f, 1177
Lard, 403
Latanoprost, 747
Latex, 1254, 1255
Laurencia cartilaginea, 260
Lauric acid, 1158t
Lavandulol, 801
Lavender oil, 339
l (levorotatory) compounds, 205, 1116
LDA. *See* Lithium diisopropylamide (LDA)
LDL particles (low density lipoproteins), 915
L-Dopa (levodopa), 7, 50–51, 1205
LDPE (low-density polyethylene), 601, 1246–1247, 1265t
Lead(II) acetate, 472
Leaving group
of alcohols, 352, 354–355, 357–362, 364–369
of carbonyl compounds, 775, 777, 778–779
of carboxylic acid derivatives, 880, 882t, 892–893
diphosphate, 616, 1171, 1173
in E1 elimination reactions, 318, 320t
in E2 elimination reactions, 312–314, 315t
in nucleophilic acyl substitution reactions, 880, 892–893
in nucleophilic substitution reactions, 261–265, 273–274, 288–289
derived from phosphorus, 291–292

good, 263–265, 264t
periodic trends in ability, 263–264
poor, 264t
reaction rate and, 283
weak bases, 263–265
sulfonate anion, 760
in terpene biosynthesis, 1171–1173
Le Châtelier's principle, 355, 900, 1127
Lecithin, 1162, 1163f
Length, units used to report wavelength, 515
Leprosy, 190
Letrozole, 916
Leucine
isoelectric point for, 1167t
metabolism of, 1003
pK_a values for, 1167t
structure of, 1165f
Leukotrienes, 1166, 1167f, 1167t
biological activity of, 380
leukotriene A_4, 380
leukotriene C_4, 99, 380, 608
synthesis of, 380
Levonorgestrel, 439
Levorotatory (*l*) compounds, 205, 1116
Lewis acid, 81–83
borane as, 419
carbocation as, 407
catalyst, 362
Lewis acid–base reactions, 82–83, 262. *See also* Nucleophilic substitution reactions
electrophilic addition, 407
S_N1 substitution reactions, 277
Lewis base, 81–83
alkene as, 407
nucleophile as, 261, 265, 277
tetrahydrofuran (THF) as, 419
Lewis structures, 13–18
of acyclic alkanes, 135–136
of benzene, 64
condensed formula conversion to, 32
drawing of, 13–15
formal charge, 16–18, 18t
isomers, 18–19
multiple bonds in, 15–16
octet rule exceptions, 19
resonance structures, 19–25
Lexan, 1243f, 1258, 1264
Ligands, in palladium-catalyzed reactions, 1188–1189
Light
polarized, 204–205
speed of, 515
symbol, 220
Lightning, 482
"Like dissolves like," 108
Limonene, 56, 483, 646, 1173
(*R*)-limonene, 400f
Linalool, 339, 801
Linalyl diphosphate, 646, 1173
Linamarin, 844
Lindlar catalyst, 472, 474f
Linear molecule, 26, 27, 29t, 49
alkynes, 435
Linezolid, 933
Linoleic acid, 597, 1158, 1158t, 1159t
as essential fatty acid, 401
melting point, 402t
sources of, 391, 401
structure, 402f, 402t
Linolenic acid, 498, 1158, 1158t
as essential fatty acid, 401
melting point, 402t
as omega-3 fatty acid, 402–403
sources of, 391, 401, 403
structure, 402f, 402t
Lipases, 906
Lipid bilayer, 114, 115f, 117f, 1163–1164
Lipids, 135, 170–171, 401–403, 1155–1184
as biomolecules, 123–124, 124f
bonding in, 1156
definition of, 170
eicosanoids, 1166–1169, 1167f, 1168t
energy in, 171, 1161
examples of, 170–171, 170f, 1156f
fat-soluble vitamins, 1165–1166, 1165f, 1166f
hydrolysis of, 905–907, 1157
hydrolyzable, 1157
metabolism of, 1161
nonhydrolyzable, 1157
origin of word, 1156
oxidation of, 596–597, 597f, 598
phospholipids. *See* Phospholipids
solubility of, 170, 1156
steroids. *See* Steroids
structure of, 170f
terpenes, 1169–1174, 1170f, 1170t, 1171f
triacylglycerols. *See* Triacylglycerols
unsaturated, oxidation of, 596–597, 597f
waxes, 1157
Lipitor. *See* Atorvastatin
Lipoic acid, 100
Lipoprotein, 915, 915f
Lipoxygenase, 377, 1167, 1167f
Lisinopril, 196
Lithium
electronegativity value of, 793
valence electrons of, 12
Lithium acetylide, 794, 798
Lithium aluminum hydride ($LiAlH_4$), 781–783, 787–791, 791t
as reducing agent, 465
reduction of aldehydes and ketones, 841
reduction of alkyl halides, 474–475, 475f
reduction of epoxides, 474–475, 475f
reduction of a nitrile, 919, 1024
Lithium diisopropylamide (LDA), 952–954, 981–982
enolate formation with, 942–945
pK_a, 942t
preparation of, 943
Lithium dimethylcuprate, 794, 1197
Lithium fluoride, 12
Lithium tri-*tert*-butoxyaluminum hydride, 787, 788, 791t
Lithium tri-*sec*-butylborohydride (L-selectride), 825
(–)-Littorine, 971
Living polymerization, 1250
Local anesthetic, 723, 735
Local mediators, 1166
Log, 66
London forces, 101. *See also* van der Waals forces
Lone pairs
in amines, 1028, 1034, 1037
in aromatic heterocycles, 664–666
Brønsted–Lowry bases and, 62–63, 62f
in carbanions, 225
in heteroatoms, 35, 92, 117
in hydrogen bonding, 103
in Lewis structures, 14–16
of negatively charged carbon atoms, 35
number of, 35
in resonance structures, 21, 23–24, 74, 617, 620
second-row elements and, 13
Long, Dr. Crawford, 348
Losartan, 1205
Low density lipoproteins (LDLs), 915
Low-density polyethylene (LDPE), 1246–1247, 1265t
Lowest unoccupied molecular orbital (LUMO), 672, 1216, 1221, 1225–1227
LSD, 951
intramolecular Friedel–Crafts acylation in the synthesis of, 698, 698f
structure of, 1021
synthesis of, 698, 698f
LUMO (lowest unoccupied molecular orbital), 672, 1216, 1221, 1225–1227
Lupin, 1036, 1062
Lupinine, 1062
Lycopene, 622, 639, 639f, 1182
Lycra. *See* Spandex
Lynen, Feodor, 1176
Lysergic acid, 648
Lysergic acid diethyl amide. *See* LSD
Lysine, 738, 773
abbreviation for, 762t
ionizable side chain of, 1166, 1167t
isoelectric point for, 1167t
pK_a values for, 1167t
structure of, 1165f
Lysozyme, 1193, 1194f
D-Lyxose, 1117, 1118f, 1135

M

M + 1 peak, in mass spectrum, 506
M + 2 peak, in mass spectrum, 508–509
Magnesium, electronegativity value of, 793
Magnetic field, 536–539, 564
Magnetic resonance imaging (MRI), 568, 568f
Mahimahi, 666
Ma huang, 187, 214
Ma'ilione, 260
Malaria, 4–6, 5f, 315
Maleic acid, 133, 632
Maleic anhydride, 632f
Malic acid, 742
Malonic acid, 90
Malonic ester synthesis, 955–959
intramolecular, 958
retrosynthetic analysis of, 958–959
steps in, 956–958
Maltose, 877, 1138
Mandelic acid, 216, 752
(*R*)-Mandelic acid, 1204
Manganese, in oxidizing agents, 476
Mannose
in Fischer proof, 1138–1139
D-mannose, 1117, 1118f, 1135
Haworth projections of, 1122–1123
Margarine, 469–470, 1048
Marijuana, 78
Markovnikov addition of water, 445
Markovnikov's rule, 408–410, 409f, 411t, 442–444, 447, 625
in cationic polymerization, 1248
Mass number, 8
Mass spectrometer
features of, 504
high-resolution, 512
schematic of, 504f
Mass spectrometry (MS)
of alkyl halides, 508–509, 509f
of amines, 1016, 1017f
analysis of unknowns using the molecular ion, 506–508
base peak, 505
of biomolecules, 514
definition of, 504
electrospray ionization (ESI), 514
features of, 504–506
fragmentation in, 505, 509–512, 510f
gas chromatography and, 513–514, 513f, 514f
high-resolution, 512
M peak, 505, 508
M + 1 peak, 506
M + 2 peak, 508–509
Mass spectrum, 505
Mass-to-charge (*m/z*) ratio, 504f, 505, 512
Matrine, 1036
Mauveine, 1050
Maytansine, 1210
MDMA (3,4-methylenedioxymethamphetamine), 1060
Mefloquine, 217
Melamine, 1274
Melanin, 640
Melmac, 1274
Melting point
of alcohols, 346t
of aldehydes and ketones, 833t
of alkanes, 149, 149t
of alkenes, 399, 401, 402t
of alkyl halides, 258t
of alkynes, 437
of amines, 1015t
of carboxylic acids, 743t
definition of, 106
of enantiomers, 205, 205t
of epoxides, 346t
of ethers, 346t
of fats, 403, 1158–1159
of fatty acids, 401, 402t
intermolecular forces and, 106–107
of monosaccharides, 1119
of oils, 403, 1158–1159
of racemic mixtures, 205, 205t
of triacylglycerols, 470f
Membrane proteins, 1164
Menthol, 56, 1169, 1170
Meperidine, 97, 967
Mercaptans, 372–373
Mercapto group, 96t, 372–373
Mercury, in Clemmensen reduction, 721
Merrifield, R. Bruce, 1188
Merrifield method, of peptide synthesis, 1188–1190
Mescaline, 1019
Mesityl oxide, 825
Meso compounds, 199–200
achiral compounds, 199, 201, 415, 416f, 632, 1196
carbene addition and, 1196
disubstituted cycloalkanes, 201
from epoxidation of alkenes, 478
plane of symmetry, 199

Mestranol, 129
Meta (prefix), 653
Metabolism, 233–235, 234f, 1111
Meta director, 703–705, 707, 709, 710f, 711–712, 714, 724, 726
Metal–carbenes, 1200
Metal catalyst, 244, 466
Metal hydride reagents
 reducing agents, 465, 474, 781–783, 787–791, 791t
 reduction of aldehydes and ketones, 781–783
 reduction of nitriles, 919
Metals
 as catalysts, 244–245
 in oxidizing agent, 476
Metathesis, 1198–1203, 1199f, 1201f
 derivation of word, 1198
 drawing products of, 1199, 1199f
 mechanism, 1200
 ring-closing (RCM), 1201–1203, 1201f
Met-enkephalin, 892
Meth- (prefix), 139
Methamphetamine, 1016f, 1019
Methanal. *See* Formaldehyde
Methanamine. *See* Methylamine
Methane
 bonding in, 36–38, 38f
 combustion of, 168
 description, 2
 halogenation of, 582, 582f
 Lewis structure, 14
 mass spectrum of, 505–507
 molecular formula, 135, 136, 137t
 molecular shape of, 27–28
 in natural gas, 147
 oxidation of, 464, 464f
 pK_a of, 67, 70
 structure of, 135
 temporary dipole, 101
 whole-number mass of, 505
Methanesulfonic acid, 760
Methanethiol, 147
Methanol
 ether formation by addition to alkenes, 413
 hybrid orbitals in, 40
 Lewis structure, 14–15
 oxidation of, 490
 as polar protic solvent, 267f
 structure of, 347f
 toxicity of, 347f
Methionine, 374
 abbreviation for, 762t
 functional groups, 97
 isoelectric point for, 1167t
 pK_a values for, 1167t
 structure of, 1165f
 synthesis of, 1170f
Methoprene, 799
p-Methoxybenzoic acid, 769
Methoxycyclohexane, conformations of, 233
Methoxy group, 344
Methoxy methyl ether, 872
p-Methoxyphenyl ester, 1207
Methyl acetate, 252
 formation of, 270–271
 NMR spectrum of, 565, 566, 567f
 structure of, 879
Methyl acrylate, 632f
Methylamine, 1013
 electrostatic potential plot of, 1012, 1012f
m-Methylaniline, 717
N-Methylaniline, NMR spectrum of, 1017, 1017f
p-Methylaniline, 717–718
 basicity of, 1032
 electrostatic potential plot of, 1033f
Methylation, 293
N-Methylbenzamide, 887t
Methylbenzene, 653. *See also* Toluene
Methyl benzoate, synthesis of, 900
(*R*)-α-Methylbenzylamine, 1171–1173
2-Methylbuta-1,3-diene, 622, 1254
2-Methylbutane, 107, 136, 147, 605
2-Methylbutanenitrile, structure of, 886f
3-Methylbutane-1-thiol, 372
2-Methylbutan-2-ol, 353
2-Methylbutanoyl chloride, structure of, 883
3-Methylbut-1-ene, 595
3-Methylbut-2-en-1-ol, 646
(*R*)-α-Methylbutyrophenone, 462
2-Methyl-CBS-oxazaborolidine, 784
Methyl chloride, electrostatic potential plot for, 49, 49f
Methyl *trans*-chrysanthemate, 1208
3-Methylcycloheptene, 396f
Methylcyclohexane, 145, 163, 163f
2-Methylcyclohexane-1,3-dione, 997
2-Methylcyclohexanethiol, 372f
trans-4-methylcyclohexanol, 288
3-Methylcyclohexanol, 342f
 cis-3-methylcyclohexanol, 369
2-Methylcyclohexanone, 447, 982
 alkylation of, 953–954
 enolates formed from, 944
1-Methylcyclohexene, 813, 1188
3-Methylcyclohexene, 189–190
5-Methylcyclopenta-1,3-diene, 680
2-Methylcyclopentanamine, in Hofmann reaction, 1040
Methylcyclopentane, 189
2-Methylcyclopentanecarboxamide, structure of, 885
(*R*)-3-Methylcyclopentanone, 842
1-Methylcyclopentene, 396f
Methylene, 1197
α-Methylene-γ-butyrolactone, 955
Methylene chloride, 259f
Methylenecyclohexane, 398f, 849
Methylene group, 137, 398
Methyl ester, 1186
3-Methylfentanyl, 508
Methyl α-D-glucopyranoside, 1128, 1133f
Methyl group, 140
 angular, 1174–1175
 as electron-donating group, 703, 708
 inductive effect of, 700
 as ortho, para director, 703, 707–708, 713, 715, 725
Methyl halides
 coupling reaction with organocuprates, 1186
 S_N2 substitution reactions, 275, 280–281, 287
6-Methylhept-6-en-2-ol, 396
4-Methylhexane-2-thiol, 372f
5-Methylhexan-3-ol, 342
5-Methylhexan-2-one, 518, 518f
2-Methylhexanoic acid, synthesis of, 959
5-Methylhex-4-en-1-yne, 437f
Methyl 5-hydroxyhexanoate, 861
Methyl ketones, 445, 448
 halogenation of, 949, 950f
 product of acetoacetic ester synthesis, 955, 959–962
Methyllithium, 793
Methylmagnesium bromide, 793
Methylmagnesium chloride, 802
Methyl 2-methylbutanoate, 891
N-Methylmorpholine *N*-oxide (NMO), 481
Methyl orange, 1049
3-Methylpentanal, 830f
4-Methylpentan-1-amine, 1014
2-Methylpentane, 183f
3-Methylpentane, 183f
4-Methyl-1,4-pentanediol, 802–804, 803f
2-Methylpentan-3-one, 832f
3-Methylpentan-2-one, 831
3-Methylpent-2-ene, 397
 (*E*)-3-methylpent-2-ene, 397
 (*Z*)-3-methylpent-2-ene, 397
4-Methylpent-3-en-2-one, 832
1-Methyl-4-phenyl-4-propionoxypiperidine, 508
N-Methylpropan-2-amine, 1013
2-Methylpropane, 136, 140
2-Methylpropan-1-ol, 296
2-Methylpropan-2-ol, 697
2-Methylpropene, 736, 1197, 1249f
2-Methylpropene oxide, 1271
2-[4-(2-Methylpropyl)phenyl]propanoic acid, 139
4-Methylpyridine, 1008
1,2-Methyl shift, 410
(*S*)-3-Methyl-3-sulfanylhexan-1-ol, 372
Methyl thioacetate, 879
Methyltriphenylphosphonium bromide, 846
β-Methylvaleraldehyde, 830f
Methyl vinyl ketone, 632f, 997
Metoprolol, 88
Micelles, 113, 113f
Michael acceptor, 995
Michael reaction, 994–996, 996f
Mifepristone, 439
Milkweed, 1185
Mineral oil, 178
Minus (–) compounds, 205
Mirror images, 181, 184–186, 198
Miscibility, 109
Misoprostol, 1168
Mixed aldol reaction. *See* Crossed aldol reactions
Molecular formula
 of acyclic alkanes, 135, 137, 137t
 of alkenes, 393
 of alkyl halides, 256
 of alkynes, 435
 of amines, 1016
 of butane, 137t
 of carbohydrates, 1110f
 of cycloalkanes, 135, 138
 of decane, 137t
 of ethane, 136, 137t
 of heptane, 137t
 of hexane, 137t
 of icosane, 137t
 of methane, 135, 136, 137t
 of nonane, 137t
 of octane, 137t
 of pentane, 137t
 of propane, 136, 137t
Molecular ion, 505–509
Molecular orbitals (MOs), 669–674, 670f, 671f
 antibonding, 669–670, 670f, 671f, 672, 674
 bonding, 669–674, 670f, 671f, 673f
 of buta-1,3-diene, 1215–1216, 1215f
 degenerate, 671
 of ethylene, 1214, 1215f
 of hexa-1,3,5-triene, 1216, 1217f
 highest occupied MO (HOMO), 672
 lowest unoccupied MO (LUMO), 672
 patterns for cyclic, completely conjugated systems, 673f
 prediction/determination of relative energies of, 672–674
Molecular orbital (MO) theory, 669, 1217
Molecular recognition, 348
Molecular shape
 bent, 29, 49
 determining
  bond angle, 26–29
  bond length, 26, 26t
 four groups around an atom, 26–29, 29t, 44t
 linear, 26, 27, 29t, 49
 tetrahedral, 26, 27–28, 29t
 three groups around an atom, 26, 27, 29t, 44t
 trigonal planar, 26, 27, 29t
 trigonal pyramid, 28
 two groups around an atom, 26, 27, 29t, 44t
Molecules
 definition of, 12
 polarity of, 49, 50f
Molina, Mario, 592
Molozonide, 482
Molybdenum, 1198
Monensin, 431, 865
Monohalogenation, 582
Monomers, 1243–1244
Monophosphate, 100, 100t
Monosaccharides, 1111–1137
 D-aldoses, family of, 1116–1118, 1118f
 as biomolecules, 120–121, 121f
 common names of, 1111, 1117
 constitutional isomers of, 1111
 conversion to esters, 1130–1131
 conversion to ethers, 1130–1131
 cyclic forms, 1119–1126
  anomers, 1121–1126, 1125f
  chair form, 1125
  drawing glucose as a cyclic hemiacetal, 1120–1122, 1122f
  furanoses, 1125–1126
  Haworth projections, 1121, 1122–1124
  three-dimensional representations for D-glucose, 1125, 1125f
 determining structure of unknown, 1136–1137
 examples of, 1111
 Fischer projection formulas of, 1112–1116, 1115f
 glycosides, 1127–1130
  formation, 1127–1128
  hydrolysis, 1128–1129
  naturally occurring, 1129–1130
 isomers (D and L) of, 1115–1116
 D-ketoses, family of, 1118, 1119f
 melting point of, 1119
 physical properties of, 1119
 reactions at the carbonyl group, 1131–1137
  addition or removal of one carbon, 1134–1137
  Kiliani–Fischer synthesis, 1134, 1135–1137

oxidation, 1132–1134, 1132f
reduction, 1132
Wohl degradation, 1134–1135, 1136
reactions at hydroxy groups, 1130–1131
solubility of, 1119, 1131
stereogenic centers in, 1112–1121, 1123–1126
D and L designations, 1115–1116
multiple, 1114–1115
structure of, 1111
sweetness of, 1119
Monosodium glutamate (MSG), specific rotation of, 207
Monoterpene, 1170, 1170t, 1172, 1173
Morphine, 62–63, 115, 531, 612, 638, 769, 1018
Morton, Dr. William, 348
Motrin. *See* Ibuprofen
M peak, in mass spectrum, 505, 508
MPPP, 508
MRI (magnetic resonance imaging), 568, 568f
MS. *See* Mass spectrometry (MS)
MSG (monosodium glutamate), specific rotation of, 207
MTBE. *See tert*-Butyl methyl ether (MTBE)
Multiple bond
in condensed structures, 31
Lewis structures, 15–16
position of, 21, 25
Multiplet, NMR spectrum, 552, 552t, 556, 556f
α-Multistriatin, 492, 492f
Muscalure, 303, 500
Muscone, 783f
Mushroom
jack-o'-lantern, 961
Psilocybe, 1020
Musk deer, 783
Mutarotation, 1121, 1127, 1138, 1139
Mycomycin, 217
Myelin sheath, 132, 1164
Myoglobin, 1163, 1199, 1199f
Myosin, 1191
Myrcene, 56, 499, 1169, 1170
Myristic acid, 1158t

N

N- (prefix), 886
Nabumetone, 962
NAD^+ (nicotinamide adenine dinucleotide), 489
NADH (nicotinamide adenine dinucleotide, reduced), 489, 786–787
NAG (*N*-acetyl-D-glucosamine), 916, 1143
Nandrolone, 1178
Naphthalene, 661, 733
1-Naphthol, 388
Naprosyn. *See* Naproxen
Naproxen, 86, 209, 953
Natta, Giulio, 1253
Natural gas, 40, 147
Natural product, 295
NBS (*N*-bromosuccinimide), 417, 594–596, 719
Neopentane, 136. *See also* 2,2-Dimethylpropane (neopentane)
Neoprene, 1255
Nepheliosyne B, 436
Neral, 832
Nerol, 433
Nerolidol, 333
Neryl diphosphate, 1173, 1183
Neuroprotectin D1, 621
Neurotoxin (hemibrevetoxin B), 95
Neurotransmitter
acetylcholine, 914
dopamine, 1019–1020, 1020f
serotonin, 1020
Neutrons, 8
Newman projection, 151–152, 152f, 154f
Niacin, 112, 787
Nickel, as catalyst, 244
Nicotinamide adenine dinucleotide (NAD^+), 489, 683
Nicotinamide adenine dinucleotide, reduced (NADH), 489, 786–787
Nicotine, 59, 115, 294, 302, 1018, 1018f, 1035
Niphatoxin B, 451, 451f
Nitration
of benzene, 688f, 691–692, 714, 724
electrophilic aromatic substitution reactions, 688f, 691–692, 714–715
Nitric oxide, 579, 592
Nitriles, 879, 916–920
acidity of, 940
addition of organometallic reagents, 920–921
dipole–dipole interactions of, 888
hydrolysis of, 917–918, 1169
infrared (IR) spectra of, 523
IR spectra of, 889
NMR spectra of, 890
nomenclature, 886–888, 886f, 887t
physical properties, 888
pK_a of, 940, 941t
products of S_N2 substitution reactions, 294, 295t
reduction of, 917, 919–920, 1024
structure and bonding of, 880, 883, 916
synthesis of, 916
p-Nitroaniline
basicity of, 1032
electrostatic potential plot of, 1033f
Nitrobenzene, 688f, 691, 692, 703, 709
reduction to aniline, 723
Nitrogen
in aromatic heterocycles, 664–668, 665f
bonding, 13f, 18t
common nucleophiles, 269t
infrared (IR) absorptions in nitrogen-containing compounds, 523–524
isotopes, 10, 512t
mass of, 512t
Nitrogen heterocycles, 1014, 1015f
Nitrogen–nitrogen double bond, 1048
Nitrogen nucleophiles, 893
Nitrogen rule, 508
Nitro group
as electron-withdrawing group, 703, 715–716
in Friedel–Crafts reactions, 711
as meta director, 703, 710, 712, 714, 724
reduction of, 723–724, 1023
Nitronium ion, 691
o-Nitrophenol, 653
p-Nitrophenol, 715
p-Nitrophenyl ester, 1207
Nitrosamines, 1041–1043
formation of
N-nitrosamines from primary amines, 1042
N-nitrosamines from secondary amines, 1043
N-Nitrosodimethylamine, 1043
Nitrosonium ion, 1041–1043
o-Nitrotoluene, synthesis from benzene, 714–715
p-Nitrotoluene, 712
Nitrous acid, reaction of amines with, 1042–1043
NMO (*N*-methylmorpholine *N*-oxide), 481
NMR. *See* Nuclear magnetic resonance (NMR) spectroscopy
NMR spectrometers, 537–538, 537f
Noble gases, stability of, 11
Node of electron density, 9
Nolvadex. *See* Tamoxifen
Nomenclature, A-3–6. *See also* Common names; IUPAC nomenclature
of acid chlorides, 883, 887t
of acyl phosphates, 883, 887t
of alcohols, 342–343, 342f
of aldehydes, 829–830, 830f, 832
of alkanes, 139–147, 146f
acyclic alkanes, 141–144, 143f
common names, 147, 147f
cycloalkanes, 144–146, 146f
IUPAC system, 141–146, 146f
substituents, 140
of alkenes, 395–398
of alkyl halides, 257–258, 258f
of alkyl substituents that contain branching, A-3
of alkynes, 436–437
of amides, 885–886, 887t
of amines, 1013–1015
of anhydrides, 887t
benzene derivatives, 653–655
of bicyclic compounds, A-5–6
of carboxylic acids, 740–742, 741t, 742f
of epoxides, 344–345
of esters, 883–884, 887t
of ethers, 343–344
E,Z system, 397–398
generic name, 139
of ketones, 830–831, 832, 832f
of monosaccharides, 1111–1112, 1117
of nitriles, 886–888, 886f, 887t
of polyfunctional compounds, A-4–5
of sulfides, 373–374
systematic name, 139
of thioesters, 883–886
of thiols, 372–373, 372f
trade name, 139
Nomex, 1270
Nonactin, 116, 116f, 117
Nonane
constitutional isomers of, 137t
molecular formula, 137t
Nonbonded interactions. *See* Intermolecular forces
Nonbonding (unshared) electrons, 13, 13f
Noncovalent interactions. *See* Intermolecular forces
Nonnucleophilic bases, 266
Nonpolar bond, 48
Nonpolar molecule
description of, 49, 50f
solubility of, 108
Nonreducing sugars, 1133, 1133f, 1140
Nonrenewable energy, 148
Nonsteroidal anti-inflammatory drugs (NSAIDs), 1168–1169
Nootkatone, 508
Noradrenaline (norepinephrine)
structure of, 1019
synthesis of, 293, 293f
Norethindrone, 111, 213, 438, 438f, 1178
Novocain. *See* Procaine
Noyori, Ryoji, 1175
NSAIDs (nonsteroidal anti-inflammatory drugs), 1168–1169
Nuclear magnetic resonance (NMR) spectroscopy, 504, 514, 535–577
^1H (proton), 536, 538–564. *See also* Proton NMR
^{13}C (carbon), 536, 564–567. *See also* Carbon NMR
basis of, 536–538
medical uses of, 568, 568f
nuclear spin in, 536–537
reference signal, 538
signal versus peak, 551
spectrometers, 537–538, 537f
variables in, 537
Nucleic acids, 122–123, 122f
Nucleophiles, 98, 265–269
acetylide anions as, 442, 449
addition to carbonyl groups, 840, 841
alkyl halide substitution reactions with, 261
ambident, 943
amines as, 1011, 1028, 1036–1038
attack of electron-deficient atoms by, 265
biological, 291–294
carbanions as, 225
common, table of, 269t
enolates, 943, 947
enols, 938
hydroxide, in alcohol synthesis, 350
Lewis base, 82
meaning of term, 82
negatively charged, 262, 269t
neutral, 262–263, 269t
nitrogen, 893
opening of epoxide rings with strong, 375–377
organometallic reagents as, 792, 793, 795–796
oxygen, 893
pi (π) bonds and, 92, 117
reactivity of, 117–118
strength, 265–269, 287–289, 290t
in substitution reactions, 261–262, 287–291. *See also* Nucleophilic substitution reactions
Nucleophile strength. *See* Nucleophilicity
Nucleophilic acyl substitution reactions, 778, 892–895, A-12
biological, 914–916, 915f
examples of, 896
general reaction of, 935
leaving groups in, 880
mechanism of, 892–893
polyester synthesis, 1256–1257

Nucleophilic acyl substitution reactions (*Continued*)
reactions of acid chlorides, 896–897, 910t
reactions of amides, 908–910, 910t
reactions of anhydrides, 897–898, 910t
reactions of carboxylic acids, 898–903, 898f, 910t
conversion to acid chlorides, 898f, 899
conversion to amides, 898f, 901–903
conversion to cyclic anhydrides, 898f, 899
conversion to esters, 898f, 900
reactions of esters, 903–905, 910t
reactivity of carboxylic acids and derivatives in, 893–894
summary of reactions, 910, 910t
Nucleophilic addition reactions
aldehydes and ketones, 777, 782, 788, 796–799, 805, 827–877, A-13
aldol reactions. *See* Aldol reactions
of amines, 1036–1038
of carbonyl compounds, 935, 956
Michael reaction, 995
with organometallic reagents, 796–798, 799f, 805, 811–812
polyurethane synthesis, 1257
reduction of acid chlorides and esters, 788, 805
Strecker synthesis, 1169–1170
α,β-unsaturated carbonyl compounds, 809–811
Nucleophilic aromatic substitution, 715–718, A-12
by addition–elimination, 715–717
by elimination–addition, 717–718
Nucleophilicity
basicity compared to, 265–266
periodic trends in, 266, 267, 268
S_N1 versus S_N2 substitution reactions, 287–288
solvent effects, 266–269
polar aprotic solvents, 268, 268f
polar protic solvents, 267, 267f
steric effects and, 266
Nucleophilic substitution reactions, A-12
acetylide anion reactions with alkyl halides, 449–451
adrenaline synthesis, 293–294, 293f
of alcohols, 351–352, 360–369, 370f
alcohol synthesis by, 349–350
aldehyde formation by, 788
of alkyl halides, 261–296
amine formation by, 1021–1023
bimolecular, 271
biological, 291–294
of carbonyl compounds, 777, 778–779, 935, 950–952
Claisen reaction, 986–989, 987f
components of, 261
crown ether use in, 348, 348f
epoxide ring opening reactions, 375–379
equilibrium and, 265
ethers with strong acids, 370–371
ether synthesis by, 349–350
features of, 261–263
first-order rate equation, 270–271
leaving groups in, 261–265, 264t, 273–274, 288–289, 291–292
mechanisms, 269–296
distinguishing between S_N1 and S_N2, 286–291, 287f, 290t
one-step, 270, 272, 275t
order of bond breaking and bond making, 269–272, 277–278
S_N1, 277–286, 277f, 278f, 279f, 281t, 287f, 290t
S_N2, 271–276, 272f, 274f, 275f, 276f, 276t, 287f, 290t, 295t
two-step, 270, 271, 277–278, 281t
nucleophile strength (nucleophilicity), 265–269, 287–289, 290t
organic synthesis by, 294–296, 295t
in polycarbonate synthesis, 1258
rate-determining step in, 270, 271
second-order rate equation, 270–271
solvent effects on, 266–269, 267f, 268f, 289, 290t
unimolecular, 271, 277
vinyl halides and, 294
Nucleotides, 119, 122–123, 122f
Nucleus, atomic, 8
Nuts, as vitamin E source, 598
Nylon
nylon 6, 1256, 1266
nylon 6,6, 1243f, 1244, 1255, 1262, 1262f
synthesis of, 1255–1256

O

-o (suffix), 257
Occidentalol, 648
(*E*)-Ocimene, 309
Octane
constitutional isomers of, 137t
gasoline and, 93, 109
molecular formula, 137t
naming, 141–142
Octanenitrile, infrared (IR) spectrum of, 524
(2*E*,4*Z*,6*E*)-Octa-2,4,6-triene, 1218–1221
Oct-4-ene, 430
Oct-1-en-3-ol, 822
Octet rule
described, 11
exceptions to, 19
Octinoxate, 33, 1007
Octylamine, infrared (IR) spectrum of, 523
Oct-1-yne, 794
Odor(s)
of acetone, 835
of acid chlorides, 896
of aldehydes, 836, 836f
of amines, 1018
of caraway, 210
of esters, 891
of spearmint, 210
Offset π stacking, 674
Ofloxacin, 1058
OH proton, NMR spectra, 559–560, 559f
-oic acid (suffix), 740, 885
Oil of ginger, 400f
Oil refinery, 148f
Oil(s), 403
crude, 134, 147–148, 148f
degree of unsaturation, 469–470
essential, 1169
fatty acid composition, 1159, 1159t
fish, 1159
hardening, 1160
hydrogenation of, 469–471, 470f, 1160
insolubility of, 150
melting points of, 403, 1158–1159
mineral, 178
oxidation of, 596
partial hydrogenation of, 469–470, 470f
rancid, 1160
unsaturated, 403
vegetable, 226, 905, 907, 1159, 1160
Oil spill, 134, 150
-ol (suffix), 342
Oleandrin, 861
Olefin metathesis, 1198–1203, 1199f, 1201f
Olefins, 392. *See also* Alkenes
Oleic acid, 525, 1158t, 1159t
melting point, 402t
structure, 402f, 402t
Olestra, 906–907
Olfactory receptors, 209, 209f
Olive oil, 403
-ol suffix, 372
Omega carbon, 741
Omega-3 fatty acids, 402, 403
-one (suffix), 830
-onitrile (suffix), 886, 887t
Open arrow, 452
Ophrys sphegodes, 136
Opium, 612
Opsin, 851, 852f
Optical activity, 203–205
of achiral compounds, 204
of chiral compounds, 204
dextrorotatory (*d*) and levorotatory (*l*) compounds, 205
enantiomeric excess, 206–207
of racemic mixtures, 205
specific rotation, 205
Optical purity, 206–207
Orajel. *See* Benzocaine
Oral contraceptives, 434, 438–439, 438f
Orbitals. *See also* Hybrid orbitals; Molecular orbitals (MOs); *specific orbitals*
description, 9–10
number of electrons in, 10
Orbital symmetry
cycloadditions and, 1224–1225
of electrocyclic reactions, 1218–1219
sigmatropic rearrangements and, 1229–1230
Order, of a rate equation, 243
Organic chemistry
definition of, 1
importance of, 1–2, 2f
malaria and, 4–6, 5f
Organic diphosphate, 291, 616
Organic halides. *See also* Alkyl halides
coupling with alkenes (Heck reaction), 1186, 1192–1194
coupling with organoboron reagents (Suzuki reaction), 1186, 1189–1192
coupling reaction with organocuprate reagents, 1186–1188
Organic molecules/compounds
aromatic compounds, 649–685
reactions of, 686–737
common features of, 4
drawing structures of, 30–36
condensed structures, 30–32, 31f
skeletal structures, 32–36, 33f
examples of, 2–4, 2f
grouping by functional group, 93
naming, 138
shape. *See* Molecular shape
solubility of, 108–109
Organic monophosphate, 291
Organic reactions, 219–254. *See also specific types of reactions*
acid–base, 63–65, 68–69, 82–83
arrows used in describing, 224, 225–227, 226t
asymmetric, 490
bimolecular, 243
bond breaking and bond making, 223–227, 225f
bond dissociation energy, 227–230, 228t
catalysts, 244, 245f
endergonic, 236
endothermic, 227–228
energy diagrams for, 236–237, 238f, 240f
energy of activation, 237–238, 238f, 240f, 244, 245f
enthalpy change in, 227–230, 235–236
entropy change in, 235–236
enzymes in, 245, 246f
equations for, 220–221, 220f
equilibrium constants for, 231–232, 232f, 232t
exergonic, 236
exothermic, 227–228, 230
free energy changes in, 231–232, 232f, 232t
general types, A-12–13
kinetics of, 241
mechanisms of. *See* Reaction mechanism
multistep, 239–241, 240f
oxidation and reduction reactions, 167–168, 168f
polar, 225, 225f
radical, 224, 225f
rate of. *See* Reaction rate
sequential, 220
thermodynamics of, 230–235, 232f, 232t
types of, 221–223. *See also specific types of reactions*
unimolecular, 243
writing, 220–221, 220f
Organic solvents, 108–109, 110t
Organic synthesis
alkenes in, 425–426
of aspirin, 295, 295f
carbon–carbon bond-forming reactions in, 1185–1211
definition of, 295
halogenation reactions, 589
halohydrin use in, 418
by nucleophilic substitution reaction, 294–296, 295t
reactions with organometallic reagents, 812–815
retrosynthetic analysis, 452–455
sodium borohydride reductions used in, 783f
Organic triphosphate, 291–292
Organoaluminum compounds, 1253
Organoborane, 447, 448
definition of, 420
in Suzuki reaction, 1186, 1189–1192
synthesis of, 1190
Organocopper reagents. *See* Organocuprate reagents

Organocuprate reagents, 793
coupling reactions of, 1186–1188
general features of, 1186–1187
hydrocarbon synthesis in, 1187–1188
stereospecificity of, 1187
preparation of, 794
reactions with acid chlorides, 806–807
reactions with α,β-unsaturated carbonyl compounds, 811
Organolithium reagents, 793
addition to nitriles, 920
in anionic polymerization, 1250
arylborane synthesis, 1190
in ethynylestradiol synthesis, 798, 799f
in lithium diisopropylamide preparation, 943
preparation of, 793
reactions
with aldehydes and ketones, 796, 842
as base, 794–795
with esters and acid chlorides, 805
with α,β-unsaturated carbonyl compounds, 811
Organomagnesium reagents, 793
Organometallic reagents, 792–812. *See also specific reagents*
acetylide anions, 794
definition of, 792
organic synthesis with, 812–815
preparation of, 793–794
protecting groups and, 802–804, 803f
reactions
with acid chlorides, 805–807
with aldehydes, 795–797, 842
as base, 793, 794–795
with carbon dioxide, 808
with carbonyl compounds, 795–799, 802–804, 803f
with carboxylic acid derivatives, 804–807
with epoxides, 808–809
with esters, 804–806
with ketones, 795–799, 799f, 842
as nucleophile, 792, 793, 795–796
rapid acid–base reactions, 802
summary of, 812
with α,β-unsaturated carbonyl compounds, 809–811
Organopalladium compounds, 1189, 1191, 1194
Organophosphorus reagents, 845–846
Organosodium reagents, 794
Orlistat, 1208
Ortho (prefix), 653
Ortho, para activator, 704–705
Ortho, para deactivator, 704
Ortho, para director, 703–705, 707–708, 710f, 712–715, 725–727, 1046
-ose (suffix), 1111
Oseltamivir, 99, 390, 926
Osmium tetroxide
as oxidizing agent, 476
in syn dihydroxylation, 481–482
Osprey, 305
Osteoporosis, 19
Oxalic acid, 742, 745
Oxaloacetate, 991
Oxaphosphetane, 847
Oxazaborolidine, 784
Oxidation, 167–168, 168f, 463–502
of alcohols, 476f, 484–490
of aldoses, 1132–1134
of alkanes, 167–169
of alkenes, 464f, 476f, 482–484
of alkyl benzenes, 721
of alkylboranes, 421–422
of alkynes, 464f, 476f, 484
biological, 489–490
of carbonyl compounds, 780, 792
carboxylic acid synthesis
from alkyl benzenes, 747
from alkynes, 747
from primary alcohols, 747
definition of, 167, 464, 779
dihydroxylation, 480–482
anti, 480, 481
syn, 480, 481–482
energy released by, 230
epoxidation, 476f, 477–480
of ethanol, 489–490
Sharpless, 490–492
stereochemistry of, 477–478
in the synthesis of disparlure, 478–480, 479f
of fatty acids, 906
general scheme for, 464f
green chemistry, 487–488
hydroboration–oxidation reactions, 405f, 419–423, 420f, 421f, 422t, 447–449
of hydrocarbons, 464, 464f
of lipids, 171
of oils, 470
oxidative cleavage
of alkenes, 476f, 482–484
of alkynes, 476f, 484
of polycyclic aromatic hydrocarbons (PAHs), 381
reduction, relationship to, 464
of substituted benzenes, 721
of unsaturated fatty acids, 1160
of unsaturated lipids, 596–597, 597f
Oxidative addition
in Heck reaction, 1194
in palladium-catalyzed reactions, 1189, 1191, 1194
in Suzuki reactions, 1191
Oxidative cleavage
of alkenes, 476f, 482–484
of alkynes, 476f, 484, 748
Oxide, alkene, 344, 345
Oxidized state, 167
Oxidizing agents, 475–476. *See also specific agents*
NAD^+, 787
peroxyacids, 475–476
Oxime, 1135
Oximene, 499
Oxiranes, 341, 344, 345. *See also* Epoxides
6-Oxoheptanal, 1005
Oxone, 488
-oxy (suffix), 344
Oxycodone, 133, 531
Oxy-Cope rearrangements, 1231–1232
Oxygen
bonding, 13f, 18t
common nucleophiles, 269t
in condensed structures, 31
formal charge, 18t
infrared (IR) absorptions in oxygen-containing compounds, 522–523
isotopes of, 512t
mass of, 512t
as radical scavenger, 581
tetrahedral, 341
Oxygen nucleophiles, 893
Oxytocin, 881, 1178–1179, 1179f, 1195
-oyl (suffix), 832
Ozone
destruction of atmospheric, 260, 592, 593f
as oxidizing agent, 475, 482–484
production of, 482
stratospheric, 3
synthesis of, 592
Ozonide, 482, 484
Ozonolysis, 482–483

P

PABA (*para*-aminobenzoic acid), 640, 734, 1051–1052
Paclitaxel, 180, 190, 214. *See also* Taxol
Padimate O, 640
PAHs. *See* Polycyclic aromatic hydrocarbons (PAHs)
Pain reliever, 962. *See also* Analgesics
Palau'amine, 535, 561
Palladium-catalyzed reactions, 244
alkene reduction, 466
alkyne reduction, 472, 474f
general features of, 1188–1189
Heck reaction, 1186, 1192–1194
hydrogenation of aldehydes and ketones, 782
oxidative addition in, 1189, 1191, 1194
in peptide synthesis, 1186–1187
reductive elimination in, 1189, 1191, 1194
Suzuki reactions, 1186, 1190–1192
Palmitic acid, 1158, 1158t
Palmitoleic acid, 1158t
Palm oil, 1159
Pantothenic acid, 112
Papaver somniferum, 612
Para (prefix), 653
Parallel π stacking, 674
Para red, 1051
Parent ion, 505. *See also* Molecular ion
Parent name, 139
Parkinson's disease, treatment of, 7, 50, 51, 1019, 1020f
Paroxetine, 88, 350
Parts per million (ppm), 538
Patchouli alcohol, 360f, 1182
PCB (polychlorinated biphenyl), 109
PCC (pyridinium chlorochromate), 476, 485–486
Peanut butter, 469, 470
PEG (polyethylene glycol), 1252
Penicillamine, 1203
Penicillin(s), 909–910
discovery of, 511, 525
penicillin G, 127, 178, 511, 525–526
Pentacycloanammoxic acid methyl ester, 1238
Pentadecane, 137t
(*E*)-Penta-1,3-diene, hydrogenation of, 624
Penta-1,4-diene, 613–614, 614f
hydrogenation of, 624
Pentalene, 678
Pentanal, 819
Pentan-2-amine, 1041, 1041f
Pentane, 506–507, 624
boiling point, 105
constitutional isomers of, 137t
melting point of, 107
molecular formula, 137t
structure, 136–137
van der Waals forces, 102, 102f, 104
Pentane-2,4-dione, 936, 940, 964, 965
Pentane-3-thiol, 372f
Pentan-3-ol, 800
Pentan-2-one, 87
mass spectrum of, 512
preparation from an acid chloride and an organocuprate reagent, 807
Pentan-3-one, boiling point of, 105, 106f
Pent-1-ene, 506–507
2-Pentylcinnamaldehyde, 980
Pent-1-yne, 506–507
Peptide bonds, 1176–1178
s-cis conformation, 1177–1178
planar geometry of, 1177–1178
s-trans conformation, 1177–1178
Peptides, 891–892, 1175–1190
constitutional isomers, 1176
interesting examples, 1178–1179, 1179f
N- and C-terminal amino acids, 1176–1177
peptide bond in, 1176–1178
sequencing, 1180–1183
amino acid analysis, 1180
Edman degradation, 1180–1181
partial hydrolysis of peptide, 1182–1183, 1182t
simple, 1176–1177
synthesis of, 1183–1190
automated, 1188–1190
protecting groups in, 1184–1188
solid phase technique, 1188–1190
Percent *s*-character, 46, 76
bond length and, 623
effect on basicity of amines, 1034
hybrid orbital in alkenes, 310
infrared (IR) absorption and, 520
Percent transmittance, 517
Perfume, 472
Pericyclic reactions, 1212–1241
defined, 1213
rules for, 1233, 1233t
types, 1213–1214
cycloaddition reactions, 1213, 1223–1227
electrocyclic reactions, 1213, 1217–1223
sigmatropic rearrangements, 1214, 1227–1233
Period, in periodic table, 9
Periodic table, 8–10
bond length trends, 26
element location and bonding type, 11
of elements common in organic chemistry, 9f
first-row elements, 9f, 10
organization of, 9
second-row elements, 9f, 10
bonding of, 12–13
exceptions to the octet rule, 19
shape arrangements, 26
third-row elements, 19
trends
acidity, 70–72
bond length, 26
electronegativity, 47, 47f

Periodic trends
in bond dissociation energies, 228
in leaving group ability, 263–264
nucleophilicity, 266, 267, 268
Periplanone B, 526, 982, 982f, 1182, 1212, 1232
Perkin, William Henry, 1050
Perlon. *See* Nylon, Nylon 6
Peroxide, as radical initiator, 580, 598–599
Peroxyacetic acid, 476–477
Peroxyacids
in epoxidation, 477–479
as oxidizing agents, 475–476
Peroxy radical, 596, 597f, 1245
Perpendicular π stacking, 674
Pesticide, 260
PET. *See* Polyethylene terephthalate (PET)
Petroleum, 134, 147–148, 148f, 150
ethylene preparation from, 399
as feedstock of polymer synthesis, 1262
Ph- (abbreviation), 654
PHAs (polyhydroxyalkanoates), 1266
PHB (polyhydroxybutyrate), 1266
PHBV (polyhydroxyvalerate), 1266
α-Phellandrene, 1170f
Phenanthrene, 661, 679, 685
Phenol(s), 653
acidity of, 750
as antioxidants, 598
in aspirin synthesis, 295, 295f
Bakelite preparation from, 1260–1261, 1261f
benzene conversion to, 177
formation from anisole, 371
homolysis reaction, 254
hydroxy group of, 340
naming, 654
pK_a of, 750
polyhalogenation of, 710
substituted, 757
synthesis by Claisen rearrangement, 1232
synthesis from aryl diazonium salts, 1044
synthesis from chlorobenzene, 717
Phenoxide, 751, 751f
Phentermine, 86
Phenylacetaldehyde, 830f, 938
Phenyl acetate, 925
Phenylacetic acid, 1006
Phenylacetonitrile, 967
Phenylalanine
abbreviation for, 762t
N-acetyl, 1175
isoelectric point for, 1167t
pK_a values for, 1167t
structure of, 1165f
synthesis of, 1169, 1175
Phenylcyclohexane
phenyl group in, 93–94
Phenylethanal, 830f
2-Phenylethanamine, 1026–1027
derivatives of, 1019–1021
in Hofmann elimination reaction, 1039
Phenyl group, 93–94, 94t, 654
Phenyl isothiocyanate, 1180–1181
(1*Z*,3*E*)-1-Phenylocta-1,3-diene, 1192
2-Phenylpropanoic acid, 218
1-Phenylpropan-1-ol, 304
N-Phenylthiohydantoin (PTH), 1180–1181, 1208
N-Phenylthiourea, 1181
Pheromones, 135, 136, 478–480, 479f, 492, 492f, 897, 982, 982f, 1191
Phomallenic acid C, 456
Phosgene, 1258
Phosphate, 291
Phosphate diester, 100
Phosphate ester, 100
Phosphatidylcholine, 1162
Phosphatidylethanolamine, 1162
Phosphine, in palladium-catalyzed reactions, 1189
Phosphoacylglycerols, 1162–1163, 1163f, 1164f
Phosphoanhydrides, 911
Phosphodiester, 100, 1162–1164
Phosphodiester linkage, 123
Phosphoenolpyruvate, 251
Phospholipids, 132
in cell membranes, 114–115, 115f, 117f, 1162–1164
phosphoacylglycerols, 1162–1163, 1163f, 1164f
sphingomyelins, 1164, 1164f
structure of, 114, 115f, 1162
Phosphoric acid, 1162
Phosphoric acid anyhydrides, 911
Phosphoric acid diester, 1162
Phosphoric acid esters, 760–761
Phosphorus
leaving groups derived from, 291–292
octet rule exceptions in, 19
organic acids containing, 760–761
Phosphorus oxychloride, dehydration of alcohols using, 353, 359, 360f
Phosphorus–oxygen (P–O) bonds, 100–101, 100t
Phosphorus tribromide, conversion of alcohols to alkyl halides with, 364, 365, 365t
Photochemical cycloaddition, 1224
Photochemical electrocyclic reactions, 1221–1222
Photons, 514–516
Photosynthesis, 347, 347f, 1110–1111
Phthalic acid, 721, 769
Phthalimide, 1022
Phylloquinone, 112, 686
Physical properties. *See* Boiling point; Melting point; Solubility
Phytyl diphosphate, 699, 699f
Pi* (π*) antibonding molecular orbital, 669–670, 670f
Pi (π) bond, 42–43, 42f, 43f, 44t
in alkene double bond, 306–307
in alkenes, 392, 393, 404–407, 424, 425
in alkynes, 326, 327f, 435–436, 440, 441f
in aromatic compounds, 93, 650, 651
of benzyne, 718
breakage in addition reactions, 222, 392, 404–407, 405f, 412, 413, 414, 416, 424, 425
bromination as test for, 413
Brønsted–Lowry bases and, 62–63, 62f
in carbonyl group, 98, 775–776, 777
in conjugated allyl carbocation, 615
in cycloaddition reactions, 1213, 1224–1227, 1227t
degrees of unsaturation and, 393, 468–469
delocalized, 615, 624
in dienes, 613–614
in electrocyclic reactions, 1213, 1218–1222, 1223t
formation in elimination reactions, 221–222, 306–307, 318, 321, 326, 327f
in functional groups, 92, 117
in Lewis base, 81, 83
radical reactions with, 580–581, 603
in resonance structure, 74
in sigmatropic rearrangements, 1214, 1229–1231
strength, 392
Pi (π) bonding molecular orbital, 669–670, 670f
Picato, 1185
Picometer (pm), 26
Pi (π) electrons
in aromatic compounds, 655, 657–660, 658–661, 659t, 672–674
in benzene, 651, 652, 687
Hückel's rule and, 657–660, 659t, 663, 673
in NMR spectroscopy, 547–548, 548t
Pinacol, 389
Pinacolone, 389
α-Pinene, 1183
Pioglitazone, 731
Piperidine
basicity of, 1034–1035, 1035t
structure, 1015f
Pi (π) stacking, 668, 668f, 674
Pitavastatin, 1009
Pitocin. *See* Oxytocin
Pivalic acid, 68
pK_a values, 66–75, A-1–2
acid strength and, 66–68
of amines, 1028, 1030–1031, 1035t
for amino acids, 1166, 1167, 1167t
of carbonyl compounds, 939, 939t, 941t
definition, 66
of leaving groups in nucleophilic substitution reactions, 264, 264t
logarithmic scale of, 68
of selected organic compounds, 67t
of strong acids, 78–79
of strong bases, 79
Planar carbocation, 411, 1127–1128
Planar double bond, 414, 420, 422
Planar molecule
benzene, 651–652
Hückel's rule, 658–659
Planar radical, 590–591
Plan B, 439
Plane of symmetry, 186, 1136–1137
in disubstituted cycloalkanes, 201
of meso compound, 199
Plane-polarized light, 204–205
Plaque, atherosclerotic, 915, 915f
Plasticizers, 109, 1261
Plastics, 2f, 94
Platinum, as catalyst, 244
Plavix, 196. *See also* Clopidogrel
β-Pleated sheet, 1192–1193, 1192f, 1193f, 1194f, 1196f
Plocoralide B, 260
Plus (+) compounds, 205
Poison ivy, 757
Polar aprotic solvents, 268, 268f, 289, 290t, 313, 315t, 944
Polar bond
carbon–halogen, 260–261, 261f
description of, 48–49, 50f
Polarimeter, 204
Polarity, of carboxylic acids, 748–749
Polarizability
boiling point and, 105, 106f
inductive effects on substituted benzenes, 700
van der Waals forces and, 102
Polarized light, 204–205
Polar molecule
description, 49, 50f
dipole–dipole interactions between, 103
solubility of, 108
Polar protic solvents, 267, 267f, 289, 290t, 319t, 320t, 945
Polar reactions, 225, 225f
Poly (prefix), 1249f
Poly(acrylic acid), 602, 1267
Polyacrylonitrile, 1249f
Polyalkylation, 711
Polyamides, 1255–1256
Polyaspartate, 1267
PolyAspirin, 1273
Polyatomic ion, 17
(*Z*)-Poly(buta-1,3-diene), 1255
Polybutyrate adipate terephthalate (PBAT), 1270
Polycarbonates, 1243f, 1258
Polychlorinated biphenyl (PCB), 109
Polycyclic aromatic hydrocarbons (PAHs), 381, 661, 661f
Polyenes, 483, 621
Poly(ester amide), 1273
Polyesters, 1256–1257
Polyethers, 348, 1251
Poly(ethyl acrylate), 609
Poly(ethyl α-cyanoacrylate), 1249f
Polyethylene, 94, 399, 400f, 601, 609, 1243f, 1246–1247, 1253
Polyethylenefuranoate (PEF), 1257
Poly(ethylene glycol) (PEG), 130, 1252
Polyethylene terephthalate (PET), 1242, 1244f, 1256–1257, 1260, 1265, 1265t
Polyfunctional compounds, nomenclature of, A-4–5
Polyhalogenation, 710
Polyhydroxyalkanoates (PHAs), 1266
Polyhydroxybutyrate (PHB), 1266
Polyhydroxyvalerate (PHBV), 1266
Polyisobutylene, 609, 1249f
Poly(lactic acid) (PLA), 1263
Polymerization, 601–603
anionic, 1249–1250, 1249f
of epoxides, 1251–1252
cationic, 1248–1249, 1249f
chain branching in, 1246–1247
coordination, 1253
definition of, 601, 1243
feedstock for, 1262
head-to-tail, 603
ionic, 1248–1250, 1249f
living, 1250
radical, 603, 1245–1246, 1246f
Ziegler–Natta, 1252–1253
Polymers, 601–603, 1242–1274
alternating, 1251
amorphous regions of, 1260–1261
atatic, 1252–1253
"Big Six," 1264–1265, 1265t
biodegradable, 1266–1267
chain-growth, 1243, 1244–1251
in consumer products, 1243, 1243f
copolymers, 1251, 1256
crystalline regions of, 1260–1261
definition of, 181, 1243

derivation of word, 1243
green synthesis of, 1261–1264, 1262f, 1263f
insoluble, 1188
isotactic, 1252–1253
in medicine and dentistry, 602t
molecular weights of, 1244
nylon, 1243f, 1244, 1255–1256, 1262, 1262f, 1266
physical properties of, 1260–1261
random, 1251
recycling of, 1242, 1264–1266, 1265t
shorthand representation of, 1244, 1244f
step-growth, 1244, 1255–1259
stereochemistry of, 1252–1253
structure of, 1243, 1260–1261
syndiotactic, 1252–1253
synthetic, 1242–1274. *See also* Synthetic polymers
Poly(methyl methacrylate) (PMMA), 609
Polynucleotides, 123
Polypropylene, 602t, 1265t
Polysaccharides, 121, 1141–1143
cellulose, 1141
glycogen, 1142
starch, 1142
Polystyrene, 109, 578, 601, 609, 1244f, 1260, 1265t
preparation from ethylene, 400f
Polystyrene derivative, 1189
Polytetrafluoroethylene (Teflon), 602t
Poly(trimethylene terephthalate) (PTT), 1263, 1263f
Polyurethanes, 1257
Poly(vinyl acetate), 602, 1249f
preparation from ethylene, 400f
Poly(vinyl alcohol), 1272
Poly(vinyl butyral), 1272
Poly(vinyl chloride) (PVC), 130, 259, 601, 602t, 1243, 1261, 1265t
preparation from ethylene, 400f
p orbitals
2*p* orbitals, 10, 36–38, 38f
in alkenes, 307–308
in alkynes, 435
description, 9
hybridization of, 36–45, 38f, 43f, 44t
hyperconjugation, 282–283
overlap of, 613–614, 620, 623, 624, 652, 652f, 657, 669–670, 671f
molecular orbitals and, 1214–1216, 1215f, 1217f
Porphyrin, 1199
Potassium *tert*-butoxide, 266
in dehydrohalogenation reactions, 306, 306t
Potassium dichromate, 476, 485–487, 487f
Potassium hydroxide, in dehydrohalogenation reactions, 306t
Potassium iodide, 11, 1044
Potassium permanganate ($KMnO_4$)
as oxidizing agent, 476, 721, 725, 792
in syn dihydroxylation, 481
Potassium peroxymonosulfate, 488
Potassium propanoate, 742f
Potassium sorbate, 746
Potassium sulfate, 488
Precor. *See* Methoprene
Prednisone, 836
Prefix, 139
cis, 396–397
cyclo-, 138, 144
D-, 1115–1116
deoxy, 1144
di-, 142, 343, 886, 1013
E, 397–398
epoxy, 344
eth-, 139
iso-, 140
L-, 1115–1116
meta-, 653
meth-, 139
N-, 886
ortho-, 653
para-, 653
poly, 1249f
R, 191–196, 200–201, 1113, 1115–1116
S, 191–196, 200–201, 1113, 1115–1116
sec-, 140
tert-, 140
tetra-, 142
trans, 396–397
tri-, 142, 1013
Z, 397–398
Pregabalin, 127
Prepolymer, 1258–1259, 1259f
Primary (1°) alcohols, 96, 423
classification as, 340
conversion to alkyl halides with HX, 360–362, 365t
conversion to alkyl halides with phosphorus tribromide, 365, 365t
conversion to alkyl halides with thionyl chloride, 364, 365t
dehydration by E2 mechanism, 355
hydrogen bonding extent, 345
from nucleophilic addition to aldehydes, 788
oxidation to carboxylic acids, 485, 486–487, 747
from reactions of aldehydes with organometallic reagents, 795–797
from reduction of acid chlorides, 787–788
from reduction of aldehydes, 469, 780, 781, 782
from reduction of carboxylic acid derivatives, 780
from reduction of carboxylic acids, 790
from reduction of esters, 787–788
from reduction of monosaccharides, 1132
Primary (1°) alkyl halides, 96
acetylide anion reactions with, 449–450
classification as, 256
E2 elimination reactions, 314, 328t, 329
S_N2 substitution reactions, 275, 287, 287f, 328t, 329
synthesis of, 360
Primary (1°) amides, 99, 880, 885, 889–890, 889t, 896, 897, 903
tautomers, 917
Primary (1°) amines, 97
addition to aldehydes and ketones, 850–852, 852f, 853f
amino acids, 1166
from direct nucleophilic substitution, 1021–1022
Gabriel synthesis, 1022–1023
IR spectra of, 1016, 1017f
nomenclature, 1013
reactions with acid chlorides and anhydrides, 1037
reactions with aldehydes and ketones, 1037
reactions with nitrous acid, 1042
from reduction of amides, 1024
from reduction of nitriles, 919, 1024
from reduction of nitro compounds, 1023
structure of, 1011
Primary carbon, 94
Primary carbon radical, 579–580, 580f
Primary hydrogen (1°), 94, 140
Primary structure, of proteins, 1191, 1196f
Pristane, 171
Procaine, 653
Procyclidine, 822
Progesterone
analogs of, 438
function in body, 1178t
structure of, 438, 1156f, 1178t
synthesis of, 985, 985f
Progestins, 1178
Proline, 772
isoelectric point for, 1167t
pK_a values for, 1167t
structure of, 1165f, 1166, 1191
Prontosil, 1051
Propanal
aldol reaction of, 975, 979
enolate formation from, 939
IR spectrum of, 833, 833f
NMR spectra of, 835, 835f
Wittig reaction, 847
Propanamide
boiling point of, 106
infrared (IR) spectrum of, 523
Propan-1-amine, 1039
Propan-2-amine, 1013
Propane
bromination of, 587–588, 588f
chlorination of, 586–589, 589f
molecular formula of, 136, 137t
naming, 140
Newman projection, 152f
as propellant, 592
structure of, 136
torsional energy of, 153
Propanedioic acid, 742
Propane-1,3-diol, 1263, 1263f
Propanenitrile, 886
Propane-1-thiol, 372, 374
Propane-1,2,3-triol, 343
Propanoic acid, NMR spectra of, 744f
Propan-1-ol
conversion to butanenitrile, 366
IR spectrum of, 517
NMR spectrum of, 566, 567f
Propan-2-ol, 80, 347f
Proparacaine, 735, 1062
Propene
borane addition to, 420
hydrohalogenation, 408
monomers in polymerization, 602t
radical halogenation, 593–594, 599–600
synthesis from propan-1-amine, 1039
Prop-2-en-1-ol, 396
Propionic acid, structure of, 741t
Propiophenone, 785
halogenation of, 948–949
Propoxyphene, 187
Propranolol, 85, 388
Propylbenzene, 253, 722
Propyl group, 140
Prostacyclins, 1166, 1167, 1167f, 1167t
Prostaglandins, 171, 241, 1166–1169
analogues of, 747, 1168
inhibitors of synthesis, 1168–1169
PGA_2, 313
PGE_1, 1168
PGE_2, 130, 1168
$PGF_{2\alpha}$, 170, 170f, 250, 432, 611, 746, 1156f, 1166, 1167f, 1181
PGG_2, 611, 746, 1167, 1167f
PGI_2, 1167f
synthesis of, 746, 1006, 1167, 1167f
Prosthetic group, 1198
Protecting groups, 802–804
acetals as, 861–862
amides for amines, 1038, 1038f
in amino acid synthesis, 1184–1188
general strategy for using, 803f
Proteins, 1163–1208, 1175. *See also* Peptides
amide linkages in, 891–892, 908
as biomolecules, 119–120, 120f
conjugated, 1198
fibrous, 1197
folding, 128
globular, 1197
important examples, 1197–1200, 1197f, 1198f, 1199f
membrane, 1164
stability of, 908
structure, 1190–1197
primary, 1191, 1196f
quaternary, 1196, 1196f
secondary, 1191–1194, 1192f, 1193f, 1194f, 1196f
tertiary, 1194–1196, 1196f
subunits, 1196
Protic solvents, polar, 267, 267f, 289, 290t, 320t, 945
Proton acceptor, 62
Protonation
in acid-catalyzed ester hydrolysis, 903, 904
of alcohols, 352, 354, 355, 357, 358, 361, 362, 363
in aldehyde/ketone nucleophilic addition reactions, 839–840, 856
of carboxylic acids, 748
of epoxides, 375, 377
of ethers, 371
in glycoside formation, 1128
in glycoside hydrolysis, 1129
in nucleophilic addition to aldehydes and ketones, 777, 782
in tautomerization, 446–447, 937
Proton donor, 62, 69
Proton NMR, 536, 538–564
of aldehydes, 835, 835f
of amines, 1017, 1017f
of aromatic compounds, 660
of benzene derivatives, 560–561, 560f, 655, 655t
of carboxylic acid derivatives, 890
of carboxylic acids, 744, 744f
characteristic absorptions, A-10–11
chemical shift
equation for, 539
predicting values of, 545–546
protons on benzene rings, 547, 548t

Proton NMR (*Continued*)
protons on carbon–carbon double bonds, 547, 548t
protons on carbon–carbon triple bonds, 548, 548t
scale, 538
values for common proton types, 546t
coupled protons, 552
of cyclohexane conformations, 560
general principles
chemical shifts, 538–539
downfield signals, 538–539
reference signal, 538
scale, 538
signal intensity, 538
upfield signals, 538–539
identification of an unknown, 561–564
integrals, 549
intensity of signals, 549–550
of ketones, 835
n + 1 rule, 555
number of signals, 539–543, 540f
equivalent protons in alkenes and cycloalkanes, 541–542
general principles, 539–540
homotopic, enantiotopic, and diastereotopic protons, 542–543
OH proton, 559–560, 559f
position of signals, 543–546
chemical shift values, 545–546, 546t
shielding and deshielding effects, 544–545, 544f, 545f, 548f, 548t
regions of spectra, 548, 548f
spin–spin splitting, 550–563
in alkenes, 557–559, 557f, 558f
common patterns of, 553t
complex examples of, 554–557
how a doublet arises, 550–551
how a triplet arises, 551–552
overview of, 550
rules and examples, 552–554, 552t
Protons
description of, 8
diastereotopic, 542–543
enantiotopic, 542–543
geminal, 557
homotopic, 542–543
number of, 8
relative acidity of, determining, 77–78
Proton transfer reaction
aspirin and, 80–81
description of, 63–65
Protosterol cation, 1176f, 1177
Proventil. *See* Albuterol
Proximity effect, 628
Prozac, 3, 208, 276, 276f
Pseudoephedrine, 72, 214, 1027, 1060
D-Psicose, 1119f
Psilocin, 1020
Psilocybe mushroom, 1020
PTH (*N*-phenylthiohydantoin), 1180–1181
PTT. *See* Poly(trimethylene terephthalate)
Puffer fish, 630, 630f
Purification procedures
extraction, 758–759, 759f
Purine, 667, 678, 1015f, 1145
Putrescine, 1018
PVC. *See* Poly(vinyl chloride)
Pyranose ring, 1120, 1125, 1138
Pyrethrin I, 127, 1195, 1208
Pyridine, 80, 403, 664–666, 736, 896, 897, 926, 927
alcohol conversion to alkyl tosylates in, 367, 369, 370f
basicity of, 1034–1035, 1035t
conversion of alcohols to alkyl halides, 364
conversion of monosaccharides to esters using, 1131
dehydration of alcohols using, 359, 360f
electrostatic potential plot of, 665, 665f
structure of, 1015f
Pyridinium chlorochromate (PCC), 476, 485–486
Pyridoxine (vitamin B_6), 129
Pyrimidine, 667, 1015f, 1145
α-Pyrone, 679
Pyrrole, 665, 679, 736
basicity of, 1034, 1035t
electrostatic potential plot of, 665, 665f
structure of, 1015f
Pyrrolidine, structure of, 1015f
Pyruvate, 753, 992
Pyruvic acid, 100, 753, 787

Q

Qinghao, 422
Quantum, 514
Quartet, NMR spectrum, 552t, 553t, 554, 559
Quaternary ammonium salt. *See also* Ammonium salt
in Hofmann elimination reaction, 1038–1041
synthesis of, 1022
Quaternary carbon, 94
Quaternary structure, of proteins, 1196, 1196f
Quiana, 1270
Quinapril, 132, 302
Quinine, 5, 216
as natural alkaloid, 1018
source of, 315
structure of, 666, 1036
synthesis of, 315, 315f, 951, 951f
Quinoline, 472, 737
o-Quinone, 1241
Quintet, NMR spectrum, 552t
Quinuclidine, 303

R

R (prefix), 191–196, 200–201
Racemic mixture (racemate), 205, 205t
of alcohols from hydroboration–oxidation of alkenes, 422
alkene halogenation, 415
of alkyl chlorides, 362
of amino acids, separation of, 1170–1173, 1171f
in carbene addition, 1196
in enolate alkylation, 946, 946f, 953
from epoxidation of alkenes, 478, 480
in epoxide ring opening reactions, 376
formation by hydrohalogenation, 410–411
halogenation reactions and, 415, 590–591
in hydride reduction of ketones, 783–784
in organometallic reagent reactions, 798
S_N1 substitution reactions and, 279–280, 279f
Racemization
at the α carbon of carbonyl compounds, 946
definition of, 279
in S_N1 substitution reactions, 279–280, 279f
Radiation
electromagnetic, 514–516, 515f
infrared, 516
RF (radiofrequency), 536–537, 568
ultraviolet, 638–640, 647
Radical anion, 473
Radical cation, 505, 509–510
Radical inhibitor, 581
Radical initiator, 580
Radical intermediates, 579, 595, 599
Radical polymerization, 1245–1246, 1246f
Radical reactions, 224, 225f, 578–611. *See also specific reactions*
of alkanes, 581–591
of alkenes, 581, 598–601
applications of
antioxidants, 597–598
oxidation of unsaturated lipids, 596–597, 597f
ozone destruction by CFCs, 592, 593f
carbon–hydrogen bond, reaction of radical with, 581
double bonds, addition to, 581, 598–601
general features of, 580–581
halogenation at the allylic carbon, 593–596
bromination, 594–596
product mixtures in, 595–596
halogenation of alkanes, 582–591
of an achiral starting material, 590, 590t
bromination, 586–588, 588ff
of a chiral starting material, 590t, 591
chlorination, 582–591, 589f
energy changes during, 584–585, 585f
mechanism of, 583–584
in organic synthesis, 589
stereochemistry of, 590–591, 590t
halogenation of alkyl benzenes, 718–720
inhibitors of, 581
polymerization, 601–603, 602t
two radicals reacting with each other, 581
Radical(s)
antioxidant reaction with, 598
bond formation from, 225
classification of, 579
definition of, 579
as electrophile, 225
as reactive intermediate, 224–225, 225f
resonance-stabilized, 595, 596
stability of, 579–580, 580f
Radical scavenger, 581
Radical substitution reaction, 582, A-12
Randomness, entropy as measure of, 235
Rate constant (*k*), 242
Rate-determining step
in chlorination of ethane, 585, 585f
E1 elimination reactions, 318, 319f, 320
electrophilic additions, 407, 407f, 412
in halogenation reactions, 585, 587–589
bromination, 588
chlorination, 589
halogenation of alkenes, 415
in a multistep mechanism, 240, 240f, 243
in nucleophilic substitution reactions, 270, 271
rate equation for, 235
S_N1 substitution reactions, 281, 283, 285–286
Rate equation, 242–244
first-order, 243, 270–271
for nucleophilic substitution reactions, 270–271
second-order, 243, 270–271
Rate law, 242–244
RCM (ring-closing metathesis), 1201–1203, 1201f
Reaction arrow, 220, 226t
curved, 224, 226t
Reaction coordinate, 236, 238f, 239, 240f
Reaction mechanism
bond cleavage, 224
bond formation, 225
concerted, 223
definition of, 223
energy diagram for a two-step, 239–241, 240f
rate equations and, 242–244
stepwise, 223, 239–241, 243
Reaction rate
carbocation stability and, 283
catalysts and, 244–245, 245f
concentration effect on, 242–244
of endothermic reactions, 238f, 284
energy of activation (E_a) and, 241
of exothermic reactions, 238f, 285
nucleophilic substitution reactions, 271, 275, 277, 283–285, 284f
rate-determining step, 240, 243
temperature effect on, 241–242
Reactions. *See* Organic reactions
Reactive intermediate
carbanion, 225, 225f
carbocation, 225, 225f, 277, 279, 356, 358
description of, 223
radicals, 224–225, 225f
synthetic intermediate compared, 426
in a two-step reaction mechanism, 239
Reactivity, functional groups and, 117–118
Reagent, 220, 220f
Rebaudioside A, 1130
(*S,S*)-Reboxetine, 500
Receptors, olfactory, 209, 209f
Recycling of polymers, 1242, 1264–1266, 1265t
Red algae, 260
Red seaweed, 260
Reduced state, 167

Reducing agents, 465. *See also specific agents*
chiral, 784
metal hydride reagents, 465, 781–783, 787–791, 791t
NADH, 786–787
Reducing sugars, 1133, 1133f, 1138, 1139
Reduction, 167–168, 168f, 463–502
of acid chlorides and esters, 787–789
of alkenes, 464, 464f, 466–469
of alkyl halides, 474–475, 475f
of alkynes, 464, 464f, 471–474, 474f
to alkanes, 464f, 471–472, 474f
to cis alkenes, 471, 472, 474f
to trans alkenes, 471, 473–474, 474f
of amides to amines, 790, 1024
of aryl ketones to alkyl benzenes, 721–722
asymmetric, 784
of carbonyl compounds, 779–791
biological, 786–787
carboxylic acids and derivatives, 787–791
by catalytic hydrogenation, 782
enantioselective, 784–787, 785f
with metal hydride reagents, 781–783, 783f, 791t
stereochemistry of, 783–784
of the carbonyl group of monosaccharides, 1132
Clemmensen, 721
definition of, 167, 464, 779
dissolving metal, 465, 473–474
of epoxides, 474–475, 475f
general scheme for, 464f
of hydrocarbons, 464, 464f
hydrogenation, 466–471
of ketones to amines, 1025–1027
of nitriles, 917, 919–920, 1024
of nitro groups, 723–724
oxidation, relationship to, 464
of polar C–X sigma (σ) bonds, 474–475
reagents for, 791t
of substituted benzenes, 721–724
of triple bonds, 471–474, 474f
Wolff–Kishner, 721
Reductive amination, of aldehydes and ketones to amines, 1025–1027
Reductive elimination
in the Heck reaction, 1194
in palladium-catalyzed reactions, 1189, 1191, 1194
in Suzuki reactions, 1191
Refining, 147–148, 148f, 150
Refrigerant, 260
Regioselectivity
of alcohol dehydrations, 353
of E1 elimination reactions, 320
of E2 elimination reactions, 316, 323
of electrophilic addition of hydrogen halides, 408, 411t
of epoxide ring opening reactions, 377–378
of halohydrin formation, 417–418, 418t
of Hofmann elimination, 1040–1041
of hydroboration, 420, 422t
Reserpine, 933, 1226
Resolution, 208
Resolution of amino acids, 1171–1174, 1171f
Resolving agent, 1171–1172
Resonance, 19–25
of allylic carbocations, 615–617
allyl type, 617
benzene, 651
examples of, 616–618
principles, 20
usage of term, 615
Resonance delocalization, 75
Resonance effects
on acidity, 74–75, 75f, 77f
on basicity of amines, 1035t
electron-donating, 700–702
electron-withdrawing, 700–702
on substituted benzenes, 700–702
Resonance hybrid, 20, 24–25, 618–620
of benzene, 651
drawing, 25
electron delocalization, hybridization, and geometry, 620
major and minor contributors, 24–25, 618–619
Resonance stabilization, of carboxylic acids, 750–751, 751f
Resonance-stabilized carbocations, 443–444, 446, 615, 615f, 625–626, 628
in electrophilic aromatic substitution reaction, 688, 690, 705–706
in glycoside formation, 1127–1128
in glycoside hydrolysis, 1129
Resonance-stabilized radicals, 595, 596
Resonance structures, 19–25, 615–620
of acetate, 750
acidity and, 74–75
of allyl radical, 594
of benzylic radical, 719
of carboxylic acid derivatives, 881
cations with a positive charge adjacent to a lone pair, 617
conjugated allyl carbocation, 615
of conjugated dienes, 623
double bond
conjugated, 617
major contributor, 618–619, 623
minor contributors, 618–619, 623
with one atom more electronegative than the other, 618
drawing, 20–24
of enolates, 938
of enols, 938
IR spectrum and, 834
isomers compared to, 20
of naphthalene, 661
peptide bond, 1177
of phenoxide, 750, 751f
stability of, 615, 619
of substituted benzenes, 700–701
three atom allyl system, 617, 618
of an α,β-unsaturated carbonyl compound, 994
of a Wittig reagent, 845
Resveratrol, 608
Retention of configuration
in chlorination, 591
in conversion of alcohols to alkyl tosylates, 367, 369
in oxidation of alkylboranes, 421
in S_N2 substitution reactions, 273
Retention time, in gas chromatography, 513
Retinal, 851–852, 852f
11-*cis*-Retinal, 398
Retinol, 111
Retro-aldol reaction, 975–976
Retro Diels–Alder reaction, 636
Retrosynthetic analysis, 425, 452–455
in acetoacetic ester synthesis, 960–961
in benzene derivative synthesis, 1046–1047
of Diels–Alder product, 635–636
of disparlure, 478–479
of Grignard products, 800–802
of Heck reaction, 1193–1194
in malonic acid synthesis, 958–959
organometallic reagent reactions, 813–815
in phenylethanamine preparation, 1026
in substituted benzene syntheses, 723–726
synthesis of *m*-bromoaniline from benzene, 723
using an aldol reaction, 977–978
Wittig reaction, 847–849
Reversible reaction, unequal equilibrium arrows and, 69
RF radiation, 536–537, 568
R group, 92, 140
identity in S_N1 substitution reactions, 280–281, 281t
identity in S_N2 substitution reactions, 275, 275t
Rhodium, 1174
Rhodopsin, 851–852, 852f
Ribbon diagrams, 1193
α-D-Ribofuranose, 1126, 1144
β-D-Ribofuranose, 1126
Ribonucleic acid (RNA), 122–123, 122f, 1144, 1145–1146, 1146f
Ribonucleosides, 1144–1145
Ribonucleotides, 1145
Ribose, 122
D-Ribose, 1117, 1118f, 1126, 1144
D-Ribulose, 1119f
Rimantadine, 1027
Ring-closing metathesis (RCM), 1201–1203, 1201f
Ring expansion reactions, 985
Ring-flipping, 161, 161f, 166
Ring formation
in aldol reaction, 984–985, 985f
in Robinson annulation, 996–1000
Ring-opening metathesis polymerization (ROMP), 1207
Ring strain, 834
Risperidone, 731
RNA, 122–123, 122f, 1144, 1145–1146, 1146f
Robinson annulation, 996–1000
Robinson, Sir Robert, 996
Rocaltrol. *See* Calcitriol
Rofecoxib, 1006, 1168–1169
Rosiglitazone, 731
Rosmarinic acid, 598
Rosuvastatin, 1002
Rotation
around carbon–carbon bonds, 150–151
barrier to, 156
conrotatory, 1219–1223, 1223t
disrotatory, 1219–1223, 1223t
restricted, 42, 308–309, 309f, 393t
Row, in periodic table, 9
Rowland, F. Sherwood, 592
RU 486, 439
Rubber, 1243f, 1254–1255, 1254f
Rubbing alcohol, 347f
Ruthenium, 1198

S

S (prefix), 191–196, 200–201
Saccharin, 1140, 1140f
Safflower oil, 403
Safrole, 861, 1060
Salicin, 80f, 745, 1129
Salicylamide, 929
Salicylates, 80
Salicylic acid, 927, 1273
Salinosporamide A, 127
Salmeterol, 379, 379f, 785f, 825
Salsolinol, 873
Salts
Brønsted–Lowry bases and, 62
of carboxylic acids, 745
formation of, 12
negatively charged nucleophile, 262
SAM (*S*-adenosylmethionine), 218, 292–293, 374
Samuelsson, Bengt, 746
Sand bee, 136
Sandmeyer reaction, 1044
Saponification, 904, 907
Saponins, 112
Saquinavir, 217
Saran, 1251
Saturated fats, 403
Saturated fatty acid, 401
Saturated hydrocarbons, 135
Saytzeff rule, 316
SBR (styrene–butadiene rubber), polymerization of, 1251
Sch38516, 1201f
s-Character, of hybrid orbitals, 76
Schiff base, 850
Schrock, Richard, 1198
s-Cis conformation, 621, 631
Scombroid fish poisoning, 666
S configuration, 1164
Scopolamine, 1010
Scurvy, 91
Seashells, chiral helical, 186
Seaweed, 259–260
Sebacic acid, 1273
Sebacoyl chloride, 1273
sec- (prefix), 140
Secondary (2°) alcohols, 96, 423
classification as, 340
conversion to alkyl halides with HX, 360–363, 365t
conversion to alkyl halides with phosphorus tribromide, 365, 365t
conversion to alkyl halides with thionyl chloride, 364, 365t
dehydration by an E1 mechanism, 354–355
hydrogen bonding extent, 345
oxidation to ketones, 485, 486
from reactions of aldehydes with organometallic reagents, 795–797
from reduction of ketones, 469, 780, 781, 782
Secondary (2°) alkyl halides, 96
acetylide anion reactions with, 449–450
classification as, 256
E1 elimination reactions, 319, 328t, 329
E2 elimination reactions, 314, 328t, 329

Secondary (2°) alkyl halides (*Continued*)
S_N1 substitution reactions, 281, 287, 287f, 328t, 329
S_N2 substitution reactions, 275, 287, 287f, 328t, 329
synthesis of, 360
Secondary (2°) amides, 99, 880, 885–886, 888, 889t, 890, 896, 897, 903
Secondary (2°) amines, 97
addition to aldehydes and ketones, 852–854, 853f
from direct nucleophilic substitution, 1021–1022
IR spectra of, 1016, 1017f
nomenclature, 1013–1014
proline, 1166
reactions with acid chlorides and anhydrides, 1037
reactions with aldehydes and ketones, 1037
reactions with nitrous acid, 1043
from reduction of amides, 1024
structure of, 1011
Secondary carbon, 94
Secondary carbon radical, 579–580, 580f
Secondary hydrogen (2°), 94, 140
Secondary structure, of proteins, 1191–1194, 1192f, 1193f, 1194f, 1196f
Second-order kinetics, 243
of E2 elimination reactions, 311
in nucleophilic substitution reactions, 271–272
Second-order rate equation, 243, 270–271
Selective serotonin reuptake inhibitors (SSRIs), 1020
Separation, of amino acids, 1170–1174
Separatory funnel, 129, 758, 759f, 1029f
Septet, NMR spectrum, 552t, 555
Serevent. *See* Salmeterol
Serine
abbreviation for, 762t
isoelectric point for, 1167t
pK_a values for, 1167t
structure of, 1165f
Serotonin, 1020
Sertraline, 699
β-Sesquiphellandrene, 631
Sesquiterpene, 1170, 1170t, 1172
Sevoflurane, 115, 348
Sex hormones, 1177–1178, 1178t. *See also specific hormones*
Sex pheromone, 982, 982f, 1191
Sextet, NMR spectrum, 552t, 556f
Shark liver oil, 171
Sharpless, K. Barry, 490
Sharpless epoxidation, 490–492, 1174
Sharpless reagent, 490–491
Shell, of electrons, 9–10
Shielding effects, in NMR, 544–545, 544t, 545f, 548f, 548t
1,2-Shifts, 356–358
Shikimic acid, 99
Sickle cell anemia, 1199, 1200
Sickle cell hemoglobin, 1199–1200
Side chain, of an amino acid, 120, 1164, 1166, 1167t
Sigma (σ) bond, 36, 38, 41–43, 42f, 43f, 44t
in alkanes, 93
in alkenes, 308, 392
in alkynes, 435–436
breakage in addition reactions, 222
carbon–oxygen, 340
in cycloaddition reactions, 1213, 1223
in electrocyclic reactions, 1213, 1217, 1219
formation in addition reactions, 404, 405f, 406
formation in elimination reactions, 221–222
functional groups that contain C–Z σ bonds, 95, 96t
hyperconjugation, 282–283
inductive effect and, 73, 281
length in conjugated dienes, 622–623
radical reactions with, 580–581
reduction of polar C–X bonds, 474–475
rotation around, 150–151
in sigmatropic rearrangements, 1214, 1227–1231
in substitution reactions, 221, 269–270
Sigmatropic rearrangement, 1227–1233
Claisen rearrangement, 1231, 1232–1233
Cope rearrangement, 1231–1232
description of, 1214
orbital symmetry and, 1229–1230
Woodward–Hoffmann rules for, 1229t
Silent Spring, 260
Silk
structure of, 1193, 1194f
Silver(I) oxide, 1130
in aqueous ammonium hydroxide, 792
as oxidizing agent, 476
Silyl ether, 803
Simmons, H. E., 1197
Simmons–Smith reaction, 1197–1198
Simmons–Smith reagent, 1198
Simvastatin, 191, 622, 622f, 1177, 1177f
Sinemat, 50
α-Sinensal, 776
Singlet, NMR spectrum, 550, 552t, 554, 559
Sirenin, 1208
Sitagliptin, 666, 1006
Six-membered ring, synthesis of
in aldol reaction, 984–985, 985f
in Dieckmann reaction, 990
in Robinson annulation, 996–1000
Skeletal structures, 32–36, 33f
with charged carbon atoms, 35
interpreting, 33, 33f
Smith, R. D., 1197
S_N1 substitution reactions, 277–286
alcohol conversion to alkyl halide, 361–363
alkyl halide identity in, 286–287, 287f
carbocation stability, 281–283
Hammond postulate and, 283–286
hyperconjugation and, 282–283
inductive effects and, 281
E1 elimination reaction compared, 321, 327–331, 328t
E2 elimination reaction compared, 327–331, 328t
energy diagram for, 278, 278f, 285, 286f
ethers with strong acids, 370–371
features of, 277, 281t
kinetics of, 277, 281t
leaving group in, 282, 288–289
nucleophile strength in, 287–288
racemization in, 279–280, 279f
rate-determining step, 281, 283, 285–286
R group in, 280–281, 281t
S_N2 reactions compared, 286–291, 287f, 290t, 327–331, 328t
solvent effects on, 289
stereochemistry, 278–280, 279f, 281t
transition state in, 278, 278f, 285, 286f, 288
two-step mechanism for, 277–278, 281t
as unimolecular, 271, 277
S_N2 substitution reactions, 271–276
in acetoacetic ester synthesis, 960
acetylide anion reactions with alkyl halides, 449–451
acetylide anion reaction with epoxides, 452
alcohol conversion to alkyl halides
with HX, 361–362
with phosphorus tribromide, 365
with thionyl chloride, 364
alcohol synthesis from alkyl halides, 349–350
alkyl halide identity in, 286–287, 287f
alkyl halide with ammonia or amine, 1021–1022
alkyl tosylate reactions, 368–369
in automated peptide synthesis, 1189
backside attack in, 273–274, 288
direct enolate alkylation, 952
E1 elimination reactions compared, 327–331, 328t
E2 elimination reactions compared, 327–331, 328t
energy diagrams for, 272, 272f
epoxide ring-opening reactions, 375–376
epoxide synthesis from halohydrins, 351
epoxy resin synthesis, 1258
ethers with strong acids, 370–371
ether synthesis from alkyl halides, 349–350
features of, 271, 275t
frontside attack in, 273
of α-halo acids with ammonia, 1167–1168, 1170f
of α-halo carbonyl compounds, 951, 951f
kinetics of, 271, 275t
leaving group in, 273–274, 288–289
in malonic ester synthesis, 957
nitrile synthesis, 916
nucleophile strength in, 287–288
one-step mechanism for, 272, 275t
organic synthesis, 294, 295t, 296
reduction of alkyl halides, 475, 475f
reduction of epoxides, 475, 475f
R group in, 275, 275t
S_N1 reactions compared, 286–290, 287f, 290t, 327–331, 328t
solvent effects on, 289
stereochemistry of, 273–274, 274f, 275t, 288
steric hindrance and, 275, 275f
transition state, 272, 272f, 274, 274f, 288
Soap, 112–114, 113f, 904, 907
Sodium, in dissolving metal reductions, 465, 473
Sodium acetate, 107, 742f
Sodium amide, for alkyne synthesis by dehydrohalogenation, 327, 327f
Sodium benzoate, 745, 758
Sodium bisulfite, 481
Sodium borohydride, 465, 781–783, 783f, 791t, 841, 1132
Sodium chloride, 11
boiling point, 105
melting point, 107
Sodium cyanoborohydride, for reductive amination of aldehydes and ketones, 1025
Sodium dichromate, 476, 485
Sodium ethoxide, 350
in dehydrohalogenation reactions, 306t
Sodium hydroxide
in dehydrohalogenation reactions, 306t, 307
use in extraction procedure, 758, 759f
Sodium hypochlorite, 488
Sodium iodide, 1044
Sodium methoxide, 350
in dehydrohalogenation reactions, 306t
Sodium salicylate, 746
Sodium trichloroacetate, 1207
Soft coral (*Capnella imbricata*), 451
Solanine, 1109, 1129
Solid phase technique, of amino acid synthesis, 1188–1190
Solubility, 107–111, 110t
of alcohols, 346t
of aldehydes and ketones, 833t
of alkanes, 149, 149t
of alkenes, 399
of alkyl halides, 258t
of alkynes, 437
of amines, 1015t
of carboxylic acids, 743t
definition of, 107
of epoxides, 346t
of ethers, 346t
intermolecular forces and, 107–111, 108f
"like dissolves like," 108
of lipids, 170, 401, 1156
of monosaccharides, 1119, 1131
soap and, 112–114, 113f
of vitamins, 111–112
Solute, 107
Solvent effects, on nucleophilic substitution reactions, 266–269, 267f, 268f, 289, 290t
Solvents, 107–108
alkyl halides, 259
in E1 elimination reactions, 320t
in E2 elimination reactions, 313, 315t
for extraction procedure, 758–759, 759f
liquid, 220
nonpolar, 108
nucleophilicity and, 266–269
organic, 108–109, 110t
polar, 108
polar aprotic, 268, 268f, 289, 290t, 313, 315t, 944
polar protic, 267, 267f, 289, 290t, 320t, 945

s orbitals
1*s* orbital, 10, 36
2*s* orbital, 10, 36–38, 38f
description, 9
hybridization of, 36–45, 38f, 42–43
Sorbitol, 201, 1132
D-Sorbose, 1119f
Sorona, 1263, 1263f
Spandex, 1257
Spearmint, odor of, 210
Specific rotation, 205–206
Spectator ions, 62
Spectrometer, NMR, 537–538, 537f
Spectroscopy, 505, 646
Spelunkers, 42
Spermaceti wax, 1157
Spermine, 97
SPF rating (sun protection factor), 640
Sphingomyelins, 132, 1164, 1164f
Sphingosine, 1164
sp hybrid orbitals, 38–39, 39t, 42–43, 43f, 44t
in alkynes, 435, 436, 440, 441
percent *s*-character, 46, 76
sp^2 hybrid orbitals, 38–39, 39t, 41–42, 42f, 44t, 340, 399
in alkenes, 307–308, 310, 392
in alkyl halides, 256, 256f
in aromatic heterocycles, 664–665, 665f
in benzene, 652, 660
in benzyne, 718
carbocation, 278, 283
carbon radical, 579, 590
carboxy group, 739
percent *s*-character, 46, 76
in vinyl halides, 294
sp^3 hybrid orbitals, 37–38, 38f, 39f, 39t, 40, 44t, 340, 341, 399
in acyclic alkanes, 135
in alkenes, 310
in alkyl halides, 256, 256f, 261, 271, 294
in nucleophilic substitution reactions, 261–262, 270–271, 294
percent *s*-character, 46, 76
Spiriva. *See* Tiotropium bromide
Spiro ring system, A-6
Splenda. *See* Sucralose
Squalene, 498, 1170f, 1171f, 1172, 1176–1177, 1176f
Squalene epoxidase, 1176f, 1177
Squalene oxide, 1176f, 1177
Stachyose, 1164
Staggered conformation, 150–157, 152f, 153f, 154f, 156f, 197, 322, 322f
Stanozolol, 683, 1178
Starch
hydrolysis of, 181f, 182, 347, 347f, 1142
structure of, 181–183, 181f, 182f, 1142
Stearic acid, 124, 972, 994, 1158, 1158t
melting point, 402t
structure, 401, 402f, 402t
Stearidonic acid, 402, 496
Step-growth polymers, 1244, 1255–1259
epoxy resins, 1258–1259, 1259f
polyamides, 1255–1256
polycarbonates, 1258
polyesters, 1256–1257
polyurethanes, 1257
Stepwise reaction, 223
Stereocenter. *See* Stereogenic centers
Stereochemistry, 150, 180–218
of addition reactions, 405, 410–412, 411t
of carbonyl reduction, 783–784
chemical properties of enantiomers, 208–210
chiral and achiral molecules, 184–186, 186f
diastereomers, 196–198, 198f, 207–208, 207f
of the Diels–Alder reaction, 632–633
of dienophiles, 632–633
of dihydroxylation, 480–482
of disubstituted cycloalkanes, 201–202
of E2 elimination reactions, 322–325, 322f
of electrocyclic reactions, 1218–1219
of electrophilic addition reactions, 410–412, 411t
of enolate alkylation, 953
of epoxidation, 477–478
of halogenation, 414–415, 416f, 590–591, 590t
of halohydrin formation, 417–418, 418t
of hydroboration–oxidation reactions, 421–422, 422t
meso compounds, 199–200, 200f
organometallic reagent reactions with carbonyl compounds, 798
physical properties of stereoisomers, 203–208, 205t, 207f
polymer, 1252–1253
of sigmatropic rearrangements, 1229
of S_N1 substitution reactions, 278–280, 279f, 281t
of S_N2 substitution reactions, 273–274, 274f, 275t, 288
of starch and cellulose, 181–183, 181f, 182f
stereogenic centers, 187–201, 189f, 193f
Stereogenic centers, 187–201
in amines, 1012–1013
in amino acids, 762, 1164, 1166
on carbon atoms not part of a ring, 187–188
carbonyl reduction, 783–784
in cyclic compounds, 189–191
cyclic hemiacetals, 863
definition of, 185
in diastereomers, 196–198
drawing enantiomers, 188–189, 189f
enantiotopic and diastereotopic protons, 542–543
formation from hydrohalogenation, 410, 411
halogenation reactions and, 590–591, 590t
inversion of configuration around, 273–275, 274f, 275t, 288
labeling with *R* or *S*, 191–196, 193f, 195f, 200–201
in meso compounds, 199–200
in monosaccharides, 1112–1121, 1123–1126
number in a molecule, 185
organometallic reagent addition, 798
priority assignment, 191–195, 193f
retention of configuration around, 273
tetrahedral, 185, 187, 188, 189, 190, 196, 199, 200
Stereoisomers, 19. *See also* Enantiomers
of alkenes, 308–310, 393t
chemical properties of, 208–210
cis, 164–167
configuration of, 183
of conjugated dienes, 621
constitutional isomers compared, 183f
of cycloalkanes, 164–167
definition of, 164
from dehydrohalogenation, 317
diastereomers, 196–198, 198f, 207–208, 207f
disubstituted cycloalkanes, 201–202
enantiomers, 185
of monosaccharides, 1114–1117, 1117f
nomenclature for alkenes, 396–398
number of, maximum, 196–197
physical properties of, 203–208, 205t, 207f
starch and cellulose, 181–183
trans, 164–167
Stereoselective reactions
alkyne reduction to a cis alkene, 472
dissolving metal reduction of an alkyne, 474
of E2 elimination reactions, 317
Stereospecific reactions
carbene addition, 1196
coupling reactions with organocuprates, 1187
definition of, 415
epoxidation, 477
halogenation of alkenes, 415
Heck reaction, 1193
Simmons–Smith reaction, 1198
Suzuki reaction, 1190
Steric effects, nucleophilicity and, 266
Steric hindrance, 266
in alcohols, 345
in hydroboration reactions, 420
S_N2 substitution reactions and, 275, 275f
Steric strain, 155, 156t
Steroids, 1174–1179
adrenal cortical, 1177, 1179
anabolic, 1178
cholesterol. *See* Cholesterol
sex hormones, 1177–1178, 1178t
structure of, 1174–1175, 1175f
synthesis in Diels–Alder reaction, 637
Stone, Reverend Edmund, 80f
Strain
angle, 157–158, 341, 352
Baeyer strain theory, 157
energy increase with, 153
steric, 155, 156t
torsional, 153, 156t, 158
s-Trans conformation, 621, 631
Strecker synthesis, 1169–1170, 1170f
Structural isomers. *See* Constitutional isomers
Strychnine, 1173
Styrene, 578, 720, 1199, 1199f, 1241
polymerization of, 1244f, 1245, 1247, 1251
structure of, 1244f, 1246f
Styrene–butadiene rubber (SBR), polymerization of, 1251
Substituents, naming, 140, 653
naming aromatic rings as, 654
Substitution reactions, A-12
alcohols, 351–352, 360–369
allylic, 595
of aromatic compounds
of benzene, 687–699
electrophilic, 687–715
nucleophilic, 715–718
of substituted benzenes, 700–715
of aryl diazonium salts, 1043–1047
of benzene, 650, 657
at the α-carbon of carbonyl compounds, 945–981
description of, 221
electrophilic aromatic. *See* Electrophilic aromatic substitution reactions
Heck reaction, 1186, 1192–1194
nucleophilic, 261–296
radical, 582
Substrate, 245, 246f
Succinaldehyde, 1006
Succinic acid, 742
Succinimide, 595
Sucralose, 1140, 1140f
Sucrose, 129, 1139–1140, 1139f
olestra synthesis from, 906–907
stereogenic centers, 190
structure of, 1139–1140, 1139f
Sudafed, 72. *See also* Pseudoephedrine
Suffix, 139
-adiene, 395
-al, 829
-aldehyde, 830
-amide, 885, 887t
-amine, 1013
-ane, 33, 137, 139, 140, 141, 145, 147, 395
-ate, 742, 887t, 888
-carbaldehyde, 829
-carbonyl chloride, 883, 887t
-carboxylic acid, 883
-dioic acid, 742
-diol, 343
-ene, 395
-enol, A-5
-enone, A-5
-enyne, A-5
-ic acid, 741, 883, 885
-ide, 257
-ine, 257
-o, 257
-oic acid, 740, 885
-ol, 342, 372
-one, 830
-onitrile, 886, 887t
-ose, 1111
-oxy, 344
-oyl, 832
-thiol, 372
-yl, 140, 344, 832
-yl chloride, 883, 887t
-ylic acid, 885
-yne, 436
Sugar, simple, 121. *See also* Monosaccharides
Sugars. *See also* Carbohydrates
amino, 1143
nonreducing, 1133, 1133f, 1140
reducing, 1133, 1133f, 1138, 1139
simple. *See* Monosaccharides
Sulfa drugs, 1051–1052, 1052f
Sulfamethoxazole, 1052f
Sulfanilamide, 1051–1052
Sulfide (functional group), 96t
Sulfides, 373–374
Sulfide (suffix), 373
Sulfisoxazole, 1052f
Sulfonate anion, 760

Sulfonation
of benzene, 688f, 691–692
electrophilic aromatic substitution reaction, 688f, 691–692
Sulfonic acids, 760
Sulfonium ion, 374
Sulfur
common nucleophiles, 269t
in fossil fuels, 79
octet rule exceptions in, 19
organic acids containing, 760–761
Sulfuric acid, 19, 78–79
dehydration of alcohols in, 353–358, 356f
in nitration and sulfonation reactions of benzene, 691–692
Sulfur nucleophile, 374, 380
Sulfur oxides, 79
Sulfur trioxide, 692
Sulfur ylides, 1208
Sunscreen, 34, 640, 734
Supercoil, 1197, 1197f
Superhelix, 1197–1198
Suprafacial bond formation, 1225–1227, 1227t
Suprafacial rearrangement, 1229–1231, 1229t
Surface area
boiling point and, 105, 106f
van der Waals forces and, 102, 102f
Suzuki, Akira, 1192
Suzuki reaction, 1186, 1188–1192, 1191f
Sweeteners, artificial, 1140, 1140f, 1179
Sweetness, of monosaccharides, 1119
Sweet'n Low. *See* Saccharin
Symbol, element, 8
Symmetrical anhydrides, 880, 883
Symmetry
melting points and, 107
plane of, 186, 199, 201
Syn additions, 405, 411, 411t, 414, 419, 421, 466, 468, 471, 477, 479, 479f, 480, 481–482, 1196
Syndiotactic polymer, 1252–1253
Syn periplanar geometry, 322, 322f
Synthetic intermediate, 426
Synthetic polymers, 601–602, 1242–1274
amorphous regions of, 1260–1261
atatic, 1252–1253
"Big Six," 1264–1265, 1265t
biodegradable, 1266–1267
chain-growth, 1243, 1244–1251
in consumer products, 1243, 1243f
crystalline regions of, 1260–1261
disposal of, 1264–1266
green synthesis of, 1261–1264, 1262f, 1263f
isotactic, 1252–1253
molecular weights of, 1244
physical properties of, 1260–1261
plasticizer addition to, 1261
recycling of, 1242, 1264–1266, 1265t
shorthand representation of, 1244, 1244f
step-growth, 1244, 1255–1259
stereochemistry of, 1252–1253
structure of, 1243, 1260–1261
syndiotactic, 1252–1253
thermoplastics, 1260
thermosetting, 1260
Systematic name, 139. *See also* IUPAC nomenclature
of amines, 1013–1014

T

2,4,5-T (2,4,5-trichlorophenoxyacetic acid herbicide), 691f
Tagamet. *See* Cimetidine
D-Tagatose, 1119f
Talose
D-talose, 1118f, 1135
Wohl degradation, 1135
Tamiflu, 99, 926. *See also* Oseltamivir
Tamoxifen, 428, 934, 954
Target compound, 453
Tartaric acid, stereoisomers of, 207, 207f
Tautomerization
in acid, 446, 937
of amides, 917–918
in base, 937
of carbonyl compounds, 917–918, 936–938, 948
mechanism of, 937–938
in sigmatropic rearrangements, 1231–1232
Tautomers
definition of, 445
enol, 445–447, 918, 936–937
keto, 445–447, 918, 936–937
Taxol, 180, 190, 779. *See also* Paclitaxel
TBDMS (*tert*-butyldimethylsilyl ether), 803, 804
Teflon, 259. *See also* Polytetrafluoroethylene (Teflon)
Telfairine, 256
Temperature, reaction rate and, 241–242
Terephthalic acid, 1242, 1244f, 1256, 1272
Terpenes, 1169–1174
biosynthesis of, 1171–1174, 1171f
classes of, 1170t
examples of, 1170f
locating isoprene units in, 1169–1171, 1170f
α-Terpinene, 1174
α-Terpineol, 433, 1173
tert- (prefix), 140
Tertiary (3°) alcohols, 96
classification as, 340
conversion to alkyl halides with HX, 360–363, 365t
dehydration by E1 mechanism, 354–355
hydrogen bonding extent, 345
from reaction of esters and acid chlorides with organometallic reagents, 804–806
from reaction of ketones with organometallic reagents, 795–797
Tertiary (3°) alkyl halides, 96
acetylide anion reaction with, 449–450
classification as, 256
E1 elimination reactions, 319, 328, 328t
E2 elimination reactions, 314, 328, 328t
S_N1 substitution reactions, 279, 281, 287, 287f, 328, 328t
synthesis of, 360, 363
Tertiary (3°) amides, 99, 880, 885–886, 896, 897, 903
enolate formation from, 940
Tertiary (3°) amines, 97
from direct nucleophilic substitution, 1021–1022
IR spectra of, 1016, 1017f
nomenclature, 1013–1014
from reduction of amides, 1024
structure of, 1011
Tertiary carbon, 94
Tertiary carbon radical, 579–580
Tertiary hydrogen (3°), 94, 140
Tertiary structure, of proteins, 1194–1196, 1196f
Testosterone, 675, 1178t
Tetra- (prefix), 142
Tetrabutylammonium fluoride, 803
Tetrachloromethane, 259. *See also* Carbon tetrachloride
Tetradecane, 137t
Tetrafluoroethylene, monomers in polymerization, 602t
Tetrahalide, 444
Tetrahedral arrangement, 26, 27–28, 29t
Tetrahedral geometry, of acyclic alkanes, 135–136
Tetrahedral oxygen, 341
Tetrahedral stereogenic center, 185, 187, 189, 190, 196, 199, 200
Tetrahedrane, 158
Tetrahydrocannabinol (THC), 130
mass spectrum of, 513, 514, 514f
structure of, 78
Tetrahydrofolate, 684
Tetrahydrofuran (THF), 344, 682, 942–945, 952–953
complexed with borane, 419
as solvent, 268f, 370
Tetrahydrogestrinone, 1178
α-Tetralone, 698
Tetramethyl α-D-glucopyranose, 1133f
Tetramethylsilane (TMS), 538, 539, 564
Tetrapeptide, 1178
Tetraterpene, 1170t
Tetravalent carbon, 4, 28, 30, 32–34
Tetrodotoxin, synthesis of, 630, 630f
Thalidomide, 190
THC. *See* Tetrahydrocannabinol (THC)
Thermodynamic enolates, 944, 945, 953, 954
Thermodynamic product, in electrophilic addition reactions of conjugated dienes, 627–629, 628f
Thermodynamics, 230–235, 232f, 232t
of hydrate formation, 855
Thermoplastics, 1260
Thermosetting polymers, 1260
THF. *See* Tetrahydrofuran (THF)
Thiamine diphosphate, 683
Thiazolinone, 1181, 1208
Thioesters, 98t, 222, 879, 882t, 883–886, 887t, 913, 914–916, 993–994
Thiol (functional group), 96t, 118–119
Thiols, 372–373, 372f
-thiol (suffix), 372
Thionyl chloride
conversion of alcohols to alkyl halides with, 364
conversion of carboxylic acids to acid chlorides, 898f, 899
Three-dimensional representation, of acyclic alkanes, 135–136
Threonine, 772
isoelectric point for, 1167t
pK_a values for, 1167t
structure of, 1165f, 1166
Threose
D-threose, 1117, 1117f, 1118f
L-threose, 1117, 1117f
Thrombin, 1164
Thromboxanes, 1166, 1167, 1167f, 1167t
Thymine, 649, 667, 1145, 1146f
Thymol, 570, 682
Ticlopidine, 873
Tiotropium bromide, 349
Titanium(IV) isopropoxide, 490–492
Toads (*Bufo*), 1020
Tobacco, 661
Tobramycin, 1143
Tollens reagent, 792, 1132–1133
Toluene, 514
in BTX mixture, 522
electrophilic aromatic substitution reaction, 703, 707
oxidation to benzoic acid, 721
structure of, 653
synthesis by Friedel–Crafts alkylation, 725
Toluene 2,6-diisocyanate, 1257
p-Toluenesulfonate. *See* Tosylate
p-Toluenesulfonic acid (TsOH), 79, 760, 857
dehydration of alcohols to alkenes in, 353, 359, 360f
p-Toluenesulfonyl chloride (TsCl), 367, 369, 370f
p-Toluidine
basicity of, 1032
electrostatic potential plot of, 1033f
Torsional energy, 153
Torsional strain, 153, 156t, 158
Tosic acid, 760
Tosylate, 367–369, 370f
Tosyl group, 367, 760
TPA (thermal polyaspartate), 1267
Trade name, 139
Trans (prefix), 397–398
Trans diaxial geometry, 323–324, 323f
Trans dihalide, 444
Trans fats, 470–471
Trans geometry
alkenes, 393, 393t, 396–397, 411, 414
cycloalkenes, 393
Trans isomers
of but-2-ene, 309–310, 309f
of cycloalkanes, 164–167
definition of, 164, 309
of disubstituted cycloalkanes, 201–202
stability, 311
Transition state
definition of, 236
drawing structure of, 237
of E1 elimination reactions, 318
of E2 elimination reactions, 312–314, 312f, 316, 322
in electrophilic addition, 407, 407f
endothermic reaction, 284, 284f
in energy diagrams, 236–240, 240f
exothermic reaction, 284–285, 285f
four-centered, 420
in halogenation reactions, 585, 585f, 587, 588f, 589, 589f
Hammond postulate and, 283–286, 284f–286f, 588
S_N1 substitution reactions, 278, 278f, 285, 286f, 288

S_N2 substitution reactions, 272, 272f, 274, 274f, 288
in two-step reactions, 239–241, 240f
Trans protons, 557, 557f
Tri- (prefix), 142, 1013
Triacylglycerols, 124f, 1158–1161
combustion of, 1161
common fatty acids in, 1158t
degrees of unsaturation in, 469–470
energy storage in, 1161
fats and oils, 403
formation of, 124
hydrogenation of, 469–471, 470f, 1160
hydrolysis of, 401, 906–907, 1160
melting point, 401, 402t, 1158–1159
mixed, 1158
olestra as a substitute for, 906–907
oxidation of, 596, 597f, 1160
saponification of, 907
saturated, 905, 906f, 1158–1159, 1158t, 1159f
simple, 1158
structure of, 401, 905, 906f, 1156f, 1158–1159, 1158t, 1159f, 1164f
unsaturated, 905, 1158–1160, 1158t, 1159f
Trialkylborane, 420, 420f
Triarylphosphine, 1192–1193
1,3,5-Tribromobenzene, synthesis of, 1045, 1045f
Tributyltin hydride, 610
Trichlorofluoromethane, 3, 592
Trichloromethane, 259. *See also* Chloroform
2,4,5-Trichlorophenoxyacetic acid herbicide (2,4,5-T), 691f
Tricyclohexylphosphine, 1189
Tridecane, 137t
Trienes, 395
Triesters, 124
Triethylamine, 80, 303, 1013, 1192–1193
1,2,4-Triethylcyclopentane, 146f
Trifluoroacetic acid, for removal of Boc protecting group, 1185
2,2,2-Trifluoroethanol, 72–73, 73f
Trifluoromethanesulfonic acid, 760
Trigonal planar carbocation, 278–279, 281t
Trigonal planar double bond, 307
Trigonal planar molecule, 26, 27, 29t
carbon radical, 579, 591
carbonyl group, 775–776, 798
carboxy group, 739
Trigonal pyramidal molecule, 28
amines, 1011
Trihalomethanes, 1195
Trimethylamine, 1018
electrostatic potential plot of, 1012, 1012f
Trimethyl borate, 1190
2,5,5-Trimethylcyclohexanecarboxylic acid, 740
1,2,2-Trimethylcyclohexanol, 389
2,5,5-Trimethylcyclohexanol, 342f
1,3,3-Trimethylcyclohexene, 410
2,3,5-Trimethylhex-2-ene, 395
2,2,4-Trimethylpentane, combustion of, 168. *See also* Isooctane
2,3,5-Trimethyl-4-propylheptane, 143f
Triols, 343
Tripeptide, 1176–1177, 1188
Triphenylene, 685
Triphenylmethyl radical, 610
Triphenylphosphine, 846, 1189, 1191
Triphenylphosphine oxide, 845, 847
Triphosphate, 100t, 291–292
Triple bonds
in alkynes, 93, 94t, 435–436
of benzyne, 717–718
bond length and bond strength, 45, 46t
components of carbon–carbon, 42–43
as functional group, 93, 94t
Lewis structure, 16
NMR spectra of, 548, 548t
reduction of, 471–474, 474f
Triple helix, of collagen, 1198, 1198f
Triplet, NMR spectrum, 550, 551–552, 552t, 553t, 556f, 559
Tristearin, 124, 1161
Triterpene, 1170t, 1172, 1176
Tri(*o*-tolyl)phosphine, 1189, 1194
Triynes, 436
Tropylium cation, 663–664, 672
Truvia. *See* Rebaudioside A
Trypsin, 1182, 1182t
Tryptophan, 773
isoelectric point for, 1167t
pK_a values for, 1167t
structure of, 1165f
TsCl (*p*-toluenesulfonyl chloride), 367, 369, 370f
TsOH. *See* *p*-Toluenesulfonic acid (TsOH)
Tuberculosis, 276
Tungsten, 1198
ar-Turmerone, 982
Two-electron bond, 12, 13, 15
Two-step reaction mechanism, 239–241, 240f
Tylenol. *See* Acetaminophen
Tyrian purple, 1050
Tyrosine
isoelectric point for, 1167t
pK_a values for, 1167t
structure of, 1165f
Tyvek, 1265

U

Ultraviolet light
absorption by conjugated dienes, 638–640, 639f, 647
skin damage from, 640
Undecane, 135, 137t
Unimolecular elimination. *See* E1 elimination reactions
Unimolecular reactions
elimination reactions, 311, 318
in nucleophilic substitution reactions, 271, 277
organic reactions, 2343
Unknown
analysis using mass spectrometry, 506–508
using ^{1}H NMR to identify an, 561–564
Unoprostone isopropyl, 1168
α,β-Unsaturated carbonyl compounds, 809–811
1,2-addition to, 809–811
1,4-addition to, 810–811
in the Michael reaction, 994–996, 996f
resonance structures, 994
in Robinson annulation, 996–1000
synthesis of
in dehydration of aldol product, 976–977, 979
from α-halo carbonyl compounds, 951
Unsaturated fatty acid. *See* Fatty acids, unsaturated
Unsaturated lipids, oxidation of, 596–597, 597f
Unsaturation, degrees of
in alkenes, 468–469
benzene, 650, 651
calculation of, 393–395, 468–469
hydrogenation data and, 468–469
in triacylglycerols, 469–470
Unsaturation, in alkynes, 435
Uracil, 1145, 1146f
Urethanes, 1257, 1272
Uridine monophosphate, 301
Urushiols, 757

V

Valdecoxib, 1168–1169
Valence bond theory, 669
Valence electrons, 10, 11–13, 18t, 36
Valence shell electron pair repulsion (VSEPR) theory, 26
Valeric acid, structure of, 741t
Valine, 762t
isoelectric point for, 1167t
pK_a values for, 1167t
structure of, 1165f
Valinomycin, 116
Valproic acid, 88
van der Waals forces, 101–102, 102f, 103t
in alkanes, 149
surface area and, 102, 102f
Vane, John, 746
Vanillin, 34, 570, 836f
Vanilloids, 4
Vasopressin, 1178–1179, 1179f, 1195
Vegetable oil, 226, 905, 907
partial hydrogenation of, 469–470, 470f
Venlafaxine, 802, 826
Ventolin. *See* Albuterol
β-Vetivone, 968, 971, 1004
Vibration of bonds, infrared spectroscopy and, 516, 518–521, 518f
Vicia faba, 7
Vicinal dibromide, 439, 440
Vicinal dihalides, 326, 413
alkyne synthesis from, 439–440
Vinegar, 745
Vinyl acetate
NMR spectrum of, 558, 558f
polymerization of, 1246, 1249f
structure of, 1246f
Vinylboranes, 1190, 1192
Vinyl bromide, 1190, 1192
Vinyl carbanion, 474
Vinyl carbocation, 442, 443, 446
Vinyl chloride, 57
monomers in polymerization, 602t
polymerization of, 1243, 1251
structure of, 1243, 1246f
1-Vinylcyclopentene, 398f
Vinyl group, 398
Vinyl halides, 256, 256f, 294
coupling reactions with alkenes, 1192–1194
organoboranes, 1190
organocuprates, 1186–1187
reactivity of, 695
synthesis in hydrohalogenation of alkynes, 442–443
Vinylidene chloride, polymerization of, 1251
Vinyl iodide, 1193
Vioxx. *See* Rofecoxib
Vitamin A, 111–112, 360f
β-carotene conversion to, 111
deficiency, 111, 1165f
electrostatic potential plot of, 1166, 1166f
role in the body, 1165f
solubility, 111–112
sources, 1165, 1165f
structure, 111, 308, 1165f, 1170
Vitamin B_3, 112, 787
Vitamin B_5, 112
Vitamin B_6, 51, 129
Vitamin C (ascorbic acid), 90, 91, 608, 933, 965
solubility of, 112
sources of, 112
structure of, 112, 891
Vitamin D_3, 622, 1222
deficiency, 1165f
role in the body, 1165f
sources of, 1165
structure of, 308, 1165f
Vitamin E, 129
as antioxidant, 597–598
deficiency, 1165f
electrostatic potential plot of, 1166, 1166f
as radical scavenger, 581
role in the body, 1165f
sources of, 1165
structure of, 597, 1165f
Vitamin K (phylloquinone), 112, 686, 699, 699f
deficiency, 1165f
role in the body, 1165f
sources of, 1165
structure of, 1165f
Vitamin(s), 111–112
definition of, 111, 1165
fat soluble, 111, 170, 1165–1166, 1166f
origin of term, 111
water soluble, 111, 112
VSEPR (valence shell electron pair repulsion) theory, 26
Vulcanization, 1254, 1254f

W

Wald, George, 852
Walden, Dr. Paul, 273
Walden inversion, 273
Warfarin, 773
Water
as an achiral molecule, 184
electrophilic addition of, 412–413, 423
electrostatic potential plot for, 50f
hybrid orbitals in, 39, 39f
hydration of alkynes, 441f, 445–447
hydrogen bonding, 103
as Lewis base, 82
Markovnikov addition of, 445
molecular shape of, 29
nucleophilicity of, 287, 288
pK_a of, 67t, 70
as a polar molecule, 49
removing from reaction mixture using Dean–Stark trap, 858, 858f
as solvent, 108–109, 112–114, 113f, 267, 267f, 289
in extraction procedure, 758, 759f

Wavelength
definition of, 515
electromagnetic spectrum and, 514–516, 515f
length units for, 515
Wavenumber, 516, 517
Waxes, 170, 170f, 1157
Wedge symbol, 28, 164, 1112–1115
Welding torch, 42
Wellbutrin. *See* Bupropion
Williamson ether synthesis, 349, 350
Willow tree, 80, 80f
Wittig, Georg, 845
Wittig reaction, 845–849
mechanism of, 846–847
retrosynthetic analysis using, 847–849
Wittig reagent, 845–849
Wohl degradation, 1134–1135, 1136
Wolff–Kishner reduction, 721
Wood alcohol, 347f. *See also* Methanol
Woodward, R. B., 933, 1005, 1214
Woodward–Hoffmann rules
for cycloaddition reactions, 1227, 1227t
for electrocyclic reactions, 1222, 1223t
for sigmatropic rearrangements, 1229t

X

Xalatan. *See* Latanoprost
m-Xylene, 713
p-Xylene, 514, 522, 606
D-Xylose, 1117, 1118f
D-Xylulose, 1119f

Y

-yl (suffix), 140, 344, 832
-yl chloride (suffix), 883, 887t
-ylic acid (suffix), 885
Ylide, 845–846
-yne (suffix), 436

Z

Z (prefix), 397–398
Zaitsev rule, 316–317, 320, 324, 353, 1041
Z configuration, double bond, 401, 1254–1255
Zidovudine, 4. *See also* AZT (azidodeoxythymidine)
Ziegler, Karl, 1253
Ziegler–Natta catalysts, 1253, 1254, 1255
Ziegler–Natta polymerization, 1252–1253
Zigzag skeletal structures, 157
Zileuton, 380
Zinc
in Clemmensen reduction, 721
conversion of an ozonide to carbonyl compounds, 482
Zinc chloride, 362
Zinc–copper couple, 1197–1198
Zingerone, 53, 983
Zingiberene, 400f, 498, 631, 1170f
Zocor. *See* Simvastatin
Zoloft. *See* Sertraline
Zolpidem, 1056
Zuclomiphene, 430
Zwitterion, 90, 120, 761, 1166
Zyban. *See* Bupropion
Zyflor. *See* Zileuton

Common Abbreviations, Arrows, and Symbols

Abbreviations

Ac	acetyl, CH_3CO-
BBN	9-borabicyclo[3.3.1]nonane
BINAP	2,2'-bis(diphenylphosphino)-1,1'-binaphthyl
Boc	*tert*-butoxycarbonyl, $(CH_3)_3COCO-$
bp	boiling point
Bu	butyl, $CH_3CH_2CH_2CH_2-$
CBS reagent	Corey–Bakshi–Shibata reagent
DBN	1,5-diazabicyclo[4.3.0]non-5-ene
DBU	1,8-diazabicyclo[5.4.0]undec-7-ene
DCC	dicyclohexylcarbodiimide
DET	diethyl tartrate
DIBAL-H	diisobutylaluminum hydride, $[(CH_3)_2CHCH_2]_2AlH$
DMF	dimethylformamide, $HCON(CH_3)_2$
DMSO	dimethyl sulfoxide, $(CH_3)_2S{=}O$
ee	enantiomeric excess
Et	ethyl, CH_3CH_2-
Fmoc	9-fluorenylmethoxycarbonyl
HMPA	hexamethylphosphoramide, $[(CH_3)_2N]_3P{=}O$
HOMO	highest occupied molecular orbital
IR	infrared
LDA	lithium diisopropylamide, $LiN[CH(CH_3)_2]_2$
LUMO	lowest unoccupied molecular orbital
m-	meta
mCPBA	*m*-chloroperoxybenzoic acid
Me	methyl, CH_3-
MO	molecular orbital
mp	melting point
MS	mass spectrometry
MW	molecular weight
NBS	*N*-bromosuccinimide
NMO	*N*-methylmorpholine *N*-oxide
NMR	nuclear magnetic resonance
o-	ortho
p-	para
PCC	pyridinium chlorochromate
Ph	phenyl, C_6H_5-
ppm	parts per million
Pr	propyl, $CH_3CH_2CH_2-$
RCM	ring-closing metathesis
ROMP	ring-opening metathesis polymerization
TBDMS	*tert*-butyldimethylsilyl
THF	tetrahydrofuran
TMS	tetramethylsilane, $(CH_3)_4Si$
Ts	tosyl, *p*-toluenesulfonyl, $CH_3C_6H_4SO_2-$
TsOH	*p*-toluenesulfonic acid, $CH_3C_6H_4SO_3H$
UV	ultraviolet